Comprehensive PHYSICS Concepts & Problems

for IIT-JEE, Engineering and Medical Entrance Examinations

Volume 2

Comprehensive
PHYSICS
Concepts & Problems

In two volumes

Contents of Volume 1

1 Systems of Units and Dimensions
2 Vectors
3 Rest and Motion: Kinematics
4 Force and Motion
5 Work, Energy and Conservation Laws
6 Centre of Mass and Rotational Motion
7 Gravitation
8 Simple Harmonic Motion
9 Fluid Mechanics
10 Mechanical Properties of Matter
11 Wave Motion
12 Wave Optics (Physical Optics)
13 Geometrical Optics
14 Optical Instruments

Comprehensive

PHYSICS

Concepts & Problems

for IIT-JEE, Engineering and Medical Entrance Examinations

Volume 2

S.L. Kakani
MSc (physics), PhD
Former Principal, SMM Government Girls PG College, Bhilwara, and Chairman, BOS and Research Board, MDS University, Ajmer
Executive Director, Institute of Technology and Management, NH 79, Chittor Road, Bhilwara Bypass, Bhilwara 311001, Rajasthan
slkakani1@yahoo.com

C. Hemrajani
MSc (physics), PhD
Former Head, Department of Physics, MLV Government PG College, Bhilwara, Rajasthan

CBS

CBS PUBLISHERS & DISTRIBUTORS

New Delhi • Bangalore • Pune (India)

Comprehensive
PHYSICS
Concepts & Problems
Volume 2
for IIT-JEE, Engineering and Medical Entrance Examinations

First Edition 2009

ISBN: 978-81-239-1680-4

Published by Satish Kumar Jain and produced by Vinod K. Jain for
CBS Publishers & Distributors,
4819/XI Prahlad Street, 24 Ansari Road, Daryaganj, New Delhi 110 002, India.
Fax: 011-23243014 e-mail: cbspubs@vsnl.com; delhi@cbspd.com
Website: www.cbspd.com

Branches:
- Seema House 2975, 17th Cross, K.R. Road, Banasankari 2nd Stage
Bangalore 560 070
Fax: 080-26771680 e-mail: cbsbng@gmail.com
- Shaan Brahmha Complex, Basement, Appa Balwant Chowk, Budhwar Peth, next to Ratan Talkies,
Pune 411 002
Fax: 020-24464059 e-mail: pune@cbspd.com

Printed at:
India Binding House, Noida, UP (India)

Preface

Present book has been prepared keeping in view the need of the students preparing for prestigious IIT-JEE and various engineering and medical competitive entrance examinations. A large number of students appear for IIT-JEE but only a small percentage of them get through. One can easily recognize that even a thorough preparation of the 10 + 2 course in physics is not at all sufficient for IIT-JEE, which relies very heavily on comprehension and application of basic principles of physics. Keeping this in view, an attempt has been made to prepare a comprehensive physics book to help students preparing for these examinations.

Each topic is self-consistent and begins with a review of basic concepts, physical laws, principles and definitions giving the important formulae and relations with SI units to be used to solve the questions and problems. The book contains good number of typical worked out problems, Short Answer Type Questions and Multiple Choice Questions with answers.

At the end of each chapter, exercises are given with hints. The work-out problems and exercises are carefully designed to test the students' comprehension and application of the basic principles of physics discussed in each chapter. At the end of Vol. II good number of model test papers with complete solutions on the latest pattern of IIT-JEE, Engineering and Medical Entrance Examinations are given.

We hope that the book in the present form shall fulfil the needs of students preparing for IIT-JEE, Engineering, Medical, etc. entrance examinations.

Any suggestions for the improvement of the book from teachers and students will be gratefully acknowledged.

S. L. KAKANI
C. HEMRAJANI

Contents

Preface v
Useful Relations and Tables xv

15. Heat and Thermodynamics 1
15.1 Introduction 1
15.2 Unit of Heat Energy 1
15.3 Zeroth Law of Thermodynamics: Concept of Temperature 2
15.4 Thermal Expansion 4
15.5 Calorimetry 6
15.6 Thermodynamics 8
15.7 Work Done in a Particular Process 10
15.8 First Law of Thermodynamics 13
15.9 Analysis of Different Processes in the Light of First Law of Thermodynamics .. 13
15.10 Limitations of the First Law of Thermodynamics 15
15.11 Second Law of Thermodynamics 15
15.12 Reversible Process and Irreversible Process 16
15.13 Heat Engines 17
15.14 Carnot Engine 18
15.15 Reversibility of Carnot's Engine; Refrigerator 22
15.16 Carnot's Theorem 23
Solved Examples 25
Problems 56
Short Answer Type Questions 62
Multiple Choice Questions 66

16. Kinetic Theory of Gases 73
16.1 Introduction 73
16.2 Assumptions of Kinetic Theory of Gases 73

16.3 Pressure of an Ideal Gas 74
16.4 Kinetic Interpretation of Temperature 76
16.5 Deduction of Gas Laws on the basis of Kinetic Theory of Gases 78
16.6 r.m.s. Speed in Terms of Temperature 80
16.7 Degrees of Freedom and Law of Equipartition of Energy 82
16.8 Specific Heats of Gases 84
16.9 Mean Free Path 85
16.10 Van der Waals Equation of State for Real Gases 86
16.11 Phase Changes: Triple Point 87
Solved Examples 88
Problems 110
Short Answer Type Questions 114
Multiple Choice Questions 119

17. Heat Transfer 126

17.1 Introduction 126
17.2 Thermal Conduction 126
17.3 Measurement of Thermal Conductivity of Metals 128
17.4 Series and Parallel Combination of Rods 129
17.5 Convection 131
17.6 Radiation 131
17.7 Some Definitions Related to Radiation 132
17.8 Krichhoff's Law 134
17.9 Prevost Theory of Exchanges 135
17.10 Stefan–Boltzmann Law 136
17.11 Newton's Law of Cooling 136
17.12 Spectral Distribution of Energy in Black Body Radiations 138
17.13 Solar Constant and Temperature of the Sun 140
Solved Examples 141
Problems 166
Short Answer Type Questions 170
Multiple Choice Questions 177

18. ELECTROSTATICS 184

18.1 Electric Charge 184
18.2 Coulomb's Law 186
18.3 Electric Field 188
18.4 Electric Lines of Force 189
18.5 Electric Flux 191
18.6 Gauss Law 191
18.7 Electric Potential 197
18.8 Electric Potential Energy 201
18.9 Electric Dipole 202
18.10 Force on a Charged Conductor 207

Solved Examples ... 207
Problems ... 244
Short Answer Type Questions ... 250
Multiple Choice Questions ... 258

19. CAPACITORS ... 269
19.1 Concept of Capacity ... 269
19.2 Capacitor or Condenser ... 271
19.3 Combination of Capacitors ... 277
19.4 Dielectrics and their Behaviour in Electric Field ... 279
19.5 A Parallel Plate Capacitor with a Dielectric ... 280
19.6 Krichhoff's Law for Capacitors ... 281
19.7 Charging and Discharging of a Capacitor through Resistance ... 281
19.8 Corona Discharge ... 283
Solved Examples ... 284
Problems ... 310
Short Answer Type questions ... 315
Multiple Choice Questions ... 327

20. ELECTRIC CURRENT AND CIRCUIT ANALYSIS ... 339
20.1 Electric Current and Current Density ... 339
20.2 Electromotive Force ... 340
20.3 Current and Drift Velocity ... 340
20.4 Classification of Substances according to their Resistivity ... 344
20.5 Electric Power and Energy Transfer in an Electrical Circuit ... 345
20.6 Krichhoff's Laws ... 346
20.7 Combination of Resistances ... 347
20.8 Grouping of Cells ... 348
20.9 Wheatstone's Bridge ... 352
20.10 Potentiometer ... 353
20.11 Measuring Instruments: Galvanometer, Ammeter and Voltmeter ... 355
Solved Examples ... 357
Problems ... 383
Short Answer Type Questions ... 388
Multiple Choice Questions ... 400

21. THERMAL AND CHEMICAL EFFECTS OF ELECTRIC CURRENT ... 417
21.1 Heating Effects of Current (Joule's Law) ... 417
21.2 Fuse and its Action ... 419
21.3 Thermoelectricity ... 419
21.4 Peltier Effect ... 422
21.5 Thomson Effect ... 422
21.6 Two Laws of Thermoelectricity ... 424
21.7 Electrolysis ... 425

21.8 Faraday's Laws of Electrolysis 427
21.9 Applications of Chemical Effects of Current 428
21.10 Primary and Secondary Cells 429
Solved examples 433
Problems 446
Short Answer Type Questions 448
Multiple Choice Questions 453

22. CLASSICAL MAGNETISM 462
22.1 Magnet and its Characteristics 462
22.2 Inverse Square Law—Coulomb's Law in Magnetostatics 463
22.3 Magnetic Field and Lines of Force 464
22.4 Magnetic Dipole 466
22.5 Dipole-Dipole Interaction 470
22.6 Terrestrial Magnetism (Earth's Magnetism) 472
22.7 Elements of Earth's Magnetic Field 473
22.8 Magnetic Properties of Matter 475
22.9 Classification of Magnetic Materials 478
22.10 Magnetic Hysteresis and B-H Curve 481
Solved Examples 483
Problems 495
Short Answer Type questions 498
Multiple Choice Questions 503

23. MAGNETIC EFFECTS OF ELECTRIC CURRENT 512
23.1 Magnetic Field 512
23.2 Motion of a Charged Particle in a Uniform Magnetic Field 514
23.3 Magnetic Force on a Current Carrying Conductor 515
23.4 Biot-Savart's Law 517
23.5 Magnetic Field due to Straight Current Carrying Conductor 518
23.6 Force between Parallel Currents 520
23.7 Magnetic Field on the Axis of a Circular Current Loop 522
23.8 Magnetic Dipole Moment of Atom 524
23.9 Ampere's Circuital Law 525
23.10 Magnetic Field due to Current through a Very Long Circular Cylinder 525
23.11 Solenoid 526
23.12 Cyclotron 528
23.13 Hall effect 529
Solved Examples 530
Problems 565
Short Answer Type Questions 571
Multiple Choice Questions 585

24. ELECTROMAGNETIC INDUCTION 598
24.1 Introduction 598
24.2 Magnetic Flux 598
24.3 Electromagnetic Induction and Faraday's Laws of Electromagnetic Induction 599
24.4 Induced emf in a Conducting Rod Moving in Uniform Magnetic Field (Motional emf) 601
24.5 emf Induced in a Conducting Rod Rotating in Magnetic Field 603
24.6 Energy Consideration: A Quantitative Study 603
24.7 Induced Electric Field and Eddy Currents 604
24.8 Types of Electromagnetic Induction and Inductance 605
24.9 Growth and Decay of Current in L-R Circuit 609
24.10 Electric Generator Dynamo 611
Solved Examples 612
Problems 645
Short Answer Type Questions 651
Multiple Choice Questions 665

25. ALTERNATING CURRENTS 680
25.1 Alternating Current and Voltage 680
25.2 Phase Relation between Current and Voltage in AC Circuits 681
25.3 L-C-R Series Circuit and Resonance 687
25.4 Power in AC Circuit and Power Factor 690
25.5 Parallel Resonance Circuit (Rejector Circuit) 691
25.6 Electromagnetic Oscillations in L-C Circuit 693
25.7 Transformer 695
Solved Examples 697
Problems 718
Short Answer Type Questions 721
Multiple Choice Questions 728

26. ELECTROMAGNETIC WAVES 737
26.1 Introduction 737
26.2 Maxwell's Displacements Current 737
26.3 Maxwell's Equations and Lorentz Force 740
26.4 Properties of Electromagnetic Waves 741
26.5 Derivation for Velocity of Electromagnetic Waves $c = \frac{1}{\sqrt{\varepsilon_0 \mu_0}}$ 743
26.6 Hertz's Experiment 745
26.7 Electromagnetic Spectrum 746
26.8 Propagation of Electromagnetic Waves in the Atmosphere 748
Solved Examples 750

Problems 765
Short Answer Type Questions 768
Multiple Choice Questionns 771

27. ATOMIC PHYSICS 776

27.1 Discharge of Electricity through Rarefied Gases 776
27.2 Determination of Specific Charge (e/m) of Electron (Cathode Rays) by J. J. Thomson Method 779
27.3 Determination of Charge on Electron by Millikan's Oil Drop Method 780
27.4 Structure of the Atoms 781
27.5 Bohr's Atomic Model and Hydrogen Spectrum 785
27.6 X-rays and X-ray Spectrum 793
27.7 LASER: Light Amplification by Stimulated Emission of Radiation 798
27.8 MASER (Microwave Amplification by Stimulated Emission of Radiation) 802
27.9 Photoelectric Effect 804
Solved Examples 813
Problems 855
Short Answer Type questions 862
Multiple Choice Questions 872

28. NUCLEAR PHYSICS 883

28.1 Introduction 883
28.2 Properties of Nucleus 883
28.3 Nuclear Forces 885
28.4 Nuclear Stability 886
28.5 Binding Energy 887
28.6 Radioactivity (Alpha, Beta and Gamma decay) 889
28.7 Laws of Radioactive Decay 895
28.8 Successive Disintegration and Radioactive Equilibrium 897
28.9 Artificial and Induced Radioactivity 899
28.10 Radioactive Carbon Dating 900
28.11 Nuclear Reactions 900
28.12 Nuclear Fission 901
28.13 Nuclear Fusion 905
Solved Examples 908
Problems 932
Short Answer Type Questions 936
Multiple Choice Questions 944

29. SEMICONDUCTORS AND DEVICES 954

29.1 Classification of Solids 954
29.2 Energy Bands in Solids: Band Theory 955
29.3 Intrinsic and Extrinsic Semiconductors 956

29.4 *p–n* Junction (Junction Diode) 961
29.5 Flow of Current through *p–n* Junction 962
29.6 *p–n* Junction Diode as a Half Wave Rectifier 965
29.7 *p-n* Junction Diode as a Full Wave Rectifier 967
29.8 Junction Transistors 970
29.9 Transistor Characteristics 973
29.10 Common Base Amplifier 974
29.11 Common-emitter Amplifier 976
29.12 Transistor as an Oscillator 978
29.13 Number Systems 979
29.14 Analog and Digital Circuits 982
29.15 Boolean Algebra 982
29.16 Logic Gates 985
29.17 Combination of Gates 988
Solved Examples 990
Problems 1007
Short Answer Type Questions 1011
Multiple Choice Questions 1019

30. PRINCIPLES OF COMMUNICATION 1031

30.1 Introduction 1031
30.2 Analog and Digital Signal 1032
30.3 Modulation 1033
30.4 Production and Detection of Amplitude Modulated Wave Production 1037
30.5 Pulse Modulation 1038
30.6 Data Transmission and Retrieval 1039
30.7 Communication Channels 1041
30.8 Satellite Communication 1043
30.9 Line Communication 1045
30.10 Optical Communication 1046
Solved Examples 1049
Problems 1054
Multiple Choice Questions 1057

Model Test Papers 1059

Model Test Paper I 1059
Model Test Paper II 1072
Model Test Paper III 1086

Model Test Papers for AIEEE Examination 1098

Model Test Paper for AIEEE Examination (I) 1098
Model Test Paper for AIEEE Examination (II) 1122

Suggested Readings 1145

Useful Relations and Tables

Algebra

$$a^{-x} = \frac{1}{a^x} \qquad a^{(x+y)} = a^x a^y \qquad a^{(x-y)} = \frac{a^x}{a^y}$$

Logarithms

$$\log a + \log b = \log(ab) \qquad \log a - \log b = \log(a/b) \qquad \log(a^n) = n \log a$$

Quadratic formula

If $ax^2 + bx + c = 0$, $\quad x = \dfrac{-b \pm \sqrt{b^2 - 4ac}}{2a}$

Power series (convergent for range of *x* shown)

$$\sin x = x - \frac{x^3}{3!} + \frac{x^5}{5!} - \frac{x^7}{7!} + \ldots \quad (\text{all } x) \qquad \cos x = 1 - \frac{x^2}{2!} + \frac{x^4}{4!} - \frac{x^6}{6!} + \ldots \quad (\text{all } x)$$

$$\tan x = x + \frac{x^3}{3} + \frac{2x^5}{15} + \frac{17x^7}{315} + \ldots \quad (|x| < \pi/2) \qquad e^x = 1 + x + \frac{x^2}{2!} + \frac{x^3}{3!} + \ldots \quad (\text{all } x)$$

$$\ln(1+x) = x - \frac{x^2}{2} + \frac{x^3}{3} - \frac{x^4}{4} + \ldots (|x| < 1)$$

Binomial theorem

$$(a+b)^n = a^n + na^{n-1}b + \frac{n(n-1)a^{n-2}b^2}{2!} + \frac{n(n-1)(n-2)a^{n-3}b^3}{3!} + \ldots$$

$$(1+x)^n = 1 + nx + \frac{n(n-1)}{2!}x^2 + \ldots \quad (\text{for } x << 1)$$

Trigonometry

In the right triangle *ABC*,

$$x^2 + y^2 = r^2$$

$$\sin\alpha = \frac{y}{r} \qquad \cos\alpha = \frac{x}{r} \qquad \tan\alpha = \frac{y}{x}$$

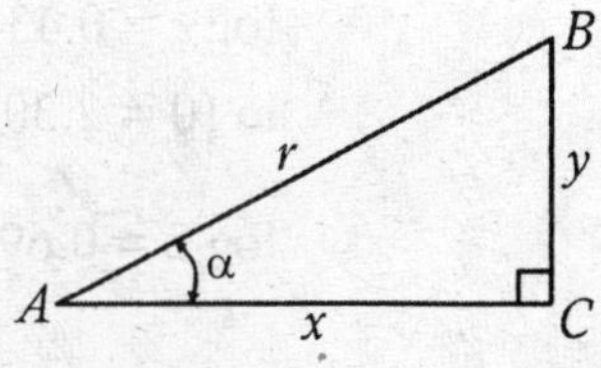

$$\sin(\alpha \pm \beta) = \sin\alpha\cos\beta \pm \cos\alpha\sin\beta \qquad \cos(\alpha \pm \beta) = \cos\alpha\cos\beta \mp \sin\alpha\sin\beta$$

$$\tan(\alpha \pm \beta) = \frac{\tan\alpha \pm \tan\beta}{1 \mp \tan\alpha\tan\beta} \qquad \cot(\alpha \pm \beta) = \frac{\cot\alpha\cot\beta \mp 1}{\cot\beta \pm \cot\alpha}$$

$$\sin\alpha = \frac{1}{\sqrt{1+\cot^2\alpha}} \qquad \cos\alpha = \frac{1}{\sqrt{1+\tan^2\alpha}}$$

$$\sin 2\alpha = 2\sin\alpha\cos\alpha \qquad \cos 2\alpha = \cos^2\alpha - \sin^2\alpha$$

$$\tan 2\alpha = \frac{2\tan\alpha}{1-\tan^2\alpha} \qquad \cot 2\alpha = \frac{\cot^2\alpha - 1}{2\cot\alpha}$$

$$\sin\alpha + \sin\beta = 2\sin\left(\frac{\alpha+\beta}{2}\right)\cos\left(\frac{\alpha-\beta}{2}\right) \qquad \sin\alpha - \sin\beta = 2\cos\left(\frac{\alpha+\beta}{2}\right)\sin\left(\frac{\alpha-\beta}{2}\right)$$

$$\cos\alpha + \cos\beta = 2\cos\left(\frac{\alpha+\beta}{2}\right)\cos\left(\frac{\alpha-\beta}{2}\right) \qquad \cos\alpha - \cos\beta = -2\sin\left(\frac{\alpha+\beta}{2}\right)\sin\left(\frac{\alpha-\beta}{2}\right)$$

$$\tan\alpha \pm \tan\beta = \frac{\sin(\alpha \pm \beta)}{\cos\alpha\cos\beta} \qquad \cot\alpha \pm \cot\beta = \pm\frac{\sin(\alpha \pm \beta)}{\sin\alpha\sin\beta}$$

$$\sin\frac{\alpha}{2} = \sqrt{\frac{1-\cos\alpha}{2}} \qquad \cos\frac{\alpha}{2} = \sqrt{\frac{1+\cos\alpha}{2}}$$

$$2\sin\alpha\sin\beta = \cos(\alpha-\beta) - \cos(\alpha+\beta) \qquad 2\cos\alpha\cos\beta = \cos(\alpha-\beta) + \cos(\alpha+\beta)$$

$$2\sin\alpha\cos\beta = \sin(\alpha-\beta) + \sin(\alpha+\beta)$$

Geometry

Circumference of circle of radius r: $C = 2\pi r$
Area of circle of radius r: $A = \pi r^2$
Volume of sphere of radius r: $V = 4\pi r^3/3$
Surface area of sphere of radius r: $A = 4\pi r^2$

NUMERICAL CONSTANTS AND APPROXIMATIONS

Numerical constants	*Approximate formulae* (for $\alpha << 1$)
$\pi = 3.1416$	$(1 \pm \alpha)^n \approx 1 \pm n\alpha$
$e = 2.7183$	$e^\alpha \approx 1 + \alpha$
$\log e = 0.4343$	$\ln(1+\alpha) \approx \alpha$
$\ln 10 = 2.3026$	$\sin\alpha \approx \alpha$
$\log_e 2 = 0.693$	$\cos\alpha \approx 1 - \frac{\alpha^2}{2}$
	$\tan\alpha \approx \alpha$

DERIVATIVES AND INTEGRALS

Functions	*Derivatives*	*Functions*	*Integrals*
x^n	nx^{n-1}	$x^n dx$	$\frac{x^{n+1}}{n+1} (n \neq -1)$
e^{nx}	ne^{nx}	$\frac{dx}{x}$	$\ln x$
a^x	$a^x \ln a$	$\sin x\, dx$	$-\cos x$
$\ln x$	$\frac{1}{x}$	$\cos x\, dx$	$\sin x$
$\sin x$	$\cos x$	$\tan x\, dx$	$\ln \cos x$
$\cos x$	$-\sin x$	$\cot x\, dx$	$\ln \sin x$
$\tan x$	$\sec^2 x$	$\frac{dx}{\cos^2 x}$	$\tan x$
$\cot x$	$-\operatorname{cosec}^2 x$	$\operatorname{cosec}^2 x\, dx$	$-\cot x$
$\sin^{-1} x$	$\frac{1}{\sqrt{1-x^2}}$	$e^x dx$	e^x
$\cos^{-1} x$	$-\frac{1}{\sqrt{1-x^2}}$	$\frac{dx}{1+x^2}$	$\tan^{-1} x$
$\tan^{-1} x$	$\frac{1}{1+x^2}$	$\frac{dx}{\sqrt{1-x^2}}$	$\sin^{-1} x$
$\cot^{-1} x$	$-\frac{1}{1+x^2}$	$-\frac{dx}{\sqrt{1+x^2}}$	$\cos^{-1} x$
$\sqrt{u}$	$\frac{u'}{2\sqrt{u}}$	$\frac{dx}{\sqrt{x^2-1}}$	$\ln(x+\sqrt{x^2+1})$
$\ln u$	$\frac{u'}{u}$	Integration by parts:	
$\frac{u}{v}$	$\frac{vu'-v'u}{v^2}$	$\int u dv = uv - \int v\, du$	

UNIT CONVERSION FACTORS

Length

1 m = 100 cm = 1000 mm = 10^6 μm = 10^9 nm
1 km = 1000 m = 0.6214 mile
1 m = 3.281 ft = 39.37 in.
1 cm = 0.3937 in.
1 ft = 30.48 cm
1in. = 2.540 cm
1 mile = 5280 ft = 1.609 km
1 Å = 10^{-10} m = 10^{-8} cm = 10^{-1} nm

Area

1 cm^2 = 0.155 $in.^2$
1 m^2 = 10^4 cm^2 = 10.76 ft^2
1 $in.^2$ = 6.452 cm^2
1 ft^2 = 144 $in.^2$ = 0.0929 m^2

Volume

1 litre = 1000 cm^3 = 10^{-3} m^3
= 0.0351 ft^3 = 61.02 $in.^3$
1 ft^3 = 0.02832 m^3 = 28.32 litre
= 7.477 gallon

Time

1 min = 60 s
1 hr = 3600 s
1 day = 86,400 s
1 yr = 3.156 × 10^7 s

Velocity

1 cm s^{-1} = 0.03281 ft s^{-1}
1 ft s^{-1} = 30.48 cm s^{-1}
1 mile min^{-1} = 60 mile hr^{-1} = 88 ft s^{-1}
1 km hr^{-1} = 5/18 m s^{-1}
1 mile hr^{-1} = 0.4470 m s^{-1}

Acceleration

1 ms^{-2} = 100 cm s^{-2} = 3.281 ft s^{-2}
1 cm s^{-2} = 0.01 m s^{-2} = 0.03281 ft s^{-2}
1 ft s^{-2} = 0.3048 m s^{-2} = 30.48 cm s^{-2}
1 mile hr^{-1} s^{-1} = 1.467 ft s^{-2}

Mass

1 kg = 10^3 g = 0.0685 slug
1 g = 6.85 × 10^{-5} slug
1 slug = 14.59 kg
1 u = 1.66 × 10^{-27} kg

Force

1 N = 10^5 dyne = 0.2247 lb
1 lb = 4.45 N = 4.45 × 10^5 dyne

Pressure

1 Pa = 1 Nm^{-2} = 1.451 × 10^{-4} lb in^{-2} = 0.209 lb ft^{-2}
1 lb in^{-2} = 6891 Pa
1 lb ft^{-2} = 47.85 Pa
1 atm = 1.013 × 10^5 Pa = 14.7 lb in^{-2} = 2117 lb ft^{-2}

Energy

1 J = 10^7 erg = 0.239 cal
1 cal = 4.186 J (based on 15° calorie)
1 ft-lb = 1.356 J
1 Btu = 1055 J = 252 cal
1 eV = 1.602 × 10^{-19} J
1 kWh = 3.600 × 10^6 J

Mass-Energy Equivalence

1 kg ↔ 8.988 × 10^{16} J
1 u ↔ 931.5 MeV
1 eV ↔ 1.073 × 10^{-19} u

Power

1 W = 1 J s^{-1}
1 hp = 746 W = 550 ft lb s^{-1}
1 Btu hr^{-1} = 0.293 W

FUNDAMENTAL PHYSICAL CONSTANTS

Name	*Symbol*	*Value*
Speed of light	c	2.9979×10^8 ms^{-1}
Charge of electron	e	1.602×10^{-19} C
Gravitational constant	G	6.673×10^{-11} Nm^{2}kg^{-2}
Planck's constant	h	6.626×10^{-34} J-s
Boltzmann's constant	k	1.381×10^{-23} J-K^{-1}
Avogadro's number	N	6.022×10^{-23} molecules mol^{-1}
Gas constant	R	8.314 J-mol^{-1}-K^{-1}
Mass of electron	m_e	9.110×10^{-31} kg
Mass of neutron	m_n	1.675×10^{-27} kg
Mass of proton	m_p	1.673×10^{-27} kg
Permittivity of free space	ε_0	8.854×10^{-12} C^{2}N^{-1}m^{-2}
	$1/4\,\pi\varepsilon_0$	8.987×10^9 Nm^{2}C^{-2}
Permeability of free space	μ_0	$4\pi \times 10^{-7}$ WbA^{-1}m^{-1}
Other Useful Constants		
Mechanical equivalent of heat	J	4.185 J cal^{-1}
Standard atmospheric pressure	1 atm	1.013×10^5 Pa
Absolute zero	0 K	–273.15 °C
Electron Volt	1 eV	1.602×10^{-19} J
Atomic mass unit	1 u	1.661×10^{-27} kg
Electron rest energy	mc^2	0.511 MeV
Energy equivalent of 1 u	Mc^2	931.5 MeV
Volume of ideal gas (0 °C and 1 atm)	V	22.4 litre mol^{-1}
Acceleration due to gravity (sea level, at equator)	g	9.78049 m s^{-2}

ASTRONOMICAL DATA

Body	*Mass (kg)*	*Radius (m)*	*Orbit radius (m)*	*Time of revolution*
Sun	1.99×10^{30}	6.95×10^8	–	–
Moon	7.36×10^{22}	1.74×10^6	0.38×10^9	27.3 day
Mercury	3.28×10^{23}	2.57×10^6	5.8×10^{10}	88.0 day
Venus	4.82×10^{24}	6.31×10^6	1.08×10^{11}	224.7 day
Earth	5.98×10^{24}	6.38×10^6	1.49×10^{11}	365.3 day
Mars	6.34×10^{23}	3.43×10^6	2.28×10^{11}	687.0 day
Jupiter	1.88×10^{27}	7.18×10^7	7.78×10^{11}	11.86 yr
Saturn	5.63×10^{26}	6.03×10^7	1.43×10^{12}	29.46 yr
Uranus	8.61×10^{25}	2.67×10^7	2.87×10^{12}	84.02 yr
Neptune	9.99×10^{25}	2.48×10^7	4.49×10^{12}	164.8 yr
Pluto	5×10^{23}	4×10^5	5.90×10^{12}	247.7 yr

DENSITY OF SUBSTANCES

Substance	Density	Substance	Density
Gases, under standard conditions, (g/cm^3)		Sodium	0.97
Hydrogen	0.09	Graphite	1.6
Methane	0.72	Porcelain	2.3
Ammonia	0.77	Aluminium	2.7
Nitrogen	1.25	Diamond	3.5
Air	1.293	Titamium	4.5
Oxygen	1.43	Zinc	7.0
Carbon dioxide	1.98	Tin	7.4
Chlorine	3.21	Iron (steel)	7.8
Liquids (g/cm^3)		Cadmium	8.65
Ether	0.72	Cobalt	8.9
Alcohol	0.79	Copper	8.9
Kerosene oil	0.80	Nickel	8.9
Benzene	0.88	Molibdenum	10.2
Castor oil	0.90	Silver	10.5
Water	1.00	Lead	11.3
Heavy water	1.10	Uranium	19.0
Glycerine	1.26	Tungsten	19.1
Mercury	13.60	Gold	19.3
Solids (g/cm^3)		Platinum	21.5
Cork	0.20		
Ice	0.916		

THERMAL EXPANSION COEFFICIENTS (AT ROOM TEMPERATURE)

Substance	*Linear expansion coefficient* α, 10^{-6} K^{-1}
Solids	
Common glass	8.5
Steel (iron)	11
Copper	16.7
Brass	18.9
Aluminium	22.9
Liquids	
Mercury	1.8
Water	2.1
Glycerine	5.0
Kerosene	10.0
Ethyl alcohol	11.0

PERMITTIVITIES (RELATIVE VALUES)

Air	1.00058 ≅ 1	Glass	6.0
Kerosene	2.0	Porcelain	6.0
Paraffin	2.0	Mica	7.5
Polyethylene	2.3	Alcohol	26
Ebonite	2.7	Water	81
Plexi glass	3.5		

RESISTIVITIES OF CONDUCTORS

Conductor	*Resistivity* (*at 20°*) ρ (nΩ m)	Temperature coefficient, α (K^{-1})
Silver	15	4.1
Copper	16	4.3
Gold	20	4.0
Aluminium	25	4.5
Tungsten	50	4.8
Iron	90	6.5
Lead	190	4.2

REFRACTIVE INDICES

Substance	*Indices*
Air	1.00029 ≅ 1
Water	1.33
Glass	1.50
Diamond	2.42

COMMONLY USED UNITS

1 year	$= 3.11 \times 10^7$ s	1 Å	$= 10^{-10}$ m
1 atm	=101.3 kPa	1 b	$= 10^{-28}$ m^2
	760 mm Hg	1 eV	$= 1.6 \times 10^{-19}$ J
1 bar	= 100 kPa		$= 1.6 \times 10^{-12}$ erg
1 mm Hg	=133.3 Pa	1 a.m.u.	$= 1.66 \times 10^{-24}$ g
1 cal	= 4.18 J		$= 9.314 \times 10^2$ MeV
1 atm	= 101.3 J	1 Ci (curie)	$= 3.70 \times 10^{10}$ dis/s

TABLE OF BASIC ELEMENTARY FUNCTIONS

1. Trigonometric functions

α°	sinα°	tanα	cotα	cosα	
0	0.0000	0.0000	–	1.000	90
1	0.0175	0.0175	57.3	1.000	89
2	0.0349	0.0349	28.6	0.999	88
3	0.0523	0.0524	19.1	0.999	87
4	0.0697	0.0699	14.3	0.998	86
5	0.0872	0.0875	11.4	0.996	85
6	0.1045	0.1051	9.51	0.995	84
7	0.1219	0.1228	8.14	0.993	83
8	0.139	0.141	7.11	0.990	82
9	0.156	0.158	6.31	0.988	81
10	0.174	0.176	5.67	0.985	80
11	0.191	0.194	5.145	0.982	79
12	0.208	0.213	4.705	0.978	78
13	0.225	0.231	4.331	0.974	77
14	0.242	0.249	4.011	0.970	76
15	0.259	0.268	3.732	0.966	75
16	0.276	0.287	3.487	0.961	74
17	0.292	0.306	3.271	0.956	73
18	0.309	0.325	3.078	0.951	72
19	0.326	0.344	2.904	0.946	71
20	0.342	0.364	2.747	0.940	70
21	0.358	0.684	2.605	0.934	69
22	0.375	0.404	2.475	0.927	68
23	0.391	0.424	2.350	0.921	67
24	0.407	0.445	2.246	0.914	66
25	0.423	0.466	2.145	0.906	65
26	0.438	0.488	2.050	0.899	64
27	0.454	0.510	1.963	0.891	63
28	0.469	0.532	1.881	0.883	62
29	0.485	0.554	1.804	0.875	61
30	0.500	0.577	1.732	0.866	60
31	0.515	0.601	1.664	0.857	59
32	0.530	0.625	1.600	0.848	58
33	0.545	0.649	1.540	0.839	57

$\alpha°$	$\sin\alpha°$	$\tan\alpha$	$\cot\alpha$	$\cos\alpha$	
34	0.559	0.675	1.483	0.829	56
35	0.574	0.700	1.428	0.819	55
36	0.588	0.727	1.376	0.809	54
37	0.601	0.754	1.327	0.799	53
38	0.616	0.781	1.280	0.788	52
39	0.629	0.810	1.235	0.777	51
40	0.643	0.839	1.192	0.766	50
41	0.656	0.869	1.150	0.755	49
42	0.669	0.900	1.111	0.743	48
43	0.682	0.933	1.072	0.731	47
44	0.695	0.966	1.036	0.719	46
45	0.707	1.000	1.000	0.707	45

2. Inverse quantities, square and cubic roots, logarithms, exponential function

x	$\frac{1}{x}$	$\sqrt{x}$	$\sqrt{10x}$	$\sqrt[3]{x}$	$\sqrt[3]{10x}$	$\sqrt[3]{100x}$	$\ln x$	e^x
1.0	1.000	1.00	3.16	1.00	2.15	4.64	0.000	2.72
1.1	0.909	1.05	3.32	1.03	2.22	4.79	0.095	3.00
1.2	0.833	1.10	3.46	1.06	2.29	4.93	0.182	3.32
1.3	0.769	1.14	3.61	1.09	2.35	5.07	0.262	3.67
1.4	0.714	1.18	3.74	1.12	2.41	5.19	0.336	4.06
1.5	0.667	1.22	3.87	1.14	2.47	5.31	0.405	4.48
1.6	0.625	1.26	4.00	1.17	2.52	5.43	0.470	4.95
1.7	0.588	1.30	4.12	1.19	2.57	5.54	0.530	5.47
1.8	0.556	1.34	4.24	1.22	2.62	5.65	0.588	6.05
1.9	0.526	1.38	4.36	1.24	2.67	5.75	0.642	6.69
2.0	0.500	1.41	4.47	1.26	2.71	5.85	0.693	7.39
2.1	0.476	1.45	4.58	1.28	2.76	5.94	0.742	8.17
2.2	0.455	1.48	4.69	1.30	2.80	6.03	0.789	9.03
2.3	0.435	1.52	4.80	1.32	2.84	6.13	0.833	9.97
2.4	0.417	1.55	4.90	1.34	2.88	6.21	0.875	11.0
2.5	0.400	1.58	5.00	1.36	2.92	6.30	0.916	12.2
2.6	0.385	1.61	5.10	1.38	2.96	6.38	0.955	13.5
2.7	0.370	1.64	5.20	1.39	3.00	6.46	0.993	14.9
2.8	0.357	1.67	5.29	1.41	3.04	6.54	1.030	16.4
2.9	0.345	1.70	5.39	1.43	3.07	6.62	1.065	18.2

x	$\frac{1}{x}$	$\sqrt{x}$	$\sqrt{10x}$	$\sqrt[3]{x}$	$\sqrt[3]{10x}$	$\sqrt[3]{100x}$	$\ln x$	e^x
3.0	0.333	1.73	5.48	1.44	3.11	6.69	1.099	20.1
3.1	0.323	1.76	5.57	1.46	3.14	6.77	1.131	22.2
3.2	0.313	1.79	5.66	1.47	3.18	6.84	1.163	24.5
3.3	0.303	1.81	5.75	1.49	3.21	6.91	1.194	27.1
3.4	0.294	1.84	5.83	1.50	3.24	6.98	1.224	30.0
3.5	0.286	1.87	5.92	1.52	3.27	7.05	1.253	33.1
3.6	0.278	1.90	6.00	1.53	3.30	7.11	1.281	36.6
3.7	0.270	1.92	6.08	1.55	3.33	7.18	1.308	40.4
3.8	0.263	1.95	6.16	1.56	3.36	7.24	1.335	44.7
3.9	0.256	1.98	6.25	1.57	3.39	7.31	1.361	49.4
4.0	0.250	2.00	6.33	1.59	3.42	7.37	1.386	54.6
4.1	0.244	2.03	6.40	1.60	3.45	7.43	1.411	60.3
4.2	0.238	2.05	6.48	1.61	3.48	7.49	1.435	66.7
4.3	0.233	2.07	6.56	1.63	3.50	7.55	1.458	73.7
4.4	0.277	2.10	6.63	1.64	3.53	7.61	1.482	81.5
4.5	0.222	2.12	6.71	1.65	3.56	7.66	1.504	90.0
4.6	0.217	2.15	6.78	1.66	3.58	7.72	1.526	99.5
4.7	0.213	2.17	6.86	1.68	3.61	7.78	1.548	110
4.8	0.208	2.19	6.93	1.69	3.63	7.83	1.569	121
4.9	0.204	2.21	7.00	1.70	3.66	7.88	1.589	134
5.0	0.200	2.24	7.07	1.71	3.68	7.94	1.609	148
5.1	0.196	2.26	7.14	1.72	3.71	7.99	1.629	164
5.2	0.192	2.28	7.21	1.73	3.73	8.04	1.649	181
5.3	0.189	2.30	7.28	1.74	3.76	8.09	1.668	200
5.4	0.185	2.32	7.35	1.75	3.78	8.14	1.686	221
5.5	0.182	2.35	7.42	1.77	3.80	8.19	1.705	244
5.6	0.179	2.37	7.48	1.78	3.83	8.24	1.723	270
5.7	0.175	2.39	7.55	1.79	3.85	8.29	1.740	299
5.8	0.172	2.41	7.62	1.80	3.87	8.34	1.758	330
5.9	0.170	2.43	7.68	1.81	3.89	8.39	1.775	365
6.0	0.167	2.45	7.75	1.82	3.92	8.43	1.792	403
6.1	0.164	2.47	7.81	1.83	3.94	8.48	1.808	446
6.2	0.161	2.49	7.87	1.84	3.96	8.53	1.825	493
6.3	0.159	2.51	7.94	1.85	3.98	8.57	1.841	545
6.4	0.156	2.53	8.00	1.86	4.00	8.62	1.856	602
6.5	0.154	2.55	8.06	1.87	4.02	8.66	1.872	665
6.6	0.152	2.57	8.12	1.88	4.02	8.71	1.887	735
6.7	0.149	2.59	8.19	1.89	4.06	8.75	1.902	812
6.8	0.147	2.61	8.25	1.90	4.08	8.79	1.918	898
6.9	0.145	2.63	8.31	1.90	4.10	8.84	1.932	992

x	$\frac{1}{x}$	$\sqrt{x}$	$\sqrt{10x}$	$\sqrt[3]{x}$	$\sqrt[3]{10x}$	$\sqrt[3]{100x}$	$\ln x$	e^x
7.0	0.143	2.65	8.37	1.91	4.12	8.88	1.946	1097
7.1	0.141	2.67	8.43	1.92	4.14	8.92	1.960	1212
7.2	0.139	2.68	8.49	1.93	4.16	8.96	1.974	1339
7.3	0.137	2.70	8.54	1.94	4.18	9.00	1.988	1480
7.4	0.135	2.72	8.60	1.95	4.20	9.05	2.001	1636
7.5	0.133	2.74	8.66	1.96	4.22	9.09	2.015	1808
7.6	0.132	2.76	8.72	1.97	4.24	9.13	2.028	1998
7.7	0.130	2.78	8.78	1.98	4.25	9.17	2.041	2208
7.8	0.128	2.79	8.83	1.98	4.27	9.21	2.054	2440
7.9	0.127	2.81	8.89	1.99	4.29	9.24	2.067	2697
8.0	0.125	2.83	8.94	2.00	4.31	9.28	2.079	2981
8.1	0.124	2.85	9.00	2.01	4.33	9.32	2.092	3294
8.2	0.122	2.86	9.06	2.02	4.34	9.36	2.104	3641
8.3	0.121	2.88	9.11	2.03	4.36	9.40	2.116	4024
8.4	0.119	2.90	9.17	2.03	4.38	9.44	2.128	4447
8.5	0.118	2.92	9.22	2.04	4.40	9.47	2.140	4914
8.6	0.116	2.93	9.27	2.05	4.41	9.51	2.152	5432
8.7	0.115	2.95	9.33	2.06	4.43	9.55	2.163	6003
8.8	0.114	2.97	9.38	2.07	4.45	9.58	2.175	6634
8.9	0.112	2.98	9.43	2.07	4.47	9.62	2.186	7332
9.0	0.111	3.00	9.49	2.08	4.48	9.66	2.197	8103
9.1	0.110	3.02	9.54	2.09	4.50	9.69	2.208	8955
9.2	0.109	3.03	9.59	2.10	4.51	9.73	2.219	9897
9.3	0.108	3.05	9.64	2.10	4.53	9.76	2.230	10938
9.4	0.106	3.07	9.69	2.11	4.55	9.80	2.241	12088
9.5	0.105	3.08	9.75	2.12	4.56	9.83	2.251	13360
9.6	0.104	3.10	9.80	2.13	4.58	9.87	2.263	14765
9.7	0.103	3.11	9.84	2.13	4.60	9.90	2.272	16318
9.8	0.102	3.13	9.90	2.14	4.61	9.93	2.282	18034
9.9	0.101	3.15	9.95	2.15	4.63	9.97	2.293	19930
10.0	**0.100**	**3.16**	**10.00**	**2.15**	**4.64**	**10.0**	**2.303**	**22026**

Formulae for taking of roots (approximate):

1. $\sqrt[n]{1+x} \approx 1 + \frac{x}{n} + \frac{1-n}{2n^2}x^2 \quad \text{for } |x| < 1$

2. $\sqrt[n]{a^n + b} \approx a\left(1 + \frac{b}{na^n} + \frac{1-n}{2n^2} \cdot \frac{b^2}{a^{2n}}\right) \quad \text{for } \left|\frac{b}{a^n}\right| < 1.$

15 Heat and Thermodynamics

15.1 INTRODUCTION

We all are well versed with terms like heat and temperature. Temperature is the measure of hotness of a body. A cup of tea is hotter than a cup of ice-cream. We also come across processes where work is converted into heat and *vice versa.* For example when we rub our palms together they become warm. When air is pumped into cycle, the pump becomes warm. In all such cases work done (in rubbing palms, pumping air, etc.) produces heat and the mechanical energy goes into the internal energy of the bodies. Thus we conclude that the cold bodies absorb energy to become hot, *i.e.* a hot-body has more internal energy than an otherwise identical cold body.

When a hot and a cold body are placed in contact, the cold body warms up and the hot body cools down. This means that the internal energy of the hot body decreases and the internal energy of cold body increases. Thus the energy is transferred from the hot body to the cold body when they are kept in contact. In this process the energy is transferred without doing the mechanical work because there are no displacements involved. This case is quite different from the transfer of mechanical energy from one body into other.

The transfer of energy from a hot body to a cold body is non-mechanical process. The energy transfer from one body to the other without any mechanical work done is called heat.

The modern concept of heat is that it is a form of energy and can be converted into other forms like mechanical work, etc. The heat energy of a body is the sum of kinetic energies of all individual molecules of the body on account of translational, vibrational and rotational motion of molecules. Thus heat possessed by a body is also called its *thermal energy.*

15.2 UNIT OF HEAT ENERGY

The most common unit of heat is a *calorie. One calorie is defined as the amount of heat energy required to raise the temperature of one gram of water through 1° C.* The rise in temperature is specified from 14.5° C to 15.5° C.

Since heat is a form of energy as such it is also measured in Joules. We have already stated that heat and mechanical work are interconvertible. Joules law states that work done W is directly proportional to heat produced, i.e.

$$W \propto H$$

or $$W = JH$$

where J is constant and is called *mechanical equivalent of heat*. J is not a physical quantity, but a conversion ratio whose value is 4.186 Joules/calorie, *i.e.* when 4.186 Joules of work is done, one calorie of heat is produced. The SI units of heat energy is Joule like all other kinds of energy.

15.3 ZEROTH LAW OF THERMODYNAMICS: CONCEPT OF TEMPERATURE

Equilibrium state in thermodynamics is defined as a state in which the macroscopic variables that characterise the system do not change with time. For example a gas inside a closed rigid container, completely insulated from surroundings, with fixed values of pressure, volume, temperature, mass and composition that do not change with time is in state of thermodynamic equilibrium. Similarly two systems are said to be in thermal equilibrium if no transfer of heat energy takes place between them when they are brought in contact with each other because in that situation the macroscopic parameters that characterise the system do not change with time.

Zeroth Law of Thermodynamics

Zeroth law of thermodynamics states that *two systems A and B in thermal equilibrium separated with system C are also in thermal equilibrium with each other.* Zeroth law was formulated in 1931 by R. H. Fowler. First and second laws of thermodynamics were stated much earlier and so this law was numbered zero.

Zeroth law gives us the concept of temperature, *i.e.* the measurement of hotness or coldness of a body. When two systems A and B are in equilibrium there must be a physical quantity that has the same value for both of them. This thermodynamic variable whose value is same for both systems in equilibrium is *called temperature* (T).

Let systems A, B and C be assigned temperatures T_a, T_b and T_c. Since system A is in thermal equilibrium with C then $T_a = T_c$. System B is also in thermal equilibrium with C then $T_b = T_c$. This impies that. $T_a = T_b$, i.e. the systems A and B are also in thermal equilibrium. Thus all bodies in thermal equilibrium are assigned equal temperature. A hotter body is assigned higher temperature than a colder body. Thus the temperatures of the two bodies decide the direction of flow of heat energy when the two are put in contact. *Heat always flows from the body at higher temperature to the body at lower temperature.*

The temperature is now regarded as one of the seven fundamental quantities like length, mass and time.

Measurement of Temperature

Zeroth law gives the concept of temperature but the most logical question is how to assign numerical values to temperatures of different bodies? This means that we have to construct a temperature scale. Thermometry deals with this basic question and the instrument which measures temperature is called thermometer.

Different types of thermometers have been developed e.g. mercury thermometers, ideal gas thermometers, electrical resistance thermometers, thermoelectric thermometers and radiation thermometers or pyrometers.

The various temperature scales in use are mentioned below:

1. **Celsius Scale (Centigrade Scale):** In this scale the melting point of ice is taken as 0° C and the boiling point of water at normal atmospheric pressure is taken as 100° C. The space between these two fixed points is divided into 100 equal parts and each part represents 1° C. This scale was given by Celsius in 1710.

2. **Fahrenheit Scale:** This scale was given by Fahrenheit in 1717. In this scale the temperature of freezing mixture was taken as zero. On this scale temperature of melting ice is 32° F and that of the boiling point of water at standard atmospheric pressure is taken 212° F. The space between these two fixed points is divided into 180 equal parts and each part is 1° F.

3. **Reaumer Scale:** This scale was designed by Reaumer in the year 1730. The melting point of ice in this scale was taken 0° R as the lowest fixed point and upper fixed point is 80° R which represents boiling point of water. The space between the two fixed points is divided into 80 equal parts. Each part represents 1° R.

If *C*, *F* and *R* represent the temperature values of a body on Celsius scale, Fahrenheit scale and Reaumer scale respectively. Then

$$\frac{C}{5} = \frac{F-32}{9} = \frac{R}{4} \qquad \text{...(15.1)}$$

Absolute Scale of Temperature: On the basis of the behaviour of gases with change in temperature led Kelvin to design *absolute scale of temperature and is also known as Kelvin scale of temperature.*

Thus ideal gas equation gives the relation between volume (V), pressure (P) and temperature (T) of the gas. The equation is

$$PV = \mu RT \qquad \text{...(15.2)}$$

Here μ is the mass of the gas in moles.

Equation (15.2) gives that for given mass of the gas if pressure remains constant, the volume of the gas is directly proportional to temperature of the gas. The variation of volume (V) with temperature T (°C) is shown in Fig. 15.1 (*a*). When this graph is extrapolated it meets the temperature axes at –273° C (Infact this value of temperature is –273.15° C). Kelvin called this value of *temperature as absolute zero.* (0 A) or (0 K). It is because at this temperature, volume of the gas would reduce to zero. Obviously pressure (P) of the gas would also become zero at this temperature. If we try to go below this temperature, volume of gas would be negative which is not possible. This means that the lowest attainable temperature is absolute zero (0 A = –273.15° C).

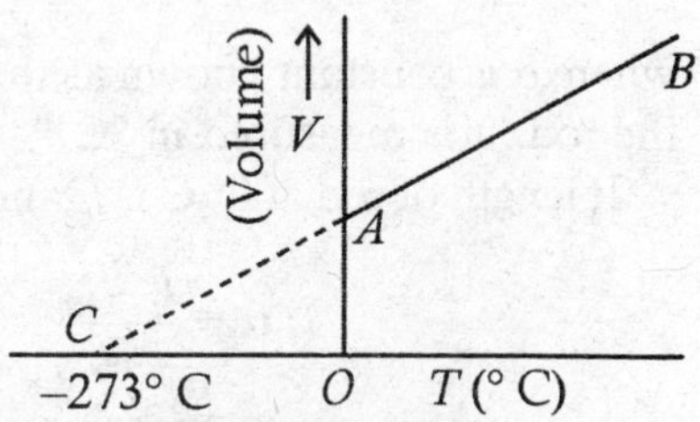

Fig. 15.1 (*a*)

Kelvin designed the scale of temperature with its zero at –273.15° C and size of each degree same as that on Celsius scale. This scale of temperature was called *Absolute scale or Kelvin scale.* Clearly

$$T\,\text{K} = 273.15 + t°\,\text{C} \qquad \text{...(15.3)}$$

This means that to obtain temperature value on Kelvin scale we add 273.15 to the temperature value on the Celsius scale. Similarly to convert temperature value on Kelvin scale to Celsius scale, we subtract 273.15, i.e.

$$t°\,\text{C} = T\,\text{K} - 273.15 \qquad \text{...(15.4)}$$

Figure 15.1 (*b*) shows the variation of volume with temperature T (K). Thus on Kelvin scale of temperature melting point of ice is 273.15 K and boiling point of water 373.15 K.

To construct the absolute temperature scale, we take the triple point of water as one fixed point. The triple point of water is the state when the solid, liquid and vapour states (phases) of water coexist in equilibrium. It is characterised by a unique temperature and pressure. That is why it is preferred over the conventional fixed points—melting point of ice and boiling point of water which are pressure dependent. Absolute zero is taken as other fixed point on this scale.

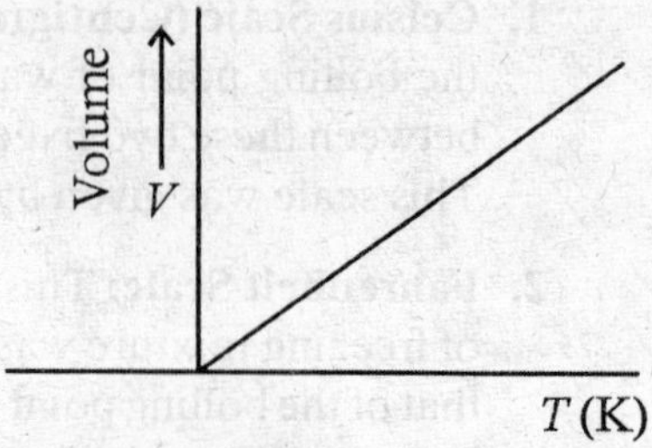

Fig. 15.1 (*b*)

The triple point of water on Celsius temperature scale in 0.01° C. If the triple point of water on absolute scale in assigned 273.16 K then the size of degree on Celsius scale becomes equal to the unit of the Kelvin absolute scale (T).

15.4 THERMAL EXPANSION

In general, most of the substances expand when heated. This phenomenon is called *thermal expansion.* The expansion may be in length, area or volume of the substance. We can study the thermal expansion of a body as a function of temperature.

Linear Expansion

A substance in the form of a long rod is heated, its length changes and the change is more easily observed. For small change in temperature ΔT, the fractional change in length $\left(\frac{\Delta l}{l}\right)$ is directly proportional to change in temperature (ΔT), i.e.

$$\frac{\Delta l}{l} \propto \Delta T$$

or
$$\frac{\Delta l}{l} = \alpha \Delta T$$

where α is constant known as the *coefficient of linear expansion* and is characteristic of the material of the rod. It is measured in °C^{-1}.

If length of rod at 0° C is l_0 and at temperature t° C (temperature measured in Celsius) is l_t, then

$$\alpha = \frac{l_t - l_0}{l_0 t} \quad \text{...(15.4)}$$

or
$$l_t = l_0(l + \alpha t) \quad \text{...(15.4 a)}$$

This equation is widely used in solving problems.

Superficial Expansion

Similarly the coefficient of superficial expansion (β) is defined as

$$\beta = \frac{\text{Change in area}}{\text{Original area at 0° C} \times \text{Rise in temperature}} = \frac{\Delta S}{S_0 t}$$

or $$\beta = \frac{S_t - S_0}{S_0 t}$$

where S_t is area at $t°$ C of the body, S_0 is the area at 0° C and t is the rise in temperature. Thus area at $t°$ C is given by

$$S_t = S_0(1 + \beta t) \quad ...(15.5)$$

Since α and β are small, hence $\beta = 2\alpha$. β is also measured in °C^{-1}.

Cubical Expansion or Volume Expansion

It we consider the fractional change in volume $\left(\frac{\Delta V}{V}\right)$ of a substance for temperature change ΔT, then coefficient of cubical expansion (or coefficient of volume expansion)

$$\gamma = \frac{\Delta V}{V \Delta T}$$

If V_0 is the volume of a body at 0° C and V_t volume at $t°$ C, then we have

$$V_t = V_0(1 + \gamma t) \quad ...(15.6)$$

where γ is a constant known as coefficient of cubical expansion and is measured in °C^{-1}.

It can be easily shown that $\gamma = 3\alpha$

Thus ratio

$$\alpha : \beta : \gamma = 6 : 3 : 2$$

The change in volume of water with temperature is slightly complicated. As the temperature is increased from 0° C to 4° C it decrease and after 4° C it increases. This is called anomalous expansion of water. Figure 15.2 shows this expansion.

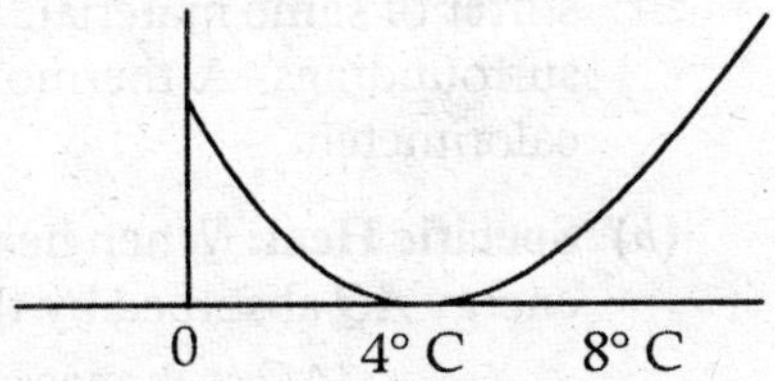

Fig. 15.2

This anomalous expansion of water has a favourable effect for animals living in water. Since the density of water is maximum at 4° C; water at the bottom of lakes remains at 4° C in winter even if it freezes at the surface. This helps marine animals to remain alive and move near the bottom.

In the study of the expansion of liquids we have to keep in mind the expansion of the container. As the temperature is raised the liquid as well as the container expand. If we consider only the expansion of liquid without the expansion of container, it is called apparent expansion and is represented by γ_a. If γ_c is the coefficient of cubical expansion of the container, then the real coefficient of cubical expansion is

$$\gamma_r = \gamma_a + \gamma_c \quad ...(15.7)$$

Molecular Theory of Thermal Expansion

Solids are made up of atoms which vibrate about their mean position. The amplitude of vibrations depends on the thermal energy of the atom which depends on the temperature of the solid. In solids the interatomic potential energy curve is asymmetric about its minimum (Fig. 15.3). As the temperature increases the amplitude of vibration also increases. As we see in Fig. 15.3 the attractive part of potential energy (BC) rises slowly compared to the repulsive part

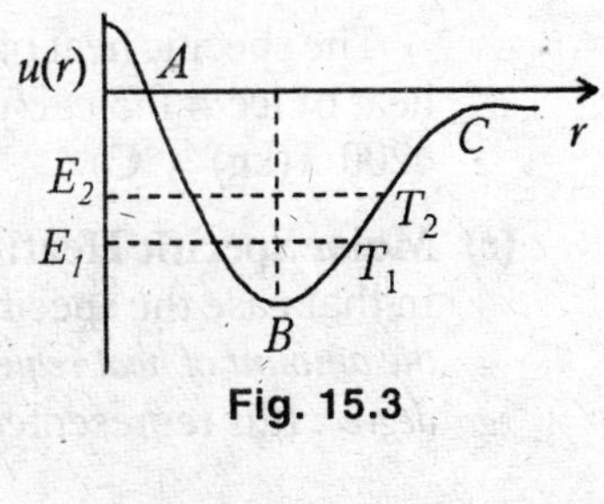

Fig. 15.3

(AB). For a given energy (*i.e.* at particular temperature) atoms vibrate with interatomic separation oscillating between its minimum and maximum values. Thus as the energy increases with temperature, the asymmetry of potential energy curve implies that its equilibrium position shifts to the right on the curve. In other words the equilibrium interatomic separation increases with temperature and thus the thermal expansion in solids takes place.

The explanation for thermal expansion given above is not rigorous, since in solids no pair of atoms can be considered separately. In case of the expansion of the gases, the molecular impacts increase with increase in temperature and hence pressure increases. To keep pressure constant, volume must increase due to this the collision frequency and hence the pressure of the gas decreases.

15.5 CALORIMETRY

Calorimetry is the method for the determination of specific heat and latent heat of a substance. The principle used for the determination is called principle of calorimetry and states that "*If objects at different temperatures are brought in contact with each other in a calorimeter, they exchange heat energy with each other and with the calorimeter. Neglecting heat lost to the surroundings, the total heat given by the hot objects equals the total heat received by the cold objects.*" Basically this is law of conservation of energy.

Here are certain definitions related to the calorimetry.

(a) Calorimeter: A simple calorimeter is a cylindrical vessel generally made of copper with stirrer of same material. The vessel is kept in a wooden box to isolate it thermally from the surroundings. A thermometer is used to measure the temperature of the contents of the calorimeter.

(b) Specific Heat: When heat is given to a body its temperature increases. The amount of heat energy ΔQ absorbed by the body

$$\Delta Q \propto m \text{ mass of the body}$$

$$\propto \Delta T \text{ the change in temperature}$$

or

$$\Delta Q = S m \Delta T \qquad \ldots(15.8)$$

Here S is a constant for a material under the given surrounding conditions. The constant S is called *the specific heat of the substance. Thus the specific heat of substance is defined as the amount of heat required to raise the temperature of unit mass of the substance through unit degree.*

$$S = \frac{\Delta Q}{m\Delta T}$$

S.I. units for specific heat are Joule $(\text{kg})^{-1}$ $(\text{Kelvin})^{-1}$

For finding out dimensions here fourth dimension taken is temperature (θ). Hence dimensional equation for specific heat is

$$S = \frac{\text{Joule}}{\text{kg}\,(\theta\text{ K})} = \frac{M^1L^2T^{-2}}{M^1\theta} = M^0L^2T^{-2}\theta^{-1} \qquad \ldots(15.9)$$

The specific heat of a substance depends on the state of the substance. For example specific heat of ice = 0.5 cal/gm/°C or 2100 J$(\text{kg})^{-1}(°\text{C})^{-1}$. Specific heat of water = 1 cal/gm/°C or 4200 J$(\text{kg})^{-1}(°\text{C})^{-1}$.

(c) Molar Specific Heat: The mass of the substance in a given body may be expressed in moles. In that case the specific heat of the substance is called molar specific heat and is defined as, *the amount of heat required to raise the temperature of one gram mole of the substance through a unit degree.* It is represented by C.

When molecular weight (M) of the substance is expressed in grams it is called 1 mole. If mass m of the substance is taken, then mass in moles is

mass moles $n = \frac{m}{M}$ or $m = nM$

Thus $$C = MS = \frac{M \cdot \Delta Q}{m\Delta T} = \frac{M\,\Delta Q}{nM\,\Delta T}$$

$$= \frac{1}{n} \cdot \frac{\Delta Q}{\Delta T} \quad \text{...(15.10)}$$

The SI units for molar specific heat are J mole^{-1}°C^{-1} or J mole^{-1}K^{-1}.

(*d*) **Heat Capacity or Thermal Capacity:** Thermal capacity of a body is defined as the amount of heat required to raise the temperature of the body through 1° C or 1 K. Thus thermal capacity is the product of specific heat and mass.

Thermal capacity = mS. ...(15.11)

Thermal capacity depends on mass and material of body. SI units are J K^{-1} or J °C^{-1}.

(*e*) **Water Equivalent:** *Water equivalent of a body is that mass of water which would absorb or evolve the same amount of heat as is done by the body in changing its temperature through the same range. It is generally represented by w.*

For a body of mass m and of specific heat S, the heat required to raise its temperature by ΔT is

$$Q = mS\Delta T$$

If W is the equivalent mass of water which when absorbs Q heat, its temperature rises by ΔT, then

$$Q = WS_w\Delta T$$

where S_W is the specific heat of water. Then

$$WS_W\Delta T = mS\Delta T$$

or $$W = m\frac{S}{S_W} \quad \text{...(15.11)}$$

In mks, SI units W is measured in kg. In cgs system S_W = 1 cal/gm/°C, hence

$$W = mS$$

and is measured in gms.

Latent Heat

When a substance is heated, apart from its rise in temperature, its state (phase) may change, i.e. solid changes into liquid or liquid to vapour. During this process of phase change (melting or vaporisation) the temperature remains constant. Whatever heat is given to the body is used in changing its state. This is called latent heat. *Thus latent heat of a substance is the amount of heat energy required to change the state of unit mass of the substance from solid to liquid or from liquid to gas/vapour without any change in temperature.*

The amount of heat required to change the state of a substance is proportional to the mass (m) of the substance, i.e.

$$Q = Lm \quad \text{...(15.12)}$$

where L is a constant for the given material and is called latent heat. When solid changes in liquid at temperature equal to melting point of solid, or liquid changes into solid, the latent heat is called *latent heat of fusion*. Similarly a liquid changes into gaseous state, at a temperature equal to boiling point of the liquid. The constant L in this case is called latent heat of vaporisation.

When a vapour condenses or a liquid solidifies heat is released to the surroundings.

In SI units the latent heat is measured is J kg^{-1}. In cgs system of units it is measured as cal g^{-1}.

Two Principal Specific Heats of a Gas

When gas is heated its volume as well as pressure both change. As such a gas has a number of (in fact infinite) specific heats depending upon the conditions under which a gas is heated. Out of all these values of specific heats of a gas two are of special significance. They are

(*a*) Specific heat of a gas at constant volume (c_V)
(*b*) Specific heat of a gas at constant pressure (c_P)

(*a*) *Specific heat of a gas at constant volume* (c_V) *is defined as the amount of heat energy required to raise the temperature of one gram of gas through 1° C or 1* K, *when volume of the gas is kept constant*. It is measured in cal $g^{-1}K^{-1}$ or J $g^{-1}K^{-1}$.

(*b*) *Specific heat of a gas at constant pressure* (c_P) *is defined as the amount of heat energy required to raise the temperature of one gram of gas through 1° C or 1* K, *when the pressure of the gas is kept constant*. It is also measured in cal $g^{-1}K^{-1}$ or J $g^{-1}K^{-1}$.

If the mass of gas is taken in gram moles, the specific heat is then called molar specific heat.

Thus molar specific heat at constant volume is the amount of heat required to raise the temperature of one gram mole of the gas through 1° C or 1 K when volume of the gas is kept constant.

It is represented by C_V and is measured in cal $mole^{-1}K^{-1}$ or J $mole^{-1}K^{-1}$.

If M is the molecular weight of the substance, then

$$C_V = Mc_V.$$

In this way molar specific heat of a gas at constant pressure is the amount of heat required to raise the temperature of one gram mole of the gas though 1° C or 1 K when pressure of the gas is kept constant. It is represented by C_P and is measured in cal $mole^{-1}K^{-1}$ or J $mole^{-1}K^{-1}$. Thus

$$C_P = Mc_P$$

The ratio of two principal specific heats of a gas is represented by γ. Hence

$$\gamma = \frac{C_P}{C_V} = \frac{c_P}{c_V} \qquad \text{...(15.13)}$$

The ratio $\frac{C_P}{C_V} = \frac{c_P}{c_V}$ (=γ) depends on the atomicity of the gas and is always greater than 1 because C_P is greater than C_V.

15.6 THERMODYNAMICS

Thermodynamics is that branch of science which deals with the study of transformation of heat energy into other forms of energy and vice versa.

We have seen that heat is just a form of energy. Mechanical work can be converted into heat. On the other hand the heat energy can be converted into mechanical energy with the help of heat engine. It is found that efficiency of heat engine is never 100%. The modern heat engines have an efficiency of the order of 55%. The basic question then arises is whether low efficiency is due to faulty design of the heat engine or it is the very nature of things that only some limited part of heat energy can be converted into mechanical energy. The study of thermodynamics attempts to answer such questions. The principles of thermodynamics have been successfully applied to physics, chemistry, electrochemistry, engineering, etc.

Some Terms Commonly Used in Thermodynamics

(*a*) **Thermodynamic State and Thermodynamic Co-ordinates:** A thermodynamic system is an assembly of an extremely large number of particles (atoms or molecules). The system is characterised by its properties like temperature, volume, pressure, etc. These are called thermodynamic parameters or co-ordinates. Thermodynamic state of a gaseous system is described by four parameters, viz., temperature (T) pressure (P) volume (V) and entropy (S) or heat content (Q). These are called thermodynamic co-ordinates. Since all these parameters are related to each other, as such only two out of these are sufficient to describe thermodynamic state of a thermodynamic system.

(*b*) **Thermodynamic Process:** *If the thermodynamic parameters (co-ordinates) of a thermodynamic system change with time, we say a thermodynamic process has taken place.* There are four important thermodynamic processes.

(*i*) **Isothermal Process:** It is the process in which the temperature remains constant. The pressure (P) and volume (V) are related to each other by *Boyle's law,* i.e. PV = constant.

(*ii*) **Adiabatic Process:** It is the process in which no heat is given to the system nor taken away from the system, i.e. the total heat contents of the system remain constant. We can say that entropy of the system remains constant. For this process the pressure and volume of the system are related by equation

$$PV^{\gamma} = \text{constant} \qquad \text{...(15.13}a\text{)}$$

(*iii*) **Isobaric Process:** In this process the pressure remains constant. The volume (V) and temperature (T) of the system are related by Charles's law, i.e.

$$V \propto T$$

or $$\frac{V}{T} = \text{Constant}$$

(*iv*) **Isochoric Process:** In this process the volume remains constant. The pressure (P) and temperature (T) are related by Gay-Lussac's law, i.e.

$$P \propto T \text{ or } \frac{P}{T} = \text{Constant}$$

Indicator (P–V) Diagram—*A graphical relation between pressure (P) and volume (V) of a thermodynamic system*: when a process is taking place is called indicator diagram. Generally volume (V) is plotted on X-axis and pressure (P) on Y-axis. One such diagram for a particular porcess is shown in Fig. 15.4. Here point $A(P_1 V_1)$ represents initial state and point $B(P_2 V_2)$ represents the final state. The intermediate curve represent the intermediate states of the system in going from A to B.

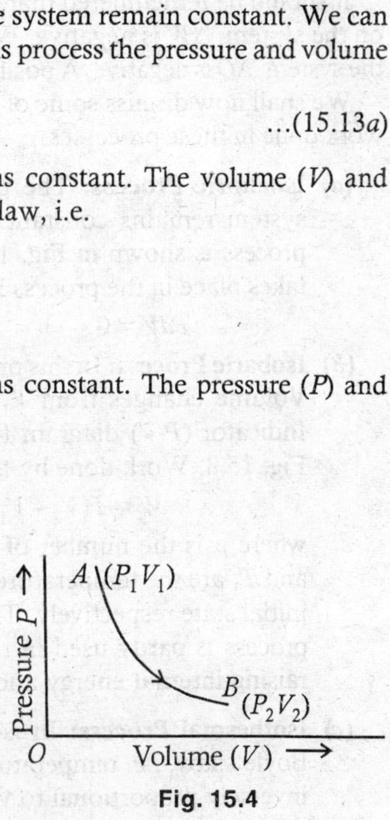

Fig. 15.4

15.7 WORK DONE IN A PARTICULAR PROCESS

Consider an ideal gas contained in a cylinder with a movable frictionless piston of cross-section A. Let the pressure, volume and temperature of the gas be P, V and T respectively. The force exerted on the piston by the gas is PA (Fig. 15.5) in outward direction. Let the gas expand a little and the piston is pushed out by a small distance Δx. The work done by the gas on the piston is

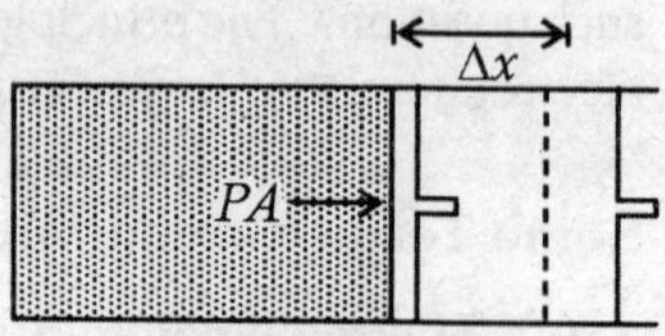

Fig. 15.5

$$\Delta W = (PA)(\Delta x) = P\Delta V$$

where $\Delta V = A\,\Delta x$ is the change in the volume of the gas. For a finite change of volume from V_1 to V_2 the pressure may not be constant. We can divide the whole process of expansion in small steps and add the work done in each step (Fig 15.6). Thus the total work done by the gas in the process is

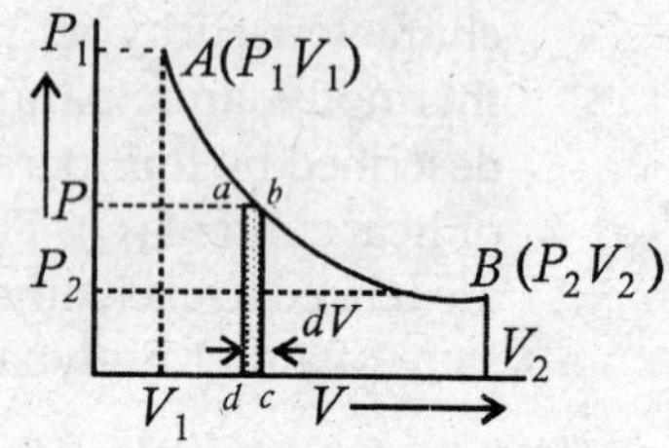

Fig. 15.6

$$W = \int_{V_1}^{V_2} PdV \qquad \text{...(15.13}b\text{)}$$

When the process is shown on P-V diagram, the work done is equal to the area bounded by the P-V curve, the V-axis and the ordinates

$$V = V_1 \text{ and } V = V_2$$

Though the equation (15.13b) is derived for a cylindrical shaped vessel, it is equally true for any shape of the vessel. It is also true for the expansion of solids and liquids and even in phase changes.

It should be remembered that when work is done by the system ΔW is positive. If work is done on the system, ΔW is negative. When heat is given to the system, ΔQ is positve. If heat is given by the system, ΔQ is negative. A positive ΔW decreases the internal energy and a positive ΔQ increases it.

We shall now discuss some of the processes mentioned above in more detail and shall calculate work done in these processes.

(*a*) **Isochoric Process:** The process in which volume of the system remains constant. The indicator diagram for the process is shown in Fig. 15.7. Since no change in volume takes place in the process hence work done is zero

$$PdV = 0 \qquad \because\ dV = 0$$

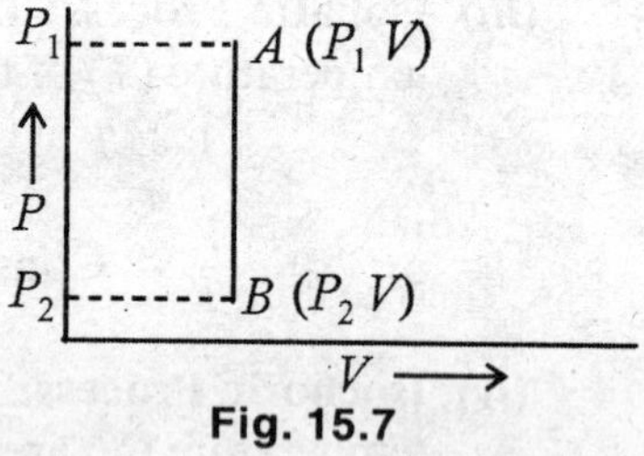

Fig. 15.7

(*b*) **Isobaric Process:** In this process pressure remains constant. Volume changes from V_1 to V_2 as temperature changes. Indicator (P-V) diagram for the process is shown in Fig. 15.8. Work done by the gas is

$$W = P(V_2 - V_1) = \mu R(T_2 - T_1) \qquad \text{...(15.14)}$$

where μ is the number of moles of the gas taken, T_2 and T_1 are the temperatures of the system in final and initial state respectively. The heat energy given in this process is partly used in raising the temperature, i.e. raising internal energy and partly to do the work.

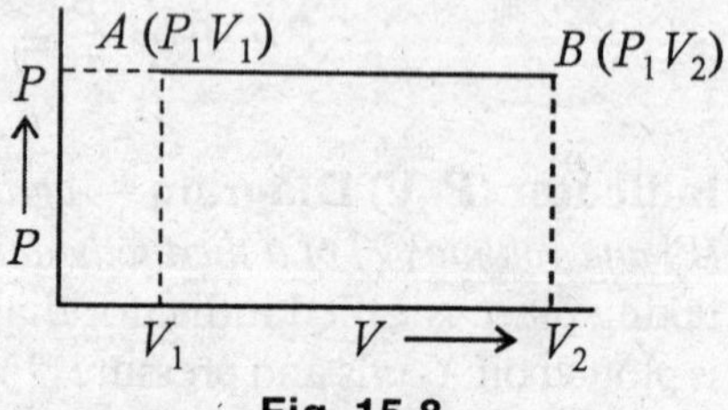

Fig. 15.8

(*c*) **Isothermal Process:** In isothermal process temperature is kept constant. The gas obeys Boyle's law, i.e. temperature remaining constant, the pressure of a given mass of the gas is inversely proportional to volume.

• $PV = $ constant

The indicator diagram is shown in Fig. 15.9. The slope of the curve at any point (P, V) is $\frac{dP}{dV}$. Differentiation gives

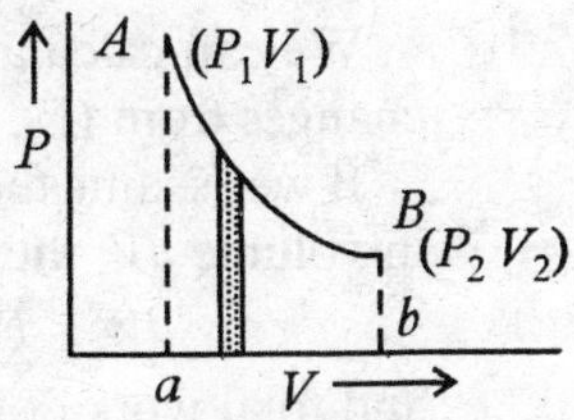

Fig. 15.9

$$PdV + VdP = 0$$

or $$\frac{dP}{dV} = -\frac{P}{V} \quad \text{...(15.15)}$$

Figure 15.9 shows an ideal gas going isothermally from state $A(P_1V_1)$ to state $B(P_2V_2)$ at temperature T. If dV is an infinitesimally small change in volume during which pressure P remains constant, the work done by the gas is

$$dW = PdV$$

and the total work done in change of state from (P_1, V_1) to (P_2, V_2) is

$$W = \int_{V_1}^{V_2} PdV \quad \text{...(i)}$$

Using ideal gas equation (eq. 15.2), we have

$$P = \frac{\mu RT}{V}$$

Substituting in (*i*), we have

$$W = \int_{V_1}^{V_2} \frac{\mu RT}{V} dV = \mu RT \log_e \frac{V_2}{V_1} \quad \text{...(15.16)}$$

This work done is equal to the area $ABbaA$ (Fig. 15.9). Changing log from base e to base 10, work done is

$$W = \mu RT(2.303)\log_{10}\frac{V_2}{V_1}$$

$$= 2.303\,\mu RT \log_{10}\frac{V_2}{V_1} = 2.303\,\mu RT \log_{10}\frac{P_1}{P_2} \quad \text{...(15.16a)}$$

(*d*) Adiabatic Process: In an adiabatic process, the system is insulated from surroundings and heat absorbed or evolved by the system is zero.

As already mentioned, the adiabatic relation between P and V for an ideal gas is (equations 15.13 and 15.13*a*)

$$PV^{\gamma} = \text{constant } (C) \quad \text{...(i)}$$

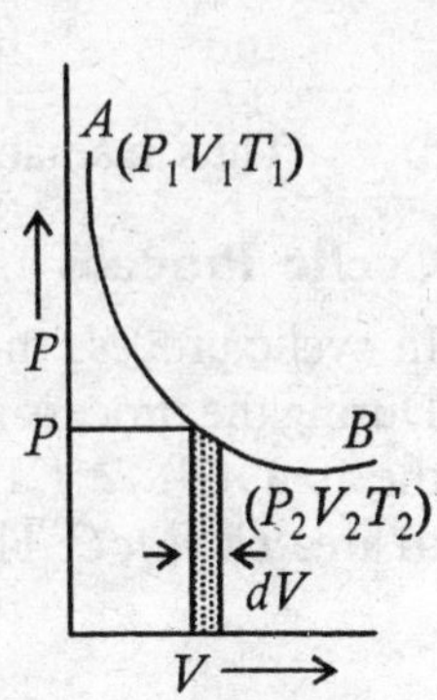

Fig. 15.10

Figure 15.10 shows the *P-V* curve (indicator diagram) for an adiabatic process. The slope of the curve at any state (P, V) is given by $\frac{dP}{dV}$. Differentiating equation (*i*) we have

$$\gamma \cdot V^{\gamma-1}dV \cdot P + V^{\gamma}dP = 0$$

or $$\frac{dP}{dV} = -\gamma\frac{P}{V} \quad \text{...(15.17)}$$

We see that the slope of *P-V* curve for adiabatic process is γ times the slope of *P-V* curve for isothermal process.

We can calculate the work done in an adiabatic process for an ideal gas when its state changes from (P_1, V_1, T_1) to (P_2, V_2, T_2).

If we assume that the pressure P remains constant during an infinitesimally small change in volume dV, then small amount of work done in this change is

$$dW = PdV$$

and total work done in adiabatic process when volume changes from V_1 to V_2 is

$$W = \int_{V_1}^{V_2} PdV \qquad ...(ii)$$

Substituting value of P from (i), we have, $P = \frac{C}{V^\gamma}$, and

$$W = \int_{V_1}^{V_2} \frac{C}{V^\gamma} dV$$

Integrating we have

$$W = \left.\frac{CV^{1-\gamma}}{1-\gamma}\right|_{V_1}^{V_2} = \frac{CV_2^{1-\gamma} - CV_1^{1-\gamma}}{1-\gamma} \qquad ...(iii)$$

From equation (i) we have

$$P_1V_1^\gamma = P_2V_2^\gamma = C$$

Substituting value of C in (iii), we get

$$W = \frac{P_2V_2^\gamma . V_2^{1-\gamma} - P_1V_1V_1^{1-\gamma}}{1-\gamma} = \frac{P_2V_2 - P_1V_1}{1-\gamma} \qquad ...(15.18)$$

If T_1 and T_2 are initial and final temperatures and μ is number of moles of gas then

$$P_1V_1 = \mu RT_1 \text{ and } P_2V_2 = \mu RT_2$$

and $$W = \frac{\mu RT_2 - \mu RT_1}{1-\gamma} = \frac{\mu R(T_2 - T_1)}{1-\gamma} \qquad ...(15.19)$$

Thus in adiabatic expansion, temperature falls and in adiabatic compression temperature rises.

Cyclic Process

In cyclic process the system returns to its initial state. One such process is shown in Fig. 15.11. During the process ABC work is done by the system and thus is positive. Work done W_1 is equal to the area $ABCcaA$. In the process CDA work W_2 is done on the system and is negative and is equal to area $CDAacC$. Thus total work done in cyclic process $ABCDA$ is

$$W = W_1 - W_2 = \text{area } ABCcaA - \text{area } CDAacC$$

$$= \text{area } ABCDA$$

If loop is traversed clockwise net work done is positive and if it is traversed anticlockwise, net work is negative.

Thus the work done in any process depends on the path traversed.

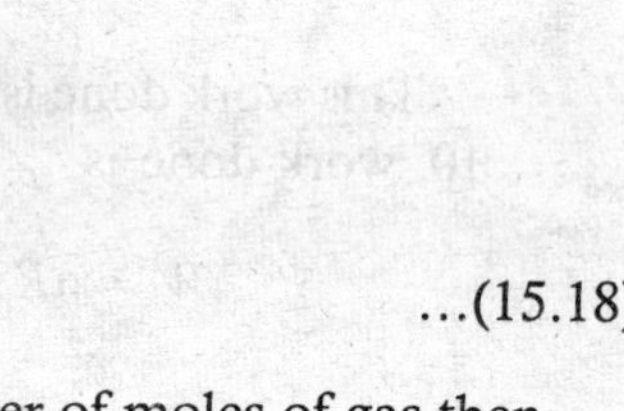

Fig. 15.11

15.8 FIRST LAW OF THERMODYNAMICS

We have seen that heat is just a form of energy and can be converted into other forms. The first law of thermodynamics states that "*heat is a form of energy.*" Since heat is a form of energy as such it must obey law of conservation of energy.

Consider an ideal gas in a cylinder with a frictionless piston attached with it. The walls of the piston are insulated and bottom is conducting. Let in a process ΔQ be the amount of heat supplied to gas. The piston will move due to expansion of gas and system will do work against external pressure. Let work done be ΔW. The total energy of the gas then must increase by $\Delta Q - \Delta W$. In thermodynamics, the kinetic energy of the system as a whole is irrelevant. Thus the random motion of the molecules increases. The sum of molecular kinetic and potential energies in the frame of reference relative to which the centre of mass of the system is at rest is called *internal energy* of the system. Thus it includes only the energy associated with the random motion of the molecules of the system. If we denote the change in internal energy by ΔU, we get

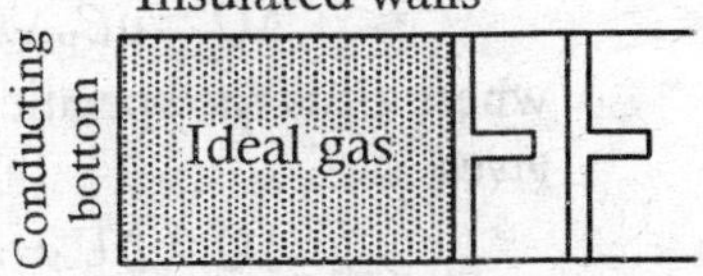

Fig. 15.12

$$\Delta U = \Delta Q - \Delta W$$

or
$$\Delta Q = \Delta U + \Delta W \qquad ...(15.20)$$

Equation (15.20) is the statement of the *first law of thermodynamics.* This equation in fact is the statement of conservation of energy and is applicable to any system, however complicated.

In an ideal monoatomic gas, the internal energy of the gas is simply the translational kinetic energy of all the gas molecules. In case of diatomic and polyatomic gases the vibrational and rotational kinetic energies of molecules may contribute to the internal energy of the gas. Potential energy corresponding to the molecular forces may as well contribute to the internal energy of the system.

It should be remembered that internal energy merely depends on temperature, i.e. an internal energy function U has a fixed value in a given state. Thus if a system is taken from initial state i to a final state f then $\Delta U = U_f - U_i$ depends only on the end states i and f and not on the process taking the system from state i to state f. Thus both ΔQ and ΔW may be different in different processes, but $\Delta Q - \Delta W$ is same for all processes taking the system from state i to state f.

15.9 ANALYSIS OF DIFFERENT PROCESSES IN THE LIGHT OF FIRST LAW OF THERMODYNAMICS

(*a*) **Isochoric Process:** In this process volume remains constant as such work done $\Delta W = 0$. From first law of thermodynamics

$$\Delta Q = \Delta U + \Delta W \qquad \Delta W = 0$$

or
$$\Delta Q = \Delta U$$

i.e. whatever the heat is given to the system is used in increasing the internal energy of the system.

If μ moles of gas are heated at constant volume and rise in temperature is ΔT, then heat given to the system is

$$\Delta Q = \mu C_V \Delta T$$

where C_V is the specific heat of the gas at constant volume. The increase in internal energy then is

$$\Delta U = \Delta Q = \mu C_V \Delta T \qquad ...(15.21)$$

(*b*) **Isobaric Process:** In this process pressure remains constant and work done when volume changes from V_1 to V_2 is

$$\Delta W = P\Delta V = P(V_2 - V_1) \quad \text{...(15.22)}$$

In the process if temperature changes from T to $T + \Delta T$, then heat given to the system is

$$\Delta Q = \mu C_P \Delta T \quad \text{...(15.23)}$$

where C_P is the specific heat of the gas at constant pressure. First law of thermodynamics gives

$$\Delta Q = \Delta U + \Delta W$$

Substituting the values from equation (15.21), (15.22) and (15.23), we have

$$\mu C_P \Delta T = \mu C_V \Delta T + P(V_2 - V_1) \quad \text{...}(i)$$

Using gas equation $PV = \mu R$ from ideal gas, we have

$$PV_1 = \mu RT$$

and $$PV_2 = \mu R(T + \Delta T)$$

or $$P(V_2 - V_1) = \mu R \Delta T$$

Substitution in (*i*) gives

$$\mu C_P \Delta T = \mu C_V \Delta T + \mu R \Delta T$$

or $$C_P - C_V = R \quad \text{...(15.24)}$$

Equation (15.24) is called *Mayor's relation.* Since R called the molar gas constant (R = 8.31 J mole^{-1}K^{-1}) is positive and it implies that $C_P > C_V$. The reason for this is that when heat is supplied at constant pressure, it is used partly in increasing the internal energy of the system and partly in doing work against the external pressure. Whereas at constant volume heat supplied merely increases the internal energy of the system. Thus $C_P > C_V$.

(*c*) **Isothermal Process**: In an isothermal process temperature remains constant as such change in internal energy is zero. First law of thermodynamics gives

$$\Delta Q = \Delta U + \Delta W \qquad \text{But } \Delta U = 0$$

i.e. $$\Delta Q = \Delta W$$

This means that heat supplied in this process is used to do work against the external pressure.

If C is the average specific heat for this process, then

$$\Delta Q = \mu C \Delta T$$

or $$C = \frac{\Delta Q}{\mu \, \Delta T}$$

but $$\Delta T = 0$$

$\therefore$ $$C = \infty$$

Thus average specific heat for this process is infinite (∞).

(*d*) **Adiabatic Process:** In an adiabatic process neither heat energy is given to the system nor is taken away from the system, i.e. $\Delta Q = 0$. The first law of thermodynamics implies that

$$\Delta Q = \Delta U + \Delta W = 0$$

or $$\Delta U = -\Delta W$$

i.e. work is done at the cost of internal energy. When a gas is compressed adiabatically, work done is negative, therefore ΔU must be positive, i.e. the internal energy of the system must increase and the gas is heated. On the other hand if the gas does work, i.e. gas expands adiabatically, ΔW is positive and the internal energy of the system will decrease and the gas is cooled.

If C is the average specific heat of the gas, then

$$\Delta Q = \mu C \Delta T = 0$$

or
$$C = \frac{0}{\mu \Delta T} = 0$$

i.e. average specific heat of the system is zero.

Thus a gas has an infinite number of specific heats and their values vary from 0 to ∞ depending on the process through which the system undergoes. Out of all these specific heats, only two (*i*) specific heat at constant volume (*ii*) specific heat at constant pressure are the principal specific heats of a gas.

15.10 LIMITATIONS OF THE FIRST LAW OF THERMODYNAMICS

The first law of thermodynamics in essentially *a law of equivalence* and only tells us about the interconvertibility of heat and mechanical work. It says nothing as to the conditions under which the conversion may occur or as to the range or limit within which it is possible. It is unable to answer the questions like:

(*i*) Why is it that heat always flows from a body at a higher temperature to the one at a lower temperature and not from the latter to the former?

(*ii*) Why is it that we are unable to convert the vast amount of heat of the earth into work, the air and water in the sea and ocean, or in general, the heat of a body or a reservoir, taken all by itself?

Of course the first law of thermodynamics is not violated in any of these two possibilities. *It merely denies creation of energy out of nothing i.e. the perpetual motion of first kind is not possible.* It also is not able to explain why *perpetual motion of second kind, i.e. full and continuous conversion into work, of heat of a body or a reservoir, taken by itself alone is not possible.*

Thus we see that the first law of thermodynamics shows the conservation of energy in heat-work conversion ($Q \rightleftharpoons W$) and flow of heat from one body to other but is unable to give the direction of flow of heat. This means that there must be law of nature other than the first law which decides, whether a given process, allowed by the first law, will actually take place or not. This law is *second law of thermodynamics.*

15.11 SECOND LAW OF THERMODYNAMICS

The idea of heat engine working in cycles is due to Sadi Carnot, a French engineer, whose outstanding work on heat engines has made the greatest single contribution towards the formulation of second law of thermodynamics.

For continuous conversion of heat into work, we need a heat engine; a thermodynamic system which can work in cycles and can renew its properties periodically. A heat engine extracts Q_1 amount of heat energy from a hot body (reservoir), converts a part of it into mechanical work and rejects the rest amount Q_2 into the cold body. The efficiency of the engine is

$$\eta = \frac{\text{Useful work done}}{\text{Total energy supplied}} = \frac{Q_1 - Q_2}{Q_1} = 1 - \frac{Q_2}{Q_1}$$

For efficiency to be 1 (i.e. 100%) Q_2 should be zero. Such an engine will not require a body at "low temperature" to which it needs to reject heat. As such it requires only one body at a single temperature from which it will extract heat and convert it completely into mechanical work. This body at single temperature can even be the surrounding and hence we will not have to burn any fuel to prepare steam or gases at high temperature to run the engine. A ship could then be run by an engine taking heat from the ocean. However all attempts to construct such an engine (efficiency 100%) failed. It is not due to the fault in fabrication of the engine but a natural phenomenon. It is due to the fact that mechanical work can be completely converted into heat but heat cannot be completely converted into mechanical work. In this respect heat and work are not equivalent but *only one sided convertibility*.

Thus we see that a heat engine with 100% efficiency is not possible and this is one form of second law of thermodynamics.

Statement of Second Law of Thermodynamics

Different workers have stated the law in different words to bring out in their own way its salient features. Few of these statements are as follows:

1. *It is not possible to design a heat engine which works in cyclic process and whose only result is to take heat from a body at a single temperature and covert it completely into mechanical work.*
   This is *Kelvin-Planck statement* of second law.
2. Oswald put it in the form: *Perpetual motion of the second kind is impossible.*
3. *It is impossible for a self acting machine unaided by any external agency, to convey heat from one body to another at a higher temperature or heat can not itself pass from a cooler to a warmer body.* This statement is by *Clausius*.

All these statements of the law are equivalent to each other.

A point worth noting is that the second law is based not on any theoretical or mathematical reasoning but on pure observation and facts of experience. Such is however, the reliance placed on it that it is considered to be simply infallible. And, indeed, there has been not one single instance so far where it has been violated or the predictions made on its basis have proved wrong.

15.12 REVERSIBLE PROCESS AND IRREVERSIBLE PROCESS

A thermodynamic process is said to be reversible if (*i*) The various stages of a process to which a system is subjected can all be reversed back in opposite direction and in reverse order such that the direct and reverse processes can be represented by the same curve with arrow inverted. Figure 15.13 gives *if* as direct process and *fi* as reverse process (*ii*) In every part of the process, the amount of energy transferred in the form of heat or work is the same in magnitude, in either direction and (*iii*) At any stage of the process an infinitesimal change in any one of its thermodynamic co-ordinates brings about its reversal.

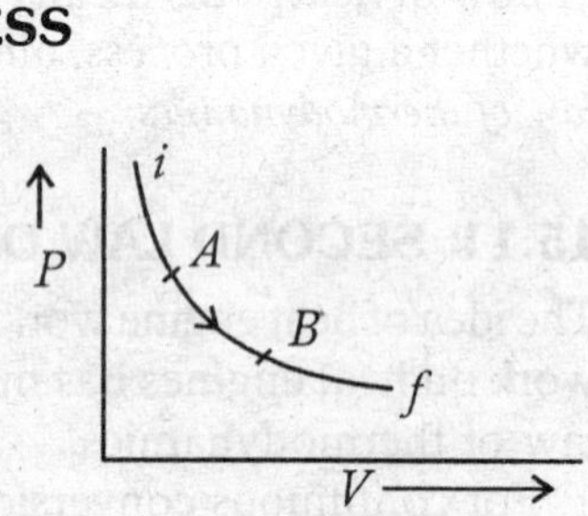

Fig. 15. 13

This implies that the process must be quasi-static consisting of a series of successive equilibrium states and hence infinitely slow. The slightest departure from any one of these conditions will result in the process becoming irreversible.

Few examples of reversible processes are mentioned below:

(*i*) A slow isothermal compression or expansion of a gas is a reversible process.
(*ii*) An infinitely slow adiabatic expansion or compression of a gas is also a reversible process.
(*iii*) An extremely slow extension or contraction of a spring is also reversible process.

However if any of the above processes be carried out rapidly, they no longer remain reversible. For them in case (*i*) and (*ii*) the temperature adjustments are not possible and part of energy is dissipated as heat and in case of (*iii*) oscillations may set in consuming source energy.

On the other hand processes involving production of heat by friction or by the passage of electric current through a conductor are irreversible. Because in these processes even if the direction of motion or the direction of electric current be reversed, the heat will be produced. Thus an irreversible process is the one in which the reversal of the process does not bring about a reversal of the thermal effects associated with it. For the same reason, Joule-Thomson effect, diffusion of gases and liquids and a number of other natural processes are in general irreversible. It is so because reversibility of a process requires rigid conditions so that it becomes more of an ideal which can, at best, be closely approached but never actually attained. But for practical purposes many processes, carried out infinitely slowly under controlled conditions, may well be taken to be reversible process.

A cyclic process in which every part of cycle can be traversed in the opposite direction and in reversed order of the associated thermal effects is called *reversible cyclic process.*

15.13 HEAT ENGINES

A heat engine *is a thermodynamic device which continuously converts heat supplied to it into mechanical work.* For this we want a process in which whatever heat is given to the system is totally converted into mechanical work and there is no increase in its internal energy. Such situation we find in isothermal expansion of an ideal gas. When an ideal gas is given heat slowly it expands and does work and if process is isothermal there in no increase in internal energy of the gas. But the pressure of the gas decreases as it expands. This gas expands till its pressure becomes equal to the atmospheric pressure. To repeat the process of work done, the system should be brought back to its original state by some other processes. In such processes some heat is absorbed, some heat is rejected and some is converted into work. Thus a heat engine has three essential parts as mentioned below.

(*a*) **Heat Source or Heat Reservoir:** A body of infinite thermal capacity, maintained at constant high temperature, so that any amount of heat drawn from it or given to it does not affect its temperature is called heat source or heat reservoir.

(*b*) **Sink:** It is a cold body also of infinite thermal capacity, maintained at a constant low temperature, so that any amount of heat given to it (or drawn from it) does not change its temperature.

(*c*) **Working Substance:** A substance which when given heat does mechanical work is called working substance.

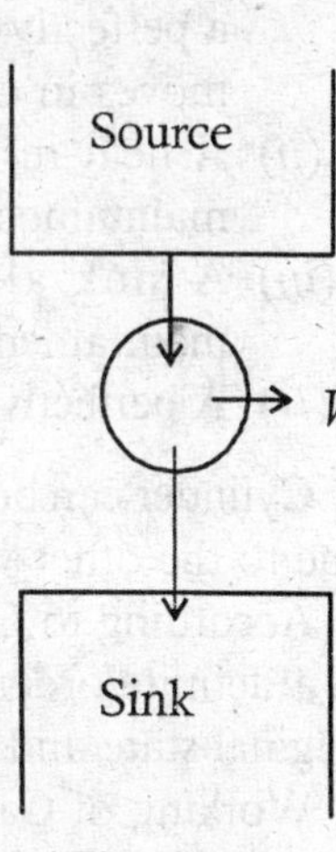

Fig. 15.14

Working substance absorbs heat Q_1 from source at higher temperature, converts a part of it into mechanical work W and rejects rest of heat Q_2 to the sink at low temperature and comes back to its initial state. The series of these processes is called a *cycle.* A system when comes back to its initial state after passing through different states is called a *cyclic process.* Continuous mechanical work is obtained by making repeated changes in working substance and the working substance periodically comes to its initial state.

Efficiency of Heat Engine

The efficiency of engine is defined as the ratio of the useful work (W) done to the total energy (Q) absorbed in doing that useful work. Thus

$$\text{Efficiency } \eta = \frac{W}{Q}$$

Since at the end of each cycle, the working substance regains its original state, its internal energy is not affected. And under ideal conditions, there in no dissipation of energy due to friction or conduction, etc., we therefore have

$$W = Q_1 - Q_2$$

$$\eta = \frac{W}{Q_1} = \frac{Q_1 - Q_2}{Q_1} = 1 - \frac{Q_2}{Q_1} \quad \text{...(15.25)}$$

Efficiency is usually expressed in percentage.

15.14 CARNOT ENGINE

Carnot engine is simple and ideal engine operating between a given source and a sink. It is ideal because

(*a*) it is free from friction and imperfections of actual engines.

(*b*) it absorbs all its heat at a constant high temperature and rejects whatever heat it has at a constant low temperature to the sink.

(*c*) it works in a cycle of operations each of which is perfectly reversible.

In fact Carnot ideal engine is an imaginary machine or device which converts heat into mechanical work and has an intimate relation with the second law of thermodynamics.

The essential parts of an ideal Carnot engine are shown in Fig. 15.15.

(*i*) It consists of a cylinder with perfectly non-conducting walls and perfectly conducting base. It is filled with an ideal gas and acts like a working substance. The cylinder is fitted with a perfectly non-conducting piston. The piston moves in cylinder without friction.

(*ii*) A heat reservoir of infinite thermal capacity maintained at a high constant temperature T_1.

(*iii*) A sink also of infinite thermal capacity maintained at constant low temperature T_2.

(*iv*) A perfectly non-conducting stand.

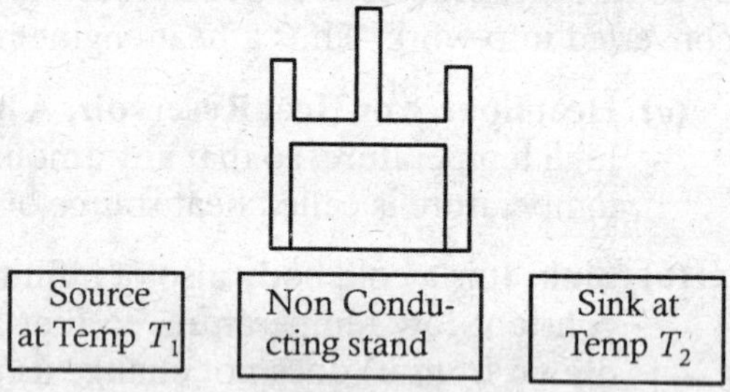

Fig. 15.15

Cylinder can be kept on any of the three—source, sink or stand and cylinder can be moved from one to the other without friction and without doing work.

According to Carnot in order to obtain continuous work from this ideal engine, it is to be run in a definite order. Engine is run in a cyclic process so that after every cycle it comes back to its original state and in this way it becomes useable for doing continuous work.

Working of Carnot engine can be explained with the help of one cycle. In this one cycle the working substance undergoes four operations, two of them are isothermal processes and the other two are adiabatic processes. This cycle is called Carnot cycle and is explained on P-V diagram in Fig.15.16.

Although any material or thermodynamic system including radiation may serve as the working substance, we shall, for the sake of convenience imagine it to be one gram mole of an ideal gas.

Four operations of a Carnot cycle are as under.

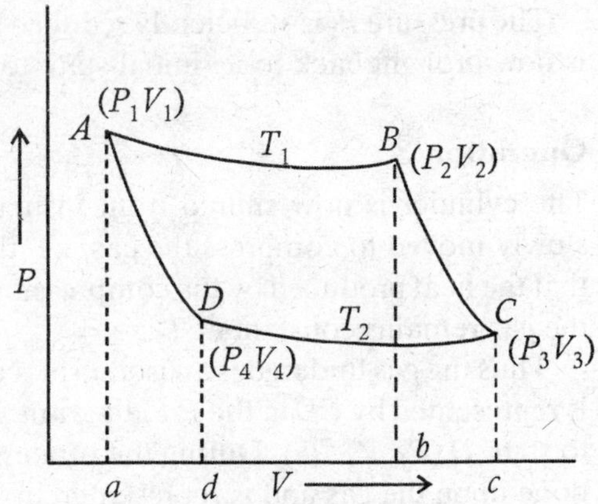

Fig. 15.16

Operation 1

The cylinder is first placed on the source, so that the gas acquires the temperature T_1 of the source. It is then allowed to undergo quasistatic expansion. As the gas expands, it does work against the external pressure and its temperature tends to fall. This fall in temperature is made up by absorbing heat from reservoir through the conducting base. Thus the gas in piston undergoes slow isothermal expansion at constant temperature T_1 by absorbing the necessary heat for maintaining constant temperature from the source. This operation is represented by isothermal *AB* on the indicator diagram (Fig. 15.16). In this process the state of the gas changes from $(P_1V_1T_1)$ at *A* to $(P_2V_2T_1)$ at *B*. In this process let Q_1 amount of heat be absorbed and work done is W_1. Since in isothermal process no change in internal energy takes place thus $Q_1 = W_1$. But in an isothermal process work done by gas, is represented by area *ABbaA*, so that

$$Q_1 = W_1 = \int_{V_1}^{V_2} PdV = \int_{V_1}^{V_2} RT_1 \frac{dV}{V} = RT_1 \log_e \frac{V_2}{V_1} = \text{area } ABbaA \quad \ldots(i)$$

Operation 2

The cylinder is now removed from the source to the non-conducting stand, so that the gas is thermally isolated from the surroundings. It is now allowed to undergo a slow adiabatic expansion performing the work at the expense of its internal energy. The gas is allowed to expand till its temperature falls to T_2, where T_2 is the temperature of the sink.

The operation is represented on *P-V* diagram by the adiabatic *BC*. In the process the state of the gas changes from (P_2, V_2, T_1) at *B* to (P_3, V_3, T_2) at *C*. In this process the work done by the gas W_2 is represented by area *BCcbB* on the indicator diagram. Thus

$$W_2 = \int_{V_2}^{V_3} PdV = \int_{V_2}^{V_3} \frac{K\,dV}{V^\gamma} \qquad \because \text{ for adiabatic change } PV^\gamma = K$$

$$= \left.\frac{KV^{1-\gamma}}{1-\gamma}\right|_{V_2}^{V_3} = \frac{KV_3^{1-\gamma} - KV_2^{1-\gamma}}{1-\gamma} = \frac{P_3V_3 - P_2V_2}{1-\gamma}$$

or
$$W_2 = \frac{R(T_2 - T_1)}{1-\gamma} = \text{area } BCcbB \quad \ldots(ii)$$

where $\gamma = \frac{C_P}{C_V}$ ratio of two specific heats of the working substance. We have used ideal gas equation $PV = RT$.

The pressure P_2 is sufficiently reduced and the gas has no capacity to do work any more. The gas is now brought back to its initial state to do work again.

Operation 3

The cylinder is now shifted from insulating stand to the sink at temperature T_2. The piston is slowly moved to compress the gas, i.e. the work is done on the gas. The process is isothermal so that the heat produced by the compression immediately passes into the sink and the temperature of the gas remains constant at T_2.

Thus the gas undergoes an isothermal compression at a constant temperature T_2. This operation is represented by CD in the P-V diagram. In this process the state of gas changes from $C(P_3, V_3, T_2)$ to state $D(P_4, V_4, T_2)$. During the process it rejects Q_2 amount of heat to the sink and W_3 work is done upon the gas and is represented by the area $CcdDE$. Thus we have

$$Q_2 = W_3 = \int_{V_3}^{V_4} P dV = \int_{V_2}^{V_4} RT_2 \frac{dV}{V} = RT_2 \log_e \frac{V_4}{V_3}$$

$$= -RT_2 \log_e \frac{V_3}{V_4} = -\text{area } CcdDE \quad \ldots(iii)$$

In this case work is done on the gas and hence work is negative.

Operation 4

The cylinder is now shifted to the non-conducting stand, thus isolating the gas thermally. The gas is further compressed but this time adiabatically. The work thus continues to be done on the gas and temperature rises. The gas is compressed till its temperature changes to T_1. In this process the gas comes back to its original state (P_1, V_1, T_1) represented by point A, thus completing one full cycle.

This process is represented by adiabatic DA and work done on the gas W_4 is represented by the area $DdaAD$. We thus have

$$W_4 = \int_{V_4}^{V_1} P dV = -\int_{V_1}^{V_4} \frac{K\,dV}{V^\gamma} = \frac{R(T_1 - T_2)}{1-\gamma} = -\text{area } DdaAD \quad \ldots(iv)$$

Thus we see that the Carnot's cycle consists of two isothermals and two adiabatic strokes one each of expansion and compression. The isothermal and adiabatic strokes alternating with each other. The cycle is repeated to obtain continuous work.

Total work done by the gas in one cycle is

$$W = W_1 + W_2 + W_3 + W_4$$

$$= RT_1 \log\left(\frac{V_2}{V_1}\right) + \frac{R(T_2 - T_1)}{1-\gamma} - RT_2 \log\left(\frac{V_3}{V_4}\right) + \frac{R(T_1 - T_2)}{1-\gamma}$$

$$= R(T_1 - T_2)\log\left(\frac{V_2}{V_1}\right) \qquad \because \frac{V_2}{V_1} = \frac{V_3}{V_4}^*$$

*B and C are on same adiabatic and $T_1V_2^{\gamma-1} = T_2V_3^{\gamma-1}$. Similarly A and D are on same adiabatic, hence $T_1V_1^{\gamma-1} = T_2V_4^{\gamma-1}$. Dividing we have $\frac{V_2}{V_1} = \frac{V_3}{V_4}$.

Hence total work done in the cycle is

$$W = Q_1 - Q_2 = R(T_1 - T_2)\log\left(\frac{V_2}{V_1}\right) = \text{area } ABCDA \qquad \ldots(15.26)$$

Efficiency of Carnot Engine

Efficiency of an engine is defined as the ratio of the heat converted into work to the heat absorbed by the engine from the source.

$$\eta = \frac{\text{Heat converted into work}}{\text{Heat absorbed from the source}} = \frac{Q_1 - Q_2}{Q_1} = 1 - \frac{Q_2}{Q_1}$$

$$\equiv \frac{R(T_1 - T_2)\log\frac{V_2}{V_1}}{R\,T_1\log\frac{V_2}{V_1}} = \frac{T_1 - T_2}{T_1} = 1 - \frac{T_2}{T_1} \qquad \ldots(15.27)$$

Equation (15.27) gives that

(*i*) Efficiency of engine depends only on the temperatures of the source and sink and does not depend on the nature of the working substance.

(*ii*) The efficiency increases with temperature difference ($T_1 - T_2$), i.e. more the temperature difference, the more is the efficiency.

(*iii*) Since source temperature T_1 is greater than sink temperature $T_2(T_1 > T_2)$ hence $\eta < 1$, i.e. efficiency of Carnot engine is always less than 1 (or 100%). In other words the efficiency of an ideal gas cannot be 100%.

(*iv*) If sink temperature is made zero ($T_2 = 0$ K), then the efficiency of an ideal engine will be maximum 1 or 100%. In this case all the heat absorbed from the source is converted into work. It is impossible to keep the sink at absolute zero temperature as such no engine can have 100% efficiency.

Note: A useful relation is obtained from equations (*i*) and (*iii*)

$$\frac{W_1}{W_2} = \frac{Q_1}{Q_2} = \frac{RT_1\log(V_2/V_1)}{RT_2\log(V_3/V_4)} \qquad \text{But } \frac{V_2}{V_1} = \frac{V_3}{V_4}$$

or

$$\frac{W_1}{W_3} = \frac{Q_1}{Q_2} = \frac{T_1}{T_2} \qquad \ldots(15.28)$$

i.e. the ratio between the two effective components W_1 and W_3 of the work done is the same as that between the temperature at which heat is absorbed and rejected respectively.

Efficiency in Terms of Adiabatic Expansion Ratio

We have seen that in adiabatic expansions

$$\frac{V_3}{V_2} = \frac{V_4}{V_1} = \rho$$

This ratio of the volumes is called adiabatic expansion ratio and is represented by ρ.

In case of adiabatic expansion, the relation between volume and temperature is

$$TV^{\gamma-1} = \text{constant}$$

Since B and C are on the same adiabatic, then

$$T_1 V_2^{\gamma-1} = T_2 V_3^{\gamma-1} \qquad \text{i.e.} \quad \frac{V_3}{V_2} = \left(\frac{T_1}{T_2}\right)^{\frac{1}{\gamma-1}}$$

Similarly D and A are on same adiabatic, hence

$$T_1 V_1^{\gamma-1} = T_2 V_4^{\gamma-1}$$

or
$$\frac{V_4}{V_1} = \left(\frac{T_1}{T_2}\right)^{1/\gamma-1}$$

$\therefore$
$$\frac{V_3}{V_2} = \frac{V_4}{V_1} = (T_1/T_2)^{1/\gamma-1} = \rho$$

or
$$\frac{T_1}{T_2} = (\rho)^{\gamma-1} \text{ and } \frac{T_2}{T_1} = \left(\frac{1}{\rho}\right)^{\gamma-1}$$

Thus efficiency of Carnot's engine is

$$\eta = 1 - \frac{T_2}{T_1} = 1 - \left(\frac{1}{\rho}\right)^{\gamma-1} \qquad ...(15.29)$$

15.15 REVERSIBILITY OF CARNOT'S ENGINE; REFRIGERATOR

The ideal Carnot's engine in real sense is a reversible engine because

(*i*) The process to which the gas or working substance is subjected in each stroke is carried out quasistatically.
(*ii*) The isothermal expansion and compression of the gas during which heat is absorbed and rejected by the gas, occur at strictly constant temperatures
(*iii*) There are no dissipative forces at work.

This is why a Carnot's engine is called reversible engine. This means that the whole cycle of operation can be made to proceed in the opposite direction or the engine can be made to run backward.

Carnot's engine, on its normal running absorbs Q_1 amount of heat from the source at temperature T_1, rejects Q_2 amount of heat to the sink at temperature T_2 and converts $(Q_1 - Q_2)$ amount of heat into work W. This engine running in reverse way would absorb Q_2 amount of heat from the sink at temperature T_2 and work W done on it, rejects Q_1 amount of heat to the source at temperature T_1 where $Q_1 = Q_2 + W$. The indicator diagram for this reverse cycle will be same as shown in Fig. 15.16 but the cycle of operations will proceed in the direction *ADCBA*. This means that gas starting from A at temperature T_1 would undergo an adiabatic expansion along *AD* untill the temperatures falls to T_2, the temperature of sink. The gas then expands isothermally along *DC* thereby absorbs heat Q_2 at temperature T_2 from the sink. The gas then undergoes adiabatic compression along *CB* till its temperature become T_1 equal to that of the source. Finally the gas undergoes an isothermal compression along *BA* rejecting heat to the source. At the end of the isothermal compression the gas regains its original state at A and has rejected Q_1 amount of heat to the source.

Thus in running in a reverse way, the engine absorbs heat from the colder reservoir and transfers it to the hotter reservoir, i.e. it works as a *refrigerator*. The refrigerating effect is Q_2.

The coefficient of performance of the refrigerator (β) is defined as the ratio of heat extracted from cold body to the work done on the system, i.e.

$$\beta = \frac{\text{Heat extracted from the cold body } (Q_2)}{\text{Work done on the system } (W)}$$

but $\qquad W = Q_1 - Q_2$

where Q_1 is the heat rejected to the reservoir

or $$\beta = \frac{Q_2}{Q_1 - Q_2} \qquad \text{...(15.30)}$$

But we know that $\frac{Q_1}{T_1} = \frac{Q_2}{T_2}$ and let $\frac{Q_1}{T_1} = \frac{Q_2}{T_2} = C$

or $\qquad Q_1 = CT_1$ and $Q_2 = CT_2$

$\therefore$ $$\beta = \frac{CT_2}{CT_1 - CT_2} = \frac{T_2}{T_1 - T_2} \qquad \text{...(15.31)}$$

But efficiency $\eta = 1 - \frac{T_2}{T_1}$ or $T_1 - T_2 = \eta T_1$ and $T_2 = (1 - \eta)T_1$

or $$\beta = \frac{(1-\eta)T_1}{\eta T_1} = \frac{1-\eta}{\eta} \qquad \text{...(15.32)}$$

Coefficient performance of a refrigerator means its efficiency. Thus higher the value of β more efficient is refrigerator. It may be noted that whereas the efficiency of a heat engine can never exceeds 1 (or 100%) the coefficient of performance of a refrigerator may exceed 1.

The equation (15.31) shows that smaller the value of $(T_1 - T_2)$, greater is the value of β, i.e smaller the temperature difference between food stuff to be cooled and atmosphere, more efficient will be the refrigerator.

The temperature T_2 goes on decreasing due to the formation of ice but T_1 is almost constant as such β decreases as refrigerator works. But when refrigerator is defrosted, T_2 increases and hence β increases. Thus for better working of refrigerator defrosting is necessary.

15.16 CARNOT'S THEOREM

Carnot's theorem consists of two parts and may be stated as follows:

(*i*) No engine can be more efficient than a reversible Carnot engine operating between the same temperatures, i.e. between the same source and sink.

(*ii*) All reversible engines operating between the same temperature limits have the same efficiency.

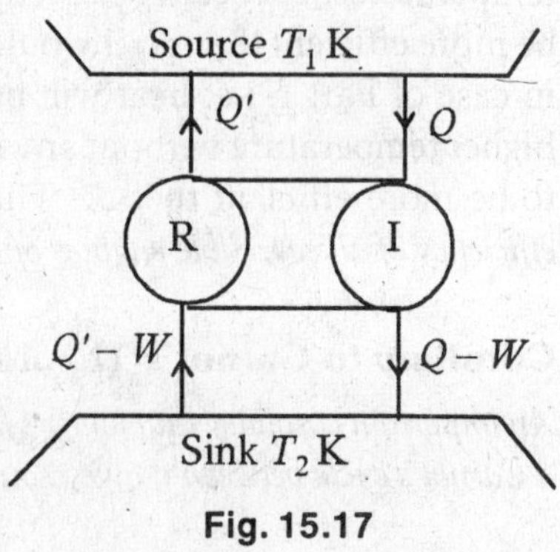

Fig. 15.17

Proof of Part I: To prove this theorem let us consider two heat engines R and I. Out of two R is reversible Carnot's engine and I some other irreversible engine (Fig. 15.17).

The amounts of working substances of these are so adjusted that the operating between the same source and sink perform same work W per cycle. Now if engine I absorbs Q units of heat per cycle from the source, the quantity of heat rejected to the sink is $(Q - W)$. And if engine R absorbs Q' units of heat per cycle the quantity of heat rejected by it to the sink is $(Q' - W)$. So that

efficiency of engine I, $\eta_I = \dfrac{W}{Q}$

and efficiency of engine R, $\eta_R = \dfrac{W}{Q'}$

Now let engine I (irreversible) be more efficient than engine R (reversible), i.e. $\eta_I > \eta_R$

or $$\frac{W}{Q} > \frac{W}{Q'} \text{ or } Q' > Q$$

This means that $(Q' - Q)$ is positive.

Let two engines be coupled together (Fig. 15.17) such that the engine I drives backwards engine R. It is obvious that engine R acts like a refrigerator driven by I. R then extracts $(Q' - W)$ units of heat from the sink and work W is being done on it by I transfers Q' units of heat to the source.

The source thus loses Q units of heat to I and gains Q' units from R and therefore gains $(Q' - Q)$ units of heat per cycle. And the sink gains $(Q - W)$ units of heat from I and loses $(Q' - W)$ units of heat to R so that it loses $(Q' - W) - (Q - W) = (Q' - Q)$ units of heat per cycle, i.e. same as gained by the source.

Thus the engines I and R combined behave like a self acting machine, requiring no external help and yet transfers $(Q' - Q)$ units of heat per cycle from the sink to the source. This transfer of heat from lower to a higher temperature without any expenditure of energy is a violation of the second law of thermodynamics. Thus we conclude that the assumption that irreversible engine is more efficient than reversible engine operating between the same source and sink is wrong. Hence we conclude that no engine operating between a given source and sink can be more efficient than a reversible engine operating between the same source and sink, i.e. *the efficiency of reversible engine operating between given source and sink is maximum.*

Proof of Part II: To prove this part of theorem, let us consider two engines R and R' both reversible and with two different working substances operating and operating between the same source and sink. Since both are reversible engines their efficiencies are maximum for the given limits of temperatures. Proceeding exactly in the same way as in Part I, we can show that if R' is supposed to be more efficient than R, it would lead to a result contrary to the second law of thermodynamics as in case of Part I, i.e. heat will be transferred from a source at lower temperature to the source at higher temperature without any expenditure of work. The same would be the result if R' is assumed to be more efficient then R. Thus no engine is more efficient than the other. *This means that the efficiency of all reversible engines operating between a given pair of source and sink must be equal.*

Corollary to Carnot's Theorem

An important corollary that follows from Carnot's theorem is that the efficiency of a reversible engine performing a Carnot's cycle between a given source and sink is quite independent of the heat absorbed from the source.

SOLVED EXAMPLES

Example 15.1: Two metal strips, each of length l and thickness d at temperature T_0 are riveted together so that their ends coincide. One strip is made of metal A having a linear coefficient of expansion α_A and other of metal B with coefficient α_B, where $\alpha_A > \alpha_B$. When this bimetal strip is heated to temperature $(T_0 + \Delta T)$ one strip becomes longer than the other and bimetallic strip bends into an arc of a circle. What is the radius of curvature of the strip?

Solution: When the temperature increases the length of rod A becomes

$$l_A = l(1 + \alpha_A \Delta T) \quad \ldots(i)$$

and that of B becomes

$$l_B = l(1 + \alpha_B \Delta T) \quad \ldots(ii)$$

or

$$l_A - l_B = l(\alpha_A - \alpha_B)\Delta T \quad \ldots(iii)$$

But $\quad l_A = r_A\theta$ and $l_B = r_B\theta$ and $r_A - r_B = d$

$$\therefore \quad l_A - l_B = r_A\theta - r_B\theta = \theta(r_A - r_B) = \theta \cdot d$$

If r is the radius of the arc, then $r\theta = l$.

Substituting the values in (*iii*), we get

$$\theta \cdot d = r\theta \cdot (\alpha_A - \alpha_B)\Delta T$$

or

$$r = \frac{d}{(\alpha_A - \alpha_B)\Delta T}$$

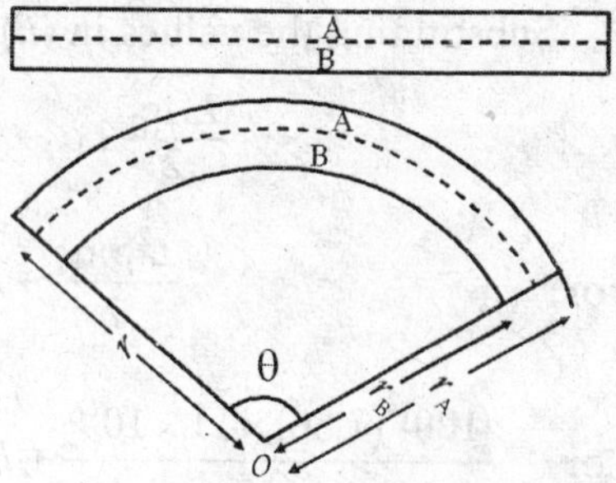

Fig. 15.18

Example 15.2: A glass tube nearly filled with mercury is attached to the bottom of an iron pendulum rod 100 cm long. How high must the mercury be in the glass tube so that the centre of mass of this pendulum will not rise or fall with changes in temperature? Given density of iron = 7.86 gm/cc, density of Hg =13.6 gm/cc γ_{Hg}=182 × 10^{-6} (°C)$^{-1}$, α_{iron} = 11 × 10^{-6} (°C)$^{-1}$, α_{glass} = 8.5 × 10^{-6}/°C.

Solution: Let L (=100 cm) and l be the lengths of iron bar and mercury column at room temperature.

The cross-sectional area of the bar and mercury column (assumed to be same) be A.

Let m_1 and m_2 be the masses of iron bar and mercury column (here we have neglected the mass of the glass tube). If ρ_1 and ρ_2 are the densities of iron and mercury, then

$$m_1 = LA\rho_1 \text{ and } m_2 = lA\rho_2$$

The centre of mass of iron bar is at its centre, i.e.

$$y_1 = L/2 \text{ from top, the support point}$$

The centre of mass of mercury is at a distance

$$y_2 = (L + l/2) \text{ from the top}$$

Thus centre of mass of the pendulum from the top is

$$y_{cm} = \frac{m_1y_1 + m_2y_2}{m_1 + m_2} = \frac{m_1 L/2 + m_2(L + l/2)}{m_1 + m_2} \quad \ldots(i)$$

When the temperature increases, the system expands in such a way that iron rod comes down and mercury rises up to keep the centre of mass fixed, i.e.

$$\frac{dy_{cm}}{dT} = 0$$

$$\therefore \quad \frac{m_1}{2}\frac{dL}{dT} + m_2\left[\frac{dL}{dT} + \frac{1}{2}\frac{dl}{dT}\right] = 0 \qquad \ldots(ii)$$

But $\dfrac{dL}{dT} = L\alpha_i$ and $\dfrac{dl}{dt} = -l\,\alpha_a$

–ve sign is due to the fact that centre of mass of iron rod comes down whereas centre of mass of mercury column goes up. Here α_i is linear coefficient of expansion of iron and α_a is the effective linear coefficient of expansion of mercury in glass given by

$$\alpha_a = \gamma_{Hg} - 2\alpha_g$$

Substituting the values in (*ii*) We have

$$\frac{LA\rho_1}{2}L\alpha_i + lA\rho_2[L\alpha_i - \frac{l}{2}\left(\gamma_{Hg} - 2\alpha_g\right) = 0$$

or
$$\frac{L^2\rho_1\alpha_i}{2} + l\rho_2\left[L\alpha_i - \frac{l}{2}\left(\gamma_{Hg} - 2\alpha_g\right)\right] = 0$$

or
$$\frac{(100)^2(7.86)\times 11\times 10^{-6}}{2} + l\times 13.6\left[100\times 11\times 10^{-6} - \frac{l}{2}(182\times 10^{-6} - 2\times 8.5\times 10^{-6}\right] = 0$$

or
$$\frac{10^4\times 7.86\times 11}{2} + 13.6\times 100\times 11\times l - \frac{l^2}{2}\times 165\times 13.6 = 0$$

or
$$0.1122\,l^2 - 1.4960\,l - 43.23 = 0$$

or
$$l = 27.4 \text{ cm.}$$

Example 15.3: Two rods each of length L_2 and coefficient of linear expansion α_2 each are connected to a third rod of length L_1 and coefficient of expansion α_1 to form an isosceles triangle. The arrangement is supported on a knife edge at the mid point of L_1 which is horizontal. What relation must exist between L_1 and L_2 so that the apex of the isosceles triangle is at constant height from the knife edge as the temperature changes?

Solution: The rods CA and CB are of equal length, i.e.

$$CA = CB = L_2$$

and
$$AB = L_1$$

The rod is supported at D so that apex C is exactly above D and let $CD = h$

For all temperature changes, CD should remain constant.

When temperature changes by ΔT, AB extends to $A'B' = L_1 + \Delta L_1$ and $CA' = CB' = L_2 + \Delta L_2$

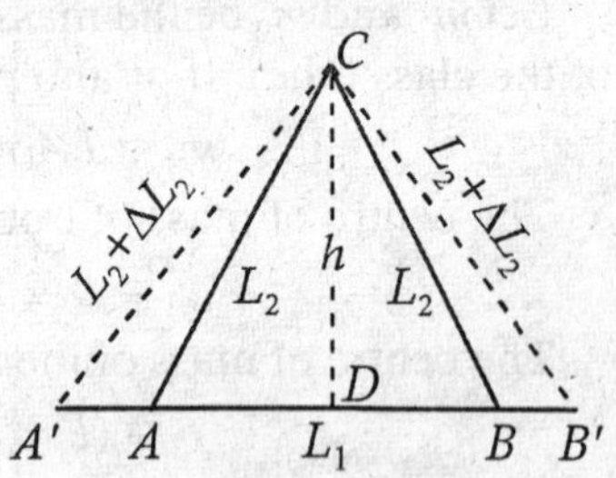

Fig. 15.19

In $\Delta\, ACD$

$$L_2^2 = h^2 + \left(\frac{L_1}{2}\right)^2$$

or
$$h^2 = L_2^2 - \frac{L_1^2}{4} \qquad \ldots(i)$$

In Δ $CA'D$, we have

$$h^2 = (L_2 + \Delta L_2)^2 - \frac{(L_1 + \Delta L_1)^2}{4}$$

$$= L_2^2 + 2L_2\Delta L_2 + (\Delta L_2)^2 - \frac{L_1^2}{4} - \frac{2 \cdot L_1 \Delta L_1}{4} - \frac{(\Delta L_1)^2}{4}$$

Since $(\Delta L_2)^2$ and $(\Delta L_1)^2$ are very small quantities hence neglecting them, we have

$$h^2 = L_2^2 - \frac{L_1^2}{4} + \left(2L_2\Delta L_2 - \frac{2L_1\Delta L_1}{4}\right) \quad \ldots(ii)$$

From (*i*) and (*ii*) we have

$$2L_2\Delta L_2 - \frac{2L_1\Delta L_1}{4} = 0$$

or $$4L_2\Delta L_2 = L_1\Delta L_1$$

But $$\Delta L_1 = L_1\alpha_1\Delta T \text{ and } \Delta L_2 = \alpha_2 L_2 \Delta T$$

$\therefore$ $$4L_2\alpha_2 L_2\Delta T = L_1 \cdot L_1\alpha_1\Delta T$$

or $$4\alpha_2 L_2^2 = L_1^2\alpha_1.$$

Example 15.4: Two rods of different metals, having the same area of cross-section A are placed end to end between two massive walls (Fig. 15.20). The first rod has length l_1, coefficient of linear expansion α_1 and Young's modulus Y_1. The corresponding quantities for second rod are l_2, α_2 and Y_2. The temperature of both rods is now raised by T. (*a*) Find the force with which the rods act on each other at the higher temperature in terms of given quantities (*b*) Also find the lengths of the rods at higher temperature. Assume that there is no change in the cross-section of the rods and rods do not bend. There is no deformation of walls.

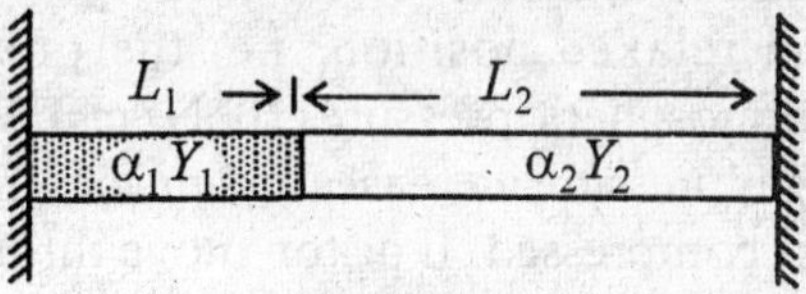

Fig. 15.20

Solution: (*a*) Due to heating, increase in length of composite rod is

$$\Delta L = (L_1\alpha_1 + L_2\alpha_2)T \quad (i)$$

and due to compressive force F from the walls due to elasticity decrease in length is

$$\Delta L = \frac{FL_1}{Y_1 A} + \frac{FL_2}{Y_2 A} = \left[\frac{L_1}{Y_1} + \frac{L_2}{Y_2}\right]\frac{F}{A} \quad (ii)$$

Equating (*i*) and (*ii*), we have

$$(L_1\alpha_1 + L_2\alpha_2)T = \left[\frac{L_1}{Y_1} + \frac{L_2}{Y_2}\right]\frac{F}{A}$$

or $$F = \frac{A(L_1\alpha_1 + L_2\alpha_2)T}{\left[\frac{L_1}{Y_1} + \frac{L_2}{Y_2}\right]}$$

(*b*) Initial length of one rod is L_1 and due to heating it increases by $\Delta L_1 = \alpha_1 L_1 T$ and due to compression decreases by $\Delta L_1' = \dfrac{FL_1}{AY_1}$. Hence its final length is

$$L_1' = (\Delta L_1 - \Delta L_1') + L_1 = L_1\left[1 + \alpha_1 T - \frac{F}{AY_1}\right]$$

Similarly final length of 2nd rod is

$$L_2' = L_2\left[1 + \alpha_2 T - \frac{F}{AY_2}\right]$$

where

$$F = \frac{A(L_1\alpha_1 + L_2\alpha_2)T}{\left[\dfrac{L_1}{Y_1} + \dfrac{L_2}{Y_2}\right]}$$

Example 15.5: An ideal monoatomic gas is confined in a cylinder by a spring-loaded piston (massless) of cross-section 8.0×10^{-3} m². Initially the gas is at 300 K and occupies volume of 2.4×10^{-3} m³ and spring is in its relaxed state. The gas is heated by an electric heater until the piston moves out slowly without friction by 0.1 m. Calculate the (*i*) final temperature of the gas and (*ii*) heat supplied (in Joule) by the heater. The force constant of the spring is 8000 N/m, atmospheric pressure is 1.0×10^5 N/m². The cylinder and piston are thermally insulated. Gas constant $R = 8.3$ J/mole-K.

Solution: (*i*) Let P_0, V_0, T_0 be the initial pressure, volume and temperature of the gas respectively. Initially the spring is in relaxed position, i.e. the pressure of the gas is P_0 = atmospheric pressure (10^5 N/m²). When the gas is heated, the gas pressure increases and pushes the piston thereby the spring is compressed. If piston move through distance Δx. The force on piston due to compression of the spring is

Fig. 15.21

$$F = k\Delta x = (8000)(0.1) = 800 \text{ N}.$$

Since A is the area of cross-section of the piston, the increase in pressure is

$$\frac{F}{A} = \frac{800}{8 \times 10^{-3}} = 10^5 \text{ N/m}^2$$

Increase in volume of the gas is

$$\Delta V = A\Delta x = 8 \times 10^{-3} \times 0.1 = 8 \times 10^{-4} \text{ m}^3$$

The final pressure of the gas

$$P = P_0 + \frac{F}{A} = 10^5 + 10^5 = 2 \times 10^5 \text{ N/m}^2$$

and final volume is

$$V = V_0 + \Delta V = 2.4 \times 10^{-3} + 8 \times 10^{-4} = 3.2 \times 10^{-3} \text{ m}^3$$

If final temperature is T, then using gas equation, we have

$$\frac{P_0V_0}{T_0} = \frac{PV}{T}$$

or $$T = \frac{PVT_0}{P_0V_0} = \frac{2\times10^5\times3.2\times10^{-3}}{10^5\times2.4\times10^{-3}}\times300 = 800\text{ K}$$

(*ii*) Heat supplied by the heater is used in compressing the spring, in raising the temperature of the gas and in moving the piston.

Work done in compressing spring

$$W_1 = \frac{1}{2}kx^2 = \frac{1}{2}\times800(0.1)^2 = 40\text{ J}$$

Heat required for raising temperature (i.e for increasing internal energy) is

$$H = nC_V\Delta T$$

where $n = \frac{P_0V_0}{RT_0}$ the number of moles of gas and $C_V = \frac{3}{2}R$ for monoatomic gas

$$\therefore \quad H = \frac{P_0V_0}{RT_0}\cdot\frac{3}{2}R(800-300)$$

$$= \frac{3}{2}\times\frac{500}{300}\times10^5\times24\times10^{-3} = 600\text{ J}$$

Work done in moving piston. Since piston moves against constant atmospheric pressure P_0 hence work done is

$$W_2 = P_0\Delta V = 10^5\times8\times10^{-3}\times0.1 = 80\text{ J}$$

Thus the total heat supplied is

$$Q = W_1 + H + W_2 = 40 + 600 + 80 = 720\text{ J.}$$

Example 15.6: An ideal gas whose ratio of specific heats is γ goes through a direct (clockwise) cycle consisting of adiabatic isobaric and isochoric process. Find the efficiency of the cycle if in adiabatic process the volume of ideal gas increases *n* fold.

Solution: Heat absorbed during the process *AB* (adiabatic process) is zero.

Heat absorbed during the process *BC* (isobaric process) is *$C_P(T_3 - T_2)$. During this process volume changes from nV_1 to V_1 at constant pressure. Thus

$$\frac{nV_1}{T_2} = \frac{V_1}{T_3} \quad \text{Charle's law}$$

$$\therefore \quad T_3 = \frac{T_2}{n} \qquad \ldots(i)$$

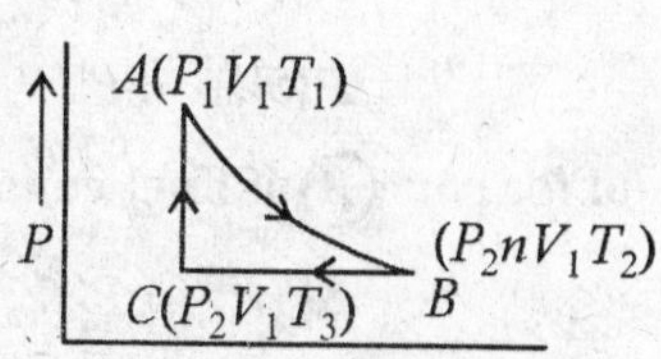

Fig. 15.21

Thus heat absorbed during isobaric process is

$$Q_1 = C_P(T_3 - T_2) = \frac{R\gamma}{\gamma-1}\left(\frac{T_2}{n} - T_2\right) = -\frac{R\gamma T_2}{n(\gamma-1)}(n-1) \qquad \ldots(ii)$$

i.e., in this process heat is rejected.

Heat absorbed during isochoric process

$$Q_2 = C_V(T_1 - T_3) = \frac{R}{\gamma-1}(T_1 - T_3) \qquad \ldots(iii)$$

---

*$C_P - C_V = R$ or $C_P\left(1 - \frac{C_V}{C_P}\right) = R$ or $C_P\left(1 - \frac{1}{\gamma}\right) = R$ or $C_P = \frac{\gamma R}{\gamma - 1}$ and $C_V = \frac{R}{\gamma - 1}$

For adiabatic process, the Poison's law is $T_1V_1^{\gamma-1} = T_2V_2^{\gamma-1}$. Thus for points A and B on the same adiabatic, we have

$$T_1V_1^{\gamma-1} = T_2(nV_1)^{\gamma-1} \text{ or } T_1 = T_2\, n^{\gamma-1} \qquad \ldots(iv)$$

Using (*i*) and (*iv*) in (*iii*), we get

$$Q_2 = \frac{R}{\gamma-1}\left[T_2\, n^{\gamma-1} - \frac{T_2}{n}\right] = \frac{RT_2}{\gamma-1}\left(\frac{n^\gamma - 1}{n}\right)$$

Hence efficiency

$$\eta = \frac{\text{Heat absorbed} - \text{Heat rejected}}{\text{Heat absorbed}}$$

or

$$\eta = \frac{Q_2 - |Q_1|}{Q_2}$$

$$\equiv \frac{\frac{RT_2}{n(\gamma-1)}(n^\gamma - 1) - \frac{R\gamma T_2}{n(\gamma-1)}(n-1)}{\frac{RT_2}{n(\gamma-1)}(n^\gamma - 1)}$$

$$\equiv \frac{(n^\gamma - 1) - \gamma(n-1)}{n^\gamma - 1} = 1 - \frac{\gamma(n-1)}{n^\gamma - 1}$$

Example 15.7: A piston divides a closed gas cylinder into two parts. Initially the piston is kept pressed such that one part has a pressure P and volume $5V$ and the other part has a pressure $8P$ and volume V. The piston is now left free. Find the new pressures and volumes for the adiabatic and isothermal process ($\gamma = 1.5$).

Solution: *For Adiabatic Process*: In equilibrium the pressures of the two parts are equal. Let this pressure be P'. Hence for part Ⓐ if final volume is V_a, then

$$P(5V)^\gamma = P'(V_a)^\gamma \qquad (i)$$

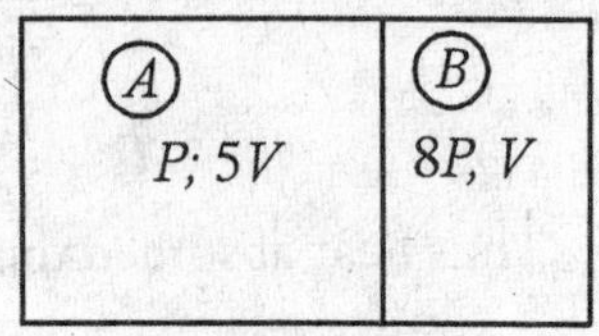

Fig. 15.22

and for part Ⓑ if final volume is V_b, then

$$8P(V)^\gamma = P'(V_b)^\gamma \qquad (ii)$$

Dividing we have

$$\frac{(5)^\gamma}{8} = \left(\frac{V_a}{V_b}\right)^\gamma$$

or

$$\frac{V_a}{V_b} = \frac{5}{(8)^{1/\gamma}} = \frac{5}{4} \qquad \because \gamma = 3/2$$

Since total volume of cylinder is $6V$, thus $V_a + V_b = 6V$.

Solving, we have $V_a = \frac{10}{3}V$ and $V_b = 8/3\, V$

Substituting in (*i*), we have

$$P' = \frac{P(5V)^\gamma}{(V_a)^\gamma} = \frac{P(5V)^\gamma}{\left(\frac{10}{3}V\right)^\gamma} = P\left(\frac{5\times 3}{10}\right)^\gamma$$

$$= P\left(\tfrac{3}{2}\right)^{3/2} = 1.837P$$

For Isothermal Change: If P' is the final pressure and V_a and V_b the volumes of A and B part respectively, then

$$P\times 5V = P'V_a \qquad (iii)$$

and $$8P\times V = P'V_b \qquad (iv)$$

Dividing, we get

$$\frac{V_a}{V_b} = \frac{5}{8} \quad \text{and} \quad V_a + V_b = 6\,V$$

Solving $$V_a = \frac{30}{13}V \quad \text{and} \quad V_b = \frac{48}{13}$$

Using in (*iii*), we have

$$P' = \frac{P\times 5V}{V_a} = P\times 5V\times\frac{13}{30V} = \frac{13}{6}P$$

Thus new pressure is = 2.16*P*.

Example 15.8: 10^{-4} kg oxygen is contained in a non-conducting cylinderical vessel of cross-sectional area 29.15×10^{-5} m^2 and volume 10^{-3} m^3. It is closed by a frictionless non-conducting piston which is free to move in a vertical direction. If the piston is slightly depressed and then released, show that the piston will execute simple harmonic motion. Find its frequency. Ignore atmospheric pressure and take $C_P/C_V = 1.4$ for oxygen.

Solution: Let the pressure and volume of the gas be P and V. Let the piston be pushed down through a small amount x, the pressure becomes $P + \Delta P$ and volume becomes $(V + \Delta V)$. Then using adiabatic equation we get

$$PV^\gamma = (P+\Delta P)(V+\Delta V)^\gamma$$

$$= PV^\gamma\left(1+\frac{\Delta P}{P}\right)\left(1+\frac{\Delta V}{V}\right)^\gamma$$

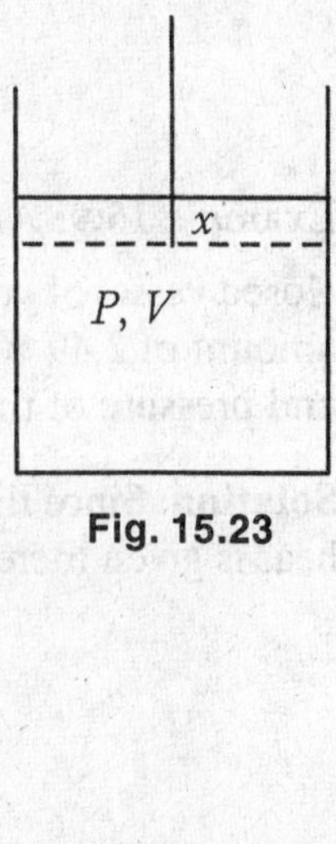

Fig. 15.23

Since $\frac{\Delta V}{V} <<1$, as such using binomial theorem

$$1 = \left(1+\frac{\Delta P}{P}\right)\left(1+\frac{\gamma\Delta V}{V}\right)$$

$$= 1+\frac{\Delta P}{P}+\frac{\gamma\Delta V}{V}+\frac{\gamma\Delta P\,\Delta V}{PV}$$

Last term $\frac{\gamma\Delta P\Delta V}{PV}$ is very small and can be neglected.

Thus $$\frac{\Delta P}{P} = -\frac{\gamma \Delta V}{V}$$

or $$\Delta P = -\gamma \frac{P}{V} \Delta V$$

Minus sign gives, when pressure increases volume decreases and *vice versa.*
If A is the area of cross-section of the piston, then $\Delta V = Ax$.
The increase in pressure of the gas will exert force F such that

$$F = \Delta P \cdot A \text{ or } \Delta P = \frac{F}{A}$$

$\therefore$ $$\frac{F}{A} = -\gamma \frac{P}{V} \cdot A \cdot x$$

or $$F = -\gamma \frac{P}{V} A^2 x$$

This force pushes the piston upwards, i.e. it is restoring force. Then acceleration of the piston is $a = \frac{F}{m} = -\frac{\gamma P A^2}{mV} x$, where m is mass of the piston. This piston executes S.H.M. with angular frequency

$$\omega = \sqrt{\frac{\gamma P A^2}{mV}}$$

or frequency f is given by ($\omega = 2\pi f$)

$$f = \frac{\omega}{2\pi} = \frac{1}{2\pi}\sqrt{\frac{\gamma P A^2}{mV}}$$

Since atmospheric pressure is neglected as such pressure P of gas is due to weight of piston, i.e.

$$P = \frac{mg}{A}$$

$\therefore$ $$f = \frac{1}{2\pi}\sqrt{\frac{\gamma mg A^2}{mAV}} = \frac{1}{2\pi}\sqrt{\frac{\gamma g A}{V}}$$

$$= \frac{1}{2\pi}\sqrt{\frac{1.4 \times 10 \times 29.15 \times 10^{-5}}{10^{-3}}}$$

$$= 0.318 \text{ Hz.}$$

Example 15.9: An ideal gas has a specific heat at constant pressure $C_P = \frac{5}{2}R$. The gas is kept in a closed vessel of volume 0.0083 m^3 at a temperature of 300 K and a pressure of 1.6×10^6 N/m^2. An amount of 2.49×10^4 Joules of heat energy is supplied to the gas. Calculate the final temperature and pressure of the gas.

Solution: Since the gas is supplied heat at constant volume as such work done is zero and whatever heat is given increases its internal energy or temperature. Thus heat given is

$$\Delta Q = nC_V \Delta T = \frac{PV}{RT} C_V \Delta T$$

where n is number of moles $n = \frac{PV}{RT_i}\cdot$ P, pressure of gas, V volume, R gas constant and T_i initial temperature

$$C_V = C_P - R = \frac{5}{2}R - R = \frac{3}{2}R \text{ specific heat at constant volume}$$

ΔT is change in temperature

Given is $\Delta Q = 2.49 \times 10^4$ J, $V = 0.0083$ m^3, $P = 1.6 \times 10^6$ N/m^2 and $T = 300$ K

Substituting we get $2.49 \times 10^4 = \dfrac{1.6 \times 10^6 \times 0.0083 \times \frac{3}{2}R \cdot \Delta T}{R \times 300}$

or $$\Delta T = \frac{2.49 \times 10^4 \times 300 \times 2}{1.6 \times 10^6 \times 0.0083 \times 3}$$

$$= \frac{2.49}{8 \times 8.3} \times 10^4 = \frac{3000}{8} = 375 \text{ K}$$

$\therefore$ Final temperature $T_f = T_i + \Delta T = 300 + 375 = 675$ K.

Since volume is constant, hence $P \propto T$.

$$\therefore \quad \frac{P_i}{P_f} = \frac{T_i}{T_f}$$

or $$P_f = T_f \frac{P_i}{T_i} = \frac{675 \times 1.6 \times 10^6}{300} = 3.6 \times 10^6 \text{ N/m}^2.$$

Example 15.10: Three moles of an ideal gas at 300 K are isothermally expanded to five times its volume and heated at this constant volume so that pressure is raised to its initial value before expansion. In the whole process 83.14 kJ heat is required. Calculate the ratio C_P/C_V of the gas (log 5 = 1.61).

Solution: During isothermal process since volume is made 5 times, the pressure becomes 1/5. If P and V are initial pressure and volume respectively then final pressure (P_f) and final volume (V_f) are

$$V_f = 5V \text{ and } P_f = P/5$$

When the gas is compressed to its initial pressure at constant volume, the temperature after the process be T_f. The initial temperature T is 300 K.

$$\therefore \quad \frac{P}{P_f} = \frac{T_f}{T}$$

or $$T_f = \frac{T \cdot P}{P/5} = 5T = 5 \times 300 = 1500 \text{ K}$$

During isothermal expansion work done is

$$W = nRT \cdot \log\frac{V_f}{V_i} \quad V_f \text{ is final volume and } V_i \text{ is initial volume}$$

$$= 3 \times 8.3 \times 300 \log 5$$

$$= 3 \times 8.3 \times 300 \times 1.6 = 11952 \text{ J}$$

Change in internal energy is $\Delta U = 0$ during isothermal process.
Change in internal energy during isochoric process

$$\Delta U = n C_V \Delta T = 3 \times C_V(1500 - 300)$$
$$= 3600\ C_V \text{ J}$$

First law of thermodynamics gives

$$\Delta Q = \Delta U + \Delta W$$

$$83.14 \times 10^3 = 3600 C_V + 11952$$

or $$C_V = \frac{83140 - 11952}{3600} = 19.77$$

and $$C_P = R + C_V = 8.3 + 19.77 = 28.07$$

or $$\gamma = \frac{C_P}{C_V} = \frac{28.07}{19.77} = 1.42$$

Example 15.11: A sample of an ideal gas has pressure P_0, volume V_0 and temperature T_0. It is isothermally expanded to twice its original volume. It is then compressed at constant pressure to have the original volume V_0. Finally the gas is heated at constant volume to get the original temperature (*a*) Show the process in a *V-T* diagram (*b*) Calculate the heat absorbed in the process.

Solution: *V-T* diagram for the process is shown in Fig. 15.24. Initial state (P_0, V_0, T_0) is shown at *A*. The gas is expanded isothermally to volume $2V_0$. During this process temperature is kept constant at T_0 and is shown as *AB*. Now the gas is compressed isobarically, i.e. volume is directly proportional to temperature (V/T = constant). This process is shown by a line *BC* which passes through the origin. At *C* the volume is V_0. Now the gas is heated at constant volume to temperature T_0. This is shown by *CA*. The final state is same as initial. (*b*) Since the process is cyclic so that the change in internal energy is zero. As such heat supplied is equal to the work done by the gas.

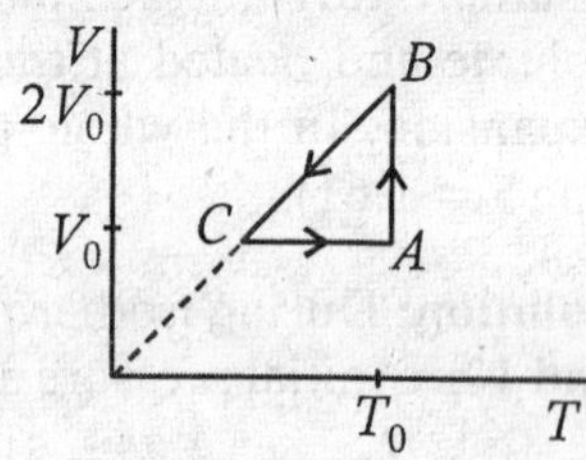

Fig 15.24

Work done during isothermal process from *A* to *B*

$$W_1 = nRT_0 \log \frac{2V_0}{V_0} = P_0V_0 \log 2 \quad \because P_0V_0 = nRT_0$$

If pressure of the gas at *B* is *P* then

$$P \times 2V_0 = P_0V_0 \text{ or } P = \frac{P_0}{2}$$

During the process *BC* pressure remains constant $P(=P_0/2)$ and volume changes from $2V_0$ to V_0. Thus work done

$$W_2 = \frac{P_0}{2}(V_0 - 2V_0) = -\frac{P_0V_0}{2}$$

During the isochoric process *CA*, volume remains constant and work done is zero.
Hence total work done is

$$W = W_1 + W_2 = P_0V_0 \log 2 - \frac{P_0V_0}{2}$$
$$= (0.693 - 0.5)P_0V_0 = 0.193P_0V_0$$

Thus heat supplied to the gas is $0.193P_0V_0$.

Example 15.12: An ideal gas having initial pressure *P*, volume *V* and temperature *T* is allowed to expand adiabatically until its volume becomes 5.66 V while its temperature falls to $\frac{T}{2}$ (*i*) How many degrees of freedom do the gas molecule have? (*ii*) Obtain the work done by the gas during the expansion as a function of initial presure *P* and volume *V*.

Solution: (*i*) In adiabatic process the relation between temperature (*T*) and volume (*V*) is

$$TV^{\gamma-1} = \text{constant}$$

If final temperature is *T'*, then

$$TV^{\gamma-1} = T'(5.66\text{ V})^{\gamma-1} \quad \text{But } T' = \frac{T}{2}$$

$$\therefore \quad TV^{\gamma-1} = \frac{T}{2}(5.66\text{ V})^{\gamma-1}$$

or $$(5.66)^{\gamma-1} = 2$$

or $$(\gamma - 1)\log 5.66 = \log 2$$

or $$\gamma = 1 + \frac{\log 2}{\log 5.66} = 1.4$$

This means that gas is diatomic and hence the degrees of freedom per molecule is 5.*

(*ii*) If change in volume is dV against presure *P*, the work done

$$dW = PdV$$

and total work done $W = \int dW = \int PdV$

In adiabatic process we have $PV^{\gamma} = C$ (constant)

$$\therefore \text{ Work done} \quad W = \int_{V_i}^{V_f} \frac{C}{V^{\gamma}} dV$$

where V_i and V_f are initial and final volumes respectively

$$\text{Integration gives} \quad W = \left.\frac{C\,V^{1-\gamma}}{1-\gamma}\right|_{V_i}^{V_f} = \frac{C\left(V_f^{1-\gamma} - V_i^{1-\gamma}\right)}{1-\gamma} \qquad (i)$$

But $$P_iV_i^{\gamma} = C = P_fV_f^{\gamma}$$

Substituting the value of *C* in (*i*), we get

$$W = \frac{P_fV_f^{\gamma}V_f^{1-\gamma} - P_iV_i^{\gamma}V_i^{1-\gamma}}{1-\gamma} = \frac{P_fV_f - P_iV_i}{1-\gamma}$$

Here P_i and P_f are initial and final pressures

But $$\frac{P_iV_i}{T_i} = \frac{P_fV_f}{T_f}$$

where T_i and T_f are initial and final temperatures of the gas.

---

*If *f* is the number of degrees of freedom, then $\gamma = \frac{f+2}{f}$. Here $\gamma = 1.4$ as such $f = 5$.

But $\quad T_i = T$ and $T_f = T/2$

$$\therefore \quad P_f V_f = \frac{P_i V_i}{2}$$

$$\therefore \quad W = \frac{\frac{P_i V_i}{2} - P_i V_i}{1-\gamma} = \frac{P_i V_i}{2(\gamma - 1)}$$

But $\quad P_i = P$ and $V_i = V$ and $\gamma = 1.4$

$$\therefore \quad W = \frac{PV}{2(1.4-1)} = \frac{PV}{0.8} = 1.25\ \text{PV}.$$

Example 15.13: One mole of a monoatomic ideal gas is taken through the cycle shown in Fig. 15.26, where *AB* is adiabatic expansion, *BC* cooling at constant volume, *CD* adiabatic compression and *DA* heating at constant volume.

The pressure and temperature at *A*, *B*, etc. are denoted by P_A, T_A, P_B, T_B, ..., etc. respectively. Given that $T_A = 1000$ K, $P_B = (2/3)P_A$ and $P_C = (1/3)P_A$. Calculate (*i*) the work done by the gas in the process $A \to B$. (*ii*) the heat lost by the gas in the process $B \to C$ (*iii*) the temperature T_D. [given. $(2/3)^{2/5} = 0.85$]

Solution: (*i*) Work done during adiabatic expansion $A \to B$ is

$$W = \frac{nR}{1-\gamma}(T_B - T_A)$$

For adiabatic change relation between presure (*P*) and temperature *T* is

$$\frac{T^\gamma}{P^{\gamma-1}} = \text{constant}$$

$$\therefore \quad \left(\frac{T_A}{T_B}\right)^\gamma = \left(\frac{P_A}{P_B}\right)^{\gamma-1}$$

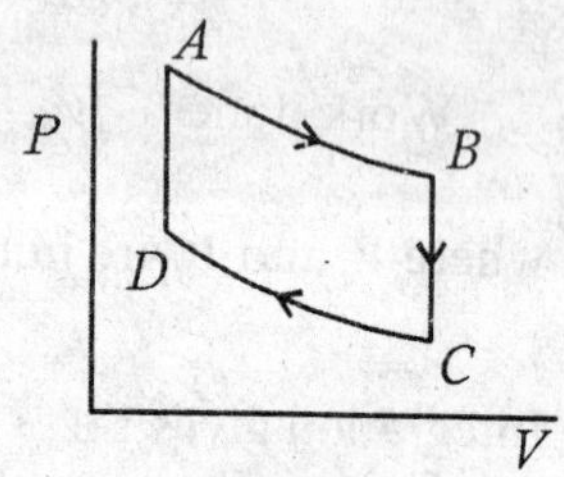

Fig.15.26

or
$$\frac{T_A}{T_B} = \left(\frac{P_A}{P_B}\right)^{\frac{\gamma-1}{\gamma}}$$

For monoatomic gas $\gamma = \frac{5}{3}$ and $\frac{\gamma-1}{\gamma} = \frac{2}{5}$

Given is $\quad \frac{P_A}{P_B} = \frac{3}{2}$

$$\therefore \quad \frac{T_A}{T_B} = \left(\frac{3}{2}\right)^{\frac{2}{5}}$$

or
$$T_B = T_A\left(\frac{2}{3}\right)^{\frac{2}{5}} = 0.85\ T_A = 0.85 \times 1000 = 850\ \text{K}$$

$$\therefore \quad W = \frac{1 \times 8.31}{1-\frac{5}{3}}(850 - 1000) \quad \text{Hence } n = 1 \text{ mol}$$

$$= 8.31 \times 150 \times 3/2 = 1869.75 \text{ Joules}$$

(*ii*) Process *BC* is isochoric and heat lost in this process is

$$\Delta Q = nC_V\Delta T = nC_V(T_C - T_B)$$

During this process we have

$$\frac{P_B}{P_C} = \frac{T_B}{T_C} \text{ or } T_C = \frac{P_C}{P_B}T_B$$

$$P_C = \frac{1}{3}P_A \text{ and } P_B = \frac{2}{3}P_A \text{ and } T_B = 850$$

$$\therefore \quad T_C = \frac{\frac{1}{3}P_A}{\frac{2}{3}P_A}\cdot 850 = \frac{850}{2} = 425 \text{ K}$$

$$\therefore \quad \Delta Q = 1\times\frac{3}{2}R(425-850) = \frac{-3\times 8.31\times 425}{2} = -5297.625 \text{ J}$$

(*iii*) Process *CD* is adiabatic and *T-P* relation is $\frac{T^\gamma}{P^{\gamma-1}}$ = constant

$$\therefore \quad \left(\frac{T_D}{T_C}\right)^\gamma = \left(\frac{P_D}{P_C}\right)^{\gamma-1}$$

$$\therefore \quad T_D = T_C\left(\frac{P_D}{P_C}\right)^{\frac{\gamma-1}{\gamma}}$$

Process *AD* is isochoric. Hence

$$\frac{P_A}{P_D} = \frac{T_A}{T_D} \text{ or } P_D = \frac{P_A\cdot T_D}{T_A}$$

and $$P_C = \frac{P_A}{3}$$

$$\therefore \quad \frac{P_D}{P_C} = \frac{P_AT_D}{T_A}\times\frac{3}{P_A} = \frac{3T_D}{T_A}$$

$$\therefore \quad T_D = T_C\left[\frac{3T_D}{T_A}\right]^{\frac{\gamma-1}{\gamma}} = T_C\left[\frac{3}{T_C}\right]^{\frac{\gamma-1}{\gamma}}\cdot T_D^{\frac{\gamma-1}{\gamma}}$$

or $$T_D^{1/\gamma} = T_C\left[\frac{3}{T_A}\right]^{\frac{\gamma-1}{\gamma}}$$

or $$T_D = T_C^\gamma\left[\frac{3}{T_A}\right]^{\gamma-1}$$

$$= (425)^{5/3}\left[\frac{3}{1000}\right]^{2/3}$$

$$= 300 \text{ K.}$$

Example 15.14: One mole of an ideal gas with heat capacity at constant pressure C_P undergoes the process $T = T_0 + \alpha V$, where T_0 and α are constants. Find (*a*) Molar heat capacity of the gas as a function of its volume (*b*) The amount of heat transferred to the gas, if volume increases from V_1 to V_2.

Solution: (*a*) If dQ is heat supplied to 1 mol of gas then molar thermal capacity of the gas is

$$C = \frac{dQ}{dT}$$

For 1 mole of ideal gas we have gas equation

$$PV = RT$$

From first law of thermodynamics, we have

$$dQ = dU + dW = dU + PdV$$

$$\therefore \quad C = \frac{dQ}{dT} = \frac{dU + PdV}{dT} = \frac{C_V dT + PdV}{dT}$$

$$= C_V + \frac{PdV}{dT} = C_V + \left(\frac{RT}{V}\right)\frac{dV}{dT}$$

Given $\quad T = T_0 + \alpha V \quad$ or $\quad \dfrac{dV}{dT} = \dfrac{1}{\alpha}$

This gives $\quad C = C_V + \dfrac{R}{V}(T_0 + \alpha V)\dfrac{1}{\alpha}$

$$= (C_V + R) + \frac{RT_0}{V\alpha}$$

But $\quad C_V + R = C_P$

$$\therefore \quad C = C_P + \frac{RT_0}{V\alpha}$$

(*b*) For ideal gas we have

$$PV = RT$$

or $\quad T = \dfrac{PV}{R} = T_0 + \alpha V$

or $\quad P = \dfrac{RT_0}{V} + \alpha R$

$$\therefore \text{ Work done } W = \int_{V_1}^{V_2} PdV = \int_{V_1}^{V_2}\left(\frac{RT_0}{V} + \alpha R\right)dV$$

$$= \left[RT_0 \log V + \alpha RV\right]_{V_1}^{V_2}$$

$$= RT_0 \log\frac{V_2}{V_1} + \alpha R(V_2 - V_1)$$

Change in internal energy

$$\Delta U = C_V \Delta T$$

$$= C_V(T_2 - T_1) = C_V[T_0 + \alpha V_2 - T_0 - \alpha V_1]$$

$$= C_V\alpha(V_2 - V_1)$$

Thus heat transferred is

$$\Delta Q = \Delta U + \Delta W$$

$$= C_V\alpha(V_2 - V_1) + RT_0 \log\frac{V_2}{V_1} + \alpha R(V_2 - V_1)$$

$$= \alpha(V_2 - V_1)(R + C_V) + RT_0 \log\frac{V_2}{V_1}$$

$$= \alpha C_P(V_2 - V_1) + RT_0 \log\frac{V_2}{V_1}$$

Example 15.15: Two moles of helium gas ($\gamma = 5/3$) are initially at temperature 27° C and occupy a volume of 20 litres. The gas is first expanded at constant pressure until the volume is doubled. Then it undergoes an adiabatic change until the temperature returns to its initial value

(*a*) Sketch a *P-V* diagram.

(*b*) What is final volume and pressure of the gas?

(*c*) What is the work done by the gas?

Solution: Using ideal gas equation

$$PV = nRT$$

$$n = 2 \text{ mol},\ T = 27°\text{ C} = 300 \text{ K},\ V = 20 \text{ litres} = 20 \times 10^{-3} \text{ m}^3.$$

$$P = \frac{nRT}{V} = \frac{2 \times 8.3 \times 300}{20 \times 10^{-3}} = 2.49 \times 10^5 \text{ N/m}^2$$

(*i*) When gas is heated at constant pressure the volume and temperature increase ($V \propto T$) if volume is doubled, the temperature is also doubled. This is shown on *P-V* diagram by *AB* (Fig. 15.27). At *B* pressure $P' = P = 2.49 \times 10^5$ N/m². $V' = 2V = 40$ litres and $T = 600$ K. The gas then cools to temperature *T* adiabetically which is shown on *P-V* diagram by *BC* (Fig. 15.27).

(*ii*) At *C* the temperature of the gas becomes $T = 300$ K. Using temperature volume relation for adiabatic change we have

$$T'V'^{\gamma-1} = T''V''^{\gamma-1}$$

or $$V''^{\gamma-1} = \frac{T'}{T''}V'^{\gamma-1}$$

or $$V'' = \left(\frac{T'}{T''}\right)^{\frac{1}{\gamma-1}} \cdot V'$$

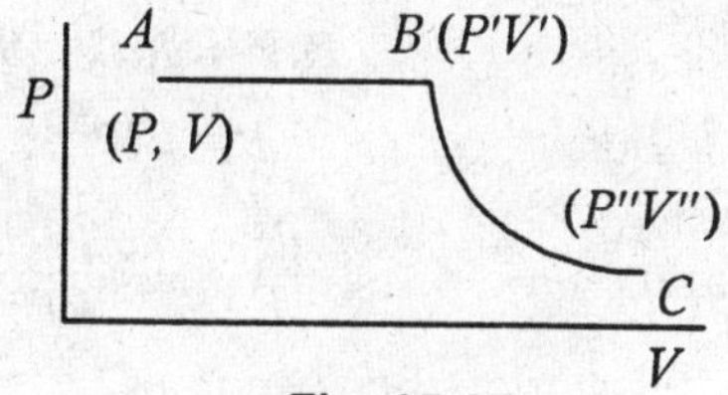

Fig. 15.27

$$= \left(\frac{600}{300}\right)^{\frac{1}{5/3-1}} \times 40 \times 10^{-3}$$

$$= (2)^{3/2} \times 40 \times 10^{-3} = 113.14 \times 10^{-3} \text{ m}^3$$

and pressure is

$$P'V'^{\gamma} = P''V''^{\gamma}$$

or $$P'' = P'\left(\frac{V'}{V''}\right)^{\gamma} = 2.49 \times 10^5 \left(\frac{T''}{T'}\right)^{\frac{\gamma}{\gamma-1}}$$

$$= 2.49 \times 10^5 \left(\frac{1}{2}\right)^{5/2}$$

$$= 0.44 \times 10^5 \text{ N/m}^2$$

(*iii*) The work done in isobaric process *AB* is

$$W_1 = P\Delta V = 2.49 \times 10^5 (40 - 20) \times 10^{-3}$$
$$= 4980 \text{ J}$$

Work done during adiabatic process *BC* is

$$W_2 = \frac{nR}{1-\gamma}(T'' - T') = \frac{2 \times 8.3}{1 - \frac{5}{3}}(300 - 600)$$
$$= 3 \times 8.3 \times 300 = 7470 \text{ J}$$

Hence total work done

$$W = W_1 + W_2 = 4980 + 7470 = 12450 \text{ J.}$$

Example 15.16: 1 mole of an ideal gas whose adiabatic exponent γ undergoes a process in which the gas pressure relates to temperature as $P = kT^{\alpha}$ where k and α are constants. Find (*a*) the work done by the gas if its temperature increases by ΔT. (*b*) The molar specific heat capacity of the gas in this process. For what value of α will it be negative.

Solution: (*a*) Work done by the gas

$$W = \int_{V_1}^{V_2} PdV \quad (i)$$

Ideal gas equation is $PV = RT$ (*ii*)

differentiating we have $PdV + VdP = RdT$

or $$PdV = RdT - VdP \quad (iii)$$

But $V = \frac{RT}{P}$ substituting in (*iii*), we get

$$PdV = RdT - RT\frac{dP}{P} \quad (iv)$$

But $P = kT^{\alpha} \quad \therefore dP = k\alpha T^{\alpha-1}dT$

and $$\frac{dP}{P} = \frac{k\alpha T^{\alpha-1}dT}{kT^{\alpha}} = \frac{\alpha}{T}dT$$

Equation then becomes

$$PdV = RdT - RT\frac{\alpha}{T}dT$$
$$= R(1-\alpha)dT$$

Hence work done is

$$W = \int_{V_1}^{V_2} PdV = \int_{T}^{T+\Delta T} R(1-\alpha)dT = R(1-\alpha)\Delta T$$

(*b*) If C is molar heat capacity for this process, then for one mole

$$\Delta Q = C\Delta T \text{ and } \Delta U = C_V \Delta T$$

Now from first law of thermodynamics we have

$$\Delta Q = \Delta U + \Delta W$$

or $$C\Delta T = C_V \Delta T + R(1-\alpha)\Delta T \qquad \text{But } C_V = \frac{R}{\gamma - 1}$$

$\therefore$ $$C = \frac{R}{\gamma - 1} + R(1-\alpha)$$

C is negative, i.e. $C < 0$. This means that

$$\frac{R}{\gamma-1} + R(1-\alpha) < 0$$

or $$1 - \alpha < -\frac{1}{\gamma - 1}$$

or $$\alpha > 1 + \frac{1}{\gamma-1} = \frac{\gamma}{\gamma-1}$$

i.e. $$\alpha > \frac{\gamma}{\gamma - 1}$$

Example 15.17: A gaseous mixture enclosed in a vessel of volume V consists of one gram mole of a gas A with $\gamma = 5/3$ and another gas with $\gamma = 7/5$ at certain temperature T. The gram molecular weight of A and B are 4 and 32 respectively. The gases A and B do not react with each other and are assumed to be ideal. The gaseous mixture follows the equation $PV^{19/13}$ = constant in adiabatic process. (*a*) Find the number of gram moles of the gas B in the gaseous mixture (*b*) If initial temperature is $T_0 = 300$ K and the mixture is compressed adiabatically to 1/5 of its initial volume V. Find the change in its adiabatic compressibility in terms of given quantities.

Solution: Let n_a and n_b be number of moles of A and B gases in mixture. For an ideal gas total internal energy remains conserved, i.e. total internal energy of mixture is equal to the sum of the internal energies of gas A and gas B. Thus

$n_a(C_V)_a T + n_b(C_V)_b T = (n_a + n_b)(C_V)_m T$, where $(C_V)_m$ is specific heat of mixture at constant volume.

But $$C_V = \frac{R}{\gamma - 1}$$

$\therefore$ $$\frac{n_a RT}{\gamma_a - 1} + \frac{n_b RT}{\gamma_b - 1} = \frac{(n_a + n_b)RT}{\gamma_m - 1}$$

or $$\frac{n_a}{\gamma_a - 1} + \frac{n_b}{\gamma_b - 1} = \frac{n_a + n_b}{\gamma_m - 1} \qquad (i)$$

But $$\gamma_a = \frac{5}{3}, \gamma_b = \frac{7}{5}, n_1 = 1 \quad \text{and} \quad \gamma_m = \frac{19}{13}$$

Substituting in (*i*), we have

$$\frac{1}{\frac{5}{3} - 1} + \frac{n_b}{\frac{7}{5} - 1} = \frac{1 + n_b}{\frac{19}{13} - 1}$$

$$\frac{3}{2}+\frac{5}{2}n_b = \frac{13}{6}+\frac{13}{6}n_b$$

or $$n_b\left(\frac{5}{2}-\frac{13}{6}\right) = \frac{13}{6}-\frac{3}{2}$$

$$\frac{n_b}{3} = \frac{2}{3} = n_b = 2$$

Thus number of moles of gas B are 2.

(*b*) Compressibility is given by $\beta = -\frac{\Delta V}{V\Delta P}$

For adiabatic change

$$PV^{\gamma} = \text{constant}$$

Differentiating, we get

$$P\gamma V^{\gamma-1}dV + V^{\gamma}dP = 0$$

or $$\gamma PdV = -VdP$$

or $$\beta = -\frac{dV}{VdP} = \frac{1}{\gamma P}$$

If P_i is initial pressure and P_f final pressure after adiabatic compression, then change in compressibility is

$$\Delta\beta = \frac{1}{\gamma P_f}-\frac{1}{\gamma P_i} = \frac{1}{\gamma}\left[\frac{1}{P_f}-\frac{1}{P_i}\right]$$

For adiabatic change PV^{γ} = constant. Thus

$$P_iV_i^{\gamma} = P_fV_f^{\gamma} \qquad \text{But } V_f = \frac{V_i}{5}$$

$\therefore$ $$P_f = P_i\left(\frac{V_i}{V_f}\right)^{\gamma} = P_i(5)^{\gamma} \text{ and } \gamma = \frac{19}{13} \text{ for mixture}$$

$\therefore$ $$\Delta\beta = \frac{1}{\gamma}\left[\frac{1}{P_i(5)^{\gamma}}-\frac{1}{P_i}\right] = \frac{1}{\gamma P_i}\left(\frac{1}{5^{\gamma}}-1\right)$$

But $$P_i = \frac{nRT_i}{V}$$

$\therefore$ $$\Delta\beta = \frac{V}{nRT_i\gamma}\left[\frac{1}{5^{\gamma}}-1\right]$$

$$= \frac{V\times 13}{3\times 8.31\times 300\times 19}\left[\frac{1}{5^{\frac{19}{13}}}-1\right]$$

$$= -8.278\times 10^{-5}V \text{ m}^2\text{N}^{-1}.$$

Example 15.18: Two moles of helium gas undergo cyclic process as shown in Fig. 15.28. Assuming the gas to be ideal, calculate the following quantities in this process: (*a*) The net change in the heat energy, (*b*) The net work done and (*c*) The net change in internal energy.

Solution: Volume of the gas at *A* is

$$V_a = \frac{nRT_a}{P_a}$$

$P_a = 2$ atm $= 2 \times 10^5$ N/m², $T_a = 300$ K, $R = 8.32$ J/mol-K, $n = 2$ mol

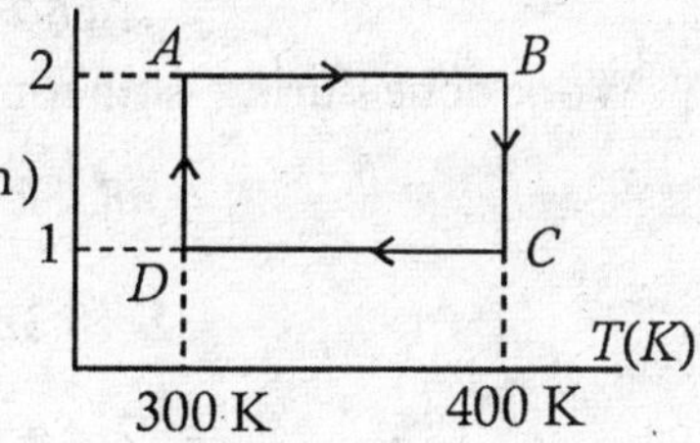

Fig 15.28

$$\therefore \quad V_a = \frac{2 \times 8.32 \times 300}{2 \times 10^5}$$

$$= 2.496 \times 10^{-2} \text{ m}^3$$

Process *AB* is isobaric at *B* since temperature is 400 K, hence volume at *B* is

$$\frac{V_a}{T_a} = \frac{V_b}{T_b}$$

$$V_b = \frac{T_b}{T_a} V_a = \frac{400}{300} \times V_a = \frac{4}{3} V_a$$

During process *BC* temperature remains constant as such the process is isothermal. At *C* pressure is half of that at *B*, hence volume at *C*

$$V_c = 2V_b = \frac{8}{3} V_a$$

Process *CD* is isobaric and since temperature at *D* is 300 K that at *C* is 400 K, hence volume at *D* is

$$V_d = V_c \frac{T_d}{T_c} = \frac{8}{3} V_a \times \frac{300}{400} = 2V_a$$

Since it is a cyclic process hence change in internal energy is

$$\Delta U = 0$$

Work done during isobaric process *AB*

$$W_1 = P\Delta V = P(V_b - V_a) = P\left(\frac{4}{3}V_a - V_a\right) = \frac{PV_a}{3}$$

$$= \frac{2 \times 10^5 \times 2.496 \times 10^{-2}}{3} = 1.664 \times 10^3 \text{ J}$$

Work done during isothermal process *BC* is

$$W_2 = nRT \log \frac{V_c}{V_b} \qquad \text{But } \frac{V_c}{V_b} = 2$$

$$= 2 \times 8.32 \times 400 \times 2.303 \log_{10} 2$$

$$= 2 \times 8.32 \times 400 \times 2.303 \times 0.3010$$

$$= 4.612 \times 10^3 \text{ J}$$

Work done during isobaric process *CD* is

$$W_3 = P\Delta V = P(V_d - V_c) = P\left(2V_a - \frac{8}{3}V_a\right)$$

$$= -P \times V_a \times 2/3$$
$$= -10^5 \times 2.496 \times 10^{-2} \times \frac{2}{3}$$
$$= -1.664 \times 10^{+3}\ \text{J}$$

Work done during isothermal process DA

$$W_4 = nRT \log_e \frac{V_a}{V_d}$$
$$= 2 \times 8.32 \times 300 \times \log_e\left(\frac{1}{2}\right)$$
$$= -2 \times 8.32 \times 300 \times 2.303 \times 0.3010$$
$$= -3.456 \times 10^3\ \text{J}.$$

Total work done

$$W = W_1 + W_2 + W_3 + W_4$$
$$= 1.664 \times 10^3 + 4.612 \times 10^3 - 1.664 \times 10^3 - 3.456 \times 10^3$$
$$= 1.156 \times 10^3\ \text{J}$$

Since there is no change in internal energy, hence net change in heat energy is equal to the total work done, i.e.

$$\Delta Q = W = 1.156 \times 10^3\ \text{J}.$$

Example 15.19: A vertical hollow cylinder contains an ideal gas. The gas is enclosed by a 5 kg movable piston with an area of cross-section 5×10^{-3} m^2. Now the gas is heated from 300 K to 350 K and piston rises by 0.1 m. The piston is now clamped at this position and the gas is cooled back to 300 K. Find the difference between the heat energy added during heating process and energy lost during cooling process.

Solution: Initial pressure of the gas in cylinder is

$$P_i = P_0 + \frac{Mg}{A}$$

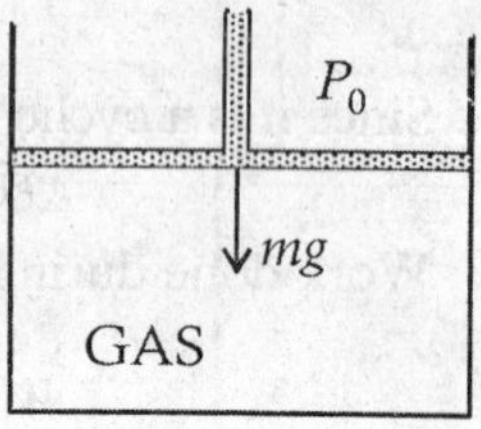

Fig. 15.29

P_0—Atmospheric pressure, $\frac{Mg}{A}$ pressure due to weight of piston

$$P_i = 10^5 + \frac{5 \times 10}{5 \times 10^{-3}} = 10^5 + 10^4 = 1.1 \times 10^5\ \text{N/m}^2$$

Since the expansion process is isobaric, hence work done is

$$W_1 = P_i \Delta V = P_i Ay \qquad \text{where } y = \text{rise in piston} = 0.1\ \text{m}$$
$$= 1.1 \times 10^5 \times 5 \times 10^{-3} \times 0.1 = 55\ \text{J}$$

If U_1 is the change in internal energy in the process, the heat supplied is

$$Q_1 = W_1 + U_1$$
$$\therefore \quad Q_1 = 55 + U_1$$

when piston is clamped it does not move and the process is isochoric. Thus work done $W_2 = 0$

Since in cooling process the original temperature is restored hence change in internal energy is again U_1 (U_1 is negative). Hence heat extracted is

$$Q_2 = W_2 + U_1 = U_1$$

Hence difference of amounts of heats in the two processes is

$$Q_1 - Q_2 = 55\ \text{J}.$$

Example 15.20: Two moles of an ideal monoatomic gas initially at pressure P_1 and volume V_1 undergo an adiabatic compression until its volume becomes V_2. Then the gas is given heat Q at constant volume V_2. (*a*) Sketch the complete *P-V* diagram, (*b*) Find the total work done by the gas, the total change in its internal energy and final temperature of the gas. Give your answer in terms of P_1, V_1, V_2, R and Q.

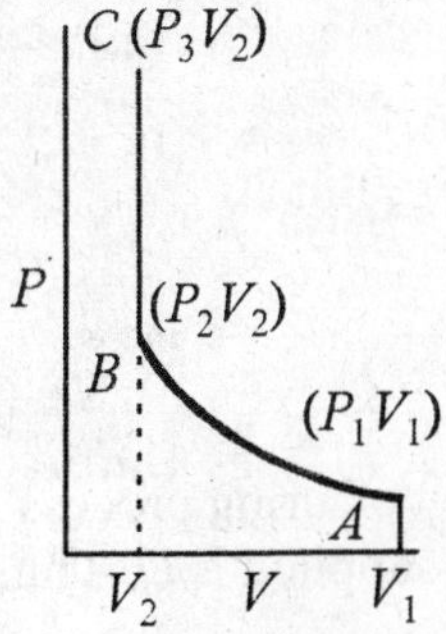

Fig.15.30

Solution: Adiabatic process is shown by *AB* in Fig. 15.30 when gas is compressed from state $A(P_1, V_1)$ to state $B(P_2, V_2)$

where $P_2 = P_1\left(\frac{V_1}{V_2}\right)^\gamma$ $\gamma = 5/3$ since gas is monoatomic

Then gas is heated at constant volume so that its pressure becomes P_3. This process is shown on *P-V* diagram by *BC*.

Work done in adiabatic process from *A* to *B* is

$$W_1 = \frac{1}{1-\gamma}(P_2V_2 - P_1V_1)$$

Substituting value of P_2, we have

$$W_1 = \frac{1}{1-\gamma}\left(P_1\left(\frac{V_1}{V_2}\right)^\gamma V_2 - P_1V_1\right)$$

$$= \frac{P_1V_1}{1-\gamma}\left(\left(\frac{V_1}{V_2}\right)^{\gamma-1} - 1\right) = \frac{P_1V_1}{1-5/3}\left(\left(\frac{V_1}{V_2}\right)^{5/3-1} - 1\right)$$

$$= \frac{3}{2}P_1V_1\left(1 - \left(\frac{V_1}{V_2}\right)^{2/3}\right)$$

During isochoric process *B* no work is done. Hence total work done in the process is

$$W = W_1 + 0 = \frac{3}{2}P_1V_1\left(1 - \left(\frac{V_1}{V_2}\right)^{2/3}\right)$$

If temperature at *A* is T_1 then temperature at *B*, T_2 is given by (using adiabatic relation between temperature and volume)

$$T_2V_2^{\gamma-1} = T_1V_1^{\gamma-1}$$

$$T_2 = T_1\left(\frac{V_1}{V_2}\right)^{\gamma-1}$$

Since process *BC* is isochoric process and heat supplied is Q. Then change in temperature ΔT is

$$Q = nC_V\Delta T \qquad n = 2 \qquad C_V = \frac{R}{\gamma - 1}$$

$$\therefore \quad \Delta T = \frac{Q}{\eta C_V} = \frac{Q(\gamma - 1)}{2 \times R} \qquad \gamma = 5/3$$

$$\Delta T = \frac{Q}{3R}$$

$\therefore$ Final temperature

$$T_f = T_2 + \Delta T = T_1\left(\frac{V_1}{V_2}\right)^{\gamma-1} + \frac{Q}{3R}$$

But $\quad T_1 = \frac{P_1V_1}{nR} = \frac{P_1V_1}{2R}$

$$\therefore \quad T_f = \frac{P_1V_1}{2R}\left(\frac{V_1}{V_2}\right)^{2/3} + \frac{Q}{3R}$$

During process AB (adiabatic process) no heat is supplied. During BC (isochoric process) heat supplied is Q. Thus total heat supplied is Q.

Total work done during the process is $W = \frac{3}{2}P_1V_1\left(1 - \left(\frac{V_1}{V_2}\right)^{2/3}\right)$.

Using first law of thermodynamics, change in internal energy is

$$\Delta U = Q - W$$

$$= Q - \frac{3}{2}P_1V_1\left(1 - \left(\frac{V_1}{V_2}\right)^{\frac{2}{3}}\right).$$

Example 15.21: During an integral number of complete cycle a reversible Carnot engine (shown by a circle) (Fig. 15.31) absorbs 1200 Joules from reservoir at 400 K and performs 200 Joules of mechanical work.

(*a*) Find quantities of heat exchanged with the other two reservoirs. State wether the reservoir absorb or lose heat.
(*b*) Find the change in entropy* of each reservoir
(*c*) What is the change in entropy of the universe.

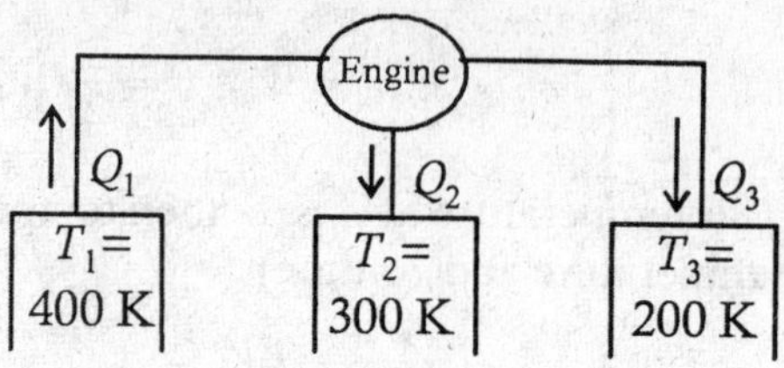

Fig 15.31

Solution:

(*a*) Heat absorbed by the working substance

$$Q_1 = 1200 \text{ J}$$

Work done by the engine $W = 200$ J

Heat rejected to sinks

$$Q_2 + Q_3 = Q_1 - W = 1200 - 200 = 1000 \text{ J} \qquad (i)$$

Using property of reversible Carnot's engine, we have

$$\left(\sum\frac{Q}{T}\right)_{\text{source}} = \left(\sum\frac{Q}{T}\right)_{\text{sinks}}$$

*Change in entropy $\Delta S = \frac{\Delta Q}{T}$. Where ΔQ is heat rejected or absorbed at temperature T. If ΔQ is positive it is positive and negative if ΔQ is negative. Its units are Joules per Kelvin.

$$\therefore \quad \frac{Q_1}{T_1} = \frac{Q_2}{T_2} + \frac{Q_3}{T_3} \qquad (ii)$$

$$\frac{1200}{400} = \frac{Q_2}{300} + \frac{Q_3}{200}$$

or $\quad 2Q_2 + 3Q_3 = 1800 \qquad (iii)$

Solving for Q_2 and Q_3 from (*i*) and (*iii*), we have

$$Q_3 = -200 \text{ J and } Q_2 = 1200 \text{ J}$$

As Q_3 is negative, it means that working substance takes heat Q_3 (= 200 J) from the reservoir and heat rejected to reservoir at temperature T_2, Q_2 = 1200 J.

(*b*) Change in entropy of first reservoir

$$\Delta S_1 = -\frac{Q_1}{T_1} = -\frac{1200}{400} = -3 \text{ J/K}$$

Negative sign indicates decrease because heat is extracted from reservoir.

Change in entropy of 2nd reservoir

$$\Delta S_2 = \frac{1200}{300} = 4 \text{ J/K} \quad \text{+ve because heat is given to this reservoir}$$

Change in entropy of third reservoir

$$\Delta S_3 = -\frac{200}{200} = -1 \text{ J/K} \quad \text{Heat is taken from the reservoir}$$

(*c*) Since there is no change in the entropy of the working substance. Hence total change in the entropy of the universe is

$$\Delta S = \Sigma_i \Delta S_i = -3 + 4 - 1 = 0$$

Example 15.22: Two moles of an ideal monoatomic gas is taken through the cycle *ABCA* as shown is *P-T* diagram (Fig. 15.32). During the process *AB*, pressure and temperature of the gas vary such that *PT* = constant if T_1 = 300 K. Calculate

(*a*) The work done on the gas in the process *AB* and

(*b*) the heat absorbed or released by the gas in each of the processes. Give answer is terms of the gas constant *R*.

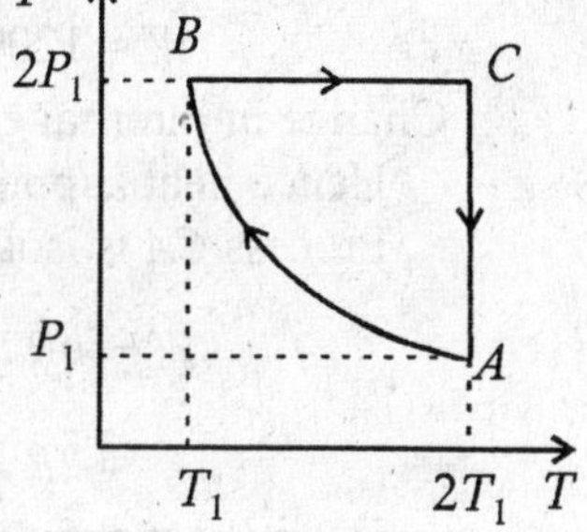

Fig 15.32

Solution:

(*a*) Work done in process *AB*

$$W = \int PdV \qquad ...(i)$$

Using gas equation $PV = nRT$ where $n = 2$

We have $\quad PV = 2RT$

$$PdV + VdP = 2RdT \qquad ...(ii)$$

The additional equation followed by the process is

$$PT = K \qquad \text{where } K \text{ is constant}$$

From (*ii*) we have

$$PdV = 2RdT - VdP\text{, But } V = \frac{2RT}{P}$$

$$\therefore \quad PdV = 2RdT - \frac{2RT}{P}dP \qquad \text{But } T = \frac{K}{P}$$

$$\therefore \quad PdV = 2RdT - \frac{2RK}{P^2}dP$$

$$W = \int PdV = 2R\int_{T_a}^{T_b} dT - 2RK\int_{P_a}^{P_b} \frac{dP}{P^2}$$

$$= 2R[T_b - T_a] - 2RK(-1)\left[\frac{1}{P_b} - \frac{1}{P_a}\right]$$

$$= 2R[T_b - T_a] - 2\,\text{R}\left[\frac{K}{P_a} - \frac{K}{P_b}\right]$$

$$\frac{K}{P_a} = T_a = 2T_1 = 600K, \ \frac{K}{P_b} = T_b = T_1 = 300K$$

$$W = 2R[300 - 600] - 2R(600 - 300)$$
$$= -1200R.$$

(*b*) During process *AB* change in internal energy

$$\Delta U = nC_V\Delta T$$
$$= 2 \cdot 3/2R \cdot (300 - 600) \text{ gas is monoatomic hence } C_V = 3/2R$$
$$= -900R$$

∴ Heat supplied

$$Q = \Delta U + \Delta W = -900R - 1200 = -2100R$$

Negative sign means heat is released by the system = 2100*R*

Process *BC* is isobaric

$$W = \int PdV = P_B(V_c - V_B)$$

$$P_BV_C = 2RT_C = 2R \times 600 = 1200R \text{ and } P_BV_B = 2RT_b = 2R \times 300 = 600R$$

$$\therefore \quad W = 1200R - 600R = 600R$$

Change in internal energy = $n\,C_V\Delta T = 2 \times 3/2R(600 - 300) = 900R$

Hence heat supplied $Q = W + \Delta U = 600R + 900R = 1200R$

Process *CA* is isothermal $\Delta U = 0$

$$W = nRT\log\frac{V_A}{V_C} \quad \text{Since pressure is halved hence } V_A = 2V_C$$

$$= 2R \times 600\log_e 2 = 2 \times 600R \times 0.693 = 831.6R$$

Hence heat supplied is

$$Q = W = 831.6R$$

Example 15.23: One gram of water (volume 1 cm^3) becomes 1671 cm^3 of steam when boiled at a pressure of one atmosphere. The latent heat of vaporisation 540 cal/gm. Compute the external work done and increase in internal energy. (1 Atmosphere = 1.013×10^5 N/m^2 and $J = 4.2$ Joule/cal)

Solution: When 1 gm of water at 100° C is heated it is converted into steam. Heat given is

$$Q = mL = 1 \times 540 = 540 \text{ cal}$$

Volume increases by $\Delta V = (1671 - 1) = 1670 \text{ cc} = 1670 \times 10^{-6} \text{ m}^3$

Work done against atmospheric pressure

$$W = P\Delta V = 1.013 \times 10^5 \times 1670 \times 10^{-6}$$

$$= 1.013 \times 167 = 169.2 \text{ J}$$

$$= \frac{169.2}{4.2} = 40.3 \text{ cal}$$

Using first law of thermodynamics

$$Q = U + W$$

or $$U = Q - W = 540 - 40.3 = 499.7 \text{ cal.}$$

Example 15.24: An ideal gas is taken through a cyclic thermodynamic process through four steps. The amounts of heat involved in these steps are $Q_1 = 5960$ J, $Q_2 = -5585$ J, $Q_3 = -2980$ J and $Q_4 = 3645$ J respectively. The corresponding works involved are $W_1 = 2200$ J, $W_2 = -825$ J $W_3 = -1100$ J and W_4 repetitively. (*i*) Find the value of W_4 (*ii*) What is the efficiency of the cycle?

Solution:

(*i*) Since the process is cyclic as such $\Delta U = 0$

Total heat absorbed = 5960 – 5585 – 2980 + 3645 = 1040 J

Total work done $W = 2200 - 825 - 1100 + W_4 = 275 + W_4$

Hence from first law

$$Q = W \qquad \because \Delta U = 0$$

or $$1040 = 275 + W_4$$

or $$W_4 = 1040 - 275 = 765 \text{ J.}$$

(*ii*) Efficiency of engine is

$$\eta = \frac{\text{Total work done}}{\text{Total heat absorbed}}$$

Total work done $W = 1040$ J

Total heat absorbed = 5960 + 3645 = 9505 J

$$\therefore \qquad \eta = \frac{1040}{9505} \times 100 = 10.83\%.$$

Example 15.25: A glass tube of length 133 cm and of uniform cross-section is filled with mercury so that the length of the tube unoccupied by mercury remains the same at all temperatures. If the coefficient of cubical expansion for glass and mercury are 0.000026/°C and 0.000182/°C respectively. Calculate the length of mercury column.

Solution: Let x_0 and x_t be the length of mercury column at 0° C and t° C respectively. A_0 and A_t be the area of cross-sections of the glass tube at 0° C and t° C respectively then

$$A_t = A_1(1 + \beta_g t) = A_0(1 + 2\alpha_g t) \qquad \ldots(i)$$

where α_g is the co-efficient of linear expansion of glass

Volume of mercury at t° C = $(V_m)_t = A_t x_t$.

Volume of glass at $t°$ C = $(V_g)_t = A_t L_t$
where L_t is the length of glass tube at $t°$ C.

Length of glass unoccupied by mercury

$$\Delta L = L_t - x_t = \frac{(V_g)_t}{A_t} - \frac{(V_m)_t}{A_t}$$

$$= \frac{(V_g)_0(1+3\alpha_g t) - (V_m)_0(1+\gamma_m t)}{A_0(1+2\alpha_g t)}$$

$$= \frac{A_0 L_0(1+3\alpha_g t) - A_0 x_0(1+\gamma_m t)}{A_0(1+2\alpha_g t)}$$

$$= L_0(1 + 3\alpha_g t)(1+2\alpha_g t)^{-1} - x_0(1+\gamma_m t)(1+2\alpha_g t)^{-1}$$

Using binomial theorem and leaving higher powers we have

$$\Delta L = L_0(1 + 3\alpha_g t)(1 - 2\alpha_g t) - x_0(1 + \gamma_m t)(1 - 2\alpha_g t)$$

$$= L_0(1 + \alpha_g t) - x_0[1 + (\gamma_m - 2\alpha_g)t]$$

ΔL is to be independent of temperature. Thus

$$\frac{d(\Delta L)}{dt} = 0 = L_0\alpha_g - x_0(\gamma_m - 2\alpha_g)$$

or

$$x_0 = \frac{L_0\alpha_g}{\gamma_m - 2\alpha_g} = \frac{133 \times (0.000026/3)}{0.000182 - 2(0.000026/3)}$$

$$= 7 \text{ cm.}$$

Example 15.26: 0.5 kg of water inside a calorimeter of thermal capacity 85 J/K is super cooled to –5° C. If water suddenly freezes, calculate (*i*) the mass of ice formed and (*ii*) latent heat of fusion of ice at –5° C? Specific latent heat of fusion at 0° C = 3.4×10^5 J/kg. Specific heat capacity of water is 4.2×10^3 J/kg-K and of ice is 2.1×10^3 J/kg-K.

Solution: (*i*) Cooling liquid below its freezing point is called super cooling. In this process molecules loose their kinetic energy as the liquid cools but it does not solidify immediately. A small disturbance such as dust particle, or stirring will solidify it. As soon as the substance solidifies it returns to its melting point (in case of water 0° C). The melting point is the only temperature at which the solid and liquid co-exist in equilibrium.

If m mass of ice melts, then heat released (latent heat) raises temperature of the super cooled liquid 0° C and of calorimeter from –5° C to 0° C. Thus

Heat released = $mL = m \times 3.4 \times 10^5$ J

Heat released to raise the temperature of water

$$Q_1 = mS\Delta T = 0.5 \times 4.2 \times 10^3 \times 5 = 1.05 \times 10^4 \text{ J}$$

Heat required to raise the temperature of calorimeter

$$Q_2 = mS\Delta T = 85 \times 5 = 425 = 0.0425 \times 10^4 \text{ J}$$

Thermal capacity = mass × specific heat
Total heat required = $(1.05 + 0.0425)\ 10^4 = 1.0925 \times 10^4$ J

$$m \times 3.5 \times 10^5 = 1.0925 \times 10^4$$

or $$m = \frac{1.0925 \times 10^4}{3.5 \times 10^5} = 0.0312 \text{ kg}$$

(*ii*) If 1 kg of water at 0° C is first freezed and then taken to –5° C, the heat extracted in this process is same as extracted if water is cooled to –5° C and then freezed.
Heat extracted in freezing 1 kg of water = $1 \times 3.4 \times 10^5$ J
Heat extracted in cooling ice from 0 °*C* to –5° C = $1 \times 2.1 \times 10^3 \times 5$ J
Total heat extracted = $3.4 \times 10^5 + 10.5 \times 10^3 = 3.505 \times 10^5$ J
Heat extracted when water is supercooled = $1 \times 4.2 \times 10^3 \times 5$ J
If L is latent heat of fusion of ice at –5° C then heat extracted in freezing supercooled water at –5° C = $mL = 1 \times L = L$

$$\therefore \quad L + 21.0 \times 10^3 = 3.505 \times 10^5$$

or $$L = 3.505 \times 10^5 - 0.21 \times 10^5 = 3.295 \times 10^5 \text{ J/kg.}$$

Example 15.27: A horizontal cylindrical vessel of length $2l$ is separated by a thin heat insulating piston into two parts each of which contains n moles of an ideal monoatomic gas at temperature T. The piston is connected to the end faces of the vessel by undeformed springs of stiffness k. When an amount of heat Q is supplied slowly to the gas in the right portion, the piston is displaced by $x = l/2$. Calculate the amount of heat Q' given away at temperature T to the thermostat with which the gas in left portion is in thermal contact at all the time.

Solution: Let instantaneous extension in one spring is x, then compression in other spring is x. If P_1 and P_2 are the pressures of the gases in two parts, then

$$P_1S - kx = P_2S + kx \qquad \ldots(i)$$

where S is the area of cross-section of the piston. If dx is the small change in the position of piston, then total work done by the system is

$$dW = (P_1 - P_2)dV = (P_1 - P_2)Sdx$$

$$\therefore \quad W = \int dW = \int_0^{l/2}(P_1 - P_2)S\,dx = \int_0^{l/2} 2\,kx\,dx = \frac{1}{4}kl^2$$

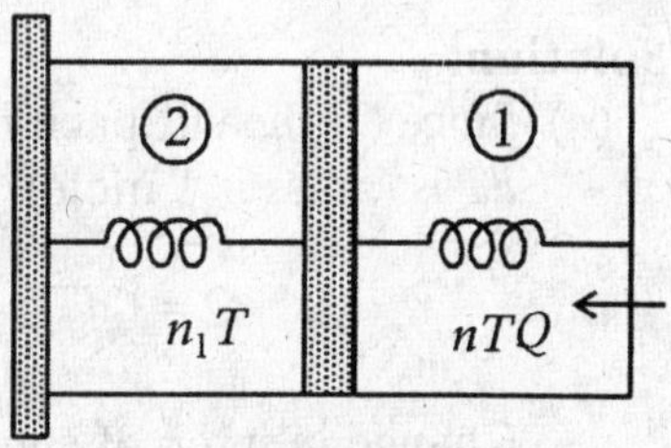

Fig. 15.33

Change in internal energy

$$\Delta U = \Delta U_1 + \Delta U_2$$

$\Delta U_2 = 0$ since heat is given to part (2) isothermally

$$\therefore \quad \Delta U = nC_V\Delta T = \frac{nR}{\gamma - 1}\Delta T = \frac{nR}{5/3 - 1}\Delta T = \frac{3}{2}nR\Delta T$$

Q is heat supplied to the gas in right portion and let Q' be transferred to left. Thus

$$Q - Q' = \Delta U + \Delta W = \frac{3}{2}nR\,\Delta T + \frac{1}{4}kl^2 \qquad \ldots(ii)$$

ΔT can be calculated from the end states of the process. In end states $x = l/2$

$$\therefore \quad P_1 = P_2 + \frac{2k \cdot l/2}{S} = P_2 + \frac{kl}{S} \qquad \ldots(iii)$$

According to gas equation

$$P_1 \times \left(l + \frac{l}{2}\right) S = nR(T + \Delta T) \quad \text{or} \quad P_1 = \frac{2}{3}\frac{nR(T + \Delta T)}{lS}$$

and $$P_2 \times \left(l - \frac{l}{2}\right) S = nRT \quad \text{or} \quad P_2 = \frac{2n\,RT}{lS}$$

Substituting values of P_1 and P_2 in (*iii*), we have

$$\frac{2}{3}\frac{nR(T + \Delta T)}{lS} = \frac{2n\,RT}{lS} + \frac{kl}{S}$$

or $$nR(T + \Delta T) = 3n\,RT + \frac{3}{2}kl^2$$

or $$nR\,\Delta T = 2n\,RT + \frac{3}{2}kl^2$$

From (*ii*) we have

$$Q - Q' = \frac{3}{2}\left\{2nRT + \frac{3}{2}kl^2\right\} + \frac{1}{4}kl^2$$

$$= 3nRT + \frac{9}{4}kl^2 + \frac{1}{4}kl^2$$

$$= 3nRT + \frac{5}{2}kl^2$$

$\therefore$ $$Q' = Q - 3nRT - \frac{5}{2}kl^2$$

Example 15.28: An ideal gas has an adiabatic exponent γ. In some process its molar heat capacity varies as $C = \frac{\alpha}{T}$, where α is a constant. Find (*a*) the work done by one mole of the gas during its heating from the temperature T_0 to ηT_0. (*b*) the equation of the process in the variables *P*, *V*.

Solution:

(*a*) Since *C* is heat capacity, hence heat supplied to the system in raising its temperature through dT is (mass is 1 mole)

$$Q = CdT = \frac{\alpha}{T}dT$$

Change in internal energy $U = C_V\,dT = \frac{R}{\gamma - 1}dT$

and if work done is dW then

$$Q = U + dW$$

or $$dW = Q - U = \frac{\alpha}{T}dT - \frac{R}{\gamma - 1}dT \qquad \ldots(i)$$

Hence work done in raising temperature from T_0 to ηT_0 is

$$W = \int_{T_0}^{\eta T_0}\left[\frac{\alpha}{T}dT - \frac{R}{\gamma - 1}dT\right]$$

$$= \left[\alpha \log T - \frac{RT}{\gamma - 1}\right]_{T_0}^{\eta T_0}$$

$$= \alpha(\log \eta T_0 - \log T_0) - \frac{R}{\gamma - 1}(\eta T_0 - T_0)$$

$$= \alpha \log \eta - \frac{RT_0}{\gamma - 1}(\eta - 1)$$

(*b*) Rewriting equation (*i*), we have

$$\frac{\alpha}{T} dT = \frac{R}{\gamma - 1} dT + PdV \qquad dW = PdV$$

Using ideal gas equation we have

$$PV = RT \qquad \ldots(ii)$$

or $PdV + VdP = RdT$ ...(*iii*)

Eliminating T and dT we have

$$\frac{R\alpha}{PV}\frac{(PdV + VdP)}{R} = \frac{R}{\gamma - 1}\frac{(PdV + VdP)}{R} + PdV$$

or $$\alpha(\gamma - 1)\frac{(PdV + VdP)}{PV} = \gamma PdV + VdP$$

or $$\alpha(\gamma - 1)\frac{(PdV + VdP)}{PV} = PV\left(\gamma \frac{dV}{V} + \frac{dP}{P}\right)$$

or $$\alpha(\gamma - 1)\frac{(PdV + VdP)}{P^2V^2} = \gamma \frac{dV}{V} + \frac{dP}{P}$$

or $$-\alpha(\gamma - 1)\, d\left(\frac{1}{PV}\right) = \gamma \frac{dV}{V} + \frac{dP}{P}$$

Integrating, we get

$$-\frac{\alpha(\gamma - 1)}{PV} = \gamma \log V + \log P + k$$

k is constant of integration. Taking $k = \log C$, we have

$$-\frac{\alpha(\gamma - 1)}{PV} = \log(CPV^{\gamma})$$

or $$CPV^{\gamma} = e^{-\alpha(\gamma - 1)/PV}$$

or $$PV^{\gamma}e^{\alpha(\gamma - 1)/PV} = \text{constant}$$

Example 15.29: Three moles of an ideal monoatomic gas perform a cycle shown in Fig. 15.34. The temperature of the gas in different states marked as *A*, *B*, *C* and *D* in figure are 400 K, 700 K, 2500 K and 1100 K respectively. Determine the work done by the gas during the process *A-B-C-D-A*.

Solution: Work done W is

$$W = \text{area } ABCDA = \text{area } OABCDO - \text{area } OADO$$

$$= \frac{1}{2} BC \times P_2 - \frac{1}{2} AD \times P_1$$

$$= \frac{1}{2}(V_3 - V_2)P_2 - \frac{1}{2}(V_4 - V_1)P_1$$

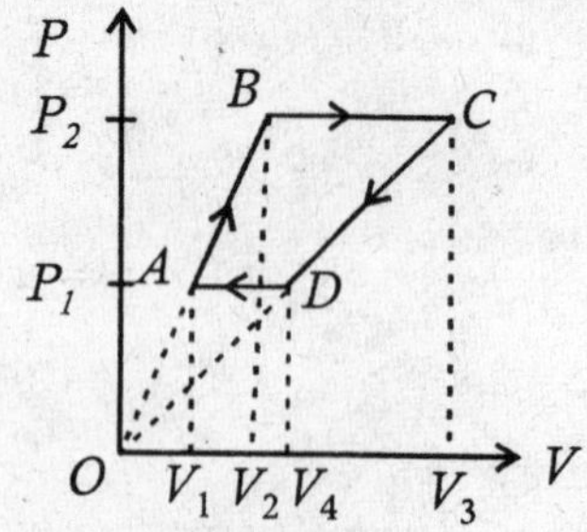

Fig. 15.34

Using ideal gas equation $PV = nRT$

$P_2V_3 = 3RT_c$, $P_2V_2 = 3RT_b$: $P_1V_4 = 3RT_d$ and $P_1V_1 = 3RT_a$

$$\therefore \quad W = \frac{3R}{2}[T_c - T_b - T_d + T_a]$$

$$= \frac{3R}{2}[2500 - 700 - 1100 + 400]$$

$$= 1650R = 1650 \times 8.4 = 1.38 \times 10^4 \text{ J.}$$

Example 15.30: An engine draws heat Q from the source and would release Q' at the sink if ideal. But in practice it also loses an amount $Q'' = 0.2Q'$ for other imperfections.

(*a*) For source at 600 K and sink at 300 K find the efficiency.
(*b*) If source temperature decreases to 500 K, find the decrease in efficiency.
(*c*) To make efficiency 60%, find the source temperature.
(*d*) If efficiency approaches zero, find to what temperature will the source approach.

Solution:

(*a*) Using the relation $\frac{Q}{T}$ = constant or $Q = CT$.

Heat extracted from source = $C(600)$

Heat rejected under ideal conditions = $C(300)$

But actual heat rejected = $C(300) + C(60) = C(360)$

Hence work done = $C(600) - C(360) = C(240)$

$$\therefore \text{ Efficiency } \eta = \frac{C(240)}{C(600)} = 40\%.$$

(*b*) When source temperature is 500 K.

Heat extracted = $C(500)$

Heat rejected = $C(300)$, since sink temperature is 300 K.

Work done $W = C(500) - C(360) = C(140)$

$$\therefore \text{ Efficiency } \eta = \frac{C(140)}{C(500)} \times 100 = 28\%$$

The efficiency decreases by 12%.

(*c*) If T is the temperature of sink for 60% efficiency, then heat extracted = $C(T)$ and heat rejected = $C(360)$

Work done = $C(T) - C(360)$

or efficiency $\eta = \dfrac{C(T) - C(360)}{C(T)} = 0.6$

or $\quad T - 360 = 0.6T$

or $\quad 0.4\,T = 360$ and $T = 900$ K.

(*d*) For efficiency zero. The heat extracted = heat rejected = $C(360)$. If T is source temperature then $C(T) = C(360)$ or $C = 360$ K.

Hence efficiency approaches zero if source temperature approaches 360 K.

Example 15.31: Two steel rods and an aluminium rod of equal length l_0 and equal cross-section are joined rigidly at their ends as shown in Fig. 15.34(*a*). All the rods are in a state of zero tension at 0° C. Find the length of the system when the temperature is raised to θ. Coefficient of linear expansion of aluminium and steel are α_a and α_s respectively. Young's modulus of aluminium is Y_a and of steel Y_s.

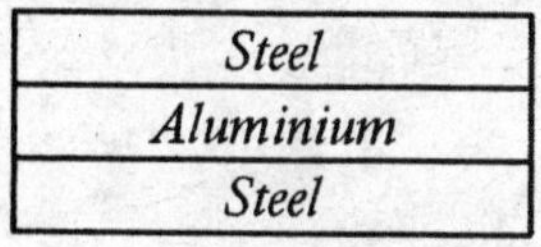

Fig. 15.34 *a*

Solution: When temperature is increased the extension in the two rods is

$$\Delta l_s = l_0\alpha_s\theta \text{ and } \Delta l_a = l_0\alpha_a\theta$$

$$\therefore \quad \Delta l_s - \Delta l_a = l_0\theta(\alpha_s - \alpha_a) \qquad ...(i)$$

Since the ends are joined rigidly as such steel rod will try to increase the length of aluminium rod and aluminium rod tries to decrease the length of steel rod. As such the tension in the the two rods is in opposite direction and that in aluminium rod tension is $2T$ when tension in steal rod is T.

Hence extension in steel rod $\Delta l_s = \dfrac{Tl_0}{AY_s}$

and compression in aluminium rod $\Delta l_a = \dfrac{2T\,l_0}{AY_a}$

$$\therefore \quad \Delta l_s - \Delta l_a = \frac{Tl_0}{A}\left(\frac{1}{Y_s} + \frac{2}{Y_a}\right)$$

Here + sign in right hand side is due to the opposite direction of the tension in two rods. Substituting $\Delta l_s - \Delta l_a$ from (*i*), we have

$$l_0\theta(\alpha_s - \alpha_a) = \frac{Tl_0}{A}\left(\frac{1}{Y_s} + \frac{2}{Y_a}\right)$$

or
$$T = \frac{\theta \cdot A(\alpha_s - \alpha_a)Y_sY_a}{Y_a + 2Y_s}$$

Hence net increase in steel rod

$$\Delta l = l_0\alpha_s\theta - \frac{Tl_0}{AY_s}$$

$$= l_0\alpha_s\theta - \frac{\theta l_0(\alpha_s - \alpha_a)Y_a}{Y_a + 2Y_s}$$

$$= l_0\theta\left[\alpha_s - \frac{(\alpha_s - \alpha_a)Y_a}{Y_a + 2Y_s}\right]$$

$$= l_0\theta\left[\frac{\alpha_s Y_a + \alpha_s 2Y_s - \alpha_s Y_a + \alpha_a Y_a}{Y_a + 2Y_s}\right]$$

$$= l_0\theta\left(\frac{\alpha_a Y_a + 2\alpha_s Y_s}{Y_a + 2Y_s}\right)$$

Hence total length of the rod is

$$L = l_0 + \Delta l = l_0\left[1 + \frac{\alpha_a Y_a + 2\alpha_s Y_s}{Y_a + 2Y_s}\theta\right].$$

PROBLEMS

1. 2.0 moles of a monoatomic ideal gas ($U = 1.5nRT$) is enclosed in an adiabatic vertical cylinder fitted with a smooth light adiabatic piston. The piston is connected to a vertical spring of spring constant 200 N/m as shown is Fig. 15.35. The area of cross-section of the cylinder is 20.0 cm^2. Initially the spring is at its natural length and temperature of the gas is 300 K. The atmospheric pressure is 100 kPa. The gas is heated slowly for some time by means of an electric heater so as to move the piston up through 10 cm. Find (*a*) the work done by the gas. (*b*) the final temperature of the gas and (*c*) heat supplied by the heater.

   [Ans. (*a*) $W = 21$ J, (*b*) $T_f = 331$ K, (*c*) $\Delta Q = 793$ J]

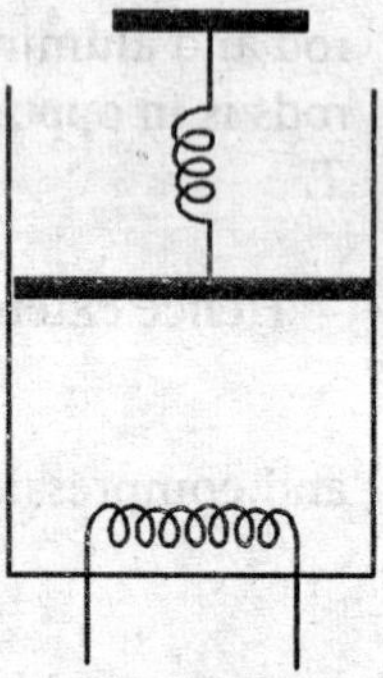

Fig. 15.35

2. A sample of 2 kg monoatomic helium (assumed ideal) is taken through the process *ABC* and another sample of 2 kg of same gas is taken through the process *ADC* (Fig. 15.36). Given molecular mass of helium = 4.

   (*i*) What is the temperature of helium in each of the states *A*, *B*, *C* and *D*?

   (*ii*) Is there any way of telling after wards which sample of helium went through the process *ABC* and which went through the process *ADC*? Write yes or no.

   (*iii*) How much is the heat involved in each of the processes *ABC* and *ADC*?

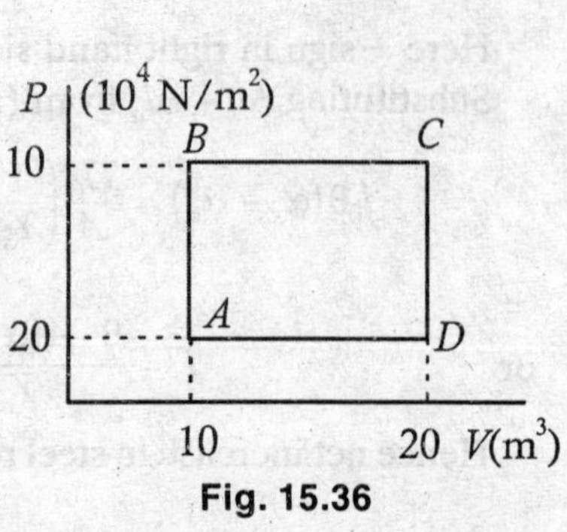

Fig. 15.36

[Ans. (*i*) $T_A = 120.3$ K, $T_B = 240.6$ K, $T_C = 481.2$ K, $T_D = 240.6$ K (*ii*) No (*iii*) $Q_{ABC} = 3.25 \times 10^6$ J, $Q_{ADC} = 2.75 \times 10^6$ J]

3. One mole of an ideal monoatomic gas is taken round the cycle ABCA as shown in Fig. 15.37. Calculate
   (*a*) the work done by the gas
   (*b*) the heat rejected by the gas in the path *CA* and the heat absorbed by the gas in the path *AB*
   (*c*) the net heat rejected or absorbed by the gas in path *BC*
   (*d*) the maximum temperature attained by the gas during the cycle.

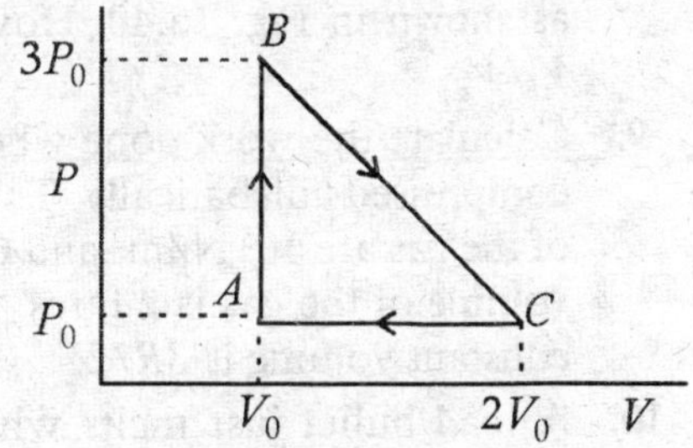

Fig. 15.37

$$\left[\textbf{Ans.}\ (a)\ W = P_0V_0\ (b)\ Q_1 = \frac{5}{2}P_0V_0,\ Q_2 = 3P_0V_0\ (c)\ Q_3 = \frac{P_0V_0}{3}\ (d)\ T = \frac{3P_0V_0}{R}\right].$$

4. In the adjoining Fig. 15.38, an ideal gas changes its state from state *A* to state *C* by two paths *ABC* and *AC* (*i*) Find the path along which the work done is the least (*ii*) The internal energy of the gas at *A* is 10 J and the amount of heat supplied to change its state to *C* through the path *AC* is 200 J. Calculate the internal energy at *C* (*iii*) The internal energy of the gas at state *B* is 10 J. Find the amount of heat supplied to gas to go from *A* to *B*.

   [**Ans.** (*i*) W_{ABC} = 60 J, W_{AC} = 40 J, thus work is least along *AC* (*ii*) U_C = 170 J (*iii*) Q_{AB} = 10 J]

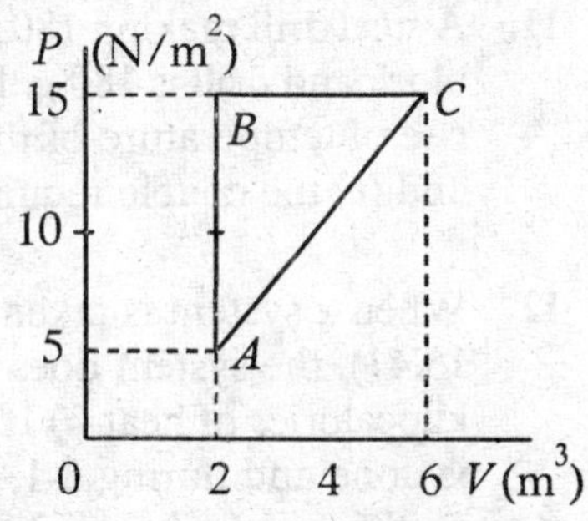

Fig. 15.38

5. At 27° C two moles of an ideal monoatomic gas occupy a volume *V*. The gas expands adiabatically to volume 2*V*. Calculate (*i*) the final temperature of the gas (*ii*) change in its internal energy and (*iii*) work done by the gas during this process.

   [**Ans.** (*i*) *T* = 189 K (*ii*) Δ*U* = –2764 J (*iii*) *W* = 2764 J]

6. Let the temperatures T_1 and T_2 of the two heat reservoirs in an ideal Carnot engine be 1500° C and 500° C respectively. Which of these, increasing T_1 by 100° C or decreasing T_2 by 100° C would result in a greater improvement in the efficiency of the engine?

   [**Ans.** In first case efficiency is 0.59 in second case it is 0.62]

7. An ideal gas is taken through a cyclic thermodynamic process through four steps starting from *A* as shown in Fig. 15.39. Volume ratios are $\frac{V_B}{V_A} = 2$ and $\frac{V_D}{V_A} = 4$. If temperature T_a at *A* is 27° C, calculate (*a*) the temperature of the gas at *B* (*b*) Heat absorbed or released by the gas in each process (*c*) the total work done by the gas during the complete cycle. Express your answer in terms of the gas constant *R*.

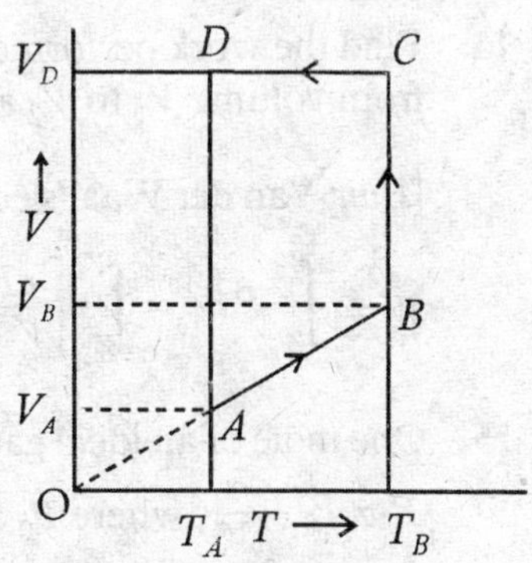

Fig. 15.39

   [**Ans.** (*a*) T_B = 600 K (*b*) Q_{AB} = 1500 R. W_{BC} = 831.8 R. U_{BC} = 0, Q_{BC} = 831.8 R, W_{CD} = 0, W_{DA} = –831.8 R. U_{DA} = 0, Q_{DA} = –831.8 R (*c*) *W* = 600 R]

8. Two different adiabatic for the same gas intersect two isothermals at T_1 and T_2 temperatures

as shown in Fig. 15.40. How does V_a/V_d compare with V_b/V_c. **[Ans.** $V_a/V_d = V_b/V_c$]

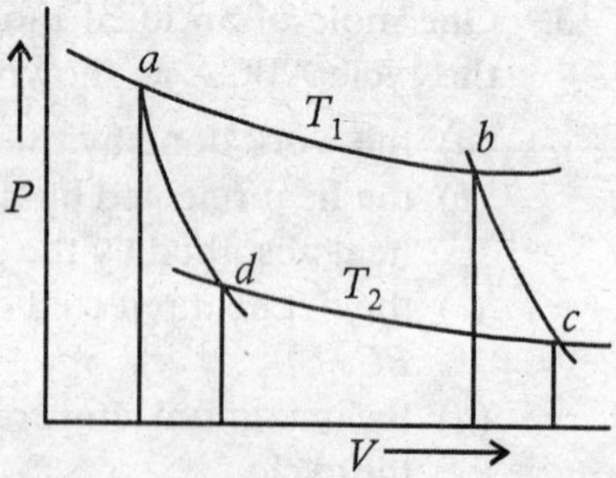

Fig. 15.40

9. Calculate the work done when 1 mole of a perfect gas is compressed adiabatically. The initial pressure and volume of the gas are 10^5 N/m^2 and 6 litres respectively. The final volume of the gas is 2 litres. Molar specific heat of gas at constant volume is $3R/2$. **[Ans.** $W = -957$ J]

10. A lead bullet just melts when stopped by an obstacle. Assuming that 25% of heat is absorbed by the obstacle. Find the velocity of the bullet if its initial temperature is 27° C. Given melting point of lead = 327° C, specific heat of lead = 0.03 cal/gm °C, latent heat of fusion of lead = 6 cal/g, J = 4.2 Joule/cal. **[Ans.** $V = 409.9$ m/s]

11. A steel drill making 180 rev/min is used to drill a hole in a block of steel. The mass of the steel block and drill is 180 g. If the entire mechanical work is used up in producing heat and rate of rise of temperature of the block 0.5° C/sec. Find (*a*) the rate of working of the drill in Watt and (*b*) the couple required to drive the drill. (Specific heat of steel = 0.1 cal/g °C).
[Ans. $P = 37.8$ Watt, $\tau = 2$ N-m]

12. When a system is taken from state 1 to 3 along path 1-2-3 (Fig. 15.41), the system does 25 kilo calories of work and absorbs 75 kilocalories of heat (*i*) If in 3-4-1 process 10 kilo calories of work is done and during 3-1, 15 kilo calories of work has to be done on the system, calculate the heat evolved on these two paths. (*ii*) If internal energy is zero at state point 1 and 45 kilo calories at state point 4. Calculate the heat on the path 3-4.
[Ans. (*i*) For process 3-4-1 $Q = -60$ k.cal for 3-1 $Q = -65$ k cal
(*ii*) process 3-4 $Q = -5$ kcal]

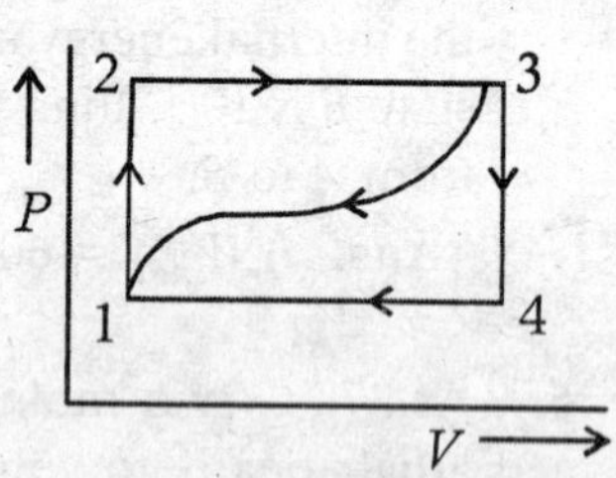

Fig. 15.41

13. The internal energy of a gas is given by $U = 1.5PV$. It expands from 100 cm^3 to 200 cm^3 against a constant pressure of 1.0×10^5 Pa. Calculate the heat absorbed by the gas in the process. **[Ans.** 25 J]

14. Find the work performed by 1 mole of a Van der Waal's gas, during its isothermal expansion from volume V_1 to V_2 at temperature T.

[*Hint*: Van der Waal's equation is $\left(P + \frac{a}{V^2}\right)(V - b) = nRT$, where a and b are constants. Hence

$$W = \int_{V_1}^{V_2} PdV = \int_{V_1}^{V_2} \left[\frac{nRT}{V-b} - \frac{a}{V^2}\right] dV = nRT \log_e \frac{(V_2 - b)}{V_1 - b} + a\left(\frac{1}{V_2} - \frac{1}{V_1}\right);$$ T is constant for isothermal process]

15. One mole of an ideal gas whose heat capacity at constant pressure is C_P: undergoes the process $P = P_0 + \frac{\alpha}{V}$, where P_0 and α are positive constants. Calculate (*a*) molar heat capacity of the gas as a function of its volume (*b*) the increase in internal energy of the gas, the work done by the gas and heat supplied to gas if it expands from volume V_1 to V_2.

$$\left[\textbf{Ans. } (a)\ C = C_P + \frac{\alpha R}{P_0 V}\quad (b)\ \Delta U = \frac{P_0}{\gamma - 1}(V_2 - V_1),\ \Delta W = P_0(V_2 - V_1) + \alpha \log_e \frac{V_2}{V_1},\right.$$

$$\left. Q = \frac{\gamma P_0}{\gamma - 1}(V_2 - V_1) + \alpha \log_e \frac{V_2}{V_1}\right].$$

16. An ideal gas whose adiabatic exponent equals γ is expanded so that the amount of heat transferred to the gas is equal to the decrease in its internal energy. Find (*a*) the molar heat capacity of the gas in this process, (*b*) the equation of process in T and V variables, (*c*) the work performed by one mole of the gas when its volume increases x times if the initial temperature of the gas is T_0.

$$\left[\textbf{Ans. } (a)\ C = -\frac{R}{\gamma-1}\quad (b)\ TV^{(\gamma-1)/2} = \text{Constant}\quad (c)\ W = \frac{2RT_0}{\gamma-1}\left\{1 - x^{(1-\gamma)/2}\right\}\right]$$

17. Figure 15.42 shows three processes for an ideal gas. One is isobaric, one is isothermal and remaining one is adiabatic. The temperature at A is 600 K, pressure 16 atm and volume 1 litre. The volume at B is 4 litres. The ratio of specific heats of the gas $\gamma = 1.5$. (*i*) Which process is adiabatic? Why? (*ii*) Calculate the pressure of the gas at B and C (*iii*) Calculate the temperature of the gas at B and C (*iv*) Calculate the volume at C.

[**Ans.** (*i*) AB is adiabatic its slope is greater. (*ii*) $P_B = P_C = 2$ atm. (*iii*) $T_C = 600$ K, $T_B = 200$ K (*iv*) $V_C = 8$ litres]

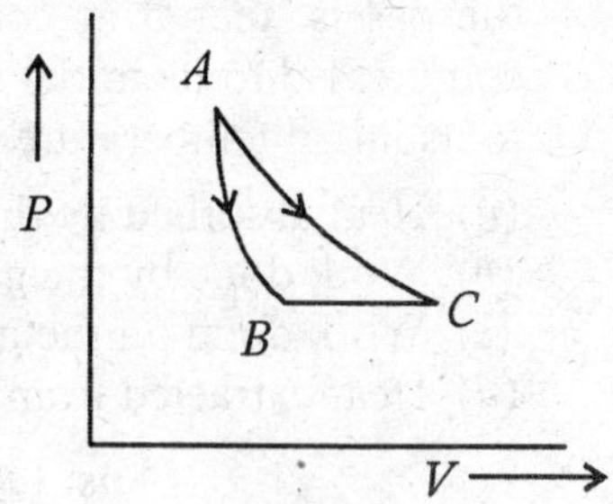

Fig. 15.42

18. Consider the cyclic process $ABCA$ shown in Fig. 15.43 performed on a sample of 2.0 moles of an ideal gas. A total of 1200 J of heat is withdrawn from the sample in the process. Find the work done by the gas during the part BC.

[**Ans.** –4520 J]

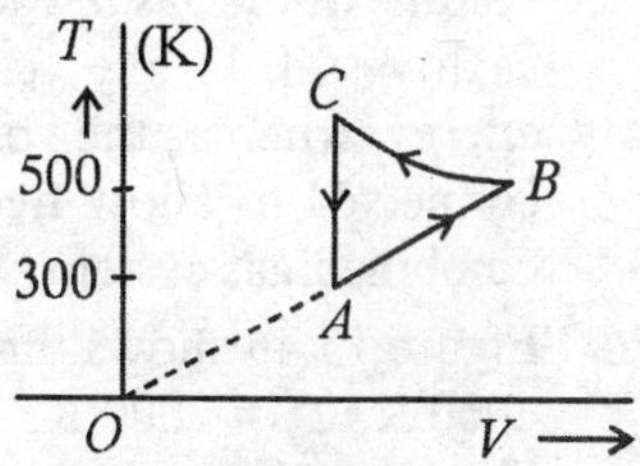

Fig. 15.43

19. The internal energy of a gas is given by $U = 1.5PV$. It expands from 100 cc to 200 cc against a constant pressure of 1.0×10^5 Pa. Calculate the heat absorbed by the gas in the process

[**Ans.** 25 J]

20. A gas is initially at pressure of 10^5 N/m^2 and its volume 2.0 m^3. Its pressure is kept constant and the volume is changed from 2.0 m^3 to 2.5 m^3. Its volume is now kept constant and the pressure is increased from 1.0×10^5 to 2.0×10^5 N/m^2. The gas is brought back to its initial state, the pressure varying linearly with its volume (*a*) whether the heat is supplied to or extracted from the gas in the complete cycle? How much heat was supplied or extracted?

[**Ans.** $Q = 25000$ J extracted]

21. A gas is enclosed is a cylindrical vessel fitted with frictionless piston. The gas is slowly heated for sometime. During the process, 10 J of heat is supplied and the piston is found to move out 10 cm. Find the increase in the internal energy of the gas. The area of cross-section of cylinder = 4 cm^2 and atmospheric pressure 100 kPa. [**Ans.** 6 J]

22. Two steel rods and an aluminium rod of equal length l_0 and equal cross-section are joined rigidly at their ends as shown in Fig. 15.44. All the rods are in a state of zero tension at 0° C. Find the length of the system when the temperature is raised to θ. Coefficient of linear expansion of aluminium and steel are α_a and α_s respectively. Young's modulus of aluminium is Y_a and of steel is Y_s.

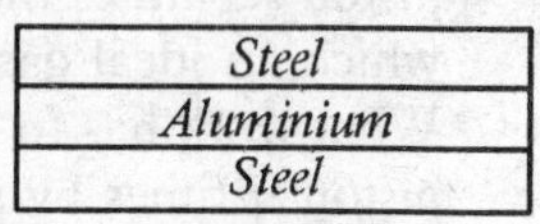

Fig. 15.44

$$\left[\textbf{Ans. } l_0\left\{1 + \frac{\alpha_a Y_a + 2\alpha_s Y_s}{Y_a + 2Y_s}\theta\right\}\right]$$

23. A steel tape 1 m long is caliberated for a temperature of 27° C. The length of a steel rod measured by this tape is found to be 63.0 cm on a hot day when the temperature is 45° C. What is the actual length of the steel rod on that day? What is the length of the same steel rod on a day, when the temperature is 27° C? Coefficient of linear expansion of steel is $1.2 \times 10^{-5}/°C$

[**Ans.** (*i*) 63.0136 cm (*ii*) 63.0 cm]

24. A mass of 8 gm of oxygen at pressure of one atmosphere and 27° C is enclosed in a cylinder fitted with a frictionless piston. The following operations are performed in the order given: (*a*) the gas is heated at constant pressure to 127° C, (*b*) then it is compressed to its initial volume isothermally, (*c*) finally it is cooled to its initial temperature at constant volume. Calculate

(*a*) Heat absorbed by the gas during operation (*a*)
(*b*) Work done by the gas in process (*a*)
(*c*) Work done on the gas in stage (*b*)
(*d*) Heat extracted from the gas in stage (*c*).

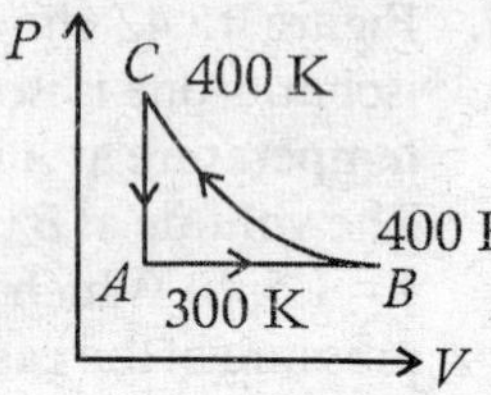

Fig. 15.45

[**Ans.** (*a*) $Q = 743.8$ J (*b*) $\Delta W = 207.8$ J (*c*) $\Delta W = -293.1$ J (*d*) $\Delta Q = -536$ J]

25. There are two vessels. Each of them contains one mole of a monoatomic ideal gas. Initially volume of the gas in each vessel is 8.3×10^{-3} m³ at 27° C. Equal amount of heat is supplied to each vessel. In one vessel, the volume of the gas is doubled without change in its internal energy, whereas the volume of the gas is held constant in second vessel. The vessels are now connected to allow free mixing of the gas. Find the final temperature and pressure of the combined gas system. [**Ans.** $T = 438.6$ K, $P = 2.45 \times 10^5$ N/m²]

26. Figure 15.46 shows three isothermals at temperatures $T_1 = 4000$ K, $T_2 = 2000$ K, $T_3 = 1000$ K. When one mole of an ideal monoatomic gas is taken through the paths $A \rightarrow B$, $B \rightarrow C$, $C \rightarrow D$ and $D \rightarrow A$. Find the change in internal energy ΔU, the work done by the gas W and Q the heat absorbed by the gas in each path. Also find these quantities for complete cycle $ABCDA$. Given $V_A = 1$ m³ and $V_B = 2$ m³.

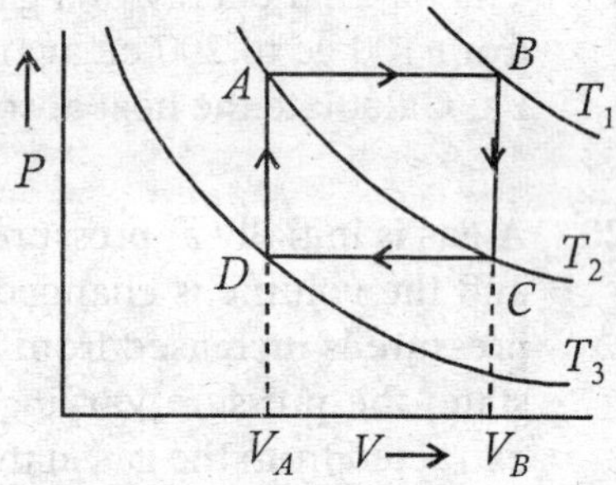

Fig. 15.46

[**Ans.** $\Delta U_{AB} = 24.93 \times 10^3$ J, $\Delta U_{BC} = -24.93 \times 10^3$ J, $\Delta U_{CD} = -12.465 \times 10^3$ J, $\Delta U_{DA} = 12.465 \times 10^3$ J, Total $\Delta U = 0$, $\Delta W_{AB} = 16.62 \times 10^3$ J, $\Delta W_{BC} = \Delta W_{DA} = 0$, $\Delta W_{CD} = 8.31 \times 10^3$ J. Total $W = 8.31 \times 10^3$ J, $\Delta Q_{AB} = 41.55 \times 10^3$ J, $\Delta Q_{BC} = -24.93 \times 10^3$ J, $\Delta Q_{CD} = -20.775 \times 10^3$ J, $\Delta Q_{DA} = 12.465 \times 10^3$ J, Total $Q = 8.31 \times 10^3$ J]

27. A piston can freely move inside a horizontal cylinder closed from both ends. Initially the piston separates the inside space of the cylinder into two equal parts each of volume V_0 in which an ideal gas is contained under the same pressure P_0 and at the same temperature. What work has to be performed in order to increase isothermally the gas volume under the piston η times by slowly raising the piston? The friction of the piston against the walls of cylinder is negligibly small.

$$\left[\textbf{Ans. } W = P_0 V_0 \log_e \left[\frac{(\eta+1)^2}{4\eta}\right]\right]$$

28. Two vessels A and B, thermally insulated contain an ideal monoatomic gas. A small tube fitted with a valve connects these vessels. Initially the vessel A has 2 litres of gas at 300 K and 2×10^5 N/m² pressure. While vessel A has 4 litres of gas at 350 K and 4×10^5 N/m² pressure. The valve is now opened and the system reaches equilibrium in pressure and temperature. Calculate the new temperature and pressure. [**Ans.** $T = 339$ K, $P = 3.33 \times 10^5$ N/m²]

29. One mole of a diatomic ideal gas ($\gamma = 1.4$) is taken through a cyclic process starting from point A. The process $A \to B$ is an adiabatic compression, $B \to C$ is isobaric expansion CD an adiabatic expansion and DA is isochoric. The volume ratios are $V_a/V_b = 16$ and $V_c/V_b = 2$ and the temperature at A is $T_a = 300$ K. Calculate the temperatures of the gas at points B and D and find the efficiency of the cycle.

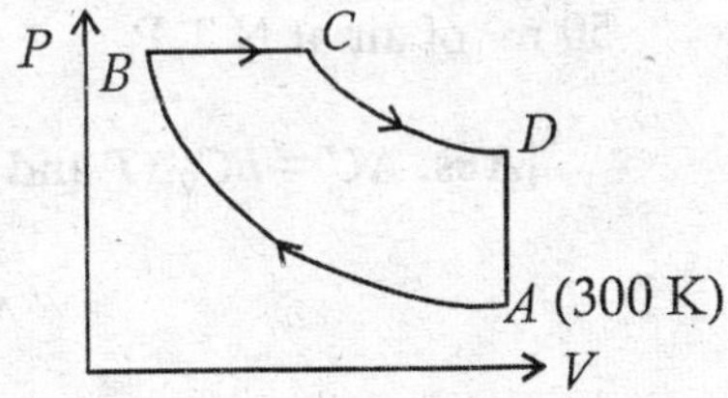

Fig. 15.47

[**Ans.** $T_b = 909$ K, $T_d = 791$ K. $\eta = 61.4\%$]

30. Consider 1 mole of an ideal gas whose pressure changes with volume as $P = \alpha V$, where α is constant. If it expands such that its volume increases η times, find change in its internal energy, work done by the gas, and heat capacity of the gas. The initial volume of the gas is V_0 and its adiabatic exponent is γ.

$$\left[\textbf{Ans. } \Delta U = \frac{\alpha(\eta^2-1)V_0^2}{\gamma-1},\ W = \frac{1}{2}\alpha V_0^2(\eta^2-1),\ C = \frac{R}{2}\frac{(\gamma+1)}{(\gamma-1)}\right]$$

31. The rectangular box shown in Fig. 15.48 has a partition which can slide without friction along the length of the box. Initially each of these two chambers of box has one mole of a monoatomic ideal gas ($\gamma = 5/3$) at a pressure P_0, volume V_0 and temperature T_0. The chamber on left is slowly heated by an electric heater. The walls of the box and partition are thermally insulated. Heat loss through lead wires of heater is negligible. The gas in left chamber expands, pushing the partition until the final pressure in both chambers become $\frac{243 P_0}{32}$. Determine (*a*) the final temperature of gas in each chamber, (*b*) work done on the gas in right chamber. [**Ans.** (*a*) $T_R = 2.25T_0$, $T_L = 12.97T_0$ (*b*) $W = -15.56T_0$]

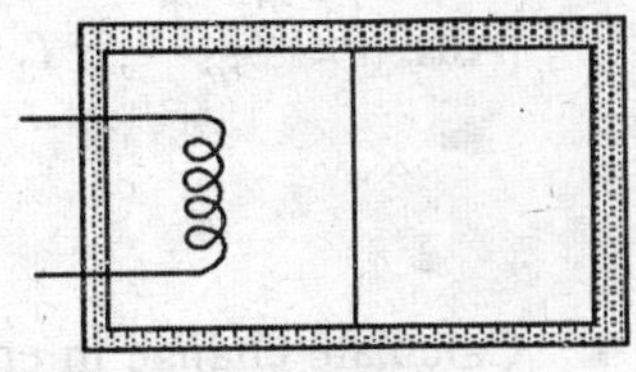
Fig 15.48

32. One mole of a certain ideal gas is contained under weightless piston of a vertical cylinder at a temperature T. The space over the piston opens into atmosphere. What work has to be performed in order to increase isothermally the gas volume under the piston η times slowly by raising the piston. The friction of the piston against the cylindrical walls is negligibly small.

[*Hint*: Work done isothermally = $RT\log\eta$ and work done on the gas by maintaining constant pressure $= \int PdV = PV(\eta-1) = RT(n-1)$. Thus total work done $= RT(n-1-\log\eta)$]

SHORT ANSWER TYPE QUESTIONS

1. Show that internal energy of an ideal gas is expressed as $\frac{PV}{\gamma - 1}$. Hence find internal energy of 50 m^3 of air at N.T.P.

[**Ans.** $\Delta U = nC_V\Delta T$ and $C_P - C_V = R$ or $C_V\left(\frac{C_P}{C_V} - 1\right) = R$ or $C_V = \frac{R}{\gamma - 1}$. $PV = nRT$. Thus

$$\Delta U = \frac{R}{\gamma - 1}\frac{PV}{R} = \frac{PV}{\gamma - 1}, \quad \Delta U = \frac{10^5 \times 50}{1.4 - 1} = \frac{50 \times 10^5}{0.4} = 1.25 \times 10^7 \text{ J}$$]

2. A heat engine works on a Carnot cycle with a heat sink at temperature 2700 K. The efficiency of the engine is 10%. Determine the temperature of heat source.

$$\left[\textbf{Ans. } \eta = \frac{T_1 - T_2}{T_1} \text{ or } 0.1 = 1 - \frac{T_2}{T_1} \text{ or } \frac{T_2}{T_1} = 1 - 0.1 = 0.9 \text{ or } T_1 = \frac{T_2}{0.9} = \frac{2700}{.9} = 3000 \text{ K}\right]$$

3. A reversible engine operates between two temperatures T_1 and $T_2 (T_1 > T_2)$, show that to increase its efficiency, it is better to decrease the temperature of sink rather than to increase temperature of source.

[**Ans.** $\eta = \frac{T_1 - T_2}{T_1}$, If T_1 is increased by x, $\eta_1 = \frac{T_1 + x - T_2}{T_1 + x}$. If T_2 is decreased by x $\eta_2 = \frac{T_1 - (T_2 - x)}{T_1}$

$$\therefore \frac{\eta_2}{\eta_1} = \frac{T_1 + x}{T_1} = 1 + \frac{x}{T_1} \quad \therefore \eta_2 > \eta_1$$]

4. Calculate change in entropy when 10 gm of ice at 0° C is converted into water at the same temperature. The latent heat of fusion of water L = cal/gm.

[**Ans.** $\Delta S = \frac{\Delta Q}{T} = -\frac{mL}{T} = -\frac{10 \times 80}{273} = -2.93$ cal-K, –ve, since heat is extracted and entropy decreases]

5. Why the internal energy of a compressed gas is less than that of a rarefield gas at the same temperature

[**Ans.** Molecules of a compressed gas experience intermolecular forces, therefore to bring molecules closer, work has to be done by the system. This causes decrease in the internal potential energy. At constant temperature the internal kinetic energy is constant. Thus net internal energy (kinetic + potential) decreases when compressed]

6. The temperature of the sun's surface is 6000 K. If we take a big lens and focus the sun rays, can we produce a temperature of 8000 K?

[**Ans.** No, second law of thermodynamics says that heat by itself cannot be transferred from cooler body to hotter body. Thus it is the violation of second law of thermodynamics to transfer heat from a body at 6000 K to 8000 K by itself]

7. A thermos bottle containing coffee is vigorously shaken. Considering the coffee as system. (*a*) Does its temperature rise? (*b*) Has heat been added to it (*c*) Has work been done on it? (*d*) Has its internal energy changed?

[**Ans.** (*a*) Yes the temperature of coffee rises, (*b*) No heat is added to coffee as thermos bottle is insulated from surroundings, (*c*) Yes, by shaking work has been done on coffee (system) against viscous forces, (*d*) Since process is adiabatic hence internal energy increases as work done is negative]

8. Water falls through a height of 250 m. Assuming that whole of the energy due to fall is converted into heat, calculate the rise in temperature of water.

[**Ans.** Energy converted into heat = $mgh = m \times 9.8 \times 250$ J. Heat produced in calories $Q = \frac{mgh}{J} = \frac{m \times 9.8 \times 250}{4.2}$ cal. If $\Delta\theta$ is rise in temperature, then $Q = mS\Delta\theta$. For water $S = 10^3$ cal/kg°C. Thus $\frac{m \times 9.8 \times 250}{4.2} = m \times 10^3 \, \Delta\theta$ or $\Delta\theta = \frac{9.8 \times 250}{4.2 \times 10^3} = 0.583°C$]

9. Calculate the heat absorbed by a system in going through the cyclic process shown in Fig. 15.49.

[**Ans.** Since the process is cyclic $\Delta U = 0$. Thus heat absorbed is equal to the work done. Work done is equal to the area of the loop. In this case area = πab, $a = 100$ cc $= 10^{-4}$ m^3, $b = 100 \times 10^3 = 10^5$ Pa. Thus $Q = \pi \times 10^{-4} \times 10^5 = 31.4$ J].

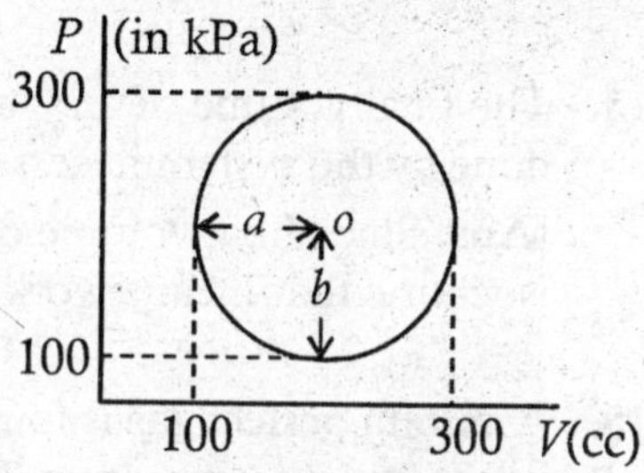

Fig. 15.49

10. A gas is taken along the path *AB* (Fig. 15.50). If 70 cal of heat is extracted from the gas in the process, calculate the change in internal energy of the system.

[**Ans.** $\Delta Q = -70$ cal $= -70 \times 4.2 = -294$ J. Work done ΔW = area enclosed by the *P-V* diagram with volume *ax* is $= -\frac{1}{2}(200 + 500) \times 10^3 \times 150 \times 10^{-6} = -52.5$ J. Negative because the work is done on the system $\therefore \Delta U = \Delta Q - \Delta W = -294 + 52.5 = -241.5$ J]

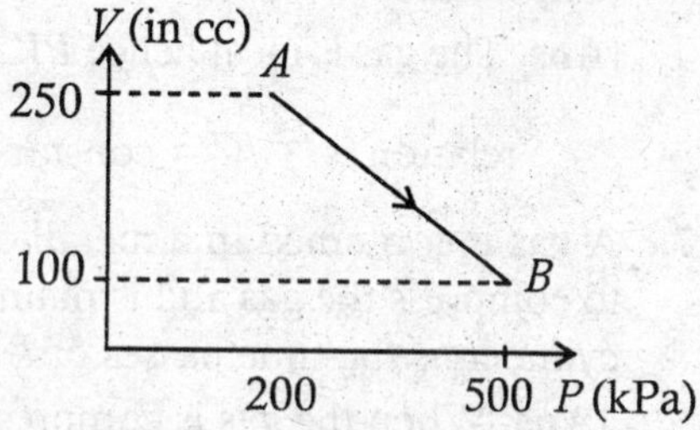

Fig. 15.50

11. An ideal gas is taken from an initial state *i* to a final state *f* in such a way that the ratio of the pressure to the absolute temperature remains constant. What will be the work done by the gas?

[**Ans.** Since $\frac{P}{T}$ is constant means volume is constant. Hence work done is zero]

12. An ungraded thermometer of uniform bore is attached to a centimeter scale and is found to read 10.3 cm in melting ice, 26.8 cm, in boiling water and 6.5 cm in freezing mixture. Calculate the temperature of freezing mixture.

[**Ans.** Size of degree $= \frac{26.8 - 10.3}{100} = \frac{16.5}{100} = 0.165°C/cm$. The temperature of freezing mixture $= \frac{(6.5 - 10.3)}{\cdot 165} = -23.03°C$]

13. Two cylinders *A* and *B* of equal capacity are connected to each other via stopcock. *A* contains a gas at standard temperature and pressure. *B* is completely evacuated. The entire system is thermally insulated. The stopcock is suddenly opened. Answer the following: (*a*) What is the final pressure of the gas in *A* and *B*? (*b*) What is the change in internal energy of the gas? (*c*) What is the change in temperature of the gas? (*d*) Do the intermediate states of the system (before settling to the final equilibrium state) lie on its *P-V-T* surface?

[**Ans.** (*a*) Since volume is doubled the final pressure is half the initial pressure i.e. 0.5 atm (*b*) Since expansion is under zero pressure hence no work is done and no heat is supplied, thus change in internal energy is zero. (*c*) Since gas is ideal hence change in temperature is zero (*d*) This process is called free expansion. The process is rapid and cannot be controlled. The intermediate states are non-equilibrium states and do not satisfy the gas equation. In due course, the gas does return to an equilibrium state which lies on its *P–V–T* surface].

14. Two identical rectangular strips, one of copper and other of steel, are rivetted together to form a bimetallic strip ($\alpha_c > \alpha_s$). On heating this strip, what shape will it take?

[**Ans.** Since $\alpha_c > \alpha_c$ as such the length of copper strip on heating will be more than that of steel. Hence the strip will bend with copper on convex side]

15. The final volume of a system is equal to the initial volume in a certain process. Is the work done by the system necessarily zero? Is it necessarily non zero?

[**Ans.** Since final volume of a system is equal to the initial volume means either the process is isochoric that means work done is necessarily zero. Or the process is cyclic in that case work done is not necessarily zero]

16. A certain perfect gas is found to obey $PV^{3/2}$ = constant during an adiabatic process. If such a gas at initial temperature *T* is adiabatically compressed to half the initial volume, find its final temperature.

[**Ans.** The gas is ideal hence $PV = nRT$ and obeys $PV^{3/2}$ = constant. Hence temperature volume relation is $T\sqrt{V}$ = constant or $T \propto \frac{1}{\sqrt{V}}$. Since *V* is halved hence temperature $= \sqrt{2}T$]

17. A gas is contained in a metallic cylinder fitted with piston. The piston is suddenly moved in to compress the gas and is maintained at this position. What will be the pressure of the gas in cylinder as the time passes on?

[**Ans.** When the gas is compressed suddenly its temperature rises. As the time passes on, the temperature decrease due to metallic walls as such the pressure decreases]

18. The density of water at 20° C is 998 kg-m^{-3} and at 40° C it is 992 kg-m^3. Find the co-efficient of cubical expansion of water.

[**Ans.** For density we have $\rho_0 = \rho_t(1 + \gamma\Delta t)$ or $\gamma = \frac{\rho_0 - \rho_t}{\rho_t \times \Delta t}$ or $\gamma = \frac{998 - 992}{992 \times 20} = 3.0 \times 10^{-4}/°C$]

19. A steel tape is caliberated at 30° C. A reading of 50 m is observed for a certain length when the temperature was 18° C (α for steel = 1.2×10^{-5}/°C). Find the correct length.

[**Ans.** Change in length when temperature changes by 12° C. $\Delta L = l\alpha\Delta T$ $\therefore$ $\Delta L = 50 \times 1.2 \times 10^{-5} \times 12 = 72 \times 10^{-4}$ m = 0.72 cm. The tape contracts. Hence correct length = 50 – 0.0072 = 49.9928 m]

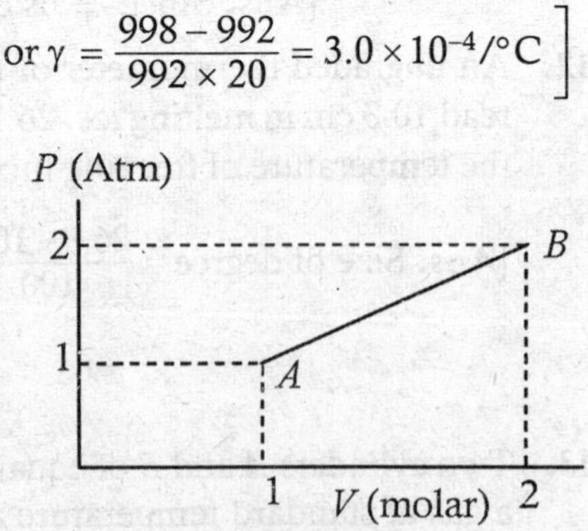

Fig. 15.51

20. Figure 15.51 gives the *P-V* indicator diagram for a monoatomic gas taken form state *A* to state *B*. The units for *P* and *V* are standard pressure and molar volume

respectively. The temperature at A is 0° C. Calculate the heat supplied in the process from A to B.
[**Ans.** PV at B is 4 times that at A. Hence temperature at B is 4×273. So change in temperature $\Delta T = 3 \times 273$. $\Delta U = nC_V\Delta T = 1 \times 3/2R \times 3 \times 273$, $n = 1$ from data at A. Work done is area under AB. $W = 3/2PV = 3/2R \times 273$ so $\Delta Q = \Delta U + \Delta W = 3/2R \times 273 \times 3 + 3/2R \times 273 = 6 \times R \times 273 = 1638R$]

21. A one litre flask contains some mercury. It is found that at all temperatures the volume of air inside the flask remains the same. If the coefficient of linear expansion of glass is $9 \times 10^{-6}/°C$ and the coefficient of volume expansion of mercury is $1.8 \times 10^{-4}/°C$; find the volume of mercury inside the flask.

[**Ans.** Volume of air = Volume of glass – volume of mercury, is constant at all temperatures.

$$\therefore \quad [V_0(1+\gamma t)]_g - [V_0(1+\gamma t)]_m = (V_0)_g - (V_0)_m \quad \text{or} \quad (V_0)_g \gamma_g t = (V_0)_m \gamma_m t \quad \text{or}$$

$$(V_0)_m = (V_0)_g \frac{\gamma_g}{\gamma_m} = \frac{1000 \times 3 \times 9 \times 10^{-6}}{1.8 \times 10^{-4}} = 150 \text{ cc}$$]

22. A gasoline engine has compression ratio 8 : 1, i.e. it expands 8 fold adiabatically during expansion. The gas is mainly air for which $\gamma = 1.4$. Find the ratio of exhaust temperature to combustion temperature.

[**Ans.** For adiabatic expansion $P_1V_1^\gamma = P_2V_2^\gamma$ and for ideal gas $\frac{P_1V_1}{T_1} = \frac{P_2V_2}{T_2}$. Eliminating pressure we have $V_1^{\gamma-1}T_1 = V_2^{\gamma-1}T_2$ or $\left(\frac{V_2}{V_1}\right)^{\gamma-1} = \frac{T_1}{T_2} \cdot \frac{V_2}{V_1} = 8$. $\therefore \frac{T_2}{T_1} = \left(\frac{1}{8}\right)^{\gamma-1} = 0.44$]

23. An ideal gas is taken through a cyclic process A→B→C→A as shown in Fig. 15.52. If total heat given to the system is 5 J, find the work done by the gas from $C \to A$.

[**Ans.** During process AB pressure is constant hence $W_{AB} = 10 \times 1 = 10$ J. Process BC is isochoric and $W_{BC} = 0$. Since the process is cyclic, the change in internal energy is zero. Thus heat given is equal to work done $\therefore W_{AB} + W_{BA} + W_{CA} = Q = 5$ or $10 + 0 + W_{CA} = 5$ or $W_{CA} = -5$ J]

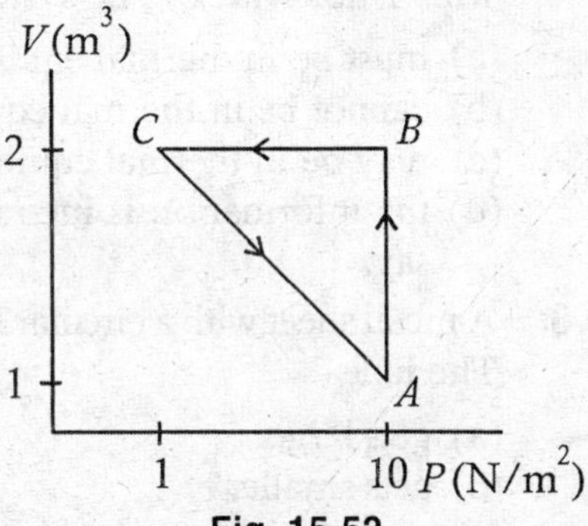

Fig. 15.52

24. The pressure and density of a diatomic gas ($\gamma = 7/5$) changes adiabatically from (P, ρ) to (P', ρ'). If $\rho'/\rho = 32$, find P'/P.

[**Ans.** For adiabatic change PV^γ = constant or $\frac{P}{\rho^\gamma}$ = constant or $\frac{P'}{P} = \left(\frac{\rho'}{\rho}\right)^\gamma = (32)^{7/5} = 128$]

MULTIPLE CHOICE QUESTIONS

PART I

In the following, a question is followed by four alternatives, of them only one is correct. Therefore, the candidate has to select the correct choice from amongst the alternatives.

1. Which of the curves in Fig. 15.53 represents the relation between Celsius and Fahrenheit temperatures?

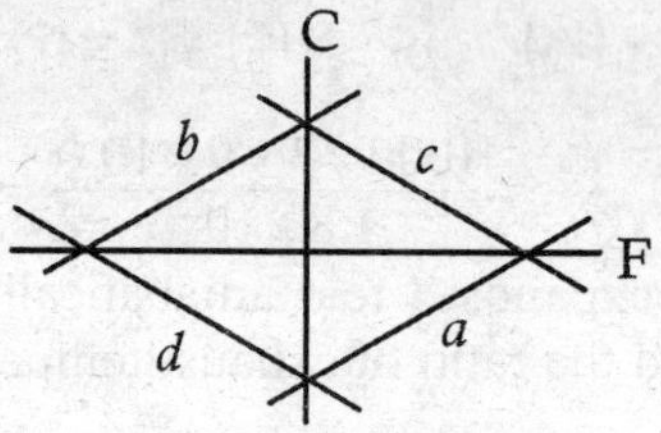

Fig. 15.53

(a) *a* (b) *b*
(c) *c* (d) *d*.

2. A system *X* is neither in thermal equilibrium with *Y* nor with *Z*. The system *Y*-*Z*
(a) must be in thermal equilibrium
(b) cannot be in thermal equilibrium
(c) may be in thermal equilibrium
(d) the information is incomplete cannot say.

3. A metal sheet with a circular hole is heated. The hole
(a) gets larger
(b) gets smaller
(c) remains of the same size
(d) gets deformed.

4. If temperature of a uniform rod is slightly increased by Δt, its moment of inertia I about a perpendicular bisector increases by
(a) zero (b) $\alpha I \Delta t$
(c) $2\alpha I \Delta t$ (d) $\frac{\alpha I \Delta t}{2}$.

5. A ideal gas is taken from state *i* to the state *f* (Fig. 15.54). The work done in the process by the gas during the process

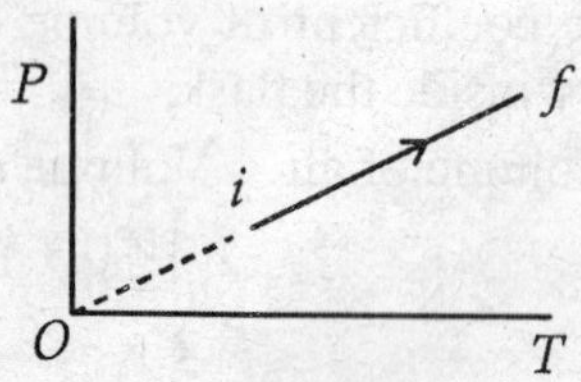

Fig. 15.54

(a) is positive (b) is negative
(c) is zero (d) cannot be obtained

6. Consider the process on a system shown in Fig. 15.55. During the process, the work done by the system

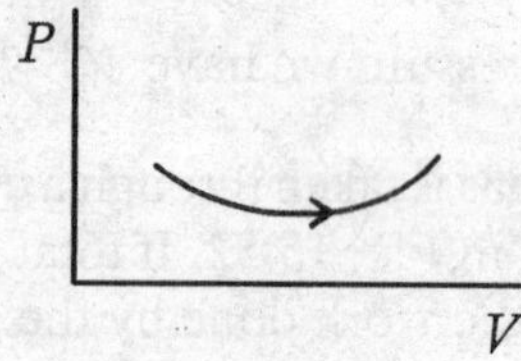

Fig. 15.55

(a) continuously increases
(b) continuously decreases
(c) first increases then decreases
(d) first increases then decreases.

7. Pressure (*P*)-volume (*V*) plots for two gases during adiabatic are shown. Plots 1 and 2 correspond to

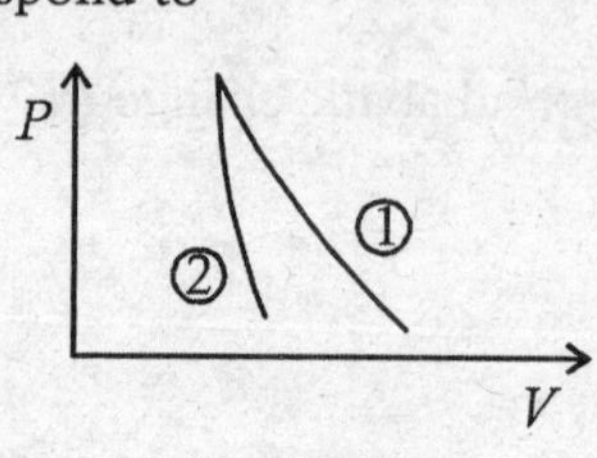

Fig. 15.56

(a) He and O_2 (b) O_2 and He
(c) He and Ar (d) O_2 and N_2.

8. In which of the following indicator diagrams do AB, BC, CA represent isothermal, isochoric and adiabatic change respectively?

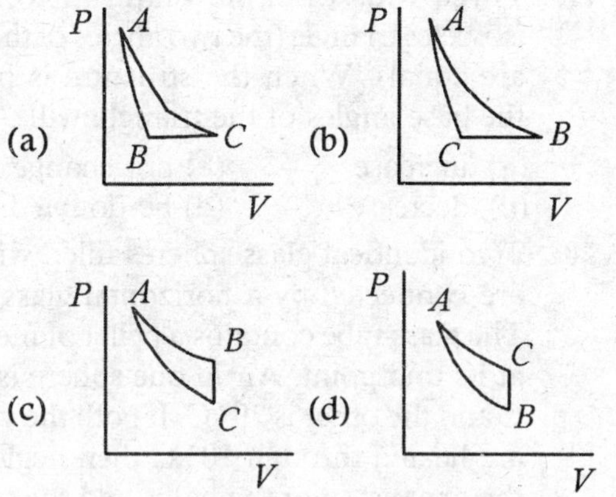

9. Equal masses of two liquids A and B contained in vessels of negligible heat capacity, are supplied heat at the same rate. The temperature-time graphs are shown in Fig. 15.57. With S for specific heat and L for latent heat, the graph tells us

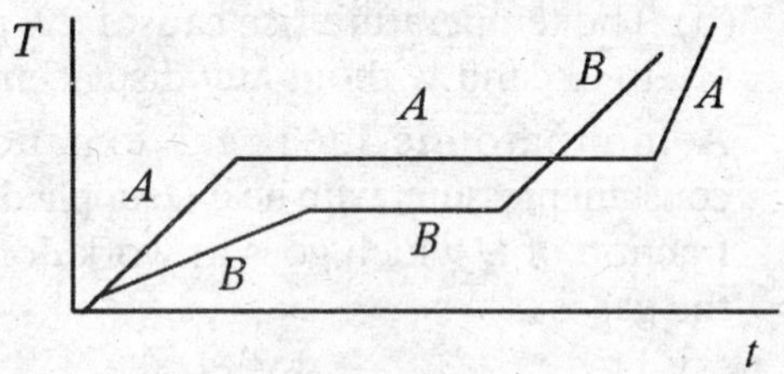

Fig. 15.57

(a) $S_A > S_B$, $L_A < L_B$
(b) $S_A > S_B$, $L_A > L_B$
(c) $S_A < S_B$ $L_A < L_B$
(d) $S_A < S_B$, $L_A > L_B$.

10. Figure 15.58. shows two processes A and B on a system. If ΔQ_1 and ΔQ_2 are the heats given to the system in process A and B respectively, then

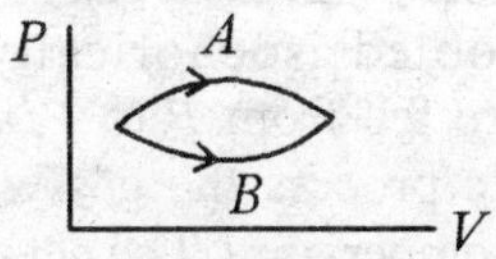

Fig. 15.58

(a) $\Delta Q_1 > \Delta Q_2$ (b) $\Delta Q_1 < \Delta Q_2$
(c) $\Delta Q_1 = \Delta Q_2$ (d) $\Delta Q_1 \leq \Delta Q_2$.

11. Consider the following two statements.
(A) If heat is added to a system, its temperature must rise.
(B) If positive work is done by the system in a thermodynamic process, its volume must increase.
(a) Both A and B are correct
(b) A is correct but B is wrong
(c) B is correct but A is wrong
(d) Both A and B are wrong.

12. A system can be taken from the initial state (P_1, V_1) to the final state (P_2, V_2) by two different methods. If ΔQ and ΔW represent the heat given to the system and work done by the system. Which of the following must be the same in both the methods?
(a) ΔQ (b) ΔW
(c) $\Delta Q + \Delta W$ (d) $\Delta Q - \Delta W$.

13. A Carnot engine is working between 127° C and 27° C. The increase in efficiency will be maximum when the temperature of
(a) the source is increased by 50° C
(b) the sink is decreased by 50° C
(c) the source is increased by 25° C and that of sink is decreased by 25° C
(d) both source and sink decreases by 25° C each.

14. A cyclic process $ABCD$ on a P-V diagram is shown is Fig. 15.59. A V-T diagram will look like

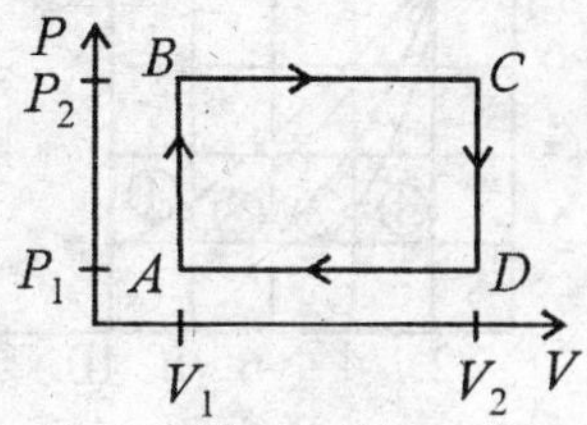

Fig. 15.59

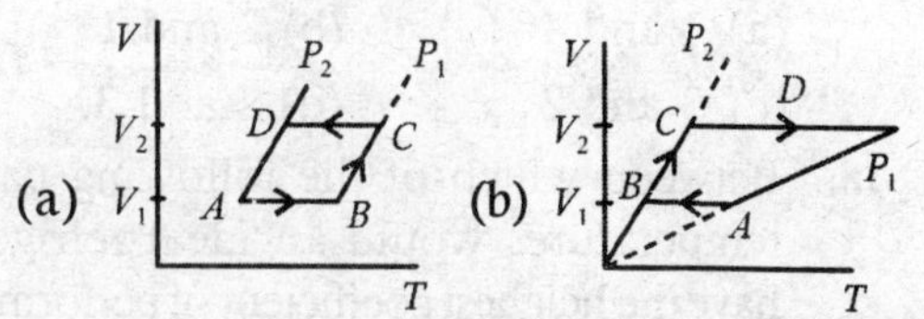

(c) 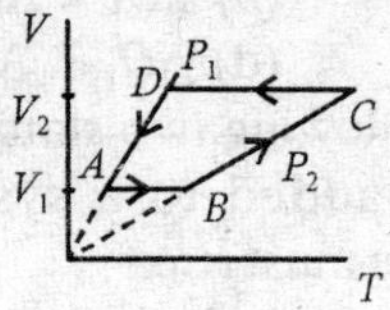

(d) 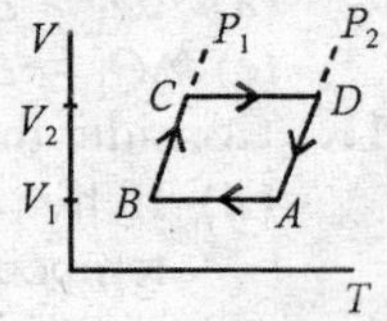

15. A copper block A of mass 1 kg at 30° C and another copper block of mass 0.1 kg at 100° C are kept in contact. Which of the following statements are true?

(a) The steady state temperature of the system is 65° C
(b) The steady state temperature of the system is 30° C
(c) Heat flows from B to A
(d) Heat flows from A to B.

16. An ideal gas is cooled from state A to state B in the following two ways: (*i*) first isobaric then isochoric process (*ii*) first isochoric then isobaric process. The amount of work done in these two processes is W_1 and W_2 respectively, then

(a) $W_1 = W_2$ (b) $W_1 = 2W_2$
(c) $W_1 < W_2$ (d) $W_1 > W_2$.

17. For a perfect gas (diatomic) four indicator diagrams are shown here nearly to scale. One of them corresponds to isothermal change and another to adiabatic change. These are respectively

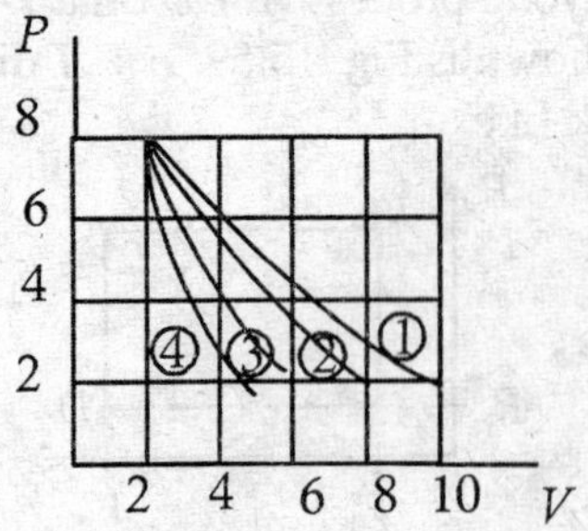

Fig. 15.60

(a) 1 and 4 (b) 2 and 1
(c) 3 and 2 (d) 2 and 3.

18. Between which of the following pairs of temperatures would an ideal refrigerator have the heighest coefficient of performance?

(a) –20° C and –10° C
(b) 0° C and 10° C
(c) 10° C and 20° C
(d) 20° C and 30° C.

19. Three rods of same material form an isosceles triangle (the two angles of the base are equal). When the structure is heated the base angles of the triangle will

(a) increase (b) not change
(c) decrease (d) be doubled.

20. Two identical glass spheres filled with air are connected by a horizontal glass tube. The glass tube contains a pellet of mercury at its mid point. Air in one sphere is at 0° C and the other is 20° C. If both the vessels are heated through 10° C then neglecting the expansion of the bulbs and the tube

(a) the mercury pellet gets displaced towards the sphere at lower temperature
(b) the mercury pellet gets displaced towards the sphere at higher temperature
(c) The mercury pellet does not get displaced at all
(d) The temperature rise causes the pellet to expand without any displacement.

21. A monoatomic ideal gas expands at constant pressure, with heat Q supplied. The fraction of Q which goes as work done by the gas is

(a) 1 (b)2/3
(c) 3/5 (d) 2/5.

22. A steam engine working between 177° C and 127° C converts 8% of the energy E drawn from the source into work. The percentage of E which is wasted due to friction and other mechanical faults is

(a) 9.2% (b) 1/2 of 92%
(c) 3% (d) 1.5%.

23. One mole of an ideal gas at temperature T was cooled isochorically till the gas pressure fell from P to P/n. Then by an isobaric process the gas was restored to initial temperature. The net amount of heat absorbed by the gas in the process is

(a) nRT (b) $\frac{RT}{n}$

(c) $RT(1 - n^{-1})$ (d) $RT(n - 1)$.

24. Which of the following temperatures is the highest?

(a) 100 K (b) −13° F

(c) −20° C (d) −23° C.

25. A gas has volume 3 litres when the thermometer reads 300 K. If the thermometer reads at 400 K, volume of the gas will be close to

(a) 4 litres (b) $\frac{9}{4}$ litres

(c) 1 litre (d) any value.

26. In an adiabatic expansion of air, the volume increases by 5% then percentage change in pressure is

(a) 7 (b) 3

(c) −5 (d) −7.

27. Some amount of gas is (*i*) heated at constant pressure to 2-fold volume (*ii*) then brought to the original volume at constant temperature, and (*iii*) finally cooled at constant volume to reach the initial state. The change in internal energy (ΔU) and work done by the gas (W) are

(a) ΔU positive in (*i*) W negative in (*ii*)
(b) ΔU negative in (*iii*) W positive in (*ii*)
(c) ΔU positive in (*i*), W negative in (*iii*)
(d) ΔU negative in (*ii*) W negative in (*i*).

28. One mole of an ideal gas at an initial temperature of T K does $6R$ Joules of work adiabatically. If the ratio of specific heats of this gas at constant pressure and constant volume is 5/3, the final temperature of the gas will be

(a) $(T + 2.4)$ K (b) $(T - 2.4)$ K

(c) $(T + 4)$ K (d) $(T - 4)$ K.

29. Two rods one of aluminium and other of steel, having initial lengths l_1 and l_2 are connected together to form a single rod of length $l_1 + l_2$. The coefficient of linear expansion for aluminium and steel are α_a and α_b respectively. If the length of each rod increases by the same amount when their temperatures are raised by t° C, then the ratio $\frac{l_1}{l_1 + l_2}$ is

(a) $\frac{\alpha_s}{\alpha_a}$ (b) $\frac{\alpha_a}{\alpha_s}$

(c) $\frac{\alpha_s}{\alpha_s + \alpha_a}$ (d) $\frac{\alpha_a}{\alpha_s + \alpha_a}$.

30. Two identical containers A and B with frictionless pistons contain the same ideal gas at the same temperature and the same volume V. The mass of the gas in A is m_a and that in B is m_b. The gas in each cylinder is now allowed to expand isothermally to the same final volume $2V$. The changes in the pressure in A and B are found to be ΔP and $1.5\Delta P$ respectively, then

(a) $4m_a = 9m_b$ (b) $2m_a = 3m_b$

(c) $3m_a = 2m_b$ (d) $9m_a = 4m_b$.

ANSWERS

1. (a)	2. (c)	3. (a)	4. (c)	5. (c)
6. (a)	7. (b)	8. (c)	9. (d)	10. (a)
11. (c)	12. (d)	13. (b)	14. (c)	15. (c)
16. (d)	17 (d)	18 (d)	19 (b)	20 (b)
21 (d)	22 (c)	23 (c)	24 (c)	25 (d)
26 (d)	27 (a)	28 (d)	29 (c)	30 (c)

PART II

In the following, a question is followed by four alternatives, of them one or more are correct. Candidate has to select the correct choice(s) from amongst the alternatives.

1. Which of the following pairs may give equal numerical values of the temperature of a body?
   (a) Fahrenheit and Kelvin
   (b) Celsius and Kelvin
   (c) Fahrenheit and Celsius
   (d) All the three.

2. In which of the following pairs of temperature scales, the size of a degree is identical?
   (a) Mercury scale and ideal gas scale
   (b) Celsius scale and mercury scale
   (c) Celsius scale and ideal gas scale
   (d) Ideal gas scale and absolute scale.

3. Which of the following statements are true is case of specific heat of a gas?
   (a) Specific heat of a gas is zero when it undergoes an adiabatic change
   (b) Specific heat of a gas is infinite when it undergoes an isothermal change
   (c) The difference between principal specific heats is the same for all gases
   (d) The ratio of principal specific heats is the same for all gases.

4. The pressure P and volume V of an ideal gas both increase in a process.
   (a) Such a process is not possible
   (b) The work done by the system is positive
   (c) The temperature of system must increase
   (d) Heat supplied to the system is equal to the change in internal energy.

5. In a process on a system, the initial pressure and volume are equal to the final pressure and volume.
   (a) The net work done by the system in the process must be zero
   (b) The initial internal energy must be equal to the final internal energy
   (c) The net heat given to the system in the process must be zero
   (d) The initial temperature must be equal to the final temperatures.

6. A certain amount of perfect gas undergoes changes in pressure and volume as shown in Fig. 15.61. During this change

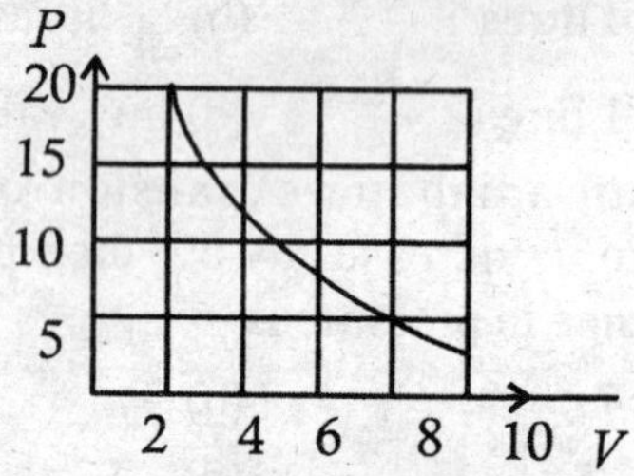

Fig. 15.61

   (a) no heat is absorbed by the gas from out side
   (b) external work is done by the gas
   (c) temperature of the gas is constant
   (d) the internal energy of the gas remains constant.

7. An ideal gas in a cylinder is compressed adiabatically to one third of its initial volume. During this process some work is done on the gas. Which of the following statements is/are true?
   (a) Change in entropy is zero
   (b) There is increase is internal energy
   (c) Pressure increases three times
   (d) Temperature of the gas decreases.

8. The principal specific heat of hydrogen at constant pressure is 3.4 cal/gm-°C and gas constant per unit mass is 4200 J/kg^2-K, then
   (a) its value is 3.4 k cal/kg°C
   (b) its value is 14280 J/kg°C
   (c) its molar specific heat is 3.4 cal/gm °C
   (d) its principal specific heat at constant volume is 1.4 cal/gm°C.

9. The internal energy of an ideal gas decreases by the same amount as work done by the system.
   (a) The process must be adiabatic
   (b) The process must be isothermal
   (c) The process must be isobaric
   (d) The temperature must decrease.

10. In Fig. 15.62, let ΔU_1 and U_2 be the change is internal energy in processes A and B respectively, ΔQ be the net heat given to the system in the process $A + B$ and ΔW be the net work done by the system in the process $A + B$.

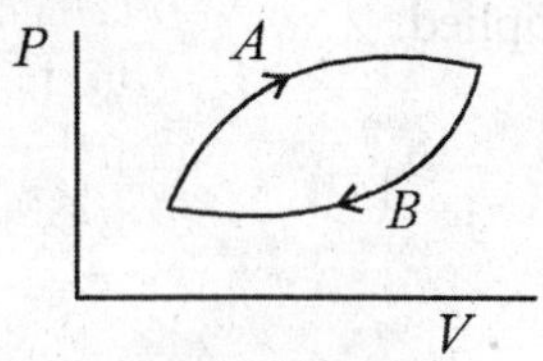

Fig. 15.62

   (a) $\Delta U_1 + \Delta U_2 = 0$ (b) $\Delta U_1 - \Delta U_2 = 0$
   (c) $\Delta Q - \Delta W = 0$ (d) $\Delta Q + \Delta W = 0$.

11. The indicator diagram (Fig. 15.63) shows ideal gas taken from state 1 to 2 through optional paths A, B, C. Let Q, W, U represents the heat supplied to, the work done by and the internal energy respectively, then

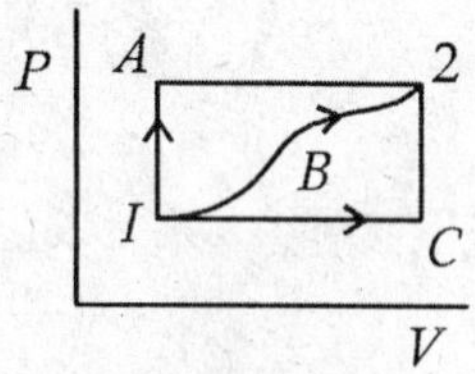

Fig. 15.63

   (a) $W_a > W_b > W_c$
   (b) $Q_a > Q_b > Q_c$
   (c) $U_2 - U_1 = Q_a - W_a$
   (d) $Q_a - W_a = Q_b - W_b = Q_c - W_c$.

12. The Fig. 15.64 shows a cyclic process starting from A for a perfect gas. The correct statements is/are

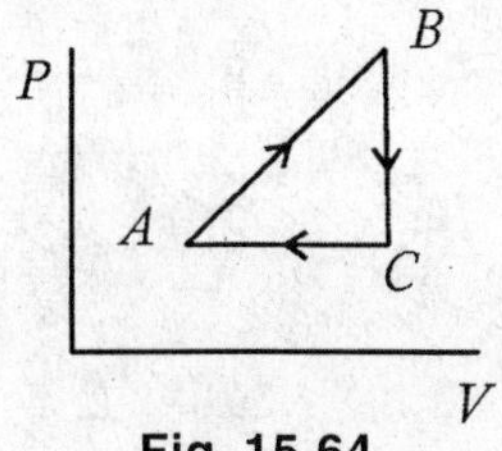

Fig. 15.64

   (a) Work is done by the gas in going through the cycle
   (b) Total net heat energy is supplied by an external agent
   (c) The internal energy of the gas increases
   (d) The temperature of the gas decreases as it goes round the cycle.

13. The temperature of an isotropic cubical solid of length l, density d and coefficient of expansion α per Kelvin is raised by 10°C, then at this temperature, to a good approximation
   (a) length is $l(1 + 10\alpha)$
   (b) total surface area is $l^2(1 + 20\alpha)$
   (c) density is $d(1 + 30\alpha)$
   (d) density is $d/(1 + 30\alpha)$.

14. Which of the following statements are true?
   (a) Rubber contracts on heating
   (b) Water expands on freezing
   (c) Water contracts on heating from 0° C to 4° C
   (d) Water expands on heating from 4° C to 40° C.

15. Select correct statements for an ideal gas
   (a) Change in internal energy in an isobaric process from temperature T_1 to T_2 is equal to $nC_V(T_2 - T_1)$ where C_V is molar specific heat at constant volume and n the number of moles
   (b) The change in internal energy of the gas and the work done by the gas are equal in magnitude in an adiabatic process
   (c) The internal energy does not change in an isothermal process
   (d) No heat is added or removed in an adiabatic process.

16. An ideal gas is taken from the state A to state B (Fig. 15.65) along a straight path. Select the correct statements.

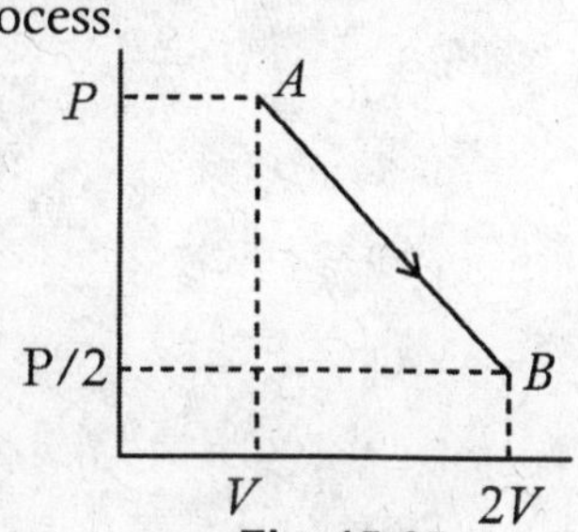

Fig. 15.65

(a) The work done by the gas in the process A to B exceeds the work that would be done by it if the system were taken from A to B along an isothermal path
(b) In T-V diagram, the paths AB becomes a part of a parabola
(c) In P-T diagram, the path AB becomes part of a parabola
(d) In going from A to B, the temperature T of the gas first increases to a maximum value and then decreases.

17. Which of the following statements are correct?
(a) Whole of work can be converted into heat
(b) Whole of heat can be converted into work
(c) Temperature of a body can be raised without heating it
(d) Temperature of a body may not change when it is heated.

18. Molar heat capacity of gas does not depend on
(a) its temperature
(b) its molecular weight
(c) its atomicity
(d) the conditions under which heat is supplied.

ANSWERS

1. (a, c)	2. (c, d)	3. (a, b)	4. (b, c)	5. (b, d)
6. (b, c, d)	7. (a, b)	8. (a, b, d)	9. (a, d)	10. (a, c)
11. (a, b, c)	12. (a, b)	13. (a, d)	14. (a, b, c, d)	15. (a, b, c, d)
16. (a, b, c, d)	17 (a, c, d)	18. (b)		

16

Kinetic Theory of Gases

16.1 INTRODUCTION

Matter is constituted of atoms and molecules, which are in constant motion. In solids they vibrate about their equilibrium position while in a gas they are free to move about in random way. The molecules of liquid are neither quite free as those in gases nor as contained as those in a solid. Of the three states of matter the molecular picture of gases is the easiest to handle. That is why theory was developed earlier and is known as *the Kinetic Theory of Gases.* This theory which was developed in nineteenth century gave successfully molecular interpretation of temperature and pressure of gas. An ideal gas is the one which obeys the perfect gas equation:

$$PV = n\,RT \qquad \ldots(16.1)$$

where P is pressure, V the volume and T, the temperature in absolute units, n is the number of moles in the gas and R is the gas constant. The real gases available to us are a good approximation of an ideal gas at high temperature and low pressures (*i.e.*, at low densities). The observed behaviour of a gas results from the detailed behaviour of its large number of molecules. The kinetic theory of gases attempts to develop a model of the molecular behaviour which should result in the observed behaviour of an ideal gas.

16.2 ASSUMPTIONS OF KINETIC THEORY OF GASES

Kinetic theory of gases is based on the molecular picture of matter. Any sample of a gas is made of large number of molecules. A molecule is a smallest unit having all the chemical properties of the sample. The molecules in a sample of gas are in incessant *random motion.* At ordinary temperature and pressure, the average distance between the molecules is very large in comparison to the size of molecule. It is roughly 10 times as such the interaction between the molecules is negligible and we assume that the molecules move freely in straight lines. Occasionally, the molecules come very close to each other and experience force (intermolecular forces) and their momentum changes. This is the simplest model, in which the molecules are thought of small hard spheres colliding incessantly with each other and walls of the container and during each collision their momentum changes in magnitude and direction. These collisions are assumed to be elastic. This picture of the gas is summarised in the form of following postulates.

1. Molecules of a gas are hard perfectly elastic spheres and are in incessant random motion. They are colliding with each other and the walls of container such that the molecular density every where is constant.
2. The size of a molecule is much smaller than the average separation between the molecules.
3. There is no force of attraction or repulsion between the molecules or between the molecules and the walls of the container except during the collision. Thus entire internal energy of the gas is kinetic.
4. All collisions between two molecules or between molecules and a wall are perfectly elastic and the time of impact is negligibly small in comparison to the time taken to travel free path.
5. The molecules obey Newton's laws of motion, i.e. between two collision's a molecule moves in a straight line with uniform velocity. The distance travelled by a molecule between two collisions is called free path. Average of all free paths is called mean free path (λ).
6. Pressure is due to the collision of the molecules on the walls of the container.

Note that the postulates are not strictly independent. Also they are strictly true for an ideal gas but only approximately true for a real gas. In case of real gas the intermolecular forces cannot be neglected at all pressures and temperatures. It is because of these intermolecular forces that we have phase transition, i.e. gases are liquified and liquids are solidified. An ideal gas with no intermolecular forces at all temperatures and pressures cannot undergo phase transition.

16.3 PRESSURE OF AN IDEAL GAS

Using the above postulates, the pressure exerted by an ideal gas can be calculated.

Consider an ideal gas enclosed in a cubical container of side l. Let the walls of the container be perfectly elastic and there are n number of molecules in the container and each molecule has mass m. Take a corner O of the container as origin and X–Y–Z axes along the edges (Fig. 16.1).

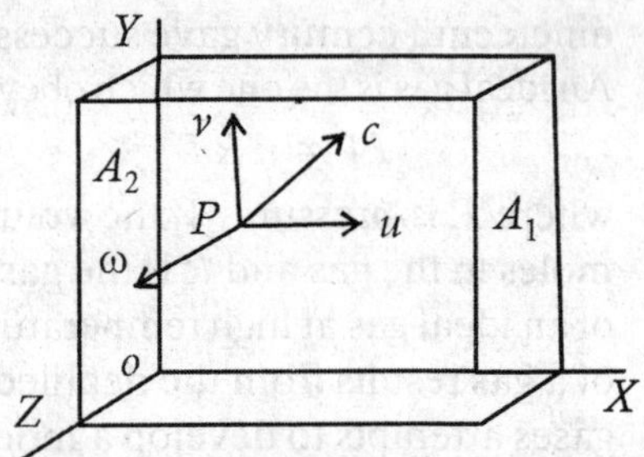

Fig 16.1

Since the molecules are constantly moving about with different velocities in different directions, they are constantly bombarding the walls of the container. Consider one molecule having velocity c at a given instant. This velocity of the molecule can be resolved into three components u, v, ω along X, Y and Z axes respectively, then

$$c^2 = u^2 + v^2 + \omega^2 \qquad ...(16.1)$$

Now consider the motion of this molecule along the X-axis, i.e. perpendicular to the faces A_1, A_2. When the molecule collides the face A_1 the X-component of the velocity (u) is reversed whereas they Y and Z-components remain unchanged (the collision being perfectly elastic). The change in the momentum of the molecule is

$$\Delta P = -mu - (+mu) = -2mu$$

As the momentum remains conserved in a collision, the change in momentum of the wall is

$$\Delta P = 2mv$$

After rebound, this molecule travels towards face A_2 with the X-component of velocity equal to $-u$. Any collision of the molecule with any other face (except for A_2) does not change the value of u. As such it travels between A_1 and A_2 with a constant X-component of velocity (= u). The collision of the molecule with other molecules can be neglected in light of the assumption that molecular density everywhere remains constant, i.e., the distribution of molecules with different velocities are independent of position, direction and time.

For each successive collision with wall A_1 the molecule must traverse the distance from A_1 to A_2 and back, i.e., a distance equal to twice the length of the cube (= $2l$). Since it covers this distance with speed u, and hence time interval between two successive collisions on wall $A_1 = \frac{2l}{u}$.

And the collision frequency, i.e. number of collisions per second $= \frac{u}{2l}$

Since momentum imparted to the wall in one collision is

$$\Delta P = 2mu$$

Hence momentum imparted to the wall A_1 per second

$$= 2mu \times \frac{u}{2l} = \frac{mu^2}{l}$$

This is the force exerted by the molecule on the wall A_1.

Hence the pressure on the wall A_1—perpendicular to the X-direction

$$= \frac{mu^2}{l \cdot l^2} = \frac{mu^2}{V} \qquad \text{where } V = l^3 \text{, volume of the gas}$$

In a similar way considering the motion of molecules along Y and Z axes the pressure exerted on the wall perpendicular to Y-axis

$$= \frac{mv^2}{V}$$

and pressure exerted on the wall perpendicular to Z-axis is

$$= \frac{m\omega^2}{V}$$

The total pressure exerted by all the molecules on the wall perpendicular to X-axis is

$$P_X = \sum \frac{mu^2}{V}$$

Similarly total pressure exerted by all the molecules on the walls perpendicular to Y-axis and Z-axis is

$$P_Y = \sum \frac{mv^2}{V}$$

and $$P_Z = \sum \frac{m\omega^2}{V}$$

Since the pressure exerted by the gas is same in all the directions, we have

$$P_X = P_Y = P_Z = P$$

Thus adding we have

$$P_X + P_Y + P_Z = 3P = \sum \frac{m}{V}(u^2 + v^2 + \omega^2)$$

or $$P = \frac{m}{3V}\Sigma\left(u^2 + v^2 + \omega^2\right) = \frac{m}{3V}\Sigma c^2$$

if $$\bar{c}^2 = \frac{c_1^2 + c_2^2 + c_3^2 + \ldots c_n^2}{n} = \frac{1}{n}\Sigma c^2$$

then $$\Sigma c^2 = n\bar{c}^2$$

where $\bar{c}^2$ is called *mean square speed* and $\bar{c}$ is called *root mean square speed* (r.m.s. speed)

Thus the pressure exerted by the gas is

$$P = \frac{1}{3}\frac{mn}{V}\bar{c}^2 \qquad \ldots(16.2)$$

Now mn = total mass of the gas and V total volume of the gas

$$\therefore \quad \frac{mn}{V} = \rho \qquad \text{where } \rho \text{ is the density of the gas}$$

$$\therefore \quad P = \frac{1}{3}\rho\bar{c}^2 \qquad \ldots(16.3)$$

Translational Kinetic Energy of Molecules

Total translational kinetic energy of all the molecules is

$$K = \sum\frac{1}{2}mc^2 = \sum\frac{1}{2}mn\frac{c^2}{n} = \frac{1}{2}mn\bar{c}^2$$

The pressure in terms of kinetic energy of molecules is

$$P = \frac{2}{3}\cdot\frac{1}{2}\frac{mn\bar{c}^2}{V}$$

But $\frac{1}{2}\frac{mn\bar{c}^2}{V}$ = kinetic energy per unit volume called energy density U

$$U = \frac{1}{2}\frac{mn\bar{c}^2}{V}$$

$$\therefore \quad P = \frac{2}{3}U \qquad \ldots(16.4)$$

i.e., pressure is equal to 2/3 of energy density.

16.4 KINETIC INTERPRETATION OF TEMPERATURE

Temperature means degree of hotness. Hotter the body larger is the internal energy. Thus higher temperature means higher internal energy and lower temperature means lower internal energy. The internal energy of an ideal gas according to the kinetic theory of gases is the total translational kinetic energy of all its molecules, i.e.

$$K = \frac{1}{2}mn\bar{c}^2$$

Thus for a given gas higher temperature means higher $\bar{c}$ and lower temperature means lower value of $\bar{c}$ (r.m.s. speed). Thus temperature must be some function of $\bar{c}$ (r.m.s. speed). Thus for a given gas

$$T = f(\bar{c})$$

If P and $\bar{c}$ be pressure of gas and r.m.s. speed of the molecules at temperature T and P_t and $\bar{c}_t$ are values at temperature 273.16 K (triple point of water). Keeping the volume V the same as that at temperature T, we have

$$PV = \frac{1}{3} mn\bar{c}^2$$

and
$$P_t V = \frac{1}{3} mn\bar{c}_t^2$$

$$\therefore \quad \frac{P}{P_t} = \frac{\bar{c}^2}{\bar{c}_t^2}$$

But we know that pressure is directly proportional to absolute temperature, i.e.

$$\frac{P}{P_t} = \frac{T}{273.16} = \frac{\bar{c}^2}{\bar{c}_t^2}$$

or
$$T = \left\{\frac{273.16}{\bar{c}_t^2}\right\}\bar{c}^2 \qquad \text{...(16.5)}$$

Since 273.16 K is triple point of water and is constant as such $\bar{c}_t$ the r.m.s. speed of molecules at this temperature and this is constant for a given gas. *Thus the absolute temperature of a given gas is proportional to the square of the r.m.s. speed of its molecules.* Consequently we can say that the temperature of a given gas is directly proportional to the total translational kinetic energy of its molecules, i.e.

$$T \propto K$$

When we use equation (16.2) along with gas equation, we have

$$PV = \frac{1}{3} mn\bar{c}^2 \qquad \text{equation (16.2)}$$

and
$$PV = \mu RT \qquad \text{Gas equation where } \mu \text{ is number of moles}$$

For $\mu = 1$ mole $n = n_0$ Avogadro number

$$\therefore \quad \frac{1}{3} mn_0\bar{c}^2 = RT$$

or
$$\frac{1}{2} m\bar{c}^2 = \frac{3}{2}\frac{R}{n_0}T = \frac{3}{2}k_B T \qquad \text{...(16.6)}$$

where $\frac{R}{n_0} = k_B$, the Boltzman's constant.

Equation (16.6) gives that mean kinetic energy of a molecule.

$$= \frac{3}{2}k_B T$$

and the energy of one gram-mole of the gas $= \frac{3}{2}n_0 k_B T = \frac{3}{2}RT$.

Thus we see that the kinetic energy of a molecule depends only upon T—the absolute temperature of the gas and is quite independent of its mass. The fact is referred to as the *Kinetic interpretation of temperature.*

Clearly at $T = 0$, $\bar{c} = 0$ indicates that at absolute zero temperature K.E. of the gas is reduced to zero, i.e. all the molecular motion ceases at this temperature and all the molecules come to rest at this temperature. But it is our common experience that all the gases get liquified or even solidified before this temperature is reached. The quantum theory predicts appreciable motion of the molecules even at this temperature (0 K). The energy possessed by the molecules at 0 K is called *zero point energy*.

Now let us consider a mixture of two gases A and B having molecules of masses m_1 and m_2 respectively. As the molecules collide with each other, they exchange energy. On an average the molecules with higher kinetic energy lose energy to those with lower kinetic energy. In thermal equilibrium, the average kinetic energy of all molecules are equal. If $\bar{c}_1$ and $\bar{c}_2$ be the r.m.s. speeds of the molecules of gas A and gas B respectively then

$$\frac{1}{2}m_1\bar{c}_1^2 = \frac{1}{2}m_2\bar{c}_2^2 \qquad \ldots(16.7)$$

We observe that for different kinds of gases it is not the r.m.s. speed but average kinetic energy of individual molecules that has a fixed value ($= 3/2k_BT$) at a given temperature. The heavier molecules move with smaller r.m.s. speed the lighter molecules move with larger r.m.s. speed.

16.5 DEDUCTION OF GAS LAWS ON THE BASIS OF KINETIC THEORY OF GASES

(*a*) **Boyle's Law:** The Boyle's law states that at a given temperature the pressure of a given mass of a gas is inversely proportional to its volume: If P is pressure and V the volume, then

$$P \propto \frac{1}{V} \quad \text{or} \quad PV = \text{constant}$$

From equation (16.2), we have

$$PV = \frac{1}{3}mn\bar{c}^2$$

But for a given gas $\bar{c}^2 \propto T$. Hence value of $\bar{c}^2$ is constant at a given temperature. For a given mass of the gas n the number of molecules are constant. Thus we have

$$PV = \text{Constant}$$

or

$$P \propto \frac{1}{V}$$

This is Boyle's law.

(*b*) **Charle's Law:** This law states that *at a given pressure, the volume of a given mass of a gas is proportional to its absolute temperature.* If V is volume T the temperature, then

$$V \propto T$$

From equation (16.2), we have

$$PV = \frac{1}{3}mn\bar{c}^2$$

Since P is constant, and mn gives mass of the gas which is also constant, hence

$$V \propto \bar{c}^2$$

But $\bar{c}^2 \propto T$ we get $V \propto T$.

(*c*) **Gay Lussac's Law:** This law states that *for a given volume of the gas, the pressure of a given mass of the gas is proportional to absolute temperature.* If P is pressure and T temperature, then

$$P \propto T$$

From equation (16.2), we have

$$PV = \frac{1}{3} mn\bar{c}^2$$

or $$P = \frac{1}{3}\frac{mn}{V}\bar{c}^2$$

If V is constant and mn the mass of gas is also constant, then

$$P \propto \bar{c}^2$$

But $$\bar{c}^2 \propto T$$

Hence $$P \propto T$$

In fact this is the definition of the absolute temperature T.

(*d*) **Avogadro's Hypothesis:** *This hypothesis states that under similar conditions of temperature and pressure, equal volumes of all gases contain equal number of molecules.*

Let n_1 and n_2 be the number of molecules in two equal volumes (V) of two gases respectively at the same temperature and pressure. Then

$$PV = \frac{1}{3} m_1 n_1 \bar{c}_1^2 = \frac{1}{3} m_2 n_2 \bar{c}_2^2 \qquad \ldots(i)$$

where m_1 and m_2 and $\bar{c}_1$ and $\bar{c}_2$ are the masses and the root mean square speeds of the molecules of the two gases respectively.

We have already proved that the average kinetic energy of a molecule is constant and depends only on temperature, i.e., under similar conditions of temperature $\frac{1}{2} m_1 \bar{c}_1^2 = \frac{1}{2} m_2 \bar{c}_2^2$

Using in (*i*) we have

$$n_1 = n_2$$

(*e*) **Dalton's Law of Partial Pressures:** *This law states that the pressure exerted by a mixture of several gases equals the sum of the pressures exerted by each gas occupying the same volume as that of the mixture.*

Kinetic theory of gases assumes that the pressure exerted by the gas on the walls of the container is due to the collision of the molecules with the walls. The total force on the wall is the sum of the forces exerted by the individual molecule.

Let $n_1, n_2, n_3, \ldots$, etc. be the number of molecules of the gas 1, 2, 3, …, etc. respectively in the mixture, then total force on the surface area A of the wall is

$$F = \text{force by } n_1 \text{ molecules of gas 1} + \text{force by } n_2 \text{ molecules of gas 2} + \ldots$$

$$= F_1 + F_2 + \ldots$$

and pressure

$$P = \frac{F}{A} = \frac{F_1}{A} + \frac{F_2}{A} + \ldots$$

If only first gas is filled in the container, then n_1 molecules will exert force F_1 and the pressure $P_1 = \frac{F_1}{A}$. Similarly for other gases. Thus

$$P = P_1 + P_2 + P_3 + \ldots$$

(*f*) **Graham's Law of Diffusion:** *When two gases at the same temperature and pressure diffuse into each other, their steady rates of diffusion are inversely proportional to the square roots of their densities. This is called Graham's law of diffusion.*

Let ρ_1 and ρ_2 be the densities of two gases diffusing into one another. Let the r.m.s. speeds of their molecules be $\bar{c}_1$ and $\bar{c}_2$ and their rates of diffusion r_1 and r_2 respectively. Then r_1 and r_2 must be proportional to their r.m.s. speeds, i.e.

$$\frac{r_1}{r_2} = \frac{\bar{c}_1}{\bar{c}_2}$$

Since pressures of the two gases are equal hence

$$P = \frac{1}{3}\rho_1\bar{c}_1^2 = \frac{1}{3}\rho_2\bar{c}_2^2$$

or

$$\frac{\bar{c}_1}{\bar{c}_2} = \sqrt{\frac{\rho_2}{\rho_1}}$$

$$\therefore \quad \frac{r_1}{r_2} = \frac{\bar{c}_1}{\bar{c}_2} = \sqrt{\frac{\rho_2}{\rho_1}}$$

(*g*) **Ideal gas equation:** We have from equation (16.2)

$$PV = \frac{1}{3}mn\bar{c}^2 \qquad \ldots(i)$$

and equation (16.3) gives the average kinetic energy per molecule, i.e.

$$\frac{1}{2}m\bar{c}^2 = \frac{3}{2}k_BT \qquad \ldots(ii)$$

Using these equations, we have

$$PV = \frac{2}{3}\left(\frac{1}{2}m\bar{c}^2\right)n = \frac{2}{3}\left(\frac{3}{2}k_BT\right)n = nk_BT$$

If the gas contains μ moles of the gas and n_0 be the Avogadro number, then

$$n = \mu n_0$$

$$\therefore \quad PV = \mu n_0 k_B T$$

But $\quad n_0 k_B = R$—the gas constant, then

$$PV = \mu RT \qquad \ldots(16.8)$$

16.6 R.M.S. SPEED IN TERMS OF TEMPERATURE

The r.m.s. speed of the molecules can be expressed in terms of absolute temperature. Equation (16.2) gives

$$PV = \frac{1}{3}mn\bar{c}^2 \qquad \ldots(i)$$

and equation (16.8) gives

$$PV = \mu RT$$

But $\mu n_0 = n$ where n is number of molecules and n_0 Avogadro number and μ number of moles.

$$\therefore \qquad PV = \frac{nR}{n_0}T = nk_BT \qquad ...(ii)$$

From these two equations, we have

$$\frac{1}{3}mn\bar{c}^2 = nk_BT$$

or

$$\bar{c} = \sqrt{\frac{3k_BT}{m}} \qquad ...(16.9)$$

This equation can be written as

$$\bar{c} = \sqrt{\frac{3n_0k_BT}{n_0m}}$$

But $\qquad n_0k_B = R$ and $n_0m = M$, the molecular weight of the gas

$$\bar{c} = \sqrt{\frac{3RT}{M}} \qquad ...(16.10)$$

The average speed

The average speed

$$< c >= \frac{\Sigma c}{n}$$

It can be shown to be equal to

$$< c >= \sqrt{\frac{8k_BT}{\pi m}} = \sqrt{\frac{8RT}{\pi m}} \qquad ...(16.11)$$

Comparing (16.10) and (16.11), we see that the average speed is somewhat smaller than the r.m.s. speed.

Maxwell's Speed Distribution Law

We have seen that molecules in a gas are incessantly moving in a random way in different directions. At a given temperature and pressure they have all possible velocities ranging from zero to infinity and along all possible directions. Such a hypothetical state of the gas obviously cannot remain steady under the incessant molecular collisions. The natural question then arises as to what then is the steady molecular feature of a gas when its bulk parameters such as pressure, temperature and volume are fixed? The answer is: it is the distribution of velocities that is steady, i.e. number of molecules dn within the speed range c and $c + dc$. Every molecule is continually changing its velocity as it collides with other molecules. Yet the velocity distribution remains fixed.

Maxwell derived an equation giving the distribution of molecules in different speeds. If dn represents the number of molecules with speeds between c and $c + dc$, then

$$dn = 4\pi n\left(\frac{m}{2\pi k_BT}\right)^{\frac{3}{2}} c^2 e^{\frac{-mc^2}{2k_BT}} dc \qquad ...(16.12)$$

The plot of $\frac{dn}{de}$ against c is shown in Fig. 16.2 at two different temperatures. We observe that there are some molecules which have speed many times greater than the mean speed. The speed c_P at which $\frac{dn}{dc}$ is maximum is called the *most probable speed*. Its value is given by

$$c_P = \sqrt{\frac{2k_BT}{m}} = \sqrt{\frac{2RT}{M}} \qquad \ldots(16.13)$$

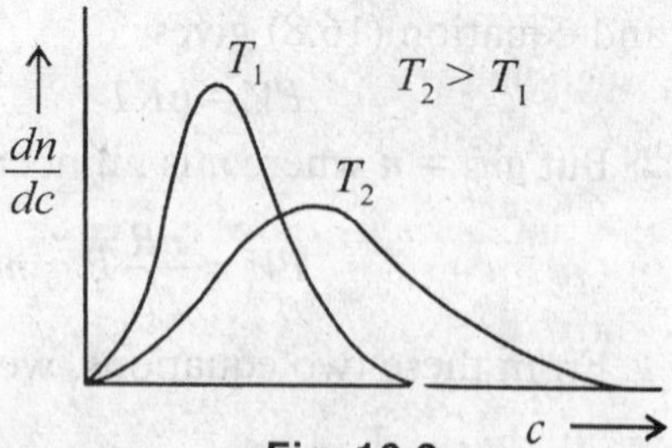

Fig. 16.2

It is also observed that the number of molecules with most probable speed decrease as the temperature increases but the number of molecules with higher speeds increase.

Notice that the Maxwell's speed distribution is not a symmetric curve. This is why $\bar{c}$, <c> and c_P are different. The most probable speed is least of all the three.

It should be remembered that the motion of molecules discussed above is truely random, i.e. the centre of mass of the gas is assumed to be at rest and any rotation about the centre of mass is assumed to be absent. Any systematic motion of a gas sample has no effect on temperature. For example a gas jar when placed in a moving train, the increase in translational kinetic energy does not increase the temperature of the gas.

16.7 DEGREES OF FREEDOM AND LAW OF EQUIPARTITION OF ENERGY

Degrees of Freedom

The number of degrees of freedom of a system is the number of mutually independent variables required to precisely define its state or position. For example, the position of a simple ideal mass point in space can be defined completely by three Cartesian co-ordinates x, y, z. It is therefore said to possess three degrees of freedom. In similar way two points in space will obviously require six co-ordinates to define their position and will therefore have six degrees of freedom.

In case of a dynamic system consisting of just a point it can at best have only three degrees of freedom (along the three axes) all of which are of translatory motion. Of course, if it is constrained to move only along a straight line, it has only one degree of freedom. If it is free to move in a plane, it has two degrees of freedom. It is only when it is free to move in space that it has three degrees of freedom.

If however, the body has a finite size and is rigid, it may have rotational motion about an axis passing through it along with translatory motion if its centre of mass. This rotational motion can also be resolved into three component rotatory motions about any three mutually perpendicular axes through a fixed point in it. In this way the body has three additional degrees of freedom of rotatory motion. *Thus the degrees of freedom of a dynamic system is the number of independent co-ordinates required to specify its position and configuration.*

It must be noted that *the number of degrees of freedom of a dynamic system depends merely on its capacity for motion and not on its actual motion and is determined purely by its geometrical rather than its physical properties.*

In case of a dynamic system with large number of molecules such as molecules of an ideal gas, the number of independent terms in the expression of energy of a molecule is called degree of freedom.

Let us now try to apply this definition of degree of freedom to various types of molecules with which alone we are concerned here.

When we assume the molecules of a gas to be like hard spheres of negligible size, the energy of each molecule may be written as

$$E = \frac{1}{2}mc_x^2 + \frac{1}{2}mc_y^2 + \frac{1}{2}mc_z^2$$

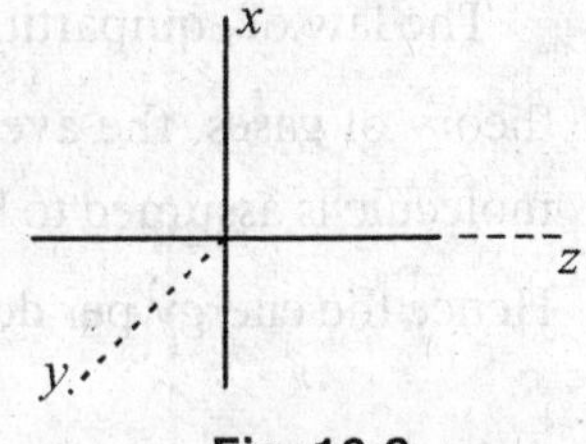

Fig. 16.3

A monoatomic gas has atoms which can be treated like spheres and the expression of energy consists of 3 independent terms. As such each molecule of monoatomic gas has 3 degrees of freedom.

In case diatomic gases, the molecules are assumed to be in the shape of dumb bells, i.e. two hard spheres of negligible size at a separation. In such case apart from translational motion, the molecule can rotate about its centre of mass and total energy is the sum of translational kinetic energy and rotational kinetic energy. If the line joining the two particles is taken as Z-axis, the total energy is then

$$E = \frac{1}{2}mc_x^2 + \frac{1}{2}mc_y^2 + \frac{1}{2}mc_z^2 + \frac{1}{2}I_x\omega_x^2 + \frac{1}{2}I_y\omega_y^2$$

where $\frac{1}{2}I_x\omega_x^2$ and $\frac{1}{2}I_y\omega_y^2$ is the rotational kinetic energy of molecule about X and Y-axis respectively. Since the size of each atom is assumed negligible, the moment of inertia I_z is zero as such the $\frac{1}{2}I_z\omega_z^2$ is not added.

Thus the number of independent terms in their expression is 5. In this situation the degrees of freedom are 5.

But if the two atoms in diatomic molecule vibrate along the line joining their centres, then the molecule has kinetic energy of vibration as well as the potential energy of the pair of atoms. This energy is

$$\frac{1}{2}\mu v^2 + \frac{1}{2}kr^2$$

where r is the instantaneous separation between the atoms and $v = \frac{dr}{dt}$, μ is the reduced mass and k the force constant of the system. Hence the total energy of a molecule has 7 components, i.e.

$$E = \frac{1}{2}mc_x^2 + \frac{1}{2}mc_y^2 + \frac{1}{2}mc_z^2 + \frac{1}{2}I_x\omega_x^2 + \frac{1}{2}I_y\omega_y^2 + \frac{1}{2}\mu v^2 + \frac{1}{2}kr^2$$

Thus in this situation the degrees of freedom are 7.

In case of triatomic and polyatomic molecules, the molecule may rotate about three mutually perpendicular axes if it is non-linear type such as H_2O, SO_2, etc. It has then three degrees of freedom of translatory motion and three degrees of freedom of rotatory motion. The number of degrees of freedom then is 6 (The vibrational motion is excluded).

If the molecule is linear like CO_2, the centre of mass lies in the central atom. It, therefore, behaves like a diatomic molecule with just 3 degrees of freedom of translation and 2 of rotation, i.e. total 5.

The number of degrees of freedom increases in general with the complexity of molecule.

Law of Equipartition of Energy

The law of equipartition of energy states that, *the total energy of a system is equally divided amongst all the degrees of freedom and the energy associated with each degree of freedom is* $\frac{1}{2}k_BT$, where k_B is Boltzmann's constant and T, the absolute temperature.

The law of equipartition of energy is the result of kinetic theory of gases. On the basis of kinetic theory of gases, the average energy of a molecule assumed to be point mass is $\frac{3}{2}k_BT$. When the molecule is assumed to be point mass it has only three degrees of freedom for translational motion. Hence the energy per degree of freedom is $\frac{1}{2}k_BT$.

16.8 SPECIFIC HEATS OF GASES

The principle of equipartition of energy has its application in the calculation of specific heats of gases. When heat is supplied to a gas, its molecules absorb in accordance with this principle, the same fraction of thermal energy for each degree of freedom. We will now take up the case of each individual type of gas separately.

Monoatomic gas

A monoatomic gas molecule has only three degrees of freedom of translational motion. So that the heat given to the gas at constant volume, increases the kinetic energy of translation of the molecules. Hence the increase in the translational kinetic energy of a molecule is equal to $\frac{3}{2}k_BT$.

Hence increase in the energy of 1 gm mole of the gas

$$U = \frac{3}{2}k_BTn_0 \qquad \text{where } n_0 \text{ is Avagadro number}$$

$$= \frac{3}{2}RT \qquad \because k_Bn_0 = R, \text{ the gas constant}$$

The molar specific heat at constant volume (C_V) is the heat required to raise the temperature of 1 mole of the gas through 1° C. Hence

$$C_V = \frac{dU}{dT} = \frac{3}{2}R \qquad \text{...(16.14)}$$

If C_P be molar specific heat at constant pressure then

$$C_P - C_V = R$$

or

$$C_P = R + C_V = R + \frac{3}{2}R = \frac{5}{2}R \qquad \text{...(16.15)}$$

and

$$\gamma = \frac{C_P}{C_V} = \frac{5}{3} = 1.67 \qquad \text{...(16.17)}$$

It is seen that expression for both C_V and C_P are independent of temperature indicating that molar specific heats of all monoatomic gases are the same at all the temperatures.

Diatomic gas

A diatomic molecule treated as a dumbbell (rigid rotator) has at ordinary temperature, five degrees of freedom, viz. 3 translational and 2 rotational (vibrational modes are not excited at ordinary temperatures).

K.E. per molecule at temperature $T = \frac{5}{2}k_BT$

and total energy per mole of the gas $U = \frac{5}{2} n_0 k_B T = \frac{5}{2} RT$

$$\therefore \quad C_V = \frac{dU}{dT} = \frac{5}{2} R \quad \text{...(16.18)}$$

and $$C_P = R + C_V = R + \frac{5}{2} R = \frac{7}{2} R \quad \text{...(16.19)}$$

and $$\gamma = \frac{C_P}{C_V} = \frac{7}{5} = 1.4 \quad \text{...(16.20)}$$

If diatomic molecule is not rigid then vibrational energy is also considered, then

$$C_V = \frac{7}{2} R,\ C_P = \frac{9}{2} R \text{ and } \gamma = \frac{9}{7} = 1.29$$

Here again the expression for C_P and C_V are independent of temperature. The molar specific heats of all diatomic gases should be same.

Triatomic and Polyatomic Gases

In triatomic gas for non-linear type molecule the number of degrees of freedom are 6. Thus the values of C_V and C_P are

$$C_V = 3R,\ C_P = 4R \text{ and } \gamma = \frac{4}{3} = 1.33$$

In general a polyatomic molecule has 3 translational and 3 rotational degrees of freedom and certain number (f) of vibrational modes. Then

$$U = \left(\frac{3}{2} k_B T + \frac{3}{2} k_B T + f k_B T\right) n_0$$

$$= (3 + f) RT$$

or $$C_V = (3 + f)R \text{ and } C_P = (4 + f)R$$

and $$\gamma = \frac{C_P}{C_V} = \frac{4 + f}{3 + f}$$

The theoretical predictions of specific heats (ignoring vibrational modes) are found to be in good agreement with experimental results. Certain discrepancies between experimental values and theoretical values of specific heats for certain gases are also observed. For example, in case of Cl_2, C_2H_6 and many other polyatomic gases the experimental values are found to be greater than the theoretical values. The agreement can be improved by including vibrational modes of motion in calculations.

16.9 MEAN FREE PATH

In postulates of kinetic theory of gases, we assumed the molecules to be mass points moving with high speeds even at ordinary temperatures. This should result in molecules all flying away in little or no time. This does not happen in practice. It is due to the fact that molecules in a gas have a finite though small size; so they are bound to undergo collisions. As a result, they cannot move straight unhindered, their paths keep getting incessantly deflected.

Between any two successive collisions, however, the molecules describe a linear path, called its free path. But as can be readily seen the length of the *free path* is not always the same. It may be greater or smaller between one pair of molecules than between another pair. We, therefore, are concerned with its average or mean value called *mean free path* (λ), which is obviously the total distance covered by the molecule divided by the number of collisions it encountered.

Let us consider molecules of the gas to be spheres of diameter *d*. Let us assume a particular molecule to be in motion where as all others are at rest. This molecule will suffer collisions with any molecule that comes within distance *d* between the centres. If *c* be the average velocity of the molecule, then it will collide with all the molecules lying in a cylinder of length *c* and area of cross-section πd^2, i.e. of volume $\pi d^2 c$ is 1 sec. If *n* be the number of molecules per unit volume in the gas, then the number of molecules in the cylinder are $\pi d^2 c n$ and hence number of collisions per sec are $\pi d^2 cn$.

The average interval between two collisions τ is thus

$$\tau = \frac{1}{\pi d^2 cn}$$

Therefore the average distance covered by the molecule between successive collisions, i.e. mean free path is

$$\lambda = \tau c = \frac{1}{\pi d^2 n} \quad \text{...(16.21)}$$

If *m* be the mass of the molecule, then density of the gas (ρ)

$$\rho = nm \text{ or } n = \frac{\rho}{m}$$

and

$$\lambda = \frac{m}{\pi d^2 \rho} \quad \text{...(16.21}a\text{)}$$

In this derivation we imagined the other molecules to be at rest. But actually all the molecules are moving and collision rate is determined by the average relative velocity of the molecule (c_r).

Maxwell on the basis of his law of distribution of velocities deduced the more accurate relation as

$$\lambda = \frac{1}{\sqrt{2}\,\pi d^2 n} = \frac{m}{\sqrt{2}\,\pi d^2 \rho} \quad \text{...(16.22)}$$

It follows that

(*i*) Mean free path (λ) is inversely proportional to density (ρ).

(*ii*) Since pressure (*P*) is directly proportional to density (ρ) and hence mean free path (λ) varies inversely as the pressure of the gas.

(*iii*) Since density (ρ) varies inversely as the absolute temperature (*T*), λ varies directly as absolute temperature (*T*) of the gas.

(*iv*) $\lambda \propto \frac{1}{d^2}$, i.e. smaller the size of the molecule, the greater the value of its mean free path (λ).

At ordinary pressure λ is of the order of 10^{-5} cm but at low pressures of about 10^{-4} cm of Hg, λ ranges between 5 cm to 10 cm.

16.10 VAN DER WAALS EQUATION OF STATE FOR REAL GASES

The equation of state for an ideal gas is given by

$$PV = \mu RT$$

where the symbols have their usual meanings. This equation is derived on the basis of simplifying assumptions made in kinetic theory of gases. Van der Waal rejected some of these assumptions in deriving the equation of state for real gas. These assumptions are (*i*) the volume occupied by the molecules is negligible compared with the total volume occupied by the gas and (*ii*) the force of attraction between the molecules is quite inappreciable.

These assumptions may be true for a gas at low pressure but they loose their validity when the pressure is high. For, however small the molecules may be, they must occupy some space, so that the volume of the gas is decreased under high pressure.

The force of cohesion in solids and liquids cannot altogether vanish; when solid or liquid passes on to gaseous state, particularly when the molecules are brought closer together under high pressure on the gas.

Taking these two facts into account Van der Waals derived the following equation of state for real gas

$$\left(P - \frac{a}{V^2}\right)(V - b) = \mu RT \quad \text{...(16.23)}$$

where a and b are positive constants. The constant a is related to the average force of attraction between the molecules and b is related to the total volume of the molecules.

Dimensions of a are of those of PV^2 and dimensional equation of a is

$$a = M^1L^5T^{-2}$$

Dimensional equation for b is $b = M^0L^3T^0$.

b represents the volume of the molecules in fact $b = 3.7 \times$ volume of all molecules.

16.11 PHASE CHANGES: TRIPLE POINT

Almost every substance known to us can exist in different forms; called its different phases. The example of water is all too familiar. It can exist in a solid, a liquid and a vaporous form or phase.

Let us consider the equilibrium between the liquid and vapour phases of the same substance. In equilibrium the vapour is saturated. The saturation vapour pressure depends on temperature. The variation of saturated vapour pressure with temperature is shown in Fig. 16.4. At pressure P and temperature T represented by point P the liquid and vapour co-exists and the vapour is saturated. Now if vapour pressure is increased by keeping the temperature constant, i.e. some vapour is added, the state is represented by the point L. Since the vapour is already saturated, the extra vapour will condense to liquid and the system will return to point P. Similarly if vapour pressure is decreased by taking out some vapour, keeping temperature constant so that the state is represented by V, the vapour becomes unsaturated and some liquid will evaporate and the system will come back to the point P.

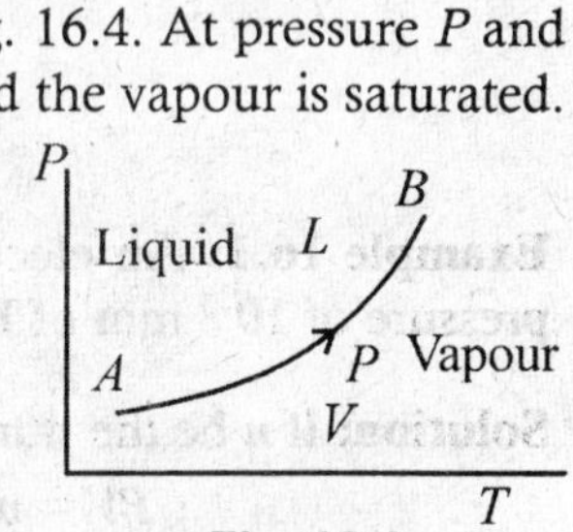

Fig. 16.4

Thus the vapour-liquid phases can co-exists only along the curve APB. At points above this curve there is pure liquid phase and at points below this curve there is pure gaseous phase.

We have defined boiling point as the temperature at which the saturated vapour pressure is equal to the external pressure. Hence curve in Fig. 16.4 gives the variation of boiling point with temperature, the temperature axis giving boiling point.

Similar curves for solid-liquid transition and for solid-vapour transition are shown in Fig. 16.5. Figure 16.5 (*a*) represents these curves for water and Fig. 16.5 (*b*) represents for CO_2. These are qualitative curves. Curve *PA* in both cases represents solid-liquid transition and *PB* represents liquid-vapour transition. Similarly *OP* represents solid-vapour transition. Solid and liquid phases can co-exist along the curve *PA* and solid and vapour phases can co-exist along the curve *OP*. Curve *PA* also represents the variation of melting point with pressure and curve *OP*, the variation of sublimation point with pressure. Thus the *P-T* space in divided into three regions called solid, liquid and vapour.

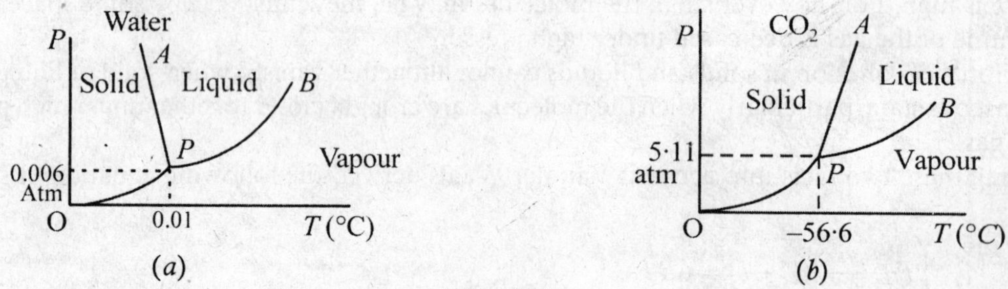

Fig. 16.5

These three curves meet at one point labelled *P*. At the pressure and temperature corresponding to point *P*, all the three phases may co-exist in equilibrium. This point is called triple point. For water the triple point occurs at the pressure 0.006 atm (= 4.58 mm of Hg) and temperature 0.01 °C (= 273.16 K).

For CO_2 the triple point temperature is –56.6° C (= 216.55 K) and at pressure 5.11 atm (Fig. 16.5 (*b*)) gives that CO_2 at one atmospheric pressure can remain either in solid phase or in vapour phase. As such when solid CO_2 open to atmosphere, is heated, it becomes vapour directly without passing through the liquid phase. It is therefore called dry ice. Notice also that the melting curves in case of CO_2 is positive while that for water is negative. That is why ice on melting to water contracts and CO_2 expand on melting.

SOLVED EXAMPLES

Example 16.1: An electric bulb of volume 250 cm³ was sealed off during manufacturing at a pressure of 10^{-3} mm of Hg at 27 °C. Find the number of molecules in the bulb.

Solution: If *n* be the number of molecules, then gas equation gives

$$PV = nk_BT$$

Given is $P = 10^{-3}$ mm of Hg $= \frac{10^{-3}}{760}$ Atmospheres $= \frac{10^{-3}}{760} \times 10^5$ Pa

$V = 250 \text{ cm}^3 = 250 \times 10^{-6} \text{ m}^3$

and $T = 27 + 273 = 300$ K

k_B is the Boltzmann's constant $= 1.4 \times 10^{-23}$ J/K

$$\therefore \quad n = \frac{PV}{k_BT}$$

$$= \frac{10^{-3} \times 10^5}{760} \times \frac{250 \times 10^{-6}}{300 \times 1.4 \times 10^{-23}}$$

$$= \frac{10^{16} \times 250}{76 \times 3 \times 1.4} = 7.8 \times 10^{15}$$

Example 16.2: 0.014 kg of nitrogen is enclosed in a vessel at a temperature of 27° C. How much heat has to be transferred to the gas to double the r.m.s. velocity of molecules?

Solution: We know that $\bar{c} = \sqrt{\frac{3RT}{M}}$

To double the $\bar{c}$, the temperature should be made 4 times, i.e. $4T$, i.e. temperature increase $\Delta T = 3T$.

Since gas is enclosed is a vessel means it is heated at constant volume. Therefore work done is zero.

Whatever heat is given increases the internal energy of the gas.

$$\therefore \quad \Delta U = nc_V \Delta T$$

$$= 0.5 \times \frac{5}{2} R \times 3T = 0.5 \times \frac{5}{2} \times 8.3 \times 3 \times 300 = 9337.5 \text{ J}$$

$$= 2223.2 \text{ cal.}$$

Example 16.3: The mass of hydrogen molecule is 3.32×10^{-27} kg. If 10^{23} hydrogen molecules strike a wall of area 2 cm² at angle 45° to the normal per second and rebound elastically with a speed of 10^3 m/s, calculate the pressure on the wall.

Solution: The molecule strike the wall along PO and rebounds along OQ. Resolving momentum along the wall and perpendicular to the wall, we have before collision

Component parallel to the wall $= mv \sin 45 = \frac{mv}{\sqrt{2}}$

Component perpendicular to the wall $= mv \cos 45 = \frac{mv}{\sqrt{2}}$

After collision,

Component parallel to the wall $= mv \sin 45 = \frac{mv}{\sqrt{2}}$

Component perpendicular to the wall $= mv \cos 45 = \frac{mv}{\sqrt{2}}$ away from the wall

Thus there is no change in momentum of the particle along the wall.

The change in momentum of the particle perpendicular to the wall

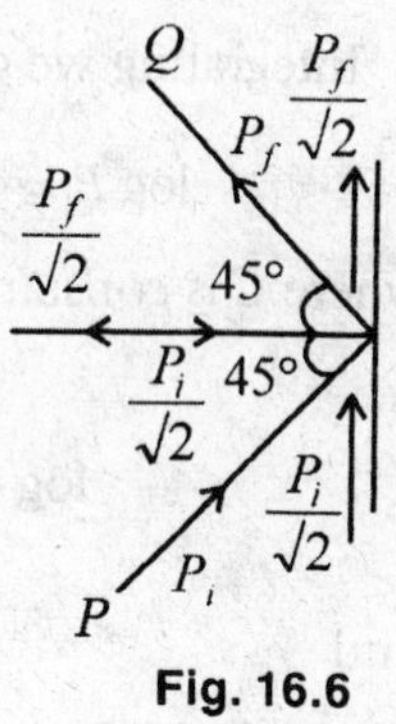

Fig. 16.6

$$\Delta P = 2\frac{mv}{\sqrt{2}} \text{ away from the wall}$$

Hence momentum imparted to the wall per collision $= \sqrt{2}\, mv$. If n particles collide per sec with the wall, the total momentum imparted to the wall per second, i.e. force on the wall is

$$F = \Delta P \cdot n = \sqrt{2}\, mvn$$

Hence pressure on the wall is

$$P = \frac{F}{A} = \frac{\sqrt{2}\,mvn}{A} = \frac{\sqrt{2} \times 3.32 \times 10^{-27} \times 10^{3} \times 10^{23}}{2 \times 10^{-4}}$$

or $$P = 2.347 \times 10^{4}\ \text{N/m}^2$$

Example 16.4: Assume that the temperature remains essentially constant in the upper part of atmosphere. Obtain an expression for pressure variation in the upper atmosphere with height. The mean molecular weight of air is M.

Solution: If the pressure at height h is P and at $h + dh$, $P + dP$ then

$$dP = -Pgdh \qquad ...(i)$$

Let us consider a small volume ΔV of air of mass Δm. Then

$$P\Delta V = nRT = \frac{\Delta m}{M}RT$$

or $$P = \frac{\Delta m}{\Delta V}\frac{R}{M}T \quad \text{But } \frac{\Delta m}{\Delta V} = \rho \text{ the density of gas}$$

or $$P = \rho\frac{RT}{M}$$

and $$\rho = \frac{PM}{RT}$$

Substituting in (*i*), we have

$$dP = -\frac{PM}{RT}g\,dh$$

or $$\frac{dP}{P} = -\frac{Mg}{RT}dh$$

Integrating we get

$$\log P = -\frac{Mg}{RT}h + C$$

where C is constant of integration. Now we have $h = 0$, let pressure be P_0, then

$$C = \log P_0$$

$$\therefore \quad \log\frac{P}{P_0} = -\frac{Mgh}{RT}$$

and $$P = P_0 e^{-\frac{Mgh}{RT}}$$

Example 16.5: A vertical cylinder closed from the both ends is equipped with an easily moving piston dividing the volume into two parts, each containing one mole of air. In equilibrium at temperature $T_0 = 300$ K, the volume of the upper part is $\eta = 4.0$ times greater than that of the lower part. At what temperature, will the ratio of these volumes be equal to $\eta' = 3.0$?

Solution: Let at temperature T_0 the upper part has pressure P_1, volume V_1 and lower part has pressure P_2 and volume V_2. Let these values change to P_1' and V_1', P_2' and V_2' respectively at temperature T.

Total volume of cylinder is constant, hence

$$V_1' + V_2' = V_1 + V_2$$

or
$$V_1'\left(1+\frac{V_2'}{V_1'}\right) = V_1\left(1+\frac{V_2}{V_1}\right)$$

or
$$\frac{V_1'}{V_1} = \frac{1+\frac{V_2}{V_1}}{1+\frac{V_2'}{V_1'}} = \frac{1+\frac{1}{\eta}}{1+\frac{1}{\eta'}} = \frac{\eta'(1+\eta)}{\eta(1+\eta')}$$

If mass of piston is m and its area of cross-section A, then in equilibrium

$$P_1 + \frac{mg}{A} = P_2 \text{ and } P_1' + \frac{mg}{A} = P_2'$$

$\therefore$
$$P_2 - P_1 = P_2' - P_1' = \frac{mg}{A}$$

Now using gas equation and remembering that number of moles of gas in each part is 1, we have

$$P_1V_1 = RT_0,\ P_2V_2 = RT_0$$

$$P_1'V_1' = RT,\ P_2'V_2' = RT$$

$\therefore$
$$P_2 - P_1 = RT_0\left(\frac{1}{V_2} - \frac{1}{V_1}\right)$$

and
$$P_2' - P_1' = RT\left(\frac{1}{V_2'} - \frac{1}{V_1'}\right)$$

As $P_2 - P_1 = P_2' - P_1'$, we have

$$T_0\left(\frac{1}{V_2} - \frac{1}{V_1}\right) = T\left(\frac{1}{V_2'} - \frac{1}{V_1'}\right)$$

or
$$\frac{T_0}{V_1}\left(\frac{V_1}{V_2} - 1\right) = \frac{T}{V_1'}\left(\frac{V_1'}{V_2'} - 1\right)$$

or
$$\frac{T}{T_0} = \frac{V_1'}{V_1}\frac{\left(\frac{V_1}{V_2} - 1\right)}{\left(\frac{V_1'}{V_2'} - 1\right)} = \frac{\eta'(1+\eta)}{\eta(1+\eta')}\frac{(\eta-1)}{(\eta'-1)}$$

or
$$\frac{T}{T_0} = \frac{3(1+4)(4-1)}{4(1+3)(3-1)} = \frac{45}{32}$$

or
$$T = T_0 \times \frac{45}{32} = 300 \times \frac{45}{32} = 421.9 \text{ K} = 148.9° \text{ C}$$

Example 16.6: One mole of oxygen at 27 °C and 1 atmosphere pressure is enclosed in a vessel. (*i*) Assuming the molecules to be moving with v_{rms} find the number of collisions per second which the molecules make with one square meter area of the wall of the vessel. (*ii*) The vessel is next thermally insulated and moved with constant speed v_0. It is then suddenly stopped. The process results in a rise in temperature of the gas by 1° C. Calculate the speed v_0 ($\gamma = 1.4$).

Solution: Gas equation gives

$$PV = nk_BT$$

or
$$P = \frac{n}{V}k_BT = n'\,k_BT$$

where n' is the number of molecules per unit volume.

$$\therefore \quad \eta' = \frac{P}{k_BT} = \frac{10^5}{1.4 \times 10^{-23} \times 300} = 2.4 \times 10^{25}/\text{m}^3$$

$$v_{rms} = \sqrt{\frac{3RT}{M}} = \sqrt{\frac{3 \times 8.4 \times 300}{32 \times 10^{-3}}} = 486.0 \text{ m/s}$$

Out of all the molecules 1/3 (= $n'/3$) molecules each travel along X, Y, and Z direction at a time in a unit volume made up of a cube of unit side length and each molecule has to travel a distance of 2 m before making a collision on the same wall. Hence the number of collisions made by molecules on one square metre area of the same wall per second

$$= \frac{n'}{3} \times \text{total number of collisions made by a molecule /sec}$$

$$= \frac{n'}{3} \times \frac{v_{rms}}{2} = \frac{1}{6}n'\,v_{rms}$$

$$= \frac{2.4 \times 10^{25} \times 486}{6} = 194.4 \times 10^{25} = 1.944 \times 10^{27}/\text{sec}$$

(*ii*) Vessel is moved with velocity v_0 and hence all molecules contained in it will move with velocity v_0 and will have an additional translational kinetic energy.
Kinetic energy of all the molecules

$$= \frac{1}{2}mv_0^2 \times n_0 \text{ where } n_0 \text{ is the Avogadro number}$$

$$= \frac{1}{2}Mv_0^2 \text{ where } M \text{ is molecular weight}$$

This kinetic energy of molecules is converted into heat energy when the vessel is stopped. This energy raises the temperature of 1 mole of the gas by 1° C. Hence heat required is

$$Q = 1 \times C_V \Delta T = \frac{R}{\gamma - 1} \qquad \because \; C_V = \frac{R}{\gamma - 1}$$

$$\therefore \quad \frac{R}{\gamma - 1} = \frac{1}{2}Mv_0^2$$

or
$$v_0 = \sqrt{\frac{2R}{M(\gamma - 1)}} = \sqrt{\frac{2 \times 8.4}{32 \times 10^{-3}(1.4 - 1)}}$$

$$= \sqrt{\frac{2 \times 8.4}{32 \times 10^{-3} \times 0.4}} = 36.2 \text{ m/s}$$

Example 16.7: We have two vessels of equal volume, one filled with hydrogen and the other with equal mass of helium. The common temperature is 27° C (*i*) What is the relative number of molecules in the two vessels. (*ii*) If pressure of hydrogen is 2 atmospheres what is the pressure of helium. (*iii*) If the temperature of helium is kept at 27° C and that of hydrogen is changed, at what temperature will its pressure become equal to that of helium? The molecular weight of hydrogen and helium are 2 and 4 respectively.

Solution: (*i*) If n_H and n_{He} be the number of molecules of hydrogen and helium respectively, and m_H and m_{He} be the molecular mass of hydrogen and helium respectively, then since their masses are equal, hence

$$n_H \cdot m_H = n_{He} \cdot m_{He}$$

or $$\frac{n_H}{n_{He}} = \frac{m_{He}}{m_H} = \frac{2}{1}$$

(*ii*) The equation of state of a gas is

$$PV = n k_B T$$

where n is the total number of molecules. For a given temperature and volume; the pressure is proportional to the number of molecules, i.e.

$$P \propto n$$

$$\therefore \quad \frac{P_H}{P_{He}} = \frac{n_H}{n_{He}} = \frac{2}{1}$$

Since $$P_H = 2 \text{ atm}$$

$$\therefore \quad P_{He} = \frac{n_H}{n_{He}} = \frac{2}{1} = 1 \text{ atm}$$

(*iii*) H_2 and He have equal volumes. For having equal pressures, we must have

$$n_H T_H = n_{He} T_{He}$$

$$\therefore \quad T_H = \frac{n_{He} T_{He}}{n_H} = \frac{1}{2} \times 300 = 150 \text{ K} = -123° \text{ C}$$

Example 16.8: Suppose the pressure P and density ρ of air are related as $\frac{P}{\rho^n}$ = constant regardless of height (n is constant here). Find the corresponding temperature gradient.

Solution: Consider a thin layer at height h and of thickness dh. Let P and $P + dP$ be the pressure on the two sides of the layer. The mass of the layer is $A\rho dh$. Considering the equilibrium of the layer (Fig. 16.7) $(P + dP)A + A\rho dhg = PA$

Here A is the area of cross-section of the layer.

or $$dP = -\rho g dh \qquad \ldots(i)$$

But $$P = C\rho^n \text{ where } C \text{ is constant} \qquad \ldots(ii)$$

$$\therefore \quad \frac{dP}{d\rho} = Cn\rho^{n-1}$$

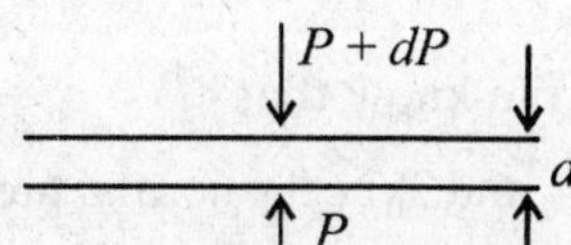

Fig. 16.7

or $$\frac{d\rho}{dP} = \frac{1}{Cn\rho^{n-1}} \qquad \ldots(iii)$$

and for ideal gas $P = \rho\dfrac{RT}{M}$ ...(iv)

Eliminating P between the two equations (ii) and (iv), we have

$$\rho\frac{RT}{M} = C\rho^n$$

or $$T = \frac{MC}{R}\rho^{n-1}$$

and $$\frac{dT}{d\rho} = \frac{MC}{R}(n-1)\rho^{n-2}$$

But $$\frac{dT}{dh} = \frac{dT}{d\rho}\cdot\frac{d\rho}{dP}\cdot\frac{dP}{dh}$$

or $$\frac{dT}{dh} = \frac{MC}{R}(n-1)\rho^{n-2}\frac{1}{Cn\rho^{n-1}}\cdot(-\rho g)$$

or $$\frac{dT}{dh} = -\frac{Mg}{R}\frac{(n-1)}{n}$$

or $$\frac{dT}{dh} = -\frac{Mg}{R}\frac{(n-1)}{n}$$

Example 16.9: A vessel of volume V = 30 litre contains ideal gas at temperature 0 °C. After a portion of the gas has been let out, pressure in the vessel is decreased by ΔP= 0.78 atm (the temperature remaining constant). Find mass of the released gas. The gas density under the normal temperature and pressure (NTP) is ρ = 1.3 g/litre.

Solution: Let m_1 and m_2 be the masses of the gas in the vessel before and after the gas is released. Hence mass of the gas released is

$$\Delta m = m_1 - m_2$$

From ideal gas equation, we have

$$P_1V = \frac{m_1 RT_0}{M} \text{ and } P_2V = \frac{m_2 RT_0}{M}$$

$\therefore$ $$(P_1 - P_2)V = \frac{RT_0}{M}(m_1 - m_2) = \Delta m\frac{RT_0}{M}$$

or $$\Delta m = \frac{(P_1 - P_2)V\cdot M}{RT_0} = \frac{\Delta P\, VM}{RT_0} \qquad \ldots(i)$$

We know that $P = \rho\dfrac{RT}{M}$

If P_0 and T_0 be the normal pressure and temperature respectively and ρ_0 density at 0° C, then

$$P_0 = \rho_0\frac{RT_0}{M}$$

$$\therefore \qquad \frac{M}{RT_0} = \frac{\rho_0}{P_0}$$

Substituting in (*i*), we have

$$\Delta m = \frac{\Delta P \cdot V \cdot \rho_0}{P_0} = 0.78 \times 30 \times 1.3 = 30.42 \text{ g}$$

Example 16.10: Find the pressure of air in a vessel being evacuated as a function of time *t*. The vessel volume is *V*, the initial pressure is P_0. The process is assumed to be isothermal and evacuation rate equal to *C* and independent of pressure.

Solution: For an ideal gas, we have

$$P = \frac{m}{M}\frac{RT}{V}$$

Thus
$$\frac{dP}{dt} = \frac{RT}{MV}\frac{dm}{dt} \qquad \ldots(i)$$

In each stroke volume *v* of the gas is ejected

$$v = \frac{V}{m_n}\left[m_n - (m_n - 1)\right]$$

In case of continuous ejection if $(m_n - 1)$ is the mass of gas in the vessel at time *t*; then m_n is the mass at time $(t + \Delta t)$, where Δt is the time in which volume *v* of the gas has come out. The ratio of evacuation therefore is $\frac{v}{\Delta t}$

$$\therefore \qquad C = \frac{v}{\Delta t} = -\frac{V}{m(t+\Delta t)}\frac{m(t+\Delta t) - m(t)}{\Delta t}$$

In the limit $\Delta t \rightarrow 0$, we get

$$C = -\frac{V}{m}\frac{dm}{dt} \qquad \ldots(ii)$$

Using (*ii*) in (*i*), we have

$$\frac{dP}{dt} = -\frac{C\,mRT}{V^2 M} = -\frac{CP}{V} \qquad \because P = \frac{mRT}{MV}$$

or
$$\frac{dP}{P} = -\frac{C}{V}dt$$

and integration gives

$$\int_{P_0}^{P}\frac{dP}{P} = -\frac{C}{V}\int_0^t dt$$

or
$$\log\frac{P}{P_0} = -\frac{C}{V}t$$

or
$$P = P_0 e^{-\frac{C}{V}t}$$

Method 2: Gas equation gives $PV = \mu RT$.

Differentiating with respect to time, we have

$$P\frac{dV}{dt} + V\frac{dP}{dt} = 0 \text{ Since process is isothermal}$$

But $$\frac{dV}{dt} = C$$

$\therefore$ $$V\frac{dP}{dt} = -PC$$

or $$\frac{dP}{P} = -\frac{C}{V}dt$$

integration gives $$\int_{P_0}^{P}\frac{dP}{P} = -\frac{C}{V}\int_0^t dt$$

or $$\log\frac{P}{P_0} = -\frac{C}{V}t$$

or $$P = P_0 e^{-\frac{C}{V}t}$$

Example 16.11: How many times is the heat capacity of oxyhydrogen gas (mixture of O_2 and H_2 in mole ratio 1 : 2) greater than that of the water vapours produced by the combustion of the gas? Solve the problem for the cases when (*i*) V = constant (*ii*) When P = constant.

Solution: Since hydrogen and oxygen both are diatomic molecules hence degrees of freedom for each of them is 5. Thus average energy per molecule is $= \frac{5}{2}k_BT$

Since there are 3 moles and total number of molecules is $3n_0$, where n_0 is Avogadro number.

Hence total energy of the gas $= \frac{5}{2}k_BT \times 3n_0 = \frac{15RT}{2}$

Total mass of the gas m = 36 gm (32 gm O_2 + 4 gm H_2)

Thus energy per gm of the system $= \frac{15RT}{2 \times 36}$

or C_V for the gas $= \frac{dQ}{dT} = \frac{15R}{72}$

For water vapour, the molecule is triatomic nonlinear. Hence degrees of freedom are 6.

Energy of each molecule $= 6 \times \frac{1}{2}k_BT = 3k_BT$

For 1 mole energy is $= 3 \times n_0 k_BT = 3RT$.

Energy per gm of water vapour $= \frac{3RT}{18}$ (water vapour $M = 18$)

$\therefore$ $$C_V = \frac{3R}{18}$$

Thus ratio is $= \frac{15R}{72 \times 3R} \times 18 = \frac{5}{4} = 1.25$

In case of constant pressure

For oxyhydrogen $C_P = \frac{7}{2}R \times \frac{3}{36} = \frac{21R}{72}$

and for water vapour $C_P = \frac{4R}{18}$

and ratio $= \frac{21R}{72} \times \frac{18}{4R} = \frac{21}{16} = 1.31$

Example 16.12: Find the maximum attainable temperature of ideal gas in each of the following processes: (*a*) $P = P_0 - \alpha V^2$ (*b*) $P = P_0 e^{-\beta V}$ where P_0, α and β are positive constants and V is the volume of one mole of gas.

Solution: (*a*) $P = P_0 - \alpha V^2$

Differentiating, we have

$$\frac{dP}{dV} = -2\alpha V \quad ...(i)$$

Gas equation gives (for 1 mole)

$$PV = RT$$

differentiating, we have

$$VdP + PdV = RdT$$

or $$V\frac{dP}{dV} + P = R\frac{dT}{dV} \quad ...(ii)$$

Substituting $\frac{dP}{dV}$ from (*i*) in (*ii*), we have

$$V(-2\alpha V) + P_0 - \alpha V^2 = R\frac{dT}{dV}$$

or $$R\frac{dT}{dV} = P_0 - 3\alpha V^2$$

for maximum temperature $\frac{dT}{dV} = 0$ gives

$$V = \sqrt{\frac{P_0}{3\alpha}}$$

Substituting the values of pressure and volume in gas equation, we have

$$T_{\max} = \frac{PV}{R} = \frac{(P_0 - \alpha V^2)V}{R}$$

$$= \frac{\left(P_0 - \alpha\frac{P_0}{3\alpha}\right)}{R}\sqrt{\frac{P_0}{3\alpha}} = \frac{2}{3}\frac{P_0}{R}\sqrt{\frac{P_0}{3\alpha}}$$

(*b*) $P = P_0 e^{-\beta V}$

Differentiating, we get $\frac{dP}{dV} = -\beta P_0 e^{-\beta V}$ ... (i)

Differentiating gas equation, we get

$$P + V\frac{dP}{dV} = R\frac{dT}{dV}$$

Eliminating P we have

$$R\frac{dT}{dV} = P_0 e^{-\beta V} + V\left(-\beta P_0 e^{-\beta V}\right)$$

$$= P_0 e^{-\beta V}(1 - \beta V)$$

For maximum temperature $\frac{dT}{dV} = 0$ or $V = \frac{1}{\beta}$

Substituting in gas equation, we have

$$T = \frac{PV}{R} = \frac{P_0 e^{-\beta\frac{1}{\beta}}}{R}\frac{1}{\beta} = \frac{P_0}{eR\beta}$$

Example 16.13: One gm of helium having r.m.s. velocity of molecules 1000 ms^{-1} and 4 gm of oxygen having r.m.s. velocity of molecules 1000 m/s are mixed in a container which is thermally isolated. What are the r.m.s. velocities of helium and oxygen molecules after the equilibrium is attained?

Solution: If n_0 is the Avogadro number 1 gm of helium has $\frac{n_0}{4}$ molecules. Helium molecule has only three degrees of freedom for translation only. Hence total energy of helium molecules

$$E_H = \frac{n_0}{4}\frac{1}{2}m_H c_H^2 \qquad c_H \text{ is r.m.s. velocity of helium molecules}$$

$$= \frac{M_H c_H^2}{8} \qquad n_0 m_H = M_H \text{ molecular weight of helium}$$

Similarly 4 gm of oxygen has $\frac{n_0}{8}$ molecules and has 3 degrees of freedom of translation and 2 degrees of freedom for rotation, i.e. total 5.

Translational K.E. of O_2 molecule $= \frac{1}{2}m_0 c_0^2$

Total energy of O_2 molecule $= \frac{1}{2}m_0 c_0^2 \times \frac{5}{3}$

and total energy of 4 gm of O_2 $= \frac{n_0}{8}\frac{1}{2}m_0 c_0^2 \times \frac{5}{3} = \frac{1}{16}M_0 c_0^2 \times \frac{5}{3}$

Hence total energy the system (mixture) is

$$E = \frac{M_H c_H^2}{8} + \frac{M_0 c_0^2}{16} \times \frac{5}{3}$$

But $\qquad c_0 = c_H = 1000$ m/s.

$$\therefore \quad E = \frac{1}{16}c_H^2\left[2M_H + \frac{5}{3}M_O\right] = \frac{1}{16}c_H^2\left[2 \times 4 + \frac{5}{3} \times 32\right] \times 10^{-3}$$

$$= \frac{(10^3)^2}{16} \times \frac{184}{3} \times 10^{-3} = \frac{10^3 \times 184}{3 \times 16} \text{ J}$$

In equilibrium the temperature remains constant and the average translational K.E. of molecules are same, i.e.

$$\frac{1}{2}m_O c_O^2 = \frac{1}{2}m_H c_H^2$$

where now c_O and c_H are r.m.s. speeds of O_2 and helium molecules when equilibrium is attained.

Energy of all helium molecules $= \frac{n_0}{4} \times \frac{1}{2}m_H c_H^2 = \frac{n_0 m_H c_H^2}{8}$

O_2 molecule has 5 degrees of freedom. Hence total energy of all O_2 molecules is

$$= \frac{n_0}{8} \times \frac{5}{3}\left(\frac{1}{2}m_O c_O^2\right) = n_0 m_O c_O^2 \times \frac{5}{16 \times 3}$$

$$= \frac{n_0 m_H c_H^2 \times 5}{16 \times 3} \text{ since } \frac{1}{2}m_H c_H^2 = \frac{1}{2}m_O c_O^2$$

∴ Total energy of the mixture is

$$= \frac{n_0 m_H c_H^2}{8} + \frac{n_0 m_H c_H^2 \times 5}{16 \times 3}$$

$$= \frac{n_0 m_H c_H^2}{16}\left[2 + \frac{5}{3}\right] = \frac{M_H c_H^2 \times 11}{16 \times 3} = \frac{4 \times 10^{-3} c_H^2}{16 \times 3} \times 11$$

Since the system is thermally isolated, total energy must remains conserved.

$$\therefore \quad \frac{4 \times 10^{-3} c_H^2 \times 11}{16 \times 3} = \frac{10^3 \times 184}{3 \times 16}$$

or $$c_H^2 = \frac{184}{44} \times 10^6$$

or $$c_H = \sqrt{\frac{184}{44}} \times 10^3 = 2.045 \times 10^3 = 2045 \text{ m/s}$$

Thus in equilibrium r.m.s. velocity of He molecules is 2045 ms^{-1}, we have

$$m_O c_O^2 = m_H c_H^2$$

or $$c_O = \sqrt{\frac{m_H}{m_O}} \cdot c_H$$

$$= \sqrt{\frac{4}{32}} \times 2045 = \frac{2045}{\sqrt{8}} = 723 \text{ m/s}$$

Thus the r.m.s. velocity of O_2 molecules in mixture is 723 m/s.

Example 16.14: Figure 16.8 shows an adiabatic cylindrical tube of volume V_0 divided into two parts by a frictionless adiabatic thin separator. Initially, the separator is kept in the middle, an ideal gas at pressure P_1 and temperature T_1 is injected to the left part and another ideal gas at pressure P_2 and temperature T_2 is injected at right part. $\frac{C_P}{C_V} = \gamma$ is same for both gases. The separator is slided slowly and is released at a position which can stay in equilibrium. Find (*a*) The volumes of the two parts, (*b*) The heat given to the gas in the left, (*c*) The final pressure of the gas (neglect the volume of the separator).

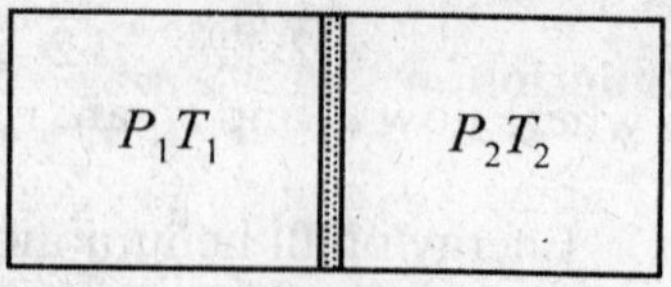

Fig. 16.8

Solution: (*a*) Initial volume of left-side $= \frac{V_0}{2}$

Initial pressure of left side $= P_1$

Initial volume of right side $= \frac{V_0}{2}$

and initial pressure of right side $= P_2$

In equilibrium the pressure be P (it will be same on both sides)
If final volume of left side in V_1 then that of right side is $V_0 - V$.
Since process is adiabatic. Then for left side

$$P_1\left(\frac{V_0}{2}\right)^\gamma = PV^\gamma \qquad \ldots(i)$$

For right side $P_2\left(\frac{V_0}{2}\right)^\gamma = P(V_0 - V)^\gamma \qquad \ldots(ii)$

Dividing, we have $\frac{P_1}{P_2} = \frac{V^\gamma}{(V_0 - V)^\gamma}$

or
$$\frac{V_0 - V}{V} = \left(\frac{P_2}{P_1}\right)^{\frac{1}{\gamma}}$$

or
$$\frac{V_0}{V} = \frac{P_2^{\frac{1}{\gamma}}}{P_1^{\frac{1}{\gamma}}} + 1 = \frac{P_1^{\frac{1}{\gamma}} + P_2^{\frac{1}{\gamma}}}{P_1^{\frac{1}{\gamma}}}$$

or
$$V = \frac{V_0 P_1^{\frac{1}{\gamma}}}{P_1^{\frac{1}{\gamma}} + P_2^{\frac{1}{\gamma}}}$$

and
$$V_0 - V = V_0 - \frac{V_0 P_1^{\frac{1}{\gamma}}}{P_1^{\frac{1}{\gamma}} + P_2^{\frac{1}{\gamma}}} = \frac{V_0 P_2^{\frac{1}{\gamma}}}{P_1^{\frac{1}{\gamma}} + P_2^{\frac{1}{\gamma}}}$$

(*b*) Since the cylinder is adiabatic hence $\Delta Q = 0$
(*c*) From (*i*) we have

$$P = P_1^{\frac{1}{\gamma}} \frac{V_0}{2V} = P_1^{\frac{1}{\gamma}} \frac{V_0\left(P_1^{\frac{1}{\gamma}} + P_2^{\frac{1}{\gamma}}\right)}{2V_0 P_1^{\frac{1}{\gamma}}}$$

$$= \frac{1}{2}\left(P_1^{\frac{1}{\gamma}} + P_2^{\frac{1}{\gamma}}\right)$$

Example 16.15: Find the minimum attainable pressure of ideal gas in the process $T = T_0 + \alpha V^2$ where T_0 and α are positive constants and V is the volume of one mole of gas. Draw the approximate P vs V plot of this process.

Solution: The process is $T = T_0 + \alpha V^2$

Since gas is ideal, hence $PV = RT$ or $V = \dfrac{RT}{P}$

$$\therefore \qquad T = T_0 + \frac{R^2 T^2 \alpha}{P^2}$$

$$\text{or} \qquad P = \frac{RT\sqrt{\alpha}}{\sqrt{T - T_0}}$$

Differentiating we have

$$\frac{dP}{dT} = \sqrt{\alpha}R\left[\frac{1}{\sqrt{T-T_0}} - \frac{1}{2}\frac{T}{(T-T_0)^{\frac{3}{2}}}\right]$$

$$= \sqrt{\alpha}R\frac{T - T_0 - \frac{1}{2}T}{(T-T_0)^{\frac{3}{2}}}$$

$$= \sqrt{\alpha}R\frac{\left(\frac{T}{2} - T_0\right)}{(T-T_0)^{\frac{3}{2}}}$$

For P to be minimum $\dfrac{dP}{dt} = 0$

$$T = 2T_0$$

$$P_{\min} = R\sqrt{\alpha}\frac{2T_0}{\sqrt{2T_0 - T_0}} = 2R\sqrt{\alpha T_0}$$

Using gas equation we have $PV = RT$. But $T = T_0 + \alpha V^2$

$$\therefore \qquad PV = R\left(T_0 + \alpha V^2\right) \text{ or } P = \frac{RT_0}{V} + R\alpha V$$

Thus P is ∞ when $V = 0$ or $V = \infty$

For P minimum $\dfrac{dP}{dV} = -\dfrac{RT_0}{V^2} + R\alpha = 0$ or $V = \sqrt{\dfrac{T_0}{\alpha}}$

Thus P-V graph is parabola with $P_{\min}$ at $V = \sqrt{\dfrac{T_0}{\alpha}}$

and $\qquad P_{\min} = 2R\sqrt{\alpha T_0}$

The plot is shown is Fig. 16.9.

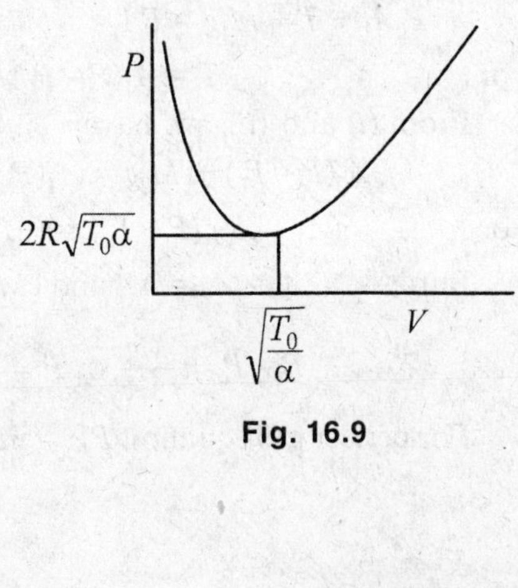

Fig. 16.9

Example 16.16: A vessel of volume $V = 5.0$ litres contains 1.4 gm of nitrogen at a temperature of 1800 K. Find the pressure of the gas if 30% of its molecules are dissociated into atoms at this temperature.

Solution: Let partial pressures of the molecular nitrogen and atomic nitrogen be P_1 and P_2 respectively. Then total pressure P is

$$P = P_1 + P_2 \qquad \ldots(i)$$

But $$P_1 = \frac{m_1}{M}\frac{RT}{V} \text{ and } P_2 = \frac{m_2}{M}\frac{RT}{V}$$

where m_1 and m_2 are the masses of molecular nitrogen and atomic nitrogen respectively.

$$m_1 = 1.4 \times 0.7 = 0.98 \text{ g} \quad \text{(70\% molecular)}$$

$$m_2 = 1.4 \times 0.3 = 0.42 \text{ g} \quad \text{(30\% atomic)}$$

$\therefore$ $$P_1 = \frac{0.98}{28} \times \frac{8.4 \times 1800}{5 \times 10^{-3}} = 1.06 \times 10^5 \text{ N/m}^2$$

Similarly $$P_2 = \frac{0.42}{14} \times \frac{8.4 \times 1800}{5 \times 10^{-3}} = 0.91 \times 10^5 \text{ N/m}^2$$

Hence pressure $$P = P_1 + P_2 = (1.06 + 0.91) \times 10^5$$
$$= 1.97 \times 10^5 \text{ N/m}^2.$$

Example 16.17: A smooth vertical tube having two different sections is open at both ends and equipped with two pistons of different areas (Fig. 16.10). Each piston slides within a respective tube section. The pistons are tied with an inextensible weightless string. One mole of an ideal gas is enclosed between them. The cross-sectional area of upper piston is $\Delta A = 10$ cm^2 greater than that of lower one. The combined mass of two pistons is equal to 5.0 kg. The external pressure is $P_0 = 1.0$ atm. By how much the gas between the pistons be heated to shift piston through $l = 5.0$ cm.

Solution: Let M_1 and M_2 be the masses of upper and lower pistons. A_1 and A_2 their areas of cross-section and T_1 tension in the string. For equilibrium of lower piston.

$$PA_2 + M_2g = P_0A_2 + T$$

or $$T = A_2(P - P_0) + M_2g \qquad \ldots(i)$$

For equilibrium of upper piston

$$P_0A_1 + T + M_1g = PA_1$$

or $$T = A_1(P - P_0) + M_1g \qquad \ldots(ii)$$

From (*i*) and (*ii*), we have

$$A_2(P - P_0) + M_2g = A_1(P - P_0) - M_1g$$

or $$(A_1 - A_2)(P - P_0) = (M_1 + M_2)g$$

But $$A_1 - A_2 = \Delta A \text{ and } (M_1 + M_2) = M$$

$\therefore$ $$P - P_0 = \frac{Mg}{\Delta A} \text{ or } P = P_0 + \frac{Mg}{\Delta A} = \text{constant} \qquad \ldots(iii)$$

For perfect gas equation $PV = nRT$.

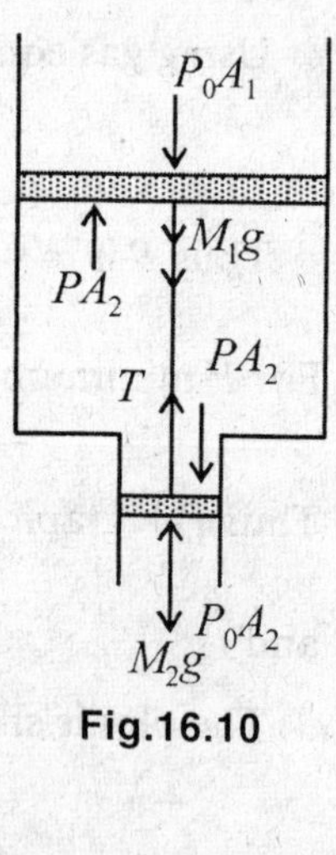

Fig.16.10

Differentiating and keeping in mind that pressure is constant, we have

$$PdV = nRdT$$

But $\quad dV = l\,\Delta A$, where l is the shift of the pistons)

$$\therefore \quad \left(P_0 + \frac{Mg}{\Delta A}\right) l\,\Delta A = nR\,dT$$

or

$$dT = \left(P_0 + \frac{Mg}{\Delta A}\right)\frac{l \cdot \Delta A}{nR}$$

$$= \left(10^5 + \frac{5 \times 10}{10 \times 10^{-4}}\right)\frac{5 \times 10^{-2} \times 10 \times 10^{-4}}{1 \times 8.4}$$

$$= \frac{1.5 \times 10^5 \times 5 \times 10^{-6} \times 10}{8.4}$$

$$= \frac{7.5}{8.4} = 0.9 \text{ K}$$

Example 16.18: An astronaut takes a cylinder of volume 10 litres containing nitrogen gas at a temperature of 27° C and pressure 50 atm. He makes a hole of area 1 sq.cm in this cylinder and places in open space. Estimate the time it would take for the cylinder to be empty.

Solution: If v is the average velocity of molecules, then according to molecular theory of leakage of gas through a hole of area a in vacuum, the mass of gas escaping out of the hole per second is $\frac{1}{4} a\rho v$

Hence volume leakage per second $= \dfrac{a}{4} v$

But

$$v = \sqrt{\frac{8k_B T}{m\pi}}$$

Thus volume leaking per second $= \dfrac{a}{4}\sqrt{\dfrac{8k_B T}{\pi m}} = a\sqrt{\dfrac{k_B T}{2\pi m}}$

Hence time taken for all gas to leak is

$$t = \frac{\text{Total volume}}{\text{Rate of leakage}}$$

$$= \frac{V}{a}\sqrt{\frac{2\pi m}{k_B T}}$$

$$= \frac{10 \times 10^{-3}}{10^{-4}}\sqrt{\frac{2\pi \times 28 \times 1.67 \times 10^{-27}}{1.4 \times 10^{-23} \times 300}}$$

$$= 10^2\sqrt{\frac{10^{-6} \times 2\pi \times 28 \times 1.67}{1.4 \times 3}}$$

$$= 8.36 \times 10^{-1} \text{ sec}$$

$$= 0.836 \text{ sec.}$$

Example 16.18: A horizontal tube of length l closed at both ends contains an ideal gas of molecular weight M. The tube is rotated at a constant angular velocity ω about a vertical axis passing through an end. Assuming the temperature to be uniform and constant, show that $P_2 = P_1 e^{M\omega^2 l^2/2RT}$, where P_1 and P_2 are the pressures of the gas at two ends of the tube.

Solution: Consider an element dx of the gas between distance x and $x + dx$ from the axis of rotation O. Let P and $P + dP$ be the pressures at the two ends of the element as shown in Fig. 16.11. The force due to the pressure difference across the element provides the necessary centripetal force for circular motion.

Force due to pressure difference = dPA

where A is area of cross-section of the tube.

If mass of the gas in element is dm then

$$dPA = dm\omega^2 x \quad \ldots(i)$$

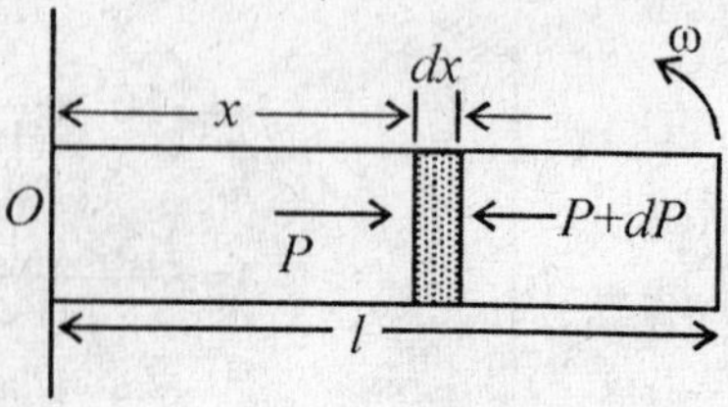

Fig. 16.11

For perfect gas, the gas equation is

$$PV = \frac{m}{M} RT$$

where M is molecular weight of the gas.

Differentiating we have

$$P\,dV = dm\frac{RT}{M}$$

or
$$dm = \frac{PM}{RT} dV$$

But
$$dV = A\,dx \;\therefore dm = \frac{PM}{RT} Adx$$

Substituting in (i), we have

$$dP \cdot A = \frac{PMA}{RT} dx \cdot \omega^2 x$$

or
$$\frac{dP}{P} = \frac{M\omega^2}{RT} dx$$

Integrating we have

$$\int_{P_1}^{P_2} \frac{dP}{P} = \frac{M\omega^2}{RT} \int_0^l x\,dx$$

or
$$\log\frac{P_2}{P_1} = \frac{M\omega^2}{2RT} l^2$$

or
$$P_2 = P_1 e^{\frac{M\omega^2 l^2}{2RT}}$$

Example 16.19: Under standard conditions, the density of the helium and nitrogen mixture equals 0.60 gm/litre. Calculate the concentration of helium atoms in the given mixture.

Solution: Let n_1 and n_2 are the number of moles and M_1 and M_2 are molar masses of helium and nitrogen respectively.

Total number of moles in mixture $n = n_1 + n_2$ or $n_2 = (n - n_1)$

Molar mass of mixture $M = \dfrac{n_1M_1 + n_2M_2}{n_1 + n_2}$

or $$M = \frac{n_1M_1 + (n - n_1)M_2}{n} = \frac{n_1}{n}(M_1 - M_2) + M_2$$

or $$\frac{n_1}{n}(M_1 - M_2) = M - M_2$$

and $$n_1 = \frac{n(M - M_2)}{M_1 - M_2}$$

Perfect gas equation is

$$PV = nRT \text{ or } n = \frac{PV}{RT}$$

$\therefore$ $$n_1 = \frac{(M - M_2)}{M_1 - M_2}\frac{PV}{RT}$$

or $$\frac{n_1}{V} = \frac{(M - M_2)}{M_1 - M_2}\frac{P}{RT}$$

For mixture

$$P = \frac{\rho RT}{M} \text{ or } M = \frac{\rho RT}{P}$$

$\therefore$ $$\frac{n_1}{V} = \frac{1}{(M_1 - M_2)}\left[\frac{MP}{RT} - \frac{M_2P}{RT}\right]$$

$$= \frac{1}{(M_1 - M_2)}\left[\rho - \frac{M_2P}{RT}\right] = \frac{1}{(M_2 - M_1)}\left(\frac{M_2P}{RT} - \rho\right)$$

Now substituting the values we have

$$\frac{n_1}{V} = \frac{1}{(28 - 4) \times 10^{-3}}\left[\frac{28 \times 10^{-3} \times 10^5}{8.4 \times 273} - 0.6\right]$$

$$= \frac{1}{24 \times 10^{-3}}[1.22 - 0.6] = \frac{0.62}{24 \times 10^{-3}}$$

$= 25.83$ moles per unit volume (moles per m^3)

$\therefore$ number of molecules per unit volume

$n = \dfrac{n_1}{V}n_0$ where n_0 is Avogadro number

$= 25.83 \times 6.02 \times 10^{23}$

$= 1.55 \times 10^{25}$ per m^3

$= 1.55 \times 10^{19}$ per cm^3

Example 16.20: A jar contains a gas and a few drops of water at T K. The pressure in the jar is 830 mm of Hg. The temperature in the jar is reduced by 1%. The saturated vapour pressure at the two temperatures are 30 mm and 25 mm of Hg. Calculate the new pressure in the jar.

Solution: The pressure in the jar is due to both, gas and saturated water-vapour. Hence initial pressure of the gas

$$P_1 = 830 - 30 = 800 \text{ mm of Hg}$$

If the final pressure in the jar is P, then pressure of the gas

$$P_2 = (P - 25) \text{ mm of Hg}$$

Now using gas equation

$$\frac{P_1 V}{T_1} = \frac{P_2 V}{T_2}$$

or $$\frac{P_2}{P_1} = \frac{T_2}{T_1} = \frac{0.99 T_1}{T_1} \quad T_2 \text{ is 1\% less than } T_1$$

∴ $$P - 25 = 0.99 \times P_1 = 0.99 \times 800 = 792$$

or $$P = 792 + 25 = 817 \text{ mm of Hg.}$$

Example 16.21: An open end wide glass tube is immersed vertically in mercury in such a way that a length 0.05 m extends above the mercury level. The open end of the tube is closed and the tube is raised further by 0.43 m. Calculate the length of the air column above mercury level in the tube.

Solution: If A is the area of cross-section of the tube then air above mercury in tube has volume $V_1 = 0.05A$ and Pressure $P_1 = P_0$ – atmospheric pressure. When tube is closed and raised let x be the length of the air column above mercury. Then volume V_2 of gas is $V_2 = xA$ and pressure P_2 then

$P_2 + (0.48 - x) = P_0$ or $P_2 = P_0 - (0.48 - x)$

Here pressure is taken in height of mercury column. $P_0 = 0.76$ m of mercury.

Using Boyle's law, we have

$$0.5A \times P_0 = xA \times \{P_0 - (0.48 - x)\}$$

or $$0.5 \times 0.76 = x[0.76 - 0.48 + x]$$

$$= 0.28x + x^2$$

or $$x^2 + 0.28x - 0.038 = 0$$

or $$x = -\frac{0.28 \pm \sqrt{(0.28)^2 + 4(0.038)}}{2}$$

$$= \frac{0.28 + 0.48}{2} \quad \text{(–ve sign is not admissible)}$$

$$= 0.1 \text{ m} = 10 \text{ cm.}$$

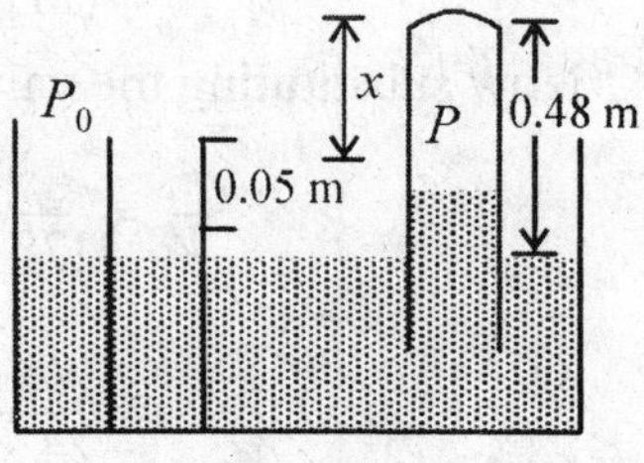

Fig. 16.12

Example 16.22: Assume air expands adiabatically when rising from lower level to higher levels in the atmosphere. If air is considered to be an ideal diatomic gas with molar mass 28.88 g. Obtain the temperature and pressure at height H (= 5 km) above see level, given that the temperature at the sea level is 27° C and pressure 1 atm, estimate the height above see level beyond which this model does not work.

Solution: Consider a slab of air at height h of thickness dh. The pressure in the slab is

$$\left(P + \frac{dP}{dh} dh\right) - P = \frac{dP}{dh} dh$$

The force due to this must balance the weight of air in the slab. If ρ is the density of air at the slab and area of cross-section of slab A, then

$$\frac{dP}{dh} A\,dh = -\rho g dh A$$

or
$$\frac{dP}{dh} = -\rho g$$

The air is considered as an ideal gas of molecular weight M, then

$$P = \frac{\rho RT}{M}$$

or
$$\frac{dP}{dh} = -\frac{Mg\,P}{RT}$$

Since expansion is adiabatic, i.e. $TP^{-\frac{(\gamma-1)}{\gamma}}$ = constant

or
$$dTP^{-\frac{(\gamma-1)}{\gamma}} - T\frac{(\gamma-1)}{\gamma}P^{-\frac{(\gamma-1)}{\gamma}-1}dP = 0$$

or
$$\frac{dT}{T} - \frac{(\gamma-1)}{\gamma}\frac{dP}{P} = 0$$

or
$$\frac{dT}{T} = \frac{\gamma-1}{\gamma}\left(-\frac{Mg\,dh}{RT}\right)$$

or
$$\frac{dT}{dh} = -\frac{\gamma-1}{\gamma}\frac{Mg}{R}$$

and integration gives

$$T = T_0 - \frac{\gamma-1}{\gamma}\frac{Mg}{R}h \quad \because \quad T = T_0 \text{ when } h = 0$$

For air $\gamma = \frac{7}{5}$ $M = 28.88g$ we have

$$T = T_0 - 9.8h \qquad \text{where } h \text{ is in km}$$

When $h = 5$ km and $T_0 = 300$ K

$$T = 300 - 9.8 \times 5 = 300 - 49.0 = 251 \text{ K}$$

We have
$$\frac{dP}{dh} = -\frac{MgP}{RT}$$

or
$$\frac{dP}{P} = -\frac{Mg\,dh}{R(T_0 - 9.8h)}$$

Integration gives

$$\log P = \frac{Mg}{9.8R}\log(T_0 - 9.8h) + C \qquad \ldots(i)$$

where C is constant of integration

Using the condition $P = P_0$ when $h = 0$, we get

$$\log P_0 = \frac{Mg}{9.8R}\log T_0 + C \qquad \ldots(ii)$$

From (*i*) and (*ii*), we have

$$\log\frac{P}{P_0} = \frac{Mg}{9.8R}\log\frac{(T_0 - 9.8h)}{T_0}$$

or

$$P = P_0\left[\frac{T_0 - 9.8h}{T_0}\right]^{Mg/9.8R}$$

Substituting the values $P_0 = 1$ atm $= 10^5$ N/m^2, $T_0 = 300$, $M = 28.88 \times 10^{-3}$ kg, $R = 8.4$ J/mole-K and $g = 10$ m/s^2 at $h = 5$ km.

We get

$$P = 0.54 \text{ atm}$$

Minimum pressure cannot go below 0. Hence model does not work above the height

$$h = \frac{T_0}{9.8} = \frac{300}{9.8} = 31 \text{ km}$$

Example 16.23: Estimate the mean free path and collision frequency of a nitrogen molecule in a cylinder containing nitrogen at 2.0 atm and temperature 17° C. Take the radius of nitrogen molecule to be 1.0 Å. Compare the collision time with the time the molecule moves freely between two successive collisions (Molecular mass of N_2 = 28.0).

Solution: Mean free path $\lambda = \dfrac{1}{\sqrt{2}\pi n d^2}$

where n is the number of molecules per unit volume $= \dfrac{P}{k_B T}$ and d is the diameter of the molecule.

$$\therefore \quad \lambda = \frac{k_B T}{\sqrt{2}\pi P d^2} = \frac{1.4\times10^{-23}\times 290}{\sqrt{2}\times\pi\times2\times10^5\times\left(2\times10^{-10}\right)^2}$$

$$= \frac{10^{-7}\times1.4\times29}{\sqrt{2}\pi\times2\times4} = 1.14\times10^{-7} \text{ m}$$

Collision frequency $f = \dfrac{v_{rms}}{\lambda} = \dfrac{1}{\lambda}\sqrt{\dfrac{3RT}{M}}$

$$= \frac{1}{1.14\times10^{-7}}\sqrt{\frac{3\times8.4\times290}{28\times10^{-3}}} = 4.48\times10^9 \text{ s}^{-1}$$

$$= 4.48\times10^9 \text{ s}^{-1}$$

Time interval between two collisions $= \dfrac{1}{f} = \dfrac{1}{4.48\times10^9} = 2.2\times10^{-10}$ sec

Collision time $= \dfrac{d}{v_{rms}} = (2\times10^{-10})\sqrt{\dfrac{28\times10^{-3}}{3\times8.4\times290}} = 3.9\times10^{-13}$ sec

Therefore

$$\frac{\text{Time interval between two collisions}}{\text{collision time}} = \frac{2.2 \times 10^{-10}}{3.9 \times 10^{-13}} = 564$$

Example 16.24: A cubical box of side 1 m contains helium gas (atomic weight 4) at a pressure of 100 N/m². During an observation time of 1 second, an atom travelling with the root-mean-square speed parallel to one of the edges of the cube was found to make 500 hits with a particular wall, without any collision with other atoms. Take $R = 25/3$ J/mol-K and $k_B = 1.35 \times 10^{-23}$ J/K.

(*a*) Evaluate the temperature of the gas
(*b*) Evaluate the average K.E. per atom
(*c*) Evaluate the total mass of helium gas in the box.

Solution: (*a*) If side of box is l and v_{rms} is the r.m.s. speed of an atom. Then time interval between successive collisions with a particular wall

$$\tau = \frac{2l}{v_{rms}}$$

Given is $l = 1$ m and $\tau = \frac{1}{500}$ sec

$$\therefore \quad v_{rms} = 2 \times 500 = 1000 \text{ m/s}$$

But $$v_{rms} = \sqrt{\frac{3RT}{M}}$$

or $$T = \frac{(v_{rms})^2 M}{3R}$$

$$= \frac{(1000)^2 \times 4 \times 10^{-3}}{3 \times \frac{25}{3}} = 160 \text{ K}$$

(*b*) $$\text{K.E.} = \frac{3}{2} k_B T = \frac{3}{2} \times 1.38 \times 10^{-23} \times 160 = 3.312 \times 10^{-21} \text{ J}$$

(*c*) Pressure $$P = \frac{1}{3} \frac{\text{mass of gas}}{\text{volume}} v_{rms}^2$$

or mass of gas $$= \frac{3PV}{(v_{rms})^2} \qquad V = 1 \text{ m}^3$$

$$= \frac{3 \times 100 \times 1}{(1000)^2} = 3 \times 10^{-4} \text{ kg}$$

Example 16.25: Figure 16.13 shows *P-T* graph for some substances. Line ① is fusion line, line ② is vaporisation line, line ③ is sublimation line. Now fill in the blanks.

(*i*) Melting point of substance................with increase in pressure.

(*ii*) Boiling point of substance................with increase in pressure.

(*iii*) At $P = 1$ atm $T = 90$ K, the substance is in ....state.

(*iv*) At $P = 1$ atm, $T = 190$ K, the substance is in ....state.

(*v*) At $P = 1$ atm, $T =$................ K, the................state and................state are in equilibrium.

(*vi*) At $T = 100$ K, $P = 2$ atm, the................state and................state are in equilibrium.

(*vii*) At $T = 100$ K, $P = 2.5$ atm, the substance is in................state.

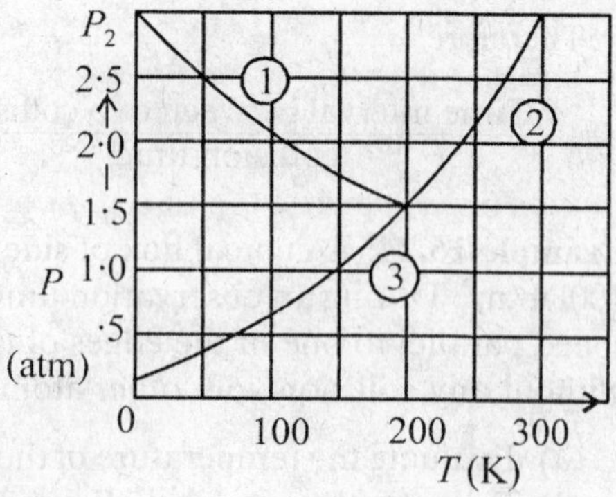

Fig. 16.13

(*viii*) At $T = 100$ K, $P = 1.5$ atm the substance is in................state.

(*ix*) At $T = 200$ K, $P = 1$ atm, the substance is in................state.

(*x*) Below $P = 1.5$ atm, the substance can exist either in................state or................state.

(*xi*) At $P = 0.5$ atm the substance is heated from 99 K to 101 K, it will go from................state to................state.

(*xii*) At $P = 2$ atm the boiling point is................K.

(*xiii*) At $T = 205$ K and $P = 2$ atm the substance is in................state.

(*xiv*) The triple point of the substance is................K.

Solutions:

(*i*) Melting point of substance decreases with increase in pressure.

(*ii*) Boiling point of substance increases with increase in pressure.

(*iii*) At $P = 1$ atm and $T = 90$ K, the substance is in solid state.

(*iv*) At $P = 1$ atm and $T = 190$ K, the substance in vapour state.

(*v*) At $P = 1$ atm and $T = 150$ K, the solid and gaseous state are in equilibrium.

(*vi*) At $T = 100$ K, $P = 2$ atm, the solid and liquid states are in equilibrium.

(*vii*) At $T = 100$ K and $P = 2.5$ atm, the substance is in liquid state.

(*viii*) At $T = 100$ K and $P = 1.5$ atm, the substance is in solid state.

(*ix*) At $T = 200$ K and $P = 1$ atm, the substance is in gaseous state.

(*x*) Below $P = 1.5$ atm, the substance can exist either in gaseous and solid state.

(*xi*) At $P = 0.5$ atm the substance is heated from 99 K to 101 K it will go from solid state to gaseous state.

(*xii*) At $P = 2$ atm, the boiling point is 250 K.

(*xiii*) At $T = 205$ K and $P = 2$ atm, the substance is in liquid state.

(*xiv*) The triple point of the substance is 200 K.

PROBLEMS

1. Consider that air is under standard conditions close to the earth's surface. Assume that the acceleration due to gravity, the temperature and molar mass of air are independent of height, find the air pressure at height 5 km over the surface and in a mine at the depth 5.0 km below the surface.

[**Hint.** $P = P_0 e^{-Mgh/RT}$ at height $h = 5000$ m, $P = 0.5$ atm. At depth $h = -5000$ m, $P = 2$ atm]

2. Two glass bulbs of equal volumes are connected by a narrow tube and are filled with a gas at 0° C and pressure 76 cm of Hg. One of the bulb is then placed in water bath maintained at 62° C. What is the new value of pressure inside the bulb? Volume of connecting tube is negligible.

[**Hint.** Total no. of molecules remain constant $n = \frac{P_0 V}{k_B T_0}$ Finally $n_1 = \frac{PV}{k_B T_0}$ and $n_2 = \frac{PV}{Tk_B}$ and $n_1 + n_2 = 2n$. $P = 83.75$ cm of Hg]

3. A closed container of volume 0.02 m³ contains a mixture of neon and argon gases at temperature of 27° C and pressure of 1×10^5 N/m². The total mass of the gas (mixture) is 28 gm. If the gm molecular weight of Ne and A are 20 and 40 respectively, find the masses of the individual gases in the container assuming them to be ideal.

[**Hint:** $m_1 + m_2 = 28$ and $PV = \left(\frac{m_{Ne}}{M_{Ne}} + \frac{m_A}{M_A}\right)RT$ give $m_{Ne} = 4$ gm. $m_A = 24$ gm]

4. Two glass bulbs of volumes 500 cm³ and 100 cm³ are connected by a narrow tube whose volume is negligible. When the apparatus is sealed off, the pressure of air inside is 70 cm of Hg and temperature 20° C. What does the pressure become if the 100 cm³ bulb is kept at 20° C and the other is heated to 100° C. [**Ans.** $P = 85.23$ cm of Hg]

5. Helium is enclosed in a closed vessel with a volume 2 litre at a temperature of 20° C and pressure 10^5 N/m².

(*a*) What amount of heat should be imparted to helium to increase its temperature by 100° C?
(*b*) What will the mean square velocity of its molecules be at new temperature?
(*c*) What will be the pressure?
(*d*) What will be the density of helium?
(*e*) What will be the energy of the thermal motion of its molecules?

[**Hint:** (*a*) $Q = nC_V\Delta T = \frac{PV}{RT}\Delta T \times \frac{3}{2}R$ Ans. 102 J. (*b*) $v_{rms} = \sqrt{\frac{3RT}{M}}$ Ans. 1.57×10^3 m/s

(*c*) $P \propto T$ Ans. 1.33×10 N/m² (*d*) $\rho = \frac{PM}{RT}$ Ans. 0.162 kg/m³ (*e*) $E = \frac{3}{2}k_B T \times$ no. of molecules.

$E = \frac{3}{2}k_B T \frac{PV}{KT} = \frac{3}{2}PV$ Ans. 4×10^2J]

6. Two vessels *A* and *B* of equal volume V_0 are connected by a narrow tube which can be closed by a valve. The vessels are fitted with pistons which can be moved to change volumes. Initially, the valve is open and the vessels contain an ideal gas $\left(\frac{C_P}{C_V} = \gamma\right)$ at atmospheric pressure P_0 and atmospheric temperature T_0. The walls of the vessels *A* are diathermic and those of *B* are adiabatic. The valve is now closed and the pitons are slowly pulled out to increase the volumes of the vessels to double the original volume.

(*a*) Find the temperatures and pressures in two vessels.
(*b*) The valve is now opened for sufficient time so that the gases acquire common temperature and pressure. Find the new values of temperature and pressure.

[**Ans.** (*a*) in *A* T_0, $\frac{P_0}{2}$, in *B* $\frac{T_0}{2^{\gamma-1}}$, $\frac{P_0}{2^\gamma}$ (*b*) T_0, $\frac{P_0}{2}$]

7. The number of molecules per cc of a gas is 2.7×10^{19} at NTP. Calculate the number of molecules per cm^3 of the gas.

(*i*) at 0° C and 10^{-6} mm of Hg pressure.

(*ii*) at 39° C and 10^{-6} mm of Hg pressure. [**Ans.** (*i*) 3.55×10^{10} (*ii*) 3.102×10^{10}]

8. How much heat energy should be added to the gaseous mixture consisting of one gm hydrogen and one gm helium to raise its temperature from 0° C to 100° C (*i*) at constant volume (*ii*) at constant pressure assuming both the gases to be ideal? ($R = 1.99$ cal/mol-K for H_2 $\gamma = 1.41$ for He $\gamma = 1.67$)

[**Hint:** C_V for mixture $= \dfrac{\frac{1}{2}\cdot\frac{5}{2}R + \frac{1}{4}\times\frac{3}{2}R}{\frac{1}{2}+\frac{1}{4}} = \dfrac{13R}{6}$ total moles $\dfrac{3}{4}, Q = nC_V\Delta T$ **Ans.** 323.375 cal

$C_P = \dfrac{19R}{6}$ $\therefore Q = 323.375 \times \dfrac{19}{13}$, $Q = 472.625$ cal]

9. A ideal gas has a molar heat capacity C_V at constant volume. Find the molar heat capacity of this gas as a function of its volume V_1 if the gas undergoes the following process.

(*a*) $T = T_0 e^{\alpha V}$ (*b*) $P = P_0 e^{\alpha V}$

[**Ans.** (*a*) $C_V + \dfrac{R}{\alpha V}$ (*b*) $C_V + \dfrac{R}{1+\alpha V}$]

10. Calculate the valve of $\gamma = \dfrac{C_P}{C_V}$ for a gaseous mixture containing V_1 moles of oxygen and V_2 moles of carbon dioxide. Assume gases to be ideal.

[**Ans.** $C_V = \dfrac{V_1 C_{1V} + V_2 C_{2V}}{V_1 + V_2}, C_P = C_V + R, \gamma = \dfrac{C_P}{C_V}$]

11. A tall cylindrical vessel with gaseous nitrogen is located in a uniform gravitational field in which the free-fall acceleration is equal to *g*. The temperature of nitrogen varies along the height *h* so that its density is the same throughout the volume. Find the temperature gradient.

[**Hint:** $dP = -dhg\rho$ and $P = \rho\dfrac{RT}{M}$ or $dP = \dfrac{\rho R}{M}dT$ $\therefore \dfrac{dT}{dh} = -\dfrac{Mg}{R}$]

12. An ideal gas of molar mass *M* is located in uniform gravitational field in which the free-fall acceleration is equal to *g*. Find the gas pressure as a function of height *h*, if $P = P_0$ at $h = 0$ and the temperature varies with height as (*a*) $T = T_0(1 - ah)$ (*b*) $T = T_0(1 + ah)$, where *a* is a positive constant.

[**Ans.** (*a*) $P = P_0(1 - ah)^n$ $h < \dfrac{1}{a}$ (*b*) $P = P_0(1 + ah)^{-n}$. Here $n = \dfrac{Mg}{RT_0}$]

13. A vertical hollow cylinder of height 1.52 m is fitted with a movable piston of negligible mass and thickness. The lower half portion of the cylinder contains an ideal gas and the upper half is filled with mercury. The cylinder is initially at 300 K. When the temperature is raised, half of the mercury comes out of the cylinder. Find this temperature assuming the thermal expansion of the mercury to be negligible. [**Ans.** $T = 337.5$ K]

14. One gm mole of oxygen at 27° C and 1 atm pressure is enclosed in a vessel. (*i*) Assuming the molecules to be moving with v_{rms} find the number of collisions per second which the molecules make with one square meter area of the wall of the vessel (*ii*) The vessel is next thermally insulated and moved with constant speed v_0. It is then suddenly stopped. The process results in a rise in temperature of the gas by 1° C. Calculate the speed v_0.

[**Ans.** No. of collisions 1.96×10^{-27}, $v = 36$ m/s]

15. How will the rate of collision of rigid diatomic molecules against the vessel's wall change if the gas is expanded adiabatically η times?

[**Hint:** Collision frequency $\propto \bar{c}\left(\frac{N}{V}\right)$, $\frac{N}{V}$ is number molecules per unit volume or $f \propto \frac{\sqrt{T}}{V}$ or

$$\frac{f_1}{f_2} = \sqrt{\frac{T_1}{V_1^2}\frac{V_2^2}{T_2}} = \sqrt{\frac{T_1}{T_2}(\eta)^2} = \sqrt{\eta^{\gamma-1}\eta^2} = \eta^{\frac{\gamma+1}{2}},\ \gamma = 7/5 \text{ for diatomic molecule } \therefore \frac{f_1}{f_2} = \eta^{\frac{6}{5}}$$]

16. Find the total random translational K.E. of the molecules of one mole of a gas at a temperature of 300 K.

17. Two vessels A and B, thermally insulated contain an ideal monoatomic gas. A small tube fitted with a valve connects these vessels. Initially the vessel A has 2 litres of gas at 300 K and 2×10^5 N/m^2 pressure. While vessel B has 4 litres of gas at 350 K and 4×10^5 N/m^2 pressure. The valve is now opened and the system reaches equilibrium in pressure and temperature. Calculate the new temperature and pressure. [**Ans.** $T = 338.7$ K $P = 3.33 \times 10^5$ N/m^2]

18. A thermally insulated vessel with gaseous nitrogen at a temperature of 27° C moves with a velocity of 100 m/s. If the vessel is suddenly stopped (*i*) What will be change in temperature of gas? (*ii*) What will be the percentage change in the pressure of the gas?

[**Ans.** $\Delta T = 6.7°$ C, $\Delta P = 2.2\%$]

19. A vessel of volume 20 litres contains a mixture of hydrogen and helium at a temperature $t =$ 20° C and pressure $P = 2.0$ atm. The mass of mixture is equal to 5.0 g. Find the ratio of the mass of hydrogen to that of helium in the given mixture. [**Ans:** 0.61]

20. Figure 16.14 represents a cylindrical tube of radius 5 cm. It is closed by a light fitting cork. The friction coefficient between the cork and the tube is 0.2. The tube contains an ideal gas at a pressure of 1 atm and a temperature of 300 K. The tube is slowly heated and it is found that the cork pops out when the temperature reaches 600 K. Let dN denotes the magnitude of normal contact force exerted by a small length dl of cork along the periphery. Assuming that the temperature of gas is inform at any instant, calculate $\frac{dN}{dt}$.

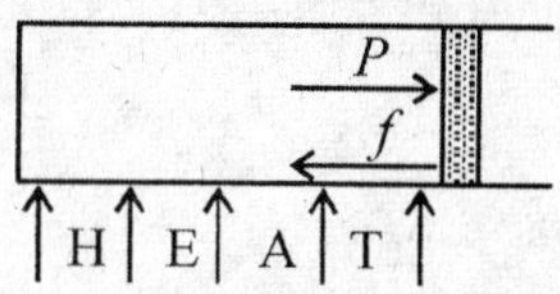

Fig. 16.14

[**Hint:** Temperature is doubled, volume is constant hence pressure is doubled or $\Delta P = 1$ atm,

$$\text{Force} = \Delta PA = \text{frictional force} = \mu\frac{dN}{dl} \times 2\pi r \text{ or } \frac{dN}{dl} = \frac{\Delta P \cdot \pi r^2}{\mu \cdot 2\pi r} = \frac{\Delta \text{Pr}}{2\mu} = 1.25 \times 10^4 \text{ N/m}$$]

21. Figure 16.15 shows a large closed cylindrical tank containing water. Initially the air trapped above the water surface has a height h_0 and pressure $2P_0$. Where P_0 is atmospheric pressure. There is a hole in the wall of the tank at depth h_1 below the top from which the water comes out. A long vertical tube is connected as shown in the figure, (*a*) Find the height h_2 of the water in the long tube above the top initially. (*b*) Find the speed with which the water comes out of the hole. (*c*) Find the height of the water in the long tube above the top when the water stops coming out of the hole.

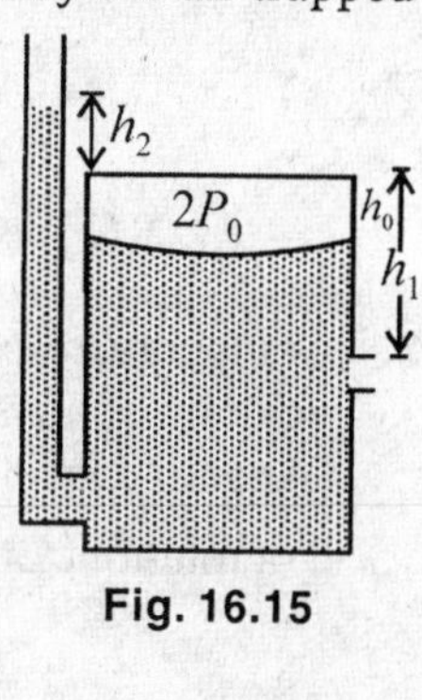

Fig. 16.15

[**Hint:** $2P_0 = P_0 + (h_2 + h_0)g\rho$ or $h_2 = \frac{P_0}{g\rho} - h_0$ (*b*) From Bernoulli's equation $2P_0 + (h_1 - h_0)\rho g = \frac{1}{2}\rho v^2 + P_0$ or $v = \sqrt{\frac{2}{P}}[P_0 + (h_1 - h_0)g\rho$ (c) When level of water in bigger tank is h_1 from top. Water stops coming out. Thus height from top is $-h_1$]

22. Oxygen is filled in a closed metal jar of volume 1.0×10^{-3} m^3 at a pressure of 1.5×10^5 Pa and temperature 400 K. The jar has a small leak in it. The atmospheric pressure is 1.0×10^5 Pa and the atmospheric temperature is 300 K. Find the mass of the gas that leaks out by the time the pressure and the temperature inside the jar equalise with the surroundings. [**Ans.** 0.16 g]

23. Figure 16.16 shows cylindrical tube with adiabatic walls and fitted with a diathermic separator. The separator can slide in the tube by an external mechanism. An ideal gas is injected in the two sides at equal pressures and temperatures. The separator remains in equilibrium at the middle. It is now slid to a position where it divides the tube in the ratio of 1 : 3. Find the ratio of the pressures in the two parts of the vessel. [**Ans.** 3 : 1]

Fig. 16.16

24. A glass tube sealed at both ends is 100 cm long. It lies horizontally with the middle 10 cm containing mercury. The two ends of the tube contain air at 27° C and at a pressure 76 cm of mercury. The air column on one side is maintained at 0° C and other side is maintained at 127° C. Calculate the length of the air column on the cooler side. Neglect the changes in the volume of the mercury and glass. [**Ans.** 36.5 cm]

25. A diving bell having a height $h = 3$ m and a constant cross-section is submerged in sea at a depth $H = 80$ m. To what hight h_0 does water rise in the bell at this depth if the temperature of water is $T_1 = 20°$ C at the surface and $T_2 = 7°$ C at the depth H = 80 m. What must be the pressure P_0 of air pumped in to the bell to remove water from the bell completely. Density of sea water is $\rho = 1.03 \times 10^3$ kg/m^3.

$$\left[\textbf{Ans. } h_0 = h\left[\frac{P_{atm}T_2}{(P_{atm} + \rho gH)T_1}\right] \approx 2.68 \text{ m}, P_0 = (P_{atm} + \rho gH) = 9.1 \times 10^5 \text{ Pa}\right]$$

SHORT ANSWER TYPE QUESTIONS

1. At what temperature the root mean square velocity is equal to escape velocity from the surface of earth for hydrogen. Given radius of earth = 6.4×10^6 m and $g = 10$ m/s^2.

$$\left[\textbf{Ans. } v_e = \sqrt{2gR} \text{ and } \frac{1}{2}mv_e^2 = \frac{3}{2}k_BT \text{ or } T = \frac{mv_e^2}{3k_B} = \frac{2}{3k_B}mgR \text{ or}\right.$$

$$\left. T = \frac{2 \times 10 \times 6.4 \times 10^6}{3 \times 1.4 \times 10^{-23}} \times \frac{2 \times 10^{-3}}{6.02 \times 10^{23}} = 1 \times 10^4 \text{ K}\right]$$

2. Temperature of a gas containing rigid diatomic molecules is 300 K. Calculate the angular velocity of a rotating molecule if the moment of inertia is = 2.1×10^{-38} kg-m^2.

[**Ans.** In case of a rigid diatomic molecule, the degrees of freedom are 5, 3 of translation and 2 of rotation. Energy associated with each degree of freedom is $\frac{1}{2}k_BT$. Hence rotational kinetic energy is $\frac{1}{2} I\omega^2 = 2 \times \frac{1}{2} k_B T$ or $\omega^2 = \frac{2 k_B T}{I}$ and $\omega = \sqrt{\frac{2k_BT}{I}}$ or $\omega = \sqrt{\frac{2 \times 1.4 \times 10^{-23} \times 300}{2.1 \times 10^{-38}}}$

$= 6.3 \times 10^8$ rad/s]

3. Calculate the root mean square velocity of molecule in a water droplet of diameter $d = 0.10$ μm suspended in air at 17° C.

[**Ans.** $\bar{c} = \sqrt{\frac{3k_BT}{m}}$ volume of the droplet $= \frac{4}{3}\pi\left(\frac{d}{2}\right)^3 = \frac{\pi d^3}{6}$ and taking density of water (ρ) = 10^3 kg/m^3; Mass m of the molecule is $m = \frac{\pi d^3}{6}\rho$ $\therefore c = \sqrt{\frac{3 k_B T \times 6}{\pi d^3 \rho}}$, $T = 273 + 17 = 290$ K

$$\therefore c = \sqrt{\frac{18 \times 1.4 \times 10^{-23} \times 290}{\pi (10^{-7})^3 \times 10^3}} = 0.152 \text{ m/s}$$]

4. 4 gm of hydrogen is mixed with 11.2 litres of He at N.T.P. in a container of volume 20 litres. If the final temperature is 300 K. Find the pressure.

[**Ans.** 4 gm of H_2 = 2 moles. 11.2 litres of He at N.T.P. = 0.5 mole. Total pressure $P = P_H + P_{He}$ $= (\mu_H + \mu_{He}) \frac{RT}{V}$ or $P = (2 + 0.5)\frac{8.4 \times 300}{20 \times 10^{-3}} = \frac{2.5 \times 8.4 \times 3}{2} \times 10^4 = 3.15 \times 10^2$ N/m^2]

5. The average translational kinetic energy of air molecules is 0.04 eV. Calculate the temperature of air.

[**Ans.** Average translational K.E. of a molecule $= \frac{3}{2} k_B T$ or $T = \frac{2}{3}\frac{E}{k_B}$ $T = \frac{2}{3}\frac{0.04 \times 1.6 \times 10^{-19}}{1.4 \times 10^{-23}}$

$= \frac{4 \times 1.6 \times 10^2}{1.4} = 457.14$ K]

6. A gas cylinder has walls that can bear a maximum pressure of 1.0×10^6 Pa. It contains a gas at 0.8×10^5 Pa and 300 K. The cylinder is steadily heated. Neglecting any change in the volume, calculate the temperature at which the cylinder will break.

[**Ans.** Cylinder will burst when final pressure becomes 10^6 Pa. The process is isochoric as such

$$P \propto T \quad \therefore \frac{P_i}{P_f} = \frac{T_i}{T_f} \text{ or } T_f = \frac{P_f}{P_i} T_i \text{ or } T_f = \frac{10^6}{0.8 \times 10^5} \times 300 = 3750 \text{ K}$$]

7. A vessel contains a mixture of 1 mole of oxygen and 2 moles of nitrogen at 300 K. Find the ratio of average rotational kinetic energy per O_2 molecule to that per H_2 molecule.

[**Ans.** Rotational kinetic energy depends only on degree of freeddom associated with the molecule. Since O_2 and nitrogen both are diatomic molecules as such have 2 digress of freedom of rotation. Hence their ratio is 1 : 1]

8. The average translational kinetic energy and rms speed of molecules in a sample of oxygen at 300 K are 6.21×10^{-21} J and 484 m/s respectively. Find the corresponding values at temp. 600 K.

[**Ans.** K.E. ∝ temp. Since temperature is doubled hence K.E. $= 2 \times 6.21 \times 10^{-21}$ or K.E. = 12.42 $\times 10^{-21}$ J. $v_{rms} \propto \sqrt{T}$, thus $v_{rms} = \sqrt{2} \times 484 = 684.4$ m/s]

9. At room temperature the r.m.s. speed of the molecules of a certain diatomic gas is found to be 1930 m/s. Find the gas.

$\left[\textbf{Ans. } v_{rms} = \sqrt{\frac{3RT}{M}} \text{ or } M = \frac{3RT}{v_{rms}^2} = \frac{3 \times 8.4 \times 300}{(1930)^2} = 2.029 \times 10^{-3}\right.$ kg or molecular weight of gas is 2. Hence gas is hydrogen]

10. Three closed vessels A, B, C are at the same temperature T and contains gases which obey Maxwellian distribution of velocities. Vessel A contains O_2, B, N_2 and C mixture of equal quantities of O_2 and N_2. If average speed of O_2 molecules in A is v_1 that of N_2 molecules in B, is v_2, then find the average speed of O_2 molecules in vessel C.

[**Ans.** The average speed $= \sqrt{\frac{2RT}{M}}$, i.e. it depends on temperature and molecular weight only as such the average speed of O_2 molecules in C is also v_1]

11. Explain why sprinkling of water reduces the temperature of a closed room?

[**Ans.** Sprinkled water vaporises and it takes latent heat of vaporisation from the room. Its latent heat of vaporisation is very high (= 540 cal/gm) as such room becomes cool]

12. If the density of a gas is 1.3 gm/litre and velocity of sound in it is 330 m/s. Find the number of degrees of freedomn of gas molecule.

[**Ans.** $v = \sqrt{\frac{\gamma P}{d}}$ or $\gamma = \frac{v^2 \cdot d}{\rho} = \frac{(300)^2 \times 1.3}{10^5} = 1.4$, i.e. the gas is diatomic and hence degrees of freedomn are 5]

13. m gm of gas of molecular weight M is flowing in an insulated tube with velocity v. If the system is suddenly stopped then find rise in temperature.

[**Ans.** When system is suddenly stopped, kinetic energy $\frac{1}{2}mv^2$ is converted into heat $\left(= \frac{1}{2}\frac{mv^2}{J}\right)$ and this raises the temperature. Thus $\frac{1}{2}\frac{mv^2}{J} = \Delta T . m . S$, S is specific heat and ΔT rise in temperature

$\therefore \Delta T = \frac{1}{2}\frac{v^2}{JS} = \frac{Mv^2}{2J(MS)}$ $MS = C_V$ molar specific heat and $C_V = \frac{R}{\gamma - 1}$. Thus $\Delta T = \frac{Mv^2(\gamma - 1)}{2JR}$]

14. Equal number of molecules of hydrogen and oxygen are contained in a vessel at one atmospheric pressure. Find the ratio of the collision frequency of hydrogen molecules to that of oxygen molecules on the container walls.

[**Ans.** Collision frequency is proportional to the r.m.s. speed. Since the gases are in equilibrium hence average kinetic energy per molecule is same, i.e. $\frac{1}{2}m_O c_O^2 = \frac{1}{2}m_H c_H^2$ or $\frac{c_H}{c_O} = \sqrt{\frac{m_O}{m_H}} = 4$]

15. Find the molar mass and the number of degrees of freedom of molecules in a gas if its heat capacities are $C_V = 0.65$ J/g-K and $C_P = 0.91$ J/g-K.

[**Ans.** $M(C_P - C_V) = R$ or $M = \frac{R}{C_P - C_V} = \frac{8.4}{91 - 0.65} = 32$. If f is the number of degrees of freedom, then $C_V = \frac{1}{2}fR$ or $f = \frac{2C_V}{R}$ $f = \frac{2C_V}{C_P - C_V} = \frac{2}{\frac{C_P}{C_V} - 1} = \frac{2}{\frac{0.91}{0.65} - 1} = \frac{2}{1.4 - 1} = \frac{2}{0.4} = 5$]

16. A tall cylindrical vessel with gaseous nitrogen is located in a uniform gravitational field in which free fall acceleration is equal to g. The temperature of nitrogen varies along the height h so that its density is throughout same. Find the temperature gradient.

[***Hint*:** $dP = -g\rho dh$, $P = \dfrac{\rho RT}{M}$ $\therefore dP = \dfrac{\rho R}{M} dT = -\rho g dh$ $\therefore \dfrac{dT}{dh} = -\dfrac{Mg}{R} = -\dfrac{28 \times 10^{-3} \times 10}{8.4}$

$= -33.0$ *k*/km]

17. The specific heat of argon at constant volume is 0.075 cal/g-K. Calculate its atomic weight and mass of an argan atom.

[**Ans.** If A is atomic weight of argan then molar specific heat $C_V = Ac_V$. But $C_v = \dfrac{3}{2}R \therefore A = \dfrac{3R}{2C_V}$

$R =$ cal/mole-K $\therefore A = \dfrac{3}{2} \times \dfrac{2}{0.075} = 40$ and mass of each atom $= \dfrac{40}{6.02 \times 10^{23}} = 6.6 \times 10^{-23} g$]

18. Determine the ratio of speed of sound v in a gas to the root mean square velocity $\bar{c}$ of this gas, if the molecules are (*a*) monoatomic (*b*) rigid diatomic.

[**Ans.** $v = \sqrt{\dfrac{\gamma P}{\rho}} = \sqrt{\dfrac{\gamma RT}{M}}$ and $\bar{c} = \sqrt{\dfrac{3RT}{M}} \therefore \dfrac{v}{\bar{c}} = \sqrt{\dfrac{\gamma}{3}}$. For monoatomic gas $\gamma = \dfrac{5}{3} \therefore \dfrac{v}{\bar{c}} = \sqrt{\dfrac{5}{9}}$

$= 0.75$ and for diatomic gas $\gamma = \dfrac{7}{5} \therefore \dfrac{v}{\bar{c}} = \sqrt{\dfrac{7}{5 \times 3}} = 0.68$]

19. An ideal gas (one mole) undergoes a process $P = \dfrac{P_0}{1 + (V/V_0)^2}$ where P_0 and V_0 are constants. Find the temperature of the gas when $V = V_0/2$.

[**Ans.** $PV = RT$ or $\dfrac{P_0 V}{1 + (V/V_0)^2} = RT$ or $\dfrac{P_0 \dfrac{V_0}{2}}{1 + \dfrac{1}{4}} = RT$ or $T = \dfrac{2 P_0 V_0}{5R}$]

20. Figure 16.17 shows a vessel partitioned by a fixed diathermic separator. Different ideal gases are filled in the two parts, The rms speed of the molecules in the left part equals the mean speed of the molecules in the right part. Calculate the ratio of the mass of a molecule in the left part to the mass of a molecule in the right part.

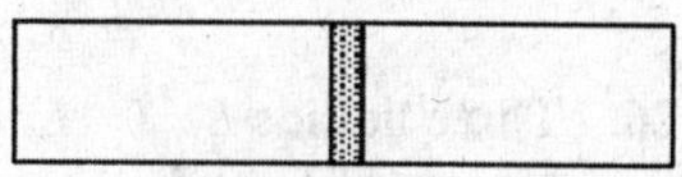

Fig. 16.17

[**Ans.** r.m.s. speed $\bar{c} = \sqrt{\dfrac{3RT}{M}}$ and mean speed $= \sqrt{\dfrac{8RT}{\pi M}} \therefore$

$\dfrac{3RT}{M_1} = \dfrac{8RT}{M_2 \pi}$ or $\dfrac{M_1}{M_2} = \dfrac{3\pi}{8} = 1.18$]

21. At the top of a mountain a thermometer read 57° C and barometer read 570 cm of Hg. At the bottom of the mountain they read 27° C and 76 cm of Hg respectively. Compare the density of the air at the top and that at the bottom.

[**Ans.** Gas equation gives $P = \rho \dfrac{RT}{M}$. Thus $\dfrac{P_t}{P_b} = \dfrac{\rho_t T_t}{\rho_b T_b}$ or $\dfrac{\rho_t}{\rho_b} = \dfrac{P_t T_b}{P_b T_t}$ or $\dfrac{\rho_t}{\rho_b} = \dfrac{70}{76} \dfrac{(273+27)}{(273+7)}$

$= \dfrac{70 \times 300}{76 \times 280} = 0.9868$]

22. The root mean square speed of oxygen molecules at a certain temperature T is v. If the temperature is doubled and oxygen gas dissociates; find the r.m.s. speed of molecules.

[**Ans.** r.m.s. speed $\bar{c} = \sqrt{\frac{3RT}{M}} = \sqrt{\frac{3RT}{32 \times 10^{-3}}}$ when the temperature is doubled it becomes $2T$.

The gas dissociates it becomes atomic, i.e. M becomes 16×10^{-3}. New $\bar{c}' = \sqrt{\frac{3R(T)}{16 \times 10^{-3}}} = 2\bar{c}$, i.e. the new r.m.s. speed becomes double]

23. A jar contains a mixture of hydrogen and helium gases in the ratio 1 : 5. What is the ratio of mean translational kinetic energy of hydrogen and helium molecules?

[**Ans.** According to kinetic theory of gases, the mean translational kinetic energy of a gas molecule is $\frac{3}{2}k_B T$ and is independent of the nature of gas and depends on its temperature. Since in the jar both the gases are at the same temperature so both will have equal translational kinetic energy or ratio is 1 : 1]

24. At what temperature does the average translational kinetic energy of a molecule in a gas becomes equal to the kinetic energy of an electron accelerated from rest through a potential difference of 1 Volt.

[**Ans.** If a charged particle (charge q) is accelerated through a potential difference of V volts its K.E. $= qV$. Average kinetic energy of a molecule $= 3/2k_B T$. Hence $3/2k_B T = qV$ or $T = \frac{2}{3}\frac{qV}{k_B}$ or

$$T = \frac{2}{3} \times \frac{1.6 \times 10^{-19} \times 1}{1.4 \times 10^{-23}} = 7.619 \times 10^3 \text{ K}$$]

25. At identical temperatures, the r.m.s. speed of hydrogen molecules is 4 times that for oxygen molecules. In a mixture of these in the mass ratio $H_2 : O_2 = 1 : 8$ the r.m.s. speed of hydrogen molecules is n times the r.m.s. speed for O_2 molecules. Find the value of n.

[**Ans.** The molecules are in the ratio of $H_2 : O_2 = 2 : 1$. This gives $(c_{rms})^2 = \frac{16 \times 2 + 1 \times 1}{2 + 1} = 11$ times the value of O_2 or $c_{rms} = \sqrt{11}$ or $n = \sqrt{11}$]

26. Three liquids L_1, L_2, L_3 are at temperature 30° C, 20° C and 10° C respectively. When equal masses of L_1 and L_2 are mixed temperature of mixture is 26° C and when equal masses at L_1 and $L_3$are mixed, the temperature of mixture is 15° C. If equal masses of L_2 and L_3 are mixed what will be the temperature of mixture?

[**Ans.** When L_1 and L_2 are mixed $m\,S_1(30 - 26) = mS_2(26 - 20)$ or $\frac{S_1}{S_2} = \frac{3}{2}$. When L_1 and L_3 are mixed $\frac{S_1}{S_3} = \frac{1}{3}$, this gives $\frac{S_2}{S_3} = \frac{2}{9}$. Hence $\frac{2}{9}(20 - \theta) = (\theta - 10)$ or $\theta = 11.8°$C]

27. Temperature of outer space is 3 K. Find the r.m.s. speed of proton in outer space. Mass of proton $= 1.67 \times 10^{-27}$ kg.

[**Ans.** Assuming 3 degrees of freedom for proton, the kinetic energy of the proton $\frac{1}{2}mv^2 = \frac{3}{2}k_B T$

$$\text{or } v = \sqrt{\frac{3k_B T}{m}} \text{ or } v = \sqrt{\frac{3 \times 1.4 \times 10^{-23} \times 3}{1.67 \times 10^{-27}}} = 275 \text{ m/s}$$]

28. In a closed vessel there is a gas. If temperature of the gas is increased by 1° C the pressure increases by 0.4%. Find the initial temperature of the gas.

[**Ans.** Since volume is constant pressure is directly proportional to temperature or $\frac{P_2}{P_1} = \frac{T_2}{T_1}$

or $\frac{P_2 - P_1}{P_1} = \frac{T_2 - T_1}{T_1}$ or $T_1 = \frac{T_2 - T_1}{\frac{P_2 - P_1}{P_1}} = \frac{1}{0.004} = 250$ K]

29. At 0° C and 1.0 atm pressure densities of air, oxygen and nitrogen are 1.293, 1.429 and 1.251 kg-m^{-3} respectively. If air is composed of only these two gases, then find the percentage of nitrogen in air.

[**Ans.** Let fraction of nitrogen in air by weight be f then fraction of oxygen in air is $(1 - f)$.

Hence $\rho_a = \rho_0(1 - f) + \rho_n f$ or $1.293 = 1.429\ (1 - f) + 1.251 f$ or $f = 0.764 = 76.4\%$].

30. Find the average magnitude of linear momentum of a helium molecule in a sample of helium gas at 0° C. Mass of helium molecule = 6.64×10^{-27} kg and Boltzman's constant = 1.38×10^{-23} J/K.

[**Ans.** Average kinetic energy E of molecule $= \frac{3}{2} k_B T$ then momentum p is $p = \sqrt{2mE}$

$= \sqrt{2m \frac{3}{2} k_B T} = \sqrt{3mk_B T}$ or $P = \sqrt{3 \times 6.64 \times 10^{-27} \times 1.38 \times 10^{-23} \times 273} = 8.66 \times 10^{-24}$ kg m s^{-1}].

MULTIPLE CHOICE QUESTIONS

PART I

In the following, a question is followed by four alternatives, of them only one is correct. Therefore, the candidate has to select the correct choice from amongst the alternatives.

1. The number of molecules in 1 cm^2 of an ideal gas at 0° C and pressure of 10^{-5} mm of Hg is

(a) 3.5×10^{11} (b) 3.5×10^{12}
(c) 7.0×10^{11} (d) 7.0×10^{12}.

2. Two moles of a monoatomic gas are mixed with three moles of a diatomic gas. The molar specific heat of the mixture at constant volume is

(a) 1.625 R (b) 2.1 R
(c) 6.25 R (d) 3.625 R.

3. The velocities of three molecules are $3V$, $4V$ and $5V$ respectively, their rms speed will be

(a) $\sqrt{\frac{50}{3}} V$ (b) $\sqrt{\frac{3}{50}} V$
(c) $\frac{50}{3} V$ (d) $\frac{3}{50} V$.

4. M kg of a material are to be kept in melted state at melting point and power required for this purpose is P. When the power source is disconnected then the sample completely solidifies in t sec. The latent heat of fusion of the material will be

(a) $\frac{M}{Pt}$ (b) $\frac{Pt}{M}$
(c) $Pt\,M$ (d) Pt.

5. A gas is filled in a container at pressure P_0. If mass of the molecules is halved and their r.m.s. speed is doubled then the resultant pressure will be

(a) P_0 (b) $4P$
(c) $2P_0$ (d) $P_0/2$.

6. From the given *P-T* graph for a fixed mass of gas, we can conclude that

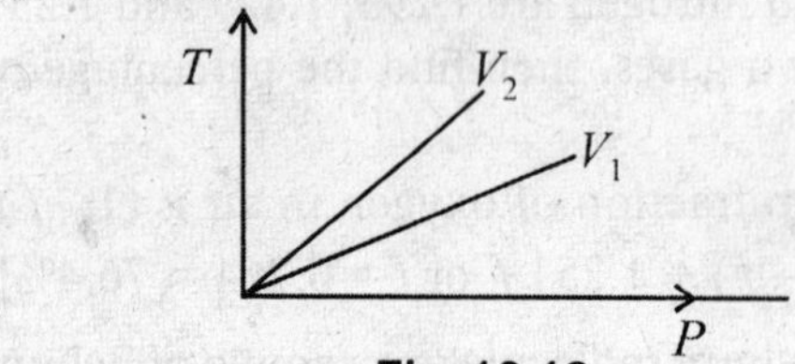

Fig. 16.18

(a) $V_2 > V_1$ (b) $V_2 = V_1$
(c) $V_2 < V_1$ (d) None of the above.

7. Two identical containers joined by a small pipe initially contain same gas at pressure P_0 and absolute temperature T_0. One container is now maintained at the same temperature while the other is heated to $2T_0$. The common pressure of the gas will be

(a) $\frac{3}{2}P_0$ (b) $\frac{4}{3}P_0$

(c) $\frac{5}{3}P_0$ (d) $2P_0$.

8. The specific heats of argon at constant pressure and constant volume are 525 J/kg°C and 315 J/kg°C respectively. Its density at NTP will be

(a) 1.77 kg/m^3 (b) 0.77 kg/m^3
(c) 1.77 gm/m^3 (d) 0.77 gm/m^3.

9. If the total translational kinetic energy of all molecules of helium when temperature is 0° C is 5×10^{-6} J; then find the temperature in a vessel which contains twice the mass of helium and total kinetic energy of all molecules equal to 10^{-5} J.

(a) 0° C (b) 273° C
(c) –136.5° C (d) 136.5° C.

10. *P-V* diagram an for a process is given by Fig. 16.19. Then

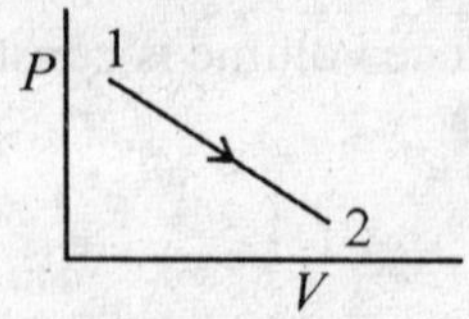

Fig. 16.19

(a) temperature remains constant throughout
(b) temperature first increases, then decreases
(c) temperature first decreases, then increases
(d) information is insufficient, cannot say.

11. Figure 16.20 depicts a process for a given mass of an ideal gas. If P_1 and P_2 represents pressure at (1) and (2) respectively of the gas, then

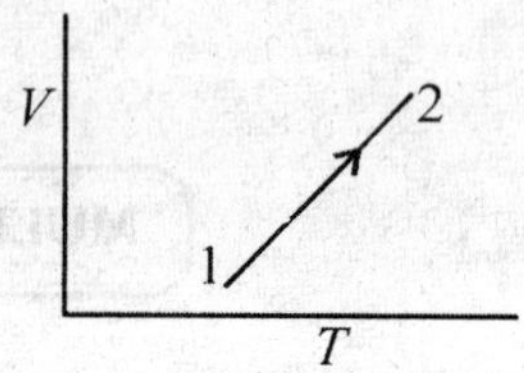

Fig. 16.20

(a) $P_1 > P_2$
(b) $P_1 = P_2$
(c) $P_1 < P_2$
(d) Pressure first increases, becomes maximum, then decreases.

12. Figure 16.21 depicts a process for a given mass of an ideal gas. If V_1 and V_2 are the volumes of the gas at (1) and (2) respectively then

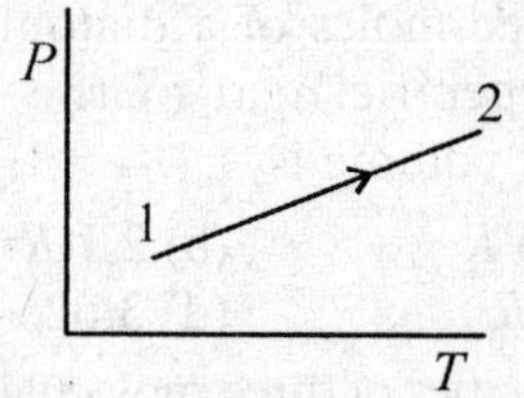

Fig. 16.21

(a) $V_1 > V_2$
(b) $V_1 = V_2$
(c) $V_1 < V_2$
(d) information is insufficient.

13. In the gas equation $PV = nRT$ the value of universal gas constant would depend on only
(a) the nature of the gas
(b) the units of measurement
(c) mass of the gas taken
(d) none of the above.

14. A vessel A has volume V and vessel B has volume $2V$. Both contain some water which has a constant volume. The pressure in space above water is P_a in vessel A and P_b in vessel B.
(a) $P_a = P_b$ (b) $P_a = 2P_b$
(c) $P_a = \frac{1}{2}P_b$ (d) $P_a = \frac{1}{4}P_b$.

15. Suppose a container is evacuated to leave just one molecule of a gas in it. Let v, $\bar{v}$ and v_p represent the average speed, the r.m.s. speed and the most probable speed of the gas, then
(a) $v > \bar{v} > v_p$ (b) $v = \bar{v} = v_p$
(c) $v_p < v < \bar{v}$ (c) $v_p > \bar{v} > v$.

16. The pressure of an ideal gas is written as $P = \frac{2}{3}\frac{E}{V}$. Here E refers to
(a) translational kinetic energy
(b) rotational kinetic energy
(c) total kinetic energy
(d) total energy.

17. Which of the following gases has maximum r.m.s. speed at a given temperature?
(a) Hydrogen (b) Nitrogen
(c) Oxygen (d) Carbon dioxide.

18. Keeping the number of moles, volume and temperature, the same, which of the following are the same for all ideal gases?
(a) r.m.s. speed of molecule
(b) density
(c) average magnitude of momentum
(d) pressure.

19. Consider the quantity $\frac{mk_BT}{PV}$ of an ideal gas, where m is mass of the gas. It depends on
(a) temperature of the gas
(b) volume of the gas
(c) pressure of the gas
(d) nature of the gas.

20. A container open to atmosphere contains air (assumed to be an ideal gas) at temperature 27° C. The temperature is now raised to 227° C. The ratio of number of atoms in the container now and at the beginning is
(a) $\frac{27}{227}$ (b) $\frac{227}{27}$
(c) $\frac{3}{5}$ (*d*) $\frac{5}{3}$.

21. Van der Waals equation of state is $(P + a/V^2)(V - b) = RT$.
Unit of constant 'a' for a given gas is
(a) cm^6 $mole^{-2}$
(b) cm^6
(c) dyne cm^4 $mole^{-2}$
(*d*) dyne cm^4-$mole^{-2}$

22. The specific heats ratio C_P/C_V for hydrogen at ordinary temperature is 7/5. As temperature is lowered sufficiently, the ratio tends to reach the value
(a) $\frac{5}{3}$ (b) $\frac{3}{2}$
(c) $\frac{9}{5}$ (d) 1.

23. Two identical glass spheres filled with air are connected by a horizontal glass tube. The glass tube contains a pellet of mercury at its mid point. Air in one sphere is at 0° C and other is at 20° C. If both the vessels are heated through 10° C, then neglecting the expansion of the bulbs and the tube
(a) the mercury pellet gets displaced towards the sphere at lower temperature
(b) The mercury pellet gets displaced towards the sphere at higher temperature

(c) The mercury pellet does not get displaced at all

(d) The temperature rise causes the pellet to expand without any displacement.

24. Van der Waals equation is $(P + a/V^2)(V - b) = RT$.
Here the constant b represents

(a) actual volume occupied by the gas molecules

(b) twice the actual volume occupied by the gas molecules

(c) four times the actual volume occupied by the gas molecules

(d) eight times the actual volume occupied by the gas molecules.

25. In phase diagram shown (Fig. 16.22) the point o corresponds to the triple point of water. The regions I, II, and III respectively correspond to phases

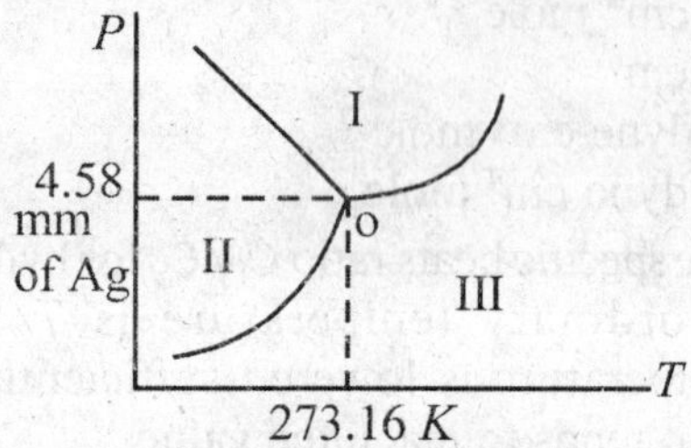

Fig. 16.22

(a) liquid, solid, vapour

(b) solid, liquid, vapour

(c) liquid, vapour, solid

(d) solid, vapour, liquid.

26. If the number of degrees of freedom of a molecule of a gas is n, then ratio C_P/C_V of the gas is

(a) $1+\dfrac{1}{n}$ (b) $1+\dfrac{1}{2n}$

(c) $1+\dfrac{2}{n}$ (d) $\dfrac{2n}{2n+1}$.

27. The ratio of root mean square speed (v_{rms}) to the most probable speed (v_m) for an ideal gas is

(a) $\sqrt{\dfrac{3}{2}}$ (b) $\sqrt{\dfrac{2}{3}}$

(c) $\dfrac{3}{2}$ (d) $\dfrac{2}{\sqrt{\pi}}$.

28. A box (thermally insulated) has two chambers separated by a membrane. One of volume V contains an ideal gas at temperature T and other of volume $\left(\dfrac{1}{2}V\right)$ is evacuated. If the membrane breaks down, the gas temperature will be

(a) $\dfrac{2T}{3}$ (b) $\dfrac{3T}{2}$

(c) T (d) data are incomplete.

29. A certain perfect gas is found to obey $PV^{3/2}$ = constant during an adiabatic process. If such a gas at initial temperature T is adiabatically compressed to half the initial volume, its final temperature will be

(a) $\sqrt{2}\,T$ (b) 2T

(c) $2\sqrt{2}\,T$ (d) $4T$.

30. Figure 16.23 shows graph between $\log v_{rms} - \log T$. Where T is temperature. The correct graph is

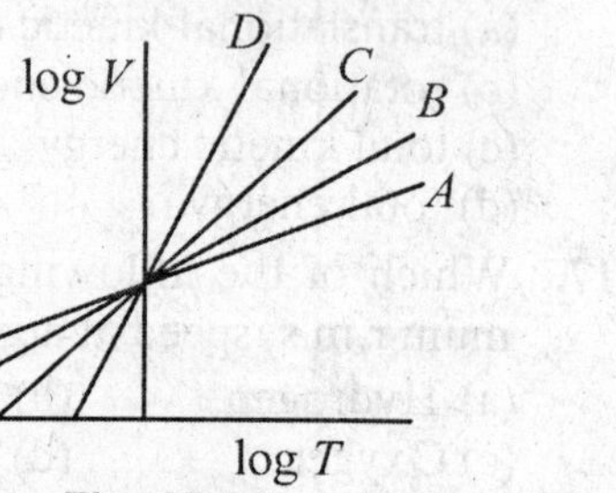

Fig. 16.23

(a) A (b) B

(c) C (d) D.

ANSWERS

1. (a)	2. (b)	3. (a)	4. (b)	5. (c)
6. (a)	7. (b)	8. (a)	9. (a)	10. (b)
11. (a)	12. (c)	13. (b)	14. (a)	15. (b)
16. (a)	17. (a)	18. (d)	19. (d)	20. (c)
21. (c)	22. (a)	23. (b)	24. (c)	25. (a)
26. (b)	27. (a)	28. (c)	29. (a)	30. (c)

PART II

In the following, a question is followed by four alternatives, of them one or more are correct. Candidate has to select the correct choice(s) from amongst the alternatives.

1. Figure 16.24 shows a cylindrical tube of cross-sectional area A fitted with two frictionless pistons. The pistons are connected to each other by a metallic wire. Initially the temperature of the gas is T_0 and pressure is P_0 which equals the atmospheric pressure. Now the temperature is increased to $2T_0$.

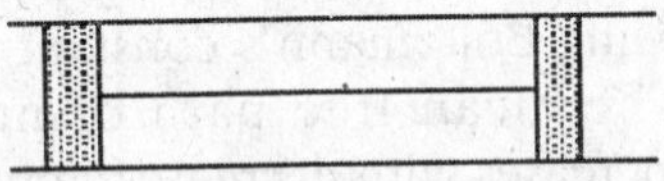

Fig. 16.24

   (a) The tension in the string when temperature is T_0 is zero
   (b) The tension in the string when temperature is $2T_0$ is $2P_0A$
   (c) The tension in the string when temperature is $2T_0$ is P_0A
   (d) The tension in the string when temperature is T_0 is P_0A.

2. A rigid container of negligible heat capacity contains one mole of an ideal gas. 10 calories of heat is required to increase its temperature by 2° C. The gas in the container may be

   (a) He (b) H_2
   (c) O_2 (d) Ne.

3. If v_a, v_r, and v_P respectively denote the average speed, root mean square speed and most probable speed of molecules in an ideal monoatomic gas at absolute temperature T. The mass of the molecule is m. Then

   (*a*) no molecule can have a speed greater than $\sqrt{2}\, v_r$
   (*b*) no molecule can have a speed less than $v_p / \sqrt{2}$
   (*c*) $v_p < v_a < v_r$
   (*d*) the average kinetic energy of molecules $\frac{3}{4} m v_p^2$.

4. An actual gas deviates from perfect gas behaviour as its molecules

   (a) have definite size
   (b) show Brownian motion
   (c) attract each other
   (d) are not spherical.

5. Consider a collision between an oxygen molecule and a hydrogen molecule in a mixture of oxygen and hydrogen kept at room temperature. Which of the following are possible?

   (a) The kinetic energies of both the molecules increase.

(b) The kinetic energies of both the molecules decrease
(c) The kinetic energy of oxygen molecules increases and that of the hydrogen molecule decreases
(d) The kinetic energy of the hydrogen molecule increases and that of oxygen molecules decreases.

6. Which of the following quantities is the same for all ideal gases at the same temperature?
(a) The kinetic energy of 1 mole
(b) The kinetic energy of 1g
(c) The number of molecules in one mole
(d) The number of molecules in 1g.

7. Which of the following quantities is the same for all the ideal gases at the same temperature?
(a) All the molecules of a gas are identical
(b) Gas molecules are just like rigid balls
(c) All gases are perfect gases
(d) Duration of molecular collision is negligible.

8. In kinetic theory of gases, it is assumed that the gas molecules
(a) do not exert any attracting force on each other
(b) collide elastically
(c) move with uniform velocity
(d) occupy negligible volume.

9. Which of the following statement(s) is/are true regarding the kinetic theory of gases?
(a) The pressure of the gas is directly proportional to the average speed of the molecules
(b) The root-mean-square speed of the molecules is directly proportional to the pressure
(c) The ratio of diffusion is directly proportional to the average speed of the molecules
(d) The average kinetic energy per molecule is directly proportional to the absolute temperature.

10. According to kinetic theory of gases which one of the following statement(s) is/are true?
(a) Real gas behaves like an ideal gas at high temperature and low pressure
(b) Liquid state of an ideal gas is impossible
(c) At any temperature and pressure ideal gas obeys the Boyle's law and Charle's law
(d) The molecules of real gas do not exert any force on one another.

11. According to kinetic theory of gases 0 K is that temperature at which
(a) pressure of ideal gas is zero
(b) volume of ideal gas is zero
(c) internal energy of ideal gas is zero
(d) ideal gas liquifies.

12. From following statements concerning ideal gas at any given temperature T, select the correct one(s).
(a) The coefficient of volume expansion at constant pressure is the same for all ideal gases
(b) The average translational kinetic energy per molecule of oxygen gas is $3k_BT$, k_B being Boltzmann's constant
(c) The mean free path of molecules increases with decrease in pressure
(d) In a gaseous mixture the average translational kinetic energy of molecules of each component is same

13. A box containing a mixture of inert gases, He, Ne, A, and is at room temperature. Identify the correct alternatives.
(*a*) The average K.E. of all the atoms is same for all gases
(*b*) v_{rms} for any atom is proportional to the temperature.
(*c*) $v_{rms}(\text{He}) > v_{rms}(\text{Ne}) > v_{rms}(\text{Ar})$
(*d*) Nothing can be said about the average energy, since their amounts are not mentions.

14. Consider Van der Waals equation for real gases. Identify the correct statement related to the constants in the equation.

(*a*) The term *a* represents the mutual attraction between the molecules of the gas

(*b*) The trem *b* represents the finite size of the molecules of the gas

(*c*) Both the terms are more important at high pressure

(*d*) $\frac{b}{aR}$ is dimensionally equivalent to temperature.

15. Mean free path of molecules of a gas is related to the following phenomenon of the gas.

(a) Liquifiction
(b) Diffusion
(c) Thermal conductivity
(d) Viscosity.

ANSWERS

1. (a, c)	2. (b, c)	3. (c, d)	4. (a, c)	5. (c, d)
6. (a, c)	7. (a, b, c)	8. (a, b, d)	9. (c, d)	10. (a, b, d)
11. (a, b, c)	12. (a, c, d)	13. (a, c)	14. (a, b, c)	15. (b, c, d)

17

Heat Transfer

17.1 INTRODUCTION

We have seen that heat transfer from one place to another or from one part of a body to another arises due to temperature difference. This transfer of heat energy may take place under three quite distinct modes. They are (*i*) *Conduction*, (*ii*) *Convection*, and (*iii*) *Radiation*.

If a metal rod is heated at one end and held at the other by the hand we feel sensation of heat. Heat travels from the hot end along the bar and produces this sensation. This process of heat transmission between adjacent part of rod without any flow of matter is called *conduction*. The power of transmission of heat energy in this manner is possessed by all substances to a varying degree.

In fluids heat is often transferred by the actual flow of matter from hotter to colder parts. This can be best illustrated by placing gently some crystals of potassium permanganate at the bottom of a beaker containing water and heating it. Heated water rises up curls down forming a closed path which is rendered visible by red colour imparted to the water. This process of heat transfer is *called* convection. The closed paths formed by rising water are called *convection currents*.

In both the above processes the intervening medium takes an active part in heat transmission. The third mode of heat transmission called radiation requires no intervening medium. If we stand near a furnace, we feel the sensation of heat. Similarly if we are in the sun, we feel warm. There is absolutely no material medium between us and the sun, still heat from the sun continues to reach us. This phenomenon is called *radiation*.

The processes of conduction and convection are relatively slow due to the action of the intervening medium. While radiation is enormously fast. In this process energy propagates with the speed of light.

17.2 THERMAL CONDUCTION

Conduction is the flow of heat between two adjacent layers of a body due to temperature difference between them. When one end of a metallic rod is put inside a furnace, the temperature of the other end gradually increases. This heat transfer takes place due to molecular collisions. The molecules at the end of the rod in furnace gain heat and their average kinetic energy increases. These molecules collide with the neighbouring molecules having less kinetic energy, the energy is shared between these two groups of molecules. The kinetic energy of these neighbouring molecules increases. These on colliding with molecule on the colder side, transfer energy to them. This way heat is transmitted along the rod from molecule to molecule with no change in the average position of the molecule and there is no mass movement of water. This process of heat transmission is called conduction.

Thus conduction is the process of heat transmission in which the average displacement of the medium particles is zero.

Thermal Conductivity

Here we shall first discuss the flow of heat inside a rod heated at one end. Consider a thin wall of the material with parallel faces such that heat flows in the direction perpendicular to the faces. Let θ_1 and θ_2 be the temperatures of the two faces, l the thickness of the wall. We also suppose the remaining surface is covered with non-conducting material so that no heat is transferred through the sides. After sufficient time, when the temperature at every point of bar is steady, i.e., *steady state* is reached, the amount of heat crossing per unit time through any cross-section (A) of the rod is equal. If ΔQ amount of heat crosses through any cross-section in time Δt, then $\frac{\Delta Q}{\Delta t}$ is called the *thermal current* (or *heat current*). It is found that the thermal current is proportional to (*i*) temperature difference $(\theta_1 - \theta_2)$ (*ii*) area of cross-section A and (*iii*) inversely proportional to length l. That is

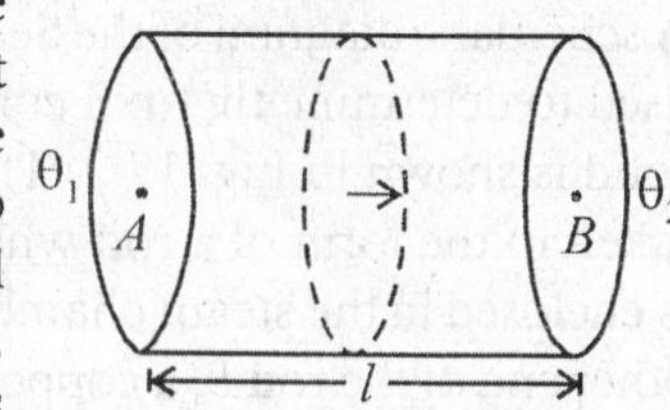

Fig. 17.1

$$\frac{\Delta Q}{\Delta t} = \frac{KA(\theta_1 - \theta_2)}{l} \qquad \text{...(17.1)}$$

where K is a constant called the *thermal conductivity* of the material. It depends upon the nature of material of the rod. Thus *conductivity of a material is defined as the quantity of heat flowing per second through a unit area of a slab of unit thickness when temperature difference between the faces is unity.*

If the area of cross-section is not uniform or the steady state is not reached, the equation can be applied to a thin layer of material perpendicular to the direction of heat flow. Let dx be a small thickness along the direction of heat flow and $d\theta$. The temperature difference across the layer, then $\frac{\theta_1 - \theta_2}{l} = -\frac{\theta_1 - \theta_2}{l} = -\frac{d\theta}{dx}$ is called *temperature gradient* at that point. The minus sign has been used before $\frac{d\theta}{dx}$ because the symbol d always stands for increase. If A is the area of cross-section at the layer, then heat current through the cross-section is

$$\frac{\Delta Q}{\Delta t} = -KA\frac{d\theta}{dx} \qquad \text{...(17.2)}$$

The quantity $\frac{1}{A}\frac{\Delta Q}{\Delta t}$ is called *thermal current density*, i.e.

$$j = \frac{1}{A}\frac{\Delta Q}{\Delta t} = -K\frac{d\theta}{dx} \qquad \text{...(17.3)}$$

The SI units of thermal conductivity are J/s-m-K or W/m-K.

In analogy with the electrical resistance, the inverse of K may be called *thermal resistivity* of the specimen and the quantity

$$\frac{l}{KA} = R \qquad \ldots(17.3a)$$

is called *thermal resistance.*

The conductivity of metals is much more than that of non-metals. It is due to the fact that in metals there are large number of free electrons which can move freely inside the body of the metal. These free electrons transmit the energy from one place to other in metal. Solids are better conductors than liquids and liquids are better conductors than gases.

17.3 MEASUREMENT OF THERMAL CONDUCTIVITY OF METALS

A schematic diagram of the Searl's apparatus used to determine thermal conductivity of a solid is shown in Fig. 17.2. The specimen is taken in the form of a rod whose one end A is enclosed in the steam chamber S. Near the other end of the rod B, a copper tube is coiled around the rod through which a steady flow of water is maintained in the copper tube. Thermometers T_3 and T_4 are provided to measure the temperatures of water at exit and entry respectively. Temperatures at two points along the bar are measured by thermometers T_1 and T_2. To ensure good contact of thermometers with rod, two holes are drilled in the rod at these places and mercury is filled in it. The whole rod is wrapped round with some non-conducting material like wood felt, etc. to prevent loss of heat from the sides.

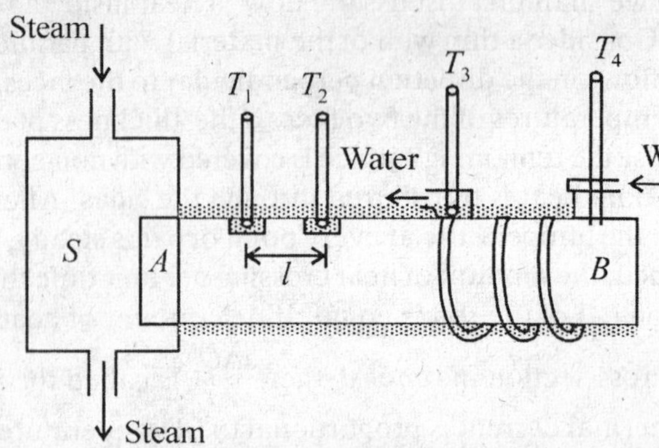

Fig. 17.2

Steam is passed in the steam chamber and a steady flow of water is maintained. The temperatures of all the four thermometers rise initially and ultimately become constant in time as the steady state is reached. In steady state let the thermometers read θ_1, θ_2, θ_3, θ_4 respectively.

The mass of water (m) that flows through, the pipes at the cold end in fixed time t, is found by collecting water in a weighed beaker. The area of cross-section (A) of the bar is determined by measuring its radius with vernier calipers. The distance between the holes (l) is measured with a scale.

From the above data, the rate of flow of heat from steam chamber by the rod is

$$\frac{\Delta Q}{\Delta t} = KA\frac{(\theta_1 - \theta_2)}{l}, \text{ where } K \text{ is thermal conductivity.}$$

As the mass of the water collected in time t is m, the heat taken by the water in this time is

$$\Delta Q = ms\,(\theta_3 - \theta_4)$$

where s is the specific heat of water.

Thus the rate at which heat is taken by water is

$$\frac{\Delta Q}{\Delta t} = \frac{ms(\theta_3 - \theta_4)}{t}$$

Since the entire rod is covered with an insulating material and the temperature of the rod does not change with time at any point, whatever heat is given by the steam chamber goes into the flowing water. Hence the rate of heat flow along the rod must be equal to the rate at which heat is taken by water. Thus equating them we have

$$KA\frac{(\theta_1 - \theta_2)}{l} = \frac{ms(\theta_3 - \theta_4)}{t}$$

or $$K = \frac{ms(\theta_3 - \theta_4)l}{At(\theta_1 - \theta_2)} \qquad ...(17.4)$$

17.4 SERIES AND PARALLEL COMBINATION OF RODS

(*a*) **Series Combination**: Consider two rods of same cross-section (A) of lengths l_1 and l_2 and of materials having thermal conductivity K_1 and K_2 respectively joined one after the other as shown in Fig. 17.3. The free ends are kept at temperatures T_1 and T_2 ($T_1 > T_2$). In steady state, the rate of flow of heat (thermal current) through the two rods is same. Such connection of rods is said to be series combination. Let the temperature of the common face be T. The heat flowing through rod A per second is

$$Q = K_1 A\frac{(T_1 - T)}{l_1} \qquad ...(i)$$

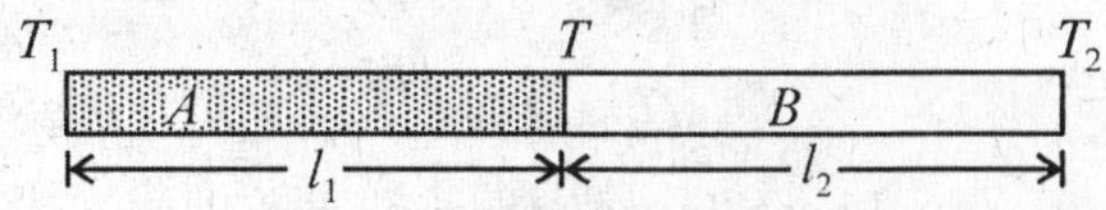

Fig. 17.3

The heat flowing per second through rod B then is

$$Q = K_2 A\frac{(T - T_2)}{l_2} \qquad ...(ii)$$

From (*i*) and (*ii*), we have

$$T_1 - T_2 = \frac{Q}{A}\left[\frac{l_1}{K_1} + \frac{l_2}{K_2}\right] \qquad ...(iii)$$

This system is equivalent to a single rod of length ($l_1 + l_2$) and of conductivity K, such that

$$Q = \frac{KA(T_1 - T_2)}{l_1 + l_2} \qquad ...(iv)$$

From (*iii*) and (*iv*), we have

$$T_1 - T_2 = \frac{Q}{A}\frac{(l_1 + l_2)}{K} = \frac{Q}{A}\left[\frac{l_1}{K_1} + \frac{l_2}{K_2}\right]$$

i.e. $$\frac{l_1 + l_2}{K} = \frac{l_1}{K_1} + \frac{l_2}{K_2}$$

or $$K = \frac{l_1 + l_2}{\left[\dfrac{l_1}{K_1} + \dfrac{l_2}{K_2}\right]} \qquad ...(17.5)$$

In general if a large number of rods are connected in this way then equivalent conductivity of the system is given by

$$K = \frac{\Sigma_i l_i}{\Sigma_i \dfrac{l_i}{K_i}} \quad \ldots(17.6)$$

If two rods of same length and conductivities K_1 and K_2 are connected in series, then equivalent conductivity of the system becomes ($l_1 = l_2 = l$)

$$K = \frac{2l}{l\left(\dfrac{1}{K_1} + \dfrac{1}{K_2}\right)} = \frac{2K_1K_2}{K_1 + K_2}$$

(b) Parallel Combinations: Now suppose two rods of area of cross-section A_1 and A_2 and of materials of thermal conductivity K_1 and K_2 and of same length are connected as shown in Fig. 17.4. Left ends of the rods are maintained at temperature T_1 and right ends at temperature T_2 so that same temperature gradient exists in the two rods. Such combination is called parallel combination. The heat flowing through rod A per second is

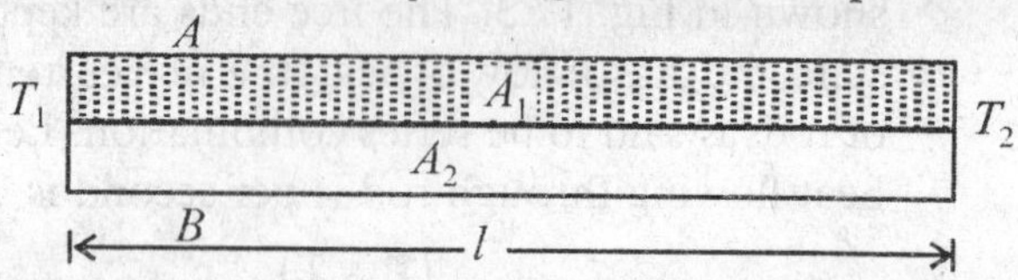

Fig. 17.4

$$Q_1 = A_1K_1\frac{(T_1 - T_2)}{l} \quad \ldots(i)$$

and heat flowing through rod B per second is

$$Q_2 = A_2K_2\frac{(T_1 - T_2)}{l} \quad \ldots(ii)$$

Total heat flowing through both rods per second is

$$Q = Q_1 + Q_2 = \left(\frac{T_1 - T_2}{l}\right)(A_1K_1 + A_2K_2) \quad \ldots(iii)$$

This system is equivalent to a single rod of length l and area of cross-section $A_1 + A_2$, of conductivity K such that

$$Q = K(A_1 + A_2)\frac{(T_1 - T_2)}{l} \quad \ldots(iv)$$

From (*iii*) and (*iv*) the equivalent conductivity of the system is

$$K(A_1 + A_2) = (K_1A_1 + K_2A_2)$$

or

$$K = \frac{K_1A_1 + K_2A_2}{A_1 + A_2} \quad \ldots(17.7)$$

In general if a number of rods are connected in parallel, then equivalent conductivity is

$$K = \frac{\Sigma_i A_iK_i}{\Sigma A_i} \quad \ldots(17.8)$$

If two rods of same area of cross-section ($A_1 = A_2$) are connected in parallel, then equivalent conductivity is given by

$$K = \frac{K_1 + K_2}{2}$$

17.5 CONVECTION

Convection is a mode of heat transmission from one place to the other by the actual motion of heated material. It is therefore possible only in fluids. It arises due to difference of densities at two places and is consequence of gravity.

When a fluid in a container is heated, the fluid at the bottom gets heated due to conduction through the bottom of the container. Its density decreases and consequently it rises up and is replaced by colder parts and thus convection currents are set-up in the fluid and the whole fluid is heated by actual movement of mass. This type of convection which results from difference in densities is called *natural convection* or *free convection*.

If the hot material is forced to move say by a blower, fan or pump, the process of heat transfer is called *forced convection*. It is found that rate of heat convection from an object is proportional to (*i*) to surface area of body exposed to fluid, i.e., the contact area and (*ii*) excess of temperature of body over its surroundings, i.e.

$$\frac{dQ}{dt} = hA(\theta - \theta_0) \qquad \ldots(17.9)$$

where h is constant of proportionality, θ is temperature of body and θ_0 that of surroundings. Constant h is called convection coefficient and depends on the properties of fluid such as density, viscosity, specific heat and thermal conductivity, etc.

If the temperature of the fluid falls at the rate of $\frac{d\theta}{dt}$, then $\frac{d\theta}{dt} = mc\frac{d\theta}{dt}$ where m is mass and c is specific heat. Then

$$mc\frac{d\theta}{dt} = -hA(\theta - \theta_0)$$

or
$$\frac{d\theta}{dt} = -\frac{hA}{mc}(\theta - \theta_0) = -K(\theta - \theta_0) \qquad \ldots(17.10)$$

where
$$K = \frac{hA}{mc}$$

Equation (17.10) is called *Newton's law of cooling*.

Regarding convection it is worthy to note that:

(*a*) Natural convection takes heat from bottom to top while forced convection takes heat in any direction.

(*b*) In case of natural convection, convection currents move warm fluid upwards and cool fluid downwards. That is why heating is done from base while cooling from top.

(*c*) Natural convection plays an important part in ventilation, in changing climates and weather and in forming land and sea breezes and trade winds.

(*d*) Natural convection cannot take place in a gravity free region.

(*e*) The heat transfer inside a human body is also an example of forced convection in which heart serves as the pump and blood as the circulating fluid.

17.6 RADIATION

Conduction and convection are the process of heat transfer which involve material particle. But the process of radiation does not need any material medium for heat transfer. The question naturally arises as to what then is the mechanism of heat transmission in this process of radiation. The answer is that heat, being also a form of energy, is very much akin to light in many ways, and, like the latter can travel through empty space. Both light and heat have come to be established as electromagnetic waves in space, different from each other only in their wavelengths—those corresponding to heat being non-luminous unlike those of visible light and very much longer.

The wavelengths immediately beyond the red form a continuous upward range and though invisible to eye, they are distinguished by the rather large amount of heat energy associated with them, though a little of it is carried by others also. The radiation in this particular range is called *infrared* and it is this that constitutes most of the heat energy radiated by a hot body and produces in us the sensation of heat. It is usually referred to as *thermal radiation*. While passing through the space, it is not accompanied by the motion of any material molecules and therefore hardly constitutes heat. It manifests itself as heat only when it falls on a material body and is absorbed by it.

The word radiation is used in two meanings. It refers to the process by which the energy is emitted by a body, is transmitted in space and falls on other body. It also refers to the energy itself which is being transmitted in space.

17.7 SOME DEFINITIONS RELATED TO RADIATION

(*a*) **Black Body:** Every hot body emits radiation from its surface which depends upon the nature of the surface, its size and nature of the surroundings. When the emitted radiation falls on water, a part is *reflected* regularly or irregularly (diffused), another part is *absorbed* while the remainder is *transmitted*. The coefficient of reflection, absorption and transmission are connected by the relation

$$r + a + t = 1 \quad \text{...(17.11)}$$

where r = coefficient of reflection, i.e. fraction of total energy reflected.

a = coefficient of absorption, i.e. fraction of total energy absorbed.

t = coefficient of transmission, i.e. fraction of total energy transmitted.

From our experience, we can define certain substances with limiting properties as regards their behaviour towards radiant energy.

A body for which $a = t = 0$ and $r = 1$, is *called perfect reflector*. Apparently it neither absorbs nor transmits the light that falls on it.

A body for which $a = r = 0$ and $t = 1$, is called *transparent body*. It neither absorbs nor reflects the light that falls on it.

A body for which $r = t = 0$ and $a = 1$, is called *perfectly black body*. Obviously, it neither reflects nor transmits the light which falls on it, absorbs all and hence appears black. But the perfectly black body is only an ideal conception for even lamp-black is found to transmit light of long wavelength. It absorbs about 96% of visible light, platinum black absorbs about 98%.

(*b*) **Absorptive Power:** Absorptive power of a surface is defined as the rate of absorption of thermal radiation per unit area of the surface, compared with that of a surface of a black body under identical conditions. It is denoted by letter '*a*'.

Thus if Q_S and Q are respective quantities of heat absorbed by unit area of surface and the black body in a given time, then

$$\text{Absorptive power of the surface, } a = \frac{Q_S}{Q}$$

As all the radiations incident on a black body is absorbed, the absorptive power of a black body is unity.

It follows that Q represents the total radiation incident on the surface in a given time. We may also, therefore, define the absorptive power of a surface as the ratio between the quantity of heat (Q_S) absorbed by it and that the total quantity of heat (Q) incident on it in given time, i.e. absorptive power is nothing but *absorptive coefficient.*

If we confine ourselves to a small range of wavelengths in the range λ and $\lambda + d\lambda$, the absorptive power of a surface for range λ to $\lambda + d\lambda$ may be defined as the ratio between the radiation of the wavelength range λ and $\lambda + d\lambda$ absorbed by the surface and the total radiation of this wavelength range falling on it. This is called as *monochromatic absorptive power.*

Note that *absorptive power is a dimensionless quantity.*

(*c*) **Emissive Power:** A surface, when heated emits thermal radiation depending on its nature, size and temperature, as also on the nature of its surroundings. This process is known as *emission.*

The amount of thermal radiations emitted per second per unit area of the surface per 1° C excess of temperature over that of the surroundings is called surface emissivity (E). Its units are Watt/m^2(°C).

The emissive power (or emittance) of a surface is defined as the rate of emission of thermal radiation per unit area of the surface at a given temperature, compared with that of the black body surface under similar conditions. It is denoted by letter (e)

It follows, therefore, that if E_S he surface emissivity and E that of the black body, then

$$\text{Emissive power of the surface, } e = \frac{E_S}{E}$$

Here the thermal radiation consists of all the wavelengths emitted by the body. That is why e is sometimes referred to as the *total emissive power of the surface.*

Many a times we can confine ourselves to a narrow range of wavelengths lying between λ and $\lambda + d\lambda$. The emissive power of a surface, then is defined as the ratio between total energy emitted per second per unit area of the surface and wavelength range $d\lambda$. It is denoted by letter e_λ and is referred to as *monochromatic emissive power of the surface.* Thus

$$e_\lambda = \frac{E}{d\lambda}$$

where E is the surface emissivity of the surface.

The reason for choosing to define emissive power in this manner is that, for a small change $d\lambda$ in wavelength, E (surface emissivity) will be proportional to $d\lambda$ so that the quotient $E/d\lambda$ remains quite independent of $d\lambda$ and is a function of only of the nature and temperature of the surface and wavelength λ. Clearly,

$$e_\lambda d\lambda = E$$

i.e. the emissive power of a surface (or a body) multiplied by wavelength $d\lambda$ gives the rate of emission of radiant energy per unit area of the surface or the body within wavelength range λ and $\lambda + d\lambda$. Obviously, therefore, the total emissivity of the surface is

$$e = \int_0^\infty e_\lambda d\lambda$$

For a perfect black body $e = 1$. It is, therefore, a perfect radiator.

17.8 KIRCHHOFF'S LAW

Kirchhoff in the year 1859 deduced an important law which may be stated as follows:

At a given temperature, the ratio of the emissive power to absorptive power is same for all substances and is equal to the emissive power of a perfectly black body.

Proof of the law

Let us consider a uniformly heated enclosure maintained at temperature T, and a body of monochromatic absorptive power a_λ is placed inside it.

Now let δQ be the quantity of radiant energy, within wavelength range λ and $\lambda + d\lambda$ incident in unit time on unit area of the body. Of this, only a portion $a_\lambda \delta Q$ will be absorbed by it and the remainder $(1 - a_\lambda)\delta Q$ will be reflected or transmitted.

If e_λ be the monochromatic emissive power of the body for the same range of wavelength λ and $\lambda + d\lambda$ at the same temperature T, the amount of energy emitted by the body per second per unit area of it will clearly be $e_\lambda d\lambda$, so that total energy given out per second per unit area of the body

$$= \text{energy reflected or transmitted} + \text{energy emitted}$$

$$= (1 - a_\lambda)\delta Q + e_\lambda d\lambda$$

Since the body is in thermal equilibrium with the enclosure, we have

$$\delta Q = (1 - a_\lambda)\delta Q + e_\lambda d\lambda$$

or $$a_\lambda \delta Q = e_\lambda d\lambda$$

In case of a perfectly black body ($a_\lambda = 1$) of emissivity E_λ, then

$$E_\lambda d\lambda = \delta Q$$

Substituting the value of δQ, we have

$$a_\lambda E_\lambda d\lambda = e_\lambda d\lambda$$

or $$\frac{e_\lambda}{a_\lambda} = E_\lambda \qquad \text{...(17.12)}$$

i.e. the ratio between the emissive power and absorptive power of a given body at a given temperature and for a given wavelength range (λ and $\lambda + d\lambda$) is a constant and is equal to the emissive power of a perfectly black body for the same temperature and wavelength range.

It may be pointed out here that, though the law is proved for a body inside a constant temperature enclosure, it has general validity under all conditions for pure temperature radiations. This is so because the absorptive power and emissive power of a body depends only upon its nature and are quite independent of the surroundings.

Some More Points to Note

(*i*) Since E_λ increases with temperature, it is clear that $e_\lambda / d\lambda$ also increases as temperature rises. It is, therefore, constant only in the sense that its value is the same for all kinds of surfaces for a given temperature and wavelength.

(*ii*) The ratio e_λ / E_λ is quite often spoken of as emissivity of the body or the surface. And since, $e_\lambda / E_\lambda = a_\lambda$, it follows that, defined in this manner, the emissivity of a surface is equal to its absorptive power.

(*iii*) For a given temperature, E_λ is a constant, so that if e_λ be large a_λ must be correspondingly large and *vice versa*. In other words, if the emissive power of surface be large, its absorptive power must also be large i.e. *good emitters must also be good absorbers.*

(*iv*) A surface for which a_λ is constant for all wavelengths (though not necessarily equal to 1), and therefore e_λ for which is exactly same fraction of E_λ for all wavelengths, (but less than unity), is called a *grey body*.

(*v*) The law also shows that a small hole in the wall of a constant temperature enclosure acts both as perfect absorber as well as the best possible emitter of radiation, i.e. it emits radiation in exactly the same manner as a perfectly black body would do, if it were maintained at the same temperature as the enclosure.

Fraunhofer Lines

Fraunhofer lines are dark lines in the spectrum of the sun and are explained on the basis of Kirchhoff's law. Fraunhofer first reported these lines and made their proper study. He mapped out them properly, measured their wavelengths and denoted the principal ones among them by letters *A*, *B*, *C*, etc. These lines, therefore, came to be known as Fraunhofer lines and were also found in the spectra of stellar light, i.e. light from the stars.

Kirchhoff explained the existence of these lines on the basis of his laws. According to him, the sun consists of a central gaseous mass at a very high temperature ($\sim 0^7$ K) called *photosphere*, which gives a continuous spectrum uncrossed by any dark lines. Around the photosphere is a comparatively cooler atmosphere, containing a number of elements in gaseous form, together with the common gases. This is called *chromosphere* and has temperature about 5000 ~ 6000 K. Now as the white light from the photosphere passes through this cooler atmosphere, these various elements present in it absorb, in accordance to Kirchhoff's law, some of the wavelengths which they emit on excitation. As a result, they are missing from the solar spectrum and we see the dark Fraunhofer lines in their place. The same is of course true for the dark lines observed in the spectra of the stars.

17.9 PREVOST THEORY OF EXCHANGES

Prior to 1792 the ideas regarding emission of radiant energy were very much confused. People used to talk of hot radiations and cold radiations. A block of ice was supposed to emit cold radiations because it produced a cold sensation. Similarly, a furnace was supposed to emit hot radiations because it produces hot sensation. Prevost put forward the theory of radiation in a systematic way which is now known as *the theory of exchanges*. According to this theory, all the bodies radiate thermal radiations irrespective of their temperatures. The amount of thermal radiations radiated per unit time depends on the nature of the emitting surface, its area and its temperature. The rate is faster at higher temperatures. Besides, a body also absorbs part of the thermal radiation emitted by the surroundings when this radiation falls on it. When a body radiates more than what it receives, the temperature falls. If a body radiates less than what it absorbs, its temperature rises.

When a body is placed in a surroundings for a long time this exchange takes place till the temperature of the body and that of the surroundings become equal. We say that the thermal equilibrium is reached. Even in thermal equilibrium the body is still radiating thermal radiations, simultaneously absorbing part of the radiations emitted by the surroundings. Thus we conclude that in thermal equilibrium it radiates at the same rate as it absorbs.

Thus if we place a hotter body in the room, it radiates at a faster rate than the rate at which it absorbs. Thus the body suffers a net loss of thermal energy in any given time and its temperature decreases. Similarly, if a colder body is kept in a warm surroundings, it radiates less to the surroundings than what it absorbs from the surroundings. Consequently, there is a net increase in the thermal energy of the body and the temperature rises.

17.10 STEFAN–BOLTZMANN LAW

The relation between the quantity of radiations from a body and its temperature has been investigated by different workers by observing the rate of cooling. The first to do so in a systematic way was Newton, with whose law of cooling we are already familiar. It refers necessarily to cooling by convection rather than by radiation.

Joseph Stefan, a Russian physicist was of the opinion that the law governing the energy radiated by body and temperature must be simple one. He interactively and empirically suggested that *the amount of energy E lost per second per unit area of the body was proportional to the 4th power of its absolute temperature T*, i.e.

$$E \propto T^4$$

or $$E = \sigma T^4 \qquad \ldots(17.13)$$

where σ is called *Stefan's constant*. The measured value of σ is 5.67×10^{-8} Watt/m²-K⁴.

Stefan had suggested this law from experimental data available on radiation and Boltzmann derived it from thermodynamical consideration. That is why this law is also known as Stefan Boltzmann law.

A body which is not a black body, emits less radiation than given by equation (17.13). It is however proportional to T^4. The energy emitted by such a body per unit time per unit area is

$$E = e\sigma T^4 \qquad \ldots(17.14)$$

where e is constant for the given surface having value between 0 and 1, e is called the emissivity of the surface. Its value is zero ($e = 0$) for perfect reflector and 1 ($e = 1$) for a black body.

Consider a body of emissivity e kept in thermal equilibrium in an enclosure at temperature T_0. The energy of radiation absorbed by it per unit time per unit area should be equal to the energy emitted by it per unit time per unit area. This is because the temperature remains constant. Thus the energy of the radiation absorbed per unit time per unit area is

$$E = e\sigma T_0^4$$

Now suppose the temperature of the body is made T but the enclosure temperature is maintained T_0, the energy of the thermal radiations emitted by the body per unit time per unit area is

$$E_e = e\sigma T^4$$

and the energy absorbed per second per unit area by the body is

$$E_a = e\sigma T_0^4$$

Thus net loss of thermal energy per unit time per unit area is

$$E = E_e - E_a = e\sigma\left(T^4 - T_0^4\right)$$

If A is the area of the body then net rate of loss of energy is

$$R = eA\sigma\left(T^4 - T_0^4\right) \qquad \ldots(17.15)$$

17.11 NEWTON'S LAW OF COOLING

We have seen that every substance emits radiation by virtue of its temperature. As such it is necessary to investigate how its emission varies with the temperature and nature of the surface. Newton was the first to study systematically the loss or gain of heat by a body placed in an enclosure. The experiment is very simple to perform. A large bulb fitted with mercury thermometer is heated and the bulb is enclosed in an evacuated flask, the walls of which are blackened inside and is maintained at fixed temperature by means of a suitable bath. The rate of fall of temperature of the thermometer gives the rate of cooling $\left(-\frac{d\theta}{dt}\right)$. If θ_0 is the temperature of the enclosure then excess of temperature is $(\theta - \theta_0)$. A plot of rate of cooling versus excess temperature is a straight line giving

$$-\frac{d\theta}{dt} = K(\theta - \theta_0)$$

i.e. the rate of cooling is proportional to the excess of temperature of the substance above the surroundings. This is *Newton's law of cooling*. The law holds when the excess temperature is not large. The experiments show considerable deviations from the law when the temperature difference was more than 40° C.

Derivation of Newton's law of Cooling from Stefan's law

Suppose a body of surface area A and emissivity e at an absolute temperature θ is kept in a surrounding having a lower temperature θ_0. The net rate of loss of thermal energy from the body due to radiation is

$$Q_1 = \sigma Ae(\theta^4 - \theta_0^4) \quad \ldots(i)$$

If temperature difference $(\theta - \theta_0)$ is small compared to θ and θ_0 then

$$\theta - \theta_0 = \Delta\theta$$

and

$$\theta = \theta_0 + \Delta\theta$$

Substituting in (*i*), we have

$$Q_1 = \sigma Ae\left\{(\theta_0 + \Delta\theta)^4 - \theta_0^4\right\}$$

$$= \sigma Ae\left\{\theta_0^4\left(1 + \frac{\Delta\theta}{\theta_0}\right)^4 - \theta_0^4\right\}$$

Since $\frac{\Delta\theta}{\theta_0}$ is very small in comparison to 1 (i.e. $\frac{\Delta\theta}{\theta_0} << 1$) hence its higher powers $\left(\frac{\Delta\theta}{\theta_0}\right)^2, \left(\frac{\Delta\theta}{\theta_0}\right)^3$, etc. are still smaller as such neglecting them and using binomial theorem, we have

$$Q_1 = \sigma Ae\left\{\theta_0^4\left(1 + \frac{4\,\Delta\theta}{\theta_0}\right) - \theta_0^4\right\}$$

$$= \sigma Ae\left(\theta_0^4 + 4\Delta\theta\,\theta_0^3 - \theta_0^4\right)$$

$$= 4\sigma Ae\,\theta_0^3\,\Delta\theta$$

$$= 4\sigma Ae\,\theta_0^3\,(\theta - \theta_0)$$

or

$$Q_1 = K_1 A(\theta - \theta_0) \quad \ldots(ii)$$

where $K_1 = 4\sigma e\theta_0^3$ is a constant. That is the rate of loss of thermal energy from body is proportional to temperature difference.

The body may also loose thermal energy due to convection in the surrounding air as discussed in section (17.5). For small temperature differences, the rate of loss of heat due to convection is

$$Q_2 = Ah(\theta - \theta_0) \qquad \text{(equation 17.9)}$$

i.e. the rate of loss of heat due to convection is also proportional to temperature difference.

The net rate of loss of thermal energy due to convection and radiation is

$$Q = Q_1 + Q_2 = K_1A(\theta - \theta_0) + Ah(\theta - \theta_0)$$
$$= A(K_1 + h)(\theta - \theta_0)$$
$$= KA(\theta - \theta_0) \qquad \text{...(17.16)}$$

where constant $K = (K_1 + h)$

If m is mass of the body and s its specific heat capacity, then rate of loss of thermal energy

$$Q = ms\left(-\frac{d\theta}{dt}\right)$$

negative sign here implies decrease of temperature with time.

Thus the rate of fall of temperature for small temperature differences between a body and its surroundings is

$$\frac{d\theta}{dt} = -\frac{Q}{ms} = -\frac{KA}{ms}(\theta - \theta_0)$$

or
$$\frac{d\theta}{dt} = -CA(\theta - \theta_0) \qquad \text{...(17.17)}$$

where $C = \dfrac{K}{ms}$ a new constant

Thus for small temperature differences between a body and its surroundings, the rate of cooling of the body is directly proportional to the temperature difference and exposed surface area (A). This is known as Newton's law of cooling.

17.12 SPECTRAL DISTRIBUTION OF ENERGY IN BLACK BODY RADIATIONS

Nature of Thermal Radiations

Thermal radiations emitted by black body at any temperature are electromagnetic waves like light. As such thermal radiations and light have identical properties.

Some of the common properties of the two are as follows:

(*i*) Thermal radiations, like light can travel through vacuum.
(*ii*) Thermal radiations travel in a straight line.
(*iii*) Thermal radiations travel with the velocity of light as such radiation is the fastest mode of heat transfer from one place to the other.
(*iv*) Thermal radiations obey inverse square law like light.
(*v*) Thermal radiations obey laws of reflection.
(*vi*) Thermal radiations suffer refraction and obey laws of refraction.
(*vii*) Thermal radiations exhibit the phenomenon of interference and diffraction.
(*viii*) Thermal radiations can also be polarised as light waves.

All these properties for thermal radiations thus provide ample evidence of its essential identity with visible light, both being transverse waves in nature, differing only in their wavelengths.

Spectral Distribution of Thermal Radiations

The important feature about the thermal radiations at any temperature is that it is not one (or a few) wavelengths but has continuous spectrum from the small to the large wavelengths. The energy contents of radiation, however, varies for different wavelengths.

The systematic study of infrared spectrum was made by Langley. He studied the radiation energy emitted by the black body at different temperatures and found out the energy emitted corresponding to each wavelength. Figure 17.5 gives the experimental plot of I (intensity of radiation) versus λ (wavelength). The relative importance of different wavelengths in a thermal radiation can be studied qualitatively from this graph. The features of the graph are:

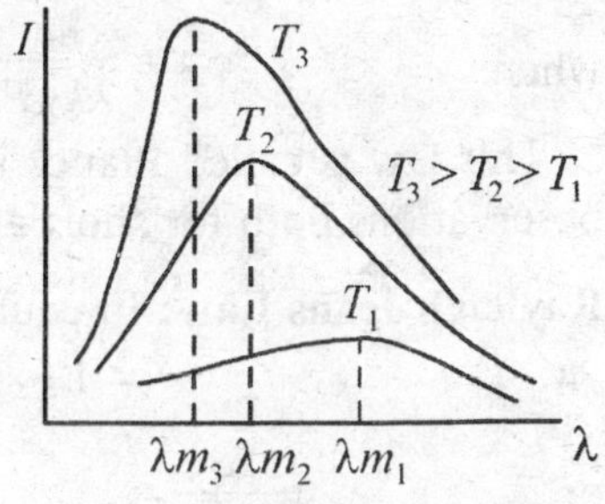

Fig. 17.5

(*i*) As wavelength decreases the intensity goes on increasing and becomes maximum for a particular wavelength λ_m and then decreases.

(*ii*) As the temperature increases, the intensity corresponding to each wavelength increases and the wavelength λ_m for which radiation energy is maximum decreases with increase in temperature. In fact, this wavelength (λ_m) is inversely proportional to the absolute temperature of radiating body. The relation between λ_m and T is given by what is called as *Wien's displacement law*:

$$\lambda_m T = \text{Constant} \quad ...(17.18)$$

Here constant is called *Wien's constant* and is represented by latter '*b*'. Thus

$$\lambda_m T = b \quad ...(17.18a)$$

For black body, the value of constant $b = 2.9 \times 10^{-2}$ mK.

(*iii*) Area under the I–λ curve is proportional to 4th power of the absolute temperature, verifying the correctness of the Stefan Boltzmann's law.

(*iv*) The maximum energy emitted (corresponding to wavelength λ_m) I_m is directly proportional to 5th power of the absolute temperature, i.e.

$$\lambda_m \propto T^5$$

or $$E_m = BT^5 \quad ...(17.19)$$

where B is a constant.

The most significant feature of the black body radiation curves is that they are *universal.* They depend only on temperature and not on the size, shape or material of the black body.

Attempts to explain black body radiation, theoretically on the basis of classical physics had predicted a very different and unrealistic wavelength distribution.

Planck's Law: Max Planck explained the distribution of energy in the spectrum of black body radiation on the basis of a novel consideration which developed into the modern quantum theory. He proposed that the exchange of electromagnetic radiation energy is not continuous but discrete. Exchange takes place (absorption or emission) in discrete bundles or packets of energy called quanta. Latter on, these quanta's of energy were called *photons.* The energy associated with a photon is given by

$$E = h\nu \quad ...(17.20)$$

where ν is the frequency of radiation and h is Planck's constant ($h = 6.6 \times 10^{-34}$ J-s).

On the basis of quantum theory of radiation Planck has shown that energy distribution in the spectrum of black body radiation is given by

$$E_\lambda = \frac{8\pi k_B T}{\lambda^4} \frac{x}{e^x - 1} \quad ...(17.21)$$

where $$x = \frac{hc}{\lambda k_B T}$$

This law is called Planck's law for radiation and is in excellent agreement with experimental observations both for short and long wavelengths.

Rayleigh Jeans Law: In equation (17.21) if x is small or λ is large, i.e. for long wavelengths

$$e^x = 1 + x \qquad \text{neglecting higher powers of } x.$$

or $$\frac{x}{e^x - 1} = 1$$

i.e. $$E_\lambda = \frac{8\pi k_B T}{\lambda^4} \qquad \text{...(17.22)}$$

This is called Rayleigh Jeans law which is true for long wavelengths.

Wien's Formula: If x is large or λ is small, i.e. for short wavelengths e^x is very large and 1 can be neglected. Thus

$$E_\lambda = \frac{8\pi k_B T}{\lambda^4}\frac{x}{e^x} = \frac{8\pi k_B T}{\lambda^4}\frac{hc}{\lambda k_B T}\frac{1}{e^{\frac{hc}{\lambda k_B T}}}$$

$$= \frac{8\pi hc}{\lambda^5} e^{-\frac{hc}{\lambda k_B T}} \qquad \text{...(17.23)}$$

This is called *Wien's formula* and is valid for short wavelengths.

This clearly shows that both Wein's and Rayleigh-Jean's formulae are in fact, special cases of Planck's formula, which thus truely is the correct formula for thermal radiation.

17.13 SOLAR CONSTANT AND TEMPERATURE OF THE SUN

The sun emits radiant energy continuously in space of which an insignificant part reaches the earth. But even of this incoming radiations, a considerable portion is lost by reflection and scattering by the terrestrial atmosphere and is sent back to the interstaller space. The best reflecting constituents of the atmosphere are water, snow and cloud. The scattering is partly due to the dust particles and partly due to the air molecules which is generally small. Further the radiation is heavily absorbed by earth's atmosphere. The total absorption varying from 20 to 40% depending upon time of the day and the season of the year and different parts of the spectrum are absorbed to a different degree.

Solar constant (S_0) is defined to be *radiant power that a unit area of the earth would receive in absence of the atmosphere when placed at right angles to the incident radiation at a distance equal to the mean distance of the earth from the sun.* Its measured value is 1340 Watt/m^2. Using this value of solar constant, the temperature of surface of the sun and earth can be easily estimated.

Temperature of the Surface of Sun

The energy radiated per second by the sun with its surface temperature T is given by Stefan-Boltzmann's law (assumed to be black body)

$$Q = \sigma 4\pi R_s^2 T^4 \qquad \text{...}(i)$$

where R_s is the radius of the sun.

If r is the mean distance of earth from the sun, the radiant power received by unit area of the earth is

$$S_0 = \frac{Q}{4\pi r^2} = \frac{\sigma\, 4\pi R_s^2 T^4}{4\pi r^2}$$

or solar constant $S_0 = \sigma \left(\frac{R_S}{r}\right)^2 T^4$ ...(17.24)

and temperature of the sun is

$$T = \left(\frac{r^2 S_0}{\sigma R_S^2}\right)^{\frac{1}{4}} \qquad ...(17.25)$$

Using known data: $R_s = 7 \times 10^8$ m, $r = 1.5 \times 10^{11}$ m, $\sigma = 5.67 \times 10^{-8}$ Watt/m^2 (K)^4 and $S_0 = 1340$ Watt/m^2. The temperature of the sun comes out to be

$$T = \left[\frac{\left(1.5 \times 10^{11}\right)^2 \times 1340}{5.67 \times 10^{-8} \times \left(7 \times 10^8\right)^2}\right]^{\frac{1}{4}} = 5740 \text{ K}$$

Temperature of Earth: If R_e is radius of earth and its temperature T. Then radiant energy radiated by the earth per second (assumed to be black body) is

$$Q = 4\pi R_e^2 \sigma T^4 \qquad ...(i)$$

The Sun's radiations fall only on one side of earth and that they are not normal. If we calculate normal area it comes out to be πR_e^2. Hence radiations absorbed by the earth in 1 sec is

$$Q = \pi R_e^2 S_0 \qquad ...(ii)$$

In equilibrium the radiant energy received by earth and radiated by the earth in unit time are equal. Equating (*i*) and (*ii*), we have

$$4\pi R_e^2 \sigma T^4 = \pi R_e^2 S_0$$

or $$T = \left(\frac{S_0}{4\sigma}\right)^{\frac{1}{4}} = \left(\frac{1340}{4 \times 5.67 \times 10^{-8}}\right)^{\frac{1}{4}}$$

$$= 278 \text{ K.}$$

SOLVED EXAMPLES

Example 17.1: Three rods of material X and three rods of material Y are connected as shown in the Fig. 17.6. All the rods are of identical length and cross-sectional area. If the end A is maintained at 60° C and the junction E at 10° C, calculate the temperature of the junction B, C and D. The thermal conductivities of X and Y are 0.92 and 0.46 SI units.

Solution: We know that total heat entering at a junction is equal to total heat leaving that junction. If T_1, T_2 and T_3 are the temperatures at B, C and D respectively. Writing the equations at these junctions, we have

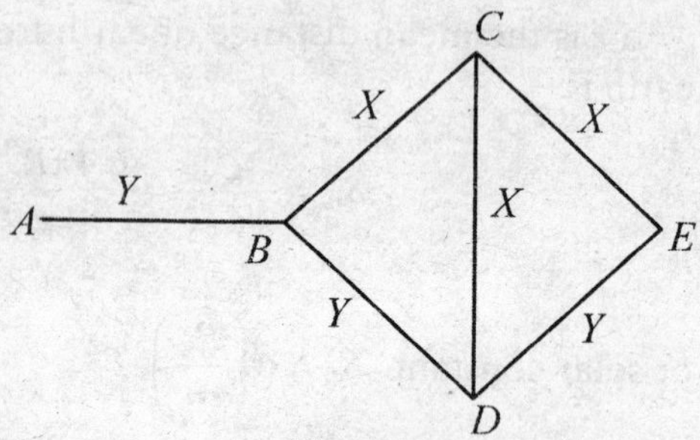

Fig. 17.6

At B

$$K_Y(60 - T_1) = K_X(T_1 - T_2) + K_Y(T_1 - T_3)$$

But $\quad K_X = 2K_Y$

or $\quad 60 - T_1 = 2(T_1 - T_2) + (T_1 - T_3)$

or $\quad 4T_1 - 2T_2 - T_3 = 60 \qquad (i)$

At C

$$K_X(T_1 - T_2) = K_X(T_2 - 10) + K_X(T_2 - T_3)$$

or $\quad T_1 - 3T_2 + T_3 = -10 \qquad (ii)$

At D

$$K_Y(T_1 - T_3) = K_X(T_3 - T_2) + K_Y(T_3 - 10)$$

or $\quad T_1 - T_3 = 2(T_3 - T_2) + (T_3 - 10)$

$$T_1 + 2T_2 - 4T_3 = -10 \qquad ...(iii)$$

From (*ii*) and (*iii*), we have $T_2 = T_3$

Now (*i*) becomes.

$$4T_1 - 3T_2 = -60 \qquad ...(iv)$$

and (*ii*) becomes

$$T_1 - 2T_2 = -10 \qquad ...(v)$$

Solving, we have

$$T_2 = 20° \text{ C and } T_1 = 30° \text{ C}$$

Thus $\quad T_2 = T_3 = 20°$ C and $T_1 = 30°$ C

Example 17.2: Show that the increase in thickness of ice on a pond from initial value x_0 to final value x is given by

$$x^2 - x_0^2 = \frac{2K\theta}{L\rho} t$$

where L is latent heat of fusion, ρ density, K thermal conductivity of ice and ($-\theta°$ C) is the temperature of air.

Solution: Consider area A and instantaneous thickness x of ice. To increase thickness of ice by dx, the heat to be extracted is

$$Q = \text{Mass} \times \text{Latent heat} = A\,dx\,\rho L = A\rho L\,dx \qquad ...(i)$$

where ρ is density of ice.

Heat conducted in time dt is

$$\text{Heat conducted} = \frac{KA(\theta_2 - \theta_1)}{x} dt$$

Ice temperature $\theta_2 = 0$ and atmospheric temperature ($-\theta$)

Hence heat conducted $= \dfrac{KA[(0)-(-\theta)]dt}{x}$

$$= \frac{KAQ\,dt}{x} \qquad ...(ii)$$

Equating (*i*) and (*ii*), we get

$$A\rho L dx = \frac{KA\theta\, dt}{x}$$

or $$\frac{\rho L}{K\theta} x\, dx = dt$$

Integrating between the limits $x = x_0$ to $x = x$, we get

$$\int_{x_0}^{x} \frac{\rho L}{K\theta} x\, dx = \int_0^t dt$$

or $$\frac{\rho L}{2K}\left(x^2 - x_0^2\right) = t$$

or $$x^2 - x_0^2 = \frac{2K\theta}{\rho L} t\,.$$

Example 17.3: Two heat reservoirs P and Q at temperatures 100° C and 35° C respectively are connected by rods as shown in Fig. 17.7. The length, area of cross-section and thermal conductivities of rods 1 and 2 are L, s and σ respectively; for rods 3, 4, 5 these are L, $2s$ and 2σ. Calculate the temperatures at B, C, and D.

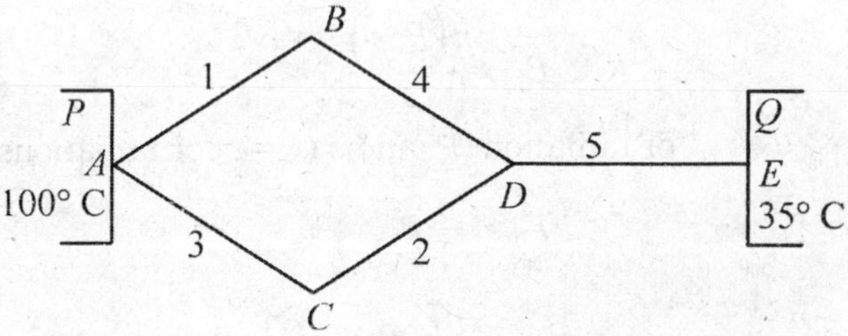

Fig. 17.7

Solution: In thermal conductivity the quantity

$$\frac{l}{\sigma s} = \frac{\text{Length}}{\text{Conductivity} \times \text{area of cross-section}} \text{ is called}$$

thermal resistance. If R is the thermal resistance of each rod 1 and 2 then 3, 5, 5 has thermal resistance $R/4$. Thus total thermal resistance from A to D is

$$X_1 = \frac{1}{2}\cdot\frac{5R}{4} = \frac{5R}{8} \qquad ...(i)$$

and thermal resistançe from D to E $X_2 = \dfrac{R}{4} = \dfrac{2R}{8}$

Thus the fall in temperature from A to D, i.e. from 100° C to 35° C is distributed as 5/7 over AD and 2/7 over DE. That gives

$$\text{Temperature at } D = 100 - \frac{5}{7}(65) = 53.6°\text{C}$$

over AB and BD the falls are in the ratio of 4/5 to 1/5.

$$\text{Temperature at } B = 100 - \frac{4}{5}(46.4) = 62.9°\text{C}$$

$$\text{Similarly temperature at } C = 100 - \frac{1}{5}(46.4) = 90.7°\text{C.}$$

Example 17.4:

(*a*) Consider the radiation trapped in a rectangular box of volume V. From kinetic theory consideration show that $PV = U/3$. Where P is the radiation pressure and U the total energy of radiation.

(*b*) Find the law governing adiabatic change of this radiation gas.

(*c*) Consider the radiation gas to be trapped in a sphere with perfectly reflecting walls. From an isotropic adiabatic expansion of the gas, show that the product λT is constant, Ignore the distribution of wavelengths for thermal radiation in cavity.

Solution:

(*a*) Let the radiation to be incident on a wall in X-Z plane. Let momentum of radiation be $\vec{P}$. The X-component changes due to reflection and imparts force. Change in momentum = $2P_x$. The number of collisions made per second on unit area of the wall is $n v_X$, where v_X is the X-component of velocity and n the number density.

Thus momentum transfer per unit area per unit time is

$$= 2n\, P_X v_X$$

and this is pressure.

Since the radiation has random distribution of momentum. Hence pressure P_0 is

$$P_0 = \frac{n\left(\vec{P}\cdot\vec{v}\right)}{3}$$

For radiation $\vec{P}$ and c ($v = c$ for radiations) are in same direction

$$\therefore \quad P_0 = \frac{n}{3} Pc$$

$$\text{or} \quad P_0 V = \frac{nV}{3} Pc = \frac{NPc}{3}$$

NPc is the total energy of the radiation.

$$P_0 V = \frac{U}{3}$$

(*b*) In the case of adiabatic change $P_0 dV = -dU$.

dU is the change in energy.

In present case $P_0 V = \dfrac{U}{3}$

$$\text{or} \quad P_0\, dV + V dP_0 = \frac{1}{3} dU$$

$$\text{or} \quad V d P_0 + \frac{4}{3} P_0\, dV = 0$$

$$\text{or} \quad \frac{dP_0}{P} + \frac{4}{3}\frac{dV}{V} = 0$$

Integration gives $PV^{4/3}$ = Constant or $P_0^3 V^4$ = Constant.

(*c*) Due to Doppler effect a ray suffers change in wavelength. Since surface is moving hence the changed wavelength is

$$d\lambda = \frac{2\lambda}{c}\frac{dr}{dt}$$

$\frac{dr}{dt}$ is the velocity of expansion of the surface. The reflection frequency is $\frac{c}{2r}$. Hence rate of change of wavelength is

$$\frac{d\lambda}{dt} = \frac{2\lambda}{c}\frac{dr}{dt} \times \frac{c}{2r} = \frac{\lambda}{r}\frac{dr}{dt}$$

or $$\frac{d\lambda}{\lambda} = \frac{dr}{r}$$

Integration gives $\frac{\lambda}{r}$ = Constant

But $P_0^3 V^4$ = constant

or $(P_0 V)^3 V$ = Constant

But $P_0 V \propto U$, U is the total internal energy when radiation is treated as a gas.

$\therefore$ $U^3 V$ = Constant

$V \propto r^3$ and $r \propto \lambda$ $\therefore$ $V \propto \lambda^3$

and $U \propto T$, where T is temperature.

This result comes from kinetic consideration.

$\therefore$ $T^3\lambda^3$ = Constant

or λT = Constant

Example 17.5: A solid body X of heat capacity c is kept in an atmosphere whose temperature T_A = 300 K. At time $t = 0$ the temperature of X is T_0 = 400 K. It cools according to Newton's law of cooling. At time t_1 its temperature is found to be 350 K.

At this time t_1 body X is connected to a large box Y at atmospheric temperature T_A through a conducting rod of length l and thermal conductivity K. The heat capacity of Y is so large that any variation in its temperature may be neglected. The cross-sectional area A of the conducting rod is small compared to surface area of X. Find the temperature of X at time $t = 3t_1$.

Solution: Initially the body cools according to Newton's law of cooling, i.e.

$$\frac{dQ}{dt} \propto \text{(temperature difference)}$$

But $$\frac{dQ}{dt} \propto \frac{dT}{dt}, \text{ i.e. rate of fall of temperature}$$

$\therefore$ $$\frac{dT}{dt} = -K'(T - T_A)$$

where K' is constant of proportionality and T the temperature of the body at any time t.

or $$\frac{dT}{T - T_A} = -K'\,dt$$

and integration gives

$$\log(T - T_A) = -K't + C$$

where C is constant of integration.

At $t = 0$, $T = 400$ K and $T_A = 300$ K gives

$$\log(400 - 300) = C \text{ or } C = \log 100$$

$$\therefore \quad \log\frac{(T - T_A)}{100} = -K't \qquad \ldots(i)$$

But at $t - t_1$ the temperature of body is $T = 350$ K

$$\therefore \quad \log\frac{50}{100} = K't_1 \text{ or } K' = \frac{\log 2}{t_1}$$

Equation (*i*) gives

$$T - T_A = 100\,e^{-K't} = 100\,e^{-t\log 2/t_1}$$

Thus temperature at any time t is

$$T = 300 + 100e^{-(\log 2)t/t_1}$$

$$= 300 + 100\exp\left(-\frac{t}{t_1}\log 2\right)$$

Now the body X is connected to body Y of infinite thermal capacity by means of a rod of length l. The body X now lose heat (*i*) by radiation as before and (*ii*) through conduction. If temperature at any time be T, then, heat lost in unit time through conduction is

$$Q_1 = -\frac{KA}{lc}(T - 300), \text{ where } c \text{ in thermal capacity.}$$

and through radiation

$$Q_2 = -K'(T - 300)$$

Net rate of loss of heat energy = $Q_1 + Q_2$. If rate of fall of temperature is $\frac{dT}{dt}$, then

$$\frac{dT}{dt} = -\frac{KA}{lc}(T - 300) - K'(T - 300)$$

$$= -(T - 300)\left(K' + \frac{KA}{lc}\right) \qquad K' = \frac{\log 2}{t_1}$$

or $$\frac{dT}{T - 300} = -\left(\frac{\log 2}{t_1} + \frac{KA}{lc}\right)dt$$

Integration gives

$$\log(T - 300) = -\left(\frac{\log 2}{t_1} + \frac{KA}{lc}\right)t + C', \qquad C' \text{ is constant of integration.}$$

When $t = 0$, $T = 350$ K

$$\therefore \quad \log 50 = C'$$

or $$\log\frac{(T - 300)}{50} = -\left(\frac{\log 2}{t_1} + \frac{KA}{lc}\right)t$$

or $$T = 300 + 50\exp\left[-\left(\frac{\log 2}{t_1} + \frac{KA}{lc}\right)t\right]$$

Hence temperature after time $3t_1$, i.e. after time $2t_1$ after conduction starts is

$$T = 300 + 50\exp\left[-2\left(\frac{\log 2}{t_1} + \frac{KA}{lc}\right)t_1\right]$$

Example 17.6: A point source of heat of power P is placed at the centre of a spherical shell of mean radius R. The material of the shell has thermal conductivity K. Calculate the thickness of the shell if temperature difference between the outer and inner surfaces of the shell in steady state is T.

Solution: Consider a spherical shell of radius r and $r + dr$ (Fig. 17.8) having temperature gradient $-\frac{d\theta}{dr}$. Negative sign gives that the temperature decreases as r increases.

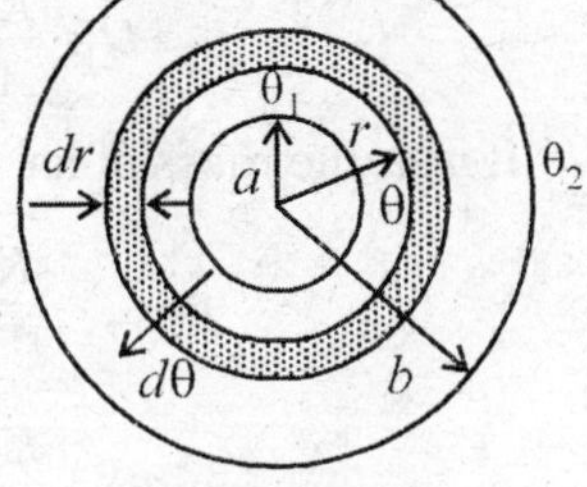

Fig. 17.8

The radial rate of flow of heat through this shell in steady state is

$$H = \frac{dQ}{dt} = -KA\frac{d\theta}{dr} \qquad \text{but } A = 4\pi r^2$$

or
$$H = -K \cdot 4\pi r^2 \cdot \frac{d\theta}{dr}$$

or
$$\frac{dr}{r^2} = -\frac{4\pi K}{H}d\theta$$

On integration, we get

$$\int_a^b \frac{dr}{r^2} = -\frac{4\pi K}{H}\int_{\theta_1}^{\theta_2} d\theta$$

or
$$-\left[\frac{1}{r}\right]_a^b = -\frac{4\pi K}{H}(\theta_2 - \theta_1)$$

or
$$\left[\frac{1}{b} - \frac{1}{a}\right] = \frac{4\pi K}{H}(\theta_2 - \theta_1)$$

and
$$H = \frac{4\pi K \cdot ab(\theta_1 - \theta_2)}{b - a} \qquad \ldots(i)$$

In steady state, no heat is absorbed, i.e., whatever energy is given by the source is all conducted away, i.e. $H = P$. Hence

$$P = \frac{4\pi K \cdot ab(\theta_1 - \theta_2)}{b - a}$$

Here $\theta_1 - \theta_2 = T$ and $a \approx b = R$

$$\therefore \qquad P = \frac{4\pi K \cdot R^2 T}{b - a}$$

or
$$b - a = \frac{4\pi K\, R^2 T}{P}$$

Example 17.7: One end of a copper rod of uniform cross-section and length 1.5 m is kept in contact with ice and the other end with water at 100° C. At what point along its length should a temperature of 200° C be maintained so that in steady state the mass of ice melted be equal to that of the steam produced in same interval of time? Assume that the whole system is insulated from the surroundings. Latent heat of fusion of ice and vaporisation of water are 80 cal/gm and 540 cal/gm respectively.

Solution: Let the point to be maintained at 200° C be at distance x from water at 100° C end.

The temperature gradient towards ice end is

$$\text{Temperature gradient} = \frac{200-0}{1.5-x} = \frac{200}{1.5-x}$$

Hence heat conducted towards ice in time t sec is

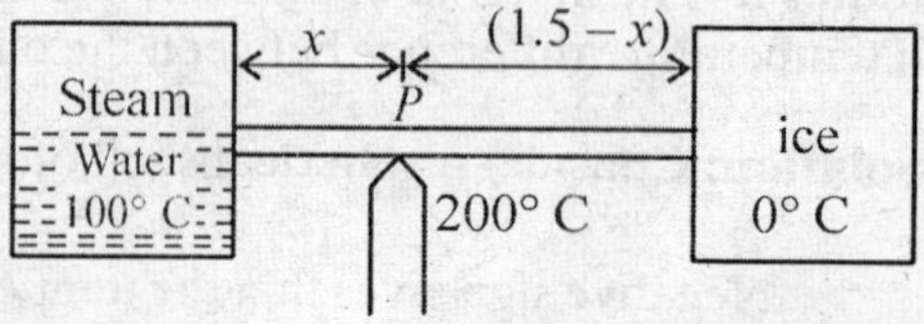

Fig. 17.9

$$Q = \frac{KA\cdot 200}{1.5-x}t$$

If m_i is the mass of ice melted then $Q = m_i L_i$

$$\therefore \quad m_i L_i = \frac{KA200}{1.5-x}t$$

$$\therefore \quad m_i = \frac{K\cdot A\times 200}{L_i(1.5-x)}t \quad \ldots(i)$$

The temperature gradient towards water end is

$$= \frac{200-100}{x} = \frac{100}{x}$$

Hence heat conducted towards water end in t sec

$$Q' = \frac{KA(100)}{x}t$$

If m_s is mass of steam formed then $Q' = m_s L_s$

$$\text{or} \quad m_s = \frac{KA(100)}{L_s\cdot x}t \quad \ldots(ii)$$

Since mass of ice melted and mass of steam formed are equal, i.e.

$$m_i = m_s$$

Then, we have

$$\frac{K\cdot A\times 200}{L_i(1.5-x)}t = \frac{KA(100)}{L_s\, x}t$$

$$\text{or} \quad \frac{200}{80(1.5-x)} = \frac{100}{540\cdot x}$$

$$\text{or} \quad 54x = 4(1.5-x)$$

$$\text{or} \quad 58x = 6$$

$$\text{or} \quad x = \frac{6}{58} = 0.1034 \text{ m} = 10.34 \text{ cm}$$

Example 17.8: Three cylindrical rods *A*, *B* and *C* of equal lengths and of equal diameters are joined in series as shown in Fig. 17.10. Their thermal conductivities are $2K$, K and $0.5K$ respectively. In steady state, if the free ends of rods *A* and *C* are at 100° C and 0° C respectively, calculate the temperature at the two junction points. Assume negligible loss through the curved surface. What will be the equivalent conductivity?

Solution: Let the temperatures at the junctions of rods *A* and *B* be T_1 and at the junctions of rods *B* and *C* be T_2.

Fig. 17.10

Temperature gradient for rod *A* is $\dfrac{100 - T_1}{L}$

Temperature gradient for rod *B* is $\dfrac{T_1 - T_2}{L}$

and that for rod *C* is $\dfrac{T_2 - 0}{L} = \dfrac{T_2}{L}$

Rate of flow of heat through rod *A* is

$$\left(\frac{dQ}{dt}\right)_A = \frac{K_A A(100 - T_1)}{L} = \frac{2KA(100 - T_1)}{L} \qquad \ldots(i)$$

Rate of flow of heat through rod *B* is

$$\left(\frac{dQ}{dt}\right)_B = \frac{K_B A(T_1 - T_2)}{L} = \frac{KL(T_1 - T_2)}{L} \qquad \ldots(ii)$$

Rate of flow of heat through rod *C* is

$$\left(\frac{dQ}{dt}\right)_C = \frac{K_C A T_2}{L} = \frac{0.5 KAT_2}{\mathrm{L}} \qquad \ldots(iii)$$

Rate of flow of heat through all the rods is same.

i.e. $$\left(\frac{dQ}{dt}\right)_A = \left(\frac{dQ}{dt}\right)_B = \left(\frac{dQ}{dt}\right)_C$$

(*i*) and (*iii*) gives

$$\frac{2\,KA(100 - T_1)}{L} = \frac{0.5\,KA\,T_2}{L}$$

or $$4T_1 + T_2 = 400 \qquad \ldots(iv)$$

(*ii*) and (*iii*) gives

$$\frac{KA(T_1 - T_2)}{L} = \frac{0.5\,KA\,T_2}{L}$$

or $$2T_1 = 3T_2 \text{ or } T_1 = \frac{3}{2}T_2$$

Substituting value of T_1 in (*iv*) we have

$$6T_2 + T_2 = 400$$

or $$T_2 = \frac{400}{7} = 57.1°\mathrm{C}$$

and $$T_1 = \frac{3}{2}T_2 = \frac{400}{7} \times \frac{3}{2} = \frac{600}{7} = 85.7°\mathrm{C}$$

Thermal resistance of rod A is $R_A = \dfrac{L}{K_A A} = \dfrac{L}{2KA}$

Thermal resistance of rod B is $R_B = \dfrac{L}{K_B A} = \dfrac{L}{\cdot 5KA} = \dfrac{2L}{KA}$

and Thermal resistance of rod C is $R_C = \dfrac{L}{K_C A} = \dfrac{L}{KA}$

Equivalent thermal resistance is

$$R = R_A + R_B + R_C = \frac{L}{2KA} + \frac{2L}{KA} + \frac{L}{KA}$$

$$= \frac{L}{KA}\left[\frac{1}{2} + 2 + 1\right] = \frac{7L}{2KA}$$

Since total length of rod is $3L$, we have

$$R = \frac{7}{6K} \cdot \frac{(3L)}{KA}$$

or equivalent conductivity is $\dfrac{6K}{7}$.

Example 17.9: Consider a cubical vessel of edge a having a small hole in one of its walls. The total thermal resistance of the walls is r. At time $t = 0$, it contains air at atmospheric pressure P_0 and temperature T_0. The temperature of the surroundings air is T_a ($> T_0$). Find the amount of the gas (in moles) in the vessel at time t. Take C_V of air to be $5R/2$.

Solution: Since there is a hole in the vessel means the gas leaks at atmospheric pressure on receiving heat from atmosphere. Let n (in moles) be the amount of the gas in the vessel at any time t and let the gas receives ΔQ amount of heat in time dt thereby its temperature increases by dT. Then

$$\Delta Q = nC_P dT \quad \ldots(i)$$

If temperature of the gas at time t is T then heat conduction rate is

$$\frac{\Delta Q}{dt} = \frac{T_a - T}{r} \quad \ldots(ii)$$

or $\quad (C_P r)\, n\, dT = (T_a - T)dt \quad \ldots(iii)$

We have $\quad P_0 a^3 = nRT \qquad \because$ volume of vessel is a^3

$$ndT + Tdn = 0 \text{ and } ndT = -Tdn$$

and $\quad T = \dfrac{P_0 a^3}{nR} \quad \ldots(iv)$

Using these results in (*iii*), we get

$$-(C_P r)\frac{P_0 a^3}{nR} dn = \left[T_a - \frac{P_0 a^3}{nR}\right]dt$$

$$\text{or}\qquad \frac{dn}{nR\left[T_a - \dfrac{P_a a^3}{nR}\right]} = -\frac{dt}{C_P r P_0 a^3}$$

on integration, we get

$$\int_{n_0}^{n} \frac{dn}{nR\left[T_a - \dfrac{P_0 a^3}{nR}\right]} = -\int_0^t \frac{dt}{C_P r P_0 a^3}$$

where $n_0 = \dfrac{P_0 a^3}{RT_0}$ is the initial mass of the gas (in moles) in the vessel. Then we have

$$\int_{n_0}^{n} \frac{dn}{nRT_a - P_0 a^3} = -\int_0^t \frac{dt}{C_P r P_0 a^3}$$

$$\text{or}\qquad \frac{1}{RT_a}\log\frac{\left(nRT_a - P_0 a^3\right)}{\left(n_0 RT_a - P_0 a^3\right)} = -\frac{t}{C_P r P_0 a^3}$$

$$\text{or}\qquad \left(nRT_a - P_0 a^3\right) = \left(n_0 RT_a - P_0 a^3\right)\exp\left\{-\frac{tRT_a}{C_P r P_0 a^3}\right\}$$

$$\text{or}\qquad nRT_a = P_0 a^3 + P_0 a^3\left(\frac{n_0 RT_a}{P_0 a^3} - 1\right)\exp\left(-\frac{tRT_a}{C_P r P_0 a^3}\right)$$

But $\dfrac{n_0 R}{P_0 a^3} = \dfrac{1}{T_0}$ and $C_P = C_V + R = 7/2R$

$$\therefore\qquad n = \frac{P_0 a^3}{RT_a}\left\{1 + \left(\frac{T_a}{T_0} - 1\right)\exp\left(-\frac{2tT_a}{7RrP_0 a^3}\right)\right\}$$

Example 17.10: A monoatomic ideal gas is contained in a rigid container of volume V with walls of total inner surface area A, thickness x and thermal conductivity K. The gas is at an initial temperature T_0 and pressure P_0. Find the pressure of the gas as a function of time if the temperature of the surroundings is T_S. All temperatures are in Kelvin.

Solution: Here volume of the gas remains constant as such to increase the temperature of the gas through dT, heat required for 1 mole of the gas is

$$\Delta Q = C_V dT \qquad \text{But } C_V = \frac{3}{2}R \text{ monoatomic gas}$$

If T be the temperature of the gas at any time, t then heat received by the gas through conduction in time Δt is

$$\frac{\Delta Q}{dt} = \frac{KA(T_s - T)}{x}$$

or $$\frac{C_V dT}{\Delta t} = \frac{KA(T_s - T)}{x}$$

or $$\frac{dT}{dt} = \frac{2KA}{3Rx}(T_s - T)$$

or $$\frac{dT}{T_s - T} = \frac{2KA}{3Rx}dt$$

Integration gives

$$\int_{T_0}^{T} \frac{dT}{T_s - T} = \int_0^t \frac{2KA}{3Rx}dt$$

or $$\log\frac{T_s - T_0}{T_s - T} = \frac{2KA}{3Rx}t$$

or $$T_s - T = (T_s - T_0)\exp\left(-\frac{2KA}{3Rx}t\right)$$

or $$T = T_s - (T_s - T_0)\exp\left(-\frac{2KA}{3Rx}t\right)$$

But pressure is directly proportional to temperature, i.e.

$$\frac{P}{T} = \frac{P_0}{T_0} \text{ or } T = \frac{P}{P_0}T_0$$

$\therefore$ $$\frac{P}{P_0}T_0 = T_s - (T_s - T_0)\exp\left(-\frac{2KA}{3Rx}t\right)$$

or $$P = \frac{P_0}{T_0}\left\{T_s - (T_s - T_0)\exp\left(-\frac{2KA}{3Rx}t\right)\right\}$$

Example 17.11: A body placed in air is cooled down according to Newton's law of cooling. The rate of decrease of temperature being K times the temperature difference from surroundings. Starting from $t = 0$, find the time in which the body will loose half the maximum heat it can loose.

Solution: If θ is the temperature of the body at any time t and θ_0 that of the surroundings. Then Newton's law of cooling gives

$$\frac{d\theta}{dt} = -K(\theta - \theta_0)$$

or $$\frac{d\theta}{\theta - \theta_0} = -K\,dt$$

and integration gives

$$\int_{\theta_1}^{\theta} \frac{d\theta}{\theta - \theta_0} = -K\int_0^t dt$$ where θ_1 is temperature of body at $t = 0$.

$$\therefore \qquad \log \frac{\theta - \theta_0}{\theta_1 - \theta_0} = -Kt$$

or $$\theta - \theta_0 = (\theta_1 - \theta_0)\, e^{-Kt} \qquad \text{...}(i)$$

The body continues to loose heat till its temperature becomes equal to that of the surroundings. In this time the total heat lost by the body is

$$\Delta Q = ms(\theta_1 - \theta_0)$$

where s is specific heat of body and m its mass. ΔQ thus is the maximum heat that the body can lose. In the time the body loses half this heat, the fall in the temperature of the body then is $\frac{(\theta_1 - \theta_0)}{2}$

Let time taken by the body to lose half of maximum heat be t_1 then temperature after time t_1, of the body is

$$\theta = \theta_1 - \frac{(\theta_1 - \theta_0)}{2} = \frac{\theta_1 + \theta_0}{2}$$

Substituting in (i), we have

$$\frac{\theta_1 + \theta_0}{2} - \theta_0 = (\theta_1 - \theta_0) e^{-Kt_1}$$

or $$\frac{1}{2} = e^{-Kt_1}$$

or $$Kt_1 = \log 2 \text{ and } t_1 = \frac{\log 2}{K}$$

Example 17.12: Two parallel plates A and B are joined together to form a compound plate (Fig. 17.11). The thickness of the plates are 4.0 cm and 2.5 cm respectively and areas of cross-section same equal to 100 cm^2. The thermal conductivities for the plates A and B are $K_A = 200$ W/m^2-°C and $K_B = 400$ W/m^2-°C. The outer surfaces of A and B are maintained at temperatures 100° C and 0° C. Find (a) the rate of heat flow through any cross-section. (b) the temperature at the interface end and (c) the equivalent thermal conductivity of the compound plate.

Solution: If θ be the temperature of the interface then temperature gradient for plate

$$A = \frac{100 - \theta}{4 \times 10^{-2}}$$

and for plate $$B = \frac{\theta}{2.5 \times 10^{-2}}$$

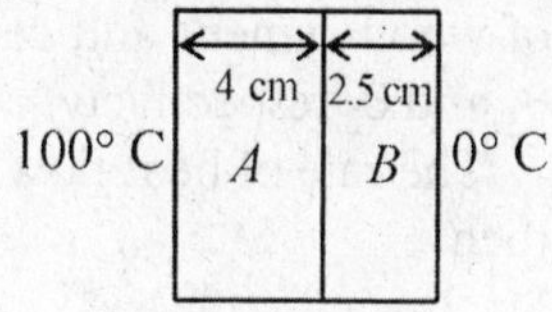

Fig. 17.11

Heat flow rate through A is

$$\left(\frac{\Delta Q}{\Delta t}\right)_A = K_A \text{ A temperature gradient}$$

$$= \frac{400 \times 100 \times 10^{-4} \times (100 - \theta)}{4 \times 10^{-2}}$$

$$= 50(100 - \theta) \qquad \text{...}(i)$$

Heat flow rate through B

$$\left(\frac{\Delta Q}{\Delta t}\right)_B = K_B \text{ A temperature gradient}$$

$$= \frac{400 \times 100 \times 10^{-4}\theta}{2.5 \times 10^{-2}}$$

$$= 160\,\theta \qquad \ldots(ii)$$

Heat flow rate through both of them is same. Hence

$$\left(\frac{\Delta Q}{\Delta t}\right)_A = \left(\frac{\Delta Q}{\Delta t}\right)_B$$

or $\quad 50(100 - \theta) = 160\theta$

or $\quad \theta = 23.8°\text{ C}$

From (*ii*) rate of heat flow

$$\frac{\Delta Q}{\Delta t} = 160\,\theta = 160 \times 23.8 = 3810 \text{ W}$$

The equivalent conductivity is given by

$$K = \frac{\Sigma_i l_i}{\Sigma_i \dfrac{l_i}{k_i}} = \frac{l_a + l_b}{\dfrac{l_a}{k_a} + \dfrac{l_b}{k_b}} = \frac{4 + 2.5}{\dfrac{4}{200} + \dfrac{2.5}{400}}$$

$$= \frac{6.5 \times 400}{10.5} = 247.6 \text{ W/m-°C}$$

Example 17.13: An electric heater is used in a room of total wall area 137 m^2 to maintain a temperature of +20° C inside it when the outside temperature is –10° C. The walls have three layers of different materials. The inner most layer is of wood of thickness 25 cm and the outer most layer is of brick of thickness 25 cm. Whereas the middle layer is of cement of thickness 1 cm. Find power of the electric heater. Assume that there is no heat loss through the floor and ceiling. The thermal conductivities of wood, cement and brick are 0.125, 1.5 and 1.0 Watt/m-0° C.

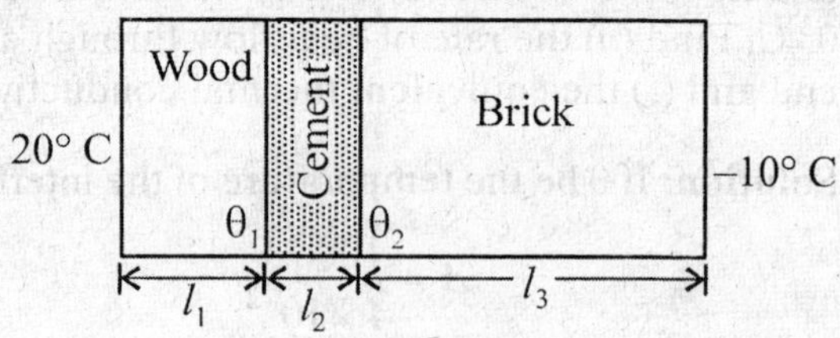

Fig. 17.12

Solution: Let l_1, l_2 and l_3 be the thickness of wood, cement and brick respectively and let temperatures of wood-cement and cement-brick contact face be θ_1 and θ_2 respectively.

The rate of heat flow through all layers is same, then

$$\frac{Q}{t} = \frac{K_1 A(20 - \theta_1)}{l_1} = \frac{K_2 A(\theta_1 - \theta_2)}{l_2} = \frac{K_3 A\{\theta_2 - (-10)\}}{l_3}$$

Substituting the values, we have

$$\frac{0.125 \times A(20 - \theta_1)}{2.5 \times 10^{-2}} = \frac{1.5 \times A(\theta_1 - \theta_2)}{1 \times 10^{-2}}$$

This gives

$$31\theta_1 - 30\theta_2 = 20 \qquad \ldots(i)$$

and $$\frac{K_1A(20-\theta_1)}{l_1} = \frac{K_3A(\theta_2+10)}{l_3}$$

Substituting the values, we have

$$\frac{0.125A(20-\theta_1)}{2.5\times10^{-2}} = \frac{1\times A(\theta_2+10)}{25\times10^{-2}}$$

This gives $5\theta_1 + 4\theta_2 = 60$ ...(*ii*)

Solving for θ_1 and θ_2, we get

$$\theta_1 = 6.86°\text{ C and }\theta_2 = 6.4°\text{ C}$$

Whatever the heat is supplied by the heater is conducted away through the wall. Hence power of the heater is

$$P = \frac{Q}{t} = \frac{K_1A(20-\theta_1)}{l_1}$$

$$= \frac{\cdot125\times137(20-6.86)}{2.5\times10^{-2}}$$

$$= 5\times137\times13.14 = 9000\text{ Watt} = 9\text{ kW}.$$

Example 17.14: A closed cubical box made of perfectly insulating material has walls of thickness 8 cm and the only way for the heat to enter or leave the box is through two solid cylindrical, metallic plugs each of cross-sectional area 12 cm^2 and length 8 cm fixed in the opposite walls of the box as shown in Fig. 17.13. Outer surface *A* is kept at 100° C while the outer surface *B* is kept at 4° C. The thermal conductivity of the material of plugs is 0.5 cal/s cm °C. A source of energy generating 36 cal s^{-1} is enclosed inside the box. Find the equilibrium temperature of the inner surface of the box assuming that it is same at all position on the inner surface.

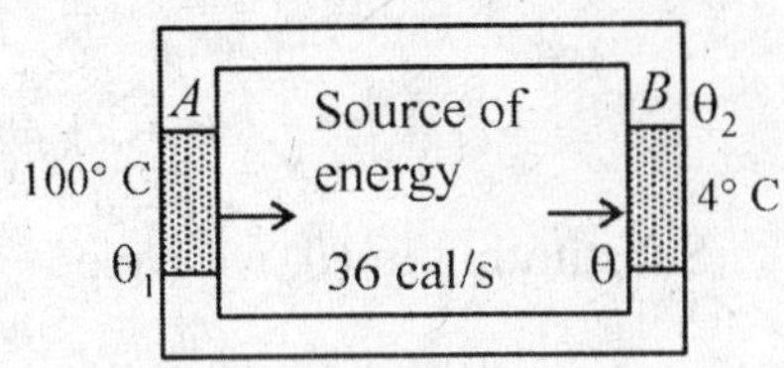

Fig. 17.13

Solution: Let Q_1 is the rate at which heat is conducted into the box from left plug. As such the heat available inside box is

$$Q_2 = Q_1 + 36$$

T_0 maintain constant temperature inside the box, this should be the rate of conduction of heat through right plug *B*.

Let θ be the equilibrium temperature, then

$$Q_1 = \frac{KA(\theta_1-\theta)}{l}$$

and $$Q_1 + 36 = \frac{KA(\theta-\theta_2)}{l}$$

or $$\frac{KA(\theta_1-\theta)}{l} + 36 = \frac{KA(\theta-\theta_2)}{l}$$

Values given are $K = 0.5$ cal/s cm °C, $A = 12$ cm², $l = 8$ cm, $\theta_1 = 100°$ C and $\theta_2 = 4$ °C. Substitution gives

$$\frac{0.5 \times 12(100-\theta)}{8} + 36 = \frac{0.5 \times 12(\theta - 4)}{8}$$

or $$3(100-\theta) + 36 \times 4 = 3(\theta - 4)$$

or $$300 - 3\theta + 144 = 3\theta - 12$$

or $$6\theta = 300 + 144 + 12 = 456$$

or $$\theta = \frac{456}{6} = 76°\text{C}$$

Example 17.15: Find the rate of heat flow through a rod of length l and having circular cross-section. The radius of cross-section varies linearly from a to b. The thermal conductivity of the material of rod is K. Temperatures are θ_1 and θ_2 ($\theta_1 > \theta_2$).

Solution: Consider a cross-section of the rod of small length dx and of radius r at a distance x from the end as shown in Fig. 17.14.

The thermal resistance of the cross-section

$$dR = \frac{dx}{KA} = \frac{dx}{K\pi r^2} \quad \ldots(i)$$

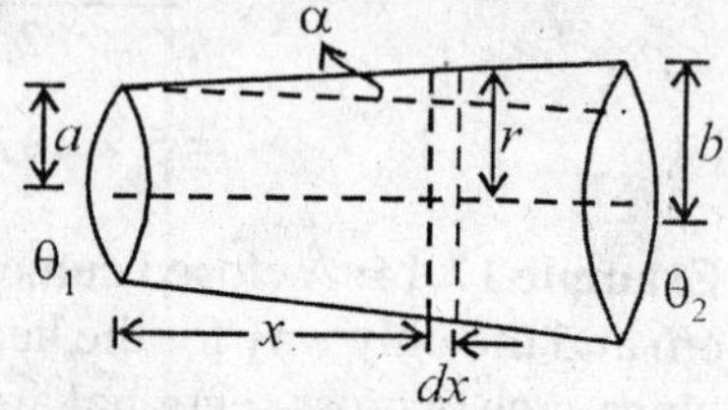

Fig. 17.14

From Fig. 17.14, we have

$$\tan\alpha = \frac{r-a}{x} = \frac{b-a}{l}$$

$$x = \frac{(r-a)}{b-a} l \text{ and } dx = \frac{l\,dr}{b-a}$$

Substituting in (*i*), we have

$$dR = \frac{l}{K\pi} \frac{dr}{(b-a)r^2}$$

Integrating, we get

$$R = \int dR = \int_a^b \frac{l}{K\pi(b-a)} \cdot \frac{dr}{r^2}$$

or $$R = -\frac{l}{K\pi(b-a)}\left[\frac{1}{r}\right]_a^b$$

$$= \frac{l}{K\pi(b-a)} \times \left[\frac{1}{a} - \frac{1}{b}\right] = \frac{l}{K\pi ab}$$

Hence rate of heat flow is

$$Q = \frac{\text{Temperature difference}}{\text{Thermal resistance}}$$

$$= \frac{\theta_1 - \theta_2}{\dfrac{l}{K\pi ab}} = \frac{K\pi ab(\theta_1 - \theta_2)}{l}$$

Example 17.16: Show that the rate of flow of heat through a spherical shell, the inner and outer walls of which have radii r_1 and r_2 and are maintained at different uniform temperatures θ_1 and θ_2 respectively is given by

$$\frac{dQ}{dt} = \frac{4K\, r_1 r_2}{r_2 - r_1}(\theta_1 - \theta)$$

where K is thermal conductivity of material of shell. Hence obtain the expression for temperature distribution.

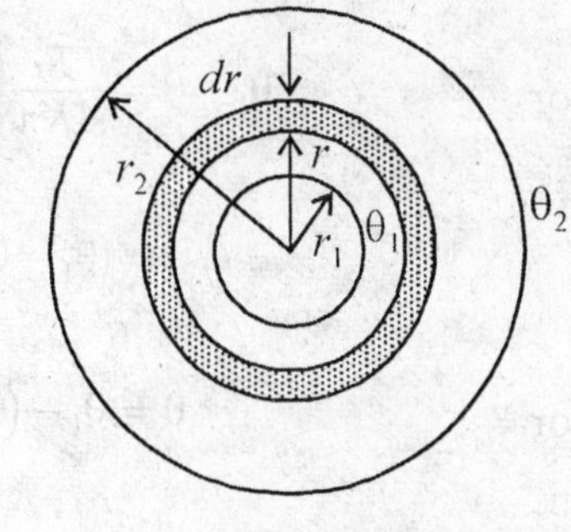

Fig. 17.15

Solution: The shell can be considered to be composed of a large number of spherical shells whose radii vary from r_1 to r_2.

Consider one such shell of radius r and thickness dr and having temperature θ and $\theta + d\theta$. The temperature gradient for this shell is $\frac{d\theta}{dr}$ and area of this shell is $4\pi r^2$

Now the rate of flow of heat through this shell is

$$\frac{dQ}{dt} = -KA\frac{d\theta}{dr} = -K\left(4\pi r^2 \frac{d\theta}{dr}\right)$$

where K is thermal conductivity of the material of shell. In steady state $\frac{dQ}{dt}$ is constant and let it be R.

$$\therefore \quad R = -K(4\pi r^2)\frac{d\theta}{dr}$$

or $$d\theta = -\frac{R \cdot dr}{4\pi K\, r^2} \qquad \ldots(i)$$

Integrating with limits $r = r_1$ to $r = r_2$ and $\theta = \theta_1$ to $\theta = \theta_2$, we get

$$\int_{\theta_1}^{\theta_2} d\theta = -\frac{R}{4\pi K}\int_{r_1}^{r_2} \frac{dr}{r^2}$$

or $$(\theta_2 - \theta_1) = \frac{R}{4\pi K}\left(\frac{1}{r_2} - \frac{1}{r_1}\right)$$

or $$(\theta_1 - \theta_2) = \frac{R}{4\pi K}\frac{(r_2 - r_1)}{r_1 r_2}$$

or $$\frac{dQ}{dt} = R = \frac{4\pi K\, r_1 r_2(\theta_1 - \theta_2)}{r_2 - r_1}$$

Temperature Distribution: Temperature distribution means the variation of temperature with r, the distance from the centre of the shell. Hence integrating equation (*i*) within limits from $r = r_1$ to r and $\theta = \theta_1$ to θ, we get

$$\int_{\theta_1}^{\theta} d\theta = -\int_{r_1}^{r} \frac{R\,dr}{4\pi K r^2}$$

$$\therefore \quad \theta - \theta_1 = \frac{R}{4K\pi}\left[\frac{1}{r} - \frac{1}{r_1}\right]$$

or $$\theta_1 - \theta = \frac{R}{4K\pi}\left(\frac{r - r_1}{r\, r_1}\right) = \frac{r_1 r_2}{r_2 - r_1}(\theta_1 - \theta_2) \times \frac{(r - r_1)}{r\, r_1}$$

$$= (\theta_1 - \theta_2)\frac{r_2 (r - r_1)}{r (r_2 - r_1)}$$

or $$\theta = \theta_1 - (\theta_1 - \theta_2)\frac{r_2 (r - r_1)}{r (r_2 - r_1)}$$

Example 17.17: Show that the radial rate of flow of heat through a uniform cylindrical shell with inner and outer radii r_1 and r_2 respectively is

$$\frac{dQ}{dt} = \frac{2\pi\, Kl\, (\theta_1 - \theta_2)}{\log \frac{r_2}{r_1}}$$

where l is length of the cylinder whose ends are maintained at temperature θ_1 and θ_2, K is the thermal conductivity of the material of the cylinder. Hence obtain the expression for temperature distribution.

Solution: Consider a cylindrical shell of radii r_1 and r_2 and length l whose inner and outer surfaces are maintained at temperature θ_1 and θ_2 so that there is radial flow of heat from axis to the outer surface.

Consider an elementary cylindrical shell of radius r and of radial thickness dr having temperatures θ and $\theta + d\theta$ (Fig. 17.16). The temperature gradient for this elementary shell is $-\frac{d\theta}{dr}$. –ve sign shows that θ decreases with increase of r.

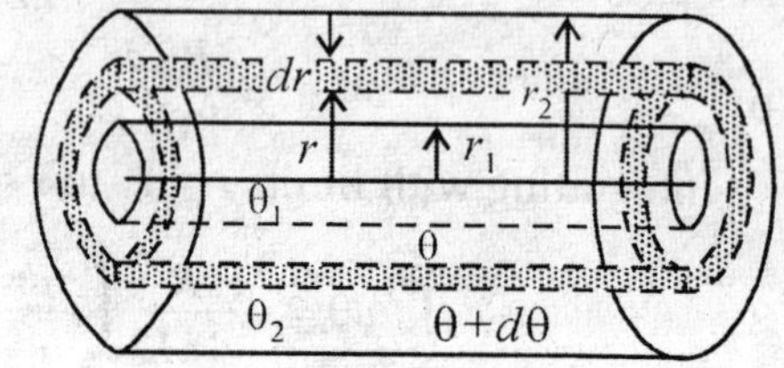

Fig. 17.16

The radial rate of flow of heat through this elementary shell is

$$\frac{dQ}{dt} = -K(2\pi r l)\frac{d\theta}{dr}$$

In steady state $\frac{dQ}{dt}$ is constant and let it be R

thus $$R = -K(2\pi r l)\frac{d\theta}{dr}$$

$$\frac{dr}{r} = -\frac{K\, 2\pi l}{R}\, d\theta \qquad \ldots(i)$$

On integration, we get

$$\int_{r_1}^{r_2} \frac{dr}{r} = -\frac{K\, 2\pi l}{R}\int_{\theta_1}^{\theta_2} d\theta$$

or $$\log \frac{r_2}{r_1} = -\frac{K\, 2\pi l}{R}(\theta_2 - \theta_1) = \frac{K \cdot 2\pi l}{R}(\theta_1 - \theta_2)$$

and radial rate of flow of heat is

$$R = \frac{dQ}{dt} = \frac{2\pi Kl(\theta_1 - \theta_2)}{\log \frac{r_2}{r_1}}$$

In order to obtain temperature distribution, integrate (*i*) within limits for $r = r_1$ to r and for $\theta = \theta_1$ to θ.

Thus we have

$$\int_{r_1}^{r} \frac{dr}{r} = -\frac{K2\pi l}{R}\int_{\theta_1}^{\theta} d\theta$$

or $$\log\frac{r}{r_1} = -\frac{K\,2\pi l}{R}(\theta - \theta_1) = \frac{K\,2\pi l}{R}(\theta_1 - \theta)$$

Substituting the value of R, we have

$$\theta_1 - \theta = \frac{R}{K\,2\pi l}\log\frac{r}{r_1} = \frac{(\theta_1 - \theta_2)}{\log\frac{r_2}{r_1}}\log\frac{r}{r_1}$$

and $$\theta = \theta_1 - \frac{(\theta_1 - \theta_2)}{\log\frac{r_2}{r_1}}\log\frac{r}{r_1}.$$

Example 17.18: A body initially at 80° C cools to 64° C in 5 minutes and to 52° C in 10 minutes. What will be its temperature after 15 minutes. Find also the temperature of the surroundings.

Solution: If a body of water equivalent w at temperature θ cools in a surroundings at temperature θ_0, then according to Newton's law of cooling, we have

$$w\frac{d\theta}{dt} = -K(\theta - \theta_0)$$

where $\frac{d\theta}{dt}$ is the rate of fall of temperature and –ve sign gives indicates that temperature decreases as time increases.

$$\therefore \quad \frac{d\theta}{\theta - \theta_0} = -\frac{K}{w}dt = -A\,dt \qquad \text{where } A = \frac{K}{w} \text{ a new constant}$$

Integration gives $\log(\theta - \theta_0) = -At + C$ ...(*i*)

Given $t = 0$, $\theta = 80°$ C and $t = 5$ min, $\theta = 64°$ C, $t = 10$ min, $\theta = 52°$ C.

Substituting, we have

$$\log(80 - \theta_0) = 0 + C \text{ or } C = \log(80 - \theta_0)$$

Thus (*i*) becomes

$$\log\frac{(\theta - \theta_0)}{(80 - \theta_0)} = -At$$

At time $t = 5$ min

$$\log(64 - \theta_0) - \log(80 - \theta_0) = -5A$$

or $$A = \frac{\log(80-\theta_0) - \log(64-\theta_0)}{5}$$

and At time $t = 10$ min

$$A = \frac{\log(80-\theta_0) - \log(52-\theta_0)}{10}$$

Equating the two values of A, we have

$$\frac{\log(80-\theta_0) - \log(64-\theta_0)}{5} = \frac{\log(80-\theta_0) - \log(52-\theta_0)}{10}$$

or $$\log(80-\theta_0) = \log\frac{(64-\theta_0)^2}{(52-\theta_0)}$$

or $$(80-\theta_0)(52-\theta_0) = (64-\theta_0)^2$$

or $$80 \times 52 - (80+52)\theta_0 + \theta_0^2 = (64)^2 - 2 \times 64\theta_0 + \theta_0^2$$

or $$4\theta_0 = 80 \times 52 - (64)^2 = 4160 - 4096$$

$$= 64$$

or $$\theta_0 = 16° \text{ C}$$

Hence room temperature is 16° C

If θ is the temperature after 15 minutes, then

$$A = \frac{\log(80-\theta_0) - \log(\theta-\theta_0)}{15}$$

Equating the values of A and substituting $\theta_0 = 16°$ C, we have

$$\frac{\log(80-16) - \log(\theta-16)}{15} = \frac{\log(80-16) - \log(64-16)}{5}$$

or $$\log 64 - \log(\theta - 16) = 3\log 64 - 3\log 48$$

or $$3\log 48 - 2\log 64 = \log(\theta - 16)$$

or $$\log\frac{(48)^3}{(64)^2} = \log(\theta - 16)$$

or $$27 = \theta - 16$$

or $$\theta - 16 = 27 \text{ or } \theta = 43° \text{ C}$$

Hence temperature after 15 minutes is 43° C.

Example 17.19: The peak emission from a black body at a certain temperature occurs at a wavelength of 9000 Å. On increasing its temperature the total radiation emitted is increased 81 times. At the initial temperature, when the peak radiation from the black body is incident on a metal surface, it does not cause any photo emission from the surface. After the increase of temperature the peak radiation from the black body caused photo emission. To bring these photoelectrons to rest, a potential equivalent to the excitation energy between $n = 2$ to $n = 3$ Bohr levels of hydrogen atom is required. Find the *work function of the metal.*

Solution: According to Stefan's law, the radiant energy emitted by a black body $E \propto T^4$.

When the energy radiated becomes 81 times i.e. $(3)^4$ times means the temperature of the body becomes 3 times. If initial temperature is T_0, then final temperature is $3T_0$

From Wien's displacement law, $\lambda_m T$ = constant, we get

$$(\lambda_m)_1 T_0 = (\lambda_m)_2 (3T_0)$$

or
$$9000 \times T_0 = (\lambda_m)_2 (3T_0)$$

or
$$(\lambda_m)_2 = \frac{9000}{3} = 3000 \text{ Å}$$

Kinetic energy of emitted photoelectrons = Excitation energy

$$= Rhc\left(\frac{1}{n_1^2} - \frac{1}{n_2^2}\right) = 13.6\left(\frac{1}{2^2} - \frac{1}{3^2}\right) = 1.89 \text{ eV}$$

$$\text{Energy of incident photon} = h\nu = \frac{hc}{\lambda} = \frac{6.62 \times 10^{-34} \times 3 \times 10^8}{3000 \times 10^{-10}}$$

$$= 6.62 \times 10^{-19} \text{ J} \equiv \frac{6.62 \times 10^{-19}}{1.6 \times 10^{-19}} \text{ eV}$$

Energy of incident photon = 4.14 eV.

According to Einstein's photoelectric equation

$$h\nu - \phi = E_K$$

where ν is frequency of incident radiation and $h\nu$ energy of the incident photon, ϕ work function of the metal and E_K, kinetic energy of the photo electrons. Thus

$$4.14 - \phi = 1.89$$

or
$$\phi = 4.14 - 1.89 = 2.25 \text{ eV}.$$

Example 17.20: Two metallic blocks with heat capacity C_1 and C_2 are interconnected by a rod of length L and cross-sectional area S and fairly low conductivity K. The whole system is thermally insulated from the surroundings. At a moment $t = 0$ the temperature difference between the two blocks of metals equals θ_0. Assuming the heat capacity of the rod to be negligible, find temperature difference between the blocks as a function of time.

Fig. 17.17

Solution: In Fig. 17.17, if θ, be the temperature difference between the blocks at anytime t, then rate of heat flow through the rod is

$$Q = \frac{KS\theta}{L}$$

Due to heat flow, the temperature of A decreases and that of B increases,

$$\text{The rate of decrease of temperature of } A = \frac{KS\theta}{LC_1}$$

$$\text{The rate of increase of temperature of } B = \frac{KS\theta}{LC_2}$$

Thus in 1 sec the temperature difference decreases by

$$\frac{d\theta}{dt} = -\frac{KS\theta}{L}\left(\frac{1}{C_1} + \frac{1}{C_2}\right) = -\alpha\theta$$

where $\alpha = \dfrac{KS}{L}\left(\dfrac{1}{C_1}+\dfrac{1}{C_2}\right) = a$ (constant)

or $$\frac{d\theta}{\theta} = -\alpha\, dt$$

and integration gives

$$\log\theta = -\alpha t + C, \quad \text{where } C \text{ is constant of integration.}$$

Initial conditions are, $t = 0$, $\theta = \theta_0$ and hence $C = \log\theta_0$

$$\therefore \quad \log\frac{\theta}{\theta_0} = -\alpha T$$

or $$\theta = \theta_0 e^{-\alpha t} \text{ with } \alpha = \frac{KS}{L}\left(\frac{1}{C_1}+\frac{1}{C_2}\right)$$

Example 17.21: A rod of length L with sides fully insulated is of a material whose conductivity K varies with temperature as $K = \alpha/T$, where α is a constant. The ends of the rod are kept at temperature T_1 and T_2. Find the function $T(x)$ where x is the distance from the end whose temperature is T_1. Find also the heat flow per unit area of cross-section per second.

Solution: End A of rod AB of length L with sides thermally insulated is at temperature T_1 and end B is at temperature T_2 ($T_1 > T_2$). Since the thermal conductivity varies with temperature hence let us consider a small strip of length dx at a distance x from end A. This strip has temperature T and temperature gradient $-\dfrac{dT}{dx}$. The rate of heat flow through this strip Q is

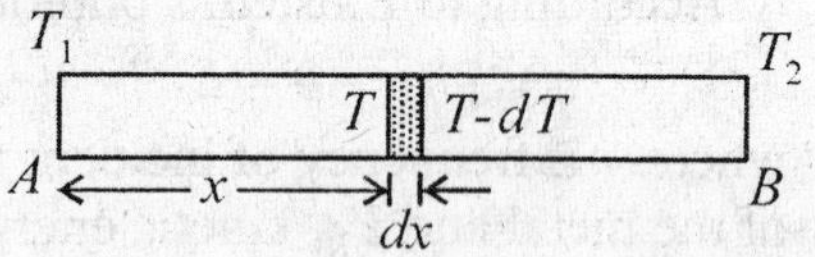

Fig. 17.18

$$Q = -KA\frac{dT}{dx}$$

where A is the area of cross-section of the rod assumed uniform.

But $$K = \frac{\alpha}{T}$$

$$\therefore \quad Q = -\frac{\alpha}{T}A\frac{dT}{dx}$$

or $$\frac{dT}{T} = -\frac{Q}{\alpha A}dx$$

Integration gives

$$\log T = -\frac{Q}{\alpha A}x + C \text{ where } C \text{ is constant of integration}$$

But when $x = 0$ $T = T_1$. This gives us

$$\log T_1 = C$$

Substituting the value of C, we get

$$\log\frac{T}{T_1} = -\frac{Q}{\alpha A}x$$

or $$T = T_1 e^{-\frac{Q}{\alpha A}x} \qquad \ldots(i)$$

We also know that $T = T_2$ when $x = L$

$$\therefore \quad T_2 = T_1 e^{-\frac{Q}{\alpha A}L}$$

or $$\frac{T_2}{T_1} = \left(e^{-\frac{Q}{\alpha A}}\right)^L$$

or $$e^{-\frac{Q}{\alpha A}} = \left(\frac{T_2}{T_1}\right)^{\frac{1}{L}}$$

Substituting the value in (*i*), we get the required function $T(x)$.

$$\therefore \quad T(x) = T_1\left(e^{-\frac{Q}{\alpha A}}\right)^x$$

$$= T_1\left(\frac{T_2}{T_1}\right)^{\frac{x}{L}} \qquad \ldots(ii)$$

Equations (*i*) and (*iii*) give us the required functions $T(x)$.
We have equation

$$e^{-\frac{Q}{\alpha A}} = \left(\frac{T_2}{T_1}\right)^{\frac{1}{L}}$$

or $$-\frac{Q}{\alpha A} = \frac{1}{L}\log\frac{T_2}{T_1}$$

$$Q = -\frac{\alpha A}{L}\log\frac{T_2}{T_1} = \frac{\alpha A}{L}\log\frac{T_1}{T_2}$$

Example 17.22: A calorimeter of negligible heat capacity contains 100 cc of water at 40° C. The water cools to 35° C in 5 minutes. The water is now replaced by K-oil of equal volume at 40° C. Find the time taken for temperature to become 35° C under similar conditions. Specific heat capacities of water and K-oil are 4200 J/kg-K and 2100 J/kg-K respectively. Density of K-oil = 800 kg/m^3.

Solution: When water is used in calorimeter, the rate of radiation is

$$R = ms\frac{d\theta}{dt} = V\rho S\frac{d\theta}{dt}$$

where m is mass, s specific heat and $\frac{d\theta}{dt}$ is the rate of fall of temperature

$$R = 100 \times 10^{-6} \times 10^3 \times 4200 \times \frac{(40-35)}{5} \text{ J/min}$$

$$= 4.2 \times 10^2 \text{ J/minute}$$

For K-oil $\quad R = 100 \times 10^{-6} \times 800 \times 2100 \dfrac{(40-35)}{t}$

$$= \frac{8 \times 21 \times 5}{t} = \frac{840}{t} \text{ J/min}$$

where t is the time taken by K-oil to cool from 40° C to 35° C.

But we know that the rate of radiation depends on (*i*) radiating area, (*ii*) surface temperature and (*iii*) temperature of the surroundings.

In present case since volumes are same as such radiating areas of the two are same.

The temperature of the body and surroundings is also same. Hence rate of radiation of the two liquids are same.

Equating rate of radiation, we have

$$4.2 \times 10^2 = \frac{840}{t}$$

or $\quad t = \dfrac{840}{4.2 \times 10^2} = 2 \text{ minutes}$

Example 17.23: A metal ball of mass 1 kg is heated by means of a 20 W heater in a room at 20° C. The temperature of the ball becomes steady at 50° C. (*a*) Find the rate of loss of heat to the surroundings when the ball is at 50° C (*b*) What is power loss when temperature is 30° C? (*c*) Assume that the temperature of the ball rises uniformly from 20° C to 30° C in 5 minutes, find the total loss of heat to the surroundings during this period (*d*) Calculate the specific heat of metal.

Solution:

(*a*) Steady state temperature of body is 50° C, i.e., whatever energy it is receiving from heater it radiates all of it in same time. Since it is receiving energy at the rate of 20 W, hence the rate of loss of heat at 50° C is 20 W.

(*b*) Newton's law says that the rate of loss of energy is proportional to the excess temperature *i.e.* $P = K(\theta - \theta_0)$

where P is power radiated, when temperature of body is θ and that of the surroundings is θ_0

When $\quad \theta = 50°\text{ C}\; P = 20 \text{ W and } \theta_0 = 20°\text{ C}$

$$\therefore \quad 20 = K(50 - 20) \text{ or } K = \frac{2}{3}$$

When $\quad \theta = 30°\text{ C}$

$$P = \frac{2}{3}(30 - 20) = \frac{20}{3} \text{ W}$$

(*c*) Power loss at 20° C is zero and at 30°C = $\dfrac{20}{3}$ W. Hence average power loss within temperature 20° C to 30° C is

$$P = \left(0 + \frac{20}{3}\right)\frac{1}{2} = \frac{20}{6} \text{ W}$$

This power is lost within 5 minutes. Hence energy radiated to the surroundings is

$$E = P \times \text{time} = \frac{20}{6} \times 5 \times 60 = 1000 \text{ J}$$

(*d*) Heater is supplying energy at constant rate of 20 W, hence the energy supplied by the heater in 5 minutes to raise the temperature of body from 20° C to 30° C is

$$E = 20 \times 5 \times 60 = 6000 \text{ J}$$

Out of 6000 J, 1000 J is radiated to the surroundings and remaining 5000 J are used in raising temperature of body of mass 1 kg. If c is the specific heat of body then

$$mc\Delta\theta = E$$

or $$1 \times c \times 10 = 5000$$

or $$c = 500 \text{ J/kg-K}$$

Example 17.24: Show that the temperature of a planet varies inversely as the square root of its distance from the sun.

Solution: If R_S is the radius of sun and T_S its temperature then energy radiated by the sun per second is

$$E = \sigma 4\pi R_S^2 T_S^4 \quad \ldots(i)$$

If d is the distance of the planet from the sun, then the energy received by unit area of planet in 1 sec is

$$Q = \frac{E}{4\pi d^2} = \frac{\sigma 4\pi R_S^2 T_S^4}{4\pi d^2} = \frac{\sigma R_S^2 T_S^4}{d^2}$$

Effective area of the planet of radius R_P which receives normal radiation is πR_P^2

Hence energy received by the planet per second $= Q\pi R_P^2$

$$E_1 = \frac{\sigma R_S^2 T_S^4}{d^2} \times \pi R_P^2$$

If temperature of the planet is T_P and is assumed to be a black body then energy radiated per second by the planet is

$$E_2 = \sigma \cdot 4\pi R_P^2 \cdot T_P^4$$

In thermal equilibrium, $E_1 = E_2$, thus

$$\sigma 4\pi R_P^2 T_P^4 = \frac{\sigma R_S^2 T_S^4 \pi R_P^2}{d^2}$$

or $$T_P^4 = \frac{R_S^2 T_S^4}{4d^2}$$

or $$T_P = \sqrt{\frac{R_S}{2d}}\, T_S \text{ or } T_P \propto \frac{1}{\sqrt{d}}$$

Example 17.25: A wire of radius r is stretched between two points and a current i is passed through it. The potential difference between its terminals is V. The wire elongates because of heating and when the steady state is reached, the increase in length in δl. The coefficient of linear expansion of the wire is α. Calculate the emission from the wire if the temperature of the surroundings is T_0 K.

Solution: In steady state whatever heat is received by the wire, it radiates in the same time. If T is the temperature of the wire in steady state then power radiated by wire is

$$P = e\sigma A\left(T^4 - T_0^4\right)$$

Here A is area = $2\pi rl$, where l is length of wire, e is emmissivity of the wire. If rise in temperature due to heat received is small then $T = T_0 + \Delta t$, then

$$P = \sigma e 2\pi rl\left[(T_0 + \Delta T)^4 - T_0^4\right]$$

$$= \sigma e 2\pi rl\left[T_0^4\left(1 + \frac{\Delta T}{T_0}\right)^4 - T_0^4\right]$$

Change in temperature Δt is small as such binomial theorem is applicable and using it, we have

$$P = e\sigma 2\pi rl\left[T_0^4\left(1 + \frac{4\Delta T}{T_0}\right) - T_0^4\right]$$

$$= e\sigma 8\pi rl \frac{\Delta T}{T_0} T_0^4 = e\sigma 8\pi rl T_0^3 \Delta T$$

The expansion in the length of wire is δl

$\therefore$ $$\delta l = l\alpha\Delta T$$

or $$\Delta T = \frac{\delta l}{l\alpha}$$

$$P = e\sigma 8\pi rl T_0^3 \frac{\delta l}{l\alpha} = e\sigma 8\pi r T_0^3 \frac{\delta l}{\alpha}$$

But power received by wire is Vi

$\therefore$ $$Vi = e\sigma 8\pi r T_0^3 \frac{\delta l}{\alpha}$$

or $$e = \frac{Vi\alpha}{\sigma 8\pi r T_0^3 \delta l}$$

PROBLEMS

1. A bar of copper of length 75 cm and a bar of steel of length 125 cm are joined together end to end. Both are of circular cross-section with diameter 2 cm. The free ends of copper and steel are maintained at 100° C and 0° C respectively. The surfaces of the bars are thermally insulated. What is the temperature of the copper steel junction? What is the heat transmitted per unit time across the junction? Thermal conductivity of copper is 9.2×10^{-2} k-cal/m °C sec. and that of steel is 1.1×10^{-2} k/cal/m °C sec. **[Ans.** Temp. = 93.31° C, $Q/t = 26 \times 10^{-2}$ cal/sec]

2. Consider the situation shown in Fig. 17.19. The frame is made of the same material and has a uniform cross-section everywhere. Calculate the amount of the heat flowing per second through a cross-section of the bent part if the total heat taken from end at 100° C per sec is 130 J.

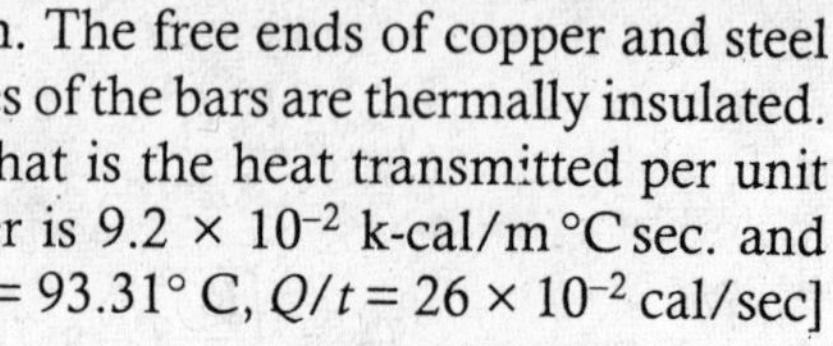

Fig. 17.19

[**Hint**. The thermal resistances of the two parts are in the ratio of their lengths. Thus $\frac{R_1}{R_2} = \frac{70}{60} = \frac{7}{6}$ and the thermal currents are inversely proportional to resistance. Hence heat flow per second through bent part $= \frac{6}{7+6} \times 130 = 60\,\text{J/sec}$]

3. Four identical rods *AB*, *CD*, *CF* and *DE* are connected as shown in Fig. 17.20. The length, cross-sectional area and thermal conductivity of each rod are *L*, *A*, *K* respectively. The free ends *A*, *E* and *F* are maintained at temperatures T_1, T_2 and T_3 respectively. Assuming no loss of heat to the atmosphere, calculate the temperature of *B*.

$$\left[\textbf{Ans. } T = \frac{2T_2 + 2T_3 + 3T_1}{7}\right]$$

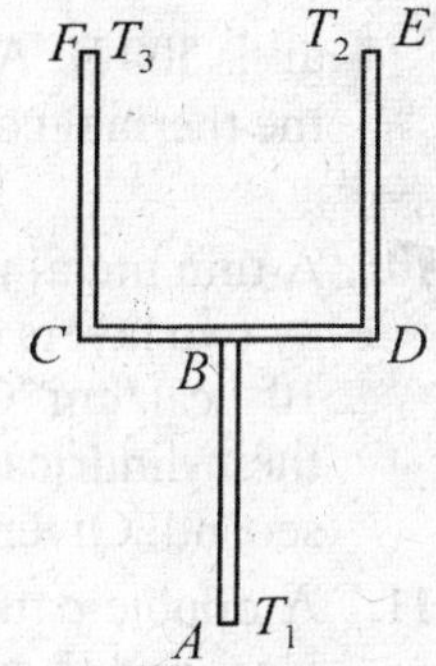

Fig. 19.20

4. A cylindrical block of length 0.4 m and area of cross-section 0.04 m² is placed co-axially on a thin metal disc of mass 0.4 kg and of the same cross-section. The upper face of the cylinder is maintained at a constant temperature of 400 K and the initial temperature of the disc is 300 K. If the thermal conductivity of the cylinder material is 10 Watt/mK and the specific heat of the material of the disc is 600 J/kg-K, how long will it take for the temperature of the disc to increase to 350 K? Assume for purpose of calculations, the thermal conductivity of the disc to be very high and the system is thermally insulated except for the upper face of the cylinder.

[**Ans.** t = 166 sec or 2 m, 46 sec]

5. Figure 17.21 shows a large tank of water maintained at constant temperature T_0 and a small vessel containing cold water of mass m at initial temperature T_1. A metal rod of length l, area of cross-section A and thermal conductivity K connects the two vessels. Find the time taken for the temperature of water in smaller vessel to become $T_2 (T_0 > T_2 > T_1)$. The specific heat capacity of water is C and all other heat capacities are negligible.

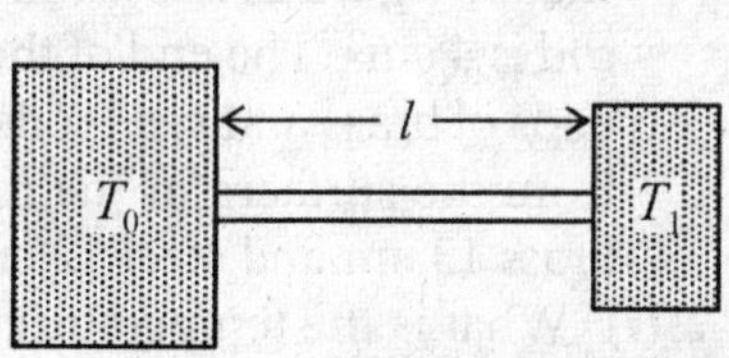

Fig. 19.21

[**Hint**. At any time if temperature of small vessel is T, then temperature gradient is $\frac{T_0 - T}{l}$ and rate of heat conduction is $\frac{dQ}{dt} = \frac{KA(T_0 - T)}{l}$ and $dQ = mcdT$ or $dT = \frac{KA(T_0 - T)}{mcl}dt$.

Integration for temperature from T_1 to T_2 gives $t = \frac{mcl}{KA}\log\left(\frac{T_0 - T_2}{T_0 - T_1}\right)$]

6. The thickness of the ice in a lake is 0.5 cm and the temperature of air is –20° C. Calculate how long will it take for the thickness of ice to be doubled. Thermal conductivity of ice = 5×10^{-4} k cal/sec m °C and latent heat of ice = 80 k-cal/kg. [**Ans.** 7 Hr. 40 min]

7. A body of mass 10 g is kept in an enclosure of temperature 27° C. If the temperature of the body is 127° C, its specific heat is 0.1 k cal/kg °C and area of emitting surface of body is 10^{-3} m², find out the rate of cooling of the body. ($\sigma = 5.72 \times 10^{-8}$ Joule m^{-2} sec^{-1}(°C)$^{-4}$

[**Ans.** 0.23° C/sec]

8. A body which has a surface area 5.0 cm² and a temperature of 727° C radiates 300 joules of energy each minute. What is emissivity? Stefan-Boltzmann's constant = 5.67×10^{-8} Watt/m² (K)⁴

[**Ans.** 0.18]

9. One end of a rod of length 0.20 m is inserted in a furnace at 800 K. The sides of the rod are covered with an insulating material and the other end emits the radiation like a black body. The temperature of this end is 750 K in the steady state. The temperature of the surrounding

air is 300 K. Assuming that energy is lost only by radiation from the open end of the rod, find the thermal conductivity of the material of the rod. Stefan's constant $\sigma = 5.7 \times 10^{-8}$ W/m^{2}K^4
[Ans. $K = 72.1$ W m^{-1}K^{-1}]

10. A thin metal pipe of 1 meter length and 1 cm radius carries steam at 100° C. This is covered by two layers of lagging. The thermal conductivity of outer layer, which is 2 cm thick is 3.6×10^{-4} cal/cm °C sec. If the outer surface of the lagging is at 30° C, find (*a*) the temperature of the cylindrical interface of two lagging materials, (*b*) the mass of the steam condensed per second. Given $\log_e 2 = 0.6931$. **[Ans.** $\theta = 47.5°$ C, $m = .0106$ gm]

11. A double-pane window used for insulating a room thermally from outside, consists of two glass sheets each of area 1m^2 and thickness 0.01 m separated by a 0.05 m thick stagnant air space. In the steady state, the room-glass interface and the glass-outdoor interface are at constant temperature of 27° C and 0° C respectively. Calculate the rate of heat flow through the window pane. Also find the temperatures of other interfaces. Given, thermal conductivities of glass and air as 0.8 and 0.08 W m^{-1} K^{-1} respectively.
[Ans. $Q = 41.54$ Watt, $\theta_1 = 0.5°$ C, $\theta_2 = 26.48°$ C]

12. Three rods of copper, brass, and steel are joined together to form a Y-shaped figure as shown in Fig. 17.22. The cross-sectional area of each rod is 4 cm^2. The end of the copper rod is maintained at 100° C and the ends of brass and steel rods at 0° C. Assume that there is no loss of heat from the surfaces of the rods. The lengths of rods are: copper 46 cm, brass 13 cm and steel 12 cm.
(*i*) What is the temperature of the junction point?
(*ii*) What is the heat current in copper rod?
Given K for copper = 0.9, for brass = 0.26 and that for steel = 0.12.
[Ans. $\theta = 40°$ C and $Q = 4.8$ cal/sec]

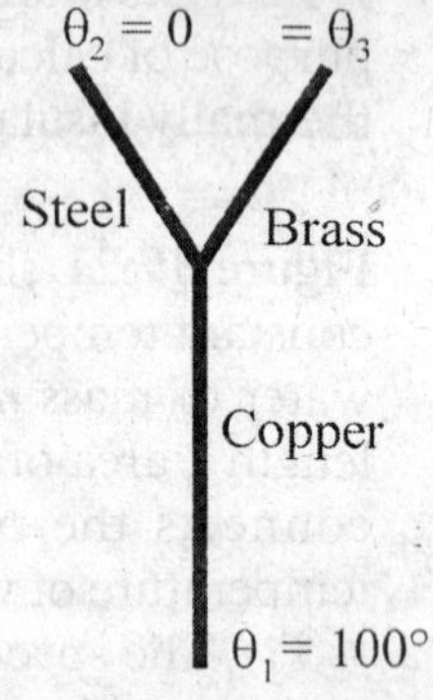

Fig. 19.22

13. An spherical tungsten piece of radius 1.0 cm is suspended in an evacuated chamber maintained at 300 K. The piece is maintained at 1000 K by heating it electrically. Find the rate at which the electrical energy must be supplied. The emissivity of tungsten is 0.3 and the Stefan's constant is 6×10^{-8} W/m^2 (K)^4. **[Ans.** 22 Watt]

14. On a winter day when the atmospheric temperature drops to –10° C, ice forms on the surface of the lake. (*a*) Calculate the rate of increase of thickness of ice when 10 cm of ice is already formed. (*b*) Calculate the total time taken in forming 10 cm of ice. Assume that the temperature of the entire water reaches 0° C before the ice starts forming. Density of water = 10^3 kg/m^3. Latent heat of fusion of ice = 3.36×10^5 J/kg and thermal conductivity of ice = 1.7 W/m °C. Neglect the expansion of water on freezing. **[Ans.** (*a*) 5×10^{-7} m/s (*b*) 27.5 hrs.]

15. A body placed in a surrounding of temperature θ_0 obeys Newton's law of cooling $\frac{d\theta}{dt} = -K(\theta - \theta_0)$. Its temperature at $t = 0$ is θ_1. The specific heat capacity of the body is s and its mass is m. Find (*a*) the maximum heat that the body can lose and (*b*) the time starting from $t = 0$ in which it will loose 90% of this maximum heat.

$$\left[\textbf{Ans.}\ (a)\ ms(\theta_1 - \theta_0)\ \ (b)\ \frac{\log 10}{K}\right]$$

16. In order to minimise heat losses from a glass container, the walls of the container are made of two sheets of glass each 2 mm thick, placed 3 mm apart, the intervening space being filled

with a poorly conducting solid. Calculate the ratio of rate of conduction of heat per unit area through this composite wall to that which would have occurred, had a single sheet of the same glass used under the same interval and external temperature conditions. Thermal conductivity of glass and the poorly conducting solid are 0.63 and 0.04 $Wm^{-1}k^{-1}$ respectively. [**Ans.** 0.05]

17. A body cools in 7 minutes from 60° C to 40° C? The temperature of the surroundings is 10° C. Assume that Newton's law of cooling holds good throughout the process, find its temperature after next 7 minutes? [**Ans.** 28° C]

18. A diode valve consists of two coaxial cylinders. The inner cathode cylinder of radius 5 mm radiates heat like a black body. The radius of the anode cylinder is large compared to that of the cathode. The cathode heater element dissipates one watt per cm length. If the steady anode temperature is 227° C, estimate the temperature of the cathode. Neglect the end effects. [**Ans.** 886 K]

19. A hollow sphere of radius R is filled by water. The temperature of atmosphere is T_0 which is below 0° C. Firstly all water comes down to 0° C and then starts freezing from the outer surface towards the centre. Assuming that the thermal conductivity of ice is K and latent heat of fusion L and the thermal conductivity of shell material to be very high, find the time taken by the whole mass of water to freeze.

$$\left[\textbf{Ans. } t = \frac{\rho L R^2}{6KT_0}\right]$$

20. A metal block of heat capacity 80 J/°C is placed in a room at 20° C, and is heated electrically. The heater is switched off when the temperature reaches 30° C. The temperature of the block rises at the rate of 2° C/s just after the heater is switched on and falls at the rate of 0.2° C/s just after the heater is switched off. Assume Newton's law of cooling to hold, (*a*) find the power of the heater, (*b*) find the power radiated by the block just after the heater is switched off, (*c*) find the power radiated by the block when the block temperature is 25° C and (*d*) assuming that the power radiated at 25° C represents the average value in heating process, find the time for which the heater was kept on. [**Ans.** (*a*) 160 W (*b*) 16 W (*c*) 8 W (*d*) 5.2 s]

21. A cubical block of mass 1.0 kg and edge 5.0 cm is heated to 227° C. It is kept in an evacuated chamber maintained at 27° C. Assuming that the block emits radiations like a black body, find the rate at which the temperature of the block will decrease. Specific heat capacity of material of the block is 400 J/kg-K.

$$\left[\textbf{Hint.} \text{ The power lost} = ms\frac{d\theta}{dt} = \sigma A\left(T^4 - T_0^4\right),\ A = 6\times5\times5\times10^{-4}\text{m}^2 \text{ or}\right.$$

$$\left.\frac{d\theta}{dt} = \frac{5.67\times10^{-8}\times150\times10^{-4}}{1\times400}\left\{(500)^4 - (300)^4\right\} = 0.12\ °\text{C/s}\right]$$

22. A spherical ball A of surface area 20 cm^2 is kept at the centre of a hollow spherical shell B of area 8 cm^2. The surface of A and the inner surface of B emit as black bodies. Assume that the thermal conductivity of the material of B is very poor and that of A is very high and that the air between A and B has been pumped out. The heat capacities of A and B are 42 J/°C and 82 J/°C respectively. Initially the temperature of A is 100° C and that of B is 20° C. Find the rate of change of temperature of A and that of B at this instant. Explain the effects of assumptions listed in the problem. [**Ans.** (*a*) = 0.05° C/s (*b*) = 0.01° C/s]

23. A compound bar 2 m long is constructed of a solid copper core 1 cm in diameter surrounded by a steel casing whose outer diameter is 2 m. The outer curved surface of bar is thermally insulated. One end of the bar is maintained at 100° C and the other at 0° C. (*a*) Find the total heat current in the bar (*b*) What percentage is carried by each material. Given K for copper = 0.9 cal cm^{-1}sec^{-1} (°C)$^{-1}$ and K for steel = 0.12 cal cm^{-1}sec^{-1}(°C)$^{-1}$.

[**Ans.** (*a*) 0.4945 cal/s (*b*) 28.58% for steel and 71.42% for Cu]

24. A cubical tank of water of volume 2 m^3 is kept at a constant temperature of 65° C by 2 kW heater. At time $t = 0$, the heater is switched off. Find the time taken by the tank to cool down to 50° C, given the temperature of room is steady at 15° C. Density of water = 10^3 kg m^{-3} and specific heat of water = 1.0 cal/gm °C. Do not assume average temperature during cooling. Take 1 kW = 240 cal/sec.

[**Hint.** In steady state power supplied by heater is equal to power lost in radiation or $P = C(65 - 15)$ or $C = \frac{2000}{50}$ Watt/°C $= \frac{48}{5}$ cal / sec-°C. Rate of loss of heat at any temperature T is $-\frac{dQ}{dt} = C(T-15)$ or $ms\frac{dT}{dt} = -C(T-15)$ integration gives time t = 20.64 hrs.]

25. An iron boiler is 1 cm thick and has a heating surface of 2 m^2. The two surfaces of the boiler are at 234° C and 100° C respectively. Find how much water can be evaporated per minute if the thermal conductivity of iron is 1.6 × 10^{-2} kcal/meter-sec-°C and latent heat of steam be 536 kcal/kg. [**Ans.** 48 kg]

SHORT ANSWER TYPE QUESTIONS

1. Two bodies A and B have thermal emissivities of 0.01 and 0.81 respectively, the outer surface areas of two bodies are the same. The two bodies emit total radiant power at the same rate. The wavelength λ_B corresponding to maximum spectral radiancy in the radiation from B is shifted from the wavelength corresponding to maximum spectral radiancy in radiation from A by 1.00 μm. If the temperature of A is 5802 K. Find the temperature of B and λ_B.

[**Ans.** Rate of radiation $= e_a A\sigma T_a^4 = e_b A\sigma\, T_b^4$ or $\frac{T_a}{T_b} = \left(\frac{e_b}{e_a}\right)^{\frac{1}{4}} = 3 \therefore T_b = \frac{5802}{3} = 1934$ K

$$\lambda_b - \lambda_a = 0.1\,\mu\text{m}.\ \frac{\lambda_a}{\lambda_b} = \frac{T_b}{T_a} = \frac{1}{3} \text{ or } \lambda_a = \frac{\lambda_b}{3} \text{ or } \frac{2}{3}\lambda_b = 1 \text{ or } \lambda_b = \frac{3}{2} = 1.5\,\mu\text{m}$$]

2. A liquid takes 5 minutes to cool from 80° C to 50° C. How much time will it take to cool from 60° C to 30° C. The temperature of the surroundings is 20° C.

[**Ans.** Newton's law of cooling gives $\frac{dT}{dt} = K(\theta - \theta_0)$ when body cools from 80° C to 50° C,

$\frac{30}{5} = K\left(\frac{80+50}{2} - 20\right)$ or $6 = K(45)$ or $K = \frac{6}{45} = \frac{2}{15}$ when body cools from 60° C to 30° C

$$\frac{30}{t} = K\left(\frac{60+30}{2} - 20\right) = K(25) \text{ or } t = \frac{30}{25K} = \frac{30\times 15}{25\times 2} = 9 \text{ minutes}$$]

3. Explain "Black body radiation in White".

[**Ans.** White radiations include all possible wavelengths. Since a black body absorbs radiations of all wavelengths, so on being heated to a suitable temperature it will emit radiations of all wavelengths, *i.e.*, black body radiation is white]

4. Water in a closed tube is heated with one arm vertically placed above the lamp. In what direction will the water begin to circulate in the tube?

[**Ans.** On heating, the liquid near the lamp will become lighter and will rise up and the heavier cooler liquid in horizontal tube will move towards left to take place of the displaced liquid. Hence liquid will circulate in clockwise direction.]

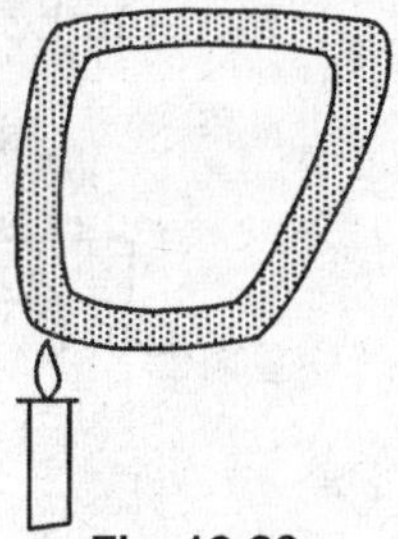

Fig. 19.23

5. Two rods *P* and *Q* are of equal length. Each rod has its end at temperatures T_1 and T_2. What is the condition that will ensure equal rates of flow of heat through the rods *A* and *B*?

[**Ans.** Rate of flow of heat $\frac{dQ}{dt} = KA\frac{(\theta_1 - \theta_2)}{l}$. For rod *P*, $\left(\frac{dQ}{dt}\right)_p = \frac{K_p A_p (T_1 - T_2)}{l_p}$. For rod *Q* $\left(\frac{dQ}{dt}\right)_Q = \frac{K_Q \cdot A_Q (T_1 - T_2)}{l_Q}$. But $l_Q = l_p$ and $\left(\frac{dQ}{dt}\right)_p = \left(\frac{dQ}{dt}\right)_Q$, $\therefore K_p A_p = K_Q A_Q$ or $\frac{K_p}{K_Q} = \frac{A_Q}{A_p}$

i.e. the ratio of the thermal conductivities of the rods should be equal to the inverse ratio of their area of cross-section.]

6. A cylinder of radius *R* made of a material of thermal conductivity K_1 is surrounded by a cylindrical shell of inner radius *R* and outer radius 2*R* made of a material of thermal conductivity K_2. The two ends of the combined system are maintained at two different temperatures. There is no loss of heat across the cylindrical shell surface and the system is in steady state. What is the effective thermal conductivity of the system?

[**Ans.** The effective conductivity of a system consisting of a number of rods in parallel is given by (Equation (17.8)) $K = \frac{\Sigma_i A_i K_i}{\Sigma A_i} = \frac{K_1 \pi R^2 + K_2(4\pi R^2 - \pi R^2)}{\pi R^2 + (4\pi R^2 - \pi R^2)} = \frac{K_1 + 3K_2}{4}$]

7. Four square pieces of insulation, of two different materials all with same thickness and area *A* are available to cover an opening area of 2*A*. This can be done in either of the two ways shown in Fig. 17.24. Which arrangement (*i*) or (*ii*) would ensure lesser heat flowing $K_1 \neq K_2$.

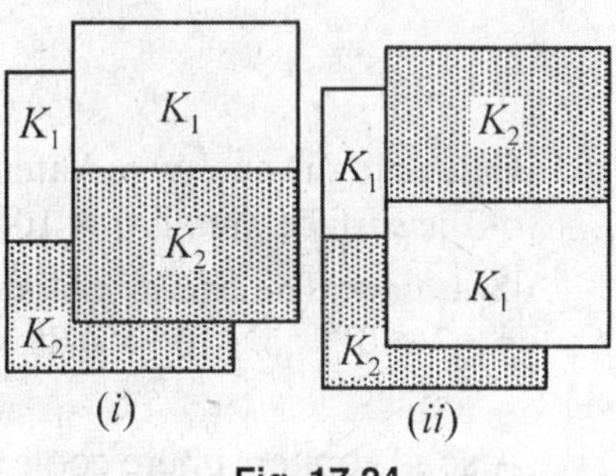

Fig. 17.24

[**Ans.** In arrangement (*i*) if R_1 and R_2 are the thermal resistance of slabs of conductivity K_1 and K_2 respectively, then $2R_1$ and $2R_2$ are in parallel and equivalent thermal resistance $R = \frac{2R_1 \times 2R_2}{2(R_1 + R_2)} = \frac{2R_1R_2}{R_1 + R_2}$ i.e. harmonic mean of R_1 and R_2. In arrangement (*ii*) $(R_1 + R_2)$ and $(R_1 + R_2)$ are in parallel and their equivalent thermal resistance $R' = \frac{R_1 + R_2}{2}$ i.e.

arithmetic mean of R_1 and R_2. Arithmetic mean is always greater than harmonic mean hence arrangement (*ii*) ensures better thermal insulation]

8. Two identical rectangular rods of metal are welded end to end as shown is Fig. 17.25 (*a*) and 10 J of heat flows through the rod in 2 min. How long would it take for 30 J of heat to flow through the rods if they are welded as shown Fig. 17.25(*b*)?

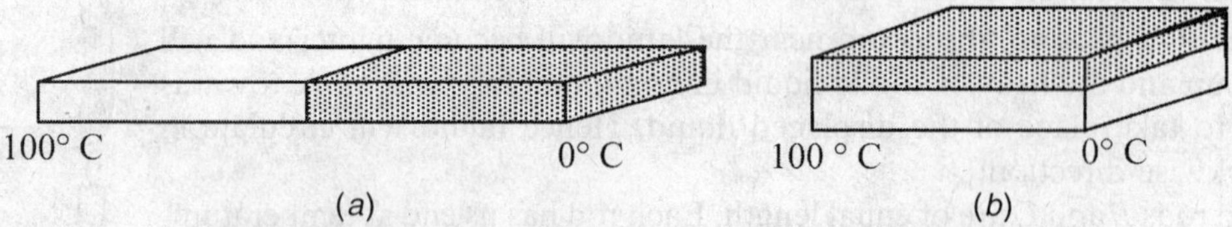

Fig. 17.25

[**Ans.** If R is thermal resistance of rod then rate of heat flow is $\frac{d\theta}{dt} = K\frac{\Delta\theta}{R}$ In case (*a*) the two thermal resistances are in series and $R_S = R_1 + R_2 = 2R$ and $\frac{10}{2} = \frac{\Delta\theta}{2R}$ and $\frac{\Delta\theta}{R} = 10$. In case (*b*) two resistances are in parallel and $R_P = R/2$ and $\frac{30}{t} = \frac{\Delta\theta}{\frac{R}{2}} = \frac{2\Delta\theta}{R}$. In both cases $\Delta\theta = (100°C - 0°C) = 100°C$ $\therefore$ $\frac{30}{t} = 2 \times 10$ or $t = 1.5$ minutes]

9. A room at 20° C is heated by a heater of resistance 20 Ω connected to 200 V mains. The temperature is uniform throughout the room and the heat is transmitted through a glass window of area 1 m² and thickness 0.2 cm. Calculate the temperature outside. Thermal conductivity of glass is 0.84 W/m²-°C.

[**Ans.** To maintain constant temperature inside a room, the rate at which the heat conducts away from the glass window should be equal to the rate at which heat is supplied by the heater. Heat supplied by heater $= \frac{V^2}{R}$ per sec and heat conducted per sec. $\equiv \frac{KA(20-\theta)}{x}$, where θ is outside temperature and x thickness of glass or $\frac{(200)^2}{20} = \frac{0.84 \times 1(20-\theta)}{0.2 \times 10^{-2}}$

or $(20-\theta) = \frac{200 \times 200}{20 \times 420} = 4.76$ or $\theta = 20 - 4.76 = 15.24°$ C]

10. Calculate the energy radiated per minute from the filament of an incandescent lamp at 2000 K of the surface area is 5×10^{-5} m² and its relative emittance is 0.85, $\sigma = 5.7 \times 10^{-8}$ mks units.

[**Solution:** The energy radiated per second by a body according to Stefan's law is $E = \sigma A e T^4 = 5.7 \times 10^{-8} \times 5 \times 10^{-5} \times 0.85 \times (2000)^4$ or $E = 5.7 \times 5 \times 0.85 \times 16 \times 10^{-1} = 387.6 \times 10^{-1}$ Joules/sec or $E = 387.6 \times 10^{-1} \times 60 = 2325.6$ J]

11. A solid copper sphere cools at the rate of 2.8° C per minute, when its temperature is 127° C. Find the rate at which another solid copper sphere of twice the radius lose its temperature at 227° C. In both cases room temperature is 27° C.

[**Ans.** Using Stefan's law we have $mc\frac{d\theta}{dt} = \sigma A(T^4 - T_0^4)$. For first sphere $mc \cdot (2.8) = \sigma A$

$\{(400)^4 - (300)^4\}$. For second sphere mass is 8 times and area is 4 times. Hence $8mc\dfrac{d\theta}{dt}$

$$= 4\sigma A\{(500)^4 - (300)^4\} \;\therefore\; \frac{d\theta}{dt} = \frac{2.8}{2}\frac{(5)^2 - (3)^4}{(4)^4 - (3)^4} = 4.35°\text{C / minute}\,]$$

12. A point source of heat of power P is placed at the centre of a spherical shell of mean radius R. The material of the shell has thermal conductivity K. If the temperature between the outer and inner surface of the shell is not to exceed T, find the thickness the shell.

[**Ans.** Rate of heat conduction = Power of the source. Hence $P = K \cdot 4\pi R^2 \dfrac{d\theta}{dx}$ for

$$d\theta = T\,dx = \frac{K\,4\pi R^2 T}{P}\,]$$

13. Earth receives 1400 Watt/m^2 of solar power. If all solar energy falling on a lens of area 0.2 m^2 is focused on to a block of ice of mass 280 gm, find the time taken to melt the ice. Latent heat of ice is 3.3×10^5 J/kg.

[**Ans.** Total energy falling on ice is $1400 \times 0.2 = 280$ J/sec. Heat required to melt 280 gm of ice

$= 280 \times 10^{-3} \times 3.3 \times 10^5$. Hence time required is $t = \dfrac{280 \times 10^{-3} \times 3.3 \times 10^5}{280} = 330\,\text{sec} = 5.5\ \text{m}$]

14. A solid copper sphere (density ρ and specific heat c) of radius r at an initial temperature 200 K is suspended inside a chamber, whose walls are at almost 0 K. Find the time required for the temperature of the sphere to drop to 100 K.

[**Ans.** If temperature of sphere at any time be θ and rate of fall of temperature $-\dfrac{d\theta}{dt}$, then

Stefan's law gives $-mc\dfrac{d\theta}{dt} = \sigma A\theta^4$ or $\dfrac{4}{3}\pi r^2 \rho c\dfrac{d\theta}{dt} = \sigma 4\pi r^2\theta^4 \;\therefore\; -\dfrac{rc\rho}{3\sigma}\dfrac{d\theta}{\theta^4} = dt$ and integration

from $\theta = 200$ K to 100 K gives $-\dfrac{rc\rho}{3\sigma}\displaystyle\int_{200}^{100}\dfrac{d\theta}{\theta^4} = \int_0^t dt$ or $t = \dfrac{rc\rho}{9\sigma}\left[\dfrac{1}{(100)^3} - \dfrac{1}{(200)^3}\right] = \dfrac{7rc \times 10^{-6}}{72\sigma}$ s]

15. A blackened metal sphere of thermal capacity 1000 cal/°C and surface area 500 cm^2 is kept inside an evacuated enclosure with blackened walls maintained at 27° C. Find the rate of cooling of the sphere when its temperature is at 127° C. Given $\sigma = 1.36 \times 10^{-12}$ cal cm^{-2}s^{-1}K^{-4}

[**Ans.** Stefan's law gives $mc\dfrac{d\theta}{dt} = \sigma A(T^4 - T_0^4) \therefore 1000\dfrac{d\theta}{dt} = 1.36 \times 10^{-12} \times 500\left[(400)^4 - (300)^4\right]$

$$\text{or } \frac{d\theta}{dt} = \frac{1.36 \times 10^{-12} \times 500 \times 175 \times 10^8}{10^3} = 0.0119°\ \text{C/s}\,]$$

16. If the power radiated by certain star is 100 times that of the sun, estimate its surface temperature assuming the temperature of sun to be 6000 K and assume both to be perfectly block bodies.

[**Ans.** Power radiated by a black body per unit area is $P = \sigma T^4 \;\therefore\; \dfrac{P_{st}}{P_{sun}} = \left(\dfrac{T_{st}}{T_{sun}}\right)^4$ or $T_{st}^4 = \dfrac{T_{st}}{T_{sun}} \times T_{sun}^4$

$$\text{or } T_{st} = \left(\frac{P_{st}}{P_{sun}}\right)^{\frac{1}{4}} T_{sun} = (100)^{\frac{1}{4}} \times 6000 \text{ or } T_{st} = \sqrt{10} \times 6000 = 18972\ \text{K}\,]$$

17. Two copper spheres of radii r and $2r$ are suspended in two evacuated chambers, each being at a temperature T_0 °C above its surroundings. Compare (*i*) the initial rates of loss of heat and (*ii*) the initial rates of fall of temperature in the two cases.

[**Ans.** Rate of loss of heat of a sphere of radius r is $Q = 4\pi r^2\sigma(\theta^4 - \theta_0^4)$ and $Q = \frac{4}{3}\pi r^3 \frac{d\theta}{dt}$ $= 4\pi^2\sigma(\theta^4 - \theta_0^4)$ or $\frac{d\theta}{dt} = \frac{3\sigma}{r}(\theta^4 - \theta_0^4)$ i.e. Rate of loss of heat is proportional to r^2 $\therefore \frac{Q_1}{Q_2} = 1:4$ and rate of fall of temperature is inversely proportional to r $\therefore \left(\frac{d\theta}{dt}\right)_1 : \left(\frac{d\theta}{dt}\right)_2 = 2:1$]

18. A copper sphere is suspended in an evacuated chamber maintained at 300 K. The sphere is maintained at a constant temperature of 500 K by heating it electrically. A total of 210 W of electric power is needed to do it. When the surface of the copper sphere is completely blackened, 700 W is needed to maintain the same temperature of the sphere. Calculate the emissivity of copper.

[**Ans.** When it radiates as copper, power radiated $= \sigma A e(T^4 - T_0^4)$ *i.e.* $210 = \sigma A e(T^4 - T_0^4)$, e is emissivity of copper. When it is radiating as black body $e = 1$ $\therefore 700 = \sigma A(T^4 - T_0^4)$

$$\therefore e = \frac{210}{700} = 0.3]$$

19. Two bodies A and B having equal surface area maintained at temperatures 10° C and 20° C. Find the ratio of thermal radiations emitted in a given time by A and B.

[**Ans.** Radiation emitted by A per sec, $R_1 = \sigma A\,(283)^4$, radiation emitted by B per sec, $R_2 = \sigma A\,(293)^4$ R. Then ratio $\frac{R_1}{R_2} = \frac{(283)^4}{(293)^4} = 0.86$]

20. A body cools from 62° C to 50° C in 10 minutes. If room temperature is 26° C, find its temperature after 10 minutes.

[**Ans.** Using Newton's law of cooling, rate of cooling $= W\frac{d\theta}{dt} = K\left(\frac{\theta_1 + \theta_2}{2} - \theta_0\right)$ using the given values we have $\frac{W(62-50)}{10} = K\left(\frac{62+50}{2} - 26\right) = 30$ K or $\frac{12W}{10} = 30$ K. If temperature after next 10 min is θ then $\frac{W(50-\theta)}{10} = K\left(\frac{50+\theta}{2} - 26\right) = \frac{K(\theta - 2)}{2}$ or $\frac{50-\theta}{12} = \frac{(\theta-2)}{2\times 30}$ solving for θ, $\theta = 30°$ C]

21. A black body at 327° C emits energy at the rate of 10 Watt/cm^2. If sun radiates at the rate of 10^5 Watt/cm^2, then find the temperature of the sun.

[**Ans.** For black body using Stefan's law $R_1 = \sigma T_1^4$ and for sun $R_2 = \sigma T_0^4$ $\therefore \frac{R_1}{R_2} = \frac{T_1^4}{T_2^4}$

$$\therefore T_2 = \left(\frac{R_2}{R_1}\right)^{\frac{1}{4}} T_1 = \left(\frac{10^5}{10}\right)^{\frac{1}{4}} \times (600) = 6000 \text{ K}]$$

22. A hollow container has a small hole. Approximately what should be its temperature so that radiations emitted by 1 m² area of it in 1 sec is 1 calorie?

[**Ans.** Using Stefan's law, radiation emitted by a black body per unit area per second $R = \sigma AT^4$, $R = 1$ calorie $= 4.2$ J and $A = 1$ m² $\therefore$ $4.2 = 5.67 \times 10^{-8} \times 1 \times T^4$ or $T = \left(\frac{4.2}{5.67} \times 10^8\right)^{\frac{1}{4}} \approx 100$ K]

23. Sun and moon emit radiations having wavelength corresponding to maximum intensity as 0.5×10^{-4} cm and 100×10^{-4} cm respectively. Find ratio of their temperatures.

[**Ans.** From Wien's displacement law, we have $\lambda_m T =$ Constant $\therefore$ $\lambda_s T_s = \lambda_m T_m$ or

$$\frac{T_s}{T_m} = \frac{\lambda_m}{\lambda_s} = \frac{100 \times 10^{-4}}{0.5 \times 10^{-4}} = 200]$$

24. 50 gm of water are put in a calorimeter of negligible water equivalent. Water takes 4 minutes to cool from 80° C to 75° C. The same calorimeter when filled with liquid (of same volume as water) of mass 40 gm it takes 130 sec to cool from 80° C to 75° C. Find the specific heat of the liquid.

[**Ans.** Rate of radiation depends on (*i*) radiating area (*ii*) temperature of body and (*iii*) temperature of surroundings. In present case all the three are same for water and liquid. Thus equating their rate of radiation, we have, for water $R = mc\frac{d\theta}{dt} = 50 \times 10^{-3} \times \frac{10^3(5)}{240} = \frac{50 \times 5}{240}$ cal/s, for liquid $R = 40 \times 10^{-3} \times C \times \frac{5}{130}$. Equating the two rates of radiation $\frac{40 \times 10^{-3} \times C \times 5}{130} = \frac{50 \times 5}{240}$

$$\text{or } C = \frac{50 \times 5 \times 130}{240 \times 40 \times 10^{-3} \times 5} = 677 \text{ cal/kg-K}]$$

25. Three rods of identical cross-section and made of the same material form the sides of an isosceles triangle ABC with B as right angle. The points A and B are maintained at temperature T and $\sqrt{2}\,T$ respectively. In steady state find the temperature of point C assuming that only heat conduction takes place.

[**Ans.** Heat flows from B to both sides A and C. From C heat flows to A. As such rate of heat flow between B and C is same as that between C and A. If θ is temperature of C then $\frac{KA(\sqrt{2}T - \theta)}{l} = \frac{KA(\theta - T)}{\sqrt{2}l}$

or $(\sqrt{2}T - \theta) = \frac{(\theta - T)}{\sqrt{2}}$. This gives $\theta = \frac{3T}{\sqrt{2}+1}$]

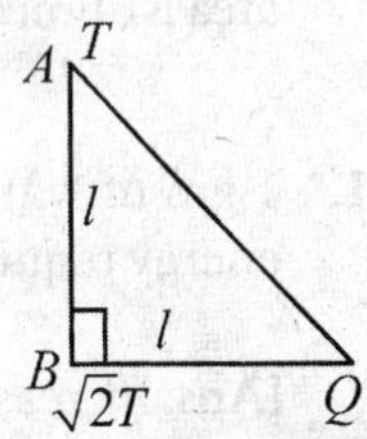

Fig. 17.26

26. Five rods of same dimensions are arranged as shown in Fig. 17.27. They have thermal conductivities and lengths as shown in figure. A and B are maintained at different temperatures. If no heat flows through central rod then find the ration between K_1, K_2, K_3, K_4 and K_5.

[**Ans.** If R_1, R_2, R_3 and R_4 are thermal resistances of rods AC, CB, AD and DB respectively, then if no thermal current flows through CD, then

$\frac{R_1}{R_2} = \frac{R_3}{R_4}$ or $R_1R_4 = R_2R_3$ or $\frac{l}{K_1A} \cdot \frac{l}{K_4A} = \frac{l}{K_2A} \cdot \frac{l}{K_3A}$ or $K_1K_4 = K_2K_3$]

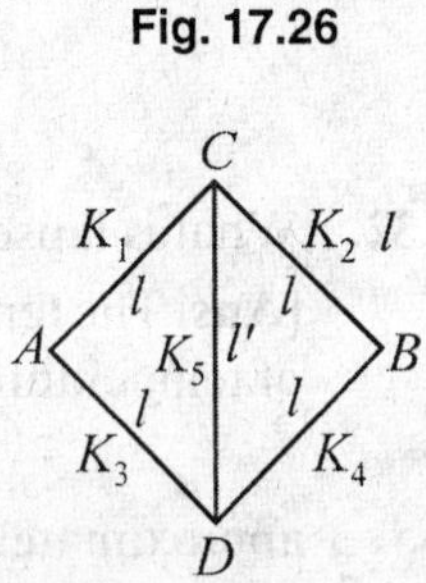

Fig. 17.27

27. The temperature of a room heated by heater is 20° C when outside temperature is –20° C and it is 10° C when the outside temperature is –40° C. Find the temperature of heater.

[**Ans.** Rate of radiation of the heater is equal to the rate of conduction. If θ is the temperature of the heater then, $\theta - 20 = K\{(20) - (-20)\} = 40$ K and $\theta - 10 = K\{(10) - (-40)\} = 50$ K. Solving $\theta = 60°$ C]

28. If the temperature of a body is increased by 50%, find percentage increase in the thermal energy radiated by the body.

[**Ans.** Initially the rate of radiation is $Q_1 = \sigma A T^4$. Now temperature is (1.5*T*) and rate of radiation is $Q_2 = \sigma A T^4 (1.5)^4$ $\therefore$ Increase in radiation energy $\Delta Q = Q_2 - Q_1 = \sigma A T^4\{(1.5)^4 - 1\}$

$\therefore$ Fractional increase is $\dfrac{\Delta Q}{Q_1} = \{(1.5)^4 - 1\} = 5.06 - 1 = 4.06$. Hence percentage increase is approximately 400]

29. A black body is at 3000 K temperature. When it cools, there is a change of 9μ in the wavelength corresponding to the maximum energy density radiation. Find now the temperature of the black body.

$$\text{[Ans. } T_1\lambda_m = b = T_2(\lambda_m + \Delta\lambda) \text{ or } T_1 = T_2\left(1 + \frac{\Delta\lambda}{\lambda_m}\right) = T_2\left(1 + \frac{\Delta\lambda\, T_1}{b}\right) \text{ or } T_2 = \frac{T_1 b}{b + \Delta\lambda T_1}$$

$$= \frac{3000 \times 3 \times 10^{-3}}{3 \times 10^{-3} + 300 \times 9 \times 10^{-6}} = 300 \text{ K]}$$

30. A copper cube of side 1 cm is kept in an evacuated enclosure to cool. Its temperature falls from 100° C to 99° C in 100 seconds. Now another cube of side 2 cm with similar surface is placed in same surroundings to cool. Its temperature falls from 100° C to 99° C. Find the time taken by this cube.

[**Ans.** In case of first cube $ms\dfrac{d\theta}{dt} = \sigma A\left(T^4 - T_0^4\right)\dfrac{d\theta}{dt} = \dfrac{1}{100}$. For second mass 8 times and surface area is 6 times $8ms\left(\dfrac{d\theta}{dt}\right)' = \sigma 4A(T^4 - T_0^4)$ or $\dfrac{8}{4}\left(\dfrac{d\theta}{dt}\right)' = \dfrac{d\theta}{dt} = \dfrac{1}{100}$ or $\left(\dfrac{d\theta}{dt}\right)' = \dfrac{1}{200}$ $\therefore$ Time taken is 200 sec $\because d\theta = 1°$ C]

31. 1 gm of water is heated from 0° F to 0° C in certain time using solar energy. Calculate the energy required.

[**Ans.** Increase in temperature is 32° F $= \dfrac{32 \times 5}{9} = \dfrac{160}{9}$ °C. Required energy $= ms\,\Delta Q = 1 \times 1 \times \dfrac{160}{9}$

$$= \frac{160}{9} \text{ cal} = \frac{160}{9} \times 4.2 = 74.6 \text{ J]}$$

32. What is lapse rate? Why is it negative during clear nights?

[**Ans.** The temperature of earth's atmosphere decreases with increase of height. The rate of fall of temperature of earth's atmosphere per unit increase in altitude is called the lapse rate. It is approximately given $\dfrac{dT}{dz} = -\dfrac{Mg}{R}\left(\dfrac{\gamma - 1}{\gamma}\right)$, where *R* is gas constant, $\gamma = \dfrac{C_p}{C_V}$, *M* = molecular weight of air. For dry air lapse rate is 10° C/km while for air saturated with water vapour it is = 6° C/km. On clear night, ground cools more quickly than the air as such heat conduct from air to earth, *i.e.* the temperature increases with altitude and hence lapse rate is negative on clear rights]

33. Two metal spheres A and B are made of same material and have got identical surface finish. The mass of A is 3 times that of B. Both the spheres are heated to the same high temperature and placed in the same room having lower temperature but are thermally insulated from each other. Find the ratio of the initial rate of cooling of A to B.

[**Ans.** From Stefan's law, we have $V\rho \frac{d\theta}{dt} = \sigma A\left(T^4 - T_0^4\right)$. Area of sphere $A \propto V^{\frac{2}{3}}$. Hence rate of fall of temperature i.e. rate of cooling is $\frac{d\theta}{dt} = \frac{\sigma V^{\frac{2}{3}}}{V\rho C}\left(T^4 - T_0^4\right)$ *i.e.* rate of cooling is inversely proportional to $(V)^{\frac{1}{3}}$ *i.e* $\frac{d\theta}{dt} \propto \frac{1}{(V)^{\frac{1}{3}}}$ $\therefore \frac{\left(\frac{d\theta}{dt}\right)_A}{\left(\frac{d\theta}{dt}\right)_B} = \left(\frac{V_B}{V_A}\right)^{\frac{1}{3}} = \left(\frac{1}{3}\right)^{\frac{1}{3}}$]

MULTIPLE CHOICE QUESTIONS

PART I

In the following, a question is followed by four alternatives, of them only one is correct. Therefore, the candidate has to select the correct choice from amongst the alternatives.

1. The plots of intensity versus wavelength for three black bodies at temperatures T_1, T_2 and T_3 respectively are shown. Their temperatures are such that

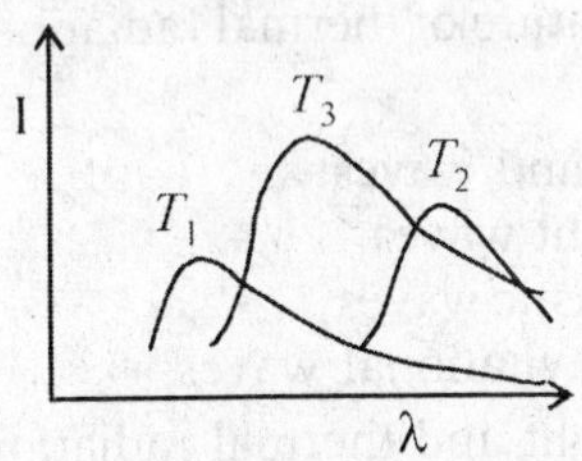

Fig. 17.28

(a) $T_1 > T_2 > T_3$ (b) $T_1 > T_3 > T_2$
(c) $T_2 > T_3 > T_1$ (d) $T_3 > T_2 > T_1$.

2. Three rods of the same material and having the same cross-section have been joined as shown in Fig. 17.29. Each rod is of the same length. The left and right ends are kept at 0° C and 90° C respectively. The temperature of the junctions of the three rods is

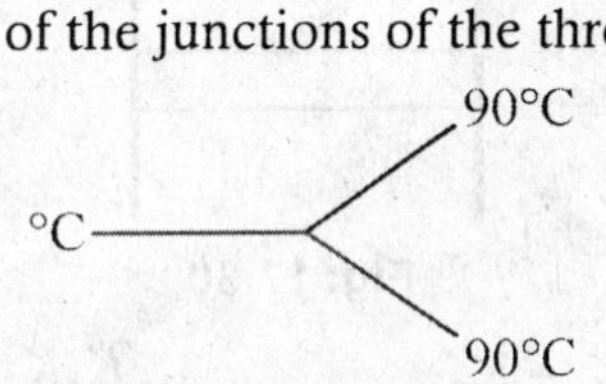

Fig. 17.29

(a) 45° C (b) 60° C
(c) 30° C (d) 20° C.

3. An ideal black body at room temperature T is thrown into a furnace. It is found that
(a) initially it appears dark and latter on bright
(b) it appears black always
(c) it is indistinguishable all the time
(d) initially it appears black and finally is indistinguishable.

4. If λ_m denotes the wavelength at which the radiative emission from black body at a temperature T K is maximum, then
(a) $\lambda_m \propto T^4$
(b) λ_m is independent of T
(c) $\lambda_m \propto T$
(d) $\lambda_m \propto T^{-1}$.

5. A black body is at a temperature of 2880 K. The energy of radiation emitted by this object with wavelength between 499 nm and 500 nm is U_1, between 999 nm and 1000 nm is U_2 and between 1499 nm and 1500 nm is U_3. The Wien's constant $b = 2.88 \times 10^6$ nm-K. Then
(a) $U_1 = 0$ (b) $U_3 = 0$
(c) $U_1 > U_2$ (d) $U_2 > U_1$.

6. A wall consists of alternating blocks with a length d and coefficient of thermal conductivity K_1 and K_2. The cross-sectional area of blocks are the same. The coefficient of thermal conductivity of the wall is

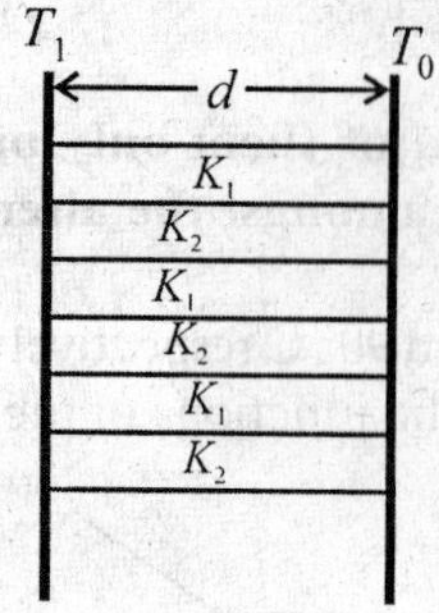

Fig. 17.30

(a) $\frac{K_1 + K_2}{2}$ (b) $\frac{K_1 K_2}{K_1 + K_2}$
(c) zero (d) infinity.

7. In a room containing air, heat can transmit from one place to another
(a) by conduction only
(b) by radiation only
(c) by convection only
(d) by all the three methods.

8. A body cools down from 70° C to 65° C in 5 minutes. It will cool down from 65° C to 60° C in
(a) 5 minutes
(b) more than 5 minutes
(c) less than 5 minutes
(d) less than or more than 5 minutes depends on the room temperature.

9. Newton's law of cooling is a special case of
(a) Wien's displacement law
(b) Kirchhoff's law
(c) Stefan's law
(d) Planck's law.

10. A hot liquid is kept in a big room. Its temperature is plotted as a function of time. Which of the following curves may represent the plot?

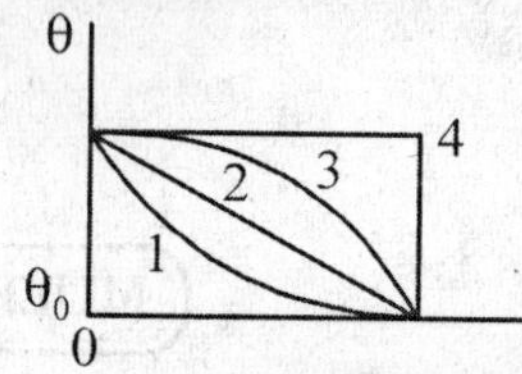

Fig. 17.31

(a) 1 (b) 2
(c) 3 (d) 4.

11. One end of a metal rod is kept in a furnace. In steady state, the temperature of the rod
(a) increases
(b) decreases
(c) remains constant
(d) is non-uniform.

12. The nature of thermal radiations is similar to
(a) sound waves
(b) light waves
(c) elastic waves
(d) gravitational waves.

13. For light and thermal radiation
(a) velocity and frequency are same
(b) velocity same but frequency different
(c) velocity and wavelength both different
(d) velocity different but wavelength same.

14. Which of the following is not the property of radiation?
(a) Transmission with velocity of light

(b) Quantum nature
(c) Electromagnetic wave nature
(d) Necessity of medium for transmission.

15. Polarisation of thermal radiation is possible. This shows that their nature is
(a) similar to transverse waves
(b) similar to electromagnetic waves
(c) similar to longitudinal waves
(d) similar to photon.

16. In sky there are stars emitting light of white, red and blue colours. If their temperatures are T_w, T_r and T_b respectively, then
(a) $T_b > T_w > T_r$
(b) $T_r > T_w > T_b$
(c) $T_w > T_b > T_r$
(d) $T_w > T_r > T_b$.

17. Four surfaces are at same temperature, same volume and similarly polished. Which of the following shape has least radiation emission rate?
(a) Cubical (b) Spherical
(c) Cylindrical (d) Disc.

18. A hollow body is at 100 K. Rate of radiation per unit area of body is
(a) 5.67 J (b) 5.67×10^2 J
(c) 5.67×10^4 J (d) 5.67×10^6 J.

19. If the temperature of metal piece is raised from 27° C to 327° C, the rate of emission will be
(a) 16 times (b) 8 times
(c) 4 times (d) 2 times.

20. Select wrong statement.
(a) Perfectly black body may not be black.
(b) At 0 K all the bodies do not emit radiations.
(c) Cold body placed in hot atmosphere does not emit radiation
(d) Thermal radiations only are in the infrared region of the electromagnetic radiation spectrum.

21. Which of the following graphs represent Newton's law of cooling?

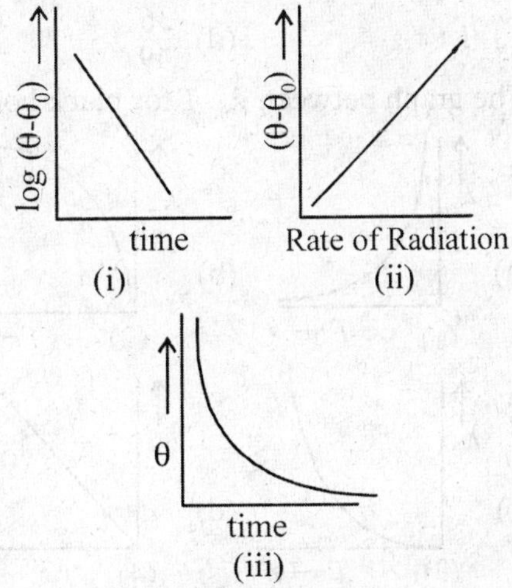

Fig. 17.33

(a) (*i*) and (*iii*) (b) (*i*) and (*ii*)
(c) (*ii*) and (*iii*) (d) all the three.

22. Equal volumes of water and glycerine are taken in identical calorimeters. They are cooled through same temperatures under identical conditions, then
(a) their rate of heat loss will be same
(b) their rate of heat loss will not be same
(c) their rate of cooling will be same
(d) their rate of cooling will not be same.

23. Rate of cooling of 150 gm water placed in calorimeter of water equivalent 10^{-2} kg at 70° C temperature is 88 cal/m. If water is replaced by oil (density 800 kgm^{-3} and specific heat 420 cal/kg-K) at 70° C, its rate of cooling is
(a) 70.4 cal/min
(b) 88 cal/min
(c) 110 cal/min
(d) information is insufficient.

24. In problem 23 the rate of fall of temperature of oil is
(a) 0.5° C/min (b) 1.5° C/min
(c) 2° C/min (d) 1° C/min.

25. Two stars *A* and *B* emit maximum intensity at wavelengths 3600 Å and 4200 Å. The ratio of their temperature is
(a) 6/7 (b) 7/6

(c) 1 : 1 (d) $\frac{36}{49}$.

26. The graph between λ_m-T for black body is

(a) 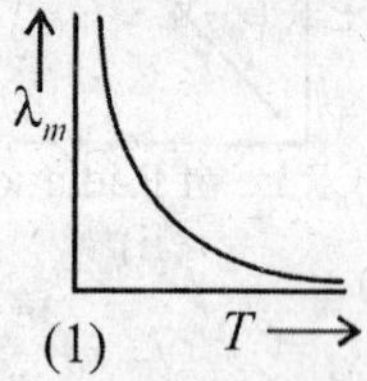

(1)

(b) 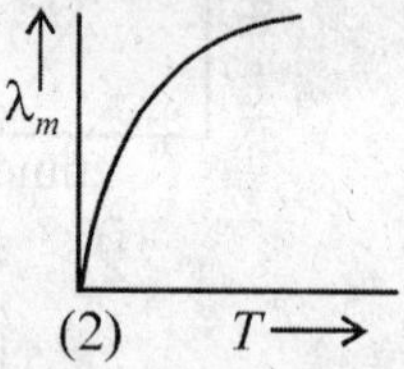

(2)

(c) 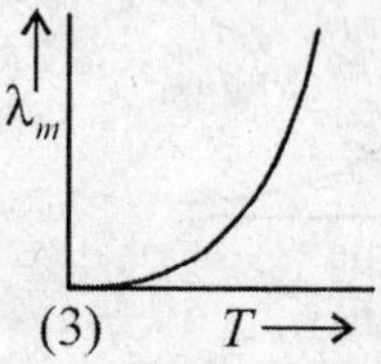

(3)

(d) 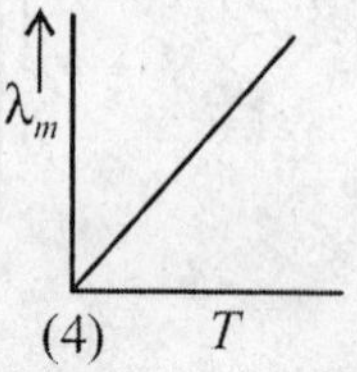

(4)

27. The presence of Fraunhoffer lines in Sun's spectrum is the result of
(a) selective absorption due to presence of elements is the outer part of sun
(b) selective absorption due to presence of elements near centre of sun
(c) selective absorption due to presence of gases in atmosphere
(d) Fraunhoffer lines are not related to the light emitted by the sun.

28. Select correct statement.
(a) Stefans law and Newton's law of cooling both are valid for all temperatures.
(b) Stefans law is valid for small temperature differences and Newton's law of cooling is valid for all temperatures.
(c) Newton's law of cooling is valid for small temperature differences and Stefan's law is valid for all temperatures.
(d) Both are valid for small temperature differences.

29. If E_m represents the maximum intensity of the emitted radiation and λ_m wavelength corresponding to maximum intensity, then on increasing temperature.
(a) E_m increases, λ_m increases
(b) E_m decreases, λ_m increases
(c) E_m increases, λ_m decreases
(d) E_m decreases, λ_m decreases.

30. The ratio of thermal conductivity of two rods of different materials is 5/4. The two rods of same area of cross-section and same thermal resistance will have lengths in ratio
(a) 5/4 (b) 1/9
(c) 9/1 (d) 4/5.

31. Below are given the graph between intensity of radiation (I) from a black body at fixed temperature and frequency (ν). The correct graph is ($T_2 > T_1$)

(a) 

(a)

(b) 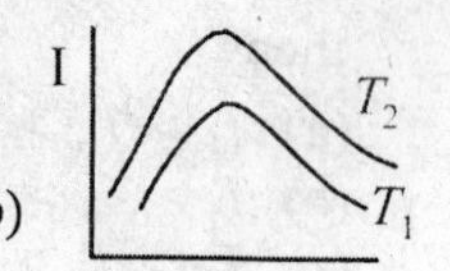

(b)

(c) 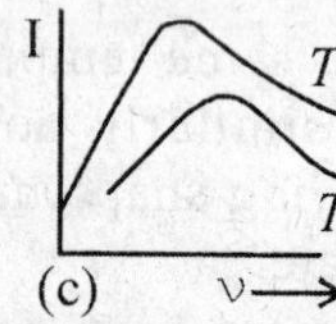

(c)

(d) 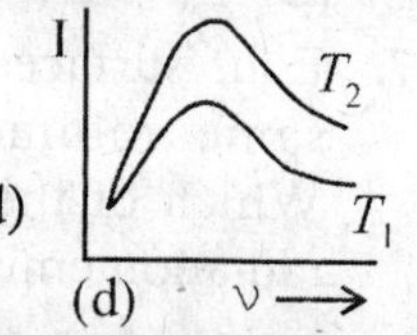

(d)

32. For a cylindrical heat source, the ratio of the thermal heat radiation intensities received at distance r and $2r$ ($r >$ radius of cylinder) is
(a) 1 : 2 (b) 2 : 1
(c) 4 : 1 (d) 1 : 4.

33. A calorimeter contains water at 100° C and is placed in a vessel containing boiling water. If now vessel is supplied heat at the rate of 540 cal/s, the rate at which water in calorimeter is converted into steam is
(a) 1 gm/s
(b) between 0 and 1 gm per sec depending on the specific heat of calorimeter
(c) zero
(d) 4.2 gm/s.

34. Cooling curves for two liquids P and Q are shown is Fig. 17.34. Same volumes of the two liquids are taken. Assuming Newton's law of cooling to be valid throughout. If S_p and S_Q are their specific heats, then

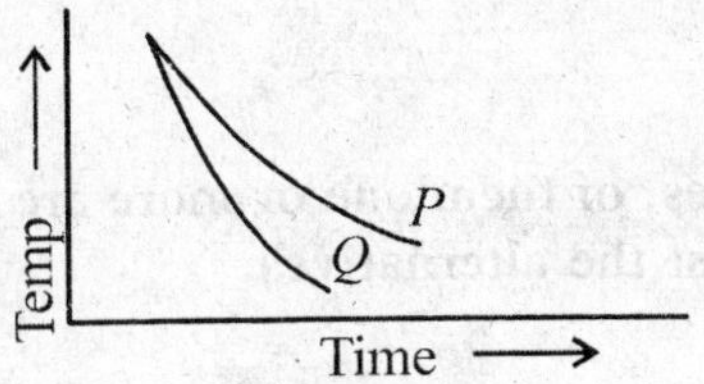

Fig. 17.34

(a) $S_P > S_Q$
(b) $S_P = S_Q$
(c) $S_P < S_Q$
(d) Information is insufficient.

35. Input power P is required to keep M mass of metal in molten state. If we switch of heater, the metal requires time t to completely solidify. If energy loss to the surroundings is constant, then specific latent heat of fusion is

(a) PMt (b) $\frac{PM}{t}$
(c) $\frac{Pt}{M}$ (d) $\frac{M}{Pt}$

36. A black foil is heated by keeping at distance r from a hot sphere. It is found that it absorbs power P. If temperature of the sphere and its distance from foil both are doubled then the foil will absorb power equal to

(a) P (b) $2P$
(c) $8P$ (d) $4P$.

37. When temperature difference between body and surroundings is small, the graph between rate of heat energy loss (E) and temperature of body θ is

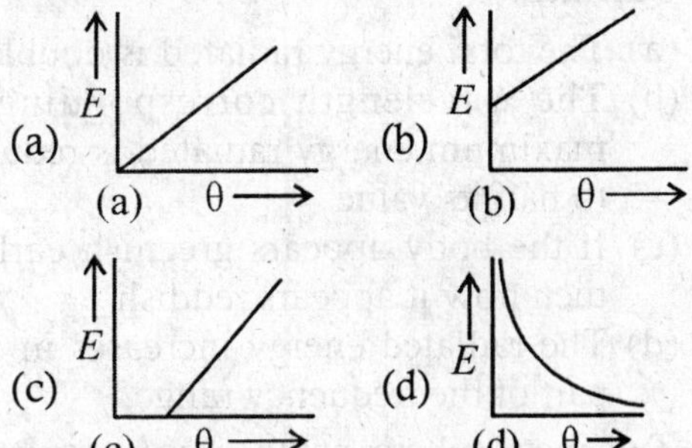

38. A body at temperature 35° C is placed in surroundings at temperature 28° C. To increase its radiant power by a factor 5 its temperature should be

(a) 175° C (b) 147° C
(c) 63° C (d) 53° C.

ANSWERS

1. (b)	2. (b)	3. (a)	4. (d)	5. (d)
6. (a)	7. (d)	8. (b)	9. (c)	10. (a)
11. (d)	12. (b)	13. (b)	14. (d)	15. (a)
16. (a)	17. (d)	18. (a)	19. (a)	20. (c)
21. (b)	22. (a)	23. (b)	24. (b)	25. (b)
26. (a)	27. (a)	28. (c)	29. (c)	30. (a)
31. (a)	32. (b)	33. (c)	34. (d)	35. (c)
36. (d)	37. (c)	38. (c)		

PART II

In the following, a question is followed by four alternatives, of them one or more are correct. Candidate has to select the correct choice(s) from amongst the alternatives.

1. The temperature of a black body is raised from T K to $2T$ K. Identify the correct statement.
   (a) The total energy radiated is doubled
   (b) The wavelength corresponding to maximum energy radiated is reduced to half its value
   (c) If the body appears greenish earlier, then now it appears reddish.
   (d) The radiated energy increases in any part of the frequency range.

2. Two metal spheres at the same temperature have their radii in the ratio 1 : 2. Then
   (a) ratio of heat contain in them is 1 : 4
   (b) ratio of heat contain in them is 1 : 8
   (c) ratio of rate of thermal radiation emitted by the two is 1 : 2
   (d) ratio of rate of thermal radiation emitted by the two is 1 : 4.

3. Two spheres A and B of same radius and of same material are at the same temperature. A is solid and B is hollow. They are placed in same surroundings.
   (a) Rate of cooling of A is greater than the rate of cooling of B
   (b) Rate of cooling of A is less than the rate of cooling of B
   (c) Rate of loss of thermal energy of A is equal to the rate of loss of heat energy of B
   (d) Rate of loss of thermal energy of A is less than the rate of loss of heat energy of B.

4. In steady state
   (a) temperature does not change with time
   (b) all parts of the body are at the same temperature
   (c) there is no flow of heat
   (d) there is no absorption of heat.

5. A hot body emits radiations which has maximum intensity near frequency ν_0. Emissivity of the material is 0.6. If the absolute temperature of the body is doubled then
   (a) the maximum intensity of the radiation will be near that of the frequency $2\nu_0$.
   (b) the maximum intensity of the radiation will be near that of the frequency $\nu_0/2$.
   (c) the maximum intensity of the radiation will increase by factor 16.
   (d) the maximum intensity of the radiation will increase by factor 9.6.

6. A piece of charcoal and a piece of shining steel of the same surface area kept for long time in an open lawn in bright sun
   (a) The steel will absorb more heat than the charcoal
   (b) The temperature of the steel will be less than that of charcoal
   (c) It both are picked up by bare hands, the steel will be felt hotter than the charcoal
   (d) If the two are picked up from the lawn and kept in a cold chamber the charcoal will loose heat at faster rate than the steel.

7. The maximum spectral radiancy of a black body corresponds to a wavelength λ. If the temperature is now changed so that the maximum spectral radiancy now corresponds to wavelength $3\lambda/4$, then
   (a) the new temperature is 4/3 times the old temperature.
   (b) the new temperature is 3/4 times the old temperature.
   (c) the power radiated by the body changes by a factor 256/81.
   (d) the power radiated by the body changes by a factor 81/256.

8. Which of the following terms represent the same in reference to radiation?

(a) Emissivity (b) Emissive power
(c) Radiancy (d) Intensity.

9. Which of the following are dimensionless quantities?

(a) Emissive power (b) Absorptic power
(c) Emissivity (d) Radiancy.

10. Which of the following statements is/are false?

(a) Thermal conductivity of the gases is zero
(b) Furcoats keep us warm by absorbing heat from out side
(c) Solids cannot transmit heat by convection
(d) If we double the length of the rod its thermal conductivity is halved.

11. Which of the following statements is/are true?

(a) A material of high conductivity can absorb large amount of heat
(b) The coefficient of thermal conductivity cannot determine the direction of propagation of heat
(c) A good conductor loses heat faster
(d) Free electrons transport heat from higher to lower temperature.

12. A wall has two layers A and B each made of different materials. The thickness of both layers is same. Thermal conductivity of A $K_A = 3K_B$. The temperature of free end of A is 20° C and of B is 0° C. In steady state

(a) the temperature difference across A is 15.0° C
(b) rate of heat transfer across A is more than that across B
(c) rate of heat transfer across both is same
(d) the temperature difference across A is 5° C.

13. A body can be cooled by

(a) conduction
(b) convection
(c) radiation
(d) any of above process.

14. The energy radiated by a body depends on

(a) area of body
(b) nature of surface
(c) mass of body
(d) temperature of body.

15. Two spheres of same material have radii 1 m and 4 m and temperature 4000 K and 2000 K respectively.

(a) The energy radiated per second by first sphere is greater than that by second
(b) The energy radiated per second by both spheres is same
(c) The rate of cooling of first sphere is 4 times than that of second
(d) The rate of cooling of first sphere is 16 times than that of second.

ANSWERS

1.	(b, d)	2.	(b, d)	3.	(b, c)	4.	(a, d)	5.	(a, c)
6.	(c, d)	7.	(a, c)	8.	(b, c, d)	9.	(b, c)	10.	(a, c, d)
11.	(b, c, d)	12.	(c, d)	13.	(a, b, c, d)	14.	(a, b, d)	15.	(b, c)

18 Electrostatics

18.1 ELECTRIC CHARGE

We know that matter is composed of certain elementary particles. Out of large number of elementary particles, the three most common elementary particles are, electrons, protons and neutrons. Two electrons (or two protons) attract each other by gravitational force. However there is a repulsive force between two electrons (or protons) which is very large in comparison to attractive gravitational force. This shows that electron has an additional property apart from their mass which is responsible for this strong electric force. We call this property *charge.* Thus electric charge is an intrinsic characteristic of the fundamental particles, it is a characteristic that automatically accompanies those particles wherever they exist. Just as masses are responsible for the gravitational force, charges are responsible for the electric force.

Two Kinds of Charges

From experiment it is inferred that there are two types of charges. A proton repels another proton and an electron repels another electron but a proton attracts an electron. We arbitrarily call the charge on an electron as *negative* charge and that on a proton a *positive* charge. This is merely a convention. One should not infer that the charge on electron is less than that on a proton.

Unit of Charge

Charge is always associated with mass, i.e. charge cannot exist without mass, though mass can exist without charge. Hence the particles whose rest mass is zero (e.g. photon, neutrino) have no charge. A charge cannot exist without mass, the presence of charge itself is a convincing proof of existence of mass.

Though it is very difficult to define a charge but it can be measured on a system by comparing it with the charge on a standard body, though we do not know what exactly it is that we want to measure.

In SI units, current is taken as a fundamental quantity and $q = it$, i.e. charge is a derived quantity with dimensions [AT] and unit (ampere × second) is called *Coulomb.* Thus SI unit of charge is Coulomb (abbreviated as C). Thus 1 Coulomb is the charge that flows through a wire in 1 second if the electric current in it is 1 A.

In CGS units, the unit of charge is static Coulomb or Franklin (Fr). The ratio of emu (electromagnetic unit) and esu (electrostatic unit) of charge is equal to velocity of light, i.e.

$$\frac{\text{emu of charge}}{\text{esu of charge}} = c = 3 \times 10^{10} \text{ cm/s}$$

$$1 \text{ Coulomb} = \frac{1}{10} \text{ emu of charge} = 3 \times 10^{9} \text{ esu of charge.}$$

Practical units of charge are ampere hour = 3600 Coulomb and

Faraday = 96500 Coulomb.

Charge is Quantised

When a physical quantity can have only discrete values rather than any fixed value, the quantity is said to be quantised. The charge is quantised means *the exchange of charge takes place in an integral multiples of a small quantity called charge quanta.* This small quantity is called *electronic charge* (e) and has the value 1.60218×10^{-19} C. The value of 'e' is so small that one can easily neglect the quantisation because the charges usually found are very large in comparison to this smallest charge. In many cases, we can assume a continuous charge.

Charge is Conserved

In an isolated system total charge remains conserved. It is possible to create or destroy charged particles but it is not possible to create or destroy net charge. It, therefore, follows that simultaneous equal quantities of positive and negative charges can appear or disappear. This is what actually happens in β-*decay process,* in *pair production* and *annihilation.*

Frictional Electricity: Induction

The study of electrical effects due to charge at rest is called *electrostatics.* This is also called '*static-electricity*'. Since charges at rest are produced due to friction between two insulating bodies which are rubbed against each other. Electrostatics is also called *frictional electricity*.

The appearance of electric charge on rubbing is due to the fact that all material bodies contain large number of electrons and equal number of protons in their normal state. When rubbed against each other some electrons from one body may pass on to the other body. The body that receives more electrons becomes negatively charged and the one donating electrons becomes positively charged because it has more protons than electrons. *Thus the mass of a negatively charged body increases on charging and that of positively charged body becomes less on charging.*

If a charged body is brought near a neutral body, the charged body will attract opposite charge and repel similar charge present in the neutral body. As a result of this one side of the neutral body becomes positively charged while the other negatively charged. This phenomenon is called *electrostatic induction.*

In case of induction it is worthwhile to note that

(*i*) Inducing body neither gains nor lose charge.

(*ii*) The nature of induced charge is always opposite to that of inducing charge.

(*iii*) induced charge can be lesser or equal to inducing charges (but never greater) its maximum value is

$$q' = -q\left(1 - \frac{1}{\varepsilon_r}\right) \quad \text{...(18.1)}$$

where q is the inducing charge and ε_r is the dielectric constant of the material of the uncharged body.

(*iv*) In case of metals $\varepsilon_r = \infty$ and $q' = -q$, i.e. in metals induced charge is equal and opposite to inducing charge.

18.2 COULOMB'S LAW

Through series of experiments, Coulomb and others came to the conclusion that the force between two point charges at rest

(*a*) is always along the line joining the charges.
(*b*) is attractive if charges are unlike and is repulsive if charges are like.
(*c*) depends on the nature of the medium between the charges.
(*d*) is directly proportional to the product of the magnitude of charges, i.e.

$$F \propto q_1 q_2 \quad \text{...(i)}$$

(*e*) is inversely proportional to the square of distance between them, i.e.

$$F \propto \frac{1}{r^2} \quad \text{...(ii)}$$

Combining (*i*) and (*ii*), we have

$$F = K\frac{q_1 q_2}{r^2} \quad \text{...(18.2)}$$

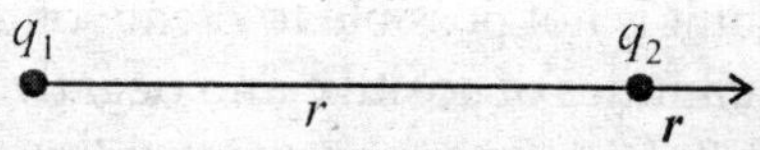

Fig. 18.1

where K is constant. In vector notation, we can write

$$\mathbf{F} = K \cdot \frac{q_1 q_2}{r^2}\hat{r} = \frac{K q_1 q_2}{r^3}\mathbf{r} \quad \text{...(18.2a)}$$

where $\mathbf{r}$ is the position vector of particle with respect to charge q_2, i.e. force experiencing particle with respect to force exerting particle. The constant $K = \dfrac{1}{4\pi\varepsilon_0} = 9 \times 10^9 \text{ N-m}^2/\text{C}^2$ (or meter per Farad). Here ε_0 is the *electric permittivity of free space* and has the value

$$\varepsilon_0 = \frac{1}{4\pi K} = 8.85 \times 10^{-12} \text{ Farad/m}$$

Thus equation (18.2) can be written

$$F = \frac{q_1 q_2}{4\pi\varepsilon_0 r^2}$$

This equation helps us to define 1 Coulomb. If $q_1 = q_2 = 1$ and $r = 1$ m, then $F = \dfrac{1}{4\pi\varepsilon_0} = 9 \times 10^9$ N.

Thus 1 Coulomb is a charge which when placed from a similar charge at a distance of 1 m in vacuum, experiences a repulsive force equal to 9×10^9 *Newton.*

Dielectric Constant or Relative Electrical Permittivity

When the charges are situated in a medium other than free space (vacuum or air), the force between them is,

$$F = \frac{q_1 q_2}{4\pi\varepsilon r^2}$$

where ε is called *absolute electrical permittivity of the medium.*

The force between these two charges when placed at the same distance in vacuum is

$$F_0 = \frac{q_1 q_2}{4\pi\varepsilon_0 r^2}$$

and the ratio of the forces is

$$\frac{F_0}{F} = \frac{\varepsilon}{\varepsilon_0} = \varepsilon_r$$

where ε_r is *called relative electric permittivity of the medium. It is also called dielectric constant of medium.*

The dielectric constant is *the ratio of the force between two charges when placed at certain distance in vacuum to the force between the same two charges when held at same distance in medium. It is also equal to the ratio of the absolute electric permittivity of medium and electrical permittivity of free space.*

The value of ε_r depends only on the nature of medium and is always greater than 1.

About Coulomb's law and Coulomb force it is worth to note that:

(*a*) The law is not derived from any physical concept. The law is based on physical observations and empirically deduced. Experiments till today reveal its universal nature.

(*b*) The force is a two body interaction, i.e. the electrical force between two point charges is independent of the presence or absence of other charges as such the principle of superposition is valid. Hence the net force on a charged particle due to a number of point charges is the vector sum of the forces due to individual point charges.

(*c*) The force is conservative, i.e. work done in moving a point charge does not depend on path followed and total work done in moving a point charge under the action of Coulomb force once round a closed path is zero.

(*d*) The law expresses the force between two point charges at rest. In case of extended bodies of finite size, we must assume that whole charge is concentrated at its centre. This assumption is true only for spherical bodies, that too for external points.

(*e*) The equilibrium of a charged particle under the action of a Coulombian force alone can never be stable. This statement is called *Earnshaw's theorem.*

(*f*) The force is action-reaction pair, i.e. the force which one charge exerts on the other is equal and opposite to the force which the other charge exerts on the first.

(*g*) The law is analogous to Newton's law of gravitation $F = \frac{G m_1 m_2}{r^2}$ with the difference:

(*i*) the electric force between charges is much stronger than gravitational force, that is why when both are present, we neglect gravitational force.

(*ii*) gravitational force is always attractive whereas electric force can be attractive as well as repulsive.

(*iii*) the electric force depends on the nature of the medium whereas gravitational force does not.

18.3 ELECTRIC FIELD

The electric field due to a charge or distribution of charges is the space around it in which its electrical effects are felt, i.e. the electrostatic force of attraction or repulsion due to this charge distribution is experienced by any other charge. Thus electric force between two charged particles can be seen as a two step process. A charge produces an electric field in space around it and this electric field exerts a force on any charge (except the source charge itself) placed in it. This electric field is there even if there is no other charge to experience force. The field takes finite time to propagate. The propagation speed of field is the speed of light (= c). This means that if a charge is displaced from its position, the field at distance r will take time $t = \frac{r}{c}$ to change.

Intensity of Electric Field

The intensity of electric field at any point in the electric filed is defined as *the force experienced by a unit positive point charge (called test charge) placed at that point*. Here by test charge, we mean a charge that exerts no force on the source charge but experiences force due to it.

If a test charge Δq in an electric field experience a force $\boldsymbol{F}$ at any point then the intensity of electric field at the point is given by

$$\boldsymbol{E} = \underset{\Delta q \to 0}{\text{Lt}} \frac{\boldsymbol{F}}{\Delta q} \quad \ldots(18.3)$$

The intensity of electric field is a vector quantity, the direction is same as that of the force and its SI units are Newton per Coulomb and dimensions $[M^1L^1T^{-3}I^{-1}]$.

From the definition of the intensity of electric field it is evident that a charge (q) either at rest or in motion in an electric field ($\boldsymbol{E}$) experiences a force

$$\boldsymbol{F} = q\boldsymbol{E}$$

in the direction of electric field in case of a positive charge and opposite to the field if it is negative charge.

The electric filed produced by the stationary charges is a conservative field. If $\boldsymbol{E}_0$ is the electric field in free space then in medium of dielectric of constant ε_r its value $\boldsymbol{E} = \boldsymbol{E}_0/\varepsilon_r$.

Electric Field Due to a Point Charge

Let a point charge q_0 be placed at a point A (Fig. 18.2) and let us find the electric field $\boldsymbol{E}$ at a point P at a distance r from q_0. For this purpose, we assume a test charge q to be placed at P. The charge q_0 creates a field at P which exerts force $q\boldsymbol{E}$ on charge q. The force on charge q according to Coulomb's law is

$$F = \frac{q_0 q}{4\pi\varepsilon_0 r^2} \quad \text{along } AP$$

Hence electric field at P is

$$E = \frac{F}{q} = \frac{q_0}{4\pi\varepsilon_0 r^2} \quad \text{along } AP$$

Fig. 18.2

In vector notation

$$\boldsymbol{E} = \frac{q_0}{4\pi\varepsilon_0 r^2}\hat{r} = \frac{q_0 \boldsymbol{r}}{4\pi r^3} \quad \ldots(18.4)$$

The electric field due to a set of charges is the vector sum of the electric fields due to individual charges.

Electric Field at any Point on the Axis of a Uniformly Charged Ring

Figure 18.3 shows a ring of radius a uniformly charged with charge q and we have to find the electric field at point P at a distance x from the centre O of the ring along the axis of ring. Let us consider a small element of the ring at point A having charge dq. The field at P due to this element is

$$dE = \frac{dq}{4\pi\varepsilon_0 (AP)^2}$$

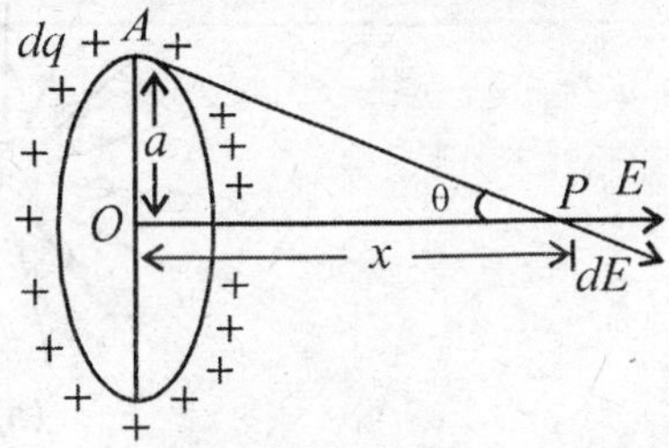

Fig. 18.3

Because of symmetry, the field at P will be along the axis because the component of electric field parallel to the plane of the ring averages out to be zero. The component of dE along the axis is $dE\cos\theta$

$$\therefore \quad dE\cos\theta = \frac{dq\cdot x}{4\pi\varepsilon_0\left(a^2+x^2\right)^{3/2}} \qquad \because AP = \sqrt{a^2+x^2}$$

Hence net field at P is

$$E = \int dE\cos\theta = \int \frac{x\,dq}{4\pi\varepsilon_0\left(a^2+x^2\right)^{3/2}} = \frac{x}{4\pi\varepsilon_0\left(a^2+x^2\right)^{3/2}}\int dq$$

$$E = \frac{xq}{4\pi\varepsilon_0\left(a^2+x^2\right)^{3/2}} \qquad \text{...(18.5)}$$

The direction is along the axis:

(*i*) When $x = 0$, $E = 0$, i.e. at the centre electric field is zero.

(*ii*) The field is maximum at $x = \pm\frac{a}{\sqrt{2}}$ and maximum field $E_{max} = \frac{q}{6\sqrt{3}\pi\varepsilon_0 a^2}$

(*iii*) The field is zero again at $x = \infty$

(*iv*) When $x \ll a$, $x^2 \ll a^2$, then the field is $E = \frac{q\cdot x}{4\pi\varepsilon_0 a^3}$

The variation of electric field with x is shown in Fig. 18.4.

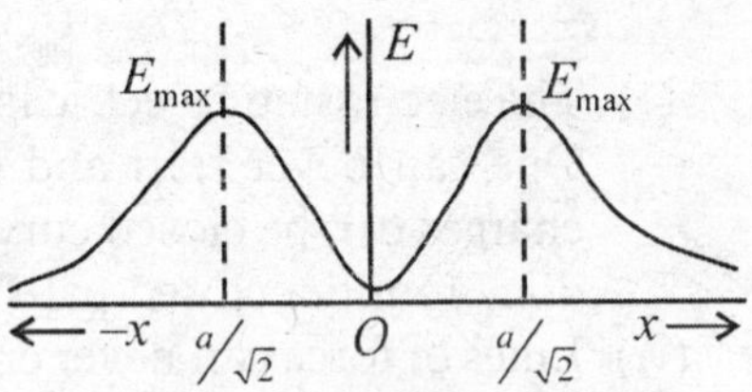

Fig. 18.4

18.4 ELECTRIC LINES OF FORCE

Michael Faraday gave the idea of lines of force and gives qualitative information about an electric field. An electric line of force is defined as a path straight or curved such that tangent to it at any point gives the direction of electric field intensity at that point. In fact lines of electric field are the graphical representation of the electric field. Some of the properties of lines of force are given below.

(*i*) A line of force in an electric field is a curve, the tangent to which at any point gives the direction of intensity at that point.

(*ii*) The number of lines of force per unit area normal to the surface surrounding that point gives the magnitude of the intensity at that point.

(*iii*) Electric lines of force start (i.e. diverge) from positive charge and terminate (converge) on negative charge or extend to

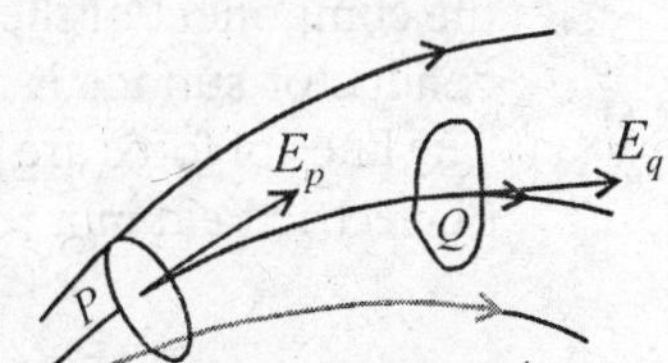

Fig. 18.5

infinity. Thus electric field due to a positive point charge is represented by straight lines originating from point (Fig. 18.6 (*a*)). The electric field due to negative point charge is represented by straight lines terminating at charge. (Fig. 18.6 (*b*)).

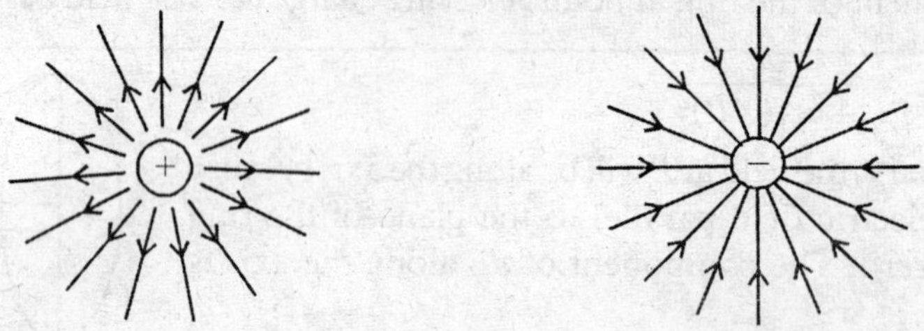

(*a*) (*b*)

Fig. 18.6

(*iv*) The lines of force for charge distribution containing more than one charge are shown in Fig. 18.7 (*a*) and (*b*), the lines may not be straight lines as one moves away from charge.

(*a*) (*b*)

Fig. 18.7

(*v*) The electrostatic electric lines of force can never be closed curves.
One can never start and end on same charge. Electric lines of force in case of induced charges can be closed curves.* In such cases the electric filed is non-conservative and work done round a closed path is not zero.

(*vi*) Lines of force can never cross each other. If they cross at a point, intensity at that point will have two directions, which is not possible.

(*vii*) The number of lines originating or terminating on a charge is proportional to the magnitude of charge. In MKS (SI) units the number of lines of force associated with unit charge (i.e. 1 Coulomb) is $\frac{1}{\varepsilon_0}(= 1.13 \times 10^{11})$.

(*viii*) The electric lines of force have a tendency to contract longitudinally like a stretched elastic string producing attraction between opposite charges and exert a lateral pressure to repel laterally resulting in repulsion between similar charges.

(*ix*) Electric lines of force end or start normally on the surface of a conductor. This means that the component of the electric field parallel to the conductor surface is zero. This is why the conductor surface is an *equipotential surface.*

(*x*) The lines of force are purely a geometrical construction which help us to visualise the nature of electric field in a region. They have no physical existence.

---

*A time varying magnetic field produces an electric field that produces eddy currents.

(*xi*) In case of field due to a set of collinear point charges (Fig. 18.7 (*c*)), *PL* is a part of line of force, then

$$q_1\cos\theta_1 + q_2\cos\theta_2 + q_3\cos\theta_3 + \ldots = \text{constant}$$

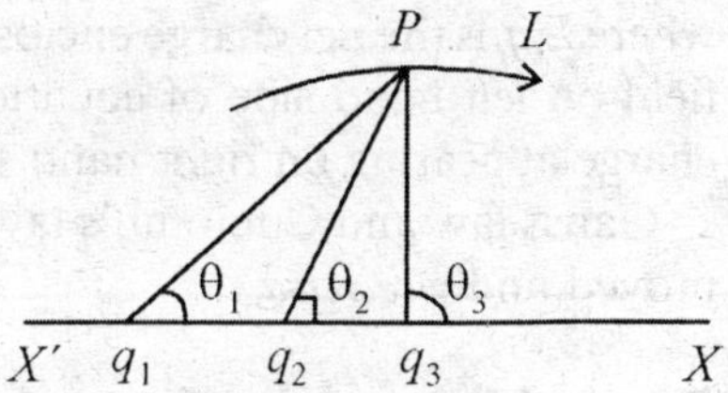

Fig. 18.7 (*c*)

18.5 ELECTRIC FLUX

If an area is considered in an electric field, then electric flux over this area is the total number of lines of force that cross this area. It is given by the product of the surface area and the component of electric field intensity normal to the area.

In electric field $\boldsymbol{E}$ a small area $\Delta \boldsymbol{s}$ is considered over which the field $\boldsymbol{E}$ is uniform (Fig. 18.8). The angle between outward drawn normal (considered positive normal) and $\boldsymbol{E}$ be θ then the quantity

$$\Delta\phi = E\Delta s\cos\theta$$

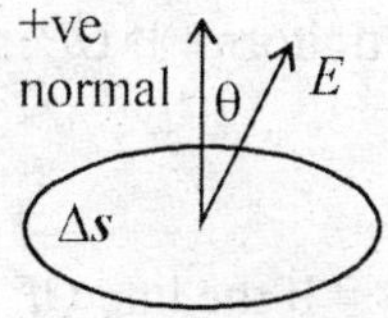

Fig. 18.8

is called the flux of the electric field through the considered surface. A vector of magnitude Δs in the direction normal to surface $\Delta \boldsymbol{s}$ is called area vector* $\Delta \boldsymbol{s}$ corresponding to Δs. Then we can write

$$\Delta\phi = \boldsymbol{E}\cdot\Delta\boldsymbol{s}$$

Flux is a scalar quantity and is added using scalar addition laws. This helps us to define the flux through surfaces which are not plane and when the field is not uniform. We can divide the given surface into smaller parts so that each part is approximately plane and the electric field over each part can be considered to be uniform then flux through each part $\Delta\phi = \boldsymbol{E}\cdot\Delta\boldsymbol{s}$ is calculated and total flux, then is

$$\phi = \Sigma\boldsymbol{E}\cdot\Delta\boldsymbol{s}$$

If $\Delta\boldsymbol{s}$ is small approaching zero, i.e. $\Delta\boldsymbol{s}\to 0$, the summation can be written in integral form. Then

$$\phi = \int_s \boldsymbol{E}\cdot d\boldsymbol{s} \qquad \ldots(18.6)$$

Here *s* means the integral is to be performed over the entire surface through which the flux is required. Right hand side of equation (18.6) is called *surface integral* of the electric field over surface *s*. Thus surface integral of the electric field over the surface *s* is the flux through the surface.

If surface under consideration is closed, i.e. it encloses volume such as spherical surface, a cubical surface with six sides, etc., the flux through closed surface is written as

$$\phi = \oint \boldsymbol{E}\cdot d\boldsymbol{s} \qquad \ldots(18.6a)$$

In case of closed surfaces an outward drawn normal is taken as positive.

Units for electric flux are Nm^2C^{-1} and dimensions for flux are $M^1L^3T^{-3}I^{-1}$.

18.6 GAUSS LAW

Gauss law states that the *total normal flux over a closed surface in vacuum is* $\frac{1}{\varepsilon_0}$ *times the total charge enclosed by the surface.* Mathematically this can be written as

$$\oint \boldsymbol{E}\cdot d\boldsymbol{s} = \frac{1}{\varepsilon_0}\Sigma_i q_i \qquad \ldots(18.7)$$

---

*The direction of the area vector is always along the normal to the corresponding surface. If $\boldsymbol{E}$ is perpendicular to the surface, it is parllel to the area vector. If $\boldsymbol{E}$ is along positive normal $\theta = 0$. If it is opposite to the positive normal then $\theta = \pi$. If electric field $\boldsymbol{E}$ is paralled to the surface, $\theta = \pi/2$.

where $\Sigma_i q_i$ is the net charge enclosed by the closed surface. It should be remembered that the electric field on left hand side of equation (18.7) is due to all the charges present in the space whereas charge appearing on right hand side includes only those which are enclosed by the surface.

Gauss law and Coulomb's law are equivalent. If we assume Coulomb's law, Gauss law can be proved and *vice versa*.

Proof of Gauss Law (Assuming Coulomb's Law)

Consider a charge q, enclosed by a closed surface (called Gaussian surface)* (Fig. 18.9). Consider a small area ds on the surface around a point P, over which the electric field is assumed to be uniform. If OP is r, then the electric field at P due to charge q is

$$E = \frac{q}{4\pi\varepsilon_0 r^2} \text{ along the line } OP$$

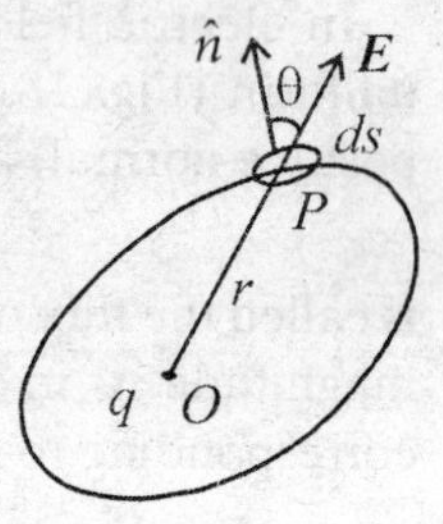

Fig. 18.9

If the line OP makes an angle θ with the outward drawn normal to ds; then the flux of electric field through ds is

$$d\phi = \mathbf{E} \cdot \mathbf{ds} = E\, ds \cos\theta$$

$$= \frac{q \cdot ds \cos\theta}{4\pi\varepsilon_0 r^2}$$

But $\dfrac{ds\cos\theta}{r^2} = d\Omega$ the solid angle subtended by area ds around P at O. Hence flux through ds is

$$d\phi = \frac{q}{4\pi\varepsilon_0} d\Omega$$

Hence flux through entire closed surface is

$$\phi = \oint d\phi = \oint \frac{q}{4\pi\varepsilon_0} d\Omega = \frac{q}{4\pi\varepsilon_0} \oint d\Omega$$

$\oint d\Omega$ is the total solid angle subtended by the closed surface and it is equal to 4π. Thus the total flux through the closed surface due to charge enclosed by it is

$$\phi = \frac{q}{4\pi\varepsilon_0} \cdot 4\pi = \frac{q}{\varepsilon_0} \qquad \ldots(18.8)$$

Flux Due to an External Charge

In case a point charge q is placed at a point outside the closed surface, the flux $d\phi$ of the electric field due to charge q through small area ds is again

$$d\phi = \frac{q}{4\pi\varepsilon_0} \frac{ds\cos\theta}{r^2} = \frac{q}{4\pi\varepsilon_0} d\Omega$$

Thus total flux through the whole closed surface is

$$\phi = \oint d\phi = \oint \frac{q}{4\pi\varepsilon_0} d\Omega = \frac{q}{4\pi\varepsilon_0} \oint d\Omega = 0$$

because the solid angle subtended by a closed surface at an external point is zero.

*A Gaussian surface is an imaginary closed surface of any shape.

Flux Due to Combination of Charges

If there are a number of charges, say q_1, q_2, q_3, ..., etc. inside the closed surface and Q_1, Q_2, Q_3, ..., etc. outside the surface; then each charge contributes to electric field at point *P* on the closed surface but the point charges outside the closed surface do not contribute to the flux. Hence using equation (18.8), we have

$$\oint \boldsymbol{E} \cdot d\boldsymbol{s} = \frac{1}{\varepsilon_0} \Sigma q_i$$

where $\boldsymbol{E}$ is the resultant electric field and Σq_i gives the total charge enclosed by the closed surface.

Gauss theorem holds good regardless of the shape of the Gaussian surface.

Applications of Gauss Theorem

(*a*) **Electric Field due to a Linear Infinite Charge Distribution**. Consider an infinite line charge of linear charge density λ. We have to calculate the electric field at point *P* at a distance *r* from the line charge (Fig. 18.10). The component of electric field parallel to the line charge is zero. If it is not then there will be force on line charge which is not possible.

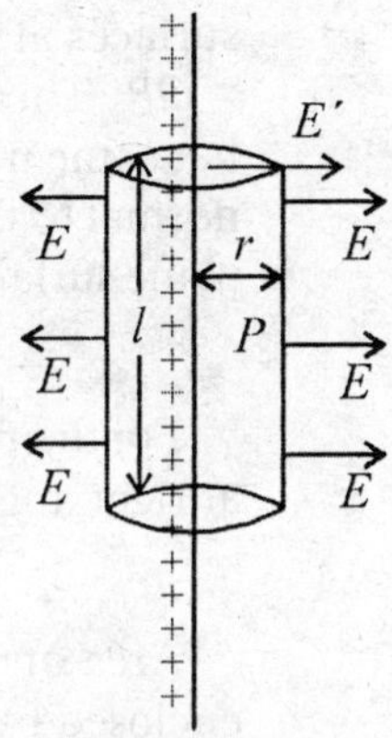

Fig. 18.10

Let us now imagine a Gaussian surface, which is a cylinder of radius *r* (*r* is the distance from line charge where the field is required), length *l* and line charge as its axis. This is shown in Fig. 18.10.

Thus the total flux through this closed surface is

$$\phi = \oint \boldsymbol{E} \cdot d\boldsymbol{s} = \int_{s_1} \boldsymbol{E} \cdot d\boldsymbol{s} + \int_{s_2} \boldsymbol{E} \cdot d\boldsymbol{s} + \int_{s_3} \boldsymbol{E} \cdot d\boldsymbol{s}$$

where s_1 and s_2 are the areas of plane surfaces and s_3 is the area of curved surface. In case of plane surfaces s_1 and s_2, the electric field vector and area vector are perpendicular to each other as such $\boldsymbol{E} \cdot d\boldsymbol{s} = 0$. Hence

$$\int_{s_1} \boldsymbol{E} \cdot d\boldsymbol{s} = \int_{s_2} \boldsymbol{E} \cdot d\boldsymbol{s} = 0$$

$$\therefore \qquad \phi = \int_{s_3} \boldsymbol{E} \cdot d\boldsymbol{s} = \int_{s_3} E ds = E \int_{s_3} ds = E \cdot 2\pi r l$$

For s_3, $\boldsymbol{E}$ and $d\boldsymbol{s}$ are parallel and $|\boldsymbol{E}|$ is same every where on the curved surface.

The charge enclosed by the Gaussian surface is λl. Using Gauss's theorem, we have

$$E \cdot 2\pi r l = \frac{1}{\varepsilon_0} \lambda l$$

or

$$E = \frac{\lambda}{2\pi\varepsilon_0 r} \qquad \text{...(18.9)}$$

In vector notation

$$\boldsymbol{E} = \frac{\lambda}{2\pi\varepsilon_0 r} \hat{n} \qquad \text{...(18.10)}$$

where $\hat{n}$ is unit vector normal to the line charge. The variation *E* with *r* is shown in Fig. 18.11. It is a rectangular hyperbola.

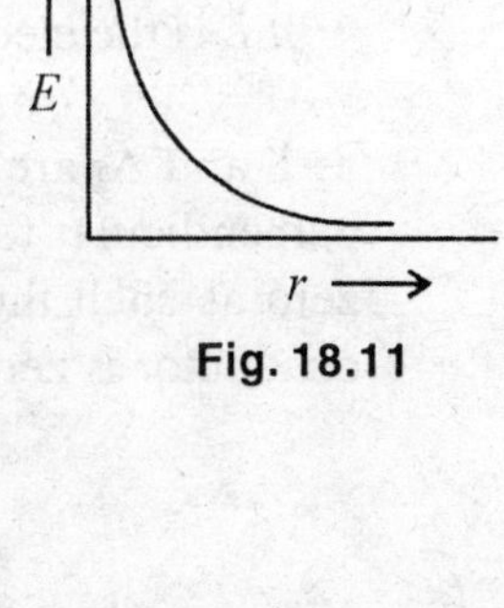

Fig. 18.11

If we take log, we have

$$\log E = \log\left(\frac{\lambda}{2\pi\varepsilon_0}\right) - \log r$$

A plot of $\log E$ and $\log r$ is a straight line with slope (= –1) and is shown in Fig. 18.12.

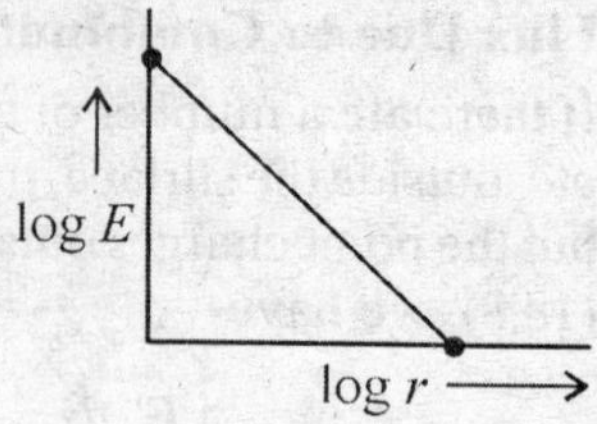

Fig. 18.12

(b) Electric Field due to Infinite Plane Sheet of Charge. In Fig. 18.13 a plane sheet of charge with surface density (charge per unit area) σ is shown. We will find an electric field at point *P* due to this charged sheet. In this case also the parallel component of electric field will be zero. The field will be along the perpendicular to the plane as shown in Fig. 18.13.

The Gaussian surface is a cylinder with two plane surfaces at *P* and *P'* parallel to the sheet and length *OP* = *OP'*. The electric field at all points on plane surfaces has same magnitude and direction is along the positive normal to the surface. Thus the flux through these two plane surfaces is

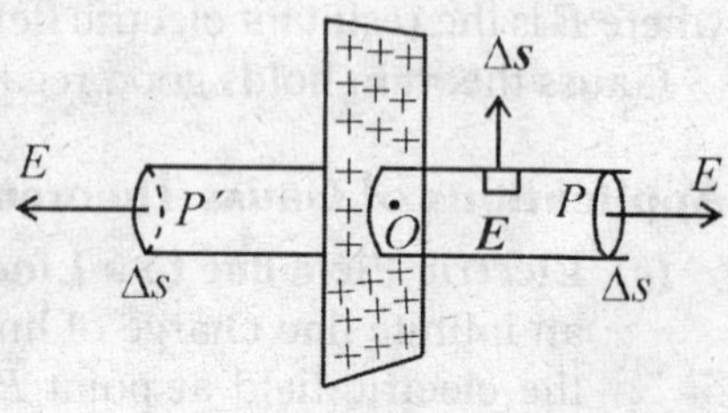

Fig. 18.13

$$\phi = 2\mathbf{E}\cdot\Delta\mathbf{s} = 2E\Delta s$$

For the points on the curved surfaces the outward drawn normal and the electric field are at right angles (θ = 90°) to each other as such $\mathbf{E}\cdot\Delta\mathbf{s} = 0$ for this surface.

$$\therefore \qquad \phi = \oint \mathbf{E}\cdot\Delta\mathbf{s} = 2E\,\Delta s + O = 2E\,\Delta s$$

Area of the charged sheet enclosed by the Gaussian surface (cylinder) is Δs. This charge enclosed by the Gaussian surface is $\sigma\Delta s$. Hence from Gauss' law, we have

$$2E\,\Delta s = \frac{1}{\varepsilon_0}\sigma\Delta s$$

or
$$E = \frac{\sigma}{2\,\varepsilon_0} \qquad \text{...(18.11)}$$

In vector notation

$$\mathbf{E} = \frac{\sigma}{2\,\varepsilon_0}\hat{n} \qquad \text{...(18.11a)}$$

where $\hat{n}$ is the unit normal vector, normal to the sheet.

We observe that the field is uniform and does not depend on distance from the charged sheet. This is true as long as the sheet is large as compared to its distance from *P*.

(c) Electric Field near a Charged Conducting Sheet. In Fig. 18.14 a large plane conducting sheet having uniform surface density of charge σ is shown. We have to find the electric field at point *P* outside this conducting sheet. In this case also the component of electric field parallel to the surface is zero and net electric field near the sheet is perpendicular to the surface. Gaussian surface in this case is a cylinder with cross-section Δs starting from point *P* and terminating in the interior of the conducting sheet.

It *E* is the electric field at *P*, the flux through the plane surface at *P* is

$$\phi = \mathbf{E}\cdot\Delta\mathbf{s} = E\Delta s$$

as **E** and $\Delta\mathbf{s}$ are parallel to each other at *P*. For curved part of the cylinder **E** and $\Delta\mathbf{s}$ are perpendicular to each other as such flux $\mathbf{E}\cdot\Delta\mathbf{s} = 0$. Since the field inside the conductor is zero as such the flux through the plane surface which terminates in the interior of the conductor is zero.

Thus the total flux through the Gaussian surface is $E\Delta s$. The charge enclosed inside the Gaussian surface is $\sigma\,\Delta S$. Hence from Gauss's theorem, we have

$$E\Delta s = \frac{1}{\varepsilon_0}\sigma\Delta s$$

or $$E = \frac{\sigma}{\varepsilon_0} \qquad ...(18.12)$$

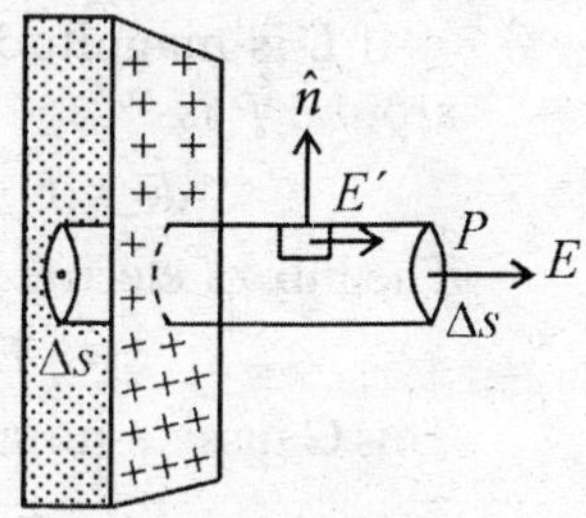

Fig. 18.14

In vector notation $$\mathbf{E} = \frac{\sigma}{\varepsilon_0}\hat{n} \qquad ...(18.12a)$$

This conducting sheet can be considered to be equivalent to two sheets of charge with surface denisty of charge σ on each side (Fig. 18.15). As such field at P outside the conducting sheet is

$$E = E_1 + E_2 = \frac{\sigma}{2\,\varepsilon_0} + \frac{\sigma}{2\,\varepsilon_0} = \frac{\sigma}{\varepsilon_0}$$

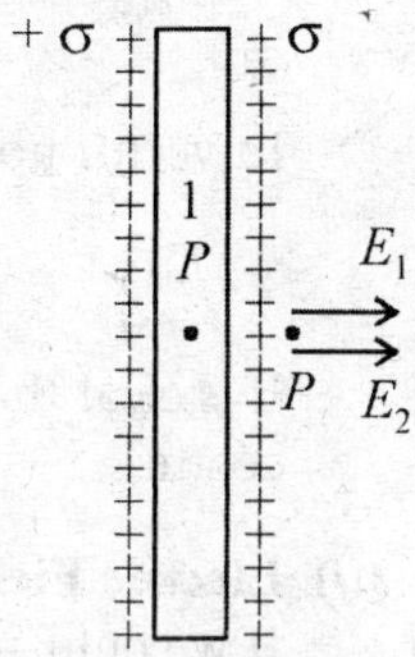

Fig. 18.15

It is same as derived from Gauss's law (Equation 18.12)

For interior points such as P' (Point inside the conductor), $\mathbf{E}_1$ and $\mathbf{E}_2$ have same magnitude $\frac{\sigma}{2\,\varepsilon_0}$ but opposite direction. The net field at P' is

$$E = E_1 - E_2 = \frac{\sigma}{2\,\varepsilon_0} - \frac{\sigma}{2\,\varepsilon_0} = 0$$

This shows that there is zero field inside the conductor. From this discussion it is apparent that electric field changes discontinuously at the surface of a conductor. Just outside the conductor, it is $\frac{\sigma}{\varepsilon_0}$ and inside it, it decreases gradually to zero in a small thickness of 4-5 atomic layers at the surface. Surface means this small thickness.

(*d*) Electric Field due to a Charged Sphere. In case of charged spheres, we will consider:

(I) A solid conducting charged sphere (total charge resides on surface)
(II) A charged spherical shell (Its thickness $t \to 0$).
(III) A uniform spherical distribution of charge.

In all the above charged spheres, we will consider the fields at points

(*i*) Outside the sphere $r > a$. Where r is the distance of the point from the centre and a is its radius.
(*ii*) On the surface of sphere $r = a$
(*iii*) Inside the surface of sphere $r < a$.

(*i*) Electric Field at Point outside the Sphere ($r > a$)

For external points all the three types of the spheres are identical.

Let us consider a sphere of radius a having charge q uniformly distributed over it. We have to find the electric field at point P at distance r from the centre of this sphere. The field in this case, will be radial by symmetry. Gaussian surface is a sphere of radius r concentric to the given sphere. The magnitude of the electric field at all the points on this Gaussian surface must be equal and outward for positive charge.

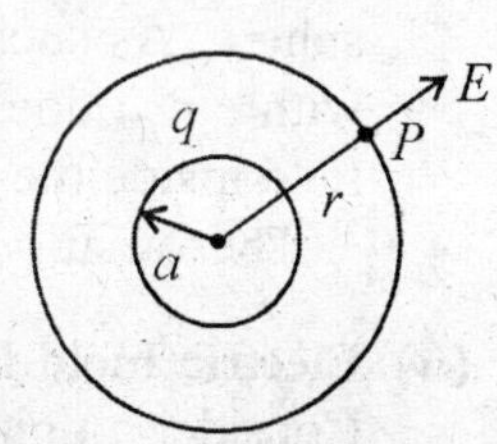

Fig. 18.16

If E is magnitude of the field at P, then the flux through a small element of area ds around P is

$$d\phi = \mathbf{E} \cdot \mathbf{ds} = E\,ds$$

The flux of electric field through this closed Gaussian surface (sphere) is

$$\phi = \oint \mathbf{E} \cdot \mathbf{ds} = \oint E\,ds = E\oint ds = E \cdot 4\pi r^2$$

This Gaussian surface encloses entire charge q on the sphere. Using Gauss theorem, we have

$$E \cdot 4\pi r^2 = \frac{1}{\varepsilon_0} \cdot q$$

or
$$E = \frac{q}{4\pi\varepsilon_0 r^2} \qquad \text{...(18.13)}$$

In vector notation

$$\mathbf{E} = \frac{q \cdot \hat{r}}{4\pi\varepsilon_0 r^2} = \frac{q\mathbf{r}}{4\pi\varepsilon_0 r^3} \qquad \text{...(18.13}a\text{)}$$

It is clear that for external points the charge on sphere behaves as if it is concentrated at the centre.

(*ii*) Electric Field on the Surface of the Sphere ($r = a$)

If we put $r = a$, we get the field on the surface. Thus

$$E = \frac{q}{4\pi\varepsilon_0 a^2} \qquad \text{...(18.14)}$$

In vector notation

$$\mathbf{E} = \frac{q\hat{n}}{4\pi\varepsilon_0 a^2} \qquad \text{...(18.14}a\text{)}$$

If we put $\dfrac{q}{4\pi a^2} = \sigma$ the surface density of charge, thus electric field

$$\mathbf{E} = \frac{\sigma}{\varepsilon_0}\hat{n}$$

The field is same as due to a charged conducting sheet having surface density of charge σ.

(*iii*) Electric Field Inside a Charged Sphere ($r < a$)

For interior points, the behaviour of uniform spherical distribution of charge is different from (*i*) the solid conducting charged sphere and (*ii*) charged spherical shell.

In case of solid conducting charged sphere and charged spherical shell there is no charge inside the sphere. As such the Gaussian surface taken as sphere with $r < a$ does not enclose any charge. Thus electric field inside the sphere is zero ($E = 0$).

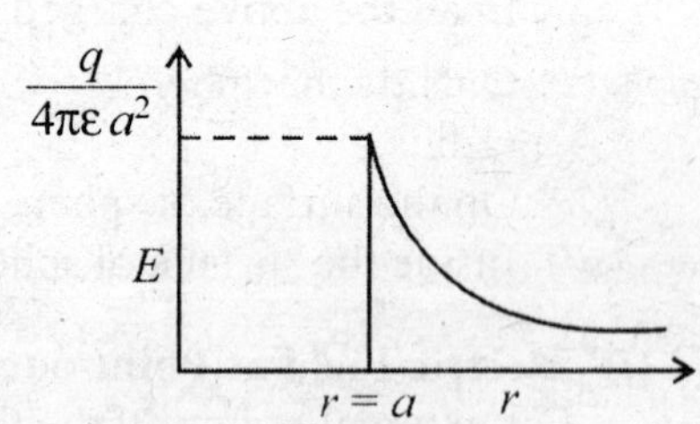

Fig. 18.17

The variation of electric field E with r is shown is Fig. 18.17.

(*iv*) Electric Field Inside a Uniform Spherical Distribution of Charge

Consider a uniform spherically distribution of charge q and radius a. We have to find the electric field at point P which is in inside at a distance r from the centre. Gaussian surface in this case is a sphere of radius r passing through point P and is concentric to the spherical distribution of charge: By symmetry all the points of this sphere are equivalent and the field is radial and has constant magnitude E. The flux through this spherical surface is

$$\phi = \oint \mathbf{E} \cdot \mathbf{ds} = \oint E ds = E \oint ds = E \cdot 4\pi r^2$$

The total charge enclosed by this Gaussian sphere is the charge within sphere of radius r.

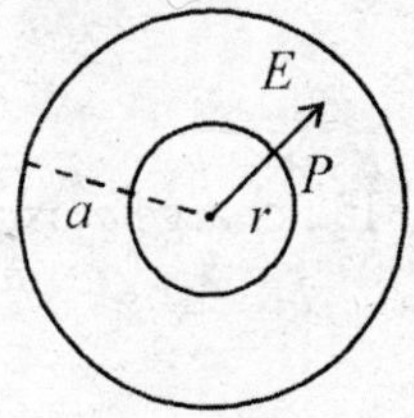

Fig. 18.18

Since volume charge density of the distribution is

$$\rho = \frac{q}{\frac{4}{3}\pi a^3} = \frac{3q}{4\pi a^3}$$

the total charge enclosed by the sphere of radius r then is

$$q' = \frac{4}{3}\pi r^3 \rho = \frac{4}{3}\pi r^3 \frac{3q}{4\pi a^3} = \frac{qr^3}{a^3}$$

Using Gauss's theorem, we have

$$E = 4\pi r^2 = \frac{1}{\varepsilon_0} \cdot q' = \frac{1}{\varepsilon_0} \cdot \frac{qr^3}{a^3}$$

or
$$E = \frac{q}{4\pi\varepsilon_0} \frac{r}{a^3} \qquad \text{...(18.15)}$$

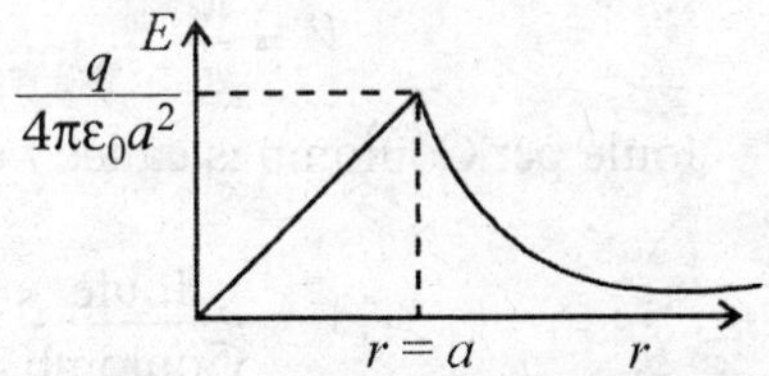

Fig. 18.19

Its direction is along the radius.

Thus the electric field at interior points is directly proportional to r, the distance from the centre. At the centre $r = 0$ and $E = 0$.

The variation of the electric field with distance from the centre of the sphere is shown in Fig. 18.19.

18.7 ELECTRIC POTENTIAL

The electric field in a region of space is described by assigning a vector quantity $\mathbf{E}$ at each point. The same field can also be described by assigning a scalar quantity V at each point. This scalar quantity is known as *electric potential.*

In an electric field, potential at any point is *defined as a physical quantity whose space rate of variation along any direction gives the intensity of electric field at that point and along that direction,* i.e.

$$\mathbf{E} = -\frac{dV}{dr}\hat{r} \qquad \text{...(15.16)}$$

where $\hat{r}$ is the unit vector in the direction $\mathbf{r}$. Negative sign is due to the fact that V decreases as r increases.

Taking dot product with $\mathbf{dr}$, we have

$$\mathbf{E} \cdot \mathbf{dr} = -\frac{dV}{dr}\hat{r} \cdot \mathbf{dr} = -dV$$

or
$$dV = -\mathbf{E} \cdot \mathbf{dr}$$

Integration gives

$$V - V_0 = -\int_{r_0}^{r} \mathbf{E} \cdot \mathbf{dr}$$

where r_0 is the reference point, when potential is V_0.

In case where inverse square law is followed generally reference point r_0 is taken at ∞ (infinity) and potential there (V_0) is assumed to be zero.

Hence potential at any point in electric field is

$$V = -\int_{\infty}^{r} \boldsymbol{E} \cdot \boldsymbol{dr}$$

If a charge q_0 is placed at point where field is $\boldsymbol{E}$ the force on this charge $\boldsymbol{F} = \boldsymbol{E} q_0$ or $\boldsymbol{E} = \dfrac{\boldsymbol{F}}{q_0}$

$$\therefore \quad V = -\int_{\infty}^{r} \frac{\boldsymbol{F}}{q_0} \cdot \boldsymbol{dr} = -\frac{1}{q_0} \int_{r=\infty}^{r} \boldsymbol{F} \cdot \boldsymbol{dr} = -\frac{W}{q_0} \qquad \ldots(18.17)$$

where W is the work done in carrying the charge q_0 from infinity to point r.

Equation (18.17) helps us to define potential at any point in an electric field.

Potential at a point in an electric field is defined as the negative of the work done by the field (or actual work done by external agency) in carrying a unit charge (q_0 = 1 Coulomb) *from infinity to that point.*

In SI, MKS system of units; potential has volt as units

$$V = \frac{W}{q_0} = \text{Joule/Coulomb} = \text{Volt}$$

i.e. Joule per Coulomb is called volt. Dimensions of potential are

$$V = \frac{\text{Joule}}{\text{Coulomb}} = \frac{[\text{M}^1\text{L}^2\text{T}^{-2}]}{[\text{IT}]} = [\text{M}^1\text{L}^2\text{T}^{-3}\text{I}^{-1}]$$

In CGS system, unit of potential is esu of potential

$$1\text{ V} = \frac{1\text{ J}}{1\text{ C}} = \frac{10^7\text{ erg}}{3 \times 10^9\text{ esu of charge}} = \frac{1}{300}\text{ esu of potential}$$

$$= \frac{10^7\text{ erg}}{\frac{1}{10}\text{ emu of charge}} = 10^8\text{ emu of potential}$$

Potential at any Point Due to a Point Charge

If $\boldsymbol{E}$ is the intensity of electric field at any point in electric field, then from definition potential at that point is

$$\frac{dV}{dr}\hat{r} = -\boldsymbol{E}$$

But electric field at any point due to a point charge q is

$$\boldsymbol{E} = \frac{q.\hat{r}}{4\pi\varepsilon_0 r^2}$$

$$\therefore \quad \frac{dV}{dr}\hat{r} = -\frac{q}{4\pi\varepsilon_0}\frac{\hat{r}}{r^2}$$

Taking dot product with $\boldsymbol{dr}$, we have

$$dV = -\frac{q}{4\pi\varepsilon_0} \cdot \frac{dr}{r^2}$$

Integration gives ,

$$\int_0^V dV = -\frac{q}{4\pi\varepsilon_0}\int_\infty^r \frac{dr}{r^2}$$

or
$$V = \frac{q}{4\pi\varepsilon_0 r} \qquad \ldots(18.18)$$

In a medium of dielectric constant ε_r, the potential becomes

$$V = \frac{q}{4\pi\varepsilon_0\varepsilon_r r} \qquad \ldots(18.18a)$$

Potential at any Point due to a Spherical Charge Distribution at an External Point (*r* > *a*)

We have already seen that charge on a sphere for external points behaves as if it is concentrated at the centre. Hence potential at a point distant r from the centre of charged sphere is

$$V = \frac{q}{4\pi\varepsilon_0 r} \qquad \ldots(18.19)$$

Thus surface potential of a charged sphere of radius a having charge q is

$$V = \frac{q}{4\pi\varepsilon_0 a} \qquad \ldots(18.20)$$

Potential Due to Charged Spherical Conductor Hollow or Solid at an Internal Point (*r* < *a*)

In case of charged conductor hollow or solid, the field inside is zero. Hence

$$\frac{dV}{dr} = -E = 0$$

or
$$dV = 0 \text{ and } V = \text{constant.}$$

Thus potential inside the charged conducting sphere, hollow or solid is same every where inside the sphere. If the potential inside is different from the potential on surface then there will be field inside the sphere. To make field zero inside, the potential inside the sphere should be same as on the surface, i.e. the potential inside is

$$V = \frac{q}{4\pi\varepsilon_0 a} \qquad \ldots(18.21)$$

The graph between potential and distance of point from the centre of the sphere is shown in Fig. 18.20.

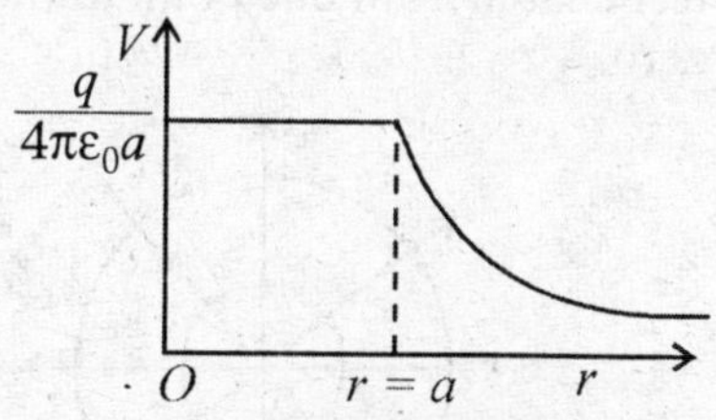

Fig. 18.20

Potential at an Interior Point (*r* < *a*) due to Uniform Spherical Distribution of Charge

In case of spherical distribution of charge, the electric field at an interior point is given by Equation 18.15.

$$\mathbf{E} = \frac{q}{4\pi\varepsilon_0}\frac{\mathbf{r}}{a^3}$$

If V is the potential at that point, then

$$\frac{dV}{dr}\hat{r} = -\mathbf{E} = -\frac{q}{4\pi\varepsilon_0}\frac{\mathbf{r}}{a^3}$$

or $$dV = -\frac{q}{4\pi\varepsilon_0}\frac{r\,dr}{a^3}$$

Integrating, we get

$$\int dV = -\frac{q}{4\pi\varepsilon_0}\int r\,dr$$

or $$V = -\frac{q}{4\pi\varepsilon_0 a^3}\frac{r^2}{2} + C \qquad C \text{ is constant of integration.}$$

But the surface potential of the spherical distribution of charge is $\frac{q}{4\pi\varepsilon_0}$, i.e. when $r = a$, $V = \frac{q}{4\pi\varepsilon_0 a}$.

This gives

$$C = \frac{3}{2}\frac{q}{4\pi\varepsilon_0 a}$$

$$\therefore \quad V = -\frac{q}{4\pi\varepsilon_0 a^3}\cdot\frac{r^2}{2} + \frac{3}{2}\cdot\frac{q}{4\pi\varepsilon_0 a}$$

$$= \frac{q}{4\pi\varepsilon_0 a}\frac{1}{2}\left(3 - \frac{r^2}{a^2}\right) \qquad \text{...(18.22)}$$

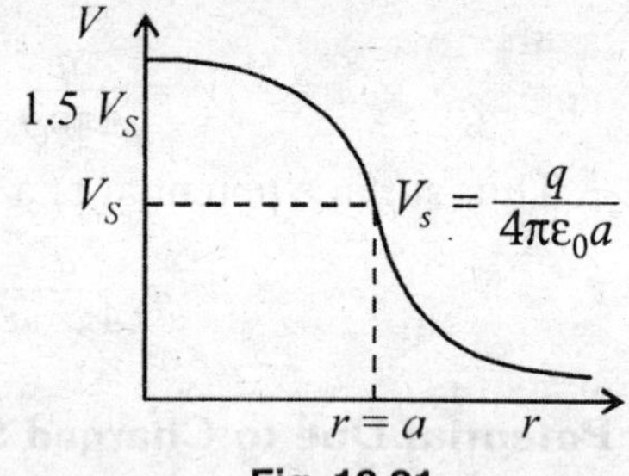

Fig. 18.21

Thus potential increases as we approach the centre of the sphere. The potential is maximum at the centre, i.e. $r = 0$.

$$V_c = \frac{3}{2}\frac{q}{4\pi\varepsilon_0 a} = \frac{3}{2}V_s \qquad \text{...(18.23)}$$

where V_s is the potential of the surface, i.e. potential at the centre is 1.5 times the surface potential.

The graph between potential and distance from the centre of the sphere is shown in Fig. 18.21.

Equipotential Surfaces

For a given charge distribution locus of all points having same potential is called equi-potential surface.

By definition, potential difference between two points is the work done in carrying a unit positive test charge from one point to the other. If the points lie on equipotential surface then work done is zero.

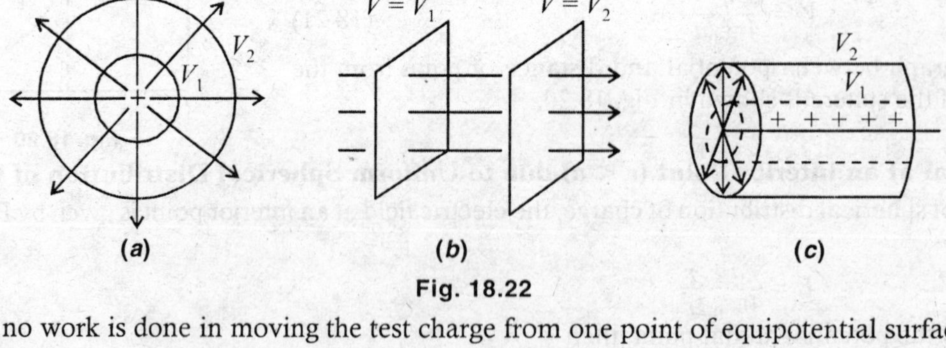

Fig. 18.22

Thus no work is done in moving the test charge from one point of equipotential surface to the other. Thus

$$dW = \mathbf{E}\cdot d\mathbf{l} = E\,dl\cos\theta = 0$$

or $\cos\theta = 0$ or $\theta = 90°$, i.e. E and dl are perpendicular, i.e. equipotential surfaces are always perpendicular to lines of force. Equipotential surfaces never cross each other.

In case of an isolated point charge, equipotential surfaces are the surfaces of concentric spheres with the charge at their centres. They are shown in Fig. 18.22 (*a*). For uniform electric field, the equipotential surfaces are parallel planes perpendicular to the field lines as shown in Fig. 18.22 (*b*). Figure 18.22 (*c*) shows equipotential surfaces for a line charge. They are coexist cylinders with line charge as axis.

18.8 ELECTRIC POTENTIAL ENERGY

The electric potential energy is the energy possessed by a system of point charges by virtue of their positions. If the position of one or more charges is changed, work may be done by the electric forces between these charges. Negative of this work done by electric forces is defined as the charge in potential energy of system of charges.

In absolute terms potential energy of the system (here system means system of charges) is defined as the work done in carrying the system to its present position from the position where potential energy is assumed to be zero.

In case of system of charges, where inverse square law is followed, the potential energy is assumed to be zero at infinity.

It should be noted that if work is done by the field, negative of the work done is the potential energy of the system and if the work is done by the external agency, this is stored in the form of potential energy.

Potential Energy of a System of Two Charges

Consider a system of two charges q_1 and q_2. Charge q_1 is fixed at A and charge q_2 is moved from point B to C along the line ABC (Fig. 18.23). Let the instantaneous position of q_2 from A be r and its distance from A changes from r to $r + dr$. The electric force on charge q_2 is

$$F = \frac{q_1 q_2}{4\pi\varepsilon_0 r^2}$$

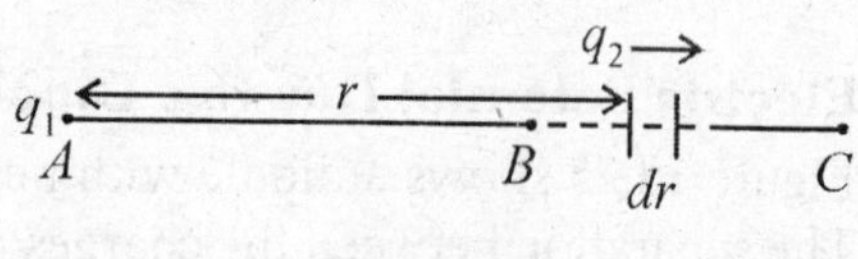

Fig. 18.23

The work done by this force in displacement dr is

$$dW = \frac{q_1 q_2}{4\pi\varepsilon_0 r^2}\, dr$$

Hence the total work done by this force when the charge moves from $B(AB = r_1)$ to $C\,(AC = r_2)$ is

$$W = \int_{r_1}^{r_2} \frac{q_1 q_2\, dr}{4\pi\varepsilon_0 r^2} = \frac{q_1 q_2}{4\pi\varepsilon_0}\left[\frac{1}{r_1} - \frac{1}{r_2}\right]$$

Since charge q_1 is fixed and is not displaced hence no work is done. Hence the change in potential energy from definition is

$$U(r_2) - U(r_1) = -W = \frac{q_1 q_2}{4\pi\varepsilon_0}\left[\frac{1}{r_2} - \frac{1}{r_1}\right] \qquad \text{...(18.24)}$$

As already stated, we take the potential energy of the two charge system to be zero when they are infinite distance apart. This means $U(\infty) = 0$. The potential energy when the separation is r is

$$r_1 = \infty,\ r_2 = r,\ U(r_2) = U(r),\quad U(\infty) = 0$$

$$\therefore \quad U(r) - 0 = \frac{q_1 q_2}{4\pi\varepsilon_0}\left(\frac{1}{r} - \frac{1}{\infty}\right) = \frac{q_1 q_2}{4\pi\varepsilon_0 r} \qquad \text{...(18.25)}$$

Equation (18.25) gives the potential energy of two charge system and depends on the separation between them and is independent of spatial location of the charged particles. Therefore equations (18.24) and (18.25) are general. Equation (18.25) gives the electric potential energy of a pair of charges. If there are three charges q_1, q_2, q_3 there are three pairs. Similarly for n particle system, there will be $\frac{n}{2}(n-1)$ pairs.

Electron Volt (eV)

It is the smallest practical unit of energy used in atomic and nuclear physics.

An electron volt is the energy acquired by an electron when accelerated by a potential difference of 1 volt.

$$1 \text{ eV} = 1.6 \times 10^{-19} \text{ J}$$

A particle of charge q when accelerated by a potential difference of V volts acquires energy

$$E = qVJ = \frac{qV}{e} \text{ eV}$$

18.9 ELECTRIC DIPOLE

An electric dipole is a combination of two equal and opposite charges $+q$ and $-q$, separated by a small distance a. The electric dipole moment of this combination is a vector defined as

$$\mathbf{P} = q\mathbf{a} \qquad \text{...(18.26)}$$

where $\mathbf{a}$ is a vector joining negative charge to the positive charge. The line joining the two charges is called the axis of the dipole. The distance in case of dipole are measured from the centre of the dipole.

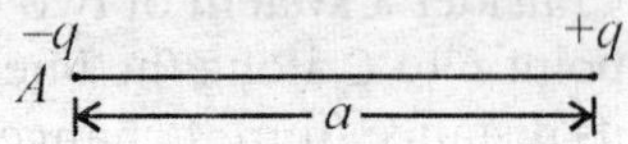

Fig. 18.24

Electric Potential Due to a Dipole

Figure 18.25 shows a dipole with negative charge $-q$ at A and the positive charge $+q$ placed at B. The separation between the charges is $AB = a$ and its mid-point is O. We have to find the potential at point P at distance r from O $(r \gg a)$ $\angle POB = \theta$.

Potential at P due to positive $(+q)$ at B is

$$V_1 = \frac{q}{4\pi\varepsilon_0} \cdot \frac{1}{BP}$$

The potential at P due to negative charge $(-q)$ at A is

$$V_2 = -\frac{q}{4\pi\varepsilon_0} \cdot \frac{1}{AP}$$

The net potential at P due to the dipole is

$$V = V_1 + V_2 = \frac{q}{4\pi\varepsilon_0}\left[\frac{1}{BP} - \frac{1}{AP}\right]$$

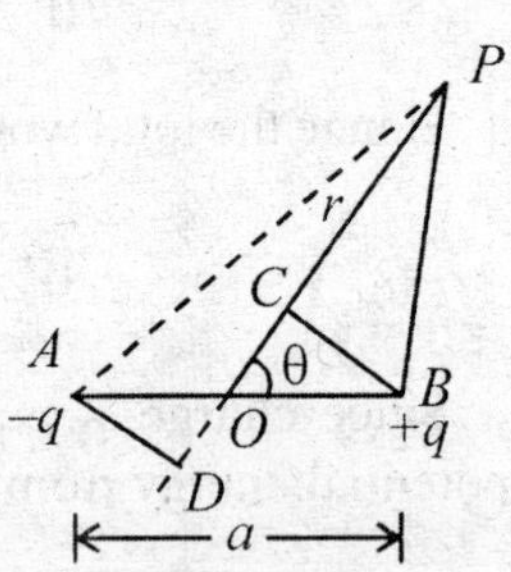

Fig. 18.25

Let BC be the perpendicular from B on OP. As $r \gg a$,

$$BP \approx PC = OP - OC = r - \frac{a}{2}\cos\theta$$

Similarly AD be perpendicular on PO produced.

$$AP \approx DP = OP + OD = r + \frac{a}{2}\cos\theta$$

Substituting values of BP and AP

$$V = \frac{q}{4\pi\varepsilon_0}\left[\frac{1}{r - \frac{a}{2}\cos\theta} - \frac{1}{r + \frac{a}{2}\cos\theta}\right]$$

$$= \frac{q}{4\pi\varepsilon_0}\frac{a\cos\theta}{r^2 - \frac{a^2}{4}\cos^2\theta} = \frac{qa\cos\theta}{4\pi\varepsilon_0 r^2}$$

Here $\frac{a^2}{4}\cos^2\theta$ being very small in comparison to r^2 and hence neglected

or $$V = \frac{P\cos\theta}{4\pi\varepsilon_0 r^2} \qquad \ldots(18.26)$$

or $$V = \frac{\boldsymbol{P}\cdot\hat{r}}{4\pi\varepsilon_0 r^2} = \frac{\boldsymbol{P}\cdot\boldsymbol{r}}{4\pi\varepsilon_0 r^3} \qquad \ldots(18.26a)$$

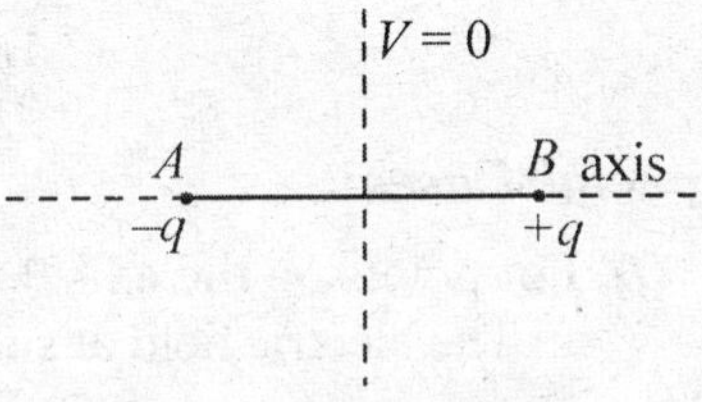

Fig. 18.26

Special cases:

For points on axis $\theta = 0$ and $V = \frac{P}{4\pi\varepsilon_0 r^2}$

For points on perpendicular bisector $\theta = \frac{\pi}{2}$ and $V = 0$.

Electric Field Due to a Dipole

We can find electric field due to a dipole using the expression for potential (equation 18.26) and the relation $E = -\frac{dV}{dr}$ (equation 18.16).

Figure 18.27 shows a dipole AB having moment $P(= qa)$. We have to find the electric field at point P, where potential is

$$V = \frac{P\cos\theta}{4\pi\varepsilon_0 r^2} \qquad \ldots(i)$$

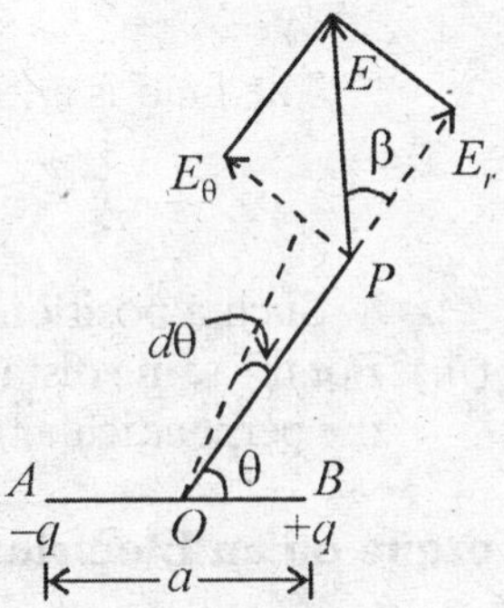

Fig. 18.27

In direction OP, θ is constant and only r varies, i.e. it is radial direction. Hence component of electric field in this direction

$$E_r = -\frac{\partial V}{\partial r} = -\frac{\partial}{\partial r}\left(\frac{P\cos\theta}{4\pi\varepsilon_0 r^2}\right)$$

$$\equiv \frac{2P\cos\theta}{4\pi\varepsilon_0 r^3} \qquad \ldots(ii)$$

In the transverse direction, i.e. perpendicular to OP, θ changes and for small change in θ (= $d\theta$), r is assumed constant, then space variation is $rd\theta$

Hence component of electric field in this direction (E_θ) is

$$E_\theta = -\frac{1}{r}\frac{\partial V}{\partial\theta} = -\frac{1}{r}\frac{\partial}{\partial\theta}\left(\frac{P\cos\theta}{4\pi\varepsilon_0 r^2}\right)$$

$$= \frac{P \sin\theta}{4\pi\varepsilon_0 r^3} \quad \ldots(iii)$$

The resultant electric field at P (Fig. 18.27), is then

$$E^2 = E_r^2 + E_\theta^2 = \frac{P^2}{(4\pi\varepsilon_0 r^3)^2}\{4\cos^2\theta + \sin^2\theta\}$$

or

$$E = \frac{P}{4\pi\varepsilon_0 r^3}\sqrt{1+3\cos^2\theta} \quad \ldots(18.27)$$

If the resultant field E at P makes an angle β with radial direction OP then

$$\tan\beta = \frac{E_\theta}{E_r} = \frac{\dfrac{P\sin\theta}{4\pi\varepsilon_0 r^3}}{\dfrac{2P\cos\theta}{4\pi\varepsilon_0 r^3}} = \frac{1}{2}\tan\theta \quad \ldots(18.28)$$

Special Cases:

(*i*) For points on the axis $\theta = 0$

The electric field at such points is

$$E = \frac{2P}{4\pi\varepsilon_0 r^3} \text{ and } \tan\beta = 0 \text{ or } \beta = 0$$

i.e., the field is in the direction of dipole. Such a position of the point P is called an *end on position* or *tan A position.*

(*ii*) For points on the perpendicular bisector of line joining charges, $\theta = 90°$

$$E = \frac{P}{4\pi\varepsilon_0 r^3} \text{ and } \beta = \frac{\tan\theta}{2} = \alpha \text{ or } \beta = 90°$$

The field is antiparallel to the direction of dipole moment, i.e.

$$\mathbf{E} = -\frac{\mathbf{P}}{4\pi\varepsilon_0 r^3}$$

Such a position of the point P is called a *broad side on or tan β position.*

(*iii*) For the same distance, the electric field on the axial point is twice in magnitude than that on the perpendicular bisector point.

Torque on an Electric Dipole Placed in an Electric Field

Consider an electric dipole consisting of two equal and opposite charges $+q$ and $-q$ separated by a small distance a (dipole moment $P = qa$) placed in a uniform electric field $\mathbf{E}$

The dipole axis at any instant makes an angle θ with the external field $\mathbf{E}$

Force F_1 on charge $+q$ at $B = qE$ along the direction of $\mathbf{E}$.

Force F_2 on charge $-q$ at $A = qE$ in direction opposite to $\mathbf{E}$.

These two forces are equal in magnitude, opposite in direction and do not act at a single point; hence constitute a

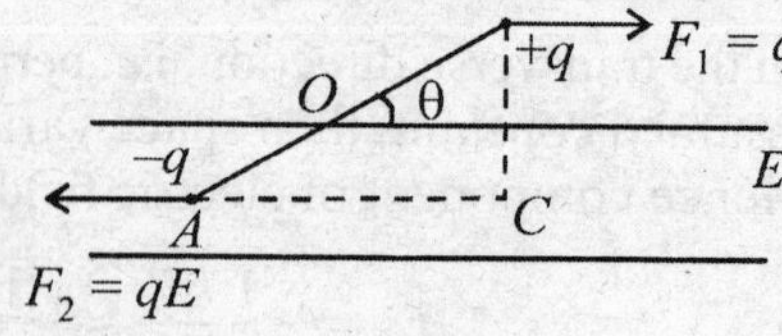

Fig. 18.28

couple, which tries to rotate the dipole to bring it in equilibrium position, i.e. parallel to the electric field. Thus torque

$$\tau = \text{Force} \times \text{perpendicular distance}$$
$$= qE(BC) \qquad \text{But } BC = a\sin\theta$$
$$= qEa\sin\theta$$
$$= PE\sin\theta \qquad \because P = qa$$

In vector form torque

$$\tau = \mathbf{P} \times \mathbf{E} \qquad \text{...(18.29)}$$

The direction of the torque is perpendicular to the plane containing the dipole axis and the electric field. In Fig. 18.28, this is perpendicular to the plane of paper and is into the paper and magnitude is $PE\sin\theta$.

Small Oscillations of an Electric Dipole in Uniform Electric Field

When an electric dipole (dipole moment $\mathbf{P}$) is placed in uniform electric field ($\mathbf{E}$), its equilibrium position is, where its axis is parallel to the field direction. Now, if the dipole is slightly displaced through a small angle θ, it experiences a torque

$$\tau = PE\sin\theta$$

which tries to bring the dipole parallel to the field direction.

If θ is small $\sin\theta \approx \theta$ and torque

$$\tau = PE\,\theta$$

If instantaneous angular acceleration is α and I moment of inertia of dipole about the axis of rotation, then

$$I\alpha = -PE\theta$$

or

$$\alpha = -\frac{PE}{I}\theta$$

Thus angular acceleration is directly proportional to angular displacement. Hence motion is S.H.M. and its time period

$$T = 2\pi\sqrt{\frac{I}{PE}} \qquad \text{...(18.30)}$$

Dipole in Non Uniform Field

In non uniform electric field some net force and torque do act on the dipole. In such case, the centre of mass of the dipole moves in the direction of increasing field and the dipole oscillates. Thus the dipole has oscillatory and translatory motions together. When dipole axis is parallel to the field the torque becomes zero and there is net force on the dipole as such the centre of mass of dipole moves linearly in the direction of the force.

Potential Energy of Dipole in a Uniform Electric Field

When an electric dipole is placed in a uniform electric field $\mathbf{E}$, a torque $\tau = \mathbf{P} \times \mathbf{E}$ acts on it. If we rotate the dipole through a small angle $d\theta$, the work done by the torque is

$$dW = \tau d\theta = -PE\sin\theta \cdot d\theta$$

Work done by the field is negative as rotation $d\theta$ is opposite to the torque. Hence work done in rotating the dipole from angle θ_1 to θ_2 is

$$W = \int_{\theta_1}^{\theta_2} -PE\sin\theta \cdot d\theta = PE(\cos\theta_2 - \cos\theta_1) \quad \text{...(18.31)}$$

Equation (18.31) gives the work done by the field in rotating the dipole from θ_1 to θ_2.

In case of dipole in electric field, potential energy is taken zero when dipole axis is perpendicular to the electric field. Hence the potential energy of a dipole in a uniform electric field is negative of the work done in rotating the dipole from $\theta_1 = 90°$ to $\theta_2 = \theta$. Thus

$$U = -W = -PE(\cos\theta - \cos 90°)$$

$$= -PE\cos\theta$$

or

$$U = -\boldsymbol{P} \cdot \boldsymbol{E} \quad \text{...(18.32)}$$

Dipole-Dipole Interaction

When a dipole (moment P_2) is placed in an electric field of another dipole (moment P_1). The dipoles experience force and torque which depends on their relative positions. In different positions their potential energy is also different. Here we take three different orientations of the dipoles and give direct formulae as the theory is beyond the scope of this book.

Two Dipoles in Same Direction

(*i*) Relative orientation for this situation is shown in Fig. 18.29.

Force $F = -\dfrac{1}{4\pi\varepsilon_0} \cdot \dfrac{6P_1P_2}{r^4}$ along r

Couple $\tau = 0$

and potential energy $U = -\dfrac{1}{4\pi\varepsilon_0} \cdot \dfrac{2P_1P_2}{r^3}$

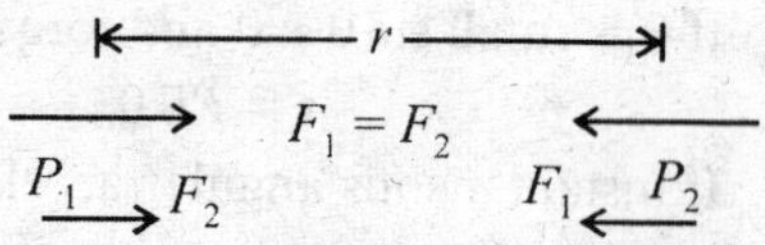

Fig. 18.29

(*ii*) Relative orientation for this situation is shown in Fig. 18.30.

Force $F = \dfrac{1}{4\pi\varepsilon_0} \cdot \dfrac{3P_1P_2}{r^4}$ along r

Couple $\tau = 0$

and potential energy $U = +\dfrac{1}{4\pi\varepsilon_0} \cdot \dfrac{P_1P_2}{r^3}$

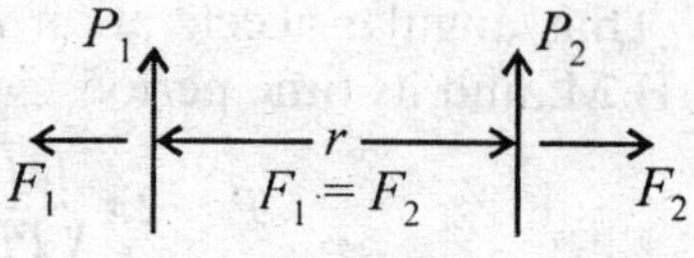

Fig. 18.30

Two Dipoles Mutually Perpendicular to Each Other

Relative orientation in this situation is shown in Fig. 18.31.

Force $= \pm\dfrac{1}{4\pi\varepsilon_0} \cdot \dfrac{P_1P_2}{r^4}$ perpendicular to r

Couple $\tau = \dfrac{1}{4\pi\varepsilon_0} \cdot \dfrac{2P_1P_2}{r^3}$ (CW) on P_1

$\tau = \dfrac{1}{4\pi\varepsilon_0} \cdot \dfrac{P_1P_2}{r^3}$ (CW) on P_2

Potential energy = 0

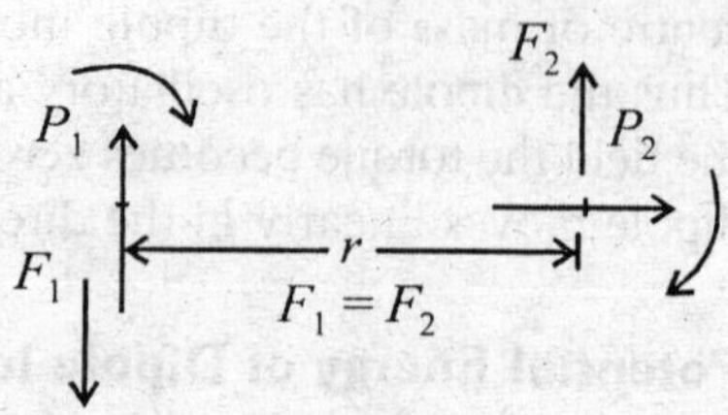

Fig. 18.31

18.10 FORCE ON A CHARGED CONDUCTOR

When a conductor is charged, it experiences a force. If σ is the surface density of charge then force per unit area on the conductor is

$$\frac{dF}{ds} = \frac{\sigma^2}{2\,\varepsilon_0} = \frac{1}{2}\varepsilon_0 E^2 \qquad \text{...(18.33)}$$

where $E = \dfrac{\sigma}{\varepsilon_0}$ the electric field near the conductor. The direction of this force is always outward normal to the conductor whether it is positively charged or negatively charged.

A soap bubble in vacuum when charged expands. In equilibrium the excess pressure inside the bubble becomes equal to electrostatic pressure. If σ is the surface density of charge on the bubble and T surface tension of soap solution and r its radius, then

$$\text{Electrostatic pressure} = \frac{1}{2}\frac{\sigma^2}{\varepsilon_0}$$

and excess pressure due to surface tension $= \dfrac{4T}{r}$

in equilibrium $\dfrac{1}{2}\dfrac{\sigma^2}{\varepsilon_0} = \dfrac{4T}{r}$ $\quad \sigma = \dfrac{q}{4\pi r^2}$, q is charge given to the bubble.

When conductor is charged, it becomes equipotential surface. Potential every where is same inside as well as on the surface. But the distribution of charge on the conductor is not uniform. The surface density of charge varies inversely as the radius of curvature of that part of the conductor, i.e. $\sigma \propto \dfrac{1}{r}$. Sharper the point, greater is the surface density σ.

The field in a cavity inside a conductor is zero resulting *electrostatic shielding*.

The energy density (Energy per unit volume) in space where electric field is E, is

$$U = \frac{1}{2}\varepsilon_0 E^2 \qquad \text{...(18.34)}$$

SOLVED EXAMPLES

Example 18.1: An α-particle passes rapidly through the exact centre of a hydrogen molecule, moving on a line perpendicular to the inter-nuclear axis. The distance between nuclei is b. Where on its path, does the α-particle experience the maximum force and find its value?

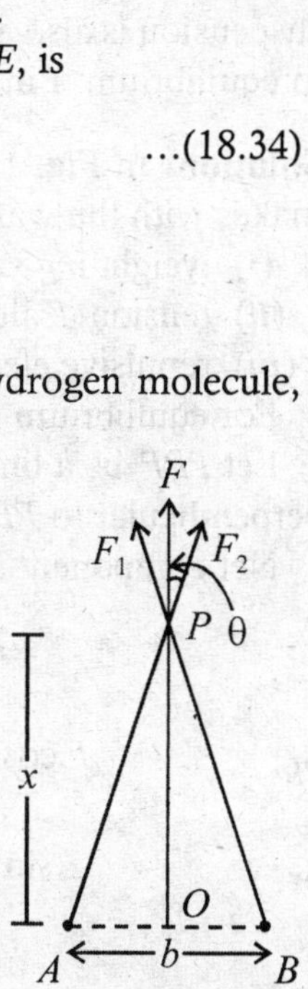

Fig. 18.32

Solution: α-particle when at distance x experience a force

$$\boldsymbol{F} = \boldsymbol{F}_1 + \boldsymbol{F}_2$$

However $\quad |\boldsymbol{F}_1| = |\boldsymbol{F}_2|$

Components of forces perpendicular to the line OP cancel each other, while along OP add up. Thus net force

$$F = 2F_1 \cos\theta$$

But $\quad F_1 = \dfrac{1}{4\pi\varepsilon_0}\dfrac{(2e)e}{(x^2+a^2)} \qquad a = \dfrac{b}{2}$ and $\cos\theta = \dfrac{x}{\sqrt{(x^2+a^2)}}$

where $2e$ is charge on α-particle and e charge on hydrogen atom.

$$\therefore \quad F = \frac{2 \times 2e^2}{4\pi\varepsilon_0 (x^2 + a^2)} \times \frac{x}{\sqrt{x^2 + a^2}}$$

$$= \frac{e^2}{\pi\varepsilon_0} \frac{x}{(x^2 + a^2)^{\frac{3}{2}}} \qquad \ldots(i)$$

For F to be maximum

$$\frac{dF}{dx} = 0, \text{ i.e. } \frac{e^2}{\pi\varepsilon_0} \cdot \frac{d}{dx}\left\{\frac{x}{(x^2 + a^2)^{\frac{3}{2}}}\right\} = 0$$

or $$\frac{1}{(x^2 + a^2)^{\frac{3}{2}}} - \frac{3}{2} \frac{x \cdot 2x}{(x^2 + a^2)^{\frac{5}{2}}} = 0$$

or $$x^2 + a^2 - 3x^2 = 0 \text{ or } x = \pm\frac{a}{\sqrt{2}} = \pm\frac{b}{2\sqrt{2}}$$

Substituting value of x in (*i*), we get maximum force

$$F_{\max} = \frac{8e^2}{3\sqrt{3}\pi\varepsilon_0 b^2}.$$

Example 18.2: A particle A having a charge q is fixed in a vertical wall. A second particle B of mass m and having equal charge is suspended by a silk thread of length l from the wall. The point of suspension is also above the particle A. Find the angle of the thread with the vertical when it stays in equilibrium. Find the angle if $q = 5 \times 10^{-7}$ Coulomb, $l = 30$ cm and m = 100 gm.

Solution: In Fig. 18.33, point O is the point of suspension and let θ is the angle that the thread makes with the wall. Forces on B are

(*i*) weight mg vertically downwards

(*ii*) tension T along the thread

(*iii*) repulsive electric force F along AB.

For equilibrium net force on B should be zero.

Let PBP' be a line perpendicular to OB. Taking components along and perpendicular to PBP'.

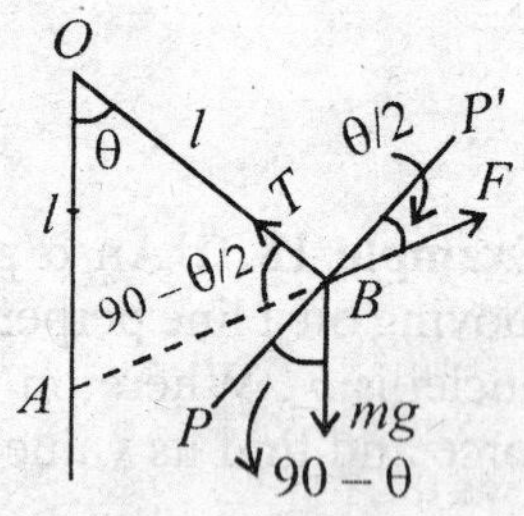

Fig. 18.33

Net component along PBP' is

$$F_1 = F\cos\frac{\theta}{2} - mg\cos(90 - \theta) = F\cos\frac{\theta}{2} - mg\sin\theta = 0$$

or $$F\cos\frac{\theta}{2} = 2mg\sin\frac{\theta}{2}\cos\frac{\theta}{2}$$

or $$\sin\frac{\theta}{2} = \frac{F}{2mg} \qquad \ldots(i)$$

Now $$F = \frac{1}{4\pi\varepsilon_0} \cdot \frac{q^2}{AB^2}$$

But $$AB = 2l\sin\frac{\theta}{2}$$

$\therefore$ $$F = \frac{1}{4\pi\varepsilon_0}\frac{q^2}{l^2\sin^2\frac{\theta}{2}\times 4}$$

Substituting in (*i*), we have

$$\sin\frac{\theta}{2} = \frac{1}{4\pi\varepsilon_0}\frac{q^2}{4l^2\sin^2\frac{\theta}{2}}\cdot\frac{1}{2mg}$$

or $$\sin^3\frac{\theta}{2} = \frac{q^2}{4\times 8\pi\varepsilon_0 l^2 mg}$$

or $$\sin\frac{\theta}{2} = \left(\frac{q^2}{4\times 8\pi\varepsilon_0 l^2 mg}\right)^{\frac{1}{3}}$$

and $$\theta = 2\cdot\sin^{-1}\left(\frac{q^2}{4\times 8\pi\varepsilon_0 l^2 mg}\right)^{\frac{1}{3}}$$

Substituting values, we have

$$\sin\frac{\theta}{2} = \left(\frac{5\times10^{-7}\times5\times10^{-7}\times9\times10^{9}}{4\times2\times0.3\times0.3\times0.1\times10}\right)^{\frac{1}{3}}$$

$$= (0.0032)^{\frac{1}{3}} = 0.15$$

or $$\theta = 17°.$$

Example 18.3: Find the electric field at a point *P* on the perpendicular bisector of a uniformly charged rod. The length of the rod is *L*, the charge on it is *q* and the distance of *P* from the centre of rod is *a*.

Solution: Charge per unit length of the rod is

$$\lambda = \frac{q}{L}$$

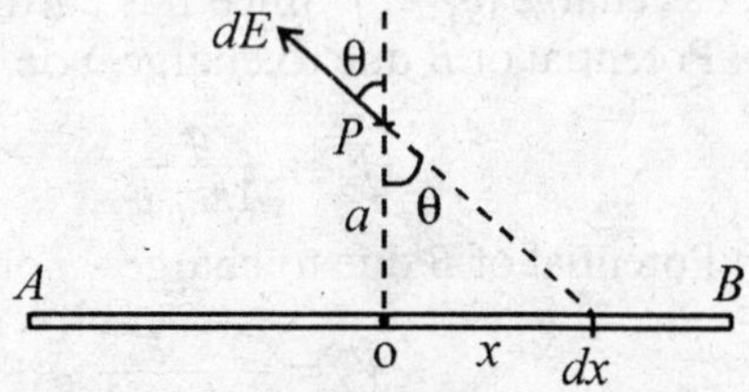

Fig. 18.34

Consider a small element of this length *dx* at a distance *x* from the centre of the rod. The charge on this element is λdx and the electric field at point *P* is given by

$$dE = \frac{\lambda\, dx}{4\pi\varepsilon_0(a^2+x^2)}$$

The component of electric field parallel to the rod will average out to zero. This can be easily proved by taking similar element on the other side of *O* on the rod. Hence component along *OP* is $dE\cos\theta$.

$$dE\cos\theta = \frac{\lambda\, dx}{4\pi\varepsilon_0(a^2+x^2)}\frac{a}{\sqrt{(a^2+x^2)}} = \frac{\lambda a\, dx}{4\pi\varepsilon_0(a^2+x^2)^{\frac{3}{2}}} \quad \ldots(i)$$

The resultant field at P is then obtained by integrating equation (i)

$$\therefore \quad E = \int dE \cos\theta = \frac{\lambda a}{4\pi\varepsilon_0} \int_{-L/2}^{+L/2} \frac{dx}{\left(a^2 + x^2\right)^{\frac{3}{2}}}$$

From Fig. 18.34, we have

$$x = a\tan\theta \text{ or } dx = a\sec^2\theta\, d\theta$$

$$\therefore \quad E = \frac{\lambda a}{4\pi\varepsilon_0} \int \frac{a\sec^2\theta\, d\theta}{a^3 \sec^3\theta} = \frac{\lambda}{4\pi\varepsilon_0 a} \int \cos\theta\, d\theta$$

$$= \frac{\lambda}{4\pi\varepsilon_0 a} \sin\theta = \frac{\lambda}{4\pi\varepsilon_0 a} \left.\frac{x}{\sqrt{x^2 + a^2}}\right|_{-L/2}^{+L/2}$$

$$= \frac{\lambda}{4\pi\varepsilon_0 a} \frac{2\frac{L}{2}}{\sqrt{\frac{L^2}{4} + a^2}} \qquad \text{But } \lambda = \frac{q}{L}$$

$$= \frac{q}{2\pi\varepsilon_0 a\sqrt{L^2 + 4a^2}}.$$

Example 18.4: Figure 18.35 shows three concentric thin spherical shells A, B and C of radii a, b and c respectively. The shells A and C are given charges $+q$ and $-q$ respectively and the shell B is earthed. Find the charges appearing on the surfaces of B and C.

Solution: The situation of the problem is shown in Fig. 18.35 (a). The charges induced on different surfaces are shown in Fig. 18.35 (b). The inner surface of B has charge $-q$ induced on it. Let the charge on outer surface of B be q_1, the inner surface of C has charge $-q_1$ and outer surface of C has charge $(q_1 - q)$. Since B is earthed means its potential is zero.

Potential of B due to charge q on sphere A is

$$V_1 = \frac{q}{4\pi\varepsilon_0 b}$$

Potential of B due to charge $-q$ on the inner surface of B is

$$V_2 = -\frac{q}{4\pi\varepsilon_0 b}$$

Potential of B due to charge q_1 on the outer of surface of B is

$$V_3 = \frac{q_1}{4\pi\varepsilon_0 b}$$

Potential of B due to charge $-q_1$ on the inner surface of C is

$$V_4 = -\frac{q_1}{4\pi\varepsilon_0 c}$$

and potential of B due to charge $q_1 - q$ on the outer surface of C is

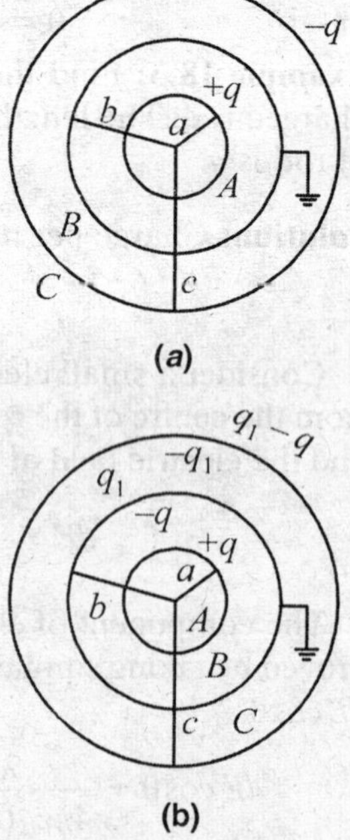

(b)

Fig. 18.35

$$V_5 = \frac{q_1 - q}{4\pi\varepsilon_0\, c}$$

Net potential of B is

$$V = V_1 + V_2 + V_3 + V_4 + V_5 = 0$$

$$= \frac{q}{4\pi\varepsilon_0\, b} - \frac{q}{4\pi\varepsilon_0\, b} + \frac{q_1}{4\pi\varepsilon_0\, b} - \frac{q_1}{4\pi\varepsilon_0\, c} + \frac{q_1 - q}{4\pi\varepsilon_0\, c} = 0$$

or $$q_1 = \frac{bq}{c}$$

Then the charges on various surfaces are

(*a*) On inner surface of B, $-q$, on outer surface of B, $q_1 = \dfrac{bq}{c}$.

(*b*) On inner surface of C, $-q_1 = -\dfrac{bq}{c}$, on outer surface of $C = q_1 - q = \dfrac{bq}{c} - q = -q\left(1 - \dfrac{b}{c}\right)$

(*c*) On A, it is $+q$.

Example 18.5: Two conducting plates A and B are placed parallel to each other. A is given a charge Q_1 and B a charge Q_2. Find the distribution of charges on the four surfaces.

Solution: Figure 18.36 (*a*) shows two parallel plates A and B given charges Q_1 and Q_2 respectively. Let us consider the Gaussian surface as shown in Fig. 18.36 (*a*). Total flux through this surface is zero, because field inside the conductor is zero and the part of the Gaussian surface outside the conductor parallel to the electric field and flux through these surfaces is also zero. Thus total flux through this Gaussian surface is zero. Hence from Gauss theorem the total charge inside it is zero. As such the charge on inner surface of A is equal and opposite to that on the inner surface of B. Let the distribution of charges on different surfaces be as shown in Fig. 18.36 (*b*).

Consider a point P inside plate A where the electric field is zero. We know that electric field due to a long sheet having surface density of charge σ is $\dfrac{\sigma}{2\varepsilon_0}$. Using this result we have for point P

Electric field due to charge $Q_1 - q$, $E_1 = \dfrac{Q_1 - q}{2A\varepsilon_0}$ (downwards)

Electric field due to charge $+q$, $E_2 = \dfrac{q}{2A\varepsilon_0}$ (upwards)

Electric field due to charge $-q$, $E_3 = \dfrac{q}{2A\varepsilon_0}$ (downwards)

and due to charge $Q_2 + q$, $E_4 = \dfrac{Q_2 + q}{2A\varepsilon_0}$ (upwards)

where A is the surface area of the plates (one side)

Net field at P

$$\mathbf{E} = \mathbf{E}_1 + \mathbf{E}_2 + \mathbf{E}_3 + \mathbf{E}_4$$

or $$E = \frac{Q_1 - q}{2A\varepsilon_0} - \frac{q}{2A\varepsilon_0} + \frac{q}{2A\varepsilon_0} - \frac{Q_2 - q}{2A\varepsilon_0} = 0$$

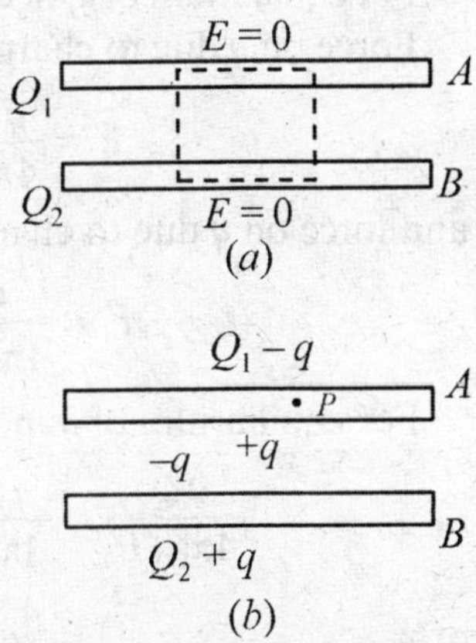

Fig. 18.36

This gives

$$Q_1 - q - q + q - Q_2 - q = 0$$

$$q = \frac{Q_1 - Q_2}{2}$$

Thus $$Q_1 - q = \frac{Q_1 + Q_2}{2}$$

and $$Q_2 + q = \frac{Q_1 + Q_2}{2}$$

This gives us an important result, i.e. when two charged conducting plates are placed parallel to each other, the two outer most surfaces get equal charges and the facing surfaces get equal and opposite charges.

Example 18.6: Two point charges $+4q$ and $+q$ are placed at a distance L apart. A third charge is so placed that all the three charges are in equilibrium. Find the location, magnitude and nature of the third charge. Discuss also, whether the equilibrium of the system is stable, unstable or neutral.

Solution: A system in equilibrium means the net force is zero. For charge to be in equilibrium, it should be placed at a point where the net electric field due to other two charges $+4q$ and $+q$ is zero. This point will always lie on the line joining the two charges. Let this be point C at a distance x from A (charge q). Field at C due to charge $+q$ is

$$E_1 = \frac{q}{4\pi\varepsilon_0 x^2} \text{ towards right}$$

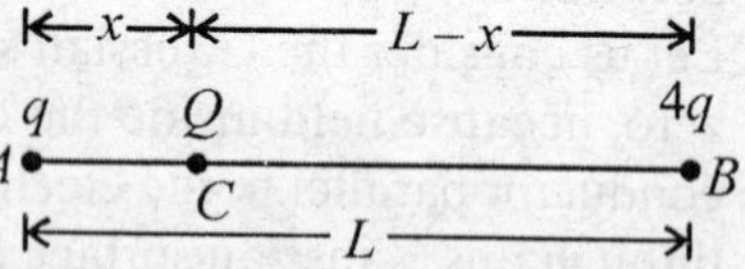

Fig. 18.37

and due to charge $+4q$, $E_2 = \dfrac{4q}{4\pi\varepsilon_0 (L-x)^2}$ towards left

for equilibrium of Q at C, we have

$$\frac{q}{4\pi\varepsilon_0 x^2} = \frac{4q}{4\pi\varepsilon_0 (L-x)^2}$$

or $$L - x = 2x \text{ or } x = L/3$$

For equilibrium of q; net force on q should be zero. For this nature of Q should be opposite that of q. Force on q due to charge $4q$ is

$$F = \frac{q \times 4q}{4\pi\varepsilon_0 L^2} \text{ towards left}$$

and force on q due to charge Q is

$$F' = \frac{Q \cdot q}{4\pi\varepsilon_0 \cdot x^2} = \frac{Qq \times 9}{4\pi\varepsilon_0 L^2} \text{ towards right}$$

For equilibrium of q, net force should be zero, i.e. $|F| = |F'|$

or $$\frac{9Qq}{4\pi\varepsilon_0 L^2} = \frac{q \times 4q}{4\pi\varepsilon_0 L^2}$$

or $$Q = \frac{4}{9} q$$

Since the nature of Q and q is opposite as such, $Q = -\frac{4}{9}q$.

If charge Q is slightly displaced towards B ($4q$ charge), the attractive force between them will increase and that between (q and Q will decrease), i.e. a net force on Q will act towards B as such equilibrium is unstable.

Example 18.7

(*a*) Two point charges each of charge 4×10^{-6} C are placed at distance 2 m apart in air, what is the force between them?

(*b*) If the medium between the above charges is filled with water of dielectric constant 81, what will be the force between them?

(*c*) If half the length of medium between them is filled with water, the remaining half length being air, what will be the force between them?

Solution:

(*a*) Force between two point charges in air is

$$F = \frac{1}{4\pi\varepsilon_0}\frac{q_1 q_2}{r^2}$$

Given $\quad q_1 = q_2 = 4 \times 10^{-6}$ C, $r = 2$ m

$$\therefore \quad F = \frac{9 \times 10^9 \times 4 \times 10^{-6} \times 4 \times 10^{-6}}{(2)^2}$$

$$= 3.6 \times 10^{-2} \text{ N}$$

(*b*) Force between two point charges in medium of dielectric constant ε_r is

$$F = \frac{1}{4\pi\varepsilon_0\varepsilon_r}\frac{q_1 q_2}{r^2}, \varepsilon_r = 81$$

$$\therefore \quad F = \frac{9 \times 10^9 \times 4 \times 10^{-6} \times 4 \times 10^{-6}}{81 \times (2)^2} = 4.4 \times 10^{-4} \text{ N}$$

(*c*) If two charges are in air, the force between them is

$$F_a = \frac{q_1 q_2}{4\pi\varepsilon_0 r^2}$$

If they are in medium of dielectric constant ε_r then if separation is r' then force is

$$F_m = \frac{q_1 q_2}{4\pi\varepsilon_0\varepsilon_r r'^2}$$

For force to be equal, we have

$$\varepsilon_r r'^2 = r^2 \text{ or } r = \sqrt{\varepsilon_r} \cdot r'$$

This shows that the force between two charges in medium of dielectric constant ε_r, when at distance r' is equivalent to distance $r = \sqrt{\varepsilon_r}\, r'$ in air.

In present case the distance between charges is

$$d = 1 \text{ m (in air)} + 1 \text{ m (in water)}$$

$$= 1 \text{ m (in air)} = \sqrt{81} \times 1 \text{ m (in air)}$$

$$= 1 + 9 = 10 \text{ m in air}$$

Hence force between charges is

$$F = \frac{q_1 q_2}{4\pi\varepsilon_0 r^2} \qquad \text{when } r = 10 \text{ m}$$

$$= \frac{9 \times 10^9 \times 4 \times 10^{-6} \times 4 \times 10^6}{(10)^2} = 1.44 \times 10^{-3} \text{ N.}$$

Example 18.8: Two similar point charges q_1 and q_2 are placed at a distance r apart in air. (*a*) If a dielectric slab of thickness t ($\ll r$) and dielectric constants ε_r is put between the charges, calculate the force of repulsion. (*b*) If the thickness of slab covers half the distance between the charges, the Coulomb repulsive force is reduced in the ratio 9 : 4. Calculate the dielectric constant of the slab.

Solution: The repulsive force in air is

$$F_a = \frac{q_1 q}{4\pi\varepsilon_0 r^2} \qquad \ldots(i)$$

Now if dielectric of constant ε_r fills the distance t, then effective distance in air is

$$r' = (r - t) + \sqrt{\varepsilon_r} \cdot t = r + t\left(\sqrt{\varepsilon_r} - 1\right)$$

The force now becomes

$$F_1' = \frac{q_1 q_2}{4\pi\varepsilon_0 \left\{r + t\left(\sqrt{\varepsilon_r} - 1\right)\right\}^2}$$

(*b*) When $t = \frac{r}{2}$, $F' = \frac{4}{9} F_a$

$$\therefore \qquad \frac{4}{9} \frac{q_1 q_2}{4\pi\varepsilon_0 r^2} = \frac{q_1 q_2}{4\pi\varepsilon_0 \left\{r + \frac{r}{2}\left(\sqrt{\varepsilon_r} - 1\right)\right\}^2}$$

or $$4\left(\frac{\sqrt{\varepsilon_r} + 1}{2}\right)^2 = 9$$

or $$\sqrt{\varepsilon_r} + 1 = 3 \text{ or } \sqrt{\varepsilon_r} = 2$$

and $$\varepsilon_r = 4.$$

Example 18.9: A charge $+10^{-9}$ C is located at the origin in free space and another charge Q at (2, 0, 0). If the X-component of the electric field at (3, 1, 1) is zero, calculate the value of Q. Is the Y-component zero at (3, 1, 1)?

Solution: The electric field at a point having position vector $\boldsymbol{r}$ with respect to point charge is

$$\boldsymbol{E} = \frac{q\boldsymbol{r}}{4\pi\varepsilon_0 r^3}$$

At origin a point charge $q = +10^{-9}$ C is placed and point B has position vector

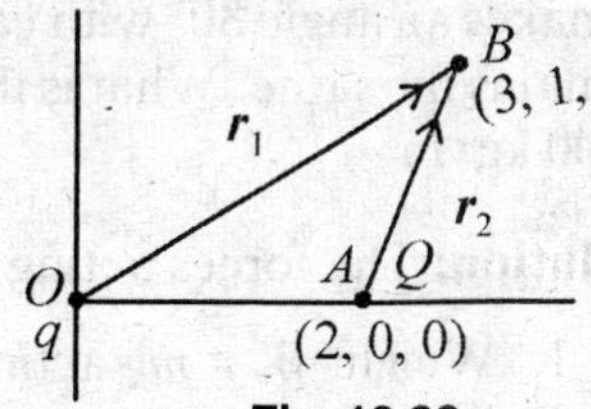

Fig. 18.38

$$\boldsymbol{r}_1 = \left(3\hat{i} + \hat{j} + \hat{k}\right)$$

and magnitude $\quad |\boldsymbol{r}_1| = \sqrt{(3)^2 + (1)^2 + (1)^2} = \sqrt{11}$ m

The position vector of B with respect to A is

$$\boldsymbol{r}_2 = \boldsymbol{r}_1 - 2\hat{i} = 3\hat{i} + \hat{j} + \hat{k} - 2\hat{i} = \hat{i} + \hat{j} + \hat{k}$$

and $\quad |\boldsymbol{r}_2| = \sqrt{1^2 + 1^2 + 1^2} = \sqrt{3}$ m

The electric field at B due to point charge q at O is

$$\boldsymbol{E}_1 = \frac{q}{4\pi\varepsilon_0}\frac{\left(3\hat{i} + \hat{j} + \hat{k}\right)}{(11)^{3/2}}$$

and electric field at B due to point charge Q at A is

$$\boldsymbol{E}_2 = \frac{Q}{4\pi\varepsilon_0}\frac{\left(\hat{i} + \hat{j} + \hat{k}\right)}{(3)^{3/2}}$$

Net field at B is $\boldsymbol{E} = \boldsymbol{E}_1 + \boldsymbol{E}_2$

$$\boldsymbol{E} = \frac{q}{4\pi\varepsilon_0}\frac{\left(3\hat{i} + \hat{j} + \hat{k}\right)}{(11)^{3/2}} + \frac{Q}{4\pi\varepsilon_0}\frac{\left(\hat{i} + \hat{j} + \hat{k}\right)}{(3)^{3/2}}$$

and x-component of the electric field at B is

$$E_x = \frac{q}{4\pi\varepsilon_0}\frac{(3)}{(11)^{3/2}} + \frac{Q}{4\pi\varepsilon_0}\frac{1}{(3)^{3/2}} = 0$$

$$\therefore \quad Q = -q\frac{(3)^{5/2}}{(11)^{3/2}} = -\frac{(3)^{5/2}}{(11)^{3/2}} \times 10^{-9} \text{ C}$$

y-component of electric field at B is

$$E_y = \frac{q}{4\pi\varepsilon_0}\frac{1}{(11)^{3/2}} + \frac{Q}{4\pi\varepsilon_0}\frac{1}{(3)^{3/2}}$$

$$= \frac{q}{4\pi\varepsilon_0}\frac{1}{(11)^{3/2}} - \frac{q}{4\pi\varepsilon_0}\frac{(3)^{5/2}}{(11)^{3/2}}\frac{1}{(3)^{3/2}}$$

$$= \frac{q}{4\pi\varepsilon_0(11)^{3/2}}[1-3] = -\frac{2q}{4\pi\varepsilon_0(11)^{3/2}} \neq 0$$

Example 18.10: Two identical charged spheres are suspended by strings of equal length. The string S makes an angle 30° with each other. When suspended in a liquid of density 800 kg/m^3, the angle remains the same. What is the dielectric constant of liquid? The density of the material of sphere is 1600 kg/m^3.

Solution: The forces acting on either sphere are shown in Fig. 18.39. They are

1. Weight $W = mg$ acting vertically downwards.
2. Electric force F between two charged spheres along BA.
3. Tension T in the string.

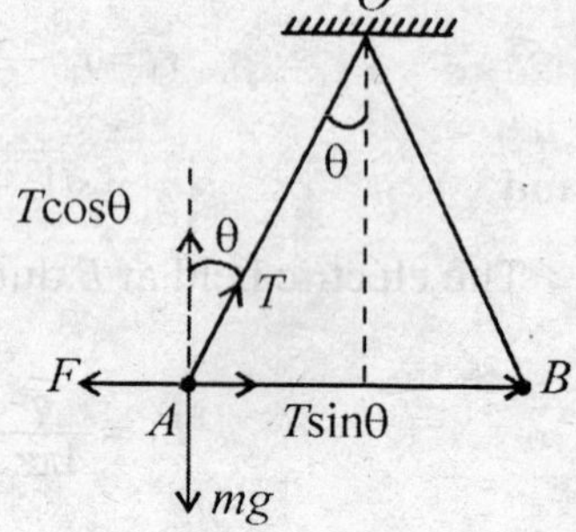

Fig. 18.39

In equilibrium the string makes an angle θ (= 15°) with vertical. For equilibrium of the spheres, we have

$$T\cos\theta = mg$$

$$T\cos\theta = F_e$$

or
$$\tan\theta = \frac{F_e}{mg}$$

When the spheres are suspended in liquid of density σ and dielectric constant ε_r, the force becomes

$$F'_e = \frac{F_e}{\varepsilon_r}$$

and weight is reduced due to upthrust. Weight is now

$$W' = W - \text{upthrust} = (mg - V\sigma g)$$

V is volume of body

$$W' = mg\left(1 - \frac{V}{m}\sigma\right) \qquad \frac{V}{m} = \frac{1}{\rho} \text{ the density of body.}$$

$$= mg\left(1 - \frac{\sigma}{\rho}\right)$$

For equilibrium of spheres in liquid, we have

$$\tan\theta = \frac{F'_e}{W'} = \frac{\dfrac{F_e}{\varepsilon_r}}{mg\left(1 - \dfrac{\sigma}{\rho}\right)}$$

$$= \frac{F_e}{mg} \cdot \frac{1}{\varepsilon_r\left(1 - \dfrac{\sigma}{\rho}\right)}$$

But
$$\tan\theta = \frac{F_e}{mg}$$

∴
$$\frac{1}{\varepsilon_r\left(1 - \dfrac{\sigma}{\rho}\right)} = 1$$

or $$\varepsilon_r = \frac{1}{1-\frac{\sigma}{\rho}} = \frac{1}{1-\frac{800}{1600}} = 2$$

or $$\varepsilon_r = 2.$$

Example 18.11: Two similar helium filled spherical balloons tied to a 5 g weight with strings each carrying an electric charge q floats in equilibrium as shown in Fig. 18.40.

(*a*) Find the magnitude of q, assuming that the charge on each balloon as if it were concentrated at its centre.

(*b*) Find the volume of each balloon (neglect the weight of the unfilled balloons and assume that the density of air $\rho = 1.29\ \text{kg/m}^3$ and density of helium $\sigma = 0.2\ \text{kg/m}^3$).

In equilibrium the separation between the balloons is 0.6 m and length of the strings is 1.0 m.

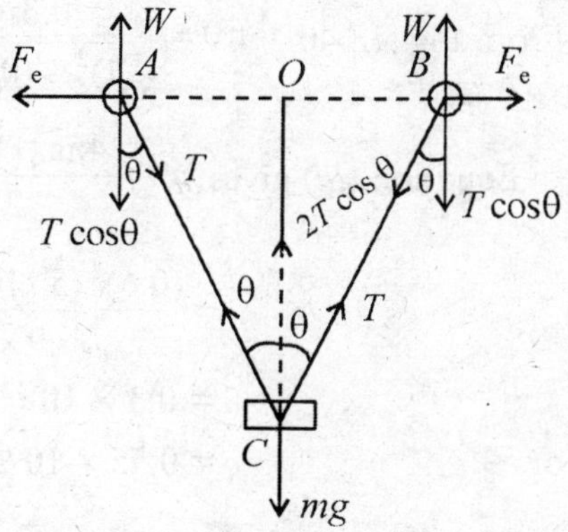

Fig. 18.40

Solution: The forces acting on each balloon are shown in Fig. 18.40.

(*i*) Effective weight of balloon W

$$W = mg - \text{upthrust} = V\rho g - V\sigma g$$

where ρ is density of air and σ density of helium and V volume of balloon.

(*ii*) Electrostatic repulsive force between charges on balloons.

$$F_e = \frac{1}{4\pi\varepsilon_0}\frac{q^2}{r^2} \text{ where } r = AB = 6 \text{ m}$$

(*iii*) Tension in string T along AC.

For equilibrium of balloon A

$$W = T\cos\theta \text{ and } F_e = T\sin\theta$$

$$\therefore \quad \tan\theta = \frac{F_e}{W} \qquad \ldots(i)$$

For equilibrium of weight C

$$2T\cos\theta = mg$$

or $$T\cos\theta = \frac{1}{2}mg \qquad \ldots(ii)$$

But $$W = T\cos\theta = \frac{1}{2}mg$$

or $$(V\rho g - V\sigma g) = \frac{1}{2}mg$$

or $$V = \frac{m}{2(\rho-\sigma)} = \frac{1}{2}\times\frac{5\times10^{-3}}{(1.29-0.2)} = \frac{5\times10^{-3}}{2\times1.09}$$

$$V = 2.293\times10^{-3}\ \text{m}^3$$

Equation (*i*) gives

$$\tan\theta = \frac{F_e}{W} \text{ and } W = \frac{1}{2}mg$$

$$= \frac{1}{4\pi\varepsilon_0}\frac{q^2}{r^2} \times \frac{2}{mg} \quad \ldots(ii)$$

From Fig. 18.40 $\tan\theta = \frac{0.3}{\sqrt{(1)^2 - (0.3)^2}} = \frac{0.3}{\sqrt{1 - 0.09}} = \frac{0.3}{\sqrt{0.91}} = 0.314$

Equation (*iii*) gives $q^2 = \frac{4\pi\varepsilon_0 r^2 \cdot mg}{2} \tan\theta$

$$= \frac{(0.6)^2 (5 \times 10^{-3} \times 10) \times 0.314}{9 \times 10^9 \times 2}$$

$$= 0.3 \times 10^{-12}$$

or $$q = 0.55 \times 10^{-6}\,\text{C} = 0.55\,\mu\text{C}.$$

Example 18.12: A particle of mass m and charge $+q$ is placed mid-way between two fixed charged particles each having a charge $+q$ and at a distance $2l$ apart. Assuming that the middle charge moves along the line joining the fixed charges, calculate the frequency of oscillations when it is displaced slightly.

Solution: The situation is shown in Fig. 18.41. When initially the third charge q is placed at O, this charge is in equilibrium. Now if it is displaced through distance x toward B, then net force on this charge due to charge at B is

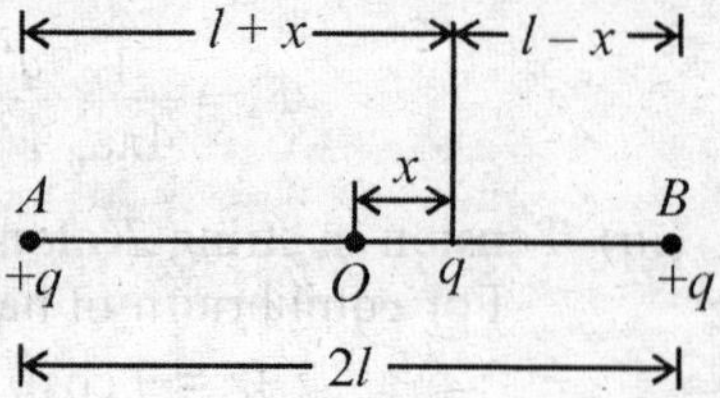

Fig. 18.41

$$F_1 = \frac{q^2}{4\pi\varepsilon_0 (l - x)^2} \text{ towards } A$$

and the force due to charge at A is

$$F_2 = \frac{q^2}{4\pi\varepsilon_0 (l + x)^2} \text{ towards } B$$

Thus net force on charge when displaced through x from O is

$$F = F_2 - F_1 = \frac{q^2}{4\pi\varepsilon_0 (l + x)^2} - \frac{q^2}{4\pi\varepsilon_0 (l - x)^2}$$

$$= \frac{q^2}{4\pi\varepsilon_0}\left[\frac{(l - x)^2 - (l + x)^2}{(l^2 - x^2)^2}\right] = -\frac{q^2}{4\pi\varepsilon_0}\frac{4lx}{(l^2 - x^2)^2}$$

Negative sign shows that displacement is towards B and net force is towards A.

If $l \gg x$, we can neglect x^2 in comparison to l^2 in denominator.

$\therefore$ $$F = -\frac{4q^2}{4\pi\varepsilon_0 l^3} x$$

$\therefore$ acceleration is

$$a = \frac{F}{m} = -\frac{4q^2}{4\pi\varepsilon_0 m l^3} x$$

Since acceleration is directly proportional to displacement as such the motion is S.H.M. with angular frequency ω

$$\omega^2 = \frac{4q^2}{4\pi\varepsilon_0 ml^3}$$

and time period $T = \frac{2\pi}{\omega} = 2\pi\sqrt{\frac{4\pi\varepsilon_0 ml^3}{4q^2}}$

or frequency $= f = \frac{1}{T} = \frac{1}{2\pi}\sqrt{\frac{q^2}{\pi\varepsilon_0 ml^3}}$.

Example 18.13: A point charge Q is distributed over two concentric hollow spheres of radii r and R ($R > r$) such that the surface densities are equal. Find the potential at the common centre.

Solution: Let q_1 and q_2 be respective charges over to concentric spheres of radii r and R such that

$$q_1 + q_2 = Q$$

and $\sigma_1 = \frac{q_1}{4\pi r^2} = \sigma_2 = \frac{q_2}{4\pi R^2}$, i.e. $\frac{q_1}{q_2} = \frac{r^2}{R^2}$

This gives $q_1 = \frac{Qr^2}{r^2 + R^2}$ and $q_2 = \frac{QR^2}{r^2 + R^2}$

Potential at the common centre due to inner sphere

$$V_1 = \frac{q_1}{4\pi\varepsilon_0 r} = \frac{Qr^2}{(r^2 + R^2)4\pi\varepsilon_0 r} = \frac{Qr}{4\pi\varepsilon_0 (r^2 + R^2)}$$

and potential at the common centre due to outer sphere

$$V_2 = \frac{q_2}{4\pi\varepsilon_0 R} = \frac{QR^2}{4\pi\varepsilon_0 R(R^2 + r^2)} = \frac{QR}{4\pi\varepsilon_0 (R^2 + r^2)}$$

Total potential at the common centre is

$$V = V_1 + V_2 = \frac{Q}{4\pi\varepsilon_0}\frac{(r + R)}{(r^2 + R^2)}.$$

Example 18.14: An inclined plane is making an angle of 30° with the horizontal electric field E of 100 V/m (Fig. 18.42). A particle of mass 1 kg and charge 0.01 C is allowed to slide down from a height of 1 m. If the coefficient of friction is 0.2, find the time the particle will take to reach the bottom.

Solution: The forces acting on the particle are

(*i*) Weight of the particle mg acting vertically downwards
(*ii*) Reaction N, perpendicular to the plane
(*iii*) Horizontal electric force qE parallel to AB
(*iv*) Frictional force μN parallel to the plane.

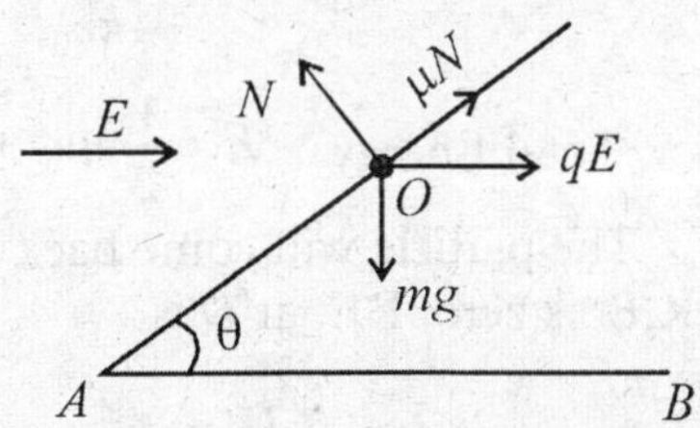

Fig. 18.42

Resolving forces parallel and perpendicular to the inclined plane, we have net force perpendicular to the plane is zero, i.e.

$$mg\cos\theta + qE\sin\theta = N \quad \ldots(i)$$

The net downward force parallel to the plane is

$$F = mg\sin\theta - qE\cos\theta - \mu N$$
$$= mg\sin\theta - qE\cos\theta - \mu(mg\cos\theta + qE\sin\theta)$$
$$= mg(\sin\theta - \mu\cos\theta) - qE(\cos\theta + \mu\sin\theta)$$

If a is the acceleration of the particle, then

$$a = g(\sin\theta - \mu\cos\theta) - \frac{qE}{m}(\cos\theta + \mu\sin\theta)$$

Substituting the values

$$\theta = 30°, \mu = 0.2, m = 1 \text{ kg}, q = 0.01 \text{ C and } g = 10 \text{ m/s}^2, E = 100 \text{ V/m}$$

$$a = 10[\sin 30° - 0.2\cos 30°] - \frac{10^{-2} \times 100}{1}(\cos 30° + 2\sin 30°)$$
$$= 10[0.5 - 0.2 \times 0.860] - [0.866 + 0.2 \times 0.5]$$
$$= 2.17 \text{ m/s}^2$$

When it comes to bottom, the distance $s = \frac{1}{\sin\theta} = \frac{1}{\sin 30} = 2$ m

Hence time taken is

$$s = v_0 t + \frac{1}{2}at^2 = \frac{1}{2} \times 2.17 \times t^2 \qquad \because v_0 = 0$$

or

$$t = \sqrt{\frac{2 \times 2}{2.17}} = 1.345 \text{ sec}$$

Example 18.15: Two fixed equal positive charges, each of magnitude q (= 5×10^{-5} C) are located at points A and B, separated by a distance of l (= 6 m). An equal and opposite charge moves towards them along the line COD; the perpendicular bisector of line AB. The moving charge when it reaches the point C at a distance d (= 4 m) from O, has a K.E. of E (= 4 J). Calculate the distance of the farthest point D at which the negative charge will reach before returning to C.

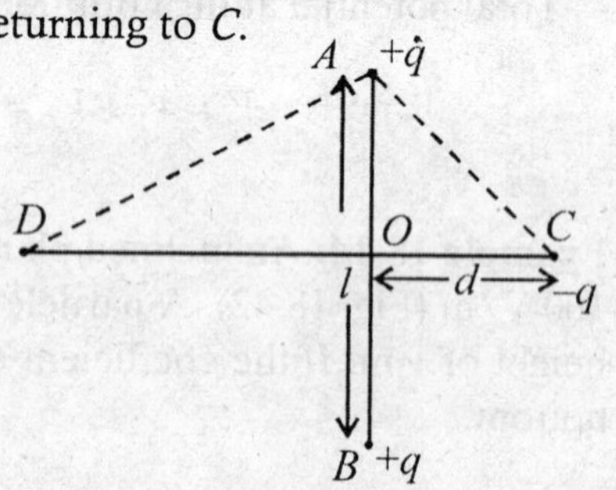

Fig. 18.43

Solution: Total energy of 3 charge system is

(*i*) K.E. of charge ($-q$) when at $C = \frac{1}{2}mv^2$

(*ii*) P.E. of 3 charge system

$$U = \frac{q^2}{4\pi\varepsilon_0 l} - \frac{q^2 \times 2}{4\pi\varepsilon_0 AC} \qquad \because AC = BC$$

$\therefore$ Total Energy $E = \frac{1}{2}mv^2 + U$

The particle will come back when its velocity becomes zero. So at point D total energy is P.E., K.E. is zero. P.E. at D is

$$U' = \frac{q^2}{4\pi\varepsilon_0 l} - \frac{q^2 \times 2}{4\pi\varepsilon_0 AD} \qquad \because AD = BC$$

Conservation of energy gives

$$\frac{q^2}{4\pi\varepsilon_0 l} - \frac{2q^2}{4\pi\varepsilon_0 AC} + \text{K.E.} = \frac{q^2}{4\pi\varepsilon_0 l} - \frac{2q^2}{4\pi\varepsilon_0 AD}$$

or $$\frac{2q^2}{4\pi\varepsilon_0}\left[\frac{1}{AC} - \frac{1}{AD}\right] = \text{K.E.}$$

or $$\frac{2q^2}{4\pi\varepsilon_0 AD} = \frac{2q^2}{4\pi\varepsilon_0 AC} - \text{K.E.}$$

Values given are $q = 5 \times 10^{-5}$ C, $AC = \sqrt{AD^2 + DC^2}$ or $AC = \sqrt{3^2 + 4^2} = 5$ m and $\frac{1}{4\pi\varepsilon_0} = 9 \times 10^9$ MKS unit, K.E. = 4 Joule. Substitution gives

$$\frac{2 \times 25 \times 10^{-10} \times 9 \times 10^9}{AD} = \frac{2 \times 25 \times 10^{-10} \times 9 \times 10^9}{5} - 4$$

or $$\frac{45}{AD} = 9 - 4 = 5$$

or $$AD = 9$$

If x is the distance of point D from O, then

$$AD^2 = x^2 + 3^2$$

or $$x^2 = 9^2 - 3^2 = 81 - 9 = 72$$

or $$x = \sqrt{72} = 8.485 \text{ m}$$

Example 18.16: Four point charges +8 μC, –1 μC, 1 μC and +8 μC are fixed at the points $-\sqrt{\frac{27}{2}}$ m, $-\sqrt{\frac{3}{2}}$ m, $+\sqrt{\frac{3}{2}}$ m and $\sqrt{\frac{27}{2}}$ m respectively on the x-axis. A particle of mass 6×10^{-4} kg and of charge 0.1 μC moves along the $-x$ direction. Its speed at $x = +\infty$ is v_0. Find the least value of v_0 for which the particle will cross the origin. Find also the kinetic energy of the particle at the origin. Assume that space is gravity free.

Solution: A particle of charge q_0 approaches from $x = \infty$ towards origin as shown in Fig. 18.44. Let at any time it is at P at a distance x from the origin. It is attracted by the charges q_1 (= –1 μC each) and repelled by charges q_2 (= 8 μC). The symmetry of the charges shows that the net force will be along x-axis and y-component is zero. Initially repulsive force is greater than the attractive force. So velocity will go on decreasing and at a point say P, force is zero and if the particle crosses this point P, the force becomes attractive. At P the net force is

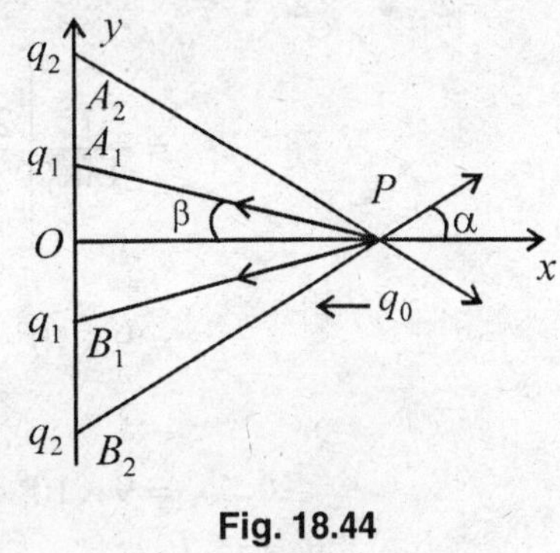

Fig. 18.44

$$F = 2\frac{1}{4\pi\varepsilon_0}\frac{q_2 q_0}{(A_2P)^2}\cos\alpha - 2 \times \frac{1}{4\pi\varepsilon_0}\frac{q_1 q_0}{(A_1P)^2}\cos\beta = 0$$

or $$\frac{q_2}{A_2P^2}\frac{x}{A_2P^2} = \frac{q_1}{(A_1P)^2}\frac{x}{(A_1P)}$$

or $$\frac{q_2}{q_1} = \frac{(A_2P)^3}{(A_1P)^3} = 8$$

or $$\frac{A_2P}{A_1P} = 2$$

$$A_2P = \sqrt{OA_2^2 + x^2} \text{ and } A_1P = \sqrt{OA_1^2 + x^2}$$

$\therefore$ $$OA_2^2 + x^2 = 4\left(OA_1^2 + x^2\right)$$

or $$3x^2 = OA_2^2 - 4OA_1^2 = \frac{27}{2} - \frac{3}{2} \times 4 = \frac{15}{2}$$

or $$x = \sqrt{\frac{5}{2}} = \sqrt{2.5} \text{ m}$$

If on reaching P the particle has same velocity, it will cross P and because of attractive force thereafter, it will reach the origin. In limiting case initial K.E. of the charge q_0 should be equal to P.E. at P

or $$\frac{1}{2}mv_0^2 \geq \frac{1}{4\pi\varepsilon_0}\left[\frac{2 \times q_0q_2}{A_2P} - \frac{2 \times q_0q_1}{A_1P}\right]$$

Now $$A_2P \geq \sqrt{\frac{27}{2} + \frac{5}{2}} = \sqrt{16} = 4 \text{ m}$$

and $$A_1P \geq \sqrt{\frac{3}{2} + \frac{5}{2}} = \sqrt{4} = 2 \text{ m}$$

$\therefore$ $$\frac{1}{2}mv_0^2 \geq \frac{1}{4\pi\varepsilon_0}\left[\frac{2 \times q_0q_2}{4} - \frac{2 \times q_0q_1}{2}\right]$$

or $$v_0^2 \geq \frac{2 \times 9 \times 10^9}{6 \times 10^{-4}}\left[\frac{2 \times 0.1 \times 10^{-6} \times 8 \times 10^{-6}}{4} - \frac{2 \times 0.1 \times 10^{-6} \times 1 \times 10^{-6}}{2}\right].$$

$$\geq \frac{2 \times 9 \times 10^9}{6 \times 10^{-4}} \times 3 \times 10^{-13} = 9$$

or $$v_0 \geq 3 \text{ m/s}$$

i.e., minimum value of v_0 is 3 m/s.

Now using conservation of energy, when the particle is at origin, if its velocity is v, then

$$\frac{1}{2}mv_0^2 - \frac{1}{2}mv^2 = \text{P.E. of } q_0 \text{ at the origin}$$

$$= \frac{1}{4\pi\varepsilon_0}\left[\frac{2 \times 0.1 \times 10^{-6} \times 8 \times 10^{-6}}{\sqrt{\frac{27}{2}}} - \frac{2 \times 0.1 \times 10^{-6} \times 1 \times 10^{-6}}{\sqrt{\frac{3}{2}}}\right]$$

$$= 9 \times 10^9 \times 10^{-13}\left[\frac{16\sqrt{2}}{3\sqrt{3}} - \frac{2\sqrt{2}}{\sqrt{3}}\right]$$

$$= 9 \times 10^9 \times 10^{-13} \times \frac{10\sqrt{2}}{3\sqrt{3}}$$

$$= \sqrt{6} \times 10^{-3}$$

or $$\frac{1}{2}mv^2 = \frac{1}{2}mv_0^2 - \sqrt{6} \times 10^{-3}$$

$$= \frac{1}{2} \times 6 \times 10^{-4} \times (3)^2 - \sqrt{6} \times 10^{-3}$$

$$27 \times 10^{-4} - 2.45 \times 10^{-3} = (27 - 24.5) \times 10^{-4}$$

$$= 2.5 \times 10^{-4} \text{ Joules.}$$

Example 18.17

(*i*) Two similar balls each of mass m are hung from silk threads of length l and carry similar charges. Prove that in equilibrium their separation will be

$$x = \left[\frac{q^2 l}{2\pi\varepsilon_0 mg}\right]^{\frac{1}{3}} \qquad \text{where } x \ll l$$

(*ii*) Find the rate $\frac{dq}{dt}$ with which the charge leaks off each sphere if their approach velocity varies as $v = \frac{a}{\sqrt{x}}$, where a is constant.

Solution: (*i*) Consider the equilibrium of ball A (Fig. 18.45). Forces acting on it are

(*i*) Weight mg vertically downwards

(*ii*) Electrostatic force $F_e = \frac{1}{4\pi\varepsilon_0}\frac{q^2}{x^2}$

(*iii*) Tension T in the string.

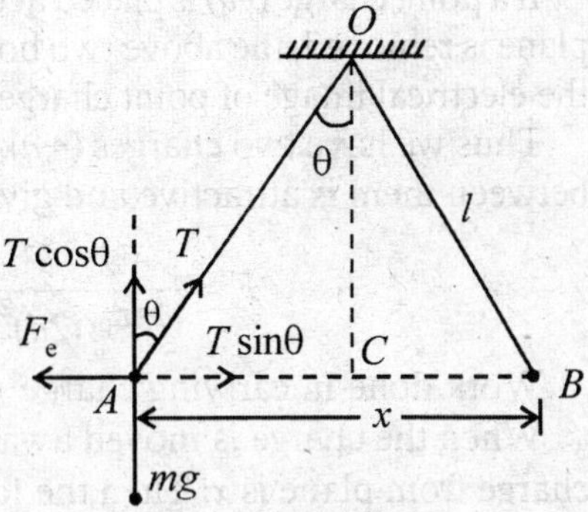

Fig. 18.45

For the equilibrium of A, we have

$$T\cos\theta = mg$$

$$T\sin\theta = F_e = \frac{1}{4\pi\varepsilon_0}\cdot\frac{q^2}{x^2}$$

$$\therefore \qquad \tan\theta = \frac{F_e}{mg} = \frac{q^2}{4\pi\varepsilon_0 x^2 mg}$$

when $x \ll l$ $\tan\theta \approx \sin\theta = \frac{\frac{x}{2}}{l} = \frac{x}{2l}$

or $$\frac{x}{2l} = \frac{q^2}{4\pi\varepsilon_0 x^2 mg} \text{ or } x^3 = \frac{q^2 \times 2l}{4\pi\varepsilon_0 mg} \qquad \ldots(i)$$

or $$x = \left\{\frac{q^2 l}{2\pi\varepsilon_0 mg}\right\}^{\frac{1}{3}}$$

(*ii*) From (*i*), we have

$$q^2 = \frac{2\pi\varepsilon_0 mg}{l}x^3$$

and $$q = \left(\frac{2\pi\varepsilon_0 mg}{l}\right)^{1/2} x^{\frac{3}{2}}$$

Differentiating with respect to x, we get

$$\frac{dq}{dt} = \left(\frac{2\pi\varepsilon_0 mg}{l}\right)^{1/2} \frac{3}{2} x^{\frac{1}{2}} \frac{dx}{dt}$$

But $$\frac{dx}{dt} = v = \frac{a}{\sqrt{x}}$$

$\therefore$ $$\frac{dq}{dt} = \left(\frac{2\pi\varepsilon_0 mg}{l}\right)^{1/2} \frac{3}{2} x^{\frac{1}{2}} \frac{a}{\sqrt{x}}$$

$$= \frac{3}{2} a \left(\frac{2\pi\varepsilon_0 mg}{l}\right)^{1/2}$$

Example 18.18: A point charge q is placed near an earthed plane conductor. Calculate the force between the charge and plane conductor. Also calculate the work done in carrying charge $+q$ to infinity.

Solution: Such type of problems can be solved by the method of electrical images. In this case a point charge q is placed near a conducting earthed plane conductor, the boundary conditions are (*i*) Potential is zero at infinity (*ii*) Potential of each point of conducting plate is zero.

If a point charge $(-q)$ is placed at distance d on opposite side of conducting plane and the conducting plane is removed, the above two boundary conditions are still satisfied. The charge $(-q)$ is then called the electrical image of point charge $+q$.

Thus we have two charges $(+q)$ and $(-q)$ at a distance $2d$ from each other placed in air. The force between them is attractive and given by

$$F = \frac{1}{4\pi\varepsilon_0} \frac{q^2}{(2d)^2} = \frac{q}{16\pi\varepsilon_0 d^2} \quad \text{(attractive)}$$

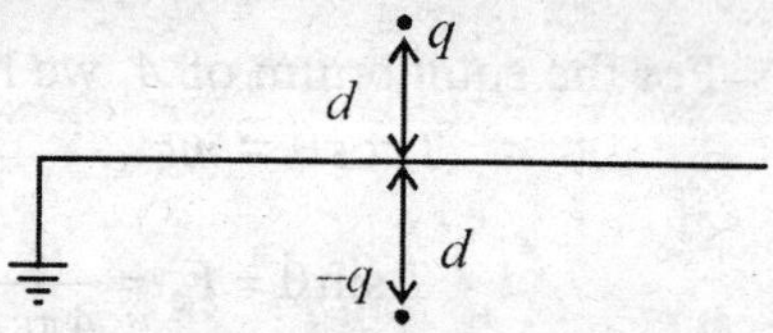

Fig. 18.46

Work done in carrying charge $+q$ to infinity:

When the charge is moved away, let at any time distance of charge from plane is x. Then the force between them is

$$F = \frac{q}{16\,\pi\varepsilon_0\, x^2}$$

and work done in displacing through distance dx is

$$dw = F\,dx = \frac{q}{16\,\pi\varepsilon_0\, x^2}\,dx$$

Then work done in moving charge from distance d to ∞ is

$$w = \int F\,dx = \int_d^\infty \frac{q}{16\pi\varepsilon_0} = \frac{dx}{x^2}$$

$$= -\frac{q}{16\,\pi\varepsilon_0}\left[\frac{1}{x}\right]_d^\infty$$

$$= \frac{q}{16\,\pi\varepsilon_0} \cdot \frac{1}{d} = \frac{q}{16\,\pi\varepsilon_0\, d}$$

Example 18.19: A line of force starts from a point charge $+q_1$ at A and terminates at point charge $-q_2$ at B. If α is the angle that the line of forces makes with AB, find the angle β at which this line of force terminates at B.

Solution: In Section (18.4) we have discussed the properties of lines of force. (property (*xi*))

For a set of collinear point charges $q_1, q_2, \ldots$, etc. if $\theta_1, \theta_2, \ldots$, etc. are the angles that the line joining point charge with any point on the line of force make with the line on which charges are placed then, we have

$$q_1\cos\theta_1 + q_2\cos\theta_2 + \ldots = \text{constant}$$

In present case, we have only two charges at points A and B, then for point A, $\theta_1 = \alpha$ and $\theta_2 = \pi$. For point B, $\theta_1 = 0$ and $\theta_2 = \pi - \beta$. Thus we have

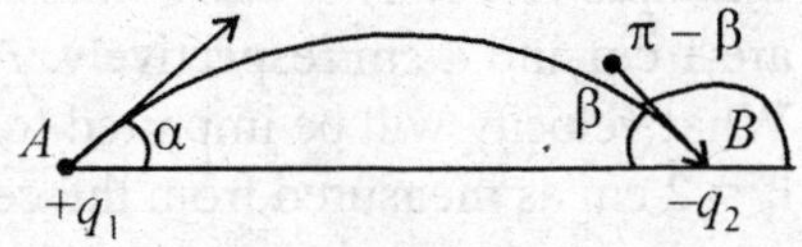

Fig. 18.47

$$q_1\cos\alpha + (-q_2)\cos\pi = q_1\cos\beta + (-q_2)\cos(\pi - \beta)$$

$$q_1\cos\alpha + q_2 = q_1 + q_2\cos\beta$$

or $$q_2(1 - \cos\beta) = q_1(1 - \cos\alpha)$$

or $$2q_2\sin^2\frac{\beta}{2} = q_1 \times 2\sin^2\frac{\alpha}{2}$$

or $$\sin^2\frac{\beta}{2} = \frac{q_1}{q_2}\sin^2\frac{\alpha}{2}$$

or $$\sin\frac{\beta}{2} = \sqrt{\frac{q_1}{q_2}}\sin\frac{\alpha}{2}$$

or $$\beta = 2\sin^{-1}\left[\sqrt{\frac{q_1}{q_2}}\sin\frac{\alpha}{2}\right]$$

Example 18.20: A uniform spherical charge distribution has density ρ. It has cavity of radius r with centre at C at a distance $OC = a$. O is the centre of charge distribution. An electron lies on the periphery of cavity at P such that PC makes an angle of 45° with OC. Find the time taken by the electron to reach the centre of cavity.

Solution: The electric field inside cavity at an arbitrary point A, is

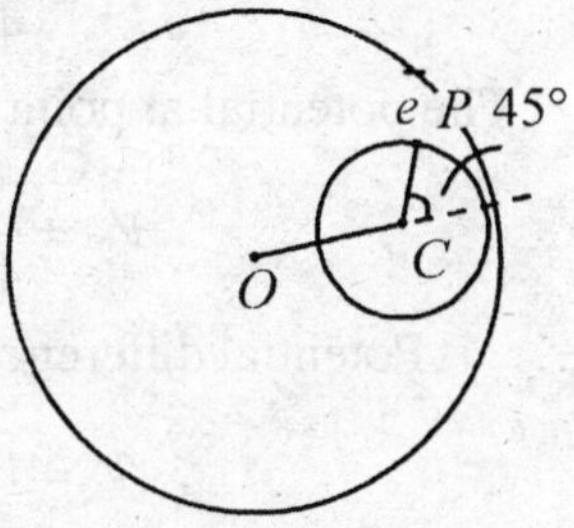

Fig. 18.48

$$\boldsymbol{E} = \frac{\rho}{3\varepsilon_0}\boldsymbol{a}$$

This relation is valid regardless of the ratio between the radii of the spheres and the distance between their centres.

The electric field has magnitude $\frac{\rho}{3\varepsilon_0}a$ and direction is along OC. Its component in the direction CP is

$$E\cos 45° = \frac{\rho}{3\varepsilon_0}a\cos 45° = \frac{\rho a}{3\sqrt{2}\,\varepsilon_0}$$

In this field electron experiences a force $F = eE\cos 45$

and has acceleration $$f = \frac{F}{m} = \frac{eE\cos 45}{m} = \frac{\rho a e}{3\sqrt{2}\,\varepsilon_0 m}$$

The electron will move in the direction PC and to reach centre of cavity C time required is t then

$$r = \frac{1}{2} f t^2 = \frac{1}{2} \frac{\rho a e}{3\sqrt{2}\,\varepsilon_0 m} t^2$$

or
$$t^2 = \frac{6\sqrt{2}\,\varepsilon_0 m r}{\rho a e}$$

or
$$t = \sqrt{\frac{6\sqrt{2}\,\varepsilon_0 m r}{\rho a e}}$$

Example 18.21: The radii of internal and external spheres of concentric spherical air core capacitor are 1 cm and 4 cm respectively. A potential difference of 3000 V is applied between the spheres. What velocity will be imparted to an electron when it approaches from a distance of r_1 = 3 cm to r_2 = 2 cm as measured from the centre of the spheres.

Solution: The potential difference between spherical conductors

$$V = \frac{1}{4\pi\varepsilon_0}\left[\frac{q}{a} - \frac{q}{b}\right]$$

Given is V = 3000 V and a = 1 cm, b = 4 cm

$\therefore$
$$q = \frac{V \times 4\pi\varepsilon_0\, ab}{b - a}$$

$$= \frac{3000 \times 10^{-2} \times 4 \times 10^{-2}}{9 \times 10^9 \times 3 \times 10^{-2}}$$

$$= \frac{4}{9} \times 10^{-8}\ \text{C}$$

The potential at point r_1 = 3 cm from the centre is

$$V_1 = \frac{q}{4\pi\varepsilon_0 r_1} - \frac{q}{4\pi\varepsilon_0 b} = \frac{q}{4\pi\varepsilon_0}\left[\frac{1}{r_1} - \frac{1}{b}\right]$$

The potential at point r_2 = 2 cm from the centre is

$$V_2 = \frac{q}{4\pi\varepsilon_0 r_2} - \frac{q}{4\pi\varepsilon_0 b} = \frac{q}{4\pi\varepsilon_0}\left[\frac{1}{r_2} - \frac{1}{b}\right]$$

$\therefore$ Potential difference

$$V = V_1 - V_2 = \frac{q}{4\pi\varepsilon_0}\left[\frac{1}{r_1} - \frac{1}{b} - \frac{1}{r_2} + \frac{1}{b}\right]$$

$$= \frac{q(r_2 - r_1)}{4\pi\varepsilon_0 r_1 r_2} = \frac{4}{9} \times \frac{10^{-8} \times (3-2) \times 10^{-2} \times 9 \times 10^9}{2 \times 3 \times 10^{-4}}$$

$$V = \frac{2}{3} \times 10^3\ \text{V}$$

and gain in K.E. of electron is

$$\text{K.E.} = eV = \frac{2}{3} \times 10^3 \times 1.6 \times 10^{-19} = \frac{3.2}{3} \times 10^{-16}\ \text{J}$$

K.E. = ½mv^2. Thus

$$\frac{1}{2}mv^2 = \frac{3.2}{3} \times 10^{-16}$$

or $$v^2 = \frac{2 \times 3.2 \times 10^{-16}}{3 \times 9.1 \times 10^{-31}} = \frac{6.4}{27.3} \times 10^{15}$$

or $$v = \sqrt{\frac{64.}{27.3} \times 10^4} = 1.54 \times 10^7 \text{ m/s}.$$

Example 18.22: A system consists of a hall of radius R carrying a spherically symmetric charge and a surrounding space filled with a charge of volume density $\rho = \frac{\alpha}{r}$, where α is a constant and r is the distance from the centre of the ball. Find the charge on the ball for which the magnitude of electric field strength is independent of r, outside the ball. What is the value of the electric field strength? The dielectric constant of ball and surrounding may be taken equal to unity.

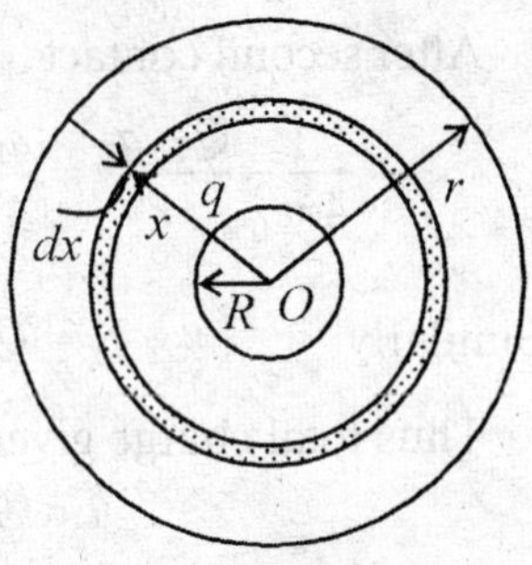

Fig. 18.49

Solution: Consider a spherical surface of radius r (> R) having the centre at the centre of the ball C. If E is the magnitude of the electric field strength at this surface then flux through the surface is

$$\phi = \oint_s \mathbf{E} \cdot \mathbf{ds} = \oint_s E ds \cos\theta = E \oint_s ds = E \cdot 4\pi r^2$$

By Gauss theorem, we have

$\phi = 4\pi r^2 E = \frac{1}{\varepsilon_0}$ total charge enclosed by this sphere.

Total charge enclosed = charge on inner sphere (spherical distribution of charge) of q + charge within hollow sphere of inner radius R and outer radius r (Q)

$$Q = \int_R^r \rho\, dv \quad dV = 4\pi x^2 dx$$

$$= \int_R^r \frac{\alpha}{x} 4\pi x^2\, dx \qquad \because\ \rho = \frac{\alpha}{r}$$

$$= 4\pi\alpha \int_R^r x\, dx = \frac{4\pi\alpha}{2}\left[r^2 - R^2\right] = 2\pi\alpha\left(r^2 - R^2\right)$$

Substituting the values, we have

$$\phi = 4\pi r^2 E = \left(q + 2\pi\alpha\,(r^2 - R^2)\right)\frac{1}{\varepsilon_0}$$

or $$E = \frac{q}{4\pi\varepsilon_0 r^2} + \frac{\alpha}{2\varepsilon_0}\frac{(r^2 - R^2)}{r^2}$$

$$= \frac{\alpha}{2\varepsilon_0} + \frac{1}{2\varepsilon_0}\left(\frac{q}{2\pi r^2} - \frac{\alpha R^2}{r^2}\right)$$

For electric field to be independent of r means second term in the expression of E should be zero.

or $$\frac{q}{2\pi r^2} - \frac{\alpha R^2}{r^2} = 0$$

i.e. $$q = 2\pi\alpha R^2$$

and electric field

$$E = \frac{\alpha}{2\varepsilon_0}$$

Example 18.23: A conducting sphere S_1 of radius r is attached to an insulating handle. Another conducting sphere S_2 of radius R is mounted on an insulating stand. S_2 is initially uncharged, S_1 is given a charge Q brought into contact with S_2 and removed. S_1 is recharged such that the charge on it is again Q and it is again brought into contact with S_2 and removed. This process is repeated n times.

(*a*) Find the electrostatic energy of S_2 after n such contacts with S_1.
(*b*) What is the limiting value of this energy as $n \to \infty$?

Solution: (*a*) Let the charges imparted to S_2 after 1st, 2nd, 3rd, etc. contacts be q_1, q_2, q_3, etc. When S_1 and S_2 are put in contact, their potentials become equal. Therefore after first contact

$$\frac{1}{4\pi\varepsilon_0}\frac{Q-q_1}{r} = \frac{q_1}{4\pi\varepsilon_0 R} \text{ or } q_1 = \frac{QR}{(R+r)}$$

After second contact

$$\frac{1}{4\pi\varepsilon_0}\frac{Q-q_2}{r} = \frac{q_1+q_2}{4\pi\varepsilon_0 R} \quad \text{or} \quad q_2 = \frac{QR - q_1 r}{R+r} = Q\left(\frac{R}{R+r}\right)^2$$

Similarly $$q_3 = Q\left(\frac{R}{R+r}\right)^3 \text{ and } q_n = Q\left(\frac{R}{R+r}\right)^n$$

Thus total charge given to S_2 in n contacts is

$$q = (q_1 + q_2 + q_3 \ldots, q_n)$$

$$= Q\left\{\frac{R}{R+r} + \left(\frac{R}{R+r}\right)^2 + \ldots\ldots + \left(\frac{R}{R+r}\right)^n\right\}$$

Electrostatic energy of S_2 after n contacts is

$$U = \frac{1}{2}qV = \frac{1}{2}q\frac{q}{4\pi\varepsilon_0 R} \text{ where } V \text{ is surface potential of } S_2$$

$$= \frac{1}{2}\frac{Q^2}{4\pi\varepsilon_0 R}\left\{\frac{R}{R+r} + \left(\frac{R}{R+r}\right)^2 + \ldots\ldots + \left(\frac{R}{R+r}\right)^n\right\}^2$$

If $$\frac{R}{R+r} = a, \text{ then}$$

$$U = \frac{Q^2}{8\pi\varepsilon_0 R}\left(a + a^2 + \ldots + a^n\right)^2$$

$$= \frac{Q^2 a^2}{8\pi\varepsilon_0 R}\left(\frac{1-a^n}{1-a}\right)^2$$

When $n \to \infty$, $a^n \to 0$, then

$$U = \frac{Q^2}{8\pi\varepsilon_0 R}\frac{a^2}{(1-a)^2} = \frac{Q^2\left(\frac{R}{R+r}\right)^2}{8\pi\varepsilon_0 R\left(1 - \frac{R}{R+r}\right)^2}$$

$$= \frac{Q^2 R}{8\pi\varepsilon_0 r^2}$$

Example 18.24: A charge q is distributed uniformly over the volume of non-conducting ball of radius R. Find (*a*) the energy stored in the ball. (*b*) the energy stored in the surrounding space. (*c*) the total energy of the system. Take dielectric constant of ball medium as 1.

Solution: Energy stored per unit volume in medium is, where electric field E is given by

$$U = \frac{1}{2}\varepsilon E^2 = \frac{1}{2}\varepsilon_0\varepsilon_r E^2 \qquad \text{but } \varepsilon_r = 1$$

$$= \frac{1}{2}\varepsilon_0 E^2$$

$\therefore$ Total energy stored is

$$U_t = \int u\,dV = \int \frac{1}{2}\varepsilon_0 E^2 dV$$

where dV is small volume element over which E is assumed constant. The electric field intensity due to a spherical charge is

$$E = \begin{bmatrix} \dfrac{1}{4\pi\varepsilon_0}\dfrac{qr}{R^3} & (r < R \text{ inside the sphere}) \\ \dfrac{1}{4\pi\varepsilon_0}\dfrac{q}{r^2} & (r > R \text{ outside the sphere}) \end{bmatrix}$$

(*a*) Energy stored in the ball

$$U_1 = \int_0^R \frac{1}{2}\varepsilon_0 E_i^2\, dV = \int_0^R \frac{1}{2}\varepsilon_0\left(\frac{1}{4\pi\varepsilon_0}\frac{qr}{R^3}\right)^2 4\pi r^2 dr$$

$$= \frac{q^2}{8\pi\varepsilon_0 R^6}\int_0^R r^4 dr = \frac{q^2}{8\pi\varepsilon_0 R^6}\frac{R^5}{5} = \frac{q^2}{40\pi\varepsilon_0 R}$$

(*b*) Energy in surrounding space

$$U_2 = \int_R^\infty \frac{1}{2}\varepsilon_0 E_0^2 dV = \int_R^\infty \frac{1}{2}\varepsilon_0\left(\frac{q}{4\pi\varepsilon_0 r^2}\right)^2 4\pi r^2 dV$$

$$= \frac{q^2}{8\pi\varepsilon_0}\int_R^\infty \frac{dr}{r^2} = \frac{q^2}{8\pi\varepsilon_0}\left[-\frac{1}{r}\right]_R^\infty$$

$$= \frac{q^2}{8\pi\varepsilon_0 R}$$

(*c*) Total energy of the system

$$U_t = U_1 + U_2 = \frac{q^2}{40\pi\varepsilon_0 R} + \frac{q^2}{8\pi\varepsilon_0 R} = \frac{3q^2}{20\pi\varepsilon_0 R}$$

Example 18.25: A non conducting disc of radius a and uniform positive surface charge density σ is placed on the ground with its axis vertical. A particle of mass m and positive charge q is dropped along its axis from height H with zero initial velocity. The particle has specific charge $\dfrac{q}{m} = \dfrac{4\varepsilon_0 g}{\sigma}$.

(*a*) Find the value of H if the particle just reaches the disc.
(*b*) Sketch the potential energy of the particle as a function of the height and find its equilibrium position.

Solution: The potential at a point at a distance x from the centre of a uniformly charged disc along the axis is

$$V = \frac{\sigma}{2\varepsilon_0}\left[\sqrt{x^2+a^2} - x\right]$$

where σ is surface density of charge on disc, a, radius of the disc.
When $x = H$, i.e. charge is at A, the electrostatic potential energy of charge q is

$$U_i = Vq = \frac{\sigma q}{2\varepsilon_0}\left[\sqrt{H^2+a^2} - H\right]$$

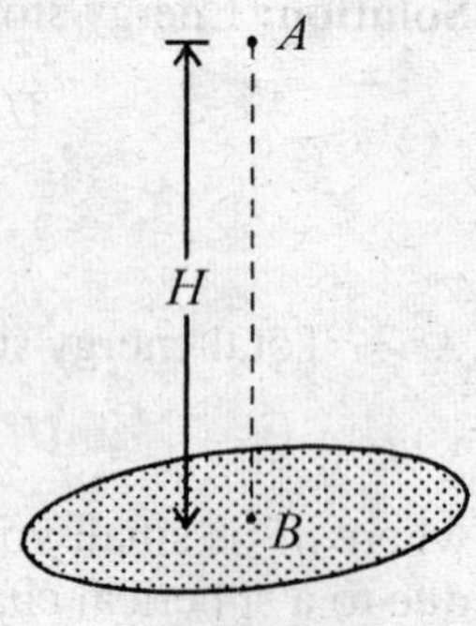

Fig. 18.50

and when at O, the centre of disc where $x = 0$.

$$U_f = Vq = \frac{\sigma q}{2\varepsilon_0} a$$

The velocity of the particle at A and O is zero, i.e. K.E. is zero. Hence conservation of energy gives

$$U_i + mgH = U_f$$

$$\frac{\sigma q}{2\varepsilon_0}\left[\sqrt{H^2+a^2} - H\right] + mgH = \frac{\sigma q}{2\varepsilon_0} a$$

or $$\frac{\sigma}{2\varepsilon_0}\frac{q}{mg}\left[\sqrt{H^2+a^2} - H\right] + H = \frac{\sigma a}{2\varepsilon_0}\frac{q}{mg}$$

But $$\frac{q}{mg} = \frac{4\varepsilon_0}{\sigma}$$

Substitution gives

$$\frac{\sigma}{2\varepsilon_0}\frac{4\varepsilon_0}{\sigma}\left[\sqrt{H^2+a^2} - H\right] + H = \frac{\sigma a}{2\varepsilon_0}\frac{4\varepsilon_0}{\sigma}$$

or $$2\sqrt{H^2+a^2} - H = 2a$$

or $$4(H^2 + a^2) = (H + 2a)^2 = H^2 + 4aH + 4a^2$$

This gives $H = \frac{4}{3}a$ ($H = 0$ is not admissible solution)

(*b*) The total potential energy at height h is

$$U = \frac{q\sigma}{2\varepsilon_0}\left(\sqrt{h^2+a^2} - h\right) + mgh$$ But $q = \frac{4\varepsilon_0 mg}{\sigma}$

$$U = \frac{4\varepsilon_0 mg}{\sigma}\frac{\sigma}{2\varepsilon_0}\left(\sqrt{h^2+a^2} - h\right) + mgh$$

$$= 2mg\left(\sqrt{h^2+a^2}-h\right)+mgh$$

$$= 2mg\sqrt{h^2+a^2}-mgh \qquad ...(i)$$

Equation (*i*) gives the variation of P.E. with h, height from the disc. Since gravitational and electric fields both are conservative and force $F = -\frac{dU}{dh}$ and for equilibrium $F = 0$, i.e. $\frac{dU}{dh} = 0$

$$\frac{dU}{dh} = 2mg\frac{1}{2}\left(h^2+a^2\right)^{-\frac{1}{2}} \times 2h - mg = 0$$

or
$$h = \frac{a}{\sqrt{3}}$$

P.E. $U = 2mg\sqrt{h^2+a^2} - mgh$

When $h = 0 \quad U = 2mga$

When $h = \frac{a}{\sqrt{3}} \quad U = \sqrt{3}\,mga$ minimum

When $h = a \quad U = 1.82mga$

The potential energy curve versus height h is shown in Fig. 18.51.

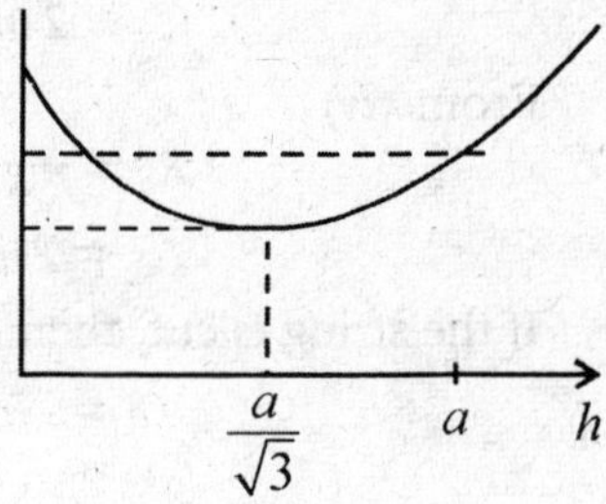

Fig. 18.51

Example 18.26: A rigid insulated wire frame in the form of a right angled triangle *ABC*, is set in a vertical plane as shown in Fig. 18.52. Two beads of equal masses m each carrying charges q_1 and q_2 are connected by a chord of length l and can slide without friction on wires. Considering the case when the beads are stationary determine (*i*) the angle α (*ii*) the tension in the chord (*iii*) normal reaction on the beads.

If the chord is now cut, what are the values of the charges for which the beads continue to remain stationary?

Solution: The forces acting on bead at P having charge q_1 are

(*i*) weight mg acting vertically downwards
(*ii*) the tension in string T along length of string PQ
(*iii*) the normal reaction R_1
(*iv*) the electrical force $F = \frac{1}{4\pi\varepsilon_0}\cdot\frac{q_1 q_2}{r^2}$ between beads along QP.

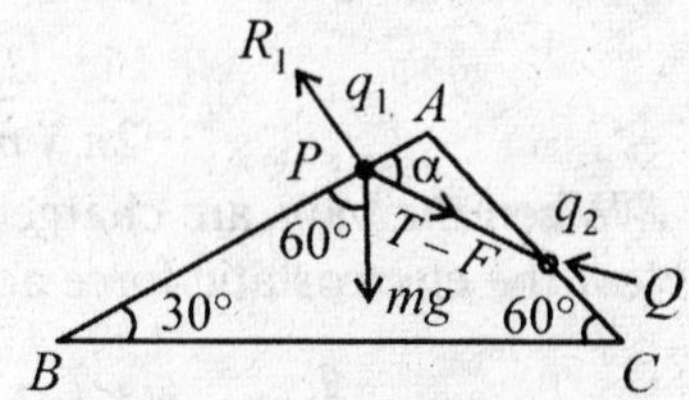

Fig. 18.52

For equilibrium of bead, the net force must be equal to zero. Resolving forces along and perpendicular to AB, we get

$$mg\cos 60° = (T - F)\cos\alpha \qquad ...(i)$$

$$R_1 = mg\cos 30° + (T - F)\sin\alpha \qquad ...(ii)$$

Similarly for the bead at Q

$$mg\sin 60° = (T - \alpha)\sin\alpha \qquad ...(iii)$$

$$R_2 = mg\cos 60° + (T - F)\cos\alpha \qquad ...(iv)$$

From (*i*) and (*iii*), we have

$$\tan\alpha = \tan 60° \text{ or } \alpha = 60°$$

Then (*i*) gives

$$mg = T - F$$

or $$T = mg + F = mg + \frac{1}{4\pi\varepsilon_0}\frac{q_1 q_2}{l^2}$$

From (*ii*) $R_1 = mg\cos 30° + (T - F)\sin\alpha$. But $T - F = mg$ and $\alpha = 60°$

$$= mg\cos 30° + mg\sin 60°$$

$$= 2\,mg\sin 60° = \sqrt{3}\,mg$$

From (*iv*)

$$R_2 = mg\cos 60° + mg\cos 60°$$

$$= 2mg\cos 60° = mg$$

If the string is cut, then the tension $T = 0$

$\therefore$ $$F = -mg$$

or $$\frac{q_1 q_2}{4\pi\varepsilon_0 l^2} = -mg$$

or $$q_1 q_2 = -4\pi\varepsilon_0 mg l^2$$

q_1 and q_2 should be of opposite nature to make positive sign.

Example 18.27: Two small identical balls *A* and *B* lying on a horizontal plane are connected by a weightless spring. The ball (*B*) is fixed at *O* and the ball (*A*) is free. The balls are charged initially. As a result the length of the spring is increased to η times. Calculate how many times the frequency of oscillations of ball *A* is charged due to charging. What will be this ratio if η = 2.0?

Solution: If *K* is the spring constant and *m* the mass of ball *A*, then before charging balls the initial frequency is

$$\nu_0 = \frac{1}{2\pi}\sqrt{\frac{K}{m}} \quad \text{...(i)}$$

When the balls are charged, the final length of the spring is ηl. Now the electrostatic force acts on the balls and in equilibrium

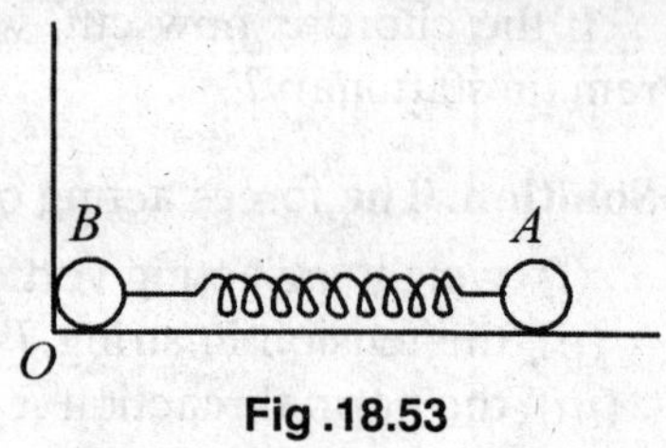

Fig.18.53

$$\frac{q^2}{4\pi\varepsilon_0(\eta l)^2} = K(\eta l - l) = Kl(\eta - 1)$$

or $$l^3 = \frac{q^2}{4\pi\varepsilon_0\eta^2(\eta - 1)K} \quad \text{...(ii)}$$

When the ball is displaced slightly by *x* from equilibrium position to the right, then net restoring force is

$$F = \frac{1}{4\pi\varepsilon_0}\frac{q^2}{(\eta l + x)^2} - K(\eta l + x - l)$$

$$= \frac{q^2}{4\pi\varepsilon_0(\eta l)^2}\left(1 + \frac{x}{\eta l}\right)^{-2} - Kl(\eta - 1) - Kx$$

Since $\frac{x}{\eta l} << 1$, hence using binomial theorem, the net force on the ball A is

$$F = \frac{q^2}{4\pi\varepsilon_0\eta^2 l^2}\left(1 - \frac{2x}{\eta l}\right) - Kl(\eta - 1) - Kx$$

$$= \frac{q^2}{4\pi\varepsilon_0\eta^2 l^2} - \frac{q^2(2x)}{4\pi\varepsilon_0\eta^3 l^3} - Kl(\eta - 1) - Kx$$

Using (*ii*), we have $F = -\left[\frac{1}{4\pi\varepsilon_0}\frac{q^2(2x)}{\eta^3 l^3} + Kx\right] = -\left(\frac{2q^2}{4\pi\varepsilon_0\eta^3 l^3} + K\right)x$

Substituting the value of l^3 from (*ii*), we have

$$\frac{2q^2}{4\pi\varepsilon_0\eta^3 l^3} = \frac{2q^2 \times 4\pi\varepsilon_0\eta^2(\eta - 1)K}{4\pi\varepsilon_0\eta^3 q^2} = \frac{2(\eta - 1)K}{\eta}$$

If the instantaneous acceleration of ball A be a, then we have

$$ma = F = -\left(\frac{2(\eta - 1)}{\eta}K + K\right)x = -\frac{3\eta - 2}{\eta}Kx$$

or $$a = -\frac{3\eta - 2}{\eta}\frac{K}{m}x$$

Acceleration is directly proportional to displacement as such the motion is S.H.M. and the angular frequency ω is given by

$$\omega = \sqrt{\frac{3\eta - 2}{\eta}\frac{K}{m}}$$

and frequency now is

$$\nu = \frac{\omega}{2\pi} = \frac{1}{2\pi}\sqrt{\frac{3\eta - 2}{\eta}\frac{K}{m}}$$

or $$\frac{\nu}{\nu_0} = \sqrt{\frac{3\eta - 2}{\eta}}$$

for $\eta = 2$, the ratio $\frac{\nu}{\nu_0} = \sqrt{\frac{6 - 2}{2}} = \sqrt{2}$.

Example 18.28: A soap bubble of radius r and surface tension T is given a potential V. Show that the new radius R of the bubble is related to initial radius by the equation

$$P(R^3 - r^3) + 4T(R^2 - r^2) - \frac{\varepsilon_0 V^2 R}{2} = 0$$

where P is atmospheric pressure.

Solution: When bubble is uncharged, the radius is r and excess pressure inside soap bubble is $\frac{4T}{r}$

Thus air inside it has pressure $P_1 = P + \frac{4T}{r}$ and volume $V_1 = \frac{4}{3}\pi r^3$.

When bubble is given a potential V, the new radius becomes R.

The electric field on the surface becomes $E = V/R$ and outward pressure on the bubble due to electrostatic field is

$$p = \frac{1}{2}\varepsilon_0 E^2 = \frac{1}{2}\varepsilon_0 \frac{V^2}{R^2}$$

Now the net pressure of the air inside bubble is

$$P_2 = P + \frac{4T}{R} - p = P + \frac{4T}{r} - \frac{1}{2}\varepsilon_0 \frac{V^2}{R^2}$$

and volume of air inside bubble is $V_2 = \frac{4}{3}\pi R^3$

Applying Boyle's law, we have

$$P_1 V_1 = P_2 V_2$$

$$\left(P + \frac{4T}{r}\right)\frac{4}{3}\pi r^3 = \left(P + \frac{4T}{R} - \frac{1}{2}\varepsilon_0 \frac{V^2}{R^2}\right)\frac{4}{3}\pi R^3$$

$$r^3 P + 4r^2 T = PR^3 + \frac{4TR^3}{R} - \frac{1}{2}\varepsilon_0 V^2 R$$

or $$P(R^3 - r^3) + 4T(R^2 - r^2) - \frac{1}{2}\varepsilon_0 V^2 R = 0.$$

Example 18.29: Two small balls A and B with charges $-q$ and $+q$ respectively are fixed on a horizontal plane at a distance d from each other. A third ball C with charge $+Q$ is suspended from a string. The string makes an angle of 30° with the vertical when the ball C is in equilibrium at a height d vertically above the ball A. When ball C is in an identical situation above ball B, find the angle which the string now makes with the vertical.

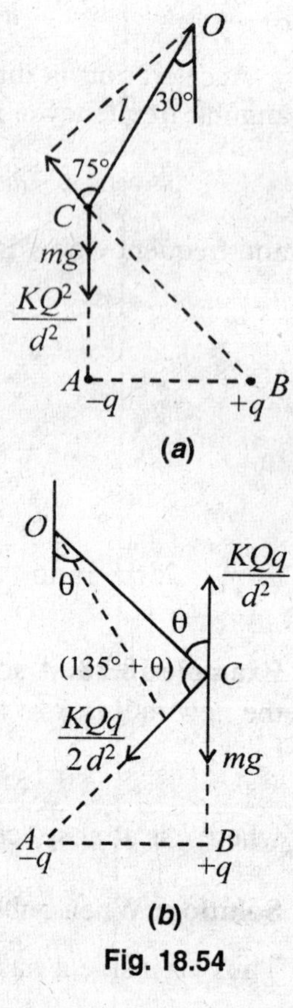

Fig. 18.54

Solution: The various forces acting on the ball C are

(*i*) Weight of the ball mg vertically downwards.

(*ii*) Electrostatic force $F_1 = \frac{KqQ}{d^2}$, attractive due to $-q$ charge at A, vertically downwards

(*iii*) Tension T along CD

(*iv*) Electrostatic repulsive force $F_2 = \frac{KQq}{2d^2}$ due to $+q$ charge at B.

Since ball C is in equilibrium, hence total torque on it about O is zero. Q is point of suspension.

$$\frac{KQq}{2d^2} l \sin 75° = \left(\frac{KqQ}{d^2} + mg\right) l \sin 30° = 0 \qquad ...(i)$$

where l is length of the string CO. Torque of tension about O is zero.

or $$\frac{KQq}{2d^2}\cos 15-\frac{KqQ}{2d^2}=\frac{1}{2}mg$$

or $$Q=\frac{mgd^2}{Kq(\cos 15-1)} \qquad \ldots(ii)$$

Similarly in second case Fig. 18.54 (*b*); if θ is the angle that the string makes with the vertical, we have

$$\left(\frac{KQq}{d^2}-mg\right)l\sin\theta=\frac{KqQ}{2d^2}l\sin(135°-\theta)$$

or $$\left(\frac{KQq}{d^2}-mg\right)\sin\theta=\frac{KqQ}{2d^2}\cos(45°-\theta) \qquad \ldots(iii)$$

Now substituting Q from (*ii*) in (*iii*), we get

$$\left(\frac{Kq}{d^2}\frac{mg\,d^2}{Kq(\cos 15-1)}-mg\right)\sin\theta=\frac{Kq}{2d^2}\frac{mg\,d^2}{Kq(\cos 15°-1)}\cos(45°-\theta)$$

or $$2\{1-(\cos 15-1)\}=\frac{\cos 45\cos\theta+\sin 45\sin\theta}{\sin\theta}$$

or $$2(2-\cos 15)=\frac{1}{\sqrt{2}}\cot\theta+\frac{1}{\sqrt{2}}$$

or $$\cot\theta=2\sqrt{2}(2-\cos 15)-1=1.92$$

or $$\theta=27.5°.$$

Example 18.30: (*a*) A charge Q Coulomb is uniformly distributed over a spherical volume of radius R. Obtain an expression for the energy of the system. (*b*) What will be the corresponding expression for the energy needed to completely dismantle the planet earth against the gravitational pull amongst its constituent particles? Assume earth to be a sphere of uniform density. Calculate this energy, given the product of mass and radius of earth to be 2.5×10^{31} kg-m and $g = 10$ m/s^2. (*c*) If the same charge of Q Coulomb as in part (*a*) is given to a spherical conductor of the same radius R, what will be the energy of the system?

Solution:

(*a*) In case of continuous distribution of charge. If dq charge is brought against potential V, work done is

$$dw = Vdq$$

and total energy

$$U=\int dw=\int_0^R Vdq$$

If at any instant radius of sphere is r then

$$V=\frac{1}{4\pi\varepsilon_0}\frac{q}{r}=\frac{1}{4\pi\varepsilon_0}\left(\frac{4}{3}\pi r^2\rho\right) \quad \text{as } q=\frac{4}{3}\pi r^3\rho$$

Since $$q=\frac{4}{3}\pi r^3\rho;\ dq=4\pi\rho r^2 dr$$

So energy is $$U=\int_0^R \frac{1}{4\pi\varepsilon_0}\left(\frac{4}{3}\pi r^2\rho\right)\left(4\pi r^2\rho\, dr\right)$$

$$=\frac{(4\pi\rho)^2}{4\pi\varepsilon_0\times 3}\int_0^R r^4\, dr=\frac{(4\pi\rho)^2}{4\pi\varepsilon_0\times 3}\times\frac{R^5}{5}$$

or $$U=\frac{1}{4\pi\varepsilon_0}\times\frac{3}{5}\left(\frac{4}{3}\pi R^3\rho\right)^2\frac{1}{R}$$

$$=\frac{3}{5}\frac{Q^2}{4\pi\varepsilon_0 R}\qquad \because\ Q=\frac{4}{3}\pi R^3\rho$$

$$=\frac{3Q^2}{20\pi\varepsilon_0 R}$$

(*b*) In case of gravitational field the self energy of a sphere is then given by

$$U=-\frac{3}{5}\frac{M^2G}{R}=-\frac{3}{5}\frac{MG}{R^2}MR=-\frac{3}{5}MgR$$

Minus sign is due to the attractive force.
Energy required to dismantle earth is

$$U=\frac{3}{5}MRg=\frac{3}{5}\times 2.5\times 10^{31}\times 10=1.5\times 10^{32}\text{ J}$$

(*c*) If the charge is distributed over the surface of the conductor then energy stored

$$U=\frac{1}{2}QV=\frac{1}{2}Q\frac{Q}{4\pi\varepsilon_0 R}=\frac{Q^2}{8\pi\varepsilon_0 R}.$$

Example 18.31: A ring of radius 0.1 m is made out of a this metallic wire of area of cross-section 10^{-6} m². The ring has a uniform charge of π Coulombs. Find the change in the radius of the ring when a charge 10^{-8} C is placed at the centre of the ring. Young's modulus of the metal is 2×10^{11} N/m².

Solution: Let T be the tension in the ring (Fig. 18.55). Consider the equilibrium of element AB of length dl of the ring.

$$2T\sin\theta=F_e$$

$$F=\frac{1}{4\pi\varepsilon_0}\frac{Q\,dq}{R^2}\quad\text{where } dq=\frac{q\,dl}{2\pi R}\ \text{charge on } dl \text{ part of the ring}$$

Since θ is very small hence $\sin\theta\approx\theta=\frac{dl/2}{R}=\frac{dl}{2R}$

$\therefore$ $$2T\cdot\frac{dl}{2R}=\frac{1}{4\pi\varepsilon_0}\cdot\frac{Qdq}{R^2}=\frac{1}{4\pi\varepsilon_0}\cdot\frac{Q}{R^2}\cdot\frac{q\,dl}{2\pi R}$$

or $$T=\frac{Qq}{8\pi^2\varepsilon_0 R^2}\qquad\ldots(i)$$

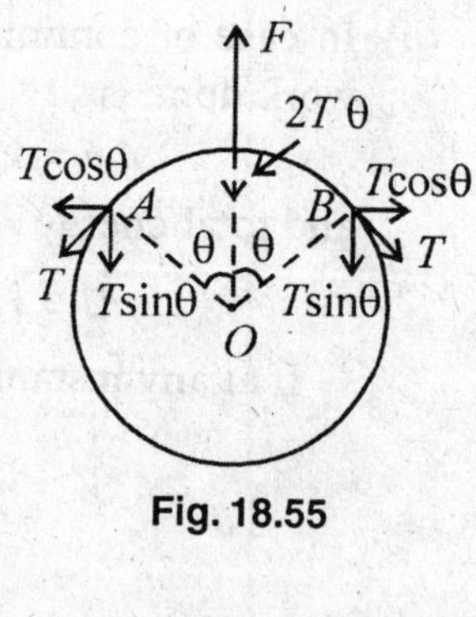

Fig. 18.55

Now if the radius of ring R changes to $R + \Delta R$ under tension then stress $= \dfrac{T}{A}$ and strain $= \dfrac{\Delta R}{R}$

Here A is area of cross-section of wire of ring.

$$\therefore \qquad Y = \frac{\text{stress}}{\text{strain}} = \frac{TR}{A\Delta R}$$

or
$$\Delta R = \frac{TR}{AY} = \frac{Q.q.R}{8\pi^2\varepsilon_0 R^2 YA}$$

$$= \frac{Qq}{8\pi^2\varepsilon_0 RYA}$$

$$= \frac{9\times10^9\times10^{-8}\times\pi}{2\times\pi\times0.1\times2\times10^{11}\times10^{-6}}$$

$$= 2.25\times10^{-3}\text{ m.}$$

Example 18.32: Two equal point charges are fixed at $x = -a$ and $x = +a$ on the x-axis. Another point charge Q is placed at the origin. Find the change in electrical energy of Q, when it is displaced by a small distance along the x-axis.

Solution: When Q is at origin the potential energy is

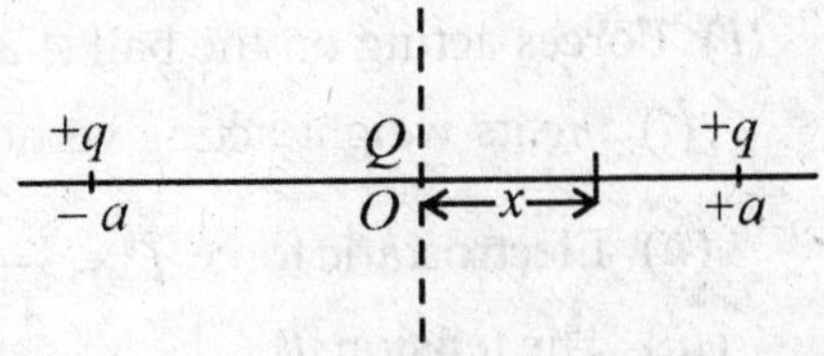

Fig. 18.56

$$U_i = \frac{1}{4\pi\varepsilon_0}\left(\frac{Qq}{a} + \frac{Qq}{a}\right) = \frac{2Qq}{4\pi\varepsilon_0}$$

When Q is displaced by x the final P.E. is

$$U_f = \frac{1}{4\pi\varepsilon_0}\left(\frac{Qq}{(a+x)} + \frac{Qq}{(a-x)}\right)$$

$$= \frac{Qq}{4\pi\varepsilon_0}\left(\frac{a-x+a+x}{a^2-x^2}\right)$$

$$= \frac{2Qqa}{4\pi\varepsilon_0(a^2-x^2)} = \frac{2Q\cdot qa}{4\pi\varepsilon_0 a^2\left(1-\dfrac{x^2}{a^2}\right)}$$

$$= \frac{2Q\cdot q\left(1-\dfrac{x^2}{a^2}\right)^{-1}}{4\pi\varepsilon_0\cdot a}$$

Since $\dfrac{x^2}{a^2} << 1$ we can use binomial theorem and we have neglecting higher powers of $\dfrac{x^2}{a^2}$.

$$U_f = \frac{2Qq}{4\pi\varepsilon_0 a}\left(1+\frac{x^2}{a^2}\right)$$

Hence change in the potential energy is

$$\Delta U = U_f - U_i = \frac{2Qq}{4\pi\varepsilon_0 a}\left(1+\frac{x^2}{a^2}\right) - \frac{2Qq}{4\pi\varepsilon_0 a}$$

$$= \frac{2Qq}{4\pi\varepsilon_0 a^3}\cdot x^2$$

Example 18.33: Two identical balls, each having a charge of 2×10^{-7} C and mass of 100 g are suspended from a common point by two insulating strings each 50 cm long. The balls are held at a separation 5.0 cm apart and then released. Find (*a*) the electric force on one of the charged balls (*b*) the components of the resultant force on it along and perpendicular to the string (*c*) the tension in the string (*d*) the acceleration of one of the balls. Answers are to be obtained only for the instant just after the release.

Solution:

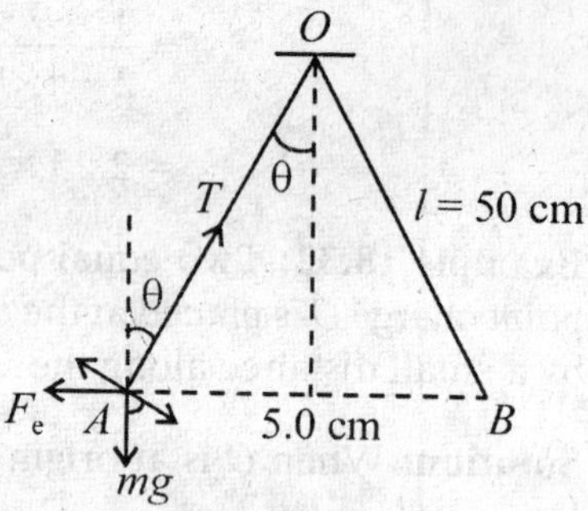

Fig. 18.57

(*a*) The electric force between the charges is

$$F_e = \frac{K\cdot q_1 q_2}{r^2} = \frac{9\times 10^9 \times 2\times 10^{-7}\times 2\times 10^{-7}}{(5\times 10^{-2})^2}$$

$$= 0.144 \text{ N}$$

(*b*) Forces acting on the ball *A* are

(*i*) *mg* its weight acting vertically downwards

(*ii*) Electrostatic force $F_e = \frac{q^2}{4\pi\varepsilon_0 r^2}$

(*iii*) The tension *T*

Since the particle do not have acceleration along the string hence net component of the force along the string is zero. The component of net force perpendicular to the string is

$$F_1 = F_e\cos\theta - mg\sin\theta$$

Since θ is small as such $\sin\theta = \frac{2.5}{50} = 0.05$

and $\cos\theta = 0.998 \approx 1$

$\therefore$ $F_1 = 0.144 - 100 \times 10^{-3} \times 10 \times 0.05$

$= 0.144 - 0.05 = 0.094$ N away from the other charge

(*c*) Tension in the string *T* is

$$T = F_e\sin\theta + mg\cos\theta$$

$$= 0.144 \times \frac{2.5}{50} + 10^{-1}\times 9.8 \times 0.999$$

$$= 0.0072 + 0.979$$

$$= 0.986 \text{ N}$$

(*d*) Net force is 0.094 N perpendicular to the string and away from the other charge. Hence net acceleration is

$a = \frac{F}{m} = \frac{0.094}{0.1} = 0.94 \text{ m/s}^2$ perpendicular to the string and away from the other charge.

Example 18.34: Three identical plates of metal with large surface areas are kept parallel to each other as shown is Fig. 18.58. The left most plate is given a charge Q, the right most a charge $-2Q$ and middle one remains neutral. Find the charge appearing on the outer surface of right most plate.

Solution: The distribution of charge on various surfaces is shown in Fig. 18.58. Let us take a point P inside plate. Net electric field at P is zero. If A be the surface area of one face of the plate then, electric field

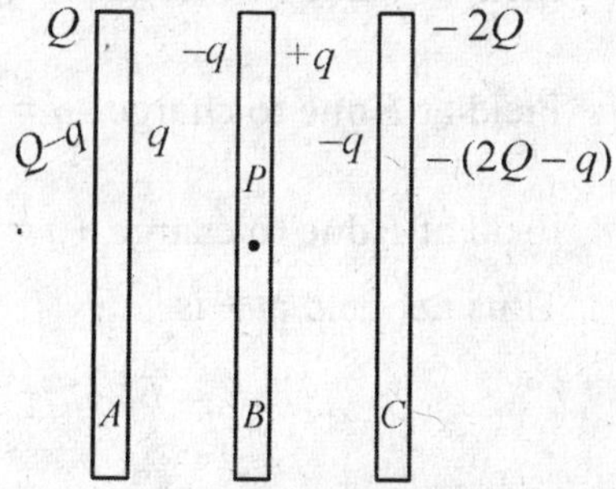

Fig. 18.58

E_1 due to charge $(Q-q)$ on A at $P = \dfrac{Q-q}{2A\varepsilon_0}$ towards right.

E_2 due to charge q on A at $P = \dfrac{q}{2A\varepsilon_0}$ towards right.

E_3 due to charge $-q$ on B at $P = \dfrac{q}{2A\varepsilon_0}$ towards left.

E_4 due to charge $+q$ on B at $P = \dfrac{q}{2A\varepsilon_0}$ towards left.

E_5 due to charge $-q$ on C at $P = \dfrac{q}{2A\varepsilon_0}$ towards right.

and E_6 due to charge $-(2Q-q)$ at $P = \dfrac{(2Q-q)}{2A\varepsilon_0}$ towards right. Then net field at P is

$$E = E_1 + E_2 - E_3 - E_4 + E_5 + E_6 = 0$$

or $$(Q-q) + q - q - q + q + (2Q-q) = 0$$

or $$2q = 3Q$$

or $$q = \frac{3}{2}Q$$

Hence the charge on the outer surface of plate $C = -(2Q-q)$ is $-Q/2$.

Example 18.35: Two conducting plates X and Y having large surface area A (on one side) are placed parallel to each other as shown in Fig. 18.59. The plate X is given a charge Q whereas the other is neutral. Find (*a*) the surface charge density at the inner surface of the plate X. (*b*) The electric field at a point in between the plates and (*d*) The electric field at a point to the right of the plates.

Solution: The distribution of charges on various plates is shown in Fig. 18.59. Using the result that charges on outer surfaces of the two plates is same, we have

$$Q - q = q \text{ or } q = Q/2$$

Since A is the area of one side.

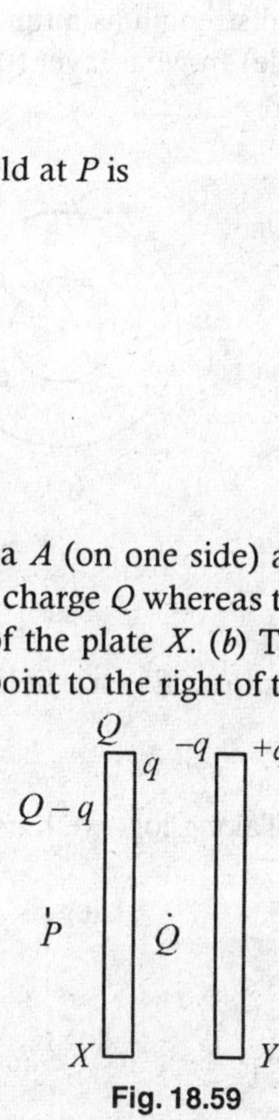

Fig. 18.59

(*a*) The surface charge density on inner surface of plate X is $\dfrac{q}{A}$ or $\sigma = \dfrac{Q}{2A}$

(*b*) Field at P

Field at P due to charge $Q - q = Q/2$ on outer side of plate $X = \frac{\sigma}{2\varepsilon_0} = \frac{Q}{2.2 A\varepsilon_0}$ or $E_1 = \frac{Q}{4 A\varepsilon_0}$ towards left.

Field at P due to charge $q = Q/2$ on inner side of plate X, $E_2 = \frac{Q}{4 A\varepsilon_0}$ towards left

Field at P due to charge $-q = -Q/2$ on inner side of plate Y, $E_3 = \frac{Q}{4 A\varepsilon_0}$ towards right

Field at P due to charge $+q = Q/2$ on outer side of plate Y, $E_4 = \frac{Q}{4 A\varepsilon_0}$ towards left

Thus net field at P is

$$E = E_1 + E_2 - E_3 + E_4 = \frac{Q}{2 A\varepsilon_0} \text{ towards left.}$$

(*c*) Similarly at Q, net field is $\frac{Q}{2 A\varepsilon_0}$ towards right

(*d*) Field at R is also $\frac{Q}{2 A\varepsilon_0}$ towards right.

Example 18.36: A point charge is surrounded by two spherical layers (Fig. 18.60) (*a*) with electric field strength as a function of distance having the form depicted in Fig. 18.60 (*b*) (on the log–log scale) In what layer (the inner or the outer) is the dielectric constant greater and by what factor?

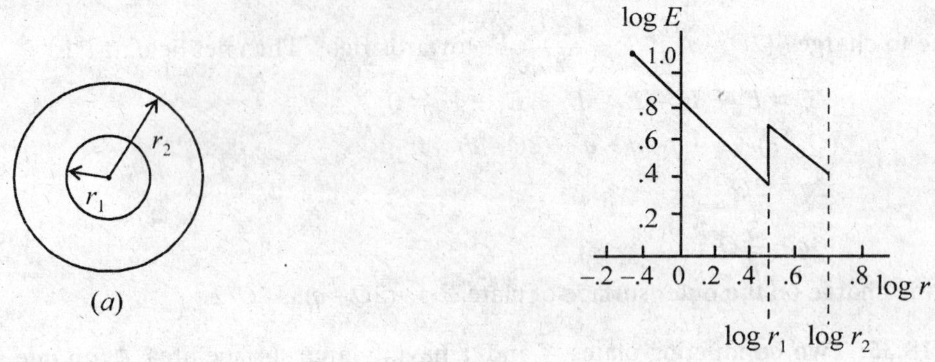

(a) **(b)**

Fig. 18.60

Solution: Electric field strength in each layer is

$$E = \frac{Q}{4\pi\varepsilon_0 \varepsilon r^2}$$

Taking log, we have

$$\log E_1 = \log\left(\frac{Q}{4\pi\varepsilon_0}\right) - \log \varepsilon_1 - 2\log r_1$$

and

$$\log E_2 = \log\left(\frac{Q}{4\pi\varepsilon_0}\right) - \log \varepsilon_2 - 2\log r_2$$

in each layer at the boundary between the layers substracting, we have

$$\log\left(\frac{E_2}{E_1}\right) = \log\left(\frac{\varepsilon_1}{\varepsilon_2}\right)$$

Hence in the inner layer the dielectric constant is higher than that in the outer layer. The difference of the logarithms of field strengths in Fig. 18.60 (*b*) is about 0.3, i.e.

$$\log\frac{E_2}{E_1} = 0.3 = \log\frac{\varepsilon_1}{\varepsilon_2}$$

or
$$\frac{\varepsilon_1}{\varepsilon_2} \approx 2.$$

Example 18.37: Two point charges Q_1 and Q_2 are positioned at a certain distance from each other. The curve in Fig.18.61 represent the distribution of potential along the straight line connecting the two charges. At which point (1, 2, and/or 3) the electric field is zero? What are the signs of charges Q_1 and Q_2 and which of the two is greater in magnitude?

Solution: The field strength vanishes only at one point, 3 because the derivative $\frac{dV}{dr}$ is zero.

Since near charge Q_2 the potential is negative while near Q_1 it is positive, we can conclude that Q_2 and Q_1 are negative and positive respectively.

The potential at every point in space is the algebraic sum of the potentials produced by all charges. To the right of Q_2 (except in the immediate vicinity of Q_2) the potential is positive. This implies that in the entire region to the right Q_2 the potential produced by Q_1 is greater in absolute terms than the potential produced by Q_2. Hence the absolute value of Q_1 is greater than that of Q_2.

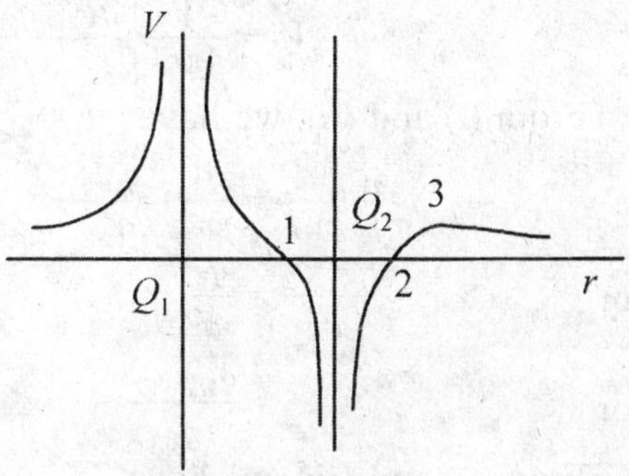

Fig. 18.61

Example 18.38: Three small identical neutral metal balls are at the vertices of an equilateral triangle. The balls are connected to a large charged sphere, held above the plane of the triangle. The first and the second ball acquire charges q_1 and q_2 respectively. How much is the value of charge q_3? Assume connecting wires to be very thin.

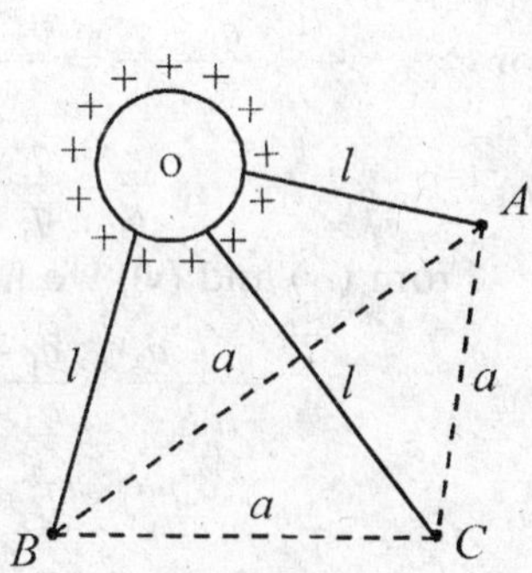

Fig. 18.62

Solution: Since the size of sphere is quite large as compared to that of the ball, as such when a ball is connected to a large charged sphere, the potential of the sphere will remain unchanged. Let l be the distance of each ball from the large sphere, r be the radius of each ball and a be the side of equilateral triangle. Let Q be the charge on the sphere and V its potential.

Let us consider the sharing of charge between the sphere and the first ball A. The potential of the ball A is

V_a = Potential due to its own charge + Potential due to charge on sphere.

$$= \frac{q_1}{4\pi\varepsilon_0 r} + \frac{Q}{4\pi\varepsilon_0 l}$$

Since ball is connected to the sphere as such its potential will be equal to that of the sphere. Thus

$$V = \frac{1}{4\pi\varepsilon_0}\left(\frac{q_1}{r} + \frac{Q}{l}\right) \quad \ldots(i)$$

Now consider the sharing of charges between (sphere + ball A) and ball B. Potential of ball B is

V_b = Potential due to its own charge + Potential due to charge on A + Potential due to charge on sphere

$$= \frac{1}{4\pi\varepsilon_0}\left[\frac{q_2}{r} + \frac{q_1}{a} + \frac{Q}{l}\right]$$

This potential is also equal to the potential of the sphere, i.e.

$$V = \frac{1}{4\pi\varepsilon_0}\left[\frac{q_2}{r} + \frac{q_1}{a} + \frac{Q}{l}\right] \quad \ldots(ii)$$

Finally consider the potential of C. If charge on it is q_3, then its potential also will be equal to that of sphere. Hence

$$V = \frac{1}{4\pi\varepsilon_0}\left[\frac{q_3}{r} + \frac{q_2}{a} + \frac{q_1}{a} + \frac{Q}{l}\right] \quad \ldots(iii)$$

From (*i*) and (*ii*), we have

$$\frac{q_1}{r} + \frac{Q}{l} = \frac{q_2}{r} + \frac{q_1}{a} + \frac{Q}{l}$$

or $$\frac{q_1 - q_2}{r} = \frac{q_1}{a}$$

or $$\frac{r}{a} = \frac{q_1 - q_2}{q_1} \quad \ldots(iv)$$

From (*i*) and (*iii*), we have

$$\frac{q_1}{r} + \frac{Q}{l} = \frac{q_3}{r} + \frac{q_2}{a} + \frac{q_1}{a} + \frac{Q}{l}$$

or $$\frac{q_1 - q_3}{r} = \frac{q_2 + q_1}{a}$$

or $$\frac{r}{a} = \frac{q_1 - q_3}{q_2 + q_1} \quad \ldots(v)$$

From (*iv*) and (*v*), we have

$$\frac{q_1 - q_2}{q_1} = \frac{q_1 - q_3}{q_2 + q_1}$$

or $$q_1q_2 - q_2^2 + q_1^2 - q_1q_2 = q_1^2 - q_1q_3$$

or $$q_3 = \frac{q_2^2}{q_1}$$

Example 18.39: Four equal small spheres each of mass m and each carrying a charge q are suspended by light strings of length l from the same point. When in equilibrium, the side of the square formed with the four spheres at its corners is a, find a. Explain why for a similar system consisting of charges Kq and string of length Kl, do the masses have still to be m only for the side of the square to be Ka?

Solution: The situation is depicted in Fig. 18.63. In this case $AB = BC = CD = DA = a$ and $PD = PA = PB = PC = l$. $ABCD$ is a square and PO is the normal from P, the point of suspension on the plane of the square. $AC = BD = a\sqrt{2}$.

Let us calculate the net electric force F_e on one of the charges at A due to the rest of the charges.

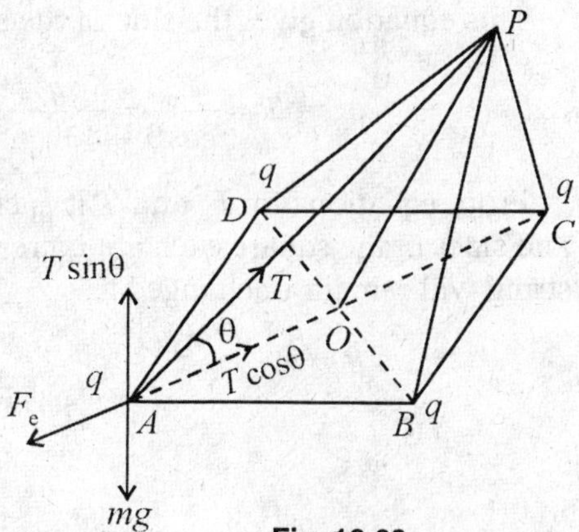

Fig. 18.63

$$\mathbf{F}_e = \frac{q^2}{4\pi\varepsilon_0}\left[\frac{\mathbf{BA}}{a^3} + \frac{\mathbf{CA}}{2\sqrt{2}a^3} + \frac{\mathbf{DA}}{a^3}\right]$$

$$= \frac{q^2}{4\pi\varepsilon_0 a^3}\left[(\mathbf{BA} + \mathbf{DA}) + \frac{\mathbf{CA}}{2\sqrt{2}}\right]$$

But from vector addition, we have $\mathbf{BA} + \mathbf{DA} = \mathbf{CB} + \mathbf{BA} = \mathbf{CA}$

or $$\mathbf{F}_e = \frac{q^2}{4\pi\varepsilon_0 a^3}\left[1 + \frac{1}{2\sqrt{2}}\right]\cdot\mathbf{CA} \qquad |\mathbf{CA}| = a\sqrt{2}$$

∴ $$F_e = \frac{q^2}{4\pi\varepsilon_0 a^2}\left(\sqrt{2} + \frac{1}{2}\right) \text{ along } CA \qquad \ldots(i)$$

Now forces acting on A are

(*i*) mg, weight acting vertically downwards
(*ii*) Electric force F_e along CA
(*iii*) Tension T along AP.

For equilibrium, we have

$$T\sin\theta = mg \quad \text{and} \quad T\cos\theta = F_e$$

or $$\tan\theta = \frac{mg}{F_e} = \frac{mg}{\dfrac{q^2}{4\pi\varepsilon_0 a^2}\left(\sqrt{2} + \dfrac{1}{2}\right)}$$

But from figure, $\tan\theta = \dfrac{\sqrt{l^2 - a^2/2}}{a/2}$

∴ $$\frac{\sqrt{l^2 - a^2/2}}{a/2} = \frac{mg}{\dfrac{q^2}{4\pi\varepsilon_0 a^2}\left(\sqrt{2} + \dfrac{1}{2}\right)} = \frac{4\pi\varepsilon_0 a^2 mg}{q^2\left(\sqrt{2} + \dfrac{1}{2}\right)}$$

or $$a^3 = 2\sqrt{l^2 - a^2/2}\,\frac{q^2\left(\sqrt{2} + \dfrac{1}{2}\right)}{4\pi\varepsilon_0 mg}$$

or $$\frac{a^3}{\sqrt{l^2 - a^2/2}} = \frac{2q^2\left(\sqrt{2} + \dfrac{1}{2}\right)}{4\pi\varepsilon_0 mg}$$

This equation gives the side of the square. Tension T in the string is

$$T = \frac{F_e}{\cos\theta} = \frac{q^2}{4\pi\varepsilon_0 a^2}\left(\sqrt{2}+\frac{1}{2}\right)\frac{\sqrt{2}l}{a} \quad \cos\theta = \frac{a/\sqrt{2}}{l}$$

From equations for F_e and T it is clear that if in a similar system of charges, lengths of strings and sides of the square each are increased K times, then also electric force F_e and tension T in the string will remain unchanged.

PROBLEMS

1. Two equally charged identical metal spheres A and B repel each other with a force 2.0×10^{-5} N. Another identical uncharged sphere C is touched to A and then placed at the mid-point between A and B. What is the net electric force on C? [**Ans.** 2×10^{-5} N along BC]
2. A copper atom consists of nucleus surrounded by 29 electrons. The atomic weight of copper is 63.5. Let us now take two pieces of copper each weighing 10 g. Let us transfer one electron from one piece to another for every 1000 atoms in that piece. What will be the Coulomb force between the two pieces after transfer of electrons if they are 1 cm apart. Avogadro number = 6×10^{23} atom/g-mole, charge on an electron = -1.6×10^{-19} Coulomb. [**Ans.** 2.057×10^{16} N]
3. It is required to hold four equal point charges ($+q$) in equilibrium at the corners a square. Find the point charge that will do it if placed at the centre of the square.

   $$\left[\textbf{Ans.}\ -q\frac{2\sqrt{2}+1}{4}\right]$$
4. A charge $+10^{-19}$ C is located at the origin in free space and another charge Q at (2, 0, 0). If the x-component of the electric field at (3, 1, 1) is zero, calculate the value of Q. Is the y-component zero at (3, 1, 1)?

   $$\left[\textbf{Ans.}\ Q = \frac{3^{5/2}}{11^{3/2}} \times 10^{-9}\ \text{C, No, } y\text{-component is not} = 0\right]$$
5. A charge Q is placed at each of two opposite corners of a square. A charge q is placed at each of the other two corners. (*a*) If the resultant electric force on Q is zero, how are Q and q related? (*b*) Should q be chosen to make the resultant force on every charge zero? [**Ans.** (*a*) $Q = -2\sqrt{2}\,q$ (*b*) No]
6. Three point charges q, $2q$, and $8q$ are to be placed on a 9 cm long straight line. Find the positions where the charges should be placed such that the potential energy of this system is minimum. In this situation what is the electric field at the position of charge q due to the other two charges.

   [**Hint.** For P.E. to be minimum the greater charge should be farthest and smaller charge in between. q should be at 3 cm from $2q$. $E = 0$]
7. A small ball of mass 2×10^{-3} kg having charge of 1 μC is suspended by a string of length 0.8 m. Another identical ball having the same charge is kept at the point of suspension. Determine the minimum horizontal velocity which should be imparted to the lower ball so that it can make complete revolution.

[**Hint.** At the highest point the net force is $T + mg - F_e = \frac{mu^2}{r}$. For minimum velocity tension T at the highest point should be zero. This gives velocity at highest point $u = \sqrt{\left(g - \frac{F_e}{m}\right) r}$.

Corresponding to this, the velocity at the lowest point is $v = \sqrt{5gl - \frac{F_e l}{m}}$. $v = 5.86$ m/s]

8. A thin fixed ring of radius 1 m has a positive charge 1×10^{-5} C uniformly distributed over it. A particle of mass m (= 0.9 g) and having a negative charge of 1×10^{-6} C is placed on its axis at a distance of 1 cm from the centre of ring. Show that the motion of the negatively charged particle is simple harmonic motion. Calculate the time period of oscillations.

$$\left[\textbf{Ans.}\ \text{acc}^n = -\frac{1}{4\pi\varepsilon_0}\frac{qq_0}{mR^3}x \text{ and } T = 2\pi\sqrt{\frac{mR^3}{qq_0} \times 4\pi\varepsilon_0} = 0.628 \text{ s}\right]$$

9. Consider the classical model of an atom such that a charge $+e$ on nucleus is uniformly distributed within a sphere of radius 2 Å. An electron of charge $-e$ at a radial distance 1 Å moves inside this sphere. Find the force attracting the electron to the centre of the sphere. Calculate the frequency with which the electron would oscillate about the centre of the sphere.

[**Hint.** Electric field at distance r from centre $E = \frac{q.r}{4\pi\varepsilon_0 a^3}$ and force on electron towards centre $F = E(-e) = -\frac{e^2 r}{4\pi\varepsilon_0 a^3}$ and acceleration $\omega = \frac{F}{m}$. Motion is S.H.M. with

$$f = \frac{1}{2\pi}\sqrt{\frac{e^2}{4\pi\varepsilon_0 ma^3}} = 9 \times 10^{14} \text{ Hz}\Bigg]$$

10. Two circular wire loops of radii 0.05 m and 0.09 m respectively are put such that their axes coincide and their centre are 0.12 m apart. Charge of 10^{-6} C is spread uniformly on each loop. Find the potential difference between the centres of the loop. [**Ans.** 7.08×10^4 V]

11. Two equal charges q are kept fixed at $-a$ and a along the x-axis. A particle of mass m and charge $q/2$ is brought to the origin and given a small displacement along (*i*) x-axis and (*ii*) y-axis. Describe the motion in the two cases.

[**Ans.** (*i*) executes S.H.M. with angular frequency $\omega = \sqrt{\frac{2q^2}{4\pi\varepsilon_0 ma^3}}$ (*ii*) Net force $F = \frac{q}{4\pi\varepsilon_0 a^3} y$.

The force is directed away from origin and particle moves away and will not come back]

12. A circular ring of radius R with uniform positive charge density λ per unit length is located in the y-z plane with its centre at the origin O. A particle of mass m and positive charge q is projected from the point $P(R\sqrt{3}, 0, 0)$ on the positive x-axis directly towards O with initial speed v. Find the smallest (non zero) value of the speed v such that the particle does not return to P.

$$\left[\textbf{Ans.}\ v_{\min} = \sqrt{\frac{\lambda q}{2\varepsilon_0 m}}\right]$$

13. An infinite number of charges each equal to q are placed along x-axis at $x = 1$, $x = 2$, $x = 4$, $x = 8$, ... and so on. (*i*) Find potential and electric field at the point $x = 0$ due to this set of charges. (*ii*) What will be the potential and electric field if in the above set up the alternate charges have opposite sign?

$$\left[\textbf{Ans. } (i)\ V = \frac{q}{2\pi\varepsilon_0}, E = \frac{q}{3\pi\varepsilon_0}\quad (ii)\ V = \frac{q}{6\pi\varepsilon_0}, E = \frac{q}{5\pi\varepsilon_0}\right]$$

14. A long charged cylinder of radius a and linear charge density λ is surrounded by an earthed hollow co-axial conducting cylinder of radius b. What is the electric field in the space between the two cylinders?

$$\left[\textbf{Ans. } E = \frac{1}{4\pi\varepsilon_0}\frac{2\lambda}{r}\right]$$

15. (*a*) Calculate the electric field strength near a long charged wire, carrying charge per unit length q.

(*b*) Two long wires carrying charges q_1 and q_2 per metre length are at separation d. Find the electrostatic force per metre length on each other.

$$\left[\textbf{Ans. } (a)\ E = \frac{1}{4\pi\varepsilon_0}\frac{2q}{r}\quad (b)\ \frac{1}{4\pi\varepsilon_0}\frac{2q_1q_2}{r}\right]$$

16. (*a*) If a small piece of radius b is removed from a charged spherical shell of radius a ($\gg b$). Calculate electric field strength at the location of aperture. Assume uniform surface charge density $= \sigma$.

(*b*) If the removed disc be placed to fill the aperture, what will be the force on this disc per unit area?

[**Hint.** The case is same as the force acting on the charged conductor per unit area. The field due to elementary part ds of the conductor is $\frac{1}{2}\frac{\sigma}{\varepsilon_0}$ and force per unit area on this conductor is $\frac{1}{2}\frac{\sigma^2}{\varepsilon_0}$]

17. The electric field in a region is radially out with magnitude $E = \alpha r$. Calculate the charge contained in a sphere of radius R centered at the origin. Calculate the charge if $\alpha = 100\ \text{V/m}^2$ and $R = 30$ m.

[**Hint.** Total flux over a Gaussian surface, a concentric sphere of radius x is $E_x 4\pi x^2$, when $x = R$,

$$E_x = E_R = \alpha R \text{ and flux } \phi = \alpha R\, 4\pi R^2 \text{ or } \phi = 4\pi\alpha R^3 = \frac{1}{\varepsilon_0}Q \text{ and } Q = 4\pi\alpha\varepsilon_0 R^3 =$$

$$\frac{1}{9\times10^9}\times 100(0.3)^3 = 3\times10^{-10}\ \text{C}$$]

18. Three point charges of 1 C, 2 C and 3 C are placed at the corners of an equilateral triangle of side 1 m. Calculate the work required to move these charges to the corners of a smaller equilateral triangle of side 0.5 m.

[**Hint.** Find potential energy of the system of charges when on the corners of triangle of side 1 m, $U_1 = 99 \times 10^9$ J. Potential energy of the system of charges when on the corners of triangle of side 0.5 = $2 \times 99 \times 10^9$ J. Work done is $U_2 - U_1 = 99 \times 10^9$ J]

19. Three point charges of 0.1 C are placed at the corners of an equilateral triangle with side L = 1 m. If this system is supplied energy at the rate of 1 kW, how much time will be required to move one of the charges on to the mid-point of the line joining the two charges?

[**Hint.** The difference of P.E. in the two configurations of the system is $\Delta U = 1.8 \times 10^8$ J. Then

$$\text{power required} = \frac{1.8 \times 10^8}{t} = 1 \times 10^3 \text{ or } t = 50 \text{ hrs}$$]

20. An electron enters a region between the plates of a parallel plate capacitor at a point equidistant from either plate. The capacitor plates are 2×10^{-2} m apart and 10^{-1} m long. A potential difference of 300 V is maintained across the plates. Assuming that the initial velocity of the electron is parallel to the capacitor plates, calculate the largest value of the velocity of the electron so that they do not fly out of the capacitor at the other end.

[**Hint.** The electron experience an acceleration $a = \frac{Ee}{m} = \frac{Ve}{dm}$ perpendicular to its velocity. If it covers $d/2$ distance in time t in which it covers distance l parallel to the plates, it will just cross the plates. The value of velocity then is, $\frac{d}{2} = \frac{1}{2}at^2$ or $t = \sqrt{\frac{d}{a}}$ and $l = vt$ or $l = v\sqrt{\frac{d}{a}}$ or $v = l\sqrt{\frac{a}{d}}$.

$$v = l\sqrt{\frac{Ve}{dm\,d}} = \frac{l}{d}\sqrt{\frac{Ve}{m}} = 3.65 \times 10^7 \text{ m/s}$$]

21. Three concentric spherical metal shells A, B and C radii a, b and c ($a < b < c$) have surface charge densities $+\sigma$, $-\sigma$ and $+\sigma$ respectively.

(*i*) Find the potentials of three shells, A, B and C.

(*ii*) If the shells A and C are at the same potential, obtain the relation between the radii a, b and c.

[**Ans.** (*i*) $V_A = \frac{\sigma}{\varepsilon_0}(a - b + c), V_B = \frac{\sigma}{\varepsilon_0}\left(\frac{a^2 - b^2}{b} + c\right)$ and $V_C = \frac{\sigma}{\varepsilon_0}\left(\frac{a^2 - b^2 + c^2}{c}\right)$

(*ii*) for $V_A = V_C$, $c = a + b$]

22. Two plane parallel conducting plates 1.5×10^{-2} m apart are held horizontally one above the other in air. The upper plate is maintained at positive potential of 1.5 kV, while the other plate is earthed. Calculate the number of electrons which must be attached to a small oil drop of mass 4.9×10^{-15} kg between the plates to maintain it at rest, assuming that the density of air is negligible in comparison to that of oil. If the potential of the above plate is suddenly changed to –1.5 kV. What is the initial acceleration of the charged drop? Also find the terminal velocity of the drop if its radius is 5×10^{-6} m and coefficient of viscosity of air is 1.8×10^{-5} N-s/m^2.

[**Hint.** For equilibrium of the drop electric force on it is equal to weight, i.e. $Eq = mg$ or $E \cdot ne = mg$, $n = \frac{mg}{Ee}$ and $E = V/d$ gives $n = 3$. When polarity of upper plate is reversed, net force downwards $F = mg + F_e$, $F = 2mg$ and acceleration = $2g$. For terminal velocity

$2mg = 6\pi\eta rv$, $v = 5.7 \times 10^{-5}$ m/s]

23. A positively charged sphere of mass 5 kg is attached to a spring of force constant $k = 10^4$ Nm^{-1}. The sphere is tied with a thread so that the spring is in its natural length. Another identical, negatively charged sphere is fixed to the floor, vertically below the positively charged sphere as shown in Fig. 18.64. Initial separation between spheres is $x_0 = 0.5$ m. Now thread is burnt and maximum elongation of the spring is 0.1 m. Calculate the charge on each sphere.

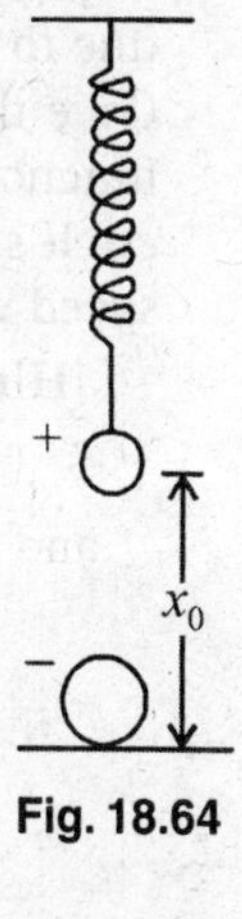

Fig. 18.64

[**Hint.** Conservation of energy gives electrostatic energy (U_e) + gravitation P.E. (U_g) + elastic potential energy (U_k) is constant. Equating these in initial and final position of the upper ball, we get $mg\,x_0 - \dfrac{q^2}{4\pi\varepsilon_0 x_0} = mg\,(x_0 - x) - \dfrac{q^2}{4\pi\varepsilon_0 (x_0 - x)} + \dfrac{1}{2}kx^2$ gives

$q = 10^{-4}$ C = 100 μC]

24. A particle of charge q and mass m moves rectilineary under the action of an electric field $E = A - Bx$. When B is a positive constant and x is a distance from the point where the particle was initially at rest, calculate

(*i*) the distance travelled by the particle till it comes to rest and

(*ii*) accel·eration at that moment.

[**Hint.** The particle will come to rest when its K.E. becomes zero, i.e. K.E.$= -\int_0^x F dx$

$= -\int_0^x q(A - Bx)\,dx = 0$. This gives $x = 0$ or $x = \dfrac{2A}{B}$. Acceleration is $F/m = \dfrac{qA}{m}$]

25. A radioactive source in the form of a metal sphere of radius 10^{-2} m emits beta particles at the rate of 5×10^{10} particles per second. The source is electrically insulated. How long will it take for its potential to be raised by 2 volt assuming that 40% of the emitted beta particles escape the source? [**Ans.** t = 700 μs]

26. Two circular loops of radius 0.05 m and 0.09 m respectively are put such that their axes coincide and their centres are 0.12 m apart. Charge of 10^{-6} Coulomb is spread uniformly on each loop. Find the potential difference between the centres of the loop. [**Ans.** 71 kV]

27. A very small sphere of mass 80 g having a charge q is held at a height 9 m vertically above the centre of a fixed conducting sphere of radius 1 m carrying an equal charge q. When released it falls and comes to rest just before it comes in contact with the sphere. Calculate the charge q. [**Ans.** q = 28 μC]

28. A particle of mass 40 μg and carrying a charge 5×10^{-9} C is moving directly towards fixed positive charge of magnitude 10^{-8} C. When it is at a distance of 10 cm from the fixed point charge it has a velocity of 50 cm/s. At what distance from the fixed charge will the particle come momentarily to rest? Is the acceleration constant during the motion?

[**Ans.** $r = 4.7 \times 10^{-2}$ m, Acceleration $\propto \dfrac{1}{r^2}$]

29. Two fixed charges $-2Q$ and Q are located at points with co-ordinates $(-3a, 0)$ and $(3a, 0)$ respectively in the x-y plane (*a*) Show that all the points in x-y plane where the electric potential due to the two charges is zero, lie on a circle. Find the radius and location of its centre. (*b*) Give the expression for the potential $V(x)$ at a general point on the x-axis and sketch the function $V(x)$ on the hole x-axis. (*c*) If a particle of charge q starts from rest at the centre of the circle show by a short quantitative argument that particle eventually crosses the circle. Find its speed when it does so.

[**Hint.** (*a*) See short exercise 22 – Locus is $(x - 5a)^2 + (4 - 0)^2 = (4a)^2$. Thus radius $r = 2a$

and center is $(5a, 0)$ (*b*) $V = \dfrac{1}{4\pi\varepsilon_0}\left(\dfrac{-2Q}{|(x+3a)|} + \dfrac{Q}{|(x-3a)|}\right)$ (*c*) $0 + \dfrac{1}{2}mv^2 = 0 + \dfrac{qQ}{4\pi\varepsilon_0 4a}$ or

$v = \sqrt{\dfrac{1}{4\pi\varepsilon_0}\dfrac{qQ}{2ma}}$]

30. A block of mass m containing a net positive charge q is placed on a smooth horizontal table which terminates in a vertical wall as shown in Fig. 18.65. The distance of the block from the wall is d. A horizontal electric field E towards right is switched on. Assuming elastic collisions (if any) find the time period of the resulting oscillations. Is it a S.H.M.?

[**Ans.** $\sqrt{\frac{8md}{qE}}$. No, motion is not S.H.M.]

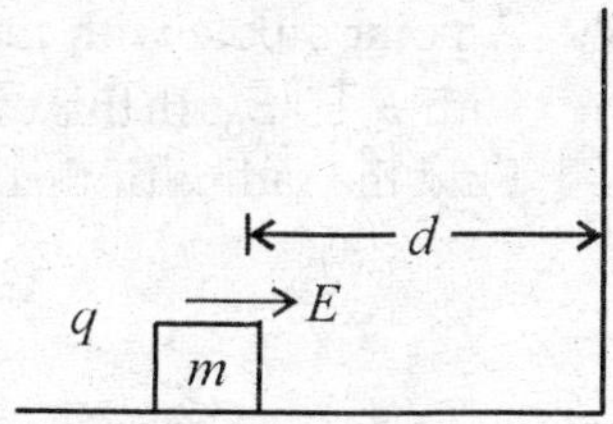

Fig. 18.65

31. A bob of simple pendulum has mass 40 g and a positive charge 40×10^{-6} C. It makes 20 oscillations in 45 s. A vertical field pointing upwards and of magnitude 2.5×10^4 N/C is switched on. How much time will it take to complete 20 oscillations? [**Ans.** 52 s]

32. A particle A having charge 2.0×10^{-6} C is held fixed on a horizontal table. A second charged particle of mass 80 g stays in equilibrium on the table at a distance of 10 cm from the first charge. The coefficient of friction between the table and this second particle is $\mu = 0.2$. Find the range within which the charge of this second particle may lie.

[**Ans.** between $\pm 8.7 \times 10^{-6}$ C]

33. A rod of length L has total charge Q distributed over its length uniformly. It is bent in the shape of a semi-circle as shown in Fig. 18.66. Find the magnitude of the electric field at the centre of curvature of the semi-circle.

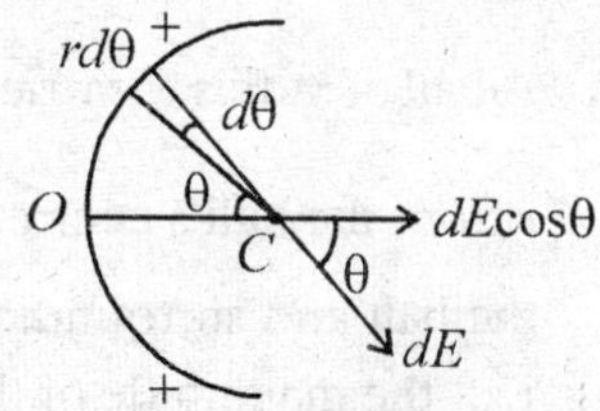

Fig. 18.66

[**Hint.** Consider small element $rd\theta$; charge on it is $\lambda r d\theta$, where $\lambda = \frac{Q}{L}$. Electric field $dE = \frac{K\lambda r d\theta}{r^2}$, its component along $OC = dE\cos\theta$ and average value of component perpendicular to OC ($= dE\sin\theta$) is zero for all the elements. Hence

$$E = \int_{-\pi/2}^{+\pi/2} dE\cos\theta = \frac{2K\lambda}{r} = \frac{Q}{2\varepsilon_0 L^2}\Bigg]$$

34. Consider a rough model of beryllium atom. The nucleus has four protons and four neutrons confined in a small volume of 10^{-15} m. The two 1s electrons make another spherical charge cloud at an average distance of 1.3×10^{-11} m, whereas the 2s electrons also make a cloud at an average distance of 5.2×10^{-11} m from the nucleus. Find the electric field at (a) a point just inside 1s cloud (b) a point just inside the 2s cloud.

[**Hint.** (a) $E = \frac{Kq}{r^2} = \frac{9 \times 10^9 \times 4 \times 1.6 \times 10^{-9}}{(1.3 \times 10^{-11})^2} = 3.4 \times 10^{13}$ N/C, it is due to only nuclear charge

(b) Field is due to 4 protons and 2, 1s electrons $E = \frac{9 \times 10^9 \times 2 \times 1.6 \times 10^{-19}}{(5.2 \times 10^{-11})^2} = 1.1 \times 10^{12}$ N/C]

35. At the moment $t = 0$, an electron leaves one plate of a parallel plate capacitor with a negligible velocity. An accelerating voltage varying as $V = kt$ is applied between the plates. The separation between the plates is l. What is the velocity of the electron at the moment it reaches the opposite plate?

[**Hint.** Acceleration of electron $a = \frac{dv}{dt} = \frac{eE}{m} = \frac{ekt}{ml}$. $\therefore\ v = \frac{1}{2}\frac{ekt^2}{ml}$, $s = \int v\,dt$ or

$$l = \int_0^t \frac{ekt^2}{2ml}dt = \frac{ekt^3}{6ml} \text{ or } t = \left(\frac{6ml^2}{ek}\right)^{1/3} \text{ and } v = \frac{1}{2}\frac{ek}{ml}\left(\frac{6ml^2}{ek}\right)^{2/3}\Bigg]$$

36. A point dipole with moment P is placed in the external field (uniform) whose strength is E_0 with $P \uparrow\uparrow E_0$. In this case one of the equipotential surfaces enclosing the dipole forms a sphere. Find the radius of this sphere.

$$\left[\textbf{Ans. } r = \left(\frac{P}{4\pi\varepsilon_0 E_0}\right)^{1/3}\right]$$

37. Find the electric field, potential at the centre of a hemisphere of radius R charged uniformly with surface density σ.

$$\left[\textbf{Ans. } V = \frac{\sigma R}{2\,\varepsilon_0} \text{ and } E = \frac{\sigma}{4\,\varepsilon_0}\right]$$

38. A dipole with an electric moment $\boldsymbol{P}$ is located a distance r from a long thread charged uniformly with a linear density λ. Find the force $\boldsymbol{F}$ acting on the dipole if the vector $\boldsymbol{P}$ is oriented (*a*) along the thread (*b*) along the radius vector (*c*) at right angles to the thread and the radius vector $\boldsymbol{r}$

$$\left[\textbf{Ans. } (a)\ F = 0\ (b)\ \boldsymbol{F} = \frac{\lambda \boldsymbol{P}}{2\pi\varepsilon_0 r^2}\ (c)\ \frac{\boldsymbol{P}\lambda}{2\pi\varepsilon_0 r^2}\right]$$

39. A ball of radius R carries a positive charge whose volume density depends only on the distance r from the ball's centre as $\rho = \rho_0\left(1 - \frac{r}{R}\right)$, where ρ_0 is a constant. Assuming permittivities of the ball and surroundings to be equal to unity, find

(*a*) the magnitude of the electric field strength as a function of the distance r both inside and outside the ball.

(*b*) the maximum intensity E_{max} and the corresponding distance.

$$\left[\textbf{Ans. } (a)\ E_r = \frac{\rho_0 R^3}{12\,r^3\varepsilon_0}\ (b)\ r_m = \frac{2R}{3} \text{ and } E_{max} = \frac{\rho_0 R}{9\,\varepsilon_0}\right]$$

40. A cone made of some non-conducting material has a total charge Q distributed uniformly over its sloping surface. Calculate the energy required to bring up small test charge q from infinity to the appex A of the cone. The cone has slope length L.

[**Hint.** Consider a small disc of slope length dx at distance x. Area is $2\pi x \sin\theta dx$. θ is semi vertical angle of the cone. Potential at apex due to this disc $dV = \frac{2\pi\sigma \cdot x \sin\theta\, dx}{4\pi\varepsilon_0 x}$ and total potential at

apex $V = \int dV = \frac{\sigma \sin\theta \cdot L}{2\,\varepsilon_0} = \frac{Q}{2\pi\varepsilon_0 L}$ work done $Vq = \frac{Qq}{2\pi\varepsilon_0 L}$]

SHORT ANSWER TYPE QUESTIONS

1. Can ever the entire or whole charge of a body be transferred to the other? If yes how and if not why?

[**Ans.** Yes, if the charged body is placed in hollow enclosure surrounded by conducting sphere and body connected to it, the whole charge will be transferred to the conducting body because the charge always lies on the outer surface of the conductor.]

2. What is the difference between charging by induction and charging by conduction?
[**Ans.** Both process of charging a conductor are different.
(*a*) In conduction two bodies are brought in contact whereas in induction they are brought near to each other.
(*b*) In conduction charges are shared between two bodies whereas in induction the charge on inducting body remains constant.
(*c*) In conduction the charge on both bodies is same in nature whereas in induction the charges on the induced body is of opposite nature to that of the inducting body.
(*d*) The charge transferred in conduction is always less than initial charge present whereas induced charge can be equal in magnitude to the inducing charge.]

3. Can two similarly charged bodies ever attract each other?
[**Ans.** Yes, when charge (Q) on body is much greater than the charge (q) on the other and are brought close to each other so that the force of attraction between Q and induced charge on the other exceeds the force of repulsion between Q and q.
However, in case of point charges, no induction takes place as such two similar point charges can never attract each other.]

4. Two identical metallic spheres of exactly equal masses are taken. One is given a positive charge (q) and the other an equal negative charge. Their masses after charging are different. Comment on the statement.
[**Ans.** Charging a body means addition or removal of electrons. In case of a positively charged conductor, electrons are removed from body as such its mass decreases. On the other hand, a negatively charged conductor is given additional electrons as such its mass increases.]

5. The electric field in a region is given by $E = \frac{A}{x^3}\hat{i}$. Write a suitable SI unit for A. Write an expression for the potential in the region assuming the potential at infinity is zero.
$\left[\textbf{Ans.}\text{ The units of } A \text{ are } A = Ex^3 = \frac{N}{C}\,\text{m}^3 \text{ or V-m}^2\text{. Potential } V = -\int_\infty^x E\,dx = -\int_\infty^x \frac{A}{x^3}dx = \frac{A}{2x^2}\right]$

6. A point charge q is at a distance d from an earthed conducting plane. How much energy is required to move the charge infinitely far from the plane.

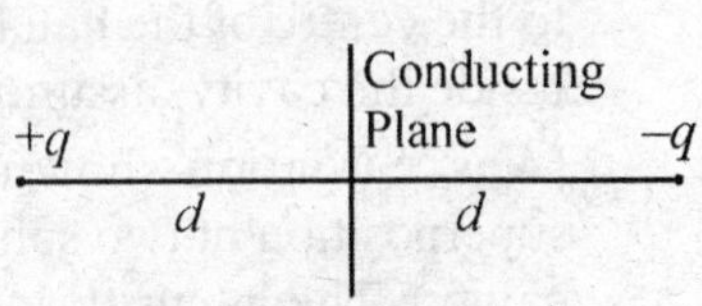

Fig. 18.67

[**Ans.** From the theory of electrical images, a point charge and a conducting plane are equivalent to two point charges $+q$ and $-q$ at a distance of $2d$, where d is the distance between point charge and conducting plane. The P.E. of the system is $U = \frac{(q)(-q)}{4\pi\varepsilon_0(2d)} = -\frac{q^2}{8\pi\varepsilon_0 d}$. Hence energy required to move the charge to infinity is $\frac{q^2}{8\pi\varepsilon_0 d}$]

7. Two point charges $+q$ and $-q$ are placed at distance d apart. What are points at which the resultant field is parallel to the line joining the charges?
[**Ans.** The two given charges constitute an electric dipole. Therefore resultant field is parallel to the dipole axis when the point (*i*) is on its axis and (*ii*) is on perpendicular bisector of the line joining the two charges.]

8. A charged particle is free to move in an electric field. Will it always move along an electric line of force?

[**Ans.** If the charged particle initially at rest in electric field and electric field is uniform or due to a point charge, it will always move along an electric line of force. But if a charged particle enters an electric field with certain velocity with its direction different from the direction of electric field, the path of the particle is not along the line of force. Also if the field is non-uniform and due to more than one charge, it will not move along line of force.]

9. If dielectric strength of air (minimum field required for ionisation of medium) is 3 MV/m. Can a metal sphere of radius 1 cm hold a charge of 1 Coulomb?

[**Ans.** When a sphere of radius r (= 1 cm) is given a charge (q = 2 Coulomb), the electric field near its surface is $E = \frac{q}{4\pi\varepsilon_0 r^2} = \frac{9\times10^9\times1}{10^{-4}} = 9\times10^{-13}$ V/m. Since this field is very large in comparison to the ionization field of air [= 3×10^6 V/m] hence the sphere cannot be given this charge.]

10. If electric field strength equals zero at a given point, must the electric potential for this point be zero?

[**Ans.** If V is potential at any point in electric field E, then $\frac{dV}{dr} = -E$. When $E = 0$, $\frac{dV}{dr} = 0$ or V = constant.]

11. A steady current passes through a cylindrical conductor. Is there an electric field inside the conductor?

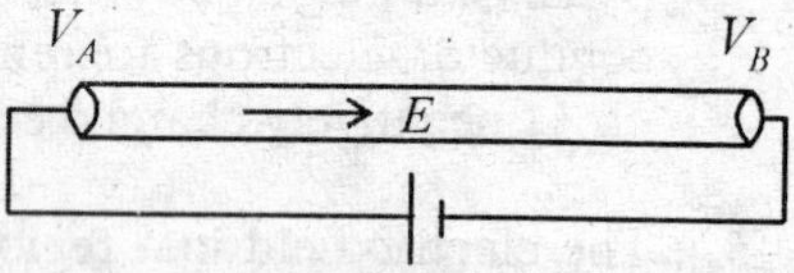

Fig. 18.68

[**Ans.** When a current i passes through a conductor there is a potential difference across the conductor, $V = iR$.

Hence there is an electric field $E = \frac{V}{l} = iR$, where l is the length of the conductor.]

12. Inside a ball charged uniformly with volume density ρ, there is a spherical cavity. The centre of cavity is displaced with respect to the centre of the ball by a distance Q. Find the field strength inside the cavity, assuming the permittivity equal to unity.

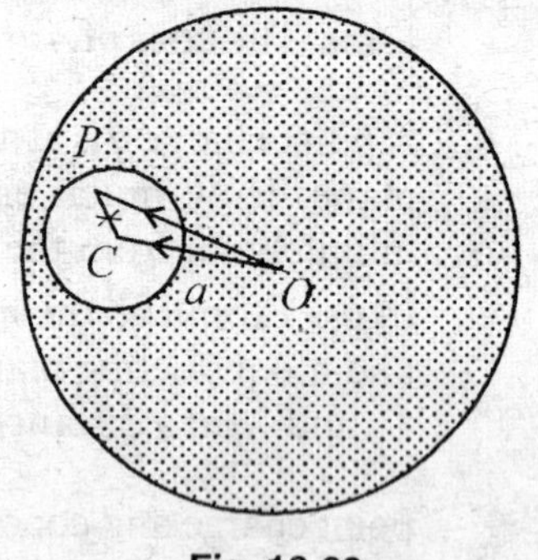

Fig. 18.69

[**Ans.** Situation shown in Fig. 18.69 can be considered as superposition of two spheres with charge density +ρ and –ρ (in cavity). The intensity at point P is due to field of bigger sphere

$$\mathbf{E}_1 = \frac{\rho}{3\varepsilon_0}\mathbf{OP} \text{ and due to } -\rho \text{ in cavity } \mathbf{E}_2 = \frac{\rho}{3\varepsilon_0}\mathbf{PC}$$

$$\mathbf{E} = \mathbf{E}_1 + \mathbf{E}_2 = \frac{\rho}{3\varepsilon_0}\mathbf{OP} + \frac{\rho}{3\varepsilon_0}\mathbf{PC} = \frac{\rho}{3\varepsilon_0}\mathbf{OC} = \frac{\rho}{3\varepsilon_0}\mathbf{a}\,]$$

13. Two spherical conductors of radii R_1 and R_2 are charged. If they are connected by a conducting wire, find the ratio of surface charge densities for both spheres.

[**Ans.** Since they are connected by a conducting wire, their potentials are equal. If V is the potential, then charges on the two spheres are $q_1 = \frac{VR_1}{K}$ and $q_2 = \frac{VR_2}{K}$, where $K = \frac{1}{4\pi\varepsilon_0}$

$$\therefore \frac{\sigma_1}{\sigma_2} = \frac{q_1}{q_2}\cdot\frac{R_2^2}{R_1^2} \quad \therefore \frac{\sigma_1}{\sigma_2} = \frac{R_1}{R_2}\times\frac{R_2^2}{R_1^2} = \frac{R_2}{R_1}\,]$$

14. A shower of protons from outer space deposits equal charges $+q$ on the earth and the moon and electrostatic repulsion then exactly counter balances the gravitational attraction. How large is q?

[**Ans.** Electrostatic force is equal in magnitude to gravitational force $\therefore \dfrac{q^2}{4\pi\varepsilon_0 r^2} = \dfrac{G M_e M_m}{r^2}$

or $q = \sqrt{G M_e M_m \times 4\pi\varepsilon_0} = \sqrt{4\pi\varepsilon_0 G M_e M_m}$]

15. A point charge q is placed at one corner of a cube, what is the flux linked with the cube?
[**Ans.** In this case consider a Gaussian surface which is a cube of side double the side of given cube with charge at the centre of the Gaussian surface. Then total flux through this cube is $\dfrac{q}{\varepsilon_0}$.

The Gaussian surface is equivalent to 8 given cubes. Hence flux through the given cube is $\dfrac{q}{8\varepsilon_0}$]

16. A hemispherical body is placed in a uniform electric field E. What is the flux linked with the curved surface if field is (*a*) parallel to the base [Fig. 18.70 (*a*)] (*b*) perpendicular to base [Fig. 18.70 (*b*)].

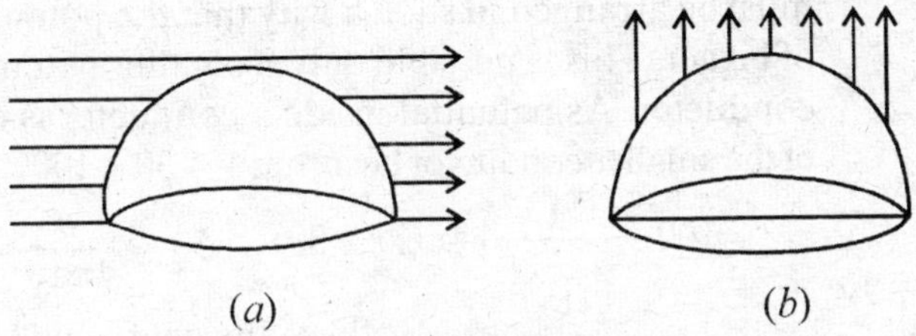

Fig. 18.70

[**Ans.** Since there is no charge inside the hemisphere hence total flux through the surface is zero, i.e. $\phi = \phi_{cs} + \phi_{ps} = 0$

In case (*i*) ϕ_{ps} flux through plane surface $\phi_{ps} = \int \mathbf{E} \cdot \mathbf{ds} = \int E\, ds \cos 90 = 0$, $\therefore \phi_{cs}$ flux through curved surface is zero. In case (*ii*) $\phi_{ps} = \int \mathbf{E} \cdot \mathbf{ds} = \int E ds = E \int ds = E \cdot \pi R^2$ and $\phi_{cs} = -\phi_{ps} = -E\pi R^2$]

17. What is the field in the cavity if a conductor having a cavity is charged? Does the result depend on the shape and size of cavity or conductor?
[**Ans.** When a conductor having cavity is charged, the charge resides on the surface as such there is no electric field inside the cavity.]

18. A solid conducting sphere having a charge Q is surrounded by an uncharged concentric hollow conducting spherical shell. The potential difference between the two spheres is V. Now outer hollow sphere is given a charge equal to $-3Q$, find the potential difference between the two spheres.
[**Ans.** When outer sphere is given additional charge and if due to this additional charge surface potential of outer sphere changes by say V_1, then potential of inner sphere will also change by V_1 and the potential difference between the two will remain unchanged, i.e. potential difference is V again.]

19. A spherical liquid drop has radius r and is given charge q when its potential is V. If two such drops coalesce, what is the potential at the surface of the drop so formed?

[**Ans.** Potential of the smaller drop $V = \dfrac{q}{4\pi\varepsilon_0 r}$ i.e. $V \propto \dfrac{q}{r}$. When two such drops coalesce, the radius of bigger drop becomes r' such that $\dfrac{4}{3}\pi r'^3 = \dfrac{4}{3}\pi r^3 \times 2$ or $r' = (2)^{1/3} r$ charge on this drop is $2q$. Thus its potential $V' = \dfrac{2q}{4\pi\varepsilon_0 (2)^{1/3} r} = (2)^{2/3} V$]

20. Calculate electric field at point P, due to infinite long charged rods having linear charge density λ [Fig. 18.71].

[**Ans.** Electric field E_y at P due to rod $AB = \dfrac{\lambda}{2\pi\varepsilon_0 y}$ and electric field E_x at P due to rod $CD = \dfrac{\lambda}{2\pi\varepsilon_0 x}$ $\therefore E = \sqrt{E_x^2 + E_y^2}$

$$= \frac{\lambda}{2\pi\varepsilon_0}\sqrt{\frac{1}{x^2}+\frac{1}{y^2}} \quad \tan\theta = \frac{E_y}{E_x} = \left(\frac{x}{y}\right)\Big]$$

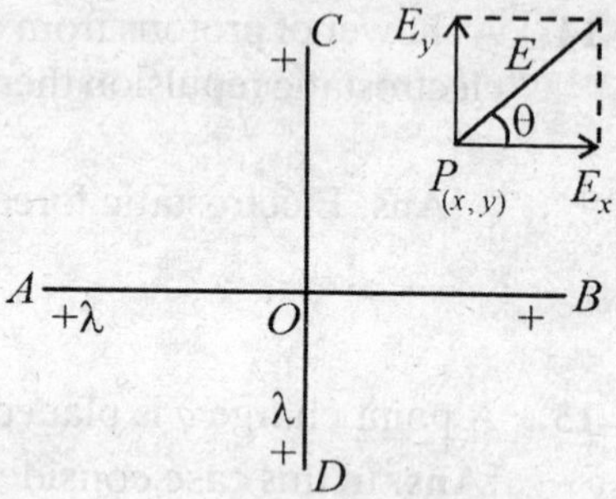

Fig. 18.71

21. Two hollow conductors are charged positively. The smaller is at 50 V and the bigger is at 100 V. How should they be arranged such that the charge flows from the smaller to the bigger conductor when connected by wire?

[**Ans.** Charge always flows from higher potential to the lower potential. The two conductors must be arranged in such a way that the potential of smaller sphere becomes higher than that of bigger. This is possible only if smaller potential conductor is surrounded by large potential conductor. As potential inside a conductor is equal to the surface value as such the potential of the smaller conductor becomes $V = 50 + 100 = 150$ V. The potential of the surface now becomes $V_s = 100 + \dfrac{1}{4\pi\varepsilon_0 R}(4\pi\varepsilon_0 r \times 50)$ $\therefore 50 = \dfrac{q}{4\pi\varepsilon_0 r} = \left(100 + 50\dfrac{r}{R}\right) < 150$. As such on connecting them the charge will flow from smaller conductor to the larger one]

22. Two electric charges q and $-2q$ are placed at a distance 6 m apart on a horizontal plane. Find the locus of points on this plane where the potential has value zero.

[**Ans.** For potential to be zero at $P(x, y)$, we have Fig. 18.72

$\dfrac{q}{r_1} = \dfrac{2q}{r_2}$ or $r_2 = 2r_1$. But $r_2^2 = (6-x)^2 + y^2$ and $r_1^2 = x^2 + y^2$ or $(6-x)^2 + y^2 = 4x^2 + 4y$ or $(x-2)^2 + (y-0)^2 = 4^2$ so locus is a circle of radius 4 m and centre at $(-2, 0)$]

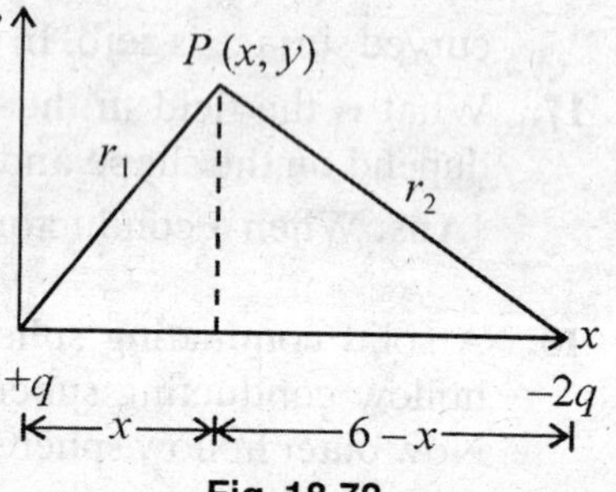

Fig. 18.72

23. Three charges are arranged on the vertices of an equilateral triangle as shown in Fig. 18.73. Find the dipole moment of combination.

[**Ans.** The system is equivalent to two dipoles one AB another CB each of magnitude $P = qa$ with direction along AB and CB respectively, i.e. the angle between the two is 60°. Dipole is a vector as such equivalent dipole moment is $P_e = \sqrt{P^2 + P^2 + 2P^2\cos 60}$ $= 2P\cos 30 = P\sqrt{3} = qa\sqrt{3}$. It is along the bisector of angle B away from the triangle.]

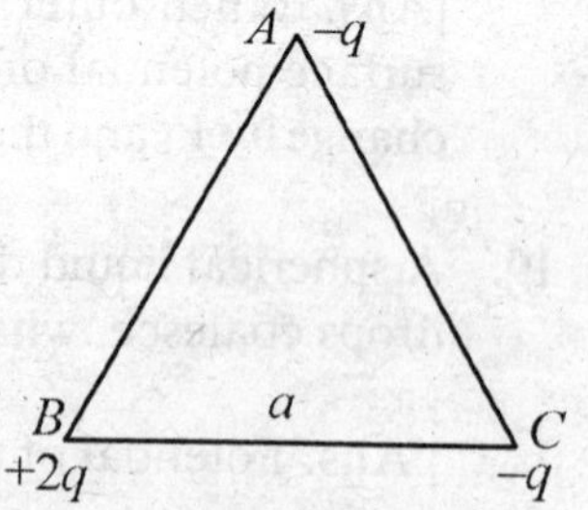

Fig. 18.73

24. An electric field $\boldsymbol{E} = 20\hat{i} + 30\hat{j}$ N/C exists in the space. If the potential at the origin is taken to be zero. Find the potential at $(2m, 2m)$.

[**Ans.** $\boldsymbol{E} = -\dfrac{dV}{dr}\hat{r}$ or $dV = -\boldsymbol{E}\cdot d\boldsymbol{r} = -\left(20\hat{i} + 30\hat{j}\right)\cdot\left(dx\,\hat{i} + dy\,\hat{j}\right)$

$\therefore dV = -(20\,dx + 30\,dy)$ $\therefore$ $\int_0^v dV = -\int(20\,dx + 30\,dy)$ or $V = -\int_0^2 20\,dx - \int_0^2 30\,dy = -100V$]

25. The electric field in a region is radially outward with magnitude $E = Ar$. Find the charge contained in a sphere of radius a centred at the origin. Take $A = 250\ \text{V/m}^2$ and $a = 10$ cm.

[**Ans.** The electric field at the surface of the sphere is $E = Ar$. Hence total flux through this sphere is $\phi = E \times 4\pi a^2 = 4\pi A a^3$. But $\phi = \frac{1}{\varepsilon_0} q$ or $q = \varepsilon_0 \phi = 8.85 \times 10^{-12} \times 4\pi \times 250\,(10^{-1})^3$ or

$q = 0.78 \times 10^{-12}$ C]

26. A thin non-conducting ring of radius R has linear charge density $\lambda = \lambda_0 \cos\phi$, where λ_0 is a constant, ϕ is the azimuthal angle. Find the magnitude of electric field strength at the centre of the ring.

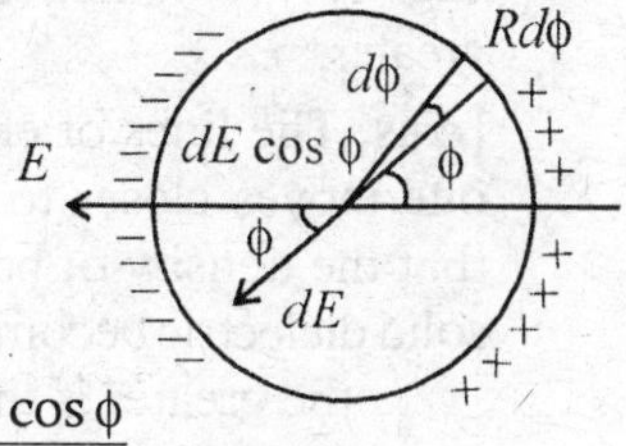

Fig. 18.74

[**Ans.** The given charge distribution is shown in Fig. 18.74. The symmetry implies that net field is towards left. dE due to elementary charge dq is, $dE = \frac{1}{4\pi\varepsilon_0}\frac{dq}{R^2}$ $dE\cos\phi = \frac{1}{4\pi\varepsilon_0}\cdot\frac{dq}{R^2}\cos\phi$

$= \frac{1}{4\pi\varepsilon_0}\cdot\frac{\lambda\cdot R\,d\phi}{R^2}\cos\phi$ $\therefore dq = \lambda R d\phi$ or $dE\cos\phi = \frac{1}{4\pi\varepsilon_0}\,\frac{\lambda_0\cos\phi\,d\phi\cdot\cos\phi}{R}$

$\because \lambda = \lambda\cos\phi$ $= \frac{\lambda_0}{4\pi\varepsilon_0 R}\cos^2\phi\,d\phi$ or $E = \int_0^{2\pi} dE\cos\phi\,d\phi$ $= \frac{\lambda_0}{4\pi\varepsilon_0 R}\int_0^{2\pi}\cos^2\phi\,d\phi$ $= \frac{\lambda_0}{4\varepsilon_0 R}$]

27. A point electric dipole with a moment P is placed in the external uniform electric field whose strength is E_0 with $P \parallel E_0$. In this case one of the equipotential surfaces enclosing the dipole forms a sphere. Find the radius of the sphere.

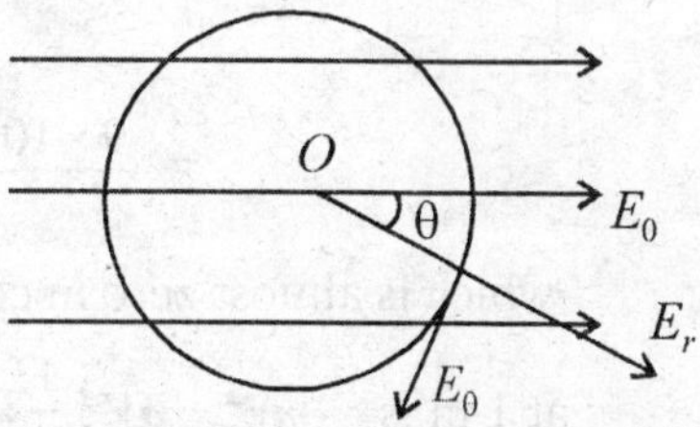

Fig. 18.75

[**Ans.** Let the dipole be placed at one of the equipotential surfaces, which is spherical (Fig. 18.75). On equipotential surface the tangential component of the field is zero. Thus

$-E_0\sin\theta + E_\theta = 0$ or $-E_0\sin\theta + \frac{P\sin\theta}{4\pi\varepsilon_0 r^3} = 0$ or $r = \left(\frac{P}{4\pi\varepsilon_0 E_0}\right)^{1/3}$]

28. A small uncharged sphere is positioned exactly in the mid-point between two charges whose absolute values are the same but whose signs are opposite. Suppose sphere is shifted some what. Will it remain in the new position or will it move in same direction?

[**Ans.** The electric field in which sphere is placed induces charges of opposite sign on the sphere. In view of this the sphere becomes a dipole. After the sphere is shifted, it finds itself in a non-uniform field, which forces it to move towards the charge to which it was shifted. Thus the equilibrium of the sphere at the midpoint between the charges is unstable.]

29. Two conducting spheres carry equal charges. The distance between the spheres is not large in comparison to the dimensions of spheres. In which case will the force of interaction between the spheres be greater (in absolute value) when they carry charges (Fig. 18.76 (*a*)) or when the carry unlike charges (Fig. 18.76 (*b*)).

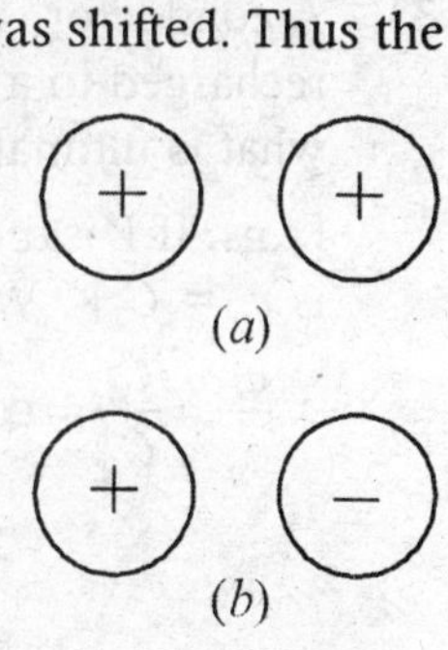

Fig. 18.76

[**Ans.** If the distance between the spheres is not very large, the charges on spheres are not evenly distributed over the surfaces. The effect of the spheres on each other results, in the case of like charges the sections of the spheres that are farthest from each other will have greater charge density. Whereas in case of unlike charges the sections of the spheres that are closest have greater charge density. For this reason the distance between the centres of charges for like charges is greater than that for unlike charges. Hence attractive force between the unlike charges will be greater than the repulsive force between like charges.]

30. The region of space between the plates of a parallel plate capacitor is filled with a liquid dielectric having dielectric constant ε_1. A solid dielectric with a dielectric constant ε_2 is immersed in the liquid. The lines of force in the liquid have the shape shown in Fig. 18.77. Which of the two dielectric constants is greater.

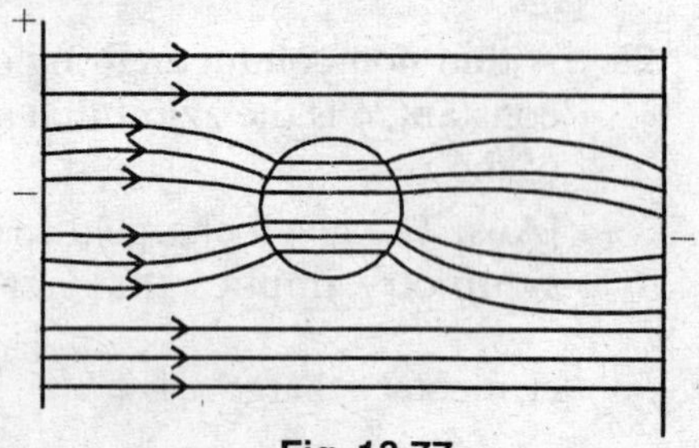

Fig. 18.77

[**Ans.** The lines of electric induction become denser as one moves closer to the solid dielectric, which means that the density of bound charges on the surface of the solid dielectric becomes enhanced. This density is higher the greater is the dielectric constant. Thus $\varepsilon_2 > \varepsilon_1$]

31. A tiny nucleus has a charge $50e$. (*i*) Find the potential V at a distance of 10^{-12} m from nucleus. (*ii*) If a proton is released from this point how fast will it move when it is 1 m from the nucleus.

$\Bigg[$**Ans.** $V = \dfrac{q}{4\pi\varepsilon_0 r} = \dfrac{9\times10^9\times50\times1.6\times10^{-19}}{10^{-12}} = 72$ kV. Potential at a distance of 1 m

$$= \frac{9\times10^9\times50\times1.6\times10^{-19}}{1} = 7.2\times10^{-8}\text{ V}$$

which is almost zero in comparison to 72 kV. If initial velocity of proton is zero then velocity at 1 m is $\frac{1}{2}mv^2 = qV$, $\frac{1}{2}\times1.67\times10^{-27}\times v^2 = 1.6\times10^{-19}\times72\times10^3$ or $v^2 = 2\times72\times10^{11}$ or

$v = \sqrt{2\times72\times10^{11}} = 3.7\times10^6$ m/s$\Big]$

32. In a Milikan oil-drop experiment, one of the drops falls at speed v without field and rises at speed $2v$ with field E applied. If the field is made $E/2$ find the speed of the drop.

[**Ans.** The field changes from zero to E makes v change from $-v$ to $+2v$, i.e. total change is $+3v$. So the change of field from 0 to $E/2$ will make velocity change by $+3v/2$, i.e. from $-v$ to $+v/2$. As such the particle will move with velocity $v/2$ upwards.]

33. A conductor is charged by repeated contacts with a metal plate. Which after each contact is recharged to a quantity of charge Q. If q is the charge of the conductor after first operation, what is ultimate charge on the conductor?

[**Ans.** If V is the common potential then charge on conductor $q_c = C_1V$ and the charge on plate is $q_p = C_2V$. Where C_1 and C_2 are the capacities of conductor and plate and are constants,

$\therefore \dfrac{q_c}{q_p} = \dfrac{C_1}{C_2}$ = constant. After first contact $q_c = q$ and $q_p = Q - q$ or $\dfrac{q_c}{q_p} = \dfrac{q}{Q-q}$. Ultimately $q_p \to Q$

$\therefore q_c \equiv \dfrac{qQ}{Q-q}\Big]$

34. An insulated conducting sphere carries a charge $+Q$. Another charge $+q$ is gradually brought towards it. The charge $+q$ experiences a repulsive or attractive force; explain?

[**Ans.** The force is repulsive, as expected at large distance from the sphere. However, the charge $+q$ induces an image charge $-q$, in the sphere. This charge exerts on $+q$, an attractive force which predominates when $+q$ is close to sphere.]

35. An isolated charge q is travelling between planar electrodes A and B which are connected through a galvanometer (Fig. 18.78). Will a current flow through the galvanometer? Justify your answer.

[**Ans.** Charge q induces negative charges q' and q'' on the electrodes A and B respectively, such that $|q|=|q'|+|q''|$. When q is nearer A $|q'|>|q''|$. As it moves towards B, q'' grows at the expense of q', i.e. negative charge flows through galvanometer from A to B or the current flows from B to A.]

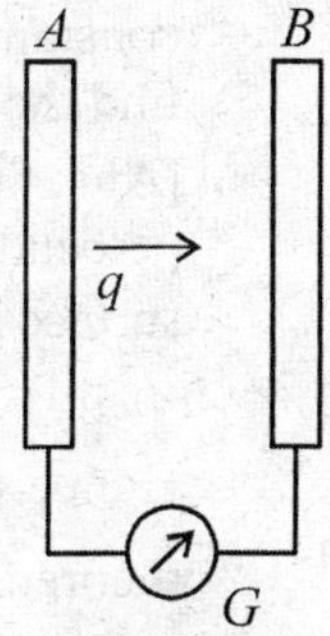

Fig. 18.78

36. An uncharged conducting ball is placed in uniform electric field of a plane capacitor. How will the form and arrangement of lines of force in the field change after the ball has been introduced and what causes this change in field. Draw the system of equipotential surfaces. Show how induced charges will appear on the ball and where will they appear.

[**Ans:** 1. Charges on sphere are of opposite nature but not uniform (Fig. 18.79). 2. Field lines are radial at the surface of sphere. 3. Equipotential lines are normal to the field lines]

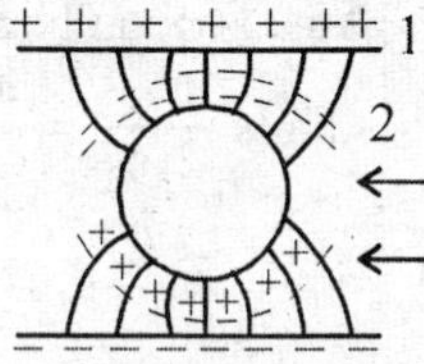

Fig. 18.79

37. Figure 18.80 shows an imaginary cube of edge $L/2$. A uniformly charged rod of length L moves towards left at a small but constant velocity v. At $t = 0$, the left end just touches the centre of the face of the cube opposite to it. Plot the graph between the electric flux through the cube and time as the rod moves through it.

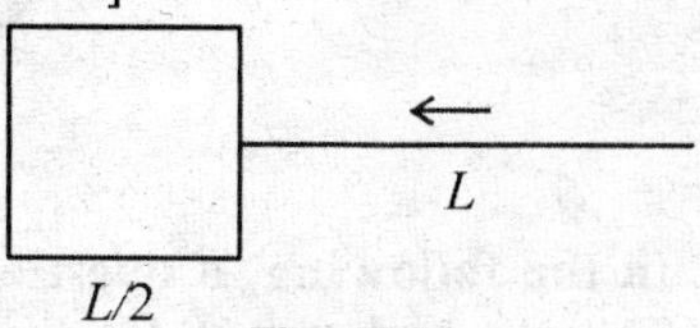

Fig. 18.80

[**Ans.** At $t = 0$ charge inside cube is zero as such $\phi = 0$ At $t = t$ length ut $(ut < L/2)$ is inside the cube and charge on it is lut. Flux increases linearly with time till $t = \frac{L}{2u}$. From $t = \frac{L}{2u}$ to $t = \frac{L}{u}$ flux is constant. After time $t = \frac{L}{u}$ flux decreases linearly with time at and becomes zero for $t = \frac{3L}{2u}$]

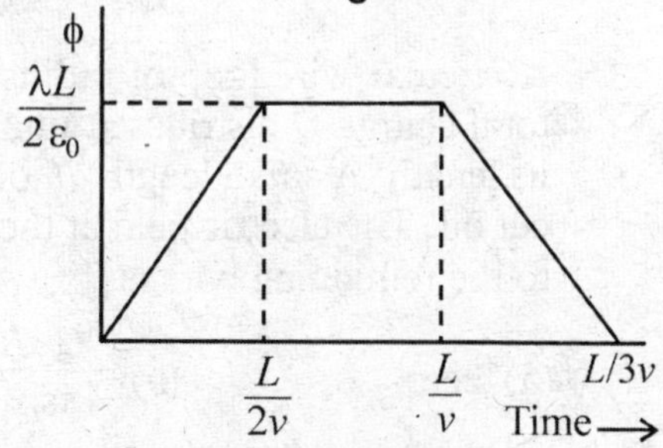

Fig. 18.81

The plot is shown in Fig. 18.81 λ is the linear charge density on the rod.

38. A spherical volume contains a uniform distributed charge of density 2.0×10^{-4} C/m³. Find the electric field at a point inside the volume at a distance 4 cm from the centre.

[**Ans.** Electric field at any point at distance r from the centre is $E = \frac{\rho r}{3\varepsilon_0} = \frac{2\times10^{-4}\times4\times10^{-2}}{3\times8.85\times10^{-12}}$

$= 3.0\times10^{5}$ N/C]

39. Figure 18.82 shows a cube of side a placed in an electric field with components $E_x = c\sqrt{x}$, $E_y = E_z = 0$ where c is a constant = 600 V/m^2. If side of the cube be 10 cm, then find the charge inside the cube.

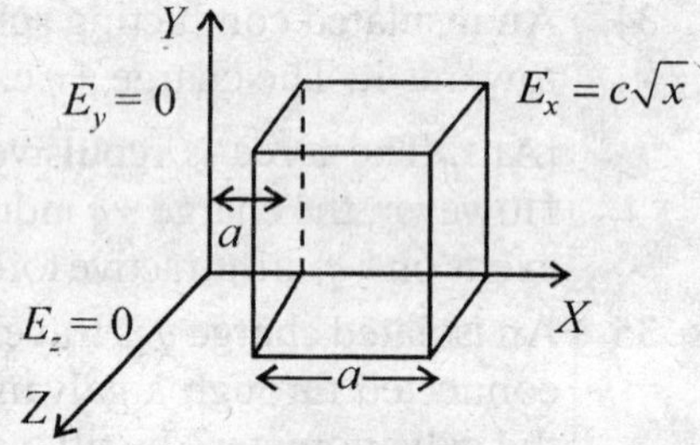

Fig. 18.82

[**Ans.** Since $E_y = E_Z = 0$ as such flux through the surfaces perpendicular to Y- and Z-axis is zero. The flux through the surface perpendicular to X-axis is $\phi = c\sqrt{2a}(a^2) - c\sqrt{a}(a^2)$ $= ca^{5/2}(\sqrt{2}-1)$. This is also the net flux through the cube. Hence charge q inside the cube according to Gauss's theorem is $q = \varepsilon_0\phi = \varepsilon_0 \cdot c\,a^{5/2}(\sqrt{2}-1) = 8.85\times10^{-12}\times600\times(10^{-1})^{5/2}\cdot(0.41) = 7\times10^{-12}$ C]

40. A soap bubble of radius R is in equilibrium when pressure inside and outside is same. If the surface tension of soap solution is T, find the potential of the bubble.

[**Ans.** For equilibrium the electrostatic pressure is equal to the excess pressure due to surface tension.

Thus $\frac{1}{2}\frac{\sigma^2}{\varepsilon_0} = \frac{4T}{R}$ or $\frac{q^2}{2\times(4\pi R^2)^2\varepsilon_0} = \frac{4T}{R}$ or $q = (128\pi^2\varepsilon_0 R^3 T)^{1/2}$ $\therefore$ Potential $V = \frac{q}{4\pi\varepsilon_0 R}$

$$V = \frac{q}{4\pi\varepsilon_0 R} = \frac{4\pi R}{4\pi\varepsilon_0 R}\cdot(8\,\varepsilon_0 RT)^{1/2} = \sqrt{\frac{8RT}{\varepsilon_0}}\ ]$$

MULTIPLE CHOICE QUESTIONS

PART I

In the following, a question is followed by four alternatives, of them only one is correct. Therefore, the candidate has to select the correct choice from amongst the alternatives.

1. A circular wire loop of radius a carries a total charge Q distributed uniformly over its length. A small length dL of the wire is cut off. The electric field at the centre due to the remaining wire is

(a) zero (b) $\frac{Q\,dL}{4\pi\varepsilon_0 a^3}$

(c) $\frac{Q\,dL}{8\pi^2\varepsilon_0 a^3}$ (d) $\frac{Q\,dL}{2\pi\varepsilon_0 a^3}$

2. The ratio of the electrical to gravitational forces between a proton and an electron in air is

(a) 10^{-40} (b) 10^{40}

(c) 10^{-39} (d) 10^{21}.

3. Charge on the outer sphere is q, and the inner sphere is grounded. Then the charge on the inner sphere q' for $b > a$ is

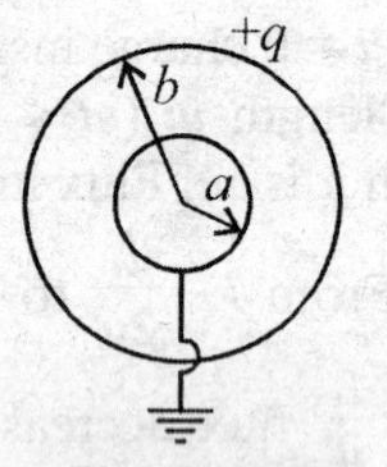

Fig. 18.83

(a) zero (b) $q' = -q$

(c) $q' = -\frac{a}{b}q$ (d) $q' = +\frac{a}{b}q$

4. Eight charged water drops each with a radius of 1 mm and a charge of 10^{-10} C merge into a single drop. The potential of bigger drop is

(a) 3600 V (b) 360 V

(c) 36 V (d) 3.6 V.

5. The variation of electric field and potential as a function of x from the centre of a uniform spherical charge distribution ($+Q$) of radius R is correctly represented by

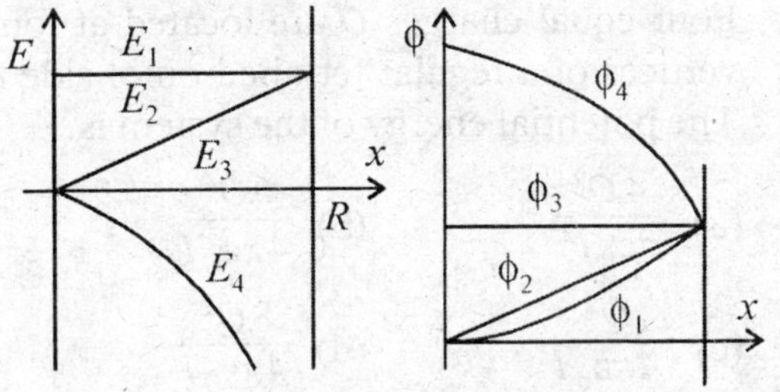

Fig. 18.84

(a) E_2, ϕ_4 (b) E_3, ϕ_4
(c) E_1, ϕ_1 (d) E_1, ϕ_2.

6. N small drops of the same size are charged to V volt each. If they coalesce to form a single drop, then its potential will be

(a) V/N (b) (VN)
(c) $VN^{1/3}$ (d) $VN^{2/3}$.

7. A metal sphere of radius r carries a charge $+Q$. The field at an external point P is E_0. If another sphere of radius $2r$ carrying charge $-2Q$ is placed at P, the field at the mid-point between the centres of the two spheres is

(a) E_0 (b) $3E_0$
(c) $12E_0$ (d) greater than $12E_0$.

8. Two conducting spheres P and Q have radii R and r respectively. They are connected by a thin wire, charged, and then separated to a large distance. Now at the surface of P and Q, the ratio of electric fields is

(a) $\frac{r}{R}$ (b) $\frac{R^2}{r^2}$

(c) $\frac{r^2}{R^2}$ (d) none of these.

9. In above question the ratio of electric potentials is

(a) $\frac{r}{R}$ (b) $\frac{R^2}{r^2}$

(c) $\frac{r^2}{R^2}$ (d) none of these.

10. In question 8, the ratio of charge densities is

(a) $\frac{r}{R}$ (b) $\frac{R^2}{r^2}$

(c) $\frac{r^2}{R^2}$ (d) none of the above.

11. Three point charges as shown in Fig. 18.85 are placed at the vertices of an isosceles right angled triangle. Which of the numbered vectors coincides in direction with the electric field at the mid-point of the hypotenuse?

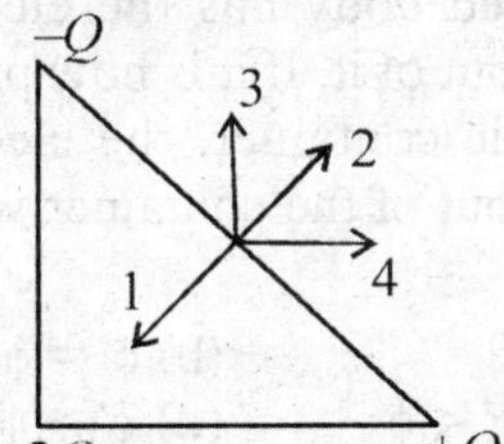

Fig. 18.85

(a) 1 (b) 2
(c) 3 (d) 4.

12. The work required to turn an electric dipole by 60° from equilibrium in uniform electric field is W. The torque required to hold it in that position is

(a) $\frac{W}{2}$ (b) $W\sqrt{3}$

(c) $\frac{\sqrt{3}}{2}W$ (d) W.

13. ABC is an equilateral triangle. Charges $+Q$ are placed at each vertex of the triangle. If r is the distance of the centroid from the vertex of the triangle, the electric intensity at the centroid is

(a) zero (b) $\frac{1}{4\pi\varepsilon_0} \cdot \frac{Q}{r^2}$

(c) $\frac{1}{4\pi\varepsilon_0} \cdot \frac{Q}{r}$ (d) $\frac{1}{4\pi\varepsilon_0} \cdot \frac{3Q}{r^2}$.

14. Five balls numbered 1 to 5 are suspended using separate threads. Pairs (1, 2), (2, 4) and (4, 1) show electrostatic attraction, while pairs (2, 3) and (4, 5) show repulsion. This implies that (1) must be

(a) positively charged
(b) negatively charged
(c) neutral
(d) uncharged metallic conductor.

15. The electric flux through a hemispherical surface of radius r placed in uniform electric field E parallel to the axis of hemisphere is

(a) $\pi r^2 E$ (b) $2\pi r^2 E$
(c) $3\pi r^2 E$ (d) $4\pi r^2 E$.

16. A charged body has the electric flux ϕ coming out of it. If it is now placed inside a metallic container, the electric flux ϕ' coming out of the container will be given by

(a) $\phi' = 0$ (b) $\phi' = \phi$
(c) $0 < \phi' < \phi$ (d) $\phi' > \phi$.

17. Work done in carrying a charge from A to B in an electric field is 2 J and from B to C is –3 J. The work done in carrying that charge from C to A will be (Fig. 18.86)

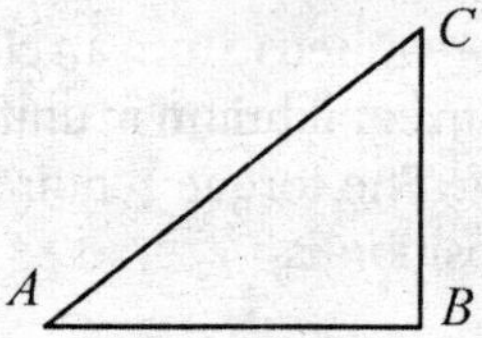

Fig. 18.86

(a) –1 J (b) 1 J
(c) $\sqrt{13}$ J (d) 5 J.

18. Three charges q_1, q_2 and q_3 are placed at the vertices of an equilateral triangle. The correct representation of the forces (in magnitude and direction) exerted by these charges on one another is given by

(a) 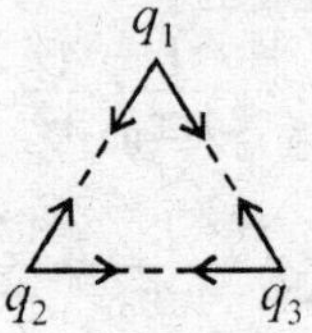

(b) 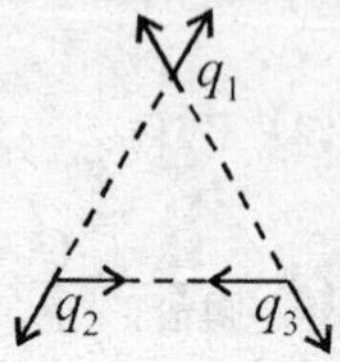

(c) 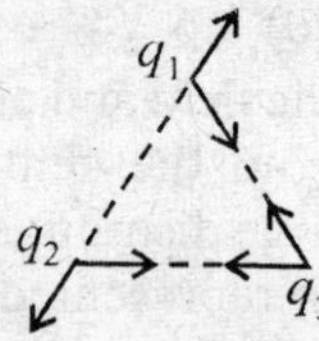

(d) 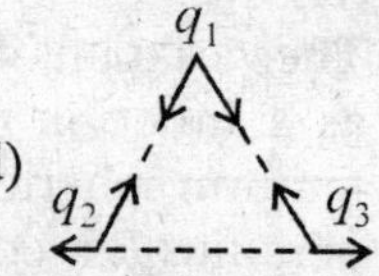

19. Four equal charges Q are located at four vertices of a regular tetrahedron of side l. The potential energy of the system is

(a) $\dfrac{4Q^2}{\pi\varepsilon_0 l^2}$ (b) $\dfrac{6Q^2}{4\pi\varepsilon_0 l}$

(c) $\dfrac{Q^2}{4\pi\varepsilon_0 l}$ (d) $\dfrac{8Q^2}{4\pi\varepsilon_0 l}$.

20. When the separation between two charges is increased, the electric potential energy of the charges

(a) increases
(b) decreases
(c) remains the same
(d) may increase or decrease.

21. If a positive charge is shifted from a low-potential region to a high-potential region, the electric potential energy

(a) increases
(b) decreases
(c) remains the same
(d) may increase or decrease.

22. An electric field E at the origin is along X-axis. A small circle is drawn with the centre at the origin cutting axes at points A, B, C and D having co-ordinates $(a, 0)$, $(0, a)$ $(-a, 0)$ and $(0, -a)$ respectively (Fig. 18.87). Out of the 4 points on the circumference of the circle, the potential is minimum at

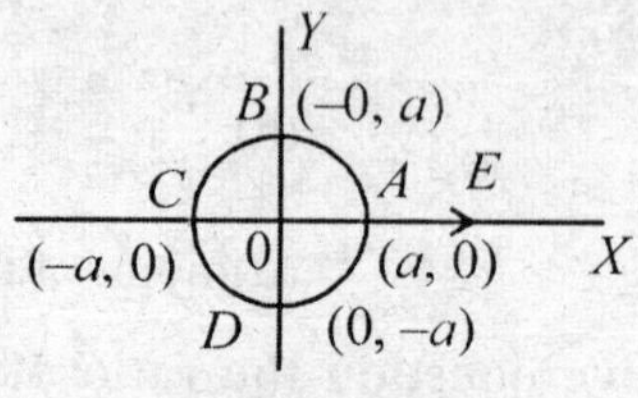

Fig. 18.87

(a) A (b) B
(c) C (d) D.

23. An electric dipole is placed in an electric field generated by a point charge
   (a) The net electric force on the dipole must be zero
   (b) The net electric force on the dipole may be zero
   (c) The torque on the dipole due to the field must be zero
   (d) The torque on the dipole due to the field may be zero.

24. A point charge q is rotated along a circle in the electric field generated by another point charge Q. The work done by the electric field on the rotating charge in one complete revolution is
   (a) zero
   (b) positive
   (c) negative
   (d) zero if Q is at the centre and non zero otherwise.

25. A body is charged by rubbing it, its weight
   (a) remains precisely constant
   (b) increases slightly
   (c) decreases slightly
   (d) may increase slightly or may decrease slightly.

26. Electric charges are distributed in a small volume. The flux of the electric field through a spherical surface of radius 1 m surrounding the total charge is 25 V-m. The flux over a concentric sphere of radius 0.5 m will be
   (a) 25 V-m (b) 50 V-m
   (c) 100 V-m (d) 12.5 V-m.

27. A charge q is placed at the centre of the open end of a cylindrical vessel (Fig. 18.88). The flux of the electric field through the surface of vessels is

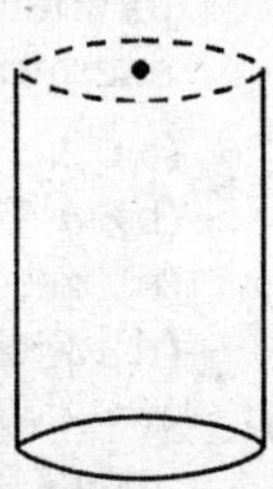

Fig. 18.88

   (a) zero (b) $\frac{q}{\varepsilon_0}$
   (c) $\frac{q}{2\varepsilon_0}$ (d) $\frac{2q}{\varepsilon_0}$.

28. Figure 18.89 shows a closed surface which intersects a conducting sphere. If positive charge is placed at point P, the flux of the electric field through the closed surface

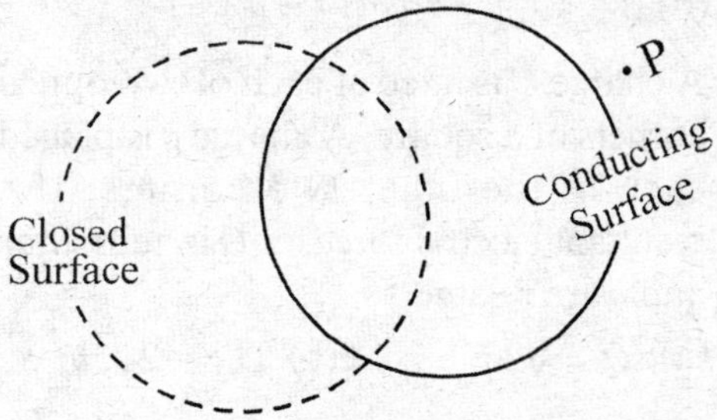

Fig. 18.89

   (a) will remain zero
   (b) will become positive
   (c) will become negative
   (d) will become undefined.

29. Consider a parallelogram $ABCD$ with angle 120° at B. A charge $+Q$ placed at corner A produces field E and potential V at corner D. If we now add charges $-2Q$ and $+Q$ at corners B and C respectively, the magnitude of field and potential at D will become respectively

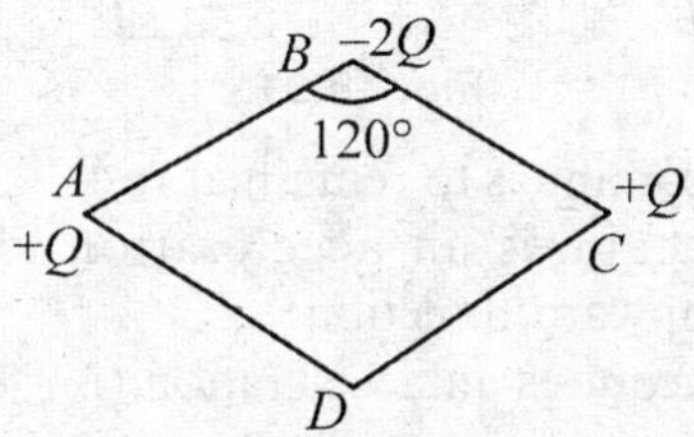

Fig. 18.90

   (a) E and 0 (b) 0 and V
   (c) $E\sqrt{2}$ and $\frac{V}{\sqrt{2}}$ (d) $\frac{E}{\sqrt{2}}$ and $\frac{V}{\sqrt{2}}$.

30. Three concentric charged metallic spherical shells A, B and C have radii a, b and c, charge densities σ, $-\sigma$, σ and potentials V_a, V_b and V_c respectively. Then which of the following relations is correct?
   (a) $V_a = (a + b + c)\frac{\sigma}{\varepsilon_0}$
   (b) $V_b = \left[\left(\frac{a^2}{b}\right) - b + c\right]\frac{\sigma}{\varepsilon_0}$

(c) $V_c = \left[\frac{(a^2 + b^2)}{c} + c\right]\frac{\sigma}{\varepsilon_0}$

(d) $V_a = V_B = V_C = (a + b + c)\frac{\sigma}{\varepsilon_0}$

31. A charge Q is fixed at each of two opposite corners of a square. A charge q is placed at each of the other two corners. If the resultant electric force on Q is zero, then Q and q are related as

(a) $Q = \sqrt{2}q$ (b) $Q = -2\sqrt{2}q$

(c) $Q = -\sqrt{2}q$ (d) $Q = 2\sqrt{2}q$.

32. In Fig. 18.91, A and B are two horizontal metal plates separated by insulating plugs. The region between A and B is evacuated. An electron of mass m remains at rest inside the chamber. If the proton of mass M replaces electron under similar conditions, then the proton

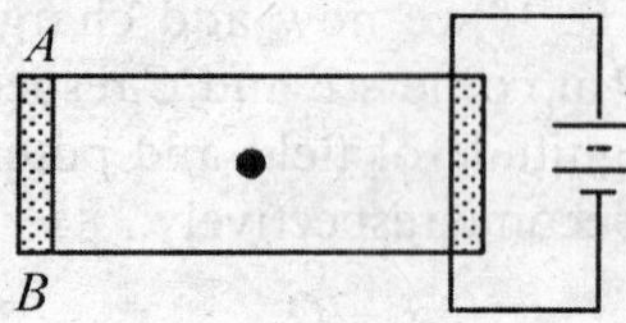

Fig. 18.91

(a) continues to remain at rest

(b) acquires an acceleration eE/M in upward direction

(c) acquires an acceleration $(Mg - eE)/M$ in the downward direction

(d) acquires an acceleration $\frac{(m+M)}{M}g$ in the downward direction.

33. A charge Q is placed 1 cm above the centre of a square of side 2 cm as shown in Fig. 18.92. The electric flux through the square is

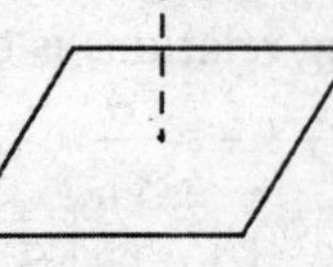

Fig. 18.92

(a) $\frac{Q}{\pi\varepsilon_0}$ (b) $\frac{4Q}{\varepsilon_0}$

(c) $\frac{Q}{\sqrt{2}\,\varepsilon_0}$ (d) $\frac{Q}{6\varepsilon_0}$.

34. The electrostatic potential on the surface of a charged conducting sphere is 100 V. Two statements are made in this regard

S_1: At any point inside the sphere, electric intensity is zero.

S_2: At any point inside the sphere, electrostatic potential is 100 V.

Which of the following is the correct statement?

(a) S_1 is true but S_2 is not true

(b) both S_1 and S_2 are wrong statements

(c) S_1 is true, S_2 is also true and S_1 is the cause of S_2.

(d) S_1 is true, S_2 is also true but the two statements are independent.

35. Three charges Q, $+q$ and $+q$ are placed at the vertices of a right angled isosceles triangle as shown in Fig. 18.93. The net electrostatic energy of configuration is zero if Q is equal to

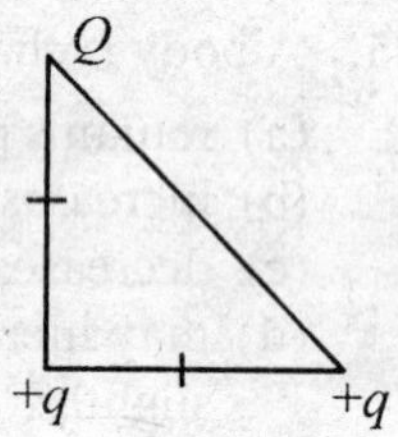

Fig. 18.93

(a) $-\frac{q}{1+\sqrt{2}}$ (b) $-\frac{2q}{2+\sqrt{2}}$

(c) $-2q$ (d) $+q$.

36. An electric dipole has the magnitude of its charge q and its dipole moment is P. It is placed in a uniform electric field E. If dipole moment is along the direction of the field, the force on it and its potential energy are respectively

(a) $2q \cdot E$ and minimum

(b) $q \cdot E$ and $P \cdot E$.

(c) zero and minimum

(d) $q \cdot E$ and maximum.

37. A point charge q is placed at the centre of circle of radius r. A point charge q_0 is carried from point A to point C and from A to B (Fig. 18.94). If work done in two cases is W_{AC} and W_{AB} respectively then

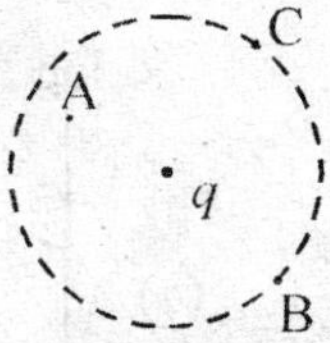

Fig. 18.94

(a) $W_{AC} > W_{AB}$
(b) $W_{AC} < W_{AB}$
(c) $W_{AC} = W_{AB} = 0$
(d) $W_{AC} = W_{AB} \neq 0$.

38. If in a region potential is given by $V = K(3x + y + z)$ then intensity of electric field depends on
(a) x-co-ordinates
(b) x and y-co-ordinates
(c) x, y and z-co-ordinates
(d) does not depend on co-ordinates.

39. An electric dipole (moment P) is placed in uniform electric field E. If angle between positive directions of P and E be θ, then the potential energy of dipole will be maximum when θ is
(a) zero (b) $\frac{\pi}{2}$
(c) π (d) $\frac{\pi}{4}$.

40. Four charges each of magnitude +2 μC are arranged on x-axis at distances x = 0, 2, 4, 8 cm. Charge at x = 2 will experience a force
(a) zero (b) 10 N
(c) 5 N (d) 10^{-3} N.

41. Two particles having charge 2 μC and 4 μC repel each other with a force of 100 N. If each particle is given an additional charge equal to –2 μC, the force between the two particles is
(a) zero (b) 400 N
(c) 200 N (d) 100 N.

42. Statement which is wrong in relation to electrostatic is
(a) charge is quantised
(b) charge is conserved
(c) for an isolated charge there is no field in the near vicinity of charge
(d) Stationary charges produce both electric and magnetic field.

43. A charge q is given to a sphere of radius r. The energy stored in the surrounding space is
(a) $\frac{q^2}{8\pi\varepsilon_0 r}$ (b) $\frac{q^2}{2\pi\varepsilon_0 r}$
(c) $\frac{q^2}{4\pi\varepsilon_0 r}$ (d) $\frac{q}{\varepsilon_0 r}$.

44. Linear charge density on a uniformly charged infinite long wire is λ. Another similarly charged wire is placed parallel to it at a distance d. The force per unit length on this wire is $\left(K = \frac{1}{4\pi\varepsilon_0}\right)$
(a) $K \cdot \frac{2\lambda}{d}$ (b) $K\frac{2\lambda^2}{d}$
(c) $K\frac{2\lambda}{d^2}$ (d) $K\frac{2\lambda^2}{d^2}$.

45. A point charge q, is displaced from point P to point S along the path P, Q, R, S in a region where uniform electric field E exists (Fig. 18.95). The co-ordinates of P, Q, R and S are (a, b, 0), (2a, 0, 0), (a, –b, 0), (0, 0, 0) respectively. In this process the work done by the field is

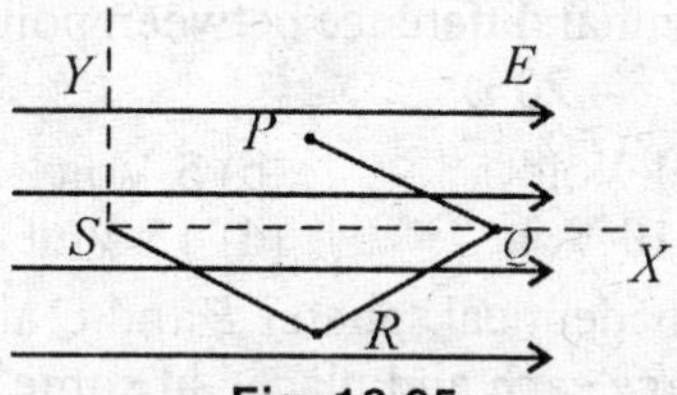

Fig. 18.95

(a) –Eaq (b) –Ebq
(c) –Eq (a + b) (d) –Eq 2(a + b).

46. A metallic bob of mass 5 g having charge 5×10^{-8} C is suspended by means of an insulated string of length 0.9 m in a region where an electric field of $98 \times 10^{+4}$ N/C is acting vertically downwards. Find the time period of the small oscillations of the bob.
(a) $\frac{\pi}{7}s$ (b) $\frac{2\pi}{7}s$
(c) $\frac{3\pi}{7}s$ (d) $\frac{4\pi}{7}s$.

47. Two particles have same charge and same mass. The gravitational force and electrostatic force of repulsion between them is same. The specific charge is

(a) $\sqrt{\frac{G}{\varepsilon_0}}$ (b) $\sqrt{4\pi\varepsilon_0 G}$

(c) $\frac{G}{\sqrt{4\pi\varepsilon_0}}$ (d) $\frac{G}{\sqrt{\varepsilon_0}}$.

48. Two spheres having charge 40 μC and –20 μC are placed at a distance r and attract each other with a force F. If they are made to touch each other and placed at the same distance, the force between them is

(a) $8F$ attractive (b) $8F$ repulsive

(c) $\frac{F}{8}$ attractive (d) $\frac{F}{8}$ repulsive.

49. The intensity of electric field is E at a distance of 2 m from a point charge q. The work done in carrying 2 Coulomb charge from infinity to that point is

(a) $2E$ (b) $4E$
(c) E (d) $E/2$.

50. In a region electric field is given by $E = \frac{100}{x^2}$, where x is in metres. The potential difference between points $x = 10$ and $x = 20$ is

(a) 1 Volt (b) 5 Volt
(c) 10 Volt (d) 15 Volt.

51. Two identical spheres P and Q are given q charge each and placed at some distance. They repel each other with force F. The third identical sphere makes a successful contact with both of them. Now the force between P and Q is

(a) $\frac{3}{4}F$ (b) $\frac{3}{8}F$

(c) $\frac{1}{4}F$ (d) $\frac{5}{8}F$

52. A thin long wire is charged. The graph between log E (E is electric field produced at distance r from the axis of wire) and log r is represented by

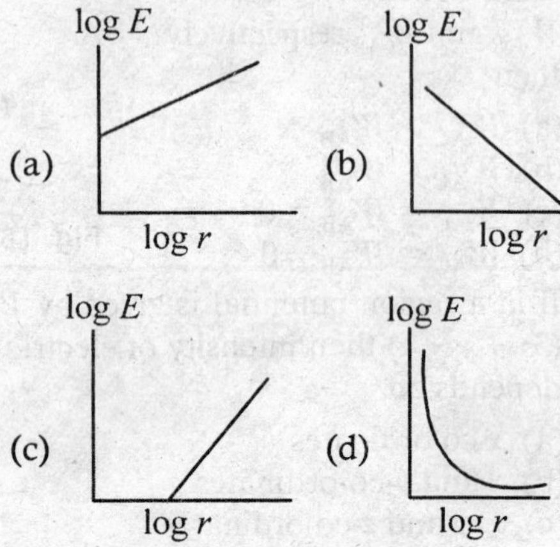

53. Three charges each of magnitude q are placed at the corners of an equilateral triangle. A fourth charge Q is placed at the centre of the triangle. For system to be in equilibrium, the charge Q is

(a) $Q = -\frac{q}{\sqrt{3}}$ (b) $Q = +\frac{q}{\sqrt{3}}$

(c) $Q = -\frac{2q}{\sqrt{3}}$ (d) $Q = +\frac{2q}{\sqrt{3}}$.

54. The force on a charge placed at a distance d from the centre of a dipole on its axis is F. If distance is made $2d$ the force on the charge becomes

(a) $F/2$ (b) $F/4$
(c) $F/8$ (d) $F/16$.

55. Two point charges 9 μC and 1 μC are placed at 10 cm from each other in air. Where a third charge of 1 μC be placed on the line joining the two charges, so that this third charge is stationary

(a) at a distance of 2.5 cm from 9 μC
(b) at a distance of 2.5 cm from 1 μC
(c) at 5 cm from each charge
(d) it can not remain in equilibrium.

ANSWERS

1. (c)	2. (b)	3. (c)	4. (a)	5. (a)
6. (d)	7. (c)	8. (a)	9. (d)	10. (a)
11. (c)	12. (b)	13. (a)	14. (d)	15. (a)
16. (b)	17. (b)	18. (c)	19. (b)	20. (d)
21. (a)	22. (a)	23. (d)	24. (a)	25. (d)
26. (a)	27. (c)	28. (b)	29. (a)	30. (b)
31. (b)	32. (d)	33. (d)	34. (c)	35. (b)
36. (c)	37. (d)	38. (d)	39. (b)	40. (b)
41. (a)	42. (d)	43. (a)	44. (b)	45. (a)
46. (c)	47. (b)	48. (d)	49. (b)	50. (b)
51. (b)	52. (b)	53. (a)	54. (c)	55. (b)

PART II

In the following, a question is followed by four alternatives, of them one or more are correct. Candidate has to select the correct choice(s) from amongst the alternatives.

1. A charge Q_1 is situated inside an uncharged conducting closed surface. Another charge Q_2 is situated outside the surface. Then

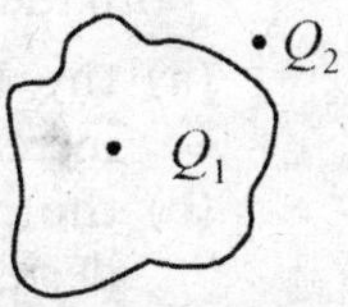

Fig. 18.96

(a) Q_1 and Q_2 experience equal and opposite forces
(b) Q_2 experiences a force but Q_1 does not
(c) there is apparent violation of Newton's third law
(d) force on Q_2 is twice the force on Q_1.

2. A dielectric slab is placed in an electric field. Mark the correct statements.

(a) All its atoms become polarised
(b) Polarisation charges appear on its faces
(c) Gauss's law can be applied to polarisation charges
(d) The electric field strength gets altered inside the dielectric.

3. The electric field and electric potential at a point are E and V respectively. Choose the wrong statement.

(a) If $E = 0$, V must be zero
(b) If $V = 0$, E must be zero
(c) If $E \neq 0$ V can not be zero
(d) If $V \neq 0$, E can not be zero.

4. Figure 18.97 shows a charge q placed at the centre of a hemisphere. A second charge Q is placed at one of the positions A, B, C and D. In which position(s) of this second charge, the flux of the electric field through the hemisphere remains unchanged?

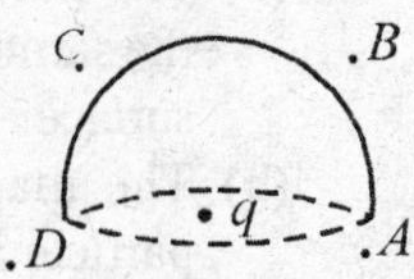

Fig. 18.97

(a) A (b) B
(c) C (d) D.

5. A closed surface S is constructed around a conducting wire connected to a battery and key (Fig. 18.98). As the key is closed, the free electrons in the wire move along the

wire. In any time interval, the number of electrons entering the closed surfaces is equal to the number leaving it. On closing the key, the flux of the electric field through the closed surface

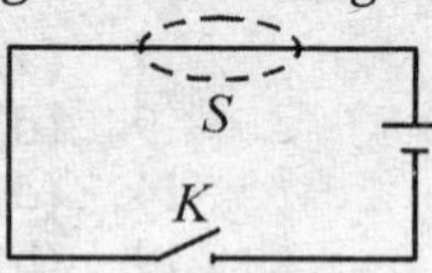

Fig. 18.98

(a) is increased
(b) is decreased
(c) remains unchanged
(d) remains zero.

6. Two charges $+2Q$ and $-2Q$ are located at co-ordinates $(+a, 0)$ and $(-a, 0)$ along x-axis. Which of the following statements are correct?

(a) The y–z plane is an equipotential surface
(b) Electric field is zero on y–z plane
(c) Electric field is constant every where also
(d) The direction of the electric field at any point on the x-axis is along positive x-direction.

7. An electric field exists in the region

(a) between the nucleus and the electron in an atom
(b) inside a current carrying conductor
(c) inside a cavity of a charged conductor
(d) inside a wooden plate held between the plates of a charged parallel plate condenser.

8. Consider a charged particle and any general set of equipotential surfaces. Which of the following is/are true?

(a) The potential energy of the particle is the same for different points on a given surface
(b) The magnitude of force on a given particle is always same at every point on a given surface
(c) The force on the particle is along the normal to the surface every where
(d) Two different equipotential surfaces never crosses one another.

9. A positively charged thin metal ring of radius R is fixed in the xy-plane with its centre at the origin O. A negatively charged particle P is released from rest at point $(0, 0, z_0)$, where $z_0 > 0$. Then the motion of P is

(a) periodic for all values of z_0 satisfying $0 < z_0 < \infty$
(b) simple harmonic for all values of z_0 satisfying $0 < z_0 \leq R$
(c) approximately simple harmonic provided $z_0 << R$
(d) such that P crosses O and continues to move along the negative z-axis toward $z = -\infty$.

10. 1 μC and –3 μC charges are placed at a distance of 1 m from each other. The points where the potential is zero is (from 1 μC charge)

(a) 0.5 m toward –3 μC
(b) 0.5 m away from –3 μC
(c) 0.25 m towards –3 μC
(d) 0.25 m away from –3 μC.

11. Two charges each of $+Q$ magnitude are brought close to each other and set free. With the passage of time

(a) the kinetic energy of the system increases
(b) the potential energy of the system decreases
(c) the potential energy of the system increases
(d) total energy of the system remains conserved.

12. Four electric charges are arranged as shown is Fig. 18.99 at four corners of a square of side a. The potential energy of the system is

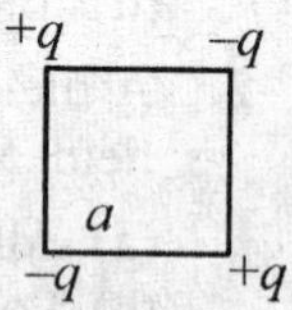

Fig. 18.99

(a) zero
(b) negative
(c) greater than $\frac{2q^2}{4\pi\varepsilon_0 a}$
(d) less than $-\frac{2q^2}{4\pi\varepsilon_0 a}$.

13. Three charged particles are in equilibrium under their electrostatic forces only

(a) the particles must be collinear

(b) all the charges cannot have the same magnitude
(c) all the charges cannot have the same sign
(d) the equilibrium is unstable.

14. Four charges of the same magnitude are placed at the four corners of a square. At the centre of the square, the potential is V and the field is E. By suitable choice of the sign of the four charges, which of the following can be obtained?
(a) $V = 0, E = 0$ (b) $V = 0, E \neq 0$
(c) $V \neq 0, E = 0$ (d) $V \neq 0, E \neq 0$.

15. A point charge Q is placed at the centre of a thin metallic spherical shell. Select the correct statement from the following.
(a) The electric field at any point outside the shell is zero
(b) The electrostatic potential at any point outside the shell is $\frac{Q}{4\pi\varepsilon_0 r}$, where r is the distance of the point from O
(c) The outer surface of the spherical shell is an equipotential shell
(d) The electric field at any point inside the shell other than O is zero.

16. An isolated metal sphere of radius R is electrically charged. It is connected by means of a thin conducting wire to a very distant uncharged metal sphere of radius r ($r < R$) then
(a) the total charge is uniformly and equally distributed over the two spheres
(b) the total charge is uniformly but unequally distributed over the two spheres
(c) the final potential energy of the system is less than the initial potential energy
(d) no amount of energy is lost in the process.

17. A, B, C are three concentric metallic shells. Shell A is inner most and shell C is outer most. A is given some charge
(a) the inner surfaces of B and C will have the same charge
(b) the inner surfaces of B and C will have the same charge density
(c) the outer surfaces of A, B and C will have the same charge
(d) the outer surfaces of A, B and C will have the same charge density.

18. A small conducting sphere of radius a mounted on an insulating handle and having a positive charge q is inserted through a hole in the walls of a hollow conducting sphere of inner radius b and outer radius c. The hollow sphere is supported on an insulating stand and is initially uncharged and the small sphere is placed at the centre of the hollow sphere. Neglect any effect of hole. Which of the following statements will be true for this system?
(a) The electric field at any point in the region between the spheres at a distance r from the centre is equal to $\left(\frac{q}{4\pi\varepsilon_0 r^2}\right)$
(b) The electric field at any point outside the hollow sphere at a distance r from the centre is equal to $\left(\frac{q}{4\pi\varepsilon_0 r^2}\right)$
(c) No work will be done in carrying a small charge from the inner conductor to the outer conductor
(d) The potential of the inner sphere with respect to the outer sphere is given by $V_{ab} = \frac{q}{4\pi\varepsilon_0}\left(\frac{1}{a} - \frac{1}{b}\right)$

19. A conducting sphere of radius R, carrying charge Q lies inside an uncharged conducting shell of radius $2R$. If they are joined by a metal wire, then
(a) $Q/3$ amount of charge will flow from the sphere to the shell
(b) $2Q/3$ amount of charge will flow from the sphere to the shell
(c) Q amount of charge will flow from the sphere to the shell

(d) $\frac{KQ^2}{4R}$ amount of heat will be produced.

20. A charged particle moves with a speed v in a circular path of radius r around a long uniformly charged conductor. If the conductor has charge per unit length λ, the particle has mass m and charge q, then

(a) $v \propto \sqrt{q}$ (b) $v \propto \sqrt{\lambda}$

(c) $v \propto \sqrt{m}$ (d) $v \propto \frac{1}{\sqrt{m}}$

ANSWERS

1. b, c	2. a, b, c, d	3. a, b, c, d	4. a, c	5. c, d
6. a, d	7. a, b, d	8. c, d	9. a, c	10. a, d
11. a, b, d	12. b, d	13. b, c, d	14. a, b, c, d	15. b, c
16. b, c	17. a, c	18. a, b, d	19. c, d	20. a, b, d

19

Capacitors

19.1 CONCEPT OF CAPACITY

When a conductor of any shape is charged, the potential of the conductor increases. It is observed that the charge (q) given to the conductor is directly proportional to the potential (V) of the conductor. Thus

$$q \propto V \quad \text{or} \quad q = CV \qquad \text{...(19.1)}$$

where C is constant and is called capacity of the conductor. It depends only on the dimensions of the conductor.

$$\text{If } V = 1, \quad C = q$$

Thus capacity of a conductor is the charge required to raise its potential through unity.

$$C = \frac{q}{V}$$

If $q = 1$ Coulomb and $V = 1$ Volt then $C = 1$ Farad, i.e. 1 Farad is the capacity of a conductor when 1 Coulomb charge increases its potential by 1 Volt.

Capacity is a scalar quantity and its dimensions are

$$C = \frac{q}{V} = \frac{q^2}{W} \qquad V = \frac{W}{q}$$

$$= \frac{I^2T^2}{M^1L^2T^{-2}} = M^{-1}L^{-2}T^4I^2$$

Capacity of a Spherical Conductor

Spherical conductor of radius r when given charge q, its surface potential becomes

$$V = \frac{q}{4\pi\varepsilon_0 r}$$

Thus its capacity is

$$C = \frac{q}{V} = 4\pi\varepsilon_0 r \qquad \text{...(19.2)}$$

The capacity of a spherical conductor is proportional to its radius. As the potential of earth is assumed to be zero as such its capacity is assumed to be ∞. Since earth's radius $r = 6.4 \times 10^6$ m, hence

$$\text{Capacity of earth} = 4\pi\varepsilon_0 r = \frac{6.4 \times 10^6}{9 \times 10^9} = 711\ \mu\text{F}$$

Energy Stored in a Charged Conductor

Any conductor when charged produces an electric field outside of it. When further similar charge is given to it, work has to be done by an external agency against the repulsive electrical force. This work is stored as the electric potential energy which resides in the electric field.

When the conductor is charged its potential increases from 0 to V. The mean potential can be taken as $\frac{0+V}{2} = \frac{V}{2}$. Against this potential the charge q is carried. Thus work done and hence the energy stored

$$U = \frac{V}{2}q = \frac{1}{2}qV = \frac{1}{2}\frac{q^2}{C} = \frac{1}{2}CV^2 \qquad \text{...(19.3)}$$

Sharing of Charges

Two charged conductors having capacities C_1 and C_2 charged to potentials V_1 and V_2 respectively when connected by a conducting wire, the charges flow from a conductor of higher potential to the conductor with lower potential. This flow continues till the potentials of the two conductors become equal. This is called common potential. In this process the total charge remains conserved.

$$\text{The common potential } V = \frac{(q_1 + q_2)}{C_1 + C_2} = \frac{C_1V_1 + C_2V_2}{C_1 + C_2} \qquad \text{...(19.4)}$$

Now the charges on the two spheres are

$$q'_1 = C_1V = \frac{C_1(C_1V_1 + C_2V_2)}{C_1 + C_2}$$

and

$$q'_2 = C_2V = \frac{C_2(C_1V_1 + C_2V_2)}{C_1 + C_2}$$

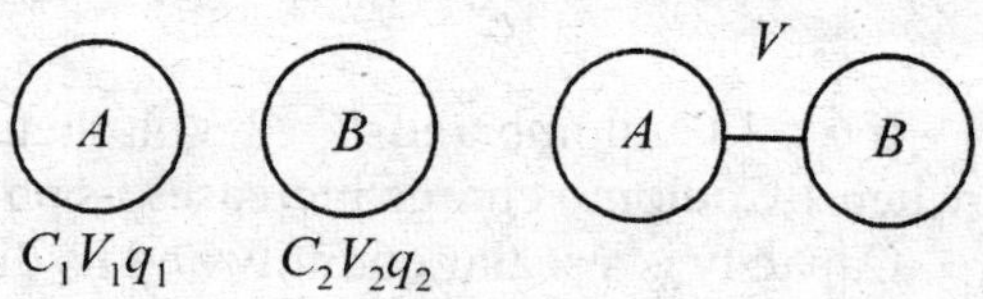

Fig. 19.1

and the charge transferred from the sphere having higher potential (say $V_1 > V_2$) to the sphere having lower potential

$$\Delta q = q_1 - q'_1 = C_1V_1 - \frac{C_1(C_1V_1 + C_2V_2)}{C_1 + C_2}$$

$$= \frac{C_1C_2}{C_1 + C_2}(V_1 - V_2)$$

Energy Loss

In the above process there will be loss of energy.

The energy stored in the system before connecting them is

$$E_i = \frac{1}{2}C_1V_1^2 + \frac{1}{2}C_2V_2^2$$

The energy stored in the system after connecting them is

$$E_f = \frac{1}{2}C_1V^2 + \frac{1}{2}C_2V^2 = \frac{1}{2}(C_1 + C_2)V^2 = \frac{1}{2}(C_1 + C_2)\frac{(C_1V_1 + C_2V_2)^2}{(C_1 + C_2)^2}$$

$$= \frac{1}{2}\frac{(C_1V_1 + C_2V_2)^2}{C_1 + C_2}$$

and the loss of energy in sharing the charges is

$$\Delta U = E_i - E_f = \frac{1}{2}C_1V_1^2 + \frac{1}{2}C_2V_2^2 - \frac{1}{2}\frac{(C_1V_1 + C_2V_2)^2}{(C_1 + C_2)^2}$$

$$= \frac{1}{2}\frac{C_1C_2}{C_1 + C_2}(V_1 - V_2)^2 \qquad \text{...(19.5)}$$

Equation (19.5) shows that there is always the loss of energy whether $V_1 > V_2$ or $V_2 > V_1$. This loss of energy is due to the flow of charges from one body to the other through the connecting wire and is lost as heat.

19.2 CAPACITOR OR CONDENSER

We have seen that each conductor has certain capacity to store charge. The capacity of a conductor is the ratio of charge given to the conductor and the potential to which conductor is raised. The capacity of a conductor depends only on its dimensions and is generally very small.

It has been observed that the potential of the charged conductor decreases if an uncharged earthed conductor is brought near to it, as such it can store more charge and its capacity is increased.

Thus a capacitor is a device by which the capacity of a conductor is increased by bringing an uncharged earthed conductor near it. This is also called *condenser.*

The capacity of a capacitor (or condenser) depends on the geometry, i.e. shape and size of the conductors, the separation between the conductors and the intervening medium (called dielectric) between them.

Capacity of a Parallel Plate Capacitor

A parallel plate capacitor consists of two large plane plates placed parallel to each other with small separation between them (Fig. 19.2). Let the area of each of the facing surface is A and the separation between the two plates is d. Also, assume that the space between the plates contains air or is vacuum. The upper plate is given charge $+q$ and the lower plate which is earthed has $-q$ charge inducted on the near surface and positive charge $(+q)$ on the other face is transferred to earth. The charge density on each of these faces has magnitude

$$\sigma = \frac{q}{A}$$

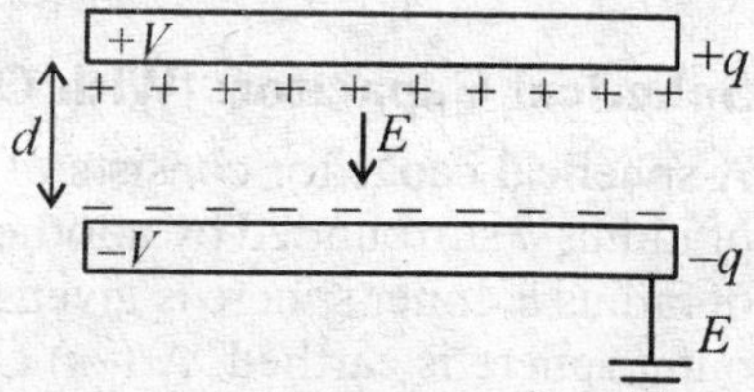

Fig. 19.2

The electric field at any point in between the plates is

$$E = \frac{\sigma}{2\varepsilon_0} - \frac{(-\sigma)}{2\varepsilon_0} = \frac{\sigma}{\varepsilon_0}$$

where ε_0 is the electric permittivity of free spaces.

But $$E = -\frac{dV}{dr}$$

$\therefore$ $$\frac{dV}{dr} = -E = -\frac{\sigma}{\varepsilon_0}$$

or $$dV = -\frac{\sigma}{\varepsilon_0} dr$$

Integration gives

$$\int_V^0 dV = -\int_0^d \frac{\sigma}{\varepsilon_0} dr$$

or $$V = \frac{\sigma}{\varepsilon_0} d = \frac{q}{A\varepsilon_0} d$$

or $$C = \frac{q}{V} = \frac{A\,\varepsilon_0}{d} \quad \text{...(19.6)}$$

If the space between the plates is filled with a medium having dielectric constant ε_r, the capacity becomes

$$C = \frac{A\,\varepsilon_r\,\varepsilon_0}{A} \quad \text{...(19.6a)}$$

It is obvious that the capacity of a parallel plate capacitor

(*i*) is directly proportional to the area of plates A, i.e.

$$C \propto A$$

(*ii*) is inversely proportional to the separation (d) between the plates, i.e.

$$C \propto \frac{1}{d}$$

(*iii*) is directly proportional to the dielectric constant ε_r of the medium between the plates, i.e.

$$C \propto \varepsilon_r$$

(*iv*) does not depend on the charge given to the plates or the potential to which the plates are raised.

Spherical Capacitor: (With Outer Sphere Earthed)

A spherical capacitor consists of a solid or a hollow spherical conductor of radius a surrounded by another concentric hollow spherical conductor of radius b. Inner sphere is given positive charge $(+q)$ and outer surface of outer sphere is earthed. A $(-q)$ charge is induced on the inner surface of outer sphere (Fig. 19.3).

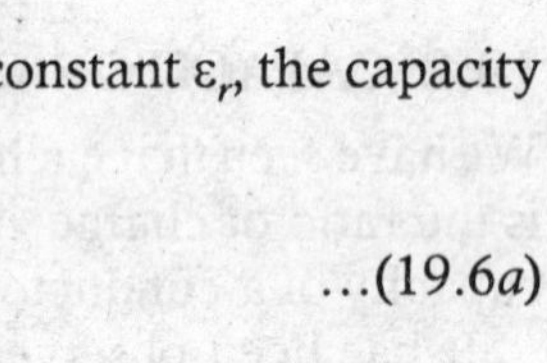

Fig. 19.3

Potential of the inner sphere is (*a*) due to its own charge $+q$ and (*b*) due to $-q$ charge induced on the outer sphere. Thus

$$V_i = \frac{q}{4\pi\varepsilon_0}\left[\frac{1}{a}-\frac{1}{b}\right]$$

The potential of outer sphere is zero as it is earthed. Hence potential difference is

$$\Delta V = \frac{q}{4\pi\varepsilon_0}\left[\frac{1}{a}-\frac{1}{b}\right] = \frac{q(b-a)}{4\pi\varepsilon_0\, ab}$$

and capacity is

$$C = \frac{q}{\Delta V} = \frac{4\pi\varepsilon_0\, ab}{(b-a)} \qquad \text{...(19.7)}$$

If the space between the two spheres is filled with medium of dielectric constant ε_r the capacity becomes

$$C = \frac{4\pi\varepsilon_0\varepsilon_r ab}{b-a} \qquad \text{...(19.7}a\text{)}$$

Equation 19.7 gives

(*i*) If $b \to \infty$ $C = 4\pi\varepsilon_0 a$, i.e. a spherical conductor is a spherical capacitor with its outer surface of infinite radius.

(*ii*) If a and b both become very large but $b - a = d$ is kept fixed and finite, we can write

$$4\pi ab = 4\pi r^2 = A$$

where r is approximately the radius of each sphere and A is area, the equation (19.7) becomes

$$C = \frac{\varepsilon_0 A}{d}$$

which is same as the expression for the capacity of a parallel plate capacitor.

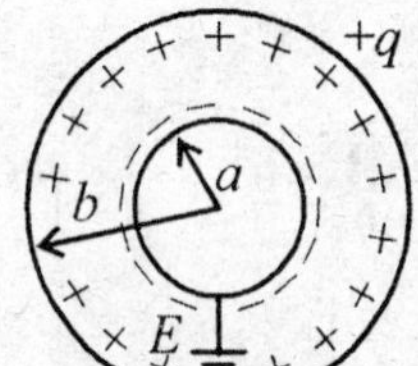

Fig. 19.4

Capacity of Spherical Capacitor with Inner Sphere Earthed

In this case the inner sphere of radius a is earthed and the outer sphere is given charge $+q$ and its radius is b (Fig. 19.4). The charge is induced on the inner sphere and is given by

$$Q = -\frac{a}{b}q$$

The potential of outer sphere is the sum of potentials due to charge on it $\frac{q}{4\pi\varepsilon_0 b}$ and due to induced charge on inner surface $-\frac{aq}{4\pi\varepsilon_0 \cdot b^2}$

Thus net potential difference is

$$V = \frac{q}{4\pi\varepsilon_0 b} - \frac{aq}{4\pi\varepsilon_0 b^2} = \frac{q}{4\pi\varepsilon_0 b^2}(b-a)$$

and capacity C is

$$C = \frac{q}{V} = \frac{4\pi\varepsilon_0 b^2}{b-a} \qquad \text{...(19.8)}$$

In this case the charges on the two spheres are not equal in magnitude as such the system is *not a capacitor.* But

$$C = 4\pi\varepsilon_0\left(\frac{b^2 - ab + ab}{b - a}\right) = 4\pi\varepsilon_0\, b + \frac{4\pi\varepsilon_0\, ab}{b - a} = C_1 + C_2$$

Here $C_1 = 4\pi\varepsilon_0 b$ the capacity of a spherical conductor of radius b and $C_2 = \frac{4\pi\varepsilon_0\, ba}{b-a}$ is the capacity of a spherical capacitor of inner sphere of radius a and outer sphere of radius b.

In this case the charge distribution on the outer sphere is in such a way that $+\frac{qa}{b}$ resides on inner surface and $\frac{q(b-a)}{b}$ resides on the outer surface as such the system behaves like two capacitors in parallel.

The Two Shells Carry Equal and Opposite Charges and are Separated by a Distance *d*

In this case the electric field at a point distant r from the centre of sphere A is

$$E = \frac{q}{4\pi\varepsilon_0 r^2} - \frac{(-q)}{4\pi\varepsilon_0 (d-r)^2}$$

$$= \frac{q}{4\pi\varepsilon_0}\left[\frac{1}{r^2} + \frac{1}{(d-r)^2}\right]$$

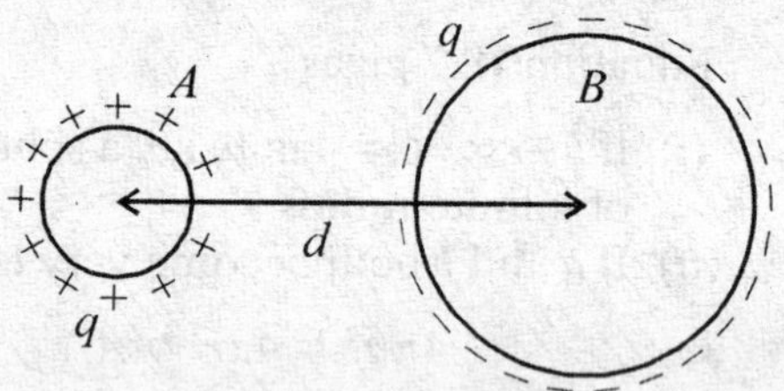

Fig. 19.5

But $$E = -\frac{dV}{dr}$$

or $$-\frac{dV}{dr} = \frac{q}{4\pi\varepsilon_0}\left[\frac{1}{r^2} + \frac{1}{(d-r)^2}\right]$$

or $$-\int_{V_A}^{V_B} dV = \frac{q}{4\pi\varepsilon_0}\int_a^{d-b}\left[\frac{1}{r^2} + \frac{1}{(d-r)^2}\right]dr$$

or $$V_A - V_B = \frac{q}{4\pi\varepsilon_0}\left[-\frac{1}{r} + \frac{1}{d-r}\right]_a^{d-b}$$

$$= \frac{q}{4\pi\varepsilon_0}\left[\frac{1}{a} + \frac{1}{b} - \frac{1}{(d-a)} - \frac{1}{(d-b)}\right]$$

$$= \frac{q}{4\pi\varepsilon_0}\left[\frac{1}{a} + \frac{1}{b} - \frac{2}{d}\right] \qquad \because d >> a \text{ and } d >> b$$

and capacity

$$C = \frac{q}{V} = \frac{4\pi\varepsilon_0}{\left(\frac{1}{a} + \frac{1}{b} - \frac{2}{d}\right)} \qquad \text{...(19.9)}$$

If $d \to \infty$

$$C = \frac{4\pi\varepsilon_0}{\frac{1}{a}+\frac{1}{b}}$$

and $$\frac{1}{C} = \frac{1}{4\pi\varepsilon_0 a} + \frac{1}{4\pi\varepsilon_0 b} = \frac{1}{C_1} + \frac{1}{C_2}$$

where $C_1 = 4\pi\varepsilon_0 a$, the capacity of conductor A and $C_2 = 4\pi\varepsilon_0 b$, the capacity of conductor B, i.e. the system is equivalent to two capacitors in series.

Cylindrical Capacitor

A cylindrical capacitor consists of a solid or a hollow cylindrical conductor (radius a) surrounded by a another coaxial hollow cylindrical conductor of radius b ($b > a$). Let the length of the cylinders be l and a positive charge q is given to the inner cylinder and outer cylinder is earthed as shown in Fig. 19.6. A charge $-q$ is induced on the inner surface of the outer cylinder.

The electric field at a point at distance r from the axis of the cylinders is

$$E = \frac{q}{2\pi\varepsilon_0 l r}$$

and the potential difference between the cylinders is

$$V_i - V_O = -\int_b^a E\,dr = -\int_b^a \frac{q}{2\pi\varepsilon_0 l}\frac{dr}{r}$$

$$= \frac{q}{2\pi\varepsilon_0 l}\log\frac{b}{a}$$

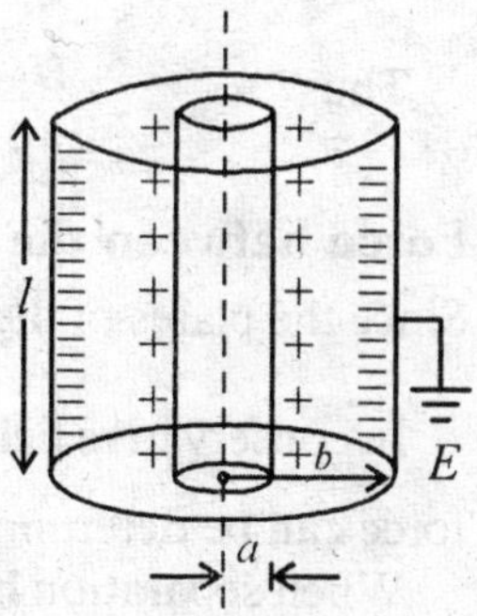

Fig. 19.6

But V_O, the potential of the outer cylinder is zero as it is earthed

$$\therefore \quad V = V_i = \frac{q}{2\pi\varepsilon_0 l}\cdot\log\frac{b}{a}$$

and capacity $$C = \frac{q}{V} = \frac{2\pi\varepsilon_0 l}{\log b/a} \qquad \text{...(19.10)}$$

And if the space between the cylinders is filled with the dielectric of constant ε_r, the capacity becomes

$$C = \frac{2\pi\varepsilon_0\varepsilon_r l}{\log b/a} \qquad \text{...(19.10}a\text{)}$$

Energy Stored in a Capacitor

When a capacitor is charged, the energy is stored in the form of electrostatic potential energy in the electric field between the plates.

If charge dq is given to capacitor when its potential is V, then work done

$$dW = Vdq = \frac{q}{C}dq$$

and total work done $$W = \int_0^q \frac{q}{C}dq = \frac{1}{2}\frac{q^2}{C} = \frac{1}{2}qV = \frac{1}{2}CV^2$$

This work done is stored in the form of electric potential energy.

Thus $$U = \frac{1}{2}CV^2$$

The energy stored per unit volume, i.e. energy density (U_d) is

$$U_d = \frac{U}{\text{Volume}} = \frac{1}{2}\frac{CV^2}{Ad}$$

where A is area of plates and d the separation between the plates. Thus

$$U_d = \frac{1}{2}\left(\frac{\varepsilon_0 A}{d}\right)\frac{V^2}{Ad} = \frac{1}{2}\varepsilon_0\left(\frac{V}{d}\right)^2 \qquad \because\ C = \frac{\varepsilon_0 A}{d}$$

and $$E = V/d$$

hence $$U_d = \frac{1}{2}\varepsilon_0 E^2 \qquad \text{...(19.11)}$$

But $E = \dfrac{\sigma}{\varepsilon_0}$ where σ is surface density of charge.

Thus $$U_d = \frac{1}{2}\frac{\sigma^2}{\varepsilon_0} \qquad \text{...(19.11a)}$$

Force between the Plates

Since the plates of the capacitor are given charge, there is force between the plates.

In conservative field $F = -\dfrac{dU}{dx}$. Since the electric field between the plates is conservative as such force can be determined using above formula.

When separation between plates is x, the energy is

$$U = \frac{1}{2}\frac{q^2}{C} = \frac{1}{2}\frac{q^2 x}{\varepsilon_0 A}$$

$$\therefore \qquad F = -\frac{d}{dx}\left(\frac{1}{2}\frac{q^2 x}{\varepsilon_0 A}\right) = -\frac{1}{2}\frac{q^2}{\varepsilon_0 A}$$

Negative sign implies that the force is attractive. The force per unit area is thus

$$\left|\frac{F}{A}\right| = \frac{1}{2}\frac{q^2}{\varepsilon_0 A^2} = \frac{1}{2}\frac{\sigma^2}{\varepsilon_0} = \frac{1}{2}\varepsilon_0 E^2 \qquad \text{...(19.12)}$$

Charging of the Capacitor

A capacitor can be charged by connecting it to a battery. If after charging battery is removed, the capacitor is said to be isolated and charge remains conserved. If the battery remains connected even after the capacitor is charged the potential difference between the plates remains constant.

(*a*) **If capacitor is isolated** (i.e. battery is removed after charging)

(*i*) the charge remains conserved.

(ii) if capacity increases due to decrease in the separation of plates or introduction of dielectric of constant (ε_r), potential difference V decreases.
(iii) the field between the plates decreases if dielectric is introduced.
(iv) the energy stored in the capacitor decreases.
(v) if the separation between the plates of an air capacitor is decreased the electric field remains unchanged.

(b) If the battery remains connected

(i) potential difference remains constant.
(ii) if capacity increases the charge on the capacitor increases.
(iii) electric field remains uncharged if the dielectric is introduced between the plates.
(iv) the energy stored increases.
(v) electric field increases if separation between plates is decreased and it decreases if separation between plates is increased.

19.3 COMBINATION OF CAPACITORS

Two or more capacitors can be connected in a number of ways. The combination should have two points which may be connected to a battery to apply potential difference. If V be the potential difference between the points and q be the magnitude of the charge supplied by the battery, then *equivalent capacitance of the combination between the two points is defined as*

$$C = \frac{q}{V}$$

If the combination is replaced by a single capacitor of this capacitance, the single capacitor will store the same amount of charge for a given potential difference as the combination does.

Replacing a combination of capacitors by a single *equivalent capacitor* is called grouping of capacitors. It simplifies the problem. Two special methods of grouping are frequently used. One is known as series combination and other as parallel combination.

Series Combination: Figure 19.7 shows three capacitors of capacities C_1, C_2, C_3 connected in series. In series combination, each capacitor has equal charges for any value of capacitance. Since potential difference $V = q/C$ hence smaller the capacitance higher is the potential difference.

Let the potential of the first plate of the first capacitor be V, the potential of 2nd plate of first capacitor and first plate of 2nd capacitor be V_1. V_2 be the potential of 2nd plate of 2nd capacitor and first plate of 3rd capacitor. V_3 be the potential of 2nd plate of 3rd capacitor. Then for first capacitor

$$q = C_1(V - V_1)$$

or

$$(V - V_1) = \frac{q}{C_1} \qquad \ldots(i)$$

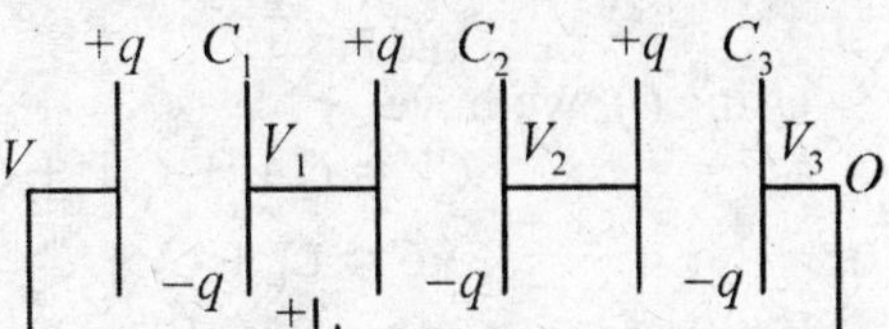

Fig. 19.7

Similarly for second capacitor

$$V_1 - V_2 = \frac{q}{C_2} \qquad \ldots(ii)$$

and for third capacitor

$$V_2 - V_3 = \frac{q}{C_3} \qquad \ldots(iii)$$

Adding, we have

$$V - V_3 = q\left(\frac{1}{C_1} + \frac{1}{C_2} + \frac{1}{C_3}\right) \quad ...(iv)$$

If the equivalent capacitance of the combination be C then from definition

$$C = \frac{q}{V - V_3} \quad \text{or} \quad V - V_3 = \frac{q}{C}$$

Then equation (*iv*) gives

$$\frac{q}{C} = q\left(\frac{1}{C_1} + \frac{1}{C_2} + \frac{1}{C_3}\right)$$

or

$$\frac{1}{C} = \frac{1}{C_1} + \frac{1}{C_2} + \frac{1}{C_3} \quad ...(19.13)$$

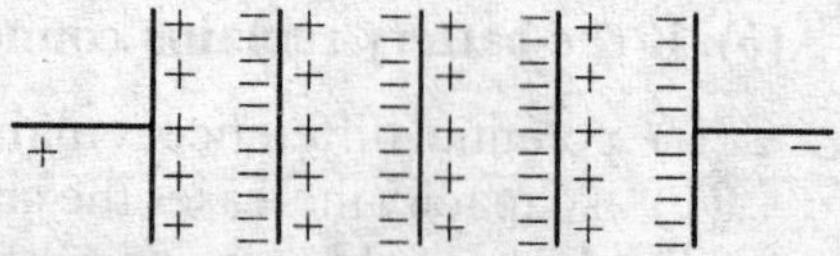

Fig. 19.8

If n plates are arranged as shown in Fig. 19.8, they form $(n - 1)$ capacitors in series each of capacity $\frac{\varepsilon_0 A}{d}$ so that equivalent capacity is

$$C_s = \frac{\varepsilon_0 A}{d(n-1)}$$

Parallel Combination: Figure 19.9 shows three capacitors of capacitance C_1, C_2 and C_3 connected in parallel. In this case the first plates of all the capacitors are connected together and connected to positive terminal of the battery and second plates of all the capacitors are connected together at negative terminal of the battery. Thus potential difference across each capacitor is same and the charge on different capacitors is different. Greater the capacitance greater is the charge on the capacitors.

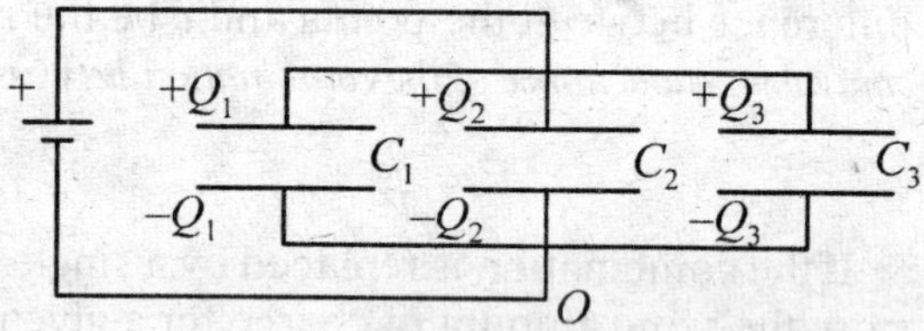

Fig. 19.9

The battery supplies a charge Q, which is distributed on the different capacitors in the ratio of their capacitance. Let the charges on these capacitors be Q_1, Q_2 and Q_3 respectively. Then

$$Q = Q_1 + Q_2 + Q_3 \quad ...(i)$$

and $\quad Q_1 = C_1V,\ Q_2 = C_2V$ and $Q_3 = C_3V$

where V is the potential of the plates connected to the positive terminal of the battery.

If equivalent capacitance be C, then from definition, we have

$$Q = CV$$

Using (*i*), we have

$$CV = C_1V + C_2V + C_3V$$

or

$$C = C_1 + C_2 + C_3 \quad ...(19.14)$$

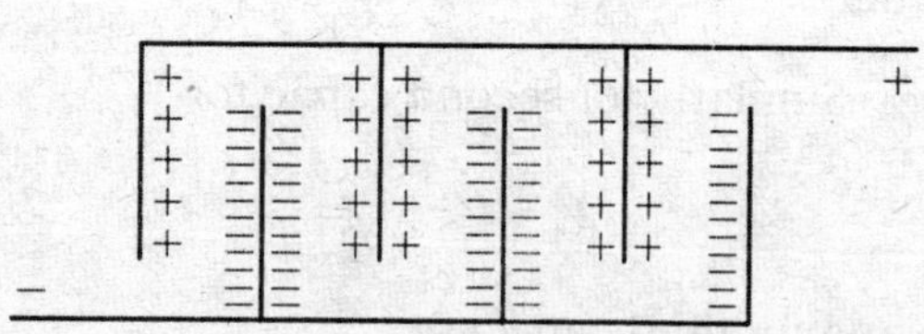

Fig. 19.10

If n plates are arranged as shown in Fig. 19.10, then they form $n - 1$ capacitors in parallel each of capacity $\frac{\varepsilon_0 A}{d}$ so that equivalent capacity is

$$C_p = (n-1)\frac{\varepsilon_0 A}{d}$$

In this case except two extreme plates, each plate is common to two adjacent capacitors.

19.4 DIELECTRICS AND THEIR BEHAVIOUR IN ELECTRIC FIELD

The dielectrics are insulating materials which transmit electric effects without conducting. In dielectric materials effectively there are no free electrons. We further know that in every atom, there is a positively charged nucleus and negatively charged electron cloud surrounding it. The two oppositely charged regions have their own centres of charge. In general, the centre of the negative charge distribution may or may not coincide with the centre of positive charge distribution. If it does not coincide, each molecule (or atom) has a permanent dipole moment ***P***. Such materials are known as *polar materials* like water, alcohol, CO_2, etc.

However different molecules have different directions of dipole moment because of random thermal agitation in the material. In any volume containing a large number of molecules, the average dipole moment is zero. If such a material is placed in an electric field, the individual dipoles experience torque due to field and they try to align along the field. On the other hand, the thermal agitation tries to randomise the orientation as such there is partial alignment. This depends on the strength of the field and temperature. As a result a net dipole moment is obtained in any volume.

On the other hand in case of materials in which the centres, of the positive charge distribution coincide with the negative charge distribution, the atoms or molecules do not possess any permanent dipole moment, such materials are called non-polar materials. They are nitrogen, oxygen, benzene, methane, etc. If such a material is placed in an electric field, the electron charge distribution is slightly shifted opposite to the electric field. This induces dipole moment in each atom or molecule and thus we get a dipole moment in any volume of the material.

Thus, when a dielectric material is placed in an electric field, dipole moment appears in any volume of it. This is *known as polarisation* of the material. The *polarisation vector* ***P*** is defined as the dipole moment per unit volume. Its magnitude $|\boldsymbol{P}|$ is referred to as the polarization.

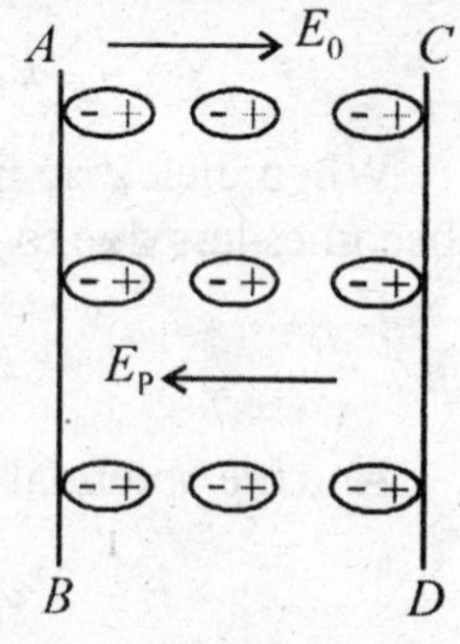

Fig. 19.11

If we consider a small volume element in the interior of the slab (Fig. 19.11), the volume charge density is zero as the net charge is zero. However a net positive charge appears on the surface *CD* and an equal net negative charge appears on the surface *AB*. These are uncancelled *induced charges*. As the induced charge appears due to a shift in the electrons bound to the nuclei, the charge is also called *bound charge*. These charges setup an electric field E_p opposite to the external field applied (E_0).

$\therefore$ Electric field inside a polarised dielectric is $E = E_0 - E_p$.

The electric field inside the dielectric is thus less than the externally applied field.

The ratio of external field applied (E_0) to the electric field (E) inside the dielectric is constant for the material and is called dielectric constant (ε_r) of the material of the dielectric i.e.

$$\varepsilon_r = \frac{E_0}{E} \qquad \text{...(19.15)}$$

Sometimes dielectric constant is also represented by letter *k*.

For conductors, electric field inside is zero, hence $\varepsilon_r = \infty$.

Electric Susceptibility

The electric field inside a polarised dielectric is

$$E = E_0 - E_p$$

or $$1 = \frac{E_0}{E} - \frac{E_p}{E}$$

The ratio of the electric field due to polarisation (E_p) to the field inside the dielectric (E) is called electric susceptibility (χ), i.e.

$$\chi = \frac{E_p}{E}$$

and $$\frac{E_0}{E} = \varepsilon_r \text{ the dielectric constant.}$$

$\therefore$ $$1 = \varepsilon_r - \chi \text{ or } \varepsilon_r = 1 + \chi \quad \ldots(19.16)$$

Dielectric Strength

If a very high electric field is created in a dielectric, the outer electrons may get detached from their parent atoms. The dielectric then behaves like a conductor. The phenomenon is known as *dielectric breakdown*. The minimum field at which the breakdown occurs is called *dielectric strength* of the material.

19.5 A PARALLEL PLATE CAPACITOR WITH A DIELECTRIC

Consider parallel plate capacitor with plate area A and plate separation d. The space between the plates is filled with a dielectric of dielectric constant ε_r upto a thickness x (<d).

Let charges $+q$ and $-q$ are given to the plates of the capacitor. The electric field E_0 produced in air is

$$E_0 = \frac{\sigma}{\varepsilon_0} = \frac{q}{A\varepsilon_0} \quad \ldots(i)$$

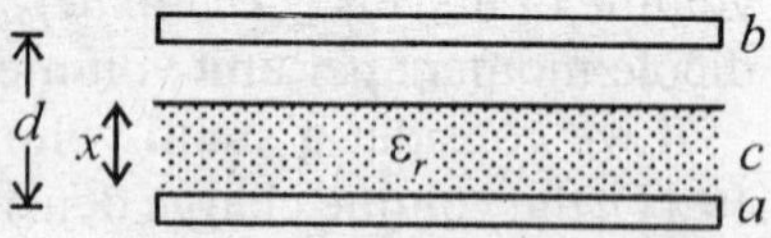

Fig. 19.12

When dielectric is introduced, the field inside the dielectric becomes less due to the polarization. The field inside the slab then becomes

$$E = \frac{E_0}{\varepsilon_r} = \frac{q}{A\varepsilon_0\varepsilon_r} \quad \ldots(ii)$$

And the potential difference between the plates becomes

$$V = E_0(d-x) + E_x = E_0(d-x) + \frac{E_0}{\varepsilon_r}x$$

$$= E_0\left[d - x\left(1 - \frac{1}{\varepsilon_r}\right)\right]$$

or $$V = \frac{q}{A\varepsilon_0}\left[d - x\left(1 - \frac{1}{\varepsilon_r}\right)\right]$$

and capacity C is

$$C = \frac{q}{V} = \frac{A\varepsilon_0}{d - x\left(1 - \frac{1}{\varepsilon_r}\right)} \quad \ldots(19.17)$$

The equation (19.17) gives that

(*i*) the introduction of the slab of thickness x ($<d$) and of dielectric constant ε_r increases the capacity as if the separation between the plates is decreased by $t = x(1 - 1/\varepsilon_r)$

(*ii*) the capacity is independent of the position of dielectric between the plates.

(*iii*) if the whole space between the plates is filled with dielectric (i.e., $x = d$) the capacity becomes

$C = \dfrac{A\varepsilon_0\varepsilon_r}{d}$, i.e. capacity increases by factor ε_r.

(*iv*) if the metal plate of thickness x is introduced instead of the dielectric the capacity becomes

$C = \dfrac{A\varepsilon_0}{d - x}$ (ε_r for metals is ∞)

In this case the capacity increases as if separation between plates is decreased by x.

(*v*) if whole space between the plates is filled with metal plate, i.e. $x = d$ and $\varepsilon_r = \infty$ the capacity becomes infinite. This means that the potential difference between the plates is zero.

(*vi*) the charge is induced on the dielectric slab. The induced charge q_i is

$$q_i = q\left(1 - \frac{1}{\varepsilon_r}\right)$$

where q is the charge on the capacitor plates.

19.6 KIRCHHOFF'S LAW FOR CAPACITORS

There are two laws. One is related to charges and other to the potential difference. The laws are

(*i*) the algebraic sum of charges meeting at any junction is zero, i.e. $\Sigma q = 0$

(*ii*) the algebraic sum of potential differences across different elements of a closed circuit containing capacitances and sources of the emf is zero, i.e. $\Sigma V = 0$.

When we apply this law, the potential fall is taken as negative and potential rise is taken as positive. When we move along a circuit, there is a potential fall in a capacitor (q/C) from positive plate to negative plate and potential rise from negative plate to positive plate. In a battery there is potential fall from positive terminal to negative terminal and potential rise from negative to positive terminal.

These laws can be used to derive condition for balanced Wheatstone's bridge for capacitors. In Fig. 19.13, five capacitors are shown and if bridge is balanced, i.e. B and D are at the same potential then capacitor C_5 will remain uncharged. The condition for this is

$$\frac{C_1}{C_2} = \frac{C_3}{C_4} \qquad \text{...(19.18)}$$

and equivalent capacity between A and C can be easily determined.

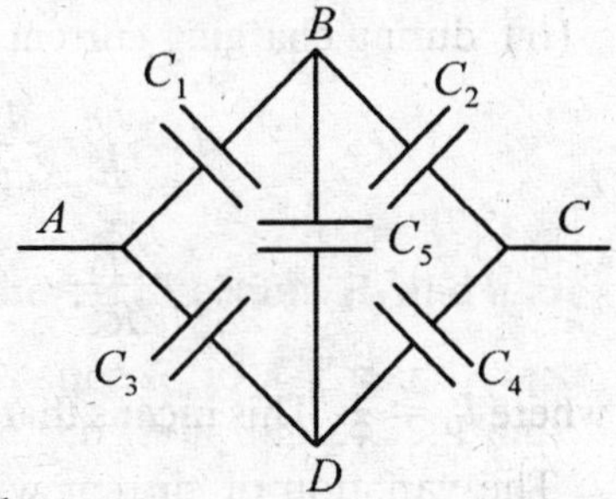

Fig. 19.13

19.7 CHARGING AND DISCHARGING OF A CAPACITOR THROUGH RESISTANCE

Figure 19.14 shows a circuit containing capacitor of capacitance C and a resistance R connected in series with a battery of emf E and a two way switch *Sab*. The battery is connected for charging of capacitor by connecting S to b and battery is disconnected for capacitor to discharge through R by connecting S to a.

Charging

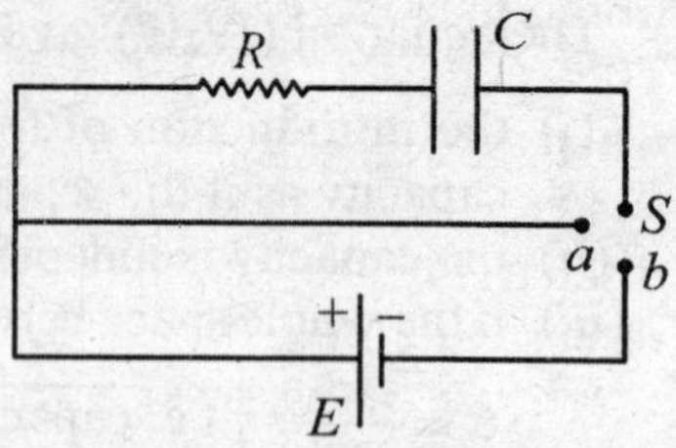

Fig. 19.14

When the capacitor is connected to a battery, the plates of capacitor will acquire equal and opposite charge and potential across it becomes equal to the end of the battery. The process of charging is not an instantaneous one but takes sometime. During this time the charge flows from the battery to the capacitor plates through resistance. During this time there is an electric current through the resistance which varies with time.

If I is the current at any time t and charge on capacitor q then, we have

$$V_C + V_R = E$$

i.e. $\frac{q}{C} + IR = E$ but $I = \frac{dq}{dt}$

and, we have

drift $\frac{dq}{dt} + \frac{q}{RC} = E$

This equation on solving gives

$$q = q_0(1 - e^{-t/RC}) \qquad \text{...(19.19)}$$

where $q_0 = CE$, the maximum charge that the capacitor will have. Equation 19.19 gives

(*i*) the charge on the capacitor increases from $q = 0$ at $t = 0$ to $q = q_0 = CE$ at $t = \infty$. The charge increases exponentially. The variation of charge with time is shown in Fig. 19.15.

(*ii*) the quantity CR has dimensions of time and is called *capacitive time constant*. The time constant gives the time in which the charge on the capacitor reaches 0.632 of its maximum value q_0 (= CR).

(*iii*) since charge q on capacitor is $q = CV$ as such the potential on capacitor also varies exponentially in the same fashion as charge varies.

$$V = \frac{q_0}{C}\left(1 - e^{-t/RC}\right) = E\left(1 - e^{-t/RC}\right)$$

(*iv*) during charging current at any time in the circuit is

$$I = \frac{dq}{dt} = \frac{d}{dt}\left[q_0\left(1 - e^{-t/RC}\right)\right]$$

$$= \frac{q_0}{RC}e^{-t/RC} = I_0 e^{-t/RC} \qquad \text{...(19.20)}$$

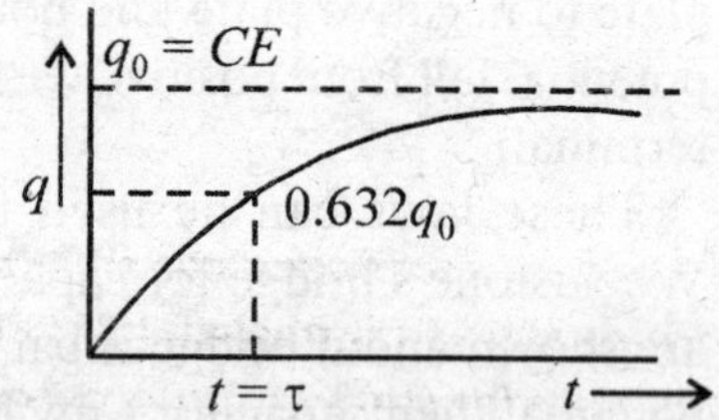

Fig. 19.15

where $I_0 = \frac{E}{R}$. This means that the current also varies exponentially from I_0 at $t = 0$ to zero at $t = \infty$

The variation of current with time is shown in Fig. 19.16. The potential difference V_R across resistance also varies in the same fashion, $V_R = IR = I_0 R e^{-t/RC} = E\, e^{-t/RC}$.

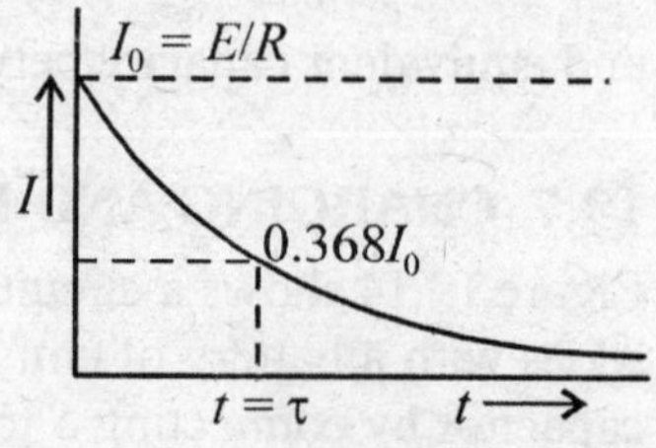

Fig. 19.16

Discharging

The capacitor is first fully charged, i.e. till it acquires charge q_0 (= CE). Now it is discharged through resistance R by throwing the switch S

from b to a (Fig. 19.14) so that the battery is disconnected. Now the capacitor gradually discharges. There is current in the circuit during the discharging of the capacitor. The discharging current stops when the potential difference across the capacitor plates is reduced to zero.

Let q be the charge on the capacitor and I is the current in the circuit at any time t during the discharging process. If V_R and V_C are the potential differences across resistance R and capacitor C respectively, then

$$V_R + V_C = 0$$

or $$RI + q/C = 0 \qquad \text{But } I = \frac{dq}{dt}$$

$\therefore$ $$\frac{dq}{dt} + \frac{q}{Rc} = 0$$

Solving the equation, we have

$$q = q_0 e^{-t/RC} \qquad \text{...(19.21)}$$

where $q_0 = EC$

And current at any time t in the circuit is

$$I = \frac{dq}{dt} = -\frac{q_0}{RC} e^{-t/RC} = -I_0 e^{-t/RC} \qquad \text{...(19.22)}$$

where $$I_0 = \frac{q_0}{RC} = \frac{EC}{RC} = \frac{E}{R}.$$

(*i*) During discharging, charge on the capacitor decreases exponentially from $q = q_0$ (= CE) at $t = 0$ to $q = 0$ at $t = \infty$. The variation of charge on capacitor with time is shown in Fig. 19.17.

(*ii*) The capacitive time constant $\tau = RC$ is the time in whch the charge on capacitor reduces to $(1/e)$, i.e. 0.368 of its initial value. The formula for time constant for charging as well as discharging ($\tau = RC$) is same but their definitions are different.

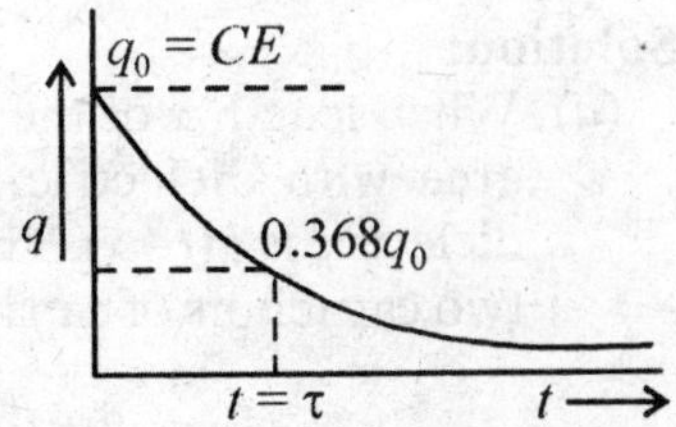

Fig. 19.17

(*iii*) The potential on capacitor also decreases exponentially with time in the same fashion as charge.

(*iv*) During discharging current also decreases exponentially from $-I_0$ at $t = 0$ to zero at $t = \infty$. The variation of current with time is shown in Fig. 19.18.

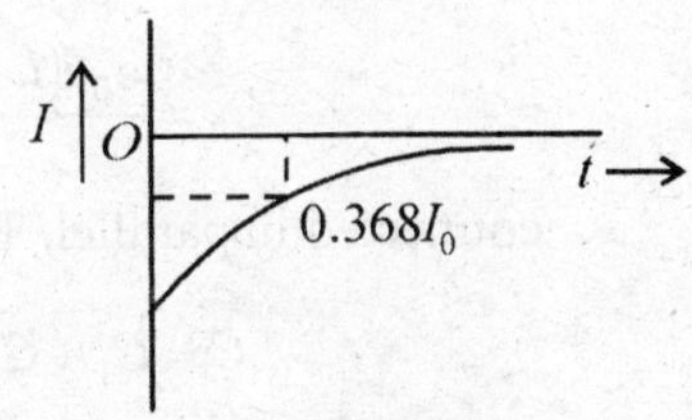

Fig. 19.18

The potential different V_R across resistance also varies in the same fashion as I (the current)

$$V_R = IR = -I_0 Re^{-t/RC} = -Ee^{-t/RC}$$

19.8 CORONA DISCHARGE

It two spheres of radii r_1 and r_2 are charged and connected to each other by a wire, their potentials will be same and charge on the two spheres will be in the ratio of their radii, i.e.

$$\frac{q_1}{q_2} = \frac{r_2}{r_1}$$

and their surface densities of charge σ_1 and σ_2 are in the inverse ratio of their radii, i.e.

$$\frac{\sigma_1}{\sigma_2} = \frac{r_2}{r_1}$$

i.e., smaller the radius larger is the charge density.

If a conductor is charged, the charge density will not be uniform on the entire surface. Where the surface is more curved the charge density will be larger.

The electric field near the surface of a charged conductor is $\frac{\sigma}{\varepsilon_0}$. Thus the electric field near the conductor with small radius of curvature (i.e. sharp points) is large compared to the field near the conductor where radius of *curvature is* large (flat part). The high electric field near the sharp points may cause dielectric *break down* in air. The charge may jump from the conductor to the air because of increased conductivity of air. Often this discharge in air is accompanied by a visible glow surrounding the pointed end. The phenomenon is called *corona discharge.*

SOLVED EXAMPLES

Example 19.1: A parallel plate capacitor with plate separation d, and plates of width b and length l is connected to a battery of emf V. The plates are rigidly clamped. A dielectric slab of thickness d and dielectric constant ε_r is slowly inserted between the plates. (*a*) Calculate the energy of the system when a length x of the slab is introduced into the capacitor. (*b*) What force should be applied on the slab to ensure that it goes slowly into the capacitor? Neglect any effect of friction or gravity.

Solution:

(*a*) When length x of the slab is introduced, the plate area with dielectric is bx and the part without dielectric is $b(l-x)$. The system is equivalent to the two capacitors of capacities

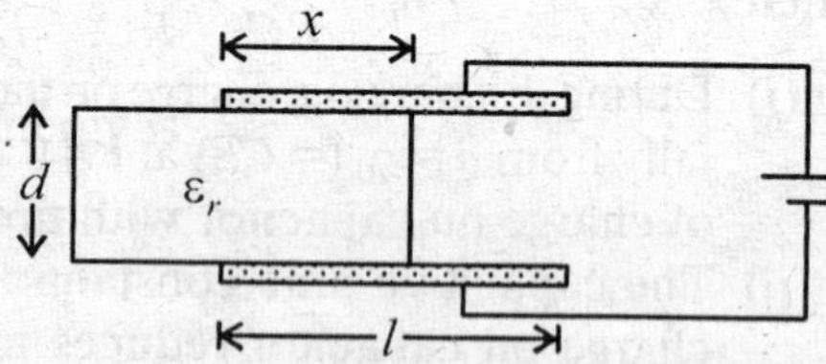

Fig. 19.19

$$C_1 = \frac{\varepsilon_r \varepsilon_0 bx}{d} \text{ and}$$

$$C_2 = \frac{\varepsilon_0 b(l-x)}{d}$$

connected in parallel. Then the capacitance of the system is

$$C = C_1 + C_2 = \frac{\varepsilon_0 b}{d}\left[\varepsilon_r x + (l-x)\right]$$

$$= \frac{\varepsilon_0 b}{d}\left[l + x(\varepsilon_r - 1)\right] \qquad \ldots(i)$$

Then the energy stored in the capacitor is

$$U = \frac{1}{2}CV^2$$

$$= \frac{1}{2}\frac{\varepsilon_0 b}{d}\left[l + x(\varepsilon_r - 1)\right]V^2$$

(*b*) If the electric field attracts the dielectric slabs with force F. An external force of same magnitude F should be applied in opposite direction so that the slab moves slowly without acceleration.

Let the dielectric moves a further distance dx inside the capacitor. The capacity increases by (dC). Since the potential difference remains constant at V, the battery supplies an additional charge

$$dq = V(dC)$$

to the capacitor. In this process work done by the battery is

$$dW = Vdq = V^2(dC)$$

The external force does the work $= -Fdx$ during the displacement dx

Thus net work done on the capacitor is

$$V^2(dC) - Fdx$$

This work done increases the energy stored in the capacitor $= \frac{1}{2}(dC)V^2$. Thus

$$\frac{1}{2}(dC)V^2 = V^2(dC) - Fdx$$

or $$F = \frac{1}{2}V^2\frac{dC}{dx}$$

From equation (*i*) $\dfrac{dC}{dx} = \dfrac{\varepsilon_0 b}{d}(\varepsilon_r - 1)$

$$\therefore \quad F = \frac{1}{2}\frac{\varepsilon_0 bV^2}{d}(\varepsilon_r - 1)$$

The electric field attracts the dielectric into the capacitor with a force $\dfrac{\varepsilon_0 bV^2}{2d}(\varepsilon_r - 1)$ and this much force should be applied in opposite direction.

Example 19.2: The space between the plates of a parallel plate capacitor is filled with a dielectric of dielectric constant ε_r. The plates are given charge of surface density σ_0.

Determine the density of bound electric charges that appear on the surface of the dielectric and the force per unit area that is exerted by the field on the dielectric.

Solution: The electric field in the dielectric medium inside the capacitor is

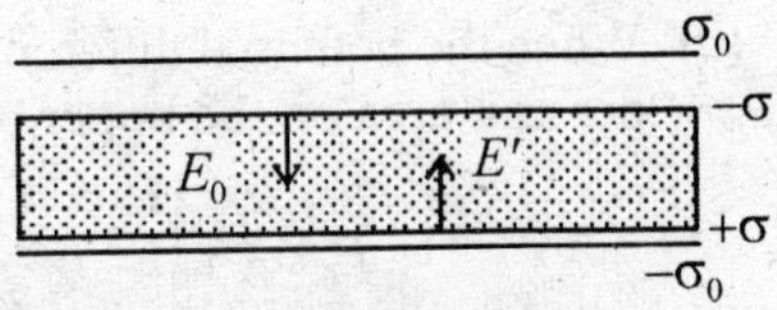

Fig. 19.20

$$E = \frac{\sigma_0}{\varepsilon_r \varepsilon_0}$$

where σ_0 is the surface charge density on the plates.

This field is due to free charges on the plates of the capacitor and due to bound charges (induced charges on the dielectric). The bound charges are on the surface of the dielectric. Let the surface density of these charges be σ'. The field created by free charges is $E_0 = \dfrac{\sigma}{\varepsilon_0}$ and that created by the bound charges is $E' = \dfrac{\sigma'}{\varepsilon_0}$.

Thus magnitude of the resultant intensity $E = E_0 - E'$

$$\frac{\sigma_0}{\varepsilon_r \varepsilon_0} = \frac{\sigma_0}{\varepsilon_0} - \frac{\sigma'}{\varepsilon_0}$$

or $$\sigma' = \sigma_0\left(1-\frac{1}{\varepsilon_r}\right)$$

To determine the force on the surface of the dielectric, we first calculate the intensity of electric field on the surface due to all the charges except the charges on the surface itself. This intensity will be equal to

$$E_1 = E_0 - \frac{E'}{2} = \frac{\sigma}{\varepsilon_0} - \frac{\sigma'}{2\varepsilon_0} = \frac{\sigma_0}{\varepsilon_0} - \sigma_0\left(\frac{\varepsilon_r - 1}{2\varepsilon_r\varepsilon_0}\right)$$

$$= \frac{\sigma_0}{2\varepsilon_0}\frac{(\varepsilon_r+1)}{\varepsilon_r}$$

Hence force per unit area acting on the upper surface of the dielectric will be directed upwards and is equal to

$$F = \sigma' E_1 = \frac{\sigma_0}{2\varepsilon_0}\left(\frac{\varepsilon_r+1}{\varepsilon_r}\right)\left(\frac{\varepsilon_r-1}{\varepsilon_r}\right)\sigma_0$$

$$= \frac{\sigma_0^2}{2\varepsilon_0}\frac{(\varepsilon_r^2+1)}{\varepsilon_r^2}$$

Example 19.3: A parallel plate capacitor is arranged horizontally. The lower plate is fixed and other is connected with a vertical spring. The area of each plate is A and in steady state the separation between the plates is d_0. When the capacitor is connected with a source of potential difference V, a new equilibrium appears with the distance between the plates d. If the mass of upper plate is m, find (*a*) spring constant K, (*b*) the maximum voltage for a given K in which an equilibrium is possible, (*c*) angular frequency of oscillating system around equilibrium value d if amplitude of oscillation is $<< d$.

Solution:

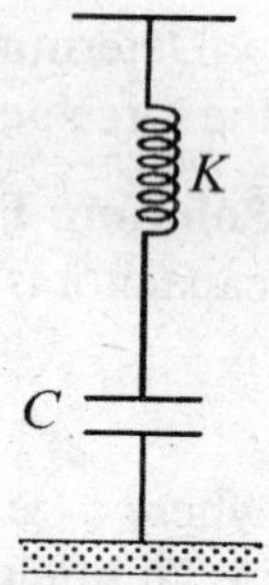

Fig. 19.21

(*a*) When the potential difference V is applied, the capacitor is charged. The force per unit area on the upper plate is

$$F_1 = \frac{1}{2}\varepsilon_0 E^2 = \frac{1}{2}\varepsilon_0\left(\frac{V}{d}\right)^2 \quad \text{as} \quad \left(E = \frac{V}{d}\right)$$

$$= \frac{1}{2}\varepsilon_0\frac{V^2}{d^2}$$

and the Coulomb force on the upper plate is

$$F = AF_1 = \frac{1}{2}\varepsilon_0\frac{V^2}{d^2}A \quad \ldots(i)$$

The upper plate goes down due to this Coulombian force. The extension in the spring then is $(d_0 - d)$ and then tension in the spring is

$$T = K(d_0 - d), \text{ where } K \text{ is spring constant.}$$

Therefore, for equilibrium of the upper plate, we have

$$K(d_0 - d) = \frac{1}{2}\varepsilon_0 \frac{V^2}{d^2} A$$

or

$$K = \frac{1}{2}\frac{\varepsilon_0 V^2}{d^2(d_0 - d)} A = \frac{\varepsilon_0 V^2 A}{2d^2(d_0 - d)} \qquad \ldots(ii)$$

(*b*) From (*ii*), we have

$$V^2 = \frac{2Kd^2(d_0 - d)}{\varepsilon_0 A}$$

or

$$V = \sqrt{\frac{2K}{\varepsilon_0 A}}\, d\,(d_0 - d)^{1/2}$$

For maximum voltage $\frac{\partial V}{\partial d} = 0$

Differentiation gives

$$\frac{\partial V}{\partial d} = \sqrt{\frac{2K}{\varepsilon_0 A}}\left[(d_0 - d)^{1/2} - \frac{1}{2}\frac{d}{\sqrt{(d_0 - d)}}\right] = 0$$

or $(d_0 - d) - \frac{d}{2} = 0$

or $d = \frac{2}{3} d_0$

and

$$V_{\max} = \sqrt{\frac{2K}{\varepsilon_0 A}\frac{2}{3} d_0}\left(d_0 - \frac{2}{3} d_0\right)^{1/2}$$

$$= \sqrt{\frac{K}{\varepsilon_0 A}} \cdot \left(\frac{2}{3} d_0\right)^{3/2}$$

(*c*) Let *y* be the instantaneous displacement of upper plate from equilibrium. Then restoring force

$$F = -\left[K(d_0 - d + y) - \frac{1}{2}\frac{\varepsilon_0 V^2 A}{(d - y)^2}\right]$$

$$= -\left[K(d_0 - d + y) - \frac{1}{2}\frac{\varepsilon_0 V^2 A}{d^2}\left(1 - \frac{y}{d}\right)^{-2}\right]$$

But $\frac{\varepsilon_0 V^2 A}{d^2} = K(d_0 - d)$ from (*ii*) and since $y << d$, using binomial theorem, we can write

$$\left(1 - \frac{y}{d}\right)^{-2} = \left(1 + 2\frac{y}{d}\right)$$

Thus restoring force becomes

$$F = -\left[K(d_0 - d + y) - K(d_0 - d)\left(1 + 2\frac{y}{d}\right)\right]$$

$$= -K\left[1 - \frac{2(d_0 - d)}{d}\right]y$$

$$= -K\left[\frac{3d - 2d_0}{d}\right]y$$

If a is the instantaneous acceleration then $F = ma$

or $$a = \frac{F}{m} = -\frac{K}{md}(3d - 2d_0)\,y$$

i.e. acceleration is proportional to displacement, i.e. the motion is simple harmonic. The angular frequency of S.H.M. is

$$\omega = \sqrt{\frac{K}{md}(3d - 2d_0)}$$

Example 19.4: A capacitor consists of two stationary plates shaped as a semicircle of radius R and a movable plate made of material of dielectric constant ε_r and capable of rotating about an axis passing through O between the stationary plates. The thickness of movable plate is equal to d which is practically the separation between the stationary plates. A potential difference V is applied across the capacitor. Calculate the magnitude of the moment of forces relative to axis O acting on movable plate in the position shown in Fig. 19.22.

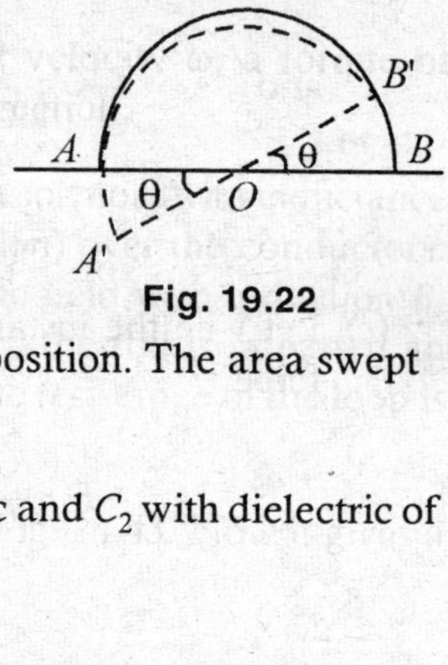

Fig. 19.22

Solution: Initial capacity of the capacitor is

$$C_0 = \frac{\varepsilon_0 \varepsilon_r \pi R^2}{2d} \qquad ...(i)$$

Let the movable plate has rotated through an angle θ from its initial position. The area swept

$$= \frac{1}{2}R(Rd\theta) = \frac{1}{2}R^2 d\theta$$

The system is now equivalent to two capacitors, C_1 with air as dielectric and C_2 with dielectric of constant ε_r. Therefore

$$C_1 = \frac{\varepsilon_0 R^2 \theta}{2d} \text{ and } C_2 = \frac{\varepsilon_0 \varepsilon_r (\pi R^2 - R^2\theta)}{2d}$$

In the position shown in Fig. 19.22, the two capacitors are in parallel and equivalent capacity is

$$C = C_1 + C_2 = \frac{\varepsilon_0 R^2}{2d}[\theta + \varepsilon_r(\pi - \theta]$$

$$= \frac{\varepsilon_0 R^2}{2d}[\pi\varepsilon_r + (1 - \varepsilon_r)\theta]$$

Initial energy of the capacitor

$$U_i = \frac{1}{2}CV^2 = \frac{1}{2}\frac{\varepsilon_0\varepsilon_r\pi R^2}{2d}V^2 = \frac{\varepsilon_0\varepsilon_r\pi R^2V^2}{4d}$$

Final energy of the capacitor

$$U_f = \frac{1}{2}CV^2 = \frac{1}{2}\frac{\varepsilon_0 R^2}{2d}[\pi\varepsilon_r + (1-\varepsilon_r)\theta]V^2$$

$$= \frac{\varepsilon_0 R^2}{4d}[\pi\varepsilon_r + (1-\varepsilon_r)\theta]V^2$$

$\therefore$ Loss in energy

$$\Delta U = U_i - U_f = \frac{\varepsilon_0 R^2}{4d}V^2[\pi\varepsilon_r - \pi\varepsilon_r - (1-\varepsilon_r)\theta]$$

$$= \frac{\varepsilon_0 R^2 V^2}{4d}(\varepsilon_r - 1)\theta$$

If τ is the torque due to force, then work done during rotation θ is $W = \tau\theta$ and this is equal to the loss in energy.

$$\therefore \qquad \tau\theta = \frac{\varepsilon_0 R^2 V^2}{4d}(\varepsilon_r - 1)\theta$$

or

$$\tau = \frac{\varepsilon_0 R^2 V^2}{4d}(\varepsilon_r - 1)$$

Example 19.5: Two identical capacitors are connected as shown in Fig. 19.23. A dielectric slab is slipped between the plates of one of the capacitor, the battery remains connected. What change if any will be observed in capacitance, the charge, the potential difference and energy stored for each capacitor?

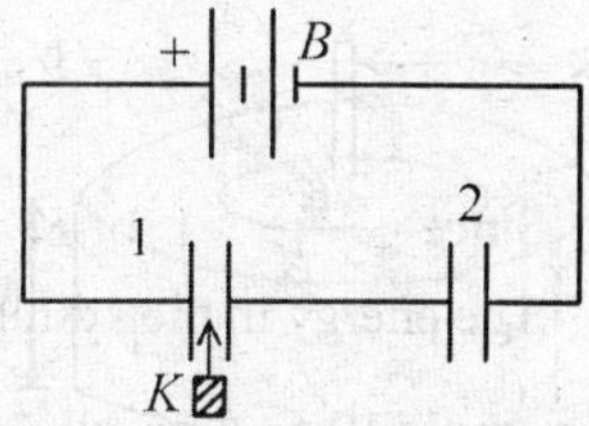

Fig. 19.23

Solution: Two capacitors each of capacity say C_0 are connected in series and their equivalent capacity is $\frac{C_0}{2}$. Battery is connected and if V_0 is emf of battery, then charge q_0 on each capacitor is $q_0 = \frac{V_0 C_0}{2}$;

Potential difference across each capacitor $= \frac{q_0}{C_0} = \frac{V_0}{2}$

and energy stored in each capacitor $U_0 = \frac{1}{2}CV^2 = \frac{1}{2}\frac{C_0V_0^2}{4} = \frac{1}{8}C_0V_0^2$

When dielectric slab is slipped between the plates of capacitor 1, its capacity becomes C_0K and the equivalent capacity of the combination is

$$C = \frac{C_1C_2}{C_1 + C_2} = \frac{C_0 \cdot C_0K}{C_0 + C_0K} = \frac{C_0K}{(1+K)} > \frac{C_0}{2}$$

i.e. the net capacitance increases. Since the applied potential difference is same and hence charge on each capacitor

$$q = CV_0 = \frac{C_0V_0K}{(1+K)} > q_0$$

i.e. charge on each capacitor increases.

Potential difference across capacitor C_1 is

$$V_1 = \frac{q}{C_0K} = \frac{V_0}{1+K} \text{ i.e. } V_1 < \frac{V_0}{2} \quad \because K > 1$$

and potential difference across capacitor C_2 is

$$V_2 = \frac{q}{C_0} = \frac{V_0K}{1+K} \text{ i.e. } V_2 > \frac{V_0}{2}$$

Thus the potential difference across C_1 decreases and that across C_2 increases. The energy stored in capacitor 1.

$$U_1 = \frac{1}{2}(C_0K)(V_1)^2 = \frac{1}{2}\frac{C_0KV_0^2}{(1+K)^2}$$

$$= \frac{1}{8}C_0V_0^2\left(\frac{4K}{(1+K)^2}\right) = U_0\frac{4K}{(1+K)^2} \text{ i.e. } U_1 < U_0$$

Since $\frac{4K}{(1+K)^2} < 1$ for $K > 1$ as such the energy stored in capacitor 1 decreases.

The energy stored in capacitor 2,

$$U_2 = \frac{1}{2}C_0V_2^2 = \frac{1}{2}\frac{C_0V_0^2K^2}{(1+K)^2} = \frac{1}{8}C_0V_0^2\frac{4K^2}{(1+K)^2}$$

$$= U_0\left(\frac{2K}{1+K}\right)^2$$

Since $\frac{2K}{1+K} > 1$ for $K > 1$ $U_2 > U_0$, i.e. energy stored in capacitor 2 increases.

The energy in the system as a whole increases.

Example 19.6: A parallel plate capacitor is placed in such a way that the plates are horizontal and the lower plate is dipped into a liquid of dielectric constant K and density ρ. The area of each plate is A. The plates are now connected to a battery which supplies a positive charge of magnitude q to the upper plate. Find the rise in the level of liquid in the space between the plates.

Solution: The charge given by the battery to the upper and lower plate be $+q$ and $-q$.

The charge induced on upper surface of liquid is q_i

$$q_i = -q\left(1 - \frac{1}{K}\right)$$

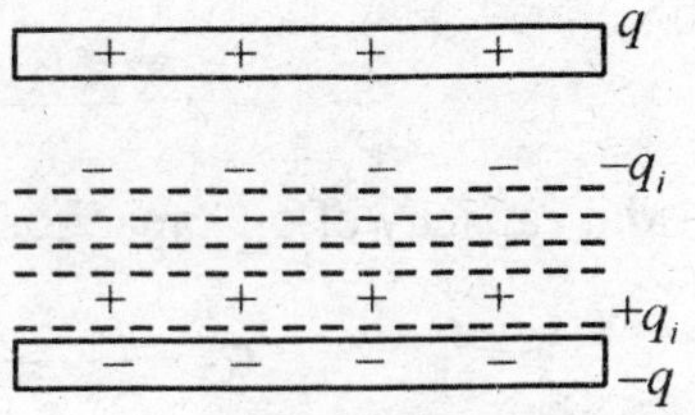

Fig. 19.24

and that on the lower surface of the liquid is $+q\left(1-\frac{1}{K}\right)$

The net charge on the lower plate is

$$q' = q\left(1-\frac{1}{K}\right) - q = -\frac{q}{K}$$

The electric field at the upper surface of liquid is E_1 due to charge on upper plate and E_2 due to charge on lower plate.

$$E_1 = \frac{q}{2\varepsilon_0 A} \text{ downwards}$$

and $$E_2 = \frac{q}{2\varepsilon_0 KA} \text{ downwards}$$

where A is area of plates. Hence net electric field is

$$E = E_1 + E_2 = \frac{q}{2\varepsilon_0 A}\left(1+\frac{1}{K}\right) = \frac{q(K+1)}{2\varepsilon_0 AK}$$

Hence the force on this surface is

$$F = Eq\left(1-\frac{1}{K}\right) \text{ upwards because of negative charge}$$

$$= \frac{q(K+1)}{2\varepsilon_0 AK}\cdot\frac{q(K-1)}{K} = \frac{q^2(K^2-1)}{2\varepsilon_0 AK^2}$$

This force pulls the liquid upward and is balanced by the weight of rise in liquid. If liquid rises by h then its weight is

$$= hA\rho g$$

Equating the two forces, we have

$$hA\rho g = \frac{q^2(K^2-1)}{2\varepsilon_0 AK^2}$$

or $$h = \frac{q^2(K^2-1)}{2\varepsilon_0 A^2 K^2 \rho g}$$

Example 19.7: Two square plates of 1 m side are kept 0.01 m apart like a parallel plate capacitor, in air, in such a way that one of their edges is perpendicular to an oil surface in tank with an insulating oil. The plates are connected to a battery of emf 500 V. The plates are then lowered vertically into the oil at a speed of 0.001 ms^{-1}. Calculate the current drawn from the battery during the process (dielectric constant of oil is 11).

Solution: The area of plates $A = l^2$ and separation between plates is d. Hence capacitance of air capacitor is

$$C_0 = \frac{\varepsilon_0 A}{d}$$

When capacitor is lowered in oil, we have two capacitors, air capacitor and oil capacitor in parallel. Let instantaneous length of the plates dipped in oil be y, then capacity of air capacitor

$$C_1 = \frac{\varepsilon_0 l(l-y)}{d}$$

and capacity of oil capacitor is

$$C_2 = \frac{\varepsilon_0 K\, ly}{d}$$

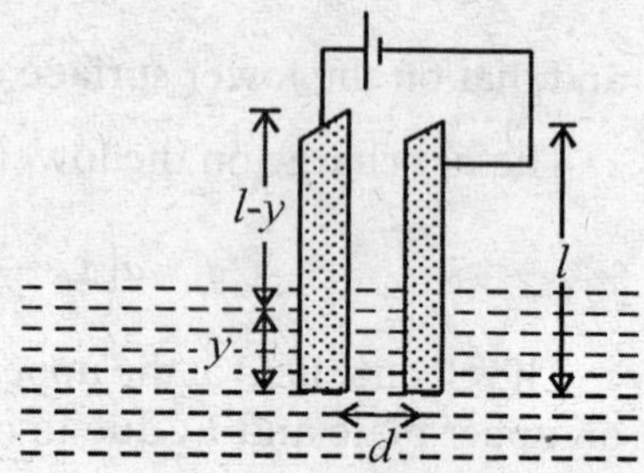

Fig. 19.25

The two are in parallel and the resultant capacity is

$$C = C_1 + C_2 = \frac{\varepsilon_0 l(l-y)}{d} + \frac{\varepsilon_0 K\, ly}{d}$$

$$= \frac{\varepsilon_0 l}{d}[l - y + Ky] = \frac{\varepsilon_0 l}{d}[l + (K-1)y]$$

Since battery has potential of 500 V, the instantaneous charge on the capacitor is

$$q = CV = \frac{\varepsilon_0 lV}{d}[l + (K-1)y]$$

And current drawn from battery is

$$i = \frac{dq}{dt} = \frac{\varepsilon_0 lV}{d}(K-1)\frac{dy}{dt} \quad \ldots(i)$$

where $\frac{dy}{dt}$ is velocity with which the plates are lowered in the oil. Given $l = 1$ m, $d = 0.01$ m, we obtain $\frac{dy}{dt} = 0.001$ ms^{-1}, $K = 11$.

Substituting these values in eq. (*i*), we have

$$i = \frac{8.85 \times 10^{-12} \times 1 \times 500}{0.01}(11-1) \times (0.001)$$

$$= 8.85 \times 5 \times 10 \times 10^{-11}$$

$$= 4.425 \times 10^{-9} \text{ A.}$$

Example 19.8: In the given circuit (Fig. 19.26), find the potential difference between the points *A* and *B*.

Solution: Let q_1, q_2, q_3 be charges on capacitors $C_1 = 4\ \mu$F, $C_2 = 2\ \mu$F and $C_3 = 1\ \mu$F respectively then using Kirchhoff's law, we get

$$q_1 + q_2 + q_3 = 0 \quad \ldots(i)$$

and
$$6 - \frac{q_1}{C_1} = 12 - \frac{q_2}{C_2} = 24 - \frac{q_3}{C_3} \quad \ldots(ii)$$

or
$$6 - \frac{q_1}{4} = 12 - \frac{q_2}{2}$$

giving
$$q_1 = 2q_2 - 24$$

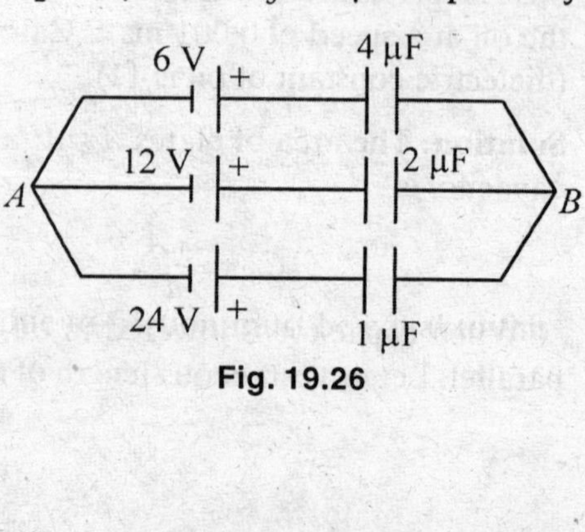

Fig. 19.26

and $\quad 12 - \frac{q^2}{2} = 24 - q_3$

giving $\quad q_3 = 12 + \frac{q_2}{2}$

Substituting in (*i*), we get

$$2q_2 - 24 + q_2 + 12 + \frac{q_2}{2} = 0$$

or $\quad \frac{7}{2}q_2 = 12$ and $q_2 = \frac{24}{7}$ μC

Now proceeding from A to B via $C_2 = 2$ μF capacitor, we have

$$V_A + 12 - \frac{q_2}{2} = V_B$$

or $\quad V_A - V_B = -12 + \frac{24}{7} \times \frac{1}{2} = -12\left(1 - \frac{1}{7}\right)$

$$= -12 \times \frac{6}{7} = -10.3 \text{ V}$$

Example 19.9: Calculate the capacity of a parallel plate capacitor with plate area A and distance between plates d. When filled with a dielectric whose dielectric constant varies as

$$\varepsilon(x) = \varepsilon_0 + \beta x \text{ for } 0 < x < d/2$$

$$\varepsilon(x) = \varepsilon_0 + \beta(d - x) \text{ for } d/2 < x < d$$

For what value of β would the capacity of the condenser be twice that of when it is without dielectric?

Solution: Let us consider a differential capacitor of width dx at a distance x from left plate, its capacity is given by

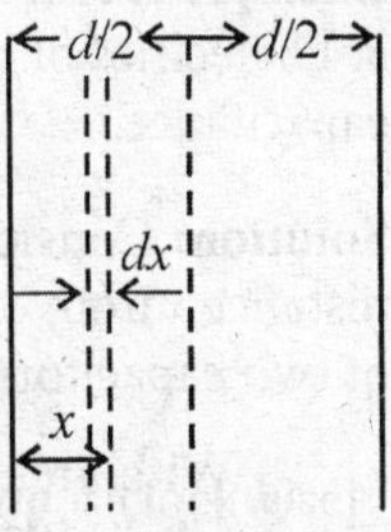

Fig. 19.27

$$dC = \frac{\varepsilon_0 \varepsilon(x) A}{dx}$$

Since all such differential capacitors are in series as such total capacity of the condenser is

$$\frac{1}{C} = \int \frac{1}{dC} = \int \frac{dx}{\varepsilon_0 \varepsilon(x) A} = \int_0^{d/2} \frac{dx}{\varepsilon_0(\varepsilon_0 + \beta x)A} + \int_{d/2}^{d} \frac{dx}{\varepsilon_0[\varepsilon_0 + \beta(d - x)]A}$$

Integration gives

$$\frac{1}{C} = \frac{1}{\varepsilon_0 A\beta} \log(\varepsilon_0 + \beta x)\Big|_0^{d/2} - \frac{1}{\varepsilon_0 A\beta} \log[\varepsilon_0 + \beta(d - x)]\Big|_{d/2}^{d}$$

$$= \frac{1}{\varepsilon_0 A\beta}\left\{\log\left(\varepsilon_0 + \frac{\beta d}{2}\right) - \log \varepsilon_0\right\} - \frac{1}{\varepsilon_0 A\beta}\left\{\log \varepsilon_0 - \log\left(\varepsilon_0 + \frac{\beta d}{2}\right)\right\}$$

$$= \frac{2}{\varepsilon_0 A\beta} \log \frac{\left(\varepsilon_0 + \frac{\beta d}{2}\right)}{\varepsilon_0} = \frac{2}{\varepsilon_0 A\beta} \log\left(1 + \frac{\beta d}{2\varepsilon_0}\right)$$

or
$$C = \frac{\varepsilon_0 A\beta}{2\log\left(1 + \frac{\beta d}{2\varepsilon_0}\right)}$$

The required capacity of the capacitor is $C = \dfrac{\varepsilon_0 A\beta}{2\log\left(1 + \frac{\beta d}{2\varepsilon_0}\right)}$

When no dielectric is introduced, capacity is $C_0 = \dfrac{\varepsilon_0 A}{d}$

for $C = 2C_0$, we get

$$\frac{\varepsilon_0 A\beta}{2\log\left(1 + \frac{\beta d}{2\varepsilon_0}\right)} = \frac{2\varepsilon_0 A}{d}$$

or
$$\beta = \frac{4}{d}\log\left(1 + \frac{\beta d}{2\varepsilon_0}\right)$$

Solving this equation, we get the value of β.

Example 19.10: A capacitor is formed by two square metal plates of edge a; separated by a distance d. Dielectrics of dielectric constant K_1 and K_2 are filled in the gap as shown in Fig. 19.28. Find the capacitance.

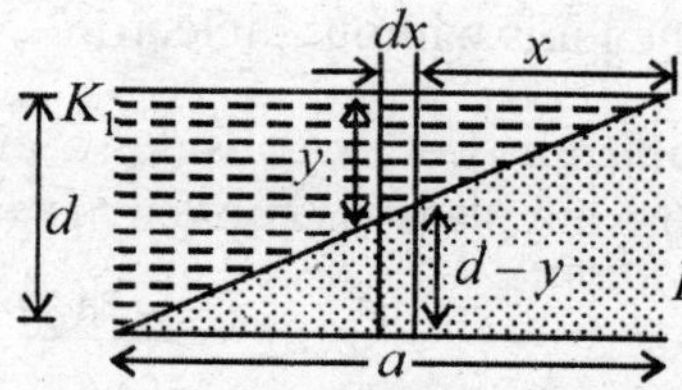

Fig. 19.28

Solution: Consider a small element of capacitor of width dx at a distance x from one end as shown in Fig. 19.28. This is composed of two capacitors, viz.

1. With area $(a\,dx)$ of plates with separation of plates y and filled with dielectric of dielectric constant K_1. Its capacity is

$$dC_1 = \frac{\varepsilon_0 K_1 a\,dx}{y} \qquad ...(i)$$

2. With area $(a\,dx)$ of plates with separation of plates $(d - y)$ and filled with dielectric of dielectric constant K_2. Its capacity is

$$dC_2 = \frac{\varepsilon_0 K_2 a\,dx}{d - y} \qquad ...(ii)$$

These two capacitors are in series as such the equivalent capacity is

$$dC = \frac{(dC_1)(dC_2)}{(dC_1) + (dC_2)}$$

$$\therefore \quad dC = \frac{\frac{\varepsilon_0 K_1 a\,dx}{y} \cdot \frac{\varepsilon_0 K_2 a\,dx}{d-y}}{\frac{\varepsilon_0 K_1 a\,dx}{y} + \frac{\varepsilon_0 K_2 a\,dx}{d-y}}$$

$$= \frac{\varepsilon_0 K_1 K_2 a\,dx}{K_1(d-y) + K_2 y} \qquad \ldots(iii)$$

From Fig. 19.28, we have $\frac{x}{y} = \frac{a}{d}$ or $x = \frac{a}{d} y$ and $dx = \frac{a}{d} dy$

Substituting in (*iii*), we get

$$dC = \frac{\varepsilon_0 K_1 K_2 a \cdot a dy}{d\left[K_1(d-y) + K_2 y\right]} = \frac{\varepsilon_0 K_1 K_2 a^2}{d} \frac{dy}{\left[K_1 d + (K_2 - K_1)y\right]}$$

Integration with respect to dy for $y = 0$ to $y = d$, we get

$$C = \int_0^d \frac{\varepsilon_0 K_1 K_2 a^2}{d} \frac{dy}{\left[K_1 d + (K_2 - K_1)y\right]}$$

$$= \frac{\varepsilon_0 K_1 K_2 a^2}{d(K_2 - K_1)} \log\left[K_1 d + (K_2 - K_1)y\right]\Big|_0^d$$

$$= \frac{\varepsilon_0 K_1 K_2 a^2}{d(K_2 - K_1)} \left\{\log\left[K_1 d + (K_2 - K_1)d\right] - \log K_1 d\right\}$$

$$= \frac{\varepsilon_0 K_1 K_2 a^2}{d(K_2 - K_1)} \left\{\log K_2 d - \log K_1 d\right\}$$

$$= \frac{\varepsilon_0 K_1 K_2 a^2}{d(K_2 - K_1)} \log \frac{K_2}{K_1}$$

Example 19.11: Consider the situation as shown in Fig. 19.29. The plates of the capacitor have plate area A and are clamped in the laboratory. The dielectric slab is released from rest with a length a inside capacitor. Neglecting any effect of friction or gravity, show that the slab will execute periodic motion and find its time period.

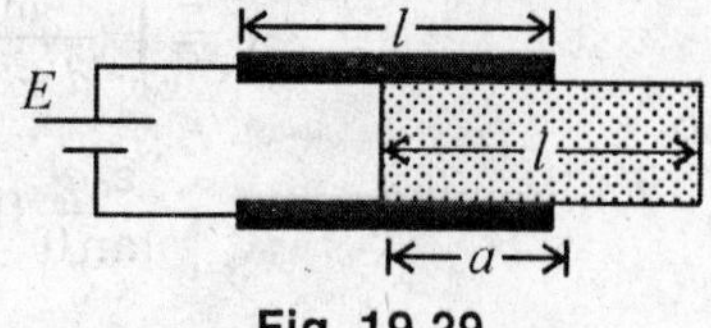

Fig. 19.29

Solution: The electric field attracts the dielectric into the capacitor with force (example 19.1)

$$F = \frac{\varepsilon_0 bE^2}{2d}(\varepsilon_r - 1)$$

where b is the breadth of the plates, d separation between plates and ε_r dielectric constant of the slab.

The acceleration of the slab is (m is mass of the slab)

$$f = \frac{F}{m} = \frac{\varepsilon_0 bE^2}{2dm}(\varepsilon_r - 1) \qquad \ldots(i)$$

Due to this acceleration, the slab moves inside and force becomes zero when whole dielectric is inside the capacitor. But the dielectric has acquired some velocity as such the slab starts moving out and the force acts on it trying to pull it inside. The slab will move out till its velocity becomes zero. Since the force is towards inside as such it will move back again. In this way, it will oscillate and execute a periodic motion.

Since the acceleration in magnitude throughout motion is constant, as such time taken by the slab to move completely inside, i.e. to move distance $(l-a)$ is

$$l-a=\frac{1}{2}ft^2=\frac{1}{2}\frac{\varepsilon_0 bE^2}{2dm}(\varepsilon_r-1)t^2$$

or

$$t^2=\frac{4(l-a)\,dm}{\varepsilon_0 bE^2(\varepsilon_r-1)}$$

and

$$t=\sqrt{\frac{4(l-a)\,dm}{\varepsilon_0 bE^2(\varepsilon_r-1)}}=\sqrt{\frac{4(l-a)l\,dm}{\varepsilon_0 AE^2(\varepsilon_r-1)}}$$

where $A = lb$ the area of plates.

The time period of the motion is $4t$

$$\therefore \qquad T=t=\sqrt{\frac{4(l-a)\,dm}{\varepsilon_0 bE^2(\varepsilon_r-1)}}=\sqrt{\frac{4(l-a)l\,dm}{\varepsilon_0 AE^2(\varepsilon_r-1)}}$$

Example 19.12: A capacitor has square plates each of side a, making an angle θ between them as shown in Fig. 19.30. Show that for small θ the capacitance is given by $C=\frac{\varepsilon_0 a^2}{d}\left\{1-\frac{\theta a}{2d}\right\}$

Solution: The given capacitor can be considered to be made up of a small elementary capacitor in parallel. One such capacitor PQ is shown in Fig. 19.30.

Let width of the elementary capacitor PQ be dx. The area of this capacitor is $a\,dx$ and separation between the plates $PQ = (d + x\tan\theta)$. The capacity of this elementary capacitor is

$$dC=\frac{\varepsilon_0(a\,dx)}{d+x\tan\theta}$$

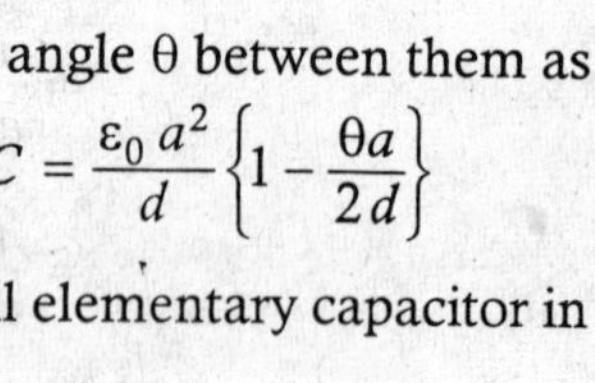

Fig. 19.30

Integration gives

$$C=\int_0^a\frac{\varepsilon_0(a\,dx)}{d+x\tan\theta}=\frac{\varepsilon_0 a}{\tan\theta}\cdot\log[d+x\tan\theta]_0^a$$

$$=\frac{\varepsilon_0 a}{\tan\theta}\{\log(d+a\tan\theta)-\log d]$$

$$=\frac{\varepsilon_0 a}{\tan\theta}\log\frac{d+a\tan\theta}{d}$$

If θ is small, $\tan\theta\approx\theta$ and

$$C=\frac{\varepsilon_0 a}{\theta}\log\left(1+\frac{a\theta}{d}\right)$$

But $\log(1+x) = x - \frac{1}{2}x^2 + \ldots$

Neglecting cubic and higher powers of $\frac{a\theta}{d}$ in the expansion of $\log\left(1+\frac{a\theta}{d}\right)$, we have

$$C = \frac{\varepsilon_0 a}{\theta}\left[\frac{a\theta}{d} - \frac{1}{2}\frac{a^2\theta^2}{d^2}\right]$$

$$= \frac{\varepsilon_0 a^2}{d}\left(1 - \frac{1}{2}\frac{a\theta}{d}\right)$$

Example 19.13: Two conductors carrying equal and opposite charges create a non-uniform field as shown in Fig. 19.31. What is the capacity of this capacitor if field along y-axis varies as $E = \frac{q}{\varepsilon_0 A}(1 + By^2)$, where B is constant?

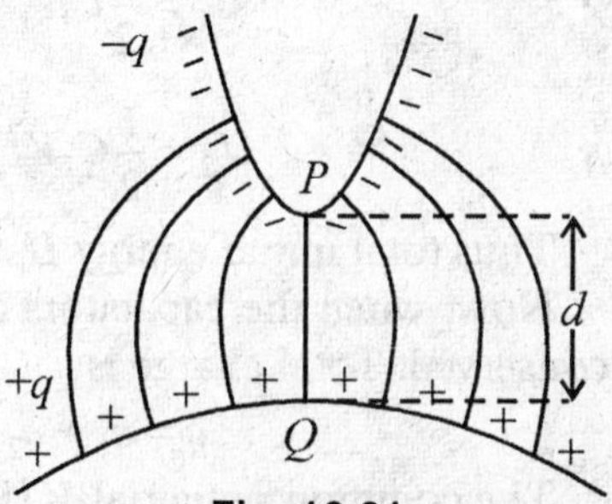

Fig. 19.31

Solution: We know that electric field is given by

$$E = -\frac{dV}{dy}$$

$$\therefore \quad \frac{dV}{dy} = -\frac{q}{\varepsilon_0 A}(1 + By^2)$$

or $$dV = -\frac{q}{\varepsilon_0 A}(1 + By^2)\,dy$$

or $$\int_{V_Q}^{V_P} dV = -\frac{q}{\varepsilon_0 A}\int_0^d (1 + By^2)\,dy$$

or $$V_P - V_Q = -\frac{q}{\varepsilon_0 A}\left(y + \frac{By^3}{3}\right)_0^d$$

or $$V_Q - V_P = \frac{q}{\varepsilon_0 A}\left(d + \frac{Bd^3}{3}\right)$$

Hence capacity

$$C = \frac{q}{V_Q - V_P} = \frac{\varepsilon_0 A}{d\left(1 + \frac{Bd^2}{3}\right)}$$

Example 19.14: Two parallel plate capacitors A and B having capacities of 1 μF and 5 μF respectively are charged separately to the same potential of 100 Volts. Now the positive plate of A is connected to the negative plate of B and the negative plate of A to the positive plate of B. Find final charges on each capacitor and total loss of electrical energy in capacitors.

Solution: Initially the capacitors are charged (Fig. 19.32 (a))

Charge on A, $q_1 = C_1V_1 = 1 \times 100 = 500$ μC.

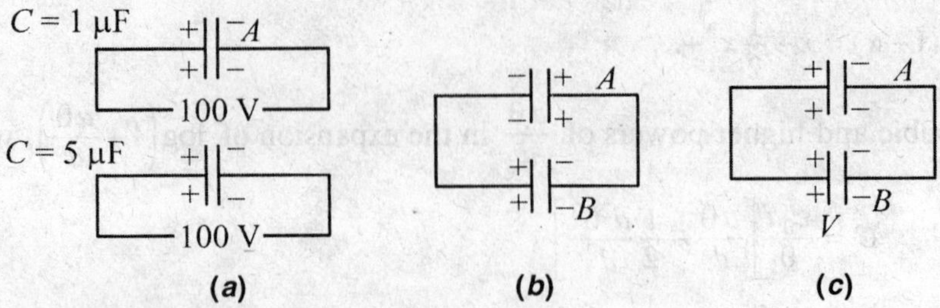

Fig. 19.32

Charge on B, $q_2 = C_2V_2 = 5 \times 100 = 500$ μC.

Energy stored in capacitor

$$A, \quad U_1 = \frac{1}{2}C_1V_1^2 = \frac{1}{2} \times 1 \times (100)^2 = 0.5 \times 10^4 \ \mu\text{J}$$

$$B, \quad U_2 = \frac{1}{2}C_2V_2^2 = \frac{1}{2} \times 5 \times (100)^2 = 2.5 \times 10^4 \ \mu\text{J}$$

Thus total initial energy $U_i = 3 \times 10^4$ μJ

Now, when the capacitors are connected as shown in Fig. 19.32 (*b*), the total charge remains conserved. Total charge is

$$q = q_1 + q_2 = |q_1| \sim |q_2| = (500 \sim 100) = 400 \ \mu\text{C}$$

The common potential V (Fig. 19.32 (*c*)) is

$$V = \frac{q_1 + q_2}{C_1 + C_2} = \frac{400}{1+5} = \frac{200}{3} \text{ V} = 66.67 \text{ V}$$

After sharing of charges, charge on each capacitor is

$$q'_1 = C_1V = 1 \times \frac{200}{3} = \frac{200}{3} \ \mu\text{C} = 66.67 \ \mu\text{C}$$

$$q'_2 = C_2V = 5 \times \frac{200}{3} = \frac{1000}{3} \ \mu\text{C} = 333.33 \ \mu\text{C}$$

The energy now stored is

$$U_f = \frac{1}{2}(C_1 + C_2)V^2 = \frac{1}{2}(1+5)\left(\frac{200}{3}\right)^2$$

$$= \frac{6}{2} \times \frac{4 \times 10^4}{9} = \frac{4}{3} \times 10^4 = 1.33 \times 10^4 \ \mu\text{J}$$

and lose of energy is

$$U_i - U_f = 3 \times 10^4 - \frac{4}{3} \times 10^4 = \frac{5}{3} \times 10^4 = 1.67 \times 10^4 \ \mu\text{J}$$

Example 19.15: Two capacitors A and B with capacities 3 μF and 2 μF are charged to a potential difference of 100 V and 180 V respectively. The plates of the capacitor are connected as shown in

Fig. 19.33 with one wire from each capacitor to an uncharged 2 μF capacitor. The upper plate of A is positive and that of B is negative. Calculate

(*i*) final charge on three capacitors,

(*ii*) the amount of electrostatic energy stored in the system before and after connecting third capacitor.

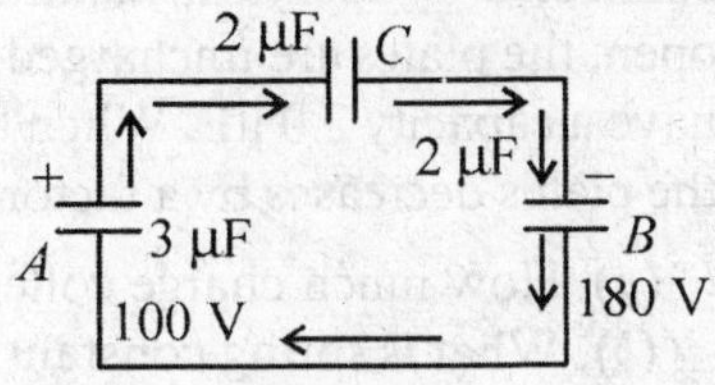

Fig. 19.33

Solution: Let charge q flows in the circuit as shown in Fig. 19.33. Using Kirchhoff's law for voltages, we have

$$\Delta V_A + \Delta V_B = \Delta V_C$$

or $$\left(100 - \frac{q}{3\times10^{-6}}\right) + \left(180 - \frac{q}{2\times10^{-6}}\right) = \frac{q}{2\times10^{-6}}$$

or $$\frac{q}{10^{-6}}\left\{\frac{1}{3}+\frac{1}{2}+\frac{1}{2}\right\} = 280$$

or $$\frac{8q}{6\times10^{-6}} = 280 \text{ or } q = \frac{280\times10^{-6}\times6}{8} = 210\times10^{-6}\,\text{C}$$

or $$q = 210\ \mu\text{C}$$

(*i*) Final charge on capacitors is

$$q_A = C_AV_A - q = 100 \times 3 \times 10^{-6} - 210 \times 10^{-6} = 90\ \mu\text{C}$$

$$q_B = C_BV_B - q = 180 \times 2 \times 10^{-6} - 210 \times 10^{-6} = 150\ \mu\text{C}$$

and $$q_C = 210\ \mu\text{c}$$

(*ii*) Electrostatic energy stored initially

$$U_i = \frac{1}{2}C_AV_A^2 + \frac{1}{2}C_BV_B^2$$

$$= \frac{1}{2}3\times10^{-6}(100)^2 + \frac{1}{2}(180)^2\times2\times10^{-6}$$

$$= 1.5 \times 10^{-2} + 3.24 \times 10^{-2}$$

$$= 4.74 \times 10^{-2}\ \text{J}$$

(*iii*) Final stored electrostatic energy

$$U_f = \frac{q_A^2}{2C_A} + \frac{q_B^2}{2C_B} + \frac{q_C^2}{2C_C}$$

$$= \frac{(90\times10^{-6})^2}{2\times3\times10^{-6}} + \frac{(150\times10^{-6})^2}{2\times2\times10^{-6}} + \frac{(210\times10^{-6})^2}{2\times2\times10^{-6}}$$

$$= \frac{10^{-6}}{2}\left[\frac{(90)^2}{3} + \frac{(150)^2}{2} + \frac{(210)^2}{2}\right]$$

$$= \frac{10^{-6}}{2}[2700 + 11250 + 22050]$$

$$= 10^{-6} \times \frac{36000}{2} = 1.8\times10^{-2}\text{J}$$

Example 19.16: The circuit shown in Fig. 19.34 consists of two identical parallel metal plates connected by identical metal springs to a 100 V battery. With switch open, the plates are uncharged, separated by a distance $d = 8.0$ mm and have a capacity 2.0 μF. When the switch is closed, the distance between the plates decreases by a factor of 0.5.

(*a*) How much charge collects on each plate?
(*b*) What is spring constant for each spring?

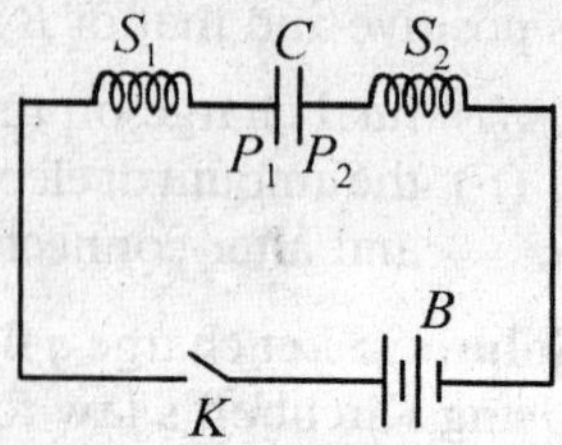

Fig. 19.34

Solution:

(*a*) Since the distance between the plates is reduced to 1/2 the capacity is doubled i.e. now capacity = 4 μF.

Since the battery remains connected as such potential difference remains constant and charge increase. Now charge on the plates is

$$q = CV = 4 \times 100 = 400 \text{ μC} \quad \ldots(i)$$

(*b*) Since the capacitor is charged, there is force between the plates and force is given by $F = \frac{q^2}{2\varepsilon_0 A}$, this force will create tension in spring. In equilibrium if x is extension in springs then tension $= Kx = F$. Equating the two forces, we have

$$Kx = \frac{q^2}{2\varepsilon_0 A}$$

or $$K = \frac{q^2}{2\varepsilon_0 A \cdot x}$$

Change in separation between the plates is $(d - d/2) = d/2$ and this is equal to the total stretch in the two springs, i.e. $2x$. Thus

$$2x = d/2 \text{ or } x = d/4$$

$$\therefore \quad K = \frac{q^2 4}{2\varepsilon_0 A d} = \frac{2}{C}\left(\frac{q}{d}\right)^2 \qquad \because C = \frac{\varepsilon_0 A}{d}$$

$$= \frac{2}{2 \times 10^{-6}}\left(\frac{4 \times 10^{-4}}{8 \times 10^{-3}}\right)^2 = \frac{2 \times 10^{-2}}{2 \times 4 \times 10^{-6}}$$

$$= 2500 \text{ N/m.}$$

Example 19.17: Two parallel plate capacitors A and B have the same separation $d = 8.85 \times 10^{-4}$ m between the plates. The plate areas of A and B are 0.04 m² and 0.02 m² respectively. A slab of dielectric constant $K = 9$ has dimensions such that it can exactly fill the space between the plates of capacitor B. (*a*) The dielectric slab is placed inside A, and is then charged to a potential difference of 110 V. Calculate the capacitance and energy stored in it. (*b*) The battery is disconnected and then the dielectric slab is removed from A. Find the work done by external agency in removing the slab from A. (*c*) The same dielectric slab is now placed in B, filling it completely. Capacitors are then connected. Calculate the energy stored in the system.

Solution: (*a*) Since dielectric completely fills capacitor B as such the area of dielectric is equal the area of B and half the area of A.

When slab is introduced in A it fills half the space. Hence capacitor A is equivalent to two capacitors with capacities $C_1 = \frac{\varepsilon_0 A}{d}$ and $C_2 = \frac{\varepsilon_0 KA}{d}$ in parallel. Here A is the area of capacitor B. Thus equivalent capacity of A is

$$C_A = \frac{\varepsilon_0 A}{d}(1+K) = \frac{8.85\times 10^{-12} \times 0.02}{8.85\times 10^{4}}(1+9)$$
$$= 2 \times 10^{-9}\text{ F}$$

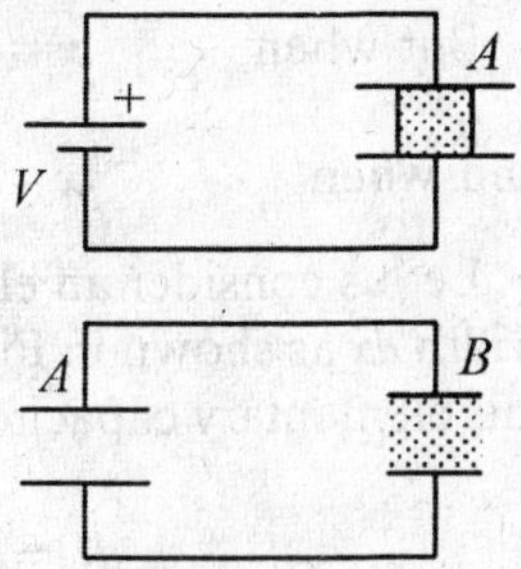

Fig. 19.35

And energy stored in capacitor A is

$$U_A = \frac{1}{2}CV^2 = \frac{1}{2}\times 2\times 10^{-9} \times (110)^2$$
$$= 1.21 \times 10^{-5}\text{ J}$$

(*b*) When battery is removed, the capacitor A is isolated and charge remains conserved. Charge on A is

$$q_A = C_A V = 2 \times 10^{-9} \times 110 = 2.2 \times 10^{-7}\text{ C}$$

Now when dielectric is removed from the capacitor, its capacity is

$$C'_A = \frac{\varepsilon_0 2A}{d} = \frac{8.85\times 10^{-12}\times 2\times 0.02}{8.85\times 10^{-4}} = 4\times 10^{-10}\text{ F}$$

Hence now energy stored in capacitor A is

$$U'_A = \frac{1}{2}\frac{q_A^2}{C'_A} = \frac{1}{2}\frac{(2.2\times 10^{-7})^2}{4\times 10^{-10}}$$
$$= 6.05 \times 10^{-5}\text{ J}$$

Hence work done in removing the dielectric is

$$W = U'_A - U_A = 6.05 \times 10^{-5} - 1.21 \times 10^{-5} = 4.48 \times 10^{-5}\text{ J}$$

(*c*) Now the two capacitors A and B are in parallel and the equivalent capacity is

$$C = \frac{\varepsilon_0(2A)}{d} + \frac{\varepsilon_0 KA}{d} = \frac{\varepsilon_0 A}{d}(2+K)$$
$$= \frac{8.85\times 10^{-12}\times 0.02}{8.85\times 10^{-4}}(2+9)$$
$$= 2.2 \times 10^{-9}\text{ F}$$

And total charge is 2.2×10^{-7} C

Hence energy stored is

$$U = \frac{1}{2}\frac{q_A^2}{C} = \frac{1}{2}\frac{(2.2\times 10^{-7})^2}{(2.2\times 10^{-9})}$$
$$= 1.1 \times 10^{-5}\text{ J.}$$

Example 19.18: A parallel plate capacitor has the space between plates filled with a medium whose dielectric constant increases uniformly with distance. If d is the distance between the plates and K_1 and K_2 are the dielectric constants of the medium at the two plates (square each of area A) respectively, show that the capacity of the capacitor is $C = \frac{\varepsilon_0 A(K_2 - K_1)}{d \log K_2/K_1}$.

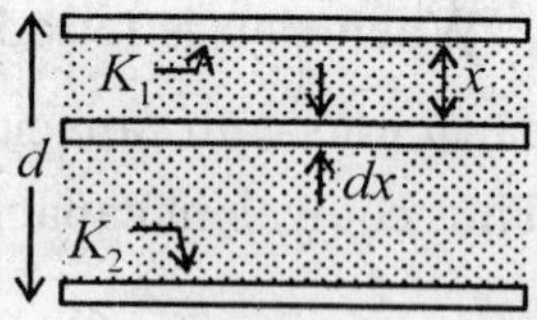

Fig. 19.36

Solution: Since the dielectric varies linearly with distance hence the variation can be written as

$$K = a + bx$$

But when $x = 0, K = K_1 \quad \therefore\ a = K_1 \qquad ...(i)$

and when $x = d, K = K_2 \quad \therefore\ b = \dfrac{K_2 - K_1}{d} \qquad ...(ii)$

Let us consider an elementary capacitor at a distance x from upper plate of area of plate A and width dx as shown in Fig. 19.36. At this point the dielectric constant $K = a + bx$. Hence capacity of this elementary capacitor is

$$dC = \frac{K\varepsilon_0 A}{dx} = \frac{\varepsilon_0 A(a+bx)}{dx}$$

or

$$d\left(\frac{1}{C}\right) = \frac{dx}{\varepsilon_0 A(a+bx)} \qquad ...(iii)$$

The given capacitor is composed of such elementary capacitors in series. Hence integration of (*iii*) for x from 0 to $x = d$ gives the reciprocal of the capacity of this capacitor.

Integration gives

$$\int d\left(\frac{1}{C}\right) = \int_0^d \frac{dx}{\varepsilon_0 A(a+bx)} = \frac{1}{\varepsilon_0 A}\int_0^d \frac{dx}{(a+bx)}$$

or

$$\frac{1}{C} = \frac{1}{\varepsilon_0 A}\frac{1}{b}\log(a+bx)\Big|_0^d$$

$$= \frac{1}{\varepsilon_0 Ab}[\log(a+bd) - \log a]$$

$$= \frac{1}{\varepsilon_0 Ab}\log\frac{(a+bd)}{a}$$

Substituting the values of a and b from (*i*) and (*ii*), we get

$$\frac{1}{C} = \frac{d}{\varepsilon_0 A(K_1 - K_2)}\log\frac{(K_1 + K_2 - K_1)}{K_1}$$

$$= \frac{d}{\varepsilon_0 A(K_2 - K_1)}\log\frac{K_2}{K_1}$$

or

$$C = \frac{\varepsilon_0 A(K_2 - K_1)}{d\log\dfrac{K_2}{K_1}}$$

Example 19.19: A capacitor is filled with two dielectrics of same dimensions but of dielectric constant 2 and 3 respectively. Find the ratio of capacities in the two possible arrangements.

Solution: Two possible arrangements are shown in Fig. 19.37. In Fig. 19.37 (*a*), the two slabs are dividing the space in two equal parts depth wise. As such two capacitors are formed with capacities

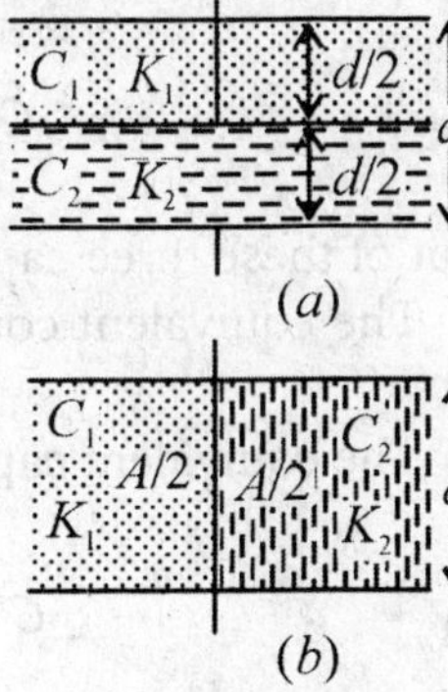

Fig. 19.37

$$C_1 = \frac{\varepsilon_0 K_1 A}{d/2} = \frac{2\varepsilon_0 K_1 A}{d}$$

and $$C_2 = \frac{\varepsilon_0 K_2 A}{d/2} = \frac{2\varepsilon_0 K_2 A}{d}$$

The two capacitors are in series. Hence equivalent capacity C is

$$\frac{1}{C} = \frac{1}{C_1} + \frac{1}{C_2} = \frac{d}{2\varepsilon_0 K_1 A} + \frac{d}{2\varepsilon_0 K_2 A}$$

$$= \frac{d}{\varepsilon_0 A}\left[\frac{1}{K_1} + \frac{1}{K_2}\right] = \frac{d(K_1 + K_2)}{2\varepsilon_0 A K_1 K_2}$$

and capacity $$C = \frac{2\varepsilon_0 A K_1 K_2}{d(K_1 + K_2)} \quad \ldots(i)$$

In Fig. 19.37 (*b*), the two slabs are dividing the space in two equal areas each of $A/2$. As such two capacitors are formed with capacities

$$C_1 = \frac{\varepsilon_0 K_1 A}{2d} \text{ and } C_2 = \frac{\varepsilon_0 K_2 A}{2d}$$

The two capacitors are in parallel and equivalent capacity C' is

$$C' = C_1 + C_2 = \frac{\varepsilon_0 K_1 A}{2d} + \frac{\varepsilon_0 K_2 A}{2d}$$

$$= \frac{\varepsilon_0 A}{2d}(K_1 + K_2) \quad \ldots(ii)$$

Hence required ratio

$$\frac{C}{C'} = \frac{2\varepsilon_0 A K_1 K_2}{2(K_1 + K_2)} \times \frac{2d}{\varepsilon_0 A(K_1 + K_2)}$$

$$= \frac{4 K_1 K_2}{(K_1 + K_2)^2}$$

But $K_1 = 2$ and $K_2 = 3$

Substituting the values, we have

$$\frac{C}{C'} = \frac{4 \times 2 \times 3}{(2+3)^2} = \frac{24}{25}$$

Example 19.20: Each of the plates shown in Fig. 19.38 has surface area $(96/\varepsilon_0) \times 10^{-12}$ F-m. On one side and the separation between the consecutive plates is 4.0 mm. The emf of the battery connected is 10 V. Find the magnitude of the charge supplied by the battery to each of the plates.

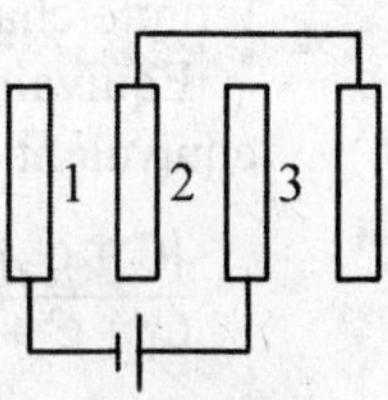

Fig. 19.38

Solution: The plates form 3 capacitors 1, 2 and 3 each of capacity

$$C = \frac{\varepsilon_0 A}{d} = \frac{\varepsilon_0 \times 96 \times 10^{-12}}{\varepsilon_0 \times 4 \times 10^{-3}} = 24 \times 10^{-9} \text{ F}$$

out of these three capacitors 2 and 3 are in parallel and 1 in series.

The equivalent combination is shown in Fig. 19.38 (*a*).

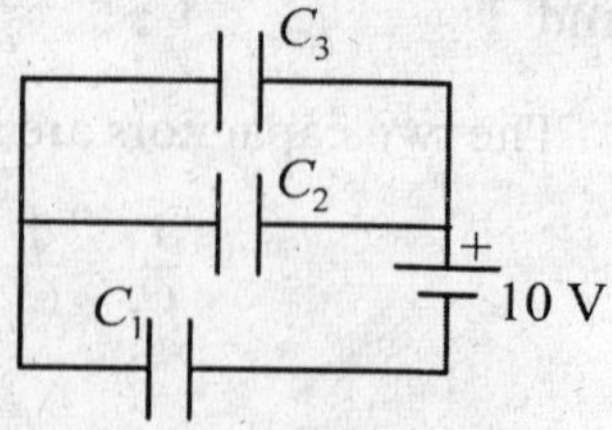

Fig. 19.38 (*a*)

The equivalent capacity $C' = \frac{C \times 2C}{C + 2C} = \frac{2}{3}C$

or $$C' = \frac{2}{3} \times 24 \times 10^{-9} = 16 \times 10^{-9} \text{ F}$$

Hence charge supplied by the battery is

$$q = CV' = 16 \times 10^{-19} \times 10 = 16 \times 10^{-8}$$
$$= 0.16 \ \mu\text{C}.$$

Example 19.21: (*a*) Find the effective capacitance between *A* and *B* of an infinite chain of capacitors joined as shown in Fig. 19.39 (*a*). (*b*) For what value of C_0 in the circuit shown in Fig. 19.39 (*b*), will the net effective capacitance between *A* and *B* be independent of the number of cells in chain?

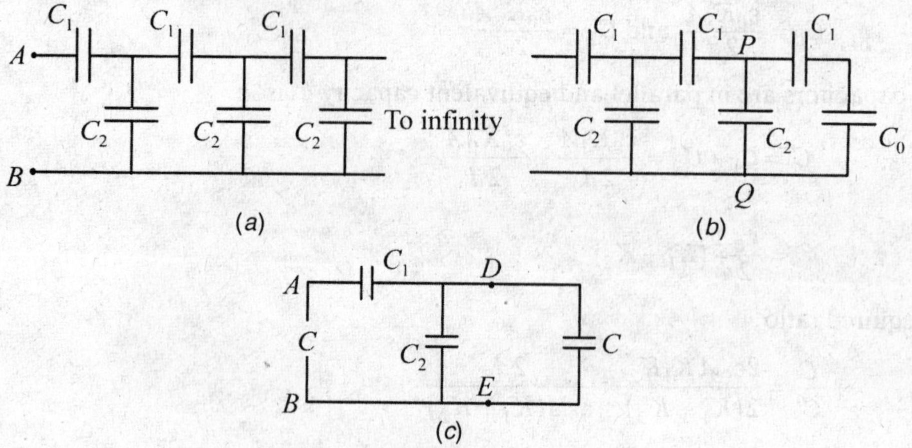

Fig. 19.39

Solution:

(*a*) Let equivalent capacitance between *A* and *B* be *C*. Since the net work is infinite, even if we remove one link from the chain, the net work will still have infinite links, i.e. effective capacitance between *DE* would still be *C* (Fig. 19.39 (*c*)). By doing so, we have reduced the infinite chain to the finite chain shown in Fig. 19.39 (*c*).

Equivalent capacity between *A* and *B*, then is *C* and C_2 is parallel with C_1 in series and equivalent capacity *C*. Thus we have

$$\frac{(C + C_2)C_1}{C_1 + C + C_2} = C$$

or $$C_1C + C_1C_2 = C_1C + C^2 + C_2C$$

or $$C^2 + C_2C - C_1C_2 = 0$$

or $$C = -\frac{C_2 \pm \sqrt{C_2^2 + 4C_1C_2}}{2}$$

Here (–) sign is not admissible as it will result C as negative.
Therefore taking +ve sign, we have

$$C = \frac{1}{2}\left\{-C_2 + C_2\sqrt{1 + 4\frac{C_1}{C_2}}\right\}$$

$$= \frac{C_2}{2}\left\{\sqrt{1 + \frac{4C_1}{C_2}} - 1\right\}$$

(*b*) In this case the net capacity across AB will be independent of the number of links if equivalent capacity across PQ (Fig. 19.39 (*b*)) becomes equal to C_0. In this case C_1 and C_0 are in series and C_2 in parallel and equivalent capacity C_0. Solving, we have

$$\frac{C_1C_0}{C_1 + C_0} + C_2 = C_0$$

or $$C_1C_0 + C_1C_2 + C_0C_2 = C_1C_0 + C_0^2$$

or $$C_0^2 - C_0C_2 - C_1C_2 = 0$$

or $$C_0 = \frac{C_2 \pm \sqrt{C_2^2 + 4C_1C_2}}{2}$$

Here negative sign is not admissible because it makes C_0 negative

$$C_0 = \frac{C}{2}\left\{1 + \sqrt{1 + \frac{4C_1}{C_2}}\right\}$$

Example 19.22: Find the equivalent capacitance of the combination of capacitors across AB as shown in Fig. 19.40.

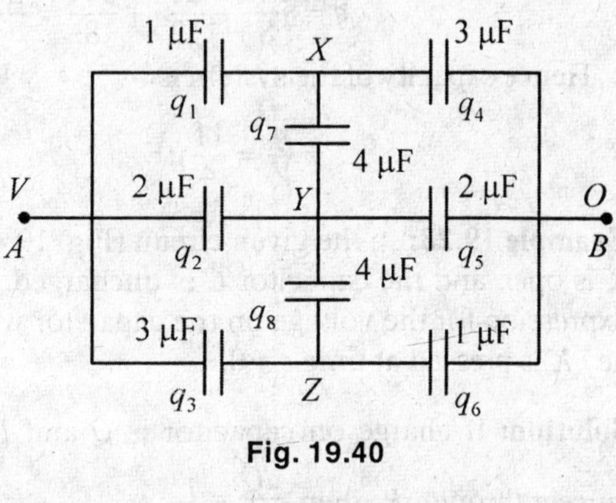

Fig. 19.40

Solution: Let a potential difference of V is applied across AB such that A is at potential V and B is at zero. Let the potentials of points X, Y and Z be x, y, z respectively. The charges on different capacitors be as shown in Fig. 19.40. From diagram,

$$q_1 = 1(V - x) = (V - x),\ q_2 = 2(V - y)$$

and $$q_3 = 3(V - z)$$

Similarly, $q_4 = 3x$, $q_5 = 2y$ and $q_6 = z$

But we know, $$q_1 + q_2 + q_3 = q_4 + q_5 + q_6$$

$\therefore$ $$(V - x) + 2(V - y) + 3(V - z) = 3x + 2y + z$$

or $$4x + 4y + 4z = 6V$$

or $$x + y + z = \frac{3}{2}V \quad \text{...(i)}$$

Now $$q_1 = q_4 + q_7$$

i.e. $$(V - x) = 4(x - y) + 3x$$

or $$V = 8x - 4y$$

or $$y = 2x - \frac{V}{4} \quad \text{...(ii)}$$

Again $$q_3 = q_6 + q_8$$

i.e. $$(V - z)\,3 = 4(z - y) + z$$

or $$3V = 8z - 4y$$

$$= 8z - (8x - V) = 8z - 8x + V$$

or $$8z = 8x + 2V \text{ or } z = x + \frac{V}{4} \quad \text{...(iii)}$$

Using (*ii*) and (*iii*) in (*i*), we get

$$x + 2x - \frac{V}{4} + x + \frac{V}{4} = \frac{3}{2}V$$

or $$4x = \frac{3V}{2} \text{ or } x = \frac{3V}{8} \quad \text{...(iv)}$$

$$y = 2x - \frac{V}{4} = \frac{3V}{4} - \frac{V}{4} = \frac{V}{2} \quad \text{...(v)}$$

and $$z = x + \frac{V}{4} = \frac{3V}{8} + \frac{V}{4} = \frac{5V}{8} \quad \text{...(vi)}$$

Total charge is

$$q = q_4 + q_5 + q_6 = 3x + 2y + z$$

$$q = \frac{9V}{8} + \frac{2V}{2} + \frac{5V}{8} = \frac{22V}{8} = \frac{11V}{4}$$

Hence capacity of the system is

$$C = \frac{q}{V} = \frac{11}{4}\,\mu\text{F}.$$

Example 19.23: In the given circuit (Fig. 19.41), key K is open and the capacitor C is uncharged. Obtain expression for the voltage on the capacitor when the key K is pressed at time $t = 0$.

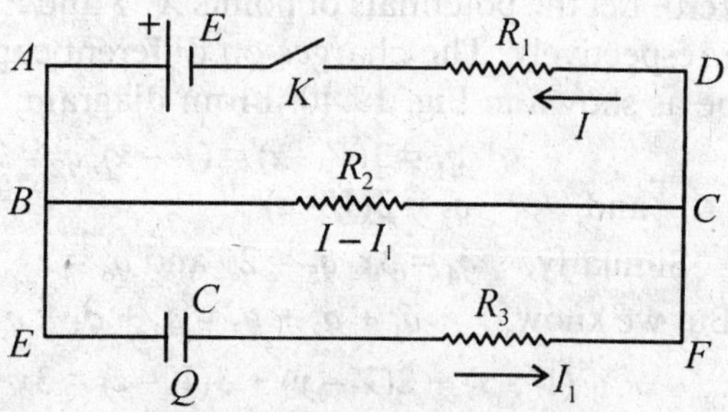

Fig. 19.41

Solution: If charge on capacitor is Q and I_1 is the current through R_3 then $\frac{dQ}{dt} = I_1$ ...(*i*)

If I is the current from battery, the current through R_2 is $I - I_1$.

Kirchhoff's law for loop $ABCDA$ gives

$$R_2(I - I_1) + IR_1 = E$$

or $$I(R_1 + R_2) = E + R_2 I_1$$

and $$I = \frac{E + R_2 I_1}{R_1 + R_2} \quad \ldots(ii)$$

Krichhoff's law for loop $AEFDA$ gives

$$\frac{Q}{C} + R_3 I_1 + R_1 I = E \quad \ldots(iii)$$

Substituting the value of I from (*ii*), we get

$$\frac{Q}{C} + R_3 I_1 + \frac{ER_1 + R_1 R_2 I_1}{R_1 + R_2} = E$$

or $$\frac{Q}{C} + I_1\left(R_3 + \frac{R_1 R_2}{R_1 + R_2}\right) = E - \frac{E\,R_1}{R_1 + R_2} = \frac{E\,R_2}{R_1 + R_2}$$

Let us put $$\alpha = R_3 + \frac{R_1 R_2}{R_1 + R_2} \text{ and } \beta = \frac{R_2}{R_1 + R_2} \text{ and } I_1 = \frac{dQ}{dt}$$

Then $$\frac{Q}{C} + \alpha \frac{dQ}{dt} = \beta E$$

and $$\frac{dQ}{dt} = \frac{\beta}{\alpha} E - \frac{Q}{\alpha C} = \frac{1}{\alpha C}[\beta C\,E - Q]$$

or $$\frac{dQ}{\beta C\,E - Q} = \frac{dt}{\alpha C}$$

Integration gives

$$-\log(\beta C\,E - Q) = \frac{t}{\alpha C} + K$$

where K is constant of integration.

But when $t = 0$, $Q = 0$ gives $K = -\log(\beta CE)$

Substitution gives

$$\log \frac{(\beta C\,E - Q)}{\beta C\,E} = -\frac{t}{\alpha C}$$

or $$\frac{\beta C\,E - Q}{\beta C\,E} = e^{-t/\alpha C}$$

and $$Q = \beta C\,E\,(1 - e^{-t/\alpha C})$$

Hence potential on capacitor at any time t is

$$V = \frac{Q}{C} = \beta\,E\left(1 - e^{-t/\alpha C}\right)$$

where $$\beta = \frac{R_2}{R_1 + R_2} \text{ and } \alpha = R_3 + \frac{R_1 R_2}{R_1 + R_2}$$

and current $$I_1 = \frac{dQ}{dt} = \frac{\beta E}{\alpha} e^{-t/\alpha C}$$

Example 19.24: Find the potential difference between points a and b shown in Fig. 19.42.

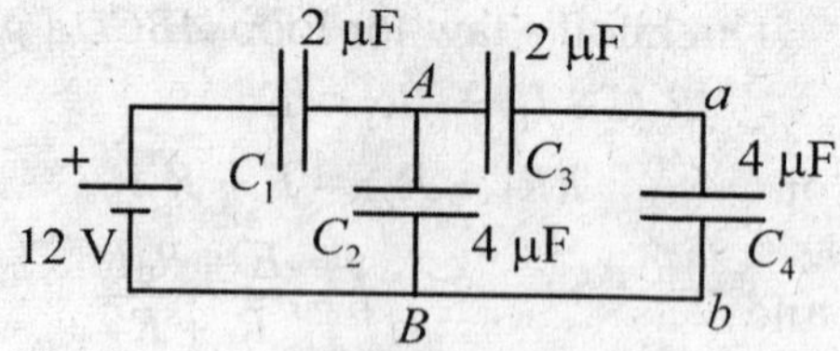

Fig. 19.42

Solution: Equivalent capacity across AB: capacitors of capacity 2 μF and 4 μF are in series and thus $C = \frac{4}{3}$ μF. Capacitor of capacity 4 μF is parallel to this. Hence equivalent capacity across $AB = 4 + \frac{4}{3} = \frac{16}{3}$ μF

The net capacity of the circuit $C = \frac{2 \times 16/3}{2 + 16/3} = \frac{16}{11}$ μF

The charge given by battery $q = \frac{16}{11} \times 12\,\mu C$

This charge divides between capacitors C_2, C_3 and C_4 combine. Hence charge on capacitors C_3 and C_4 which are in series is

$$q_1 = \frac{q \times \frac{4}{3}}{4/3 + 4} = \frac{16}{11} \times \frac{12 \times \frac{4}{3}}{16/3} = \frac{12 \times 4}{11}$$

Hence potential difference across $ab = \frac{q_1}{4} = \frac{12 \times 4}{11 \times 4} = \frac{12}{11}$ V.

Example 19.25: Eleven capacitors each of capacity C are arranged to form eleven sides of a cube as shown in Fig. 19.43. Find the equivalent capacity across AD.

Solution: Let a potential difference V is applied across AD. A charge $2q_1$ enters the network which is distributed amongst various capacitors as shown in Fig. 19.43. The equivalent capacity C_e will be

$$C_e = \frac{2q_1}{V} \quad \ldots(i)$$

Applying Kirchhoff's law to the mesh $HGFEH$, we have

$$\frac{q_2}{C} + \frac{2q_2}{C} + \frac{q_2}{C} - \frac{(q_1 - q_2)}{C} = 0$$

or $\quad 5q_2 = q_1$ or $q_2 = \frac{q_1}{5} \quad \ldots(ii)$

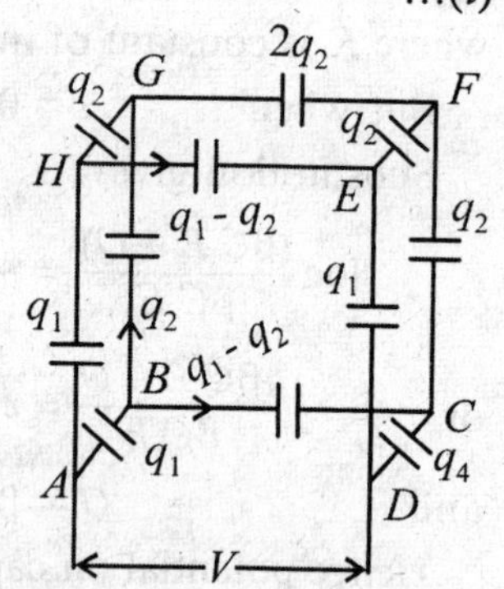

Fig. 19.43

Applying Kirchhoff's law to mesh $ABCD$, we get

$$\frac{q_1}{C} + \frac{q_1 - q_2}{C} + \frac{q_1}{C} = V$$

or $\quad 3q_1 - q_2 = CV$

Substituting $\quad q_2 = \frac{q_1}{5}$ we have

$$3q_1 - \frac{q_1}{5} = CV$$

or $\quad 14q_1 = 5CV$

or $\quad C_e = \frac{2q_1}{V} = \frac{5C}{7}$

Example 19.26: What amount of heat will be generated in the circuit shown in Fig. 19.44 after the switch S is shifted from position 1 to position 2. $C_0 = 2\ \mu F$, $C = 1\ \mu F$ and $E = 12$ V.

Solution: When switch is in position 1, capacitors a and b are in parallel while capacitor c is in series.

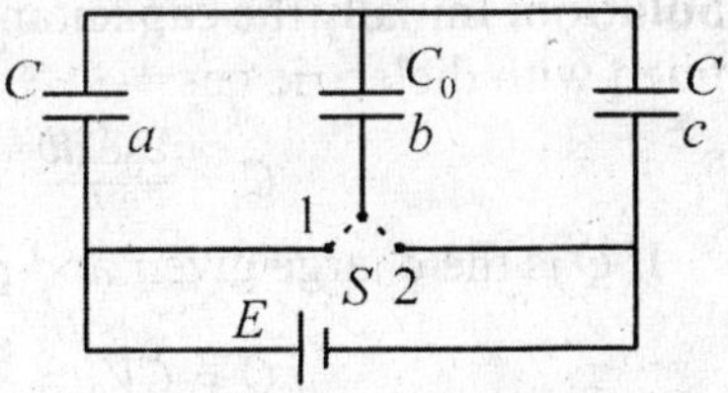

Fig. 19.44

Equivalent capacity of network is $C' = \dfrac{C(C+C_0)}{2C+C_0}$

and charge supplied by the battery is $q = C'E = \dfrac{C(C+C_0)}{2C+C_0}E$

The charge on capacitor C is $q = \dfrac{C(C+C_0)}{2C+C_0}E$

Potential difference across capacitors a and b is

$$V = \frac{q}{C+C_0} = \frac{C(C+C_0)E}{(2C+C_0)(C+C_0)} = \frac{CE}{2C+C_0}$$

Charge on capacitor $a = VC = \dfrac{EC^2}{2C+C_0}$ ...(*i*)

Energy stored in the system is $= \dfrac{1}{2}C'E^2 = \dfrac{1}{2}\dfrac{C(C+C_0)}{2C+C_0}E^2$ ...(*ii*)

When switch is changed to position (2), the capacitor (b) and (c) are in parallel and (a) is in series.

Now charge on capacitor (a) is $q = \dfrac{C(C+C_0)E}{2C+C_0}$

and charge on capacitor (c) is $= VC = \dfrac{EC^2}{2C+C_0}$

Thus extra charge on (a) or (c) is

$$\Delta Q = \frac{EC(C+C_0)}{2C+C_0} - \frac{EC^2}{2C+C_0} = \frac{EC_0C}{2C+C_0}$$

The energy stored in the system remains unchanged.
Thus heat produced = Energy drawn from battery.

$$H = \Delta Q \cdot E = \frac{E^2C_0C}{2C+C_0}$$

Substituting the values, we have

$$H = \frac{(12)^2(2\times10^{-6})(1\times10^{-6})}{2\times1\times10^{-6}+2\times10^{-6}}$$

$$= 72 \times 10^{-6}\ \text{J}$$

Example 19.27: The space between the rectangular plates (with sides a and b) of a parallel plate capacitor (the separation between plates is l) is filled with a solid dielectric whose dielectric constant is ε. The capacitor is charged to a potential V_0 and disconnected from voltage source. After that the dielectric is slowly removed out of the capacitor, which means that the section x not filled with the dielectric gradually increases in size. How will the potential difference between the plates and the surface charge densities on both parts of the capacitor (with and without the dielectric) change in the process?

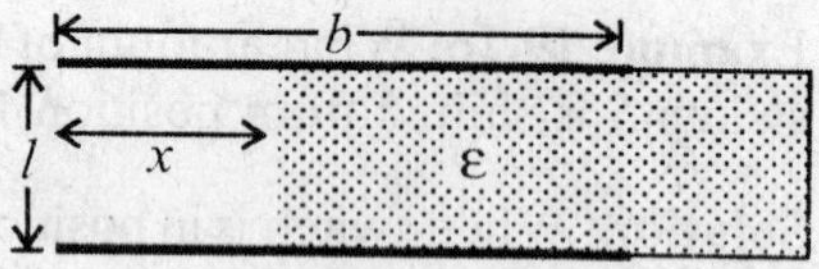

Fig 19.45

Solution: Initially the capacitance of the capacitor completely filled with dielectric ε is

$$C = \frac{\varepsilon_0 \varepsilon\, ab}{l}$$

If Q is the charge given and potential difference V_0 then

$$Q = CV_0 = \frac{\varepsilon_0 \varepsilon\, ab}{l} V_0 \qquad \ldots(i)$$

When dielectric is moved out by a distance x, the capacity now becomes,

$$C' = \frac{\varepsilon_0\, a\left[x + \varepsilon(b - x)\right]}{l}$$

Since the total charge remains conserved and hence potential now is

$$V = \frac{Q}{C'} = \frac{\varepsilon_0 \varepsilon\, ab\, V_0}{l} \times \frac{l}{\varepsilon_0\, a\left[x + \varepsilon(b - a)\right]}$$

$$= \frac{\varepsilon\, bV_0}{\varepsilon\, b - (\varepsilon - 1)x} \qquad \ldots(ii)$$

The field strength will also increase in the same proportion. The charge density in the part without the dielectric is

$$\sigma_1 = \varepsilon_0 E = \frac{\varepsilon_0 V}{l} = \frac{\varepsilon_0 \varepsilon\, b\, V_0}{\{\varepsilon\, b - (\varepsilon - 1)x\}l} = \frac{Q}{a\{cb - (\varepsilon - 1)x\}}$$

While on the part with the dielectric is

$$\sigma_2 = \varepsilon_0 \varepsilon E = \frac{\varepsilon_0\, \varepsilon^2\, bV_0}{l\{(\varepsilon b - (\varepsilon - 1)x\}} = \frac{\varepsilon Q}{a\{\varepsilon b - (\varepsilon - 1)x\}}$$

PROBLEMS

1. The capacitors A and B, two batteries and a resistor are connected as shown in Fig. 19.46. Find the charge on capacitor A.

   [Ans. 20 μC]

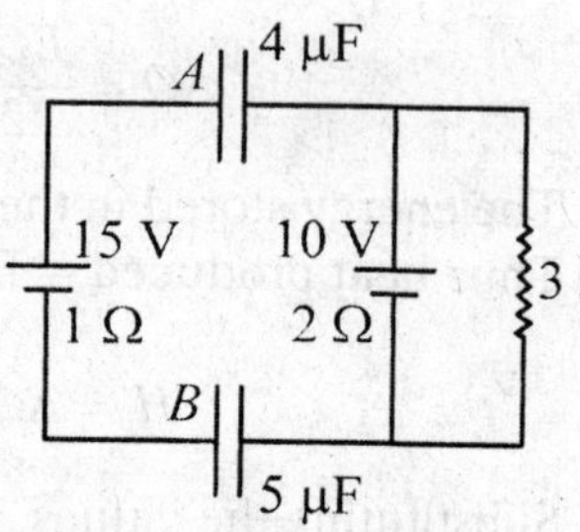

Fig. 19.46

2. A capacitor is made of a flat plate of area A and a second plate having a stair structure as shown in Fig. 19.47. Width of stair is a and height b. Find the capacitance of the assembly.

   $$\left[\textbf{Ans.}\ \frac{\varepsilon_0 A(3d^2 + 6\,bd + 2b^2)}{3d(d + b)(d + 2b)}\right]$$

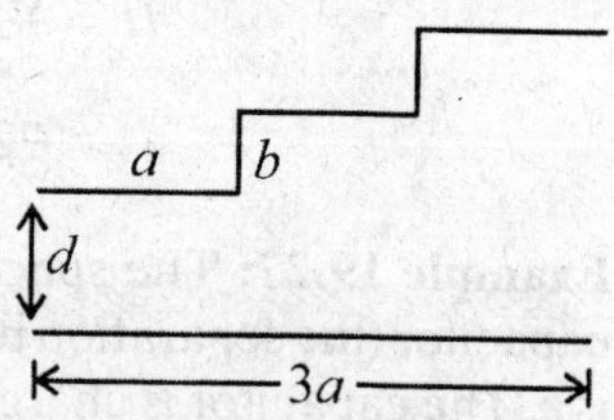

Fig. 19.47

3. The capacity of a variable radio capacitor can be charged from 50 pF to 950 pF by turning the dial from 0° to 180°. With the dial set at 180°, the capacitor is connected to a 400 V battery. After charging, the capacitor is disconnected from the battery and the dial is turned to 0°.

(*a*) What is the potential difference across the capacitor when the dial reads 0°?

(*b*) How much work is required to turn the dial if friction is neglected?

[**Ans.** (*a*) $V = 7600$ V, (*b*) $W = 1.368 \times 10^{-3}$ J]

4. *N* drops of mercury of equal radii and possessing equal charges combine to form a big drop. Compare the charge; capacitance, potential of bigger drop and energy stored by bigger drop with the corresponding quantities of individual drop.

[**Ans.** $Q = Nq$, $V = N^{2/3}v$; $U = N^{5/3}u$]

5. Find the capacitance of an isolated ball shaped conductor of radius R_1 surrounded by a concentric layer of dielectric constant *K* and outside radius R_2.

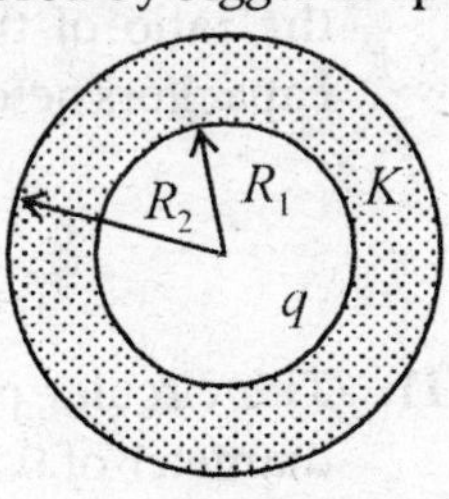

Fig. 19.48

$$\left[\textbf{Ans. } C = \frac{4\pi\varepsilon_0 K R_1 R_2}{(K-1)R_1 + R_2}\right]$$

6. What amount of heat will be generated in the circuit shown in Fig. 19.49 after the switch *S* is shifted from position (1) to position (2).

$$\left[\textbf{Ans. } \frac{1}{2}C_1E_1^2 + \frac{1}{2}C_2E_2^2\right]$$

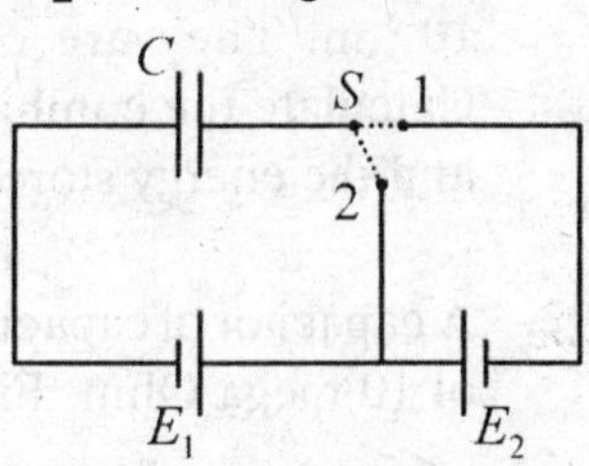

Fig. 19.49

7. The capacitance of all the capacitors shown in Fig. 19.50 are in μF. (*a*) What is the equivalent capacitance between *A* and *B* (*b*) If the charge on 5 μF capacitor is 120 μC, what is the potential difference between *A* and *B*?

[**Ans.** (*a*) $C = 5$ μF, (*b*) $V = 64$ V]

8. A parallel plate capacitor of plate area 0.2 m² and separation 10^{-2} m is charged to 10^3 V and is then disconnected from the battery. How much work is required if the plates are pulled apart to double the plate spacing? Calculate the final voltage of the capacitor.

[**Ans.** $W = 8.8 \times 10^{-5}$ J, $V = 2 \times 10^3$ V]

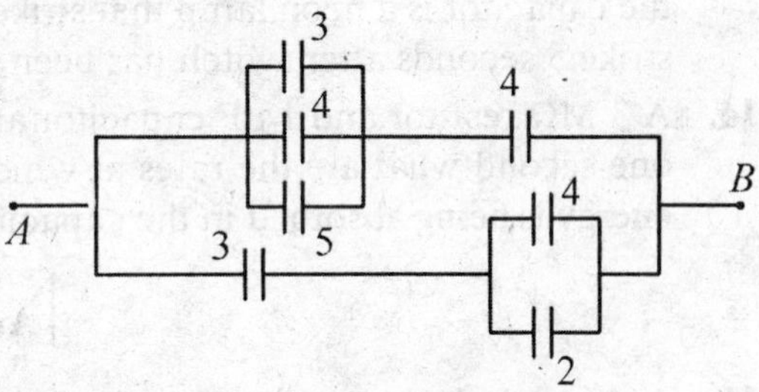

Fig. 19.50

9. (*i*) A 1 μF capacitor and a 2 μF capacitor are connected in series across a 100 V supply line. Find the charge on each capacitor and voltage across each.

(*ii*) The charged capacitors are disconnected from the line and reconnected with each other, with their positive plates together and their negative plates together. No external voltage being applied. Find the charge and potential difference across each capacitor.

(*iii*) The charged capacitors of (*i*) are disconnected from the line and reconnected with plates of opposite polarity connected together. No external voltage being applied. Find the charge and potential difference across each capacitor.

(*iv*) The charged capacitor of case (*ii*) are disconnected and reconnected with plates of opposite polarity together. Find the charge and potential difference across each capacitor.

$$\left[\textbf{Ans. } (i)\ q = 200\ \mu\text{C},\ V_1 = 200\ \text{V},\ V_2 = 100\ \text{V}\ (ii)\ q_1 = \frac{400}{3}\mu\text{C},\ q_2 = \frac{800}{3}\mu\text{C},\ V = \frac{400}{3}\text{V}\right.$$

$$\left[(iii)\ q_1 = q_2 = 0,\ V = 0\ (iv)\ V = \frac{400}{9}\ \text{V},\ q_1 = \frac{400}{9}\ \mu\text{C},\ q_2 = \frac{800}{9}\ \mu\text{C}\right]$$

10. Figure 19.51 shows two identical capacitors connected with a battery and the switch S which is initially closed. The switch is now opened and the free space between the plates of the capacitors is filled with a dielectric of constant $K = 3$. Find the ratio of the total electrostatic energy stored in both capacitors before and after the introduction of the dielectric.

$$\left[\textbf{Ans. } \frac{U_i}{U_f} = \frac{3}{5}\right]$$

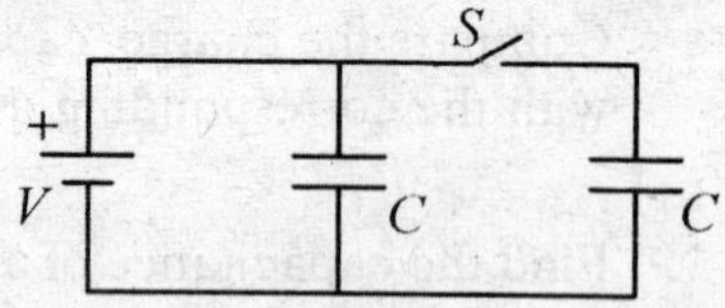

Fig. 19.51

11. The circular plates of a parallel plate capacitor C_1 have diameter of 0.1 m and are 2×10^{-3} m apart. The plates of capacitor C_2 have a diameter of 0.12 m and separation 3×10^{-3} m. They are connected as shown in Fig. 19.52. Calculate the combined capacitance of the arrangement and the energy stored in it.

[**Ans.** $C = 17$ PF, $U = 1.224 \times 10^{-7}$J]

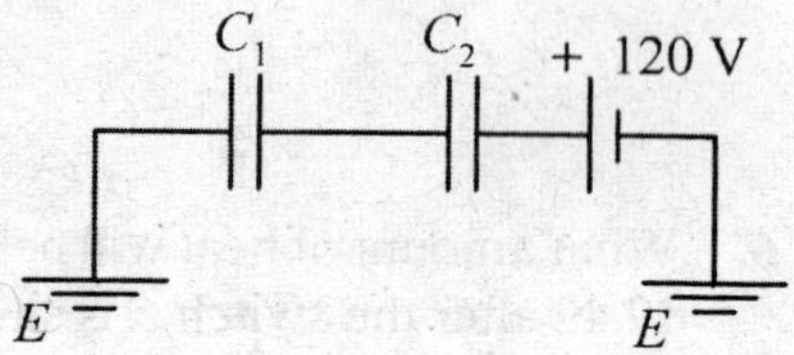

Fig. 19.52

12. A capacitor of capacity 0.1 micro Farad is first charged and then discharged through a resistance of 10 mega Ohm. Find the time in which the potential will fall of half its original value.

[**Ans.** $t = \log_e 25$]

13. A resistance R and a 2 μF condenser in series are connected to a 200 V direct supply. Across the capacitor is a neon lamp that strikes at 120 V. Calculate the value of R to make the lamp strike 5 seconds after switch has been closed. [**Ans.** $R = 2.729$ MΩ]

14. A 3 MΩ resistor and 1 μF capacitor are connected in series with a battery of 4 V emf. After one second what are the rates at which (*i*) the charge on the capacitor is increasing (*ii*) the energy is being absorbed in the capacitor.

$$\left[\textbf{Ans. } \frac{dq}{dt} = 9.75 \times 10^{-7}\ \text{Coulomb/S},\ \frac{dE}{dt} = 1.1 \times 10^{-6}\ \text{W}\right]$$

15. A parallel plate capacitor is to be constructed using rubber as a dielectric, having dielectric constant ($K = 3$) and a dielectric strength of 2×10^7 V/m. The capacitor is to have a capacitance of 0.15 μF and must be able to withstand potential difference of 6000 V. What is the maximum area of the plates of capacitor? [**Ans.** 1.69 m^2]

16. A dielectric slab of dielectric constant ($K = 7$) fills half of space between the plates of a parallel plate condenser of area 100 cm^2. Separated by a distance of 1.0 cm and charged to a potential of 100 V. What percentage of energy is stored in the air gap? [**Ans.** 87.5%]

17. The paper dielectric in a paper and tin foil capacitor is 0.005 cm thick. If dielectric constant is 2.5 and dielectric strength is 5×10^7 V/m

(*a*) what area of paper and of tin foil is required for 0.1 μF capacitor?

(*b*) if the electric field in the paper is not to exceed one-half the dielectric strength, what is the maximum potential difference that can be applied across the capacitor?

[**Ans.** (*a*) 0.226 m^2 (*b*) 1250 V]

18. Two parallel plate capacitors A and B having capacities of 1 μF and 5 μF are charged separately to the same potential of 100 V. Now a positive plate of A is connected to the negative plate of B and the negative plate of A to the positive plate of B. Find the final charges on each condenser and the total loss of electrical energy in condensers.

$$\left[\textbf{Ans. } q_1 = \frac{200}{3}\,\mu\text{C},\ q_2 = \frac{1000}{3}\,\mu\text{C},\ \Delta U = 1.62\times10^{-2}\,\text{J}\right]$$

19. Find the equivalent capacitance of a given network of capacitors between A and B (Fig. 19.53). Each capacitor has capacitance $C = 1$ μF.

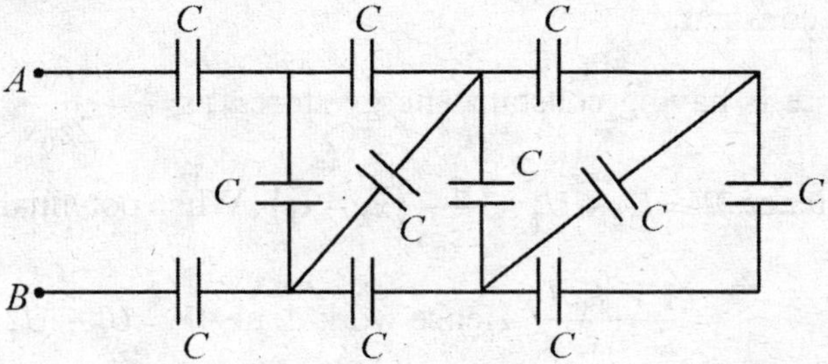

Fig. 19.53

$$\left[\textbf{Ans. } \frac{155}{144} = 0.38\,\mu\text{F}\right]$$

20. Twelve capacitors each having a capacitance C are connected to form a cube (Fig. 19.54). Find the equivalent capacity between

(*a*) diagonally opposite corners AF $\left[\textbf{Ans. } \frac{6}{5}C\right]$

(*b*) Corners AB $\left[\textbf{Ans. } \frac{12}{7}C\right]$

(*c*) Corners AG $\left[\textbf{Ans. } \frac{4}{3}C\right]$

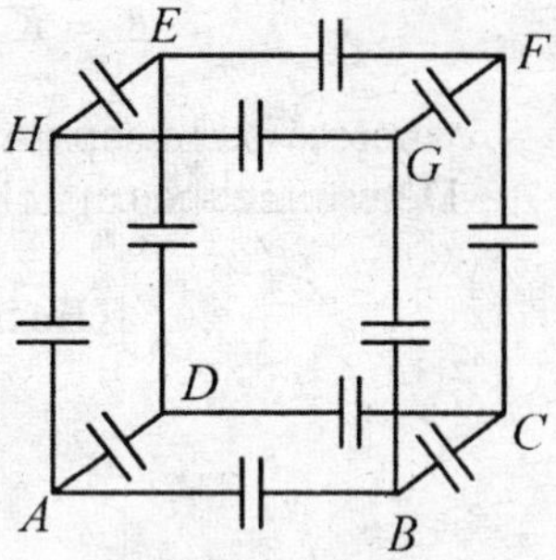

Fig. 19.54

21. In Fig. 19.55, find the value of capacitance C if the equivalent capacitance between points A and B to be 1 μF.

[**Ans.** 32/25 μF]

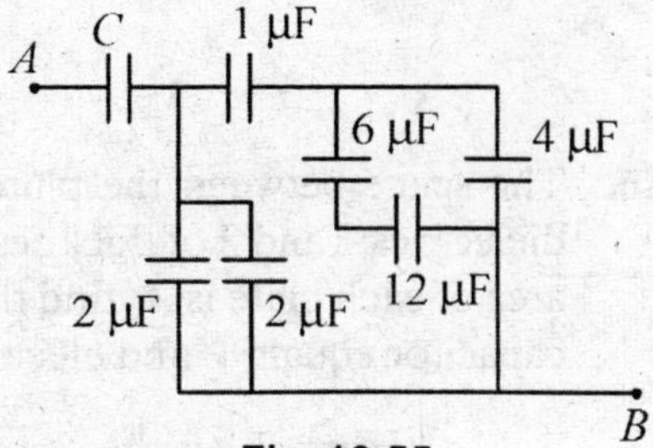

Fig. 19.55

22. Find the equivalent capacitance between P and Q in Fig. 19.56

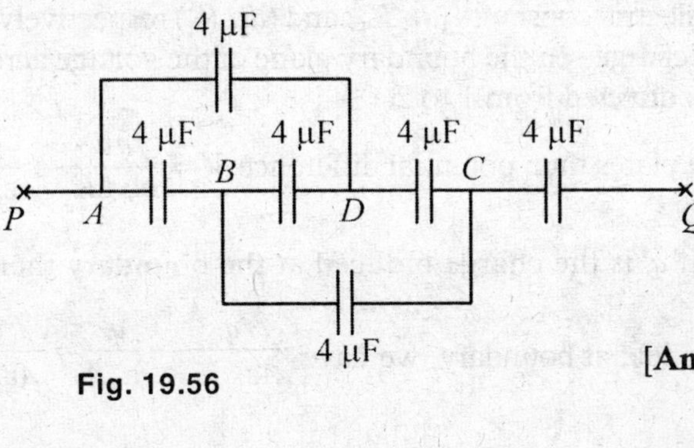

Fig. 19.56

[**Ans.** 2 μF]

23. Find the capacitance of a system of three parallel plates, each of area A separated by d_1 and d_2. The space between them is filled with dielectric of dielectric constant K_1 and K_2.

$$\left[\textbf{Ans. } \frac{\varepsilon_0 A}{\frac{d_1}{K_1}+\frac{d_2}{K_2}}\right]$$

24. Find the work done by an agent against the electric force to increase the distance between the plates of a parallel plate capacitor of area s from x_1 to x_2, if

(*a*) the charge of the capacitor is constant

(*b*) its voltage V is constant.

[**Hint**: When charge is having constant energy stored as $U = \frac{q^2 x}{2\varepsilon_0 s}$, i.e. energy is directly proportional to x. Hence $W = U_2 - U_1 = \frac{q^2}{2\varepsilon_0 s}(x_2 - x_1)$. When potential is constant $U = \frac{1}{2}CV^2$

$= \frac{1}{2}\frac{V^2 \varepsilon_0 A}{x}$. Hence work done $W = U_2 - U_1 = \frac{1}{2}V^2\varepsilon_0 A\left(\frac{1}{x_2} - \frac{1}{x_1}\right)$]

25. The gap between the plates of a plane parallel plate capacitor is filled with an isotropic insulator whose dielectric constant varies in the direction perpendicular to the plates according to the law

$$K = K_1\left(1 + \sin\frac{\pi}{d}x\right)$$

where d is the separation between the plates and K_1 is constant. The area of the plates is A. Determine the capacitance of the capacitor.

[**Hint:** Elementary capacitor at x of separation dx is $dC = \frac{\varepsilon_0 KA}{dx}$ $\therefore \frac{1}{C} = \int \frac{1}{dC}$

$$= \frac{1}{\varepsilon_0 A}\int \frac{dx}{K_1\left(1+\sin\frac{\pi x}{d}\right)} = \frac{1}{\varepsilon_0 A K_1}\int_0^d \frac{\sec^2\frac{\pi x}{2d}\,dx}{\left(1+\tan\frac{\pi x}{2d}\right)^2} = \frac{2d}{\pi}\frac{1}{\varepsilon_0 A K_1}$$

and $C = \frac{\pi\varepsilon_0 A K_1}{2d}$]

26. The space between the plates of a parallel plate capacitor is filled consecutively with two dielectrics 1 and 2 of thickness and dielectric constants (d_1, K_1) and (d_2, K_2) respectively. If the area of each plate is A_1 find the bound charges on the boundary plane of the voltage across the capacitor equals V and electric field is directed from 1 to 2.

[**Hint:** If q is the charge on the plates then potential difference $V = \frac{q d_1}{\varepsilon_0 A K_1} + \frac{q d_2}{\varepsilon_0 A K_2}$

$= \frac{q}{\varepsilon_0 A}\frac{(d_1 K_2 + d_2 K_1)}{K_1 K_2}$. If q' is the charge induced at the boundary then from continuity of electric field at boundary, we have $\frac{q}{A\varepsilon_0 K_2} - \frac{q'}{\varepsilon_0 A} = \frac{q}{A\varepsilon_0 K_1}$ or

$$q' = q\left(\frac{1}{K_2} - \frac{1}{K_1}\right) = q\frac{(K_1 - K_2)}{K_1K_2} \text{ or } q' = \frac{V\varepsilon_0 A K_1 K_2}{K_1 d_2 + d_1 K_2} \times \frac{K_1 - K_2}{K_1 K_2} = \frac{V\varepsilon_0 A(K_1 - K_2)}{K_1 d_2 + d_1 K_2}\Bigg]$$

27. A parallel plate capacitor is filled with a dielectric whose permittivity varies with applied voltage according to the law $K = aV$. The same (but containing no dielectric) capacitor charged to voltage $V_0 = 156$ V is connected in parallel to the first nonlinear uncharged capacitor. Find the common potential (Given $a = 1\ \text{V}^{-1}$).

[**Hint:** When voltage is V, capacity is $CK = CaV$, where C is capacity without dielectric. Combined capacity $= C(1 + aV)$ and charge $q = CV(1 + aV) =$ charge initial $= CV_0$ or

$$aV^2 + V - V_0 = 0 \text{ or } V = -\frac{1 \pm \sqrt{4aV_0 + 1}}{2a} = 12\text{ V}\Bigg]$$

28. Four identical plates are arranged as shown in Fig. 19.57. Area of each plate is s and separation between neighbouring plates is d. The outer plates A and B are connected together. Find the capacitance of this system between points 1 and 2.

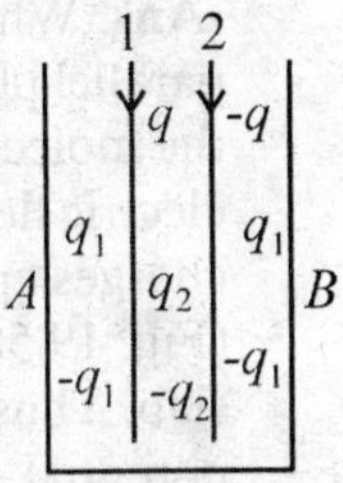

Fig. 19.57

[**Hint:** Potential difference between plates A and B is zero, i.e. $-\frac{q_1}{C_0} + \frac{q_2}{C_0} - \frac{q_1}{C_0} = 0$

where $C_0 = \frac{\varepsilon_0 s}{d}$. Potential difference between plates 1 and 2 $V = \frac{q_2}{C_0}$. Hence

$$\text{capacity } C = \frac{q}{V} = \frac{q_1 + q_2}{q_2/C_0} = \frac{3}{2}C_0 = \frac{3}{2}\frac{\varepsilon_0 s}{d}\Bigg]$$

29. A cylindrical layer of a homogenous dielectric constant K is introduced into a cylindrical capacitor so that the layer fills the gap of width d between the plates. The mean radius of the plates is R such that $R >> d$. The capacitor is connected to a source of permanent voltage V. Find the force pulling the dielectric inside the capacitor.

$$\left[\textbf{Ans. } F = \frac{\varepsilon_0 (K-1)\pi R V^2}{d}\right]$$

30. Find the work done against the electric force in order to remove a dielectric plate from a parallel charged capacitor. It is assumed that the charge q of the capacitor remains unchanged and that the dielectric fills the entire space between the capacitor plates. The capacitance of the capacitor in absence of dielectric is C.

$$\left[\textbf{Ans. } W = \frac{q^2}{2C}\left(1 - \frac{1}{K}\right)\right]$$

SHORT ANSWER TYPE QUESTIONS

1. A capacitor of capacitance C is charged by connecting it to a battery of emf E. The capacitor is now disconnected and connected to the battery with the polarity reversed. Calculate the heat developed in the connecting wires.

[**Ans.** When a capacitor is connected to the battery it gets charge $Q = CE$ on one plate and on other $-Q(= -EC)$ appears. When the polarity is reversed, a charge $-Q$ appears on the plate with $+Q$ charge initially and second plate get $+Q$. Thus the battery supplies a charge $2Q$ and battery does work $W = (2Q)E = 2CE^2$ in the process. The energy stored in the capacitor is same and the work done by the battery appears as heat in the connecting wires. The heat produced is therefore $2CE^2$]

2. Can we give any desired charge to a conducting sphere?

[**Ans**. No. When charge on sphere is increased, the electric field near it increases $\left(E = \frac{Kq}{a^2}\right)$. At a certain stage dielectric strength of medium (air strength ~10^7 V/m) will break. This will result in a spark and the charge will leak to earth]

3. Explain why the introduction of a dielectric between the plates of a capacitor changes its capacity?

[**Ans**. When a dielectric slab is placed between the plates of a parallel plate capacitor, the slab gets electrically polarised and the molecular dipoles tends to orient themselves parallel to the electric field between the plates. The net effect is that negative charges appear on one face and positive charges on other face (Fig. 19.58). This causes an opposing electric field within the slab. Thus net field within the slab reduces and consequently potential difference between plates reduces. Thus increasing the capacity.]

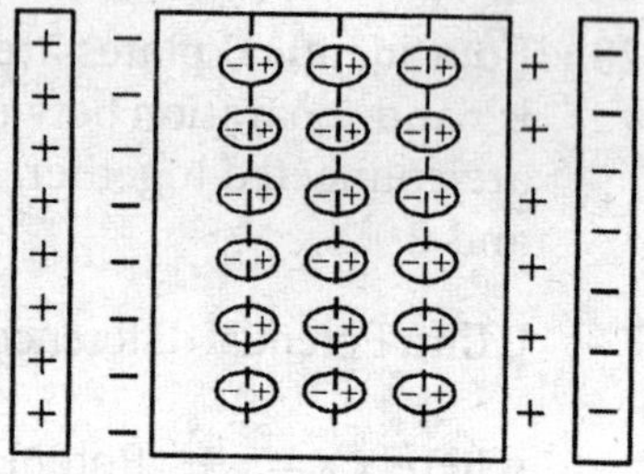

Fig. 19.58

4. Why should circuits containing capacitor be handled cautiously even when there is no current?

[**Ans**. When the circuit containing capacitor is connected to a source of high voltage, the capacitor gets charged to a high potential. On removing the source of voltage, the capacitor does not itself discharge. So handling the charged capacitor without due precautions may cause to a severe shock to handle.]

5. Five identical capacitor plates each of area A are arranged such that adjacent plates are at a distance d apart. The plates are connected to a source of emf V as shown in Fig. 19.59. What is magnitude and nature of charge on plate 1 and 4 respectively?

[**Ans**. These five plates form 4 capacitors in parallel and each having capacity $= \frac{\varepsilon_0 A}{d}$. Since potential difference is V as such charge on each capacitor is $= \frac{\varepsilon_0 A}{d} V$. Since plate 1 is connected to positive terminal hence charge on it is $+\frac{\varepsilon_0 AV}{d}$. Plate 4 is connected to negative terminal of the battery and is part of two capacitors. Hence charge on it is $-\frac{2\varepsilon_0 AV}{d}$]

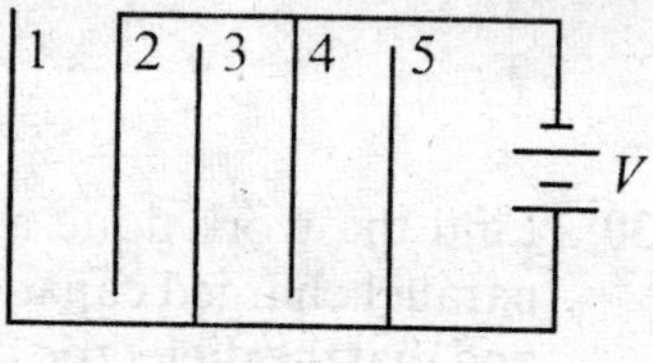

Fig. 19.59

6. A 10 μF capacitor is charged through a resistance of 0.1 MΩ by a battery of 1.5 V. Calculate the time required for capacitor to get charged to 0.75 V for situations as shown in Fig. 19.60 (*a*) and (*b*).

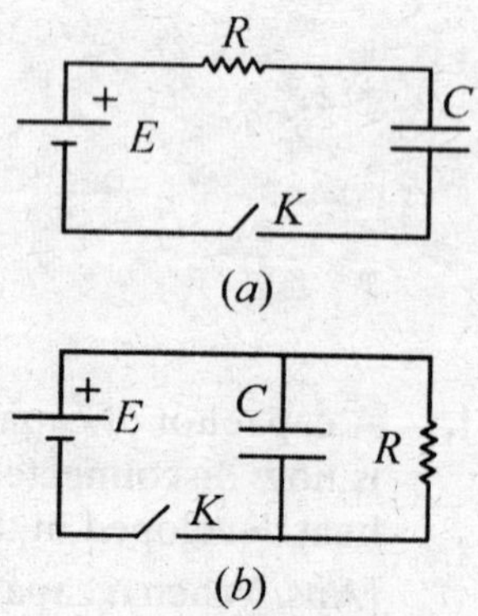

Fig. 19.60

[**Ans**. (*a*) $q = q_0(1 - e^{-t/RC})$ but $\frac{q}{q_0} = \frac{V}{V_0} = \frac{1}{2}(1 - C^{-t})$. Since $RC = 10$

$\times 10^{-6} \times 0.1 \times 10{-}6 = 1$ or $e{-}t = 1/2$ or $t = \log 2 = 0.693$ sec. (*b*) The capacitor is connected directly to the battery as such charging time (time constant) is zero as such it will charge instantaneously, i.e. $t = 0$.]

7. Capacitor C_1 of capacity 1 μF and capacitor C_2 of capacity 2 μF are separately charged fully by a common battery. The two capacitors are separately allowed to discharge through equal resistances. At $t = 0$ (*a*) in which circuit the value of current is more (*b*) in which circuit the capacitor loses 50% of its initial charge sooner?

[**Ans.** (*a*) Current during discharging of capacitor is $i = i_0 e^{-t/RC}$ and $i_0 = \dfrac{E}{R}$. Thus current at $t = 0$ is $i = \dfrac{E}{R}$ independent of the capacity of capacitor. As such current in both will be same.

(*b*) Instantaneous charge on capacitor $q = q_0 e^{-t/RC}$ and $t = CR \log \dfrac{q_0}{q}$, i.e. time depends on the capacity. C_1 looses 50% charge sooner $\dfrac{t_1}{t_2} = \dfrac{C_1}{C_2} = \dfrac{1}{2}$]

8. A leaky parallel plate capacitor is filled completely with a material having dielectric constant $K = 5$ and electrical conductivity $\sigma = 7.4 \times 10^{-12}\ \Omega^{-1}\text{m}^{-1}$. If the charge on plate at instant $t = 0$ is $q = 8.85$ μC, then calculate the leakage current at the instant $t = 125$.

[**Ans.** Leakage resistance $R = \dfrac{\rho d}{A} = \dfrac{d}{\sigma A}$ and time constant τ is $\tau = RC = \dfrac{d}{\sigma A} \cdot \dfrac{\varepsilon_0 KA}{d} = \dfrac{\varepsilon_0 K}{\sigma}$ and

$$\text{leakage current is } i = \frac{q_0}{RC} e^{-t/\tau} = \frac{8.85 \times 10^{-6} \times 7.4 \times 10^{-12}}{8.85 \times 10^{-12} \times 5} e^{-12/6} \left(\because \frac{\varepsilon_0 K}{\sigma} = 6 \right)$$

$$\text{or } i = \frac{7.5 \times 10^{-6}}{5} e^{-2} = 0.25\,\mu\text{A}$$]

9. Two parallel plate capacitors of capacity C and $2C$ are connected in parallel and charged to a potential difference V. The battery is then disconnected and space between the plates of capacitor with capacity C is filled with material of dielectric constant K. What is the final potential difference across the capacitors?

[**Ans.** Initially the equivalent capacity is $3C$ and total charge is $q = 3CV$. When battery is disconnected, the system is isolated and charge is conserved. Now the equivalent capacity after introducing the dielectric is $(CK + 2C) = C(2 + K)$. Hence potential difference across the capacitors is $V' = \dfrac{3CV}{C(2+K)} = \dfrac{3V}{2+K}$]

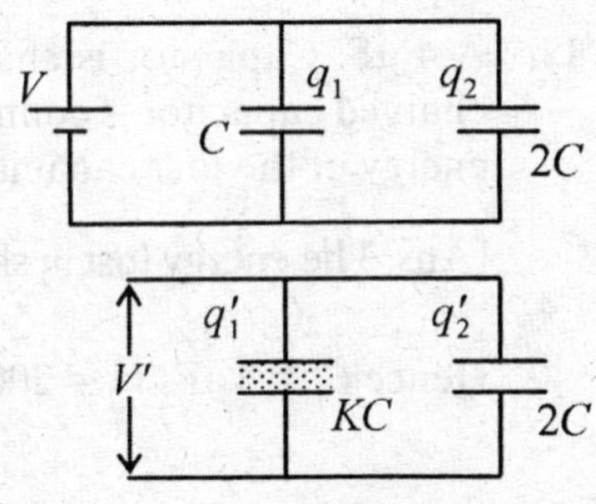

Fig. 19.61

10. A parallel plate capacitor contains a mica sheet (thickness 10^{-3} m) and a sheet of fibre (thickness 0.5×10^{-3} m). The dielectric constant of mica is 8 and that of fibre is 2.5. Assuming that the fibre breaks down when subjected to an electric field 6.4×10^6 V/m. Find the maximum safe voltage that can be applied to the capacitor.

[**Ans.** The maximum potential difference that can be given to fibre capacitor is $V = E \times d = 6.4 \times 10^6 \times 0.5 \times 10^{-3} = 3.2 \times 10^3$ V. Since the given capacitor is equivalent to two capacitors in series as such the charge on the two capacitors, is same. As such the maximum potential difference across mica capacitor is $V_m = \frac{C_f V_f}{C_m} = \left(\frac{K_f}{K_m} \cdot \frac{dm}{df}\right) V_f = \frac{2.5}{8} \times \frac{10^{-3}}{0.5 \times 10^{-3}} \times 3.2 \times 10^3 = 2 \times 10^3$ V.

Hence total safe voltage is $V = 3.2 \times 10^3 + 2 \times 10^3 = 5.2 \times 10^3$ V = 5.5 kV.]

11. Two parallel wires each having a radius R are separated by distance d (Fig. 19.62). Find the expression for the capacity per unit length of the system assuming $d >> R$.

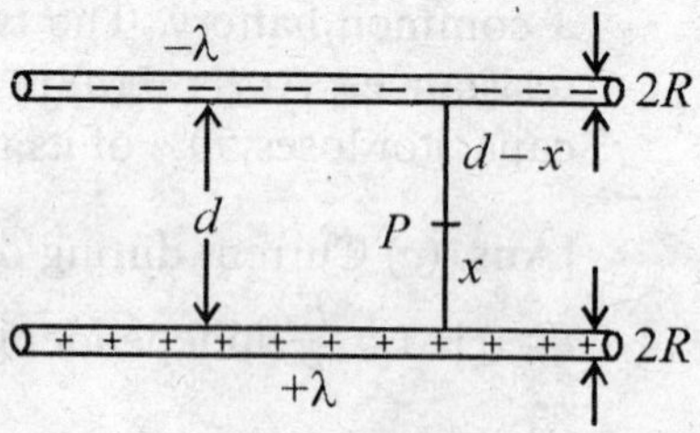

Fig. 19.62

[**Ans.** Electric field at P (Fig. 19.62). is $E = \frac{\lambda}{2\pi\varepsilon_0}\left(\frac{1}{x} + \frac{1}{d - x}\right)$

and $\frac{dV}{dx} = -E = -\frac{\lambda}{2\pi\varepsilon_0}\left(\frac{1}{x} + \frac{1}{(d - x)}\right)$

or $dV = -\frac{\lambda}{2\pi\varepsilon_0}\left(\frac{1}{x} + \frac{1}{d - x}\right)dx$

Integration gives

$$\int_0^V dV = -\frac{\lambda}{2\pi\varepsilon_0}\int_{d-R}^{R}\left(\frac{1}{x} + \frac{1}{(d - x)}\right)dx$$

or $V = -\frac{\lambda}{2\pi\varepsilon_0}\left[\log x - \log(d - x)\right]_{d-R}^{R}$

$$= \frac{\lambda}{2\pi\varepsilon_0}\left[\log\frac{d - x}{x}\right]_{d-R}^{R} = \frac{\lambda}{2\pi\varepsilon_0}\left[\log\frac{d - R}{R} - \log\frac{R}{d - R}\right]$$

$= \frac{\lambda}{\pi\varepsilon_0}\log\frac{d - R}{R} \approx \frac{\lambda}{\pi\varepsilon_0}\log d/R$ as $d >> R$. Hence capacity $C = \frac{\lambda}{V} = \frac{\pi\varepsilon_0}{\log d/R}$]

12. A 4 μF. Capacitor is charged by a 200 V supply. The supply is then disconnected and the charged capacitor is connected to another uncharged 2 μF capacitor. How much electrostatic energy of the first capacitor is lost in the process of attaining the steady situation?

[**Ans.** The energy lost in sharing of charges between the capacitors is $\Delta E = \frac{1}{2}\frac{C_1 C_2}{C_1 + C_2}(V_1 - V_2)^2$

Hence $C_1 = 4$ μF, $V_1 = 200$ V, $C_2 = 2$ μF, $V_2 = 0$ ∴ $\Delta E = \frac{1}{2} \times \frac{4 \times 2}{4 + 2}(200 - 0)^2 = \frac{4 \times 2 \times 4 \times 10^4}{2 \times 6}$

$= \frac{8}{3} \times 10^4 = 2.67 \times 10^4$ μJ $= 0.0267$ J]

13. A parallel plate capacitor is to be designed with a voltage rating 1 kV using a dielectric constant 3 and dielectric strength about 10^7 V m^{-1}. For safety, we would like the field never to exceed say, 10% of the dipole strength. What minimum area of the plates is required to have a capacitance of 50 PF?

[**Ans.** The safe field in the capacitor is 10^6 Vm^{-1}. Potential difference between plates is 10^3 V. Hence d is the separation between the plates is $E = \frac{V}{d}$ or $d = \frac{V}{E} = \frac{10^3}{10^6} = 10^{-3}$ m. Capacity of a parallel plate capacitor is $C = \frac{\varepsilon_0 KA}{d}$ or $A = \frac{Cd}{K\varepsilon_0} = \frac{50 \times 10^{-12} \times 10^{-3}}{3 \times 8.85 \times 10^{-12}} = 1.88 \times 10^{-3}$ m^2]

14. A spherical capacitor is made of two conducting spherical shells of radii a and b. The space between the shells is filled with a dielectric constant K up to a radius c as shown in Fig. 19.63, calculate the capacitance.

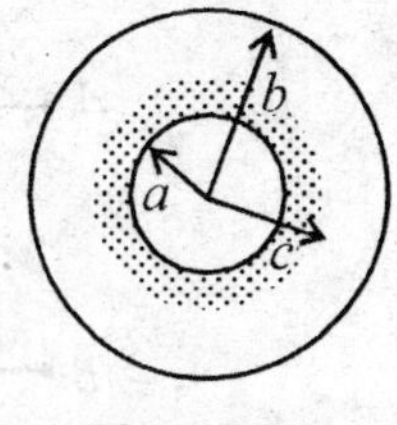

Fig. 19.63

[**Ans.** The system is equivalent to two spherical capacitors in series with capacities $C_1 = \dfrac{4\pi\varepsilon_0 K\, ac}{c-a}$ and $C_2 = \dfrac{4\pi\varepsilon_0\, bc}{b-c}$. Hence equivalent capacity

$$C_s \text{ is } \frac{1}{C_S} = \frac{1}{C_1} + \frac{1}{C_2} = \frac{(c-a)}{4\pi\varepsilon_0 K\, ac} + \frac{b-c}{4\pi\varepsilon_0\, bc} = \frac{1}{4\pi\varepsilon_0 c}\left[\frac{c-a}{aK} + \frac{b-c}{b}\right]$$

$$= \frac{1}{4\pi\varepsilon_0 c}\left\{\frac{b(c-a) + Ka(b-c)}{a\,Kb}\right\} \text{ or } C_S = \frac{4\pi\varepsilon_0\, ab\,cK}{Ka\,(b-c) + b\,(c-a)}$$]

15. Consider the situation shown in Fig. 19.64. The width of each plate is b and separation d. The capacitor plates are rigidly clamped in the laboratory and connected to a battery of emf E. All surfaces are frictionless. Calculate the value of M for which the dielectric slab will stay in equilibrium.

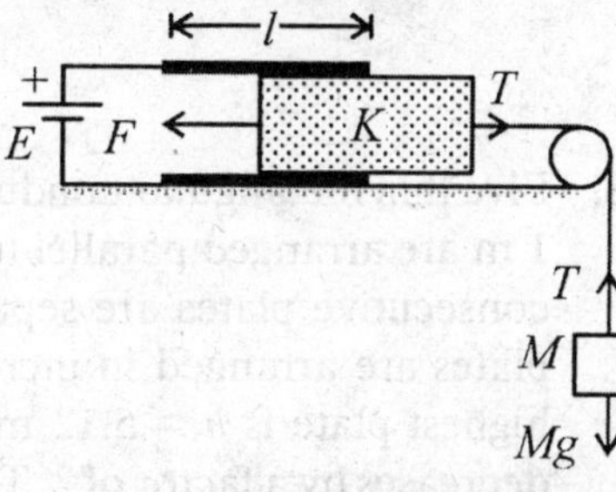

Fig. 19.64

[**Ans.** The dielectric in the plates experiences the force F, where $F = \dfrac{\varepsilon_0\, b E^2 (K-1)}{2d}$ (Example 19.1). For equilibrium of the slab $F = T = Mg$ or $M = \dfrac{F}{g} = \dfrac{\varepsilon_0\, bE^2(K-1)}{2dg}$]

16. Figure 19.65 shows two parallel plates capacitors with fixed plates and connected to two batteries. The separation between the plates is the same for two capacitors. The plates are rectangular in shape with width b and lengths l_1 and l_2. The left half of the dielectric slab has a dielectric constant K_1 and the right half K_2. Neglecting any friction, find the ratio of the emf of the left battery to that of right battery for which the dielectric slab may remain in equilibrium.

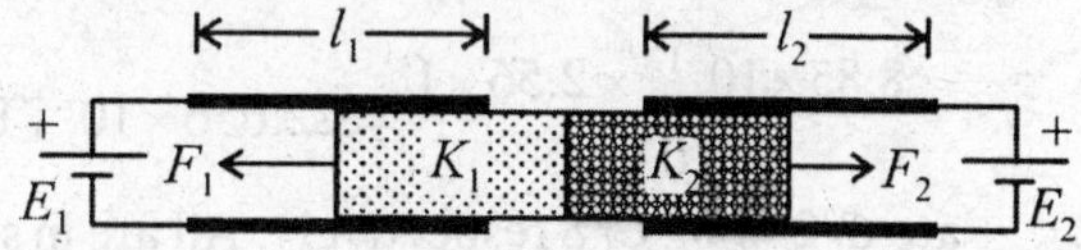

Fig. 19.65

[**Ans.** For equilibrium the force F_1 on dielectric slab K_1 should be equal in magnitude and opposite in direction to the force F_2 on dielectric slab K_2. Force $F_1 = \dfrac{\varepsilon_0\, bE_1^2(K_1-1)}{2d}$ and $F_2 = \dfrac{\varepsilon_0 bE_2^2(K_2-1)}{2d}$. Equating the two forces, we have $\dfrac{\varepsilon_0\, bE_1^2(K_1-1)}{2d} = \dfrac{\varepsilon_0\, bE_2^2(K_2-1)}{2d}$

or $\dfrac{E_1}{E_2} = \sqrt{\dfrac{K_2-1}{K_1-1}}$]

17. Find the equivalent capacitance of the infinite lader shown in Fig. 19.66 between the points A and B. Given $C_1 = 2\ \mu\text{F}$ and $C_2 = 1\ \mu\text{F}$.

[**Ans.** Since lader is infinite as such the capacity between PQ will be same as the capacity between AB. Let it be C (Fig. 19.66 (*a*)). The equivalent circuit is shown in Fig. 19.66 (*b*).

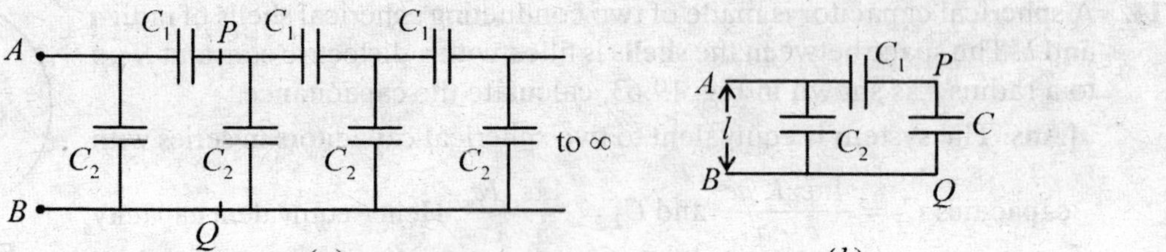

Fig. 19.66

Hence $\frac{C_1C}{C_1+C}+C_2=C$ or $C_1C+C_1C_2+CC_2=C_1C+C^2$ or $C^2-CC_2-C_1C_2=0$

or $C=\frac{C_2\pm\sqrt{C_2^2+4C_1C_2}}{2}$. Negative sign is not admissible. Hence

$$C=\frac{C_2}{2}\left[1+\sqrt{1+\frac{4C_1}{C_2}}\right]=\frac{1}{2}\left[1+\sqrt{1+8}\right]=2\,\mu\text{F}]$$

18. Five thin rectangular conducting plates with each of constant breadth 1 m are arranged parallel to each other as shown in Fig. 19.67. Two consecutive plates are separated by equal distance d = 1 mm. The plates are arranged in increasing order of height. The height of the highest plate is h = 5.12 m and the height of each successive plate decreases by a factor of 2. Then find the equivalent capacitance of the system between the first and the last plate, assuming that air is present in between the plates.

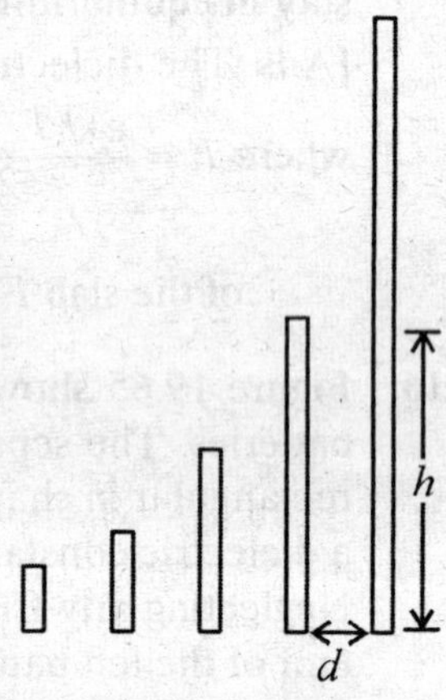

Fig. 19.67

[**Ans.** The first capacitor on right extreme has the capacity $C=\frac{\varepsilon_0 A}{d}$

$=\frac{8.85\times10^{-12}\times2.56\times1}{10^{-3}}=22.656\times10^{-9}$ F. The capacities of others are $C/2$, $C/4$, $C/8$ respectively. All are in series. Equivalent capacity is

$$C_S=\frac{C}{15}=1.5\times10^{-9}\text{ F}]$$

19. In the circuit shown in Fig. 19.68 switch is turned on at t = 0, then

(*i*) find the current supplied by the battery at t = 0

(*ii*) find the current through R_3 at t = 0

(*iii*) find the current supplied by the battery in steady state.

(*iv*) find the steady state current in R_3.

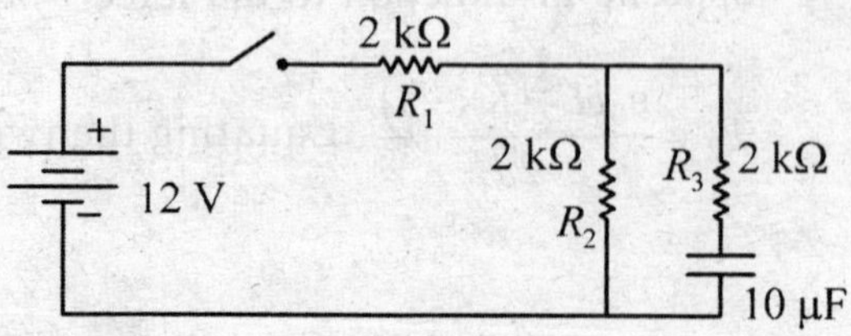

Fig. 19.68

[**Ans.** (*i*) At t = 0 the capacitor acts as short circuit. Hence total resistance is 3 kΩ and current in circuit is 4 mA. (*ii*) Hence current through R_3 = 2 mA. (*iii*) In steady state, the capacitor acts as open circuit. Hence total resistance is 4 kΩ and current in circuit is 3 mA and (*iv*) current through R_3 = 0.]

20. In circuit shown in Fig. 19.69, $R_1 = R_2 = 10\ k\Omega$, $C_1 = 10$ pF and $C_2 = 20$ pF. The switch is turned on at $t = 0$. Calculate the output voltage V_o at the moment $t = 0$ and after a long time.

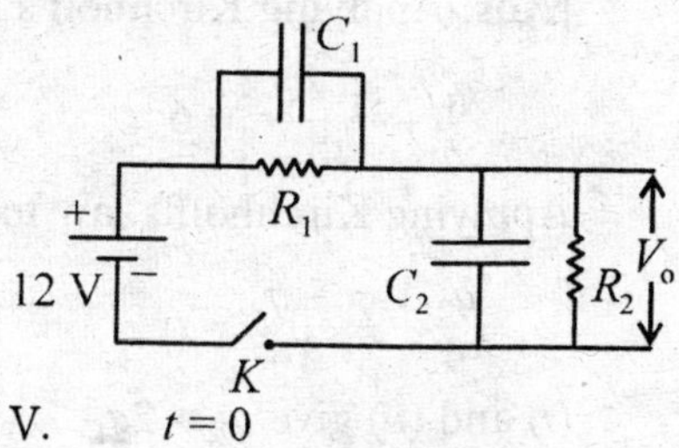

Fig. 19.69

[**Ans**. When switch is turned on, the capacitors have very low effective resistance as such the shunting effect of R_1 and R_2 can be neglected. Hence $V_o(0) = \frac{C_1}{C_1 + C_2} V = \frac{10}{30} \times 10 = 4$ V.

After a long time $t \to \infty$, each capacitor behaves like an open circuit and output is decided by resistors. Hence $V_o(\infty) = \frac{R_2}{R_1 + R_2} \times V = \frac{10}{20} \times 12 = 6$ V]

21. For the parallel plate system shown in Fig. 19.70 deduce the electric field in medium. Given $V = 200$ V and dielectric constant K of medium = 5.

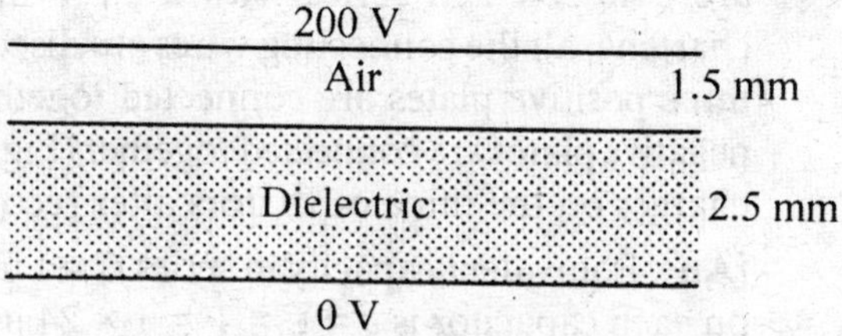

Fig. 19.70

[**Ans**. If electric field in air is E then electric field in medium $E' = \frac{E}{K}$. Let t_1 be air thickness and t_2 that of medium then potential difference

$$V = Et_1 + E't_2 \therefore V = Et_1 + \frac{E}{K}t_2 \text{ or } 200 = E\left(1.5 + \frac{2.5}{2}\right) \times 10^{-3} \text{ or } E = \frac{200}{2.0 \times 10^{-3}} = 10^5 \text{ V/m}$$

$$E' = \frac{E}{5} = 2 \times 10^4 \text{ V/m}]$$

22. Figure 19.71 shows a battery of emf 120 V connected through a 4.0 MΩ resistor to a 0.50 micro farad capacitor. A neon bulb N which strikes at 100 V is connected as shown. It shows repetitive glow. Find the period of this glow.

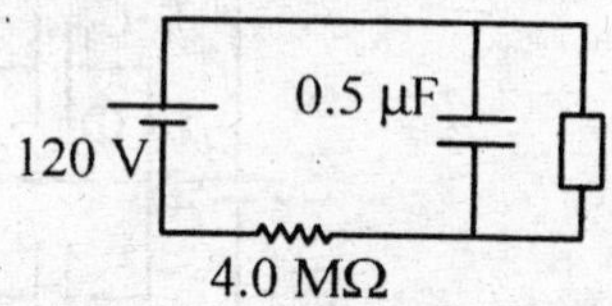

Fig. 19.71

[**Ans**. Voltage variation across capacitor in RC circuit is $V = V_0(1 - e^{-t/RC})$. Here $RC = 4 \times 10^6 \times 0.5 \times 10^{-6} = 2$ sec and $V_0 = 120$ V. Time for voltage to become 100 volts is

$$\frac{100}{120} = 1 - e^{-t/2} \text{ or } e^{-t/2} = \frac{1}{6} \text{ and } t = 2 \log 6 \text{ sec}]$$

23. An isolated charge q is travelling between conducting plates A and B which are connected through galvanometer. Will a current flow through the galvanometer? Justify your answer.

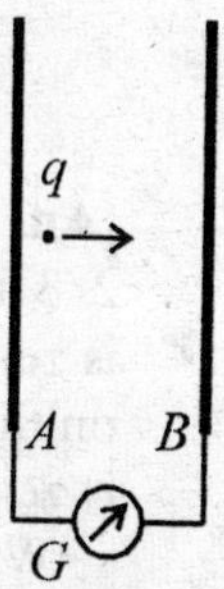

19.72

[**Ans**. Charge q induces negative charges q_1 and q_2 on plates A and B respectively such that $|q| = |q_1| + |q_2|$. When q is near A, $|q_1| > |q_2|$, as it moves towards B, q_2 grows at the expense of q_1, i.e. negative charge flows through the galvanometer from A to B or the current flows from B to A.]

24. Find the charge on the three capacitors shown in Fig. 19.73. Given $C_1 = 2\ \mu$F, $C_2 = 4\ \mu$F and $C_3 = 5\ \mu$F. Each battery has emf 6 V.

[**Ans**. Applying Kirchhoff's law to mesh *ABCDA*, we get

$$\frac{q_1}{2} + \frac{q_1 + q_2}{5} = 6 \quad \ldots(i)$$

Applying Kirchhoff's law to mesh *EFCDE*, we get

$$\frac{q_2}{4} + \frac{q_1 + q_2}{5} = 6 \quad \ldots(ii)$$

(*i*) and (*ii*) give $q_2 = 2q_1$

Solving, we get $q_1 = \frac{60}{11} = 5.45\ \mu C$, $q_2 = 10.9\ \mu C$ and $q_1 + q_2 = 16.35\ \mu C$]

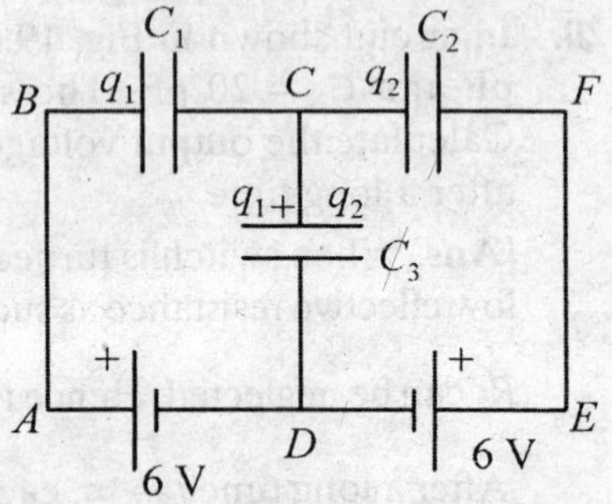

Fig. 19.73

25. Three capacitors of capacitance, 2 μF, 3 μF and 6 μF are connected in series with a 24 V battery. After charging, all the connecting wires are disconnected and three positive plates are connected together and three negative plates are connected together (Fig. 19.74). Find charges on the three capacitors after reconnection.

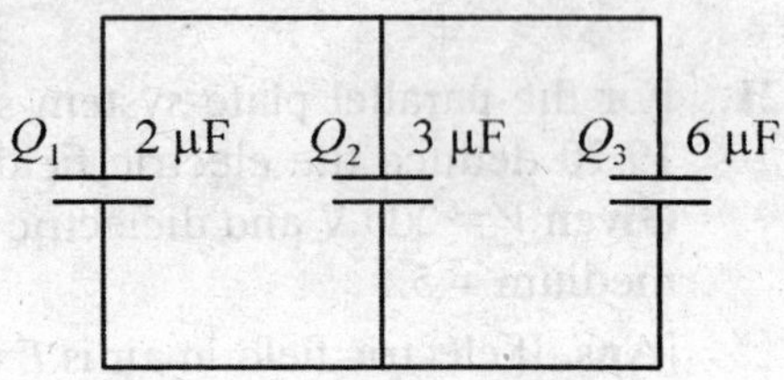

Fig. 19.74

[**Ans.** Equivalent capacity in series $C = 1\ \mu F$ and charge on each capacitor is $q = C \times V = 1 \times 24\ \mu C$. Now they are connected in parallel and equivalent capacity $C = 11\ \mu F$. Hence common potential is $V = \frac{24 \times 3}{11} = \frac{72}{11}$ Volts. Hence charge $q_1 = \frac{72 \times 2}{11} = \frac{144}{11} = 13.1\ \mu C$, $q_2 = \frac{72 \times 3}{11} = 19.6\ \mu C$ and $q_3 = \frac{72}{11} \times 6 = 39.2\ \mu C$]

26. In Fig. 19.75, the connections are established with switch *S* open. Find the charge that will flow through the switch if it is closed. Given $C_1 = 1\ \mu F$, $C_2 = 2\ \mu F$ and $V = 24$ Volts.

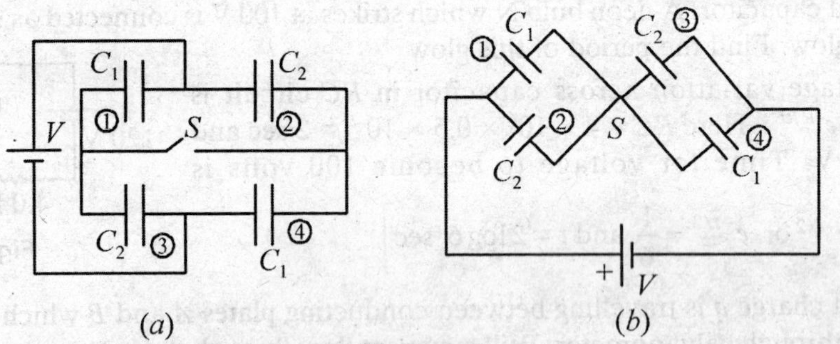

Fig. 19.75

[**Ans**. Before switch *S* is closed, capacitors (1) and (3) are in series having equivalent capacity 2/3 μF and charge is 16 μC. The total charge on two plates connected to the switch therefore is zero. Similarly equivalent capacity of series capacitors (2) and (4) is 2/3 μF. Total charge on these capacitors is also 16 μF. The total charge on the two plates connected to the switch is zero.

When switch is closed, the situation is as shown in Fig. 17.75 (*b*). Capacitors (1) and (2) are in parallel with equivalent capacity 3 μF. Similarly capacitors (3) and (4) are in parallel with equivalent capacity 3 μF. This combination is in series with combination (1) and (2) giving equivalent capacity 1.5 μF and charge on each parallel combination is 24 × 1.5 = 36 μC. Thus charge on capacitor (1) is 24 μC and on (2) is 12 μC, on (3) is 12 μC and on (4) is 24 μC. Before switch was on, total charge on plates of (1) and (3) connected to switch was zero and now it is 24 μC and 12 μC. Hence 12 μC charge flows through the switch.]

27. Find the energy stored in the electric field produced by a metal sphere of radius R containing charge Q.

[**Ans**. Consider a concentric spherical shell at infinity. This forms a spherical capacitor of capacity $4\pi\varepsilon_0 R$. Hence energy stored in this capacitor is the energy stored in electric field. The energy stored in the capacitor is $E = \frac{1}{2}\frac{q^2}{C} = \frac{1}{2}\frac{Q^2}{4\pi\varepsilon_0 R} = \frac{Q^2}{8\pi\varepsilon_0 R}$]

28. A capacitor of capacitance C is charged by connecting it to a battery of emf E. The capacitor is now disconnected and reconnected to the battery with polarity reversed. Calculate the heat developed in the connecting wires.

[**Ans**. When a capacitor is connected to the battery, a charge $q = CE$ appears on one plate and $-q$ on the other. When the polarity is reversed, a charge $-q$ appears on first plate and $+q$ on the second, i.e. a charge $2q$, passes through the battery from negative to the positive terminal. Thus the battery does work equal to $2CE^2$.

In this process, the energy stored in the capacitor is the same in two cases. Thus the work done by the battery appears as heat in the connecting wires. Heat produced in the wires therefore is $2CE^2$]

29. A charge 1 μC is given to one plate of a parallel plate capacitor of capacitance 0.1 μF and a charge of 2 μC is given to the other plate. Find the potential difference developed between the plates.

[**Ans**. When charge q is given to the plate of one side surface area A, the surface density of charge $\sigma = \frac{q}{2A}$ and electric field near it becomes $E = \frac{\sigma}{\varepsilon_0} = \frac{q}{2A\varepsilon_0}$ and potential difference between two plates separated by distance d is $V = Ed = \frac{qd}{2A\varepsilon_0} = \frac{q}{2C}$ where C is capacity of the capacitor. If second plate is given charge q', then potential difference becomes $V' = \frac{q'}{2C}$. If both charges are positive, then net potential difference is $V' - V = \frac{1}{2C}(q' - q)$. In present case $q' = 2$ μC,

$q = 1$ μC, C = 0.1 μF. Hence $V' - V = \frac{1}{2 \times 0.1 \times 10^{-6}}(2-1) \times 10^{-6} = 5$ V]

30. A large conducting plane has a surface charge density 2×10^{-5} cm^{-2}. Find the electrostatic energy stored in a cubical volume of edge 1.0 cm in front of the plate.

[**Ans**. If at a point electric field is E then energy density in the region is $\frac{1}{2}\varepsilon_0 E^2$. Hence energy within volume V will be $= \frac{1}{2}\varepsilon_0 E^2 V = \frac{1}{2}\varepsilon_0\left(\frac{\sigma}{\varepsilon_0}\right)^2 V \left(\because E = \frac{\sigma}{\varepsilon_0}\right) = \frac{1}{2}\frac{\sigma^2}{\varepsilon_0}V = \frac{1}{2}\frac{\left(2 \times 10^{-5}\right)^2}{8.85 \times 10^{-12}} \times 10^{-6}$

$= 2.26 \times 10^{-5}$ J]

31. Consider an assembly of three conducting concentric spherical shells of radii a, b, c as shown in Fig. 19.76. Find the capacitance of the assembly between points A and B.

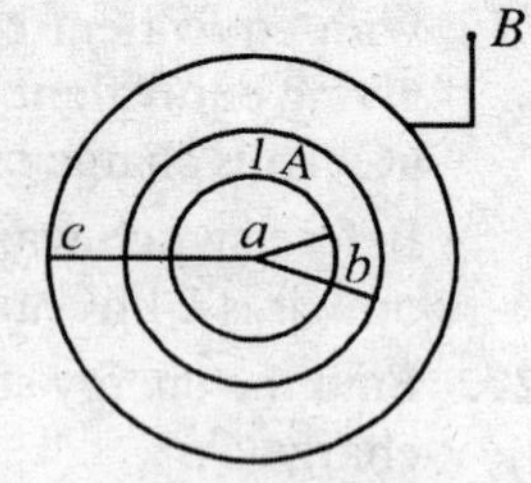

Fig. 19.76

[**Ans.** The system is equivalent to two capacitors with capacitance $C_1 = \frac{4\pi\varepsilon_0 ab}{b-a}$ and $C_2 = \frac{4\pi\varepsilon_0 bc}{c-b}$ in series. Hence equivalent capacity

$$C_S \text{ is given by } \frac{1}{C_S} = \frac{1}{C_1} + \frac{1}{C_2} = \frac{b-a}{4\pi\varepsilon_0 ba} + \frac{c-a}{4\pi\varepsilon_0 bc}$$

$$= \frac{1}{4\pi\varepsilon_0 b}\left\{\frac{bc-ca+ac-ab}{ca}\right\} = \frac{c-a}{4\pi\varepsilon_0 ca} \text{ or } C_S = \frac{4\pi\varepsilon_0 ca}{c-a}\Bigg]$$

32. An air-filled parallel plate capacitor is to be constructed which can store 12 μC of charge when operated at 1200 V. Find the minimum plate area of the capacitor? The dielectric strength of air is 3×10^6 V/m.

[**Ans.** Since maximum electric field that can be produced between the plates of the capacitor is 3×10^6 V/m. Hence separation d between the plates is $E = \frac{V}{d} = 3\times10^6$ or $d = \frac{1200}{3\times10^6} = 4\times10^{-4}$ m.

Capacity of capacitor C should be $C = \frac{q}{V} = \frac{12}{1200}10^{-2}\ \mu\text{F} = 10^{-8}$ F. But $C = \frac{\varepsilon_0 A}{d}$ or $A = \frac{cd}{\varepsilon_0}$

$$= \frac{10^{-8}\times4\times10^{-4}}{8.85\times10^{-12}} = 0.45 \text{ m}^2\Bigg]$$

33. A parallel plate capacitor is maintained at a certain potential difference. When a 3 mm thick slab is introduced between the plates in order to maintain same potential difference, the distance between the plates is increased 2.4 mm. Find the dielectric constant of the slab.

[**Ans.** When *a* dielectric slab of constant K and thickness t is introduced between the plates, the capacity becomes $C' = \frac{\varepsilon_0 A}{d - t(1-1/K)}$, where initial capacity is $C = \frac{\varepsilon_0 A}{d}$. The effect of introduction of the slab is that the capacity increases as if the separation between the plates is reduced by $t\left(1-\frac{1}{K}\right)$. To maintain the same potential difference, i.e. to make its capacity equal to its original value, the separation should be increased by $X = t(1-1/K)$. Given is $X = 2.4$ mm, $t = 3$ mm or

$$2.4 = 3\left(1-\frac{1}{K}\right) \text{ or } 0.8 = 1-\frac{1}{K} \text{ or } \frac{1}{K} = 1-0.8 = 0.2 \text{ and } K = 5]$$

34. A parallel plate capacitor has capacity of 2 μF. A dielectric slab of constant $K = 5$ is inserted between the plates and the capacitor is charged to 100 V and then isolated (*a*) What is the new potential difference if the dielectric is removed? (*b*) How much work is required to remove the dielectric slab?

[**Ans.** (*a*) When dielectric slab is removed, the capacity becomes C/K and charge remains conserved. Hence potential becomes VK, i.e. new potential = 100 × 5 = 500 V. (*b*) Energy stored

$$\text{is } E_i = \frac{1}{2}\frac{q^2}{C} \text{ and } E_f = \frac{1}{2}\frac{q^2K}{C}. \text{ Hence work done} = \frac{1}{2}\frac{q^2}{C}(K-1) = \frac{1}{2}CV^2(K-1)$$

$$= \frac{1}{2}\times5\times2(100)^2(5-1)\times10^{-6} = 0.2 \text{ J}\Bigg]$$

35. In the circuit shown in Fig. 19.77 with steady current, find the potential drop across the capacitor.

[**Ans**. In steady state no current flows through the capacitor. Hence current in the circuit is $i = \dfrac{V}{3R}$. Hence potential difference across $AB = 2V - 2\,iR = \dfrac{4V}{3}$. If potential drop across capacitor is V', then $V + V' = \dfrac{4V}{3}$ or $V' = V/3$]

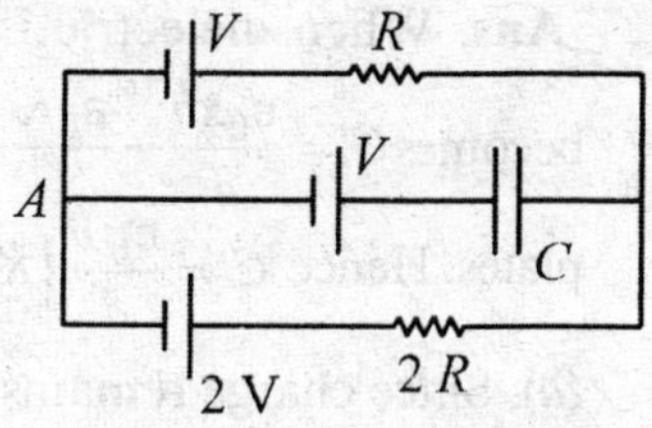

Fig. 19.77

36. In the arrangement shown in Fig. 19.78, each capacitor has capacity 1 μF. Find equivalent capacity across *AB*.

[**Ans**. First row has capacity *C*. Second row has *C*/2. Third row has *C*/4, etc. All these rows are in parallel. Hence total capacity is $C_t = C + C/2 + C/4 + \ldots = C\left(\dfrac{1}{1-1/2}\right) = 2C$

$\therefore\ C_t = 2\ \mu\text{F}$]

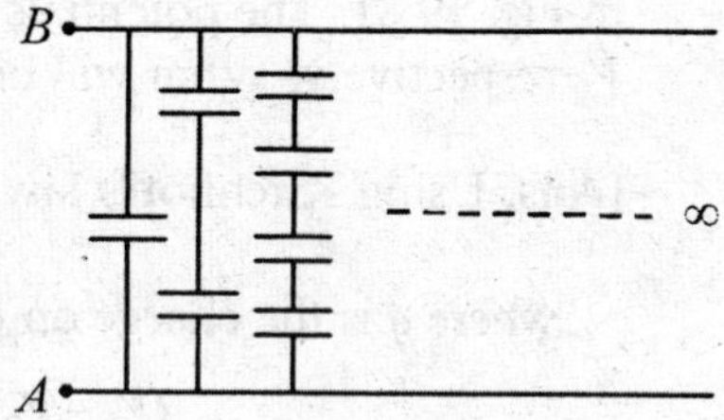

Fig. 19.78

37. In the circuit shown in Fig. 19.79 three batteries and three capacitors are shown. Find the potential difference across capacitor *A*.

[**Ans**. If V_A, V_B and V_C be the potential difference on capacitor *A*, *B* and *C* respectively, then $15 - V_A = 7 - V_B = V_C - 11$ and charge conservation gives $3V_A + 4V_B = 0.5V_C$. Solving, we have $V_A = 6$ V, $V_B = -2$ V and $V_C = 20$ V]

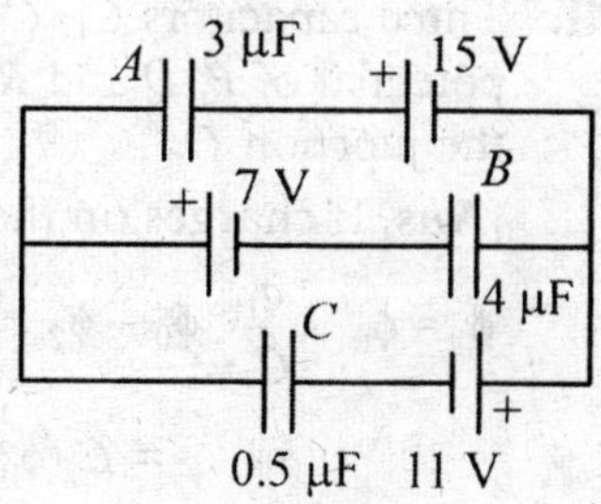

Fig. 19.79

38. If the current charging a capacitor is kept constant, show that the potential difference is directly proportional to the charging time.

[**Ans**. If *q* is the instantaneous charge on the capacitor then current $i = \dfrac{dq}{dt} = K$ (constant) or $q = Kt$. Since constant of integration is zero, i.e. $q = 0$ at $t = 0$. Hence $q \propto t$]

39. Between the plates of a parallel plate capacitor a dielectric slab is introduced just to fill the space between the plates. The capacitor is charged and then disconnected from the battery. The dielectric plate is slowly drawn out of the capacitor. Plot the graph between capacity of the capacitor and the length of dielectric drawn out. Plot also the graph between potential difference.

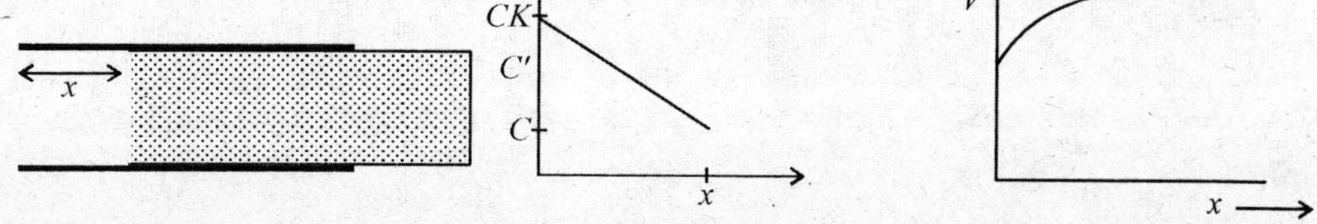

(*a*) (*b*)

Fig. 19.80

[**Ans**. When dielectric is drawn out through x (Fig. 19.80). The capacity of the capacitor becomes $C' = \frac{\varepsilon_0 xb}{d} + \frac{\varepsilon_0 K\,b(l-x)}{d}$ where d is separation between plates and b is the breadth of the plates. Hence $C' = \frac{\varepsilon_0 b}{d}\{Kl - x(K-1)\}$. The plot of C' vs x is a straight line as shown in Fig. 19.80 (*a*). Since charge remains conserved and potential V is given by $V = \frac{q}{C'} = \frac{qd}{\varepsilon_0 b\{Kl - x(K-1)\}}$.

It is hyperbola as shown in Fig. 19.80 (*b*)]

40. Two capacitors C_1 and C_2 in a circuit are connected as shown in Fig. 19.81. The potentials of the points A and B are V_1 and V_2 respectively. What will be the potential of point C?

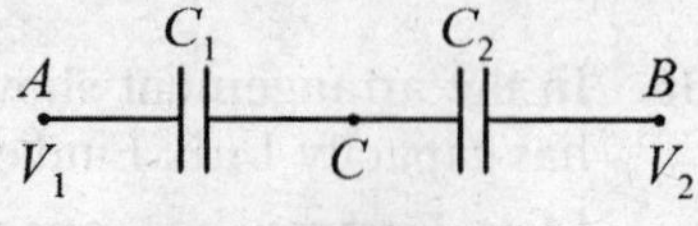

Fig. 19.81

[**Ans**. Using Kirchhoff's law, we have $V_1 - \frac{q}{C_1} - \frac{q}{C_2} = V_2$,

where q is the charge on each capacitor which are in series. Thus $q\left(\frac{C_2 + C_1}{C_1 C_2}\right) = V_1 - V_2$ or

$q = \frac{(V_1 - V_2)\,C_1 C_2}{C_1 + C_2}$. And potential of point D is $V_d = V_1 - \frac{q}{C_1} = \frac{C_1 V_1 + C_2 V_2}{C_1 + C_2}$]

41. Three capacitors C_1, C_2, C_3 are connected as shown in Fig. 19.82. The potential of P, Q and R are ϕ_1, ϕ_2, ϕ_3 respectively. Find the potential at the junction O.

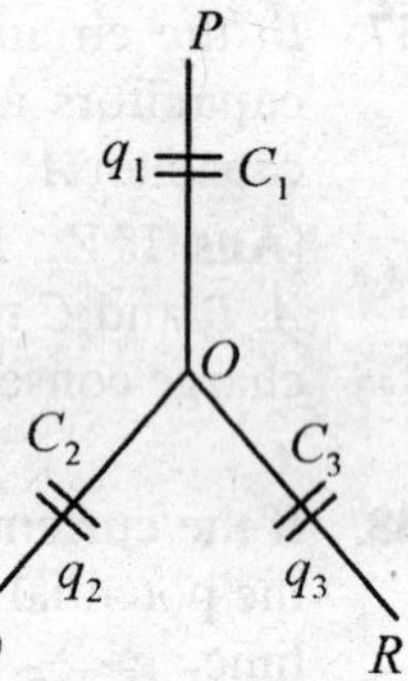

Fig. 19.82

[**Ans**. If charges on the three capacitors be q_1, q_2, q_3 respectively then

$\phi_1 - \phi_0 = \frac{q_1}{C_1}$, $\phi_0 - \phi_2 = \frac{q_2}{C_2}$, $\phi_0 - \phi_3 = \frac{q_3}{C}$. But $q_1 = q_2 + q_3$. Hence $C_1(\phi_1 - \phi_0)$

$= C_2(\phi_0 - \phi_2) + C(\phi_0 - \phi_3)$ or $\phi_0 = \frac{\phi_1 C_1 + \phi_2 C_2 + \phi_3 C_3}{C_1 + C_2 + C_3}$]

42. Find the capacitance of a system of two identical metal balls of radius a if the distance between their centres is equal to b ($b >> a$). The system is located in a uniform dielectric with permittivity (K).

[**Ans**. From symmetry of the problem $V_+ - V_- = 2|V|$, where V is the potential of any one ball. But

$|V| = \frac{q}{4\pi\varepsilon_0 Ka} + \frac{-q}{4\pi\varepsilon_0 K(b-a)} = \frac{q(b-2a)}{4\pi\varepsilon_0 Ka(b-a)}$

Hence $C = \frac{q}{2|V|} = \frac{2\pi\varepsilon_0 Ka(b-a)}{(b-2a)}$. If $b >> a$ then

$C = 2\pi\varepsilon_0 Ka$]

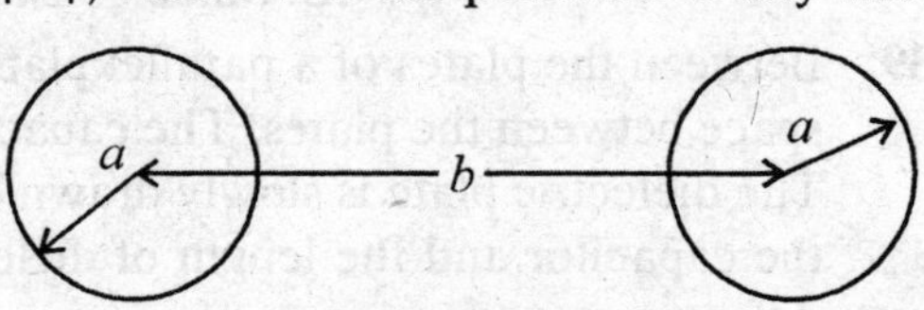

Fig. 19.83

MULTIPLE CHOICE QUESTIONS

PART I

In the following, a question is followed by four alternatives, of them only one is correct. Therefore, the candidate has to select the correct choice from amongst the alternatives.

1. With reference to the arrangement of capacitors shown in Fig. 19.84, the effective capacitance between points A and B is

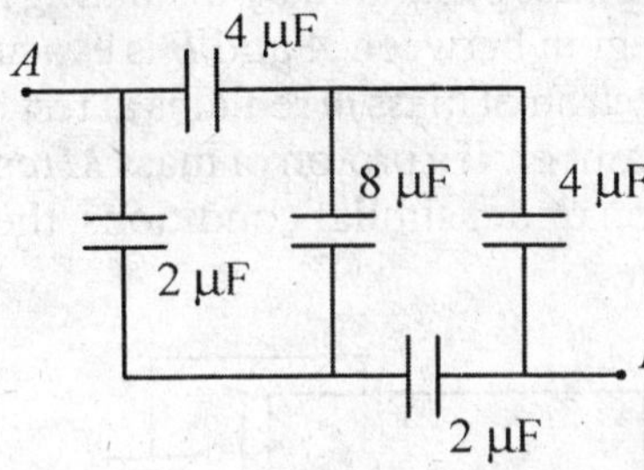

Fig. 19.84

(a) 3 μF (b) 8/3 μF
(c) 20 μF (d) 12 μF.

2. Electrical energy stored in the given network (Fig. 19.85) is

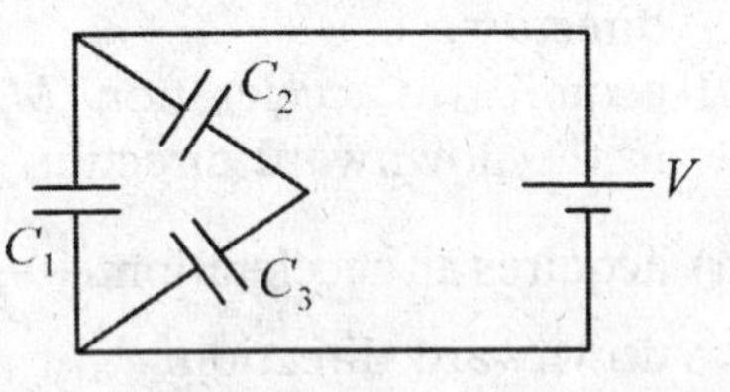

Fig. 19.85

(a) $\frac{1}{2}(C_1 + C_2 + C_3)V^2$

(b) $\frac{1}{2}\frac{(C_1C_2 + C_2C_3 + C_3C_1)}{(C_2 + C_3)}V^2$

(c) $\frac{C_1C_2C_3V^2}{2(C_1 + C_2 + C_3)}$

(d) $\frac{(C_1C_2 + C_2C_3 + C_3C_1)}{2(C_1 + C_2 + C_3)}V^2$

3. When the given circuit (Fig. 19.86) is switched on, the graph of voltage V versus time is as shown in

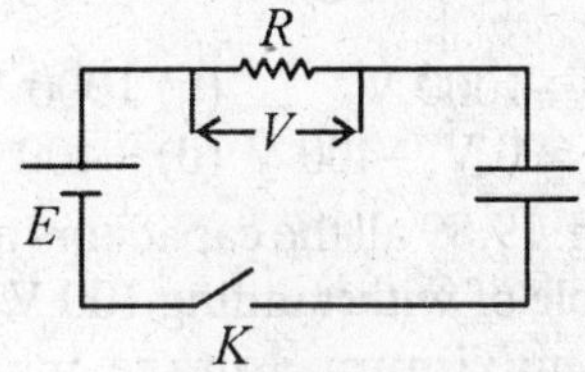

Fig. 19.86

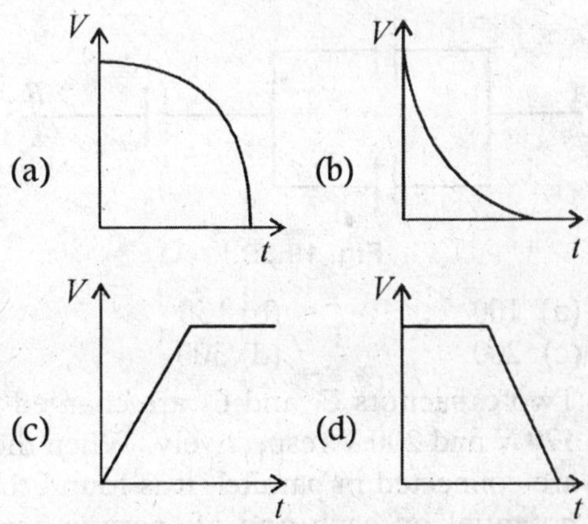

4. A network made up of six identical capacitors, each of value C is made as shown in Fig. 19.87. Equivalent capacitance between points A and B is

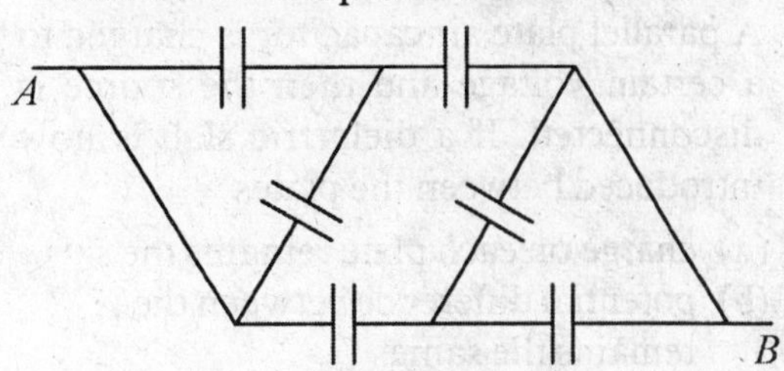

Fig. 19.87

(a) $C/4$ (b) $3C/4$
(d) $4C/3$ (d) $3C$.

5. Figure 19.88 shows a circuit with E as the earthing of the common plate. The potentials at P and Q are

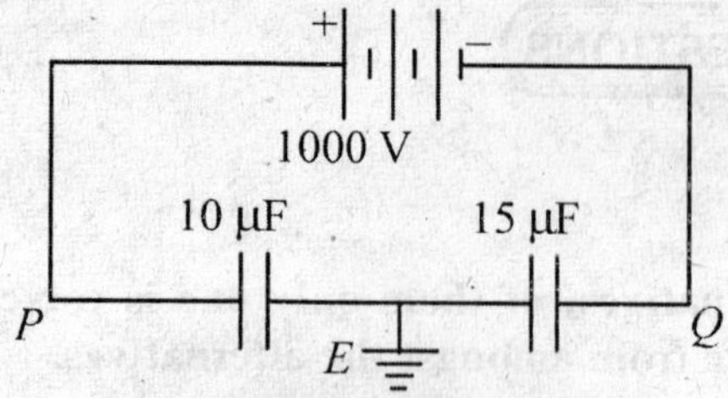

Fig. 19.88

(a) 0, –1000 V (b) 1000 V, 0 V
(c) +600 V, –400 V (d) +400 V, –600 V.

6. In Fig. 19.89 all the capacitors are identical, capable of withstanding 100 V maximum. The maximum voltage which can be applied across terminals A, B (in volts) is

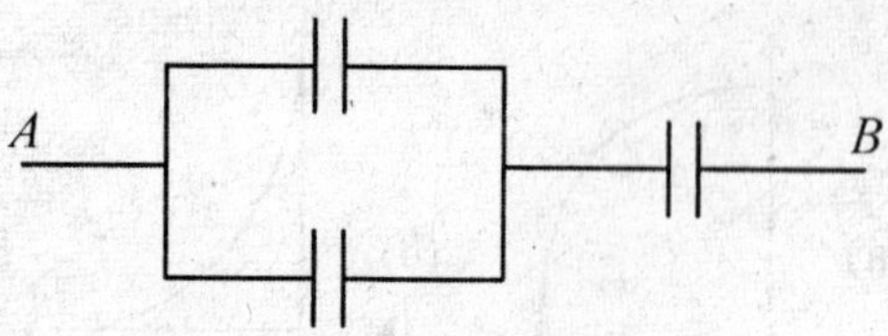

Fig. 19.89

(a) 100 (b) 150
(c) 200 (d) 300.

7. Two capacitors C_1 and C_2 are charged to 120 V and 200 V respectively. When they are connected in parallel, it is found that potential on each one of them is zero. Therefore,

(a) $5C_1 = 3C_2$ (b) $3C_1 = 5C_2$
(c) $3C_1 + 5C_2 = 0$ (d) $9C_1 = 4C_2$.

8. A parallel plate air capacitor is charged to a certain voltage and then the source is disconnected. If a dielectric slab is now introduced between the plates

(a) charge on each plate remains the sa...
(b) potential difference between the p... remains the same
(c) energy of the capacitor remains th... same
(d) capacitance of the capacitor remains the same.

9. In Fig. 19.90 the steady state ...ltages V_1 and V_2 are respectively

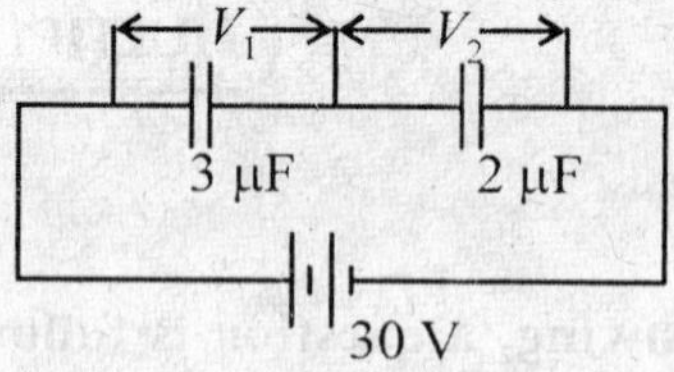

Fig. 19.90

(a) 12 V, 18 V (b) 18 V, 12 V
(c) 15 V, 15 V (d) 30 V, 30 V.

10. In Fig. 19.91 A and B are two horizontal metal plates separated by insulating plugs. The region between A and B is evacuated. An electron of mass m remains at rest inside the chamber. If a proton of mass M replaces electron under similar conditions, then the proton

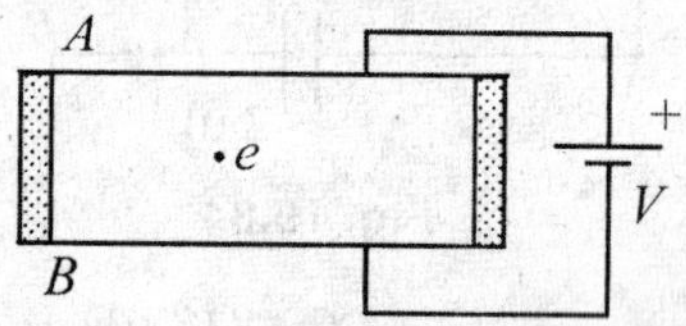

Fig. 19.91

(a) continues to remain at rest
(b) acquires an acceleration $\frac{eE}{M}$ in upward directio...
(c) acquire... an acceleration $(Mg - eE)/M$ in the downward direction
(d) acquires an acceleration $\frac{(m + M)}{M} g$ in downward direction.

11. In the circ...t shown in Fig. 19.92, capacitors A and ... identical geometry, but a mate... ...ctric constant 3 is present ... es of B. Potential difference ... nd B respectively are

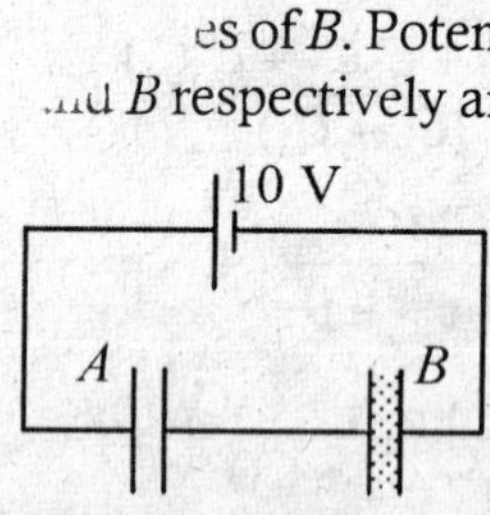

Fig. 19.92

(a) 2.5 V, 7.5 V (b) 2 V, 8 V
(c) 7.5 V, 2.5 V (d) 8 V, 2 V.

12. In the circuit shown in Fig. 19.93, assume the capacitors are initially uncharged and the switch is off. When the switch is turned on, then the out put voltage V_0

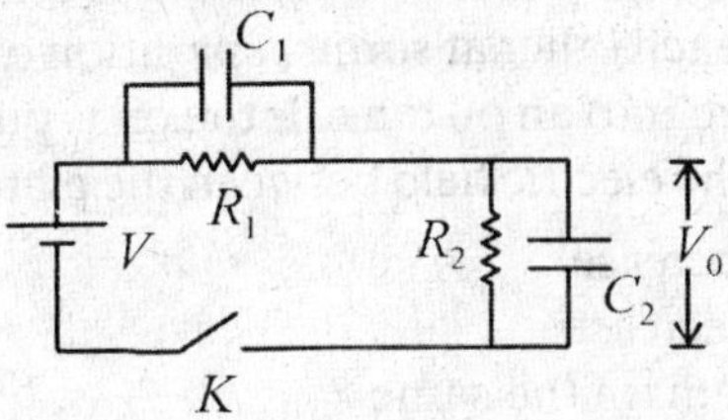

Fig. 19.93

(a) remains constant to the value $\frac{R_2}{R_1 + R_2} V$

(b) remains constant to the value $\frac{C_1}{C_1 + C_2} V$

(c) changes linearly from $\frac{C_1}{C_1 + C_2} V$ to $\frac{R_2}{R_1 + R_2} V$

(d) changes exponentially from $\frac{C_1}{C_1 + C_2} V$ to $\frac{R_2}{R_1 + R_2} V$.

13. In the circuit shown in Fig. 19.94, C is connected first to A. It charges capacitor of 4 μF. Now connection of C is switched to B. The charge on the 4 μF capacitor thereby changes by a factor

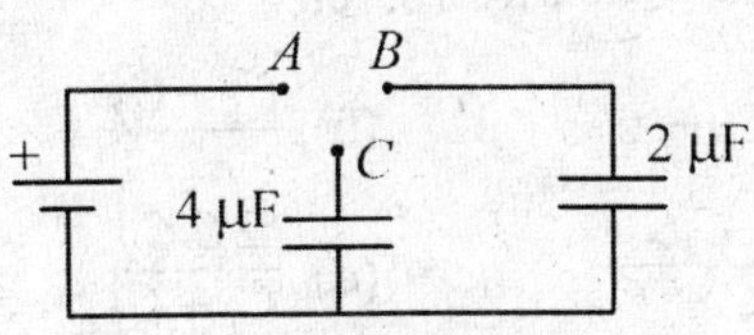

Fig. 19.94

(a) 1 (b) 2/3
(c) 3/4 (d) 1/3.

14. In the network shown in Fig. 19.95, all the capacitors are identical. When a voltage is applied between A, B one of the capacitors remains uncharged. That capacitor is

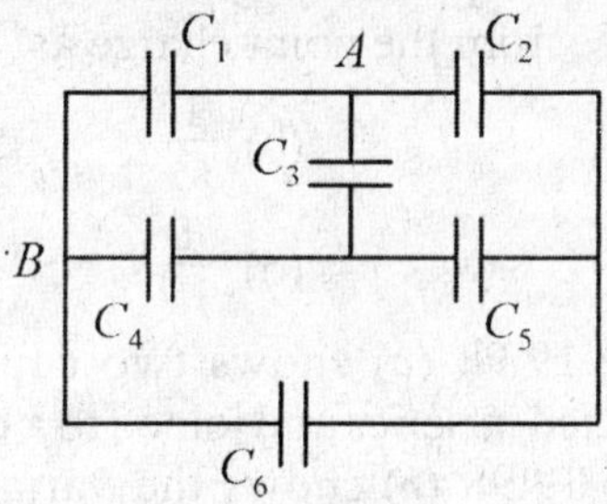

Fig. 19.95

(a) C_1 (b) C_2
(c) C_5 (d) C_6.

15. In Fig. 19.96, separation between the plates of a parallel plate capacitor is 2 mm, a dielectric slab of 1.6 mm thickness is introduced. The dielectric constant of this material is 10. If a potential difference 24 V is applied at the plates; the electric field in the slab is (in V/m)

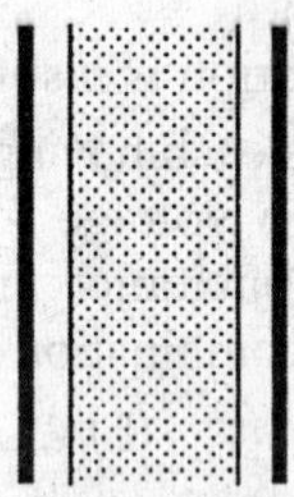

Fig. 19.96

(a) 4.3×10^4 (b) 4.3×10^3
(c) 4.3×10^2 (d) 4.3×10^1.

16. Circuit shown in Fig. 19.97 contains 4 capacitors each of 20 μF. The capacitance across the terminals 1 and 2 is

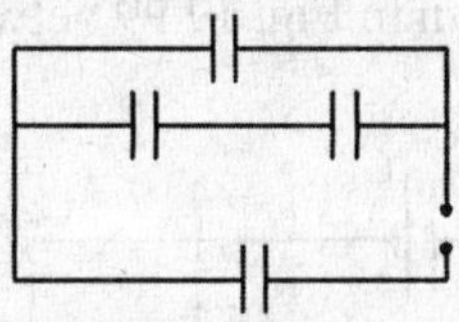

Fig. 19.97

(a) 12 μF (b) 15 μF
(c) 20 μF (d) 30 μF.

17. The energy density in the electric field created by a point charge falls off with the distance from the point charge as

(a) $\frac{1}{r}$ (b) $\frac{1}{r^2}$

(c) $\frac{1}{r^3}$ (d) $\frac{1}{r^4}$.

18. Figure 19.98 (*a*) shows two capacitors connected in series and joined to a battery. Figure 19.98 (*b*) shows the variation in potential with distance from *A* as one moves towards *B*. Then

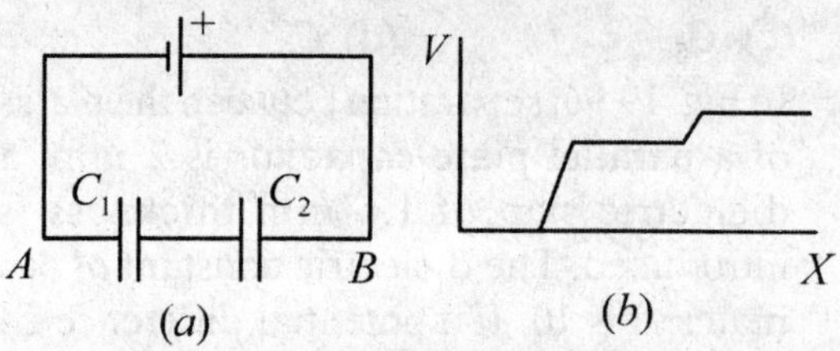

Fig. 19.98

(a) $C_1 > C_2$ (b) $C_1 = C_2$
(c) $C_1 < C_2$
(d) the information is insufficient.

19. Two capacitors each having capacitance *C* and breakdown voltage *V* are joined in series. The capacitance and breakdown voltage of the combination will be

(a) $2C, 2V$ (b) $C/2, V/2$
(c) $2C, V/2$ (d) $C/2, 2V$

20. If the capacitors in question number 19 are joined in parallel, the capacitance and breakdown voltage of the combination will be

(a) $2C, 2V$ (b) $C, 2V$
(c) $2C, V$ (d) C, V.

21. The equivalent capacitance of the combination shown in Fig. 19.99 between *A* and *B* is

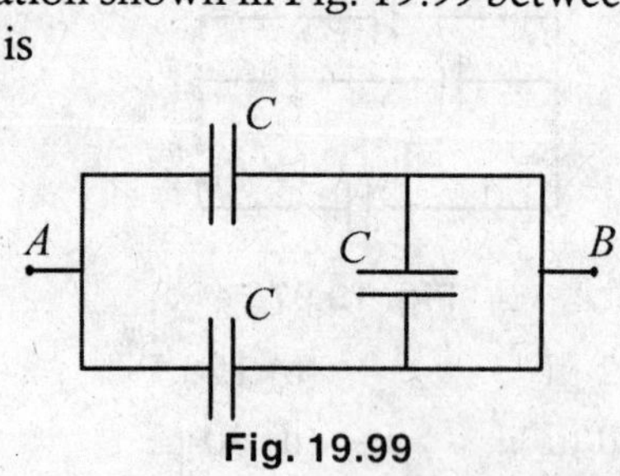

Fig. 19.99

(a) C (b) $2C$
(c) $C/2$ (d) $\frac{2C}{3}$.

22. Two metal plates having charges *Q* and $-Q$ face each other at some separation and are dipped into an oil tank. If the oil is pumped out, the electric field between the plates will

(a) increase
(b) decrease
(c) remain the same
(d) become zero.

23. A dielectric slab is inserted between the plates of an isolated capacitor. The force between the plates will

(a) increase
(b) decrease
(c) remain unchanged
(d) become zero.

24. The equivalent capacitance between *A* and *B* in network shown in Fig. 19.100 is

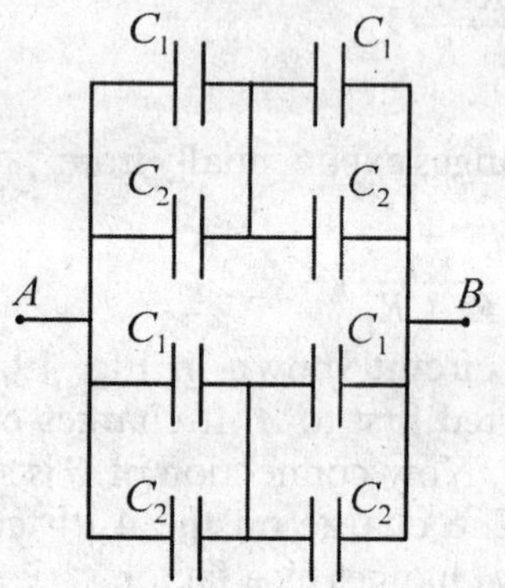

Fig. 19.100

(a) $C_1 + C_2$ (b) $\frac{C_1 + C_2}{2}$

(c) $\frac{C_1C_2}{C_1 + C_2}$ (d) $\frac{2C_1C_2}{C_1 + C_2}$.

25. Capacitor stores 50 μC charge when connected across a battery. When the gap between the plates is filled with a dielectric, a charge of 100 μC flows through the battery. The dielectric constant of material is

(a) 2 (b) 3
(c) 2.5 (d) 1.5.

26. A 5.0 μF capacitor is charged to 12 V. The positive plate of this capacitor is now connected to the negative terminal of a 12 V battery and *vice versa*. The heat developed in connecting wires is (in mJ)

(a) 1.44 (b) 0.72
(c) 0.36 (d) 2.88.

27. The potential difference between points *P* and *Q* in the circuit shown in Fig. 19.101 is

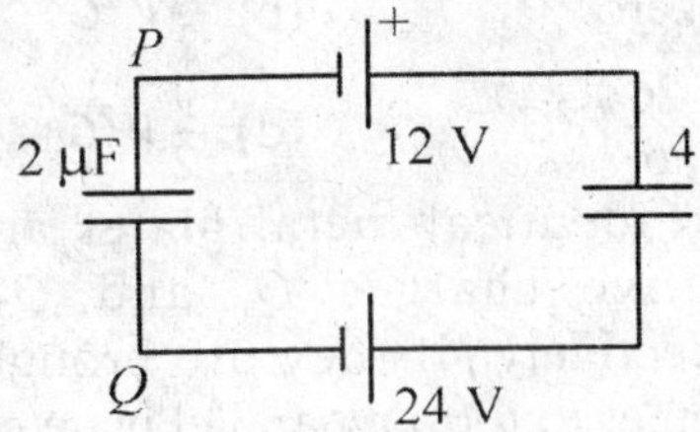

Fig. 19.101

(a) +8 V (b) –8 V
(c) +12 V (d) –12 V.

28. Potential difference $V_a - V_b$ between the points *a* and *b* shown in Fig. 19.102 is

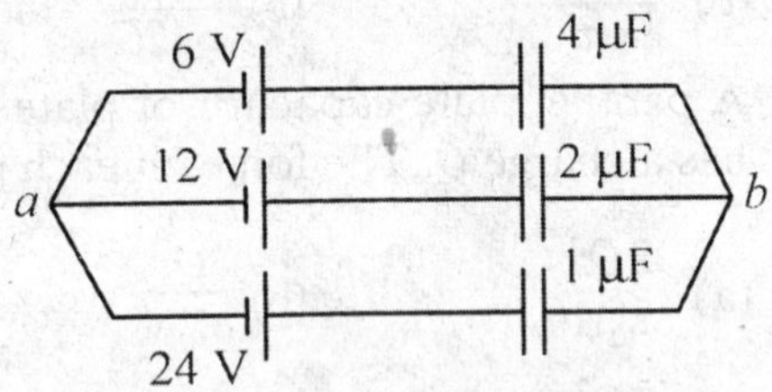

Fig. 19.102

(a) –10.3 V (b) –15.6 V
(c) –21.3 V (d) zero.

29. A charge of 20 μC is placed on the positive plate of an isolated parallel plate capacitor of capacitance 10 μF. The potential difference developed between the plates is

(a) 2 V (b) 1 V
(c) 0.5 V (d) 1.5 V.

30. The capacitance between the adjacent plates shown in Fig. 19.103 is 50 nF. A charge of 1.0 μC is placed on middle plate. The charge on the outer surface of the upper plate (*q*) and potential difference (*V*) developed between the upper and middle plates respectively is

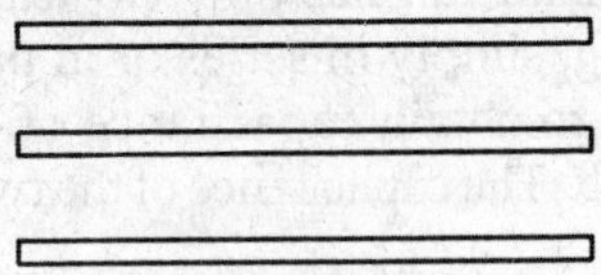

Fig. 19.103

(a) $q = 1$ μC, $V = 10$ V
(b) $q = 0.5$ μC, $V = 10$ V
(c) $q = -1$ μC, $V = 5$ V
(d) $q = -0.5$ μC, $V = 5$ V.

31. A capacitor with stored energy 4.0 J is connected with an identical capacitor with no electric field in between them. The energy stored in the two capacitors is

(a) 4.0 J (b) 2.0 J
(c) 1.0 J (d) zero.

32. A dielectric slab is inserted between the plates of an isolated charged capacitor. Which of the following quantities will remain same?

(a) The electric field in the capacitor
(b) The charge on the capacitor
(c) The potential difference between the plates
(d) The stored energy in the capacitor.

33. A dielectric slab is introduced between the plates of a capacitor. The charge on the capacitor is *Q*. The magnitude of the induced charge is

(a) $Q\left(1-\frac{1}{K}\right)$ (b) $Q\left(1+\frac{1}{K}\right)$
(c) Q/K (d) *K*.

34. A parallel plate capacitor is connected to a battery. A metal sheet of negligible thickness is placed between the plates. The sheet remains parallel to the plates of the capacitor.

(a) The battery will supply more charge
(b) The capacitance will increase
(c) The potential difference between the plates will increase

(d) Equal and opposite charges will appear on the two faces of the metal plate.

35. A technician has only two capacitors. By using, simply in series or in parallel, he is able to obtain capacitance of 4, 5, 20 and 25 μF. The capacitance of the two capacitors is

(a) 4, 25 μF (b) 5, 20 μF
(c) 4, 20 μF (d) 5, 25 μF.

36. A parallel plate capacitor of area A, plate separation d and capacitance C is filled with three dielectric materials having dielectric constant K_1, K_2, K_3 as shown in Fig. 19.104. If a single dielectric material is to be used to have same capacitance C then the dielectric will have constant K given by

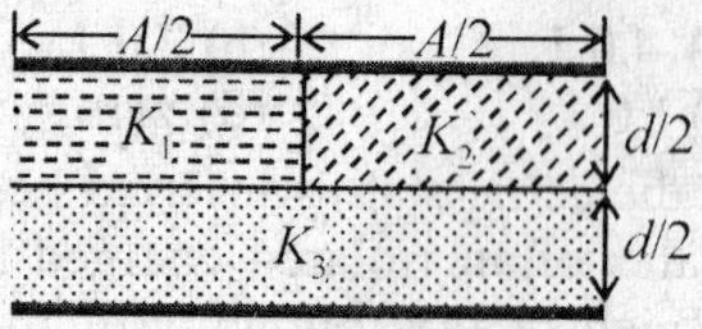

Fig. 19.104

(a) $\frac{1}{K} = \frac{1}{K_1} + \frac{1}{K_2} + \frac{1}{K_3}$

(b) $\frac{1}{K} = \frac{1}{K_1 + K_2} + \frac{1}{2K_3}$

(c) $K = \frac{K_1 K_2}{K_1 + K_2} + 2K_3$

(d) $K = K_1 + K_2 + 2K_3$.

37. Four capacitors are connected as shown in Fig. 19.105. The equivalent capacitance between points A and B is

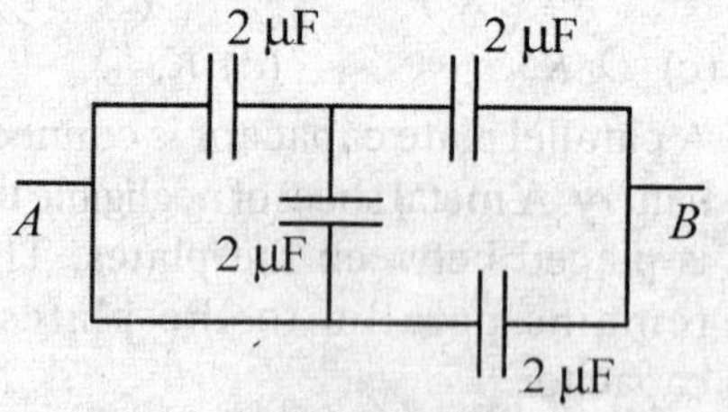

Fig. 19.105

(a) $\frac{28}{9}$ μF (b) 4 μF
(c) 5 μF (d) 18 μF.

38. A parallel plate capacitor of capacitance C is connected to a battery and is charged to a potential difference V. Another capacitor of capacitance $2C$ is similarly charged to a potential 2 V. The charging battery is then disconnected and the capacitors are connected in parallel to each other in such a way that the positive terminal of one is connected to the negative terminal of the other. The final energy of the configuration is

(a) zero (b) $\frac{3}{2}V^2C$
(c) $\frac{26}{6}V^2C$ (d) $\frac{9}{2}V^2C$.

39. Two identical metal plates are given positive charges Q_1 and Q_2 ($<Q_1$) respectively. If they are brought close together to form a parallel plate capacitor with capacitance C, the potential difference between them is

(a) $\frac{Q_1 + Q_2}{2C}$ (b) $\frac{Q_1 + Q_2}{C}$
(c) $\frac{Q_1 - Q_2}{C}$ (c) $\frac{Q_1 - Q_2}{2C}$.

40. A parallel plate capacitor of plate area A has a charge Q. The force on each plate is

(a) $\frac{2Q^2}{\varepsilon_0 A}$ (b) $\frac{Q^2}{\varepsilon_0 A}$
(c) $\frac{Q^2}{2\varepsilon_0 A}$ (d) zero.

41. In case of a spherical capacitor, the radius of inner sphere is kept fixed and the radius of outer sphere is increased. The capacity of the capacitor

(a) will increase
(b) will decrease
(c) will remain unchanged
(d) depends on the nature of the spheres.

42. A 100 μF capacitor is charged upto 200 V potential difference. Now it is discharged through a resistance of 2 Ω. Heat developed in resistance is

(a) 2 J (b) 4 J
(c) 0.2 J (d) 0.4 J.

43. Capacity of a parallel plate capacitor is C when plate separation is d and medium between plates is air. If a slab of thickness t ($<d$) and of dielectric constant K is introduced in the plates, the capacity now becomes

(a) $\dfrac{Cd}{d+t(1-1/K)}$ (b) $\dfrac{Cd}{d-t(1-1/K)}$

(c) $\dfrac{Cd}{d-t(1+1/K)}$ (d) $\dfrac{Cd}{d+t(1+1/K)}$.

44. A radio capacitor of variable capacity is formed using n plates of area A and separation d. Alternate plates are connected together. One group of alternate plates is fixed whereas other group is movable. The maximum capacity of this capacitor is

(a) $\dfrac{n\varepsilon_0 A}{d}$ (b) $\dfrac{(n-1)\,\varepsilon_0 A}{d}$

(c) $\dfrac{\varepsilon_0 A}{nd}$ (d) $\dfrac{\varepsilon_0 A}{(n-1)\,d}$.

45. A capacitor of capacity 500 μF is given the charge at the rate of 100 μC/s. The time in which its potential will be 20 V is

(a) 5 s
(b) 10 s
(c) 20 s
(d) none of the above.

46. Figure 19.106 shows a network of capacitors and resistors. Equivalent capacity across A and B is

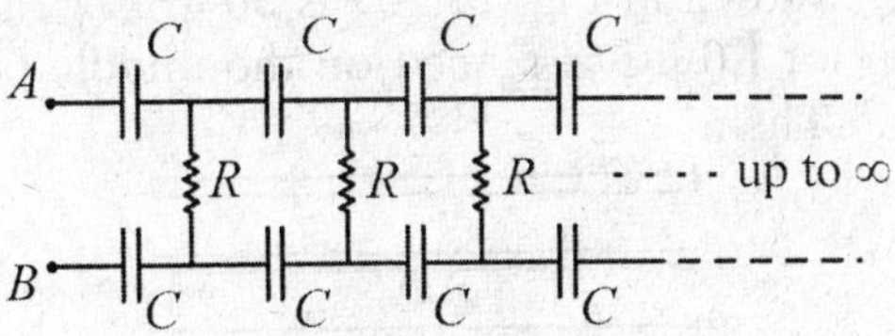

Fig. 19.106

(a) C (b) $2C$
(c) $C/2$ (d) infinity.

47. Work done in increasing the potential difference of a capacitor from 15 V to 30 V is W. The work done in raising its potential from 30 to 60 V is

(a) $2W$ (b) $3W$
(c) $4W$ (d) $8W$.

48. A parallel plate capacitor has capacity C, potential difference V and plate separation d. The force of attraction between the plates is

(a) $\dfrac{1}{2}\dfrac{CV^2}{d}$ (b) $\dfrac{1}{2}\dfrac{C^2V}{d}$

(c) $\dfrac{1}{2}\dfrac{C^2V^2}{d}$ (d) $\dfrac{1}{2}C^2V\,d$.

49. Seven capacitors each having capacity 2 μF are connected such that their equivalent capacity is $\dfrac{10}{11}$ μF. The possible arrangement is

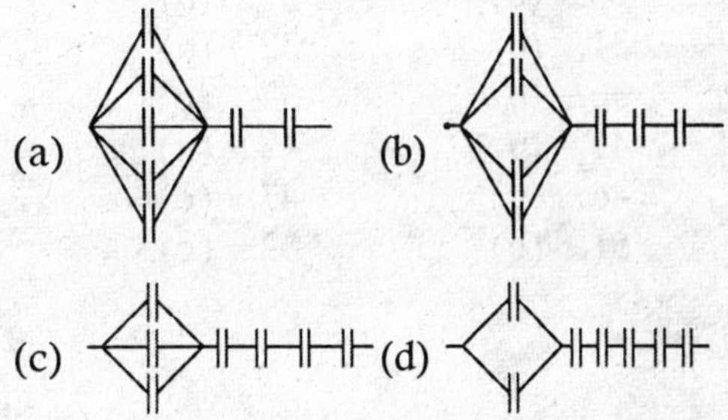

50. Two spherical conductors A and B of radii a and b ($b > a$) are placed concentrically in air. The two are connected by wire as shown in Fig. 19.107. The capacity of the system is

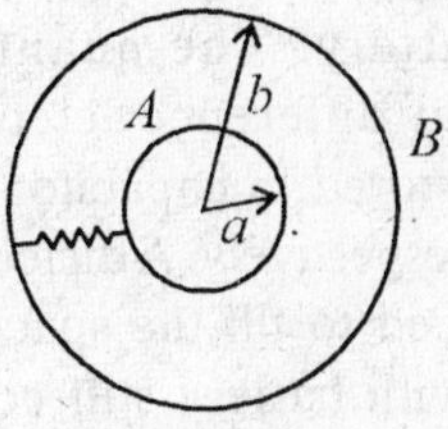

Fig. 19.107

(a) $\dfrac{4\pi\varepsilon_0\, ab}{b-a}$ (b) $4\pi e_0(a+b)$

(c) $4\pi\varepsilon_0 b$ (d) $4\pi\varepsilon_0 a$.

51. Force acting on a charged particle kept between the plates of a charged condenser is F. If one of the plates of the condenser is removed, the force acting on the same particle becomes

(a) 0 (b) $F/2$
(c) F (d) $2F$.

52. The equivalent capacitance of the network shown in Fig. 19.108 across AB is

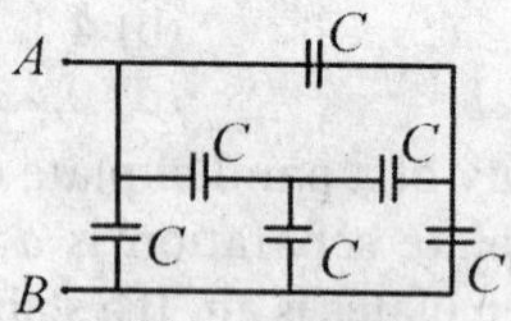

Fig. 19.108

(a) 2.5 C (b) C
(c) 2 C (d) $\frac{4}{3}$C

ANSWERS

1. (a)	2. (b)	3. (b)	4. (c)	5. (c)
6. (b)	7. (b)	8. (a)	9. (a)	10. (d)
11. (c)	12. (d)	13. (b)	14. (c)	15. (b)
16. (a)	17. (d)	18. (c)	19. (d)	20. (c)
21. (b)	22. (a)	23. (c)	24. (a)	25. (b)
26. (a)	27. (b)	28. (a)	29. (b)	30. (b)
31. (b)	32. (b)	33. (a)	34. (d)	35. (b)
36. (b)	37. (c)	38. (b)	39. (d)	40. (c)
41. (b)	42. (a)	43. (b)	44. (b)	45. (d)
46. (c)	47. (c)	48. (a)	49. (a)	50. (c)
51. (c)	52. (c).			

PART II

In the following, a question is followed by four alternatives, of them one or more are correct. Candidate has to select the correct choice(s) from amongst the alternatives.

1. A parallel plate air capacitor is connected to a battery. The quantities charge, potential difference, electric field and energy stored in capacitor are Q_0, V_0, E_0 and U_0 respectively. A dielectric slab is now introduced to fill the space between the plates with battery still connected. The corresponding quantities now given by Q, V, E and U are related to the initial values as

(a) $Q > Q_0$ (b) $V > V_0$
(c) $E > E_0$ (d) $U > U_0$.

2. The capacitance between the adjacent plates shown in Fig. 19.109 is 50 μF. A charge of 1.0 μC is placed on the middle plate. Then

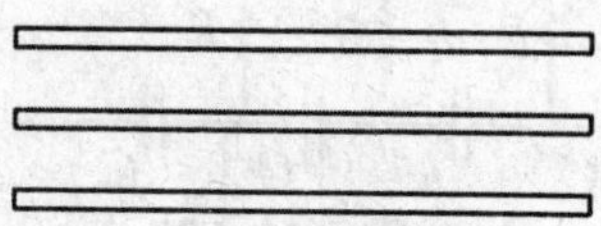
Fig. 19.109

(a) the potential difference between upper and middle plate is 10 V
(b) the potential difference between upper and middle plate is 20 V

(c) the potential difference between middle and lower plate is 10 V

(d) the potential difference between middle and lower plate is 20 V.

3. Each capacitor in Fig. 19.110 has capacitance 10 μF. The emf of the battery is 100 V.

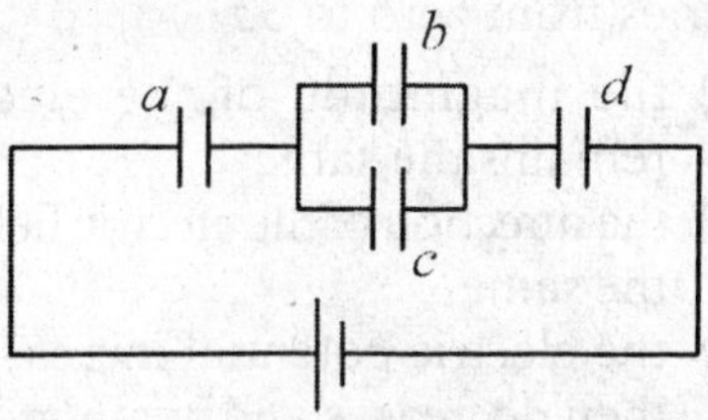

Fig. 19.110

(a) Energy stored in *b* is 8 mJ

(b) Energy stored in *d* is 8 mJ

(c) Energy stored in *c* is 2 mJ

(d) Energy supplied by the battery is 40 mJ.

4. A parallel plate capacitor with the plate area 100 cm^2 and the separation between the plates 1.0 cm is connected to the battery of emf 2.4 Volts.

(a) The force of attraction between the plates is 2.5×10^{-7} N

(b) The energy stored in the capacitor is 2.5×10^{-9}J

(c) The energy supplied by the battery is 2.5×10^{-9}J

(d) Heat produced in the connecting wires is 2.5×10^{-9}J.

5. Each plate of a parallel plate capacitor has a charge *q* on it. The capacitor is now connected to the battery. Now

(a) the facing surfaces of the capacitor have equal and opposite charges

(b) the two plates of the capacitors have equal and opposite charges

(c) the battery supplies equal and opposite charges to the plates

(d) the outer surfaces of the plates have equal charges.

6. The separation between the plates of a charged parallel plate capacitor is increased. Which of the following quantities will change?

(a) Charge on the capacitor

(b) Energy of the capacitor.

(c) Potential difference across the capacitor

(d) The energy density between the plates.

7. The following operations, can be performed on a capacitor

A – Connect the capacitor to a battery of emf *E*.

B – Disconnect the battery.

C – Reconnect the battery with reversed polarity.

D – Insert a dielectric in the capacitor.

(a) In ABC operation the energy stored remains unchanged and no thermal energy is developed

(b) After the operation ADB, the charge appearing on the capacitor is more than the charge appearing after operation ABD

(c) After the operation DAB, the electric energy stored in the capacitor is more than that after operation ABD

(d) The electric field between the capacitor plates after the operation AD is the same as that after operation DX.

8. Using all the three identical capacitors of capacity 6 μF each the possible values of capacity that can be obtained are

(a) 18 μF (b) 9 μF

(c) 4 μF (d) 2 μF.

9. In the circuit shown in Fig. 19.111.

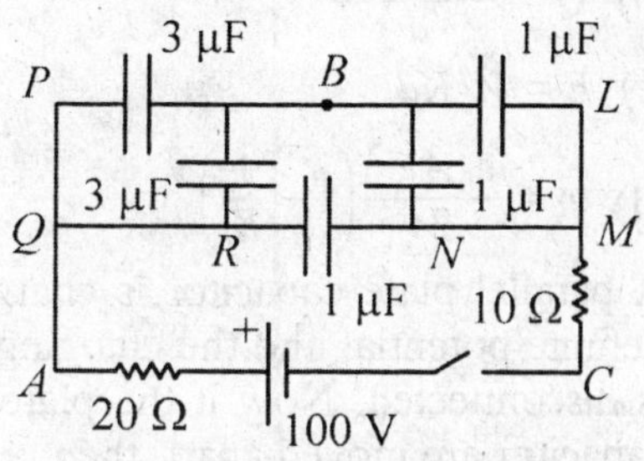

Fig. 19.111

(a) the potential difference between *A* and *B* is 25 V

(b) the potential difference between B and C is 75 V
(c) the energy stored in the capacitor between R and N is 5 mJ
(d) the charge on 3 μF capacitor is 75 μC.

10. A positive charge Q is located at the centre of a thin metallic spherical shell. Select the correct statement.
(a) The electric field at any point outside the shell is zero
(b) The electric potential at any point outside the shell is $Q/4\pi\varepsilon_0 r$, where r is the distance of the point from the centre
(c) The outer surface of the spherical shell is equipotential surface
(d) The electric field at any point inside the shell other than the centre is zero.

11. A parallel plate capacitor of plate area A and plate separation d is charged to potential difference V and then the battery is disconnected. A slab of dielectric constant K is then inserted between the plates of the capacitor so as to fill the space between the plates. If Q, E and W denote respectively the magnitude of charge on each plate, the electric field between the plates (after inserting the slab), and work done on the system in question, in the process of inserting the slab, then
(a) $Q = \frac{\varepsilon_0 AV}{d}$
(b) $Q = \frac{\varepsilon_0 K AV}{d}$
(c) $E = V/Kd$
(d) $\omega = \frac{\varepsilon_0 AV^2}{2d}\left(1 - \frac{1}{K}\right)$

12. A parallel plate capacitor is charged to a definite potential and the charging battery is disconnected. Now if the plates of the capacitor are moved apart, then
(a) the stored energy of the capacitor increases
(b) charge on the capacitor increases
(c) voltage of the capacitor increases
(d) the capacitance increases.

13. A dielectric slab of thickness d is inserted in a parallel plate capacitor whose negative plate is at $x = 0$ and positive plate is at $x = 3d$. The slab is equidistant from the plates. The capacitor is given some charge. As x varies from zero to $3d$
(a) the magnitude of the electric field remains the same
(b) the direction of the electric field remains the same
(c) the electric potential increases at first, then decreases and again increases
(d) the electric potential increases continuously.

14. Two capacitors of capacity 1 μF and 2 μF are separately charged by a common battery. The two capacitors discharge through equal resistances at $t = 0$.
(a) The current in each of the two discharging circuits is zero at $t = 0$
(b) The current in the two discharging circuits at $t = 0$ are equal but not zero
(c) The currents in two discharging circuits are equal at all times
(d) Capacitor of 1 μF capacity loses 50% of its charge earlier than capacitor of capacity 2 μF.

15. In the circuit shown in Fig. 19.112,

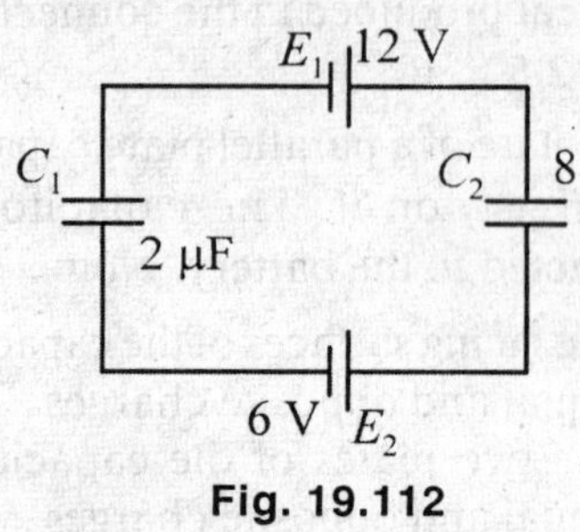

Fig. 19.112

(a) charge on C_2 is greater than on C_1
(b) energy stored in C_1 is more than that in C_2
(c) potential difference across C_1 equals to that across C_2
(d) charge on both capacitors is same.

16. A capacitor of capacitance 500 μF is connected to a battery through a 10 kΩ resistor. The charge stored on the capacitor in first 5 sec is larger than the charge stored in the next

(a) 5 sec (b) 50 sec
(c) 500 sec (d) 5000 sec.

17. A parallel plate capacitor is charged and then disconnected from the source of steady emf. The plates are then drawn apart further. Again it is connected to the same source. Then

(a) the potential difference across the plates increases, while the plates are being drawn apart
(b) the potential difference across the plates decreases while the plates are being drawn apart
(c) the charge from the capacitor flows into the source when the capacitor is re-connected
(d) the electric intensity between the plates remains constant during the drawing apart of the plates.

18. The dielectrics K_1 and K_2 are filled between the plates of a capacitor as shown in Fig. 19.113. The dielectric can be filled in two ways shown in Fig. (*a*) and (*b*).

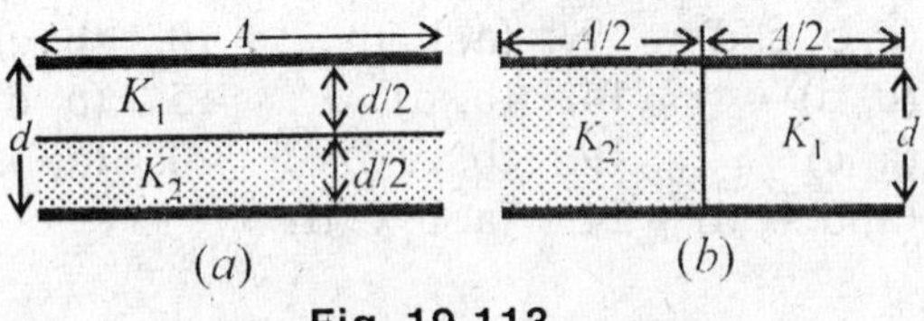

Fig. 19.113

(a) The ratio of the capacities in two systems is $\frac{C_a}{C_b} = \frac{4K_1K_2}{(K_1+K_2)^2}$
(b) If both capacitors are connected to the battery of emf E then charge on (*a*), (q_a) is greater than charge on (*b*), (q_b), i.e. $q_a > q_b$
(c) When the battery is connected the energy stored in (*a*) E_a is less than the energy stored in (*b*) E_b, i.e. $E_a < E_b$
(d) If $K_1 = K_2 = K$ then $C_a < C_b$.

19. If a capacitor of capacitance C is charged to potential difference V, then units of CV^2 are

(a) Watt (b) Joule
(c) erg (d) electron Volt.

20. In the circuit shown in Fig. 19.114, all the capacitors have same capacitance (= 10 μF)

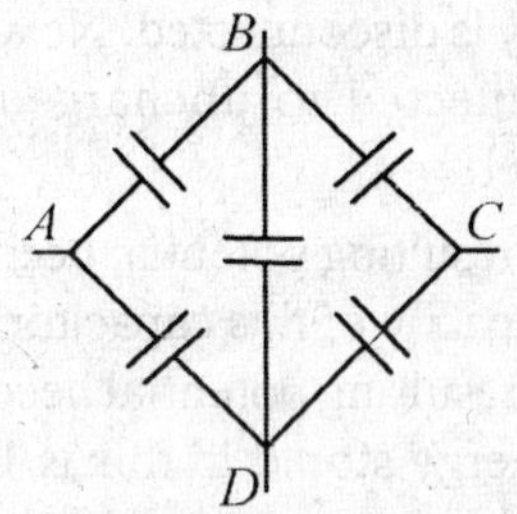

Fig. 19.114

(a) Equivalent capacity between B and D is 20 μF
(b) Equivalent capacity between A and D is 6 μF
(c) Equivalent capacity between D and C is 10 μF
(d) Equivalent capacity between A and C is 10 μF.

21. The capacitance of two capacitors are C_1 and C_2. They have equal energies stored in them. Then

(a) the ratio of their potential $\frac{V_1}{V_2} = \sqrt{\frac{C_2}{C_1}}$
(b) the ratio of charges on them $\frac{q_1}{q_2} = \sqrt{\frac{C_2}{C_1}}$
(c) the ratio of their potential $\frac{V_1}{V_2} = \sqrt{\frac{C_1}{C_2}}$
(d) the ratio of charges on them $\frac{q_1}{q_2} = \sqrt{\frac{C_1}{C_2}}$

22. A capacitor of capacitance 400 pF has plate separation 2 mm. It is charged upto 1500 V and battery removed.

(a) If plate separation is doubled the potential becomes 3000 V
(b) Energy required to double the plate separation is 4.5×10^{-4} J
(c) Energy stored in the capacitor before increasing the separation is 4.5×10^{-4} J
(d) The electric field between the plates after separation is doubled the initial field.

23. A 10 μF capacitor is charged to 100 V and battery is disconnected. Now this capacitor is connected to uncharged capacitor in parallel.
(a) If resulting potential becomes 40 V, the capacity of this capacitor is 15 μF
(b) If resulting potential becomes 40 V, the energy stored in this is 12 mJ.
(c) If resulting potential is 40 V, the charge from 10 μF capacitor transferred to this capacitor is 600 μC.
(d) If resulting potential is 40 V the energy converted in heat in the process is 30 mJ.

24. In Fig. 19.115, all the three plates have same area and mid plate is given charge q.

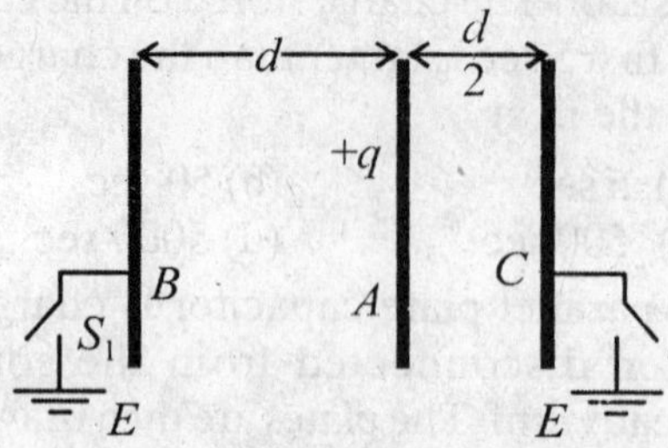

Fig. 19.115

(a) If switch S_1 is closed charge q goes to earth (S_2 is open)
(b) If switch S_2 is closed charge q goes to earth (S_1 is open)
(c) If S_1 and S_2 both are closed charge $2q/3$ goes to earth through S_1 and charge $q/3$ goes to earth through S_2
(d) If S_1 and S_2 both are closed, the potential difference between A and B is same as potential difference between A and C.

ANSWERS

1. (a, d)	2. (a, c)	3. (b, c, d)	4. (a, b, d)	5. (a, c, d)
6. (b, c)	7. (b, c, d)	8. (a, b, c, d)	9. (a, b, c)	10. (b, c)
11. (a, c, d)	12. (a, c)	13. (b, d)	14. (b, d)	15. (b, d)
16. (a, b, c, d)	17. (a, c, d)	18. (a, c)	19. (b, c, d)	20. (a, b, d)
21. (a, d)	22. (a, b, c)	23. (a, b, c, d)	24. (a, b, c, d)	

20

Electric Current and Circuit Analysis

20.1 ELECTRIC CURRENT AND CURRENT DENSITY

The flow of charge from one side of an area to the other is called an *electric current*. The direction of moving positive charges is taken as the direction of current. In fact, in conductors it is the electrons (–ve charge carriers) which carry the charge from one side of area to the other and as such the current direction is opposite to the flow of these negative charges.

In case of liquids *positively and negatively charged ions are the current carriers*, whereas in gases *the positive ions and electrons are the current carriers.*

The time rate of flow of charge through any cross-section is called current. If a charge ΔQ crosses an area in time Δt, we define average electric current as

$$i = \frac{\Delta Q}{\Delta t}$$

and instantaneous current is defined as

$$i = \underset{\Delta t \to 0}{\text{Lt}} \frac{\Delta Q}{\Delta t} = \frac{dQ}{dt} \qquad \ldots(20.1)$$

Practical unit of current is *ampere.* If one Coulomb of charge crosses an area in one second, we say the current is one ampere.

Current is taken as a fundamental quantity with ampere as unit in mks SI units and is represented by I or A. The cgs unit of current is emu (electro magnetic unit) of current and is called as biot (Bi).

$$1\text{ A} = \frac{1\text{ C}}{\text{s}} = \frac{(1/10)\text{ emu of charge}}{\text{s}} = \frac{1}{10}\text{ Bi}$$

Conventionally a direction is associated with current (the direction of motion of positive charges or direction of applied electric field) yet it is treated as a scalar quantity because for its addition, law of scalar addition is applicable and not the law of vector addition.

We know that charge remains conserved and the current is the rate of flow of charge, this means that the charge entering at one end of the conductor per second is equal to the charge leaving the other end of the conductor per second. This implies that the current does not change with change in cross-section.

The conductor remains uncharged when the current flows through it.

Current Density

Current density at a point in a conductor is the current that flows per unit area of the conductor around that point, the area is taken perpendicular to the direction of flow of current.

If Δi is the current through the area Δs around a point, then average current density is

$$j = \frac{\Delta i}{\Delta s}$$

The current density at the point is then

$$j = \underset{\Delta t \to 0}{\text{Lt}} \frac{\Delta i}{\Delta s} = \frac{di}{ds}$$

For a current i uniformly distributed over an area s and is perpendicular to i then

$$j = \frac{i}{s} \qquad \text{...(20.2)}$$

Current density is a *vector quantity* and has the direction same as the direction of current, i.e. along the moving positive charges.

Current is also defined as flux of current density, i.e.

$$i = \int \boldsymbol{j} \cdot \boldsymbol{ds} \qquad \text{...(20.3)}$$

20.2 ELECTROMOTIVE FORCE

Current is due to motion of charge carriers. A continuous electric current in a closed circuit is possible if there is an external source of force which makes the charge carrier to move in a definite direction. This external force which compels the charge carries to move in a definite direction is called *electromotive force (emf). The electromotive force in fact is not a force but energy* and is defined as *the work done in carrying a charge once around the closed circuit.* In order to have a continuous current in a closed circuit, a device is required which supplies the energy for charges to move around a closed circuit. Such a device is called a source of emf. It may be a cell, a battery, a generator, etc.

For a cell emf is defined as the potential difference between its electrodes when no current is drawn from the cell, i.e. when the cell is in open circuit.

SI unit of emf is volt.

20.3 CURRENT AND DRIFT VELOCITY

In conductors the valance electrons are loosely bound. As such they are free to move inside the metal but cannot escape from it. These are called *free electrons.* The free electrons are moving randomly whereas the relatively heavy positive ions called lattice keep on vibrating about their mean position. The randomly moving free electrons collide with the lattice. The speed and direction of the electrons changes randomly at each such collision. As a result the electron moves in a zig zag way. Since there is very large number of free electrons having random motion as such in absence of any electric field the number of electrons crossing any area from one side very nearly equals the number crossing the other side in any given time. As a result the net current through a cross-section is zero.

In the presence of electric field inside the conductor, a force acts on each electron in the direction opposite to the direction of electric field. As a result of this force the electrons drift slowly in the direction of force, i.e. in the direction opposite to the direction of electric field. The net result of this is that the electron acquires a small velocity towards the positive end of the conductor called *drift velocity*. Due to this the drift, the random motion of the electrons gets modified and there is a net transfer of electrons across a cross-section resulting in current. Thus when an electric field is applied inside the conductor, a dynamic equilibrium is achieved due to two processes, collision and acceleration, and the electron gets average drift velocity.

Expression for Drift Velocity

If $\boldsymbol{E}$ is the electric field applied, then electron in this field experiences a force

$$\boldsymbol{F} = (-e)\boldsymbol{E}$$

and acquires acceleration

$$a = \frac{F}{m} = -\frac{eE}{m}$$

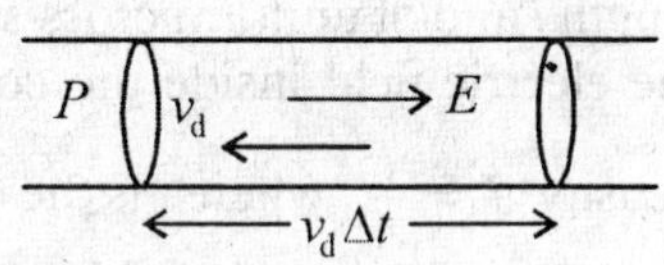

Fig. 25.1

I' - be the average time between two successive collisions, then drift velc ;

$$v_d = a\tau = -\frac{e\tau}{m}\boldsymbol{E} \qquad \text{...(20.4)}$$

ıve sign merely gives that direction of drift velocity is opposite to the direction of ıd.

observe that drift velocity of electrons depends on the applied electric field. Drift velocity per ıt electric field is called mobility (μ), i.e.

$$\mu = \frac{v_d}{E} = \frac{e\tau}{m} \qquad \text{...(20.5)}$$

Expression for Current in Terms of Drift Velocity

Consider a conductor of cross-sectional area A in which an electric field E is applied (Fig. 20.1). Consider a length $v_d\Delta t$ of this conductor where v_d is the drift velocity of the electrons. Volume of this portion of conductor is $Av_d\Delta t$. All the electrons within this volume of the conductor will cross cross-section ρ in time Δt. If n be the free electrons per unit volume in the conductor, then $nAv_d\Delta t$ electrons cross-section ρ in time Δt and each carries charge e. Hence current

$$i = \frac{nAv_d\Delta te}{\Delta t} = nAv_d e \qquad \text{...(20.6)}$$

and in terms of relaxation time $\left(v_d = \frac{e\tau}{m}E\right)$

$$i = \frac{nAe^2E\tau}{m} \qquad \text{...(20.7)}$$

Then the current density is given by

$$J = \frac{i}{A} = nev_d = \frac{ne^2E\tau}{m} \qquad \text{...(20.8)}$$

Ohm's Law: Equation 20.8 gives

$$\frac{J}{E} = \frac{ne^2\tau}{m} = \sigma \qquad \ldots(20.9)$$

where $\sigma = \dfrac{ne^2\tau}{m}$ is a constant and depends only on the material of the conductor and its temperature. This constant is called electrical conductivity of the material. This is known as Ohm's law.

Reciprocal of the conductivity is called *resistivity* (ρ), i.e.

$$\rho = \frac{1}{\sigma} = \frac{m}{ne^2\tau} \qquad \ldots(20.10)$$

Let us apply a potential difference V across a conductor of length l and of uniform cross-sectional area A (Fig. 20.2). Then the electric field inside the conductor is $E = V/l$ and current density $J = \dfrac{i}{A}$, where i is the current through the conductor as a result of the electric field applied. Then

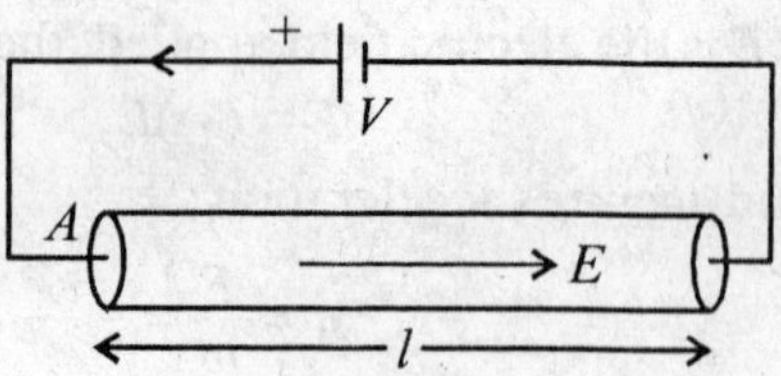

Fig. 20.2

$$\frac{i}{A} = \sigma\frac{V}{l}$$

or

$$V = \frac{1}{\sigma}\frac{l}{A}i = \rho\frac{li}{A} = Ri \qquad \ldots(20.11)$$

where $R = \dfrac{\rho l}{A}$ and is called resistance of the given conductor. Equation (20.11) is also a form of Ohm's law generally used in circuit analysis.

The unit of resistance is Ohm

$$1 \text{ Ohm} = \frac{1 \text{ Volt}}{1 \text{ Amp}}$$

The dimensions of Ohm are $M^1L^2T^{-3}I^{-2}$

The unit of resistivity ρ is Ohm-m and that of conductivity σ is $(\text{Ohm-m})^{-1}$ or mhos-m^{-1}.

More about Current and Current Density

When steady current flows in a metallic conductor of non-uniform cross-section, then

1. the current density J will not be constant and will vary with area A as $J = \dfrac{I}{A}$, where I is steady current.
2. $J = \sigma E$ or $E = \dfrac{J}{\sigma} = \dfrac{I}{A\sigma}$, this implies that electric field will not be constant and will vary inversely as area (A)
3. $J = nev_d$ or $v_d = \dfrac{I}{neA}$ this also means that drift velocity v_d will not be constant and will vary inversely as area (A)
4. Equation (20.8) gives $v_d = \dfrac{e\tau}{m}E = \dfrac{e\tau}{m}\left(\dfrac{V}{l}\right)$

   (*a*) $v_d \propto V$ (V is the applied voltage)

(*b*) $v_d \propto \frac{1}{l}$ (l is the length of the conductor)

(*c*) In case of constant potential difference applied across the conductor, the drift velocity is independent of cross-section area.

Dependence of Resistance

(*a*) Ohm's law states that the potential difference applied is directly proportional to current, i.e.

$$V \propto I \text{ or } V = IR$$

Such resistances are called *Ohmic resistances* and I versus V curve for them is a straight line. Resistance of such conducting devices is independnet of magnitude and direction of current passing through them.

(*b*) Though resistance of Ohmic-conductors is independent of magnitude and direction of flow of current, still for a given device (conductor) it is not unique and depends on l and A, i.e. how the potential difference is applied, e.g. for conductor shown in Fig. 20.3, the resistance (R_1) between faces $ABCD$ and $EFGH$ is different from the resistance (R_2) between faces $ABGH$ and $EFCD$. Resistance $R_1 > R_2$.

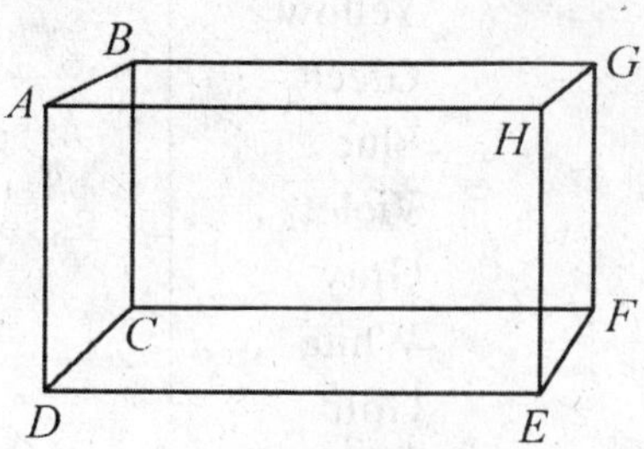

Fig. 20.3

(*c*) The resistance of a conductor depends on temperature.

If R_0 is the resistance at 0° C and R_t is resistance at t° C, then change in resistance

$$R_t - R_0 = R_0 \alpha t$$

where α is constant called *temperature co-efficient of resistivity*. Thus resistance at any temperature t is

$$R_t = R_0(1 + \alpha t) \quad \text{...(20.12)}$$

α is a positive constant and has order $\sim 10^{-5}$/°C.

(*d*) The resistivity ρ of the conductor also varies with temperature according to relation

$$\rho_t = \rho_0 (1 + \alpha t + \beta t^2 + \ldots)$$

where t is rise in temperature and α and β are constants. For small rise in temperature (i.e., t is small)

$$\rho_t = \rho_0(1 + \alpha t)$$

(*e*) In case of conductors the rise in temperature increases the thermal speed of the free electrons and amplitude of lattice vibrations. The collision frequency of free electrons with lattice will increase thus decreasing the relaxation time and increasing resistivity.

Colour Code for Carbon Resistances

The high resistors widely used in electronic circuits are made from carbon by molding it with some binding agent. The value of the resistance is indicated by four coloured bands marked on the surface. The meanings of the 4 positions of the bands are shown in Fig. 20.4 and the colour code is given in Table 20.1.

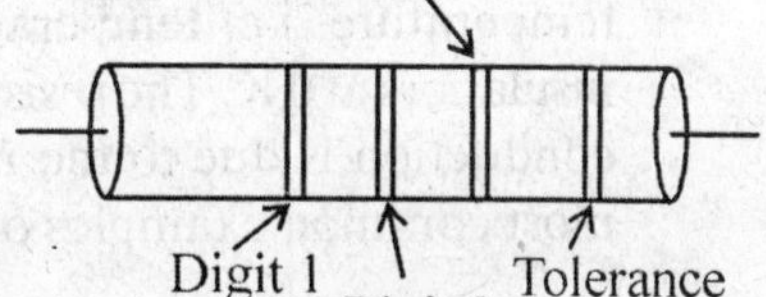

Fig. 20.4

Suppose the colour bands on the resistor shown are, red, green, blue and silver as red from left. The resistance is

$$(25 \pm 2.5) \times 10^6 \ \Omega$$

Sometimes, the tolerance band is missing from the code so that there are only three bands. This means 20% tolerance

Table 20.1

Colour	*Digit*	*Multiplier*	*Tolerance*
Black	0	10^0	
Brown	1	10^1	
Red	2	10^2	
Orange	3	10^3	
Yellow	4	10^4	
Green	5	10^5	
Blue	6	10^6	
Violet	7	10^7	
Gray	8	10^8	
White	9	10^9	
Gold		10^{-1}	5%
Silver		10^{-2}	10%

20.4 CLASSIFICATION OF SUBSTANCES ACCORDING TO THEIR RESISTIVITY

All the substances have been classified into following four classes depending upon their resistivity and its variation with temperature. Their behaviour is explained on the basis of band theory.

1. **Insulators:** These have very high resistivity ($\sim 10^{15}$ Ω–m) and the band gap is of the order 5 to 6 eV. These have no free electrons. The ideal insulator has infinite resistivity and it decreases with rise in temperature. Examples are rubber, glass porcelain, fused quartz, etc.

2. **Conductors:** In conductors there are a large number of free electron ($\sim 10^{28} - 10^{29}$ m^{-3}). As such their resistivity is small ($\sim 10^{-8}$ Ω-m). The band gap is zero. In fact the conduction band and valance band overlap. The resistivity of an ideal conductor is zero and increases with increase in temperature.
   Examples are copper, aluminimum, silver, etc. In case of alloys of metals such as nichrome, manganin, and constantan resistivity is more than that of constituent metals and varies slowly with temperature than pure metals.

3. **Semi-conductors:** The resistivity of semi conductors is of the order of 10^{-1} to 10^{+1} Ω-m. The band gap also varies from 0.6 eV to 1.1 eV. They have few conduction electrons (10^{15}–10^{16} m^{-3}) compared to conductors. The resistivity of semi-conductors decreases with temperature, i.e. temperature co-efficient of resistance (α) is negative. They are perfect insulatiors at 0 K. The resistivity of semi-conductors decreases with impurities. The electrical conduction is due to the motion of both electrons and holes. Silicon and germanium are most common examples of semi-conductors.

4. **Super-conductors:** For certain materials the resistivity suddenly drops to zero below a certain temperature called *critical temperature* (T_c). Above critical temperature the resistivity follows

the trend of a normal metal (Fig. 20.5). Kamerlingh Onnes discovered the phenomenon of super-conductivity in 1911. He observed that resistance of Hg become zero below 4.2 K.

An electrical current once set up in super-conductors will persist indefinitely without any applied emf. It is also observed that steady currents persist in super-conducting loops without any appreciable decrease. This is why super-conductors are used to construct very strong magnets and as a power line for transmission of electrical energy without any loss.

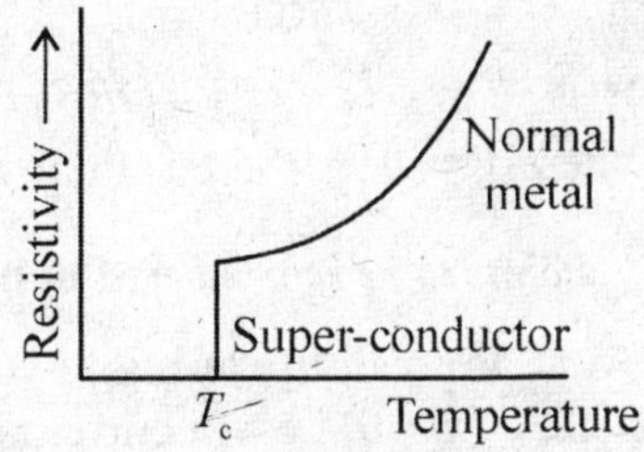

Fig. 20.5

However, the requirement of low temperature makes its application expensive and impracticable. At present cuperate super-conductors around 133 K are reported and efforts are on to develop compounds and alloys which would be super-conducting at room temperature (300 K).

20.5 ELECTRIC POWER AND ENERGY TRANSFER IN AN ELECTRICAL CIRCUIT

A simple circuit consisting of a resistor R connected to a battery CD of emf E is shown in Fig. 20.6. The connecting wires AC and BD are assumed to be of zero resistance. This means the potential difference across CA and DB is zero even if the current is flowing. This implies that the potential difference across resistor is the same as that across the battery. If i is the current in the circuit then this potential difference

$$V = V_A - V_B = V_C - V_D = iR$$

The charge that passes through the circuit in time t is $q = it$. As the charge goes from A to B, the electric potential energy decreases by

$$U = qV = (it)(iR) = i^2Rt \qquad \text{...(20.13)}$$

This loss in electric potential energy appears as thermal energy of the resistor, which heats up the resistor.

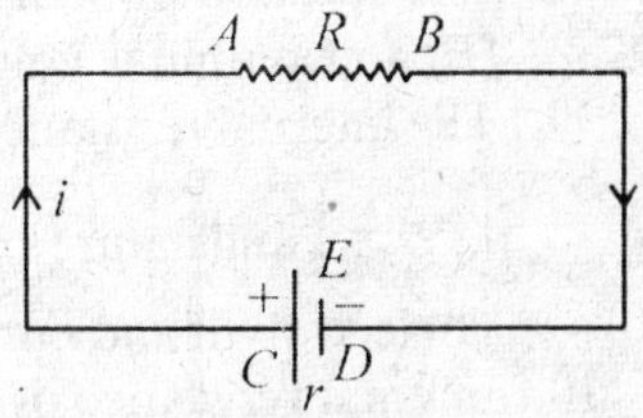

Fig. 20.6

Thus a current for a time t through a resistor R produces thermal energy equal to i^2Rt and the power developed is

$$P = \frac{U}{t} = i^2R$$

But $\quad V = iR$

$$\therefore \quad P = i^2R = Vi = \frac{V^2}{R} \qquad \text{...(20.14)}$$

Internal Resistance of Battery

As the charge $q = it$ goes through the battery from the negative terminal D to positive terminal C work is done against the resisting force offered by the electrolyte. This opposition of electrolyte to the flow of current through it is called *internal resistance*. The battery develops thermal energy as current passes through it and potential difference between terminals is smaller than the emf. This potential difference is called *terminal voltage*.

Relation between *E*, *V* and *r*

In Fig. 20.6 if r be the internal resistance of the battery, then current i is

$$i = \frac{E}{R + r}$$

or $\quad iR + ir = E$

or $\quad V = iR = E - ir$ ...(20.15)

where $V = iR$ is the terminal voltage.

Points to be remembered in connection with internal resistance

1. If $r = 0$, the terminal voltage is equal to emf (E) of the battery.
2. If $i = 0$, i.e. no current is drawn from the battery (in open circuit) terminal voltage is equal to emf.
3. In closed circuit, the terminal voltage is less than emf if current in battery goes from negative terminal to positive terminal.
4. If the current in battery goes from positive to negative terminal the terminal voltage $V = E + ir$, i.e. terminal voltage is greater than emf.
5. A battery without internal resistance is called an ideal battery. It is represented by symbol shown in Fig. 20.7 (*a*).
6. A battery with internal resistance is called a non ideal battery and is represented by Fig. 20.7 (*b*).
7. The thermal energy developed in time t is i^2rt.
8. The addition of the resistance r accounts for difference in emf (E) and terminal voltage (V).
9. The internal resistance depends on the distance between electrodes ($r \propto d$), area of electrodes $\left(r \propto \frac{1}{A}\right)$ and nature and concentration of electrolyte ($r \propto c$). It also depends on the temperature of electrolyte and current drawn.
10. Internal resistance of different types of cells is different.

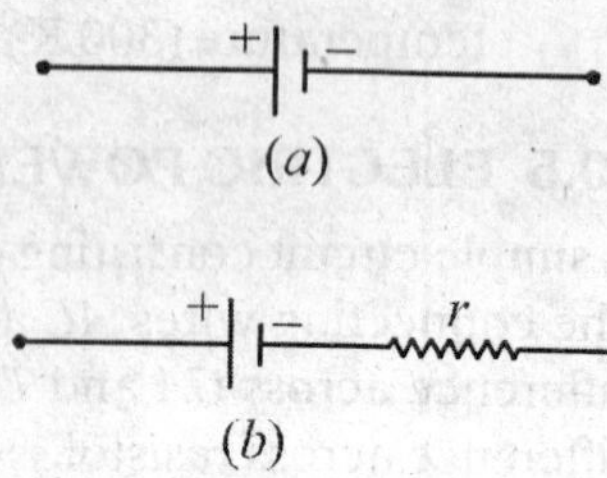

Fig. 20.7

20.6 KIRCHHOFF'S LAWS

In certain circuits such as unbalanced Wheatstones bridge, disimilar cells connected in parallel or a loop containing capacitor, Ohm's law is insufficient to analyse such circuits. For analysis of such complex DC circuits, we use Kirchhoff's laws. There are two laws:

Current Law or the Junction Law

This law states that the algebraic sum of all the currents directed towards a point is zero, i.e.

$$\Sigma i = 0$$

This implies that sum of the currents directed towards a point in a circuit is equal to the sum of all the currents directed away from the point. This is illustrated in Fig. 20.8.

This law is the consequence of law of conservation of charge which means that net charges coming towards the point should be equal to the charges going away from the point in the same time.

If we take the current approaching a point as positive then the current directed away from the point is negative.

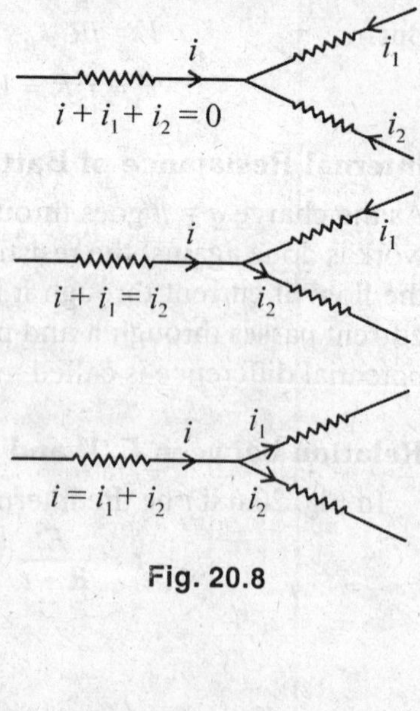

Fig. 20.8

This law is also applicable to the capacitors also using the concept of displacement current $\left(i_d = \varepsilon_0 \frac{d\phi_e}{dt}\right)$ treating the resistance of capacitor zero during charging and discharging and infinite in steady state.

Potential Law or Loop Law

This law is related to the potential differences across all the elements of the closed mesh.

This law states that the algebraic sum of the potential differences (product of current and resistance) along a closed mesh in a circuit is zero, i.e.

$$V = \Sigma iR = 0$$

The potential drop is taken to be positive and rise in potential is taken to be negative.

In Fig. 20.9 for mesh *ABDA*, we have

$$i_1R_1 + i_3R_5 - i_2R_2 = 0$$

For mesh *ABCEFA*, we have

$$i_1R_1 + (i_1 - i_3)R_3 - E = 0$$

and for mesh *ABCDA*, we have

$$i_1R_1 + (i_1 - i_3)R_3 - (i_2 + i_3)R_4 - i_2R_2 = 0$$

(*i*) This law is consequence of law of conservation of energy. In this whatever the energy is given to the charge by the source of emf is lost in passing through the resistance.

(*ii*) This law can be applied to mesh containing a capacitor by treating the capacitor as a source of emf (q/C) in open circuit in steady state and short circuited in transient state.

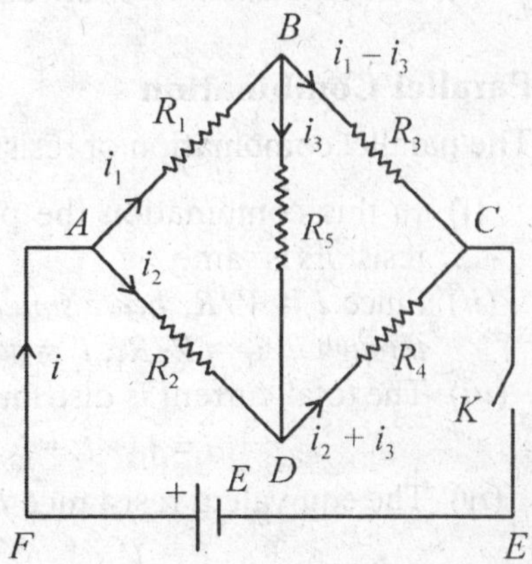

Fig. 20.9

20.7 COMBINATION OF RESISTANCES

The combination of several resistances is called a network. If we replace this network by a single resistance such that it draws the same current when same potential difference (V) is applied to it as was applied to the network, then this resistance is called *equivalent resistance*, i.e.

$$R_{eq} = \frac{V}{i}$$

We have two types of combinations—*series* and *parallel* combination.

Series Combination

Series combination of resistances is shown in Fig. 20.10.

(*i*) In series combination same current flows through all the resistors.

(*ii*) Potential difference is different across different resistors. Greater the resistance greater is the potential difference ($V = iR$)

$$V_1 = iR_1,\ V_2 = iR_2,\ V_3 = iR_3$$

and total potential difference $V = V_1 + V_2 + V_3 = iR_1 + iR_2 + iR_3$

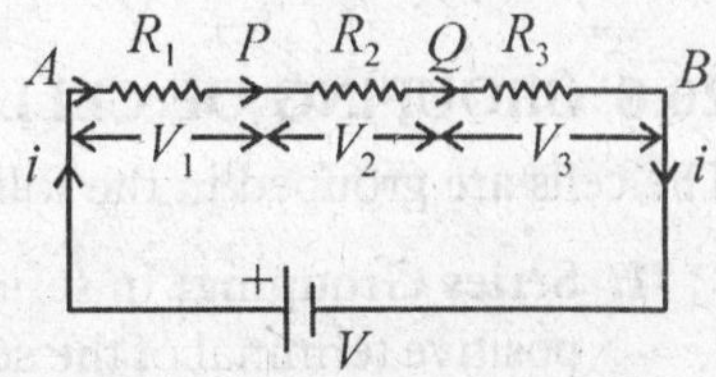

Fig. 20.10

(*iii*) The equivalent resistance R_S will draw the same current i when same potential difference V is applied to it, i.e.

$$V = iR_S$$

Equating the values of V, we have

$$iR_S = iR_1 + iR_2 + iR_3 \qquad \because i \neq 0$$

Hence we have

$$R_S = R_1 + R_2 + R_3$$

The argument can be extended to any number of resistors connected in series. Thus

$$R_S = R_1 + R_2 + R_3 + \ldots \qquad \ldots(20.16)$$

(*iv*) If n resistances each of value r are connected in series the equivalent resistance

$$R_S = nr$$

(*v*) The equivalent resistance is greater than the highest resistance in the network.

Parallel Combination

The parallel combination of resistances is shown in Fig. 20.11.

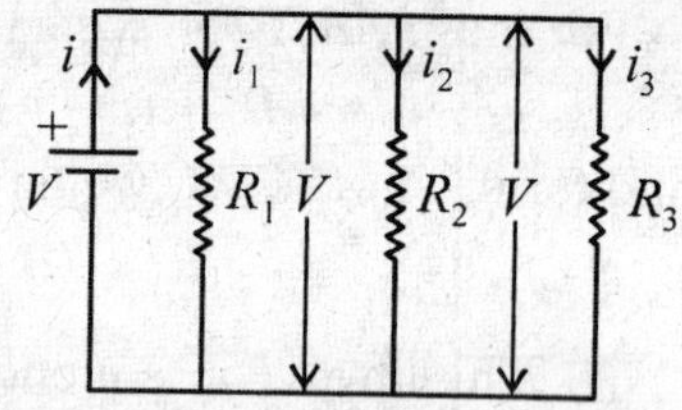

Fig. 20.11

(*i*) In this combination the potential difference across all the resistors is same.

(*ii*) Since $i = V/R$, *hence smaller the resistance more is the current through it.* $i_1 = V/R_1$, $i_2 = V/R_2$ and $i_3 = V/R_3$

(*iii*) The total current is distributed among resistors such that

$$i = i_1 + i_2 + i_3$$

(*iv*) The equivalent resistance R_P will draw current i when potential difference V is applied, i.e.

$$i = \frac{V}{R_P}$$

Thus we have

$$\frac{V}{R_P} = \frac{V}{R_1} + \frac{V}{R_2} + \frac{V}{R_3}$$

and

$$\frac{1}{R_P} = \frac{1}{R_1} + \frac{1}{R_2} + \frac{1}{R_3} \qquad \ldots(20.17)$$

The argument can be extended to any number of resistances.

(*v*) For n resistances each of value r connected in parallel, the equivalent resistance is

$$R_P = \frac{r}{n}$$

(*vi*) The equivalent resistance is less than the least resistance in the network.

20.8 GROUPING OF CELLS

The cells are grouped in the following three ways:

1. **Series Grouping:** In series connection the negative terminal of first cell is connected to the positive terminal of the second cell. The negative terminal of second cell is connected to the positive terminal of third cell and so on. n cells each of emf E and internal resistance r

connected in series is shown in Fig. 20.12 (*a*). This combination will behave like a single cell of emf nE and internal resistance nr. The equivalent circuit is shown in Fig. 20.12 (*b*).

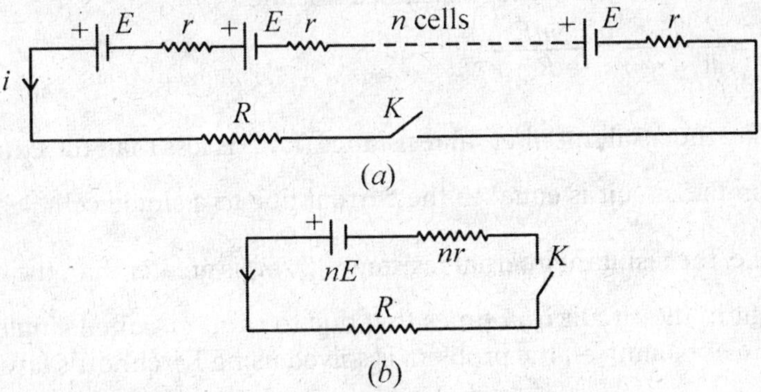

Fig. 20.12

The current through the external resistance R is

$$i = \frac{nE}{R + nr}$$

(*i*) If $nr << R$, i.e. the resultant internal resistance is less than the external resistance, then

$i = \frac{nE}{R}$, i.e. current is n times the current due to single cell.

(*ii*) If $nr >> R$, i.e. the resultant internal resistance is greater than the external resistance, then $i = E/r$, i.e. current in the circuit is equal to that of a single short circuited cell.

(*iii*) If cells in series are not identical, then

$$i = \frac{\Sigma E_j}{R + \Sigma r_j}$$

(*iv*) If out of these n cells s cells are reversed then equivalent emf $= (n - s)E - sE = (n - 2s)E$

and current $i = \frac{(n - 2s)E}{R + nr}$.

2. Parallel Groupings: If positive terminals of all the cells are connected at one junction and negative at the other, the cells are said to be in parallel. m cells each of emf E and internal resistance r are shown connected in parallel in Fig. 20.13 (*a*). This combination is equivalent

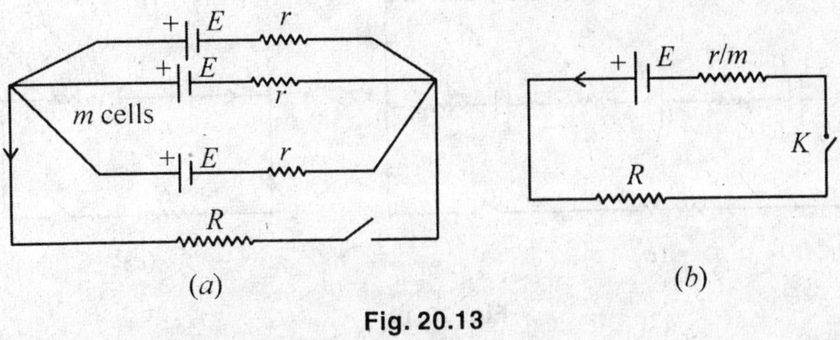

Fig. 20.13

to a single cell of emf E and internal resistance r/m. The equivalent circuit is shown in Fig. 20.13 (*b*), R is the external resistance.

The current through the external resistance R is then

$$i = \frac{E}{R + r/m} = \frac{mE}{mR + r}$$

(*i*) If $\frac{r}{m} << R$, i.e. the resultant internal resistance (r/m) is less than the external resistance R, the current in the circuit is equal to the current due to a single cell.

(*ii*) If $\frac{r}{m} >> R$, i.e. the resultant internal resistance (r/m) is greater than the external resistance R, the current in the circuit is m times that due to short circuited single cell.

(*iii*) If the cells are not identical, the problem is solved using Kirchhoff's laws and the terminal voltage V is

$$E_{\min} < V < E_{\max}$$

where $E_{\min}$ and $E_{\max}$ respectively are the minimum and maximum emf of the cells arranged in parallel.

Thus to get maximum current, cells are to be connected in series if net internal resistance is small in comparision to the external resistance and in parallel if net internal resistance is greater than the external resistance.

3. Mixed Grouping: Here we have n identical cells (each of emf E and internal resistance r) in a single row and there are m such rows, i.e. total number of cells are $n \times m$ (Fig. 20.14 (*a*)).

We can treat each row as a single cell of emf nE and internal resistance nr. As such we have in all m such cells in parallel. This is shown in Fig. 20.14 (*b*).

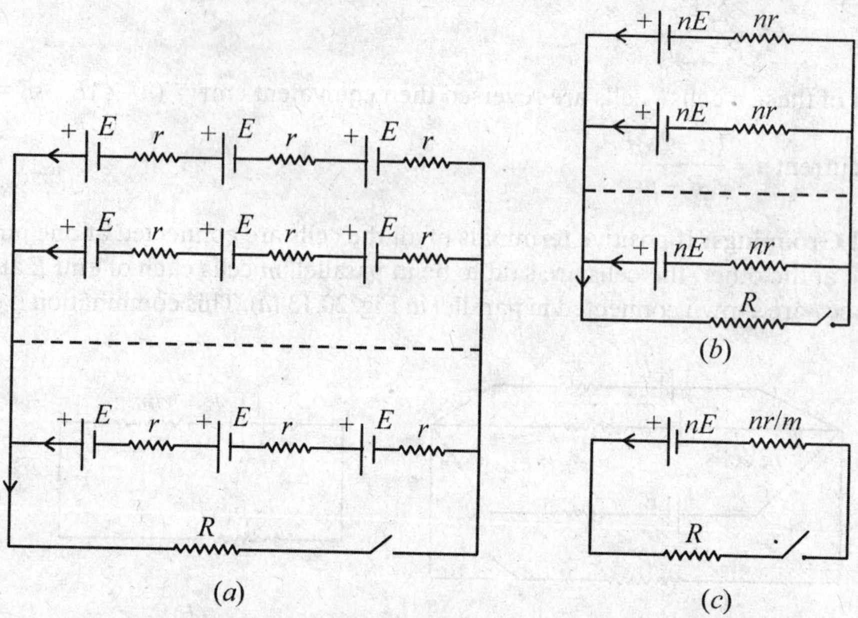

Fig. 20.14

This is then equivalent to a single cell of emf nE and internal resistance $\frac{nr}{m}$ (Fig. 20.14 (*c*)).

The current in the circuit through external resistance R, then is

$$i = \frac{nE}{\left(R + \frac{nr}{m}\right)} = \frac{mnE}{mR + nr} = \frac{mnE}{\left\{\left(\sqrt{mR} - \sqrt{nr}\right)^2 + \sqrt{2mnrR}\right\}}$$

for current to be maximum

$$\sqrt{mR} - \sqrt{nr} = 0$$

or $$R = \frac{n}{m}r \qquad \text{...(20.18)}$$

i.e. for maximum current the total internal resistance must be equal to external resistance.

Maximum Power Transferred to the Load

Power transfer is maximum when load resistance is equal to the internal resistance. This is called maximum power transfer theorem. Hence power transferred to external resistance is

$$P = \frac{n^2E^2R}{\left(R + \frac{nr}{m}\right)^2} \qquad \text{But} \quad R = \frac{nr}{m}$$

$$\therefore \quad P_{max} = \frac{n^2E^2R}{4R^2} = \frac{n^2E^2}{4R}$$

Parallel Combination of Two Cells of Different emf

Figure 20.15 shows two cells of emf, E_1 and E_2 and internal resistance r_1 and r_2 connected in parallel. The currents in different branches are also shown. Applying Kirchhoff's loop law to the loop containing E_1, R and r_1, we have

$$R(i_1 + i_2) + r_1i_1 - E_1 = 0$$

or $$i_1(r_1 + R) + i_2R = E_1 \qquad \text{...}(i)$$

Similarly applying Kirchhoff's loop law to the loop containing E_2, R and r_2, we get

$$i_1R + (r_2 + R)i_2 = E_2 \qquad \text{...}(ii)$$

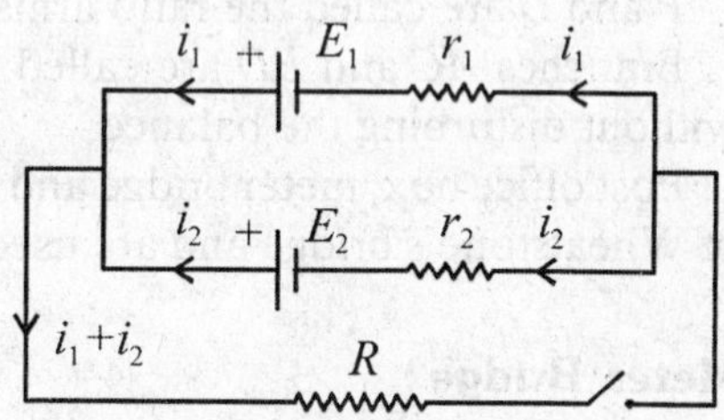

Fig. 20.15

Multiply (*i*) by r_2 and (*ii*) by r_1 and adding, we have

$$(i_1 + i_2)\{r_1r_2 + R(r_1 + r_2)\} = E_1r_2 + E_2r_1$$

$$(i_1 + i_2) = \frac{E_1r_2 + E_2r_1}{r_1r_2 + R(r_1 + r_2)}$$

or $i = i_1 + i_2$ the current through external resistance R is

$$i = \frac{\frac{E_1r_2 + E_2r_1}{r_1 + r_2}}{R + \frac{r_1r_2}{r_1 + r_2}} = \frac{E}{R + r}$$

i.e., for external resistance the combination behaves as a single cell of emf

$$E_0 = \frac{E_1 r_2 + E_2 r_1}{r_1 + r_2}$$

and internal resistance

$$r_0 = \frac{r_1 r_2}{r_1 + r_2}$$

and when $E_1 = E_2$ and $r_1 = r_2 = r$
then $E_0 = E$ and $r_0 = r/2$.

20.9 WHEATSTONE'S BRIDGE

A network of 4 resistances with galvanometer shown in Fig. 20.16 is called Wheatstone's bridge network. The current through various branches is also shown.

Applying loop rule to mesh *ABDA*, we have

$$i_V P + i_g G - i_2 R = 0 \qquad \ldots(i)$$

For mesh *BDCB*, we get

$$i_g G + (i_2 + i_g)\, S - (i_1 - i_g) Q = 0 \qquad \ldots(ii)$$

The bridge is said to be balanced if *B* and *D* are at the same potential, i.e. no current flows through branch *BD*, i.e. $i_g = 0$ and galvanometer shows no deflection.

Now when bridge is balanced, equations (*i*) and (*ii*) become

$$i_1 P - i_2 R = 0 \quad \text{or} \quad i_1 P = i_2 R \qquad \ldots(iii)$$

and

$$i_2 S - i_1 Q = 0 \quad \text{or} \quad i_1 Q = i_2 S \qquad \ldots(iv)$$

Dividing, we get

$$\frac{P}{Q} = \frac{R}{S} \qquad \ldots(20.19)$$

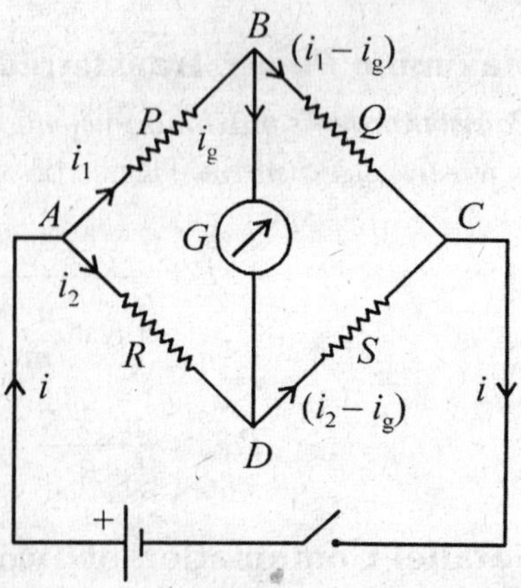

Fig. 20.16

This is the condition when the Wheatstone's bridge is balanced. Now if three resistances are known, then 4th can be determined.

P and *Q* are called the ratio arms. *R* is known resistance and *S* is unknown.

Branches *AC* and *BD* are called conjugate arms, i.e. these two branches can be interchanged without disturbing the balance.

Post office box, meter bridge and Carey Foster bridge are the instruments based on this principle of Wheatstone's bridge and are used to measure the unknown resistance and specific resistance.

Meter Bridge

Meter bridge in balanced condition is shown in Fig. 20.17.

In this uniform cross-section wire of 1 meter long (AC) is used. In balanced conditions *B* and *D* are at same potential. Then

$$P = \rho l \text{ and } Q = \rho(100 - l)$$

when ρ is the resistance per unit length of wire, then

$$\frac{P}{Q} = \frac{\rho l}{\rho(100 - l)} = \frac{l}{100 - l}$$

Then $\qquad \dfrac{P}{Q}=\dfrac{R}{S}$ or $S=\dfrac{R\cdot Q}{P}=\dfrac{R(100-l)}{l}$

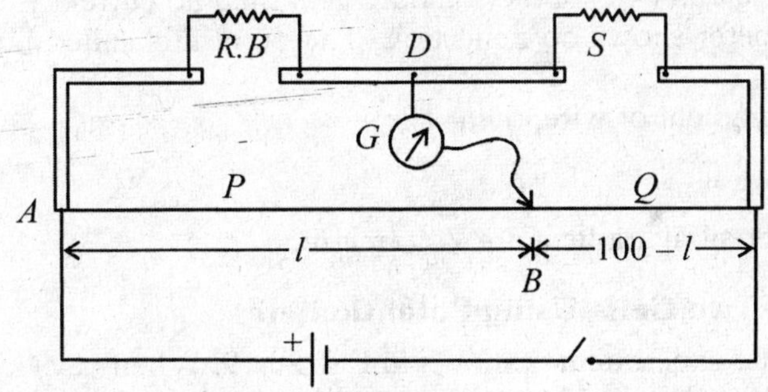

Fig. 20.17

20.10 POTENTIOMETER

Potentiometer is an ideal voltmeter because it does not draw any current from the circuit and still measures potential difference. In this sense it is voltmeter with infinite resistance.

A potentiometer consists of a long uniform wire generally made of nichrome, or constantan and is stretched on a wooden board. The length of wire varies from instrument to instrument (usually 5 to 10 m). There is Jockey with the help of which the contact can be made at any point on the wire.

Primary circuit of potentiometer is shown in Fig. 20.18 in which *AB* is an stretched wire connected in series with a cell of constant emf *E*, a key *K* and a rheostate (Rh).

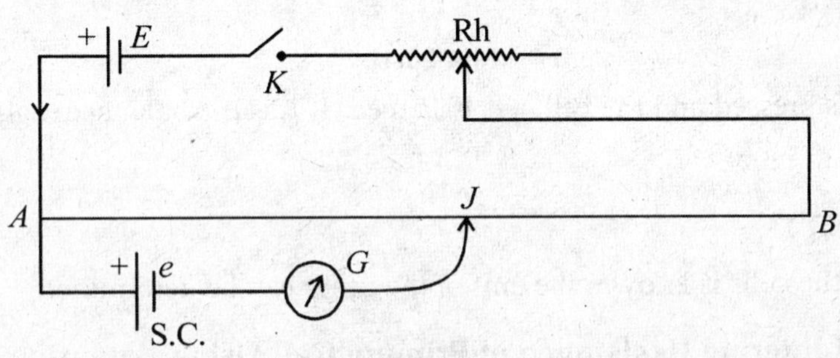

Fig. 20.18

As the current passes through the potentiometer wire *AB* the potential drops along *AB*. Point *A* is at higher potential than point *B*. The fall of potential per unit length along the wire is called potential gradient. If *V* is the potential difference across *AB* and *L* is the length of potentiometer wire then

$$\text{Potential gradient } x=\frac{V}{L} \qquad \text{...(20.20)}$$

If i is the current through the wire and R its resistance then $V = iR$, and

$$\text{Potential gradient } x=\frac{iR}{L}=i\rho \qquad \text{...(20.21)}$$

when ρ is the resistance per unit length of the potentiometer wire.

Potential gradient is determined using standard cell (S.C.) of emf e.

Point J on the wire is located using Jockey such that the potential difference across AJ length ($= l$) of wire is equal to the emf ($= e$) of the standard cell, then no current is drawn from standard cell (S.C.) and galvanometer shows no deflection. The point J is called null point and length l balancing length.

Now if x is potential gradient of wire, then

$$xl = e \quad \text{or} \quad x = \frac{e}{l} \qquad \ldots(20.22)$$

The SI mks units of potential gradient are Volt/metre.

Comparison of emf of Two Cells Using Potentiometer

The circuit diagram for the experiment is shown in Fig. 20.19. When key (1, 2) is pressed the cell e_1 is balanced. If balance length is l_1 and x is potential gradient, then

$$e_1 = xl_1 \qquad \ldots(i)$$

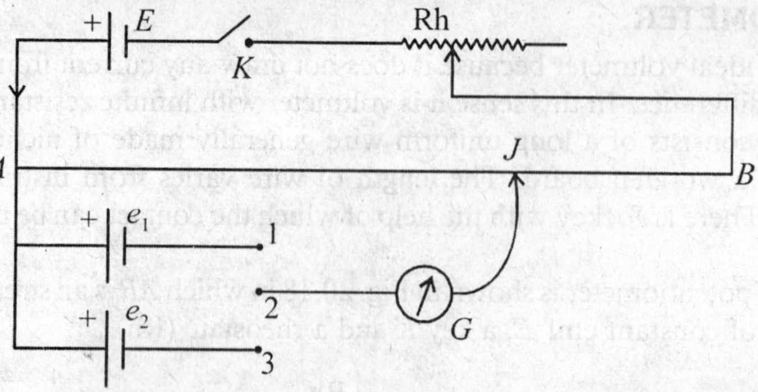

Fig. 20.19

Now key (2, 3) is pressed and the cell e_2 is balanced. If l_2 is the balance length now, then

$$e_2 = xl_2 \qquad \ldots(ii)$$

dividing, we have $\frac{e_1}{e_2} = \frac{l_1}{l_2}$ ...(20.23)

If emf of one of the cells is known, the emf of the other can be determined.

Determination of Internal Resistance of Primary Cell Using Potentiometer

The circuit diagram is shown in Fig. 20.20. The emf of the cell is e and internal resistance r. A known resistance R and key K are connected across the cell.

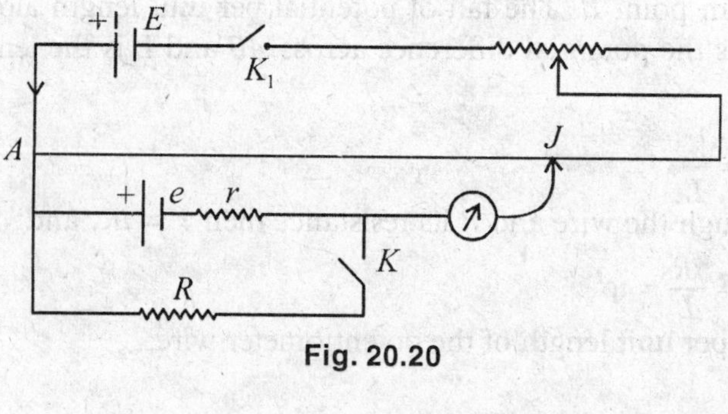

Fig. 20.20

The balance point J is searched on the potentiometer wire AB with the help of jockey when the circuit is open (key K open). In this case no current is drawn from cell as such potential difference across length $AJ\ (= l_1)$ is equal to the emf e of the cell. Then

$$e = xl_1 \qquad \ldots(i)$$

where x is the potential gradient of the potentiometer wire AB.

Now the key K is closed and the new balance point is searched. Let it be at length l_2. Then potential difference across length l_2 on the potentiometer wire will be equal to the terminal voltage (V) across the cell. Then

$$V = xl_2 \qquad \ldots(ii)$$

$$\therefore \quad \frac{e}{V} = \frac{l_1}{l_2} \qquad \ldots(iii)$$

when key K is closed the current i is

$$i = \frac{e}{R+r}$$

and terminal voltage $V = iR = \dfrac{eR}{R+r}$ or $\dfrac{e}{V} = \dfrac{R+r}{R}$

$$\therefore \quad \frac{R+r}{R} = \frac{l_1}{l_2}$$

and solving r we have

$$r = \frac{R(l_1 - l_2)}{l_2} \qquad \ldots(20.24)$$

20.11 MEASURING INSTRUMENTS: GALVANOMETER, AMMETER AND VOLTMETER

Throughout our discussion in this chapter, we have talked about current and potential difference. But so far we have not mentioned as to how these quantities are measured. The basic instruments for measuring current and potential difference is galvanometer. A galvanometer is a device which gives deflection whenever current passes through it, i.e. it detects the flow of current in circuit.

Galvanometer*

In DC circuit usually moving coil galvanometers are used. The deflection θ in the galvanometer coil is proportional to current, i.e.

$$I \propto \theta \text{ or } I = K\theta$$

where K is galvanometer constant. A sensitive galvanometer can detect a current of the order of 10^{-9} A passing through it.

The galvanometer coil has some resistance called galvanometer resistance and is represented by letter G.

The current required for its full scale deflection is represented by i_g and is called *current for full scale deflection.*

The current sensitivity of the galvanometer is defined as the deflection for unit current. If θ is the deflection when i current passes through it, then

$$\text{Current sensitivity by } = \frac{\theta}{i} = \frac{1}{K}$$

*The detailed construction and working of galvanometer will be discussed in chapter on magnetic fields.

The reciprocal of the current sensitivity is called *figure of merit*. Thus figure of merit is the current required for unit deflection. Hence

$$\text{Figure of merit} = \frac{i}{\theta} = K$$

Ammeter

A *shunt* is a low resistance connected in parallel to the galvanometer. *An ammeter is a shunted galvanometer.* A shunted Galvanometer is shown in Fig. 20.21. If i is the main current then current through galvanometer is

$$i_g = \frac{Si}{S+G}$$

and through shunt $i_s = \dfrac{Gi}{S+G}$

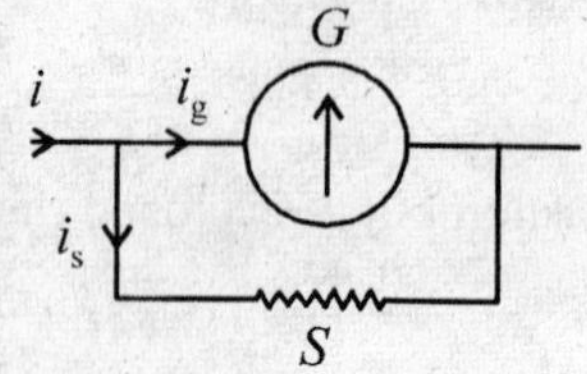

Fig. 20.21

If nth part of the main current is to pass through galvanometer, then

$$\frac{i_g}{i} = \frac{1}{n} = \frac{S}{S+G}$$

and
$$S = \frac{G}{n-1} \qquad \text{...(20.25)}$$

The shunt required to convert galvanometer into ammeter reading up to i amps is

$$S = \frac{Gi_g}{i - i_g} \qquad \text{...(20.26)}$$

Voltmeter

A voltmeter is a device to measure potential difference. It is a galvanometer with high resistance in series. Figure 20.22 illustrates the principle of voltmeter. If voltmeter is of range V, the resistance required to be connected in series is R given by

$$i_g(R+G) = V$$

$$R = \frac{V}{i_g} - G \qquad \text{...(20.27)}$$

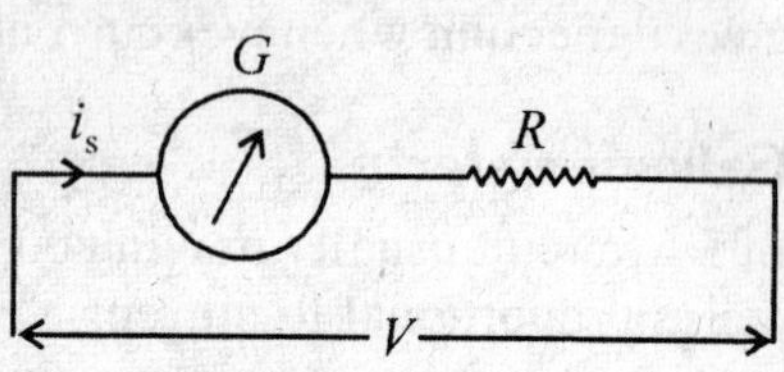

Fig. 20.22

The Calibration of ammeter and voltmeter is done using sensitive potentiometer.

SOLVED EXAMPLES

Example 20.1: A uniform copper wire of mass 2.33×10^{-3} kg carries a current of 1 A when 1.7 V is applied across it. Calculate its length and area of cross-section. If the wire is uniformly stretched to double its length, calculate the new resistance (Density of copper is 8.92×10^3 kg m^{-3} and resistivity is 1.7×10^{-8} Ω-m).

Solution: When voltage applied is 1.7 V the current is 1 A. Hence resistance of wire R is

$$R = \frac{V}{i} = \frac{1.7}{1} = 1.7\,\Omega \qquad ...(i)$$

If L is the length of the wire and A area of cross-section, then

$$R = \rho \cdot \frac{L}{A}$$

or
$$\frac{L}{A} = \frac{R}{\rho} = \frac{1.7}{1.7 \times 10^{-8}} = 10^8 \qquad ...(ii)$$

If mass is m then density $d = \frac{m}{V} = \frac{m}{AL}$

$$\therefore \quad AL = \frac{m}{d} = \frac{2.23 \times 10^{-3}}{8.92 \times 10^3} = \frac{10^{-6}}{4} \qquad ...(iii)$$

From (*ii*) and (*iii*), we have

$$L^2 = \frac{10^2}{4} \quad \text{or} \quad L = 5 \text{ m}$$

and
$$A = 5 \times 10^{-8} \text{ m}^2$$

when the wire is stretehed to double its length, the area of cross-section becomes $A/2$ and resistance becomes

$$R' = \rho \frac{2L}{A/2} = 4\frac{\rho L}{A} = 4R = 4 \times 1.7 = 6.8\ \Omega.$$

Example 20.2: Find the current in R_3 and the energy stored in the capacitor in circuit shown in Fig. 20.23.

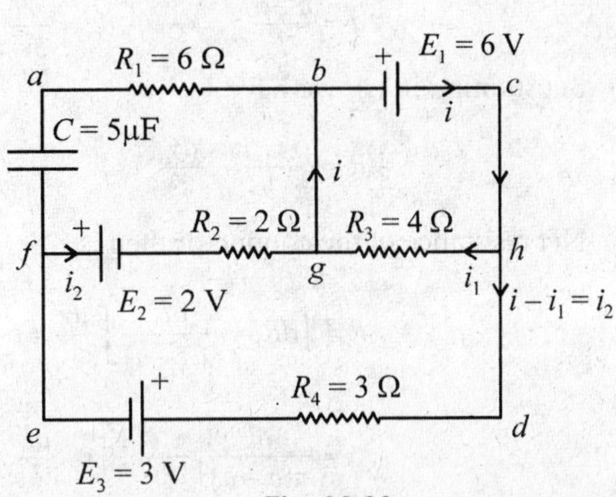

Fig. 20.23

Solution: Figure 20.23 shows the distribution of current through various elements of the circuit. It is to be remembered that no current flows through the capacitor in steady state. Applying Kirchhoff's law to mesh *bchg*, we have

$$4i_1 = 6$$

or
$$i_1 = \frac{6}{4} = 1.5 \text{ A} \qquad ...(i)$$

i.e. the current through resistor R_3 is 1.5 A.

Applying Kirchhoff's law to mesh *fghde*, we have

$$3i_2 + 3 + 2 + 2i_2 - 4i_1 = 0 \qquad \text{where } i_2 = i - i_1$$

or $\quad 4i_1 - 5i_2 = 5$

or $\quad 4i_1 - 5(i - i_1) = 5$

or $\quad -5i + 9i_1 = 5 \quad$ or $\quad i = 1.7\text{ A} \qquad \because i_1 = 1.5\text{ A}$

$\therefore \quad i_2 = i - i_1 = 0.2\text{ A}$

Consider the mesh *afgba* (*ii*)

$$V_a - V_f + 2 + 2 \times 0.2 = 0$$

or $\quad V_f - V_a = 2.4\text{ V}$

This is the potential difference across the capacitor *C*.
Hence energy stored in the capacitor is

$$E = \frac{1}{2}CV^2 = \frac{1}{2} \times 5 \times 10^{-6} \times (2.4)^2 = 14.4 \times 10^{-6}\text{ J}.$$

Example 20.3: A resistor is in the shape of a truncated right cone. The end radii are *a* and *b* (*b* > *a*). The specific resistance of material is ρ and altitude is *l*. If the taper is small, then find the resistance between the plane faces of the resistor.

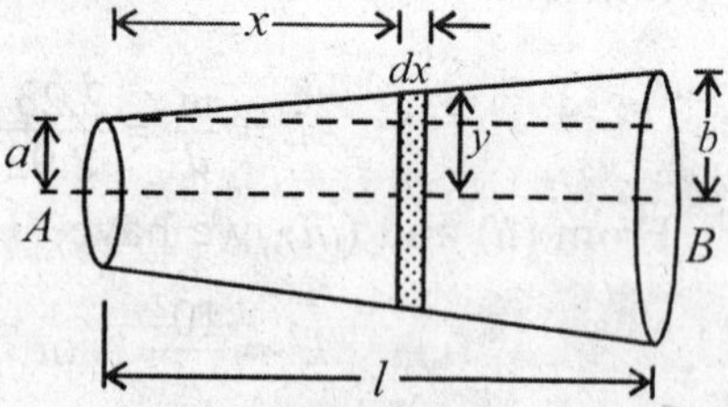

Fig. 20.24

Solution: Consider a small element of length *dx* at a distance of *x* from end *A*. Its radius be say *y*. Then its resistance is

$$dR = \rho \cdot \frac{dx}{\pi y^2} \qquad \ldots(i)$$

From the Fig. 20.24, it is clear that

$$\frac{b-a}{y-a} = \frac{l}{x}$$

or $$y - a = \frac{b-a}{l}x \quad \text{or} \quad dx = \frac{l}{b-a}dy$$

Substituting in (*i*), we have

$$dR = \frac{\rho}{\pi}\frac{l}{b-a}\frac{dy}{y^2}$$

Net resistance of the conductor then is

$$R = \int dR = \frac{\rho l}{\pi(b-a)}\int_a^b \frac{dy}{y^2} = \frac{\rho l}{\pi(b-a)}\left[-\frac{1}{y}\right]_a^b$$

$$= \frac{\rho l}{\pi(b-a)}\left[\frac{1}{a} - \frac{1}{b}\right] = \frac{\rho l}{\pi ab}.$$

Example 20.4: A long round conductor of cross-sectional area A is made of a material whose resistivity depends only on the distance r from the axis of the conductor as $\rho = \frac{\alpha}{r^2}$, where α is a constant. Find (*a*) the resistance per unit length of such a conductor, (*b*) the electric field strength in the conductor due to which current I flows through it.

Solution:

(*a*) Consider a cylindrical conductor of length 1 m and radius a. The conductor may be supposed to be formed of a large number of thin ring shaped conductors each of length 1 m. Consider such a conductor of radius r and thickness dr. The resistance of this conductor is

$$dR = \frac{\rho \times 1}{2\pi r dr} = \frac{\rho}{2\pi r dr}$$

but $$\rho(r) = \frac{\alpha}{r^2}$$

$$\therefore \quad dR = \frac{\alpha}{2\pi r^3 dr}$$

or $$\frac{1}{dR} = \frac{2\pi r^3 dr}{\alpha}$$

All these ring shaped conductors are in parallel and net resistance R is given by integration.

$$\therefore \quad \frac{1}{R} = \int d\left(\frac{1}{R}\right) = \int_0^a \frac{2\pi r^3 dr}{\alpha} = \frac{2\pi a^4}{4\alpha} = \frac{\pi a^4}{2\alpha}$$

$$\therefore \quad R = \frac{2\alpha}{\pi a^4}$$

But $$\pi a^2 = A \quad \text{or} \quad a^2 = \frac{A}{\pi}$$

$$\therefore \quad R = \frac{2\alpha}{\pi \frac{A^2}{\pi^2}} = \frac{2\pi\alpha}{A^2}$$

(*b*) Let required electric field be E then

$$E = \frac{V}{L} = \frac{IR}{L} = \left(\frac{R}{L}\right) I$$

Here V is potential difference across length L and $\frac{R}{L}$ is resistance per unit length. Hence required field is

$$E = \frac{2\pi\alpha}{A^2} I$$

Example 20.5: An infinite ladder network of resistances is constructed with 1 Ω and 2 Ω resistances (Fig. 20.25 (*a*)). The 6 V battery between A and B has negligible resistance. (*i*) Show that the equivalent resistance between A and B is 2 Ω. (*ii*) What is the current that passes through the 2 Ω resistance nearest to the battery.

Solution

(*i*) Let equivalent resistance between A and B be R. Since the network consists of an infinite identical links each having two resistances 1 and 2 Ω. We can reduce one link and still the net resistance will be R. Thus the resistance between P and Q should also be R. Then the given network can be represented as shown in Fig. 20.25 (b). Hence equivalent resistance across AB then is

$$\frac{2 \times R}{2+R} + 1$$

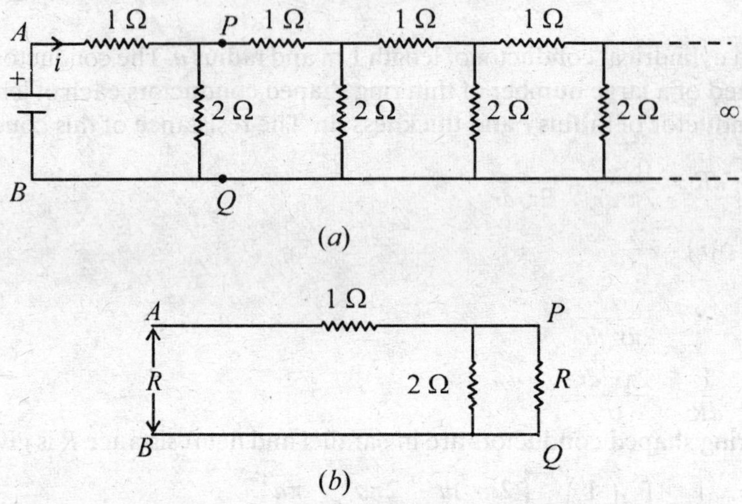

Fig. 20.25

But equivalent resistance across AB is R, then

$$R = \frac{2R}{2+R} + 1$$

or $\quad 2R + R^2 = 2R + 2 + R$

or $\quad R^2 - R - 2 = 0$

or $(R-2)(R+1) = 0$

This gives $R = 2$.

(ii) Since 6 Volt battery is connected, hence current drawn from the battery is

$$i = \frac{E}{R} = \frac{6}{2} = 3 \text{ amp}$$

Resistance between P and Q is also 2 Ω as such current 3 amp is divided equally and the current through 2 Ω resistance near battery is $3/2 = 1.5$ amp.

Example 20.6: Consider an infinite ladder of network shown in Fig. 20.26 (a). If the voltage is halved after each section, find the ratio $\frac{R_1}{R_2}$. Suggest a method to terminate it after a few sections without introducing much error in its attenuation.

Solution: If applied voltage across AB is V then voltage across PQ is $V/2$ and the voltage across R_1 is also $V/2$. If i is the current through R_1, then

$$R_1 i = V/2$$

and current through nearest R_2 is $i/2$, then

$$R_2 i_1/2 = \frac{V}{2} = R_1 i$$

$$\therefore \quad \frac{R_1}{R_2} = \frac{1}{2}$$

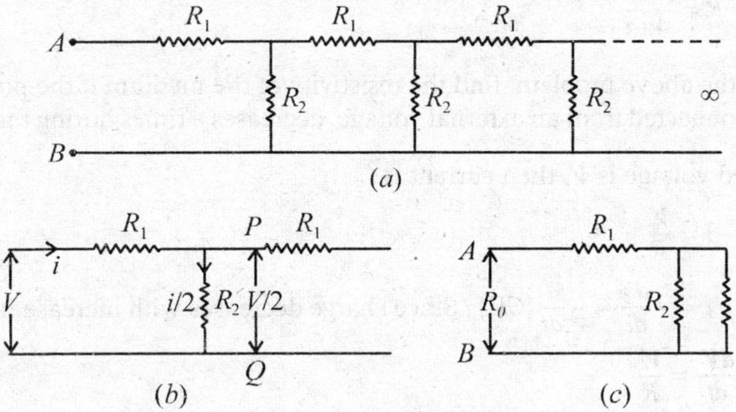

Fig. 20.26

Now the attenuation produced by circuit on termination by a resistance will not be affected if equivalent resistance becomes independent of number of links in the circuit. This is possible if the terminating resistance R_0 is itself equal to equivalent resistance (Fig. 20.26 (*c*)), i.e.

$$R_0 = R_1 + \frac{R_0 R_2}{R_0 + R_2}$$

or $$R_0^2 + R_0 R_2 = R_1 R_0 + R_1 R_2 + R_0 R_2$$

or $$R_0^2 - R_1 R_0 - R_1 R_2 = 0$$

or $$R_0 = \frac{R_1 \pm \sqrt{R_1^2 + 4R_1 R_2}}{2} = \frac{R_1}{2}\left[1 + \sqrt{1 + \frac{4R_2}{R_1}}\right]$$

Thus the circuit may be terminated after any number of links if resistance R_0 is connected is parallel (Fig. 20.26 (*c*)).

Example 20.7: A metal sphere of radius *a* is surrounded by a then concentric metal shell of radius *b*. The space between these electrodes is filled up with a poorly conducting homogeneous medium of resistivity ρ. Find the resistance of interelectrode gap. What is the result if $b \to \infty$.

Solution: Consider a spherical shell of radius *x* and thickness *dx* such that $a < x < b$. Surface area of shell is

$$\text{area} = 4\pi x^2$$

and resistance of the shell is

$$dR = \rho \frac{dx}{4\pi x^2}$$

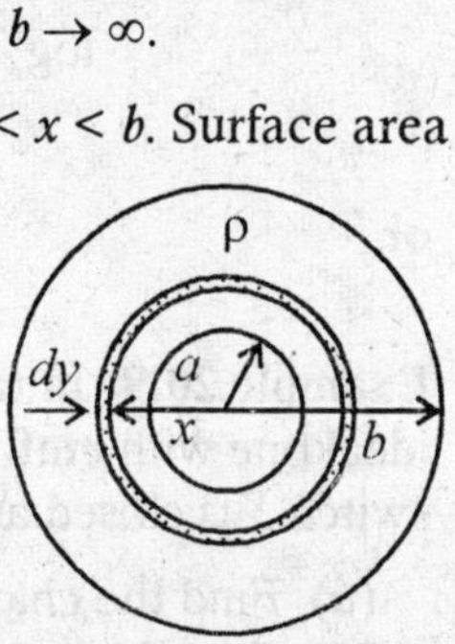

Fig. 20.27

All such shells are in series. Hence total resistance is

$$R = \int dR = \int_a^b \frac{\rho dx}{4\pi x^2} = \frac{\rho}{4\pi}\int_a^b \frac{dx}{x^2} = \frac{\rho}{4\pi}\left[-\frac{1}{x}\right]_a^b$$

$$= \frac{\rho}{4\pi}\left(\frac{1}{a} - \frac{1}{b}\right) = \frac{\rho(b-a)}{4\pi ab}$$

when $b \to \infty$

$$R = \frac{\rho}{4\pi a}$$

Example 20.8: In the above problem, find the resistivity of the medium if the potential difference when they are disconnected from an external voltage, decreases n times during the time internal Δt.

Solution: If applied voltage is V, then current is

$$i = \frac{V}{R}$$

and $$i = -\frac{dq}{dt} = -\frac{d}{dt}(CV)$$. Since charge decreases with increase in time.

or $$-C\frac{dV}{dt} = \frac{V}{R}$$

or $$\frac{dV}{V} = -\frac{dt}{RC}$$

Integration gives

$\log_e V = -\frac{t}{RC} + K$, where K is constant of integration.

At $t = 0$ voltage is V and after time Δt voltage is V/n. Thus we get

$$\log_e V = 0 + K$$

$$\log_e V/n = -\frac{\Delta t}{RC} + K$$

or $$\log_e V - \log_e V/n = \frac{\Delta t}{RC}$$

or $$\log_e n = \frac{\Delta t}{RC}$$

Substituting the value of $R = \frac{\rho(b-a)}{4\pi ab}$, we get

$$\log_e n = \frac{\Delta t\, 4\pi ab}{\rho(b-a)C}$$

or $$\rho = \frac{4\pi ab\Delta t}{(b-a)C\log_e n}$$

Example 20.9: In the circuit shown in Fig. 20.28 the battery is ideal one with emf V. The capacitor is initially uncharged. The switch S is closed at time $t = 0$.

(*a*) Find the charge Q on the capacitor at time t.

(*b*) Find the current in BE at time t. What is limiting value at $t = \infty$? (IIT 98)

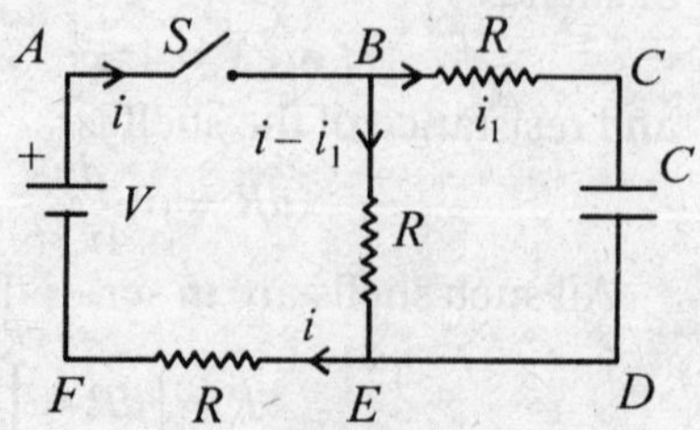

Fig. 20.28

Solution: The distribution of current in different branches of the circuit are shown in Fig. 20.28 and let charge on capacitor be q then $i_1 = \frac{dq}{dt}$. Applying Kirchhoff's law to mesh $ABEFA$, we have

$$R(i - i_1) + iR - V = 0$$

or
$$V = R(2i - i_1) \qquad ...(i)$$

Applying Kirchhoff's law to mesh $ABCDEFA$, we get

$$i_1R + \frac{q}{C} + iR - V = 0$$

or
$$V = R(i + i_1) + \frac{q}{C} \qquad ...(ii)$$

Eliminating i between (i) and (ii), we get

$$3Ri_1 = V - \frac{2q}{C} \qquad \text{But} \quad i_1 = \frac{dq}{dt}$$

$$\therefore \quad 3R\frac{dq}{dt} = V - \frac{2q}{C}$$

or
$$\frac{dq}{dt} = \frac{1}{3RC}[VC - 2q] = \frac{2}{3RC}\left[\frac{VC}{2} - q\right]$$

or
$$\frac{dq}{\left(\frac{VC}{2} - q\right)} = \frac{2\,dt}{3RC}$$

Integrating, we get

$$\int \frac{dq}{\left(\frac{VC}{2} - q\right)} = \int \frac{2}{3RC}\,dt$$

or
$$-\log_e\left(\frac{VC}{2} - q\right) = \frac{2t}{3RC} + K, \; K \text{ is constant of integration.}$$

But at $t = 0$, $q = 0$

$$\therefore \quad K = -\log_e \frac{VC}{2}$$

and
$$-\log_e\left(\frac{VC}{2} - q\right) = \frac{2t}{3RC} - \log_e \frac{VC}{2}$$

or
$$\log_e\left(\frac{VC}{2} - q\right) - \log_e \frac{VC}{2} = -\frac{2t}{3RC}$$

and
$$\frac{\frac{VC}{2} - q}{VC/2} = e^{-2t/3RC}$$

and
$$q = \frac{VC}{2}\left(1 - e^{-2t/3RC}\right)$$

and $$i_1 = \frac{dq}{dt} = \frac{VC}{2} \times \frac{2}{3RC} e^{-2t/3RC}$$

$$= \frac{V}{3R} e^{-2t/3RC}$$

and the current in branch BE is

$$i - i_1 = i - \frac{V}{3R} e^{-2t/3RC}$$

But equation (i) gives

$$\left(\frac{V}{R} + i_1\right) = 2i \quad \text{or} \quad i = \frac{V}{2R} + \frac{i_1}{2}$$

$$\therefore \qquad i - i_1 = \frac{V}{2R} + \frac{i_1}{2} - i_1 = \frac{V}{2R} - \frac{i_1}{2}$$

$$= \frac{V}{2R} - \frac{V}{6R} e^{-2t/3RC}$$

$$= \frac{V}{2R}\left(1 - \frac{1}{3} e^{-2t/3RC}\right)$$

At $t = \infty$, $$i - i_1 = \frac{V}{2R}$$

Example 20.10: Find the resistance R_{ab} of the frame made of thin wire assuming that the number of successive embedded equilateral triangles (with sides decreasing to half) tends to infinity. Side AB is equal to a and resistance per unit length of wire is ρ.

Solution: Let equivalent resistance across AB (Fig. 20.29 (a)) is x. since the resistance is directly proportional to length as such the equivalent resistance between A_1B_1 will be $x/2$. The equivalent circuit for the given network is shown in Fig. 20.29 (b). If AB length has resistance $2r$, then resistance of $AA_1 = A_1C = BB_1 = B_1C = r$.

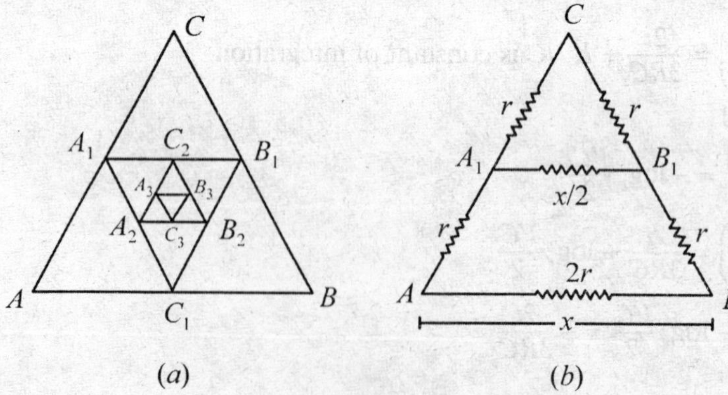

Fig. 20.29

From figure, the effective resistance between A_1 and B_1 is $2r$ in series and $x/2$ in parallel. Thus

$$R_1 = \frac{2r \times x/2}{2r + x/2} = \frac{2rx}{4r + x} \qquad \ldots(i)$$

R_1 is in series with AA_1 and B_1B. Therefore effective resistance is

$$R_2 = R_1 + 2r = \frac{2rx}{4r+x} + 2r \quad \ldots(ii)$$

Now R_2 and $2r$ are in parallel and give equivalent resistance x. Thus

$$x = \frac{R_2 \times 2r}{R_2 + 2r} = \frac{\left(\frac{2rx}{4r+x} + 2r\right)2r}{\frac{2rx}{4r+x} + 2r + 2r}$$

$$= \frac{4r^2x + 4r^2(4r+x)}{2rx + 4r(4r+x)}$$

Solving, we get

$$3x^2 + 4rx - 8r^2 = 0$$

or $$x = \frac{-4r \pm \sqrt{16r^2 + 4 \times 3 \times 8r^2}}{6}$$

Since x cannot be negative, hence

$$x = \frac{-4r + r\sqrt{112}}{6} = \frac{4\sqrt{7} - 4}{6} r$$

$$= \frac{2\sqrt{7} - 2}{3} r$$

But $$r = \rho \frac{a}{2}$$

Hence $$x = \frac{2\sqrt{7} - 2}{3} \frac{\rho a}{2} = \frac{\sqrt{7} - 1}{3} \rho a = 0.55 \text{ Pa}$$

Example 20.11: Calculate the relaxation time and mean free path in copper at room temperature (300 K), if the number of free electrons per unit volume is $8.5 \times 10^{28}/\text{m}^3$ and resistivity $\rho = 1.7 \times 10^{-8}$ Ω-m.

Solution: We know that relaxation time τ is

$$\tau = \frac{m}{ne^2\rho} = \frac{9.1 \times 10^{-31}}{8.5 \times 10^{28}(1.6 \times 10^{-19})^2 \times 1.7 \times 10^{-8}}$$

$$= 2.5 \times 10^{-14} \text{ sec}$$

On the basis of kinetic theory of gases, the rms speed of free electrons is given by

$$v_{rms} = \sqrt{\frac{3kt}{m}} = \sqrt{\frac{3 \times 1.38 \times 10^{-23} \times 300}{9.1 \times 10^{-31}}}$$

$$= 1.2 \times 10^5 \text{ m/s}$$

Mean free path

$$\lambda = v_{rms} \cdot \tau = 1.2 \times 10^5 \times 2.5 \times 10^{-14}$$

$$= 3 \times 10^{-9} \text{ m} = 30 \text{ Å}.$$

Example 20.12: In the circuit shown in Fig. 20.30 potential of point P and Q are +20 V and +10 V respectively. Current 2 A flows in the branch CD and terminal E is earthed. Find (*a*) the potential of point A, (*b*) the value of resistance R.

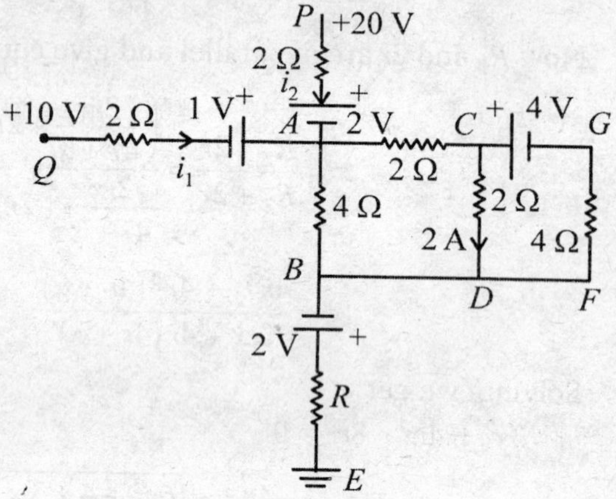

Fig. 20.30

Solution: Potential difference

$$V_C - V_D = 2 \times 2 = 4V$$

As such no current flows through $CGFD$.

Applying Kirchhoff's law to mesh $ACDBA$, we have

$2 \times 2 + 2 \times 2 - 4 \times i = 0$ where i is the current in branch AB or i = 2A.

If i_1 and i_2 be the currents in branches QA and PA respectively, then for branch QA, we can write

$$V_Q - 2i_1 + 1 = V_A \quad \ldots(i)$$

and for branch PA we have

$$V_P - 2i_2 - 2 = V_A \quad \ldots(ii)$$

Adding (*i*) and (*ii*) we get

$$(V_P + V_Q) - 2(i_1 + i_2) - 1 = 2V_A$$

But $\quad V_P + V_Q = 30$ V $\quad$ and $\quad i_1 + i_2 = 4$ A

or $\quad 30 - 2 \times 4 - 1 = 2V_A$

or $\quad 2V_A = 21$

or $\quad V_A = \dfrac{21}{2} = 10.5$ V

Applying Kirchhoff's law to the branch ABE, we get

$$V_A - 4 \times 2 + 2 - 4R = 0$$

or $\quad 4R = 10{\cdot}5 - 8 + 2 = 4{\cdot}5$

or $\quad R = \dfrac{4.5}{4} = 1.125\,\Omega$.

Example 20.13: In Fig. 20.31 two cells of emf e_1 and e_2 ($e_2 > e_1$) are connected. If the polarity of e_1 is reversed then potential difference across AB increases by 10 V. Find the value of e_1.

Solution: Let in first position, the current distribution be as shown in Fig. 20.31.

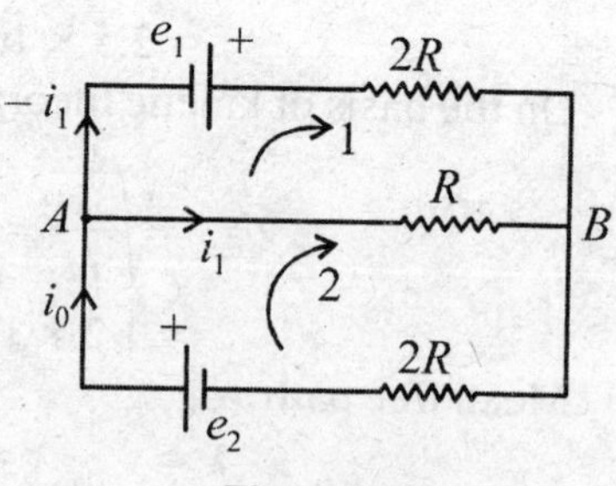

Fig. 20.31

Applying Kirchhoff's voltage law to loop (1), we have

$$2R(i_0 - i_1) - i_1R = e_1$$

or $\quad 2Ri_0 - 3i_1R = e_1 \quad \ldots(i)$

Applying Kirchhoff's voltage law to loop (2), we have

$$i_1R + i_0 2R = e_2 \quad \ldots(ii)$$

Subtracting (*i*) from (*ii*), we get

$$4Ri_1 = e_2 - e_1$$

or $\quad Ri_1 = (e_2 - e_1)/4$

Hence potential difference between A and B is

$$V_{AB} = Ri = \frac{e_2 - e_1}{4} \qquad \ldots(iii)$$

When the polarity of e_1 is changed the potential difference between A and B becomes

$$V'_{AB} = \frac{e_2 + e_1}{4}$$

But $\quad V'_{AB} - V_{AB} = 10 \text{ V}$

$$\therefore \quad \frac{e_2 + e_1}{4} - \frac{e_2 - e_1}{4} = 10$$

or $\quad 2e_1 = 40 \quad \text{or} \quad e_1 = 20 \text{ V}.$

Example 20.14: An uncharged capacitor of capacity C is connected as shown in Fig. 20.32. Find the charge on the capacitor as a function of time after switch is closed at $t = 0$.

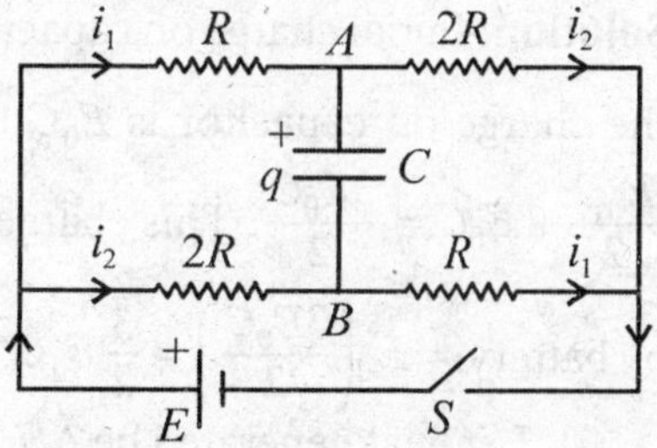

Fig. 20.32

Solution: The current flowing through various branches is shown in Fig. 20.32. Using Kirchhoff's law, we get

$$i_1R + 2Ri_2 = E \qquad \ldots(i)$$

and potential difference between A and B is

$$\frac{q}{C} = i_2 2R - i_1 R \qquad \ldots(ii)$$

where q is charge on capacitor after time t after the switch is closed, solving for i_1 and i_2, we have

$$i_1 = \frac{1}{2}\left(\frac{E}{R} - \frac{q}{RC}\right) \quad \text{and} \quad i_2 = \frac{1}{4}\left(\frac{q}{RC} + \frac{E}{R}\right)$$

$$\therefore \quad i_1 - i_2 = \frac{dq}{dt} = \frac{1}{2}\left(\frac{E}{R} - \frac{q}{RC}\right) - \frac{1}{4}\left(\frac{q}{RC} + \frac{E}{R}\right)$$

$$= \frac{E}{4R} - \frac{3q}{4RC} = \frac{1}{4RC}(EC - 3q)$$

or $$\frac{dq}{(EC - 3q)} = \frac{dt}{4RC}$$

Integrating, we get

$$\int \frac{dq}{(EC - 3q)} = \int \frac{dt}{4RC}$$

or $$-\frac{1}{3}\log_e(EC - 3q) = \frac{t}{4RC} + k$$

where k is constant of integration

Initial conditions are $t = 0 \quad q = 0 \qquad \therefore \; k = -\frac{1}{3}\log_e EC$

$$\therefore \quad -\frac{1}{3}\log_e(EC - 3q) = \frac{t}{4RC} - \frac{1}{3}\log_e EC$$

or $\quad \frac{1}{3}\log_e \frac{(EC-3q)}{EC} = -\frac{t}{4RC}$

or $$\frac{EC-3q}{EC} = e^{-\frac{3t}{4RC}}$$

or $$q = \frac{EC}{3}\left(1 - e^{-\frac{3t}{4RC}}\right)$$

Example 20.15: A capacitor of capacitance C is charged to a potential difference of $\frac{E_0}{2}$ and after that it is connected to a circuit with the polarity shown in Fig. 20.33. At $t = 0$ switch S is closed.

(*a*) Find heat energy generated through the resistance after a long time.
(*b*) Draw a schematic curve of charge on the capacitor with respect to time t.

Solution: Initial charge on capacitor is $\frac{E_0C}{2}$. After connecting with the battery in opposite direction the charge on capacitor is E_0C but with opposite polarity. Hence charge supplied by battery = $\frac{E_0C}{2} + E_0C = \frac{3E_0C}{2}$. This charge is supplied under potential difference E_0. Hence energy supplied by battery = $E_0\left(\frac{3E_0C}{2}\right) = \frac{3}{2}E_0^2C.$

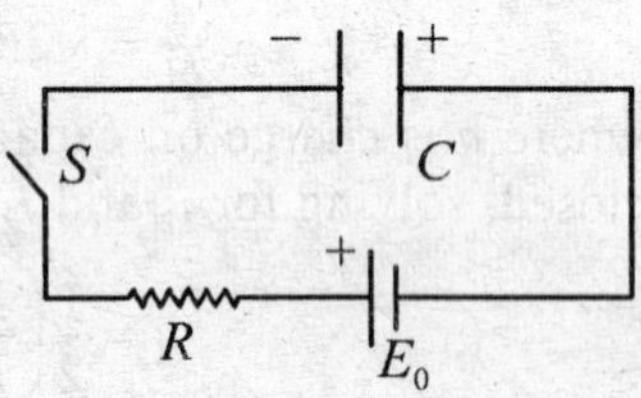

Fig. 20.33

(*i*) Let heat generated be ΔH

(*ii*) When the capacitor is fully charged energy stored = $\frac{1}{2}C\,E_0^2$

(*iii*) Initially energy stored in capacitor = $\frac{1}{2}C\left(\frac{E_0}{2}\right)^2 = \frac{CE_0^2}{8}$

Conservation of energy gives

$$\frac{3}{2}E_0^2C + \frac{E_0^2C}{8} = \Delta H + \frac{1}{2}CE_0^2$$

or $$\Delta H = \frac{9}{8}CE_0^2$$

(*b*) If now charge on capacitor at any time t be q and current i then

$$Ri + \frac{q}{C} = E_0 \quad \text{with } t = 0 \quad q = -\frac{CE_0}{2}$$

Since $i = \frac{dq}{dt}$, then equation can be written as

$$R\frac{dq}{dt} + \frac{q}{C} = E_0$$

or $$R\frac{dq}{dt} = \left(E_0 - \frac{q}{C}\right)$$

or $$\frac{dq}{dt} = \frac{1}{RC}(E_0C - q)$$

or $$\frac{dq}{E_0C - q} = \frac{dt}{RC}$$

and integration gives

$$-\log_e(E_0C - q) = \frac{t}{RC} + k, \text{ where } k \text{ is constant of integration}$$

Using initial conditions one obtains $k = -\log_e \frac{3E_0C}{2}$

Substitution gives

$$\log \frac{E_0C - q}{3/2E_0C} = -\frac{t}{RC}$$

or $$q = CE_0 - 3/2CE_0e^{-t/RC} = CE_0\left(1 - \frac{3}{2}e^{-t/RC}\right)$$

This gives at $t = \infty$ $q = CE_0$ and $q = 0$ when

$$\frac{3}{2}e^{-t/RC} = 1$$

or $$e^{-t/RC} = 2/3$$

and $$t = RC\log_e 3/2$$

The plot of charge q versus time is shown in Fig. 20.34.

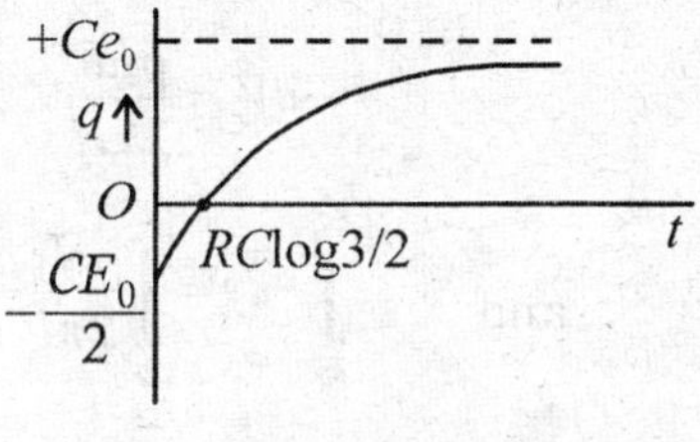

Fig. 20.34

Example 20.16: A cylindrical tube of length l has inner radius a and outer radius b. Find the resistance of the tube between (*a*) its ends (*b*) its inner and outer surface (The resistivity of the material is ρ).

Solution:

(*a*) In this case the current and field are along the axis of the tube. The tube can be considered to be made up of large number of co-axial annular discs. Consider such a disc of thickness dx at a distance x from one end (Fig 20.35), then

$$E = -\frac{dV}{dx} \quad \text{and} \quad J = \frac{I}{\pi(b^2 - a^2)} \quad [\text{area of cross-section} = \pi(b^2 - a^2)]$$

But $$\frac{J}{E} = \sigma \text{ (conductivity)} = \frac{1}{\rho} \text{ (}\rho\text{ is resistivity)}$$

or $$E = \rho J$$

$$\therefore \quad -\frac{dV}{dx} = \frac{\rho I}{\pi(b^2 - a^2)}$$

and $$-dV = \frac{\rho I dx}{\pi(b^2 - a^2)}$$

$$\int_{V_1}^{V_2} -dV = \frac{\rho I}{\pi(b^2 - a^2)} \int_0^{\rho} dx$$

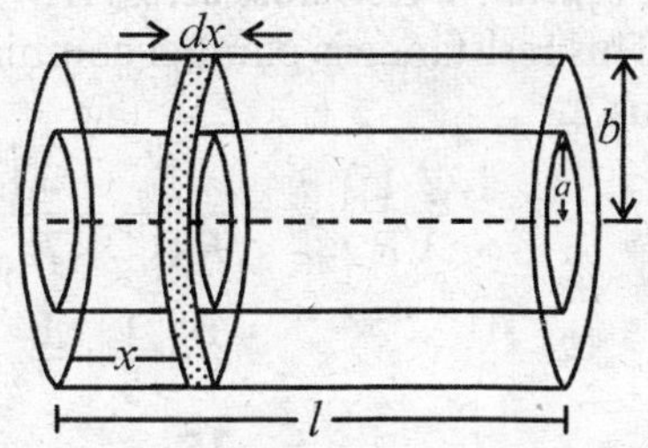

Fig 20.35

or $$V_1 - V_2 = \frac{\rho I l}{\pi(b^2 - a^2)}$$

or Resistance $$R = \frac{V_1 - V_2}{I} = \frac{\rho l}{\pi(b^2 - a^2)}$$

(*b*) Here field is radial. Consider the tube to be made up of large number concentric cylindrical shells. Now consider a shell of radius r and thickness dr, we have

$$E = -\frac{dV}{dr} \quad \text{and} \quad J = \frac{I}{2\pi rl}$$

Here area of cross-section is $2\pi rl$

Now $$-\frac{dV}{dr} = \frac{\rho I}{2\pi rl}$$

or $$-dV = \frac{\rho I dr}{2\pi rl}$$

and $$\int_{V_1}^{V_2} -dv = \int_a^b \frac{\rho I}{2\pi l}\frac{dr}{r}$$

Integration gives $V_1 - V_2 = \frac{\rho I}{2\pi l}\log_e \frac{b}{a}$

$\therefore$ Resistance $R = \frac{V_1 - V_2}{I} = \frac{\rho}{2\pi l}\log_e b/a$

Example 20.17: A nichrome wire of uniform cross-section is bent to form a rectangular loop *ABCD*. Another nichrome wire of the same cross-section is connected to form the diagonal *AC*. Find the ratio of resistances across *BD* and *AC* if $AB = 0.4$ m and $BC = 0.3$ m. (Roorkee 2000)

Solution: Since wire is of uniform cross-section as such the resistance is directly proportional to length. Hence resistances of wires can be taken as

$AB = CD = 4$ units

$BC = AD = 3$ units and $AC = 5$ units.

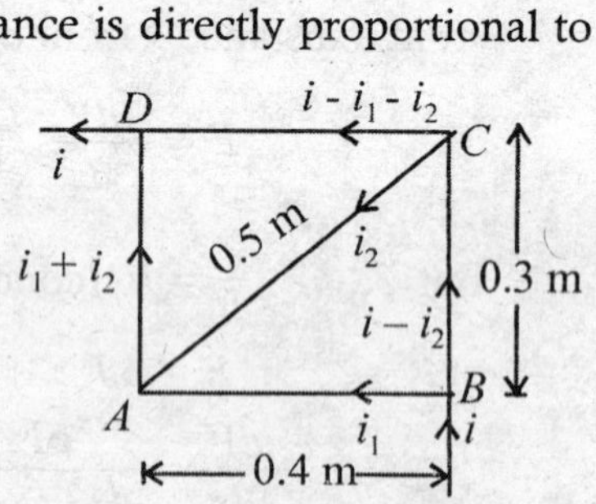

Fig. 20.36

The equivalent resistance across AC

This resistance is parallel combination of, R_{ABC}, R_{AC} and R_{ADC}.

Thus

$$\left(\frac{1}{R}\right)_{AC} = \frac{1}{R_{ABC}} + \frac{1}{R_{AC}} + \frac{1}{R_{ADC}}$$

$$= \frac{1}{7} + \frac{1}{5} + \frac{1}{7} = \frac{17}{35}$$

or $$R_{AC} = \frac{35}{17} \text{ units} \qquad \ldots(i)$$

The equivalent resistance across *BD* can be determined using Kirchhoff's law.

Let a potential difference V is applied between *BD* such that *B* has potential V and *D* has zero. The distribution of current in various branches is shown in Fig. 20.36.

Applying Kirchhoff's law to mesh $ABCA$, we get

$$3(i - i_1) + 5i_2 - 4i_1 = 0$$

i.e. $$3i - 7i_1 + 5i_2 = 0 \qquad \ldots(ii)$$

and from mesh $ADCA$, we get

$$3(i_1 + i_2) - 4(i - i_1 - i_2) + 5i_2 = 0$$

or $$7i_1 + 12i_2 - 4i = 0 \qquad \ldots(iii)$$

Solving for i_1 and i_2 in terms of i, we get

$$17i_2 = i \quad \text{or} \quad i_2 = \frac{i}{17}$$

and $$i_1 = \frac{8i}{17}$$

If R_{BD} is effective resistance across BD, then

$$iR_{BD} = 4i_1 + 3(i_1 + i_2)$$

$$= 7i_1 + 3i_2 = \frac{7 \times 8i}{17} + \frac{3i}{17}$$

$$= \frac{59i}{17}$$

or $$R_{BD} = \frac{59}{17}$$

Thus ratio $$\frac{R_{BD}}{R_{AC}} = \frac{59}{17} \times \frac{17}{35} = \frac{59}{35}.$$

Example 20.18: A battery has an emf E and internal resistance r. A variable resistor R is connected across the terminals of the battery. Find the value of R such that (*a*) the current in the circuit is maximum. (*b*) The potential difference across the terminals is maximum. (*c*) The power delivered to the resistor (called load) is maximum.

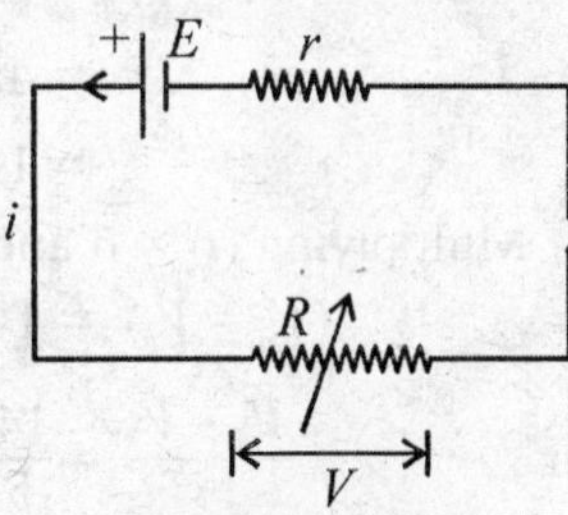

Fig. 20.37

Solution:

(*a*) Using Ohm's law, we have

Current in the circuit $i = \dfrac{E}{R + r}$

Now i is maximum when $R + r$ = minimum

i.e., R = minimum = 0

i.e., current is maximum when cell is short circuited.

In this situation $i_{max} = E/r$, $V = 0$ and Power $P_L = 0$

i.e., power delivered to load is zero and potential difference across load is also zero.

(*b*) Potential difference across load V is

$$V = iR = \frac{ER}{R + r}$$

this also is the terminal voltage.

Now V is maximum when $1 + r/R$ is minimum, i.e., $R \to \infty$

In this situation $i = 0$, $V = E$ and Power $P_L = 0$.

i.e., terminal voltage in maximum (= E) when the cell is in open circuit.

(*c*) Power delivered to the load is

$$P_L = i^2 R = \frac{E^2 R}{(R+r)^2}$$

Power delivered is maximum when $R\,(R + r)^{-2}$ is maximum, i.e.

$$\frac{d}{R}\left[R(R+r)^{-2}\right] = 0$$

i.e., $(R + r)^{-2} + R(-2)\,(R + r)^{-3} = 0$

or $R + r - 2R = 0$ or $R = r$

In this situation $i = \dfrac{E}{2r}$, $V = E/2$ and $P_L = \dfrac{E^2}{4r}$

i.e., power transferred to the load R is maximum when external resistance is equal to internal resistance.

Example 20.19: Find the equivalent resistance between *a* and *b* of the circuit shown in Fig. 20.38.

Solution: Suppose current i enters the circuit at the point *a*, i_1 goes through 4 Ω resistance and $i - i_1$ through 2 Ω resistance. By symmetry, the current i coming out from point *b* will be composed of a part i_1 from 4 Ω resistance and $i - i_1$ from 2 Ω resistance.

Applying Kirchhoff's law, the current through resistance 2 Ω between *cd* is $(2i_1 - i)$. Now we have

$$V_a - V_b = (V_a - V_c) + (V_c - V_b)$$

$$= 4i + 2(i - i_1)$$

or $V_a - V_b = 2i_1 + 2i$ ...(*i*)

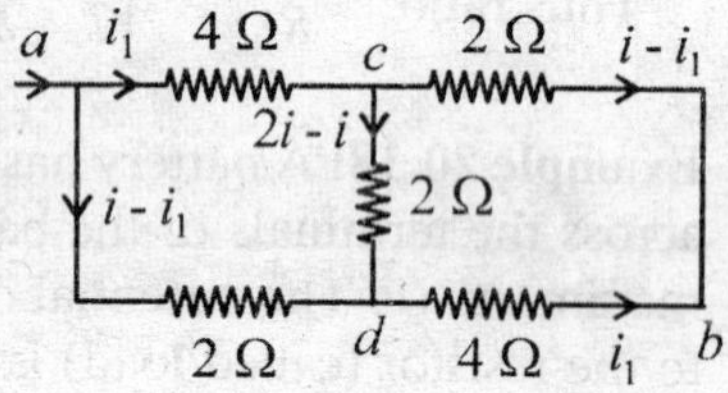

Fig. 20.38

and for path *acdb*, we have

$$V_a - V_b = (V_a - V_c) + (V_c - V_d) + (V_d - V_b)$$

$$= 4i_1 + 2(2i_1 - i) + 4i_1$$

$$= 12i_1 - 2i \qquad ...(ii)$$

Multiplying (*i*) × 6 and subtracting (*ii*) from it, we eliminate i_1 and get

$$5(V_a - V_b) = 14i$$

or $\dfrac{V_a - V_b}{i} = \dfrac{14}{5} = 2 \cdot 8\,\Omega$

Thus the equivalent resistance between *a* and *b* is 2.4 Ω.

Example 20.20: Find the equivalent resistance of the network shown in Fig. 20.39 between the points *a* and *b*.

Solution: Suppose the current i enters the network at point *a* and the same current leaves it at *b*. Let the currents in *ac*, *ad* and *ae* are i_1, i_2 and i_3 respectively (Fig. 20.39 (*a*)). Because of symmetry the distribution of current at *b* will be similar. The current i leaves the point *b* composed of i_1 from *db*, i_2 from *eb* and i_3 from *ab*.

Since the current in *ae* is equal to that in *eb* the current in *ce* is same as the current in *ed* using junction law. This netwrok acts as if the branches *ced* and *aeb* do not physically touch at *e*, as such we can represent branch *aeb* by a single resistance of 4 Ω connected between *a* and *b*. Similarly branches *ce* and *ed* can be replaced by a single resistance of 2 Ω as shown in Fig. 20.39 (*b*). By assuming so we are not changing the distribution of the current.

This circuit shown in Fig. 20.39 (*b*) is equivalent to the circuit shown in Fig. 20.38 with which a resistance 4 Ω is connected in parallel.

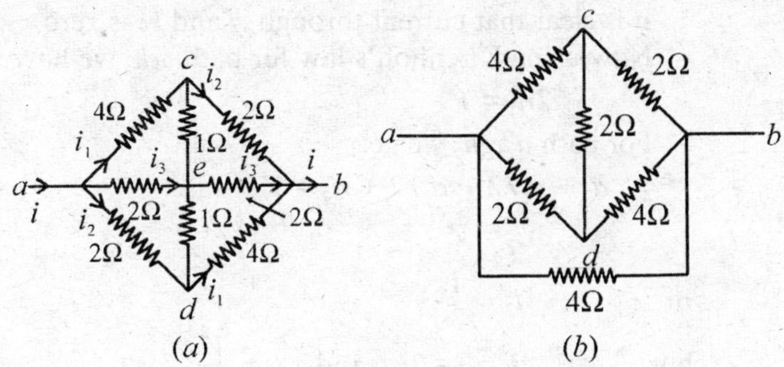

Fig. 20.39

Thus equivalent resistance of whole network is

$$R = \frac{4 \times 2.8}{4 + 2.8} = 1.65\,\Omega$$

Example 20.21: A capacitor having a charge of 20 μC is discharged through a wire of resistance 5 Ω. Find the heat dissipated in the wire between 25 to 50 μs after the connections are made.

Solution: After the connections are made, the charge on the capacitor at anytime t is

$$Q = Q_0 e^{-t/RC}$$

or

$$i = \frac{dQ}{dt} = -\frac{Q_0}{RC} e^{-t/RC}$$

Hence heat dissipated during the time t_1 to t_2 is

$$H = \int_{t_1}^{t_2} i^2 R dt = \int_{t_1}^{t_2} \frac{Q_0^2}{RC^2} \cdot e^{-2t/RC} dt$$

$$= -\frac{Q_0^2}{2C}\left[e^{-2t/RC}\right]_{t_1}^{t_2}$$

$$= \frac{Q_0^2}{2C}\left[e^{-2t_1/RC} - e^{-2t_2/RC}\right]$$

$$RC = 5 \times 5 \times 10^{-6} = 25\ \mu\text{s} \quad \text{and} \quad t_1 = 25\ \mu\text{s} \quad t_2 = 50\ \mu\text{s}$$

$$\therefore \quad H = \frac{(20 \times 10^{-6})^2}{2 \times 5 \times 10^{-6}}(e^{-2} - e^{-4})$$

$$= 4.7\ \mu\text{J}.$$

Example 20.22: Twelve wires each of resistance r are joined to form a cube. Find the equivalent resistance between the ends of (*a*) face diagonal (*b*) the ends of an edge (*c*) the body diagonal.

Solution:

(*a*) When a potential difference V is applied between a and h, current i enters at a and the same current leaves the circuit at h. The current distribution is shown in Fig. 20.40 (*a*). By symmetry the path *abh* and *aeh* carry the same current i_1 and the path *ad* carries the current

$$i_2 = i - 2i_1 \quad \ldots(i)$$

It is clear that current through *ef* and *bc* is zero.

Now using Kirchhoff's law for path *aeh*, we have

$$2ri_1 = V \quad \text{...(ii)}$$

For path *adcgh*, we get

$$ri_2 + ri_2/2 + ri_2/2 + ri_2 = V$$

$$3ri_2 = V$$

or $$i_2 = \frac{V}{3r} \quad \text{...(iii)}$$

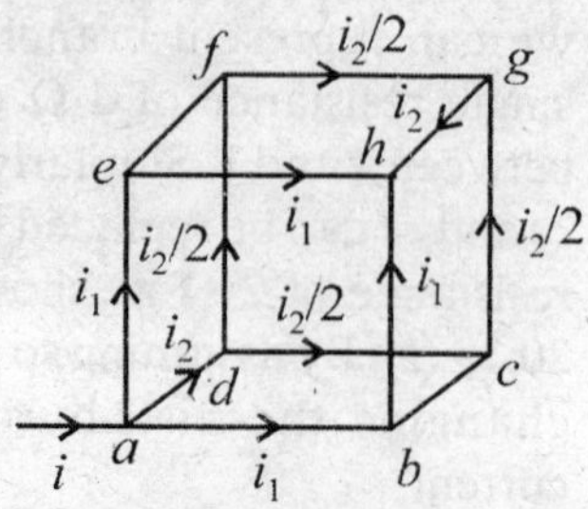

Fig. 20.40 (*a*)

but $$i_2 = i - 2i_1 \quad \text{and} \quad i_1 = \frac{V}{2r}$$

$\therefore$ $$\frac{V}{3r} = i - \frac{V}{r}$$

or $$i = \frac{V}{3r} + \frac{V}{3} = \frac{4V}{3r}$$

and equivalent resistance

$$R = \frac{V}{i} = \frac{3r}{4}$$

(*b*) Suppose the current *i* enters the circuit at the point *a* and the same current leaves *b* when a potential difference *V* is applied between *ab* the edge of the cube. The paths through *ae* and *ad* are equivalent and carry equal current i_1 and the current through *ab* then is $i - 2i_1$. The same distribution of current holds will be at junction *d* and *e*. The further distribution of current is shown in Fig. 20.40 (*b*).

Using Kirchhoff's law for path *ab*, we have

$$r(i - 2i_1) = V \quad \text{or} \quad i - 2i_1 = \frac{V}{r} \quad \text{...(i)}$$

Using Kirchhoff's law for path *efghe*, we have

$$r(i_1 - i_2) + 2(i_1 - i_2)r + (i_1 - i_2)r - i_2 r = 0$$

or $$5i_2 = 4i_1$$

or $$i_2 = \frac{4}{5}i_1 \quad \text{...(ii)}$$

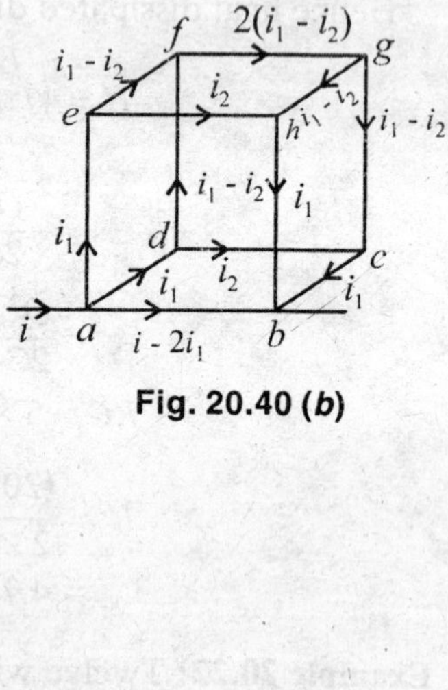

Fig. 20.40 (*b*)

Using Krichhoff's law for path *adcb*, we get

$$i_1 r + i_2 r + i_1 r = V$$

or $$2i_1 + i_2 = \frac{V}{r} \quad \text{or} \quad 2i_1 + \frac{4}{5}i_1 = \frac{V}{r} \quad \text{or} \quad i_1 = \frac{5V}{14r}$$

Substituting the value in (*i*), we have

$$i - 2i_1 = V/r$$

or $$i - \frac{5V}{7r} = \frac{V}{r}$$

$$i = \left(\frac{5}{7} + 1\right)V/r = \frac{12}{7}V/r$$

and equivalent resistance

$$R = \frac{V}{i} = \frac{7r}{12}$$

(c) Let i current enters the network at a and leaves at g when a potential difference V is applied between a and g. The current distribution in various branches is shown in Fig. 20.40 (c). Because of symmetry the current at a is equally divided between branches ab, ad and ae, each carrying current $i/3$. Again at junction b, d and e the current is further divided equally between branches de and df, eh and ef and bc and ch, each carrying current $i/6$.

Now using Kirchhoff's law for path $abcg$, we have

$$r\frac{i}{3} + r\frac{i}{6} + r\,i/3 = V$$

$$\frac{5ir}{6} = V$$

or equivalent resistance is

$$R = \frac{V}{i} = \frac{5r}{6}$$

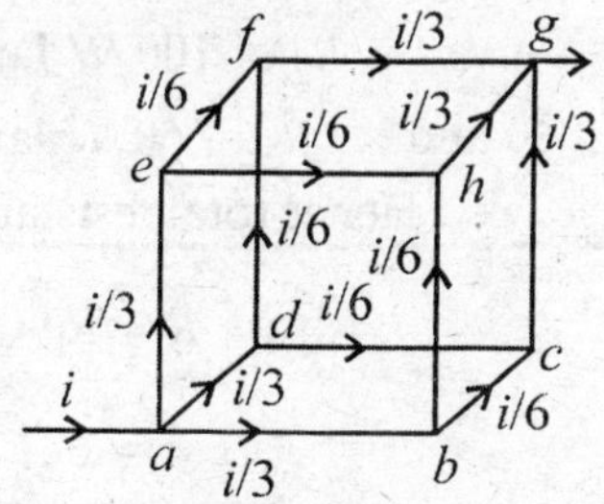

Fig. 20.40 (c)

Example 20.23: Three bulbs of rating 100 W, 250 W and 60 W, 250 V are connected as shown in Fig. 20.41. Neglecting change of resistance with temperature

(i) Find the maximum safe voltage that can be applied to the circuit.
(ii) Which bulb glows brightest?
(iii) Total cost of consumption of energy in 6 hrs at the rate of Rs. 2.50 per unit.
(iv) What would happen if 500 V are applied to the circuit.

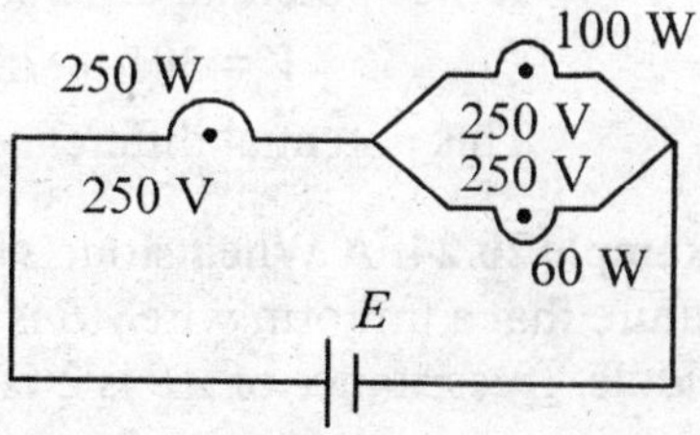

Fig. 20.41

Solution:

(i) 100 W and 60 W lamps must be given maximum required voltage of 250 V because their total power is 160 W, which is less than 250 W power of the third bulb. Total current in these two lamps is

$$i = \frac{160}{250} = 0.64 \text{ Amp} \quad \ldots(i)$$

The same current will pass through the third lamp, whose resistance is

$$R = \frac{V^2}{P} = \frac{250 \times 250}{250} = 250\,\Omega \quad \ldots(ii)$$

Hence potential difference across 250 Watt lamp is

$$V = iR = 0.64 \times 250 = 160 \text{ V}$$

Hence total maximum safe voltage that can be applied is

$$V = 250 + 160 = 410 \text{ V}$$

(ii) Power consumed in 250 W lamp is

$$P = i^2R = (0.64)^2 \times 250$$
$$= 102.4 \text{ W}$$

The power consumed in other two bulbs is 100 W and 60 W hence 250 W lamp glows brighter.

(iii) Total power consumed = 102.4 + 100 + 60 = 262.4 W

Hence cost $= \dfrac{262.4}{1000} \times 6 \times 2.50 = \text{Rs. } 3.94$

(*iv*) Resistance of each bulb is $\dfrac{V^2}{W}$

R_1 of 250 W lamp = 25 Ω

R_2 of 100 W lamp = 625 Ω

and R_3 of 60 W lamp = 1042 Ω

Hence total resistance

$$R = R_1 + \frac{R_1R_2}{R_1 + R_2} = 250 + \frac{625 \times 1042}{625 + 1042} = 250 + 391$$

$$= 641\ \Omega$$

and current in the circuit when 500 V potential difference is applied is

$$i = \frac{500}{641} = 0.78 \text{ Amp}$$

Hence potential difference across 100 W and 60 W bulbs is

$V = 391 \times .78 = 305$ V

This potential difference is greater then 250 V as such both these bulbs will fuse.

Example 20.24: A Wheatstone's bridge (Fig. 20.42) is set up with arms *P, Q, R, X* and an additional feature that a uniform wire *AB* is connected between *X* and *R*. It has 50 divisions with zero at the middle. If resistance of *AB* is 2 Ω and *P, Q, R* are respectively 10, 20 and 30 Ω and *X* is a variable unknown, then deduce:

(*a*) The value of *X* when null point is at zero on *AB*
(*b*) The value of *X* when null point is at +8 div. on *AB*
(*c*) The least count in Ohms if 1 div. shift on *AB* can be clearly seen on galvanometer
(*d*) The minimum sensitivity of galvanometer in *mV*/div. if the battery voltage is 2 V and 1 div. shift on *AB* is to show at least 1/5 div. deflection on the galvanometer.

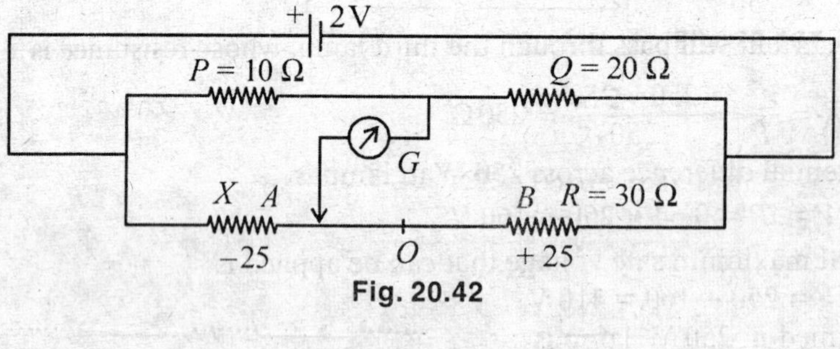

Fig. 20.42

Solution:

(*a*) Using principle of Wheatstone's bridge, when null point is at 0 of the scale, then

$$\frac{10}{20} = \frac{X + 1}{30 + 1}$$

this gives $x = 14.5\ \Omega$

(*b*) The resistance per unit division of wire = $\frac{2}{50}$

Hence resistance of 8 divisions = $\frac{2 \times 8}{50} = 0.32\ \Omega$

This makes $R' = R + (1 - 0.32) = 30 + 0.68 = 30.68\ \Omega$

and $X' = X + (1 + 0.32) = X + 1.32$

Now, again using conditions for balanced Wheatstone's bridge, we get

$$\frac{P}{Q} = \frac{R'}{X'} \quad \text{or} \quad \frac{10}{20} = \frac{30.68}{X + 1.32}$$

This gives $x = 14.02\ \Omega$

(*c*) Here we note that wire *AB* has resistance 0.04 Ohm/div. but least count comes 3/2 times, i.e. 0.06 Ohm/div. Why ? It is because 1 div. added to one arm, 1/2 div. is reduced from other arm due to ratio $P/Q = 1/2$.

(*d*) Order of current *X* is

$$i = \frac{2}{30 + 15} = 0.04 \text{ amp}$$

Mean resistance of *X* is taken 15 Ω because when null point is at *A*, $X = 16$ and when it is at *B*, $X = 13$.

The potential difference across 1 div. = $0.04 \times 0.04 = 0.0016$ V

We want at least $1/5 = 0.2$ div. deflection

Hence sensitivity is = $\frac{0.0016}{0.2} = 8 \times 10^{-3}$ V/div = 8 mV/div

Example 20.25: *N* cells each of emf *E* and internal resistance *r* are arranged in series forming a closed circuit. Two points including *n* cells on one side and (*N* – *n*) cells on the other are connected to a resistor *R*. Calculate the current through *R*.

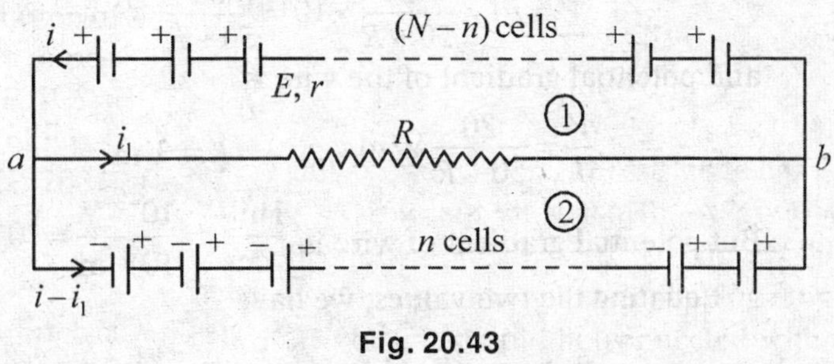

Fig. 20.43

Solution: The given network is shown in Fig. 20.43.

The current in different branches is shown. Using Kirchhoff's law for loop (1), we have

$$Ri_1 + (N - n)\,ri = (N - n)\,E \quad \ldots(i)$$

Using Kirchhoff's law for loop (2), we have

$$(i - i_1)nr - i_1R = nE$$

or $$nri - i_1(nr + R) = nE \quad \ldots(ii)$$

Multiplying (*i*) by *n* and (*ii*) by (*N* – *n*), we get

$$nRi_1 + n\,(N - n)\,ri = (N - n)\,nE \quad \ldots(iii)$$

$$-i_1(nr + R)(N - n) + n(N - n)ri = (N - n)nE \quad \ldots(iv)$$

(*iii*) – (*iv*) gives

$$i_1\,[nR + (nr + R)\,(N - n)] = 0$$

or $$i_1 = 0$$

Thus no current flows in branch ab and hence potential difference across ab is zero.

Note that $i \neq 0$, $i = E/r$, as such terminal voltage of each cell is zero. The current flows through the cells.

Example 20.26: Figure 20.44 shows a potentiometer, which has 4 coils of 10.0 Ω each and an uniform wire PQ of length 100.0 cm and resistance 10 Ω with 2.0 V battery (negligible internal resistance). R is adjusted to get a potential gradient 1.0 mV/cm on the wire PQ. (*i*) What is the value of resistor R? (*ii*) What is the voltage when connected across ab and shows balance?

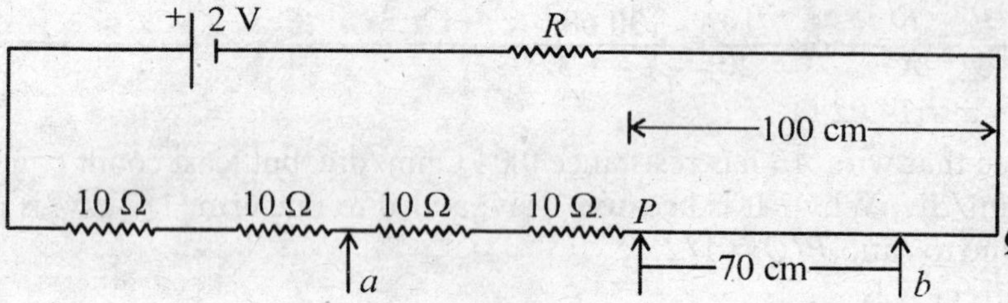

Fig. 20.44

Solution:

(*i*) Total resistance of the circuit is (50 + R)

Hence current in the circuit is

$$i = \frac{2}{50 + R}$$

Hence potential difference across wire PQ is

$$V = iR = \frac{2}{50+R} \times 10 = \frac{20}{50+R}\ \text{V}$$

and potential gradient of the wire is

$$\frac{V}{L} = \frac{20}{50+R}\ \text{V/m} \quad \because \quad L = 1\ \text{m}$$

But potential gradient of wire is $\frac{1\,\text{mV}}{\text{cm}} = \frac{10^{-3}\ \text{V}}{10^{-2}\ \text{m}} = 10^{-1}\ \text{V/m}$

Equating the two values, we have

$$\frac{20}{50 + R} = \frac{1}{10}$$

or $\quad 50 + R = 200 \quad$ and $\quad R = 150\ \Omega$.

(*ii*) Resistance of 70 cm wire is 7 Ω.

Hence total resistance between a and b is

10 + 10 + 7 = 27 Ω

current in circuit

$$i = \frac{2}{50+150} = \frac{2}{200} = 0.01\ \text{amp}$$

Hence potential difference across ab is

$$V = i \times 27 = 27 \times 0.01 = 0.27\ \text{V}$$

Hence a voltage of 0.27 V connected between ab shows balance.

Example 20.27:

(*a*) To measure the resistance of a coil R (Fig. 20.45) two students made connections X and Y. Ammeter A has resistance 0.50 Ω and voltmeter V has resistance of 1000 Ω. If reading in case of X comes to 0.20 amp and 6.10 V, deduce the voltmeter reading in case Y, if the ammeter reading is set again at 0.2 amp.

(*b*) Discuss the effect in case Y, if the voltmeter has resistance 10,000 Ω or 100,000 Ω.

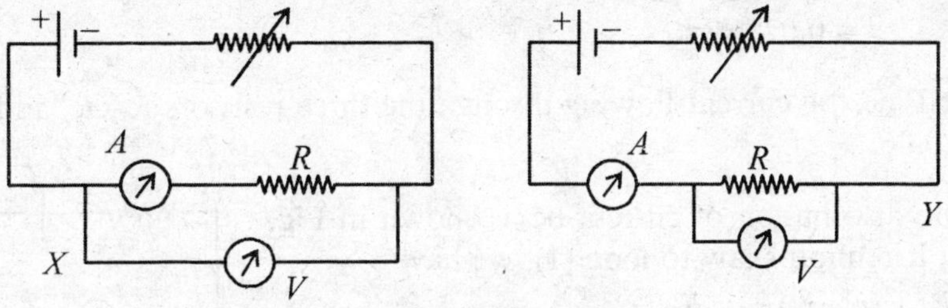

Fig. 20.45

Solution:

(*i*) For circuit X

$(R + 0.5)\,(0.2) = 6.10$

and this gives $R = 30\ \Omega$

For circuit Y, we have potential difference across R is

$$V' = \frac{0.2\,R \times 1000}{R + 1000} \qquad \because \text{ current through } R = \frac{iR_1}{R_1 + R}$$

where R_1 is resistance of voltmeter.

or $$V' = \frac{0.2 \times 30 \times 1000}{1030} = 5.28\ \text{V}$$

(*b*) If R is unknown, the circuit Y would give it as

$$R' = \frac{5.82}{0.2} = 29.1\ \Omega$$

The difference arises due to the factor $\frac{R_1}{R + R_1}$, larger the value of R_1, makes the factor closer to 1.

Example 20.27: Calculate the drift velocity of electrons when 1 amp current passes through a wire of radius 1 mm of copper. Given atomic weight of copper = 64, density of copper = 7.5×10^3 kg/m^3 and assume one free electron per atom.

Solution: The current in terms of drift velocity

$$i = nAeV_d \qquad \ldots(i)$$

where n is number density of free electrons.

$$n = \frac{n_0 d}{W}$$

where n_0 is Avogadro number = 6.02×10^{23}, W is atomic weight and d is density.

$$\therefore \quad n = \frac{6.02\times10^{23}\times7.5\times10^{3}}{64\times10^{-3}} = 0.7\times10^{29}\ /\text{m}^3$$

Using (*i*) the drift velocity V_d is

$$V_d = \frac{i}{nAe} = \frac{1}{0.7\times10^{29}\times\pi\left(10^{-3}\right)^2\times1.6\times10^{-19}}$$

$$= 2.8\times10^{-5}\ \text{m/s}$$

$$= 0.028\ \text{mm/s}.$$

Example 20.29: Find the current flowing through the three resistors R_1, R_2 and R_3 in the circuit shown in Fig. 20.46.

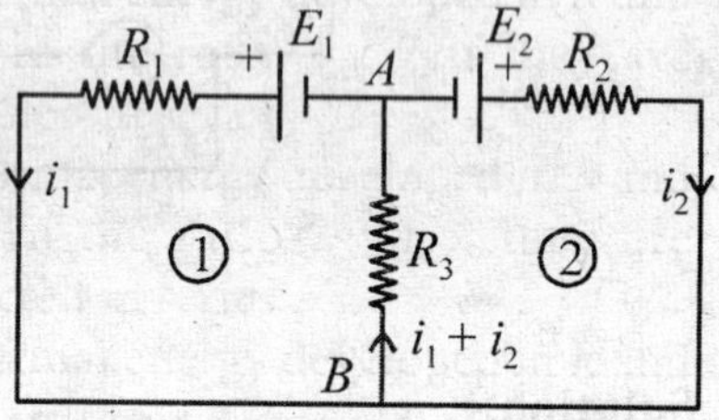

Fig. 20.46

Solution: Let the distribution of current be as shown in Fig. 20.46. Applying Kirchhoff's law to loop (1), we have

$$i_1R_1 + (i_1+i_2)R_3 = E_1$$

or $$i_1(R_1+R_3) + R_3 i_2 = E_1 \qquad \ldots(i)$$

Applying Kirchhoff's law to loop (2), we have

$$i_2R_2 + (i_1+i_2)R_3 = E_2$$

or $$i_1R_3 + (R_2+R_3)i_2 = E_2 \qquad \ldots(ii)$$

To eliminate i_2 multiply (*i*) by (R_2+R_3) and (*ii*) by R_3 and subtract. This gives

$$i_1 = \frac{E_1(R_2+R_3) - E_2R_3}{(R_1+R_2)(R_2+R_3) - R_3^2}$$

$$= \frac{E_1(R_2+R_3) - E_2R_3}{R_1R_2 + R_2R_3 + R_3R_1} \qquad \ldots(iii)$$

Similarly eliminating i_2 from (*i*) and (*ii*), we get

$$i_2 = \frac{E_2(R_2+R_3) - E_1R_3}{R_1R_2 + R_2R_3 + R_3R_1} \qquad \ldots(iv)$$

and $$i_1 + i_2 = \frac{E_1R_2 + E_2R_1}{R_1R_2 + R_2R_3 + R_3R_1} \qquad \ldots(v)$$

Example 20.30: A copper wire of length (l) and radius r is nickel plated till its final radius is (R) and length l. If the specific resistance of nickel and copper be ρ_n and ρ_c respectively, find the conductivity of this composite wire.

Solution: The resistance of copper wire $R_c = \dfrac{\rho_c l}{\pi r^2}$

where ρ_c is resistivity of copper.

Since cross-section of the nickel coating is $\pi(R^2 - r^2)$ as such resistance of this part of wire is

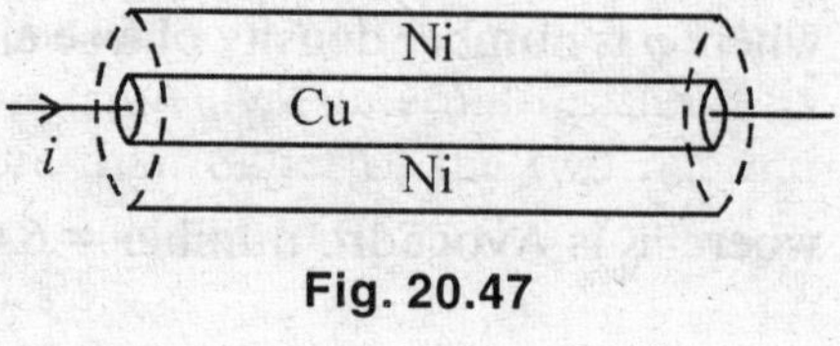

Fig. 20.47

$$R_n = \frac{\rho_n l}{\pi\left(R^2 - r^2\right)}$$

The two resistances are in parallel. Hence equivalent resistance is

$$\frac{1}{R} = \frac{1}{R_c} + \frac{1}{R_n} = \frac{\pi r^2}{\rho_c l} + \frac{\pi\left(R^2 - r^2\right)}{\rho_n l} = \frac{\pi}{l}\left\{\frac{r^2}{\rho_c} + \frac{\left(R^2 - r^2\right)}{\rho_n}\right\}$$

If this composite rod is replaced by a single rod of radius R and length l, then conductivity of composite rod σ is

$$\frac{1}{R} = \frac{\sigma\pi R^2}{l} = \frac{\pi}{l}\left[\frac{r^2}{\rho_c} + \frac{\left(R^2 - r^2\right)}{\rho_n}\right]$$

or
$$\sigma = \frac{1}{R^2}\left[\frac{r^2}{\rho_c} + \frac{\left(R^2 - r^2\right)}{\rho_n}\right]$$

Example 20.31: Find the current flowing through the resistance R_1 of the circuit as shown in Fig. 20.48, if $R_1 = 10\ \Omega$, $R_2 = 20\ \Omega$, $R_3 = 30\ \Omega$ and the potential of points 1, 2, 3 are $V_1 = 10$ V, $V_2 = 6$ V and $V_3 = 5$ V respectively.

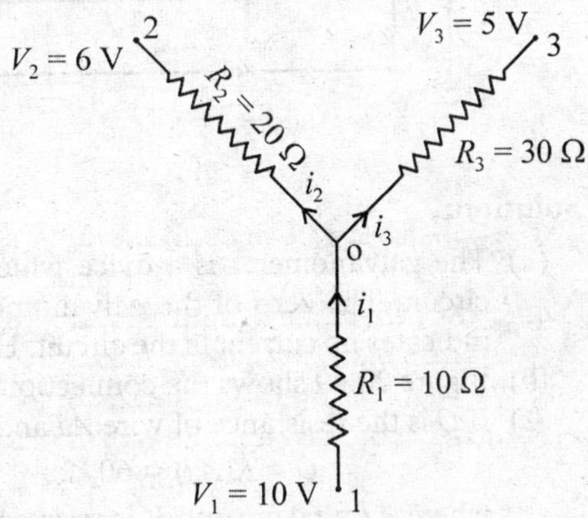

Fig. 20.48

Solution: The distribution of currents in different branches is shown in Fig. 20.48.

Kirchhoff's law at junction o gives

$$i_1 = i_2 + i_3 \qquad \ldots(i)$$

Using Kirchhoff's II law across junction o, we get for points 1 and 2

$$V_1 - i_1R_1 - i_2R_2 = V_2$$

or $$10i_1 + 20i_2 = 10 - 6 = 4$$

or $$5i_1 + 10i_2 = 2 \qquad \ldots(ii)$$

Similarly applying Kirchhoff's II law for junction o for points 1 and 3, we get

$$V_1 - i_1R_1 - i_3R_3 = V_3$$

or $$10i_1 + 30i_3 = 10 - 5 = 5$$

or $$2i_1 + 6i_3 = 1 \qquad \ldots(iii)$$

Equation (*ii*) gives $i_2 = \dfrac{2 - 5i_1}{10}$

and equation (*iii*) gives $i_3 = \dfrac{1 - 2i_1}{6}$

Using (*i*), we get

$$i_1 = \frac{2 - 5i_1}{10} + \frac{1 - 2i_1}{6}$$

or $$30i_1 = 6 - 15i_1 + 5 - 10i_1$$

or $$55i_1 = 11 \quad \text{or} \quad i_1 = \frac{11}{55} = 0.2 \text{ A}$$

Example 20.32: A thin uniform wire *AB* of length 1 m, an unknown resistance *X* and a resistance of 12 Ω are connected by a thick conducting strip as shown in Fig. 20.49. A battery and a galvanometer (with a sliding jockey) connected to it, are also available. Connections are to be made to measure the unknown resistance *X*, using principle of Wheatstone's bridge. Answer the following questions:

(*a*) Are there positive and negative terminals on the galvanometer?

(*b*) Copy the figure in your answer book and show the battery and the galvanometer (with Jockey) connected at approximate points.

(*c*) After appropriate connections are made, it is found no deflection takes place in the galvanometer when the sliding jockey touches the wire at a distance of 60 cm from *A*. Obtain the value of resistance *X*.

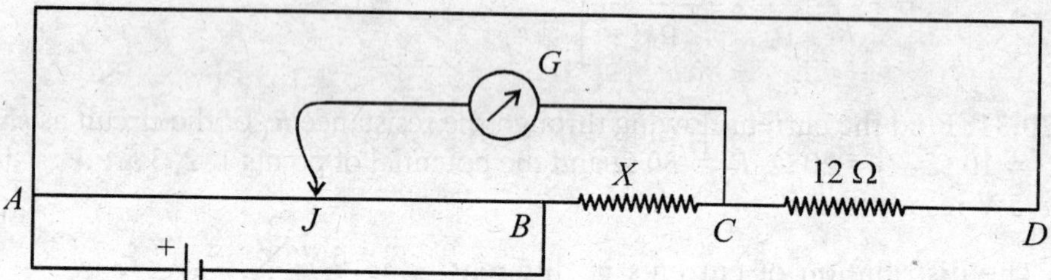

Fig. 20.49

Solution:

(*a*) The galvanometer is a dvice which is used for detecting (not measuring) current in the circuit. The zero of the galvanometer scale is in the middle of the scale. Null deflection indicates no current in the circuit. Hence galvanometer has no positive or negative terminal.

(b) Figure 20.49 shows the connections of galvanometer and battery.

(*c*) If *Q* is the resistance of wire *AJ* and *P* that of *JP*.

$$\therefore \qquad Q = K(AJ) = 60\,K$$

where $AJ = 60$ cm and *K* is resistance per unit length of the wire, then

$$P = K(JB) = 40K$$

This is the case of balanced Wheatstone's bridge, hence

$$\frac{Q}{P} = \frac{12}{X}$$

$$\text{or} \qquad \frac{60}{40} = \frac{12}{X} \quad \text{or} \quad X = \frac{12 \times 40}{60} = 8\ \Omega$$

PROBLEMS

1. In the circuit shown in Fig. 20.50, find the resistance in which power consumed is maximum. If the battery emf is 29 V. Find the maximum heat generated. **[Ans.** 4 Ω, 100 J]
2. A conductor with rectangular cross-section dimensions ($a \times 2a \times 4a$) as shown in Fig. 20.51. Resistance across *AB* is *x*, across *CD* is *y* and *EF* is *Z*. Then prove that $Z^2 = xy$.

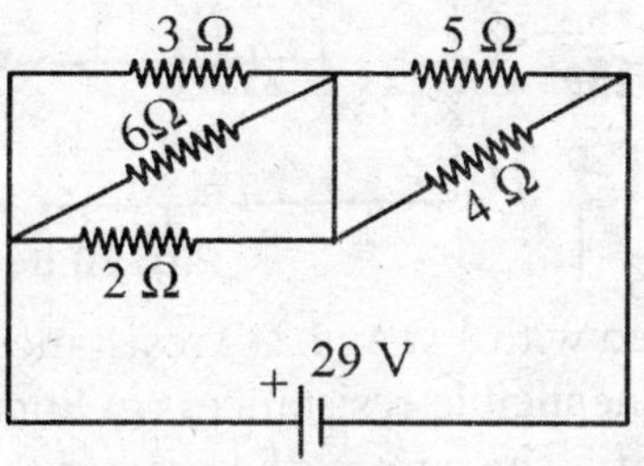

Fig. 20.50

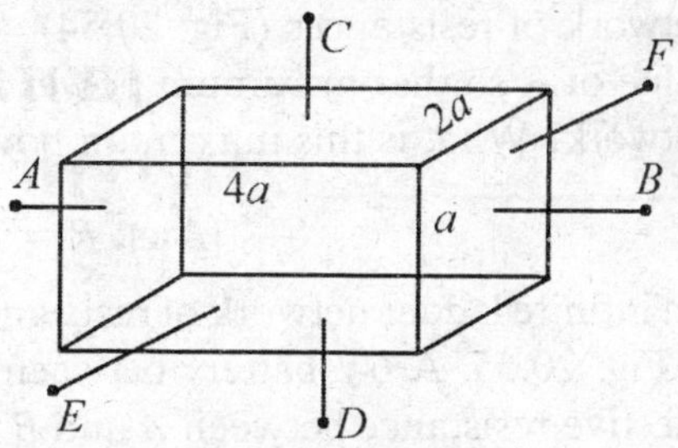

Fig. 20.51

3. The electric current of 5 A is divided into three branches forming parallel combination. The lengths of the wire in three cases are in the ratio 2 : 3 : 4. While their diameters are respectively in the ratio 3 : 4 : 5. Find current in each branch if wires are of the same material. **[Ans.** $i_1 = 1.4$ A, $i_2 = 1.66$ A, $i_3 = 1.94$ A]
4. A thick spherical shell has inner radius *a* and outer radius *b*. The material has resistivity ρ. Show that when *a* potential difference is applied between the inner and outer surface its resistance is $R = \dfrac{\rho(b-a)}{4\pi ab}$.
5. Find the effective resistance between points *A* and *B* for network shown in Fig. 20.52.

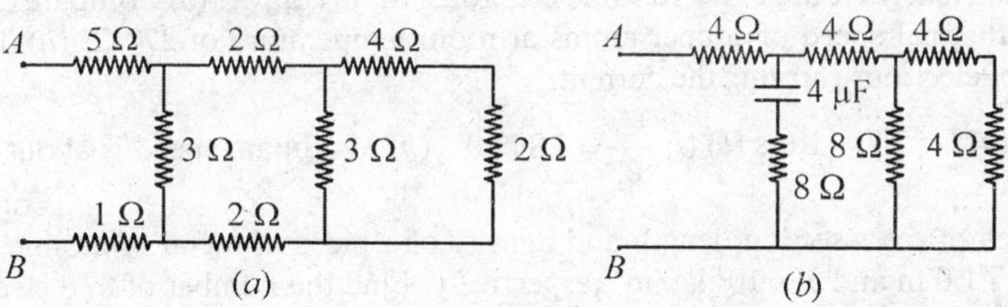

Fig. 20.52

[Ans. (*a*) 8 Ω (*b*) 12 Ω]

6. For what value of load, the power transfer is maximum when two identical cells each of emf *E* and internal resistance *r* are connected (*a*) in series (*b*) in parallel.

$$\left[\textbf{Ans. } (a)\ r_{eq} = 2r\left(P_{max}\right) = \frac{E^2}{2r} \quad (b)\ r_{eq} = \tfrac{r}{2}\left(P_{max}\right) = \frac{E^2}{2r}\right]$$

7. What is the potential difference between the points *M* and *N* for the circuit shown in Fig. 20.53. **[Ans.** $V = 8.4$ V]

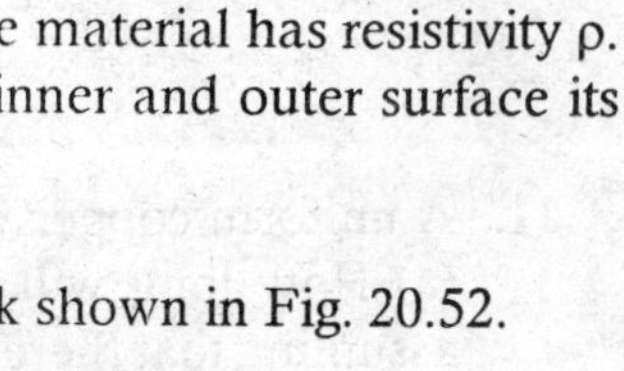

Fig. 20.53

8. A 30 V storage battery is charged from a 120 V direct current supply mains with a resistor being connected in series with battery to limit the charging current to 15 amp. Find the rate at which energy is wasted (if any) in heat in the circuit. If all this heat produced, could be made available in heating water, how long will it take to bring 1 kg of water from 15° C to the boiling point. Neglect the internal resistance of the battery.

[**Ans.** $P = 1350$ W, Time $t = 4.4$ minutes]

9. A battery of internal resistance 4 Ω is connected to the network of resistances (Fig. 20.54). What must be the value of R so that maximum power is delivered to the network? What is this maximum power? (IIT 95)

$$\left[\textbf{Ans. } R = 2\ \Omega\ \ P_{max} = \frac{E^2}{16}\right]$$

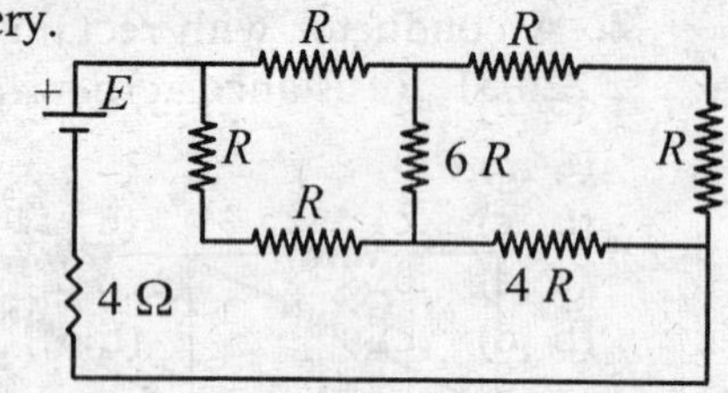

Fig. 20.54

10. An infinite ladder network of resistances is connected with 1 Ω and 2 Ω resistance as shown in Fig. 20.55. A 6 V battery between A and B has negligible resistance. (*a*) Show that the effective resistance between A and B is 2 Ω (*b*) What is the current that passes through 2 Ω resistance nearest to the battery?

[**Ans.** $i = 1.5$ A]

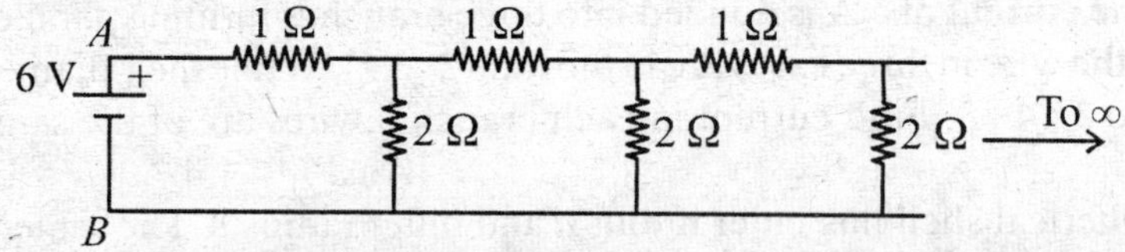

Fig. 20.55

11. A uniform copper wire of length 3.0 m and diameter 0.050 cm carries a current of 1.0 A. (*a*) How long will an electron take to drift from one end of the wire to the other end assuming that there are 8.4×10^{28} free electrons/m³ in copper? (*b*) Compare drift speed with (*i*) the thermal speed of copper atoms at room temperature of 27° C. (*ii*) The maximum speed of electrons carrying the current.

[**Ans.** (*a*) $t = 7.7 \times 10^3$ s (*b*) (*i*) $\frac{V_d}{V_t} = 5.8 \times 10^{-7}$ (*ii*) Maximam speed is about 10^9 times that of drift velocity]

12. The area of cross-section, length and density of a piece of metal of atomic weight 60 are 10^{-6} m², 1.0 m and 5×10^3 kg/m³ respectively. Find the number of free electrons per unit volume if every atom contributes one free electron. Also find the drift velocity of electrons in the metal when a current of 16 A passes through it (Given Avogadro no. $n_0 = 6 \times 10^{23}$ mol⁻¹).

[**Ans.** $n = 5 \times 10^{28}$/m³, $V_d = 2$ mm s⁻¹].

13. A copper wire is stretched to make it 0.1% longer. Find the percentage change in its resistance.

[**Ans.** 2 %]

14. Calculate the steady state current in the 2 Ω resistor shown in Fig. 20.56. The internal resistance of the battery is negligible and the capacity of the capacitor is 0.5 μF.

[**Ans.** $i = 0.9$ A]

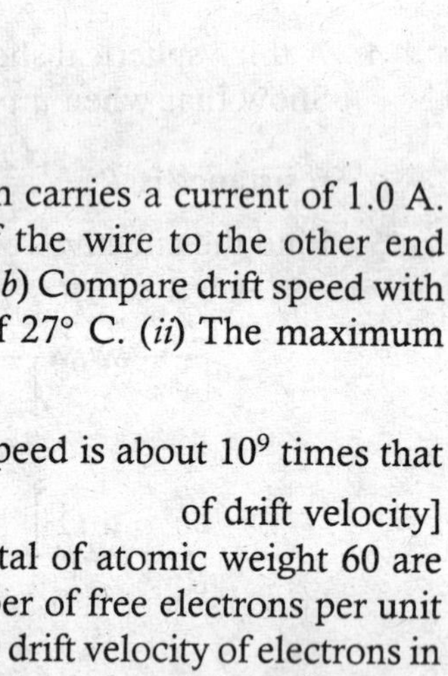

Fig. 20.56

15. In the circuit shown in Fig. 20.57, $E_1 = 3$ V, $E_2 = 2$ V, $E_3 = 1$ V and $R = r_1 = r_2 = r_3 = 1$ Ω. (*a*) Find the current through each

branch. (*b*) If r_2 is short circuited and the point A is connected to B, find the current through cells E_1, E_2, E_3 and through the resistor R.
[**Ans.** (*a*) $i_1 = 1$ A, $i_2 = 0$ and $i_3 = 1$ A (*b*) $i_1 = i_3 = 1$ A, $i_2 = 2$ A and current through R is zero]

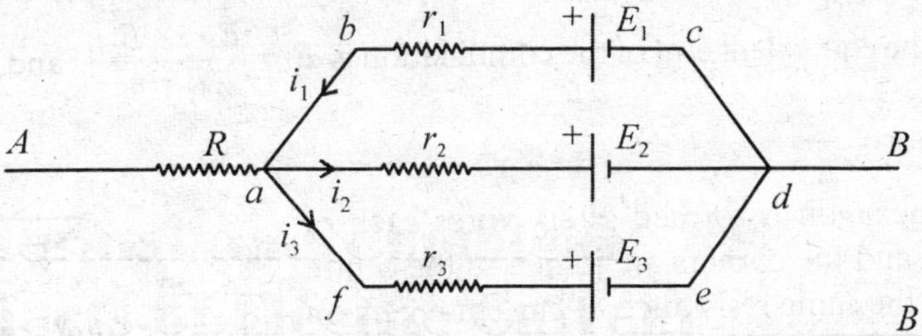

Fig. 20.57

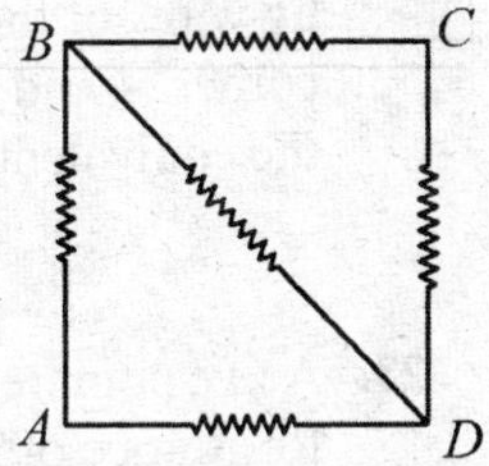

Fig. 20.58

16. A wire of resistance 0.1 $\Omega\,cm^{-1}$ is bent to form a square *ABCD* of side 10 cm. A similar wire is connected between the corners B and D to form the diagonal *BD*. Find the effective resistance of this combination between corners A and C. If a 2 V battery of negligible internal resistance is connected across A and C, calculate the total power dissipated. [**Ans.** $R = 1\,\Omega$ $P = 4$ W]

17. (*a*) A galvanometer having a coil resistance of 100 Ω gives a full scale deflection when a current of 1 mA is passed through it. What is the value of resistance which can convert this galvanometer into ammeter giving a full scale deflection for a current of 10 A?
(*b*) A resistance of the required value is available but it will get burnt if the energy dissipated in it is greater than 1 Watt. Can it be safely used for conversion of the given galvanometer?
(*c*) This modified galvanometer is connected across the terminals of a battery, it shows a current of 4 A. The current drops to 1 A, when a resistance of 1.5 Ω is connected in series with the modified galvanometer. Find the emf and internal resistance of the battery.

[**Ans.** (*a*) $S = \dfrac{1}{99.99}\,\Omega$ (*b*) $P = 0.9999$ W which is less than 1 Watt hence shunt can be used safely (*c*) $E = 2$ V, $r = 0.49\,\Omega$]

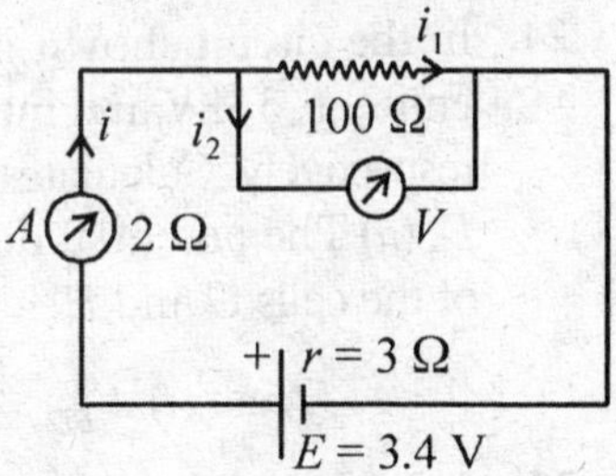

Fig. 20.59

18. A cell of emf 3.4 V and internal resistance 3 Ω is connected to an ammeter having resistance of 100 Ω. When a voltmeter is connected across the 100 Ω resistance the ammeter reading is 0.04 A. Find the voltage reading by the voltmeter and its resistance. Had the voltmeter been an ideal one, what would have been its reading?
[**Ans.** $V = 3.2$ V, $R = 400\,\Omega$, for ideal voltmeter $V = 3.24$ V].

19. A galvanometer of resistance 20 Ω gives full scale deflection when a current of 0.001 A is passed through it. Calculate the value of the shunt that must be used to convert it into an ammeter that can read current up to 1.0 A. When this ammeter is put in the circuit does it read (slightly) less or more than the actual current in the original circuit?
[**Ans.** $R_s = 0.02\,\Omega$, it reads slightly less]

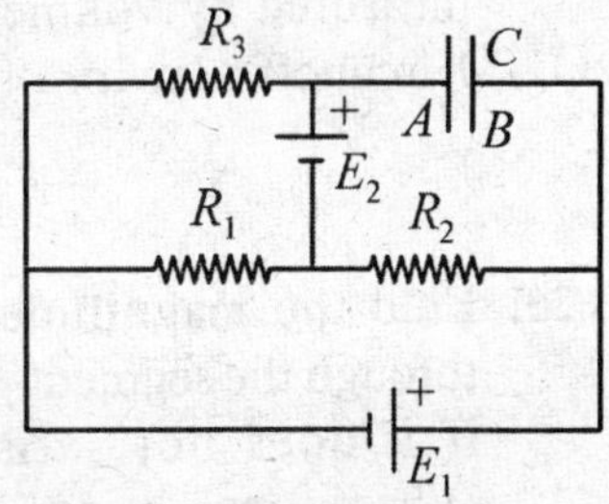

Fig. 20.60

20. Find the potential difference $V_A - V_B$ between the plates of a capacitor C in the circuits shown in Fig. 20.60. Given emf of sources $E_1 = 4.0$ V, $E_2 = 1.0$ V, $R_1 = 20\ \Omega$, $R_2 = 10\ \Omega$, $R_3 = 30\ \Omega$. The internal resistance of sources is negligible.

[**Ans.** $V_A - V_B = 1$ V]

21. Two cells of emfs E_1 and E_2 and internal resistances r_1 and r_2 are connected in parallel. Show that the equivalent emf of the combination is $E = \dfrac{E_1 r_2 + E_2 r_1}{r_1 + r_2}$ and the effective internal resistance is $r = \dfrac{r_1 r_2}{r_1 + r_2}$.

22. A regular hexagon is formed of six wires each of resistance r and the corners are joined to the centre by wires of the same resistance. If current enters on corner A and leaves at the opposite corner B, find the equivalent resistance of the conductor.

$\left[\textbf{Ans. } R_{eq} = \dfrac{4}{5} r\right]$

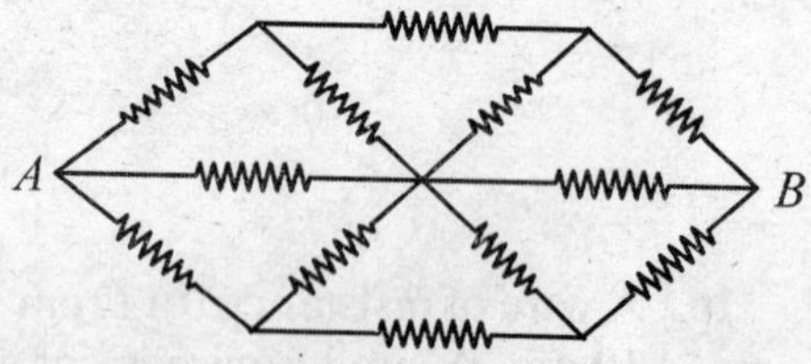

Fig. 20.61

23. A five point regular star has been soldered together from a uniform wire. Resistance of $AK = KB = LC = CF = FD = DG = GH = EH = HA = r$.

(*a*) Find the resistance of section KBL.

(*b*) What is equivalent resistance across the terminals A and F?

[**Hint**: If resistance of $KB = r$, then of $KL = 0.6r$. Hence R_{eq} for section $KBL = \dfrac{0.6r \times 2r}{0.6r + 2r} = \dfrac{1.2}{2.6} r = 0.46r\ \Omega$. Resistance of $AKLF$ section $= (0.46 + 0.46 + 1)r = 1.92r$. Hence net resistance between AF $= \dfrac{1.92r}{2} = 0.96r$]

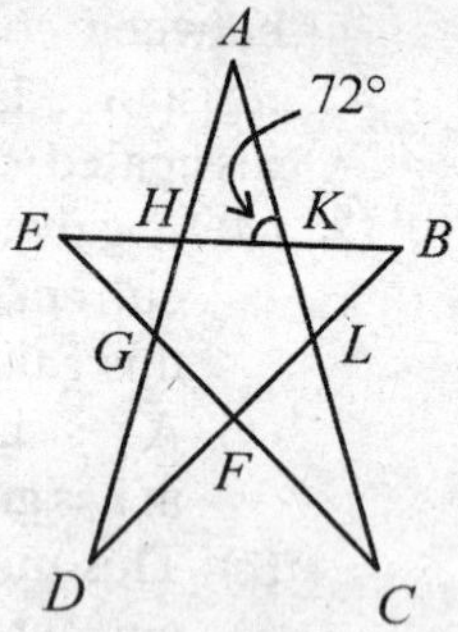

Fig. 20.62

24. In the circuit shown in Fig. 20.63, E, F, G, H are cells of emf 2, 1, 3, 1 V and internal resistances, 2, 1, 3 and 1 Ohms respectively. Calculate (*i*) Potential difference between B and D. (*ii*) The potential difference across the terminals of each of the cells G and H.

[**Ans.** (*i*) $V_{BD} = \dfrac{2}{13}$ V (*ii*) $V_G = 1.61$ V, $V_H = 1.61$ V]

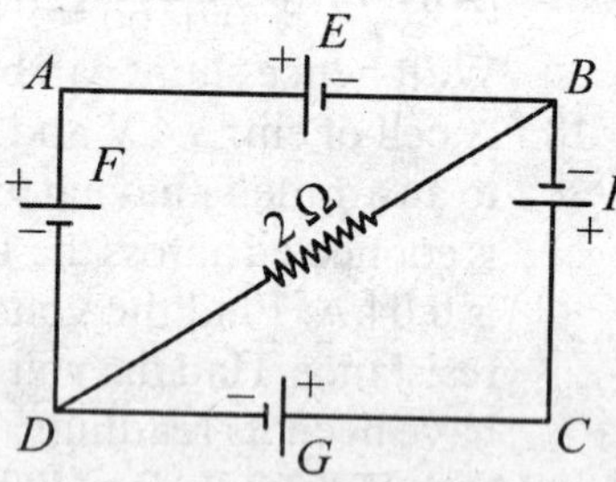

Fig. 20.63

25. An electric circuit is shown in Fig. 20.64. Calculate the potential difference across the resistor of 400 Ω as will be measured by voltmeter V of resistance 400 Ω either by Kirchhoff's law or otherwise.

$\left[\textbf{Ans. } V = \dfrac{20}{3} \text{ V}\right]$

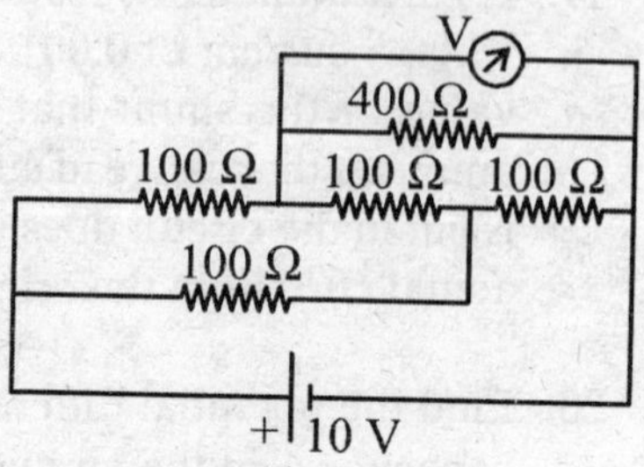

Fig. 20.64

26. Find the magnitude and direction of current flowing through the segment CD of the circuit shown in Fig. 20.65, if it does not satisfy the condition of balance of Wheatstone's bridge.

[**Ans.** $i = \dfrac{V(R_2R_3 - R_1R_4)}{R_1R_2(R_3 + R_4) + R_3R_4(R_1 + R_2)}$ from C to D if $\dfrac{R_1}{R_2} < \dfrac{R_3}{R_4}$]

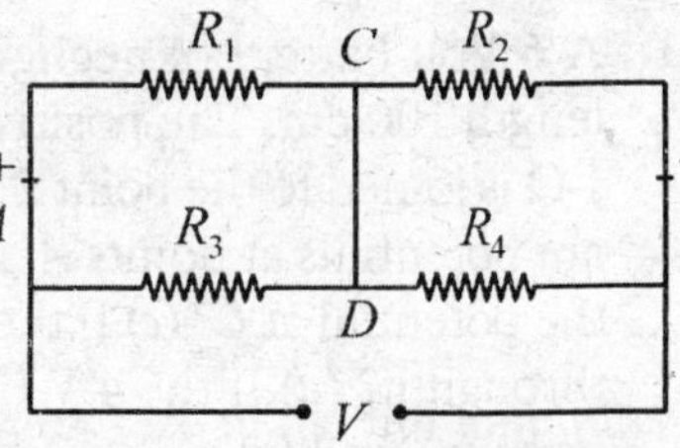

Fig. 20.65

27. Suppose the switch S has been closed sufficiently long for the capacitor to become fully charged. Find (*a*) the steady state current through each resistor, (*b*) the charge on the capacitor, (*c*) the switch S is now opened at $t = 0$. Find the time it takes for the charge on the capacitor to fall to half its initial value, (*d*) write an equation for the current through R_2 as a function of time.

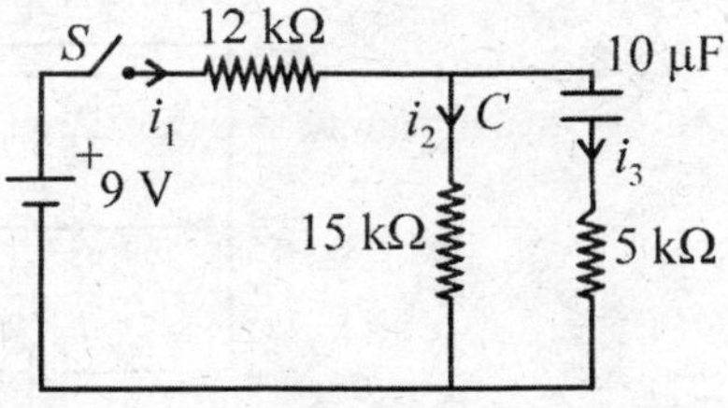

Fig. 20.66

[**Ans.** (*a*) $i_1 = i_2 = \dfrac{1}{3}$ mA, $i_3 = 0$ (*b*) $q = 50\ \mu C$ (*c*) $t = 0.1386$ s (*d*) $i = 0.25e^{-5t}$]

28. A copper wire of radius 0.1 mm and resistance 1 kΩ is connected across a power supply of 20 V. (*a*) How many electrons are transferred per second between the supply and the wire at one end? (*b*) Write down the current density in the wire.

[**Ans.** (*a*) 1.25×10^{17} (*b*) 6.37×10^5 A/m²]

29. Find the charges on four capacitor of capacitance 1 μF, 2 μF, 3 μF, and 4 μF shown in Fig. 20.67.

[**Ans.** 2 μC, 8 μC, 9 μC and 12 μC]

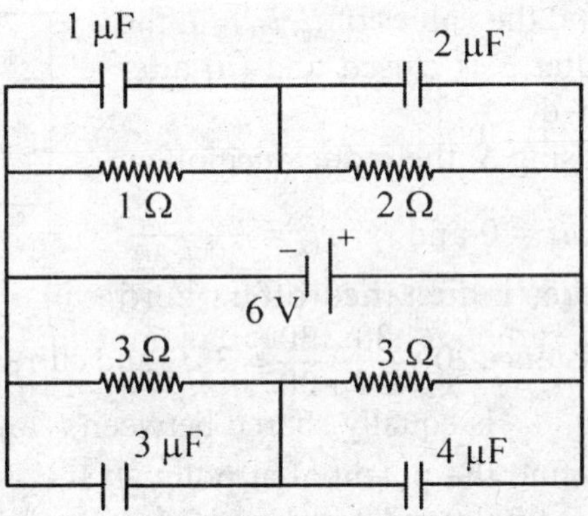

Fig. 20.67

30. The charge supplied by source varies with time as $Q = at - bt^2$, find the total heat produced in resistance $2R$.

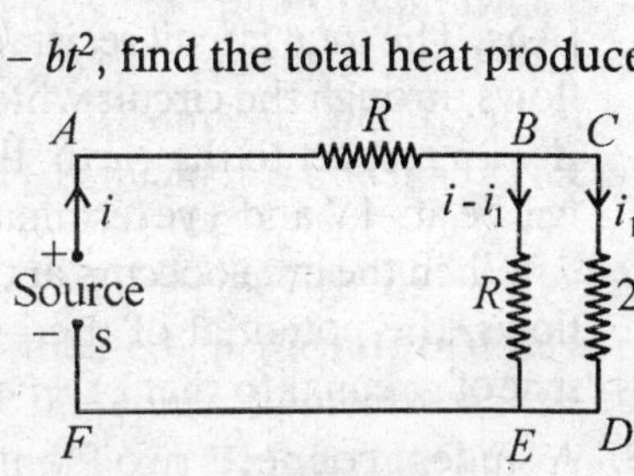

Fig. 20.68

[**Hint.** $i = \dfrac{dQ}{dt} = a - 2bt$ and $i_1 = i/3 = \dfrac{a - 2bt}{3}$

$$\therefore H = \int_0^{t_0} i_1^2(2R)dt = \int_0^{t_0} \frac{(a - 2bt)^2}{9} \times 2R\,dt$$

where $t_0 = a/2b$ integration gives $\dfrac{a^3R}{27b}$]

31. A 6 Volt battery of negligible internal resistance is connected across a uniform wire *AB* of length 100 cm. The positive terminals of another battery of emf 4 V and internal resistance 1 Ω is joined to the point *A* as shown in Fig. 20.69. Take the potential at *B* to be zero, (*a*) what are potentials at points *A* and *C*? (*b*) At what point *D* of the wire *AB*, the potential is equal to the potential at *C*? (*c*) If the points *C* and *D* are connected by a wire, what will be the current through it? (*d*) If the 4 V battery is replaced by 7.5 V battery, what would be the answer of parts (*a*) and (*b*)?

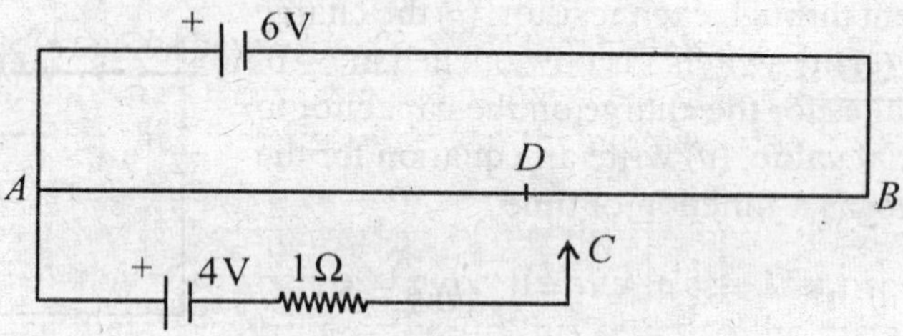

Fig. 20.69

[**Ans.** (*a*) $V_A = 6$ V, $V_C = 4$ V (*b*) $AD = 66.7$ cm (*c*) Zero (*d*) $V_A = 6$ V, $V_C = -1.5$ V no such point *D* exists]

SHORT ANSWER TYPE QUESTIONS

1. In circuit shown in Fig. 20.70 switch *S* is initially open. If switch is now closed, find the values of i_1, i_2, i_3 at the instant (*i*) immediately after *S* is closed and (*ii*) after steady values are established.

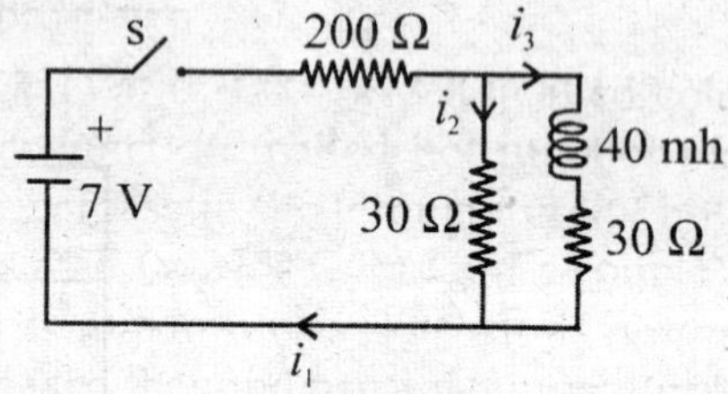

Fig. 20.70

[**Ans.** Immediately on closing *S*, the inductance offers infinite resistance. Hence $i_3 = 0$ and $i_1 = i_2 = \frac{7}{20 + 30}$ = 0.14 A. In steady state, inductance offers zero resistance. Hence total resistance $20 + \frac{30 \times 30}{30 + 30} = 35\,\Omega$ and current $i_1 = \frac{7}{35} = 0.2$ amp and this is equally shared between i_2 and i_3. Thus $i_2 = i_3 = 0.1$ amp]

2. In the circuit (Fig. 20.71) find the potential at point *A*. If the connections breaks at the point indicated by the arrow, find the new potential at *A*.

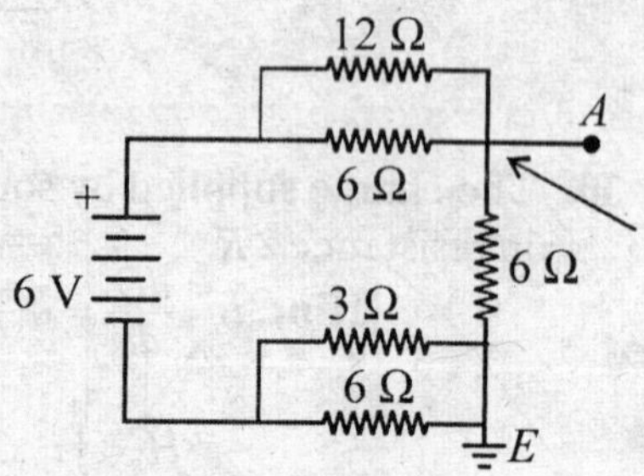

Fig. 20.71

[**Ans.** The total circuit resistance is 12 Ω. A current of 0.5 A flows through the circuit which gives the potential of 3 V at *A* with respect to the earth, the –ve terminal of the battery will be at –1V and +ve terminal of the battery will be at +5 V.

When the break occurs at the indicated place, no current flows, the potential of the –ve terminal will be zero and that of *A* equal to that of +ve terminal, i.e. 6 V.]

3. A student connects two filament lamps rated 100 W and 50 W (for 220 V) in series to a 440 V supply. He finds that one of them fuses. Discuss which lamp will fuse.

[**Ans.** Since bulbs are in series, the same current will pass through both of them. The one which has greater resistance will have greater potential difference across it. 50 W bulb will have greater resistance $\left(R = \frac{V^2}{W}\right)$ hence potential difference across this bulb will be more than 220 V. Hence 50 W lamp will fuse.

4. In the circuit shown in Fig. 20.72, $R_1 = 2\ \Omega$, $R_2 = 6\ \Omega$, $R_3 = 5\ \Omega$; $E_1 = 2$ V and $E_2 = 1$ V. Find potential difference between P and Q.

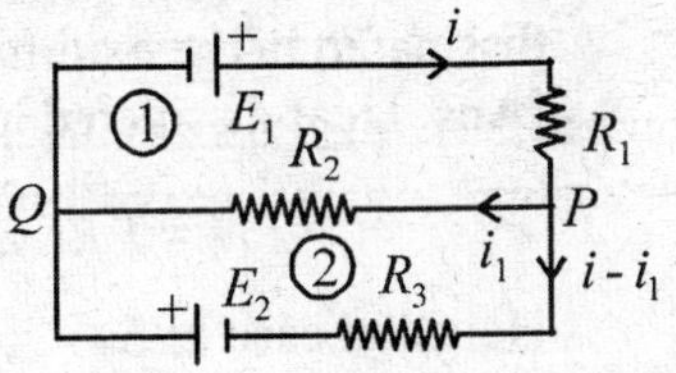

Fig. 20.72

[**Ans.** Using Kirchhoff's law for loop (1), we have

$$2i + 6i_1 = 2 \text{ or } i + 3i_1 = 1 \qquad \ldots(i)$$

For loop (2), we get

$$-6i_1 + 5(i - i_1) = 1 \text{ or } 5i - 11i_1 = 1 \qquad \ldots(ii)$$

$5 \times (i) - (ii)$ gives

$$26i_1 = 4 \text{ or } i_1 = 2/13$$

Hence potential difference between

$$PQ = 6i_1 = 6 \times \frac{2}{13} = \frac{12}{13} \text{ V}\Big]$$

5. Three resistors shown in Fig. 20.73 satisfy the relation $R_2 = 2R_3$ and $R_2^{-1} + R_3^{-1} = R_1^{-1}$. If P_i stands for the power dissipated in ith resistor, then find the ratio $P_1 : P_2 : P_3$.

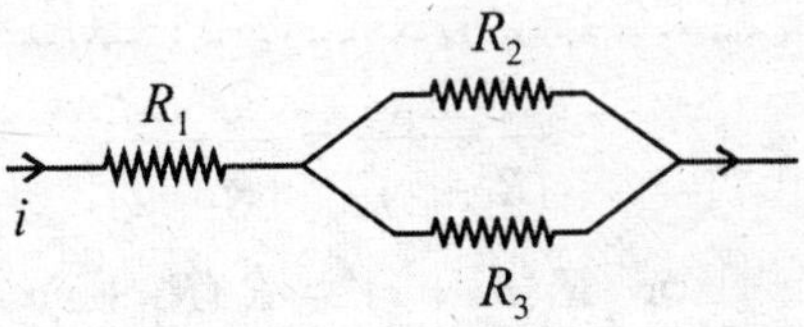

Fig. 20.73

[**Ans.** If current through R_2 is i then current through $R_3 = 2i$ and that through R_1 is $3i$.

$$\text{Resistance } R_1 = \frac{R_2 R_3}{R_2 + R_3} = \frac{2}{3} R_3$$

$$\therefore \quad P_1 = (3i)^2 R_1 = \frac{2}{3} R_3 9i^2 = 6i^2 R_3$$

$$P_2 = i^2 R_2 = i^2 2R_3 = 2i^2 R_3$$

$$\text{and} \quad P_3 = (2i)^2 R_3 = 4i^2 R_3 \qquad \therefore \quad P_1 : P_2 : P_3 = 3 : 1 : 2.]$$

6. Figure 20.74 shows a Wheatstone's net, with $P = 1000\ \Omega$, $Q = 10\ \Omega$, R (unknown), S variable and near 150 Ω for balance. If the connection across AC and BD are interchanged, discuss the error range in R determination.

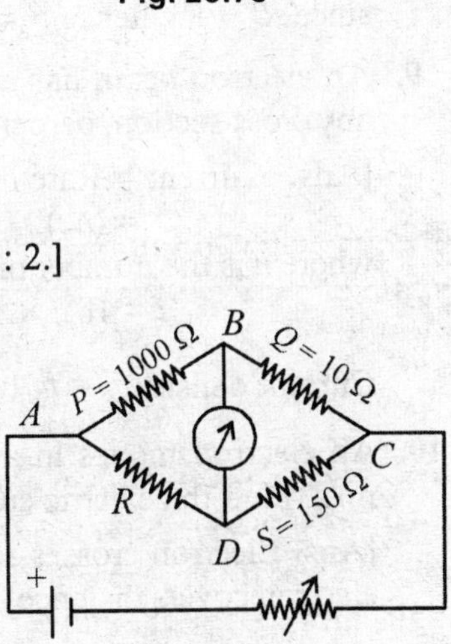

Fig. 20.74

[**Ans.** R is of the order of 15000 Ω. The junction of highest two and lowest two resistances are A and C and for better sensitivity the galvanometer should be between these junctions. Hence error will decrease with interchange suggested.

Estimation: With the given connections $\Delta S = 1\ \Omega$ will cause at D a voltage change $\frac{150}{15150}$ to $\frac{151}{15151}$. Fraction of $V(AC)$ with connections interchanged the corresponding change is $\frac{150}{160}$ to $\frac{151}{161}$ that is about 6 fold.]

7. Two cells of emf E_1 and E_2 ($< E_1$) having zero internal resistances are connected in parallel. What will be the terminal voltage?

[**Ans.** When a cell is short circuited, the terminal voltage drops to zero. However, here one cell (source of smaller emf) is short circuiting the other, the terminal voltage will be equal to the smaller emf, i.e. $V_1 = V_2 = E_2$.]

8. An accumulator is connected first to an external resistance R_1 and then to another external resistance R_2. For what value of the internal resistance of the accumulator will the heat dissipated in the external resistances be the same in the two cases?

[**Ans.** Heat dissipated in the external resistance is

$$H = i^2Rt \text{ and } i = \frac{E}{R+r}$$

Hence in case of R_1

$$H_1 = \frac{E^2R_1t}{(R_1+r)^2}$$

and in case of R_2, $H_2 = \dfrac{E^2R_2t}{(R_2+r)^2}$. For $H_1 = H_2$, we have

$$\frac{E^2R_1t}{(R_1+r)^2} = \frac{E^2R_2t}{(R_2+r)^2}$$

or $R_1(R_2 + r)^2 = R_2(R_1 + r)^2$ or $(R_2 - R_1)(r^2 - R_1R_2) = 0$

since $R_1 \neq R_2$, hence $r^2 = R_1R_2$ and $r = \sqrt{R_1R_2}$]

9. An electron beam has an aperture 1.0 mm^2. A total number of 10^{20} electrons go through any cross-section, perpendicularly in 1 sec. Find the current and current density in the beam.

[**Ans.** Current = Rate of flow of charge.
$= (ne)$

where n is the number of electrons crossing cross-section in 1 sec.

$\therefore \quad i = 10^{20} \times 1.6 \times 10^{-19} = 16$ amp

Current density $j = i/A = \dfrac{16}{10^{-6}} = 16 \times 10^6$ amp/m^2.]

10. An electron moves in a circular orbit of radius 0.5 Å with a constant speed of 3×10^6 m/s. Find the electric current at a point on the circle.

[**Ans.** Electron crosses any point on the circle once in every revolution. In one revolution electron covers distance $2\pi r$

Hence its time period $T = \dfrac{2\pi r}{v}$

or frequency $= \dfrac{1}{T} = \dfrac{v}{2\pi r}$

This means the electron passes $\dfrac{v}{2\pi r}$ times in 1 s through this point; carrying change e.

Hence current $= e \times$ frequency $= 1.6 \times 10^{-19} \dfrac{v}{2\pi r}$

$$= \frac{1.6 \times 10^{-19} \times 3 \times 10^6}{2\pi \times .5 \times 10^{-10}}$$

$$= \frac{10^{-3} \times 1.6 \times 3}{\pi} = 1.5 \times 10^{-3} \text{ amp} \Big]$$

11. Find the equivalent resistance between the points *a* and *b* of the network shown in Fig. 20.75. Each resistance is equal to *r*.

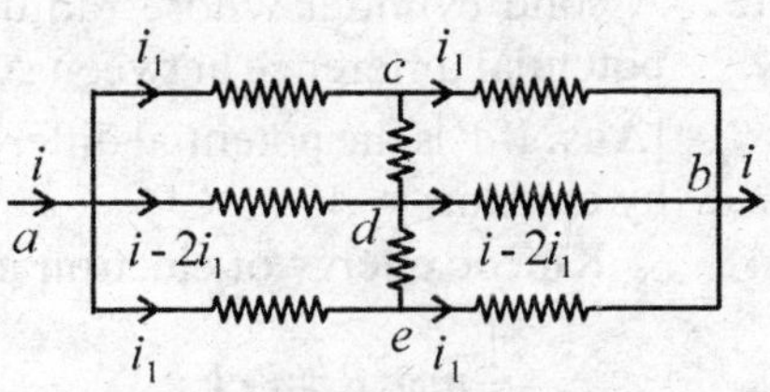

Fig. 20.75

[**Ans.** When a potential difference V is applied between *a* and *b*, *i* current enters at *a* and leaves at *b*. Because of symmetry current in *ac* and *ae* will be equal $(= i_1)$ and current in *ad* $= i - 2i_1$. Similarly currents from *cb*, *db* and *eb* combine at *b* to give the total current *i*. Since the situation at *b* is equivalent to that at *a*, by symmetry the currents in *cb*, *eb* will be i_1 and that in *db* will be $i - 2i_1$.

Using junction law at *c*, we see that the current in *cd* is zero. Similarly the current in *dc* is also zero. Thus the circuit is equivalent to three resistances each of value $2r$, in parallel. The equivalent resistance is $\frac{2r}{3}$]

12. 3.2 A current is flowing through a conductor. Find the number of electrons crossing any cross-section in 1 sec.

[**Ans.** If *n* is the number of electrons crossing any cross-section in 1 sec., then total charge crossing it in 1 sec is *ne*. Hence

$$i = ne \quad \text{or} \quad n = \frac{i}{e} = \frac{3.2}{1.6 \times 10^{-19}} = 2 \times 10^{19} \Big]$$

13. A wire of resistance R is stretched so that its length increases by 10%. Find the increase in the resistance of the wire.

[**Ans.** Volume of the wire will remain constant on stretching. If initial length of the wire be l and cross-section A and volume $V = Al$. On stretching 10% length becomes $1.1l$ and cross-section $A' = A/1.1$. Hence new resistance $R' = \frac{\rho \cdot (1.1)l}{A/1.1} = \frac{\rho l}{A}(1.1)^2 \cdot R$

or $\quad \frac{R'-R}{R} = (1.1)^2 - 1 = 0.21$

∴ Percentage increase in resistance is 21%]

14. V–I graph for a conductor at temperatures T_1 and T_2 are shown in Fig. 20.76. Show that $(T_1 - T_2)$ is proportional to $\cot 2\theta$.

[**Ans.** Let R_0, R_1 and R_2 be the resistances at 0° C, T_1° C and T_2° C, then

$$R_1 = \tan\theta = R_0(1 + \alpha T_1) \quad \ldots(i)$$

$$R_2 = \cot\theta = R_0(1 + \alpha T_2) \quad \ldots(ii)$$

Subtracting we get $\tan\theta - \cot\theta = R_0\alpha(T_1 - T_2)$

or $\quad R_0\alpha(T_1 - T_2) = \frac{\cos^2\theta - \sin^2\theta}{\sin\theta\cos\theta} = -\frac{2\cos 2\theta}{\sin 2\theta} = -2\cot 2\theta$

or $\quad T_1 - T_2 = -\frac{2}{R_0\alpha}\cot 2\theta \Big]$

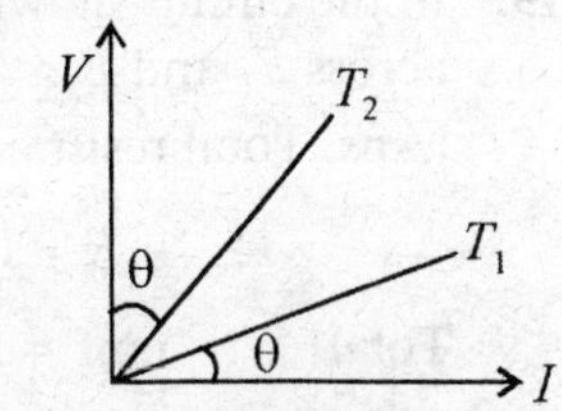

Fig. 20.76

15. A solid cylinder whose radius is R, rotates with a constant angular velocity ω. Find the potential difference between surface of cylinder and the axis.

[**Ans.** If V is the potential difference between axis and surface of cylinder then energy acquired by electron $= eV$.

Kinetic energy of electron increases from 0 to $1/2mu^2$ when $u = \omega R$.

$$\therefore \quad \frac{1}{2} m\omega^2 R^2 = eV$$

or $$V = \frac{m\omega^2 R^2}{2e} \Big]$$

16. A rectangular wire mesh of infinite extent in a plane has 1 amp of current fed into it at point P and 1 amp taken out of it at point Q (Fig. 20.77). Find the current in wire PQ.

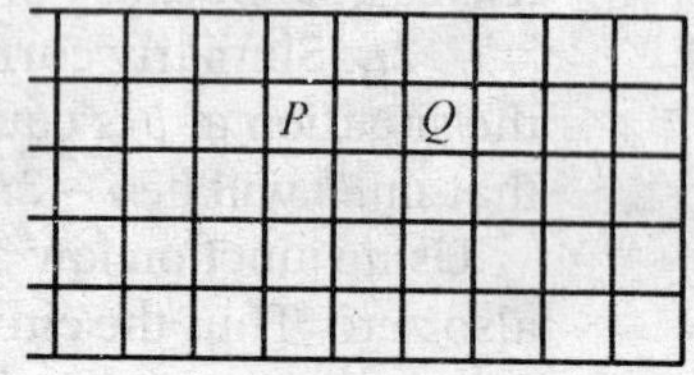

Fig. 20.77

[**Ans.** If one ampere current is fed into P and taken out at infinity, then from symmetry $\frac{1}{4}$ amp will flow in PQ. Similarly, if 1 amp is taken from Q and fed at infinity, $\frac{1}{4}$ amp will flow in PQ. By superposition the current in wire PQ will be 1/2 A.]

17. In a circuit (Fig. 20.78), find the potential difference between points B and D ($V_B - V_D$).

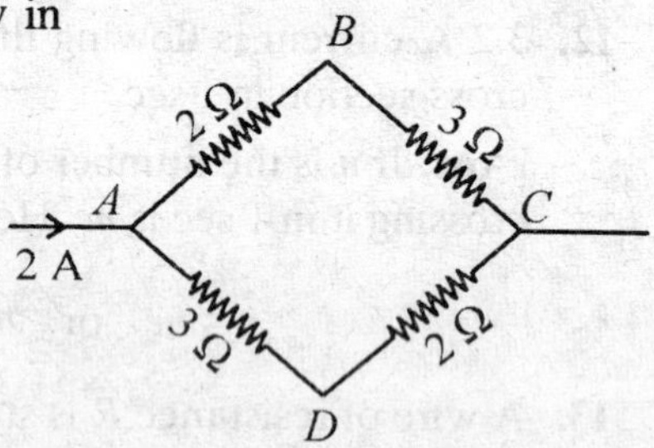

Fig. 20.78

[**Ans.** Resistances in the two branches ABC and ADC are same, hence 1 amp current flows through each branch. Hence

Potential difference $V_A - V_B = 2 \times 1 = 2$

Potential difference $V_A - V_D = 3 \times 1 = 3$

$\therefore$ Potential difference $V_B - V_D = 1$ V]

18. Figure 20.79 shows the graph between terminal voltage (V) across the cell terminals and the current (I) drawn. Find the emf of the cell and internal resistance of the cell.

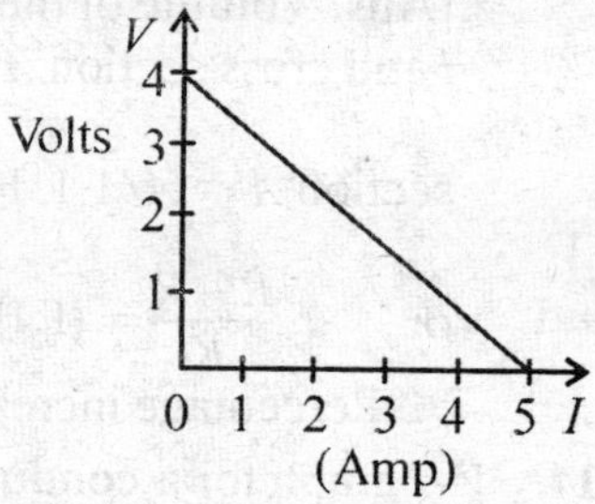

Fig. 20.79

[**Ans.** Terminal voltage V, emf E, current i and internal resistance r are related by equation

$$V = E - ir$$

Thus $E = 4$ V and $r = \frac{E - V}{i} = \frac{4 - 0}{5} = 0.8\Omega$]

19. In the circuit shown in Fig. 20.80, find the potential difference across E_1 and E_2.

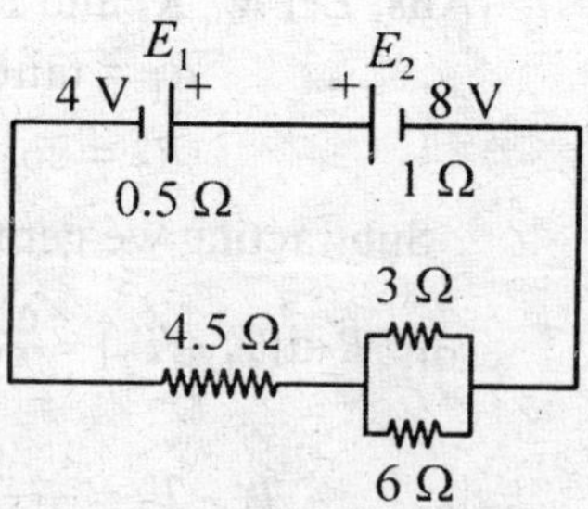

Fig. 20.80

[**Ans.** Total resistance of the circuit is

$$R = 1 + 0.5 + 4.5 + \frac{3 \times 6}{6 + 3} = 8\Omega$$

Total $\quad$ emf $= E_2 - E_1 = 8 - 4 = 4$ V

Hence current in circuit $= \frac{4}{8} = 0.5$ amp.

Terminal voltage across E_1, $V_1 = E_1 + ir_1 = 4 + 0.5 \times 0.5 = 4.25$ V and terminal voltage across E_2 is $V_2 = E_2 - ir_2 = 8 - 1 \times 0.5 = 7.5$ V]

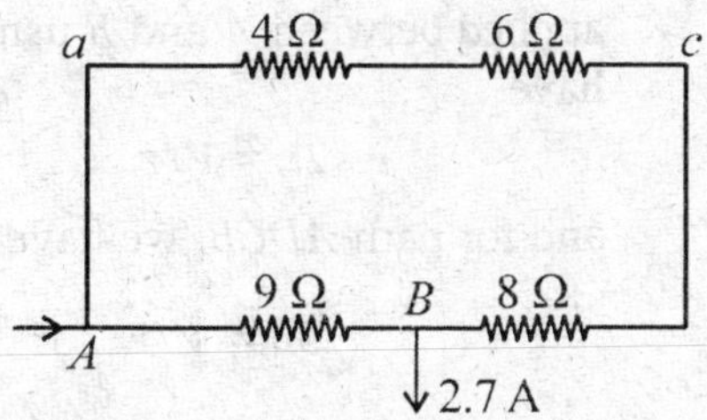

Fig. 20.81

20. In Fig. 20.81, find the potential difference between A and B.

[**Ans.** Resistance of branch $AacB$ (=18 Ω) is twice the resistance (= 9 Ω) of branch AB. Hence current through $AacB$ is 0.9 amp and through AB = 1.8 amp. Hence potential difference between A and B is $V = 9 \times 1.8 = 1.6$ V]

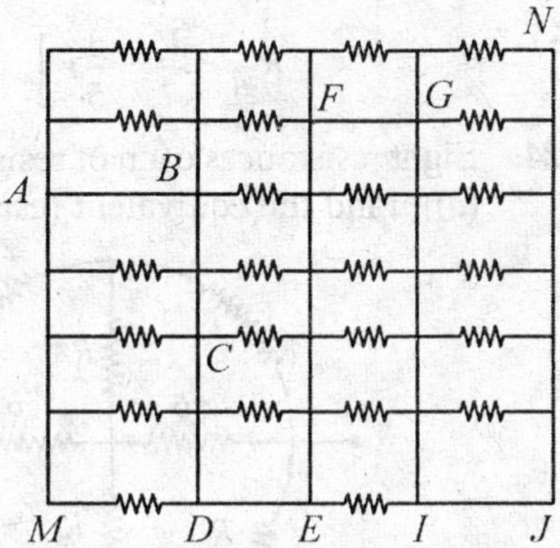

Fig. 20.82

21. In the circuit shown in Fig. 20.82, find the resistance between points M and N. Each resistance of network is r_0.

[**Ans.** When a battery is connected between points M and N, then it is clear that

$$V_M = V_A = V_B = V_D = V_E = V_F = V_G = V_I = V_J = V_N$$

$$\therefore \quad V_M - V_N = 0$$

If equivalent resistance is R then

$$iR = V_M - V_N = 0 \text{ or } R = 0]$$

22. Nine resistances each of resistance R are connected in circuit as shown in Fig. 20.83 (*a*). Find the equivalent resistance between A and B.

[**Ans.** It is clear from figure that 3 resistances each of resistance R are connected in parallel at AE and BF. Similarly other three resistances each of value R are connected between CG and AE and other three between CG and FB. Equivalent diagram is shown in Fig. 20.83 (*b*). Thus equivalent resistance between AB is $R/3$ and $R/3$ in series $\left(= \frac{2R}{3}\right)$ which is in parallel with $R/3$.

$$\therefore \quad R_{eq} = \frac{R/3 \times 2R/3}{R/3 + 2R/3} = \frac{2}{9}R \,]$$

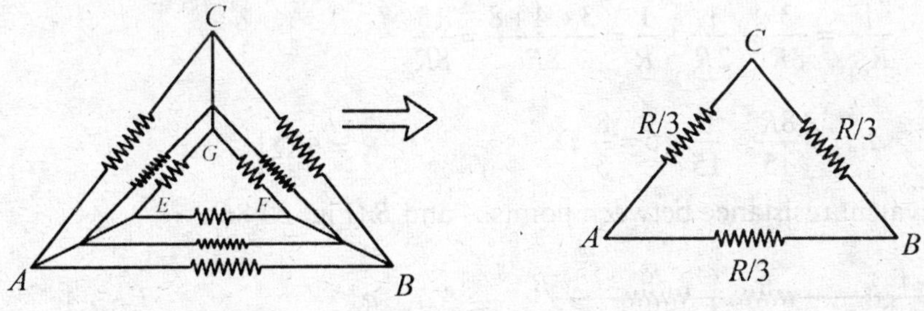

Fig. 20.83

23. In the circuit shown in Fig. 20.84 all wires have equal resistance r. Find the equivalent resistance between A and B.

[**Ans.** Points e and E are symmetrically situated with respect to A and B and hence are at same potential and no current flows through CE. Similarly no current flows through DF. The current in different branches is then shown in Fig. 20.84. When a potential difference V is applied between A and B using Kirchhoff's loop law for path AB, we have

$$i - 2i_1 = V/r \qquad \ldots(i)$$

and for path $ADCB$, we have

$$3i_1 = V/r \quad \text{or} \quad i_1 = \frac{V}{3r} \qquad \ldots(ii)$$

(i) and (ii) gives $i = \dfrac{V}{r} + \dfrac{2V}{3r} = \dfrac{5}{3}\dfrac{V}{r}$

$$\therefore \quad R_{eq} = \frac{V}{i} = \frac{3}{5}r \Big]$$

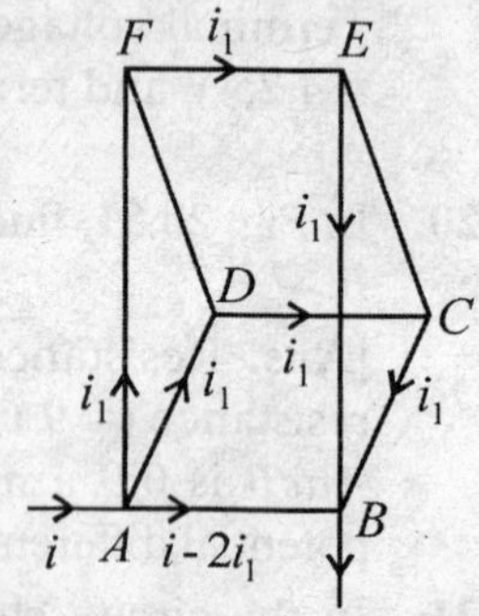

Fig. 20.84

24. Eight resistances each of resistance R (= 5 Ω) are connected as shown in the circuit (Fig. 20.85 (a)). Find the equivalent resistance between A and B.

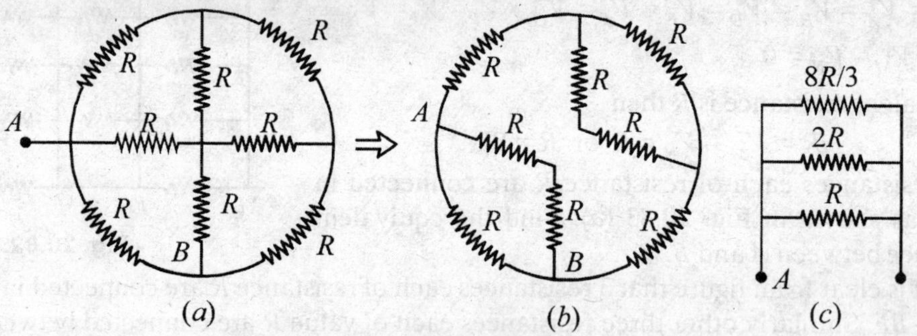

Fig. 20.85

[**Ans.** The equivalent circuit is shown in Fig. 20.85 (b). This is equivalent to three resistances $R_1 = \left(\dfrac{8R}{3}\right)$, $R_2 = 2R$ and $R_3 = (R)$ all in parallel. This is shown in Fig. 20.85 (c). Thus equivalent resistance R_{eq} is

$$\frac{1}{R_{eq}} = \frac{3}{8R} + \frac{1}{2R} + \frac{1}{R} = \frac{3+4+8}{8R} = \frac{15}{8R}$$

or
$$R_{eq} = \frac{8R}{15} = \frac{8}{15} \times 5 = \frac{8}{3}\ \Omega \qquad \because \ R = 5\ \Omega \Big]$$

25. Find equivalent resistance between points A and B (Fig. 20.86).

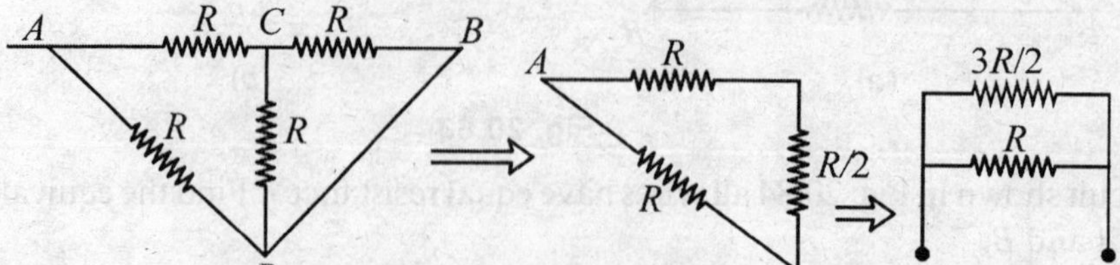

Fig. 20.86

[**Ans.** Points B and B_1 are same, hence resistances CB and CB_1 are in parallel. The equivalent circuit then is shown in Fig. 20.86 (*b*). Which gives resistance R and $R/2$ in series $(3R/2)$ which is again in parallel to R. Hence equivalent resistance is

$$R_{eq} = \frac{R \times 3R/2}{R + 3R/2} = \frac{3}{5}R\Big]$$

26. In the circuit shown in Fig. 20.87, potential difference between points A and B is 16 V. Find the current passing through 2 Ω resistance.

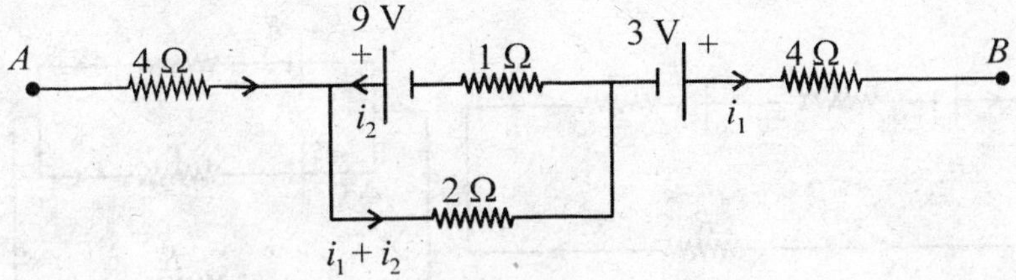

Fig. 20.87

[**Ans.** We have $V_A - V_B = 16$ V ...(*i*)

Using Kirchhoff's law between A and B, we have

$$4i_1 + 2(i_1 + i_2) - 3 + 4i_1 = 16 \quad ...(ii)$$

For closed loop, we have

$$2(i_1 + i_2) + 1i_2 = 9 \quad ...(iii)$$

Solving i_1 and i_2, we have $i_1 = 1.5$ A and $i_2 = 2$ A

∴ Current through 2 Ω resistor is = 2 + 1.5 = 3.5 amp.]

27. In the circuit shown in Fig. 20.88, the voltage across 15 Ω resistance is 30 V having the polarity as shown. Find the value of R.

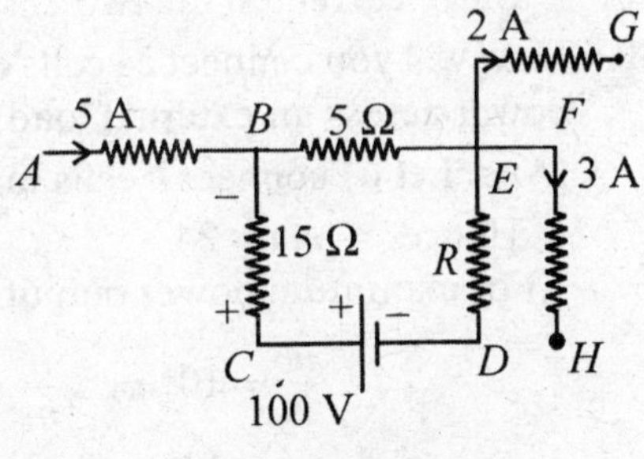

Fig. 20.88

[**Ans.** Current in 15 Ω resistance $= i = \frac{30}{15} = 2$ amp

Current through 5 Ω resistance 5 + 2 = 7 amp

Potential difference across BE = 5 × 7 = 35 V. Hence potential difference across R = 100 – (35 + 30) = 35 V and current = 2 amp

$$\therefore \quad R = \frac{35}{2} = 17.5\,\Omega\Big]$$

28. In the circuit shown in Fig. 20.89, switch S_1 is closed and S_2 is open, the ideal voltmeter shows a reading of 18 V. When switch S_2 is closed and S_1 is open the voltmeter reads 24 V. What will be the reading of voltmeter if both S_1 and S_2 are closed?

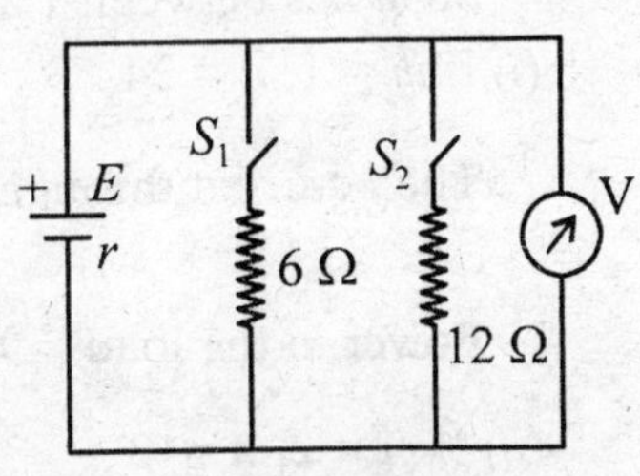

Fig. 20.89

[**Ans.**

$$18 = \frac{E \times 6}{6 + r} \text{ or } 18 + 3r = E \quad ...(i)$$

$$24 = \frac{E \times 12}{12 + r} \text{ or } 24 + 2r = E \quad ...(ii)$$

Solving we get $r = 6\ \Omega$ and $E = 36$ V

When both S_1 and S_2 are closed equivalent resistance $= \dfrac{12 \times 6}{12+6} = 4\,\Omega$

Hence voltmeter reading, $V = \dfrac{36 \times 4}{4+6} = 14.4$ V]

29. The two cells each of emf E and internal resistance r are connected to external resistance in two ways shown in Fig. 20.90 (*a*) and (*b*). Find the internal resistance so that the current in two situations through resistance R is same.

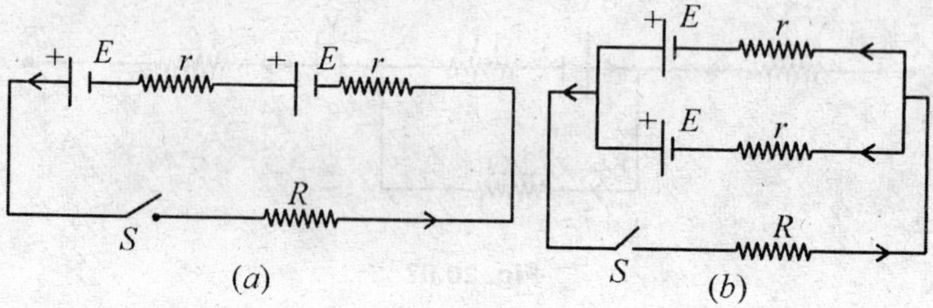

Fig. 20.90

[**Ans.** In situation (*a*) the current through R is

$$i = \frac{2E}{R+2r} \qquad \ldots(i)$$

In situation (*b*) the current through R is

$$i' = \frac{E}{R+r/2} = \frac{2E}{2R+r} \qquad \ldots(ii)$$

Since current in the two cases should be same, means. $i = i'$ gives $R = r$]

30. How will you connect 24 cells each of emf E and internal resistance $1\ \Omega$ so as to get maximum power across an external load of $10\ \Omega$? Find that the maximum power, if $E = 10$ V.

[**Ans.** Let us connect n cells in series and have m such rows.

Hence $\quad mn = 24 \qquad \ldots(i)$

For maximum power output, the total internal resistance should be equal to external load.

$$\therefore \quad \frac{nr}{m} = 10 \text{ or } \frac{n}{m} = 10 \qquad r = 1\ \Omega \qquad \ldots(ii)$$

$$\therefore \quad n = 10m$$

or $10m^2 = 24$ or $m = \sqrt{2.4} = 1.56$

So m lies between 1 and 2.

(*i*) If $m = 1$, $n = 24$

Then current through $10\ \Omega$ load $= \dfrac{24E}{10+24} = \dfrac{24 \times 10}{34}$

Power in the load $= i^2R = \left(\dfrac{24 \times 10}{34}\right)^2 \times 10 = 498$ Watt

(*ii*) If $m = 2$, $n = 12$

The current through 10 Ω load $i = \dfrac{12E}{10 + \dfrac{12r}{2}} = \dfrac{12 \times 10}{16}$

Power in the load $= i^2R = \left(\dfrac{12 \times 10}{16}\right)^2 \times 10 = 562.5$ W

Hence we must connect 12 cells in series with two such rows to get maximum power in load]

31. A constant potential difference V is applied to a copper wire of diameter d and length l. What is the effect on electron drift velocity of doubling (*i*) V (*ii*) l and (*iii*) d?

[**Ans.** Average drift velocity when potential difference V is applied across conductor of length l is given by $v_d = \dfrac{eV\tau}{2ml}$ where τ is relaxation time and m is mass of electron.

(*i*) If V is doubled the drift velocity is also doubled.
(*ii*) If l is doubled the drift velocity is halved.
(*iii*) If d is doubled the drift velocity remains unchanged.]

32. Eleven equal wires each of resistance 2 Ω form the edges of an incomplete skeleton cube. Find the total resistance between the points a and d.

[**Ans.** When a potential difference V is applied between points a and d, the current distribution is shown in Fig. 20.91.

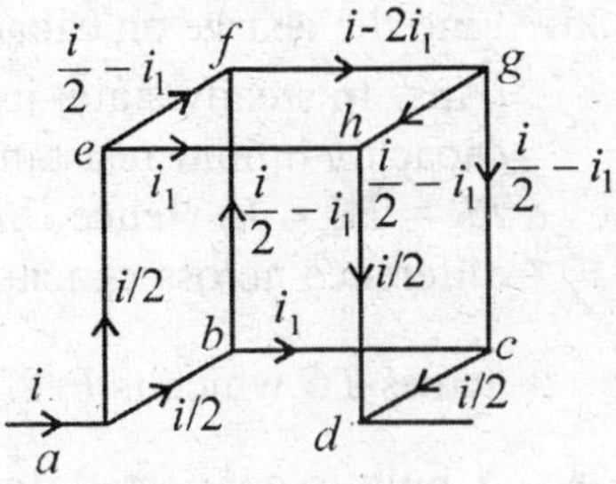

Fig. 20.91

For loop *efgh*, we have

$$r(i/2 - i_1) + r(i - 2i_1) + r(i/2 - i_1) - i_1 r = 0$$

or $\quad 2i - 5i_1 = 0$ or $i_1 = \dfrac{2i}{5}$

For path *abcd*, we have

$$i/2r + i_1 r + i/2r = V$$

or $\quad ir + \dfrac{2}{5}ir = V$ or $\dfrac{7}{5}ir = V$

and equivalent resistance $R = \dfrac{V}{i} = \dfrac{7}{5}r$

and since $r = 2$ Ω, hence $R = \dfrac{7}{5} \times 2 = \dfrac{14}{5} = 2.8\Omega$]

33. In the given circuit, with steady current find the potential drop across the capacitor (Fig. 20.92).

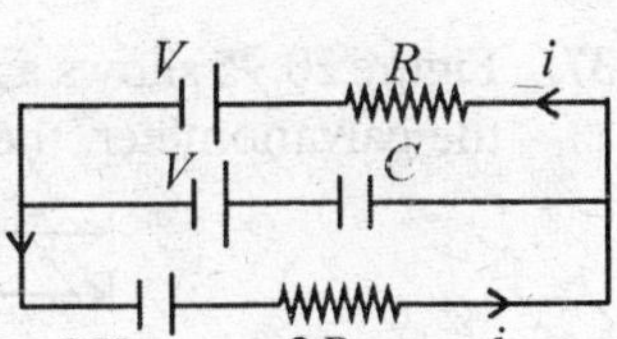

Fig. 20.92

[**Ans.** In steady state no current flows through the capacitor. Hence if i is the current in the circuit, then

$$2iR + iR = 2V - V = V$$

or $\quad i = V/3R$

Hence terminal voltage of cells is $2V - i(2R) = 2V - \dfrac{2V}{3} = \dfrac{4V}{3}$

Hence potential drop across capacitor $= \dfrac{4V}{3} - V = V/3$]

34. In a standard meterbridge experiment to measure the specific resistance of wire, the following data are found. Length $L = (1 \pm 0.01)$ m measured by meter scale, radius of wire (1 ± 0.01)

mm measured by screw gauge and resistance of wire $R = (5 \pm 0.01)\ \Omega$. Find the maximum possible error in the measurement of specific resistance. (You may use the formula $\rho = \frac{\pi r^2 R}{L}$).

[**Ans.** Specific resistance $\rho = \frac{\pi r^2 R}{L}$.

Taking logs, we have $\log \rho = \log \pi + 2 \log r + \log R - \log L$

Differentiation gives $\frac{d\rho}{\rho} = 2\frac{dr}{r} + \frac{dR}{R} - \frac{dL}{L}$.

$$\therefore \quad \frac{d\rho}{\rho} = 2\frac{(0.01)}{1} + \frac{0.01}{5} + \frac{0.01}{1} \text{ (+ sign because errors are added).}$$

$$= 0.032$$

But $\rho = \frac{\pi r^2 R}{L} = \frac{\pi (10^{-3})^2 \times 5}{1} = 5\pi \times 10^{-6}\ \Omega\text{-m}$

Hence maximum error is $d\rho = 0.032 \times \rho = 0.032 \times 5\pi \times 10^{-6} = 0.502 \times 10^{-6}$

or $d\rho = 0.502\ \mu\Omega\text{-m}$]

35. Find the charge on capacitors A and B in the steady state.

[**Ans.** In steady state, inductance offers zero resistance and capacitor infinite resistance. Hence potential difference across $PR = PQ = E$. Hence charge on capacitor A is CE. Potential difference across cpacitor B is same as potential difference across TQ which is $E/2$. Hence charge on it is $\frac{CE}{2}$]

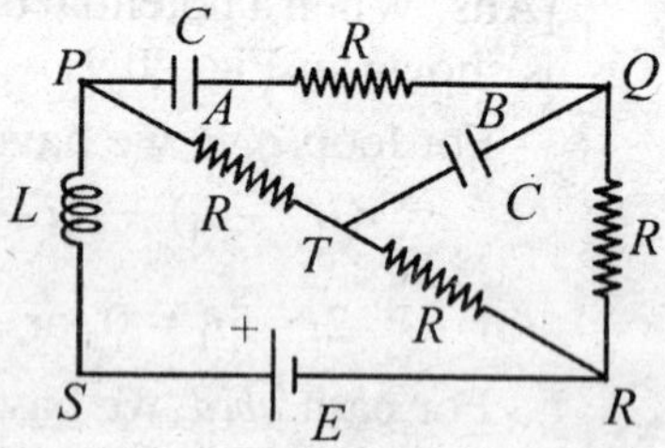

Fig. 20.93

36. A bulb is connected across a voltage source. A resistor R is connected in parallel to the bulb through a switch S. If switch S is closed what is the effect on the brightness of the source?

[**Ans.** When switch S is closed, even then the potential difference across bulb is V and the current thorugh it is same hence its brightness is unaffected and remains the same.]

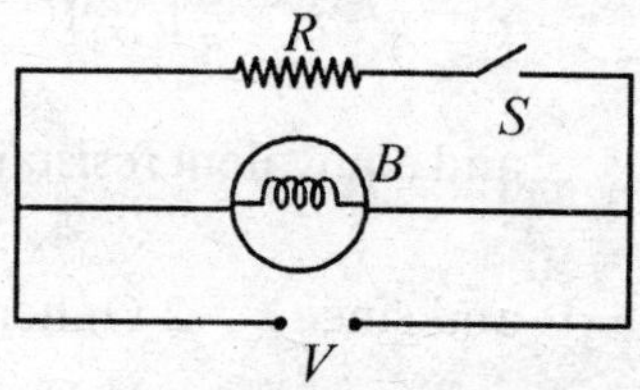

Fig. 20.94

37. Figure 20.95 shows a meter bridge. If there is no current through the galvanometer, then find the length l.

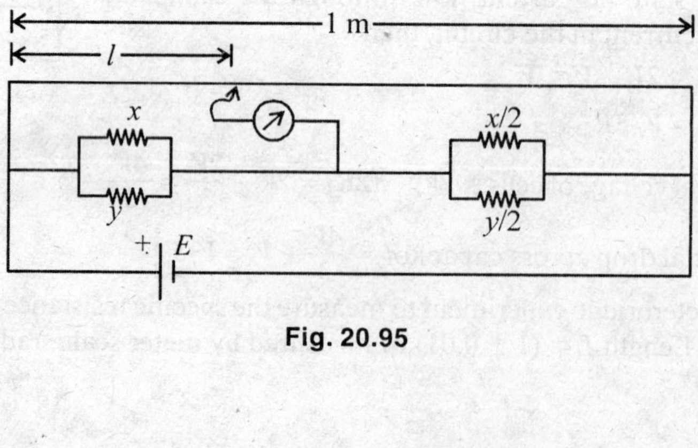

Fig. 20.95

[**Ans.** Since the bridge is balanced hence

$$\frac{l}{100-l} = \frac{\text{Resistance in left gap}}{\text{Resistance in right gap}}$$

$$\text{Resistance in left gap} = \frac{xy}{x+y} \text{ and in right gap } = \frac{x/2 \cdot y/2}{x/2+y/2} = \frac{xy}{2(x+y)}$$

$$\therefore \quad \frac{l}{100-l} = \frac{xy}{x+y} \cdot \frac{2(x+y)}{xy} = 2$$

or $\quad l = 200 - 2l \quad$ or $\quad l = \frac{200}{3}$ cm]

38. The potential difference across the resistor as measured by a voltmeter is 60 ± 1.5 V and the current through the resistor as measured by ammeter is (2 ± 0.01) A. Find the maximum % error in the measurement of R?

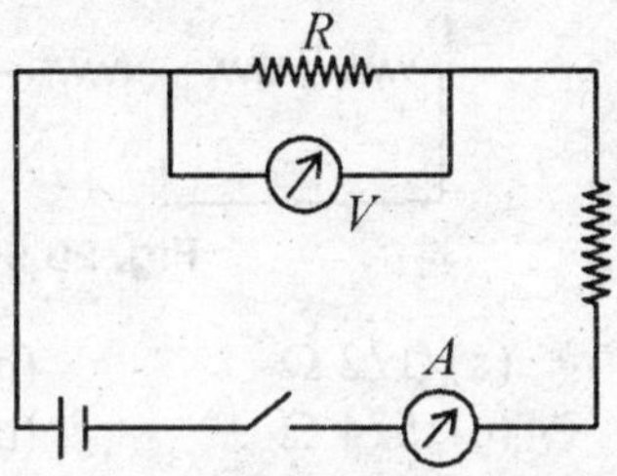

Fig. 20.96

[**Ans.** Ohms law gives $i = \frac{V}{R}$ or $R = \frac{V}{i}$

Taking log, we have $\log R = \log V - \log i$

Differenciation gives $\frac{dR}{R} = \frac{dV}{V} - \frac{di}{i}$

The errors are additive, hence

$$\frac{dR}{R} = \frac{1.5}{60} + \frac{0.01}{2} = \frac{1.8}{60}$$

$$\therefore \quad \% \text{ error} = \frac{1.8}{60} \times 100 = 3\%]$$

39. In the given network (Fig. 20.97), find the equivalent resistance between A and B.

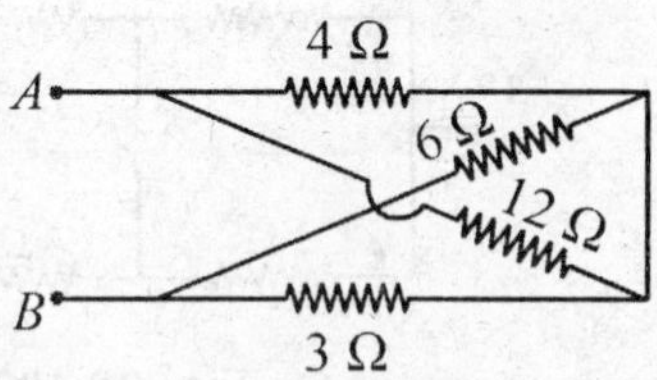

Fig. 20.97

[**Ans.** 4 Ω and 12 Ω resistances are in parallel. Their combined resistance is 3 Ω. Similarly 6 Ω and 3 Ω resistances are also in parallel and their combined resistances is 2 Ω. Now the 3 Ω and 2 Ω resistances are in series. So the net resistance between A and B is 5 Ω.]

40. Find the condition under which current passing through the resistance R can be increased by short circuiting the battery of emf E_2 (Fig. 20.98). The internal resistances of the two batteries are r_1 and r_2 respectively.

[**Ans.** The current through the resistance R before battery E_2 is short circuited is $i_1 = \frac{E_1+E_2}{R+r_1+r_2}$

Current through R after battery E_2 is short circuited is

$$i_2 = \frac{E_1}{R+r_1}$$

for $\quad i_2 > i_1 \quad \frac{E_1}{R+r_1} > \frac{E_1+E_2}{R+r_1+r_2}$

This gives $E_1 r_2 > E_2(R + r_1)$]

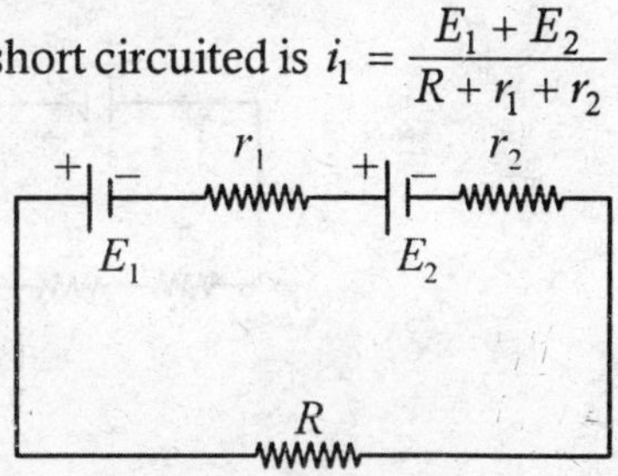

Fig. 20.98

MULTIPLE CHOICE QUESTIONS

PART I

In the following, a question is followed by four alternatives, of them only one is correct. Therefore, the candidate has to select the correct choice from amongst the alternatives.

1. Resistors each of value 1 Ω are arranged as shown in Fig. 20.99. The equivalent resistance between points A and B is

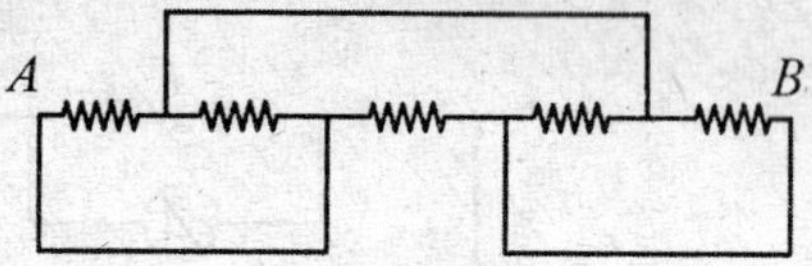

Fig. 20.99

(a) 1/2 Ω (b) 1/5 Ω
(c) 5/4 Ω (d) none of the above.

2. What is the current in amperes flowing across the switch S when it is closed (Fig. 20.100)?

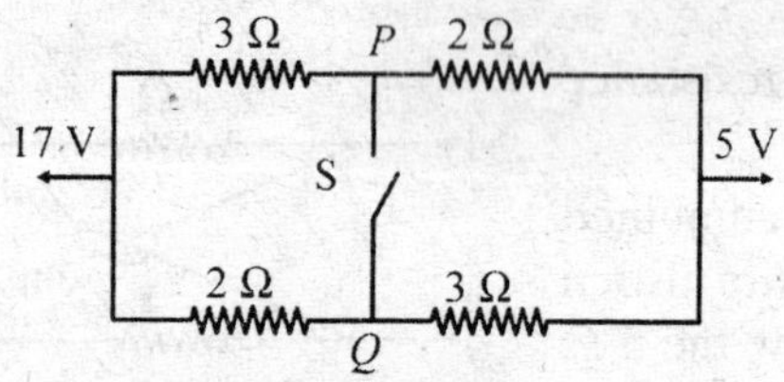

Fig. 20.100

(a) 1 amp from Q to P
(b) 1 amp from P to Q
(c) 2 amp from P to Q
(d) 2 amp from Q to P.

3. In the network of resistors shown in Fig. 20.101 the current through the 2 Ω resister is I_0. That through the 20 Ω one is

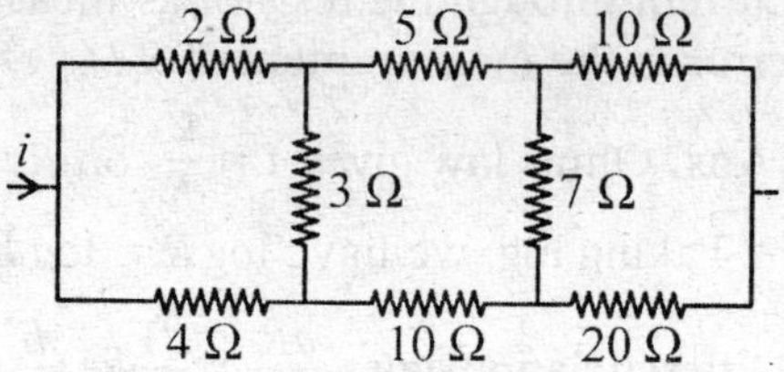

Fig. 20.101

(a) $I_0/2$ (b) $2I_0$
(c) I_0 (d) $\frac{20I_0}{17}$.

4. A potentiometer circuit is shown in Fig. 20.102. Find the correct statement.

(a) The sensitivity (mm/Volt) would double if R_1 is made two fold
(b) The sensitivity would double if R_2 is made half
(c) If the voltage E_2 is balanced as shown with S = 30.1 cm, E_2 = 18.01 mV
(d) The potential gradient on the wire QR is affected if S is moved a way from balanced point.

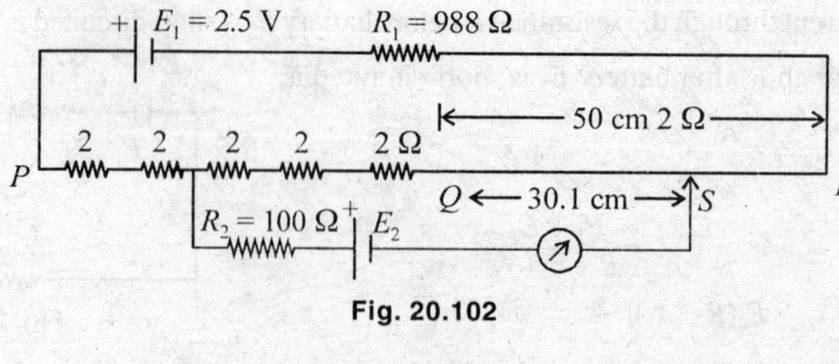

Fig. 20.102

5. In the circuit shown in Fig. 20.103,

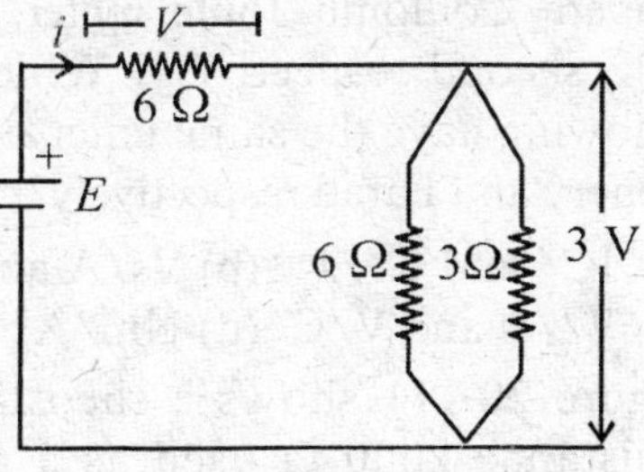

Fig. 20.103

(a) $E = 6$ V and $i = 1$ amp
(b) $E = 12$ V and $V = 6$ V
(c) $E = 12$ V and $V = 9$ V
(d) $E = 9$ V and $V = 6$ V.

6. In the network shown in Fig. 20.104, equivalent resistance between A and B is

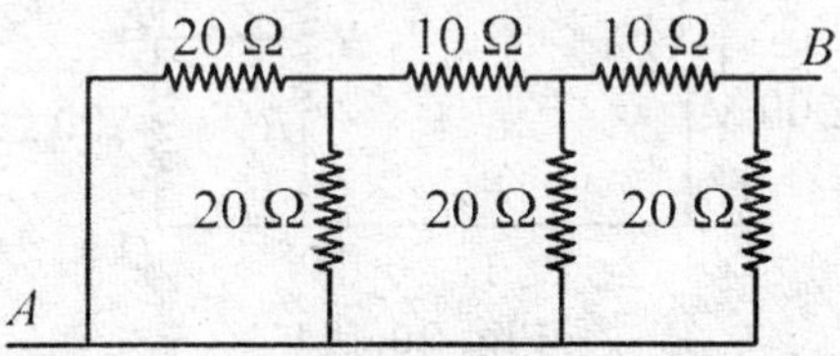

Fig. 20.104

(a) 10 Ω (b) 20 Ω
(c) 40 Ω (d) 80 Ω.

7. A voltmeter of resistance 9800 Ω has a least count of 0.2 V and range (0-10 V). Instead of connecting the voltmeter in parallel with one of 100 Ω resistor, by mistake a student connected it in series as shown in Fig. 20.105. Which of the statement is correct?

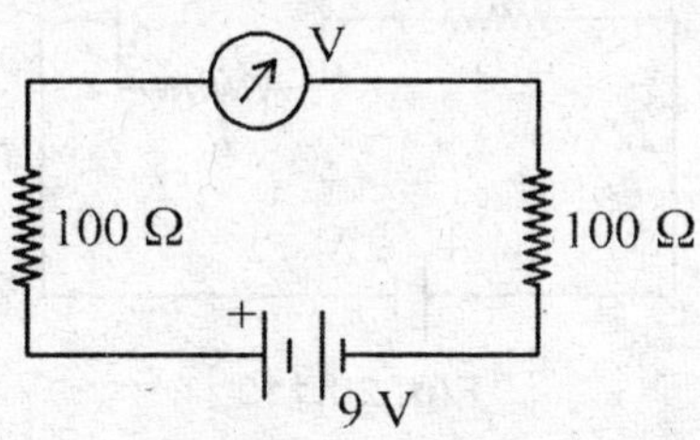

Fig. 20.105

(a) The voltmeter will get damaged
(b) The voltmeter will read zero
(c) The voltmeter will read 9 V
(d) The voltmeter will read 8.8 V.

8. For the network shown in Fig. 20.106, the effective resistance between points P and Q is

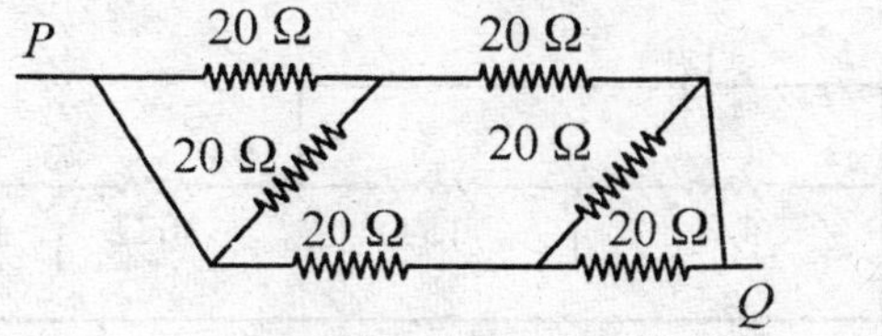

Fig. 20.106

(a) 15 Ω (b) 40 Ω
(c) 60 Ω (d) 80 Ω.

9. An ammeter has range 1 amp without shunt. The range can be varied by using different shunt resistances. The graph between shunt resistance and range will have the nature.

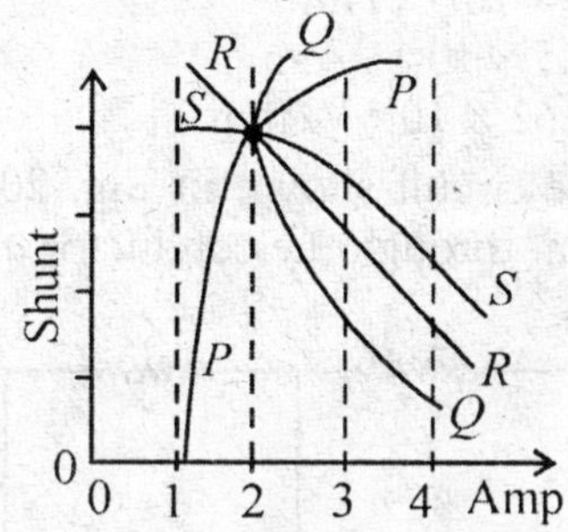

Fig. 20.107

(a) P (b) Q
(c) R (d) S.

10. In the circuit shown in Fig. 20.108, every resistance is of 24 Ω. The resistance between terminals P and Q is

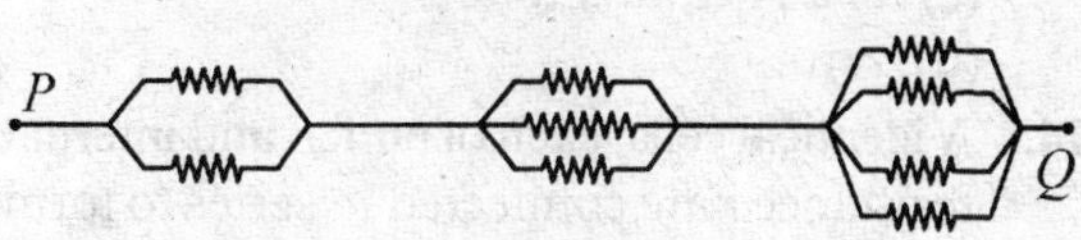

Fig. 20.108

(a) 26 Ω (b) 32 Ω
(c) 36 Ω (d) 42 Ω.

11. Four resistances are connected as shown in Fig. 20.109 and a 6 V battery of negligible

resistance is connected across terminals of A and C. The potential difference across terminals $B\ D$ will be

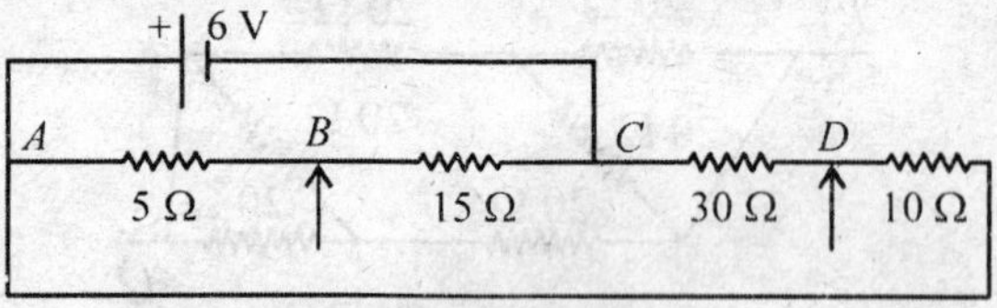

Fig. 20.109

(a) 0 V (b) 1.5 V
(c) 2 V (d) 3 V.

12. A potentiometer wire has length 10 m and resistance 20 Ω. A 2.5 V battery of negligible internal resistance is connected across the wire with an 80 Ω series resistance. The potential gradient on the wire will be
(a) 1.0×10^{-5} V/mm
(b) 5×10^{-5} V/mm
(c) 2.5×10^{-4} V/cm
(d) 0.62×10^{-4} V/mm.

13. In the circuit shown in Fig. 20.110, the current through the conductor ab is

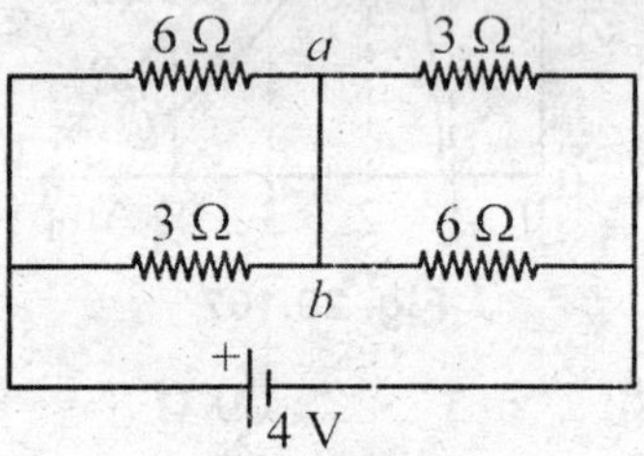

Fig. 20.110

(a) 1/3 A from a to b
(b) 1/3 A from b to a
(c) 2/3 A from a to b
(d) zero.

14. N identical cells, each of emf E and internal resistance r are connected in series to form a closed circuit. An ideal voltmeter connected across n cells will read
(a) nE (b) $(N - n)E$
(c) zero (d) $n\left(1-\frac{n}{N}\right)E$.

15. With A, C, J, m, N, V, s standing for ampere, Coulomb, Joule, meter, Newton, volt, second respectively. Which of the following have the same units as those of Henery and Farad respectively?
(a) JC^2 and C^2/J (b) Vs/A and C/V
(c) V/As and V/C (d) Nm/A^2 and 1/J.

16. Figure 20.111 shows a rheostat AC of resistance 2000 Ω used as a potential divider, with B as the sliding point to vary R'. The load across ED is 200 Ω and source resistance is negligible. For $R' = 20$ Ω, 200 Ω and 1000 Ω, the voltage across the load will respectively be nearest to

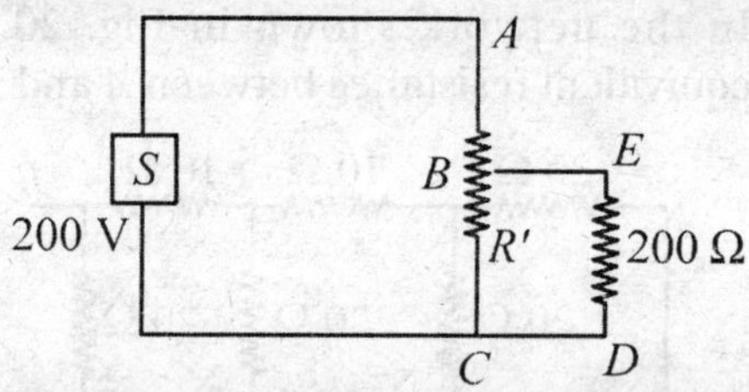

Fig. 20.111

(a) 2 V, 10.5 V, 29 V (b) 2 V, 20 V, 100 V
(c) 16 V, 20 V, 33 V (d) 22 V, 40 V, 120 V.

17. In the circuit shown in Fig. 20.112, when R is removed an additional resistance of 72 Ω must be introduced in series with the battery in order to keep the current through 30 Ω resistance unaltered. Hence R is

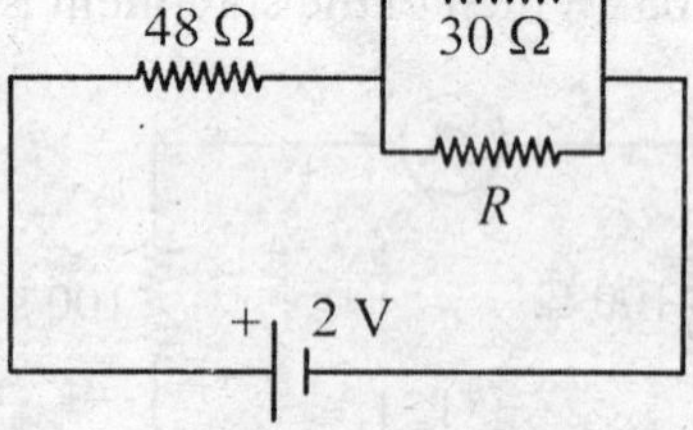

Fig. 20.112

(a) 15 Ω (b) 18 Ω
(c) 20 Ω (d) 21 Ω.

18. A potential difference of 220 V is maintained across a 12000 Ω reheostat (Fig. 20.113). The voltmeter has a resistance of 6000 Ω

and point c is at out fourth of the distance from a to b. The reading of the voltmeter is

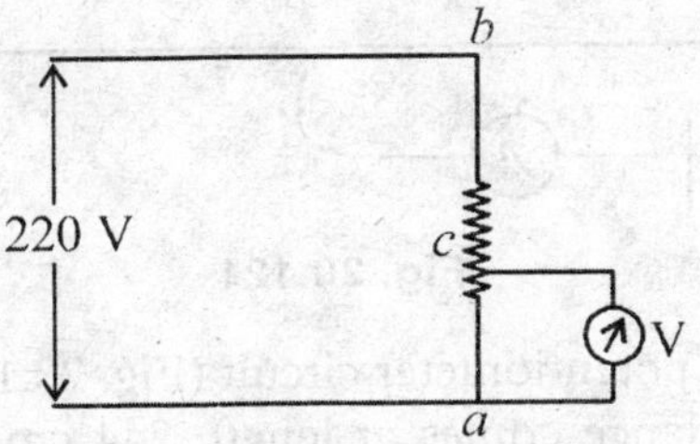

Fig. 20.113

(a) 32 V (b) 36 V
(c) 40 V (d) 42 V.

19. Fig. 20.114 shows a balanced Wheatstone's net. Now if it is disturbed by changing P to 11 Ω, which of the following steps will not bring the bridge to balance again?

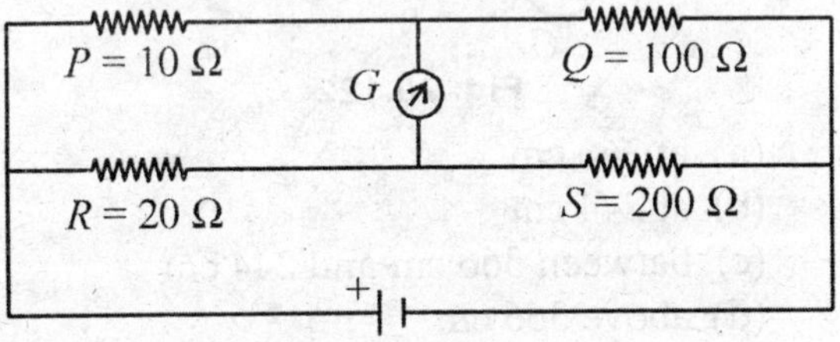

Fig. 20.114

(a) Increasing R by 2 Ω
(b) increasing S by 20 Ω
(c) Increaseing Q by 10 Ω
(d) Making product $RQ = 22000\,(\Omega)^2$.

20. Just as electricity is supplied at 220 V for domestic use in India, it is supplied at 110 V in USA. If resistance of a 60 W bulb for use in India is R that of a 60 W bulb for use in USA will be

(a) $R/4$ (b) $R/2$
(c) R (d) $2R$.

21. Two resistances 300 Ω and 400 Ω are connected in series with 60 V supply. A voltmeter connected across 400 Ω reads 30 V. The same voltmeter when connected across the other resistance will read

(a) 22.5 V (b) 50 V
(c) 30 V (d) 10 V.

22. In the circuit shown in Fig. 20.115 current in branch AB is

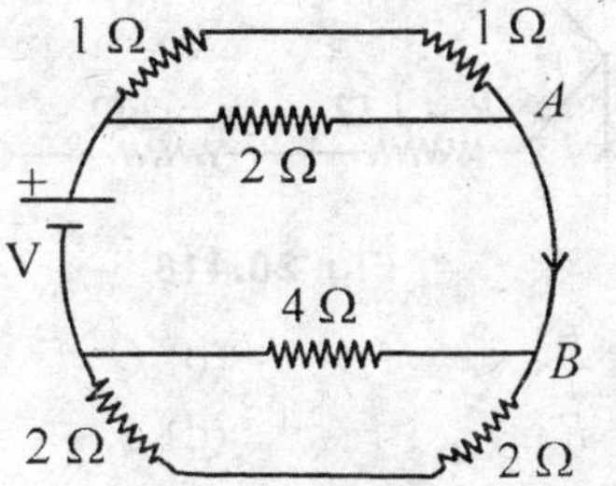

Fig. 20.115

(a) 1 A (b) 2 A
(c) 1.5 A (d) 3 A.

23. In the circuit (Fig. 20.116) the potential difference across PQ will be nearest to

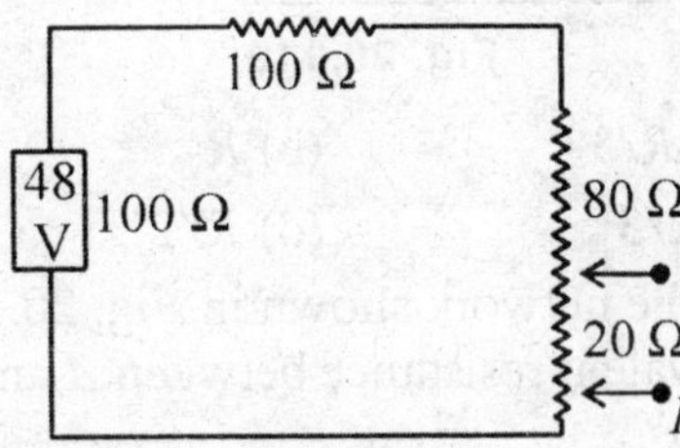

Fig. 20.116

(a) 9.6 V (b) 6.6 V
(c) 4.8 V (d) 3.2 V.

24. In the circuit shown in Fig. 20.117, the resistances R_1 and R_2 respectively are

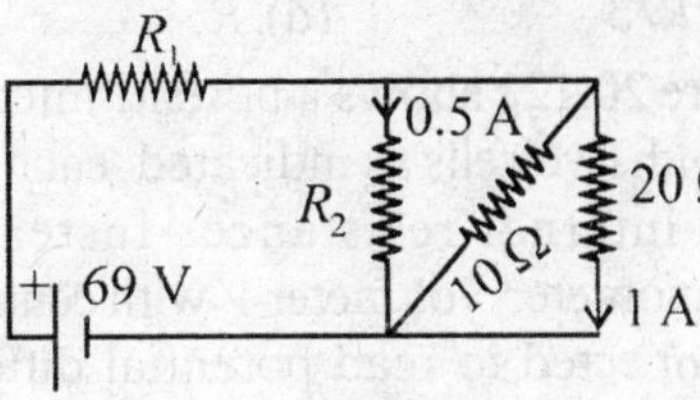

Fig. 20.117

(a) 14 Ω and 40 Ω
(b) 40 Ω and 14 Ω
(c) 40 Ω and 30 Ω
(d) 14 Ω and 30 Ω.

25. Which of the four resistances (Fig. 20.118) generate the maximum amount of heat when a current flows from A to B?

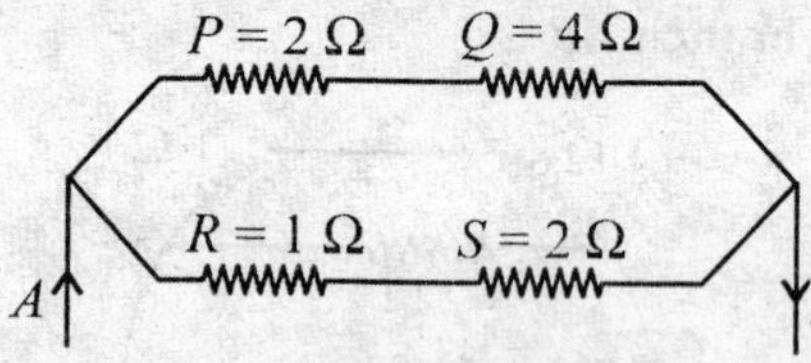

Fig. 20.418

(a) P (b) Q
(c) R (d) S.

26. What is the effective resistance between points A and B in network as shown in Fig. 20.119?

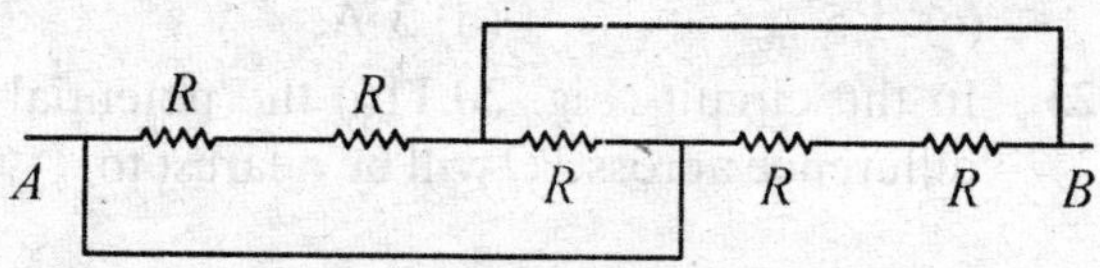

Fig. 20.119

(a) $4R/3$ (b) R
(c) $R/3$ (d) $R/2$.

27. For the network shown in Fig. 20.120 the equivalent resistance between A and B is

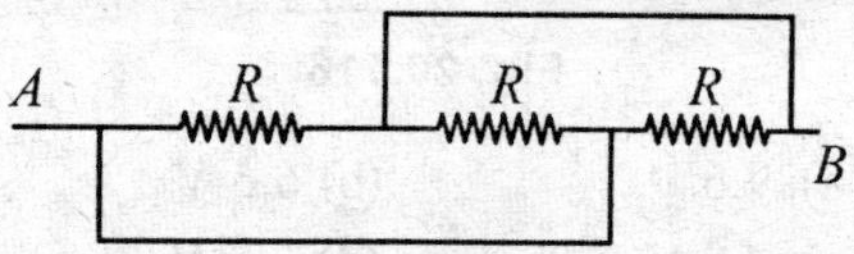

Fig. 20.120

(a) $3R/2$ (b) $R/3$
(c) $2R/3$ (d) R.

28. Figure 20.121 shows a potentiometer wire AB with two cells as indicated, each having zero internal resistance. Instead of a galvanometer, voltmeter V with central zero is connected to read potential difference. As the jockey point P is slided from A to B, its reading will vary from

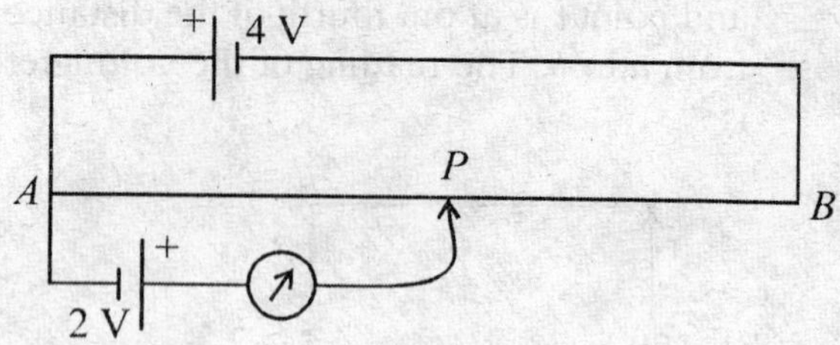

Fig. 20.121

(a) +2 V to 0 and back to +2 V
(b) +2 to 6 V
(c) 0 to 4 V
(d) +2 to −2 V.

29. In potentiometer circuit (Fig. 20.122), the balance comes at length 244 cm. If R is made 3/2 fold and R' is made 2 fold, the balance will occur

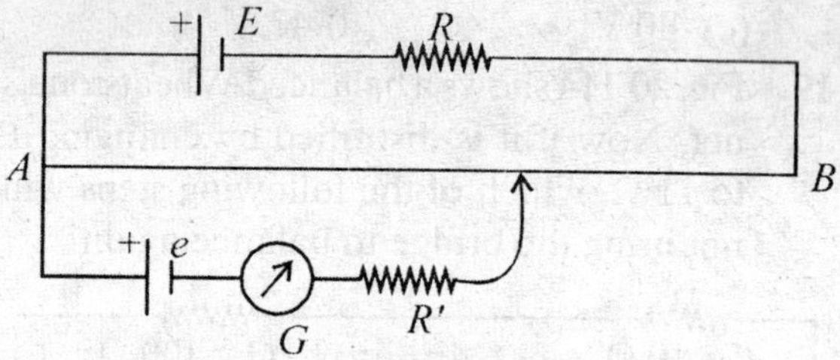

Fig. 20.122

(a) at 366 cm
(b) at 244 cm
(c) between 366 cm and 244 cm
(d) above 366 cm.

30. In the circuit (Fig. 20.123), cells A and B have emf 10 V each and internal resistance is 5 Ω for A and 2 Ω for B. For what value of R will the potential difference across cell A be zero?

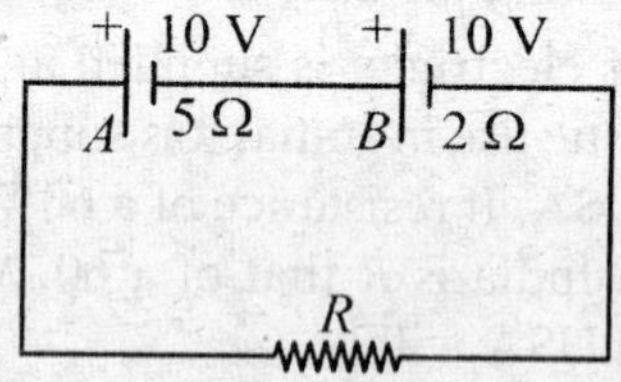

Fig. 20.123

(a) Zero (b) 1 Ω
(c) 2 Ω (d) 3 Ω.

31. In the potentiometer circuit (Fig. 20.124) V and R are set to give a potential gradient of 10 mV/cm on the wire MN. If $R' = 2.0$ Ω and the null point occurs at $MP = 82.8$ cm. The current through R' is

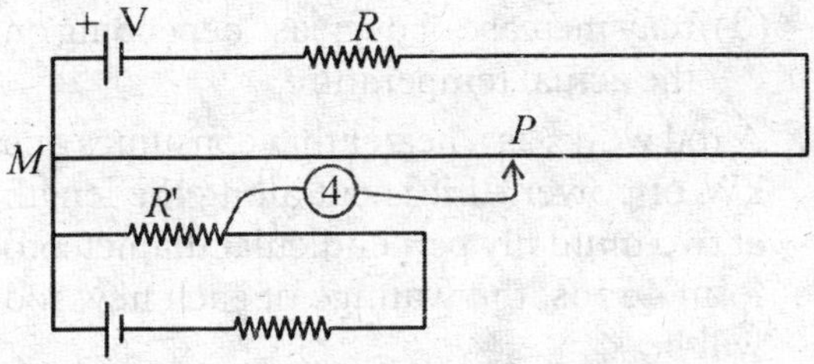

Fig. 20.124

(a) 0.414 A (b) 0.828 A
(c) 1.656 A (d) 0.207 A.

32. Figure 20.125 shows a Wheatstone's net with $P = 1000\ \Omega$, $Q = 10\ \Omega$ and R (unknown), S variable near 150 Ω for balance. If the connections across AC and BD are interchanged, the error in R determination would

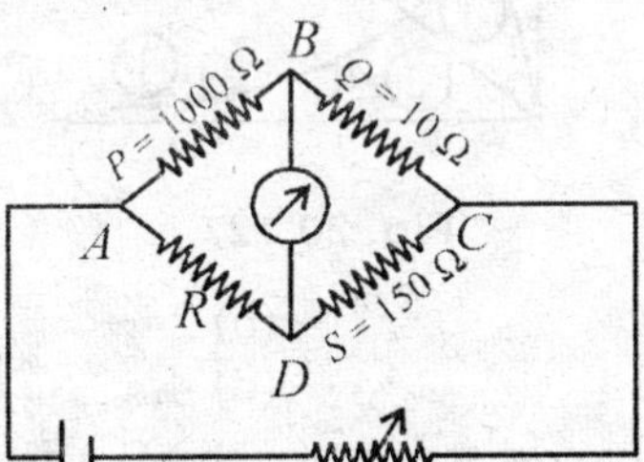

Fig. 20.125

(a) remain unaffected
(b) increase substantially
(c) increase marginally
(d) decrease substantially.

33. Two resistors A and B have resistances R_A and R_B respectively with $R_A < R_B$. Resistivity of their material is ρ_A and ρ_B, then

(a) $\rho_A > \rho_B$
(b) $\rho_A = \rho_B$
(c) $\rho_A < \rho_B$
(d) information is insufficient to find relation between ρ_A and ρ_B.

34. A metallic resistor is connected across a battery. If the number of collisions of free electrons with the lattice is somehow decreased in the resistor, the current will

(a) increase (b) decrease
(c) remain constant (d) become zero.

35. Product of resistivity and conductivity of a cylindrical conductor depends on

(a) temperature
(b) nature of material
(c) area of cross-section
(d) is always 1.

36. In an electric circuit containing a battery, the charge (assumed positive) inside the battery

(a) always goes from the positive terminal to the negative terminal
(b) may go from the positive terminal to the negative terminal
(c) always goes from the negative terminal to the positive terminal
(d) does not move.

37. A steady current flows in a metallic conductor of non-uniform cross-section. The quantity constant along the length of the conductor is

(a) current, electric field and drift speed
(b) current only
(c) drift speed only
(d) current and drift speed only.

38. A resistor R is connected to an ideal battery. If the value of R is decreased, the power dissipated in the resistor will

(a) increase
(b) decrease
(c) remain unchanged
(d) may increase or decrease depends on the emf of the battery.

39. A current passes through a resistor. K_1 and K_2 represent the average kinetic energy of the conduction electrons and metal ions respectively then

(a) $K_1 < K_2$
(b) $K_1 = K_2$
(c) $K_1 > K_2$
(d) Any of these many occur.

40. The ratio of the lengths, densities, masses and resistivities of two wires A and B are 1 : 2, 1 : 2, 1 : 1 and 1 : 4 respectively. The ratio of their resistances is

(a) 1 : 32 (b) 1 : 16
(c) 8 : 1 (d) 4 : 1.

41. The electrons are emitted by a hot filament and are accelerated by an electric field E (Fig. 20.126). The two stops at the left ensure that the electrons beam has a uniform cross-section.

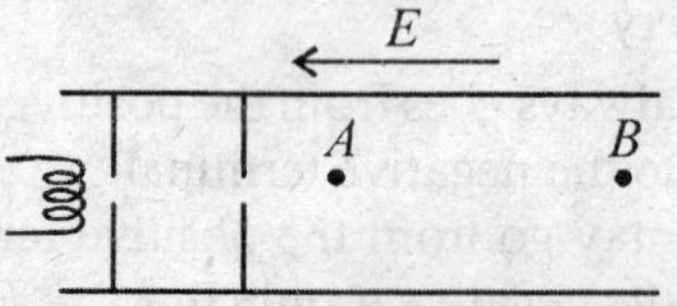

Fig. 20.126

(a) The speed of electron is more at B than at A
(b) The electric current is from left to right
(c) The magnitude of the current at A is larger than at B
(d) The current density is more at B than at A.

42. A capacitor with no dielectric is connected to a battery at $t = 0$. Consider a point A in the connecting wires and a point B in between the plates.
(a) There is no current through A
(b) There is a current through B
(c) There is a current through A as long as the charging is not complete
(d) There is a current through B as long as the charging is not complete.

43. Which of the following quantities does not change when a resistor connected to a battery is heated due to the current?
(a) Drift speed
(b) Resistivity
(c) Resistance
(d) Number of free electrons.

44. As the temperature of conductor increases its resistivity and conductivity change. The ratio of the conductivity to resistivity
(a) increases
(b) decreases
(c) remains constant
(d) may increase or decrease depending on the actual temperature.

45. A rod works as a heater that consumes one kW of power. If it is cut along the length at two mutually perpendicular diameter to form 4 rods, the wattage of each new rod will be
(a) 1 kW (b) 0.25 kW
(c) 4 kW (d) 16 kW.

46. In Fig. 20.127, voltage versus current graphs are shown for different devices. Which one represents the ohmic resistance?

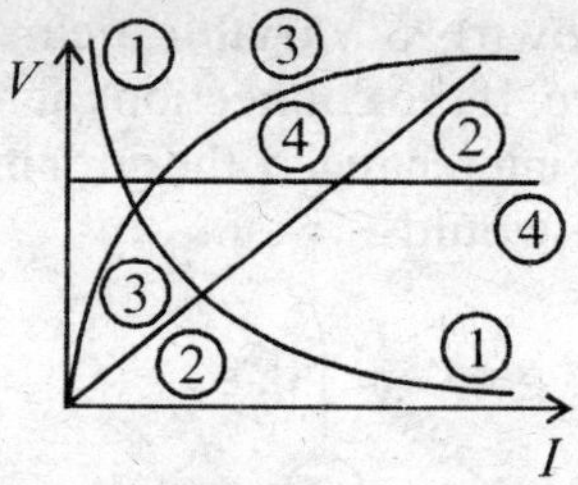

Fig. 20.127

(a) 1 (b) 2
(c) 3 (d) 4.

47. Twelve resistances each of resistance 1 Ω are connected as shown in Fig. 20.128. The equivalent resistance between points A and H is

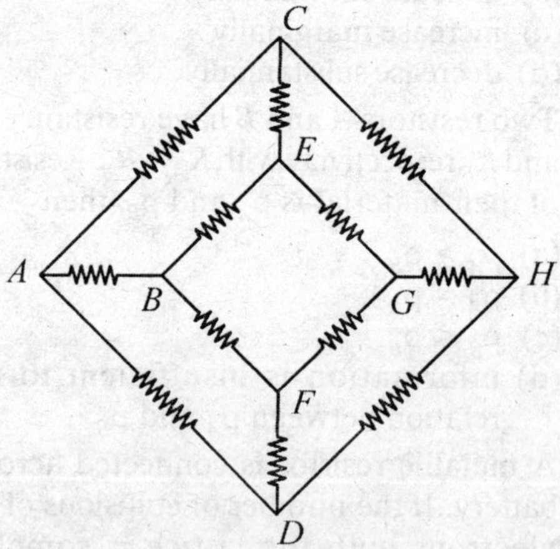

Fig. 20.128

(a) 5/3 Ω (b) 1 Ω
(c) 3/4 Ω (d) 7/6 Ω.

48. n identical cells are joined in series with two cells A and B with reversed polarities. Emf of each cell is E and internal resistance is r. Potential difference across cell A or B is $(n > 4)$

(a) $\frac{2E}{n}$ (b) $2E\left(1-\frac{1}{n}\right)$

(c) $\frac{4E}{n}$ (d) $2E\left(1-\frac{2}{n}\right)$.

49. The length of a potentiometer wire is l. A cell of emf E is balanced at a length $l/3$ from the positive end of the wire. If the length of the wire is increased by $l/2$. Now the same cell will give the balance point at

(a) $\frac{2l}{3}$ (b) $l/2$

(c) $l/6$ (d) $\frac{4l}{3}$.

50. The potential difference between points A and B is (Fig. 20.129)

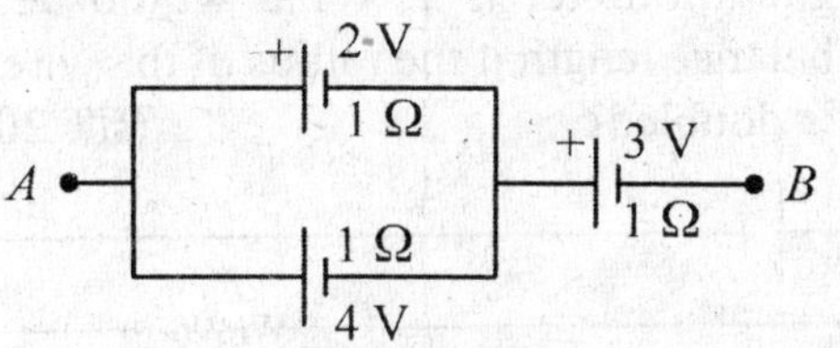

Fig. 20.129

(a) 2 V (b) 6 V

(c) 4 V (d) 3 V.

51. In the circuit shown in Fig. 20.130, the heat developed per second between A and B through C is

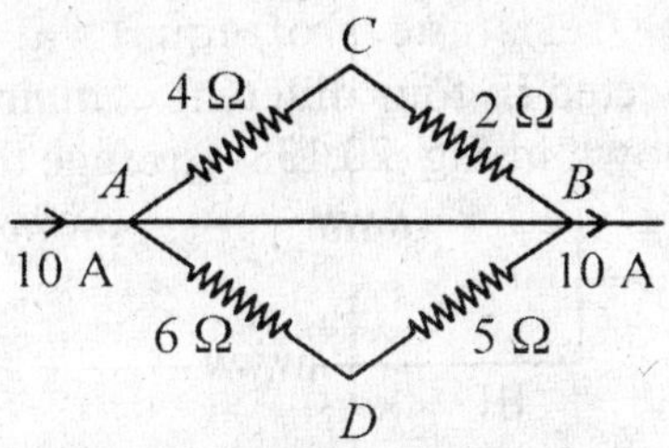

Fig. 20.130

(a) 650 J/s (b) 500 J/s

(c) 900 J/s (d) 0 J/s.

52. 5 identical lamps each having resistance R = 110 Ω are connected as shown in Fig. 20.131. The reading of an ideal ammeter is

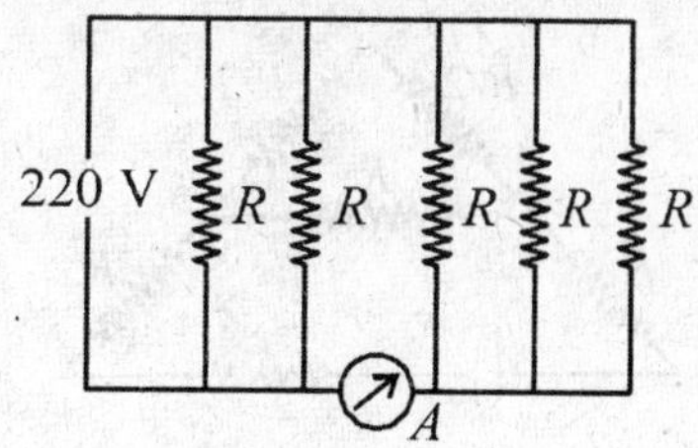

Fig. 20.131

(a) 1 A (b) 2 A

(c) 4 A (d) 6 A.

53. n resistors of resistance R are joined with capacitors of capacity C each and a battery of emf E (Fig. 20.132). In steady state condition, ratio of charges stored in the first and last capacitor is

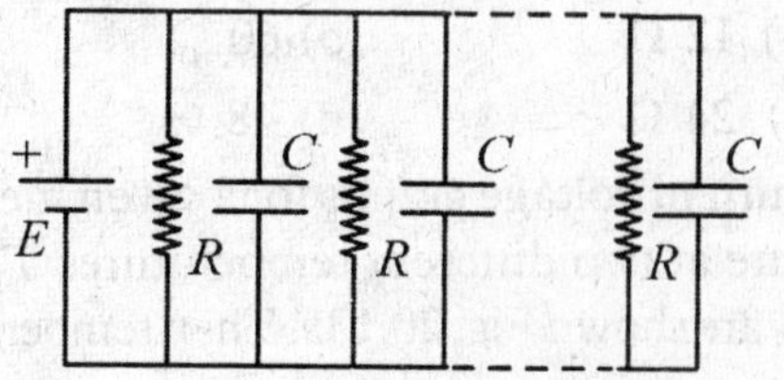

Fig. 20.132

(a) $n : 1$

(b) $(n-1) : n$

(c) $(n^2 + 1) : (n^2 - 1)$

(d) 1 : 1.

54. The amount of heat generated in 500 Ω resistance, when key is thrown over from contact 1 to 2 (Fig. 20.133) is

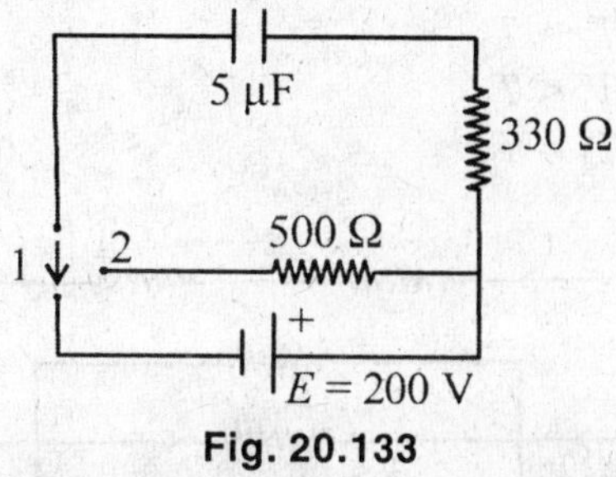

Fig. 20.133

(a) 40×10^{-3} J (b) 50×10^{-3} J

(c) 60×10^{-3} J (d) 30×10^{-3} J.

55. A battery of emf E is connected across the circuit. The value of R for which heat generated in the circuit is maximum

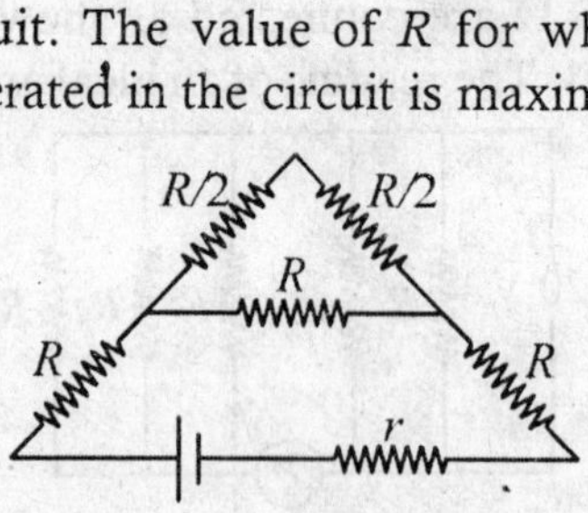

Fig. 20.134

(a) $R = \frac{r}{4}$ (b) $R = \frac{5r}{2}$

(c) $R = \frac{r}{5}$ (d) $\frac{2r}{5}$.

56. The deflection in a moving coil galvanometer falls from 50 div. to 10 div., when a shunt of 12 Ω is applied. The resistance of the galvanometer is

(a) 12 Ω (b) 60 Ω

(c) 24 Ω (d) 48 Ω.

57. Current voltage graphs for a given metallic wire at two different temperatures T_1 and T_2 are shown Fig. 20.135. Then temperature

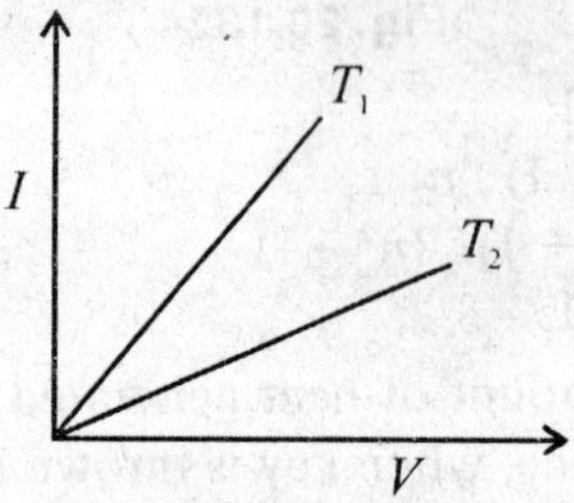

Fig. 20.135

(a) $T_1 > T_2$
(b) $T_1 < T_2$
(c) $T_1 = T_2$
(d) cannot be decided from given data.

58. The effective resistance between points P and Q of the electrical circuit shown in Fig. 20.136 is (IIT 2002)

(a) $\frac{2Rr}{R+r}$ (b) $\frac{8R(R+r)}{3R+r}$

(c) $2r + 4R$ (d) $\frac{5R}{2} + 2r$.

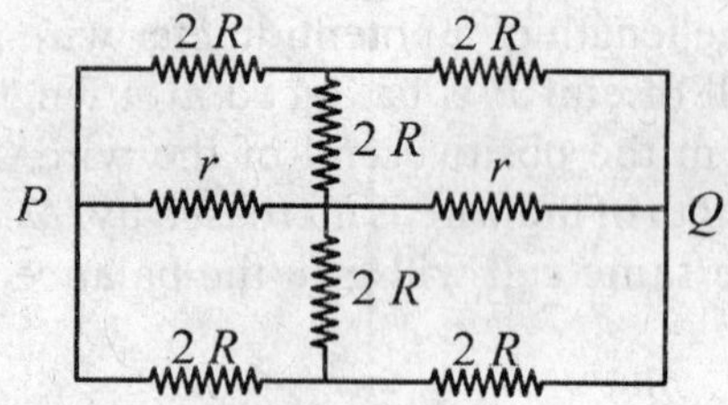

Fig. 20.136

59. In meter bridge experiment shown in Fig. 20.137, the balance length AC corresponding to the null deflection of the galvanometer is x. What would be the balance length if the radius of the wire AB is doubled? (IIT 2003)

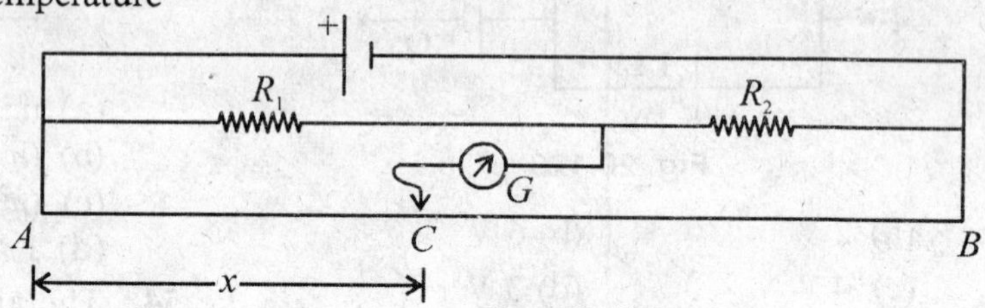

Fig. 20.137

(a) $x/2$ (b) x
(c) $2x$ (d) $4x$

60. Three resistances of equal value are connected in four different combinations as shown in Fig. 20.138. Arrange them in

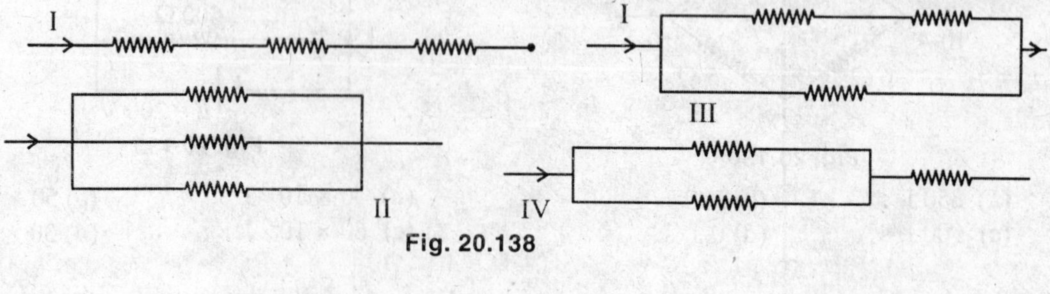

Fig. 20.138

increasing order of power dissipated. (IIT 2003)

(a) III < II < IV < I
(b) II < III < IV < I
(c) I < IV < III < II
(d) I < III < II < IV.

61. Six equal resistances are connected between points P, Q and R (Fig. 20.139). The net resistance will be maximum between (IIT 2004)

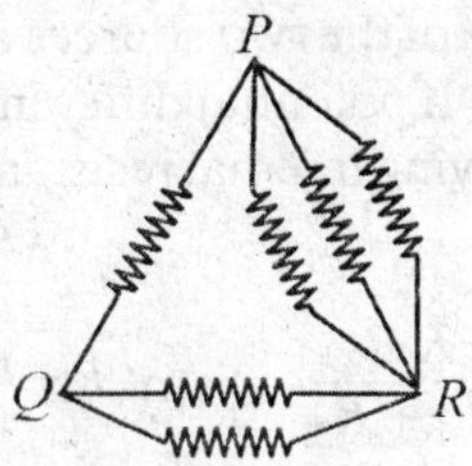

Fig. 20.139

(a) P and Q
(b) Q and R
(c) P and R
(d) is same between any two points.

62. The resistance of the series combination of two resistances is S. When they are joined in parallel, the total resistance is P. If $S = nP$ then minimum possible value of n is (AIEEE 2004)

(a) 4 (b) 1
(c) 2 (d) 3.

63. An electric current is passed through a circuit containing two wires of same material, connected in parallel. If the lengths and radii of the wires are in the ratio $\frac{4}{3}$ and $\frac{2}{3}$, then the ratio of the currents passing through the wires will be. (AIEEE 2004)

(a) 3 (b) 2
(c) 8/9 (d) 1/3.

64. In meterbridge experiment null point is obtained at 20 cm from one end of the wire when resistance X is balanced against an other resistance Y. If $X < Y$ then where will be the new position of the null point from same end, if one decides to balance a resistance of 4xXagainst Y? (AIEEE 2005)

(a) 50 cm (b) 70 cm
(c) 40 cm (d) 80 cm.

65. Five equal resistances each of resistance R are connected as shown in Fig. 20.140. A battery of V volts is connected between A and B. The current flowing in $ASRQB$ is (CBSE PMT 2004)

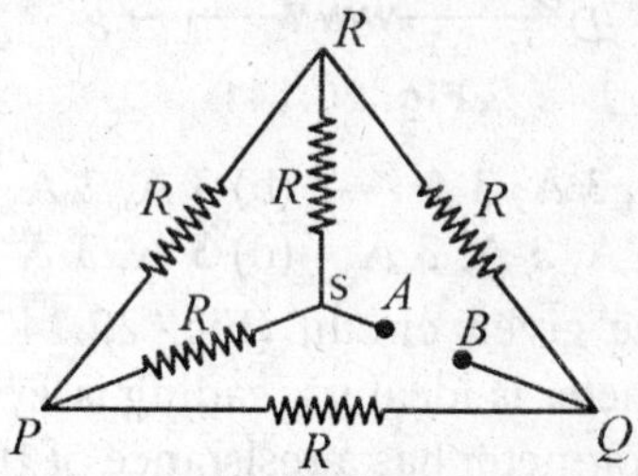

Fig. 20.140

(a) $\frac{3V}{R}$ (b) $\frac{V}{R}$
(c) $\frac{V}{2R}$ (d) $\frac{2V}{R}$.

66. A galvanometer of 50 Ω resistance has 25 div. A current of 4×10^{-4} A gives a deflection of one division. To convert this galvanometer into voltmeter having range of 25 volts, it should be connected with a resistance of (CBSE PMT 2004)

(a) 2500 Ω as shunt
(b) 2450 Ω as a shunt
(c) 2550 Ω in series
(d) 2450 Ω in series.

67. When three identical bulbs of 60 Watt, 200 V rating are connected in series to a 200 V supply, the power drawn by them will be (CBSE PMT 2004)

(a) 60 Watt (b) 180 Watt
(c) 10 Watt (d) 20 Watt.

68. In the given network (Fi.g 20.141) if a battery of emf 36 V and negligible internal resistance is connected between A and B. Currents through the 6 Ω resistance between (*i*) O and D (*ii*) O and B and (*iii*) A and C respectively are

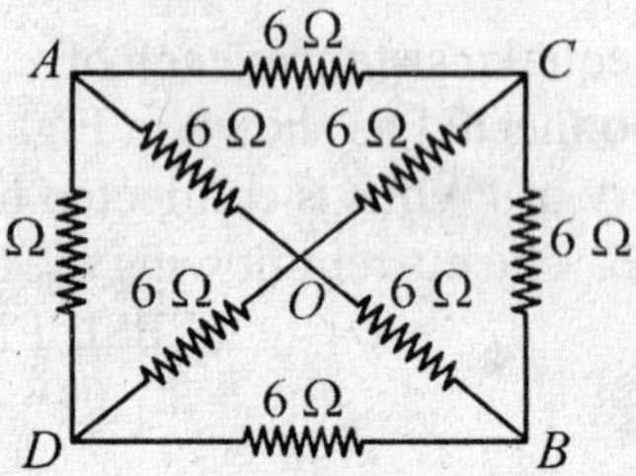

Fig. 20.141

(a) 0, 3 A, 3 A (b) 3 A, 3 A, 3 A
(c) 6 A, 3 A, 3 A (d) 3 A, 3 A, 6 A.

69. In the given circuit (Fig. 20.142) if the voltmeter is ideal its reading is V_1 volts. If the voltmeter has a resistance of 100 Ω the reading is V_2 volts, then $|V_2 - V_1|$ is equal to (IIT 2005)

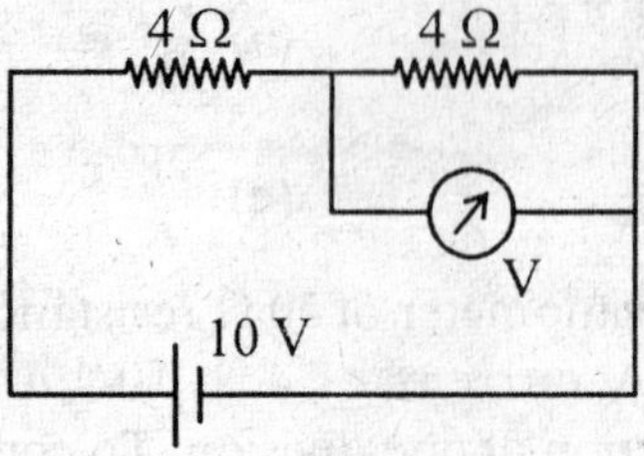

Fig. 20.142

(a) zero (b) 0.3 V
(c) 0.2 V (d) 0.1 V.

70. In the Fig. 20.143, the current through 2 Ω resistor is (IIT 2005)

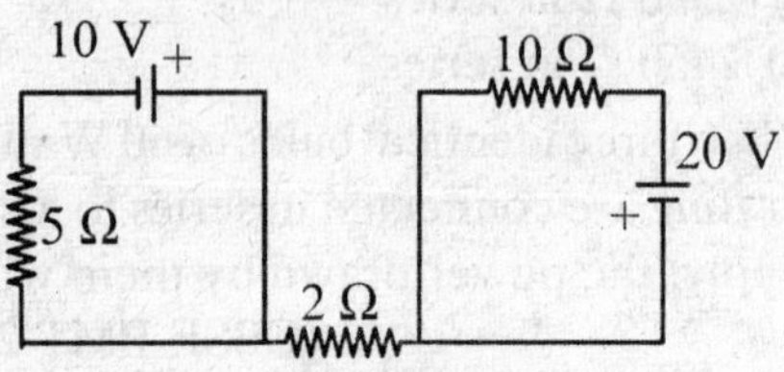

Fig. 20.143

(a) 2 A (b) 0 A
(c) 4 A (d) 6 A.

71. A galvanometer has resistance 100 Ω and it requires current 100 μA for full scale deflection. A resistor 0.1 Ω is connected to make it an ammeter. The smallest current required to produce the full scale deflection is (IIT 2005)

(a) 1000.1 mA (b) 1.1 mA
(c) 10.1 mA (d) 100.1 mA.

72. Two sources of equal emf are connected to an external resistance R. The internal resistance of the two sources are R_1 and R_2 ($R_2 > R_1$). If potential difference across the source having internal resistance R_2 is zero, then (AIEEE 2005)

(a) $R = \dfrac{R_1 R_2}{R_1 + R_2}$ (b) $R = \dfrac{R_1 R_2}{R_2 - R_1}$

(c) $R = R_2 - R_1$ (d) $R = \dfrac{R_2(R_1 + R_2)}{R_2 - R_1}$

73. In the circuit (Fig. 20.144), the galvanometer shows zero deflection. If the batteries A and B have negligible internal resistance, the value of resistor R is (AIEEE 2005)

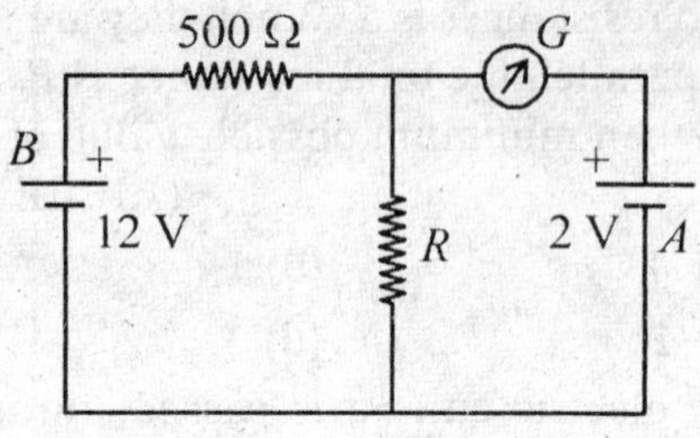

Fig. 20.144

(a) 1000 Ω (b) 500 Ω
(c) 100 Ω (d) 200 Ω.

74. A moving coil galvanometer has 150 equal divisions. Its current sensitivity is 10 divisions per milliampere and voltage sensitivity 2 division per millivolt. In order that each division reads 1 volt the resistance in Ohms needed to be connected in series with the coil will be (AIEEE 2005)

(a) 9995 Ω (b) 99995 Ω
(c) 10^5 Ω (d) 10^3 Ω.

75. The resistance of hot tungsten filament is about 10 times the cold resistance. What will be the resistance of 100 W and 200 V lamp when not in use? (AIEEE 2005)

(a) 200 Ω (b) 400 Ω
(c) 20 Ω (d) 40 Ω.

76. In a potentiometer experiment, the balancing length with a cell is at 240 cm. On shunting the cell with resistance of 2 Ω, the balancing length becomes 120 cm. The internal resistance of the cell is (AIEEE 2005)

(a) 2 Ω (b) 4 Ω
(c) 0.5 Ω (d) 1 Ω.

ANSWERS

1. (a)	2. (a)	3. (a)	4. (c)	5. (c)
6. (a)	7. (d)	8. (a)	9. (b)	10. (a)
11. (a)	12. (b)	13. (b)	14. (c)	15. (b)
16. (a)	17. (c)	18. (c)	19. (b)	20. (a)
21. (a)	22. (b)	23. (d)	24. (a)	25. (d)
26. (d)	27. (b)	28. (b)	29. (c)	30. (d)
31. (a)	32. (d)	33. (d)	34. (a)	35. (d)
36. (b)	37. (b)	38. (a)	39. (c)	40. (a)
41. (a)	42. (c)	43. (d)	44. (b)	45. (b)
46. (b)	47. (c)	48. (d)	49. (b)	50. (b)
51. (d)	52. (d)	53. (d)	54. (c)	55. (b)
56. (d)	57. (b)	58. (a)	59. (b)	60. (c)
61. (a)	62. (a)	63. (d)	64. (a)	65. (c)
66. (d)	67. (d)	68. (a)	69. (d)	70. (b)
71. (d)	72. (c)	73. (c)	74. (a)	75. (d)
76. (d)				

PART II

In the following, a question is followed by four alternatives, of them one or more are correct. Candidate has to select the correct choice(s) from amongst the alternatives.

1. A micro ammeter has a resistance 100 Ω and full scale deflection current of 50 μA. It can be used as a voltmeter or as a higher range ammeter provided a resistance is added. Choose the correct range and resistance combinations from the following

(a) 50 V range, 10 kΩ resistance in series
(b) 10 V range, 200 kΩ resistance in series
(c) 5 mA range and 1 Ω resistance in parallel
(d) 10 mA range and 1 Ω resistance in parallel.

2. In the circuit elements given in Fig. 20.145, all individual resistances are identical. The resistance between terminals P and Q in different cases is

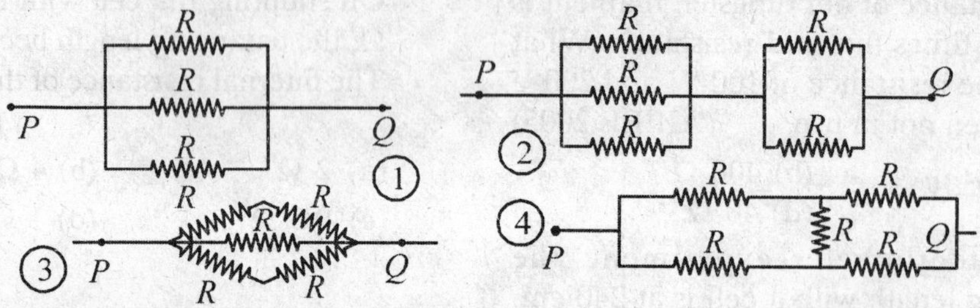

Fig. 20.145

(a) in (1) it is $3R$ (b) in (2) it is $5R/6$
(c) in (3) it is $R/2$ (d) in (4) it is R.

3. In the Fig. 20.146

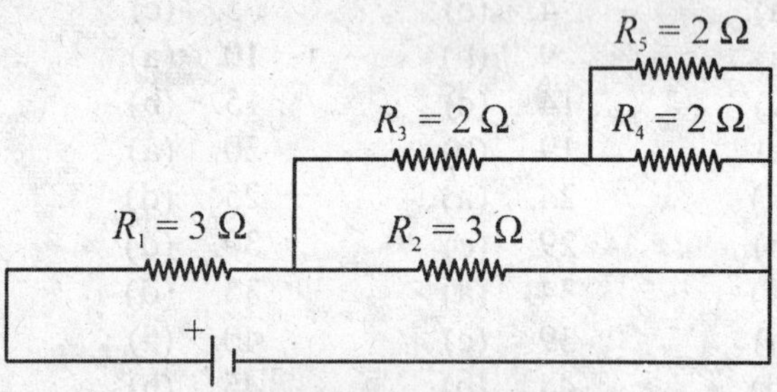

Fig. 20.146

(a) current in R_4 and R_5 are equal
(b) current in R_2 and R_3 are equal
(c) current in R_1 is equal to the sum of currents in R_3, R_4 and R_5
(d) current in R_1 is equal to the sum of currents in R_2, R_4 and R_5.

4. Mark the correct option(s).

(a) In absence of an electric field, electrons move in straight lines between collisions
(b) Relaxation time is of the order of 10^{-14} sec
(c) Resistivity is inversely proportional to relaxation time when number of electrons per unit volume of material remains constant
(d) When current passes through the wire of non-uniform cross-section, then product of drift velocity and area of cross-section is constant

5. Two resistances R and $2R$ are connected in a circuit with a battery, then.

(a) the thermal energy developed in R and $2R$ is in the ratio 1 : 2 if they are connected in series
(b) the thermal energy developed in R and $2R$ is in the ratio 2 : 1 if they are connected in series
(c) the thermal energy developed in R and $2R$ is in the ratio 2 : 1 if they are connected in parallel
(d) the thermal energy developed in R and $2R$ is in the ratio 1 : 2 if they are connected in parallel.

6. Choose the correct statements.

(a) Kirchhoff's junction law follows from conservative nature of electric field
(b) Kirchhoff's junction law follows from conservation nature of charge
(c) Kirchhoff's loop law follows from conservation nature of electric field
(d) Kirchhoff's loop law follows from conservation of charge.

7. Two non ideal batteries are connected together, then

(a) their equivalent internal resistance is less than either of the two internal resistances, when they are connected in series
(b) their equivalent internal resistance is less than either of the two internal resistances, when they are connected in parallel
(c) their equivalent internal resistance is more than either of the two internal resistances when they are connected in series

(d) their equivalent internal resistance is less than either of the two internal resistances when they are connected in series.

8. Two non ideal batteries with emfs E_1 and E_2 $(E_1 > E_2)$ are connected together. If their combined emf is E, then

(a) $E > E_1, E > E_2$ if they are connected in parallel

(b) $E > E_1, E > E_2$ if they are connected in series

(c) $E_1 > E > E_2$ if they are connected in series

(d) $E_1 > E > E_2$ if they are connected in parallel.

9. When no current is passed through a conductor

(a) the free electrons do not move

(b) the average speed of a free electron over a large period of time is zero

(c) the average velocity of a free electron over a large period of time is zero

(d) the average of velocities of all the free electrons at an instant is zero.

10. In Fig. 20.147 PQ is a potentiometer wire. If for a long range of jockeys no balance point is obtained, then

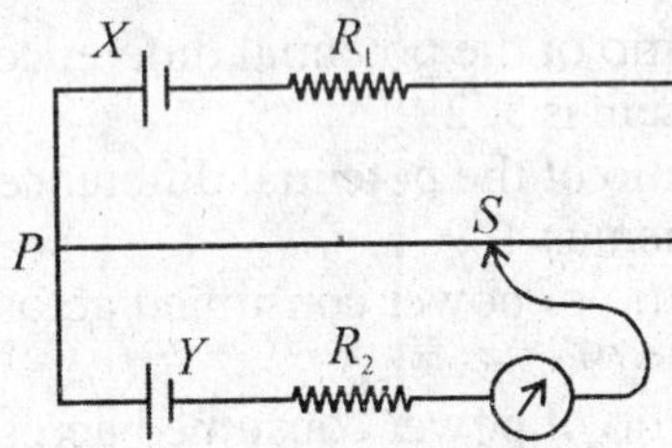

Fig. 20.147

(a) resistance R_1 may be very large

(b) resistance R_2 may be very large

(c) cell Y is discharged

(d) cell X is discharged.

11. In Fig. 20.148, when potential difference is measured with an ideal voltmeter, the reading is

(a) 2 V when connected across cell A

(b) 4 V when connected across B

(c) 6 V when connected across resistance 3 Ω

(d) 12 V when connected across resistance 3 Ω.

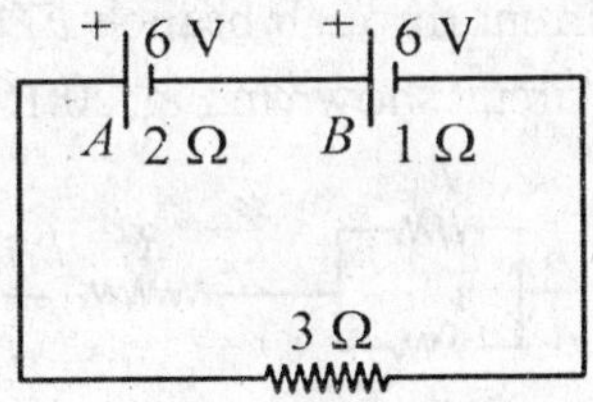

Fig. 20.148

12. In the circuit shown in Fig. 20.149

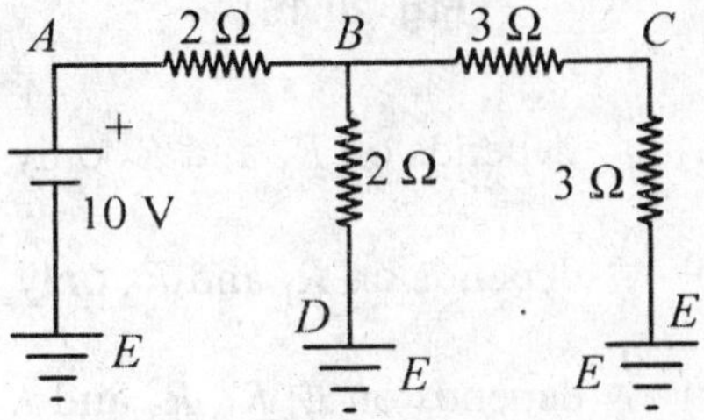

Fig. 20.149

(a) current through 3 Ω resistance is 5/7 amp

(b) current through battery is 15/7 amp

(c) current through resistance 2 Ω (BD) is 15/7 amp

(d) current through resistance 2 Ω (AB) is 20/7 amp.

13. In the circuit shown in Fig. 20.150, seven resistances each of value 5 Ω are connected. The potential difference between A and B is 7 V, then

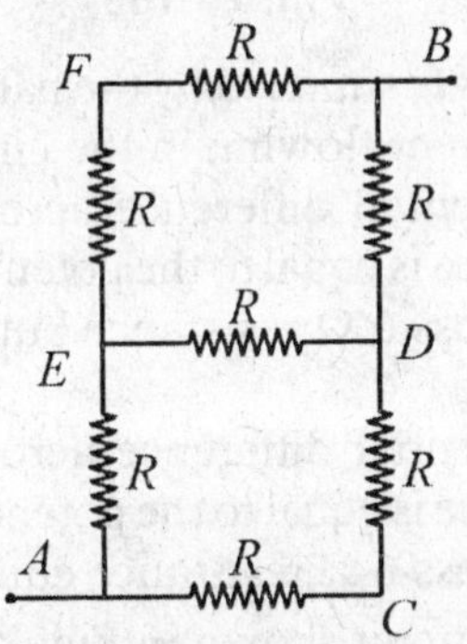

Fig. 20.150

(a) current through branch AE is 0.6 amp
(b) current through branch ED is 0.2 amp
(c) current through branch ACD = current through branch EFB
(d) current through branch ED is zero.

14. In the circuit shown in Fig. 20.151, the ratio

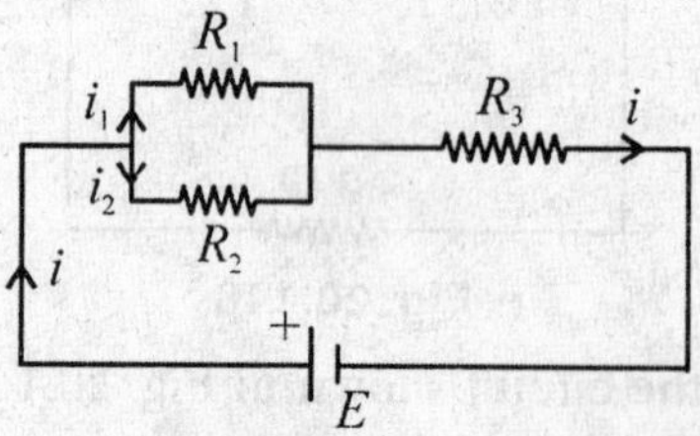

Fig. 20.151

(a) $\frac{i_1}{i_2}$ depends on R_1 and R_2 only
(b) $\frac{i}{i_1}$ depends on R_1 and R_2 only
(c) $\frac{i}{i_1}$ depends on E, R_1, R_2 and R_3
(d) $\frac{i_1}{i_2}$ depends on E, R_1 and R_2.

15. In the circuit shown in Fig. 20.152

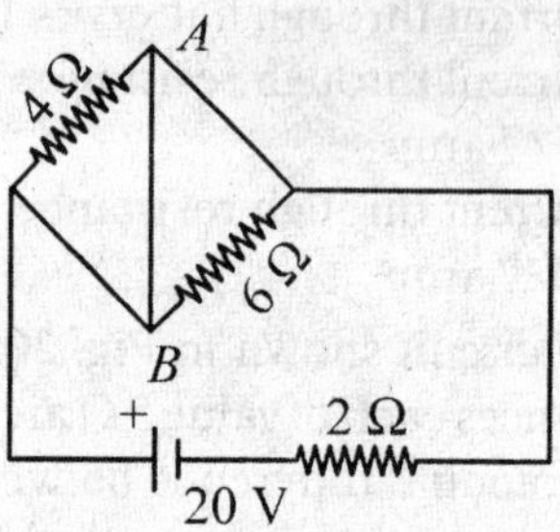

Fig. 20.152

(a) power supplied by the battery is 200 W
(b) current flowing in the circuit is 5 A
(c) potential difference across 4 Ω resistance is equal to the potential difference across 6 Ω resistance but not equal to zero
(d) potential difference across 4 Ω resistance is equal to the potential difference across 6 Ω resistance equal to zero.

16. In the circuit shown in Fig. 20.153 if R_3 = 10 Ω, then

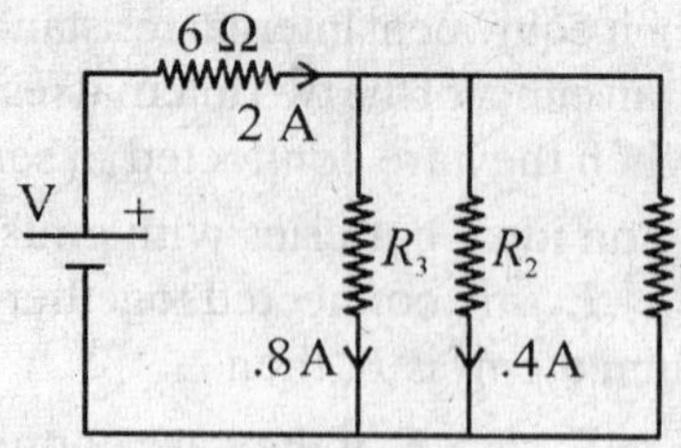

Fig. 20.153

(a) V = 10 V
(b) R_1 = 10 Ω
(c) R_2 = 20 Ω
(d) net resistance of circuit is 10 Ω.

17. Potential difference across the terminals of a non ideal battery is

(a) zero when it is short circuited
(b) zero when no current is drawn from it
(c) less than its emf when current flows from negative terminal to positive terminal inside the battery
(d) greater than its emf when current flows from positive terminal to negative terminal inside the battery.

18. Two bulbs consume power when operated at 200 V and 300 V respectively when these bulbs are connected in series across a DC source of 500 V, then

(a) ratio of the potential difference across them is 3/2
(b) ratio of the potential difference across them is 4/9
(c) ratio of power consumed across them is 4/9
(d) ratio of power consumed across them is 2/3.

19. In the circuit shown in Fig. 20.154

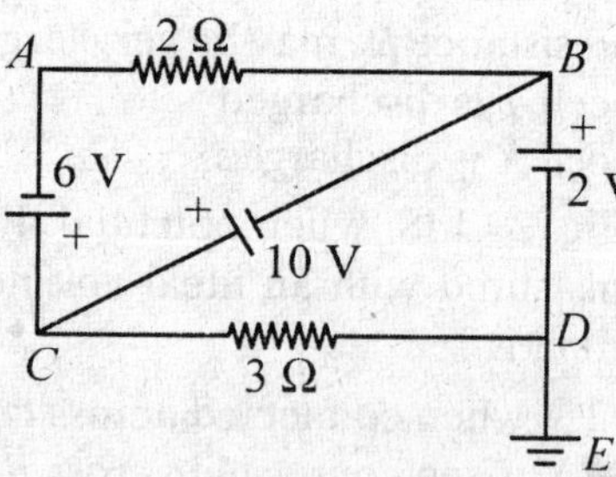

Fig. 20.154

(a) current passing through 2 Ω resistance is 2 A
(b) current passing through 3 Ω resistance is 4 A
(c) current in wire DE is zero
(d) potential of point A is 10 V.

20. A voltmeter reads the potential difference across the terminals of an old battery as 1.40 V. While a potentiometer reads its voltage to be 1.55 V. The voltmeter resistance is 280 Ω. Then

(a) the emf of the battery is 1.4 V
(b) the emf of the battery is 1.55 V
(c) the internal resistance r of the battery is 30 Ω
(d) the internal resistance r of the battery is 5 Ω.

21. In the circuit shown in Fig. 20.155, switch S is closed at $t = 0$. The correct statements are

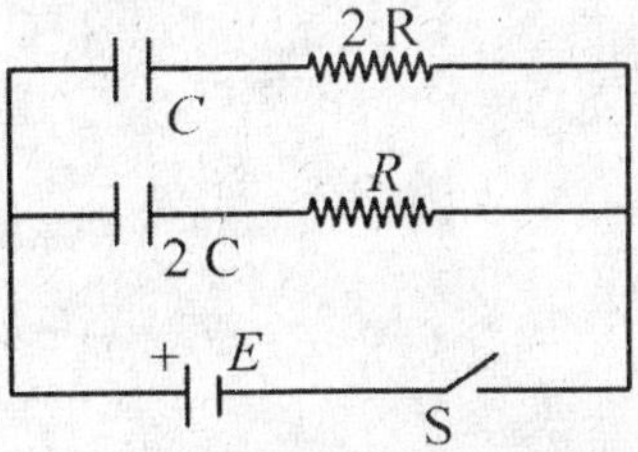

Fig. 20.155

(a) rate of increase of charge is same in both capacitors
(b) rate of charges stored in capacitors C and $2C$ at anytime t would be 1 : 2
(c) time constants of both capacitors are same
(d) steady state charges in capacitors C and $2C$ are in the ratio 1 : 2.

22. In the circuit shown in Fig. 20.156

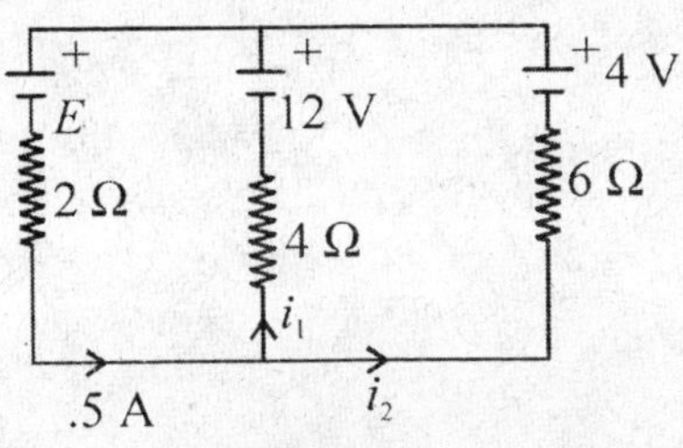

Fig. 20.156

(a) $E = 6.6$ V (b) $i_1 = 1.1$ A
(c) $i_2 = 0.6$ A (d) $i_2 = -0.6$ A.

23. A battery of emf E is connected to a net of resistances (Fig. 20.157)

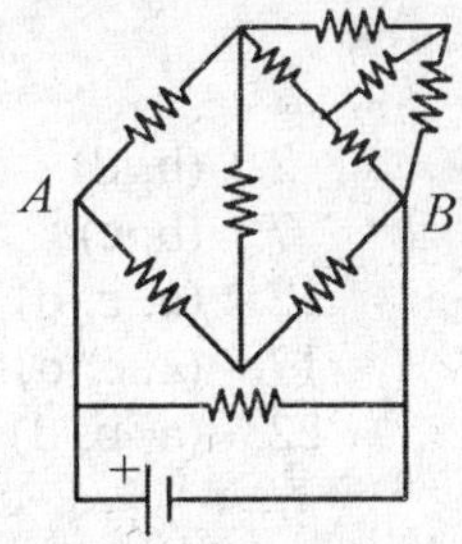

Fig. 20.157

(a) current through the battery is $\frac{E}{R}$
(b) current through the battery is $\frac{2E}{R}$
(c) equivalent resistance between A and B is $2R$
(d) equivalent resistance between A and B is $R/2$.

24. A capacitor C_1 of capacitance 1 μF and a capacitor C_2 of capacitance 2 μF are separately charged by a common battery for a long time. The two capacitors are then separately discharged through equal resistors. Both discharge circuits are connected at $t = 0$

(a) the current in each of two discharging circuits is zero at $t = 0$
(b) the currents in the two discharging circuits at $t = 0$ are equal but not zero
(c) the currents in two discharging circuits at $t = 0$ are unequal
(d) C_1 loses 50% of initial charge sooner than C_2 looses 50% of its initial charge.

25. In the circuit shown in Fig. 20.158, the ammeter A_1 reads 2.4 amp. Neglect the resistance of ammeters, then

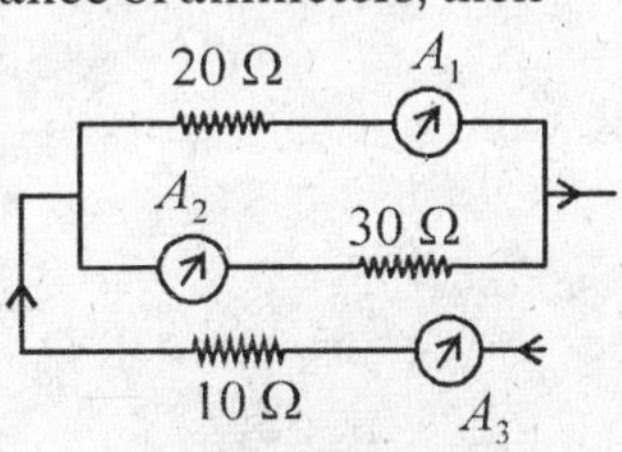

Fig. 20.158

(a) the ammeter A_2 reads 3.6 amps
(b) the ammeter A_2 reads 1.6 amps
(c) the ammeter A_3 reads 3.6 amps
(d) the ammeter A_3 reads 4.6 amp.

ANSWERS

1. (b, c)	2. (b, d)	3. (a, b, c, d)	4. (a, b, c, d)	5. (a, c)
6. (b, c)	7. (b, c)	8. (b, d)	9. (c, d)	10. (a, d)
11. (a, b, c)	12. (a, c, d)	13. (a, b, c)	14. (a, b)	15. (a, d)
16. (b, c, d)	17. (a, c, d)	18. (b, c)	19. (a, b, c)	20. (b, c)
21. (b, c, d)	22. (a, b, d)	23. (b, d)	24. (b, d)	25. (b, d)

21

Thermal and Chemical Effects of Electric Current

In last chapter we studied about electric current. When electric current flows through different media various interesting phenomena take place. For example, when electric current passes through a resistor, heat is produced. There are other situations in which electric current can produce or absorb thermal energy. Here we will study these thermal effects and the chemical effects such as electrolysis, etc.

21.1 HEATING EFFECTS OF CURRENT (JOULE'S LAW)

When an electric current passes through a resistor it becomes hot, i.e. its thermal energy increases. This shows that the electrical energy is converted into heat energy. This phenomenon is called *heating effect of the current or Joule heating effect.*

When a charge dq flows through a resistor R, across which a potential difference V is maintained, the work done by the electric field on the charge is

$$dW = Vdq$$

If the charge flows in time dt then current $i = \frac{dq}{dt}$ or $dq = idt$

$$\therefore \quad dW = Vidt$$

and total work done in time t is

$$W\int_0^t Vidt = Vit \qquad \text{...(21.1)}$$

But Ohm's law gives $V = iR$. Where R is the resistance of the conductor.

$$\therefore \quad W = Vit = \frac{V^2}{R}t = i^2Rt \qquad \text{...(21.2)}$$

This work appears as thermal energy is the resistor through the collision of accelerated electrons with lattice. This thermal energy is referred to as *the heat produced in resistor.*

This relation was first verified experimentally by Joule. That is why it is called *Joule heating*.

If i is in amperes and V in volts, t in seconds, then heat produced $H = i^2Rt$ is in Joules. To convert it in calories, we divide it with J (= 4.2 Joules/cal) the mechanical equivalent of heat. This heat produced in calories is

$$H = \frac{i^2Rt}{J} = \frac{V^2t}{RJ} \qquad \ldots(21.3)$$

Regarding heating effects of current, it follows from equation (21.2) that

(*a*) the heat produced in a given resistor in a given time is proportional to the square of the current through it, i.e.

$$H \propto i^2 \qquad \ldots(i)$$

(*b*) the heat produced in a given resistor by a given current is proportional to the time for which the current passes through it, i.e.

$$H \propto t \qquad \ldots(ii)$$

(*c*) the heat produced in a resistor by a given current in given time is proportional to its resistance, i.e.

$$H \propto R \qquad \ldots(iii)$$

These are known as *Joule's laws.* It must be remembered that

(*i*) Joule heating effect is *irreversible process,* i.e., on reversing the direction of current in a resistor, heating is produced not the cooling.

(*ii*) Due to Joule heating, power dissipated in the form of heat is

$$P = \frac{H}{t} = i^2R = \frac{V^2}{R} = Vi \qquad \ldots(21.4)$$

This means that if resistances are connected in series (same current through all). Power consumed in higher resistance will be more. On the other hand if resistances are connected in parallel (Potential difference same) current and power consumed is more in smaller resistance.

This means that more power is consumed in larger resistance if resistances are connected in series and in smaller resistance, they are connected in parallel.

(*iii*) Under a potential difference V power dissipated in resistance R is

$$P = V^2/R$$

So if resistance is changed from R to R/n, keeping V same the power dissipated is

$$P' = V^2/{R/n} = \frac{nV^2}{R} = nP$$

(*iv*) This means that if n equal resistances are connected in series with a voltage source, the power consumed is

$$P_s = \frac{V^2}{nR} = \frac{P}{n}$$

and if they are connected in parallel with same voltage source

$$P_p = \frac{nV^2}{R} = nP$$

i.e. $$\frac{P_p}{P_s} = n^2$$

It means if V remains the same, the power consumed by n equal resistors is parallel in n^2, times the power consumed in series.

Electric Energy

The total work done or energy supplied by the source of emf in maintaining the current in an electric circuit for a given time is called electric energy consumed in the circuit.

$$\text{Electric Energy} = Vit = P \cdot t$$
$$= \text{electric power} \times \text{time}.$$

S.I. unit of electric energy is Joule.

$$1 \text{ Joule} = 1 \text{ amp} \times 1 \text{ volt} \times 1 \text{ sec} = 1 \text{ Watt} \times 1 \text{ sec}$$

The commercial unit of electric energy is 1 kilo Watt-hour (kWh).

$$1 \text{ kwh} = 1 \text{ kilo Watt} \times 1 \text{ hour}$$
$$= 1000 \text{ Watt} \times 60 \times 60 \text{ sec} = 3.6 \times 10^6 \text{ J}$$

This is also called *Board of Trade Unit (B.O.T.U.)* or *Unit of electricity*.

21.2 FUSE AND ITS ACTION

Fuse is used with the main electrical circuit for the safety of electrical appliances. A fuse wire must have a high resistance and low melting point. It is made of tin-lead alloy.

For the safe use of fuse in steady state of temperature heat produced per second must be equal to heat radiated by it per second.

Consider a fuse wire of length l, cross-section a, of material of resistivity ρ and resistance R.

Heat produced in fuse wire per second is

$$H = i^2R = \frac{i^2\rho l}{a} = \frac{i^2\rho l}{\pi r^2} \qquad \ldots(i)$$

It e is emissivity of fuse material and T its excess temperature above surroundings, then according to Newton's law of cooling

$$\text{energy radiated per second} = e2\pi rl\,T \qquad \ldots(ii)$$

Equating (*i*) and (*ii*), we have

$$e2\pi rlT = \frac{i^2\rho l}{\pi r^2}$$

or
$$T = \frac{i^2\rho}{2\pi^2 er^3}$$

This means that steady state temperature is independent of fuse length.

For a given material of a fuse wire

$$i^2 \propto r^3 \qquad \ldots(21.5)$$

21.3 THERMOELECTRICITY

In last section, we have seen that current passing through a conductor produces heat, i.e., electrical energy is converted into heat energy in an irreversible process. From principle of physical symmetry it is natural to expect a phenomenon in which heat energy can be converted into energy. It now is possible to convert heat energy into electrical energy by a very important reversible effect called thermoelectric effect. This effect has entirely different explanation from resistance heating effect. The thermoelectric effect consists of three related phenomenon, e.g. (*i*) *Seebeck effect* (*ii*) *the Peltier effect* and (*iii*) *Thomson effect*. We will discuss them in detail.

The Seebeck Effect

Seebeck in 1882 observed that when two junctions of two dissimilar metals are maintained at two different temperatures an electric current flows in closed circuit. The phenomenon is known as the *Seebeck effect* or *thermo electric effect*. The two metals forming the circuit are known as *thermo couple*. The existence of a current in circuit implies that there is an emf in the circuit. This emf is called *thermoelectric emf*.

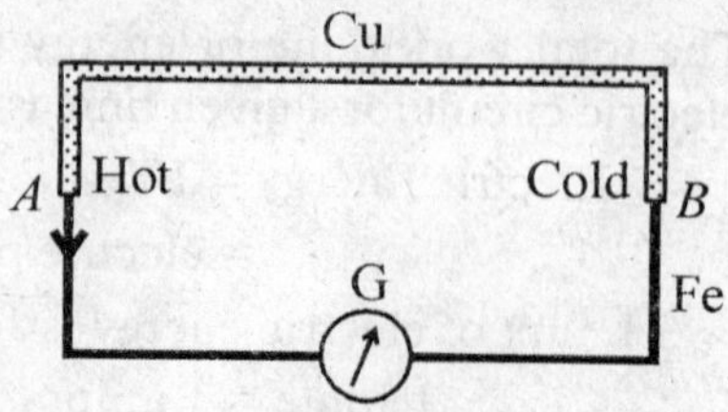

Fig. 21.1

Seebeck studied the behaviour of many pairs of metals. He found that he could arrange them in series. This series is such that if any pair of the metals was selected for use in a thermocouple, the current flowed from the metal occurring earlier in this series to the metal occurring later in the series across the hot junction. Further its magnitude depended upon the extent of separation of the metals in the series. The Seebeck series is Bi, Ni, Co, Pd,U, Cu, Mn, Ti, Hg, Pb, Sn, Cr, Mo, Rh, Ir, Au, Ag, Zn, W, Cd, Fe, As, Sb.

The series shows that a bismuth-antimony thermocouple is the most effective thermocouple and the current in this thermocouple will flow from bismuth to antimony across the hot junction. Similarly in a copper iron thermocouple, the current will flow from copper to iron across the hot junction.

The Variation of Seebeck emf (Thermo emf) with Temperature

Suppose we keep one junction of Cu-Fe thermocouple (Fig. 21.1) at 0° C (= 273 K) called cold junction and measure the emf as the temperature of the other junction (called hot junction) is gradually increased. It is found that the emf increases with an increase in the temperature of hot-junction. It reaches a maximums value when the hot junction is at T_n(= 275° C for Cu-Fe couple). This temperature of hot junction at which the thermo-current is maximum is known as *neutral temperature* for the couple. When the temperature is further increased (beyond neutral temperature) the current decreases till the temperature of hot junction becomes T_i(= 550° C for Cu-Fe couple). The current at this temperature becomes zero. If hot junction is heated further, the direction of current is reversed and the magnitude iscreases. This temperature of hot junction at which the current in couple is zero and reversal is about to begin is called *the temperature of inversion*.

The value of neutral temperature is constant for a couple and depends on the nature of metals forming the couple. It is independent of the temperature of cold junction.

The temperature of inversion depends on the nature of metals forming the couple and the temperature of cold junction. The temperature of inversion (T_i) is as much above the neutral temperature (T_n) as the neutral temperature is above cold junction (T_c), i.e.

$$T_i - T_n = T_n - T_c$$

or
$$T_n = \frac{T_i + T_c}{2} \quad \text{...(21.6)}$$

The graph of thermo emf against temperature difference of the junctions is shown in Fig. 21.2. For majority of thermocouples, the curve is a parabola to a first approximation.

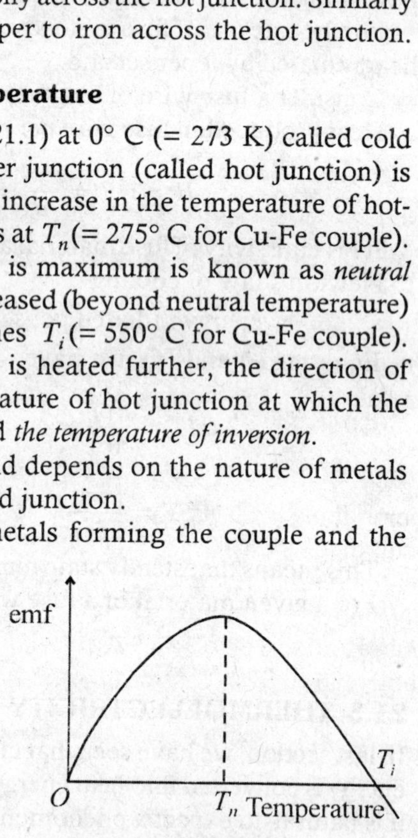

Fig. 21.2

If the cold junction is at 0° C and the hot junction at T° C, the dependence of thermo emf on temperature of hot junction is

$$E = \alpha T + \frac{1}{2}\beta T^2 \qquad ...(21.7)$$

when α and β are constants for the pair of metals. Instead of plotting E vs T, it is convenient to plot a quantity called the *thermo-electric power* (P)—against the temperature of the hot junction.

The thermo-electric power is represented by $\frac{dE}{dT}$. Where dE is the small emf generated in thermocouple whin its junctions are at the temperatures T and $T + dT$ respectively. *The thermo-electric power at any temperature can be defined as the change in thermo emf per degree change in temperature of hot junction.* Differentiating equation (21.7), we get

$$P = \frac{dE}{dT} = \alpha + \beta T \qquad ...(21.8)$$

Thus we find that thermo-electric power has a linear dependence on temperature and as such graph between thermo-electric power and temperature is a straight line.

From equation (21.8), we see that, when $T = 0$, $P = \alpha$, it shows that thermo-electric power is independent of the temperature of the cold junction.

At neutral temperature (T_n) thermo emf in the couple is maximum, i.e. at neutral temperature of the hot junction

$$\frac{dE}{dT} = 0 \quad \text{i.e.}$$

$$\alpha + \beta T_n = 0 \quad \text{or} \quad T_n = -\frac{\alpha}{\beta}$$

and if cold junction is at 0° C then temperature of inversion T_i is

$$T_i = -\frac{2\alpha}{\beta}$$

It should be remembered that thermo-electric power is positive if temperature of hot junction lies in between temperature of cold junction and neutral temperature and is negative for temperature between neutral temperature and temperature of inversion.

Origin of Seeback Effect

The origin of Seebeck effect can be explained in terms of free electron theory of metals. According to this theory, the free electrons in a metals can be considered to move at random within the lattice and that the free electron density (free electrons per unit volume) is different in different metals.

When a junction is formed the free electrons tend to diffuse from the metal with higher concentration to the other with lower electron concentration. As a result of this transfer of electrons a potential difference is set up between the metals at each junction. This is called *contact potential.* When both junctions are at the same temperature the contact potentials at the two junctions will be the same resulting no current in the couple. But if one junction is heated up, the rate of diffusion of free electrons at this junction will increase. As a result of this the contact potentials at the two junctions will be different and there will be a net emf acting in the circuit. This is called *thermo emf.*

21.4 PELTIER EFFECT

Peltier in 1834 observed an important phenomenon which was converse of Seebeck effect. He observed that if a current is passed through the junction of two dissimilar metals, heat is either evolved or absorbed at that junction. Peltier found that if current is passed through thermocouple from B to A (Fig. 21.3) heat is absorbed at A (cooling is produced) and heat is evolved at B (heating is produced). On comparing Fig. 21.1 and Fig. 21.3, we find that the junction A is cooled which was heated in order to get thermo-current in the same direction.

Thus Peltier concluded that *whenever current is passed through the couple, heat is absorbed at one junction and evolved at the other junction. He further observed that the heat is absorbed at the junction which is heated to give the current in the same direction. This is known as Peltier effect.*

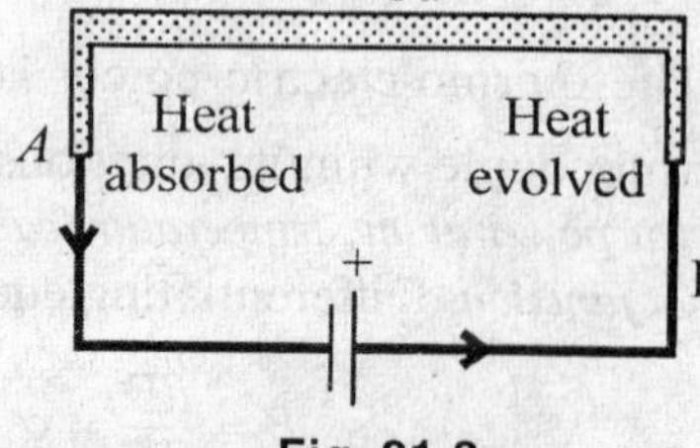

Fig. 21.3

About Peltier Effect

1. The rate at which the Peltier heat is transferred is proportional to the first power of the current, i.e. $H = \pi it$. The quantity π is called the Peltier coefficient and is equal to the heat energy absorbed or evolved at a junction when unit current flows for one second. In S.I. units the Peltier coefficient is expressed in Joule/Coulomb, i.e. Volt.
2. The Peltier effect is reversible, i.e. when the direction of the current is reversed, the Peltier heat is the same but in opposite direction.
3. The Peltier coefficient depends on the temperature and material of the junction and is independent of the temperature of the other junction.
4. The Peltier coefficient π (between metals A and B) is taken as positive when current passing from metal A to metal B causes an absorption of heat at that junction.

Distinction between Joule Heating Effect and Peltier Effect

(*a*) The Peltier effect takes place at a junction only but the Joule heating effect is distributed along the entire length of the conductor.

(*b*) Peltier effect is reversible whereas the Joule heating is irreversible.

(*c*) The rate of heat generation (or absorption) in Peltier effect is directly proportional to the current whereas the rate at which heat is generated in Joule effect is proportional to the square of the current.

(*d*) In Joule effect heat is always evolved whereas in Peltier effect heat is evolved at one junction and absorbed at the other.

21.5 THOMSON EFFECT

Thomson discovered that when a temperature gradient exist along the conductor and a current is allowed to pass through the conductor, the absorption or evolution of heat takes place all along the conductor itself. This phenomenon is also reversible and is known as Thomson effect.

Let us consider a copper rod ABC (Fig. 21.4 (*a*)) in which a symmetric temperature gradient is maintained along the length by heating its central point B to a temperature T and keeping the ends A and C at a lower temperature T_1. The height of ordinates at each point is a measure of temperature of rod at that point. Now if a current is passed along the copper rod in the direction ABC, it is observed heat is absorbed in part AB, i.e. when the current flows from cold to hot parts, whereas heat is evolved when the current flows from B to C, i.e. from hot to cold part. Thus we find that there is a transfer of heat due to flow of current.

Metals like silver, zinc, antimony and cadmium show a similar behaviour. These metals are said to possess positive Thomson's effect. On the other hand, metals like iron, bismuth, cobalt, platinum and nickel possess a negative Thomson's effect. In these metals heat is evolved when current flows from cold to hot end and heat is absorbed when current flows from hot to cold end. In both the above cases, Thomson's effect gets reversed when the direction of the current is reversed.

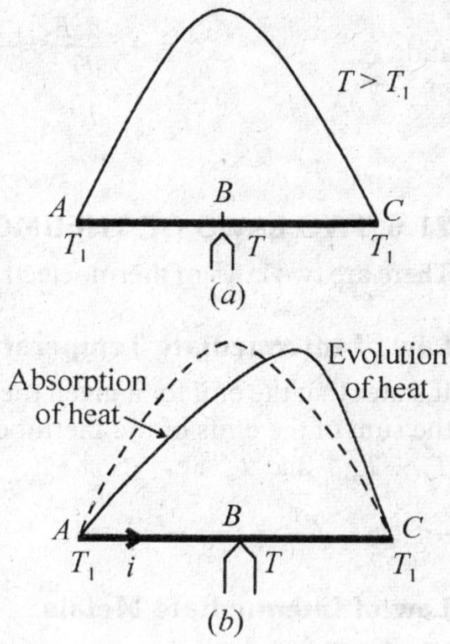

Fig. 21.4

About Thomson's Effect

1. The rate at which Thomson heat is transferred into a small region of conductor carrying current i and having a temperature difference dT is proportional to idT and hence

$$H = \sigma i dT$$

where σ is known as Thomson's coefficient. Thomson's coefficient is numerically equal to the energy absorbed or evolved when a current of one ampere flows for one second between two points of conductor which differ in temperature by 1 K. σ is measured in Joules and sometimes is called the specific heat of electricity.

2. The Thomson coefficient is taken positive when a current opposite to the direction of temperature gradient causes an absorption of heat by the conductor.
3. Thomson's effect is reversible if the direction of the current is reversed heat is absorbed in the region where it was evolved previously.
4. Thomson's coefficient is characteristic of a particular metal under consideration. For metals like copper it is positive whereas for metals like iron it is negative. Thomson's coefficient for lead is zero. For a particular metal Thomson's coefficient is not constant but a function of temperature.

If σ is the Thomson's coefficient, then energy absorbed (or evolved) by element of the conductor with temperature T and $T + dT$ at the ends when unit charge flows through it is

$$dH = \sigma dt$$

and the total heat absorbed or evolved by the conductor is

$$H = \int_{T_1}^{T_2} \sigma dT$$

where T_1 and T_2 are the temperatures of the ends of the conductor.

Relation between Thermoelectric Power P, Peltier Coefficient π and Thomson's Coefficient σ

From thermodynamics we can establish

$$\pi = T\frac{dE}{dT} = TP \qquad \text{...(21.9)}$$

and $$\sigma = -T\frac{d^2E}{dT^2} = -T\frac{d}{dT}\left(\frac{dE}{dT}\right)$$

$$= -T\frac{dP}{dT} \qquad ...(21.10)$$

21.6 TWO LAWS OF THERMOELECTRICITY

There are two laws of thermoelectricity which are established experimentally. They are

Law of Intermediate Temperatures

It states that the emf for a given thermocouple with junctions at temperatures T_1 and T_n is equal to the sum of the emfs of the thermocouples of the same metals with junctions at T_1 and T_2, T_2 and T_3... T_{n-1} and T_n, i.e.

$$E_{T_1}^{T_n} = E_{T_1}^{T_2} + E_{T_2}^{T_3} + E_{T_3}^{T_4} + \dots + E_{n-1}^{T_n} \qquad ...(21.11)$$

Law of Intermediate Metals

If a number of metals are in successive contacts so as to form a chain, then the effective emf between extreme metals when placed in direct contact at a given temperature is equal to the sum of the individual emfs between adjacent metals provided all the junctions are at the same temperature.

If A and D are the extreme metals, B and C are intermediate metals, then

$$E_A^D = E_A^B + E_B^C + E_C^D \qquad ...(21.12)$$

We have

$$E_A^D = \alpha_{AD}T + \frac{1}{2}\beta_{AB}T^2$$

$$E_A^B = \alpha_{AB}T + \frac{1}{2}\beta_{AB}T^2$$

$$E_B^C = \alpha_{BC}T + \frac{1}{2}\beta_{BC}T^2$$

$$E_C^D = \alpha_{CD}T + \frac{1}{2}\beta_{CD}T^2$$

and using equation (21.12), we can find the values of α and β for any pair of metals.

Applications of Thermoelectric Effect

Important applications of thermoelectric effect are as follows:

(*i*) **To measure temperature, using thermoelectric thermometers:** The thermocouples of different metals are constructed depending upon the temperature to be measured. The range of temperature measurement for different couples is shown in the table below.

Thermocouple	Range of temperature
(*a*) Copper and gold-iron alloy	1 K to 50 K
(b) Copper and constantan	50 K to 400 K
(*c*) Platinum and Platinum-iridium alloy	400 K to 1300 K
(*d*) Platinum and Rhodium Platinum alloy	1300 K to 2000 K

(*ii*) To detect heat radiation using thermopile
(*iii*) Thermoelectric refrigerator
(*iv*) Thermoelectric generator.

21.7 ELECTROLYSIS

The liquids are classified into three classes according to their electric behaviour. They are:

1. **Good Conductors:** These are the liquids through which the current can pass without dissociation into ions. For example, mercury (liquid metal at ordinary temperature).
2. **Insulators:** These are the liquids through which current does not pass. For example distilled water, vegetable oil.
3. **Electrolytes:** These are the liquids through which current can pass and dissociate into ions during the passage of current through them. For example the solutions of salts, acids and bases.

Thus phenomenon of electrolysis, i.e. the passage of current through conducting liquid solutions, is different from the conduction of electricity in conductors in two important respects. First, of course is the occurrence of chemical reaction and here also the chemical reactions at the anode and cathode are different from each other. Second, unlike metallic conduction here matter is actually transported through the solution. The phenomenon of electrolysis is, therefore, due to mechanism different from the one responsible for conduction of electricity through metals.

Theory of Electrolysis

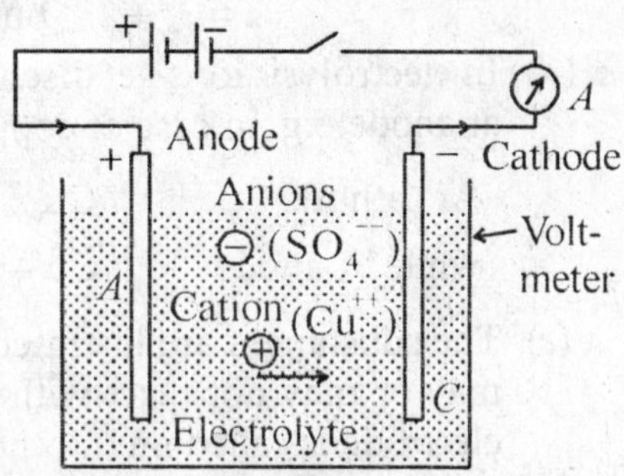

Fig. 21.5

Electrolysis is the process of decomposition of electrolyte solution into ions on passing the current through it. The vessel in which the electrolysis is carried out is called voltameter (Fig. 21.5). We have already seen that the mechanism of electric conduction in electrolytes is quite different from that in conductors (metals). In order to understand the process let us take an specific example of $CuSO_4$.

Let us take a solution of copper sulphate ($CuSO_4$) in electrolytic cell (voltameter) and immerse two copper rods called electrodes. Electric current is passed through the solution by connecting the rods to an external battery. The current enters the solution through the electrode called anode (*A*) and leaves through the electrode called cathode (*C*). With the passage of time we observe that copper gets deposited on the Cathode (*C*) and an equal amount of copper is removed from the anode (*A*).

When $CuSO_4$ is dissolved in water, then according to Arrhenius it dissociates into Cu^{++} and SO_4^{--} ions. Each of these ions have electric charge. The ions with positive charge is called cation (here Cu^{++}) and an ion with negative charge is called anion (here SO_4^{--}). These ions move freely in the the solution. When the electrodes are connected to the battery, an electric field is setup in the solution from anode towards the cathode. Thus the cations (Cu^{++}) move towards cathode and anions (SO_4^{--}) move towards anode. The ions give up their charges at the electrodes and substance making up the ions is liberated. The liberated substance may get deposited on the electrodes or may take part in some secondary chemical reaction.

In case of $CuSO_4$ the Cu^{++} ions move to the cathode, the charge is given up there and copper atoms are deposited on the cathode. In fact a Cu^{++} ions receives two electrons from the cathode to become neutral Cu atom.

The SO_4^{--} (anion) ion moves to the anode and gives the two extra electrons to it and is converted to SO_4 and reacts with copper atom of the anode to form $CuSO_4$ which goes into the solution. This way copper is continuously removed from anode and is deposited on cathode and the concentration of the electrolyte remains the same. In this process, the movement of cations (positive charges) towards the cathode and that anions (negative charges) towards the anode makes the current through the electrolyte.

The electrons given up by an anion to the anode are passed through the battery back to the cathode which has supplied the electrons to the cations. In this way the potential difference is maintained between the electrodes. The number of electrons passed through the battery is equal to the number of electrons absorbed or released at each electrode, i.e. the charge passed through the circuit is equal to the charge released at each electrode.

Here it is worth while to mention that in electrolysis:

(*a*) It takes place for direct current and low frequency alternating current. At higher frequency *AC*, the ions can not follow the frequency of *AC* due to inertia.

(*b*) In this process electrical energy is converted into chemical energy while in a cell chemical energy is converted into electrical energy.

(*c*) In this process the charge carriers are ions. The current due to positive and negative ions are not equal because the masses of cations and anions are not equal as such their mobilities are different, i.e.

$$i = i_+ + i_- \quad \text{but} \quad i_+ \neq i$$

(*d*) In electrolysis ions get discharged at the electrodes with reduction at cathode and oxidation at anode e.g. in case of copper voltameter with copper anode and $CuSO_4$ electrolyte.

At cathode $\quad Cu^{+2} + 2e^- \rightarrow Cu$ (reduction)

while at anode $\quad Cu \rightarrow Cu^{+2} + 2e^-$ (oxidation)

(*e*) The substances are liberated at the electrodes due to decomposition of electrolytes, which may or may not react with them. If the liberated mass does not react with electrodes the electrode is called inert, otherwise soluble. In case of water voltameter platinum electrodes are inert whereas copper anode in copper voltameter is soluble.

(*f*) If electrodes are soluble, electrolysis obeys Ohm's law. In case of inert electrodes, electrolysis does not obey Ohm's law. Fig. 21.6 (*a*) shows *i-V* curve for inert electrodes and (*b*) shows for soluble electrodes.

The conductivity of electrolytes is poor in comparision to that of metals (of the order 10^{-5} to 10^{-6}). The conductivity of electrolytes increases with temperature. The reasons for poor conductivity are

(*i*) no of ions per unit volume is small.

(*ii*) ions are massive and hence less mobile.

(*iii*) liquid is dense and disordered, i.e. viscosity affects adversely to the mobility.

The conductivity increases with temperature because viscosity decreases and degree of ionisation increases.

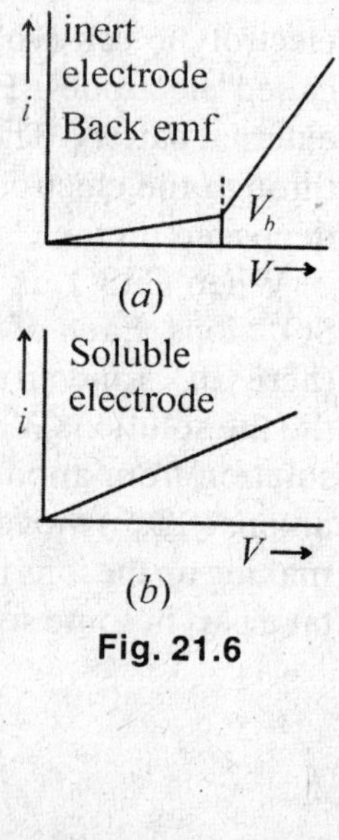

Fig. 21.6

21.8 FARADAY'S LAWS OF ELECTROLYSIS

After systematic study of electrolysis Faraday in 1834 discovered the following two laws.

(*a*) *The mass of the substance deposited or liberated at an electrode is directly proportional to the total charge passing through the electrolyte.*

If an electric current i is passed through an electrolyte for a time t, the amount of charge passed is $Q = it$. Then if mass deposited on the electrode be m, then

$$m \propto Q$$

or $$m = ZQ = Zit \qquad \ldots(21.13)$$

where Z is a constant for the substance liberated and is called *electrochamical equivalent* (*ECE*).

Electrochemical equivalent is the characteristic of ions and is equal to the mass deposited or liberated at an electrode when one Coulomb of charge is passed through the electrolyte. S.I. units of *ECE* is kg/C and dimensions are $M^1L^0T^{-1}I^{-1}$.

(*b*) Second law states that *if same charge is passed through different voltameters, the masses of the substances deposited or liberated are directly proportional to their chemical equivalents.* (The chemical equivalent is equal to atomic mass of substance divided by valency), i.e.

$$\frac{m_1}{m_2} = \frac{E_1}{E_2} \qquad \textit{if} \qquad Q_1 = Q_2$$

here E_1 and E_2 are the chemical equivalents of the substances deposited or liberated and m_1 and m_2 their masses respectively.

From first law, we have

$$\frac{m_1}{m_2} = \frac{Z_1}{Z_2} \qquad \because Q_1 = Q_2$$

$$\therefore \qquad \frac{E_1}{E_2} = \frac{Z_1}{Z_2}$$

or $$\frac{E_1}{Z_1} = \frac{E_2}{Z_2} \ldots = \frac{E}{Z} = \text{constant}$$

i.e. in case of electrolysis, the ratio of chemical equivalent (*CE*) and electrochemical equivalent (*ECE*) is constant and same for all electrolytes, *i.e.* is independent of nature of electrolyte.

From first law, we have if mass deposited is m, then charge required is

$$Q = \frac{m}{Z}$$

If $m = E$, i.e. 1 gm equivalent (Equivalent weight expressed in grams) then charge required

$$Q = \frac{E}{Z}$$

This means that ratio E/Z represents the charge required to liberate or deposit 1 gram equivalent of substance through electrolysis. This quantity of charge is called 1 *Faraday* and is represented by letter F, i.e.

$$F = \frac{E}{Z}$$

Since Faraday is the charge on 1 gm equivalent ions, so if p is the valance of ions and N_0 is the Avogadro number, the charge on 1 gm equivalent ions

$$F = \text{(charge on ion) (no. of ions in 1 g equivalent)}$$

$$= p \cdot e \cdot \left(\frac{N_0}{p}\right) = eN_0 \qquad \text{...(21.14)}$$

$$= (6.022 \times 10^{23})(1.602 \times 10^{-19})$$

i.e. $\quad 1\ F = 96487\ C = 96500\ C$

i.e. Faraday is a universal constant having value 96500 C.

21.9 APPLICATIONS OF CHEMICAL EFFECTS OF CURRENT

Some of the important applications of chemical effects of current are given below.

1. **It is Used to Measure Charge or Current:** From first law, we have

$$Q = \frac{m}{Z} \quad \text{or} \quad i = \frac{Q}{t} = \frac{m}{Zt}$$

   Because of the high accuracy attainable, silver voltameters are used in laboratories for caliberation and standardisation.

2. **Electroplating:** The process of depositing a thin uniform layor of a metal on any conducting surface is called electroplating. The article to be electroplated is made cathode and the metal to be deposited made anode. The electrolyte used is same salt of the metal to be deposited. When current is passed a thin layer of uniform thickness of the metal is deposited on the article. Using this technique silver is electroplated with gold, brass with nickel and iron with zinc.

3. **In Determining the equivalent weight and atomic weight** (or valance): From second law, we have

$$\frac{m_1}{m_2} = \frac{E_1}{E_2} \quad \text{when} \quad Q_1 = Q_2$$

   If equivalent weight of one metal is known then equivalent weight of second can be determined and hence atomic weight can be determined if valance is known and *vice versa.*

4. **In Manufacturing Non Metals:** Non metals like hydrogen, oxygen and chlorine, etc. are obtained by electrolysis.

5. **In Manufacturing Compounds:** Compounds like NaOH, NO_2CO_3, $KClO_3$ and $KMnO_4$ etc. are manufactured by electrolysis.

6. **In Constructing Electrolytic Capacitors:** An electrolytic capacitor consists of two aluminium electrodes dipping in a mixture of boric acid, glycerine and ammonia water. When current is passed, a thin film of aluminium oxide is formed on anode due to electrolysis which acts as dielectric. As the thickness of oxide layer is very small, the capacitor has large capacity.

7. **It is Used to Ascertain the Polarity of the Battery:** In electrolysis the mass of cathode increases and that of anode decreases. Hence knowing the mass of the electrodes before and after the electrolysis, the polarity of the battery can be ascertained.

8. **In Electrorefining of Metals:** Metals like copper, silver, gold, aluminium, tin, etc. can be refined using electrolysis.

9. **In Electro-metallurgy:** The metals like sodium, potassium, manganese and calcium, etc. are obtained by electrolysis of fused electrolytes.

21.10 PRIMARY AND SECONDARY CELLS

A cell is a device which gives us an steady current. When a metal electrode is dipped in an electrolyte, negative or positive ions tend to go from electrolyte to electrode. Thus a potential difference is produced between the electrolyte and the electrode. If two electrodes of different metals are dipped in an electrolyte, the potentials produced at the two electrodes are different and there is a net potential difference (called emf) between the electrodes. This forms a *voltaic cell* also called simply a *cell*. The electrode at higher potential is called the *positive electrode* and that at lower potential is called the *negative electrode*. This is the basic theory of voltaic cell in which electrolysis is used to produce an emf between two terminals.

A cell is called *primary* if chemical energy is directly converted into electrical energy. This cell is used only for discharge. The current leaves the cell at the positive terminal, goes through the external circuit and enters the cell at negative terminal.

In secondary cells the electrical energy is stored in the form of chemical energy by charging it, and this stored chemical energy is converted into electrical energy at the time of discharge. Thus a secondary cell can be charged as well as discharged. The current can go both ways in the cell.

Primary Cells

(*a*) **Daniel Cell:** In the simplest form a Daniel cell consists of a copper vessel containing copper sulphate solution. The vessel itself acts like the positive electrode of the cell. A porous pot of porcelain or fired clay containing zinc rod in dilute H_2SO_4 is placed in the copper sulphate solution (Fig. 21.7). The porous pot allows ions to pass from one solution to the other but prevents the two solutions to mix. The zinc surface is amulgamated to avoid the local effects of impurities.

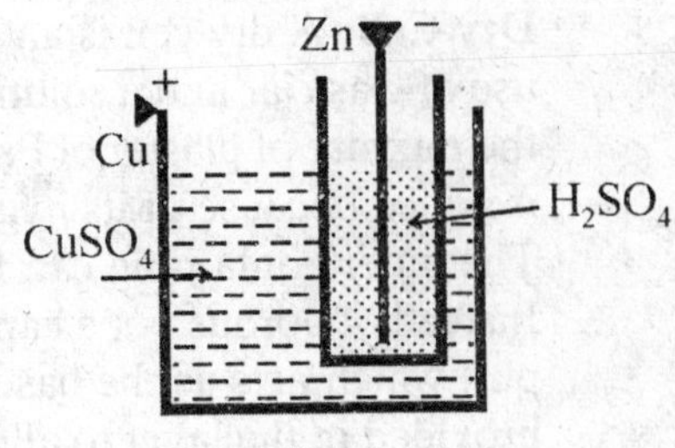

Fig. 21.7

The H_2SO_4 solution consists of H^+ ions and SO_4^{--} ions. SO_4^{--} ions move to the zinc electrode and form $ZnSO_4$. The negative charge is given to the zinc electrode. The H^+ ions move out through the porous pot. These H^+ ions combine with SO_4^{--} ions in $CuSO_4$ solution to form H_2SO_4. The Cu^{++} ion move towards the walls of the outer vessel where they are deposited and the positive charge is given to the vessel which acts like a positive electrode. In this process hydrogen is prevented from collecting near the anode. Any hydrogen collected near the anode may stop the function of the cell. This is called *polarisation of the cell.*

Due to the chemical reactions described above the positive charge is accumulated on the copper electrode and negative charge on the zinc rod. The zinc electrode acts as a negative electrode and copper as positive electrode.

The emf of Daniel cell is 1.08 V and internal resistance 1 Ω. This gives continuous current.

(*b*) **Leclanche Cell:** Leclanche cell consists of a glass vessel containing saturated solution of NH_4Cl which acts as electrolyte. The carbon electrode packed in a porous pot containing manganese dioxide and charcoal powder is immersed in ammonium chloride solution

(Fig. 21.8). Since MnO_2 which acts as a depolariser is a poor conductor of electricity, as such charcoal powder is added to improve the conductivity. In the outer glass vessel a zinc rod is immersed in electrolytic solution. The zinc electrode forms the negative terminal and the carbon forms the positive terminal. When an external circuit is connected across the cell a current passes through the cell. The Cl^- ions combine with Zn forming $ZnCl_2$ and NH_4^+ ions move towards carbon electrode. The NH_4^+ reduces to ammonia (NH_3) and hydrogen (H^+). The H^+ ions move towards carbon electrode through porous pot. Positive charge is given to the carbon electrode. The hydrogen reacts with MnO_2 to form Mn_2O_3 and water. The reactions is

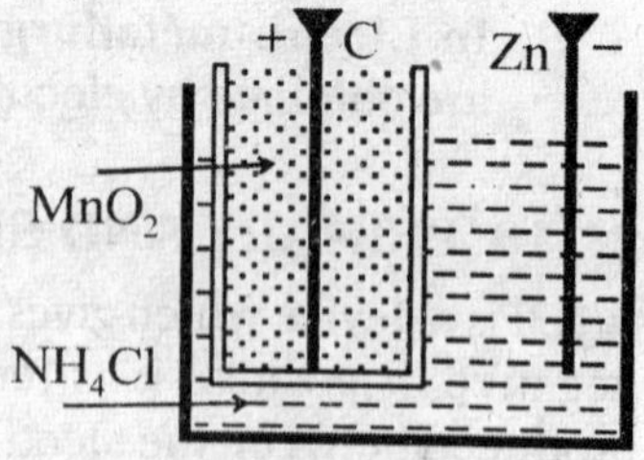

Fig 21.8

$$2MnO_2 + 2H^+ \rightarrow Mn_2O_3 + H_2O + 2e^+$$

Thus, MnO_2 prevents hydrogen from collecting there on the anode, which could otherwise stop the cell's function. In this way it acts like a *depolariser.*

The depolarising action by the solid MnO_2 is quite slow. Thus if the current is drawn continuously from the Leclanche cell, hydrogen starts collecting at the carbon electrode and thereby the cell stops functioning. When external circuit is switched off the hydrogen gas escapes. The cell regains its emf and becomes ready for use. Thus the Leclanche cell is used where intermittent currents are required. The emf of the cell is about 1.5 V.

Dry Cell: A dry cell is an easily portable compact form of the Leclanche cell in which the use of glass jar and a solution of ammonium chloride is avoided. A moist paste made from the mixture of plaster of Paris, ammonium chloride and zinc chloride is used. Zinc chloride is added because of its hygroscopic nature. It helps to maintain the moistness of the paste. The outer containing can is made up of zinc which itself serves as the negative electrode. Instead of porous pot a canvas wrapping is used. At the top of the cell a layer of saw dust is put, which acts as the base for the top layer of bitumin used for ceiling purpose. A vent is provided in the layer to allow the gases formed in the chemical reaction to escape. The emf of the dry cell is about the same as that of the Leclanche cell but the internal resistance is much less.

The dry cells are among the most widely used form of cells because of the advantages of compactness of size and easily portability.

Secondary Cell

(*a*) **Lead Acid Accumulator:** The lead acid accumulator is based on the fact that (*i*) the oxide of lead is highly electronegative with respect to lead and (*ii*) lead and lead oxide form a reversible cell in sulphuric acid. It was developed by Gaston Plante in 1859 and since then, it has undergone considerable commercial development.

Lead accumulator consists of electrodes made of PbO_2 and spongy Pb immersed in a dilute Sulphuric acid (H_2SO_4) having specific gravity of 1.20 to 1.28. The electrodes are in the form of a number of plates connected together. This is done to increase the capacity of the cell. To prevent plates of opposite polarity from coming into contact, certain suitable insulators are used.

When a new cell is to be used its plates are first allowed to *form.* This *forming of plates* is achieved by passing a current through a new cell first for some time in one direction and then for some time in the reverse direction. This makes the PbO_2 layer thicker and at the same time spongy. Also the lead plate connected to the positive of the charging source turns brown while the other plate stays grey.

When a fully charged cell is used for supplying current to an external load, it is said to be getting discharged. The chemical reactions taking place during the discharging and charging processes are as follows:

(*i*) ***Discharging Process*:** Sulphuric acid dissociates as

$$H_2SO_4 \rightarrow 2H^+ + SO_4^{--}$$

SO_4^{--} ions move towards Pb electrode (negative electrode) while H^+ ions move towards PbO_2 electrode (positive electrode).

At negative electrode (at cathode)

$$Pb + SO_4^{--} \rightarrow PbSO_4 + 2e^-$$

and at positive electrode (at anode)

$$PbO_2 + 2H^+ + H_2SO_4 \rightarrow PbSO_4 + 2H_2O + 2e^+$$

Thus during the discharging process whitish $PbSO_4$ is formed on both the electrodes. As the sulphuric acid is used up in discharging, the specific gravity of the acid decreases. When the specific gravity falls below 1.18 the cell is considered to be fully discharged and any further current drawn from it, may permanently damage the electrodes.

(*ii*) **Charging Process:** Charging process is reverse of discharging. A current is forced from the positive to the negative electrode inside the cell. Now the H^+ ions move towards the cathode and SO_4^{--} ions move towards anode.

At cathode

$$PbSO_4 + 2H^+ \rightarrow Pb + H_2SO_4 + 2e^+$$

and at anode

$$PbSO_4 + SO_4^{--} + 2H_2O \rightarrow PbO_2 + 2H_2SO_4 + 2e^-$$

Thus the $PbSO_4$ deposited at the two electrodes is dissolved. Pb is deposited at the negative electrode (cathode) and PbO_2 at the positive electrode (anode). When fully charged emf of the lead accumulator is 2.1 to 2.2 V and a very low ($\approx$ 0.01 to 0.02 Ω) internal resistance.

The capacity of an accumulator is measured in ampere-hour. Thus a 100 ampere-hour accumulator can supply 20 A current for 5 hours and 1 A current for 100 hours.

(*b*) **Alkali Accumulator: Ni-Fe cell:** Alkali accumulator was developed by Edison and has the following constructional features:

In Ni-Fe cell, the positive electrode is made up of nickel hydroxide [$Ni(OH)_2$] mixed with finely flaked nickel and the negative one of finely divided iron and yellow mercury oxide. These materials are packed into pocket provided in the nickel steelgrids. The electrolyte is 20% solution of potassium hydroxide (KOH) and 1% lithium hydroxide (LiOH). This helps to improve the reaction at the positive plate. The electrolyte has specific gravity 1.17 which remains practically unchanged during the charging or discharging.

Chemical Reactions During Charging

$$KOH \rightarrow K^+ + (OH)^-$$

The K^+ ions go towards cathode, $(OH)^-$ ions go towards anode. Hence at cathode

$$FeO + 2K^+ + H_2O \rightarrow Fe + 2KOH + 2e^-$$

and at anode

$$2Ni(OH)_2 + 2(OH)^- \rightarrow 2Ni(OH)_4 + 2e^+$$

Chemical Reactions During Discharging

Now the K^+ ions go towards anode and $(OH)^-$ towards cathode

at cathode

$$Fe + 2(OH)^- \rightarrow F(OH)_2 + 2e^-$$

and at anode

$$2Ni(OH)_4 + 2K^+ \rightarrow 2Ni(OH)_2 + 2KOH + 2e^+$$

The reactions explains that the specific gravity is practically same. The emf of alkali accumulator is 1.3 V when fully charged. Alkali accumulator is costlier than the acid accumulator but has several advantages over acid accumulators. The advantages are:

(*i*) low weight/capacity ratio
(*ii*) has capability to withstand over charging, over discharging and idleness.
(*iii*) It is not damaged if subjected to low temperatures or freezing and plates are not spoiled even if allowed to stand for long periods in the electrolyte in the discharged conditions.

A new type of accumulators that are better than either the acid or alkali type described above are the silver-cadmium secondary cells. The characteristics of these is very long life and a very compact size. They are widely used in missiles, satellites and under sea cameras.

Solid State Cells

The cells discussed above have some disadvantages. They are

(*i*) leakage on long term storage.
(*ii*) corrosion due to the use of liquid acid/alkaline solution
(*iii*) low wattage per kg mass of the cell
(*iv*) short life
(*v*) limit on miniaturisation.

Recently some solid state electrolytes in the form gels, polymers, campsites, polycrystalline solids, thin solid film have been developed in which ions can move. These solid electrolytes are used in fabricating *solid state cells.* One such cell is shown in Fig. 21.9. with mobile cation A^+ and anion B^-. In such cells either one or both ions can move.

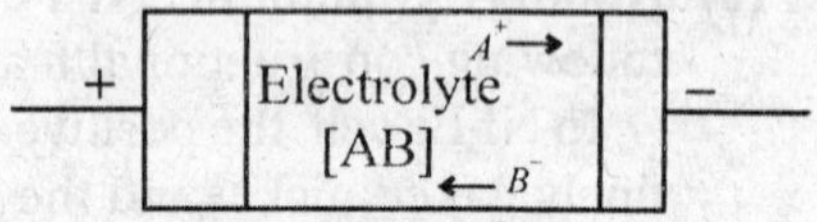

Fig. 21.9

Some electrolyte is also mixed in cathode or anode to decrease polarisation. Recently many solid state materials have been developed which are used as cathodes, anode or electrolyte in solid state cells.

One such cell in use is lithium solid state cell in which basic reaction is

$$2Li^+ + I_2 \rightarrow 2LiI + 2e^+ \qquad ...(21.14)$$

Lithium batteries using solid state electrolyte such as ($LiI + Al_2O_3$) are in use. They are used in mobile phones and heart pacer batteries. Polymer Li-batteries and H^+–batteries are in advanced stage of development for electrical cars.

Some electrode materials like doped $LiCoO_2$ or $LiMnO_2$ have provided excellent rechargeability to these cells.

SOLVED EXAMPLES

Example 21.1: An electric kettle has two coils. When one coil is switched on, it takes 15 minutes to boil water and when the second coil is switched on it takes 30 minutes. How long will it take to boil water when both the coils are used in (*a*) series and (*b*) in parallel.

Solution: According to Joule's law, heat produced in resistor R in time t when connected to a potential difference V is

$$H = \frac{V^2}{R}t$$

For 1st coil $$H = \frac{V^2}{R_1} \times 15 \qquad ...(i)$$

and for second coil $$H = \frac{V^2}{R_2} \times 30 \qquad ...(ii)$$

i.e. $$\frac{15}{R_1} = \frac{30}{R_2} \quad \text{or} \quad R_2 = 2R_1 \qquad ...(iii)$$

(*a*) When both are used in series

$$R_S = R_1 + R_2 = 3R_1$$

and heat produced is

$$H = \frac{V^2}{R_S}t = \frac{V^2}{3R_1}t$$

Equating with (*i*), we have

$$\frac{V^2}{R_1} \times 15 = \frac{V^2}{3R_1}t$$

or $$t = 45 \text{ minutes}$$

(*b*) When both are in parallel

$$R_p = \frac{R_1R_2}{R_1 + R_2} = \frac{R_1 \times 2R_1}{3R_1} = \frac{2}{3}R_1$$

and heat produced is

$$H = \frac{V^2}{R_p}t = \frac{3V^2}{2R_1}t$$

Equating with (*i*), we get

$$\frac{V^2}{R_1} \times 15 = \frac{3V^2}{2R_1} t$$

or $$t = \frac{15 \times 2}{3} = 10 \text{ minutes.}$$

Example 21.2: Find the heat developed in each of resistances shown in Fig. 21.10 in 1 minute.

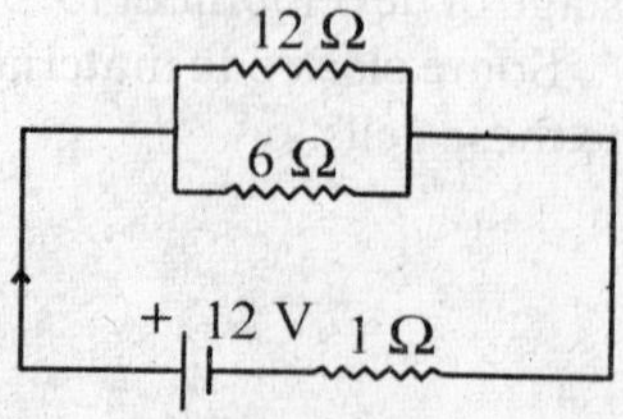

Fig. 21.10

Solution: The equivalent resistance of 6 Ω and 12 Ω in parallel is

$$= \frac{12 \times 6}{12 + 6} = 4\,\Omega$$

This is connected in series with 1 Ω resistor. Hence equivalent resistance of the circuit is

$$R = 4 + 1 = 5\,\Omega$$

The current through the battery is

$$i = \frac{12}{5} = 2.4 \text{ amp}$$

The current through 1 Ω resistor is 2.4 amp and the heat developed in this resistor is

$$H = i^2Rt = (2.4)^2(1)(60) = 345.6 \text{ Joules}$$

The current through 6 Ω resistor is

$$= \frac{2.4 \times 12}{12 + 6} = 1.6 \text{ A}$$

The heat developed is

$$H = i^2Rt = (1.6)^2 \times 6 \times 60 = 921.6 \text{ Joules}$$

The current through 12 Ω resistor is = 2.4 – 1.6 = 0.8 A
The heat developed in it

$$H = i^2Rt = (0.8)^2 \times 12 \times 60 = 460.8 \text{ Joules.}$$

Example 21.3: A room heater is rated 500 W; 220 V. (*a*) Find the resistance of its coil (*b*) If the supply voltage drops to 200 V. What will be power consumed? (*c*) If an electric bulb rated 100 W 220 V is connected in series with this heater, what will be the power consumed by the heater and by the bulb when the supply is at 220 V?

Solutim: (*a*) Power consumed by a coil of resistance *R* when connected across a supply *V* is

$$P = \frac{V^2}{R}$$

$$\therefore \quad R = \frac{V^2}{P} = \frac{(220)^2}{500} = 96.8\,\Omega$$

(*b*) When the supply voltage drops to 200 V, the power consumed will be

$$P = \frac{V^2}{R} = \frac{(200)^2}{96.8} = 413 \text{ W}$$

(c) The resistance of the 100 Watt, 220 V bulb is

$$R = \frac{(220)^2}{100} = 484\,\Omega$$

If this is connected in series with the heater of 96.8 Ω, the current will be

$$i = \frac{220}{484 + 96.8} = 0.379$$

Thus the power consumed by heater

$$= i^2R \times (0.379)^2 \times 96.8 = 13.9 \text{ W}$$

and that by the bulb = $i^2 \times 484 = (0.379)^2 \times 484 = 69.5$ W.

Example 21.4: Two tungsten lamps with resistance R_1 and R_2 respectively are connected first in parallel and then in series in a lighting circuit of negligible internal resistance. If $R_1 > R_2$, (a) Which lamp will glow more brightly when they are connected is parallel. (b) If the lamp of resistance R_1 now burns out how will the net illumination produced change? (c) Which lamp will glow more brightly when they are connected in series? (d) If the lamp of resistance R_2 now burns out and the lamp R_1 alone is plugged in, will the net illumination increase or decrease?

Solution:

(a) In parallel combination, the potential difference across each resistor is same and power consumed $P = \frac{V^2}{R}$. Then the lamp with lesser resistance, i.e. R_2 will glow more brightly.

(b) When both the bulbs are operating in parallel, total power consumed

$$P = \frac{V^2}{R_1} + \frac{V^2}{R_2}$$

when lamp of resistance R_1 burns out, lamp of resistance R_2 only will give light. Now power consumed is

$$P' = \frac{V^2}{R_2} < P$$

Hence the net illumination decreases.

(c) In series power consumed is i^2R. Since i is constant, hence bulb with greater resistance, i.e. R_1 will glow more brightly.

(d) When both the bulbs are operating in series, total power consumed

$$P = \frac{V^2}{R_1 + R_2}$$

when R_1 is alone in operation, power consumed

$$P' = \frac{V^2}{R_1} > P$$

So net illumination increases.

Example 21.5: A series circuit consists of three identical lamps, connected to a battery as shown in Fig. 21.11. When the switch S is closed, what happens (a) to the intensities of lamp A and B. (b) to

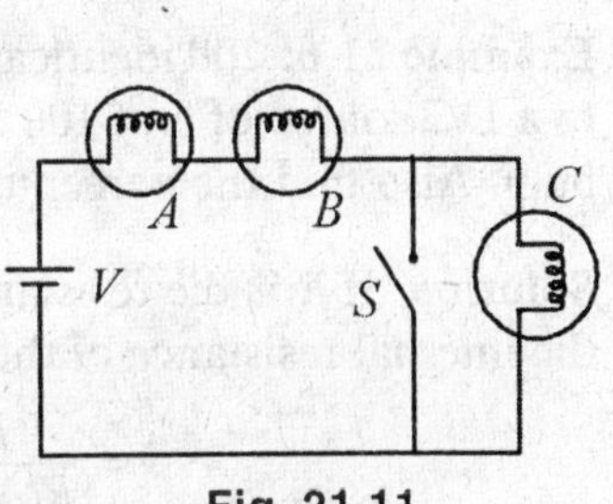

Fig. 21.11

the intensity of lamp C (*c*) to the current in the circuit and (*d*) to the voltage drop across the three lamps? Does the power dissipated in the circuit increase; decrease or remain the same?

Solution: The three bulbs are in series and identical. When switch is open

$$V_a = V_b = V_c = \frac{1}{3}V$$

and hence power consumed in each bulb is same, i.e.

$$P_a = P_b = P_c = \left(\frac{V}{3}\right)^3 \frac{1}{R} = \frac{V^2}{9R} \qquad \ldots(i)$$

where R is the resistance of each bulb.

Now when switch is closed

(*a*) The bulb C is short circuited and hence potential difference across it is zero and potential difference across A and B is

$$V_a' = V_b' = \frac{V}{2}$$

and power consumed in each bulb is

$$P_a' = P_b' = (1/2)^2 \frac{1}{R} = \frac{V^2}{4R} \qquad \ldots(ii)$$

Hence intensities of each bulb will increase and become 2.25 times of their initial value.

(*b*) Since potential difference across C is zero and hence no current passes through C and will give no light, i.e. $P_c' = 0$.

(*c*) Before switch is closed total resistance of circuit is $3R$ and current $i = \frac{V}{3R}$. After switch is closed, the total resistance changes to $2R$ and current changes to $i' = \frac{V}{2R}$, i.e. current in the circuit becomes 1.5 times.

(*d*) Before switch is closed the potential difference across each bulb is $V/3$. After switch is closed the potential difference across bulbs A and B become $V/2$ (each) i.e. across bulbs A and B potential difference increases but across C it reduces from $V/3$ to zero.

Since the total resistance of circuit changes from $3R$ to $2R$ hence power dissipated in circuit increases from $P = \frac{V^2}{3R}$ to $P' = \frac{V^2}{2R}$, i.e. power dissipated is 1.5 times the power dissipated before switch is on.

Example 21.6: 200 identical electrical bulbs each having resistance 400 Ω are connected in parallel to a DC source of emf 100 V and internal resistance 0.1 Ω. What is the power consumed by each bulb. Also find the percentage change in power consumed by each bulb if one bulb burns out.

Solution: If R is the resistance of each bulb then resistance of N bulbs in parallel is R/N and if r is the internal resistance of the cell, then current i is

$$i = \frac{E}{R/N + r}$$

This current is equally divided between N bulbs. Thus current through each bulb is

$$i' = \frac{i}{N} = \frac{1}{N}\frac{E}{R/N + r} = \frac{E}{R + Nr}$$

and power consumed in each bulb is

$$P = i'^2 R = \frac{E^2 R}{(R + Nr)^2}$$

$$= \frac{(100)^2(400)}{(400 + 200 \times 0.1)^2} = 0.227 \text{ W}$$

With one bulb burnt out, total number of bulbs are $(N - 1)$. Now power consumed is

$$P' = \frac{E^2 R}{\{R + (N-1)r\}^2}$$

Change in power consumed is

$$\Delta P = P' - P = \frac{E^2 R}{\{R + (N-1)r\}^2} - \frac{E^2 R}{\{R + Nr\}^2}$$

and

$$\frac{\Delta P}{P} = \left\{\frac{(R + Nr)^2}{\{R + (N-1)r\}^2} - 1\right\} = \left\{\frac{1}{\left\{1 - \frac{r}{R + Nr}\right\}^2} - 1\right\}$$

$$= \left\{\left(1 - \frac{r}{R + Nr}\right)^{-2} - 1\right\}$$

Since r is very small hence $\frac{r}{R + Nr} << 1$. Hence using binomial theorem, we have

$$\frac{\Delta P}{P} = 1 + \frac{2r}{R + Nr} - 1 = \frac{2r}{R + Nr} = \frac{2 \times 0.1}{400 + 200 \times 0.1} = 4.8 \times 10^{-4}$$

Hence percentage change $= \frac{\Delta P}{P} \times 100 = 4.8 \times 10^{-4} \times 10^2 = 0.048\%$.

Example 21.7: The circuit shown in Fig. 21.12 is closed at time $t = 0$. Calculate the total amount of heat generated in R_2 during the time in which the capacitor is charged to a voltage of 20 V.

Solution: The circuit is an RC circuit and charge on the capacitor any time t is

$$q = CE(1 - e^{-t/RC})$$

where R is the total resistance of circuit and

$$R = \frac{R_1 R_2}{R_1 + R_2} = \frac{20 \times 40}{60} = \frac{40}{3}\Omega$$

and current at any time t in the circuit is

$$i = \frac{dq}{dt} = \frac{E}{R}e^{-t/RC}$$

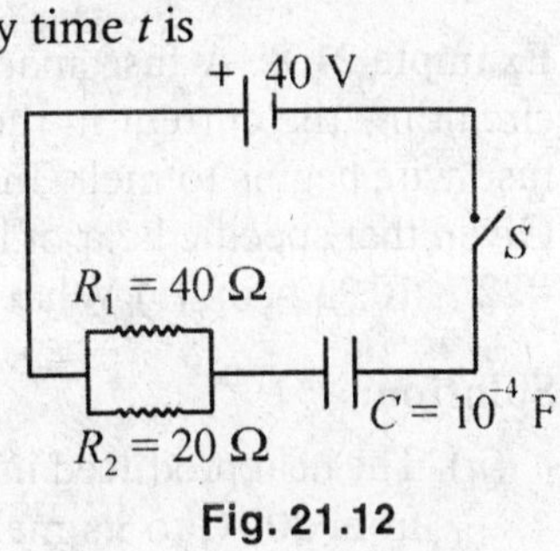

Fig. 21.12

Hence current through R_2 is

$$i_2 = \frac{iR_1}{R_1 + R_2} = \frac{i \times 40}{40 + 20} = \frac{2i}{3}$$

$$\therefore \quad i_2 = \frac{2}{3}\frac{E}{R}e^{-t/RC} = \frac{2}{3} \times \frac{40 \times 3}{40}e^{-t/RC} = 2\,e^{-t/RC}$$

Hence the heat generated in R_2 within time t and $t + dt$ is

$$dH = i_2^2 R_2 dt$$
$$= 4\,e^{-2t/RC} dt \times 20$$
$$= 80\,e^{-2t/RC}$$

Hence heat generated in resistance R_2 in time $t = 0$ to $t = t_0$ is

$$H = \int dH = \int_0^{t_0} 80\,e^{-2t/RC} = -\frac{80\,RC}{2}\left[e^{-2t/RC}\right]_0^{t_0}$$

$$= \frac{80RC}{2}\left[1 - e^{-2t_0/RC}\right]$$

where t_0 is the time in which capacitor acquires a voltage of 20 V

Charge on capacitor at any time t is

$$q = CE\,(1 - e^{-t/RC})$$

and potential $\quad V = \frac{q}{C} = E(1 - e^{-t/RC})$

But $E = 40$ V and $V = 20$ V or

$$\therefore \quad \frac{1}{2} = 1 - e^{-2t_0/RC}$$

or $\quad e^{-t_0/RC} = \frac{1}{2}$ or $t_0 = RC\log 2$

$\therefore$ Heat generated is

$$H = \frac{80 \times 40 \times 10^{-4}}{2 \times 3}\left[1 - e^{-2\log 2}\right]$$

But $\quad e^{-2\log 2} = \frac{1}{4} \qquad \therefore \; H = \frac{80 \times 40 \times 10^{-4}}{2 \times 3} \times \frac{3}{4} = 4 \times 10^{-2}$ J.

Example 21.8: A fuse made up of lead wire has an area of cross section 0.2 sq.mm. On short circuiting, the current in the fuse wire reaches 30 A. (*a*) How long after the short circuiting, the fuse wire begins to melt. Initial temperature of wire is 20° C and melting point of lead is 327° C. Given that specific heat of lead = 0.032 cal/g/°C, density of lead = 11.34 g per cc and resistivity =22 × 10^{-6} Ω-cm, (*b*) What will this time be if the length of the fuse wire is doubled?

Solution:

(*a*) The heat produced in the fuse wire due to current should be sufficient to raise its temperature from 20° C to its melting point (=370° C).

Then heat required is

$$H = mc\Delta\theta = (\pi r^2 Ld)\, C(\theta_m - \theta_r) \qquad \ldots(i)$$

Here r is radius of wire, L—its length and d density. θ_m is melting point, θ_r room temperature.

This heat is generated when current passes through the fuse wire. If i is current and R resistance of wire, then heat generated is

$$H = \frac{i^2 Rt}{J} = \frac{i^2 \rho L}{J\pi r^2} t \qquad \ldots(ii)$$

Equating the two equations, we obtain

$$\frac{i^2 \rho L}{J\pi r^2} t = \pi r^2 LdC(\theta_m - \theta_r)$$

or

$$t = \frac{\pi^2 r^4 dC(\theta_m - \theta_r)J}{i^2 \rho} \qquad \ldots(iii)$$

where J is mechanical equivalent of heat = 4.2 J/cal

given is $\pi r^2 = 0.2 \times 10^{-2} = 2 \times 10^{-3}$ sq.cm, $d = 11.34$ g/cc, $C = 0.032$ cal/g/°C, $Q_m = 370°$ C, $Q_r = 20°$ C, $i = 30$ amp and $\rho = 22 \times 10^{-6}\ \Omega$-cm

Substituting the values, we have

$$t = \frac{(2\times10^{-3})^2 \times 11.34 \times 0.032 \times (370°-20°) \times 4.2}{(30)^2 \times 22 \times 10^{-6}}$$

$$= \frac{4 \times 11.34 \times 32 \times 10^{-3} \times 350 \times 4.2}{9 \times 10^2 \times 22} = 0.11 \text{ sec}$$

(*b*) From equation (*iii*), we see that time is independent of the length of fuse wire. If length is doubled the time remains unchanged.

Example 21.9: The walls of a closed cubical box of edge 50 cm are made of a material of thickness 1 mm and thermal conductivity 4×10^{-4} cal s^{-1} cm^{-1} (°C)$^{-1}$. The interior of the box is maintained 100° C above the outside temperature by a heater placed inside the box and connected across 400 V DC. Calculate the resistance of the heater.

Solution: The rate of heat transmission through the wall is

$$Q = \frac{H}{t} = \frac{kA(\theta_2 - \theta_1)}{d}$$

where A is total surface area = 6 × area of each face = 6 × 50 × 50 cm^2 = 1.5×10^4 cm^2, $k = 4 \times 10^{-4}$ cal s^{-1}cm^{-1}(°C)$^{-1}$. $\theta_2 - \theta_1 = 100°$ C, $d = 1$ mm = 0.1 cm.

$$\therefore \qquad Q = \frac{4 \times 10^{-4} \times 1.5 \times 10^4 \times 100}{0.1} = 6000 \text{ cal}$$

To maintain constant temperature difference between inside and outside the box, the electric heater must produce the heat at the same rate. Heat produced by the heater in 1 sec is

$Q = i^2 R = \dfrac{V^2}{R}$. If V is in volts and R in Ω, Q is in Joules

$$\therefore \qquad Q = \frac{V^2}{JR} \text{ cal}$$

$$\therefore \quad \frac{400 \times 400}{4.2 \times R} = 6000$$

$$\text{or} \quad R = \frac{400 \times 400}{4.2 \times 6000} = 6.35\ \Omega.$$

Example 21.10: Two uniform wires of same material, each weighing 1 gm but one having the double length than the other are connected in series. They carry a current of 10 A, the length of the larger wire is 20 cm. Calculate the rate of consumption of energy in each of the two wires, which one will get hotter? Density of material of wire = 11.34 gm/cc, specific resistance of the material = 20.6×10^{-8} Ω-m.

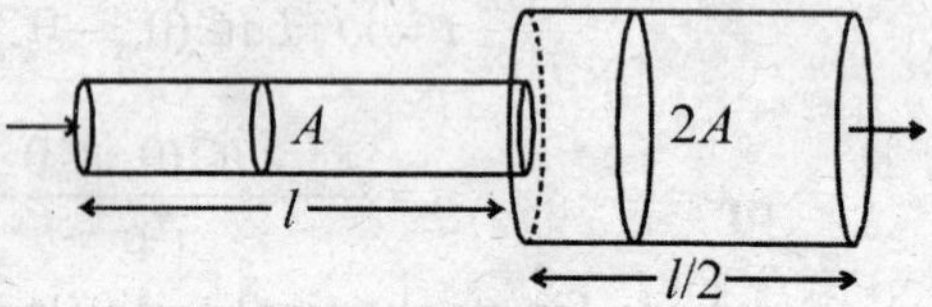

Fig. 21.13

Solution: Mass of wire = Volume × density

$$\therefore \quad m = Ald \quad \text{or} \quad A = \frac{m}{ld}$$

where A is area of cross-section, l, length and d density

and resistance of wire $R = \dfrac{\rho l}{A} = \dfrac{\rho l^2 d}{m}$.

The resistance of longer wire

$$R_1 = \frac{20.6 \times 10^{-8} \times (0.2)^2 (11.34 \times 10^3)}{10^{-3}} = 9.34 \times 10^{-2}\ \Omega$$

For given mass $R \propto l^2$. Hence resistance of shorter wire

$$R_2 = \frac{R_1}{4} = \frac{9.34 \times 10^{-2}}{4} = 2.34 \times 10^{-2}\ \Omega$$

Rate of consumption of energy in longer wire

$$P_1 = i^2 R_1 = (10)^2 \times (9.34 \times 10^{-2}) = 9.34 \text{ Watt}$$

Rate of consumption of energy in shorter wire

$$P_2 = i^2 R_2 = (10)^2 \times (2.34 \times 10^{-2}) = 2.34 \text{ Watt.}$$

Example 21.11: A 10 m long nichrome wire having 80 Ω resistance has current carrying capacity of 5 A. What is the maximum power which can be obtained as heat by the wire from a 200 V mains supply? How should the wire be connected to obtain maximum power?

Solution: When the wire is connected directly across the battery, the current in the wire is

$$i = \frac{V}{R} = \frac{200}{80} = 2.5 \text{ A}$$

and power obtained $P = i^2 R = (2.5)^2 \times 80 = 500$ Watt

The wire can carry maximum current of 5 A. Therefore to double the current, resistance should be halved. Thus the wire if divided in two equal parts (each having resistance 40 Ω) and two parts if connected in parallel, then current through each part will be 5 A.

Hence power consumed in each wire

$$P = i^2 R = (5)^2 \times 40 = 1000 \text{ W}$$

Hence total power consumed in both wires is 2 × 1000 = 2000 W.

Example 21.12: In a circuit shown in Fig. 21.14 (*a*), the capacitance of each capacitor is equal to C and resistance equal to R. One of the capacitor was connected to voltage V_0 by closing switch S_1. Now at $t = 0$ the switch S_1 is opened and S_2 is closed. Calculate (*a*) the current i in the circuit as a function of time t (*b*) the amount of heat generated in the resistance R.

Solution:

(*a*) Initial charge on capacitor 1 is $Q = CV_0$. When switch S_2 is closed, let charge q flows in time t The charge on two capacitors after time t is $(Q - q)$ and q (Fig. 21.14 (*b*)).

According to Kirchhoff's second law

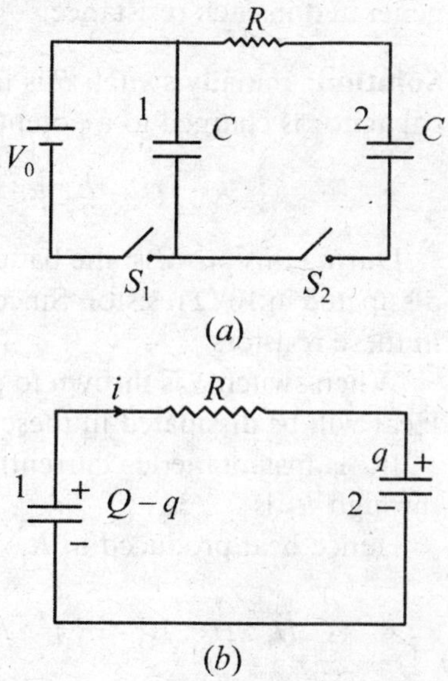

Fig. 21.14

$$\frac{Q-q}{C} - iR - \frac{q}{C} = 0$$

where i is the current at time t and $i = \dfrac{dq}{dt}$

or $$iR = \frac{Q-2q}{C}$$

and $$R\frac{dq}{dt} = \frac{Q-2q}{C} \quad \text{or} \quad \frac{dq}{Q-2q} = \frac{dt}{RC}$$

Integrating we get

$$\int_0^q \frac{dq}{Q-2q} = \frac{1}{RC}\int_0^t dt$$

or $$-\frac{1}{2}\log(Q-2q)\Big|_0^q = \frac{t}{RC}$$

$$\log\frac{(Q-2q)}{Q} = -\frac{2t}{RC}$$

and $$q = \frac{Q}{2}(1 - e^{-2t/RC})$$

But $Q = CV_0$, we have charge on capacitor 2 at any time t as

$$q = \frac{CV_0}{2}(1 - e^{-2t/RC})$$

and current i through the resistor R is

$$i = \frac{dq}{dt} = \frac{CV_0}{2}\frac{2}{RC}e^{-2t/RC} = \frac{V_0}{R}e^{-2t/RC}$$

(*b*) Heat produced

$$dH = i^2 R \cdot dt$$

and total heat produced

$$H = \int_0^\infty i^2 R dt = \frac{V_0^2}{R^2} R \int_0^\infty e^{-4t/RC} dt$$

$$= \frac{V_0^2}{R} e^{-4t/RC} \times \left(-\frac{RC}{4}\right)\Bigg|_0^{\infty}$$

$$= \frac{1}{4} V_0^2 C \qquad \because e^{-\infty} = 0 \text{ and } e^0 = 1.$$

Example 21.13: Two way switch S is used in the circuit as shown in Fig. 21.15. First the capacitor is charged by putting the switch S in position 1. Now switch is thrown in position 2. Find the heat generated in each resistance.

Solution: Initially switch S is in position 1. The battery charges the capacitor. In steady state the capacitor is charged to a potential difference of 60 V and the energy stored is

$$U = \frac{1}{2} CV^2 = \frac{1}{2} \times 0.1 \times 60 \times 60 = 180 \text{ J}$$

During this process, the battery has supplied the energy = 2 × 180 = 360 J. Energy 180 J is then dissipated in 10 Ω resistor. Since no current flows through R_1, R_2 and R_3, hence no heat is dissipated in these resistors.

When switch S is thrown to position 2, the capacitor discharges through resistors R_1, R_2 and R_3. Heat will be dissipated in these three resistors and not in 10 Ω resistor.

If i is instantaneous current, in the circuit, then current through R_1 is i, through R_2 is $i/3$ and through R_3 is $2i/3$.

Hence heat produced in R_1, R_2 and R_3, is in the ratio

$$H_1 : H_2 : H_3 = R_1 i^2 : R_2 (i/3)^2 : R_3 \left(\frac{2i}{3}\right)^2$$

$$= 4 : \frac{2}{3} : \frac{4}{3} = 6 : 1 : 2.$$

But total heat dissipated in all the resistors is 180 J.
Hence heat dissipated in R_1 = 120 J
Heat dissipated in R_2 = 20 J
and heat dissipated in R_3 = 40 J.

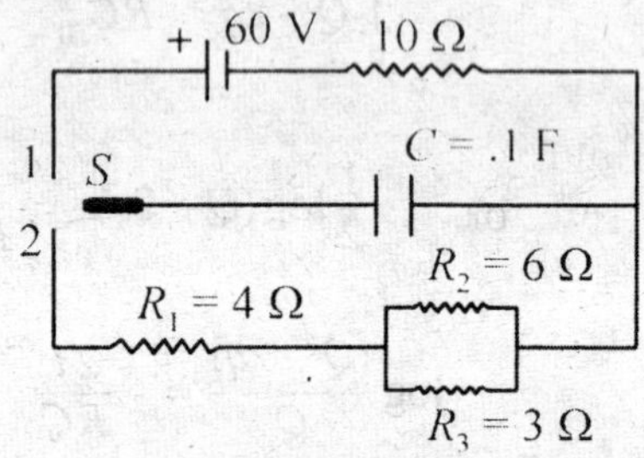

Fig. 21.15

Example 21.14: A 2 Ω resistor shown in Fig. 21.16 is dipped in to a calorimeter containing water. The heat capacity of the calorimeter together with water is 2000 J/K (*a*) If the circuit is active for 15 minutes what would be the rise in temperature of the water? (*b*) Suppose 6.0 Ω resistor gets burnt, what would be the rise in the temperature of the water in the next 15 minutes.

Solution:

(*a*) In the circuit resistance 2 Ω and 6 Ω are in parallel and their equivalent resistance is

$$= \frac{2 \times 6}{2 + 6} = 1.5\ \Omega$$

and total resistance of the circuit = 1.5 + 1 = 2.5 Ω and current in circuit is

$$i = \frac{6}{2.5} = 2.4 \text{ A}$$

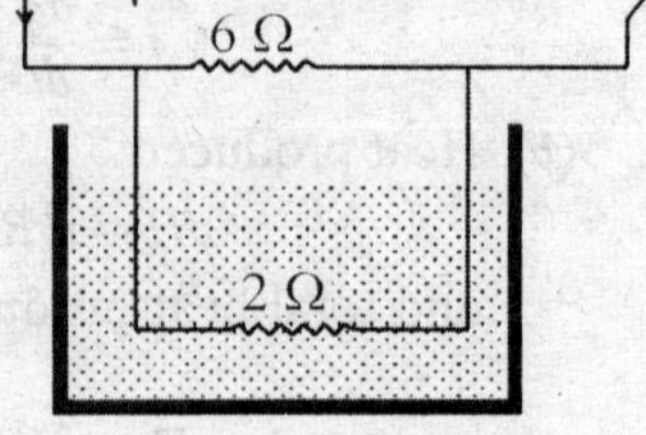

Fig. 21.16

and the current through 2 Ω resistor is

$$i = \frac{6i}{6+2} = \frac{6}{8} \times 2.4 = 1.8 \text{ A}$$

Hence heat generated in 15 minutes in 2 Ω resistor due to current 1.8 A through it is

$$H = i^2Rt = (1.8)^2\,(2)\,(15 \times 60) = 5832.0 \text{ J} \qquad \ldots(i)$$

This heat raises the temperature of water and calorimeter. If θ is the rise in temperature then heat required is

$$H = C\theta = 2000 \times \theta \text{ J} \qquad \ldots(ii)$$

Equating (*i*) and (*ii*), we get

$$2000\theta = 5832.0$$

or $$\theta = \frac{5832}{2000} = 2.916°\text{C (2.916 K rise in temperature)}$$

(*b*) When 6 Ω resistor burns, the total resistance becomes 3 Ω and a current of $i = \frac{6}{3} = 2$ A passes through resistor 2 Ω producing heat in it. Heat produced in 15 minutes is

$$H = i^2Rt = (2)^2(2)(15 \times 60) = 7100 \text{ J} \qquad \ldots(iii)$$

If now θ is further rise in temperature, then heat required is

$$H = C\theta = 2000\ \theta \text{ J}$$

Equating with (*iii*), we have

$$2000\ \theta = 7200$$

or $$\theta = 3.6°\text{ C.}$$

Example 21.15: A plate of area 10 cm^2 is to be electroplated with copper (density = 9 × 10^3 kg/m^3) to a thickness of 10 micrometers on both sides using a cell of 12 V. Calculate the energy spent by the cell in the process of deposition. If this energy is used to heat 100 gm water, calculate the rise in temperature of water. ECE of copper = 3 × 10^{-7} kg/C and specific heat of water = 4200 J/kg-K.

Solution: Volume of the copper deposited on one side

$$V = At = 10 \times 10^{-4} \times 10 \times 10^{-6} = 10^{-8} \text{ m}^3$$

Hence mass of this electroplated copper $m = V\rho = 10^{-8} \times 9 \times 10^3 = 9 \times 10^{-5}$ kg. Since it is to be electroplated on both the sides, hence total mass deposited is

$$= 2m = 2 \times 9 \times 10^{-5} = 1.8 \times 10^{-4} \text{ kg}$$

Using Faradays first law, we have

mass deposited = ZQ

Where Z is ECE of copper and Q is the charge passed through the voltameter

Hence $$Q = \frac{\text{mass deposited}}{Z}$$

$$= \frac{1.8 \times 10^{-4}}{3 \times 10^{-7}} = 0.6 \times 10^3 \text{ C}$$

∴ Energy required is QV where V is the potential difference across the voltameter

$\therefore$ Energy $E = QV = 0.6 \times 10^3 \times 12 = 7.2$ kJ

If this energy is used to heat 100 gm of water and rise in temperature be θ then

$$\text{Energy} = \text{mass} \times \text{sp. heat} \times \text{rise in temperature}$$

$$\therefore \quad 7.2 \times 10^3 = 10^{-1} \times 4200 \times \theta$$

or
$$\theta = \frac{7.2 \times 10^3}{4.2 \times 10^2} = 17.14 \text{ K}.$$

Example 21.16: The coil of an electric bulb takes 40 Watts to start glowing. If more than 40 W is supplied 60 % of the extra power is converted into light and remaining into heat. The bulb consumes 100 W at 220 V. Find the percentage drop in the light intensity at a point if the supply voltage changes from 220 V to 200 V.

Solution: The bulb consumes 100 W at 220 V.

Hence resistance of coil

$$R = \frac{V^2}{W} = \frac{220 \times 220}{100} = 484\,\Omega$$

when 220 V is supplied extra power over 40 watt is 60 W.

Hence intensity I is proportional to 60×0.6

i.e. $\quad I = k(60 \times 0.6) \quad$ ...(*i*)

When voltage drops to 200 V the power consumption is

$$P = \frac{200 \times 200}{484} = 82.6 \text{ W}$$

and extra power over 40 Watt = 82.6 – 40 = 42.6 W

and intensity I' now is proportional to 42.6×0.6

i.e. $\quad I' = k(42.6 \times 0.6) \quad$ ...(*ii*)

Hence percentage change in intensity is

$$\frac{I - I'}{I} \times 100 = \frac{k(160 \times 0.6) - k(42.6 \times 0.6)}{k(60 \times 0.6)} \times 100 = \frac{60 - 42.6}{60} \times 100 = 29\%.$$

Example 21.17: Along a metal rod of cross-section 2 cm square, there is a uniform temperature gradient of 1° C/cm. The specific resistance of the material of the rod is 160×10^{-6} Ω-cm. The temperature gradient remains unaltered when a current of 0.05 A is sent from the hot end to the cold end. Calculate the Thomson's coefficient for this metal.

Solution: When current flows through the conductor, the temperature gradient remains constant. This implies that the heat produced by Joule's effect is equal to the heat absorbed due to Thomson effect.

Resistance of 1 cm length of metal conductor $R = \dfrac{\rho l}{A}$

or
$$R = \frac{160 \times 10^{-6} \times 1}{2 \times 2} = 40 \times 10^{-6}\Omega$$

Hence potential difference across 1 cm length of the conductor

$$V = iR = 0.05 \times 40 \times 10^{-6} = 2 \times 10^{-6} \text{ V}$$

But by definition of Thomson's coefficient, i.e potential difference between two points having temperature difference of 1° C is numerically equal to the Thomson's coefficient. Since here temperature gradient is 1° C/cm. Hence potential difference across 1 cm length of conductor gives Thomson's coefficient. Hence $\sigma = 2 \times 10^{-6}$ V/°C.

Example 21.18: A thermocouple used with one junction at 0° C has an emf given by equation $E = \alpha\theta + \frac{1}{2}\beta\theta^2$. Where θ is in °C, temperature of hot junction. If the emf has the value 4.05 and 9.85 mV for θ = 100° C and θ = 300° C respectively, what is the emf at θ = 500° C?

Solution: Equation for thermo emf is $E = \alpha\theta + \frac{1}{2}\beta\theta^2$

when θ = 100° C, E = 4.05 mV. The equation then becomes

$$4.05 = 100\,\alpha + \frac{\beta}{2}(100)^2 \qquad \ldots(i)$$

and when θ = 300° C, E = 9.85 mV, the equation then becomes

$$9.85 = 300\,\alpha + \frac{\beta}{2}(300)^2 \qquad \ldots(ii)$$

equation (*i*) × 3 gives

$$12.15 = 300\,\alpha + \frac{3\beta}{2}(100)^2 \qquad \ldots(iii)$$

(*ii*)-(*iii*) gives

$$-2.3 = 3\,\beta \times 10^4 \text{ or } \beta = -\frac{2.3}{3} \times 10^{-4}\ \text{mV}/(°\text{C})^2$$

Substituting in (*i*), we get

$$4.05 = 100\,\alpha - \frac{2.3}{6}$$

or

$$\alpha = 4.43 \times 10^{-2}\ \text{mV/°C}$$

Hence thermo emf at 500° C is

$$E = \alpha\theta + \frac{1}{2}\beta\theta^2$$

or

$$E = 4.43 \times 10^{-2} \times 500 - \frac{2.3}{6} \times 10^{-4}\,(500)^2$$

$$= 22.15 - 9.58$$

$$= 12.57\ \text{mV}.$$

Example 21.19: A heating coil of 2000 W is immersed in an electric kettle. How much time will it take in raising the temperature of 1 litre of water from 4° C to 100° C? Only 80% of the thermal energy produced is used in raising the temperature of water.

Solution: Volume of water = 1000 cc, and mass of water 1000 g (since density of water is 1 g/cc)

Rise in temperature = 100 – 4 = 96° C

Specific heat of water = 1 cal/g/°C

Hence heat energy required to raise temperature from 4° C to 100° C is

$$H = mS\Delta\theta = 1000 \times 1 \times 96 = 96000 \text{ cal}$$

$$= 96000 \times 4.2 \text{ Joules}$$

Heat generated in coil in t sec is = Power × time

$$= 2000 \times t \text{ J}$$

80% of this heat is used to raise the temperature, i.e. heat energy = $2000 \times t \times 0.8$ J is used to raise the temperature. Equating the two we have

$$2000t \times 0.8 = 96000 \times 4.2$$

or $$t = \frac{96000 \times 4.2}{2000 \times 0.8} = 252 \text{ sec.}$$

Example 21.20: A house is fitted with 20 lamps of 60 watt each, 10 fans consuming 0.5 A each and an electric kettle of resistance 110 Ω. If the power supply is at 220 V and costs Rs 3/- per kWh, calculate the bill for November while running these appliances for 6 hours a day.

Solution: Here n_1 = 20 lamps, P_1 = 60 W

Total power consumed by all bulbs = 20 × 60 = 1200 W

n_2 = 10 fans consuming current of 0.5 A at 220 V

Total power consumed by all fans = 10 × 0.5 × 220 = 1100 W

$$n_3 = 1 \text{ kettle consuming power } = \frac{V^2}{R} = \frac{(220)^2}{110} = 440 \text{ W}$$

Hence total power consumption = 1200 + 1100 + 440

= 2740 W

All these operate for 6 hours a day, hence the energy consumed per day = 2740 × 6 Joule-hr = 2.740 × 6 kWh.

November month has 30 days, hence total energy consumed is

$$E = 2.740 \times 6 \times 30 \text{ kWh}$$

Hence cost is = 2.740 × 6 × 30 × 3 = 1479.6, i.e. Rs 1480.00.

PROBLEMS

1. A copper wire having cross-sectional area 0.5 mm^2 and a length of 0.1 m is initially at 25° C and is thermally insulated from the surrounding. If a current of 10 A is setup in this wire,
   (*i*) find the time in which the wire will start melting. The change in resistance with temperature may be neglected,
   (*ii*) what will be the time taken if the length of the wire is doubled? Given for copper, the density = 9×10^3 kg/m^3, specific heat = 9×10^{-2} kcal kg^{-1}(°C)$^{-1}$, melting point = 1075° C and specific resistance = 1.6×10^{-8} Ohm-meter

   [**Ans.** (*i*) t = 588.14 sec (*ii*) time is independent of length]

2. A piece of metal weighing 200 g is to be electroplated with 5% of its weight in gold. If the strength of the available current is 2 amperes, how long would it take to deposit the required amount of gold? *ECE* of hydrogen is 0.1044×10^{-4} g/C, atomic weight of gold is 197.1 and that of hydrogen is 1.008. **[Ans.** 2.04 hr]

3. Silver and copper voltameters are connected in parallel to a 12 V battery of negligible internal resistance. If in 30 minutes 1.0 g silver and 1.8 g of copper are deposited at what rate energy is being supplied by the battery? *ECE* of silver and copper are respectively 11.2×10^{-4} g/C and 6.6×10^{-4} g/C. **[Ans.** $W = 24.156$ W]

4. Figure 21.17 shows an electrolyte of AgCl through which a current is passed. It is observed that 2.68 g of silver is deposited in 10 minutes on the cathode. Find the heat developed in the 20 Ω resistor during this period. Atomic weight of silver is 108. **[Ans.** 190 kJ]

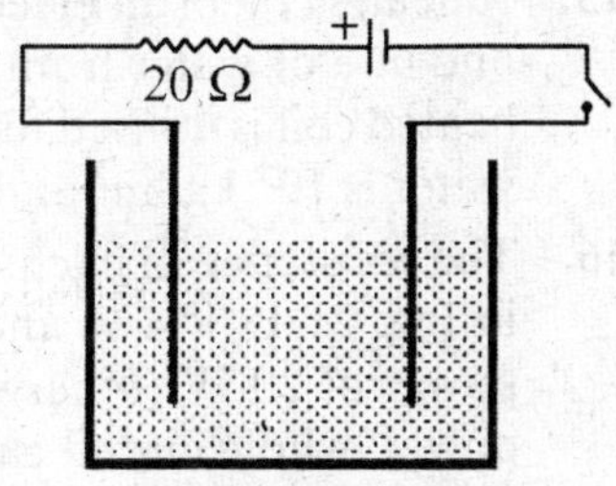

Fig. 21.17

5. An electroplating unit, plates 3.0 g of silver on a brass plate in 3.0 minutes. Find the current used by the unit. *ECE* of silver is 1.12×10^{-3} kg/C. **[Ans.** $i = 15$ A]

6. The temperature of the junctions of a bismuth-silver thermocouple are maintained at 0° C and 0.001° C. Find the thermo emf (Seebeck emf) developed. For bismuth-silver, $a = -46 \times 10^{-6}$ V/deg and $b = -0.48 \times 10^{-6}$ V/deg^2. **[Ans.** -4.6×10^{-8} V]

7. Find the neutral temperature and temperature of inversion of copper-iron couple if the reference junction is kept at 0° C. Given $a_{Cu} = 2.76$ μV/°C, $b_{Cu} = 0.012$ μV/°C. $a_{Fe} = 16.6$ μV/°C, $b_{Fe} = -0.03$ μV/(°C)^2. All the values are with respect to lead. **[Ans.** $T_n = 330°$ C, $T_i = = 660°$ C]

8. A current of 1 A is passed through a dilute solution of sulphuric acid for sometime to liberate 1 g of oxygen. How much hydrogen is liberated during this period? How long was the current passed? Faraday constant is 96500 C/mole. **[Ans.** $m = 0.125$ g, $t = 3$ hr, 20 min]

9. In a copper-constantan thermocouple, the thermo emf E and temperatures θ of the junction (with cold junction at 0° C) are found to satisfy the equation $E = \alpha\theta + \beta\theta^2$ where E is in μV, θ in °C, $\alpha = 41$ μV/°C and $\beta = 0.041$ μV/(°C)^2. What is the temperature of the hot junction when thermo emf is measured as 5.5 mV. **[Ans.** 120° C]

10. The thermo emf of a thermocouple is 40 μV per degree Kelvin at room temperature. A galvanometer of resistance 100 Ω capable of detecting current as low an 10 μA is connected with one such thermocouple. Calculate the smallest temperature that can be detected by such a thermocouple. **[Ans.** 25 k]

11. A thermocouple of resistance 1.6 Ω connected in series with a galvanometer of 8 Ω resistance. The thermo couple develops an emf of 10 μV per degree temperature difference between two junctions. When one junction is kept at 0° C and the other in molten metal, the galvanometer records 8 mV. Find the temperature of molten metal when emf varies linearly with temperature difference. **[Ans.** 960° C]

12. The mass of hydrogen evolved per Coulomb of electricity passed through water voltameter is 1.05×10^{-8} kg/Coul. The corresponding figures for copper deposited in copper voltameter and silver deposited in silver voltameter are 3.29×10^{-7} kg/C and 1.12×10^{-6} kg/C respectively. What conclusion is implied by these results. Predict the amount of iron deposited per Coulomb in voltameter with Fe electrodes in $FeCl_3$. Relative atomic mass of hydrogen, coppers, silver and iron are 1.008, 63.54, 107.9 and 55.85 respectively. **[Ans.** 1.93×10^{-7} kg/C]

13. A steady current deposits 0.15 g of copper in 15 minutes from $CuSO_4$ solution. An ammeter in series with the voltameter reads 0.6 A. What correction is to be applied to the ammeter reading? Given *ECE* of copper is 3.3×10^{-4} kg/C. **[Ans.** –0.1 A]

14. Twenty one electric bulbs are connected in series with mains of a 220 V supply. After one bulb is fused, the remaining 20 bulbs are again connected in series across the same mains. By what percentage will the illumination of (*i*) a bulb change (*ii*) all the bulbs change? **[Ans.** (*i*) 10.25% (*ii*) 5%]

15. A heater coil of an electric kettle is rated 2000 W at 200 V. How much time will it take to heat one litre of water from 20° C to 100° C. Assuming that the entire energy liberated from the heated coil is utilised for heating the water? Also calculate the resistance of the coil. Density of water is 10^3 kg/m^3 and specific heat of water 10^3 cal/kg(K)$^{-1}$. **[Ans.** (*i*) 168 S, (*ii*) 20 Ω]

16. The resistance of each of the three wires joined an shown in Fig. 21.18 is 4 Ω and each one can have a maximum power of 20 W (otherwise it will melt). What maximum power, whole circuit can take? **[Ans.** 30 W]

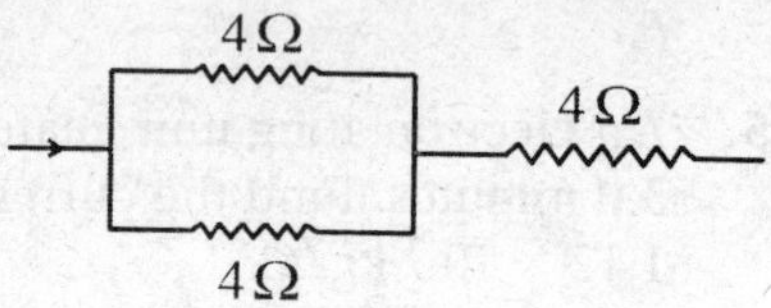

Fig. 21.18

17. A series of 6 lead accumulators of emf 2.0 V and internal resistance 0.5 Ω is charged by a 100 V DC supply. (*a*) What series resistance should be used in the changing circuit in order to limit the current to 8.0 A? Using the required resistor, obtain (*b*) the power supplied by the DC source (*c*) the power dissipated as head (*d*) the chemical energy stored in the battery in 15 min. **[Ans.** (*i*) R = 80 Ω (*b*) 8000 Watt (*c*) 707 Watt (*d*) 86400 J]

18. A copper kettle weighing 1 kg contains 900 g water at 20°C. It takes 12 minutes to raise the temperature to 100° C. If electric energy is supplied at 210 V, calculate the strength of current, assuming that 10% of heat is lost in radiation. Specific heat of copper is 0.1 cal/g-°C. **[Ans.** i = 1.47 A]

19. An electric kettle boils some water in 16 minutes. Due to some defect 10% turns of heating coil of the kettle are removed. Find the time taken by the coil to boil the same mass of water. **[Ans.** 14.4 minutes]

20. A metal rod of cross-section 10^{-4} m^2 and specific resistance of material of rod is 150×10^{-8} Ω-m has a uniform temperature gradient of 1°C/m. When a current of 0.05 A is sent through hot junction to cold, temperature gradient remains unaltered. Find the Thomson-coefficient of the material of rod. **[Ans.** 7.5×10^{-4} V]

SHORT ANSWER TYPE QUESTIONS

1. The cold junction of a thermocouple is maintained at 10° C. No thermo emf is developed when the hot junction is maintained at 530°C. Find the neutral temperature.
[Ans. Since thermo emf at temperature 530° C is zero means temperature of inversion T_i = 530° C. Temperature of cold junction T_c = 10° C].

Hence t_2 the neutral temperature $T_n = \dfrac{T_i + T_c}{2} = \dfrac{530 + 10}{2} = 270°$ C

2. Find coefficients a and b for a bismuth-iron couple. Given is the coefficient a and b for metals with lead. For bismuth a = – 43.7 μV/°C, b = –0.47 μV/(°C)^2 and for iron a = 16.6 mV/°C and b = – 0.03 mV/(°C)^2.

[**Ans.** Using the law of intermediate metals, we have, $a_{\text{Bi-Fe}} = a_{\text{Bi-Pb}} - a_{\text{Fe-Pb}} = -43.7 - 16.6 = -60.3\ \mu\text{V/°C}$ and $b_{\text{Bi-Fe}} = b_{\text{Bi-Pb}} - b_{\text{Fe-Pb}} = -0.47 + 0.03 = -0.44\ \mu\text{V/°C}$].

3. Two heater coils made of the same material are connected in parallel across the mains, the length and diameter of one coil is double of the other. Which of them will produce more heat?

[**Ans.** Rate at which heat is produced is $H = \frac{V^2}{R}$ but $R = \frac{\rho l}{\pi r^2}$, $\therefore H = \frac{V^2 \pi r^2}{\rho l}$, $\frac{H_1}{H_2} = \frac{V^2 \pi r_1^2 \rho l_2}{\rho l_1 V^2 \pi r_2^2}$ $= \frac{l_2}{l_1}\left(\frac{r_1}{r_2}\right)^2 = \frac{2\,l_1}{l_1}\left(\frac{r_1}{2\,r_1}\right)^2 = \frac{1}{2}$, $\therefore H_2 = 2H_1$. Thus heat produced in thicker wire is more than the thinner wire]

4. The n resistors of equal value are connected in series across a source of emf. The power dissipated is P. Find the power dissipated if the same resistors are connected in parallel across the same source.

[**Ans.** In series total resistance $R_s = nR$ where R is the resistance of each resistor. Hence power dissipated $P_s = \frac{V^2}{nR}$, where V is the voltage of source. In parallel total resistance $R_p = R/n$ and power dissipated $= \frac{V^2}{R/n}$, i.e. $P_p = \frac{nV^2}{R}$, Hence $\frac{P_s}{P_p} = \frac{1}{n^2}$ or $P_p = n^2 P_s$]

5. Two electric bulbs of 100 W and 200 W are given. Which one will glow brighter when they are connected to the mains (*a*) in series (*b*) in parallel. The operating voltage for both bulbs is same.

[**Ans.** (*a*) In series same current passes through both the bulbs. Hence power dissipated $P = i^2R$. Higher the resistance, greater the power consumption and higher glow. Hence in series 100 Watt bulb will glow brighter as its resistance is more (double) than that of 200 Watt bulb.

(*b*) In parallel, potential difference across both the bulbs is same and power consumed $P = \frac{V^2}{R}$. Lower the resistance, greater is the power consumption and higher the glow. Hence 200 Watt bulb will glow brighter].

6. Calculate the electric current required to deposit 0.972 of chromium in 1 hour. *ECE* of chromium is 0.00018 g/C.

[**Ans.** We know that mass deposited, $m = Zit$ or $i = \frac{m}{Zt} = \frac{0.972}{0.00018 \times 60 \times 60} = \frac{0.972}{1.8 \times 0.36}$ $= 1.5$ A]

7. A current of 30 A is registered when the terminals of a dry cell of emf 1.5 V are connected through an ammeter. The ammeter has a resistance of 0.01 Ω. Find the amount of heat produced in the battery in 10 seconds.

[**Ans.** If r is the internal resistance of the battery then total resistance $R = r + 0.01$. Current $i = \frac{E}{R}$ or $30 = \frac{1.5}{r + 0.01}$ or $r + 0.01 = 0.05$ or $r = 0.04$. Hence heat generated in battery $= i^2rt$ $= (30)^2(0.04) \times 10 = 360$ J]

8. If two bulbs of 25 W and 100 W rated at 220 V are connected in series across a 440 V supply. Will both bulbs fuse? If not which one?

[**Ans.** Resistance (R_1) of 25 W bulb is 4 times the resistance (R_2) of 100 W bulb hence potential difference across 25 W bulbs is 4 times (=352 V) than the potential difference across 100 W bulb (= 88 V). Hence 25 W bulbs will fuse]

9. A standard 50 W electric bulb in series with a room heater is connected across the mains. If the 50 W bulb is replaced by 100 W bulbs, will the heater output be larger, smaller or remain the same?

[**Ans.** Since heater and bulb are is series, hence same current flows through them (i) Hence power consumed by heater = i^2R, where R is resistance of heater coil. If 50 W bulb is replaced by a 100 W bulb, the total resistance of circuit decreases and the current in the circuit (i') increases. Hence power consumed by heater ($=i'^2R$) increases. Hence heater output increases]

10. What will happen to the light of the bulb (Fig. 21.19) after the switch is closed?

[**Ans.** In R-C circuit at $t = 0$ capacitor behaves as if its resistance is zero. Hence current is maximum $i = E/R$. Where R is resistance of bulb and at $t = \infty$ it behaves like an infinite resistance and current is zero. Thus intensity of bulb decreases exponentially from maximum to zero].

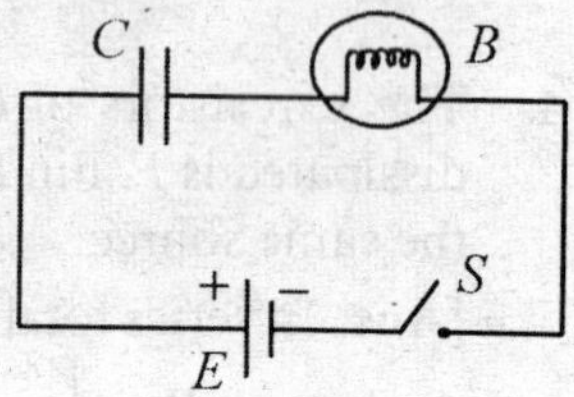

Fig. 21.19

11. Two resistors R_1 and R_2 may be connected in series or in parallel across a bettery of zero internal resistance. It is required that the Joule heating for parallel combination be 5 times that for the series combination. If $R_1 = 100\ \Omega$ find R_2.

[**Ans.** Joule heat in parallel combination $= \dfrac{V^2(R_1 + R_2)}{R_1R_2}$ and series combination $= \dfrac{V^2}{(R_1 + R_2)}$

$$\therefore \frac{R_1 + R_2}{R_1R_2} = \frac{5}{R_1 + R_2} \text{ or } (R_1 + R_2)^2 = 5\,R_1R_2 \text{ or } R_1^2 + R_2^2 - 3\,R_1R_2 = 0 \text{ or } R_2 = \frac{3R_1 \pm \sqrt{9R_1^2 - 4R_1^2}}{2}$$

$$\text{or } R_2 = \frac{3R_1 \pm \sqrt{5}\,R_1}{2}. \text{ Since } R_1 = 100\ \Omega, \text{ hence } R_2 = \frac{300 \pm 100\sqrt{5}}{2} = 50\left(3 \pm \sqrt{5}\right)\Omega\ ]$$

12. In the circuit shown in Fig. 21.20. If heat generated in 5 Ω resistor is H per sec, find the heat generated in 4 Ω resistor in 1 sec.

[**Ans.** If i is the current through 5 Ω resistor then heat generated in it $H = 5\,i^2$. The current through 4 Ω resistor then is $i/2$. Hence heat generated in 1 sec is $H' = 4(i/2)^2$

$$= \frac{4\,i^2}{4} = i^2 = H/5\,]$$

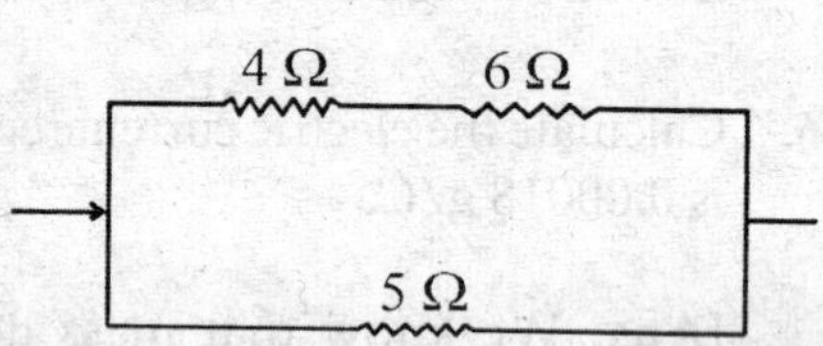

Fig. 21.20

13. Three equal resistors each of $R\ \Omega$ are connected as shown in Fig. 21.21. A battery of 2 V and internal resistance 0.1 is connected across the circuit. Calculate the value of R for which the heat generated in the circuit is maximum.

[**Ans.** The three resistors are connected in parallel, hence equivalent resistance is $R/3$. For maximum power delivered to load the external resistance should be equal to internal resistance, i.e. $\dfrac{R}{3} = 0.1$ or $R = 0.3\ \Omega$].

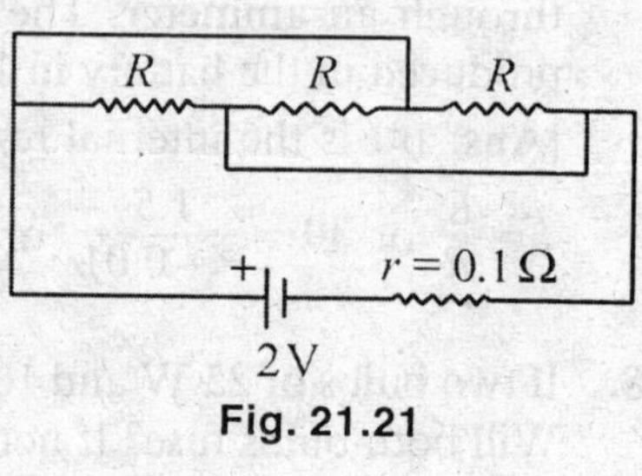

Fig. 21.21

14. An electric bulb rated 500 W at 100 V is used in a circuit having a 200 V supply. What resistance R must be put in series with the bulb so that bulb delivers 500 W?

[**Ans.** The bulb has resistance $R = \dfrac{V^2}{W} = \dfrac{100 \times 100}{500} = 20\ \Omega$. When connected with supply of 200 V, the resistance should be such that potential difference across it should also be 120 V. Hence resistance required to be added in series is 20 Ω].

15. Two fuse wires are made of the same material. The radius of first and second wires are $r_1 = 1$ mm and $r_2 = 4$ mm. The first wire blows when a current of 10 A passes through it. Find the current required to blow the second wire.

[**Ans.** For a fuse wire of a given material $i^2 \propto r^3$, i.e. $\left(\dfrac{i_2}{i_1}\right)^2 = \left(\dfrac{r_2}{r_1}\right)^3$ or $\dfrac{i_2}{i_1} = \left(\dfrac{r_2}{r_1}\right)^{3/2}$. or

$$i_2 = i_1\left(\frac{r_2}{r_1}\right)^{3/2} = 10\left(\frac{4}{1}\right)^{3/2} = 80\ \text{A}\Big]$$

16. In producing chlorine through electrolysis 100 kW power at 125 V is being consumed. How much chlorine per minute is liberated [*ECE* for chlorine is 0.367×10^{-6} kg/C.

[**Ans.** Current in the circuit is $i = W/V = \dfrac{100 \times 10^3}{125} = 800$ A. Faraday's first law gives $m = Zit$

$$= 0.367 \times 10^{-6} \times 800 \times 60 = 17.6 \times 10^{-3}\ \text{kg/C}]$$

17. Compute the time required for a current of 3 A to decompose electrolytically 18 g of water. Equivalent weight of hydrogen is 1.0 and of oxygen is 8.0 and 1 F = 96500 C.

[**Ans.** During decomposition of water $H_2O \rightarrow H_2 + \frac{1}{2}O_2$, one mole i.e. 18 g of water on decomposition give 2 g of hydrogen and 16 g of oxygen. From first law $m = Zit$. But $Z = \dfrac{E}{F}$,

E is equivalent weight, hence $t = \dfrac{m}{Zi} = \dfrac{mF}{Ei} = \dfrac{2 \times 96500}{1 \times 3} = 17.87$ Hrs]

18. A current passing through copper plating bath and a silver plating bath in series deposits 0.4 g of copper in 10 minutes. How much silver is deposited. Atomic weights of silver and copper are 108 and 64 respectively.

[**Ans.** Faraday's second law gives that mass deposited is proportional to the equivalent weight.

Hence $\dfrac{m_c}{m_s} = \dfrac{E_c}{E_s}$ or $m_s = m_c\dfrac{E_s}{E_c} = \dfrac{0.4 \times 108}{32} = 1.35$ g. (Equivalent weight = Atomic weight/ Valance)]

19. Thermo emf E of a copper constantan thermocouple and the temperature θ of hot junction (with cold junction at 0° C) are found to satisfy approximately the relation $E = \alpha\theta + \beta\theta^2$, where E is in μV and θ in °C, $\alpha = 41\ \mu\text{V}°\text{C}^{-1}$ and $\beta = 0.041\ \mu\text{V}(°\text{C})^{-2}$. Find the temperature of hot junction when the thermo emf is measured to be 5.5 mV.

[**Ans.** $E = \alpha\theta + \beta\theta^2$, $E = 5.5$ mV $= 5.5 \times 10^{+3}$ μV. Substituting the values, we have 5.5×10^3

$= 41\theta + 0.041\theta^2$ or $0.041\theta^2 + 41\theta - 5.5 \times 10^3 = 0$ or $\theta = \dfrac{-41 \pm \sqrt{(41)^2 + 4(5.5 \times 10^3) \times 0.041}}{2 \times .041}$.

Solving θ = 120° C]

20. A nichrome heating element across 230 V supply consumes 1.5 kW of power and heats up to a temperature of 750° C. A tungsten bulb across the same supply operates at much higher temperature of 1600° C in order to be able to emit light. Does it mean that the tungsten bulb necessarily consumes more power?

[**Ans.** No. The steady state temperature acquired by a resistor depends not only on the power consumed, but also on the characteristics such as surface area, emissivity, etc. the factors that determine its power loss due to radiation]

21. The charge q flowing through resistance R varies with time t as $q = at - bt^2$. Find the total heat produced in the resistance R.

[**Ans.** The instantaneous current through resister R is $i = \frac{dq}{dt} = a - 2\,bt$ and the current will be zero at time $t = \frac{a}{2\,b}$. The heat generated in time interval t and $t + dt$ is $dH = i^2Rdt = (a - 2bt)^2Rdt$.

Hence total heat generated is $H = \int dH = \int_0^{a/2b} (a - 2bt)^2 Rdt = R\int_0^{a/2b} (a^2 - 4abt + 4b^2t^2)\,dt$

$$= R\left[a^2t - \frac{4abt^2}{2} + \frac{4b^2t^3}{3}\right]_0^{a/2b} \qquad H = R\left[a^2\frac{a}{2\,b} - 2\,ab\left(\frac{a}{2\,b}\right)^2 + \frac{4b^2}{3}\left(\frac{9}{2b}\right)^2\right]$$

$$= R\left(\frac{a^3}{2\,b} - \frac{a^3}{2\,b} + \frac{a^3}{6\,b}\right) = \frac{a^3R}{6\,b}\Bigg]$$

22. Calculate the smallest temperature difference that can be measured with a combination of a thermocouple of thermo emf 30 μV per degree and a galvanometer of 50 Ω resistance capable of measuring a minimum current of 3×10^{-7} A.

[**Ans.** Smallest potential difference that the given galvanometer can measure is $V = 30 \times 10^{-7} \times 50 = 15 \times 10^{-6}$. If θ is the temperature difference between two junctions of a thermo couple, then voltage developed is $30 \times \theta$ μV $= 30 \times \theta \times 10^{-6}$ V. Equating the two, $\theta = 0.5°$ C]

23. A fuse wire with a circular cross-sectional radius of 0.02 mm blows with a current of 5 A. Find the current required for another wire of same material but of cross-section 0.04 mm to blow it.

$\Bigg[$**Ans.** Radii of the two fuse wires are in ratio $\frac{r_1}{r_2} = \sqrt{\frac{1}{2}}$. But $\left(\frac{i_1}{i_2}\right)^2 = \left(\frac{r_1}{r_2}\right)^3$ or $i_2 = i_1\left(\frac{r_2}{r_1}\right)^{3/2}$

$$= 5\left(\sqrt{2}\right)^{3/2} = 8.4\text{A}\Bigg]$$

24. Explain why an electric bulb becomes dim when an electric heater in parallel circuit is made on. Why brightness increases after sometime?

[**Ans.** Electric heater consumes more power than an electric bulb hence its resistance is less. When heater connected in parallel is switched on it draws more current and the current through the bulb decrease hence is becomes dim. After sometime the heater coil becomes hot its resistance increases and begins to draw less current resulting increase in the current through the bulb and increase in its brightness]

25. What is the power transferred per unit volume into Joule heat in a resistor?

[**Ans.** Power $P = i^2R = \frac{i^2\rho l}{\pi r^2}$, where ρ is resistivity, l, length and πr^2 is area of cross-section of the wire. Volume of resistor is $\pi r^2 l$. Thus power transferred per unit volume is $\frac{P}{V} = \frac{i^2 \rho l}{\pi r^2 . \pi r^2 l}$

$$= \rho\left(\frac{i}{\pi r^2}\right)^2 = \rho J^2, \text{ when } J \text{ is current density}]$$

MULTIPLE CHOICE QUESTIONS

PART I

In the following, a question is followed by four alternatives, of them only one is correct. Therefore, the candidate has to select the correct choice from amongst the alternatives.

1. A fuse with a circular cross-sectional radius of 0.15 mm blows at 15 A. What should be the radius of cross-section of fuse made of same material which blows at 30 A?

(a) 0.15 mm (b) 0.24 mm
(c) 0.30 mm (d) 0.075 mm.

2. A 1 kW heater and a 100 W tungsten bulb are marked 230 V.

(a) Resistance of tungsten bulb > resistance of heater
(b) Resistance of tungsten bulb = Resistance of heater
(c) Resistance of tungsten bulb < Resistance of heater
(d) No such relation can be determined from the data given above.

3. Which of the plots (Fig. 21.22) may represent the thermal energy produced in a resistor in a given time as a function of electric current?

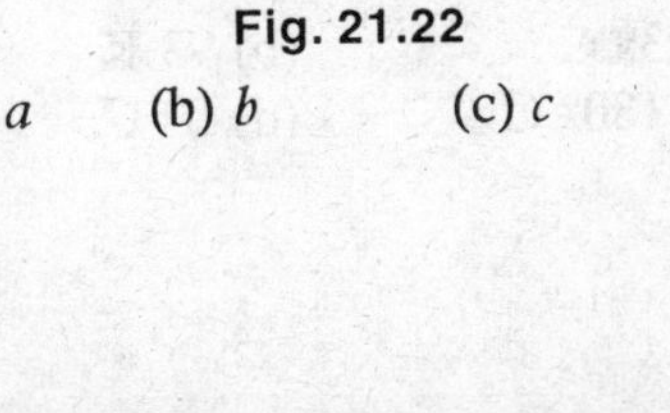

Fig. 21.22

(a) a (b) b (c) c (d) d.

4. A constant current i is passed through a resistor. Taking the temperature coefficient of resistance into account, indicate which of the graphs in Fig. 21.23 best represents the rate of production of thermal energy.

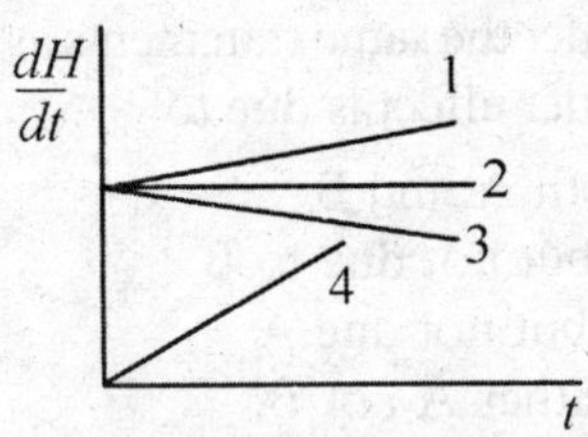

Fi.g 21.23

(a) 1 (b) 2
(c) 3 (d) 4.

5. Faraday constant

(a) depends on the charge passed through the electrolyte
(b) depends on the current on the electrolyte
(c) depends on the electrolyte concentration
(d) is a universal constant.

6. In an electrolysis experiment, the experiment is stopped and the terminals of the battery are reversed.

(a) Electrolysis will stop
(b) The rate of liberation of material at the electrodes will increase

(c) The rate of liberation of material will remain same
(d) Heat will be generated at greater rate.

7. Heat generated in a system is proportional to the current through it
(a) It can not be Thomson effect
(b) It can not be Peltier effect
(c) It can not be Joule heat effects
(d) It can be any of the three heats mentioned above.

8. Consider the following two statements
(*i*) Free-electron density is different in different metals
(*ii*) Free-electron density in metals depends on temperature. Seebeck effect is caused
(a) due to both A and B
(b) due to A but not due to B
(c) due to B but not due to A
(d) neither due to A nor due to B.

9. Consider the same statements as in question 8. Peltier effect is due to
(a) both A and B
(b) A but not due to B
(c) B but not due A
(d) neither A nor B.

10. Consider the same statements as in question 8. Thomson effect is due to
(a) both A and B
(b) A but not due to B
(c) B but not due to A
(d) neither A nor B.

11. A plate of area 10 cm^2 has to be electroplated with silver to a thickness 0.002 cm on both sides using a battery of 54 V. Energy spent in the process is, [silver is monovalent with atomic mass = 108 and density 10 g/cc. ECE of hydrogen is 10^{-5} g/C].
(a) 2×10^4 J (b) 4.5×10^4 J
(c) 6×10^4 J (d) 9.5×10^4 J.

12. Time taken by a 836 W heater to heat one litre of water from 10° C to 40° C is (AIEEE 2004)
(a) 50 S (b) 200 S
(c) 150 S (d) 100 S.

13. Thermo emf of a thermocouple varies with temperature θ of the hot junction as $E = a\theta + b\theta^2$ in volt. When the ratio a/b is 700° C. If the cold junction is kept at 0° C, then neutral temperature is
(a) 700° C
(b) 1400° C
(c) 350° C
(d) no neutral temperature is possible for this thermocouple.

14. The electrochemical equivalent of a metal is 3.3×10^{-7} kg/C. The mass of the metal liberated at the cathode when a 3A current is passed for 2 s will be
(a) 19.8×10^{-7} kg (b) 1.1×10^{-7} kg
(c) 6.6×10^{-7} kg (d) 9.9×10^{-7} kg.

15. One junction of a certain thermocouple is at temperature T_C and the other junction at temperature T. The emf generated is represented by $E = k(T - T_C)\left[T_0 - \frac{1}{2}(T + T_C)\right]$ where T_0 is constant. The thermoelectric power at temperature $T = T_0/2$ is
(a) kT_0 (b) $\frac{1}{2}kT_0$
(c) $2kT_0$ (d) $2k(T_0 - T_C)$.

16. Thermo emf of a Cu-Fe thermocouple varies with the temperature θ of hot junction (cold junction is at 0° C) as $E(\mu V) = 14\theta - 0.02\theta^2$, then neutral temperature is
(a) 700° C
(b) 350° C
(c) 175° C
(d) There is no neutral temperature for the given couple.

17. The emf in thermocouple changes sign at 600 K. If the neutral temperature is 210° C, then temperature of cold junction is
(a) 93° C (b) 93 K
(c) –180° C (d) 0° C.

18. The thermo emf E of copper-constantan thermocouple and the temperature θ of the hot junction (with cold junction at 0° C) are found to satisfy the relation

$$E = \alpha\theta + \frac{1}{2}\beta\theta^2$$

where E is in μV, θ in °C and α = 41 μV/°C and β = 0.041 μV/(°C)^2. Then the temperature of inversion for this couple is

(a) –1000° C (b) –2000° C
(c) 1000° C (d) 2000° C.

19. Consider a thermocouple made of iron and constantan. If the thermo emf of iron and constantan against platinum are 800 and –1700 μV per 50° C, difference of temperature between the junction is 10° C. The thermo emf developed is

(a) –180 μV (b) 500 μV
(c) 900 μV (d) 2500 μV.

20. If the current in an electric bulb decreases by 0.5%, the power in the bulb decreases by approximately

(a) 0.25% (b) 0.5%
(c) 1% (d) 2%.

21. When a current is passed through a conductor, rise in temperature is θ° C. If the current is doubled and passed for the same times, rise is temperature will be

(a) θ° C (b) 2θ° C
(c) 4θ° C (d) $\frac{\theta°}{4}$C.

22. The emf of the battery in a thermocouple is doubled. The rate of heat generation at one junction

(a) becomes half
(b) remains unchanged
(c) becomes double
(d) becomes four times.

23. Two heater wires of equal length, area of cross-section and of same material are first connected in parallel and then in series, the ratio of the heat produced in the two cases is

(a) 2 : 1 (b) 1 : 2
(c) 4 : 1 (d) 1 : 4.

24. How many calories of heat will be developed in a 210 W electric bulb in 5 minutes?

(a) 1050 (b) 1500
(c) 6300 (d) 80,000.

25. If potential difference across a conductor of specific resistance ρ, remains constant, then heat produced in the conductor is proportional to

(a) $\frac{1}{\sqrt{\rho}}$ (b) $\frac{1}{\rho}$
(c) ρ (d) ρ^2.

26. In order to liberate 22.4 litres of hydrogen at N.T.P. by passing electricity through acidulated water, the charge required is

(a) 1 C (b) 1.6×10^{-16} C
(c) 96500 C (d) 19300° C.

27. The thermo emf E (in Volts) of thermo-couple is found to vary with θ (in °C) as per relation $E = 40\theta - \frac{1}{10}\theta^2$. The neutral temperature of the thermocouple is

(a) 100° C (b) 200° C
(c) 300° C (d) 400° C.

28. In a copper voltameter, mass deposited in 30 sec is m grams. If the time-current graph is as shown in Fig. 21.24. Where current is in mA (milli amperes). *ECE* of copper is

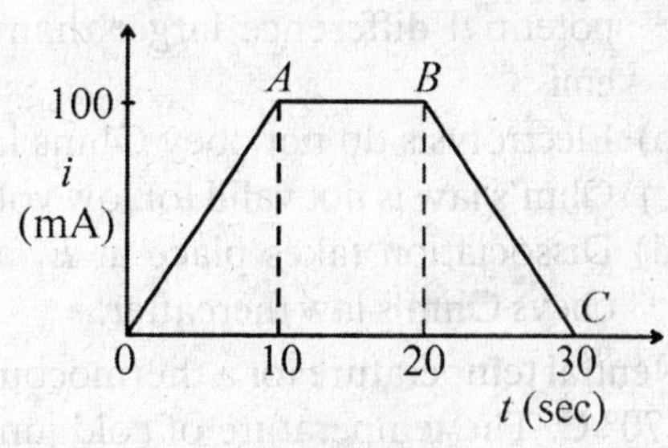

Fig. 21.24

(a) 0.1 m (b) 0.5 m
(c) 0.2 m (c) 0.6 m.

29. If R_1 and R_2 are the filament resistances of 200 W and a 100 W bulb respectively. Both designed to run at the same voltage, then

(a) $R_1 = 4R_2$ (b) $R_1 = \frac{1}{4}R_2$
(c) $R_1 = 2R_2$ (d) $R_1 = \frac{1}{2}R_2$.

30. The emf of a copper nickel thermocouple in μV per °C is given by $E = 16.34t - 0.02t^2$. The thermoelectric power at 100° C in μV/°C/s will be
(a) 10.5 (b) 12.34
(c) 20.5 (d) 1.234.

31. A heater of 200 V heats a volume of water in 5 minutes time. A heater of 110 V and of same resistance will heat the same volume of water in
(a) 5 minutes (b) 10 minutes
(c) 20 minutes (d) 40 minutes.

32. Figure 21.25 represents the current voltage behaviour of a water voltameter. The correct statement is

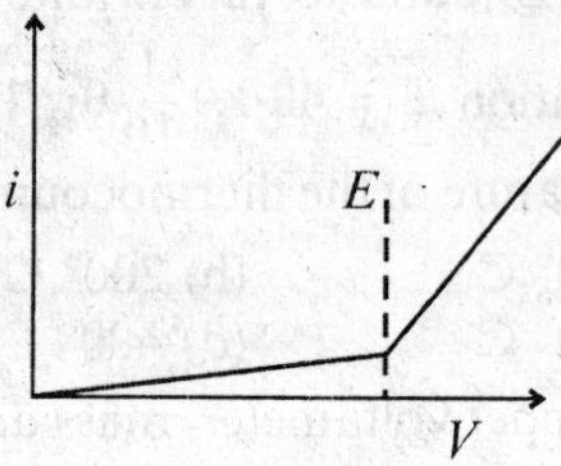

Fig. 21.25

(a) Ohm's law is obeyed but only for potential difference larger than back emf
(b) Electrolysis do not obey Ohms law
(c) Ohm's law is not valid for low voltages
(d) Dissociation takes place at E_1 and it obeys Ohm's law thereafter.

33. Neutral temperature for a thermocouple is 270° C. The temperature of cold junction is 15° C. The temperature of inversion is
(a) 255° C (b) 285° C
(c) 575° C (d) 525° C.

34. Nearly 10^5 C liberate 1 g equivalent of aluminium, then the amount of aluminium (equivalent weight = 9) deposited through electrolysis in 20 minutes by a current of 50 A is
(a) 0.09 g (b) 0.6 g
(c) 5.4 g (d) 10.8 g.

35. Two electric bulbs whose resistances are in the ratio 1 : 2 are connected in parallel to a constant voltage source. The powers dissipated in them have the ratio
(a) 2 : 1 (b) 1 : 2
(c) 1 : 1 (d) 1 : 4.

36. A certain charge liberates 0.8 g of oxygen. The same charge will liberate silver equal to
(a) 108 g (b) 10.8 g
(c) 0.8 g (d) 108 × 0.8 g.

37. The electric current passing through a metallic wire produces heat because of
(a) collision of conduction electrons with each other
(b) the energy released in the ionisation of the atoms of the metal
(c) collision of the atoms of the metal with each other
(d) collision of the conduction electrons with the atoms of the metallic wire.

38. How many lamps each of 50 W and 100 V can be connected in parallel across a 120 V battery of internal resistance 10 Ω so that each glows to full power?
(a) 2 (b) 4
(c) 6 (d) 8.

39. Fifty electric bulbs are connected in series across a supply of 220 V. The total illumination is I_0. Five bulbs are fused and remaining forty five bulbs are connected in series, the illumination produced is I, which of the following is true?
(a) $I_0 < I$ (b) $I_0 > I$
(c) $I_0 = I$ (d) none of the above.

40. In above question by what percentage the total illumination will change?
(a) Increase by 10%
(b) Decrease by 10%

(c) Increase by 11%
(d) decrease by 11%.

41. In question 39 percentage change in illumination of each bulb is

(a) increase by 10%
(b) decrease by 19%
(c) increase by 23.5%
(d) increase by 24.5%.

42. An electric kettle taking 3 A at 200 V brings 1 litre of water from 20° C to the boiling point in 10 minutes. Its efficiency is

(a) 33.30% (b) 66.6%
(c) 87.7% (d) 93.3%.

43. The emf of a thermocouple, one junction of which is kept at 0° C is given by equation

$$E = 400\theta - \frac{1}{10}\theta^2$$

The temperature of inversion of thermocouple is

(a) 400° C (b) 200° C
(c) −200° C (d) −100° C.

44. An ammeter reads 0.9 A. When connected in series with a silver voltameter. A steady current passed for 40 minutes deposits 2.6 g of silver on cathode. Take *ECE* of silver = 1.118×10^{-3} g/C, then percentage error in the ammeter reading is

(a) 6.66% (b) 7.3%
(c) −6.66% (d) −7.3%.

45. A current of 1.6 A is passed through a solution of $CuSO_4$. How many Cu^{++} ions are liberated in one minute?

(a) 3×10^{20} (b) 3×10^{10}
(c) 6×10^{20} (d) 6×10^{10}.

ANSWERS

1. (b)	2. (a)	3. (a)	4. (a)	5. (d)
6. (c)	7. (c)	8. (a)	9. (b)	10. (c)
11. (a)	12. (c)	13. (d)	14. (a)	15. (b)
16. (b)	17. (a)	18. (b)	19. (b)	20. (c)
21. (c)	22. (c)	23. (c)	24. (b)	25. (b)
26. (d)	27. (b)	28. (b)	29. (d)	30. (b)
31. (c)	32. (a)	33. (d)	34. (c)	35. (a)
36. (b)	37. (d)	38. (b)	39. (a)	40. (c)
41. (c)	42. (d)	43. (a)	44. (c)	45. (a)

PART II

In the following, a question is followed by four alternatives, of them one or more are correct. Candidate has to select the correct choice(s) from amongst the alternatives.

1. 1 kWh electric heater is meant for use at 200 V supply,
   (a) Current through the heater is 5 A
   (b) Heat produced per second in calories is 238 cal/s
   (c) If supply is 100 V, the power consumed is 500 W
   (d) If supply is 100 V the current through the heater is 5 A.

2. In the circuit shown in Fig. 21.26. If heat dissipated in 9 Ω resistor per sec is H then

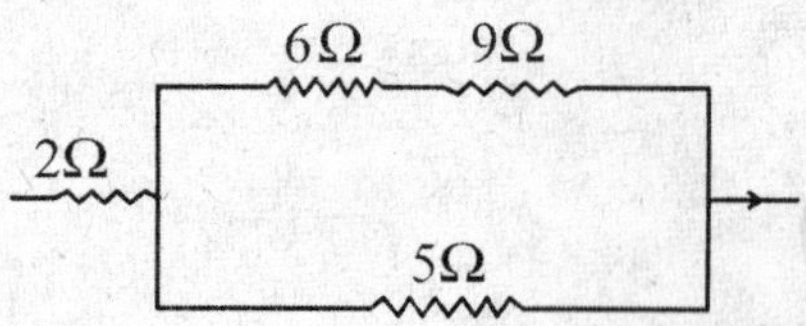

Fig. 21.26

   (a) heat dissipated in resistance 5 Ω is 5 H/sec
   (b) heat dissipated in resistance 2 Ω is 2 H/sec
   (c) heat dissipated in resistance 6 Ω is 2/3 H/sec
   (d) heat dissipated in the circuit is approximately 10 H/sec.

3. Two electric bulbs each designed to operate with a power of 500 W in 220 V line are put in series in 110 V line
   (a) power consumed by each bulb is $\frac{500}{4}$ W
   (b) power consumed by each bulb is $\frac{500}{16}$ W
   (c) The current through each bulb is 0.57 A
   (d) The current through each bulb is 1.15 A.

4. A 200 W and a 100 W bulb, both meant for operation at 220 V are connected to a supply
   (a) total power consumed is 300 W, if they are connected in series
   (b) total power consumed is 300 W, if they are connected in parallel
   (c) total power consumed is 66.6 W, if they are connected in parallel
   (d) total power consumed is 66.6 W, if they are connected in series.

5. Three 60 W, 120 V bulbs are connected across a 120 V supply as shown in Fig. 21.27

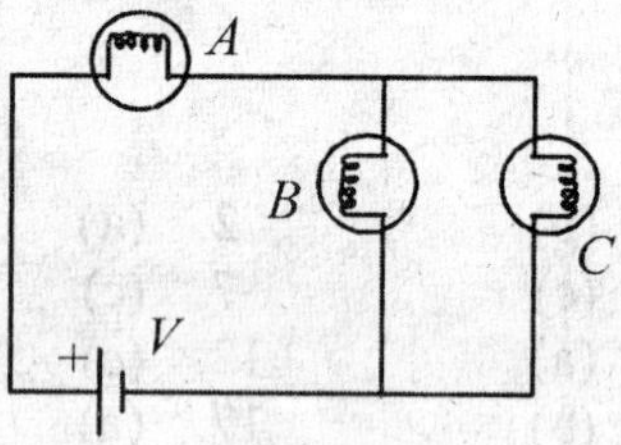

Fig. 21.27

   (a) voltage across bulb A is 40 V and that across B is 80 V
   (b) power consumed by bulb A is double the power consumed by B
   (c) total power consumed is 3 times, the combined power consumed in B and C
   (d) total power consumed is 1.5 times the power consumed only in A.

6. A charge d capacitor of capacity 5×10^{-2} F is discharged through a resistance (=20 Ω) and a copper voltameter of internal resistance 30 Ω connected in series. 4.62×10^{-6} kg of copper is deposited. Take ECE of copper as 3.3×10^{-7} kg/C, then

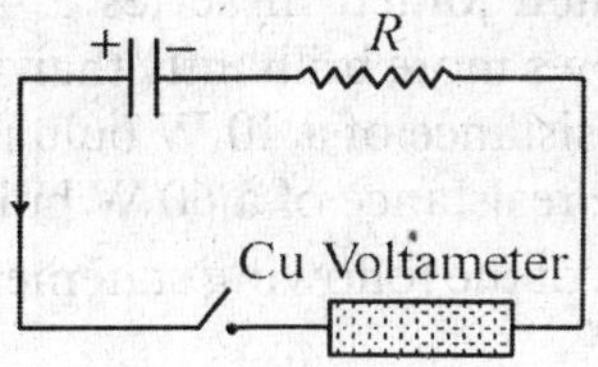

Fig. 21.28

(a) charge on the capacitor before discharging was 14 C
(b) heat dissipated in resistor R is 784 J
(c) the potential difference to which capacitor was initially charged was 300 V
(d) total heat developed is 1960 J.

7. A dry cell of emf 1.5 V and internal resistance 0.1 Ω connected across a resistor in series with a very low resistance ammeter. When the circuit is switched on, the ammeter reading in steady state is 2.0 A
(a) rate of energy dissipated in resistor is 2.6 W
(b) rate of chemical energy consumption of the cell is 3 W
(c) power output of cell is 2.6 W
(d) potential difference across resistor is 1.5 V.

8. The constants a and b for the pair silver-lead are $a = 2.5\ \mu V/°C$ and $0.012 \mu V/(°C)^2$ respectively. For silver-lead thermocouple cold junction at 0° C
(a) there will be no neutral temperature
(b) there will be no inversion temperature
(c) there will not be any thermo emf, even if the junctions are kept at different temperatures
(d) there will be no current in the thermo couple even if the junctions are kept at different temperature.

9. A copper strip AB and an iron strip AC are joined at A. The junction A is maintained at 0° C and free ends B and C are maintained at 100° C. There is potential difference between

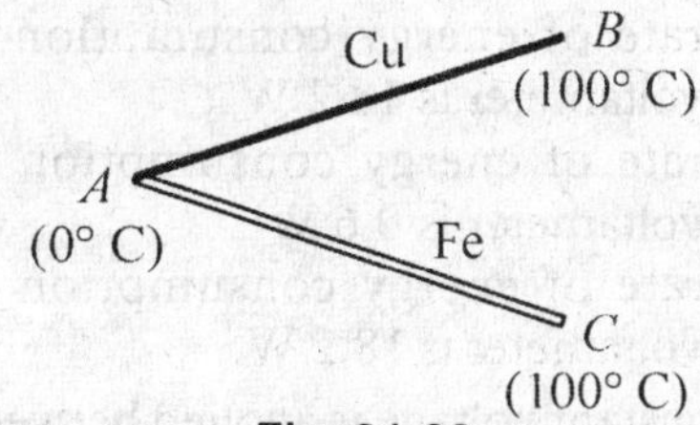

Fig. 21.29

(a) two ends of the copper strip
(b) two ends of iron strip
(c) copper end and iron end at the junction
(d) the free ends B and C.

10. Two resistors having equal resistances are joined in series and current is passed through the combination. Neglect any variation in resistance as the temperature changes. In a given time interval
(a) equal amounts of thermal energy must be produced in the resistors
(b) unequal amounts of thermal energy may be produced
(c) the temperature must rise equally in resistors
(d) the temperature may rise equally in the resistors.

11. The thermo emf of a thermocouple is given by $E = 2164t - 6.2t^2$. Where t is temperature of hot junction and cold junction is at 0° C. Then (here temperature is in °C and emf is in micro volts)
(a) Peltier coefficient at 100° C is 9.24×10^{-4} V/°C
(b) Thomson coefficient at 100° C is 1.24×10^{-3} V/°C
(c) Thomson coefficient at 100° C is 1.24×10^{-3} V/°C^2
(d) Thomson coefficient at 100° C is 1.24×10^{-4} V/°C^2.

12. Silver and copper voltameter are connected in parallel with a battery of emf 12 V. 1 gm of silver and 1.8 gm of copper are liberated. If $Z_{Cu} = 6.6 \times 10^{-4}$ g/C and $Z_{Ag} = 11.2 \times 10^{-4}$ g/C; then
(a) rate of energy consumption in Ag voltameter is 0.6 W

(b) rate of energy consumption is Cu voltameter is 18.2 W
(c) rate of energy consumption in Cu voltameter is 0.6 W
(d) rate of energy consumption in Ag voltameter is 18.2 W.

13. A constant voltage is applied between two ends of a uniform metallic wire some heat is developed in it. The heat developed is doubled if
(a) length of the wire is halved
(b) length and radius of the wire are doubled
(c) radius of the wire is halved
(d) length and radius of the wire are halved.

14. Two electric bulbs of 40 Watt 100 Volts are connected together
(a) power consumed when connected in series with a 100 Volt source is 20 W
(b) power consumed when connected in parallel with a 100 V source is 20 W
(c) power consumed when connected in parallel with a 100 V source is 80 W
(d) power consumed when connected in series with a 100 V source in 10 W.

15. Two identical batteries each of emf 2 V and internal resistance 1 Ω are available to produce Joulean heat by passing the current through a 0.5 Ω resistor.
(a) Minimum heat generated in resistor is 2 W
(b) Maximum heat generated in resistor is 2 W
(c) Maximum power delivered by the batteries is 6.4 W
(d) Maximum power dissipated in the batteries is 5.12 W.

16. Which of the following statements are correct?
(a) Resistivity of electrolytes, decreases on increasing temperature
(b) Resistance of mercury falls on decreasing its temperature
(c) When joined in series a 40 W bulb glows more brilliantly than 60 W bulb
(d) Resistance of a 40 W bulb is less than the resistance of a 60 W bulb.

17. Which of the following statements are not correct?
(a) Both Peltier and Joule effects are reversible
(b) Both Peltier and Joule effects are irreversible
(c) Joule effect is reversible whereas Peltier effect is irreversible
(d) Joule effect is irreversible whereas Peltier effect is reversible.

18. A 100 W and 500 W bulbs are given and connected to a fixed mains
(a) 100 W bulb will glow more than 500 W if connected in series
(b) 100 W bulb will glow more than 500 W if connected in parallel
(c) 500 W bulb will glow more than 100 W if connected in parallel
(d) 500 W bulb will glow more than 100 W if connected in series.

19. A resistance coil is made by connecting two coils each of resistance R in parallel. They are connected to a battery of emf V for t sec, then
(a) total power consumed is $\frac{2V^2}{R}t$
(b) total power consumed is $\frac{V^2}{2R}t$
(c) current through each coil is V/R
(d) current through each coil is $\frac{V}{2R}$.

20. Three resistors each of resistance R, and having a maximum power P are connected as shown in Fig. 21.30

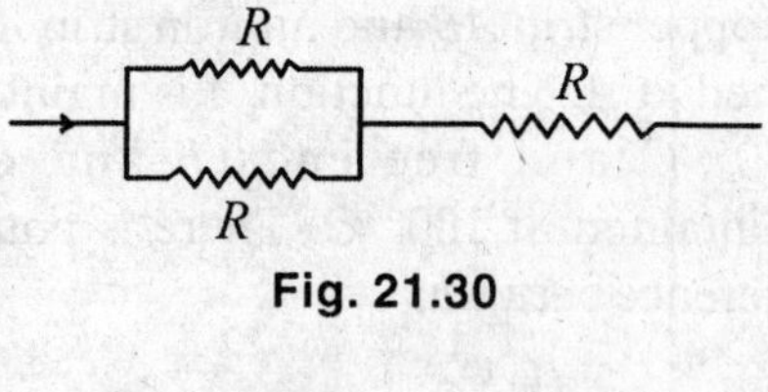

Fig. 21.30

(a) The maximum voltage that can safely be applied across the circuit is $2\sqrt{\frac{W}{R}}$

(b) The maxim voltage that can safely be applied across the circuit is $\frac{3}{2}\sqrt{\frac{W}{R}}$

(c) Total power consumed is 3 W

(d) Total power consumed is $\frac{3W}{2}$.

ANSWERS

1. (a, b)	2. (a, c, d)	3. (b, c)	4. (b, d)	5. (c, d)
6. (a, b, d)	7. (a, b, c)	8. (a, b)	9. (a, b, c, d)	10. (a, d)
11. (a, c)	12. (a, b)	13. (a, b)	14. (c, d)	15. (b, c, d)
16. (a, b, c)	17. (a, b, c)	18. (a, c)	19. (a, c)	20. (b, d)

22

Classical Magnetism

22.1 MAGNET AND ITS CHARACTERISTICS

Certain stones, which are a form of iron ore (Fe_3O_4), called *magnetite*, are found to possess attractive as well as directive properties, i.e. they attract small pieces of iron placed near them and point in north-south direction when suspended freely. These stones are called *natural magnets*. The study of magnets and their properties is called magnetism.

Natural magnets such as *lead stone* have odd shape and weak attractive power as such they are rarely used. Now-a-days pieces of iron and other materials of suitable shapes and sizes are made magnets by either rubbing them with natural magnets or by passing direct current through a wire wound around them. The magnets so formed are called *artificial magnets* and are available in various shapes and sizes.

Fundamental Properties of Magnets

1. **Attractive Property and Poles:** Magnets attract small pieces of iron or steel. The attraction is maximum near the two ends of the magnet. These regions of maximum attraction are called *poles*. The place (near centre) of minimum attraction is called *neutral region*. The line joining the poles of a magnet is called *magnetic axis*.

2. **Directive Property and N-S Poles:** A magnet when suspended freely always points in same direction—its length always becomes parallel to Geographical N-S direction. The end pointing towards north is called *north pole (N-pole)* and the end pointing towards south is called *south pole (S-pole). The north and south poles of a magnet are always of equal strength and opposite nature.* North pole is also called as *magnetic positive charge* and south pole as *magnetic negative charge* from analogy with electrostatics.

   The two poles of a magnet cannot be separated. If a magnet is broken into a number of pieces, each piece becomes a magnet with two equal and opposite poles. This implies that an isolated pole (called *monopole*) does not exist.

3. **Consequent Pole and no-Pole:** A monopole does not exist in magnetism but there can be a magnet with two poles of equal strength at the ends and a pole of double the strength and

of opposite nature at the centre. The pole at the centre is called *consequent pole* (Fig. 22.1 (*b*)). These magnets are formed due to faulty magenetisation and poles developed are temporary.

Figure 22.1 (*a*) shows a magnetised ring. This is the example of magnet with no poles. Torroid or solenoid of infinite length has properties of magnet but no poles.

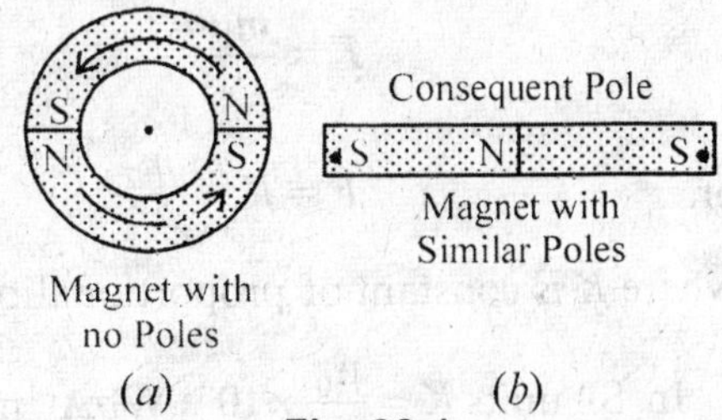

Fig. 22.1

4. **Magnetic Length:** The distance between two poles along the axis of a magnet is called *effective* or *magnetic length of magnet*. As poles are not exactly at the ends the effective length is lesser than the actual length of the magnet. The ratio of the magnetic length and geometrical length as shown in Fig. 22.2 is 0.84.

$$\frac{\text{Magnetic length}}{\text{Geometrical length}} = 0.84$$

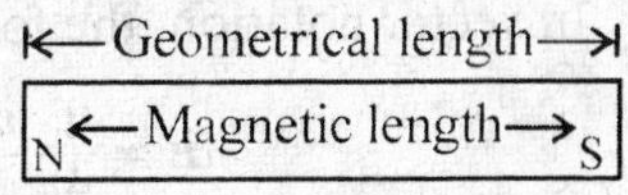

Fig. 22.2

5. **Magnetic Meridian:** The line joining the two poles of a magnet is called magnetic axis. The vertical plane passing through the axis of a freely suspended magnet is called *magnetic meridian*.

6. **Pole Strength and Magnetic Moment:** Pole strength of both poles is same and is represented by m. North pole is taken as positive pole (pole strength $+m$) and south pole is taken as negative pole (pole strength $-m$). Pole strength is a scalar quantity. It is anologous to charge in electrostatics. The pole strength depends on the nature of material of the magnet, the state of magnetisation (with maximum magnetisation called saturation) and area of cross-section. When a magnet is divided into two parts, each having same cross-section, the each will be a magnet with same pole strength and if a magnet is divided into two equal parts such that the cross-section of each part is half that of the initial, each part will behave like a magnet of pole strength half the initial pole strength. The SI units of pole strength is Amper-meter (A-m).

The product of pole strength (m) and the magnetic length ($2l$) of magnet is called *magnetic moment* $M (= m \times 2l)$. Magnetic moment is a vector quantity with magnitude $M (= 2ml)$ and direction is always from south pole to north pole. Its unit in SI system of units is Ampere-(meter)2 (= A-m^2). This unit is equivalent to Joule per Tesla (J/T). Thus if a magnet is cut into two equal parts either along its length or perpendicular to its length, magnetic moment becomes half of the original value (Fig. 22.3). The magnetic moments of the two magnets will be added or subtracted according to the laws of vector addition.

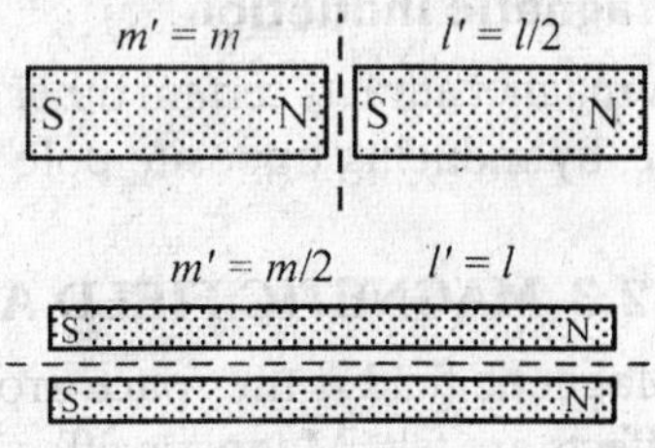

Fig. 22.3

22.2 INVERSE SQUARE LAW—COULOMB'S LAW IN MAGNETOSTATICS

The force between magnetic poles was first investigated quantitatively by Coulomb in 1785. He observed that like poles repel each other and unlike poles attract each other.

Coulomb further observed that the force of attraction or repulsion between two magnetic poles of strength m_1 and m_2 separated by a distance r is directly proportional to the product of pole strengths and inversely proportional to the square of the distance between their centres, i.e.

$$F \propto \frac{m_1 m_2}{r^2}$$

or $$F = K\frac{m_1 m_2}{r^2}$$

where K is constant of proportionality.

In SI units $K = \frac{\mu_0}{4\pi} \times 10^{-7}$ $\text{WbA}^{-1}\text{m}^{-1}$ (or Henry per meter), μ_0 is called absolute magnetic permeability of free space (i.e. vacuum)

$$\therefore \quad F = \frac{\mu_0}{4\pi}\frac{m_1 m_2}{r^2} \quad \text{...(22.1)}$$

In vector notation, the force is written as

$$\boldsymbol{F} = \frac{\mu_0}{4\pi}\frac{m_1 m_2}{r^2}\hat{r} = \frac{\mu_0}{4\pi}\frac{m_1 m_2 \boldsymbol{r}}{r^3} \quad \text{...(22.2)}$$

Equation (22.1) helps us to define unit pole.
If $m_1 = m_2 = 1$ and $r = 1$ m, then

$$F = \frac{\mu_0}{4\pi} = 10^{-7} \text{ N}$$

Thus a unit pole is a pole which when placed from similar pole at a distance of 1 m in vacuum experiences a force of 10^{-7} N. This is called SI Sommerfeld version and units of m are A-m.

Equation (22.2) represents Coulomb's law. This is in reality approximation to actual conditions for it assumes point pole which do not exist.

In cgs system of units $K = 1$, when r is in cms and F in dynes, m is in emu (electro magnetic units).

In SI (Kenelly) units $K = \frac{1}{4\pi\mu_0}$, when r is in meters and F in Newtons m is in Webers (Wb).

Magnetic Induction

A magnet attracts certain other magnetic substances through the phenomenon of magnetic induction, i.e. by inducing opposite pole in magnetic material on the side facing it.

22.3 MAGNETIC FIELD AND LINES OF FORCE

Magnetic field is the space around a magnet or current carrying conductor in which the magnetic effects are felt. Magnetic effects means a moving charge experiences a force.

In a magnetic field the field strength at any point is generally represented by *vector* $\boldsymbol{B}$. Vector $\boldsymbol{B}$ is called *magnetic field vector, or magnetic induction or magnetic flux density*. In terms of pole concept, *it is defined as the force experienced by a unit test north-pole supposed to be placed at that point*. This means that if a pole of strength m_0 placed at point P in magnetic field experiences a force $\boldsymbol{F}$, then field vector at point P is

$$\boldsymbol{B} = \frac{\boldsymbol{F}}{m_0} \text{ or } \boldsymbol{F} = m_0\boldsymbol{B} \quad \text{...(22.3)}$$

In SI units if m_0 is in *ampere-meter* and F in Newton then B has units *Tesla*. Thus *one tesla is the magnetic induction in which a pole of strength one ampere-meter experiences a force of 1 Newton*. The dimensions of magnetic vector (magnetic induction) are $M^1L^0L^{-2}I^{-1}$.

In cgs system of units, the unit used for magnetic field vector $\boldsymbol{B}$ is Gauss (G) 1 Gauss = 10^{-4} Tesla.

If the field is produced by a pole of strength m then the field at point P at distance r from pole in vacuum is, by inverse square law

$$\boldsymbol{B} = \frac{\mu_0}{4\pi}\frac{m}{r^2}\hat{r} \qquad \ldots(22.4)$$

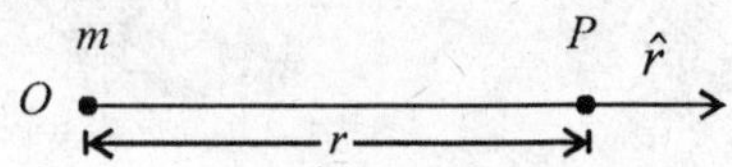

Fig. 22.4

where $\hat{r}$ is a unit vector in the direction of r (Fig. 22.4).

Magnetic Lines of Force and their Characteristics

In order to show magnetic field graphically, Michael Faraday introduced the concept of *lines of force*. He defined the line of force as *an imaginary curve traced by a free north pole placed in magnetic field. It is the curve the tangent to which at any points gives the direction of the magnetic field vector B at that pint.*

Properties of Magnetic Lines of Force

(*a*) Magnetic lines of force are closed curves, i.e. they appear to converge or diverge at the poles. Outside the magnet, lines of force go from north pole to south pole and inside they go from south to north pole.

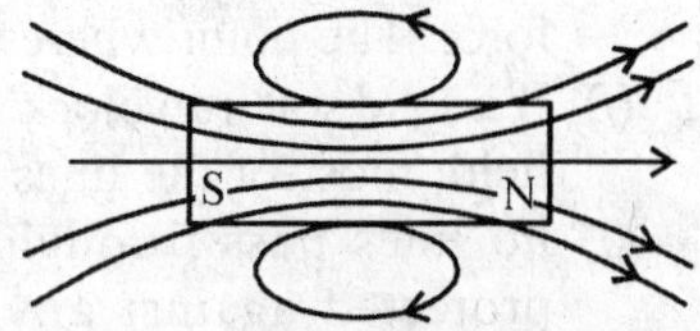

Fig. 22.5

(*b*) The number of magnetic lines of force originating or terminating on a pole are proportional to its strength, i.e. number of lines of force from (or to) a pole-n is

$$n \propto m \text{ or } n = \mu_0 m \qquad \ldots(22.5)$$

where μ_0 is the magnetic permeability of free space.

Equation (22.5) gives that the number of lines of force associated with a unit pole is μ_0. This means that a closed surface enclosing pole of strength m, the total lines of force linked with the closed surface will be $\mu_0 m$. This is called *magnetic flux. The magnetic flux is thus defined as the total magnetic lines of force linked with the surface.*

The magnetic flux through a surface of area s in magnetic field of vector $\boldsymbol{B}$ is given by

$$\phi = \int_s \boldsymbol{B} \cdot d\boldsymbol{s} \qquad \ldots(22.6)$$

In SI units the magnetic flux has unit, Tesla (meter)^2 or Weber and has dimensions $M^1L^2T^{-2}I^{-1}$.

(*c*) Magnetic lines of force never cross each other because if they intersect at a point, the magnetic field vector $\boldsymbol{B}$ will have two directions and that is not possible.

(*d*) Number of lines of force per unit area, normal to the area at a point, represents the magnitude of field at that point; as such the regions with crowded lines will represent strong field regions and regions with far off lines represent weak field regions. If the lines of force are equidistant and straight, the field is uniform, otherwise non-uniform.

(*e*) Magnetic lines of force have tendancy to contract longitudinally like a stretched elastic string (causing attraction between opposite poles (Fig. 22.6 (*a*))) and repel each other latterally (causing repulsion between similar poles (Fig. 22.6 (*b*))).

(*f*) Magnetic lines of force exist inside every magnetised material.

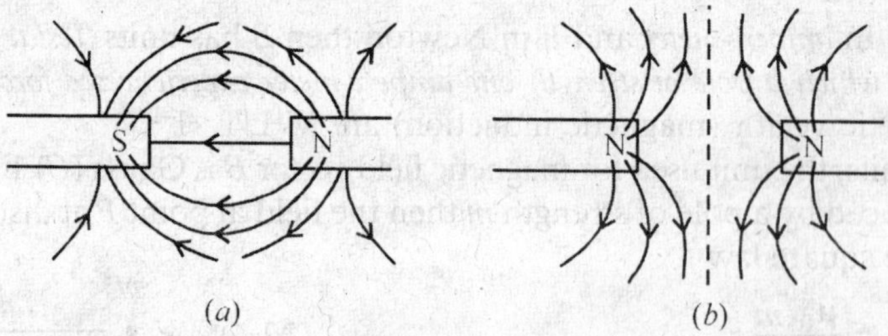

Fig. 22.6

(g) Since isolated pole (mono pole) does not exist. Hence the total magnet flux over a closed surface is always zero, i.e.

$$\oint_S \boldsymbol{B} \cdot \boldsymbol{ds} = \mu_0(0) = 0 \qquad \text{...(22.7)}$$

Equation (22.7) is called *Gauss' law* in magentostatics.

(h) Magnetic lines of force though imaginary can be mapped by using iron fillings or using a compass needle.

(i) In a region where the magnetic field due to magnet and earth are equal in magnitude and opposite in direction, resultant magnetic filed is zero in that region. There are no lines of force. The point where the two fields cancel each other is called *neutral point.*

(j) If a solid soft iron block with a cavity is placed in magnetic field, most of the lines pass through the iron block but no lines pass through the cavity. The cavity is thus protected against any external magnetic field. This phenomenon is called *magnetic shielding or magnetic screening*. This is used to protect costly instruments from magnetic field. This is done by placing them in soft iron boxes.

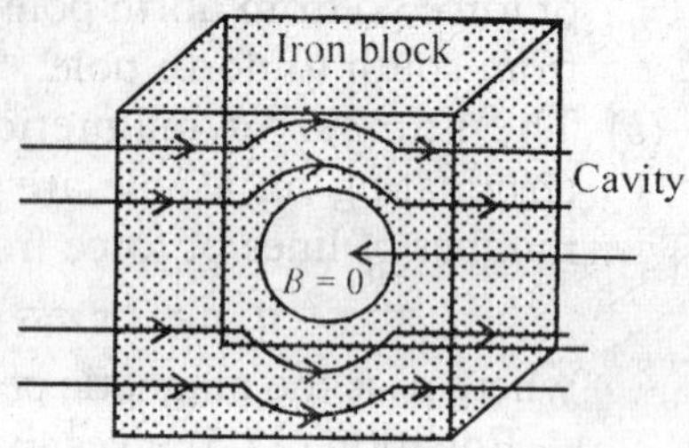

Fig. 22.7

22.4 MAGNETIC DIPOLE

The unlike poles of equal pole strength and separated by a small distance constitute a dipole. Magnetic dipole moment is then defined as the product of pole strength (m) and the distance between the two poles ($2l$), i.e.

$$\boldsymbol{M} = m(2l)\,\hat{n} \qquad \text{...(22.8)}$$

where $\hat{n}$ is a unit vector along the axis of magnet from south to north.

The SI unit of magnetic moment is ampere (meter)^2. Since magnetic monopole does not exist as such the source of magnetic fields are usually dipoles.

A current carrying closed loop behaves like a magnetic dipole for distant points with magnetic dipole moment

$$M = IA \qquad \text{...(22.9)}$$

where I is current through loop and A is area of the loop. In this case the direction of magnetic dipole moment is perpendicular to the plane of the loop (i.e. in the direction of vector area). Looking to the loop if current is anticlockwise, it is north pole and if it is clockwise it is south pole.

The earth for its magnetic effects behaves a dipole with its N pole in the geographical southern hemisphere and S pole in geographical northern hemisphere.

(a) Magnetic Field Due to a Magnetic Dipole

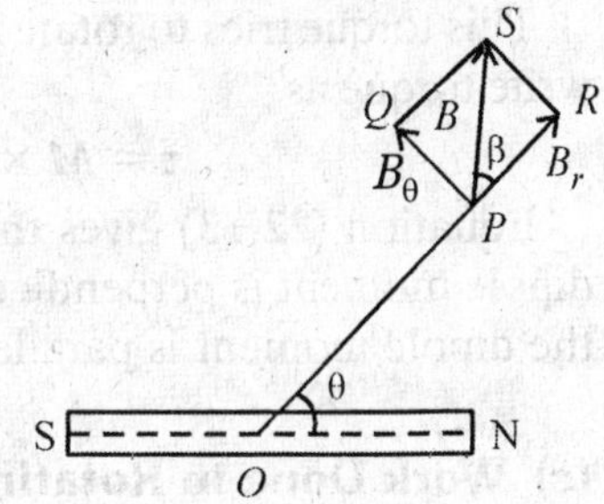

Fig. 22.8

Magnetic analogue of electric dipole is the magnetic dipole. Hence following the treatment similar to electric dipole, it can be shown that the field due to magnetic dipole of moment M at point (r, θ) is given by

$$B_r = \frac{\mu_0}{4\pi}\frac{2M\cos\theta}{r^3} \text{ and } B_\theta = \frac{\mu_0}{4\pi}\frac{M\sin\theta}{r^3} \quad ...(22.10)$$

where B_r is the magnetic field in the direction of r and B_θ in the direction perpendicular to r.

Hence resultant magnetic field B is

$$B = \sqrt{B_r^2 + B_\theta^2} = \frac{\mu_0}{4\pi}\frac{M}{r^3}\sqrt{1+3\cos^2\theta} \quad ...(22.11)$$

with $$\tan\beta = \frac{B_\theta}{B_r} = \frac{1}{2}\tan\theta \quad ...(22.12)$$

where β is the angle that the resultant field makes with the direction of r.

(a) If $\theta = 0$, B is maximum $B = \frac{\mu_0}{4\pi}\frac{2M}{r^3}$. In this case point is on the axis of dipole. Point P in Fig. 22.9. This position is usually referred to as *axial, end-on or tan A position.*

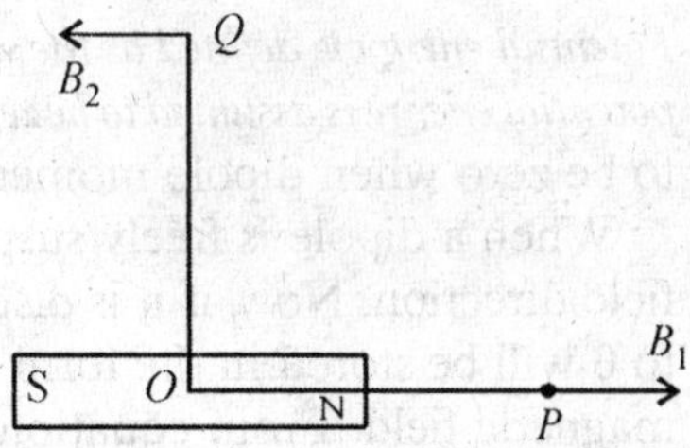

Fig. 22.9

(b) If $\theta = 90°$, B is minimum $B = \frac{\mu_0}{4\pi}\frac{M}{r^3}$. In this case the point is on the equatorial line of dipole. This position is usually referred to as *equatorial, broad-side-on* or tan$B$ position. This is shown as point Q in Fig. 22.9.

(c) For constant r, θ charges from 0° to 90°, B decreases from $\frac{\mu_0}{4\pi}\cdot\frac{2M}{r^3}$ to $\frac{\mu_0}{4\pi}\cdot\frac{M}{r^3}$, i.e. for same distance field on axial point is twice the field on equitorial point.

(d) For constant θ, the field due to dipole is inversely proportional to the cube of distance (i.e. varies as $1/r^3$). Thus the field due to dipole is never zero except at infinity.

(b) Torque on a Bar Magnet Placed in a Magnetic Field

When a dipole is placed in uniform field, (Fig. 22.10) a force equal to mB (m is pole strength of dipole) acts on N pole in the direction of magnetic field and force mB acts on S pole in the direction opposite to the field. Thus net force on the dipole will be zero. As such the centre of mass of dipole will not move.

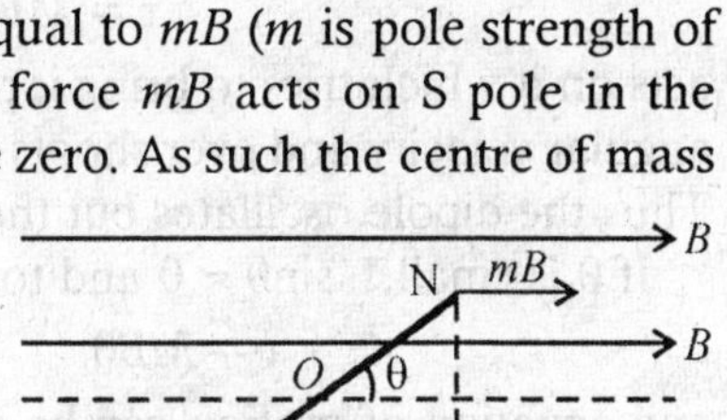

Fig. 22.10

But the two forces are not acting at the same point hence a torque will act on it. Perpendicular distance between two forces is $NP = 2l\sin\theta$, where $2l$ is the magnetic length of the dipole and θ is the angle between magnetic field and dipole moment. Hence

$$\text{Torque} = mB\cdot 2l\sin\theta = M\cdot B\sin\theta \quad ...(22.13)$$

when $M = m\cdot 2l$, the dipolc moment of magnet.

This torque tries to rotate magnet so as to align it with the field. Hence in vector notation we can write torque as

$$\tau = \boldsymbol{M} \times \boldsymbol{B} \quad \text{...(22.14)}$$

Equation (22.13) gives that torque acting on a dipole is maximum = MB, when $\theta = 90°$, i.e. dipole moment is perpendicular to the field direction and minimum equal to zero when $\theta = 0$, i.e. the dipole moment is parallel to the field direction.

(c) Work Done in Rotating Dipole in Magnetic Field

If an external agent rotates magnet slowly, the agent has to exert torque $MB\sin\theta$ opposite to that exerted by field. The work done by the agent in changing the angle from θ to $\theta + d\theta$ is $dW = MB\sin\theta \cdot d\theta$. Hence total work done in rotating the dipole from angle θ_0 to θ is

$$W = \int_{\theta_0}^{\theta} MB\sin\theta\, d\theta = MB(\cos\theta_0 - \cos\theta) \quad \text{...(22.15)}$$

Now if the dipole is rotated from field direction $\theta_0 = 0$ to position θ, the work done is

$$W = MB(1 - \cos\theta) \quad \text{...(22.16)}$$

This equation gives that maximum work done (= $2MB$) when the dipole is rotated through $\theta = 180°$ from the parallel position.

(*d*) Potential Energy

Potential energy is defined as the work done in bringing the body to its present position from a position where potential energy is assumed to be zero. In case of dipole in magnetic field the potential energy is assumed to be zero when dipole moment is perpendicular to the field direction.

When a dipole is freely suspended in a uniform magnetic field it will align itself parallel to the field direction. Now, if it is displaced through an angle θ, then work done in rotating from $\theta_0 = 90°$ to θ will be stored in the form of potential energy. This is called the potential energy of dipole in magnetic field. From equation (22.15),

$$\text{P.E.} = MB(\cos 90° - \cos\theta) = -MB\cos\theta = -\boldsymbol{M} \cdot \boldsymbol{B} \quad \text{...(22.17)}$$

(e) Motion of a Magnetic Dipole in Uniform Magnetic Field

When a magnetic dipole is freely suspended in uniform magnetic field it aligns itself parallel to the magnetic field. If it is displaced through an angle θ, a torque

$$\tau = MB\sin\theta$$

acts on it which tries to bring it parallel to the field where the torque becomes zero. But it acquires angular velocity and over shoots its equilibrium position. Now, the direction of the torque reverses. Thus the dipole oscillates but the oscillations are not simple harmonic.

If θ is small, $\sin\theta \approx \theta$ and torque

$$\tau = MB\theta$$

and equation of motion can be written as

$$I\alpha = -MB\theta$$

when I is moment of inertia of the dipole about the axis of rotation and α its instantaneous angular acceleration.

$$\therefore \quad \alpha = -\frac{MB}{I}\theta = -\omega^2\theta \text{ where } \omega^2 = \frac{MB}{I}$$

i.e. angular acceleration is directly proportional to angular displacement, hence motion is SHM with angular frequency ω.

Thus the dipole executes SHM with time period

$$T = \frac{2\pi}{\omega} = 2\pi\sqrt{\frac{I}{MB}} \qquad \text{...(22.18)}$$

Vibration magnetometer is based on this principle and is used to compare two fields or magnetic moments of two magnets.

In non-uniform magnetic field, the dipole experiences a net force in the direction of increasing field. Hence the centre of mass moves in that direction. Simultaneously a torque also acts on it as such that the dipole oscillates also. Thus the dipole has two motions simultaneously one translatory and other oscillatory.

(f) Tangent Law

If a dipole (say a magnetic needle) of magnetic dipole moment M is free to rotate in two mutually perpendicular fields B_0 and B (Fig. 22.11) then in equilibrium if dipole moment $\boldsymbol{M}$ makes an angle θ with B_0, then ratio of the two fields is equal to the tangent θ, i.e.

$$\frac{B}{B_0} = \tan\theta \qquad \text{...(22.19)}$$

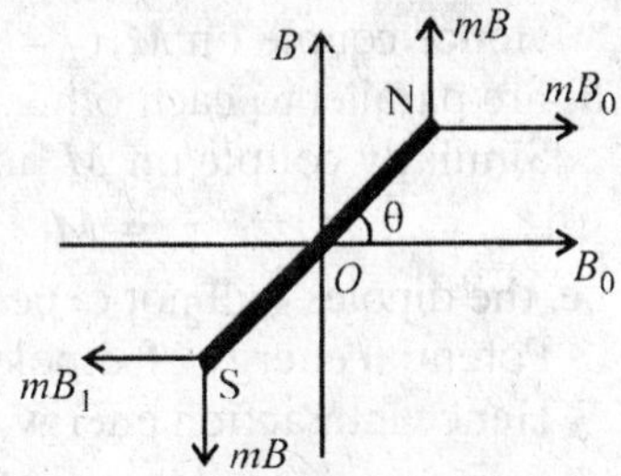

Fig. 22.11

This is called *tangent law*. Deflection magnetometer used to compare the magnetic moments of two magnets are based on this principle. In deflection magnetometer B_0 generally is *the horizontal component of earth's magnetic field* while B is produced by magnetic dipole.

(g) Magnetic Scalar Potential Due to Magnetic Dipole

Figure 22.12 shows a magnetic dipole NS of length $2l$ and of pole strength m and magnetic moment equal to $2ml = M$. Scalar potential at point $P(r, \theta)$ is

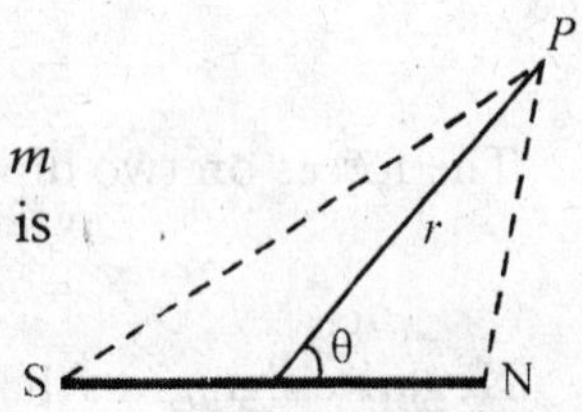

Fig. 22.12

$$V = \frac{\mu_0}{4\pi}\frac{m(2l\cos\theta)}{r^2 - l^2\cos^2\theta} \quad \text{for } r >> l, \text{ we get}$$

$$V = \frac{\mu_0}{4\pi}\frac{M\cos\theta}{r^2} = \frac{\mu_0}{4\pi}\frac{M\cdot\hat{r}}{r^2} \qquad \text{...(22.20)}$$

Using equation (22.20) for scalar magnetic potential, we can find magnetic field at point P. Magnetic field along r is

$$B_r = -\frac{\partial V}{\partial r} = \frac{\mu_0}{4\pi}\frac{2M\cos\theta}{r^2} \qquad \text{...(i)}$$

and magnetic field in the direction θ increasing

$$B_\theta = -\frac{1}{r}\frac{\partial V}{\partial \theta} = \frac{\mu_0}{4\pi}\frac{M\cos\theta}{r^3} \qquad \text{...(ii)}$$

These equations are same as equations (22.9) and (22.10).

22.5 DIPOLE-DIPOLE INTERACTION

When a dipole is placed in the field of another dipole, it usually experiences a couple, or force or both and has potential energy. Depending on the orientations of the two dipoles relative to each other, three situations given below are possible:

(a) Two Dipoles Placed along the Line Joining their Centres

When two dipoles have their opposite poles facing each other (Fig. 22.13 (*a*)), there will be attractive force between them.

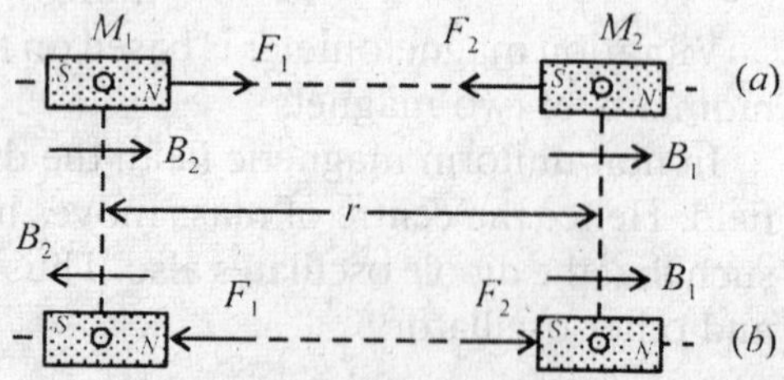

Fig. 22.13

The magnetic field at the centre of M_2 due to dipole of moment M_1 is

$$\boldsymbol{B}_1 = \frac{\mu_0}{4\pi}\frac{2\boldsymbol{M}_1}{r^3}$$

Hence couple on $\boldsymbol{M}_2\tau_2 = \boldsymbol{M}_2 \times \boldsymbol{B}_1 = 0$. Since M_2 and B_1 are parallel to each other.

Similarly couple on $\boldsymbol{M}_1$ in the field due to $\boldsymbol{M}_2$ is

$$\tau_1 = \boldsymbol{M}_1 \times \boldsymbol{B}_2 = 0 \text{ as } M_1 \text{ and } B_2 \text{ are parallel to each other.}$$

i.e. the dipoles will not experience any couple.

Potential energy of dipole of moment $\boldsymbol{M}$ in magnetic field $\boldsymbol{B}$ is $U = -\boldsymbol{M}\cdot\boldsymbol{B}$

Hence interaction energy of system

$$U = -\boldsymbol{M}_1\cdot \boldsymbol{B}_2 = -\boldsymbol{M}_2\cdot \boldsymbol{B}_1 = -\frac{\mu_0}{4\pi}\frac{2M_1M_2}{r^3}$$

and the force on dipole is

$$F = -\frac{\partial u}{\partial r} = -\frac{\partial}{\partial r}\left[-\frac{\mu_0}{4\pi}\cdot\frac{2M_1M_2}{r^3}\right]$$

$$= -\frac{\mu_0}{4\pi}\cdot\frac{6M_1M_2}{r^2}$$

The forces on two dipoles will be equal in magnitude and opposite in direction. Hence

$$F_1 = F_2 = -\frac{\mu_0}{4\pi}\frac{6M_1M_2}{r^4} \qquad \text{...(22.21)}$$

Negative sign gives that the force is attractive and varies inversely as the 4th power of the distance $(1/r^4)$.

If similar poles face each other (Fig. 12.13 (*b*)), the angle between the two dipoles is 180°, U and hence F will be positive, i.e. force will be one of repulsive.

(b) Two Dipoles Perpendicular to the Line Joining their Centres

In Fig. 22.14 (*a*) the two dipoles having their similar poles facing each other are shown. The magnetic field due to dipole of moment $\boldsymbol{M}_1$ at the centre of dipole of moment M_2 is $\boldsymbol{B}_1$. Then

$$\boldsymbol{B}_1 = -\frac{\mu_0}{4\pi}\cdot\frac{M_1}{r^3}$$

i.e. the angle between $\boldsymbol{B}_1$ and $\boldsymbol{M}_1$ is 180°. Hence couple τ_2 on $\boldsymbol{M}_2$ in field $\boldsymbol{B}_1$ is

$$\tau_2 = \boldsymbol{M}_2 \times \boldsymbol{B}_1 = 0$$

and similarly couple τ_1 on $\boldsymbol{M}_1$ due to field $\boldsymbol{B}_2$ is $\tau_2 = \boldsymbol{M}_1 \times \boldsymbol{B}_2 = 0$, i.e. torque is zero.

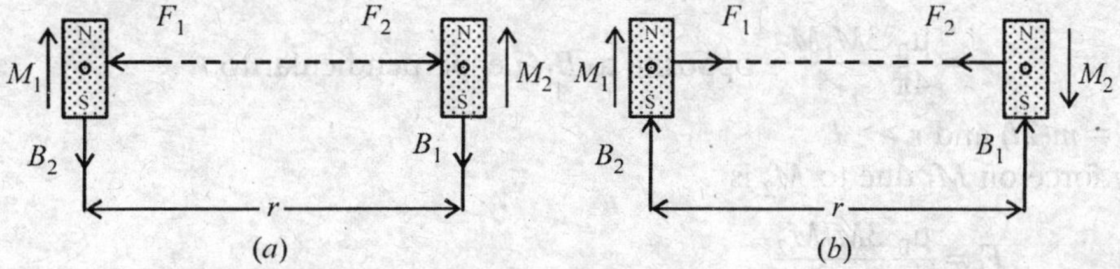

Fig. 22.14

Interaction potential energy of the system is

$$U = -\boldsymbol{M}_2 \cdot \boldsymbol{B}_1 = -\boldsymbol{M}_1 \cdot \boldsymbol{B}_2 = \frac{\mu_0}{4\pi}\frac{M_1 M_2}{r^3}$$ ($\because$ angle between $\boldsymbol{B}_1$ and $\boldsymbol{M}_2$ or between $\boldsymbol{B}_2$ and $\boldsymbol{M}_1$ is 180°)

And force F is

$$F = -\frac{dU}{dr} = \frac{\mu_0}{4\pi}\frac{3M_1 M_2}{r^4} \quad \text{...(22.22)}$$

Since force is positive, i.e. the force is repulsive and varies inversely as the 4th power of r. The forces F_1 and F_2 on two dipoles are equal in magnitude and opposite in direction.

If opposite poles of the two dipoles face each other (Fig. 22.14 (*b*)) the angle between $\boldsymbol{B}_1$ and $\boldsymbol{M}_2$ as well as between $\boldsymbol{B}_2$ and $\boldsymbol{M}_1$ is zero, hence U and F will be negative, i.e. force becomes one of attraction.

(c) Axes of the Two Dipoles are Mutually Perpendicular to Each Other

In this case M_2 is in broad side on position relative to M_1 and M_1 in end-on position relative to M_2. Hence magnetic field due to M_1 at the centre of M_2 is

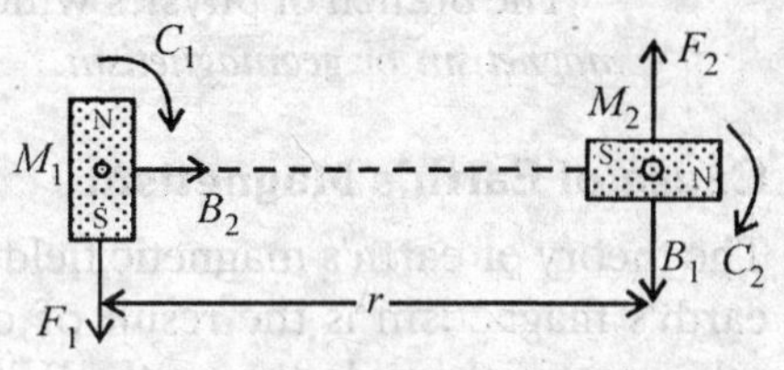

Fig. 22.15

$$B_1 = \frac{\mu_0}{4\pi}\cdot\frac{M_1}{r^3}$$

M_2 and B_1 are perpendicular to each other. Hence torque on M_2 due to magnetic field B_1 is

$$\tau_2 = \boldsymbol{M}_2 \times \boldsymbol{B}_1 = \frac{\mu_0}{4\pi}\frac{M_1 M_2}{r^3}$$ (M_2 and B_1 are perpendicular to each other)

The torque on $\boldsymbol{M}_1$ due to field $\boldsymbol{B}_2$ and due to $\boldsymbol{M}_2$ is

$$\tau_1 = \boldsymbol{M}_1 \times \boldsymbol{B}_2 = \frac{\mu_0}{4\pi}\frac{2M_1 M_2}{r^3}$$ (M_1 and B_2 are perpendicular to each other)

The interaction potential energy

$$U = -\boldsymbol{M}_1 \cdot \boldsymbol{B}_2 = -\boldsymbol{M}_2 \cdot \boldsymbol{B}_1 = 0$$

Since interaction potential energy is zero hence force cannot be calculated using $F = -\frac{\partial U}{\partial r}$ relation.

In this case the force is calculated by finding out the field at north pole and south pole separately and force is equal to field multiplied by pole strength. Hence net force on M_2 is

$$F_2 = \frac{\mu_0}{4\pi}\left[\frac{mM_1}{(r-l)^3} - \frac{mM_2}{(r+l)^3}\right]$$

$$= \frac{\mu_0}{4\pi}\frac{3M_1M_2}{r^4} \text{ opposite to } B_1, \text{ i.e. perpendicular to } r.$$

Here $M_2 = m(2l)$ and $r >> l$.
Similarly force on M_1 due to M_2 is

$$F_1 = \frac{\mu_0}{4\pi}\frac{3M_1M_2}{r^3}$$

These forces are shown in Fig. 22.15.

22.6 TERRESTRIAL MAGNETISM (EARTH'S MAGNETISM)

The idea that earth itself behaves like a huge magnet and is a natural source of magnetic field was suggested by Dr. William Gilbert at the end of sixteenth century. His statement was based on experimental facts as follows:

1. A bar magnet freely suspended near the earth's surface always comes to rest along *geographical north-south direction*. This behaviour of magnet shows as if there is a magnetic field every where near earth directed from *geographical south to geographical north*.
2. When magnetic field lines for a magnet are drawn, we come across neutral points. *Neutral points* are the points where the magnetic field due to magnet is exactly equal in magnitude and opposite in direction to that due to earth.
3. If a bar of soft iron is burried under the surface of earth in north-south direction, it is found to acquire magnetic properties after sometime.

The branch of physics which deals with the study of magnetism of earth is called *terrestrial magnetism or geomagnetism*.

Cause of Earth's Magnetism

The theory of earth's magnetic field is not yet well developed. But the common opinion is that the earth's magnetism is the result of circulating electric currents induced in the molten liquid and other material inside the earth. With the rotation of earth, charged metallic fluid in the core of earth (approximate radius 3500 km) also rotates resulting in the currents in the core of earth. This is equivalent to a current carrying closed loop which behaves like a magnetic dipole.

Besides this the outer layers of earth's atmosphere (ionosphere) are in the ionised state. As earth rotates, strong electric currents are set up due to movement of these charges. This can also be one of the causes of the earth's magnetism.

Most of the magnetic properties of earth can be easily explained if we assume a giant magnetic dipole (of moment about 8.0×10^{22} J/T) placed at the centre of the earth, and the axis of the dipole inclined at a small angle (= 11.5°) with the earth's axis of rotation (Fig. 22.16). The magnetic north pole lies in southern hemisphere of earth and the magnetic south pole lies in northern hemisphere. This conclusion is based on the fact that north pole of the freely suspended magnet should be attracted by the south pole of assumed dipole of earth.

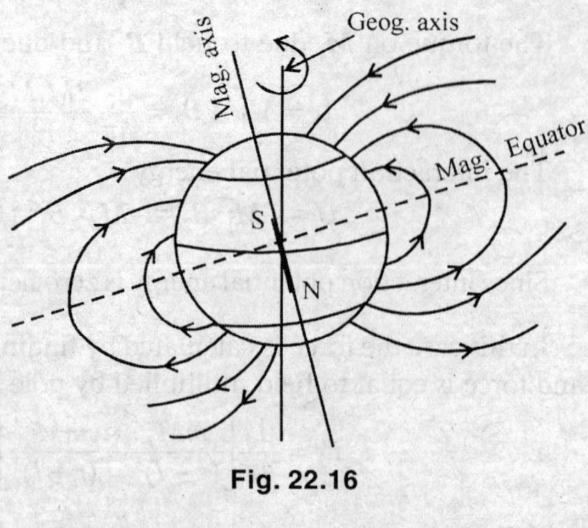

Fig. 22.16

The points where the axis of this assumed magnetic dipole cuts the earth surface are called *magnetic poles*. The strength of earth's magnetic field is of the order of 10^{-4} tesla (= 1 Gauss). The earth's magnetic field changes both in magnetic and direction as time passes on. On geological time scale (10^5~10^6 yrs) the direction of earth's magnetic field has *even reversed itself*. It is believed that once in 10^5~10^6 yrs, electric currents deep in earth's core slow down; come to a stop and then pick up in opposite direction. This might be the cause of direction reversal of earth's magnetic field.

22.7 ELEMENTS OF EARTH'S MAGNETIC FIELD

The earth's magnetic field at a point on its surface is usually characterized by three quantities: (*a*) declination (*b*) inclination or dip and (*c*) horizontal component of the earth's magnetic field. These are known as the *elements of the earth's magnetic field*. Before discussing these elements of earth's magnetic field, we define certain terms:

(*a*) **Magnetic Meridian:** *It is vertical plane passing through the axis of a freely suspended magnet*, i.e. it is a plane containing the point *P* (the point of suspension) and the magnetic axis.

(*b*) **Geographical Meridian:** *It is the vertical plane passing through the geographical north-south*, i.e. it is a plane containing the axis of rotation of earth at the place.

(*c*) **Magnetic Equator:** It is a great circle on earth's surface, with centre at earth's centre and is perpendicular to the magnetic axis. The magnetic equator divides earth in to two hemisphere. The geographical northern hemisphere contains magnetic south polarity and the geographical southern hemisphere contains magnetic north polarity.

(*d*) The magnetic field of earth is not constant and changes irregularly from place to place on the surface of earth. The lines of force of earth's magnetic field are shown in Fig. 22.16. They are parallel to the earth's surface near the equator and perpendicular to it near poles. It also varies with time at a given place.

Declination or Variation

The declination at a given place is *the angle between the geographical meridian and magnetic meridian*. The declination at a place is expressed as θ° W or θ° E depending upon the north pole of the compass needle lies to the west (left) or to the east (right) of the geographical north-south direction. Figure 22.17 shows angle θ ($\angle BAB_1$), the declination at the place. Navigators use magnetic compass needle.

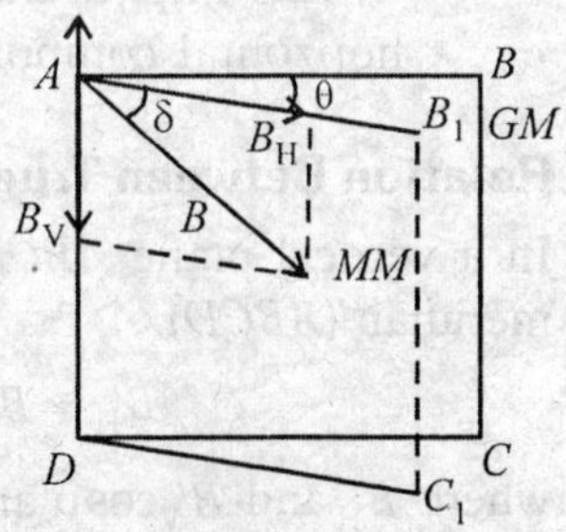

Fig. 22.17

Inclination or Angle of Dip

The angle made by the earth's magnetic field with the horizontal direction in the magnetic meridian, is called the inclination or dip angle at that point. Actually it is the angle that the axis of a freely suspended magnet (up or down) subtends with the horizontal in magnetic meridian at a given place. Figure 22.17 shows the direction of *B* (the magnetic field of earth) making angle δ with the horizontal. Hence δ gives the dip angle at that place.

We have already mentioned that geographical northern hemisphere contains magnetic south polarity as such north pole of a freely suspended magnet (or pivoted compass needle) will dip downwards, i.e. north pole of needle dips towards the earth, while opposite will take place in the southern hemisphere.

Angle of dip at a place is measured by the instrument called dip-circle. In *dip circle* a magnetic needle is free to rotate in a vertical plane, which can be set in any vertical plane.

The knowledge of declination and inclination completely specifies the direction of earth's magnetic field.

Horizontal Component of Earth's Magnetic Field

As the name suggests, *the horizontal component means the component of the earth's magnetic field in the horizontal direction of the magnetic meridian. This direction is towards magnetic north.*

Figure 22.17 shows all the three elements. If at a place magnetic field of earth is B and the angle of dip δ, then the horizontal component B_H and vertical component B_V is

$$B_H = B\cos\delta : B_V = B\sin\delta$$

hence $$B = \sqrt{B_V^2 + B_H^2} \quad \ldots(22.23)$$

and $$\tan\delta = \frac{B_V}{B_H} \quad \ldots(22.24)$$

The values of B_V, B_H and δ are different at difference places.

(*a*) At equator earth's magnetic field B is parallel to earth's surface, i.e. horizontal. Hence

$$B_H = B,\ \delta = 0 \text{ and } B_V = 0$$

This implies that at equator the freely suspended magnet will be horizontal and has no vertical component.

(*b*) At a place on poles, earth's magnetic field is perpendicular to the surface of earth, i.e. vertical. Hence $\delta = 90°$.

and $$B_V = B \text{ and } B_H = 0$$

This implies that at poles a freely suspended magnet will become vertical and has no horizontal component.

Relation Between True Dip and Apparent Dip

In a vertical plane $ABCD$ (Fig. 22.18) at an angle α to the magnetic meridian ($ABCD$).

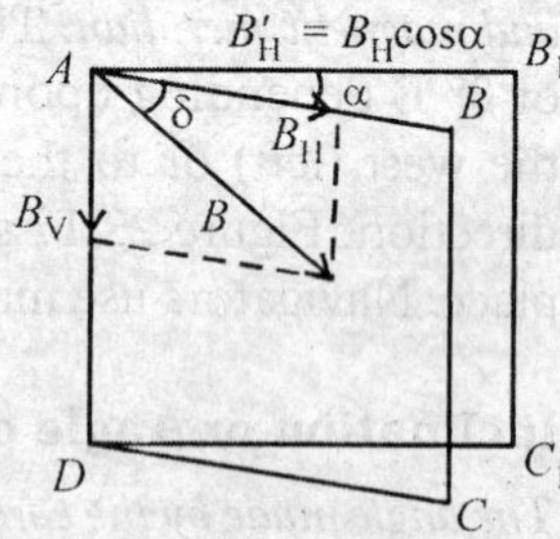

Fig. 22.18

$$B'_H = B_H\cos\alpha \text{ and } B'_V = B_V$$

where B'_H and $B'_V\cos\alpha$ are horizontal and vertical components of earth's magnetic field in plane $ABCD$.

Hence the angle of dip (δ_1) in this plane $ABCD$ is called the apparent dip,

$$\tan\delta_1 = \frac{B'_V}{B'_H} = \frac{B_V}{B_H\cos\alpha} = \frac{\tan\delta}{\cos\delta} \quad \ldots(22.25)$$

where $\tan\delta = \dfrac{B_V}{B_H}$ is the *true dip* at that place.

The equation (22.25) implies that in a plane other than magnetic meridian $\alpha > 0$, $\delta_1 > \delta$, i.e. the angle of dip is more than that in magnetic meridian. In the plane perpendicular to the magnetic meridian $\alpha = 90°$ and $\tan\delta_1 = \infty$ or $\delta_1 = 90°$, i.e. in this plane needle will remain vertical.

In a plane perpendicular to the plane *ABCD*, i.e. plane making an angle of (90 + α) with magnetic meridian, the apparent dip is δ_2, then

$$\tan\delta_2 = \frac{B'_V}{B'_H} = \frac{B_V}{B_H\cos(90+\alpha)} = -\frac{\tan\delta}{\sin\alpha} \quad ...(22.26)$$

Equations (22.25) and (22.26) give

$$\cot^2\delta = \cot^2\delta_1 + \cot^2\delta_2 \quad ...(22.27)$$

Relation Between Angle of Dip (δ) at a Place and its Latitude (λ)

Consider the magnetic moment of earth's magnetic dipole be *M*. Then B_r and B_θ are

$$B_r = \frac{2M\cos\theta}{r^3} \text{ and } B_\theta = \frac{M\sin\theta}{r^3}$$

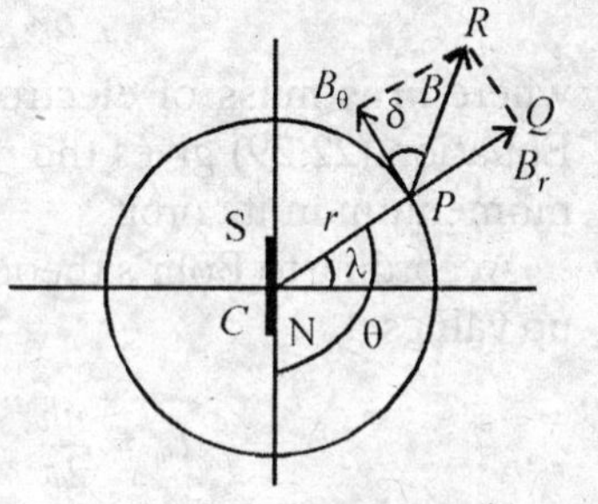

Fig. 22.19

and $$\tan\delta = -\frac{B_r}{B_\theta} = -2\cot\theta$$

But from diagram θ = 90 + λ

∴ $$\tan\delta = -2\cot(90+\lambda) = 2\tan\lambda \quad ...(22.28)$$

Magnetic Maps

We have defined the three elements of earth's magnetic field. They are declination, dip and horizontal component. They are found to be different at different places on the surface of earth. Magnetic maps showing the variation of these magnetic elements from place to place have been drawn. Generally lines are drawn (on geographical maps) joining all places having same value of an element. Such maps are called *magnetic maps.*

The lines joining places of equal declination are called *isogonic lines.*

The lines joining places of zero declination are called *agonic lines.*

The lines joining places of equal dip (inclication) are *called isoclinic lines.*

The lines joining places of zero dip are called *aclinic lines.*

The lines joining places of equal values of horizontal component (B_H) are called *isodynamic lines.*

However, earth's magnetic field at a given place is found to change with time, so magnetic maps are revised from time to time.

22.8 MAGNETIC PROPERTIES OF MATTER

(*a*) Atom as a Magnetic Dipole

Langevin for the first time proposed that magnetic behaviour of materials is primarily due to motion of the electrons. The electron revolving round the nucleus in a circular orbit is equivalent to a current-carrying circular loop and therefore possesses an orbital magnetic dipole moment. When electron revolves in anticlockwise direction, the current is clockwise. Therefore upper face of the electron loop acts as a south pole and lower face acts as north pole (Fig. 22.20).

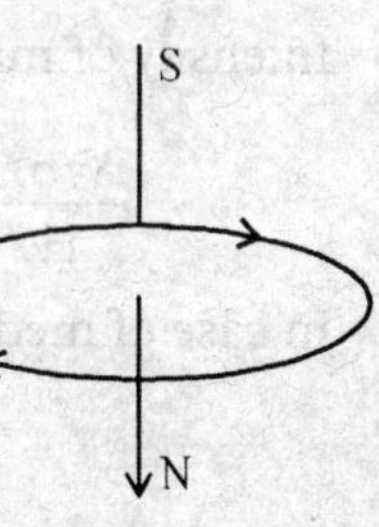

Fig. 22.20

If electron (charge e) is revolving in nth orbit of radius r_{n_1}, round the nucleus, with frequency f; then it is equivalent to current

$$i = ef$$

and area of the orbit is $A_n = \pi r_n^2$

Then magnetic moment of atom, due to electron in nth orbit is

$$M_n = iA = ef.\pi r_n^2$$

$$= \frac{e}{2m}(2\pi f)mr_n^2 = \frac{e}{2m}\cdot\omega_n mr_n^2 \quad \omega_n = 2\pi f$$

$$M_n = \frac{e}{2m}L_n \qquad \text{...(22.29)}$$

where m is mass of electron, $L_n = m\omega_n r_n^2$ is angular momentum of the electron in nth orbit. Equation (22.29) gives the relation between magnetic dipole moment of electron and its angular momentum in its orbit.

According to Bohr's theory, the angular momentum of electron in an orbit is quantified. It takes up values

$$L_n = \frac{nh}{2\pi} \qquad \text{where } n = 1, 2, 3, 4, \ldots, \text{etc.}$$

Hence magnetic moment of atomic dipole is

$$M_n = \frac{e}{2m}\frac{nh}{2\pi} = \frac{eh}{4\pi m}n \qquad \text{...(22.30)}$$

This shows that the magnetic moment of an atom is quantised and least value of atomic dipole moment is $\mu_B = \frac{eh}{4\pi m}$. This is called *Bohr Magneton*: and has value 9.27×10^{-24} Am^2.

Bohr magneton can be defined as the magnetic dipole moment associated with an atom due to orbital motion an electron in the first orbit of hydrogen atom.

(*b*) Intensity of Magnetising Field *H*

The degree to which a magnetic field can magnetise a material is represented in terms of magnetising force or intensity of magnetising field (***H***). In magnetic field the ratio of magnetising field $\boldsymbol{B}_0$ to the permeability of free space μ_0 is called intensity of magnetising field, i.e.

$$\boldsymbol{H} = \frac{\boldsymbol{B}_0}{\mu_0} \text{ or } \boldsymbol{B}_0 = \mu_0\boldsymbol{H} \qquad \text{...(22.31)}$$

Intensity of magnetising field (***H***) is a vector in the direction of magnetic field and has unit

$$\frac{\text{Wb/m}^2}{\text{H/m}} = \frac{\text{Ampare}}{\text{meter}} \text{ and dimensions } [\text{IL}^{-1}]$$

In case of medium of magnetic permeability μ; the intensity of magnetising field

$$\boldsymbol{H} = \frac{\boldsymbol{B}}{\mu} \qquad \text{...(22.32)}$$

Here B is the magnitude of total field inside the material.

(c) Intensity of Magnetisation (*I*)

When a magnetic material is magnatised by placing in a magnetising field, *the induced dipole moment per unit volume in the specimen is called intensity of magnetisation*, i.e.

$$\boldsymbol{I} = \frac{\boldsymbol{M}}{V} \quad \text{...(22.33)}$$

If m is pole strength induced and L the length of magnet, a be area of cross-section perpendicular to the direction of magnetisation then, $M = mL$ and $V = aL$.

or
$$I = \frac{mL}{aL} = \frac{m}{a} \quad \text{...(22.34)}$$

Thus the intensity of magnetisation of a magnetic material can be defined as *the pole strength per unit area of cross-section of the material.* Its unit is (A/m) and dimension (IL^{-1}).

(d) Magnetic Susceptibility: (χ_m)

It is a property which shows how easily a specimen can be magnetised. *It is defined as the ratio of the intensity of magnetisation (I) induced in the material to the magnetising force (H) applied.* Thus

$$\chi_m = \frac{I}{H} \quad \text{...(22.35)}$$

Since I is magnetic moment per unit volume, χ_m *is usually called volume susceptibility of material. The susceptibility of unit mass of specimen is called specific susceptibility* and *the susceptibility of 1 mole of specimen is called molar susceptibility*.

Susceptibility is a scalar quantity and is a dimensionless quantity.

(e) Magnetic Permeability (μ)

When a magnetic material is placed in a magnetising field, *the ratio of magnitude of total field inside the material to that of intensity of magnetising field is called permeability*.

$$\mu = \frac{B}{H} \text{ or } B = \mu H \quad \text{...(22.36)}$$

It measures the degree to which a magnetic material can be penetrated or permeated by magnetising field and is a scalar quantity. Its SI units are Henery per meter, $\left(\frac{H}{m}\right)$ and has dimensions $[M^1L^1T^{-2}I^{-2}]$.

The ratio of permeability of medium (μ) to that of free space (μ_0) is called relative permeability (μ_r), i.e.

$$\mu_r = \frac{\mu}{\mu_0} = \frac{\mu H}{\mu_0 H} = \frac{B}{B_0} \quad \text{...(22.37)}$$

Thus *relative permeability is also equal to the ratio of the magnitude of total field inside the material (B) to the magnetising field (B_0).*

It has no units and is a dimensionless quantity.

(f) Relation Between Permeability and Susceptibility

When a magnetic material is placed in a magnetising field for its magnetisation, the field inside the magnetic material is the resultant of the magnetising field $\boldsymbol{B}_0$ and the induced field $\boldsymbol{B}_i$, i.e.

$$\boldsymbol{B} = \boldsymbol{B}_0 + \boldsymbol{B}_i$$

But by definition $B_0 = \mu_0 H$ and $B_i = \mu_0 I$.* Hence

$$B = \mu_0 H + \mu_0 I = \mu_0(H + I)$$

or $$\frac{B}{H} = \mu_0\left(1 + \frac{I}{H}\right) \qquad \text{But } \frac{I}{H} = \chi_m \text{ –magnetic susceptibility}$$

and $$\frac{B}{H} = \mu$$

∴ $$\mu = \mu_0(1 + \chi_m)$$

or $$\frac{\mu}{\mu_0} = \mu_r = 1 + \chi_m \qquad \text{...(22.38)}$$

22.9 CLASSIFICATION OF MAGNETIC MATERIALS

On the basis of the behaviour of the substances in magnetic field, Faraday classified all magnetic substances into the three classes:

(a) Diamagnetic Substances

The substances which when placed in a magnetising field get feebly magnetised in a direction opposite to the magnetising field are called diamagnetic substances. These are the substances whose atoms/molecules/ions do not possess magnetic moment of their own. The resultant field in such materials is, therefore, smaller than the applied field. Examples of such substances are, bismuth, antimony, copper, gold, quartz, mercury, water, alcohol, air, hydrogen, etc.

In fact magnetic moments are induced in all materials whenever a magnetic field is applied. Thus diamagnetism is the property of all materials. However, if there is permanent atomic magnetic moment, paramagnetism or ferromagnetism is much stronger than diamagnetism and the material does not show diamagnetic properties.

Properties of Diamagnetic Substances

1. The diamagnetic substances are weakly repelled by the field as such when placed in non-uniform magnetic field have tendency to move from regions of strong field to the regions of weak field, because of this
   (a) A diamagnetic rod when free to rotate between the pole pieces of a magnet set itself perpendicular to the field. It is due to the fact that the field is strong at poles (Fig. 21.21).
   (b) A diamagnetic liquid in U-tube depresses in the bimb which between the poles of strong magnets (Fig. 22.22). In this situation, the field is strong between the poles.

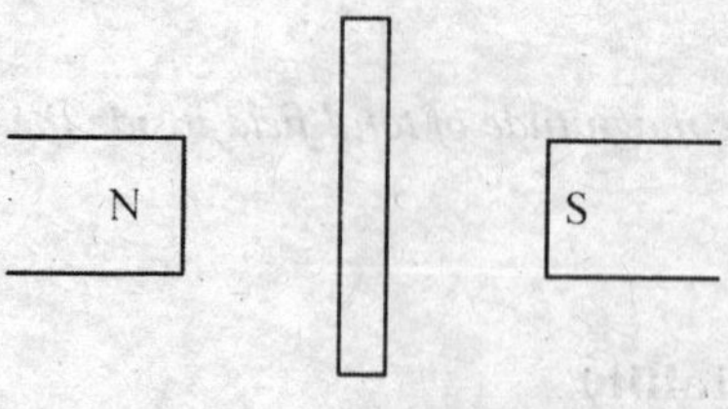

Fig. 22.21

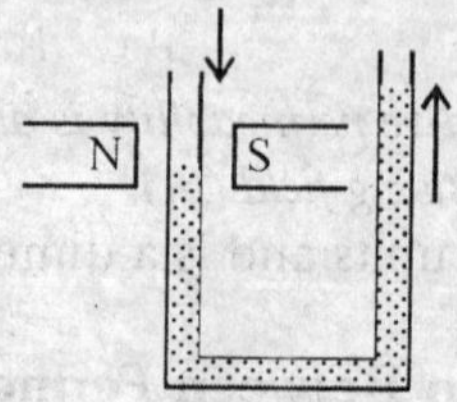

Fig. 22.22

*With unit pole lines of force associated are $\mu_0 I$ where I is the pole strength induced per unit area. Hence lines of force per unit area associated with the specimen are $\mu_0 I$, i.e. induced field is $\mu_0 I$.

2. Intensity of magnetisation I is very small, negative and directly proportional to magnetising field (Fig. 22.23 (*a*)).

3. Magnetic susceptibility $\chi_m \left(= \frac{I}{H}\right)$ is small and negative ($\approx 10^{-5}$) and is independent of temperature (Fig. 22.23 (*b*)).

Fig. 22.23

4. Relative permeability $\mu_r \left(= \frac{\mu}{\mu_0}\right)$ is slightly less than unity, i.e. $\mu < \mu_0$ as such $\chi_m = \mu_r - 1$ is small and negative. This implies that lines of force and hence B is less than the magnetic field B_0 outside the specimen. Thus a diamagnetic substance is placed in magnetic field has tendency to expell lines of force (Fig. 22.24 (*a*)).

Superconductors exhibit perfect diamagnetism, i.e. field inside is zero (Fig. 22.24 (*b*)). This was first of all discovered in 1933 by Meissner and Ochsenfeld and is called Meissner effect. For superconductors $B = 0$ and $I = -H$ and $\chi_m = -1$ and $\mu_r = 0$.

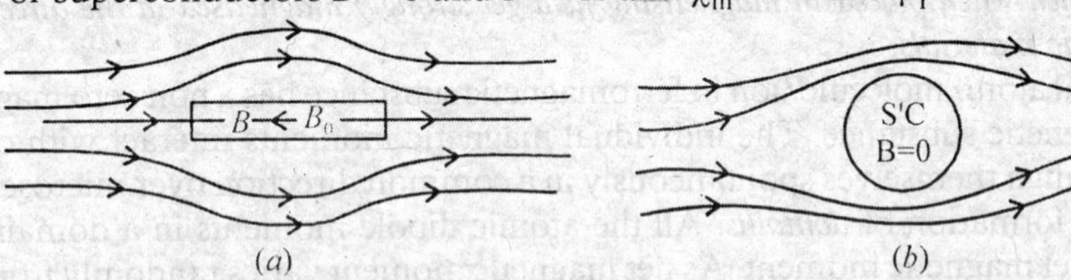

Fig. 22.24

5. Diamagnetism is exhibited by those specimen in which spins are paired, i.e. the atoms which do not have any permanent magnetic moment.

(*b*) Paramagnetic Substances

Paramagnetic substances are those in which each individual atom/molecule/ion has a net non zero magnetic moment of its own. When such a substance is placed in an external magnetic field of induction B, it tries to align individual dipole moments in the direction of the field. For strong values of B there is a net average magnetic dipole moment in the direction of the field B. Substances like, aluminium, platinum, chromium, manganese, copper sulphate, sodium, potassium are all paramagnetic substances.

Properties of Paramagnetic Substances

1. Paramagnetic substances are weakly attracted by the field. Paramagnetic substance when placed in non-uniform magnetic field have a tendency to move from weak to strong field.

Hence a paramagnetic rod sets itself parallel to the field if free to rotate between the poles of a strong bar magnet. Similarly a paramagnetic liquid in U-tube ascends in the limb which is between the poles of a strong magnet.

2. Intensity of magnetisation I is very small, positive and directly proportional to the magnetising field.
3. Magnetic susceptibility χ_m (= I/H) is small and positive (~10^{-3} to 10^{-5}). It is independent of the field but varies inversely with temperature

$$\chi_m = \frac{C}{T} \quad ...(22.39)$$

where C is constant and equation (22.39) is called Curie law.

The plot of $1/\chi_m$ versus T (Temperature in Kelvin) is a straight line (Fig. 22.25).

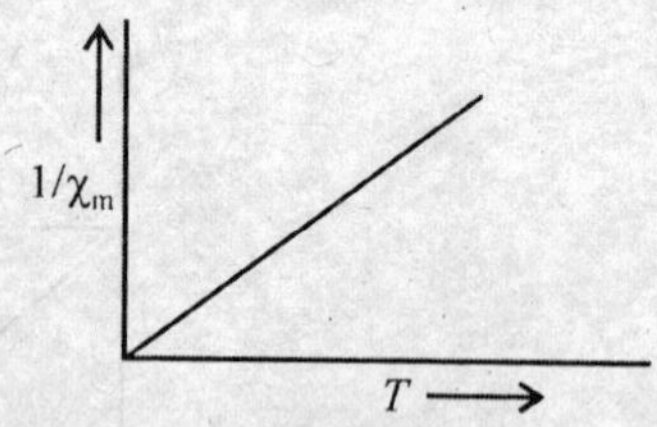

Fig. 22.25

4. Relative permeability μ_r (= μ/μ_0) is slightly greater than unity, i.e. $\mu > \mu_0$ hence $\chi_m = \mu_r - 1$ is positive but small. This means that the lines of force inside the specimen are more than outside or magnetic field inside specimen (B) is slightly greater than the magnising field outside (= B_0).
5. Paramagnetism is due to partial alignment of randomly oriented atomic dipoles along the field. So atoms of paramagnetic substances have permanent dipole moment, i.e. unpaired spin. The paramagnetism is adversely affected due to rise in temperature which randomises the orientation of dipoles.

(c) Ferromagnetic Substances

The substances which when placed in magnetising field get strongly magnetised in the direction of field are called ferromagnetic materials.

Each individual atom/molecule/ion in ferromagnetic substance has a non zero magnetic moment like in a paramagnetic substance. The individual magnetic moments interact with one another in such a way as to align themselves spontaneously in a common direction over microscopic volumes. This leads to the formation of *domains*. All the atomic dipole moments in a domain are lined up leading to some net magnetic moment. As net magnetic moment varies randomly from one domain to the other, there is no bulk magnetic moment of ferromagnetic substance. When a ferromagnetic material is placed in an external magnetic field of induction B, magnetic moments of different domains are aligned and material gets strongly magnetised. Examples of ferromagnetic substances are, iron, nickel, cobolt and a number of their alloys.

Properties of Ferromagnetic Substances

1. Ferromagnetic substances are strongly attracted by the field hence in non-uniform field, these stick to the poles because the field there is strongest.
2. Intensity of magnetisation I is very large, positive and varies non-linearly with magnetising field intensity or temperature (Fig. 22.26). However in each case it has an upper limit *called saturation magnetisation* (I_s). Magnetic saturation depends on the nature of the material.
3. Susceptibility χ_m of the ferromagnetic substances is very large (~10^3-10^5). It is dependent on temperature and varies in a complex way with temperature, for temperature $T < T_C$; the variation in magnetic susceptibility (χ_m) obeys Curie-Weiss law, i.e.

$$\chi_m \propto \frac{1}{T - T_C} \text{ or } \chi_m = \frac{C}{T - T_C} \quad ...(22.40)$$

where T_C is constant called *Curie temperature*. It depends on the nature of material. For iron it is 1043 K (770° C). C is also a constant. Above *Curie temperature* ($T > T_C$) *the ferromagnetic material behaves as a paramagnetic substance.* Figure 22.27 gives the variation of $1/\chi_m$ with temperature.

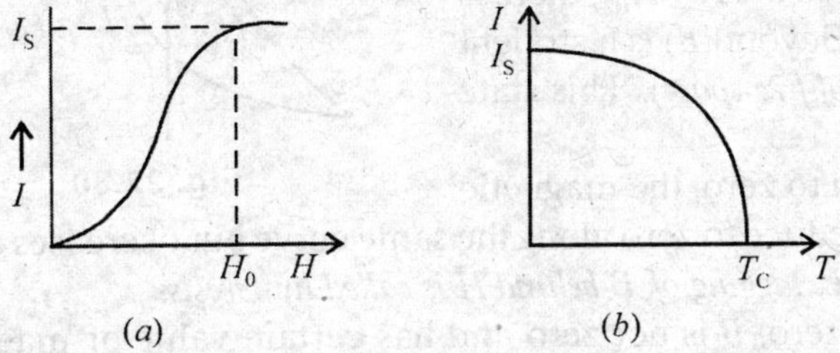

(a) (b)

Fig. 22.26

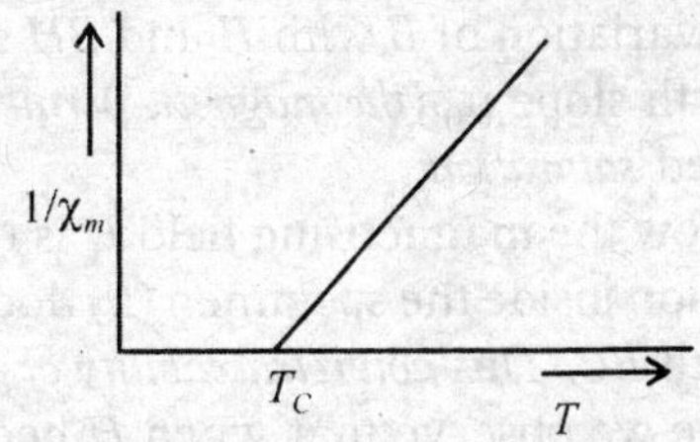

Fig. 22.27

4. Relative permeability $\mu_r \left(= \dfrac{\mu}{\mu_0}\right)$ is very large (~10^3-10^5). $\chi_m = \mu_r - 1$ is also very large. μ_r varies non-linearly with magnetising field (Fig. 22.28). Large value of μ_r implies that the field inside the specimen (B) is much stronger than the magnetising field B_0.

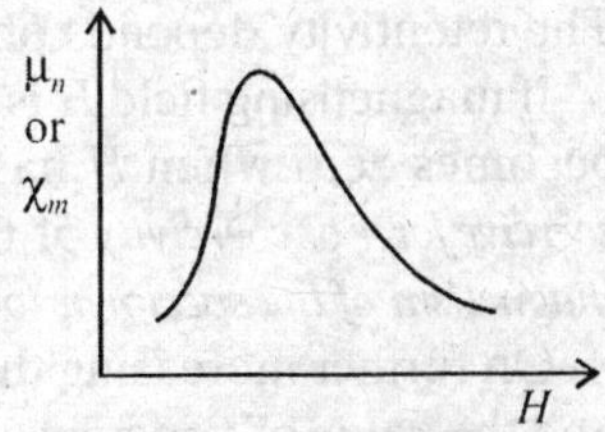

Fig. 22.28

5. In contrast to dia and paramagnetic substances, ferromagnetic substances can have a net magnetisation even in the absence of an external field (*called spontaneous magnetisation*) and are always crystalline solids.
6. Ferromagnetism is due to *spin orbit interaction*. According to Weiss, because of spin interaction of electrons of one atom with that of a neighbouring atoms, ferromagnetic materials get divided into small regions called *domains*; each of which is at all times completely magnetised, but the direction of magnetisation varies from domain to domain so the resultant magnetisation of the material is zero. When a magnetising field is applied, the magnetic material gets magnetised either by domain growth through the wall displacement (in weak fields) or by domain rotation (in strong fields) (Fig. 22.29).

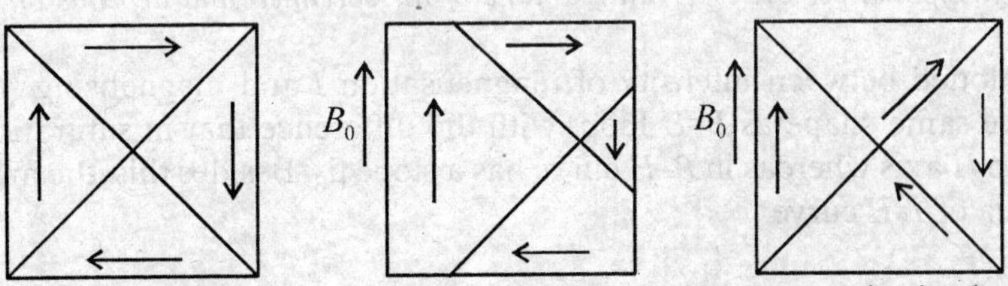

Unmagnetised material | Magnetisation by Domain growth | Magnetisation by Domain rotation

Fig. 22.29

22.10 MAGNETIC HYSTERESIS AND *B-H* CURVE

Some peculiar properties of ferromagnetic materials are commonly studied from *B-H curve or hysteresis curve*. As already stated the magnetic field (***B***) inside the material increases non-linearly with magnetising field ***H***.

When $\boldsymbol{H}$ is increased from O the magnetic induction B also increases. This variation is shown by graph OAB in Fig. 22.30. After a definite increase in magnetising field $H (= H_0)$, there is no further magnetisation of the specimen, i.e. intensity of magnetisation (I) remains constant as such beyond this value of $H (= H_0)$, there is linear variation of B with H and BH curve (Beyond B) is a straight line with slope μ_0 (*the magnetic permeability of free space*). This state is called *saturation*.

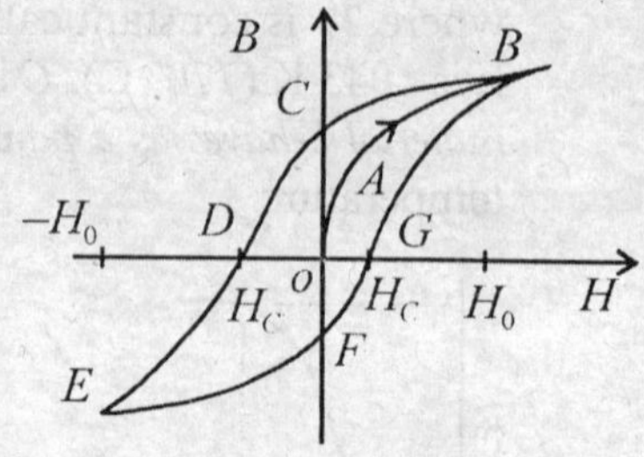

Fig. 22.30

If now the magnetising field H is reduced to zero, the magnetic induction inside the specimen (B) does not reduce to zero along the same curve but decreases along the path BC. *This non retraceability of path*, i.e. *lagging of B behind H is called hysteresis.*

Here we observe that when H becomes zero, B is not zero and has certain value of magnetic induction equal to OC. This remaining field ($= B_r$) in the specimen in the absence of any external field is called *the residual magnetism, retentivity or remanence.* This retentivity of a specimen is a measure of the magnetic field remaining in the specimen when the magnetising field is removed. The retentivity depends on the nature of the material.

If magnetising field H is increased in the reversed direction, the value of B further decreases and becomes zero when H has the value equal to OD ($= H_C$). This value of magnetic field is called *coercive force or coercivity* of the specimen. *The coercivity in the magnetic field required to reduce the residual magnetism of the specimen to zero. It also depends on the nature of the specimen.*

On further increasing the magnetising field, the specimen gets magnetised in opposite direction and the value of B increases with H upto saturation state E. Now when H is reduced to demagnetise the specimen, the curve is not retraced but we get $EFGB$. In this situation also, to reduce residual magnetism to zero, we require mangetising field H equal to OG ($= H_C$). Now to magnetise a specimen to saturation again we have to increase H but now the curve obtained is GB. In this way starting from B, one cycle of magnetisation and demagnetisation is represented by cycle $BCDEFGB$. This loop is called *B-H loop* or *hysteresis loop*.

In this process of taking a ferromagnetic substance through a cycle of magnetisation and demagnetisation involves expenditure of energy which is converted into heat and cannot be recovered. This loss of energy is called *hysteresis loss. The area of B–H curve (hysteresis loop) is the measure of energy dissipated per cycle per unit volume of the specimen and depends on the nature of the material.*

If a graph is plotted between intensity of magnetisation I and magnetising field H; the loop obtained will have same shape as B–H loop; with the difference that in saturation state the I–H curve is parallel to H-axis whereas in B–H curve has a slope μ_0. Besides this, the area of B–H curve is μ_0 times the area of I–H curve.

Classification of Ferromagnetic Substances

The ferromagnetic substances have been divided into two groups depending on their retentivity, coercivity and hysteresis loss. They are:

(*a*) **Soft Magnetic Materials:** These are the substances which have low coercivity, low retentivity and low hysteresis loss. Soft iron, mu-metal and stalloy are the examples of soft ferromagnetic materials. These are suitable for making electromagnets, cores of transformers, motors and generators.

(*b*) **Hard Magnetic Materials:** Those ferromagnetic substances which have high retentivity, high coercivity and high hysteresis loss come under this class. Steel, alnico alcomax and ticonal are the examples of hard ferromagnetic substances. These materials are used for making permanent magnets and various kinds of electric meters, loudspeakers, etc.

Note: For magnetic tape the specimen must possess, high retentivity and low coercivity, so that signal recorded on it can be changed at will. Soft iron and special alloys are used for this.

SOLVED EXAMPLES

Example 22.1: A roland ring of ferromagnetic substance has 10^3 turns per metre. A current of 2.0 A passes through the coil. This produce a magnetic field of 10 Weber/m^2. Calculate (*i*) magnetising force, (*ii*) intensity of magnetisation (*iii*) relative permeability of the specimen ($\mu_0 = 4\pi \times 10^{-7}$ Henry/m).

Solution: (*i*) In roland ring (or torroid)

$$B = \mu_0 ni \quad \text{or} \quad H = \frac{B}{\mu_0} = ni$$

where n is number of turns per unit length and i is current.

$\therefore$ Magnetising field $H = ni = 10^3 \times 2 = 2000$ A/m.

(*ii*) In an specimen

$$B = \mu_0(H + I)$$

or
$$I = \frac{B}{\mu_0} - H = \frac{10}{4\pi \times 10^{-7}} - 2000$$

$$= 7.96 \times 10^6 - 0.002 \times 10^6$$

$$= (7.962 - 0.002) \times 10^6 = 7.96 \times 10^6 \text{ A/m}.$$

(*iii*)
$$\mu_r = \frac{\mu}{\mu_0} = \frac{1}{\mu_0}\left(\frac{B}{H}\right)$$

or
$$\mu_r = \frac{1 \times 10}{4\pi \times 10^{-7} \times 2 \times 10^3}$$

$$= \frac{10^5}{8\pi} = 3.98 \times 10^3$$

Example 22.2: The atomic magnetic moment of iron is 1.8×10^{-23} A-m^2. Calculate the saturation magnetic moment of iron specimen, 5 cm long and 1 sq.cm cross-section. Atomic weight of iron is 56 and density 7.6 g/cc.

Solution: In saturation state all the atomic dipoles are aligned parallel to the magnetising field. If n is the number of atoms and each atom has magnetic moment μ then total magnetic moment of specimen is

$$M = \mu n \quad \ldots(i)$$

Volume of rod is $V = 5 \times 1 = 5$ cc

Mass of rod $m = V \cdot d = 5 \times 7.6 = 38$ g

Since atomic weight of iron is 56, hence 56 g of iron contain n_0 (= Avogadro) number of atoms. Hence $m(= 38$ g) mass of iron contains

$$n = \frac{n_0 \times 38}{56} \text{ number of atoms } = \frac{6.02 \times 10^{23} \times 38}{56} \text{ atoms}$$

where $n_0 = 6.02 \times 10^{23}$.

Hence net dipole moment of specimen is

$$M = \mu n = \frac{1.8 \times 10^{-23} \times 6.02 \times 10^{23} \times 38}{56}$$

$$= 7.353 \text{ A-m}^2.$$

Example 22.3: A magnetising field of 2×10^3 A/m produces a magnetic flux 6.0×10^{-4} Weber in an iron rod. The area of cross-section of rod is 2×10^{-5} m^2. Find the relative magnetic permeability and intensity of magnetisation of the specimen.

Solution: If B is magnetic induction, then flux

$$\phi = BA \text{ or } B = \frac{\phi}{A} = \frac{6.0 \times 10^{-4}}{2 \times 10^{-5}} = 30 \text{ T}$$

Given $H = 2 \times 10^3$

$$\therefore \quad \mu = \frac{B}{H} \text{ and } \mu = \mu_0 \mu_r$$

$$\therefore \quad \mu_r = \frac{\mu}{\mu_0} = \frac{B}{\mu_0 H} = \frac{30}{4\pi \times 10^{-7} \times 2 \times 10^3}$$

$$= 1.19 \times 10^4.$$

Intensity of magnetisation

$$I = \frac{B}{\mu_0} - H \qquad [\because B = \mu_0(H + I)]$$

$$= \frac{30}{4\pi \times 10^{-7}} - 2 \times 10^3$$

$$= 2.39 \times 10^7 \text{ A/m}.$$

Example 22.4: Two short magnets A and B of moments 0.72 A-m^2 and 0.34 A-m^2 are placed side by side with their axes parallel to magnetic meridian separated by a distance of 0.2 m. If north pole of A points geographical north and that of B points geographical south, find the resultant field at a point midway between them if horizontal component of earth's field is 0.38×10^{-4} T.

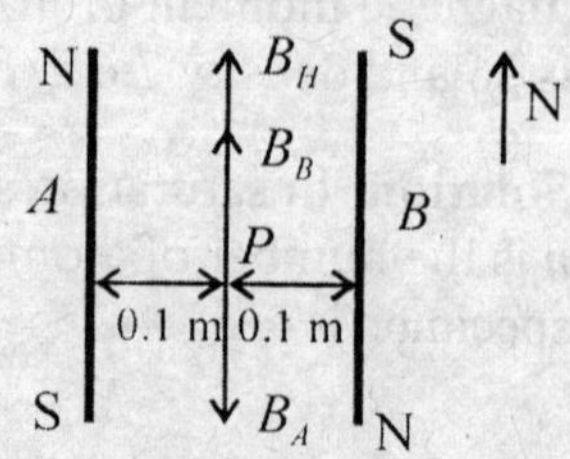

Fig. 22.3

Solution: The situation is as shown in Fig. 22.31. P is the point where net field is to be determined. For both magnets the point is in tan-B position. Hence field at P due to magnet A is

$$B_A = \frac{\mu_0}{4\pi} \frac{M}{r^3} = \frac{10^{-7} \times 0.72}{(0.1)^3} = 0.72 \times 10^{-4} \text{ T}$$

The direction of B_A is towards geographical south.

Similarly the field at P due to magnet B is

$$B_B = \frac{\mu_0}{4\pi}\frac{M}{r^3} = \frac{10^{-7} \times 0.34}{(0.1)^3} = 0.34 \times 10^{-4}\ \text{T}$$

The direction B_B is towards geographical north.

Magnitude field due to earth is

$$B_H = 0.38 \times 10^{-4}\ \text{T towards geographical north.}$$

Hence resultant field P is

$$B = B_H + B_B - B_A$$
$$= 0.38 \times 10^{-4} + 0.34 \times 10^{-4} - 0.72 \times 10^{-4} = 0$$

Hence point P is a neutral point.

Example 22.5: A magnet is suspended such that it oscillates in a horizontal plane. When at a place where the angle of dip is 30°, it gives 20 oscillations per minute and where angle of dip is 60° it gives 15 oscillations per minute. Compare the total field strength of earth's magnetic field at the two places.

Solution: When a magnet oscillates in earth's magnetic field, the time period is given by

$$T = 2\pi\sqrt{\frac{I}{MB_H}}$$

where I is moment of inertia of the magnet, M its magnetic moment and B_H is horizontal component of earth's magnetic field.

If B is total field strength of earth's magnetic field and δ angle of dip at the place then $B_H = B\cos\delta$.

Hence time period of oscillation is

$$T = 2\pi\sqrt{\frac{I}{MB\cos\delta}}$$

If T_1 and T_2 are the time periods at two places where earth's total fields are B_1 and B_2 and angles of dip δ_1 and δ_2, then

$$T_1 = 2\pi\sqrt{\frac{I}{MB_1\cos\delta_1}} \quad \text{and} \quad T_2 = 2\pi\sqrt{\frac{I}{MB_2\cos\delta_2}}$$

$$\therefore \quad \frac{T_1}{T_2} = \sqrt{\frac{B_2\cos\delta_2}{B_1\cos\delta_1}}$$

$$\text{or} \quad \frac{T_1^2}{T_2^2} = \frac{B_2}{B_1}\frac{\cos\delta_2}{\cos\delta_1}$$

$$\text{or} \quad \frac{B_1}{B_2} = \frac{T_2^2}{T_1^2}\frac{\cos\delta_2}{\cos\delta_1}$$

$$T_1 = \frac{60}{20} = 3\ \text{s} \quad \text{and} \quad T_2 = \frac{60}{15} = 4\ \text{s}$$

$$\therefore \quad \frac{B_1}{B_2} = \left(\frac{4}{3}\right)^2\frac{\cos 60°}{\cos 30°} = \frac{16}{9}\cdot\frac{1/2}{\sqrt{3}/2} = \frac{16}{9\sqrt{3}}$$

Example 22.6: Find the force of attraction or repulsion between two short magnets of moments 10 A-m^2 and 20 A-m^2 with their centres 0.2 m apart and their axis in the same direction and along the same line.

Solution: Magnetic field at distance r from a magnetic dipole of moment M_1 along its axis is

$$B = \frac{\mu_0}{4\pi}\frac{2M}{r^3}$$

When another magnetic dipole of moment M_2 is placed in this magnetic field, the potential energy is

$$U = -\boldsymbol{M}\cdot\boldsymbol{B}$$

$$= -M_2\frac{\mu_0}{4\pi}\frac{2M_1}{r^3} = -\frac{\mu_0}{4\pi}\frac{2M_1M_2}{r^3}$$

and force $F = -\dfrac{dU}{dr} = \dfrac{\mu_0}{4\pi}\dfrac{6M_1M_2}{r^4}$

$$= 10^{-7} \times \frac{6\times 10\times 20}{(0.2)^4}$$

$$= 7.5 \times 10^{-2}\text{ N.}$$

Example 22.7: A short bar magnet of magnetic moment 2×10^{-7} Wb-m lying in a horizontal plane with its north pole pointing 60° east of north. Find the net horizontal magnetic field at a point north of magnet 0.2 m away from it. Horizontal component of earth's magnetic field = 0.3×10^{-4} T.

Solution: The situation is shown in Fig. 22.32. The components of magnetic field due to bar magnet at point P are

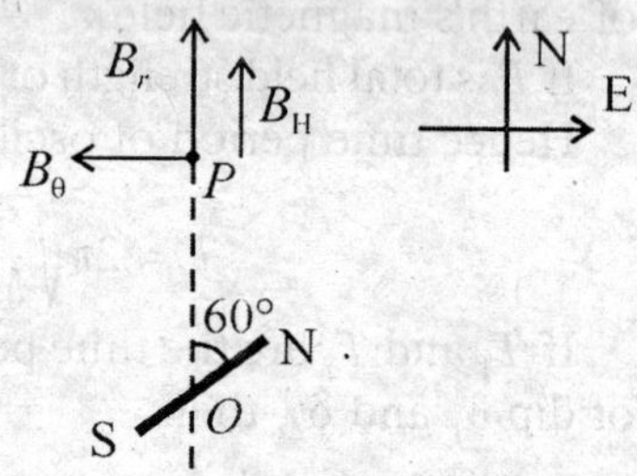

Fig. 22.32

$$B_r = \frac{\mu_0}{4\pi}\frac{2M\cos\theta}{r^3} \quad\text{and}\quad B_\theta = \frac{\mu_0}{4\pi}\frac{M\sin\theta}{r^3}$$

in formulae given above M—magnetic moment is in A-m^2. But we are given magnetic moment in Wb-m, hence

$$M = \frac{2\times 10^{-7}}{\mu_0} = \frac{2\times 10^{-7}\times 10^7}{4\pi} = \frac{2}{4\pi}\text{ Amp-m}^2$$

$$\therefore \quad B_r = 10^{-7}\times 2\times\frac{2}{4\pi}\frac{\cos(60°)}{(0.2)^3} = \frac{10^{-7}\times 2}{4\pi\times 10^{-3}\times 8} = \frac{10^{-4}}{16\pi}$$

$$= 0.02 \times 10^{-4}\text{ T} = 2 \times 10^{-6}\text{ T}$$

and $$B_\theta = 10^{-7}\times\frac{2}{4\pi}\frac{\sin(60°)}{(0.2)^3} = \frac{10^{-7}\times\sqrt{3}}{10^{-3}\times 4\pi\times 8}$$

$$= \frac{\sqrt{3}}{32\pi}\times 10^{-4} = 1.7\times 10^{-6}\text{ T}$$

The horizontal component of earth's magnetic field = 30×10^{-6} is in the direction of B_r. Hence net magnetic field in direction r is

$$B_r' = B_r + B_H = 2\times 10^{-6} + 30\times 10^{-6} = 32\times 10^{-6}\text{ T}$$

Hence net field at P is

$$B = \sqrt{B_r'^2 + B_\theta^2} = 10^{-6}\sqrt{(32)^2 + (1.7)^2} = 32.05 \times 10^{-6}\ \text{T}$$

The angle that this resultant B makes with OP is

$$\tan\theta = \frac{B_\theta}{B_r} = \frac{1.7 \times 10^{-6}}{32 \times 10^{-6}} = \frac{1.7}{32} = 0.053$$

or $$\theta = \tan^{-1}(0.053) = 3°.$$

Example 22.8: A 6 cm long bar magnet possessing magnetic moment 0.3 A-m^2 is placed vertically on a horizontal wooden table. The north pole of the magnet touches the table. A neutral point is found on the table at a distance of 8 cm south of the magnet. Find the horizontal component of earth's magnetic field. (Roorkee 2001)

Solution: If m is the pole strength of the magnet and L its length, the magnetic moment $M = mL$

In Fig. 22.33, the magnetic field at P due to isolated N-pole N is

$$B_\text{N} = \frac{\mu_0}{4\pi}\frac{m}{NP^2} = \frac{\mu_0}{4\pi}\frac{m}{d^2} \text{ along } NP$$

Field at P due to south pole is

$$B_\text{S} = \frac{\mu_0}{4\pi}\frac{m}{SP^2} = \frac{\mu_0}{4\pi}\frac{m}{(d^2 + L^2)} \text{ along } PS$$

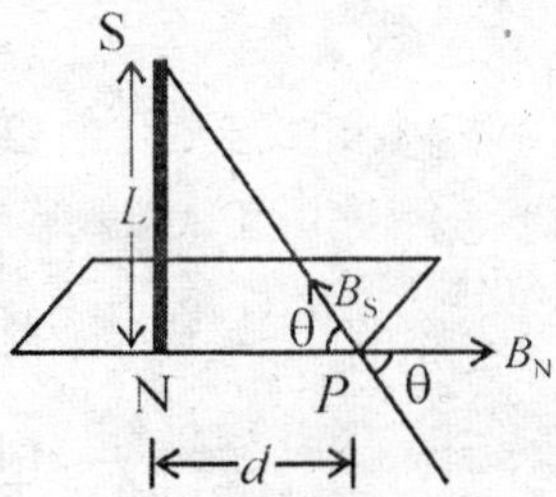

Fig. 22.33

Its component along $PN = B_\text{S}\cos\theta = \dfrac{\mu_0 m}{4\pi(d^2 + L^2)}\cos\theta$

$$\cos\theta = \frac{d}{\sqrt{d^2 + L^2}}$$

Hence component of field due to S-pole along PN

$$= \frac{\mu_0 m d}{4\pi(d^2 + L^2)^{3/2}}$$

Hence resultant horizontal field at P is

$$B_\text{H} = \frac{\mu_0 m}{4\pi}\left[\frac{1}{d^2} - \frac{d}{(d^2 + L^2)^{3/2}}\right]$$

or $$B_\text{H} = \frac{\mu_0 M}{4\pi L}\left\{\frac{1}{d^2} - \frac{d}{(d^2 + L^2)^{3/2}}\right\}$$

Since point P is a neutral point. Hence B_H is horizontal component of earth's magnetic field. Hence

$$B_\text{H} = \frac{10^{-7} \times 0.3}{6 \times 10^{-2}}\left[\frac{1}{(8 \times 10^{-2})^2} - \frac{8 \times 10^{-2}}{(10 \times 10^{-2})^3}\right]$$

$$= 3.81 \times 10^{-5}\ \text{T}.$$

Example 22.9: A small bar magnet having a magnetic moment of 9×10^{-9} Wb-m is suspended at its centre of gravity by a light torsion-less string at distance of 10^{-2} m. Vertically above a long straight horizontal wire carrying a current of 1.0 A. Find the frequency of oscillation of magnet about its equilibrium position assuming that the motion is undamped. The moment of inertia of magnet is 6×10^{-9} kgm^2.

Solution: Magnetic field produced by a current carrying wire at distance r from it is

$$B = \frac{\mu_0 i}{2\pi r} \quad \text{...}(i)$$

In this magnetic field the magnet oscillates and time period is given by

$$T = 2\pi\sqrt{\frac{I}{MB}}$$

or frequency $$f = \frac{1}{T} = \frac{1}{2\pi}\sqrt{\frac{MB}{I}} \qquad \left(\because B = \frac{\mu_0 i}{2\pi r}\right)$$

$$= \frac{1}{2\pi}\sqrt{\frac{M \cdot \mu_0 i}{2\pi r \cdot I}} \qquad \frac{\mu_0}{4\pi} = 10^{-7} \text{ H/m}$$

$$= \frac{1}{2\pi}\sqrt{\frac{2 \times 10^{-7} \times 9 \times 10^{-9} \times 1}{10^{-2} \times 6 \times 10^{-9}}}$$

$$= \frac{1}{2\pi}\sqrt{3.0 \times 10^{-5}}$$

$$= 8.7 \times 10^{-4} \text{ Hz.}$$

Example 22.10: Figure 22.34 shows two identical magnetic dipoles a and b of magnetic moments M each placed at a separation d with their areas perpendicular to each other. Find the magnetic field at the point P midway between the dipoles.

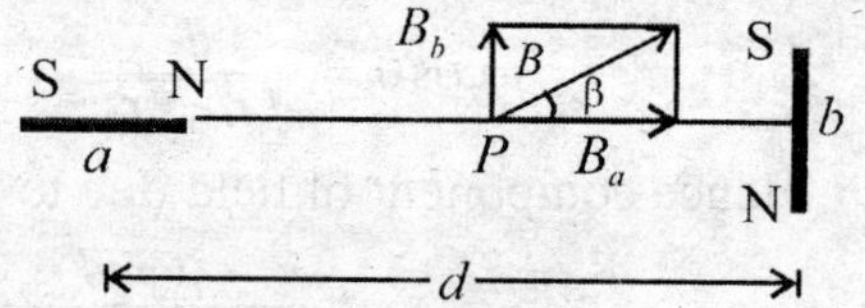

Fig. 22.34

Solution: The point P is in end on position for the dipole a and in broad side on position for the dipole b. The magnetic field at P due to a is

$$B_a = \frac{\mu_0}{4\pi}\frac{2M}{(d/2)^3} \quad \text{along the axis of } a$$

and that due to b is

$$B_b = \frac{\mu_0}{4\pi}\frac{M}{(d/2)^3} \quad \text{parallel to the axis of } b \text{ (Fig. 22.34).}$$

Resultant field at P is therefore

$$B = \sqrt{B_a^2 + B_b^2} = \frac{\mu_0}{\pi}\frac{2M}{d^3}\sqrt{1 + (2)^2}$$

$$= \frac{2\sqrt{5}\,\mu_0 M}{\pi d^3}$$

The direction of this field makes an angle β with B_a such that $\tan\beta = \frac{B_b}{B_a} = \frac{1}{2}$.

Example 22.11: The earth's magnetic field at geomagnetic poles has a magnitude 6.2×10^{-5} T. Find the magnitude and direction of the field at a point on the earth's surface where the radius makes an angle of 135° with the axis of the earth's assumed magnetic dipole. What is the inclination (dip) at this point?

Solution: Earth's magnetic field is assumed to be due to a dipole placed at the centre of earth with N-pole in Southern hemisphere (Fig. 22.35). The field at geomagnetic poles is 6.2×10^{-5} T and these poles are in end on position. If M is magnetic moment of earth's magnetic dipole then field B_P at poles is

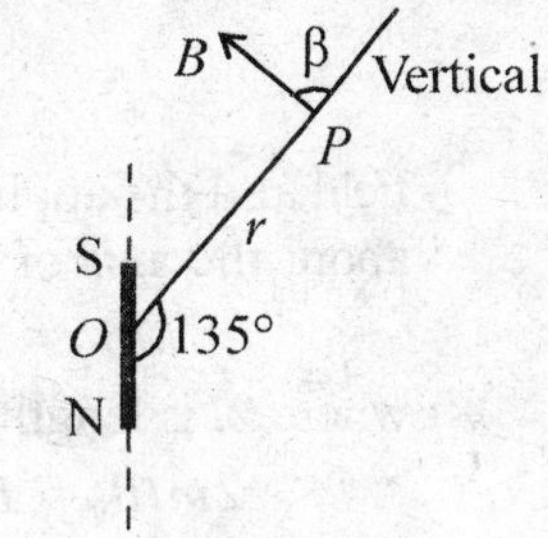

Fig. 22.35

$$B_P = \frac{\mu_0}{4\pi}\frac{2M}{r^3} \quad \text{where } r \text{ is radius of earth.}$$

$$\therefore \quad \frac{\mu_0 M}{4\pi r^3} = \frac{B_P}{2} = \frac{6.2 \times 10^{-5}}{2} = 3.1 \times 10^{-5} \text{ T}$$

The magnetic field at point P is

$$B = \frac{\mu_0}{4\pi}\frac{M}{r^3}\sqrt{1 + 3\cos^2\theta} \quad \text{where } \theta = 135°$$

$$= 3.1 \times 10^{-5}\sqrt{1 + 3\cos^2 135°} \quad \because \cos 135° = -\frac{1}{\sqrt{2}}$$

$$= 3.1 \times 10^{-5} \times \sqrt{2.5}$$

$$= 4.9 \times 10^{-5} \text{ T.}$$

This field is in direction making an angle β with the radial direction

Such that $\tan\beta = (\tan\theta)/2$

$$= \frac{1}{2}\tan(135°) = -0.5$$

or $$\beta = 153°$$

The inclination (dip) is the angle made by earth's magnetic field with the horizontal plane. Hence the dip angle.

$$\delta = 153° - 90° = 63° \text{ below the horizontal.}$$

Example 22.12: A magnetic needle of length 10 cm suspended at its middle point through a thread, stays at an angle of 45° with the horizontal. The horizontal component of the earth's magnetic field is 18 μT. (*a*) Find the vertical component of this field. (*b*) If the pole strength of the needle is 1.6 A-m. What vertical force should be applied to an end so as to keep it in horizontal position.

Solution: (*a*) When no field is applied, the needle stays in equilibrium in the direction of resultant magnetic field of earth. The needle in absence of field makes an angle of 45° with the horizontal. Hence the angle of dip δ = 45°. From definition of dip, we have

$$\tan\delta = \frac{B_V}{B_H} \quad \text{or} \quad B_V = B_H = 18\ \mu\text{T}$$

(*b*) When force F is applied, the needle stays horizontal, this means the torque due to magnetic

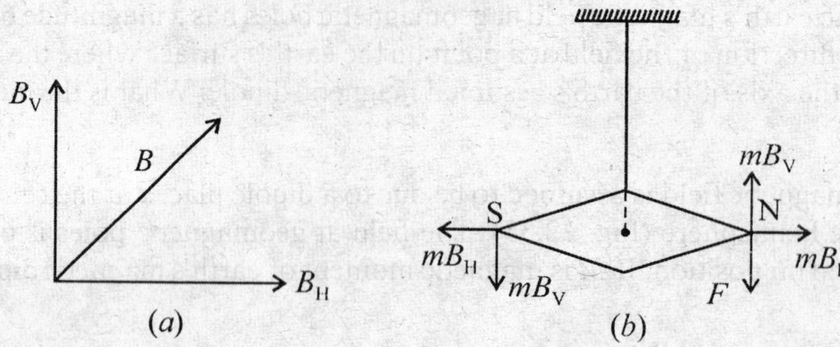

Fig. 22.36

field and the applied force are equal in magnitude and opposite in direction. Taking torques about the axis of rotation, we have (Fig. 22.36 (*b*))

$$MB_V = Fl, \quad \text{but} \quad M = 2ml$$

where $2l$ is length of the magnetic needle and m its pole strength.

$$\therefore \quad 2mlB_V = Fl$$

or $$F = 2mB_V = 2 \times 1.6 \times 18 \times 10^{-6}$$

$$= 5.76 \times 10^{-5} \text{ N.}$$

Example 22.13: A torroid has a mean radius R equal to $20/\pi$ cm and a total of 400 turns of wire carrying a current of 2.0 A. An aluminium ring at temperature 280 K inside the torroid provides the core.

(*a*) If the magnetisation I is 4.8×10^{-2} A/m, find the susceptibility of aluminium at 280 K.
(*b*) If the temperature of aluminium ring is raised to 320 K, what will be the magnetisation?

Solution: (*a*) Magnetic intensity H in the core is

$$H = ni$$

where n is number of turns per unit length $n = \dfrac{400}{2\pi R} = \dfrac{400}{2\pi \times \dfrac{20}{\pi} \times 10^{-2}} = 1000$

and $$H = 1000 \times 2 = 2000 \text{ A/m}$$

and susceptibility $\chi_m = \dfrac{I}{H} = \dfrac{4.8 \times 10^{-2}}{2 \times 10^3} = 2.4 \times 10^{-5}$

(*b*) In case of paramagnetic substances $\chi_m = \dfrac{C}{T}$, where T is temperature

$$\therefore \quad \frac{\chi_2}{\chi_1} = \frac{T_1}{T_2} \quad \text{and} \quad \chi_2 = \frac{\chi_1 T_1}{T_2}$$

Hence susceptibility at temperature 320 K is

$$\chi_2 = 2.4 \times 10^{-5} \times \frac{280}{320} = 2.1 \times 10^{-5}$$

Hence magnetisation at 320 K is

$$I = \chi_2 H = 2.1 \times 10^{-5} \times 2000$$

$$= 4.2 \times 10^{-2} \text{ A/m.}$$

Example 22.14: A bar magnet of length 1 cm and cross-sectional area 1 cm^2 produces a magnetic field of 1.5×10^{-4} T at a point in end on position at a distance 15 cm away from the centre. (*a*) Find the magnetic moment M of the magnet. (*b*) Find the magnetisation I of the magnet. (*c*) Find the magnetic field B at the centre of the magnet.

Solution: (*a*) A bar magnet of magnetic moment M produces a field in end on position at a distance r from centre

$$B = \frac{\mu_0}{4\pi}\frac{2M}{r^3}$$

given is $B = 1.5 \times 10^{-4}$, and $r = 15$ cm $= 1.5 \times 10^{-1}$ m

$$\therefore \quad 1.5\times10^{-4} = \frac{10^{-7}\times 2M}{(1.5\times10^{-1})^3}$$

or

$$M = \frac{(1.5)^4}{2} = 2.5 \text{ A-m}^2$$

(*b*) Magnetisation I is magnetic moment per unit volume.

Volume of magnet is $V = 1 \times 1 = 1$ cc $= 10^{-6}$ m^3

$$\therefore \quad I = \frac{M}{V} = \frac{2.5}{10^{-6}} = 2.5\times10^{-6} \text{ A/m}$$

It is from south pole to north pole of the magnet.

(*c*) The magnetic field at the centre is

$$B = \mu_0(H + I)$$

H at centre is

$$H = \frac{2\times m}{4\pi r^2} = \frac{2M}{4\pi r^2 l}$$

where m is pole strength and r distance of centre from pole. It is directed from north to south pole of magnet. Hence

$$B = \mu_0(H + I) = 4\pi\times10^{-7}\left(2.5\times10^{-6} - \frac{2\times2.5}{4\pi(10^{-2})(0.5\times10^{-2})^2}\right)$$

$$= 2.5\times4\pi\times10^{-1} - \frac{5\times10^{-7}}{25\times10^{-8}}$$

$$= \pi - 2 = 1.14 \text{ T.}$$

Example 22.15: Show that the magnetic field at a point due to magnetic dipole is perpendicular to the magnetic axis if the line joining the point with the centre of the dipole makes an angle of $\tan^{-1}(\sqrt{2})$ with the magnetic axis.

Solution: Figure 22.37 shows a magnet NS and field at point $P(r, \theta)$ is B and is perpendicular to the magnetic axis. We know that if β is the angle that the direction of B makes with position vector r (OP line) then

$$\tan\beta = \frac{1}{2}\tan\theta$$

From Fig. 22.37, $\beta = 90 - \theta$

$\therefore \quad \tan(90 - \theta) = \frac{1}{2}\tan\theta$

or $\quad \cot\theta = \frac{1}{2}\tan\theta$

or $\quad \tan^2\theta = 2$ and $\tan\theta = \sqrt{2}$ or $\theta = \tan^{-1}(\sqrt{2})$

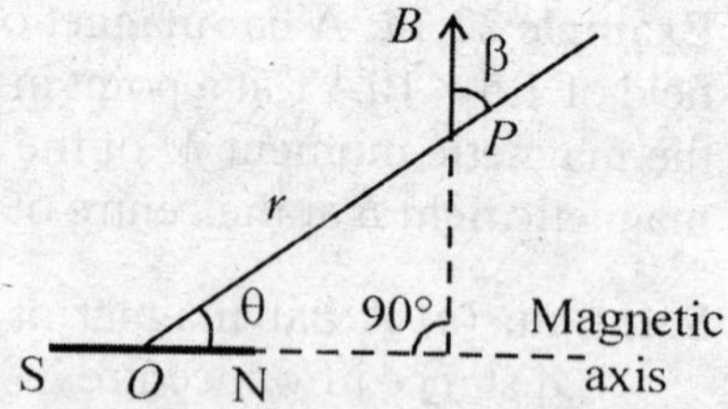

Fig. 22.37

Example 22.16: A magnetic dipole of magnetic moment $0.72\sqrt{2}$ A-m^2 is placed horizontally with the north pole pointing towards east. Find the position of the neutral point if the horizontal component of the earth's magnetic field is 18 μT.

Solution: Neutral point is obtained at a point where the magnetic field is equal in magnitude to the horizontal component of earth's magnetic field and opposite in direction. The situation is shown in Fig. 22.38. At point P the magnetic field is perpendicular to the axis of magnet and $\theta = \tan^{-1}(\sqrt{2})$. From example 22.15

The magnetic field at B due to magnetic dipole is

$$B = \frac{\mu_0}{4\pi}\frac{M}{r^3}\sqrt{1 + 3\cos^2\theta}$$

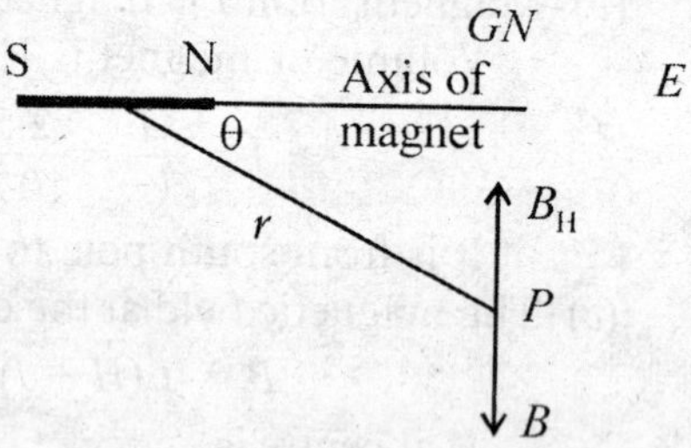

Fig. 22.38

Since $\tan\theta = \sqrt{2}$ hence $\cos\theta = \frac{1}{\sqrt{3}}$

$$\therefore \quad 18 \times 10^{-6} = 10^{-7} \times \frac{0.72\sqrt{2}}{r^3}\sqrt{1 + 3/3}$$

or $$r^3 = \frac{10^{-7} \times 0.72\sqrt{2} \times \sqrt{2}}{18 \times 10^{-6}}$$

$$= 8 \times 10^{-3}$$

or $$r = 2 \times 10^{-1} \text{ m} = 20 \text{ cm}$$

Hence neutral point is obtained at 20 cm from the dipole at an angle $\theta = \tan^{-1}(\sqrt{2})$ south of east.

Example 22.17: A short bar magnet placed with its axis at 30° experiences a torque of 0.016 N-m in an external field of 800 G. (*a*) What is the magnetic moment of the magnet? (*b*) What is work done by an external force in moving it from its most stable to most unstable position? (*c*) What is the work done by the force due to the external magnetic field in the process mentioned in part (*b*)? (*d*) The bar magnet is replaced by a solenoid of cross-sectional area 2×10^{-4} m^2 and 1000 turns but the same magnetic moment. Determine the current flowing through the coil.

Solution: (*a*) A magnet in magnetic field experiences a torque

$$\tau = MB\sin\theta$$

Given $\theta = 30°$, $B = 800$ G $= 800 \times 10^{-4}$ T and $\tau = 0.016$ N-m.

$$\therefore \quad M = \frac{\tau}{B\sin\theta} = \frac{0.016}{800 \times 10^{-4} \times \sin 30} = \frac{16 \times 10^{-3} \times 2}{8 \times 10^{-2}}$$

$$= 0.4 \text{ A-m}^2$$

(*b*) When a magnetic dipole is rotated in magnetic field from position θ_1 to θ_2, the work done is

$$W = MB(\cos\theta_1 - \cos\theta_2)$$

Most stable position means $\theta_1 = 0°$ and most unstable position means $\theta_2 = 180°$. Hence work done by external force is

$$W = 0.4 \times 800 \times 10^{-4} (\cos 0° - \cos 180°)$$
$$= 0.4 \times 800 \times 10^{-4} \times 2$$
$$= 6.4 \times 10^{-2} \text{ J.}$$

(*c*) Work done by the field is negative of the work done by the external force.

Work done by the field = -6.4×10^{-2} J.

(*d*) A current carrying loop is equivalent to a dipole of moment

$$M = \text{Area} \times \text{Current} = NiA$$

or

$$i = \frac{M}{NA}$$

$$i = \frac{0.4}{1000 \times 2 \times 10^{-4}} = 2 \text{ A.}$$

Example 22.18: A domain in ferromagnetic iron is in the form of a cube of side length 1 mm. Estimate the number of iron atoms in the domain. The molecular mass of iron is 55 g/mole and its density 7.9 g/cc. Assume that each atom has a dipole moment of 9.27×10^{-24} A-m².

Solution: Volume of the domain $V = (10^{-3})^3 = 10^{-9}$ m³

$$\text{Mass} = \text{Volume} \times \text{density} = 10^{-9} \times 7.9 \times 10^3 = 7.9 \times 10^{-6} \text{ kg.}$$

Molar mass of iron is 55 g and it contains 6.02×10^{23} atoms of iron. Hence number of iron atoms in 7.9×10^{-6} kg is

$$n = \frac{6.02 \times 10^{23}}{55 \times 10^{-3}} \times 7.9 \times 10^{-6}$$
$$= 8.65 \times 10^{19}$$

Each iron atom has dipole moment = 9.27×10^{-24} A-m²

Hence maximum possible dipole moment

$$M = 9.27 \times 10^{-24} \times 8.65 \times 10^{-19} = 8.0 \times 10^{-4} \text{ A-m}^2$$

Hence intensity of magnetisation

$$I = \frac{M}{V} = \frac{8 \times 10^{-4}}{10^{-9}} = 8 \times 10^5 \text{ A/m}$$

Example 22.19: Figure 22.39 shows a small magnetised needle *A* placed at centre. The arrow shows the direction of its magnetic moment. The other arrows show different positions (and orientations of the magnetic moment) of another identical magnetised needle *B*.

(*a*) In which configurations is the system not in equilibrium.

(*b*) In which configurations is the system (*i*) in stable (*ii*) in unstable equilibrium.

(*c*) Which configuration corresponds to the lowest potential energy among all the configurations shown.

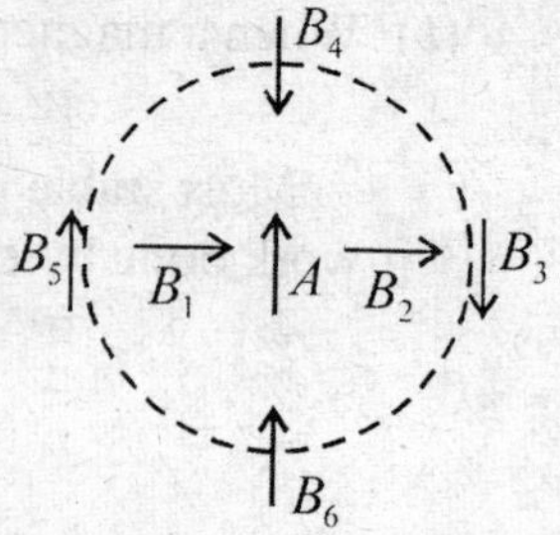

Fig. 22.39

Solution: For equilibrium the torque acting on magnetic dipole must be zero.

(*a*) In configuration B_1 and B_2 the magnetic moment B and field due to A are perpendicular to each other and hence torque is maximum. In rest of configuration the two are either parallel to each other or anti-parallel as such torque is zero. Hence in configuration AB_1 and AB_2 the system is not in equilibrium.

(*b*) The system is in stable equilibrium if potential energy is minimum. In case of a dipole in magnetic field, the potential energy is minimum when the two are parallel. In configuration AB_3 and AB_6, the magnetic field due to A and magnetic moment of B are parallel. Hence configuration AB_3 and AB_6 gives stable equilibrium.
Equilibrium is unstable if magnetic moment and magnetic field are anti-parallel (angle = 180°).
In configuration AB_4 and AB_5 the two vectors are anti-parallel. Hence configuration AB_4 and AB_5 are in unstable equilibrium.

(*c*) As already stated in (*b*) the configurations AB_3 and AB_6 are the configurations of minimum potential energy.

Example 22.20: A sample of paramagnetic salt contains 2.0×10^{24} atomic dipoles each of dipole moment 1.5×10^{-23} JT^{-1}. The sample is placed under a homogeneous magnetic field of 0.84 T and cooled to a temperature of 4.2 K. The degree of magnetic saturation achieved is equal to 15%. What is the total dipole moment of the sample for a magnetic field of 0.98 T and a temperature of 2.8 K? (Assume Curie Law).

Solution: Sample has total atomic dipoles = 2.0×10^{24} and each has dipole moment = 1.5×10^{-23} J T^{-1}.

If all the atomic dipoles are aligned parallel to the magnetic field the total magnetic moment will be = $1.5 \times 10^{-23} \times 2 \times 10^{24} = 30$ J T^{-1}. But only 15% of the magnetisation takes place. Hence the magnetic moment of the sample is

$$M = 30 \times 1.5 = 4.5 \text{ J T}^{-1}$$

Magnetisation is directly proportional to magnetic field B and inversely proportional to temperature (T in Kelvin), i.e.

$$M \propto \frac{B}{T}$$

$$\therefore \quad \frac{M_1}{M_2} = \frac{B_1}{T_1} \times \frac{T_2}{B_2}$$

or $$M_2 = \frac{M_1 T_1 B_2}{B_1 T_2} = \frac{4.5 \times 4.2 \times 0.98}{0.84 \times 2.8} = 7.87 \text{ JT}^{-1}$$

Example 22.21: A magnetic dipole is under the influence of two magnetic fields. The angle between the field directions is 60° and one of the fields has magnitude 1.2×10^{-2} T. If the dipole comes to stable equilibrium at an angle 15° with this field, what is the magnitude of the other field?

Solution: When a magnetic dipole of moment M is placed in magnetic field B, it experiences a torque given by

$$\tau = MB\sin\theta$$

where θ is the angle between vectors M and B.

In present case dipole in equilibrium makes an angle α (= 15°) with magnetic field B_1 (= 1.2 × 10^{-2} T). Hence torque due to this field is

$$\tau_1 = MB_1 \sin\alpha \qquad \text{(clockwise)}$$

Magnetic field B_2 makes an angle of 60° with field B_1 hence in equilibrium magnetic dipole makes an angle of (60° – α) with the field B_2 and the torque acting on magnetic dipole is

$$\tau_2 = MB_2 \sin(60 - \alpha) \qquad \text{(anticlockwise)}$$

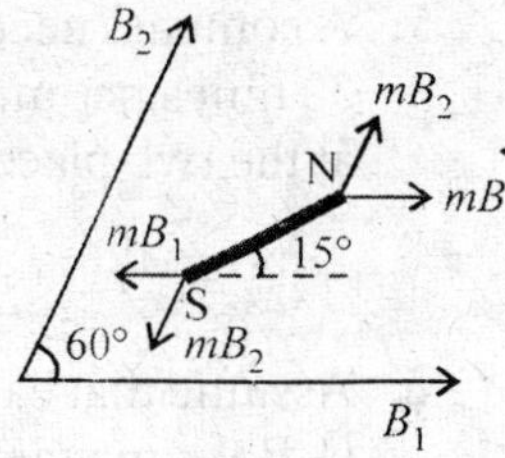

Fig. 22.40

In equilibrium, the two torques must be equal in magnitude.

$$\therefore \qquad MB_1 \sin\alpha = MB_2 \sin(60 - \alpha)$$

or

$$B_2 = \frac{B_1 \sin\alpha}{\sin(60 - \alpha)} \qquad (\alpha = 15° \text{ and } B_1 = 1.25 \times 10^{-2}\ \text{T})$$

$$= \frac{1.25 \times 10^{-2} \sin 15°}{\sin 45°} = 4.6 \times 10^{-3}\ \text{T}.$$

PROBLEMS

1. A small magnet of magnetic moment M_2 is placed at broad-side on position of a magnet of magnetic moment M_1 of length $2l_1$, in such a way that the axis of the former coincides the perpendicular bisector of the latter. The separation between their centres is d. Calculate the nature of interaction (force and couple) among them. What is limiting value when d becomes very large? **[Hint:** This is the case of dipole-dipole interaction and is solved article 22.5]
2. Two identical dipoles each of magnetic moment 1.0 Am² are placed at a separation of 2 m with their axes perpendicular to each other. What is the magnetic field at a point midway between the dipoles. **[Ans.** $B = \sqrt{5} \times 10^{-7}$ T $\theta = 26.6°$]
3. Two short magnets of magnetic moment M_1 and M_2 are placed on a table as shown in Fig. 22.41. Find the direction and magnitude of the magnetic field produced by these magnets at point P. M_1 = 2700 cgs units, M_2 = 3200 cgs units, d_1 = 30 cm and d_2 = 40 cm.

   [Hint: In cgs system of units the value of constant $\left(\frac{\mu_0}{4\pi}\right) = 1$.

   Hence field in tan A position $= \frac{2M}{d^3}$ and in tan $B = \frac{N}{d^3}$ and $H = B$, i.e. 1 Gauss = 1 Oersted. Hence Ans: B = 0.112 G = 1.12 × 10^{-5} T in the direction θ = 26° with PX]

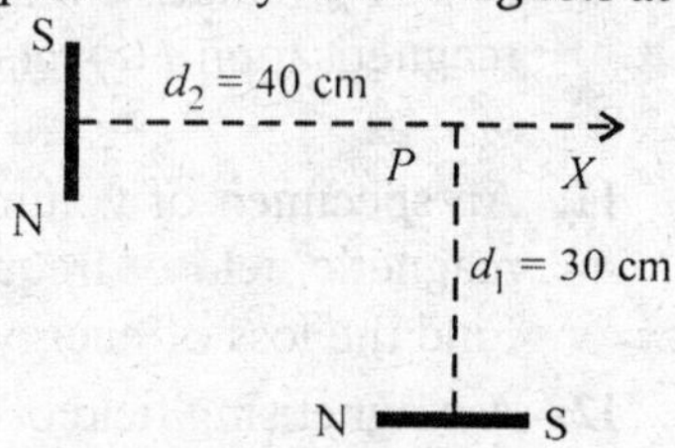

Fig. 22.41

4. Two short bar magnet of magnetic moment M_1 = 2.7 Am² and M_2 = 4.0 Am² are placed in the magnetic meridian (Fig. 22.42 (*a*)). What should be the distance d such that point P is a neutral point? Indicate on diagram N-S poles of magnet. Horizontal component of earth's magnetic field is 0.3 Gauss. **[Ans.** (*i*) B_1 is in same direction B_H, d = 20 cm (*ii*) B_1 in direction opposite to B_H, d = 34.2 cm]

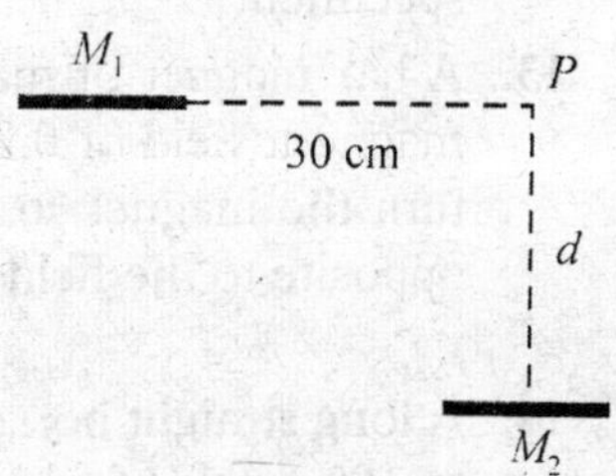

Fig. 22.42

5. A compass needle oscillates 20 times per minute at a place where the angle of dip is 45° and 30 times per minute when the dip is 30°. Compare the total magnetic field due to the earth at the two places.

[**Ans.** $= \dfrac{B_2}{B_1} = 1.83$]

6. Assume that each iron atom has a permanent magnetic moment equal to 2 Bohr magnetons (1 Bohr magneton equals 9.27×10^{-24} A-m^2). The density of atoms in iron is 8.52×10^{28} atoms/m^3. (*a*) Find the maximum magnetisation *I* in a long cylinder of iron. (*b*) Find the maximum magnetic field *B* on the axis inside the cylinder.

[**Ans.** (*a*) 1.58×10^6 A/m (*b*) 2.0 T]

7. The magnetic field intensity *H* at the centre of a long solenoid carrying a current of 2.0 A is found to be 1500 A/m. Find the number of turns per centimeter of the solenoid. Now a rod is inserted as a core in the current carrying solenoid, (*a*) What is the magnetic intensity *H* at the centre? (*b*) If the magnetisation *I* of the core is found to be 0.12 A/m, find the susceptibility of the material of the rod. (*c*) Is the material paramagnetic, diamagnetic or ferromagnetic.

[**Ans.** $n = 75$ (*a*) 1500 A/m (*b*) 8.0×10^{-5} (*c*) Paramagnetic]

8. A closely wound solenoid of 2000 turns and area of cross-section 1.6×10^{-4} m^2, carrying a current of 4.0 A is suspended through its centre allowing it to turn in a horizontal plane.
(*a*) What is the magnetic moment associated with the solenoid?
(*b*) What is the force and torque on the solenoid if a uniform horizontal magnetic field of 7.5×10^{-2} T is set up at an angle of 30° with the axis of the solenoid?

[**Ans.** (*a*) 1.28 A-m^2 (*b*) Force = 0 torque = 0.048 N-m]

9. A magnetising field of 50 A/m produces a flux of 4×10^{-6} W in an steel rod. Area of cross-section of rod is 0.5 cm^2. Calculate (*a*) flux density (*b*) absolute permeability and (*c*) intensity of magnetisation.

[**Ans.** (*a*) $B = 8 \times 10^2$ T (*b*) $\mu = 1.6 \times 10^{-3}$ H/m (*c*) $I = 6.34 \times 10^4$ A/m]

10. A roland ring of a ferromagnetic substance has 10^3 number of turns per metre. If a current of 2 A produces a magnetic field of 1 T, then find (*a*) magnetising field *H*. (*b*) intensity of magnetisation *I* (*c*) relative magnetic permeability.

[**Ans.** (*a*) $H = 200$ A/m (*b*) $I = 7.94 \times 10^5$ A/m (*c*) $\mu_r = 397.8$]

11. An specimen of volume 10^{-3} m^3 of a ferromagnetic substance is placed in an alternating magnetic field of frequency 50 Hz; the area of hysteresis loop obtained is 0.1 mks units. Find the loss of energy per second in the specimen. [**Ans.** 5×10^{-3} J]

12. A magnetising field of 2×10^3 A/m produces a magnetic flux of 6.28×10^{-4} Wb in an iron rod of area of cross-section 2×10^{-5} m^2. Find the relative magnetic permeability of the specimen. [**Ans.** $\mu_r = 1.25 \times 10^4$]

13. A bar magnet of magnetic moment 1.5 JT^{-1} lies aliened with the direction of a uniform magnetic field of 0.22 T. (*a*) What is the amount of work required by an external torque to turn the magnet so as to align its magnetic moment (*i*) normal to the field direction (*ii*) opposite to the field direction. (*b*) What is the torque on the magnet in cases (*i*) and (*ii*)?

[**Ans.** (*a*) (*i*) 0.33 J (*ii*) 0.66 J (*b*) (*i*) $\tau = 0.33$ J (*ii*) zero]

14. A long straight horizontal cable carries a current of 2.5 A in the direction 10° south of west to 10° north of east. The magnetic meridian of the place happens to be 10° west of the geographical meridian. The earth's magnetic field at location is 0.33 G and the angle of dip is zero. Locate the line of neutral points (Ignore the thickness of the cable).

[**Hint:** $B = \frac{\mu_0 i}{2\pi r} = 0.33 \times 10^{-4}$, $r = 1.5$ cm, direction should be opposite to the earth's field hence point is above the cable. The locus of such points is parallel to the cable]

15. A compass needle free to turn in a horizontal plane is placed at the centre of circular coil of 30 turns and radius 12 cm. The coil is in a vertical plane making an angle of 45° with the magnetic meridian. When the current in the coil is 0.35 A the needle points east-west.

(*a*) Determine the horizontal component of earth's magnetic field at the location.

(*b*) The current in the coil is reversed and the coil is rotated about vertical axis by an angle of 90° in the anticlockwise sense looking from above. Predict the direction of needle. Take magnetic declination at the place to be zero.

[**Ans.** (*a*) $B_H = \frac{\mu_0 ni}{2r} \cos 45 = 3.9 \times 10^{-5}$ T (*b*) then needle will merely reverse its direction]

16. A telephone cable at a place has four long horizontal straight wires carrying a current of 1.0 A in the same direction east to west. The earth's magnetic field at the place is 0.39 G and the angle of dip is 35°. The magnetic declination is nearly zero. What are the resultant magnetic fields 4.0 cm below and above the cable?

[**Hint:** B due to current through wire $B = \frac{4 \times \mu_0 i}{2\pi r} = 0.2\,\text{G}$ below cable $B_H = 0.39 \cos 35° - 0.2 = 0.12$ G. $B_V = 0.39 \sin 35° = 0.22$ G, $B = \sqrt{B_H^2 + B_V^2} = 0.25$ G and $\theta = \tan^{-1} \frac{B_V}{B_H} = 62°$. Above the cable $B_H = 0.39 \cos 35° + 0.2 = 0.52$ G, $B_H = 0.22$ G and $B = 0.57$ G and $\theta = 23°$]

17. A short bar magnet of magnetic moment 5.25×10^{-2} JT^{-1} is placed with its axis perpendicular to the earth's field direction. At what distance from the centre of the magnet the resultant field is inclined at 45° with the earth's field on (*a*) its normal bisector (*b*) its axis. Magnitude of the earth's field at the place is 0.42 G. Ignore the length of the magnet in comparision to the distance involved. [**Ans.** (*a*) $r = 5.0$ cm, (*b*) $r = 6.3$ cm]

18. A short bar magnet placed in a horizontal plane has its axis aligned along the magnetic north-south direction. Null points are found on the axis of the magnet at 14 cm from the centre of the magnet. The earth's magnetic field at the place is 0.36 G and the angle of dip is zero. What is the total magnetic field on the normal bisector of the magnet at the same distance as the null point (i.e. 14 cm) from the centre of the magnet?

If now bar magnet is turned around by 180°, where will the new null-points be located?

[**Ans.** $B = 0.54$ G in the direction of earth's field. $r = 11.1$ cm on the normal bisector.]

19. A magnetic needle free to rotate in a vertical plane parallel to the magnetic meridian has its north pole pointing down at 22° with the horizontal. The horizontal component of the earth's magnetic field at the place is known to be 0.35 G. Determine the magnitude of the earth's magnetic field at the place. [**Ans.** $B = 0.38$ G]

20. A circular coil of 16 turns and radius 10 cm carrying a current of 0.75 A rests with its plane normal to an external field of magnitude 5.0×10^{-2} T. The coil is free to turn about an axis in its plane perpendicular to the field direction. When the coil is turned slightly and released, it oscillates about its stable equilibrium with frequency 2.0 per sec. What is the moment of inertia of coil about its axis of rotation? [**Ans.** $I = 1.2 \times 10^{-4}$ kg m^2]

SHORT ANSWER TYPE QUESTIONS

1. An specimen of volume 30 cc is placed in a magnetising field of intensity 5 Oersted. The magnetic moment induced is 6 A-m^2. Calculate intensity of magnetisation and magnetic induction.

[**Ans.** $I = \frac{M}{V} = \frac{6}{30 \times 10^{-6}} = 2 \times 10^5$ A/m

$B = \mu_0(H + I)$

H is 5 oersted. Oersted is the unit of magnetising field in cgs system. In cgs system 1 Oersted = 1 Gauss.

$$\therefore \; 1 \text{ Orested} = \frac{10^{-4}\text{ T}}{\mu_0} = \frac{10^{-4}}{4\pi \times 10^{-7}} = \frac{10^3}{4\pi} \text{ A/m}$$

$$\therefore \quad B = 4\pi \times 10^{-7}\left(\frac{5 \times 10^3}{4\pi} + 2 \times 10^5\right) = 5 \times 10^{-4} + 8\pi \times 10^{-2}$$

$= 0.2518$ T]

2. The volume magnetic susceptibility of an specimen is 30×10^{-4}. Find its absolute and relative magnetic permeability.

[**Ans.** $\chi_m = 30 \times 10^{-4}$ and $\mu_0 = 12.57 \times 10^{-7}$ Henry/m

$\mu = \mu_0(1 + \chi_m)$ and $\mu_r = 1 + \chi_m$

$\therefore \quad \mu_r = 1 + 0.003 = 1.003$

$\mu = \mu_0(1 + \chi_m) = 4\pi \times 10^{-7} \times 1.003 = 12.61 \times 10^{-7}$ H/m]

3. A magnet of mass 75 g has magnetic moment equal to 9×10^{-7} A-m^2. If density of the material of magnet is 7.5×10^3 kg/m^3, then find its intensity of magnetisation.

[**Ans.** $I = \frac{M}{V}$ and $V = \frac{\text{Mass}}{\text{Density}} = \frac{m}{d}$

$\therefore \quad I = \frac{Md}{m}$ where M is magnetic moment

$$\therefore \quad I = \frac{9 \times 10^{-7} \times 7.5 \times 10^3}{75 \times 10^{-3}} = 9 \times 10^{-2} = 0.09 \text{ A/m}]$$

4. A ferromagnetic specimen has mass 0.6 kg and density 7.8×10^3 kg/m^3 when this specimen is placed in magnetic field of frequency 50 Hz, the area of the hysteresis loop in mks units is 0.0722. Find the energy loss in specimen per second.

[**Ans.** Volume of specimen $= \frac{\text{Mass}}{\text{Density}} = \frac{0.6}{7.8 \times 10^3} = 7.7 \times 10^{-5}$ m^3

Energy loss per unit volume per cycle = area of the loop

Energy loss in 1 sec = area of loop × volume × frequency

$= 0.0722 \times 7.7 \times 10^{-5} \times 50$

$= 27.8 \times 10^{-5}$ Joules]

5. Two short magnets of equal dipole moment M are fastened perpendicular to each other as shown in Fig. 22.43. Find the magnetic field at point P at a distance r from centre O.
[**Ans.** Resolving magnetic moments along OP, we have

$$= M\sqrt{2}$$

Point P is on its axis at distance r. Hence magnetic field is

$$B = \frac{\mu_0}{4\pi}\frac{2\sqrt{2}M}{r^3} \text{ in the direction of } OP.$$

Component of magnetic moment perpendicular to OP is zero.

Hence magnetic field perpendicular to OP is zero. Hence net field is $\frac{\mu_0}{4\pi}\frac{2\sqrt{2}M}{r^3}$]

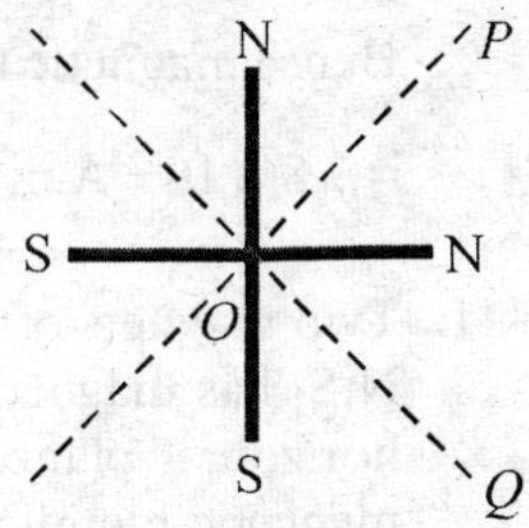

Fig. 22.43

6. In above problem calculate the field at Q.
[**Ans.** For point Q axial component of magnetic moment, i.e. along OQ is zero. Hence B is in direction OQ is zero.

Component of magnetic moment along OP is $\sqrt{2}M$ and for this component, point P is in broad side on position (tan$B$ position). Hence magnetic field is $\frac{\mu_0}{4\pi}\frac{\sqrt{2}M}{r^3}$]

7. Calculate the moment of the couple required to keep a bar magnet of moment 200 cgs units in uniform field of 0.39 oersted at an angle of 30° with the direction of the field.
[**Ans.** Moment of the couple acting on the magnet in field is

$$\tau = MB\sin\theta = 200 \times 0.39 \times \sin 30$$
$$= 39 \text{ dyne cm}]$$

8. What is magnetostriction effect?
[**Ans.** When a rod is magnetised its length increases in the direction of magnetisation due to alignment of spins parallel to the field. This effect is called *magnetostriction effect*. This effect is used to produce ultrasonic waves by magnetising and demagnetising an iron rod at high frequency by the current in a wire wound over it.]

9. Two identical bar magnets each of length L and pole strength m are placed at right angles to each other with N-pole of one touching the south pole of the other. Find the magnetic moment of the system.
[**Ans.** Magnetic moment is a vector having direction from south to north. Vector sum of the magnetic moments is along SN (Fig. 22.44). Since magnetic moments are equal hence $\theta = 45°$ and magnitude $= M\sqrt{2} = \sqrt{2}ml$]

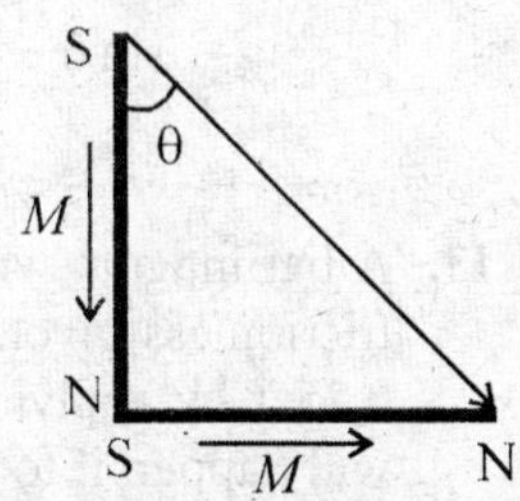

Fig. 22.44

10. Earth's magnetic field be supposed to be due to small bar magnet located at the centre of the earth. If the magnetic field at a point on the magnetic equator is 0.3×10^{-4} T. Find the magnetic moment of such a bar magnet? Find also the magnetic field at the north pole. Radius of earth is 6.4×10^6 m.
[**Ans.** Figure 22.45 shows the small bar magnet placed at the centre of earth and the point on magnetic equator is the broad-side on position.

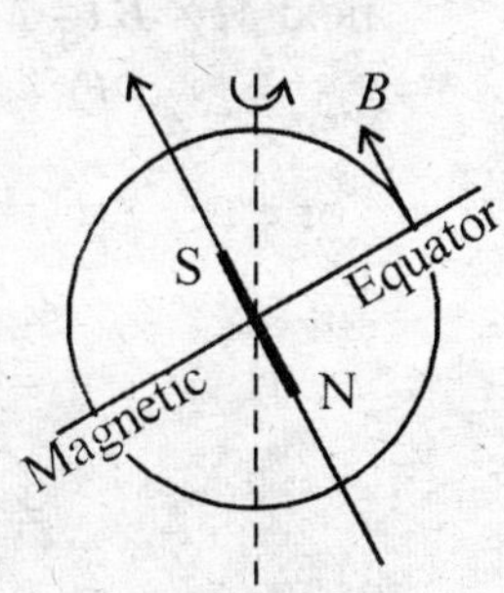

Fig. 22.45

Hence magnetic field here is $B = \frac{\mu_0}{4\pi}\frac{M}{r^3}$ or $M = B \cdot r^3 \times \frac{4\pi}{\mu_0} = 0.3 \times 10^{-4} \times (6.4 \times 10^6)^3 \times 10^7$ $= 7.8 \times 10^{22}$ A-m^2. North pole is on end on posiotion and B here is twice that on broad side on position. Hence $B = 0.6 \times 10^{-4}$ T.]

11. Two magnets of equal mass are joined at right angles to each other (Fig. 22.46). Magnet N_1S_1 has magnetic moment 3 times that of N_2S_2. This arrangement is free to rotate in the horizontal plane. Find the angle that the magnet N_1S_1 makes with magnetic meridian.

[**Ans.** For equilibrium of system torque on M_1 and M_2 due to B_H must be equal in magnitude, i.e.

$$|\boldsymbol{M}_1 \times \boldsymbol{B}_H| = |\boldsymbol{M}_2 \times \boldsymbol{B}_H|$$

$$M_1 B_H \sin\theta = M_2 B_H \cos\theta$$

or $\tan\theta = \frac{M_2}{M_1} = \frac{1}{3}$ or $\theta = \tan^{-1}(1/3)$]

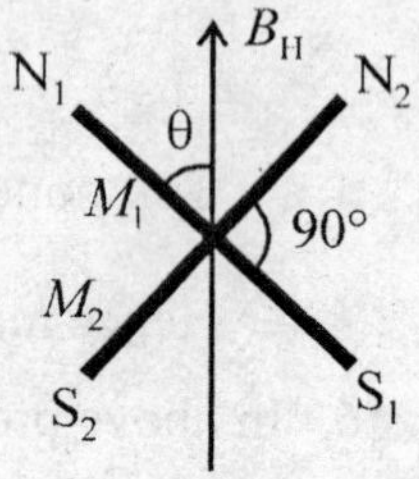

Fig. 22.46

12. A compass needle whose magnetic moment is 60 A-m^2 is pointing geographical north at a certain place, where horizontal component of earth's magnetic field is 40 μT. It experiences a torque 1.2×10^{-3} N-m. Find the declination of place.

[**Ans.** In equilibrium a compass needle free to rotate in horizontal plane rests parallel to the magnetic meridian (MM). It is held in parallel to geographical meridian (GM). If θ is the angle between MM and GM then torque

$$\tau = MB\sin\theta$$

and by definition θ is declination at the place. Hence

$$\sin\theta = \frac{\tau}{MB} = \frac{1.2 \times 10^{-3}}{60 \times 40 \times 10^{-6}} = \frac{1.2 \times 10}{24} = \frac{1}{2}$$

or $\theta = 30°$]

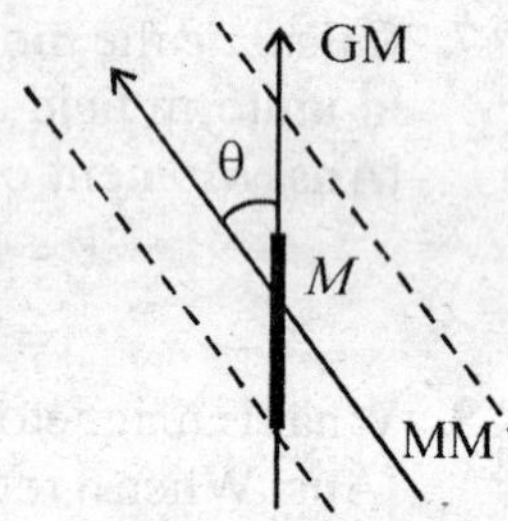

Fig. 22.47

13. A bar magnet with poles 20 cm apart and pole strength 15 A-m^2 rests with its centre on a frictionless pivot. It is held in equilibrium at 60° to a uniform magnetic field of induction 0.25 T by applying a force F at right angles to the axis 10 cm from its pivot. Find F. What will happen if force is removed?

[**Ans.** In equilibrium the torque due to force F (= Fr) and magnetic field $\boldsymbol{M} \times \boldsymbol{B}$ (= $MB\sin\theta$) should be equal in magnitude, i.e.

$$Fr = MB\sin\theta$$

or $F = \frac{MB\sin\theta}{r} = \frac{15 \times 0.2 \times 0.25 \times \sin 60°}{10}$ ($\because M = ml$)

$$= \frac{15 \times 0.2 \times 0.025}{10 \times 10^{-2}}\frac{\sqrt{3}}{2}$$

$$= 6.5 \text{ N}$$

If force is removed the bar magnet will come to rest parallel to magnetic field.]

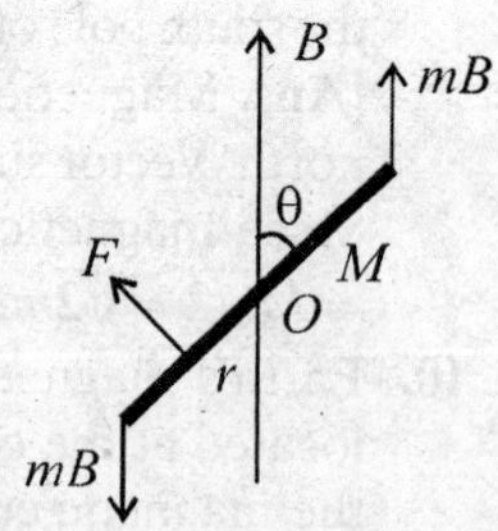

Fig. 22.48

14. A thin rectangular magnet suspended freely has a period of oscillation of 4 sec. Find the period of oscillation if the magnet is broken into two halves (each having half the original length) and only one of the pieces is made to oscillate in the same field? Assume breadth of magnet to be negligible with respect to length.

[**Ans.** Time period of oscillation is $T = 2\pi\sqrt{\frac{I}{MB}}$

When the magnet is broken into two pieces its magnetic moment becomes $M/2$ and moment of intertia $= \frac{I}{8}\left(I = \frac{ml^2}{12}\right)$. Thus new time period $T' = 2\pi\sqrt{\frac{I/8}{(M/2)B}} = 2\pi\sqrt{\frac{I}{4MB}} = \frac{T}{2} = 2$ sec]

15. For a ferromagnetic substance, what is the main difference between, *B-H* and *I-H* curves.
[**Ans.** Area of *B-H* loop is μ_0 times the area of *I-H* curve. In saturation state *I-H* curve is parallel to *H*-axis (*I* is constant) whereas in *B-H* it becomes a straight time of slope μ_0]

16. Find the magnetic field due to a dipole of magnetic moment 1.2 A-m² at a point 1 m away from it in a direction making an angle of 60° with the dipole axis.
[**Ans.** Magnetic field due to a dipole at point (r, θ) is

$$B = \frac{\mu_0}{4\pi}\frac{M}{r^3}\sqrt{1 + 3\cos^2\theta}$$

$$= \frac{10^{-7}\times 1.2}{(1)^3}\sqrt{1 + 3\times\frac{1}{4}} = 1.6\times 10^{-7}\text{ T}$$

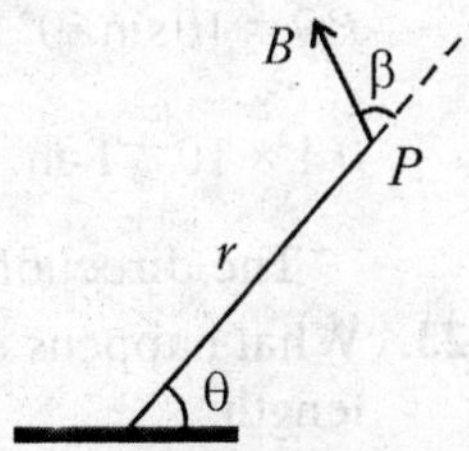

Fig. 22.49

The direction of the field β with radial line is

$$\tan\beta = \frac{1}{2}\tan\theta = \frac{\sqrt{3}}{2}\,]$$

17. Find the magnetic intensity *H* at the centre of a long solenoid having *n* turns per unit length and carrying a current *i* (*a*) when no material is kept in it and (*b*) when a long copper rod is inserted in the solenoid.
[**Ans.** (*a*) When there is no rod, the magnetic intensity is $H = ni$ where *n* is number of turns per unit length and *i* current, and (*b*) Since solenoid and rod are long as such there is no effect on the magnetic intensity at the centre. Its value in both cases is $H = ni$]

18. Find percentage increase in the magnetic field *B* when the space within a current carrying torroid is filled with aluminium having susceptibility $\chi_m = 2.1 \times 10^{-5}$.
[**Ans.** In absence of specimen (Aluminium) $B_0 = \mu_0 H$
In presence of specimen $B = \mu H = \mu_0\mu_r H = \mu_0(1 + \chi_m) H$

$$\text{Hence percentage increase} = \frac{B - B_0}{B_0}\times 100 = \frac{\mu_0(1+\chi_m)H - \mu_0 H}{\mu_0 H}\times 100$$

$$= \chi_m \times 100 = 2.1 \times 10^{-3}\,]$$

19. Maximum value of the permeability of an alloy (65% Ni, 17% Fe, 10% Cu and 8% Cr) is 0.126 T-m/A. Find maximum relative permeability and susceptibility.

[**Ans.** $\mu_r = \frac{\mu}{\mu_0} = \frac{0.126}{4\pi\times 10^{-7}} = 1.0\times 10^5$ and susceptibility $\chi_m = \mu_r - 1 \approx 1.0 \times 10^5$]

20. The coercive force for a certain permanent magnet is 4×10^4 A/m. This magnet is placed inside a long solenoid of 40 turns/cm and a current is passed in the solenoid to demagnetise it completely. Find the current?

[**Ans.** If current is passed through the solenoid, the reverse field produced is

$$H = ni = 40 \times 10^2 \times i$$

to demagnetise the magnet completely. This magnetising field should be equal to coercive force. Hence $40 \times 10^2 \times i = 4 \times 10^4$ or $i = 10$ A]

21. The susceptibility of annealed iron at saturation is 5500. Find the permeability of annealed iron at saturation [**Ans.** $\mu = \mu_0\mu_r = \mu_0(1 + \chi_m) = 4\pi \times 10^{-7}(1 + 5500) = 6.9 \times 10^{-3}$]

22. Figure 22.50 shows some of the equipotential surfaces of the magnetic scalar potential. Find the magnetic field B at a point in the region.

[**Ans.** If V is the scalar potential at any point then vector field $B = -\dfrac{dV}{dr}$. Here all the lines are equipotential lines and the perpendicular distance between the two lines is $PQ = 10\sin 60° = 5\sqrt{3}$ cm and potential difference $\Delta V =$ 0.1×10^{-4} T-m. Hence $B = \dfrac{0.1\times 10^{-4}}{5\sqrt{3}\times 10^{-2}} = 1.15\times 10^{-4}$ T.

The direction of the field is from P to Q]

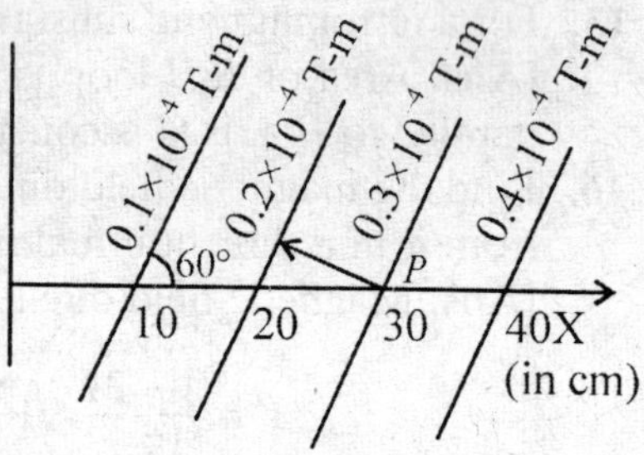

Fig. 22.50

23. What happens if a bar magnet is cut into two pieces: (*i*) transverse to its length (*ii*) along its length.

[**Ans.** In each case we get two magnets each of magnetic moment half the magnetic moment of original magnet. In case (*i*) the pole strength remains same and length is halved. In second case pole strength is halved and length remains same]

24. A magnetic needle in a uniform magnetic field experiences a torque but no net force. An iron nail near a bar magnet, however experiences a force of attraction in addition to a torque. Why?

[**Ans.** In uniform field a magnetic needle experience no force because force on south pole and north pole is equal in magnitude and opposite in direction. Hence net force is zero. If the point of application of force is not same a torque acts.

In case of a bar magnet and nail, bar magnet produces non uniform magnetic field. A nail is magnetised by induction and the forces on S-pole and N-pole are not equal in magnitude hence net force acts. If the point of application is not same a torque also acts]

25. A roland ring of mean radius 15 cm has 3500 turns of wire wound on a ferromagnetic core of relative permeability 800. What is the magnetic field B in the core for a magnetising current of 1.2 A?

[**Ans.** $B = \mu_0 \mu_r ni$ where n is number of turns per unit length.

$$= 4\pi \times 10^{-7} \times 800 \times \frac{3500}{2\pi \times 0.15} \times 1.2$$

$$= 4.48 \text{ T}$$]

MULTIPLE CHOICE QUESTIONS

PART I

In the following, a question is followed by four alternatives, of them only one is correct. Therefore, the candidate has to select the correct choice from amongst the alternatives.

1. A magnetic needle of length l and magnetic moment M is bent in the form of a semicircle. Its magnetic moment is

(a) M (b) $\frac{M}{\pi}$

(c) $\frac{2M}{\pi}$ (d) $\frac{M}{2\pi}$.

2. Which of the following are vectors?

(*i*) Pole-strength
(*ii*) Magnetic moment
(*iii*) Magnetisation
(*iv*) Permeability of free space.

(a) only (*i*)
(b) only (*ii*) and (*iii*)
(c) only (*iii*) and (iv)
(d) only (*i*) and (*ii*).

3. It is found that neutral points lie along the axis of a magnet placed on the table, the orientation of the magnet is

(a) along the magnetic meridian with N-pole pointing geographical north
(b) along the magnetic meridian with N-pole pointing geographical south
(c) perpendicular to the magnetic meridian with N-pole pointing west
(d) perpendicular to the magnetic meridian with N-pole pointing west.

4. A bar magnet is placed with its north pole pointing north and south pole pointing south. The neutral points lie

(a) on its axis
(b) on its equitorial line
(c) no where (neutral points are not obtained)
(d) on the bisector of angle between axis and equitorial line.

5. The work done in deflecting a magnet of moment M through an angle θ from the direction of uniform field of strength B is

(a) $MB(1 - \cos\theta)$ (b) $-MB\cos\theta$
(c) $MB\cos\theta$ (d) $MB(\cos\theta - 1)$.

6. The work done in deflecting a magnet of moment M through 60° from the direction of uniform magnetic field of strength B is W. Then torque required to keep this magnet in equilibrium in this position is

(a) W (b) $\sqrt{3}W$
(c) $\frac{1}{\sqrt{3}}W$ (d) $2W$.

7. The horizontal component of earth's magnetic field is 35 μT. The vertical component of earth's magnetic field at equator is

(a) 35 μT (b) 70 μT
(c) 17.5 μT (d) zero.

8. A magnet is held vertical on a horizontal board, the number of neutral points obtained are

(a) one (b) two
(c) zero (d) infinite.

9. A bar magnet has a magnetic moment 2.5 J/T and is placed in a magnetic field of 0.2 T. The work done in turning the magnet from parallel to anti parallel position relative to field direction is

(a) 1 J (b) 0.5 J
(c) 2 J (d) zero.

10. A thin rectangular magnet suspended freely has a period of oscillation T. The magnet is now broken into two equal halves. Now both together are made to oscillate in same field with their N-poles in same direction.

The time period now becomes (neglect breadth of magnet)

(a) T (b) $2T$
(c) $T/2$ (d) a (infinite).

11. α_1 and α_2 are the angles of dip in two vertical planes at right angles to each other and α is true dip, then

(a) $\tan^2\alpha = \tan^2\alpha_1 + \tan^2\alpha_2$
(b) $\tan^2\alpha = \tan^2\alpha_1 \sim \tan^2\alpha_2$
(c) $\cot^2\alpha = \cot^2\alpha_1 + \cot^2\alpha_2$
(d) $\cot^2\alpha = \cot^2\alpha_1 \sim \cot^2\alpha_2$

12. A ship is sailing due east according to mariner's compass. The declination of the place is 10° east, the true direction of motion of the ship is

(a) 10° south of east
(b) 10° north east
(c) 80° south of east
(d) 80° north of east.

13. A dip circle is placed in such a way that the magnetic needle is vertical. The circle is then rotated through angle $\alpha°$ about the vertical axis and the dip measured in this position is found to be δ. The true dip at the place is

(a) $\phi = \tan^{-1}(\sin\alpha \tan\delta)$
(b) $\phi = \tan^{-1}(\cos\alpha \tan\delta)$
(c) $\phi = \tan^{-1}(\cos\alpha \cot\delta)$
(d) $\phi = \tan^{-1}(\sin\alpha \cot\delta)$.

14. At a place the angle of dip is δ and latitude is λ. The correct relation is

(a) $\tan\lambda = 2\tan\delta$ (b) $\tan\delta = 2\tan\lambda$
(c) $\cos\delta = 2\cos\lambda$ (d) $\cos\lambda = 2\cos\delta$.

15. A uniform magnetic field parallel to the plane of paper exists in the space directed from left to right. When a bar of soft iron is placed in the field parallel to it, the lines of force passing through it will be represented by

(a) 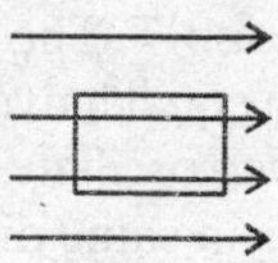 (b) 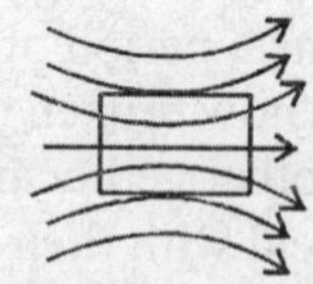

(c)  (d) 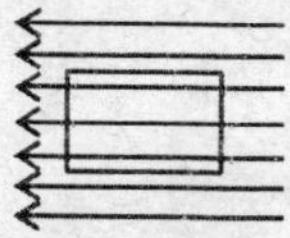

16. A rod when placed in strong magnetic field tries to orient perpendicular to the lines of force. The substance is

(a) paramagnetic (b) ferromagnetic
(c) nonmagnetic (d) diamagnetic.

17. In small magnetising fields, intensity of magnetisation (I) and magnetising field (H) bear a relation

(a) $I \propto \frac{1}{H}$ (b) $I \propto H°$
(c) $I \propto H$ (d) $I \propto H^2$.

18. The magnetic susceptibility of paramagnetic substance varies with temperature. The correct relation is

(a) It is independent
(b) $\chi \propto T$
(c) $\chi \propto \frac{1}{T - T_C}$
(d) $\chi \propto \frac{1}{T}$

19. The magnetic permeability of a substance is

(a) $\mu = \frac{I}{H}$ (b) $\mu = \frac{M}{V}$
(c) $\mu = \frac{B}{H}$ (d) $\mu = \frac{M}{H}$

where symbols have their usual meanings.

20. The relation between magnetic susceptibility (χ_m) and relative permeability μ_r is

(a) $\chi_m = \mu_r + 1$
(b) $\chi_m = \mu_r - 1$
(c) $\chi_m = \mu_r$
(d) there is no relation between them.

21. The magnetic moment of a magnetised rod is 8×10^{-7} A-m^2. Its mass is 75 g and specific gravity is 7.5. Its intensity of magnetisation is

(a) 8×10^{-8} A/m (b) 8×10^{-5} A/m
(c) 8×10^{-2} A/m (d) 8.0 A/m.

22. A magnetising field of 2000 A/m produces a flux of 6.28×10^{-4} Wb. Area of cross-section of rod is 2×10^{-6} m^2. Its relative permeability is

(a) 0.25 (b) 0.75
(c) 10^3 (d) 1.25×10^4.

23. In a tube light choke, a soft iron core is used. It is used to provide

(a) large permeability and low resistance
(b) large permeability and low inductance
(c) large permeability and high inductance
(d) large permeability and low hysteresis loss.

24. The magnetic susceptibility of a substance is 7.54×10^{-3}. Its relative permeability and nature of substance, repectively are

(a) 8.54×10^{-3}, diamagnetic
(b) 1.0075, paramagnetic
(c) 6.0×10^3, ferromagnetic
(d) none of the above.

25. An iron rod *B* when slipped over another rod of iron *A*, *B* is always attracted by *A*. This means

(a) both rods are magnetised
(b) both are not magnetised
(c) rod *A* is magnetised but not *B*
(d) rod *B* is magnetised but not *A*.

26. At Curie temperature

(a) paramagnetic substance is converted into diamagnetic substance
(b) ferromagnetic substance is demagnetised
(c) magnetisation of ferromagnetic substance is lost spontaneously
(d) diamagnetic substance is converted into paramagnetic substance.

27. Which of the following statement is wrong?

(a) Diamagnetism is the property of all substances
(b) At Curie temperature ferromagnetic substance is converted into paramagnetic
(c) Those atoms whose inner orbits are not filled become paramagnetic
(d) The magnetic susceptibility of ferromagnetic substances at Curie temperature is zero.

28. A paramagnetic material is placed in a magnetic field. Consider the following statements.

(*i*) If the magnetic field is increased, the magnetisation is increased
(*ii*) If the temperature is increased, the magnetisation is increased.

(a) Both (*i*) and (*ii*) are correct
(b) (*i*) is correct but (*ii*) is incorrect
(c) (*i*) is incorrect but (*ii*) is correct
(d) Both (*i*) and (*ii*) are incorrect.

29. A paramagnet material is kept in a magnetic field. The field is increased till the magnetisation becomes constant. If the temperature is now decreased the magnetisation

(a) will increase
(b) decrease
(c) remain constant
(d) may increase or decrease.

30. A long straight wire carries a current *i*. The magnetising field intensity *H* is measured at a point *P* close to the wire. A long, cylindrical iron rod is brought close to the wire so that the point *P* is at the centre of the rod. The value of *H* at *P* will

(a) increase
(b) decrease
(c) remain almost constant
(d) become zero.

31. A ferromagnetic specimen is placed in an external magnetic field. The magnetic domains

(a) increase in size
(b) decrease in size
(c) may increase or decrease in size
(d) have no relation with the field.

32. When a current in a circular loop is equivalently replaced by a magnetic dipole

(a) the pole strength *m* of each pole is fixed
(b) the distance *d* between the poles is fixed

(c) the product is fixed
(d) none of the above is fixed.

33. A compass needle which is allowed to move in a horizontal plane is taken to geomagnetic pole. It will

(a) stay in north-south direction only
(b) stay in east-west direction only
(c) become rigid showing no movement
(d) stay in any position.

34. A dip circle is taken to geomagnetic equator. The needle is allowed to move in a vertical plane perpendicular to the magnetic meridian. The needle will stay in

(a) horizontal direction only
(b) vertical direction only
(c) any direction except horizontal and vertical
(d) any direction.

35. The permanent magnetic moment of the atoms of a material is not zero. The material

(a) must be paramagnetic
(b) must be diamagnetic
(c) must be ferromagnetic
(d) may be paramagnetic.

36. Permanent magnetic moment of the atoms of a material is zero. The material

(a) must be paramagnetic
(b) must be diamagnetic
(c) must be ferromagnetic
(d) may be paramagnetic.

37. The susceptibility of an specimen at 300 K is 1.2×10^{-5}. The temperature at which its susceptibility is 1.8×10^{-5} is

(a) 300 K (b) 200 K
(c) 400 K (d) 450 K.

38. A uniform magnetic field of 2.0×10^{-4} T exists in a region. The change in scalar potential as one moves through 50 cm along the field is

(a) -1×10^{-4} T-m (b) $+1 \times 10^{-4}$ T-m
(c) -4×10^{-2} T-m (d) $+4 \times 10^{-2}$ T-m.

39. In the magnetic meridian of a certain place, the horizontal component of earth's magnetic field is 0.26 G and the angle of dip is 60°. In this location the magnetic field of the earth is

(a) 0.52 G (b) 0.13 G
(c) $\frac{0.26}{\sqrt{3}}$ G (d) $0.26\sqrt{3}$ G.

40. Mass of an iron rod is 80 gm and its magnetic moment 10 A-m^2. If density of iron is 8 g/cc, then intensity of magnetisation is (in A/m)

(a) 10^3 (b) 10^4
(c) 10^5 (d) 10^6.

41. If magnetic lines of force are drawn by keeping magnet vertical, the number of neutral points obtained will be

(a) one (b) two
(c) zero (d) infinite.

42. The ratio between total intensity of magnetic field at equator to poles is

(a) 1 : 1 (b) 1 : 2
(c) 2 : 1 (d) 1 : 4.

43. The force acting between two small magnets, placed in end on position 0.1 m apart from their centre is (Given the magnetic moment of each magnet is 5 A-m^2)

(a) 0.6 N (b) 0.8 N
(c) 0.15 N (d) 0.2 N.

44. If the angular momentum of an electron is J, then the magnitude of the magnetic moment will be

(a) $\frac{eJ}{m}$ (b) $\frac{eJ}{2m}$
(c) $2eJm$ (d) $\frac{2m}{eJ}$.

45. An atom is paramagnetic if it has

(a) an electric dipole moment
(b) no magnetic moment
(c) a magnetic moment
(d) no electric dipole moment.

46. If the distance between two dissimilar poles held d distance apart be doubled, then the force of attraction between them

(a) increases to four times
(b) decreases by half
(c) remains unchanged
(d) decreases to one fourth.

47. The magnetic field on the axis of a short bar magnet at a distance of 10 cm is 0.2 Oersted. What will be the field at a point 5 cm on the line perpendicular to the axis passing through the magnet?

(a) 0.025 Oersted (b) 0.2 Oestered
(c) 0.4 Oersted (d) 0.80 Oersted.

48. Points A and B are situated along the extended axis of a 2 cm long bar magnet at a distance x and $2x$ respectively from the pole nearer to the point. The ratio of the magnetic field at A and B will be

(a) exactly 8 : 1 (b) slightly <8 : 1
(c) slightly >8 : 1 (d) exactly 4 : 1.

49. Two short magnets placed along the same axis with their like poles facing each other will repel with force which varies inversely as

(a) distance (r)
(b) square of distance (r^2)
(c) cube of distance (r^3)
(d) fourth power of distance (r^4).

50. A magnet of magnetic moment M is rotated through 360° in magnetic field B, the work done will be

(a) MB (b) $2MB$
(c) $4MB$ (d) zero.

51. Work done in turning a magnet of magnetic moment M by an angle 90° from the meridian is n times the corresponding work done to turn it through an angle of 60°, where n is

(a) 1/2 (b) 2
(c) 1/4 (d) 1.

52. A magnet of magnetic moment 4.0 A-m^2 is held in a uniform magnetic field 5.0×10^{-4} T with magnetic moment vector making an angle 30° with the field. The work done in increasing the angle from 30° to 45° is about

(a) 3.2×10^{-4} J (b) 4.3×10^{-4} J
(c) 7.3×10^{-4} J (d) 1.6×10^{-4} J.

53. A magnetic needle suspended by a silk thread is vibrating in the earth's magnetic field. If the temperature of the needle is increased by 500° C

(a) the time period decreases
(b) the time period increases
(c) the time period remains unchanged
(d) the needle stops vibrating.

54. Two bar magnets having magnetic moments M_1 and M_2 are joined together pole for pole and suspended by a string. The time of oscillation of this assembly in magnetic field is 3 sec. If, now the polarity of one magnet is changed and combination again made to oscillate in the same field. The time of oscillation is 6 sec. The ratio $\frac{M_1}{M_2}$ is

(a) 5/3 (b) 2
(c) 8/3 (d) 11/3.

55. A small piece of an unmagnetised substance gets repelled when it is brought near a powerful magnet. The substance can be

(a) paramagnetic (b) ferromagnetic
(c) non-magnetic (d) diamagnetic.

56. Magnetic moment of atomic neon is equal to

(a) zero (b) $\frac{1}{2}\mu_B$
(c) μ_B (d) $\frac{3}{2}\mu_B$.

57. Which of the following exhibits perfect diamagnetism?

(a) Conductor (b) Superconductor
(c) Insulator (d) Semiconductor.

58. The temperature above which a ferromagnetic substance becomes paramagnetic is called

(a) critical temperature
(b) Boyle's temperature
(c) Debye's temperature
(d) Curie temperature.

59. Ferromagnetic substance is
(a) only crystalline solid
(b) only amorphous soild
(c) solid and liquid
(d) solid, liquid and gas.

60. Soft iron is used to manufacture electromagnets because their
(a) magnetic saturation limit is high and retentivity and coercive force are small
(b) retentivity is high
(c) coercive force is high
(d) area of hysteresis curve is large.

ANSWERS

1. (c)	2. (b)	3. (b)	4. (b)	5. (a)
6. (b)	7. (b)	8. (a)	9. (a)	10. (c)
11. (c)	12. (a)	13. (a)	14. (b)	15. (b)
16. (d)	17. (c)	18. (d)	19. (c)	20. (b)
21. (c)	22. (d)	23. (d)	24. (b)	25. (c)
26. (c)	27. (d)	28. (b)	29. (c)	30. (c)
31. (c)	32. (c)	33. (d)	34. (d)	35. (d)
36. (b)	37. (b)	38. (a)	39. (a)	40. (d)
41. (a)	42. (b)	43. (c)	44. (b)	45. (c)
46. (d)	47. (d)	48. (b)	49. (d)	50. (d)
51. (b)	52. (a)	53. (b)	54. (a)	55. (d)
56. (a)	57. (b)	58. (d)	59. (a)	60. (a)

PART II

In the following, a question is followed by four alternatives, of them one or more are correct. Candidate has to select the correct choice(s) from amongst the alternatives.

1. A uniformly magnetised specimen has magnetic susceptibility χ_m (= 3.14×10^{-4}) is placed in a magnetising field of H (= 6×10^4 A/m).
(a) Intensity of magnetisation of specimen is 18.84 A/m
(b) Intensity of magnetisation of specimen is 1.9×10^8 A/m
(c) Relative magnetic permeability of specimen is 1.0003
(d) Relative magnetic permeability of specimen is 1.003.

2. SI (Sommerfeld) unit of
(a) intensity of magnetisation is Ampere per meter
(b) magnetic flux is Wb/m^2
(c) absolute permeability is H/m.
(d) magnetic susceptibility is Ampere-meter.

3. The angle of dip
(a) is 90° in a plane perpendicular to the magnetic meridian
(b) is 0° in a plane perpendicular to the magnetic meridian

(c) is 0° at the poles
(d) is 90° at the poles.

4. A magnetic dipole is free to rotate in a uniform magnetic field; the orientation of magnet with respect to field
(a) for which potential energy is maximum is zero
(b) for which potential energy is maximum is 180°
(c) rate of change of torque with deflection is maximum, is 0°
(d) rate of change of torque with deflection is maximum, is 90°.

5. A bar magnet has pole strength 3.6 A-m magnetic length 12 cm and area of cross-section 0.9 cm^2.
(a) Magnetic intensity H at the centre of rod is 159.2 A/m towards N-pole
(b) Magnetic intensity H at the centre of rod is 159.2 A/m towards S-pole
(c) Magnetic field B at the centre is 50 mT towards N-pole
(d) Magnetic field B at the centre is 50 mT towards S-pole.

6. To measure the magnetic moment of a bar magnet, one may use
(a) a tangent galvanometer
(b) a deflection magnetometer if the earth's horizontal field is known
(c) a vibration magnetometer if the earth's horizontal field is known
(d) both deflection and vibration magnetometer if the earth's horizontal field is not known.

7. A horizontal circular loop carries a current that looks clockwise when viewed from the above. It is replaced by an equivalent magnetic dipole consisting of south pole S and a north pole N.
(a) the line SN should be along a diameter of the loop
(b) the line SN should be perpendicular to the plane of the loop
(c) the S-pole should be below the loop
(d) the N-pole should be below the loop.

8. Consider a magnetic dipole kept in the north-south direction. Let A, B, C, D be four points at the same distance from the dipole towards north, south, east and west. The direction of the magnetic field due to the dipole are the same at
(a) A and B (b) C and D
(c) A and C (d) B and D.

9. In the situation described in problem 8, the direction of the magnetic field due to the dipole are opposite at
(a) A and B (b) C and D
(c) A and C (d) B and D.

10. Pick up the correct options.
(a) Magnetic field is produced by electric charges only
(b) Magnetic poles are only mathematical assumptions having no real existance
(c) A north pole is equivalent to a clockwise current and a south pole is equivalent to an anticlockwise current
(d) A bar magnet is equivalent to a long, straight current carrying conductor.

11. Pick the correct statement.
(a) All the electrons have magnetic moment
(b) All the protons have magnetic moment
(c) All the nuclei have magnetic moment
(d) All the atoms have magnetic moment.

12. Which of the following pair have quantities of same dimensions?
(a) Magnetic field B and magnetising field H
(b) Magnetic field B and intensity of magnesation I
(c) Magnetising field intensity H and intensity of magnetisation I
(d) Magnetic susceptibility and longitudinal strain.

13. When a ferromagnetic material goes through a hysteresis loop, the magnetic susceptibility

(a) has fixed value (b) may be zero
(c) may be infinity (d) may be negative.

14. Mark out the correct options.

(a) Diamagnetism results from partial alignment of permanent magnetic moments
(b) The magnetic field of induced magnetic moment is opposite to the applied field
(c) The magnetising field intensity H is always zero in free space
(d) Diamagnetism occurs in all materials.

15. The magnetic field B and intensity of magnetising field H in a material are found to be 1.6 T and 1000 A/m respectively.

(a) Relative permeability of specimen is 1.3×10^3
(b) Magnetic susceptibility of specimen is 1.3×10^3
(c) Intensity of magnetisation of the specimen is nearly 1.3×10^6 A/m
(d) Intensity of magnetisation of the specimen is nearly 1.3×10^3 A/m.

16. The magnetic field at a point 10 cm away from a magnetic dipole is found to be 2.0×10^{-4} T. Then magnetic moment of the dipole is

(a) 1.0 A-m^2 if the point is in end-on position
(b) 2.0 A-m^2 if the point is in end-on position
(c) 1.0 A-m^2 if the point is in broad-side-on position
(d) 2.0 A-m^2 if the point is in broad-side-on position.

17. A solenoid has 500 turns/m and carries a current of 3 A. Its core is made of iron having relative permeability $\mu_r = 5000$.

(a) The intensity of magnetisation of core is 7.5×10^6 A-m^{-1}
(b) The intensity of magnetisation of core is 9.4×10^6 A-m^{-1}
(c) Magnetic field inside the specimen is 9.4 T
(d) Magnetic field inside the core is 7.5 T.

18. A copper ring is placed in a magnetising field H (= 3000 A/m). The magnetic susceptibility of copper χ_m (= -1.18×10^{-11}).

(a) The intensity of magnetisation in copper ring is 3.54×10^{-8} A/m
(b) The intensity of magnetisation in copper ring is 3.7×10^{-3} A/m
(c) The magnetic flux density in ring is 3.7×10^{-3} Wb/m^2
(d) The magnetic flux density in ring is 3.54×10^{-8} Wb/m^2.

19. A short bar magnet of magnetic moment M = 0.32 JT^{-1} is placed in uniform magnetic field of 0.15 T. If the bar is free to rotate in the plane of the field,

(a) in stable equilibrium $\boldsymbol{M}$ and $\boldsymbol{B}$ are parallel with P.E. = -4.8×10^{-2} J
(b) in stable equilibrium $\boldsymbol{M}$ and $\boldsymbol{B}$ are perpendicular with P.E. = 0
(c) in instable equilibrium $\boldsymbol{M}$ and $\boldsymbol{B}$ are parallel with P.E. = 4.8×10^{-2} J
(d) in instable equilibrium $\boldsymbol{M}$ and $\boldsymbol{B}$ are antiparallel with P.E. = 4.8×10^{-2} J.

20. An iron rod is subjected to cycles of magnetisation at the rate of 50 Hz. Given the density of the rod is 8×10^3 kg/m^3 and specific heat is 0.11×10^3 cal/kg°C. The area enclosed by B-H loop corresponds to energy 10^{-2} J.

(a) The rise in temperature of specimen in 1 minute is 8.1×10^{-6} °C
(b) The rise in temperature of specimen in 1 minute is 8.1×10^{-6} K
(c) The rise in temperature of specimen in 1 minute is 34.02×10^{-6} °C
(d) The rise in temperature of specimen in 1 minute is 34.02×10^{-6} K.

ANSWERS

1. (a, c)	2. (a, c)	3. (a, d)	4. (b, d)	5. (b, c)
6. (b, c, d)	7. (b, d)	8. (a, b)	9. (a, b)	10. (a, b)
11. (a, b)	12. (c, d)	13. (b, c, d)	14. (b, d)	15. (a, b, c)
16. (a, d)	17. (a, c)	18. (a, c)	19. (a, d)	20. (a, b).

23

Magnetic Effects of Electric Current

23.1 MAGNETIC FIELD

A *magnetic field* is the space in which magnetic effects are felt. By magnetic effects we mean that in this field a moving charge experiences a force. This field is different from *electric field* which always exerts a force on a charged particle whether it is at rest or in motion. The magnetic field is denoted by letter ***B*** and the force exerted by *magnetic field* is called *magnetic force.*

Definitions of Magnetic Field *B*

In order to define magnetic field ***B***, let us study the force experienced by a moving charge in a magnetic field ***B***.

(*i*) When a charged particle is projected in different directions in a magnetic field, it is found that there is a definite direction of projection along which the *charged particle experiences no magnetic force. This direction is defined as the direction of magnetic field.**

(*ii*) If the particle is projected with speed v in magnetic field in a direction making an angle θ with the direction of magnetic field (defined in (*i*), the magnitude of the force is directly proportional to $|v\sin\theta|$, i.e.

$$F \propto v\sin\theta$$

(*iii*) The force is proportional to the magnitude of the charge (q) and its direction is opposite for positive and negative charges.

$$F \propto q$$

(*iv*) The magnitude of the force is directly proportional to the magnitude of magnetic field (B) i.e. $F \propto B$

(*v*) The direction of the magnetic force is perpendicular to the plane containing ***B*** and ***v***, i.e. it is perpendicular to both magnetic field vector ***B*** and velocity ***v***.

---

*This direction of magnetic field is not uniquely defined; because there are two opposite directions along any line.

Combining all these, we have

$$F = k \cdot qvB\sin\theta \quad ...(23.1)$$

where k is constant of proportionality.

In mks SI system of units, when q is in Coulomb, v in meters per second, F in Newton, the unit of B is so chosen that k becomes 1. This unit of B is called Tesla. In these units magnetic force on a charged particle moving in magnetic field is

$$F = qvB\sin\theta \quad ...(23.2)$$

In vector notations, the magnetic force is written as

$$\boldsymbol{F} = q\boldsymbol{v} \times \boldsymbol{B} \quad ...(23.3)$$

Equation (23.3) uniquely determines the direction of $\boldsymbol{B}$ using rules of vector product.

Equation (23.2) defines the magnitude of the magnetic field. If $\theta = 90°$, $\sin\theta = +1$, $v = 1$ m/s and $q = 1$ Coulomb, then

$$F = B$$

Thus *the magnetic field induction* at any point is defined as the force experienced by a unit charge* (= 1 *Coulombs*) *when moving with unit velocity* (= 1 *m/s*) *at right angles to the magnetic field direction.* If this force experienced is equal to 1 Newton then B is called one tesla (T).

Thus one tesla is defined as the magnetic field induction in which 1 *Coulomb charge moving with velocity of* 1 *m/s perpendicular to the direction of field, experiences a force of* 1 *Newton.* The unit Weber/m^2 is also used for magnetic field and is same as tesla.

Another unit commonly used is gauss. 1 gauss = 10^{-4} T.

The dimensions of B are $= \dfrac{M^1LT^{-2}}{Q \cdot LT^{-1}} = M^1L^0T^{-1}Q^{-1}$ and in terms of current as 4th dimension $Q = IT$. We have dimensions of $B = M^1L^0T^{-2}I^{-1}$.

As we have already studied that static charge is source of electric field, similarly moving charges or currents produce a magnetic field in the surrounding space.

It should be remembered that *stationary charges produce only electric field but not magnetic field. A moving charge produce both electric and magnetic fields. On the other hand, a current carrying wire produces only magnetic field but not electric field.***

In fact the electric field and magnetic field are not basically independent. They are the two aspects of the same entity which we call *electromagnetic field.* Whether the electromagnetic field will show up as an electric field or a magnetic field or combination of both, depends on the frame of reference from which we are observing the field. From a particular frame of reference we can observe these two fields as separate entities.

The direction of magnetic field, when applied perpendicular to the plane of paper, is diagramatically shown as in Fig. 23.1. When the magnetic field B is perpendicular to the paper and coming out of the paper, by convention is shown as in Fig. 23.1 (*a*). Figure 23.1 (*b*) shows the direction of magnetic field perpendicular to the plane of paper going into it. The circle around the cross or dot is quite often omitted.

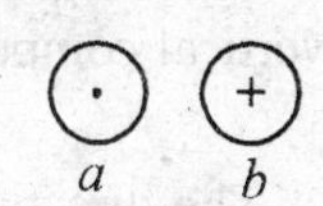

Fig. 23.1

---

*Magnitude of magnetic field is called intensity of magnetic field or magnetic field induction or magnetic flux density.

**In fact in case of current carrying wire, a small charge density is there on the surface of wire. Which does produce an electric field near the wire. This field is small and is generally neglected.

23.2 MOTION OF A CHARGED PARTICLE IN A UNIFORM MAGNETIC FIELD

When a charge q is projected with speed v in the plane of paper where the magnetic field B exists perpendicular to the plane of paper going into it (Fig. 23.2) a force $\boldsymbol{F} = q\boldsymbol{v} \times \boldsymbol{B}$ acts on it which is perpendicular to both v and B. In present case since $\boldsymbol{v}$ and $\boldsymbol{B}$ are perpendicular to each other hence magnitude of force is

$$F = qvB$$

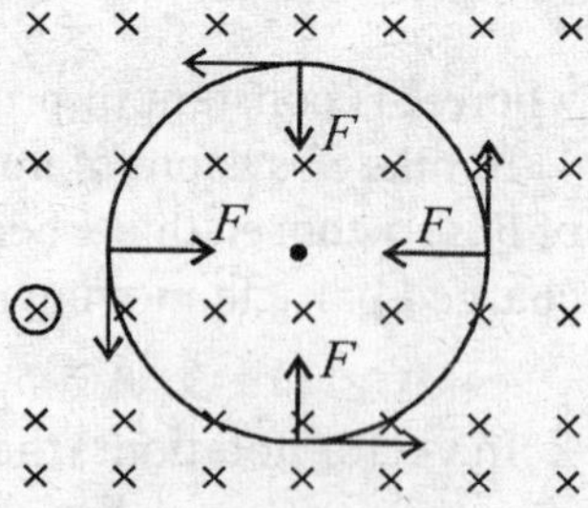

Fig. 23.2

This force will deflect the particle without changing the speed and the particle will move along a circle perpendicular to the field. The magnetic force provides the centrepetal force. If r be the radius of the circle, then

$$qvB = \frac{mv^2}{r}$$

or $$r = \frac{mv}{qB} \quad ...(23.4)$$

Time period, i.e. the time taken by the particle to complete the circle is

$$T = \frac{2\pi r}{v} = \frac{2\pi m}{qB} \quad ...(23.5)$$

It is clear that time period is independent of the speed v. If particle moves faster, the radius is larger, it has to move a longer distance to complete the circle so that the time taken is same.

The frequency of revolution is

$$\nu = \frac{1}{T} = \frac{qB}{2\pi m} \quad ...(23.6)$$

This frequency is called *cyclotron frequency*.

As the magnetic force on a particle is perpendicular to the velocity, it does not do any work on the particle. Hence, the kinetic energy or the speed of the particle does not change due to magnetic force.

Motion when Charge Velocity is not Perpendicular to the Magnetic Field

Let a particle of mass m and charge q enters a uniform magnetic field induction $\boldsymbol{B}$ at O with velocity $\boldsymbol{v}$ making an angle θ with the direction of magnetic field acting in the plane of paper (Fig. 23.3). Resolving $\boldsymbol{v}$ into two perpendicular components, we have $\boldsymbol{v}\cos\theta$ acts in the direction of magnetic field, and $\boldsymbol{v}\sin\theta$ acts perpendicular to the direction of magnetic field.

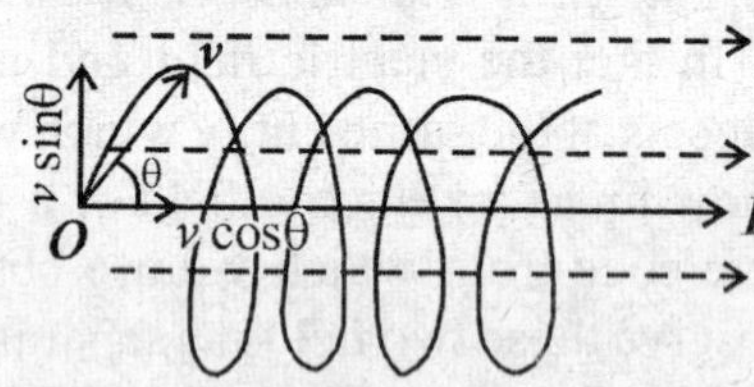

Fig. 23.3

The magnetic force acts on the charged particle due to its vertical component ($v\sin\theta$) and has magnitude.

$$F = qvB\sin\theta \quad ...(23.7)$$

The direction of this force is perpendicular to the plane of paper. Hence the charged particle moves in a circular path. This force provides the necessary centrepetal force $\left(\frac{mv^2\sin^2\theta}{r}\right)$ for the circular motion

i.e. $$qvB\sin\theta = \frac{mv^2\sin^2\theta}{r}$$

or $$r = \frac{mv\sin\theta}{qB} \qquad ...(23.8)$$

If angular velocity $\omega = \frac{v\sin\theta}{r} = \frac{qB}{m}$

The frequency of rotation called cyclotron frequency will be

$$\nu = \frac{\omega}{2\pi} = \frac{qB}{2\pi m} \qquad ...(23.9)$$

Equation (23.6) and (23.9) are same.

Since the component $v\cos\theta$ is parallel to magnetic field as such there will be no force on the particle in the direction of magnetic field. The charge particle will move with constant velocity (= $v\cos\theta$) in the direction of magnetic field.

The combined effect of the two motions, i.e. the resultant path is helix with axis parallel to the direction of the magnetic field.

The linear distance covered by the charged particle in the magnetic field in time equal to one revolution of its circular path is called *pitch of helix*.

Pitch of helix $= v\cos\theta \cdot T$ But $T = \frac{2\pi m}{qB}$

$$= v\cos\theta \cdot \frac{2\pi m}{qB}$$

$$= \frac{2\pi mv}{qB} \cdot \cos\theta \qquad ...(23.10)$$

Note:

1. If charged particle having charge q is in magnetic field, it experience force.
   $F = qvB\sin\theta$
2. If particle is at rest $v = 0$, $F = 0$, i.e. it experiences no force.
3. If charged particle is moving parallel to the direction of $\boldsymbol{B}$, i.e. $\theta = 0$ or 180°, $\sin\theta = 0$ and $F = 0$, i.e. it experiences no force.
4. If charged particle is moving perpendicular to the direction of $\boldsymbol{B}$, i.e. $\theta = 90°$, force is maximum (= qvB), the force is perpendicular to both $\boldsymbol{v}$ as well as $\boldsymbol{B}$ and it describes a circular motion of radius
   $$r = \frac{mv}{qB}$$
5. The force experienced by a charged particle moving in both electric and magnetic fields is called *Lorentz force*. If $\boldsymbol{E}$ is electric field $\boldsymbol{B}$ magnetic field and $\boldsymbol{v}$ velocity of the charged particle having charge q, then force
   $$\boldsymbol{F} = q(\boldsymbol{E} + \boldsymbol{v} \times \boldsymbol{B}) \qquad ...(23.11)$$

23.3 MAGNETIC FORCE ON A CURRENT CARRYING CONDUCTOR

Let a conductor carrying current i is placed in magnetic field $\boldsymbol{B}$ (Fig. 23.4). Consider a small element of length dl of this conductor. The current through the conductor in terms of drift velocity v_d is

$$i = nev_dA$$

where n is number of free electrons per unit volume, A is area of cross-section and e electronic charge.

Each electron then is moving with velocity v_d in magnetic field $\boldsymbol{B}$ and hence experiences magnetic force

$$\boldsymbol{f} = -e\boldsymbol{v}_d \times \boldsymbol{B}$$

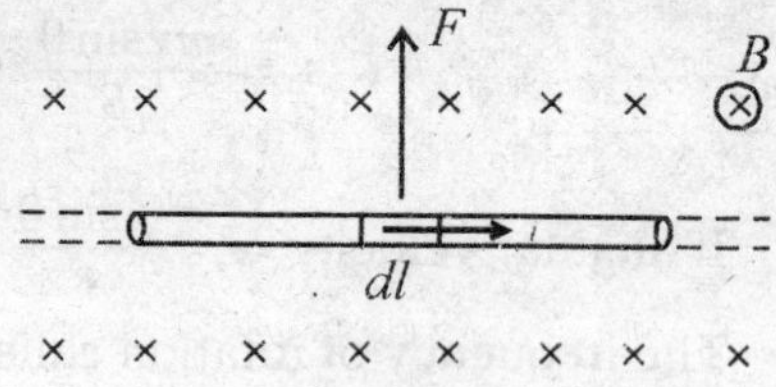

Fig. 23.4

The number of free electrons in small element dl of the conductor considered is $nAdl$. Thus the magnetic force on the element dl of the conductor considered is

$$d\boldsymbol{F} = (nAdl)\,(-e\boldsymbol{v}_d \times \boldsymbol{B})$$

If we denote the length dl along the current by $\boldsymbol{dl}$, the above equation can be written as

$$d\boldsymbol{F} = nAev_d(\boldsymbol{dl} \times \boldsymbol{B})$$
$$= (i\,\boldsymbol{dl} \times \boldsymbol{B}) = (i\boldsymbol{dl})^* \times (\boldsymbol{B}) \quad ...(23.11)$$

For a conductor of length l, the force is found by integration. For an straight current carrying conductor of length l in uniform magnetic field, magnetic force is

$$\boldsymbol{F} = (i\boldsymbol{l}) \times \boldsymbol{B} \quad ...(23.12)$$

The direction of the force can be obtained using laws of vector product.

Torque on a Current Loop

A current carrying loop when placed in a uniform magnetic field experiences a torque. It does not experience net force. For simplicity we will consider rectangular loop but the expression obtained is true for any shape of the loop.

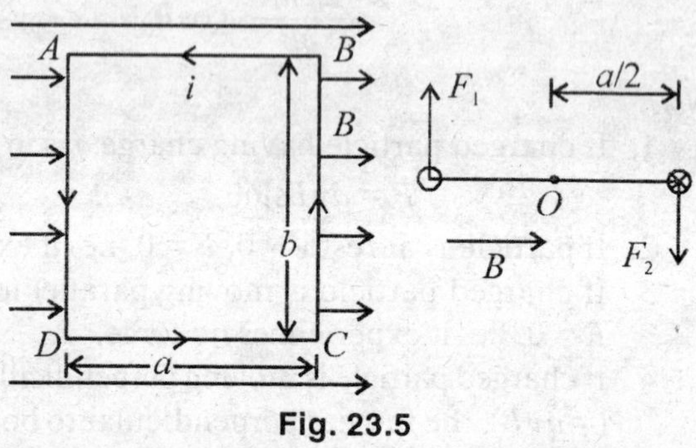

Fig. 23.5

We first consider the simple case in which the rectangular loop is placed with its plane parallel to the direction of the uniform magnetic field $\boldsymbol{B}$ (Fig. 23.5). In this situation magnetic field is parallel to sides AB and CD. Hence field does not exert any force on these arms of the loop. The field is perpendicular to the arms AD and BC. The force on arm AD is $F_1 = ibB$ is directed out of the plane of the loop and the force on arm BC, $F_2 = ibB$ directed into the plane of the loop. Thus net force on the loop is zero. The points of application of the two forces are not same hence a torque acts on the loop. The side Fig. 23.5 gives the sectional view of the loop. The torque has the magnitude

$$\tau = F_1 a/2 + F_2 a/2 = iabB = iAB \quad ...(23.13)$$

where $A = ab$ the area of the loop. Equation (23.13) is a particular case when the field direction $\boldsymbol{B}$ and vector area $\boldsymbol{A}$ are perpendicular to each other.

For arbitrary angle θ (other than 90°) between magnetic field vector $\boldsymbol{B}$ and area vector $\boldsymbol{A}$, the situation is shown in Fig. 23.6. The force on arms AB and CD is same (ilB) as before, the net torque now

$$\tau = iAB\sin\theta \quad ...(23.14)$$

*In fact product ($i\boldsymbol{dl}$) is taken as vector called curent element. It has direction as the direction of current and magnitude equal to the product (idl).

We have in last chapter stated that a current carrying loop is equivalent to a magnetic dipole whose magnetic moment is current multiplied by area of loop, i.e.

$$\boldsymbol{M} = \boldsymbol{A}i$$

and its direction is same as the direction of vector area. Hence torque is

$$\tau = MB\sin\theta$$

$$\boldsymbol{\tau} = \boldsymbol{M} \times \boldsymbol{B} \qquad \ldots(23.15)$$

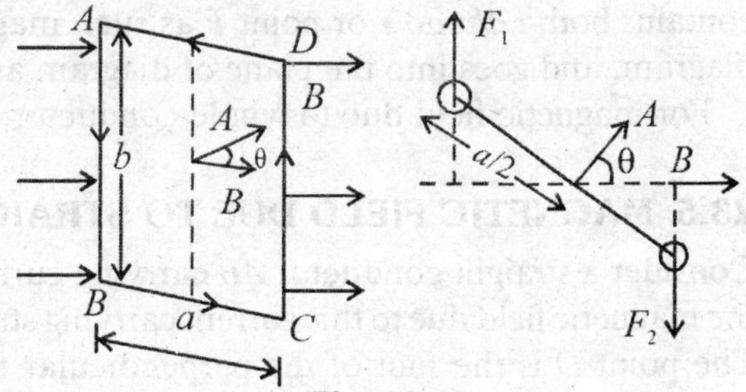

Fig. 23.6

This equation is analogous to electrostatic case $\tau = \boldsymbol{P} \times \boldsymbol{E}$

If there are n turns in the loop, each turn experiences a torque. The net torque then is

$$\tau = ni\boldsymbol{A} \times \boldsymbol{B}$$

$$= \boldsymbol{M} \times \boldsymbol{B}$$

In this case magnetic moment is defined as $\boldsymbol{M} = ni\boldsymbol{A}$

23.4 BIOT-SAVART'S LAW

We have already defined magnetic field in terms of the force it exerts on a moving charge. Here we shall discuss how magnetic field can be produced. As already stated a magnetic field can be produced by moving charges or electric currents. The basic equation for magnetic field (***dB***) due to an electrical current carrying element (***idl***) is the *Biot-Savart's* law. Figure 23.7 shows a finite conductor *AB* carrying current *i*. Consider an infinitesimal current element ***idl***. The field due to this element at point *P* at distance *r* from the element is given by Biot-Savart's law. According to this law

$dB \propto idl$ (Current element)

$\propto \frac{1}{r^2}$ (r distance of point P from current element)

$\propto \sin\theta$ (θ angle between ***idl*** and ***r***)

Combining we have

$$dB = \frac{\mu_0}{4\pi} \cdot \frac{i\,dl\sin\theta}{r^2} \qquad \ldots(23.16)$$

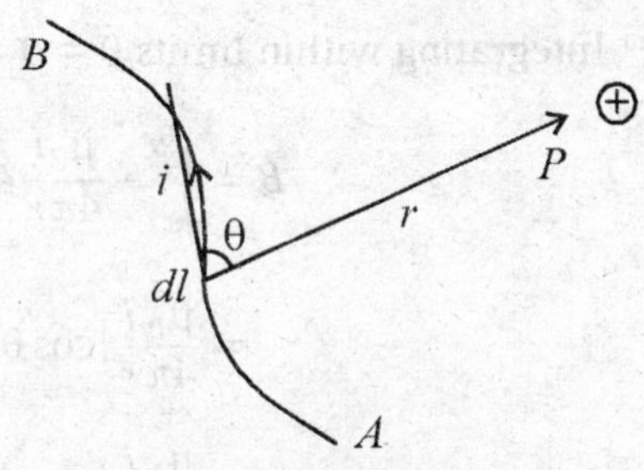

Fig. 23.7

where $\frac{\mu_0}{4\pi}$ is constant of proportionality. Its value is 10^{-7} Henry per meter. μ_0 is called magnetic permeability of free space. Since dB (magnetic field) is vector, hence equation (23.16) can be written in vector notation as

$$\boldsymbol{dB} = \frac{\mu_0}{4\pi} \cdot \frac{i\boldsymbol{dl} \times \hat{r}}{r^2}$$

$$= \frac{\mu_0}{4\pi} \cdot \frac{(i\,\boldsymbol{dl}) \times \boldsymbol{r}}{r^3} \qquad \ldots(23.17)$$

The direction of the field is perpendicular to the plane containing the current element and the point P according to the rules of vector product. In present case (Fig. 23.7), the plane of paper contains both $i\,dl$ and r or point P as such magnetic field $\boldsymbol{B}$ is then perpendicular to the plane of diagram, and goes into the plane of diagram as depicted in the figure.

For magnetic field due to whole conductor, we have to integrate equation (23.17).

23.5 MAGNETIC FIELD DUE TO STRAIGHT CURRENT CARRYING CONDUCTOR

Consider a straight conductor AB carrying current i. Let point P is at a distance r ($OP = r$). Where the magnetic field due to this current carrying straight conductor, of finite length is to be determined. The point O is the foot of the perpendicular from P to the conductor. Let us consider a small element dx of the conductor at distance x from the point O. The element dx makes an angle θ with line CP. The magnetic field dB at P due to this current element (idx), using Biot-Savart's law is

$$dB = \frac{\mu_0}{4\pi}\frac{i\,dx\sin\theta}{CP^2} = \frac{\mu_0}{4\pi}\frac{i\,dx\sin\theta}{x^2+r^2} \qquad \ldots(i)$$

The direction of the field is determined by vector product $(\boldsymbol{idx}) \times \boldsymbol{CP}$. It is perpendicular to the plane of diagram going into it (Fig. 23.8). The direction of the field is the same for all elements of the wire and hence the net field due to the current carrying conductor AB is obtained by integrating equation (i) under proper limits.

From the Fig. 23.8, we have

$$x = r\cot\theta \text{ or } dx = -r\,\text{cosec}^2\theta \cdot d\theta$$

or $$x^2 + r^2 = r^2(1 + \cot^2\theta) = r^2\text{cosec}^2\theta$$

under this transformation equation (i) becomes

$$dB = \frac{\mu_0}{4\pi}\frac{i(-r\,\text{cosec}^2\theta d\theta)\sin\theta}{r^2\text{cosec}^2\theta} = -\frac{\mu_0 i}{4\pi r}\sin\theta\cdot d\theta$$

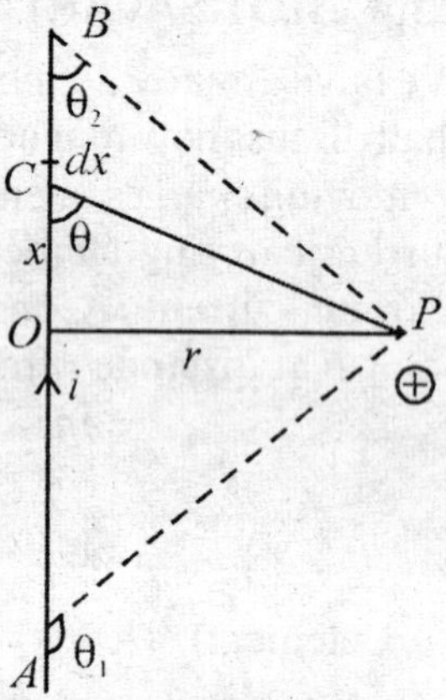

Fig. 23.8

Integrating within limits $\theta = \theta_1$ to $\theta = \theta_2$ we have

$$B = \int_{\theta_1}^{\theta_2} -\frac{\mu_0 i}{4\pi r}\sin\theta\cdot d\theta$$

$$= \frac{\mu_0 i}{4\pi r}\left[\cos\theta\right]_{\theta_1}^{\theta_2}$$

$$= \frac{\mu_0 i}{4\pi r}(\cos\theta_2 - \cos\theta_1) \qquad \ldots(23.18)$$

If α and β are the angles that the two ends of a current carrying conductor make at point P with perpendicular PO, then

$$\theta_2 = 90 - \alpha \text{ and } \theta_1 = 90 + \beta$$

Then magnetic field at P is

$$\beta = \frac{\mu_0 i}{4\pi r}\left[\cos(90-\alpha) - \cos(90+\beta)\right]$$

$$= \frac{\mu_0 i}{4\pi r}(\sin\alpha + \sin\beta) \qquad \ldots(23.19)$$

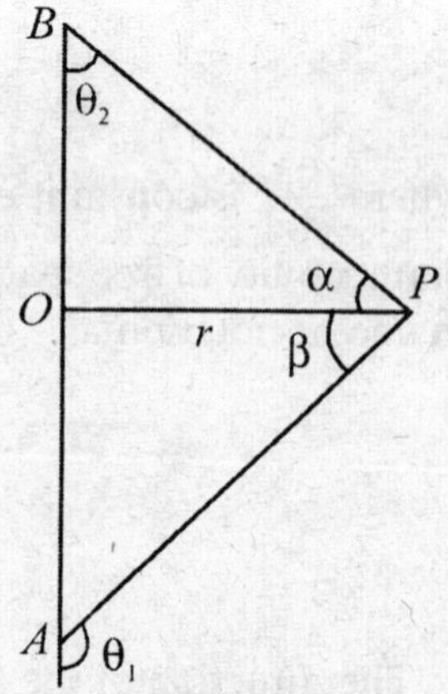

Fig. 23.9

Some Special Cases

Conductor infinite long (Fig. 23.10) gives $\alpha = \frac{\pi}{2}$ and $\beta = \frac{\pi}{2}$. Hence

$$B = \frac{\mu_0 i}{4\pi r}\left(\sin\frac{\pi}{2} + \sin\frac{\pi}{2}\right)$$

$$= \frac{\mu_0 i}{2\pi r} \qquad \text{...(23.20)}$$

The direction of magnetic field at P can also be determined by *right hand thumb rule.* This rule says that if we *stretch the thumb of our right hand along the direction of current in infinite long conductor and curl our fingers to pass through the point P, the direction of the fingers at P gives the direction of the magnetic field there* (Fig. 23.11).

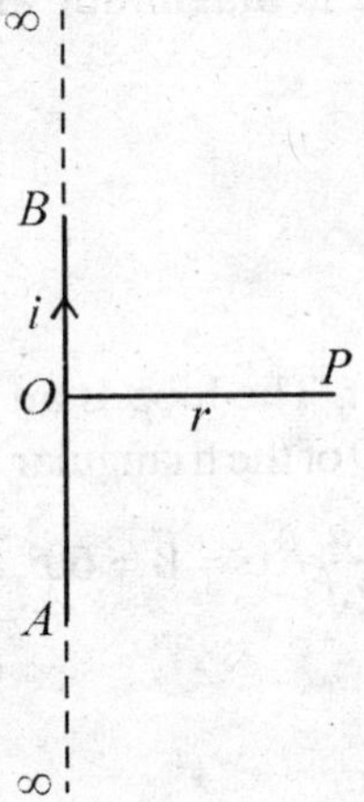

Fig. 23.10

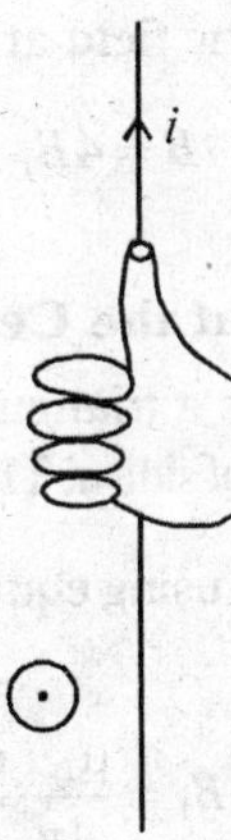

Fig. 23.11

Conductor infinite long with one end in the plane of point P

The situation is showing in Fig. 23.12. In this case $\alpha = 0$ and $\beta = \frac{\pi}{2}$. Hence magnetic field B at P is

$$B = \frac{\mu_0 i}{4\pi r}\left(\sin 0 + \sin\frac{\pi}{2}\right) = \frac{\mu_0 i}{4\pi r} \qquad \text{...(23.21)}$$

In these cases the magnetic field lines can be drawn on the pattern of electric field lines. A tangent to a magnetic field line gives the direction of magnetic field existing at that point. For a long straight current carrying conductor field lines are circles with their centres on the conductor (Fig. 23.13).

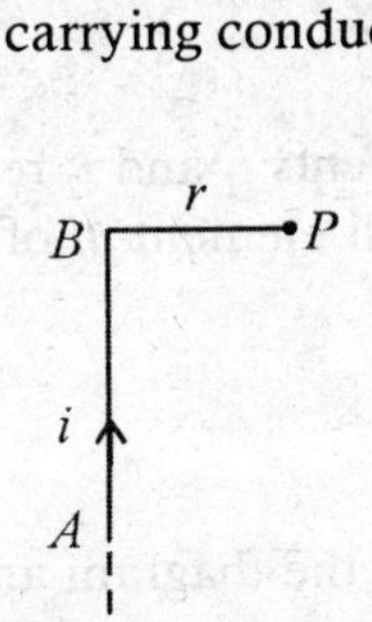

Fig. 23.12

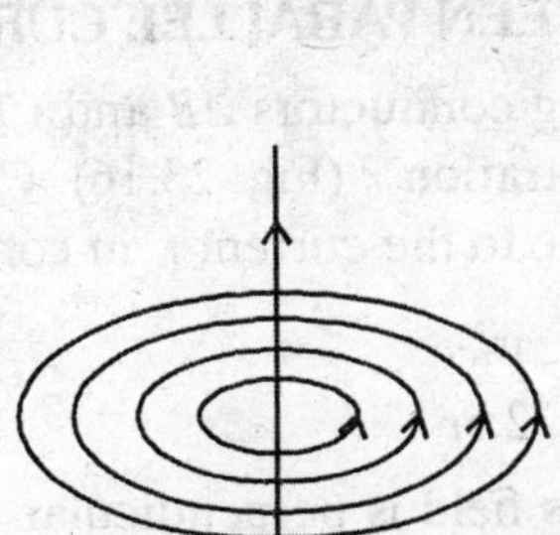
Fig. 23.13

Magnetic Field at the Centre of a Square Loop

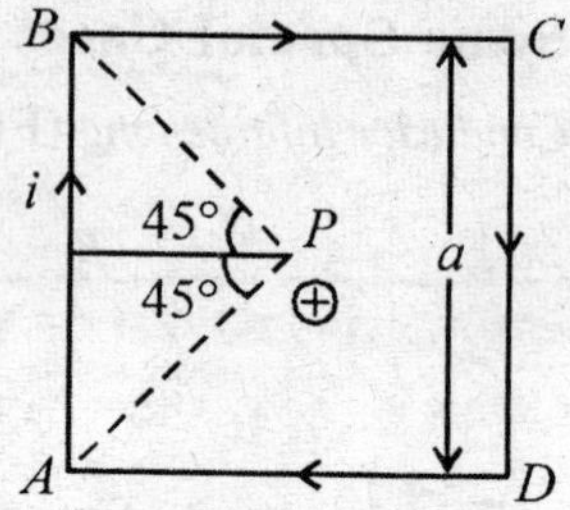

Fig. 23.14

Figure 23.14 shows a current carrying loop of side a. Current i flows through the loop. Magnetic field at P due to current i in AB is found using equation (23.19). Here $r = a/2$, $\alpha = 45°$, $\beta = 45°$, hence

$$B_1 = \frac{\mu_0 i}{4\pi} \frac{(\sin 45° + \sin 45°)}{a/2}$$

$$= \frac{\mu_0 i}{\sqrt{2}\pi a}$$

The direction of the field as obtained using right hand thumb rule, is perpendicular to the plane of loop into the paper (Fig. 23.14).

The magnetic field due to other three sides is also same in magnitude and parallel in direction. Hence net magnetic field at P due to the loop is

$$B = 4B_1 = \frac{2\sqrt{2}\mu_0 i}{\pi a} \tag{23.22}$$

Magnetic Field at the Centre of a Triangular Loop

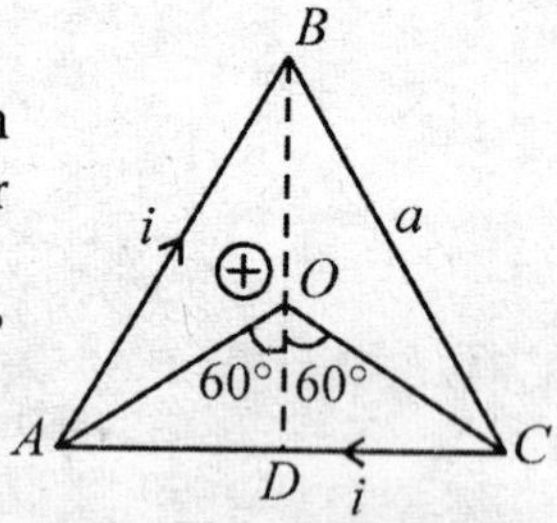

Fig. 23.15

Figure 23.15 shows a triangular loop ABC carrying current i. The loop is an equilateral triangle of side a. The magnetic field at the centre O of the triangular loop is determined using equation (23.19). Here $r = OD = \frac{a}{2\sqrt{3}}$, $\alpha = \beta = 60°$, hence

$$B_1 = \frac{\mu_0 i}{4\pi} \frac{(\sin 60° + \sin 60°)}{a/2\sqrt{3}}$$

$$= \frac{3\mu_0 i}{2\pi a}$$

The direction is perpendicular to the plane of loop into the paper (Fig. 23.15).

The magnetic field due to other two sides is also same in magnitude and parallel in direction. Hence net magnetic field at P due to the loop is

$$B = 3B_1 = \frac{9\mu_0 i}{2\pi a} \quad \text{...(23.23)}$$

23.6 FORCE BETWEEN PARALLEL CURRENTS

Consider two very long conductors AB and CD carrying currents i_1 and i_2 respectively. They are placed parallel at separation r (Fig. 23.16). Consider a small element dl of conductor CD. The magnetic field at dl due to the current i_1 in conductor AB is

$$B = \frac{\mu_0 i}{2\pi r} \quad \text{...(i)}$$

The direction of this field is perpendicular to the plane of the diagram and going into it. The magnetic force on the element $i_2 dl$ is

$$d\mathbf{F} = (i_2 \mathbf{dl}) \times \mathbf{B}$$

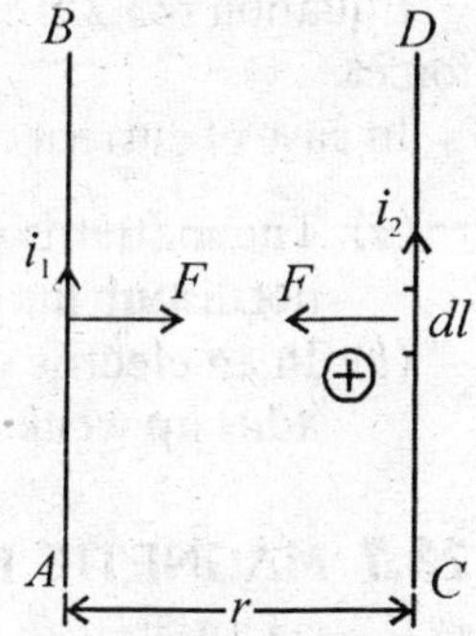

Fig. 23.16

or $$dF = i_2\, dl \frac{\mu_0\, i_1}{2\pi r} = \frac{\mu_0\, i_1 i_2\, dl}{2\pi r}$$

and has a direction $(i\, d\mathbf{l}) \times (\mathbf{B})$, i.e. towards the conductor *AB*. Hence force per unit length of the conductor *CD* due to the current through *AB* is

$$\frac{dF}{dl} = \frac{\mu_0\, i_1 i_2}{2\pi r} \qquad \text{...(23.24)}$$

If we take an element *dl* in the conductor *AB* and calculate the magnetic force per unit length of the conductor *AB* due to current through *CD* it is same in magnitude with direction towards conductor *CD*. Thus the two conductors carrying parallel currents attract each other.

If these conductors carry currents in opposite directions, the conductors repel each other.

Definition of Ampere

Equation (23.24) can be used to define an '*ampere*'. If $i_1 = i_2$ 1 A and $r = 1$ m, then force per unit length

$$\frac{dF}{dl} = 2 \times 10^{-7} \text{ N}$$

Thus 1 *ampere is the current passing through two parallel long wires kept* 1 *m apart in vacuum, exert an attractive force of* 2×10^{-7} *per meter of each wire.*

Comparison of Electrical and Magnetic Forces

Consider the instantaneous situation of the two charges shown in Fig. 23.17. The electrostatic repulsive force between the charges is

$$F_e = \frac{1}{4\pi\varepsilon_0} \cdot \frac{q_1 q_2}{r^2} \qquad \text{...}(i)$$

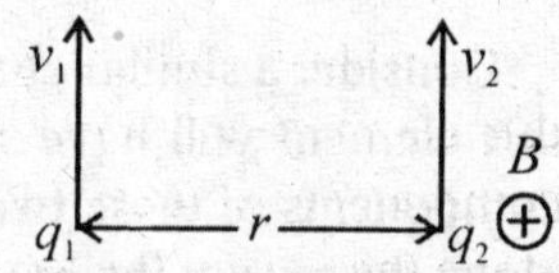

Fig. 23.17

Charge q_1 moving with velocity v_1 produces magnetic field at the site of charge q_2

$$B = \frac{\mu_0}{4\pi} \cdot \frac{q_1\, dl}{dt} \cdot \frac{1}{r^2} = \frac{\mu_0}{4\pi}\, q_1 \frac{v_1}{r^2} \qquad \text{...}(ii)$$

Since a charge q_1 moving with velocity v_1 is equivalent to current element $(q_1 v_1)$. This field is perpendicular to the plane of paper and into the plane (Fig. 23.17).

Charge q_2 moving with velocity v_2 in this magnetic field experiences an attractive magnetic force

$$F_m = q_2 v_2 B = \frac{\mu_0}{4\pi} \frac{q_1 q_2\, v_1 v_2}{r^2} \qquad \text{...}(iii)$$

Comparing (*i*) and (*iii*), we get

$$\frac{F_e}{F_m} = \frac{1}{v_1 v_2\, \varepsilon_0\, \mu_0} = \frac{c^2}{v_1 v_2} \qquad \text{...(23.25)}$$

where $c^2 = \dfrac{1}{\varepsilon_0\, \mu_0}$ and c is velocity of light in vacuum.

Equation (23.25) shows that the magnetic forces are very weak in comparison to electrostatic forces.

In case of current carrying conductors, electric forces never dominate magnetic forces because

(*a*) The matter is electrically neutral to very high degree of accuracy. The flowing charges do not disturb the neutrality of the matter. Thus the Coulomb's electric forces are rarely observed.

(*b*) In an electric current the large number of electrons are drifting in the same direction. This adds up weak magnetic fields which are observable.

23.7 MAGNETIC FIELD ON THE AXIS OF A CIRCULAR CURRENT LOOP

Figure 23.18 shows a circular loop of radius a carrying a current i. We have to find the magnetic field at a point P on the axis of the loop at distance x from its centre O. In figure the loop is perpendicular to the plane of the figure, while its axis is in the plane of the figure. The current comes out of the plane at A and goes into it at B. Consider a current element $i\,dl$ at point A. The magnetic field at P due to this current element is

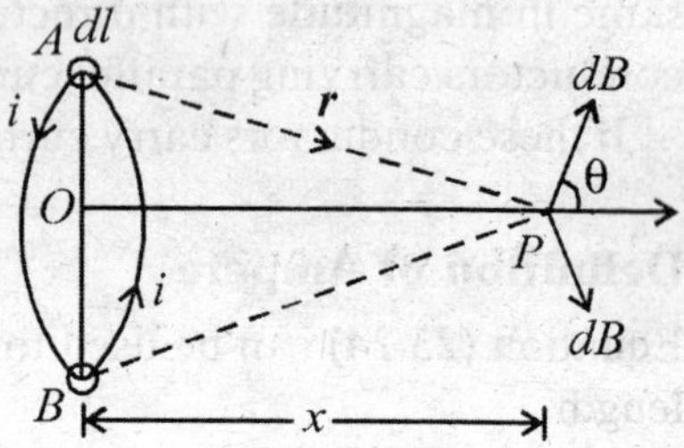

Fig. 23.18

$$\boldsymbol{dB} = \frac{\mu_0}{4\pi}\frac{(i\,\boldsymbol{dl})\times\boldsymbol{r}}{r^3}$$

where $\boldsymbol{r}$ is vector joining the current element to the point P.

Since $(i\boldsymbol{dl})$ is perpendicular to the plane of the figure as such $(i\boldsymbol{dl})\times\boldsymbol{r}$ must be in the plane of figure. The direction of $\boldsymbol{dB}$ is shown in figure as per rules of vector product. The magnitude of the field is

$$dB = \frac{\mu_0\, i\, dl}{4\pi r^2} = \frac{\mu_0 i}{4\pi}\frac{dl}{(a^2+x^2)} \qquad \ldots(i)$$

Consider a similar current element at the diametrically opposite point at B. The field at P due to this element will have same magnitude but now direction is perpendicular to the line BP. The components of these two fields perpendicular to the axis cancel each other. They have resultant along the axis of the loop. Dividing the loop in such pairs of diametrically opposite elements, we conclude that the resultant field at P must be along the axis of the loop. The resultant field at P can therefore be obtained by integrating the component of dB along the axis, i.e. $dB\cos\theta$

$$dB\cos\theta = \frac{\mu_0 i}{4\pi}\frac{dl}{(a^2+x^2)}\frac{a}{\sqrt{a^2+x^2}} = \frac{\mu_0\, ia\, dl}{4\pi(a^2+x^2)^{3/2}}$$

$\therefore$

$$B = \int dB\cos\theta = \int \frac{\mu_0\, ia\, dl}{4\pi(a^2+x^2)^{3/2}} = \frac{\mu_0\, ia}{4\pi(a^2+x^2)^{3/2}}\int dl$$

But $\int dl$ = total length of coil = $2\pi an$

where n is number of turns in the loop.

$\therefore$

$$B = \frac{\mu_0\, ia}{4\pi(a^2+x^2)^{3/2}}\times 2\pi\, an$$

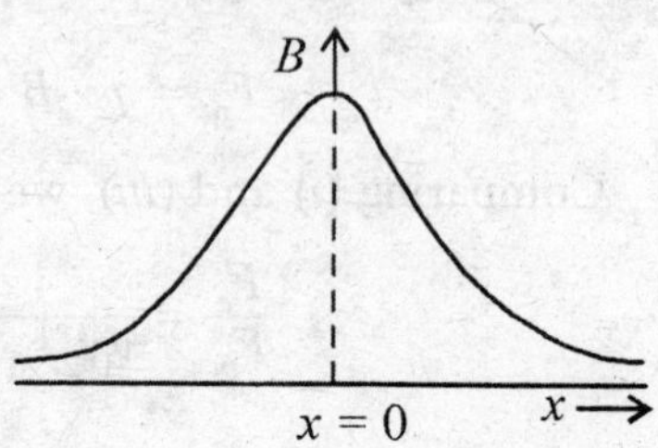

Fig. 23.19

$$= \frac{\mu_0 ia^2 n}{2(a^2 + x^2)^{3/2}} \qquad ...(23.26)$$

The graph between B and x is shown in Fig. 23.19.

Special Case

(i) Field at the Centre of the Coil

At the centre $x = 0$; equation (23.26) then becomes

$$B = \frac{\mu_0 in}{2a} \qquad ...(23.27)$$

The magnetic field lines due to circular coil form closed loops and are given in Fig. 23.20. The direction of the magnetic field is given by *right hand thumb rule* already explained.

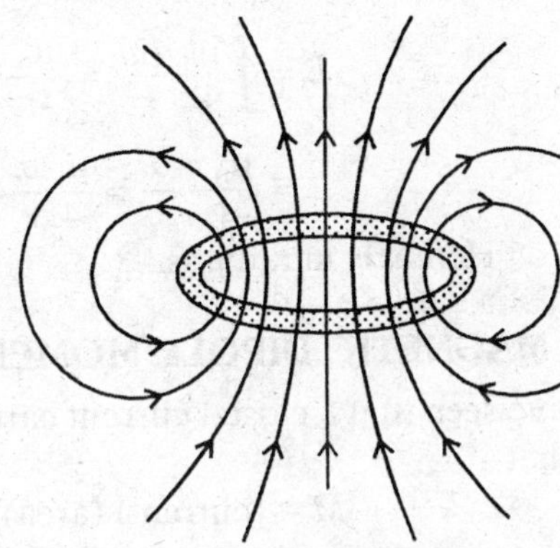

Fig. 23.20

(ii) Field at Infinity: At Infinity, i.e. $x = \infty$

$$B = 0$$

This shows that magnetic field is maximum at the centre and goes on decreasing as we move away along the axis and becomes zero at infinity.

(iii) Field at a Very Large Distance from Centre ($x >> a$)

When $x >> a$, a^2 can be neglected in comparison to x^2 in equation (23.26). We then get

$$B = \frac{\mu_0 ia^2 n}{2x^3} \qquad ...(23.28)$$

This equation can be written as

$$B = \frac{2\mu_0}{4\pi} \frac{(in)(\pi a^2)}{x^3} \qquad ...(23.28a)$$

We have proved in previous chapter (classical magnetism), that magnetic field due to a magnetic dipole of moment M at a distance x from the centre along its axis is

$$B = \frac{\mu_0}{4\pi} \cdot \frac{2M}{x^3}$$

Comparing this equation with equation (23.28*a*) we observe that a current carrying loop is equivalent to a magnetic dipole of moment M given by

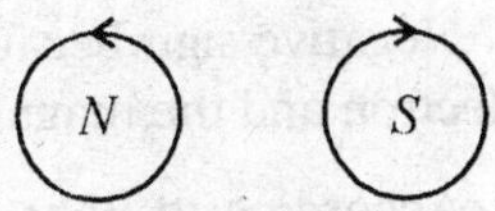

Fig. 23.21

$$M = (in)(\pi a^2) = \text{(total current) (area of loop)} \qquad ...(23.28b)$$

Polarity of the dipole is given as follows. Looking towards the coil, if the current is anticlockwise, it is North pole. If the current in clockwise, it is South pole (Fig. 23.21).

(iv) Field Due to a Segment of Current Loop at its Centre

Consider a segment AB of a loop of radius r carrying current i. The segment makes an angle α at the centre and length $AB = l$. Hence $l = r\alpha$. Consider small element dl of this segment

(Fig. 23.22). Magnetic field at the centre of loop due to this current element idl, using Biot-Savart's law is

$$B = \frac{\mu_0}{4\pi}\frac{i\,dl}{r^2}$$

The direction is perpendicular to the plane of paper going into the paper plane.

Hence field due to the whole segment is

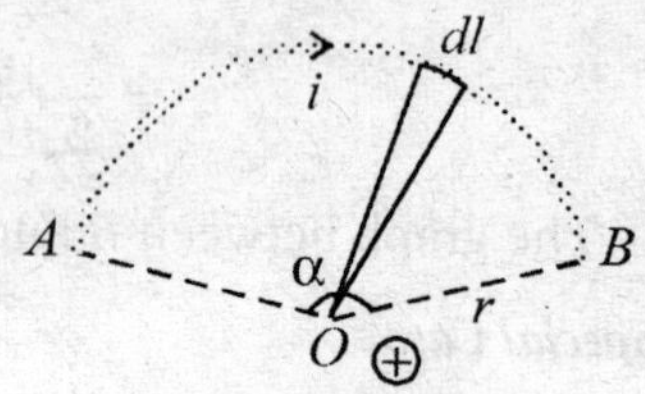

Fig. 23.22

$$B = \int \frac{\mu_0}{4\pi}\frac{i\,dl}{r^2} = \frac{\mu_0\, il}{4\pi r^2} \qquad \ldots(23.29)$$

$$= \frac{\mu_0\, ir\,\alpha}{4\pi r^2} = \frac{\mu_0 i\alpha}{4\pi r} \qquad \ldots(23.29a)$$

Here a is in radians.

23.8 MAGNETIC DIPOLE MOMENT OF ATOM

We have seen that a closed current carrying loop is equivalent to a magnetic dipole of moment M such that

$$M = \text{(current) (area)} = iA$$

The direction of the magnetic moment is the direction of the vector area of the loop.

An electron revolving round a nucleus can be treated as a current carrying closed loop.

If f_n is the frequency of revolution of electron in nth orbit of radius r_n; then it is equivalent to a current loop of radius r_n carrying current ef_n. Hence magnetic moment of atom is

$$M_n = ef_n \cdot \pi r_n^2 = \frac{e}{2m}(2\pi f_n)mr_n^2$$

where m is mass of electron, $2\pi f_n = \omega_n$, angular velocity of electron and $mr_n^2\omega_n$ is angular momentum L_n of the electron in nth orbit

$$\therefore \quad M_n = \frac{e}{2m}L_n \qquad \ldots(23.30)$$

Thus we see that angular momentum is related to the magnetic moment. In vector notation the equation (23.30) can be written as

$$\mathbf{M}_n = -\frac{e}{2M}\mathbf{L}_n \qquad \ldots(23.30a)$$

Negative sign because of negative charge of electron, shows that the angular momentum of electron and the magnetic moment associated with it are opposite in nature.

According to Bohr atomic theory, angular momentum is quantised and has value $L_n = \frac{nh}{2\pi}$ where n is an integral number.

$$\therefore \quad M_n = \frac{e}{2m}\frac{nh}{2\pi} = \frac{eh}{4\pi m}n \qquad \ldots(23.31)$$

This means that magnetic moment of atom is also quantised and takes up integral values of

$$\mu_B = \frac{eh}{4\pi m} \qquad \ldots(23.32)$$

μ_B is called Bohr magneton $\mu_B = 9.27 \times 10^{-24}$ A-m^2

23.9 AMPERE'S CIRCUITAL LAW

The Ampere's law is another from of Biot-Savart's law. This law states that *the line integral of magnetic field over a closed path is equal to μ_0 times the total current enclosed by the path.* The closed path is called an Amperean loop (analogous to Gaussian surface in electrostatics). It is a geometrical entity and not a real loop. Mathematically, it is written as

$$\oint \boldsymbol{B} \cdot \boldsymbol{dl} = \mu_0 \Sigma i \qquad \ldots(23.33)$$

Biot-Savart's law is an inverse-square law and is the magnetic analogue of Coulomb's law $\boldsymbol{E} = \frac{1}{4\pi\varepsilon_0}\int \frac{dq\,\boldsymbol{r}}{r^3}$. While Ampere's law is analogous to Gauss' law $\oint_s \boldsymbol{E} \cdot \boldsymbol{ds} = \frac{1}{\varepsilon_0}\Sigma q$

Ampere's law may be derived from Biot-Savart's law and Biot-Savart's law may be derived from Ampere's law. As such in scientific content the two laws are equivalent.

Though these laws are general and hold good for any assembly of currents, Ampere's law finds its application only in situation where, by symmetry of problem $\boldsymbol{B}$ has constant value both in magnitude and direction, either parallel or perpendicular to the length elements along some path of integration.

23.10 MAGNETIC FIELD DUE TO CURRENT THROUGH A VERY LONG CIRCULAR CYLINDER

Consider an infinite long cylinder of radius *a* carrying current *i*. A magnetic field is set up around the cylinder in the form of circular magnetic lines of force with their centres on the axis of the cylinder (Fig. 23.23). These lines of force are perpendicular to the length of the cylinder. We will find the magnetic field due to this current carrying cylinder.

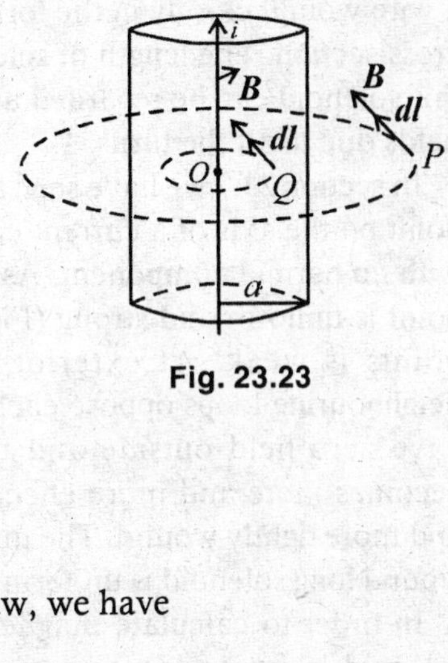

Fig. 23.23

Case (*i*) Point *P* Lies Outside the Cylinder (*r* > *a*)

Through point *P* let us take a closed path—which is circle of radius *r* with its centre on the axis of the cylinder. Line integral of magnetic field *B* over this closed path is

$$\oint \boldsymbol{B} \cdot \boldsymbol{dl} = \oint B\,dl = 2\pi r B$$

Since *B* and *dl* are parallel to each other at all point.

The closed path encloses the current *i*. Hence using Ampere's law, we have

$$\oint \boldsymbol{B} \cdot \boldsymbol{dl} = 2\pi r B = \mu_0 i$$

or
$$B = \frac{\mu_0 i}{2\pi r}$$

A result which we have already proved (Equation (23.20)).

Case (*ii*) Point *P* Lies on the Surface the Cylinder (*r* = *a*)

Magnetic field *B* is obtained by putting *r* = *a*

∴
$$B = \frac{\mu_0 i}{2\pi a}$$

Case (*iii*) Point *P* Lies inside the Cylinder ($r < a$)

Let point Q is at distance r from the axis (Fig. 23.23). Let us take closed path as a circle of radius r through Q with its centre on the axis. The line integral of the magnetic field over this closed path, i.e. circle is

$$\oint \boldsymbol{B} \cdot \boldsymbol{dl} = \oint B\, dl = 2\pi r B$$

If J is the current density then current enclosed by the path is

$$i' = J \cdot \pi r^2 \text{ But } J = \frac{i}{\pi a^2} \therefore i' = \frac{i}{\pi a^2} \times \pi r^2 = \frac{i r^2}{a^2}$$

Using Ampere's law, we get

$$2\pi r B = \frac{\mu_0 i r^2}{a^2}$$

or

$$B = \frac{\mu_0 i}{2\pi} \cdot \frac{r}{a^2} \qquad \text{...(23.24)}$$

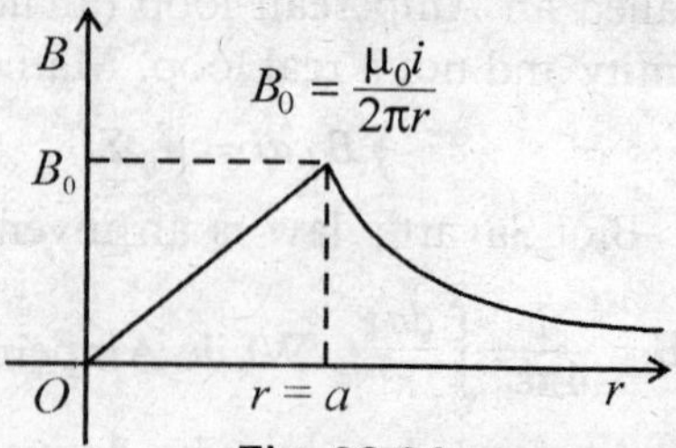

Fig. 23.24

If we plot a graph between magnetic induction B and distance r from the axis of cylinder for current flowing through the cylinder, we get a curve as shown in Fig. 23.24.

23.11 SOLENOID

A solenoid is a device in which the current always flows perpendicular to the length. It consists of a wire wound closely in the form of helix over a long tube of circular cross-section or rectangular cross-section. The length of solenoid is large compared to its transverse dimensions. Each turn of this solenoid can be regarded as a closed loop and the net magnetic field is the vector sum of the fields due to all the turns.

In section 23.7 we have seen that the magnetic field at any point on the axis of a current carrying coil is along the axis with no normal component. As such the field at the interior point is uniform and strong (Fig. 23.25) and at the exterior points is weak. At exterior points, the field due the neighbouring loops oppose each other. These tendencies to have zero field outside and uniform strong field inside becomes more and more effective as the solenoid is more and more tightly wound. The magnetic field inside a very tightly wound long solenoid is uniform everywhere and is zero outside.

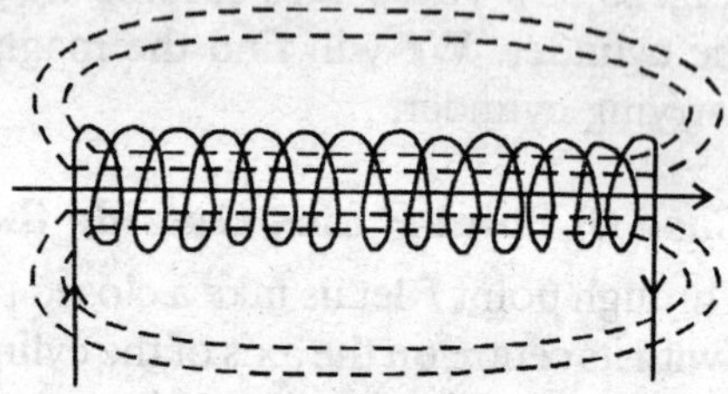
Fig. 23.25

In order to calculate magnetic field at a point P inside the solenoid, let us consider a rectangular closed path $ABCD$ as shown in Fig. 23.26. This path called Amperean path has line AD parallel to the solenoid and hence parallel to magnetic field $\boldsymbol{B}$.

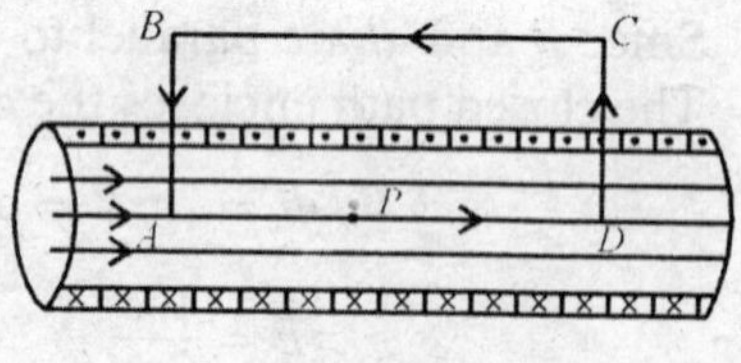

Fig. 23.26

The line integral of magnetic field over closed path $ABCD$ is

$$\oint \boldsymbol{B} \cdot \boldsymbol{dl} = \int_{AD} \boldsymbol{B} \cdot \boldsymbol{dl} + \int_{DC} \boldsymbol{B} \cdot \boldsymbol{dl} + \int_{CB} \boldsymbol{B} \cdot \boldsymbol{dl} + \int_{BA} \boldsymbol{B} \cdot \boldsymbol{dl}$$

For last three integrals $\boldsymbol{B} \cdot \boldsymbol{dl}$ is zero as $\boldsymbol{B}$ is either zero (outside the solenoid or perpendicular to $\boldsymbol{dl}$ (inside the solenoid). Thus the line integral over the closed path is

$$\oint \boldsymbol{B} \cdot \boldsymbol{dl} = \int_{AD} \boldsymbol{B} \cdot \boldsymbol{dl} = B(AD)$$

Let N be the number of turns within distance AD of solenoid, the total current enclosed by the closed path is Ni, where i is the current through the solenoid wire. From Ampere's law we have

$$\oint \mathbf{B} \cdot d\mathbf{l} = \mu_0 i$$

$$\therefore \quad B(AD) = \mu_0 N i$$

or
$$B = \mu_0 \frac{N}{DA} i = \mu_0 n i \qquad \ldots(23.25)$$

where $n = \frac{N}{DA}$ is the number of turns per unit length on the solenoid.

The direction of the magnetic field is given by the right hand thumb rule. The solenoid is commonly used to obtain a uniform magnetic field.

If a soft iron core of magnetic permeability μ_r is introduced in the solenoid, the magnetic field inside the solenoid becomes

$$B = \mu_0 \mu_r n i \qquad \ldots(23.25a)$$

General formula for magnetic field at any point P due to a current. carrying solenoid: In Fig. 23.26 (*a*), the magnetic field at P is

$$B = \frac{\mu_0 n i}{2}(\cos\theta_1 - \cos\theta_2)$$

where n is number of turns per unit length on the solenoid.

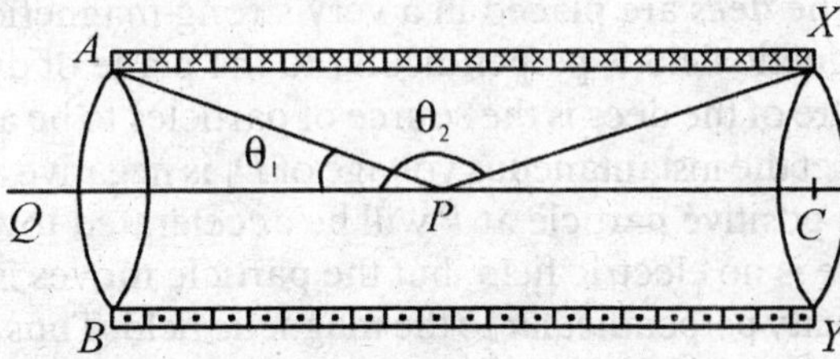

Fig. 23.26 (*a*)

For a very long solenoid $\theta_1 = 0°$ and $\theta_2 = 180°$ and

$$B = \frac{\mu_0 n i}{2}[\cos 0° - \cos 180°] = \mu_0 n i$$

For point at the end of solenoid (point C) $\theta_1 = 0$ and $\theta_2 = 90°$ and

$$B = \frac{\mu_0 n i}{2}, \text{ i.e. half the field at centre}$$

The direction is given by right hand thumb rule.

Toroid: *A toroid is an endless solenoid.* A solenoid bent in a circular shape with ends joined is called toroid. Thus toroid is a hollow circular ring on which a large number of turns of a wire are closely wound. A toroid carrying current i is shown in Fig. 23.27. As already explained in case of solenoid, the magnetic field in the open space inside (point P) and exterior to the toroid (Point Q) is zero. The field B inside the toroid is constant in magnitude in case of an ideal toroid of closely wound turns.

The magnetic field in such a toroid can be obtained in similar way as we obtained in case of long solenoid using Ampere's law. The magnetic field inside the toroid also has the same expression.

$$B = \mu_0 n i$$

where n is the number of turns per unit length in the toroid. If r is the mean radius of toroid and N total number of turns on the toroid, then

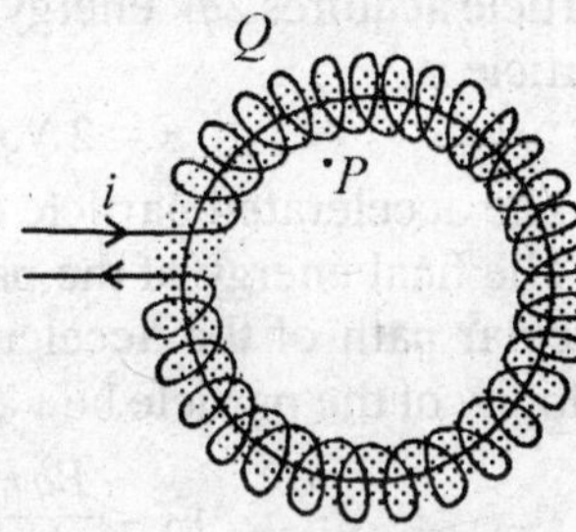

Fig. 23.27

$$n = \frac{N}{2\pi r}$$

and magnetic field B inside the toroid is

$$B = \frac{\mu_0 Ni}{2\pi r} \quad ...(23.26)$$

In an ideal toroid the coils are circular. In fact the turns of the toroidal coil forms a helix and there is always a small magnetic field outside the toroid. Toroids are used in equipments for plasma confinement in fusion power reactors called *tokamak*.

23.12 CYCLOTRON

Cyclotron is a device used to accelerate charged particles or ions to high energies. The cyclotron uses both electric and magnetic fields to generate energetic particles.

It consists of two D-shaped hollow evacuated metal chambers D_1 and D_2 called the dees. These dees are placed horizontally with their diametric edges parallel and slightly separated from each other. The dees are connected to a high frequency oscillator (H.F.O.) of potential difference of the order of 10^4 V at the cyclotron frequency of the particle to be accelerated.

The dees are placed in a very strong magnetic field (~1 to 10 T). The magnetic field is perpendicular to the plane of dees (Fig. 23.28). P at the centre of the dees is the source of particles to be accelerated.

Let the instantaneous voltage of D_1 is negative and that of D_2 is positive. The positive particle at P will be accelerated towards dee D_1. Inside D_1, there is no electric field, but the particle moves in magnetic field with its velocity perpendicular to the magnetic field. Thus the particle will describe circular motion with frequency

$$f = \frac{1}{2\pi}\frac{Bq}{m}$$

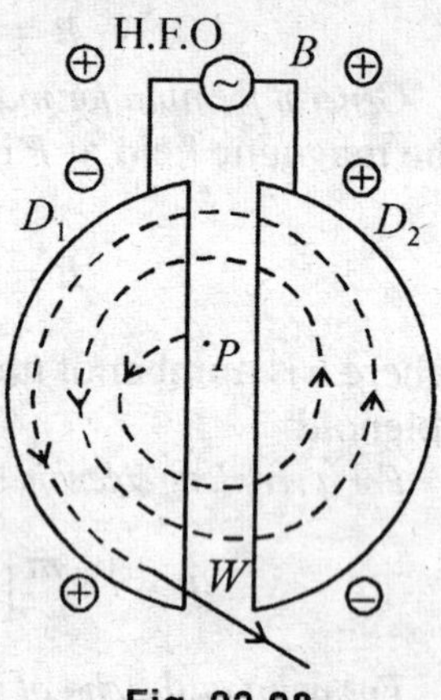

Fig. 23.28

This frequency is called cyclotron frequency.

Since the frequency of the alternating voltage applied between dees is also equal to the cyclotron frequency as such, the time taken by the charged particle to describe semi-circle inside D_1 and reaches the gap between the dees, the oscillator completes half cycle and hence the polarity of the dees reverses, i.e. D_2 becomes negative and D_1 positive as such the positively charged particle is accelerated towards D_2 where it moves with greater speed and in circle of greater radius. Every time the charged particle reaches the gap between the dees exactly at the time when the polarity of the dees is reversed. Thus the charged particle goes on accelerating every time it comes into the gap between the dees. If q is the charge on the particle and V potential difference between dees, then every time it passes the gap it acquires an energy = qV. Thus during one revolution in dees the particle acquires $2qV$ energy. If it makes N revolutions in the dees, then total energy acquired by the particle is

$$E = 2NqV \quad ...(23.27)$$

The accelerated particle can be removed from dees through window W.

The final energy of the particle can also be calculated as follows. The maximum radius of the circular path of the accelerated particle will be equal to the radius of dee (= r_0). If maximum velocity of the particle be v_0, then

$$v_0 = \frac{Bqr_0}{m}$$

and the kinetic energy of the particle

$$E = \frac{1}{2}mv_0^2 = \frac{1}{2}\frac{B^2q^2r_0^2}{m} \qquad \text{...(23.27}a\text{)}$$

Equations (23.27) and (23.27*a*) are identical and give the final energy of the particle accelerated.

Limitations of Cyclotron

1. Cyclotron cannot accelerate uncharged particles such as neutrons.
2. Cyclotron cannot be used to accelerate light particles like electrons. In case of light particles, the velocity becomes comparable to that of light and the mass increases

$$m = \frac{m_0}{\sqrt{1 - v^2/c^2}}$$

m_0 is rest mass, m is mass of particle when its speed in v and c is speed of light.

As mass increases the cyclotron frequency of the particle decreases. As a result the charged particle will not reach the gap between dees at the exact time when the polarity of the dees is reversed. Hence the cyclotron is suitable only for heavy particles like protons, deutrons, α-particle, etc.

23.13 HALL EFFECT

When a current carrying metal strip is placed in magnetic field such that the current direction and magnetic field are perpendicular to each other an electric field is developed at right angles to both the current and magnetic field. This transverse field developed is called Hall field. This effect is called Hall effect. The voltage responsible for Hall field is called *Hall voltage.*

Figure 23.29 shows a rectangular metal strip through which current i flows in Y-direction and magnetic field is applied in X-direction. It is found that a potential difference is developed between the faces C and D and can be measured using sensitive galvanometer. This potential difference is called Hall voltage.

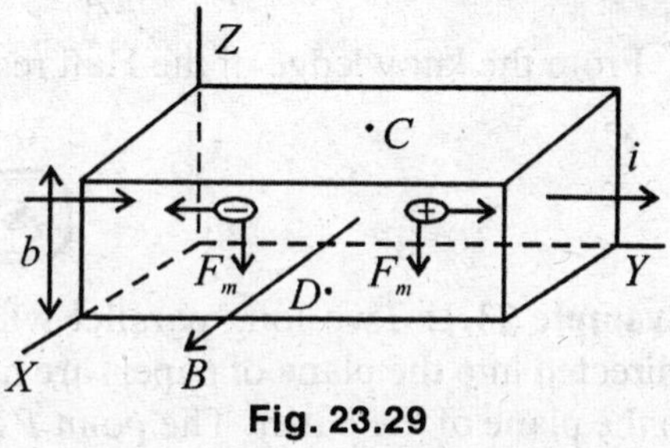

Fig. 23.29

If charge carriers are electrons, the conventional current flows in OY direction and electron drifts with velocity $\boldsymbol{v}_d$ towards negative Y-direction. Magnetic force on drifting electron ($-e$ charge) is then

$$\boldsymbol{F}_m = -e(\boldsymbol{v}_d \times \boldsymbol{B})$$

This face is towards negative Z-direction. Thus electrons will collect at face D making its potential negative and face c will be at positive potential. This creates a potential difference between CD called Hall potential difference.

On the other hand if charge carriers are positive charges (as holes in P-type semi conductor) the current is due to their movement in positive Y-direction. Hence force on positive charge due to magnetic field is

$$\boldsymbol{F}_m = e(\boldsymbol{v}_d \times \boldsymbol{B})$$

This force is towards the face D. Thus in this case the positive charges will collect at face D and making it positive and making the face C as negative. Thus a potential difference is developed between C and D called Hall potential difference. Thus we infer that

(*i*) *if the face D is negative the charge of the current carriers is negative*
(*ii*) *if the face D is positive the charge of the current carriers is positive.*

Magnitude of Hall Voltage

The magnetic force on the charge carrier (Charge *e*) is

$$F_m = ev_d B$$

If V_H is the Hall Voltage developed between faces *C* and *D* and *b* the distance between the two faces then Hall field developed is

$$E_H = \frac{V_H}{b}$$

This electric field will exert force eE_H on the current carrying charges in the direction opposite to the force due to magnetic field. In equilibrium the two forces are equal. Thus

$$\frac{eV_H}{b} = ev_d B$$

or

$$V_H = v_d B d b \qquad \text{...(23.28)}$$

If *i* is the current through the strip and *A* area of cross-section of the strip (perpendicular to the direction of flow of current) then

$$i = neAv_d$$

where *n* is the current carrier concentration (number of charge carriers per unit volume)

Substituting value of v_d in (23.28), we get

$$V_H = \frac{Bb \cdot i}{neA} \qquad \text{...(23.29)}$$

We define R_H *the Hall resistance* (or Hall coefficient) as

$$R_H = \frac{V_H}{i} = \frac{Bb}{neA} \qquad \text{...(23.30)}$$

From the knowledge of the Hall resistance we can know the nature of charge carriers.

SOLVED EXAMPLES

Example 23.1: Two long parallel wires carrying currents 2.5 A and *i* A in the same direction (directed into the plane of paper) are held at *P* and *Q* respectively such that they are perpendicular to the plane of the paper. The point *P* and *Q* are located at a distance of 5 m and 2 m respectively from a collinear point *R* (Fig. 23.30).

(*a*) An electron moving with velocity 4×10^5 m/s along positive *X*-direction experiences a force of magnitude 3.2×10^{-20} N at point *R*. Find the value of *i*.

(b) Find all the positions at which a third wire carrying a current of magnitude 2.5 A may be placed so that the magnetic induction at *R* is zero.

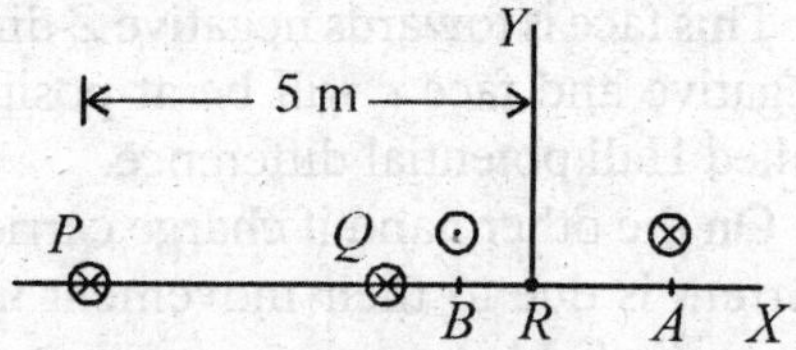

Fig. 23.30

Solution:

(*a*) Magnetic field at *R* due to current carrying wire through *P* is

$$B_1 = \frac{\mu_0 i}{2\pi r}$$

Here $i = 2.5$ A and $r = 5$ m, $\frac{\mu_0}{2\pi} = 2 \times 10^{-7}$ H/m

$$\therefore \quad B_1 = \frac{2 \times 10^{-7} \times 2.5}{5} = 10^{-7} \text{ T}$$

It is in the negative Y-direction

Magnetic field at R due to current carrying wire through Q is

$$B_r = \frac{\mu_0 i}{2\pi r} \quad \text{where } r = 2 \text{ m}$$

$$= \frac{2 \times 10^{-7} \times i}{2} = 10^{-7} i \text{ T}$$

This field is also in negative Y-direction.

Hence total field at R is $B = B_1 + \vec{B}_2 = (1 + i) \times 10^{-7}$ T in negative Y-direction.

An electron at R moving in X-direction experiences a force equal to 3.2×10^{-20} N.

Since magnetic field and velocity are perpendicular to each other, hence magnetic force on electron is

$$F_m = q\mathbf{v} \times \mathbf{B} = evB$$

or $3.2 \times 10^{-20} = 1.6 \times 10^{-19} \times 4 \times 10^5 \times (1 + i) \times 10^{-7}$

or $\quad 1 + i = \frac{10 \times 3.2}{1.6 \times 4} = 5$

or $\quad i = 4$ A

(*b*) If the distance of point R from third current carrying wire is x, then resultant field at R should be zero.

Magnetic field due to this third current carrying wire at R is

$$\frac{\mu_0}{2\pi} \times \frac{2.5}{x} = \frac{5 \times 10^{-7}}{x} \text{ T}$$

Magnetic field due to wires P and Q is $= (1 + i) \times 10^{-7} = 5 \times 10^{-7}$ T

$$\therefore \quad \frac{5 \times 10^{-7}}{x} + 5 \times 10^{-7} = 0$$

Now there are two possibilities.

(*i*) If the current in third wire is into the page, then third wire must be placed at A so that the magnetic field at R due to this third wire is in opposite direction to that due to P and Q wires. Hence

$$\frac{-5 \times 10^{-7}}{x} + 5 \times 10^{-7} = 0 \text{ or } x = 1 \text{ m}$$

(*ii*) If the current in third wire is out of the page as in B then also the field at R due to this third wire is in opposite direction to that due to wires P and Q. Hence

$x = 1$ m.

Example 23.2: Two long straight parallel wires are 2 m apart; perpendicular to the plane of the paper. The wire A carries a current of 9.6 A directed into the plane of the paper. The wire B carries a current such that the magnetic field of induction at the point P at a distance 10/11 m from wire B is zero (Fig. 23.31). Find (*a*) the magnitude and direction of the current in B, (*b*) the magnitude of the magnetic field at the point S and (*c*) the force per unit length on the wire B.

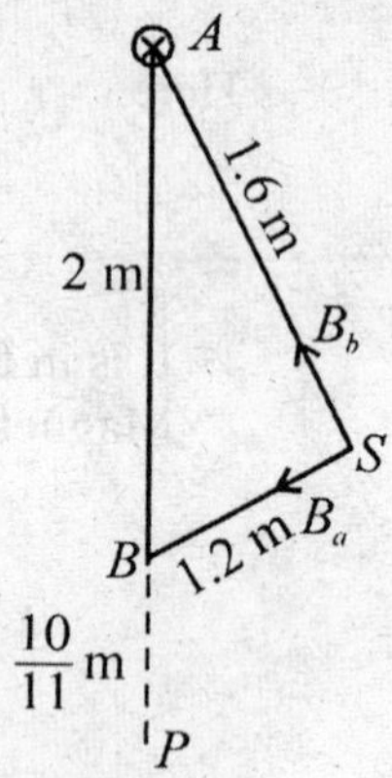

Fig. 23.31

Solution: (*a*) Net magnetic field at P is zero. This means the direction of magnetic field at P due to current carrying wire A is opposite to that due to wire A. This gives that the current in B is directed out of the plane of the paper.

Magnetic field due to a long current carrying wire at a point distance r from it is given by $B = \frac{\mu_0 i}{2\pi r}$ so field at P will be

$$B = B_a - B_b = \frac{\mu_0}{4\pi}\left(\frac{2i_a}{r_a} - \frac{2i_b}{r_b}\right) = 0$$

or

$$\frac{i_a}{r_a} = \frac{i_b}{r_b}$$

and

$$i_b = \frac{r_b \cdot i_a}{r_a} = \frac{10 \times 9.6}{11\left(2 + \frac{10}{11}\right)} = \frac{10 \times 9.6}{32} = 3 \text{ A}$$

(*b*) Magnetic field at S due to current carrying wire A (B_a) has direction perpendicular to AS (using right hand thumb rule). While that due to current carrying wire B (B_b) has direction perpendicular to BS. From geometry of the Fig. 23.31, the magnetic fields B_a and B_b are perpendicular to each other. The magnitudes of the two fields are

$$B_a = \frac{\mu_0}{4\pi}\frac{2ia}{r_a} = \frac{10^{-7} \times 2 \times 9.6}{1.6} = 12 \times 10^{-7} \text{ T}$$

and

$$B_b = \frac{\mu_0}{4\pi}\frac{2i_b}{r_b} = \frac{10^{-7} \times 2 \times 3}{1.2} = 5 \times 10^{-7} \text{ T}$$

Hence net magnetic field at S is

$$B = \sqrt{B_a^2 + B_b^2} = \sqrt{(12 \times 10^{-7})^2 + (5 \times 10^{-7})^2}$$

$$= 13 \times 10^{-7} \text{ T} = 1.3 \times 10^{-6} \text{ T}$$

(*c*) Force per unit length between two parallel current carrying wires is

$$\frac{dF}{dL} = \frac{\mu_0}{4\pi}\frac{2i_1i_2}{r}$$

$$= \frac{10^{-7} \times 2 \times 9.6 \times 3}{2} = 28.8 \times 10^{-7} \text{ N/m}$$

$$= 2.88 \times 10^{-6} \text{ N/m}$$

Since the currents are anti-parallel as such forces are repulsive.

Example 23.3: Two straight infinitely long and thin parallel wires are spaced 0.1 m apart and carry a current of 10 A each. Find the magnetic field at a point distance 0.1 m from both wires in two cases when the currents are in the (*a*) same (*b*) opposite direction.

Solution:

(*a*) In case the current through wires is same; the situation is as shown in Fig. 23.32 (*a*). B_a is perpendicular to *AP* and B_b is perpendicular *BP*. The angle between the fields due to *A* and *B* is 60°. Since the distances are same, hence magnetic fields in magnitudes are also equal. Thus

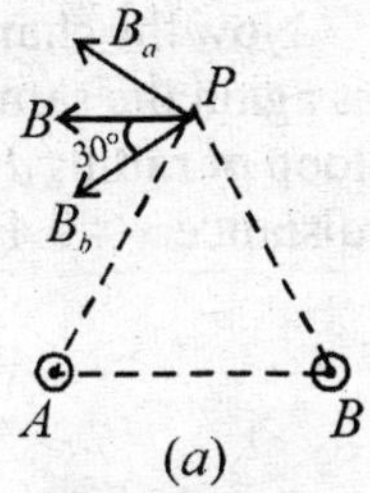

Fig. 23.32 (*a*)

$$B_a = B_b = \frac{\mu_0 i}{2\pi r} = \frac{2 \times 10^{-7} \times 10}{10^{-1}} = 2 \times 10^{-7} \text{ T}$$

Hence resultant magnetic field at *P* due to two current carrying wires is

$$B = 2B_a \cos 30 = 2 \times 2 \times 10^{-7} \times \frac{\sqrt{3}}{2} = 2\sqrt{3} \times 10^{-7} \text{ T}$$

along the line parallel to *BA*.

(*b*) Figure 23.32 (*b*) shows situation *B* in which currents through the two wires are in opposite direction. Field B_a due to wire *a* is perpendicular to line *AP* and due to B_b is perpendicular to *BP*. The angle between the two directions is 120°. Magnitudes of the two fields again is same

$$B_a = B_b = 2 \times 10^{-7} \text{T}$$

Hence resultant magnetic field at *P* is

$$B = 2B_a \cos 60 = 2 \times 2 \times 10^{-7} \times \frac{1}{2} = 2 \times 10^{-7} \text{ T}$$

along the line perpendicular to *AB*.

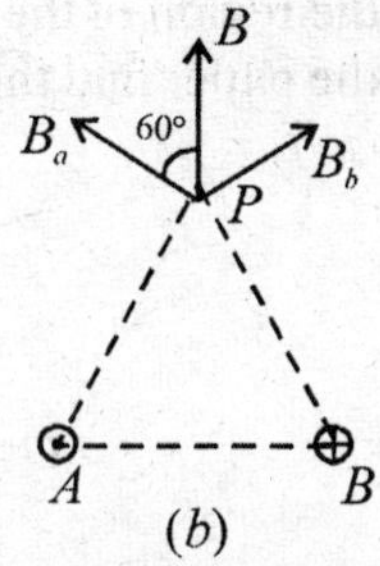

Fig. 23.32 (*b*)

Example 23.4: A charge of 1 Coulomb is placed at one end of a non-conducting rod of length 0.6 m. The rod is rotated in a vertical plane about a horizontal axis passing through the other end of the rod with angular frequency $\pi \times 10^4$ rad/sec. Find the magnetic field at a point on the axis of rotation at a distance of 0.8 m from the centre of the path.

Now half of the charge is removed from one end and placed on the other end. The rod is rotated in a vertical plane about a horizontal axis passing through the mid-point of the rod with the same angular frequency. Calculate the magnetic field at a point on the axis at a distance of 0.4 m from the centre of the rod.

Solution: A revolving charge is equivalent to a current carrying loop. The current

$$i = qt = \frac{1 \times \pi \times 10^4}{2\pi} = 5 \times 10^3 \text{ A}$$

Hence magnetic field at a distance *x* from centre on the axis is

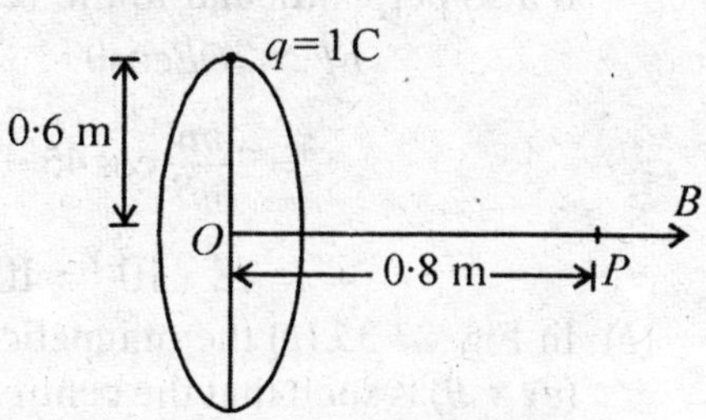

Fig. 23.34

$$B = \frac{\mu_0 i a^2}{2(a^2 + x^2)^{3/2}}$$

$$= \frac{4\pi \times 10^{-7} \times 5 \times 10^3}{2(0.6^2 + 0.8^2)^{3/2}}(0.6)^2 = 2\pi \times 10^{-4} \times 5 \times 36 \times 10^{-2}$$

$$= 1.13 \times 10^{-3} \text{ T}$$

Now the charge is equally distributed between the two ends of the rod, i.e. total charge revolving is again the same and frequency is same hence now the system is equivalent to a current carrying loop of radius $a = 0.3$ m and current $i = 5 \times 10^3$ A. Now we have to find out the magnetic field at distance $x = 0.4$ m from the centre. Hence

$$B = \frac{\mu_0 i a^2}{2(a^2 + x^2)^{3/2}} = \frac{4\pi \times 10^{-7} \times 5 \times 10^3 \times (0.3)^2}{2(0.3^2 + 0.4^2)^{3/2}}$$

$$= \frac{20\pi \times 10^{-4} \times 0.09}{2 \times (0.5)^3} = 2.26 \times 10^{-3} \text{ T}$$

Example 23.5: A particle of mass $m = 1.6 \times 10^{-27}$ kg and charge $q = 1.6 \times 10^{-19}$ C moves at a speed of 1.0×10^7 m/s. It enters a region of uniform magnetic field at a point E (Fig. 23.35). The field has a strength of 1.0 T (*a*) Magnetic field is directed into the plane of the paper. The particle leaves the region of the field at point F. Find the distance EF and angle θ. (*b*) If the field is coming out of the paper find the time spent by the particle in the region of the magnetic field after entering it at E.

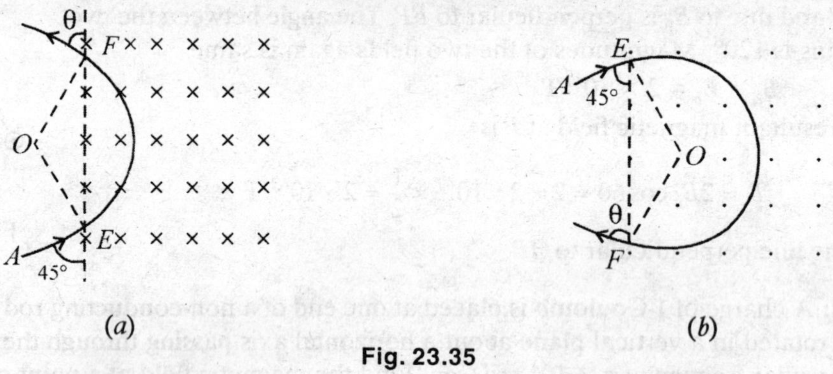

Fig. 23.35

Solution:

(*a*) As the particle enters the magnetic field, a magnetic force ($= q\boldsymbol{v} \times \boldsymbol{B}$) acts on it and it describes circular path of radius $\left(r = \dfrac{mv}{qB}\right)$. The centre of the path lies on the line perpendicular to its velocity. In Fig. 23.35 (*a*), the centre of the circle will be outside the field. Since AE is tangent to the circle hence $\angle AEO = 90°$ (OE is radius). $\angle OEF = \angle OFE = 45°$. OF (radius) is also perpendicular to the velocity at F, so that $\theta = 45°$. Hence from triangle OEF

$$EF = 2OE\cos\theta$$

$$= 2\frac{mv}{qB}\cos 45 = \frac{\sqrt{2} \times 1.6 \times 10^{-27} \times 10^7}{1.6 \times 10^{-19} \times 1.0}$$

$$= \sqrt{2} \times 10^{-1} = 10 \text{ cm}$$

(*b*) In Fig. 23.35 (*b*) the magnetic field is coming out of the paper. The direction of the force ($q\boldsymbol{v} \times \boldsymbol{B}$) is such that the centre (O) of the circular path θ will be inside the field region. Here again $\theta = 45°$. This means that the particle describes 3/4 of the circle inside the magnetic field. Speed v is uniform. Hence time spent inside the field is

$$t = \frac{3}{4} \cdot \frac{2\pi r}{v} = \frac{3}{4} \cdot \frac{2\pi m}{qB}$$

$$= \frac{3\pi m}{2qB} = \frac{3 \times \pi \times 1.6 \times 10^{-27}}{2 \times 1.6 \times 10^{-19} \times 1} = 4.7 \times 10^{-8} \text{ sec}$$

Example 23.6: An electron is released from the origin at a place where a uniform electric field *E* and a uniform magnetic field *B* exists along the negative *Y*-axis and negative *Z*-axis respectively. Find the displacement of the electron along the *Y*-axis when its velocity becomes perpendicular to the electric field for the first time.

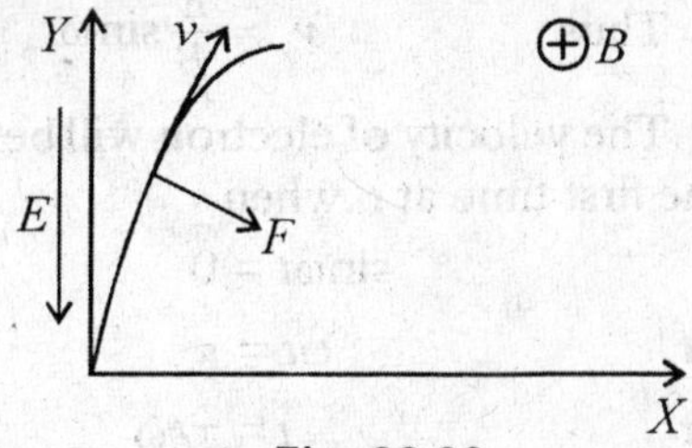

Fig. 23.36

Solution: Electric and magnetic fields can be written as

$$\boldsymbol{E} = -E\,\hat{j} \text{ and } \boldsymbol{B} = -B\hat{k}$$

Let instantaneous velocity of electron be $\boldsymbol{v}$. Thus

$$\boldsymbol{v} = v_x\hat{i} + v_y\hat{j}$$

The force on the electron in magnetic and electric field is

$$\boldsymbol{F} = -e(\boldsymbol{E} + \boldsymbol{v} \times \boldsymbol{B})$$

$$= -e\left[-E\hat{j} + (v_x\hat{i} + v_y\hat{j}) \times (-B\hat{k})\right]$$

$$= eE\hat{j} - ev_x \cdot B\hat{j} + ev_yB\hat{i}$$

$$= ev_yB\hat{i} + e(E - v_xB)\hat{j}$$

Hence the force acts in *X-Y* plane and F_x and F_y, *x* and *y* components of force are

$$F_x = ev_yB \text{ and } F_y = e(E - v_xB)$$

If *m* is the mass of electron then components of acceleration of electron are

$$a_x = \frac{F_x}{m} = \frac{dv_x}{dt} = \frac{ev_yB}{m} \qquad \ldots(i)$$

and $$a_y = \frac{F_y}{m} = \frac{dv_y}{dt} = \frac{e}{m}(E - v_xB) \qquad \ldots(ii)$$

$$\therefore \quad \frac{d^2v_y}{dt^2} = -\frac{e}{m}B\frac{dv_x}{dt} = -\frac{eB}{m}\frac{eB}{m} \cdot v_y$$

$$= -\frac{e^2B^2}{m^2}v_y = -\omega^2 v_y \text{ where } \omega^2 = \frac{e^2B^2}{m^2}$$

This equation is similar to the differential equation for simple harmonic motion. Hence its solution is

$$v_y = v_0\sin(\omega t + \theta) \qquad \ldots(iii)$$

where v_0 and θ are constants of integration and can be determined from initial conditions.

Differentiating (*iii*), we have

$$\frac{dv_y}{dt} = v_0\omega\cos(\omega t + \theta) \qquad \ldots(iv)$$

at $t = 0$ $v_y = 0$ and $\frac{dv_y}{dt} = \frac{F_y}{m} = \frac{eE}{m}$

Substituting these values in (*iii*) and (*iv*), we get

$$\theta = 0,\ v_0 = \frac{eE}{m\omega} = \frac{E}{B}$$

Thus $$v_y = \frac{E}{B}\sin\omega t \qquad \ldots(v)$$

The velocity of electron will be perpendicular to the *Y*-axis when $v_y = 0$. This will be the case for the first time at *t*, when

$$\sin\omega t = 0$$

or $$\omega t = \pi$$

or $$t = \pi/\omega$$

or $$t = \frac{\pi m}{eB}$$

Now $$v_y = \frac{dy}{dt} = \frac{E}{B}\sin\omega t$$

$\therefore$ $$\int_0^y dy = \int_0^t \frac{E}{B}\sin\omega t \cdot dt$$

or $$y = \frac{E}{B\omega}(1 - \cos\omega t) \qquad \text{But } t = \frac{\pi m}{eB} = \frac{\pi}{\omega}$$

$$= \frac{E}{B\omega}\left(1 - \cos\omega\frac{\pi}{\omega}\right) = \frac{2E}{B\omega}$$

Thus displacement along *Y*-axis is

$$y = \frac{2E}{B\omega} = \frac{2Em}{B \cdot Be} = \frac{2E}{B^2 e}$$

Example 23.7: Electrons emitted with negligible speed from electron gun are accelerated through a potential difference *V* along the *X*-axis. These electrons emerge from a narrow hole into a uniform magnetic field *B* directed along this axis. However some of the electrons emerging from the hole make slightly divergent angle as shown in Fig. 23.37. Show that these paraxial electrons are re-focused on the *X*-axis at a distance $\sqrt{\frac{8\pi^2 mV}{eB^2}}$.

Solution: Electron accelerated by a potential difference of *V*, its energy becomes $E = eV$ and its velocity

$$v = \sqrt{\frac{2E}{m}} = \sqrt{\frac{2eV}{m}} \qquad \ldots(i)$$

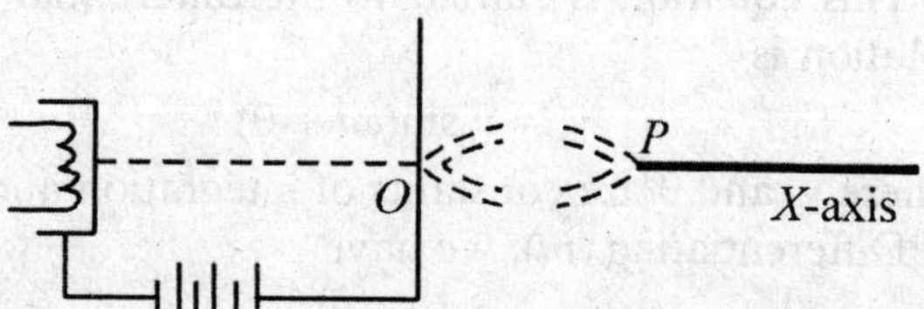

Fig. 23.37

Since the electrons are paraxial, they will have helical path. The cyclotron frequency of the electrons is

$$f = \frac{qB}{2\pi m} = \frac{eB}{2\pi m}$$

and time period $T = \frac{2\pi m}{eB}$ ...(ii)

All the electrons coming out at O will be refocused at P if OP is equal to the pitch of the helix d.

$$d = vT = \sqrt{\frac{2eV}{m}} \cdot \frac{2\pi m}{eB}$$

$$= \sqrt{\frac{8\pi^2 mV}{eB^2}}$$

Example 23.8: Two particles each having a mass m are placed at a distance d in a uniform magnetic field B (Fig. 23.38). They have opposite charges of equal magnitude q. At time $t = 0$ the particles are projected towards each other, each with speed v. Suppose the Coulomb force between the charges is switched off (*a*) Find the maximum value of v_m of the projection speed that the two particles do not collides. (*b*) What would be minimum and maximum separation between the particles if $r = v_m/2$? (*c*) At what instant will a collision occur between the particles if $v = 2v_m$? (*d*) Suppose $v = 2v_m$ and the collision between the particles is completely inelastic? Describe the motion after the collision.

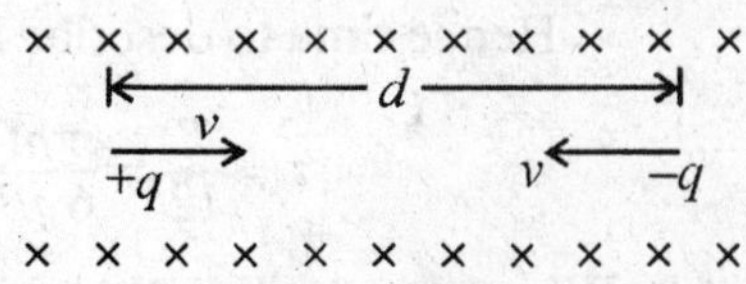

Fig. 23.38

Solution:

(*a*) For no collision between the charged particles the radius of the circular paths described by the charged particles should be less than $d/2$, i.e. $r < d/2$ or $2r < d$. The situation is shown in Fig. 23.38 (*a*). Here for just collision v should be such that $r = d/2$. For a particle of charge q moving in magnetic field B the radius of circular path described by particle is

$$r = \frac{mv}{qB}$$

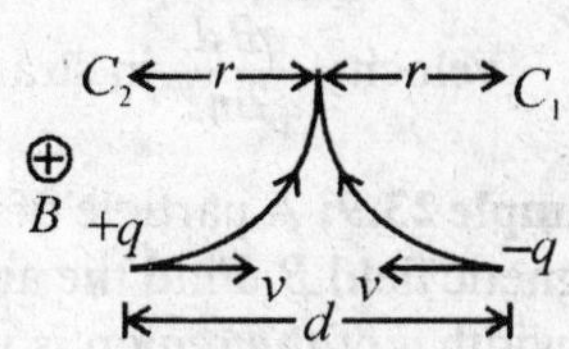

Fig. 23.38 (*a*)

In present case for $r = d/2$ $v = v_{max}$

$$\therefore \quad \frac{d}{2} = \frac{mv_{max}}{qB}$$

or $$v_{max} = \frac{qB\,d}{2m}$$

(*b*) If $v = \frac{v_m}{2} = \frac{qBd}{4m}$, the circular path has radius $r = \frac{mv}{qB} = \frac{d}{4}$

Then minimum separation $= d - 2r = d - \frac{d}{2} = \frac{d}{2}$

The maximum separation $= d + 2r = d + \frac{d}{2} = \frac{3d}{2}$

(*c*) For $v = 2v_{max}$ the radius of the circular path is

$$r = \frac{2mv_{max}}{qB} = d$$

This means the radii of circular paths of the two particles are *d*. The collision of the two particles is shown in Fig. 23.38 (*b*). The situation shown in Fig. 23.38 (*b*) shows that the particles will collide after moving through an angle of 30°.

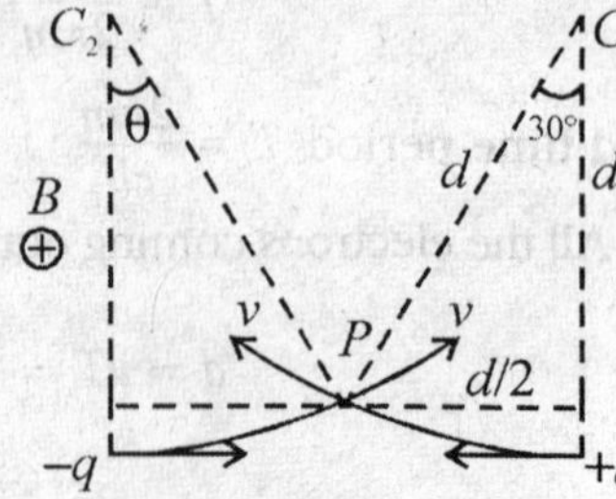

Fig. 23.28 (*b*)

The time of the revolution of particle in magnetic field is

$$T = \frac{2\pi r}{v} = \frac{2\pi d}{2v_{max}} = \frac{2\pi\, dm}{qB\, d}$$

$$= \frac{2\pi\, m}{qB}$$

Hence time to describe angle θ = 30° = *T*/12. Hence the particles will collide after time

$$t = \frac{T}{12} = \frac{\pi m}{6qB}$$

(*d*) When they collide their velocities make an angle of 60° with *PQ* which is perpendicular bisector of the line joining to the starting points *A* and *B* of the particles (Fig. 23.38 (*c*)). If the collision is inelastic, then momentum of the combined particles after collision $= 2m(2v_{max})\cos 60$

momentum $P = 2mv_{max}$

and the velocity of the combined mass is $v_{max} = \frac{qB\,d}{2m}$

Now since total charge is zero, the system will move with velocity $\frac{qB\,d}{2m}$ in straight line along *PQ*.

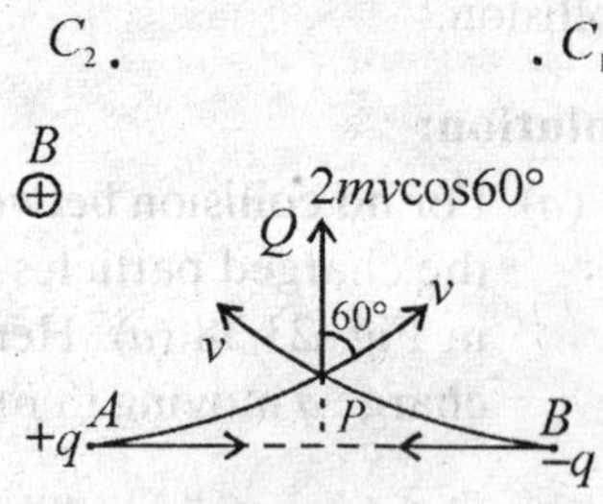

Fig. 23.38 (*c*)

Example 23.9: A particle of mass *m* and charge *q* is projected into a region having a perpendicular magnetic field *B*. Find the angle of deviation of the particle as it comes out of the magnetic field if the width *d* of the region is very slightly greater than

(*a*) $\frac{mv}{qB}$ (*b*) $\frac{mv}{2qB}$ (*c*) $\frac{2mv}{qB}$

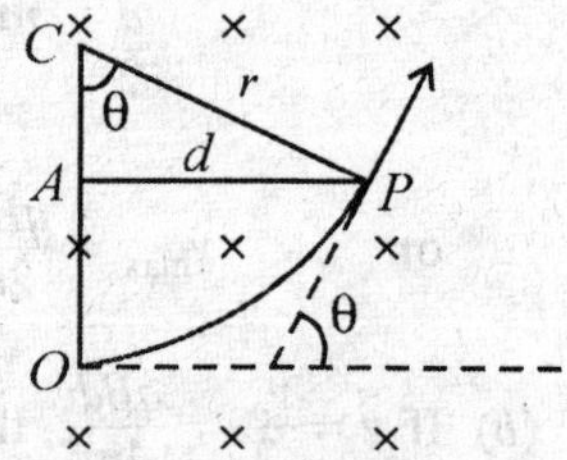

Fig. 23.39

Solution: When a particle enters a magnetic field, it describes a circular motion of radius

$$r = \frac{mv}{qB}$$

At the end of the magnetic field the particle will come out of the magnetic field. Its velocity will be tangential to the trajectory it has followed in the magnetic field. The situation is shown in Fig.

23.39. The particle enters the magnetic field at O and emerges out at P. The deviation suffered by the particle is θ such that

$$\sin\theta = \frac{d}{r} = \frac{d\,qB}{mv}$$

(a) If $d = \frac{mv}{qB}$ i.e. $d = r$ then $\sin\theta = 1$ or $\theta = \pi/2$

(b) If $= \frac{mv}{2qB}$ i.e. $d = r/2$ then $\sin\theta = \frac{r}{2r} = \frac{1}{2}$ or $\theta = \pi/6$

(c) $d = \frac{2mv}{qB} = 2r$

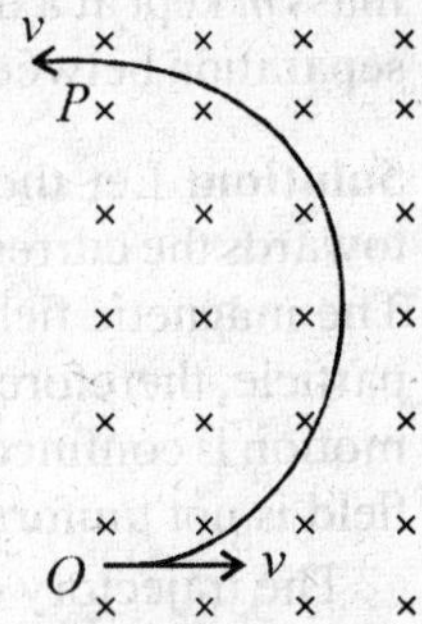

Fig. 23.39 (a)

The situation is shown in Fig. 23.39 (a). The particle completes a semi-circle and comes out at P if it enters at O. Thus the angle of deviation is π

Example 23.10: Fig. 23.40 shows a cross-section of a large metal sheet carrying current along its surface. The current in a strip of width dl is $k\,dl$ where k is constant. Find the magnetic field at a point P at a distance x from the metal sheet.

Solution: Let us assume the current carrying strip be made up of a large number of strips (Fig 23.40). Consider two such strips A and B symmetrically situated on two sides of point P. The magnetic field at P due to strip A is B_a perpendicular to AP and due to strip B be B_b, perpendicular to BP. The resultant of these two is parallel to the width AB of the sheet. In this way the whole width of the current carrying sheet can be divided in such symmetric pairs and the net field due the whole sheet will be parallel to the width of the sheet. Let this field be B.

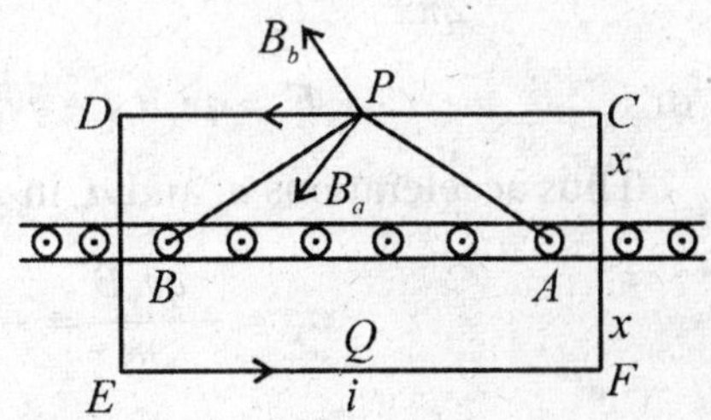

Fig. 23.40

Similarly the field at point Q opposite to P at the same distance on the other side of sheet is same in magnitude but opposite in direction.

Let us apply Ampere's law to the closed loop $CDEF$ which is a rectangle. The line integral of magnetic field for the closed path

$$\oint \mathbf{B}\cdot d\mathbf{l} = \int_{CD} \mathbf{B}\cdot d\mathbf{l} + \int_{DE} \mathbf{B}\cdot d\mathbf{l} + \int_{EF} \mathbf{B}\cdot d\mathbf{l} + \int_{FC} \mathbf{B}\cdot d\mathbf{l}$$

$$\int_{DE} \mathbf{B}\cdot d\mathbf{l} = \int_{FC} \mathbf{B}\cdot d\mathbf{l} = 0 \text{ as } \mathbf{B} \text{ and } d\mathbf{l} \text{ are perpendicular to each other}$$

for $\int_{CD} \mathbf{B}\cdot d\mathbf{l} = \int_{CD} B\,dl = Bl$ as B and dl

Similarly $\int_{EF} \mathbf{B}\cdot d\mathbf{l} = Bl$

$\therefore$ $\oint \mathbf{B}\cdot d\mathbf{l} = 2Bl$

Total current enclosed by the loop $i = \int_0^l k\,dl = kl$

$\therefore$ $2Bl = \mu_0 kl$

or $B = \frac{\mu_0 k}{2}$

We see that the magnetic fields is independent of x.

Example 23.11: A long straight wire carries a current i. A particle having a positive charge q and mass m kept at a distance x_0 from the wire is projected towards it with a speed v. Find the minimum separation between the wire and the particle.

Solution: Let the particle be projected from point P. The particle be projected with velocity v towards the current carrying wire. Let us assume the wire as Y-axis and line OP extended as X-axis. The magnetic field B on the right of the wire is along negative Z-axis. The magnetic force on the particle, therefore is in the X-Y plane. Since there is no initial velocity in the Z-direction, hence the motion is confined in X-Y plane. The speed of the particle will remain unchanged. As the magnetic field is not uniform, the particle does not describe circular motion.

The trajectory of the particle is as shown in Fig. 23.41. Initially at $t = 0$ velocity $\boldsymbol{v} = -v\,\hat{i}$. At the time of closest approach X-component of the velocity will be zero and velocity will be $\boldsymbol{v} = -v_x\hat{j}$.

Let the velocity of the particle at any time t be $\boldsymbol{v} = v_x\hat{i} + v_y\hat{j}$. The force due to magnetic field on the particle is

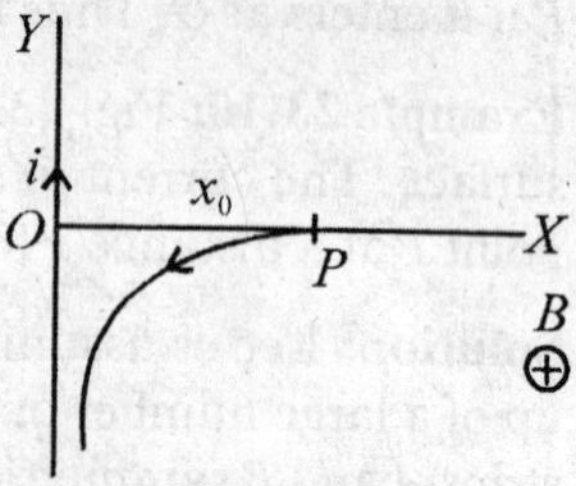

Fig. 23.41

$$\boldsymbol{F} = q(v_x\hat{i} + v_y\hat{j})(-B\hat{k})$$

where $B = \dfrac{\mu_0 i}{2\pi x}$, x is the distance of the particle from the wire.

or
$$\boldsymbol{F} = qv_xB\,\hat{j} - qv_y \cdot B\,\hat{i}$$

Thus accelerations a_x and a_y in X and Y-direction are

$$a_x = -\frac{qv_yB}{m} = -\frac{q\mu_0\, i}{2\pi\, x}\cdot\frac{v_y}{m} = -P\cdot\frac{v_y}{x} \quad \text{...(i)}$$

where
$$P = \frac{\mu_0\, qi}{2\pi m}$$

But
$$a_x = \frac{dv_x}{dt} = \frac{dv_x}{dx}\cdot\frac{dx}{dt} = v_x\frac{dv_x}{dx} \quad \text{...(ii)}$$

As
$$v_x^2 + v_y^2 = v^2 \text{ (constant)}$$

$\therefore$
$$2v_x\,dv_x + 2v_y\,dv_y = 0$$

or
$$v_x\,dv_x = -\,v_y\,dv_y \quad \text{...(iii)}$$

But
$$a_x = v_x\cdot\frac{dv_x}{dx} = -v_y\frac{dv_y}{dx} = -P\cdot\frac{v_y}{x}$$

or
$$\frac{dx}{x} = \frac{dv_y}{P}$$

Initially $x = x_0$ and $v_y = 0$. At minimum separation from the wire $v_x = 0$ and $v_y = -v$. Hence integrating with these limits, we have

$$\int_{x_0}^{x}\frac{dx}{x} = \int_0^{-v}\frac{dv_y}{P}$$

$$\log \frac{x}{x_0} = -\frac{v}{P}$$

or $$x = x_0 e^{-\frac{v}{P}} = x_0 e^{-\frac{2\pi m v}{\mu_0 q i}}$$

Example 23.12: Figure 23.42 shows a square loop *ABCD* with edge length *a*. The resistance of the wire *ABC* is *R* and that of *ADC* is 2*R*. Find the magnetic field *B* at the centre of the loop; assume uniform width.

Solution: The situation in Fig. 23.42 shows that the current through branch $ABC = \frac{2i}{3}$ and through *ADC* is $i/3$.

Magnetic field at *O* due to *AB* part is

$$B_1 = \frac{\mu_0 \, 2i/3 (\sin 45 + \sin 45)}{4\pi \, a/2}$$

$$= \frac{2\mu_0 i}{3\pi a\sqrt{2}} \oplus$$

Hence field at *O* due to part *ABC* of the loop is

$$2B_1 = \frac{2\sqrt{2}\mu_0 i}{3\pi a} \oplus \qquad \ldots(i)$$

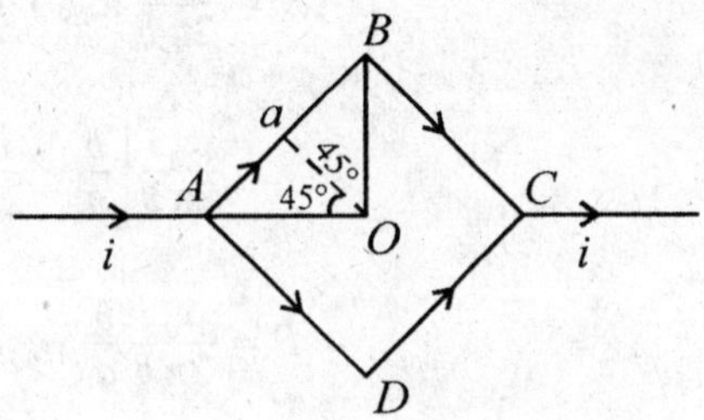

Fig. 23.42

Similarly the magnetic field at *O* due part *ADC* of the loop is

$$2B_2 = \frac{\sqrt{2}\mu_0 i}{3\pi a} \odot$$

Hence the net field at *O* due to the current through the loop is

$$B = 2B_1 - 2B_2 = \frac{2\sqrt{2}\mu_0 i}{3\pi a} - \frac{\sqrt{2}\mu_0 i}{3\pi a}$$

$$= \frac{\sqrt{2}\mu_0 i}{3\pi \, a} \oplus$$

Example 23.13: A very long thin strip of metal of width *b* carries a current *i* along its length as shown in Fig. 23.43. Find the magnitude of magnetic field in the plane of the strip at a distance *a* from the edge nearest to the point.

Solution: Consider the strip to be made up of a large number of elementary strips parallel to its length. Let us consider one such strip of width *dx* at a distance of *x* from the point *P* where the magnetic field due to whole strip carrying current is to be determined. This strip can be treated as a very long current carrying conductor, through which current $i' = \frac{i}{b} dx$ is passing. Hence magnetic field at point *P* due to this current carrying conductor is

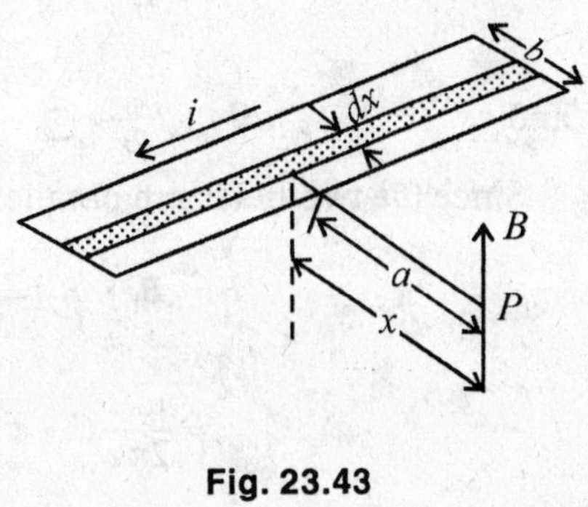

Fig. 23.43

$$dB = \frac{\mu_0 i'}{2\pi x} = \frac{\mu_0 i \, dx}{2\pi \, xb}$$

Hence magnetic field due to whole strip is

$$B = \int dB = \int_a^{(a+b)} \frac{\mu_0 \, i \, dx}{2\pi \, bx}$$

$$= \frac{\mu_0 \, i}{2\pi \, b} \log x \Big|_a^{(a+b)}$$

or $$B = \frac{\mu_0 i}{2\pi b} \log \frac{(a+b)}{a}$$

For a strip of small width, i.e. $b << a$

$$B = \frac{\mu_0 \, i}{2\pi \, b} \log\left(1 + \frac{b}{a}\right) \quad \text{for } b << a$$

$$= \frac{\mu_0 \, i}{2\pi \, b}\left[\frac{b}{a} - \frac{b^2}{2a^2} + \ldots\right]$$

$$B = \frac{\mu_0 \, i}{2\pi \, b}\frac{b}{a} \text{ neglecting higher powers of } b/a$$

$$= \frac{\mu_0 \, i}{2\pi \, a}$$

i.e. for a distant point, strip behaves like a current carrying wire.

Example 23.14: A long wire is bent into the shape as shown in Fig. 23.44 without cross contact at P. Determine the magnitude and direction of field at O. Now the circular loop is rotated through an angle 60° without distortion about its diameter perpendicular to the straight portion of the wire. Determine the field at O if point A of the loop goes into the paper on rotation.

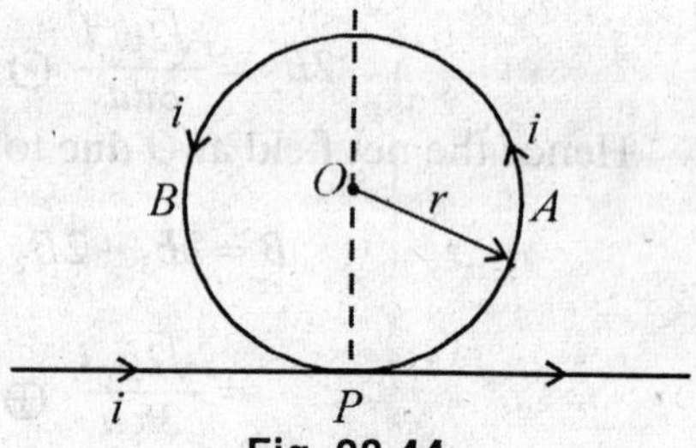

Fig. 23.44

Solution: The field at O, the centre of the loop will be the vector sum of magnetic field $\boldsymbol{B}_1$ due to straight infinite wire and magnetic field $\boldsymbol{B}_2$ due to the circular loop

$$\boldsymbol{B}_1 = \frac{\mu_0 \, i}{2\pi \, r} \odot \quad \ldots(i)$$

and $$\boldsymbol{B}_2 = \frac{\mu_0 \, i}{2r} \odot \quad \ldots(ii)$$

Since the two fields are parallel out of the paper plane. Hence net field at O is

$$\boldsymbol{B} = \boldsymbol{B}_1 + \boldsymbol{B}_2 = \left[\frac{\mu_0 \, i}{2\pi \, r} + \frac{\mu_0 \, i}{2r}\right] \odot$$

$$= \frac{\mu_0 \, i}{2\pi \, r}(1 + \pi) \odot$$

(*b*) In this position the magnetic field at O due to straight wire is perpendicular to the plane of diagram out of the page and B_2 the field due to the coil is along the axis and the angle between the directions of the two fields is 60° (Fig. 23.45). Magnitude of these fields are same as given by equation (*i*) and (*ii*). Hence resultant field B at O is

$$B = \sqrt{B_1^2 + B_2^2 + 2B_1B_2 \cos 60°}$$

$$= \frac{\mu_0 i}{2\pi r}\sqrt{1+\pi+\pi^2}$$

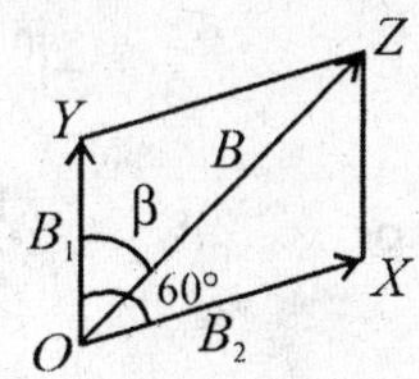

Fig. 23.45

If β is the angle that the direction of B makes with the vertical, then

$$\tan\beta = \frac{B_2 \sin 60°}{B_1 + B_2 \cos 60°}$$

$$\equiv \frac{\frac{\mu_0 i}{2r}\sqrt{3/2}}{\frac{\mu_0 i}{2\pi r} + \frac{\mu_0 i}{2r}\cdot\frac{1}{2}} = \frac{\sqrt{3/2}}{\frac{1}{\pi}+\frac{1}{2}}$$

$$= \frac{\sqrt{3}\pi}{2+\pi} = 1.05$$

or $\beta \approx 45°$.

Example 23.15: A constant current i flows through a metal rod of length l and mass m that slides on a frictionless rails as shown in Fig. 23.46. If the initial speed of the rod is v_0 and a magnetic field B acts vertically up, find (*a*) the speed of the rod as a function of time (*b*) the total distance moved by the rod before coming to rest.

Solution: The magnetic force on the rod carrying current in is $F_m = ilB$ and is in the direction opposite to the motion of the rod. Hence retardation $= \frac{F_m}{m} = \frac{ilB}{m}$ where m is mass of the rod.

If instantaneous velocity is v, then retardation $= -\frac{dv}{dt}$

$\therefore$ $$-\frac{dv}{dt} = \frac{ilB}{m} \quad \text{...(}i\text{)}$$

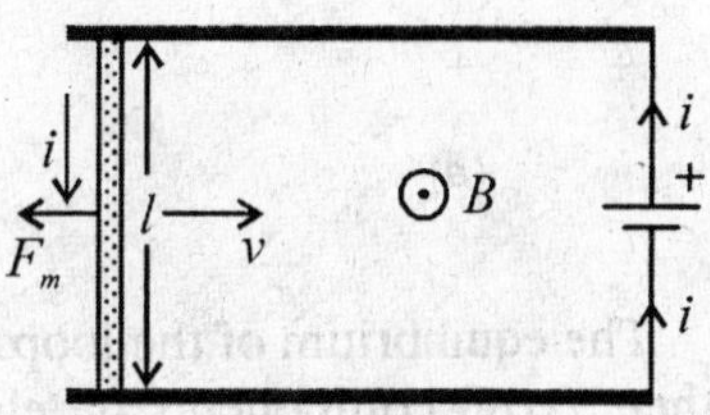

Fig. 23.46

or $$dv = -\frac{ilB}{m}dt$$

and integration gives

$$\int_{v_0}^{v} dv = -\frac{ilB}{m}\int_0^t dt$$

or $$v - v_0 = -\frac{ilB}{m}t \text{ or } v = v_0 - \frac{ilB}{m}t \quad \text{...(}ii\text{)}$$

Equation (*i*) can be written as

$$-\frac{dv}{ds}\cdot\frac{ds}{dt} = \frac{ilB}{m}$$

where s is instantaneous displacement.

or $\qquad -v\,dv = \dfrac{i\,l\,b}{m}\,ds$

integration gives

$$-\int_{v_0}^{0} v\,dv = \int_0^s \frac{i\,l\,B}{m}\,ds$$

or $\qquad \dfrac{1}{2}v_0^2 = \dfrac{i\,lB}{m}\,s$

or $\qquad s = \dfrac{mv_0^2}{2i\,l\,B}$

Thus rod will come to rest after moving a distance $= \dfrac{mv_0^2}{2i\,lB}$

Example 23.16: A flexible conducting wire of length 0.5 m lies in a magnetic field of 1.0 tesla perpendicular to the plane of the loop. Show that when a current is passed through the loop it opens into a circle. Also calculate the tension developed in the wire if a current of 1.57 amperes is passed through it.

Solution: When a current carrying loop is placed in a magnetic field with its plane perpendicular to the field, every part of it will experience a force which is always perpendicular to both the current element (*idl*) and the field. The loop will change into a circle in such a way that all the three (*idl*), ***B*** and ***F*** are mutually perpendicular to each other. This is shown in Fig. 23.47 (*b*).

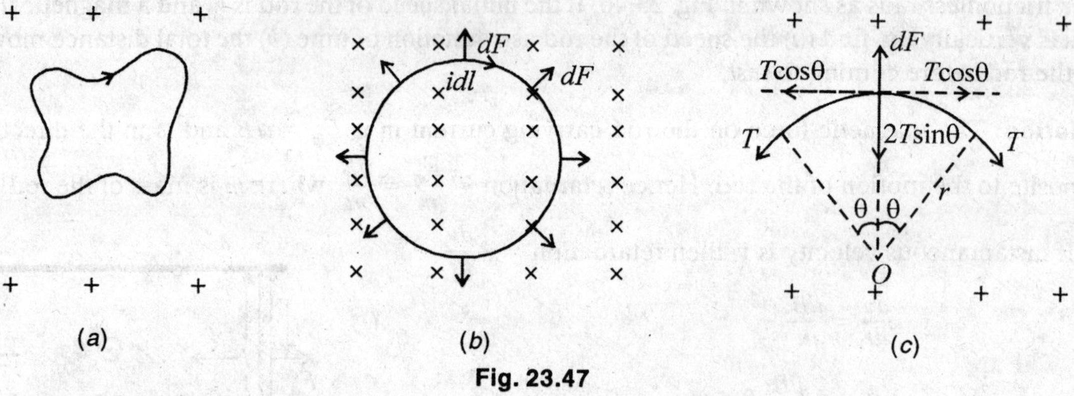

Fig. 23.47

The equilibrium of the loop is shown in Fig. 23.47 (*c*). If *T* is the tension developed in the loop, then *T*cosθ components on element (*idl*) cancel each other and component 2*T*sinθ balance the magnetic field *dF*. Thus in equilibrium

$$2T\sin\theta = dF = idlB$$

Since θ is small, hence $\sin\theta \approx \theta = \dfrac{dl}{2r}$, where *r* is radius of the loop.

∴ $\qquad 2T\dfrac{dl}{2r} = idl\,B$

or $\qquad T = iBr$

Thus a current carrying loop when placed in magnetic field takes a circular shape and tension developed = iBr.

In given problem $i = 1.57$ A, $r = \frac{l}{2\pi} = \frac{0.5}{2\pi} m$ and $B = 1.0$ T

$$\therefore \quad T = \frac{1.57 \times 1 \times 0.5}{2\pi} = 0.125 \text{ N}$$

Example 23.17: A long horizontal wire AB which is free to move in a vertical plane and carries a current of 20 A is in equilibrium at height of 0.01 m above another parallel wire CD which is fixed in a horizontal plane and carries a current of 30 A as shown in Fig. 23.48. Show that when AB is slightly depressed and released, it executes simple harmonic motion. Find the period of oscillation.

Solution: Force per unit length between two parallel current carrying conductors separated by a distance h is

$$\frac{dF}{dl} = \frac{\mu_0}{2\pi}\frac{i_1 i_2}{h}$$

If currents are in opposite direction, the force is repulsive. For equilibrium of wire AB, we have

$$\frac{mg}{l} = \frac{\mu_0 \, i_1 i_2}{2\pi \, h} \qquad \ldots(i)$$

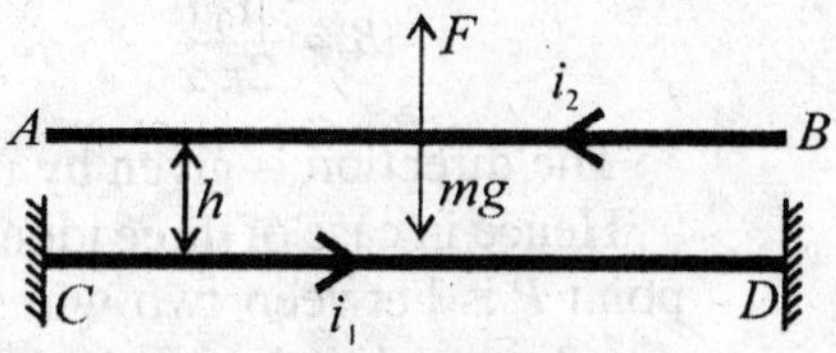

Fig. 23.48

Now if the upper wire is displaced by y towards the lower one, the magnetic restoring force will be

$$F = \frac{\mu_0}{2\pi} i_1 i_2 l\left(\frac{1}{h-y} - \frac{1}{h}\right) = \frac{\mu_0 \, i_1 i_2 \, l y}{2\pi \, h(h-y)} \text{ vertically upwards}$$

$$= -\frac{\mu_0 \, i_1 i_2 \, l \cdot y}{2\pi \, h^2}$$

Since $h >> y$ and force and y are in opposite direction, hence negative sign. Under the conditions specified above, the restoring force is directly proportional to displacement. Hence motion is S.H.M. with instantaneous acceleration

$$a = \frac{F}{m} = -\frac{\mu_0 \, i_1 i_2 l y}{2\pi \, m h^2}$$

The angular frequency ω is

$$\omega = \sqrt{\frac{\mu_0 \, i_1 i_2 l}{2\pi \, m h^2}}$$

and time period $T = \frac{2\pi}{\omega} = 2\pi\sqrt{\frac{2\pi \, m h^2}{\mu_0 \, i_1 i_2 \, l}}$

But $\quad \frac{\mu_0 \, i_1 i_2 \, l}{2\pi \, h} = mg \qquad$ from (i)

$$\therefore \quad T = 2\pi\sqrt{\frac{h}{g}}$$

Values given are, $h = 0.01$ m and $g = 10$ m/s^2

$$\therefore \quad T = 2\pi\sqrt{\frac{10^{-2}}{10}} = 2\pi\sqrt{\frac{1}{10^3}} \approx \frac{2}{10} = 0.2 \text{ sec}$$

Example 23.18: Three infinite long thin wires each carrying current in the same direction are in the X-Y plane of a gravity free space. The central wire is along Y-axis while the other two are along $x = \pm d$. (*a*) Find the locus of the points for which the magnetic field is zero (*b*) If the central wire is displaced along the Z-axis by a small amount and released, show that it will execute simple harmonic motion. If the linear density of the wire is λ, find the frequency of oscillation.

Solution:

(*a*) Magnetic field due to an infinite long current carrying wire at distance x from wire is given by

$$B = \frac{\mu_0 i}{2\pi x}$$

The direction is given by right hand thumb rule.

Hence in case of three identical current carrying wires the resultant can be zero only if the point P is between two wires otherwise B due to all the three wires will be in the same direction and so the resultant B cannot be zero. Now let the point P where the net magnetic field is zero is at distance x from the central wire between wires B and C (Fig. 23.49), then $\mathbf{B}$ at P is

$$\mathbf{B} = \frac{\mu_0 i}{2\pi}\left[\frac{1}{x} + \frac{1}{d+x} - \frac{1}{d-x}\right](-\hat{k})$$

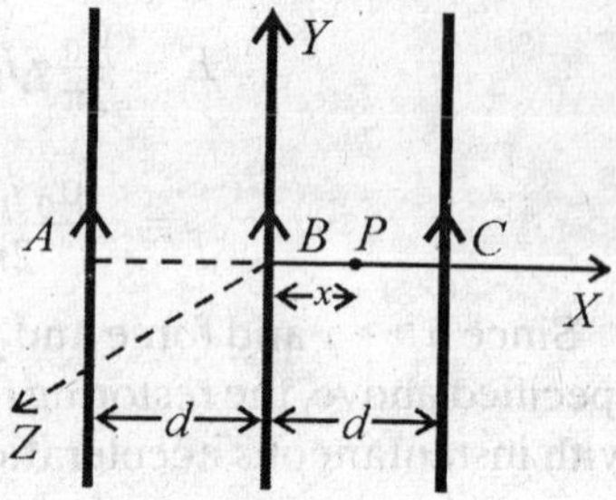

Fig. 23.49

For $\qquad \mathbf{B} = 0$ we have

$$(d-x)(d+x) + x(d-x) - x(d+x) = 0$$

This gives

$$x = \pm\frac{d}{\sqrt{3}}$$

Hence locus of points where resultant field is zero are straight lines parallel to Y-axis at distance $= \pm\dfrac{d}{\sqrt{3}}$

(*b*) Force between two current carrying parallel wires per unit length at distance r is

$$f = \frac{dF}{dl} = \frac{\mu_0 i_1 i_2}{2\pi r}$$

Hence if middle wire is displaced through distance z along Z-axis then restoring force per unit length acting on this wire is

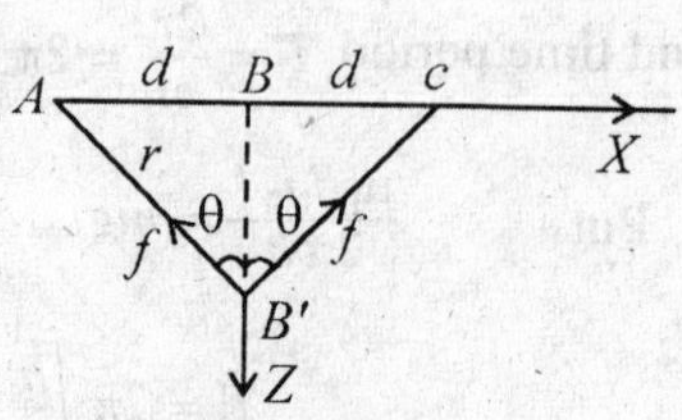

Fig. 23.50

$$F = 2f\cos\theta = \frac{2\mu_0 i^2}{2\pi r}\frac{z}{r} \qquad \left(\because \cos\theta = \frac{z}{r}\right)$$

$$= \frac{2\mu_0 i^2 z}{2\pi r^2} = \frac{2\mu_0 i^2 z}{2\pi(d^2+z^2)} \qquad (\because r^2 = d^2 + z^2)$$

Since $d >> z$, hence

Restoring force per unit length on the wire is

$$F = -\frac{\mu_0 i^2 z}{\pi d^2}$$

Hence force on length l of the wire is

$$Fl = -\frac{\mu_0 i^2 zl}{\pi d^2} = -\frac{\mu_0 i^2 lz}{2\pi d^2}$$

Negative sign because force and displacement are in opposite direction. Since force is directly proportional to displacement, hence the motion is simple harmonic motion. If m is mass of the wire, then instantaneous acceleration is

$$a = \frac{Fl}{m} = -\frac{\mu_0 i^2 lz}{\pi d^2 m} \qquad \text{But } \frac{m}{l} = \lambda$$

$$= -\frac{\mu_0 i^2}{\pi d^2 \lambda} z$$

and angular frequency

$$\omega = \sqrt{\frac{\mu_0 i^2}{\pi d^2 \lambda}}$$

$$\text{Hence frequency } f = \frac{\omega}{2\pi} = \frac{1}{2\pi}\sqrt{\frac{\mu_0 i^2}{\pi d^2 \lambda}} = \frac{i}{2\pi d}\sqrt{\frac{\mu_0}{\pi \lambda}}$$

Example 23.19: A thin isolated wire forms a plane spiral consisting of a large number n of closely packed turns through which a direct current i is flowing. The radii of internal and external loops are a and b respectively. Find the magnetic induction B as the centre of the spiral and the magnetic moment of the spiral for the given current.

Solution: Magnetic field due to a circular current carrying loop of radius r at its centre is given by

$$B_r = \frac{\mu_0 i}{2r}$$

The given spiral is made up of concentric loops having different radii varying from a to b. If dN is the number of turns in interval r and $r + dr$, then the field at the centre is

$$dB = B_r \cdot dN$$

$$= \frac{\mu_0 i}{2r} dN$$

But $$dN = \frac{N}{b-a} dr$$

Hence $$dB = \frac{\mu_0 i}{2r}\frac{N}{b-a} dr$$

and the total magnetic field at the centre due to whole spiral is

$$B = \int_a^b \frac{\mu_0 i}{2r}\frac{N}{b-a} dr$$

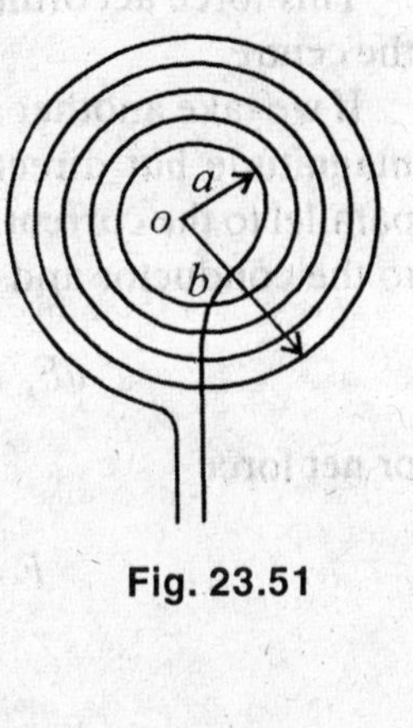

Fig. 23.51

$$= \frac{\mu_0 iN}{2(b-a)} \log r \Big|_a^b$$

$$= \frac{\mu_0 iN}{2(b-a)} \log \frac{b}{a}$$

Magnetic moment is equal to current multiplied by area.

Hence magnetic moment of turn of radius r, having dN number of turns in interval r and $r + dr$ is

$$dM = \pi r^2 i dN$$

Hence the total magnetic moment

$$M = \int_a^b \pi r^2 i dN = \pi i \int_a^b \frac{r^2 N}{b-a} dr$$

or
$$M \equiv \frac{\pi iN}{b-a} \cdot \frac{r^3}{3}\Big|_a^b = \frac{\pi iN(b^3 - a^2)}{3(b-a)}$$

Example 23.20: A circular conductor of radius r carrying a current i is coplanar with long straight conductor carrying a current i_0 the distance of the centre of the circular conductor being a from the straight current. Find the magnetic interaction force between the conductors.

Solution: The magnetic induction due to straight current carrying wire is directed normally into the page and is non uniform (Fig. 23.52). The distances of different circular current carrying elements are different and the magnetic field depends on the distance of the element from the straight conductor.

Consider a small element dl of the circular current. Its distance from current carrying wire is $a + r\cos\theta$ and magnetic field at the site of element dl is

$$B = \frac{\mu_0}{2\pi} \frac{i_0}{(a + r\cos\theta)}$$

and force on element $dl = rd\theta$ is

$$\mathbf{dF} = i\mathbf{dl} \times \mathbf{B}$$

or
$$dF = \frac{\mu_0 i_0 i \, rd\theta}{2\pi(a + r\cos\theta)}$$

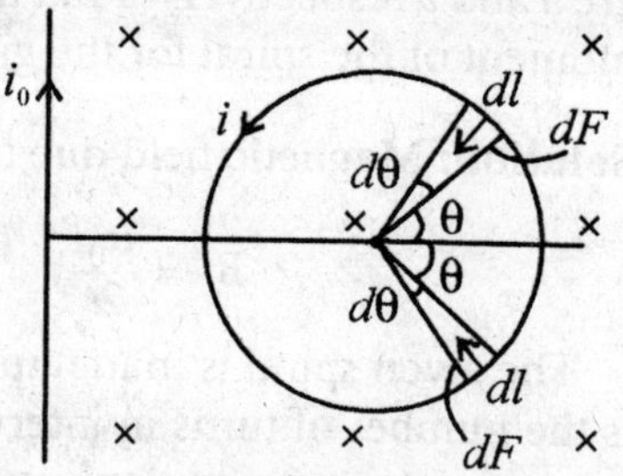

Fig. 23.52

This force according to vector product law is directed towards the centre.

If we take another symmetric element as shown in Fig. 23.52. It will experience a force of same magnitude but direction towards centre as shown in Fig. 23.52. The component of the force parallel to the current carrying conductor cancel each other. The only component is perpendicular to the conductor and towards it, i.e. negative X-direction. Hence

$$dF_x = dF\cos\theta = -\frac{\mu_0 i_0 i \, r\cos\theta d\theta}{2\pi(a + r\cos\theta)}$$

or net force

$$F = 2\int dF_x = -\frac{\mu_0 i_0 i}{2\pi} 2\int_0^{\pi} \frac{r\cos\theta \, d\theta}{a + r\cos\theta}$$

$$= -\frac{\mu_0 i_0 i}{\pi} \int_0^{\pi} \frac{\{(a + r\cos\theta) - a\}\, d\theta}{a + r\cos\theta}$$

or
$$F = \frac{\mu_0 i_0 i}{\pi} \int_0^{\pi} \frac{a\, d\theta}{a + r\cos\theta} - \frac{\mu_0 i_0 i}{\pi} \int_0^{\pi} d\theta$$

Integration gives

$$F = \mu_0 i_0 i \left[\frac{a}{\sqrt{a^2 - r^2}} - 1 \right]$$

Example 23.21: A non conducting thin disc of radius r charged uniformly over one side with the charge q rotates about its axis with angular velocity ω. Find the magnetic moment of the disc and magnetic field at the centre.

Solution: The disc can be treated as composed of an infinite number of rings with radii varying from $x = 0$ to $x = r$. Consider one such ring with inner radius x and outer radius $x + dx$. The charge on this ring element is

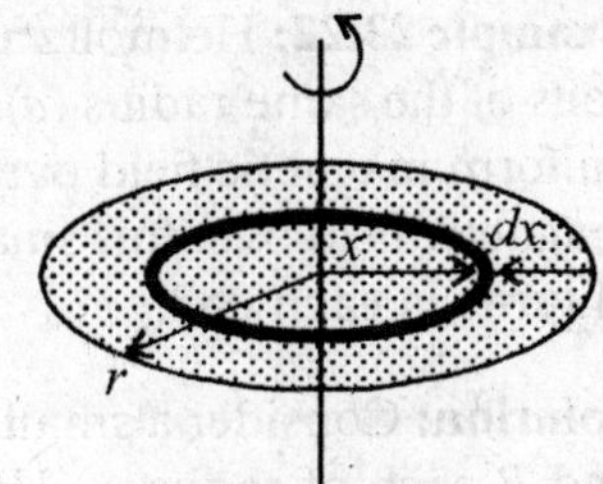

Fig. 23.53

$$dq = \frac{q}{\pi r^2} \times 2\pi x\, dx = \frac{2q\, x\, dx}{r^2}$$

Since it is rotating with angular velocity ω hence it is equivalent to a current loop of current

$$di = \frac{\omega}{2\pi} dq = \frac{q\omega\, x\, dx}{\pi r^2} \qquad \ldots(i)$$

This current carrying loop is equivalent to a magnetic dipole of moment dM.

dM = current × area of the loop

$$= \frac{q\omega\, x\, dx}{\pi r^2} \cdot \pi x^2 = \frac{q\omega}{r^2} x^3 dx$$

Hence the disc is a magnetic dipole of moment M and

$$M = \int dM = \int_0^r \frac{q\omega}{r^2} x^3 dx = \frac{q\omega\, r^4}{4r^2}$$

$$= \frac{q\omega\, r^2}{4}$$

Note: Angular momentum of this rotating disc is $L = \frac{mr^2}{2}\omega$. Where m is mass of disc.

$$\therefore \qquad \frac{M}{L} = \frac{q}{2m}$$

and in vector notation $\boldsymbol{M} = \frac{q}{2m} \boldsymbol{L}$ ...(ii)

Thus the angular momentum and magnetic moment are related by above given equation.

(*b*) **Magnetic Field**: Since a small width loop of radii x and $x + dx$ carries a current, equation (*i*)

$$di = \frac{q\omega\, x\, dx}{\pi r^2}$$

Hence magnetic field due to this current carrying loop is

$$dB = \frac{\mu_0\, di}{2x} = \frac{\mu_0}{2}\frac{q\omega\, x\, dx}{\pi r^2 x}$$

$$= \frac{\mu_0\, q\omega}{2\pi r^2}\, dx$$

Hence magnetic field of the centre

$$B = \int dB = \int_0^r \frac{\mu_0\, q\omega}{2\pi r^2}\, dx = \frac{\mu_0\, q\omega}{2\pi r}$$

Example 23.22: Helmoltz an eminent scientist of the 19th century, showed that two narrow circular coils of the same radius (a) and carrying the same current (i) in the same direction can produce a uniform magnetic field over a small region around a point midway between them. Show that the magnetic field over this small region is given approximately by $B = 0.72\mu_0 ni$ where n is the number of turns in each coil.

Solution: Consider a small region $PQ = 2d$ near the point C which is midway between two coils A and B each of radius a. Hence $CP = CQ = d$ and $OC = CO' = a/2$ (Fig. 23.54). Let us find the magnetic field at point P due to the two coils. Magnetic field at P due to coil A is

$$B_a = \frac{\mu_0\, ni\, a^2}{2} \times \frac{1}{\{(a/2 - d)^2 + a^2\}^{3/2}} \quad \ldots(i)$$

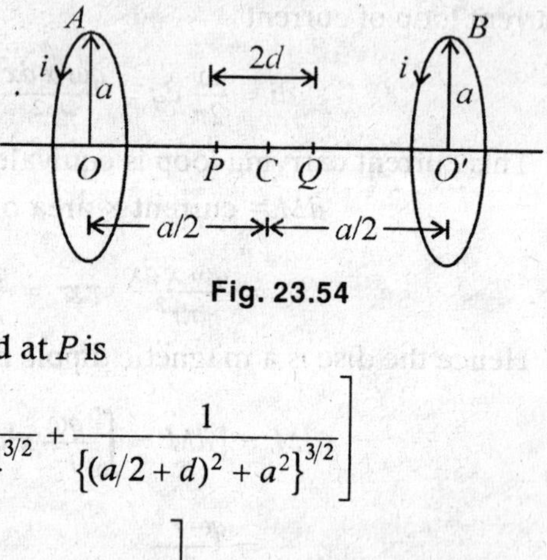

Fig. 23.54

Similarly magnetic field at P due to coil B is

$$B_b = \frac{\mu_0\, ni\, a^2}{2} \times \frac{1}{\{(a/2 + d)^2 + a^2\}^{3/2}} \quad \ldots(ii)$$

Since currents in the two coils are in the same direction, the two fields at P are in the same direction. Hence net field at P is

$$B = B_a + B_b = \frac{\mu_0\, ni\, a^2}{2}\left[\frac{1}{\{(a/2 - d)^2 + d^2\}^{3/2}} + \frac{1}{\{(a/2 + d)^2 + a^2\}^{3/2}}\right]$$

$$= \frac{\mu_0\, ni\, a^2}{2}\left[\frac{1}{\left(\frac{5a^2}{4} - ad + d^2\right)^{3/2}} + \frac{1}{\left(\frac{5a^2}{4} + ad + d^2\right)^{3/2}}\right]$$

Since d^2 is very small, hence neglecting d^2, we have

$$B = \frac{\mu_0\, ni\, a^2}{2}\frac{1}{\left(\frac{5a^2}{4}\right)^{3/2}}\left[\left(1 - \frac{4d}{5a}\right)^{-3/2} + \left(1 + \frac{4d}{5a}\right)^{-3/2}\right]$$

$$= \frac{\mu_0 ni}{2a}(4/5)^{3/2}\left(1+\frac{6d}{5a}+1-\frac{6d}{5a}\right) \text{ (Using binomial theorem)}$$

$$= \frac{\mu_0 ni}{a}(4/5)^{3/2} = 0.72\frac{\mu_0 ni}{a}$$

Example 23.23: A wire loop carrying a current i is placed in the X-Y plane (Fig. 23.55). (*a*) If a particle with charge q and mass m is placed at the centre P and given a velocity v along NP. Find its instantaneous acceleration (*b*) If an external magnetic field, $\boldsymbol{B} = B\hat{i}$ is applied, find the force and the torque acting on the loop due to this field.

Solution: Arc MN subtends an angle of 120° at the centre P. Hence magnetic field at P due to this arc is 1/3 that is produced by the loop. Let this field be B_1. Thus

$$B_1 = \frac{\mu_0 i}{6a}$$

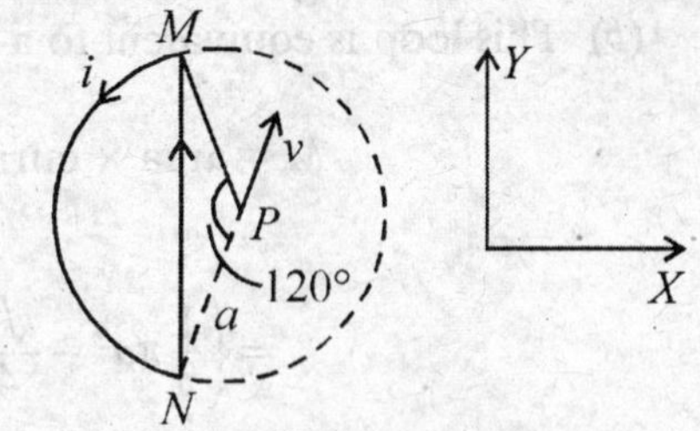

Fig. 23.55

This field is directed along positive Z-axis.

Magnetic field at P due to straight part NM is

$$B_2 = \frac{\mu_0 i}{4\pi}\frac{(\sin\alpha + \sin\beta)}{x}$$

Figure 23.55 (*a*) shows $x = OP = a\cos 60° = a/2$, $\alpha = \beta = 60°$

$$\therefore \quad B_2 = \frac{\mu_0 i}{4\pi} \times \frac{2\sqrt{3}}{2a/2} = \frac{\sqrt{3}\mu_0 i}{2\pi a}$$

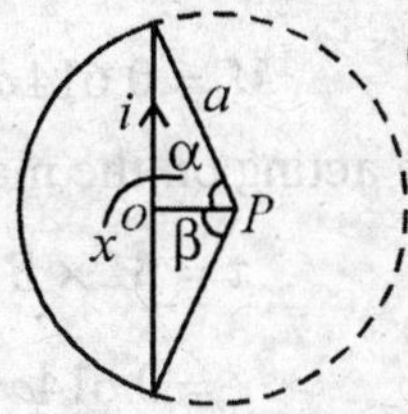

Fig. 23.55 (*a*)

This field is directed along the negative Z-axis.

Hence net magnetic field induction at P due to the loop is

$$B = B_1 - B_2 = \frac{\mu_0 i}{6a} - \frac{\sqrt{3}\mu_0 i}{2\pi a} = -\frac{0.11\,\mu_0 i}{a}$$

The negative sign shows that B is directed along the negative Z-direction. Hence in vector notation we have

$$\boldsymbol{B} = -0.11\frac{\mu_0 i}{a}\hat{k} \qquad \ldots(i)$$

The particle is given velocity v at P making an angle of 60° with X-axis. Hence

$$\boldsymbol{v} = v\cos 60°\,\hat{i} + v\sin 60°\,\hat{j} \qquad \ldots(ii)$$

This particle in magnetic field experiences force $\boldsymbol{F} = q\boldsymbol{v} \times \boldsymbol{B}$

$$\therefore \quad \text{Acceleration of the particle } \boldsymbol{a} = \frac{\boldsymbol{F}}{m}$$

or

$$\boldsymbol{a} = \frac{q}{m}\left(\frac{v}{2}\hat{i} + \frac{\sqrt{3}v}{2}\hat{j}\right) \times \left(-0.11\frac{\mu_0 i}{a}\hat{k}\right)$$

$$= \frac{qv}{2m}\left(\frac{0.11\mu_0 i}{a}\right)\left(-\sqrt{3}\,\hat{i} + \hat{j}\right)$$

Thus magnitude of the acceleration is

$$a = \sqrt{a_x^2 + a_y^2} = \frac{qv}{2m}\left(\frac{0.11\mu_0 i}{a}\right)\sqrt{3+1}$$

$$= \frac{0.11\mu_0\, i\, qv}{m}$$

The acceleration is directed along the angle θ such that

$$\tan\theta = \frac{a_y}{a_x} = -\frac{1}{\sqrt{3}}$$

i.e., $\theta = 30$ with negative *X*-axis.

(*b*) This loop is equivalent to a magnetic dipole whose dipole moment is

$$M = \text{area} \times \text{current} = \left(\frac{1}{3}\pi a^2 - 2\times\frac{1}{2}a\cos 60°\times a\sin 60°\right)i$$

$$= \left(\frac{1}{3}\pi a^2 - \frac{\sqrt{3}}{4}a^2\right)i = 0.614\,a^2 i$$

This magnetic moment is directed perpendicular to the plane of the diagram

$$\boldsymbol{M} = 0.614\,a^2 i\hat{k}$$

The torque acting on the magnetic dipole in magnetic field is

$$\tau = \boldsymbol{M}\times\boldsymbol{B} = (0.614\,a^2 i\,\hat{k})\times(B\hat{i})$$

$$= 0.614a^2 iB\,\hat{j}$$

This torque has magnitude $0.614a^2iB$ and is directed along the positive *Y*-axis.

Example 23.24: A particle of charge $+q$ and mass m moving under the influence of a uniform electric field $E\hat{i}$ and a magnetic field $B\hat{k}$ follows a trajectory from *P* to *Q* as shown in Fig. 23.56. The velocities at *P* and *Q* are $v\hat{i}$ and $-2v\,\hat{j}$ respectively, calculate

(*a*) the magnitude of electric field.
(*b*) rate of work done by electric field at *P* and *Q*
(*c*) rate of work done by magnetic field at *P* and *Q*.

Solution:

(*a*) In electric field the charged particle experiences a force

$$\boldsymbol{F} = qE\hat{i}$$

At *P* the position vector of the particle is $\boldsymbol{r}_1 = a\,\hat{j}$ and at *Q* its position vector $\boldsymbol{r}_2 = 2a\hat{i}$. Hence displacement of the particle is

$$\boldsymbol{r} = \boldsymbol{r}_2 - \boldsymbol{r}_1 = (2a\hat{i} - a\,\hat{j})$$

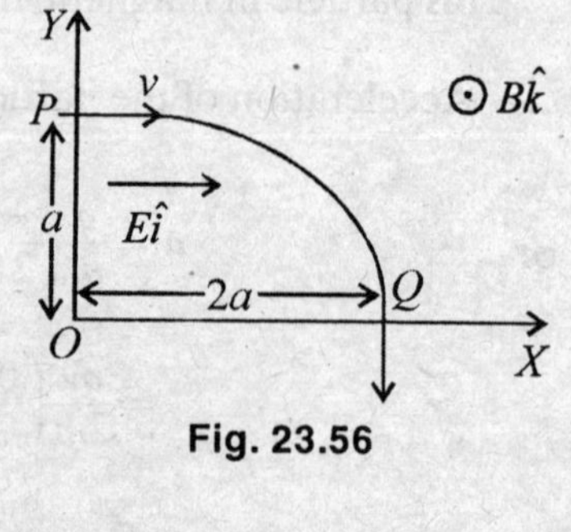

Fig. 23.56

Hence work done $= \boldsymbol{F}\cdot\boldsymbol{r} = qE\hat{i}\cdot(2a\hat{i} - a\,\hat{j}) = 2qEa \quad \ldots(i)$

Since no work is done by the magnetic force on the charge. Hence this work done must be equal to the change in kinetic energy of the particle

$$\Delta K = \frac{1}{2}m(2v)^2 - \frac{1}{2}m(v)^2 = \frac{3}{2}mv^2 \qquad \ldots(ii)$$

Equating (*i*) and (*ii*), we get

$$2q\,Ea = \frac{3}{2}mv^2$$

or

$$E = \frac{3}{4}\frac{mv^2}{qa}$$

(*b*) At *P* force on charged particle is $\boldsymbol{F} = qE\hat{i}$. Its velocity is $\boldsymbol{v} = v\hat{i}$. Hence rate of work done by electric field at *P* is

$$P_a = \boldsymbol{F}\cdot\boldsymbol{v} = qEi\cdot v\hat{i} = qEv = \frac{3}{4}\frac{mv^3q}{qa} = \frac{3}{4}\frac{mv^3}{a}$$

At *Q* force on charged particle is $\boldsymbol{F} = qE\hat{i}$. Its velocity is $\boldsymbol{v} = -2v\,\hat{j}$. Hence rate of work done at *Q* by the electric field is

$$P_q = \boldsymbol{F}\cdot\boldsymbol{v} = qEi(-2v\hat{j}) = 0$$

(*c*) Force on charged particle in magnetic field is $\boldsymbol{F} = q(\boldsymbol{v}\times\boldsymbol{B})$. Hence rate of work done by magnetic field is

$$P = \boldsymbol{F}\cdot\boldsymbol{v} = q(\boldsymbol{v}\times\boldsymbol{B})\cdot\boldsymbol{v} = 0$$

Hence work done by magnetic field at *P* and at *Q* is zero.

Example 23.25: A straight line segment *OC* (of length *L* meters) of a circuit carrying a current *i* amperes is placed along the *X*-axis as shown in Fig. 23.57. Two infinitely long straight wires *A* and *B* each extending from $-\infty$ to $+\infty$ fixed at $y = -a$ and $y = +a$ meters. If the wires *A* and *B* carry current *i* amperes into the plane of the paper, obtain the expression for the force on the segment *OC*. What will be the force on *OC* if the current in wire *B* is reversed?

Solution: To find force on the segment *OC* of the current carrying conductor, let us consider a small element *dx* of the conductor at *C′* at distance *x* from the origin *O* (Fig. 23.57). Magnetic field at *C′* due to wire *A* is

$$B_a = \frac{\mu_0 i}{2\pi(AC')} \text{ directed perpendicular to line } AC'$$

Magnetic field at *C′* due to wire *B* is

$$B_b = \frac{\mu_0 i}{2\pi(BC')} \text{ directed perpendicular to line } BC'.$$

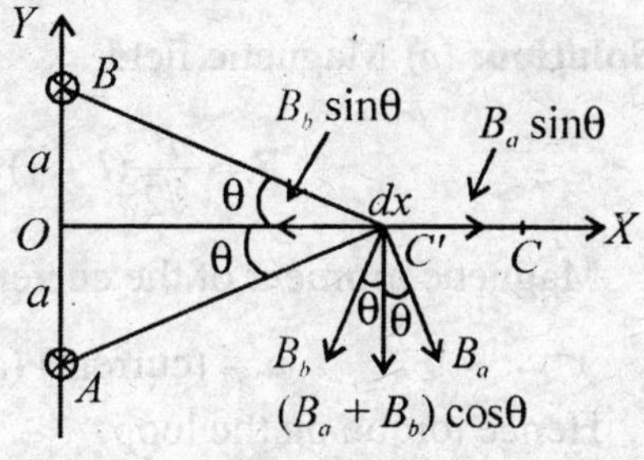

Fig. 23.57

From Fig. 23.57 $AC' = BC' = \sqrt{x^2 + a^2}$

Resolving B_a and B_b along and perpendicular to *OC*, we see that component $B_a\sin\theta = B_b\sin\theta$ (since $|B_a| = |B_b|$) and cancel each other. The net magnetic field at point *C′* is

$$B = 2B_a\cos\theta = \frac{\mu_0 i\cos\theta}{\pi\sqrt{x^2 + a^2}} = \frac{\mu_0 i\cdot x}{\pi(a^2 + x^2)}$$

Hence force on element dx of the segment OC is

$$dF = iB\,dx = \frac{\mu_0 i^2 x\,dx}{\pi(a^2 + x^2)}$$

Hence total force on the whole current carrying segment OC is

$$F = \int_0^L \frac{\mu_0 i^2 x\,dx}{\pi(a^2 + x^2)} = \frac{\mu_0 i^2}{2\pi} \log (a^2 + x^2)\Big|_0^L$$

$$= \frac{\mu_0 i^2}{2\pi} \log \frac{(a^2 + L^2)}{a^2}$$

The direction of this force is determined by Fleming's left hand rule. This force is directed towards negative Z-axis. Hence

$$\boldsymbol{F} = -\frac{\mu_0 i^2}{2\pi} \log \frac{(a^2 + L^2)}{a^2}\,\hat{k}$$

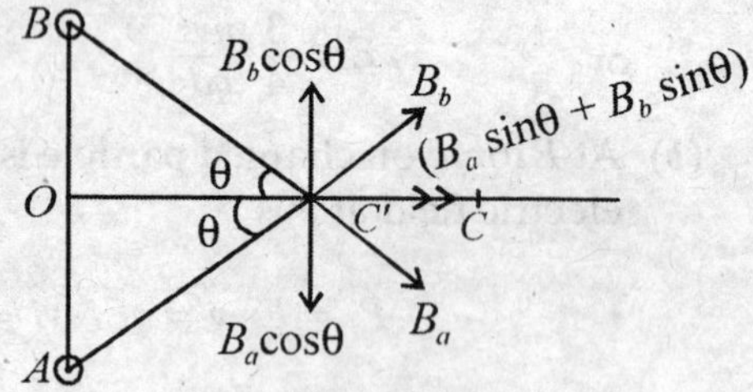

Fig. 23.58

If the direction of the current in wire B is reversed; the direction of the magnetic field is perpendicular to BC' as shown in Fig. 23.58. Hence the components of the field perpendicular to the current carrying segment cancel each other. The only component parallel to the field is there. Thus the field and current segment are parallel. Hence force on the conductor is zero.

Example 23.26: A uniform constant magnetic field B is directed at an angle of 45° to the X-axis in X-Y plane (Fig. 23.59). $PQRS$ is a rigid square wire frame carrying a steady current I_0 with its centre at the origin 0. At time $t = 0$, the frame is at rest in the position shown in the figure with its sides parallel to the X and Y axes. Each side of the frame is of mass M and length L.

(*a*) What is the torque τ about O acting on the frame due to the magnetic field B?

(*b*) Find the angle by which the frame rotates under the action of this torque in a short interval of time Δt and the axis about which the rotation occurs. (Δt is so short that any variation in the torque during this time interval may be neglected). Given moment of inertia of the frame about an axis through its centre perpendicular to its plane is $\frac{4}{3} ML^2$.

Solution: (*a*) Magnetic field

$$\boldsymbol{B} = \frac{B}{\sqrt{2}}(\hat{i} + \hat{j}) \qquad \ldots(i)$$

Magnetic moment of the current carrying loop is

$$\mu = \text{(current) (area)} = I_0 L^2\,\hat{k} \qquad \ldots(ii)$$

Hence torque on the loop

$$\tau = \mu \times \boldsymbol{B} = I_0 L^2 \hat{k} \times \frac{B}{\sqrt{2}}(\hat{i} + \hat{j})$$

$$= \frac{I_0 L^2 B}{\sqrt{2}}(-\hat{i} + \hat{j}) \qquad \ldots(iii)$$

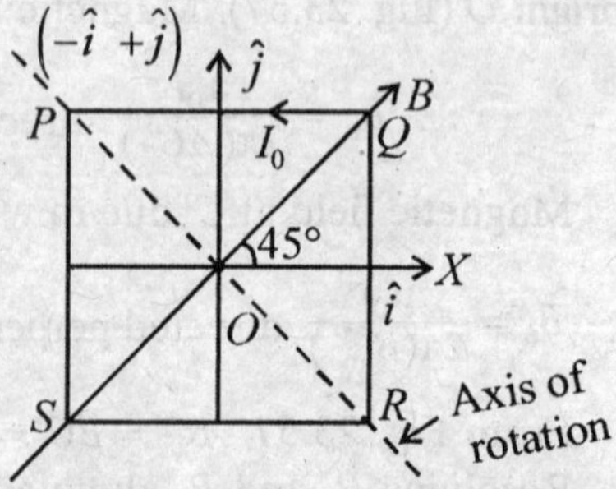

Fig. 23.59

This means that the direction of the torque is along the vector $(-\hat{i} + \hat{j})$, i.e. parallel to PR and is directed from R to P (Fig. 23.59).

Magnitude of the torque $\tau = I_0 L^2 B$ ...(iv)

(b) This torque will produce an angular acceleration in the frame about the axis of rotation RP. Hence equation of motion is

$$\tau = I\alpha \qquad \alpha \text{ is angular acceleration and } I \text{ moment of inertia}$$

or
$$\alpha = \frac{\tau}{I}$$

But
$$I_{RP} = \frac{1}{2} I_{OZ} = \frac{1}{2}\left(\frac{4}{3} ML^2\right) = \frac{2\,ML^2}{3}$$

$$\therefore \quad \alpha = \frac{I_0 L^2 B \times 3}{2\,ML^2} = \frac{3 I_0 B}{2\,M}$$

Hence the rotation of the frame in time Δt, (since initial velocity is zero) is

$$\theta = \frac{1}{2}\alpha(\Delta t)^2 = \frac{1}{2}\frac{3\,I_0 B}{2\,M}(\Delta t)^2 = \frac{3\,I_0 B}{4\,M}(\Delta t)^2$$

Example 23.27: A particle of mass m and charge q is moving in a region where uniform constant electric and magnetic fields $\boldsymbol{E}$ and $\boldsymbol{B}$ are parallel to each other. At time $t = 0$ the velocity v_0 of the particle is perpendicular to E. Find the velocity v of the particle at time t. Express your answer in terms of t, q, m the vectors v_0, $\boldsymbol{E}$ and $\boldsymbol{B}$ and their magnitudes. Assume $v << c$.

Solution: The Lorentz force acting on a particle of charge q moving with velocity $\boldsymbol{v}$ in electric and magnetic field $\boldsymbol{E}$ and $\boldsymbol{B}$ is given by

$$\boldsymbol{F} = q(\boldsymbol{E} + \boldsymbol{v} \times \boldsymbol{B})$$

If m is mass of particle and $\boldsymbol{v}$ instantaneous velocity, then

$$m\frac{d\boldsymbol{v}}{dt} = \boldsymbol{F} = q(\boldsymbol{E} + \boldsymbol{v} \times \boldsymbol{B})$$

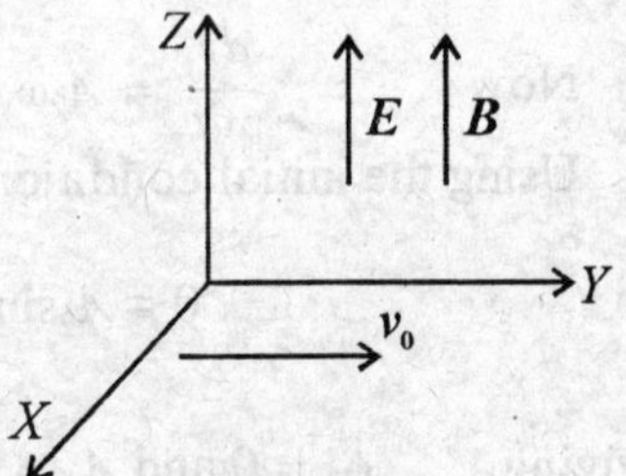

Fig. 23.60

Let the direction of $\boldsymbol{E}$ and $\boldsymbol{B}$ be along Z-axis.

Hence
$$\boldsymbol{E} = E\hat{k} \text{ and } \boldsymbol{B} = B\hat{k}$$

and
$$\boldsymbol{v} = v_x\hat{i} + v_y\hat{j} + v_z\hat{k}$$

$\therefore$

$$m\frac{dv_x}{dt}\hat{i} + m\frac{dv_y}{dt}\hat{j} + m\frac{dv_z}{dt}\hat{k} = q\left[E\hat{k} + (v_x\hat{i} + v_y\hat{j} + v_z\hat{k}) \times (B\hat{k})\right]$$

$$= q\left[E\hat{k} - v_x B\hat{j} + v_y B\hat{i}\right]$$

Hence we have

$$\frac{dv_x}{dt} = \frac{qv_y B}{m} \qquad \text{...(i)}$$

$$\frac{dv_y}{dt} = -\frac{qv_x B}{m} \text{ and } \frac{dv_z}{dt} = \frac{qE}{m} \qquad \text{...(ii)}$$

Differentiating (*i*) we have

$$\frac{d^2v_x}{dt^2} = \frac{qB}{m} \cdot \frac{dv_y}{dt} = -\frac{q^2B^2}{m^2} v_x = -\omega^2 v_x \qquad \ldots(iii)$$

Similarly differentiation of $\frac{dv_y}{dt}$ gives

$$\frac{d^2v_y}{dt^2} = -\frac{qB}{b}\frac{dv_x}{dt} = -\left(\frac{qB}{m}\right)^2 v_y = -\omega^2 v_y \qquad \ldots(iv)$$

where $\omega = \frac{qB}{m}$. Equations (*iii*) and (*iv*) are the equation representing simple harmonic variations. The solution of equation (*iii*) is

$$v_x = A_1 \sin(\omega t + \phi_1) \qquad \ldots(v)$$

$$\therefore \quad \frac{dv_x}{dt} = A_1\omega \cos(\omega t + \phi_1) \qquad \ldots(vi)$$

It is given that at $t = 0$ $v_x = v_0^*$ and since v_0 is constant hence at $t = 0$, $\frac{dv_x}{dt} = 0$. Using these conditions, we have

$$v_0 = A_1 \sin\phi_1 \text{ and } 0 = A_1\omega\cos\phi_1$$

giving $\phi_1 = \pi/2$ and $A_1 = v_0$

Using these values of constants equation (*v*) becomes

$$v_x = v_0 \sin(\omega t + \pi/2) = v_0 \cos\omega t \qquad \ldots(vii)$$

Similarly solution of equation (*iv*) can be written as

$$v_y = A_2 \sin(\omega t + \phi_2) \qquad \ldots(viii)$$

The initial conditions are $v_y = 0$ at $t = 0$ $\frac{dv_y}{dt} = -\frac{qB\,v_0}{m}$

Now $\frac{dv_y}{dt} = A_2\omega \cos(\omega t + \phi_2)$

Using the initial conditions, we have

$$0 = A_2 \sin\phi_2 \text{ and } -\frac{qB\,v_0}{m} = \omega A_2 \cos\phi_2$$

giving $\phi_2 = 0$ and $A_2 = -\frac{qB\,v_0}{m\omega}$

$$\therefore \quad v_y = -\frac{qB\,v_0}{m\omega} \sin\omega t = -v_0 \sin\omega t \qquad \ldots(ix)$$

Solution of *Z*-equation is $v_z = \frac{qE}{m} t$

Hence $\quad v = (v_0 \cos\omega t)\hat{i} - (v_0 \sin\omega t)\hat{j} + \frac{qE}{m} t\hat{k} \qquad \ldots(x)$

---

*It is given that v_0 is peperdicular to electric field *E*. Since *E* is taken along *Z*-direction and direction of v_0 is taken along *X*-axis (Fig. 23.60).

Now putting $\omega \equiv \frac{qB}{m}$, we get

$$\boldsymbol{v} = \cos\frac{qB}{m}t\,\boldsymbol{v}_0 + \sin\frac{qB}{m}t\frac{\boldsymbol{v}_0 \times \boldsymbol{E}}{E} + \frac{qt}{m}\boldsymbol{E}$$

or
$$\boldsymbol{v} = \cos\frac{qB}{m}t\,\boldsymbol{v}_0 + \sin\frac{qBt}{m}\frac{\boldsymbol{v}_0 \times \boldsymbol{B}}{B} + \frac{qt}{m}\boldsymbol{E}$$

Example 23.28: The region between $x = 0$ to $x = L$ is filled with a uniform steady magnetic field $B_0\hat{k}$. A particle of mass m, charge $+q$ and velocity $v_0\hat{i}$ travels along X-axis and enters the magnetic field. Neglecting gravity, (*a*) find the value of L if the particle emerges from region of magnetic field with its final velocity at an angle of 30° its initial velocity, (*b*) find the final velocity of the particle and the time spent by it in the magnetic field, if the magnetic field now extends upto $2.1L$.

Solution: A charged particle in magnetic field experiences a force

$$\boldsymbol{F} = q(\boldsymbol{v} \times \boldsymbol{B}) = q(v_0\hat{i}) \times (B_0\hat{k})$$

$$= -qv_0B_0\hat{j} \quad \ldots(i)$$

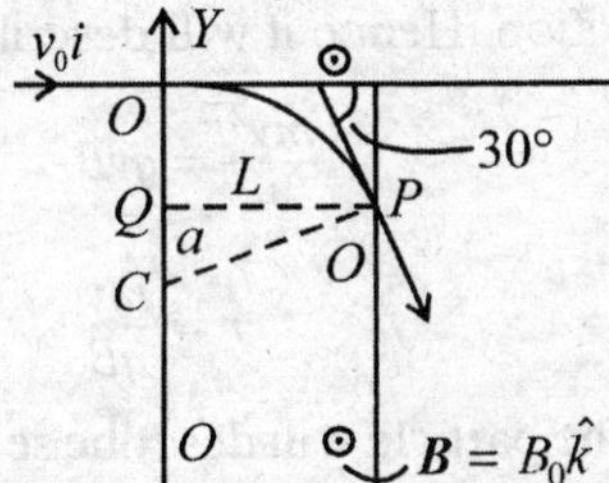

Fig. 23.61

Since force is always perpendicular to the velocity hence it describes circular motion. The radius of circular path r is given by

$$\frac{mv_0^2}{r} = qv_0B_0$$

$$r = \frac{mv_0}{qB_0} \quad \ldots(ii)$$

(*a*) Since the particle emerges from magnetic field with velocity vector making an angle of 30° with the initial vector, hence from triangle PQC.

$$\frac{L}{r} = \sin 30° \quad \text{or } r = 2L$$

or
$$L = r\sin 30° = \frac{mv_0}{2qB_0} \quad \ldots(iii)$$

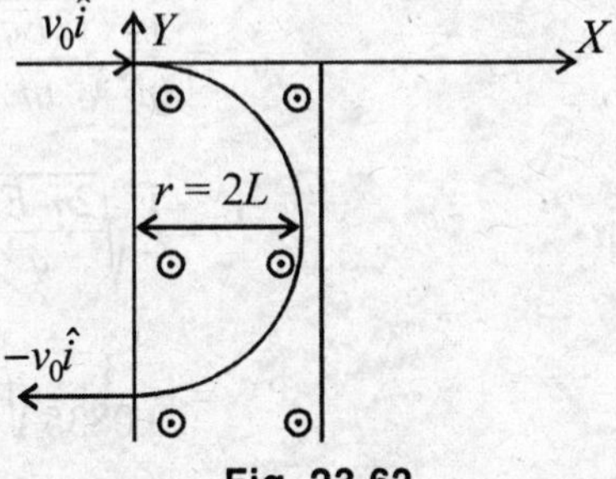

Fig. 23.62

(*b*) It is apparent that $r = 2L$. If now the magnetic field extends up to $x = 2.1L$, the particle will continue to move in a circular path. Its path is shown in Fig. 23.62. It completes semi-circle and emerges out of the magnetic field with velocity $-v_0\hat{i}$. Thus

Path travelled by the particle in magnetic field is

$$S = \frac{2\pi r}{2} = \frac{\pi m v_0}{qB_0}$$

Since this path is traversed by particle with velocity v_0, thus time taken by the particle is

$$t = \frac{S}{v_0} = \frac{\pi m}{qB_0}$$

Example 23.29: A constant homogeneous electric field E (= 100 V/m) exists in a region $x = 0$ to $x = 0.167$ m pointing in positive X-direction. A constant magnetic field B also exists in region from

$x = 0.167$ m to $x = 0.334$ m pointing in positive Z-direction. A proton at rest at $x = 0$, $y = 0$ is released in the positive X-direction. Find the minimum strength of magnetic field B so that the proton is deflected back at $x = 0$, $y = 0.167$ m.

Solution: The proton when released in electric field will get accelerated. The acceleration is $a = \frac{Eq}{m}$. If electric field extends upto distance x, then at the end of the electric field its velocity v is

$$v^2 = 2ax = 2 \cdot \frac{Eq}{m} \cdot x$$

$$v = \sqrt{\frac{2Eq}{m}x} \quad \ldots(i)$$

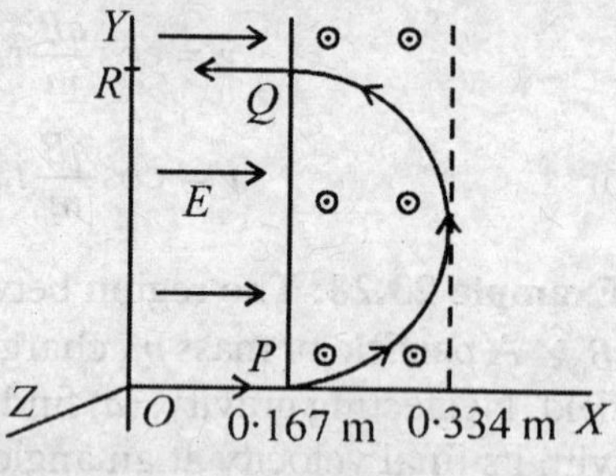

Fig. 23.63

Now the proton with its velocity v in X-direction enters a region of magnetic field B in Z-direction. Hence it will describe circular notion of radius r, such that

$$\frac{mv^2}{r} = qvB$$

or $$r = \frac{mv}{qB} \quad \ldots(ii)$$

The particle will describe semi circle if the distance upto which magnetic field extends is slightly greater than the radius of the circular path, i.e., $d > r$.

Since the particle has to come out at co-ordinates $x = 0$ and $y = 0.167$ m means particle must describe circular path of diameter 0.167 m or radius $r = \frac{0.167}{2} = 0.0835$ m.

Using (i) and (ii), we have

$$r = \frac{m}{qB}\sqrt{\frac{2Eq}{m}x} = \sqrt{\frac{2mEx}{q}}\,\frac{1}{B}$$

or $$B = \frac{1}{r}\sqrt{\frac{2mEx}{q}}$$

$$= \frac{1}{0.0835}\sqrt{\frac{2 \times 1.67 \times 10^{-27} \times 100 \times 0.167}{1.6 \times 10^{-19}}}$$

$$= 7.07 \times 10^{-3}\ \text{T}$$

Note: Here magnetic field extends upto distance $d = 0.334 - 0.167 = 0.167$ m. Hence particle will describe semi circle.

Particle enters electric field at $x = 0$. Leaves electric field at P. $x = 0.167$ m where it enters magnetic field. Leaves magnetic field at Q where $x = 0.167$ and $y = 0.167$. Here again the particle enters the electric field and leaves electric field at R having coordinates $x = 0$ and $y = 0.167$ m.

Example 23.30: A circular loop of radius R is bent along a diameter and given a shape as shown in Fig. 23.64. One of the semi-circles (KNM) lies in the X-Z plane and the other one (KLM) in the Y-Z plane with their centres at the origin. Current i flows through each of the semi-circles as shown in Fig. 23.64.

(a) A particle of charge q is released at the origin with velocity $v = -v_0\hat{i}$. Find the instantaneous force f on the particle. Assume that space is gravity free.

(b) If an external uniform magnetic field $B\hat{j}$ is applied, determine the forces F_1 and F_2 on the semi-circles KLM and KNM due to this field and the net force F on the loop.

Solution: Magnetic field at the centre of current carrying semi-circular wire loop of radius R is

$$B = \frac{\mu_0 i}{4R}$$

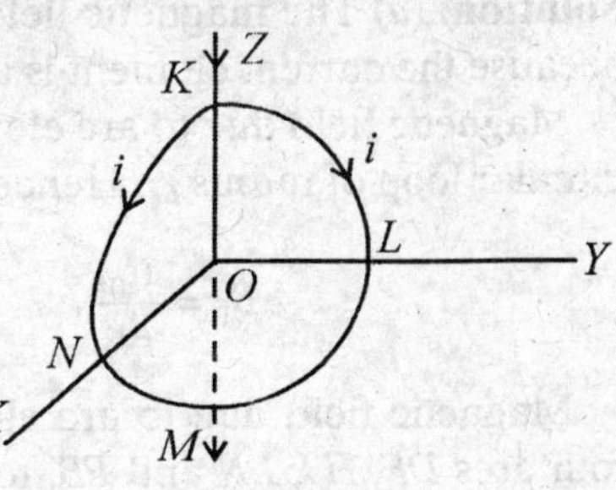

Fig. 23.64

and is directed along axis of loop perpendicular to the plane of loop. Hence the magnetic field at O due to loop KLM. B_1 is

$$B_1 = \frac{\mu_0 i}{4R}(-\hat{i}) \text{ where } \hat{i} \text{ is unit vector along } X\text{-axis.}$$

Similarly magnetic field at O due to loop KNM is

$$B_2 = \frac{\mu_0 i}{4R}(\hat{j}) \text{ where } \hat{j} \text{ is unit vector along } Y\text{-axis.}$$

Hence total magnetic field at the centre is

$$\boldsymbol{B} = \frac{\mu_0 i}{4R}(-\hat{i} + \hat{j}) \quad \text{...(i)}$$

(a) If a charged particle (charge = q) is released at the centre with velocity $v = -v_0\hat{i}$; the instantaneous force on the particle is

$$\boldsymbol{F} = q\boldsymbol{V} \times \boldsymbol{B} = q(-v_0\hat{i}) \times \frac{\mu_0 i}{4R}(-\hat{i} + \hat{j})$$

$$= -\frac{\mu_0 q v_0 i}{4R}\hat{k} \text{ where } \hat{k} \text{ is unit vector along } Z\text{-axis}$$

Hence the force on the charged particle is $\frac{\mu_0 q v_0 i}{4R}$ in magnitude and is directed along negative Z-axis.

(b) When a current (i) carrying semi-circular loop of radius r is placed in magnetic field (B) it experiences a force

$$F = 2riB$$

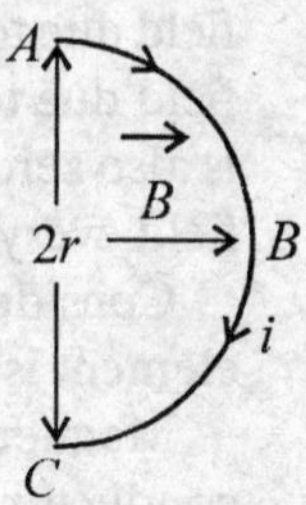

Fig. 23.65

The direction is given by Fleming's left hand rule. Hence force on the loop KLM is

$$F_1 = 2RiB\,\hat{i}$$

Similarly force on the loop KNM is

$$F_2 = 2RiB\,\hat{i}$$

Hence total force on the loop is $\boldsymbol{F} = \boldsymbol{F}_1 + \boldsymbol{F}_2 = 4iRB\,\hat{i}$.

Example 23.31: A current of 10 A flows around a closed path in a circuit which is in the horizontal plane as shown in Fig. 23.66. The circuit consists of eight alternate areas of radii r_1 and r_2 ($r_1 = 0.08$ m and $r_2 = 0.12$ m). Each arc subtends the same angle at the centre (a) Find the magnetic field produced by this circuit at the centre. (*b*) An infinitely long straight wire carrying a current of 10 A is passing through the centre of the above circuit vertically with the direction of the current being into the plane of the circuit. What is the force acting on wire at the centre due to the current in the circuit? What is the force acting on the arc *AC* and the straight segment *CD* due to the current at the centre.

(IIT 2001)

Solution: (*a*) The magnetic field at the centre *O* due to 8 straight portions *AB, CD, ..., OP* is zero because the current element is along radius vector for each of them. Hence ($id\boldsymbol{l} \times \boldsymbol{r}$) is zero.

Magnetic field due to arc elements of radius r_1. The four arcs *AC, EG, IK* and *MO* form a semi-circular loop of radius r_1. Hence magnetic field at the centre is

$$\boldsymbol{B}_1 = \frac{\mu_0 i}{4r_1}\hat{k} \qquad ...(i)$$

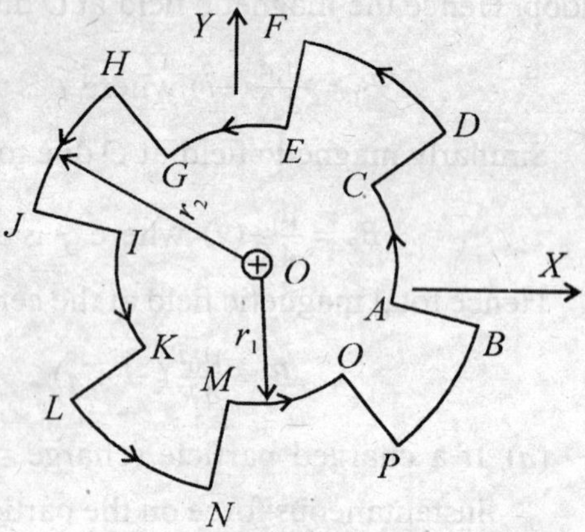

Fig. 23.66

Magnetic field due to arc elements of radius r_2. The four arcs *DF, HJ, LN* and *PB*, form semi-circular loop of radius r_2. Hence magnetic field at the centre is

$$\boldsymbol{B}_2 = \frac{\mu_0 i}{4r_2}\hat{k} \qquad ...(ii)$$

Hence net magnetic field at the centre of the loop is

$$\boldsymbol{B} = \boldsymbol{B}_1 + \boldsymbol{B}_2 = \frac{\mu_0 i}{4}\left(\frac{1}{r_1} + \frac{1}{r_2}\right)\hat{k}$$

Given $\mu_0 = 4\pi \times 10^{-7}$ H/m; $i = 10$ A, $r_1 = 0.08$ m and $r_2 = 0.12$ m. Substituting, we get

$$\boldsymbol{B} = \frac{4\pi \times 10^{-7} \times 10}{4}\left(\frac{1}{0.08} + \frac{1}{0.12}\right)\hat{k} = 6.54 \times 10^{-5}\hat{k}\text{ T}$$

(*b*) The force on a current carrying conductor in magnetic field is given by

$$\boldsymbol{F} = (id\boldsymbol{l}) \times \boldsymbol{B}$$

In present situation the current element $id\boldsymbol{l}$ for straight conductor at centre and magnetic field due to loop are parallel hence force is zero. Similarly the force exerted by the magnetic field due to current through straight conductor at the centre on the element *AC* of the loop is also zero as the field is tangential to the arc *AC*. But the force on part *CD* is not zero as the field everywhere is perpendicular to length *CD* and in the plane of the paper (Fig. 23.67).

Consider a small element *dx* of straight part *CD* at a distance *x* from *O*. Force on this element is ($\boldsymbol{i'dx}$) × ($\boldsymbol{B}$)

Magnetic field $\boldsymbol{B}$ at the distance *x* from current carrying conductor is

$$B = \frac{\mu_0 i}{2\pi x}$$

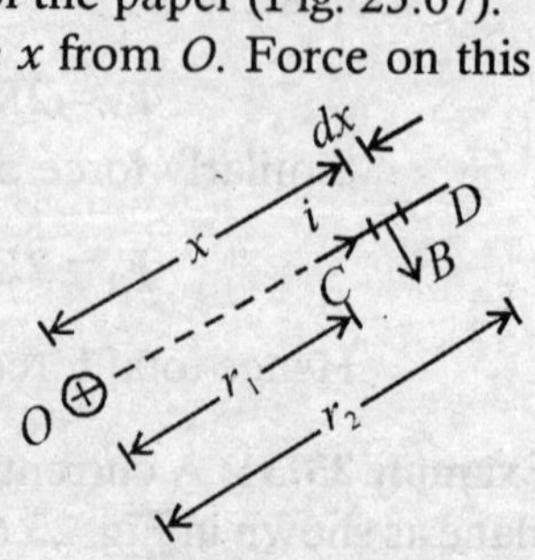

Fig. 23.67

Hence magnitude of the force on element *dx* is

$$dF = \frac{\mu_0 i i'}{2\pi}\frac{dx}{x}$$

$\therefore$ Total force $F = \dfrac{\mu_0 i i'}{2\pi}\displaystyle\int_{r_1}^{r_2}\dfrac{dx}{x} = \dfrac{\mu_0 i i'}{2\pi}\log\dfrac{r_2}{r_1}$

The direction of this force using Fleming's left hand rule is negative *Z*-direction. Hence

$$\boldsymbol{F} = \frac{\mu_0 i i'}{2\pi}\log\frac{r_2}{r_1}(-\hat{k})$$

Substituting the values, we get

$$\boldsymbol{F} = 2\times10^{-7}\times10\times10\times\log_e\frac{0.12}{0.08}(-\hat{k})$$
$$= -8.11\times10^{-6}(\hat{k})$$

Example 23.32: A rectangular loop *PQRS* made from a uniform wire has length *a* and width *b* and mass *m*. It is free to rotate about arm *PQ*. Which remains hinged along horizontal line taken as the *Y*-axis (Fig. 23.68). Take the vertically upward direction as *Z*-axis. A uniform magnetic field $\boldsymbol{B} = (3\hat{i} + 4\hat{k})B_0$ exists in the region. The loop is held in *X*-*Y* plane and a current *i* is passed through it. The loop is now released and is found to stay in the horizontal position in equilibrium. (*a*) What is the direction of the current *i* in *PQ*? (*b*) Find the magnetic force on the arm *RS* (*c*) Find the expression for *i* in terms of B_0, *a*, *b* and *m*. (IIT 2002)

Solution: (*a*) Referring to the Fig. 23.68 the weight of the loop acts in negative *Z*-direction. Part *PQ* of the loop is hinged parallel to *Y*-axis. Hence force on part *RS* must be in positive *Z*-direction. Hence the current in part *SR* must be from *R* to *S*. (From Fleming's left hand rule). Hence current in *PQ* must be along positive $y(=+\hat{j})$ direction.

(*b*) Since the current in part *RS* is along $(-\hat{j})$ direction, hence force on this is

$$\boldsymbol{F} = (i\boldsymbol{l})\times\boldsymbol{B} = ib(-\hat{j})\times(3\hat{i}+4\hat{k})B_0$$
$$= 3\,ibB_0\,\hat{k} - 4\,ibB_0\hat{i} = ibB_0(3\hat{k}-4\hat{i})$$

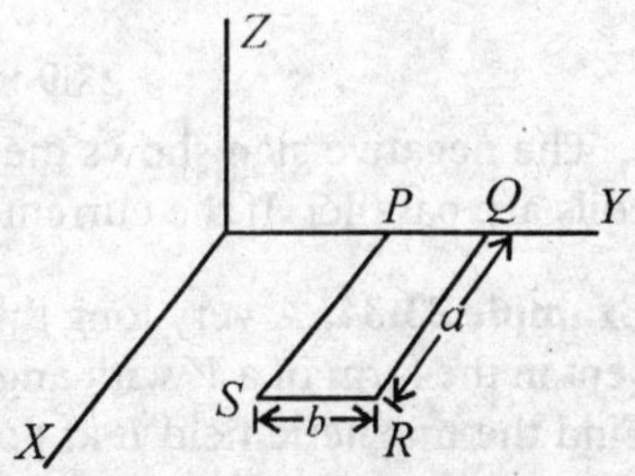

Fig. 23.68

(*c*) The torque τ_1 due to this force about the *Y*-axis (hinge) is

$$\tau_1 = \boldsymbol{r}\times\boldsymbol{F}$$

where $r = a\hat{i}$ is the position vector of arm *SR*.

$$\therefore \quad \tau_1 = (a\hat{i})\times ibB_0(3\hat{k}-4\hat{i}) = -3\,iabB_0\hat{j}$$

Now torque due to the weight of the loop is

$$\tau_2 = \boldsymbol{r}\times\boldsymbol{F}$$

Here $\quad r = \dfrac{a}{2}\hat{i}$ and $\boldsymbol{F} = mg(-\hat{k})$

$$\therefore \quad \tau_2 = \frac{a}{2}\hat{i}\times mg(-k) = \frac{mga}{2}\hat{j}$$

For equilibrium $\tau_1 = \tau_2$ in magnitude.

$$3\,iabB_0 = \frac{mag}{2} \text{ or } i = \frac{mg}{6\,B_0 b}$$

Example 23.33: Two coaxial loops of radii 0.5 m and 0.05 m are separated by a distance 0.5 m and carry currents 2 A and 1 A respectively. What is the force between the loops?

Solution: As shown in Fig. 23.69, the magnetic field at the centre of coil A (O_2) due to current in coil B is

$$B = \frac{\mu_0 \times 2\pi b^2 i_1}{4\pi(b^2 + x^2)^{3/2}}(\hat{i})$$

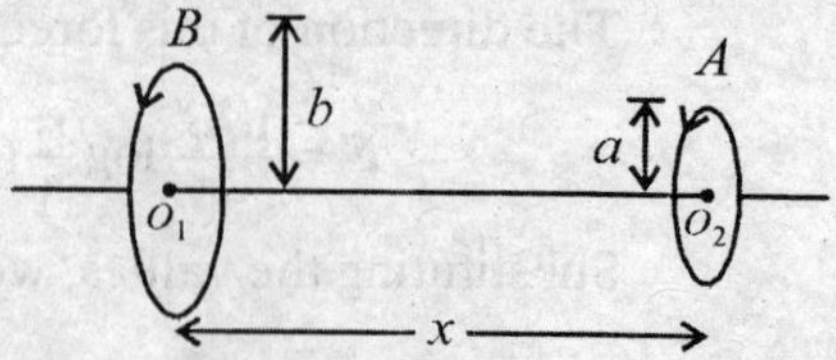

Fig. 23.69

and the magnetic moment of coil A is $M = \pi a^2 i_2(\hat{i})$

It is the case of a dipole of moment M placed in magnetic field B. Hence potential energy of the system is

$$U = -\boldsymbol{M}\cdot\boldsymbol{B} = -\frac{\mu_0 \cdot 2\pi b^2 i_1 \cdot \pi a^2 i_2}{4\pi(b^2 + x^2)^{3/2}}(\hat{i}\cdot\hat{i}) \qquad \hat{i}\cdot\hat{i} = 1$$

$$= -\frac{\mu_0 2\pi^2 i_1 i_2 a^2 b^2}{4\pi(b^2 + x^2)^{3/2}}$$

Hence the force between the coils is

$$F = -\frac{du}{dx} = -\frac{\mu_0}{4\pi}\frac{2\pi^2 \cdot i_1 i_2 a^2 b^2}{(b^2 + x^2)^{5/2}} \times 3x$$

$$= -\frac{\mu_0}{4\pi}\cdot\frac{6\pi^2 i_1 i_2 a^2 b^2}{(b^2 + x^2)^{5/2}}x$$

Substituting the values, we get

$$F = -\frac{10^{-7} \times 6\pi^2(2)(1)(.5)^2(.05)^2(.5)}{\left[(0.5)^2 + (0.5)^2\right]^{5/2}}$$

$$= -2.09 \times 10^{-7}\text{ N}$$

The negative sign shows the force is attractive. It is due to the fact that the currents in the two coils are parallel. If the current in the two coils is opposite the force will be one of the repulsion.

Example 23.34: A very long thin wire carrying a current i is bent in the form of a V with angular half span α (Fig. 23.70). Find the magnetic field B at points P and Q.

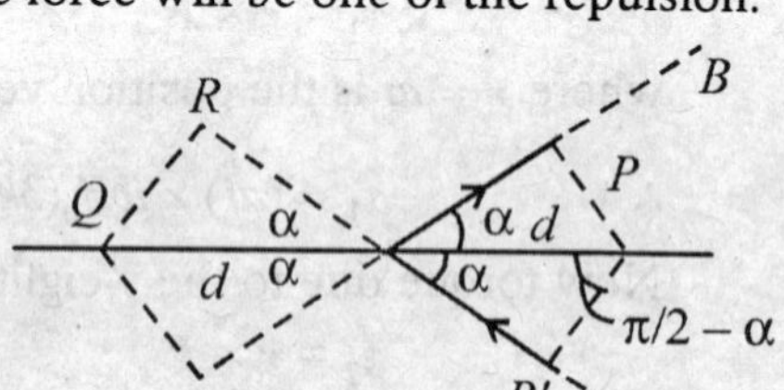

Fig. 23.70

Solution: For point P for symmetry reasons, the total field is twice that generated by each leg and has still the same direction.

For a finite length of current carrying wire, the magnetic field at distance a from the wire is (Fig. 23.71)

$$B = \frac{\mu_0 i}{4\pi}\frac{(\sin\theta_1 + \sin\theta_2)}{a}$$

In Fig. 23.70 for portion AO, $a = d\sin\alpha$, $\theta_1 = \frac{\pi}{2} - \alpha$ and $\theta_2 = \frac{\pi}{2}$

$$\therefore \quad B = \frac{\mu_0 i}{4\pi} \frac{\left[\sin\frac{\pi}{2} + \sin\left(\frac{\pi}{2} - \alpha\right)\right]}{d \sin\alpha} = \frac{\mu_0 i}{4\pi} \frac{(1 + \cos\alpha)}{d \sin\alpha}$$

$$= \frac{\mu_0 i}{4\pi d} \cot\frac{\alpha}{2}$$

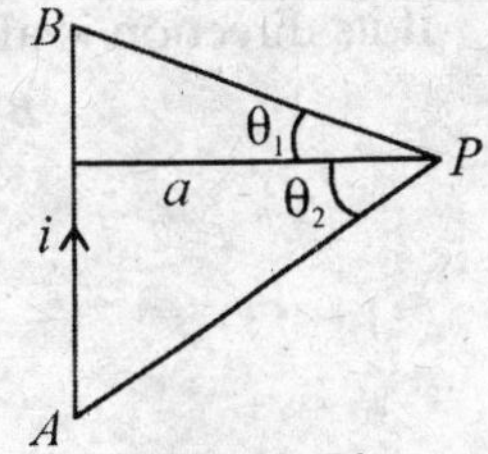

Fig. 23.71

Hence B at P due to both legs AO and OB is $= \dfrac{\mu_0 i}{2\pi d} \cot\dfrac{\alpha}{2}$

If the plane of diagram is taken as X-Y plane, then this field is in negative Z-direction.

For point Q, magnetic field due to portion OA is equal to the field due to AR minus the field due to portion OR.

Field at Q due to AR is, $\theta_1 = 0$, $\theta_2 = \pi/2$, $a = d\sin\alpha$

$$B_1 = \frac{\mu_0 i}{4\pi} \frac{(\sin 0 + \sin\frac{\pi}{2})}{d \sin\alpha} = \frac{\mu_0 i}{4\pi d \sin\alpha}$$

and is in positive Z-direction and for portion OR, $\theta_1 = 0$, $\theta_2\ \pi/2 - \alpha$ and $a = d\sin\alpha$

$$B_2 = \frac{\mu_0 i}{4\pi} \frac{\left[\sin 0 + \sin\left(\frac{\pi}{2} - \alpha\right)\right]}{d \sin\alpha} = \frac{\mu_0 i}{4\pi} \frac{(\cos\alpha)}{d \sin\alpha}$$

This has also the direction, positive Z-direction. Hence net field at Q due to leg AO is

$$B_1 - B_2 = \frac{\mu_0 i}{4\pi d \sin\alpha} - \frac{\mu_0 i \cos\alpha}{4\pi d \sin\alpha} = \frac{\mu_0 i}{4\pi d} \frac{(1 - \cos\alpha)}{\sin\alpha}$$

$$= \frac{\mu_0 i}{4\pi d} \tan\alpha/2 \text{ in positive } Z\text{-direction.}$$

For symmetry reasons the total field is twice that generated by each leg and has still the same direction. Hence B at Q is

$$B = \frac{\mu_0 i}{2\pi d} \tan\alpha/2 \text{ in positive } Z\text{-direction}$$

Example 23.35: Infinite number of straight wires each carrying a current i are equally placed as shown in Fig. 23.72. Adjacent wires have current in opposite direction. Find the net magnetic field at point R.

Solution: Magnetic field due to a current carrying straight wire is

$$B = \frac{\mu_0 i}{4\pi d} (\sin\theta_1 + \sin\theta_2)$$

Referring to Fig. 23.73 $d = a\cos 30° = \dfrac{\sqrt{3}a}{2}$

and $\quad B = \dfrac{\mu_0 i}{4\pi d} (\sin 30° + \sin 30°) = \dfrac{\mu_0 i}{4\pi d}$

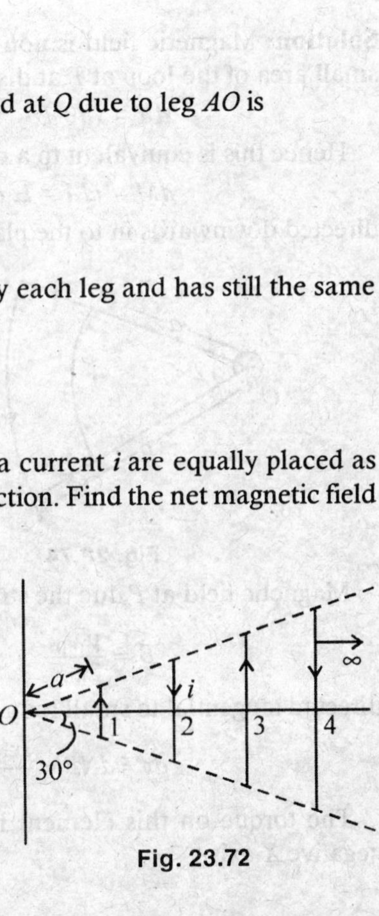

Fig. 23.72

If its direction is along positive Z-direction, net field at O due to all wires is

$$\mathbf{B} = \mathbf{B}_1 + \mathbf{B}_2 + \mathbf{B}_3 + \ldots$$

$$= \frac{\mu_0 i}{4\pi d}\left(\hat{k} - \frac{\hat{k}}{2} + \frac{\hat{k}}{3} - \frac{\hat{k}}{4} + \ldots + \infty\right)$$

$$= \frac{\mu_0 i}{4\pi d}\left(1 - \frac{1}{2} + \frac{1}{3} - \frac{1}{4} + \ldots + \infty\right)\hat{k}$$

$$= \frac{\mu_0 i}{4\pi d}\log_e 2$$

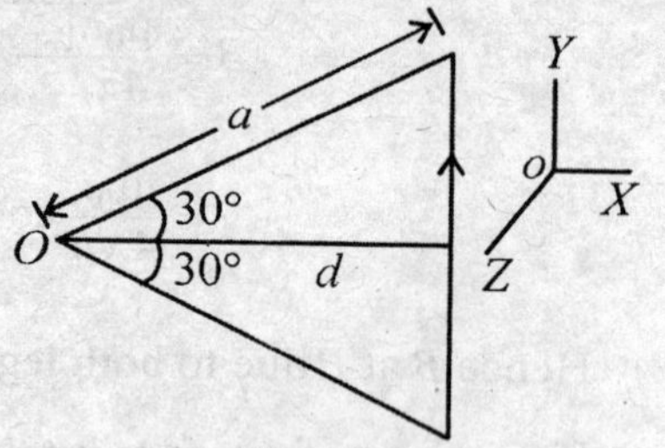

Fig. 23.73

Substituting the value of d, we get

$$\mathbf{B} = \frac{\mu_0 i}{4\pi \frac{\sqrt{3}}{2} a}\log 2 = \frac{\mu_0 i \log_e 2}{2\sqrt{3}\pi a}$$

Example 23.36: A loop with current i is in the field of a long straight wire with current i_0. The plane of the loop is perpendicular to the straight wire (Fig. 23.74). Find the moment of the Ampere's forces acting on the loop.

Solution: Magnetic field is non-uniform and varies from point to point on the loop. Consider a small area of the loop at P at distance x from O. Referring to Fig. 23.75, the area of the loop is

$$dA = x d\theta\, dx$$

Hence this is equivalent to a dipole of moment

$$dM = idA = ix\, d\theta\, dx \qquad \ldots(i)$$

directed downwards in to the plane of paper.

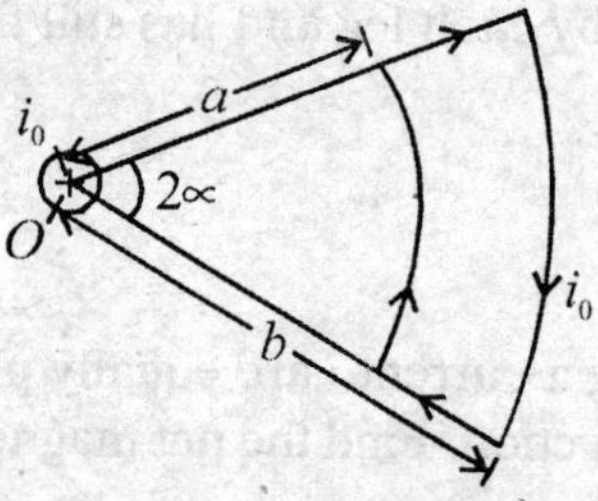

Fig. 23.74

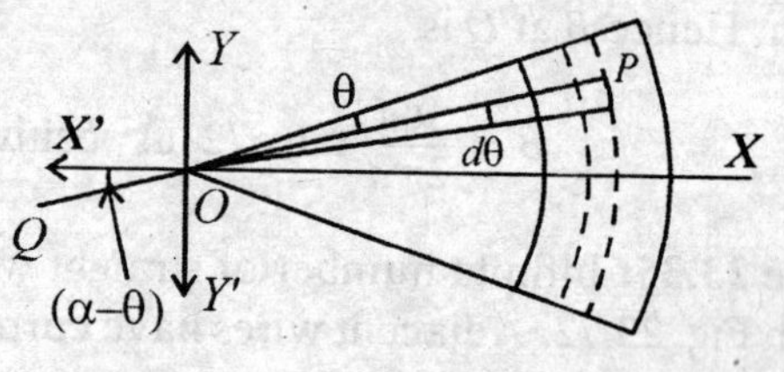

Fig. 23.75

Magnetic field at P due the straight current carry conductor at O is

$$B = \frac{\mu_0 i_0}{2\pi x} \qquad \ldots(ii)$$

directed tangential to small loop at P. Hence torque on magnetic dipole is

$$d\tau = dNB = \frac{ix\, d\theta\, dx\, \mu_0 i_0}{2\pi x} = \frac{\mu_0 i i_0}{2\pi} d\theta \cdot dx$$

The torque on this element is directed along OQ (Fig 23.75) making an angle $\alpha - \theta$ with negative X-axis.

Component of torque along Y-axes averages out to zero because of the symmetry of loop about $X'OX$. The only component is

$$d\tau \cos(\alpha - \theta) = \frac{\mu_0 i i_0}{2\pi} \cos(\alpha - \theta)\, d\theta \cdot dx$$

Hence total torque on the loop is

$$\tau = \int d\tau \cos(\alpha - \theta) = \int_a^b \int_0^{2\alpha} \frac{\mu_0 i i_0}{2\pi} \cos(\alpha - \theta)\, d\theta \cdot dx$$

$$= \frac{\mu_0 i i_0}{2\pi} \left[x\right]_a^b \int_0^{2\alpha} \cos(\alpha - \theta)\, d\theta = \frac{\mu_0 i i_0}{2\pi}(b - a) \int_0^{2\alpha} \cos(\alpha - \theta)\, d\theta$$

$$= \frac{\mu_0 i i_0}{2\pi}(b - a)\left[-\sin(\alpha - \theta)\right]_0^{2\alpha}$$

$$= \frac{\mu_0 i i_0}{2\pi}(b - a)\left[-\sin(\alpha - 2\alpha) + \sin(\alpha - 0)\right]$$

$$= \frac{\mu_0 i i_0 (b - a)}{2\pi} \times 2\sin\alpha = \frac{\mu_0 i i_0 (b - a)\sin\alpha}{\pi}$$

PROBLEMS

1. A positive charge q and mass m kept at a distance x_0 from a fixed very long straight current carrying wire is projected normally away from it with a speed of v. Find the maximum separation between the wire and the particle. **[Ans.** $x_m = x_0 e^{2\pi m v/\mu_0 q i}$]

2. A non-conducting hollow sphere of radius charged uniformly with surface density σ rotates with an angular velocity ω about the axis passing through its centre. Find the magnetic induction at its centre.

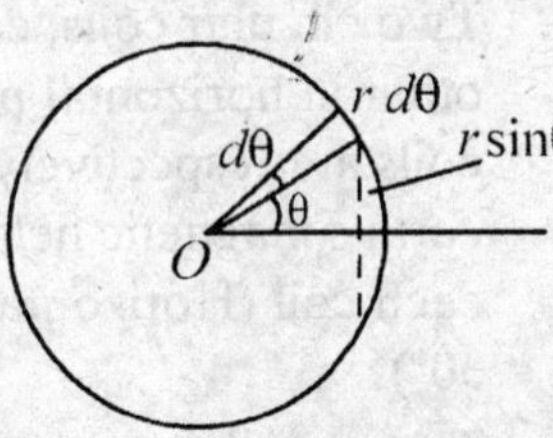

Fig. 23.76

[**Hint:** Hollow sphere can be treated as made up of rings. One such ring of radius $r\sin\theta$ and width $rd\theta$ is shown in Fig. 23.76. Charge on it is $\sigma 2\pi \sin\theta\, r d\theta = 2\pi r^2 \sigma \sin\theta\, d\theta$. Since it is rotating with frequency $\frac{\omega}{2\pi}$ hence it is equivalent to current loop of current

$i = \frac{\omega}{2\pi} dq = r^2 \sigma \omega \sin\theta\, d\theta$. Field at the centre is $dB = \frac{\mu_0 i a^2}{2(a^2 + x^2)^{3/2}} = \frac{\mu_0 i r^2 \sin^2\theta}{2r^3}$ or

$dB = \frac{\mu_0 r^4 \sigma\omega \sin^3\theta \cdot d\theta}{2r^3} = \frac{\mu_0 r \sigma\omega}{2} \sin^3\theta\, d\theta$. Integeration gives $B = 2/3 \mu_0 \sigma \omega r$]

3. Two wires P and Q each of length 64 cm and carry a current of 5.0 A. Wire P is bent into a circle and wire Q into a square. Find the magnitude of the magnetic field (a) at the centre of circle and (b) at the centre of the square which current loop produces a greater magnetic field at the centre?

[**Hint:** A circular loop of radius a produces magnetic field at the centre, $B_c = \frac{\mu_0 i}{2r}, 2\pi r = l$.

This gives $B_c = 3.08 \times 10^{-5}$ T. For square loop of side a, $B_s = 4 \cdot \frac{\mu_0 i}{4\pi} \frac{(\sin 45° + \sin 45°)}{a/2} = \frac{2\sqrt{2}\mu_0 i}{\pi a}$.

Here $a = 0.16$ m. Thus $B_s = 3.53 \times 10^{-5}$ T. The magnetic field at the centre of the square is greater]

4. A small rectangular loop of sides 5.0 cm and 3.0 cm carries a current of 7.0 A. It is placed with its longer side parallel to a long straight conductor of length 5.0 cm at a distance of 2.0 cm from it. If the current in the long conductor is 20 A, find the net force on the loop. Is the loop attracted towards, or repelled away from the long conductor?

[**Ans.** $F = 4.\dot{2} \times 10^{-5}$ N, attracted towards the long conductor]

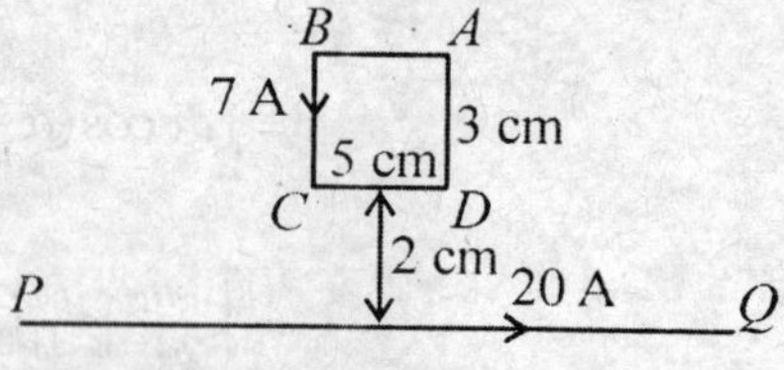

Fig. 23.77

5. An electron of energy 10,000 eV describes a circular path in a plane at right angles to uniform magnetic field 0.01 T. (*i*) What is the radius of circular orbit? (*ii*) What is the cyclotron frequency? (*iii*) What is the period of revolution? (*iv*) What is the direction of revolution as viewed by an observer in the direction of the field.

[**Ans.** (*i*) $r = 3.4 \times 10^{-2}$ m (*ii*) $f = 2.8 \times 10^{8}$ Hz (*iii*) $\tau = 3.6 \times 10^{-9}$ sec (*iv*) The electron revolves in a clockwise sense]

6. A particle of mass 1×10^{-26} kg and charge $+1.6 \times 10^{-19}$ C travelling with velocity 1.28×10^{6} m s^{-1} in positive X-direction enters a region in which a uniform electric field E and uniform magnetic field of induction B are present such that $E_x = E_y = 0$ and $E_z = -102.4$ kVm^{-1} and $B_x = B_z = 0$, $B_y = 8 \times 10^{-2}$ T. The particle enters this region at time $t = 0$. Determine the location (X, Y and Z co-ordinates) of the particle at $t = 5 \times 10^{-6}$ s. If the electric field is switched off at this instant (with magnetic field still present). What will be the position of the particle at $t = 7.45 \times 10^{-6}$ s?

[**Ans.** (*a*) $x = 6.4$ m, $y = 0$, $z = 0$. (*b*) $x = 6.4$ m, $y = 0$, $z = 2.0$ m]

7. Two circular coils, each of 100 turns, are held such that one lies in the vertical plane and the other in horizontal plane, with their centres coinciding. The radii of vertical and horizontal coils are respectively 20 cm and 30 cm. The directions of current in them are such that the earth's magnetic field at the centre of the coils is exactly neutralized. Calculate the current in each coil (Horizontal component of earth's magnetic field = 27.8 Am^{-1} and angle of dip = 30°).

[**Ans.** $i_1 = 0.1112$ A, $i_2 = 0.0963$ A]

8. Two straight and infinitely long and then parallel wires are spaced 0.1 m apart and carry a current of 10 A. Find the magnetic field at a point 0.1 m from both wires in the two cases when the currents are in the (*a*) same direction (*b*) opposite direction.

[**Ans.** (*a*) $B = 3.46 \times 10^{-5} (-\hat{j})$ T (*b*) $B = 2 \times 10^{-5}$ $(\hat{i})$ T]

9. An electron gun G emits electrons of energy 2 eV travelling in the positive X-direction. The electrons are required to hit the spot S where $GS = 0.1$ m and the line GS makes an angle $\theta = 60°$ with X-axis (Fig. 23.78). A uniform magnetic field B parallel to GS exists in the region outside the electron gun. Find the minimum value of B needed to make electron hit S.

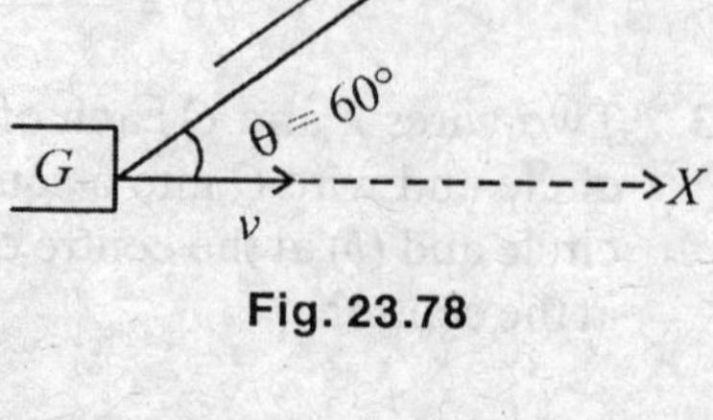

Fig. 23.78

[**Hint:** The electron moves along helical path with line GS as axis. The electron hits the target at S_1 if $v\cos\theta\ T$ (pitch of the helix) = GS, where T is the period of revolution of the electron.

$T = \dfrac{2\pi m}{eB}$ and $v = \sqrt{\dfrac{2k}{m}}$. Substituting values $B = 7.547 \times 10^{-9}$ T]

10. A beam of charged particles, having K.E. = 10^3 eV containing particles of masses 8×10^{-27} kg and 1.6×10^{-26} kg emerges from one end of an accelerator tube. There is a plate at a distance 10^{-2} m from the end of the tube and placed perpendicular to the beam. Calculate the magnitude of the smallest magnetic field which can prevent the beam from striking the plate.

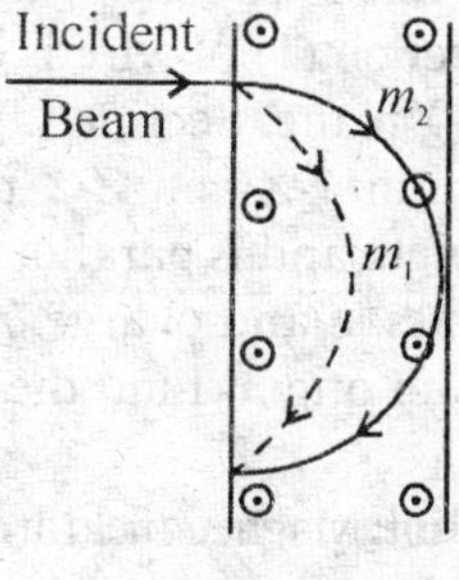

Fig. 23.79

[**Hint:** Radius of circular path is $r = \dfrac{mv}{qB} = \dfrac{\sqrt{2mE}}{qB}$. Greater the mass greater is radius and smaller is the field required. For mass (8×10^{-27} kg) if radius $r (= 10^{-2}$ m) the separation between plates,

then required $B = \sqrt{2mE}\Big/qr = 1.41$ T. This is the minimum field required]

11. Centres of two circular coils P and Q having the same number of turns are located at coordinates (3, 0, 0) and (0, 4, 0) such that the planes of the coils are perpendicular to X and Y axes respectively and areas of the coils Q and P are in the ratio 4 : 3. Coil P has $9\sqrt{3}$ A current in anticlockwise and coil Q has a current 16 A in clockwise direction as seen from the origin. A small compass needle is placed at the origin. Find the deflection of the needle assuming the earths magnetic field to be negligible and radii of the two coils very small compared to their distances from the origin.

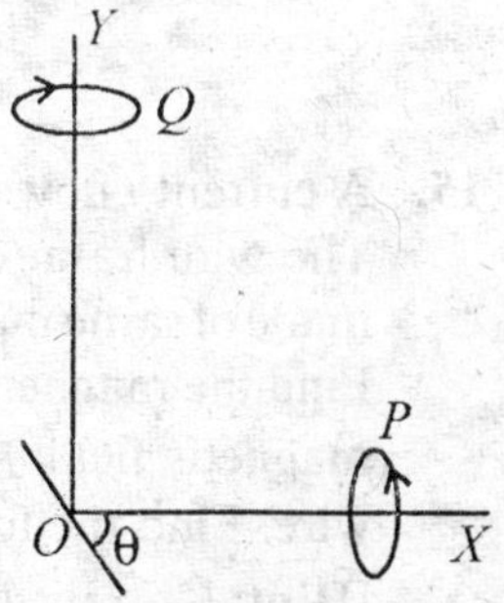

Fig. 23.80

[**Hint:** $\tan\theta = \dfrac{B_p}{B_q}$, $\boldsymbol{B}_p = \dfrac{\mu_0 N i_p r_p^2}{2x^3}(-\hat{i})$ and $B_q = \dfrac{\mu_0 N i_q r_q^2}{2y^3}(\hat{j})$

$$\therefore \quad \tan\theta = \frac{i_p r_p^2}{x^3}\frac{y^3}{i_q (r_q)^2} = \frac{9\sqrt{3}}{(3)^2}\left(\frac{3}{4}\right)\frac{(4)^3}{16} = \sqrt{3} \text{ or } \theta = 60^\circ$$]

12. A wire loop consists of a straight segment AB and a circular arc ACB of radius r. The segment AB subtends an angle 60° at the centre O of the circular arc. The wire loop carries a current i in the clockwise direction (Fig. 23.81). Find the magnitude and direction of the magnetic field at O due to the current in wire loop.

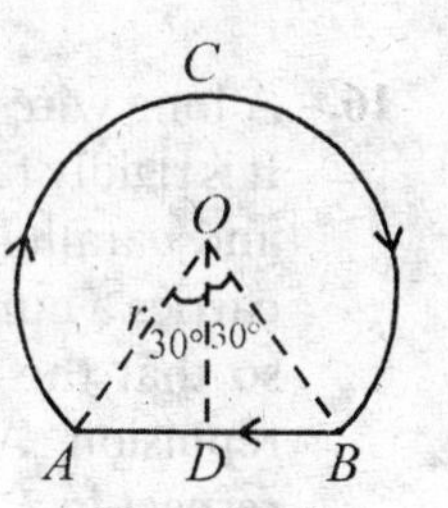

Fig. 23.81

[**Hint:** Magnetic field at O due to straight portion $B_1 = \dfrac{\mu_0 i}{4\pi}\dfrac{(\sin 30 + \sin 30)}{OD}$.

But $OD = r\cos 30$. Hence $B_1 = \dfrac{\mu_0 i}{4\pi r\sqrt{3}/2} = \dfrac{\mu_0 i}{2\sqrt{3}\pi r} \oplus$. Field at O due to

circular arc $B_2 = \dfrac{\mu_0 i}{2r}\dfrac{5}{6} = \dfrac{5\mu_0 i}{12r} \oplus$. Hence total field $= B = B_1 + B_2 = \dfrac{\mu_0 i}{12r}\left[5 + 2\sqrt{3}\right] \oplus$]

13. A particle of mass 10^{-26} kg and charge 1.6×10^{-19} C is travelling with velocity 1.28×10^6 m/s along the X-direction. It enters at $t = 0$, a region in which uniform electric and magnetic fields are present with $E_x = E_y = 0$ and $E_z = -102.4$ kV/m and $B_x = B_y = 0$ and $B_y = 8 \times 10^{-2}$ T. Determine the θ-ordinates of the particle at $t = 5 \times 10^{-6}$ s. If the electric field is switched off at this instant. What will be the position of the particle at $t = 7.45 \times 10^{-6}$ s?

[Ans. (6.4, 0, 0), (6.4, 0, 2)]

14. A fixed wire carrying current i has shape shown in Fig. 23.82. Section (1) of wire is of infinite length and is parallel to X-axis in X-Y plane. Section (2) and (3) are quarter circular of radius r and are in Y-Z and Z-X plane respectively. Section (4) is of infinite length and is parallel to Y-axis in X-Y plane. A charged particle of mass m and charge q is projected with velocity v_0 along $+X$-axis from origin. Find the instantaneous acceleration at O.

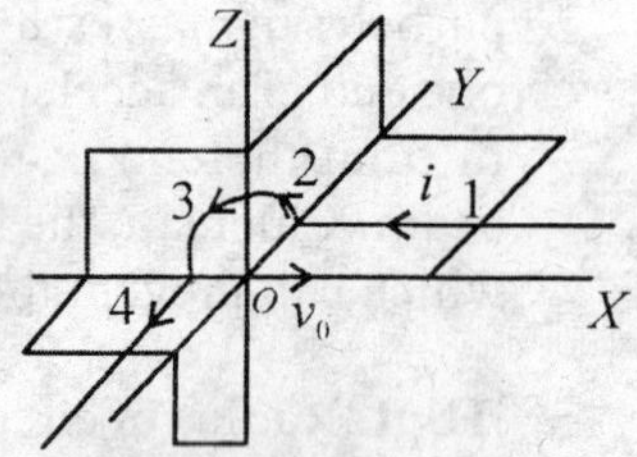

Fig. 23.82

[Hint: Magnetic field at origin is $\boldsymbol{B} = \frac{\mu_0 i}{4\pi r}\hat{k} + \frac{\mu_0 i}{8r}\hat{i} + \frac{\mu_0 i}{8r}(-\hat{j}) + \frac{\mu_0 i}{4\pi r}\hat{k}$.

$\boldsymbol{B} = \frac{\mu_0 i}{2\pi r}\left[\frac{\pi}{4}\hat{i} - \frac{\pi}{4}\hat{j} + \hat{k}\right]$ and $\boldsymbol{v} = v_0\hat{i}$. Hence $\boldsymbol{F} = q\boldsymbol{v} \times \boldsymbol{B}$ ∴ Acceleration $= F/m = \frac{q}{m}\boldsymbol{v} \times \boldsymbol{B}$

$$= -\frac{q\mu_0 i v_0}{2\pi r m}\left(\hat{j} + \frac{\pi}{4}\hat{k}\right)\Bigg]$$

15. A current i flows through a wire as shown in Fig. 23.83. The wire frame lies in the X-Y plane. All the wires are made of same material and have same cross-section area. Find the magnetic field at C (center) due to $N'Q'$. (*ii*) If a magnetic field $B_0\hat{k}$ is switched on in space enclosed by wire. Find the force on section NPQ.

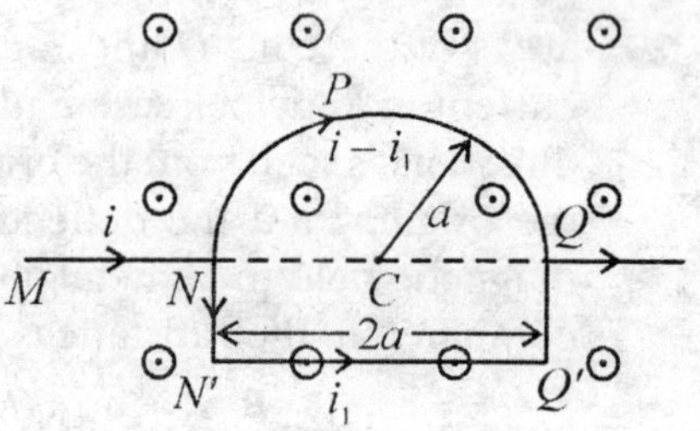

Fig. 23.83

[Hint: Current i at N divides such that $\pi a(i - i_1) = 4ai_1$

or $i_1 = \frac{\pi}{\pi + 4}i$. Hence magnetic field at C due to $N'Q'$

$= \frac{\mu_0 i_1}{4\pi a}(\sin 45° + \sin 45°) = \frac{\sqrt{2}\mu_0}{4a(\pi + 4)}\hat{k}$. Current through $NPQ = \frac{4}{4 + \pi}i$

and Force $= 2aiB = \frac{8aiB}{4 + \pi}(-\hat{j})\Big]$

16. A long wire P placed horizontally carries a current of 50 A. It is rigidly fixed. Another fine wire Q is placed directly above and parallel to P. The weight of wire Q is 0.075 Nm^{-1} and carries a current of 25 A. Find the position of wire Q from P so that the wire Q remains suspended due to magnetic repulsion. Also indicate the direction of current in Q with respect to P. **[Ans.** $d = 1/3 \times 10^{-2}$ m]

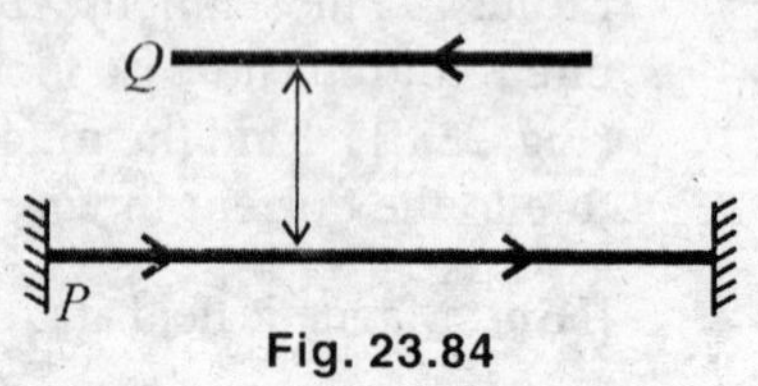

Fig. 23.84

17. A small magnet of magnetic moment $\pi \times 10^{-10}$ A-m^2 is placed on the Y-axis at a distance 0.1 m from the origin with its axis parallel to the X-axis. A coil having 169 turns and radius 0.05 m is placed on the X-axis at a distance of 0.12 m from the origin with the axis of the coil coinciding with the axis. Find the magnitude and direction of the current in the coil for compass needle placed at the origin to point in north-south direction.

[Ans. $i = 2.6 \times 10^{-10}$ A]

18. In Fig. 23.85, a coil of single turn is wound on a sphere of radius a and mass m. The plane of the coil is parallel to the plane and lies in the equatorial plane of the sphere. Current in the coil is i. Find the value of B if the sphere is in equilibrium.

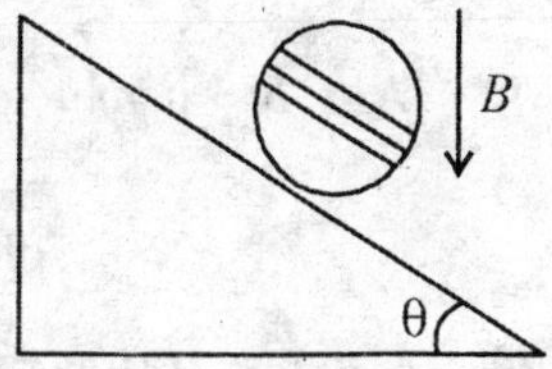

Fig. 23.85

$$\left[\textbf{Ans. } B = \frac{mg}{\pi i R}\right]$$

19. Find the magnitude of the magnetic induction B of a magnetic field generated by a system of thin conductor along which a current i is flowing at a point A (O, R, O). That is the centre of circular conductor of radius R.

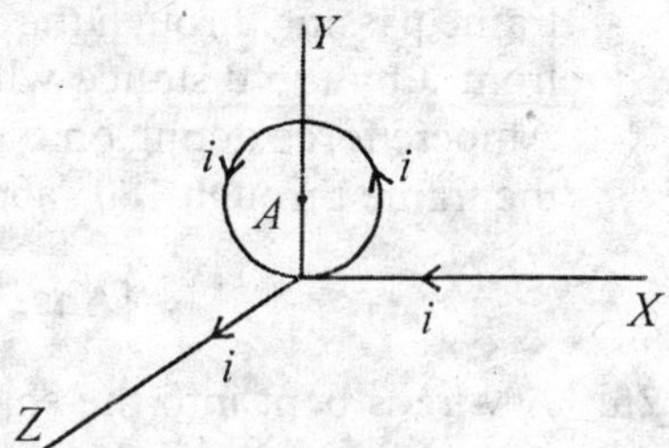

Fig. 23.86

$$\left[\textbf{Ans. } B = \frac{\mu_0 i}{4\pi R}\sqrt{2(2\pi^2 - 2\pi + 1)}\right]$$

20. Three conductors carrying currents are perpendicular to the plane of the drawing. They intersect the plane at three points that lie on a single straight line; with the distance from the middle conductor to the other two being equal. The currents in the outer conductors flow away from the reader, while the currents in the middle conductor flows towards the reader. How is the magnetic field vector directed at the point P on the straight line that is perpendicular to the straight line passing through the three conductors in the plane of the drawing and is separated from the middle conductor by a distance equal to the distances between that conductor and outer conductors. All three currents are equal is magnitude. [**Ans.** $B = 0$]

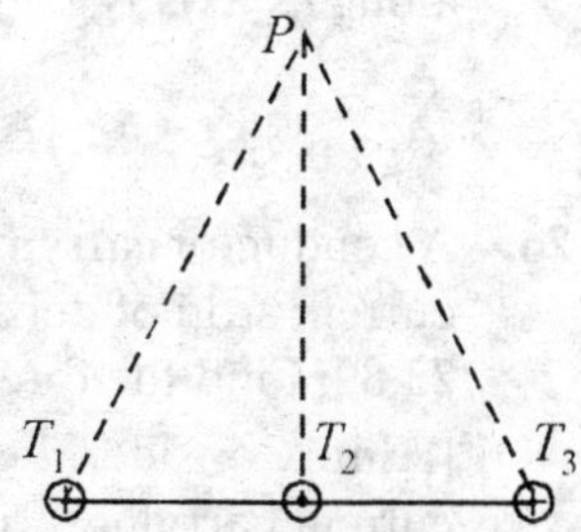

Fig. 23.87

21. A current $i = 6.28$ A flows in a circuit having the shape of an isosceles trapezium. The ratio of the bases of the trapezium is 2. Find the magnetic induction B at point O in the plane of the trapezium. The length of the smaller base of the trapezium is $l = 100$ mm, the distance $b = 50$ mm. [**Ans.** $B = 8.9\pi$ T]

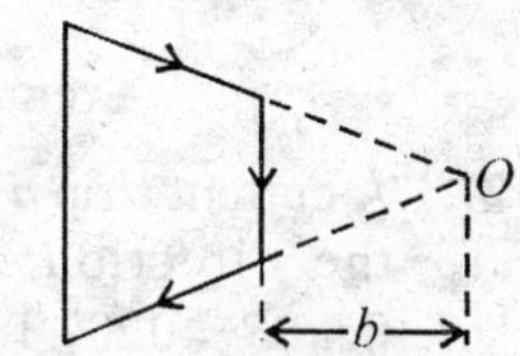

Fig. 23.88

22. A single layer coil (solenoid) has length l and ceross-section radius R. A number of turns per unit length is equal to n. Find the magnetic field induction at the centre of the coil when a current i flows through it.

$$\left[\textbf{Ans. } B = \frac{\mu_0 n i}{\sqrt{1 + \left(\frac{2R}{l}\right)^2}}\right]$$

23. A very long straight solenoid has a cross-section radius R and n turns per unit length. A direct current i flows through the solenoid. Suppose x is the distance from the end of the solenoid measured along its axis. Find

(*a*) the magnetic induction B on the axis as a function of x, draw an appropriate plot of B vs the ratio x/R.

(*b*) the distance x_0 to the point on the axis at which the value of B differs by $\eta = 1\%$ from that in the middle section of the solenoid.

[**Ans.** $B = \mu_0 ni\left(1 - \frac{x}{\sqrt{x^2+R^2}}\right)$ for $x > 0$, for point outside, B falls with x. $x < 0$ point lies inside the solenoid, B increases with x and eventually becomes $\mu_0 ni$ $|x_0| \equiv \frac{(1-2\eta)R}{2\sqrt{\eta(1-\eta)}}$]

24. A square frame carrying a current $i = 0.90$ A is located in the same plane as a long straight wire carrying a current $i_0 = 5.0$ A. The frame side has a length $a = 8.0$ cm. The axis of the frame passing through the midpoints of opposite sides is parallel to the wire and is separated from it by the distance which is $\eta = 1.5$ times greater than the side of the frame. Find (*a*) Ampere force acting on the frame (*b*) the mechanical work to be performed in order to turn the frame through 180° about its axis, with the currents maintained constant.

[**Ans.** Net force $= \frac{2\mu_0 ii_0}{\pi(4\eta^2-1)} = 0.4$ N ; $W = \frac{\mu_0 ii_0 a}{\pi}\log\left(\frac{2\eta+1}{2\eta-1}\right) = 0.1$ μJ]

25. A wire is bent into the shape of a square of edge length l. Show that when the current i is passed in the loop, the magnetic field at point P a distance x from the centre of the square along its axis is given by

[**Ans.** $B = \frac{\mu_0 il^2}{2\pi(x^2+l^2/4)(x^2+l^2/2)^{1/2}}$]

26. A conductor carrying a current i is placed parallel to the current strip of density J_0 and width d as shown in Fig. 23.89. Find the force per unit length on the conductor.

[**Hint:** Consider the strip to be made up of elementary current carrying strips. Consider one such strip at x and of width dx. Current through this is jdx. Due to symmetry X-component of force becomes zero only Z-component remains. Net force is $= \frac{\mu_0 i\, j_0}{\pi}\tan^{-1}\frac{d}{2\pi}(-\hat{k})$]

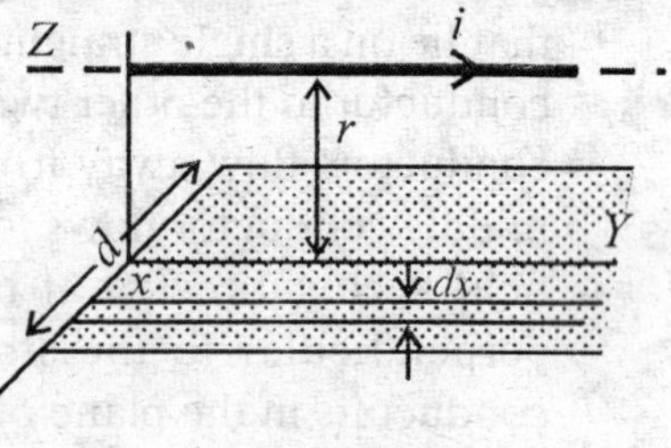

Fig. 23.89

27. A current strip $d = 2$ cm and carrying current $i = 15$ A in the direction of X-axis is placed in a magnetic field $\mathbf{B} = 0.20\hat{j}$ T as shown in Fig. 23.90. Find the force per unit length on the strip. [**Ans.** $\mathbf{F} = 3.0\hat{k}$ N/m]

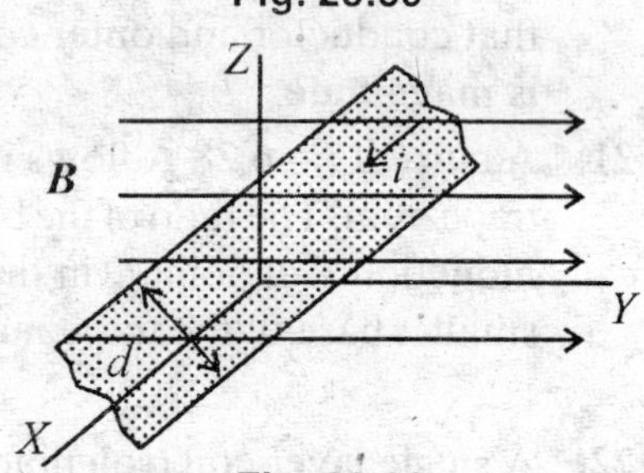

Fig. 23.90

28. Figure 23.91 shows one quarter of a single circular loop of wire carrying a current $i = 14$ A. Its radius $a = 5$ cm. A uniform magnetic field $B = 300$ G and is directed in the $+X$-direction. Find the torque on the entire loop and the direction in which it will rotate.

[**Hint:** The loop is equivalent to a dipole of moment $M = \frac{i\pi a^2}{4}$ perpendicular to the plane of the loop. The direction makes an angle of 30° with Z-axis and 90° with Y-axis and then torque $\mathbf{M} \times \mathbf{B} = \frac{i\pi a^2}{4} B\sin 60°$

$$\tau = \frac{14 \times \pi(25\times10^{-4})\times 300\times10^{-4}}{4}\frac{\sqrt{3}}{2} = 7.14\times10^{-4} \text{ N-m}$$]

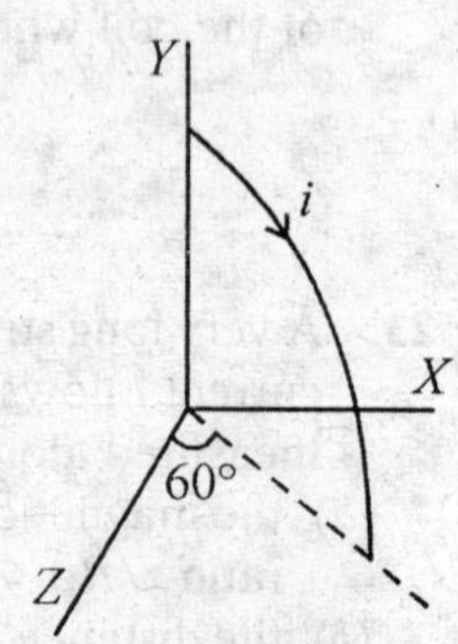

Fig. 23.91

29. Inside long straight uniform wire of round cross-section, there is a long round cylindrical cavity whose axis is parallel to the axis of wire and displaced from the latter by $\boldsymbol{l}$ (Fig. 23.92). A direct current of density j flows along the wire. Find the magnitude of magnetic induction inside the cavity. Consider the particular case $\boldsymbol{l} = 0$.

[**Hint:** We can think of given current which will be assumed uniform, as arising due to a negative current, flowing in the cavity, superimposed on true current everywhere including cavity $\boldsymbol{B} = \frac{1}{2}\mu_0 \boldsymbol{j} \times (\boldsymbol{OP} + \boldsymbol{PO'}) = \frac{1}{2}\mu_0 \boldsymbol{j} \times \boldsymbol{l}$ for $\boldsymbol{l} = 0$, the cavity is concentric with the conductor. Hence $B = 0$]

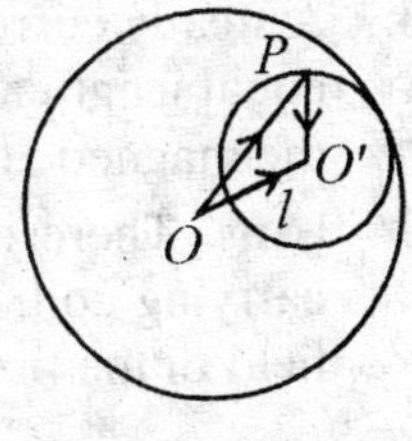

Fig. 23.92

30. A ring of mass m and radius r is rotated in a magnetic field B which is perpendicular to the plane of the loop with constant angular velocity ω_0. Find the net Ampere force on the ring and the tension developed in the ring if there is a current in the ring.

[**Hint:** (Component of tension – magnetic force) = centrepetal force $\int_0^{\pi} 2T d\theta - \int_0^{2\pi} iB \cdot r d\theta = m\omega_0^2 r$ or $2\pi T - iBr \cdot 2\pi = \omega_0^2 mr$ or

$$T = \frac{r}{2\pi}(m\omega_0^2 + 2\pi iB)]$$

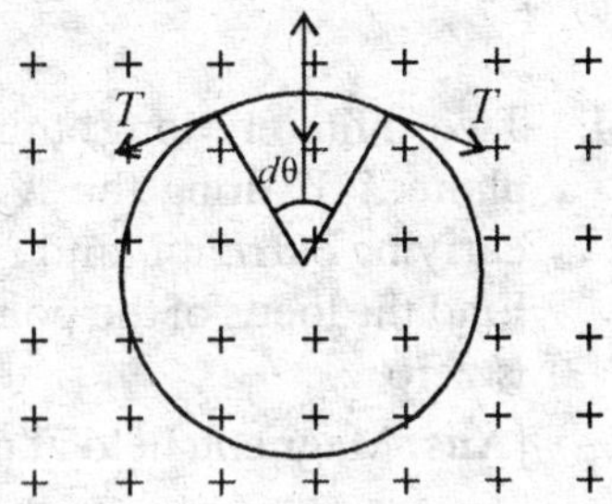

Fig. 23.93

SHORT ANSWER TYPE QUESTIONS

1. A vertical straight long wire carries a current vertically upwards. Points P and Q lie respectively to the east and west of the conductor at the same distance from it. Is the field at P less or more than that Q?

[**Ans.** Field B at points P and Q due to the current through the wire is same in magnitude but opposite in direction. B_H earth's horizontal component at both points is same (Fig. 23.94). Hence net field at Q is less than that at P]

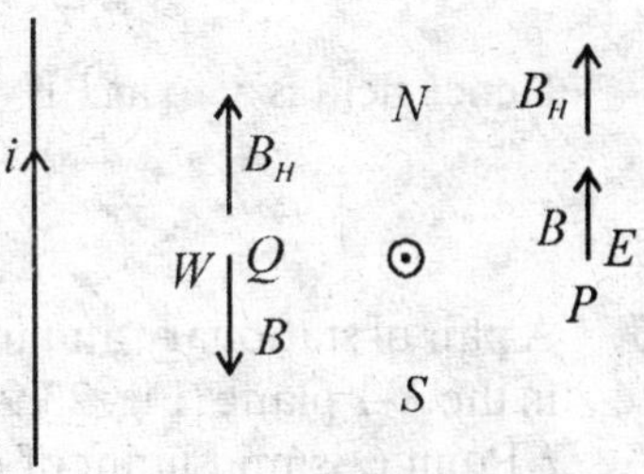

Fig. 23.94

2. Equal currents are flowing through two infinitely long parallel wires. Point P is exactly half way between the wires. What will be the magnetic field at point P when the currents through them are (*a*) in same direction (*b*) in opposite direction?

[**Ans.** Figure 23.95 (*a*) shows two current carrying wires, with currents in same direction. Magnetic field at P due to the current through two wires is same in magnitude $\left(= \frac{\mu_0 i}{2\pi d}\right)$ and opposite in direction. Hence net field at P is zero. Fig. 23.95 (*b*) shows the situation in which currents are in opposite direction. The fields at B are in same direction downwards and magnitudes $\left(= \frac{\mu_0 i}{2\pi d}\right)$ are equal. Hence net field is $\frac{2 \times \mu_0 i}{2\pi d} = \frac{\mu_0 i}{\pi d}$]

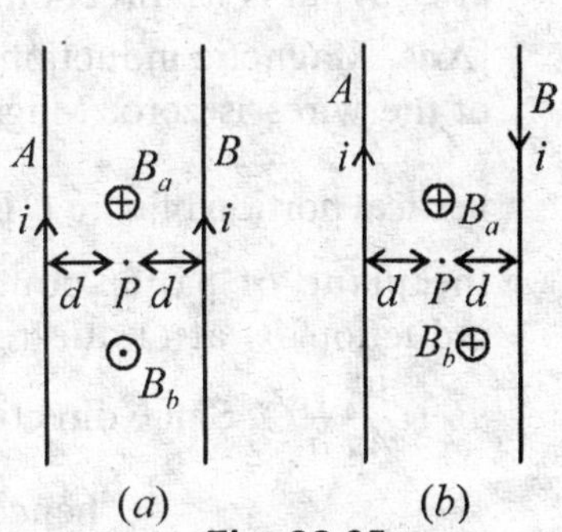

Fig. 23.95

3. A steady current is set up in a network composed of wires of equal length and resistances as shown in Fig. 23.96. What is the magnetic field at the centre *P*?

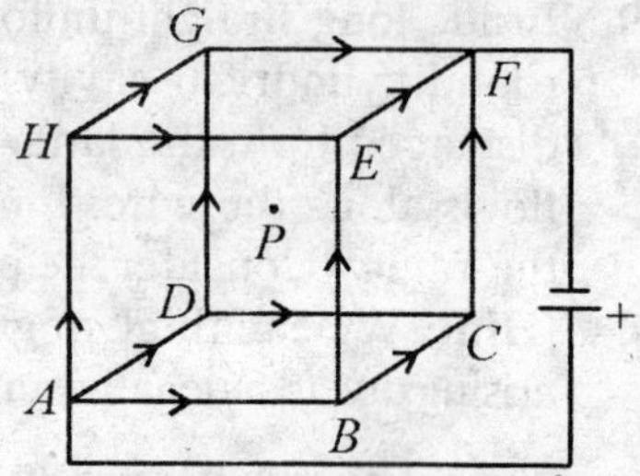

Fig. 23.96

[**Ans.** There is symmetry in the problem. For each current carrying coil, there is another wire such that it cancels the field of the first, e.g., the field at *P* due to *AB* is cancelled by the field due to *GF* because product *il*, $\left[B = \dfrac{\mu_0\, il(\sin\alpha + \sin\beta)}{4\pi r^2}\right]$ is same for both of them. Thus the resultant field at *P* due to the given network is zero]

4. Two infinitely long, thin, insulated straight wires lie in the plane *X*-*Y* along the *X* and *Y*-axis respectively. The wires carrying current i_1 and i_2 respectively as shown in Fig. 23.97. Find the locus of the points in this plane where magnetic field is zero.

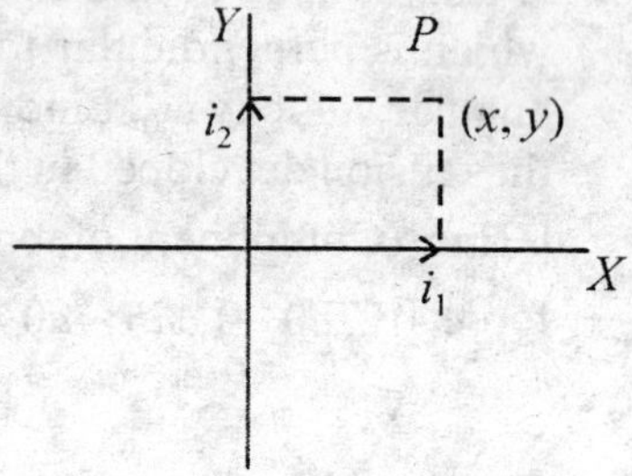

Fig. 23.97

[**Ans.** Magnetic field at point *P* due to i_1 current carrying wire is $B_1 = \dfrac{\mu_0 i_1}{2\pi y}$ in the direction perpendicular to the plane of diagram upwards. Magnetic field at point *P* due to i_2 current carrying wire is $B_2 = \dfrac{\mu_0\, i_2}{2\pi x}$, in the direction perpendicular to the plane of diagram downwards.

Hence field is zero at *P* if $\dfrac{\mu_0\, i_1}{2\pi\, y} = \dfrac{\mu_0\, i_2}{2\pi\, x}$ or $y = \dfrac{i_1}{i_2} x$. Thus locus is a straight line passing through origin and having slope $= \dfrac{i_1}{i_2}$]

5. A pair of stationary and infinite long bent wires are placed in the *X*-*Y* plane (Fig. 23.98). Each wire carry same current *i*. Point *O* is mid-point of *PQ*. Find the magnetic induction at *O*. What is its direction?

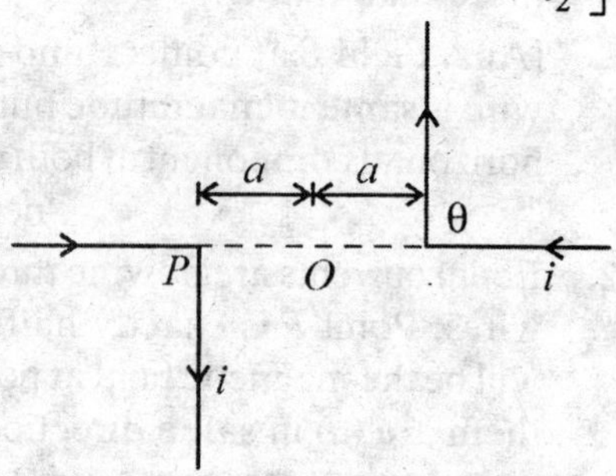

Fig. 23.98

[**Ans.** Magnetic induction at *O* due to horizantal portion of the wires is zero. Magnetic induction B_1 at *O* due to vertical portion of wire *Q* is $B_1 = \dfrac{\mu_0}{4\pi}\dfrac{i}{a} \odot$ perpendicular to the plane of paper coming out. Similarly magnetic induction B_2 at *O* due to vertical portion of wire *P* is $B_2 = \dfrac{\mu_0}{4\pi}\dfrac{i}{a} \odot$. Since direction of the two fields are parallel,

hence total field at *O* is $B = B_1 + B_2 = \dfrac{\mu_0 2i}{4\pi a} \odot$]

6. A current of 1 ampere is flowing in a triangular loop of side 4.5 × 10^{-2} m. Find the magnetic field at the centeroid of the triangle.

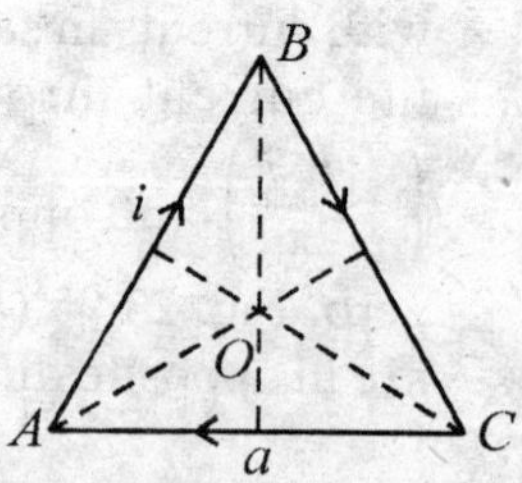

Fig. 23.99

[**Ans.** We have seen that the magnetic field at the centre of the current carrying loop is $B = \frac{9\mu_0 i}{2\pi a}$ Here $i = 1$ A and $a = 4.5 \times 10^{-2}$ m $\therefore$ $B = \frac{9 \times 2 \times 10^{-7} \times 1}{4.5 \times 10^{-2}} = 4 \times 10^{-5}$ T.

Since the current is clock-wise hence direction is perpendicular to the plane into it (= ⊕)]

7. In the Bohr model of hydrogen atom, the electron circulate around the nucleus in the path radius of 5.1×10^{-11} m at a frequency of 6.8×10^{15} rev/s. Calculate the magnetic induction B at the centre of the orbit. What is the equivalent dipole moment?

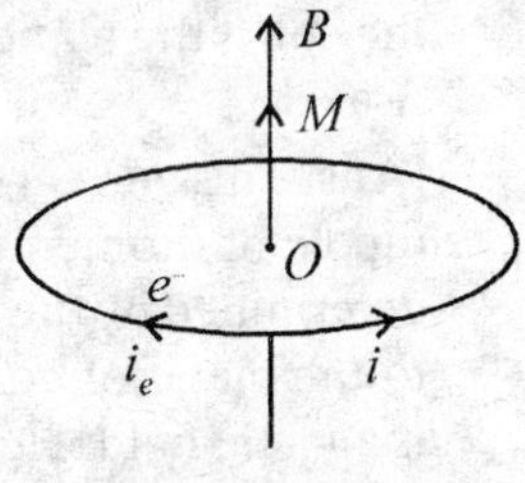

Fig. 23.100

[**Ans.** A revolving electron is equivalent to a current loop. Current i = charge × frequency = $1.6 \times 10^{-19} \times 6.8 \times 10^{15} = 1.088 \times 10^{-3}$ A.

Hence the field at the centre due to a current carrying loop is $B = \frac{\mu_0 i}{2a}$

$= \frac{4\pi \times 10^{-7} \times 1.088 \times 10^{-3}}{2 \times 5.1 \times 10^{-11}} = 13.4$ T. A current loop is equivalent to a dipole of moment M. M = current × area = $1.088 \times 10^{-3} \times \pi(5.1 \times 10^{-11})^2 = 8.88 \times 10^{-24}$ A-m^2.

The direction of the magnetic field and dipole moment are shown in Fig. 23.100]

8. A charge q (= 2.0 μC) moves with a speed of 2.0×10^6 m/s along the positive X-axis. A magnetic field $\boldsymbol{B}$ of strength $(0.2\hat{j} + 0.4\hat{k})$ T exists in space. Find the magnetic force acting on the charge.

[**Ans.** Force on charge $\boldsymbol{F} = q\boldsymbol{v} \times \boldsymbol{B}$, $\boldsymbol{v} = 2.0 \times 10^6 \hat{i}$. Hence force $\boldsymbol{F} = 2 \times 10^{-6} \times 2.0 \times 10^6 (\hat{i}) \times (0.2\hat{j} + 0.4\hat{k}) = (0.2\hat{k} - 0.4\hat{j}) = (0.8\hat{k} - 1.6\hat{j})$ N]

9. A proton, a deutron and an alpha particle moving with equal kinetic energies enter perpendicularly into a magnetic field. If r_p, r_d and r_α are their radii of the circular paths respectively, find the ratio $r_p : r_d : r_\alpha$.

[**Ans.** A charged particle (charge q) moving in magnetic field describes a circular path of radius $r = \frac{mv}{qB} = \frac{\sqrt{2mE}}{qB}$. Where E is kinetic energy of panticle. For proton $r_p = \frac{\sqrt{2m_pE}}{qB} = \frac{\sqrt{2m_pE}}{eB}$.

For deutron $r_d = \frac{\sqrt{2m_dE}}{qB} = \frac{\sqrt{4m_pE}}{eB}$ $m_d = 2m_p$ and For α-particle $r_\alpha = \frac{\sqrt{2m_\alpha E}}{qB} = \frac{\sqrt{8m_pE}}{2eB}$

$m_\alpha = 4m_p \therefore r_p : r_d : r_\alpha = 1 : \sqrt{2} : 1$]

10. A particle having a charge 20 μC and mass 20 μg moves along a circle of radius 5.0 cm under the action of magnetic field $B = 1.0$ T. When the particle is at a point P a uniform electric field E is switched on and it is found that the particle continues on the tangent through P with a uniform velocity, find the electric field.

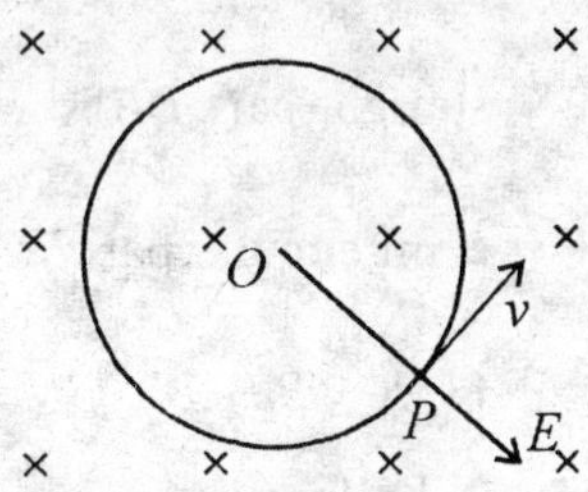

Fig. 23.101

[**Ans.** When a particle moves along a circle in the magnetic field it experiences a radial force = qvB inward. If electric field of proper magnitude is switched on which is directed radially outward, the particle may experience no net force. Under such conditions the particle will move along a straight line with constant velocity. This happens when $qE = qvB$ or $E = vB$. But radius of circle described by the particle in magnetic field is $r = \frac{mv}{qB}$ or $v = \frac{rqB}{m}$ $\therefore E = \frac{rqB^2}{m}$

$= \frac{5 \times 10^{-2} \times 20 \times 10^{-6} \times (1)^2}{20 \times 10^{-9}} = 50$ N/C. The field should be directed radially outward]

11. A rectangular coil of size 3 cm × 4 cm and having 100 turns is pivoted about the Z-axis (Fig. 23.102). The coil carries an electric current of 2.0 A and magnetic field of 1.0 T is present along Y-axis. Find the torque acting on the coil if the plane of the coil makes on angle of 37° with X-axis.

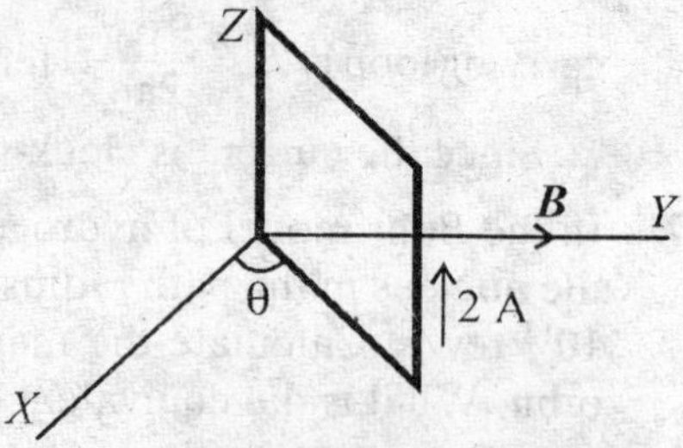

Fig. 23.102

[**Ans.** This current carrying coil is equivalent to a magnetic dipole of moment $\boldsymbol{M} = i\boldsymbol{A}$. Direction of area vector $\boldsymbol{A}$ is determined by the direction of current. It is perpendicular to the plane of coil and makes an angle θ (Fig. 23.102) (= 37) with Y-axis, the direction of magnetic field. Hence torque on the dipole is $\tau = \boldsymbol{M} \times \boldsymbol{B} = 100 \times 2 \times 12 \times 10^{-4} \times 1 \sin 37 = 0.14$ N-m. The torque is along positive Z-axis]

12. A long solenoid is formed by winding 20 turns/cm. What current is necessary to produce a magnetic field of 20 mT inside the solenoid?

[**Ans.** Magnetic field inside the solenoid is $B = \mu_0 ni$. Here $B = 20 \times 10^{-3}$ T, $n = 20$ turns/cm $= 2000$ turns/m and $\mu_0 = 4\pi \times 10^{-7}$ H/m $\therefore 20 \times 10^{-3} = 4\pi \times 10^{-7} \times 2000 \times i$ or

$$i = \frac{20 \times 10^{-3}}{4\pi \times 10^{-7} \times 2000} = \frac{100}{4\pi} = 8.0 \text{ A}$$]

13. Two long wires carrying currents i_1 and i_2 are placed perpendicular to each other, in such a way that they just avoid a contact. Find the magnetic force on a small length dl of the second wire situated at distance l from the first wire.

[**Ans.** Magnetic field at the element dl of second wire due to the current i_1 through first wire is $B = \frac{\mu_0 i_1}{2\pi l}$. This field is perpendicular to the plane of paper going into it. Hence the magnetic force on the dl element of second conductor which carries a current of i_2 is $F = i_2 B\, dl = \frac{i_2 \mu_0 i_1\, dl}{2\pi l} = \frac{\mu_0 i_1 i_2\, dl}{2\pi l}$. This force is parallel to the current i_1]

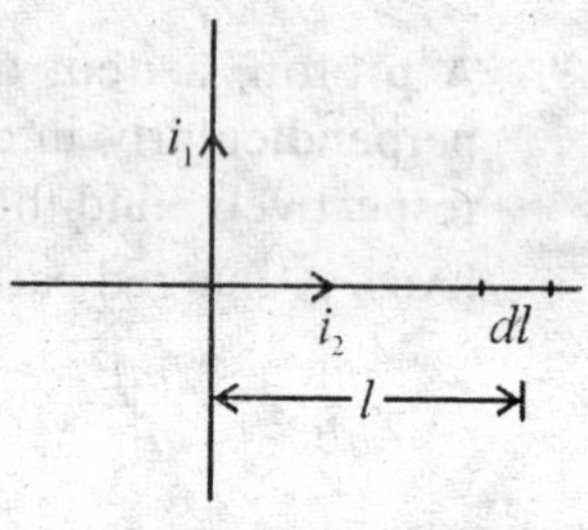

Fig. 23.103

14. Calculate the magnetic field at the centre of a semi-circular wire of radius r in situation shown in Fig. 23.104. Assume the straight part of the wire to be infinite.

[**Ans.** Magnetic field B_1 at O due to infinite long straight part is $B_1 = \frac{\mu_0 i}{2\pi r} \odot$. Similarly magnetic field B_2 at O due to infinite long straight part DE is $B_2 = \frac{\mu_0 i}{2\pi r} \odot$. Magnetic field B_3 at O due to semi-circular part BCD at O is $B_3 = \frac{\mu_0 i}{2r} \oplus$. Hence net field B at O is $B = B_1 + B_2 - B_3 = \frac{\mu_0 i}{\pi r} - \frac{\mu_0 i}{2r}$ or $B = \frac{\mu_0 i}{2\pi r}(2 - \pi)$ or $B = \frac{\mu_0 i}{2\pi r}(\pi - 2) \oplus$]

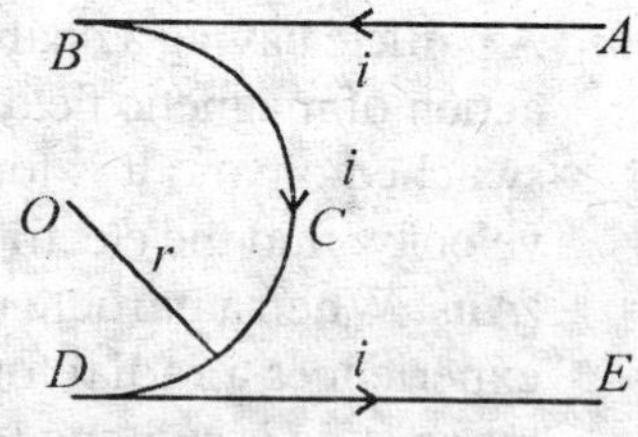

Fig. 23.104

15. A ballery is connected between two points A and B on the circumference of a uniform conducting ring of radius r and resistance R as shown in Fig. 23.105. One of the arc AB of the ring subtends an angle θ at the centre. What is the field at the centre due the current in the ring?

[**Ans.** We know that a current i through an arc of radius r produces a magnetic field at its centre $B = \dfrac{\mu_0 i\theta}{4\pi r}$ and $\theta = l/r$ where l is arc length or $B = \dfrac{\mu_0 il}{4\pi r^2}$. In present case if l_1 is arc length ADB and l_2 is arc length ACB through which the current i_1 and i_2 respectively is passing, then magnetic fields at O due to the two parts of the arc are $B_1 = \dfrac{\mu_0 i_1 l_1}{4\pi r^2} \odot$ and $B_2 = \dfrac{\mu_0 i_2 l_2}{4\pi r^2} \oplus$. But ADB and ACB are in parallel. Hence $\dfrac{i_1}{i_2} = \dfrac{R_2}{R_1} = \dfrac{l_2}{l_1}$ or $i_1 l_1 = i_2 l_2$. Hence magnitude of B_1 and B_2 at O is same and direction opposite.

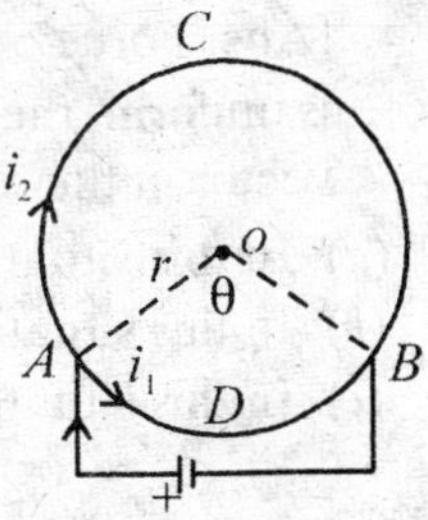

Fig. 23.105

Hence net field at O is zero]

Note: Field at the centre is zero irrespective of the value of θ. Hence wherever the points A and B may be on the ring, the field is always zero.

16. A solenoid of length 0.4 m and diameter 0.6 m consists of a single layer of 1000 turns of fine wire carrying a current of 5×10^{-3} A. Calculate the magnetic field on the axis at the middle and at the end of the solenoid.

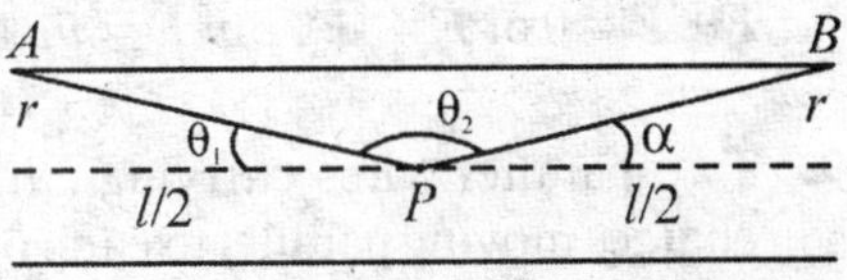

Fig. 23.106

[**Ans.** The magnetic field at any point P on the axis of a solenoid is $B = \dfrac{\mu_0 ni}{2}(\cos\theta_1 - \cos\theta_2)$. Where n is number of turns per unit length and θ_1 and θ_2 are shown in Fig. 23.106. For given solenoid $\theta_2 = (180° - \alpha)$, $B = \dfrac{\mu_0 ni}{2}(\cos\theta_1 + \cos\alpha)$, $\cos\theta_1 = \cos\alpha = \dfrac{l/2}{\sqrt{l^2/4 + r^2}} = \dfrac{0.2}{\sqrt{(0.2)^2 + (0.3)^2}} = 0.55$

or $B = \mu_0 ni \times 0.55 = 4\pi \times 10^{-7} \times \dfrac{1000}{0.4} \times 5 \times 10^{-3} \times 0.55$

$= 8.6 \times 10^{-6}$ T. For point at the end $\cos\theta_1 = \dfrac{l}{\sqrt{r^2 + l^2}}$

$= \dfrac{0.4}{\sqrt{(0.3)^2 + (0.4)^2}} = 0.8$ and $\cos\theta_2 = \cos 90° = 0$ $\therefore$

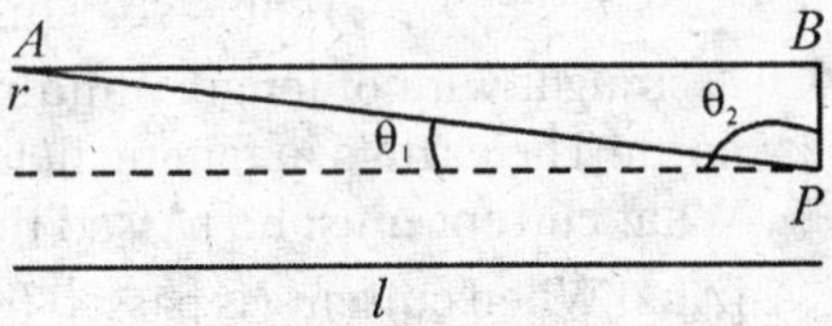

Fig. 23.107

$B = \dfrac{\mu_0 ni}{2} \times 0.8 = \dfrac{4\pi \times 10^{-7}}{2} \times \dfrac{1000}{0.4} \times 0.8 \times 5 \times 10^{-3} = 6.28 \times 10^{-6}$ T]

17. What is the force experienced by a semi-circular wire of radius r when it carries a current i and is placed in magnetic field of induction B as shown in Fig. 23.108.

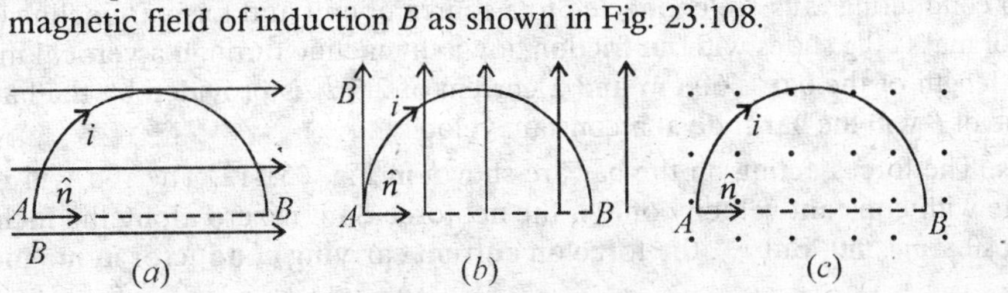

Fig. 23.108

[**Ans.** Force on a current element $\boldsymbol{idl}$ in magnetic field B is given by $\boldsymbol{dF} = \boldsymbol{idl} \times \boldsymbol{B}$. If the field is uniform, the face on the conductor is $\boldsymbol{F} = \int \boldsymbol{idl} \times \boldsymbol{B} = i[\int \boldsymbol{dl}] \times \boldsymbol{B}$. But $\int \boldsymbol{dl} = 2r\hat{n}$ where $\hat{n}$ is unit vector in the direction AB $\therefore$ $\boldsymbol{F} = 2ir(\boldsymbol{n} \times \boldsymbol{B})$

(*a*) In this situation $\boldsymbol{B}$ and $\boldsymbol{n}$ are parallel, $\theta = 0$ or $F = 0$

(*b*) In this situation $\boldsymbol{B}$ is perpendicular to $\hat{n}$. $\theta = \pi/2$ or $F = 2irB$ $\odot$

(*c*) In this situation also $\boldsymbol{B}$ is perpendicular to $\hat{n}$, $\theta = \pi/2$ or $F = 2irB$

The force is directed perpendicular to AB, downwards.]

18. A wire $ABCDEF$ (with each side L) bent as shown in Fig. 23.109 and carrying a current i is placed in a uniform magnetic induction B parallel to Y-direction. What is the force experienced by the wire?

[**Ans.** If we join A to F by a straight conductor, it becomes a closed current carrying loop in magnetic field. Hence the net force on this loop in uniform magnetic field is zero. If $\boldsymbol{F}$ is force on the given net work and $\boldsymbol{F'}$ force on wire AF, then $\boldsymbol{F} + \boldsymbol{F'} = 0$ or $\boldsymbol{F} = -\boldsymbol{F'}$. But $\boldsymbol{F'} = ilB$ along negative Z-direction.

Hence $\boldsymbol{F} = ilB$ along Z-axis]

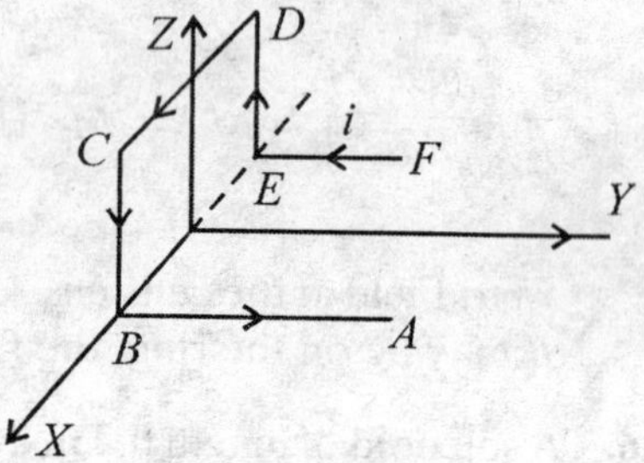

Fig. 23.109

19. Two parallel wires carrying current in same direction attract each other while two similar charges moving parallel to each other in same direction repel each other. How can this apparent contradiction be explained?

[**Ans.** In case of parallel wires carrying current only a magnetic force acts between them. This force is $F = \dfrac{\mu_0 i_1 i_2}{2\pi r}$ and is attractive if the currents are in same direction. In case of two moving charged particles, two forces are acting. One is magnetic force and other is electrostatic force. Magnetic force is attractive and very weak. Electrostatic force is repulsive force and very strong. Hence the net interaction between two similar charges moving parallel to each other is repulsive. This is why a beam of moving charged particles always diverge]

20. A straight wire of length l and mass m lies in the direction θ east of north. The earth's magnetic field is horizontal and has magnitude B. What current must be passed through it so that it floats in air?

[**Ans.** When current i is passed through conductor placed in magnetic field it experiences a force $\boldsymbol{il} \times \boldsymbol{B} = ilB\sin\theta$. In Fig. 23.110, if this force becomes equal to the weight of the wire, it floats in air. Hence $ilB\sin\theta$

$$= mg \text{ or } i = \frac{mg}{lB\sin\theta}\Big]$$

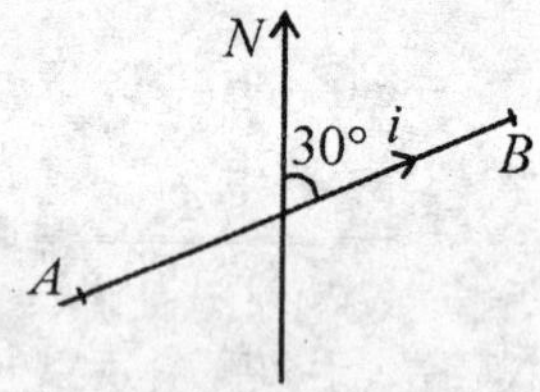

Fig. 23.110

21. Two conducting rails are connected to a source of emf and form an incline (Fig. 23.111). A bar of mass 50 g slides without friction down the incline through a vertical magnetic field B. The length of the bar is 50 cm and a current of 2.5 A is provided by the battery. For what value of B will the bar slide at a constant velocity?

[**Ans.** The forces acting on the bar are shown in Fig. 23.112. The rod will move down the plane with constant velocity only if the net force on it is zero along the inclined plane, i.e. $F_m\cos\theta = mg\sin\theta$. But F_m, the force on current carrying conductor in magnetic field $= ilB$,

$$\therefore ilB\cos\theta = mg\sin\theta \text{ or } B = \frac{mg}{il}\tan\theta = \frac{50\times10^{-3}\times10}{2.5\times0.5}\tan 37° \left(\because \tan 37° = \frac{3}{4}\right) = 0.3 \text{ T}]$$

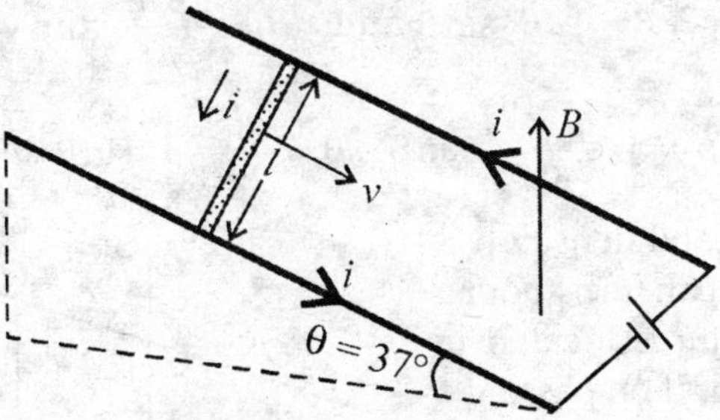

Fig. 23.111

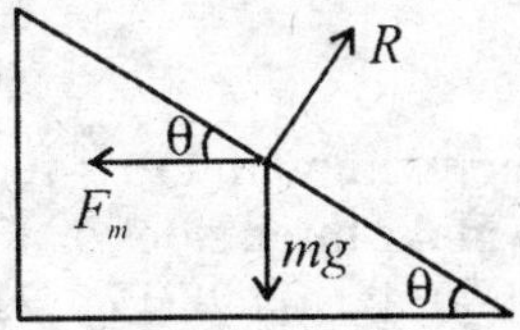

Fig. 23.112

22. A metal rod of mass 10 g and length 25 cm is suspended on two springs as shown in Fig. 23.113. The springs are extended by 4 cm. When a 20 Ampere current passes through the rod it rises by 1 cm. Determine the magnetic field assuming acceleration due to gravity to be 10 cm/s^2.

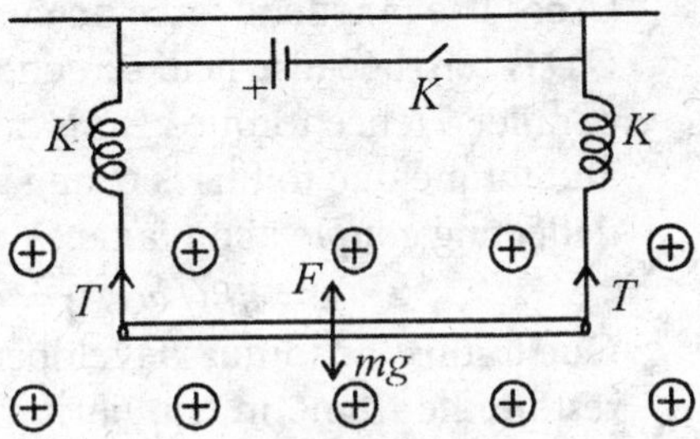

Fig. 23.113

[**Ans.** For initial equilibrium of the rod $2T_0 = mg$ and $T_0 = kx_0$ when current is passed through the rod, it experiences a force $F = ilB$ vertically upwards. If tension in spring is T now then for equilibrium $2T + iBl = mg$

where $T = kx$ $\therefore$ $\dfrac{T_0}{T} = \dfrac{mg}{mg - ilB} = \dfrac{x_0}{x}$ or $mgx = mgx_0 - iBlx_0$

$$\text{or } B = \frac{mg(x_0 - x)}{x_0 il} = \frac{10 \times 10^{-3} \times 10(4-3) \times 10^{-2}}{4 \times 10^{-2} \times 20 \times 0.25} = 5 \times 10^{-3}\ \text{T}$$]

23. Find the magnetic induction at the point O if the wire carries a current $i = 10$ A has the shape shown in Fig. 23.114. The radius of the curved part of the wire is 10 cm; the linear parts of the wire are very long.

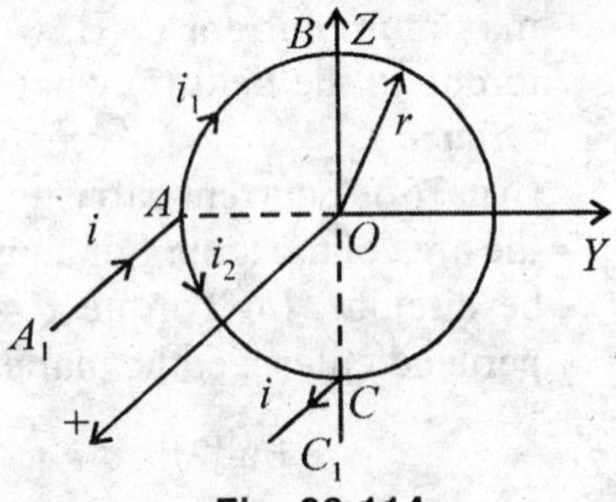

Fig. 23.114

[**Ans.** Magnetic field at O due to straight part A_1A is $\boldsymbol{B}_1 = \dfrac{\mu_0 i}{4\pi r}(-\hat{k})$ and due to part CC_1 $\boldsymbol{B} = \dfrac{\mu_0 i}{4\pi r}(-\hat{j})$. Magnetic field at O due to circular part ABC and AC are equal in magnitude and opposite in direction. Hence it is zero. Hence net magnetic field at O' is $\boldsymbol{B} = \boldsymbol{B}_1 + \boldsymbol{B}_2 = -\dfrac{\mu_0 i}{4\pi r}(\hat{i} + \hat{j})$ or $|B| = \dfrac{\mu_0 i}{4\pi r}\sqrt{2}$. This is in Y-Z plane inclined at an angle of 45° with Z-axis in 3rd quadrant.

24. Current flows in a long straight conductor having the shape of a groove with a cross-section in the form of a thin half-ring of radius r (Fig. 23.115). The current is directed from the reader behind the plane of diagram. Find the magnetic induction on the axis O.

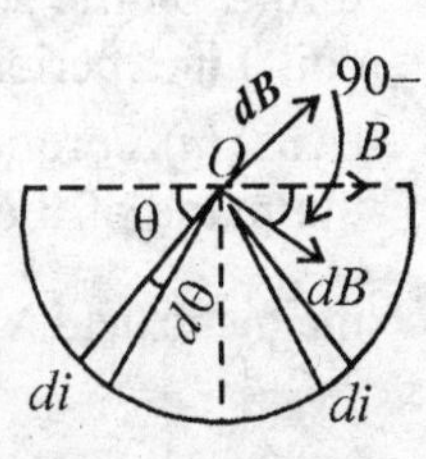

Fig. 23.115

[**Ans.** Let us find the direction of the magnetic field at O. For this purpose we divide the entire conductor into elementary fragments with current di. It is obvious that the sum of any two symmetric fragments gives vector $\boldsymbol{dB}$ as shown in Fig. 23.115. Consequently the resultant vector $\boldsymbol{B}$

will also be directed towards right. Hence $B = \int dB \cdot \sin\theta$ and $dB = \frac{dB\, di}{2\pi r}$ and $di = \frac{i}{\pi} d\theta$

$$\therefore\ B = \int_0^{\pi} \frac{\mu_0}{2\pi r} \cdot \frac{i}{\pi} \sin\theta \cdot d\theta = \frac{\mu_0 i}{2\pi^2 r} \int_0^{\pi} \sin\theta \cdot d\theta = \frac{\mu_0 i}{\pi^2 r}\ ]$$

25. A copper wire with cross-sectional area α and density ρ is bent to make three sides of a square, which can turn about a horizontal axis OO' (Fig. 23.116). The wire is located in uniform magnetic field in vertical direction. If on passing a current i the frame deflects through an angle θ, then find the magnetic field applied.

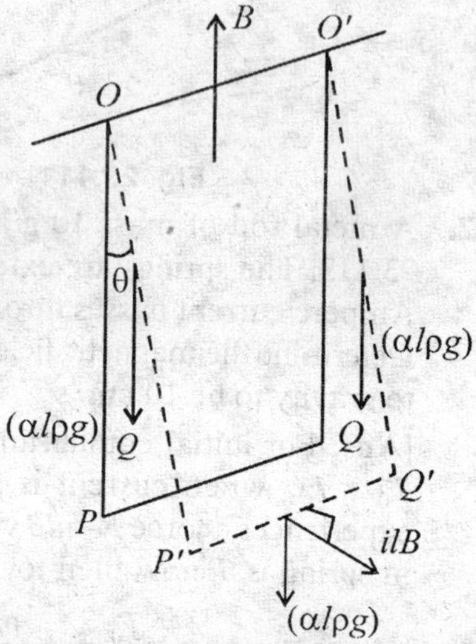

Fig. 23.116

[**Ans.** The magnetic force on current carrying wires OP and $O'Q$ is zero because field direction and current direction are parallel. Hence torque also is zero. Force on wire PQ is ilB and torque due to this is trying to deflect the frame. Hence deflecting couple about axis OO' is

$$\tau = ilB(l\cos\theta) = il^2B\cos\theta$$

Due to this the torque developed due to the weight tries to restore the frame in equilibrium position. Hence restoring torque is $\tau = (\alpha l\rho g)\ (l/2\sin\theta) + (\alpha l\rho g)\ (l/2\sin\theta) + (\alpha l\rho g) l\sin\theta$. For equilibrium the two torques must be equal in magnitude. Hence $il^2B\cos\theta = 2\alpha l^2\rho g \sin\theta$

or $$B = \frac{2\alpha\rho g}{i} \tan\theta\ ]$$

26. A uniform field $B = 0.01$ T exists in a region of space. A wire of 50 cm length can carry maximum current $i = 15$ A. How a loop is formed so that (*a*) it experiences a maximum torque in the field (*b*) what should be the orientation of the loop (*c*) what is the maximum torque?

[**Ans.** (*a*) A current carrying loop behave like a magnetic dipole. Dipole moment depends on the area of the loop. For a given perimeter, the circle has maximum area. Hence loop should be circular. (*b*) Torque $\tau = M \cdot B\sin\theta$. Hence for maximum torque **B** and **M** should be perpendicular, i.e. the plane of the loop should be parallel to the **B**. (*c*) $M = i\pi r^2$ or $\tau = i\pi r^2 B$

$$\text{But } 2\pi r = l \therefore\ \tau = \frac{i\pi l^2}{4\pi^2} B \text{ or } \tau = \frac{il^2B}{4\pi} = \frac{15 \times (0.5)^2 \times (0.01)}{4\pi} = 2.98 \times 10^{-3} \text{ N-m}\ ]$$

27. A circular coil of 50 turns and mean diameter 20 cm is placed in a uniform magnetic field of 200 G normal to the plane of the coil. The current in the coils is 5.0 A. Find (*a*) the total torque on the coil (*b*) the total force on the coil and (*c*) average drift speed of electron in the coil if it experiences a force of 5×10^{-25} N.

[**Ans.** (*a*) Torque $\tau = iBA\sin\theta$. Where θ is the angle between the direction of B and normal to the plane of the coil. Since $\theta = 0$, hence $\tau = 0$. (*b*) Net force on the planer current loop in magnetic field is zero. (*c*) Force on a moving charge in magnetic field $F = q\mathbf{v} \times \mathbf{B} = qvB$ or $v = \frac{F}{qB}$

$$= \frac{5.0 \times 10^{-25}}{200 \times 10^{-4} \times 1.6 \times 10^{-19}} = 1.56 \times 10^{-4} \text{ m/s}\ ]$$

28. A potential difference of 600 V is applied across the plates of a parallel plate condenser. The separation between the plates is 3 mm. An electron projected vertically parallel to the plates, with velocity of 2×10^6 m s^{-1} moves undeflected between the plates. Find the magnitude and direction of the magnetic field in the region between the plates.

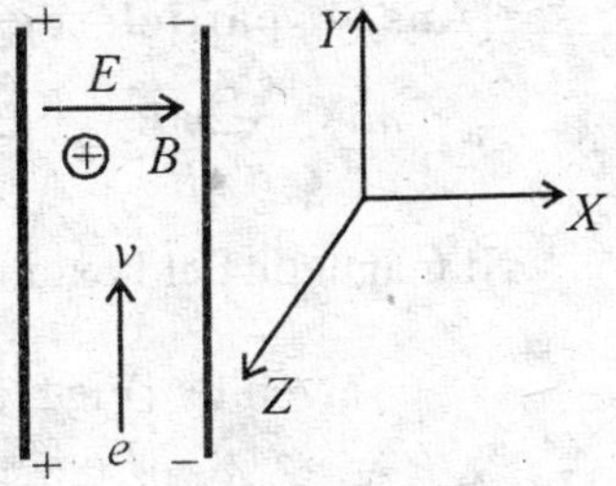

Fig. 23.117

[**Ans.** Electric field between the plates $E = \frac{V}{d} = \frac{600}{3 \times 10^{-3}} = 2 \times 10^5$ V/m. Electron in electric field experiences a force Ee in the direction opposite to the field, i.e. in negative X-direction (Fig. 23.117). If magnetic field B is vertically downward (i.e., perpendicular to the plane of Figure), then electron experiences a force evB in the positive X-direction. If magnitudes of the two forces are equal then electron goes undeviated, $\therefore eE = evB$ or $B = \frac{E}{V} = \frac{2 \times 10^5}{2 \times 10^6}$
$= 10^{-1} = 0.1$ T]

Note: A charged particle charge q, moving with velocity $v\hat{i}$, where electric field $\mathbf{E} = E_0\hat{j}$ and magnetic field $\mathbf{B} = B_0\hat{k}$ will go undeviated if velocity $v = \frac{E_0}{B_0}$]

29. A metal wire of mass m slides without friction on two rails spaced a distance d apart. The track lies in a vertical uniform magnetic field B. A constant current i flows along one rail across the wire and back (Fig. 23.118). Find the velocity (magnitude and direction) of the wire as a function of the time (t). Assume it to be at rest initially.

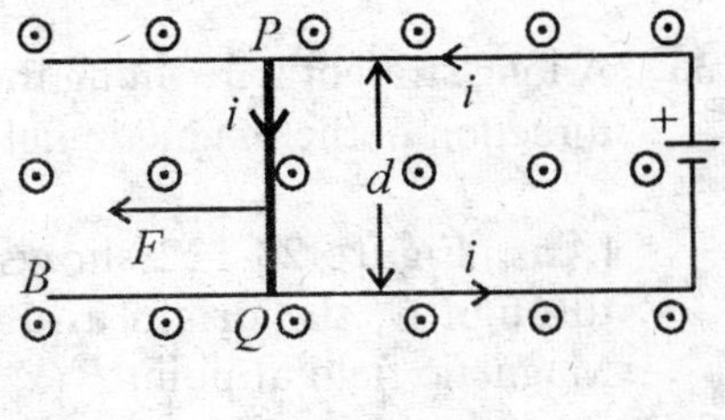

Fig. 23.118

[**Ans.** Since the current carrying wire PQ is placed in magnetic field it experiences a force $F = idB$ towards left.

Hence its acceleration $a = \frac{dv}{dt} = F/m = \frac{idB}{m}$. Since acceleration is constant (independent of time) and initial velocity is zero, hence velocity at anytime t is $v = at = \frac{idB}{m}t$ towards left]

30. A circular coil of 100 turns has an effective radius of 0.05 m and carries a current of 0.1 A. How much work is required to turn it in an external magnetic field of 1.5 Wb m^{-2} through 180° about an axis perpendicular to the magnetic field? The plane of the coil is initially perpendicular to the magnetic field.

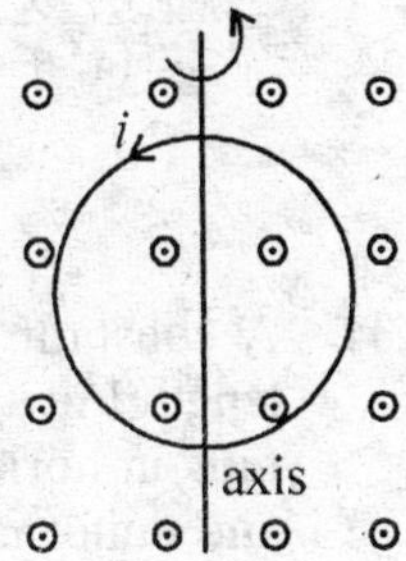

Fig. 23.119

[**Ans.** This current carrying coil is equivalent to a magnetic dipole of moment $M = iAn = i\pi r^2 n$ with its direction perpendicular to the plane of coil. Work done in rotating a magnetic dipole in magnetic field from angle θ_1 to θ_2 is $W = MB(\cos\theta_1 - \cos\theta_2)$. In present problem $\theta_1 = 0$ and $\theta_2 = 180°$, then $W = 2MB = 2 \times i\pi r^2 n \times B = 2 \times 0.1 \times \pi\,(0.05)^2 \times 100 \times 1.5 = 0.2356$ J]

31. An α-particle is accelerated by a potential difference of 10^4 V. Find the change in its direction of motion, if it enters normally in a region in which magnetic induction of 0.1 T is acting upto length 0.1 m mass of α-particle $= 6.4 \times 10^{-27}$ kg.

[**Ans.** α-particle describes a circular motion of radius

$$r = \frac{mv}{qB} = \frac{\sqrt{2mE}}{qB} = \sqrt{\frac{2mV}{qB^2}} = \sqrt{\frac{2 \times 6.4 \times 10^{-27} \times 10^4}{2 \times 1.6 \times 10^{-19} \times 10^{-2}}} = 0.2 \text{ m}.$$

If magnetic field extends upto distance $l < r$, the deflection θ is given by (Fig. 23.120) $\sin\theta = \frac{l}{r} = \frac{0.1}{0.2} = \frac{1}{2}$ or $\theta = 30°$]

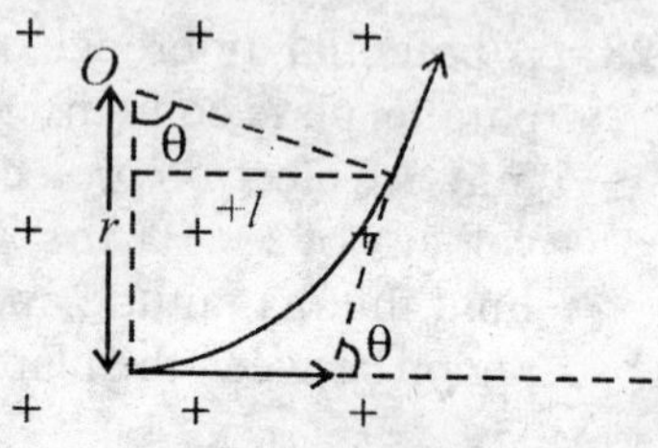

Fig. 23.120

32. A wire consists of two linear portions *AB* and *DE* each of length *L* and a semi-circular loop *BCD* of radius *r*. The wire carries a current *i* as shown in Fig. 23.121. Find the magnetic field at the centre *O* of the circular loop due to the current in entire wire.

[**Ans.** We know that the magnetic field due to a straight current carrying current is given by $B = \frac{\mu_0 i(\sin\alpha + \sin\beta)}{4\pi r}$. In situation shown in Fig. 23.120 for part *AB* and *DE* $\alpha = \beta = 0$. Hence value of *B* at centre due to these straight parts is zero. Hence field at *O* is only due to semi-circular loop $= \frac{\mu_0 i}{4r}$. The field is perpendicular to the plane of diagram going into the page (+)]

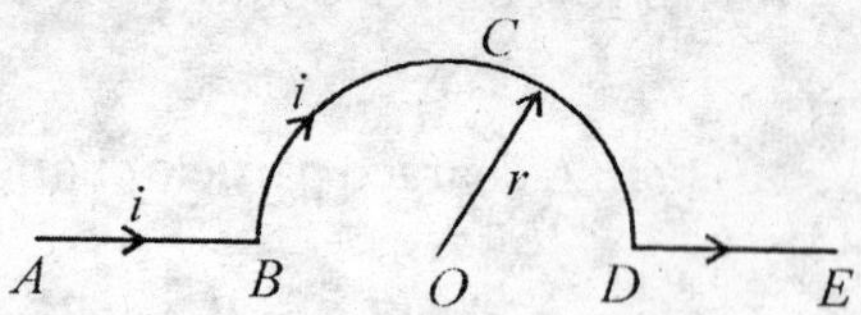

Fig. 23.121

33. A long straight wire along the *Z*-axis carries a current *i* in the negative *Z*-direction. Find the direction of the magnetic field vector at a point having co-ordinates (*x*, *y*) in the $Z = 0$ plane. (IIT 2002)

[**Ans.** Figure 23.122 shows current carrying wire passing through *O*, the origin and current *i* flows into the plane. Magnetic field at point *P* (*x*, *y*, *O*) is perpendicular to line *OP* $\left(= \sqrt{x^2 + y^2}\right)$. Magnetic field at $P = \frac{\mu_0 i}{2\pi\, OP} = \frac{\mu_0 i}{2\pi\sqrt{x^2 + y^2}}$.

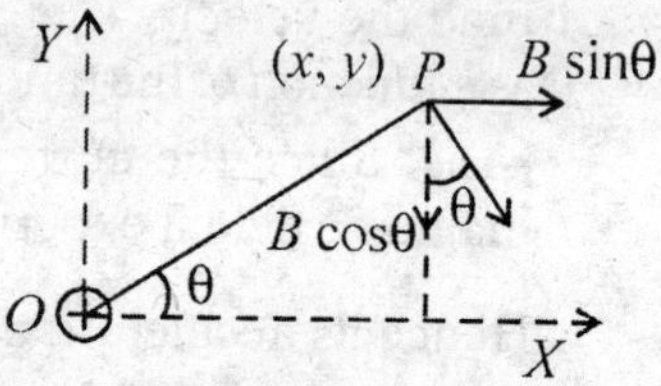

Fig. 23.122

Resolving *B* along *X* and *Y* directions, we have $B_x = B\sin\theta$

$$= \frac{\mu_0 i}{2\pi}\frac{y}{(x^2 + y^2)} \quad \because \sin\theta = \frac{y}{\sqrt{x^2 + y^2}} \text{ and } B_y = -B\cos\theta = -\frac{\mu_0 i}{2\pi}\frac{x}{(x^2 + y^2)} \quad \because \cos\theta = \frac{x}{\sqrt{x^2 + y^2}}$$

Hence $\mathbf{B} = \frac{\mu_0 i}{2\pi(x^2 + y^2)}(y\hat{i} - x\hat{j})$]

34. A constant current 2 A flow through a metal rod of length 1 m and mass 0.5 kg. Which slides on frictionless rails in horizontal plane. Initial speed of rod is 2 m/s and uniform magnetic field of 0.5 T is acting vertically upwards (Fig. 23.123). Find the distance moved by the rod before coming to rest.

[**Ans.** The current carrying rod experiences force in magnetic field. Force on rod is $F = ilB$ in direction opposite to the velocity. Hence retardation of rod is

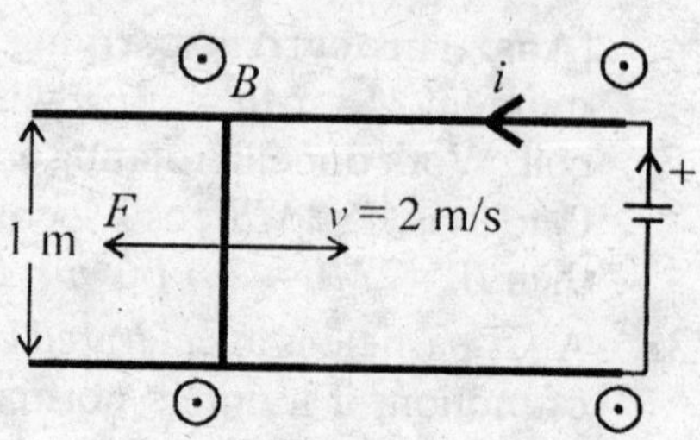

Fig. 23.123

$a = \frac{F}{m} = \frac{ilB}{m} = \frac{2 \times 1 \times 0.5}{0.5} = 2\ \text{m/s}^2$. It will stop after moving distance $s = \frac{v^2}{2a} = \frac{2 \times 2}{2 \times 2} = 1\ \text{m}$]

35. Two long parallel wires carry currents of equal magnitude but in opposite direction. These wires are suspended from rod PQ by four chords of same length L as shown in Fig. 23.124. The mass per unit length of the wires is λ. Determine the value of θ assuming it to be small.

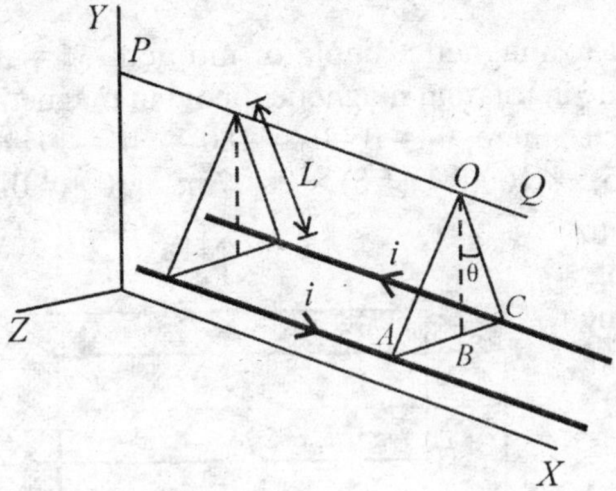

Fig. 23.124

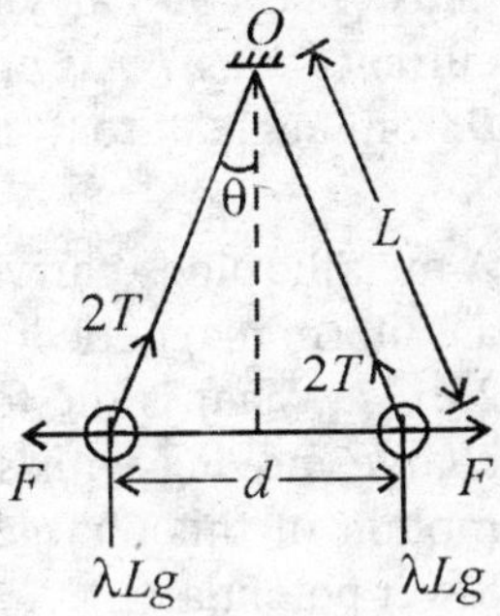

Fig. 23.125

[**Ans.** Force per unit length on the current carrying conductors is $\frac{dF}{dL} = \frac{\mu_0 i_1 i_2}{2\pi d}$ or $F = \frac{\mu_0 i_1 i_2}{2\pi d} dL$. Where d is separation between wires. Equilibrium of wires is shown in Fig. 23.125. This gives

$$2T\cos\theta = \lambda dLg \text{ and } 2T\sin\theta = F \therefore \tan\theta = \frac{F}{\lambda Lg} = \frac{\mu_0 i^2 dL}{2\pi dL\, g\lambda} = \frac{\mu_0 i^2}{2\pi dg\,\lambda} \text{ or } \theta = \tan^{-1}\left(\frac{\mu_0 i^2}{2\pi dg\,\lambda}\right)$$]

36. For a given length L of a wire carrying a current i, how many turns would produce the maximum magnetic moment and of what value?

[**Ans.** Magnetic moment of a current carrying loop is $M = i \times$ area. For a given perimeter area of circle is maximum. Hence a circular loop will give maximum dipole moment. If r is radius and n number of turns then $2\pi rn = L$ or $r = \frac{L}{2\pi n}$ $\therefore M = i \times \pi r^2 n = i\pi n \cdot \frac{L^2}{4\pi n^2} = \frac{iL^2}{4\pi n}$. Hence for maximum moment n should be minimum, i.e. $n = 1$. Hence maximum dipole moment

$$M = \frac{iL^2}{4\pi}$$]

37. A coil in the shape of an equilateral triangle of side 0.02 m is suspended from a vertex such that it is hanging in a vertical plane between the pole pieces of a permanent magnet producing a horizontal magnetic field of 5×10^{-2} T. Find the couple acting on the coil when a current of 0.1 A is passed through it.

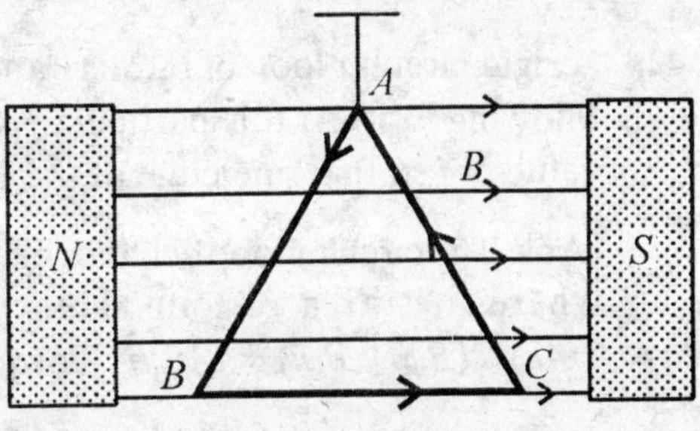

Fig. 23.126

[**Ans.** Area of coil $= \frac{1}{2}a(a\cos 30°) = \frac{\sqrt{3}}{4}a^2$. It carries current i and is equivalent to a magnetic dipole of moment $M = i$ Area $= \frac{\sqrt{3}}{4}a^2 i$ perpendicular to the plane of coil. Hence torque on it is $\tau = \boldsymbol{M} \times \boldsymbol{B} = MB$

$\because$ M and B are perpendicular or $\tau = \frac{\sqrt{3}}{4} a^2 i \ B = \frac{\sqrt{3}}{4}(0.02)^2(0.1)(5\times10^{-2}) = 5\times\sqrt{3}\times10^{-7}$ N-m]

38. A circular coil has 100 turns and effective radius 0.05 m. It carries a current of 0.1 A. How much work is required to turn it in external magnetic field of 1.5 Web/m^2 through 180° about an axis perpendicular to the magnetic field? Plane of the coil initially is perpendicular to the magnetic field.

[**Ans.** The coil carrying current is equivalent to magnetic dipole of moment M = area $\times$ current $\times$ $\pi r^2 n i$, n is number of turns. Work done in rotating magnetic dipole in magnetic field B from angle θ_1 to θ_2 is $W = MB = (\cos\theta_1 - \cos\theta_2)$, here, $\theta_1 = 0°$, $\theta_2 = 180°$ $\therefore$ $W = \pi r^2 niB \times 2$ $= \pi(0.05)^2 \times 100(0.1)(1.5) \times 2 = 75\pi \times 10^{-3} = 0.236$ J]

39. A metallic block carrying current i is subjected to a uniform magnetic induction $\boldsymbol{B}$ as shown in Fig. 23.127. What is the force $\boldsymbol{F}$ experienced by the charge moving with speed v and as a result of motion of this charge which face will acquire lower potential?

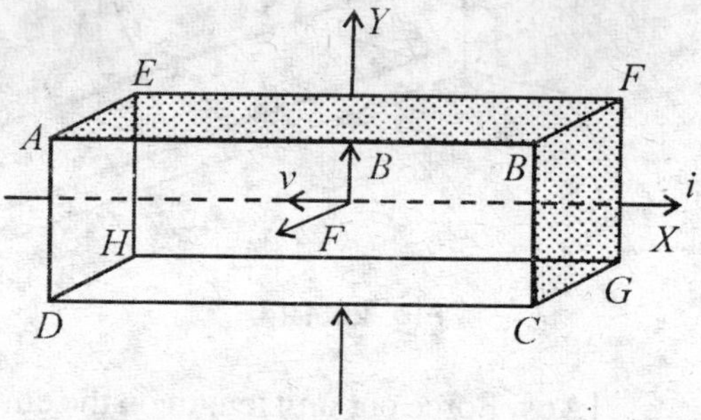

Fig. 23.127

[**Ans.** If the direction of the current is $\hat{i}$ (X-axis) then the velocity of free electrons is $(-v\hat{i})$. Let magnetic field be in Y-direction. Hence $\boldsymbol{B} = B\hat{j}$. The electron then experiences a force of $(q\boldsymbol{v} \times \boldsymbol{B})$

$\boldsymbol{F} = (-e)(-v\hat{i}) \times (B\hat{j} = evB\hat{k})$ positive Z-direction. Thus force on the electrons is towards face $ABCD$. Hence this face will be at lower potential]

40. Can a magnetic field set a resting electron in motion?

[**Ans.** Yes. We know that a time varying magnetic field produces an electric field. Hence an electron at rest in time varying magnetic field will be set in motion due to this induced electric field]

41. A particle of charge q and mass m starts moving from the origin under the action of an electric field $\boldsymbol{E} = E_0\hat{i}$, $\boldsymbol{B} = B_0\hat{i}$ with velocity $\boldsymbol{v} = v_0\hat{i}$. Find the time when the speed of the particle becomes $2v_0$.

[**Ans.** $\boldsymbol{E}$ is parallel to $\boldsymbol{B}$ and $\boldsymbol{v}$ is perpendicular to both. Hence the path of the particle is helix with increasing pitch. Speed of particle at any time t is $v = \sqrt{v_x^2 + v_y^2 + v_z^2}$. But $v_x^2 + v_y^2 = v_0^2$ and

$$v_x^2 = \left(\frac{qE}{m}t\right)^2 \text{ and } v = 2v_0 \therefore \frac{qE}{m}t = \sqrt{3}v_0 \text{ or } t = \frac{\sqrt{3}v_0 m}{qE}$$]

42. A rigid circular loop of radius r and mass m lies in X-Y plane on a flat table and has a current i flowing in it. At this particular place the earth's magnetic field is $\boldsymbol{B} = B_x\hat{i} + B_z\hat{k}$. Find the value of i so that one edge of the loop lifts from the table.

[**Ans.** The circular loop behaves like a magnetic dipole with magnetic dipole moment $\boldsymbol{M} = iA\hat{k}$ where A is area of loop. Hence deflecting torque in magnetic field is $\boldsymbol{\tau} = \boldsymbol{M} \times \boldsymbol{B} = (iA\hat{k}) \times (B_x\hat{i} + B_z\hat{k}) = iAB_x\hat{j}$. Torque due to weight $\tau' = mgr(-\hat{j})$. For equilibrium, the two torques should be equal in magnitude, $\therefore$ $iAB_x = mgr$ or $i = \frac{mgr}{AB_x} = \frac{mgr}{\pi r^2 B_z} = \frac{mg}{\pi r B_x}$]

43. A metalic wire is folded to form a square loop of side a. It carries a current i and is kept perpendicular to uniform magnetic field. If the shape of loop is changed from square to a circle without changing the length and current, find the work done in doing so.

[**Ans.** Square loop has magnetic moment $M = a^2 i$. Its potential energy in magnetic field is $U_i = -\boldsymbol{M} \cdot \boldsymbol{B} = -a^2 iB$. In case of circular loop, $2\pi r = 4a$ or $r = \frac{4\theta}{2\pi}$. The dipole moment of circular loop $= \pi r^2 i = \pi i \frac{16a^2}{4\pi^2} = \frac{4}{\pi} a^2 i$. Its potential energy in magnetic field U_f is $U_f = -M \cdot B = -\frac{4}{\pi} a^2 iB$.

Hence work done $W = U_f - U_i = a^2 iB\left(1 - \frac{4}{\pi}\right)$]

44. A conducting rod of mass m and length l is placed over a smooth horizontal table. A uniform magnetic field B is acting perpendicular to the rod. Charge q is suddenly passed through the rod and it acquires an initial velocity u on the surface, find q.

[**Ans.** If i is the current then force $F = ilB$. Then impulse $= Fdt$. $\therefore \int F dt =$ change in momentum

$\therefore \int ilB dt = mv$ or $qlB = mv$ or $q = \frac{mv}{lB}$]

45. Equal currents $i = 1$ A are flowing through the wires parallel to Y-axis located at $x = +1$ m, $x = +2$ m, $x = +4$ m, etc. but in opposite directions as shown in Fig. 23.128. Find the magnetic field at origin (in tesla).

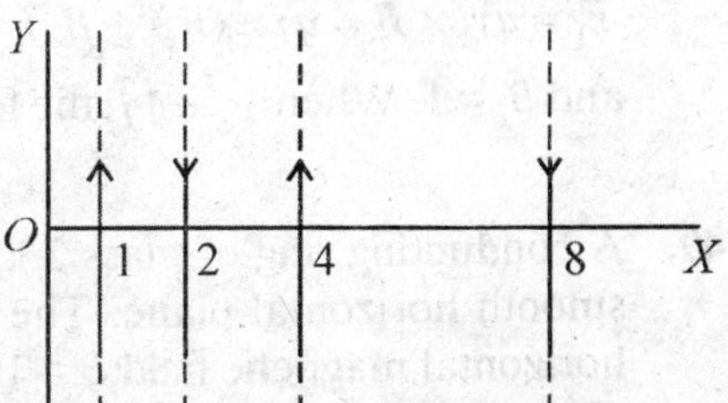

Fig. 23.128

[**Ans.** Magnetic field at the origin due to different currents is

$$B = \frac{\mu_0 i}{2\pi(1)} - \frac{\mu_0 i}{2\pi(2)} + \frac{\mu_0 i}{2\pi(4)} - \frac{\mu_0 i}{2\pi(8)} + \ldots + \infty$$

$$= \frac{\mu_0 i}{2\pi}\left(1 + \frac{1}{4} + \frac{1}{16} + \ldots\right) - \frac{\mu_0 i}{2\pi}\left(\frac{1}{2} + \frac{1}{8} + \frac{1}{32} + \ldots\right)$$

$$= \frac{\mu_0 i}{2\pi}\left(1 + \frac{1}{2^2} + \frac{1}{2^4} + \ldots\right) - \frac{\mu_0 i}{2\pi} \cdot \frac{1}{2}\left(1 + \frac{1}{2^2} + \frac{1}{2^4} + \ldots\right)$$

$$= \frac{\mu_0 i}{2\pi(2)}\left(1 + \frac{1}{2^2} + \frac{1}{4^2} + \ldots\right) = \frac{\mu_0 i}{4\pi} \frac{1}{1 - (1/2)^2} = \frac{\mu_0 i}{4\pi} \frac{4}{3}$$

$= \frac{\mu_0 i}{3\pi} = 1.33 \times 10^{-7}$ T. The direction of of the field is +ve Z-direction, i.e. $B = 1.33 \times 10^{-7} \hat{k}$.]

46. A charged particle of mass m and charge q is moving in a uniform magnetic field $\boldsymbol{B}$ which acts into the plane of motion. The plane is rough having coefficient of friction μ. The speed of the charged particle just before entering into the region is v_0. Find the radius of curvature of the path after time $\frac{v_0}{2\mu g}$.

[**Ans.** Here centripetal force is provided by magnetic force and the tangential acceleration is provided by friction. If v is the velocity after time $t = \frac{v_0}{2\mu g}$, then $\frac{mv^2}{r} = qvB$ or $r = \frac{mv}{qB}$.

Tangential acceleration $a = -\frac{\mu mg}{m} = -\mu g$. Hence velocity after time $t = \frac{v_0}{2\mu g}$ $v = v_0 - at$

$= v_0 - \frac{v_0}{2\mu g}(\mu g) = v_0/2 \ \therefore r = \frac{mv_0}{2qB}$]

47. A charged particle moving in a uniform magnetic field and losses 4% of its K.E. The radius of curvature of the path changes by how much?

[**Ans.** Radius of path is given by $r = \frac{mv}{qB} = \frac{\sqrt{2mE}}{qB}$, where E is K.E. of the particle $\therefore r \propto \sqrt{E}$

$\therefore \frac{dr}{r} = \frac{1}{2}\frac{dE}{\sqrt{E}} \times \frac{1}{\sqrt{E}}$ or $\frac{dr}{r} = \frac{1}{2}\frac{dE}{E}$ or $\frac{dE}{E} = 2\frac{dr}{r}$. Since $\frac{dE}{E} \times 100 = 4\%$ then $\frac{dr}{r} \times 100 = 2\%$]

48. A charge q is moving with a velocity $v_1 = 1\hat{i}$ m/s at a point in a magnetic field and experiences a force $F_1 = q[-1\hat{j} + 1\hat{k}]$ N. If the charge is moving with velocity $v_2 = 1\hat{j}$ m/s at the same point, it experiences a force $F_2 = q(1\hat{i} - 1\hat{k})$ N. Find the magnetic induction at that point.

[**Ans.** Let magnetic field at that point be $\boldsymbol{B} = B_x\hat{i} + B_y\hat{j} + B_z\hat{k}$. When $v_1 = 1\hat{i}$ the force is $F_1 = qv_1 \times \boldsymbol{B} = q\hat{i} \times (B_x\hat{i} + B_y\hat{j} + B_z\hat{k}) = q(B_y\hat{k} - B_z\hat{j})$ But force $F_1 = q(-1\hat{j} + 1\hat{k})$ giving $B_y = 1$ and $B_z = 1$. When $v_2 = 1\hat{j}$ the force is $F_2 = q\boldsymbol{v}_2 \times \boldsymbol{B} = q\hat{j} \times (B_x\hat{i} + B_y\hat{j} + B_z\hat{k}) = q(-B_x\hat{k} - B_z\hat{i})$,

equating with $F_2 = q(1\hat{i} - 1\hat{k}), B_x = 1$. Hence $\boldsymbol{B} = (1\hat{i} + 1\hat{j} + 1\hat{k})$]

49. A conducting ring of mass 2 kg and radius 0.5 m is placed on a smooth horizontal plane. The ring carries a current $i = 4$ A. A horizontal magnetic field $B = 10$ T is switched on at the time $t = 0$ (Fig. 23.129). Find the initial angular acceleration of the ring.

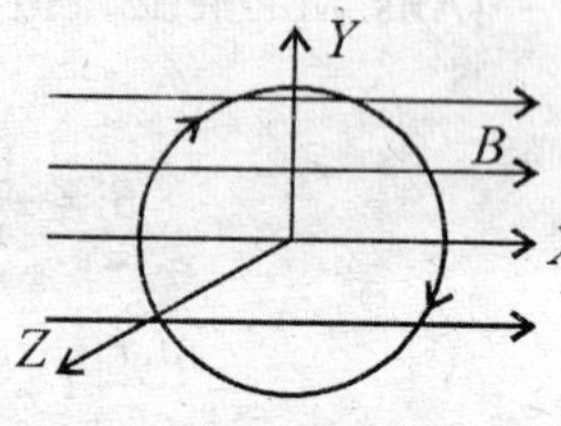

Fig. 23.129

[**Ans.** A current carrying loop is equivalent to a magnetic dipole of moment $M = -iA\hat{k}$. In magnetic field $\boldsymbol{B} = B\hat{i}$ experiences a torque $\tau = \boldsymbol{M} \times \boldsymbol{B}$ or $\tau = -(iA\hat{k}) \times B(\hat{i}) = -iAB\hat{j}$. But torque $\tau = I\alpha$. Since torque is along Y-axis which is diameter of the ring. M.I I of ring about its diameter is

$\frac{mr^2}{2}$ $\therefore iAB = I\alpha$ or $\alpha = \frac{iAB}{I} = \frac{i \cdot \pi r^2 \cdot B \times 2}{mr^2}$ or $\alpha = \frac{4 \times \pi \times 10 \times 2}{2} = 40\pi$ rad/s^2]

50. In the phase shown in Fig. 23.130, find the magnetic force on q_1 by q_2 and on q_2 by q_1. Could you justify answer according to Newton's third law?

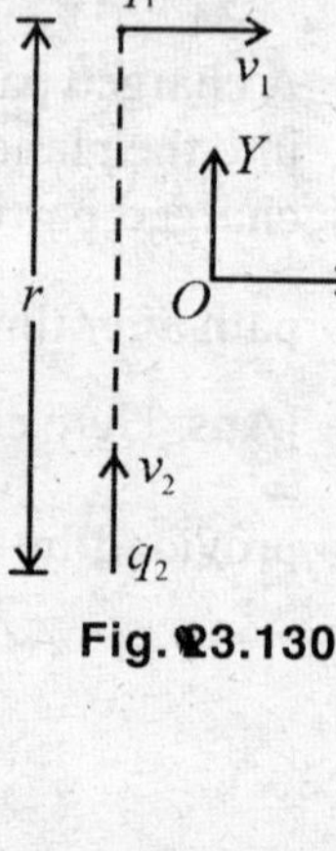

Fig. 23.130

[**Ans.** A charge q moving with velocity v is equivalent to current element qv. Using Biot Savarts law, magnetic field at q_2 due to moving charge q_1 is $\boldsymbol{B}_1 = \frac{\mu_0}{4\pi}\frac{(q_1\boldsymbol{v}_1) \times \boldsymbol{r}}{r^3} = \frac{\mu_0 q_1 v_1}{4\pi r^2}(-\hat{k})$ Hence force on charge q_2 $\boldsymbol{F} = (q_2\boldsymbol{v}_2) \times \boldsymbol{B}$

$= \frac{\mu_0 q_1 q_2 v_1 v_2}{4\pi r^2}(-\hat{i})$. Magnetic field due to moving charge q_2 at q_1 is

$\boldsymbol{B}_2 = \frac{\mu_0}{4\pi}\frac{(q_2\boldsymbol{v}_2) \times \boldsymbol{r}}{r^2} = 0$. Hence force on $q_1 = 0$. Newton's third law is violated in this case]

MULTIPLE CHOICE QUESTIONS

PART I

In the following, a question is followed by four alternatives, of them only one is correct. Therefore, the candidate has to select the correct choice from amongst the alternatives.

1. Three identical long solenoids A, B, and C are connected to each other as shown in Fig. 23.131. If the magnetic field at the centre of A is B; the field at the centre of B is (Assume that the field due to any solenoid is confined within the volume of the solenoid only)

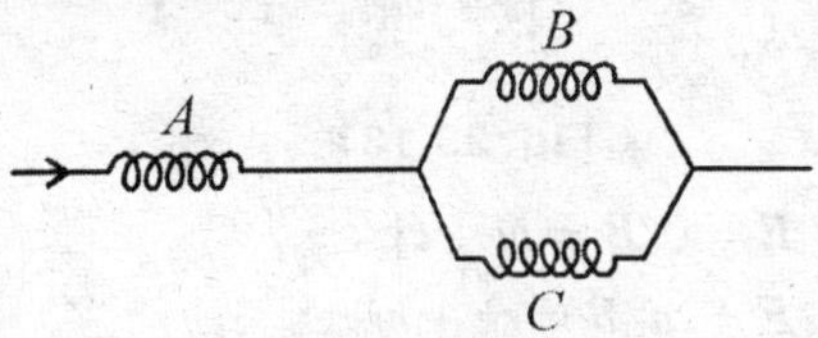

Fig. 23.131

(a) B (b) $B/2$
(c) $2B$ (d) $B/4$.

2. Magnetic field at the centre of semi-circular wire (Fig. 23.132) of radius r is (assume straight portion of wire to be infinite)

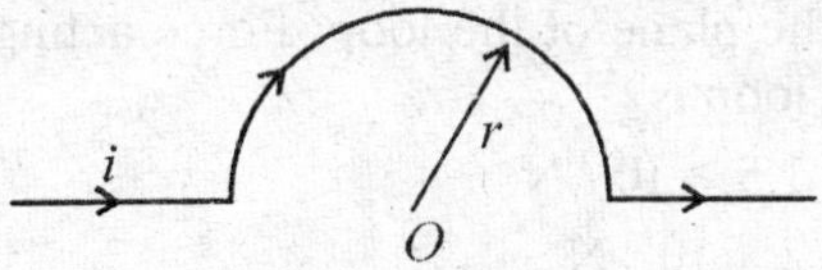

Fig. 23.132

(a) $\frac{\mu_0 i}{4r} \odot$ (b) $\frac{\mu_0 i}{4r} \oplus$
(c) $\frac{\mu_0 i}{2r} \odot$ (d) $\frac{\mu_0 i}{2r} \oplus$.

3. In the situation shown in Fig. 23.133, magnetic field at the centre of semi-circle of radius r is (assume straight part of the wire to be infinite long)

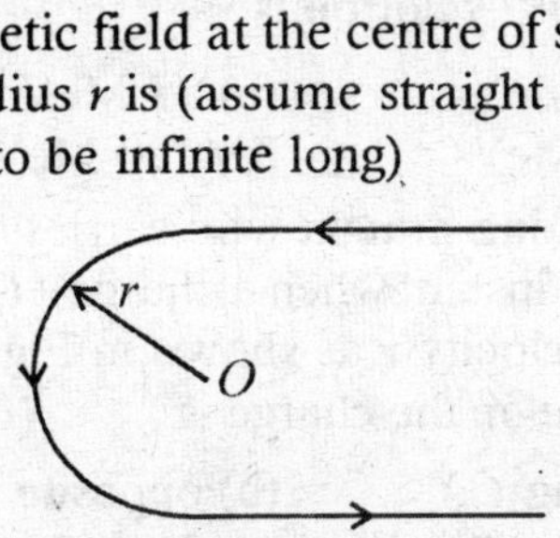

Fig. 23.133

(a) $\frac{\mu_0 i}{4\pi r}(\pi+2) \odot$ (b) $\frac{\mu_0 i}{4\pi r}(\pi-2) \odot$
(d) $\frac{\mu_0 i}{4\pi r}(\pi+2) \oplus$ (d) $\frac{\mu_0 i}{4\pi r}(\pi-2) \oplus$.

4. Figure 23.134 shows a closed loop made of wire. A current i passes through the loop as shown. The magnetic field at the centre of the loop is

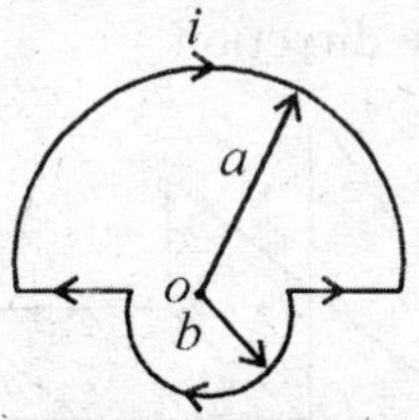

Fig. 23.134

(a) $\frac{\mu_0 i}{4\pi}\left(\frac{1}{a}+\frac{1}{b}\right)$ (b) $\frac{\mu_0 i}{4}\left(\frac{1}{a}+\frac{1}{b}\right)$
(c) $\frac{\mu_0 i}{4\pi}\left(\frac{1}{b}-\frac{1}{a}\right)$ (d) $\frac{\mu_0 i}{4}\left(\frac{1}{b}-\frac{1}{a}\right)$.

5. A solid sphere of radius r is given charge q which is uniformly distributed. It is rotated with angular velocity ω about its diameter. Its equivalent magnetic moment is

(a) $\frac{2q\omega r^2}{5}$ (b) $\frac{q\omega r^2}{5}$
(c) $\frac{2}{3}q\omega r^2$ (d) $\frac{1}{3}q\omega r^2$.

6. Two particles A and B of masses m_a and m_b respectively and having the same charge are moving in a plane. A uniform magnetic field exists perpendicular to this plane. The speeds of the particles are v_a and v_b respectively and trajectories are as shown in Fig. 23.135, then (IIT 2001)

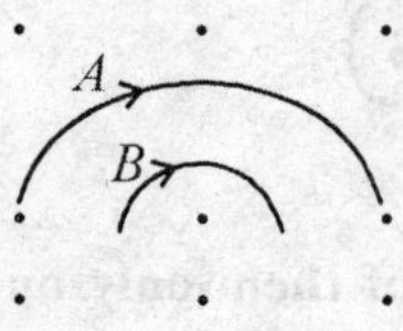

Fig. 23.135

(a) $m_a v_a < m_b v_b$
(b) $m_a v_a > m_b v_b$
(c) $m_a < m_b$ and $v_a < v_b$
(d) $m_a = m_b$ and $v_a = v_b$.

7. A non-planar loop of conducting wire carrying a current i is placed as shown in Fig. 23.136. Each of the straight sections of the loop is of length $2a$. The magnetic field due to this loop at point O (0, 0 0) points in the direction (IIT 2001)

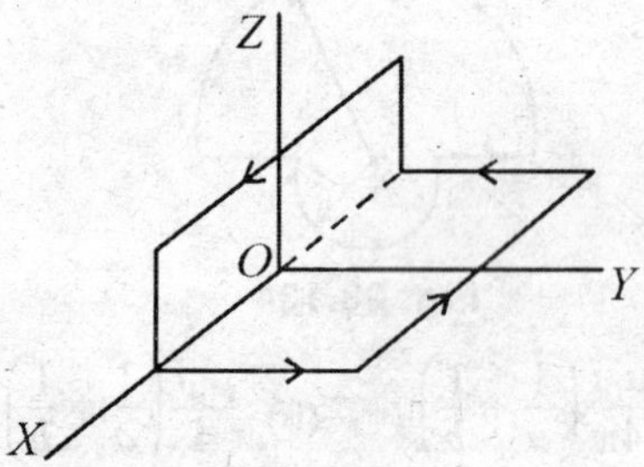

Fig. 23.136

(a) $\frac{1}{\sqrt{2}}(\hat{i}+\hat{j})$ (b) $\frac{1}{\sqrt{2}}(\hat{j}+\hat{k})$
(c) $\frac{1}{\sqrt{2}}(\hat{k}+\hat{i})$ (d) $\frac{1}{\sqrt{3}}(\hat{i}+\hat{j}+\hat{k})$.

8. A charged particle of mass m and charge q moves with a constant velocity v_1 along the positive X-direction. It enters a region containing a uniform magnetic field $\mathbf{B}$ directed along the negative Z-direction extending from $x = a$ to $x = b$. The minimum value of v required so that the particle can just enter the region $x > b$ is

(a) $\frac{qbB}{m}$ (b) $\frac{q(b-a)B}{m}$
(c) $\frac{qaB}{m}$ (d) $\frac{q(b+a)B}{2m}$.

9. An electron is moving in X-Y plane along the positive X-axis. There is a sudden change in its path due to its presence of electric and/or magnetic field beyond P as shown in Fig. 23.137. The curved path lies in X-Y plane and is found to be non-circular. Which of the following combinations is possible?

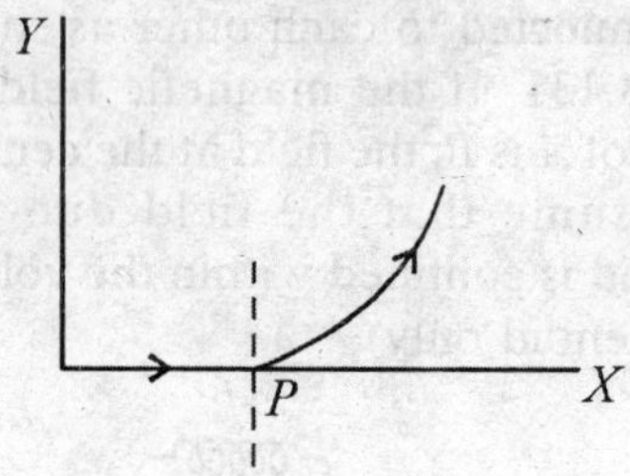

Fig. 23.137

(a) $\mathbf{E} = 0, \mathbf{B} = b\hat{i} + c\hat{k}$
(b) $\mathbf{E} = ai, \mathbf{B} = c\hat{k} + b\hat{i}$
(c) $\mathbf{E} = 0, \mathbf{B} = c\hat{j} + b\hat{k}$
(d) $\mathbf{E} = a\hat{i}$ $\mathbf{B} = c\hat{k} + b\hat{j}$.

10. A rectangular loop 5 cm × 4 cm and carrying a current 1.5 A is placed in a uniform magnetic field $B = 0.5$ T such that the direction of magnetic field is parallel to the plane of the loop. Force acting on the loop is

(a) 1.5×10^{-3} N
(b) 2×10^{-3} N
(c) 4.5×10^{-3} N
(d) None of the above.

11. Maximum energy of protons coming out of a cyclotron is 30 MeV. Maximum energy of deutrons which can be obtained from this cyclotron is

(a) 15 MeV (b) 75 MeV
(c) 60 MeV (d) 120 MeV.

12. A very long straight wire carries a current i. At the instant when a charge $+Q$ at point P has velocity v as shown in Fig. 23.138, the force on the charge is (IIT 2005)

(a) along OX (b) opposite OY
(c) along OY (d) opposite OX.

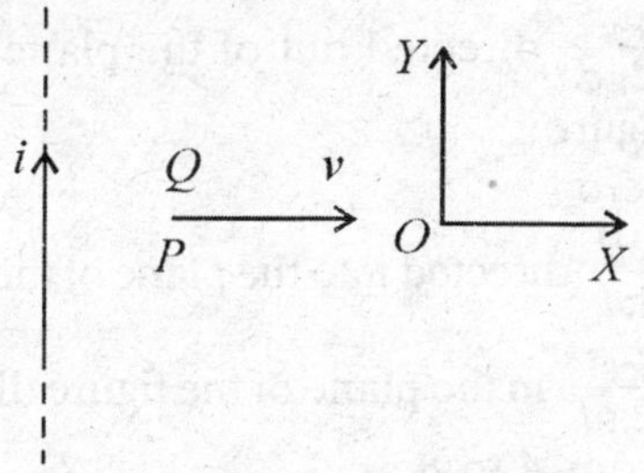

Fig. 23.138

13. A coil in the shape of an equilateral triangle of side l is suspended between the pole pieces of a permanent magnet such that B is in the plane of triangle. If due to current i in the triangle, a torque τ acts on it, the side l of the triangle is (CBSE PMT 2005)

(a) $\frac{2}{\sqrt{3}}\left(\frac{\tau}{Bi}\right)$ (b) $\frac{1}{\sqrt{3}}\frac{\tau}{Bi}$

(c) $2\left(\frac{\tau}{\sqrt{3}Bi}\right)^{1/2}$ (d) $\frac{2}{\sqrt{3}}\left(\frac{\tau}{Bi}\right)^{1/2}$.

14. A square loop of side $4R$ ($ABCD$) and a square loop of side $2R$ ($EFGH$) carry current i. Magnetic field at the centre O of the loop is

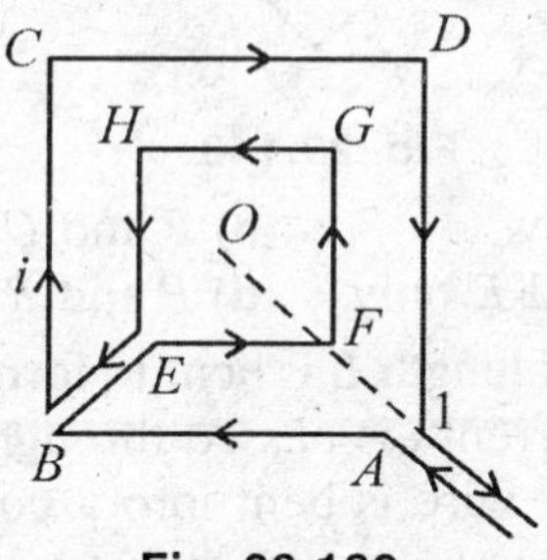

Fig. 23.139

(a) $\frac{\mu_0 i}{2\pi R}\odot$ (b) $\frac{\mu_0 i}{2\pi R}\oplus$

(c) $\frac{\mu_0 i}{\sqrt{2}\pi R}\oplus$ (d) $\frac{\mu_0 i}{\sqrt{2}\pi R}\odot$.

15. Two concentric coils each of radius equal to 2π are placed at right angles to each other. 3 A and 4 A are the currents flowing in each of coils respectively. The magnetic inductions in Web/m^2 at the centre of the coils will be (AIEEE 2005)

(a) 7×10^{-5} (b) 5×10^{-5}

(c) 10^{-5} (d) 12×10^{-5}.

16. Uniform electric and magnetic fields are acting along the same direction in a certain region. If an electron is projected along the direction of the field with a certain velocity, then (AIEEE 2005)

(a) it will turn towards left of direction of motion

(b) it will turns towards right of direction of motion

(c) its velocity will increase

(d) its velocity will decrease.

17. The earth has a magnetic dipole moment of 7.8×10^{22} A-m^2 . What current would have to be set up in a single turn of wire going around the earth at its magnetic equator if we wish to set up such a dipole through magnetic effects of current? (radius of earth is 6.4×10^3 m)

(a) 6.06×10^8 A (b) 3.03×10^8 A

(c) 6.06×10^4 A (d) 3.03×10^4 A.

18. An electric current i is passed through a circular loop of folded copper wire (Fig. 23.140). The magnetic induction at the centre is

Fig. 23.140

(a) zero (b) $\frac{2\mu_0 i}{r}$

(c) $\frac{\mu_0 i}{r}$ (d) $\frac{\mu_0 i}{2r}$.

19. Figure 13.141 shows two long conducting wires placed at right angles. They carry currents as indicated. Assume that 1 A current in a long wire produces magnetic field B at distance d from wire. The strength

and direction of the field at point X will be

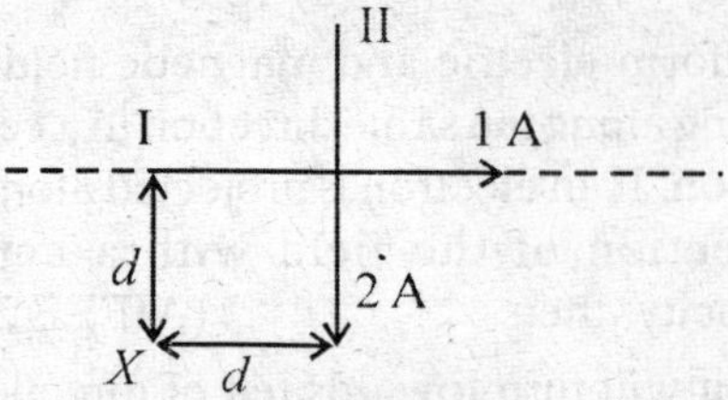

Fig. 23.141

(a) $3B$ into the plane of paper
(b) B into the plane of paper
(c) B out of the plane of paper
(d) $3B$ out of the plane of paper.

20. In Fig. 23.141, suppose wire II were free to rotate about its point of intersection with wire I. The wire II will have stable position
(a) as it is
(b) after turning 180° in plane of the paper
(c) after turning 90° clockwise in the plane of paper
(d) after turning 90° anticlockwise in the plane of paper.

21. Two short bar magnets have magnetic moments 1.2 A-m^2 and 1.0 A-m^2 respectively. They are placed horizontally on table paralled to each other with their north poles pointing towards the south. They have a common magnetic equator and are separated by a distance of 20.0 cm. The value of the resultant horizontal magnetic induction at the mid point O of the line joining their centres is (horizontal component of the earth's magnetic induction is 3.6×10^{-5} Wb/m^2)

(a) 3.6×10^{-5} Wb/m^2
(b) 2.56×10^{-4} Wb/m^2
(c) 3.50×10^{-5} Wb/m^2
(d) 5.8×10^{-4} Wb/m^2.

22. A and B are long wires carrying a current i each (Fig. 23.142). The magnetic field at point P midway between the wires is

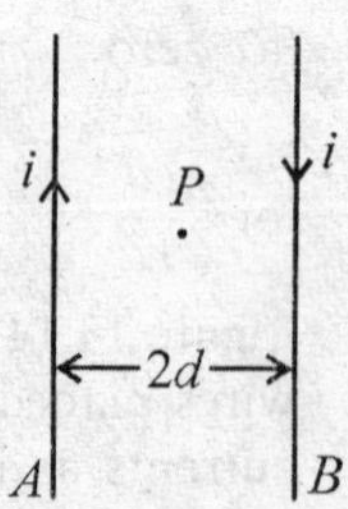

Fig. 23.142

(a) $\frac{\mu_0 i}{2\pi d}$ directed out of the plane of the figure
(b) zero
(c) $\frac{\mu_0 i}{\pi d}$ directed into the plane of the figure
(d) $\frac{\mu_0 i}{2\pi d}$ in the plane of the figure directed from A to B.

23. A circular coil of radius r carries a current i. The magnetic field at a point on the axis of the coil at a distance x from the centre is 1/8th of the field at the centre of the coil. Then x is

(a) $2\sqrt{2}r$ (b) $0.5r$
(c) $r^{3/2}$ (d) $\sqrt{3}r$.

24. Two thick wires and two thin wires, all of the same material and same length, form a square in three different ways, P, Q, R as shown in Fig. 23.143. With the current connections shown, the magnetic field at the centre of the square is zero in cases

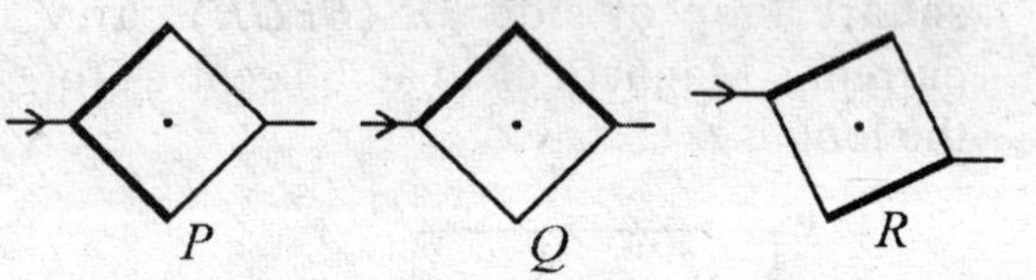

Fig. 23.143

(a) P only (b) P and Q only
(c) Q and R only (d) P and R only.

25. A wire of length L is bent to form a circle and a current i is passed through it. Now the same wire is bent into a coil of two turns and the same current is passed through it. The magnetic induction at the centre now as compared to its value in the first case is

(a) one quarter (b) one half
(c) twice (d) four times.

26. The magnetic induction associated with current flowing in a hollow copper tube will be

(a) only out side
(b) only inside

(c) both inside and outside
(d) neither inside nor outside.

27. The magnetic field at the centre O of the arc shown in the Fig. 23.144 is

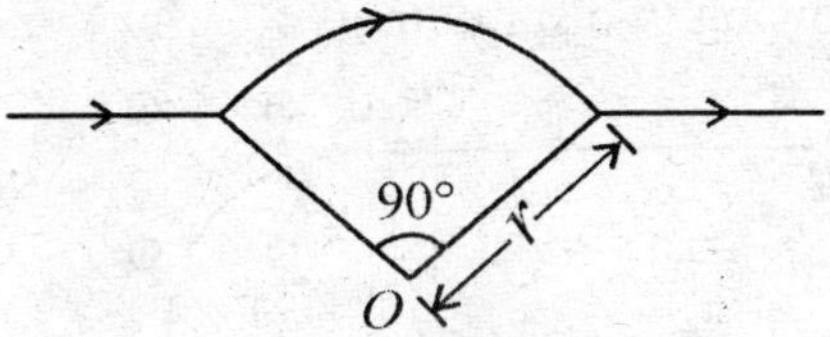

Fig. 23.144

(a) $2i(\sqrt{2} + \pi) \times 10^{-7}/r$

(b) $2i(\sqrt{2} + \pi/4) \times 10^{-7}/r$

(c) $i(\sqrt{2} + \pi) \times 10^{-7}/r$

(d) $i(\sqrt{2} + \pi/4) \times 10^{-7}/r$.

28. P, Q, R are long straight wires in air carrying currents as shown in Fig. 23.145. Force on Q is directed towards

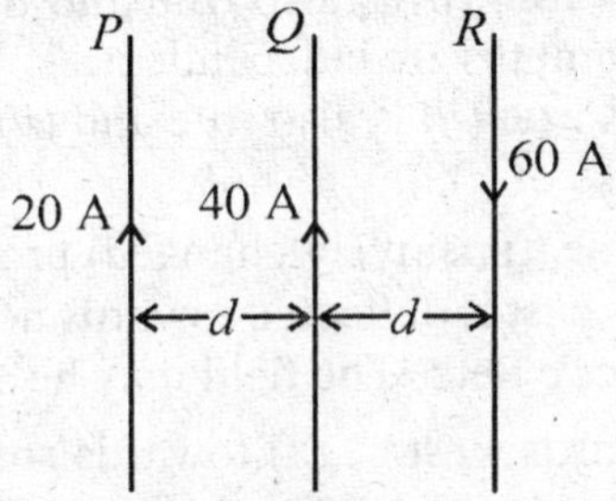

Fig. 23.145

(a) left
(b) right
(c) perpendicular to the plane of diagram
(d) along the current in Q.

29. The graph showing variation of magnetic field induction B with distance r in case of a long straight wire carrying current is

(a) 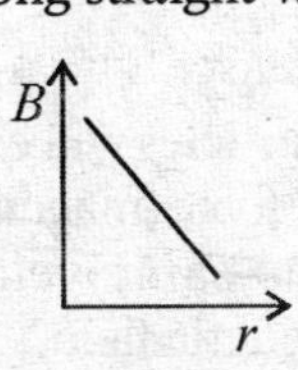

(b) 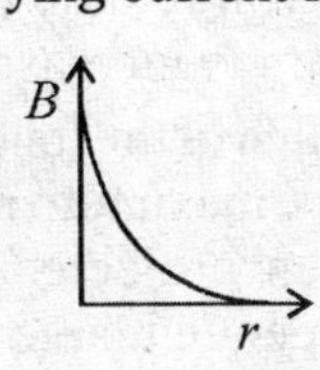

(c) 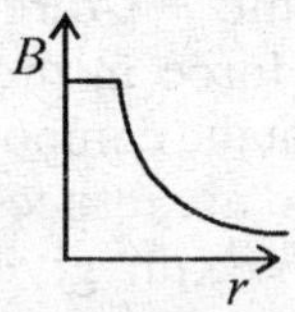

(d) 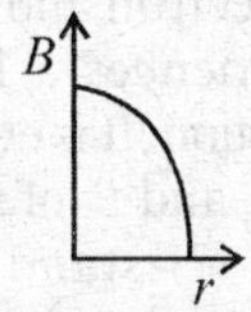

30. Which of the following relations between Weber (Wb), second (s), Ampere (A) volt (V) and Coulomb (C) is correct?

(a) Wb × s = V (b) Wb × s^{-1} = V
(c) Wb × s^{-1} = A (d) Wb × s = C.

31. Figure 23.146 shows three configurations of current and an observation point P. Values of i and r same in all cases. The straight portion extends to infinitly and the curved ones are circular. The magnetic induction at P in cases a, b and c have the ratio

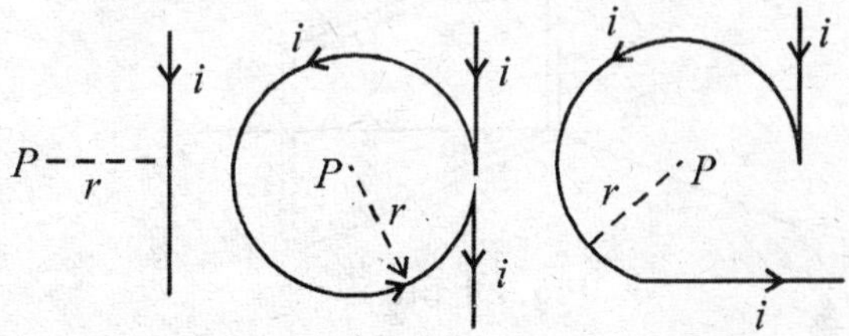

(a) (b) (c)

Fig. 23.146

(a) 1 : 1 : 1
(b) 1 : $(1 - \pi)$: $(-3\pi/4)$
(c) 1 : $(1 + \pi)$: $(3\pi/4)$
(d) 1 : $(\pi - 1)$: $(1 - 3\pi/4)$.

32. A straight wire, infinite long ABC (Fig. 23.147) carrying a current i would create a magnetic field B_0 at point P. If the wire is bent in the form $ABC'DE$ such that $BC'D$ forms a semicircle with P as centre and DE is parallel to AB, the magnetic field at P will be

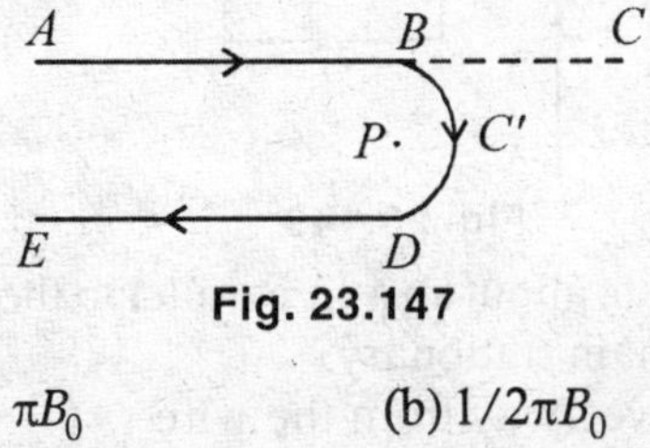

Fig. 23.147

(a) πB_0 (b) $1/2\pi B_0$
(c) $(1 - \pi/2)B_0$ (d) $(1 + \pi/2)B_0$.

33. A electron moving in the $+X$-direction experiences a Lorentz force in the +Z-direction. It is evident that the components B_x, B_y and B_z of the B field are respectively with (?) standing for "Expt. gives no information."

(a) ?, +ve, 0 (b) 0, +ve, ?
(c) 0, –ve, 0 (d) ?, –ve, 0.

34. Figure 23.148 shows a rectangular frame of side a = 40 cm, b = 30 cm. It has 80 turns of thin wire would on it; and carries a current 8.0 A. If there exists a uniform magnetic field of 40 mT directed along the X-axis, the magnitude of the torque experienced by the frame is nearest to

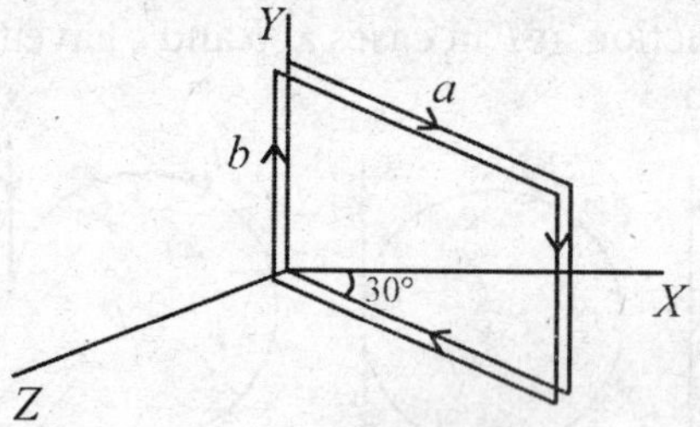

Fig. 23.148

(a) 2.7 N-m (b) 1.5 N-m
(c) 3.1 N-m (d) 2.5 N-m.

35. A rectangular loop carrying a current i is situated near a long straight wire carrying a steady current i such that the wire is parallel to one of the sides of the loop and is in the plane of the loop. Then current loop will

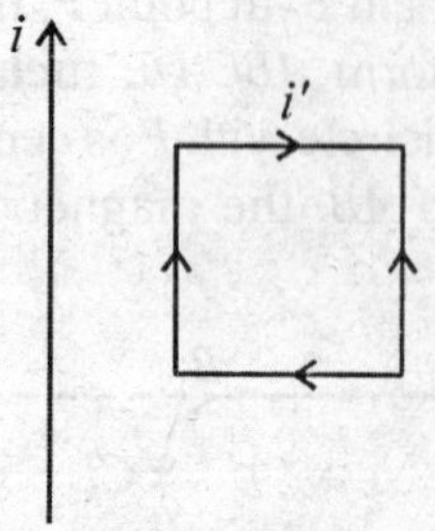

Fig. 23.149

(a) rotate about an axis parallel to the wire
(b) remain stationary
(c) move away from the wire
(d) move towards the wire.

36. When a charged particle moving with velocity v enters a region containing a perpendicular magnetic field, it moves along a semi-circle path of radius r as shown in Fig. 23.150. Consider the following two statements

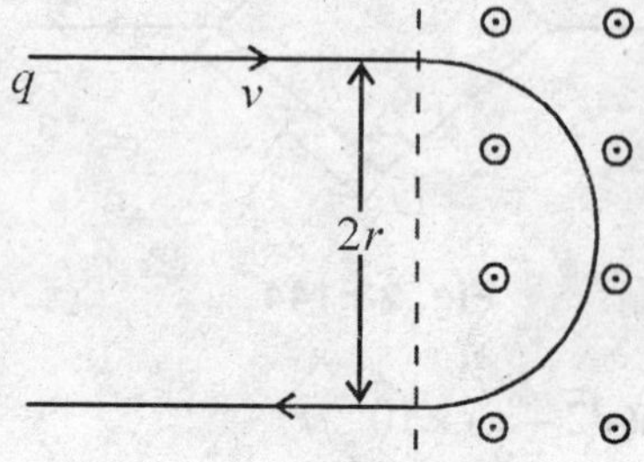

Fig. 23.150

(*i*) The radius 'r' of the circular path is proportional to the initial velocity
(*ii*) The time required for the particle to traverse the semi-circle is independent of v then

(a) (*i*) is true but (*ii*) is incorrect
(b) (*i*) is incorrect while (*ii*) is correct
(c) (*i*) is true (*ii*) is also true and the two statements are independent
(d) (*i*) is true (*ii*) is also true and (*ii*) is the cause of (*i*).

37. A particle, (positively changed) projected towards east is deflected towards north by a magnetic field. The field may be

(a) towards west (b) towards south
(c) upward (d) downward.

38. A charged particle is whirled in a horizontal circle on a frictionless table by attaching it to string fixed at one point. If a magnetic field is switched on in vertical direction, the tension in the string

(a) will increase
(b) will decrease
(c) will remain the same
(d) may increase or decrease.

39. Which of the following particles will experience maximum magnetic force (magnitude) when projected with the same velocity perpendicular to the magnetic field?

(a) electron (b) proton
(c) He^+ (d) Li^{++}.

40. A circular loop of area 1 cm^2 carrying a current of 10 A is placed in magnetic field of 0.1 T perpendicular to the plane of the loop. The torque on the loop due to the magnetic field is
(a) zero (b) 10^{-4} N-m
(c) 10^{-2} N-m (d) 1 N-m.

41. A vertical wire carries a current in upward direction. An electron beam sent horizontally towards the wire will be deflected
(a) towards right (b) towards left
(c) upwards (d) downwards.

42. A circular loop is kept in a vertical plane which contains the north south direction. It carries a current that is towards north at top most point. Let A be a point on the axis of circle to east of it and B a point on this axis to the west of it. The magnetic field due to the loop
(a) is towards east at A and towards west at B
(b) is towards west at A and towards east at B
(c) is towards east at A and B both
(d) is towards west at A and B both.

43. Two particles X and Y having equal charges after being accelerated through the same potential difference enter a region of uniform magnetic field and describe circular paths of radii R_1 and R_2 respectively. The ratio of mass of X to that of Y is
(a) $\left(\frac{R_1}{R_2}\right)^{1/2}$ (b) $\left(\frac{R_1}{R_2}\right)$
(c) $\left(\frac{R_1}{R_2}\right)^2$ (d) R_1R_2.

44. Two parallel long wires carry currents i_1 and i_2 with $i_1 > i_2$. When the currents are in the same direction, the magnetic field at a midway between them is 10 μT. If the direction of i_2 is reversed, the field becomes 30 μT. The ratio i_1/i_2 is
(a) 4 (b) 3
(c) 2 (d) 1.

45. Consider a long, straight wire of cross-sectional area A carrying a current i. Let there be n free electrons per unit volume. An observer places himself on a trolley moving in the direction opposite to the current with speed $v = \frac{i}{nAe}$ and separated from the wire by a distance r. The magnetic field seen by the observer is very nearly
(a) $\frac{\mu_0 i}{2\pi r}$ (b) zero
(c) $\frac{\mu_0 i}{\pi r}$ (d) $\frac{2\mu_0 i}{\pi r}$.

46. A current is uniformally distributed over the cross-section of a long hollow cylindrical wire of inner and outer radii R_1 and R_2 respectively. Magnetic field B varies with distance r from the axis of cylinder as

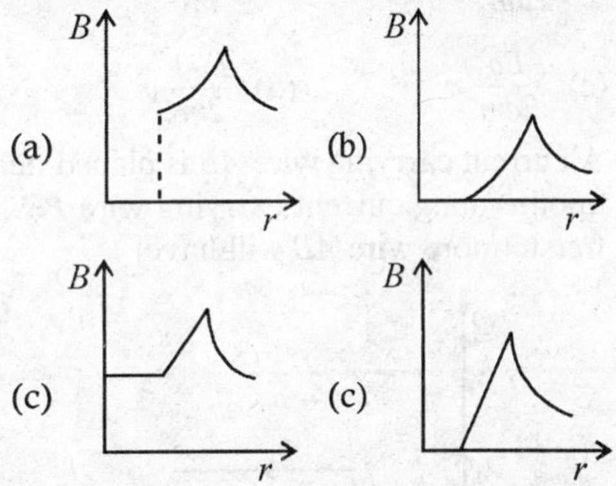

47. In Fig. 23.151 the net magnetic field at the centre O due to a current carrying loop is (θ < 180°)

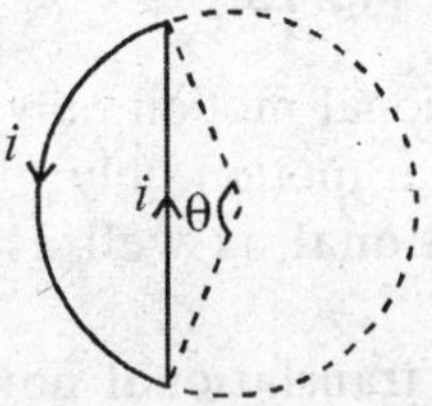

Fig. 23.151

(a) zero
(b) perpendicular to paper inwards
(c) perpendicular to paper outwards
(d) is perpendicular to paper inwards if $\theta \leq 90°$ and perpendicular to paper outwards if $90° \leq \theta < 180°$.

48. Magnetic field at the centre of circular loop of area A is B. Then the magnetic moment of the loop will be

(a) $\frac{BA^2}{\mu_0\pi}$ (b) $\frac{BA}{\mu_0}\sqrt{A}$

(c) $\frac{BA\sqrt{A}}{\mu_0\pi}$ (d) $\frac{2BA}{\mu_0}\sqrt{\frac{A}{\pi}}$.

49. A uniform magnetic field $\boldsymbol{B} = B_0\hat{j}$ exists in space. A particle of mass m and charge q is projected towards negative X-axis with speed v from a point $(d, 0, 0)$. The maximum value of v for which the particle does not hit the Y-Z plane is

(a) $\frac{2Bq}{dm}$ (b) $\frac{Bqd}{m}$

(c) $\frac{Bq}{2dm}$ (d) $\frac{Bqd}{2m}$.

50. A current carrying wire AB is placed near another long current carrying wire PQ. If free to more wire AB will have

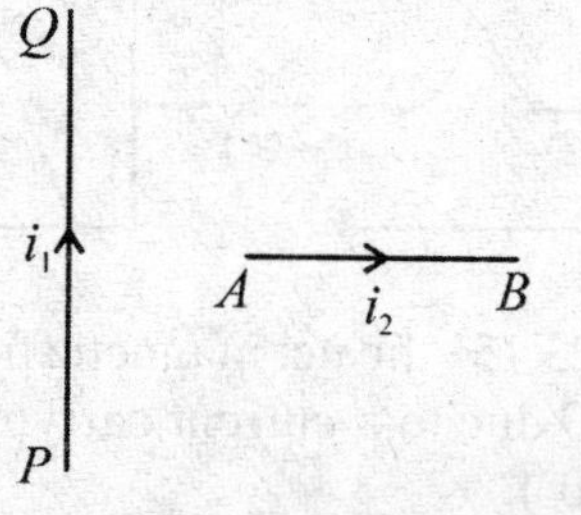

Fig. 23.152

(a) translational motion only
(b) rotational motion only
(c) translational as well as rotational motion
(d) neither translational nor rotational motion.

51. A current carrying loop is placed near an infinite long current carrying wire (Fig. 23.153). The torque acting on the loop is

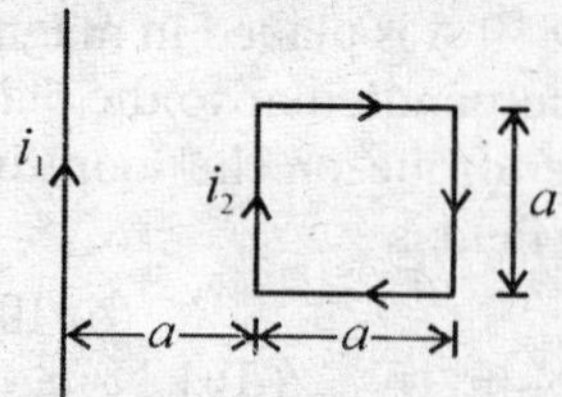

Fig. 23.153

(a) $\frac{\mu_0}{2\pi}\left(\frac{i_1 i_2 a}{2}\right)$ (b) $\frac{\mu_0 i_1 i_2 a}{2\pi}$

(c) $\frac{\mu_0 i_1 i_2 a}{2\pi}\log$ (d) zero.

52. Figure 23.154 shows a straight wire of length l carrying a current i. The magnitude of magnetic field produced by the current at point P is

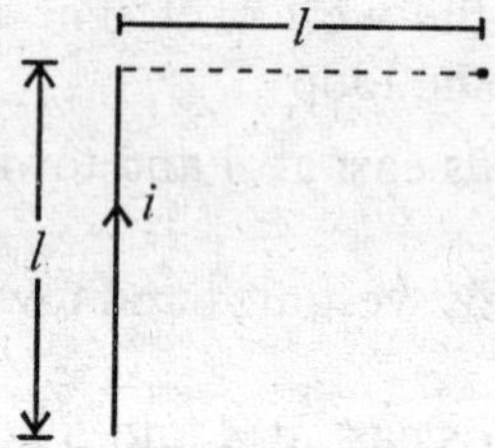

Fig. 23.154

(a) $\frac{\sqrt{2}\,\mu_0 i}{\pi l}$ (b) $\frac{\mu_0 i}{4\pi l}$

(c) $\frac{\sqrt{2}\,\mu_0 i}{8\pi l}$ (d) $\frac{\mu_0 i}{2\sqrt{2}\pi l}$.

53. Equal currents i flow in two segments of a circular loop in the direction shown (Fig. 23.155). Radius of the loop is a. Magnetic field at the centre of the loop is

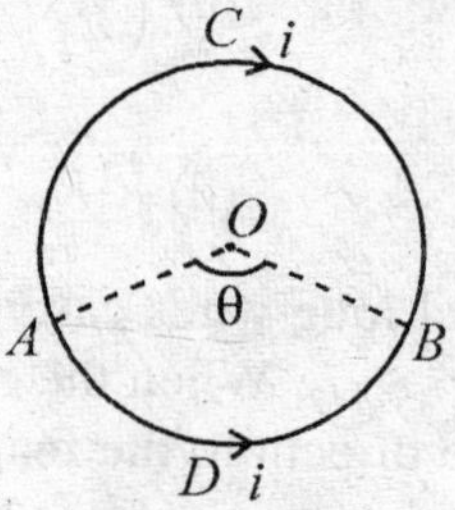

Fig. 23.155

(a) zero (b) $\left(\frac{\pi-\theta}{\pi}\right)\frac{\mu_0 i}{2a}$

(c) $\left(\frac{2\pi-\theta}{\pi}\right)\frac{\mu_0 i}{2a}$ (d) $\frac{\theta}{2\pi}\frac{\mu_0 i}{2a}$.

54. Same current $i = 2$ A is flowing in a wire frame as shown in Fig. 23.156. The frame is combination of two equilateral triangles ACD and CDE of side 1 m. It is placed in uniform magnetic field $B = 4$ T acting perpendicular to the plane of frame. The magnitude of the magnetic force acting on the frame is

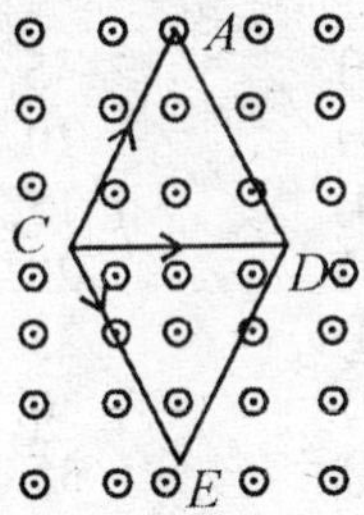

Fig. 23.156

(a) 24 N (b) zero
(c) 16 N (d) 8 N.

55. A charged particle enters a uniform magnetic field with velocity vector at an angle of 45° with the magnetic field. The pitch of the helical path followed by the particle is P. The radius of the helix is

(a) $\frac{P}{\sqrt{2}\pi}$ (b) $\sqrt{2}P$

(c) $\frac{P}{2\pi}$ (d) $\frac{\sqrt{2}P}{\pi}$.

56. Figure 23.157 shows a square current carrying loop of side 2 m and current 0.5 A. The magnetic moment of the loop is

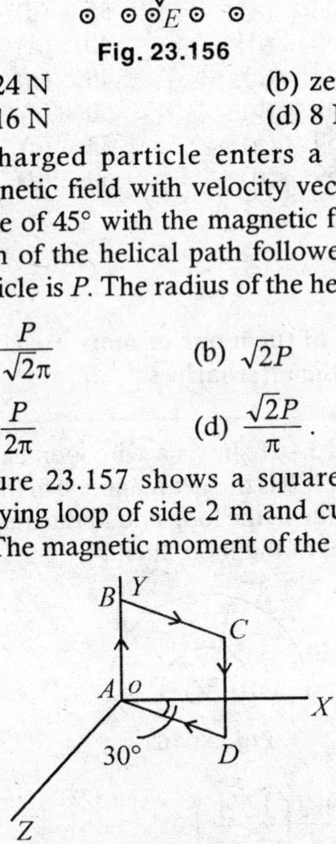

Fig. 23.157

(a) $(\hat{i}-\sqrt{3}\hat{k})$ A-m² (b) $(\hat{j}-\hat{k})$ A-m²
(c) $(\sqrt{3}\hat{i}+\hat{k})$ A-m² (d) $(\hat{j}+\hat{k})$ A-m².

57. Two identical particles having the same mass m and charge $+q$ and $-q$ separated by a distance d enter in a uniform magnetic field B directed perpendicular to the paper inwards with speed v_1 and v_2 as shown in Fig. 23.158. The particles will not collide if

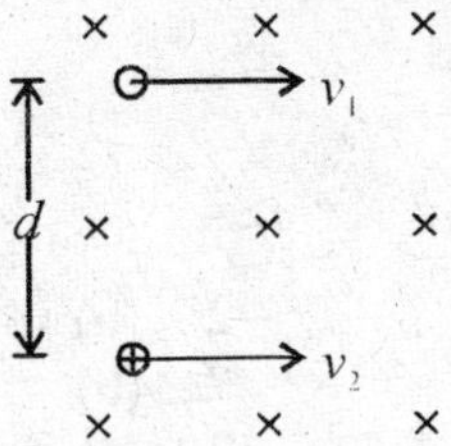

Fig. 23.158

(a) $d > \frac{m}{Bq}(v_1+v_2)$ (b) $d < \frac{m}{Bq}(v_1+v_2)$

(c) $d > \frac{2m}{Bq}(v_1+v_2)$ (c) $v_1 = v_2$.

58. An equilateral triangular loop ACD of side l carries a current i in the directions shown in Fig. 23.159. The loop is kept in a uniform magnetic field B as shown in Fig. 23.159. Net force on the loop is

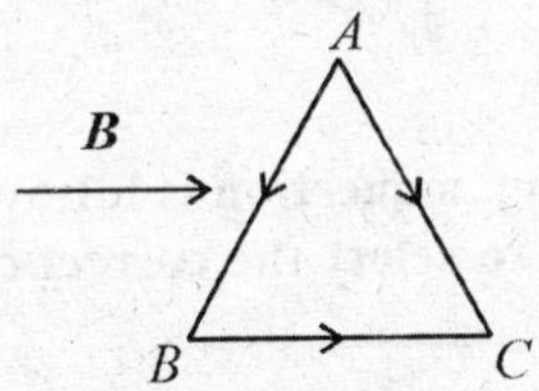

Fig. 23.159

(a) zero
(b) $\frac{\sqrt{3}}{2}ilB$
(c) perpendicular to paper inward
(d) perpendicular to paper outward.

59. Two protons are moving with same velocity in magnetic field of same magnitude then
(a) magnetic force on protons may be zero
(b) magnetic force on protons must be zero
(c) magnetic force on both may or may not be same due to each other

(d) both (*a*) and (*c*) are correct.

60. A circular coil having mass m is kept above the ground (X-Z plane) at some height. Coil carries a current i in the direction shown in Fig. 23.160.

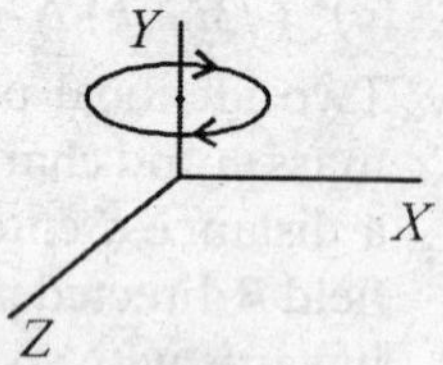

Fig. 23.160

In which direction a uniform magnetic field $\boldsymbol{B}$ be applied so that the magnetic field force balances the weight of the coil?

(a) positive X-direction
(b) negative X-direction
(c) positive Z-direction
(d) none of these.

ANSWERS

1. (b)	2. (b)	3. (a)	4. (b)	5. (b)
6. (b)	7. (b)	8. (b)	9. (b)	10. (d)
11. (a)	12. (c)	13. (c)	14. (d)	15. (b)
16. (d)	17. (a)	18. (a)	19. (a)	20. (d)
21. (b)	22. (c)	23. (d)	24. (d)	25. (d)
26. (a)	27. (b)	28. (a)	29. (b)	30. (b)
31. (b)	32. (d)	33. (d)	34. (a)	35. (d)
36. (d)	37. (d)	38. (d)	39. (d)	40. (a)
41. (c)	42. (d)	43. (c)	44. (c)	45. (a)
46. (b)	47. (c)	48. (d)	49. (b)	50. (c)
51. (d)	52. (c)	53. (b)	54. (a)	55. (c)
56. (a)	57. (c)	58. (d)	59. (d)	60. (d)

PART II

In the following, a question is followed by four alternatives, of them one or more are correct. Candidate has to select the correct choice(s) from amongst the alternatives.

1. An electron moves with a constant speed v along a circle of radius r.

(a) It is equivalent to a current loop with current $i = \dfrac{ev}{\pi r}$
(b) It is equivalent to magnetic dipole of moment $\mu = \dfrac{evr}{2}$
(c) Its angular momentum is $L = \dfrac{2m\mu}{e}$
(d) Magnetic field at the centre is $B = \dfrac{\mu_0\, ev}{4\pi r^2}$.

2. Figure 23.161 shows a wire loop carrying current. If B is the net magnetic field at the centre due to the loop and M the magnetic moment of the loop, then

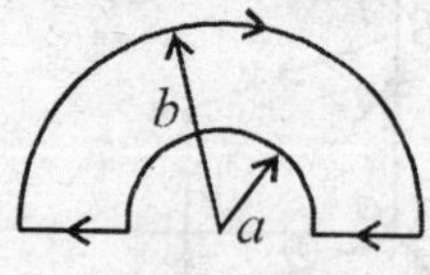

Fig. 23.161

(a) $B = \dfrac{\mu_0 i}{4}\left(\dfrac{1}{a} - \dfrac{1}{b}\right) \odot$

(b) $B = \frac{\mu_0 i}{4}\left(\frac{1}{a}+\frac{1}{b}\right)\oplus$

(c) $M = \frac{1}{2}\pi i (b^2 - a^2)\oplus$

(d) $M = \frac{1}{2}\pi i (b^2 - a^2)\odot$.

3. Two infinitely long parallel wires carry equal currents in the same direction (Fig. 23.162)

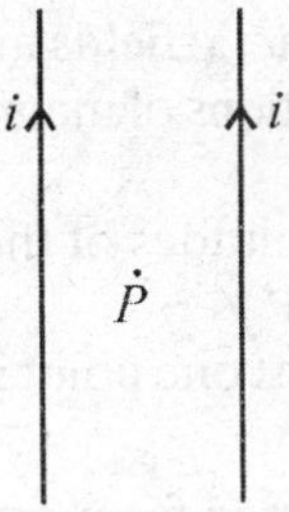

Fig. 23.162

(a) The magnetic field at point P equidistant from both wires is equal to twice the field due to each wire

(b) If the distance between the wires is doubled the force between the wires is halved

(c) If the current through the wires is doubled the force between then becomes four times

(d) The locus of the points of zero magnetic field is a straight line parallel to two wires.

4. A cyclotron oscillator has frequency 10 MHz

(a) Operating magnetic field to accelerate protons is approximately 0.6 T

(b) Operating magnetic field to accelerate protons is 1.0 T.

(c) If radius of dees is 0.6 m; then maximum energy of protons is approximately 15 MeV.

(d) If radius of dees is 1.0 m then maximum energy of protons is approximately 40 MeV.

5. Figure 23.163 shows a rectangular loop carrying current placed 2 cm away from a long straight current carrying current.

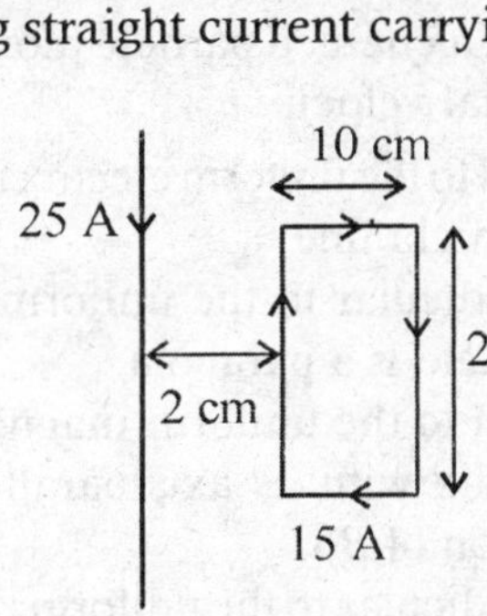

Fig. 23.163

(a) The net force on the loop is 3.125×10^{-5} N repulsive

(b) Net torque is 10^{-5} N-m anticlockwise

(c) Net torque is zero

(d) Net force is zero.

6. A solenoid of length 0.4 m and having 500 turns of wire carries a current of 3 A. A thin coil having 10 turns of wire and of radius 0.01 m carries a current of 0.4 A.

(a) The magnetic dipole moment of coil is $4\pi \times 10^{-5}$ A-m^2

(b) The magnetic dipole moment of coil is $4\pi \times 10^{-4}$ A-m^2

(c) Torque required to hold the coil in its position is 6×10^{-6} N-m

(d) Torque required to hold the coil in its position is 6×10^{-5} N-m.

7. Two particles of masses m_1 and m_2 with same charge enter a uniform magnetic field at right angles to the magnetic field. The ratio of radii of their path is

(a) $\frac{m_1}{m_2}$ if their velocities are same

(b) $\sqrt{\frac{m_1}{m_2}}$ if their velocities are same

(c) = 1 if their momentas are same

(d) $\frac{m_1}{m_2}$ if their kinetic energies are same.

8. Select the correct statement(s) from the following.

The path of a charged particle moving with an initial velocity u

(a) parallel to the uniform electrostatic field is a straight line
(b) perpendicular to the uniform electrostatic field is a parabola.
(c) parallel to the uniform magnetic field is a helix with its axis parallel to the direction of B
(d) perpendicular to the uniform magnetic field is a circle in plane normal to the field.

9. If a charged particle goes unaccelerated in a region containing electric and magnetic fields

(a) $\boldsymbol{E}$ must be perpendicular to $\boldsymbol{B}$
(b) $\boldsymbol{v}$ must be perpendicular to $\boldsymbol{E}$
(c) $\boldsymbol{v}$ must be perpendicular to $\boldsymbol{B}$
(d) E must be equal to vB.

10. Two ions have equal masses but one is singly ionized and the other doubly ionised. They are projected from the same place in a uniform magnetic field with the same velocity perpendicular to the field

(a) Both ions will go along circles of equal radii
(b) The circle described by the signly-ionized ion will have a radius double that of the other circle
(c) The two circles do not touch each other
(d) The two circles touch each other.

11. The magnetic field at the origin due to a current element (idl) placed at a position $\boldsymbol{r}$ is

(a) $\frac{\mu_0 (idl) \times \boldsymbol{r}}{4\pi r^3}$ (b) $-\frac{\mu_0 \boldsymbol{r} \times (idl)}{4\pi r^3}$

(c) $\frac{\mu_0 \boldsymbol{r} \times (idl)}{4\pi r^3}$ (d) $-\frac{\mu_0 (idl) \times \boldsymbol{r}}{4\pi r^3}$

12. Consider three quantities $x = \frac{E}{B}$, $y = \sqrt{\frac{1}{\varepsilon \mu_0}}$ and $z = \frac{l}{RC}$. Here l is length, R, resistance and C capacitance. All the other symbols have their usual meaning.

(a) x, y have the same dimensions
(b) x, z have the same dimensions
(c) z, x have the same dimensions
(d) None of the three pairs have same dimensions.

13. A long straight wire carries a current along the Z-axis. One can find two points in the X-Y plane such that

(a) the magnetic fields are equal
(b) the directions of magnetic fields are the same
(c) the magnitudes of the magnetic fields are equal
(d) the field at one point is opposite to that at the other.

14. A proton is fired from origin with velocity $\boldsymbol{v} = v_0 \hat{j} + v_0 \hat{k}$ in a uniform magnetic field $\boldsymbol{B} = B_0 \hat{j}$. In the subsequent motion of the proton

(a) its z co-ordinates can never be negative
(b) its x co-ordinates can never be positive
(c) its x and z co-ordinates cannot be zero at the same time
(d) its y co-ordinate will be proportional to its time of flight.

15. Two concentric circular coils of radii a and b ($a << b$) carry currents i_1 and i_2 respectively. If the smaller coil is rotated slightly about one of its diameter it starts oscillating. Then which of the following statement(s) is/are correct?

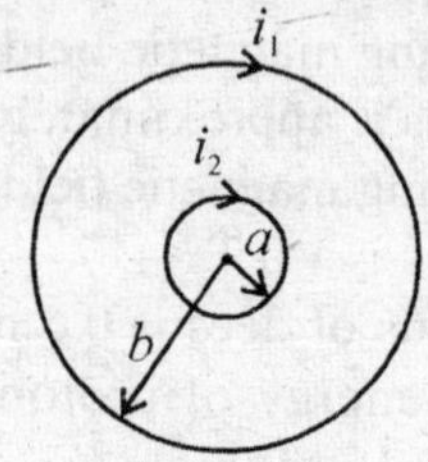

Fig. 23.164

(a) The oscillations are simple harmonic in nature

(b) The frequency of oscillation is proportional to product $i_1 i_2$.
(c) The frequency of oscillation is proportional to square root of b
(d) The frequency is independent of radius r.

16. A proton enters in a uniform electric ($\boldsymbol{E}$) and magnetic ($\boldsymbol{B}$) fields. Velocity of proton is $\boldsymbol{v}$. All the three vectors are mutually perpendicular. Proton is deflected along positive X-axis, when either of the fields or both are switched on simultaneously. Which of the statement(s) is/are correct?

(a) $\boldsymbol{v}$ may be along positive X-axis
(b) $\boldsymbol{E}$ is along positive X-axis
(c) $\boldsymbol{B}$ may be along positive Z-axis
(d) $\boldsymbol{B}$ may be along negative Z-axis.

17. A charged particle with velocity $v = x\hat{i} + y\hat{j}$ moves in magnetic field $\boldsymbol{B} = y\hat{i} + x\hat{j}$. The magnitude of magnetic force acting on the particle is F. Which one of the following statement(s) is/are correct?

(a) No force will act on particle if $x = y$
(b) $F \propto (x^2 - y^2)$ if $x > y$
(c) The force will act along Z-axis if $x > y$
(d) The force will act along Y-axis if $y > x$.

18. A rectangular loop as shown in Fig. 23.165 carries a current i. A uniform magnetic field $\boldsymbol{B} = B_0\hat{i}$ exists, then

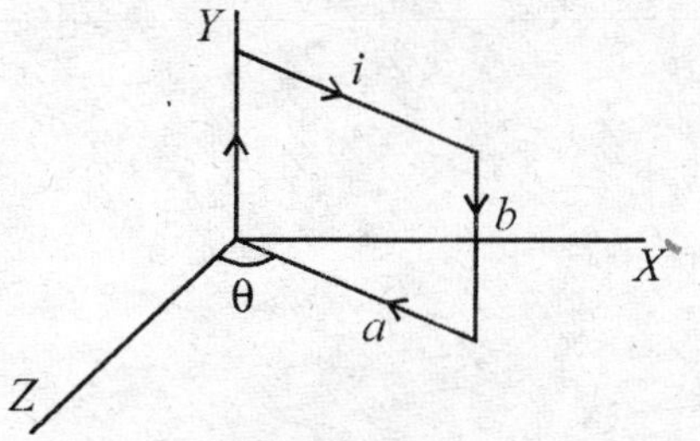

Fig. 23.165

(a) a torque $iabB_0 \sin\theta$ acts on the loop
(b) torque on the loop is in negative Y-direction
(c) if allowed to rotate, the loop turns so as to increase θ
(d) if allowed to rotate, the loop turns so as to decrease θ.

19. In a region where both non zero uniform electric field and magnetic field co-exist, the path of a charged particle

(a) cannot be a circle
(b) may be a circle
(c) may be a straight line
(d) may be a helix.

20. Magnetic field at the centre of a circular coil of radius R and carrying a current i is

(a) $\dfrac{\mu_0 i}{2R}$ (b) $\dfrac{i}{2C^2\varepsilon_0 R}$

(c) $\dfrac{\mu_0 i}{2\pi R}$ (d) $\dfrac{iC^2}{2\varepsilon_0 R}$

ANSWERS

1. (b, c, d)	2. (a, c)	3. (b, c, d)	4. (a, c, d)	5. (a, b)
6. (b, c)	7. (a, c, d)	8. (a, b, d)	9. (a, b)	10. (b, d)
11. (c, d)	12. (a, b, c)	13. (b, c, d)	14. (b, d)	15. (a, d)
16. (a, b, c)	17. (a, b, c)	18. (b, d)	19. (a, c, d)	20. (a, b)

24

Electromagnetic Induction

24.1 INTRODUCTION

In last chapter, we studied the magnetic effects of current and saw that the magnetic fields are produced by electric currents due to moving charges. These magnetic effects were first observed by Oersted in 1820. Since then attempts were made by scientists to observe the reverse effect. Michael Faraday and Joseph Henry observed that an emf is produced across the ends of conductor when the number of magnetic lines (called magnetic flux) of force linked with the conductor changes. The emf developed lasts till this change lasts. This phenomenon by virtue of which an emf is generated by changing the magnetic flux linked with the conductor is called electromagnetic induction. The emf developed in this way is called induced emf. If the circuit is complete, current flows in the circuit. This is called *induced current.*

24.2 MAGNETIC FLUX

Magnetic flux ϕ through any surface in a magnetic field $\boldsymbol{B}$ is measured as the total number of magnetic lines of force crossing the surface. Mathematically magnetic flux ϕ through an area $\boldsymbol{ds}$ in a magnetic field $\boldsymbol{B}$ is defined as

$$d\phi = \boldsymbol{B} \cdot \boldsymbol{ds} \qquad \ldots(24.1)$$

and to find flux through any surface of area s, we integrate equation (24.1) for the whole surface s. Then

$$\phi = \int_S \boldsymbol{B} \cdot \boldsymbol{ds}$$

The lines of force are imaginary but flux is a real scalar physical quantity.

Units and Dimensions of Magnetic Flux

Magnetic flux $\quad \phi = [B][s] = \left(\frac{F}{il}\right)s \qquad (\because F = ilB)$

$$= \frac{M^1L^1T^{-2}}{I\,L^1}L^2 = M^1L^2T^{-2}I^{-1}$$

$M^1L^2T^{-2}$ are the dimensions of energy. Hence

$$\text{Magnetic flux } \phi = \frac{\text{Energy}}{\text{Ampere}} = \frac{\text{Joule}}{\text{Ampere}} = \frac{\text{Joule-sec}}{\text{Coulomb}} \qquad \left(\because i = \frac{Q}{\text{Time}}\right)$$

$$= \text{Volt-sec} \qquad \left(\because \frac{\text{Joule}}{\text{Coul.}} = \text{Volt}\right)$$

$$= \text{Weber.}$$

Since units of $\boldsymbol{B}$ are Tesla (T). Hence units of ϕ are also (T-m^2)

In cgs system of units flux is measured in Maxwell (Mx).

$$1 \text{ Weber} = 1 \text{ Volt} \cdot \text{sec} = 10^8 \text{ emu of potential difference} \times \text{sec} = 10^8 \text{ Mx.}$$

Since magnetic flux is the dot product of magnetic field induction $\boldsymbol{B}$ and vector area $\boldsymbol{S}$, i.e.

$$\phi = \boldsymbol{B} \cdot \boldsymbol{S} = BS\cos\theta$$

where θ is the angle between magnetic field induction $\boldsymbol{B}$ and vector area $\boldsymbol{s}$. Hence $\phi = 0$ if $\theta = 90°$, i.e. $\boldsymbol{B}$ is parallel to the plane of the area and maximum when it is perpendicular to the plane of the area (Fig. 24.1).

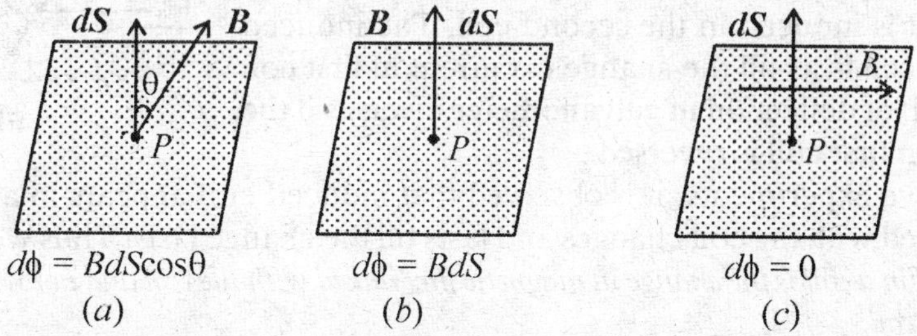

Fig. 24.1

In case of closed surface in magnetic field either uniform or non-uniform, outward flux is taken positive while incoming flux is taken as negative. Fig. 24.2 shows negative as well as positive flux.

If $\boldsymbol{B} = 0$, $\phi = \int_s \boldsymbol{B} \cdot d\boldsymbol{S} = 0$. However the converse may or may not be true. This is because if $\boldsymbol{B} \neq 0$, flux for some surface element may be positive and for other may be negative such that the net flux through the surface may be zero. For a closed surface in uniform magnetic field total flux is zero (Fig. 24.2). Similarly flux through a closed surface enclosing a magnetic dipole moment is zero. Total flux through closed surface is given by

$$\oint_S \boldsymbol{B} \cdot d\boldsymbol{S} = 0 \qquad \ldots(24.2)$$

This is called *Gauss's law for magnetism.*

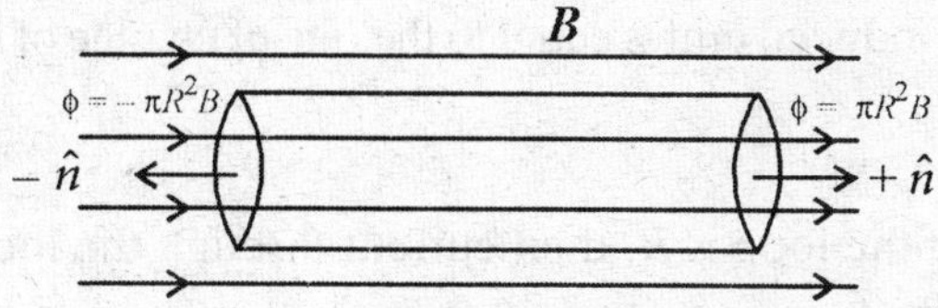

Fig. 24.2

24.3 ELECTROMAGNETIC INDUCTION AND FARADAY'S LAWS OF ELECTROMAGNETIC INDUCTION

(a) Faraday's Experiments

With the discovery of magnetic effects by electric currents, a link between electricity and magnetism was thought, i.e. electric current could be produced by magnetic fields. Faraday performed a series of experiments to produce induced currents.

Experiment 1: Figure 24.3 shows a circular loop with galvanometer in series. NS is a magnet placed along the axis of the loop. If the magnet is moved towards the loop, there is deflection in the galvanometer, i.e. there is current in the loop. The current is there as long as there is relative motion between magnet and loop. The deflection in the galvanometer is reversed if the direction of motion is reversed. It is also observed that deflection is more if the magnet is moved faster and less when the magnet is moved slowly.

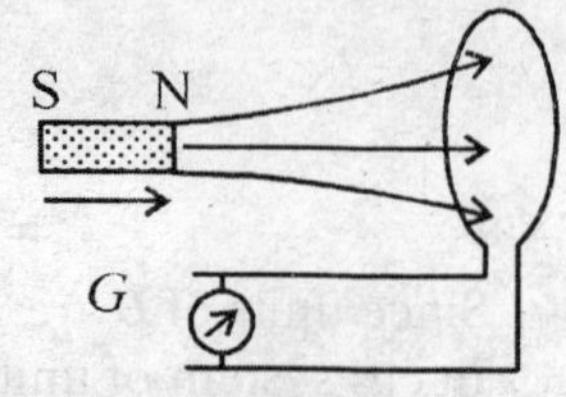

Fig. 24.3

Experiment 2: Figure 24.4 shows experimental set up of the experiment. The two coils are arranged so that a steady current flows in one and other has a galvanometer in series to detect induced current if any. It is observed that if current in first coil changes, a current is induced in the second coil. The induced current in second coil lasts till the change in current in first coil lasts. The direction of deflection in galvanometer reverses if the current direction in first coil is reversed.

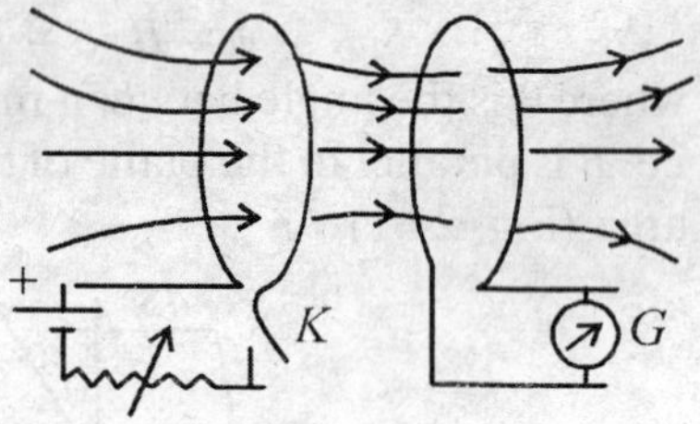

Fig. 24.4

In both the above experiments, it is observed that induced emf appears in a coil whenever the magnetic flux linked with the coil changes and lasts till the change lasts. Thus we conclude that the *cause of induced emf in a coil is the change in magnetic flux linked with the coil and not with the total magnetic flux linked with the coil.*

One can change the magnetic flux in a number of ways. Flux can be changed by changing the magnitude of the magnetic field $\boldsymbol{B}$ at the site of the loop, area of the loop, or the angle between the area vector $\boldsymbol{ds}$ and the magnetic field $\boldsymbol{B}$. In any case as long as the flux keeps on changing, the emf is present. The emf so produced is the cause of current in the loop.

(b) Faraday's Laws of Electromagnetic Induction

On the basis of results of the above described experiments, Faraday concluded that:

(*i*) Whenever there is change in magnetic flux linked with the area bounded by a closed conducting loop an emf is developed in it. The phenomenon is called electromagnetic *induction* and the *emf, induced emf.* This emf lasts till the change in magnetic flux lasts.

If the circuit is closed a current flows in the circuit called *induced current.*

(*ii*) The magnitude of the induced emf is equal to the rate of change of flux linked with the loop, i.e.

$$|e| = \frac{d\phi}{dt} \qquad \text{...(24.3)}$$

If the resistance of the loop is R, then current through the loop is

$$i = \frac{1}{R}\frac{d\phi}{dt} \qquad \text{...(24.4)}$$

and the charge through the loop

$$q = idt = \frac{1}{R}d\phi = \frac{1}{R}\text{ (change in flux)} \qquad \text{...(24.4}a\text{)}$$

The above two laws are called *Faraday's laws of electromagnetic induction.*

(c) Lenz's Law

This law gives us the direction of current induced in the circuit. According to *Lenz's law, the effect of the induced current is such as to oppose the very cause that produces it.*

If we combine Lenz's law with Faraday's second law the equation (24.3) can be written as

$$e = -\frac{d\phi}{dt} \quad \text{...(24.5)}$$

The negative sign shows that, if emf is produced due to increase in the flux linked with loop, the induced current produces the magnetic field and hence flux in such a way that it reduces the flux linked with the loop and *vice versa.*

In case of electromagnetic induction whenever flux linked changes an *emf* $|e| = \frac{d\phi}{dt}$ always exists, whether the circuit is closed or open. But the current exists only if the circuit is closed.

(d) Lenz's Law and Energy Conservation

Lenz's law is in accordance with the law of conservation of energy. The current is induced in the coil (loop) by changing magnetic flux. The induced current itself in the coil will produce a magnetic field and hence a magnetic flux.

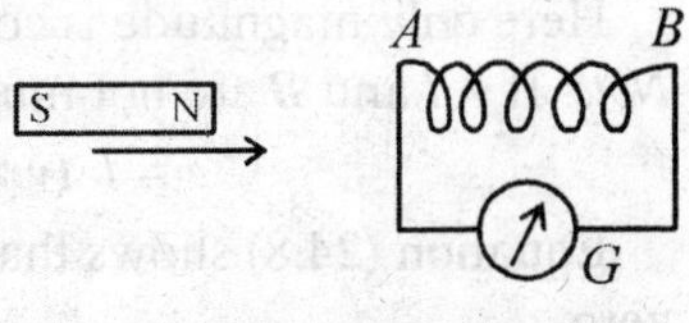

Fig. 24.5

Now suppose induced emf is produced by approaching N-pole of a magnet, then according to Lenz's law face *A* of coil acquires north polarity. As such work has to be done against the force of repulsion to bring the magnet closer to the coil. Similarly when N-pole moves away, South polarity developed on face *A* (Fig. 24.5). Therefore work has to be done against force of attraction in taking away the magnet from the coil.

In this process mechanical work is done in moving the magnet with respect to the coil that is converted into electrical energy producing induced current. In this way energy is merely transferred from mechanical to electrical energy. When the magnet is not moved no flux changes hence no induced current is produced. Thus Lenz's law can be treated as a consequence of the principle of energy conservation.

24.4 INDUCED EMF IN A CONDUCTING ROD MOVING IN UNIFORM MAGNETIC FIELD (MOTIONAL EMF)

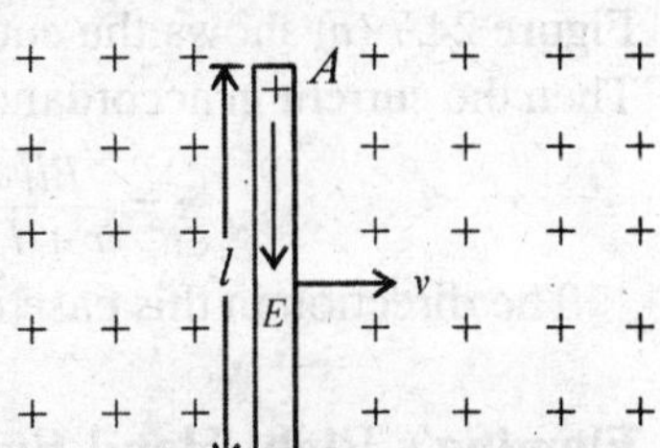

Fig. 24.6

Let us consider a conductor *AB* moving with uniform velocity *v* in uniform and time independent magnetic field (Fig. 24.6). For simplicity we have taken length (l) magnetic field (***B***) and velocity (***v***) mutually perpendicular to each other. The free electrons in the conductor also move with velocity ***v*** along with the random thermal velocity in the conductor. Since the average of random velocity is zero hence the net force on each electron due to magnetic field is $(-e)\ \mathbf{v} \times \mathbf{B}$. As the directions are shown in Fig. 24.6, this force is from *A* to *B*, i.e. the negative charge is accumulated at *B* and positive charge at *A*. As a result an electrostatic field (***E***) is developed within the conductor from *A* to *B*. This field exerts a force (= $(-e)\ \mathbf{E}$) on each free electron. The charges keep on accumulating till the two forces (magnetic force and electrostatic force) are equal in magnitude, i.e.

$$(-e)\,\boldsymbol{v} \times \boldsymbol{B} = (-e)\boldsymbol{E} \qquad \text{(since } \boldsymbol{v} \text{ and } \boldsymbol{B} \text{ are perpendicular)}$$

or
$$E = vB \qquad \ldots(24.6)$$

After this there is no resultant force on the free electrons. The potential difference between the ends A and B (= V) is then

$$E = \frac{V}{l} = vB$$

or
$$V = vlB \qquad \ldots(24.7)$$

As the emf is produced due to the motion of the conductor is called *motional emf.*

The *emf* developed in the conductor moving in magnetic field can be explained using Faraday's law by electromagnetic induction.

Conductor AB of length l is moved in magnetic field which is perpendicular to both l and v. If conductor moves a distance dx time dt, it sweeps area ldx and flux through the loop is $\phi = Bldx$ and the induced *emf*

$$e = -\frac{d\phi}{dt} = -Bl\frac{dx}{dt} = Blv$$

Here only magnitude is considered.

Note: If $\boldsymbol{v}$, $\boldsymbol{l}$ and $\boldsymbol{B}$ are not mutually perpendicular then induced *emf* developed is given by

$$e = \boldsymbol{l} \cdot (\boldsymbol{v} \times \boldsymbol{B}) \qquad \ldots(24.8)$$

Equation (24.8) shows that if any two of $\boldsymbol{l}$, $\boldsymbol{v}$ and $\boldsymbol{B}$ are parallel, the induced emf developed is zero.

Current in the Loop and Equivalent Circuit

If the ends A and B of the conductor are connected to an external resistance Fig. 24.7 (*a*), a current is established in the circuit. The current flows as shown in Fig. 24.7 (*a*).

Thus we can replace the moving conductor AB by a battery of emf vBl with positive terminal at A and the negative at B. The resistance of the conductor AB (= r) is taken to be the internal resistance of the battery. Figure 24.7 (*b*) shows the equivalent circuit. Then the current in accordance with Ohm's law

$$i = \frac{Blv}{r + R} \qquad \ldots(24.9)$$

The direction in this case is clockwise.

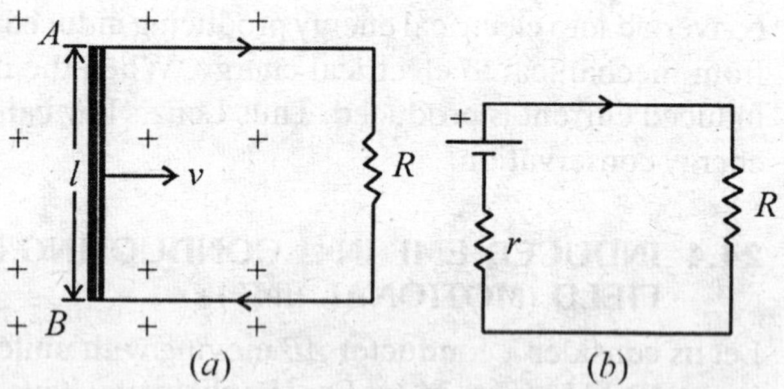

Fig. 24.7

Fleming's Right Hand Rule

Though in general, the direction of induced emf or current is determined by Lenz's law in case of moving conductor in magnetic

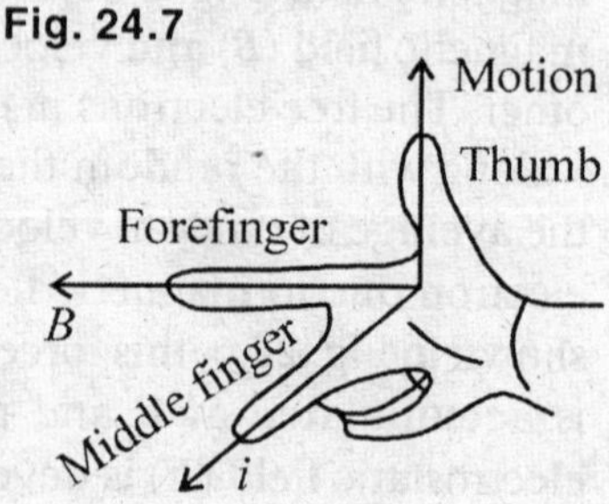

Fig. 24.8

field, it can also be determined by so called Fleming's right hand rule. According to this rule if we stretch forefinger, middle finger and thumb of right had mutually perpendicular to each other, such that the forefinger points in the direction of field and the thumb in the direction of motion then the middle finger will point in the direction of the induced current (Fig. 24.8).

24.5 EMF INDUCED IN A CONDUCTING ROD ROTATING IN MAGNETIC FIELD

Consider a conducting rod of length l rotating in magnetic field B with angular velocity ω. Magnetic field is perpendicular to the plane of diagram (Fig. 24.9) outward and the axis of rotation is parallel to the magnetic field. The induced emf in a small element of length dx at distance x from centre is

$$de = B \cdot \omega x \, dx$$

Hence total emf in the rod is

$$e = \int de = \int_0^l B\omega x \, dx = \frac{1}{2} B\omega l^2 \qquad \ldots(24.10)$$

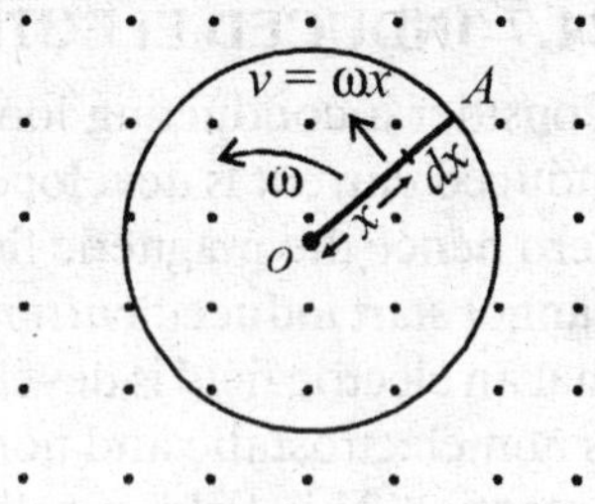

Fig. 24.9

In this way if a disc of radius r is made to rotate in magnetic field B, the induced emf developed between the centre and axes is

$$e = \frac{1}{2} B\omega r^2 = \frac{1}{2} B(2\pi f) r^2$$

where f is frequency of rotation

$$\therefore \qquad e = Bf(\pi r^2) = Bfs \qquad \ldots(24.11)$$

where $\pi r^2 = s$ the area of disc.

24.6 ENERGY CONSIDERATION: A QUANTITATIVE STUDY

Consider smooth railings PQ and RS placed in magnetic field B, directed perpendicular to the plane of diagram inwards. The railings are connected to resistance R. Over these railings a conductor AB is free to slide. If the conductor moves with velocity v towards right (Fig. 24.10) an induced emf equal to Blv is developed. Hence current through the conductor and resistance is

$$i = \frac{Blv}{R} \qquad \ldots(i)$$

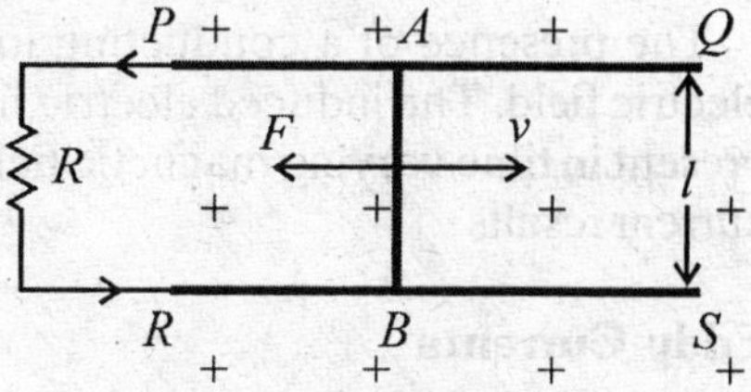

Fig. 24.10

The direction of the current is determined using Fleming's right hand rule. In present situation it is anticlockwise.

Since conductor AB carries current i and is moving in magnetic field B hence experiences force $F = ilB$ and the direction is opposite to the direction of motion according to Lenz's law.

If we want the conductor to continue to move with constant velocity v, an external force F must be applied in the direction of motion. Hence power supplied by the external agency to keep conductor moving is

$$P = \mathbf{F} \cdot \mathbf{v} = Fv = ilBv = \frac{Blv}{R} lBv$$

$$= \frac{B^2 l^2 v^2}{R} \qquad \ldots(ii)$$

Since the current passes through the conductor and resistor, heat produced due to Joule effect per second is

$$P_j = i^2 R = \frac{B^2 l^2 v^2}{R^2} R = \frac{B^2 l^2 v^2}{R} \qquad \ldots(iii)$$

This heat lost is exactly equal to the power supplied for conductor to move with velocity v, by the external agency.

Thus mechanical energy required to move conductor AB is converted into electrical energy and then to thermal energy. Thus Lenz's law is consistent with the conservation of energy.

24.7 INDUCED ELECTRIC FIELD AND EDDY CURRENTS

Consider a conducting loop at rest in magnetic field $\boldsymbol{B}$. If the magnetic field is time varying, an induced current is developed in the loop. Since the average thermal velocity of the free electrons is zero hence the magnetic field cannot exert force on free electrons at rest. Thus the magnetic field cannot start induced current. But the current is induced in the conducting loop, hence we conclude that an electric field is developed due to a time varying magnetic field. The electric field so produced is non-electrostatic and non conservative* in nature. Corresponding to this field, we cannot define potential. This field is called *induced electric field.* The field lines of force in this field are closed curves as such there are no starting and terminating points.

If $\boldsymbol{E}$ is the electric field induced due to time varying magnetic field the force on charge q is $q\boldsymbol{E}$. Hence work done in moving a unit charge through $\boldsymbol{dl}$ is $\boldsymbol{E} \cdot \boldsymbol{dl}$. The emf developed in the loop then is

$$e = \oint \boldsymbol{E} \cdot \boldsymbol{dl}$$

And using Faraday's law of electromagnetic induction

$$e = -\frac{d\phi}{dt}$$

$$\therefore \qquad \oint \boldsymbol{E} \cdot \boldsymbol{dl} = -\frac{d\phi}{dt} \qquad \ldots(24.12)$$

The presence of a conducting loop in time varying magnetic field is not necessary to induce electric field. The induced electric field is always present in time varying magnetic field. If a loop is present in time varying magnetic field, the free electrons start drifting and consequently an induced current results.

Eddy Currents

When a solid conducting plate is placed in changing magnetic field so that flux linked with it changes, *circulating currents called eddy current* (because they resemble in shape with eddies in a fluid) are induced in the plate. As the resistance of the bulk conductor is usually small, eddy currents have usually large magnitude and heat up the conductor.

Let a solid metallic plate on which a loop is drawn enters a region where a magnetic field exists (Fig. 24.11 *a*). As the plate moves, the magnetic flux through the area bounded by the loop changes

---

* This electric field is different from field produced by charged particles according to Coulomb's law or Gauss's law.

and hence a current is induced. There may be a number of such loops on the plate and hence currents are induced on the surface along a variety of paths. These currents are called eddy currents.

In fact we do not have a definite loop to guide the induced currents. The system itself looks for the loops on the surface along which eddy currents are induced. As already stated, heat is developed due to these currents. This thermal energy comes at the cost of the kinetic energy of the plate and the plate slows down. This is called electromagnetic damping. To reduce electromagnetic damping, one can cut slots in the plate (Fig. 24.11 (*b*)) this reduces the possible paths of the eddy currents considerably.

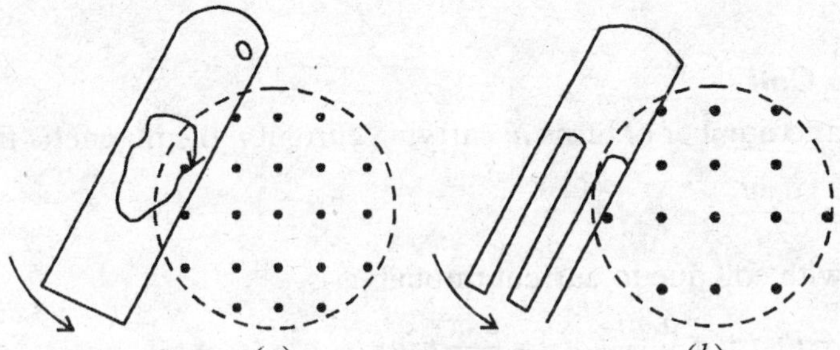

Fig. 24.11

24.8 TYPES OF ELECTROMAGNETIC INDUCTION AND INDUCTANCE

There are two types of electromagnetic inductions: (*a*) Self-induction (*b*) Mutual induction.

(*a*) Self-induction

When an electric current passing through a closed conducting loop, changes, the magnetic flux linked with the loop also changes. As a result an emf is induced in the loop. This phenomenon is called self-induction.

The magnetic field at any point due to current is proportional to the current. Hence the magnetic flux through the area bounded by a current carrying loop is, therefore proportional to the current. Hence we can write

$$\phi = Li \qquad \text{...(24.13)}$$

where L is constant called coefficient of self inductance of the loop and depends on the geometrical construction of the loop. The induced emf then is

$$e = -\frac{d\phi}{dt} = -L\frac{di}{dt} \qquad \text{...(24.14)}$$

Coefficient of self induction can be defined using equations (24.13) and (24.14).

In equation (24.13) if $i = 1$ then $\phi = L$

Thus *coefficient of self induction of a coil is numerically equal to the flux linked with it when unit current passes through it.*

If a current of 1A produces a flux equal to 1 Wb (T-m²), the coefficient of induction of such coil or loop is said to be 1 Henry.

In equation (24.14), if $\frac{di}{dt}$ = unity, then $e = L$

So *coefficient of self induction of a coil is numerically equal to the emf induced in the coil or loop when the current in the loop varies at unit rate.*

If current in the loop varies at the rate of 1 A/s and induced emf in the loop is 1 V, then inductance of the coil is said to be 1 Henry (abbreviated as H).

Dimensions of L are

$$L = \frac{e}{\frac{di}{dt}} = \frac{e\,dt}{di} = \frac{\text{Joule Sec}}{\text{Coul. Amp}}$$

$$= \frac{M^1L^2T^{-2}T^1}{ITI} = M^1L^2T^{-2}I^{-2} \quad \text{...(24.15)}$$

Self Inductance of a Coil

For a coil of radius a and number of turns n, carrying current i, the magnetic field at the centre is

$$B = \frac{\mu_0 n i}{2a}$$

Hence flux linked with coil due to current through it is

$$\phi = B\pi a^2 n = \frac{\mu_0 n i}{2a}\pi a^2 n = \frac{\mu_0 \pi}{2} n^2 a i$$

and from definition self inductance of this coil is

$$L = \frac{\phi}{i} = \frac{\mu_0 \pi}{2} n^2 a \quad \text{...(24.16)}$$

Self Inductance of a Solenoid

In case of a solenoid having n number of turns per unit length and of length l carrying current i, magnetic field inside the solenoid is

$$B = \mu_0 n i$$

If cross-sectional area of solenoid is S, then flux linked with it is

$$\phi = B \cdot (nl) S \qquad (\because \text{ total no. of turns} = nl)$$

$$= \mu_0 n i \cdot nl \cdot S = \mu_0 n^2 l S i$$

$$\therefore \quad L = \frac{\phi}{i} = \mu_0 n^2 l S \quad \text{...(24.17)}$$

Energy Stored in the Coil

If the current through the coil is changing at the rate $\frac{di}{dt}$, when instantaneous current is i, then induced emf is

$$e = L\frac{di}{dt}$$

and the power required to pass current i against induced emf is

$$P = L\frac{di}{dt} i$$

and energy required (work done) is

$$dW = Pdt = Lidi$$

Hence total work done in establishing current i is

$$W = \int_0^i Li\,di = \frac{1}{2}Li^2 \qquad \text{...(24.18)}$$

This work is stored as '*magnetic potential energy*' which is not localized but is distributed in the field associated with the current carrying coil.

Thus magnetic potential energy

$$U_m = W = \frac{1}{2}Li^2 \qquad \text{...(24.18a)}$$

From equation (24.18*a*) if $i = 1$ A, $L = 2$ W, i.e. the coefficient of *self inductance of a coil is also numerically equal to twice the work required to establish a unit current in the coil.*

Coils in Series and Parallel

If two coils having inductances L_1 and L_2 are connected in series, the equivalent inductance L_S is (Fig. 24.12 *a*)

$$L_S = L_1 + L_2 \qquad \text{...(24.18b)}$$

If the two coils are connected in parallel then equivalent inductance L_P is (Fig. 24.12 *b*)

$$\frac{1}{L_P} = \frac{1}{L_1} + \frac{1}{L_2} \qquad \text{...(24.19)}$$

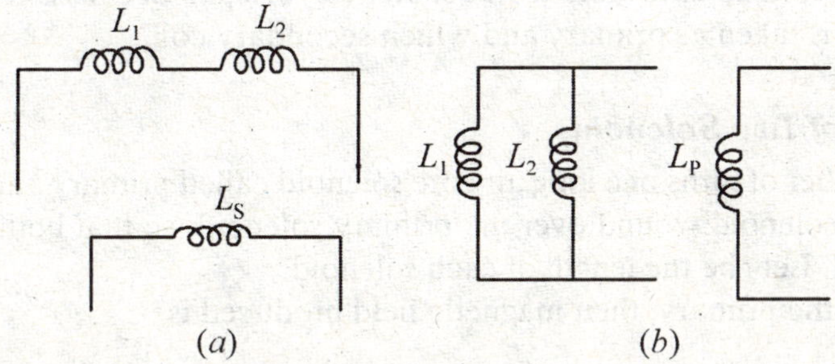

Fig. 24.12

(*b*) Mutual Induction

Mutual induction is the property of two coils by virtue of which each opposes any change in the strength of current flowing through the other by developing an induced emf.

Consider two closed loops placed close to each other and current i is passed in one. This current produces a magnetic field and this field has a flux ϕ linked through the area bounded by the other loop. Since the magnetic field at any point is proportional to the current producing it, hence the flux linked with the other loop can be written as

$$\phi = Mi \qquad \text{...(24.20)}$$

where M is constant called *mutual inductance* of the given pair of loops. It depends on the geometrical shapes of the two loops and their relative orientation. Here the loop through which

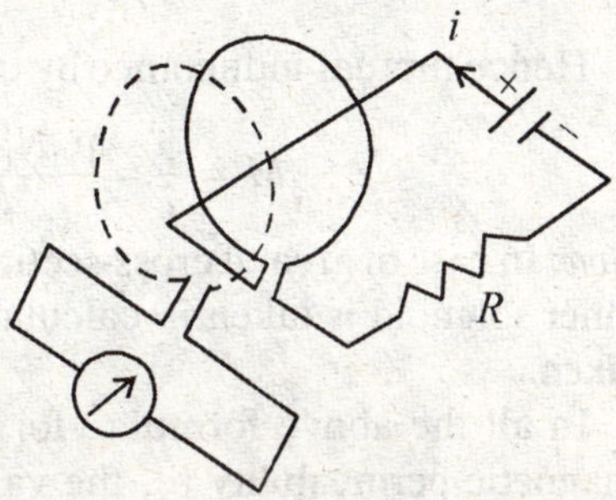

Fig. 24.13

current is passed is called *primary coil* (*or loop*) and the loop in which the current is induced is called *secondary coil* (*or loop*).

If the current i changes with time in primary, the flux through the secondary also changes and an emf is induced in the secondary coil. Induced emf is

$$e = -\frac{d\phi}{dt} = -M\frac{di}{dt} \qquad ...(24.21)$$

Here also like self inductance, the mutual inductance of two coils can be defined using equations (24.20) and (24.21).

In equation (24.20), if $i = 1$, $\phi = M$.

Thus coefficient of mutual inductance of two coils is numerically equal to the flux linked with one coil (secondary), when unit current passes through primary.

If a current of 1 amp. in primary coil produces a flux of 1 T-m^2 through the secondary then mutual inductance of the two coils is said to be 1 Henry.

In equation (24.21), if $\frac{di}{dt} = 1$, then $e = -M$

i.e. coefficient of mutual induction is numerically equal to emf induced in one coil (secondary) when the rate of change of current in primary is unity. If a current change at the rate of 1 ampere/sec in primary produces an induced emf of 1 Volt in secondary, the mutual inductance of pair of coils is said to be 1 Henry.

It should be remembered that the mutual inductance (M) of a pair of coils is constant, irrespective of the fact, which coil is taken as primary and which secondary coil.

Mutual Inductance of Two Solenoids

Let N_1 is the total number of turns of a long air core solenoid called primary and N_2 be the number of turns in secondary solenoid wound over the primary solenoid, so that both have nearly same area of cross-section A. Let l be the length of each solenoid.

If i is the current in the primary, then magnetic field produced is

$$B_1 = \mu_0 \frac{N_1}{l} i \qquad ...(i)$$

Hence the magnetic flux linked with the secondary is

$$\phi = B_1 A N_2$$

$$= \frac{\mu_0 N_1}{l} i A N_2 = \frac{\mu_0 N_1 N_2 A}{l} i$$

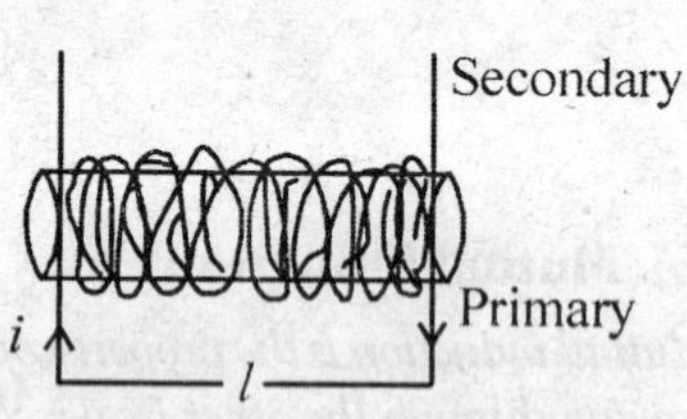

Fig. 24.14

Hence mutual inductance by definition is

$$M = \frac{\phi}{i} = \frac{\mu_0 N_1 N_2 A}{l} \qquad ...(24.22)$$

Note: In case of area of cross-section of the two solenoids are different, then, area of cross-section of inner solenoid is taken in calculating M for the solenoids. Similarly length of smaller solenoid is taken.

In all the above formulae for inductance (L) and mutual inductance (M) the medium is of magnetic permeability μ_r, the value of inductance L and mutual inductance M is increased by factor μ_r.

The mutual inductance of two coils having self inductance L_1 and L_2 is given by

$$M = k\sqrt{L_1 L_2} \qquad \ldots(24.23)$$

where k is constant called *coefficient of coupling (or coupling constant)*. If the coils are wound over each other, the coupling is said to be '*tight*' otherwise '*loose*'. For tight coupling $k = 1$ and $M = \sqrt{L_1 L_2}$ while loose coupling $0 < k < \sqrt{L_1 L_2}$.

An inductance is said to be ideal if it has no resistance. In practice due to finite resistance of conductors an inductance always has a resistance, i.e. one cannot have inductance without resistance. However converse may or may not be true, i.e. one can have a resistance with or without having inductance. A resistance without inductance is called non inductive resistance.

24.9 GROWTH AND DECAY OF CURRENT IN L-R CIRCUIT

(*a*) Growth of Current

Inductance L and resistance R are connected in series to a battery through a key. Initially when key is open there is no current in the circuit (Fig. 24.15). At $t = 0$, the key ab is closed, the current in the circuit will grow with time. If i be the current in the circuit at any time t and varies at the rate $\frac{di}{dt}$; then a self induced emf $\left(= -L\frac{di}{dt}\right)$ is produced. Hence using Kirchhoff's law

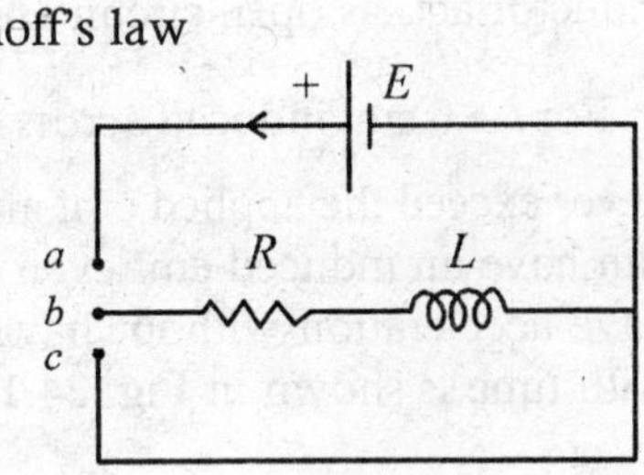

Fig. 24.15

$$E - L\frac{di}{dt} = Ri \qquad \ldots(i)$$

where E is the emf of the battery.

Now rearranging the terms, we get

$$L\frac{di}{dt} = E - Ri = R\left(\frac{E}{R} - i\right) \qquad \text{If } \frac{E}{R} = i_0$$

Then $\dfrac{di}{i_0 - i} = \dfrac{R}{L}dt$

Integration gives

$$-\log(i_0 - i) = \frac{R}{L}t + C, \text{ where } C \text{ is constant of integration}$$

At $t = 0$, $i = 0$, gives $C = -\log i_0$

or $$\log\frac{i_0 - i}{i_0} = -\frac{R}{L}t \qquad \text{i.e.} \qquad \frac{i_0 - i}{i_0} = e^{-\frac{R}{L}t}$$

or $$i = i_0\left(1 - e^{-\frac{R}{L}t}\right) \qquad \ldots(24.24)$$

Differentiating equation (24.24), we get

$$\frac{di}{dt} = \frac{i_0 R}{L}e^{-\frac{R}{L}t} = \frac{E}{L}e^{-\frac{R}{L}t} \qquad (\because i_0 = E/R) \qquad \ldots(24.25)$$

The current in circuit grows exponentially with time from $i = 0$ at $t = 0$ to i_0 ($= E/R$) at $t = \infty$ as shown in Fig. 24.16 (*a*). The constant L/R has dimensions that of time and is called *time constant* of the *L-R* circuit. After time $t = \tau$ (after one time constant)

$$i = i_0\left(1 - \frac{1}{e}\right) = 0.63\, i_0$$

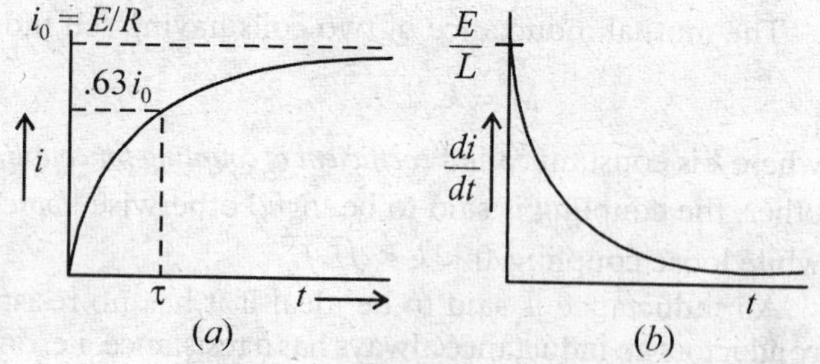

Fig. 24.16

Thus the time constant of L-R circuit is numerically equal to the time in which the current in circuit grows to 0.63 (or 63%) of the maximum value in the circuit.

The time constant tells us how fast will the current grow. If time constant is small, the growth is steep. In principle it takes infinite time for the current to attain its maximum value. In practice however a small number of time constants may be sufficient for the current to reach almost the maximum value.

Since at $t = 0$, $i = 0$ hence potential difference across *R*, the resistance is zero and since $\frac{di}{dt} = \frac{E_0}{L}$, hence potential difference across *L*, the inductance is *E*, i.e. immediately after closing the key inductor acts as open circuit and after a long time, i.e. in steady state it behaves as a short circuit.

For $t \geq 0$ emf induced across *L* is $|e| = L\frac{di}{dt} = Ee^{-\frac{R}{L}t}$ i.e. $e \leq E$, i.e. induced emf in the circuit can never exceed the applied emf and is independent of the current. We can conclude that an inductor can have an induced emf even in the absence of current through it. This is similar as a body can have acceleration without having velocity. The variation of potential difference across inductor with time is shown in Fig. 24.16 (*b*).

(*b*) Decay of Current

Let us consider the arrangement as shown in Fig. 24.15. Once the steady current i_0 is maintained through the circuit by pressing key *ab*; the key then is moved to connect *b* to *c* at $t = 0$. This disconnects the battery and completes the circuit through wire. The current in the circuit does not drop to zero abruptly because of the opposition by the induced emf across the inductance and decays to zero overtime.

The induced emf is $L\frac{di}{dt}$ and the potential difference across resistance is Ri. Hence from Kirchhoff's law, we have

$$L\frac{di}{dt} + Ri = 0$$

or
$$\frac{di}{i} = -\frac{R}{L}dt$$

Integration gives $\log i = -\frac{R}{L}t + C$, C is constant of integration

At $t = 0$, $i = i_0$ gives $C = \log i_0$

$$\therefore \quad \log\frac{i}{i_0} = -\frac{R}{L}t$$

or $$i = i_0 e^{-\frac{R}{L}t} \quad \text{...(24.26)}$$

and $$\frac{di}{dt} = -\frac{i_0 R}{L} e^{-\frac{R}{L}t} = -\frac{E}{L} e^{-\frac{R}{L}t} \quad \text{...(24.27)}$$

Here $t = L/R$ is *time constant* of the circuit.

The plot of i vs t and $\frac{di}{dt}$ vs t are shown in Fig. (24.17 *a*) and (24.17 *b*) respectively. The decay is also exponential. The current falls to $1/e$th of the maximum current in one time constant. After $t = \tau = \frac{L}{R}$ $i = i_0 e^{-1} = \frac{i_0}{e} = 0.37\ i_0$.

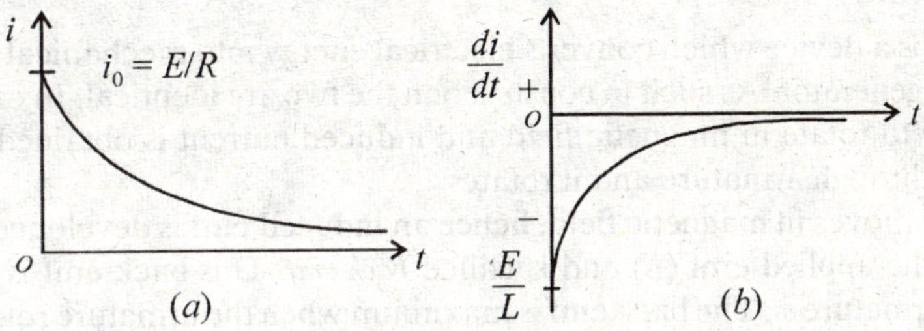

Fig. 24.17

Thus the current falls to 0.37 (or 37%) of the maximum current in one time constant.

24.10 ELECTRIC GENERATOR DYNAMO

Generator is a device which converts mechanical energy (rotational kinetic energy) into electric energy. This basically consists of a coil wound over an iron core called armature, which is made to rotate in a strong magnetic field. The dynamos are based on the principle of electromagnetic induction.

When the armature is made to rotate, the magnetic flux linked with the coil changes and an induced emf in the coil is developed. If the two ends of the coil rotate, in slip-rings an alternating current (voltage) is developed and if it rotates in split ring commutator, a direct current (voltage) is developed.

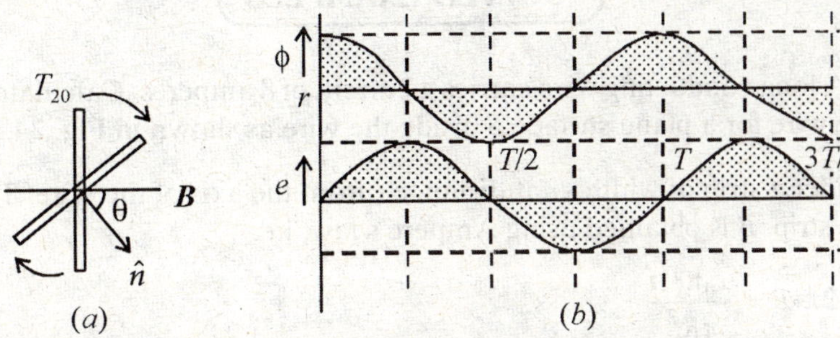

Fig. 24.18

If initially the plane of coil is perpendicular to the field, the flux linked with it at any time t is

$$\phi = Bs\cos\theta = nsB\cos\omega t \qquad (\theta = \omega t;\ n \text{ total number of turns})$$

and induced emf developed is

$$e = -\frac{d\phi}{dt} = nsB\omega \sin \omega t$$

$$= e_0 \sin\omega t \qquad \ldots(24.28)$$

where $e_0 = nsB\omega$

It is clear that induced emf lags behind the flux in phase by $\pi/2$, i.e. when flux is maximum, i.e. plane of coil is perpendicular to the magnetic field induced emf is zero and when flux is zero, i.e. plane of coil is parallel to the magnetic field emf is maximum. This is shown in Fig. 24.18 (*b*).

Back emf in Motors

An electric motor is a device which converts electrical energy into mechanical energy. This device is just reverse of a generator. As such in construction the two are identical. In case of generator, the armature is made to rotate in magnetic field and induced current is obtained whereas in motor, current is passed through armature and it rotates.

Since armature moves in magnetic field, hence an induced emf is developed. This induced emf (e) is opposite to the applied emf (E) and is called *back emf.* This back emf is proportional to the angular speed of armature ω. The back emf is maximum when the armature rotates with maximum speed. If R is the total resistance of motor, then current is given by

$$i = \frac{E - e}{R} \quad \text{with } e \propto \omega \qquad \ldots(24.29)$$

When the motor is turned on the armature is at rest and there is no back emf. Thus the start up current can be quite large. To reduce this current a resistance called 'starter' is put in series with the armature. This remains in the circuit for short period when the motor is started and remains in the circuit till motor acquires full speed.

The efficiency of the motor is given by

$$\eta = \frac{\text{Induced emf}}{\text{Applied emf}} = \frac{e}{E} \times 100 \qquad \ldots(24.30)$$

SOLVED EXAMPLES

Example 24.1: A long conducting wire carries a current of 8 amperes. Calculate the magnetic flux per meter of the wire for a plane surface S inside the wire as shown in Fig. 24.19.

Solution: Consider a strip of width dr at distance r from the axis of the wire. The magnetic field, at the site of the strip, B is obtained using Ampere's law, i.e.

$$2\pi rB = \frac{\mu_0 \pi r^2 i}{\pi R^2}$$

i.e. $$B = \frac{\mu_0 ir}{2\pi R^2}$$

where i is the current through the wire.

The direction is perpendicular to the plane surface S.
Flux linked with unit length of this element is

$$d\phi = B \cdot dS = \frac{\mu_0 ir}{2\pi R^2} \cdot dr \quad (dS = 1 \times dr)$$

Hence the total flux linked with the surface is

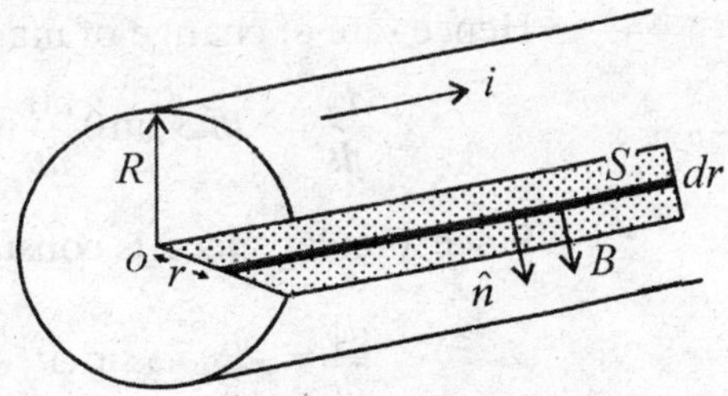

Fig. 24.19

$$\phi = \int_0^R \frac{\mu_0 ir}{2\pi R^2} dr = \frac{\mu_0 i}{2\pi R^2} \int_0^r r dr$$

$$= \frac{\mu_0 i}{2\pi R^2} \cdot \left[\frac{r^2}{2}\right]_0^R = \frac{\mu_0 i}{4\pi}$$

or $\qquad \phi = 10^{-7} \times 8 = 8 \times 10^{-7}$ Wb.

Example 24.2: A coil of n turns and area S is situated in a uniform magnetic field of induction B with its plane perpendicular to the field as shown in Fig. 24.20. Calculate the rate of change of flux linked with the coil if it is rotated uniformly with angular velocity ω (a) about an axis through its centre and perpendicular to its plane, (b) about a diameter through 180°, (c) about a diameter continuously.

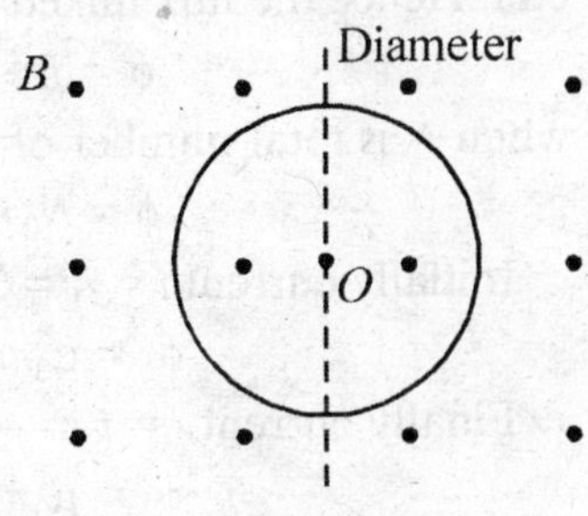

Fig. 24.20

Solution:

(a) The flux linked with the coil at any instant is

$$\phi = BnS\cos\theta = BnS \qquad \text{since } \theta = 0$$

Since all the terms B, n and S are constant, hence

$$\frac{d\phi}{dt} = 0$$

(b) In this situation (Fig. 24.20), the normal to the surface and magnetic field are parallel. Hence flux

$$\phi_i = BnS \qquad \ldots(i)$$

After it is rotated through 180°, the upper face goes down, i.e. normal to the surface now is making an angle of 180° with the magnetic field. Hence flux now is

$$\phi_2 = BnS\cos 180° = -BnS \qquad \ldots(ii)$$

Hence change in magnetic flux is

$$\Delta\phi = \phi_2 - \phi_1 = -BnS - BnS = -2BnS$$

This change has taken place in time $t = \pi/\omega$

Hence rate of change of magnetic flux is $\dfrac{\Delta\phi}{t} = \dfrac{-2BnS\omega}{\pi}$

(c) If in time t plane of coil has turned through an angle θ with respect to its initial position, the normal to the coil will also turn by the same angle θ. Hence angle between $\boldsymbol{B}$ and vector are $\boldsymbol{S}$ is θ at time t. Hence flux linked with coil at any time t is

$$\phi = BnS\cos\theta$$

Hence rate of change of magnetic flux linked with the coil is

$$\frac{d\phi}{dt} = -BnS\sin\theta\frac{d\theta}{dt}$$

Since angular speed is constant hence $\frac{d\phi}{dt} = \omega$ and $\theta = \omega t$.

$$\therefore \qquad \frac{d\phi}{dt} = -BnS\omega\sin\omega t$$

The variation of flux with time is sinusoidal.

Example 24.3: Along solenoid of diameter 0.1 m has 2×10^4 turns per metre. At the centre of solenoid a 100 turn coil of radius 0.01 m is placed with its axis coinciding with that of the solenoid. The current in the solenoid is decreasing at constant rate from +2 A to –2 A in 0.05 sec. Find the emf induced in the coil. Also find the total charge flowing through the coil during this time if the resistance of the coil is $10\,\pi^2$ Ohms.

Solution: The magnetic field at the centre of a current carrying solenoid is

$$B = \mu_0 ni \quad n \text{ is number of turns per unit length of solenoid}$$

The direction of this field at the centre when the coil is placed is parallel to the vector area of the coil. Hence the flux linked with it where current i is passing is

$$\phi = B \cdot N\pi r^2$$

when N is total number of turns in coil and r its radius.

$$\therefore \qquad \phi = N_0 ni \cdot N\pi r^2 = \mu_0 \pi nNr^2 i$$

Initially current $i = i_1 = 2$ A. Hence flux is

$$\phi_1 = \mu_0\pi nNr^2 i_1$$

Finally current $i = i_2 = -2$ A. Hence flux is

$$\phi_2 = \mu_0\pi nNr^2 i_2$$

Hence change in flux is

$$\Delta\phi = \phi_2 - \phi_1 = \mu_0\pi nNr^2(i_2 - i_1)$$

This change in flux takes place in time $\Delta t = 0.05$ sec.

Hence induced emf

$$e = -\frac{\Delta\phi}{\Delta t} = -\frac{\mu_0\pi nNr^2(i_2 - i_1)}{\Delta t}$$

Substituting the values $n = 2 \times 10^4$, $N = 100$, $r = 0.01$ m, we have

$$e = -\frac{4\pi \times 10^{-7} \times \pi \times 2 \times 10^4 \times 100 \times (10^{-2})^2(-2-2)}{0.05}$$

$$= \frac{16\pi^2 \times 2 \times 10^{-5}}{5 \times 10^{-2}} = \frac{32\pi^2}{5} \times 10^{-3} = 63.1 \text{ mV}$$

(*b*) If R is the resistance of coil, then current i is

$$i = \frac{e}{R} = \frac{dq}{dt}$$

Hence charge flowing through the coil is

$$dq = \frac{e}{R}dt = \frac{63.1 \times 10^{-3} \times 0.05}{10\pi^2}$$

$$= 32\ \mu C.$$

Example 24.4: A rectangular frame *ABCD* made of a uniform metal wire has a straight connection between *E* and *F* made of the same wire as shown in Fig. 24.21. *AEFD* is a square of side 1 m and $EB = FC = 0.5$ m. The entire circuit is placed in a steady increasing uniform magnetic field directed into the plane of the paper and normal to it. The rate of change of the magnetic field is 1 Ts^{-1}. The resistance per unit length of wire is 10 Ωm^{-1}. Find the magnitude and direction of currents in segments *AE*, *BE* and *EF*.

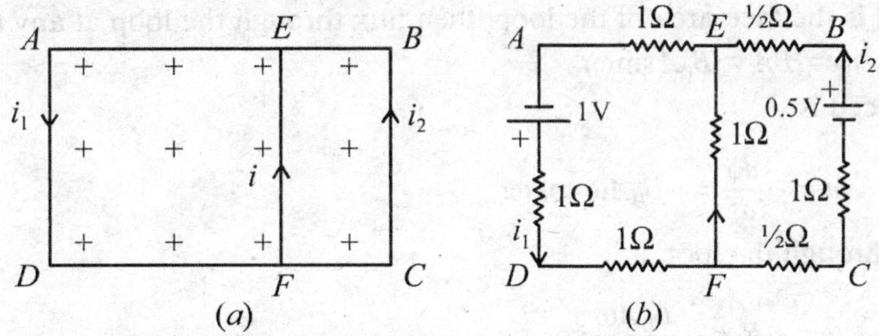

Fig. 24.21

Solution: Induced emf in a loop placed in time varying magnetic field is

$$e = \frac{d\phi}{dt} = \frac{d}{dt}(BS) = S\frac{dB}{dt}$$

Hence in loop *AEFD* emf induced is $e_1 = (1 \times 1)\ 1 = 1$ V

and in loop *EBCF* emf induced is $e_2 = \left(1 \times \frac{1}{2}\right)1 = 0.5$ V

The currents are induced in the two loops in such a way that the magnetic field due to induced currents opposes the change in flux with the two loops. If magnetic field is increasing downwards then the direction of the current in the two loops is as shown in Fig. 24.21 (*a*).

The current through $FE = i = i_1 - i_2$.

Equivalent circuit diagram is shown in Fig. 24.21 (*b*).

Applying Kirchhoff's law to loop *AEFD*, we have

$$i_1 \times 1 + i_1 \times 1 + (i_1 - i_2) \times 1 + i_1 \times 1 = 1$$

or $$4i_1 - i_2 = 1 \quad \ldots(i)$$

Applying Kirchhoff's law to loop *EFBE*, we have

$$0.5i_2 - (i_1 - i_2) \times 1 + 0.5\ i_2 + i_2 \times 1 = 0.5$$

$$-i_1 + 3i_2 = 0.5$$

or $$-2i_1 + 6i_2 = 1 \quad \ldots(ii)$$

Solving (*i*) and (*ii*) for i_1 and i_2, we get

$$i_1 = \frac{7}{22} \text{ A and } i_2 = \frac{6}{22} \text{ A}$$

Hence current in segment AE is $i_1 = 7/22$ A from E to A
Current in segment $BE = i_2 = 6/22$ A from B to E
Current in segment $EF = i = i_1 - i_2 = 1/22$ A from F to E.

Example 24.5: A conducting circular loop of face area 2.5×10^{-3} m² is placed perpendicular to a magnetic field which varies as $B = B_0 \sin\omega t$. (*a*) Find the charge flowing through any cross-section during the time $t = 0$ to $t = 40$ ms (*b*) If the resistance of the loop is 10 Ω, find the thermal energy developed in the loop in this period. Take $B_0 = 0.25$ T and $\omega = 50\pi$ rad/s.

Solution: (*a*) If A is the face area of the loop, then flux through the loop at any time t is

$$\phi = BA = B_0 A \sin\omega t$$

and the emf induced is

$$e = -\frac{d\phi}{dt} = -B_0 A\omega \cos\omega t$$

and the current through the loop is

$$i = e/R = -\frac{B_0 A\omega}{R}\cos\omega t$$

Hence charge that flows through the loop in time dt is

$$dq = idt = -\frac{BA\omega}{R}\cos\omega t\, dt$$

The current varies sinusoidally with time period $T = \frac{2\pi}{\omega} = \frac{2\pi}{50\pi} = 40$ ms.

Hence the charge that passes through the loop in time $t = 0$ to $= 40$ ms, i.e. in one time period is

$$Q = \int dq = -\frac{BA\omega}{R}\int_0^T \cos\omega t\, dt$$

$$= -\frac{BA}{R}\sin\omega t\Big|_0^T = 0$$

(*b*) The thermal energy produced in time dt is $dH = i^2 R dt$

i.e.
$$dH = \frac{RB_0^2 A^2 \omega^2}{R^2}\cos^2\omega t \cdot dt$$

$$= \frac{B_0^2 A^2 \omega^2}{2R}(1 + \cos 2\omega t)\, dt$$

Hence total thermal energy produced within time $t = 0$ to $t = 40$ ms is

$$H = \int_0^T \frac{B_0^2 A^2 \omega^2}{2R}(1 + \cos 2\omega t)\, dt$$

$$= \frac{B_0^2 A^2 \omega^2}{2R} \int_0^T dt + \frac{B_0^2 A^2 \omega^2}{2R} \int_0^T \cos 2\omega t dt$$

$$= \frac{B_0^2 A^2 \omega^2}{2R} T \qquad \because \int_0^T \cos 2\omega t dt = 0$$

or
$$H = \frac{B_0^2 A^2 \omega^2}{2R} \frac{2\pi}{\omega} = \frac{\pi B_0^2 A^2 \omega}{R}$$

$$= \frac{\pi \times (0.25)^2 (2.5 \times 10^{-3})^2 \times (50\pi)}{10}$$

$$= 1.93 \times 10^{-5} \text{ J.}$$

Example 24.6: A wire of mass m and length l can freely slide on a pair of parallel, smooth, horizontal rails placed in a vertical magnetic field B (Fig. 24.22). The rails are connected to a capacitor of capacitance C. The electric resistance of the rails and the wire is zero. If a constant force F acts on the wire as shown in the figure, find the acceleration of the wire.

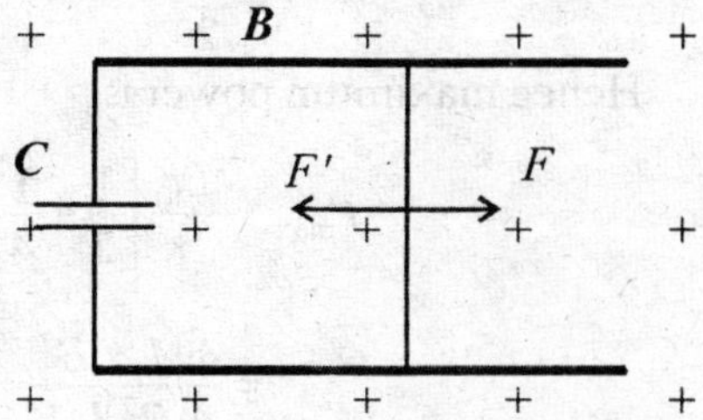

Fig. 24.22

Solution: Let the velocity of the wire at any time t be v, then the induced emf is $e = vlB$.

Hence charge on the capacitor $q = Ce = CvlB$.

The current in the circuit is $i = \frac{dq}{dt} = ClB\frac{dv}{dt} = ClBa$

where a is the acceleration of the wire at time t.

Because of this current, the wire will experience force due to magnetic field. This magnetic force will have direction opposite to F and magnitude

$$F = ilB = Cl^2B^2a$$

Hence the net force on the conductor is

$$F - F' = F - Cl^2B^2a$$

According to Newton's law, this must be equal to mass × acceleration

$$\therefore \quad F - Cl^2B^2a = ma$$

or
$$a(m + Cl^2B^2) = F$$

and
$$a = \frac{F}{m + Cl^2B^2}$$

Example 24.7: An inductance L and a resistance R are connected in series with a battery of emf E. Find the maximum rate at which the energy is stored in the magnetic field.

Solution: In L–R circuit, the current at any time is i, given by

$$i = i_0\left(1 - e^{-t/\tau}\right)$$

where $i_0 = \frac{E}{R}$ and τ is time constant $= \frac{L}{R}$.

The rate at which the energy stored is

$$P = \frac{du}{dt} = \frac{d}{dt}\left(\frac{1}{2}Li^2\right) = \frac{d}{dt}\left(\frac{1}{2}L\right)\left\{i_0\left(1-e^{-t/\tau}\right)\right\}^2$$

$$= \frac{1}{2}Li_0^2\frac{d}{dt}\left(1-e^{-t/\tau}\right)^2 = \frac{1}{2}Li_0^2 \times 2\left(1-e^{-t/\tau}\right)\left(\frac{e^{-t/\tau}}{\tau}\right)$$

$$P = \frac{Li_0^2}{\tau}\left(e^{-t/\tau} - e^{-2t/\tau}\right)$$

This rate will be maximum when $\frac{dP}{dt} = 0$

i.e. $$\frac{d}{dt}\left\{\frac{Li_0^2}{\tau}\left(e^{-t/\tau} - e^{-2t/\tau}\right)\right\} = 0$$

or $$\frac{Li_0^2}{\tau}\left[\frac{-e^{-t/\tau}}{\tau} + \frac{2e^{-2t/\tau}}{\tau}\right] = 0$$

$$e^{-t/\tau} = \frac{1}{2}$$

Hence maximum power is

$$P_{max} = \frac{Li_0^2}{\tau}\left[\frac{1}{2} - \frac{1}{4}\right] = \frac{Li_0^2}{\tau}\frac{1}{4}$$

$$= \frac{1}{4}\frac{L \times E^2}{R^2\, L/R} = \frac{E^2}{4R}$$

Example 24.8: Figure 24.23 shows a metallic frame of edge a in vertical plane. A uniform magnetic field B exists in the space in a direction perpendicular to the plane of the figure. Two boys pull the opposite corners of the square to deform it into rhombus. They start pulling the corners at $t = 0$ and displace the corners at a uniform speed v. (*a*) Find the induced emf in the frame at the instant when the angles at these corners reduces to 60°. (*b*) Find the induced current in the frame at this instant if the total resistance of the frame is R. (*c*) Find the total charge which flows through a side of the frame by the time the square is deformed into a straight line.

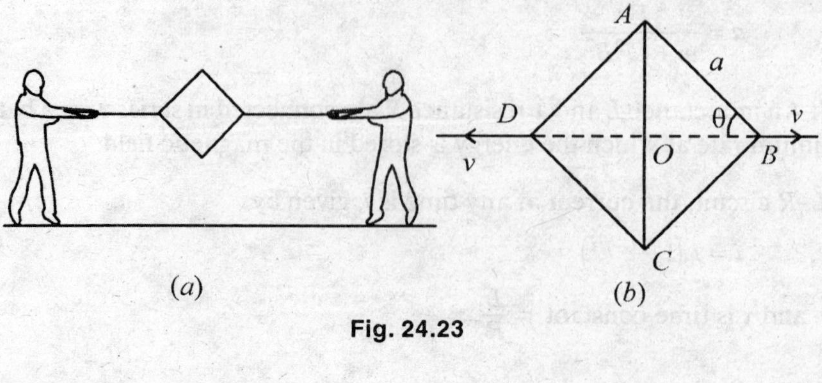

Fig. 24.23

Solution:

(*a*) Let at any time *t* the shape of the frame is as shown in Fig. 24.23 (*b*). Diagonal *DB* makes an angle θ with side *AB*. Hence area of the frame is

$$A = 4 \times \frac{1}{2} a \sin\theta \times a \cos\theta$$
$$= a^2 \sin 2\theta.$$

and flux linked with the frame is

$$\phi = BA = Ba^2 \sin 2\theta \qquad \because \textbf{\textit{B}} \text{ and vector area } \textbf{\textit{A}} \text{ are parallel.}$$

Hence induced emf

$$e = -\frac{d\phi}{dt} = -2Ba^2 \cos^2\theta \frac{d\theta}{dt} \qquad \ldots(i)$$

At this moment semi length of diagonal *OB* is

$$x = a\cos\theta.$$

and $$\frac{dx}{dt} = -a\sin\theta\frac{d\theta}{dt} \qquad \ldots(ii)$$

But $\frac{dx}{dt} = v$ because the boy is pulling corner *B* with velocity *v*.

$$\therefore \qquad \frac{d\theta}{dt} = -\frac{v}{a\sin\theta} \qquad \ldots(iii)$$

Using (*iii*) in (*i*), we get

$$e = -2Ba^2 \cos 2\theta \left(-\frac{v}{a\sin\theta}\right)$$

$$= 2Bav\frac{\cos 2\theta}{\sin\theta}$$

Hence induced emf when θ = 30° (because angle at *B* is 2θ = 60°)

$$\therefore \qquad e = 2Bav\frac{\cos 60^\circ}{\sin 30^\circ} = 2Bav$$

(*b*) Since total resistance is *R* and induced emf, *e*, then current at this instant is

$$i = \frac{e}{R} = \frac{2Bav}{R}$$

(*c*) Initially the area of the frame is a^2 and the flux linked is

$$\phi_i = Ba^2$$

when the frame is deformed into a straight line, the flux linked is

$$\phi_f = 0$$

Hence change in flux $\Delta\phi = \phi_f - \phi_i = -Ba^2$

Charge that passes through the side is $q = \dfrac{\text{Change in flux}}{\text{Resistance}}$

$$\therefore \qquad q = \frac{Ba^2}{R}$$

Example 24.9: A uniform magnetic field B exists in a cylindrical region, shown in Fig. 24.24 by dotted line. The magnetic field increases at constant rate $\frac{dB}{dt}$. Consider a circle of radius r coaxial with cylindrical region. (*a*) Find the magnitude of the electrical field E at a point on the circumference of the circle. (*b*) Consider a point P on the side of the square circumscribing the circle. Show that the component of the induced electric field at P along ba is same as the magnitude found in part (*a*).

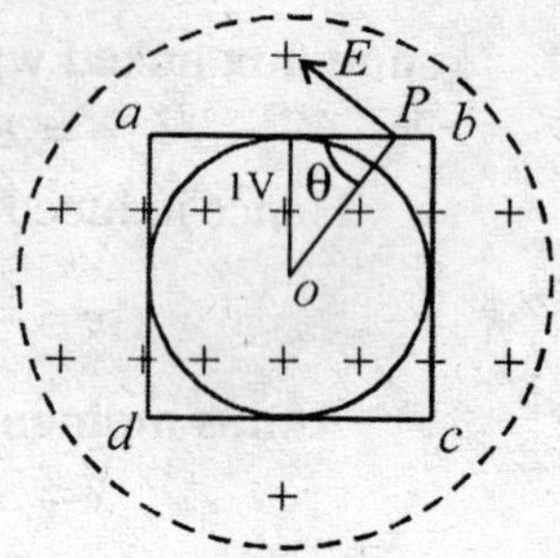

Fig. 2.24

Solution:

(*a*) If E is the electric field developed due to time varying magnetic field, then

$$\oint \mathbf{E} \cdot d\mathbf{l} = \frac{d\phi}{dt} \qquad \ldots(i)$$

for circle everywhere E and dl are parallel and E is same at all points on the circumference. Hence

$$\oint \mathbf{E} \cdot d\mathbf{l} = \oint E dl \cos\theta = E \oint dl = E 2\pi r$$

where r is the radius of the circle.

If at any time magnetic field is B, then flux linked with the surface is $\phi = B \cdot \pi r^2$

and $$\frac{d\phi}{dt} = \frac{d}{dt}(B\pi r^2) = \pi r^2 \frac{dB}{dt}$$

Then from (*i*), we have

$$E 2\pi r = \pi r^2 \frac{dB}{dt}$$

or $$E = \frac{r}{2}\frac{dB}{dt}$$

(*b*) Let distance OP be x. Then electric field induced at P due to time varying magnetic field is from (*a*)

$$E = \frac{x}{2}\frac{dB}{dt}$$

Its direction is perpendicular to OP. From Fig. 24.24.

$x \sin\theta = r$ $\qquad$ r is radius of the inscribed circle, then

$$E = \frac{r}{2\sin\theta}\frac{dB}{dt}$$

or $$E \sin\theta = \frac{r}{2}\frac{dB}{dt}$$

But $E \sin\theta$ is component of electric field induced at E, along ba. Hence the result.

Example 24.10: A square loop of side a is carrying a current i in the direction as shown in Fig. 24.25. All the sides are uniform and mass of each side is m. There exists a uniform magnetic field B parallel to Z-axis. If square loop is released from the position as shown in Fig. 24.25 and rotated about an axis passing through the centre of mass and parallel to Y-axis, then calculate its initial angular acceleration. Neglect friction.

Solution: The current carrying square loop is equivalent to a magnetic dipole having magnetic moment

$$M = ia^2 \qquad \ldots(i)$$

The direction is perpendicular to the plane of coil. The magnetic field B is along Z-axis.

Hence the angle between B and M is $(90 - \theta)$.

Hence torque on the loop $= MB\sin(90 - \theta) = ia^2B\cos\theta$.

This torque acts about an axis parallel to Y-axis passing through the centre of mass of the coil.

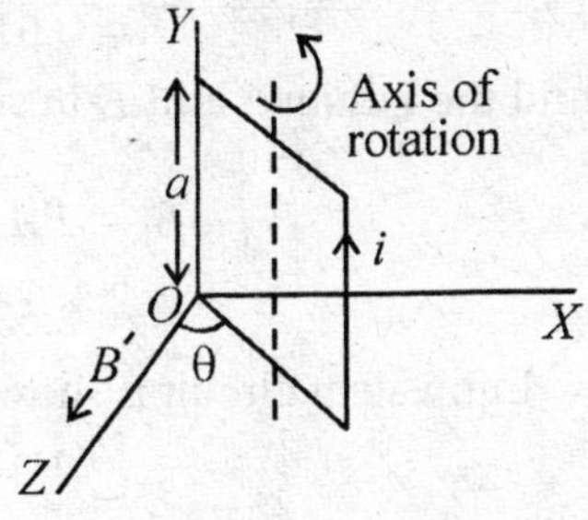

Fig. 24.25

Moment of inertia about the axis of rotation is

$$I = 2m\left(\frac{a}{2}\right)^2 + 2m\left(\frac{a^2}{12}\right) = \frac{ma^2}{2} + \frac{ma^2}{6} = \frac{4ma^2}{6} = \frac{2}{3}ma^2$$

If α is the angular acceleration, then

$$\alpha = \frac{\tau}{I} = \frac{ia^2B\cos\theta}{2/3\,ma^2} = \frac{3}{2}\frac{iB\cos\theta}{m}$$

Example 24.11: A circular coil of radius R is revolving around a vertical wire PQ with a constant angular speed ω. Magnetic field B is acting perpendicular to plane of loop and into the plane. The resistance per unit length of circular loop is $1\ \Omega/\text{m}$, while that of wire PQ is $\pi R\ \Omega$. Find the current in wire PQ when direction of magnetic field is along the plane of the loop.

Solution: When the plane of the loop is parallel to the field, then θ, the angle between normal to the loop and magnetic field is $90°$.

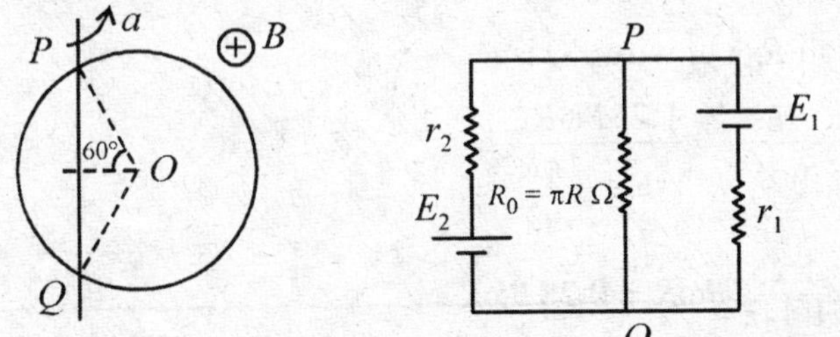

Fig. 24.26

Let any time t the normal to the plane of the loop makes an angle θ, the flux linked is $\phi = BA\cos\theta$ and the emf induced in the loop is

$$e = -\frac{d\phi}{dt} = -\frac{d}{dt}(BA\cos\theta) = BA\sin\theta\frac{d\theta}{dt}$$

or $$e = BA\omega\sin\theta, \text{ where } \omega = \frac{d\theta}{dt} = \text{angular velocity.}$$

when $\theta = 90°$, the induced emf E_2 in left loop is

$$E_2 = BA_2\omega \quad \text{where } A_2 = \frac{\pi R^2}{3} - \frac{\sqrt{3}}{4}R^2 = \left(\frac{\pi}{3} - \frac{\sqrt{3}}{4}\right)R^2 = 0.61R^2$$

$$= 0.61B\omega R^2$$

and the induced emf E in right loop is

$$E_1 = BA_1\omega \quad \text{where } A_1 = \frac{2\pi R^2}{3} + \frac{\sqrt{3}}{4}R^2 = \left(\frac{2\pi}{3} + \frac{\sqrt{3}}{4}\right)R^2$$

$$= 2.53B\omega R^2.$$

Equivalent circuit is shown in Fig. 24.26. The two cells behave like a single cell of emf E given by

$$E = \frac{E_1 r_2 + E_2 r_1}{r_1 + r_2} \quad \text{with } r_1 = \frac{4\pi R}{3} \text{ and } r_2 = \frac{2\pi R}{3}$$

With internal resistance $\frac{r_1 r_2}{r_1 + r_2}$

Hence

$$E = \frac{2.53\,B\omega R^2 \times \frac{2\pi R}{3} + 0.61\,B\omega R^2 \times \frac{4\pi R}{3}}{\frac{4\pi R}{3} + \frac{2\pi R}{3}}$$

$$= B\omega R^2 \frac{(2.53 + 1.21)}{3} = B\omega R^2 \times \frac{3.74}{3}$$

$$= 1.25\,B\omega R^2.$$

and internal resistance

$$r = \frac{\frac{4\pi R}{3} \times \frac{2\pi R}{3}}{\frac{4\pi R}{3} + \frac{2\pi R}{3}} = \frac{4\pi R}{9}$$

Hence current through R_0 i.e. wire PQ is

$$i = \frac{E}{R_0 + r} = \frac{1.25\,B\omega R^2}{\pi R + \frac{4\pi R}{9}}$$

$$= \frac{1.25}{1.44\pi}B\omega R = 0.28\,B\omega R$$

Example 24.12: A thin ring of mass 0.1 kg and diameter 0.5 m carrying charge q is free to rotate about an axis as shown in Fig. 24.27. At $t = 0$ time dependent field of $B = 8t$ tesla is applied vertically down on the ring. If the ring is at rest before applying magnetic field and induced magnetic effects due to rotation of the ring are neglected, the power developed by the force acting on the ring after 10 sec is found to be 6.4 mW. Find the charge on the ring.

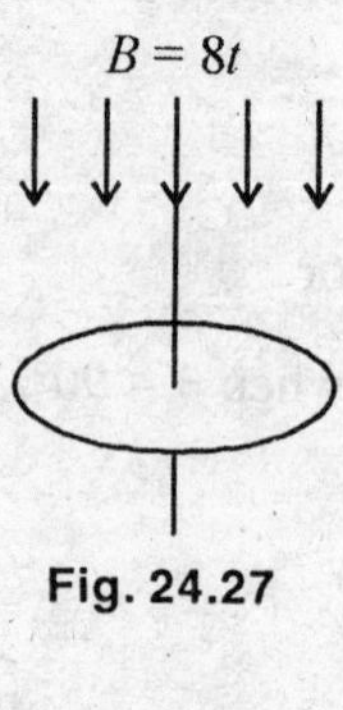

Fig. 24.27

Solution: Due to time varying magnetic field, an electric field is developed at the ring. If E is induced electric field and r radius of the ring, then

$$\oint \boldsymbol{E} \cdot \boldsymbol{dl} = \pi r^2 \frac{dB}{dt}$$

Since E and dl are parallel hence

$$\oint \mathbf{E} \cdot d\mathbf{l} = 2\pi r E = \pi r^2 \frac{dB}{dt}$$

and $$E = \frac{r}{2}\frac{dB}{dt} = \frac{r}{2} B_0 \quad \text{where } B_0 = 8 \text{ TS}^{-1}.$$

In this electric field the ring experiences a force

$$F = qE \qquad \text{where } q \text{ is the charge on the ring.}$$

This force is tangential to the ring. Hence torque on the ring is

$$\tau = Fr = qEr$$

If α is angular acceleration and I moment of inertia of ring about its own axis about which it rotates, we have

$$\tau = I\alpha = qEr$$

or $$\alpha = \frac{qEr}{I}$$

Its angular velocity after time t is $\omega = \alpha t$. Hence power

$$P = \tau\omega = qEr\frac{qEr}{I}t = \frac{q^2E^2r^2}{I}t$$

But $E = \frac{r}{2}B_0$ and $I = mr^2$. Hence

$$P = \frac{q^2r^2B_0^2r^2}{4mr^2}t = \frac{q^2r^2B_0^2}{4m}t$$

and $$q^2 = \frac{1}{rB_0}\sqrt{\frac{4mP}{t}}$$

$$= \frac{1}{(0.25)\times 8}\sqrt{\frac{4\times .1\times 6.4\times 10^{-3}}{10}}$$

$$= \frac{1}{2}\times 10^{-3}\times 2\times 8 = 8\times 10^{-3} \text{ C}.$$

Example 24.13: A magnetic field in the vertical region is varying at a constant rate $\alpha = \frac{dB}{dt}$ and is zero at $t = 0$. A horizontal concentric tube of radius r is fixed in the cylindrical region. At $t = 0$ a particle of mass m and charge q is placed inside the tube at position P as shown in Fig. 24.28. Draw the graph of horizontal component of reaction of groove on the particle vs angle θ made by the line joining centre to particle with OP.

Solution: At the instant when line joining centre to the particle makes an angle θ with OP, let the horizontal component of reaction be N.

Due to time varying magnetic field an electric field E is induced at the site of the tube. Since radius of tube is r, then induced field E is given by

$$2\pi r E = \pi r^2 \frac{dB}{dt} \quad \text{or} \quad E = \frac{r}{2}\frac{dB}{dt} = \frac{r\alpha}{2}$$

In this field the particle experiences a force $F = qE = \dfrac{qr\alpha}{2}$

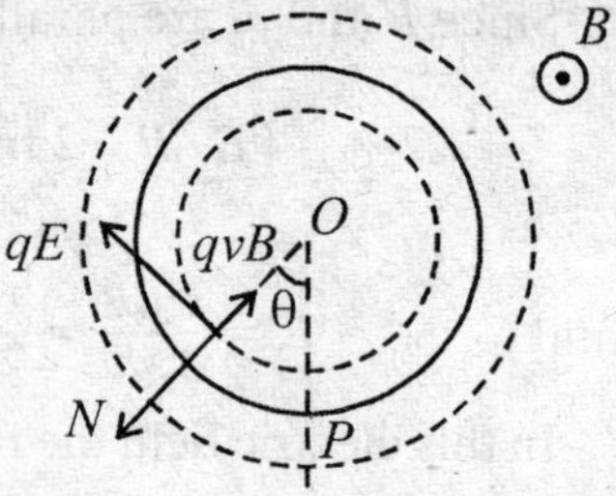

Fig. 24.28

Hence acceleration of particle $a = \dfrac{F}{m} = \dfrac{qr\alpha}{2m}$

If instantaneous velocity is v then $a = \dfrac{dv}{dt}$

or $$\frac{dv}{dt} = \frac{qr\alpha}{2m}$$

and $$v = \frac{qr\alpha}{2m}t \text{ and } B = \alpha t$$

Since the particle moves in a circle of radius r, then required centripetal force

$$\frac{mv^2}{r} = qvB - N$$

$$N = qvB - \frac{mv^2}{r}$$

$$= q\frac{qr\alpha}{2m}t\alpha t - \frac{m}{r}\frac{q^2r^2\alpha^2}{4m^2}t^2$$

$$= \frac{rq^2\alpha^2t^2}{4m}$$

Since tangential acceleration of the particle is $a = \dfrac{qr\alpha}{2m}$

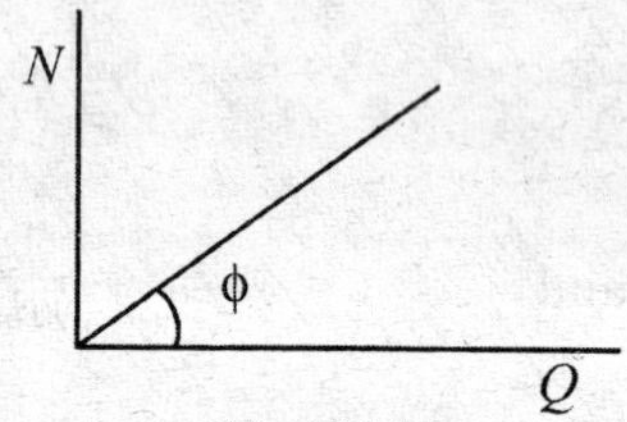

Fig. 24.29

Hence its angular acceleration $\dfrac{a}{r} = \dfrac{q\alpha}{2m}$

Hence angular displacement in time t is

$$\theta = \frac{1}{2}\left(\frac{q\alpha}{2m}\right)t^2 = \frac{q\alpha}{4m}t^2$$

$$\therefore \quad N = qr\alpha\theta$$

Hence graph between N and θ is a straight line shown in Fig. 24.29 with slope $\tan\phi = qr\alpha$.

Example 24.13: There is a small metallic ring of radius r_0 placed perpendicular to a constant magnetic field B_0. One end of a rod is hinged at the centre of ring O and other end is placed on the ring. Now rod is rotated with constant angular velocity ω_0 by some external agent and circuit is connected as shown in Fig. 24.30. If switch S in closed at $t = 0$, then calculate heat loss from the resistor R_2 from $t = 0$ to the instant when voltage across capacitor becomes V_0. (Assume plane of ring to be horizontal and friction to be absent at all times).

Solution: Rod is rotating with angular velocity ω_0. Hence emf induced across its ends is $V = \frac{1}{2}B_0\omega_0 r_0^2$. Charge on capacitor $q = CV_0$ when its potential is V_0. Since charge q flows under potential difference V. Hence energy supplied is $E = qV$. This energy is supplied by external agency in rotating the rod.

This energy is dissipated in heat across resistance R_1 and R_2 and stored in capacitor as electrostatic potential energy.

If H_1 and H_2 are the heat energies lost in resistance R_1 and R_2 respectively, then

$$H_1R_1 = H_2R_2$$

or
$$H_1 = \frac{R_2}{R_1}H_2$$

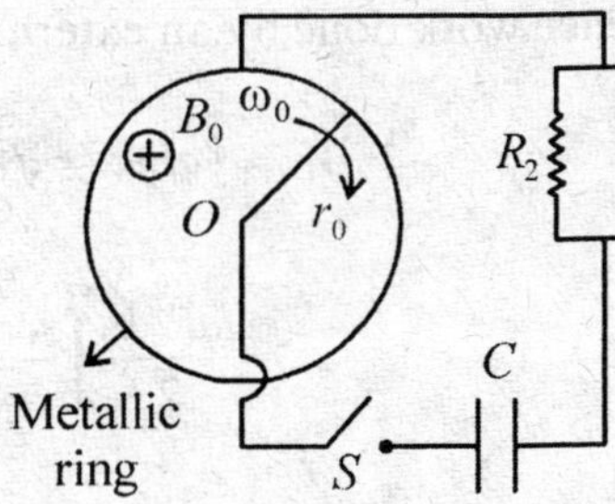

Fig. 24.30

Conservation of energy gives

$$qV = \frac{1}{2}CV_0^2 + H_1 + H_2$$

or
$$CV_0\frac{1}{2}B_0\omega_0r_0^2 = \frac{1}{2}CV_0^2 + \frac{R_2}{R_1}H_2 + H_2$$

or
$$H_2\frac{(R_1+R_2)}{R_1} = \frac{1}{2}CV_0(B_0\omega_0r_0^2 - V_0)$$

or
$$H_2 = \frac{1}{2}CV_0\frac{R_1}{R_1+R_2}(B_0\omega_0r_0^2 - V_0)$$

Example 24.14: In a cylindrical region of radius R, there exists a time varying magnetic field B such that $\frac{dB}{dt} = k(>0)$. A charged particle having charge q is placed at the point P at a distance d (>R) from its centre O. Now the particle is moved in the direction perpendicular to OP by an external agent up to infinity. Find the work done by an external agent, so that there is no gain in kinetic energy of the charged particle.

Solution: At any time let the particle be at A at distance x from P as shown in Fig. 24.31 (*b*)

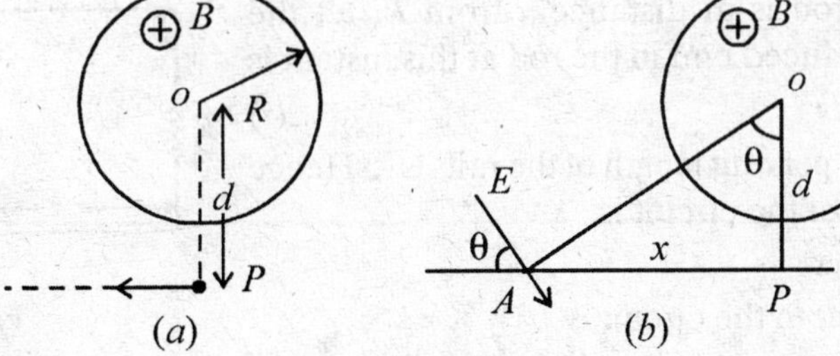

Fig. 24.31

The electric field E induced at A is given by

$$\oint \mathbf{E}\cdot d\mathbf{l} = A\frac{dB}{dt}$$

or
$$E \times 2\pi\sqrt{x^2+d^2} = \pi R^2\frac{dB}{dt}$$

or
$$E = \frac{R^2k}{2\sqrt{x^2+d^2}} \qquad \ldots(i)$$

Hence work done by an external agency is

$$W = \int_0^{\infty} q\boldsymbol{E}d\boldsymbol{x} = q\int_0^{\alpha} E\cos\theta dx$$

$$= \frac{qR^2k}{2}\int_0^{\infty}\frac{\cos\theta\cdot dx}{\sqrt{x^2+d^2}} \qquad \text{But } \cos\theta = \frac{d}{\sqrt{x^2+d^2}}$$

$$\therefore \quad W = \frac{qR^2kd}{2}\int_0^{\infty}\frac{dx}{x^2+d^2}$$

$$= \frac{qR^2kd}{2}\left[\frac{1}{d}\tan^{-1}\frac{x}{d}\right]_0^{\infty} = \frac{qR^2kd}{2d}\frac{\pi}{2}$$

$$= \frac{qR^2k\pi}{4} = \frac{\pi}{4}qR^2k$$

This work done is independent of d.

Example 24.15: Two long parallel horizontal rails, a distance d apart and each having a resistance λ per unit length, are joined at one end by a resistance R. A perfectly conducting rod MN of mass m is free to move along the rails without friction (Fig. 24.32). There is a uniform magnetic field B normal to the plane of the paper and directed into the paper. A variable force F is applied to the rod MN such that as the rod moves, a constant current I flows through R. (*a*) Find the velocity of the rod and the applied force F as a function of distance x of the rod from R. (*b*) What fraction of work done per second by F is converted into heat?

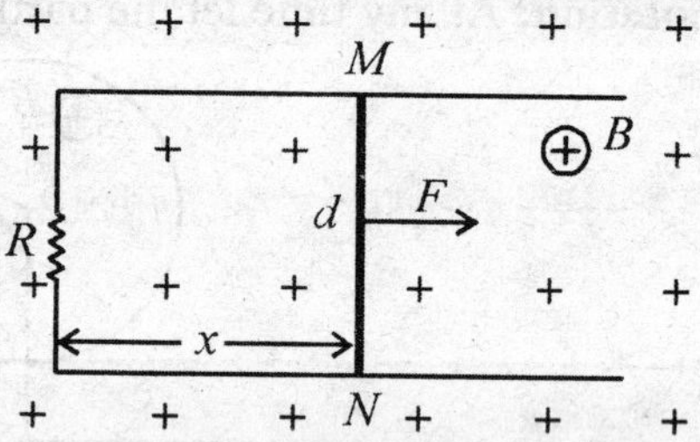

Fig. 24.32

Solution: (*a*) When the rod is at distance x from R, let the velocity of the rod be v. Induced emf in the rod at this instant is

$$e = Bdv \qquad \ldots(i)$$

Since the resistance per unit length of the rails is λ. Hence the total resistance of the circuit is

$$R + 2\lambda x$$

Hence the current in the circuit is

$$I = \frac{Bdv}{R+2\lambda x} \qquad \ldots(ii)$$

The magnetic force on the rod then is

$$F_m = IdB$$

and is directed towards left. If F is the external force on the rod at this instant then net force on the rod is

$$f = F - F_m = F - IdB$$

Then equation of motion is

$f = m \cdot a$, where a is instantaneous acceleration

$$\therefore \quad f = ma = \frac{mdv}{dt} = mv \cdot \frac{dv}{dx}$$

But from (*ii*) $v = \dfrac{I(R + 2\lambda x)}{Bd}$ ...(*iii*)

Hence $\quad f = \dfrac{mI(R + 2\lambda x)}{Bd} \dfrac{d}{dx}\left[\dfrac{I(R + 2\lambda x)}{Bd}\right]$

$$= \frac{2mI^2\lambda}{B^2d^2}(R + 2\lambda x)$$

But the applied force $F = f + F_m$

$$\therefore \quad F = \frac{2mI^2\lambda}{B^2d^2}(R + 2\lambda x) + BId \qquad \text{...(iv)}$$

Equation (*iii*) and (*iv*) give us, the velocity and force as a function of *x* respectively.

(*b*) Work done per second by the external force is

$$P = Fv = \frac{FI(R + 2\lambda x)}{Bd}$$

Heat produced per second in resistance is

$$H = I^2 \quad \text{Resistance} = I^2(R + 2\lambda x)$$

$$\therefore \quad \frac{H}{P} = \frac{I^2(R + 2\lambda x)}{FI(R + 2\lambda x)} Bd = \frac{IBd}{F}$$

$$= \frac{IBdB^2d^2}{2mI^2\lambda(R + 2\lambda x) + IB^3d^3} = \frac{B^3d^3}{2mI\lambda(R + 2\lambda x) + B^3d^3}$$

or $\quad \dfrac{H}{P} = \dfrac{B^3d^3}{2mI\lambda(R + 2\lambda x) + B^3d^3}$

Example 24.16: A copper rod of length 0.18 m is moving with uniform velocity of 10 ms^{-1} parallel to a long straight wire carrying a current 5 A. The rod itself is perpendicular to the wire with its ends at distances 0.01 m and 0.2 m from wire, calculate the emf induced in the rod.

Solution: As the rod *AB* moves in the magnetic field produced by the current carrying conductor an induced emf is developed in the rod. Consider a small element *dr* of the rod at a distance *r* from the current carrying wire.

The magnetic field due to current *i* through the wire at the element *dr* is

$$B = \frac{\mu_0 i}{2\pi r}$$

The emf induced in the element *dr* of the rod is

$$de = Bvdr = \frac{\mu_0 iv}{2\pi r} dr$$

Hence the emf induced in the whole rod is

$$e = \int_a^b \frac{\mu_0 iv}{2\pi r} dr = \frac{\mu_0 iv}{2\pi} \int_a^b \frac{dr}{r} = \frac{\mu_0 iv}{2\pi} \log_e \frac{b}{a}$$

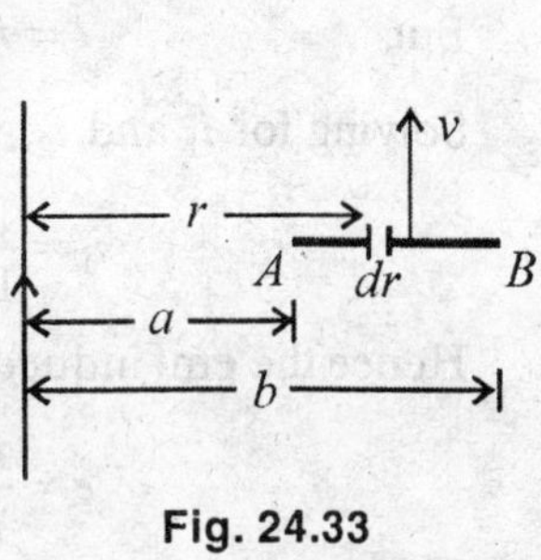

Fig. 24.33

The values given are $\frac{\mu_0}{2\pi} = 2 \times 10^{-7}$ H/m. $v = 10$ m/s, $a = 0.01$ m, $b = 0.2$ m and $i = 5$ A.

Substitution gives

$$e = 2 \times 10^{-7} \times 5 \times 10 \times \log\left(\frac{0.2}{0.01}\right)$$

$$= 10^{-5} \log_e 20 = 3 \times 10^{-5} \text{ V}$$

$$= 30 \ \mu\text{V}.$$

Example 24.17: Two parallel vertical rails *AB* and *CD* are separated by 1 m. They are connected at the two ends by resistances R_1 and R_2 as shown in Fig. 24.34. A horizontal metallic bar *L* of mass 0.2 kg slides without friction vertically down the rails under the action of gravity. There is a uniform horizontal magnetic field of 0.6 T perpendicular to the plain of the rails. It is observed that when the terminal velocity is attained the powers dissipated in R_1 and R_2 are 0.76 W and 1.2 W respectively. Find the terminal velocity of bar *L* and the values of R_1 and R_2.

Solution: When the rod moves with terminal velocity, the net force on it is zero. At this instant the forces on the rod are (*i*) its weight *mg* vertically downwards and (*ii*) magnetic force = *Bid* vertically upwards, where *i* is the induced current through the rod and *d* its length. Hence

$$Bid = mg$$

or $$i = \frac{mg}{Bd} = \frac{0.2 \times 10}{0.6 \times 1} = \frac{10}{3} \text{ A} \quad \ldots(i)$$

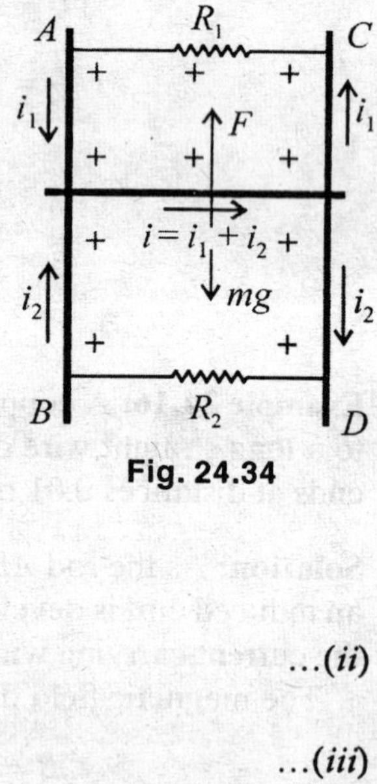

Fig. 24.34

If i_1 and i_2 is the current through resistances R_1 and R_2 (Fig. 24.34), then

$$i = i_1 + i_2 \text{ and } e = i_1 R_1 = i_2 R_2$$

Powers dissipated in the two resistances are

$$P_1 = i_1^2 R_1 = 0.76 \quad \text{and} \quad P_2 = i_2^2 R_2 = 1.2$$

$$\therefore \quad \frac{P_1}{P_2} = \frac{i_1^2 R_1}{i_2^2 R_2} = \frac{i_1}{i_2} = \frac{0.76}{1.2} \qquad \because i_1 R_1 = i_2 R_2$$

or $$\frac{i_1}{i_2} = \frac{19}{30} \quad \ldots(ii)$$

But $$i = i_1 + i_2 = \frac{10}{3} \quad \ldots(iii)$$

Solving for i_1 and i_2, we have

$$i_1 = \frac{19}{15} \text{ A} \quad \text{and} \quad i_2 = 2 \text{ A}$$

Hence the emf induced across the rod is

$$e = \frac{P_1}{i_1} = \frac{0.76}{19/15} = 0.6 \text{ V}$$

If terminal velocity of rod is v, then emf induced e is

$$e = Bvd$$

or $$v = \frac{e}{Bd} = \frac{0.6}{0.6 \times 1} = 1 \text{ m/s}$$

$\because$ $$e = i_1 R_1 = i_2 R_2$$

Hence $$R_1 = \frac{e}{i_1} = \frac{0.6 \times 15}{19} = 0.47\ \Omega$$

and $$R_2 = \frac{e}{i_2} = \frac{0.6}{2} = 0.3\ \Omega.$$

Example 24.18: Space in Fig. 24.35 is divided by the line *AD* into two regions. Region I is field free and region II has a uniform magnetic field *B* directed into the plane of the paper. *ACD* is a semi circular loop of radius *r* with centre at *O* and the plane of the loop is in the plane of paper. The loop is now made to rotate with a constant angular velocity ω about an axis passing through *O* and perpendicular to the plane of the paper. Effective resistance of the loop is *R*. (*i*) Obtain an expression for the magnitude of the induced current in the loop. (*ii*) Show the direction of the current when the loop is entering into the region II. (*iii*) Plot a graph between the induced emf and time of rotation for two periods of rotation.

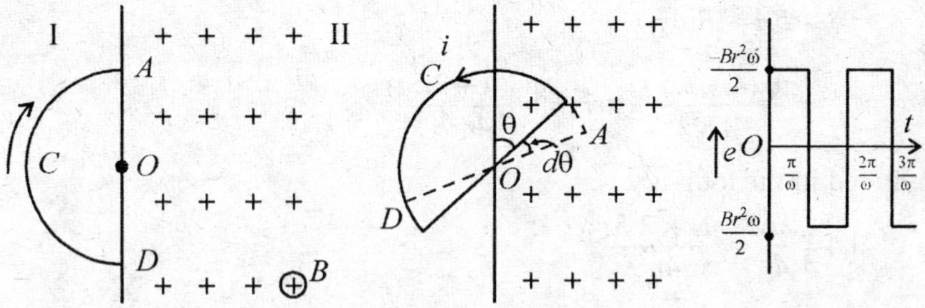

Fig. 24.35

Solution: When the loop has entered through an angle θ in the magnetic field, the area of the loop in magnetic field is $A = \frac{1}{2} r^2 \theta$. Hence the flux linked with the loop at this instant is

$$\phi = BA = B\frac{1}{2} r^2 \theta$$

and the emf induced in the loop is

$$e = -\frac{d\phi}{dt} = -\frac{1}{2} Br^2 \frac{d\theta}{dt} = -\frac{1}{2} Br^2 \omega$$

Since effective resistance of the loop is R, hence current is

$$i = \frac{e}{R} = \frac{Br^2\omega}{2R}$$

The induced current is produced due to increase in flux downwards, hence Lenz's law gives that induced current in loop should be anticlockwise when it enters the field and clockwise when it comes out.

The time period of rotation is $\frac{2\pi}{\omega}$. The graph between the induced emf and the time of rotation (t) is shown in Fig. 24.35.

Example 24.19: Two concentric coplanar circular loops made of wire with resistance per unit length $10^{-4}\ \Omega m^{-1}$, have diameters 0.2 m and 2 m. A time varying potential difference $V = (4 + 2.5t)$ Volt is applied to the longer loop. Calculate the current in the smaller loop.

Solution: Let r_1 and r_2 be the radii of the two loops with ($r_1 > r_2$).

Hence the resistance of the outer loop

$$R_1 = 2\pi r_1 \lambda, \qquad \text{where } \lambda \text{ is resistance per unit length} = 10^{-4}\ \Omega m^{-1}.$$

Hence the current in the loop is

$$i_1 = \frac{V}{R_1} = \frac{(4+2.5\,t)}{2\pi r_1 \lambda}$$

and the magnetic field at the centre of this current carrying loop is

$$B = \frac{\mu_0 i_1}{2r_1} = \frac{\mu_0(4+2.5t)}{4\pi r_1^2 \lambda}$$

The smaller loop has a radius much smaller than that of the bigger one hence magnetic field B at centre of the smaller loop is considered to be uniform. Hence flux with smaller loop at any time t is

$$\phi = BA = B\pi r_2^2$$

$$= \frac{\mu_0(4+2.5\,t)}{4\pi r_1^2 \lambda}\pi r_2^2 = \frac{\mu_0(4+2.5t)}{4r_1^2\lambda}r_2^2$$

and the emf induced in the loop is

$$|e| = \frac{d\phi}{dt} = \frac{\mu_0 \times 2.5 r_2^2}{4r_1^2 \lambda}$$

$$= \frac{4\pi \times 10^{-7} \times 2.5 \times (0.1)^2}{4\,(1)^2 \times 10^{-4}} = 2.5 \times \pi \times 10^{-5}\ \text{V}$$

Resistance of the smaller loop is $R_2 = 2\pi r_2 \lambda = 2 \times \pi\,(0.1) \times 10^{-4} = 2\pi \times 10^{-5}\ \Omega$

Hence current induced in the loop is

$$i = \frac{e}{R_2} = \frac{2.5\pi \times 10^{-5}}{2\pi \times 10^{-5}} = 1.25\ \text{A}.$$

Example 24.20: A pair of parallel horizontal conducting rails of negligible resistance shorted at one end is fixed on the table. The distance between the rails is l. A conducting massless rod of resistance R can slide on the rails without friction. The rod is tied to a massless string which passes over a pulley fixed to the edge of the table. A mass m is tied to the other end of the string hangs vertically. A constant magnetic field B exists perpendicular to the table. If the system is released from rest, calculate (a) the terminal velocity acquired by the rod and (b) the acceleration of the mass at the instant when the velocity of the rod is half the terminal velocity.

Solution: Let the instantaneous velocity of the rod be v along Y-axis (Fig. 24.36). The magnetic field is along Z-axis and the table lies in X-Y plane. The emf induced then is

$$e = Blv$$

If R is the resistance then current through the rod is

$$i = \frac{e}{R} = \frac{Blv}{R} \quad \text{...(i)}$$

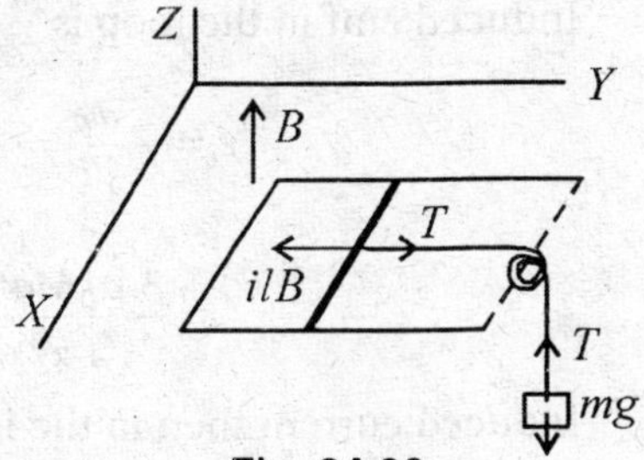

Fig. 24.36

Because of this current the rod experiences a force $F = ilB$ in the direction opposite to the velocity. For the rod to move with terminal velocity ($= v_0$) the net force on the rod should be zero. Figure 24.36 shows this magnetic force must be equal to the weight suspended, i.e.

$$ilB = mg \quad \text{or} \quad i = \frac{mg}{lB} \quad \text{...(ii)}$$

But from (*i*) current when velocity is v_0 is $i = \frac{Blv_0}{R}$

Equating with (*ii*), we have

$$\frac{Blv_0}{R} = \frac{mg}{lB}$$

or $$v_0 = \frac{mgR}{B^2l^2}$$

When velocity of rod is $v = v_0/2$ half the terminal velocity. The magnetic force

$$F = ilB \qquad \text{But } i = \frac{Blv_0}{2R}$$

$$\therefore \quad F = \frac{B^2l^2v_0}{2R}$$

Now net force on mass m is

$$mg - F$$

and if a is acceleration, then $mg - F = ma$

or $$a = g - \frac{F}{m} = g - \frac{B^2l^2v_0}{2Rm} \quad \text{But } v_0 = \frac{mgR}{B^2l^2}$$

$$\therefore \quad a = g - g/2 = g/2.$$

Example 24.21: An infinitesimally small bar magnet of dipole moment M is pointing and moving with a speed v in the X-direction. A small closed circular conducting loop of radius a and negligible self inductance lies in the Y-Z plane with its centre at $x = 0$ and its axis coinciding with the X-axis. Find the force opposing the motion of the magnet, if the resistance of the loop is R. Assume that the distance x of the magnet from the centre of the loop is much greater than a.

Solution: The magnetic field at distance x from the magnet of magnetic moment M on its axis is given by

$$B = \frac{\mu_0}{4\pi}\frac{2M}{x^3} \quad \text{...(i)}$$

Hence flux linked with the loop is

$$\phi = BA = \frac{\mu_0}{4\pi}\frac{2M}{x^3}(\pi a^2) = \frac{\mu_0 Ma^2}{2x^3}$$

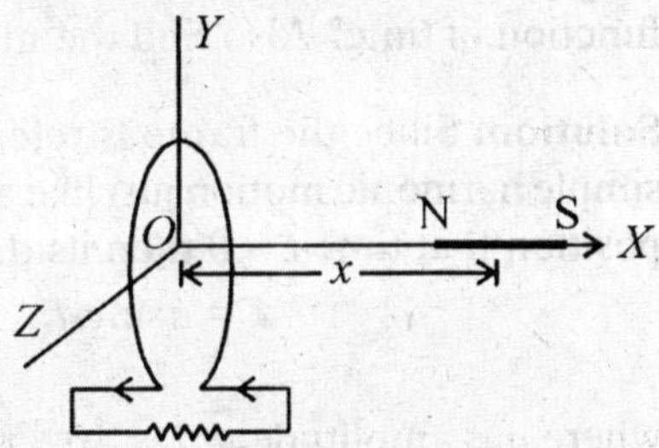

Fig. 24.37

Induced emf in the loop is

$$e = -\frac{d\phi}{dt} = -\frac{d}{dt}\left(\frac{\mu_0 M a^2}{2x^3}\right) = \frac{3}{2}\frac{\mu_0 M a^2}{x^4}\frac{dx}{dt}$$

$$= \frac{3}{2}\frac{\mu_0 M a^2 v}{x^4} \qquad \because \frac{dx}{dt} = v, \text{ speed of the magnet}$$

Induced current then in the loop is

$$i = \frac{e}{R} = \frac{3}{2}\frac{\mu_0 M a^2 v}{R x^4}$$

This loop behaves like a magnetic dipole of moment

$$M' = \text{Current} \times \text{Area}$$

$$= \frac{3}{2}\frac{\mu_0 M a^2 v}{R x^4}(\pi a^2)$$

$$= \frac{3}{2}\frac{\pi \mu_0 M a^4 v}{R x^4}$$

The force between two magnetic dipoles of magnetic moments M and M' at a distance x is given by

$$F = \frac{\mu_0}{4\pi}\frac{6MM'}{x^4}$$

$$= \frac{\mu_0}{4\pi}\frac{6M}{x^4}\frac{3}{2}\frac{\pi \mu_0 M a^4 v}{R x^4}$$

$$= \frac{9\mu_0^2 M^2 a^4 v}{4 R x^8}$$

This force is caused by the moving magnet and from Lenz's law this force opposes the motion of the magnet.

Example 24.22: A wire frame of area 3.92×10^{-4} m^2 and resistance 20 Ω is suspended freely from 0.392 m long thread. There is a uniform horizontal magnetic field of 0.784 tesla and the plane of the wire frame is perpendicular to the magnetic field. The frame is made to oscillate under the force of gravity by displacing it through 2×10^{-2} m from its initial position along the direction of the magnetic field and does not rotate about it. What is the induced emf in the wire frame as a function of time? Also find the maximum current in the frame.

Solution: Since the frame is released after displacing it through 2×10^{-2} m $(= a)$, it will execute simple harmonic motion just like a simple pendulum. If we assume the frame to be at its equilibrium position P at time $t = 0$ then its displacement (x) at any time t, is given by

$$x = a \sin \omega t \qquad \ldots(i)$$

where a is amplitude $= 2 \times 10^{-2}$ m and $\omega = \sqrt{\frac{g}{l}} = \sqrt{\frac{10}{0.392}} = 5$ rad/s

The position of the loop at any time t is shown in Fig. 24.38. At this time the vector area makes an angle θ with the direction of magnetic field B, and the flux linked with the loop is

$$\phi = BA\cos\theta$$

and induced emf e is

$$e = -\frac{d\phi}{dt} = BA\sin\theta\frac{d\theta}{dt}$$

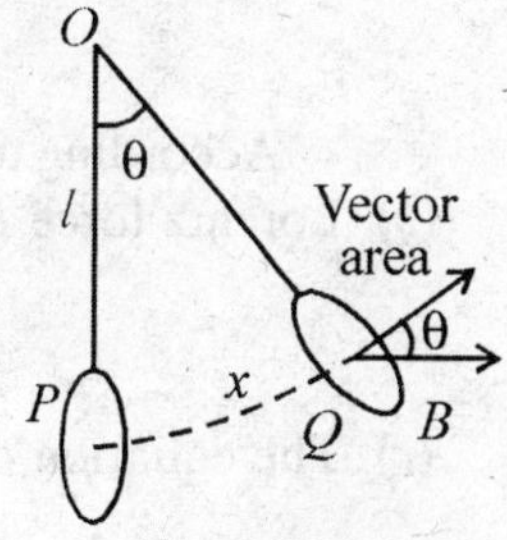

Fig. 24.38

From diagram $\theta = \frac{x}{l}$ and since θ is small $\sin\theta \approx \theta$

$\therefore$ Induced emf is

$$e = BA\theta\frac{d\theta}{dt} = BA\frac{x}{l}\frac{d}{dt}\left(\frac{x}{l}\right)$$

or
$$e = \frac{BAx}{l^2}\frac{dx}{dt}$$

But equation (i) gives displacement x of the loop at any time t. Then, we have

$$e = \frac{BA}{l^2}a\sin\omega t\frac{d}{dt}(a\sin\omega t)$$

$$= \frac{BAa^2\omega}{l^2}\sin\omega t\cos\omega t$$

Substituting the values, we have

$$\frac{BAa^2\omega}{2l^2} = \frac{0.78\times 3.92\times 10^{-4}\times(2\times 10^{-2})^2\times 5}{2\times(0.392)^2} = 2\times 10^{-6}\text{ V}$$

$$\therefore \quad e = 2\times 10^{-6}\sin 10t\text{ V} \qquad \because \omega = 5\text{ rad/s}$$

This gives the induced emf as a function of time. Maximum value of induced emf then is

$$e_{max} = 2\times 10^{-6}\text{ V}$$

Hence maximum induced current is

$$I_{max} = \frac{e_{max}}{R} = \frac{2\times 10^{-6}}{20} = 10^{-7}\text{ A} = 0.1\,\mu\text{A}$$

Example 24.23: A magnetic field $B = \left(\frac{B_0 y}{a}\right)\hat{k}$ is into the paper in the $+Z$ direction. B_0 and a are positive constants. A square loop $EFGH$ of side a mass m and resistance R, in X-Y plane, starts falling under the influence of gravity. The directions of X and Y are shown in Fig. 24.39. Find,

(a) the induced current in the loop and indicate its direction
(b) the total Lorentz force acting on the loop and indicate its direction and
(c) an expression for the speed of the loop $v(t)$ and its terminal value.

Solution:

(a) Let at $t = 0$, $y = 0$, i.e. EF coincides with X-axis and at, $t = t$, $y = y$ and v is the velocity of the loop vertically downwards. The emf induced in segment $EF = B_2va$ and induced emf in segment $GH = B_4va$

Hence induced emf in loop $e = (B_4 - B_2)\, va$

But $\boldsymbol{B}_4 - \boldsymbol{B}_2 = B_0\,\hat{k}$

$\therefore \quad e = vaB_0$ and induced current $i = \dfrac{e}{R} = \dfrac{vaB_o}{R}$

According to Lenz's law, the direction of current is anticlockwise.

(*b*) Lorentz force $F = (B_4 ia - B_2 ia) = (B_4 - B_2) ia$

$$= B_0 \frac{B_0 va}{R} a = \frac{B_0^2 a^2 v}{R} \text{ upward (Fleming's right hand rule)}$$

(*c*) The equation of falling loop is

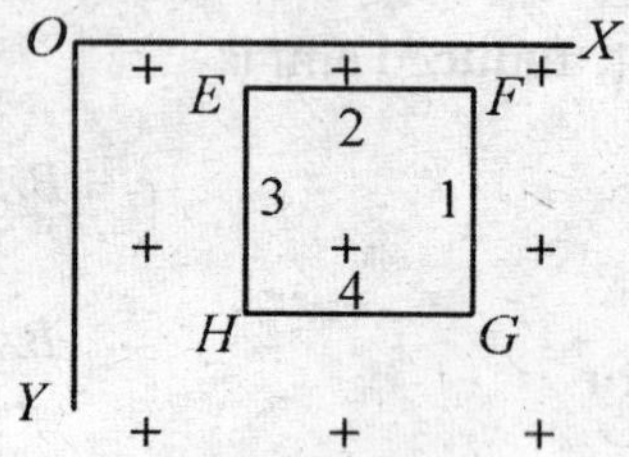

Fig. 24.39

$$m\frac{dv}{dt} = mg - \frac{B_0^2 a^2 v}{R}$$

or

$$dt = \frac{m\,dv}{mg - \dfrac{B_0^2 a^2 v}{R}}$$

integrating, we get

$$t = -\frac{mR}{B_0^2 a^2} \log\left(mg - \frac{B_0^2 a^2 v}{R}\right) + C$$

where C is constant of integration.

At $t = 0$, $v = 0$ gives $C = \dfrac{mR}{B_0^2 a^2} \log mg$

$$\therefore \quad t = -\frac{mR}{B_0^2 a^2}\left[\log\left(mg - \frac{B_0^2 a^2 v}{R}\right) - \log mg\right]$$

$$= -\frac{mR}{B_0^2 a^2} \log \frac{\left(mg - \dfrac{B_0^2 a^2 v}{R}\right)}{mg}$$

$$= -\frac{mR}{B_0^2 a^2} \log\left(1 - \frac{B_0^2 a^2 v}{mgR}\right)$$

or

$$\log\left(1 - \frac{B_0^2 a^2 v}{mgR}\right) = -\frac{B_0^2 a^2}{mR} t$$

or

$$1 - \frac{B_0^2 a^2 v}{mgR} = e^{-\frac{B_0^2 a^2}{mR} t}$$

and

$$v = \frac{mgR}{B_0^2 a^2}\left(1 - e^{-\frac{B_0^2 a^2}{mR} t}\right)$$

Terminal velocity means velocity at $t = \infty$. Hence

$$v_t = \frac{mgR}{B_0^2 a^2}$$

Example 24.24: A rectangular conducting loop in the vertical X-Z plane has length l, width w mass m and resistance R. It is dropped length wire from rest. At $t = 0$ bottom of the loop is at a height h above the horizontal X-axis. There is a uniform magnetic field B, perpendicular to the X-Z plane; below the X-axis. The bottom and the top of the loop cross this axis at $t = t_1$ and $t = t_2$ respectively. Obtain the expression for the velocity of the loop for the time $t_1 \le t \le t_2$. [Roorkee 2000]

Solution: The loop falls freely under gravity for time t_1 and falls distance h. Hence its velocity at $t = t_1$ is

$$v_1 = \sqrt{2gh} \qquad \ldots(i)$$

During the time $t_1 \le t \le t_2$ the flux linked with the loop changes and the emf indued is

$$e = -\frac{d\phi}{dt} = -Bv\omega \qquad \ldots(ii)$$

where v is the velocity of loop at any instant t.

The induced current in the loop is

$$i = -\frac{e}{R} = -\frac{Bv\omega}{R}$$

and magnetic force $F = i\omega B = -\dfrac{B^2\omega^2 v}{R}$ in upward direction

Hence equation of motion of the loop is

$$m\frac{dv}{dt} = mg - \frac{B^2\omega^2 v}{R}$$

or $$\frac{mdv}{\left(mg - \dfrac{B^2\omega^2 v}{R}\right)} = dt$$

or $$\frac{dv}{g - \dfrac{B^2\omega^2 v}{mR}} = dt$$

and integration gives

$$-\frac{mR}{B^2\omega^2}\log\left(g - \frac{B^2\omega^2 v}{mR}\right) = t + C$$

where C is constant of integration. At $t = t_1$ $v = gt_1$

$$\therefore \qquad C = -t_1 - \frac{mR}{B^2\omega^2}\log\left(g - \frac{B^2\omega^2 gt_1}{mR}\right)$$

Substituting the value of C, we get

$$-\frac{mR}{B^2\omega^2}\log\left(g - \frac{B^2\omega^2 v}{mR}\right) = t - t_1 - \frac{mR}{B^2\omega^2}\log\left(g - \frac{B^2\omega^2 gt_1}{mR}\right)$$

or $$\frac{mR}{B^2\omega^2}\log\left[\frac{g - \dfrac{B^2\omega^2 v}{mR}}{g - \dfrac{B^2\omega^2 gt_1}{mR}}\right] = -(t - t_1)$$

or $$g-\frac{B^2\omega^2 v}{mR}=\left(g-\frac{B^2\omega^2 g t_1}{mR}\right)e^{-\frac{B^2\omega^2}{mR}(t-t_1)}$$

or $$\frac{B^2\omega^2 v}{mR}=\left[g-\left(g-\frac{B^2\omega^2 g t_1}{mR}\right)e^{-\frac{B^2\omega^2}{mR}(t-t_1)}\right]$$

or $$v=\frac{mgR}{B^2\omega^2}=\left[1-\left(1-\frac{B^2\omega^2 t_1}{mR}\right)e^{-\frac{B^2\omega^2}{mR}(t-t_1)}\right]$$

This equation gives the velocity v of the loop within time interval $t_1 \le t \le t_2$.

Example 24.25: A coil of inductance 1 henry and resistance 10 Ω is connected to a battery of emf 50 V and zero internal resistance at $t = 0$. Calculate the ratio of the rate at which magnetic energy is stored in the coil to the rate at which the energy is supplied by the battery at $t = 0.1$ s.

Solution: In an $L–R$, circuit the growth of the current is given by

$$i = i_0(1 - e^{-t/\tau})$$

where $\tau = \frac{L}{R} = \frac{1}{10} = 0.1$ sec

$$\frac{di}{dt} = \frac{i_0}{\tau}e^{-t/\tau} \text{ and } i_0 = \frac{50}{10} = 5 \text{ A}$$

Energy stored in the coil in the form of magnetic field energy is

$$U = \frac{1}{2}Li^2$$

Hence the rate of increase of magnetic field energy is

$$\frac{dU}{dt} = Li\frac{di}{dt}$$

$$= Li_0\left(1-e^{-t/\tau}\right)\times\frac{i_0}{\tau}e^{-t/\tau}$$

$$= \frac{Li_0^2}{\tau}\left(e^{-t/\tau}-e^{-2t/\tau}\right)$$

$$= \frac{1\times(5)^2}{(0.1)}\left(e^{-1}-e^{-2}\right)$$

$$= 250\frac{(e-1)}{e^2} = 58.1 \text{ J}$$

Rate of heat produced

$$\frac{dH}{dt} = i^2R = i_0^2R\left(1-e^{-t/\tau}\right)^2$$

$$= 25\times 10\left(1-\frac{1}{e}\right)^2 = 250\left(1-\frac{1}{e}\right)^2 = 99.89 \text{ J}$$

Hence energy supplied by the battery is $= \frac{dU}{dt} + \frac{dH}{dt}$

$$= 58.1 + 99.89 = 157.99 \text{ J}$$

Hence required ratio $= \dfrac{58.1}{157.99} = 0.37.$

Example 24.26: A uniformly wound solenoid coil of self inductance L (= 1.8×10^{-4} H) and resistance R(= 6 Ω) is broken up into two identical coils. These identical coils are then connected in parallel across a 12 V battery of negligible resistance. Find the time constant for the current in the circuit and the steady state current through the battery.

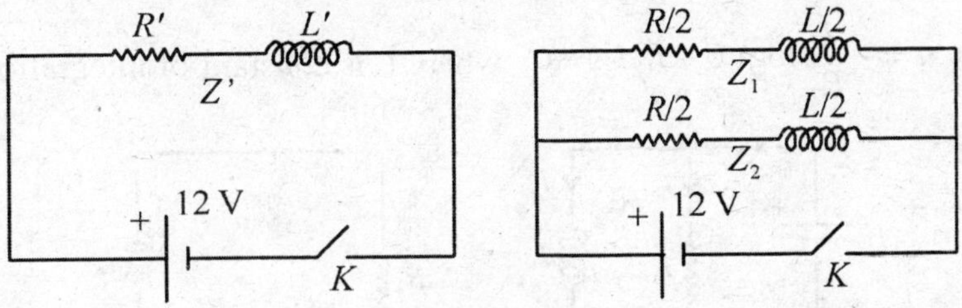

Fig. 24.40

Solution: When the coil is broken up into two identical parts, each part will have resistance $(R/2)$ and inductance $(L/2)$ and the problem then reduces to that of two impedances in parallel, i.e.

$$Z' = R' + j\omega L' = \frac{Z_1 Z_2}{Z_1 + Z_2}$$

But in present problem

$$Z_1 = Z_2 = \frac{R}{2} + j\omega\frac{L}{2}$$

i.e. $$R' + j\omega L' = \frac{R}{4} + j\omega\frac{L}{4}$$

or $$R' = \frac{R}{4} \text{ and } L' = \frac{L}{4}$$

As such the equivalent circuit is as shown in Fig. 24.40.

Hence time constant of the circuit is $\tau = \dfrac{L'}{R'} = \dfrac{L/4}{R/4} = \dfrac{L}{R}$

or $$\tau = \frac{L}{R} = \frac{1.8 \times 10^{-4}}{6} = 30\ \mu\text{s}$$

In steady state inductance behaves like a short circuit and current through battery is

$$i = \frac{E}{R/4} = \frac{12}{6/4} = 8\ \text{A}.$$

Example 24.27: An inductor of inductance $L = 400$ mH and resistors of resistance $R_1 = 2\ \Omega$ and $R_2 = 2\ \Omega$ are connected to a battery of emf $E = 12$ V as shown in Fig. 24.41. The internal resistance of the battery is negligible. The switch S is closed at time $t = 0$.

What is the potential drop across L as a function of time? After the steady state is reached, the switch is opened. What is the direction and magnitude of the current through R_1 as a function of time? [IIT 2001]

Solution: During the growth of the current let the currents through different branches be as shown in Fig. 24.41(*a*). Applying Kirchhoff's potential law to circuit containing E, L and R_2, we get

$$E - L\frac{di_2}{dt} = R_2 i_2$$

or
$$L\frac{di_2}{dt} = E - R_2 i_2 \quad \text{and} \quad dt = \frac{L di_2}{E - i_2 R_2}$$

or
$$t = -\frac{L}{R_2}\log(E - R_2 i_2) + C, \text{ where } C \text{ is constant of integration}$$

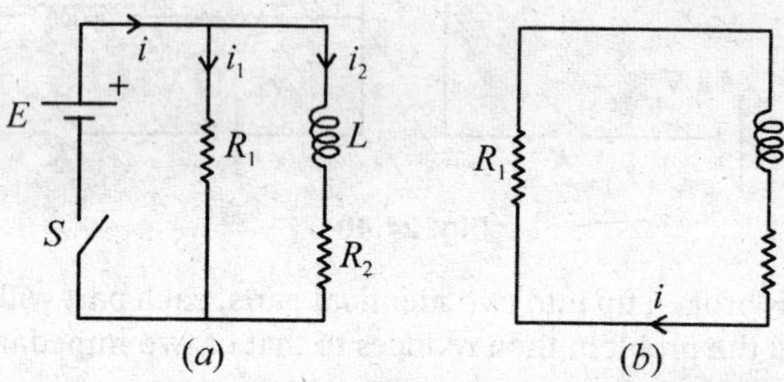

Fig. 24.41

Initial conditions $t = 0$ $i_2 = 0$ give $C = \frac{L}{R_2}\log E$

Substitution gives

$$t = -\frac{L}{R_2}[\log(E - R_2 i_2) - \log E]$$

and rearranging the terms, we get

$$i_2 = \frac{E}{R_2}\left(1 - e^{-\frac{R_2}{L}t}\right)$$

The potential drop across the inductor as a function of time will be

$$V = L\frac{di_2}{dt} = \frac{LE}{R_2}\frac{R_2}{L}e^{-\frac{R_2}{L}t} = Ee^{-\frac{R_2 t}{L}}$$

But $R_2 = 2\ \Omega$ and $L = 400$ mH. Hence $\frac{R_2}{L} = \frac{2}{400 \times 10^{-3}} = 5\ \text{sec}^{-1}$

$\therefore$ $\quad V = 12\, e^{-5t}$ V $\quad \because E = 12$ V

In steady state there is no potential drop across inductor as current becomes constant.

When the switch is opened after a long time, the battery is disconnected and the current flows through R_1, L and R_2. At the time of breaking the circuit, i.e. at $t = 0$, current through R_1 becomes zero and through R_2 and L decays exponentially through R_1 and R_2.

At $t = 0$ current through R_2 is $i_0 = 6$ A. Now the time constant is $\frac{L}{R_1 + R_2}$

i.e.
$$\tau = \frac{400 \times 10^{-3}}{2 + 2} = 0.1\ \text{sec}$$

Hence the current at any time t in the circuit is

$$i = i_0 e^{-t/\tau} = 6\,e^{-0.1t} \text{ A}$$

The direction through R_1 now is shown in Fig. 24.41 (*b*).

Example 24.28: A metal rod OA of mass m and length l is kept rotating with a constant angular speed ω in a vertical plane about a horizontal axis at the end O. The free end A is arranged to slide without friction along fixed conducting circular ring in the same plane as that of rotation. A uniform and constant magnetic induction B is applied perpendicular and into the plane of rotation as shown in Fig. 24.42 (*a*). An inductor L and an external resistance R are connected through a switch S between the point O and C on the ring to form an electrical circuit. Neglect the resistance of the ring and the rod. Initially the switch is open.

(*a*) What is the induced emf across the terminals of the switch?

(*b*) The switch S is closed at time $t = 0$.

(*i*) Obtain an expression for the current as a function of time.

(*ii*) In the steady state, obtain the time dependence of the torque required to maintain the constant angular speed; given that the rod OA was along the positive X-axis at $t = 0$.

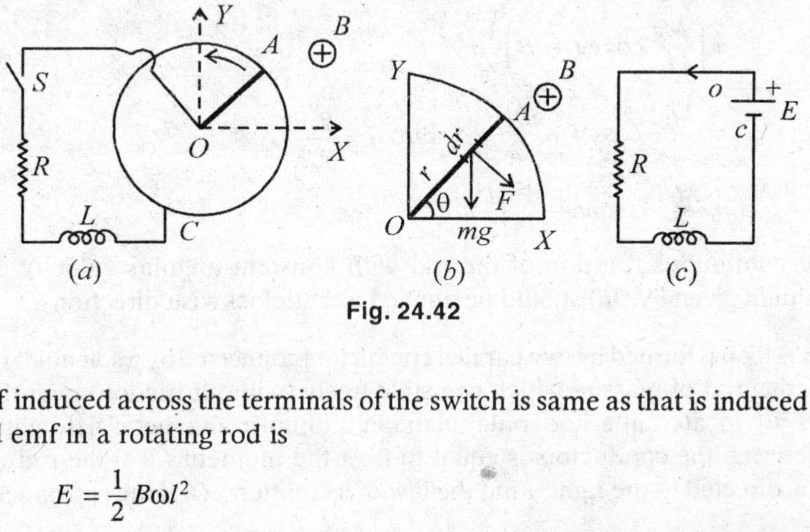

Fig. 24.42

Solution:

(*a*) The emf induced across the terminals of the switch is same as that is induced in the rod. An induced emf in a rotating rod is

$$E = \frac{1}{2} B\omega l^2 \qquad \text{...(i)}$$

The direction of the induced current in rod according to Fleming's right hand rule is from A to O as such O is at higher potential.

(*b*)

(*i*) The equivalent circuit is shown in Fig. 24.42 (*c*). It is just like an L-R circuit connected a battery of emf $E = \frac{1}{2} B\omega l^2$

Hence maximum current $i_0 = \frac{E}{R} = \frac{1}{2}\frac{B\omega l^2}{R}$ and the current at any time is given by

$$i = i_0\left(1 - e^{-\frac{R}{L}t}\right)$$

$$= \frac{B\omega l^2}{2R}\left(1 - e^{-\frac{R}{L}t}\right) \qquad \text{...(ii)}$$

(*ii*) In steady state the current is independent of time and $i = i_0 = \dfrac{B\omega l^2}{2R}$. Since the rod through which current is passing rotates in magnetic field and hence experiences a force. Referring to Fig. 24.42 (*b*), the magnetic force on an element dr at a distance r from O is

$$dF = iBdr$$

and its torque about O is $= rdF = iBrdr$ clockwise. In addition to it a torque $= mg\,r\cos\theta$ also acts in clockwise direction due to the weight of the element. Hence net torque is

$$d\tau = mg\,r\cos\theta + iBrdr$$

If M is mass of rod then $m = \dfrac{M}{l}dr$ and $\theta = \omega t$ as at $t = 0 \quad \theta = 0$

$$\therefore \quad d\tau = \frac{Mg}{l}\cos\theta\, rdr + iBr\,dr = \left(\frac{Mg}{l}\cos\theta + iB\right)rdr$$

Hence total torque on the rod is

$$\tau = \left(\frac{Mg}{l}\cos\omega t + iB\right)\int_0^l rdr$$

$$= \left(\frac{Mg}{l}\cos\omega t + iB\right)\frac{1}{2}l^2$$

$$\tau = \frac{Mgl}{2}\cos\omega t + \frac{iBl^2}{2} \qquad \text{But } i = \frac{B\omega l^2}{2R}$$

$$= \frac{Mgl}{2}\cos\omega t + \frac{B^2\omega l^4}{4R} \qquad \ldots(iii)$$

For continuous rotation of the rod with constant angular velocity ω, a torque of magnitude given by (*iii*) should be applied in anticlockwise direction.

Example 24.29: A loop is formed by two parallel conductors connected by a solenoid with inductance L and a conducting rod of mass m which can slide freely (without friction) over the conductors. The conductors are located in a horizontal plane in a uniform magnetic field with induction B. The distance between the conductors is equal to l. At the moment $t = 0$ the rod is imparted an initial velocity v_0 directed to the right. Find the law of its motion $x(t)$ if the resistance of the loop is negligible.

Solution: Let the position of the conducting rod at $t = 0$ is as shown in Fig. 24.43. Now it gives a velocity v_0 towards right. Let at time $t = t$, velocity of rod be v then

$$L\frac{di}{dt} = Blv$$

or $$Ldi = Bldx$$

and integration gives $\int_0^i Ldi = \int_0^x Bldx$

or $$Li = Blx \text{ and } i = \frac{Blx}{L}$$

Since current through rod is i. Hence magnetic force on the rod is

$$F = ilB \text{ in the direction opposite to the velocity}$$

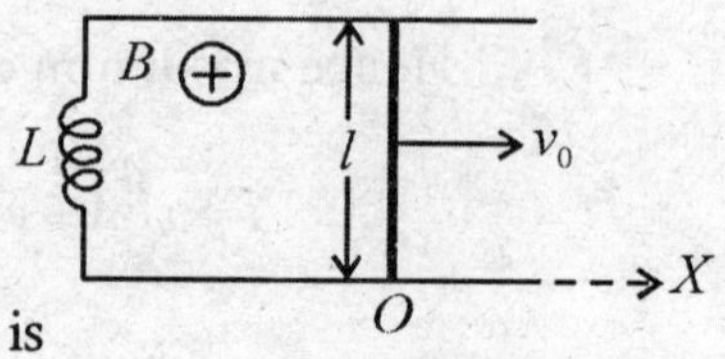

Fig. 24.43

If a is the acceleration at any time t then equation of motion is

$$F = ma = -ilB = -\frac{B^2l^2}{L}x$$

or
$$a = -\frac{B^2l^2}{mL}x \qquad \ldots(i)$$

Negative sign is due to the fact that acceleration (a) and displacement (x) are in opposite direction.

Since acceleration is directly proportional to displacement hence motion is simple harmonic and solution is

$$x = A\sin(\omega t + \delta) \text{ where } \omega^2 = \frac{B^2l^2}{mL}$$

At $t = 0$, $x = 0$ gives $\delta = 0$

Hence
$$x = A\sin\omega t \qquad \ldots(ii)$$

Then
$$\frac{dx}{dt} = A\omega\cos\omega t$$

But at $t = 0$
$$\frac{dx}{dt} = v_0$$

$\therefore$
$$v_0 = A\omega \quad \text{or} \quad A = \frac{v_0}{\omega}$$

Hence equation of motion is

$$x(t) = \frac{v_0}{\omega}\sin\omega t \text{ with } \omega = \frac{Bl}{\sqrt{mL}}$$

Example 24.30: A long wire is arranged along the symmetry axis of a toroidal coil of rectangular cross-section, whose dimensions are given in Fig. 24.44. The number of turns of coil is N and permeability of the surrounding medium is unity. Find the amplitude of the emf induced in this coil if the current $i = i_0\cos\omega t$ flows along the straight wire.

Solution: Consider a small strip area in the solenoid at distance r from the current carrying wire (i.e. axis of the solenoid). The area of the strip is hdr (Fig. 24.44). Magnetic field at this strip due to the current carrying wire is

$$B = \frac{\mu_0 i}{2\pi r} \qquad \ldots(i)$$

Vector area hdr and B are parallel. Hence flux through this area is

$$d\phi = Bhdr = \frac{h\mu_0 i}{2\pi}\frac{dr}{r} \qquad \ldots(ii)$$

Hence total flux linked with the toroid is

$$\phi = \int d\phi = \int_a^b \frac{h\mu_0 i}{2\pi}\frac{dr}{r}$$

$$= \frac{\mu_0 ih}{2\pi}\log(b/a) = \frac{\mu_0 h}{2\pi}\log(b/a)\, i_0\cos\omega t$$

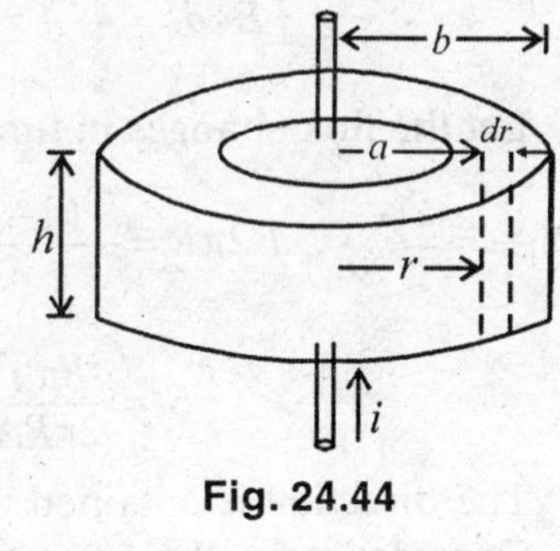

Fig. 24.44

and induced emf developed is

$$e = -\frac{d\phi}{dt} = \frac{\mu_0 h i_0 \omega}{2\pi}\log(h/a)\sin\omega t$$

Since there are N number of turns, hence total induced emf is

$$E = Ne = \frac{\mu_0 h i_0 \omega N}{2\pi} \log(b/a) \sin \omega t$$

The emf induced is alternating emf with peak value

$$E_0 = \frac{\mu_0 h i_0 \omega N}{2\pi} \log(b/a)$$

Example 24.31: A line charge λ per unit length is passed uniformly on to the rim of a wheel of mass M and radius R. The wheel has light non-conducting spokes and it is free to rotate about a vertical axis as shown in Fig. 24.45. A uniform magnetic field extends over a radial region given by

$$\boldsymbol{B} = B_0 \hat{k} \; (r \le a < R)$$
$$= 0 \text{ otherwise.}$$

What is the angular velocity of the wheel when this field is suddenly switched off?

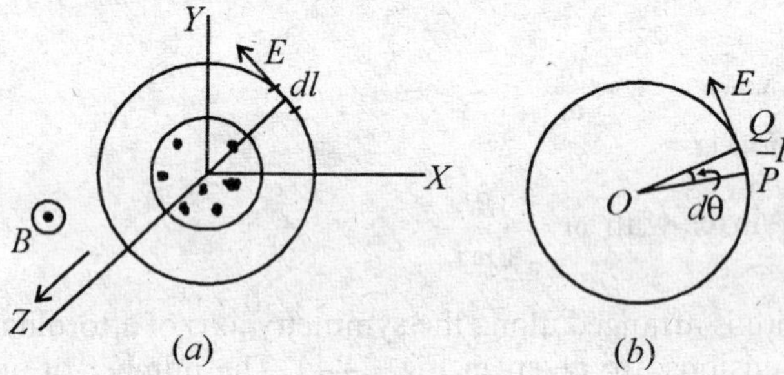

Fig. 24.45

Solution: Since the magnetic flux linked with wheel changes from $B\ \pi a^2$ to zero. Hence the electric field induced at the periphery of the wheel is given by

$$\oint \boldsymbol{E} \cdot \boldsymbol{dl} = -\frac{d\phi}{dt}$$

Let the flux changes in time Δt then

$$E 2\pi R = -\frac{(0 - B\pi a^2)}{\Delta t}$$

$$E = \frac{B\pi a^2}{2\pi R \Delta t} = \frac{Ba^2}{2R\Delta t} \qquad \ldots(i)$$

The direction is obtained using Lenz's law and it is anticlockwise (Fig. 24.45 (*a*)).

Consider a small element $PQ = Rd\theta$ of the wheel (Fig. 24.45 (*b*)) charge on this element $dq = Rd\theta\, \lambda$, where λ is charge per unit length.

Hence force on this element is

$$dF = Edq = \frac{Ba^2}{2R\Delta t} Rd\theta\lambda = \frac{Ba^2 \lambda d\theta}{2\Delta t}$$

Hence this force exerts torque on the wheel. Hence torque about Z-axis is

$$d\tau = RdF = \frac{Ba^2\lambda R d\theta}{2\Delta t}$$

Hence total torque on the wheel due to induced electric field is

$$\tau = \int d\tau = \int_0^{2\pi} \frac{Ba^2\lambda R d\theta}{2\Delta t} = \frac{B\pi a^2 \lambda R}{2\Delta t} \qquad \ldots(ii)$$

and the angular impulse is

$$\tau\Delta t = B\pi a^2\lambda R \qquad \ldots(iii)$$

This must be equal to change in angular momentum. Initial angular momentum is zero. If wheel starts rotating with angular velocity ω the angular momentum is

$L = I\omega$ where I is M.I. of the wheel about Z-axis.

Equating the two, we have

$$B\pi a^2\lambda R = I\omega - 0 \qquad \because I = MR^2$$

we have $$\omega = \frac{B\pi a^2\lambda R}{MR^2} = \frac{B\pi a^2\lambda}{MR}$$

Since it is along Z-axis hence

$$\omega = \frac{B\pi a^2\lambda}{MR}\hat{k}$$

Example 24.32: A rectangular loop of sides a and b has a resistance R and lies at a distance C from an infinite straight wire carrying current i_0. The current decreases to zero in time t_0. $i = i_0\left(\frac{t_0 - t}{t_0}\right)$ $0 < t < t_0$. Find the charge flowing through the rectangular loop.

Solution: Let us consider a small strip of the loop at distance x from the current carrying wire and of width dx. Fig. 24.46.

Magnetic field at the strip when the current through the conductor is i is

$$B = \frac{\mu_0 i}{2\pi x}$$

and the flux linked with the strip at any time t is

$$d\phi = Bb\,dx = \frac{\mu_0 ib}{2\pi x}dx$$

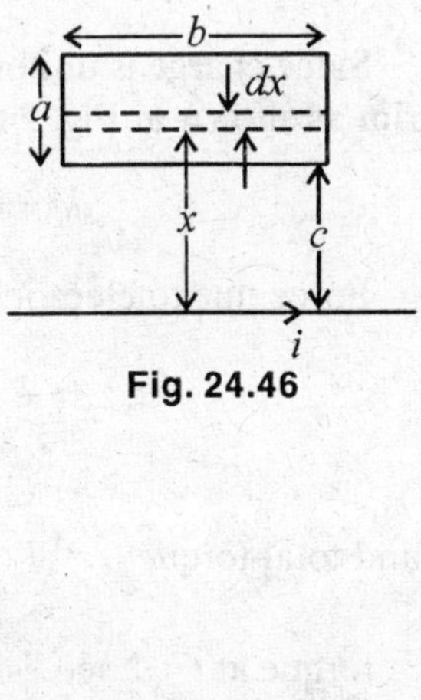

Fig. 24.46

Hence the total flux with the loop at any time t is

$$\phi = \int d\phi = \int_c^{c+a} \frac{\mu_0 ib}{2\pi}\frac{dx}{x} = \frac{\mu_0 ib}{2\pi}\log\left(\frac{c+a}{c}\right)$$

But current at time $t = i_0\left(\frac{t_0 - t}{t_0}\right)$

$$\therefore \quad \phi = \frac{\mu_0 b}{2\pi}\log\left(\frac{c+a}{c}\right) i_0 \left(\frac{t_0 - t}{t_0}\right)$$

Hence the change in flux within time dt is

$$d\phi = \frac{\mu_0 b i_0}{2\pi} \log\left(\frac{c+a}{c}\right)\left(\frac{-dt}{t_0}\right)$$

or the total change in flux in time $t = 0$ to $t = t_0$ is

$$\Delta\phi = \int d\phi = -\frac{\mu_0 b i_0}{2\pi t_0} \log\frac{c+a}{c} \int_0^{t_0} dt$$

$$= -\frac{\mu_0 b i_0}{2\pi} \log\left(\frac{c+a}{c}\right)$$

Since resistance of the loop is R. Hence charge flowing through the rectangular loop is

$$q = \frac{|\Delta\phi|}{R} = \frac{\mu_0 b i_0}{2\pi} \log\left(1 + \frac{a}{c}\right)$$

Example 24.33: A non conducting ring having charge q uniformly distributed over its circumference is placed on a rough horizontal surface. A vertical time varying magnetic field $B = 4t^2$ is switched on at time $t = 0$. Mass of the ring is M and radius R. The ring starts rotating after 2 seconds. Find the coefficient of friction between the ring and table.

Solution: Due to time varying magnetic field an electric field is developed at the rim of the ring. If E is the electric field, then

$$\oint \mathbf{E} \cdot d\mathbf{l} = -\frac{d\phi}{dt}$$

$$E 2\pi R = -\pi R^2 \frac{dB}{dt}$$

or magnitude of electric field which in present case is anticlockwise is

$$E = \frac{R}{2}\frac{dB}{dt} \quad \text{But } B = 4t^2 \text{ or } \frac{dB}{dt} = 8\,\text{t}$$

$$E = \frac{R}{2} 8\text{t} = 4Rt \qquad \ldots(i)$$

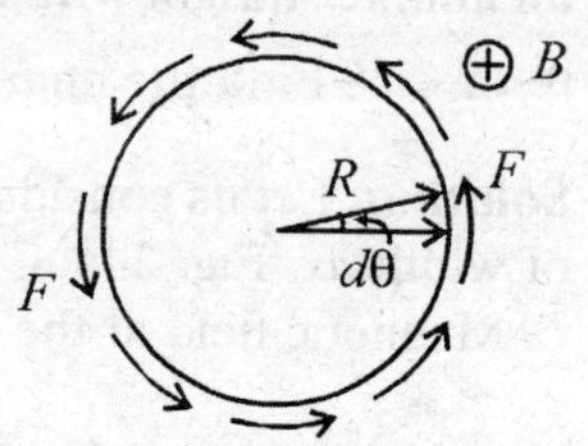

Fig. 24.47

Since charge is uniformly distributed over it hence a force acts on the rim as shown in Fig. 24.47. Change on small arc length $Rd\theta$ is

$$dq = \frac{qRd\theta}{2\pi R} = \frac{q}{2\pi} d\theta$$

Force due to electric field on this charge is $dF = Edq = \frac{4Rqt}{2\pi} d\theta$ and torque due to this force is

$$d\tau = \frac{4R^2 qt}{2\pi} d\theta$$

and total torque $\quad \tau = \int d\tau = \int_0^{2\pi} \frac{4R^2 qt}{2\pi} d\theta = 4R^2 qt$

torque at $t = 2$ sec. is $\tau = 8R^2 q$.

Since ring starts rotating under this torque, hence torque due to friction must be equal to this torque due to electric field.

If μ is the coefficient of friction, then torque due to frictional force is

$$\tau = \mu mgR$$

Equating the two torques, we get

$$\tau mgR = 8R^2 q$$

or

$$\mu = \frac{8Rq}{mg}$$

PROBLEMS

1. A long solenoid of radius 2 cm has 100 turns/cm and is surrounded by a 100 turn coil of radius 4 cm having a total resistance of 20 Ω. The coil is connected to galvanometer. If the current in the solenoid is changed from 5 A in one direction to 5 A in opposite direction, find the charge which flows through the galvanometer.

[**Ans.** 800 μC]

2. A square metal wire loop of side 10 cm and resistance 1 Ω is moved with constant velocity v_0 in a uniform magnetic field of induction $B = 2\ \text{Wbm}^{-2}$ as shown in Fig. 24.48. The magnetic field lines are perpendicular to the plane of the loop. The loop is connected to network of resistors each of value 3 Ω. The resistance of lead wires *PB* and *QD* are zero. What should be the speed of the loop so as to have a steady current of 1 mA in the loop? Give the direction of the current in the loop.

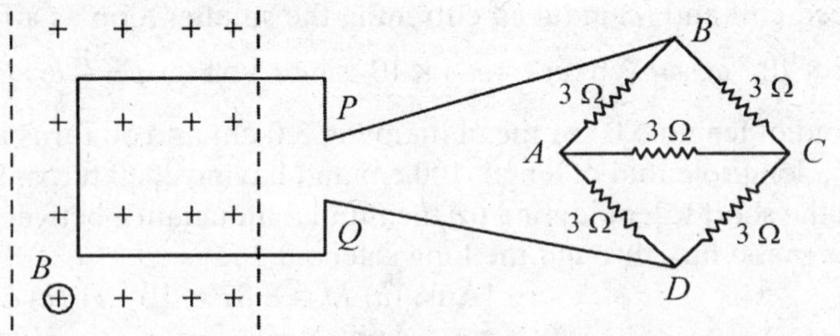

Fig. 24.48

[**Ans.** v_0 = 2 m/s. Induced current in loop is clockwise]

3. Figure 24.49 shows a metal rod *PQ* resting on rails *A* and *B* and positioned between the poles of a permanent magnet. The rails, the rod and the magnetic field are in three mutually perpendicular directions. A galvanometer *G* connects the rails through a switch *K*. Length of the rod is 15 cm, $B = 0.5$ T, resistance of the closed loop containing the rod is 9.0 mΩ. Answer the following questions:

(*a*) Suppose *K* is open and the rod moves with a speed of 12 cm s^{-1} in the direction shown. Give the polarity and magnitude of induced emf.

(*b*) Is there an excess charge built up at the ends of rod when *K* is open? What if *K* is closed.

(*c*) With *K* open and the rod moving uniformly, there is no net force on the electron in the rod *PQ* even though they do experience magnetic force due to the motion of the rod. Explain.

(*d*) What is the retarding force when *K* is closed.

(*e*) How much power is required (by an external agent) to keep the rod moving at the same speed (= 12 m s^{-1}) when *K* is closed.

(*f*) How much power is dissipated as heat in the closed circuit? What is the source of this power?

(*g*) What is the induced emf in moving rod when the permanent magnet is rotated to a vertical position so that the field is parallel to the rails?

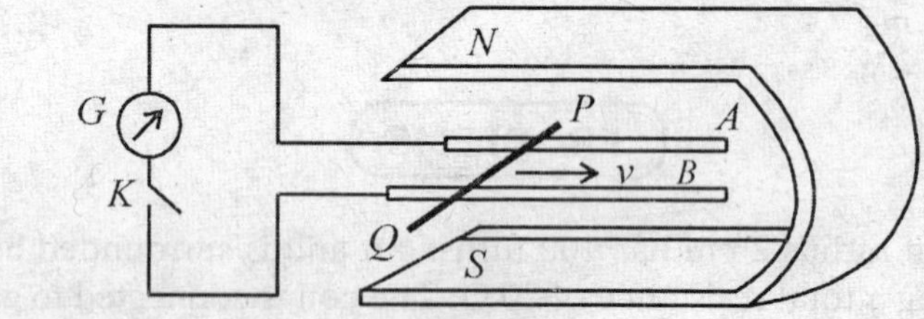

Fig. 24.49

[**Ans.** (*a*) $e = 9.0$ mV with *P* possible (*b*) Yes. When *K* is pressed induced current maintains the excess charge (*c*) electric force *Ee* is equal to magnetic force *evB* (*d*) $F = 7.5 \times 10^{-2}$ N (*e*) $P = 9.0$ mW (*f*) $H = 9.0$ mW, external agent (*g*) In this case **B** and ***v*** are parallel hence $e = 0$]

4. A very small circular loop of area 5×10^{-4} m^2 and resistance 2 Ohm and negligible inductance is initially coplanar and concentric with a much larger fixed circular loop of radius 0.1 m. A constant current of 1 ampere is passed in the bigger loop and smaller loop is rotated at an angular velocity ω rad/s about a diameter. Calculate (*a*) the flux linked with the smaller loop (*b*) induced emf and (*c*) induced current in the smaller loop as a function of time.

[**Ans.** (*a*) $\phi = \pi \times 10^{-9} \cos\omega t$ Wb (*b*) $e = \pi\omega \times 10^{-9} \sin\omega t$ volt (*c*) $I = \frac{\pi}{2}\omega \times 10^{-9} \sin\omega t$ ampere]

5. A short solenoid of length 5.0 cm mean diameter 3.0 cm and 50 turns is placed inside and on the axis of a long solenoid of length 100 cm and having 2000 turns. If a current of 4.0 A flows through the short solenoid, find (*a*) the mutual inductance between the two solenoids and (*b*) the magnetic flux through the long solenoid.

[**Ans.** (*a*) $M = 8.87 \times 10^{-5}$ H (*b*) $\phi = 3.55 \times 10^{-4}$ Wb]

6. Two coaxial circular loops of radii 0.5 m and 5×10^{-2} m are separated by a distance 0.5 m and carry currents 2 A and 1 A respectively (*a*) Calculate the mutual inductance (*b*) What is the force between the loops?

[**Ans.** (*a*) $M = 3.47 \times 10^{-9}$ H (*b*) $F = 1.18 \times 10^{-7}$ N]

7. A circular coil of radius 8.0 cm and 20 turns rotates about its vertical diameter with an angular speed of 50 rad/s in a uniform horizontal magnetic field of strength 3.0×10^{-2} T (*a*) Find the maximum and average emf induced in the coil. (*b*) If the coil forms a closed loop of resistance 10 Ω how much power is dissipated as heat? What is the source of this power?

[**Ans.** (*a*) $e_{av} = 0$, $e_{max} = 0.603$ V (*b*) $P_{av} = 0.018$ W. Source is the external agent which rotates the coil]

8. A 10 Ω coil of mean area 500 cm^2 and having 1000 turns is held perpendicular to a uniform field of 0.4 Gauss. The coil is turned through 180° in 0.1 sec. Calculate (*a*) the change in flux. (*b*) the average induced emf (*c*) average induced current (*d*) the total induced charge.

[**Ans.** (*a*) $\phi = 4$ mWb (*b*) $V = 40$ mV (*c*) $i = 4$ mA (*d*) $q = 400$ μC]

9. Two long parallel wires of zero resistance are connected to each other by a battery of 1.0 V. The separation between the wires is 0.5 m. A metallic bar, which is perpendicular to the wires and resistance 10 Ω moves on the wires when the magnetic field of 0.02 tesla is acting perpendicular to the plane containing the bar and the wires. Find the velocity of the bar as a function of time if mass of the bar is 0.002 kg. Find the steady state velocity of the bar also.

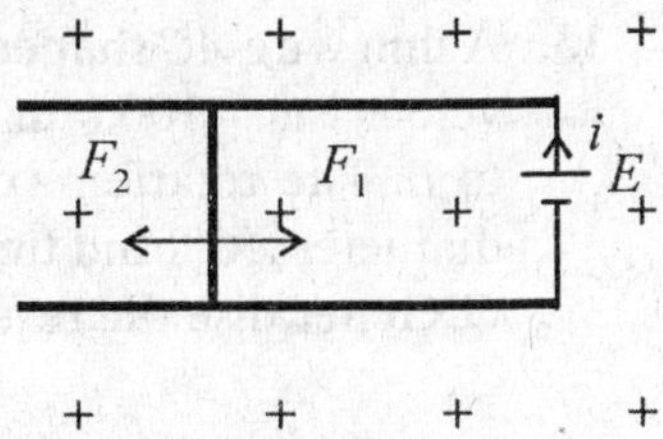

Fig. 24.50

[**Ans.** If v is the instantaneous velocity of the rod then current i in the circuit is $i = \frac{E - Blv}{R}$ and the net force on the rod is $F = F_1 - F_2 = ilB$. Hence equation of motion is $m\frac{dv}{dt} = F = ilB$ $= \frac{E - Blv}{R} lB$. The solution of the equation is $v = \frac{E}{Bl}(1 - e^{-t/\tau})$ where $\tau = \frac{mR}{(Bl)^2}$. Hence

$$v = 100\left(1 - e^{-t/200}\right) \text{ ms}^{-1} \text{ and } v_T = 100 \text{ m/s}\Big]$$

10. A solenoid has an inductance of 10 H. and a resistance of 2 Ω. It is connected to a 10 V battery. How long will it take for the magnetic energy to reach 1/4 of its maximum value?

[**Ans.** t = 3.465 sec]

11. A square wire frame with side a and a long straight conductor with a direct current i_0 lie in the same plane (Fig. 24.51). The inductance of the frame is L and its resistance is R. The frame has been rotated through 180° around the axis OO' and then stopped. Find the amount of charge that has passed through the frame. The distance between the axis OO' and the straight wire b is known.

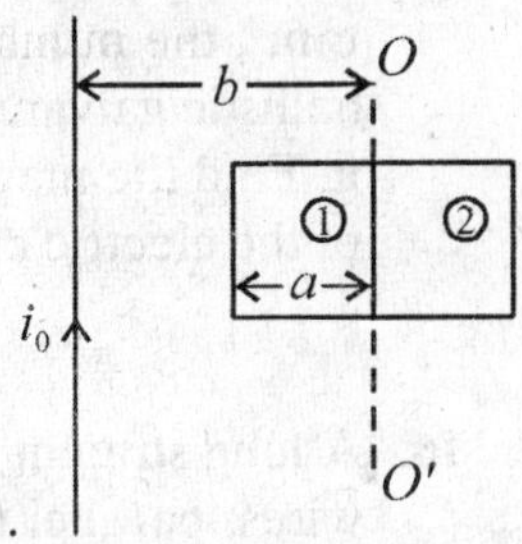

Fig. 24.51

[**Hint:** $q = \frac{1}{R}$(change in flux). Flux $\phi_1 = \int_{b-a}^{b} \frac{\mu_0 i}{2\pi r} adr = \frac{\mu_0 ai}{2\pi} \log\left(\frac{b}{b-a}\right)$.

Flux $\phi_2 = -\int_{b}^{b+a} \frac{\mu_0 ai}{2\pi} \frac{dr}{r} = -\frac{\mu_0 ia}{2\pi} \log\left(\frac{b+a}{b}\right)$. Hence $\Delta\phi = -\frac{\mu_0 ai}{2\pi} \log\frac{(b+a)}{(b-a)}$. Hence $q = \frac{\Delta\phi}{R}$

$$= \frac{\mu_0 ia}{2\pi R} \log\frac{(b+a)}{(b-a)} \text{ neglecting negative sign}]$$

12. An infinitely long straight conductor lies in the plane of a square loop with an ohmic resistance R and a side length a at a distance r_0 parallel to one of the loops side (Fig. 24.52). The current flowing in the conductor varies according to the law $i = \alpha t^3$. Where α is constant. Find the current in the loop at time t = 10 s.

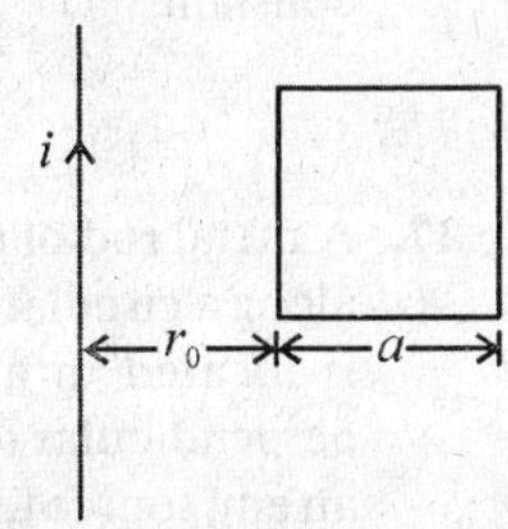

Fig. 24.52

$$\Big[\textbf{Ans. } i = \frac{3\mu_0 a\alpha \log(1 + r_0/a)}{2\pi R} t^2 \text{ for } t = 10 \text{ sec}$$

$$i = \frac{150\mu_0 a\alpha \log(1 + r_0/a)}{2\pi R}\Big]$$

13. A thin wire AC shaped as a semicircle of diameter $d = 20$ cm rotates with a constant angular velocity $\omega = 100$ rad/s in a uniform magnetic field of induction $B = 5.0$ mT with ω parallel to B. The rotation axis passes through the end A of the wire and is perpendicular to the diameter AC. Find the value of the line integral $\int \mathbf{E} \cdot d\mathbf{r}$ along the wire from point A to point C. Generalise the result so obtained.

$$\left[\textbf{Ans. } \int_A^C \mathbf{E} \cdot d\mathbf{r} = -\frac{1}{2}\omega B d^2 = -10 \text{ mV}\right]$$

14. A wire shaped as a semicircle of radius a rotates about an axis OO' with an angular velocity ω in a uniform magnetic field of induction B (Fig. 24.53). The rotation axis is perpendicular to the field direction. Total resistance of the circuit is R. Neglecting the magnetic field of the induced current, find the mean amount of thermal power being generated in the loop during a rotation period.

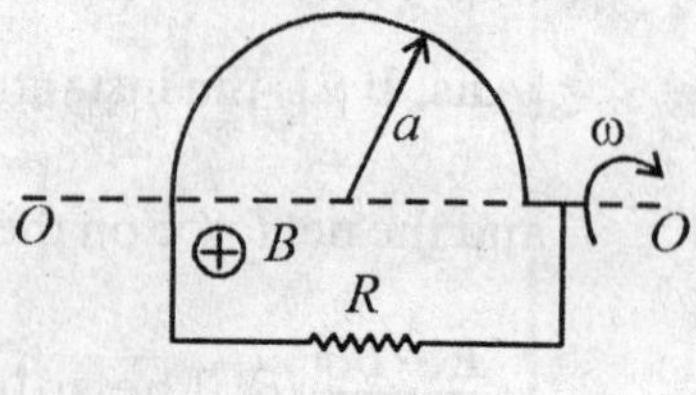

Fig. 24.53

$$\left[\textbf{Ans. } H = \frac{1}{2R}\left(\frac{\pi\omega a^2 B}{2}\right)^2\right]$$

15. A small coil is introduced between the plates of an electromagnet so that its axis coincides with the magnetic field direction. The cross-sectional area of the coil is equal to $S = 3.0$ mm^2, the number of turns $N = 60$. When the coil turns through 180° about its diameter, a ballistic galvanometer connected to the coil indicates a charge $q = 4.5$ μC flowing through it. Find the magnetic induction magnitude between the poles, provided the total resistance of the electric circuit equals $R = 40\ \Omega$.

$$\left[\textbf{Ans. } B = \frac{1}{2}\frac{qR}{NS} = 0.5 \text{ T}\right]$$

16. A long straight wire carries a current I_0. At a distance a and b from it there are two other wires; parallel to the former one, which are connected by a resistance R (Fig. 24.54). A conductor slides without friction along the wires with a constant velocity v. Assuming the resistances of the wires, the conductor, the sliding contacts and the self-inductance of the frame to be negligible, find (*a*) the magnitude and the direction of the current induced in the conductor. (*b*) the force required to maintain the conductor's velocity constant.

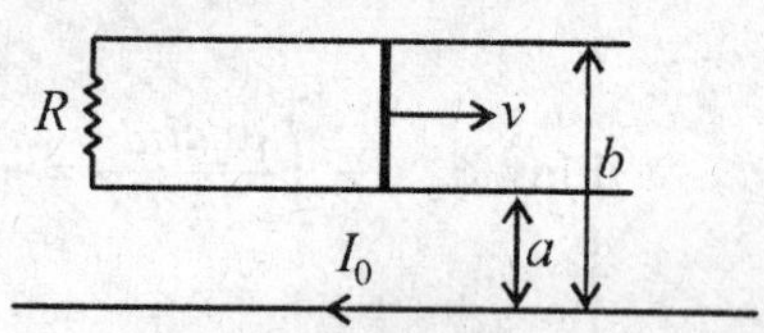

Fig. 24.54

$$\left[\textbf{Ans. } (a)\ I = \frac{\mu_0 I_0 v}{2\pi R}\log\frac{b}{a} \quad (b)\ F = \frac{v}{R}\left(\frac{\mu_0 I_0}{2\pi}\log\frac{b}{a}\right)^2\right]$$

17. A metal rod of mass m can rotate about a horizontal axis O. Sliding along a circular conductor of radius a (Fig. 24.55). The arrangement is located in a uniform magnetic field of induction B directed perpendicular to the ring plane. The axis and ring are connected to an emf source to form a circuit of resistance R. Neglecting the friction, circuit inductance, and ring resistance, find the law according to which the source emf must vary to make the rod rotate with a constant angular velocity.

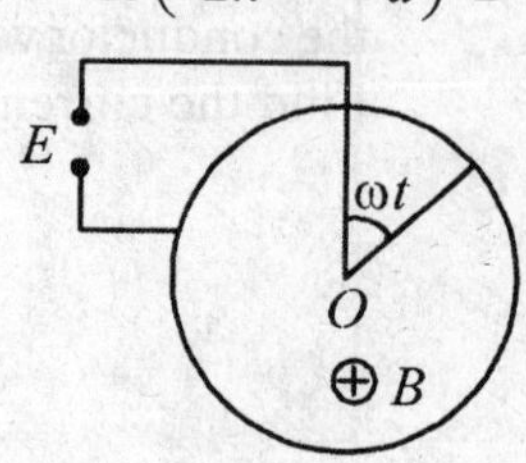

Fig. 24.55

$\left[\textbf{Ans. } E(t) = \frac{1}{2}(\omega a^3 B^3 + 2mg\sin\omega t)/aB\right]$

18. A square frame with side a and a long straight wire carrying a current I are located in the same plane as shown in Fig. 24.56. The frame translates to the right with a constant velocity v. Find the emf induced in the loop as a function of x.

$\left[\textbf{Ans. } E(x) = \frac{\mu_0}{4\pi}\frac{2Ia^2v}{x(x+a)}\right]$

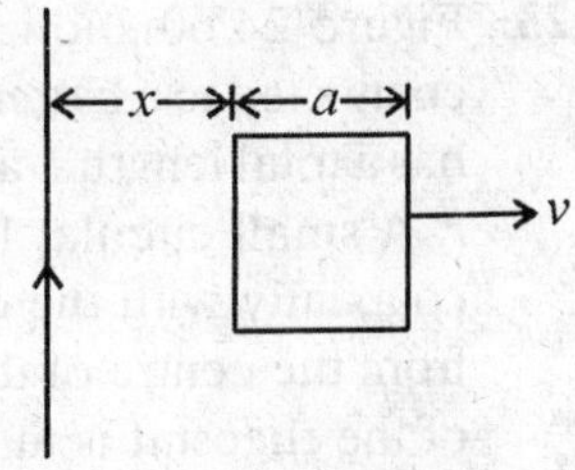

Fig. 24.56

19. (*a*) Obtain an expression for the mutual inductance between a long straight wire and a square loop of side a as shown in Fig. 24.57. (*b*) Evaluate induced emf in the loop if the wire carries a current of 50 A and the loop has an instantaneous velocity v = 10 m/s at the location x = 0.2 m. Take a = 0.1 m and assume that the loop has large resistance.

$\left[\textbf{Ans. } (a)\ M = \frac{\mu_0 a}{2\pi}\log\left(1+\frac{x}{a}\right)\ (b)\ E = 17\ \mu V\right]$

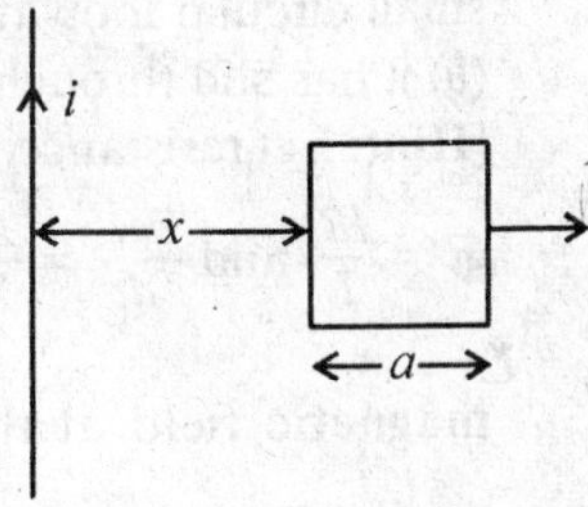

Fig. 24.57

20. An air core solenoid with length 30 cm, area of cross-section 25 cm^2 and number of turns 500 carries a current of 2.5 A. The current is suddenly switched off in a brief time of 10^{-3} s. How much is the back emf induced across the ends of open switch in the circuit? Ignore the variation in magnetic field near the ends of the solenoid. [**Ans.** E = 6.5 V]

21. A wire AB of length l, mass m and resistance R slides on a smooth thick pair of metallic rails joined at the bottom as shown in Fig. 24.58. The plane of the rails makes an angle θ with the horizontal. A vertical magnetic field B exists in the region. If the wire slides on the rails at a constant speed v show that

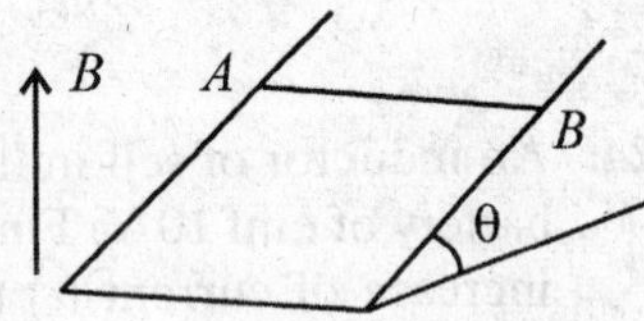

Fig. 24.58

$$B = \sqrt{\frac{mgR\sin\theta}{vl^2\cos^2\theta}}$$

22. The rectangular wire frame, shown in Fig. 24.59 has a width d, mass m, resistance R and large length. A uniform magnetic field B exists there to the left of frame. A constant force F starts pushing the frame into the magnetic field at t = 0. (*a*) Find the acceleration of the frame when its speed has increased to v. (*b*) Show that after sometime the frame will move with a constant velocity till the whole frame enters into magnetic field. Find this velocity v_0. (*c*) Find also its velocity at time t.

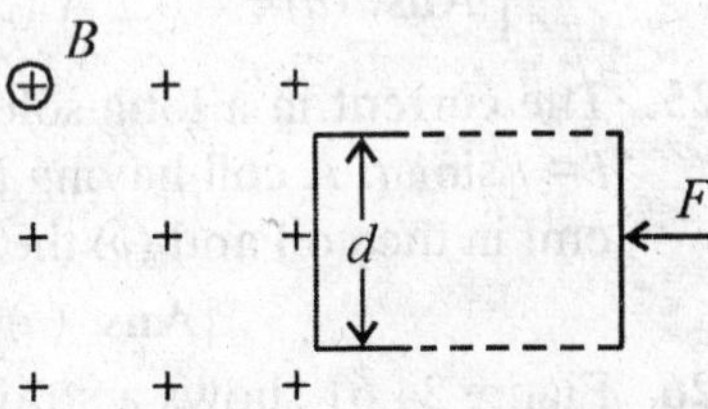

Fig. 24.59

[**Hint:** (*a*) When velocity is v, current in the frame $i = \frac{Blv}{R}$. Hence magnetic force $F_1 = ilB$

$= \frac{B^2l^2v}{mR}$. Hence acceleration $a = \frac{F - F_1}{m}$ or $a = \frac{1}{m}\left[F - \frac{B^2l^2v}{R}\right] = \frac{FR - B^2l^2v}{mR} = \frac{dv}{dt}$. ...(*i*)

Velocity will be constant when $a = 0$ or $FR = B^2l^2v_0$ or $v_0 = \frac{FR}{B^2l^2}$ integration of (*i*) under condition $t = 0$ $v = 0$ give $v = v_0\left(1 - e^{-Ft/mv_0}\right)$]

23. Figure 24.60 shows a circular coil of N turns and radius a connected to a battery of emf E. through a rheostat. The rheostat has a total length L and resistance R. The resistance of the coil is r. A small circular loop of radius a' and resistance r' is placed co-axially with the coil. The centre of the loop is at distance x from the centre of the coil. In the beginning the sliding contact of the rheostat is at the left end and then onwards it is moved towards right at constant speed v. Find the emf induced in the small circular loop at the instant (*a*) contact begins to slide and (*b*) it has slid through half the length of the rheostat.

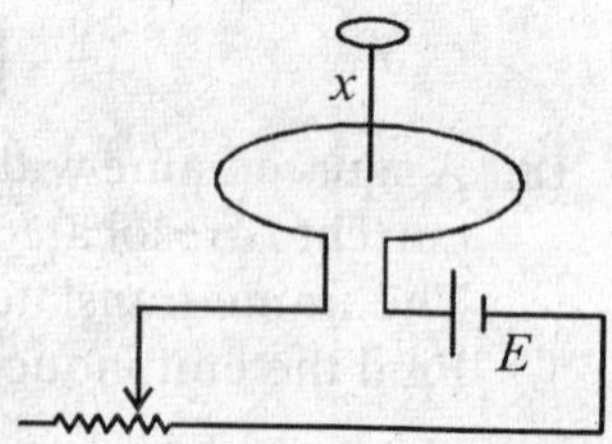

Fig. 24.60

[**Hint:** Let resistance in rheostat be R' when the sliding contact has moved a distance l. Then $R' = \frac{lR}{L}$ and $\frac{dR'}{dt} = \frac{R}{L}\frac{dl}{dt} = \frac{R}{L}v$. At this moment current through the coil is $i = \frac{E}{R'+r}$ and magnetic field at the centre of the loop is $B = \frac{\mu_0 i N a^2}{2(a^2+x^2)^{3/2}} = \frac{\mu_0 E N a^2}{2(a^2+x^2)^{3/2}} \times \frac{1}{R'+r}$.

And the magnetic flux linked with the loop is $\phi = B\pi a'^2 = \frac{\pi\mu_0 E N a^2 a'^2}{2(a^2+x^2)^{3/2}} \times \frac{1}{(R'+r)}$ and

induced emf $e = -\frac{d\phi}{dt} = \frac{\pi\mu_0 E N a^2 a'^2}{2(a^2+x^2)^{3/2}(R'+r)^2}\frac{dR'}{dt}$ or $e = \frac{\pi\mu_0 E N a^2 a'^2 R v}{2(a^2+x^2)^{3/2}(R'+r)^2 L}$. For (*a*) $R' = R$ and for (*b*) $R' = R/2$]

24. An inductor of self inductance 5 H and a resistance of 100 Ω are connected in series to a battery of emf 10 V. Find (*a*) the steady state current in the circuit (*b*) the maximum rate of increase of current (*c*) time constant (*d*) the value of current after time $t = 0.5$ T (*e*) the potential drop across the inductor at $t = 0.5$ T.

[**Ans.** (*a*) $i = 0.1$ A (*b*) $\left(\frac{di}{dt}\right)_{max} = 2$ A/s (*c*) $T = 0.05$ s (*d*) $i = 0.049$ A (*e*) $V = 5.1$ V]

25. The current in a long solenoid of radius R and having n turns per unit length is given by $i = i_0 \sin\omega t$. A coil having N turns is wound around it near the centre. Find (*a*) the induced emf in the coil and (*b*) the mutual inductance between the solenoid and the coil.

[**Ans.** (*a*) $\pi\mu_0 i_0 n N\omega R^2 \cos\omega t$ (*b*) $\pi\mu_0 n N R^2$]

26. Figure 24.61 shows a straight long wire carrying a current i and a rod of length l coplanar with the wire and perpendicular to it. The rod moves with a constant velocity v in a direction parallel to the wire. The distance of the wire from the centre of the rod is x. Find the motional emf induced in the rod.

[**Ans.** $E = \frac{\mu_0 i v}{2\pi}\log\left(\frac{2x+l}{2x-l}\right)$]

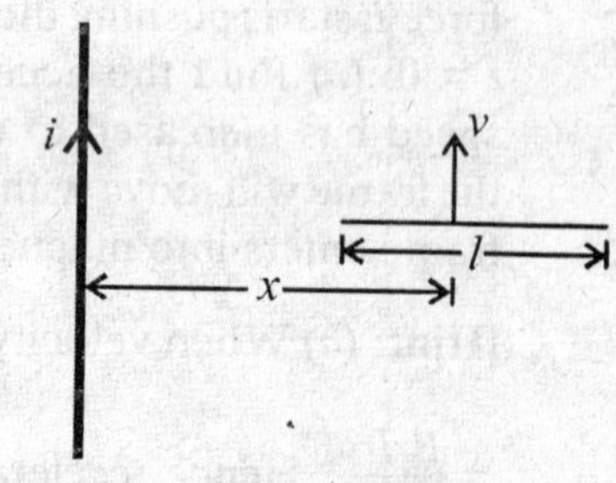

Fig. 24.61

27. Consider the situation depicted in problem 26 except that the ends of the rod slide on a pair of thick metallic rails laid parallel to the wire. At one end, the rails are connected by resistor of resistance R (Fig. 24.62 (*a*)). What force is needed to keep the rod sliding at a constant speed? (*b*) In this situation what is the current in the resistor R.

(*c*) Find the rate of heat developed in the resistor. (*d*) Find the power delivered by the external agent exerting the force on the rod.

$$\left[\textbf{Ans.}\ (a)\ F = \frac{v}{R}\left(\frac{\mu_0 i}{2\pi}\log\frac{2x+l}{2x-l}\right)^2 \quad (b)\ I = \frac{\mu_0 iv}{2\pi R}\log\frac{(2x+l)}{(2x-l)}\right.$$

$$\left.(c)\ P = \frac{1}{R}\left(\frac{\mu_0 iv}{2\pi}\log\frac{2x+l}{2x-l}\right)^2 \quad (d)\ \text{Same as } (c)\right]$$

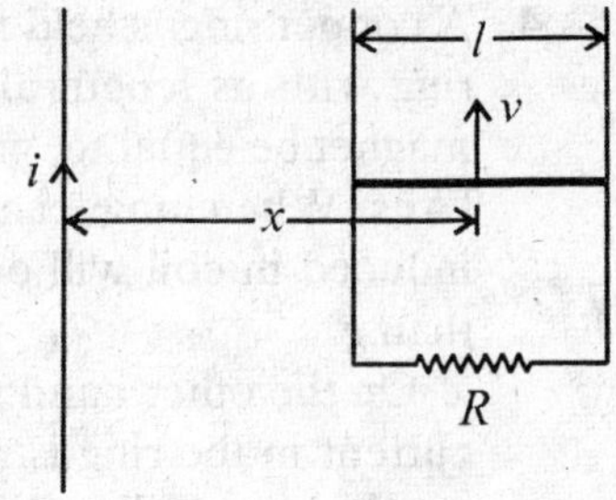

Fig. 24.62

SHORT ANSWER TYPE QUESTIONS

1. At a given place, horizontal and vertical components of earth's magnetic field B_H and B_V are along X and Y axes respectively as shown in Fig. 24.63. What is the total flux of earth's magnetic field with area S, if the area S is in (*a*) X-Y plane (*b*) Y-Z plane (*c*) Z-X plane?

[**Ans.** Magnetic field $\boldsymbol{B} = B_H\hat{i} - B_V\hat{j}$ and area is $\boldsymbol{S}$, then flux $\phi = \boldsymbol{B}\cdot\boldsymbol{S}$.

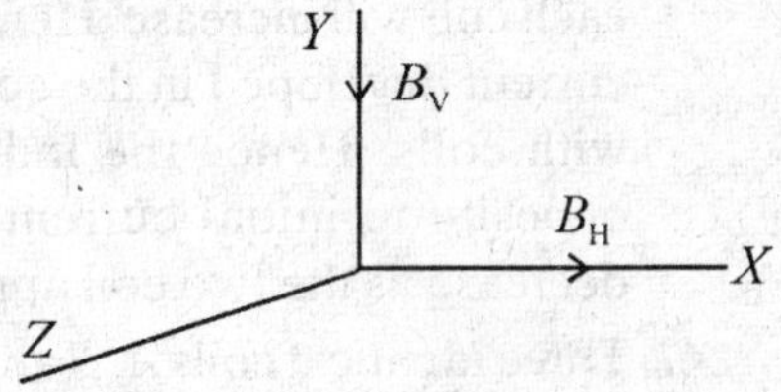

Fig. 24.63

(*a*) For area in X-Y plane $\boldsymbol{S} = S\hat{k}\phi = \left(B_H i - B_V\hat{j}\right)\cdot S\hat{k} = 0$

(*b*) For area in Y-Z plane $\boldsymbol{S} = S\hat{i}, \phi = \left(B_H\hat{i} - B_V\hat{j}\right)\cdot S\hat{i} = B_H S$

(*c*) For area in Z-X plane $\boldsymbol{S} = S\hat{j}, \phi = \left(B_H\hat{i} - B_V\hat{j}\right)\cdot S\hat{j} = -B_V S$]

2. Calculate the rate at which the flux linked with the area generated changes with time when a rod of length l is translated as in Fig. 24.64.

[**Ans.** We have to find out the rate of change of flux means induced emf in rod AB as it moves with velocity $\boldsymbol{v}$ in magnetic field $\boldsymbol{B}$. Since $\boldsymbol{v}$ makes an angle θ with the length, then induced emf is

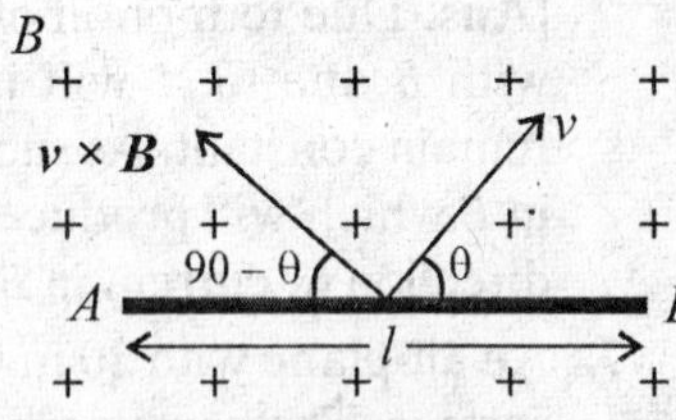

Fig. 24.64

$$e = \boldsymbol{l}\cdot(\boldsymbol{v}\times\boldsymbol{B})$$

$\boldsymbol{v}\times\boldsymbol{B} = vB$ in the direction shown in Fig. 24.64. Hence

$$e = lvB\cos(90-\theta) = lvB\sin\theta]$$

3. A rod of length 15×10^{-2} m rotates about an axis passing through one end with a uniform angular velocity of 60 rad/s. A uniform magnetic field of 0.1 tesla exists in the direction of rotation. Calculate the emf induced between the ends of the rod.

[**Ans.** Induced emf between the ends of the rod is given by

$$e = \frac{1}{2}Bl^2\omega$$

$$= \frac{1}{2}\times 0.1\times(15\times 10^{-2})^2\times 60$$

$$= 3\times 225\times 10^{-5} = 67.5 \text{ mV}]$$

4. A copper ring is held horizontally and a bar magnet is dropped through the ring with its length along the axis of ring. Will the acceleration of falling magnet be equal to, greater than or less than that due to gravity?
[**Ans.** When magnet approaches the coil, according to Lenz's law, a current induced in coil will oppose its motion. Hence it's acceleration will be less than g.

On the other hand when magnet has passed the ring, the direction of the current in the ring is reversed, pulling the magnet towards the coil. Hence acceleration will be less than g]

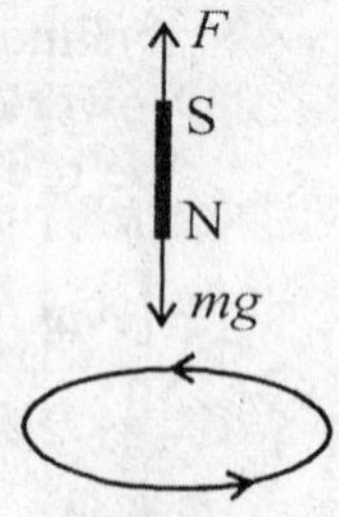

Fig. 24.65

5. Two identical coaxial circular loops carry equal currents circulating in the same direction. What will happen to the current in each loop if the loops approach each other?
[**Ans.** When the coils approach each other, the flux linked with each coil will increase. Hence according to Lenz's law induced current developed in the coils will try to reduce the flux linked with coils. Hence the induced current in each coil will be opposite to initial current. Hence current in each coil will decrease as the two coil approach each other]

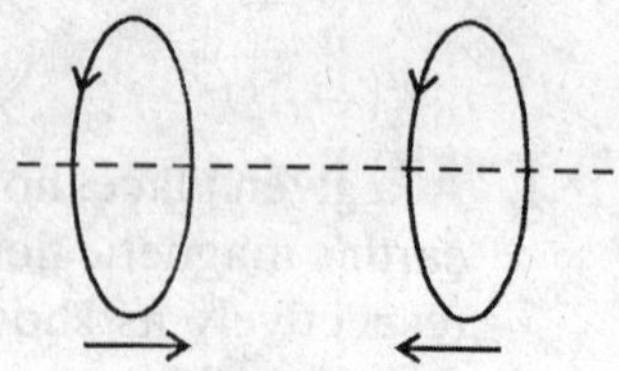

Fig. 24.66

6. Three identical coils A, B and C are placed with their planes parallel to one another. Coils A and C carry equal currents as shown in Fig. 24.67. Coils B and C are fixed in position and coil A is moved towards B with uniform motion. Is there any induced current in B? If no, give reason, if yes mark the direction of induced current.
[**Ans.** Due to motion of coil A towards B, the flux linked with B due to A will increase while that due to C will remain constant. As such an induced current is developed in B which will produce the effect such that it opposes the approach of A towards B. Hence direction of current in B is opposite to that in A, because opposite currents repel each other]

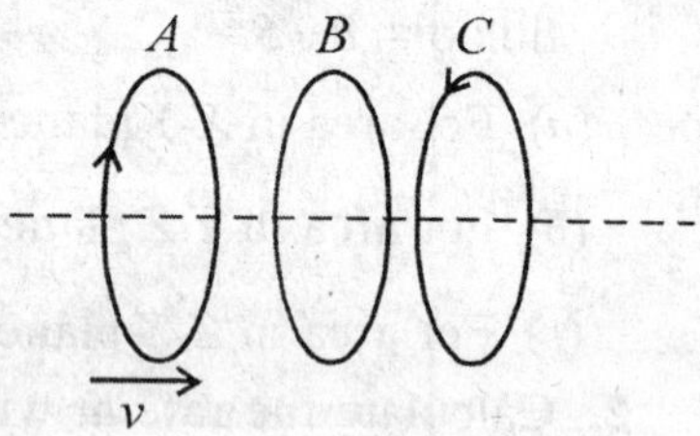

Fig. 24.67

7. An air-plane with 20 m wing spread is flying at 250 m/s, straight south parallel to the earth's surface. The earth's magnetic field has a horizontal component of 2×10^{-5} Wb/m² and dip angle 60°. Calculate the induced emf between the plane wings.
[**Ans.** If a conductor of length l moves in a magnetic field B with velocity v, the induced emf $e = Blv$.

The wings of the plane can be considered as a conductor. Since it is moving due south, hence its velocity is parallel to horizontal component 'of earth's magnetic field as such B_H cannot induce emf in rod. It is vertical component (B_V) of earth's field is cut and will produce emf, but $\frac{B_V}{B_H} = \tan\delta$. Where δ is angle of dip or $B_V = B_H \tan\delta$

Hence emf induced is

$$e = B_H \tan\delta \cdot lv$$
$$= 2 \times 10^{-5} \tan 60° \times 20 \times 250$$
$$= \sqrt{3} \times 10^{-1} = 0.173 \text{ V}]$$

8. A conducting circular loop is placed in a uniform magnetic field $B = 0.025$ T with its plane perpendicular to the field. Some how, the radius of the loop starts shrinking at a constant rate of 1.5 mm/s. Find the induced emf in the loop at the instant when the radius is 2.5 cm.
[**Ans.** Let at any time *t*, the radius of the loop be *r*. The magnetic flux linked with the loop is $\phi = B \cdot \pi r^2$ and the induced emf is

$$e = \frac{d\phi}{dt} = 2\pi r B \frac{dr}{dt}$$

or $$e = 2\pi \times (2.5 \times 10^{-2})\,(0.025)\,(1.5 \times 10^{-3})$$

$$= 5.89\ \mu\text{V}]$$

9. A square loop of edge *a* having *n* turns is rotated with a uniform angular velocity ω about one of its diagonals which is kept fixed in a horizontal position (Fig. 24.68). A uniform magnetic field *B* exists in the vertical direction. Find the emf induced in the coil.

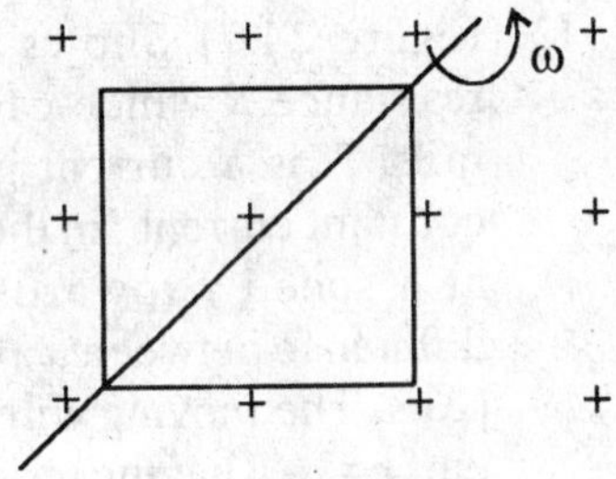

Fig. 24.68

[**Ans.** The area of the square loop is $A = a^2$. Let the normal to the surface area of frame makes an angle $\theta = 0$ at $t = 0$ with the magnetic field, then the angle θ, that the normal makes with magnetic field at time *t* is $\theta = \omega t$. Hence at this time the flux linked with the coil is

$$\phi = nBA\cos\theta = nBa^2\cos\omega t$$

Hence emf induced is $e = -\frac{d\phi}{dt} = nBa^2\omega \sin \omega t = e_0 \sin \omega t$, where $e_0 = na^2B\omega$, i.e. an alternating emf is developed]

10. Figure 24.69 shows a wire of length *l* which can slide on a *U* shaped rail of negligible resistance. The resistance of the wire is *R*. The wire is pulled to the right with a constant speed *v*. Draw an equivalent circuit diagram representing the induced emf by a battery. Find the current in the wire using this diagram.

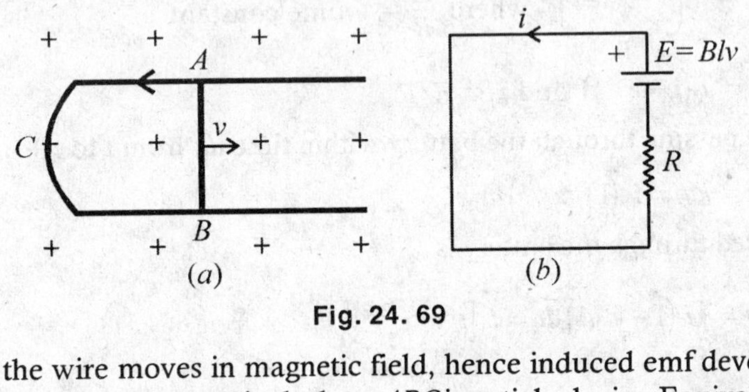

Fig. 24. 69

[**Ans.** Since the wire moves in magnetic field, hence induced emf developed = Blv. Using Lenz's law the induced current in the loop *ABC* is anticlockwise. Equivalent circuit is shown in Fig. 24.69 (*b*) with battery having emf $e = Blv$ and current $i = \frac{Blv}{R}$]

11. Figure 24.70 shows a horizontal magnetic field which is uniform above the dotted line and is zero below it. A long rectangular conducting loop of width *l* mass *m* and resistance *R* is placed partly above and partly below the dotted line with the lower edge parallel to it. With what velocity should it be pushed downwards so that it may continue to fall without any acceleration?

[**Ans.** If the loop is pushed with velocity v the induced emf $= Blv$ and the current in the loop is $i = \frac{e}{R} = \frac{Blv}{R}$.

Hence force on side AB is $F = ilB = \frac{B^2l^2v}{R}$.

For no acceleration net force should be zero, i.e.

$$\frac{B^2l^2v}{R} = mg \text{ or } v = \frac{mgR}{B^2l^2}\Big]$$

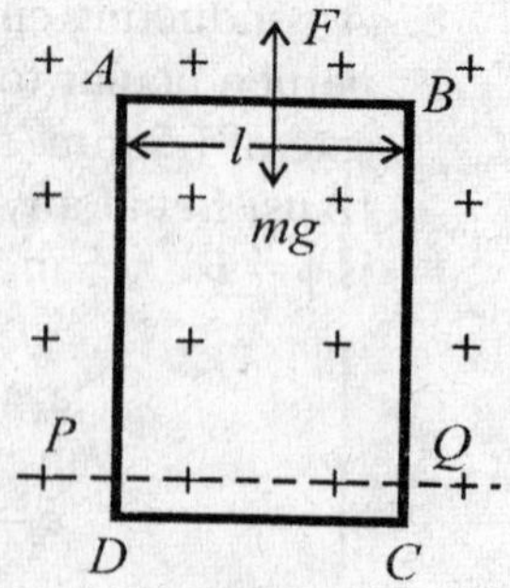

Fig. 24.70

12. Figure 24.71 Shows a wire ab of length l and resistance R which can slide on a smooth pair of rails. I_g is a current generator which supplies a constant current i in the circuit. If the wire ab slides at a speed v towards right, find the potential difference between a and b.

[**Ans.** The moving wire is equivalent to a battery of emf $e = ulB$ having resistance R. If it moves towards right and the magnetic field is going into the plane of paper, the magnetic force will push the positive charges towards a ($\mathbf{F} = q(v \times \mathbf{B})$). The equivalent circuit is shown in Fig. 24.71. The potential difference between a and b then is

$$V_a - V_b = vBl - iR]$$

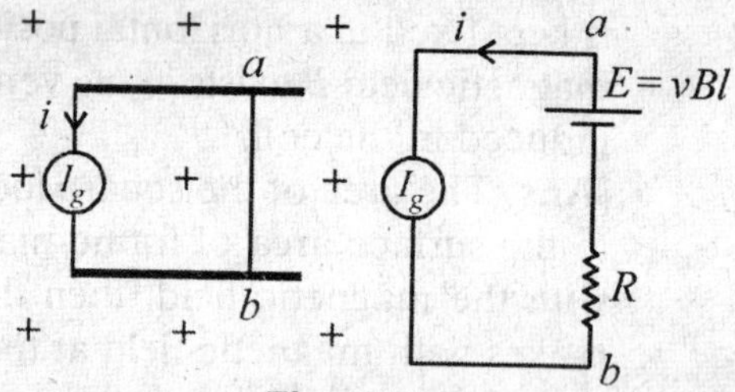

Fig. 24.71

13. An inductor-resistance-battery circuit is switched on at $t = 0$. If the emf of the battery is E, find the charge which passes through the battery in one time constant.

[**Ans.** In L–R circuit, current at any time is given by

$$i = i_0\left(1 - e^{-\frac{R}{L}t}\right) \quad \text{where } \frac{L}{R} = \tau \text{ time constant}$$

$$= i_0(1 - e^{-t/\tau}) \text{ and } i_0 = E/R$$

The charge passing through the battery within time dt, from t to $t + dt$ is

$$dq = idt = i_0\left(1 - e^{-t/\tau}\right)dt$$

$\therefore$ charge passed through the battery is

$$q = \int_O^\tau i_0(1 - e^{-t/\tau})\,dt = i_0\left[t + \tau e^{-t/\tau}\right]_O^\tau$$

$$= i_0\left[\tau + \tau e^{-1} - \tau\right] = i_0\frac{\tau}{e}\Big]$$

14. A right angled triangle abc, made from a metallic wire, moves at a uniform speed v in its plane as shown in Fig. 24.72. A uniform magnetic field B exists in the perpendicular direction. Find the induced emf (a) in loop abc (b) in the segment bc (c) in the segment ac (d) in the segment ab.

[**Ans.** (*a*) Since no flux changes in the loop during its motion hence induced emf is zero. (*b*) Segment *bc* is perpendicular to *B* and *v*, hence induced emf $e = Bv(bc)$. From Lenz's law end *c* is positive. (*c*) Segment *ac* moves with length parallel to *v* hence emf induced *e* is zero. (*d*) emf induced in segment $(cb + ba + ac) = 0$

$\therefore$ emf induced in segment ba = –emf induced in segment bc.

$= Bv(bc)$ with *a* positive]

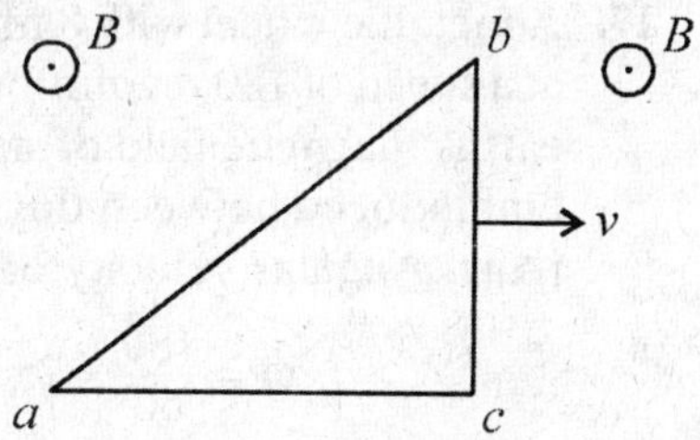

Fig. 24.72

15. Consider the circuit shown in Fig. 24.73 (*a*). Find the current through the battery a long time after the switch *S* is closed. (*b*) Suppose the switch is again opened at $t = 0$. What is the time constant of the discharging circuit? (*c*) Find the current through the inductor after one time constant.

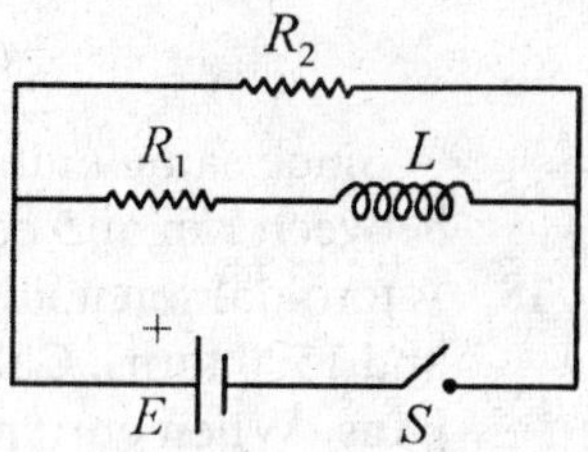

Fig. 24.73

[**Ans.** (*a*) In steady current state inductance behaves like a short circuit. Hence current through R_1 and R_2 respectively is

$$i_1 = \frac{E}{R_1} \text{ and } i_2 = \frac{E}{R_2}.$$

Hence current through battery is

$$i = i_1 + i_2 = E\left(\frac{1}{R_1} + \frac{1}{R_2}\right) = E\frac{R_1 + R_2}{R_1 R_2}.$$

(*b*) When switch is open, the current decays through resistance $R_1 + R_2$. Hence time constant is $\frac{L}{R_1 + R_2}$. (*c*) At $t = 0$ current through inductor is

$$i_1 = \frac{E}{R_1}$$

Hence current after one time constant is

$$i = \frac{i_1}{e} = \frac{E}{R_1 e}\Big]$$

16. The magnetic field exists in a cylindrical region of radius $r = 3$ m as shown in Fig. 24.74. What is the induced emf in the segment *AB* if the field changes at uniform rate of $\frac{2}{\pi}$ Ts^{-1}.

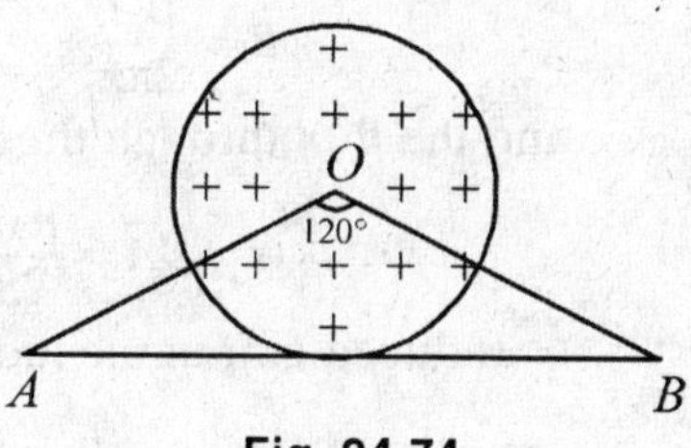

Fig. 24.74

[**Ans.** If *B* is magnetic field at any time, then flux linked with loop *OAB* is

$$\phi = \frac{1}{3}\pi R^2 B \text{ and induced emf } e = \frac{d\phi}{dt}$$

or $$e = \frac{1}{3}\pi R^2 \frac{dB}{dt} = \frac{1}{3}\pi R^2 \times \frac{2}{\pi} = \frac{2}{3}(3)^2 = 6 \text{ V} \quad \because R = 3 \text{ m}$$

Induced emf along $OA = \int_{OA} \mathbf{E} \cdot d\mathbf{l}$. Since *E* and *dl* along *OA* are perpendicular to each other hence induced emf in segment $OA = 0$. Similarly induced emf in segment *OB* is also zero. Hence emf induced in segment $AB = 6$ V]

17. A metallic wheel with 8 metallic spokes each 0.40 m long is rotated at a speed of 180 revolutions per minute in a plane perpendicular to earths magnetic field of strength 0.32 G. Find the magnitude of the emf induced between the axle and the rim of the wheel.

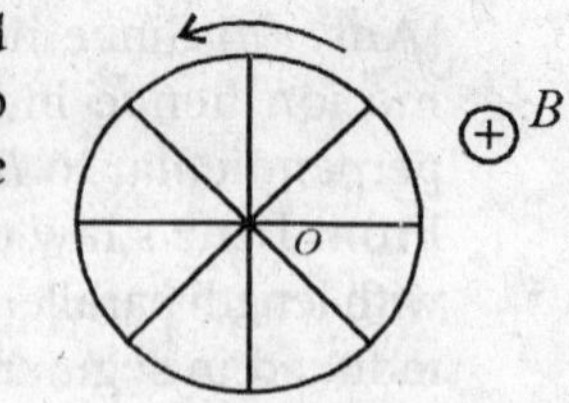

Fig. 24.75

[**Ans.** Angular velocity of the wheel is

$$\omega = \frac{180}{60} \times 2\pi$$

or $\qquad \omega = 6\pi$ radius/s

If a rod of length l rotates in magnetic field, the emf induced is

$$e = \frac{1}{2}l^2\omega B = \frac{1}{2}(0.4)^2\,(6\pi)(0.32\times10^{-4}) = 4.8\times10^{-5}\text{ V}$$

Since same emf is induced in all the spokes and all are in parallel. Hence emf induced between rim and centre = 4.8×10^{-5} V]

18. A toroidal solenoid with an air core has a mean radius of 50 m, area of cross-section 15 cm^2 and 1500 turns. Calculate the self-inductance of the solenoid.

[**Ans.** When current i flows, the magnetic field

$$B = \mu_0 \frac{N}{2\pi R} i$$

Flux linked with the toroid $\phi = BAN = \dfrac{\mu_0 AN^2 i}{2\pi R}$

and $\qquad L = \dfrac{\phi}{i} = \dfrac{\mu_0 AN^2}{2\pi R} = \dfrac{4\pi\times10^{-7}\times15\times10^{-4}\times(1500)^2}{2\pi\times0.5} = 1.35\times10^{-3}\text{ H}$]

19. A current $i = 3.36(1 + 2t) \times 10^{-2}$ A increases at a steady rate in a long straight wire. A small circular lop of radius 10^{-3} m has its plane parallel to the wire and is placed at a distance of 1 m from the wire. The resistance of the loop is $8.4 \times 10^{-4}\ \Omega$. Find the magnitude and direction of the induced current in the loop.

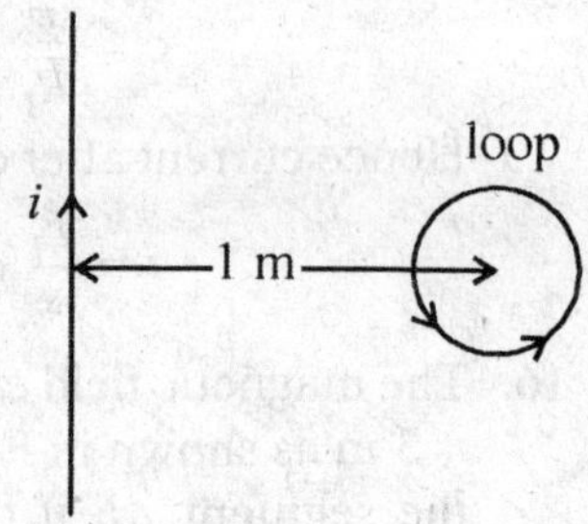

Fig. 24.76

[**Ans.** Magnetic field at the centre of the loop due to current through straight wire

$$B = \frac{\mu_0 i}{2\pi x}$$

and the flux through the loop is

$$\phi = BA = \frac{\mu_0 iA}{2\pi x}$$

Induced emf in the loop

$$e = -\frac{d\phi}{dt} = -\frac{\mu_0 A}{2\pi x}\frac{di}{dt}$$

and the current

$$i = \frac{e}{R} = -\frac{\mu_0 A}{2\pi Rx}\frac{di}{dt}$$

Given is $\mu_0 = 4\pi \times 10^{-7}$ H/m, $A = \pi r^2 = (\pi \times 10^{-6})$ m^2, $R = 8.4 \times 10^{-4}\ \Omega$ $x = 1$ m and $i = 3.36(1 + 2t)10^{-2}$ A or $\dfrac{di}{dt} = 2\times3.36\times10^{-2}$ A/s.

Substitution gives

$$i = \frac{4\pi \times 10^{-7} \times \pi \times 10^{-6} \times 2 \times 3.36 \times 10^{-2}}{2\pi \times 8.4 \times 10^{-4} \times 1} = 1.6\pi \times 10^{-11}\text{ A}$$

Due to current in wire flux increases in downward direction, hence Lenz's law gives the direction of current anticlockwise]

20. A wire in the form of a circular loop of radius 10 cm lies in a plane normal to a magnetic field of 100 T. If this wire is pulled to take a square shape in the same plane in 0.1 s. Find the average induced emf in the loop. (Roorkee 2000)

[**Ans.** Initially the flux linked with the circular loop is

$$\phi_1 = B \cdot \pi r^2$$

Now if square loop has side l, then $4l = 2\pi r$ or $l = \frac{\pi r}{2}$ and area of the loop $l^2 = \frac{\pi^2 r^2}{4}$ and the flux linked with square loop is

$$\phi_2 = Bl^2 = \frac{B\pi^2 r^2}{4}$$

Hence change in flux $\Delta\phi = \phi_2 - \phi_1 = \frac{B\pi^2 r^2}{4} - B\pi r^2$

or $$\Delta\phi = \frac{B\pi r^2}{4}(\pi - 4)$$

Hen induced emf is $e = -\frac{\Delta\phi}{\Delta t} = \frac{B\pi r^2}{4\Delta t}(4 - \pi)$

or $$e = \frac{100 \times \pi \times (0.1)^2 (4 - \pi)}{4 \times 10^{-1}} = 6.74\text{ V}\Big]$$

21. The network shown in Fig. 24.77 is a part of a complete circuit. What is the potential difference $V_B - V_A$ when the current i (= 5 A) and is decreasing at the rate of 10^3 (A/s)?

[**Ans.** Using Krichhoff's law for potential difference for the circuit, we get

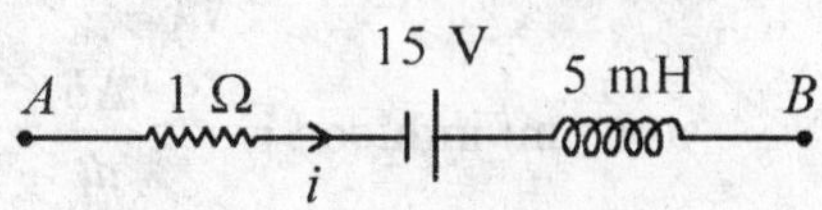

Fig. 24.77

$$V_A - iR + E - L\frac{di}{dt} = V_B$$

or $$V_B - V_A = -iR + E - L\frac{di}{dt} = -5 \times 1 + 15 - 5 \times 10^{-3}(-10^3)$$

$$= 15\text{ V}\,]$$

22. The current in a coil of self inductance 2.0 H is increasing according to $i = 2\sin t^2$ A. Find the amount of energy spent during the period when the current changes from 0 to 2 A.

[**Ans.** When the current changes in the coil an induced emf $e = L\frac{di}{dt}$ is produced. To maintain current i through inductor, work done in time dt is

$$dW = iL\frac{di}{dt}\,dt = Li\,di \qquad (W = Vi\,dt)$$

$$\therefore \qquad W = \int_0^2 Li\,di = \frac{1}{2}Li^2\Big|_0^2 = 2L = 4 \text{ J} \;\because L = 2 \text{ H}\Big]$$

23. A constant current 2 A flows through a metal rod of length 1 m and mass 0.5 kg which slides on frictionless rails in a horizontal plane. Initial speed of rod is 2 m/s and uniform magnetic field 0.5 T is acting vertically upwards as shown in Fig. 24.78. Calculate the distance moved before coming to rest.

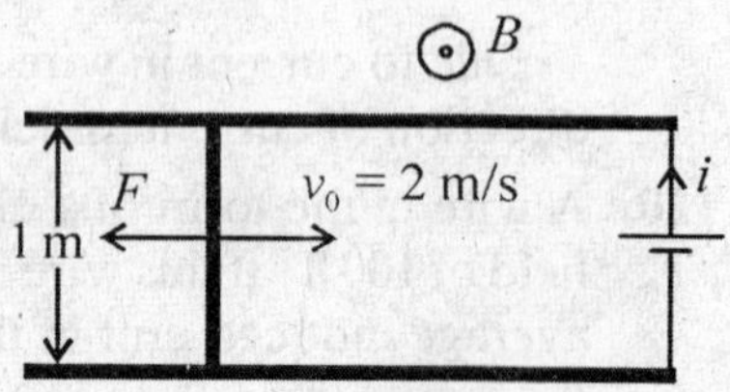

Fig. 24.78

[**Ans.** Since current carrying conductor is in magnetic field hence experiences force $F = ilB$. As shown in Fig. 24.78, the force is in opposite direction to the motion hence retardation $a = \frac{F}{m} = \frac{ilB}{m}$. Hence it will move a distance $S = \frac{v_0^2}{2a}$ before coming to rest, i.e.

$$S = \frac{v_0^2}{2a} = \frac{v_0^2 m}{2\,ilB} = \frac{(2)^2 \times 0.5}{2 \times 2 \times 1 \times 0.5} = 1 \text{ m}\Big]$$

24. A wire in the shape of a parabola $y = kx^2$ is in uniform magnetic field B perpendicular to the plane X-Y. A jumper translates without initial velocity and at an acceleration a (constant) from apex of the parabola. Find the emf induced in the formed loop as a function of y.

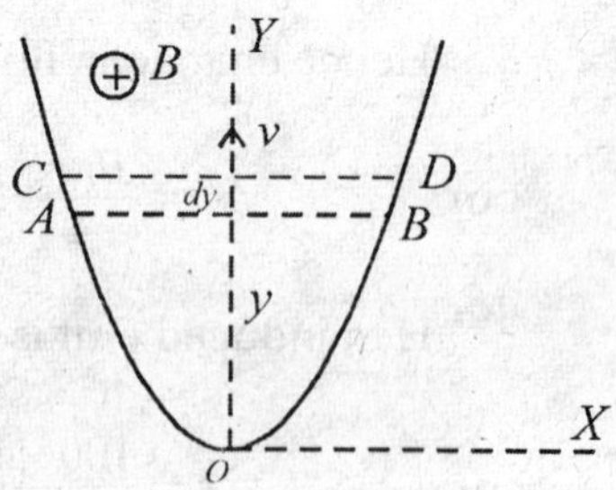

Fig. 24.79

[**Ans.** At any instant if the jumper is at AB (Fig. 24.79) and moves to positions CD in time dt then change in flux $\Delta\phi = B \cdot 2x\,dy$.

But $\qquad y = kx^2$ or $x = \sqrt{\frac{y}{k}}$

Hence $\quad \Delta\phi = 2B\sqrt{\frac{y}{k}}dy$

and emf induced is $e = \frac{\Delta\phi}{dt} = 2B\sqrt{\frac{y}{k}}\frac{dy}{dt} \qquad \frac{dy}{dt} = v = \sqrt{2ay}$

$$\therefore \qquad e = 2B\sqrt{\frac{y}{k}}\sqrt{2ay} = By\sqrt{\frac{8a}{k}}\;\Big]$$

25. A metal disc of radius a rotates with a constant angular velocity ω about its axis. Find the potential difference between the centre and the rim. Assume no external magnetic field is there.

[**Ans.** As the metal disc rotates free electrons also rotate with it. The electrons at distance r from the centre experience centripetal force $= m\omega^2 r$ towards the centre. A charged particle experiences force in electric and magnetic fields. Since there is no magnetic field hence we conclude that there is an electric field E. Hence

$$-Ee = m\omega^2 r \text{ or } E = -\frac{m\omega^2 r}{e}$$

But $\qquad E = -\frac{dV}{dr} = -\frac{m\omega^2 r}{e}$ or $dV = \frac{m\omega^2}{e}r\,dr$

and potential difference between centre and rim is

$$\int_{V_c}^{V_r} dV = \int_O^a \frac{m\omega^2}{e} r\,dr \quad \text{or} \quad V_r - V_c = \frac{m\omega^2 a^2}{2e}\Big]$$

26. A wire is bent into three circular segments of radius r = 10 cm as shown in Fig. 24.80. Each segment is a quadrant of circle *ab* lying in *X-Y* plane; *bc* in *Y-Z* plane and *ca* in *Z-X* plane (*a*) If a magnetic field *B* points in the positive *X*-direction, what is the magnitude of the emf developed in the wire when *B* increases at the rate of 3 mT per second? (*b*) What is the direction of the current in the segment *bc*?

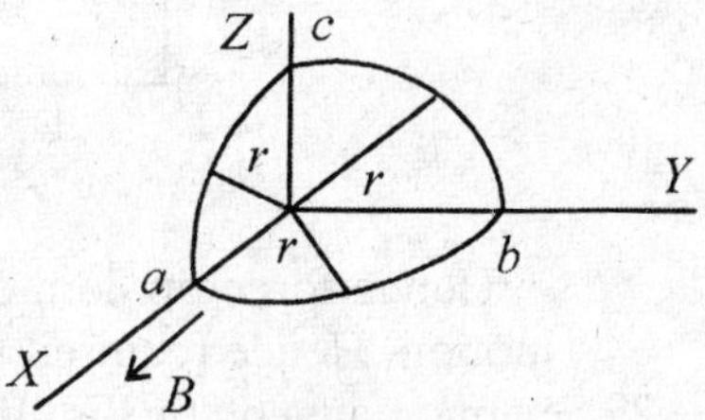

Fig. 24.80

[**Ans.** Since vector areas of segments *ab* and *ac* are perpendicular to the direction of magnetic field, hence flux through them is zero. Area of segment $CB = \frac{1}{4}\pi r^2$ and is parallel to the magnetic field. Hence flux through it is $\phi = \frac{1}{4}\pi r^2 B$ and induced emf $e = -\frac{d\phi}{dt} = -\frac{1}{4}\pi r^2 \frac{dB}{dt}$. Hence $e = -\frac{1}{4}\pi(10^{-1})^2 \times 3 \times 10^{-3} = -2.4 \times 10^{-5}$ V. Lenz's law gives the direction of the current from *c* to *b*]

27. A straight wire with a resistance λ per unit length is bent in the form of an angle 2α as shown in Fig. 24.81. A rod of the same wire perpendicular to the bisector of the angle 2α forms a closed triangular loop with the bent wire. The loop is placed in uniform magnetic field with the induction *B* perpendicular to the plane of triangle. Find the direction and magnitude of the current flowing in the loop, when the rod moves at the constant speed *v*.

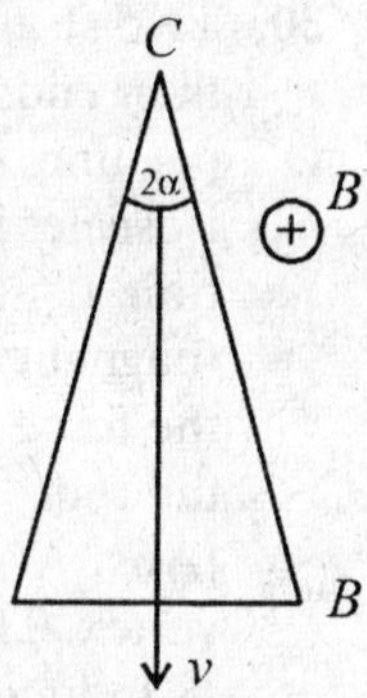

Fig. 24.81

[**Ans.** Due to the fall of the frame in magnetic field emf induced in the wires *AC* and *BC* are equal and opposite. Hence the net emf induced is that in wire *AB*. If length of *AC* is *l*, then length of $AB = 2l\sin\alpha$ and emf induced is

$$e = B\cdot(2l\sin\alpha)v$$

Total length of the frame is $(2l + 2l\sin\alpha)$ and resistance of the frame is $\lambda 2l(1+\sin\alpha)$. Hence current

$$i = \frac{e}{R} = \frac{2lB\sin\alpha \cdot v}{2l\lambda(1+\sin\alpha)} = \frac{Bv\sin\alpha}{\lambda(1+\sin\alpha)}\Big]$$

28. A rectangular loop with a sliding connector of length $l = 1.0$ m is situated in a uniform magnetic field $B = 2$ T perpendicular to the plane of the loop. Resistance of connector is $r = 2\ \Omega$. Two resistances 6 Ω and 3 Ω are connected as shown in Fig. 24.82. Find the external forces required to keep the connector moving with constant velocity $v = 2$ m/s.

[**Ans.** When connector moves induced emf developed $e = Blv = 2 \times 1 \times 2 = 4$ V. Equivalent circuit is shown in Fig. 24.82 (*b*). Total resistance of the circuit is 4 Ω. Hence current in the connector is $i = \frac{e}{R} = 1$ amp.

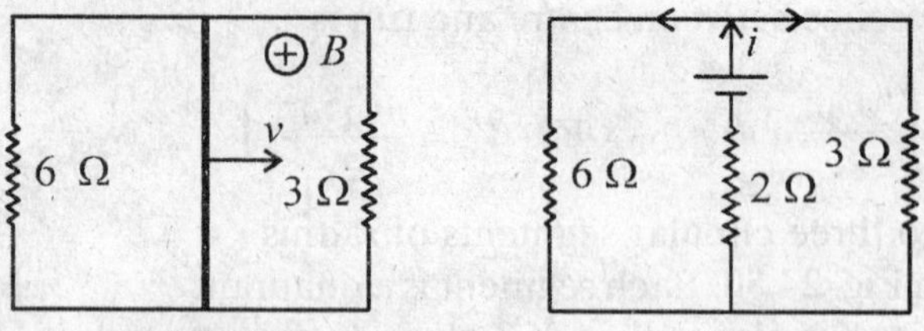

(*a*) (*b*)

Fig. 24.82

Hence force on connector $F = ilB = 1 \times 1 \times 2 = 2$ N in the direction opposite to the motion. Hence required force is 2 N in the direction of velocity]

29. Some magnetic flux is changed from a coil of resistance 10 Ω. As a result an induced current is developed in it, which varies with time as shown in Fig. 24.83. Find the magnitude of the change in flux through the coil.

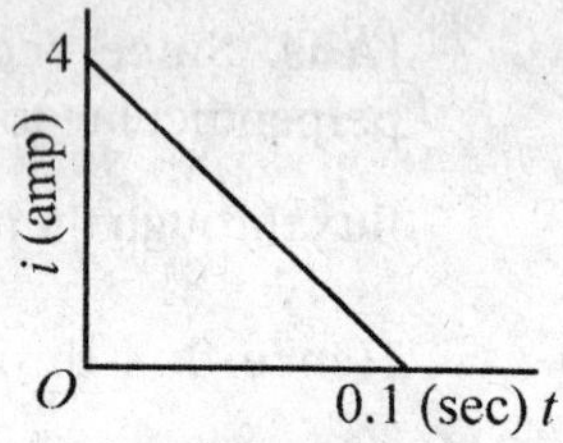

Fig. 24.83

[**Ans.** Charge passed through the coil $q = \int i dt$

i.e. charge = area under i–t curve

$$\therefore \quad q = \frac{1}{2} \times 4 \times 0.1 = 0.2 \text{ C}$$

Change flux $\Delta\phi = qR = 0.2 \times 10 = 2$ Wb]

30. Two straight superconducting rails form an angles θ when their ends are joined. A conducting bar having λ resistance per unit length is in contact with the rails and forms an isosceles triangle with them. The bar starts at the vertex at time $t = 0$ and moves with constant velocity v to right. A magnetic field B is present into the region (Fig. 24.84). Find the force exerted by external agent to maintain velocity of the rod.

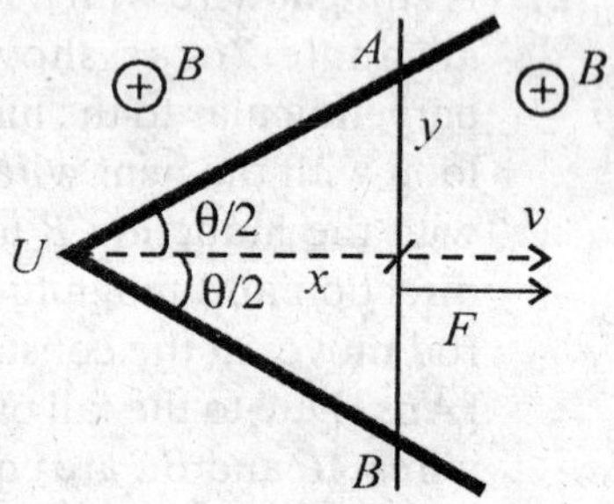

Fig. 24.84

[**Ans.** At any instant t let the rod be at distance x from the vertex. Then length of the rod is $2y = 2x \tan\theta/2$.

Induced emf in the rod then is $e = 2y \cdot vB = 2xvB \tan\theta/2$.

Resistance of the rod then is $R = 2y\lambda$. Hence current in the rod is

$$i = \frac{e}{R} = \frac{2xvB \tan\theta/2}{2y\lambda} = \frac{2xvB \tan\theta/2}{2x \tan\theta/2\ \lambda} = \frac{vB}{\lambda}$$

Hence magnetic force on the conductor $= i2y \cdot B$

$$F = \frac{vB}{\lambda} 2x \tan\theta/2\, B$$

$$= \frac{2vB^2x}{\lambda} \tan\theta/2$$

But $x = v$. Hence

$$F = \frac{2v^2B^2t}{\lambda} \tan\theta/2$$

This force acts in the direction opposite to the velocity and hence to maintain velocity this force is to be applied in the direction of the velocity]

31. Two concentric and coplanar coils have radii a and b ($>>a$) as shown in Fig. 24.85. Resistance of the inner coil is R, current in the outer coil is increased from 0 to i, then find total charge circulating the inner coil.

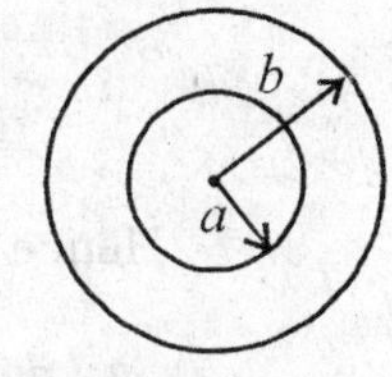

Fig. 24.85

[**Ans.** Due to the current in outer coil the magnetic field induction at the centre is

$$B = \frac{\mu_0 i}{2b}$$

The centres of the two coils are coincident and inner coil radius is small. Hence flux linked with inner coil $\phi = \pi a^2 B = \frac{\pi\mu_0 i a^2}{2b}$.

Since current varies from 0 to i hence flux linked with inner coil changes by

$$\Delta\phi = \frac{\pi\mu_0 i a^2}{2b}$$

Hence charge circulating through it is $q = \frac{\Delta\phi}{R} = \frac{\pi\mu_0 i a^2}{2bR}$]

32. The magnetic field in the a region is given by $\boldsymbol{B} = B_0\left(\frac{x}{a}\right)\hat{k}$. A square loop of side d is placed with its edges along the X and Y axis. The loop is moved with constant velocity $\boldsymbol{v} = v_0\,\hat{i}$. Find the emf induced in the loop.

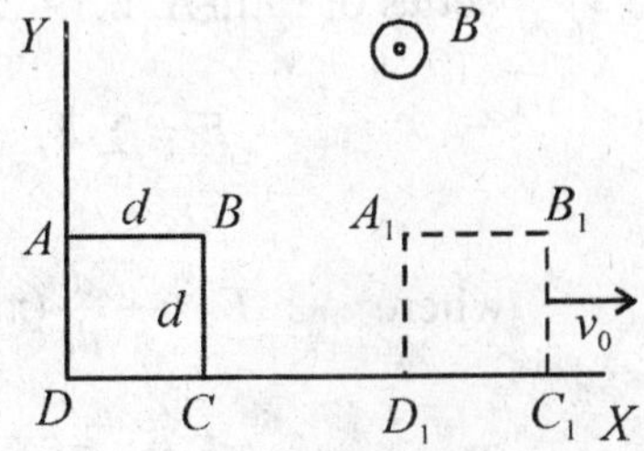

Fig. 24.86

[**Ans.** Let at any instant, loop is at $A_1B_1C_1D_1$ so that side A_1D_1 is at distance x from Y-axis. Hence distance of $B_1C_1 = x + d$.

Magnetic field at A_1D_1 is $B_1 = \frac{B_0 x}{a}$ and at

$$B_1C_1 = \frac{B_0(x+d)}{a}.$$

Induced emf in A_1D_1 and B_1C_1 is $e_1 = B_1 d v_0 = \frac{B_0 v_0 d}{a} x$ and $e_2 = \frac{B_0 v_0 d}{a}(x+d)$. Hence net emf induced in the loop is $e_2 - e_1 = \frac{B_0 v_0 d^2}{a}$. Since the emf induced in side A_1B_1 and D_1C_1 is zero. It is due to the fact that l and v are parallel]

33. A square loop of side a is placed in the same plane as a long straight wire carrying a current i. The centre of the loop is at a distance r ($r >> a$) from the wire. The loop is moved away from the wire with constant velocity v. Find the emf induced in the loop.

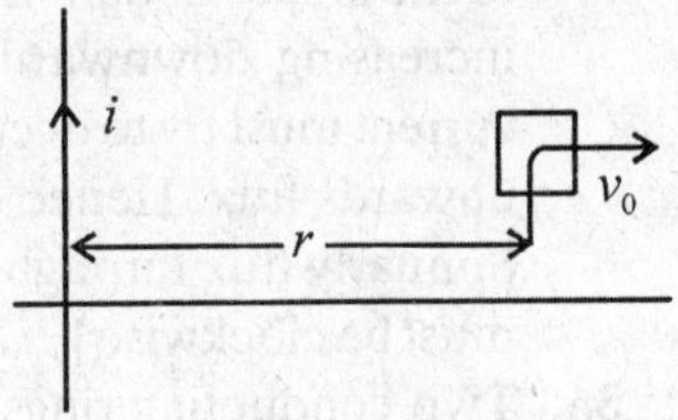

Fig. 24.87

[**Ans.** Since $r >> a$ hence magnetic field at all points on the loop can be considered to be constant and uniform. Magnetic field at the centre of the loop due to current through the conductor is

$$B = \frac{\mu_0 i}{2\pi r}$$

The vector area of the loop a^2 and B vector are parallel. Hence flux through the loop is

$$\phi = Ba^2 = \frac{\mu_0 i a^2}{2\pi r}$$

Hence emf induced in the loop is $e = -\frac{d\phi}{dt} = -\frac{d}{dt}\left(\frac{\mu_0 i a^2}{2\pi r}\right) = \frac{\mu_0 i a^2}{2\pi r^2}\frac{dr}{dt}$.

But $\frac{dr}{dt} = v_0$ the velocity of the loop. Hence

$$e = \frac{\mu_0 i a^2}{2\pi r^2} v \Big]$$

34. A conducting wire forms a spiral in a plane starting with radius R; with the radius decreasing by Δ per turn till it completes n turns (Fig. 24.88). A spatially uniform magnetic field $B = B_0 \sin \omega t$ is applied in direction perpendicular to the plane of the spiral. Determine the emf induced between the two ends of the spiral ($R = n\Delta$ and n is very large).

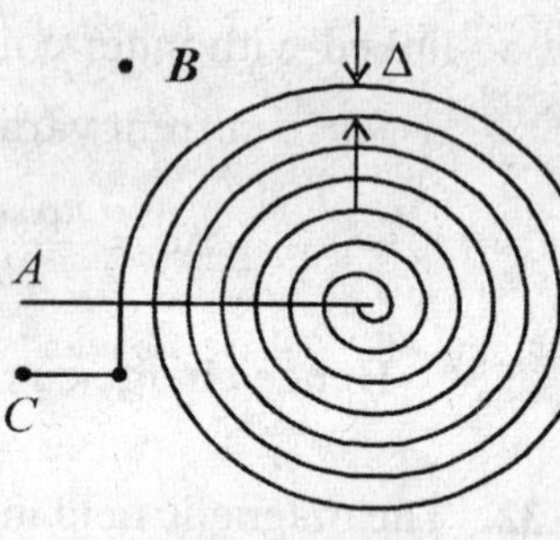

Fig. 24.88

[**Ans.** The circuit is approximately equivalent to n circuits in series of radii $\Delta, 2\Delta, ..., i\Delta, ..., n\Delta$. The emf induced is

$$E = \sum_{i=1}^{n} E_i$$

where $E_i = -\frac{d}{dt}(\pi r_i^2 B_0 \sin \omega t) = -\pi r_i^2 B_0 \omega \cos \omega t$

or $E = -\pi B_0 \sum r_i^2 \cos \omega t = -\pi B_0 \omega \left(\sum \Delta^2\right) \cos \omega t$

$$= -\pi B_0 \omega \frac{n(n+1)(2n+1)}{6} \cos \omega t \Big]$$

35. What is the direction of induced current in coils P and Q. P is entering in the magnetic field and Q is leaving the magnetic field.

[**Ans.** Magnetic field is directed downwards normal to the loops. Loop P is entering the field means flux is increasing downwards. Hence flux due to induced current must try to decrease downward flux or increase upwards flux. Hence current must be anticlockwise. Similarly flux through Q is decreasing. Hence current must be clockwise]

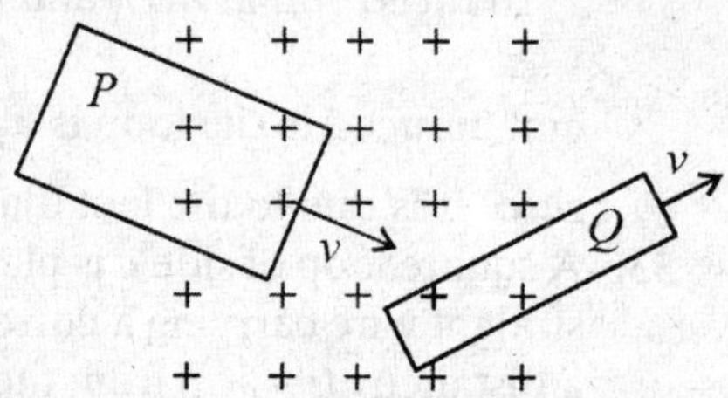

Fig. 24.89

36. Two conducting rings of radii a and $2a$ move in opposite direction with velocity $2v$ and v respectively on a conducting surface XY. There is a uniform magnetic field of induction B perpendicular to the plane of rings. Find the potential difference between higher points E and F of the rings.

[**Ans.** Induced emf in ring (E) $E_1 = B(2v)(2a) = 4vaB$

Induced emf in ring (F) $E_2 = B(v)(4a) = 4vaB$.

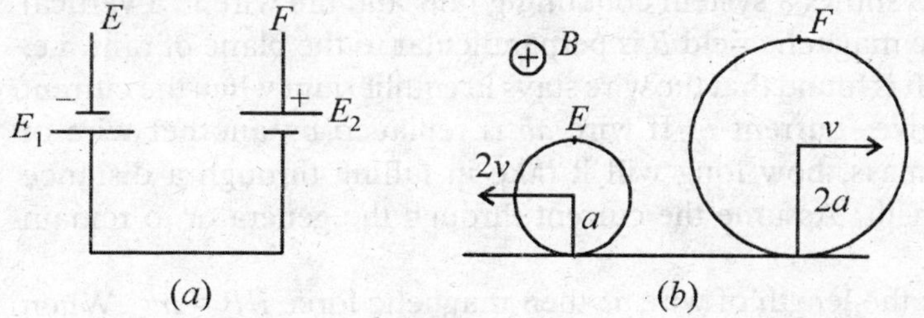

Fig. 24.90

The equivalent circuit is shown in Fig. 24.90 (*b*). As the rings are moving in opposite direction, thus net potential difference between *E* and *F* is

$$E = 4vaB + 4vaB = 8vaB]$$

37. A long wire carries a current *i*. A rod *AB* of length *l* is moving parallel to wire with velocity *v* in the direction of the current. Find the magnitude of induced emf across the rod.

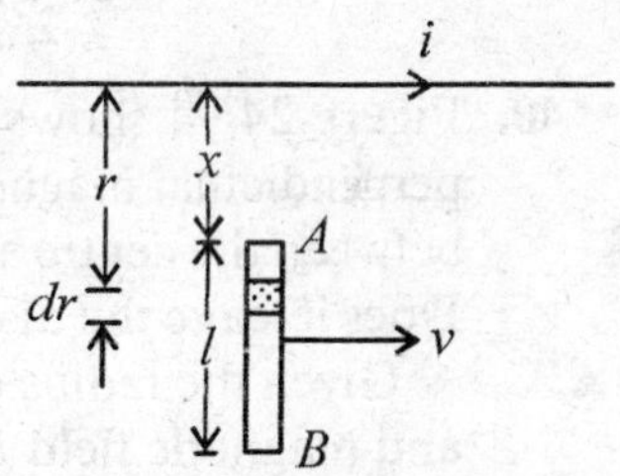

Fig. 24.91

[**Ans.** Let us consider a small element of the rod *dr* at distance *r* from the current carrying wire. Magnetic field at the element is

$$B = \frac{\mu_0 i}{2\pi r}$$

and the induced emf *dE* in this element is

$$dE = Bv\,dr = \frac{\mu_0 iv}{2\pi r}\,dr$$

Hence total emf induced $E = \int dE \int_x^{l+x} \frac{\mu_0 iv dr}{2\pi r} = \frac{\mu_0 iv}{2\pi} \log r\Big|_x^{l+x}$

or $$E = \frac{\mu_0 iv}{2\pi}\log\left(1 + \frac{l}{x}\right)]$$

38. Consider the situation shown in Fig. 24.92. The wire *OC* has zero resistance and the circular loop has resistance *R* uniformly distributed along its length. The rod *OA* is made to rotate with uniform angular speed ω about the direction of magnetic field as axis. Find the current in the rod when $\angle AOC = 90°$.

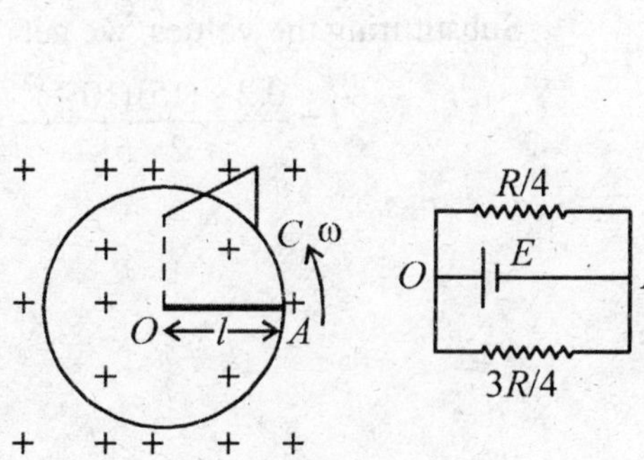

Fig. 24.92

[**Ans.** When rod rotates induced emf in rod *OA* is $E = \frac{1}{2}B\omega l^2$ with *O* as positive. When rod is rotated through 90°, the two parts of the coil have resistance $\frac{R}{4}$ and $\frac{3R}{4}$. The equivalent circuit is shown in Fig. 24.92. Total resistance then is $\frac{3R}{16}$ and the current through the rod is $i = \frac{E \times 16}{3R} = \frac{1}{2}\frac{B\omega l^2 \times 16}{3R}$ or $i = \frac{8}{3}\frac{B\omega l^2}{R}$]

39. Figure 24.93 shows a system containing rails and the wire in a vertical plane where magnetic field B is perpendicular to the plane of rails, i.e. horizontal. It is found that the wire stays in equilibrium when the current generator gives current i_g. If wire ab is replaced by another wire of double its mass, how long will it take in falling through a distance equal to length? Assume the current through the generator to remain unchanged.

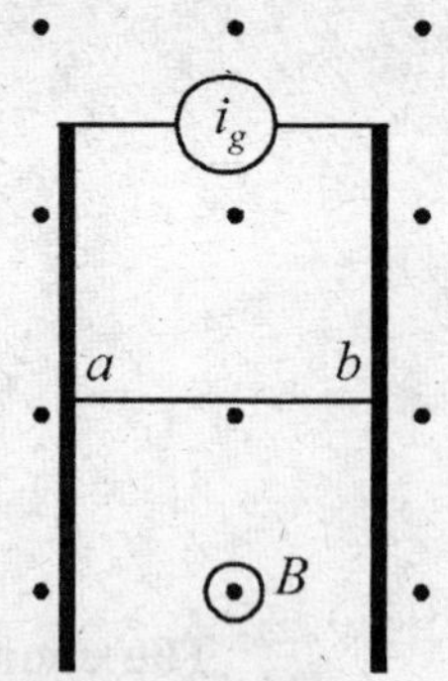

Fig. 24.93

[**Ans.** If l is the length of wire ab then magnetic force $ilB = mg$. When wire of double the mass is placed then $2mg - ilB = 2ma$, i.e. the wire will fall with acceleration $a = g/2$. Hence time taken to fall through distance l is

$$l = \frac{1}{2}at^2 \text{ or } t = \sqrt{\frac{2l}{a}} = \sqrt{\frac{2l}{g/2}} = \sqrt{\frac{4l}{g}} = 2\sqrt{\frac{l}{g}}\ ]$$

40. Figure 24.94 shows a conducting disc rotating about its axis in a perpendicular magnetic field B. A resistor of resistance R is connected between the centre and the rim. Calculate the current in the resistor. Does it leave the disc or enter at the centre?

Given the radius of disc $a = 10.0$ cm, angular speed $\omega = 15$ rad/s and magnetic field $B = 0.2$ T, Resistance $R = 5\ \Omega$.

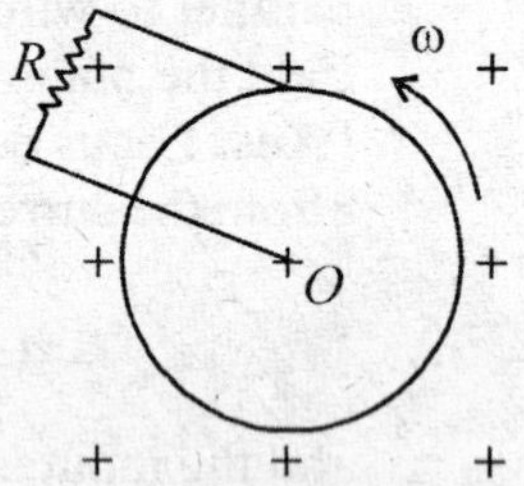

Fig. 24.94

[**Ans.** When disc rotates, the emf induced between rim and centre is

$$E = \frac{1}{2}B\omega a^2$$

and the current through resistor $i = \dfrac{E}{R} = \dfrac{B\omega a^2}{2R}$.

Since the disc is rotating in anticlockwise direction hence centre becomes negative and rim positive. Hence the current leaves the disc at centre.

Substituting the values, we get

$$i = \frac{0.2 \times (15)(10^{-1})^2}{2 \times 5} = 0.3 \times 10^{-2} = 3 \text{ mA}\ ]$$

MULTIPLE CHOICE QUESTIONS

PART I

In the following, a question is followed by four alternatives, of them only one is correct. Therefore, the candidate has to select the correct choice from amongst the alternatives.

1. An angle ABC made of conducting wire moves along its bisector through a magnetic field B as shown in Fig. 24.95. If the magnetic field is perpendicular to the plane of angle, the potential difference between A and B is

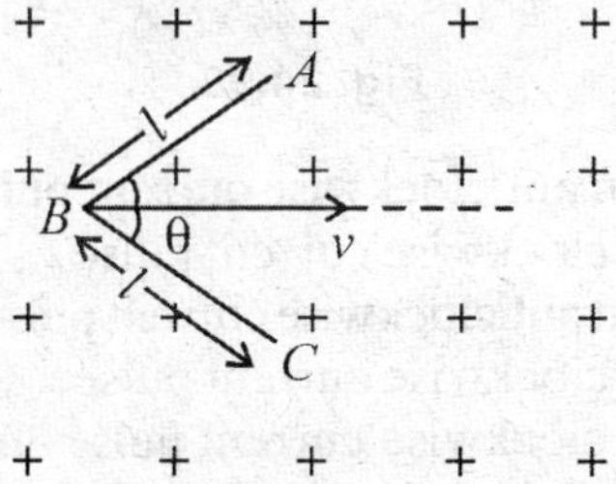

Fig. 24.95

(a) $2Blv\sin\theta/2$ (b) $2Blv\sin\theta$
(c) $Blv\sin\theta/2$ (d) $Blv\sin\theta$.

2. One end of conducting U tube can slide inside another as shown in Fig. 24.96 maintaining electrical contacts between the tubes. The magnetic field B is perpendicular to the plane of the figure. If each tube moves towards the other at constant speed v, then the emf induced in the circuit in terms of B, l, and v, where l is the width of each tube, will be (AIEEE 2005)

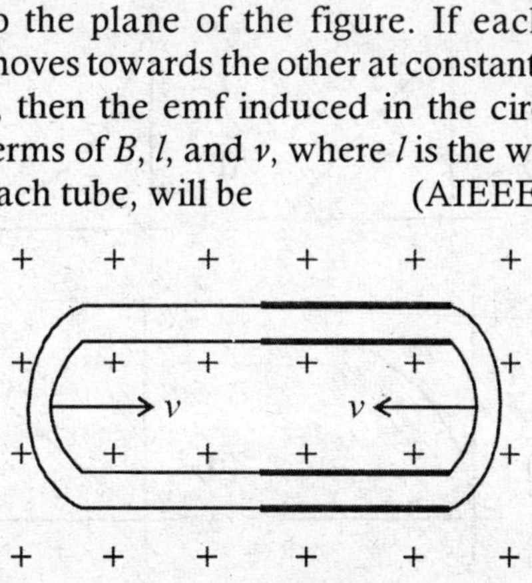

Fig. 24.96

(a) $2lBv$ (b) zero
(c) $-Blv$ (d) Blv.

3. A coil of inductance 300 mH and resistance 2 Ω is connected to a source of 2V. The current reaches half its steady state value in
(a) 0.3 s (b) 0.15 s
(c) 0.1 s (d) 0.05 s.

4. A conducting square loop of length l is placed in a magnetic field of induction B_0. End A and C are moved away from each other, so that finally the square loop becomes a straight line. The total charge flowing through the loop, if its resistance is 'R', then is

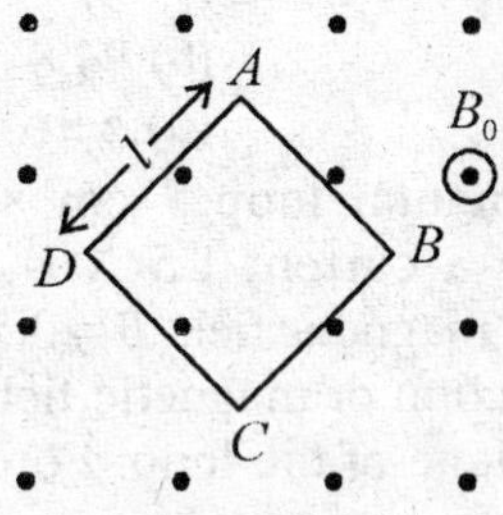

Fig. 24.97

(a) $\frac{B_0 l^2}{\pi R}$ (b) $< \frac{B_0 l^2}{R}$
(c) $\frac{B_0 l^2}{R}$ (d) $> \frac{B_0 l^2}{R}$.

5. There exists a constant magnetic field in a triangular region ABC. A wire with a small galvanometer is bent as shown in Fig. 24.98 (*b*). During the process
A. The galvanometer shows deflection.
R. Initially flux through the wire was zero but in final condition there is some flux linked with wire loop, then
(a) Both A and R are correct and R is correct explanation of A

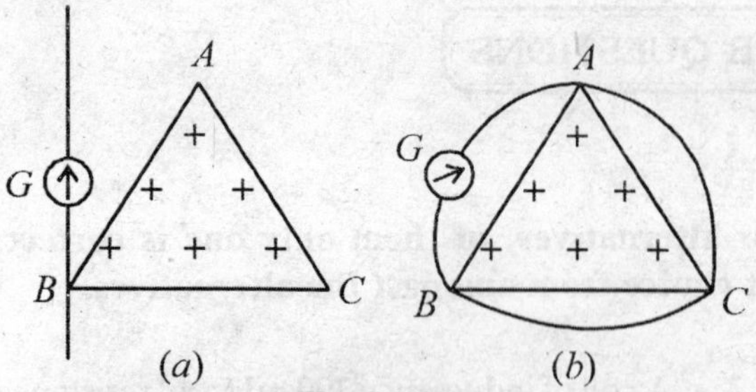

Fig. 24.98

(b) Both A and R are correct but R is not correct explanation of A
(c) A is correct but R is wrong
(d) Both A and R are wrong.

6. An ideal solenoid of diameter 0.02 m has 600 turns per metre. At the centre of this solenoid, another coil of 100 turns is wrapped closely around it.

If the current in the coil changes from 0 to 2A in 1 ms, the induced emf in the second coil is (in mV)

(a) 47.4 (b) 94.8
(c) 23.7 (d) zero.

7. A rectangular loop 5 cm × 4 cm and carrying a current 1.5 A is placed in a uniform magnetic field $B = 0.5$ T such that the direction of magnetic field is parallel to the plane of the loop. Force acting on the loop is

(a) 1.5×10^{-3} N (b) 3×10^{-3} N
(c) 4.5×10^{-3} N (d) None of the above.

8. Two solenoids of the same length and practically the same cross-section are completely inserted into one another. The inductances of solenoids are L_1 and L_2, their mutual inductance is

(a) $M = \sqrt{L_1 L_2}$ (b) $M > \sqrt{L_1 L_2}$
(c) $M < \sqrt{L_1 L_2}$ (d) $M < \sqrt{L_1 + L_2}$.

9. A rod of length l rotates with a small but uniform angular velocity ω about its perpendicular bisector. A uniform magnetic field B exists parallel to the axis of rotation. The potential difference between the centre and end of the rod is

(a) zero (b) $\frac{1}{8}\omega B l^2$
(c) $\frac{1}{2}\omega B l^2$ (d) $B\omega l^2$.

10. Consider the situation shown in Fig. 24.99. If the switch is closed and after sometime it is opened again the closed loop will show

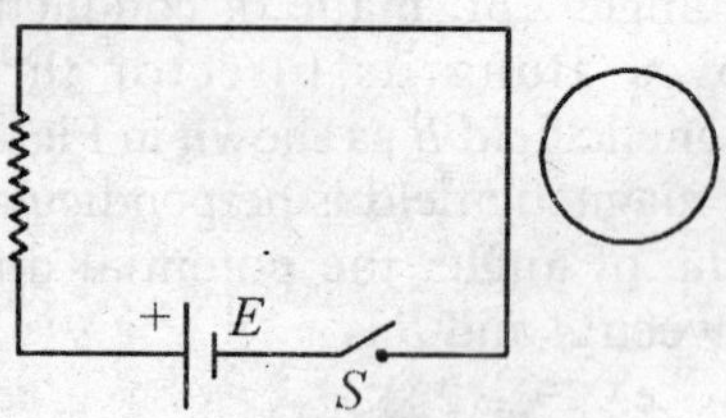

Fig. 24.99

(a) an anticlockwise current pulse
(b) a clockwise current pulse
(c) an anticlockwise current pulse and then a clockwise current pulse
(d) a clockwise current pulse and then an anticlockwise current pulse.

11. In an L-R circuit connected to a battery of emf E and switch S. S is closed at time $t = 0$. If e denotes the induced emf across inductor and i the current in the circuit at any time t, then which of the following graphs shows the variation of e with i?

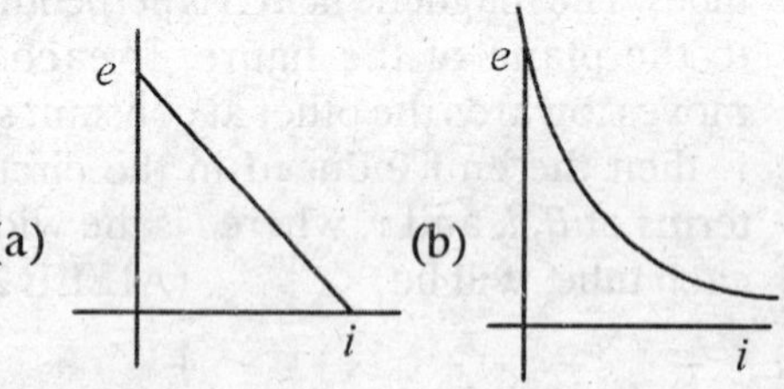

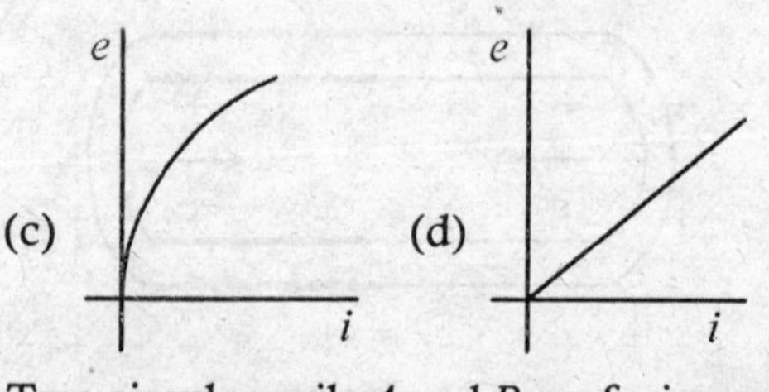

12. Two circular coils A and B are facing each other as shown in Fig. 24.100. The current i through A can be altered

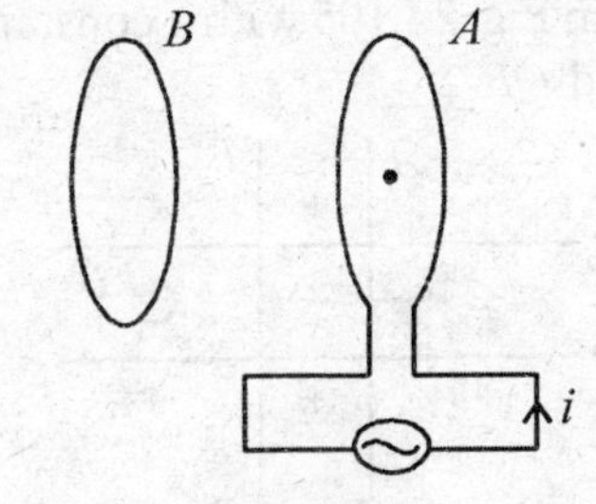

Fig. 24.100

(a) there will be repulsion between A and B if i is increased
(b) there will be attraction between A and B if i is increased
(c) there will be neither attraction nor repulsion when i is changed
(d) attraction or repulsion between A and B depends on the direction of the current. It does not depend whether the current is increased or decreased.

13. In an L-R circuit connected to a battery the rate at which energy is stored in the inductor is plotted against time during the growth of the current in the circuit. Which of the following curves gives the correct representation?

(a) 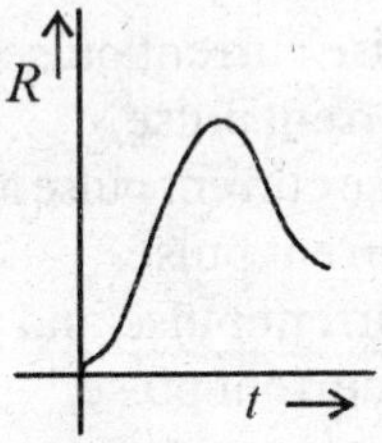

(b) 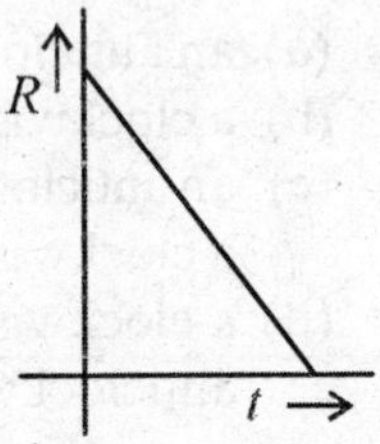

(c) 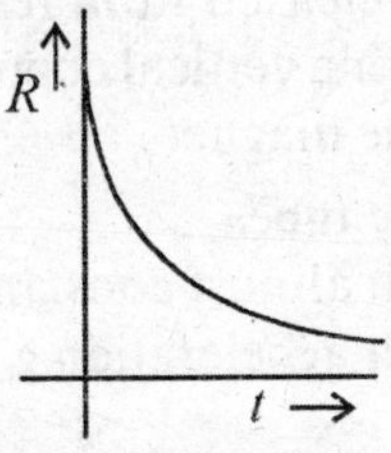

(d) 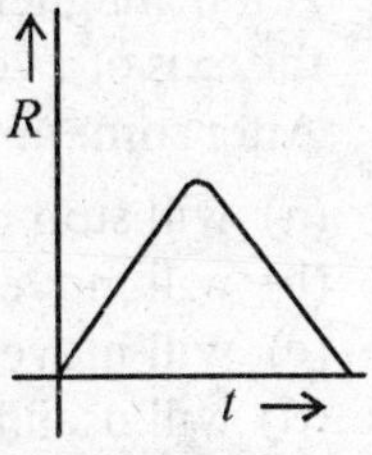

14. Three resistances each of magnitude R are connected in the form of an equilateral triangle of side a. The combination is placed in a magnetic field $B = B_0 e^{-\lambda t}$ perpendicular to the plane. The induced current in the circuit is

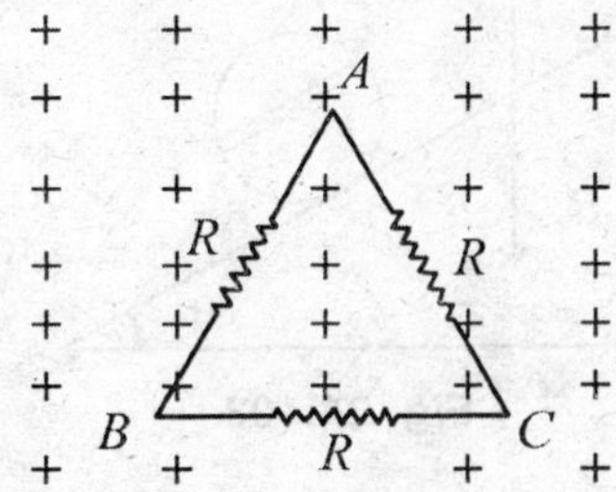

Fig. 24.101

(a) $\left[\dfrac{a^2\lambda}{2\sqrt{3}R} B_0\right] e^{-\lambda t}$

(b) $\left[\dfrac{a^2\lambda}{4\sqrt{3}R} B_0\right] e^{-\lambda t}$

(c) $\left[\dfrac{a^2 B_0}{4\sqrt{3}\lambda R}\right] e^{-\lambda t}$

(d) $\left[\dfrac{a^2 B_0 R}{4\sqrt{3}\lambda}\right] e^{-\lambda t}$.

15. A metal rod AB of length l is rotated with a constant angular velocity ω about an axis passing through O and normal to length. Potential difference between ends of the rod in absence of external field is

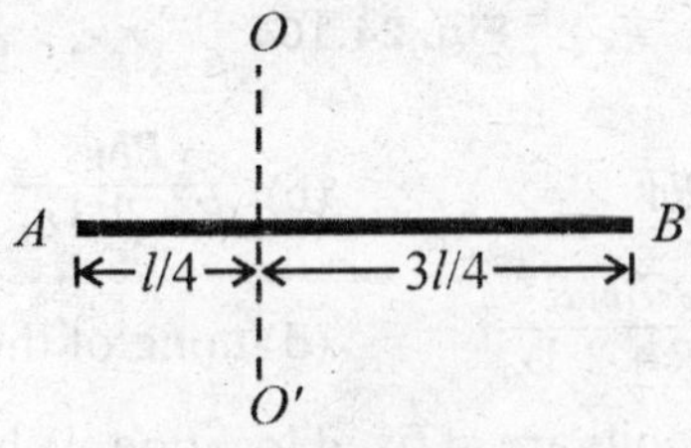

Fig. 24.102

(a) zero (b) $\dfrac{m\omega^2 l^2}{4e}$

(c) $\dfrac{m\omega^2 l^2}{2e}$ (d) $\dfrac{m\omega^2 l^2}{8e}$.

16. In the Figure 24.103 shown a coil of single turn is wound on a sphere of radius r and mass m. The plane of the coil is parallel to the inclined plane of the sphere. If the coil carries current i and sphere is in rotational equilibrium, the value of B then is

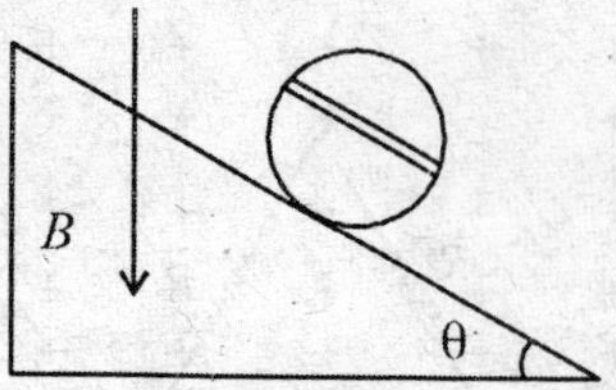

Fig. 24.103

(a) $\frac{mg}{\pi r i}$ (b) $\frac{mg \sin \theta}{\pi i}$

(c) $\frac{mgr \sin \theta}{\pi i}$ (d) none of these.

17. A wire of length 1.5 b slides at speed v along the rails separated by a distance b. The resistance per unit length of the wire is r_0. The potential difference between ends of the rod is

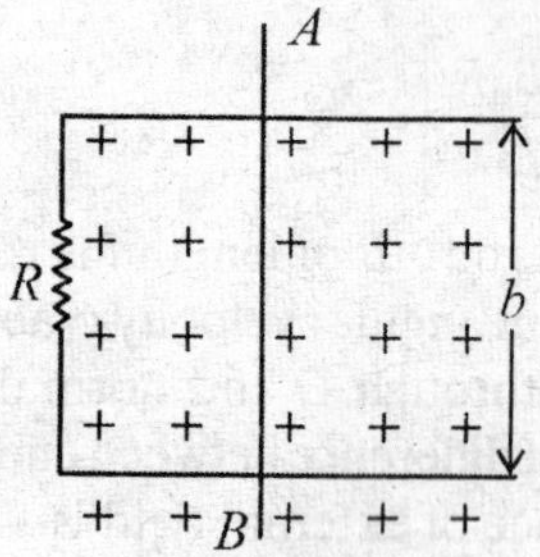

Fig. 24.104

(a) Bbv (b) $\frac{Bbv}{R + 2/3\, r_0}$

(c) $\frac{BbvR}{R + 2/3\, r_0}$ (d) none of these.

18. Two coils are at fixed location. When coil 1 has no current and current in coil 2 increases at the rate 15.0 A/s, the emf in coil 1 is 25.0 mV. When coil 2 has no current and coil 1 has a current of 3.6 A, the flux linked in coil 2 is

(a) 16 mWb (b) 10 mWb

(c) 4.00 mWb (d) 6 mWb.

19. Two identical conductors A and B are placed on two frictionless rails C and D in a uniform magnetic field is directed into the plane. If A is moved in the direction shown in Fig. 24.105 with a constant speed, then rod QB

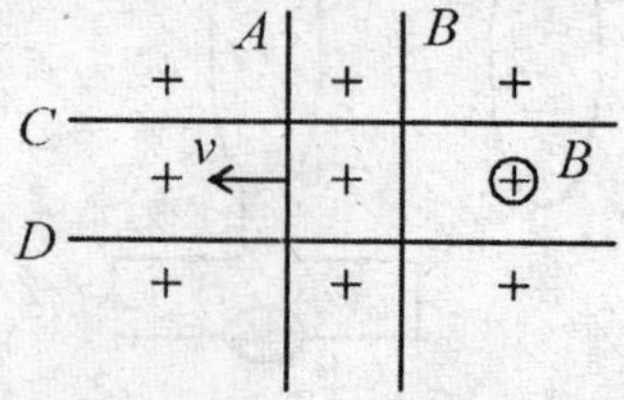

Fig. 24.105

(a) will be attracted towards A
(b) will be repelled away from A
(c) will remain stationary
(d) may be repelled or attracted towards A.

20. Consider the situation shown in Fig. 24.106. If the switch is closed and after some time it is open again; the closed loop will show

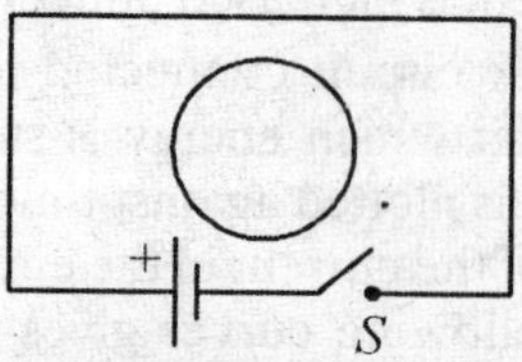

Fig. 24.106

(a) an anticlockwise current pulse
(b) a clockwise current pulse
(c) an anticlockwise current pulse and then a clockwise current pulse
(d) a clockwise current pulse and then an anticlockwise current pulse.

21. A bar magnet is released from rest along the axis of a very long vertical copper tube. After sometime the magnet

(a) will stop in the tube
(b) will move with almost constant speed
(c) will move with acceleration g
(d) will oscillate.

22. A conducting rod is moved with a constant velocity v in a magnetic field. A potential difference appears across the two ends

(a) if $\mathbf{v}$ is parallel to $\mathbf{l}$
(b) if $\mathbf{v}$ is parallel to $\mathbf{B}$

(c) if l is parallel to $\mathbf{B}$
(d) none of these.

23. A rod AB moves with a uniform velocity v in a uniform magnetic field as shown in Fig. 24.107.

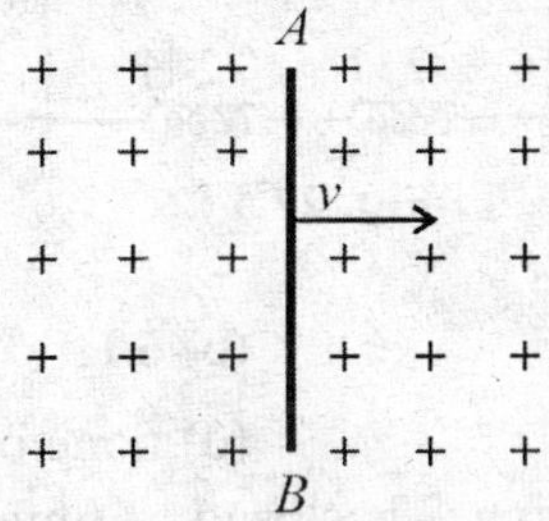

Fig. 24.107

(a) The rod becomes electrically charged
(b) The end A becomes positively charged
(c) The end B becomes positively charged
(d) The rod becomes hot because of joule heating.

24. An L–R circuit with a battery is connected at $t = 0$. Which of the following quantities is not zero just after the connection?
(a) Current in the circuit
(b) Magnetic field energy in the inductor
(c) Power delivered by the battery
(d) emf induced in the inductor.

25. A conducting square loop of side a and resistance R moves in its plane with a uniform velocity v perpendicular to one of its sides. A uniform magnetic field B exists along the perpendicular to the plane of the loop as shown in Fig. 24.108. The current induced in the loop will be

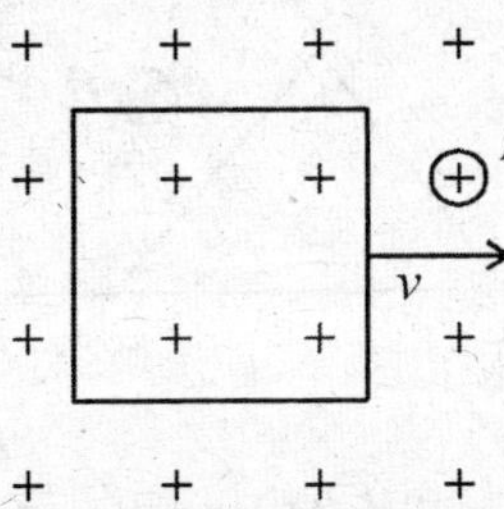

Fig. 24.108

(a) $\frac{Bav}{R}$ clockwise
(b) $\frac{Bav}{R}$ anticlockwise
(c) $\frac{2Bav}{R}$ anticlockwise
(d) zero.

26. Figure 24.109 shows a conducting loop being pulled out of a magnetic field with velocity v. Which of the following plots may represent the power delivered by the pulling agent as a function of speed v?

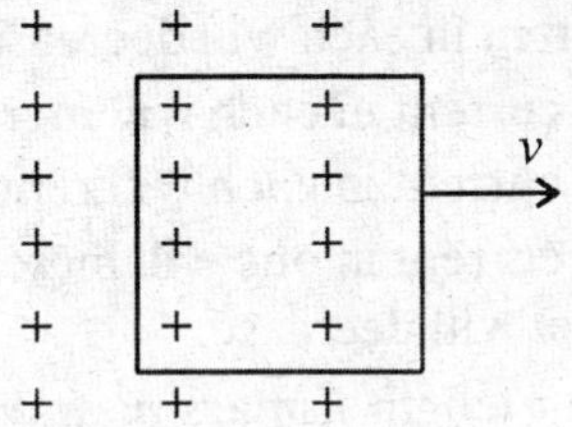

Fig. 24.109

(a)
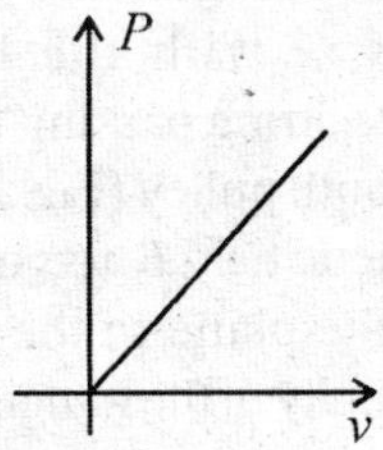

(b)
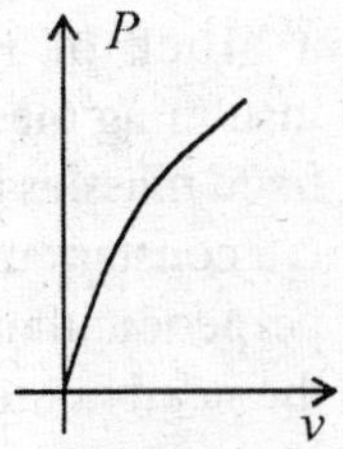

(c)
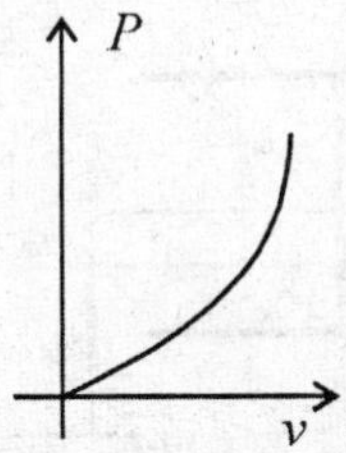

(d)
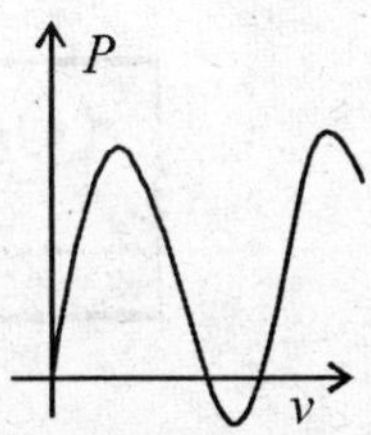

27. Figure 24.110 shows a wire sliding on two parallel conducting rails placed at separation l. A magnetic field B exists in a direction perpendicular to the plane of the rails, if R is the resistance of wire then force necessary to keep the wire moving at constant speed is

(a) $\frac{B^2l^2v}{R}$ (b) $\frac{B^2l^2v^2}{R}$
(c) $\frac{Bl^2v^2}{R}$ (d) zero.

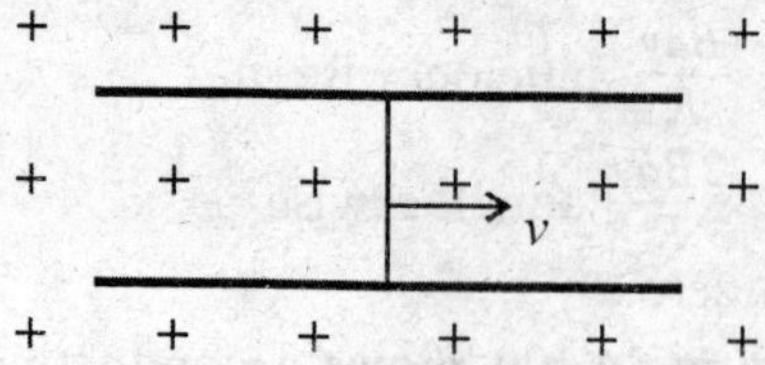

Fig. 24.110

28. Two coaxial circular loops carry a current i each circulating in the same direction. If the loops approach each other

(a) current in each will decrease

(b) the current in each will increase

(c) the current in each will remain same

(d) the current in one will increase and in other will decrease.

29. A rod of length l, mass m and negligible resistance slides on two horizontal frictionless rails of negligible resistance by having a block of mass m with the help of insulating massless string passing through fixed massless smooth pulley (Fig. 24.111). If a constant magnetic field B acts upwards perpendicular to the plane of the Figure, the steady state velocity of the hanging mass is

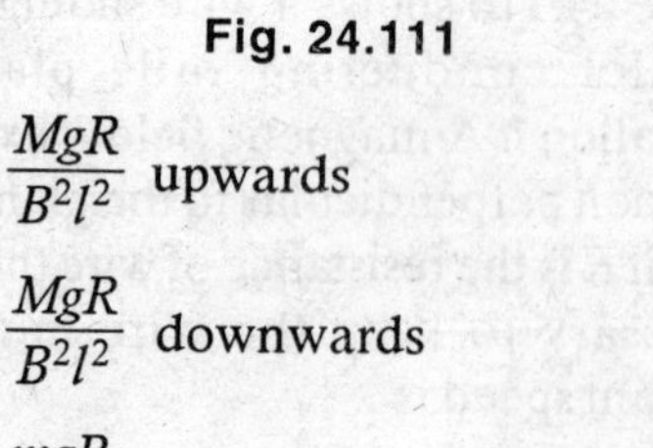

Fig. 24.111

(a) $\dfrac{MgR}{B^2l^2}$ upwards

(b) $\dfrac{MgR}{B^2l^2}$ downwards

(c) $\dfrac{mgR}{B^2l^2}$ upwards

(d) $\dfrac{mgR}{B^2l^2}$ downwards.

30. Three pure inductors each of 2 H are connected as shown in Fig. 24.112. The equivalent inductance of the circuit is

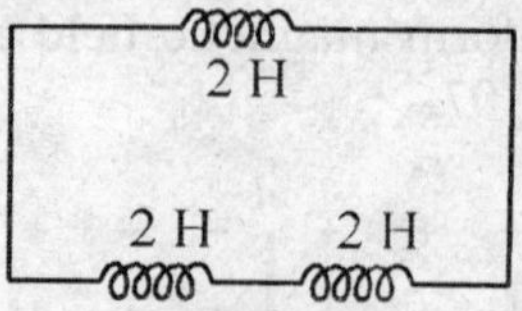

Fig. 24.112

(a) $\dfrac{8}{6}$ H (b) 6 H

(c) 2 H (d) none of these.

31. The switch S is closed at time $t = 0$, the current through the battery at $t = 0$ and $t = \infty$ will be

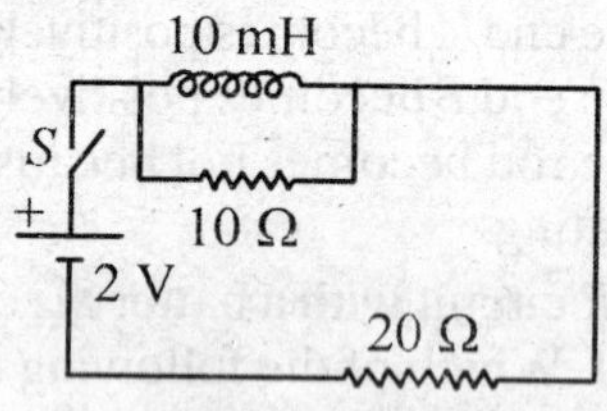

Fig. 24. 113

(a) $\dfrac{1}{15}$ A, $\dfrac{1}{15}$ A (b) $\dfrac{1}{10}$ A, $\dfrac{1}{15}$ A

(c) $\dfrac{2}{15}$ A, $\dfrac{1}{10}$ A (d) $\dfrac{1}{15}$ A, $\dfrac{2}{20}$ A.

32. A long conducting wire AB is moved over a conducting triangular wire PQR with a constant velocity v in uniform magnetic field $\boldsymbol{B}$ directed into the paper. The resistance per unit length of each wire is λ, then

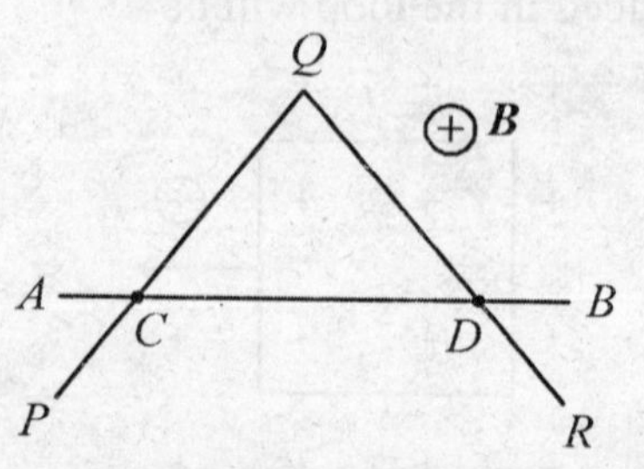

Fig. 24. 114

(a) a constant clockwise induced current will flow in closed loop

(b) an increasing anticlockwise induced current will flow in the closed loop
(c) a decreasing anticlockwise induced current will flow in the closed loop
(d) a constant anticlockwise induced current will flow in the closed loop.

33. An equilateral triangular loop *PQR* having some resistance is pulled with a constant velocity *v* out of a uniform magnetic field directed into the paper. At time $t = 0$ side *QR* of the loop is at the edge of the magnetic field. The induced current (i) versus time (t) graph will be

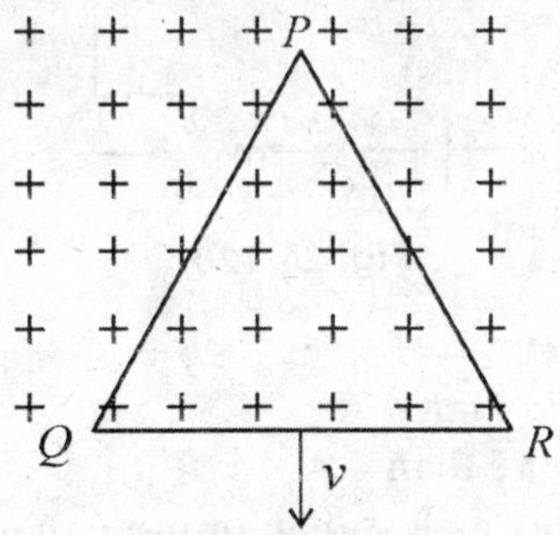

Fig. 24.115

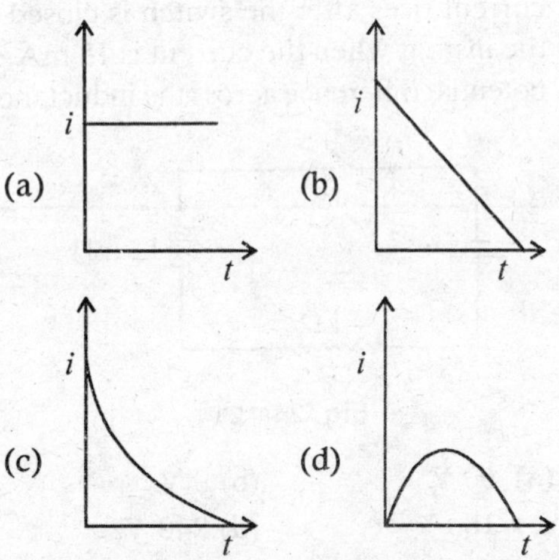

34. A cylindrical space of radius *a* is filled with a uniform magnetic field of induction *B* parallel to the axis of the cylinder. If *B* changes at constant rate, the graph showing the variation of induced electric field with distance *r* from the axis is

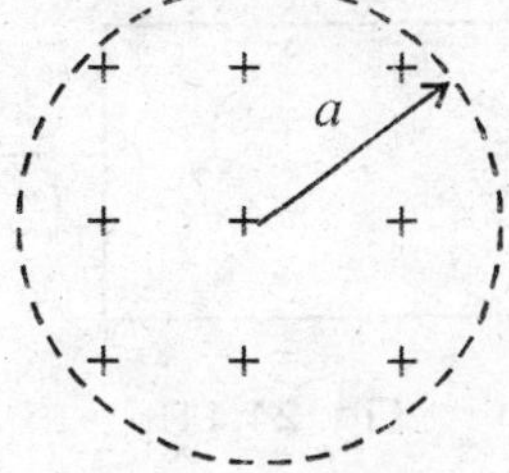

Fig. 24.116

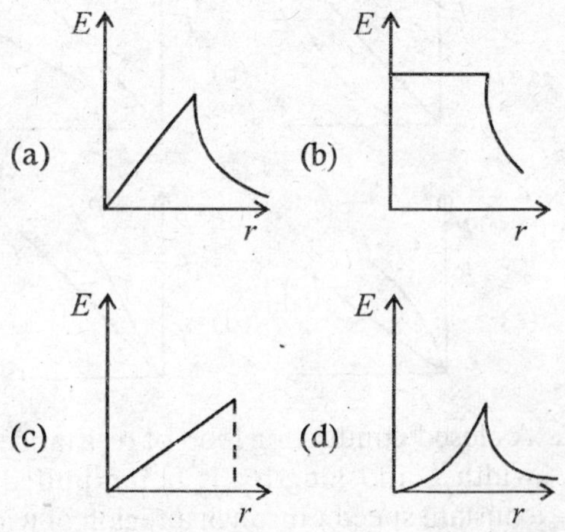

35. In the circuit shown in Fig. 24.117, switch *S* is closed at $t = 0$. The charge which passes through the battery in one time constant is

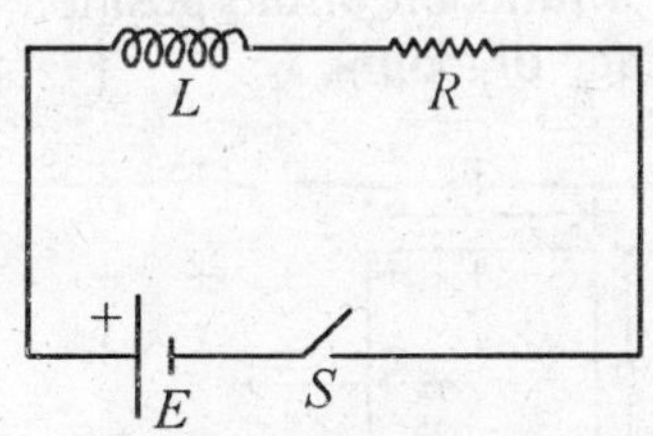

Fig. 24.117

(a) $\frac{eR^2E}{L}$ (b) $E\left(\frac{L}{R}\right)$

(c) $\frac{EL}{eR^2}$ (d) $\frac{eL}{ER}$

36. Figure 24.118 shows a rectangular loop being pulled out in magnetic field *B* with constant speed *v*, then force and power by external agent vary with speed *v* as

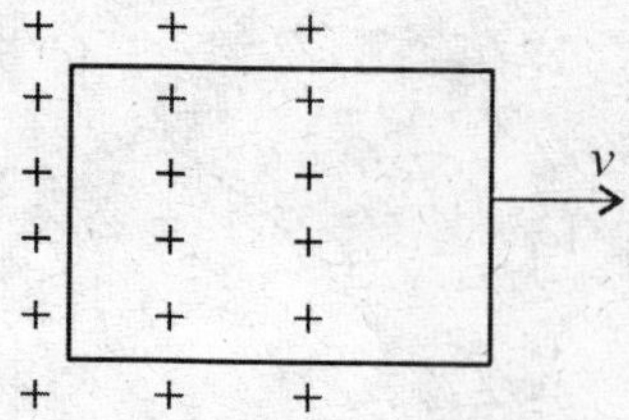

Fig. 24.118

(a) 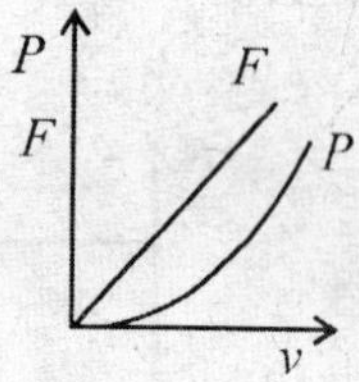

(b) 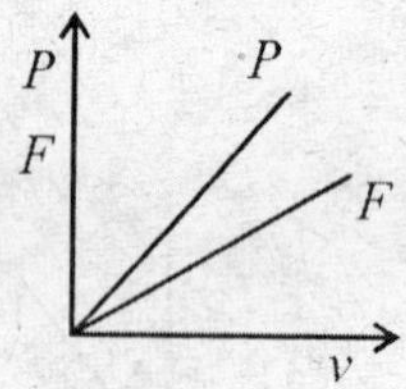

(c) 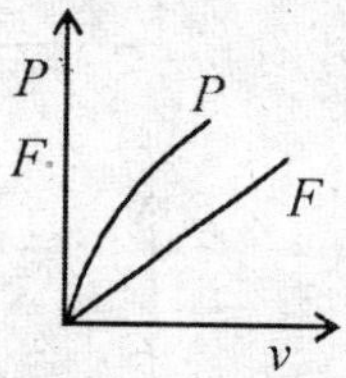

(d) 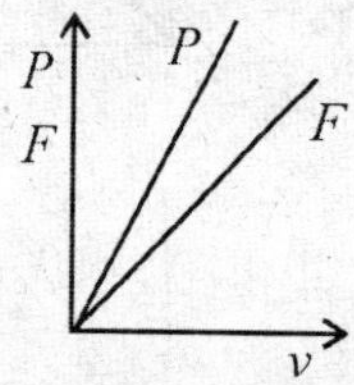

37. A closed conducting loop of resistance R, width b and length l is being pulled at constant speed v through a region of width d ($> l$) in which a uniform magnetic field B is present (Fig. 24.119). Plot of the rate of production of internal energy (P) in the log as a function of the position x of the right edge of loop is

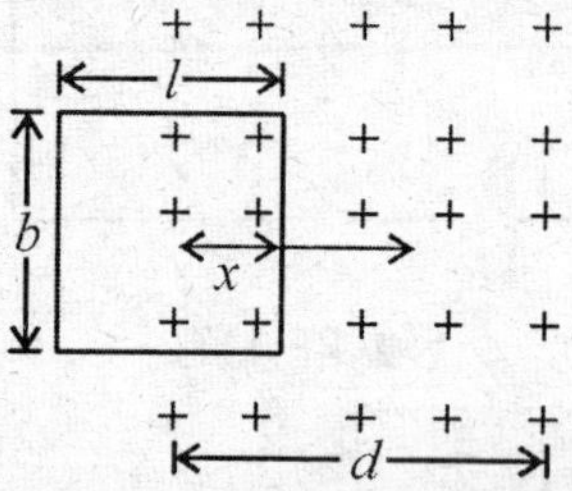

Fig. 24.119

(a) 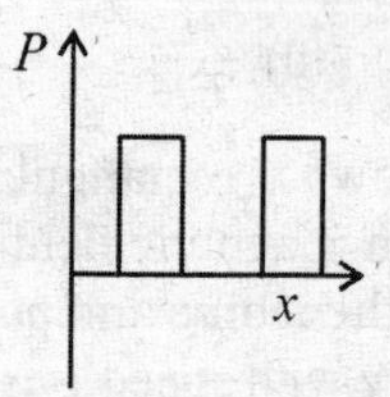

(b) 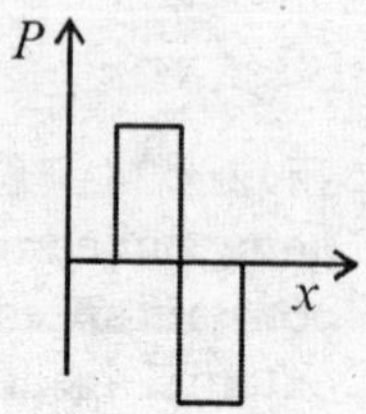

(c) 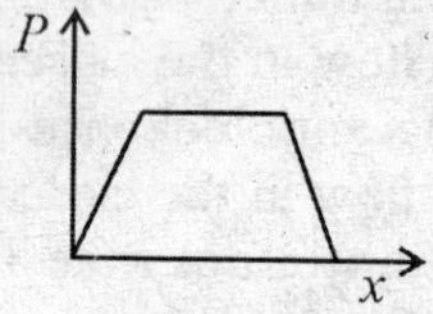

(d) none of these.

38. In the circuit shown in Fig. 24.120 the jockey J is being pulled towards right so that the resistance in the circuit is increasing. Its value at some instant is 5 Ω. The current in the circuit at this instant will be

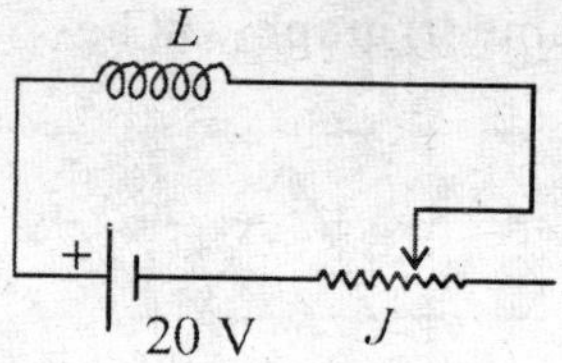

Fig. 24.120

(a) 4 A
(b) less than 4 A
(c) more than 4 A
(d) may be less than or more than 4 A.

39. In the circuit shown in Fig. 24.121, the current rises after the switch is closed. At the instant when the current is 15 mA, the potential difference across the inductance is

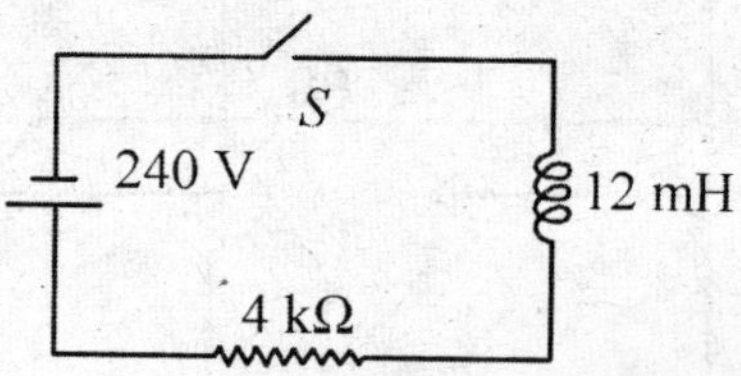

Fig. 24.121

(a) 60 V (b) 0 V
(c) 180 V (d) 240 V.

40. A thick walled hallow metal cylinder has its axis vertical. A bar magnet is dropped into it from a large height with its axis vertical. The shape of the velocity v of the magnet versus time t as it passes through the cylinder is best represented by

(a) 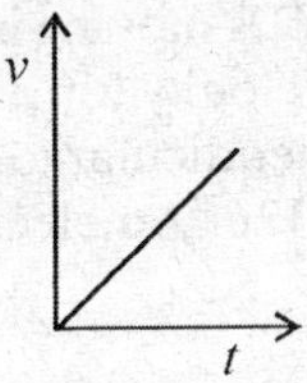

(b) 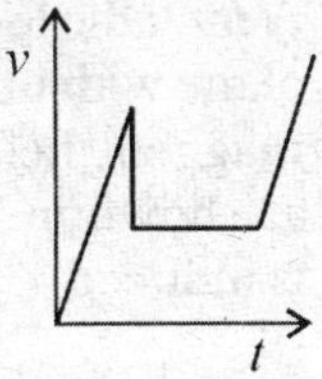

(c) 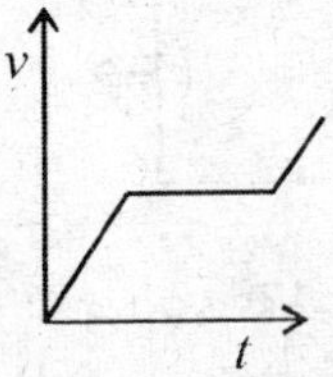

(d) 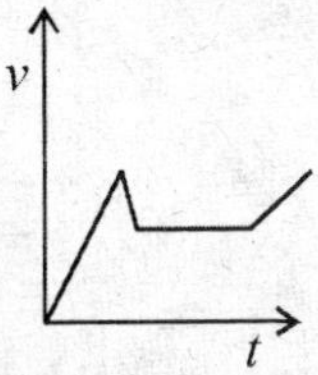

41. In the circuit (Fig. 24.122) the time taken by the current to reach 3/4 of its final value after switching on is (in ms)

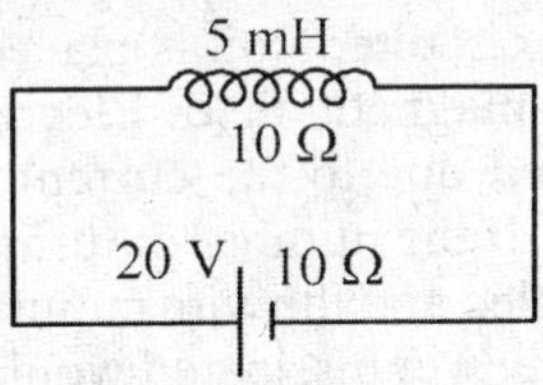

Fig. 24.122

(a) $1/2 \log 2$ (b) $2 \log 2$
(c) $5 \log 2$ (d) $\log(3/4)$.

42. Dimensions of magnetic flux are

(a) $M^1L^2T^{-2}I^{-1}$ (b) $M^1L^1T^{-2}I^1$
(c) $M^1L^2T^{-1}I^{-1}$ (d) $M^1L^2T^{-2}I^1$.

43. A rod *PQ* is connected to the capacitor plates (Fig. 24.123). There is a magnetic field (*B*) directed downwards perpendicular to the plane of the paper. If the rod is pulled out of the magnetic field with velocity *v* as shown,

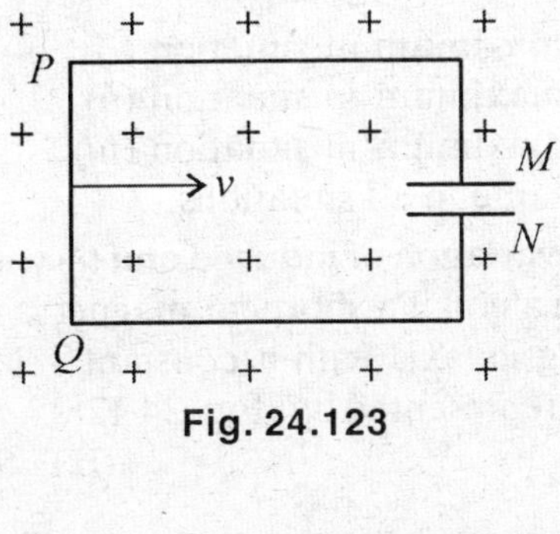

Fig. 24.123

(a) plate *M* will be positive
(b) plate *N* will be positive
(c) both plates will be similarly charged
(d) no charge will be collected on the plates.

44. A disc has radius *R*. It is kept in a non-uniform magnetic field *B* perpendicular to the plane of the disc. The emf induced between its centre and end is

(a) zero
(b) constant but non-zero
(c) BR
(d) $B\pi R^2$.

45. A metal rod moves at a constant velocity in a direction perpendicular to its length. A constant uniform magnetic field exists in space in direction perpendicular to the rod as well as its velocity. Which of the following is correct statement?

(a) The entire rod is at the same potential
(b) There is an electric field in the rod
(c) The electric potential is highest at the centre
(d) The electric potential is lowest at the centre.

46. A thin semicircular conducting ring of radius *R* is falling with its plane vertical in a horizontal magnetic induction *B*. At the position *MNQ* (Fig. 24.124), the speed of the ring is *v*. Then potential difference developed across the ring is

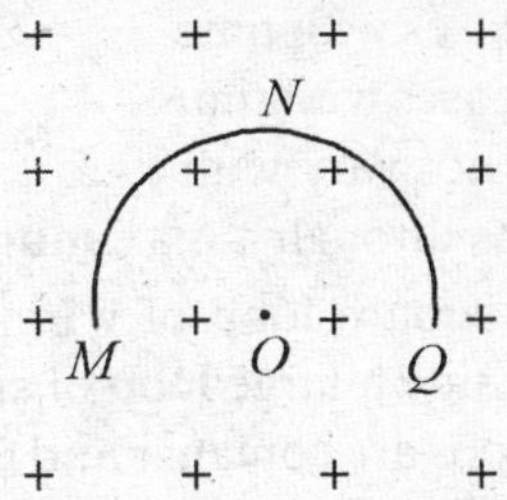

Fig. 24.124

(a) zero
(b) $\frac{1}{2} Bv\pi R^2$ and *M* at higher potential
(c) πRBv and *Q* at higher potential
(d) $2RBv$ and *Q* at higher potential.

47. A uniform but time varying magnetic field $B(t)$ exists in a circular region of radius a and is directed into the plane as shown in Fig. 24.125. Magnitude of the induced electric field at a point P at a distance r from the centre of the circular region is

(IIT 2000)

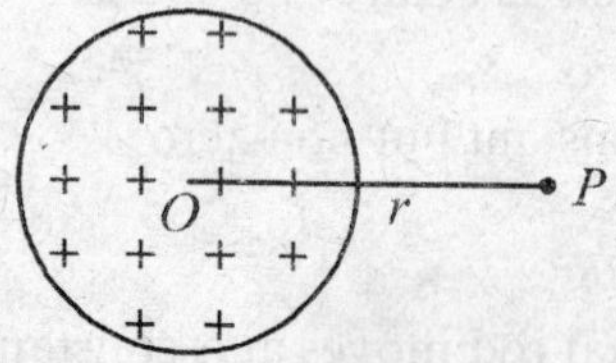

Fig. 24.125

(a) zero (b) decreases as $\frac{1}{r}$

(c) increases as $\frac{1}{r}$ (d) decreases as $\frac{1}{r^2}$.

48. A coil of wire having finite inductance and resistance has a conducting ring placed co-axially within it. The coil is connected to a battery at time $t = 0$, so that a time dependent current $I_1(t)$ starts flowing through the coil. If $I_2(t)$ is the current induced in the ring and $B(t)$ is the magnetic field at the axis of the coil due to $I_1(t)$, then as the function of time t ($t > 0$) the product $I_2(t)\,B(t)$

(IIT Screening 2000)

(a) increases with time
(b) decreases with time
(c) does not vary with time
(d) passes through a maximum.

49. A small square loop of wire of side l is placed inside a large loop of side L ($L >> l$). The loops are coplanar and their centres coincide. The mutual inductance of system is proportional to

(a) $\frac{l}{L}$ (b) $\frac{l^2}{L}$

(c) $\frac{L}{l}$ (d) $\frac{L^2}{l}$.

50. A metallic loop $ABCD$ is moving in its own plane with uniform velocity v in a uniform magnetic field B perpendicular to its plane as shown in Fig. 24.126. An electric field is induced

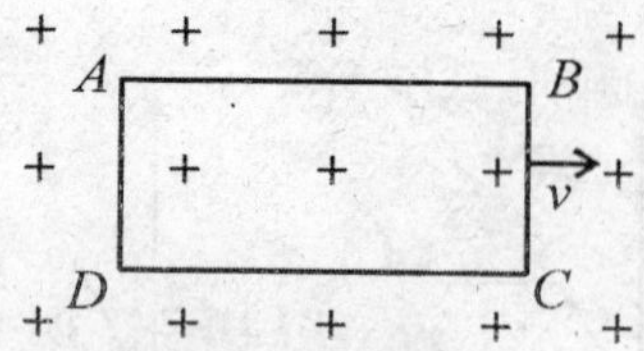

Fig. 24.126

(a) in AD but not in BC
(b) in BC but not in AD
(c) neither in AD nor in BC
(d) in both AD and BC.

51. A short circuited coil is placed in a time varying magnetic field. Electric power is dissipated due to the current induced in the coil. If the number of turns were to be quadrupled and the wire radius halved, the electric power dissipated would be

(IIT S 2002)

(a) halved (b) the same
(c) doubled (d) quadrupled.

52. Two circular coils can be arranged in any of the three situations as shown in Fig. 24.127. Their mutual inductance will be

(IITS 2001)

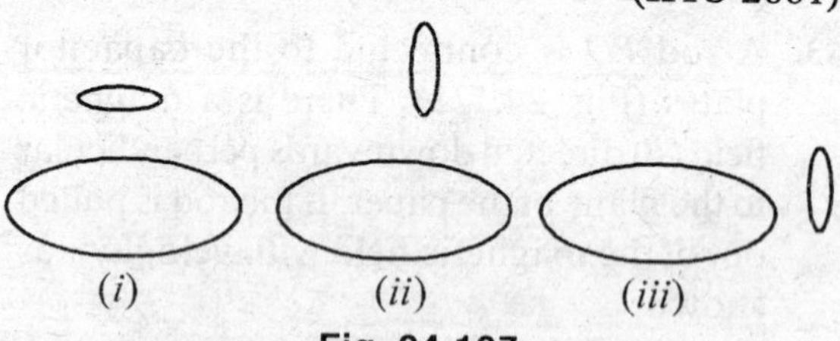

Fig. 24.127

(a) maximum in situation (*i*)
(b) maximum in situation (*ii*)
(c) maximum in situation (*iii*)
(d) same in all situations.

53. The variation of induced emf (e) with time (t) in a coil if a short bar magnet is moved along its axis with a constant velocity is best represented by (Fig. 24.128)

(IIT S 2004)

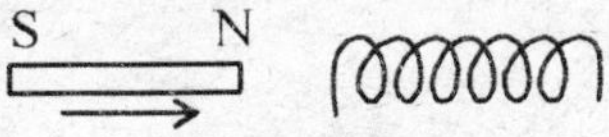

Fig. 24.128

(a) 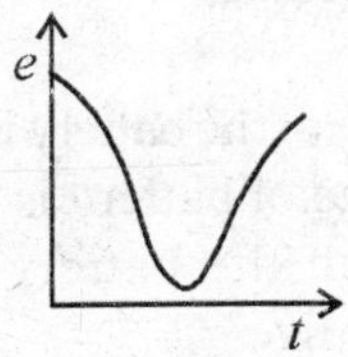

(b) 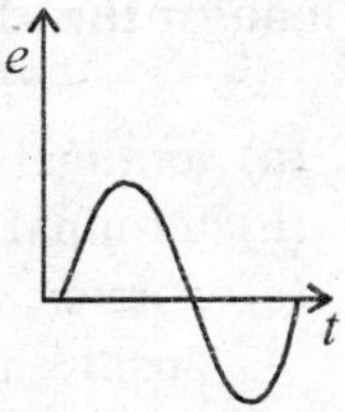

(c) 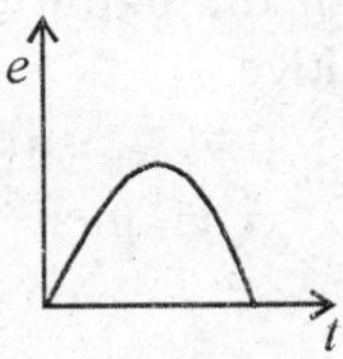

(d) 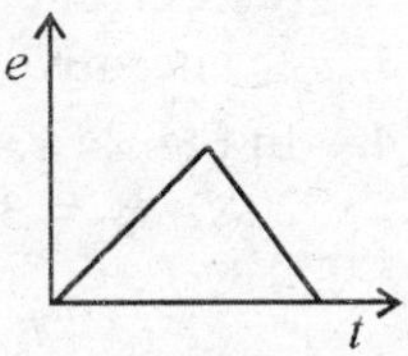

54. As shown in Fig. 24.129 P and Q are two coils placed co-axially at some distance. When switch S is closed a clockwise current i_P flows in P (as seen by eye) and an induced current i_Q flows in Q. The switch S remains closed for a long time. When switch S is opened, a current i'_Q flows in Q, then the direction of i_Q and i'_Q are (IIT S 2002)

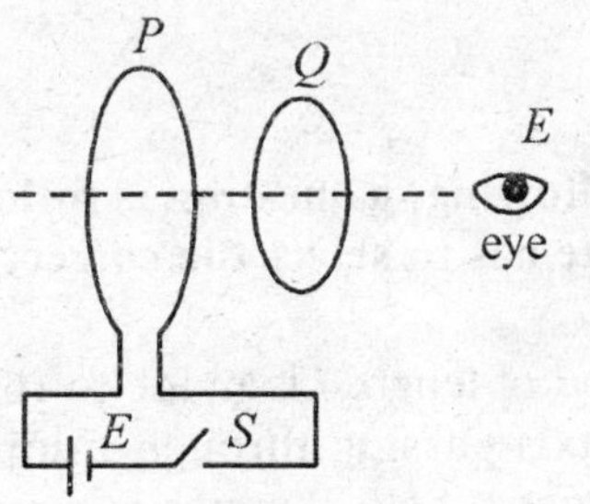

Fig. 24.129

(a) respectively clockwise and anticlockwise
(b) both clockwise
(c) both anticlockwise
(d) respectively anticlockwise and clockwise.

55. A rectangular loop of wire of resistance R is placed in a uniform magnetic field B acting normally on the plane of the loop. If we attempt to pull (l is length of the loop and b the breadth) it out of the field with velocity v, the power required is

(a) $\dfrac{B^2v^2l^2}{R}$ (b) $\dfrac{B^2l^2v}{R}$

(c) $\dfrac{B^2l^2v^2}{R}$ (d) $\dfrac{1}{2}\dfrac{B^2l^2v^2}{R}$.

ANSWERS

1. (a)	2. (b)	3. (b)	4. (c)	5. (a)
6. (a)	7. (d)	8. (a)	9. (b)	10. (d)
11. (a)	12. (a)	13. (a)	14. (b)	15. (b)
16. (a)	17. (d)	18. (d)	19. (a)	20. (c)
21. (b)	22. (d)	23. (b)	24. (d)	25. (d)
26. (c)	27. (d)	28. (a)	29. (b)	30. (a)
31. (d)	32. (d)	33. (b)	34. (a)	35. (c)
36. (a)	37. (a)	38. (b)	39. (c)	40. (b)
41. (a)	42. (a)	43. (a)	44. (a)	45. (b)
46. (d)	47. (b)	48. (d)	49. (b)	50. (d)
51. (d)	52. (a)	53. (b)	54. (d)	55. (a).

PART II

In the following, a question is followed by four alternatives, of them one or more are correct. Candidate has to select the correct choice(s) from amongst the alternatives.

1. A rod of length l is made to rotate about an axis passing through midpoint O in magnetic field B which is parallel to the axis of rotation. If angular speed is ω, then

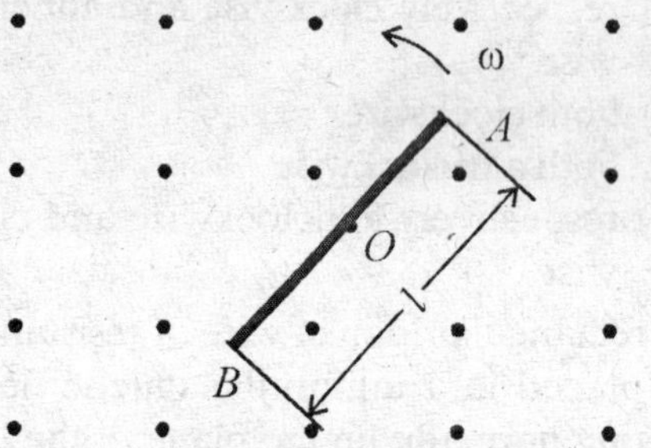

Fig. 24.130

(a) induced emf e between A and O is $\frac{l^2}{2} B\omega$
(b) induced emf e between A and O is $\frac{l^2}{8} B\omega$
(c) potential diffference between A and B is zero
(d) potential diffference between A and B is $l^2 B\omega$.

2. Inductance of a solenoid becomes
(a) double if number of turns are doubled
(b) four times if number of turns are doubled
(c) becomes double if both number of turns and length is doubled
(d) becomes half if length is doubled.

3. A square wire loop with 2 m sides is perpendicular to a uniform magnetic field with half area of the loop in the field (Fig. 24.131). The loop contains 20 V battery with negligible internal resistance. The magnitude of the field varies with time according to $B = 0.042 - 0.87t$, then

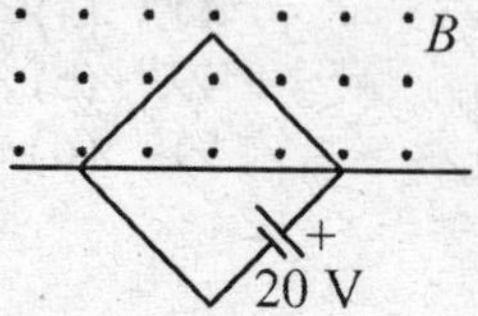

Fig. 24.131

(a) terminal voltage of the battery is 21.74 V
(b) terminal voltage of battery is 18.26 V
(c) current through the battery is from positive to negative
(d) current through the battery is from negative to positive.

4. In Fig. 24.132, $E = 100$ V, $R_1 = 10\ \Omega$, $R_2 = 20\ \Omega$, $R_3 = 30\ \Omega$, $L = 2$ H, then

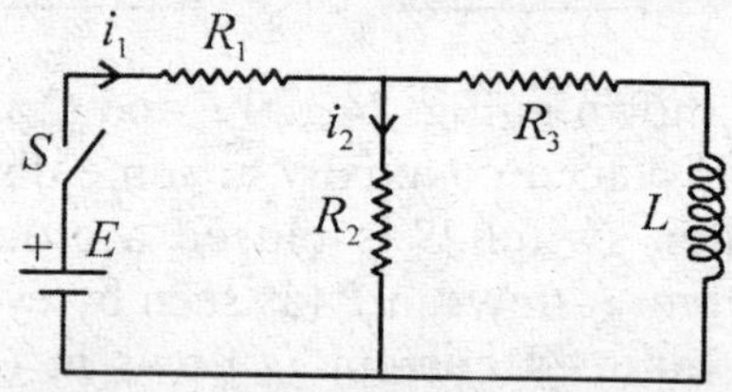

Fig. 24.132

(a) current i_2 immediately after switch S is closed is $\frac{10}{3}$ A
(b) current i_1 after long time switch S closed is $\frac{50}{11}$ A
(c) current i_2 immediately after switch S is opened is $\frac{20}{11}$
(d) current i_2 a long time after (c) is zero.

5. Consider the situation shown in Fig. 24.133. The wire PQ has a negligible resistance and is made to slide on the rails with constant speed of 5 m/s

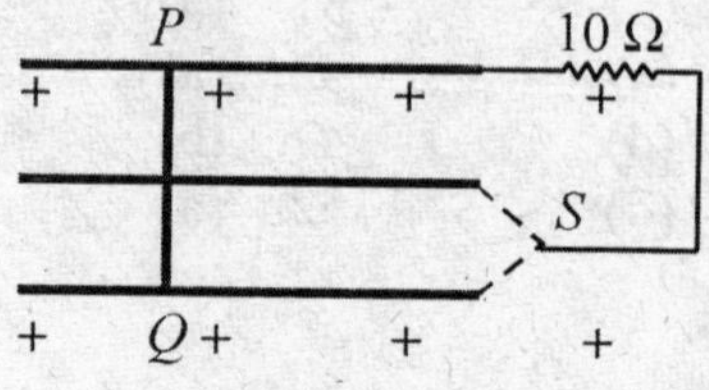

Fig. 24.133

(a) current through 10 Ω resistor is 0.1 mA when switch S is thrown to middle rail
(b) current through 10 Ω resistor is 0.05 mA when switch S is thrown to middle rail
(c) current through 10 Ω resistor is 0.1 A when switch S is thrown to bottom rail
(d) current through 10 Ω resistor is 0.2 A when switch S is thrown to bottom rail.

6. Consider the situation shown in Fig. 24.134. The wires AB and CD are made to slide on the rails with same speed 5 m/s. The current through the 19 Ω resistor is, B = 1.0 T,

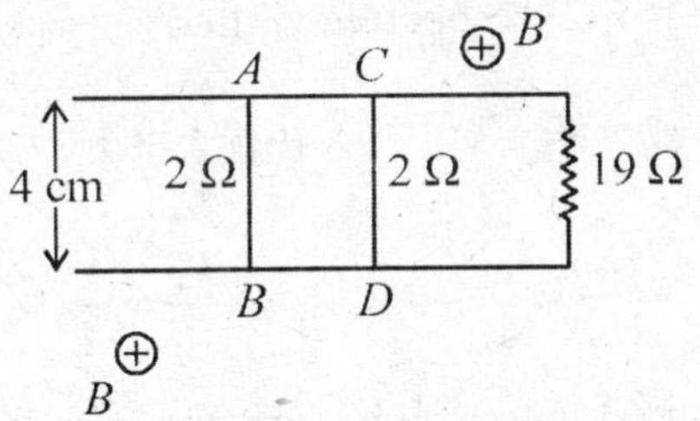

Fig. 24.134

(a) zero if both wires move towards right
(b) 0.1 mA if AB moves towards left and CD towards right
(c) zero if AB moves towards left and CD towards right
(d) 0.1 mA if both wires move towards right.

7. A bar magnet is moved along the axis of a copper ring placed far away from the magnet looking from the side of the magnet, an anticlockwise current is found to be induced in the ring. Which of the following may be true?
(a) The south pole faces the ring and the magnet moves towards it
(b) The north pole faces the ring and the magnet moves towards it
(c) The south pole faces the ring and the magnet moves away from it
(d) The north pole faces the ring and the magnet moves away from it,

8. A conducting rod of length l is hinged at point O. It is free to rotate in a vertical plane. There exists a uniform magnetic field $\boldsymbol{B}$ in horizontal direction. The rod is released from position shown in Fig. 24.135. The potential difference between two ends of the rod is proportional to

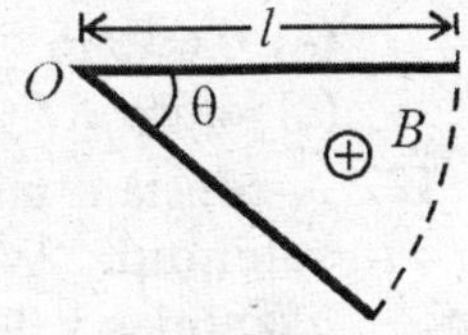

Fig. 24.135

(a) $l^{3/2}$ (b) l^2
(c) $\sin\theta$ (d) $(\sin\theta)^{1/2}$.

9. Which of the following statement(s) is/are correct regarding the electric field produced by the changing magnetic field?
(a) It is non-electrostatic in nature
(b) It is non-conservative in nature
(c) Potential can be defined corresponding to this field
(d) The lines of this field are closed curves.

10. A conducting loop is placed in uniform magnetic field with its plane perpendicular to the field. An emf is induced in the loop if
(a) it is translated
(b) it is rotated about its axis
(c) it is rotated about its diameter
(d) it is deformed.

11. A square loop $PQRS$ of side 0.2 m is placed in a magnetic field $B = 3t^2$. Here t is in seconds and B in tesla. The magnetic field is out of the paper. At time t = 2 sec, induced electric field in SR is

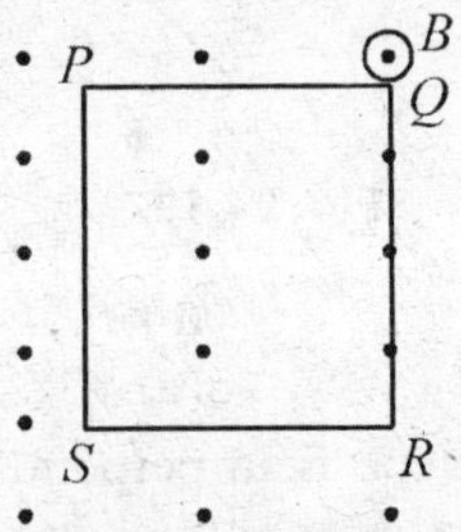

Fig. 24.136

(a) 0.6 V/m (b) along SR
(c) along RS (d) 0.3 V/m.

12. A constant current i is maintained in a solenoid. Which of the following quantities will increase if an iron rod is inserted in the solenoid along its axis?

(a) Magnetic field at the centre
(b) Magnetic flux linked with the solenoid
(c) Self inductance of the solenoid
(d) Rate of joule heating.

13. A metal sheet is placed in front of a strong magnetic pole. A force is needed to

(a) hold the sheet there if the metal is magnetic
(b) hold the sheet there if the metal is non-magnetic
(c) move the sheet away from the pole with uniform velocity if metal is magnetic
(d) move the sheet away from the pole with uniform velocity if metal is non-magnetic.

14. Two straight conducting rails form a right angle where their ends are joined. A conducting bar in contact with the rails starts at the vertex at time $t = 0$ and moves with constant velocity v along them as shown in Fig. 24.137. A magnetic field B is directed into the page. The induced emf in the circuit at any time t is proportional to

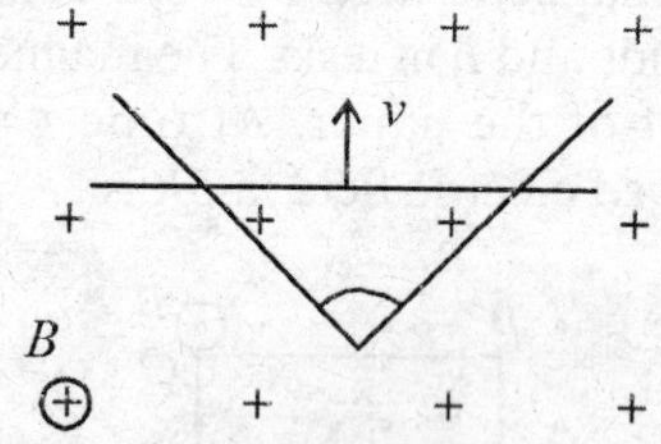

Fig. 24.137

(a) t (b) t^2
(c) v^2 (d) v^o

15. The magnetic field perpendicular to the plane of a conducting ring of radius a changes at the rate of λ, then

(a) the emf induced in the ring is $\pi\lambda a^2$
(b) the induce electric field $E = \frac{\lambda a}{2}$
(c) the electric field intensity at any point on the ring is zero
(d) all the points on the ring are at the same potential.

16. Two different coils have self inductance $L_1 = 8$ mH and $L_2 = 2$ mH. Current in the two coils is increased at the same constant rate. At a certain instant of time the power given to the two coils is also the same. At this instant the current, the induced voltage and energy stored in the coils is i_a, V_a, U_a ($a = 1, 2$) respectively, then

(a) $\frac{i_1}{i_2} = \frac{1}{4}$ (b) $\frac{i_1}{i_2} = 4$

(c) $\frac{U_1}{U_2} = \frac{1}{4}$ (d) $\frac{V_1}{V_2} = \frac{1}{4}$.

17. Two solenoids have identical geometrical construction but one is made of thick wire and the other of thin wire. Which of the following quantities are different for the two solenoids?

(a) Self inductance
(b) Rate of joule heating if the same current passing through them
(c) Magnetic field energy if the same current passing through them
(d) Time constant if one solenoid is connected to one battery and the other is connected to another battery.

18. A magnetic field in the cylindrical region as shown in Fig. 24.138 increases at constant rate of 20.0 mT/s. Each side of the square $ABCD$ and $DEFA$ has a length of 1.00 cm and a resistance of 4 Ω. The current in the wire AD is

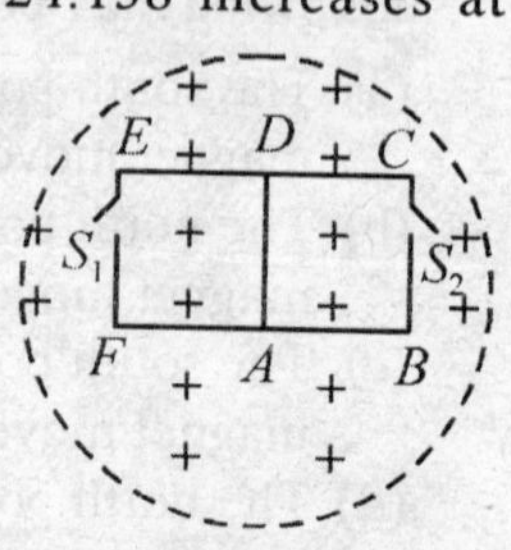

Fig. 24.138

(a) 1.25×10^{-7} A from A to D if the switch S_1 is closed and S_2 is open
(b) 1.25×10^{-7} A from D to A if the switch S_1 is open and S_2 is closed
(c) zero if both switches S_1 and S_2 are closed
(d) 2.50×10^7 A from D to A if the switch S_1 and S_2 are both closed.

19. A conducting circular loop of radius a and resistance R is kept on a horizontal plane. A vertical time varying magnetic field $B = 2t$ is switched on at time $t = 0$, then
(a) flow of charge per unit time from any section of the coil is constant
(b) total charge passed through any section between time $t = 0$ to $t = 2$ sec is $\left(\frac{4\pi a^2}{R}\right)$
(c) power generated in the coil at any time t is constant
(d) none of the above is correct.

20. Figure 24.139 shows a conductor located in magnetic field along the inward normal to the plane of the figure. Magnetic field starts diminishing, then the induced current

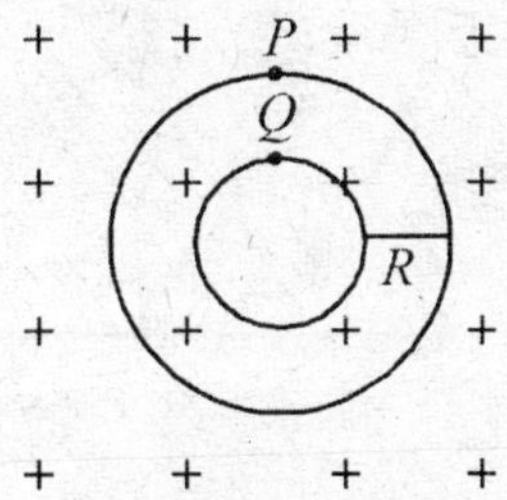

Fig. 24.139

(a) at point P is clockwise
(b) at point Q is anticlockwise
(c) at point Q is clockwise
(d) at point R is zero.

21. The SI unit of inductance 'henry' is the same as
(a) Weber/ampere (b) Volt sec/ampere
(c) Joule/(amp)^2 (d Ohm-sec.

22. A circular loop of radius R carrying current I lies in X-Y plane with its centre at origin. The total magnetic flux through X-Y plane is (IIT 1999)
(a) directly proportional to I
(b) directly proportional to R
(c) directly proportional to R^2
(d) zero.

ANSWERS

1. (a, b, c) 2. (b, c, d) 3. (a, d) 4. (a, b, c, d) 5. (a, d)
6. (c, d) 7. (b, c) 8. (a, d) 9. (a, b, d) 10. (c, d)
11. (a, b) 12. (a, b, c) 13. (a, c, d) 14. (a, c) 15. (a, b, d)
16. (a, c) 17. (b, d) 18. (a, b, c) 19. (a, b, c) 20. (a, b, d)
21. (a, b, c, d) 22. (a, b)

25

Alternating Currents

25.1 ALTERNATING CURRENT AND VOLTAGE

Of the three circuit elements, resistance R, capacitance C and inductance L, we have so far discussed the series combination RC (Chapter 19) and R-L (in Chapter 24). In these two types of circuits, we found that the charge, current and potential difference grow and decay exponentially. The time scale of the growth or decay is given by time constant τ $(= L/R$ or $RC)$. In all these circuits, direct currents were used, for which primary source of emf is a battery. In steady states such currents are *time independent*.

The currents (or voltages) which in steady state vary with time are called time *varying currents*. The time varying currents may be periodic or non periodic. The simplest type of periodic variation is 'Sine' or 'Cosine' function of time. *Such currents or voltages which vary sinusoidally with time are called alternating currents or voltages.* Their variation with time is shown in Fig. 25.1. The main characteristics of alternating current are:

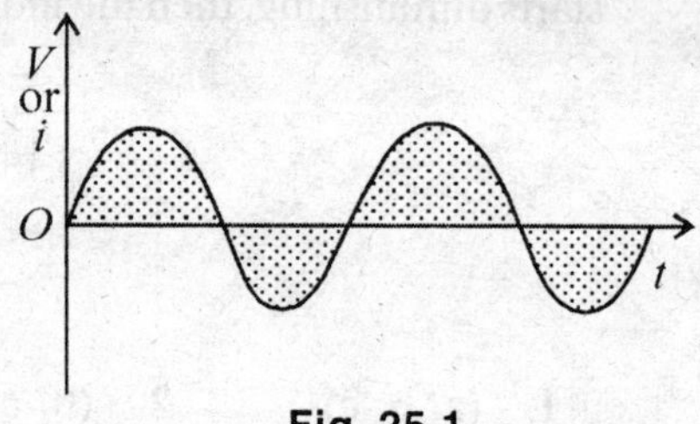

Fig. 25.1

(*a*) a constant amplitude. Amplitude in alternating currents (abbreviated as AC) is also called *peak value*.

(*b*) alternate half cycle is positive and half negative.

(*c*) they are represented mathematically like

$$i = i_0 \sin \omega t \quad \text{or} \quad V = V_0 \sin \omega t \qquad \text{...(25.1)}$$

because in circuits they also vary with time as given by equation (25.1). Here i and V are called instantaneous values of current and potential difference, ω is called angular frequency, such that

$$\omega = 2\pi f$$

where f is called frequency i_0 and V_0 are called amplitudes or *peak values* of current and voltage respectively.

Average and rms (Root Mean Square) Values of Alternating Currents

The average value of alternating currents for a cycle is defined as

$$\langle i \rangle = \frac{\int_0^T i dt}{\int_0^T dt} = \frac{\int_0^T i_0 \sin \omega t \cdot dt}{T}$$

$$= -\frac{i_0}{T\omega} \cos \omega t \Big|_0^T = 0 \qquad \because \omega T = 2\pi$$

The average value of alternating current for half cycle is similarly defined as

$$\langle i \rangle = \frac{\int_0^{T/2} i dt}{\int_0^{T/2} dt} = \frac{2}{T} \int_0^{T/2} i_0 \sin \omega t \, dt$$

$$= -\frac{2i_0}{T\omega} \cos \omega t \Big|_0^{T/2} = \frac{2i_0}{\pi} \qquad \ldots(25.1a)$$

Its value for positive half cycle is positive and negative for negative cycle.

Alternating currents are measured using *hot wire instruments* (ammeter and voltmeter). These instruments are based on the Joule heating effects, i.e. square of the current and voltage applied to a conductor. Hence average of i^2 for a cycle is

$$\langle i^2 \rangle = \frac{\int_0^T i^2 \sin \omega t \, dt}{\int_0^T dt} = \frac{1}{T} \int_0^T i_0^2 \sin^2 \omega t \, dt$$

$$= \frac{i_0^2}{2}$$

under-root of this average is called rms value of AC, i.e.

$$i_{\text{rms}} = \sqrt{\langle i^2 \rangle} = \frac{i_0}{\sqrt{2}} \qquad \ldots(25.2)$$

This value is also called *effective or virtual value of alternating current.*

All the *AC* instruments measure this value. If an *AC* voltmeter measures 200 V means $V_{\text{rms}} = 200$ V and the peak value of this alternating voltage $V_0 = \sqrt{2} \times 200 = 283$ V.

The ratio of i_{rms} and i_{av}^* for half cycle is called *form factor*. Form factor for sinusoidal alternating current is

$$\text{Form factor} = \frac{i_{\text{rms}}}{i_{\text{av}}} = \frac{i_0}{\sqrt{2}} \times \frac{\pi}{2i_0} = \frac{\pi}{2\sqrt{2}}$$

Physical Significance of rms Value

rms value of an AC is that value of DC which produces the same amount of heat in same time when passed through a conductor as produced by AC in the same conductor in same time.

25.2 PHASE RELATION BETWEEN CURRENT AND VOLTAGE IN AC CIRCUITS

In DC circuits if transient effects are excluded only resistance R can be used to control the current because inductor behaves like a short circuit while capacitor as open circuit. But in case of AC

---

*Average value <*i*> is also written as i_{av}.

circuit, all three elements or any combination of these can be used to control the current in the circuit. In such cases it is just possible that an AC current through an element may not be in phase with potential difference across that element. By phase difference, we mean that the current and potential difference may not be zero or maximum at the same time. In order to study this phase relation we will consider the following cases.

Pure Resistive Circuit

When an alternating voltage $E = E_0 \sin\omega t$ is applied across a resistor R as shown in Fig. 25.2, the instantaneous current through resistor is

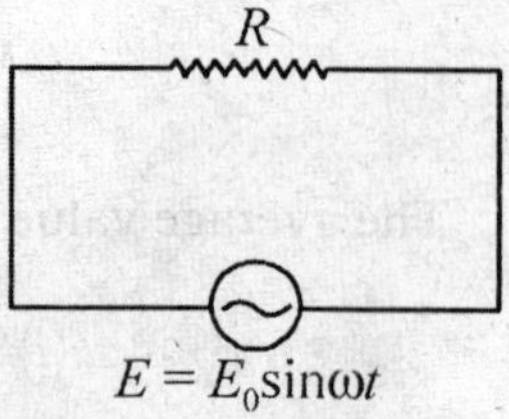

Fig. 25.2

$$i = \frac{E}{R} = \frac{E_0}{R}\sin\omega t = i_0 \sin\omega t \qquad \text{...(25.3)}$$

where $i_0 = \frac{E_0}{R}$, from equation (25.3), we have

(*a*) the current and potential difference across resistor R are in same phase. This is shown in Fig. 25.3 (*a*)

(*b*) Figure 25.3 (*b*) shows the vector diagram for the current and potential difference across resistor.

(*c*) R the resistance of the resistor is independent of the frequency.

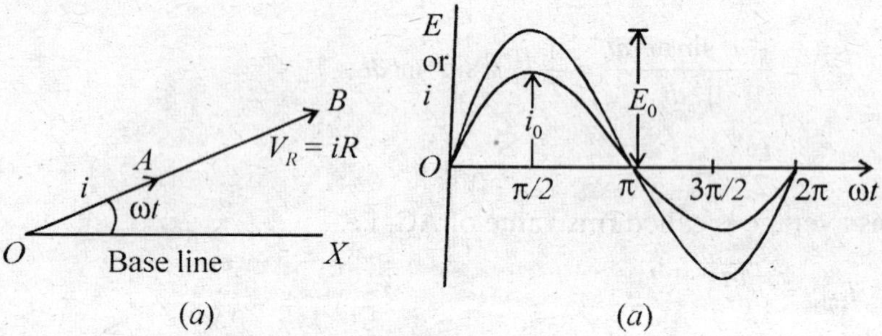

Fig. 25.3

Pure Inductive Circuit

When an alternating voltage $E = E_0 \sin\omega t$ is applied across the inductance L, a current flows through the inductor. If i is instantaneous value of current and it varies at the rate $\frac{di}{dt}$, then we have

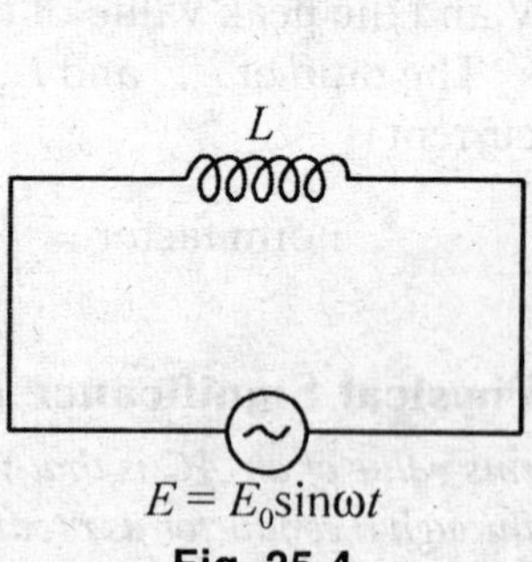

Fig. 25.4

$$L\frac{di}{dt} = E = E_0 \sin\omega t$$

or
$$di = \frac{E_0}{L}\sin\omega t\, dt$$

On integration, we get

$$i = -\frac{E_0}{\omega L}\cos\omega t + C$$

where C is constant of integration. Since C is independent of time, hence represents DC. Hence $C = 0$

$$\therefore \qquad i = -\frac{E_0}{\omega L}\cos\omega t = i_0 \sin(\omega t - \pi/2) \qquad \text{...(25.4)}$$

where $i_0 = \dfrac{E_0}{\omega L}$. From equation (25.4), the following can be concluded.

(*a*) The current through inductor and potential difference across inductor are not in phase. The current through the inductor lags behind the potential difference by $\pi/2$. The graph between E (and i) versus ωt is shown in Fig. 25.5 (*a*). When potential difference leads the current in phase, the phase is said to be positive. The phase diagram for pure inductive circuit is shown in Fig. 25.5 (*c*).

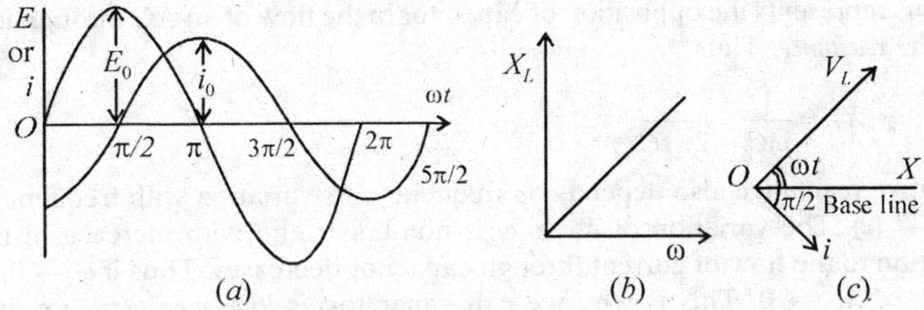

Fig. 25.5

(*b*) The quantity ωL has the dimensions of resistance because

$$[\omega L] = \left(\frac{\text{rad}}{S} H\right) = \left(\frac{\text{rad}}{S} \times \Omega \times S\right) = [\Omega]$$

This quantity is called *inductive reactance* and is represented by X_L and represents opposition (or resistance) of the coil to the flow of AC. Thus

$$X_L = \omega L = 2\pi f \qquad \text{...(25.5)}$$

(*c*) Unlike resistance, the inductive reactance depends on frequency and its variation with frequency is shown in Fig. 25.5 (*b*).

If $\omega \to 0$, $X_L \to 0$ and when $\omega \to \infty$, $X_L \to \infty$. Because of this an inductor is called *low pass filter*, i.e. it allows low frequency currents to pass through it and offers greater opposition to high frequency currents. For DC, i.e. $\omega = 0$ or $X_L = 0$, the opposition of an inductance is zero. This means that an inductor acts like a simple conducting wire in steady state in DC circuits.

Pure Capacitive Circuit

When an alternating emf $E = E_0 \sin\omega t$ is applied across the capacitor (Fig. 25.6) the instantaneous potential difference across capacitor is E and charge q on it is

$$q = CE = CE_0 \sin\omega t$$

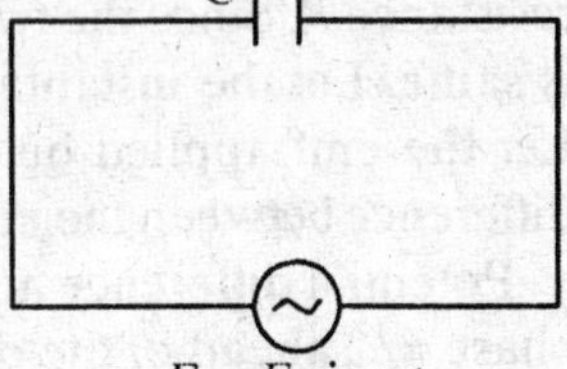

Fig. 25.6

Hence the current through the capacitor $i = \dfrac{dq}{dt}$

$$\therefore \quad i = CE_0\omega\cos\omega t = i_0 \sin(\omega t + \pi/2) \qquad \text{...(25.6)}$$

where $i_0 = C\omega E_0$. From equation (25.6), the following can be concluded.

(*a*) The current and voltage across capacitor are not in phase. In this case the current leads the emf in phase by $\pi/2$. The plot of E and i versus time ($= \omega t$) is shown in Fig. 25.7 (*a*). Since the emf lags behind the current as such this phase is taken negative. The phase diagram for pure capacitive circuit is shown in Fig. 25.7 (*b*).

(*b*) $i_0 = E_0 C\omega = \frac{E_0}{X_C}$, where $X_C = \frac{1}{\omega C}$ has the dimensions of resistance as

$$\frac{1}{\omega C} = \left[\frac{1}{\text{rad}/S} \times \frac{\text{Volt}}{\text{Coulomb}}\right] = \left[\frac{\text{Volt}}{\text{Ampere}}\right] = \Omega$$

It thus represents the opposition of capacitor to the flow of an AC through it and is called *capacitive reactance.* Thus

$$X_C = \frac{1}{\omega C} = \frac{1}{2\pi fC} \quad \ldots(25.7)$$

(*c*) Capacitive reactance also depends on frequency. Its variation with frequency is shown in Fig. 25.7 (*c*). The variation of X_C vs ω is non-linear and with increase in frequency the opposition to the flow of current through capacitor decreases. Thus if $\omega \to 0$, $X_C \to \infty$ and if $\omega \to \infty$, $X_C \to 0$, This is why we call capacitor as *high pass filter*, i.e. it allows high frequency currents to pass through it and offers greater opposition to low frequency currents. For DC, i.e. $\omega = 0$, $X_C = \infty$, the opposition of capacitor is infinite and acts as open circuit in DC circuits in steady state.

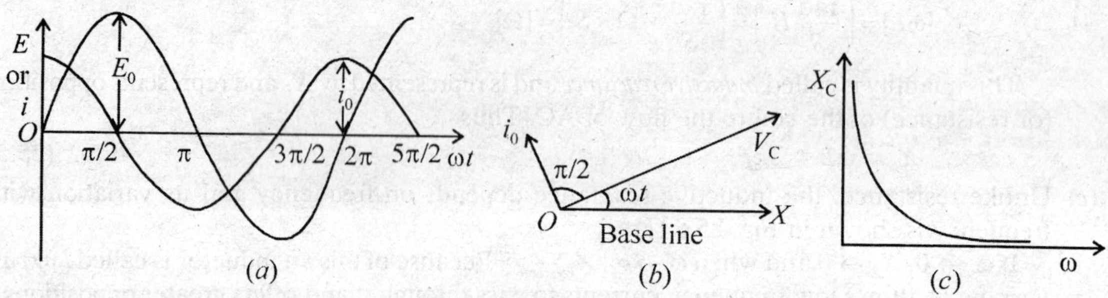

Fig. 25.7

Circuit Containing Inductance and Resistance (L-R Series Circuit)

Figure 25.8 shows an AC circuit containing inductance L and resistance R. Since the two are in series hence current through them is same. Let the instantaneous value of the current be $i = i_0 \sin\omega t$. Let the emf applied be $E = E_0 \sin(\omega t + \theta)$ where θ is the phase difference between the current and applied emf.

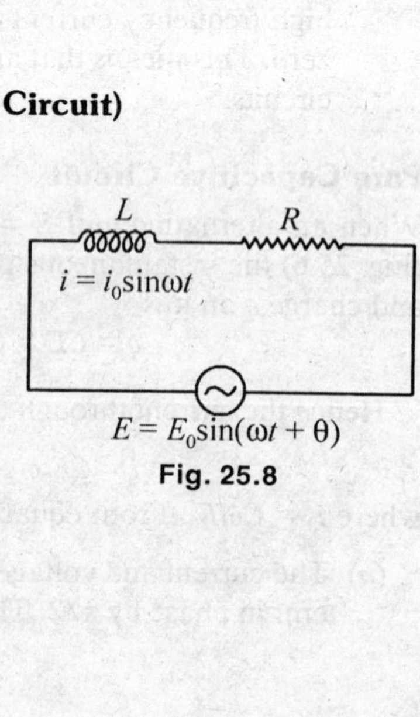

Fig. 25.8

Potential difference across inductance then is $V_L = i\omega L$ and is in phase $\pi/2$ ahead of the current through it.

Similarly, the potential difference across resistance R is $V_R = iR$ and is in phase with the current. The vector resultant of V_L and V_R must give the applied voltage E. The vector diagram is shown in Fig. 25.9. From diagram, we have

$$E^2 = V_L^2 + V_R^2 = i^2\omega^2 L^2 + i^2 R^2 = i^2 (\omega^2 L^2 + R^2)$$

or $$E = i\sqrt{\omega^2 L^2 + R^2} = iZ \qquad \ldots(25.8)$$

with $Z = \sqrt{\omega^2 L^2 + R^2}$ and $\tan\theta = \dfrac{V_L}{V_R} = \dfrac{i\omega L}{iR} = \dfrac{\omega L}{R}$ ...(25.9)

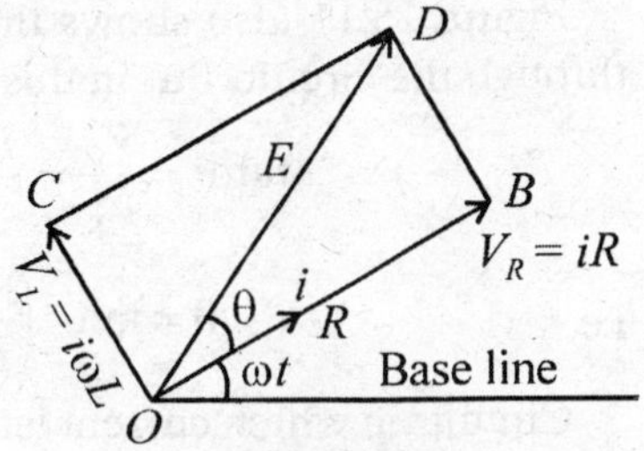

Fig. 25.9

Equation (25.8) shows that here Z plays the part of resistance for the circuit and is called *impedance of the circuit*. Thus

$$Z = \sqrt{\omega^2 L^2 + R^2} \qquad \ldots(25.10)$$

impedance is also the function of the frequency of applied AC.

Equation (25.9) gives that emf (potential difference applied) is leading the current in phase by angle $\theta = \tan^{-1}\left(\dfrac{\omega L}{R}\right)$. Such circuits in which emf leads the current in phase are called *inductive circuits.*

Circuit Containing Capacitor and Resistor (R-C Series Circuit)

Figure 25.10 shows an AC circuit containing capacitor (C) and resistor (R). Since the two are in series hence current through both of them is same.

Let the instantaneous value of the current be $i = i_0\sin\omega t$. The applied emf be $E = E_0\sin(\omega t - \theta)$ where θ is the phase difference between the current and applied emf.

Potential difference across capacitor then is $V_C = iX_C = \dfrac{i}{\omega C}$ and is in phase $\pi/2$ behind the current through it.

Similarly the potential difference across resistor R is $V_R = iR$ and is in phase with the current. The vector resultant of V_C and V_R must give the applied voltage E, the vector diagram is shown in Fig. 25.11.

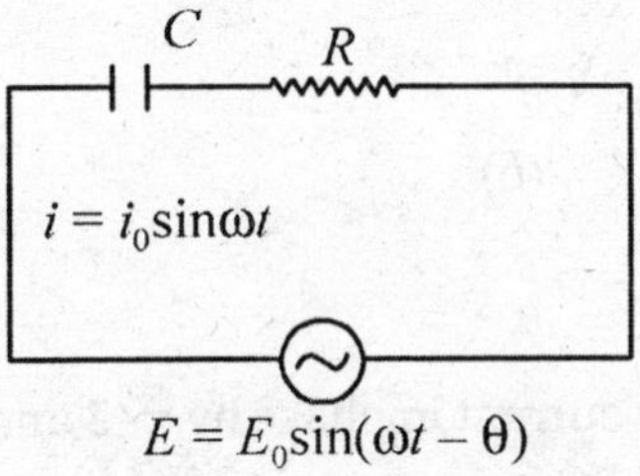

Fig. 25.10

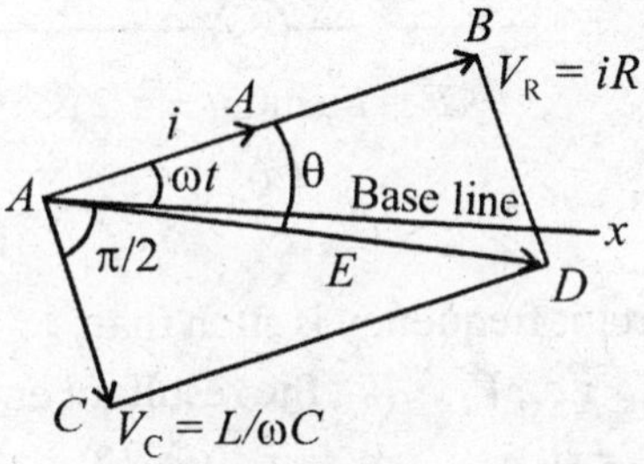

Fig. 25.11

From vector diagram, we have

$$E^2 = V_C^2 + V_R^2 = \frac{i^2}{\omega^2 C^2} + i^2 R^2 = i^2\left(\frac{1}{\omega^2 C^2} + R^2\right)$$

or $$E = i\sqrt{\frac{1}{\omega^2 C^2} + R^2} = iZ \qquad \ldots(25.10)$$

where $$Z = \sqrt{\frac{1}{\omega^2 C^2} + R^2} \qquad \ldots(25.11)$$

Equation (25.10) shows that here Z plays the part of resistance for the circuit and is called *impedance of the circuit*. Here also the impedance is function of frequency of applied AC.

Figure 25.11 also shows that there is phase difference between the applied emf and current through the circuit. But in this case the emf lags behind the current by an angle given by

$$\tan\theta = \frac{V_C}{V_R} = \frac{i/\omega C}{iR} = \frac{1}{\omega CR} \quad \text{...(25.12)}$$

i.e. $$\theta = \tan^{-1}\left(\frac{1}{\omega CR}\right) \quad \text{...(25.12 a)}$$

Circuits in which current leads the emf are called capacitive circuits.

Circuit Containing Inductance and Capacitance (*L-C* Circuit)

Figure 25.12 (*a*) shows AC circuit containing inductance L and capacitance C. Since the two are in series hence current through them is same and let its instantaneous value be $i = i_0 \sin\omega t$.

Potential difference across inductance $V_L = iX_L$

or $V_L = i\omega L$ and is $\pi/2$ ahead in phase with current.

Similarly potential difference across capacitor $V_C = iX_C$

$$V_C = \frac{i}{\omega C}$$

and is lagging in phase by $\pi/2$ with current. The vector diagram is shown in Fig. 25.12 (*b*). Resultant of V_C and V_L should give us the applied voltage.

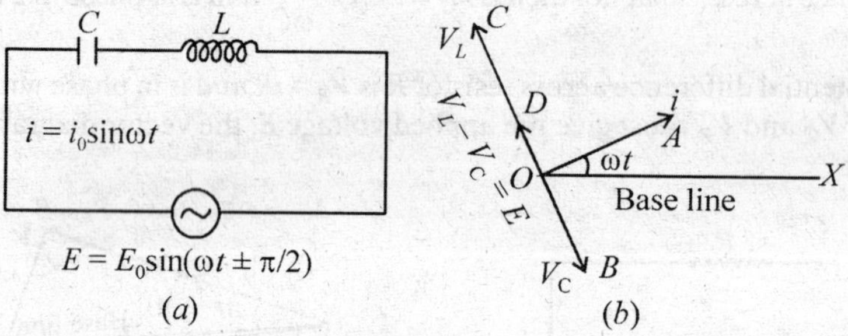

Fig. 25.12

If in the circuit frequency is such that

(*i*) $X_L > X_C$ i.e., $V_L > V_C$; the resultant emf E, leads the current in phase by $\pi/2$ and we say that the circuit behaves as inductive circuit. This is the situation when $\omega > \omega_0$ where $\omega_0 = \frac{1}{\sqrt{LC}}$.
The instantaneous value of emf is given by $E = E_0 \sin(\omega t + \pi/2)$ and impedance is given by $Z = \omega L - \frac{1}{\omega C}$.

(*ii*) $X_L < X_C$ i.e., $V_L < V_C$; the resultant emf E lags behind the current in phase by $\pi/2$ and we say that the circuit behaves as capacitive circuit. This is the situation when $\omega < \omega_0$. The instantaneous value of emf is given by $E = E_0 \sin(\omega t - \pi/2)$ and impedance is given by $Z = \frac{1}{\omega C} - \omega L$.

(*iii*) $X_L = X_C$ i.e., $V_L = V_C$ then $Z = 0$ and the current in the circuit is infinity. This happens for frequency ω_0 so that

$$\omega_0 L = \frac{1}{\omega_0 C} \quad \text{or} \quad \omega_0 = \frac{1}{\sqrt{LC}} \quad \text{...(25.12 b)}$$

or frequency $f_0 = \frac{\omega_0}{2\pi} = \frac{1}{2\pi}\sqrt{\frac{1}{LC}}$

This situation is called *electrical resonance.* In practice the current is prevented from becoming infinite due to presence of some resistance in circuit. The variation of current with frequency and impedance with frequency is shown in Fig. 25.13. The frequency f_0 (or ω_0) is called *resonance frequency*.

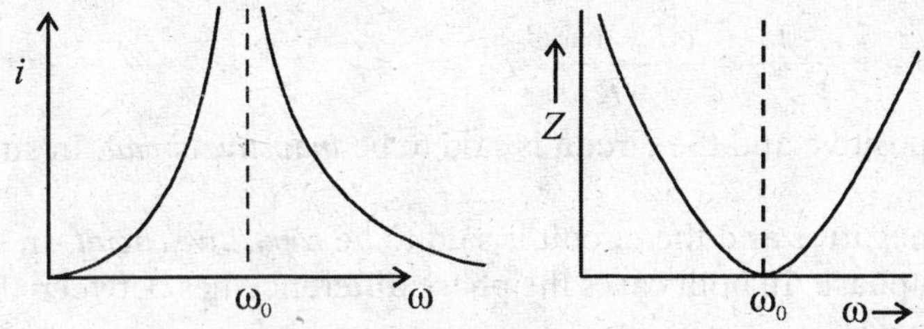

Fig. 25.13

25.3 L-C-R SERIES CIRCUIT AND RESONANCE

Figure 25.14 (*a*) shows *L-C-R* AC series circuit with applied AC voltage E (instantaneous value). Since all the three elements are in series hence the current through all of them is same. Let the instantaneous value of current be $i = i_0 \sin\omega t$.

(*a*) Potential difference across inductance $V_L = iX_L = i\omega L$ and is in phase $\pi/2$ ahead of current.

(*b*) Potential difference across capacitance $V_C = iX_C = \frac{i}{\omega C}$ and is in phase $\pi/2$ lagging behind current.

(*c*) Potential difference across resistance $V_R = iR$ and is in phase with current.

The vector sum of all the three is equal to the applied voltage E. The vector diagram is shown in Fig. 25.14 (*b*).

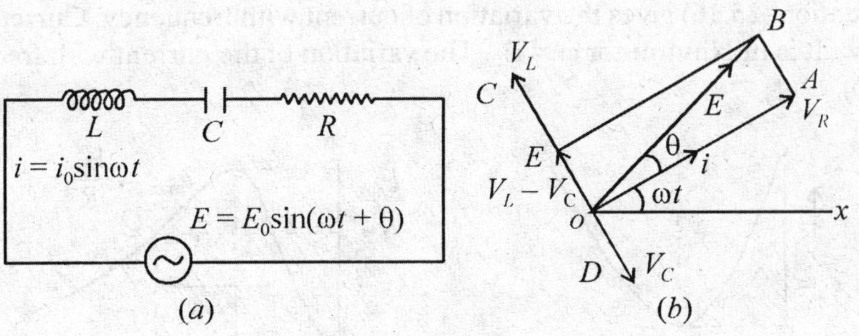

Fig. 25. 14

Since V_L and V_C are opposite in phase as such their resultant is $V_L - V_C$ (= OE). Here we have taken $V_L > V_C$. Thus

$$E^2 = (V_L - V_C)^2 + V_R^2$$

or $$E^2 = (i\omega L - i/\omega C)^2 + i^2R^2$$

$$= i^2\{(\omega L - 1/\omega C)^2 + R^2\}$$

or $$E = i\sqrt{(\omega L - 1/\omega C)^2 + R^2} = iZ \qquad ...(25.13)$$

where $Z = \sqrt{(\omega L - 1/\omega C)^2 + R^2}$ and is called the *impedance* of the circuit. Impedance gives the opposition to the flow of current, i.e., plays the part of resistance and

$$\tan\theta = \frac{V_L - V_C}{V_R} = \frac{\omega L = 1/\omega C}{R} \qquad ...(25.14)$$

When $V_L > V_C$, θ is positive and the circuit is said to be *inductive circuit.* In such circuits emf leads the current in phase.

When $V_L < V_C$, θ is negative and the circuit is said to be *capacitive circuit.* In such circuit emf lags behind the current in phase. In both cases the phase difference lies between 0 and $\pi/2$.

Equation (25.13) gives

$$i = \frac{E}{Z} \quad \text{where} \quad Z = \sqrt{(\omega L - 1/\omega C)^2 + R^2} \qquad ...(25.15)$$

or Peak value of current $$i_0 = \frac{E_0}{\sqrt{(\omega L - 1/\omega C)^2 + R^2}} \qquad ...(25.16)$$

Equation (25.15) gives that impedance is also function of frequency. When frequency $\omega \to 0$, $Z \to \infty$ and when $\omega \to \infty$, $Z \to \infty$. The impedance becomes minimum when $\omega L - \frac{1}{\omega C} = 0$, i.e. impedance is minimum for frequency

$$\omega_0 = \frac{1}{\sqrt{LC}} \quad \text{or} \quad f = \frac{1}{2\pi}\frac{1}{\sqrt{LC}} \qquad ...(25.17)$$

The minimum value of impedance $Z = R$ (the resistance). The variation of impedance with frequency is shown in Fig. 25.15 (*a*).

Similarly equation (25.16) gives the variation of current with frequency. Current $i_0 \to 0$ for both $\omega \to 0$ or $\omega \to \infty$. It is maximum for $\omega = \omega_0$. The variation of the current with frequency is shown in Fig. 25.15 (*b*).

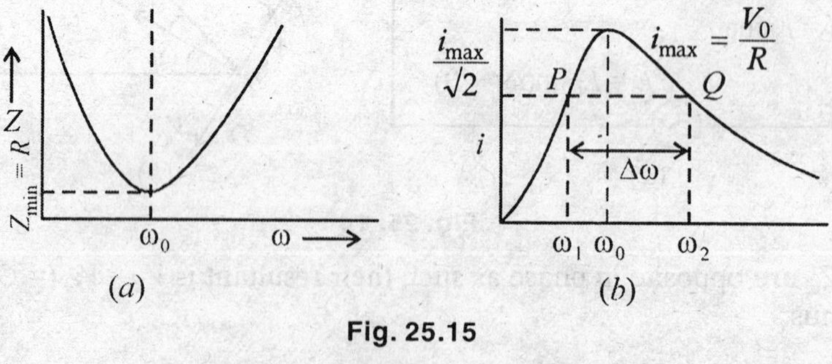

Fig. 25.15

As stated earlier $\omega_0 = \frac{1}{\sqrt{LC}}\left(\text{or } f_0 = \frac{1}{2\pi\sqrt{LC}}\right)$ is called *resonance frequency*. At this particular frequency the *L-C-R* series circuit is called *resonance circuit.*

At resonance frequency

(*i*) $X_L = X_C$, i.e., inductive reactance is equal to the capacitive reactance.

(*ii*) $X = X_L - X_C = 0$, i.e. reactance of the circuit is zero.

(*iii*) $Z = R$, i.e. impedance of the circuit is equal to resistance and is minimum at this frequency.

(*iv*) $i_0 = \frac{V_0}{Z} = \frac{V_0}{R}$ = maximum , i.e. current in the circuit is maximum.

(*v*) All the applied voltage comes across resistance and net potential difference across L and C together is zero.

(*vi*) For frequency $\omega < \omega_0$, θ is negative, i.e. current leads the emf in phase and circuit is called capacitive circuit.

(*vii*) For frequency $\omega > \omega_0$, θ is positive, i.e. current lags behind the applied emf in phase and circuit is said to be inductive circuit.

(*viii*) The power factor of the circuit $\cos\theta = \frac{R}{Z} = +1$ (maximum) as such power consumed is maximum

$$P = V_{rms} I_{rms} \cos\theta = \frac{1}{2} V_0 I_0$$

or
$$P = \frac{1}{2} R I_0^2 = \frac{1}{2}\frac{V_0^2}{R} \qquad ...(25.18)$$

(*ix*) Frequencies corresponding to which the current in the circuit is $\frac{i_{max}}{\sqrt{2}}$ are called *half power points*. In Fig. 25.15 (*b*) P and Q are half power points. The difference of these frequencies $\Delta\omega = \omega_2 - \omega_1$ is called *width of the resonance*.

(*x*) The series resonant circuit is called '*acceptor circuit*' because at resonance its impedance is minimum and it most readily accepts signal out of many signals whose frequency is equal to its natural frequency. In radio or TV tuning we receive the desired station by making the frequency of the circuit equal to that of desired station.

(*xi*) At resonance $\omega_0 = \frac{1}{\sqrt{LC}}$ and $i_0 = \frac{E_0}{R}$ and

$$V_L = V_C = i_0\omega_0 L = \frac{\omega_0 L}{R} E_0 = \frac{i_0}{\omega_0 C} = \frac{E_0}{R\omega_0 C} = QE_0 \qquad ...(25.19)$$

with
$$Q = \frac{\omega_0 L}{R} = \frac{1}{R}\sqrt{\frac{L}{C}} \qquad ...(25.20)$$

Q is called quality factor (or Q factor) of the circuit: Thus at resonance the potential difference across inductance (L) or capacitance (C) is Q times the applied voltage. Q is a dimensionless number and gives the '*voltage magnification*.'

Note: Q-factor also gives *sharpness of the resonance*. Sharpness is defined *as the change in i_{max} at resonance with deviation $\delta\omega$ in frequency from resonance frequency*. For a small change in frequency ($\delta\omega$) at resonance if change in i_{max} (δi_{max}) is large, the resonance is called sharp. Sharpness and flatness of resonance is relative. For same value of L and C, smaller the value of R sharper is the resonance. The Q is also given by

$$Q = \frac{\text{Frequency at resovance}(\omega_0)}{\text{Width of resovance }(\Delta\omega)} = \frac{\omega_0}{\omega_2 - \omega_1}$$

$$= \frac{f_0}{f_2 - f_1} \qquad \text{...(25.21)}$$

25.4 POWER IN AC CIRCUIT AND POWER FACTOR

When an AC voltage $E = E_0 \sin\omega t$ is applied to a circuit and the current in the circuit is $i = i_0 \sin(\omega t - \theta)$, then instantaneous power consumed by the circuit P_i is

$$P_i = iE = i_0 E_0 \sin\omega t \sin(\omega t - \theta)$$

$$= \frac{i_0 E_0}{2}[\cos\theta - \cos(2\omega t - \theta)] \qquad \text{...(25.22)}$$

It is clear that the power consumed by the circuit is function of time. Hence the average power consumed is

$$P_{av} = \frac{\int_0^T P_i dt}{\int_0^T dt} = \frac{1}{T}\int_0^T P_i dt$$

or $$P_{av} = \frac{1}{T}\int_0^T \frac{i_0 E_0}{2}[\cos\theta - \cos(2\omega t - \theta)]dt$$

$$= \frac{i_0 E_0}{2T}\cos\theta \int_0^T dt - \frac{i_0 E_0}{2T}\int_0^T \cos(2\omega t - \theta)\,dt$$

$$= \frac{i_0 E_0 \cos\theta}{2} \qquad \because \text{ Second integral is zero.}$$

Thus average power consumed is

$$P_{av} = \frac{i_0}{\sqrt{2}} \cdot \frac{E_0}{\sqrt{2}} \cos\theta = i_{rms} E_{rms} \cos\theta \qquad \text{...(25.23)}$$

Here $\cos\theta$ is called *power factor* of the circuit.

In *L-C-R* series circuit

$$\tan\theta = \frac{\omega L - 1/\omega C}{R}$$

and from Fig. 25.16, we have

$$\cos\theta = \frac{R}{Z} \qquad \text{...(25.24)}$$

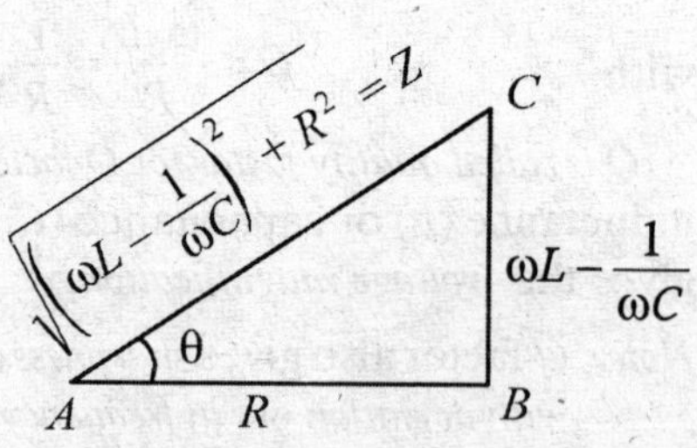

Fig. 25.16

Hence $$P_{av} = \frac{i_0}{\sqrt{2}} \frac{E_0}{\sqrt{2}} \frac{R}{Z} \qquad \text{but } E_0 = i_0 Z$$

$$= \frac{i_0^2 R}{2} = (i_{rms})^2 R$$

$$= \frac{E_0^2 R}{2Z^2} = (E_{rms})^2 R/Z^2 \qquad \text{...(25.25)}$$

Triangle shown in Fig. 25.16 is called *impedance triangle.*

From equation (25.24), i.e. $\cos\theta = \frac{R}{Z} = \frac{R}{\sqrt{(\omega L - 1/\omega C)^2 + R^2}}$, we infer

(*i*) For pure resistive circuit $X = \omega L - 1/\omega C = 0$. Hence $\cos\theta = +1$ and the power consumed is maximum, $P_{av} = \frac{1}{2} I_0 E_0$.

(*ii*) In case of pure inductive (or capacitive) circuit, $R = 0$

and $P_{av} = 0 \quad \because \cos\theta = 0$

i.e., average power consumed in such circuits is zero. Such currents are called *wattless currents.*

(*iii*) In actual practice, an inductor or capacitor has large reactance and a finite small resistance as such phase difference between E and I is large but less than 90° and so the power consumed by inductor or capacitor is not zero but very small.

(*iv*) Figure 25.17 gives the vector diagram for E and I with phase difference θ. If instantaneous value of current is I and phase difference θ, then component of the wattless current

Wattless current = $I\sin\theta$ ...(25.26)

and its peak value is = $I_0\sin\theta$ ...(25.26*a*)

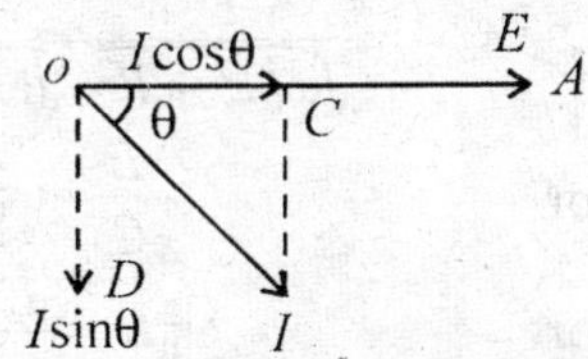

Fig. 25.17

Choke Coil

In direct current circuits the current through circuit is controlled using rheostat (resistance). This causes the loss of energy i^2R per second in the form of heat.

The current in an AC circuit can be reduced by means of devices which involves very small amount of energy loss. This is called *choke coil.* Coil consists of copper wire wound over a core. For low frequency AC coil is wound over a soft iron laminated core and for high frequency AC it is air core wound coil. The coil is put in series with the circuit in which the current is to be reduced.

This circuit is L-R circuit and the current in the circuit is

$$i = \frac{E}{Z} \quad \text{where } Z = \sqrt{(R+r)^2 + (\omega L)^2}$$

Because of the large inductance (L) of the coil, the current is reduced considerably. However small resistance of the coil (r) the powerloss is very small in the coil

$$P = V_{rms} i_{rms} \cos\theta \rightarrow 0$$

$$\cos\theta = \frac{r}{Z} \rightarrow 0$$

25.5 PARALLEL RESONANCE CIRCUIT (REJECTOR CIRCUIT)

An alternating voltage $E = E_0 \sin\omega t$ is applied across an ideal capacitor (C) and inductor of inductance L and resistance R connected in parallel (Fig. 25.18).

Impedance of the coil $= \sqrt{\omega^2 L^2 + R^2}$ and the current through it

$$i_L = \frac{E}{\sqrt{\omega^2 L^2 + R^2}}$$

and lags behind in phase by θ such that $\tan\theta = \frac{\omega L}{R}$, to the applied voltage.

The current through capacitor is

$$i_C = \frac{E}{X_C} = \omega CE$$

and is 90° ahead in phase with the applied voltage.

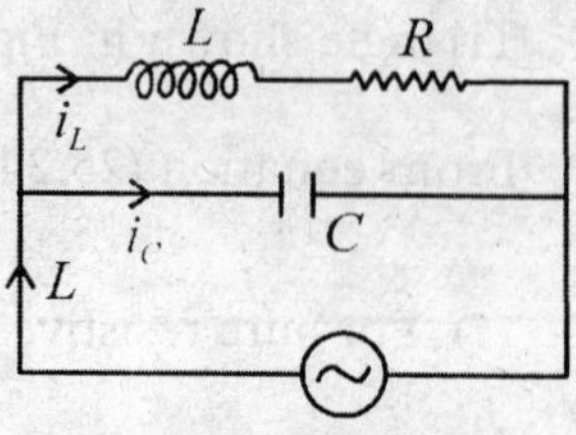

$E = E_0 \sin\omega t$

Fig. 25.18

The circuit is said to be parallel resonance circuit if the applied voltage and the current i in the circuit is in same phase.

Thus we have for resonance, from Fig. 25.18 (*a*)

$$i_L \sin\theta = i_C \qquad \text{But } \sin\theta = \frac{\omega_0 L}{\sqrt{\omega_0^2 L^2 + R^2}}$$

Here ω_0 is the resonance frequency

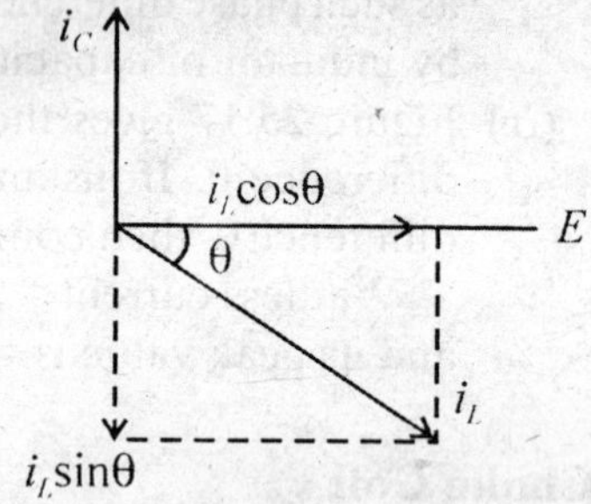

Fig. 25.18 (*a*)

$$\therefore \quad \frac{E}{\sqrt{\omega_0^2 L^2 + R^2}} \frac{\omega_0 L}{\sqrt{\omega_0^2 L^2 + R^2}} = \omega_0 CE$$

or $$\frac{L}{C} = \omega_0^2 L^2 + R^2$$

or $$\omega_0^2 L^2 = \frac{L}{C} - R^2 \quad \text{or} \quad \omega_0^2 L^2 = \frac{1}{LC} - \frac{R^2}{L^2}$$

and $$\omega_0 = \sqrt{\frac{1}{LC} - \frac{R^2}{L^2}} \qquad \text{...(25.27)}$$

ω_0 is called the resonance frequency for parallel circuit.

Equation (25.27) gives that $R < \sqrt{L/C}$ *for parallel resonance.*

If R is very small, i.e. $\frac{R^2}{L^2} << \frac{1}{LC}$ the resonance frequency is

$$\omega_0 = \frac{1}{\sqrt{LC}} \qquad \text{...(25.28)}$$

$$f_0 = \frac{1}{2\pi\sqrt{LC}} \qquad \text{...(25.28 a)}$$

At resonance, the net current in the circuit is $i = i_L \cos\theta$, i.e.

$$i = \frac{E}{\sqrt{R^2 + \omega_0^2 L^2}} \frac{R}{\sqrt{R^2 + \omega_0^2 L^2}} = \frac{ER}{R^2 + \omega_0^2 L^2}$$

$$i = \frac{ER}{R^2 + L^2\left(\frac{1}{LC} - \frac{R^2}{L^2}\right)} = \frac{ER}{L/C} = \frac{ERC}{L} \qquad \text{...(25.29)}$$

Equation (25.29) gives that $i \propto R$. Since R is very small hence current i in the circuit is very small. If $R = 0$, $i = 0$, impedance in circuit is

$$Z = \frac{E}{i} = \frac{L}{RC} \qquad \text{...(25.30)}$$

and when R is small (negligible) then Z impedance becomes infinite.

Thus at the resonance frequency ω_0 the current in each branch i_L and i_C is large in comparison to the main current. Thus in this case we have current amplification.

In this circuit the current amplification

$$= \frac{i_L}{i} = \frac{i_C}{i} = \frac{\omega_0 L}{R} = \frac{1}{\omega_0 CR} = Q$$

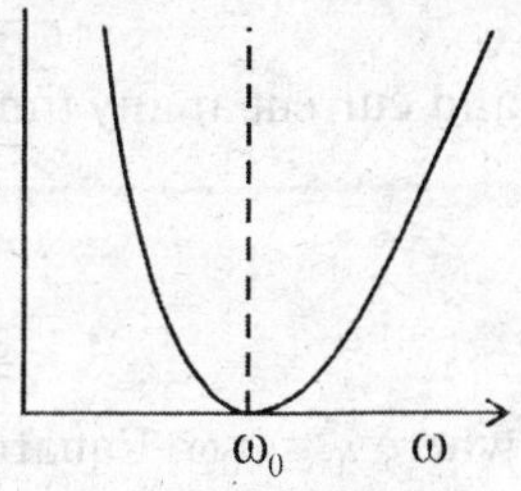

Fig. 25.19

This is the same as voltage amplification in series resonance circuit?

For $R = 0$, the current in circuit (i) and frequency of the source (ω_0) graph is shown in Fig. 25.19, which shows $i = 0$ for $\omega = \omega_0$.

This circuit does not allow the AC of particular frequency (ω_0) to pass through it. Hence this is also called filter circuit. This also is called reflector circuit.

25.6 ELECTROMAGNETIC OSCILLATIONS IN L-C CIRCUIT

We have studied that capacitor and inductor can store electrical energy ($=1/2CV^2$) and magnetic energy ($=1/2\ Li^2$) respectively. The capacitor stores energy in the electric field between the plates and inductor stores energy in magnetic field in its core. When a charged capacitor is discharged through an inductor of zero resistance, the charge on the capacitor and the current through the inductor exhibit the phenomenon of electrical oscillations similar to oscillation in mechanical system (like simple pendulum).

Let a capacitor C, an inductance L of negligible resistance and a battery be connected as shown in Fig. 25.20. When switch (1-2) is put on, the capacitor is charged. Let charge on capacitor when it is fully charged be q_0. The electrical energy, $U_e = \frac{1}{2}\frac{q_0^2}{C}$ is stored between the plates of capacitor and the magnetic energy in the inductor $U_B = 0$, because no current passes through the inductor.

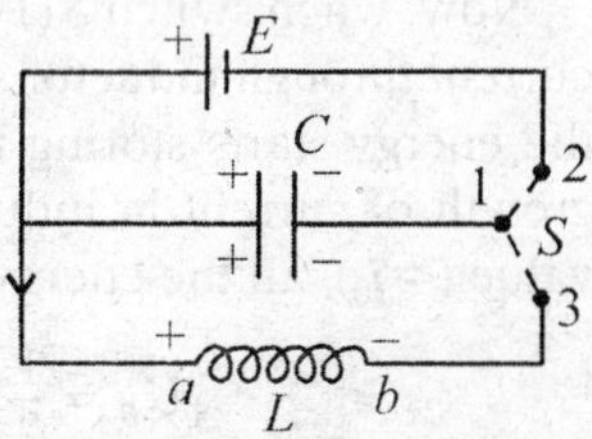

Fig. 25.20

Now, the capacitor is connected to inductor by pulling on switch (1-3) after disconnecting switch (1-2). Now the charge on the capacitor starts decreasing giving rise to the current in the circuit though the inductor. Let q and i be the charge and current in the circuit at any instant t then from Kirchhoff's loop law, we have

$$\frac{q}{C} + L\frac{di}{dt} = 0 \qquad \ldots(i)$$

But $i = \frac{dq}{dt}$, therefore

$$\frac{d^2q}{dt^2} + \frac{q}{LC} = 0 \qquad \ldots(ii)$$

The equation is similar to that for simple harmonic oscillator $\frac{d^2x}{dt^2} + \omega_0^2 x = 0$. The charge thus oscillates with natural frequency

$$\omega_0 = \frac{1}{\sqrt{LC}} \qquad \ldots(25.31)$$

and varies sinusoidally with time as

$$q = q_0 \sin(\omega t + \theta) \qquad \ldots(25.32)$$

where q_0 is the maximum charge on the capacitor and θ is initial phase. Using the condition that when $t = 0$, $q = q_0$, $\theta = \pi/2$. Hence

$$q = q_0 \sin(\omega t + \pi/2) = q_0 \cos\omega t \quad \text{...(25.33)}$$

and current at any time t is

$$i = \frac{dq}{dt} = -q_0\omega \sin \omega t$$

$$= i_0 \sin(\omega t - \pi/2) \quad \text{...(25.34)}$$

where $i_0 = q_0\omega$. Equations (25.33) and (25.34) give us that the current and charge are not in same phase. The current lags behind the change in phase by $\pi/2$.

Let us now visualize how the oscillations take place in the circuit. Capacitor with initial charge q_0 has electrical energy $U_e = \frac{1}{2}\frac{q_0^2}{C}$ stored in it. This is now connected to an ideal inductor. Since there is no current in the circuit, hence energy in the inductor is zero. Thus total energy in the L-C circuit is

$$U = U_e + U_B = \frac{1}{2}\frac{q_0^2}{C} \quad \text{...(25.35)}$$

Now, when switch $S(1\text{-}3)$ is closed the capacitor starts discharging through inductor as such the current through inductor starts increasing and sets up a magnetic field in the inductor and thereby the energy starts storing in the inductor. An induced emf developed in the circuit opposed the growth of current in inductor (L) and hence delays it. When the current reaches its maximum value ($= i_0$), all the energy is stored in the magnetic field: This energy is

$$U_B = \frac{1}{2} L i_0^2$$

At this stage capacitor has no charge and hence no energy. Thus

$$U = U_B + U_e = \frac{1}{2} L i_0^2$$

or

$$\frac{1}{2} L i_0^2 = \frac{1}{2}\frac{q_0^2}{C} \quad \text{...(25.36)}$$

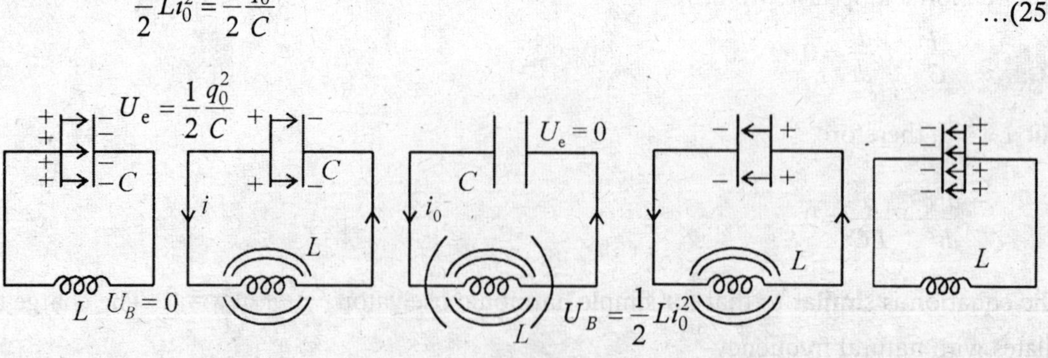

Fig. 25.21

As soon as the discharge of capacitor is complete, current stops and magnetic field linked with inductor (L) starts decreasing. This produces the induced current in opposite direction which starts charging the capacitor in opposite direction. The recharging is also opposed and delayed. When the capacitor is recharged completely, the magnetic field energy in inductor (L) reappears in the form of electric field energy between the plates of capacitor.

The whole process just described will now repeat itself till the system reverts to its original state. Thus the energy in the system oscillates between the capacitor and inductor. All this is shown in Fig. 25.21.

The L-C oscillations are similar to the mechanical oscillations of a simple pendulum or block attached to an ideal spring. In these oscillations, the kinetic energy and potential of the bob or block oscillates.

If the circuit has no resistance, no loss of energy would occur. The oscillations produced will be of constant amplitude. These are called *undamped oscillations*. In actual practice, however there is some loss of energy. Hence the amplitude of oscillations goes on decreasing. These are called *damped oscillations*.

25.7 TRANSFORMER

A transformer is an electrical device which transforms an alternating voltage from one to another of greater or smaller value. This is done using the principle of mutual induction. There are two types of transformers:

(*a*) **Step up Transformer:** *It transforms low voltage high current into high voltage low current.*

(*b*) **Step Down Transformer:** *It transforms high voltage low current into low voltage high current.*

Transformer consists of two coils wound on the same core. Core in general is rectangular in shape and made up of insulated strips of soft iron or silicon steel. The coil which is connected to the source is called primary while other through which output is taken is called secondary. The alternating current passing through the primary produces a continuously changing flux through the core. This changing flux linked with secondary induces an alternating emf in the secondary. The value of this induced emf depends on the number of turns in the secondary. We will consider an ideal transformer in which the primary has negligible resistance and all the flux links both primary and secondary windings. As magnetic lines of force are closed curves, the flux per turn of primary must be equal to the flux per turn of the secondary.

When a voltage V_p is applied to the primary, let ϕ be the flux in each turn. Then induced emf E_S in the secondary with turns N_S is

$$E_S = -N_S \frac{d\phi}{dt}$$

The alternating flux ϕ also induces an emf called back emf in the primary. This is

$$E_p = -N_p \frac{d\phi}{dt}$$

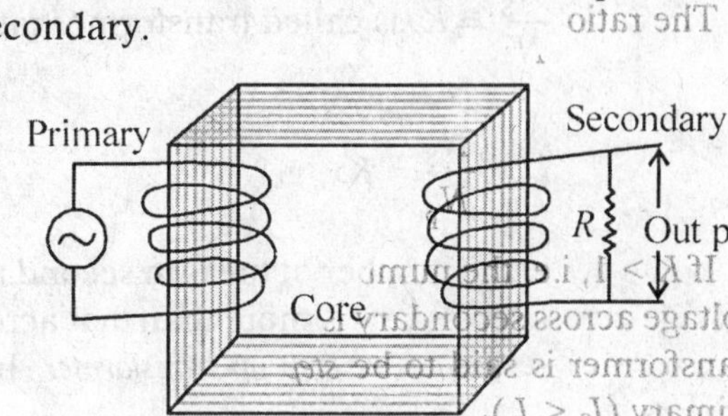

Fig. 25.22

Since we have taken primary resistance to be zero, hence $E_p = V_p$, otherwise the primary current would be infinite. If the secondary is an open circuit or current taken from it is small, then we can take $E_S = V_S$, where V_S is the voltage across the secondary. Thus we have

$$\frac{V_S}{V_p} = \frac{E_S}{E_p} = \frac{N_S}{N_p} \qquad ...(25.37)$$

The equation (25.37) is obtained under the following three assumptions

(*i*) The primary resistance and current are small.
(*ii*) The same flux links both the primary and secondary and very little flux escapes from the core.
(*iii*) The secondary current is very small.

If the transformer is to be assumed ideal, i.e. 100% efficient (no loss of energy), then power input must be equal to the power output. If I_p and I_S are currents in primary and secondary respectively, then

$$I_S V_S = I_p V_p$$

or
$$\frac{V_S}{V_p} = \frac{I_p}{I_s} = \frac{N_S}{N_p} \qquad \ldots(25.37)$$

If Z_p and Z_S are the impedances of primary and secondary respectively, then

$$V_s = I_S Z_S \quad \text{and} \quad V_p = I_p Z_p$$

or
$$\frac{Z_S}{Z_p} = \frac{V_S I_p}{V_p I_S} = \frac{N_S^2}{N_p^2}$$

or
$$\frac{N_S}{N_p} = \sqrt{\frac{Z_S}{Z_p}}$$

Thus equation (25.38) can be written as

$$\frac{V_S}{V_p} = \frac{I_P}{I_S} = \sqrt{\frac{Z_S}{Z_p}} = \frac{N_S}{N_p} \qquad \ldots(25.39)$$

The ratio $\frac{N_S}{N_p} = K$ is called transformer ratio, i.e.

$$\frac{N_S}{N_p} = K \qquad \ldots(25.40)$$

If $K > 1$, i.e. the number of turns in secondary coil are more than that in primary ($N_S > N_p$), the voltage across secondary is more than that across primary ($V_S > V_p$), i.e. voltage is stepped up. The transformer is said to be *step up transformer*. In this the current in secondary is less than that in primary ($I_S < I_p$).

If $K < 1$, i.e. the number of turns in secondary coil are less than that in primary ($N_S < N_p$), the voltage across secondary is less than that across primary ($V_S < V_p$), i.e., voltage is stepped down. The transformer is said to be *step down transformer*. In this the current in secondary is more than that in primary ($I_S > I_p$).

All the above equations are derived assuming ideal transformer (without any energy losses). But in actual transformer some energy losses do occur. Following are the major sources of energy loss in transformer.

(*a*) **Copper Loss:** Energy is lost in the form of heat in copper coils of transformer. This is due to Joule heating produced by the current through the copper windings.

(*b*) **Iron Loss:** This is the energy loss in the form of heat in the iron core of transformer. This is due to eddy currents induced in the iron core. As such transformer cores are made up of laminated strips to minimise this loss.

(*c*) **Magnetic Flux Leakage:** All the flux due to primary does not pass through the secondary due to poor design of the core or the air gaps in the core. This causes leakage of some flux. It can be reduced by winding the primary and secondary coils one over the other.

(*d*) **Hysteresis Loss:** This is the loss of energy due to repeated magnetisation and demagnetisation of iron core by the alternating magnetic field. The resulting energy loss in the core appears as heat and is kept to minimum by using a magnetic material which has a low hysteresis loss.

(*e*) **Magnetostriction:** It is found that when an specimen is magnetised its length changes. The phenomenon is called *magnetostriction.* The alternating magnetic field causes changes in the length of the core which produces humming noise and some energy given to transformer is lost in the form of sound energy.

The transformers are used in the large scale transmission and distribution of electrical energy over long distances. Using step up transformer the voltage output of the generator is stepped up thereby the current is reduced which reduces heat losses (i^2R) through the transmission lines. This voltage is then stepped down at the substation near the consumers.

In India for the domestic consumption the voltage is 220 V and frequency is 50 Hz.

SOLVED EXAMPLES

Example 25.1: A series AC circuit contains an inductance (= 20 mH) a capacitor (= 100 μF), a resistor (= 50 Ω) and AC source of 12 V, 50 Hz. Find the energy dissipated in the circuit in 1000 s.

Solution: The impedance of the circuit is

$$Z = \sqrt{R^2 + (\omega L - 1/\omega C)^2}$$

We have $\omega L = 2\pi f L = 2\pi \times 50 \times 20 \times 10^{-3} = 2\pi\ \Omega$

and $$\frac{1}{\omega C} = \frac{1}{2\pi \times 50 \times 100 \times 10^{-6}} = \frac{100}{\pi}\ \Omega$$

and $$R = 20\ \Omega$$

$\therefore$ $$Z = \sqrt{(50)^2 + \left(2\pi - \frac{100}{\pi}\right)^2} = \sqrt{3150}\ \Omega$$

Average power dissipated per second

$$P_{av} = i_{rms} V_{rms} \cos\theta \quad \text{where } \theta \text{ is phase difference}$$

But $$\tan\theta = \frac{\omega L - 1/\omega C}{R} \quad \text{or} \quad \cos\theta = \frac{R}{Z}$$

$$\therefore \quad P_{av} = i_{rms} V_{rms} \frac{R}{Z} \quad \text{but} \quad i_{rms} = \frac{V_{rms}}{Z}$$

$$\therefore \quad P_{av} = V_{rms}^2 \frac{R}{Z^2}$$

Hence energy dissipated in 1000 sec is

$$E = V_{rms}^2 \frac{R}{Z^2} \times 100 = \frac{(12)^2 \times 50}{3150} \times 1000$$

$$= 2.3 \times 10^3 \text{ J.}$$

Example 25.2: Find (*a*) the average and (*b*) rms value for the saw-tooth voltage of peak value V_0 as shown in Fig. 25.23.

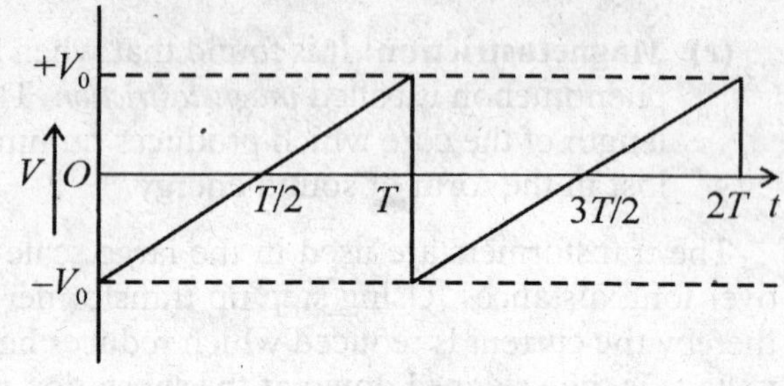

Fig. 25.23

Solution: Figure 25.23 gives the slope of the curve as slope $= \dfrac{2T}{V_0}$ and at $t = 0$, $V = -V_0$. Hence the equation of the saw tooth wave can be written as

$$V = \frac{2V_0}{T} t - V_0$$

(*a*) Average value of voltage for half cycle is

$$V_{av} = \frac{\int_0^{T/2} V dt}{\int_0^{T/2} dt} = \frac{2}{T} \int_0^{T/2} \left(\frac{2V_0}{T} t - V_0 \right) dt$$

$$= \frac{2V_0}{T} \left[\frac{t^2}{T} - t \right]_0^{T/2} = \frac{2V_0}{T} \left[\frac{T^2}{4T} - \frac{T}{2} \right]$$

$$= -\frac{V_0}{2}$$

for positive half cycle for $t = T/2$ to $t = T$. $V_{av} = +V_0/2$.

Hence average value for half cycle

$$V_{av} = \left| \frac{V_0}{2} \right|$$

and for full cycle $V_{av} = 0$

$$(b) \quad \langle V^2 \rangle = \frac{\int_0^T V^2 dt}{\int_0^T dt} = \frac{1}{T} \int_0^T \left(\frac{2V_0 t}{T} - V_0 \right)^2 dt$$

$$= \frac{V_0^2}{T}\int_0^T \left(\frac{4t^2}{T^2} - \frac{4t}{T} + 1\right) dt$$

$$= \frac{V_0^2}{T}\left[\frac{4}{3}\frac{t^3}{T^2} - \frac{2t^2}{T} + t\right]_0^T$$

$$= \frac{V_0^2}{T}\left[\frac{4}{3}T - 2T + T\right] = \frac{V_0^2}{3}$$

$$\therefore \quad V_{\text{rms}} = \sqrt{\langle V^2 \rangle} = \frac{V_0}{\sqrt{3}}$$

Example 25.3: Find the effective value of the current obtained due to superposition of a direct current of value *a* and an alternating current of instantaneous value $i = b\sin\omega t$.

Solution: Due to superposition, the instantaneous value of the current is

$$I = a + i = a + b\sin\omega t$$

Hence average value of I^2 for one cycle is

$$\langle I^2 \rangle = \frac{\int_0^T I^2 dt}{\int_0^T dt} = \frac{1}{T}\int_0^T (a + b\sin\omega t)^2 dt$$

$$= \frac{1}{T}\int_0^T \left(a^2 + 2ab\sin\omega t + b^2\sin^2\omega t\right) dt$$

$$= \frac{1}{T}\left[a^2 T + \frac{b^2}{2}T\right]$$

$$\because \quad \int_0^T \sin\omega t\, dt = 0 \text{ and } \int_0^T \sin^2\omega t\, dt \int_0^T \frac{(1-\cos^2\omega t)}{2} dt = \frac{T}{2}$$

$$\therefore \quad \langle I^2 \rangle = \left(a^2 + \frac{b^2}{2}\right)$$

or rms value of current $I_{\text{rms}} = \sqrt{a^2 + \dfrac{b^2}{2}}$

Example 25.4: A small town with a demand of 800 kW of electric power of 220 V is situated 15 km away from an electric plant generating 440 V. The resistance of the two wire line carrying power is 0.5 Ω per km. The town gets power from the line through a 4000-200 V step down transformer at a substation in the town.

(*a*) Estimate the line power loss in the form of heat.

(*b*) How much power must the plant supply, assuming there is no loss in power due to leakage?

(*c*) Characterise the step up transformer at the plant.

Solution: The line resistance = 2 × 15 × 0.5 = 15 Ω

Since the town gets power = 800 × 1000 = 8 × 10^5 Watts and voltage V = 4000 V. Hence the current i in the line is

$$i = \frac{W}{V} = \frac{8 \times 10^5}{4000} = 200 \text{ A}$$

(*a*) Hence power loss in line = $i^2R = (200)^2(15)$ = 600 kW

(*b*) Hence power supplied by the plant = 800 + 600 = 1400 kW

(*c*) Voltage drop on the line = 200 × 15 = 3000 V

Hence voltage at the plant is 4000 + 3000 = 7000 V

Power is generated at voltage 440 V

Hence transformer at the plant is step up 440 – 7000 V.

Example 25.5: A coil of inductance 0.5 H and resistance 100 Ω is connected to a 240 V, 50 Hz, AC supply. (*a*) What is the maximum current in the coil? (*b*) What is the time lag between the voltage maximum and current maximum?

Solution:

(*a*) Impedance of the circuit is

$$Z = \sqrt{R^2 + \omega^2 L^2} = \sqrt{(100)^2 + (2\pi \times 50 \times 0.5)^2}$$

$$= 10^2\sqrt{3.5} = 1.87 \times 10^2 \ \Omega$$

Hence current $i = \dfrac{V}{Z} = \dfrac{240}{1.87 \times 10^2} = 1.28$ A

and Peak value is $i_0 = i\sqrt{2} = 1.28 \times \sqrt{2} = 1.81$ A.

(*b*) Phase difference between the current and voltage is

$$\tan\theta = \frac{\omega L}{R} = \frac{2\pi \times 50 \times 0.5}{100} = \frac{\pi}{2} = 1.57$$

or $$\theta = \tan^{-1}(1.57) = 57.5° = \frac{57.5 \times \pi}{180}$$

and time interval $= \dfrac{\theta}{\omega} = \dfrac{57.5 \times \pi}{180 \times 2\pi \times 50} = 3.2 \times 10^{-3}$ sec = 3.2 ms

Example 25.6: Figure 25.24 shows a series *L-C-R* circuit connected to a variable frequency 230 V source. L = 5.0 H, C = 80 μF and R = 40 Ω.

(*a*) Determine the source frequency which derives current in the circuit at resonance

(*b*) Obtain the impedance of the circuit and the amplitude of current at the resonance frequency

(*c*) Determine the rms potential drop across the three elements of the circuit

(*d*) Show that the potential drop across the *LC* combination is zero at the resonance frequency.

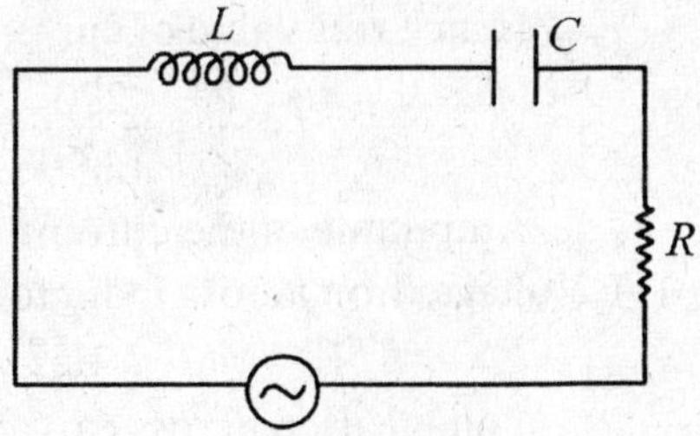

Fig. 25.24

Solution:

(*a*) Resonance angular frequency

$$\omega_0 = \frac{1}{\sqrt{LC}} = \frac{1}{\sqrt{5 \times 80 \times 10^{-6}}}$$

$$= \frac{100}{2} = 50 \text{ rad/s}$$

and the resonance frequency $f = \frac{50}{2\pi} = \frac{25}{\pi}$ Hz

(*b*) Impedance $Z = \sqrt{R^2 + (\omega L - 1/\omega C)^2}$

at resonance frequency $\omega L = 1/\omega C$ and $Z = R = 40\ \Omega$

(*c*) Current at resonance is

$$i = \frac{V}{Z} = \frac{230}{40} = 5.75 \text{ A}$$

and current amplitude $i_0 = i\sqrt{2} = 5.75 \times \sqrt{2} = 8.13$ A

(*d*) Potential difference across L at resonance is

$$V_L = i\omega L = 5.75 \times 50 \times 5$$

$$= 1437.4 \text{ V}$$

This voltage is $\pi/2$ ahead in phase with the current.

The potential difference across C at resonance is

$$V_C = \frac{i}{\omega C} = \frac{5.75}{50 \times 80 \times 10^{-6}}$$

$$= 10^3 \times \frac{5.75}{4} = 1.4371 = 1437.4 \text{ V}$$

This voltage lags behind the current in phase by $\pi/2$.

The total potential drop across L and C combined = 0.

Example 25.7: A circuit consists of a series combination of a 50 mH inductor and a 20 μF capacitor connected to 220 V, 50 Hz supply. The circuit has negligible resistance

(*a*) Find the amplitude and rms value of the current

(*b*) Find the rms value of voltage across inductor and capacitor

(*c*) What is the average power transferred to (*i*) the inductor and (*ii*) to the capacitor?

(*d*) What is the total power absorbed by the circuit?

Solution:

(*a*) Impedance of the circuit is $Z = \sqrt{R^2 + (\omega L - 1/\omega C)^2}$

Since $R = 0$, hence $Z = (\omega L \sim 1/\omega C) = 2\pi \times 50 \times 50 \times 10^{-3} \sim \frac{1}{2\pi \times 50 \times 20 \times 10^{-6}}$

$$= 5\pi \sim \frac{1000}{2\pi} = 143.4\ \Omega$$

Hence rms value of current is

$$i = \frac{V}{Z} = \frac{220}{143.4} = 1.53 \text{ A}$$

Amplitude of the current $i_0 = \sqrt{2}i = \sqrt{2} \times 1.53 = 2.17$ A

(*b*) Voltage drop across inductor is

$$V_L = i\omega L = 1.53 \times 5 \times \pi = 24.0 \text{ V}$$

Voltage drop across capacitor is

$$V_C = i\omega C = 1.53 \times \frac{500}{\pi} = 243.5 \text{ V}$$

(*c*) In inductor the current lags behind the voltage in phase by $\pi/2$, i.e. power factor $\cos\theta = \cos\pi/2 = 0$. Hence average power transferred to inductor is zero. Similarly in capacitor the current leads the voltage in phase by $\pi/2$, i.e. power factor is $\cos\pi/2 = 0$. Hence average power transferred to capacitor is also zero.

(*d*) Total power absorbed

= Power absorbed by inductor + Power absorbed by capacitor

= 0.

Example 25.8: An *L-C-R* series circuit with 100 Ω resistance is connected to an AC source of 200 V and angular frequency 300 rad/s. When only the capacitor is removed, the current lags behind the voltage by 60°. When only the inductance is removed, the current leads the voltage by 60°. Calculate the current and the power dissipated in the *L-C-R* circuit.

Solution: When capacitor is removed, the circuit contains inductance resistance (*L-R*) only. In *L-R* circuit the current lags behind the voltage in phase by θ given by

$$\tan\theta = \frac{\omega L}{R}$$

Since θ = 60° $\quad\because\ \tan 60° = \frac{\omega L}{R} = \sqrt{3}$ ...(*i*)

When inductor is removed, the circuit contains capacitance and resistance (*C-R*) only. In *R-C* circuit, the current leads the voltage in phase by θ such that

$$\tan\theta = \frac{1}{\omega CR}$$

Since θ = 60° $\quad \tan 60° = \frac{1}{\omega CR} = \sqrt{3}$ ...(*ii*)

From (*i*) and (*ii*), we have

$$\frac{\omega L}{R} = \frac{1}{\omega CR} \quad \text{or} \quad \omega = \frac{1}{\sqrt{LC}}$$

This means that *L-C-R* series circuit is resonance circuit. Its impedance $Z = R$.

Hence current in the circuit $= \frac{V}{Z} = \frac{200}{100} = 2$ A

The power dissipated in the circuit is

$$P = i^2R = (2)^2 \times 100 = 400 \text{ Watt.}$$

Example 25.9: An *L-C* circuit contains a 50 mH inductor and 80 μF capacitor. Initially (at $t = 0$) the charge on the capacitor is 20 mC. The resistance of the circuit is negligible. The circuit is closed at $t = 0$

(*a*) What is the natural frequency of *L-C* oscillation of the circuit?
(*b*) What is the total energy stored initially? Is this energy conserved during the *L-C* oscillations?
(*c*) At what values of time *t* is the energy stored (*i*) completely electrical (*ii*) completely magnetic?
(*d*) At what values of *t* is the energy shared equally between the inductor and the capacitor?

Solution:

(*a*) The nature of frequency of *L-C* oscillations is

$$f = \frac{1}{2\pi\sqrt{LC}} = \frac{1}{2\pi\sqrt{50\times10^{-3}\times80\times10^{-6}}}$$

$$= \frac{1000}{4\pi} = 79.6 \text{ Hz}$$

and time period $T = \frac{1}{f} = \frac{1}{79.6} = 1.26\times10^{-2}$ sec

(*b*) Initially the entire energy is stored in the capacitor as there is no current in the inductor. Hence total energy

$$E = \frac{1}{2}\frac{q_0^2}{C} = \frac{1}{2}\frac{(20\times10^{-3})^2}{80\times10^{-6}}$$

$$= \frac{1}{2}\frac{4\times10^{-4}}{8\times10^{-5}} = 2.5 \text{ J}$$

Since there is no resistance in the circuit, this energy is conserved in *L-C* oscillations.

(*c*) The charge on the capacitor at any time *t* is given by

$$q = q_0 \sin(\omega t + \theta) \text{ where } \omega = \frac{1}{\sqrt{LC}}$$

and from initial conditions, i.e. $q = q_0$ at $t = 0$, we have $\theta = \pi/2$.

$$\therefore \quad q = q_0\sin(\omega t + \pi/2) = q_0\cos\omega t$$

(*i*) The energy will be completely electrical, i.e. stored in capacitor when $q = \pm q_0$, i.e. at times when $\cos\omega t = \pm1$.

or $\quad \omega t = 0, \pi, 2\pi, \ldots, n\pi$, where $n = 0, 1, 2, 3, \ldots$

$$\therefore \quad t = \frac{n\pi}{\omega} = \frac{nT}{2}, \text{ where } T = 1.26\times10^{-2} \text{ sec}$$

or $\quad t = 0, \frac{T}{2}, T, \frac{3T}{2}, 2T, \ldots$ etc.

(*ii*) The energy will be entirely magnetic when $q = 0$, i.e. when $\cos\omega t = 0$.

or $\quad \omega t = (2n-1)\pi/2$ where n 1, 2, 3, …, etc.

or $\quad t = \frac{(2n-1)\pi}{2\omega} = \frac{(2n-1)T}{4}$

or $\quad t = \frac{T}{4}, \frac{3T}{4}, \frac{5T}{4}, \ldots$

(*d*) The energy will be equally shared between *L* and *C* at values of time (*t*). When the electrical energy stored in capacitor becomes half the total energy. Let the charge at this instant be *q* then

$$\frac{1}{2}\frac{q^2}{C} = \frac{1}{2}\cdot\frac{1}{2}\frac{q_0^2}{C} \quad \text{or} \quad q = \pm\sqrt{2}q_0$$

i.e. $\quad \cos\omega t = \pm\frac{1}{\sqrt{2}}$

or $\quad \omega t = (2n-1)\frac{\pi}{4} \quad$ where $n = 1, 2, 3, \ldots$

or $\quad t = (2n-1)\frac{\pi}{4\omega} = (2n-1)\frac{T}{8}$

or $\quad t = \frac{T}{8}, \frac{3T}{8}, \frac{5T}{8}, \ldots$

Example 25.10: A thermocole vessel contains 0.5 kg of distilled water at 30° C. A metal coil of area 5×10^{-3} m^2, number of turns 100, mass 0.06 kg and resistance 1.6 Ω is lying horizontally at the bottom of the vessel. A uniform time varying magnetic field is setup to pass vertically through the coil at time $t = 0$. The field is first increased from zero to 0.8 T, at constant rate between 0.0 sec to 0.2 s and then decrease to zero at the same rate between 0.2 s to 0.4 s. This cycle is repeated 12000 times. Make sketch of current through the coil and the power dissipated in the coil as a function of time for the first two cycles. Clearly indicate the magnitudes of the quantities on the axes. Assume that no heat is lost to the vessel or surroundings. Determine the final temperature of the water under equilibrium. Specific heat capacity of the metal = 500 J kg^{-1}(K)$^{-1}$ and the specific heat capacity of water = 4200 J kg^{-1}K^{-1}. Neglect inductance of the coil. (IIT 2000)

Solution: Since the magnetic field is changing with time at the rate

$$\frac{dB}{dt} = \frac{0.8}{0.2} = 4\ \text{Ts}^{-1}$$

hence the magnitude of the induced emf is

$$e = nA\frac{dB}{dt} = 100 \times 5 \times 10^{-3} \times 4 = 2\ \text{V}$$

Hence induced current in the coil is

$$i = \frac{2}{1.6} = 1.25\ \text{A}$$

As the magnetic field changes at a constant from $t = 0$ to $t = 0.2$ sec, current *i* is constant at 1.25 A during this half cycle. From $t = 0.2$ s to $t = 0.4$ s field decreases at the same rate. Hence the magnitude of current remains constant at 1.25 A during this second half cycle but its direction is reversed. Figure 25.25 shows the sketch of the current through the coil as a function of time for the first two cycles.

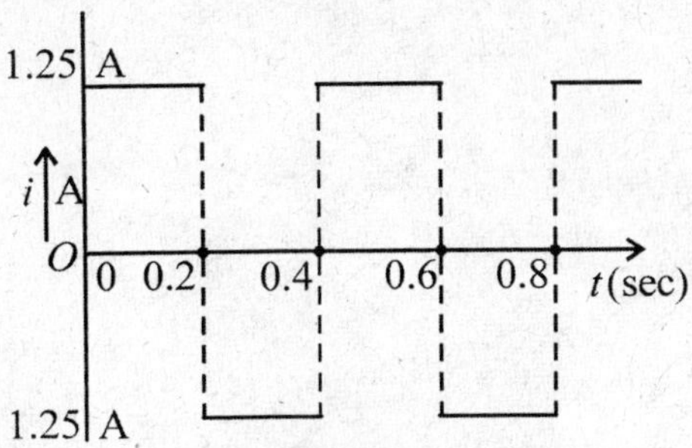

Fig. 25.25

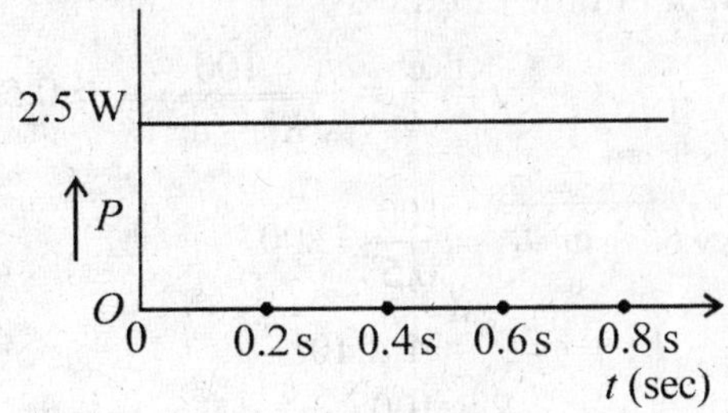

Fig. 25.25 (a)

Power dissipated in the coil is

$$P = i^2R = (1.25)^2 \times 1.6 = 2.5 \text{ W}$$

Since power is proportional to the square of the current ($P \propto i^2$) the power dissipated remains constant at 2.5 W. Plot of power dissipated with time is shown in Fig. 25.25 (*a*).

Heat produced in the coil during 0.2 sec is

$$H = i^2Rt = (1.25)^2 \times 1.6 \times 0.2 = 0.5 \text{ J}$$

During the next 0.2 sec (i.e. from 0.2 s to 0.4 s) also heat generated is

$$H = 0.5 \text{ J}$$

Hence heat generated during one cycle = 0.5 + 0.5 = 1 J
and heat generated in 12000 cycles is = 1 × 12000 = 12000 J.

If C_w and C_m represent specific heats of water and metal of the coil respectively, the rise in temperature be $\Delta\theta$ then heat required is

$$H = (m_w C_w + m_m C_m)$$

$$\therefore \quad \Delta\theta = \frac{H}{m_w C_w + m_m C_m}$$

$$= \frac{12000}{0.5 \times 4200 + 0.06 \times 500} = 5.61 \text{ K}$$

Since size of degree on Kelvin and Celsius scale is same. Hence rise in temperature = 5.61° C.
Hence final temperature = 30° + 5.61° C = 35.61° C.

Example 25.11: When 100 V DC is applied across a coil, a current of 1 amp flows through it. When 100 V AC of 50 Hz is applied to the same coil, only 0.5 A flows. Calculate the resistance and inductance of the coil.

Solution: For DC the coil behaves like a resistance. If resistance of coil is R, then

$$i = \frac{V}{R} \quad \text{or} \quad R = \frac{V}{i}$$

$$\therefore \quad R = \frac{100}{1} = 100\ \Omega$$

For AC it is like *L-R* circuit.
Impedance of the circuit then is

$$Z = \sqrt{R^2 + \omega^2 L^2}$$

and current i in the circuit is

$$i = \frac{E}{Z} = \frac{100}{\sqrt{R^2 + \omega^2 L^2}} = 0.5$$

or $$\sqrt{R^2 + \omega^2 L^2} = \frac{100}{0.5} = 200$$

or $$R^2 + \omega^2 L^2 = 4 \times 10^4$$

But $$R = 100$$

$\therefore$ $$\omega^2 L^2 = 4 \times 10^4 - 10^4 = 3 \times 10^4$$

or $$\omega L = 100\sqrt{3}\ \Omega$$

or $$L = \frac{100\sqrt{3}}{\omega} = \frac{100\sqrt{3}}{2\pi \times 50} = \frac{\sqrt{3}}{\pi}\ \text{H} = 0.55\ \text{H}.$$

Example 25.12: An AC source of angular frequency ω is fed across a capacitor C and *a* resistor R in series. The current registered is I. If now the frequency of the source is changed to $\omega/3$ (but maintaining the same voltage) the current in the circuit is found to be halved. Calculate the ratio of reactance to the resistance at the original frequency.

Solution: Since circuit is R-C circuit. Hence impedance Z of the circuit at frequency ω is

$$Z = \sqrt{R^2 + 1/\omega^2 C^2}$$

and current $$I = \frac{V}{Z} = \frac{V}{\sqrt{R^2 + \frac{1}{\omega^2 C^2}}}$$

when frequency is changed to $\omega/3$, the impedance now is Z'

$$Z' = \sqrt{R^2 + \frac{9}{\omega^2 C^2}}$$

and current is $$I' = \frac{V}{Z'} = \frac{V}{\sqrt{R^2 + 9/\omega^2 C^2}}$$

But $$I' = I/2$$

$\therefore$ $$\frac{2V}{\sqrt{R^2 + 9/\omega^2 C^2}} = \frac{V}{\sqrt{R^2 + 1/\omega^2 C^2}}$$

or $$4\left(R^2 + 1/\omega^2 C^2\right) = R^2 + \frac{9}{\omega^2 C^2}$$

or $$3R^2 = \frac{5}{\omega^2 C^2}$$

But reactance at frequency ω is $X_C = \dfrac{1}{\omega C}$

$\therefore$ $$3R^2 = 5X_C^2$$

or $\left(\frac{X_C}{R}\right)^2 = \frac{3}{5}$ or $\frac{X_C}{R} = \sqrt{\frac{3}{5}}$

Example 25.13: What is the rms current in the circuit shown in Fig. 25.26 (*a*). Given $C = \frac{100}{\pi}$ μF, $L = 1/\pi\ H$ and $E = 200\sin(100\pi t)$.

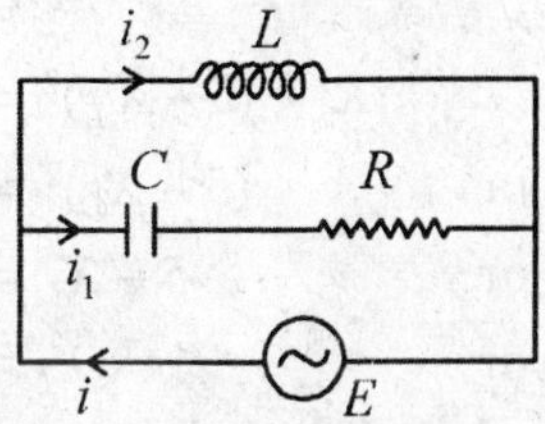

Fig. 25.26 (*a*)

Solution: Impedance of the *R-C* circuit is

$$Z_1 = \sqrt{R^2 + \frac{1}{\omega^2 C^2}} = \sqrt{(100)^2 + \frac{1}{\left(100\pi \times \frac{100}{\pi} \times 10^{-6}\right)^2}}$$

$$= \sqrt{(100)^2 + (100)^2} = 100\sqrt{2}\ \Omega$$

Hence current through this circuit $= i_1 = \frac{E_0}{Z} = \frac{220}{100\sqrt{2}} = 1.1\sqrt{2}$ A. This is the peak value and is in phase ($= \theta$) ahead of voltage where

$$\tan\theta = \frac{1}{\omega RC} = \frac{1}{100\pi \times 100 \times \frac{100}{\pi} \times 10^{-6}} = 1$$

or $\theta = 45°$

Reactance of the inductance $X_L = \omega L = 100\pi \times \frac{1}{\pi} = 100\ \Omega$ and the peak value of the current through inductor is

$$i_2 = \frac{220}{100} = 2.2\ \text{A}$$

This current lags behind the emf in phase by $\pi/2$.

The vector diagram is shown in Fig. 25.26 (*b*).

Component of i_2 in the direction $OB = 2.2\cos 45 = 1.1\sqrt{2}$ A and hence cancels with i_1. Hence net peak value of the current is

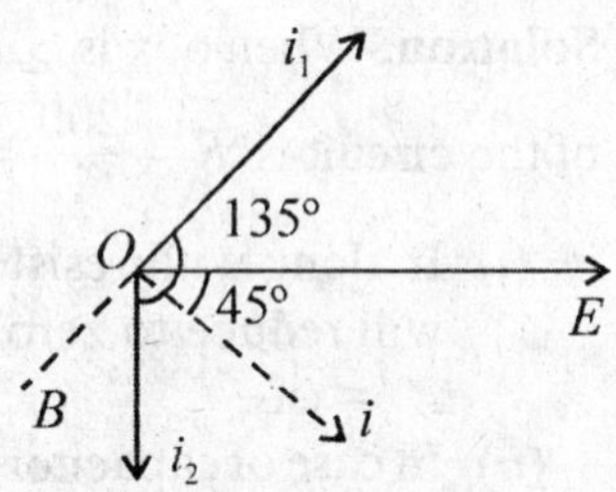

Fig. 25.26 (*b*)

$$i = i_1 \sin 45 = 2.2 \times \frac{1}{\sqrt{2}} = 1.1\sqrt{2}\text{A}$$

or rms value of current $= \frac{1.1\sqrt{2}}{\sqrt{2}} = 1.1$ A

and lags behind the emf *E* in phase by 45°.

Example 25.14: In the circuit shown in Fig. 25.27, ammeter reading is 3 A; $R = 3\ \Omega$, reactance of capacitor is 4 Ω and AC supply voltage is 15 V, 100 Hz. If the box consists of one capacitor of reactance 1 Ω and an inductor. Find the arrangement of inductor and capacitor in box and also find the value of the inductance of the inductor.

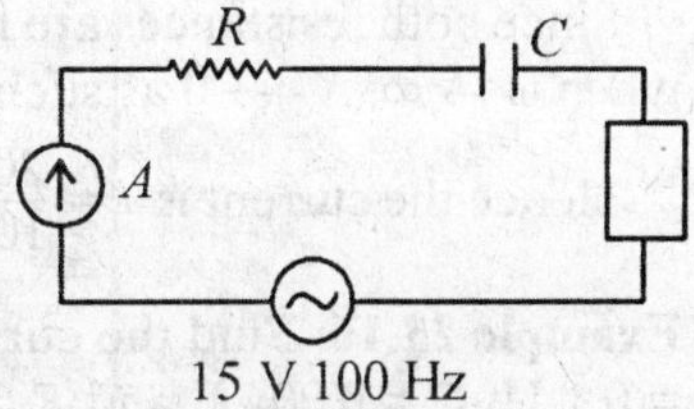

Fig. 25.27

Solution: Impedance of the circuit without box is

$$Z = \sqrt{4^2 + 3^2} = 5\ \Omega$$

Voltage across box = 15 – iZ = 15 – 3 × 5 = 0 V.

Hence arrangement in the box can be possible with series combination of L and C.

$$\therefore \quad Z = \sqrt{R^2 + (4 + 1 - X_L)^2} = 5$$

$$\therefore \quad R^2 + (5 - X_L)^2 = 25$$

or $\quad (5 - X_L)^2 = 25 - R^2 = 25 - 9 = 16$

or $\quad 5 - X_L = \pm 4$

or $\quad X_L = 1 \text{ or } 9\ \Omega$

when $X_L = 1$ then

$$\omega L = 1 \quad \text{or} \quad L = \frac{1}{\omega} = \frac{1}{2\pi \times 100} = 1.59 \text{ mH}$$

when $X_L = 9$ then

$$\omega L = 9 \quad \text{or} \quad L = \frac{9}{\omega} = \frac{9}{2\pi \times 100} = 14.31 \text{ mH}$$

Example 25.15: A box contains two identical resistors and either an inductor or a capacitor. When the box is connected to a DC source of emf 200 V, the current drawn from the source is 1 A and when it is connected to an AC source it is observed that the current increases to a finite value as the frequency of the source is increased. (*a*) Identify the element in the box and draw the possible circuit arrangement of the box. Find the maximum possible value of the rms current drawn from the AC source whose rms emf is 200 V.

Solution: When box is connected to DC source of 200 V the current is 1 A. Hence total resistance of the circuit is $R = \frac{200}{1} = 200\ \Omega$. When connected to AC

(*i*) If along with resistor an inductor is there, the current will reduce to zero as frequency is increased because $X_L = \omega L$.

(*ii*) In case of capacitor the reactance reduces as frequency increases and ultimately the net impedance will be equal to net resistance and current will attain finite value.

(*iii*) Hence box contains two resistances and a capacitance arranged as shown in Fig. 25.28.

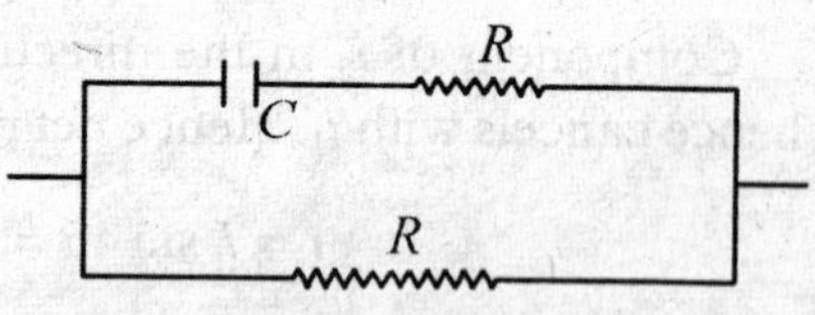

Fig. 25.28

Since both resistances are identical hence each resistance is $R = 200\ \Omega$.
when $\omega \to \infty$, $X_C \to 0$ as such the total resistance of the circuit is $R/2 = 100\ \Omega$.

Hence the current is $i = \frac{200}{100} = 2$ A.

Example 25.16: Find the current at any time in the circuit shown (Fig. 25.29). Given $R = 5\ \Omega$, $L = 0.1$ H, $C = 0.004$ F and $E = 100\sin(50t)$ V.

Solution: Impedance of LR circuit is

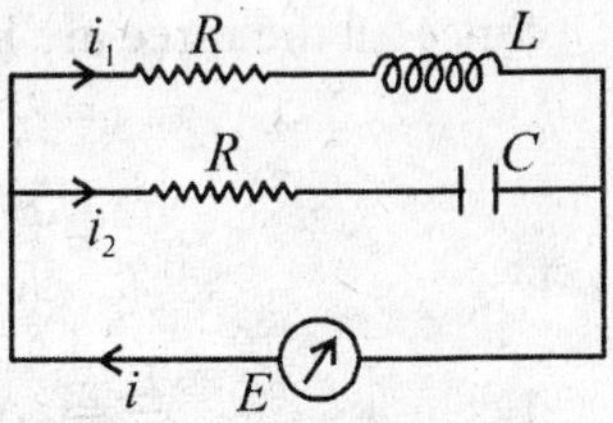

Fig. 25.29

$$Z = \sqrt{R^2 + (\omega L)^2} = \sqrt{(5)^2 + (50 \times 0.1)^2}$$

$$= 5\sqrt{2}\,\Omega$$

current $$i_1 = \frac{E}{Z} = \frac{100}{5\sqrt{2}} = \frac{20}{\sqrt{2}} \text{ A (Peak value)}$$

It is in phase θ lagging behind the applied emf where

$$\tan\theta = \frac{\omega L}{R} = 1 \qquad \because \omega L = R = 5\,\Omega$$

or $$\theta = 45°$$

Similarly, impedance of R–C circuit is

$$Z = \sqrt{R^2 + \left(\frac{1}{\omega C}\right)^2} = 5\sqrt{2}\,\Omega \qquad \because \frac{1}{\omega C} = \frac{1}{50 \times 4 \times 10^{-3}} = 5\Omega$$

Hence peak value of the current i_2 in this circuit is

$$i_2 = \frac{E}{Z} = \frac{100}{5\sqrt{2}} = \frac{20}{\sqrt{2}} \text{ A}$$

leading is in phase by θ given by

$$\tan\theta = \frac{1}{\omega CR} = 1 \quad \text{or} \quad \theta = 45°$$

The vector diagram of the two currents is shown in Fig. 25.29 (*a*).

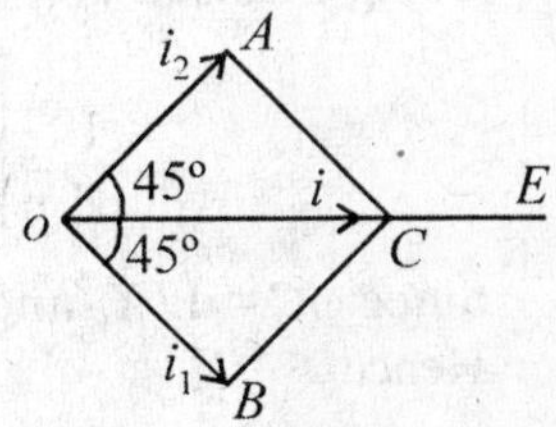

Fig. 25.29 (*a*)

Hence the peak value of the current $i = \sqrt{i_1^2 + i_2^2} = 20$ A and is in phase with the applied emf E. Hence instantaneous value of the current is

$$i = 20\sin(50t) \text{ A}.$$

Example 25.17: For the circuit shown in Fig. 25.30, find the expression for the impedance of the circuit and phase of current.

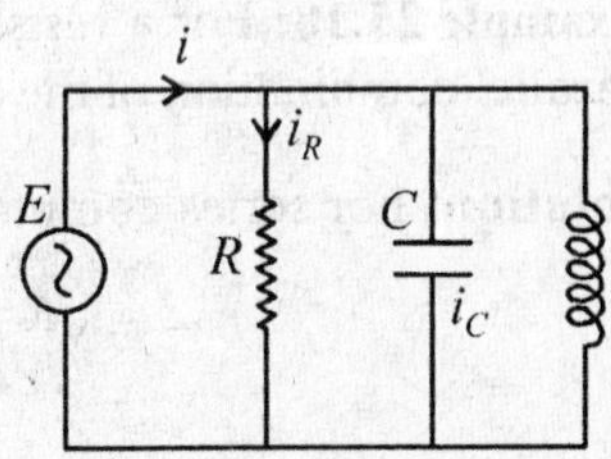

Fig. 25.30

Solution: Here L-C and R are connected in parallel to the same source, as such potential difference across each element is same but the current through them is different in magnitude as well as in phase.

Current through inductor lags behind voltage by an angle $\pi/2$. Hence

$$i_L = \frac{E}{\omega L}\sin(\omega t - \pi/2) \qquad \ldots(i)$$

Current through capacitor leads the voltage in phase by $\pi/2$. Hence

$$i_C = \frac{E}{X_C}\sin(\omega t + \pi/2) = \omega CE \sin(\omega t + \pi/2) \qquad \ldots(ii)$$

Current through resistor is in phase with the potential difference. Hence

$$i_R = \frac{E}{R}\sin\omega t \qquad \ldots(iii)$$

Since all the three are in parallel, hence total current is

$$i = i_L + i_C + i_R$$

$$= \frac{E}{R}\sin\omega t + E\left(\omega C - \frac{1}{\omega L}\right)\cos\omega t \qquad \ldots(iv)$$

Let $\frac{E}{R} = i_0 \cos\theta$ and $E\left(\omega C - \frac{1}{\omega L}\right) = i_0 \sin\theta$

then $i = i_0 \sin(\omega t + \theta)$

when $$\tan\theta = \frac{\omega C - 1/\omega L}{1/R} = R(\omega C - 1/\omega L) \qquad \ldots(v)$$

and $$i_0^2 = \frac{E^2}{R^2} + E^2\left(\omega C - \frac{1}{\omega L}\right)^2 = E^2\left[\frac{1}{R^2} + \left(\omega C - \frac{1}{\omega L}\right)^2\right]$$

and $$i_0 = E\left[\frac{1}{R^2} + \left(\omega C - \frac{1}{\omega C}\right)^2\right]^{1/2}$$

Using Ohms law $i_0 = \frac{E}{Z}$, where Z is the impedance of the circuit, then

$$\frac{1}{Z} = \left[\frac{1}{R^2} + \left(\omega C - \frac{1}{\omega L}\right)^2\right]^{1/2}$$

Since $\omega C = 1/X_C$ and $\omega L = X_L$

Hence

$$\frac{1}{Z} = \left[\frac{1}{R^2} + \left(\frac{1}{X_C} - \frac{1}{X_L}\right)^2\right]^{1/2}$$

Example 25.18: For a resistance R and capacitance C in series the impedance is twice that of a parallel combination of the same elements. What is the frequency of the applied emf?

Solution: For series combination (Fig. 25.31 (*a*)) the impedance is

$$Z = \sqrt{R^2 + 1/\omega^2 C^2} \qquad \ldots(i)$$

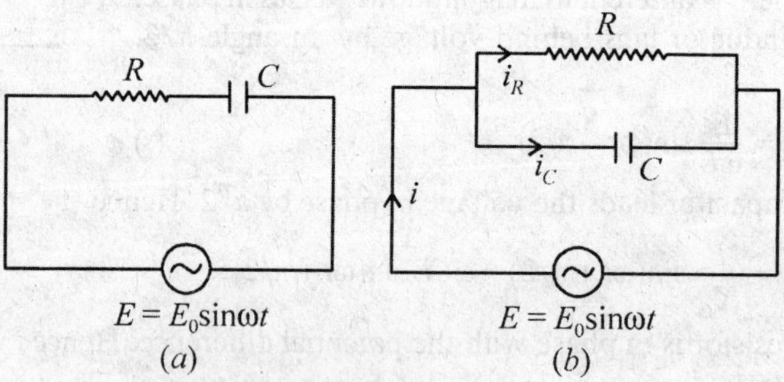

Fig. 24.31

For parallel combination current through resistor R, i_R is in phase with the applied emf, i.e.

$$i_R = \frac{E}{R}\sin\omega t$$

The current through capacitor i_C is $\pi/2$ ahead in phase with applied emf.
Hence

$$i_C = \omega CE\sin(\omega t + \pi/2) = \omega CE\cos\omega t$$

and total current $$i = i_R + i_C = \frac{E}{R}\sin\omega t + \omega CE\cos\omega t$$

or $$i = i_0\sin(\omega t + \theta) \qquad \text{...(ii)}$$

where $$\tan\theta = \frac{\omega CE}{E/R} = \omega CR \qquad \text{...(iii)}$$

and $$i_0^2 = \frac{E^2}{R^2} + (\omega CE)^2 = E^2\left[\frac{1}{R^2} + (\omega C)^2\right]$$

$$i_0 = E\left[\frac{1}{R^2} + (\omega C)^2\right]^{1/2}$$

and from Ohms law we have impedance of the circuit is Z_1 such that

$$\frac{1}{Z_1} = \left[\frac{1}{R^2} + (\omega C)^2\right]^{1/2} \qquad \text{...(iv)}$$

But as per problem $Z = 2Z_1$

i.e. $$Z_1 = \frac{1}{2}Z = \frac{1}{2}\sqrt{R^2 + \omega^2 L^2}$$

and then from (*iv*), we have

$$\frac{2}{\sqrt{R^2 + 1/\omega^2C^2}} = \left[\frac{1}{R^2} + (\omega C)^2\right]^{1/2}$$

or $$\frac{4}{R^2 + 1/\omega^2C^2} = \frac{1}{R^2} + (\omega C)^2 = \frac{1 + R^2\omega^2C^2}{R^2}$$

or $$4R^2 = (1 + R^2\omega^2C^2)(R^2 + 1/\omega^2C^2)$$

or $$4R^2 = \left(1 + R^2\omega^2C^2\right)R^2\left(1 + \frac{1}{\omega^2C^2R^2}\right)$$

or $$4 = 1 + R^2\omega^2C^2 + \frac{1}{\omega^2C^2R^2} + 1$$

or $$4 = R^2\omega^2C^2 + 2 + \frac{1}{\omega^2C^2R^2} = \left(R\omega C + \frac{1}{R\omega C}\right)^2$$

$$\text{or} \qquad R\omega C + \frac{1}{R\omega C} = 2$$

$$\text{or} \qquad R^2\omega^2C^2 + 1 = 2R\omega C$$

$$\text{or} \qquad R^2\omega^2C^2 - 2R\omega C + 1 = 0$$

$$\text{or} \qquad R\omega C - 1 = 0 \quad \text{or} \quad \omega = \frac{1}{RC}$$

$$\text{or} \qquad f = \frac{1}{2\pi RC}$$

Example 25.19: A box contains L, C and R. When 250 V DC is applied to the terminals of the box, a current of 1.0 A flows in the circuit. When an AC source of 250 V rms at 2250 rad/s is connected a current of 1.25 A rms flows. It is observed that the current rises with frequency and becomes maximum at 4500 rad/s. Find the values of L, C and R. Draw the circuit diagram.

Solution: Since the circuit works on DC and all the elements L, C and R cannot be in series. Further with AC supply current changes with frequency, increases becomes maximum and then decreases. This happens only in L-C series circuit. Arrangement is as shown in Fig. 25.32. With DC supply of 250 V, current is 1.0 A

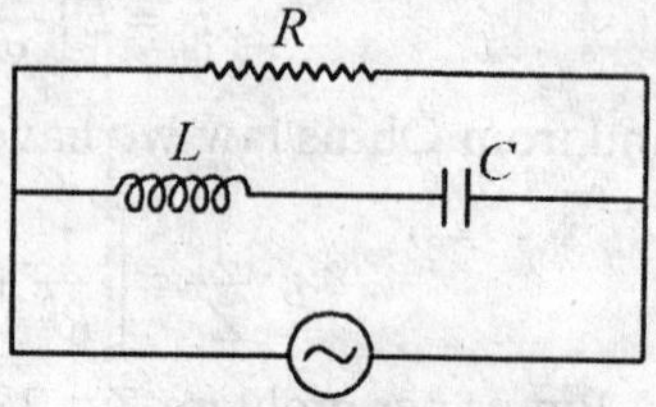

Fig. 25.32

$$\therefore \qquad R = \frac{V}{I} = \frac{250}{1} = 250\ \Omega$$

With AC current is maximum for frequency $\omega = 4500$ rad/s, i.e.

$$\omega^2 = \frac{1}{LC} \quad \text{or} \quad LC = \frac{1}{\omega^2} = \frac{1}{(4500)^2} \qquad \ldots(i)$$

For AC at frequency $\omega = 2250$ rad/s the current is 1.25. The current through resistance is

$$i_R = \frac{250}{250} = 1\text{ A; in phase with the applied voltage.}$$

If reactance is X then current through L-C is ($X = \omega L \sim 1/\omega C$)

$$i_X = \frac{250}{X} \text{ and in phase } \pi/2 \text{ with applied voltage.}$$

Hence phase difference between i_R and i_x is $\pi/2$. Thus

$$i^2 = i_R^2 + i_X^2$$

$$\text{or} \qquad (1.25)^2 = 1^2 + \left(\frac{250}{X}\right)^2$$

$$\text{or} \qquad \left(\frac{250}{X}\right)^2 = (1.25)^2 - 1^2 = (0.25)(2.25)$$

$$\text{or} \qquad \frac{250}{X} = 0.5 \times 1.5$$

$$\text{or} \qquad X = \frac{250}{0.5 \times 1.5} = \frac{1000}{3}\ \Omega$$

But $$x = \omega L \sim \frac{1}{\omega C}$$

or $$2250\, L \sim \frac{1}{2250\, C} = \frac{1000}{3}$$

But $$LC = \frac{1}{(4500)^2} \quad \text{or} \quad L = \frac{1}{(4500)^2\, C}$$

$$\therefore \quad \frac{2250 \times 1}{4500 \times 4500\, C} \sim \frac{1}{2250\, C} = \frac{1000}{3}$$

or $$\frac{1}{C}\left[\frac{1}{9000} \sim \frac{1}{2250}\right] = \frac{1000}{3}$$

or $$\frac{3}{9000\, C} = \frac{1000}{3} \quad \text{or} \quad C = 10^{-6}\text{ F} = 1\ \mu\text{F}$$

We know that $LC = \dfrac{1}{(4500)^2}$

or $$L = \frac{1}{(4500)^2\, C} = \frac{1}{(4.5)^2 \times 10^6 \times 10^{-6}} = \frac{1}{(4.5)^2} = 0.049\text{ H}$$

Example 25.20: An *L-C-R* circuit has L = 10 mH, R = 3 Ω and C = 1 μF connected in series to a source of $15 \cos \omega t$ V. Calculate the current amplitude and the average power dissipated per cycle at a frequency that is 10% less than the resonance frequency.

Solution: Resonance frequency $\omega_0 = \dfrac{1}{\sqrt{LC}} = \dfrac{1}{\sqrt{10 \times 10^{-3} \times 10^{-6}}} = 10^4$ rad/s

Frequency 10% less than resonance frequency is

$$\omega = \omega_0 - \frac{\omega_0}{10} = \frac{9}{10}\omega_0 = 9 \times 10^3 \text{ rad/s}$$

At this frequency inductive reactance

$$X_L = \omega L = 9 \times 10^3 \times 10 \times 10^{-3} = 90\ \Omega$$

Capacitance reactance is $X_C = \dfrac{1}{\omega C} = \dfrac{1}{9 \times 10^3 \times 10^{-6}} = \dfrac{1000}{9} = 111.11\,\Omega$

Hence net reactance $X = X_L \sim X_C = 111.11 - 90 = 21.11\ \Omega$

and impedance of the circuit is

$$Z = \sqrt{R^2 + X^2} = \sqrt{(3)^2 + (21.11)^2} = 21.32\ \Omega$$

Hence peak value of the current is ($E = 15 \cos \omega t$ V)

$$i_0 = \frac{E_0}{Z} = \frac{15}{21.32} = 0.704\text{ A}$$

Average power dissipated is

$$P_{av} = \frac{i_0 E_0 \cos\theta}{2} = \frac{i_0 E_0}{2} \cdot \frac{R}{Z}. \quad \text{But } E_0 = i_0 Z$$

$\therefore$

$$P_{av} = \frac{1}{2} i_0^2 R = \frac{1}{2}(0.704)^2 \times 3 = 0.74 \text{ W}$$

Frequency is $f = \frac{\omega}{2\pi} = \frac{9 \times 10^3}{2\pi} \cdot \frac{\text{Cycles}}{\text{Sec}}$

Hence

$$\frac{P_{av}}{\text{cycle}} = \frac{0.74}{9 \times 10^3} \times 2\pi = 5.16 \times 10^{-4} \text{ J/S}.$$

Example 25.21: A series *L-C-R* circuit containing a resistance of 120 Ω has angular resonance frequency 4 × 10⁵ rad/s. At resonance the voltages across resistance and inductance is 60 V and 40 V respectively. Find the values of *L* and *C*. At what frequency the current in the circuit lags the voltage by 45°?

Solution: At resonance $X = 0$ and all the voltage comes across resistance and current i is

$$i = \frac{V_R}{R} = \frac{60}{120} = 0.5 \text{ A}$$

Potential difference across inductor V_L is

$$V_L = i\omega L$$

or

$$L = \frac{V_L}{i\omega} = \frac{40}{0.5 \times 4 \times 10^5} = 0.2 \text{ mH}$$

The resonance frequency $\omega_0 = \frac{1}{\sqrt{LC}}$

or

$$C = \frac{1}{L\omega_0^2} = \frac{1}{4 \times 0.2 \times 10^{-3} (4 \times 10^5)^2}$$

$$= \frac{10^{-6}}{32} = 0.03 \times 10^{-6} \text{ F}$$

$$= 0.03 \text{ μF}$$

In case of series *L-C-R* series circuit, the phase difference θ between current and voltage is

$$\tan\theta = \frac{\omega L \sim 1/\omega C}{R}$$

Since θ = 45°. Hence

$$R = \omega L - \frac{1}{\omega C}$$

or

$$\omega CR = \omega^2 LC - 1$$

or

$$\omega^2 - \frac{R}{L}\omega - \frac{1}{LC} = 0$$

or

$$\omega = \frac{\frac{R}{L} \pm \sqrt{\frac{R^2}{L^2} + 4/CL}}{2}$$

–ve sign is not admissible because this will make frequency negative.

$$\frac{R}{L} = \frac{120}{0.2\times10^{-3}} = 6\times10^5 \text{ and } \frac{4}{CL} = \frac{4\times32\times10^6}{0.2\times10^{-3}} = 64\times10^{10}$$

$$\therefore \quad \frac{R^2}{L^2} + \frac{4}{CL} = 36\times10^{10} + 64\times10^{10} = 10^{12}$$

$$\therefore \quad \omega = \frac{1}{2}\left[6\times10^5 + 10\times10^5\right] = \frac{16\times10^5}{2} = 8\times10^5 \text{ rad/s.}$$

Example 25.22: A current of 4 A flows in a coil when connected to a 12 V DC source. If the same coil is connected to a 12 V, 50 rad/s AC source a current of 2.4 A flows in the circuit. Determine the inductance of the coil. Also find the power developed in the circuit if a 2500 μF capacitor is connected in series with the coil.

Solution: Since 4 A current flows through the coil when connected to a 12 V DC. Hence its resistance is

$$R = \frac{12}{4} = 3\ \Omega$$

Since inductance offers zero resistance with AC currents 2.4 A. Hence impedance of the coil is

$$Z = \frac{12}{2.4} = 5\Omega$$

But $\quad Z = \sqrt{R^2 + \omega^2 L^2}$

or $\quad R^2 + \omega^2 L^2 = Z^2$

and $\quad \omega^2 L^2 = Z^2 - R^2 = (5)^2 - (3)^2 = 16$

or $\quad \omega L = 4$

and $\quad L = \frac{4}{\omega} = \frac{4}{50} = 0.08 \text{ H}$

when the capacitor is connected in series to the above circuit capacitive reactance

$$X_C = \frac{1}{\omega C} = \frac{1}{50\times2.5\times10^{-3}} = \frac{10^3}{125}\ \Omega = 8\ \Omega$$

Now impedance of the circuit is

$$Z = \sqrt{R^2 + (X_L - X_C)^2} = \sqrt{(3)^2 + (4-8)^2} = 5\ \Omega$$

and hence $\quad i = \frac{V}{12} = \frac{12}{5} = 2.4$

and average power $P_{av} = i^2 R = (2.4)^2 \times (3) = 17.28$ W

Hence i is rms value.

Example 25.23: An inductor 20×10^{-3} H, a capacitor 100 μF, and a resistor 50 Ω are connected in series across a source of emf $V = 10\sin 314t$. Find the energy dissipated in the circuit in 20 minutes. If the resistance is removed from the circuit and the value of inductance is doubled then find the variation of current with time in new circuit. (Roorkee 1999)

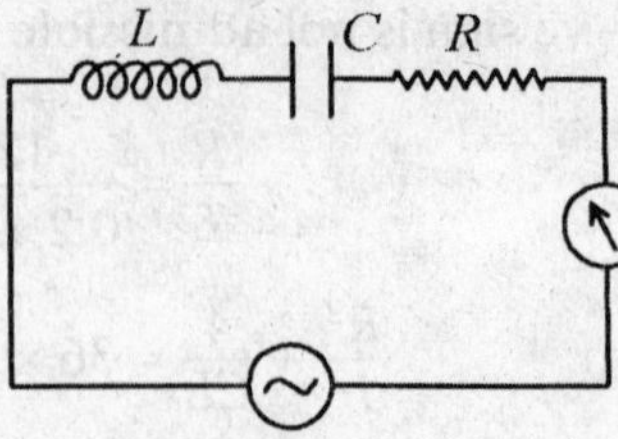

Fig. 25.33

Solution: Time for 1 cycle is

$$T = 1/50 = 0.02 \text{ s}$$

We have to find the average energy dissipated for time 20 minutes, i.e. for time much larger than T (0.02 s).

Average energy consumed in time t second is

$$E = \frac{1}{2} i_0 E_0 \frac{R}{Z} t = \frac{1}{2} E_0^2 \frac{R}{Z^2} t \qquad \because\ i_0 = \frac{E_0}{Z}$$

where Z is impedance and is given by

$$Z = \sqrt{R^2 + (\omega L - 1/\omega C)^2}$$

But $\quad \omega L = 314 \times 20 \times 10^{-3} = 6.28\ \Omega$

and $\quad \dfrac{1}{\omega C} = \dfrac{1}{314 \times 100 \times 10^{-6}} = \dfrac{10^4}{314} = 31.8\ \Omega$

Hence $\quad Z = \sqrt{(50)^2 + (6.28 - 31.8)^2} = 56\ \Omega$

Hence energy dissipated is

$$E = \frac{1}{2}(10)^2 \times \frac{50}{(56)^2} \times 20 \times 60$$

$$= \frac{3 \times 10^6}{(56)^2} = 956.6 \text{ J}$$

When resistance is removed, if new impedance is Z', then power factor

$$\cos\theta = \frac{R}{Z} = 0 \qquad \text{i.e. } \theta = \pi/2$$

Now impedance is $Z' = \dfrac{1}{\omega C} - \omega L = \dfrac{1}{314 \times 10^{-4}} - 314 \times 40 \times 10^{-3}$

$$= 31.8 - 12.56 = 19.22\ \Omega.$$

and current $\quad i = \dfrac{E_0}{Z'} \sin(\omega t + \theta)$

$$= \frac{10}{19.22} \sin(314t + \pi/2)$$

$$= 0.52\cos(314t).$$

Example 25.24: A box P and a coil Q are connected in series with an AC source of variable frequency. The emf of the source is constant at 10 V. Box P contains a capacitance of 1 μF in series with a resistance of 32 Ω. Q has a coil of self inductance 4.9 mH and a resistance of 68 Ω in series. The frequency is adjusted so that the maximum current flows in P and Q. Find the impedance of P and Q at this frequency. Also find the voltage across P and Q respectively. (Roorkee 99)

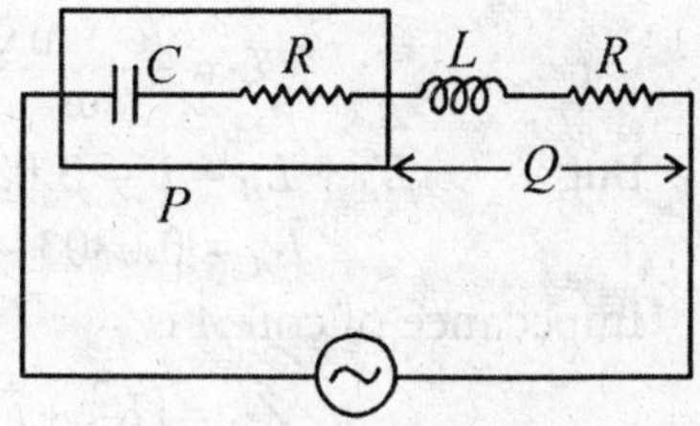

Fig. 25.34

Solution: The resonance frequency is

$$\omega_0 = \frac{1}{\sqrt{LC}} = \frac{1}{\sqrt{(4.9\times10^{-3})(10^{-6})}} = \frac{10^5}{7} \text{ rad/s}$$

Since the circuit is resonance circuit hence all the applied potential is across resistance R. Hence current in the circuit is

$$i = \frac{E}{R} = \frac{10}{(32+68)} = \frac{10}{100} = 0.1 \text{ A}$$

Impedance of P is

$$Z_P = \left[R_1^2 + \left(\frac{1}{\omega C}\right)^2\right]^{1/2} = \left(32^2 + \left(\frac{7}{10^5\times10^{-6}}\right)^2\right)^{1/2} = \sqrt{5924} = 77\ \Omega$$

and $$Z_Q = (R_2^2 + \omega^2 L^2) = \left[(68)^2 + \left(\frac{7}{10^5}\times 4.9\times10^{-3}\right)^2\right]^{1/2}$$

$$\sqrt{(9527)} = 97.6\,\Omega$$

$\therefore$ $$V_P = iZ_P = 0.1 \times 77 = 7.7 \text{ V}.$$

and $$V_Q = iZ_Q = 0.1 \times 97 = 9.7 \text{ V}.$$

Example 25.25: Two coils A and B are connected in series to an AC source of 240 V 50 Hz. A has resistance 5 Ω and B has inductance 0.02 H. The power dissipated in the circuit is 3 kW and power factor is 0.75. Find the resistance of coil B and inductance of coil A. Find also the voltage across coil A.

Solution: Power consumed is $EI\cos\theta$

$\therefore$ $$EI\cos\theta = P$$

or $$240 \times I \times 0.75 = 3 \times 10^3$$

or $$I = \frac{3\times10^3}{240\times0.75} = \frac{50}{3} \text{ A}$$

Hence impedance $Z = \frac{E}{I} = \frac{240}{50}\times 3 = 14.4\ \Omega$

But power factor $= \frac{R}{Z} = 0.75$

$\therefore$ $$R = 0.75 \times 14.4 = 10.8\ \Omega$$

and $$R = R_A + R_B$$

$$R_B = R - R_A = 10.8 - 5 = 5.8\ \Omega$$

If phase difference is θ, then

$$\tan\theta = \frac{L\omega}{R}$$

Since $\cos\theta = 3/4$ then $\tan\theta = \frac{\sqrt{7}}{3}$

$\therefore \qquad L = \frac{R \cdot \tan\theta}{\omega} = \frac{10.8 \times \sqrt{7}}{3 \times 2\pi \times 50} = 3.03 \times 10^{-2}$ H

But $\qquad L_A + L_B = L = 3.03 \times 10^{-2}$ but $L_B = 0.02$ H

$\therefore \qquad L_A = 0.0303 - 0.02 = 0.0103$ H $= 10.3$ mH.

Impedance of coil A is

$$Z_A = \left(R_A^2 + \omega^2 L_A^2\right)^{1/2} = \left[5^2 + (2\pi \times 50 \times 0.0103)^2\right]^{1/2}$$

$$= [25 + 10.5]^{1/2} = 6\,\Omega$$

Hence potential difference across coil A is

$$V_A = Z_A I = 6 \times \frac{50}{3} = 100 \text{ V}$$

PROBLEMS

1. A small town with a demand of 800 kW of electric power at 220 V is situated at 15 km away from an electric plant generating power at 440 V. The resistance of the two wire line carrying current is 0.5 Ω per km. The town gets power from the line through a 40000-220 V step down transformer at substation in the town.
   (*a*) Estimate power loss in the line in the form of heat.
   (*b*) How much power must the plant supply, assuming there is negligible power loss due to leakage?
   (*c*) Characterize the step up transformer at the plant.
   [**Ans.** (*a*) Line loss = 6 kW (*b*) Power supplied = 806 kW (*c*) Voltage drop V = 300 V, transformer is 440 V 40300 V]
2. A circuit containing a 80 mH inductor and a 60 μF capacitor in series is connected to a 230 V 50 Hz supply. The resistance of the circuit is negligible
   (*a*) Obtain current amplitude and rms values.
   (*b*) Obtain the rms value of potential drops across each element.
   (*c*) What is the average power transferred to the inductor?
   (*d*) What is the average power transferred to the capacitor?
   (*e*) What is the total average power absorbed by the circuit?
   [**Ans.** (*a*) $i_{rms} = 8.24$ A (*b*) $V_L = 207$ V, $V_C = 437$ V (*c*) $P_{av} = 0$ (*d*) $P_{av} = 0$ (*e*) $P_{av} = 0$]
3. At hydro electric power plant, the water pressure head is at a height of 300 m and the water flow available is 100 m^3/s. If the turbine generator efficiency is 60% estimate the electric power available from the plant. [**Ans.** 176 Watt]
4. A series *L-C-R* circuit with $L = 0.12$ H, $C = 480$ μF and $R = 23\,\Omega$ is connected to a 230 V variable frequency supply (*a*) What is the source frequency for which the current amplitude is maximum? Obtain this maximum current amplitude. (*b*) What is the source frequency for which the average power absorbed by the circuit is maximum? Obtain the value of this maximum power. (*c*) For which frequencies of the source, is the power transferred to the circuit half the power at resonance frequency? What is the current amplitude at these frequencies.
   [**Ans.** (*a*) $f_r = 663$ Hz, $i_0 = 14.1$ A (*b*) $P_{max} = 2300$ W (*c*) $f_1 = 648$ Hz, $f_2 = 678$ Hz, $i_0 = 10$ A]

5. Show that the average power transferred from the AC source to a circuit consisting of L-C-R is given by $P = V_{rms} L_{rms} \frac{R}{Z}$.

6. An inductor coil, a capacitor and an AC source of rms value 24 V are connected in series. When the frequency of source is varied, a maximum rms current of 6 A is observed. If this inductor coil is connected to a battery of 12 V and internal resistance 4.0 Ω. What will be current?

[**Hint:** Maximum current means reactance zero. Hence $R = \frac{24}{6} = 4\,\Omega$ and current when DC is connected is $i = \frac{12}{4+4} = \frac{12}{8} = 1.5$ A]

7. In a series L-C-R circuit with an AC source, $R = 300\,\Omega$, $C = 25\,\mu F$, $E_0 = 50$ V and $f = \frac{50}{\pi}$ Hz. Find (*a*) the rms current in the circuit and (*b*) the rms potential diference across the capacitor, the resistor and inductor.

[**Ans.** (*a*) 0.1 A (*b*) $V_C = 50$ V, $V_L = 10$ V, $V_R = 30$ V]

8. A choke coil is needed to operate an arc lamp at 160 V (rms) and 50 Hz. The arc lamp has an effective resistance of 5 Ω when running at 10 A (rms). Calculate the inductance of the choke coil. If the same arc lamp is to be operated on 160 V (DC) what additional resistance is required? Compare the power losses in both cases.

[**Ans.** $L = 4.84 \times 10^{-2}$ H, $P = 500$, P(DC) = 1100 W]

9. A circuit draws a power of 550 W from a 220 V 50 Hz source. Power factor of the circuit is 0.8 and the current lags behind the voltage in phase. In order to make the power factor 1.0 find the value of the capacitor to be connected in the circuit.

[**Ans.** $C = 75\,\mu F$]

10. An AC voltage $V = 10\sqrt{2}\sin(200t)$ is connected across a coil of inductance 1 H and a resistance of 200 Ω. Find (*a*) Impedance of the circuit (*b*) Phase difference between voltage and current (*c*) Peak value of the current (*d*) Voltage across R and (*e*) Voltage across L.

[**Ans.** (*a*) $Z = 200\sqrt{2}\,\Omega$ (*b*) $\theta = 45°$ (*c*) $i_0 = 0.05$ A (*d*) $V_R = 7.07$ V and (*e*) $V_L = 7.07$ V]

11. A coil draws a current of 1.0 A and 100 Watt power from a source of 110 V, 50 Hz. Find the resistance and inductance of the coil.

[**Ans.** $R = 100\,\Omega$, $L = 0.146$ H]

12. A 1.5 μF capacitor is charged by a battery of 57 V. It is then connected to a coil of inductance 12 mH, so that the charge oscillates. If there is no resistance in the circuit then find the maximum current passing through the coil.

[**Ans.** 0.637 A]

13. A non inductive resistance of 12 Ω and pure inductance of 0.05/π henry are connected in series with a 130 V 50 Hz AC source. Find the current in the resistance and potential difference across resistance and inductance.

[**Ans.** $i = 10$ A, $V_R = 120$ V, $V_L = 50$ V]

14. In a primary of a transformer the number of turns are 500 and in secondary it is 10 (*a*) If $V_P =$ 120 V then find V_S. (*b*) If resistance of secondary is 15 Ω, then find the current in primary and secondary coil.

[**Ans.** $V_S = 2.4$ V, $i_S = 0.16$ A, $i_P = 3.2$ mA]

15. In an *L-C-R* series circuit an inductance ($L = 1$ H), a capacitance ($C = 0.2\ \mu$F) and a resistance ($R = 200\ \Omega$) are connected with an AC source of 200 V and $\frac{1000}{2\pi}$ Hz frequency. Find (*i*) impedance of the circuit (*ii*) phase difference between current and emf (*iii*) potential difference across coil, capacitor and resistor.

[**Ans.** (*i*) $Z = 5000\ \Omega$ (*ii*) $\theta = \tan^{-1}(1.33) = 53°$ (*iii*) $V_L = 40$ V, $V_C = 200$ V, $V_R = 120$ V]

16. Figure 25.35 shows a series circuit in which inductance $\left(L = \frac{2}{5\pi}\ \text{H}\right)$ and resistance ($R = 8\ \Omega$). The applied voltage is 100 V at 45 Hz, then draw the vector diagram and hence find the power factor of the circuit.

[**Ans.** $\cos\theta = 0.2$]

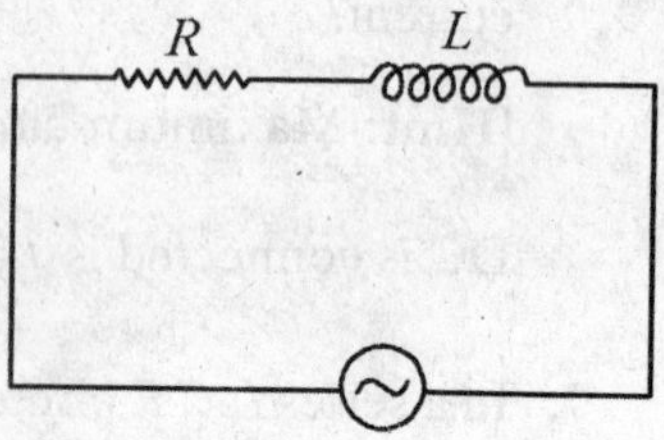

Fig. 25.35

17. In an *L-C-R* series circuit $R = 100\ \Omega$ is connected to a 200 V, 300 rad/s source. When only capacitor is removed, the current lags behind the emf in phase by 60°. If only inductance is removed the current leads the emf in phase by 68. Calculate the current in the circuit and power dissipated.

[**Ans.** $i = 2$ A, $P = 400$ W]

18. Figure 25.36 consists of a resistor *R* and an unknown section *X*. An *AC* is passing through it. Using only one voltmeter how will you find *X* (*i*) is inductor (*ii*) is pure resistor (*iii*) is resistanceless and inductanceless conductor.

[**Hint:** Find potential difference across *AB*, *BC* and *AC* using voltmeter. (*i*) If $V_{AB} + V_{BC} > V_{AC}$ then *X* is inductance (*ii*) if $V_{AB} + V_{BC} = V_{AC}$, then *X* is pure resistor (*iii*) $V_{BC} = 0$ then *X* is a conductor]

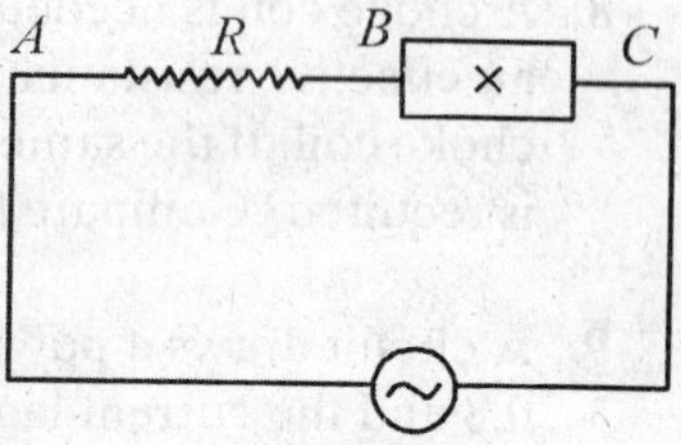

Fig. 25.36

19. In *L-C-R* series circuit shown in Fig. 25.37 resistance has value 60 Ω then find (*i*) current through capacitor (*ii*) If the frequency of AC is 100 rad/s, then find the value of inductance (*iii*) Capacity of the capacitor.

[**Ans.** (*i*) = 0.25 A (*ii*) $L = 0.8$ H (*iii*) $C = 250\ \mu$F]

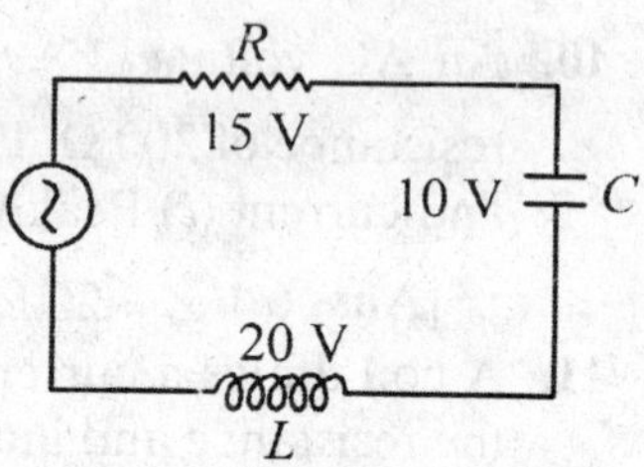

Fig. 25.37

20. A bulb is rated 100 V, 50 Hz. How will you operate it on 200 V AC mains with frequency 50 Hz.

[**Ans.** We will connect a coil of reactance $200\sqrt{3}\ \Omega$ in series with the bulb]

21. Instantaneous voltage of an AC is $E = 110 \sin(\omega t + \pi/6)$ and current is given by $i = 5\sin(\omega t - \pi/6)$. Find the phase difference between *E* and *i*. What is the impedance of the circuit and power dissipated ?

[**Ans.** $\phi = \pi/3$, $Z = 22\ \Omega$, $P = 137.5$ W]

22. In an AC circuit, there is a coil and a resistance. A potential of 130 V is applied across this combination. If potential drop across resistance is 50 V then find the potential drop across choke coil.

[**Ans.** 120 V]

23. In a circuit the emf and current are represented by $V = 28\sin(628t - 30°)$ and $i = 3.5\sin(628t + 30°)$ A. Then find (*i*) peak value of current (*ii*) rms value of current (*iii*) time period (*iv*) phase difference between current and emf.

[**Ans.** (*i*) 3.5 A (*ii*) 2.48 A (*iii*) $T = 10^{-2}$ sec. (*iv*) $\phi = 60°$]

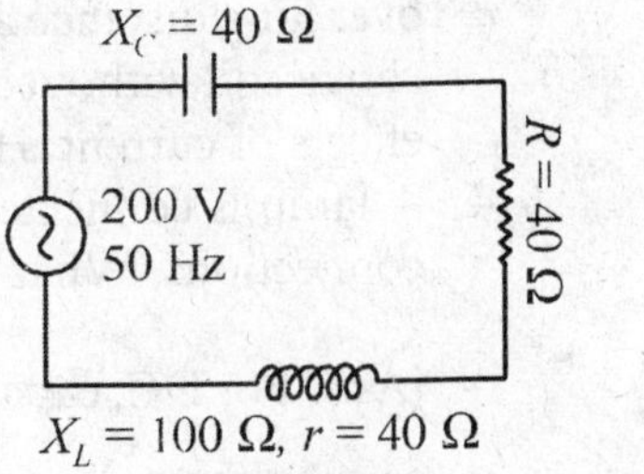

Fig. 25.38

24. Find in Fig. 25.38 (*i*) What is the power factor (*ii*) Potential difference across the coil (*iii*) Power loss in the coil.

[**Hint**: (*i*) $\cos\theta = \frac{R}{Z} = \frac{40+40}{\sqrt{(80)^2 + (100-40)^2}} = 0.8$ (*ii*) $V_L = iZ_L$,

$i = \frac{200}{100} = 2$ A. $Z_L = \sqrt{(40)^2 + (100)^2} = 108\,\Omega$ and $V_L = 2 \times 108$ $= 216$ V (*iii*) $P = i^2 r = 2 \times 2 \times 40 = 160$ W]

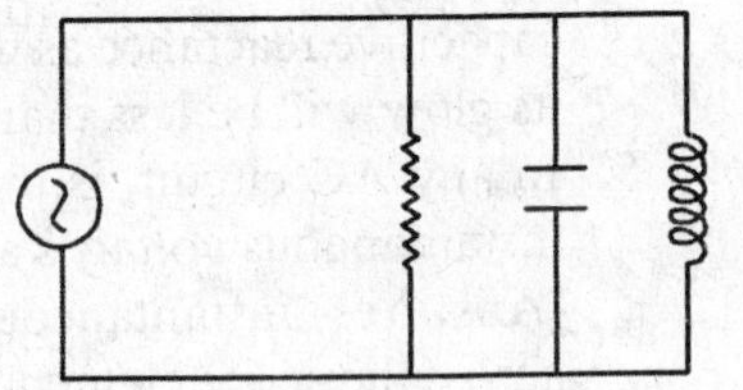

Fig. 25.39

25. In Fig. 25.39, $R = 100\ \Omega$, $L = 0.1$ H; $C = 0.1\ \mu$F and the peak value of applied source is 100 V. Find (*a*) resonance frequency (*b*) resonance angular frequency (*c*) maximum current through resistor at resonance (*d*) maximum current through the circuit at resonance (*e*) maximum current through inductance at resonance (*f*) maximum current through capacitor at the resonance (*g*) maximum energy stored in inductor and capacitor at resonance.

[**Ans.** (*a*) $f = 1592$ (*b*) $\omega = 10^4$ rad/s (*c*) $i_0 = 0.1$ A (*d*) $i_0 = 1$ A (*e*) $i_0 = 0.1$ A (*f*) $i_0 = 0.1$ A (*g*) $E = 5.0 \times 10^{-4}$ J, 5×10^{-4} J]

SHORT ANSWER TYPE QUESTIONS

1. The peak value of an alternating voltage is 300 V and its frequency is 60 Hz. Find its rms value. How long will the voltage take to reach the peak value starting from zero?

[**Ans.** $V_{rms} = \frac{V_0}{\sqrt{2}} = \frac{300}{\sqrt{2}} = 212$ V

Time period $T = \frac{1}{f} = \frac{1}{60}$ s

The voltage takes 1/4 of the time period to reach the peak value starting from zero. Thus time required is $t = \frac{1}{4 \times 60} = \frac{1}{240}$ s.]

2. An inductor ($L = 200$ mH) is connected to an AC source of peak value 210 V and frequency 50 Hz. Calculate the peak current. What is the instantaneous voltage of the source when the current is at its peak value?

[**Ans.** Reactance of inductor $X_L = \omega L = 2\pi f L = 2\pi \times 50 \times 200 \times 10^{-3} = 62.8\ \Omega$. Hence peak value of current $i_0 = \frac{V_0}{X_L} = \frac{210}{62.8} = 3.3$ A.

Since circuit is inductive as such the voltage leads the current in phase by $\pi/2$. Thus the voltage is zero when the current has its peak value.]

3. Give relative advantages of AC over DC.

[**Ans.** (*a*) AC is cheaper than DC (*b*) AC can be easily converted into DC with the help of a rectifier. It is very difficult to convert DC into AC. A DC can be converted into AC using inverter. (*c*) It can be stepped up or stepped down using a transformer. (*d*) It can be transmitted over long distance at high voltage without much power loss. (*e*) It can be regulated by using choke coil without any significant loss of energy. (*f*) AC cannot be used for study of chemical effects of current. Hence for electrolysis and electroplating only DC is used]

4. A lamp is connected in series with a capacitor. Predict your observations for DC and AC connections. What happens in each case if the capacity is reduced.

[**Ans.** For DC, capacitor offers infinite reactance $\left(X_C = \frac{1}{\omega C}\right)$, i.e. it behaves like open circuit as such the bulb will not glow at all even if capacity (C) is reduced. For AC the capacitor offers finite reactance and current flows in circuit hence the bulb glows. Reducing C increases capacitive reactance as well as impedance, thereby reducing the current in the circuit. Hence its glow will be less than before]

5. In any AC circuit, is the applied instantaneous voltage equal to the algebraic sum of the instantaneous voltages across the series elements of the circuit? Is that true for rms values?
[**Ans.** Yes, instantaneous voltage equals the algebraic sum of the instantaneous voltages across the series elements of the circuit. The same is not true for rms values, because voltage across different elements may not be in same phase]

6. An applied voltage signal consists of a superposition of a DC voltage and an AC voltage with high frequency. The circuit consists of an inductor and a capacitor in series. Show that the DC signal will appear across C and the AC signal across L.
[**Ans.** For DC; impedance of L is negligible and of C very high (infinite) so the DC signal appears across C. For high frequency AC, impedance of L is high and that of C is low. As such the AC signal appears across L]

7. A choke coil in series with a lamp is connected to a DC line. The lamp is seen to shine brightly. In section of an iron core in the choke causes no change in the lamp's shining. Predict the corresponding observation if the connection is to be an AC.
[**Ans.** For steady state DC inductance (L) has no effect even if it is increased by iron core. For AC the shining of the bulb will decrease as the impedance increases which reduces the current. Hence the introduction of iron core reduces the shine of the bulb]

8. For circuits used for transporting electric power, a low power factor implies large power loss in transmission.
[**Ans.** To supply a given power, low power factor means high current ($P = L_{rms} V_{rms} \cos\theta$). This causes high heat losses due to i^2R]

9. Power factor can often be improved by the use of a capacitor of appropriate value in the circuit.

[**Ans.** Power factor $\cos\theta = \frac{R}{Z}$. Many AC machines have inductive reactance. A capacitance of appropriate value reduces the net reactance so that Z approaches R.]

10. A power transmission line feeds input power at 2300 V to step down transformer with its primary windings having 4000 turns. What should be the number of turns in the secondary in order to get output power at 230 V?
[**Ans.** If E_S is potential difference across secondary with number of turns n_s and E_P is potential difference across primary with number of turns n_P, then

$$\frac{E_s}{E_p} = \frac{n_s}{n_p}$$

or $\dfrac{230}{2300} = \dfrac{n_s}{4000}$ $\therefore n_s = 400$]

11. Obtain the resonance frequency and Q factor of a series L-C-R circuit with $L = 3.0$ H $C = 27$ μF and $R = 7.4\ \Omega$. It is desired to improve the sharpness of resonance of circuit by reducing its full width at half maximum by a factor 2. Suggest a suitable way.

[**Ans.** Resonance frequency $f = \dfrac{1}{2\pi}\dfrac{1}{\sqrt{LC}} = \dfrac{1}{2\pi}\cdot\dfrac{1}{\sqrt{3\times 27\times 10^{-6}}}$

or $$f = \frac{1}{2\pi}\times\frac{10^3}{9} = 17.7 \text{ Hz}$$

$$Q = \frac{\omega_0 L}{R} = \frac{1}{R}\sqrt{\frac{L}{C}} = \frac{1}{7.4}\sqrt{\frac{3}{27\times 10^{-6}}} = \frac{10^3}{7.4\times 3} = 45$$

Since sharpness depends on Q factor and Q factor $\left(= \dfrac{\omega_0 L}{R}\right)$ depends on resistance. Hence to double Q factor R should be made 1/2 i.e., $R = 3.7\ \Omega$]

12. A 100 μF capacitor in series with a 40 Ω resistance is connected to a 110 V, 60 Hz supply. (*a*) What is the maximum current in the circuit. (*b*) What is the time lag between the current maximum and voltage maximum.

[**Ans.** (*a*) Impedance of the circuit is $Z = \sqrt{R^2 + \dfrac{1}{\omega^2 C^2}}$. Hence the current in the circuit is

$$i = \frac{V}{Z} = \frac{V}{\sqrt{R^2 + \dfrac{1}{\omega^2 C^2}}} = \frac{110}{\sqrt{(40)^2 + \left(\dfrac{1}{2\pi\times 60\times 100\times 10^{-6}}\right)^2}}$$

or $$i = \frac{110}{\sqrt{(40)^2 + (26.5)^2}} = 2.3 \text{ A}$$

Hence maximum current $i_0 = i\sqrt{2} = 2.3\sqrt{2} = 3.24$ A.

(*b*) Phase difference between current and voltage is

$$\tan\phi = \frac{1}{\omega CR} = \frac{1}{2\pi 60\times 100\times 10^{-6}\times 40} = 0.66$$

or $$\phi = 33.5°$$

Then time lag $= \dfrac{\phi}{\omega} = \dfrac{33.5}{360}\times\dfrac{1}{60} = 1.5$ ms]

13. State whether the following statements are true or false giving reasons in brief.

(*a*) DC is more dangerous than AC of same value.

(*b*) The average value of AC is defined over half cycle.

[**Ans.** (*a*) False: It is a common misconception that DC is more dangerous than AC of same value as due to it an electrocuted person sticks to the line while AC repels away from the line. As a matter of fact in case of 220 V AC the peak value $V_0 = 220\sqrt{2}$ and as 311 V will cause more harm to the human body than 230 V DC. AC is more dangerous than DC of the same value.

(*b*) False: No doubt the average value of AC is defined over a half cycle (as over full cycle its value is zero), the half cycle must be either positive or negative. If in a half cycle, one quarter is positive and the other is negative the average value for the half cycle will be zero]

14. The current in the discharging *L-R* circuit is given by $i = i_0 e^{-t/\tau}$, where τ is the time constant of the circuit. Calculate the rms current for the period $t = 0$ to $t = \tau$.

[**Ans.** The average value of the current i^2 is

$$\langle i^2 \rangle = \frac{\int_0^\tau i^2 dt}{\int_0^\tau dt} = \frac{1}{\tau}\int_0^\tau \left(i_0 e^{-t/\tau}\right)^2 dt$$

$$= \frac{i_0^2}{\tau}\int_0^\tau e^{-2t/\tau} dt = \frac{i_0^2}{\tau}\left(-\frac{\tau}{2}\right) e^{-2t/\tau}\Big|_0^\tau$$

$$= -\frac{i_0^2}{2}\left[e^{-2} - 1\right] = \frac{i_0^2}{2}(1 - e^{-2})$$

$$= \frac{i_0^2}{e^2}\left(\frac{e^2 - 1}{2}\right)$$

$$\therefore \quad i_{rms} = \sqrt{\langle i^2 \rangle} = \frac{i_0}{e}\sqrt{\frac{e^2 - 1}{2}}\ \Big]$$

15. An alternating voltage $E = 200\sqrt{2}\sin(100t)$ V is connected to 1 μF capacitor through an AC ammeter. What will be the reading of the ammeter?

[**Ans.** Reactance of the circuit $X_C = \frac{1}{\omega C} = \frac{1}{100 \times 10^{-6}} = 10^4\ \Omega$

Hence peak value of current $i_0 = \frac{E_0}{X_C} = \frac{200\sqrt{2}}{10^4} = 2\sqrt{2} \times 10^{-2}$ A

Ammeter records rms value. Hence ammeter reading is

$$i = \frac{i_0}{\sqrt{2}} = 0.02 \text{ A}\ \Big]$$

16. A 200 km long telegraph wire has capacity of 0.014 μF/km. If it carries an alternating current of frequency 5 Hz, what should be the value of an inductance required to be connected in series so that the impedance is minimum?

[**Ans.** Total capacity of the cable = $C = 0.014 \times 200 = 28$ μF

For minimum impedance $\omega L = 1/\omega C$ or $\omega = \frac{1}{\sqrt{LC}}$ and $\omega = 2\pi f$

$$\therefore \quad L = \frac{1}{\omega^2 C} = \frac{1}{(2\pi f)^2 C}$$

or

$$L = \frac{1}{(2\pi \times 5 \times 10^3)^2} \times \frac{1}{2.8 \times 10^{-6}}$$

$$= \frac{1}{100\pi^2 \times 2.8} = 0.036 \times 10^{-2}\ \text{H}$$

$$= 0.36\ \text{mH}]$$

17. Find the power factor of the circuit as shown in Fig. 25.40. Given is $R = 20\ \Omega$, $C_1 = \frac{20}{\pi}\ \mu\text{F}$, $L_1 = \frac{5}{\pi}$ H, $X_{C_2} = 20\ \Omega$, $X_{L_2} = 30\ \Omega$. Applied voltage is 220 V, 50 Hz.

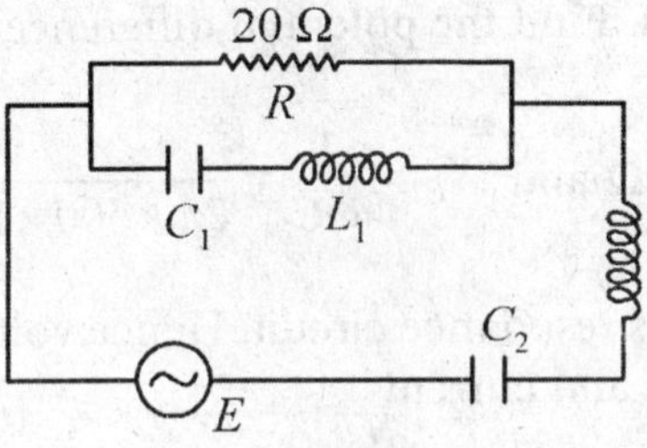

Fig. 25.40

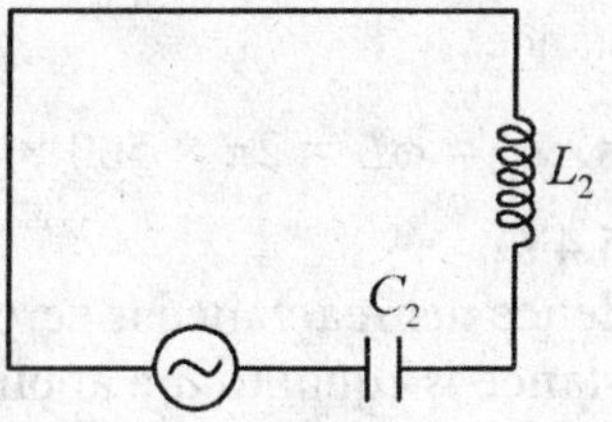

Fig. 25.40 (*a*)

[**Ans.** At this frequency

$$X_{C_1} = \frac{1}{\omega C_1} = \frac{1}{2\pi \times 50 \times \frac{20}{\pi} \times 10^{-6}} = 500\ \Omega$$

and $$X_{L_1} = \omega L_1 = 2\pi \times 50 \times \frac{5}{\pi} = 500\ \Omega$$

Hence net reactance X_1 is zero and the circuit is as shown in Fig. 25.40 (*a*). Power factor $= \frac{R}{Z} = 0$, since $R = 0$]

18. Consider the situation shown in Fig. 25.41. The current in the segment containing the capacitor leads the voltage source by phase angle $\pi/3$. Find the reactance of the capacitor.

[**Ans.** In *R-C* circuit the current leads the emf in phase by angle θ, such that

$$\tan\theta = \frac{X_C}{R} \qquad \because \theta = \pi/3$$

Hence $\tan \pi/3 = \frac{X_C}{50} = \sqrt{3}$ or $X_C = 50\sqrt{3}\ \Omega$]

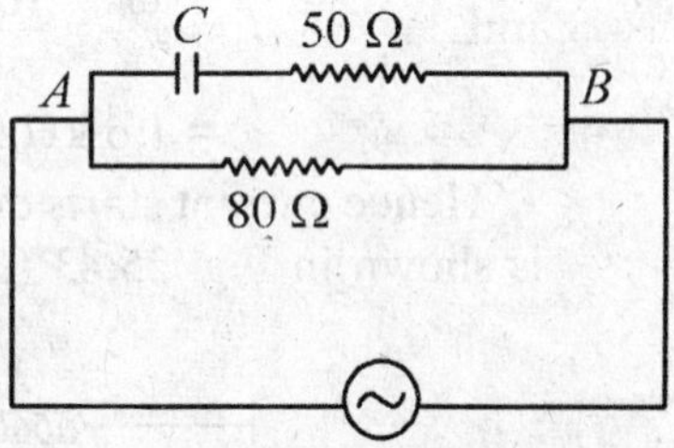

$V = 200 \sin(100\pi t)$

Fig. 25.41

19. For circuit shown in Fig. 25.42, current in inductance is 0.8 A while in capacitance is 0.6 A. What is the current drawn from the source?

[**Ans.** When an AC, $E = E_0 \sin\omega t$ is applied, the current in inductor lags behind the emf in phase by $\pi/2$. Hence current through inductor is

$$i_L = \frac{E_0}{X_L}\sin(\omega t - \pi/2) = -0.8\cos\omega t$$

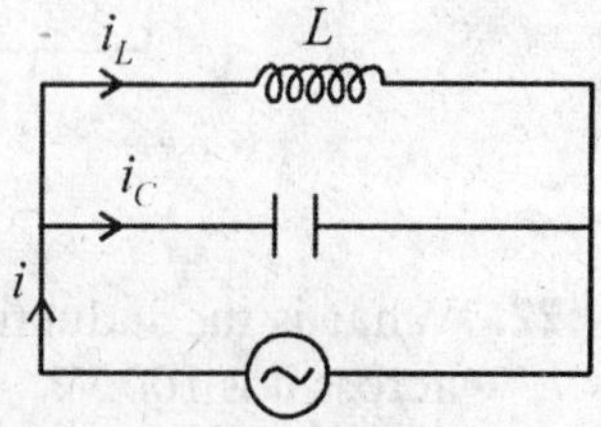

Fig. 25.42

and the current through capacitor leads the emf in phase by $\pi/2$. Hence current in the capacitor is

$$i_C = \frac{E_0}{X_C}\sin(\omega t + \pi/2) = 0.6\cos\omega t$$

So the current drawn = $i = -0.2\cos\omega t$. Negative sign means the current lags behind the emf in phase by $\pi/2$]

20. A 100 volt AC source of frequency 500 Hz is connected to an L-C-R circuit with $L = 8.1$ mH, $C = 12.5$ μF and $R = 10\ \Omega$ all connected in series. Find the potential difference across the resistance.

[**Ans.** $X_L = \omega L = 2\pi \times 500 \times 8.1 \times 10^{-3} = 25.40\ \Omega$ and $X_C = \frac{1}{\omega C} = \frac{1}{2\pi \times 500 \times 12.5 \times 10^{-6}}$ $= 25.4\ \Omega$.

Hence net reactance is zero means the circuit is resonance circuit. Hence voltage across resistance is equal to the applied voltage or 100 V and current

$$i_R = \frac{100}{10} = 10\text{ A}\,]$$

21. In an L-C circuit inductance 0.01 H and capacity 1 μF is connected to a variable frequency source as shown in Fig. 25.43 (*a*). Draw a rough sketch of the current variation as the frequency is changed from 1 kHz to 2 kHz.

[**Ans.** Here L and C are connected in parallel and acts as a rejecter circuit with rejecter frequency (resonance frequency)

$$\omega_0 = \frac{1}{\sqrt{LC}} = \frac{1}{\sqrt{(0.01)(10^{-6})}} = 10^4\text{ rad/s}$$

and $$f = \frac{\omega}{2\pi} = \frac{10 \times 10^3}{2\pi} = 1.6\text{ kHz}$$

Hence current starts decreasing and becomes zero at 1.6 kH and then increases. The plot is shown in Fig. 25.43 (*b*)]

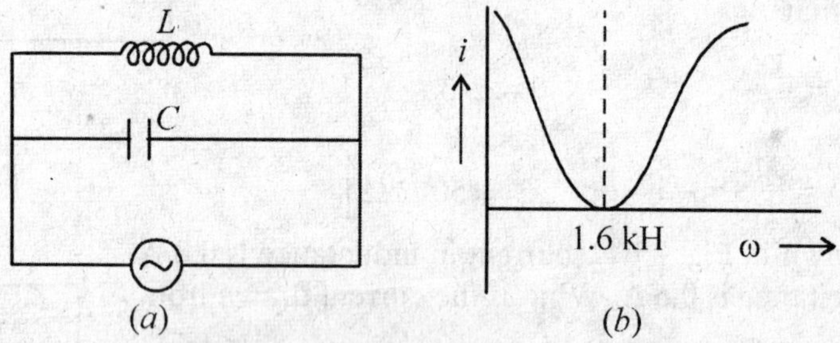

Fig. 25.43

22. What is the inductive reactance of a coil if the current through it is 20 mA and voltage across it is 100 V?

$$[\textbf{Ans.}\ X_L = \frac{V}{I} = \frac{100}{20 \times 10^{-3}} = 5 \times 10^3\ \Omega = 5\text{ k}\Omega\,]$$

23. A voltage of 10 V and frequency 1000 Hz is applied to a 0.1 μF capacitor in series with a resistor of 500 Ω. Find the power factor of the circuit and power dissipated.
[**Ans.** Impedance of the circuit is

$$Z = \sqrt{R^2 + \left(\frac{1}{\omega C}\right)^2} = \sqrt{(500)^2 + \left(\frac{1}{2\pi \times 1000 \times 10^{-7}}\right)^2}$$

or $$Z = \sqrt{(500)^2 + \left(\frac{5000}{\pi}\right)^2} = 500\sqrt{1 + \left(\frac{10}{\pi}\right)^2} = 500 \times 3.34\ \Omega$$

and power factor $\cos\theta = \dfrac{R}{Z} = \dfrac{500}{500 \times 3.34} = 0.3$

$$\text{Power dissipated} = \frac{V^2 R}{Z^2} = \frac{(10)^2 \times 500}{(500 \times 3.34)^2} = \frac{100}{500 \times (3.34)^2} = 0.018\ \text{W}\ ]$$

24. A group of electric lamps having a total power rating of 1000 watt is supplied by an AC voltage $E = 200\sin(310t + 60°)$ then find the rms value of the current in the circuit ?

[**Ans.** Power $= \dfrac{i_0 E_0}{2} \cos\theta$

Since total power is 1000 watt. Hence

$$1000 = \frac{i_0 \times 200}{2} \cos 60 = i_0 \times \frac{200}{2 \times 2}$$

or $$i_0 = \frac{1000 \times 4}{200} = 20\ \text{A}$$

Hence rms value of the current is $\dfrac{i_0}{\sqrt{2}} = \dfrac{20}{\sqrt{2}} = 10\sqrt{2}\ \text{A}$]

25. A coil of negligible resistance has reactance 22 Ω at 200 Hz. It is connected to a 200 V – 50 Hz supply. Find the inductance of the coil and the current in the circuit.
[**Ans.** If L is inductance then $\omega L = 22$ where $f = 200$ Hz.

$$\therefore \quad L = \frac{22}{\omega} = \frac{22}{2\pi \times 200} = 1.75 \times 10^{-2}\ \text{H} = 17.5\ \text{mH}$$

Reactance of the coil at frequency 50 Hz is

$$X_L = 2\pi \times 50 \times 17.5 \times \times 10^{-3} = \frac{22}{4} = 5.5\ \Omega$$

Hence current in the coil is $i = \dfrac{200}{5.5} = 36.3\ \text{A}$]

MULTIPLE CHOICE QUESTIONS

PART I

In the following, a question is followed by four alternatives, of them only one is correct. Therefore, the candidate has to select the correct choice from amongst the alternatives.

1. The peak power consumed by a resistive coil when connected to an AC source is 80 W. The energy consumed by the coil in 100 sec which is many times larger than the time period of the source, is
   (a) 8.0 kJ (b) 4.0 kJ
   (c) 2 kJ (d) zero.

2. A transformer has 50 turns in the primary and 100 in the secondary. If the primary is connected to a 220 V DC supply; the voltage across the secondary is
   (a) zero (b) 440 V
   (c) 110 V (d) non of the above.

3. Which of the following plots (Fig. 25.44) may represent the reactance of a series LC combination?

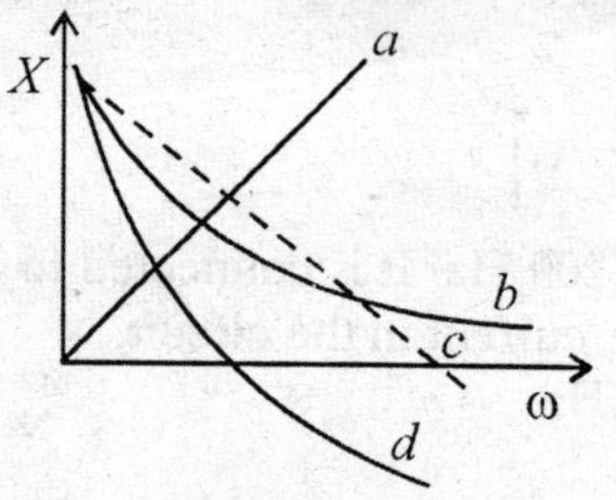

Fig. 25.44

4. An AC source producing emf $E = E_0[\cos(100\pi t) + \cos(500\pi t)$ is connected in series with a capacitor and a resistor. The steady state current in the circuit is found to be $i = i_1\cos[(100\pi t) + \phi_1] + i_2\cos[500\pi t + \phi_2]$, then
   (a) $i_1 > i_2$
   (b) $i_1 = i_2$
   (c) $i_1 < i_2$
   (d) information is insufficient to find the relation between i_1 and i_2.

5. An alternating current is given by $i = i_1\cos\omega t + i_2\sin\omega t$, the rms current is given by
   (a) $\dfrac{i_1 + i_2}{\sqrt{2}}$ (b) $\dfrac{|i_1 + i_2|}{\sqrt{2}}$
   (c) $\dfrac{\sqrt{i_1^2 + i_2^2}}{2}$ (d) $\sqrt{\dfrac{i_1^2 + i_2^2}{2}}$.

6. An alternating current having peak value 10 A is used to heat a metal wire. To produce the same heating effect a constant current i can be used; where i is
   (a) 10 A (b) 14 A
   (c) 7 A (d) 20 A.

7. An AC source is rated 220 V, 50 Hz. The average voltage is calculated in time internal of 0.01 s, it
   (a) must be zero (b) may be zero
   (c) is never zero (d) is $\dfrac{220}{\sqrt{2}}$ V.

8. The time required for a 50 Hz alternating current to change its value from zero to rms value is (in milli seconds)
   (a) 2 (b) 2.5
   (c) 1 (d) 4.

9. An AC source $V = V_0\sin\omega t$ is connected to a circuit as shown in Fig. 25.33. If $-\dfrac{1}{\sqrt{LC}}$ the reading of ammeter is
   (a) $\dfrac{V_0}{R}$ (b) $\dfrac{V_0}{\sqrt{2}R}$
   (c) $\dfrac{V_0}{\sqrt{2}\omega L}$ (d) $\dfrac{V_0}{\sqrt{2}}\omega C$.

10. The phase difference between the alternating current and emf is $\pi/2$. Which of

the following can not be the constituent of the circuit? (AIEEE)

(a) L alone (b) L-C
(c) L-R (d) C alone.

11. A circuit has a resistance of 12 Ω and an impedance of 15 Ω. The power factor of the circuit will be (AIEEE)

(a) 0.125 (b) 1.25
(c) 0.4 (d) 0.8.

12. Power in an AC circuit is given by $P = E_{rms}\, i_{rms} \cos\theta$. The value of power factor $\cos\phi$ in series L-C-R circuit at resonance is

(a) zero (b) 1
(c) 1/2 (d) $1/\sqrt{2}$.

13. A 40 Ω electric heater is connected to a 200 V, 50 Hz main supply. The peak value of electric current flowing in the circuit is approximately

(a) 2.5 A (b) 5.0 A
(c) 7 A (d) 10 A.

14. The average power dissipated in AC circuit is, (where ϕ is phase difference between emf and current in the circuit)

(a) $\frac{1}{2} E_0 I_0 \sin\phi$ (b) $\frac{1}{2} E_0 I_0 \cos\phi$

(c) $\frac{1}{2} E_0 I_0 \tan\phi$ (d) $\frac{1}{2} E_0 i_0\, \phi$.

15. Which of the following curves represents capacitive reactance versus angular frequency graph?

(a) 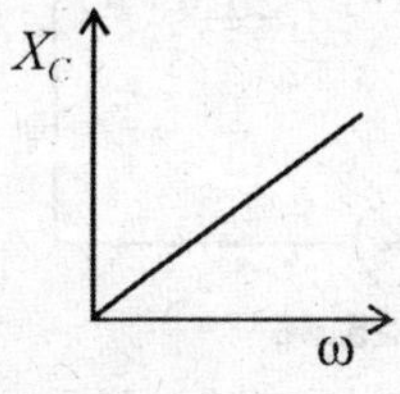

(b) 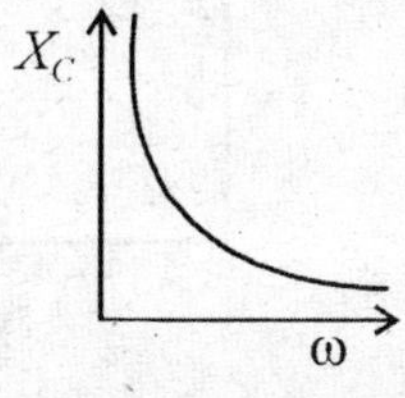

(c) 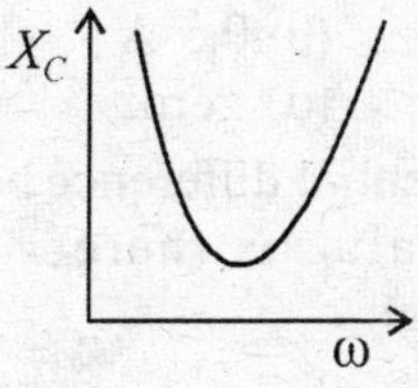

(d) 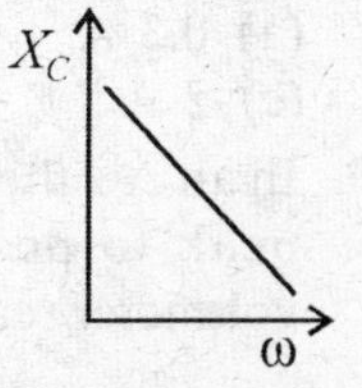

16. At resonance in the circuit (Fig. 25.45)

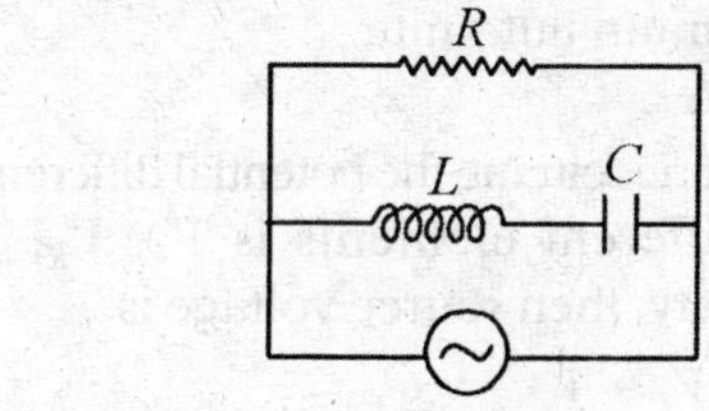

Fig. 25.45

(a) the power factor is zero
(b) the current through the AC source is zero
(c) the current through the AC source is maximum
(d) the current through L and R are equal.

17. An inductor L, a capacitor C and ammeters A_1, A_2 and A_3 are connected to an oscillator in the circuit shown in Fig. 25.46. When the frequency of the oscillator is changed, then at resonance frequency, the ammeter reading is zero in the case of

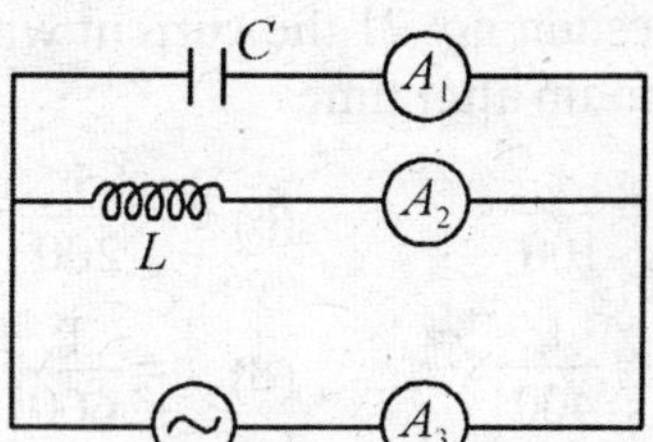

Fig. 25.46

(a) Ammeter A_1 (b) Ammeter A_2
(c) Ammeter A_3 (d) all the three.

18. A circuit is made up of a resistance 1 Ω and inductance 0.01 H and alternating emf 200 V at 50 Hz. The phase difference between the current and the emf in the circuit is

(a) $\tan^{-1}\pi$ (b) $\tan^{-1}(\pi/2)$
(c) $\tan^{-1}(\pi/4)$ (d) $\tan^{-1}(\pi/3)$.

19. Current in resistor R at resonance is

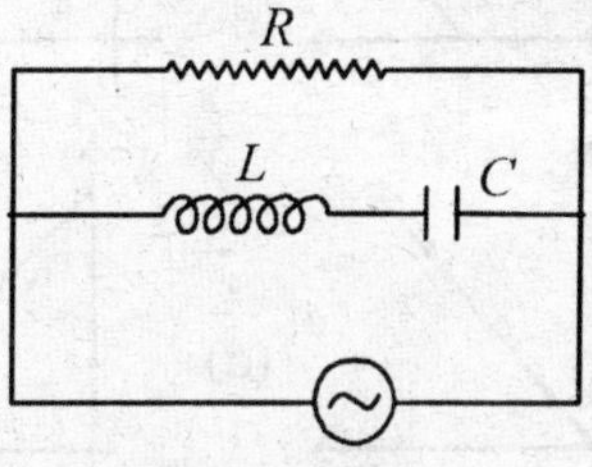

Fig. 25.47

(a) zero (b) minimum but finite
(c) maximum but finite
(d) infinite.

20. In *L-C-R* series circuit, the potential difference across different elements is V_L, V_R, V_C respectively, then source voltage is

(a) $V_R + V_L + V_C$
(b) $\left(V_R^2 + V_L^2 + V_C^2\right)^{1/2}$
(c) $\left[V_R^2 + (V_L + V_C)^2\right]^{1/2}$
(d) $\left[V_R^2 + (V_L - V_C)^2\right]^{1/2}$

21. In an AC circuit instantaneous value of current is $i = 2\sin(100\pi t + \pi/3)$ A. The value of current at time $t = 0$ is

(a) $2\sqrt{3}$ A (b) $\sqrt{3}$ A
(c) $\frac{\sqrt{3}}{2}$ A (d) zero.

22. In question no. 21 the current will be first maximum after time

(a) $t = \frac{1}{100}$ s (b) $t = \frac{1}{200}$ s
(c) $t = \frac{1}{400}$ s (d) $t = \frac{1}{600}$ s.

23. A bulb is rated 220 V, 100 Watt. The resistance of its filament is

(a) 565 Ω (b) 704 Ω
(c) 968 Ω (d) 1210 Ω.

24. In an AC circuit the reactance of coil varies with frequency as

(a) 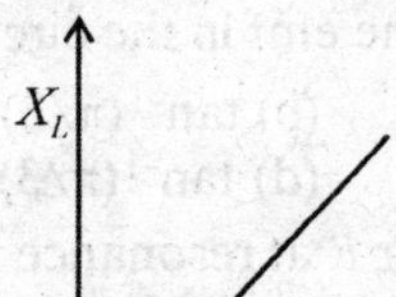

(b) 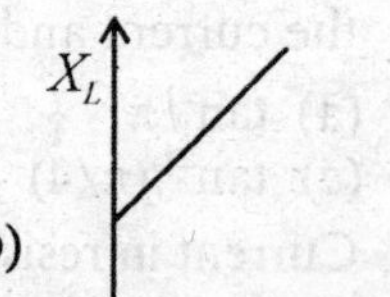

(c) 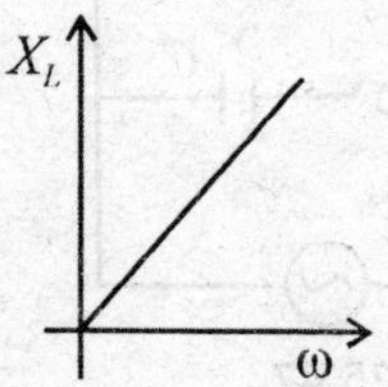

(d) 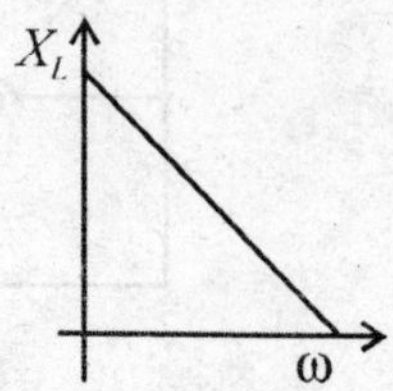

25. Which of the following do not have dimensions of frequency?

(a) $\frac{1}{RC}$ (b) $\frac{R}{L}$
(c) $\frac{1}{\sqrt{LC}}$ (d) $\frac{C}{L}$.

26. In an *L-C-R* series circuit power factor is 1 means

(a) $R = 0$ (b) $L = C$
(c) $\omega L = \frac{1}{\omega C}$ (d) $\omega L - \frac{1}{\omega C} = R$.

27. An AC circuit has resonance frequency 100 Hz. If the frequency is made 1 kHz, then impedance

(a) increases 10 times
(b) remains unchanged
(c) increases and becomes inductive
(d) increases and becomes capacitive.

28. For resonance

(a) current maximum and $\pi/2$ phase difference between i and V
(b) current maximum and 0 phase difference between i and V
(c) voltage maximum and $\pi/2$ phase difference between i and V
(d) voltage maximum and 0 phase difference between i and V.

29. In *L-C-R* series circuit, if the resistance of the circuit is 40 Ω, then current through inductance is

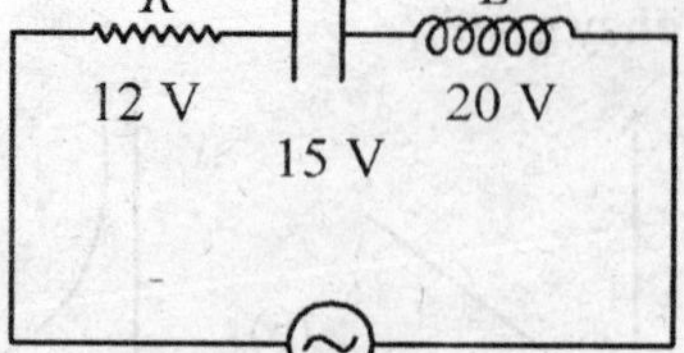

Fig. 25.48

(a) 0.3 A (b) 0.5 A
(c) 2 A (d) zero.

30. In an AC the potential difference between peak to peak value is (here V is the voltmeter reading)

(a) $\frac{V}{\sqrt{2}}$ (b) $\sqrt{2}$ V

(c) $2\sqrt{2}$ V (d) $\frac{V}{2\sqrt{2}}$.

31. The correct phase diagram between current i and voltage V in R-C AC circuit is

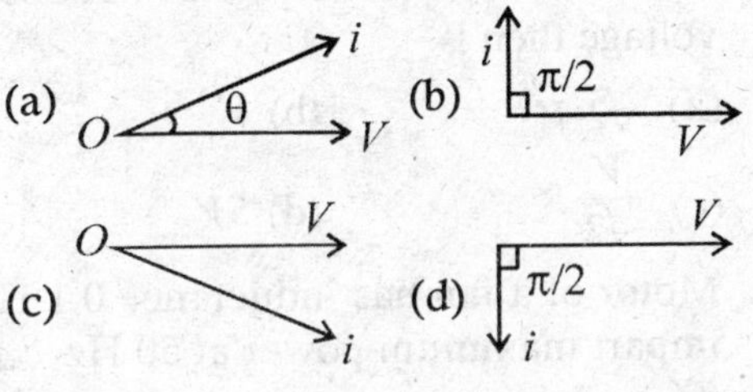

32. In an L-C-R series circuit, resistance is 40 Ω and resonance frequency is 500 Hz. If resistance is made 10 Ω, the resonance frequency will become

(a) 1000 Hz (b) 500 Hz
(c) 250 Hz (d) will not resonate.

33. An inductor, a resistor and a capacitor are joined in series with an AC source. As the frequency of the source is slightly increased from a very low value, the reactance

(a) of the inductor increases
(b) of the resistance increases
(c) of the capacitor increases
(d) of the circuit increases.

34. A circuit has resistance R (= 50 Ω) and the phase difference between current and emf is $\pi/3$ then reactance of the circuit is

(a) 50 Ω (b) $25\sqrt{3}$ Ω
(c) 25 Ω (d) none of the above.

35. In above question the impedance of the circuit is

(a) 50 Ω (b) 100 Ω
(c) $50\sqrt{3}$ Ω (d) 75 Ω.

36. Form factor for a sinusoidal voltage is

(a) $\pi\sqrt{2}$ (b) $2\sqrt{2}\,\pi$

(c) $\frac{\pi}{\sqrt{2}}$ (d) $\frac{\pi}{2\sqrt{2}}$.

37. The dimensions of LC are

(a) $M^0L^0T^0A^0$ (b) $M^0L^1T^2A^0$
(c) $M^0L^0T^2A^0$ (d) $M^0L^0T^4A^0$.

38. In L-C-R series circuit, if L is decreased by 20% and C is increased by 25% the resonance frequency

(a) will increase by 10%
(b) will decrease by 10%
(c) will remain unchanged
(d) will increase by 2.5%.

39. On superposition of currents $i = i_0 \sin\omega t$ and $i' = i_0 \cos\omega t$, the rms value of the current becomes

(a) i_0 (b) $\frac{i_1}{\sqrt{2}}$

(c) $\sqrt{2}i_0$ (d) zero.

40. In Fig. 35.49 the capacitor is charged to voltage V and then the key is pressed. The charge on the capacitor oscillates with frequency ω. To make frequency double

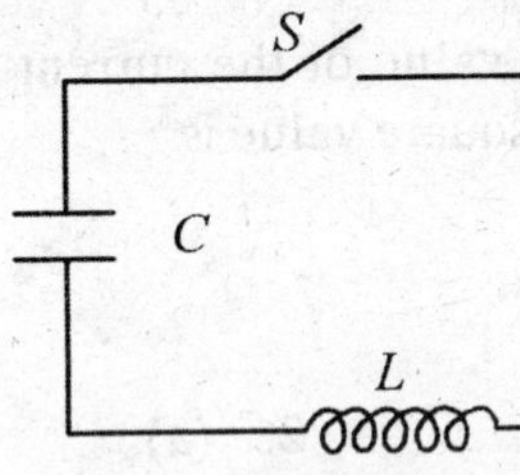

Fig. 25.49

(a) voltage V should be doubled
(b) both L and C should be halved
(c) both L and C should be doubled
(d) L should be doubled keeping C same.

41. The Fig. 25.50 shows variation of X_C and X_L with frequency f. The point showing resonance frequency is

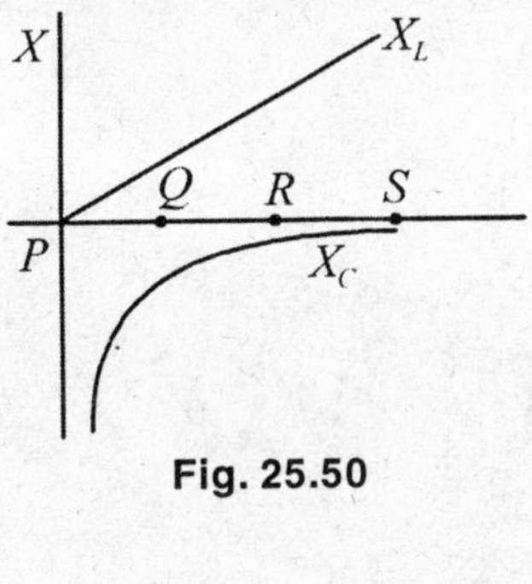

Fig. 25.50

(a) P (b) Q
(c) R (d) S.

42. In the *R-C* circuit shown in Fig. 25.51, the potential difference across capacitor is 3 V. The potential difference across R_L is

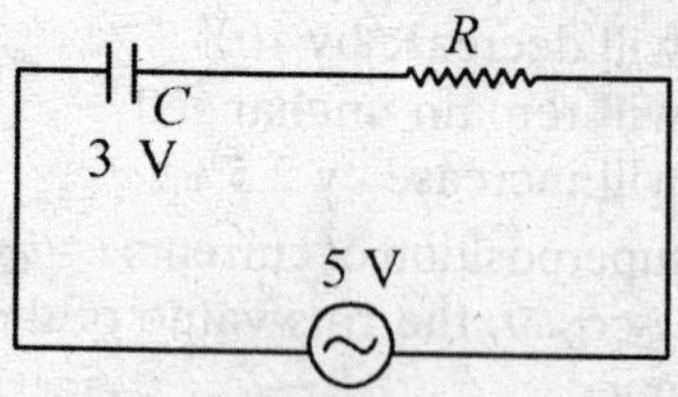

Fig. 25.51

(a) 4 V (b) 2 V
(c) 1 V (d) 0 V.

43. If in a circuit, impedance is given by $Z = 100\ L30°\ \Omega$. The resistance in the circuit then is

(a) 50 Ω (b) 100 Ω
(c) $50\sqrt{3}\ \Omega$ (d) $\frac{100}{\sqrt{3}}\ \Omega$.

44. If the peak value of the current is $\sqrt{2}$ A then mean square value is

(a) 1 A^2 (b) $\frac{1}{\sqrt{2}}\ A^2$
(c) 2 A^2 (d) $2\sqrt{2}\ A^2$.

45. In *L-C-R* series circuit if voltage across resistor is V, across capacitor is $V_C = 2V$ and across inductor is $V_L = 3V$. The source voltage then is

(a) $\sqrt{2}\ V$ (b) V
(c) $\frac{V}{\sqrt{2}}$ (d) $5V$.

46. Motor of a fan has inductance 0.1 H. To impart maximum power at 50 Hz

(a) a 30 Ω resistance must be connected in series
(b) a 100 μF capacitor must be connected in series
(c) a 50 μF capacitor must be connected in series
(d) a 30 Ω resistance should be connector in series.

ANSWERS

1. (b)	2. (a)	3. (d)	4. (c)	5. (d)
6. (c)	7. (b)	8. (b)	9. (b)	10. (c)
11. (d)	12. (b)	13. (c)	14. (b)	15. (b)
16. (c)	17. (c)	18. (a)	19. (a)	20. (d)
21. (b)	22. (d)	23. (c)	24. (c)	25. (d)
26. (c)	27. (c)	28. (b)	29. (a)	30. (c)
31. (a)	32. (b)	33. (a)	34. (d)	35. (b)
36. (d)	37. (c)	38. (c)	39. (a)	40. (b)
41. (b)	42. (a)	43. (c)	44. (a)	45. (a)
46. (b).				

PART II

In the following, a question is followed by four alternatives, of them one or more are correct. Candidate has to select the correct choice(s) from amongst the alternatives.

1. A resistor R, an inductor L and a capacitor C are connected in series with an AC source of variable frequency. The voltmeters, V_1, V_2 and V_3 are connected as shown in Fig. 25.52. Then at resonance frequency

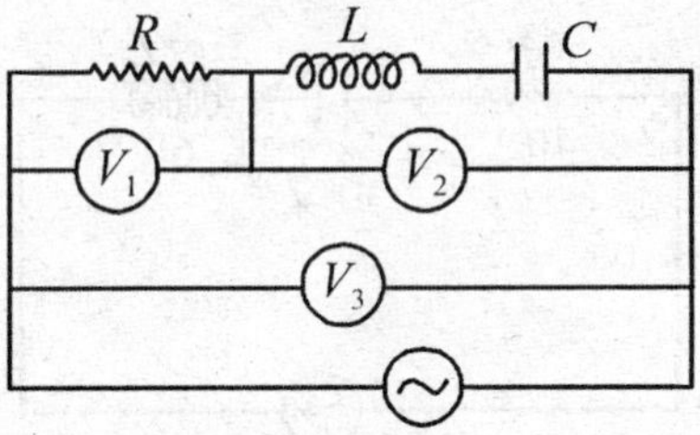

Fig. 25.52

(a) V_1 will show reading equal to that shown by V_3
(b) V_2 and V_3 will show same reading
(c) V_2 will show zero reading
(d) V_1 will show zero reading.

2. In L-C-R series circuit shown in Fig. 25.53 $LC = 10^{-5}$, then

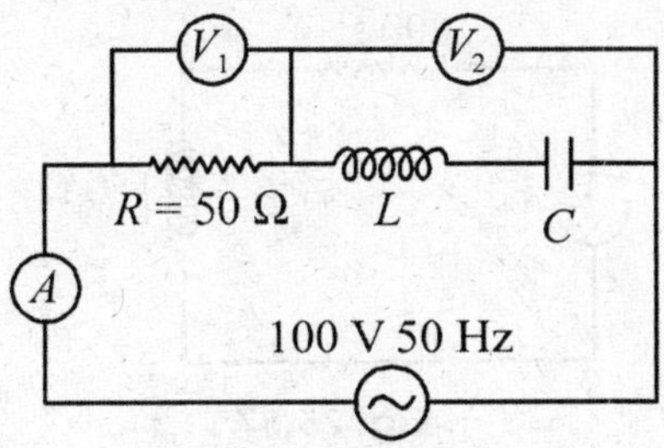

Fig. 25. 53

(a) voltmeter V_1 reads 100 V
(b) ammeter A reads 2 A
(c) voltmeter V_2 reads zero
(d) the power factor of the circuit is one.

3. In an AC circuit, the instantaneous value of current is given by $i = 2\sin(100\pi t + \pi/3)$, then

(a) the value of current at 0.5 s is $\sqrt{3}$ A
(b) the value of current at 0.2 s is $\sqrt{3}$ A
(c) the phase angle of current at 0.2 s is $\frac{61\pi}{3}$ rad
(d) the phase angle of current at 0.5 s is $\frac{61\pi}{3}$ rad.

4. Which of the following have dimensions of frequency?

(a) $\frac{1}{RC}$ (b) $\frac{R}{L}$
(c) $\sqrt{\frac{L}{C}}$ (d) $\frac{C}{L}$.

5. In a circuit, the instantaneous value of current and emf is given by $i = 4\sin(100\pi t + \pi/3)$ and $E = 12\sin(100\pi t)$, then

(a) the power factor is 0.5
(b) impedance of the circuit is 3 Ω
(c) the resistance in circuit is 1.5 Ω
(d) the instantaneous power at $t = 0.25$ sec is zero.

6. In a circuit the current and emf is given by $i = \sin(314t + \pi/3)$ and $E = 200\sin 314t$, then

(a) possible elements in circuit are L-C-R is series
(b) the circuit may contain L-R in series
(c) the circuit may contain R-C in series
(d) the circuit must contain L-C in series.

7. In a circuit, the current and emf are given by $i = \sin\omega t$ and $E = 100\sin(\omega t - \pi/3)$, then, (the frequency $f = 50$ Hz)

(a) vector diagram is (Fig. 25.54)
(b) resistance of the circuit is 50 Ω
(c) ratio of reactance and resistance is $1/\sqrt{3}$

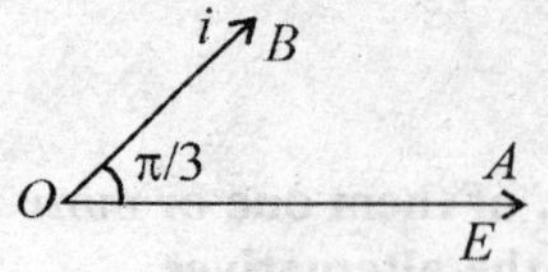

Fig. 25.54

(d) possible elements of the circuit are *L-R*.

8. In an AC circuit, instantaneous current $i = i_0 \sin\omega t$ and emf $e = e_0 \sin(\omega t + \pi/3)$

(a) Peak value of wattless current is $i_0/2$

(b) Peak value of wattless current is $i_0 \cdot \frac{\sqrt{3}}{2}$

(c) Average power consumed is $\frac{i_0 e_0}{4}$

(d) Average power consumed is $\frac{i_0 e_0 \sqrt{3}}{4}$.

9. In an *L-C-R* series circuit voltage across resistance is 40 V, across inductance 90 V and across capacitor is 60 V

(a) the power factor in the circuit is 0.8

(b) the ratio of reactance and resistance is 0.75

(c) applied voltage is 70 V

(d) ratio of reactance and impedance is 0.6.

10. In an AC circuit there is only pure inductance *L* and current of peak value i_0 at frequency ω is passing then

(a) power consumed is zero

(b) Power consumed is $\frac{1}{2} I_0^2 \omega L$

(c) If in addition to *L* a resistance *R* is also connected in series, then power consumed is $\frac{1}{4} I_0^2 R$.

(d) If in addition to *L* pure capacitor of capacitance *C* is connected then power consumed is zero.

11. In Fig. 25.55, *L-C-R* series circuit is shown. If resistance in the circuit is 40 Ω, then

(a) inductive reactance is 67 Ω

(b) current through capacitor 0.3 A

(c) applied voltage is 13 V

(d) current and emf are in same phase.

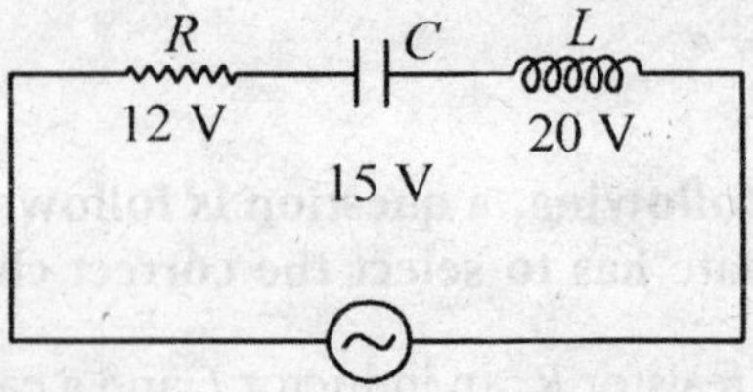

Fig. 25.55

12. In *L-C-R* series circuit shown in Fig. 25.56. If the resistance in the circuit is 40 Ω and the angular frequency of the applied emf is 100 rad/s, then

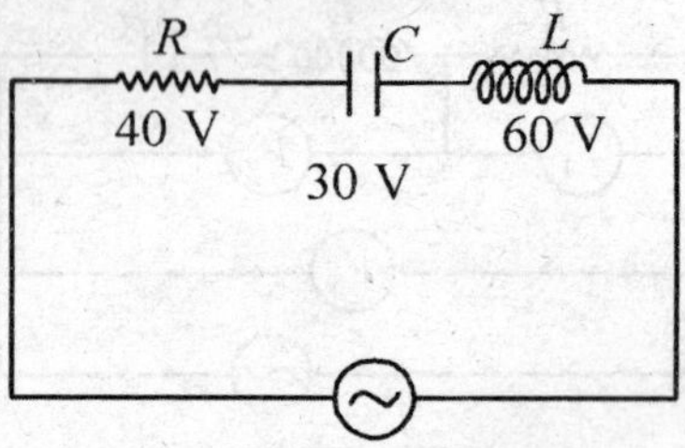

Fig. 25.56

(a) inductance is 0.6 H

(b) current leads the emf in phase by an angle $\theta = \tan^{-1}(0.75)$

(c) the capacity of capacitor is $\frac{1}{3}$ mF

(d) power dissipated in the circuit is 40 W.

13. In the circuit shown in Fig. 25.57

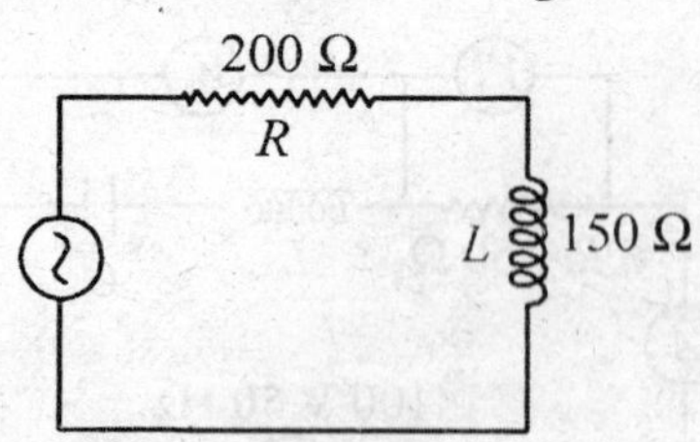

Fig. 25.57

(a) emf is $\tan^{-1}(0.75)$ ahead of current in phase

(b) emf lags behind current in phase by $\tan^{-1}(0.75)$

(c) impedance of the circuit is 250 Ω

(d) impedance of the circuit is 350 Ω.

14. An AC source of voltage 100 V is connected to pure ohmic resistance of value 20 Ω.

(a) Average power consumed is 500 Watt
(b) Average power consumed is 1000 Watt
(c) Wattless current is 5 A
(d) Wattless current is zero.

15. Inductive reactance becomes
(a) zero for DC
(b) finite for AC
(c) depends on frequency of AC
(d) finite for DC.

16. The reactance of a circuit is zero. It is possible that the circuit contains
(a) an inductor and a capacitor
(b) an inductor but no capacitor
(c) a capacitor but no inductor
(d) neither an inductor nor a capacitor.

17. In an AC series circuit, the instantaneous current is zero when the instantaneous voltage is maximum. Connected to the source may be a
(a) pure inductor
(b) pure capacitor
(c) pure resistor
(d) combination of an inductor and a capacitor.

18. An inductor coil having some resistance is connected to an AC source. Which of the following quantities have zero average value over a cycle?
(a) Current
(b) Induced emf in the inductor
(c) Joules heat
(d) Magnetic energy stored in the inductor.

19. An AC source rated 100 V supplies a current of 10 A to a circuit. The average power delivered to the source
(a) must be 1000 W
(b) may be 1000 W
(c) may be greater than 1000 W
(d) may be less than 1000 W.

20. In *L-C-R* sereis circuit, emf and current are given by $i = i_0 \sin(\omega t + \phi)$ and $E = E_0 \sin\omega t$, then ϕ can have value
(a) $\phi < \pi/2$ (b) $\phi = 0$
(b) $\phi > \pi/2$ (d) $-\pi/2 < \phi < \pi/2$.

21. An *L-C-R* series circuit is connected to a 100 V AC source. If $X_L = 20\ \Omega$, $X_C = 20\ \Omega$ and $R = 5\ \Omega$, then
(a) potential difference across *L* and *C* combine is zero
(b) potential difference across inductance is 400 V
(c) wattless current is zero
(d) wattless current 20 A.

22. An *L-C-R* series circuit has resistance *R* (= 100 Ω), and is connected to an AC voltage of 200 V and frequency 300 rad/s. If inductance is removed from the circuit; the current leads the emf by 60°. If only capacitor is removed, the current lags behind the emf in phase by 60°, then
(a) current in circuit is 2 A
(b) power loss in the circuit is 400 W
(c) inductive reactance in the circuit is $100\sqrt{3}\ \Omega$
(d) capacitive reactance in the circuit is $100\sqrt{3}\ \Omega$.

23. Select the correct statement.
(a) In *L-C-R* series, resonance circuit power factor is 1
(b) In *C-R* series circuit the power factor is $\dfrac{\omega RC}{\sqrt{[1+(R\omega C)^2]}}$
(c) In *L-R* series circuit power factor $\dfrac{R}{\sqrt{R^2+L^2}}$
(d) In *L-C* circuit power factor is 1.

24. A bulb is connected with 200 V, 50 Hz supply. The bulb draws a current of 10 A. When connected to DC source
(a) inductance of the bulb filament is 55 mH
(b) inductance of the bulb filament is 75 mH
(c) power factor for the circuit is 0.2
(d) power factor for the circuit is 0.5.

25. In *L-C-R* series resonance circuit, reactance of coil is 400 Ω and resistance is 50 Ω. If applied voltage is *E* then

(a) potential difference across coil is 8*E*
(b) potential difference across resistance is *E*/8
(c) power factor of the circuit resistance is 1/8
(d) wattless current is approximately by *E*/50 A.

26. In a circuit, current flowing is given by $i = 2\sin 4\pi t$ A, then

(a) average value of current from time $t = 0.25$ s to 0.75 s is zero
(b) average value of current from time $t = 0$ to $t = T/2$ is $4/\pi$ A
(c) average value of current from time $t = T/2$ to $t = T$ is zero
(d) average value of current for one cycle is 1 A.

27. In Fig. 25.58 log*f* (*f* is the frequency of the applied AC) of constant voltage and is plotted on *X*-axis then

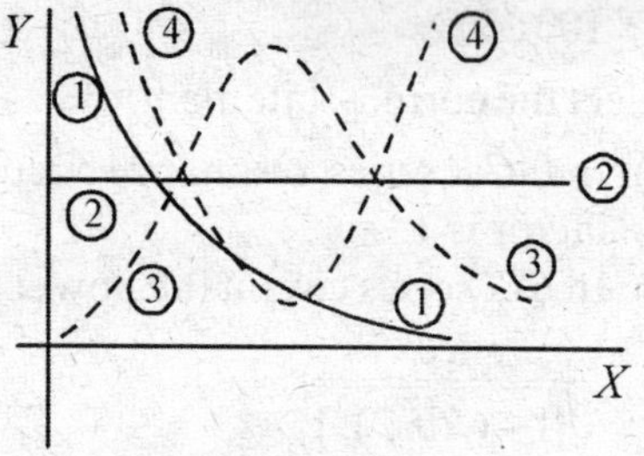

Fig. 25.58

(a) If $|X_C|$ is plotted on *Y*-axis, the graph is (1)
(b) If *Z* is plotted on *Y*-axis, the graph is (4)
(c) If current *i* is plotted on *Y*-axis, the graph is (3)
(d) If resistance is plotted on *Y*-axis, the graph is (2).

28. Figure 25.59 shows variation of X_C, X_L and *R* with

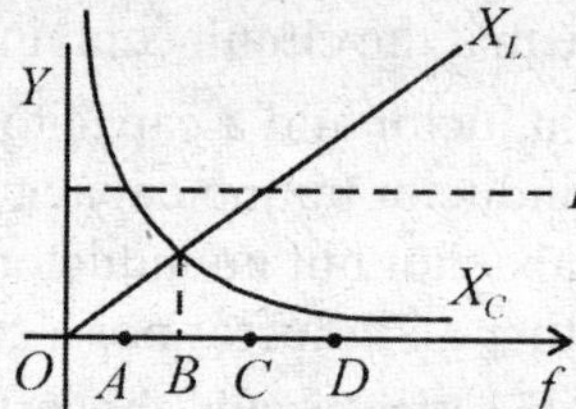

Fig. 25.59

(a) the frequency corresponding to which the circuit is capacitive corresponds to point *A*
(b) the frequency corresponding to which the circuit is pure resistive corresponds to point *B*
(c) the frequency corresponding to which the circuit is inductive corresponds to point *C*
(d) the frequency corresponding to which the circuit resonates corresponds to point *D*.

ANSWERS

1. (a, c)	2. (a, b, c, d)	3. (a, b, c)	4. (a, b)	5. (a, b, c, d)
6. (a, c)	7. (a, b)	8. (a, c)	9. (a, b, d)	10. (a, c, d)
11. (a, b, c)	12. (a, c, d)	13. (a, c)	14. (a, d)	15. (a, b, c)
16. (a, d)	17. (a, b, d)	18. (a, b)	19. (b, d)	20. (a, b)
21. (a, b, c)	22. (a, b, c, d)	23. (a, b, c)	24. (a, d)	25. (a, c, d)
26. (a, b)	27. (a, b, c, d)	28. (a, b, c)		

26

Electromagnetic Waves

26.1 INTRODUCTION

We have seen that magnetic fields are produced by electric currents as well as by moving charges and that two current carrying conductors exert force on each other. We have also studied that a time varying magnetic field produces an electric field. James Clark Maxwell was of the opinion that a time varying electric field must produce a magnetic field. We have also seen that in certain situations light may be treated as a wave. Maxwell keeping these facts in mind showed that a beam of light is a travelling wave of electric and magnetic fields—*an electromagnetic wave* and thus optics, the study of visible light, is a branch of electromagnetism. He formulated a set of equations involving electric and magnetic fields and their source, the charges and current densities. These equations are known as *Maxwell's equations*. Together with the Lorentz force formula, they expressed mathematically all the basic laws of electromagnetism.

The most striking result of Maxwell's equations is the existence of electromagnetic waves, i.e. coupled time varying electric and magnetic fields propagate in space. They do not require any medium for their propagation and the speed of waves according to these equations comes out to be very close to the speed of light ($= 3 \times 10^8\ \text{m s}^{-1}$). This led to the important conclusions that light is an electromagnetic wave. We will give brief discussion of this theory here.

26.2 MAXWELL'S DISPLACEMENT CURRENT

According to Ampere's law, the line integral of the magnetic field over a closed path is equal to μ_0 times the current enclosed by the path, i.e.

$$\oint \boldsymbol{B} \cdot \boldsymbol{dl} = \mu_0 i \qquad \ldots(26.1)$$

This equation is valid for non-time varying field. This law tells us that an electric current produces magnetic field and gives a method to calculate the field.

Ampere's law, as it is, (Eq. 26.1) is not valid if the field at the surface is time varying. To show this let us consider a capacitor being charged by the battery (Fig. 26.1). If we place a compass needle in the space between the plates, the needle, in general deflects; showing that a magnetic field exists in this region. Figure 26.1 shows a closed loop C, completely inside the plates. The plane surface of this loop is parallel to the plates.

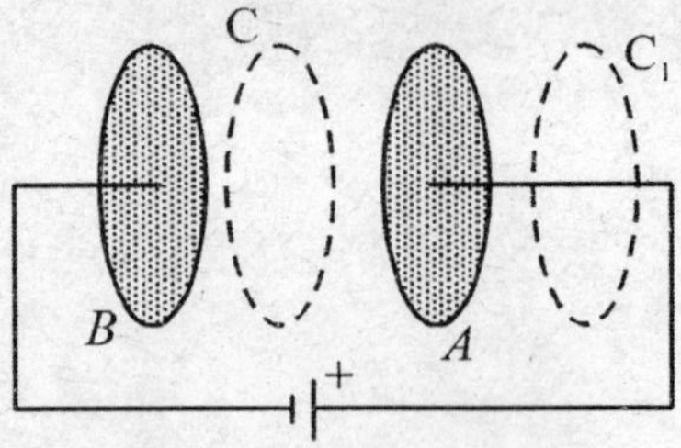

Fig. 26.1

During charging, the electric current flows through the wires and charge accumulates on the plates. Due to this there is a time varying electric field at the surface of the loop. As already stated a magnetic field exists there at the points on the loop. Hence the line integral of magnetic field is non zero, i.e.

$$\oint_c \boldsymbol{B} \cdot \boldsymbol{dl} \neq 0$$

Since no charge crosses through the loop, the electric current enclosed by the loop is zero. Hence

$$\oint_c \boldsymbol{B} \cdot \boldsymbol{dl} \neq \mu_0 i \qquad \ldots(26.2)$$

Now we can use *Biot-Savart's law* to deduce *Ampere's law.* Let the magnetic field due to current elements, using Biot-Savart's law be B' then its line integral for the closed path C is

$$\oint_c \boldsymbol{B'} \cdot \boldsymbol{dl} = 0$$

This shows that the actual field $\boldsymbol{B}$ is different from the field $\boldsymbol{B'}$ produced by the electric currents only. This means there is some other source of magnetic field. Thus other source is nothing but the changing electric field between the plates. As the capacitor gets charged the electric field between the plates changes and this changing electric field produces magnetic field.

We know that a changing magnetic field produces electric field. Faraday's law gives the relation between the two, i.e.

$$\oint \boldsymbol{E} \cdot \boldsymbol{dl} = -\frac{d\phi_m}{dt}$$

where ϕ_m is the magnetic flux through the loop around which the line integral of $\boldsymbol{E}$, the electric field is calculated $\phi_m = \int \boldsymbol{B} \cdot \boldsymbol{ds}$. Now we have come to the conclusion that a time varying electric field produces a magnetic field. The relation between the changing electric field $\boldsymbol{E}$ and the magnetic field $\boldsymbol{B}$ resulting from it is given by

$$\oint \boldsymbol{B} \cdot \boldsymbol{dl} = \mu_0 \varepsilon_0 \frac{d\phi_e}{dt} = \mu_0 \varepsilon_0 \int_s \boldsymbol{E} \cdot \boldsymbol{ds} \qquad \ldots(26.3)$$

Here ϕ_e is the electric flux through the area bounded by the closed path along which the line integral of B is calculated.

Equation (26.1) gives the magnetic field resulting from an electric current due to flow of charges, whereas equation (26.3) gives the magnetic field due to changing electric field. If there is an electric current as well as a changing electric field, then the resultant magnetic field is given by

$$\oint \boldsymbol{B} \cdot \boldsymbol{dl} = \mu_0 i + \mu_0 \varepsilon_0 \frac{d\phi_e}{dt}$$

or

$$\oint \boldsymbol{B} \cdot \boldsymbol{dl} = \mu_0 i + \mu_0 i_d \qquad \ldots(26.4)$$

where $i_d = \varepsilon_0 \frac{d\phi_e}{dt}$, Maxwell called this i_d as *displacement* current and current due to flow of charges is called *conduction current* and is denoted by i_c

Maxwell generalized the Ampere's law from equation (26.1) to equation (26.4).

Continuity of Current

Maxwell's modified form of Amperes law gives

$$\oint \boldsymbol{B} \cdot \boldsymbol{dl} = \mu_0(i_c + i_d)$$

where $i_d = \varepsilon_0 \dfrac{d\phi_c}{dt}$ called displacement current, i_c is the conduction current and ϕ_e is the electric flux through the closed loop C. *The sum of the currents ($i_c + i_d$) has an important property of continuity along any closed path although individually they may not be continuous.*

To prove this consider two loops C and C_1 (Fig. 26.1) having boundaries exactly like the boundaries of the plates of capacitor. Loop C is within the plates and loop C_1 is just outside the plates of the capacitor.

Let at any time t the charge on the capacitor be q and is increasing at the rate $\dfrac{dq}{dt}$. Hence current through the lead wires is

$$i_c = \frac{dq}{dt}$$

and the instantaneous electric field between the plates is

$$E = \frac{\sigma}{\varepsilon_0} = \frac{q}{A\varepsilon_0}$$

where A is area of the plates.

Using Ampere's law for loop C_1 we have

$$\oint_{C_1} \boldsymbol{B} \cdot \boldsymbol{dl} = \mu_0(i_c + i_d)$$

But since at C_1 there is no electric flux hence i_d is zero and we have

$$\oint_{C_1} \boldsymbol{B} \cdot \boldsymbol{dl} = \mu_0 i_i = \mu_0 \frac{dq}{dt} \qquad \ldots(i)$$

For loop C we have

$$\oint_C \boldsymbol{B} \cdot \boldsymbol{dl} = \mu_0(i_c + i_d) = \mu_0 i_d \qquad \ldots(ii)$$

Since within the plates $i_c = 0$

But $\qquad i_d = \varepsilon_0 \dfrac{d\phi_e}{dt}$

The electric flux with loop C is $\phi_e = \int_A \boldsymbol{E} \cdot \boldsymbol{ds}$

or
$$\phi_e = \int_A \frac{q}{A \cdot \varepsilon_0} ds = \frac{q}{A\varepsilon_0} \int_A ds = \frac{q \cdot A}{A\varepsilon_0} = \frac{q}{\varepsilon_0}$$

Hence
$$i_d = \varepsilon_0 \frac{d\phi_e}{dt} = \varepsilon_0 \frac{d}{dt}\left(\frac{q}{\varepsilon_0}\right) = \frac{dq}{dt} = i_c$$

Thus we conclude that the conduction current and displacement currents taken together are continuous, though individually they may not be continuous. Here the conduction current starts from battery and flows up to plate A. The displacement current starts from plate A and ends on plate B and the conduction current starts from plate B and ends at the battery.

26.3 MAXWELL'S EQUATIONS AND LORENTZ FORCE

In absence of any dielectric or magnetic material, the four Maxwell's equations are given below:

1. **Gauss law in electrostatics** is mathematically expressed as

$$\oint_s \boldsymbol{E} \cdot d\boldsymbol{s} = \frac{q}{\varepsilon_0} \quad \ldots(i)$$

This gives the total electric flux through the closed surface in terms of total charge enclosed by it. This law predicts that electric lines of force do not form continuous closed path and existence of the isolated charge.

2. **Gauss law in magnetostatics** is stated as

$$\oint_s \boldsymbol{B} \cdot d\boldsymbol{s} = 0 \quad \ldots(ii)$$

and predicts the non-existence of a monopole and that the magnetic lines of force always form a closed path.

3. **Faraday's law of electromagnetic induction:** This law predicts that a time varying magnetic field produces an electric field. This law gives the relation between induced electric field (E) and changing magnetic flux. It is stated as

$$\oint \boldsymbol{E} \cdot d\boldsymbol{l} = -\frac{d\phi_m}{dt} = -\frac{d}{dt}\int_s \boldsymbol{B} \cdot d\boldsymbol{s} \quad \ldots(iii)$$

This law shows that the line integral of electric field over any closed path (i.e. the emf) is equal to time rate of change of magnetic flux through the surface bounded by the closed path.

4. **Ampere's-Maxwell's Modified Law:** This gives the line integral of magnetic field over a closed path in terms of the total current enclosed by the path, i.e.

$$\oint \boldsymbol{B} \cdot d\boldsymbol{l} = \mu_0(i_c + i_d) \quad \ldots(iv)$$

i_c is the conduction current i_d is the displacement current $\left(i_d = \varepsilon_0 \frac{d\phi_e}{dt}\right)$

It predicts that a time varying electric field produces a magnetic field.

These equations are collectively known as Maxwell's equations.

5. **Lorentz Force:** This gives the force acting on a moving charge in electric and magnetic field.

$$\boldsymbol{F} = q(\boldsymbol{E} + \boldsymbol{v} \times \boldsymbol{B}) \quad \ldots(v)$$

The above five equations give a complete description of all electromagnetic interactions.

The solution of the Maxwell's equations is obtained which gives the wave equation for propagation of electromagnetic waves in X-direction as

$$E = E_0 \sin\omega\left(t - \frac{x}{c}\right) \quad \ldots(26.5)$$

and

$$B = B_0 \sin\omega\left(t - \frac{x}{c}\right) \quad \ldots(26.6)$$

The electric field E and magnetic field B oscillate sinusoidally at position x at time t. The constant c in these equations is the speed of light in vacuum ($c = 3 \times 10^8$ m s^{-1}). The two oscillate mutually perpendicular and perpendicular to the direction of propagation (Fig. 26.2).

Such a combination of mutually perpendicular electric and magnetic fields is referred to as *an electromagnetic wave.*

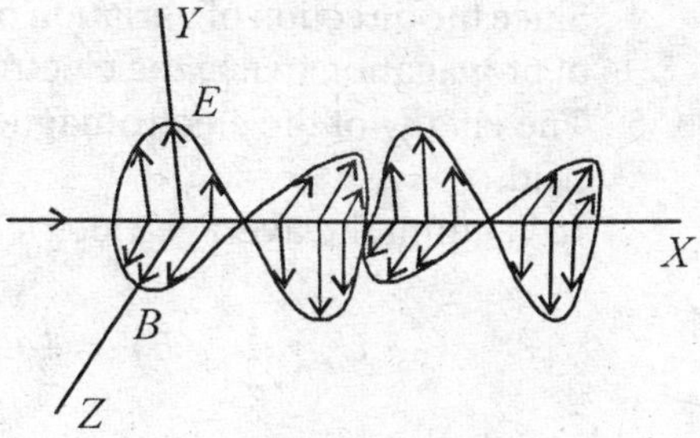

Fig. 26.2

Production of Electromagnetic Waves

We know that an stationary charge produces only electric field around it. A moving charge produces both the electric and magnetic fields. If the charge is moving with constant velocity, the electric and magnetic fields will not change with time and no electromagnetic waves can be produced. But if the charge is moving with non zero acceleration, both electric and magnetic fields so produced will be time varying as well as space dependent, it then produces electromagnetic waves. Thus we come to the conclusion that *accelerated charges produce electromagnetic waves.*

We have already seen that an L-C circuit is an oscillatory circuit. In that the charge oscillates between the plates of a capacitor through an inductor as such it emits electromagnetic waves of the same frequency as that of the oscillating charge.

26.4 PROPERTIES OF ELECTROMAGNETIC WAVES

1. The electromagnetic waves are produced by accelerated charge or oscillating charge. They do not require material medium for their propagation. They can propagate through vacuum also.

   The velocity of these waves depends on the electric and magnetic properties of the medium. In vacuum, velocity is

$$c = \frac{1}{\sqrt{\varepsilon_0 \mu_0}} = 3 \times 10^8 \text{ ms}^{-1} \qquad \text{...(26.7)}$$

   In medium it is given by

$$c_m = \frac{1}{\sqrt{\varepsilon \mu}} \qquad \text{...(26.8)}$$

   where ε and μ are *absolute electric permittivity and absolute magnetic permeability* of medium respectively.

   But $\varepsilon = \varepsilon_0 \varepsilon_r$ and $\mu = \mu_0 \mu_r$

   ε_r and μ_r are *relative electric permittivity (dielectric constant) of the medium and relative magnetic permeability of medium* respectively

$$\therefore \quad n = \frac{c}{c_m} = \sqrt{\varepsilon_r \mu_r} \qquad \text{...(26.9)}$$

   where n is the refractive index of the medium.

3. The equation (26.5) and (26.6), we see that electric and magnetic fields oscillate sinusoidally and are in same phase. The ratio

$$\frac{E}{B} = \frac{E_0}{B_0} = c \qquad \text{...(26.10)}$$

   i.e., the ratio of the magnitudes of electric and magnetic fields at any time is constant and equal to the speed of light in vacuum.

4. Since the direction of variation of electric and magnetic fields is perpendicular to the direction of propagation hence the electromagnetic waves are transverse waves.
5. The energy of the electromagnetic waves is equally divided between electric and magnetic field.

   In vacuum the average electric energy density is given by

$$U_e = \frac{1}{2}\varepsilon_0\langle E\rangle^2 = \frac{1}{2}\varepsilon_0\left(\frac{E_0}{\sqrt{2}}\right)^2 = \frac{1}{4}\varepsilon_0 E_0^2 \qquad \ldots(26.11)$$

   and the average magnetic energy density is given by

$$U_B = \frac{\langle B^2\rangle}{2\mu_0} = \frac{1}{2\mu_0}\left(\frac{B_0}{\sqrt{2}}\right)^2 = \frac{1}{4}\frac{B_0^2}{\mu_0} \qquad \ldots(26.12)$$

   Since $\frac{E_0}{B_0} = c = \frac{1}{\sqrt{\varepsilon_0\mu_0}}$, we have

$$U_e = \frac{1}{4}\varepsilon_0(cB_0)^2 = \frac{1}{4}\frac{c_0B_0}{\varepsilon_0\mu_0} = \frac{B_0^2}{4\mu_0} = U_B$$

   Thus $U_e = U_B$, i.e. the average electric energy density and average magnetic energy density are equal.

   Total energy density

$$U = U_e + U_B = \frac{1}{4}\varepsilon_0 E_0^2 + \frac{1}{4}\frac{B_0^2}{\mu_0}$$

$$= \frac{1}{2}\varepsilon_0 E_0^2 = \frac{1}{2}\frac{B_0^2}{\mu_o} \qquad \ldots(26.13)$$

6. The electric vector is responsible for the optical effects of an electromagnetic wave and is called light vector.
7. The electromagnetic waves are unaffected by the electric and magnetic field.
8. **Intensity of Electromagnetic Waves:** Intensity is defined as the energy crossing unit area in unit time; area is taken perpendicular to the direction of propagation. Consider a cylinder with area of cross-section A and length $c\Delta t$ (Fig. 26.3). Energy contained in this cylinder will cross area A in time Δt. Here c is wave propagation speed.

   Volume of this cylinder is

$$V = c\Delta t \cdot A$$

   Hence energy crossing this area in time Δt is

$$E = U \cdot V$$

$$= \frac{1}{2}\varepsilon_0 E_0^2 c\Delta t \cdot A$$

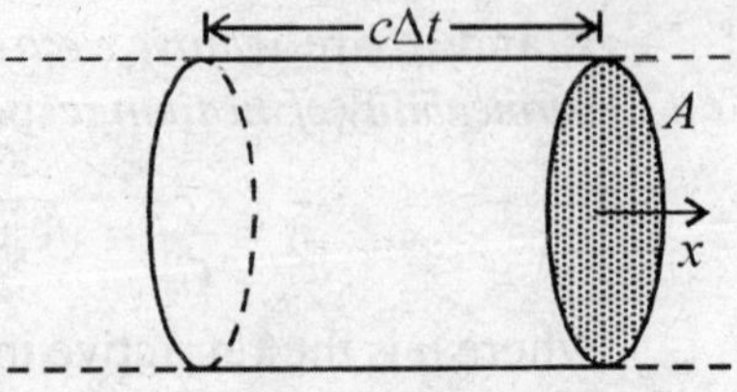

Fig. 26.3

   The intensity then is

$$I = \frac{E}{\Delta t A} = \frac{1}{2}\varepsilon_0 E_0^2 c \qquad \ldots(26.14)$$

9. **Momentum:** Electromagnetic wave also carries linear momentum with it. The linear momentum carried by the portion of wave having energy E is given by

$$p = \frac{E}{c} \qquad \ldots(26.15)$$

Thus if the wave incident on the material surface is totally reflected, the momentum delivered is $\frac{2E}{c}$ because the momentum of the wave changes from p to $-p$. Similarly if the incident wave is absorbed completely by the surface on which it is incident, then momentum transferred is E/c and energy delivered is E. It follows that the electromagnetic wave exerts force on the surface on which they are incident.

10. **Poynting Vector:** The rate of flow of energy in an electromagnetic wave is described by a vector $\boldsymbol{S}$ called the Poynting vector; defined by the equation

$$\boldsymbol{S} = \frac{1}{\mu_0}(\boldsymbol{E} \times \boldsymbol{B}) \qquad \ldots(26.16)$$

The magnitude of the Poynting vector gives the rate at which the energy flows through a unit surface area perpendicular to the direction of wave propagation. Thus it represents power per unit area. Obviously the direction of vector $\boldsymbol{S}$ is along the direction of propagation of the wave. The SI units of Poynting vector are W/m^2. For a plane electromagnetic wave, the magnitude $\boldsymbol{E} \times \boldsymbol{B} = EB$. Then

$$S = \frac{EB}{\mu_0} = \frac{E^2}{\mu_0 c} = \frac{c}{\mu_0} B^2 \qquad \ldots(26.17)$$

This equation for S represent the instantaneous rate at which energy is passing through unit area.

If time average of S is taken over a cycle, it is called intensity. Thus we have

$$I = S_{av} = \frac{E_0 B_0}{2\mu_0} = \frac{E_0^2}{2\mu_0 c} = \frac{cB_0^2}{2\mu_0} \qquad \ldots(26.18)$$

26.5 DERIVATION FOR VELOCITY OF ELECTROMAGNETIC WAVES $c = \frac{1}{\sqrt{\varepsilon_0 \mu_0}}$

The whole subject of electricity and magnetism may be described mathematically with the help of four fundamental equations called Maxwell's equations for electromagnetic waves. We have already discussed them in article (26.3).

In vacuum there are no charges and hence no conduction currents. Under this condition Faraday's law and Ampere's law can be stated as

Faraday's law $\oint \boldsymbol{E} \cdot d\boldsymbol{l} = -\frac{d\phi_B}{dt}$ ...(*i*)

and Ampere's law $\oint \boldsymbol{B} \cdot d\boldsymbol{l} = \mu_0 \varepsilon_0 \frac{d\phi_e}{dt}$ ...(*ii*)

Solution of these equations as plane progressive electromagnetic waves is given by

$$E = E_y = E_0 \sin \omega\left(t - \frac{x}{c}\right) \qquad \ldots(26.19a)$$

$$B = B_z = B_0 \sin \omega\left(t - \frac{x}{c}\right) \qquad \ldots(26.19b)$$

The waves described in these equations propagates along the positive X-direction, the electric field oscillates along Y-direction and the magnetic field oscillates along Z-direction. Amplitude of the fields is E_0 and B_0 as such their magnitude oscillates between $\pm E_0$ and $\pm B_0$ respectively. It is linearly polarised wave; polarised along the Y-axis.

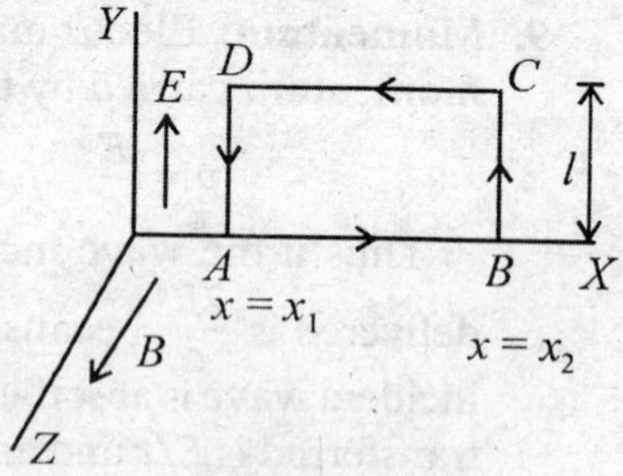

Fig. 26.4 (*a*)

Let us consider a rectangular path $ABCD$ in the X-Y plane as shown in Fig. 26.4 (*a*). Let us find out the line integral of electric field along this path as the electromagnetic wave propagates.

$$\oint \boldsymbol{E}\cdot d\boldsymbol{l} = \int_A^B \boldsymbol{E}\cdot d\boldsymbol{l} + \int_B^C \boldsymbol{E}\cdot d\boldsymbol{l} + \int_C^D \boldsymbol{E}\cdot d\boldsymbol{l} + \int_D^E \boldsymbol{E}\cdot d\boldsymbol{l}$$

$$0 = +E(x_2)l + 0 + E(x_1)(-l)$$

$$= E_0 l\,[\sin\omega\,(t - x_2/c) - \sin\omega\,(t - x_1/c)] \quad \ldots(iii)$$

Now let us calculate the flux ϕ_B of magnetic field through the same $ABCD$ [Fig. (26.4*b*)]. The flux through an strip of width dx at x is

$$B(x)l\,dx = B_0 \sin\omega(t - x/c)l\,dx$$

Hence the flux through the rectangular $ABCD$ is

$$\phi_B = \int_{x_1}^{x_2} B_0 l \sin\omega(t - x/c)dx$$

$$= +\frac{B_0 lc}{\omega}[\cos\omega(t - x/c) - \cos\omega(t - x/c)]$$

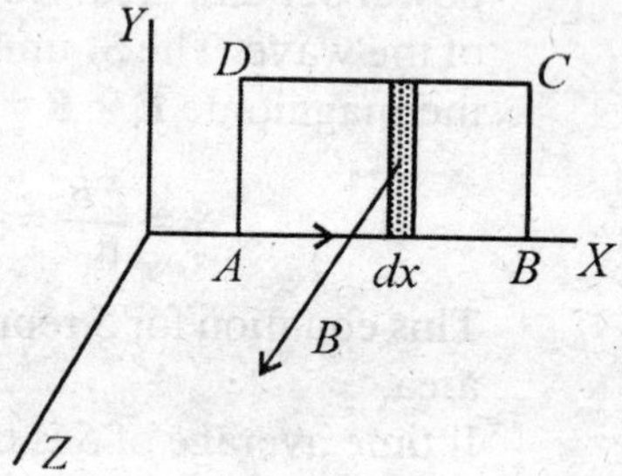

Fig. 26.4 (*b*)

Now from Faraday's law

$$\oint \boldsymbol{E}\cdot d\boldsymbol{l} = -\frac{d\phi_B}{dt} \text{ we have}$$

$$E_0 l\left[\sin\omega\left(t - \frac{x_2}{c}\right) - \sin\omega\left(t - \frac{x_1}{c}\right)\right]$$

$$= B_0 lc\left[\sin\omega\left(t - \frac{x_2}{c}\right) - \sin\omega\left(t - \frac{x_1}{c}\right)\right]$$

which gives $\quad E_0 = B_0 c$ or $\dfrac{E_0}{B_0} = c \quad \ldots(26.20)$

Again let us consider rectangle $EFGH$ [Fig. 26.5 (*a*)]. The line integral of magnetic field B over the path $EFGH$ is

$$\oint \boldsymbol{B}\cdot d\boldsymbol{l} = \int_{EF} \boldsymbol{B}\cdot d\boldsymbol{l} + \int_{FG} \boldsymbol{B}\cdot d\boldsymbol{l} + \int_{GH} \boldsymbol{B}\cdot d\boldsymbol{l} + \int_{HE} \boldsymbol{B}\cdot d\boldsymbol{l}$$

$$= B(x_1)d + 0 - B(x_2)d + 0$$

$$= B_0 d\left[\sin\omega\left(t - \frac{x_1}{c}\right) - \sin\omega\left(t - \frac{x_2}{c}\right)\right]$$

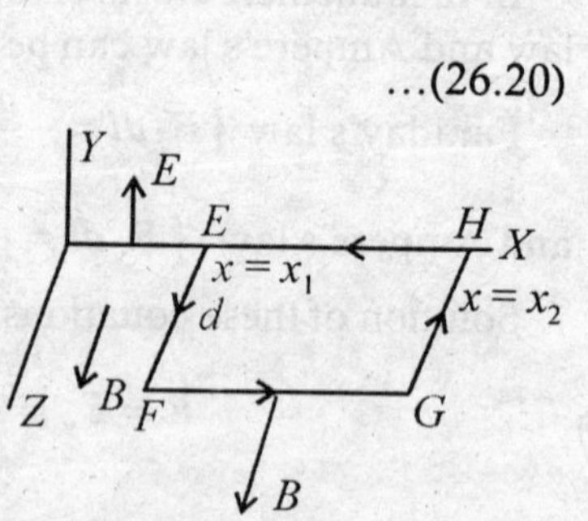

Fig. 26.5 (*a*)

The flux of the electric field through the same rectangle *EFGH* [Fig. 26.5 (*b*)] is

$$\phi_e = \oint \boldsymbol{E} \cdot \boldsymbol{dl} = \int_{x_1}^{x_2} E(x) d \cdot dx$$

$$= E_0 d \int_{x_1}^{x_2} \sin \omega \left(t - \frac{x}{c} \right) dx$$

$$\equiv \frac{cE_0 d}{\omega} \left[\sin \omega \left(t - \frac{x_2}{c} \right) - \sin \omega \left(t - \frac{x_1}{c} \right) \right]$$

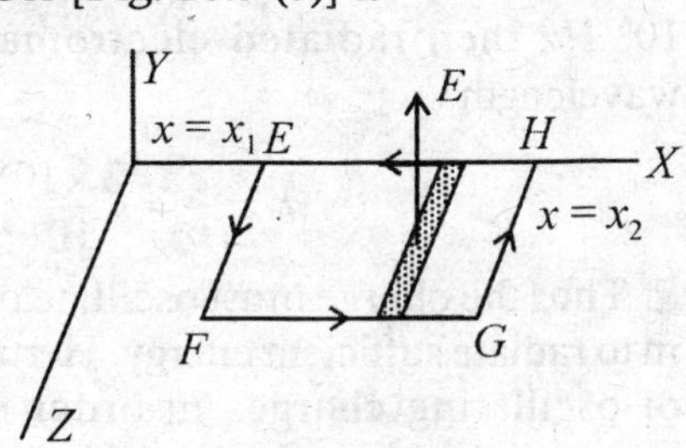

Fig. 26.5 (*b*)

and using Ampere's law

$$\oint \boldsymbol{B} \cdot \boldsymbol{dl} = \mu_0 \varepsilon_0 \frac{d\phi_e}{dt}, \text{ we have}$$

$$B_0 d \left[\sin \omega \left(t - \frac{x_2}{c} \right) - \sin \omega \left(t - \frac{x_1}{c} \right) \right]$$

$$= cE_0 d \mu_0 \varepsilon_0 \left[\sin \omega \left(t - \frac{x_2}{c} \right) - \sin \omega \left(t - \frac{x_1}{c} \right) \right]$$

or $$B_0 = cE_0 \mu_0 \varepsilon_0 \text{ and } E_0 = cB_0$$

$$B_0 = c^2 B_0 \mu_0 \varepsilon_0 \text{ or } c^2 = \frac{1}{\mu_0 \varepsilon_0}$$

and $$c = \frac{1}{\sqrt{\mu_0 \varepsilon_0}} \quad \text{...(26.21)}$$

The above discussion shows that Maxwell's equations have a solution giving a plane electromagnetic wave of the form given by equations (26.19) with $E_0 = cB_0$ and $c = \frac{1}{\sqrt{\mu_0 \varepsilon_0}}$

It may be remembered that the speed of electromagnetic waves is the same as that of the speed of light.

26.6 HERTZ'S EXPERIMENT

Faraday from his experimental study of electromagnetic induction concluded that a time varying magnetic field produces electric field and a time varying electric field produces a magnetic field. This idea led Maxwell to conclude that the variation in electric and magnetic field vectors perpendicular to each other leads to the production of electromagnetic disturbances in space. These disturbances have the properties of a wave and can travel in space without any material medium. He called these waves as *electromagnetic waves.* Maxwell's idea was conclusively demonstrated experimentally by Heinrich Hertz in 1887.

Hertz experiment was based on the fact that an oscillating electric charge radiates electromagnetic waves and that these waves carry energy which is being supplied at the cost of kinetic energy of the oscillating charge. The detailed study revealed that the electromagnetic radiation is significant only if the distance to which the charge oscillates is comparable to the wavelength of the radiation.

For example if the charge oscillates with a frequency 10^6 Hz then radiated electromagnetic waves have wavelength

$$\lambda = \frac{c}{\nu} = \frac{3 \times 10^8}{10^6} = 300 \text{ m}$$

Thus the charge must oscillate over a distance of 300 m to radiate sufficient energy. Hertz developed a system of oscillating charges in order to produce electromagnetic waves in the laboratory.

Figure 26.6 shows the experimental set up used by Hertz. In this set up two large metal spheres S and S' are connected to two large metal plates PP'. A high potential difference of several thousand volts is applied across the spheres using induction coil. Due to high potential difference across S and S', the air between the spheres gets ionized and provides the path for the discharge of plates. Due to it a spark is produced between S and S' and electromagnetic waves of high frequency are radiated. Here the two plates act as a capacitor of small capacity C and the connecting wires provide the two inductance L and the high frequency of oscillations is given by

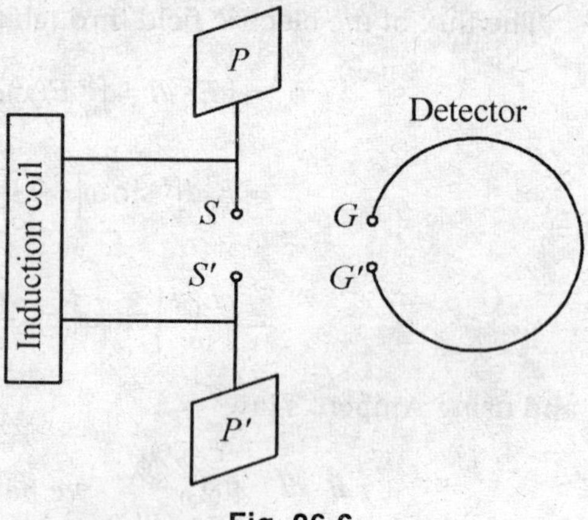

Fig. 26.6

$$f = \frac{1}{2\pi\sqrt{LC}} \qquad \text{...(26.22)}$$

To detect these waves, Hertz designed a detector which consists of a single loop of wire connected to two spheres GG'. It has its own effective inductance, capacitance and natural frequency of oscillation. The electromagnetic waves reaching the gap of detector had an electric field strong enough to establish a high potential difference in the gap GG' and cause a spark. Hertz observed tiny sparks jumping between the gap GG'—thereby providing detection of electromagnetic waves. Hertz also found that when the detector gap is parallel to the source gap SS' the spark between GG' is maximum. If the two gaps are perpendicular to each other no spark is observed in gap GG', i.e., electromagnetic waves are not detected. When the two gaps are parallel, i.e. in a direction perpendicular to the direction of propagation of the radiation the spark is maximum because the electric vector of radiation is parallel to the two gaps showing that the electromagnetic waves are transverse in nature.

In a series of experiments, Hertz also showed that the radiation generated by his spark experiment device exhibited the wave properties of interference, diffraction, reflection, refraction and polarization all of which are the properties exhibited by light. Thus it became evident that the radio-frequency waves had properties similar to light waves and differed only in frequency and wavelength.

26.7 ELECTROMAGNETIC SPECTRUM

Maxwell's equations are applicable for electromagnetic waves of all wavelengths. Visible light has wavelengths roughly in range 380 nm to 780 nm. At present we are familiar with electromagnetic waves having wavelengths as small as 30 fm (1 fm = 10^{-15} m) to as large as 30 km.

The orderly distribution of electromagnetic radiations according to their wavelength or frequency is called electromagnetic spectrum.

The whole spectrum has been classified into different parts and sub-parts in order of increasing wavelength; according to their type of excitation. The boundaries separating different regions of spectrum are not sharply defined and there is overlapping of different regions of spectrum.

The different parts of the electromagnetic spectrum with wavelength range, frequency range and names of the sources of the various electromagnetic radiations is given in Table 26.1.

Table 26.1: Characteristics of different parts of electromagnetic spectrum

S.N.	*Name*	*Frequency Range* (Hz)	*Wavelength Range* (m)	*Source*
1.	Gamma rays	5×10^{22} to 3×10^{19}	6×10^{-14} to 1×10^{-11}	Nuclear origin
2.	*X*-rays	3×10^{19} to 1×10^{16}	1×10^{-11} to 3×10^{-8}	Sudden deceleration of high energy electron
3.	Ultraviolet	5×10^{17} to 8×10^{14}	6×10^{-10} to 4×10^{-7}	Excitation of atom, spark and arc lamp
4.	Visible	8×10^{14} to 4×10^{14}	4×10^{-7} to 8×10^{-7}	Excitation of valance electron
5.	Infrared	4×10^{14} to 1×10^{13}	8×10^{-7} to 3×10^{-5}	Excitation of atoms and molecules
6.	Thermal radiations	3×10^{13} to 3×10^{9}	10^{-5} to $\times 10^{-1}$	Hot bodies
7.	Microwaves	3×10^{11} to 1×10^{9}	10^{-3} to 0.3	Oscillating current in special vacuum tube
8.	Ultra high frequency	3×10^{9} to 3×10^{8}	1×10^{-1} to 1	Oscillating circuit
9.	Very high radio frequency	3×10^{8} to 3×10^{7}	1 to 10	Oscillating circuit
10.	Radio frequencies	3×10^{7} to 3×10^{4}	10 to 10^{4}	Oscillating circuit
11.	Power frequencies	60 to 50	5×10^{6} to 6×10^{6}	Weak radiations from AC circuit

Here a brief description of various waves of electromagnetic spectrum is given.

Radio Waves

These are generally in the frequency range of 500 kHz to 1710 kHz. The amplitude modulated (AM) band has range 530 kHz to 1710 kHz. Higher frequencies upto 54 MHz are used for short wave bands. TV waves range from 54 MHz to 890 MHz. The frequency modulated (FM) radio band extends from 88 MHz to 108 MHz. Cellular phones use radio waves to transmit voice communication in the ultra high frequency (4 HF) band. How these waves are transmitted and received is discussed in chapter on communication.

Microwaves

They are short wavelength radio waves. They are used for the radar system and for studying the atomic and molecular properties of matter. Microwave owens represent an interesting domestic application of these waves. In this wavelength $\lambda = 0.122$ m is used.

Infrared Waves

These are used (*a*) in revealing the secret writings on the ancient walls (b) in green houses to keep the plants warm. (c) in warfare for looking through haze, fog or mist as these radiations can pass through them.

Ultraviolet Radiations

The sun is an important source of ultraviolet light, which is main cause of sun burns. Ultraviolet light from the sun is absorbed by ozone (O_3) molecules in the earths upper atmosphere in a layer called the stratosphere. This ozone shield converts lethal high-energy UV radiations to infrared radiations. Absorption of ultraviolet radiation results in warming the atmosphere. As a result of the chemicals emitted from aerosol spray cans and refrigerants there is the possibility of depletion of the protective *ozone layer*.

Visible Light

The component of electromagnetic waves can be detected by human eye. Photons of light are emitted by the deexcitation of atoms and molecules. The various wavelengths of visible light correspond to different colours. The range of colours lies between red ($\lambda = 7 \times 10^{-7}$ m) and violet ($\lambda = 4 \times 10^{-7}$ m). The sensitivity of human eye depends on wavelength and is maximum for wavelength of about 5.5×10^{-7} m which lies in the yellow region.

X-Rays

The most common source of *X*-rays is the deceleration of high energy electrons bombarding a metal target. *X*-rays are used as a diagnostic tool in medicine and as a treatment for certain forms of cancer. Since *X*-rays damage or destroy living tissues and organisms, care must be taken to avoid unnecessary exposure or overexposer. *X*-rays are also used in the study of crystal structure because their wavelength is comparable to the atomic separation distances ($\approx$1 Å).

Gamma Rays

These are the electromagnetic radiations emitted by radioactive nuclei and during certain nuclear reactions. High energy γ-rays are a component of cosmic rays entering the earth's atmosphere from space. Their wavelength range is 10^{-10} m to less than 10^{-14} m. They are highly penetrating and produce serious damage when absorbed by living tissues. They can be absorbed by thick layers of lead.

26.8 PROPAGATION OF ELECTROMAGNETIC WAVES IN THE ATMOSPHERE

The gaseous envelope surrounding the earth is called earth's atmosphere. Before we discuss the propagation of em waves through atmosphere, it is necessary to know a few things about the atmosphere and its various layers. Because of gravitational attraction it always remains with the earth. It mainly consists of nitrogen (78%) and oxygen (21%) along with little amount of water-vapour, argon, carbon dioxide, dust particles, hydrocarbonates, etc. As we go up the air density and pressure goes on decreasing. The earth's atmosphere has been divided into various regions which have no sharp boundaries. The different layers are known by different names. The different regions of atmosphere with their heights from the surface of earth is shown in Fig. 26.7. These are mentioned below:

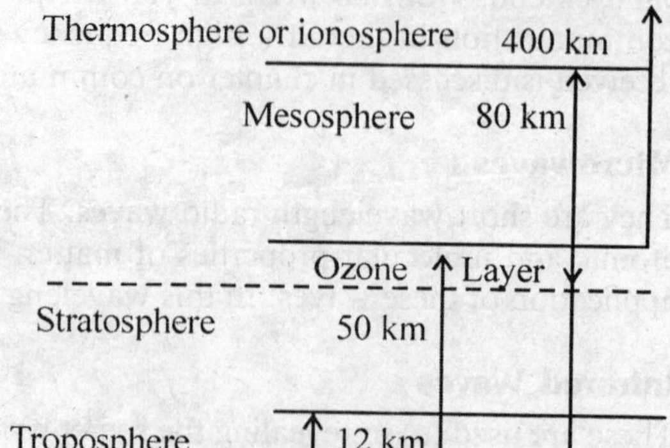

Fig. 26.7

(*a*) **Troposphere:** This region extends upto 12 km from the surface of earth. The temperature and density decreases with height. The temperature varies from 290 K to 220 K and density falls to $\frac{1}{10}$th of that on the earth. This layer is responsible for all the important weather phenomenon affecting our environment.

(*b*) **Stratosphere:** This layer extends from 12 km to 50 km from the surface of earth. The temperature increases from 220 K to 280 K and density varies from 10^{-1} to 10^{-3} of that on the earth. The upper part of this layer (from 30 km to 50 km) is called ozone layer because most of the ozone in the atmosphere is concentrated in this layer. This absorbs most of the ultraviolet radiations emitted by the sun.

(*c*) **Mesosphere:** It extends from 50 to 80 km from the surface of earth. The temperature here falls from 280 K to 180 K and the density of the air falls from 10^{-3} to 10^{-5} of that on the earth.

(*d*) **Ionosphere:** It extends from 80 km to a height of 400 km above the surface of earth. The temperature here varies from 180 K to 700 K. The density varies as 10^{-5} to 10^{-10} that of the density on earth. As the temperature in this region rises to very high temperature hence it is called *Thermosphere*. The ionosphere is mainly composed of electrons and positive ions; the rest of the atmosphere is mostly composed of neutral molecules.

The atmosphere is transparent to visible radiations and we can see the sun and the stars through it clearly. However most infrared radiations are unable to pass through it, as they are absorbed by the atmosphere. Low lying clouds in the atmosphere also prevent infrared radiations from passing through. The ozone layer blocks the passage of ultraviolet radiation from the sun. Practically all the radiations of wavelength less than 3×10^{-7} m are absorbed by the ozone layer.

The energy from sun obviously heats the earth which like any other hot body radiates the energy. This radiation travels mostly in the infrared region unlike solar radiation. This radiation from earth however is unable to cross the lower atmosphere which reflects it back. As such the earth's atmosphere is rich in infrared radiations which is sometime called heat radiations. These infrared radiations keep the earth's surface warm at night. This phenomenon is called *the green house effect*.

Propagation of Radio Waves

The electromagnetic waves of frequency range from a few kilo Hertz to about a few mega Hertz (i.e., waves of wavelength 10^{-3} and higher are called radio waves). The behaviour of these waves in propagating through atmosphere is considered important in all modern forms of communication such as radio, television, etc.

A signal emitted by an antenna from a certain point can be received at another point on the surface in two possible ways. The wave can travel directly following the surface of the earth. These are known as *ground waves.* They can also reach after being reflected back from the ionosphere. These are called as *the sky waves.*

Electromagnetic waves of frequency less than 30 MHz (or wavelength 10 m or more) form *amplitude modulated band* (A.M. band). Those waves having frequency upto 1500 KHz (or wavelength more than 200 m) are called *ground waves or surface waves* because they travel directly following the surface of earth. This region of amplitude modulated wave is referred to as *medium wave band*. The ground wave transmission becomes weaker as frequency increases. For the radio communication of frequencies more than 1500 kHz upto 40 MHz (i.e., wavelength between 7.5 m to 200 m) *sky waves* can only be used. They are received after being reflected from the ionosphere. These waves are normally referred to as *short wave band.*

Beyond a certain frequency (i.e., above 40 MHz) the ionosphere bends any incident electromagnetic wave but does not reflect it back towards the earth. The electromagnetic waves of frequency range 80 MHz to 200 MHz form frequency modulated band (FM band).

Television signals have frequency range of 100 MHz to 200 MHz. These waves neither follow curvature of earth nor get reflected by ionosphere. Their communication via sky waves or ground waves is not possible. The communication of these waves is possible (*i*) by using communication geostationary satellites which reflects the television signals back to earth (*ii*) by using tall receiver antenna which may intercept directly the signal. In order to have greater television coverage transmission has to be made by using tall antenna. This is called *space wave propagation.*

Radio waves with frequencies higher than television signals are called *the microwaves. Microwaves are electromagnetic waves of wavelength of the order of few millimetres* (i.e. frequencies of the order of $10^4 \sim 10^5$ MHz). On account of smaller wavelengths the microwaves can be transmitted as beam signals in a particular direction much better than the radio waves because microwaves do not spread or bend around the corners of any obstacle coming in their way. The signals in this range from the broadcasting stations are beamed towards a geostationary satellite, which in turn broadcasts it back to earth. In this way signals can be propagated over a large part of the earth's surface.

The detailed study of radio and TV communication is given in chapter on communication.

SOLVED EXAMPLES

Example 26.1: Figure 26.8 shows a capacitor made of two circular plates each of radius 12 cm and separated by 5.0 mm. The capacitor is being charged by an external source. The charging current is constant and equal to 0.15 A

(*a*) Calculate the capacitance of the capacitor and the rate of change of potential difference between the plates.

(*b*) Obtain the displacement current between the plates.

(*c*) Is the Kirchoff's first law valid at each plate?

(*d*) Calculate the magnetic field between the plates at a point (*i*) on the axis (*ii*) 6.5 cm from the axis (*iii*) 15 cm from the axis.

(*e*) At what distance from the axis is the magnetic field due to displacement currents maximum? Obtain the maximum value of the field.

Solution: (*a*) $C = \frac{\varepsilon_0 A}{d} = \frac{\varepsilon_0 \pi r^2}{d} = \frac{8.85\times10^{-12}\times\pi\times(12\times10^{-2})^2}{5\times10^{-3}}$

$= 80 \times 10^{-12}$ F = 80 pF

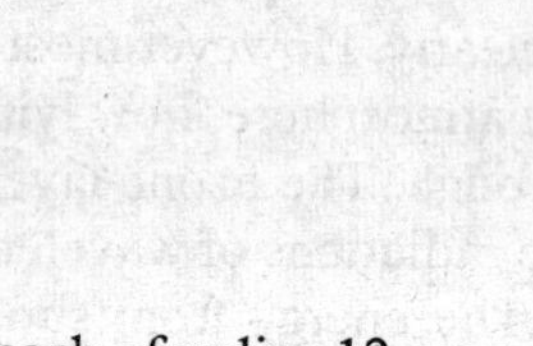

Fig. 26.8

Now $q = CV$ or $\frac{dq}{dt} = \frac{CdV}{dt}$ but $\frac{dq}{dt} = i$ (constant)

$\therefore \quad \frac{dV}{dt} = \frac{i}{C} = \frac{0.15}{80\times10^{-12}} = 1.87\times10^9$ V/s (*b*) Since displacement current and conduction currents when taken together are continuous hence displacement current

$$i_d = i_c = 0.15 \text{ A}$$

(*c*) Yes Kirchhoff's first law is valid at each plate of the capacitor provided we take the current to be the sum of conduction and displacement currents.

(*d*) For calculating magnetic field within the plates let us consider a closed loop of radius r between two circular plates coaxial to the plates. Then using Ampere's circuital law, we have

$$\oint \boldsymbol{B} \cdot \boldsymbol{dl} = u_0(i_c + i_d)$$

Since within the plates i_c is zero and $i_d = \varepsilon_0 \frac{d\phi_e}{dt} = 0.15$ A in present situation.

$$\therefore \qquad 2\pi r B = u_0 i'_d$$

Here i'_d in the current enclosed by the loop $i'_d = \frac{i_d \cdot \pi r^2}{\pi R^2} = \frac{i_d \cdot r^2}{R^2}$ where R is the radius of the plates.

$$\therefore \qquad B = \frac{u_0 i'_d}{2\pi r} = \frac{\mu_0 \cdot i_d \cdot r^2}{2\pi r R^2} = \frac{\mu_0 i_d r}{2\pi R^2} \text{ for } r < R$$

(*i*) On the axis $r = 0 \therefore B = 0$

(*ii*) For $r = 6.5$ cm $B = \frac{\mu_0 i_d r}{2\pi R^2} = \frac{2 \times 10^{-7} \times 0.15 \times (6.5 \times 10^{-2})}{(12 \times 10^{-2})^2}$

or $\qquad B = 1.35 \times 10^{-7}$ T

(*iii*) For $r = 15$ cm For $r > R$

$$B.2\pi r = \mu_0 i_d \text{ and } B = \frac{\mu_0 i_d}{2\pi r}$$

$$\therefore \qquad B = \frac{2 \times 10^{-7} \times 0.15}{15 \times 10^{-2}} = 2 \times 10^{-7} \text{ T}$$

(*e*) Field is maximum when $r = 12$ cm

$$B_{\max} = \frac{\mu_0 i_d}{2\pi r} = \frac{2 \times 10^{-7} \times 0.15}{12 \times 10^{-2}} = 2.5 \times 10^{-7} \text{ T}$$

Example 26.2: A parallel plate capacitor made of circular plates each of radius $R = 6.0$ cms has a capacity $C = 100$ PF. The capacitor is connected to a 230 V AC supply with angular frequency of 300 rad/s.

(*a*) What is the rms value of conduction current?
(*b*) Is the conduction current equal to the displacement current?
(*c*) Determine the magnitude of B at a point 3.0 cm from the axis between the plates.

Solution:

(*a*) The reactance of the capacitor $X_c = \frac{1}{\omega C} = \frac{1}{100 \times 10^{-12} \times 300} = \frac{10^8}{3} \ \Omega$

Hence conduction current $i_c = \frac{E}{X_c} = \frac{230 \times 0.3}{10^8} = 6.9 \times 10^{-6}$ A

$i_c = 6.9\ \mu A$

(*b*) Yes $i_d = i_c$ whether i_c is a steady DC or AC. This can be shown as below

$$i_d = \varepsilon_0 \frac{d\phi_e}{dt} = \varepsilon_0 \frac{d}{dt}(EA)$$

$$= \varepsilon_0 A \frac{d}{dt}(E) = \varepsilon_0 A \frac{d}{dt}\left(\frac{q}{A\varepsilon_0}\right) = \frac{dq}{dt} = i_c$$

(*c*) For $r < R$ the value of B is given by

$$B = \frac{\mu_0 i_d r}{2\pi R^2}$$

Since $i_d = i_c$ then $B = \dfrac{\mu_0 i_c r}{2\pi R^2}$

B will be maximum (B_{max}) when $i_c = i_0$ (peak value) $= i_c\sqrt{2}$

$$\therefore \quad B_{max} = \frac{\mu_0 \cdot i_c\sqrt{2} r}{2\pi R^2} = \frac{2\times10^{-7}(3\times10^{-2})\times6.9\times10^{-6}\times\sqrt{2}}{(6\times10^{-2})^2}$$

$$= 1.63 \times 10^{-11}\ \text{T}$$

Example 26.3: An AC voltage is applied directly across 8 μF capacitor. The frequency of the source is 3 kHz and the voltage amplitude is 30 V. Find the displacement current between the plates of the capacitor.

Solution: Angular frequency of the source is $\omega = 2\pi f = 2\pi(3 \times 10^3)$

or $\omega = 6\pi \times 10^3$ rad/s.

Hence the instantaneous voltage applied is

$$V = V_0 \sin\omega t$$

and the instantaneous charge on the capacitor is

$$q = CV = CV_0 \sin\omega t$$

and the displacement current

$$i_d = \frac{dq}{dt} = \frac{d}{dt}(CV_0 \sin\omega t)$$

or $i_d = CV_0\omega\cos\omega t$

$$= (8 \times 10^{-6})(30)(6\pi \times 10^3)\cos\omega t$$

$$= 4.52\cos\omega t\ \text{A, where } \omega = 6\pi \times 10^3\ \text{rad/s}$$

The displacement current varies sinusoidally with time and has maximum value 4.52 A.

Example 26.4: A plane electromagnetic wave of frequency 40 MHz travels in free space in *X*-direction. At some point and at some instant, electric field ***E*** has its maximum value of 750 N/C and is along *Y*-axis.

(*a*) Determine the wavelength and period of the wave.

(*b*) Calculate the magnitude and direction of the magnetic field ***B*** when ***E*** $= 750\,\hat{j}$ N/C.

(*c*) Write an expression for the space time variation of the electric and magnetic field components of this plane wave.

(*d*) Find the total average energy density of the electromagnetic wave.

Solution:

(*a*) For wave motion if λ is wavelength and f frequency then

$\lambda f = c$ where c is the wave velocity

$$\therefore \quad \lambda = c/f = \frac{3\times 10^8}{40\times 10^6} = 7.5 \text{ m}$$

If time period is T, then

$$T = \frac{1}{f} = \frac{1}{40\times 10^6} = 2.5\times 10^{-8}\text{ sec}$$

(*b*) If B is the magnitude of magnetic field, then

$$B = \frac{E}{c} = \frac{750}{3\times 10^8} = 2.5\times 10^{-6}\text{ T}$$

Since E and B are perpendicular to each other and perpendicular to the direction of propagation, hence B is along positive Z-direction. Hence

$$\boldsymbol{B}_{\text{max}} = 2.5 \times 10^{-6}\,\hat{k}\text{ T}$$

(*c*) $\quad E = E_0\sin(\omega t - kx) = 750 \sin(\omega t - kx)$ N/C

and $\quad B = B_0\sin(\omega t - kt) = 2.5 \times 10^{-6} \sin(\omega t - kx)$ T

when $\quad \omega = 2\pi f = 2\pi \times 40 \times 10^6 = 8\pi \times 10^7$ rad/s

and $\quad k = \dfrac{2\pi}{\lambda} = \dfrac{2\pi}{7.5} = 0.84\text{ m}^{-1}$

(*d*) Total energy density $U = \dfrac{1}{2}\varepsilon_0 E_0^2 = \dfrac{8.85\times 10^{-12}}{2}\times (750)^2$

$= 2.5 \times 10^{-6}$ J/m^3.

Example 26.5: Calculate the electric and magnetic fields produced by the radiation coming from a 100 watt bulb at a distance of 3 m. Assume the efficiency of the bulb is 2.5% and it is a point source.

Solution: Since the bulb is treated as a point source, hence intensity at a distance of 3 m is

$$I = \frac{\text{Power}}{\text{Area}} = \frac{100\times 0.025}{4\pi r^2} = \frac{2.5}{4\pi(3)^2} = 0.022\ \text{W/m}^2$$

The intensity of electromagnetic waves is given by

$$I = \frac{1}{2}\varepsilon_0 E_0^2 c$$

$$\therefore \quad E_0^2 = \frac{2I}{\varepsilon_0 c} = \frac{2\times 0.022}{8.85\times 10^{-12}\times 3\times 10^8} = \frac{2\times 220}{3\times 8.85} = 16.57$$

or $\quad E_0 = 4.07$ V/m

Magnetic field B is given by

$$B = \frac{E_0}{c} = \frac{4.07}{3\times 10^8} = 1.34\times 10^{-8}\text{ T}$$

Example 26.6: An isotropic point source having radiation power P is located on the axis of an ideal mirror plate. The distance between the source and the plate exceeds the radius of the plate n-fold. In terms of photons theory, find the force that light exerts on the plate.

Solution: Intensity of photons at distance r from the point source is

$$I = \frac{P}{4\pi r^2}$$

If the distance of the point source is l and radius of the mirror R, then $\frac{l}{R} = n$

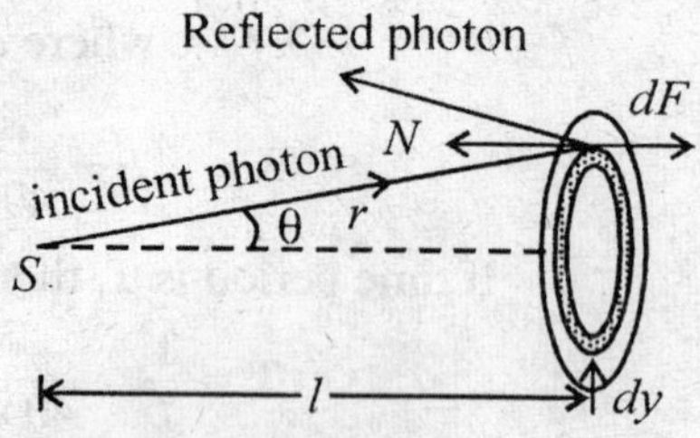

Fig. 26.9

Consider a ring on the mirror of radius y and $y + dy$ which is at distance r from the source (Fig. 26.9) inclined at an angle θ. The energy falling in one second normally on this arc is

$$dE = I\,dA\cos\theta \qquad ...(i)$$

where I is the intensity. Then momentum is $\frac{dE}{c}$ and the momentum transferred normally to the area is $\frac{dE}{c}\cos\theta$

Hence the momentum transferred normally to the area on striking it per second is

$$\left(\frac{dP}{dt}\right) = \frac{dE}{c}\cos\theta = \frac{I\,dA\cos^2\theta}{c} \qquad ...(ii)$$

Momentum transferred on reflection is $\left(\frac{dP}{dt}\right) = -\frac{\rho I\,dA\cos^2\theta}{c}$ (iii)

Here ρ is coefficient of reflection.

Hence rate of momentum transfer to the mirror by the photons is

$$\frac{dP}{dt} = \frac{IdA\cos^2\theta}{c} + \frac{\rho IdA\cos^2\theta}{c} = \frac{(1+\rho)IdA\cos^2\theta}{c}$$

Since mirror is a perfect reflector hence ρ = 1. Hence the rate of change of momentum is

$$\frac{dP}{dt} = \frac{2IdA\cos^2\theta}{c}$$

and from Newton's second law of motion this is the force on the annular ring of the mirror considered

$$\therefore \quad dF = \frac{2IdA\cos^2\theta}{c}$$

But $dA = 2\pi y\,dy$ and $\tan\theta = \frac{y}{l}$ or $dy = l\sec^2\theta \cdot d\theta$

Hence force on annular ring of mirror is

$$dF = \frac{2I}{c}2\pi y\,dy\cos^2\theta$$

$$= \frac{2I}{c}2\pi(l\tan\theta)(l\sec^2\theta d\theta)\cdot\cos^2\theta$$

$$= \frac{4\pi l^2 I}{c}\tan\theta\cdot d\theta \qquad \because\ I = \frac{P}{4\pi r^2}$$

$$= \frac{4\pi l^2}{c}\left(\frac{P}{4\pi r^2}\right)\tan\theta \cdot d\theta$$

$$= \frac{P}{c}\left(\frac{l^2}{r^2}\right)\tan\theta \cdot d\theta \qquad \text{But } \frac{l}{r} = \cos\theta$$

$$\therefore \quad dF = \frac{P}{c}\cos^2\theta\tan\theta \cdot d\theta = \frac{P}{c}\sin\theta \cdot \cos\theta \cdot d\theta$$

Hence force F is $F = \int_0^{\theta_0} \frac{P}{c}\sin\theta \cdot \cos\theta \cdot d\theta$ where $\sin\theta_0 = \frac{R}{\sqrt{l^2 + R^2}}$

$$= \frac{P}{2c}\sin^2\theta_0 = \frac{P}{2c}\frac{R^2}{(l^2 + R^2)} \equiv \frac{P}{2c}\frac{1}{(1 + l^2/R^2)}$$

$$F = \frac{P}{2c} \cdot \frac{1}{(1 + n^2)}$$

Example 26.7: A plane wave of intensity $I = 0.70\ \text{W/cm}^2$ illuminates a sphere with ideal mirror surface. The radius of the sphere is $R = 5.0$ cm. From the stand point of photon theory, find the force that light exerts on the sphere.

Solution: Consider the sphere to be made of a large number of rings. Consider one such ring of radius R and thickness $ds = Rd\theta$ which subtends an angle θ at the centre.

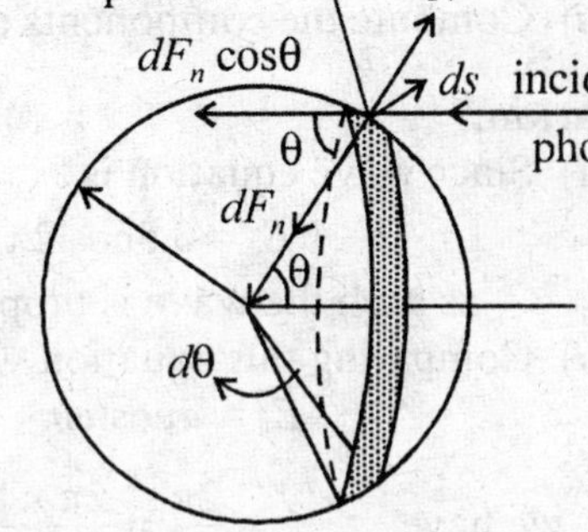

Fig. 26.10

Momentum per second of incident photon normal to surface

$$\left(\frac{dP}{dt}\right)_{in} = \frac{I}{c}dA\cos\theta$$

Since the surface of mirror is ideal, i.e. reflection coefficient is one. The photons suffer momentum change only in normal direction. Hence net rate of change of momentum is

$$\frac{dP}{dt} = -\frac{2I}{c}dA\cos^2\theta$$

Hence rate of momentum transfer to the ring is

$$\frac{dP}{dt} = \frac{2I}{c}dA\cos^2\theta$$

And this rate of change of momentum is the force on the ring

$$dF_n = \frac{2I}{c}dA\cos^2\theta$$

Resolving this force in horizontal and vertical components, we see that the vertical component for whole sphere averages out to zero. The horizontal component is

$$dF_n\cos\theta = \frac{2I}{c}dA\cos^3\theta$$

and total force is

$$F = \int dF_n\cos\theta = \int \frac{2I}{c}dA\cos^3\theta$$

But $$dA = (2\pi R\sin\theta)\,(Rd\theta)$$

$$\therefore \quad F = \int_0^{\pi/2} \frac{2I}{c}(2\pi R^2)\cos^3\theta \cdot \sin\theta \cdot d\theta$$

$$\equiv \frac{4\pi R^2 I}{c}\int_0^{\pi/2} \cos^3\theta \cdot \sin\theta \cdot d\theta$$

$$\equiv \frac{\pi R^2 I}{c}$$

The values given are $I = 0.7\ \text{W/cm}^2 = 0.7 \times 10^4\ \text{W/m}^2$

$R = 5\ \text{cm} = 5 \times 10^{-2}\ \text{m},\ c = 3 \times 10^8\ \text{m/s}$

$$\therefore \quad F \equiv \frac{\pi(5\times10^{-2})^2 \times 0.7 \times 10^4}{3\times10^8}$$

$$= 0.18\ \mu\text{N}.$$

Example 26.8: The electric field of a plane electromagnetic wave in vacuum is represented by $E_x = 0$, $E_y = 0.5\cos[2\pi \times 10^8\,(t - x/c)]$, $E_z = 0$.

(*i*) What is the direction of propagation of the electromagnetic wave?
(*ii*) Determine the wavelength of the wave
(*iii*) Compute the components of associated magnetic wave.

Solution:

(*i*) Since wave equation is

$$E_y = 0.5\cos[2\pi \times 10^8(t - x/c)$$

as such the wave is propagating in positive X-direction.

(*ii*) Comparing this equation with standard wave equation

$$E_y = a\cos(\omega t - kx)$$

we have $$k = \frac{2\pi}{\lambda} = \frac{2\pi \times 10^8}{c}$$

or $$\lambda = \frac{c}{10^8} = \frac{3\times10^8}{10^8} = 3\ \text{m}$$

Magnetic field is perpendicular to both, electric field and direction of propagation. Hence

$$B_x = 0,\ B_y = 0 \text{ and } B_z = \frac{E_y}{c} = \frac{0.5}{3\times10^8}\cos[2\pi\times10^8(t - x/c)]$$

Example 26.9 (*a*) Determine the average intensity of the wave propagating in positive X-direction with $E_0 = 120$ N/C and frequency $\nu = 50.0$ MHz. (*b*) Write an expression of its Poynting vector.

Solution:

(*a*) Average intensity $I = Uc$ where U is energy density.

$$\therefore \quad I = \frac{1}{2}\varepsilon_0 E_0^2 \cdot c$$

$$= \frac{1}{2}(8.85 \times 10^{-12}) \times (120)^2(3 \times 10^8)$$

$$= 19.1 \text{ W/m}^2$$

(*b*) $$I = \frac{1}{2}\varepsilon_0 E_0^2 c = \frac{1}{2}\varepsilon_0 B_0 E_0 c^2 = \frac{1}{2}\frac{B_0 E_0}{\mu_0}$$

$$\therefore \quad \frac{E_0 B_0}{\mu_0} = 2I$$

Hence Poynting vector $\boldsymbol{S}$ of this wave is given by expression

$$\boldsymbol{S} = 2I \sin^2(kx - \omega t)\,\hat{s}$$

The magnitude of this vector is the wave intensity and its direction is the direction in which the energy is propagating. Hence

$$\boldsymbol{S} = 38.2 \sin^2(kx - \omega t)\,\hat{i}$$

$$\text{where } k = \frac{2\pi}{\lambda} = \frac{2\pi\nu}{c} = \frac{2\pi \times 50 \times 10^6}{3 \times 10^8} = 1.05 \text{ m}^{-1}$$

and $$\omega = 2\pi\nu = 2\pi \times 50 \times 10^6 = 3.14 \times 10^8 \text{ rad/s.}$$

Example 26.10: Consider an electromagnetic wave travelling in air whose dielectric constant K is 1.006. (*a*) Determine the speed of visible light in air (*b*) Given that speed of light in glass is 2.0×10^8 m/s. Determine ε_r for glass.

Solution:

(*a*) Speed of light in medium is given by

$$c_m = \frac{1}{\sqrt{\varepsilon\mu}} = \frac{1}{\sqrt{\varepsilon_0 \varepsilon_r \mu_0 \mu_r}}$$

The value of μ_r for dielectrics is nearly 1. Hence

$$c_m = \frac{c}{\sqrt{\varepsilon_r}}$$

where $c = \dfrac{1}{\sqrt{\varepsilon_0 \mu_0}}$ is velocity of light in vacuum and ε_r is the dielectric constant of the medium.

$$\therefore \quad c_m = \frac{3 \times 10^8}{\sqrt{1.006}} = 2.99 \times 10^8 \text{ m/s}$$

(*b*) For glass

$$\varepsilon_r = \frac{c^2}{(c_g)^2} \text{ where } c_g \text{ is velocity of light in glass}$$

$$= \left(\frac{3 \times 10^8}{2 \times 10^8}\right)^2 = (1.5)^2 = 2.25.$$

Example 26.11: Figure 26.11 shows a section of long straight wire of radius b and resistance R carrying a uniformly distributed current i. Find the magnitude and direction of the Poynting vector that describes energy flux into the wire and use this result to evaluate power supplied to the wire.

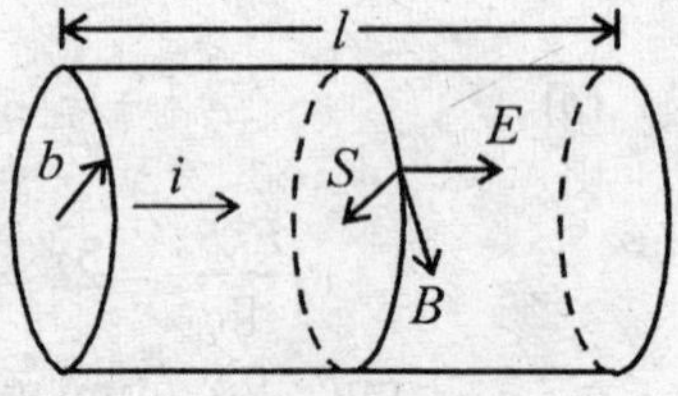

Fig. 26.11

Solution: Electric field E is along the length of the wire and is uniform. The magnetic lines of induction circulate around the axis. On the surface of wire, the magnetic field is tangential to the surface. Poynting vector is given by

$$S = \frac{\boldsymbol{E} \times \boldsymbol{B}}{\mu_0}$$

Thus the direction of Poynting vector is towards the axis as shown in Fig. 26.11.

Electric field $E = \frac{iR}{l}$

From Ampere's law the magnetic field on the surface of the current carrying wire is

$$B = \frac{\mu_0 i}{2\pi b} \text{ where } b \text{ is the radius of wire.}$$

Hence the magnitude of the Poynting vector is

$$S = \frac{iR \cdot \mu_0 i}{l\,\mu_0 \times 2\pi b} = \frac{i^2 R}{2\pi\, bl}$$

This Poynting vector is everywhere normal to the periphery of wire. By definition of S, i.e. it is power per unit area normal to S. Hence power

$$P = S \times 2\pi bl = i^2 R$$

and i^2R is the power supplied to the wire.

Example 26.12: The energy from the sun falls on the earth at a rate of 1353 W/m^2, which is known as solar constant. Find the rms values of the electric and magnetic field in the sunlight reaching the top of the atmosphere.

Solution: For electromagnetic wave of sinusoidal form

$$\boldsymbol{E} = E_0 \cos(kx - \omega t)\,\hat{j} \text{ and } \boldsymbol{B} = B_0 \cos(kx - \omega t)\,\hat{k}$$

The energy flux is given by Poynting vector and the magnitude of the Poynting vector is

$$S = \frac{E_0 B_0}{\mu_0} \cos^2(kx - \omega t)$$

Hence average value is $S = \frac{E_0^2}{2\mu_0 c}$ $\quad \because B_0 = \frac{E_0}{c}$ and $< \cos^2\theta > = \frac{1}{2}$

But $\frac{E_0^2}{2} = E_{rms}^2$ and average energy flux is intensity. Hence

$$I = \frac{E_{rms}^2}{\mu_0 c}$$

$$\therefore \quad E_{rms} = \sqrt{I \mu_0 c}$$

$$= \sqrt{1353 \times 4\pi \times 10^{-7} \times 3 \times 10^{8}}$$

$$= 714 \text{ V/m}$$

and $$B_{rms} = \frac{E_{rms}}{c} = \frac{714}{3 \times 10^{8}} = 2.381 \times 10^{-6} \text{ T}$$

Example 26.13: A parallel plate capacitor having plate area A and plate separation d is joined to a battery of emf E and internal resistance R at $t = 0$. Consider a plane surface of area $A/2$ parallel to the plates and situated symmetrically between them. (*a*) Find the displacement current through this surface as a function of time. (*b*) If displacement resistance is defined as $R_d = \frac{V}{i_d}$ (of the space between the plates, then show that R_d varies with time as $R_d = R(e^{t/RC} - 1)$ where $C = \frac{\varepsilon_0 A}{d}$.

Solution: Since the circuit is R–C charging circuit hence charge at any time t is

$$q = q_0(1 - e^{-t/RC}) \text{ where } q_0 = EC$$

The electric field between the plates at any time t is

$$\varepsilon = \frac{q}{A\varepsilon_0}$$

and the electric flux through area $A/2$ is

$$\phi_e = \varepsilon A/2 = \frac{q}{2\varepsilon_0}$$

and the displacement current through this area is

$$i_d = \varepsilon_0 \frac{d\phi_e}{dt} = \varepsilon_0 \frac{d}{dt}\left(\frac{q}{2\varepsilon_0}\right)$$

$$= \frac{1}{2}\frac{d}{dt}(q) = \frac{1}{2}\frac{d}{dt}\{q_0(1 - e^{-t/RC})\}$$

$$= \frac{1}{2} q_0 \cdot e^{-t/RC} \times \frac{1}{RC}$$

$$i_d = \frac{E}{2R} e^{-t/RC}$$

Since charge at any time t on the capacitor is

$$q = q_0(1 - e^{-t/RC}) = EC(1 - e^{-t/RC})$$

Hence the potential difference across capacitor at any time t is

$$V = \frac{q}{C} = E(1 - e^{-t/RC}) \qquad \ldots(i)$$

and displacement current at that time is

$$i_d = \frac{dq}{dt} = \frac{EC\, e^{-t/RC}}{RC} = \frac{E}{R} e^{-t/RC} \qquad \ldots(ii)$$

Hence displacement resistance at time t is

$$R_d = \frac{V}{i_d} = \frac{RE(1-e^{-t/RC})}{E\,e^{-t/RC}} = R(e^{t/RC}-1)$$

Example 26.14: The electric field in an electromagnetic wave is given by

$$E = (50)\sin\omega(t - x/c)\ \text{N/C}$$

(*a*) Find the energy contained in a cylinder of cross-section 100 cm^2 and length 1.5 m along the X-axis.

(*b*) Find its intensity also.

Solution:

(*a*) Energy density $U = \frac{1}{2}\varepsilon_0 E_0^2$

Hence energy within a cylinder of cross-section A and length l is

Energy = Energy density × volume

$$= \frac{1}{2}\varepsilon_0 E_0^2 Al$$

$$= \frac{1}{2}\times 8.85\times 10^{-12}(50)^2\times(100\times 10^{-4})(1.5)$$

$$= 1.66\times 10^{-10}\ \text{J}$$

(*b*) Intensity $\quad I = Uc = \frac{1}{2}\varepsilon_0 E_0^2 c$

$$= \frac{1}{2}\times 8.85\times 10^{-12}\times(50)^2\times 3\times 10^8$$

$$= 3.3\ \text{W/m}^2.$$

Example 26.15: A plane electromagnetic wave falls at right angles to the surface of a plane parallel plate of thickness l. The plate is made of nonmagnetic substance whose permittivity decreases exponentially from a value ε_1 at the front surface down to a value ε_2 at the near one. How long does it take a given wave phase to travel across this plate?

Solution: Since electric permittivity varies exponentially hence the permittivity at distance x from front surface then is

$$\varepsilon(x) = \varepsilon_1 e^{-(x/l)\log \varepsilon_1/\varepsilon_2} \qquad \ldots(i)$$

Hence local velocity of light in the plate is

$$v(x) = \frac{c}{\sqrt{\varepsilon(x)}} = \frac{c}{\sqrt{\varepsilon_1}}e^{\frac{x}{2l}\log\frac{\varepsilon_1}{\varepsilon_2}}$$

The time taken by the wave phase to travel distance dx is

$$dt = \frac{dx}{v(x)}$$

and the required time is

$$t = \int_0^l \frac{dx}{v(x)} = \frac{\sqrt{\varepsilon_1}}{c}\int_0^l e^{-\frac{x}{2l}\log\frac{\varepsilon_1}{\varepsilon_2}}dx$$

$$= -\frac{\sqrt{\varepsilon_1}}{c}\frac{2l}{\log \varepsilon_1/\varepsilon_2}\left[e^{-\frac{x}{2l}\log\frac{\varepsilon_1}{\varepsilon_2}}\right]_0^l$$

$$= \frac{\sqrt{\varepsilon_1}}{c}\frac{2l}{\log \varepsilon_1/\varepsilon_2}\left[1 - e^{-\frac{1}{2}\log\frac{\varepsilon_1}{\varepsilon_2}}\right]$$

or
$$t = \frac{\sqrt{\varepsilon_1}}{c}\frac{2l}{\log \varepsilon_1/\varepsilon_2}\left[1 - \frac{\sqrt{\varepsilon_2}}{\sqrt{\varepsilon_1}}\right]$$

$$= \frac{2l}{c}\frac{\sqrt{\varepsilon_1} - \sqrt{\varepsilon_2}}{\log \varepsilon_1 - \log \varepsilon_2}$$

Example 26.16: An alternating sinusoidal current of frequency $\omega = 1000$ rad/s flows in the windings of a straight solenoid whose cross-sectional radius is equal to $R = 6.0$ cm. Find the ratio of peak values of electric and magnetic energies within the solenoid.

Solution: The current $i = i_0 \sin\omega t$

If n is the number of turns per unit length of the solenoid, and l is length of the solenoid, then its, inductance

$$L = \mu_0 \pi n^2 R^2 l$$

and the peak magnetic energy is

$$U_m = \frac{1}{2}Li_0^2 = \frac{1}{2}\mu_0 \pi n^2 R^2 l \cdot i_0^2 \qquad \ldots(i)$$

A changing magnetic field produces an electric field. If E is the electric field then using Faraday's law, we get

$$\oint \boldsymbol{E}\cdot d\boldsymbol{l} = -\frac{d}{dt}\int \boldsymbol{B}\cdot d\boldsymbol{s} \qquad B = \mu_0 ni = \mu_0 ni_0 \sin\omega t$$

or
$$E\, 2\pi r = -\mu_0 ni_0 \omega \cos\omega t\,(\pi r^2)$$

or
$$E = -\frac{1}{2}\mu_0 nr\omega i_0 \cos\omega t$$

The energy density $= \frac{1}{2}\varepsilon_0 E^2$. Hence if we consider a small volume dV then associated electric energy is

$$U_e = \int \frac{1}{2}\varepsilon_0 E^2 dV$$

$$dV = 2\pi r l\, dr$$

$$\therefore \quad U_e = \int_0^R \frac{1}{2}\varepsilon_0\left(\frac{1}{4}\mu_0^2 n^2 r^2 \omega^2 i_0^2 \cos^2\omega t\right)(2\pi r l)dr$$

$$= \frac{1}{8}\varepsilon_0 \mu_0^2 n^2 i_0^2 \omega^2 \cos^2\omega t(2\pi l)\int_0^R r^3 dr$$

$$= \frac{1}{8}\varepsilon_0\mu_0^2 n^2 i_0^2 \omega^2 \cos^2 \omega t \frac{\pi R^4 l}{2}$$

and peak value of electric energy is

$$U_e = \frac{1}{16}\varepsilon_0\mu_0^2 n^2 i_0^2 \omega^2 \pi R^4 l$$

and the ratio of peak values is

$$\frac{U_e}{U_m} = \frac{1}{16}\frac{\varepsilon_0\mu_0^2 n^2 i_0^2 \omega^2 \pi R^4 l \times 2}{\mu_0 \pi n^2 R^2 l i_0^2}$$

$$= \frac{1}{8}\varepsilon_0\mu_0\omega^2 R^2 \quad \text{But } \varepsilon_0\mu_0 = \frac{1}{c^2}$$

$$\therefore \quad \frac{U_e}{U_m} = \frac{1}{8}\left(\frac{\omega R}{c}\right)^2 = 5 \times 10^{-15}$$

Example 26.17: A parallel plate air capacitor whose electrodes are shaped as discs of radius $R = 6.0$ cm is connected to a source of alternating sinusoidal voltage with frequency $\omega = 1000$ rad/s. Find the ratio of peak values of magnetic and electric energies within the capacitor.

Solution: Applied voltage is $V = V_0 \sin\omega t$

Inside capacitor electrical energy stored $= \frac{1}{2}CV^2$, $C = \frac{\varepsilon_0 \pi R^2}{d}$

Hence peak value of electric energy stored is

$$U_e = \frac{1}{2}CV_0^2 = \frac{1}{2}\frac{\varepsilon_0 \pi R^2}{d}V_0^2 \quad \ldots(i)$$

A changing electric field produces displacement current and displacement current density

$$j_d = \frac{1}{A}\frac{\partial}{\partial t}(q) = \frac{1}{A}\frac{\partial}{\partial t}(VC) = \frac{\varepsilon_0 A}{Ad}\frac{d}{dt}(V_0 \sin \omega t)$$

$$\equiv \frac{\varepsilon_0 V_0 \omega \cos \omega t}{d}$$

This gives rise to a magnetic field. If B is the magnetic field at distance r from the axis of the plates, then

$$\oint \boldsymbol{B} \cdot \boldsymbol{dl} = \mu_0 i_d = \mu_0 \pi r^2 j_d$$

or $\quad 2\pi r B = \mu_0 \pi r^2 j_d$

and $\quad B = \frac{1}{2}\mu_0 r j_d = \frac{1}{2}\frac{\mu_0 \varepsilon_0 V_0 \omega \cdot r \cos \omega t}{d}$

Energy density of magnetic field $= \frac{1}{2}\frac{B^2}{\mu_0}$. Hence magnetic energy with in volume dV is

$$dU = \frac{1}{2}\frac{B^2}{\mu_0}dV \quad \text{and} \quad dV = 2\pi r d dr$$

$$\therefore \quad dU = \frac{1}{2}\frac{1}{4}\frac{u_0^2\varepsilon_0^2V_0^2\omega^2}{d^2\mu_0}r^2\cos^2\omega t\cdot 2\pi r d\,dr$$

$$= \frac{1}{8}\frac{\mu_0\varepsilon_0^2V_0^2\omega^2(2\pi d)\cos^2\omega t\cdot r^3dr}{d^2}$$

Hence total magnetic energy is

$$U = \int_0^R \frac{1}{8}\frac{\mu_0\varepsilon_0^2\cdot V_0^2\omega^2(2\pi)}{d}\cos^2\omega t\cdot r^3dr$$

$$= \frac{\pi}{16}\frac{\mu_0\varepsilon_0^2\omega^2\cdot R^4}{d}V_0^2\cos^2\omega t$$

Hence the maximum magnetic field is

$$U_m = \frac{\pi}{16}\frac{\mu_0\varepsilon_0^2\omega^2R^4V_0^2}{d} \quad \ldots(ii)$$

$$\therefore \quad \text{ratio } \frac{U_m}{U_e} = \frac{\pi}{16}\frac{\mu_0\varepsilon_0^2\omega^2R^4V_0^2\times 2d}{\varepsilon_0\pi R^2V_0^2d}$$

$$= \frac{1}{8}(\mu_0\varepsilon_0\omega^2R^2) \text{ But } \mu_0\varepsilon_0 = \frac{1}{c^2}$$

$$= \frac{1}{8}\left(\frac{\omega R}{c}\right)^2$$

Substituting the given values, we have

$$\frac{U_m}{U_e} = \frac{1}{8}\left(\frac{1000\times 6\times 10^{-2}}{3\times 10^8}\right)^2 = 5\times 10^{-15}$$

Note: These results are valid under the approximation $\omega R << c$.

Example 26.18: A plane electromagnetic wave $E = E_0\cos(\omega t - kx)$ propagating in vacuum induces the emf e_i in a square frame with side l. The orientation of the frame is shown in Fig. 26.12. Find the amplitude value of e_i, if $E_0 = 50$ mV/m, the frequency is 5.0 MHz and $l = 50$ cm.

Solution: Consider a small element dx at distance x (Fig. 26.12)

The flux through this area is

$$d\phi = Bl\,dx$$

But $\quad B = B_0\cos(\omega t - kx) \quad \text{where } B_0 = \frac{E_0}{C}$

$$\therefore \quad \phi = \int_0^l B_0 l\cos(\omega t - kx)dx$$

$$= -\frac{B_0 l}{k}[\sin(\omega t - kx)]_0^l$$

$$= \frac{B_0 l}{k}[\sin\omega t - \sin(\omega t - kl)]$$

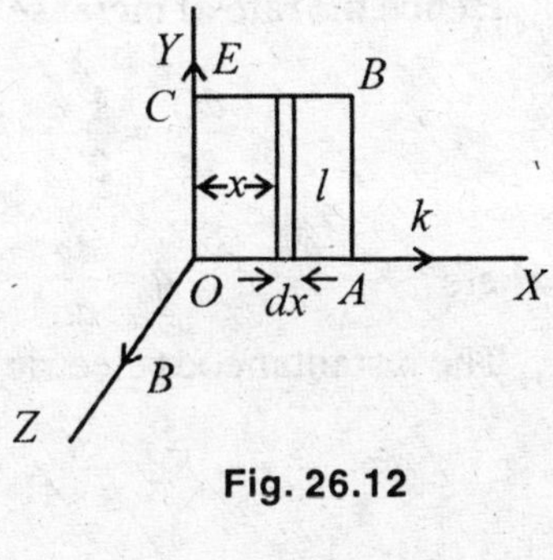

Fig. 26.12

$$= \frac{2B_0 l}{k} \sin\frac{kl}{2} \cos\left(\omega t - \frac{kl}{2}\right)$$

Hence emf induced is

$$e_i = -\frac{d\phi}{dt} = \frac{2B_0 l\omega}{k} \sin\frac{kl}{2} \sin\left(\omega t - \frac{kl}{2}\right)$$

$$\frac{\omega}{k} = c \text{ and } B_0 c = E_0. \text{ Hence}$$

$$e_i = 2E_0 l \sin\frac{kl}{2} \sin\left(\omega t - \frac{kl}{2}\right)$$

Hence amplitude of e_i is

$$e_0 = 2E_0 l \sin\frac{kl}{2}$$

Since $\frac{kl}{2}$ is very small hence $\sin\frac{kl}{2}$ can be approximated to $\frac{kl}{2}$

$$\therefore \quad e_0 = \frac{2E_0 l^2 k}{2} = E_0 l^2 k = E_0 l^2 \frac{2\pi\nu}{c}$$

$$= \frac{2\pi\nu l^2 E_0}{c}$$

$$\therefore \quad e_0 = \frac{2\pi \times 5 \times 10^6 \times (0.5)^2 \times 50 \times 10^{-3}}{3 \times 10^8}$$

$$= 13 \text{ mV.}$$

Example 26.19: A parallel plate capacitor whose electrodes are shaped as round discs is charged slowly. Demonstrate that the flux of the Poynting vector across the capacitors lateral surface is equal to the increment of the capacitor's energy per unit time. The dissipation of the field at the edge is to be neglected in calculations.

Solution: Let the charge on the capacitor at any time be q. Hence energy stored at this instant is

$$U = \frac{1}{2}\frac{q^2}{C}$$

Hence the rate of increase of energy of capacitor is

$$\frac{dU}{dt} = \frac{1}{2}\frac{d}{dt}\left(\frac{q^2}{C}\right) = \frac{q\dot{q}}{C}$$

where $\dot{q} = \frac{dq}{dt}$

The instantaneous electric field between the plates is

$$E = \frac{q}{\varepsilon_0 A} \quad \text{wher } A = \pi R^2 \text{ area of the plates.}$$

and the displacement current

$$i_d = \varepsilon_0 \frac{d}{dt}(\phi_e) = \varepsilon_0 \frac{d}{dt}\left(\frac{q}{\varepsilon_0 A} A\right) = \frac{dq}{dt} = \cdot(\dot{q})$$

This produces magnetic field at distance r such that

$$\oint \boldsymbol{B} \cdot \boldsymbol{dl} = \mu_0 i_d \frac{\pi r^2}{\pi R^2} = \mu_0 i_d \frac{r^2}{R^2}$$

and $$B 2\pi r = \mu_0 q \frac{r^2}{R^2} \text{ or } B \equiv \frac{\mu_0 (\dot{q}) r}{2\pi R^2}$$

and Poynting vector is

$$\boldsymbol{S} = \frac{1}{\mu_0} \boldsymbol{E} \times \boldsymbol{B} \text{ Since } \boldsymbol{E} \text{ and } \boldsymbol{B} \text{ are perpendicular}$$

Hence magnitude $S = \frac{EB}{\mu_0} = \frac{q}{\varepsilon_0 A} \cdot \frac{\mu_0 (\dot{q}) r}{2\pi R^2} \cdot \frac{1}{\mu_0}$

$$= \frac{q(\dot{q})\, r}{\varepsilon_0 A 2\pi R^2}$$

Hence the inward Poynting vector at the edge is ($r = R$)

$$S = \frac{q(\dot{q})}{\varepsilon_0 A 2\pi R}$$

Hence total flux flow = $S 2\pi R d$ where d is separation between plate

$$= \frac{q\dot{q} 2\pi R d}{\varepsilon_0 A 2\pi R} = \frac{q\dot{q}}{\varepsilon_0 A/d} = \frac{q\dot{q}}{c} = \frac{dU}{dt}$$

Hence the result.

PROBLEMS

1. A sinusoidal electromagnetic plane wave travels through empty space in the X-direction. The electric wave is polarised in the Y-direction. Frequency of wave is 1.5 GHz. At certain point P in space, the oscillating electric field E_y attains a maximum magnitude $E_0 = 63$ V/m. (*a*) What is the wavelength of the wave? (*b*) What is the maximum magnitude of B_0? What is the direction of B? (*c*) Express the electric and magnetic fields as function of x and t in the vicinity of P.

   [**Ans.** (*a*) $\lambda = 0.2$ m (*b*) $B_0 = 2.1 \times 10^{-7}$ T along Z-direation (*c*) Here $c = 3 \times 10^8$ ms^{-1} $\boldsymbol{E_y} = 63\cos[31(x - ct)]\, \hat{j}$ V/m, $\boldsymbol{B_z} = 2.1 \times 10^{-7}\cos[31(x - ct)\, \hat{k}$ T]

2. A nozzle throws a stream of gas against a wall with a velocity v much larger than thermal agitation of the molecules. The wall deflects the molecules without changing the magnitude

of their velocity. Also assume that the force exerted on the wall by molecules is perpendicular to wall. Find the pressure exerted on the wall. [**Ans.** $P = 2nmv^2\cos^2\theta$]

3. A plane wave of light of intensity $I = 0.2$ W/cm^2 falls on surface of a plane mirror with reflection coefficient $\rho = 0.8$. The angle of incidence is 45°. In terms of corpuscular theory, find the magnitude of the normal pressure exerted on that surface. [**Ans.** $P = 0.5$ N/cm^2]

4. Find the maximum K.E. of photoelectrons liberated from the surface of lithium by electromagnetic radiation whose electric component varies with time as $E = a(1 + \cos\omega t)\cos\omega_0 t$, where a is constant, $\omega = 6.0 \times 10^{14}$ s^{-1} and, $\omega_0 = 3.60 \times 10^{15}$ s^{-1}, work function of lithium = 2.39 eV.

   [**Hint.** $\boldsymbol{E} = a(1 + \cos\omega t)\cos\omega_0 t$ is superposition of three waves which are $E_1 = a\cos\omega_0 t$, $E_2 = \frac{a}{2}\cos(\omega - \omega_0)t$ and $E_3 = a/2\cos(\omega + \omega_0)t$. Maximum frequency is $(\omega + \omega_0)$ and from Einstein's photoelectric equation $h\nu = W + (\text{K.E.})_{max}$. $(\text{K.E.})_{max} = \dfrac{h(\omega + \omega_0)}{2\pi} - W = 0.36$ eV]

5. An isotropic point source emits light of wavelength $\lambda = 589$ nm. The radiation power of the source is $P = 10$ W. Find (*a*) the mean density of flow of photons at a distance $r = 2.0$ m from the source (*b*) the distance between source and the point at which the mean concentration of photons is equal to $n = 100$ cm^{-3}. [**Ans.** $r = 9$ cm]

6. Light of wavelength λ from a small 0.5 mW He-Ne laser source is used in the school laboratory shines a spacecraft of mass 1000 kg. Explain why this causes the spacecraft to accelerate. Estimate the time needed for spacecraft to reach a velocity of 1.0 km/s from rest. The momentum P of a photon of wavelength λ is given by $P = h/\lambda$; where h is Plank's constant.

   [**Ans.** $t = 6 \times 10^{17}$ sec]

7. The magnetic field in a plane electromagnetic wave is given by $B_z = 2 \times 10^{-7}\sin(0.5 \times 10^3 x + 1.5 \times 10^{11} t)$ T. (*a*) What is the wavelength and frequency of the wave? (*b*) Write an expression for the electric field vector.

   [**Ans.** $\lambda = 1.26$ cm, $\nu = 23.6$ GHz (*b*) $E_y = 60\sin(500x + 1.5 \times 10^{11} t)$ V/m]

8. Show that the instantaneous magnitude of the Poynting vector in free space may be written in the form $S = \dfrac{c}{2}\left(\varepsilon_0 E^2 + \dfrac{B^2}{\mu_0}\right)$, where the symbols have usual meanings.

9. At a distance of 6 m from a monochromatic point source the amplitude of the electric field is 10 V/m. Find (*a*) the amplitude of the magnetic field. (*b*) The average power output of the source. [**Ans.** (*a*) $B_0 = 3.33 \times 10^{-8}$ T (*b*) $P = 60$ W]

10. A cylindrical coil used as an AM antenna has 250 turns of diameter 1.5 cm. Find the peak emf induced by a 10^4 W station (treated as a point source) broadcasting at 800 kHz at a distance of 2 km. The axis of the coil is parallel to the direction of the magnetic field of the wave. [**Ans.** $V_0 = 2.87 \times 10^{-4}$ V]

11. A leaky capacitor with circular plates of radius $R = 12$ cm has a capacitance of 5 μF and an effective resistance of 4×10^5 Ω. At $t = 0$ the potential difference across the plate is zero but is increasing at constant rate of 2000 V/s (*a*) Find the displacement current I_d (*b*) At what time is I_d equal to the conduction current?

[**Hint:** $I_d = \varepsilon_0 \frac{d\phi_e}{dt} = \varepsilon_0 \frac{d}{dt}(E \cdot A) = \frac{\varepsilon_0 A}{d}\frac{dV}{dt} = C\frac{dV}{dt} = 5 \times 10^{-6} \times 2000 = 0.01$ A I_d will be equal to I_c after a long time, i.e. after a time constant = RC = 2 s]

12. A standing electromagnetic wave $E = E_0 \cos kx \cos \omega t$ is sustained along the X-axis in vacuum. Find the projection of the Poynting vector on the X-axis $S_x(x, t)$ and the mean value of that projection averaged over an oscillation period.

$$\left[\textbf{Ans. } S_x(x, t) = \frac{1}{2}\varepsilon_0 c E_0^2 \sin 2kx \sin^2 \omega t, < S_x > = 0\right]$$

13. The circular plates of parallel plate capacitor have a radius 2 cm and are separated by 4 mm. They are connected to 60 Hz AC source with peak potential difference 120 V. Find the peak magnetic field half way from the centre to the edge of plate. [**Ans.** $B_0 = 6.28 \times 10^{-13}$ T]

14. A plane electromagnetic wave with intensity 220 W/m^2 is incident normal to a flat plate of radius 30 cm. If the plate absorbs 60% and reflects 40% of the incident radiations, what is the momentum transferred to it in 5 m? [**Ans.** $P = 8.7 \times 10^{-5}$ kgms^{-1}]

15. A plane electromagnetic wave varies sinusoidally at 90 MHz as it travels along the +X-direction. The peak value of electric field is 2m V/m and it is directed along the ±Y-direction. (*a*) Find the wavelength, the period and the peak value B_0 of the magnetic field. (*b*) Write expression in SI units for the space and time variation of the electric field and of the magnetic field. Include numerical values as well as subscripts to indicate coordinate directions. (*c*) Find the average power per unit area that this wave propagates through space. (*d*) Find the average energy density in the radiation (mJ/m^3). (*e*) What radiation pressure would this wave exert up on a perfectly reflecting surface at normal incidence?

[**Ans.** (*a*) λ = 3.3 m, T = 11.1ns, B_0 = 6.67 PT (*b*) $E = 2.0 \cos 2\pi\left(\frac{x}{\lambda} - \frac{t}{T}\right)\hat{j}$ V/m

$B = B_0 \cos 2\pi\left(\frac{x}{\lambda} - \frac{t}{T}\right)\hat{k}$ PT (*c*) <P> = 5.31×10^{-9} W/m^2

(*d*) <U> = 1.77×10^{-14} J/m^3 (*e*) $P_r = 3.54 \times 10^{-14}$ Pa]

16. A plane electromagnetic wave has an energy flux of 750 W/m^2. A flat rectangular surface of dimensions 50 cm × 100 cm is placed perpendicular to the direction of the plane wave. If the surface absorbs half of the energy and reflects half (*i.e.*, a 50% reflecting surface). Calculate (*a*) the total energy absorbed by the surface in a time/minute. (*b*) momentum absorbed by this time. [**Ans.** (*a*) E = 11.25 kJ (*b*) $P = 11.25 \times 10^{-5}$ kgms^{-1}]

17. A linearly polarized wave of wavelength 1.5 cm is directed along the positive X-axis. The electric field vector has maximum value of 175 V/m. and vibrates in the X-Y plane. (*a*) Assume that the magnetic field component of the wave can be written in the form $B = B_0 \sin(kx - \omega t)$ and give values of B_0, k and ω. Also determine in which plane the magnetic field vector vibrates. (*b*) Calculate the Poynting vector for this plane wave. (*c*) What radiation pressure would this wave exert if directed at normal to incidence to a perfectly reflecting sheet? What acceleration would be imparted to a 500 g sheet (perfectly reflecting and at normal incidence) with dimension 1 m × 0.75 m?

[**Ans.** (*a*) B_0 = 583 nT, k = 419 rad/m, $\omega = 1.26 \times 10^{11}$ rad/s, X-Z plane (*b*) 40.6 W/m^2 average value (*c*) 271 nPa. (*d*) 406 nm/s^2]

18. A rectangular surface of dimensions 120 cm × 40 cm is parallel to and 4.4 m from a very large conducting sheet in which there is a sinusoidally varying surface current which has maximum value of 10 A/m. (*a*) Calculate the average power incident on smaller sheet. (*b*) What power per unit area is radiated by the current carrying sheet?

[**Ans.** (*a*) $<P>$ = 2.26 kW (*b*) P/A = 4.71 kW/m^2]

19. The average energy density in a sinusoidal electromagnetic wave is 10^{-7} J/m^3. Find the amplitude of (*a*) the electric field (*b*) the magnetic field.

[**Ans.** (*a*) E_0 = 150 V/m (*b*) B_0 = 0.5 μT]

20. A current flowing in the windings of a long straight solenoid is increased at a sufficient slow rate. Demonstrate that the rate at which the energy of the magnetic field in the solenoid increases is equal to the flux of the Poynting vector across the lateral surface of the solenoid.

SHORT ANSWER TYPE QUESTIONS

1. Using $B = \mu_0 H$, find the ratio E_0/H_0 for a plane electromagnetic wave propagating through vacuum. Find the dimensions of E_0/H_0.

[**Ans.** For plane electromagnetic wave $B_0 = \frac{E_0}{c}$ $\therefore \mu_0 H_0 = \frac{E_0}{c}$ or $\frac{E_0}{H_0} = \mu_0 c = 4\pi \times 10^{-7} \times 3 \times 10^8$ $= 337\ \Omega$. This is called the *characteristic impedance of free space.* The ratio $\frac{E_0}{H_0}$ has dimensions.

$E_0 = \frac{\text{Newton}}{\text{Coulomb}} = \frac{M^{-1}L^1T^{-2}}{IT} = M^1L^1T^{-3}I^{-1}$, $H_0 = \frac{\text{Ampere}}{\text{meter}} = M^0L^{-1}T^0I^1$ $\therefore \frac{E_0}{H_0} = \frac{M^1L^1T^{-3}I^{-1}}{M^0L^{-1}T^0I^1}$ $= M^1L^2T^{-3}I^{-2}$. Resistance $R = \frac{V}{I} = \frac{\text{Joule}}{\text{constant } I} = \frac{M^1L^2T^{-2}}{ITI} = M^1L^2T^{-3}I^{-2}$ Hence $\frac{E_0}{H_0}$ has the dimensions of resistance]

2. A point charge is moving along a straight line with constant velocity v. Consider a small area A perpendicular to the direction of motion of the charge (Fig. 26.13). Find the displacement current through the area when its distance from the charge is x. The value x is not large so that the electric field at any instant is given by Coulomb's law.

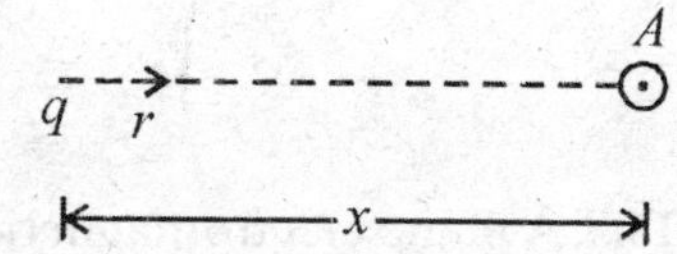

Fig. 26.13

[**Ans.** When q is at distance x the electric field at A is $E = \frac{1}{4\pi\varepsilon_0}\frac{q}{x^2}$. Its direction is normal to the plane of area. The flux through area A is $\phi_e = \int_A \boldsymbol{E} \cdot d\boldsymbol{s} = EA = \frac{qA}{4\pi\varepsilon_0 x^2}$. Hence displacement current $i_d = \varepsilon_0 \frac{d\phi_e}{dt}$ or $i_d = \varepsilon_0 \frac{qA}{4\pi\varepsilon_0} \cdot \frac{d}{dt}\frac{1}{x^2} = -\frac{qA}{4\pi}\frac{1}{x^3}\frac{dx}{dt}$. But $\frac{dx}{dt} = v$, hence magnitude of the displacement is $i_d = \frac{qAv}{4\pi x^3}$]

3. The magnetic field in a plane electromagnetic wave is given by B = (200 μT)sin $4 \times 10^{15}(t - x/c)$. Find the maximum electric field and the average energy density corresponding to the electric field.

[**Ans.** We have B_0 = 200 μT. If E_0 is the maximum electric field then $E_0 = CB_0$ $\therefore E_0 = 3 \times 10^8 \times 200 \times 10^{-6} = 6 \times 10^4$ V/m. If E_0 is the maximum electric field, then the average energy density of the medium is $U = \frac{1}{2}\varepsilon_0 E_0^2 = \frac{1}{2} \times 8.85 \times 10^{-12} \times (6 \times 10^4)^2 = 0.16$ J/m^2]

4. A laser beam has intensity 2.5×10^{14} W/m^2. Find the amplitudes of electric and magnetic fields in the beam.

[**Ans.** If U is energy density then intensity $I = Uc = \frac{1}{2}\varepsilon_0 E_0^2 \cdot c$. Where E_0 is the amplitude of the electric field. Then $E_0^2 \equiv \frac{2I}{\varepsilon_0 c} = \frac{2 \times 2.5 \times 10^{14}}{8.85 \times 10^{-12} \times 3 \times 10^8} = 18.83 \times 10^{16}$ or $E_0 = 4.3 \times 10^8$ V/m

and $B_0 = \frac{E_0}{c} = \frac{4.3 \times 10^8}{3 \times 10^8} = 1.4$ T]

5. Find the energy stored in 12 m length of a laser beam operating at 3 mW.

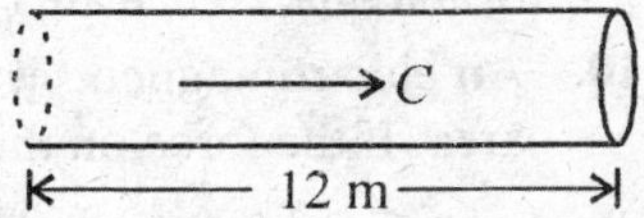

Fig. 26.14

[**Ans.** Time taken by laser beam to cross 12 m length is $t = \frac{12}{3 \times 10^8}$ $= 4 \times 10^{-8}$ sec. Since the power of beam is 3 mW, hence energy passing through any cross-section in 4×10^{-8} sec is $E = P \times t = 3 \times 10^{-3} \times 4 \times 10^{-8} = 12 \times 10^{-11}$ J]

6. An electromagnetic wave of frequency $\nu = 3.0$ MHz passes from vacuum into a nonmagnetic medium with permittivity $\varepsilon = 4.0$. Find the increment in its wavelength.

[**Ans.** Velocity of light in medium $c_m = \frac{c}{\sqrt{\varepsilon}}$. Frequency remains unchanged. Hence wavelength in air $\lambda_a = \frac{c}{\nu}$ and in medium $\lambda_m = \frac{c_m}{\nu} = \frac{c}{\sqrt{\varepsilon}\,\nu}$. Hence change in wavelength $\lambda_a - \lambda_m$

$= \frac{c}{\nu}\left(1 - \frac{1}{\sqrt{\varepsilon}}\right) = \frac{3 \times 10^8}{3 \times 10^6}\left(1 - \frac{1}{\sqrt{4}}\right) = 10^2 \times \frac{1}{2} = 50$ m. Hence increase in wavelength = –50 m]

7. A plane electromagnetic wave of frequency $\nu = 10$ MHz propagates in a poorly conducting medium with conductivity $\sigma = 10$ m mhos/m and permittivity $\varepsilon_r = 9$. Find the ratio of amplitudes of conduction and displacement current densities.

[**Ans.** If E is electric field then conduction current density is $j_c = \sigma E$, where σ is conductivity. and if $E = E_0 \sin(\omega t - kx)$ is the plane electromagnetic wave, then amplitude of conduction current density is $j_c = \sigma E_0$. Displacement current $i_d = \varepsilon \frac{d\phi_e}{dt}$ where $\varepsilon = \varepsilon_0 \varepsilon_r$. $\phi_e = AE = AE_0 \sin$ $(\omega t - kx)$ $\therefore\ i_d = \varepsilon_0 \varepsilon_r A E_0 \frac{d}{dt}\sin(\omega t - kx)$ and displacement current density $j_d = \frac{i_d}{A}$

$= \varepsilon_0 \varepsilon_r E_0 \omega \cos(\omega t - kx)$ and displacement current density amplitude is $j_d = \varepsilon_0 \varepsilon_r E_0 \omega$ $\therefore \frac{j_c}{j_d} = \frac{\sigma}{\varepsilon_0 \varepsilon_r \omega}$

$= \frac{10 \times 10^{-3}}{8.85 \times 10^{-12} \times 9 \times 2\pi \times 10^7} = 2$]

8. A ball of radius $R = 50$ cm is located in a nonmagnetic medium with permittivity $\varepsilon = 4.0$. In that medium a plane electromagnetic wave propagates, the strength amplitude of whose electric component is equal to $E_0 = 200$ V/m. Find the amount of energy reaching the ball during time interval $t = 1.0$ minute.

[**Ans.** Poynting vector's magnitude is $S = \frac{1}{2}\frac{E_0 B_0}{\mu_0} = \frac{1}{2}\frac{E_0^2}{\mu_0 c_m}$, c_m is velocity of light in medium

$= \frac{1}{2}\frac{E_0^2\sqrt{\varepsilon_0\varepsilon_r\mu_0}}{\mu_0} = \frac{1}{2}E_0^2\sqrt{\frac{\varepsilon_0\varepsilon_r}{\mu_0}} = \frac{1}{2}(200)^2\sqrt{\frac{8.85\times10^{-12}\times4}{4\pi\times10^{-7}}} = 106\ \text{W/m}^2$. Hence normal area of sphere on which energy falls is πr^2. Hence energy reaching the ball in time t is $= S\pi r^2 \cdot t$

$= 106 \times \pi\ (0.5)^2 \times 60 = 5$ kJ]

9. A TV tower has a height 100 m. Find the area covered by the TV broadcast, if the radius of earth is 6400 km.

[**Ans.** The ground waves cover distance $d = \sqrt{2Rh}$, where R is radius of earth and h is the height of antenna. Hence area covered is $A = \pi d^2 = \pi(2Rh) = 2\pi Rh = 2\pi \times 6.4 \times 10^6 \times 100 = 4.2 \times 10^9\ \text{m}^2$]

10. An electromagnetic wave with Poynting vector $5\ \text{W/m}^2$ is absorbed by a surface of some area. If the force on the surface is 10^{-7} N, then find the area.

[**Ans.** We have if P is pressure exerted on the source and I is intensity then $P = \frac{I}{c}$ and force

$$F = P \times \text{area} = \frac{I}{c}\cdot A \ \therefore\ A = \frac{Fc}{I} = \frac{10^{-7}\times3\times10^8}{5} = 6\ \text{m}^2$$]

11. The incident intensity on a horizontal surface at sea level from sun is about $1\ \text{kW/m}^2$. (*a*) Assume 50% this intensity is reflected and 50% is absorbed, determine the radiation pressure on the horizontal surface. (*b*) Find the ratio of this pressure to atmospheric pressure $P_0(= 10^5\ P_a)$ at sea level.

[**Ans.** If I is intensity then radiation pressure due to (*i*) reflected light $P_r = \frac{2I}{c}$, where c is velocity of light and (*ii*) absorbed light $P_a = \frac{I}{c}$. In present case $P_r = \frac{I}{c}$. Since $I/2$ intensity is reflected and $P_a = \frac{I}{2c}$. Hence total radiation pressure $P = P_r + P_a = \frac{3}{2}\frac{I}{c} = \frac{3}{2}\times\frac{10^3}{3\times10^8} = 5\times10^{-6}\ \text{P}_\text{a}$ and

ratio $\frac{P}{P_0} = \frac{5\times10^{-6}}{10^5} = 5\times10^{-11}$]

12. Figure 26.15 illustrates a segment of a double line carrying direct current whose direction is indicated by the arrows. Taking into account that the potential $V_2 > V_1$ and making use of the Poynting vector establish on which side (left or right) the source of the current is located.

[**Ans.** Since $V_2 > V_1$ the directions of electric field E and magnetic field B are as shown in Fig. 26.15(*b*). Hence Poynting vector $\boldsymbol{S} = \boldsymbol{E} \times \boldsymbol{B}$ is parallel to the wires and towards right.

Hence source must be on the left]

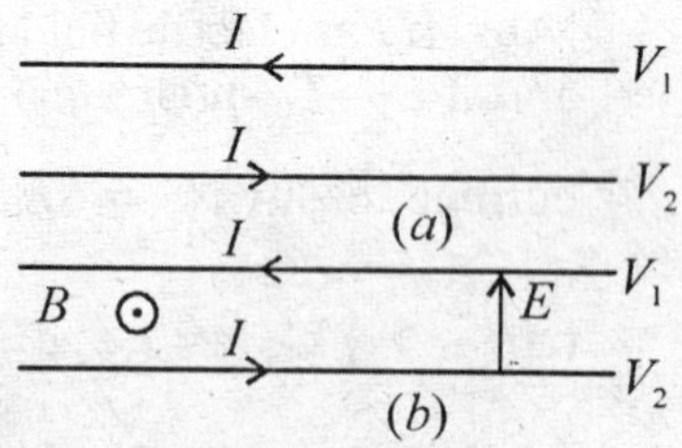

Fig. 26.15

13. A plane harmonic electromagnetic wave with plane polarisation propagates in vacuum. The electric component of the wave has a strength amplitude $E_0 = 50$ mV/m and the frequency is $\nu = 100$ MHz. Find (*a*) the effective value of the displacement current density (*b*) the mean energy flow density averaged over an oscillation period.

[**Ans.** Displacement current $i_d = \varepsilon_0\frac{d\phi_e}{dt} = \varepsilon_0\frac{d}{dt}\boldsymbol{E}\cdot\boldsymbol{A} = \varepsilon_0 A\frac{d}{dt}E_0\sin(\omega t - kx)$ $\therefore$ Current density $j_d = \frac{i_d}{A} = \varepsilon_0 \cdot E_0 \cdot \omega\cos(\omega t - kx)$. Hence the amplitude of displacement current density

$j_0 = \varepsilon_0 E_0 \omega$, $j_0 = 8.85 \times 10^{-12} \times 50 \times 10^{-3} \times 2\pi \times 100 \times 10^6 = 10^{-5} \times 8.85 \times \pi = 2.78 \times 10^{-4}$ A/m^2. Hence effective value is $\frac{j_0}{\sqrt{2}} = \frac{2.78 \times 10^{-4}}{\sqrt{2}} = 0.2\ mA/m^2$ (*b*) Mean energy density flow

$$< S > = \frac{1}{2}\frac{E_0 \cdot B_0}{\mu_0} = \frac{1}{2}\frac{E_0^2}{\mu_0 c} = \frac{1}{2}\sqrt{\frac{\varepsilon_0}{\mu_0}}E_0^2 = \frac{1}{2}\sqrt{\frac{8.85 \times 10^{-12}}{4\pi \times 10^{-7}}} \times (50 \times 10^{-3}) = 3.3\ \mu W/m^2 \Big]$$

14. Find the mean Poynting vector $<S>$ of a plane electromagnetic wave $E = E_0\cos(\omega t - kr)$ if the wave propagates in vacuum.

[**Ans.** Let the wave propagates in X-direction then $\boldsymbol{E} = E_0\cos(\omega t - kx)\,\hat{j}$ and magnetic field vector $\boldsymbol{B} = B_0\cos(\omega t - kx)\,\hat{k}$ where $B_0 = \frac{E_0}{c}$. Hence Poynting vector $\boldsymbol{S} = \frac{\boldsymbol{E} \times \boldsymbol{B}}{\mu_0}$ $= \frac{E_0^2}{\mu_0 c}\cos^2(\omega t - kx)\hat{i}$. But average value of $\cos^2(\omega t - kx) = 1/2$ $\therefore$ $< S > = \frac{E_0^2}{2\mu_0 c}\hat{i}$ $= \frac{E_0^2 \cdot \varepsilon_0}{2\mu_0\varepsilon_0 c}\hat{i} = \frac{1}{2}\varepsilon_0 E_0^2 c\,\hat{i}$. Since we have assumed direction $\hat{i}$ to be direction of propagation or direction of k. Then $< S > = \frac{1}{2}E_0^2\varepsilon_0\frac{c \cdot \boldsymbol{k}}{k} = \frac{1}{2}\sqrt{\frac{\varepsilon_0}{\mu_0}} \cdot E_0^2 \cdot \frac{\boldsymbol{k}}{k}$]

15. It is predicted that a nuclear war on the earth would be followed by a severe nuclear winter with devastating effects on life on the earth. What might be the cause of prediction?

[**Ans.** The clouds produced by the nuclear war, besides causing severe devastating effects, would perhaps cover substantial parts of the sky preventing solar light from reaching many parts of the globe. This would cause a winter.

16. Using Wien's displacement law ($\lambda_m T = b$) obtain the part of electromagnetic spectrum which belongs to 2.7 K. Take $b = 3 \times 10^{-3}$ mK.

[**Ans.** $\lambda_m = \frac{b}{T} = \frac{3 \times 10^{-3}}{2.7} = \frac{1}{9} \times 10^{-2}\ m = 0.11\ cm$. This belongs to microwave region (1 mm to 30 mm)]

MULTIPLE CHOICE QUESTIONS

PART I

In the following, a question is followed by four alternatives, of them only one is correct. Therefore, the candidate has to select the correct choice from amongst the alternatives.

1. An electromagnetic wave going through vacuum is described by $E = E_0\sin(\omega t - kx)$. Which of the following is independent of the wavelength?
   (a) k (b) ω
   (c) k/ω (d) $k\omega$.
2. Speed of electromagnetic waves is the same
   (a) for all wavelengths
   (b) in all media
   (c) for all intensities
   (d) for all frequencies.
3. The energy contained in a small volume through which an electromagnetic wave is passing oscillates with

(a) zero frequency
(b) the frequency of the wave
(c) half the frequency of the wave
(d) double the frequency of the wave.

4. An electromagnetic wave going through vacuum is described by $E = E_0\sin(kx - \omega t)$; $B = B_0\sin(kx - \omega t)$, then

(a) $E_0k = B_0\omega$
(b) $E_0B_0 = \omega k$
(c) $E_0\omega = B_0k$
(d) No such relation exists.

5. Dimensions of $\frac{1}{\varepsilon_0\mu_0}$ is

(a) L^1T^{-1} (b) T^1L^{-1}
(c) L^2T^{-2} (d) T^2L^{-2}.

6. The maximum electric field in a plane electromagnetic wave is 900 N/C. The wave is propagating in the X-direction and the electric field in the Y-direction. The maximum magnetic field in the wave is

(a) 3×10^{-6} T in Y-direction
(b) 3×10^{-6} T in Z-direction
(c) 2.7×10^{-6} T in X-direction
(d) 2.7×10^{-6} T in Z-direction.

7. A parallel plate capacitor with plate area A and separation between the plates d, is charged by a constant current i. Consider a plane surface of area $A/2$ parallel to the plates and drawn symmetrically between the plates. The displacement current through this area is

(a) i (b) $2i$
(c) $i/2$ (d) zero.

8. An object is placed at some distance from a radio station. If the interval between transmission and reception of pulse is 2.66 $\times 10^{-2}$ sec, then the distance of the reflector is

(a) 4000 km (b) 2000 km
(c) 3000 km (d) 2500 km.

9. The wavelength of a radio wave of frequency 1 MHz is

(a) 400 m (b) 300 m
(c) 350 m (d) 200 m.

10. A compass needle is placed in the gap of a parallel plate capacitor. The capacitor is connected to a battery through a resistance. The compass needle

(a) does not deflect
(b) deflects for a very short time and then comes back to the original position
(c) deflects and remains deflected as long as the battery is connected
(d) deflects and gradually comes to the original position in a time which is large compared to the time constant.

11. Electromagnetic waves are produced by

(a) a static charge
(b) a moving charge
(c) an accelerating charge
(d) a neutral moving particle.

12. Some radio waves of frequency 1.5×10^9Hz were received by a radio telescope from a distant star. If the speed of the waves is 3×10^8 m/s, then wavelength of the wave will be

(a) 0.1 m (b) 0.6 m
(c) 0.2 m (d) 0.46 m.

13. A radio wave of intensity I is reflected by a surface. The intensity of reflected wave is I. If the pressure exerted on the surface is 2×10^{-8} N/m^2, then I is (in W/m^2)

(a) 3 (b) 4
(c) 6 (d) 7.

14. If a bulb emits 20 W of electromagnetic radiation uniformly in all directions, then average power per unit area at a distance of 2 m is

(a) 0.69 W/m^2 (b) 0.56 W/m^2
(c) 0.78 W/m^2 (d) 0.39 W/m^2.

15. A plane electromagnetic wave is incident on a material surface. The wave delivers momentum P and energy E, then

(a) $P = 0$ $E \neq 0$ (b) $P \neq 0$ $E = 0$
(c) $P \neq 0$ $E \neq 0$ (d) $P = 0$, $E = 0$.

16. Consider the following two statements regarding a linearly polarized, plane electromagnetic wave

(*i*) The electric field and magnetic field have equal average values.
(*ii*) The electric energy and magnetic energy have equal average values.

(a) both (*i*) and (*ii*) are true
(b) (*i*) is false (*ii*) is true
(c) (*i*) is true (*ii*) is false
(d) both (*i*) and (*ii*) are false.

17. If at a certain instant, the magnetic induction of the electromagnetic wave in vacuum is 6.7×10^{-12} T, then the magnitude of electric field intensity is

(a) 2×10^{-3} N/C (b) 3×10^{-3} N/C
(c) 4×10^{-3} N/C (d) 1×10^{-3} N/C.

18. A free electron is placed in the path of a plane electromagnetic wave. The electrons will start moving

(a) along the electric field
(b) along the magnetic field
(c) along the direction of propagation of the wave
(d) in a plane containing the magnetic field and the direction of propagation.

19. An electric field **E** and magnetic field **B** exists in a region. The fields are not perpendicular to each other

(a) This is not possible
(b) No electromagnetic wave is passing through the region
(c) An electromagnetic wave may be passing through the region
(d) An electromagnetic wave is certainly passing through the region.

20. If the speed of gamma rays is ($= v_g$), that of *X*-ray is ($= v_x$) and of microwaves is ($= v_m$) then correct relation is

(a) $v_g > v_x > v_m$ (b) $v_g < v_x < v_m$
(c) $v_g = v_x = v_m$ (d) $v_g > v_x < v_m$.

21. The idea of displacement current was introduced by

(a) Maxwell (b) Hertz
(c) Marconi (d) Bose.

22. The average value of electric energy density in an electromagnetic wave is (E_0 is peak value)

(a) $\frac{1}{2}\varepsilon_0 E_0^2$ (b) $\frac{E_0^2}{2\varepsilon_0}$
(c) $\varepsilon_0 E_0^2$ (d) $\frac{1}{4}\varepsilon_0 E_0^2$.

23. The relation between electric field *E* and magnetic field *H* in an electromagnetic wave is

(a) $E = H$ (c) $E = \frac{\mu_0}{\varepsilon_0}H$
(c) $E = \sqrt{\frac{\mu_0}{\varepsilon_0}}H$ (d) $E = \sqrt{\frac{\varepsilon_0}{\mu_0}}H$.

24. A lamp radiates power P_0 uniformly in all directions, the amplitude of electric field strength E_0 at a distance *r* from it is

(a) $E_0 = \frac{P_0}{2\pi\varepsilon_0 c r^2}$ (b) $E_0 = \sqrt{\frac{P_0}{2\pi r^2 \varepsilon_0 c}}$
(c) $E_0 = \sqrt{\frac{P_0}{4\pi r^2 \varepsilon_0 c}}$ (d) $\sqrt{\frac{P_0}{\pi r^2 \varepsilon_0 c}}$.

25. An electromagnetic wave propagates along *Z*-axis. Which of the following pairs of space time varying fields would generate such fields?

(a) E_x, B_y (b) E_y, B_x
(c) E_z, B_x (d) E_y, B_z.

26. In an electromagnetic wave the phase difference between oscillating electric and magnetic fields have phase difference of

(a) 0 (b) $\pi/2$
(c) π (d) $\pi/4$.

27. The wavelength 21 cm emitted by atomic hydrogen in interstallar space belongs to

(a) radio waves (b) infrared waves
(c) microwaves (d) γ-rays.

28. Which of the following has the least wavelength? (AIEEE 2003)

(a) *X*-rays (b) microwaves
(c) ultra violet rays (d) radio waves.

29. Green house effect is due to

(a) Absorption of ultraviolet rays by ozone layer

(b) Transmission of visible light by atmosphere
(c) Reflection of infrared rays emitted by earth atmosphere
(d) Radiations from sun.

30. Radio waves and visible light wave
(a) same wavelength but different speed in vacuum
(b) same speed in vacuum but different wavelengths
(c) different speeds in vacuum and different wavelengths
(d) same speed in vacuum and same wavelength.

31. Displacement current arises due to
(a) movement of positive and negative ions
(b) motion of electrons
(c) time varying magnetic fields
(d) time varying electric fields.

32. To which region of spectrum the energy 14.4 keV belongs
(a) visible region
(b) ultra violet region
(c) *X*-rays
(d) γ-rays.

33. Frequency modulated radio band extends from
(a) 550 KHz to 1710 KHz
(b) 54 MHz to 890 MHz
(c) 88 MHz to 108 MHz
(d) 890 MHz to 1080 MHz.

34. If the energy of the photon is increased by the factor of 4, its momentum
(a) does not change
(b) decreases by a factor 4
(c) increases by a factor 4
(d) decreases by a factor 2.

35. In which one of the following regions of electromagnetic spectrum will the vibrational motion of molecules give rise to absorption?
(a) Ultraviolet (b) Microwaves
(c) Infrared (d) Radio waves.

ANSWERS

1. (b)	2. (c)	3. (d)	4. (a)	5. (c)
6. (b)	7. (c)	8. (a)	9. (b)	10. (d)
11. (c)	12. (c)	13. (a)	14. (d)	15. (c)
16. (b)	17. (a)	18. (a)	19. (c)	20. (c)
21. (a)	22. (d)	23. (c)	24. (b)	25. (a)
26. (a)	27. (a)	28. (a)	29. (c)	30. (b)
31. (d)	32. (c)	33. (c)	34. (c)	35. (b).

PART II

In the following, a question is followed by four alternatives, of them one or more are correct. Candidate has to select the correct choice(s) from amongst the alternatives.

1. The intensity of the sun light reaching the earth is 1380 W/m^2. Assume this light to be a plane monochromatic wave
(a) the amplitude of electric field is 1.02×10^3 V/m
(b) the amplitude of magnetic field is 3.40×10^{-6} T
(c) the amplitude of magnetic field is 3.0×6^{-11} T
(d) the amplitude of electric field is 9.18×10^{-3} V/m.

2. The displacement current goes through the gap between the plates of a capacitor when the charge of the capacitor
(a) increases
(b) decreases
(c) does not change
(d) is zero.

3. Which of the following have zero average value in a plane electromagnetic wave?
(a) Electric field (b) Magnetic field
(c) Electric energy (d) Magnetic energy.

4. An electromagnetic wave going through vacuum is described by $E = E_0 \sin(\omega t - kx)$, which of the following are dependent on the wavelength?
(a) k (b) ω
(c) ω/k (d) $k\omega$.

5. A magnetic field can be produced by
(a) a moving charge
(b) a changing electric field
(c) an stationary charge
(d) all of them.

6. A light beam travelling in the X-direction is described by the electric field $E_y = 300 \sin\omega(t - x/c)$ V/m. An electron is constrained to move along the Y-direction with a speed of 2.0×10^{-17} m/s.
(a) Maximum electric force on electron is 4.8×10^{-17} N
(b) Maximum electric force on electron is 3.2×10^{-17} N
(c) Maximum magnetic force on electron is 4.8×10^{-18} N
(d) Maximum magnetic force on electron is 3.2×10^{-18} N.

7. A small bulb emits 10.0 W of electromagnetic radiations uniformly in all directions
(a) the average power per unit area at a distance of 1 m is 0.8 W/m^2
(b) the maximum energy density in the electric field is 2.65×10^{-9} J/m^3
(c) the maximum energy density in magnetic field is 2.65×10^{-9} J/m^3
(d) the average power per unit area at a distance of 2 m is 0.4 W/m^2.

8. In electromagnetic waves, if E_0 is amplitude of electric field and B_0 amplitude of magnetic field and assume the wave to be propagating in vacuum, then
(a) $E_0 = cB_0$ (b) $E_0 = \dfrac{B_0}{c}$
(c) $\sqrt{\varepsilon_0}E_0 = \dfrac{B_0}{\sqrt{\mu_0}}$ (d) $\sqrt{\mu_0}E_0 = \dfrac{B_0}{\sqrt{\varepsilon_0}}$.

9. Which of the following statements about electromagnetic waves are not true?
(a) X-rays in vacuum travel faster than light waves in vacuum
(b) the energy of X-ray photon is greater than that of a light photon
(c) the light can be polarized but infrared rays cannot
(d) infrared rays and X-rays travel with same speed in vacuum.

ANSWERS

1. (a, b)	2. (a, b)	3. (a, b)	4. (a , d)	5. (a, b)
6. (a, d)	7. (a, b, c)	8. (a, c, d)	9. (a, c)	

27

Atomic Physics

27.1 DISCHARGE OF ELECTRICITY THROUGH RAREFIED GASES

Gases in general are the bad conductors of electricity. It is due to the fact that they do not have free charged particles, which may carry charge in an applied electric field. However at low pressures and under high electric fields (of the order of 10^5 V/m) the discharge is found to pass through air.

In order to study the discharge through gases, the experimental set-up required consists of a long tube of length of about 30 cm and diameter of about 4 cm. This tube is called discharge tube. This tube is fitted with two metal electrodes. A side tube is used to pump out the air inside the tube. The electrodes are connected to the secondary of an induction coil so that a high voltage (of the order of 30 to 40 kV) may be applied across the gas. The electrode connected to the positive terminal is called anode (A) and the one connected to negative is called cathode (C).

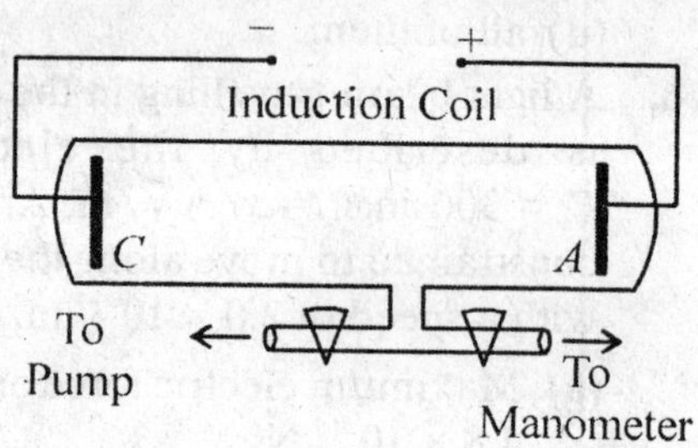

Fig. 27.1

As the potential difference between the electrodes is gradually increased, at a certain stage the sparking occurs between the electrodes. This is called *sparking potential*. The sparking potential depends on the pressure of the gas (P) and separation between the plates (d). Paschen empirically showed that sparking potential is some function of product Pd, i.e.

$$V = f(Pd) \qquad \ldots(27.1)$$

This is called *Paschen's law*

Keeping potential difference between the electrodes constant and lowering the pressures different phenomena are observed at different pressures. The observed phenomena are (for a tube of length 30 cm and potential difference of 10000 to 15000 V):

1. At the atmospheric pressure, no discharge passes through the discharge tube.
2. As the pressure is decreased to 10 mm of mercury column, a cracking sound is heard and luminous streaks of light (*called blue streamers*) are found to pass through the gas between the electrodes [Fig. 27.2 (*a*)].

3. As the pressure is reduced to about 5 mm of Hg column, the discharge widens and fills entire tube. This luminous column is called as *positive column*. The colour of this positive column depends on the nature of the gas in the column. It is red for nitrogen, pink for helium, green for chlorine and blue for hydrogen [Fig. 27.2 (*b*)].
4. At the pressure of about 0.5 mm of mercury column the positive column detaches from the cathode and contracts towards anode. A dark space appears between cathode and positive column called *Faraday's dark space* (*FDS*). Now the cathode begins to glow with blue colour. This glow is called *negative glow* (*NG*) [Fig. 27.2 (*c*)].
5. When the pressure is further reduced to 0.1 mm of Hg column, the negative glow detaches from the cathode and moves towards anode creating another dark space between cathode and negative glow. This is called *Crooke's Dark Space* (*CDS*) [Fig. 27.2 (*d*)].
6. With further reduction in pressure, the positive column further reduces and at a pressure of about 0.05 mm of mercury column it splits into dark and bright discs of light called *striations*. Another glow appears at the cathode called *cathode glow* [Fig. 27.2 (*e*)].
7. With further reduction in pressure, the striations and the negative glow begins to faint and the Crooke's dark space widens. At a pressure of about 0.01 mm, it fills the entire tube. At this stage the whole tube itself glows with a faint greenish light. This glow is due to fluorescence of the glass tube produced by rays coming from the cathode. These invisible rays coming from cathode are called *cathode rays* [Fig. 27.2 (*f*)].
8. As the pressure is further reduced, it becomes more and more difficult for the discharge to pass through the tube and finally when the pressure becomes 10^{-4} mm of mercury column no discharge takes place.

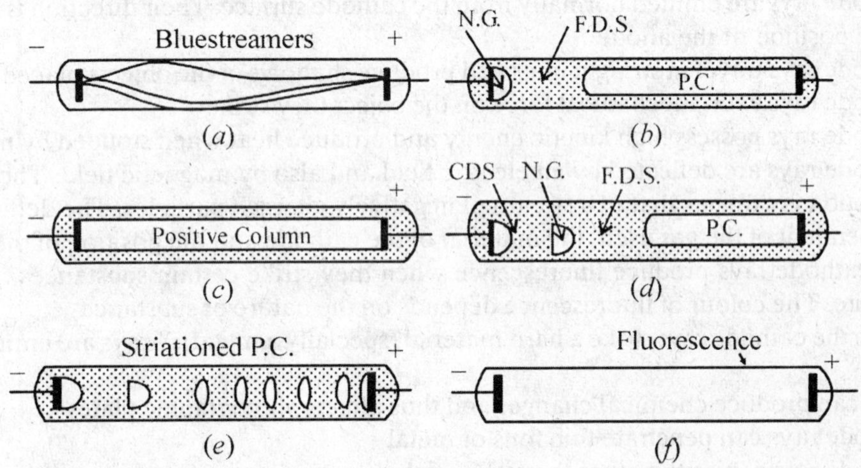

Fig. 27.2

Explanation of Discharge Phenomenon

Due to cosmic rays and other phenomenon, some ions are always present in the gas. These ions after being accelerated through a high potential difference in the discharge tube, strike the cathode with high kinetic energy. The collision liberates electrons from the cathode. These free electrons can further liberate ions from gas molecules through collisions. The positive ions move towards cathode and negative electrons move towards anode. *The negative glow is formed due to the collision between neutral molecules and electrons near the cathode.*

The electrons liberated from cathode due to collision of positive ions, are considerably slowed down due to collision with comparatively massive positive ion. Thus ionising capacity of electrons is considerably reduced. They acquire sufficient kinetic energy only after being accelerated through a certain distance. As such no ionisation takes place during this period. This causes the *Faraday's dark space.*

The breaking up of the continuous positive column in striations is due to the fact that after collision the electron losses its energy and has to travel a certain distance before it can ionise the molecules further.

The Faradays dark space is several times longer than the Crooke's dark space because the electric field here is not as intense as it is near the cathode.

As the pressure is further lowered, the mean free path of electrons increases and hence the length of Crooke's dark space increases. At extremely low pressures, when the mean free path becomes longer than the length of the tube, the Crooke's dark space fills the entire tube and no cathode glow or positive column is observed. The electrons coming out of the cathode strike the walls of the tube causing fluorescence.

Properties of Cathode Rays

Cathode rays are the stream of negatively charged particles. The charge on them is found to be equal to the charge on electron (= 1.6×10^{-19} C). Their mass is also equal to the mass of electron (= 9.1×10^{-31} kg). Their specific charge q/m is 1.76×10^{11} C/kg. This shows that cathode rays are nothing but stream of electrons. Their properties are as follows:

1. Cathode rays are emitted normally from the cathode surface. Their direction is independent of the position of the anode
2. Cathode rays travel in straight lines and produce shadows of the objects placed in their path
3. Cathode rays exert mechanical force on the object they strike
4. Cathode rays possess high kinetic energy and produce heat when stopped by matter
5. Cathode rays are deflected by an electric field and also by magnetic field. The direction of deflection is same as that of a stream of negatively charged particles. The deflection thus is independent of the gas used, the material of the cathode and the position of the anode, etc.
6. The cathode rays produce fluorescence when they strike certain substances, such as zinc sulphite. The colour of fluorescence depends on the nature of substance
7. When the cathode rays strike a hard material (specially a metal) *X*-rays are emitted from the target
8. They can produce chemical changes and thus affects photographic plates
9. Cathode rays can penetrate thin foils of metal
10. Cathode rays ionise the gases through which they pass.

Positive Rays or Canal Rays

If cathode in the discharge tube is perforated and the pressure is around 1 mm of mercury column, streams of faint luminous glow come out from each hole on the back side of the cathode. These are called *canal rays or positive rays.* The origin of the canal rays can easily be understood. The molecules near the code are ionized by the collision with electrons from cathode. These ionized molecules passing through the holes constitute the positive rays or canal rays. Since they are having positive charge as such are deflected by electric magnetic fields.

27.2 DETERMINATION OF SPECIFIC CHARGE (*e*/*m*) OF ELECTRON (CATHODE RAYS) BY J. J. THOMSON METHOD

Principle

The fact that electrons having negative charge are deflected by electric and magnetic fields. This fact was used by J.J. Thomson to determine specific charge (i.e., charge mass ratio) of the electrons.

We know that a charged particle (charge q) moving in electric field ($\boldsymbol{E}$) and magnetic field ($\boldsymbol{B}$), with velocity $\boldsymbol{v}$ experiences a force, $\boldsymbol{F}$, such that

$$\boldsymbol{F} = q(\boldsymbol{E} + \boldsymbol{v} \times \boldsymbol{B}) \qquad ...(i)$$

If $\boldsymbol{E}$ and $\boldsymbol{B}$ are adjusted such that the net force on particle is zero then a beam of charged particles can be made to pass through the two fields undeviated.

In case of electrons ($q = -e$) if $\boldsymbol{v}$ is along X-direction, $\boldsymbol{E}$ along Y-direction and $\boldsymbol{B}$ along Z-direction, the electrons go undeviated if

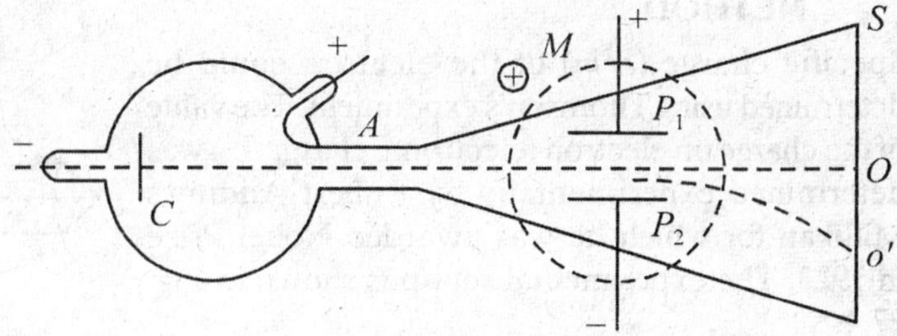

Fig. 27.3

$$v = \frac{E}{B} \qquad ...(27.2)$$

This means that if a beam of electrons is made to pass through mutually perpendicular electric and magnetic field, then those electrons whose velocity is equal to ratio (E/B) will go undeviated. Thus knowing the velocity, specific charge can be determined.

Apparatus

Thomson essentially used a cathode ray tube for this purpose which is shown in Fig. 27.3. It has a cathode C and anode A. Anode is in the form of a cylindrical tube having a fine bore along the axis of the discharge tube. P_1 and P_2 are two plates between which a suitable uniform electric field can be applied.

A magnetic field is also set up in the same region. The magnetic field is perpendicular to the electric field. This is shown by dotted circle. The magnetic field is perpendicular to the plane of paper going into it. A fluorescent screen S is used to visually observe the spot when the electrons strike the screen.

Theory and Working

If V is the potential difference between anode A and cathode C, then kinetic energy of the electrons coming out of anode tube is eV and velocity of electrons is v such that

$$\frac{1}{2}mv^2 = eV \qquad ...(ii)$$

These electrons pass through the region between plates P_1 and P_2 where simultaneous electric and magnetic fields are applied. The fields are so adjusted that the visual spot of the electrons on the screen remains unshifted whether the fields are on or off. This gives that the velocity of electrons is

$$v = \frac{E}{B}$$

Then from (*ii*) we get

$$\frac{1}{2}m\left(\frac{E}{B}\right)^2 = eV$$

or
$$\frac{e}{m} = \frac{E^2}{2VB^2} \qquad \ldots(27.3)$$

The experimental value of specific charge of electron obtained is $e/m = 1.759 \times 10^{11}$ C/kg.

27.3 DETERMINATION OF CHARGE ON ELECTRON BY MILLIKAN'S OIL DROP METHOD

Specific charge (e/m) of the electron could be determined using Thomson's experiment. The value of the charge on electron (electronic charge 'e') was determined experimentally by Robert Andrews Millikan for which he was awarded Nobel Prize in 1923. The experimental set up is shown in Fig. 27.4.

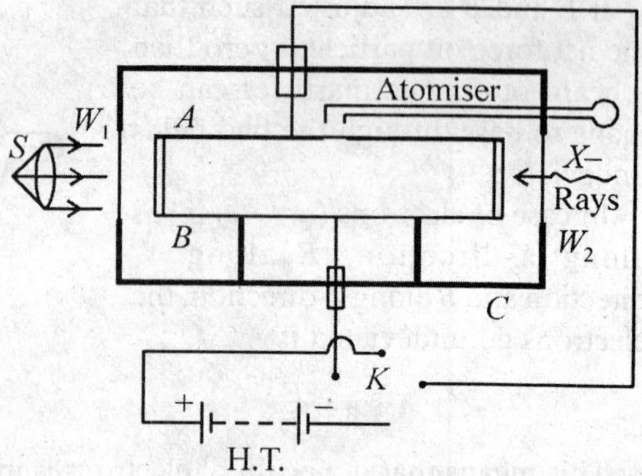

Fig. 27.4

A and *B* are two parallel metal plates separated by a small distance of the order of few millimetres. The plates are enclosed in a chamber with glass windows W_1, W_2. An electric field E is applied between the plates by connecting them to a high tension battery (H.T..). The field is applied vertically downwards and its direction can be changed by reversing key. Small droplets of oil are sprayed in the region between the plates (with the help of atomiser) through a hole in the upper plate *A*. Some of the droplets acquire electric charge due to friction with air. (This can also be achieved by passing *X*-rays through the air in chamber or putting some radioactive material in the chamber). The chamber is illuminated by passing light through it (Fig. 27.4). The droplets can be seen by using a telescope placed perpendicular to the light beam. These droplets appear like a bright shining star moving either up or down. If a droplet has a mass m and charge q, the various forces acting on it are (*a*) its weight $W = mg$ (*b*) electric force qE (*c*) buoyancy B and (*d*) the viscous force f.

Working and Theory

A single drop carrying negative charge is chosen for observation. Initially the field is put off and the droplet is allowed to fall under gravity. When the viscous force which is velocity dependent ($= 6\pi\eta rv$) becomes just equal to the resultant weight, the droplet moves with a constant velocity v_1 called terminal velocity. This can be easily measured with the help of micrometer eyepieces attached with the telescope. Then $F = W - B$.

$$6\pi\eta r v_1 = mg - B = \frac{4}{3}\pi r^3 \rho g - \frac{4}{3}\pi r^3 \sigma g \qquad \ldots(i)$$

where ρ is the density of the material of droplet and σ is the density of air. Solving we have

$$r^2 = \frac{9}{2}\frac{\eta v_1}{(\rho - \sigma)g}$$

or $$r = \left[\frac{9}{2}\frac{\eta v_1}{(\rho - \sigma)g}\right]^{1/2} \quad ...(ii)$$

Now the electric field between the plates is switched on such that the electric force on the droplet acts upwards. Soon it starts moving with terminal velocity v_2 which can easily be measured.

Now $$qE = mg - B + f$$

or $$qE = mg - B + 6\pi\eta r v_2$$

But from (*i*) we have $mg - B = 6\pi\eta r v_1$

$\therefore$ $$qE = 6\pi\eta r(v_1 + v_2)$$

or $$q = \frac{6\pi\eta r(v_1 + v_2)}{E}$$

If V is the potential difference between the plates and d the separation between the plates then $E = V/d$ and radius of droplet is given by equation (*ii*). Substitution gives

$$q = \frac{6\pi\eta d(v_1 + v_2)}{V}\left[\frac{9}{2}\frac{\eta v_1}{(\rho - \sigma)g}\right]^{1/2} \quad ...(27.4)$$

Millikan repeated his experiments large number of times and made the measurements on thousands of droplets. He found that the charge q calculated for each drop was some integral multiple of the elementary charge e (= 1.602×10^{-19} C) within the accuracy of the experiment. From this he concluded that $e = 1.6 \times 10^{-19}$ C is the charge on an electron. If one electron is attached to the droplet, the charge on it is e. If two electrons are attached the charge on the droplet is $2e$ and so on. This experiment provides a conclusive evidence of the quantisation of charge.

27.4. STRUCTURE OF THE ATOMS

The structure of matter has been a subject of study since long. Scientists began to think that matter despite its appearance of being continuous possesses a definite structure on a microscopic level. The first contribution came from Dalton who postulated that matter is made up of small particles. He called them atoms which are indivisible. The experiments on electric discharge through gases carried out by J.J. Thomson revealed that atoms of different elements contain negatively charged particles (electrons) that are completely identical. The electrical neutrality of atom therefore tells that atom must also contain some positive charge. By that time electrons had been identified but not much was known about the nature of positive charge. Even then J. J. Thomson in 1808 proposed an atomic model. According to him an *atom is basically a spherical cloud of positive charges with electrons embedded in it like seeds of watermelon.* This model though was taken seriously but electrostatically it constituted an unstable system.

It could not explain the origin of spectral series of hydrogen and other atoms observed experimentally.

α-Particle Scattering and Rutherford's Nuclear Model of Atom

An important experiment to find out what was inside the atom was performed by Geiger and Marsden on the suggestion from Rutherford. They studied the scattering of α-particles spontaneously emitted by certain radioactive elements. An α-particle is helium nucleus containing 2 protons and 2 neutrons as such the mass of α-particles is 4 units of mass and two units of positive charge. Many radioactive elements emit α-particles. The experimental set up used by Rutherford and his co-workers is shown in Fig. 27.5.

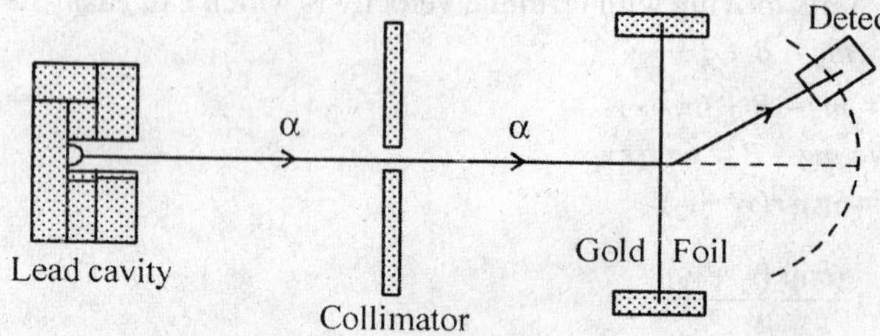

Fig. 27.5

α-emitting radioactive source such as Polonium or Thorium is placed in a lead cavity with a small opening. The α-particles emitted by the source are collimated into a narrow beam with the help of a lead slit (collimator). This collimated beam is allowed to fall on a thin foil of gold of thickness 10^{-7} m. The scattered α-particles were observed through a movable detector consisting of zinc sulphite screen and a microscope. The scattered α-particles on striking the screen produce scintillations. These scintillations could be viewed through the microscope and counted at different angles from the direction of the incident beam.

A graph plotted between the total number of α-particles scattered at different angles in a given interval of time is shown in Fig. 27.6. In the graph dots represent the exacts observed points (experimentally) and the solid curve is the theoretical prediction based on the assumption that the atom has a small, massive positively charged *nucleus*.

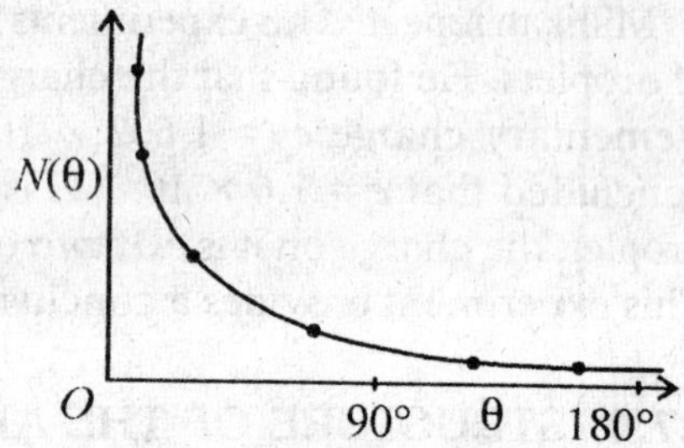

Fig. 27.6

The actual observation made by Geiger and Marsden were, that most of the α-particles indeed emerged without deviation (θ < 90°) some were scattered through very large angle (θ > 90°). A few were even scattered in the backward direction, i.e. through 180° (Fig. 27.7).

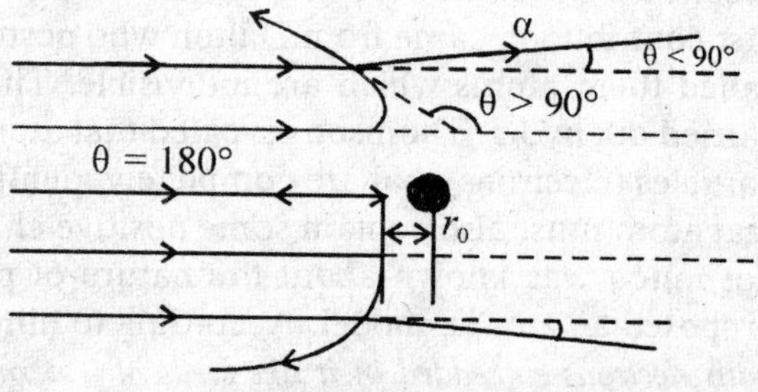

Fig. 27.7

Explanation: Since α-particles are relatively heavy (over 7000 times more massive than electrons) and those used in this experiment travelled at high speed, it was clear that strong forces must be acting on them to cause such marked deflections. To explain these results Rutherford was forced to postulate that in atom entire positive charge and the whole mass of it is concentrated in a very tiny central core called *nucleus*.

Considering an atom as largely empty space, it is easy to see why most α-particles go right through a thin foil. When an α-particle approaches a nucleus, it encounters an intense electric field and is likely to be scattered through a considerable angle. Rarely an α-particle may be aimed just at the nucleus. The force of repulsion will slow down the particle which will stop and then repel it along the direction of its approach. The electron in the atom are so light, that they do not appreciably affect the motion of incident α-particles.

Thus the scattering experiment by Rutherford leads to the following nuclear atom model.

(*i*) An atom consists of a positively charged and massive central core called the nucleus

(*ii*) The nucleus occupies a very small space (~10^{-14} m) as compared with the size of the atom (~10^{-10} m)

(*iii*) The nucleus is surrounded by a proper number of electrons whose total negative charge is equal to the positive charge of the nucleus to make atom electrically neutral.

Nuclear Dimensions (Distance of Closest Approach) and Impact Parameter

In order to calculate nuclear dimensions, Rutherford made some simplifying assumptions. They are

(*a*) Since gold foil is very thin as such α-particles will encounter only one scattering during their passage through the foil. Hence the computation of the trajectory of an α-particle scattered by a single nucleus is sufficient

(*b*) The nucleus of gold being 50 times heavier than an α-particle, it is reasonable to assume that it remains stationary throughout the scattering process

(*c*) The nucleus and the α-particle both are taken as point charges having no dimensions

(*d*) The scattering is due to elastic collision between nucleus and α-particle.

Using these assumptions the trajectory of an α-particle can be computed using Newton's second law of motion and the Coulomb's law for electrostatic force of repulsion between the α-particles and the positively charged nucleus.

Rutherford experiment also permits us to determine an upper limit to nuclear dimensions. Let us compute the distance of closest approach r_0 of the most energetic α-particle used in the experiment. If E (= $1/2mv^2$) is the kinetic energy of the α-particles then at the closest distance between the nucleus and α-particle, the potential energy

$$U = \frac{1}{4\pi\varepsilon_0}\frac{(Ze)(2e)}{r_0}$$

Here Ze is the charge (+ve) on the gold nucleus.

At this distance of closest approach all the initial K.E. is converted into electrostatic potential energy due to Coulomb's repulsion between the two. Thus

$$\frac{2Ze^2}{4\pi\varepsilon_0 r_0} = E$$

or $$r_0 = \frac{2Ze^2}{4\pi\varepsilon_0 E} \quad ...(27.5)$$

The maximum value of E found in α-particles of natural origin is 8 MeV and using knowing values of e and Z we have

$$r_0 = \frac{2 \times 79 \times (1.6 \times 10^{-19})^2 \times 9 \times 10^9}{8 \times 10^6 \times 1.6 \times 10^{-19}}$$

$$= 2.84 \times 10^{-14} \text{ m}.$$

Thus the radius of gold nucleus (~2.84×10^{-14} m) is much under (10^{-10} m) the radius of the atom as a whole.

The radius of nucleus is generally expressed in Fermi (f) and 1 f = 10^{-15} m.

Impact Parameter

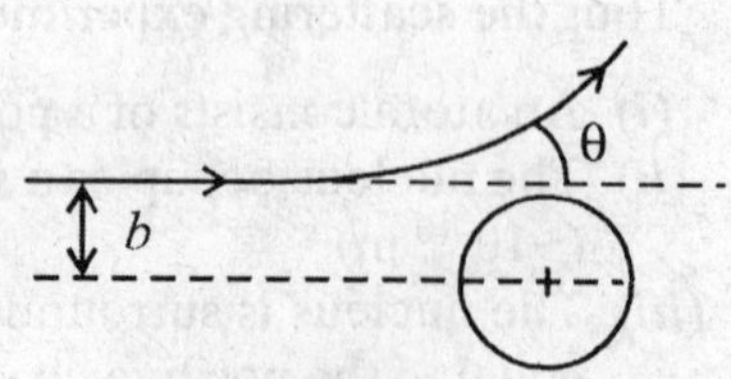

Fig. 27.8

The trajectory followed by an α-particle depends on its distance of closest approach from nucleus or an equivalent length called the impact parameter 'b'. It is defined as the distance of the initial velocity vector of the α-particle from the centre of the nucleus (Fig. 27.8). For large impact parameters the force on α-particles is weak because of its inverse square law character. It is observed that for a large impact parameter (Fig. 27.9) the α-particle goes practically undeviated and for small impact parameters it suffers large scattering. For head on approach, the impact parameter is zero and the α-particle rebounds back.

Rutherford calculated analytically the relation between the impact parameter (b) and scattering angle θ. This relation is

$$b = \frac{1}{4\pi\varepsilon_0} \frac{Ze^2 \cot \theta/2}{E} \quad \ldots(27.6)$$

Here E is the kinetic energy of the α-particle.

Success and Limitations of Rutherford Model

(*i*) It could explain large angle scattering of α-particles through thin foils

(*ii*) It justified the classification of elements in the Periodic Table on the basis of their atomic number instead of atomic weight.

Limitations

The objections against the nuclear model proposed by Rutherford were based on its *stability*. If the electrons are assumed to be at rest, the electrostatic force of attraction between positively charged nucleus and negatively charged electrons should make them fall into the nucleus; thus destroying the stable structure of the atom.

To overcome this, Rutherford assumed electrons to be revolving round the nucleus in various orbits (as the earth revolves round the sun) at such a speed that the necessary centripetal force is provided by the electrostatic force of attraction between the electrons and the nucleus, thus maintaining the stability.

But according to the laws of classical electrodynamics, an *accelerated charge always radiates energy in the form of electromagnetic waves.* Electron revolving round the nucleus is subjected to a constant acceleration and would continuously radiate energy in the form of electromagnetic radiations. This energy will come only from the energy of the system (electron-nucleus). Thus energy of the system will continuously decrease. The electron then will approach the nucleus along a spiral path. In the process it will give out radiations of constant increasing frequency. This does not agree with the observed emission of spectral lines of fixed frequency. Moreover, the system would last for a very short duration. But the atoms show stability and do not have tendency to collapse. Thus Rutherford model led to certain consequences which were in sharp contrast to experimental observations.

27.5. BOHR'S ATOMIC MODEL AND HYDROGEN SPECTRUM

Rutherford model though could explain scattering results successfully but had two serious drawbacks in explaining stability of atom and experimentally observed spectra of hydrogen atom. Neils Bohr thought over the matter and came to the conclusion that inspite of the success of electromagnetic theory in explaining large-scale phenomenon, it could not be applied to process at the atomic scale. Thus it became evident that to understand the structure of atoms and the relation of atomic structure to atomic spectra, a fairly radical departure from the established principles of classical mechanics and electromagnetism would be needed.

Neils Bohr solved the problem of Rutherford model by combining classical and early quantum concepts (by Max Planck) in the form of postulates. Basic postulates of Bohr's atomic model are:

1. Every atom consists of a central core called nucleus, in which entire positive charge and almost entire mass of the atom are concentrated. A suitable number of electrons (having as much negative charge as the positive charge on nucleus to make atom electrically neutral) revolve around the nucleus in circular orbits. The necessary centripetal force is provided by the electrostatic force of attraction between positively charged nucleus and negatively charged electron.

   If m is mass of electron moving in circular orbit of radius r with speed v then required centripetal force is

   $$F = \frac{mv^2}{r}$$

   If (Ze) is positive charge on nucleus (Z is atomic number), then electrostatic force of attraction is

   $$F_e = \frac{1}{4\pi\varepsilon_0}\frac{(Ze)(e)}{r^2}$$

   Thus $$\frac{mv^2}{2} = \frac{Ze^2}{4\pi\varepsilon_0 r^2}$$

   or $$mv^2 = \frac{ze^2}{4\pi\, \epsilon_0\, r} \qquad \ldots(i)$$

2. The electrons in an atom revolve in certain stable orbits without the emission of radiant energy. This means that so long the electron is revolving in its orbit, it does not radiate energy contrary to the laws of electrodynamics. Bohr called these orbits as *stationary radiationless orbits*.
3. Stationary orbits are those in which the angular momentum of orbiting electron is an integral multiple of $h/2\pi$ $(=\hbar)$ where h is Plank's constant. Angular momentum of revolving electron is mvr. Hence for permitted orbits

   $$mvr = \frac{nh}{2\pi} \qquad \ldots(ii)$$

   where n is a positive integer, $n = 1, 2, 3, \ldots$. n is also called the principal quantum number. This condition restricts the number of allowed orbits.
4. The electron can make a transition from one stationary radiationless orbit to another one. When it does so a single photon is emitted (or absorbed) having energy equal to the energy

difference between the initial and final state. The frequency of the emitted (or absorbed) photon is then given by

$$E_i - E_f = h\nu \qquad \text{...(27.7)}$$

when h is Planck's constant, E_i and E_f are the energies of initial and final states. This is also called Bohr's frequency condition.

5. Total number of electrons in an orbit are $2n^2$, where n is the quantum number of the orbit.

Bohr's Theory of Hydrogen Like Atoms

Let us consider an electron (mass m and charge e) orbiting in nth orbit has radius r_n and speed v_n. We can now use Bohr's postulates to calculate radius of this stationary orbit, speed of electron revolving in this orbit, its total energy in the orbit and we can also explain the spectrum of hydrogen atom experimentally observed.

Radius of Bohr's Stationary Orbit

Equation (*ii*) gives

$$mv = \frac{nh}{2\pi r_n} \quad \text{or} \quad m^2v_n^2 = \frac{n^2h^2}{4\pi^2 r_n^2}$$

and equation (*i*) gives

$$m^2v_n^2 = \frac{Ze^2m}{4\pi\varepsilon_0 r_n}$$

Equating the two equations, we have

$$\frac{Ze^2m}{4\pi\varepsilon_0 r_n} = \frac{n^2h^2}{4\pi^2 r_n^2}$$

or

$$r_n = \frac{n^2h^2\varepsilon_0}{Ze^2m\pi} = \frac{\varepsilon_0 n^2h^2}{\pi Ze^2m} = 0.51\frac{n^2}{Z}\ \text{Å} \qquad \text{...(27.8)}$$

Equation (27.8) gives

(*i*) For a given atom the radius of orbit is directly proportional to the square of the principal quantum number (n)

(*ii*) For a given n, the radius of stationary orbit is inversely proportional to the atomic number of the atom

(*iii*) For hydrogen atom ($Z = 1$) the radius of first stationary orbit ($n = 1$) $r_n = 0.51$ Å.

Orbital Velocity of Electron

Equation (*ii*) gives

$$v_n = \frac{nh}{2\pi m r_n}$$

Substituting the value of r_n from equation (27.8), we have

$$v_n = \frac{nh \cdot Ze^2m\pi}{2\pi m \cdot n^2h^2\varepsilon_0} = \frac{e^2Z}{2\varepsilon_0 nh} = 2.2 \times 10^6 \frac{Z}{n}\ \text{m/s} \qquad \text{...(27.9)}$$

in terms of velocity of light

$$v_n = \frac{c}{137} \cdot \frac{Z}{n}\ \text{m/s} \qquad \text{...(27.9}a\text{)}$$

(*i*) The orbital velocity is directly proportional to atomic number (Z)
(*ii*) For a given atom, orbital velocity is inversely proportional to the principal quantum number.
(*iii*) For hydrogen ($Z = 1$), orbital velocity of electron in first orbit ($n = 1$) is $v_1 = \frac{c}{137}$ or $\frac{v_1}{c} = \frac{1}{137} = \alpha$. Here this ratio $\frac{v_1}{c} = \alpha$ is called *fine structure constant.*

Orbital Frequency

The orbital frequency of electron in nth orbit is

$$f_n = \frac{v_n}{2\pi r_n} = \frac{1}{2\pi}\left(\frac{e^2 Z}{2\varepsilon_0 nh}\right) \times \frac{\pi Z e^2 m}{\varepsilon_0 n^2 h^2}$$

$$= \frac{mZ^2 e^4}{4\varepsilon_0^2 n^3 h^3} = 6.6 \times 10^{15} \frac{Z^2}{n^3} \text{ Hz} \quad \text{...(27.10)}$$

and time period

$$T_n = \frac{1}{f_n} = \frac{4\varepsilon_0^2 n^3 h^3}{mZ^2 e^4} \quad \text{...(27.11)}$$

(*i*) Frequency f_n is proportional to the square of atomic number (Z)
(*ii*) Frequency is inversely proportional to the cube of principal quantum number (n)
(*iii*) For hydrogen ($Z = 1$) frequency in first orbit $n = 1$ is

$$f_n = 6.6 \times 10^{15} \text{ Hz}$$

Energy of the Electron

Total energy of electron (E_n) in an stationary orbit is the sum of its kinetic energy K_n and potential energy U_n. Kinetic energy of electron is

$$K_n = \frac{1}{2} m v_n^2 = \frac{1}{2} m \cdot \left(\frac{e^2 Z}{2\varepsilon_0 nh}\right)^2$$

$$= \frac{1}{8} \frac{Z^2 e^4 m}{\varepsilon_0^2 h^2 n^2} \quad \text{...(27.12)}$$

Potential energy of electron is

$$U_n = \frac{1}{4\pi\varepsilon_0} \frac{(Ze)(-e)}{r_n} = -\frac{Ze^2}{4\pi\varepsilon_0 r_n}$$

$$= -\frac{1}{4} \frac{Z^2 e^4 m}{\varepsilon_0^2 h^2 n^2} \quad \text{...(27.13)}$$

Total energy of electron in nth stationary orbit is

$$E_n = U_n + K_n = -\frac{1}{8} \frac{Z^2 e^4 m}{\varepsilon_0^2 h^2 n^2} = -\frac{13.6 Z^2}{n^2} \text{ eV} \quad \text{...(27.14)}$$

The total energy can be written as

$$E_n = -Rhc \frac{Z^2}{n^2} \cdot J = -\frac{Z^2}{n^2} \text{ Rydberg} \quad \text{...(27.15)}$$

where $R = \dfrac{me^4}{8\varepsilon_0^2 ch^3} = 1.097 \times 10^7\ m^{-1}$ and is called Rydberg's constant. Rhc joule i.e., 13.6 eV is called one Rydberg and its dimensional equation is $[M^1L^2T^{-2}]$.

Note: Here $R = \dfrac{me^4}{8\varepsilon_0^2 ch^3}$ is written assuming the mass of the nucleus to be infinite in comparison to the mass of electron. If it is not so, then m is replaced by reduced mass of the system.

Equation $E_n = -\dfrac{13.6\,Z^2}{n^2}$ eV gives

(*a*) For hydrogen ($Z = 1$) $E_n = -\dfrac{13.6}{n^2}$ eV

(*b*) Total energy of electron in stationary orbits is negative means the system is *bound system*, i.e. electron is not free but bound to the nucleus. On giving the atom energy equal to $\left(+\dfrac{13.6\,Z^2}{n^2}\text{ eV}\right)$ 13.6 eV to the hydrogen atom, the electron will come out of the atom. Hence this energy is called *binding energy of electron*. For hydrogen binding energy of electron is 13.6 eV.

(*c*) Energy of atom is not continuous but discrete. As n increases the numerical value of E_n decreases. Since total energy is negative hence with increase in n, the total energy increases and becomes maximum (= zero) at infinity.

(*d*) For a given value of n, $E_n \propto Z^2$, i.e. as Z increases the total energy becomes more negative.

(*e*) $|K_n| = |E_n| = \dfrac{1}{2}|U_n|$, i.e. kinetic energy of electron is equal to the magnitude of total energy and half the magnitude of potential energy. Kinetic energy is always positive and, potential energy and total energy are negative. With increase in radius, the kinetic energy decreases whereas potential energy and total energy increases.

Emission Transition and Formula for Emitted Wavelength

When an atom acquires energy from external source, its electron jumps from lower energy state n_1 to the higher energy state n_2 we say *atom is in excited state*. These excited states are short lived (lifetime ~10^{-8} sec) as such the electron cannot remain there for long time and jumps to the lower energy state. The difference of the energies of two levels is emitted in the form of radiation energy according to the Bohr's frequency condition (equation 27.7). Thus the frequency of the radiation emitted is

$$h\nu = E_{n_2} - E_{n_1}$$

$$= -\frac{RhcZ^2}{n_2^2} - \left(-\frac{RhcZ^2}{n_1^2}\right)$$

or
$$\nu = RcZ^2\left(\frac{1}{n_1^2} - \frac{1}{n_2^2}\right)$$

or
$$\frac{c}{\lambda} = RcZ^2\left(\frac{1}{n_1^2 - n_2^2}\right) \qquad \because \nu = \frac{c}{\lambda}$$

Hence frequency in wave numbers $\bar{\nu} = 1/\lambda$ is

$$\bar{\nu} = 1/\lambda = RZ^2\left(\frac{1}{n_1^2} - \frac{1}{n_2^2}\right) \qquad \ldots(27.16)$$

Equation (27.16) is the general equation which helps to explain the spectrum of atom in general. For hydrogen $Z = 1$, then

$$\bar{\nu} = 1/\lambda = R\left(\frac{1}{n_1^2} - \frac{1}{n_2^2}\right) \qquad \ldots(27.17)$$

Since both n_1 and n_2 are integers, this gives immediately the result that in transition between different atomic levels, light is radiated in various discrete frequencies.

Explanation of Hydrogen Spectrum

The various spectral series of hydrogen are explained on the basis of quantum mechanics. According to quantum mechanics ionized state of hydrogen atom is taken as zero energy state, then according to Bohr's theory, the energies of different energy states of atom can be expressed by equation

$$E_n = -\frac{Rhc}{n^2} = -\frac{13.6}{n^2}\text{ eV} \quad \text{where } n = 1, 2, 3, \ldots \text{ an integer}$$

When electron is in its lowest energy state, i.e. $n = 1$, it is called its *normal state or ground state.* When atom gets energy from outside source (either by heating or collision), the electron jumps from lowest energy state to higher energy state. This is called excitation. As already stated these excited states are shortlived (life time ~10^{-8} sec) as such the electron returns to its ground state either directly or through different lower energy states. In the process the atom emits light or other electromagnetic radiations. Using equation (27.17), we can explain the existence of all the lines in all the series of hydrogen spectrum.

Lyman Series: When the atom jumps from high energy excited state to ground state, the wavelength of the different emitted spectral lines is obtained by putting $n_1 = 1$ and $n_2 = 2, 3, 4, \ldots$, etc. in equation (27.17). Thus

$$1/\lambda = R\left(\frac{1}{1^2} - \frac{1}{n_2^2}\right)$$

The largest wavelength of this series is (putting $n_2 = 2$)

$$\frac{1}{\lambda} = R\left(\frac{1}{1^2} - \frac{1}{4}\right) = \frac{3}{4}R$$

$$\lambda = \frac{4}{3R} = 1215\text{ Å}$$

The smallest wavelength of this series is (putting $n_2 = \alpha$)

$$\frac{1}{\lambda} = R\left(\frac{1}{1} - \frac{1}{\alpha}\right) = R$$

or

$$\lambda = \frac{1}{R} = 912\text{ Å}$$

This is also called series limit. Emitted all lines of this series lie in ultraviolet region. This series is called Lyman series, because Lyman for the first time photographed these lines in 1916.

Balmer Series: When atom de-excites to second energy state ($n_1 = 2$) from higher energy state ($n_2 = 3, 4, 5, \ldots$, etc.), then wavelengths of different lines are obtained by equation

$$\frac{1}{\lambda} = R\left(\frac{1}{2^2} - \frac{1}{n_2^2}\right) \quad n_2 = 3, 4, 5, \ldots, \text{etc.}$$

The longest wavelength of this series is

$$\frac{1}{\lambda} = R\left(\frac{1}{4} - \frac{1}{9}\right) = \frac{5R}{36}$$

or $$\lambda = \frac{36}{5R} = 6563 \text{ Å}$$

and the shortest wavelength of this series is ($n_2 = \infty$)

$$\frac{1}{\lambda} = R\left(\frac{1}{4} - \frac{1}{\infty}\right) = R/4$$

$$\lambda = \frac{4}{R} = 3646 \text{ Å}$$

These lines were first observed by Balmer in 1885 and is called Balmer series. This series lies in visible region.

Paschen Series: This series is obtained when atom deexcites from higher energy state ($n_2 = 4, 5, \ldots$, etc.) to the third energy state ($n_1 = 3$). The maximum wavelength of this series is

$$\frac{1}{\lambda} = R\left(\frac{1}{3^2} - \frac{1}{4^2}\right) = R\,\frac{7}{144}$$

or $$\lambda = \frac{144}{7R} = 18752 \text{ Å}$$

and the minimum wavelength of this series is ($n_2 = \infty$)

$$\frac{1}{\lambda} = R\left(\frac{1}{9} - \frac{1}{\infty}\right) = \frac{9}{R} = 8204 \text{ Å}$$

This series is called Paschen series. All the lines of this series lie in infrared region.

Brackett Series: When the atom from higher energy state ($n_2 = 5, 6, 7, \ldots$, etc.) de-excites to fourth energy state ($n_1 = 4$), the series so obtained is called Brackett series. The longest wavelength is

$$\frac{1}{\lambda} = R\left(\frac{1}{4^2} - \frac{1}{5^2}\right) = R\frac{9}{400}$$

or $$\lambda = \frac{16}{9R} = 40515 \text{ Å}$$

and the shortest wavelength of this series is

$$\frac{1}{\lambda} = R\left(\frac{1}{4^2} - \frac{1}{\infty}\right) = \frac{R}{16}$$

or $$\lambda = \frac{16}{R} = 14585 \text{ Å}$$

All the lines of this series also live in infrared region.

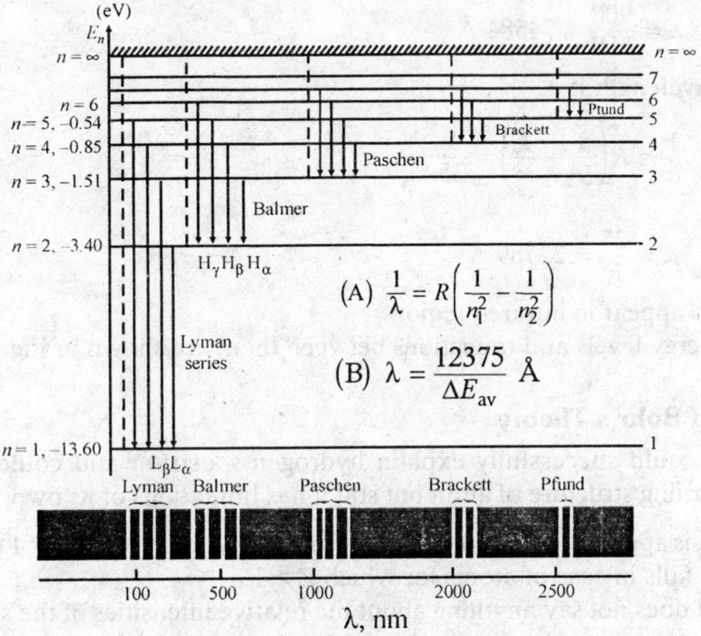

(a)

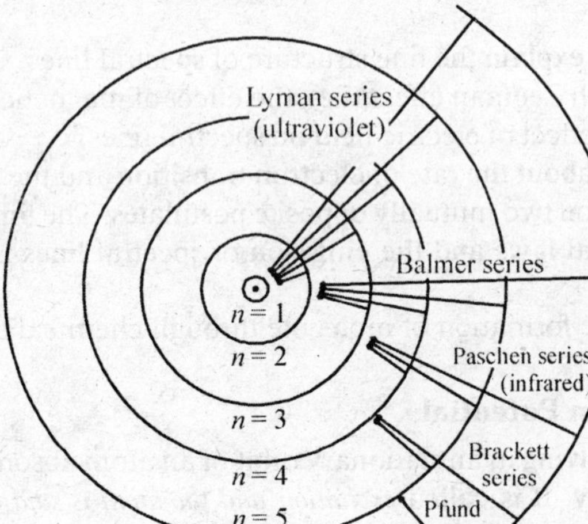

(b)

Fig. 27.9

Pfund Series: The atom when de-excites to fifth energy state ($n_1 = 5$) from higher energy states ($n_2 = 6, 7, \ldots$, etc.), the spectral; series so obtained is called Pfund series. The maximum wavelength of this series is

$$\frac{1}{\lambda} = R\left(\frac{1}{5^2} - \frac{1}{6^2}\right) = R\frac{11}{900}$$

or $$\lambda = \frac{900}{11R} = 74584 \text{ Å}$$

and minimum wavelength is

$$\frac{1}{\lambda} = R\left(\frac{1}{5^2} - \frac{1}{\infty}\right) = \frac{R}{25}$$

or $$\lambda = \frac{25}{R} = 22789 \text{ Å}$$

These lines also appear in infrared region.

The various energy levels and transitions between them are shown in Fig. 27.9.

Shortcomings of Bohr's Theory

The Bohr model could successfully explain hydrogen spectrum and could provide important informations regarding structure of atom but still it has limitations of its own. The limitations are:

(*i*) This theory is applicable only to simple atoms like hydrogen with $Z = 1$ (e.g. He^+, Li^{++}, etc.). The theory fails in case of atoms for which $Z > 1$

(*ii*) This model does not say anything about the relative intensities of the spectral lines

(*iii*) It does not state about the distribution of the electrons in the atom

(*iv*) The model does not explain why the orbits are circular, whereas the elliptical orbits are also possible

(*v*) The model does not explain the fine structure of spectral lines

(*vi*) It is unable to explain Zeeman effect (i.e., the effect of magnetic field on spectral lines) and Stark effect (i.e. the effect of electric field on spectral lines)

(*vii*) This theory is silent about the rate of electron transition and the section rules for transitions

(*viii*) This theory is based on two mutually opposite postulates. The force equilibrium is explained according to classical laws and the emission of spectral lines is explained on the basis of quantum laws

(*ix*) It cannot explain the formation of molecule through chemical bond between the atoms.

Excitation and Ionisation Potential

Whenever an electron revolving in an stationary orbit of an atom absorbs energy it may jump over to an orbit of higher energy. It is called *excitation and the atom is said to be in the excited state. The minimum accelerating potential which provides an electron energy sufficient to jump from the inner most orbit (ground state) to one of the outer orbits is called excitation potential or resonance potential.*

For hydrogen the energy of electron in nth orbit is $E_n = -\frac{13.6}{n^2}$ eV

Thus

$$E_1 = -13.6 \text{ eV}$$

$$E_2 = -\frac{13.6}{2^2} = -3.40 \text{ eV}$$

$$E_3 = -\frac{13.6}{3^2} = -1.51 \text{ eV}$$

$$E_4 = -\frac{13.6}{4^2} = -0.85 \text{ eV}$$

and $$E_\infty = -\frac{13.6}{(\infty)^2} = 0$$

The first excitation energy for hydrogen is $E_2 - E_1$

i.e. $$E_2 - E_1 = -3.40 - (-13.6) = 10.2 \text{ eV}$$

Hence first excitation potential of hydrogen is 10.2 V

Similarly 2nd excitation energy is

$$E_3 - E_1 = -1.51 - (-13.6) = 12.09 \text{ eV}$$

and second excitation potential is 12.09 V.

In this way we can find different excitation potentials.

If sufficient energy is given to the atom so that it is removed from the atom (i.e. $n = \infty$), the process is said *ionisation. The minimum accelerating potential which would provide an electron energy just sufficient to remove it from the atom is called ionisation potential.*

Since ionisation energy is

$$E_\infty - E_1 = 0 - (-13.6) = 13.6 \text{ eV}$$

Hence ionisation potential is 13.6 V.

27.6. X-RAYS AND X-RAY SPECTRUM

When energetic electrons (energy of the order of keV) bombard a hard material, *X*-rays are emitted. Hard material must have following properties.

(1) High atomic number (2) High melting point (3) High electrical conductivity and (4) Low vapour pressure at high temperatures.

The materials used as targets are, copper, tungsten, etc. They were first observed by Wilhelm. K. Roentgen in 1895 and were originally called *Roentgen rays.*

X-rays are electromagnetic waves of short wavelength. The *X*-rays wavelength range is from 0.001 nm to 1 nm. A device to produce *X*-rays was designed by Coolidge and is known as Coolidge tube to produce *X*-rays, commonly called *X*-ray tube. A typical Coolidge tube is shown in Fig. 27.10. It consists of a filament *F* and a target *T*. Target is made up of copper or tungsten. They are enclosed in a highly evacuated enclosure *C*. The filament is heated electrically and emits electrons

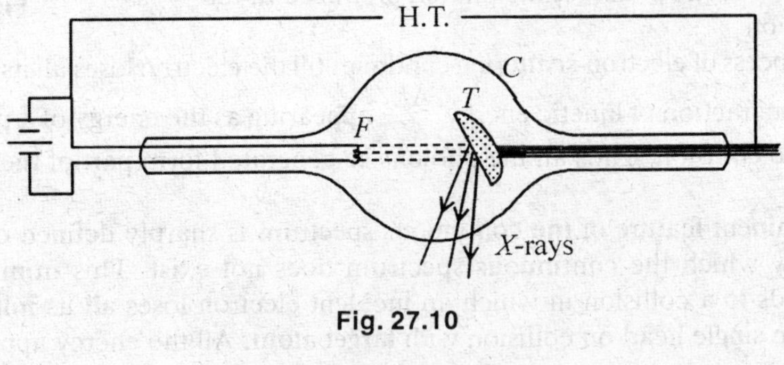

Fig. 27.10

by thermionic effect. These electrons are accelerated by applying a high potential difference (H-T) between the two electrodes and hit the target with a very high speed. The X-rays are emitted from the anode surface (target) as a result of the bombardment of energetic electrons. In the process, large amount of heat is developed, as such an arrangement is made to cool down the tube by continuously running cold water.

Continuous and Characteristic X-rays and their Origin

The wavelength spectrum of X-rays produced when a beam of about 40 keV electrons are bombarded on a molybdenum target is shown in Fig. 27.11. The graph shows the intensity of different wavelength components emitted by the target. It is observed that there is a minimum wavelength (λ_{min}) below which no X-ray is emitted. This is called *the cut off wavelength* or *threshold wavelength*. It is observed that the intensity varies gradually with wavelength except at sharply defined wavelength (K_α, K_β) where intensity of X-rays is very large. Intensity never falls to zero. Thus we conclude that the X-ray spectrum consists of *continuous X-rays* on which *the line spectrum* is superposed. This line spectrum is called *characteristic spectrum*. This line spectrum is the characteristics of the target element.

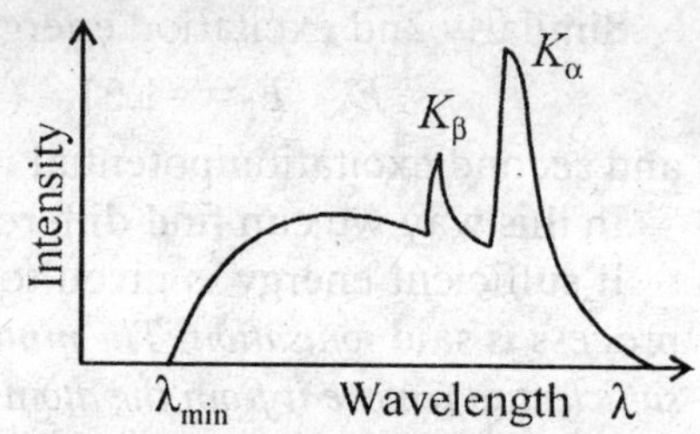

Fig. 27.11

Origin of X-ray Spectrum

If V is the potential difference between target and filament, then the kinetic energy acquired by the electron is $K = eV$ before it strikes the target. As the electron enters the target it loses its energy and is brought to rest inside the target material. Electron before finally being stopped suffers several collisions with the atoms in the target. At each collision one of the following two processes may take place.

(*i*) The electron may lose some of its kinetic energy ΔK on collision with one of the target atoms. This loss of energy will appear in the form of X-ray photon.

$$h\nu = \Delta K$$

and X-ray photon of frequency ν is radiated away from the sight of interaction. The recoil energy transferred to the atom is converted into heat as the atom cannot move being very massive. The scattered electron (Fig. 27.12) with less energy may have a second collision with another atom generating second photon whose energy in general will be different from that of the photon produced in the first collision.

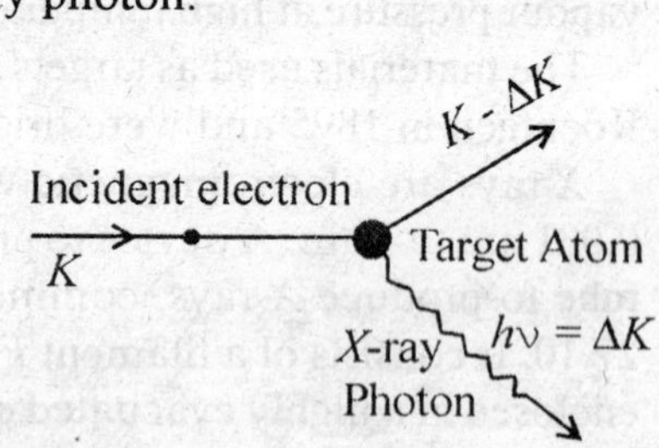

Fig. 27.12

This process of electron-scattering continues till the electron loses all its energy and comes to rest. The fraction of kinetic energy $\frac{\Delta K}{K}$ appearing as the energy of a photon varies from collision to collision. Thus all the photons so generated form part of the continuous X-ray spectrum.

A prominent feature of the continuous spectrum is sharply defined cut off wavelength (λ_m) below which the continuous spectrum does not exist. This minimum wavelength corresponds to a collision in which an incident electron loses all its initial kinetic energy ($K = eV$) in single head-on collision with target atom. All the energy appears as the energy of a single photon whose associated wavelength (the minimum possible X-ray wavelength λ_{min}) is given by

$$h\nu = \frac{hc}{\lambda_{min}} = K = eV$$

or $$\lambda_{min} = \frac{hc}{eV} \quad ...(27.18)$$

$$= \frac{12.4 \times 10^3}{V} \text{ Å} \quad ...(27.19)$$

The equation gives the λ_{min}, here V is in volts. We see that the cut off wavelength (λ_{min}) depends only on the accelerating voltage (V) applied between the target and the filament. It does not depend on the material of the target.

(*ii*) The electron may knockout an inner electron (K- or L-electron) of the atom with which it collides, it will create a vacancy in this shell. Suppose the incident electron knocks out an electron from the K-shell (n = 1). This means that there is only one electron with n = 1 whereas two could be accommodated by Pauli's exclusion principle. As such an electron from higher energy state (with n = 2, 3, ..., etc.) may make a transition to this vacant state. In the process the energy difference (ΔE) of the two states is emitted in the from of X-ray photon of wavelength

$$\lambda = \frac{hc}{\Delta E} \quad ...(27.20)$$

X-rays emitted due to electromic transition from a higher energy state to a vacancy created in the K-shell are called K-X-rays. If the electron that fills the K (n = 1) shell vacancy jumps from L-shell (n = 2) the emitted radiation is the K_α-line (Fig. 27.11). If it jumps from M-shell (n = 3) it produces K_β line and so on. An electron from still farther out in the atom will fill the vacancy in the L or M-shell. In a similar way if the electron jumps to L-shell will give L_α, L_β, etc. lines. A simplified energy level diagram of these transitions is shown in Fig. 27.13. The base line E = 0 represents the neutral atom in its ground state. The levels marked E_K, E_L, E_M, etc. represent the energy of atom with a vacancy in K, L, M shell respectively. The emission of K_α, K_β, L_α, L_β, etc lines is also shown.

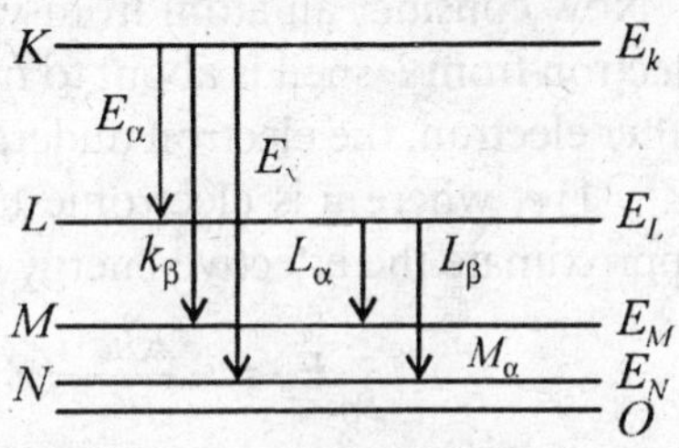

Fig. 27.13

Energies E_K, E_L, ..., etc. are characteristic energies of the material. For different materials the values of these energies are different. The values $E_K - E_L$, $E_L - E_N$, etc. also have definite values for a given material. The wavelengths of the X-rays emitted corresponding to these transitions are:

For K_α $$\lambda = \frac{hc}{E_K - E_L}$$

For K_β $$\lambda = \frac{hc}{E_K - E_M}$$

For K_α $$\lambda = \frac{hc}{E_L - E_N}$$

etc. These wavelengths, therefore have definite values for a particular material. The X-rays emitted in this way are characteristic X-rays shown in Fig. 27.11. The characteristic X-rays thus constitute a line spectrum and are characteristic of the target element. They are so named because their wavelengths may be used to identify the element from which they originate.

Moseley's Law

H. J. G. Moseley a British physicist studied the characteristic x-rays for many elements, using them as targets. He used them as targets in x-ray tube of his own design. Moseley measured the frequencies of characteristic X-rays from a large number of elements and plotted square root of frequency $\sqrt{\nu}$ against its position number (atomic number Z) in the periodic table. He observed that the plot is very close to a straight tine Fig. 27.14. Where $\sqrt{\nu}$ of K_α x-rays is plotted against the position number in the periodic table. From this linear relation Moseley concluded that *there must be a fundamental property of the atom which increases by regular steps as one moves from one element to the other. This quantity was later on identified to be the number of protons in the nucleus and was referred to as the atomic number (z).*

The result obtained by Moseley can be mathematically expressed as

$$\sqrt{\nu} = a(Z - b) \qquad \ldots(27.21)$$

where a and b are constants. The relation is known as *Moseley law*. Moseley's law can be qualitatively understood from Bohr's atomic model.

The energy of an electron in nth orbit of element of atomic number = Z is

$$E_n = -\frac{RhcZ^2}{n^2} \qquad \ldots(i)$$

Now consider an atom from which an electron from the K shell has been knocked out. Let an electron from L shell is about to make a transition to the vacant site. Because of the presence of the other electron, the electron undergoing transition sees an effective nuclear charge of approximately $(Z - 1)\,e$, where e is electronic charge and Z the atomic number of the element. We can thus approximate the effective energy of the atom by replacing Z with $(Z - b)$ with $b = 1$. This gives

$$E_n = -\frac{Rhc}{n^2} \cdot (Z - 1)^2$$

Now for K_α emission electron jumps from $N_2 = 2$ to $N_1 = 1$ (from L shell to K-shell). Hence change in energy is

$$\Delta E = Rhc\,(Z - 1)^2 \left[\frac{1}{1^2} - \frac{1}{2^2}\right] = \frac{3}{4} Rhc\,(Z - 1)^2$$

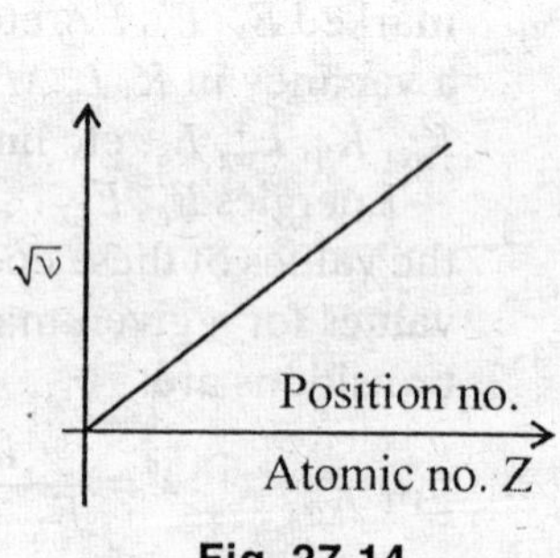

Fig. 27.14

If ν is the frequency of the radiation (x-rays) emitted then

$$h\nu = \Delta E = \frac{3}{4} Rhc\,(Z - 1)^2$$

or

$$\nu = \frac{3}{4} Rc\,(Z - 1)^2$$

and

$$\sqrt{\nu} = \sqrt{\frac{3}{4} Rc}\,(Z - 1)^2 \qquad \ldots(27.22)$$

In general the equation then can be written as equation (27.21) with $a = \sqrt{\frac{3}{4} Rc}$

Bragg's Law

X-rays are electromagnetic waves of short wavelengths. As such can be diffracted by suitable diffracting centres. But the diffraction effects are appreciable only when the diffracting apertures are of the order of wavelength. In case of diffraction of *X*-rays the apertures should be of the order of 0.1 nm. In practice it is difficult to construct slits with such small widths.

In solid crystals, the arrangement is fairly regular with interatomic distance of the order of 0.1 nm. Crystals can act as natural three dimensional gratings for the *X*-rays diffraction.

The structure of a solid can be viewed as a series of parallel planes of atoms separated by a distance d (Fig. 27.15). Suppose an *X*-ray beam is incident on solid with grazing angle θ. These *X*-rays are reflected by different atoms and the reflected rays interfere. The direction in which the different rays interfere constructively we have strong *X*-rays intensity along that direction. For constructive interference, the path difference between the two rays should be $n\lambda$ where n is an integer (n = 1, 2, 3, ..., etc.)

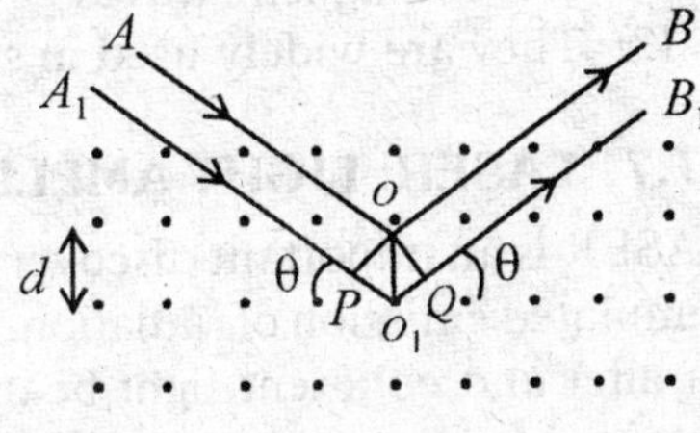

Fig. 27.15

From Fig. 27.15, the path difference between the two rays AOB and $A_1O_1B_1$ is

$$\Delta = PO_1 + O_1Q = 2d\sin\theta$$

Hence for constructive interference

$$2d\sin\theta = n\lambda \qquad \ldots(27.23)$$

Equation (27.23) is known as Bragg's law. For specific grazing angles there will be strong intensity in reflected beam otherwise they are absorbed. By using monochromatic *X*-rays beam and noting the angles of reflection, the interplanar spacing d and informations regarding structure of the solid can be obtained.

Properties and Uses of X-rays

As already stated *X*-rays are electromagnetic waves. As such it has many properties common to light. Some of the properties of *X*-rays are

1. *X*-rays are invisible, uncharged electromagnetic waves with wavelength range 0.1 Å to 100 Å ($\nu = 3 \times 10^{19}$ Hz to 3×10^{17} Hz). Their speed is 3×10^8 m/s, i.e. same as that of light. They travel in a straight line.
2. *X*-rays are diffracted by crystals and obey Bragg's law.
3. They are not deflected by electric and magnetic field.
4. *X*-rays affect a photographic plate more strongly than visible light.
5. They obey all the laws obeyed by light such as reflection, refraction and exhibit the phenomena like interference, diffraction and polarisation.
6. They excite fluorescence in certain substances such as barium-platinocyanide, zinc sulphide and cadmium tungestate.
7. When passed through gases they ionise the gas molecules.
8. When incident on living beings, their tissues and white blood cells are destroyed.
9. *X*-rays can penetrate into several metals and other materials. Thus they can pass through small thickness of aluminium, wood, plastics, human flesh, etc. They are stopped by materials of high density and high atomic number.

10. They are absorbed by substances also. When passed through a substance of thickness X, the intensity of the emergent X-rays (I) is given by

$$I = I_0 e^{-\mu x} \quad \dots(27.24)$$

where I_0 is the intensity of incident X-rays and μ is absorption coefficient. The value of μ depends on atomic number (Z) of the absorber, and wavelength (λ) of the incident X-rays. $\mu = AZ^4\lambda^3$, where A is constant.

11. They produce photoelectric effect and Compton effect, which proves the quantum nature of electromagnetic waves.
12. They are widely used in surgery and radiotherapy.

27.7. LASER: LIGHT AMPLIFICATION BY STIMULATED EMISSION OF RADIATION

LASER is an important discovery of quantum physics and is abbreviation of light amplification by stimulated emission of radiation. It is a process by which a highly monochromatic, almost perfectly parallel and coherent light beam is obtained. The word Laser is also used for the light beam obtained by this process. All the photons in the beam, emitted by different atoms at different instances are in phase that is why it is highly coherent. The spread in wavelength $\Delta\lambda$ is very small of the order of 10^{-6} nm. Which is even 10^{-3} times smaller than the spread in ^{86}Kr light. A laser beam can go to moon and come back to earth without much loss of intensity and a very small deviation. This shows that the laser beam almost perfectly parallel can be obtained. To understand the process involved in laser we will define certain terms.

Stimulated Absorption of Light Photon

In Bohr atomic model, we have seen that an isolated atom can exist either in the ground state (state of lowest energy) whose energy is say E_1 or in a state of higher energy (an excited state) whose energy may be E_2. If a photon of energy $(E_2 - E_1)$ is incident on the atom which is in the lowest energy state, the atom may absorb the photon and jump to the higher energy state E_2 (Fig. 27.16 (*a*)). This process is called stimulated absorption of light photon. If the frequency of the photon absorbed is ν then

$$h\nu = E_2 - E_1 \text{ where } h \text{ is Planck's constant.}$$

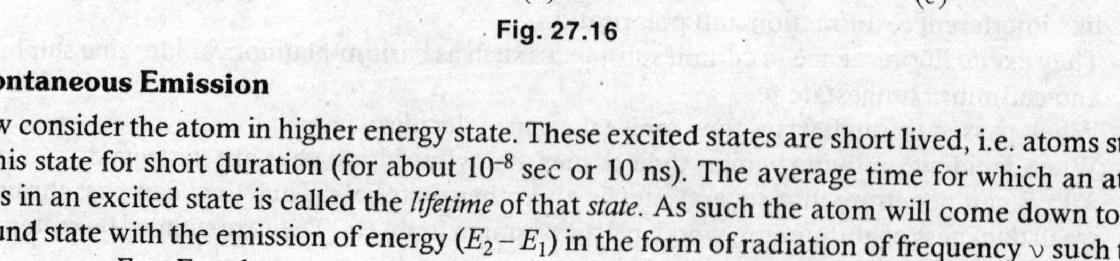

Fig. 27.16

Spontaneous Emission

Now consider the atom in higher energy state. These excited states are short lived, i.e. atoms stays in this state for short duration (for about 10^{-8} sec or 10 ns). The average time for which an atom stays in an excited state is called the *lifetime* of that *state*. As such the atom will come down to the ground state with the emission of energy $(E_2 - E_1)$ in the form of radiation of frequency ν such that

$$E_2 - E_1 = h\nu$$

This process is called *spontaneous emission* (Fig. 27.16 (*b*)). It is spontaneous because it is not triggered by any outside influence. In this process atoms emit photon of light independently of each other. As a result, different photons have different phases and the light as a whole becomes incoherent. Besides the energy of transition differs slightly from photon to photon so that the wavelength is not strictly, uniquely defined. There is a spread $\Delta\lambda$ in the wavelength λ. The direction is also different for different transitions so that we do not get strictly parallel beam of light.

Stimulated Emission

There are atoms which have certain excited states with mean lifetime as much as 10^5 times longer than the usual lifetime. These long lifetime states are called *metastable states.* Now suppose the atom is in the excited state (energy E_2) but this time it is in metastable state. Now radiation with frequency ν such that

$$h\nu = E_2 - E_1$$

is also simultaneously present. This photon can stimulate the atom to move to its ground state. A fresh photon is emitted in the process. This process is different from spontaneous emission in which the atom makes transition to lower energy state on its own. In the present case, the process is stimulated by the external photon. *Thus stimulated emission is the process in which an atom due to its interaction with a photon incident on it emits a photon. The emitted photon has exactly the same energy, phase and direction as the incident photon.* (Fig. 27.16 *c*).

Population Inversion

When the higher energy state in an atomic system gets occupied due to induced absorption to a greater extent than the lower energy state, the phenomenon is called *inverted population of energy levels* or simply *population inversion*. However the population inversion does not sustain in normal atomic systems, as the atoms tend to return to lower energy state by spontaneous emission.

The process of induced absorption for producing the population inversion of energy levels is known as pumping.

Principle of Production of Laser Beam

For producing a laser beam an atomic system with metastable state is chosen. The system may be gas or a liquid in cylindrical tube or a solid in the shape of cylindrical rod. Figure (27.17) shows one such system with metastable state of energy E_2 and lower energy state E_1. Now suppose by some technique, the number of atoms in the metastable state E_2 is made much larger than those in energy state E_1. When a photon of energy $E_2 - E_1$ is incident on the system with atoms in metastable state, the atom will de-excite to the lower energy state E_1 with emission of a photon of energy same as that of the incident photon. This emitted photon is in phase with incident photon and is emitted in same direction. The two photons then interact with two more atoms in the metastable state. In this way the photons go on multiplying. All these photons possess the same energy, phase, and direction. This process is termed as amplification of light. For this process to continue smoothly, following two requirements must be fulfilled.

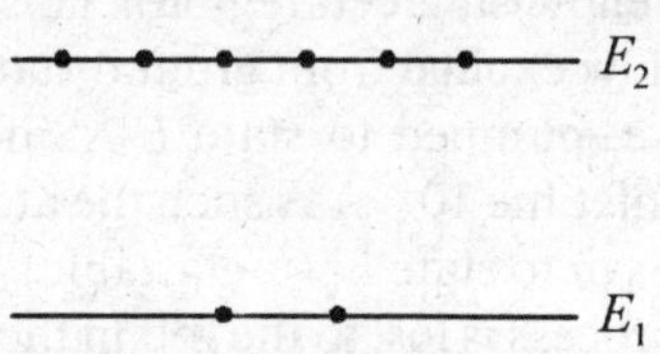

Fig. 27.17

(*a*) The metastable state should all the times be densely populated than the lower energy state. This means that atoms from the lower energy state E_1 must be continuously pumped to metastable state E_2 to maintain population inversion.

(*b*) The photons emitted due to stimulated emission should stimulate other atoms to multiply the photons inside the system. This means that the stimulated photons should spend enough time in the system.

Production of Laser Beam

The stimulated emission is the key to operation. Einstein introduced this concept in 1917. But the first successful laser was fabricated by T. H. Maiman in 1960. He used ruby rod and is referred as *ruby laser*. Since then a variety of lasers have been developed. Of these a few laser systems are three energy level systems while other are four energy level systems.

Ruby Laser

Ruby is a crystal of aluminimum oxide (Al_2O_3) doped with 0.05% chromium oxide. Chromium oxide imparts red colour to the crystal. The crystal is in the form of a rod of length 5 cm and diameter 0.5 cm. The two ends of crystal are made optically plane and coated with silver. One end of the rod is made reflector by thick coating and the other end is made partially transparent by light coating. It is this face through which the laser beam emerges out of ruby rod. A xenon flash lamp is wound round the ruby rod which provides the necessary energy for pumping to produce population inversion. The crystal is kept cool by circulating liquid nitrogen around it (Fig. 27.18).

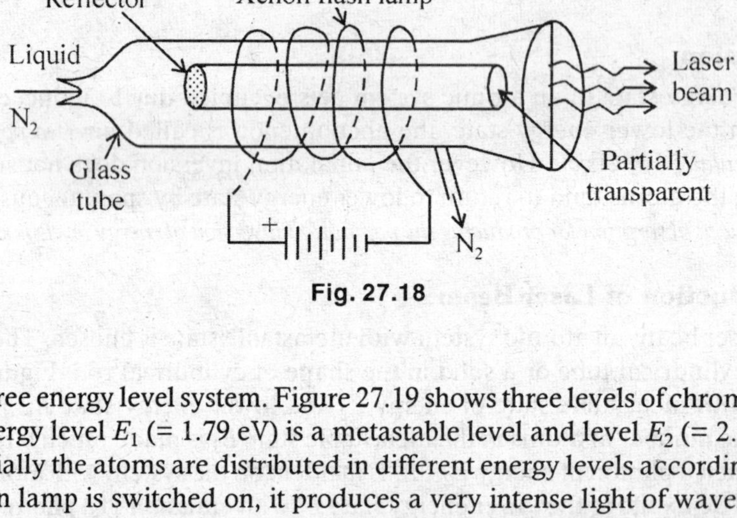

Fig. 27.18

Theory

Ruby laser is a three energy level system. Figure 27.19 shows three levels of chromium ion. E_0 is the ground state. Energy level E_1 (= 1.79 eV) is a metastable level and level E_2 (= 2.26 eV) is a higher energy level. Initially the atoms are distributed in different energy levels according to Boltzmann's law. When xenon lamp is switched on, it produces a very intense light of wavelength 5500 Å (in yellowish green region). This light acts as a pumping agent for chromium atoms. The atoms are now excited from ground state to higher energy state E_2, i.e. pumped to state E_2. Since this level is short lived (lifetime 10^{-8} s) as such the atoms de-excite either to state E_0 or to state E_1 (metastable). The energy released in this process is lost to the rod in the form of heat. The crystal is continuously cooled by circulating liquid nitrogen to prevent rise in temperature. In a short time the population inversion takes place, i.e. metastable state E_1 has more atoms than the ground state E_0.

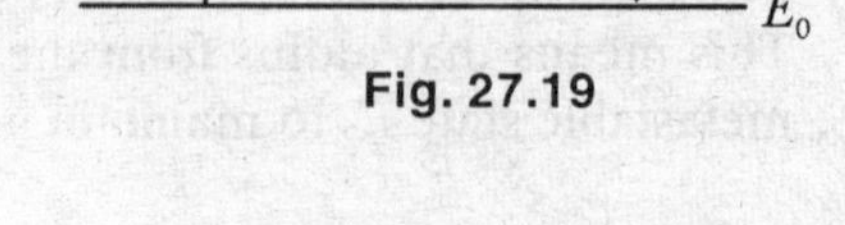

Fig. 27.19

At this stage due to an spontaneous transition from metastable state E_1 to ground state E_0 produces a photon of energy $E_1 - E_0$. This photon now stimulates the other atoms to de-excite to ground state thereby producing photons of same energy, phase and direction. The emitted photons suffer multiple reflections between the two silvered ends of ruby rod. The process of amplification continues till quite an intense beam of laser emerges out from the partially transparent end of the rod. The wavelength (λ) of the laser beam obtained can be calculated from the known energies of the different levels. Hence

$$\frac{hc}{\lambda} = E_1 - E_0 \quad \text{or} \quad \lambda = \frac{hc}{E_1 - E_0}$$

using the values $E_1 = 1.79$ eV, $E_0 = 0$, $\lambda = 6943$ Å

Helium-Neon Laser

The experimental arrangement consists of a cylindrical glass tube containing mixture of helium (= 90%) and neon (= 10%) at low pressures. Two parallel mirrors M_1 and M_2 are fixed at the two ends of the glass tube. The mirror M_1 is thickly silvered while the mirror M_2 is semi-silvered. A high voltage across the electrodes A and B is applied to establish a strong electric field in the tube (Fig. 27.20).

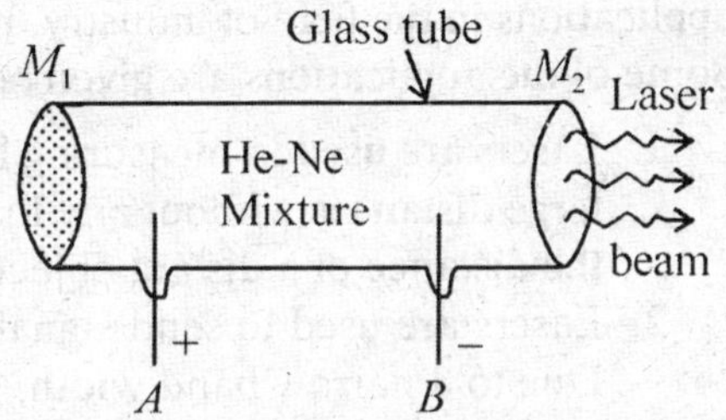

Fig. 27.20

Theory

The relevant energy levels of helium and neon are shown in Fig. 27.21. The lasing action takes place between the state at energy $E_2 = 20.66$ eV and the state $E_1 = 18.70$ eV of neon atoms. Helium has a metastable state at $E_3 = 20.61$ eV which happens to be close to the state E_2 of neon. Helium is used to pump the neon atoms to the state E_2 from where they may be stimulated to come down to the state E_1 with emission of energy

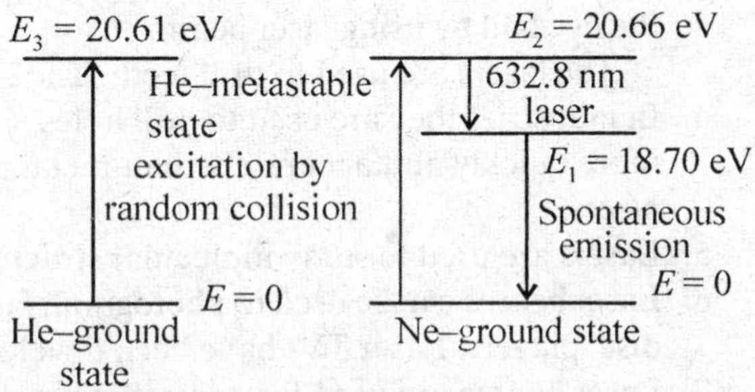

Fig. 27.21

$$E_2 - E_1 = 20.66 - 18.70 = 1.96 \text{ eV}$$

So that the wavelength of helium neon laser is

$$\lambda = \frac{hc}{E_2 - E_1} = \frac{6.6 \times 10^{-34} \times 3 \times 10^8}{1.96 \times 1.6 \times 10^{-19}} = 632.8 \text{ nm}$$

Working

When the electric field is established in the tube some of the atoms of mixture get ionized. The electrons freed in this process are accelerated by the high electric field in the tube. These electrons collide with helium atoms. The helium atoms are then excited to the metastable state E_2 (= 20.61 eV). These atoms collide with neon atoms and transfer the energy to neon atoms which are then excited to energy state E_2 = (20.60 eV). As a result, the helium atom comes back to ground state. This process continues so that the neon atoms are continuously pumped to the state at energy E_2 to keep the population of this state large.

Note that this higher energy state E_2 of neon is not itself a metastable. But the metastable state of helium accumulates atoms at higher energy which take neon atoms to the level E_2 by means of collisions.

Now the stimulated emission takes place between the states having energies E_2 and E_1. The state at energy E_1 has small life time (~10 ns) these atoms de-excite spontaneously to the ground state rapidly. This way the population of the state at energy E_1 is always very small.

He-Ne laser is a four energy level system and in contrast to ruby laser—a three level system, requires much less pumping power.

Applications of Lasers

The first laser was fabricated in 1960. Since then the laser technology has greatly advanced. Lasers have proved to be a boon to the growth of modern scientific world. Lasers have wide spread applications in the field of industry, medical science, telecommunication, scientific research, etc. Some of the applications are given below:

1. Lasers are used in measuring long distances. Being highly coherent, laser beams can travel large distances without any loss in their intensity. As such a laser beam is used to measure the distance of a distant object such as moon.
2. Lasers are used to send signals over long distance communications through optical fibre. Due to a narrow band width, thousands of signals can be sent simultaneously.
3. Lasers are used to perform bloodless surgery. As the laser beam cuts through tissue, it 'heats seals' blood vessels and this technique is used in removing tumour from the body, the diseased part of the liver, etc. In the eye a detached retina can be 'welded' back to the wall of the eye ball by using laser beam.

   The lasers are used to drill teeth. The skin blemishes are also decoloured using laser.
4. In industries they are used to drill holes, cut the metal sheets. Laser does these types of jobs more quickly and accurately than metal drills and saws. They are also used to weld metal pieces.
5. Lasers are used to study molecular structure and nature of chemical reactions.
6. Laser beams can be used to photograph fast moving objects. They are also used in 'compact disc' players. Laser TVs have been developed to produce large TV pictures on walls.
7. Laser beams are used for controlling and guiding rockets, missiles, satellites and locating planes. Intense laser beams can be used to destroy tanks and planes.
8. Laser beams are used to obtain clear pictures of clouds and wind movements for forecasting weather.
9. Lasers are used in nuclear fusion research which is likely to be the ultimate source of energy for us.
10. An interesting application of laser is to produce holograms, which record a 3 D image of an object. When the hologram is viewed, again with a laser, the same 3 D perception is achieved as it is with the actual object.

27.8 MASER (MICROWAVE AMPLIFICATION BY STIMULATED EMISSION OF RADIATION)

MASER stands for microwave amplification by stimulated emission of radiation. Laser produces highly monochromatic coherent narrow and an intense beam in visible region, whereas MASER produces similar beam in microwave (invisible) region. Like laser, maser is produced by creating population inversion of excited atoms in medium and allowing the microwave of appropriate frequency to interact with medium causing stimulated emission. The frequency of the electromagnetic waves used in this case is less than 10^{11} Hz. The first maser was fabricated by J.P. Gordon and C.H. Townes in 1954. They used ammonia gas as medium. This maser is known as ammonia maser.

Ammonia Maser

The device employees ammonia gas. The ammonia molecule has the shape of a pyramid (Fig. 27.22). The three hydrogen atoms in ammonia (NH_3) form the base triangle while the nitrogen atom is at the apex of the pyramid. The N-atom is able to move through the base at right angles to it and can turn this pyramid upside down as shown by dotted line.

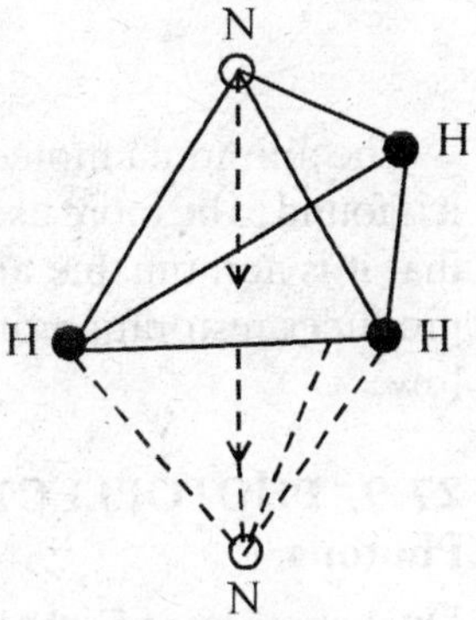

Fig. 27.22

Classically, the nitrogen atom is pictured as flipping back and forth at the rate of 23870 MHz and at any given instant it is on one or the other side of the plane of hydrogen atom. Quantum mechanically, N atom has a finite probability of being on both sides simultaneously. In this sense it is on both sides partly. The potential energy function of nitrogen atom is shown in Fig. 27.23. Moreover, the molecules as a whole has distinct energy states and their energy difference equals the energy of a photon of frequency 23870 MHz.

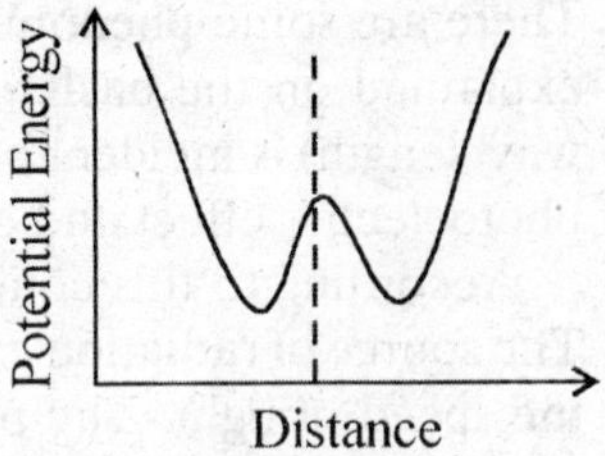

Fig. 27.23

The first requirement of the maser system is that the lower energy moelcules be separated from the upper energy molecules. At any temperature the population in the two energy levels can be represented by the Boltzmann's formula

$$N_2 = N_1 e^{-(E_2-E_1)/k_BT} = N_1 e^{-h\nu/k_BT} \quad \ldots(27.25)$$

where N_2 is the number of molecules in the level E_2 and N_1 is the number of molecules in the level E_1. The above equation suggests that under normal conditions $N_2 < N_1$. Population inversion is obtained by segregating the lower energy molecules from the higher energy molecules by an ingeneous device. The difference in the dipole moments of the ammonia molecule in the two energy states enables one to separate the higher energy molecules by applying a non-homogeneous electric field. In an electric field the perturbation increases the energy of the upper state and depresses the lower energy states. Since the force on the molecules is the negative gradient of the potential energy $\left(F = -\dfrac{dV}{dr}\right)$ the upper state molecules will be repelled away from the centre while the others will be attracted towards the walls of the separator which consists of a cylinder of electrically charged rods as shown in Fig. 27.24 out of the far end of this cylinder only the ammonia molecules in the upper state E_2 emerge.

If these molecules of the ammonia in the upper state E_2 are now irradiated with microwaves of frequency 23870 Hz only downward transition will occur. Energy will be given up by ammonia molecules to the microwaves which will get amplified. As a photon emerges from the downward transition of the ammonia molecule from the upper level E_2 it interacts with another molecule and so on till a self-sustaining chain comes into existence.

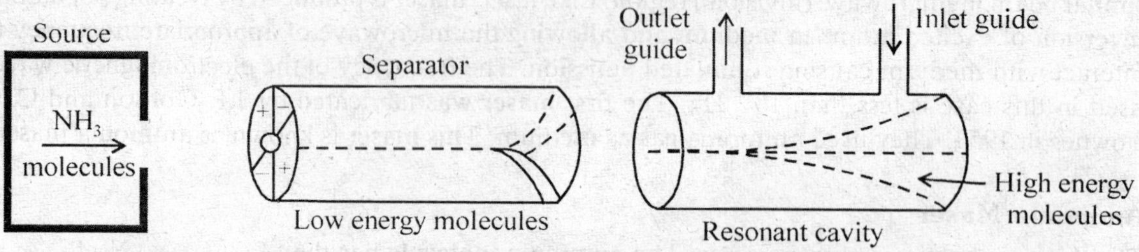

Fig. 27.24

The ammonia molecule gives extraordinary stable and monochromatic oscillations and hence it is found to be more useful as a frequency standard. However ammonia maser has the disadvantage that it is non-tunable and low powered ($\sim 10^{-9}$ watt). To prevent inter molecular collisions which produces restoration of the thermal equilibrium, the pressure of the ammonia gas has to be kept low.

27.9. PHOTOELECTRIC EFFECT

Photons

Dual character of light is well known, i.e. light exhibits wave character as well as particle character. Depending on the interaction it exhibits one of the two characteristics. While passing through double slit, interference is observed which can only be explained on the basis of wave character. There are some phenomenon like, photoelectric effect, Compton effect, etc., which can only be explained on the basis of particle character. When light of sufficient high frequency (i.e. low wavelength) is incident on the metal surface electrons are emitted. This phenomenon is called the photoelectric effect and can be understood in terms of *particle nature of light*.

According to this concept the energy in electromagnetic wave is not continuous but discrete. The source of radiation emits *energy packets* (or energy particles), which travel in a straight line with the speed of light. The particles of light are called *photons.* These particles of light have several properties in common with the material particles and several other properties which are different from the material particles. Some of the important properties of photons are given below.

1. According to Planck's quantum theory, a source of radiation emits energy in the form of photons which travel in straight line at a speed $c = 3.0 \times 10^8$ m/s in vacuum. This is true for any frame of the reference, used, to observe the photon.
2. Each photon has a definite energy and definite momentum. For radiation of frequency ν and wavelength λ, the energy E and momentum P of photon is
$$E = h\nu = \frac{hc}{\lambda} \quad \text{and} \quad P = \frac{E}{c} = \frac{h}{\lambda} \qquad \ldots(27.26)$$
where h is Planck's constant and has value 6.628×10^{-34} J-s $= 4.136 \times 10^{-15}$ eV-s. Thus all the photons of light of a particular wavelength (λ) have the same energy (E) and same momentum (P).
3. The mass of photon is not defined in the sense of Newtonian mechanics. This concept is not applicable to photons. Relativity gives the mass of particle in motion as
$$m = \frac{m_0}{\sqrt{1 - v^2/c^2}}$$
m_0 is rest mass of particle and m is mass when it is moving with speed v.

Since photon moves with speed of light as such $m_0 = 0$ and we say rest mass of photon is zero this means that photon cannot exist at rest.

The mass of photon of frequency ν is given by

$$m = \frac{h\nu}{c^2} \quad \text{...(27.27)}$$

4. Photons are electrically neutral as such are not deflected by electric and magnetic fields.
5. The speed of photon is different in different media. This is due to change in wavelength. The frequency of radiation does not change with medium, i.e. the energy of the photon does not change with medium.
6. A photon when collides with a material particle, the total energy and total momentum remain conserved in such a collision. The photon may get absorbed and/or a new photon may be emitted. Thus the number of photons may not be conserved.
7. Under suitable conditions, the stream of photons may show the phenomenon of differaction.
8. If the intensity of light of a given wavelength is increased, there is an increase in the number of photons crossing a given area in a given time. The energy of each photon remains the same.

The energy of light of a given wavelength (λ) is an integral multiple of $\frac{hc}{\lambda}$, as such it can be varied only in steps (quanta) of $\frac{hc}{\lambda}$. The photon theory is therefore, also called as the *quantum theory of light.*

Photoelectric Effect

When light of suitable wavelength is incident on a metal surface, the electrons are ejected from the metal. The phenomenon is called *the photoelectric effect.* The emitted electrons are called *photo-electrons* and the current so produced is called *photoelectric current.*

Different materials emit photo-electrons when they are exposed to radiation of suitable frequencies or wavelengths. For example alkali metals (like Na, K, Cs, etc.) show photoelectric effect with visible light, whereas the metals like zinc, cadmium, magnesium, etc. are sensitive only to ultraviolet radiations.

Photoelectric effect was discovered by Hallwachs in 1888, experimentally verified by Heinrich Hertz in 1894 and successfully explained theoretically by Einstein in 1905.

Experimental Study of Photoelectric Effect

Figure 27.25 gives the experimental arrangement for the study of photoelectric effect. It consists of an evacuated quartz tube having photo sensitive plate C which acts as a cathode and another plate A which acts as an anode. The plate A can be given a desired positive or negative potential with respect to C using potential divider method shown in Fig. 27.25. Voltmeter V measures the potential difference between and plates and μA (micro-ammeter) measures the current in the circuit.

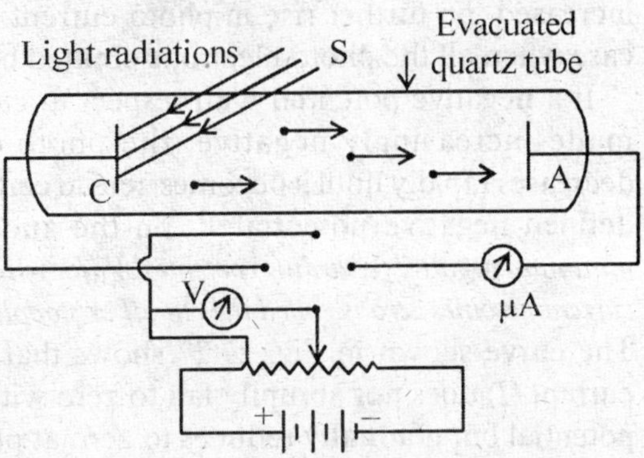

Fig. 27.25

When monochromatic radiations of suitable frequency fall on the plate C, the photo-electrons are emitted which get accelerated towards the plate A. These electrons cause the current in the outer circuit. This current is called *photoelectric current* and is measured by the μA in the circuit.

Using the experimental arrangement shown in Fig. 27.25, we can study the variation of photoelectric current with (*a*) intensity of incident radiation (*b*) frequency of incident radiation (*c*) the potential difference between the electrodes A and C and (*d*) the nature of the material of cathode plate C. The intensity of light can be varied by changing the distance of the light source from the emitter and the frequency can be varied by using filters of different wavelengths.

The experiments revealed that for a given metallic surface, there is a smallest value of frequency (ν_0) for which the incident light can eject photo-electrons from the metal surface. Light of frequency smaller than this particular value cannot eject electrons, no matter for how long it falls on the surface or how much high intense it is. This frequency is called *threshold frequency*. Light of frequency greater than this critical frequency ejects electrons of different velocities. The maximum velocity with which the electrons are ejected depends upon the frequency of the incident light and not on the intensity of incident light.

The variation of photoelectric current with various parameters mentioned above is given below.

Effect of Intensity of Incident Radiation

Maintaining constant potential difference between cathode (C) and anode (A), the intensity of incident radiation of constant frequency (greater than the threshold frequency ν_0) can be varied. It is found that the photoelectric current increases linearly with intensity (Fig. 27.26). It means that number of photo-electrons emitted per second from the photo sensitive plate (C) is directly proportional to the intensity (no. of photons per second) of incident radiation.

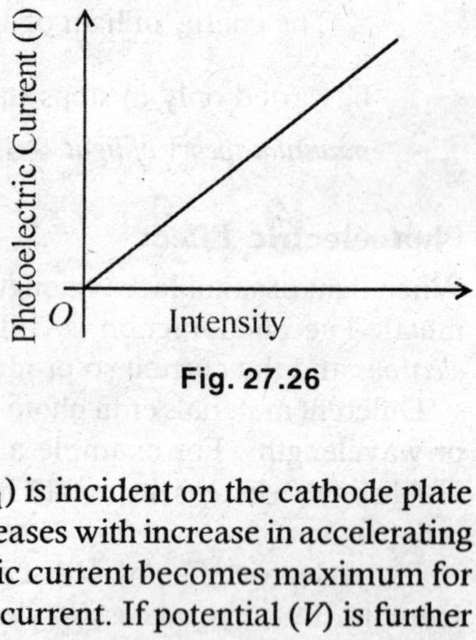

Fig. 27.26

Effect Potential

When light of fixed frequency ($> \nu_0$) and of constant intensity (E_1) is incident on the cathode plate C, and the potential of anode A is varied, the photo current (I) increases with increase in accelerating potential (V) of anode A. A stage comes at which the photoelectric current becomes maximum for a certain positive potential of anode A. This is called saturation current. If potential (V) is further increased, no further rise in photo current is observed. This *saturation current* corresponds to the case when all the photo-electrons emitted by the cathode reach the anode (Fig. 27.27).

If a negative potential with respect to cathode (C) is given to the anode (A) and is gradually made increasingly negative, the photo-current (I) decreases rapidly until it becomes zero at certain sharply defined negative potential V_0 on the anode A. *This minimum negative (retarding) potential V_0 for which the photo-current becomes zero is called the cut off or stopping potential.* The curve shown in Fig. 27.27 shows that the photo-current (I) does not abruptly fall to zero with retarding potential but gradually reduces to zero at potential V_0. This shows that the photo-electrons emitted from the

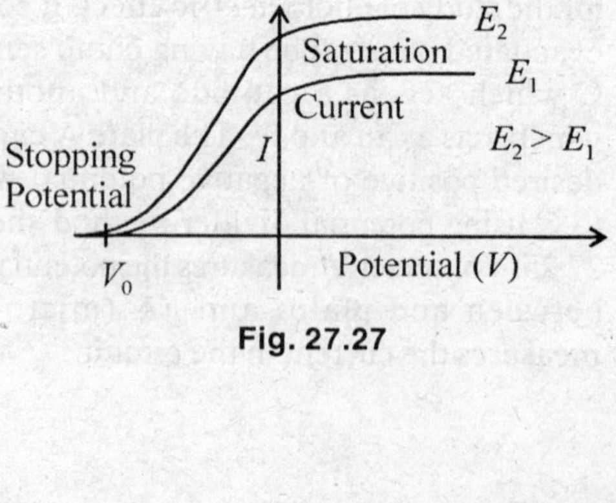

Fig. 27.27

metal do not all have the same energy. Since current is zero at potential V_0, means the maximum kinetic energy (K_m) of electrons emitted is

$$K_m = eV_0 \qquad \ldots(27.28)$$

If the experiment is repeated with the radiation of same frequency but of greater intensity ($E_2 > E_1$), it is observed that the saturation current is greater (proportional to intensity) but the stopping potential is same (Fig. 27.27). Thus for a given frequency of the incident radiation, *the stopping potential is independent of its intensity.* Since the stopping potential is related to the maximum kinetic energy (K_m) of the electrons, hence *the maximum kinetic energy of photo-electrons is independent of intensity of incident radiation.*

Effect of Frequency of Incident Radiation on Stopping Potential

When the cathode is irradiated with light of different frequencies (ν_1, ν_2, etc.) but of the same intensity, the following observations are made:

(*i*) The saturation current is same for radiations of different frequencies. The resulting variation is shown in Fig. 27.28.

(*ii*) The stopping potential V_0 varies linearly with the frequency of incident radiation for a given photo-sensitive material. This means that more the frequency more is stopping potential and more the maximum kinetic energy of the photo-electrons.

(*iii*) There is a certain minimum cut-off frequency ν_0 for which the stopping potential is zero.

(*iv*) For frequencies ν less than cut-off frequency ν_0 no photoelectric emission is possible even if the intensity is large enough.

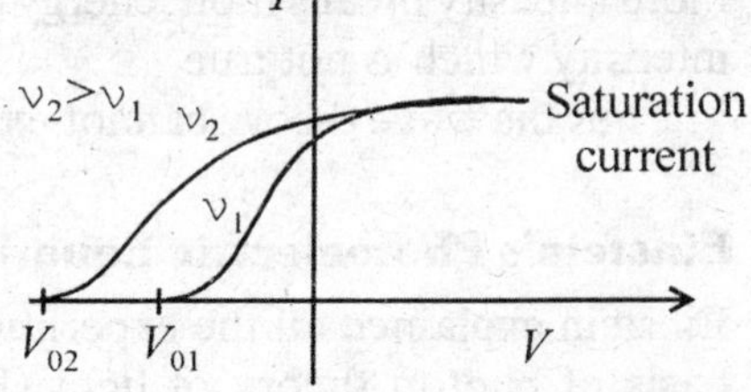

Fig. 27.28

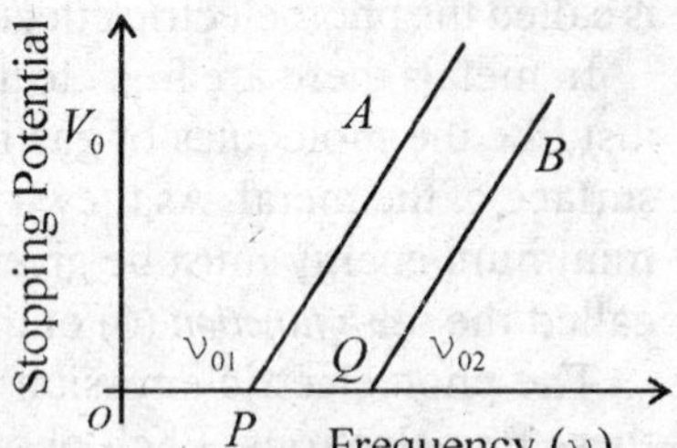

Fig. 27.29

A plot of stopping potential (V_0) and frequency of the incident radiation is a straight line (Fig. 27.29), i.e.

$$V_0 \propto \nu$$

There is a certain minimum frequency ν_0 of the incident radiation for which the stopping potential is zero. This frequency is called *threshold frequency*.

For two different photo-sensitive metals, the curves are parallel, i.e. their slopes are same. The two curves cut the frequency axis at two different points P and Q. This intercepts $OP (= \nu_{01})$ and OQ $(= \nu_{02})$ gives the threshold frequency for the two photo-sensitive metals.

In addition to all these observations, the photoelectric emission is found to be an instantaneous phenomenon. This means that emission is without any apparent time log ($\sim 10^{-9}$ sec) even when the incident radiation is made exceedingly dim.

Failure of Wave Theory of Light to Explain the Photoelectric Effect

The electromagnetic wave theory was well established by the time photoelectric effect was discovered. This theory could very well explain the phenomenon like interference, diffraction and polarisation. Let us now see if the wave theory of light can explain the observations on photoelectric emission discussed above.

According to wave theory, when light falls on metal surface, energy is continuously distributed over the surface. All the free electrons at the surface receive light energy. An electron may be ejected if the wave incident on it imparts energy more than the minimum energy needed to get out of the metal. If we use a low intensity, it may take hours before an electron acquires this much energy from the light. In this duration there will be many collisions and any extra energy accumulated so far will be shared with the remaining metal. This will result in no photoelectric emission.

Secondly the energy of a wave depends on the intensity. Hence an intense beam of incident radiation should be able to cause photoelectric emission even if the frequency is less than threshold frequency. However these facts are not found correct experimentally. If the frequency of incident radiation is greater than the threshold frequency photo-electrons are ejected how small the intensity may be and that too without any appreciable delay of time.

The maximum kinetic energy of a photo-electron does not depend on the intensity of incident light. This fact also can not be understood on the basis of wave theory. According to this theory, more intensity means more energy and the maximum kinetic energy must increase with increase in intensity which is not true.

Thus the wave theory cannot explain the photoelectric emission.

Einstein's Photoelectric Equation

Einstein explained all the experimental observations related to the photoelectric emission on the basis of photon theory of light (Planck's quantum theory). When light of sufficiently small wavelength is incident on a metal surface, electrons are ejected from the metal. The phenomenon is called the photoelectric effect. The electrons so ejected from the metal are called photo-electrons.

In metals there are free electron which move throughout the body due to their thermal velocity just like the molecules of gas in a container.* However, these electrons are not free to leave the surface of the metal. As they try to come out of the metal, the metal attracts them back. As such a minimum energy must be given to them to bring them out of the metal. This minimum energy is called the *work function* (ϕ) of the metal.

The photoelectric emission is the result of the elastic collision between the light photons and these free electrons. Each photon of light has energy $h\nu_1$, where h is Plank's constant and ν the frequency of light. During collision between electron and photon of energy ($h\nu$) the electron absorbs the photon energy. If this energy absorbed exceeds the minimum energy needed for the electron to come out (work function) of the metal, the kinetic energy of the emitted electron then is

$$K = h\nu - \phi \quad \text{...(27.29)}$$

Different electrons in the metal have different energies and therefore need different minimum energy ϕ to be ejected from the metal. Work function ϕ_0 is the least value of ϕ. Thus the maximum kinetic energy of photo-electrons is given by

$$K_m = h\nu - \phi_0 \quad \text{...(27.30)}$$

If V_0 is the stopping potential, then maximum kinetic energy of electron $K_m = eV_0$. Then equation (27.30) can be written as

$$eV_0 = h\nu - \phi_0 \quad \text{or} \quad V_0 = \frac{h\nu}{e} - \frac{\phi_0}{e} \quad \text{...(27.31)}$$

*Some times this is referred to as free electron gas.

This equation is known as *Einstein's photoelectric equation*. This equation satisfactorily explains all the experimental observations on photoelectric effect.

(*a*) Equation (27.30) shows the linear dependence of maximum kinetic energy (K_m) of photo-electron, on the frequency (ν) of the incident radiation and is independent of the intensity of radiation. This is in agreement with experimental observation. It is because Einstein assumed the photoelectric effect to be due to the absorption of a single photon (one quantum of energy) by a single electron. As such intensity of radiation, i.e. the number of photons per second is irrelevant to this basic process.

(*b*) The maximum kinetic energy of electron must be a positive quantity. This implies that

$$h\nu_0 = \phi_0$$

$$\nu_0 = \frac{\phi_0}{h} \qquad ...(27.32)$$

Thus there exists a threshold frequency ν_0 ($= \phi_0/h$) below which no photoelectric emission is possible whatever the intensity of radiation may be.

(*c*) According to this concept, greater the number of photons, greater is the number of photo-electrons or greater photo current. Thus if frequency is greater than the threshold frequency ($\nu > \nu_0$) the photoelectric current is proportional to the intensity.

(*d*) Since the photoelectric emission process involves the absorption of a photon, as such the process must be instantaneous whatever the intensity may be. Because it is the energy of the photon and not the number of photons which is involved in the process of ejection.

(*e*) Equation (27.31) predicts that V_0 versus ν curve is a straight line with slope = (h/e) and intercept $= -\dfrac{\phi_0}{e}$

Since the slope is a constant hence the curves for two different metals are parallel. The intercept on frequency axis gives the threshold frequency (ν_0) of the metal surface and intercept on potential axis gives the work function (ϕ_0) of the metal surface.

Photo Cell

Photo cell is a device which converts light energy into an electrical energy. It consists of an evacuated glass or quartz bulb (Fig. 27.30) containing a semi-cylindrical cathode. It consists of a metal coated with photo-sensitive material like alkali metal (Na, K, Cs, etc.). This acts like a photo-cathode. Along the axis of the semi-cylinder, a straight wire is put which acts like an anode. A positive potential is applied to the anode through a battery. When light of appropriate frequency is allowed to fall on the cathode, photo-electrons are emitted which are collected by the anode. The photoelectric current produced in the circuit is measured by the microammeter (μA) connected in the circuit. When the irradiation of the cathode is stopped, the photo-current in the circuit stops.

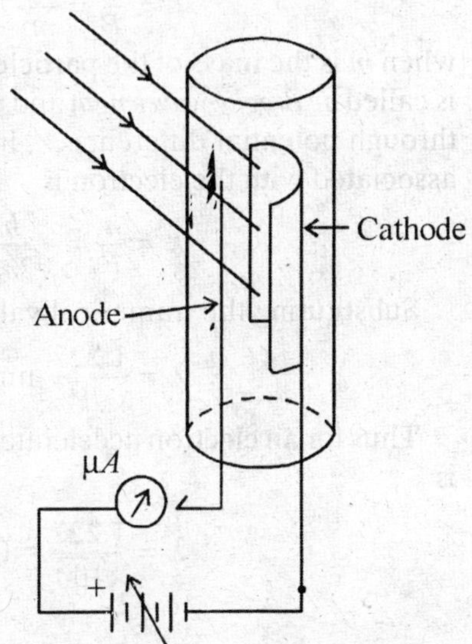

Fig. 27.30

The current in the vacuum type of cell is weak. Therefore gas filled type of photo cells are used. The gas used is either argon or helium and when bombarded by photo-electrons, the gas molecules are ionized. A current then flows which will be evidently greater, but still small so that it cannot be used for practical purposes. Hence the photoelectric current is usually amplified.

Uses of Photoelectric Cells

1. They are used in reproduction of sound recorded on films along with pictures in movies.
2. They are used in television cameras for telecasting scene and photographs.
3. They are used to measure the temperature of stars and to study the spectrum of heavenly bodies.
4. They are used in photometry to compare the illuminating power of two sources.
5. They are used to control the temperature of chemical reactions.
6. They are used as photoelectric sorters to sort out the materials of different shades.

Matter Waves: Wave Nature of Particles

We have studied the phenomena like interference, diffraction and polarisation of light which involves only energy transfer.

In such phenomena, wave nature of light shows up. On the other hand in phenomena like photoelectric effect, Compton effect which involve both energy and momentum transfer, the particle nature of light is involved. Whether a particle or wave description is exhibited in a phenomenon depends on the nature of interaction involved in the experiment. So far no single experiment has been devised which shows both wave and particle behaviour of light (radiation) at the same time.

In 1924 the French physicist Louis Vietorde Broglie put forward the bold hypothesis that *moving particles (electrons, protons, etc.) of matter should display wave like properties under suitable conditions.* His argument was based on the fact that nature was symmetrical and the two basic physical quantities—matter and energy must have symmetrical character. If radiation shows dual aspects so should matter. He proposed that the wavelength (λ)—characteristics of wave property, associated with a particle of momentum (P)—characteristics of particle property is given as*

$$\lambda = \frac{h}{P} = \frac{h}{mv} \qquad \ldots(27.33)$$

when m is the mass of the particle, v its velocity and h is called Planck's constant. This wavelength is called *de Broglie wavelength* and the relation is called *de Broglie relation.* For an electron accelerated through potential difference V, has energy $E = Ve$ and momentum $P = \sqrt{2mE}$. Thus wavelength associated with the electron is

$$\lambda = \frac{h}{P} = \frac{h}{\sqrt{2mE}} = \frac{h}{\sqrt{2meV}}$$

Substituting the numerical values of h, e and m, we get

$$\lambda = \frac{1.227}{\sqrt{V}} \text{ nm} \qquad \ldots(27.34)$$

Thus for an electron accelerated by a potential difference of $V = 100$ V, the wavelength associated is

$$\lambda = \frac{1.227}{\sqrt{100}} = 0.1227 \text{ nm}$$

---

*Infact this relation is valid for photon $P = \frac{h\nu}{c} = \frac{h}{\lambda}$ but de Broglie extended it to material particles also.

The de Broglie wave equation (27.33) gives

(*i*) If $v = 0$ $\lambda = \infty$ and if $v = \infty$, $\lambda = 0$

This means that the wave character is exhibited only by material particles in motion. It does not depend on the charge of the particle. This implies that the *de Broglie waves cannot be electromagnetic in nature because electromagnetic waves are produced by accelerated charged particles.*

(*ii*) Since the position of the wave cannot be located exactly, therefore the wave nature of material particle introduces an uncertainty of the position (location) of the particle. The wavelength in fact defines a region of uncertainty in which the location of particle is unknown.

The wave packet description of an electron is shown in Fig. 27.31. The square of the instantaneous amplitude gives the probability of location of the electron at that particular point. The wave packet corresponds to the spread of wave around some central location and gives uncertainty in position (Δx) and uncertainty in momentum (ΔP).

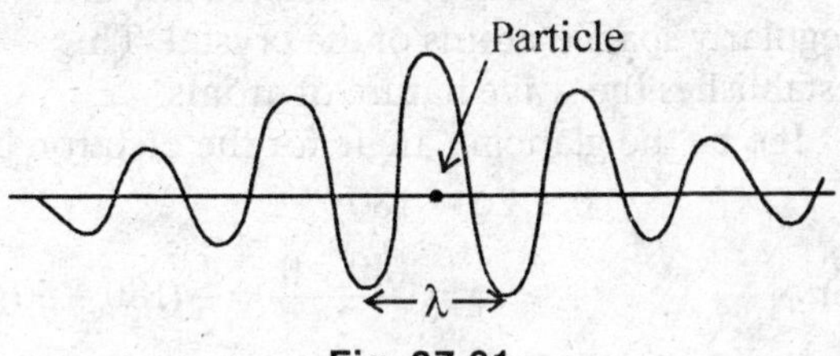

Fig. 27.31

Experimental Demonstration of Wave Nature of Electron. The Davison and Germer Experiment

The first experimental verification for de Broglie's bold hypothesis came from experiments by C.J. Davison and L.H. Germer in 1927. Later on in 1928 G.P. Thomson observed diffraction effects with beams of accelerated electrons scattered by crystals.

The experimental setup used by Davison and Germer is shown Fig. 27.32. It consists of tungsten filament (F) coated with barium oxide and heated by low tension (L.T.) battery. Electrons are emitted from filament on heating and accelerated to a known velocity by applying a suitable potential from high tension (H.T.) battery. These electrons are collimated to a fine beam by allowing them to pass through a cylinder with five holes along its axis. All this arrangement is known as electron gun. This collimated beam of electrons is made to fall on the surface of a nickel crystal. The crystal acts as a three dimensional grating and the electrons are scattered in all directions by the atoms of the crystal. The intensity of the electron beam scattered in a given direction is measured by movable detector. All this arrangement is enclosed in an evacuated chamber. By rotating the detector on circular scale, the intensity of the scattered electron beam is measured for different scattering angles (θ) and for different accelerating voltages.

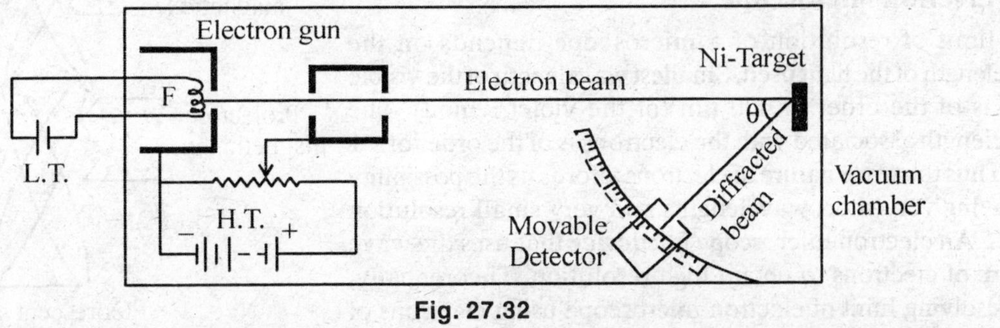

Fig. 27.32

In Davison and Germer experiment, the accelerating voltage was varied from 44 V to 68 V and for each voltage a polar graph between intensity (I) of the scattered beam and scattering angle θ was plotted. Figure 27.33 gives some such plots. Figure 27.33 (*b*) shows a strong peak corresponding to a sharp diffraction maxima in electron distribution at an accelerating potential of 54 V and scattering angle θ = 50°. The appearance of the peak in a particular direction is due to the constructive interference of electrons scattered from different layers of the regularly spaced atoms of the crystal. This establishes the wave nature of atoms.

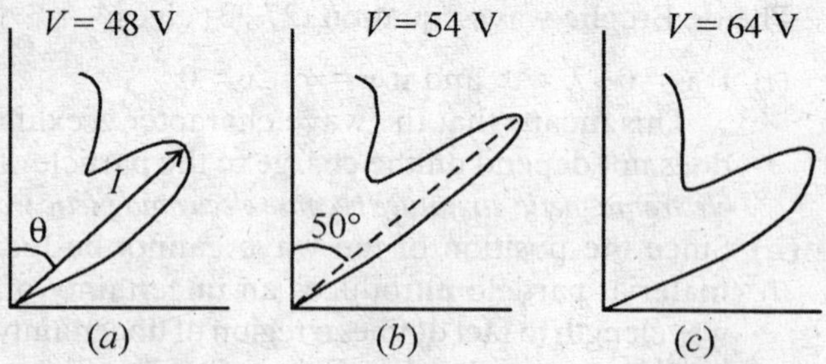

Fig. 27.33

If ϕ be the glancing angle for the electron beam in the crystal, then we have

$$\theta + 2\phi = 180$$

or
$$\phi = \frac{180-\theta}{2} = \frac{1}{2}(180-50) = 65°$$

According to Bragg's law for first order maxima $n = 1$

$$2d\sin\phi = n\lambda$$

we have
$$2d\sin\phi = \lambda$$

where d is the spacing between parallel planes of the diffracting crystal. For nickel $d = 0.91$ Å.

$$\therefore \quad \lambda = 2d\sin\phi = 2 \times 0.91 \times \sin 65° = 1.65 \text{ Å}$$

According to de Broglie hypothesis, the wavelength of the wave associated with electron is given by

$$\lambda = \frac{12.27}{\sqrt{V}} = \frac{12.27}{\sqrt{54}} = 1.66 \text{ Å}$$

There is thus an excellent agreement between theoretical value and the experimentally obtained value of de Broglie wavelength. *This proves the existence of de Broglie waves for the electron in motion.*

The de Broglie hypothesis has led to the development of modern quantum mechanics and field of electron optics. The wave properties of electrons have been used to design electron microscope which has higher resolution (~4000) over the optical microscope.

The Electron Microscope

The limit of resolution of a microscope depends on the wavelength of the light used. Smallest wavelength of the visible light is of the order of 400 nm (of the violet region). The wavelength associated with the electrons is of the order of 0.1 nm. Thus the wave nature of electrons affords us the possibility of having very short wavelengths and very small resolution limits. An electron microscope is a device that uses the wave nature of electrons to obtain high resolution. Theoretically, the resolving limit of electron microscope using electrons of 50 kV would be 5.5×10^{-3} nm about 10^{-5} times that of optical microscope.

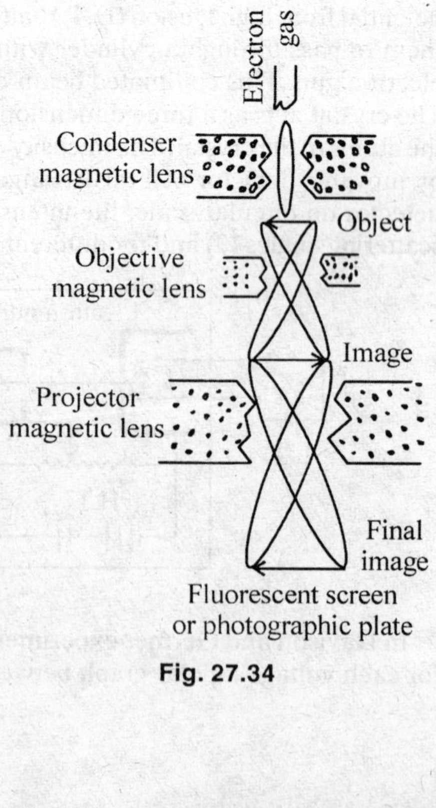

Fig. 27.34

In electron microscope electric and magnetic fields are used to focus the electrons. In this sense they act much like a lens. An schematic diagram of electron microscope is shown in Fig. 27.34.

Bohr's Quantum Condition from de Broglie Relation

de Broglie's concept of waves associated with moving material particles has one simple application, i.e. deduction of the quantum condition for electron orbit in the hydrogen atom.

According to de Broglie, only those orbits can be stationary for which the total length of the orbit ($2\pi r$) is an integral multiple of de Broglie wavelength (λ). This gives, for nth orbit

$$2\pi r_n = n\lambda_n = n\frac{h}{P_n}$$

where P_n is the linear momentum of the electron in the nth orbit. The angular momentum L_n is given by

$$L_n = P_n r_n = \frac{nh}{2\pi} \qquad \text{...(27.35)}$$

This equation is same as equation (*ii*) of section (27.5), the Bohr's quantisation condition.

In Bohr's theory the orbits were quantised through the adhoc quantisation rule for angular momentum (L_n). In de Broglie procedure the orbits are quantised naturally in the same manner as the vibration frequencies of a sonometer wire are quantised through standing wave conditions.

We shall not discuss details about the nature and properties of matter waves. In particular there may be difficulty of how the waves go around a circle. The important point is that quantisation of states come through the consideration *that stationary waves are the only acceptable solutions in the bound system*.

In case of stationary waves no transfer of energy takes place. This also explains the fact the electron so long is moving in an orbit does not radiate any energy, Bohr called these orbits as radiationless, stationary orbits.

SOLVED EXAMPLES

Example 27.1: Consider ionisation energy of a hydrogen atom as W_i. Which of the following; electron, hydrogen ion and helium ion is more effective in ionising the hydrogen atom?

Solution: Let us assume a perfectly inelastic collision between incident particle (mass m) and hydrogen atom (mass m_H). If v_0 is the velocity of incident particle and v the velocity after collision, then conservation of momentum gives

$$mv_0 = (m + m_H)v \quad \text{or} \quad v = \frac{mv_0}{m + m_H}$$

Initial kinetic energy $W_0 = 1/2\, mv_0^2$

and final kinetic energy $W = \frac{1}{2}(m + m_H)v^2$

$$= \frac{1}{2}\frac{m^2 v_0^2}{(m + m_H)} = W_0 \frac{m}{m + m_H}$$

Decrease in kinetic energy is

$$W_0 - W = W_0 - \frac{W_0 m}{m + m_H} = W_0 \frac{m_H}{m + m_H}$$

This decrease in kinetic energy must be equal to ionization energy, i.e.

$$W_i = W_0 \frac{m_H}{m + m_H}$$

This equation suggests that greater the mass of projectile particle the smaller is the fraction of the incident kinetic energy that can be used for ionization.

when an electron is used $m_e << m_H$, i.e. $W_i \approx W_0$

This means that in case of electron the initial kinetic energy of the electron is almost completely used for ionization.

When hydrogen atom is used $W_i = \frac{W_0}{2}$ or $W_0 = 2W_i$

This shows that initial kinetic energy of projectile hydrogen atom must be double that of electron.

For helium atom $W_i = \frac{W_0}{5}$ or $W_0 = 5W_i$

This means that initial kinetic energy of projectile helium atom must be five times that of the electron.

Example 27.2: Estimate the average kinetic energy of hydrogen atoms (or molecules) at room temperature and use the result to explain why nearly all hydrogen atoms are in the ground state at room temperature (= 300 K) and hence emit no light.

Solution: According to the kinetic theory of gases, the average kinetic energy of molecules at temperature T (Kelvin) is

$$E_k = 3/2\, k_B T$$

when k_B is Boltzmann's constant and T temperature.

Using the values $k_B = 1.38 \times 10^{-23}$ J/K and $T = 300$ K.

$$E_k = \frac{3}{2} \times 1.38 \times 10^{-23} \times 300 = 6.2 \times 10^{-21} \text{ J}$$

$$= 0.04 \text{ eV}.$$

Thus the average kinetic energy is very small compared to the energy difference between the ground state and the next higher energy state (13.6 – 3.4 = 10.2 eV). Any atom in excited state emits light and eventually falls to the ground state. Once in ground state, collision with other atoms can transfer energy of 0.04 eV on the average. A small fraction of atoms can have much more energy but even the kinetic energy that is 100 times the average is not enough to excite atom to the next higher energy state. Thus at room temperature, nearly all the atoms are in ground state.

Example 27.3: A monochromatic point source S radiating light of wavelength 6000 Å with power 2 watt. An aperture AB of diameter 0.1 m and a large screen Sc are placed as shown in Fig. 27.35. A photo emissive detector D of surface area 0.5 cm^2 is placed at the centre of the screen. Efficiency of the detector is 0.9.

(*a*) Calculate the photon flux at the centre of the screen and the photo-current in the detector.

(*b*) If a concave lens L of focal length 0.6 m is inserted in the aperture, find the new value of photon flux and photo-current. Assume uniform average transmission of 80% from the lens.

(*c*) If work function of the photo emissive surface is 1 eV. Calculate the values of stopping potential in the two cases (without and with the lens in aperture).

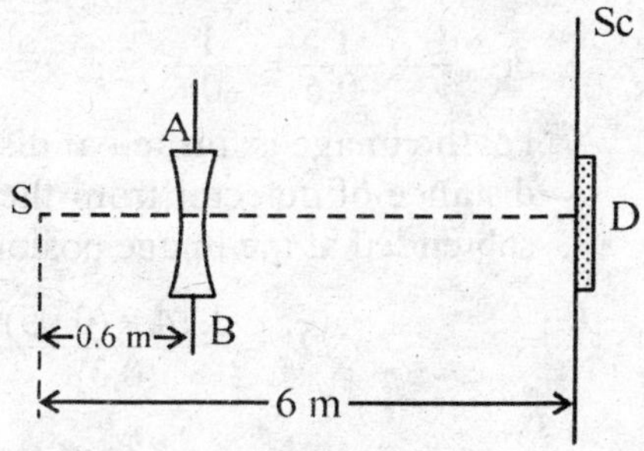

Fig. 27.35

Solution (*a*): The number of photons (n) of wavelength $\lambda = 6000$ Å emitted per second by the source of 2 W is

$$n\frac{hc}{\lambda} = \text{Power}$$

or

$$n = \frac{\text{Power} \times \text{Wavelength}}{hc} = \frac{2 \times 6 \times 100^{-7}}{6.6 \times 10^{-34} \times 3 \times 10^{8}}$$

$$= 6.03 \times 10^{18}$$

Radius of aperture AB, $r = 0.05$ m and its distance from the source is 0.6 m. Hence the solid angle subtended by the aperture at the source is

$$\Omega_a = \frac{\text{Area}}{(\text{Distance})^2} = \frac{\pi r^2}{(\text{SB})^2} = \frac{\pi (0.05)^2}{(0.6)^2} = 2.18 \times 10^{-2} \text{ SR}$$

Similarly the solid angle subtended by the detector at the source is

$$\Omega_d = \frac{0.5 \times 10^{-4}}{(6)^2} = 1.39 \times 10^{-6} \text{ SR (steradian)}$$

Since $\Omega_d << \Omega_a$ all the photons reaching the detector are those which are transmitted by the aperture. Hence photon flux ϕ (i.e. the number of photons reaching per unit area per second) at the centre of screen is

$$\phi = \frac{n}{4\pi d^2}$$

where d is the source screen distance $d = 6$ m.

$$\therefore \quad \phi = \frac{6.03 \times 10^{18}}{4\pi (6)^2} = 1.33 \times 10^{16} \text{ per (m}^2\text{ sec)}$$

Since efficiency of the detector is 90% hence 90% of the photons falling on the detector eject photoelectrons from the detector. Thus the number of electrons emitted from the detector per second is

$$n' = 0.9 \times \text{number of photons incident on the detector per second}$$
$$= 0.9 \times \text{photon flux } (\phi) \times \text{area of detector}$$
$$= 0.9 \times 1.33 \times 10^{16} \times 0.5 \times 10^{-4} = 5.98 \times 10^{11} \text{ per sec.}$$

Hence photo-current

$$i = n' \times \text{electronic charge}$$
$$= 5.98 \times 10^{11} \times 1.6 \times 10^{-19} = 9.6 \times 10^{-8} \text{ A.}$$

(*b*) When concave lens is inserted in the aperture as shown in Fig. 27.35 the source is at the focus of the lens, i.e. the object distance $u = -0.6$ m, focal length $f = -0.6$ m. The image distance v is obtained using the relation

$$\frac{1}{v} - \frac{1}{u} = \frac{1}{f}$$

or $\frac{1}{v} - \frac{1}{-0.6} = \frac{1}{-0.6}$ or $v = -0.3$ m

i.e. the image is formed at distance 0.3 m from lens toward left of the lens. Hence now the distance of detector from the image is $d' = 6 - 0.3 = 5.7$ m. Hence the new solid angles subtended at the image position are

$$\Omega'_a = \frac{3.14 \times (0.05)^2}{(0.3)^2} = 8.72 \times 10^{-2} \text{ Steradian}$$

and $$\Omega'_d = \frac{0.5 \times 10^{-4}}{(5.7)^2} = 1.54 \times 10^{-6} \text{ Steradian}$$

Now the number of photons reaching the aperture per second is

$$n_1 = n \times \frac{\text{Area of aperture}}{4\pi (SB)^2}$$

$$= \frac{6.03 \times 10^{18} \times \pi (0.05)^2}{4\pi (0.6)^2} = 1.05 \times 10^{16} \text{ per sec}$$

Since the lens transmits 80% of the photons, hence number of photons reaching the detector per second is

$$n_2 = 0.8 \times n_1 \times \frac{\text{Solid angle made by detector at image } \Omega'_d}{\text{Solid angle made by aperture at image } \Omega'_a}$$

$\therefore$ $$n_2 = 0.8 \times 1.05 \times 10^{16} \times \frac{1.54 \times 10^{-6}}{8.72 \times 10^{-2}}$$

$$= 1.48 \times 10^{11} \text{ per sec}$$

Hence the photon flux at the detector is

$$\phi_1 = \frac{n_2}{\text{Detector area}} = \frac{1.48 \times 10^{11}}{0.5 \times 10^{-4}} = 2.96 \times 10^{15} \text{ per } (\text{m}^2\text{s})$$

And now the photo-current is

$i' = 0.9 \times e \times$ number of photons reaching the detector per second

$= 0.9 \times 1.6 \times 10^{-19} \times 1.48 \times 10^{11}$

$= 2.1 \times 10^{-8}$ A

(*c*) The stopping potential V_0 is independent of the photo-current. Hence in both cases (with or without lens) V_0 is given by

$$eV_0 = \frac{hc}{\lambda} - \phi_0 \qquad \text{where } \phi_0 \text{ is the work function}$$

or $$V_0 = \frac{hc}{e\lambda} - \phi_0/e \qquad \because \frac{\phi_0}{e} = 1 \text{ V}$$

$$= \frac{6.6 \times 10^{-34} \times 3 \times 10^8}{1.6 \times 10^{-19} \times 6 \times 10^{-7}} - 1$$

$= 2.06 - 1 = 1.06$ V.

Example 27.4: In a photoelectric set up, a point source of light of power 3.2×10^{-3} W emits monoenergetic photons of energy 5.0 eV. The source is located at a distance of 0.8 m from the centre of stationary metallic sphere of work function 3.0 eV and of radius 8.0×10^{-3} m. The efficiency of photoelectric emission is one for every 10^6 incident photons. Assume that the sphere is isolated and initially neutral and the photoelectrons are instantly swept away after emission.

(*a*) Calculate the number of photoelectrons emitted per second.

(*b*) Find the ratio of the wavelength of incident light to the de Broglie wavelength of the fastest electrons emitted.

(*c*) It is observed that the photo-electron emission stops at certain time '*t*' after the light source is switched on. Why?

(*d*) Evaluate this time '*t*'.

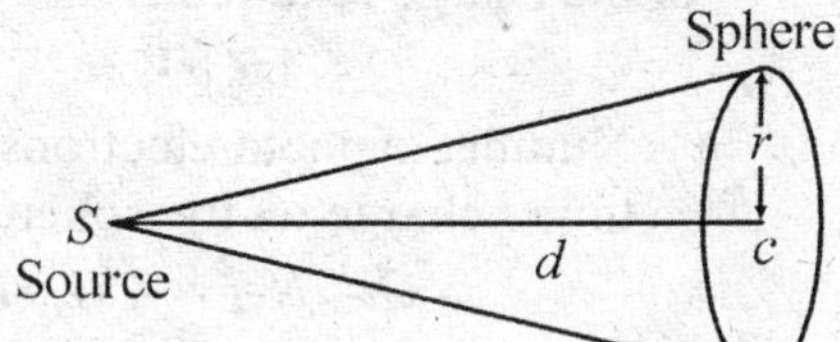

Fig. 27.36

Solution: (*a*) Power of the source is $P = 3.2 \times 10^{-3}$ W

Energy of each photon $= E = 5$ eV $= 5 \times 1.6 \times 10^{-19}$ J.

Hence number of photons emitted by the source per second

$$n = \frac{P}{E} = \frac{3.2 \times 10^{-3}}{5 \times 1.6 \times 10^{-19}} = 4 \times 10^{15} \text{ per second}$$

If r is the radius of the sphere, then normal area of the sphere receiving photons is πr^2. If d is the source sphere distance, then number of photons reaching the sphere per second is

$$n_1 = \frac{n \cdot \pi r^2}{4\pi d^2} = \frac{nr^2}{4d^2} = \frac{4 \times 10^{15} \times (8 \times 10^{-3})^2}{4 \times (0.8)^2}$$

$$= 10^{11} \text{ s}^{-1}$$

Since 10^6 photons eject one photo-electron, hence number of photo-electrons (N) ejected are

$$N = \frac{n_1}{10^6} = \frac{10^{11}}{10^6} = 10^5 \text{ s}^{-1}$$

(*b*) Since work function of the sphere is 3.0 eV. Hence maximum kinetic energy of photo-electrons is

$$K_{max} = 5.0 - 3.0 = 2.0 \text{ eV}$$

If E_P is energy of photon then its wavelength $\lambda_P = \frac{hc}{E_P}$ and E_e is the energy of electron its momentum $= \sqrt{2m E_e}$ and the de Broglie wavelength associated with it is

$$\lambda_e = \frac{h}{\sqrt{2m E_e}} \qquad \text{where } E_e = K_m$$

$$\therefore \quad \frac{\lambda_P}{\lambda_e} = \frac{hc}{E_P} \times \frac{\sqrt{2m E_e}}{h} = \frac{c\sqrt{2m E_e}}{E_P}$$

$$= \frac{3 \times 10^8 \sqrt{2 \times 9.1 \times 10^{-31} \times 2 \times 1.6 \times 10^{-19}}}{5 \times 1.6 \times 10^{-19}}$$

$$= 2.86 \times 10^2$$

(*c*) Due to the emission of photo-electrons, the metallic sphere acquires a positive charge and it will oppose the ejection of photo-electrons due to attractive force exerted by the positive charge of the sphere on the electrons. The photoelectric emission will stop when the sphere acquires a positive potential equal to the stopping potential.

(*d*) The photoelectric emission will stop when the sphere has acquired a positive potential of 2.0 V due to emission of the electrons. If Q is the charge on the sphere, then potential

$$V = \frac{Q}{4\pi\varepsilon_0 r}$$

or the charge required is

$$Q = 4\pi\varepsilon_0 rV$$

Number of photo-electrons ejected per second is N and each electron carries charge e and in time t charge on the sphere will be Net.

$$\therefore \quad Q = Net = 4\pi\varepsilon_0 rV$$

or $$t = \frac{4\pi\varepsilon_0 rV}{Ne} = \frac{8\times10^{-3}\times2}{9\times10^{9}\times10^{5}\times1.6\times10^{-19}}$$

$$= 111 \text{ sec} = 1 \text{ m } 51 \text{ sec.}$$

Example 27.5: A doubly ionized lithium atom is hydrogen like with atomic number $Z = 3$. Find the wavelength of the radiation required to excite the electron in Li^{2+} from the first to the third Bohr orbit. Given the ionization energy of hydrogen atom as 13.6 eV.

Solution: The energy electron in *n*th orbit of hydrogen like atom is given as

$$E_n = -\frac{13.6\,Z^2}{n^2}$$

For Li^{2+} atom $Z = 3$ the energies of electron in the first and third orbit are

$$E_1 = -\frac{13.6\times9}{1} = -122.4 \text{ eV}$$

and $$E_3 = -\frac{13.6\times9}{9} = -13.6 \text{ eV}$$

Thus energy required to excite the electron from first to third orbit is

$$\Delta E = E_2 - E_2 = 122.4 - 13.6 = 108.8 \text{ eV}$$

Hence the wavelength required is

$$\frac{hc}{\lambda} = \Delta E$$

or $$\lambda = \frac{hc}{\Delta E} = \frac{6.6\times10^{-34}\times3\times10^{8}}{108.8\times1.6\times10^{-19}}$$

$$= 0.1137 \text{ m} = 113.7 \text{ Å.}$$

Example 27.6: A moving hydrogen atom makes a head-on inelastic collision with a stationary hydrogen atom. Before collision both atoms are in the ground state and after collision they move together. What is the minimum velocity of the moving hydrogen atom if one of the atoms is to be given the minimum excitation energy after the collision.

Solution: The minimum excitation energy for hydrogen atom is

$$E_i = E_2 - E_1 = [-3.4 - (-13.6)] = 10.2 \text{ eV}$$

If m_t and m_p are the masses of target particle and projectile particle E_p and E_i are the initial kinetic energy of projectile particle and excitation energy of target particle then from example 27.1, we have

$$E_i = E_p \cdot \frac{m_t}{m_p + m_t}$$

In present problem $m_t = m_p = m_H$ and $E_i = 10.2$ eV

$$\therefore \quad E_p = E_i \times 2 = 20.4 \text{ eV} = 20.4 \times 1.6 \times 10^{-19} \text{ J}$$

If v is the velocity of projectile hydrogen atom, then $E_p = \frac{1}{2} m_H v^2$

$$\therefore \quad \frac{1}{2} m_H v^2 = 20.4 \times 1.6 \times 10^{-19} \qquad \because m_H = 1.67 \times 10^{-27} \text{ kg}$$

or

$$v^2 = \frac{20.4 \times 1.6 \times 10^{-19} \times 2}{1.67 \times 10^{-27}} = 10^8 \times 39.1$$

or

$$v = \sqrt{10^8 \times 39.1} = 6.25 \times 10^4 \text{ m/s}.$$

Example 27.7: A particle of charge equal to that of an electron (e^-) and mass 208 times the mass of electron (called the μ-meson) moves in a circular orbit around a nucleus of charge $+3e$ (Take mass of the nucleus to be infinite). Assuming that the Bohr model of the atom is applicable to the system.

(*a*) Derive an expression for the radius of the nth Bohr orbit.
(*b*) Find the value of n for which the radius of orbit is approximately the same as that of the first Bohr orbit for the hydrogen atom.
(*c*) Find the wavelength of the radiation emitted when the μ-messon jumps from the third orbit to the first orbit.

Solution:

(*a*) Radius of Bohr's nth orbit is, (equation 27.8)

$$r_n = \frac{n^2 h^2 \varepsilon_0}{Z e^2 m_e \pi}$$

For our model we have $Z = 3$ and $m = 208\, m_e$, we get

$$r_n = \frac{n^2 h^2 \varepsilon_0}{3e^2 (208 m_e) \pi} = \frac{n^2 h^2 \varepsilon_0}{624\, e^2 m_e \pi}$$

(*b*) For r_n (μ-mesonic atom) $= r_1$ (hydrogen atom), we have

$$\frac{n^2 h^2 \varepsilon_0}{624 \pi m_e e^2} = \frac{h^2 \varepsilon_0}{e^2 m_e \pi}$$

or $\quad n^2 = 624$

or $\quad n = 25$

(*c*) Energy of the electron in *n*th orbit in hydrogen atom is (equation 27.14)

$$E_n = -\frac{Z^2 e^4 m_e}{8\varepsilon_0^2 h^2 n^2}$$

For our model $Z = 3$ and $m = 208\ m_e$

$$E_n = -\frac{9e^2(208\,m_e)}{8\varepsilon_0^2 h^2 n^2} = -\frac{1872\,m_e e^4}{8\varepsilon_0 h^2 n^2}$$

$$= -1872\left(\frac{m_e e^4}{8\varepsilon_0 h^3 c}\right)\frac{hc}{n^2} = -1872\frac{Rhc}{n^2}$$

where $R = \frac{m_e e^4}{8\varepsilon_0 h^3 c}$ is the Rydberg constant, assuming mass of nucleus to be infinite.

When the μ-meson jumps from the third orbit to the first orbit, the difference in energy is radiated as a photon of frequency ν such that

$$h\nu = E_3 - E_1$$

or $$\frac{hc}{\lambda} = E_3 - E_1 = 1872\,Rhc\left[\frac{1}{1^2} - \frac{1}{3^2}\right]$$

or $$\frac{1}{\lambda} = 1872\,R \times \frac{8}{9}$$

or $$\lambda = \frac{9}{1872 \times 8 \times R} = \frac{9}{1872 \times 8 \times (1.097 \times 10^7)}$$

$$= 0.5478 \times 10^{-10}\text{ m}$$

$$= 0.5478\text{ Å}.$$

Example 27.8: A hydrogen like atom (atomic no. *Z*) is in a higher excited state of quantum number *n*. This excited atom can make a transition to the first excited state by successively emitting two photons of energies 10.20 eV and 17.00 eV respectively. Alternatively the atom from the same excited state can make a transition to the second excited by successively emitting two photons of energies 4.25 eV and 5.95 eV respectively. Determine the values of *n* and *Z* (ionization energy of hydrogen atom is 13.6 eV).

Solution: The change in energy of hydrogen like atom in transition from state n_2 to state n_1 is given by

$$\Delta E = 13.6\,Z^2\left(\frac{1}{n_1^2} - \frac{1}{n^2}\right)$$

For transition from higher state $n_2 = n$ to first excited state $n = 2$ the total energy released is ΔE = (10.2 + 17.0), i.e. 27.2 eV. Substituting we have

$$27.2 = 13.6\,Z^2\left(\frac{1}{4} - \frac{1}{n^2}\right) \qquad \ldots(i)$$

For the transition to second excited state $n_1 = 3$ the total energy released is (4.25 + 5.95) = 10.2 eV. Substitution gives

$$10.2 = 13.6\,Z^2\left(\frac{1}{9} - \frac{1}{n^2}\right) \qquad \ldots(ii)$$

$$\therefore \quad \frac{27.2}{10.2} = \frac{n^2-4}{n^2-9}\cdot\frac{9}{4}$$

Solving we get $n^2 = 36$ or $n = 6$

Substituting n in (*i*), we get

$$27.2 = 13.6\,Z^2\left(\frac{1}{4}-\frac{1}{36}\right)$$

or $$Z^2 = 9 \quad \text{or} \quad Z = 3.$$

Example 27.9: A single electron orbits around a stationary nucleus of charge $+Ze$, where Z is a constant and e is magnitude of electronic charge. It requires 47.2 eV to excite the electron from the third to the fourth Bohr orbit. Find

(*a*) the value of Z.

(*b*) the energy required to excite the electron from the third to the fourth Bohr orbit.

(*c*) the wavelength of electromagnetic radiation required to remove the electron from the first Bohr orbit to infinity.

(*d*) the kinetic energy, potential energy and angular momentum of the electron in the first Bohr orbit.

(*e*) the radius of the first Bohr orbit.

[The ionization energy of hydrogen atom is 13.6 eV, Bohr radius = 5.3×10^{-11} m, speed of light = 3×10^8 m/s Plank's constant = 6.63×10^{-34} Js].

Solution: The energy of electron in nth orbit of hydrogen like atom is given by

$$E_n = -Z^2\left(\frac{me^4}{8\varepsilon_0^2ch^3}\right)\frac{hc}{n^2} = -\frac{Z^2Rhc}{n^2}$$

when $R = \dfrac{me^4}{8\varepsilon_0^2ch^3} = 1.097\times10^7\ \text{m}^{-1}$ is Rydberg constant.

(*a*) The energy required to excite the electron from $n = 2$ to $n = 3$, Bohr orbit is

$$E_3 - E_2 = -Z^2Rhc\left(\frac{1}{3^2}-\frac{1}{2^2}\right) = Z^2Rhc\times\frac{5}{36}$$

Given is ionization energy of hydrogen = Rhc = 13.6 eV and $E_3 - E_2$ = 47.2 eV

$$\therefore \quad 47.2 = Z^2\times13.6\times\frac{5}{36}$$

or $$Z^2 = \frac{47.2\times36}{13.6\times5} = 25$$

or $$Z = 5.$$

(*b*) The energy required to excite the electron from $n = 3$ to $n = 4$, Bohr orbit is

$$E_4 - E_3 = Z^2Rhc\left[\frac{1}{3^2}-\frac{1}{4^2}\right] = 5^2\times13.6\times\frac{7}{144}$$

$$= 16.53 \text{ eV}.$$

(*c*) The ionization energy of the atom is the energy required to remove the electron from the first ($n = 1$) Bohr orbit to infinity ($n = \infty$). Thus

$$\text{Ionisation energy} = E_\infty - E_1 = Z^2Rhc\left(\frac{1}{1^2} - \frac{1}{\infty}\right) = Z^2Rhc$$

Ionization energy = $(5)^2$ (13.6) = 340 eV

The wavelength λ of electromagnetic radiation corresponding to this energy is given by

$$\frac{hc}{\lambda} = 340\,\text{eV} = 340 \times 1.6 \times 10^{-19}\ \text{J}$$

or
$$\lambda = \frac{hc}{340 \times 1.6 \times 10^{-19}} = \frac{6.63 \times 10^{-34} \times 3 \times 10^8}{340 \times 1.6 \times 10^{-19}} = .03656 \times 10^{-7}\ \text{m}$$
$$= 36.06\ \text{Å}.$$

(*d*) Kinetic energy of electron in first Bohr orbit is $-E_1 = 340$ eV

Potential energy of electron in the first Bohr orbit is = –2 (K.E.)

= –680 eV.

The angular momentum in nth orbit $L_n = \dfrac{nh}{2\pi}$

$$\therefore \quad L_1 = \frac{h}{2\pi} = 1.06 \times 10^{-34}\ \text{Js}$$

(*e*) The radius of the nth Bohr orbit of atom is

$$r_n = \frac{\varepsilon_0 h^2 n^2}{\pi m Z e^2}$$

For $n = 1$
$$r_1 = \frac{\varepsilon_0 h^2}{\pi m Z e^2} = \frac{\text{Radius of first Bohr orbit of hydrogen}}{Z}$$
$$= \frac{5.3 \times 10^{-11}}{5} = 1.06 \times 10^{11}\ \text{m}.$$

Example 27.10: Electrons in hydrogen like atom ($Z = 3$) make transition from the fifth to the fourth orbit and from fourth to third orbit. The resulting radiations are incident normally on metal plate and eject photo-electrons. The stopping potential for the photo-electrons ejected by the shorter wavelength is 3.95 V. Calculate the work function of the metal and the stopping potential for the photo-electrons ejected by the longer wavelength. (Rydberg constant = 1.094×10^7 m^{-1}).

Solution: The energy of electron in nth orbit of a hydrogen like atom is

$$E_n = -\frac{RhcZ^2}{n^2} \qquad \text{since } Rhc = 13.6\ \text{eV}$$

For $Z = 3$
$$E_n = -\frac{9 \times 13.6}{n^2} = -\frac{122.4}{n^2}\,\text{eV}$$

$$\therefore \quad E_5 = -\frac{122.4}{25} = -4.896\,\text{eV}$$

$$E_4 = -\frac{122.4}{16} = -7.650\ \text{eV}$$

$$E_3 = -\frac{122.4}{9} = -13.6\ \text{eV}$$

Hence wavelength of radiation emitted by transition from 5th to 4th state is

$$\frac{hc}{\lambda} = E_5 - E_4$$

or $$\frac{hc}{\lambda} = -4.896 - (-7.650) = 2.754 \text{ eV}$$

and λ' the wavelength of radiation emitted by transition from 4th to 3rd orbit is

$$\frac{hc}{\lambda'} = -7.650 - (-13.6) = 5.95 \text{ eV}$$

It is clear that $\lambda' < \lambda$. Hence photo-electrons ejected by radiation of wavelength λ' have maximum kinetic energy = 3.95 eV. Now using Einstein's photoelectric equation, we have

$$\frac{hc}{\lambda} - \phi_0 = K_{max}$$

where ϕ_0 is work function of the metal and K_{max} is the maximum kinetic energy of the ejected photo-electrons. Substitution gives

$$5.95 - \phi_0 = 3.95 \quad \text{or} \quad \phi = 5.95 - 3.95 = 2 \text{ eV}$$

If the stopping potential for electrons ejected by the radiations of shorter wavelength (λ) then K_{max} for these electrons is eV_0 Joules or V_0 eV. Thus we have

$$\frac{hc}{\lambda} - \phi_0 = K_{max}$$

$$2.754 - 2 = V_0$$

or $$V_0 = 0.754 \text{ V}.$$

Example 27.11: Hydrogen atom in its ground state is excited by means of monochromatic radiation of wavelength 975 Å. How many different lines are possible in the resulting spectrum? Calculate the longest wavelength among them. You may assume the ionization energy for hydrogen atom as 13.6 eV.

Solution: The energy of incident photon in eV is

$$E = \frac{hc}{\lambda} = \frac{6.63 \times 10^{-34} \times 3 \times 10^8}{975 \times 10^{-10} \times 1.6 \times 10^{-19}} = 12.75 \text{ eV}$$

Energy of electron in nth orbit in hydrogen atom is

$$E_n = -\frac{13.6}{n^2}$$

Hence energy required to excite the electron from 1st orbit ($n = 1$) to nth orbit is

$$\Delta E = E_n - E_1 = 13.6\left(1 - \frac{1}{n^2}\right)$$

The incident photon of energy 12.75 eV will excite this atom to nth orbit if $\Delta E = 12.75$. Equating, we have

$$13.6\left(1 - \frac{1}{n^2}\right) = 12.75$$

or $$1-\frac{1}{n^2}=\frac{12.75}{13.6}$$

or $$\frac{1}{n^2}=1-\frac{12.75}{13.6}=\frac{0.85}{13.6}$$

or $$n^2=\frac{13.6}{0.85}=16$$

or $$n=4$$

The incident photon excite the electron to 4th orbit. The possible transitions are shown in Fig. 27.37. In all there will be six transitions and hence six spectral lines. The longest wavelength radiation will have minimum energy, i.e. transition $n = 4$ to $n = 3$ will give us the radiation of longest wavelength.

$$\lambda_{max}=\frac{hc}{\Delta E_{min}}$$

$$\Delta E_{min}=E_4-E_3=13.6\left(\frac{1}{9}-\frac{1}{16}\right)=\frac{13.6\times 7}{144}\text{ eV}$$

$$=\frac{13.6\times 7\times 1.6\times 10^{-19}}{144}\text{ J}$$

∴ $$\lambda_{max}=\frac{hc}{\Delta E_{min}}=\frac{6.6\times 10^{-34}\times 3\times 10^{8}\times 144}{13.6\times 7\times 1.6\times 10^{-19}}$$

$$=18.718\times 10^{-7}\text{ m}=18718\text{ Å}.$$

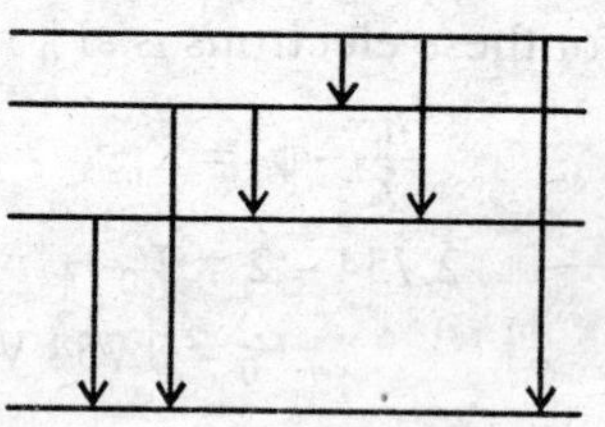

Fig. 27.37

Example 27.12: The ionization energy of a hydrogen like Bohr atom is 4 Rydberg:

(*a*) What is the wavelength of radiation emitted when the electron jumps from the first excited to the ground state?

(*b*) What is radius of the first orbit of this atom? Given that the Bohr radius of hydrogen atom $= 5\times 10^{-11}$ m and 1 Rydberg $= 2.2\times 10^{-18}$ J.

Solution: Energy of electron in *n*th orbit of hydrogen like atom in terms of Rydberg is

$$E_n=-\frac{RZ^2}{n^2}\qquad \text{where } R=2.2\times 10^{-18}\text{ J.}$$

The ionization energy of atom is

$$\Delta E=E_\infty-E_1=-RZ^2\left(\frac{1}{\infty}-\frac{1}{1}\right)=RZ^2$$

But $$\Delta E=4\text{ Rydberg}$$

∴ $$4R=RZ^2$$

or $$Z=2$$

(*a*) The energy of the radiation emitted when the electron jumps from the first excited state (n = 2) to the ground state (n = 1) is

$$\Delta E=RZ^2\left(\frac{1}{1^2}-\frac{1}{2^2}\right)=\frac{3}{4}RZ^2$$

$$= 3R = 3 \times 2.2 \times 10^{-18}\ \text{J}$$

Wavelength of the radiation is then

$$\frac{hc}{\lambda} = 3R$$

$$\lambda = \frac{hc}{3R} = \frac{6.6\times10^{-34}\times3\times10^{8}}{3\times2.2\times10^{-18}} = 3\times10^{-8}\ \text{m}$$

$$= 300\ \text{Å}.$$

(*b*) Radius of 1st Bohr orbit is

$$r = \frac{(r_1)_H}{Z} = \frac{5\times10^{-11}}{2} = 2.5\times10^{-11}\ \text{m}.$$

Example 27.13: Suppose the potential energy between the electron and the proton at a distance *r* is given by $-\frac{ke^2}{3r^3}$. Use Bohr's theory to obtain energy levels of such a hypothetical hydrogen atom.

Solution: The potential energy is

$$U = -\frac{ke^2}{3r^3}$$

∴ Force between electron and proton is

$$F = -\frac{du}{dr} = -\frac{ke^2}{r^4}$$

The negative sign shows the attractive force. Hence electron in *n*th circular orbit (r_n) has velocity v_n such that

$$\frac{mv_n^2}{r_n} = \frac{ke^2}{r_n^4} \quad \text{or} \quad mv_n^2 = \frac{ke^2}{r_n^3} \qquad ...(i)$$

Using Bohr's quantum condition, we have

$$mv_nr_n = \frac{nh}{2\pi} \qquad ...(ii)$$

From (*i*) and (*ii*), we have $r_n = \frac{4\pi^2mke^2}{n^2h^2}$ ...(*iii*)

Kinetic energy of electron in *n*th orbit is

$$E_n = \frac{1}{2}mv_n^2 = \frac{ke^2}{2r_n^3}$$

and the potential energy of the electron in *n*th orbit is

$$E_p = -\frac{ke^2}{3r_n^3}$$

Hence total energy in *n*th orbit is

$$E_n = E_K + E_p = \frac{ke^2}{2r_n^3} - \frac{ke^2}{3r_n^3} = \frac{ke^2}{6r_n^3}$$

Substituting the value of r_n from (*iii*) we have

$$E_n = \frac{ke^2}{6}\left(\frac{n^2h^2}{4\pi^2mke^2}\right)^3$$

$$= \frac{n^6h^6}{384\pi^6m^3k^2e^4}$$

Note: Since total energy is positive as such a system is unstable and physically not possible.

Example 27.14: A gas of identical hydrogen like atoms has some atoms in the lowest (ground) energy level A and some atoms in a particular upper (excited) energy level B and there are no atoms in any energy level. The atoms of the gas make transition to higher energy level by absorbing monochromatic light photon of energy 2.7 eV. Subsequently the atoms emit radiations of only six different photon energies. Some of the emitted photon have energy 2.7 eV, some have energy more and some have less than 2.7 eV.

(*a*) Find the principal quantum number of the initially excited level B.
(*b*) Find the ionization energy of the gas atoms.
(*c*) Find the maximum and minimum energies of the emitted photons.

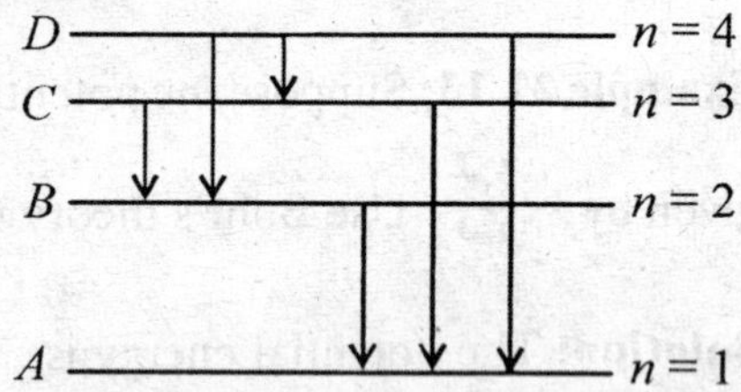

Fig. 27.38

Solution: The energy of an electron in nth orbit of hydrogen like atom is given by

$$E_n = -\frac{RhcZ^2}{n^2} = -\frac{13.6\,Z^2}{n^2}\text{eV} \qquad ...(i)$$

(*a*) When this atom is given energy equal to 2.7 eV it is excited to the energy state. Let this state be D (Fig. 27.38). Since the atom now emits six different energy photons, i.e. there are six different transitions. The six different transitions are possible only if D has $n = 4$. These transitions are shown in Fig. 27.38. Since energy of some of the emitted photons is greater than 2.7 eV as such n can not be one (i.e. $n \neq 1$). Hence n must be either 2 or 3. If $n = 3$ it cannot emit photon of energy less than 2.7 eV. Thus $n = 2$. Hence initial excited state of some of the atoms is corresponds to $n = 2$.

(*b*) We are given $E_4 - E_2 = 2.7$ eV. Hence from (*i*), we have

$$E_4 - E_2 = 2.7 = -13.6Z^2\left(\frac{1}{4^2} - \frac{1}{2^2}\right)$$

or $\qquad Z = 1.$

The ionization energy corresponds to transition $n = \infty$ to $n = 1$.
Thus the ionization energy of atom is

$$E = -(13.6)Z^2\left(\frac{1}{\infty} - \frac{1}{1^2}\right) = +13.6 \text{ eV } (Z = 1)$$

(*c*) The maximum energy of the emitted radiation corresponds to the transitions $n = 4$ to $n = 1$ and this is

$$E_{\text{max}} = -(13.6)\left(\frac{1}{4^2} - \frac{1}{1^2}\right) = -13.6 \times \left(-\frac{15}{16}\right)$$

$$= 12.75 \text{ eV}$$

The minimum energy of the emitted radiation corresponds to the transition $n = 4$ to $n = 3$ and this is

$$E_{min} = -(13.6)\left(\frac{1}{16} - \frac{1}{9}\right) = -13.6 \times \frac{7}{144} = 0.67 \text{ eV}$$

Example 27.15: A beam of light has three wavelengths 4144 Å, 4972 Å and 6216 Å. With total intensity 3.6×10^{-3} Wm^{-2} equally distributed amongst the three wavelengths. The beam falls normally on an area 1.0 cm^2 of clean metallic surface of work function 2.3 eV. Assume that there is no loss of light by reflection and that each energetically capable photon ejects one electron. Calculate the number of photo-electrons liberated in two seconds.

Solution: Threshold wavelength (λ_0) for the surface is

$$\lambda_0 = \frac{hc}{\phi_0} \qquad \text{where } \phi_0 \text{ is work function of metal}$$

$$\therefore \quad \lambda_0 = \frac{6.6 \times 10^{-34} \times 3 \times 10^8}{2.3 \times 1.6 \times 10^{-19}} \text{ m}$$

$$= 5.380 \times 10^7 \text{ m} = 5380 \text{ Å}.$$

This means that light of wavelength 6216 Å is incapable of ejecting photo-electrons.

The intensity of each of these wavelengths is $I = \dfrac{\text{Intensity incident}}{3}$

$$I = \frac{3.6 \times 10^{-3}}{3} = 1.2 \times 10^{-3} \text{ W/m}^2$$

Since the area of metal surface $A = 1.0$ cm^2 = 10^{-4} m^2 hence the incident energy of each wavelength per second falling on the metal surface is

$$E = I \times A = 1.2 \times 10^{-3} \times 10^{-4} = 1.2 \times 10^{-7} \text{ W}$$

Hence the number of photons of wavelength 4144 Å incident on the metal surface per second is

$$n_1 = \frac{E}{h\nu_1} = \frac{E\lambda_1}{hc} = \frac{1.2 \times 10^{-7} \times 4144 \times 10^{-10}}{6.6 \times 10^{-34} \times 3 \times 10^8}$$

$$= 2.5 \times 10^{11} \text{ per second.}$$

Similarly, the number of photons of wavelength $\lambda_2 = 4972$ Å incident per second on the metal surface is

$$n_2 = \frac{E\lambda_2}{hc} = \frac{1.2 \times 10^{-7} \times 4972 \times 10^{-10}}{6.6 \times 10^{-34} \times 3 \times 10^8}$$

$$= 3.0 \times 10^{11}.$$

Hence total photons incident on the metal surface per second is

$$n = n_1 + n_2 = 5.5 \times 10^{11} \text{ per sec}$$

Hence number of photo-electrons ejected per second is n. Since each photon ejects one electron. Hence total photo-electrons ejected in 2 seconds is

$$N = 2n = 5.5 \times 2 \times 10^{11} = 1.1 \times 10^{12}.$$

Example 27.16: Radiation of wavelength 180 nm ejects photo-electrons from a metal plate whose work function is 2.0 eV. If a uniform magnetic field of flux density 5.0×10^{-5} T (or Wb m^{-2}) is applied parallel to the plate, what should be the radius of the path followed by electrons ejected normally from the plate with maximum energy?

Solution: Einstein's photoelectric equation gives the maximum energy of the photo-electrons emitted from the surface as

$$E = \frac{hc}{\lambda} - \phi = \frac{6.63 \times 10^{-34} \times 3 \times 10^{8}}{180 \times 10^{-9}} - 2 \times 1.6 \times 10^{-19}$$

$$= 7.85 \times 10^{-19} \text{ J.}$$

Hence the velocity of these electrons v is

$$\frac{1}{2} mv^2 = E$$

or

$$v = \sqrt{\frac{2E}{m}} = \sqrt{\frac{2 \times 7.85 \times 10^{-19}}{9.1 \times 10^{-31}}}$$

$$= 1.31 \times 10^{6} \text{ m/s.}$$

Since the electron now moves in magnetic field B experiences a force $F (= evB)$ which provides the necessary centrepetal force $\left(= \frac{mv^2}{r}\right)$ to the electron for circular motion in orbit of radius r. Thus

$$evB = \frac{mv^2}{r}$$

$$r = \frac{mv}{eB} = \frac{9.1 \times 10^{-31} \times 1.31 \times 10^{6}}{1.6 \times 10^{-19} \times 5 \times 10^{-5}}$$

$$= 0.149 \text{ m.}$$

Example 27.17: A 40 W ultraviolet light source of wavelength 2480 Å illuminates a magnesium (Mg) surface placed 2 m away. Determine the number of photons emitted from the surface per second and the number incident on a unit area of Mg surface per second. The work function of Mg is 3.68 eV. Calculate the kinetic energy of the fastest electrons ejected from the surface. Determine the maximum wavelength for which the photoelectric effect can be observed with Mg surface.

Solution: The energy of emitted photon is $E = \frac{hc}{\lambda}$. Power of the source $P = 40$ W. Therefore the number of photons emitted per second by the source is

$$n = \frac{P}{E} = \frac{P\lambda}{hc} = \frac{40 \times 2480 \times 10^{-10}}{6.6 \times 10^{-34} \times 3 \times 10^{8}} = 5 \times 10^{+19} \text{ s}^{-1}$$

If d is the distance of the metal surface then number of photons falling on unit area of the metal surface per second is

$$N = \frac{n}{4\pi d^2} = \frac{5 \times 10^{19}}{4\pi (2)^2} = 10^{18} \text{ s}^{-1}$$

Kinetic energy of the fastest electron is given by

$$E_{max} = \frac{hc}{\lambda} - \phi_0 = \frac{6.6\times10^{-34}\times3\times10^8}{2480\times10^{-10}} - 3.68\times1.6\times10^{-19}$$

$$= 8.0 \times 10^{-19} - 3.68 \times 1.6 \times 10^{-19}\ \text{J}$$

$$= \frac{8.0}{1.6} - 3.68\ \text{eV}$$

$$= 5 - 3.68 = 1.32\ \text{eV}.$$

The maximum (threshold) wavelength for photoelectric emission from Mg surface is

$$\lambda_0 = \frac{hc}{\phi_0} = \frac{6.6\times10^{-34}\times3\times10^8}{3.68\times1.6\times10^{-19}}$$

$$= 3.362 \times 10^{-7}\ \text{m} = 3362\ \text{Å}.$$

Example 27.18: A stationary He^+ ion emitted a photon corresponding to the first line of the Lyman series. That photon liberated a photoelectron from a stationary hydrogen atom in the ground state. Find the velocity of the photoelectron.

Solution: When an electron makes a transition from state n_2 to state n_1, the energy liberated by the hydrogen like atom of atomic number Z is

$$\Delta E = RchZ^2\left(\frac{1}{n_1^2} - \frac{1}{n_2^2}\right) \qquad ...(i)$$

In present problem this energy is shared between the photon (energy $h\nu$) and helium ion ($= E_r$). Thus from conservation of energy, we have

$$\Delta E = h\nu + E_r \qquad ...(ii)$$

E_r is kinetic energy of recoiling helium atom.

Conservation of momentum gives

$$P_{Ph} = -P_{He} \qquad ...(iii)$$

Here –ve sign indicates that the two particles move in opposite direction. Since energy of photon is $h\nu$, hence momentum is $P_{Ph} = \frac{h\nu}{c} = P_{He}$

Hence recoil kinetic energy of $He^+ E_r = \frac{P_{He}^2}{2m'} = \frac{h^2\nu^2}{2c^2m'}$. Where m' is the mass of He^+ ion.

From (*ii*), we have

$$\Delta E = h\nu + \frac{h^2\nu^2}{2c^2m'}$$

or
$$h^2\nu^2 + (2c^2m')h\nu - (2c^2m')\Delta E = 0$$

or
$$h\nu = \frac{-2m'c^2 \pm \sqrt{(2c^2m')^2 + (8c^2m')\Delta E}}{2}$$

$$= -m'c^2 + m'c^2\sqrt{1 + \frac{2\Delta E}{m'c^2}}$$

Negative sign is not admissible and makes ν (frequency) negative.

or $$h\nu = m'c^2\left\{\sqrt{1+\frac{2\Delta E}{m'c^2}}-1\right\}$$

Multiplying and dividing with $\left[\sqrt{1+\frac{2\Delta E}{m'c^2}}+1\right]$, we have

$$h\nu = m'c^2\frac{\left[1+\frac{2\Delta E}{m'c^2}-1\right]}{\sqrt{1+\frac{2\Delta E}{m'c^2}}+1} = \frac{2\Delta E}{\sqrt{1+\frac{2\Delta E}{m'c^2}}+1}$$

Since ΔE is less than the ionisation energy (=4 × 1.36 = 54.2 eV) and $m'c^2 \approx \frac{4\times1.67\times10^{-27}\times9\times10^{16}}{1.6\times10^{-19}}$

$\approx 4\times10^9$ eV i.e. 4 GeV. Hence in denominator in under root $\frac{2\Delta E}{m'c^2}$ can be neglected, this gives

$$h\nu = \Delta E \qquad ...(iv)$$

For first line of Lyman series $n_2 = 2$ and $n_1 = 1$. ΔE from (*i*) is

$$\Delta E = RchZ^2\left(\frac{1}{1}-\frac{1}{4}\right) \quad \text{and } Z = 2$$

$$= 3\,Rch \qquad ...(v)$$

Now the photons of energy 3 *Rch* are incident on hydrogen atom. The ionization energy of hydrogen is *Rch*. Hence the difference of energy 2 *Rch* is given to the ejected electron. Hence kinetic energy of ejected electron is 2 *Rch*.

i.e. $$\frac{1}{2}mv^2 = 2Rch = 2\times13.6\text{ eV} = 2\times13.6\times1.6\times10^{-19}\text{ J}$$

or $$v = \left[\frac{2\times2\times13.6\times1.6\times10^{-19}}{9.1\times10^{-31}}\right]^{1/2}$$

$$= 3.1\times10^6\text{ m/s.}$$

Example 27.19: A sample of hydrogen gas in its ground state is irradiated with photons of 10.02 eV energies. The radiation from the above sample is used to irradiate two other samples of excited ionized He^+ and excited ionized Li^{+2} respectively. Both the ionized samples absorb the incident radiation.

(*i*) How many lines are obtained in the He^+ and Li^{+2} emission spectra respectively?
(*ii*) What are the smallest and largest wavelengths in their spectra?

Solution: In hydrogen atom the energy of electron in *n*th orbit is

$$E_n = -\frac{13.6}{n^2}\text{ eV}$$

∴ $E_1 = -13.6$ eV and $E_2 = -3.4$ eV. Hence difference is $E_2 - E_1 = 10.2$ eV. Thus hydrogen atom after absorbing this energy is excited to the 1st excited state. Hence on de-excitation, it will release energy equal to 10.2 eV.

Energy of electron in nth orbit of He^+ ion is

$$E_n = -\frac{13.6 Z^2}{n^2} = -\frac{13.6 \times (2)^2}{n^2} = -\frac{54.4}{n^2}$$

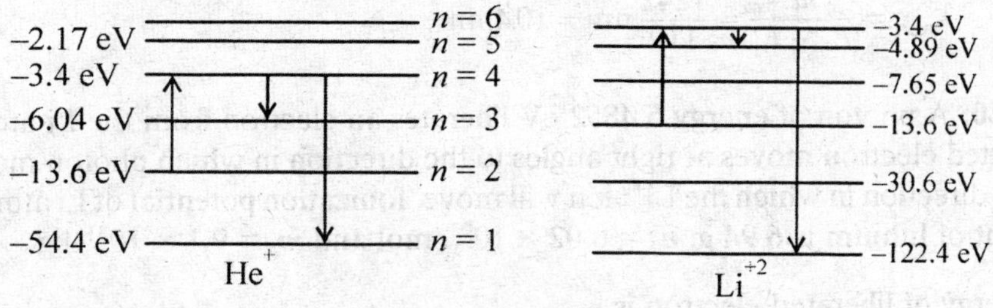

Fig. 27.39

The different energy levels are shown in Fig. 27.39. Since it absorbs energy equal to 10.2 eV, hence the He^+ ion must be in first excited state and after absorbing 10.2 eV energy, must be excited to state with $n = 4$. Hence in all six lines will be emitted. The minimum energy difference between levels is

$$E_4 - E_3 = -3.4 - (-6.04) = 2.64 \text{ eV}$$

Thus maximum wavelength is

$$\frac{hc}{\lambda_{max}} = 2.64 \times 1.6 \times 10^{-19}$$

or
$$\lambda_{max} = \frac{hc}{2.64 \times 1.6 \times 10^{-19}} = \frac{6.6 \times 10^{-34} \times 3 \times 10^8}{3.64 \times 1.6 \times 10^{-19}}$$

$$= 4.6875 \times 10^{-7} \text{ m} = 468.75 \text{ nm.}$$

The maximum energy difference between levels is

$$E_4 - E_1 = 54.4 - 3.4 = 51 \text{ eV}$$

Hence minimum wavelength is

$$\lambda_{min} = \frac{hc}{E_4 - E_1} = \frac{6.6 \times 10^{-34} \times 3 \times 10^8}{51 \times 1.6 \times 10^{-19}} = 0.2426 \times 10^{-7} \text{ m}$$

$$= 24.26 \text{ nm.}$$

In case of lithium, energy of electron in nth orbit is

$$E_n = -\frac{13.6 Z^2}{n^2} = -\frac{13.6 \times 9}{n^2} = -\frac{122.4}{n^2} \text{ eV}$$

The different energy levels are shown in Fig. 27.39. The Li^{+2} must be in second excited state and on absorbing 10.2 eV energy, it must be excited to $n = 6$ level and on deexcitation must give us 15 lines ($^6C_2$). Minimum energy difference between levels is

$$E_6 - E_5 = -3.4 - (-4.89) = 1.49 \text{ eV}$$

Hence maximum wavelength is

$$\lambda_{max} = \frac{hc}{E_6 - E_5} = \frac{1242}{1.49} \text{ nm} = 833.5 \text{ nm}$$

Maximum energy difference between levels is

$$E_6 - E_1 = -3.4 - (-122.4) = 119 \text{ eV}$$

and the minimum wavelength is

$$\lambda_{min} = \frac{hc}{E_6 - E_1} = \frac{1242}{119} \text{ nm} = 10.4 \text{ nm}$$

Example 27.20: A photon of energy 5.4852 eV liberates an electron from the Li atom initially at rest. The emitted electron moves at right angles to the direction in which photon moves. Find the speed and the direction in which the Li^{+2} ion will move. Ionization potential of Li atom is 5.3918 V. Atomic weight of lithium is 6.94 g. $n_0 = 6.02 \times 10^{23}$/mol and $m_e = 9.1 \times 10^{31}$ kg.

Solution: Energy of liberated electron is

$$\frac{1}{2}mv^2 = 5.4852 - 5.3918 = 0.0934 \text{ eV}$$

$$= 0.0934 \times 1.6 \times 10^{-19} \text{ J.}$$

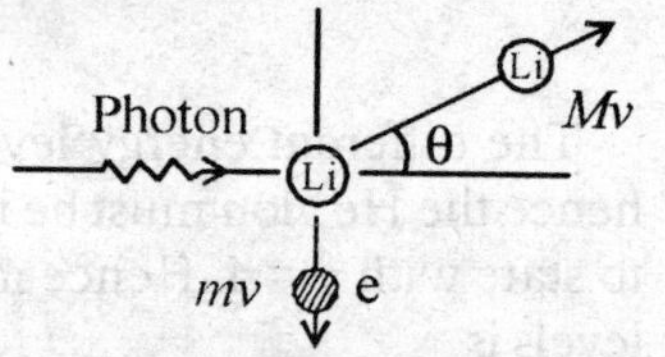

Fig. 27.40

$$\therefore \quad v = \sqrt{\frac{2 \times 0.0934 \times 1.6 \times 10^{-19}}{9.1 \times 10^{-31}}} = 0.18 \times 10^6 \text{ m/s}$$

If Li ion (mass M) moves with velocity V, then conservation of momentum gives (Fig. 27.40).

$$MV\sin\theta = mv \qquad \ldots(i)$$

and

$$MV\cos\theta = \frac{h\nu}{c} \qquad \ldots(ii)$$

$$\therefore \quad \tan\theta = \frac{mvc}{h\nu} = \frac{9.1 \times 10^{-31} \times 0.18 \times 10^6 \times 3 \times 10^8}{5.4852 \times 10^{-19} \times 1.6}$$

$$= 55.99.$$

or

$$\theta = 89°$$

and

$$V = \frac{mv}{M\sin\theta}$$

$$M = \frac{6.94 \times 10^{-3}}{6.02 \times 10^{23}}$$

$$\therefore \quad V = \frac{mv}{M\sin\theta} = \frac{9.1 \times 10^{-31} \times 0.18 \times 10^6 \times 6.02 \times 10^{23}}{6.94 \times 10^{-3} \times 0.999}$$

$$= 14.2 \text{ m/s.}$$

Example 27.21: Light from a discharge tube containing hydrogen atoms falls on the surface of a piece of sodium. The kinetic energy of fastest photo-electrons emitted from sodium is 0.73 eV and the work function of sodium is 1.82 eV. Find

(*a*) the energy of the photons causing the photoelectric emission.
(*b*) the quantum numbers of the two levels involved in the emission of these photons.
(*c*) the change in the angular momentum of the electron in hydrogen atom in above transition.
(*d*) the recoil speed of the emitting atom assuming it to be at rest before the transition.

(Ionization potential of hydrogen is 13.6 V and mass of hydrogen atom is 1.67×10^{-27} kg).

Solution:

(*a*) Einstein's photoelectric equation gives

$$h\nu = E_{max} + \phi_0$$

$h\nu$ is the energy of incident photon, E_{max} is the maximum kinetic energy of the photo-electrons ejected and ϕ_0 is work function of the metal surface.

$$\therefore \quad h\nu = 0.73 + 1.82 = 2.55 \text{ eV} = 2.55 \times 1.6 \times 10^{-19} \text{ J}$$

$$= 4.08 \times 10^{-19} \text{ J.}$$

(*b*) The energy of electron in *n*th orbit of atom (atomic number = *Z*) is

$$E_n = -\frac{13.6 Z^2}{n^2} \text{ eV} \quad \text{for Hydrogen } Z = 1.$$

and different energies are

$$E_1 = -13.6 \text{ eV}, \; E_2 = -3.4 \text{ eV}, \; E_3 = -1.51 \text{ eV}, \; E_4 = -0.85 \text{ eV, etc.}$$

the difference of levels given energy 2.55 eV is $(E_4 - E_2)$. Hence the quantum number for which, the transition gives photon of energy 2.55 eV are $n = 2$ to $n = 4$.

(*c*) The transition in hydrogen atom takes place from $n = 4$ to $n = 2$. Hence angular momentum in 4th orbit is $L_n = \dfrac{4h}{2\pi}$ and in 2nd orbit is $L_2 = \dfrac{2h}{2\pi}$

Thus change in angular momentum is

$$\Delta L = L_4 - L_2 = \frac{4h}{2\pi} - \frac{2h}{2\pi} = \frac{h}{\pi} = 2.11 \times 10^{-34} \text{ Js}$$

(*d*) The conservation of momentum gives that the momentum of emitted photon is equal to the recoil momentum of the atom. Thus

$$P = \frac{h\nu}{c}$$

If *M* is mass of recoil atom and *v* its velocity then

$$P = Mv = \frac{h\nu}{c}$$

or

$$v = \frac{h\nu}{Mc} \qquad \text{But } h\nu = 4.08 \times 10^{-19} \text{ J}$$

$$= \frac{4.08 \times 10^{-19}}{1.67 \times 10^{-27} \times 3 \times 10^8}$$

$$= \frac{4.08}{1.67 \times 3} = 0.81 \text{ ms}^{-1}$$

Example 27.22: A neutron of kinetic energy 65 eV collides inelastically with a singly ionized helium atom at rest. It is scattered at an angle of 90° with respect to its original direction.

(*a*) Find the allowed values of the energy of the neutron and that of the atom after the collision.

(*b*) If the atom gets de-excited subsequently by emitting radiation, find the frequencies of the emitted radiation.

[Given mass of the He atom = 4 × (mass of neutron), ionization energy of H atom = 13.6 eV].

Solution: Figure 27.41 depicts the scattering of the neutron. If m is mass of neutron, then mass of He atom is $4\,m$. The velocities of different particles are shown in Fig. 27.41.

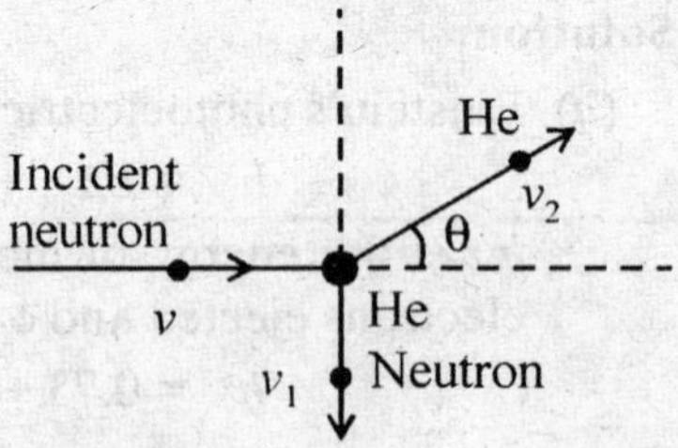

Fig. 27.41

Conservation of linear momentum gives

$$mv = 4mv_2\cos\theta \quad \text{or} \quad v = 4v_2\cos\theta \qquad \ldots(i)$$

$$mv_1 = 4mv_2\sin\theta \quad \text{or} \quad v_1 = 4v_2\sin\theta \qquad \ldots(ii)$$

and conservation of energy gives

$$\frac{1}{2}mv^2 = \frac{1}{2}mv_1^2 + \frac{1}{2}4mv_2^2 + Q$$

or
$$v^2 = v_1^2 + 4v_2^2 + \frac{2Q}{m} \qquad \ldots(iii)$$

where Q is excitation energy, equation (*i*) and (*ii*) gives

$$v^2 + v_1^2 = (4v_2)^2$$

Substituting value of $(4v_2)^2$ in (*iii*), we have

$$v^2 = v_1^2 + \frac{v^2}{4} + \frac{v_1^2}{4} + \frac{2Q}{m}$$

or
$$5v_1^2 = 3v^2 - \frac{8Q}{m}$$

If $K_1 = \frac{1}{2}mv^2$ and $K_2 = \frac{1}{2}mv_1^2$, the kinetic energies of neutron before and after scattering respectively then

$$5\left(\frac{1}{2}mv_1^2\right) = 3\left(\frac{1}{2}mv^2\right) - 4Q$$

or
$$5K_2 = 3K_1 - 4Q$$

or
$$K_2 = \frac{1}{5}(3K_1 - 4Q) \qquad \ldots(iv)$$

The energy of electron in nth orbit of the atom is given by

$$E_n = -\frac{13.6\,Z^2}{n^2} \quad \text{and } Z = 2$$

Hence excitation energy for transition from state n_1 to state n_2 is

$$Q = (13.6)(2)^2\left(\frac{1}{n_1^2} - \frac{1}{n_2^2}\right)\text{eV} = (54.4)\left(\frac{1}{n_1^2} - \frac{1}{n_2^2}\right)$$

We have $K_1 = 65$ eV, then from (*iv*)

$$K_2 = \frac{1}{5}\left[(3\times 65) - 4\times(54.4)\left(\frac{1}{n_1^2} - \frac{1}{n_2^2}\right)\right]$$

$$= \frac{1}{5}\left[195 - 217.6\left(\frac{1}{n_1^2} - \frac{1}{n_2^2}\right)\right] \text{ eV}$$

The energy of neutron after collision depends on the excitation energy of the electron in helium atom.

(*i*) For transition from $n_1 = 1$ to $n_2 = 2$, $\frac{1}{n_1^2} - \frac{1}{n_2^2} = 1 - \frac{1}{4} = \frac{3}{4}$

$$\therefore \quad K_2 = \frac{1}{5}\left[195 - 217.6 \times \frac{3}{4}\right] = 6.36 \text{ eV}$$

(*ii*) For transition from $n_1 = 1$ to $n_2 = 3$, $\frac{1}{n_1^2} - \frac{1}{n_2^2} = 1 - \frac{1}{9} = 8/9$

$$\therefore \quad K_2 = \frac{1}{5}\left[195 - 217.6 \times \frac{8}{9}\right] = 0.32 \text{ eV}$$

The transition from $n_1 = 1$ to $n = 4$ gives $\frac{1}{n_1^2} - \frac{1}{n_2^2} = 1 - \frac{1}{16} = \frac{15}{16}$ and $217.6 \times \frac{15}{9} = 20$ eV this gives K_2 negative which is not possible.

The kinetic energy of He atom after collision are

$$K_{He} = K_1 - K_2 - Q(1 \to 2)$$

$$Q(1 \to 2) = 54.4\left(1 - \frac{1}{4}\right) = 54.4 \times \frac{3}{4} = 40.8 \text{ eV}$$

$$\therefore \quad K_H = 65 - 6.36 - 40.8 = 17.84$$

In this case when He atom de-excites the frequency of radiation emitted is

$$\nu_1 = \frac{Q(1 \to 2)}{h} = \frac{40.8 \times 1.6 \times 10^{-19}}{6.6 \times 10^{-34}} = 9.89 \times 10^{15} \text{ Hz}$$

For Q (1 → 3) transition $Q = 54.4 \times \frac{8}{9} = 48.36$ eV

Hence kinetic energy of He atom is

$$K_{He} = K_1 - K_2 - Q - (1 \to 2) = 65 - 0.32 - 48.36 = 16.32 \text{ eV}$$

and the frequency of radiation emitted is

$$\nu_2 = \frac{48.36 \times 1.6 \times 10^{-19}}{6.6 \times 10^{-34}} = 1.17 \times 10^{16} \text{ Hz}.$$

Example 27.23: The peak emission from a black body at a certain temperature occurs at a wavelength of 9000 Å. On increasing the temperature, the total radiation is 81 times. At the initial temperature when the peak radiation from the black body is incident on a metal surface, it does not cause any photo emission from the surface. After the increase of temperature, the peak radiation from the black body caused photo emission. To bring these photo-electrons to rest, a potential equivalent to the excitation energy between $n = 2$ to $n = 3$, Bohr levels of hydrogen atom is required. Find the work function of the metal.

Solution: Since on raising temperature, the total radiation becomes 81 times. Using Stefans law, we have

$$\frac{R_1}{R_2} = \left(\frac{T_1}{T_2}\right) = \frac{1}{81}$$

∴ Final temperature $T_2 = 3T_1$

Using Wien's displacement law $\lambda_m T$ = constant. Since temperature is made 3 times, hence the wavelength of peak emission now becomes 1/3 of the initial one, i.e. 3000 Å.

Using Einstein's photoelectric equation, we have

$$\frac{hc}{\lambda} - \phi_0 = E_{max} \quad E_{max} \text{ is maximum kinetic energy.}$$

Excitation energy of transition $n = 2$ to $n = 3$ in Bohr model is

$$E_{ex} = 13.6\left(\frac{1}{2^2} - \frac{1}{3^2}\right) \text{eV} = 13.6 \times \frac{5}{36} = 1.89 \text{ eV}$$

$$\frac{hc}{\lambda} = \frac{6.6 \times 10^{-34} \times 3 \times 10^8}{3000 \times 10^{-10}} \text{ J} = \frac{6.6 \times 10^{-34} \times 3 \times 10^8}{3000 \times 10^{-10} \times 1.6 \times 10^{-19}} \text{ eV}$$

$$= 4.13 \text{ eV.}$$

The excitation energy is equal to the maximum kinetic energy of photo-electrons emitted, i.e. $E_{max} = 1.89$.

$$\therefore \qquad \phi_0 = \frac{hc}{\lambda} - E_{max} = 4.13 - 1.89 = 2.24 \text{ eV.}$$

Example 27.24: The radiation emitted when an electron jumps from $n = 3$ to $n = 2$ orbit of hydrogen atom falls on a metal to produce photo-electrons. The electrons emitted from the metal surface with maximum kinetic energy are made to move perpendicular to a magnetic field of $\frac{1}{320}$ *T* in a radius of 10^{-3} m. Find

(*a*) the kinetic energy of the electrons

(*b*) the work function of the metal, and

(*c*) the wavelength of the radiation.

Solution:

(*a*) When a particle of charge q moves in magnetic field it describes a circle. If r is the radius of the circle, then

$$\frac{mv^2}{r} = qvB \quad \text{or} \quad mv = qBr$$

and its kinetic energy is

$$E_K = \frac{1}{2}mv^2 = \frac{q^2B^2r^2}{2m}$$

$$= \frac{(1.6 \times 10^{-19} \times 1/320 \times 10^{-3})^2}{2 \times 9.1 \times 10^{-31}} = 0.86 \text{ eV}$$

(*b*) The energy of the photon emitted when an electron jumps from $n = 3$ to $n = 2$ orbit of hydrogen atom is

$$E = h\nu = 13.6\left(\frac{1}{2^2} - \frac{1}{3^2}\right) = 13.6 \times \frac{5}{36} = 1.89 \text{ eV}$$

From Einstein's photoelectric equation, we have

$$E = h\nu = E_K + \phi_0 \qquad \phi_0 \text{ is work function}$$

$$\therefore \qquad \phi_0 = E - E_K = 1.89 - 0.86 = 1.03 \text{ eV}$$

(*c*) $E = \dfrac{hc}{\lambda} = 1.89$ eV or $\lambda = \dfrac{hc}{1.89 \times 1.6 \times 10^{-19}}$

$$\therefore \qquad \lambda = \frac{6.6 \times 10^{-34} \times 3 \times 10^8}{1.89 \times 1.6 \times 10^{-19}} = 6.548 \times 10^{-7} \text{ m} = 6548 \text{ Å}$$

Example 27.25: Assume that the de Broglie wave associated with an electron can form a standing wave between the atoms arranged in a one-dimensional array with nodes at each of the atomic sites. It is found that one such standing wave is formed if the distance *d* between the atoms of array is 2 Å. A similar standing wave is again formed if *d* is increased to 2.5 Å but not for any intermediate value of *d*. Find the energy of the electrons (in electron volt) and the least value of *d* for which the standing wave of the type described above can form ($h = 6.6 \times 10^{-34}$ Js).

Solution: In standing waves the distance between two consecutive nodes is $\lambda/2$. If *d* is the distance and *n* is the mode of vibration, then

$$n\lambda/2 = d \quad \text{or} \quad n\lambda = 2d \qquad \ldots(i)$$

when *d* is increased to *d'* so that $(n + 1)$th mode is observed then

$$(n + 1)\lambda = 2d' \qquad \ldots(ii)$$

$$\therefore \qquad \frac{d}{d'} = \frac{n}{n+1} \quad \text{given is } d = 2 \text{ Å and } d' = 2.5 \text{ Å}$$

$$\therefore \qquad \frac{2}{2.5} = \frac{n}{n+1} \quad \text{or} \quad n = 4$$

Using (*i*), we have $\lambda = \dfrac{2d}{n} = \dfrac{2 \times 2}{4} = 1$ Å

If *P* is the momentum of the electron then its energy is

$$E = \frac{P^2}{2m}$$

but $P = \dfrac{h}{\lambda}$. Hence $E = \dfrac{h^2}{2m\lambda^2}$

$$\therefore \qquad E = \frac{(6.6 \times 10^{-34})^2}{2 \times 9.1 \times 10^{-31} \times (10^{-10})^2} = 2.4 \times 10^{-17} \text{ J} = 150 \text{ eV}$$

The least value of *d* is obtained by putting $n = 1$ in (*i*)

$$\therefore \qquad d_{min} = \lambda/2 = 0.5 \text{ Å}.$$

Example 27.26: Photo-electrons are emitted when 400 nm radiation is incident on a surface of work function 1.9 eV. These photo-electrons pass through a region containing α-particles. A maximum energy electron combines with an α-particle to form a He^+ ion, emitting a single photon in this process. He^+ ions thus formed are in their fourth excited state. Find the energies (in eV) of the photons, lying in the 2 to 4 eV range; that are likely to be emitted during and after the combination. Take $h = 4.14 \times 10^{-15}$ eV.

Solution: The energy of the incident photon is

$$E_{ph} = h\nu = \frac{hc}{\lambda} = \frac{4.14\times 10^{-15}\times 3\times 10^{8}}{400\times 10^{-9}} = 3.1 \text{ eV}$$

The maximum kinetic energy of electrons (E_{max}) from Einstein's photoelectric equation is

$$E_{max} = E_{ph} - \phi_0 \quad \text{where } \phi_0 \text{ is work function of surface}$$

$$= 3.1 - 1.9 = 1.2 \text{ eV}$$

We are given that

$$E_{max} + \text{He} \rightarrow \underset{\text{(in 4th excited state)}}{\text{He}^+} + \text{Photon}$$

Fourth excited state corresponds to $n = 5$. For helium, the energy of electron in nth orbit is

$$E_n = -\frac{13.6\,Z^2}{n^2} \text{ eV} \qquad \ldots(i)$$

and

$$E_5 = -\frac{13.6\times(2)^2}{(5)^2} = -2.18 \text{ eV}$$

The photon emitted in the process has energy

$$E = E_{max} - E_5 = 1.2 - (-2.18) = 1.2 + 2.18 = 3.38 \text{ eV}$$

This energy is within the range 2 to 4 eV.

After the recombination the helium ion (He^+) can deexcite to different levels. The energies in these levels are (from (*i*))

$$E_4 = -\frac{13.6\times 4}{16} = -3.4 \text{ eV}$$

$$E_3 = -\frac{13.6\times 4}{9} = -6.04 \text{ eV}$$

$$E_2 = -\frac{13.6\times 4}{4} = -13.6 \text{ eV}$$

$$E_1 = -\frac{13.6\times 4}{1} = -54.4 \text{ eV}$$

The difference of energy between ($E_5 - E_n$) where n = 4, 3, 2, 1, which lie between 2 to 4 eV is

(*i*) $E_4 - E_3 = -3.4 - (-6.04) = 2.64$ eV
(*ii*) $E_5 - E_3 = -2.18 - (-6.04) = 3.86$ eV

Thus the energies of photons that are likely to be emitted with energies in the range 2 to 4 eV are 2.64 eV, 3.38 eV and 3.86 eV.

Example 27.27: A gas of hydrogen like ion is prepared in such a way that the ions are only in the ground state and the first excited state. A monochromatic light of wavelength 1216 Å is absorbed by the ions. The ions are lifted to higher excited states and emit radiations of six different wavelengths; some higher and some lower than the incident wavelength. Find the principal quantum number of all the excited states. Identify the nuclear charge on the ions. Calculate the values of maximum and minimum wavelengths.

Solution: Since the ions are excited to levels from where on de-excitation emit six different spectral lines. Hence the highest excited level must have $n = 4$. The different transitions are shown in Fig. 27.42.

Since initially some of the ions are in ground state ($n = 1$) and first excited state ($n = 2$), on absorption of light, the excited states of the atom are $n = 3$ and $n = 4$.

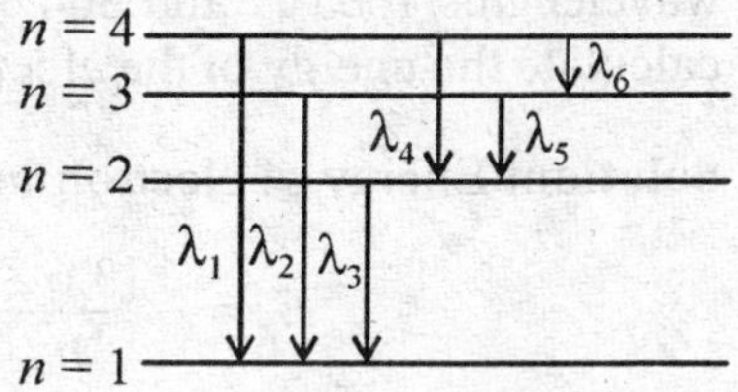

Fig. 27.42

Energy of the electron in the nth state is

$$E_n = -\frac{13.6\,Z^2}{n^2} \qquad \ldots(i)$$

Radiation of wavelength $\lambda = 1216$ Å has photons of energy

$$E_{ph} = \frac{hc}{\lambda} = \frac{6.6\times10^{-34}\times3\times10^{8}}{1216\times10^{-10}}\ \text{J} = 16.3\times10^{-19}\ \text{J}$$

$$= \frac{16.3\times10^{-19}}{1.6\times10^{-19}}\ \text{eV} = 10.2\ \text{eV}$$

This photon excites the ion from $n = 2$ to $n = 4$. Hence

$$13.6\,Z^2\left(\frac{1}{2^2}-\frac{1}{4^2}\right) = 10.2$$

or

$$Z^2 = \frac{10.2\times16}{3\times13.6} = 4 \quad \text{or} \quad Z = 2$$

The nucleus contains 2 positive charges, i.e. it has 2 protons.

The maximum wavelength is emitted in transition from $n = 4$ to $n = 3$.

$$\therefore \qquad \frac{hc}{\lambda_{max}} = 13.6\times1.6\times10^{-19}\times(2)^2\left(\frac{1}{3^2}-\frac{1}{4^2}\right)$$

$$= \frac{7}{144}\times13.6\times1.6\times10^{-19}\times4$$

or

$$\lambda_{max} = \frac{hc\times144}{4\times7\times13.6\times1.6\times10^{-19}}$$

$$= \frac{6.6\times10^{-34}\times3\times10^{8}\times144}{4\times7\times13.6\times1.6\times10^{-19}} = 4.679\times10^{-7}\ \text{m}$$

$$= 4679\ \text{Å}.$$

The minimum wavelength is emitted in transition from $n = 4$ to $n = 1$

$$\therefore \qquad \frac{hc}{\lambda_{min}} = 13.6\times1.6\times10^{-19}\times(2)^2\left(\frac{1}{1^2}-\frac{1}{4^2}\right)$$

or $$\lambda_{min} = \frac{hc \times 16}{13.6 \times 1.6 \times 4 \times 10^{-19} \times 15}$$

$$= \frac{6.6 \times 10^{-34} \times 3 \times 10^{8} \times 16}{13.6 \times 1.6 \times 4 \times 10^{-19} \times 15} = 0.243 \times 10^{-7} \text{ m}$$

$$= 243 \text{ Å}.$$

Example 27.28: A 100 eV electron collides with a stationary helium ion (He^+) in its ground state and excites to a higher level. After the collision He^+ ion emits two photons in succession with wavelengths 1085 Å and 304 Å. Find the principal quantum number of the excited state. Also calculate the energy of the electron after the collision.

Solution: Energy of electron in *n*th orbit of He^+ ion is given by

$$E_n = -\frac{13.6\, Z^2}{n^2} = -\frac{54.4}{n^2} \text{ eV} \qquad \ldots(i)$$

The photon of wavelength λ = 1085 Å has energy.

$$E_1 = -\frac{12470}{\lambda(\text{Å})} \text{ eV} = \frac{12370}{1080} = 11.5 \text{ eV} \qquad \ldots(ii)$$

and the photon of wavelength λ = 304 has energy,

$$E_2 = \frac{12370}{304} = 40.7 \text{ eV} \qquad \ldots(iii)$$

Total energy emitted in the form of photons is

$$E = E_1 + E_2 = 11.5 + 40.7 = 52.2 \text{ eV}$$

If n be the principal quantum number of the excited state, then transition from this level to n = 1 will release energy

$$\Delta E = 54.4\left(\frac{1}{1^2} - \frac{1}{n^2}\right) \text{eV}$$

and this energy must be equal to 52.2 eV. Hence equating, we have

$$52.2 = 54.4\left(1 - \frac{1}{n^2}\right)$$

or $$1 - \frac{1}{n^2} = \frac{52.2}{54.4} = 0.96$$

or $$\frac{1}{n^2} = 1 - 0.96 = 0.04 \quad \text{or} \quad n = 5$$

Energy of incident electron 100 eV. Energy given to He^+ ion for excitation is 52.2 eV. The energy left with the electron for collision is

$$E = 100 - 52.2 = 47.8 \text{ eV}$$

Example 27.29: A particle of mass m, atomic number Z', initial velocity v and impact parameter b is scattered by a heavy nucleus of atomic number Z. Use principle of conservation of angular momentum and energy to obtain a relation between the minimum distances of the particle from the nucleus in terms of Z, Z', v and b. Show that for $b = 0$, s reduces to the distance of closest approach r_0 given by

$$r_0 = \frac{1}{4\pi\varepsilon_0}\frac{2ZZ'e^2}{mv^2}$$

Solution: Particle of charge $q_1 (=Z'e)$ is used for scattering. Figure 27.43 shows the trajectory of the particle scattered by the nucleus of charge $q_2 = Ze$. The impact parameter is b.

When particle is far away from the nucleus its energy is entirely kinetic and is

$$E_K = \frac{1}{2}mv^2 \quad ...(i)$$

When it approaches the nucleus the repulsive force of nucleus acts on it as such kinetic energy decreases and the potential energy increases. At the least distance s from the nucleus, let its velocity be v'. v' will be tangential to the trajectory. The total energy will be

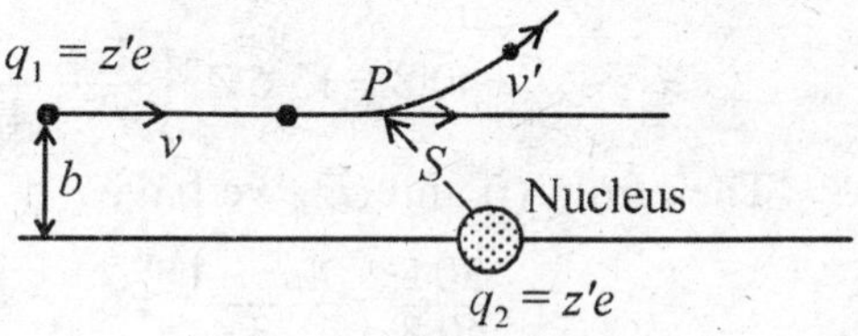

Fig. 27.43

$$E = \text{K.E.} + \text{P.E.} = \frac{1}{2}mv'^2 + \frac{1}{4\pi\varepsilon_0}\cdot\frac{(Z'e)(Ze)}{s}$$

and the conservation of energy gives

$$\frac{1}{2}mv^2 = \frac{1}{2}mv'^2 + \frac{Z'Ze^2}{4\pi\varepsilon_0 s} \quad ...(ii)$$

At P the point of least distance velocity is tangential to the trajectory and is perpendicular to s—the radius vector. Conservation of angular momentum gives

$$mvb = mv's \quad \text{or} \quad v' = \frac{vb}{s}$$

Substitution in (*ii*) gives

$$\frac{1}{2}mv^2 = \frac{1}{2}m\cdot\frac{v^2b^2}{s^2} + \frac{Z'Ze^2}{4\pi\varepsilon_0 s}$$

or

$$s^2 = b^2 + \frac{2Z'Ze^2 s}{4\pi\varepsilon_0 mv^2}$$

For head on collision $b = 0$, and this gives

$$r_0 = s = \frac{2Z'Ze^2}{4\pi\varepsilon_0 mv^2}$$

Example 27.30: A hydrogen like atom of the atomic number Z is in an excited state of quantum number $2n$. It can emit a maximum energy photon of 204 eV. If it makes a transition to quantum state n, a photon of energy 40.8 eV is emitted. Find n, Z and the ground state energy (in eV), for this atom. Also calculate the minimum energy (in eV) that can be emitted by this atom during de-excitation. Ground state energy of hydrogen atom is –13.6 eV. (IIT 2000)

Solution: The energy of electron in nth orbit of hydrogen like atom of atomic number Z is

$$E_n = -\frac{13.6\,Z^2}{n^2}$$

For energy transition from $n_2 = 2n$ to $n_1 = 1$, the energy of the emitted photon will be maximum. Thus

$$E_{max} = 13.6\,Z^2\left(\frac{1}{1} - \frac{1}{4n^2}\right) \text{eV}$$

and this $E_{max} = 204$ eV

$$\therefore \qquad 204 = 13.6\,Z^2\left(1 - \frac{1}{4n^2}\right) = \frac{13.6\,Z^2\,(4n^2 - 1)}{4n^2} \qquad \ldots(i)$$

For transition from $n_2 = 2n$ to $n_1 = n$; the photon of energy 40.8 eV is emitted. Hence

$$40.8 = 13.6\,Z^2\left(\frac{1}{n^2} - \frac{1}{4n^2}\right) = \frac{3 \times 13.6\,Z^2}{4n^2} \qquad \ldots(ii)$$

Then from (*i*) and (*ii*), we have

$$\frac{204}{40.8} = \frac{4n^2 - 1}{3} = 5$$

or $\qquad 4n^2 - 1 = 15$ or $4n^2 = 16$ and $n = 2$.

Using this value of n in (*ii*), we get

$$40.8 = \frac{3 \times 13.6 Z^2}{4n^2} \qquad \text{for } n = 2$$

or $$Z^2 = \frac{40.8 \times 4 \times 4}{3 \times 13.6} = 16$$

or $\qquad Z = 4$

The ground state energy is obtained by putting $n = 1$.

$$\therefore \qquad E_1 = -\frac{13.6\,Z^2}{n^2} = -\frac{13.6\,(4)^2}{1} = 217.6 \text{ eV}$$

Since $n = 2$, the given excited state of atom is $2n = 4$.

For minimum energy photon emission, the transition must take place from $n_2 = 4$ to $n_1 = 3$. Thus

$$E_{min} = -13.6\,Z^2\left(\frac{1}{4^2} - \frac{1}{3^2}\right)$$

$$= +13.6\,(4)^2\left(\frac{1}{9} - \frac{1}{16}\right) = \frac{7}{144} \times 217.6 = 10.58 \text{ eV}$$

Example 27.31: When a beam of 10.6 eV photons of intensity 2.0 W/m^2 falls on a platinum surface of area 1.0×10^{-4} m^2 and work function 5.6 eV, 0.53% of the incident photons eject photo-electrons. Find the number of photo-electrons emitted per second and their energies (in eV).

(IIT 2000)

Solution: Energy of photons beam striking the surface in 1 sec is

$$E = \text{intensity} \times \text{area of the surface} = 2 \times 10^{-4} \text{ J}.$$

Energy carried by each photon = 10.6 eV = $10.6 \times 1.6 \times 10^{-19}$ J.
And the number of photons striking the surface per second is

$$n = \frac{2 \times 10^{-4}}{10.6 \times 1.6 \times 10^{-19}} = 1.18 \times 10^{14}$$

It is given that 0.53% of the incident photons eject photo-electrons. Hence number of photo-electrons emitted per second is

$$n_e = \frac{n \times 0.53}{100} = \frac{1.18 \times 10^{14} \times 0.53}{100} = 6.25 \times 10^{11}$$

Since work function of platinum is 5.6 eV, the minimum energy of the emitted photo-electrons is zero and maximum energy is

$$E_{max} = 10.6 - 5.6 = 5 \text{ eV}.$$

Example 27.32: A hydrogen like atom (described by Bohr model) is observed to emit six wavelengths, originating from all possible transitions between a group of levels. The levels have energies between –0.85 eV and –0.544 eV (including with these values)

(*a*) Find the atomic number of the atom.

(*b*) Calculate the smallest wavelength emitted in these transitions. (Take hc = 1240 eV nm ground state energy of hydrogen atom is –13.6 eV). (IIT 2002)

Solution: Four energy levels are required for six transitions. Figure 27.44 shows these transitions.
The energy of electron in nth orbit of hydrogen like atom of atomic number Z is

$$E_n = -\frac{13.6 Z^2}{n^2} \text{ eV} \quad \ldots(i)$$

$$\therefore \quad E_{n+3} = -\frac{13.6 Z^2}{(n+3)^2} \text{ eV} \quad \ldots(ii)$$

$$\therefore \quad \frac{E_n}{E_{n+3}} = \frac{-0.85}{-0.544} = \frac{(n+3)}{n^2}$$

$$\text{or} \quad \frac{n+3}{n} = \sqrt{\frac{85}{54.4}} = 1.25$$

$$\text{or} \quad n = 12$$

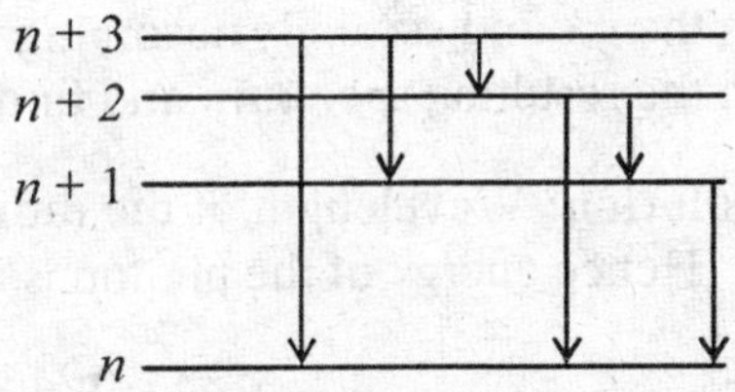

Fig. 27.44

Using (*i*) and substituting $n = 12$, $E_n = -0.85$ eV, we get

$$-0.85 = -\frac{13.6 Z^2}{(12)^2}$$

$$\text{or} \quad Z^2 = \frac{0.85 \times (12)^2}{13.6} = 9$$

$$\text{or} \quad Z = 3$$

The smallest wavelength corresponds to the maximum energy difference.

$$E_{max} = 0.85 - 0.544 = 0.306 \text{ eV}$$

$$\text{and} \quad \lambda_{min} = \frac{hc}{E_{max}} = \frac{1240}{0.306} = 4052 \text{ nm}$$

Example 27.33: An insulated conducting shell is exposed to a radiation of wavelength λ. The threshold wavelength of the conducting shell is $\lambda_0 > \lambda$.

(*a*) Write the condition when emission of photo-electrons will stop. Initially switch S is open.

(*b*) By closing the switch, a capacitor is connected to the sphere (Fig. 27.45). Find the charge on the capacitor after a long time if it was uncharged initially.

Solution: The incident radiation has wavelength (λ) less than the threshold wavelength (λ_0), the photo-electrons are emitted from the surface of the shell and surface gets positively charged. The emitted electrons are then attracted towards the sphere. When the potential of the sphere becomes equal to the stopping potential of the photo-electrons emitted, the emission of the photo-electrons will stop. If V is the stopping potential then

$$\frac{hc}{\lambda} = \frac{hc}{\lambda_0} + eV$$

or

$$V = \frac{hc}{e}\left[\frac{1}{\lambda} + \frac{1}{\lambda_0}\right]$$

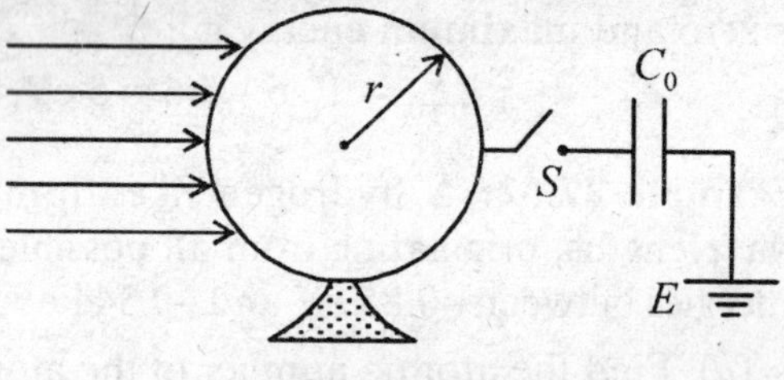

Fig. 27.45

The charge required to make the potential V of the sphere of radius r is

$$Q = 4\pi\varepsilon_0 rV$$

When the capacitor is connected to the sphere, charge on the capacitor is

$$q = C_0V = \frac{C_0hc}{e}\left[\frac{1}{\lambda} - \frac{1}{\lambda_0}\right]$$

Example 27.34: A monochromatic radiation of wavelength 975 Å is incident on a hydrogen atom in the ground state. By absorbing this radiation the atom is excited. Find the total number of lines in the resulting spectrum and find the longest wavelength in spectrum.

Solution: Wavelength of the radiation is 975 Å.

Hence energy of the photon is

$$E_{Ph} = \frac{hc}{\lambda} = \frac{12460\,(\text{eV. Å})}{975\,(\text{Å})} = 12.72 \text{ eV}$$

For hydrogen atom, the energy of electron in nth orbit is

$$E_n = -\frac{13.6}{n^2} \text{ eV}$$

Hence in order to excite this atom to nth level, energy required is

$$E = E_n - E_1 = 13.6\left[\frac{1}{1^2} - \frac{1}{n^2}\right]\text{eV}$$

But energy absorbed is 12.72 eV

$$\therefore \quad 12.72 = 13.6\left[\frac{1}{1^2} - \frac{1}{n^2}\right]$$

gives $\quad n = 4.$

As such 6 lines will be observed when this atom de-excites. Longest wavelength is observed when energy difference between levels is least. Hence

$$\frac{hc}{\lambda_{max}} = E_4 - E_3 = 13.6\left(\frac{1}{9} - \frac{1}{16}\right) = 13.6 \times \frac{7}{144} \text{ eV}$$

or
$$\lambda_{max} = \frac{hc \times 144}{13.6 \times 7 \times 1.6 \times 10^{-19}} = \frac{6.6 \times 10^{-34} \times 3 \times 10^8 \times 144}{13.6 \times 7 \times 1.6 \times 10^{-19}}$$

$$= 18.72 \times 10^{-7} \text{ m} = 18720 \text{ Å}.$$

Example 27.35: How many photons having wavelength $\lambda = 5 \times 10^{-7}$ m must fall per second on a blackened metallic plate so that a force of magnitude 10^{-5} N is produced on the plate? What is the rate of rise in temperature if mass of plate is 1.5 g and specific heat 0.1 k-cal/kg°C.

Solution: Momentum of each photon $P = \frac{h}{\lambda}$. Since plate is a black body as such photon will be absorbed by the plate. Hence each photon transferred momentum h/λ to the plate. If n photons fall on the plate per second, then total momentum transferred to the plate per second is $h/\lambda n$. From Newtons second law this is the force.

$$\therefore \quad F = \frac{h}{\lambda} n$$

or
$$n = \frac{F \cdot \lambda}{h} = \frac{10^{-5} \times 5 \times 10^{-7}}{6.6 \times 10^{-34}} = 7.6 \times 10^{21} \text{ /s}$$

Each photon has energy $h\nu = \frac{hc}{\lambda}$. Hence total energy absorbed is

$$E = \frac{nhc}{\lambda}$$

This energy is used in raising the temperature of the plate and if rise in temperature is θ then heat energy required is

$$H = ms\theta.$$

when m is mass of plate, s its specific heat.

$$H = 1.5 \times 10^{-3} \times 0.1 \times \theta \text{ k-cal} \qquad \text{But 1 cal} = 4.2 \text{ J}$$

$$= 1.5 \times 10^{-3} \times 0.1 \times \theta \times 10^3 \times 4.2 = 1.5 \times 0.1 \times 4.2\,\theta$$

$$= 0.63\theta \text{ J}$$

Equating the two, we have

$$\frac{nhc}{\lambda} = 0.63\theta$$

or
$$\theta = \frac{nhc}{\lambda \times 0.63} = \frac{7.6 \times 10^{21} \times 6.6 \times 10^{-34} \times 3 \times 10^8}{5 \times 10^{-7} \times 0.63}$$

$$= 4.8 \times 10^3 \text{ °C/s}.$$

Example 27.36: Consider a singly ionized helium atom.

(*a*) Compute the four lowest energy levels and construct the energy level diagram.

(*b*) What is the minimum energy of photon that can be absorbed by such atom in the state $n = 2$?

(*c*) What wavelengths (in Å) can be emitted when these atoms in the ground state are bombarded by electrons that have been accelerated through a potential difference of 50 V?

(*d*) If these atoms are in the ground state, can they absorb radiation having a wavelength of 350 Å?

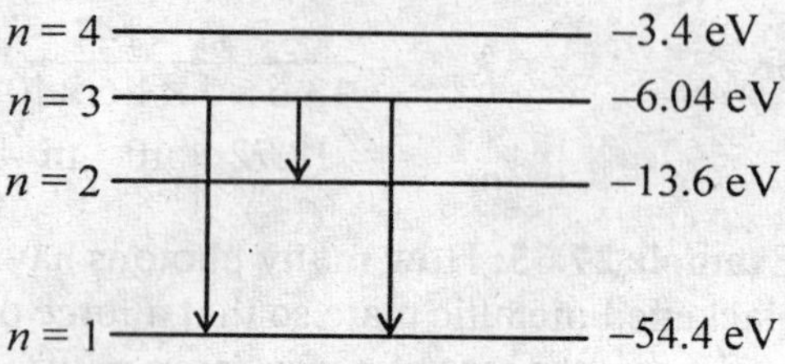

Fig. 27.46

Solution:

(*a*) Energy of electron in *n*th orbit of hydrogen like atom of atomic number *Z* is

$$E_n = -\frac{13.6\,Z^2}{n^2}\ \text{eV} = -\frac{54.4}{n^2}\ \text{eV} \quad Z = 2 \text{ for He}^+$$

The lowest energy levels are

$$E_1 = -54.4 \text{ eV},\ E_2 = -13.6 \text{ eV},\ E_3 = 6.04 \text{ eV and } E_4 = -3.4 \text{ eV}$$

The energy level diagram is shown in Fig. 27.46.

(*b*) The He^+ in state $n = 2$ has energy −13.6 eV and state with $n = 3$ has energy −6.04. Hence the minimum energy of photon that can be absorbed by this atom is

$$E_3 - E_2 = -6.04 - (-13.6) = 7.56 \text{ eV}$$

(*c*) When this atom is bombarded by an electron accelerated by a potential difference of 50 V (energy = 50 eV), then it can be excited to $n = 2$ and $n = 3$ states. It cannot be excited to $n =$ 4 state as $E_4 - E_1 = 51$ eV (>50 eV). The possible transitions on de-excitation are shown in Fig. 27.46. The wavelengths of the lines shown are given by

$$\lambda = \frac{12400}{\text{Energy difference in eV}}\ \text{Å}$$

Thus different wavelengths are

$$\lambda_1 = \frac{12400}{54.4 - 6.04} = \frac{12400}{48.36} = 256.5 \text{ Å}$$

$$\lambda_2 = \frac{12400}{54.4 - 13.6} = \frac{12400}{40.8} = 304 \text{ Å}$$

$$\lambda_3 = \frac{12400}{13.6 - 6.04} = \frac{12400}{7.56} = 1640 \text{ Å}$$

(*d*) A photon of wavelength 350 Å has energy

$$E = \frac{12400}{350} = 35.43 \text{ eV}$$

Since minimum energy required for excitation is 54.4 − 13.6 = 40.8 eV as such this photon cannot be absorbed by this atom.

Example 27.37: A light source of frequency ν illuminates a metallic surface and ejects photo-electrons. The photo-electrons having maximum energy are just able to excite the hydrogen atoms in the ground state. When the whole experiment is repeated with an incident radiation of frequency $\frac{6}{5}\nu$, the photo-electrons so emitted are about to excite the hydrogen atoms which then emit a radiation of six different wavelengths.

(*a*) Find the work function of the metal.
(*b*) What is the frequency of radiation?

Solution: The photo-electrons emitted are just able to excite the hydrogen atoms in ground state. Thus energy required is 13.6 – 3.4 = 10.2 eV. Hence for photoelectric emission, we have

$$h\nu - \phi_0 = E_{max} = 10.2 \text{ eV} \quad \ldots(i)$$

The radiations of frequency $\frac{6}{5}\nu$ emits photo-electrons which excite the hydrogen atom from $n = 1$ to $n = 4$ because on de-excitation it gives 6 different wavelengths. This means that the photo-electrons now emitted have maximum kinetic energy

$$E_{max} = E_4 - E_1 = 13.6\left(1 - \frac{1}{16}\right) = 13.6 \times \frac{15}{16} = 12.75 \text{ eV}$$

So we have for photoelectric emission

$$\frac{6}{5}h\nu - \phi_0 = E_{max} = 12.75 \text{ eV} \quad \ldots(ii)$$

Subtracting (*i*) from (*ii*), gives

$$\frac{1}{5}h\nu = 12.75 - 10.2 = 2.55 \text{ eV} = 2.55 \times 1.6 \times 10^{-19} \text{ J}$$

or

$$\nu = \frac{5 \times 2.55 \times 1.6 \times 10^{-19}}{6.6 \times 10^{-34}} = 3.09 \times 10^{15} \text{ Hz}$$

This gives $h\nu = 2.55 \times 5 = 12.75$ eV

Using in (*i*) we get

$$12.75 - \phi_0 = 10.2$$

or

$$\phi_0 = 12.75 - 10.2 = 2.55 \text{ eV}.$$

Example 27.38: Photons of energy 3 eV fall on a photo sensitive metal of work function 1 eV. The de Broglie wavelength of the most energetic ejected electron is found to be $(1.09)^2\sqrt{10}$ times the wavelength K_α, *X*-ray coming from a certain element *A* when it is bombarded by fast moving electron. Find the atomic number (*Z*) of the element A. Rydberg's constant = 1.09×10^7 m^{-1}.

Solution: Photo-electrons ejected from the photo sensitive surface have maximum energy

$$E_{max} = 3 - 1 = 2 \text{ eV} = 2 \times 1.6 \times 10^{-19} \text{ J}$$

and its wavelength (de Broglie wavelength)

$$\lambda = \frac{h}{P} = \frac{h}{\sqrt{2mE_{max}}}$$

If atomic number of element is *Z*, then the wavelength of the K_α line (λ_α) is

$$\frac{1}{\lambda_\alpha} = R(Z-1)^2\left(\frac{1}{1^2} - \frac{1}{2^2}\right) = \frac{3}{4}R(Z-1)^2$$

$$\therefore \quad \lambda_\alpha = \frac{4}{3R(Z-1)^2}$$

But $\lambda = (1.09)^2\sqrt{10}\ \lambda_\alpha$. Equating we have

$$\frac{h}{\sqrt{2mE_{max}}} = \frac{(1.09)^2\sqrt{10} \times 4}{3R(Z-1)^2}$$

$$(Z-1)^2 = \frac{(1.09)^2 \times \sqrt{10} \times 4 \times \sqrt{2mE_{max}}}{3R \times h}$$

$$= \frac{(1.09)^2 \times \sqrt{10} \times 4 \times \sqrt{2 \times 9.1 \times 10^{-31} \times 2 \times 1.6 \times 10^{-19}}}{3 \times 1.09 \times 10^7 \times 6.6 \times 10^{-34}}$$

$$= 5.314 \times 10^2$$

or $\quad Z - 1 = 23$

or $\quad Z = 24.$

Example 27.39: The radiation emitted, when an electron jumps from $n = 4$ to $n = 3$ in a lithium atom, falls on a metal surface to produce photo-electrons. When the photo-electrons with maximum K.E. are made to move perpendicular to the magnetic field 4×10^{-4} T, it traces out a circular path of radius 1.68 cm.

(*a*) Find the wavelength of radiation falling on the atom

(*b*) Find kinetic energy of electrons and work function of the metal.

Solution:

(*a*) Electron in *n*th orbit of lithium ($Z = 3$) has energy

$$E_n = -\frac{13.6\,Z^2}{n^2} = -\frac{13.6 \times 9}{n^2}$$

Hence radiation emitted due to transition from $n = 4$ to $n = 3$ has energy

$$\Delta E = E_4 - E_3 = 13.6 \times 9\left(\frac{1}{9} - \frac{1}{16}\right) = \frac{13.6 \times 7}{16}$$

$$= 5.95 \text{ eV}$$

Wavelength of the radiation is

$$\frac{hc}{\lambda} = 5.95 \times 1.6 \times 10^{-19} \text{ J}$$

or $$\lambda = \frac{hc}{5.95 \times 1.6 \times 10^{-19}} = \frac{6.6 \times 10^{-34} \times 3 \times 10^8}{5.95 \times 1.6 \times 10^{-19}}$$

$$= 2.08 \times 10^{-7} \text{ m} = 208 \text{ nm}.$$

(b) When electron enters the magnetic field it describes circular motion and the necessary centrepetal force $\left(= \frac{mv^2}{r}\right)$ is provided by magnetic force (qvB). Thus

$$\frac{mv^2}{r} = qvB \quad \text{or} \quad mv = qBr$$

and $$K.E. = \frac{1}{2}mv^2 = \frac{q^2B^2r^2}{2m}$$

$$= \frac{(1.6 \times 10^{-19})^2 \times (4 \times 10^{-4})^2 \times (1.68 \times 10^{-2})^2}{2 \times 9.1 \times 10^{-31}} \text{ J}$$

$$= \frac{1.6\times10^{-19}\times16\times10^{-8}\times(1.68)^2\times10^{-4}}{2\times9.1\times10^{-31}}\text{ eV}$$

$$E_n = 3.97\text{ eV}$$

Work function is

$$\phi_0 = \Delta E - E_k = 5.95 - 3.97 = 1.8\text{ eV}$$

Example 27.40: A gas having hydrogen like atoms, emits radiation in the Lyman series. If the light waves corresponding to the first and second lines of the Lyman series fell on a photo sensitive plate of work function ϕ_0 the de Broglie wavelengths of most energetic electrons are found to be 2.033 Å and 1.852 Å respectively. Find the atomic number of hydrogen like atom and the work function of the photo sensitive plate.

Solution: The energy levels in a hydrogen like atom of atomic number Z are given by

$$E_n = -\frac{13.6\,Z^2}{n^2}$$

Hence energy of first Lyman line is

$$\Delta E_1 = E_2 - E_1 = 13.6\,Z^2\left(\frac{1}{1^2}-\frac{1}{2^2}\right) = \frac{13.6\times3\,Z^2}{4}\text{ eV}$$

Similarly energy of second Lyman line is

$$\Delta E_2 = E_3 - E_1 = 13.6\,Z^2\left(\frac{1}{1^2}-\frac{1}{3^2}\right) = \frac{13.6\times8\,Z^2}{9}\text{ eV}$$

If the de Broglie wavelength of electron is λ then momentum is

$$P = \frac{h}{\lambda}$$

and energy is $$E = \frac{P^2}{2m} = \frac{h^2}{2m\lambda^2}$$

The first line gives electrons of maximum energy.

$$K_1 = \frac{h^2}{2m\lambda_1^2} = \frac{(6.6\times10^{-34})^2}{2\times9.1\times10^{-31}\times(2.033\times10^{-10})^2}\text{ J}$$

$$= 36\text{ eV}$$

and similarly the second line gives the electrons of maximum kinetic energy

$$K_2 = \frac{h^2}{2m\lambda_2^2} = \frac{(6.6\times10^{-34})^2}{2\times9.1\times10^{-31}\times(1.852\times10^{-10})^2}\text{ J}$$

$$= 43.6\text{ eV}$$

Using photoelectric equation

$$\frac{13.6\times3}{4}Z^2 = \phi_0 + 36 \qquad \ldots(i)$$

and $$\frac{13.6\times 8}{9}Z^2 = \phi_0 + 43.6 \qquad ...(ii)$$

$$\therefore \quad 13.6\,Z^2\left(\frac{8}{9}-\frac{3}{4}\right) = 7.6$$

or $$Z = 2$$

Using equation (*i*), we have

$$\phi_0 = \frac{13.6\times 3}{4}(2)^2 - 36$$
$$= 40.8 - 36.0 = 4.8 \text{ eV}.$$

Example 27.41: A photoelectric plate is initially exposed to a spectrum of hydrogen gas excited to second energy level. Later when the same photoelectric plate is exposed to a spectrum of some unknown hydrogen like gas, excited to second energy level, it is found that the de Broglie wavelength of the photo-electrons now ejected has decreased $\sqrt{6.1}$ times. For this new gas difference of energies of first Lyman line and Balmer series limit is found to be two times the ionization potential of hydrogen atom. Detect the atom and determine the work function of the photoelectric plate. For minimum de Broglie wavelength of electron, the momentum and hence K.E. of electron must be maximum.

Solution: The energy of hydrogen spectrum excited to second energy level ($n = 2$) is $E_2 - E_1 = 13.6 - 3.4 = 10.2$ eV.

If ϕ_0 is work function and K_1 maximum kinetic energy of photo-electrons emitted from the plate then

$$\phi_0 + K_1 = 10.2 \qquad ...(i)$$

Let atomic number of unknown hydrogen like atom be Z, then the energy levels are given by

$$E_n = -\frac{13.6\,Z^2}{n^2}$$

The energy of radiation emitted due to excitation of this atom to second energy level is

$$E_2 - E_1 = 13.6\,Z^2\left(\frac{1}{1^2}-\frac{1}{2^2}\right) = \frac{13.6\times 3Z^2}{4} \text{ eV}$$

This emits photo-electrons from plate with maximum kinetic energy K_2 (say). Then

$$\phi_0 + K_2 = 10.2\,Z^2 \qquad ...(ii)$$

An electron with kinetic energy K has de Broglie wavelength

$$\lambda = \frac{h}{P} = \frac{h}{\sqrt{2mK}}$$

Hence the photo-electrons due to two radiations have wavelengths

$$\lambda_1 = \frac{h}{\sqrt{2mK_1}} \quad \text{and} \quad \lambda_2 = \frac{h}{\sqrt{2mK_2}}$$

but $$\lambda_1 = \sqrt{6.1}\,\lambda_2$$

$$\therefore \quad \frac{h}{\sqrt{2mK_1}} = \sqrt{6.1}\frac{h}{\sqrt{2mK_2}}$$

or $$K_2 = 6.1K_1 \qquad \ldots(iii)$$

For this atom the first Lyman line has energy

$$E_2 - E_1 = 13.6\,Z^2\left(1-\frac{1}{4}\right) = \frac{13.6\times 3}{4}Z^2$$

and the Balmer series limit is

$$E_2 - E_\infty = 13.6\,Z^2\left(\frac{1}{4}-\frac{1}{\infty}\right) = \frac{13.6}{4}Z^2$$

It is given that the difference of these is twice the ionization energy of hydrogen atom, i.e. 2 × 13.6 eV.

$$\therefore \quad \frac{13.6\times 3}{4}Z^2 - \frac{13.6}{4}Z^2 = 2\times 13.6$$

or $$Z^2 = 4 \text{ or } Z = 2$$

i.e. atom is helium.

Using (*iii*) in (*i*) and (*ii*) with $Z = 2$, we get

$$\phi_0 + K_1 = 10.2$$

and $$\phi_0 + 6.1K_1 = 40.8$$

$$\therefore \quad 5.1K_1 = 30.6 \quad \text{or} \quad K_1 = 6 \text{ eV}$$

$$\therefore \quad \phi_0 = 10.2 - K_1 = 10.2 - 6 = 4.2 \text{ eV}$$

Example 27.42: A photo cell is operating in saturation mode with a photo current 4.8 mA. When a monochromatic radiation of wavelength 3000 Å and power 1 W is incident. When another monochromatic radiation of wavelength 1650 Å and power 5 mW is incident, it is observed that maximum velocity of photoelectrons increases to two times. Assuming efficiency of photoelectrons generation per incident photon to be same for both the cases, calculate

(*a*) threshold wavelength for the cell
(*b*) saturation current in second case
(*c*) efficiency of photo-electron generation per incident photon.

Solution:

(*a*) Radiation of wavelength $\lambda_1 = 3000$ Å has energy

$$E_1 = \frac{hc}{\lambda_1} = \frac{12400}{3000} = 4.13 \text{ eV}$$

Similarly radiation of wavelength $\lambda_2 = 1650$ has energy

$$E_2 = \frac{hc}{\lambda_2} = \frac{12400}{1650} = 7.52 \text{ eV}$$

If radiation of wavelength λ_1, ejects electrons of maximum K.E. K_1 then radiation of wavelength λ_2 ejects photo-electrons of maximum kinetic energy $K_2 = 4K_1$ (double the velocity). If ϕ_0 is the work function of the surface, then

$$E_1 = \phi_0 + K_1 \quad \text{or} \quad \phi_0 + K_1 = 4.13 \text{ eV}$$

$$E_2 = \phi_0 + 4K_1 \quad \text{or} \quad \phi_0 + 4K_1 = 7.52 \text{ eV}$$

Solving $\phi_0 = 3$ eV

Hence threshold wavelength is $\lambda_0 = \dfrac{12400}{3} = 4133.3$ Å

(*b*) Photon of wavelength 3000 Å has energy $E_1 = \dfrac{hc}{\lambda_1}$

Hence source of 1 W emits number of photons per second $n_1 = \dfrac{P}{E_1} = \dfrac{\lambda_1}{hc}$

These photons produce a saturation current $i_1 = 4.8 \times 10^{-3}$ A

Photon of wavelength 1650 has energy $E_2 = \dfrac{hc}{\lambda_2}$

Hence source of 5 mW emits number of photons per second $n_2 = \dfrac{P}{E_2} = \dfrac{5\times10^{-3}\times\lambda_2}{hc}$

Hence they produce saturation current $i_2 = \dfrac{i_1 n_2}{n_1}$

or
$$i_2 = \frac{4.8\times10^{-3}\times5\times10^{-3}\times\lambda_2\times hc}{hc\times\lambda_1}$$

$$= \frac{4.8\times10^{-3}\times5\times10^{-3}\times1650}{3000} = 13.2\times10^{-6}\text{A} = 13.2\ \mu\text{A}$$

(*c*) When n_1 number of photons fall on the surface per second, they produce current equal to 4.8×10^{-3} A. This means that the number of photo-electrons emitted per second from the surface is

$$n_e = \frac{4.8\times10^{-3}}{1.6\times10^{-19}} = 3\times10^{16} \text{ per second}$$

Hence efficiency $\eta = \dfrac{n_e}{n_1}\times100$

$$n_1 = \frac{\lambda_1}{hc} = \frac{3000\times10^{-10}}{6.6\times10^{-34}\times3\times10^8} = 1.5\times10^{18} \text{ per second}$$

and efficiency
$$\eta = \frac{n_e}{n_1} = \times100 = \frac{3\times10^{16}}{1.5\times10^{18}}\times100 = 2\% .$$

Example 27.43: Two metallic plates *A* and *B*, each of area 5×10^{-4} m^2 are placed parallel to each other at a separation of 1 cm. Plate *B* carries a positive charge of 33.7×10^{-12} C. A monochromatic beam of light with photons of energy 5 eV each, starts falling on plate *A* at $t = 0$ so that 10^{16} photons fall on it per square meter per second. Assume that one photo-electrons is emitted for energy 10^6 incident photons. Also assume that all the emitted photo-electrons are collected by plate *B* and the work function of plate *A* remains constant at the value 2 eV. Determine

(*a*) the number of photo-electrons emitted up to 10 sec.
(*b*) the magnitude of the electric field between the plates *A* and *B* al $t = 10$ sec and
(*c*) the kinetic energy of the most energetic photo-electron emitted at $t = 10$ sec. When it reaches plate *B*

Neglect the time taken by the photo-electron to reach plate *B*. Take $\varepsilon_0 = 8.85 \times 10^{-12}$ Farad/m.

Solution:

(*a*) The number of photons falling on plate A, per unit area per second = 10^{16}

Hence number of photons falling on plate A per second

$$= 10^{16} \times 5 \times 10^{-4} = 5 \times 10^{12}$$

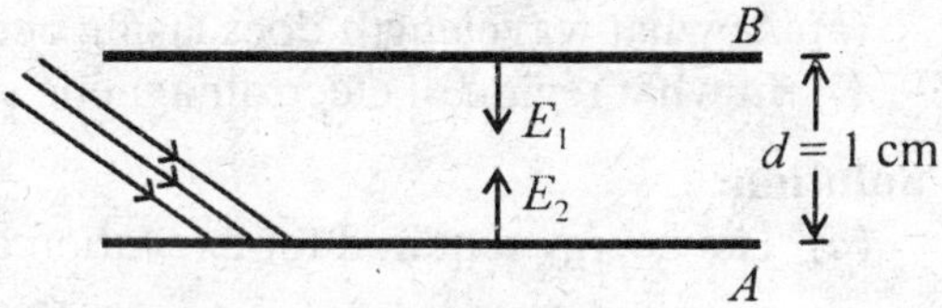

Fig. 27.47

Hence total number of photons falling on plate A in 10 sec is

$$n_1 = 5 \times 10^{12} \times 10 = 5 \times 10^{13}$$

Since 10^6 photons emit one photo-electron, hence total photo-electrons emitted are

$$N_e = \frac{n_1}{10^6} = 5 \times 10^7 \text{ photo-electrons.}$$

(*b*) The charge carried by these photo-electrons $q = 1.6 \times 10^{-19} \times 5 \times 10^7$ coul.

or $\quad q = 8 \times 10^{-12}$ C

Hence charge on plate B is reduced by q, i.e. it becomes

$$q_a = (33.7 - 8) \times 10^{-12} = 25.7 \times 10^{-12}$$

It produces an electric field E_1 between the plates A and B

$$E_1 = \frac{q_a}{2 \times \text{Area} \times \varepsilon_0} = \frac{25.7 \times 10^{-12}}{2 \times 5 \times 10^{-4} \times 8.85 \times 10^{-12}}$$

$$= 2.9 \times 10^3 \text{ V/m towards plate } A$$

Because of emission of photo-electrons plate A becomes positively charge with charge

$$q_b = 8 \times 10^{-12} \text{ C}$$

It produces an electric field E_2 between the plate A and B

$$E_2 = \frac{q_b}{2 \times \text{Area} \times \varepsilon_0} = \frac{8 \times 10^{-12}}{2 \times 5 \times 10^{-4} \times 8.55 \times 10^{-12}}$$

$$= 0.9 \times 10^3 \text{ N/C toward plate } B$$

Hence net field between plates is

$$E = E_1 - E_2 = 2.9 \times 10^3 - 9 \times 10^3 = 2000 \text{ V/m}$$

The net field is towards plate B.

The electrons are emitted with maximum kinetic energy

$$= 5 - 2 = 3 \text{ eV}$$

Since electric field between the plates is 2000 V/m and separation is 1 cm. Thus potential difference between the plates is

$$V = E \times d = 2000 \times 10^{-2} = 20 \text{ V.}$$

Additional kinetic energy acquired by photo-electrons in moving through space between plates is 20 eV. Hence maximum kinetic energy of the electrons reaching plate B is 20 + 3 = 23 eV.

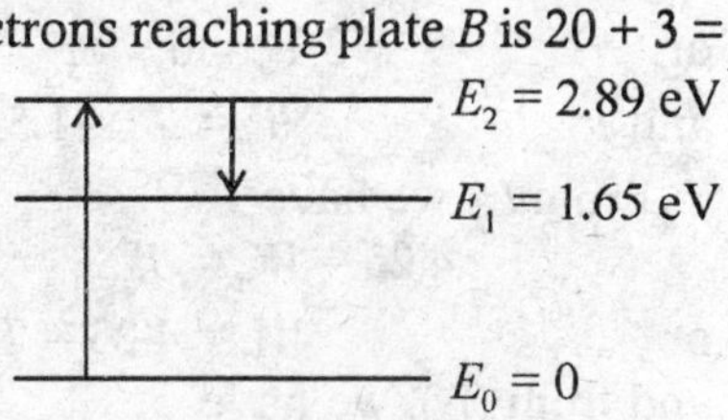

Fig. 27.48

Example 27.44: Sunlight can produce natural laser action in carbon dioxide molecules. The energy levels involved in lasing action are shown in Fig. 27.48.

(*a*) What wavelength of sunlight excites the molecules in laser action?

(*b*) At what wavelength does lasing occur?
(*c*) In what region of electromagnetic spectrum do the excitation and lasing wavelength lie.

Solution:

(*a*) The energy required for excitation of the of the molecule is $E = 2.89 - 0 = 2.89$ eV.

$$\text{Hence wavelength } \lambda = \frac{hc}{E} = \frac{12400\,(\text{eV}\cdot\text{Å})}{2.89\,(\text{eV})} = 4291\,\text{Å}$$

(*b*) The energy difference for lasing action is

$$E_2 - E_1 = 2.89 - 1.65 = 1.24 \text{ eV}$$

$$\text{Hence required wavelength } \lambda = \frac{hc}{E_2 - E_1} = \frac{12400}{1.24} = 10000\,\text{Å}$$

(*c*) Thus the wavelength of sunlight, which excites lasing action is 4291 Å and the lasing occurs at 10000 Å. The exciting radiation lies at the violet end and the laser light lies at the infrared end of the electromagnetic spectrum.

Example 27.45: When a photon of energy 4.25 eV strike the surface of a metal *A*, the ejected photoelectrons have maximum kinetic energy T_a eV and de Broglie wavelength λ_a. The maximum kinetic energy of photo-electrons liberated from another metal *B* by photons of energy 4.70 eV is $T_b = (T_a - 1.50)$ eV. If the de Broglie wavelength of these photo-electrons is $\lambda_b = 2\lambda_a$, then find (*a*) Work functions of *A* and *B* and (*b*) T_a and T_b.

Solution: From Einstein's photoelectric equation, we have

$$E - W = K_{max}$$

for metal *A*, we have

$$4.25 - W_a = T_a \qquad \ldots(i)$$

$$\therefore \quad T_b = T_a - 1.50 = 4.25 - W_a - 1.50 = 2.75 - W_a \qquad \ldots(ii)$$

But for metal *B*, we get

$$T_b = 4.7 - W_b \qquad \ldots(iii)$$

Thus from (*ii*) and (*iii*), we get

$$W_b - W_a = 1.95 \qquad \ldots(iv)$$

If *K* is the kinetic energy, then de Broglie wavelength is

$$\lambda = \frac{h}{\sqrt{2mK}}$$

$$\therefore \quad \frac{\lambda_b}{\lambda_a} = \sqrt{\frac{T_a}{T_b}} = \sqrt{\frac{T_a}{T_a - 1.5}} = 2$$

or $\quad T_a = 4(T_a - 1.5)$

or $\quad 3\,T_a = 6 \quad$ or $\quad T_a = 2$ eV.

and $\quad T_b = T_a - 1.5 = 2 - 1.5 = 0.5$ eV.

From (*i*) we have

$$4.25 - W_a = T_a.$$

or $\quad W_a = 4.25 - T_a = 4.25 - 2 = 2.25$ eV

and from (*iv*), we have

$$W_b = W_a + 1.95 = 2.25 + 1.95 = 4.20 \text{ eV}.$$

PROBLEMS

1. Two hydrogen like atoms A and B are of different masses and each atom contains equal number of protons and neutrons. The difference in the energies between the first Balmer lines emitted by A and B is 5.667 eV. When the atom A and B moving with the same velocity, strike a heavy target they rebound back with the same velocity. In the process atom B imparts twice the momentum to the target than that imparted by A. Identify the atoms A and B.

[**Hint:** $E_n = -\frac{13.6 Z^2}{n^2}$ eV. For Balmer lines $n_2 = 3$, $n_1 = 2$, Hence $E_a = 13.6 Z_a^2 \times \frac{5}{36}$, $E_b = 13.6 Z_b^2 \times \frac{5}{36}$ $\therefore$ $13.6 \times \frac{5}{36}(Z_b^2 - Z_a^2) = 5.66$ $Z_b^2 - Z_a^2 = 3$. Hence $Z_a = 1$ and $Z_b = 2$ and $A = {}_1H^2, B = {}_2H^4$]

2. The energy levels of a hypothetical one electron atom are given by $E_n = -\frac{18.0}{n^2}$ eV, where $n = 1, 2, 3, \ldots$

(*a*) Compute the four lowest energy levels and construct the energy level diagram.
(*b*) What is the excitation potential of state $n = 2$?
(*c*) What wavelengths (Å) can be emitted when these atoms in the ground state are bombarded by electrons that have been accelerated through a potential difference of 16.2 V?
(*d*) If these atoms are in the ground state, can they absorb radiation having a wavelength of 2000 Å?
(*e*) What is the photoelectric threshold wavelength of this atom?

[**Ans.** (*a*) –18.0 eV, –4.5 eV, –2.0 eV, –1.125 eV (*b*) 13.5 V (*c*) 4970 Å, 777 Å, 920 Å (*d*) No (*e*) 690 Å]

3. A μ-meson moves in a circular orbit around a very heavy nucleus (of infinite mass) of charge +3 *e*. Assuming Bohr's model is applicable to this system,

(*a*) derive an expression for the radius of *n*th orbit.
(*b*) Find *n* for which the radius of the orbit is approximately same as that of 1st Bohr orbit for hydrogen atom.
(*c*) Find the wavelength of radiation emitted when μ-meson jumps from 3rd orbit to 1st orbit [μ-meson is a particle, whose charge is equal to that of an electron and mass = 208 times that of an electron]

[**Ans.** (*a*) $r_n = \frac{n^2 h^2 \varepsilon_0}{624 \pi m_e e^2}$ (*b*) $n \approx 25$ (*c*) $\lambda = 55$ pm]

4. If potential energy between an electron and a proton at a separation of r is given by $U = -k/r^4$ when k is a positive constant, then for such a hypothetical hydrogen atom, calculate the radius of *n*th Bohr's orbit and energy levels.

[**Ans.** $r_n = \frac{4\pi\sqrt{km}}{nh}$; $E_n = \left(\frac{3h^4}{256\, k\pi^4 m^2}\right) n^4$]

5. A metal surface is irradiated with the light of wavelength 1980 Å. Photo current disappears if a retarding potential greater than 2.25 V is applied. Find

(*a*) work function of the surface and cut-off wavelength.

(*b*) If the surface is irradiated with the largest wavelength of Lyman series of hydrogen spectrum, find the maximum energy of the emitted photo-electron.

[**Ans.** (*a*) $\phi_0 = 4.0$ eV, $\lambda_0 = 3100$ Å (*b*) $E_{max} = 6.2$ eV]

6. A small plate of a metal of work function $\phi = 1.24$ eV is placed at a distance of 2 m from a monochromatic light source of wavelength 520 nm and power 1.2 watt. The light falls normally on the plate. Find the number of photons striking the metal plate per square cm per sec. If a constant magnetic field of induction 5×10^{-4} T is applied parallel to the metal surface, find the largest circular path traced by the photo-electron.

[**Ans.** $n = 6.27 \times 10^{+15}$ $r = 14.44 \times 10^{-3}$ m]

7. An electron of charge $(-e)$ and mass m is moving around a very heavy nucleus of charge $(+e)$ in a circular orbit of radius r. Assuming that Bohr's postulates are valid, calculate the frequency of radiation emitted when the electron makes a transition from the nth to $(n-1)$th orbit. Compare the above frequency with the frequency of revolution of the electron in the nth orbit. Show that the two frequencies become equal when $n \to \infty$.

[**Hint:** Energy of electron in nth orbit $E_n = -\dfrac{me^4}{8\varepsilon_0^2 n^2 h^2}$. Hence frequency ν for the transition n to $(n-1)$ is $h\nu = \dfrac{me^4}{8\varepsilon_0^2 h^2}\left[\dfrac{1}{(n-1)^2} - \dfrac{1}{n^2}\right] = \dfrac{me^4\,(2n-1)}{8\varepsilon_0^2 h^2 n^2 (n-1)^2}$ or $\nu = \dfrac{me^4\,(2n-1)}{8\varepsilon_0^2 h^3 n^2 (n-1)^2}$. For $n \to \infty$

$\nu = \dfrac{me^4}{4\varepsilon_0^2 h^3 n^3}$. Frequency of electron in nth orbit $f = \dfrac{v}{2\pi r} = \dfrac{e^2}{2\varepsilon_0 nh} \times \dfrac{1}{2\pi} \times \dfrac{\pi m e^4}{\varepsilon_0 n^2 h^2} = \dfrac{me^4}{4\varepsilon_0^2 h^3 n^3} = \nu$]

8. The path of photo-electrons emitted due to electromagnetic radiation incident on a sample material A is found to have a maximum bending radius of 0.1 m in magnetic field of $\dfrac{\sqrt{2}}{3} \times 10^{-4}$ T. When the radiation is incident normally on a double slit having a slit separation of 1.0 mm, it is observed that there are 10 fringes in width of 0.31 cm on a screen placed at a distance of 1 m from the double slit. Find the work function of the material and the corresponding threshold wavelength. What should be the wavelength of the incident light so that the bending radius is one half of what it was before? Given that $hc = 12400$ eV·Å.

[**Hint:** Momentum $(P = mv)$ of photo-electrons moving in magnetic field is $P = qBr$ and energy $E = \dfrac{P^2}{2m} = \dfrac{q^2B^2r^2}{2m} = 1.95$ eV. $\left(q = 1.6 \times 10^{-19}\,\text{C},\ B = \dfrac{\sqrt{2}}{3} \times 10^{-4}\,\text{T},\ r = 0.1\,\text{m}\right)$ wavelength of radiation is $\lambda = \dfrac{\beta d}{D} = 3100$ Å ($\beta = 3.1 \times 10^{-4}$ m, $d = 10^{-3}$ m, $D = 1$ m). Energy of incident radiation $\dfrac{hc}{\lambda} = 4$ eV. Hence work function $\phi_0 = 4 - 1.95 = 2.05$ eV and $\lambda_0 = \dfrac{hc}{\phi_0} = 6200$ Å. For radius half energy of electron should be 1/4, i.e. photo-electron must have energy $\dfrac{1.95}{4} = 0.5$ eV. Work function is 2 eV. Hence incident radiation has energy 2.5 eV and wavelength $= \dfrac{12400}{2.5} = 4960$ Å]

9. In a photo cell collector and emitter plates are separated by a distance 15 cm. The photo cell is connected with ammeter in series. If light of wavelength between 400 nm-700 nm fall on the plate whose work function is 2.1 eV. Calculate the minimum value of magnetic field and its direction, present between the plates of the cell so that the ammeter reading falls to zero.

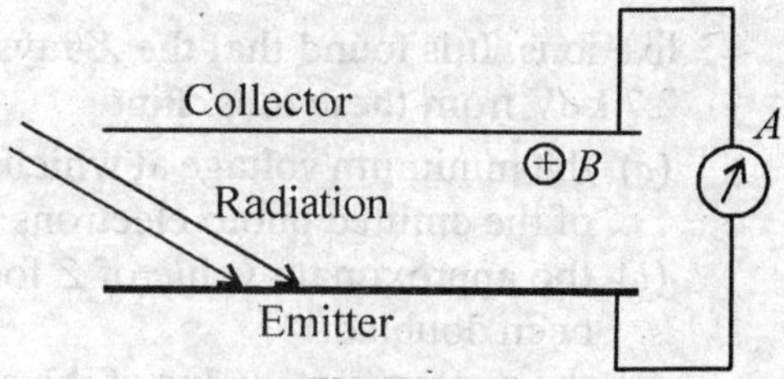

Fig. 27.49

[**Hint:** Photo electrons emitted must not reach the collector for no current to flow, i.e. radius of circular path should be 15 cm. Maximum energy of radiation is $E = \dfrac{12400}{4000} = 3.1$ eV and that of the electrons is $E_e = 1$ eV. Hence required field is $B = \dfrac{mv}{qr} = \dfrac{\sqrt{2mE_e}}{qr} = 2.25 \times 10^{-5}$ T and the direction should be perpendicular to the velocity]

10. A neutron with K.E. 120 eV collides inelastically with a doubly ionized lithium ion in ground state at rest. The neutron is scattered at an angle of 90° with respect to its original direction. If after the collision the Li ion gets excited, find the allowed values of the energy of the neutron and that of the ion after the collision (Given mass of Li = 6 × mass of neutron).

[**Ans.** 7.03 eV. 21.17 eV]

11. A gas of identical hydrogen like atoms has some atoms in the ground state and some atoms in a particular excited state and there are no atoms in any other energy level. The atoms of the gas make transition to a higher energy state by absorbing monochromatic light of wavelength 304 Å. Subsequently, the atoms emits radiation of only six different photon energies. Some of emitted photons have wavelength 304 Å. Some have wavelength more and some have wavelength less than 304 Å, then

(*a*) find the principal quantum number of the initially excited state
(*b*) identify the gas
(*c*) find the ground state energy (in eV)
(*d*) find the maximum and minimum energies of emitted photons.

[**Ans.** (*a*) $n = 2$ (*b*) $Z = 4$ (c) –217.6 eV (*d*) 204 eV, 10.6 eV]

12. A mixture of hydrogen atoms (in their ground state) and hydrogen like ion (in their first excited state) is being excited by electrons which have been accelerated by a potential difference V. After excitation when both atoms come into their ground state, the wavelengths emitted are found in the ratio 5 : 1, then

(*a*) for what minimum value of V both the atoms get excited after collision with electrons?
(*b*) find the other hydrogen like ion and the energy of emitted light.

[**Hint:** For hydrogen like atom of atomic number Z, energy difference in first excited state and ground state is $\dfrac{3}{4} \times 13.6\,Z^2 = 10.2\,Z^2$ eV. Hence $(10.2\,Z^2 + V) = \dfrac{hc}{\lambda_1}$ and for hydrogen atom $V = \dfrac{hc}{\lambda_2} = \dfrac{hc}{5\lambda_1}$. This gives $4V = 10.2\,Z^2$. Minimum excitation voltage for hydrogen is 10.2 V, which gives $Z = 2$. Thus the energy of the radiation emitted is $10.2 \times (2)^2 + 10.2 = 51$ eV]

13. X-rays are emitted by a tube containing the element Niobium ($Z = 41$) as the anti-cathode, and the K_α lines emitted are allowed to be incident on an unknown gas containing hydrogen like ions. It is found that the *X*-rays cause the emission of photo-electrons with an energy of 2.7 keV from these ions. Find

(*a*) the minimum voltage at which the *X*-ray tube should be operated so that the momentum of the emitted photo-electrons is doubled.
(*b*) the approximate value of *Z* for the target of the anti cathode after the momentum has been doubled.
(*c*) the approximate value of the atomic number of the atoms of gas.
(*d*) the ionization in the gas caused by *X*-rays, if the intensity of *X*-ray is 100 mW/m² and 1% of the *X*-rays cause ionization.

[**Hint:** (*a*) K_α —*X*-ray photon of Niobium has energy $E_1 = 13.6(Z-1)^2 \times \frac{3}{4}$ or $E_1 = 16.320$ keV. Maximum K.E. of electron $K = 2.7$ k eV. Hence work function of gas $= E_1 - K = 13.52$ k eV. For momentum double energy must be 4 times, i.e. $K = 2.7 \times 4 = 10.8$ keV. Hence incident photon must have energy $= 13.52 + 10.8 = 24.32$ keV. Hence minimum voltage required is 24.32 keV. (*b*) If atomic number is *Z* then $13.6\,(Z-1)^2 \times \frac{3}{4} = 24.32 \times 10^3$ gives $Z = 50$. (*c*) Since work function is 13.52 eV and if *Z* is atomic number of gas then ionization energy is $13.6\ Z^2 = 13.52 \times 1000$ or $Z \approx 32$. (*d*) No. of photons falling on unit area per second

$$n = \frac{\text{Intensity}}{\text{Energy}} = \frac{10^{-1}\ \text{W/m}^2}{24.32\ \text{keV}} \text{ or } n = 2.6 \times 10^{+13} \text{ per sec.}$$

Since 1% of photons cause the ionization as such 2.6×10^{11} photons cause ionization.

$$\text{Hence total ionization} = \frac{2.6 \times 10^{11}}{3 \times 10^8} = 0.867 \times 10^3 = 867/\text{m}^3\ ]$$

14. When only the Paschen series of wavelengths emitted by a gas (*A*) containing hydrogen like particles (ions or atoms) is incident on metal *B*, it is observed that photo-electrons are just emitted. If the light from the Bracket series of transition of the hydrogen like gas (*C*) is incident on the same metal (*B*), the tracks of the photo-electrons emitted have radius of 1.47 cm when the photo-electrons enter a region of uniform magnetic field $\sqrt{10} \times 10^{-4}$ T perpendicularly. It is known that the atomic number of *C* is 1 unit higher than that of *A*. Find the atomic number of the gases *A* and *C* and work function of metal *B*.

[**Hint**: For A atomic number *Z* for Paschen series maximum energy is $E_1 = \frac{13.6\,Z^2}{9}$. This must be work function of *B*. For *C* energy of Bracket series is $E_2 = \frac{13.6(Z+1)^2}{16}$. This emits photo-electrons of energy $E_e = \frac{q^2B^2r^2}{2m} = 2$ eV $\therefore\ 13.6\left(\frac{Z+1}{4}\right)^2 - 13.6\left(\frac{Z}{3}\right)^2 = 2$ or

$$\left(\frac{Z+1}{4}\right)^2 - (Z/3)^2 = 2/13.6 \text{ gives } Z = 1 \text{ and work function } \phi_0 = E_1 = \frac{13.6 \times 1}{9} = 1.5 \text{ eV}\ ]$$

15. A low intensity ultra violet light of wavelength 2271 Å irradiates a photo cell made of Molybdenum metal. If the stopping potential is –1.3 V, find the work function of the metal. Will the photo cell work if it is irradiated by a high intensity red light of wavelength 6328 Å (Given $h = 6.62 \times 10^{-34}$ Js) ? [**Ans.** $\phi_0 = 4.16$ eV; No since $\nu < \nu_0$]

16. Monochromatic light of wavelength 640.2 nm irradiates a photo cell made of cesium on tungsten. The stopping voltage is measured to be 0.5 V. What will be the new stopping voltage of the cell if it is irradiated by monochromatic light of wavelength 427.2 nm?

[**Ans.** $V = 1.51$ Volt]

17. When a surface is irradiated with light of wavelength $\lambda = 4050$ Å, a photo current appears which vanishes if a retarding potential greater than 0.6 V is applied across the photo tube. When a different source of light is used; it is found that the critical retarding potential is changed to 1.1 V. What is the work function of the surface and the wavelength of the second source? If the photo-electrons (after emission from the surface) are subjected to a magnetic field of 10 tesla what changes will be observed in the above two readings of potentials?

[**Ans.** $\phi_0 = 1.9$ eV, $\lambda = 4125$ Å. No change]

18. Doubly ionized lithium atom is hydrogen like with $Z = 3$.

(*a*) Find the wavelength of the radiation required to excite the electron in Li^{++} from the first to the third Bohr orbit (Ionization energy of hydrogen equals 13.6 eV).

(*b*) How many spectral lines are observed in the emission spectrum of the above excited system?

[**Ans.** (*a*) $\lambda = 114.3$ Å, (*b*) 3 lines]

19. An electron of energy 20 eV collides with a hydrogen atom in the ground state. As a result of the collision, the atom is excited to a higher energy state and the electron is scattered with reduced velocity. The atom subsequently returns to its ground state with emission of radiation of wavelength 1.216×10^{-7} m. Find the velocity of the scattered electron.

[**Ans.** $v = 1.86 \times 10^6$ ms^{-1}]

20. Suppose a moving hydrogen atom makes a head on elastic collision with a stationary hydrogen atom. Before collision both atoms are in the ground state and after collision they move together. What is the maximum velocity of the moving hydrogen atom if one of the atoms is to be given the minimum excitation energy after the collision. Given mass of hydrogen atom $= 1.0078 \times 1.66 \times 10^{-27}$ kg and Rydberg constant $R = 1.097 \times 10^7$ m^{-1}.

[**Ans.** $v = 6.25 \times 10^4$ m/s]

21. An electron in the ground state of hydrogen atom is revolving in anticlockwise direction in circular orbit of radius R.

(*a*) Obtain an expression for the orbital magnetic dipole moment of the electron.

(*b*) The atom is placed in a uniform magnetic induction B such that the plane-normal to the electron orbit makes an angle of 30° with the magnetic induction. Find the torque experienced by the orbiting electron.

$$\left[\textbf{Ans. } (a)\ M = \frac{eh}{4\pi m}\hat{n}\ \text{Amp m}^2\ \ (b)\ \tau = \frac{ehB}{8\pi m}\ \text{N-m}\right]$$

22. An energy of 680 eV is required to excite a hydrogen like atom from its second Bohr orbit to the third. The nuclear charge is Ze. Find the value of Z, the kinetic energy of the electron in the first Bohr orbit and the wavelength of the electromagnetic radiation required to eject the electron from the first Bohr orbit to infinity. [**Ans.** $Z = 6$, K.E. = 489.6 eV, $\lambda = 25.3$ Å]

23. (*a*) A stopping potential of 0.82 V is required to stop the emission of photo-electrons from the surface of a metal by light of wavelength 4000 Å. For light of wavelength 3000 Å, the stopping potential is 1.85 V. Find the value of Plank's constant.

(*b*) At stopping potential, if the wavelength of incident light is kept fixed at 4000 Å but the intensity of light is increased two times, will the photoelectric current be obtained? Give the reason for your answer. [**Ans.** (*a*) $h = 6.6 \times 10^{-34}$ Js (*b*) No]

24. In a Millikan's oil-drop experiment, a charged oil drop of mass density 880 kgm^{-3} is held stationary between two parallel plates 6.0 mm apart held at a potential difference of 10^3 V. When the electric field is switched off, the drop is observed to fall a distance of 2.00 mm in 35.7 s. (*a*) What is the radius of the drop? (*b*) Estimate the charge of the drop. How many excess electrons does it carry? Viscosity of air = 1.8×10^{-15} Nsm^{-2}, density of air = 1.29 kg/m^3.
[**Ans.** (*a*) $r = 7.26 \times 10^{-7}$ m (*b*) $q = 8.05 \times 10^{-19}$ C, 5 excess electrons]

25. A point source of light is placed at the centre of curvature of a hemispherical surface. The radius of curvature is r and the inner surface is completely reflecting. Find the force on the hemisphere due to the light falling on it if the source emits a power W.

[**Hint:** Energy falling on area dA at P in unit time is $\dfrac{W\,dA}{4\pi r^2}$. As the light is reflected, the momentum transferred to area dA in unit time is $\Delta P = \dfrac{2W\,dA}{4\pi r^2 c}$. Hence component of the force parallel to OX (Fig. 27.50) is $dF\cos\theta = \dfrac{2W\,dA}{4\pi r^2 c}\cos\theta$. $dA\cos\theta$ is the projection of area dA in the plane containing rim. Hence total force is $F = \dfrac{2W\cdot\pi r^2}{4\pi r^2 c} = \dfrac{W}{2c}$]

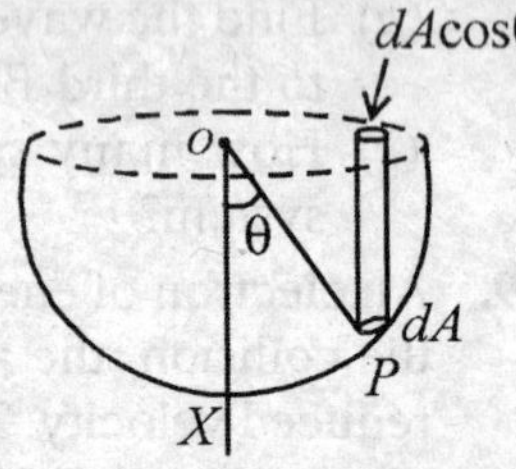

Fig. 27.50

26. A neutron moving with speed V makes a head-on collision with a hydrogen atom in ground state kept at rest. Find the minimum kinetic energy of the neutrons for which inelastic (completely or partially) collision may take place. The mass of neutron = mass of hydrogen = 1.67×10^{-27} kg. [**Ans.** E_{min} = 20.4 eV]

27. A small particle of mass m moves in such a way that the potential energy $U = 1/2m^2\omega^2r^2$ where ω is constant and r is the distance of the particle from the origin. Assume Bohr's model of quantization of angular momentum and circular orbits, show that the radius of the nth allowed orbit is proportional to $\sqrt{n}$.

28. An *X*-ray tube operates at 40 kV. Suppose the electron converts 70% of its energy into a photon at each collision. Find the lowest three wavelengths emitted from the tube. Neglect the energy imparted to the atom with which it collides.
[**Ans.** λ_1 = 44.3 pm, λ_2 = 148 pm, λ_3 = 493 pm]

29. A uniform magnetic field exists in a region. An electron projected perpendicular to the field goes in a circle. Assuming Bohr's quantisation rule for angular momentum. Calculate (*a*) the smallest possible radius of the electron (*b*) the radius of *n*th orbit and (*c*) the minimum possible speed of the electron.

[**Hint:** $\dfrac{mv^2}{r} = qvB$ or $mv = qBr$ and $mvr = \dfrac{nh}{2\pi}$. Solving $r_n = \sqrt{\dfrac{nh}{2\pi eB}}$ gives $r_{min} = \sqrt{\dfrac{h}{2\pi eB}}$

and $v_n = \sqrt{\dfrac{nheB}{2\pi m^2}}$ and $v_{min} = \sqrt{\dfrac{heB}{2\pi m^2}}$]

30. A totally reflecting small plane mirror placed horizontally faces a parallel beam of light (Fig. 27.51). The mass of the mirror is 20 g. Assume that there is no absorption in the lens and that

30% of the light emitted by source goes through the lens. Find the power of the source needed to support the weight of the mirror.

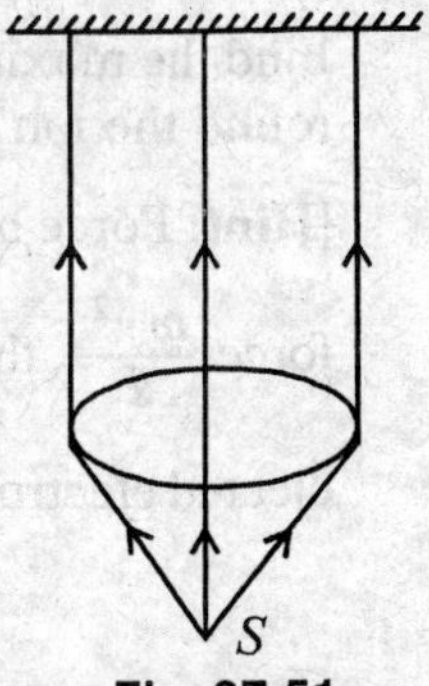

Fig. 27.51

[Hint: Momentum transferred by each photon to the mirror is $\frac{2h\nu}{c}$. If n is number of photons falling per second then $\frac{2h\nu}{c}n = wt = 20\times10^{-3}\times10$

or $h\nu n = \frac{20\times10^{-3}\times10\times c}{2}$. If P is the power then $0.3P = h\nu n = 3\times10^{+7}$

or $P = \frac{3\times10^7}{0.3} = 10^8$ watt = 100 MW]

31. In the arrangement shown in Fig. 27.52, $y = 0.1$ mm $d = 0.24$ mm and $D = 1.2$ m. The work function of the material of the emitter is 2.2 eV. Find the stopping potential needed to stop the photo-electrons.

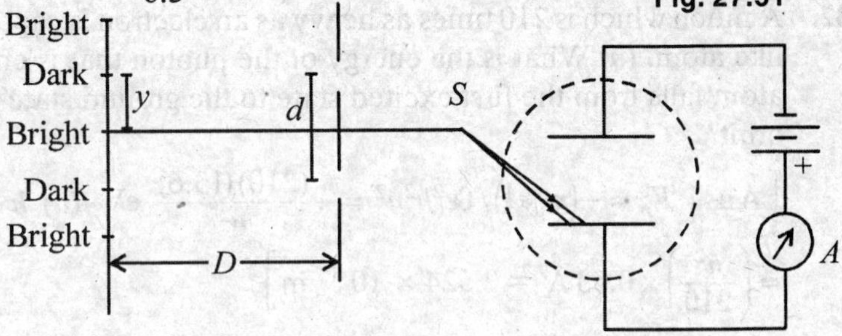

Fig. 27.52

[**Ans.** $V = 0.9$ V]

32. A horizontal cesium plate ($\phi = 1.9$ eV) is moved vertically downward at a constant speed v in a room full of radiation of wavelength 250 nm and above. What should be minimum value of v so that the vertically upward component of velocity is non positive for each photo-electron.
[Hint: Velocity of plate should be equal to the velocity of photo-electrons emitted $v = 1.04\times10^6$ m/s]

33. A light beam of wavelength 400 nm is incident on a metal plate of work function 2.2 eV. (*a*) A particular electron absorbs a photon and makes two collisions before coming out of the metal. If in each collision it looses 10% of its energy, find the kinetic energy of this electron as it comes out of the metal, (*b*) Under the same assumption, find the maximum number of collisions the electron can suffer before it becomes unable to come out of the metal.

[**Ans.** (*a*) $E = 0.31$ eV (*b*) number of collisions 4]

34. In an experiment on photoelectric emission, it was found that the stopping potential decreases from 1.85 V to 0.82 V as the wavelength of the incident light is varied from 300 nm to 400 nm. Calculate the value of the Planck's constant from these values.

[**Ans.** $h = 4.12\times10^{-16}$ eV]

35. When the sun is directly overhead, the surface of the earth receives 1.4×10^3 W/m^2 of sunlight. Assume that the light received is monochromatic with average wavelength 500 nm and that no light is absorbed in between the sun and earth's surface. The distance between earth and the sun is 1.5×10^{11} m. (*a*) Calculate the number of photons falling per second on each square meter of earth's surface directly below the sun. (*b*) How many photons are there in each cubic meter near the earth's surface at any instant? (*c*) How many photons does the sun emit per second?

[**Ans.** (*a*) 3.5×10^{21} (*b*) 1.2×10^{13} (*c*) 9.9×10^{44}]

36. A small metal plate (work function ϕ) is kept at a distance d from singly ionized, fixed ion. A monochromatic light beam is incident on the metal plate and photo-electrons are emitted.

Find the maximum wavelength of the light beam so that some of the photo-electrons may go round the ion along circle.

[**Hint:** Force between ion and electron is $F = \frac{ke^2}{d^2}$. If this provides the necessary centrepetal force $\frac{mv^2}{d}$, then photo-electrons will move round the ion, where v is the velocity (max) of the ejected electrons. Photoelectric equation gives $\frac{hc}{\lambda} - \phi = \frac{1}{2}mv^2$ and $\frac{mv^2}{d} = \frac{ke^2}{d^2}$. Solving the two equations $\lambda = \frac{8\pi\varepsilon_0 dhc}{e^2 + 8\pi\varepsilon_0 \phi d}$]

37. A muon which is 210 times as heavy as an electron is captured by a proton to form a hydrogen like atom. (*a*) What is the energy of the photon that is emitted when the muon of the above atom falls from the first excited state to the ground state? (*b*) What is the radius of first Bohr orbit?

[**Ans.** $E_n = -\left(m_\mu e^4\right)/8\varepsilon_0^2 h^2 n^2 = -\frac{(210)(13.6)}{n^2}$ eV (*a*) $h\nu = 2142$ eV (*b*) $r_n = \frac{\varepsilon_0^2 h^2 n^2}{\pi(210 m_e e^2)}$
$= \left(\frac{n^2}{210}\right) \times 0.53$ Å $= 2.524 \times 10^{-13}$ m]

SHORT ANSWER TYPE QUESTIONS

1. Find the ratio of minimum to maximum wavelength of radiation emitted by electron in ground state of Bohr's hydrogen atom.

[**Ans.** Energy of radiation corresponding to the transition between two energy levels, n_1 and n_2 is

$$\Delta E = 13.6\left(\frac{1}{n_1^2} - \frac{1}{n_2^2}\right) \text{eV}$$

ΔE is minimum when $n_1 = 1$ and $n_2 = 2$ or $\Delta E_{\text{min}} = 13.6\left(\frac{1}{1} - \frac{1}{4}\right)$

or $\Delta E_{\text{min}} = \frac{3}{4} 13.6$ eV

ΔE is maximum when $n_1 = 1$ and $n_2 = \infty$ $\Delta E_{\text{max}} = 13.6$ eV

$\therefore \frac{\Delta E_{\text{min}}}{\Delta E_{\text{max}}} = \frac{3}{4} = \frac{\lambda_{\text{min}}}{\lambda_{\text{max}}}$ $\because \Delta E = \frac{hc}{\lambda}$]

2. What is the angular momentum of an electron in Bohr's hydrogen atom whose energy is –3.4 eV?

[**Ans.** Energy of Bohr atom in nth orbit is $E_n = -\frac{13.6}{n^2}$ eV

$\therefore$ $-3.4 = -\frac{13.6}{n^2}$ or $n = 2$

the angular momentum in nth orbit is $L_n = \frac{nh}{2\pi}$

$\therefore$ $L_2 = \frac{2h}{2\pi} = \frac{h}{\pi}$ Js]

3. A proton approaches another proton at rest along the line joining the two particles with velocity v_0. Find the distance of closest approach.

[**Ans.** Since the two will move due to their mutual repulsive interaction. Hence the velocity of first particle will decrease and that of second increases. The distance of closest approach will be when their relative velocity is zero, i.e. both move with same velocity (v). Conservation of momentum gives $mv_0 = mv + mv = 2mv$ or $v = v_0/2$ and conservation of energy gives $\frac{1}{2}mv_0^2 = 2\frac{1}{2}m(v_0/2)^2 + \frac{e^2}{4\pi\varepsilon_0 r}$, where r is the distance of closest approach $\therefore\ r = \frac{e^2}{\pi\varepsilon_0 mv_0^2}$]

4. Find the frequency of light which ejects electrons from a metal surface, fully stopped by a retarding potential of 3 V. The photoelectric effect begins at a frequency of 6×10^{14} Hz. Find the work function of the metal.

[**Ans.** Einstein's photoelectric equation gives $K_{max} = h\nu - h\nu_0$, but $K_{max} = eV_0$ or $\nu = \frac{eV_0}{h} + \nu_0$

$= \frac{1.6\times10^{-19}\times3}{6.6\times10^{-34}} + 6\times10^{14} = 7.27 \times 10^{14} + 6 \times 10^{14} = 1.327 \times 10^{15}$ Hz. Work function

$\phi_0 = h\nu_0 = 6.6 \times 10^{-34} \times 6 \times 10^{14}$ J $= 3.96 \times 10^{-19}$ J $= 2.475$ eV]

5. Ultraviolet light of wavelengths 830 Å and 700 Å when allowed to fall on hydrogen atoms in their ground state is found to liberate electrons with kinetic energies 1.8 eV and 4.0 eV respectively. Find the value of Planck's constant.

[**Ans.** For hydrogen atom if E_i is the ionization energy and E is the kinetic energy of electrons, and λ the wavelength of incident radiation then $\frac{hc}{\lambda_1} - E_i = E\ \therefore\ hc\left(\frac{1}{\lambda_1} - \frac{1}{\lambda_2}\right) = E_1 - E_2$ or

$$h = \frac{(E_1 - E_2)\lambda_1\lambda_2}{c(\lambda_2 - \lambda_1)} = \frac{(4-1.8)\times1.6\times10^{-19}\times830\times700\times10^{-20}}{3\times10^8(830-700)\times10^{-10}} = \frac{2.2\times1.6\times830\times700\times10^{-39}}{3\times130\times10^{-2}}$$

$= 5.2\times10^{-34}$ Js]

6. The mean free path of the electron in a discharge tube is 20 cm. The tube itself is 10 cm long. What is the length of Crookes dark space?

[**Ans.** The mean free path of the electrons is much longer than the length of the tube. Thus the electrons in general, do not collide in between. As such no ionization takes place and hence no light is emitted. The Crookes dark space fills the entire tube and its length is 10 cm]

7. Consider a cylindrical tube closed at one end and filled with a conducting movable piston at the other end. A cathode is fixed in the tube near the closed end and an anode is fixed with piston. A gas is filled in the tube at pressure P. Using Paschen equation $V = f(Pd)$ show that the sparking potential does not change as the piston is slowly moved in or out. Assume that the temperature does not change in the process.

[**Ans.** As the piston is moved the volume and pressure both change. Since temperature does not change, hence the product (Pd) remains constant—Boyle's law as such the sparking potential does not change as the piston is moved]

8. Consider energy level diagram of a hydrogen atom (Fig. 27.53). How will the kinetic energy and potential energy of electron vary if the electron moves from a lower level to higher level?

[**Ans.** Kinetic energy of electron in hydrogen atom ($Z = 1$) in nth orbit is $E_K = \frac{me^4}{8\varepsilon_0^2h^2}\cdot\frac{1}{n^2}$ and potential energy is $E_p = -\frac{me^4}{4\varepsilon_0^2 h^2}\cdot\frac{1}{n^2} = -2E_K$. Potential energy is negative. As n increases magnitude of potential energy decreases but it itself increases. It becomes maximum at infinity (equal to zero). Whereas kinetic energy is a positive quantity it decreases as n increases and becomes zero at ∞]

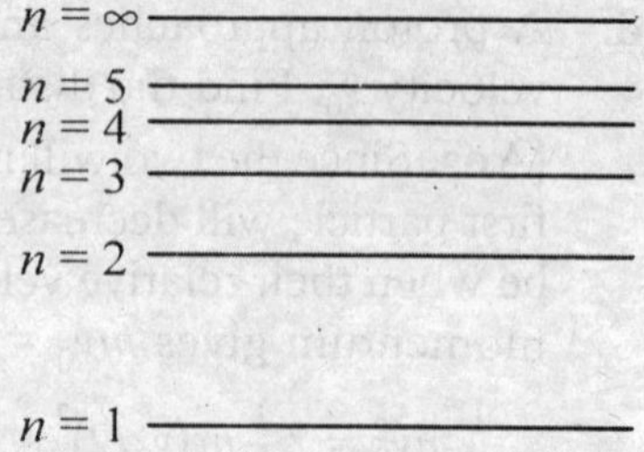

Fig. 27.53

9. Use the Bohr model to determine the ionization energy of the He^+ ion which has a single electron. Also calculate the minimum wavelength a photon must have to cause ionization.

[**Ans.** According to Bohr model, the energy of electron in nth orbit of atom is $E_n = -\frac{RhcZ^2}{n^2}$ Rhc = 13.6 eV. Hence ionization energy of singly ionized helium (He^+) is (Z = 2, n = 1) $E = -\frac{13.6(2)^2}{1} = -13.6\times 4 = -54.4$ eV. The photon must have this much minimum energy to cause ionization. Hence wavelength of this photon is $\frac{hc}{\lambda} = E$ or $\lambda = \frac{hc}{E}$. But $hc = 12.4\times 10^{-7}$ eV

$$\therefore\ \lambda = \frac{12.4\times 10^{-7}}{54.4} = 0.228\times 10^{-7}\text{ m} = 22.8\text{ nm}$$]

10. (*a*) An *X*-ray tube produces a continuous spectrum of radiation with the short wavelength end at 0.45 Å. What is the maximum energy of a photon in the radiation (*b*) from your answer to (*a*), guess what order of accelerating voltage (for electrons) is required in such a tube?

[**Ans.** For continuous *X*-rays the minimum wavelength $\lambda_m = 0.45$. Hence the photons corresponding to this wavelength has maximum energy. Thus $E_{max} = \frac{hc}{\lambda_{min}} = \frac{6.6\times 10^{-34}\times 3\times 10^{+8}}{0.45\times 10^{-10}}$ $= 44\times 10^{-16}\text{ J} = \frac{44\times 10^{-16}}{1.6\times 10^{-19}} = 27.5\times 10^3\text{ eV} = 27.5\text{ k eV}$. Thus the order of voltage is 27.5 kV]

11. The electron in a hydrogen atom makes a transition from $n_1 \to n_2$, where n_1 and n_2 are the principal quantum numbers of the two states. Assume the Bohr model to be valid. The time period of the electron in the initial state is eight times that in the final state. What are the possible values of n_1 and n_2?

[**Ans.** This period of electron in nth orbit is $T_n = \frac{2\pi r_n}{v_n} = \frac{4\varepsilon_0^2 h^3 n^3}{mZ^2e^4}$, i.e. $T_n \propto n^3$ or $\left(\frac{T_1}{T_2}\right) = \left(\frac{n_1}{n_2}\right)^3 = 8$ or $\frac{n_1}{n_2} = 2$. Thus possible values n_1 and n_2 are $n_1 = 2$, $n_2 = 1$, or $n_1 = 4$, $n_2 = 2$ or $n_1 = 6$, $n_2 = 3$ and so on ...]

12. Taking into account the motion of nucleus of hydrogen atom, find the expression for the electron's binding energy in the ground state and for the Rydberg constant. How much (in per cent) do the binding energy and the Rydberg constant, differ from this more accurate values then obtained without taking the motion of the nucleus into account?

[**Ans.** When the motion of the nucleus is not taken into account, the binding energy E_b of the hydrogen atom in ground state and Rydberg constant R_∞ is given by $E_b = \frac{me^4}{8\varepsilon_0^2 h^2}$ and

$R_\infty = \dfrac{me^4}{8\varepsilon_0^2 ch^3}$. Now if motion of nucleus is taken into account, mass of electron is to be replaced by reduced mass $\mu = \dfrac{mM}{M+m}$, where M is the mass of nucleus. Hence new binding energy E and Rydberg constant R are $E = \dfrac{Mme^4}{(M+m)8\varepsilon_0^2 h^2}$ and $R = \dfrac{mM}{M+m}\dfrac{e^4}{8\varepsilon_0^2 ch^3}$. Hence difference in energy $\Delta E = E_b - E = \dfrac{e^4}{8\varepsilon_0^2 ch^3} \times \dfrac{m^2}{M+m}$ $\therefore \dfrac{\Delta E}{E_b} = \dfrac{m}{M+m} \approx \dfrac{m}{M} = 0.055\%$. Similarly

$$\frac{\Delta R}{R} = \frac{m}{M+m} \approx \frac{m}{M} = 0.055\%\,]$$

13. An electron in a hydrogen like atom is in excited state. It has a total energy of –3.4 eV. Calculate (*a*) the kinetic energy (*b*) the de Broglie wavelength of the electron.

[**Ans.** (*a*) In hydrogen like atom kinetic energy of electron is equal to negative total energy. Thus K.E. = –(–3.4) = 3.4 eV (*b*) The de Broglie wavelength $\lambda = \dfrac{h}{P}$ where momentum $P = \sqrt{2m E_k}$, E_k is kinetic energy $\therefore \lambda = \dfrac{h}{\sqrt{2mE}} = \dfrac{6.6 \times 10^{-34}}{\sqrt{2 \times 9.1 \times 10^{-31} \times 3.4 \times 1.6 \times 10^{-19}}} = 0.7$ nm]

14. *X*-rays and a beam of electrons (accelerated by an appropriate voltage have same wavelength λ (= 1 Å). Which one has greater energy?

[**Ans.** X-ray photon has energy $E_{ph} = \dfrac{hc}{\lambda} = \dfrac{6.6 \times 10^{-34} \times 3 \times 10^8}{10^{-10}} = 1.98 \times 10^{-15}\ \text{J} = 12.37 \times 10^3$ eV.

The electron has momentum $P = h/\lambda$ and energy $E_e = \dfrac{P^2}{2m}$ $\therefore E_e = \dfrac{h^2}{2m\lambda^2} = \dfrac{(6.6 \times 10^{-34})^2}{2 \times 9.1 \times 10^{-31} \times 10^{-20}}$

$= 2.4 \times 10^{-17}\ \text{J} = 1.5 \times 10^{+2}$ eV. Definitely $E_{ph} > E_e$]

15. A neutron has the same kinetic energy (=150 eV) as that of electron (K.E. = 150 eV). Will a neutron beam be equally suitable for crystal diffraction experiment, explain? Mass of neutron is 1.67×10^{-27} kg.

[**Ans.** Electron has wavelength $\lambda_e = 1$ Å. Whereas neutron has wavelength $\lambda_n = \dfrac{h}{\sqrt{2m E_n}}$ $= \dfrac{6.6 \times 10^{-34}}{\sqrt{2 \times 1.67 \times 10^{-27} \times 150 \times 10^{-19} \times 1.6}} = 0.233 \times 10^{-11}$ m = 0.0233 Å. Bragg's law gives $2d \sin\theta = n\lambda$ or $\sin\theta = \dfrac{n\lambda}{2d}$. If $\lambda << d$ the angle of diffraction θ does not exist unless n is very large. For large values of n the intensity of diffracted waves become very small and diffraction effects are not observable. Hence we conclude that unlike an electron beam, a neutron beam of same energy is not suitable for crystal diffraction]

16. The wavelength of the characteristic *X*-ray K_α line emitted from zinc ($Z = 30$) is 1.415 Å. Find the wavelength of the K_α-line emitted from molybdenum ($Z = 42$)

[**Ans.** Moseley's law gives for *k*-lines, $\nu \propto (Z-1)^2$. But $\nu = \dfrac{c}{\lambda}$ $\therefore \dfrac{1}{\lambda} = k(Z-1)^2$ where k is a constant,

$$\therefore \frac{\lambda_{Zn}}{\lambda_{Mo}} = \left(\frac{Z_{Mo}-1}{Z_{Zn}-1}\right)^2 \quad \therefore \lambda_{Mo} = \lambda_{Zn}\left(\frac{Z_{Zn}-1}{Z_{Mo}-1}\right)^2 = 1.415\left(\frac{30-1}{42-1}\right)^2 = 1.415 \times \left(\frac{29}{41}\right)^2 = 0.708 \text{ Å}\,\Big]$$

17. If the series limit of Balmer series for hydrogen is 3644 Å, then find the atomic number of element which gives the X-rays wavelength down to 1 Å. Identify the element.

[**Ans.** For hydrogen, Balmer series has shortest wavelength when $n_1 = 2$ and $n_2 = \infty$. Thus $\frac{1}{\lambda} = R\left(\frac{1}{2^2} - \frac{1}{\infty}\right) = \frac{R}{4}$ or $R = \frac{4}{\lambda} = \frac{4}{3644}(\text{Å}^{-1})$. The wavelengths of K-lines in X-ray characteristic spectrum is given by $\frac{1}{\lambda_k} = R(Z-1)^2\left(\frac{1}{1^2} - \frac{1}{n^2}\right)$. The shortest wavelength correspond to $n = \infty$ $\therefore \left(\frac{1}{\lambda_k}\right)_{\min} = R(Z-1)^2$. But $(\lambda_k)_{\min} = 1\text{ Å}$ or $(Z-1)^2 = \frac{1}{R} = \frac{3644}{4} = 911$ or $Z - 1 = 30$ or $Z = 31$ and element is gallium]

18. The wavelength of the characteristic X-ray k_α line emitted by a hydrogen like element is 0.32 Å. Calculate the wavelength of k_β line emitted by the same element.

[**Ans.** For hydrogen like atom, the wavelength of k lines is given by $1/\lambda_k = Z^2R\left(\frac{1}{1^2} - \frac{1}{n^2}\right)$.

For k_α line $n = 2$ $\therefore \frac{1}{\lambda_\alpha} = Z^2R\left(1 - \frac{1}{4}\right) = \frac{3Z^2R}{4}$ for k_β line $n = 3$ $\therefore \frac{1}{\lambda_\beta} = Z^2R\left(1 - \frac{1}{9}\right) = \frac{8Z^2R}{9}$

$\therefore \frac{\lambda_\beta}{\lambda_\alpha} = \frac{3}{4} \times \frac{9}{8}$, $\lambda_\alpha = 0.32$ Å $\therefore \lambda_\beta = \frac{27}{32} \times \lambda_\alpha = \frac{27}{32} \times 0.32 = 0.27$ Å]

19. A material whose K absorption edge is 0.2 Å is irradiated by X-rays of wavelength 0.15 Å. Find the maximum energy of the photo-electrons that are emitted from the K-shell.

[**Ans.** The binding energy of k shell electron is $E_k = \frac{hc}{\lambda} = \frac{6.6 \times 10^{-34} \times 3 \times 10^8}{0.2 \times 10^{-10}} = 9.9 \times 10^{-15}$ J.

The energy of incident photon is $E_p = \frac{hc}{\lambda_p} = \frac{6.6 \times 10^{-34} \times 3 \times 10^8}{0.15 \times 10^{-10}} = 13.2 \times 10^{-15}$ J. Hence maximum K.E. of the photo-electrons is $E = E_p - E_x = 3.3 \times 10^{-15}$ J $= 2.06 \times 10^4$ eV $= 20.6$ eV]

20. Frequency of a photon emitted due to transition of electron from L to K shell is found to be 4.2×10^{18} Hz. Using Moseley's law, find the atomic number of element, given that Rydberg's constant $R = 1.1 \times 10^7$ m^{-1}.

[**Ans.** Mosley's law gives relation between frequency ν and atomic number of the element as $\sqrt{\nu} = a(Z - b)$ where $a = \sqrt{\frac{3}{4}RC}$ and for electron transition from L to K shell $b = 1$. Hence

$$\sqrt{\nu} = \sqrt{\frac{3}{4}RC}\,(Z-1) \text{ or } Z - 1 = \sqrt{\frac{4\nu}{3RC}} = \left(\frac{4 \times 4.2 \times 10^{18}}{3 \times 1.1 \times 10^7 \times 3 \times 10^8}\right)^{1/2} = 41 \text{ or } Z = 42\,]$$

21. In a photo cell, photons of energy 8×10^{-19} J are incident on the cathode having work function 3 eV. For intensity $I_1 = 10^{-4}$ Wm^{-2} the saturation current is $i_1 = 2$ μA. Plot the current-voltage graph of the photo cell. How will the graph change if the intensity of the incident light is changed to $I_2 = 2 \times 10^{-4}$ Wm^{-2}.

[**Ans.** Energy of incident photon $= 8 \times 10^{-19}$ J $= 5$ eV. Work function $\phi_0 = 3$ eV. Hence stopping potential is $V = 5 - 3 = 2$ V. Figure 27.54 shows the graph (i-V) for the photo cell.

Saturation current is proportional to the incident intensity. The stopping potential is independent of the incident intensity. If intensity is doubled the saturation current is doubled]

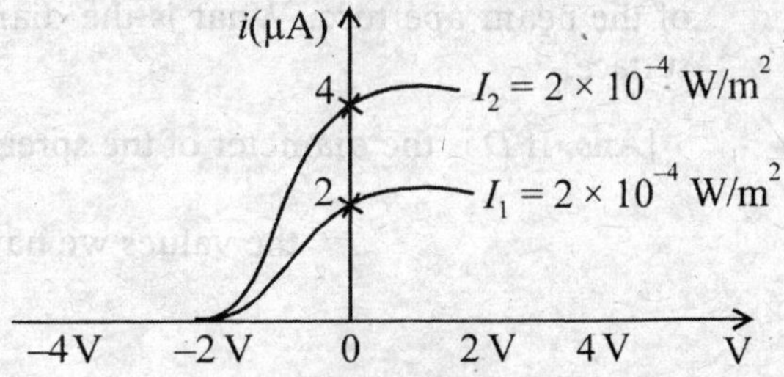

Fig. 27.54

22. An electron beam is generated when a potential difference of 40 kV is applied across an *X*-ray tube. A particular electron losses 50% of its energy in a collision with the metal anode. Calculate the wavelength of *X*-ray generated due to its collision?

[**Ans.** Since electron imparts 50% of its energy on collision hence excitation energy for atom is 20 k eV. Wavelength of the *X*-ray is $\lambda = \frac{hc}{eV} = \frac{12400}{V} \text{Å} = \frac{12400}{20 \times 10^3} = 0.62 \text{ Å}$]

23. If the *K*, *L*, *M* energy, levels of platinum involved in the emission of K-series of *X*-rays of this element are 78000 eV, 12000 eV and 3000 eV respectively. Calculate the wavelengths of K_α, and K_β lines for platinum.

[**Ans.** K_α line is obtained by transition of electron from *L*-shell to *K*-shell. Hence energy released in transition is $E = E_K - E_L = 78000 - 12000 = 66000$ eV. Hence $\lambda_\alpha = \frac{hc}{E_K - E_L} = \frac{12400 \text{ (eV} \cdot \text{Å)}}{66000 \text{ eV}}$

$= 0.188$ Å . Similarly K_β line is emitted. When electron jumps from *M*-shell to *K*-shell. Hence energy difference is $E = 78000 - 3000 = 75000$ eV and $\lambda_\beta = \frac{hc}{E_K - E_M} = \frac{12400 \text{ (eV} \cdot \text{Å)}}{75000 \text{ (eV)}} = 0.165 \text{ Å}$]

24. An *X*-ray tube with copper target is found to be emitting wavelength other than those due to copper. The K_α lines of copper is found to have a wavelength of 1.540 Å and the other two K_α lines observed because of the presence of other impurity, have wavelengths of 0.7092 Å and 1.6578 Å. Identify the impurities (Atomic number of copper = 29).

[**Ans.** Mosley's law gives, for K_α emission, $\nu = K(Z-1)^2$, where ν is frequency, *Z* atomic number of element and *K* is constant. $\therefore \frac{hc}{\lambda} = K(Z-1)^2 \; \therefore \frac{hc}{\lambda_x} = \frac{(Z_x - 1)^2}{(Z_C - 1)^2}$ where *C* stands for copper and *x* for unknown element. $\therefore (Z_x - 1)^2 = \frac{(Z_C - 1)^2 \cdot \lambda_C}{\lambda_x} = (29-1)^2 \times \frac{1.5404}{0.7092} \; \therefore Z_x - 1$

$= 28 \times 1.47 = 41.16$ or $Z_x = 42$. Similarly for other impurity $Z'_x - 1 = (Z_C - 1)\sqrt{\frac{1.5404}{1.6578}} = 27$ or $Z'_x = 28$]

25. A high powered laser beam ($\lambda = 650$ nm) with a beam diameter of 0.1 m is aimed at the moon. 3.5×10^8 m distant. The beam spreads only because of diffraction. The angular location of the edge of the central diffraction disc is given by $\sin\theta = \frac{1.22\lambda}{d}$, where *d* is the diameter

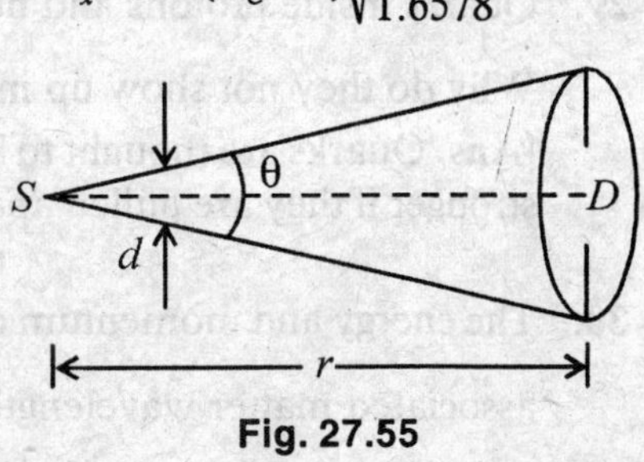

Fig. 27.55

of the beam aperture. What is the diameter of the central diffraction disc on the moon's surface?

[**Ans.** If D is the diameter of the spread on the moon than $\sin\theta \approx \tan\theta = \frac{D}{r}$. Substituting the values we have $d\sin\theta = 1.22\lambda$ or $\frac{dD}{r} = 1.22\lambda$ or $D = \frac{1.22\lambda r}{d}$ $\therefore$

$$D = \frac{1.22 \times 650 \times 10^{-9} \times 3.8 \times 10^{8}}{10^{-1}} = 3013.4 \text{ m}]$$

26. Monochromatic light of frequency 6.0×10^{14} Hz is produced by a laser. The power emitted is 2.0 mW (*a*) What is the energy of a photon in the light beam? (*b*) How many photons per second on the average, are emitted by the source?

[**Ans.** Energy of photons $= h\nu = 6.6 \times 10^{-34} \times 6 \times 10^{14}$ J $= 3.96 \times 10^{-19}$ J $= \frac{3.96 \times 10^{-19}}{1.6 \times 10^{-19}} = 2.5$ eV.

Number of photons emitted by the source is $N = \frac{\text{Power}}{\text{Energy of photon}} = \frac{2 \times 10^{-3}}{3.96 \times 10^{-19}}$

$$= 5 \times 10^{15} \text{ per sec}]$$

27. A particle is moving three lines as fast as an electron. The ratio of the de Broglie wavelength of the particle to that of the electron is 1.813×10^{-4}. Calculate the particle's mass and identify the particle.

[**Ans.** If v is the velocity of electron then velocity of particle is $3v$. Let masses be m and m_e respectively. Then $\lambda = \frac{h}{m(3v)}$ and $\lambda_e = \frac{h}{m_e v}$ or $\frac{\lambda}{\lambda_e} = \frac{m_e v}{3mv} = \frac{m_e}{3m} = 1.813 \times 10^{-4}$ or

$$m = \frac{m_e}{3 \times 1.813 \times 10^{-4}} = \frac{9.1 \times 10^{-31}}{3 \times 1.813 \times 10^{-4}} = 1.67 \times 10^{-27} \text{ kg}. \text{ Particle is proton}]$$

28. In an accelerator experiment on high energy collisions of electrons with positrons, a certain event is interpreted as annihilation of an electron-positron pair of total energy 10.2 BeV into two γ-rays of equal energy. What is the wavelength associated with each γ-ray? (1 BeV = 10^9 eV)

[**Ans.** Energy of each γ-ray photon is $\frac{10.2}{2} = 5.1$ BeV $= 5.1 \times 10^9$ eV. Wavelength $\lambda = \frac{hc}{\text{energy}}$

$$= \frac{6.6 \times 10^{-34} \times 3 \times 10^{8}}{5.1 \times 10^{9} \times 1.6 \times 10^{-19}} = 2.43 \times 10^{-16} \text{ m}]$$

29. Quarks inside protons and neutrons are thought to carry fractional charges $\left[\left(+\frac{2e}{3}\right), \left(-\frac{e}{3}\right)\right]$. Why do they not show up in Millikans oil-drop experiment?

[**Ans.** Quarks are thought to be confined within a proton or neutron by forces which becomes stronger if they are pulled apart. It is therefore concluded that fractional charges though may exist in nature yet observable charges are integral multiples of e]

30. The energy and momentum of an electron are related to the frequency and wavelength of the associated matter wavelength by relation $E = h\nu$ and $P = \frac{h}{\lambda}$. But while the value of λ is physically significant, the value of ν and therefore the value of phase speed $\nu\lambda$, has no physical significance. Why?

[**Ans.** The absolute value of energy E of any particle is arbitrary but the momentum is not. Because for energy an additive constant is necessary. As such λ has physical significance whereas frequency ν of a matter wave does not have. Similarly phase speed ($= \nu\lambda$) is not physically significant. The group speed (v_g) $v_g = \frac{\partial\omega}{\partial K} = \frac{\partial\nu}{\partial(1/\lambda)} = \frac{dE}{dp} = \frac{d}{dp}\left(\frac{p^2}{2m}\right) = \frac{p}{m}$ is physically meaningful]

31. Every metal has a definite work function. Why do all photo-electrons not come out with the same energy if incident radiation is monochromatic? Why is there an energy distribution of photo electrons?

[**Ans.** Work function merely is the minimum energy required for the electron in the highest level of the conduction band to get out of the metal. Not all electrons in the metal belong to this level. They occupy a continuous band of levels. Consequently, for the same incident radiation electrons knocked off from different levels come out with different energies]

32. A cobalt target is bombarded with electrons and the wavelengths of its characteristic X-ray spectrum are measured. There is also a second fainter characteristic spectrum which is due to an impurity in cobalt. The wavelengths of K_α lines are 178.9 pm (cobalt) and 143.5 pm (impurity). Using these data, identify the impurity.

[**Ans.** Mosley's law gives $\nu \propto (Z-\sigma)^2$, where ν is frequency of radiation, Z atomic number and σ screening constant. σ for k-lines is equal to 1. $\therefore \frac{hc}{\lambda} = K(Z-\sigma)^2$. For cobalt $\frac{hc}{\lambda_c} = K(Z_c-1)^2$ and for impurity $\frac{hc}{\lambda_i} = K(Z_i-1)^2$ Hence $\frac{\lambda_i}{\lambda_C} = \frac{(Z_C-1)^2}{(Z_i-1)^2}$ $Z_C = 27$. $\therefore \frac{143.5}{178.9} = \frac{(27-1)^2}{(Z_i-1)^2} = 0.8$

or $\frac{27-1}{Z_i-1} = 0.9$ or $Z_i - 1 = \frac{26}{0.9} = 29$ or $Z_i = 30$]

33. A helium atom consists of two electrons orbiting round a nucleus of charge $Z = 2$. But the electrons do not see the full charge $Z = 2$ of the nucleus. Each electron sees the nucleus slightly screened by other electron so that the effective charge Z_{eff} seen by each electron is less than 2. The ionization potential for a He atom in its ground state is measured experimentally to be 24.46 V. Estimate the effective charge of the nucleus as seen by each electron in the helium ground state.

[**Ans.** If each electron see schange Z_{eff} then each electron has energy in ground state as $-13.6\ Z_{eff}^2$ and the energy of helium atom in ground state $= -2 \times 13.6\ Z_{eff}^2 = -27.2\ Z_{eff}^2$. Energy of helium in ion (He^+) in ground state $= -54.4$ eV (13.6×4) for He^+ there is no screening. Ionization potential then is $V = -54.4 - (-27.2\ Z_{eff}^2) = 24.46$ (Experimental value). This gives $Z_{eff} = 1.7$]

34. A positronium atom is a bound state of an electron (e^-) and its antiparticle, the positron (e^+) revolving round their centre of mass. In which part of the electromagnetic spectrum does the system radiate when it de-excites from its first excited state to the ground state?

[**Ans.** In an atom nucleus is much heavier than electron as such its motion is ignored. In positronium the mass of the two particles is equal as such the two will move about their centre of mass. Such a system can be reduced to a single particle system with mass equal to the reduced mass of the two particles. Which in present case is $m_e/2$ (where m_e is mass of electron). Since in hydrogen atom energy in nth orbit is given by $E_n = -\dfrac{me^4}{8\varepsilon_0^2 h^2 n^2} = -\dfrac{13.6}{n^2}$ eV and for present system it will be $E_n = -\dfrac{me^4}{16\varepsilon_0^2 h^2 n^2} = -\dfrac{6.8}{n^2}$ eV. Hence transition from $n = 2$ to $n = 1$ will release energy $= 6.8 \times \dfrac{3}{4} = 5.1$ eV and the wavelength of the radiation $\lambda = \dfrac{hc}{E} = \dfrac{12400}{5.1} = 2431$ Å]

35. A molecule of gas, filled in a discharge tube, gets ionized when an electron is detached from it. An electric field of 5.0 kV/m exists in the vicinity of the event. (*a*) Find the distance travelled by the free electron in 1 μs assuming no collision. (*b*) If mean free path of the electron is 1.0 mm, estimate the time of transit of the free electron between successive collisions.

[**Ans.** (*a*) Acceleration of electron $a = \dfrac{eE}{m}$. Hence distance travelled by electron in time $t =$ 1 μs is $S = \dfrac{1}{2}at^2 = \dfrac{1}{2}\dfrac{eE}{m}t^2 = \dfrac{1}{2}\dfrac{1.6 \times 10^{-19} \times 5 \times 10^3 \times (10^{-6})^2}{9.1 \times 10^{-31}} = 440$ m (*b*) Time for $S = 1.0$ mm $= 10^{-3}$ m is $t = \sqrt{\dfrac{2S}{a}} = \sqrt{\dfrac{2 \times 10^{-3} \times 9.1 \times 10^{-31}}{1.6 \times 10^{-19} \times 5 \times 10^3}} = 1.5 \times 10^{-9}$ s $= 1.5$ ns]

36. When the electron in hydrogen atom makes transition from $n = 3$ to $n = 2$ in 10^{-8} s. Find the order of torque acting on the electron in this period.

[**Ans.** When the electron makes transition from $n = 3$ to $n = 2$ the change in angular momentum

$$\Delta L = \frac{3h}{2\pi} - \frac{2h}{2\pi} = \frac{h}{2\pi}. \text{ The torque } \tau = \frac{\Delta L}{\Delta t} = \frac{h}{2\pi \times 10^{-8}} = \frac{6.6 \times 10^{-34}}{2\pi \times 10^{-8}} = 10^{-26} \text{ N-m} \Big]$$

37. Find the magnetic field at the centre (nucleus) of hydrogen like atoms (atomic number Z) due to motion of electron in nth orbit.

[**Ans.** If frequency of electron in nth orbit is f_n then it is equivalent to a current loop with current $i_n = ef_n$. Hence magnetic field at the centre is $B = \dfrac{\mu_0 i_n}{2r_n} = \dfrac{\mu_0 ef_n}{2r_n}$ or $B \propto \left(\dfrac{f_n}{r_n}\right)$, but $f_n = \dfrac{v_n}{2\pi r_n}$.

$\therefore\ B \propto \left(\dfrac{v_n}{r_n^2}\right)$ But $v_n \propto Z/n$ and $r_n \propto n^2/Z$ or $B \propto \dfrac{(Z/n)}{(n^2/Z)^2}$ or $B \propto \dfrac{Z^3}{n^5}$ and constant of proportionality is $\dfrac{\pi\mu_0 m^2 e^7}{8\varepsilon_0^3 h^5}$]

38. A hydrogen atom is in an excited state of principal quantum number n. It emits a photon of wavelength λ when returns to the ground state. Find the value of n.

[**Ans.** When electron makes transition from state n to ground state the wavelength emitted is $1/\lambda = R\left(\dfrac{1}{1^2} - \dfrac{1}{n^2}\right)$ when R is Rydberg constant. $\therefore\ 1 - \dfrac{1}{n^2} = \dfrac{1}{R\lambda}$ or $\dfrac{1}{n^2} = 1 - \dfrac{1}{R\lambda} = \dfrac{R\lambda - 1}{R\lambda}$ or

$$n^2 = \frac{R\lambda}{R\lambda - 1} \quad \text{or} \quad n = \sqrt{\frac{R\lambda}{R\lambda - 1}} \Big]$$

39. Find the ratio of the maximum wavelength of Lyman series in hydrogen spectrum to the maximum wavelength in Paschen series.

[**Ans.** In hydrogen spectrum wavelength is given by $\frac{1}{\lambda} = R\left(\frac{1}{n_1^2} - \frac{1}{n_2^2}\right)$. For Lymann series $n_1 = 1, n_2 = 2, 3, \ldots$, etc. For maximum wavelength $n_2 = 2$. $\therefore \frac{1}{\lambda_L} = R\left(\frac{1}{1^2} - \frac{1}{4}\right) = \frac{3R}{4}$ or $\lambda_L = \frac{4}{3R}$.

For Paschen series $n_1 = 3$ and $n_2 = 4, 5, \ldots$, etc. For maximum wavelength $n_2 = 4$. $\therefore \frac{1}{\lambda_P} = R\left(\frac{1}{9} - \frac{1}{16}\right)$ $= \frac{7R}{144}$ or $\lambda_p = \frac{144}{7R}$ and ratio $\frac{\lambda_L}{\lambda_P} = \frac{4}{3R} \times \frac{7R}{144} = \frac{7}{108}$]

40. A neutron moving with speed v makes a head-on collision with a hydrogen atom in ground state kept at rest. Find the minimum kinetic energy of neutron for which inelastic collision will take place. Assume the mass of neutron to be nearly equal to that of proton.

[**Ans.** Assuming collision to be perfectly inelastic, the velocity after collision is $v/2$ and kinetic energy of the system is $\frac{1}{2}(2m)\frac{v^2}{4}$ or $\frac{mv^2}{4}$. Hence loss in kinetic energy due to collision is $\Delta E = \frac{1}{2}mv^2 - \frac{1}{4}mv^2 = \frac{1}{4}mv^2 = \frac{1}{2}E$, where E is initial K.E. of neutron. For excitation of hydrogen atom this must be 10.2 eV or E = 20.4 eV]

41. Find the maximum angular speed of electron of hydrogen atom in a stationary orbit.

[**Ans.** Maximum angular speed will be in its ground state or $\omega = \frac{v_1}{r_1}$, where $v_1 = \frac{e^2}{2\varepsilon_0 h} =$ 2.2×10^6 m/s and $r_1 = \frac{\varepsilon_0 h^2}{\pi m e^2} = 0.51 \times 10^{-10}$ m. $\therefore \omega = \frac{2.2 \times 10^6}{6.51 \times 10^{-10}} = 4.1 \times 10^{16}$ rad/s]

42. Find the ratio of the acceleration of electron in singly ionized helium atom and hydrogen atom, both in ground state.

[**Ans.** Acceleration is $a = \frac{v^2}{r}$ where velocity in ground state of atom of atomic number Z is

$v = \frac{e^2 Z}{2\varepsilon_0 h}$ and $r = \frac{\varepsilon_0 h^2}{\pi m e^2 Z}$ $\therefore a = \frac{v^2}{r} = \left(\frac{e^2 Z}{2\varepsilon_0 h}\right)^2 \times \left(\frac{\pi m e^2 Z}{\varepsilon_0 h^2}\right)$ or $a \propto Z^3$ $\therefore \frac{a_1}{a_2} = \left(\frac{Z_1}{Z_2}\right)^3 = 8$]

43. The ground state and first excited state energies of hydrogen atom are –13.6 eV and –3.4 eV respectively. If the potential energy in ground state is taken to be zero; then find (*i*) potential energy in first excited state (*ii*) kinetic energy in first excited state (*iii*) total energy in ground state.

[**Ans.** Under our model total energy is negative of kinetic energy and potential energy is twice total energy. Thus in ground state kinetic energy is 13.6 eV, potential energy is –27.2 eV. Now when potential energy is taken zero in ground state as such potential energy in each orbit is increased by 27.2 eV and K.E. remains unchanged. Thus in 1st excited state P.E. = –2 × 3.4 + 27.2 = 20.4 eV and K.E. = 3.4 eV and total energy = 20.4 + 3.4 = 23.8 eV. In ground state P.E. = 0, K.E. = 13.6 and total energy = 13.6 eV]

44. Binding energy of an electron in the ground state of He is equal to 24.6 eV. Find the energy required to remove both the electrons.

[**Ans.** When one electron is removed, the remaining atom is hydrogen like atom whose ionization energy is $E_1 = (13.6)Z^2 = 13.6 \times 4 = 54.4$ eV. Hence to remove second electron additional energy required is 54.4 eV. Thus to remove both electrons total energy required is

$$E = 24.6 + 54.4 = 79 \text{ eV}]$$

45. In hydrogen atom, an electron makes a de-excitation transition from $n = 5$ such that change of radius of the orbit is 21 times the radius of that orbit in which the binding energy of the electron is -13.6 eV. Find the percentage change in angular momentum due to transition.

[**Ans.** Let the electron de-excites to level p. The radius of orbit is proportional to square of the principal quantum number. This means due to de-excitation change in radius is $\Delta r \propto (5^2 - P^2)$ or $\Delta r = K(5^2 - P^2)$. This change is 21 times the radius of first orbit of the hydrogen atom, i.e. $\Delta r = k1^2$. Equating $P = 2$. Hence angular momentum changes by $\Delta L = \frac{5h}{2\pi} - \frac{2h}{2\pi} = \frac{3h}{2\pi}$.

Thus percentage change is $3/5 \times 100 = 60\%$.]

MULTIPLE CHOICE QUESTIONS

PART I

In the following, a question is followed by four alternatives, of them only one is correct. Therefore, the candidate has to select the correct choice from amongst the alternatives.

1. The de Broglie wavelength associated with an electron accelerated by a potential difference of 100 V is

(a) 0.123 nm (b) 1.23 nm
(c) 12.3 nm (d) 0.0123 nm.

2. Cathode rays are passing through a discharge tube. In the tube there is

(a) an electric field but no magnetic field
(b) a magnetic field but no electric field
(c) an electric as well as magnetic field
(d) neither an electric field nor a magnetic field.

3. Cathode rays constitute a stream of

(a) electrons (b) protons
(c) positive ions (d) negative ions.

4. Two discharge tubes have identical material structure and the same gas is filled in them. The length of one tube is 10 cm and that of the other tube is 20 cm. Sparking starts in both the tubes when the potential difference between the cathode and anode is 100 V. If the pressure in the shorter tube is 1.0 nm of mercury, the pressure in the bigger tube is (in mm of mercury)

(a) 1.0 (b) 0.5
(c) 4 (d) 0.25.

5. Two photons having

(a) equal wavelengths have equal linear momenta
(b) equal energies have equal linear momenta
(c) equal frequencies have equal linear momenta
(d) equal linear momentum have equal wavelengths.

6. Let P and E denote the linear momentum and energy of a photon. If the wavelength is decreased

(a) both P and E increase
(b) P increases and E decreases
(c) P decreases and E increases
(d) both P and E decrease.

7. Planck's constant has the same dimensions as
(a) Force × distance
(b) Force × speed
(c) Force × distance × time
(d) Force × time.

8. A hydrogen atom absorbs 10.2 eV energy when in its ground state, orbital angular momentum of the electron increases by
(a) 1.05×10^{-34} J (b) 2.11×10^{-34} Js
(c) 3.16×10^{-34} Js (d) 4.22×10^{-34} Js.

9. Which of the following parameters are the same for all hydrogen like atoms and ions in their grounds state?
(a) Radius of the orbit
(b) Speed of the electron
(c) Energy of the atom
(d) Orbital angular momentum of the electron.

10. The minimum orbital angular momentum of the electron in hydrogen atom is
(a) h (b) $h/2$
(c) $\frac{h}{2\pi}$ (d) $\frac{h}{\lambda}$.

11. In a laser tube all the photons
(a) have same energy
(b) move with same speed
(c) move in same direction
(d) have same wavelength.

12. A red bulb and a blue bulb of equal power when lighted emit photons. If n_r and n_b represent the number of photons emitted by red and blue bulb respectively in a given time, then
(a) $n_r < n_b$
(b) $n_r > n_b$
(c) $n_r = n_b$
(d) the information is insufficient to get relation between n_r and n_b.

13. The frequency and the intensity of a light source are both doubled. Consider the following statements
(A) The saturation photo current remains almost same
(B) The maximum kinetic energy of the photo-electrons is doubled
(a) Both A and B are true
(b) Both A and B are false
(c) A is true but B is false
(d) A is false but B is true.

14. A chromatic light is used in an experiment on photoelectric effect. The stopping potential
(a) is related to mean wavelength
(b) is related to longest wavelength
(c) is related to shortest wavelength
(d) is not related to wavelength.

15. If the frequency of light in a photoelectric experiment is doubled the stopping potential will
(a) be doubled
(b) be halved
(c) become more than double
(d) become less than double.

16. Which of the following curves may represent the speed of the electron in a hydrogen atom as a function of the principal quantum number?

(a) 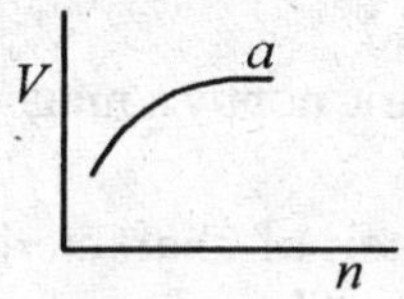

(b) 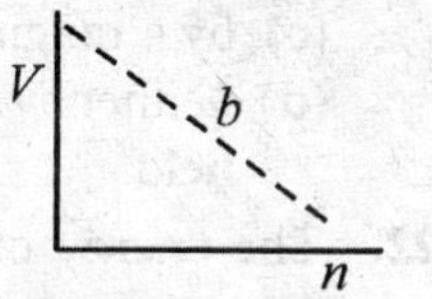

(c) 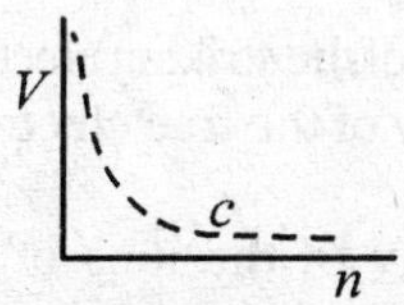

(d) 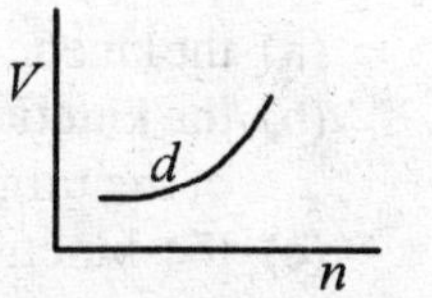

17. Three photons coming from excited atomic-hydrogen sample are picked up. Their energies are 12.1 eV, 10.2 eV and 1.9 eV. These photons must come from
(a) a single atom
(b) two atoms
(c) three atoms
(d) either two or three atoms.

18. In which of the following systems will the radius of the first orbit (n = 1) be maximum?

(a) Hydrogen atom
(b) Deuterium atom
(c) Singly ionized helium atom
(d) Doubly ionized lithium atom.

19. In which of the following systems will the wavelength corresponding to $n = 2$ to $n = 1$ be minimum?
(a) Hydrogen atom
(b) Singly ionized helium
(c) Deuterium atoms
(d) Doubly ionized lithium.

20. If the potential difference applied to the tube is doubled and the separation between the filament and the target is also doubled, the cut-off wavelength
(a) will remain unchanged
(b) will be doubled
(c) will be halved
(d) will become four times the original one.

21. *X*-rays can be deflected
(a) by an electric field
(b) by an electric as well as by magnetic field
(c) by a magnetic field
(d) neither by an electric nor by a magnetic field.

22. The energy of a photon of characteristic *X*-ray from a Coolidge tube comes from
(a) the kinetic energy of the striking electron
(b) the kinetic energy of the free electrons of the target
(c) the kinetic energy of the ions of the target
(d) an atomic transition in the target.

23. According to Bohr's theory of hydrogen atom, the product of the binding energy of the electron in the *n*th orbit and its radius in the *n*th orbit
(a) is proportional to n^2
(b) is inversely proportional n^3
(c) has a constant value of 102 eV-Å
(d) has a constant value of 7.2 eV-Å.

24. An electron and a photon have same wavelength. If P is the momentum of electron and E the energy of photon. The magnitude of P/E in SI units is
(a) 3.0×10^8 (b) 3.33×10^{-9}
(c) 9.1×10^{-31} (d) 6.64×10^{-34}.

25. In *X*-ray tube when the accelerating voltage V is halved, the difference between the wavelength of k_α-line and minimum wavelength of continuous *X*-ray spectrum
(a) remains constant
(b) becomes more than two times
(c) becomes half
(d) becomes less than two times.

26. The de Broglie wavelength of electron in ground state of an hydrogen atom is
(a) 1.06 Å (b) 1.52 Å
(c) 0.53 Å (d) 3.33 Å.

27. A potential difference of 10^3 V is applied across an *X*-ray tube. The ratio of the de Broglie wavelength of incident electrons to the shortest wavelength of the *X*-rays emitted is
(a) $\frac{1}{20}$ (b) $\frac{1}{100}$
(c) 1 (d) $\frac{1}{10^4}$.

28. When stopping potential is applied in an experiment on photoelectric effect, no photo- current is observed. This means that
(a) the emission of photo-electrons is stopped
(b) the photo-electrons are emitted but are re-absorbed by the emitter metal
(c) the photo-electrons are accumulated near the collector
(d) the photo-electrons are dispersed from the sides of the apparatus.

29. A point source causes photoelectric effect from a small metal plate. Which of the following curves may represent the saturation photo current as a function of the distance between the source and metal?

(a) 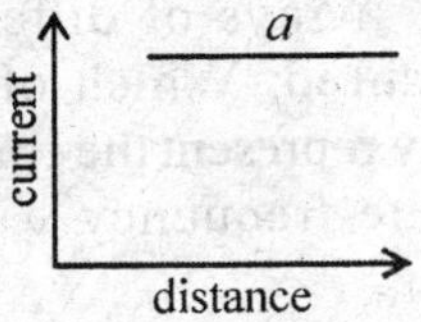

(b) 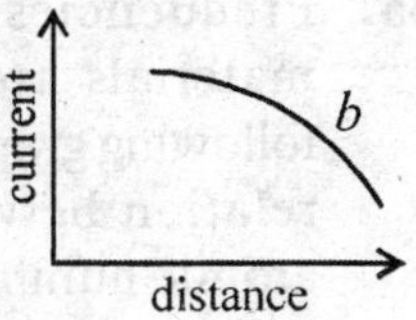

(c) 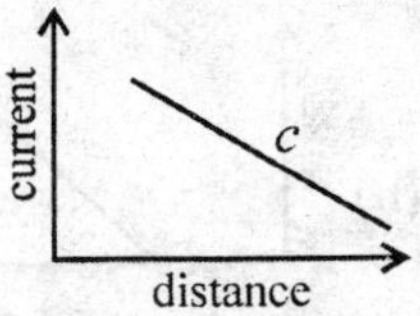

(d) 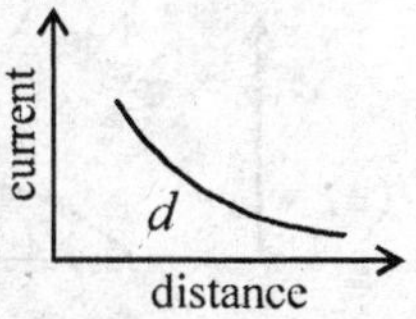

30. A proton and an electron are accelerated by the same potential difference. If λ_e and λ_p are the de Broglie wavelengths of electron and proton respectively, then

(a) $\lambda_e = \lambda_p$ (b) $\lambda_e > \lambda_p$
(c) $\lambda_e < \lambda_p$
(d) depends on the accelerating voltage.

31. The equation $E = Pc$ is valid (c speed of light)

(a) for an electron as well as for a photon
(b) for an electron but not for a photon
(c) for a photon but not for an electron
(d) neither for an electron nor for a photon.

32. The photo current in an experiment on photoelectric effect increases if

(a) the intensity of the source is increased
(b) the exposure time is increased
(c) the intensity of the source is decreased
(d) the exposure time is decreased.

33. The collector plate in an experiment on photoelectric effect is kept vertically above the emitter plate. Light source is put on and a saturation photo current is recorded. An electric field is switch on which has a vertically downward direction

(a) the photo current will increase
(b) the kinetic energy of photo-electrons will increase
(c) the stopping potential will decrease
(d) the threshold wavelength will increase.

34. A photon of energy $h\nu$ is absorbed by a free electron of a metal having work function $\phi < h\nu$

(a) the electron is sure to come out
(b) the electron is sure to come out with kinetic energy $h\nu - \phi$
(c) either the electron does not come out or it comes out with a kinetic energy $h\nu - \phi$
(d) it may come out with a kinetic energy less than $h\nu - \phi$.

35. An electron with kinetic energy 5 eV is incident on a hydrogen atom in its ground state. The collision

(a) must be elastic
(b) may be partially elastic
(c) must be completely inelastic
(d) may be completely inelastic.

36. Ionization energy of a hydrogen like ion A is greater than that of another hydrogen like ion *B*. Let *r*, *v*, *E* and *L* represent the radius of the orbit, speed of the electron, energy of the atom and orbital angular momentum of the electron respectively. In the ground state

(a) $r_A > r_B$ (b) $v_A > v_B$
(c) $E_A > E_B$ (d) $L_A > L_A$.

37. Consider a photon of *X*-ray coming from a Coolidge tube. Its energy comes from

(a) the kinetic energy of the striking electron
(b) the kinetic energy of the free electrons of the target
(c) the kinetic energy of the target
(d) the atom transition in the target.

38. If the current in the circuit for heating the filament is increased, the cut-off wavelength

(a) will increase
(b) will decrease
(c) will remain unchanged
(d) will change.

39. The minimum kinetic energy of an electron, hydrogen ion, helium ion required for ionization of a hydrogen atom is E_1 in case electron is collided with hydrogen atom. It is E_2 if hydrogen ion is collided and E_3 when the helium ion is collided. Then

(a) $E_1 = E_2 = E_3$ (b) $E_1 > E_2 > E_3$
(c) $E_1 < E_2 < E_3$ (d) $E_1 > E_3 > E_2$.

40. In hydrogen and hydrogen like atoms the ratio difference $E_{4n} - E_{2n}$ and $E_{2n} - E_n$ varies with atomic number Z and principal quantum number n as

(a) $\frac{Z^2}{n^2}$ (b) $\frac{Z^4}{n^4}$
(c) $\frac{Z}{n}$ (d) none of the above.

41. Angular momentum (L) and radius (r) of a hydrogen atom are related as
(a) Lr = constant (b) Lr^2 = constant
(c) Lr^4 = constant (d) none of the above.

42. A particle of mass $3m$ at rest decays into two particles of mass m and $2m$ having non-zero velocities. The ratio of the de Broglie wavelengths of the particles (λ_1/λ_2) is
(a) 1/2 (b) 1/4
(c) 2 (d) none of these.

43. The radius of second orbit of an electron in hydrogen atom is 2.116 Å. The de-Broglie wavelength associated with this electron in this orbit is
(a) 6.64 Å (b) 1.058 Å
(c) 2.116 Å (d) 13.28 Å.

44. Figure 27.56 shows the intensity wavelength relations for X-rays coming from two different Coolidge tubes. In (A) potential difference V_a is applied and target material has atomic number Z_b. The quantities for B at Z_b and V_b. Then

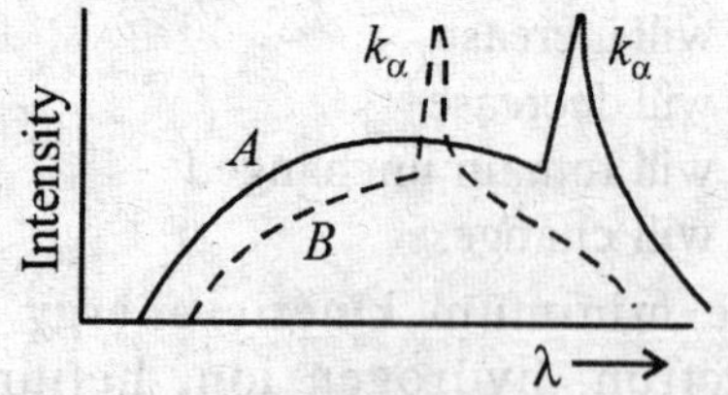

Fig. 27.56

(a) $V_a > V_b$, $Z_a > Z_b$
(b) $V_a > V_b$, $Z_a < Z_b$
(c) $V_a < V_b$, $Z_a > Z_b$
(d) $V_a < V_b$, $Z_a < Z_b$.

45. Frequencies of K_α X-rays of different materials are measured. Which of the following graphs may represent the correct relation between the frequency ν and atomic number Z.

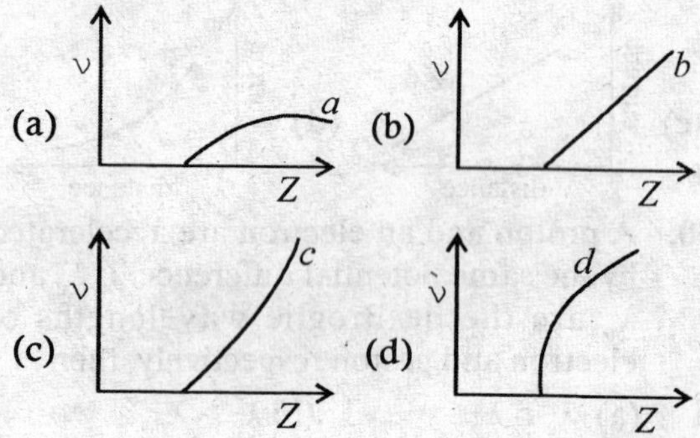

46. Consider a photon of continuous X-ray and a photon of characteristic X-ray of same wavelength. Which of the following is different for the two photons?
(a) Frequency
(b) Energy
(c) Penetrating power
(d) Method of creation.

47. A metallic surface ejects photo-electrons (n per sec) when exposed to green colour light of certain intensity I. The long wavelength limit for the surface being 560 nm. If the surface is exposed to same intensity I of green, yellow and red light simultaneously. Then the number of electrons emitted per sec will be
(a) n (b) $2n$
(c) $3n$ (d) $9n$.

48. A photon of energy 10.5 eV is allowed to collide with a hydrogen atom in its ground state. Then
(a) the photon is completely absorbed by the hydrogen atom
(b) the photon cannot excite the H-atom and comes out with energy 10.5 eV
(c) the photon transfers 10.2 eV energy to H-atom exciting it to first excited state
(d) none of these.

49. An X-ray tube is operating at 150 kV and 10 mA. If only 1% of the electric power is converted into X-rays, the rate at which the target is heated in calories per second is

(a) 3.55 (b) 35.5
(c) 355 (d) 3550.

50. A photon having wavelength 6000 Å moves from vacuum to a medium of refractive index 1.5, the values of photon energy in vacuum and in medium are

(a) 2.06 eV and 3.09 eV
(b) 2.06 eV and 1.37 eV
(c) 2.06 eV and 2.06 eV
(d) 1.37 eV and 2.06 eV.

51. According to Bohr model, magnetic field at the centre of a hydrogen atom due to the motion of electron in nth orbit is proportional to

(a) $1/n^3$ (b) $1/n^5$
(c) n^5 (d) n^3.

52. A hydrogen sample is prepared in a particular excited state. Photons of energy 2.55 eV get absorbed into the sample to take some of the electrons to a further excited state B. The quantum numbers corresponding to the state A and B respectively are

(a) 2 and 4 (b) 3 and 5
(c) 3 and 6 (d) 4 and 7.

53. The angular momentum of an electron in an orbit is quantized because it is a necessary condition for the compatibility with

(a) the wave nature of electron
(b) the particle nature of electron
(c) Pauli's exclusion behaviour
(d) none of the above.

54. The shortest wavelength of the Bracket series of a hydrogen like atom (atomic number Z) is the same as the shortest wavelength of the Balmer series of hydrogen atom. The value of Z is

(a) 2 (b) 3
(c) 4 (d) 5.

55. When an electron in hydrogen atom in ground state absorbs a photon of energy 12.1 eV its angular momentum

(a) decreases by 2.11×10^{-34} J-s
(b) decreases by 1.055×10^{-34} J-s
(c) increases by 2.11×10^{-34} J-s
(d) increases by 1.055×10^{-34} J-s.

56. An ionized gas contains both positive and negative ions. If it is subjected simultaneously to an electric field along the +ve X-direction and a magnetic field along +ve Z-direction, then (IIT 2000)

(a) positive ions deflect towards $+Y$ direction and negative ions towards –ve Y-direction
(b) all ions deflect towards $+Y$ direction
(c) all ions deflect towards $-Y$ direction
(d) positive ions deflect towards $-Y$ direction and negative ions towards +ve Y-direction.

57. The electron in a hydrogen atom makes a transition from excited state to ground state. Which of the following statement is true? (IIT 2000)

(a) Its kinetic energy increases and its potential and total energy decreases
(b) Its kinetic energy decreases, potential energy increases and total energy remains the same
(c) Its kinetic energy and total energy decrease and its potential energy increases.
(d) Its kinetic, potential and total energy decreases.

58. The intensity of X-rays from a Coolidge tube is plotted against wavelength λ as shown in Fig. 27.57. The minimum wavelength is λ_c and the wavelength of the K_α is λ_k. As the accelerating voltage is increased

(a) $\lambda_k - \lambda_c$ increases (b) $\lambda_k - \lambda_c$ decreases
(c) λ_k increases (d) λ_k decreases.

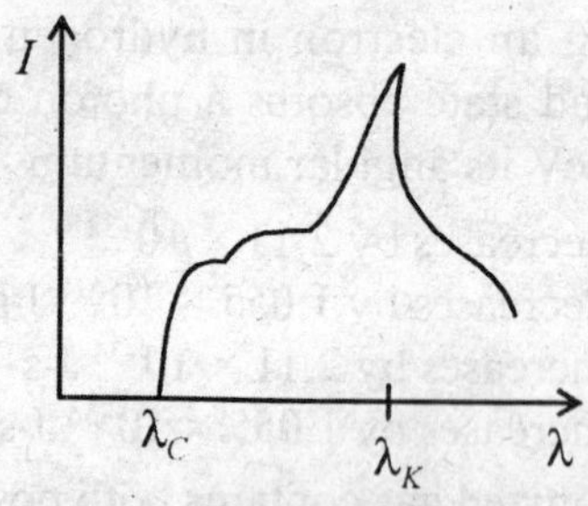

Fig. 27.57

59. The transition from the state $n = 4$ to $n = 3$ in a hydrogen like atom results in ultra-violet radiation. Infrared radiation will be obtained in the transition

(a) $2 \to 1$ (b) $3 \to 2$
(c) $4 \to 2$ (d) $5 \to 4$.

60. The potential difference applied to *X*-ray tube is 5 kV and the current through the tube is 3.2 mA. Then the number of electrons striking the largest per second is

(a) 2×10^{16} (b) 5×10^{6}
(c) 1×10^{17} (d) 4×10^{15}.

61. A hydrogen like atom and a Li^{++} ion are both in the second excited state. If l_H and l_{Li} are their respective electronic angular momenta and E_H, E_{Li} their respective energies, then (IIT 2000)

(a) $l_H > l_{Li}$, $|E_H| > |E_{Li}|$
(b) $l_H = l_{Li}$, $|E_H| < |E_{Li}|$
(c) $l_H = l_{Li}$, $|E_H| > |E_{Li}|$
(d) $l_H < l_{Li}$, $|E_H| < |E_{Li}|$.

62. The energy of a photon is equal to the kinetic energy of a proton. The energy of photon is E. Let λ_1 be the de Broglie wavelength of the proton and λ_2 be the wavelength of the photon. Then λ_1/λ_2 is proportional to

(a) E^0 (b) $E^{1/2}$
(c) E^{-1} (d) E^{-2}.

63. The Fig. 27.58 shows the variation of photo current with anode potential for a photosensitive surface for three different radiations. Let I_a, I_b and I_c be the intensities and f_a, f_b, f_c be the frequencies for curves *a*, *b*, and *c* respectively, then (IIT 2000)

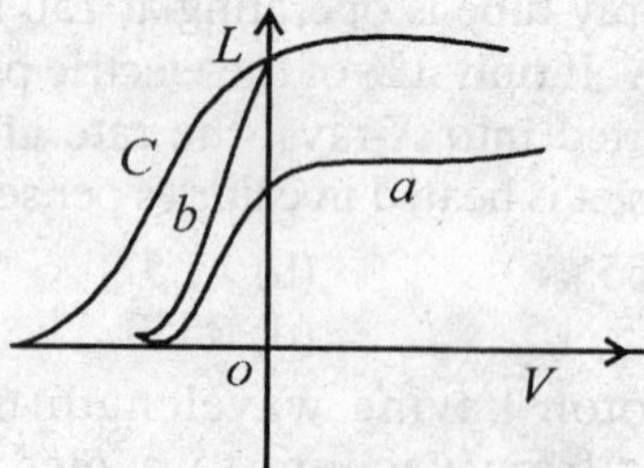

Fig. 27.58

(a) $f_a = f_b$ and $I_a \neq I_b$
(b) $f_a = f_c$ and $I_a = I_c$
(c) $f_a = f_b$ and $I_a = I_b$
(d) $f_b = f_c$ and $I_b = I_c$.

64. The wavelength of K_α *X*-ray of an element having atomic number $Z = 11$ is λ. The wavelength of another element of atomic number Z' is 4λ. Then Z' is

(a) 11 (b) 44
(c) 6 (d) 4.

65. A photon of energy 10.2 eV collides with a hydrogen atom in ground state inelastically. After a few micro seconds one more photon of energy 15 eV collides with the same hydrogen atom. Then what can be detected by suitable detector? (IIT 2005)

(a) One photon of 10.2 eV and an electron of energy 1.4 eV
(b) Two photons of energy 10.2 eV
(c) Two photons of energy 3.4 eV
(d) One photon of energy 3.4 eV and one electron of 1.4 eV.

66. The work function of metals A, B and C are respectively 1.92 eV, 2.0 eV and 5.0 eV. According to Einstein's equation, the metals which will emit photo-electrons for a radiation of wavelength 4100 Å is (CBSE PMT 2005)

(a) none
(b) A only
(c) A and B only
(d) all the three metals.

67. A photo sensitive metallic surface has work function $h\nu_0$. If photon of energy $2h\nu_0$ fall on the surface, the electrons come out with maximum velocity of 4×10^6 m/s. When the photon energy is increased to $5h\nu_0$, then the velocity (maximum) of photo electrons will be (CBSE PMT 2005)

(a) 2×10^7 m/s (b) 2×10^6 m/s
(c) 8×10^5 m/s (d) 8×10^6 m/s.

68. The total energy of an electron in first excited state of hydrogen atom is about –3.4 eV. Its kinetic energy in this state is (CBSE PMT 2005)

(a) –6.8 eV (b) 3.4 eV
(c) 6.8 eV (d) –3.4 eV.

69. When light of intensity I_1 and frequency ν_1 is incident on surface, photoelectric emission takes place and variation of current (i) with accelerating/retarding potential (V) is shown in Fig. 27.59 as curve 1. However if light of intensity I_2 and frequency ν_2 falls on the same metal surface, the i vs V graph is as shown 2. Which of the following is correct?

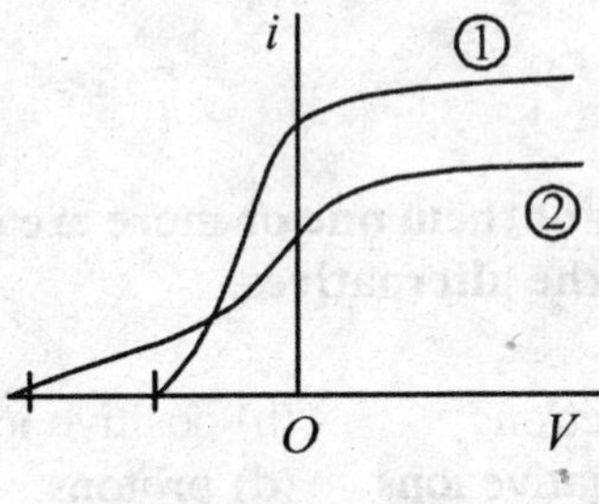

Fig. 27.59

(a) $I_1 = I_2, \nu_1 > \nu_2$ (b) $I_1 > I_2, \nu_1 > \nu_2$
(c) $I_1 > I_2, \nu_1 < \nu_2$ (d) $I_1 < I_2, \nu_1 < \nu_2$.

70. Photons of energy 5 eV fall on the surface of a metal X resulting in emission of photo-electrons having maximum kinetic energy E (eV) and de Broglie wavelength λ. Y is another metal on the surface of which photons of energy 6 eV are incident; the photo-electron of maximum kinetic energy $(E - 2)$ eV are emitted and has de Broglie wavelength $\sqrt{3}\,\lambda$. Work function of metals X and Y are in ratio

(a) 3 : 2 (b) 2 : 5
(c) 1 : 3 (d) 3 : 5.

71. Difference in the frequencies of series limit of Lyman series and Brackett series is equal to the frequency of

(a) second line of Balmer series
(b) first line of Paschen series
(c) third line of Lyman series
(d) first line of Brackett series.

72. Wavelength of characteristic X-ray K_α-line emitted by a hydrogen like atom is 1.28 Å. Wavelength of K_β line for this atom will be

(a) 1 Å (b) 1.46 Å
(c) 1.08 Å (d) 0.8 Å.

73. A non monochromatic light is used in an experiment on photoelectric effect. The stopping potential is

(a) related to the mean wavelength
(b) related to the longest wavelength
(c) related to the shortest wavelength
(a) not related to any of the wavelength.

74. A particle of rest mass m_0 is moving at a speed such that its kinetic energy is twice its rest mass energy. De Broglie wavelength of the particle is (h—Planck's constant, c—speed of light)

(a) $\frac{h}{2.8\, m_0 c}$ (b) $\frac{h}{4.6\, m_0 c}$
(c) $\frac{h}{2\, m_0 c}$ (d) $\frac{h}{5.2\, m_0 c}$.

75. A photo cell is illuminated by a small bright source placed 1 m away. When the same source of light is placed 1/2 m away, the number of electrons emitted by the photo cell would (AIEEE 2002)

(a) increase by a factor of 2
(b) decrease by a factor of 2
(c) increase by a factor of 4
(d) decrease by a factor of 4.

76. Which of the following is the minimum kinetic energy required by a hydrogen atom in ground state to produce a photon after collision with another identical hydrogen atom at rest?

(a) 6.8 eV (b) 3.4 eV
(c) 20.4 eV (d) 2.04 eV.

ANSWERS

1. (a)	2. (c)	3. (a)	4. (b)	5. (d)
6. (a)	7. (c)	8. (a)	9. (d)	10. (c)
11. (b)	12. (b)	13. (b)	14. (c)	15. (c)
16. (c)	17. (d)	18. (a)	19. (d)	20. (c)
21. (d)	22. (d)	23. (d)	24. (b)	25. (d)
26. (d)	27. (c)	28. (b)	29. (d)	30. (c)
31. (c)	32. (a)	33. (b)	34. (d)	35. (a)
36. (b)	37. (a)	38. (c)	39. (c)	40. (d)
41. (d)	42. (d)	43. (a)	44. (b)	45. (c)
46. (d)	47. (a)	48. (b)	49. (c)	50. (c)
51. (b)	52. (a)	53. (a)	54. (a)	55. (c)
56. (c)	57. (a)	58. (a)	59. (d)	60. (a)
61. (b)	62. (b)	63. (a)	64. (c)	65. (a)
66. (c)	67. (d)	68. (b)	69. (c)	70. (b)
71. (c)	72. (c)	73. (c)	74. (c)	75. (c)
76. (c).				

PART II

In the following, a question is followed by four alternatives, of them one or more are correct. Candidate has to select the correct choice(s) from amongst the alternatives.

1. Work function of caesium is 2.14 eV

(a) If stopping potential is 1 V then the wavelength of incident radiation is approximately 4000 Å
(b) Threshold wavelength of Caesium is approximately 5800 Å
(c) The threshold frequency for caesium is 5×10^{16} Hz
(d) Wavelength 5000 Å of incident radiation cannot produce photo electric effect.

2. Electric conduction takes place in a discharge tube due to movement of

(a) electrons (b) positive ions
(c) negative ions (d) protons.

3. Which of the following are true for cathode rays?

(a) It travels along a straight line
(b) It is an electromagnetic wave
(c) It emits *X*-rays when strikes a metal
(d) It is not deflected by magnetic field.

4. In which of the following situations the heavier of the two particles has smaller de Broglie wavelength? The two particles

(a) move with same speed
(b) move with same linear momentum

(c) move with same kinetic energy
(d) have fallen through same height.

5. When the intensity of light source is increased
(a) the number of photons emitted by the source in unit time increases
(b) the total energy of the photons emitted per unit time increases
(c) more energetic electrons are emitted
(d) faster photons are emitted.

6. Photoelectric effect supports quantum nature of light because
(a) there is a minimum frequency below which no photo-electrons are emitted
(b) the maximum kinetic energy of photo-electrons depends only on the frequency of light and not on its intensity
(c) even when the metal surface is faintly illuminated the photo-electrons leave the surface immediately
(d) electric charge of the photo-electrons is quantized.

7. Which of the following products in a hydrogen atom are independent of the principal quantum number *n*? The symbols have their usual meaning.
(a) vn (b) Er
(c) En (d) vr.

8. Let A_n be the area enclosed by the *n*th orbit in a hydrogen atom. The graph of log (A_n/A_1) against log (n)
(a) will pass through origin
(b) will be a straight line with slope 4
(c) will be a monotonically non linear curve
(d) will be a circle.

9. When a photon stimulates the emission of another photon, the two photons have
(a) same energy (b) same direction
(c) same phase (d) same wavelength.

10. For harder *X*-rays?
(a) the wavelength is higher
(b) the intensity is higher
(c) the frequency is higher
(d) the photon energy is higher.

11. Cut-off wavelength of *X*-rays coming from a Coolidge tube depends on
(a) target material
(b) accelerating voltage
(c) separation between the target and the filament
(d) temperature of the filament.

12. When an electron strikes the target in a Coolidge tube, its entire kinetic energy
(a) is converted into a photon
(b) may be converted into photon
(c) is converted into heat
(d) may be converted into heat.

13. *X*-rays incident on a material
(a) exert a force on it
(b) transfer energy to it
(c) transfers momentum to it
(d) transfers impulse to it.

14. The wavelength of K_α *X*-rays of lead isotopes Pb^{208}, Pb^{206} and Pb^{204} are λ_1, λ_2 and λ_3 respectively. Then
(a) $\lambda_1 = \lambda_2 = \lambda_3$ (b) $\lambda_1 > \lambda_2 > \lambda_3$
(c) $\lambda_1 < \lambda_2 < \lambda_3$ (d) $\lambda_2 = \sqrt{\lambda_1 \lambda_3}$.

15. The wavelengths and frequencies of photons in transition 1, 2 and 3 (Fig. 27.60) for hydrogen like atom are λ_1, λ_2, λ_3, ν_1, ν_2 and ν_3 respectively. Then

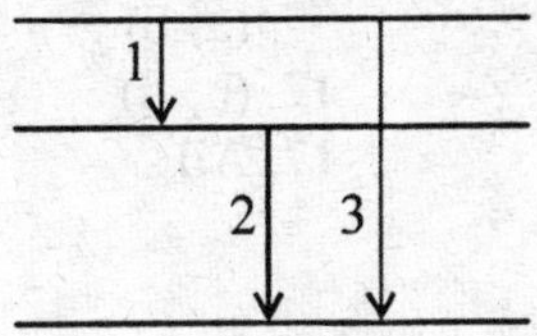

Fig. 27.60

(a) $\nu_3 = \nu_1 + \nu_2$ (b) $\nu_3 = \frac{\nu_1 \nu_2}{\nu_1 + \nu_2}$
(c) $\lambda_3 = \lambda_1 + \lambda_2$ (d) $\lambda_3 = \frac{\lambda_1 \lambda_2}{\lambda_1 + \lambda_2}$.

16. An electron in hydrogen atom makes transition from second excited state to first

excited state and then from 1st excited state to ground state. Let the ratio of wavelength, momentum and energy of photons emitted in these two cases be a, b, c respectively, then

(a) $c = 1/a$ (b) $a = 9/4$
(c) $b = 5/27$ (d) $c = 5/27$.

17. The energy, the magnitude of linear momentum and orbital radius of an electron in a hydrogen atom corresponding to the quantum numbers are E, P and r respectively. Then according to Bohr's theory of hydrogen atom

(a) $EPr \propto \frac{1}{n}$
(b) $P/E \propto n$
(c) Er = constant for all orbits
(d) $Pr \propto n$..

18. Which of the following transitions in He^+ ion will give rise to spectral line which has the same wavelength as some spectral lines in the hydrogen atom?

(a) $n = 4$ to $n = 2$ (b) $n = 6$ to $n = 3$
(c) $n = 6$ to $n = 2$ (d) $n = 8$ to $n = 4$.

19. An electron is excited from a lower energy state in hydrogen atom, which of the following quantities decrease in the excitation?

(a) Potential energy
(b) Angular speed
(c) Kinetic energy
(d) Angular momentum.

20. An electron in a hydrogen atom makes a transition from $n_1 \rightarrow n_2$. Where n_1 and n_2 are the principle quantum numbers of the two states. Assume the Bohr model to be valid. The time period of the electron in initial state is eight times that in the final state. The possible values of n_1 and n_2 are

(a) $n_1 = 4$, $n_2 = 2$ (b) $n_1 = 3$, $n_2 = 2$
(c) $n_1 = 8$, $n_2 = 1$ (d) $n_1 = 6$, $n_2 = 3$.

21. In which of the following cases the heavier of the two particles has a smaller de Broglie wavelength? The two particles

(a) move with same speed
(b) move with same linear momentum
(c) move with same kinetic energy
(d) have the same change of potential energy in conservative field.

ANSWERS

1. (a, b, c)	2. (a, b, c)	3. (a, c)	4. (a, c, d)	5. (a, b)
6. (a, b, c)	7. (a, b)	8. (a, b)	9. (a, b, c, d)	10. (c, d)
11. (b)	12. (b, d)	13. (a, b, c, d)	14. (a, d)	15. (a, d)
16. (a, c, d)	17. All	18. (a, d)	19. (b, c)	20. (a, d)
21. (a, c, d).				

28

Nuclear Physics

28.1 INTRODUCTION

Rutherford in order to explain the scattering of α-particles suggested an atomic model. According to this model, in every atom the positive charge is concentrated at the centre of atom called *nucleus*. The overall dimensions of a nucleus are much smaller than those of the atom but nevertheless contains most of its total mass. The scattering experiments revealed that the diameter of the nucleus is of the order of 10^{-15} m (= 1 Fermi). This is smaller than the diameter of the atom by a factor of about 10^4. This gives that atom is mostly empty space. The electrons are confined to only a limited part of this empty space.

The basic question then arises as to what is the structure of the nucleus? Is it a continuous mass distribution or does it have discrete constituents? If so what is the nature of these constituents and how are they held together? The data suggest that positive charges in an atom are concentrated in small volume which we call nucleus and that protons and neutrons exist in the nucleus. In this chapter we shall try to answer these questions and discuss various properties of atomic nucleus.

28.2 PROPERTIES OF NUCLEUS

(*i*) **Charge:** Number of protons (Z) in a nucleus is called atomic number. The number of electrons in different orbits of atom are also Z. This makes atom electrically neutral. Since each proton carries change $+e$ as such total charge on nucleus is $+Ze$.

(*ii*) **Constituents of a Nucleus:** A nucleus is made of protons and neutrons. A nucleus $_ZX^A$ has Z protons (called atomic number) and A is called mass number. Then number $(A - Z)$ gives the number of neutrons in the nucleus. This is represented by N.

Thus $\quad N = A - Z$ or $A = N + Z$

A is also called the total number of *nucleons*. A neutron is a neutral particle and carries no charge. Mass of neutron is slightly greater than that of proton. The masses of a proton and a neutron are

$$m_p = 1.6726231 \times 10^{-27} \text{ kg} = 1.0073 \text{ amu} = 938.3 \text{ MeV}$$

and $$m_n = 1.6749286 \times 10^{-27} \text{ kg} = 1.0087 \text{ amu} = 939.6 \text{ MeV}$$

where amu is another unit of mass *called atomic mass unit*. This atomic mass unit is obtained by assuming mass of carbon as 12 amu. Mass is also expressed in terms of energy*. We have

$$1 \text{ amu} = 1.660559 \times 10^{-27} \text{ kg} = 931.5 \text{ MeV}$$

(iii) **Size of Nucleus:** The nucleus is a small region as such sharp boundaries for it cannot be defined. For such dimensional particles, quantum mechanical description is given in terms of the wave function. Classically nucleus is considered *probate sphereoid* but deformation (eccentricity $e \approx 1$) is so small that practically it is taken as sphere. Experiments show that the average radius R of the nucleus may be written as

$$R = R_0 A^{1/3} \quad \text{...(28.1)}$$

where $R_0 = 1.1 \times 10^{-15}$ m = 1.1 fm is a constant and A is the mass number

The volume of the nucleus then is

$$V = \frac{4}{3}\pi R^3 = \frac{4}{3}\pi R_0^3 A$$

and mass of nucleus of mass number A is $M = A \times 1.67 \times 10^{-27}$ kg.

Hence density of nucleus is

$$\rho = \frac{M}{V} = \frac{A \times 1.67 \times 10^{-27} \times 3}{4\pi(1.1 \times 10^{-15})^3 A}$$

$$= 3 \times 10^{17} \text{ kg/m}^3.$$

This gives that the nuclear density is independent of mass number and is very high (roughly 3×10^{14} times that of water). Thus we conclude that matter in nucleus is very densely packed.

(iv) **Nuclear Spin:** The protons and neutrons inside a nucleus move in well defined quantum states and thus have orbital angular momenta. Apart from this, protons and neutrons also have internal spin angular momentum. The spin quantum number m_s is $+1/2$ or $-1/2$. The total angular momentum of the nucleus is the resultant of all the spin and orbital angular momenta of the individual nucleons. This total angular momentum of a nucleus is called the nuclear spin of the nucleus.

Since the spin of the proton and neutron is 1/2, as such they are called *Fermions* and obey *Fermi-Dirac statistics*.

The hyper fine structure is explained on the basis of *spin-orbit interaction of nucleons.*

(v) **Magnetic Moment:** When electron revolves round nucleus it is equivalent to a current loop and has magnetic moment (M)

$$M = \frac{e}{2m} L$$

where L is angular momentum of electron and m mass of electron and for $n = 1$ ($L = nh/2\pi$). Magnetic moment is called Bohr magneton

$$\mu_e = \frac{eh}{4\pi m}$$

---

*Mass is expressed in terms of energy using Einstein's mass energy relation $E = mc^2$.

Similarly due to spin angular momentum of nucleons, nucleus has magnetic moment. The magnetic moment due to proton is

$$\mu_p = \frac{eh}{4\pi m_p} = 5.05 \times 10^{-27} \text{ A-m}^2$$

It is clear that magnetic moment of electron is approximately 2000 times that of proton. Nuclear magnetic moment is measured in terms of *nuclear magneton* (μ_N) and

$$\mu_N = \frac{eh}{4\pi m_p} \quad \text{where } m_p \text{ is mass of proton}$$

For free proton magnetic moment = 2.7928 μ_N and for neutron it is –1.9135 μ_N. Thus the magnetic moment of nucleus is very small in comparison to that of electron.

28.3 NUCLEAR FORCES

Nuclear forces are responsible for stability of the nucleus. Repulsive forces between the protons inside nucleus would tear nucleus apart. Since nucleus is stable, there must be an attractive force which keeps nucleons together. The nuclear forces are strong, attractive and short range and very complex in nature. Some of the characteristics of these forces are given below.

(*a*) The nuclear forces are the strongest forces in nature. Their magnitude is roughly 100 times that of electrostatic forces and 10^{38} times that of gravitational forces between nucleons.

(*b*) Nuclear forces are independent of charge. The nuclear forces between two protons is same as that between two neutrons or between a proton and a neutron.

(*c*) The nuclear forces are very short range forces. They are most effective upto a distance of the order of a few fermi. The nuclear force decreases rapidly with the separation between nucleus and becomes negligible at separation of about 10 fm. The range upto which the nuclear force is effective is called *nuclear range.*

(*d*) The variation of nuclear forces with distance is not known exactly. However the nuclear interaction may be represented as shown in Fig. 28.1. This gives that beyond a certain distance (~10 fm) the force is zero. At very short distances (< 0.5 fm) force is repulsive.

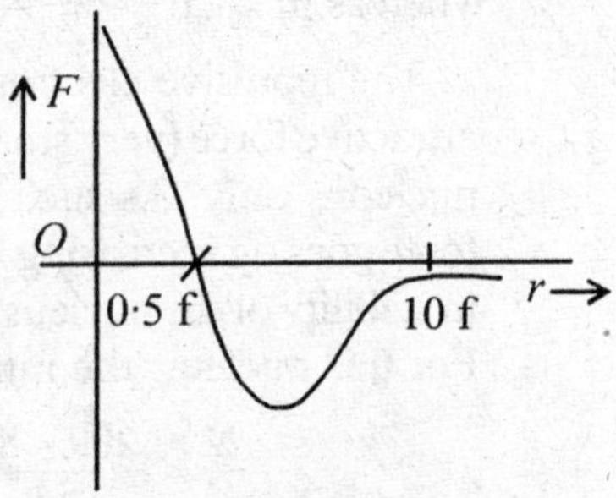

Fig. 28.1

(*e*) The nuclear forces show saturation properties *i.e.* each nucleon interacts with its immediate neighbours only, rather than with all the other nucleons in the nucleus.

(*f*) The nuclear forces are not central forces. The force between a pair of nucleons is not solely determined by the distance between the nucleons but also by the relative orientation of the spins of nucleons, i.e. the nuclear force depends on the direction of the spins of the nucleons. The force is stronger if spins of the nucleons are parallel and is weaker if the spins are anti-parallel.

(*g*) These forces are not conservative forces.

According to Yukawa, nuclear forces can be treated as exchange forces, which result from the continual exchange of a particle called π meson between a proton and neutron.

A neutron becomes a proton inside the nucleus by emitting π^- meson. Similarly a proton absorbs π^- meson and is converted to a neutron.

$$n \rightarrow p + \pi^- \text{ and } p + \pi^- \rightarrow n$$
$$p \rightarrow n + \pi^+ \text{ and } n + \pi^+ \rightarrow p$$

The transfer of momentum associated with the shift of pions is equivalent to the action of force. Thus, nuclear force is the result of the exchange of pions between the nucleons.

28.4 NUCLEAR STABILITY

A nucleus ${}_Z^AX$ contains Z number of protons and $N = A - Z$ number of neutrons. All are confined to a sphere of radius 10^{-15} m. The stability of nucleus is decided by the relative number of protons and neutrons in the nucleus.

Some Points with Regards to the Stability of Nucleus

(*i*) More than 1000 nuclides have been identified but not all are stable. An unstable nucleus emits some kind of particle and changes its constitution. A stable nucleus maintains its constitution all the time.

(*ii*) In light element (except hydrogen) atoms, the number of neutrons and number of protons in nucleus are equal or nearly equal. For atom $A \leq 40$ (or $Z \leq 20$) $A - Z \approx Z$ (number neutrons $\approx$ number of protons). In such nuclei, the nuclear force of attraction between nucleons, completely overpowers the repulsive electrostatic force between protons, as such light elements are stable.

(*iii*) As we move towards heavy elements ($A > 40$ or $Z > 20$) the number of neutrons and protons in the nucleus increase but the number of neutron increase more in comparison to the number of protons, e.g. in ${}_{26}Fe^{56}$ the ratio $\frac{N}{Z}$ is

$$\frac{N}{Z} = \frac{56 - 26}{26} = 1.15$$

whereas in ${}_{29}Cu^{63}$ $\frac{N}{Z} = \frac{63 - 29}{29} = 1.17$

The repulsive electrostatic force between a pair of protons always acts whereas nuclear attractive force (very small range force and due to saturation) acts between nearest neighbour nucleons only. As such with increase in atomic number (Z) the net electrostatic repulsive force goes on increasing in comparison to the nuclear attractive force. This results in decrease in stability of the nucleus. The maximum nuclear force between nucleon is for ${}_{83}Bi^{209}$ nucleus. For this nucleus, the ratio

$$\frac{N}{Z} = \frac{209 - 83}{83} = 1.52$$

This is the heaviest stable nucleus.

(*iv*) For nuclei having $Z > 82$, the net electrostatic repulse force between the protons is more than the net attractive nuclear force between the nucleons. As such all the nuclei with $Z > 82$ are unstable and they spontaneously continue to emit α, β and γ–rays. After emission they are converted to light stable nuclei*.

(*v*) Mostly even-even (atomic number Z and neutron number N both even) nuclei are stable. 60% of the stable nuclei have even number of protons and neutrons.

(*vi*) Some of nuclei with Z even and N odd as well as Z odd and N even are also found to be stable. But a few of nuclei with Z odd and N odd (odd–odd nuclei) are found to be stable.

* ${}^{209}_{85}Bi$ is an exception.

Out of 1000 known nuclei only 0.5% odd–odd nuclei are found to be stable. These are ${}^{2}_{1}H$, ${}^{6}_{3}Li$, ${}^{10}_{5}B$, ${}^{14}_{7}N$ and ${}^{180}_{73}Ta$.

(*vii*) Like atom, there are different energy states in nuclei also. In each energy state, two protons or two neutrons of opposite spin, give more stability to the nucleus because in this configuration the energy is minimum.

(*viii*) In all stable nuclei (except ${}^{1}_{1}H$ and ${}^{3}_{2}H_e$), the number of neutrons (N) is more than the number of protons (Z). That is why ${}^{11}_{5}B$ is stable whereas ${}^{11}_{6}C$ is unstable.

Isotopes

Isotopes of an element are the atoms which have the same atomic number (Z) but different mass number (A). For example:

${}^{1}_{1}H$, ${}^{2}_{1}H$, and ${}^{3}_{1}H$ are the isotopes of hydrogen.

${}^{3}_{2}He$, ${}^{4}_{2}He$ and ${}^{5}_{2}He$ are the isotopes of helium.

${}^{10}_{6}C$, ${}^{11}_{6}C$, ${}^{12}_{6}C$, ${}^{13}_{6}C$, ${}^{14}_{6}C$ are the isotopes of carbon.

All the known elements have one or more isotopes. The relative abundance of different isotopes differs from element to element. The atomic weight of an element is the weighted average of the masses of all its isotopes.

Isobars

Isobars are the atoms of different elements which have the same mass number (A) but different atomic number (Z).

Examples are ${}^{22}_{11}Na$ and ${}^{22}_{10}Ne$, ${}^{40}_{20}Ca$ and ${}^{40}_{18}Ar$, ${}^{37}_{17}Cl$ and ${}^{37}_{16}S$

The chemical properties of isobars are widely different, their physical properties may be identical. The isobars occupy different places in the Periodic Table.

Isotones

Isotones are the nucleides which contain the same number of neutrons, i.e. $(A - Z) = N$ is same for these. Examples are ${}^{37}_{17}Cl$ and ${}^{39}_{19}K$.

28.5 BINDING ENERGY

We have seen that the nucleus is made up of neutrons and protons (called nucleons). These nucleons in every nucleus are bound together with short range forces called nuclear forces. Therefore energy must be supplied to the nucleus to separate the constituent nucleons to large distance. The amount of energy needed to do this is called the *binding energy* of the *nucleus*.

The origin of the nuclear binding energy is explained on the basis of Einstein's theory of mass energy equivalence. It is found that the rest mass energy of a nucleus is always less than the rest mass energy of its constituent nucleons in free state. The difference of the two energies is the binding energy. If m_p, m_n and M are the masses of free proton, free neutron and nucleus, then their rest mass energies respectively are m_pc^2, m_nc^2 and Mc^2, where c is the velocity of light in vacuum. For a nucleus ${}^{A}_{Z}X$, containing Z protons and $N = (A - Z)$ neutrons, the binding energy is

$$B = [Zm_p + Nm_n - M]c^2 \qquad \text{...(28.2)}$$

where M is mass of nucleus and c is the velocity of light in vacuum.

The quantity

$$\Delta m = Zm_p + Nm_n - M \quad \ldots(28.3)$$

is called *mass defect*. This shows that certain mass disappears in the formation of nucleus and this difference appears in the form of *binding energy* (according to mass energy equivalence) and is responsible for binding the nucleons together in nucleus.

The binding between the constituents of nucleus, to show how well nucleons are held together is better measured in terms of what we call as *binding energy per nucleon*. This is defined as *the ratio of binding energy (B) of nucleus to the number of the nucleons (A) in the nucleus (= B/A).*

Thus the binding energy per nucleon is the average energy required to separate nucleus into its individual nucleons. It gives the measure of the strength of the force, which binds the nucleons together inside a nucleus. The physical significance of this parameter (B/A) is better understood if we study its variation with mass number of nucleus. The variation of binding energy per nucleon (B/A) with mass number of nucleus (A) is shown in Fig. 28.2.

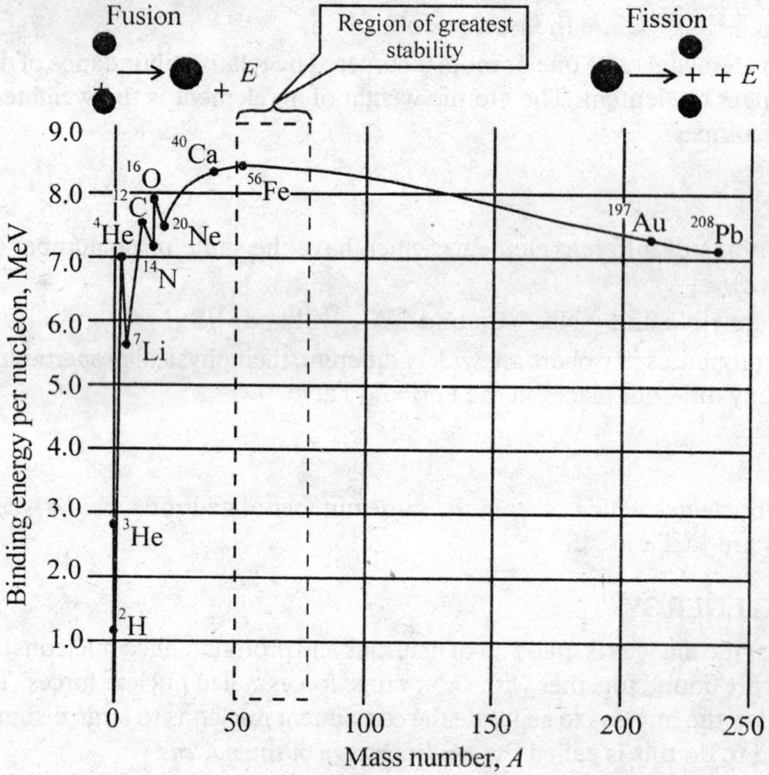

Fig. 28.2

The curve depicts the following important information.

(*a*) Average binding energy per nucleon for light nuclei like ^1_1H, ^2_1H, ^3_1H is small. The interesting feature in this range is that the binding energies per nucleon is ^1_1H and ^6_3Li are lower than in ^4_2He. Similarly binding energy per nucleon in $^{11}_5B$ and $^{14}_7N$ are lower than in $^{12}_6C$. This shows the existence of shell structure at nuclear level, very similar to that at the atomic level.

(*b*) The binding energy per nucleon is approximately equal to 8 MeV for nuclei of middle mass (A = 30 to A = 120). The maximum value of this is 8.8 MeV/nucleon for $^{56}_{26}Fe$. This is the most stable nucleus. This shows that binding energy per nucleon is independent of the size of nucleus. This gives the relative strength of the nuclear force.

(*c*) If each nucleon interacted with every other nucleon, there would be $(A - 1)$ interactions for each nucleon, and the binding energy would be proportional to $(A - 1)$ rather than constant. Instead there is a fixed number of interactions per nucleon because each nucleon attracts to its nearest neighbours. Such a situation leads to a constant nuclear density. If the binding energy per nucleon were instead proportional to the number of nucleons then the radius would be approximately constant as in case of atom.

(*d*) The binding energy per nucleon is lower for both light nuclei ($Z < 10$) and very heavy nuclei ($Z > 70$). The decrease in the binding energy for heavy nuclei ($A > 100$) shows the contribution of the increasing Coulomb interaction (repulsive) between the protons inside the nucleus.

(*e*) Dropping of binding energy curve at high mass number tells us that nucleons are more tightly bound when they are assembled into two middle mass nuclei rather than into a single high mass nucleus. High mass number nuclei are thus prone to a process called *spontaneous fission*. Energy is released in the nuclear fission of a single massive nucleus into two smaller fragments.

(*f*) The dropping of the binding energy curve at low mass number gives that the energy will be released if two nuclei of small mass number combine to form a single nucleus of higher mass number. The new nucleus has more binding energy per nucleon. For example, if two ^2_1H (deuterium) nuclei combine to form ^4_2He nucleus, over 23 MeV energy is released. Such a process is called nuclear fusion.

Packing Fraction

Packing fraction of a nucleus is defined as the mass excess per nucleon.

$$\text{Packing fraction} = \frac{\text{Mass excess}}{\text{Mass no.}} = \frac{M - A}{A} \qquad \ldots(28.4)$$

where M is the atomic weight and A is mass number of the nucleus. The value of the packing fraction can be positive, negative or zero, according as M is greater than A, less than A or equal to A. A is a whole number always whereas M is not necessarily so.

It is observed that the very light and heavy nucleide ($A < 10$ and $A > 180$) have an average mass per nucleon more than unity. Whereas those with mass number between 20 and 180 have a mass per nucleon less than unity. A positive value of packing fraction generally implies a tendency towards instability in the nucleus. A negative packing fraction suggests that some mass has been converted into energy when that isotope is constituted. In general a lower value of packing fraction implies greater stability. Transient elements or iron family have lower values of packing fractions, they show maximum stability.

28.6 RADIOACTIVITY (ALPHA, BETA AND GAMMA DECAY)

We have seen that some nucleides have definite Z, N combination and they are stable. Nuclides with other Z, N combination are also found in nature or may be be prepared in the laboratory. However these nuclides are unstable and they decay into other nuclides by various processes. *This spontaneous disintegration of heavy nucleus without any external provocation is called radioactivity. The disintegration process is spontaneous means, i.e. the process cannot be started, stopped, accelerated or decelerated by any external means.*

The phenomenon was discovered by a French physicist, Henry Becquerel in 1896. He observed that Uranium salts possessed a peculiar property of affecting a photographic plate even when it was in light proof package. He was of the opinion that this was due to certain active radiations emitted by Uranium salt. He called these radiations "*Becqueral rays*". This phenomenon of emission of active radiation was termed as *Radioactivity* and the elements exhibiting this property were called *radioactive element.*

The experiments have revealed that the radiations emitted by radioactive elements consist of three kinds. The one with least penetrating power and named as *Alpha particles.* Other kind with a comparatively larger penetrating power are named as *Beta particles* and the highly penetrating are called *Gamma rays.* It should be remembered that all the three types of radiations need not necessarily be emitted by one radioactive element.

Properties of α-Particles and α-Decay

(*i*) Each α-particle contains 2 protons and two neutrons. As such mass of α-particle is nearly 4 amu and charge $+2e$. Thus α-particle is nothing but a helium nucleus, i.e. a doubly ionized helium atom. It's specific charge q/m is

$$\frac{2e}{4\,\text{amu}} = \frac{3.2\times10^{-19}}{4\times1.67\times10^{-27}} = 4.8\times10^{7}\ \text{Coulomb/kg}$$

(*ii*) The speed of α-particles ranges between 1.4×10^7 ms^{-1} to 2.1×10^7 ms^{-1} depending upon the source emitting it.

(*iii*) Because of large mass, the penetrating power of α-particles is least. Approximately $\frac{1}{100}$ of that of β-particle and $\frac{1}{10000}$ of that γ-rays.

(*iv*) Because of large mass and high velocity, α-particles have large ionizing power (100 times that of β-particles and 10,000 times that of γ-rays).

(*v*) Being charged particle (made of 2 protons and 2 neutrons) they can be deflected by an electric field as well as by a magnetic moment.

(*vi*) α-particles coming from a particular decay scheme have same energy (mono energetic).

(*vii*) The range of α-particles in air (distance through which they can travel in air) depends upon the radioactive source producing it. If λ is the decay constant of the element and E the energy of emitted α-particles then

$$\log\lambda = a + b\log E \qquad \text{...(28.5)}$$

where a and b are constants. This is called *Geiger Nuttal law.*

The range (R) of α-particle depends on energy (E) as

$$R \propto E^{3/2} \qquad \text{...(28.6)}$$

Combining (28.5) and (28.6), we get

$$\log\lambda = A + B\log R \qquad \text{...(28.7)}$$

where A and B are new constants.

(*viii*) α-particles produce fluorescence in certain materials such as barium platino cynide. They produce scintillations on these materials.

(*ix*) α-particles affect photographic plate.

(*x*) α-particles cause burns on human body.

(*xi*) α-particles produce heat when stopped.

(*xii*) α-particles are scattered while passing through thin metal foils. Most of the α-particles are scattered at small angles but a few of them are scattered at an angle greater than 90° also.

α-Decay

It is the process of emission of an α-particle from a radioactive nucleus. The unstable nucleus when emits an α-particle its proton number Z and neutron number N each reduces by 2. The α-decay process is represented as

$$^{A}_{z}X \rightarrow {}^{A-4}_{z-2}Y + {}^{4}_{2}\text{He} \qquad \ldots(28.8)$$

Since atomic number Z is changed, the element itself is changed and hence the chemical symbol of residual nucleus is different from that of the original nucleus. The nucleus before the decay ($^{A}_{z}X$) is called the *parent nucleus and* the product nucleus ($^{A-4}_{z-2}X$) after the decay is called *daughter nucleus*. Example of α-decay is

$$^{238}_{92}\text{U} \rightarrow {}^{234}_{90}\text{Th} + {}^{4}_{2}\text{He and } {}^{212}_{83}\text{Bi} \rightarrow {}^{208}_{81}\text{Tl} + {}^{4}_{2}\text{He}$$

Energy is released in this process in the form of kinetic energy of α-particles and daughter nucleus Y. It can be calculated using Einstein's mass energy equivalence relation, i.e.

$$Q = (m_X - m_Y - m_\alpha)c^2 \qquad \ldots(28.9)$$

where m_X, m_Y and m_α are masses of nucleus X, Y and α-particle respectively.

If K_Y and K_α are the kinetic energies of nucleus Y and α-particle respectively, then, assuming nucleus X in stable state,

$$Q = K_Y + K_\alpha$$

and conservation of momentum gives

$$\mathbf{p}_Y + \mathbf{p}_\alpha = 0$$

where $\mathbf{p}_Y$ and $\mathbf{p}_\alpha$ are the momentum of nucleus Y and α-particle respectively. Hence magnitude $P_Y = P_\alpha$. Then

$$K_\alpha = \frac{m_Y}{m_Y + m_\alpha} Q = \frac{A-4}{A} Q \qquad \ldots(28.10)$$

where A is mass number of parent nucleus (X). It shows that $K_\alpha >> K_Y$.

For spontaneous α-decay process Q is always positive. In α-decay process the energy of α-particles is of the order of 4–9 MeV but it has to cross potential barrier of 25 to 40 Mev to come out of nucleus. Thus classically α-decay from nucleus is not possible. Gamow Germi and Condon first presented theoretical explanation of α-decay on the basis of quantum mechanics.

According to them the independent existence of α-particles in heavy nuclei is possible. The α-particle is always in motion and continuously strikes the potential barrier. In each impact there is small but a finite propability, that inspite of small energy of α-particles, they can cross the high energy potential barrier as if it tunnels through the nuclear barrier*. This is called *tunnel effect*.

If n is the number of impacts on nuclear barrier per second and P is the probability of tunneling through potential barrier then decay constant (λ) is

$$\lambda = np \qquad \ldots(28.11)$$

*In reality there is no tunnel.

If energy of the α-particles is large the tunneling probability is high, i.e. probability of α-decay is large and half life of sample is less.

Properties of β-particles and β-decay

(*i*) β-particles are negatively charged particles with charge equal to charge on electron (1.6×10^{-19} C) and mass equal to that of an electron ($m = 9.1 \times 10^{-31}$ kg) and their specific charge

$$\frac{q}{m} = \frac{e}{m} = \frac{1.6 \times 10^{-19}}{9.1 \times 10^{-31}} = 1.76 \times 10^{11} \text{ Coulomb/kg}$$

Thus it is an stream of electrons coming from the nuclei. Then the properties of β-rays, cathode ray, thermions, photoelectrons, etc. are all identical except for their origin. β-particles are created at the time of nuclear transformation, whereas, in cathode ray, thermions, etc. the electrons are already present and get ejected.

(*ii*) The velocity of β-particles ranges from 33% to 99% of the velocity of light.

(*iii*) Because of small mass their penetrating power is large and their ionization power is small, only 1/100 that of α-particles. The range of β-particles in air is several meters.

(*iv*) Beta rays produce fluorescence in some substances like barium platinocyanide and zinc sulphide, i.e. they produce scintillations when they fall on these materials.

(*v*) They affect the photographic plate.

(*vi*) They being negatively charged particles are affected by electric and magnetic fields.

(*vii*) The energy of beta particles coming from the same decay scheme are not equal. The energy spectrum of β-decay is shown in Fig. 28.3. This is because the available energy is shared by antinutrino. The energy of β-particles thus varies between zero and a maximum (E_0) called *end point energy*.

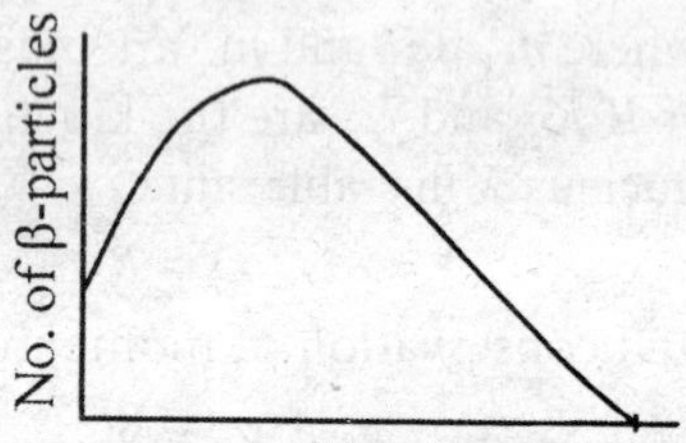

Fig. 28.3

Beta (β) Decay

A nucleus that decays spontaneously by emitting an electron or a positron is said to undergo *beta decay*. Like alpha decay, this is a spontaneous process, with a definite disintegration energy and half life. In β-decay either a neutron is converted into a proton or proton is converted into a neutron as such the ratio N/Z is altered. If a nucleus is formed with more number of neutrons than needed for stability, a neutron will convert itself into a proton to move towards stability. Similarly, if a nucleus is formed with more number of protons than needed for stability a proton will convert itself into a neutron. Such transformations take place because of weak forces operating within a neutron or a proton. When a neutron is converted into a proton, an electron and a new particle called anti-neutrino are created and emitted from the nucleus, i.e.

$$\text{n} \rightarrow \text{p} + \beta^- + \bar{\nu} \quad \text{...(28.12)}$$

$\bar{\nu}$ is called anti *neutrino*. The neutrino is assigned the following properties:

(*i*) It is a neutral particle.

(*ii*) Its rest mass is practically zero (rest mass = 7 eV/c²).

(*iii*) It spin angular momentum is $\pm\frac{\hbar}{2} = \pm\frac{1}{2}\left(\frac{h}{2\pi}\right)$

(*iv*) The magnetic moment is finite but small.
(*v*) Its anti-particle is $\bar{\nu}$ (anti neutrino).
(*vi*) It does not interact with matter.
(*vii*) It was postulated by Pauli and experimentally detected by Rains and Coven.

The electron emitted from the nucleus is called a beta particle ($\bar{\beta}$). A stream of such beta particles coming from a bulk of unstable nuclei is called *Beta ray*. The β-decay process may be represented by

$$ {}_{z}^{A}X \rightarrow {}_{z+1}^{\;\;A}Y + \beta^{-} + \bar{\nu} + Q \qquad \text{...(28.13)}$$

It is observed that in β decay the atomic number of daughter nucleus increases by 1 and mass number remains same. Q is the energy released in β decay. The energy released can be calculated as

$$ Q = [m({}_{z}^{A}X) - m({}_{z+1}^{\;\;A}Y)]c^2 \qquad \text{...(28.14)}$$

This energy appears in the form of kinetic energy of the product particles. Because of the large mass of the residual nucleus ${}_{z+1}Y^A$ it does not share appreciable kinetic energy. Thus the energy Q is shared by the anti-neutrino and the beta particle. Depending on the fraction taken away by the anti-neutrino, the kinetic energy of the beta particle can be anything between zero and maximum value Q (= E_0 – called end point energy). The energy carried by anti-neutrino is

$$ E_{\nu} = E_0 - E_{\beta} \qquad \text{...(28.15)}$$

If unstable nucleus has excess protons than needed for stability, a proton converts itself into a neutron in the process a positron and a neutrino are created and emitted from the nucleus.

$$ \mathrm{p} \rightarrow \mathrm{n} + \beta^{+} + \nu \qquad \text{...(28.16)}$$

β^+ is called position and has positive charge equal to electronic charge (e^+) and has mass equal to the mass of electron. It is anti particle of electron. Due to their annihilation, energy in the form of γ-ray photon is produced

$$ \beta^{-} + \beta^{+} \rightarrow 2h\nu \qquad \text{...(28.17)}$$

β^+ decay process is represented as

$$ {}_{Z}X^{A} \rightarrow {}_{Z-1}^{\;\;A}Y + \beta^{+} + \nu + Q \qquad \text{...(28.18)}$$

Note: An isolated proton cannot decay into neutron emitting positron and neutrino as shown in equation (28.15). This is due to the fact that mass of neutron is greater than that of proton and Q value of such process is negative.

Electron Capture Process

In certain nuclei a similar process known as *electron capture* takes place. In this process, the nucleus captures one of the atomic electron (say from *K*-shell). A proton in the nucleus combines with the electron and converts itself into a neutron. A neutrino is created in the process and emitted from the nucleus.

$$ \mathrm{p} + e^{-} \rightarrow \mathrm{n} + \nu $$

The process may be represented as

$$ {}_{Z}^{A}X + e^{-} \rightarrow {}_{Z-1}^{\;\;A}Y + \nu \qquad \text{...(28.19)}$$

and the Q value of the process is

$$ Q = [m({}_{Z}^{A}X) - m({}_{z-1}^{\;\;A}X)]c^2 \qquad \text{...(28.20)}$$

Since an atomic electron is captured a vacancy is created in the atomic shell and X-rays are emitted as a result of the capture.

The daughter nucleus formed as a result of an alpha decay or a beta decay may not be stable and undergo another alpha or beta decay. Thus a series of decays proceed till a stable nucleus is formed. Such a series always ends at $^{206}_{82}Pb$.

Properties of γ-rays and γ-decay

(*i*) Gamma rays is an electromagnetic radiation of short wavelength. Its wavelength in general is smaller than X-rays. As such they are unaffected by electric as well as magnetic fields.

(*ii*) It has the largest penetrating power and the least ionizing power.

(*iii*) All the photons coming from a particular gamma decay scheme have the same energy.

(*iv*) γ-rays can produce fluorescence in a substance like willimite. They also affect the photographic plate.

(*v*) γ-rays can eject photo-electrons from the surface of metal on which they fall.

(*vi*) γ-rays can produce nuclear reactions.

(*vii*) The intensity of γ-rays after penetrating a distance x in some material is given by

$$I = I_0 e^{-\mu x} \qquad \text{...(28.21)}$$

where I_0 is the intensity of incident γ-rays and μ absorption co-efficient. The absorption co-efficient μ depends on wavelength ($\mu \propto \lambda^3$) and nature of absorber.

γ-decay

γ-decay is the process of emission of gamma ray photon from a radioactive nucleus. The nucleons (protons and neutrons) inside a nucleus move in discrete quantum states with definite energies. The ground state is the one which occupied by nucleons makes the total energy of the nucleus minimum. However higher energy states are also available to the nucleus. As a result of very strong nuclear interaction the excitation energies are very high (of the order MeV). When an excited nucleus makes transition to lower energy states, energy in the form of electromagnetic radiation is released. Since energy difference is very large (MeV) wavelength of emitted radiation is small (< 0.01 Å).

The energy differences in allowed energy levels of a nucleus are generally large, it is therefore difficult to excite the nucleus to higher energy levels by usual methods of supplying energy (like heating, collision by accelerated electrons, etc.). However when an α-decay or β-decay takes place, the daughter nucleus is generally formed in one of its excited state. This daughter nucleus by single transition or sometimes by successive transitions reaches the ground state by emitting one or more gamma rays, Figure 28.4 shows $^{60}_{27}Co$ nucleus transforming into $^{60}_{28}Ni$ when a β-decay takes place. This process leaves $^{60}_{28}Ni$ nucleus in excited state. When this nucleus de-excites it emits γ-rays of energy 1.17 MeV (λ = 0.0106 Å) and 1.33 MeV (λ = 0.009 Å).

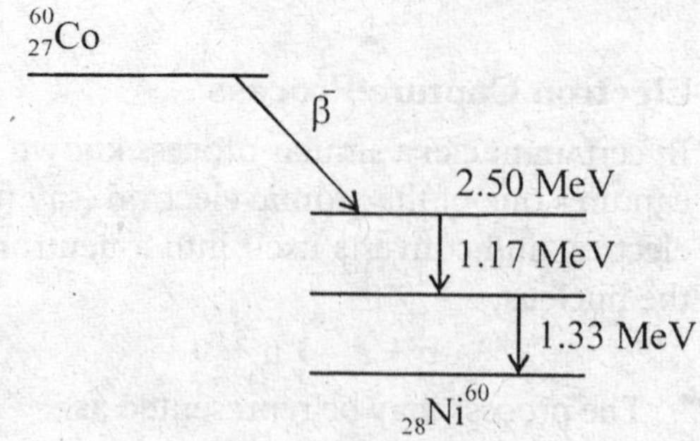

Fig. 28.4

This process of a nucleus de-exciting to lower energy state by emitting a γ-ray photon is called γ-decay. The γ-decay can then be represented as

$$^{A}_{Z}X \rightarrow ^{A}_{Z}X + \gamma \qquad ...(28.22)$$

28.7 LAWS OF RADIOACTIVE DECAY

Rutherford and Soddy on the basis of their experimental study formulated certain laws regarding radioactive disintegration. They are called laws of radioactive disintegration.

(*a*) Unstable nuclei disintegrate spontaneously emitting α-particles, β-particles and γ-rays. The process of disintegration cannot be accelerated, decelerated nor stopped by any physical or chemical means.

(*b*) It is a random process, i.e. it is impossible to predict which particular atom of radioactive substance will disintegrate in a given time interval.

(*c*) During disintegration of nucleus either α-particle or a β-particle is emitted. Both of these particles are never emitted simultaneously. Also at a time only one α or one β particles is emitted by an atom.

(*d*) α-emission changes parent nucleus into daughter nucleus with charge number reduced by 2 and mass number reduced by 4.

(*e*) Emission of a β-particle increases the charge number of daughter nucleus by 1 and mass number remains same.

(*f*) The rate disintegration (number of atoms disintegrated per second) at any instant is directly proportional to the number of active atoms present at that moment.

If at any time (t) there are N active atoms and at time $t + dt$ reduces to N-dN, then rate of disintegration at instant t is $\left(-\frac{dN}{dt}\right)$. Hence

$$-\frac{dN}{dt} \propto N \quad \text{or} \quad \frac{dN}{dt} = -\lambda N \qquad ...(28.23)$$

where λ is constant called *disintegration constant*. It is a constant for a particular element. Rearranging equation (28.23), we get

$$\frac{dN}{N} = -\lambda\, dt$$

Integration gives

$$\log_e N = -\lambda t + C$$

where C is constant of integration and can be determined from initial conditions. If at $t = 0$ the active number of nuclei are N_0 then

$$C = \log_e N_0$$

Substituting the value of c, we get

$$\log_e N = -\lambda t + \log_e N_0$$

or $\qquad \log_e N/N_0 = -\lambda t$

or $\qquad N = N_0 e^{-\lambda t} \qquad ...(28.24)$

Equation 28.24 shows that the number of nuclei in a sample decays exponentially with time.

Since equation 28.24 gives the number of active atoms in sample at any time t. Hence the number of atoms decayed in time t is

$$(N_0 - N) = N_0(1 - e^{-\lambda t}) \quad \text{...(28.25)}$$

The decay rate, i.e. the number of disintegration per unit time is called *activity*. Activity R of a radioactive sample is obtained by differentiating equation 28.24 with respect to time

$$|R| = \left|\frac{dN}{dt}\right| = N_0\lambda e^{-\lambda t} = \lambda N = R_0 e^{-\lambda t} \quad \text{...(28.26)}$$

where R_0 is the activity at $t = 0$ and $N\lambda$ is activity at any time t. The SI unit of activity is Becquerel.

1 Becquerel = 1 disintegration per second.

Other units in use are

1 Curie = 3.7×10^{10} disintegration per second

1 Rutherford = 10^6 disintegration per second.

Equations (28.24) and (28.26) give that both R and N decrease exponentially with time. The variation of N with time is shown in Fig. 28.5.

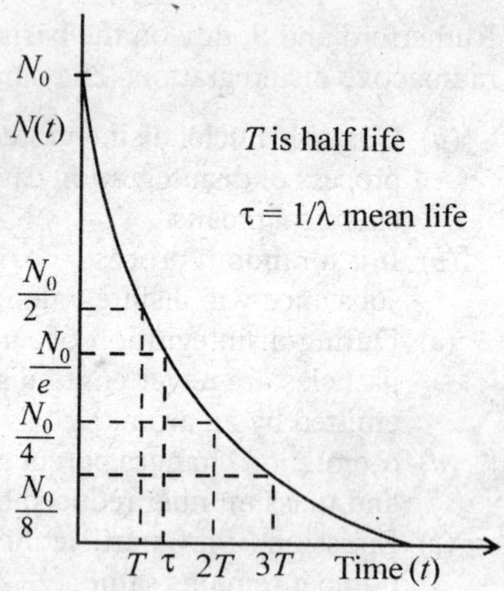

Fig. 28.5

Half Life

Half life of a radioactive substance is the time it takes in which half of a given number of radioactive nuclei to decay.

Setting $N = N_0/2$ and $t = T$, we have from equation (28.24).

$$\frac{N_0}{2} = N_0 e^{-\lambda T}$$

or $$\lambda T = \log_e 2$$

or $$T \equiv \frac{\log_e 2}{\lambda} = \frac{0.693}{\lambda} \quad \text{...(28.27)}$$

In terms of half life T, of a radioactive substance, equation 28.24 can be written as

$$N = N_0 e^{-\lambda T} = N_0 e^{-\frac{0.693t}{T}} = N_0(2)^{\frac{t}{\tau}} \quad \text{...(28.28)}$$

Mean Lifetime

Mean lifetime (average lifetime) τ of a radioactive substance is defined as average time the nucleus survives before it decays. Mean life can be determined by calculating total life time of all the nuclei initially present (N_0) and dividing it by total number of nuclei. Let the number of nuclei that decay in time interval t and $t + dt$ be dN. The life time of each of these atoms is t. Thus the total life time of these nuclei is $t\,dN$. Hence total life time of all the active atoms of the sample is

$$\text{Total life time} = \int_0^{N_0} t\, dN$$

and by definition mean life time τ is

$$\tau = \frac{\text{Total life time}}{N_0} = \frac{1}{N_0}\int_0^{N_0} t\,dN$$

But $\quad N = N_0 e^{-\lambda t} \therefore dN = -N_0 \lambda e^{-\lambda t} dt$

Substituting the value of dN in the integral, we have

$$\tau = \frac{1}{N_0}\int_{\infty}^{0} -\lambda N_0 e^{-\lambda t} \cdot t \cdot dt$$

The limits of integration in terms of time are when $N = N_0$, $t = 0$ and when $N = 0$, $t = \infty$

Thus we have $\quad \tau = \lambda \int_0^{\infty} te^{-\lambda t} dt$

Integrating by parts, we have

$$\tau = \lambda \left[\frac{te^{-\lambda t}}{-\lambda}\right]_0^{\infty} - \lambda \int_0^{\infty} \frac{e^{-\lambda t}}{-\lambda} dt$$

$$= \int_0^{\infty} e^{-\lambda t} dt = -\frac{1}{\lambda}\left[e^{-\lambda t}\right]_0^{\infty} = \frac{1}{\lambda}$$

or $$\tau = \frac{1}{\lambda} \qquad \ldots(28.29)$$

Hence average life time of a radioactive element is reciprocal of the decay constant of the element.

$$\lambda = \frac{0.693}{T}$$

or $$\tau = \frac{1}{\lambda} = \frac{T}{0.693} = 1.44T \qquad \ldots(28.30)$$

i.e., the average life period of a radioactive element is 1.44 times the half life of the element.

28.8 SUCCESSIVE DISINTEGRATION AND RADIOACTIVE EQUILIBRIUM

A radioactive nucleus (A) when disintegrates into (B) and B disintegrates into (C) and so on. Such decays are called *successive disintegrations*. Here A will be called parent nucleus and B daughter nucleus of A. Any two adjacent nuclei may be considered as parent and daughter nuclei.

Suppose N_a and N_b are the active number of nuclei of A and B at anytime t. They undergo decay and their decay constants be λ_a and λ_b respectively. Thus

Rate of disintegration of $A = \frac{dN_a}{dt} = -\lambda_a N_a$ which is also the rate of formation of B.

Rate of disintegration of $B = \frac{dN_b}{dt} = -\lambda_b N_b \qquad \ldots(i)$

Thus net rate of formation of B is $\frac{dN_b}{dt} = \lambda_a N_a - \lambda_b N_b \qquad \ldots(ii)$

The rate of increase of the (stable) nuclei C is equal to the rate of decay of nuclei B. Thus

$$\frac{dN_c}{dt} = N_b \lambda_b \qquad \ldots(iii)$$

If initially at $t = 0$ the number of nuclei of A were N_0 then number of nuclei at any time t of A are

$$N_a = N_0 e^{-\lambda at} \qquad \ldots(iv)$$

Substituting N_a in (ii), we get

$$\frac{dN_b}{dt} + N_b\lambda_b = \lambda_a N_0 e^{-\lambda_a t}$$

Multiplying both sides by $e^{\lambda_b t}$, we get

$$\left(\frac{dN_b}{dt} + N_b\lambda_b\right)e^{\lambda_b t} = \lambda_a N_0 e^{-(\lambda_a - \lambda_b)t}$$

or
$$\frac{d}{dt}\left(N_b e^{\lambda_{bt}}\right) = \lambda_a N_0 e^{-(\lambda_a - \lambda_b)t}$$

Integration gives

$$N_b e^{\lambda_b t} = \frac{\lambda_a N_0}{\lambda_b - \lambda_a} e^{-(\lambda_a - \lambda_b)t} + C$$

where C is the constant of integration.

At
$$t = 0\ N_b = 0 \text{ gives } C = -\frac{\lambda_a N_0}{\lambda_b - \lambda_a}$$

Substituting the value of C, we get

$$N_b = \frac{\lambda_a N_0}{\lambda_b - \lambda_a}\left(e^{-\lambda_a t} - e^{-\lambda_b t}\right)^* \qquad \ldots(28.31)$$

Equation (28.31) gives the number of daughter atoms at anytime t. Thus N_b depends not only on its own decay constant λ_b but also on, parents decay constant λ_a. Substituting the value of N_b in (iii) and using the condition $N_c = 0$ at $t = 0$

$$N_c = N_0\left[1 + \frac{\lambda_a}{\lambda_b - \lambda_a}e^{-\lambda_b t} - \frac{\lambda_b}{\lambda_b - \lambda_a}e^{-\lambda_a t}\right] \qquad \ldots(28.32)$$

After time t, the ratio of activities is

$$\frac{\lambda_b N_b}{\lambda_a N_a} = \frac{\lambda_b}{\lambda_b - \lambda_a}\left[1 - e^{-(\lambda_b - \lambda_a)t}\right] \qquad \ldots(28.33)$$

Secular and Transient Radioactive Equilibrium

(*a*) **Secular equilibrium:** In successive disintegration when the parent has very-very long half life, a state is reached when the daughter nuclei are formed at the same rate as they decay. At this stage the proportions of the different radioactive atoms in the mixture becomes constant, i.e. do not change with time. In this situation the parent is said to be in *secular radioactive equilibrium with its daughter products.*

Let the parent atom (A) has much longer half life ($T_a >> T_b$) than any of the decay products (B). Then $\lambda_a \approx 0$ and $\lambda_a << \lambda_b$, $\lambda_a << \lambda_c$, etc. Then substitution in (iv) gives

$$N_a = N_0 \text{ and } N_b = \frac{\lambda_a}{\lambda_b}N_0\left(1 - e^{-\lambda_b t}\right)$$

*Equation 28.31 is based on the asumption that parent nuclei is more lived than daughter nuclei, i.e. $\lambda_a < \lambda_b$.

or $$N_b = \frac{\lambda_a N_a}{\lambda_b}\left(1 - e^{-\lambda_b t}\right)$$

For $t \to \infty$ $$N_b = \frac{\lambda_a N_a}{\lambda_b}$$

or $$\lambda_b N_b = \lambda_a N_a$$

Thus after a sufficient time, the activities of parent and daughter become equal. This condition is known as *secular equilibrium*.

(*b*) **Transient Equilibrium:** The parent has long average life (longer-lived) than the daughter ($\tau_a > \tau_b$), but the half life of the parent is not very long ($T_a \approx T_b$).

After *t* becoming sufficient large, $e^{-\lambda_b t}$ becomes negligible compared with $e^{-\lambda_a t}$ so that the number of daughter atoms is given by equation (28.31).

$$N_b = \frac{\lambda_a}{\lambda_b - \lambda_a} N_0 e^{-\lambda_a t}$$

This gives that daughter eventually decay with the same half life as the parent. Comparing with (*iv*)

$$N_a = N_0 e^{-\lambda_a t}$$

$$N_b = \frac{\lambda_a}{\lambda_b - \lambda_a} N_a$$

or $$\frac{N_a}{N_b} = \frac{\lambda_b - \lambda_a}{\lambda_a} \approx \frac{\lambda_b}{\lambda_a} \quad \text{for } \lambda_a << \lambda_b$$

After sufficient time the ratio of parent atoms to the daughter atoms becomes constant and both eventually decay. This condition is called *transient equilibrium*.

(*c*) When the parent has a shorter half life time than the daughter ($T_a < T_b$ or $\lambda_a < \lambda_b$) no state of equilibrium is attained. If initially there were only the parent atoms, then as the parent atoms decay, the daughter atoms increase in number, pass through a maximum and eventually decay with their own half life.

28.9 ARTIFICIAL AND INDUCED RADIOACTIVITY

Spontaneous disintegration of nuclei is known as natural radioactivity. It is also observed that certain elements which are otherwise stable, can be made to disintegrate by bombarding them with suitably accelerated particles. *The phenomenon of disintegration of otherwise stable nuclei by bombarding them with suitable accelerated particles is called artificial radioactivity.* Usually, the disintegration stops as soon as the bombardment is over. However, if the bombard element continues to disintegrate even after the bombardment is stopped, the phenomenon is called *induced radioactivity*.

Curie and Juliot observed that boron, when bombarded by α-particles continued to disintegrate even after the bombardment is stopped. It is represented as

$${}^{10}_{5}B + {}^{4}_{2}He \to {}^{14}_{7}N^* \to {}^{13}_{7}N^* + {}_0n^1$$

$${}^{13}_{7}N^* \to {}^{13}_{6}C + \beta^+$$

The half life of $^{13}_{7}N$ is 11 minutes and it is converted into $^{13}_{6}C$ emitting a positron.

Other examples are

(a) $^{23}_{11}Na + ^{2}_{1}H \rightarrow ^{24}_{11}Na^{*} + ^{1}_{1}H$

$$^{24}_{11}Na^{*} \rightarrow ^{24}_{12}Mg + \beta^{-} + \gamma$$

(b) $^{27}_{13}Al + ^{4}_{2}He \rightarrow ^{31}_{15}P^{*} \rightarrow ^{30}_{15}P^{*} + ^{1}_{0}n$

$$^{30}_{15}P^{*} \rightarrow ^{30}_{14}Si + \beta^{+}$$

28.10 RADIOACTIVE CARBON DATING

The half life of a radioactive element is helpful in estimating the age of earth, rocks, mineral deposits, etc. One of the methods used is carbon dating.

Cosmic rays coming from interstellar space on colliding with atoms of upper atmosphere form a number of radioactive atoms. One of such isotope of carbon is ($^{14}_{6}C$), formed by bombardment of high energy neutron with nitrogen.

$$^{1}_{0}n + ^{14}_{7}N \rightarrow ^{14}_{6}C + ^{1}_{1}H$$

This isotope of carbon is radioactive and has half life of $T = 5730$ years. This oxidises in isotopic carbon oxide ($^{14}_{6}CO_2$). In atmosphere this isotopic carbon dioxide is present with normal carbon dioxide ($_6C^{12}O_2$). In atmosphere their relative abundance is in the ratio $1.3 \times 10^{12} : 1$ and is constantly maintained.

Carbon dioxide is always absorbed by plants. As such the ratio of $^{12}_{6}C$ to $^{14}_{6}C$ in plants and human beings remains constant equal to $1.3 \times 10^{12} : 1$. This ratio remains there till they are alive. After death the amount of ordinary carbon ($^{12}_{6}C$) remains constant whereas the radioactive carbon ($^{14}_{6}C$) continues to decay.

$$^{14}_{6}C \rightarrow ^{14}_{7}N + \beta^{-1} + \bar{\nu}$$

Knowing the activity of the dead specimen and half life of the radioactive $^{14}_{6}C$ (= 5730 yrs), the age of the specimen can be determined.

Let R_0 be the activity of live specimen and R the activity of the dead specimen. If T is half life, and t the life of specimen, then

$$R = R_0 e^{-\lambda t} = R_0 (2)^{-\frac{t}{T}}$$

or
$$\log \frac{R_0}{R} = \frac{t}{T} \log 2$$

or
$$t = T \frac{\log(R_0/R)}{\log 2} = 3.32 T \log_{10}(R_0/R)$$

This process of finding out the age of archaeological objects is called carbon dating.

28.11 NUCLEAR REACTIONS

When a nucleus is bombarded with various particles (such as neutron, proton, α-particle, etc.), it may cause scattering of projectile particle and it may lead to change in structure of target nuclei. Collisions in which identity of nuclei is changed (i.e., target nucleus is changed into a new nucleus) is called a nuclear reaction.

Symbolically this may be written as

$$^{A}_{Z}X + ^{4}_{2}He \rightarrow ^{A+4}_{Z+2}C \rightarrow ^{A+3}_{Z+1}Y + ^{1}_{1}H + Q$$

where $^{A}_{Z}X$ the target nucleus, bombarded by an α-particle (called the projectile), C is compound nucleus which is unstable. It disintegrates to give a product nucleus Y and a product particle (proton). The reaction is named (α, p) reaction after the name projectile particle (α) and product particle (proton p). In nuclear reaction, Q is the total energy change in the reaction. It is called nuclear reaction energy or Q value of the reaction.

In all the nuclear reactions, following conservation laws are obeyed.

(i) **Conservation of Charge:** This means that total charge of reactants is equal to the total charge of product particles.

(ii) **Conservation of Energy:** Total energy of the system is conserved. Mass is also taken as energy (= mc^2).

(iii) **Conservation of Momentum:** Total momentum of the reactants before reaction is equal to total momentum of the product particles after reaction.

(iv) **Conservation of Mass Number (Nucleons):** Total number of nucleons in reaction remain conserved.

(v) **Conservation of Angular Momentum:** Angular momentum of all the particles taking part in the reaction remains conserved.

Note: All the above laws are also obeyed in natural as well as artificial radioactive disintegration.

Rutherford was the first to perform an experiment in 1919 on artificial transmutation of elements. He observed the reaction

$$^{14}_{7}N + ^{4}_{2}He \rightarrow ^{18}_{9}F \rightarrow ^{17}_{8}O + ^{1}_{1}H \qquad (\alpha, p) \text{ reaction}$$

Some other reactions are

$$^{7}_{3}Li + ^{1}_{1}H \rightarrow ^{8}_{4}Be^{*} \rightarrow ^{4}_{2}He + ^{4}_{2}He \qquad (p, \alpha) \text{ reaction}$$

$$^{11}_{5}B + ^{1}_{1}H \rightarrow ^{12}_{6}C^{*} \rightarrow ^{11}_{6}C + ^{1}_{0}n \qquad (p, n) \text{ reaction}$$

$$^{12}_{6}C + ^{1}_{1}H \rightarrow ^{13}_{7}N^{*} \rightarrow ^{13}_{7}N + \gamma \qquad (p, \gamma) \text{ reaction}$$

$$^{14}_{7}N + ^{1}_{0}n \rightarrow ^{15}_{7}N^{*} \rightarrow ^{14}_{6}C + ^{1}_{1}H \qquad (n, p) \text{ reaction}$$

$$^{2}_{1}H + \gamma \rightarrow ^{1}_{1}H + ^{1}_{0}n \qquad \text{(Photo disintegration)}$$

Q value of nuclear reaction is calculated using Einstein's mass energy equivalence ($E = \Delta mc^2$). It may be positive or negative. If Q value is positive i.e. energy is released it is called *exoenergetic* reaction and if Q value is negative, i.e. energy is supplied, it is called *endoenergetic.*

The most important nuclear reactions have been investigated are *fission and fusion.*

28.12 NUCLEAR FISSION

Nuclear fission is the phenomenon of splitting of a heavy nucleus (A > 230) into two or more lighter nuclei.

In fission process certain mass disappears, i.e. the total mass of the reactant components is slightly greater than the total mass of product components. This difference called mass defect (Δm) is released in the form of energy according to Einsteins mass energy relation ($E = \Delta mc^2$).

In 1939 Ottohann and Strassman observed that, when high energy neutrons (~10^6 eV) are bombarded on $_{92}U^{238}$, it splits up into two nearly equal light nuclei $_{56}Ba^{148}$ and $^{88}_{36}Kr$. In this fission 3 neutrons along with 200 MeV of energy is released. The reaction is

$$^{238}_{92}U + ^{1}_{0}n \rightarrow ^{148}_{56}Ba + ^{88}_{36}Kr + 3^{1}_{0}n + Q$$

Mass defect in this reaction comes out to be 0.2153 amu and the energy released is (1 amu = 931 MeV).

$$Q = \Delta m \times 931 = 0.2153 \times 931 = 200.4 \text{ MeV}$$

Thus in the fission of a single nucleus ($^{238}_{92}U$) an energy of about 200 MeV is released and such a huge energy in about 10^{-9} sec. This energy appears in the form of kinetic energy of fragments of fission, released neutrons and γ-rays. It must be remembered that the fission fragments are not always Ba and Kr. Other nucleides have also been obtained. In all more than 100 isotopes of more than 20 elements are obtained whose mass number varies from 75 to 160, but the energy released in each fission of a nucleus ($^{238}_{92}U$) is same in all the cases. This phenomenon is the bases of atomic bomb and atomic reactors.

Natural Uranium has two isotopes $^{235}_{92}U$ and $^{238}_{92}U$. In this naturally occurring Uranium, 99.3% is Uranium-238 and 7% is Uranium-235. Both are fissionable. Uranium-238 can be fissioned only by fast neutron (energy not less than 1 MeV) whereas fission of Uranium-235 is possible by thermal neutrons (slow neutron of energy 1eV). Thus for fission process Uranium -235 is more useful. For atom bomb Uranium-235 is used.

Explanation of Nuclear Fission

The fission process is similar to the oscillations of a liquid drop. ^{235}U nucleus absorbs a thermal neutron (Fig. 28.6 (*a*)) forms compound nucleus ^{236}U with excess energy and oscillates violently (Fig. 28.6 (*b*)). This forms a neck (Fig. 28.6 (*c*)), and the Coulomb repulsive forces further stretches (Fig. 28.6 (*d*)) and the oscillations become violent and the nucleus split into two. The fragments separate (Fig. 28.6 (*e*)) and neutrons are emitted. *This is called liquid drop model of nucleus.* The critical energy (E_c) needed by ^{236}U nucleus to undergo fission can be calculated. (^{236}U is the compound nucleus formed by capture of neutron by ^{235}U).

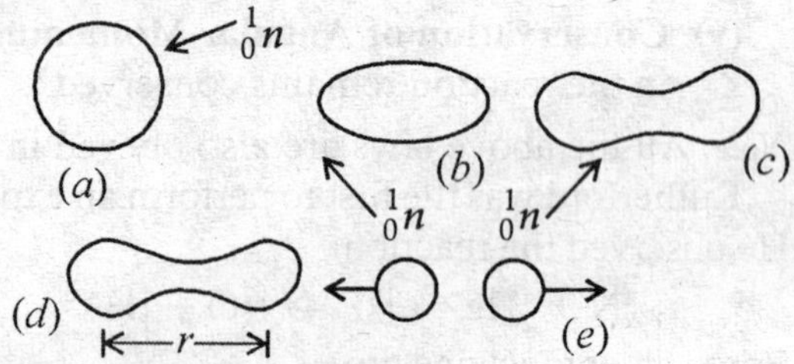

Fig. 28.6

Critical energy (E_c) is the energy required by fragments to overcome Coulomb force. For this nucleus (^{236}U) the excitation energy produced when ^{235}U captures a neutron is 6.5 MeV. The critical energy is about 6.2 MeV (less than the excitation energy). Therefore in the critical state the resulting ^{236}U nucleus has enough energy to break apart. Whereas the critical energy for fission of ^{239}U nucleus is 5.9 MeV. After capturing a neutron ^{238}U nucleus produces an energy of only 5.2 MeV. Therefore when ^{238}U captures a neutron to form ^{239}U, the excitation energy is not enough for fission to occur unless the neutron is very fast one. Consequently excited ^{239}U nucleus de-excites by γ or α emission.

Difference Between Radioactive Decay and Nuclear Fission

Though radioactive decay and nuclear fission both are nuclear phenomenon but they are quite different phenomenon. The main differences are:

(*a*) Radioactive disintegration is as spontaneous phenomenon. Whereas the nuclear fission is not an spontaneous phenomenon. In fission a heavy nucleus is bombarded with neutrons.

(*b*) In radioactive disintegration particles like α-particles and β-particles are emitted by the nucleus and the energy is released in the form of their kinetic energy and γ-rays. This energy is not very large of the order of 6 to 10 MeV. Whereas in nuclear fission, the nucleus breaks in two nearly equal heavy nuclei and the energy released is very high (order of 200 MeV per nucleus).

(*c*) In radioactive disintegration the atomic number changes by either 1 (β-decay) or 2 (α-decay) and mass number changes by either zero (β-decay) or 4 (α-decay). In nuclear fission the atomic number Z and mass number A both change forming nuclei of nearly equal Z and A.

(*d*) Radioactive disintegration cannot be controlled whereas nuclear fission rate can be controlled by using moderators and arresters.

Chain Reaction in Nuclear Fission

When $^{235}_{92}U$ is bombarded with thermal neutrons then each nucleus breaks in two nearly equal parts. In this process of fission, three new neutrons and a large amount of energy is released. These three neutrons in favourable conditions bring about the fission of three more $^{235}_{92}U$ nuclei producing 27 neutrons and so on. Thus a continuous chain of reactions starts. This chain of reactions is called *chain reaction*. Thus in a very short time (~10^{-6} to 10^{-7} sec) a huge amount of energy is released. This chain reaction is shown in Fig. 28.7. Here FP represents fission product. If we have a control over this reaction, the energy released can be used for peaceful purposes. This leads to the construction of *nuclear reactor* or *nuclear pile*. However if the reaction cannot be controlled, the energy released will bring a disastrous effects. This is what we call as *atom bomb*.

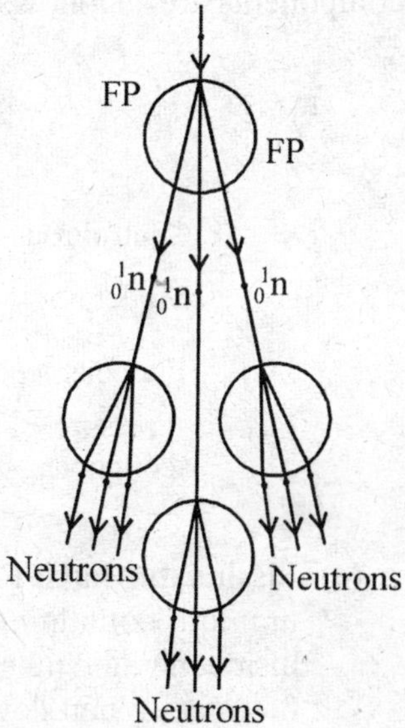

Fig. 28.7

For a sustained chain reaction one of the neutrons (on an average) emitted in fission of ^{235}U must be captured by another U^{235} nucleus and cause its fission. Thus for sustained nuclear chain reaction the leakage of secondary neutrons should be prevented. There should be no absorption of neutrons by nonfissionable impurities. This can be achieved using enriched Uranium-235.

The chain reaction once started will, remain steady, accelerate or retard depends on *neutron production factor (K)*. This is defined as the *ratio of rate of production of neutrons to the rate of loss of neutrons due to leakage and absorption, i.e.*

$$K = \frac{\text{Rate of production of neutrons}}{\text{Rate of loss of neutrons}}$$

The maximum possible value of K is about 2.5. Due to escape of neutrons from the region containing fissionable nuclei and capture of some of the neutrons (may be by non-fissionable nuclei) K is reduced below 2.5. If K is exactly 1, the reaction will be self sustained and steady. The size of fissionable material used is said to be the critical size and its mass the critical mass. If it is less than 1 the reaction will die out. The size of material used is called *sub critical.* If K is significantly greater than 1, the reaction rate will increase rapidly resulting in an explosion. The size of material used in this case is *super critical.*

The energy of neutrons emitted in fission is of the order of 1 MeV or higher. The neutron capture probability by ^{235}U is high at low energies (for thermal neutrons ~1 eV). For sustained reaction the neutrons are slowed down using *moderator*. Generally, heavy water, graphite, deuterium, paraffin wax, etc. are used as moderators.

For peaceful use of atomic energy, while fabricating the atomic reactor, the availability of neutrons for fission is regulated by using a series of rods made of a neutron absorbing material (B, Hf, Cd) are used. It maintains control over the number of neutrons (as well as *K*).

Nuclear Fission Reactor

A nuclear reactor is based on *the controlled nuclear chain reaction* and is a device to produce nuclear energy for constructive purposes. The schematic block diagram is shown in Fig. 28.8. The main components are as follows:

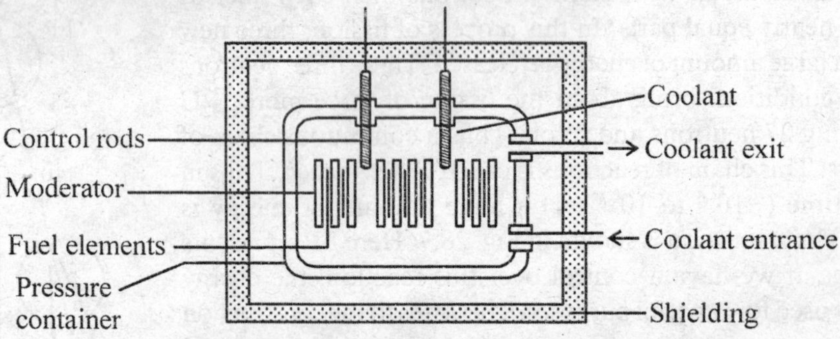

Fig. 28.8

(*a*) **Fissionable Fuel:** Fuel used is fissionable material like ^{235}U, ^{233}U, ^{239}Pu. Generally enriched uranium containing 90% $^{235}_{92}U$ is used. The fuel in the form of uranium oxide is placed inside metallic tubes and are called fuel elements (or rods). Cluster of such rods constitute fuel core. When slow neutrons interact with fuel, the fission starts and $^{235}_{92}U$ content decreases. Part of $^{235}_{92}U$ is converted into $^{239}_{94}Pu$ and other isotopes of Pu. As such the fuel rods need replacement from time to time.

(*b*) **Moderator:** It slows down the fast neutrons produced during the fission. It is a substance of low nuclear mass which surrounds the fuel rods and does not absorb neutrons. The moderator can be water if the enriched uranium is used, heavy water or graphite if the uranium is natural. Moderator slows down the secondary neutrons to low energy neutrons called *thermal neutrons*.

(*c*) **Control Rods:** A series of rods made of neutron absorbing material (B, Hf, Cd) are used to control the availability of neutrons for fission. These rods are adjusted from outside the reactor. They control the chain reaction from becoming violent.

(*d*) **Coolant:** To remove the heat produced in the core of the reactor, and transfer it to heat exchanger a fluid is used. This is called coolant. At ordinary temperatures water or heavy water are used as coolants but at high temperatures liquid sodium is used as a coolant. This prevents the excessive rise in temperature.

(*e*) **Shielding:** Thick concrete walls (about 2 to 2.5 m) surround the core chamber of the reactor. This prevents the emitted radiations during reactions, producing harmful effects on the persons working on the reactor.

Breader Reactor

The fission of ^{235}U although produces a large amount of energy but its less abundance in nature and limited capacity of enrichment, makes its use in reactors, uneconomical in long term. The fission can easily be produced in ^{239}Pu. But ^{239}Pu is not a naturally occurring isotope. However ^{238}U can capture a neutron to produce ^{239}Pu, i.e.

$$^{238}_{92}U + ^{1}_{0}n \rightarrow ^{239}_{92}U \xrightarrow{\beta^-} ^{239}_{93}Np \xrightarrow{\beta^-} ^{239}_{94}Pu$$

$^{239}_{94}Pu$ so produced can be used as a fuel.

In order to produce $^{239}_{94}Pu$ by the above process, used uranium rods which contain only ^{238}Pu are placed in or around uranium-reactor core. The geometry of reactor is made such that on an average out of 2.4 neutrons produced in a fission, one neutron is absorbed by ^{238}U nucleus in these rods resulting in ^{239}Pu. In this way we produce more fuel in the form of ^{239}Pu than what we consume. Thus apart from nuclear energy, such reactors give us fresh nuclear fuel which often exceeds the nuclear fuel use. Such reactors are called *breader reactors.*

$^{239}_{94}Pu$ is fissionable and undergoes α-decay like

$$^{239}_{94}Pu \xrightarrow{\alpha} ^{235}_{92}U$$

The half life of $^{239}_{94}Pu$ is 24000 yrs to produce $^{235}_{92}U$. Thus using an unfissionable material $^{238}_{92}U$, we produce fissionable material $^{239}_{94}Pu$ which can be fissioned even by fast neutrons and therefore do not require moderator.

28.13 NUCLEAR FUSION

Nuclear fusion is the process in which two light nuclei combine to form a single large nucleus. The binding energy curve Fig. 28.2 shows that energy can be released in such a process. Few examples of nuclear fusion are:

$$^{2}_{1}H + ^{2}_{1}H \rightarrow ^{3}_{1}H + ^{1}_{1}H + 4.0 \text{ MeV} \quad \ldots(a)$$

$$^{1}_{1}H + ^{1}_{1}H \rightarrow ^{2}_{1}H + \beta^{+} + \nu + 4.2 \text{ MeV} \quad \ldots(b)$$

$$^{2}_{1}H + ^{2}_{1}H \rightarrow ^{4}_{2}He + 24 \text{ MeV} \quad \ldots(c)$$

$$^{2}_{1}H + ^{3}_{1}H \rightarrow ^{4}_{2}He + ^{1}_{0}n + 32.6 \text{ MeV} \quad \ldots(d)$$

In all the above reaction we observe that energy released in fusion per unit mass is much larger than the energy released per unit mass in the process of fission. Secondary in all these reactions two positively charged particles combine to form a larger nucleus. As such, this process is hindered by the Coulomb repulsion that acts to prevent the two positively charged particles from getting close enough to be within the range of their attractive nuclear forces and thus fusing. The height of this Coulomb barrier depends on the charges and the radii of the two interacting nuclei.

For two protons (radius r) having kinetic energy K each will fuse if their distance is $2r$ or less. Then

$$2K = \frac{e^2}{4\pi\varepsilon_0(2r)} \text{ or } K = \frac{e^2}{16\pi\varepsilon_0 r}$$

For $r = 1 \text{ f} = 10^{-15}$ m, $K = 400$ keV

i.e., for two protons, the barrier height is 400 keV. The barrier height for more highly charged nuclei is higher.

For fusion to occur in bulk matter an energy of atleast 400 keV is needed. For particles to acquire this much energy the temperature of the material must be raised to $10^8 \sim 10^9$ K (temperature $10^8 = 10$ keV).

The process of fusion of two light particles, by raising the temperature so high, so that the particles can penetrate Coulomb potential barrier is called *thermonuclear fusion.*

Besides high temperature and pressure, high density of fusing particles is necessary for fusion to occur; so that collisions between light nuclei occur quite frequently. J. D. Lawson showed that in order to get an energy output greater than the energy input a fusion reactor should achieve

$$n\tau > 10^{14}\ \text{g/cm}^3$$

where n is the density of the interacting particles and τ is confinement time. The quantity $n\tau$ is called *Lawson number.*

The ratio of the energy output to energy input is known as Q of the fusion machine. For a viable fusion machine, Q should be greater than 1.

Stellar Energy

For thermonuclear fusion to take place, extreme conditions of temperature and pressure are required. These conditions are met within the interior of the sun and stars and cannot be arranged easily in laboratory.

Stellar energy is the energy obtained continuously from the sun and stars. According to rough estimation sun is radiating energy at the rate of 3.8×10^{26} J/s. Modern view for this vast source of energy is *the fusion of light nuclei.* 90% part of the sun is made of hydrogen and helium and the rest 10% part consists of other elements in which majority are light elements. The temperature of the interior part (called photosphere) has a temperature of the order of 10^7 K. At this high temperature, the electrons of all the elements are ejected from their orbits and the electrons and nuclei (positive ions) are present in large number. *This state of matter in which electrons and positive ions are present in large number is called Plasma state.* In plasma state the elements are in nuclear form. Due to high temperature when four protons (hydrogen nuclei) fuse to from α-particle (helium nuclei), a vast amount of energy is released in the process. This fusion energy is the source of solar energy.

Two types of thermonuclear reactions have been postulated in the stars for fusion of four hydrogen nuclei to form a helium nucleus and release vast amount of energy. They are

(*i*) **Carbon-Nitrogen Cycle:** According to American Scientist Bethe, in stars and sun, fusion of 4-hydrogen nuclei in a helium nucleus is not direct but through a cycle of thermonuclear reactions. These thermonuclear reactions are very rapid. In these reactions carbon acts like a catalyst. These reactions are

$${}_1^1H + {}_6^{12}C \rightarrow {}_7^{13}N + \gamma + Q(= 1.93\ \text{MeV})$$

$${}_7^{13}N \rightarrow {}_6^{13}C + {}_1\beta^0 + \nu + Q(= 1.20\ \text{MeV})$$

$${}_1^1H + {}_6^{13}C \rightarrow {}_7^{14}N + \gamma + Q(= 7.6\ \text{MeV})$$

$${}_1^1H + {}_7^{14}N \rightarrow {}_8^{15}O + \gamma + Q(= 7.39\ \text{MeV})$$

$${}_8^{15}O \rightarrow {}_7^{15}N + {}_1\beta^0 + \nu + Q(= 1.71\ \text{MeV})$$

$${}_1^1H + {}_7^{15}N \rightarrow {}_6^{12}C + {}_2^{14}He + Q(= 4.99\ \text{MeV})$$

Adding all the above, we get

$${}_1^1H + {}_1^1H + {}_1^1H + {}_1^1H \rightarrow {}_2^4He + 2\,{}_1\beta^0 + 2\nu + \gamma + Q(= 24.82\ \text{MeV})$$

Thus in C–N cycle, 4 hydrogen nuclei fuse together to form ${}^4_2$He (helium) nucleus and two positrons ($2\,{}_1\beta^0$) and 24.82 MeV energy is released. These two positrons annihilate with two electrons to release an energy of about 2 MeV. Thus total energy released in carbon-nitrogen cycle is 26.82 MeV.

(*ii*) **Proton-Proton Cycle:** Bethe's principle is valid for stars with very high temperatures. For stars with less temperatures the C–N cycle probability is very small as such in these stars p–p cycle probability is more. This cycle is

$$ {}^1_1\text{H} + {}^1_1\text{H} \rightarrow {}^2_1\text{H} + {}_1\beta^0 + \nu + Q\,(= 0.42 \text{ MeV}) $$

$$ {}^2_1\text{H} + {}^1_1\text{H} \rightarrow {}^3_2\text{He} + \gamma + Q\,(= 5.49 \text{ MeV}) $$

$$ {}^3_2\text{He} + {}^3_2\text{He} \rightarrow {}^4_2\text{He} + {}^1_1\text{H} + {}^1_1\text{H} + Q\,(= 12.86 \text{ MeV}) $$

Multiplying first two by 2 and adding to 3rd gives

$$ {}^1_1\text{H} + {}^1_1\text{H} + {}^1_1\text{H} + {}^1_1\text{H} \rightarrow {}^4_2\text{He} + 2\,{}_1\beta^0 + 2\nu + 2\gamma + Q\,(= 24.68 \text{ MeV}) $$

Here also 2 positrons ($2\,{}_1\beta^0$) annihilate with 2 electrons to release 2.04 MeV energy thus total energy released is 26.72 MeV.

In the sun both proton-proton and carbon-nitrogen cycles occur with equal probability. In stars whose temperatures are less than that of the sun proton-proton cycle dominates the energy generation. On the other hand in stars whose interior temperatures are more than that of the sun, the energy generation is mainly due to carbon, nitrogen cycle.

Life of Sun

Sun radiates energy at the rate of 3.8×10^{26} J/s. Using Einstein's mass energy relation, it looses its mass at the rate of

$$ \Delta m = \frac{E}{c^2} = \frac{3.8 \times 10^{26}}{9 \times 10^{16}} = 4.1 \times 10^9 \text{ kg/s} $$

Thus sun is loosing its mass rapidly. Total mass of sun is estimated to be 2×10^{30} kg. According to these data the age of the sun is

$$ T = \frac{2 \times 10^{30}}{4.1 \times 10^9 \times 3.2 \times 10^7} \qquad (1\text{yrs} = 3.2 \times 10^7\text{s}) $$

$$ \approx 10^{13} \text{ yrs.} $$

Controlled Thermonuclear Fusion

On November 1, 1952, USA exploded a fusion device, generating energy equivalent to 10 million tons of TNT. (1 TNT, on explosion releases 2.6×10^{22} MeV of energy). This was the first uncontrolled thermonuclear reaction on the earth in the form of hydrogen bomb. But a controlled sustained thermonuclear reaction is difficult to achieve. The main reasons for this are

(*a*) **High Particle Density:** The density of interacting particles deuterium, tritium (D–T) must be very high so that the collision frequency becomes large. At high temperatures deuterium (${}^2_1$H) is completely in ionized state (D^+, e^-). This is called plasma. Due to repulsive forces between positive ions it is very difficult to increase the density.

(*b*) **High Plasma Temperature:** For fusion, the temperature of the plasma must be high. In laboratory a temperature of 4×10^8 K can be produced. This temperature is approximately 20 times the temperature of the inner core of the sun.

(c) **Long Confinement Time:** To keep this hot plasma (temperature 4×10^8) into a container for a long time, so that it does not touch the walls of the container is a very difficult problem. In one of the processes plasma is placed in a very intense magnetic field so that the plasma moves in circles and is confined in a definite region. This confinement is achieved for short duration (≈ few micro seconds) fusion thus proceeds in bursts or pulses.

Due to a number of difficulties (some of them as mentioned above) the controlled thermonuclear chain reaction has not been possible in laboratory. Efforts are being made world over to achieve thermonuclear fusion in laboratory. When that happens we will get an unlimited source and unpolluted energy.

Comparison Between Fusion Reactor and Fission Reactor

Fusion reactors have the following advantages over the fission reactors:

(i) In fusion reactors neither radioactive material is used nor it is produced as a result of fusion. As such there is no problem of atmospheric pollution nor there is a problem of disposal of radioactive wastes. At present the radioactive waste from fission reactors is sealed in steel containers and dumped into the sea or in desert.

(ii) The persons working in fission reactors have to protect themselves from hazardous radiations coming from the reactor. Whereas in fusion reactors no such problem is there.

(iii) The fuel used in fusion reactors is available in abundance and will be available for long time to come. Whereas the fuel used in fission reactors have limited source and will not last for more than 100 years at the present rate of consumption.

SOLVED EXAMPLES

Example 28.1: A radioactive nucleus can decay by two different processes. The half life for the first process is T_1 and that for the second process is T_2. Find the effective value of half-life T of the nucleus.

Solution: The decay constant for the first process is $\lambda_1 = \dfrac{\log 2}{T_1}$ and for the second process is $\lambda_2 = \dfrac{\log 2}{T_2}$

The probability that an active nucleus decays by first process in a time interval dt is $\lambda_1 dt$. Similarly the probability that it decays by the second process is $\lambda_2 dt$. The probability, that either decays by first process or by the second process is $\lambda_1 dt + \lambda_2 dt$. If the effective decay constant is λ, this probability is also equal to λdt. Therefore

$$\lambda dt = \lambda_1 dt + \lambda_2 dt$$

or

$$\lambda = \lambda_1 + \lambda_2$$

and

$$\frac{\log 2}{T} = \frac{\log 2}{T_1} + \frac{\log 2}{T_2}$$

$$\therefore \quad \frac{1}{T} = \frac{1}{T_1} + \frac{1}{T_2} \text{ or } T = \frac{T_1 T_2}{T_1 + T_2}$$

Example 28.2: A nucleus X initially at rest, undergoes α-decay according to the equation

$$^{A}_{92}X \rightarrow ^{228}_{Z}Y + \alpha$$

(*a*) Find the value of A and Z in the above process.

(*b*) The α-particle in the above process is found to move in a circular track of radius 0.11 m in uniform magnetic field of 3 T. Find the energy (in MeV) released during the process and the binding energy of the parent nucleus X. Given that

$$m_y = 228.03 \text{ u} \quad m_\alpha = 4.003 \text{ u} \quad m_n = 1.009 \text{ u} \quad m_p = 1.0084 \text{ u}$$

Solution:

(*a*) α-Decay process is

$$^{A}_{92}X \rightarrow ^{238}_{Z}Y + ^{4}_{2}\text{He}$$

Hence the conservation of mass number gives

$$A = 228 + 4 = 232$$

and the conservation of charge number gives

$$92 = Z + 2 \text{ or } Z = 90$$

(*b*) Since α-particle moves in circular orbit in the magnetic field. Thus

$$qv_\alpha B = \frac{m_\alpha v_\alpha^2}{r}$$

or $$v_\alpha = \frac{qBr}{m_\alpha} = \frac{2\times1.6\times10^{-19}\times3\times0.11}{4.003\times1.67\times10^{-27}}$$

∴ $$q = 2e = 2 \times 1.6 \times 10^{-19}\text{C and } m_\alpha = 4.0034 = 4.003 \times 1.67 \times 10^{-27} \text{ kg}$$

$$v_\alpha = 1.58 \times 10^7 \text{ m/s}$$

Conservation of momentum gives that momentum of α-particle and Y nucleus are equal in magnitude and opposite in direction. Hence total energy released is

$$E = E_\alpha + E_y = \frac{p^2}{2m_\alpha} + \frac{p^2}{2m_y} = \frac{p^2}{2}\left(\frac{m_y + m_\alpha}{m_\alpha m_y}\right)$$

But $$p = m_\alpha v_\alpha$$

∴ $$E = \frac{m_\alpha v_\alpha^2}{2}\left(\frac{m_y + m_\alpha}{m_y}\right)$$

$$= \frac{1}{2} \times \frac{4.003\times1.67\times10^{-27}\times(1.58\times10^7)^2}{1.6\times10^{-13}}\left(\frac{228.03 + 4.003}{228.03}\right)$$

$$= 5.342 \text{ MeV}$$

This energy is equivalent to mass $= \dfrac{5.342}{931.5} = 0.00574$ amu

Applying the principle of conservation of energy, we get

$$^{232}_{92}X \rightarrow m_y + m_\alpha + 0.0057$$

$$= 228.03 + 4.003 + 0.0057 = 232.03874$$

Since $^{232}_{92}X$ nucleus contains 92 protons, 140 neutrons then

Binding energy = mass defect

$$= 92(1.008) + 140(1.009) - 232.0387 = 1.95734$$

or B.E. $= 1.9573 \times 931.5 = 1823$ MeV.

Example 28.3: A radioactive nuclide A_1 with decay constant λ_1 transforms into a radioactive nuclide A_2 with decay constant λ_2. Assuming that at the initial moment the preparation contained only the radioactive nuclide A_1, find

(*a*) the equation describing accumulation of the radioactive nuclide A_1 with time,

(*b*) the time interval after which the activity of radioactive nuclide A_2 reaches the maximum,

(*c*) what happens if $\lambda_1 = \lambda_2 = \lambda$?

Solution:

(*a*) Let N_1 and N_2 be the number of active nuclides of A_1 and A_2 respectively and at $t = 0$ there were N_0 active atoms of A_1. Then

$$\frac{dN_1}{dt} = -\lambda_1 N_1 \qquad ...(i)$$

$$\frac{dN_2}{dt} = \lambda_1 N_1 - \lambda_2 N_2 \qquad ...(ii)$$

and $$N_1 = N_0 e^{-\lambda_1 t} \qquad ...(iii)$$

From (*ii*), we have

$$\frac{dN_2}{dt} + \lambda_2 N_2 = \lambda_1 N_1 = \lambda_1 N_0 e^{-\lambda_1 t}$$

Multiplying both sides by $e^{\lambda_2 t}$, we get

$$\left(\frac{dN_2}{dt} + \lambda_2 N_2\right)e^{\lambda_2 t} = \lambda_1 N_0 e^{-(\lambda_1 - \lambda_2)t}$$

or $$\frac{d}{dt}\left(N_2 e^{\lambda_2 t}\right) = \lambda_1 N_0 e^{-(\lambda_1 - \lambda_2)t}$$

Integration with respect to time gives

$$N_2 e^{\lambda_2 t} = \frac{\lambda_1 N_0}{\lambda_2 - \lambda_1} e^{-(\lambda_1 - \lambda_2)t} + C$$

where C is constant of integration.

Now at $t = 0$ $N_2 = 0$ gives $C = -\frac{\lambda_1 N_0}{\lambda_2 - \lambda_1}$

Thus $$N_2 = \frac{\lambda_1 N_0}{\lambda_2 - \lambda_1}\left[e^{-(\lambda_1 - \lambda_2)t} - 1\right]e^{-\lambda_2 t}$$

$$= \frac{\lambda_1 N_0}{\lambda_2 - \lambda_1}\left[e^{-\lambda_1 t} - e^{-\lambda_2 t}\right]$$

(*b*) The activity of A_2 at any time t is $N_2\lambda_2$. This is maximum when N_2 is maximum, *i.e.* $\frac{dN_2}{dt} = 0$

i.e. $\lambda_1 e^{-\lambda_1 t} - \lambda_2 e^{-\lambda_2 t} = 0$

or $e^{(\lambda_2-\lambda_1)t} = \lambda_2/\lambda_1$

or $$t = \frac{\log(\lambda_2/\lambda_1)}{\lambda_2 - \lambda_1}$$

(*c*) When $\lambda_1 = \lambda_2 = \lambda$, let $\lambda_2 = \lambda_1 + \varepsilon$ where $\varepsilon \to 0$. Then

$$N_2 = \frac{\lambda_1 N_0}{\lambda_1 + \varepsilon - \lambda_1}\left[e^{-\lambda_1 t} - e^{-(\lambda_1+\varepsilon)t}\right]$$

$$= \frac{\lambda_1 N_0}{\varepsilon}\left[1 - e^{-\varepsilon t}\right]e^{-\lambda_1 t}$$

$$= \frac{\lambda_1 N_0}{\varepsilon}\left[1 - 1 + \varepsilon t + \text{....higher power of } \varepsilon\right]e^{-\lambda_1 t}$$

$$N_2 = \lambda_1 N_0 e^{-\lambda_1 t} \qquad \text{where } \varepsilon \to 0$$

$$N_2 = \lambda N_0 e^{-\lambda t}$$

N_2 is maximum when

$$e^{(\lambda_2-\lambda_1)t} = \frac{\lambda_2}{\lambda_1}$$

or $$e^{\varepsilon t} = \frac{\lambda + \varepsilon}{\lambda} = 1 + \frac{\varepsilon}{\lambda}$$

or $t = 1/\lambda$ when $\varepsilon \to 0$.

Example 28.4: A nucleus A decays to B with constant λ_a and B decays to C with decay constant λ_b. Initially at $t = 0$, the number of nuclei A and B are $2N_0$ and N_0 respectively. At some instant $t = t_0$ number of nuclei of B stop changing. Find t_0, if at this instant number of nuclei of B are $\frac{3N_0}{2}$.

Solution: $A \xrightarrow{\lambda_a} B \xrightarrow{\lambda_b} C$

At $t = 0$ $N_a = 2N_0$ and $N_b = N_0$

At any instant t let N_a and N_b be the number of active atoms, then

$$\frac{dN_a}{dt} = -\lambda_a N_a$$

$$\frac{dN_b}{dt} = +\lambda_a N_a - \lambda_b N_b$$

at $$t = t_0,\ \frac{dN_b}{dt} = 0 \text{ and } N_b = \frac{3N_0}{2} \text{ and } N_a = 2N_0 e^{-\lambda_a t_0}$$

$$0 = 2N_0\lambda_a e^{-\lambda_a t_0} - \lambda_b \frac{3}{2} N_0$$

or $$e^{-\lambda_a t_0} = \frac{3\,\lambda_b}{4\,\lambda_a}$$

or $$-\lambda_a t_0 = \log\left(\frac{3\,\lambda_b}{4\,\lambda_a}\right)$$

and $$t_0 = \frac{1}{\lambda_a}\log\left(\frac{4\,\lambda_a}{3\,\lambda_b}\right)$$

Example 28.5: A radioactive nucleide with half life T = 14.3 days is produced in a reactor at a constant rate n = 2.7 × 10⁹ nuclei per second. How soon after the beginning of production of daughter nucleide will its activity be equal to A = 1.0 × 10⁹ dis/s?

Solution: Let A be instantaneous activity. The number of nuclei produced in time $dt = n\,dt$. Thus the number of nuclei decayed in time dt is

$$-dN = (A - n)dt$$

But $A = N\lambda$ where N is number of active nuclei and λ is decay constant.

∴ $$-dN = (N\lambda - n)dt$$

or $$\frac{dN}{\frac{n}{\lambda} - N} = \lambda dt$$

Integration gives

$$-\log\left(\frac{n}{\lambda} - N\right) = \lambda t + C$$

where C is constant of integration.

But at $t = 0$, $N = 0$ and $C = -\log\frac{n}{\lambda}$

Substitution gives

$$-\log\left(\frac{n}{\lambda} - N\right) = \lambda t - \log\frac{n}{\lambda}$$

or $$-\log\left(\frac{n}{\lambda} - N\right) + \log\frac{n}{\lambda} = \lambda t$$

or $$\log\left(\frac{n}{\lambda} - N\right) - \log\frac{n}{\lambda} = -\lambda t$$

or $$\log\left(1 - \frac{N\lambda}{n}\right) = -\lambda t$$

But $N\lambda = A$ the activity. Hence

$$t = -\frac{1}{\lambda}\log\left(1 - \frac{A}{n}\right)$$

But $$\lambda = \frac{\log 2}{T},\ A = 1.0 \times 10^9 \text{ dis/s} \quad n = 2.7 \times 10^9 \text{ Nuclei/s}$$

$$\therefore \quad t = -\frac{T}{\log 2}\log\left(1-\frac{10^9}{2.7\times 10^9}\right)$$

$$= -\frac{T}{\log 2}\log\left(1-\frac{1}{2.7}\right) = -\frac{T(-0.46)}{\log 2}$$

$$= \frac{0.46T}{\log 2} = 0.67T$$

But $T = 14.3$ days.

$\therefore$ $T = 0.67 \times 14.3 = 9.5$ days.

Example 28.6: To investigate the β-decay of Mg^{23} radionucleide a counter was activated at the moment $t = 0$. It registered N_1 beta particles by a moment $t_1 = 2.0$ s and by a moment $t_2 = 6$ sec. The number of registered β-particles was 2.66 times greater. Find the mean life time of the given radionucleide.

Solution: If N_0 is the number of nuclei active at $t = 0$. Then nuclei active at time t are $N = N_0 e^{-\lambda t}$ where λ is disintegration constant. Hence the nuclei decayed in time t are

$$n = N_0 - N_0 e^{-\lambda t} = N_0(1 - e^{-\lambda t})$$

At time $t = t_1 = 2$ sec, counter recorded N_1 disintegration

$$\therefore \quad N_1 = N_0(1 - e^{-2\lambda})$$

and at time $t = t_2 = 6$ sec, it registered N_2 disintegration

$$\therefore \quad N_2 = N_0\left(1 - e^{-6\lambda}\right)$$

or

$$\frac{N_2}{N_1} = \frac{1-e^{-6\lambda}}{1-e^{-2\lambda}} = 2.66$$

Let $e^{-2\lambda} = x$

$$\therefore \quad \frac{1-e^{-6\lambda}}{1-e^{-2\lambda}} = \frac{1-x^3}{1-x} = 1 + x + x^2$$

Thus $x^2 + x + 1 = 2.66$

or $x^2 + x - 1.66 = 0$

and

$$x = \frac{-1 \pm \sqrt{1+4\times 1.66}}{2} = 0.882$$

–ve sign before under root is not admissible as it will make x –ve

$\therefore$ $e^{-2\lambda} = 0.882$

or $-2\lambda = -0.126$ or $\lambda = 0.063$

or mean life $\tau = 1/\lambda = 16$ sec.

Example 28.7: Nuclei of radioactive element A are being produced at a constant rate α. The element has a decay constant λ. At time $t = 0$ there are N_0 nuclei of the element A.

(*a*) Calculate the number N of the nuclei of A at time t.

(*b*) If $\alpha = 2N_0\lambda$, calculate the number of nuclei of A after one half-life of A and also the limiting value of N as $t \to \infty$.

Solution:

(*a*) The rate of production of radioactive element *A* is

$$\frac{dN_a}{dt} = \alpha$$

and the rate of decay of element *A* is $-\frac{dN_a}{dt} = \lambda N_a$

or $$\frac{dN_a}{dt} = -\lambda N_a$$

Thus net rate of element *A* is

$$\frac{dN_a}{dt} = \alpha - \lambda N_a$$

or $$\frac{dN_a}{\alpha - \lambda N_a} = dt$$

and integration gives

$$-\frac{1}{\lambda}\log(\alpha - \lambda N_a) = t + C$$

where *C* is constant of integration. Initial conditions $t = 0$ $N_a = N_0$ gives

$$C = -\frac{1}{\lambda}\log(\alpha - \lambda N_0)$$

or $$-\frac{1}{\lambda}\log(\alpha - \lambda N_a) = t - \frac{1}{\lambda}\log(\alpha - \lambda N_0)$$

$$-\frac{1}{\lambda}\log\frac{\alpha - \lambda N_a}{\alpha - \lambda N_0} = t$$

or $$\frac{\alpha - \lambda N_a}{\alpha - \lambda N_0} = e^{-\lambda t}$$

or $$N_a = \frac{1}{\lambda}\left[\alpha - (\alpha - \lambda N_0)e^{-\lambda t}\right]$$

(*b*) If $\alpha = 2N_0\lambda$ then

$$N_a = \frac{1}{\lambda}\left[2N_0\lambda - (2N_0\lambda - N_0\lambda)e^{-\lambda t}\right]$$

$$= N_0\left(2 - e^{-\lambda t}\right) \quad \text{But } \lambda = \frac{\log 2}{T}, \text{ where } T \text{ is half life}$$

$$\therefore \quad N_a = N_0\left[2 - (e^{\log 2})^{-t/T}\right] = N_0\left\{2 - (2)^{-t/T}\right\}$$

for $t = T$

$$N_a = N_0\{2 - (2)^{-1}\} = N_0(2 - 0.5) = 1.5N_0$$

when $t \to \infty$

$$N_a = N_0(2) = 2N_0$$

Example 28.8: The element curium $^{248}_{96}Cm$ has a mean life of 10^{13} sec. Its primary decay modes are spontaneous fission and α-decay. The former with probability of 8% and the latter with a probability of 92%. Each fission releases 200 MeV of energy. The masses involved in α-decay are as follows: $^{248}_{96}Cm$ = 248.072220 u $^{244}_{94}Pu$ = 244.064100 u and $^{4}_{2}He$ = 4.002603 u. (1 u = 931 MeV). Calculate the power output from sample of 10^{20} Cm atoms.

Solution: α-decay gives

$$^{248}_{96}Cm \rightarrow {}^{244}_{94}Pu + {}^{4}_{2}He$$

In α-decay mass defect is

$$\Delta m = (244.064100 \text{ u} + 4.002603 \text{ u}) - 248.072220 = -0.0055166 \text{ u}$$

∴ Energy released per α-decay $E_d = 0.0055166 \times 931 = 5.136$ MeV

Energy released per fission E_f is 200 MeV

Since probability of α-decay is 92% and that of fission is 8%. Hence the total energy released in transformation is

$$E = 0.92E_d + 0.08E_f$$

$$= 0.92 \times 5.136 + 0.08 \times 200$$

$$= 4.725 + 16 = 20.725 \text{ MeV}$$

Since there are 10^{20} Cm atoms, hence total energy released in transformation is

$$E_{total} = 20.725 \times 10^{20} \text{ MeV}$$

$$= 20.725 \times 10^{20} \times 1.6 \times 10^{-13} \text{ J}$$

This energy is released in mean life time (= 10^{13} sec) of curium. Hence power generated is

$$P = \frac{E_{total}}{\text{life time}} = \frac{20.725 \times 10^{20} \times 1.6 \times 10^{-13}}{10^{+13}}$$

$$= 3.32 \times 10^{-5} \text{ W.}$$

Example 28.9: In nuclear reactor, ^{235}U undergoes fission liberating 200 MeV of energy. The reactor has a 10% efficiency and produces 1000 MW power. If the reactor is to function for 10 yrs. Find the total mass of the uranium required. [IIT 2001]

Solution: Energy liberated in each fission = 200 MeV.

$$= 200 \times 1.6 \times 10^{-13} \text{ J}$$

$$= 3.2 \times 10^{-11} \text{ J}$$

1000 MW power means = $10^3 \times 10^6 = 10^9$ J/s

To run this reactor for 10 yrs, total energy required = $10^9 \times 3.2 \times 10^7 \times 10$

(1 year = 3.2×10^7 sec and 10 years = $3.2 \times 10^7 \times 10$)

∴ Total energy required = 3.2×10^{17} J

Energy released in one fission = 3.2×10^{-11} J

and efficiency of reactor is 10%, i.e. useful energy in the form of power per fission is

$$= \frac{3.2 \times 10^{-11}}{10} = 3.2 \times 10^{-12} \text{J}$$

Hence required number of uranium atoms for fission is

$$= \frac{3.2 \times 10^{17}}{3.2 \times 10^{-12}} = 10^{29} \text{ atoms}$$

Since 235 gm of uranium–235 contain Avogadro no. (= 6.2 × 10^{23}) of atoms. Hence mass of 10^{29} atoms is

$$m = \frac{235}{6.2 \times 10^{23}} \times 10^{29} = 3.79 \times 10^7 \text{ g} = 3.79 \times 10^4 \text{ kg}.$$

Example 28.10: A nucleus at rest undergoes decay emitting α-particle of de Broglie wavelength λ = 5.76 × 10^{-15} m. If the mass of daughter nucleus is 223.610 amu and that of α-particle is 4.002 amu, determine the total kinetic energy in the final state. Hence obtain the mass of the parent nucleus in amu (1 amu = 931.47 MeV/C). [IIT 2001]

Solution: α-decay of parent nucleus $_ZX^A$ into daughter element Y is represented as

$$^{A}_{Z}X \rightarrow \, ^{A-4}_{Z-2}Y + \, ^{4}_{2}\text{He}$$

de Broglie wavelength of α-particles is λ = 5.76 × 10^{-15} m. Hence its momentum

$$p_\alpha = \frac{h}{\lambda} = \frac{6.6 \times 10^{-34}}{5.76 \times 10^{-15}} = 1.15 \times 10^{-19} \text{ kg m s}^{-1}$$

Since parent nucleus was at rest, hence the magnitude of the moment of daughter nucleus and α-particles is same. Hence total kinetic energy released is

K = Kinetic energy of α-particle + kinetic energy of Y

$$= \frac{p_\alpha^2}{2m_\alpha} + \frac{p_\alpha^2}{2m_Y}$$

$$= \frac{p_\alpha^2}{2}\left[\frac{1}{m_\alpha} + \frac{1}{m_Y}\right]$$

$$\because \quad 1 \text{ amu} = 931.47 \frac{\text{MeV}}{c^2} = 1.66 \times 10^{-27} \text{ kg}$$

$$\therefore \quad m_\alpha = 4.002 \times 1.66 \times 10^{-27} \text{ and } m_Y = 223.61 \times 1.66 \times 10^{-27} \text{ kg}$$

$$\therefore \quad K = \frac{(1.15 \times 10^{-19})^2}{2 \times 1.66 \times 10^{-27}}\left[\frac{1}{4.002} + \frac{1}{223.61}\right]$$

$$= \frac{(1.15)^2 \times 10^{-11}}{2 \times 1.66}[0.25 + 0.0047]$$

$$= \frac{(1.15)^2 \times 10^{-11} \times 0.2547}{2 \times 1.66} = 1.01 \times 10^{-12} \text{J}$$

This kinetic energy is equivalent to mass m equal to K/c^2

$$\therefore \quad m = \frac{1.01 \times 10^{-12}}{(3 \times 10^8)^2} = 1.1 \times 10^{-29} \text{kg}$$

$$= \frac{1.1\times10^{-29}}{1.66\times10^{-27}} = 0.0066\,\text{amu}$$

Hence mass of the parent nucleus *X* is

$$m_X = m_Y + m_\alpha + K$$
$$= 223.60 + 4.002 + 0.0066$$
$$= 227.6086 \text{ amu.}$$

Example 28.11: A radioactive nucleus *X* decays to a nucleus *Y* with a decay constant $\lambda_X = 0.1\ \text{s}^{-1}$. *Y* further decays to a stable nucleus *Z* with a decay constant $\lambda_y = \frac{1}{30}\ \text{s}^{-1}$. Initially there are only *X* nuclei and their number is $N_0 = 10^{20}$. Set up the rate equations for the population of *X*, *Y* and *Z*. The population *Y* nucleus as a function of time is given by $N_Y(t) = \frac{N_0\lambda_X}{\lambda_X - \lambda_Y}\left[e^{-\lambda_Y t} - e^{-\lambda_X t}\right]$. Find the time at which N_Y is maximum and determine the population of *X* and *Z* at that instant.

[IIT 2001]

Solution: If N_X and N_Y are the number of atoms of *X* and *Y* respectively at any time, then

Rate of decay of *X* is $\frac{dN_X}{dt} = -\lambda_X N_X$ ...(*i*)

Rate of formation of *Y* is $\frac{dN_Y}{dt} = -\lambda_Y N_Y + \lambda_X N_X$ ...(*ii*)

Since *Z* is an stable atom hence rate at which *Z* is produced is equal to rate at which *Y* is decaying.

$$\therefore \quad \frac{dN_Z}{dt} = N_Y\lambda_Y \qquad ...(iii)$$

Integrating (*i*) and remembering at $t = 0$ $N_X = N_0 = 10^{20}$, we have

$$\int_{N_0}^{N_X} \frac{dN_X}{N_X} = -\lambda_X dt$$

$$N_X = N_0 e^{-\lambda_X t} \qquad ...(iv)$$

Given $$N_Y = \frac{N_0\lambda_X}{\lambda_X - \lambda_Y}\left[e^{-\lambda_Y t} - e^{-\lambda_X t}\right] \qquad ...(v)$$

N_Y will be maximum at time *t* when $\frac{dN_Y}{dt} = 0$

$$\therefore \quad 0 = \frac{N_0\lambda_X}{N_X - N_Y}\left[-\lambda_Y e^{-\lambda_Y t} + \lambda_X e^{-\lambda_X t}\right]$$

or $$\frac{\lambda_X}{\lambda_Y} = \frac{e^{-\lambda_Y t}}{e^{-\lambda_X t}} = e^{(\lambda_X - \lambda_Y)t}$$

or $$t = \frac{1}{\lambda_X - \lambda_Y} \log \frac{\lambda_X}{\lambda_Y}$$

Substituting $\lambda_X = 0.1\ s^{-1}$ $\lambda_Y = \frac{1}{30} s^{-1}$, we get

$$t = \frac{1}{0.1 - \frac{1}{30}} \log(0.1 \times 30)$$

$$= 15 \log 3 = 16.48 \text{ sec}$$

(*iv*) gives $$N_X = N_0 e^{-\lambda_X t} = 10^{20} e^{-(0.1)(16.48)}$$

$$= 10^{20} e^{-1.648} = 1.92 \times 10^{19}$$

At $$\frac{dN_Y}{dt} = 0$$

$$-\lambda_Y N_Y + \lambda_X N_X = 0$$

or $$N_Y = \frac{\lambda_X N_X}{\lambda_Y} = \frac{0.1}{\frac{1}{30}} \times 1.92 \times 10^{19}$$

$$= 3 \times 1.92 \times 10^{19} = 5.76 \times 10^{19}$$

The population of Z is

$$N_Z = N_0 - N_X - N_Y$$

$$= 10^{20} - 10^{19}(1.92 + 5.76)$$

$$= 2.32 \times 10^{19}.$$

Example 28.12: A laboratory has got 1.49 μg of pure ${}^{13}_{7}N$ which has a half life of 10.0 minutes. (*a*) How many nuclei are present initially? (*b*) What is the activity initially? (*c*) What is the activity after 1.0 hr? (*d*) After approximately how long will activity drop to less than one per second?

Solution:

(*a*) 1.49 μg of ${}_7N^{13}$ contain number of atoms

$$N_0 = \frac{6.02 \times 10^{23}}{13} \times 1.49 \times 10^{-6} = 6.9 \times 10^{16} \text{ nuclei}$$

(*b*) Activity at $t = 0$, i.e. initial activity is

$$R_i = N_0 \lambda = N_0 \frac{(0.693)}{T}$$ where T is half life = 10 min = 600 sec

$$= \frac{6.9 \times 10^{16} \times 0.693}{600} = 8 \times 10^{13} \text{ disintegrations/s}$$

$$= \frac{8 \times 10^{13}}{3.7 \times 10^{10}} = 215.4 \text{ Curie}$$

(*c*) Activity after 1 hour means activity after 6 half lives. In one half life activity becomes half.

Hence activity after 6 half lives becomes $\left(\frac{1}{2}\right)^6$.

$$\therefore \quad R = R_i\left(\frac{1}{2}\right)^6 = \frac{8\times10^{13}}{(2)^6} = 1.25\times10^{12} \text{ disintegrations/s}$$

$$= 33.7 \text{ Curie}$$

(*d*) Activity reduces to 1 per second, i.e. $R = 1$ disint/s $= N\lambda$ and $R_i = N_0\lambda = 8 \times 10^{13}$ disint/s.

$$\therefore \quad \frac{R}{R_i} = \frac{1}{8\times10^{13}} = \frac{N}{N_0} = e^{-\lambda t} = (2)^{-t/T}$$

$$\therefore \quad \frac{1}{8\times10^{13}} = (2)^{-t/T}$$

and taking log, we have

$$-13.91 = -\frac{t}{T}(0.3010)$$

or
$$t = \frac{13.91\times T}{0.3010} \qquad T = 10 \text{ m} = 600 \text{ sec}$$

$$t = \frac{13.91\times 600}{0.3010} = 2.71\times10^4 \text{ sec.}$$

Example 28.13: In the fission of $_{94}Pu^{239}$ by a thermal neutron, two fission fragments of equal masses and sizes are produced and 4 neutrons are emitted. Find the force between the two fission fragments at the momentum they are produced. Given $R_0 = 1.1$ fermi.

Solution: Initial total mass no. = 239 + 1 = 240

In fission, 4 neutrons are ejected and the two products have same mass.

Hence mass of each fission product $(A) = \frac{240-4}{2} = 118$

Atomic number of each product $= \frac{94}{2} = 47$

The charge number of each fission product is 47.

Charge on each nucleus $= 47 \times 1.6 \times 10^{-19}$ C

Radius of each nucleus $= R_0(A)^{1/3} = 1.1 \times 10^{-15}(118)^{1/3}$

The distance between the two fragments $= 2 \times 1.1 \times 10^{-15}\,(118)^{1/3}$

The force between the two fragments then is

$$F = \frac{1}{4\pi\varepsilon_0}\cdot\frac{q_1q_2}{r^2}$$

$$= \frac{9\times10^9\times47\times1.6\times10^{-19}\times47\times1.6\times10^{-19}}{(2\times1.1\times10^{-15})^2\,(118)^{2/3}}$$

$$= \frac{9\times10^9\times(7.52\times10^{-18})^2}{(10.79\times10^{-15})^2} = 4.37\times10^3 \text{ N.}$$

Example 28.14: The deuterium–tritium fusion reaction (call D–T reaction) is most likely to be the basic reaction in a future thermonuclear fusion reaction

$${}^{2}_{1}\text{H} + {}^{3}_{1}\text{H} \rightarrow {}^{4}_{2}\text{He} + {}^{1}_{0}\text{n} + Q$$

(*a*) Calculate the amount of energy released in the reaction, given $m({}_1\text{H}^2) = 2.014102$ amu, $m({}^3_1\text{H}) = 3.016090$ amu, $m_n = 1.008665$ amu, $m({}^4_2\text{He}) = 4.002603$ amu.

(*b*) Find the kinetic energy needed to overcome Coulomb repulsion. Assume the radius of both deuterium and tritium to be approximately 1.5×10^{-15} m.

(*c*) To what temperature must the gases be heated to initiate the fusion reaction? Boltzmann's constant $K = 1.38 \times 10^{-23}$ J K^{-1}.

Solution:

(*a*) Mass defect in the reaction given above is

$$\Delta m = m({}^2_1\text{H}) + m({}^3_1\text{H}) - m({}^4_2\text{He}) - m_n$$

$$= 2.014102 + 3.016090 - 4.002603 - 1.008665$$

$$= 0.018924 \text{ amu}$$

$$1 \text{ amu} = 931.5 \text{ MeV}$$

$$\therefore \quad Q = 0.018924 \times 931.5 = 17.63 \text{ MeV}$$

(*b*) In order to fuse (*D* and *T*) nuclei, they must touch each other, i.e. separation between the two nuclei must be twice the radius of the nucleus, i.e.

$$r = 2 \times \text{radius} = 2 \times 1.5 \times 10^{-15} = 3 \times 10^{-15} \text{ m}$$

The potential energy of nuclei at distance *r* then is

$$U = \frac{K \cdot q_1 q_2}{r} \qquad q_1 = 1e = 1.6 \times 10^{-19} \text{ C} = q_2$$

$$= \frac{9 \times 10^9 \times 1.6 \times 10^{-19} \times 1.6 \times 10^{-19}}{3 \times 10^{-15}} = 7.68 \times 10^{-14} \text{ J}$$

Thus the kinetic energy required to overcome Coulomb repulsion between these nuclei is

$$E_K = U = 7.68 \times 10^{-14} \text{ J}$$

(*c*) The energy required by each nucleus to overcome the Coulombian repulsion then is

$$\frac{1}{2} E_K = \frac{7.68}{2} \times 10^{-14} = 3.84 \times 10^{-14} \text{ J}$$

If *T* is the temperature of gas then energy of each gas molecule is 3/2 *KT*.

$$\therefore \quad \frac{3}{2} KT = 3.84 \times 10^{-14}$$

or

$$T = \frac{2}{3} \times \frac{3.84 \times 10^{-14}}{K} = \frac{2}{3} \times \frac{3.84 \times 10^{-14}}{1.38 \times 10^{-23}}$$

$$= 1.86 \times 10^9 K.$$

Example 28.15: There is a stream of neutrons with kinetic energy of 0.0327 eV. If the half life of neutrons is 700 second what fraction of neutrons will decay before they have travelled a distance of 10 km? Given mass of neutron = 1.676×10^{-27} kg.

Solution: If v is the velocity of neutrons, then

$$\frac{1}{2}mv^2 = E_K = 0.03273 \text{ eV} = 0.0327 \times 1.6 \times 10^{-19} \text{ J}$$

or
$$v^2 = \frac{2 \times 0.327 \times 1.6 \times 10^{-19}}{1.676 \times 10^{-27}} = 0.62 \times 10^8$$

or
$$v = 2.5 \times 10^3 \text{ m/s}$$

Time taken by neutron to travel a distance of 10 km is

$$t = \frac{10 \times 10^3}{2.5 \times 10^3} = 4 \text{ sec}$$

Decay equation is

$$N = N_0 e^{-\lambda t} = N_0 (2)^{-t/T} \qquad T \text{ is half life} = 700 \text{ s.}$$

Hence fraction remained undecayed is

$$\frac{N}{N_0} = (2)^{-4/700} = (0.5)^{4/700} = (0.5)^{1/175}$$

$$\therefore \quad \log\frac{N}{N_0} = \frac{1}{175}\log(0.5) = \frac{1}{175}(-0.3010)$$

$$= -1.72 \times 10^{-3}$$

or
$$\frac{N}{N_0} = \text{anti}\log(-1.72 \times 10^{-3}) = 0.996$$

$\therefore$ Fraction decayed $= 1 - \dfrac{N}{N_0} = 1 - 0.996 = 0.004.$

Example 28.16: Find the energy of the reaction $^{14}N\ (\alpha, p)\ ^{17}O$ if the kinetic energy of the incoming α-particle is $K_\alpha = 4.0$ MeV and the proton outgoing at an angle $\theta = 60°$ to the direction of motion of α-particle has kinetic energy $K_p = 2.09$ MeV.

Solution: The reaction is

$$^{14}_{7}N + {}^{4}_{2}He \rightarrow {}^{17}_{8}O + {}^{1}_{1}H + Q$$

Initial kinetic energy of α-particle $K_\alpha = 4.0$ MeV

The momentum of incident α-particle is

$$\boldsymbol{p}_\alpha = \sqrt{2m_\alpha K_\alpha}\,\hat{i} = \sqrt{2\eta_\alpha m_0 K_\alpha}\,\hat{i}$$

and the momentum of outgoing proton is

$$\boldsymbol{p}_p = \sqrt{2m_p K_p}\,(\cos\theta\hat{i} + \sin\theta\hat{j})$$

$$= \sqrt{2n_p m_0 K_p}\,(\cos\theta\hat{i} + \sin\theta\hat{j})$$

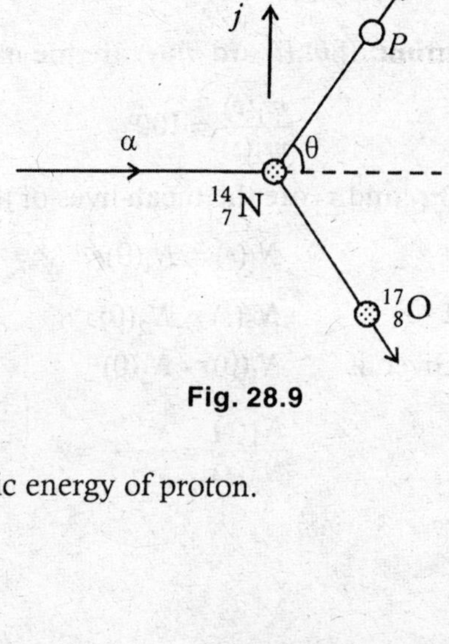

Fig. 28.9

where $\eta_\alpha = \dfrac{m_\alpha}{m_0}$, $\eta_p = \dfrac{m_p}{m_0}$ and m_0 is mass of O^{17}. K_p is kinetic energy of proton.

Conservation of momentum gives, momentum of ^{17}O (p_0) as

$$p_0 = p_\alpha - p_p = \left(\sqrt{2\eta_\alpha m_0 K_\alpha} - \sqrt{2\eta_p m_0 K_p}\cos\theta\right)\hat{i} - \sqrt{2\eta_p m_0 K_p}\sin\theta\hat{j}$$

Nucleus ^{14}N was initially at rest.
Conservation of energy (including rest mass energy) gives

$$M_{14}c^2 + M_\alpha c^2 + K_\alpha = M_p c^2 + K_p + M_0 c^2 + \text{K.E. of } ^{17}O\ (K_O)$$

and $$K_O = \frac{P_O^2}{2m_O} = \frac{1}{2m_0}\left[2\eta_\alpha m_0 K_\alpha + 2\eta_p m_0 K_p \cos^2\theta - 2\sqrt{4\eta_\alpha m_0^2 K_\alpha K_p \eta_p}\cos\theta + 2\eta_p m_0 K_p \sin^2\theta\right]$$

$$= \eta_\alpha K_\alpha + n_p K_p - 2\sqrt{\eta_\alpha \eta_p K_p K_\alpha}\cos\theta$$

Hence by definition of the energy of reaction (Q), we have

$$Q = M_{14}c^2 + M_\alpha c^2 - M_0 c^2 - M_p c^2$$

But conservation of energy gives

$$M_{14}c^2 + M_\alpha c^2 - M_0 c^2 - M_p c^2 = K_p + K_0 - K_\alpha$$

$$\therefore \quad Q = K_p + K_0 - K_\alpha = K_p + n_\alpha K_\alpha + n_p K_p - 2\sqrt{\eta_p K_p K_\alpha \eta_\alpha}\cos\theta - K_\alpha$$

$$= K_p(1+n_p) - K_\alpha(1-\eta_\alpha) - 2\sqrt{\eta_p K_p K_\alpha \eta_\alpha}\cos\theta$$

Given is $K_p = 2.09$ MeV $= K_\alpha = 4.0$ MeV, $\theta = 60°$ $\eta_\alpha = \frac{4}{17} = 0.24$ and $\eta_p = \frac{1}{17} = 0.059$

Substituting the values, we get

$$Q = 2.09(1+0.059) - 4(1-0.24) - 2\sqrt{(0.059)(2.09)(4)(0.24)} \times \frac{1}{2}$$

$$= 2.21 - 3.04 - 0.344 = -1.17 \text{ MeV}.$$

Example 28.17: In the chemical analysis of a rock, the mass ratio of two radioactive isotopes is found to be 100 : 1. The mean lives of the two isotopes are 4×10^9 yrs and 2×10^9 yrs respectively. If it is assumed that at the time of formation of rock, the atoms of the two isotopes were equal in proportion, estimate the age of the rock. The ratio of the atomic weights of the two isotopes is 102 : 1.

Solution: If $m_1(t)$ and $m_2(t)$ are the masses of the two radioactive isotopes at any time t, then

$$\frac{m_1(t)}{m_2(t)} = 100$$

If τ_1 and τ_2 are the mean lives of the two isotopes, then number of active atoms at time t is

$$N_1(t) = N_1(0)e^{-t/\tau_1} \qquad \therefore \lambda = 1/\tau$$

and $$N_2(t) = N_2(0)e^{-t/\tau_2}$$

Given is $$N_1(0) = N_2(0)$$

$$\therefore \quad \frac{N_1(t)}{N_2(t)} = \frac{e^{-t/\tau_1}}{e^{-t/\tau_2}} = e^{-t\left(\frac{1}{\tau_1} - \frac{1}{\tau_2}\right)}$$

For an element of atomic weight A, the number of atoms in mass m is

$$N = \frac{n_0\, m}{A} \quad \text{where } n_0 \text{ is Avogadro number}$$

$$\therefore \quad \frac{N_1(t)}{N_2(t)} = \frac{m_1}{A_1} \cdot \frac{A_2}{m_2} \qquad \text{But } \frac{m_1}{m_2} = 100 \text{ and } \frac{A_1}{A_2} = \frac{1.02}{1}$$

$$= 100 \times \frac{1}{1.02}$$

$$\therefore \quad \frac{100}{1.02} = e^{-t\left(\frac{1}{\tau_1} - \frac{1}{\tau_2}\right)}$$

Taking logs, we get

$$\log_e\left(\frac{100}{1.02}\right) = -t\left(\frac{1}{\tau_1} - \frac{1}{\tau_2}\right) \qquad \tau_1 = 4 \times 10^9 \text{ yrs}\ ;\ \tau_2 = 2 \times 10^9 \text{ yrs}$$

or
$$4.58 = -t\left[\frac{-1}{4 \times 10^9}\right]$$

or
$$t = 4.58 \times 4 \times 10^9 \text{ yrs} = 1.832 \times 10^{10} \text{ yrs.}$$

Example 28.18: A small quantity of solution containing ^{24}Na radionuclide (half life $T = 15$ hours) of activity 1.2 micro Curie is injected into the blood of a person. A sample of the blood of volume 1 cc taken after 5 hours shows an activity of 296 disintegration per minute. Determine the total volume of blood in the body of the person. Assume that the radioactive solution mixes uniformly in the blood of the person (1 Curie = 3.7×10^{10} disintegrations/s).

Solution: Initial activity is $R_0 = 1$ m Curie $= 1 \times 3.7 \times 10^{10} \times 10^{-6} = 3.7 \times 10^4$ disint./s.

$R_0 = N_0\lambda$, where N_0 is the initial number of active nucleides.

$$\therefore \quad N_0 = \frac{R_0}{\lambda} = \frac{3.7 \times 10^4 \times 15 \times 60 \times 60}{0.693} \qquad \lambda = \frac{0.693}{T},\ T = 15 \text{ hours}$$

$$= 2.88 \times 10^9$$

Hence number of active nuclides (N) after 5 hours are

$$N = N_0^{-\lambda t} = N_0(2)^{-t/T}$$

$$= 2.88 \times 10^9 (2)^{-5/15} = \frac{2.88 \times 10^9}{8} = 3.6 \times 10^8$$

Activity of 1 cc of blood is found to be 296 disint./min $= \dfrac{296}{60}$

Hence number of active atoms in 1 cc are

$$N' = \frac{R}{\lambda} = \frac{296 \times 15 \times 60 \times 60}{60 \times 0.693} = 6.41 \times 10^4$$

Total number of active atoms are 3.6×10^8

1 cc contain 6.4×10^4 active atoms

$$\text{Hence volume of blood is } = \frac{3.6 \times 10^8}{6.4 \times 10^4} = 0.5625 \times 10^4 \text{ cc}$$

$$= 5.625 \text{ litres.}$$

Example 28.19: A polonium nucleus $^{210}_{84}Po$ emits an α-particle ^4_2He and is converted into lead nucleus ($^{206}_{82}Pb$). This reaction is used to produce electric power in space mission. Po–210 has a half life of 138.6 days. Assuming an efficiency of 10%, how much Po–210 is required to produce 1.2×10^7 J of electric energy per day at the end of 693 days? Also find the initial activity of the material. Given mass of nuclei $^{210}_{84}Po = 209.98264$ amu, $^{206}_{82}Pb = 205.97440$ amu and $^4_2He = 4.00260$ amu. 1 amu = 931 MeV and Avogadro number = 6×10^{23} mol^{-1}.

Solution: The reaction is

$$^{210}_{84}Po \rightarrow {}^{206}_{82}Pb + {}^4_2He + Q$$

Q the energy of the reaction is equal to mass defect.

$$\Delta m = 209.98264 - 205.97440 - 4.00260$$

$$= 0.00564 \text{ amu}$$

$$= 0.00564 \times 931 = 5.25 \text{ MeV}$$

$$= 5.25 \times 1.6 \times 10^{-13} = 8.4 \times 10^{-13} \text{ J}$$

Since efficiency is 10% hence useful available energy per nucleon is

$$E = 8.4 \times 10^{-14} \text{ J}$$

Required energy per day is 1.2×10^7 J

Hence required nucleon reactions, i.e. disintegration are

$$n = \frac{1.2 \times 10^7}{8.4 \times 10^{-14}} = \left(\frac{1}{7}\right) \times 10^{21} \text{ per day}$$

$$\text{Decay constant is } \lambda = \frac{0.693}{T} \qquad T \text{ is half life}$$

$$= \frac{0.693}{138.6} = 5 \times 10^{-3} \text{ per day}$$

For an activity of n per day, if N is the number of active atoms required, then

$$N = \frac{\text{activity}}{\lambda} = \frac{(1/7 \times 10^{21})}{5 \times 10^{-3}} = \frac{1}{35} \times 10^{24}$$

If N_0 is the number of atoms initial, i.e. 693 days earlier, then

$$N = N_0 e^{-\lambda t} \qquad \text{where } t = 693 \text{ days}$$

$$\therefore \qquad N_0 = N e^{\lambda t} = \frac{10^{24}}{35} e^{5 \times 10^{-3} \times 693}$$

$$= \frac{10^{24}}{35} (e^{.693})^5 = \frac{10^{24}}{35} (2)^5 = \frac{32}{35} \times 10^{24}$$

Thus mass of polonium required is

$$m = \frac{A}{n_0} \times N_0 \qquad \text{where } n_0 \text{ is Avogadro number}$$

and A is mass number

$$\therefore \qquad m = \frac{210}{6 \times 10^{23}} \times \frac{32}{35} \times 10^{24} = 320 \text{ g}$$

Initial activity (R_i) was

$$R_i = N_0 \lambda = \frac{32}{35} \times 10^{24} \times 5 \times 10^{-3}$$

$$= 4.57 \times 10^{21} \text{ disint./day.}$$

Example 28.20: In nature, a decay chain series starts with $^{232}_{90}Th$ and finally terminates at $^{208}_{82}Pb$. A thorium ore sample was found to contain 8×10^{-5} mL of helium at NTP and 5×10^{-7}g of ^{232}Th. Find the age of the ore sample assuming the source of helium to be only due to decay of ^{232}Th. Also assume complete retention of helium within the ore. Half life of $^{232}Th = 1.39 \times 10^{10}$ yr.

Solution: Number of helium atoms in 8×10^{-5} mL is

$$N = \frac{V \times n_0}{22400} \qquad n_0 \text{ is Avogadro number}$$

$$= \frac{8 \times 10^{-5} \times 6.02 \times 10^{23}}{22400} = 2.15 \times 10^{15} \text{ atoms}$$

Thorium–232 decay to Pb–208 and in the process gives 6 atoms of helium. Hence number of thorium atoms used in the process are (N_{th})

$$N_{th} = \frac{N}{6} = \frac{2.15 \times 10^{15}}{6} = 3.585 \times 10^{14} \text{ atoms}$$

and the number of thorium atoms undecayed, i.e. active atoms is

$$N'_{th} = \frac{5 \times 10^{-7} \times 6.02 \times 10^{23}}{232} = 1.29 \times 10^{15} \text{ atoms}$$

The number of thorium atoms at the start of decay was

$$N_0 = N'_{th} + N_{th} = 1.29 \times 10^{15} + 3.585 \times 10^{14}$$

$$= 1.6485 \times 10^{15} \text{ atoms}$$

Using decay formula

$$N'_{th} = N_0 e^{-\lambda t} = N_0 (2)^{-t/T} \qquad T \text{ is half life time}$$

$$\frac{1.29 \times 10^{15}}{1.6485 \times 10^{15}} = (2)^{-t/T}$$

$$-0.1065 = -\frac{t}{T}(0.3010)$$

$$\therefore \qquad t = \frac{0.1065}{0.3010} T = \frac{0.1065}{0.3010} \times 1.39 \times 10^{10} = 4.9 \times 10^{9} \text{ yrs}$$

Example 28.21: Show that ^{55}Fe may capture electron but cannot decay β^+. Given $M(^{55}_{26}Fe)$ = 54.938298 u, $M(^{55}_{25}Mn)$ = 54.938050 u and m_e = 0.0005494 u.

Solution: The possible reactions are

Electron capture $^{55}_{26}Fe + e \rightarrow {}^{55}_{25}Mn + \nu$ ...(*i*)

β^+ decay $^{55}_{26}Fe \rightarrow {}^{55}_{25}Mn + \beta^+ + \nu$ ...(*ii*)

Q value for reaction (*i*)

$$\Delta m = M(^{55}_{26}Fe) + m_e - M(^{55}_{25}Mn)$$
$$= 54.938298 + 0.000549 - 54.938050$$
$$= 7.97 \times 10^{-4} \text{ u}$$

and $$Q = \Delta m \times 931 = 7.97 \times 10^{-4} \times 931 = 0.742 \text{ MeV}$$

Positive value Q implies that electron capture is possible.

Q value for reaction (*ii*)

$$\Delta m = M(^{55}_{26}Fe) - M(^{55}_{25}Mn) - m_e$$
$$= 54.938298 - 54.9380504 - 0.000549$$
$$= -2.924 \times 10^{-4} \text{ u}$$

and $$Q = \Delta m \times 931 = -2.92 \times 10^{-4} \times 931 = -0.272 \text{ MeV}$$

–ve Q value of reaction implies that positron decay is not possible.

Example 28.22: Show that the nucleide ^8Be has a positive binding energy but is unstable with respect to decay into two α-particles. Given is m_p = 1.007825 u, m_n = 1.008665 u, and $M(^8Be)$ = 8.005305 u.

Solution: Mass defect when 4 protons an 4 neutrons combine to form ^8Be is

$$\Delta m = 4(m_p) + 4(m_n) - M(^8Be)$$
$$= 4(1.007825) + 4(1.008665) - 8.005305$$
$$= 0.060655 \text{ u.}$$

and binding energy $E = \Delta m \times 931 = 56.5$ MeV

Thus the nucleus ^8Be has positive binding energy.

Now let us calculate the binding energy of the decay of Be^8 into two α-particles.

$$^8_4Be \rightarrow 2\,{}_2He^4$$

Mass defect is

$$\Delta m = 2M(^4_2He) - M(^8Be)$$
$$= 2 \times 4.002603 - 8.005305$$
$$= -9.9 \times 10^{-5} \text{ u}$$

Hence binding energy for the reaction is

$$E = -9.9 \times 10^{-5} \times 931 = -0.092 \text{ MeV}$$

Since binding energy is negative hence ^8_4Be is unstable against decay to two α-particles. This means that ^8_4Be does spontaneously decay into two α-particles. Because of the instability of ^8_4Be, it is difficult for nuclei to join together to make heavier nuclei. This is the reason why most stars consist of mostly hydrogen and helium.

Example 28.23

(*a*) Find the energy needed to remove a neutron from the nucleus of the calcium isotope $^{42}_{20}$Ca.
(*b*) Find the energy needed to remove a proton from this nucleus.
(*c*) Why are these energies different?

Given [$^{41}_{20}$Ca] = 40.962278 u, m_n = 1.008665 u, m_p = 1.007825 u,

$M(^{42}_{20}\text{Ca})$ = 41.958622 and $M(^{41}_{19}\text{K})$ = 40.961828.

Solution:

(*a*) On removing a neutron form $^{42}_{20}$Ca, we get isotope of calcium $^{41}_{20}$Ca

i.e. $^{42}_{20}\text{Ca} \rightarrow\, ^{41}_{20}\text{Ca} + \,^{1}_{0}\text{n}$

Mass defect in the reaction is

$$\Delta m = M(^{42}_{20}\text{Ca}) - M(^{41}_{20}\text{Ca}) - m_n$$
$$= 41.958622 - 40.962278 - 1.008665$$
$$= -0.012321 \text{ u}$$

and energy = 0.012321 × 931 = 11.47 MeV

–ve sign means when neutron was in $^{42}_{20}$Ca its BE was 11.47 MeV, i.e. this energy is required to remove neutron from $^{42}_{20}$Ca.

(*b*) When proton is removed atomic number decreases by 1 and mass number also decreases by 1.

$^{42}_{20}\text{Ca} \rightarrow\, ^{41}_{19}\text{K} + \,^{1}_{1}\text{H}$

Mass defect is

$$\Delta m = M[^{42}_{20}\text{Ca}] - M[^{41}_{19}\text{K}] - m_p$$
$$= 41.958622 - 40.961828 - 1.007825$$
$$= -0.011031 \text{ u}$$

and energy = 0.011031 × 931 = 10.27 MeV

–ve sign here means energy is required to remove proton from $^{42}_{20}$Ca and it must be the binding energy of proton in $^{42}_{20}$Ca

(*c*) Neutron and proton have different energies because only attractive nuclear forces act on the neutron, whereas the proton is also acted upon by repulsive electric forces that decrease its binding energy.

Example 28.24: Calculate the excitation energy of the compound nuclei produced when U^{235} and U^{238} absorb thermal neutrons.

Solution: The two reactions are

$$^{235}_{92}\text{U} + \,_0\text{n}^1 \rightarrow\, ^{236}_{92}\text{U}^* \quad \ldots(i)$$
$$^{238}_{92}\text{U} + \,_0\text{n}^1 \rightarrow\, ^{293}_{92}\text{U}^* \quad \ldots(ii)$$

We can find excitation energy using mass defect method. Here we will neglect thermal energy of neutrons (= 0.04 eV) in comparison to excitation energy (~ MeV)

In reaction (*i*) mass defect is

$$\Delta m = M(^{235}_{92}\text{U}) + m_n - M(^{236}_{92}\text{U})$$

$$= 235.0439 + 1.0087 - 236.0456$$
$$= 7 \times 10^{-3}\text{ u} = 6.5\text{ MeV}$$

This energy excites the nucleus. Thus excitation energy is 6.5 MeV. In reaction (*ii*) mass defect is

$$\Delta m = M(^{238}_{92}\text{U}) + m_n - M(^{239}_{92}\text{U})$$
$$= 238.0508 + 1.0087 - 239.0543$$
$$= 5.2 \times 10^{-3}\text{ u} = 4.84\text{ MeV}$$

Thus excitation energy of ^{238}U is 4.84 MeV

Thus ^{236}U has more excitation energy than ^{239}U. When both are produced by thermal neutron absorption. This is why ^{235}U easily undergoes thermal neutron fission.

Example 28.25: The nuclear reaction $^9_4\text{Be} + ^4_2\text{He} \rightarrow ^{12}_6\text{C} + ^1_0\text{n}$ has Q value equal to 5.7 MeV. If this reaction is initiated by an α-particle with kinetic energy 5.3 MeV. What would be the kinetic energy of the neutron if it moves at right angles to the initial direction of α-particles? Take m_α = 4.031882 u, m_n = 1.008665 u and m_C = 12.095646 u.

Solution: Let K_α, K_n and K_C be the kinetic energies of α-particle neutron and carbon nucleus.

Then referring to Fig. 28.10, conservation of momentum gives

$$p_\alpha = p_C \cos\theta \text{ or } \sqrt{2m_\alpha K_\alpha} = \sqrt{2K_C m_C}\cos\theta \quad ...(i)$$

and

$$p_n = p_C \sin\theta \text{ or } \sqrt{2m_n K_n} = \sqrt{2K_C m_C}\sin\theta \quad ...(ii)$$

Conservation of energy gives

$$K_\alpha + Q = K_C + K_n \quad ...(iii)$$

Eliminating θ from (*i*) and (*ii*), we get

$$m_\alpha K_\alpha + m_n K_n = K_C m_C \quad ...(iv)$$

Substituting the value of K_C from (*iii*), we get

$$m_\alpha K_\alpha + m_n K_n = m_C(K_\alpha + Q - K_n)$$

or

$$K_n(m_n + m_C) = m_C(K_\alpha + Q) - m_\alpha K_\alpha$$

or

$$K_n \equiv \frac{m_C(K_a + Q) - m_\alpha K_\alpha}{m_n + m_C}$$

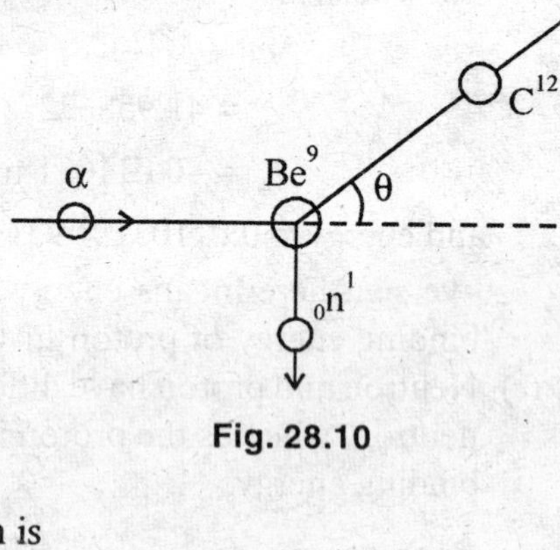

Fig. 28.10

Substituting the given values; the kinetic energy of neutron is

$$K_n = \frac{12.095646(5.3 + 5.7) - 4.031882 \times 5.3}{1.008665 + 12.095646}$$
$$= 8.5\text{ MeV}$$

Example 28.26: In the nuclear reaction

$$^1_1\text{H} + ^{15}_7\text{N} \rightarrow ^A_Z X + ^1_0\text{n}$$

(*a*) Find A, Z and identify the nucleus X.

(*b*) Find the Q value of the reaction.

(*c*) If proton were to collide with $^{15}_7$N at rest, find the minimum kinetic energy needed by the proton to initiate the above reaction.

(*d*) If proton has twice the energy in (*c*) and the outgoing neutron emerges at an angle of 90° with the direction of the incident proton, find the momentum of the nucleus *X*.

$m_p = 1.007825$ u , $m(^{15}N) = 15.000$ u, $m_n = 1.008665$ u, $m(^{15}O) = 15.0031$ u.

Solution:

(*a*) Conservation of mass number gives

$$1 + 15 = A + 1 \text{ or } A = 15$$

Z increases by 1 hence $Z = 8$. Hence ${}^{A}_{Z}X = {}^{15}_{8}O$

(*b*) Mass defect of the reaction is

$$\Delta m = m_p + m(^{15}N) - m(^{15}O) - m_n$$
$$= 1.007825 + 15.000 - 15.0031 - 1.008665$$
$$= -0.00394 \text{ u}$$

and Q value is $= -0.00394 \times 931 = -3.67$ MeV

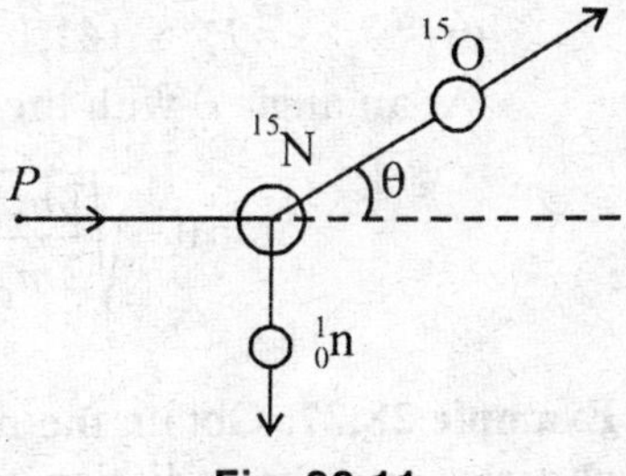

Fig. 28.11

(*c*) Threshold energy is given by

$$K_{th} = \frac{m+M}{M}|Q|$$

where m is mass of projectile particle and M is mass of target particle.

$$K_{th} = \frac{m_p + m(^{15}N)}{m(^{15}N)}|Q|$$

$$= \frac{1.007825 + 15.000}{15.000} \times 3.67 = 3.92 \text{ MeV}$$

(*d*) If proton has energy $K_p = 2K_{th} = 2 \times 3.92 = 7.84$ MeV then let K_O be the kinetic energy of oxygen atom and K_n be the kinetic energy of neutron. Then from Fig. 28.11, conservation of momentum gives

$$p_O \cos\theta = p_p \text{ i.e. } \sqrt{2m_O K_O}\cos\theta = \sqrt{2m_p K_p} \qquad \ldots(i)$$

and $$p_O \sin\theta = p_n \text{ i.e. } \sqrt{2m_O K_O}\sin\theta = \sqrt{2m_n K_n} \qquad \ldots(ii)$$

and conservation of energy gives

$$K_p + Q = K_O + K_n$$

or $$K_O + K_n = 7.84 - 3.67 = 4.17 \text{ MeV} \qquad \ldots(iii)$$

From (*i*) and (*ii*), we have

$$2m_O K_O = 2m_p K_p + 2m_n K_n$$

or $$m_O K_O = m_p K_p + m_n K_n = m_p K_p + m_n (K - K_O)$$

where $K = 4.17$ MeV (From (*iii*))

or $$K_O (m_O + m_n) = m_p K_p + m_n K$$

or $$K_O = \frac{m_p K_p + m_n K}{m_O + m_n} = \frac{(1.007825)(7.84) + (1.008665)(4.17)}{15.000 + 1.008665}$$

$$= 1.25 \text{ MeV}$$

$\therefore \quad K_n = 4.17 - K_O = 4.17 - 1.25 = 2.92$ MeV

(*i*) and (*ii*) give

$$P_0^2 = 2m_pK_p + 2m_nK_n$$

$$= (2 \times 1.007285 \times 7.84 + 2 \times 1.008665 \times 2.92)\, 931$$

$\because$ mass is substituted in MeV

or $\quad P_O = 141.1$ MeV/c

At an angle θ with the incident direction of the proton such that

$$\tan\theta = \sqrt{\frac{2m_nK_n}{2m_pK_p}} = 0.61 \text{ or } \theta = \tan^{-1}(0.61) = 31.4°.$$

Example 28.27: Obtain the maximum kinetic energy of β-particles and the radiation frequencies of γ-decays in the following decay scheme. You are given $M[^{198}\text{Au}]$ = 197.968233 u, $M(^{198}\text{Hg})$ = 197.966760 u.

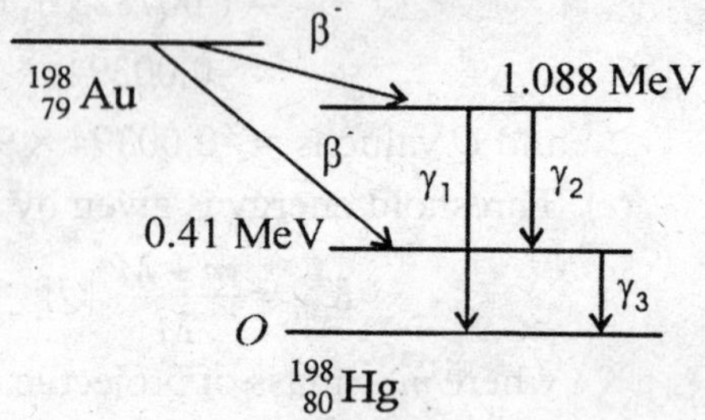

Fig. 28.12

Solution: Radiation frequency is

$$\nu = \frac{\text{Energy}}{h}$$

$$\therefore \quad \nu_1 = \frac{(1.088 - 0)\text{ MeV}}{h} = \frac{1.088 \times 1.6 \times 10^{-13}}{6.6 \times 10^{-34}} = 2.6 \times 10^{20} \text{ Hz}$$

$$\nu_2 = \frac{(1.088 - 0.41)\text{ MeV}}{h} = \frac{0.678 \times 1.6 \times 10^{-13}}{6.6 \times 10^{-34}} = 1.64 \times 10^{20} \text{ Hz}$$

and

$$\nu_3 = \frac{(0.41 - 0)\text{ MeV}}{h} = \frac{0.41 \times 1.6 \times 10^{-13}}{6.6 \times 10^{-34}} = 9.9 \times 10^{19} \text{ Hz}$$

Maximum kinetic energy of β-particles is

$$K_{max} = \left[M\left(^{198}_{79}\text{Au}\right) - M(^{198}_{80}\text{Hg})\right]c^2 \text{ — Excitation energy}$$

$$\therefore \quad K_1 = [197.968233 - 197.966760]\, c^2 - (1.088) \text{ MeV}$$

$$= (1.473 \times 10^{-3} \text{ amu})\, c^2 - (1.088 \text{ MeV})$$

$$= 1.37 \text{ MeV} - 1.088 \text{ MeV} = 0.283 \text{ MeV}$$

Similarly

$$K_2 = (1.473 \times 10^{-3} \text{ amu})\, c^2 - (0.41 \text{ MeV})$$

$$= 1.37 - 0.41 = 0.96 \text{ MeV}.$$

Example 28.28: Protons striking a stationary lithium target, activate a reaction $^7_3$Li (p, n)$^7_4$Be. At what value of the protons kinetic energy can the resulting neutron be stationary? $M(^7\text{Li})$ = 7.01601 u $M[^7\text{Be}]$ = 7.01693 u, m_n = 1.00867 u and m_p = 1.00783 u.

Solution: The reaction is

$$^7_3\text{Li} + \text{p} \rightarrow {}^7_4\text{Be} + {}^1_0\text{n}$$

The mass defect is

$$\Delta m = M(^7\text{Li}) + m_p - M(^7\text{Be}) - m_n$$
$$= 7.01601 + 1.00783 - 7.01693 - 1.00867$$
$$= -1.76 \times 10^{-3}\,\text{u} = -1.64\ \text{MeV}$$

Thus Q value of the reaction is = –1.64 MeV
From conservation of momentum gives

$$p_p = p_{Be}$$

Since initially Li^7 and finally neutron are at rest.
Conservation of energy gives

$$\frac{p_p^2}{2m_p} = \frac{p_{Be}^2}{2m_{Be}} + 1.64$$

or $$\frac{p_p^2}{2m_p}\left(1 - \frac{m_p}{m_{Be}}\right) = 1.64$$

or $$\frac{p_p^2}{2m_p} \times \frac{6}{7} = 1.64$$

∴ Kinetic energy of proton is

$$\frac{p_p^2}{2m_p} = \frac{7}{6} \times 1.67 = 1.91\ \text{MeV}$$

Example 28.29: A particle of mass m strikes a stationary nucleus of mass M and activates an endoenergetic reaction. Demonstrate that the threshold (minimal) kinetic energy required to initiate this reaction is

$$K_{th} = \frac{m+M}{M}|Q|, \quad \text{where } Q \text{ is energy of reaction.}$$

Solution: Energy required is minimum when the reaction products all move in the direction of the incident particle with the same velocity, so that the combination is at rest in the centre of mass frame. Then conservation of momentum gives

$$\sqrt{2mK_{th}} = (m+M)v$$

Conservation of energy gives

$$K_{th} - |Q| = \frac{1}{2}(m+M)v^2 = \frac{mK_{th}}{m+M}$$

or $$K_{th}\frac{M}{m+M} = |Q| \text{ or } K_{th} = \frac{m+M}{M}|Q|$$

Example 28.30: A proton with kinetic energy K_p = 1.5 MeV is captured by a deutron (^2_1H). Find the excitation energy of the formed nucleus.

Solution: The reaction is $^1_1H + ^2_1H \rightarrow ^3_2He$

The excitation energy of $_2He^3$ is just the energy available in centre of mass. Using conservation of momentum, we have

$$p_p = p_{He}$$

or $$\sqrt{2m_pK_p} = m_{He}v = 3m_pv \qquad m_{He} = m_p + m_d \quad m_d = 2m_p$$

and v is the velocity of He^3 atom

$$\therefore \quad v = \frac{1}{3}\sqrt{\frac{2K_p}{m_p}} \qquad \ldots(i)$$

Hence kinetic energy of ^3He nucleus is

$$K_{He} = \frac{1}{2}m_{He}v^2 = \frac{1}{2}3m_p\frac{1}{9}\frac{2K_p}{m_p}$$

$$= \frac{1}{3}K_p = 0.5 \text{ MeV} \quad \text{Since } K_p = 1.5 \text{ MeV}$$

Q value of the reaction is obtained from mass defect

$$\Delta m = m_p + m_d - m_{He}$$
$$= 1.00783 + 2.01410 - 3.01603$$
$$= 5.9 \times 10^{-3} \text{ amu}$$

and energy $$Q = \Delta m \times 931 = 5.9 \times 10^{-3} \times 931 = 5.5 \text{ MeV}$$

Conservation of energy gives

$$K_p + Q = K_{He} + \text{excitation energy}$$

$\therefore$ excitation energy = 1.5 + 5.5 – 0.5 = 6.5 MeV.

PROBLEMS

1. A ^{210}Bi radioactive nucleus decays according to the following chain $^{210}Bi \xrightarrow{\lambda_a} {}^{210}Po \xrightarrow{\lambda_b} {}^{206}Pb$(stable). Where decay constants are λ_a and λ_b. Calculate the α and β activities of m_0 gm of Bi^{210} after time t of its preparation.
[**Hint:** In Art. 28.8 equation (28.31) gives the number of active atoms at any time t. $N_b = \frac{\lambda_a N_0}{\lambda_b - \lambda_a}\left(e^{-\lambda_a t} - e^{-\lambda_b t}\right)$. Here $N_0 = \frac{m_0}{210}n_0$ where n_0 is Avogadro number. Hence $R_b = N_b\lambda_b = \frac{m_0 n_0}{210}\frac{\lambda_a\lambda_b}{(\lambda_b - \lambda_a)}\left(e^{-\lambda_a t} - e^{-\lambda_a t}\right)$. Active atoms of A at any time t are $N_a = N_0 e^{-\lambda_a t}$.
Its activity is $R_a = N_a\lambda_a = \frac{\lambda_a m_0 n_0}{210}e^{-\lambda_a t}$]

2. Find the binding energy and the binding energy per nucleon of the nucleus $^{16}_{8}O$. Given atomic mass of $^{16}_{8}O$ = 15.994915 amu and mass of proton = 1.007825 amu; mass of neutron = 1.008665 and 1 amu = 931.5 MeV.

$$\left[\textbf{Ans. } \text{B.E.} = 127.62 \text{ MeV}, \ \frac{\text{B.E.}}{A} = 7.976 \text{ MeV}\right]$$

3. How much energy is required to remove a neutron from the nucleus of ${}^{41}_{20}Ca$? Given m_n = 1.008665 amu, atomic mass of ${}^{41}_{20}Ca$ = 40.962278 amu and atomic mass of ${}^{40}_{20}Ca$ = 39.962591 amu. **[Ans.** E = 8.362 MeV]
4. The ${}^{230}_{90}Th$ nucleus is unstable against α-decay with half life of 7.6 × 10^3 yrs. Write down the equation of the decay and estimate the kinetic energy of the emitted α-particles from the following data: ${}^{230}_{90}Th$ (m) = 230.0381 amu, ${}^{226}_{88}Ra$ (m) = 226.0254 amu and ${}^{4}_{2}He$ (m) = 4.0026 amu.
[Ans. ${}^{230}_{90}Th \to {}^{226}_{88}Ra + {}^{4}_{2}He + Q$, K_α = 9.25 MeV]
5. ${}^{239}_{94}Pu$ is undergoing α-decay according to the equation ${}^{239}_{94}Pu \to {}^{235}_{92}Pu + {}^{4}_{2}He$. The process is accompanied by an emission of γ-rays of wavelength 0.125 Å. Find the speed of the emitted α-particles. Given, mass of ${}^{239}_{94}Pu$ = 239.05122 amu, mass of ${}^{235}_{92}Pu$ = 235.04209 amu, mass of ${}^{4}_{2}He$ = 4.00260 amu and 1 amu = 931 MeV. **[Ans.** $v = 1.57 \times 10^7$ ms^{-1}]
6. A nuclear reactor using ^{235}U generates 250 MW of electrical power. The efficiency of reactor is 25%. What is the amount of ^{235}U used in the reactor per year? Thermal energy released per fission of ^{235}U is 200 MeV. Avogadro number is 6.023 × 10^{23} mol^{-1}.
[Ans. Mass = 384.4 kg]
7. What is the power output of a ${}^{235}_{92}U$ reactor, if it takes 30 days to use up 2 kg of fuel and if each fission gives 185 MeV of useable energy? **[Ans.** P = 58.6 MW]
8. It is proposed to use the nuclear fusion reaction ${}^{2}_{1}H + {}^{2}_{1}H \to {}^{4}_{2}He$ in a nuclear reactor of 200 MW rating. If the energy from the above reaction is used with a 25% efficiency in the reactor, how many grams of deuterium fuel will be needed per day? (The masses of ${}^{2}_{1}H$ and ${}^{4}_{2}He$ are 2.0141 and 4.0026 amu respectively). **[Ans.** mass = 121.3 g]
9. A nuclear reactor generates power at 50% efficiency of by fission of ${}^{235}_{92}U$ into two equal fragments ${}^{116}_{46}Pd$ with the emission of two γ-ray photons of energy 5.2 MeV each and three neutrons. The average binding energy per nucleon of U–235 and Pd–116 is 7.2 MeV and 8.2 MeV respectively. Calculate the energy released in one fission event. Also estimate the amount of U–235 consumed per hour to produce 1600 megha watt power.
[Ans. Energy per fission = 200 MeV, Mass = 140.5 g]
10. A count rate meter is used to measure the activity of a given sample. At one instant the meter shows 4750 counts per minute. Five minutes later it showed 2700 counts per minute. Find the half life of the sample. Also find the decay constant. **[Ans.** T = 6.1 min, λ = 0.113 min^{-1}]
11. 1 g of Cesium–137 (${}^{137}_{55}Cs$) decays by β-emission with half life of 30 yrs. If the initial activity of caesium is 1.0 mci and atomic mass of caesium is 137 kg/kmol. What is (*a*) the resulting isotope? (*b*) the number of atoms left after 5 yrs? (*c*) the activity of caesium after 5 yrs?
[Ans. (*a*) ${}^{137}_{55}Cs \to {}^{137}_{56}Ba + \beta^-$ (*b*) $N = 3.907 \times 10^{21}$ (*c*) R = 0.89 mci]
12. The normal activity of a living matter containing carbon is found to be 15 decays per minute per gram of carbon. An archaeological specimen gives 6 decays per minute per gram of carbon. If the half life of carbon is 5730 years, estimate age of the specimen.
[Ans. Age = 7576 yrs]
13. In an ore containing uranium, the ratio of U–238 to Pb–206 is 3. Calculate the age of the ore, assuming that all the lead present in the ore is the final stable product of U–238. Take the half life of U–238 to be 4.5 × 10^9 yrs.
[Ans. Age = 1.867 × 10^9 yrs]

14. A bone suspected to have originated during the period of Ashoka the great, was found in Bihar. Accelerator techniques gave its $^{14}C : ^{12}C$ ratio as 1.1×10^{-12}. Is the bone old enough to have belonged to that period? Take initial ratio of ^{14}C to $^{12}C = 1.2 \times 10^{-12}$.

[**Ans.** It is 7194 old. The bone does not belong to Ashoka times]

15. A radioactive isotope X has a half life of 3 second. Initially a given sample of this isotope contains 8000 atoms. Calculate (*a*) its decay constant (*b*) the time t, when 1000 atoms of the isotope X remain in the sample, and (*c*) the number of decay per second in the sample at $t = t_1$.

[**Ans.** (*a*) $\lambda = 0.231\ s^{-1}$ (*b*) $t_1 = 9$ s (*c*) 231 s^{-1}]

16. A piece of burnt wood of mass 20 g is found to have a ^{14}C activity of 4 Bq. If half life of ^{14}C is 5730 yrs, find its age. [**Ans.** Age = 1842 yrs]

17. 20 mg of the non-radioactive isotope $^{40}_{20}Ca$ are mixed with 5 mg of the radio isotope $^{45}_{20}Ca$ How much will the specific activity of preparation decrease? Half life of $_{20}Ca^{45}$ = 164 days and $n_0 = 6.02 \times 10^{23}$ per mole. [**Ans.** 5.234×10^{14} disint./s]

18. Find whether α-decay or any of the β-decay are allowed for $^{226}_{89}Ac$.

[**Ans.** (*a*) α-decay is allowed $Q = 5.50$ MeV > 0. (*b*) β^- decay is allowed $Q = 1.12$ MeV > 0 (*c*) β^+ decay is not allowed $Q = -0.38$ MeV < 0 (*d*) electron capture is allowed Q = 0.64 MeV > 0]

19. Calculate the minimum kinetic energy of protons incident on $^{13}_{6}C$ nuclei at rest in the laboratory frame that will produce the endothermic reaction $^{13}_{6}C$ (p, n) $^{13}_{7}N$. Given $M(C^{13}) = 13.003355$ u, $M(^{13}N) = 13.005738\ u$. [**Ans.** $E_{min} = 3.23$ MeV]

20. The nuclear reaction $^{10}_{5}B + ^{1}_{0}n \rightarrow ^{7}_{3}Li + ^{4}_{4}He$ is observed to occur when slow neutrons strike a boron atom at rest. For a particular reaction in which $K_n = 0$ (kinetic energy of neutron) the helium is observed to have a speed of 9.30×10^6 m/s. Determine (*a*) the kinetic energy of lithium and (*b*) the Q value of the reaction. Given $M(^{7}_{3}Li) = 7.0160$ u and $M(^{4}_{2}He) = 4.0026$ u.

[**Ans.** (*a*) $K_{Li} = 1.02$ MeV (*b*) $Q = 2.82$ MeV]

21. Electrons emitted by some radioactive substance A and α-particles emitted by another radioactive substance B in a particular radioactive sample are absorbed and the kinetic energy is converted to electricity. The efficiency of conversion for the β-decay and α-decay are 10% and 40% respectively. The initial power generated is 20 MW. Half life period of A and B are 2 hours and 4 hours respectively. If average K.E./mole for each of α and β particles emitted in the process are 1 J and 2 J respectively, find the total number of moles of radioactive species initially. [**Ans.** $M = 1.039 \times 10^3$ moles]

22. Natural uranium is a mixture of three isotopes $^{234}_{92}U$, $^{235}_{92}U$, $^{238}_{92}U$ with the share of 0.0006%, 0.71% and 99.2804% respectively. The half life of three isotopes are 2.5×10^5 yrs, 7.1×10^8 yrs and 4.5×10^9 yrs respectively. Determine the share of radioactivity of each isotope into the total activity of the natural uranium.

[**Ans.** $^{234}_{92}U \rightarrow 9.9\%$, $^{235}_{92}U \rightarrow 3.9\%$, $^{238}_{92}U \rightarrow 86.4\%$]

23. A sample contains two radioactive nuclei X and Y with half life 2 hr and 1 hr respectively. The nucleus X decays into the nucleus Y and Y decays into an stable nucleus Z. At time $t = 0$ the activities of the components in the sample were equal. Find the ratio of the number of active nuclei of Y at $t = 4$ hr to the number at $t = 0$. [**Ans.** 716]

24. An insulated solid radioactive material decays by means of α-emission with a decay constant λ. If the molecular weight of the material is M and the specific heat of the parent material is s (the molar specific heat of the material does not change due to decay), find the variation of temperature of the material as a function of time. Assume that the energy released during the decay of each nucleus is Q and the entire energy is converted into heat and the initial temperature of the material is T_0.

$$\left[\textbf{Ans. } T = T_0 + \frac{\lambda Q n_0}{Ms}\left(1 - e^{-\lambda t}\right),\ n_0 \text{ is Avogadro number}\right]$$

25. Suppose a nucleus initially at rest undergoes α-decay according to equation

$${}^{225}_{92}X \rightarrow {}^{221}_{90}Y + {}^{4}_{2}He$$

At $t = 0$, the emitted α-particle enters in a region of space where a uniform magnetic field $\boldsymbol{B} = B_0\hat{i}$ and electric field $\boldsymbol{E} = E_0\hat{i}$ exists. The α-particle enters in the region with velocity $\boldsymbol{v} = v_0\hat{j}$ from $x = 0$. At time $t = \sqrt{3} \times 10^6 \dfrac{M_\alpha}{q_\alpha E_\alpha}$, the particle was observed to have speed twice the initial v_0, then find

(a) the initial velocity of α-particles at time t.

(b) the initial velocity v_0 of the α-particle.

(c) the binding energy per nucleon of α-particle.

Given $m_n = 1.009$ u, $m_p = 1.008$ u and $m_\alpha = 4.003$.

$$\left[\textbf{Ans. } (a)\ v = \sqrt{v_0^2 + \left(\frac{E_0 q}{m_\alpha}t\right)^2}\quad (b)\ v_0 = 10^6 \text{ m/s (c) B.E.} = 7.21 \text{ per nucleon}\right]$$

26. A hypothetical radionuclide ${}^{225}_{92}X$ with disintegration constant $\lambda = 1\ hr^{-1}$ is produced in a reactor at constant rate $K = 1 \times 10^{17}$ nuclei per second. Its primary decay modes are spontaneous fission and α-decay. The former with probability of $P_1 = 10\%$ and latter with probability $P_2 = 90\%$. Each fission releases $E = 200$ MeV energy. The masses involved in α-decay are as follows ${}^{225}_{92}X = 225.052$ u, ${}^{221}_{90}Y = 221.03$ u and ${}^{4}_{2}He = 4.002$ u. 20% of the energy released is utilized in increasing the temperature of water. Find the increase in temperature of 1000 kg of water in 1 hr. Specific heat of water is 4.2×10^3 J/kg°C. Assume there is no loss of energy through water surface. **[Ans.** $\Delta T = 37.0°$ C]

27. Use the following given data $M({}^{238}_{92}U) = 288.05079$ u, $M({}^{4}_{2}He) = 4.00260$ u, $M({}^{234}_{90}Th) = 234.04363$ u, $M({}^{1}_{1}H) = m_p = 1.00783$ u, $M({}^{237}_{91}Pa) = 237.0512$ u.

(*a*) Calculate the energy released during the α-decay of ${}^{238}_{92}U$

(*b*) Show that ${}^{238}_{92}U$ cannot spontaneously emit a proton.

[Ans. (*a*) $Q = 4.25$ MeV (*b*) $Q = -7.68$ MeV and cannot emit ${}^{1}_{1}H$]

28. Find the disintegration energy Q for the fission event shown below.

$${}^{235}_{92}U + n \rightarrow {}^{236}_{92}U \rightarrow {}^{140}_{54}Xe + {}^{94}_{38}Sr + 2\,{}^{1}_{0}n$$

where ${}^{140}_{54}Xe \rightarrow {}^{140}_{58}Ce$ and ${}^{94}_{38}Sr \rightarrow {}^{94}_{40}Zr$

Take into account the decay of fission fragments as shown. Given is $M({}^{235}_{92}U) = 235.0439$ u, $m_n = 1.00867$ u, $M({}^{140}_{58}Ce) = 139.9054$ u, $M({}^{94}_{40}Zr) = 93.9063$ u.

[Ans. $Q = 208$ MeV]

29. Consider the D–T (deuterium–tritium) reaction given below

$${}^{2}_{1}H + {}^{3}_{1}H \rightarrow {}^{4}_{2}He + {}^{1}_{0}n + 17.59 \text{ MeV}$$

(*a*) Calculate the energy released in MeV in this reaction from the following data

$M(_1^2\text{H}) = 2.014102$ u, $M(_1^3\text{H}) = 3.016049$ u

$M(_2^4\text{He}) = 4.002603$ u, $m_n = 1.00867$ u.

(*b*) Consider the radius of both deuterium to be approximately 1.5 fm. What is the kinetic energy needed to overcome the Coulomb repulsion? To what temperature must the gases be heated to initiate?

[**Ans.** (*a*) $Q = 17.59$ MeV (*b*) $K = 0.48$ MeV, $T = 3.7 \times 10^9$ K]

30. Protons with kinetic energy $K = 1.0$ MeV striking a lithium target induce a nuclear reaction $^7\text{Li} + \text{p} \rightarrow {}_2^4\text{He}$. Find the kinetic energy of each α-particle and the angle of their divergence provided their motion directions are symmetrical with respect to that of incoming protons.

[**Ans.** $K = 9.18$ MeV, $\theta = 170.5°$]

31. A neutron with kinetic energy $K = 10$ MeV activates a nuclear reaction ^{12}C (n, α) ^9Be whose threshold is $K_{th} = 6.17$ MeV. Find the kinetic energy of the α-particles outgoing at right angles to the incoming neutron's direction.

[**Ans.** $K_\alpha = 2.21$ MeV]

SHORT ANSWER TYPE QUESTIONS

1. The activity of a radionuclide is found to decrease by 4% every hour. Find the decay constant and the mean life of nuclide.

[**Ans.** Activity $R = N\lambda$ ∴ $\frac{R}{N} \times 100 = 4$, activity decreases by 4%. ∴ $\frac{R}{N} = \lambda = 0.04 \text{ hours}^{-1}$.

Mean life $= \frac{1}{\lambda} = \frac{1}{0.04} = 25 \text{ hours}$]

2. Neon–23 decays in the following way

$$^{23}_{10}\text{Ne} \rightarrow {}^{23}_{11}\text{Na} + \beta^- + \overline{\nu}$$

Find the minimum and maximum kinetic energies that the β-particle can have. Atomic masses of ^{23}Ne and ^{23}Na are 22.9945 u and 22.9898 u respectively.

[**Ans.** Mass defect is $= (m)_{\text{Ne}} - (m)_{\text{Na}} = 22.9945 - 22.9898$, $\Delta m = 0.0047$ u. But 1 amu = 931 MeV. Hence $\Delta m = 0.0047 \times 931 = 4.4$ MeV. This energy is shared between β-particle and neutrino. Hence maximum energy of β-particle is 4.4 MeV and minimum energy of β-particle is 0]

3. A nucleus of ^{60}Ni is in excited state and decays to its ground state by emitting a photon of energy 1.33 MeV. What is the recoil energy and recoil speed of the ^{60}Ni nucleus?

[**Ans.** From conservation of momentum, we have the momentum of photon $\boldsymbol{p}_1$ and momentum of nucleus $\boldsymbol{p}_2$ are equal in magnitude, i.e. $\boldsymbol{p}_1 + \boldsymbol{p}_2 = 0$ $|\boldsymbol{p}_1| = |\boldsymbol{p}_2| = p = h/\lambda =$ (momentum of photon). Hence recoil energy of nucleus is $E = \frac{p^2}{2m} = \frac{h^2}{2m\lambda^2}$. $E = \left(\frac{hc}{\lambda}\right)^2 \cdot \frac{1}{2mc^2}$. Now $\frac{hc}{\lambda} = 1.33$ MeV and $mc^2 = 60 \times 931$ MeV. Since $m = 60$u ∴ $E = \frac{(1.33)^2}{2 \times 60 \times 931} = 1.58 \times 10^{-5}$ MeV $= 15.8$ eV. $\frac{1}{2}mv^2 = 15.8 \text{ eV} = 15.8 \times 1.6 \times 10^{-19}$ J ∴ $v = \sqrt{\frac{2 \times 15.8 \times 1.6 \times 10^{-19}}{60 \times 1.67 \times 10^{-27}}} = 7.1 \times 10^3$ m/s]

4. A radioactive substance emits n beta particles in the first 2 sec and $0.5n$ beta particles in next 2 sec. What is the mean life of the sample?

[**Ans.** Using radioactive decay formula we have $n = N_0(1 - e^{-2\lambda})$ and $1.5n = N_0(1 - e^{-4\lambda})$

$$\therefore 1.5 = \frac{(1-e^{-4\lambda})}{1-e^{-2\lambda}} = \frac{(1-e^{-2\lambda})(1+e^{-2\lambda})}{1-e^{-2\lambda}} = 1+e^{-2\lambda} \text{ or } e^{-2\lambda} = 1.5 - 1 = 0.5 = \frac{1}{2}, \text{ or } -2\lambda = -\log 2$$

$$\therefore \text{Mean life } \tau = \frac{1}{\lambda} = \frac{2}{\log 2} = 2.88 \text{ sec}$$]

5. The binding energies per nucleon for deuterium ($^2_1$H) and helium ($^4_2$He) are 1.1 MeV and 7.0 MeV respectively. Calculate the energy released when two denterium nuclei fuse to form helium nucleus.

[**Ans.** $^2_1\text{H} + ^2_1\text{H} \rightarrow ^4_2\text{He} + Q$. Conservation of energy gives $\{2(m_p) + 2(m_n)\}c^2 - 2\text{ B.E.}(^2_1\text{H}) = \{2(m_p) + 2(m_n)\}c^2 - \text{B.E.}(_2\text{He}^4) + Q$, where m_p and m_n are the masses of proton and neutron respectively $\therefore Q = \text{B.E.}(^4_2\text{He}) - 2\text{B.E.}(^2_1\text{H}) = 4 \times 7 - 2 \times (2 \times 1.1) = 28 - 4.4 = 23.6$ MeV]

6. How many α- and β-particles are emitted when uranium ($_{92}$U^{238}) decays to lead ($_{82}$Pb^{206})?

[**Ans.** When α-particle is emitted mass no. changes by 4. If n is no. of α-particles emitted then $238 - 4n = 206$ or $n = 8$. When β-particle is emitted charge no. increases by 1. If m is no. of β-particles emitted then charge no. increase by m. When α-particle is emitted charge no. decreases by 2. Due to emission of 8 α-particles, total charge no. decreases by $2 \times 8 = 16$ $\therefore 82 = 92 - 16 + m$ or $m = 6$]

7. The binding energy of $^{35}_{17}$Cl nucleus is 298 MeV. Find its atomic mass. Given mass of proton $(m_p) = 1.007825$ amu and mass of neutron $(m_n) = 1.008665$ amu and 1 amu = 931.5 MeV.

[**Ans.** Mass defect in nucleus $\Delta m = \dfrac{\text{B.E.(in MeV)}}{\text{(1 amu)(in MeV)}} = \dfrac{298}{931.5} = 0.32$ amu. The nucleus has 17 protons and 18 neutrons. Hence total mass is $m = 17(m_p) + 18(m_n) = 17(1.007825) + 18(1.008665) = 35.289$ amu. Atomic mass of $_{17}$Cl^{35} = 35.289 – 0.32 = 34.969 amu]

8. The mean life times of a radioactive substance are 1620 yrs and 405 yrs for α-emission and β-emission respectively. Find out the time during which three-forths of a sample will decay if it is decaying by both α-emission and β-emission simultaneously.

[**Ans.** Decay constant $\lambda = \lambda_a + \lambda_b = \dfrac{1}{\tau_a} + \dfrac{1}{\tau_b} = \dfrac{1}{1620} + \dfrac{1}{405} = \dfrac{1}{324}$ yer^{-1}. If t is the time in which 3/4th of the sample decays means active nuclei left are 1/4 $\therefore \dfrac{1}{4} = e^{-\lambda t} = e^{-t/324}$ here time is in years. Taking logs we have $-2\log_e 2 = -\dfrac{t}{324}$ or $\dfrac{t}{324} = 2 \times 0.693\ (\because \log_e 2 = 0.693)$

or $t \times 2 \times .693 \times 324 = 449.064$ yrs]

9. The half life of radon is 3.8 days. After how many days will only one twentieth of radon sample be left over?

[**Ans.** Decay equation is $N = N_0 e^{-\lambda t} = N_0(2)^{-t/T}$. T is half life. $\dfrac{N}{N_0} = \dfrac{1}{20}$ $\therefore \dfrac{1}{20} = (2)^{-t/T}$. Taking logs we have $-1.3010 = -\dfrac{t}{T}(0.3010)$ $\therefore t = \dfrac{1.3010}{0.3010}T = \dfrac{1.3010 \times 3.8}{0.3010} = 16.4$ days]

10. Write missing symbols denoted by x in the following nuclear reactions (*a*) ^{10}B (x, α) ^{8}Be (*b*) ^{17}O (d, n)x (*c*) ^{23}Na (p, x) ^{20}Ne (*d*) x (p, n) ^{37}Ar.

[**Ans.** (*a*) Here x must carry two nucleons and a unit of positive charge. The reaction is

$$^{10}_{5}B + ^{2}_{1}H \rightarrow ^{8}_{4}Be + ^{4}_{2}He \text{ (d, } \alpha) \text{ reaction}$$

(*b*) It is ($_1H^2$, $_0n^1$) reaction hence x must contain a proton in addition to the constituents of O^{17}. Thus reaction is $^{17}_{8}O + ^{2}_{1}H \rightarrow ^{18}_{9}F + ^{1}_{0}n$

(*c*) In this reaction x must have 4 nucleons, i.e. 2 protons and 2 neutrons. Thus reaction is $^{23}_{11}Na + ^{1}_{1}H \rightarrow ^{20}_{10}Ne + ^{4}_{2}He$ (p, α) reaction

(*d*) The particle x must have mass number 37 and must have one unit less of positive charge. The reaction then is $^{37}_{17}Cl + ^{1}_{1}H \rightarrow ^{37}_{18}Ar + ^{1}_{0}n$.]

11. Find the binding energy of a nucleus consisting of equal numbers of protons and neutrons and having the radius one and a half times smaller than that of ^{27}Al nucleus.

[**Ans.** Mass number and radius of nucleus are related as $r = R_0(A)^{1/3}$. Hence mass number of nucleus is $= 27/(3/2)^3 = 8$. Thus the nucleus is $^{8}_{4}Be$. Hence binding energy is $E_b = 4(m_n) + 4(m_p) - 8 = 4(1.00867) + 4(1.00783) - 8 = 0.06069$ amu = 56.5 MeV]

12. What is the α-activity of a 10 kg sample of ^{235}U that is used in a nuclear reactor? ^{235}U has a half life for emitting α-particles of $T = 7.04 \times 10^8$ yrs.

[**Ans.** Number of atoms in 10 kg of sample are $N = \frac{Mn_0}{A}$ where M is mass of sample, n_0 is Avogadro number and A atomic weight. ∴ $N = \frac{10 \times 10^3 \times 6.02 \times 10^{23}}{235} = 2.56 \times 10^{25}$

$$\therefore \text{ Activity } R = N\lambda = 2.56 \times 10^{25} \frac{\log 2}{T} \quad T = \text{half life} = 7.04 \times 10^8 \text{ yrs}$$

$$= \frac{2.56 \times 10^{25} \times 0.693}{7.04 \times 10^8 \text{(yrs)}} \quad 1 \text{ yr} = 3.2 \times 10^7 \text{ s}$$

$$= \frac{2.56 \times 10^{25} \times 0.693}{7.04 \times 10^8 \times 3.2 \times 10^7} = 7.9 \times 10^8 \text{ disint./s}$$

or

$$R = \frac{7.9 \times 10^8}{3.7 \times 10^{10}} = 2.14 \times 10^{-2} \text{ curie} = 21.4 \text{ mCi}.$$

13. A sample of ^{210}Po is an α-emitter with half life T = 138 days. It is observed to have 2000 disint./s (2000 Bq). (*a*) What is the activity in μCi of this source? (*b*) What is the mass of the ^{210}Po sample?

[**Ans.** (*a*) Activity $R = 2000 \text{ Bq} = \frac{2000}{3.7 \times 10^{10}} \text{ Ci} = 0.54 \times 10^{-7} \text{ Ci} = 0.054 \times 10^{-6} \text{ Ci} = 0.054 \, \mu\text{Ci}$

(*b*) No. of Po nuclei in sample $N = \frac{R}{\lambda} = \frac{2000}{\log 2 / T} = \frac{2000}{\log 2} \times 138 \times 24 \times 3600$. Mass $= \frac{A}{n_0} \times N$. where A is Atomic weight and n_0 is Avogadro number.

$$m = \frac{210}{6.02 \times 10^{23}} \times \frac{2000}{\log_e 2} \times 138 \times 24 \times 3600 = 1.1 \times 10^{-11} \text{ g}$$]

14. Calculate the binding energy per nucleon for $^{20}_{10}Ne$. Given is $m_p = 1.007825$ amu, $m_n = 1.008665$ amu and $M_{20} = 19.992440$ amu.

[**Ans.** Nucleus $^{20}_{10}N$ contains 10 protons and 10 neutrons. Thus mass defect is $\Delta m = 10(m_p) + 10(m_n) - M_{20}$ or $\Delta m = 10 \times 1.007825 + 10 \times 1.008665 - 19.992440 = 0.17246$ amu $= 0.17246 \times 931 = 160.56$ MeV. Hence binding energy per nucleon $= \dfrac{160.56}{20} = 8.028$ MeV]

15. Calculate the ground state Q value of the induced fission reaction

$$^{235}_{92}U + {}^1_0n \rightarrow {}^{236}_{92}U \rightarrow {}^{99}_{40}Zr + {}^{134}_{52}Te + 3\,{}^1_0n$$

The reacting neutron is thermal neutron. Given $M(^{235}U) = 235.0439$ u, $M(^{99}Zr) = 98.9165$ u, $M(^{134}Te) = 133.9115$ u.

[**Ans.** Mass defect in the reaction is $\Delta m = M(^{235}U) + m_n - M(^{99}Zr) - M(^{134}Te) - 3m_n = 235.0439 - 98.9165 - 133.915 - 2 \times 1.0087 = 0.1985$ u. Energy $Q = \Delta m \times 931 = 184.8$ MeV.

Note: Here we have neglected the initial K.E. of thermal neutron. Thermal neutron has K.E. $E = \dfrac{3}{2}kT = .04\,\text{eV}$. This is negligible in comparison to Q]

16. Show that $^{230}_{92}U$ does not decay by emitting a neutron or a proton.

[**Ans.** Decay equations are (*i*) $^{230}_{92}U \rightarrow {}^{229}_{92}U + {}^1_0n$ (*ii*) $^{230}_{92}U \rightarrow {}^{229}_{91}Pa + p({}^1_1H)$. Using the standard value of masses, equation (*i*) gives mass defect $\Delta m = M(^{230}_{92}U) - M(^{229}_{92}U) - m_n = 230.033927 - 229.033496 - 1.008665 = -8.234 \times 10^{-3}\,u = -7.7$ MeV]. Since $Q < 0$ (negative) hence neutron decay is not possible. (*ii*) gives $\Delta m = M({}_{92}U^{230}) - M({}_{91}Pa^{229}) - m_p = 230.033927 - 229.032089 - 1.0078251 = -5.9871 \times 10^{-3}\,u = -5.57$ MeV. This energy is also –ve as such proton decay is not possible]

17. Find the Q value of the reaction $^7_3Li + {}^1_1H \rightarrow 2\,{}^4_2He$. The atomic masses of $^1_1H = 1.007825$ u, ${}_2He^4 = 4.002603$ u and ${}_3Li^7 = 7.016004$ u.

[**Ans.** Mass defect of the reaction is $\Delta m = M(^7_3Li) + m_p = 2M({}_2He^4) = 7.016004 + 1.007825 - 2(4.002603) = 0.018623$ u and Q value of the reaction is $Q = \Delta m \times 931 = 17.35$ MeV. Since energy is released in the reaction, hence reaction is exoenergetic]

18. What is the activity of a container holding 125 cm³ of tritium (^3_1H) half life $T = 12.3$ yrs at a pressure of 5.0×10^5 Pa at 300 K? Express activity in Curies.

[**Ans.** No. of active atoms in container N are $N = \dfrac{PV}{k_BT} = \dfrac{5 \times 10^5 \times 125 \times 10^{-6}}{1.38 \times 10^{-23} \times 300} = 1.51 \times 10^{22}$ atoms. Activity $R = N\lambda = N\dfrac{.693}{T}$ $(T = 12.3 \text{ yrs}) = \dfrac{1.51 \times 10^{22} \times .693}{12.3 \times 3.2 \times 10^7}$ (1 yr $= 3.2 \times 10^7$ sec)

$= 2.66 \times 10^{13}$ disint./s (Bq) $= \dfrac{2.66 \times 10^{13}}{3.7 \times 10^{10}} = .72 \times 10^3$ Curie $= 720$ Curie]

19. Plutonium isotope $^{239}_{94}Pu$ is α-radioactive. Its decay process is as follows.

$$^{239}_{94}Pu \rightarrow {}^{235}_{92}U + {}^4_2He$$

The decay results in the energy released that consists mostly of the kinetic energy of α-particle. A part of energy is however released by Uranium by emitting γ-rays. Determine the velocity of α-particles emitted in the above decay if γ-rays have energy 0.09 MeV. Given is $M[^{239}_{94}Pu] = 239.05122$ u, $M(^{235}_{92}U) = 235.04299$ u and $m_\alpha = 4.00260$ u.

[**Ans.** Mass defect $\Delta m = M(\text{Pu}) - M(\text{U}) - m_\alpha = 239.05122 - 235.04299 - 4.00260$. $= 5.63 \times 10^{-3}$ amu = 5.24 MeV. This energy is shared by α-particle (K.E.) and U–235 (excitation energy). U–235 on de-excitation releases energy = 0.09 MeV. Hence K.E. of a-particle = 5.24 – 0.09 = 5.15 MeV $\therefore 1/2 m_\alpha v_\alpha^2 = 5.15$ MeV $= 5.15 \times 1.6 \times 10^{-13}$ J or $v_\alpha^2 = \dfrac{2 \times 5.15 \times 1.6 \times 10^{-13}}{4.00260 \times 1.67 \times 10^{-27}}$

$$= 2.5 \times 10^{14} \text{ or } v_\alpha = \sqrt{2.5 \times 10^{14}} = 1.57 \times 10^7 \text{ m/s}\,]$$

20. In reaction ${}^{11}_{5}\text{B} + {}^{4}_{2}\text{He} \rightarrow {}^{14}_{7}\text{N} + {}^{1}_{0}\text{n}$ the masses of boron, nitrogen atoms and α-particles are 11.0128 amu, 14.00752 amu and 4.0037 amu respectively. If the incident α-particle had 5.25 MeV of energy and the resultant neutron and nitrogen atoms had energies 3.26 MeV and 2.139 MeV respectively. What is the mass of neutron?

[**Ans.** If m_n is the mass defect, then $\Delta m = M(^{11}\text{B}) + M(^{4}\text{He}) - M(^{14}\text{N}) - m_n$ ...(*i*)
Energy released in the reaction = 3.26 + 2.136 – 5.25 = 0.146 MeV and energy in terms of mass Δm = 0.000157 amu. Substituting in (*i*), we get 0.000157 = 11.0128 + 4.0037 – 14.00752 – m_n. m_n = 1.00898 – 0.000157 = 1.008823 amu]

21. The Fig. 28.13 shows the potential well diagram for a proton in nucleus. Find the kinetic and potential energies of the proton in level *A*.

[**Ans.** The kinetic energy is measured from the bottom of the well. Thus K.E. = 9 MeV and P.E. = –8 MeV.

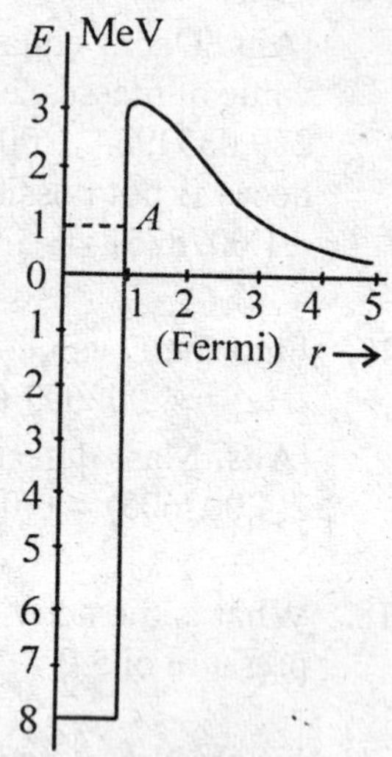

Fig. 28.13

22. The binding energy per nucleon for deuterium (${}^{2}_{1}\text{H}$) and helium (${}^{4}_{2}\text{He}$) are 1.1 MeV and 7.0 MeV respectively. Find the energy released when two deuterium nuclei fuse to form a helium nucleus.

[**Ans.** B.E. of deuterium = 1.1 × 2 = 2.2 MeV. B.E. of helium = 4 × 7 = 28.0 MeV. The reaction is ${}^{2}_{1}\text{H} + {}^{2}_{1}\text{H} = {}^{4}_{2}\text{He} + Q$, Q is energy released. Conservation of energy gives $(2m_p + 2m_n)c^2 - 2\text{B.E.}({}^{2}_{1}\text{H}) = (2m_p + 2m_n)c^2 - \text{B.E.}({}^{4}_{2}\text{He}) + Q \therefore Q = \text{B.E.}({}^{4}_{2}\text{He}) - 2\text{B.E.}({}^{2}_{1}\text{H}) = 28 - 4.4 = 23.6$ MeV]

23. Elements *A*, *B*, and *C* show a cyclic nuclear reaction. Find the possible particle and their number added in the process *C* to *A*.

[**Ans.** If nucleus *C* is ${}^{A}_{Z}X$ then *C* is formed when *B* emits 4β particles, i.e. *B* must have atomic number less by 4 and same mass number. Thus for *B* (${}^{A}_{z-4}\text{B}$) B is formed due to emission of 2 α-particles. Hence *A* must have atomic number greater by 4 and mass number by 8. Hence *A* is (${}^{A+8}_{Z}A$). Hence *A* is formed by adding 8 neutrons to *C*]

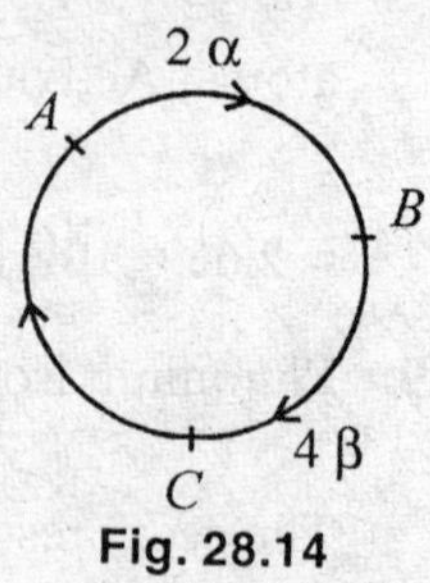

Fig. 28.14

24. A radioactive sample contains 2.2 mg of pure ${}^{11}_{6}\text{C}$ which has half life period of 1224 seconds. Calculate

(*i*) the number of atoms present initially.

(*ii*) the activity when 5 μg of the sample well be left.

[**Ans.** (*i*) 11 gm of ${}^{11}_{6}\text{C}$ contain Avogadro number (n_0) of atoms. Hence no. of atoms in 2.2 mg are

$$= \frac{n_0 \cdot m}{A} \text{ or } N = \frac{6.02 \times 10^{23} \times 2.2 \times 10^{-3}}{11} 1.204 \times 10^{20} \text{ Atoms}$$

(*ii*) 5 μg contain N' no. of atoms, i.e. $N' = \frac{6.02 \times 10^{23} \times 5 \times 10^{-6}}{11} = 2.74 \times 10^{17}$ Atoms

$$\text{Activity} = N'\lambda = N'\frac{\log 2}{T} = \frac{2.74 \times 10^{17} \times 0.693}{1224} = 1.55 \times 10^{14} \text{ disint./sec}$$

$\therefore$ Activity = 1.55×10^{14} Bq = 4.189×10^{3} Curie]

25. When N_0 molecules of a radioactive nuclide are taken at $t = 0$ the activity reduces to A in time t_0. When $3N_0$ molecules of the same nuclide are taken, find the time when activity becomes $3Ae$.

[**Ans.** After time t_0 activity $A = \lambda N_0 e^{-\lambda t_0}$ ...(*i*)

When $3N_0$ atoms are taken, number of atoms after time t are $N = 3N_0 e^{-\lambda t}$ and Activity becomes $3Ae = N\lambda = 3\lambda N_0 e^{-\lambda t}$ ...(*ii*)

From (*i*) and (*ii*), we get $\frac{A}{3Ae} = \frac{\lambda N_0 e^{-\lambda t_0}}{3\lambda N_0 e^{-\lambda t}}$ or $e^{-1} = \frac{e^{-\lambda t_0}}{e^{-\lambda t}} = e^{-\lambda(t_0 - t)}$ or $\lambda(t_0 - t) = 1$ or $t_0 - t = \frac{1}{\lambda}$ and $t = t_0 - 1/\lambda$]

26. Suppose the daughter nucleus in a nuclear decay is itself radioactive. Let λ_p and λ_d be decay constants of parent and the daughter nuclei. Also let N_p and N_d be the number of parent and daughter nuclei at time t. Find the condition for which the number of daughter nuclei becomes constant.

[**Ans.** The rate of decay of parent nuclei is $R_p = \lambda_p N_p$. This is also the rate at which daughter nuclei are produced. If N_d is number of daughter nuclei, then its rate of decay is $R_d = \lambda_d N_d$. For daughter nuclei to be constant, its decay rate and production rate should be same. Thus the conditions is $N_d \lambda_d = N_p \lambda_p$]

27. A radioactive substance is being produced at constant rate of 200 nuclei/s. The decay constant of the substance is 1 s^{-1}. After what time the number of radioactive nuclei will become 100. Initially there are no nuclei.

[**Ans.** Let N be the number of nuclei at any time t. They are produced at constant rate $A = 200$ nuclei/s and their decay rate is $N\lambda$. Hence the net rate at which they are produced is $\frac{dN}{dt} = A - N\lambda$ or $\frac{dN}{A - N\lambda} = dt$. Integration gives $-\frac{1}{\lambda}\log(A - N\lambda) = t + C$. C is constant of integration. Initially at $t = 0$, $N = 0$ gives $C = -\frac{1}{\lambda}\log(A)$. This gives $\log\frac{(A - N\lambda)}{A} = -\lambda t$ or $N = \frac{A}{\lambda}(1 - e^{-\lambda t})$. Substituting the given values $N = \frac{200}{1}(1 - e^{-t})$ where $N = 100$ $\therefore \frac{1}{2} = 1 - e^{-t}$ or $e^{-t} = \frac{1}{2}$ $\therefore t = \log 2$ sec]

28. A radioactive isotope is being produced at constant rate X. Half life of radioactive substance is Y. After sometime the number of radioactive nuclei becomes constant. Find the value of this constant.

[**Ans.** Number of nuclei becomes constant when its rate of production (X) becomes equal to its decay rate ($N\lambda$) $\therefore X = N\lambda$ or $N = \frac{X}{\lambda}$, $\lambda = \frac{\log 2}{T} = \frac{\log 2}{Y}$ $\therefore N = \frac{XY}{\log 2}$]

29. There are two radioactive substances A and B. Decay constant of B is twice that of A. Initially both have equal number of nuclei. After n half lives of A, rate of disintegration of both are equal. What is the value of n?

[**Ans.** Let N_0 be the initial number of active nuclei of each. If $\lambda_a = \lambda$ then $\lambda_b = 2\lambda$. After n half lives of A number of atoms of A are $N_a = \frac{N_0}{(2)^n}$. Since $\lambda_b = 2\lambda$ means half life of B is half the life of A. Therefore after $2n$ half lives of B, the number of active atoms of B are $N_b = \frac{N_0}{(2)^{2n}}$

Activities of the two at this time are equal. Hence $N_b\lambda_b = N_a\lambda_a$. $\frac{N_0}{(2)^{2n}} 2\lambda = \frac{N_0}{(2)^n}\lambda$ or $2^{n+1} = 1$

or $n = 1$]

30. A radioactive element X converts into another stable element Y. Half life of X is 2 hours. Initially only X is present. After time t, the ratio of atoms of X and Y is found to be 1 : 4. Find the value of t.

[**Ans.** Let at $t = 0$ there are N_0 atoms of X. After 4 hours 2 half lives $N_x = \frac{N_0}{4}$ and $N_y = \frac{3N_0}{4}$

$\therefore \frac{N_x}{N_y} = \frac{1}{3}$. After 6 hrs, i.e. 3 half lives $N_x = \frac{N_0}{8}$ and $N_y = \frac{7N_0}{8}$ $\therefore \frac{N_x}{N_y} = \frac{1}{7}$. The ratio 1/4 lies between 1/3 and 1/7. Hence t lies between 4 and 6 hrs.

Aliter: Let N_0 be the no. of atoms of X at $t = 0$. After time t, $N_x = N_0 e^{-\lambda t}$ and $N_y = N_0(1 - e^{-\lambda t})$

and $\frac{N_x}{N_y} = \frac{e^{-\lambda t}}{1 - e^{-\lambda t}} = \frac{1}{4}$ or $4e^{-\lambda t} = 1 - e^{-\lambda t}$ or $e^{-\lambda t} = \frac{1}{5}$ or $(2)^{-t/T} = \frac{1}{5}$ $\therefore$ $-\frac{t}{T}\log 2$

$= \log\left(\frac{1}{5}\right) = -\log 5$ $\therefore$ $t = \frac{\log 5}{\log 2} \times T = 4.64$ Hrs]

31. $\frac{15}{16}$ of a radioactive substance decays in 80 days. A given nucleus, by chance survives and remains undecayed. (*i*) What is the probability of this nucleus to survive 20 more days. (*ii*) If this nucleus survives these 20 more days, what is its probability to survive the next 20 days?

[**Ans.** $\frac{15}{16}$ of a radioactive substance decays in 80 days. Hence active substance after 20 days is $\frac{1}{16} = \left(\frac{1}{2}\right)^4$, i.e. time is 4 half lives or half life $T = 20$ days. In next 20 days its decay probability is 1/2. The probability of its decay in next 20 days further halved, i.e. 1/4]

32. A star having 10^{42} deuterons produces energy through the process

$${}_1H^2 + {}_1H^2 \rightarrow {}_1H^3 + {}_1H^1$$

$${}_1H^2 + {}_1H^3 \rightarrow {}_2He^4 + {}_0n^1$$

masses of nuclei being $M({}_1H^2) = 2.014$ amu, $M({}_2He^4) = 4.001$ amu, $m_p = 1.007$ amu, $m_n = 1.008$ amu. If the star radiates an average power of 10^{18} W, deutrons of the star will be exhausted in time t. Find this time.

[**Ans.** The reaction can be written as

$$3\,{}_1H^2 \rightarrow {}_2He^4 + {}_1H^1 + {}_0n^1 + Q$$

Mass defect is $\Delta m = 3M(^2_1\text{H}) - M(^4_2\text{He}) - m_p - m_n = 3 \times 2.014 - 4.001 - 1.007 - 1.008 = 0.026$ amu.

Energy released $Q = 0.026 \times 931 = 24.206$ MeV.

3 deuterons combine to give us 24.206 MeV energy = $24.206 \times 1.6 \times 10^{-13}$ J. Hence total energy released when all the deuterons are exhausted is

$$E = \frac{24.206 \times 1.6 \times 10^{-13} \times 10^{42}}{3}$$

Energy radiated per second is 10^{18} J. Hence time for which all the deuterons will survive is

$$t = \frac{E}{\text{power}} = \frac{24.206 \times 1.6 \times 10^{-13} \times 10^{42}}{3 \times 10^{18}} = 1.3 \times 10^{12} \text{ sec}\Big]$$

33. What is the difference between the balancing of nuclear reaction and chemical reaction?

[**Ans.** A chemical equation is balanced in the sense that the number of atoms of each element is the same on both sides of the equation. A chemical reaction merely alters the original combination of atoms. In a nuclear reaction, elements may be transmitted. Thus the number of atoms of each element is not necessarily conserved in a nuclear reaction. In this total charge is conserved and total number of nucleons (baryon number) is conserved.]

34. If both number of protons and the number of neutrons are conserved in each nuclear reaction, in what way is mass converted into energy (or *vice versa*) in nuclear reaction?

[**Ans.** It is due to the fact that the masses of neutron and proton in free state are different from their masses in bound state in nucleus. Since proton number and neutron number are conserved in nuclear reaction as such the total rest mass is conserved but the total binding energy (due to mass defect in bound state) on two sides may be different. The difference of these binding energies appears as energy released or absorbed in a nuclear reaction. The difference in total mass of nuclei on the two sides get converted into energy and *vice versa.* In fact nuclear reaction is an example of mass-energy inter conversion]

35. To activate the reaction (n, α) with stationary B^{11} nuclei, neutrons must have the threshold kinetic energy $K_{th} = 4.0$ MeV. Find the energy of reaction.

[**Ans.** If Q is energy of reaction, then threshold energy is

$$K_{th} = \frac{M+m}{M}|Q| \text{ or } |Q| = \frac{K_{th}M}{M+m} = \frac{4 \times 11}{11+1} = 3.67 \text{ MeV}\Big]$$

36. 1 MeV positron and 1 MeV electron collide directly. The perfect inelastic collision causes two γ-ray photons. Mass of electron is 0.000549 amu. Find the energy of each γ-ray photon.

[**Ans.** Using energy conservation, we have

$[m_0c^2 + \text{K.E.}(\bar{e}) + m_0c^2 + \text{K.E.}(e+)] = 2 \times$ energy of photon

2 (rest mass energy + K.E.) = 2 × energy of photon

∴ Energy of photon = rest mass energy + K.E.

$= 0.000549 \times 931 + 1$ MeV

$= 0.51 + 1 = 1.51$ MeV]

MULTIPLE CHOICE QUESTIONS

PART I

In the following, a question is followed by four alternatives, of them only one is correct. Therefore, the candidate has to select the correct choice from amongst the alternatives.

1. The activity of a radioactive sample, reduces to 1/8th of the initial activity in time which is equal to

(a) one half life (b) two half lives
(c) three half lives (d) four half lives.

2. A heavy nucleus at rest, disintegrates by emitting β radiation and γ-ray such that daughter nucleus is at rest, then which of the following is true?

(a) Directions of β radiation and γ-rays is same
(b) Directions of β-radiation and γ-rays is opposite
(c) β-radiation and γ-rays are inclined at 90°
(d) Nothing can be said about mutual directions of γ-rays and β-radiations.

3. The count rate of radioactive nuclei falls from 992 counts per minute to 62 counts per minute in 10 hours. The half life of the element is

(a) 1.0 hour (b) 2.5 hours
(c) 5 hours (d) 6 hours.

4. Four nuclei *P*, *Q*, *R*, *S* participate in processes involving α and β decays as shown in Fig. 28.15. The mass number and atomic number of *P* are 210 and 83 respectively. Which of the following statement is true?

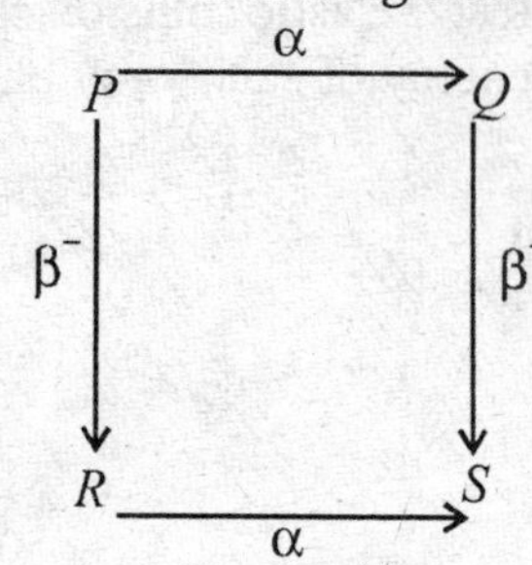

Fig. 28.15

(a) Mass number of *R* is 210 and atomic number 82
(b) Mass number of *Q* is 206 and atomic number 84
(c) Mass number of *S* is 206 and atomic number 82
(d) The energy released in the resultant process *P* → *S* via *Q* and via *R* is the same.

5. A $_{92}U^{238}$ nucleus emits eight α and sixteen β particles, the resulting nucleus is

(a) $^{206}_{92}U$ (b) $^{206}_{94}U$
(c) $^{238}_{88}U$ (d) $^{210}_{92}U$.

6. Suppose a parent nucleus decays into a daughter nucleus with decay constant of λ_p and the daughter nucleus of λ_d. At any instant the ratio of parent to daughter nulei is observed to be 2. What should be the ratio λ_p/λ_d if the number of daughter nuclei is constant?

(a) 2 (b) 1/2
(c) 1/4 (d) 4.

7. The half life of a radioactive substance is 2 hrs. What fraction of original activity will remain after 10 hrs.?

(a) 1/10 (b) 1/20
(c) 1/32 (d) 1/16.

8. A nuclear transformation is denoted by *X* (n, α) $^{7}_{3}Li$. Which of the following is the nucleus of element *X*? [AIEEE 2005]

(a) $^{11}_{4}Be$ (b) $^{9}_{5}B$
(c) $^{10}_{5}B$ (d) $^{12}_{6}C$.

9. Helium nuclei combine to form oxygen nucleus. The binding energy per nucleon of oxygen nucleus is (Given m_O = 15.834 amu and m_α = 4.0026 amu) [IIT 2005]

(a) 10.24 MeV (b) 0 MeV
(c) 5.24 MeV (d) 4 MeV.

10. If R is radius of nucleus and A its mass number, then graph between $\log R/R_0$ and $\log A$ is ($R_0 = 1.1 \times 10^{-15}$ fm)

(a) a straight line (b) a parabola
(c) an ellipse (d) none of them.

11. If F_{pp}, F_{pn} and F_{nn} denote the magnitudes of the nuclear force by a proton on proton, by a proton on a neutron and by a neutron on a neutron respectively when separation is 1 fm, then

(a) $F_{pp} > F_{pn} = F_{nn}$ (b) $F_{pp} = F_{pn} = F_{nn}$
(c) $F_{pp} > F_{pn} > F_{nn}$ (d) $F_{pp} < F_{pn} = F_{nn}$.

12. Let F_{pp}, F_{pn}, F_{nn} denote the magnitudes of the net force by a proton on proton, by a proton on a neutron and by a neutron on neutron respectively. Neglect gravitational force. When the separation is 1 fm, then

(a) $F_{pp} > F_{pn} > F_{nn}$ (b) $F_{pp} > F_{pn} = F_{nn}$
(c) $F_{pp} = F_{pn} = F_{nn}$ (d) $F_{pp} < F_{pn} = F_{nn}$.

13. Free ^{238}U nuclei kept in a train emit α-particles. When the train is stationary and a uranium nucleus decays. A passenger measures that the separation between the α-particle and the recoiling nucleus becomes x in time t after decay. If a decay takes place when the train is moving at uniform speed v, the distance between the α-particle and the recoiling nucleus at a time t after the decay, as measured by the passenger will be

(a) $x + vt$
(b) $x - vt$
(c) x
(d) depends on the direction of the train.

14. During a nuclear fission reaction

(a) a heavy nucleus breaks into two fragments by itself
(b) a light nucleus bombarded by thermal neutrons breaks up
(c) a heavy nucleus bombarded by thermal neutrons breaks up
(d) two light nuclei combine to give a heavier nucleus and possibly other products.

15. A radioactive substance X decays into another radioactive substance Y. Initially only X was present. λ_X and λ_Y are the disintegration constants of X and Y. N_X and N_Y are the number of nuclei of X and Y at any time t. Number of nuclei N_Y will be maximum when

(a) $\dfrac{N_Y}{N_X - N_Y} = \dfrac{\lambda_Y}{\lambda_X - \lambda_Y}$

(b) $\dfrac{N_X}{N_X - N_Y} = \dfrac{\lambda_X}{\lambda_X - \lambda_Y}$

(c) $\lambda_Y N_Y = \lambda_X N_X$

(d) $\lambda_Y N_X = \lambda_X N_Y$.

16. There are two radioactive nuclei A and B. A is α-emitter and B β-emitter. Their disintegration constants are in the ratio 1 : 2. What should be the ratio of number of atoms of A and B at any time so that probabilities of getting α and β particles are same at that instant?

(a) 2 : 1 (b) 1 : 2
(c) e (d) e^{-1}.

17. Half life of a radioactive substance A is two times the half life of another radioactive substance B. Initially the number of nuclei A and B are N_A and N_B respectively. After three half lives of A, the number of nuclei of both are equal. Then ratio $\dfrac{N_A}{N_B}$ is

(a) 1/4 (b) 1/8
(c) 1/3 (d) 1/6.

19. Mass number of a nucleus is

(a) always less than its atomic number
(b) always more than its atomic number
(c) equal to its atomic number
(d) sometimes more than and sometimes equal to its atomic number.

20. A freshly prepared radioactive source of half life $2h$ emits radiations of intensity which is 64 times the permissible safe level. The minimum time after which it would be possible to work safely with this source is

(a) 6*h* (b) 12*h*
(c) 24*h* (d) 128*h*.

21. Ten grams of ^{57}Co kept in an open container, β-decays with half life of 270 days. The weight of the material inside the container after 540 days will be very nearly

(a) 10 g (b) 5 g
(c) 25 g (d) 1.25 g.

22. An α-particle is bombarded on ^{14}N. As a result, a O^{17} nucleus is formed and a particle is emitted. This particle is a

(a) neutron (b) proton
(c) electron (d) positron.

23. $^{238}_{92}U$ changes to $^{210}_{85}Al$ by a series of α and β decays. The number of α and β decays undergone is

(a) 7 and 5 (b) 7 and 7
(c) 5 and 7 (d) 7 and 9.

24. Number of nuclei of radioactive nucleus at time $t = 0$ are 1000 and 900 at time $t = 2$ sec. Then number of nuclei at time $t = 4$ sec will be

(a) 800 (b) 810
(c) 790 (d) 700.

25. Decay constant of a radioactive sample is λ. The half life and the average life of the sample are respectively

(a) 1/λ, log2/λ (b) log2/λ, 1/λ
(c) λ/log2, 1/λ (d) λ/log2, 1/λ.

26. In one average life

(a) half the entire nuclei decay
(b) less than half the active nuclei decay
(c) more than half the active nuclei decay
(d) all the nuclei decay.

27. In a radioactive decay, neither the atomic number, nor the mass number changes. Which of the following particles is emitted in the decay?

(a) Electron (b) Proton
(c) Photon (d) Neutron.

28. β-particles emitted by a radioactive element is

(a) an electromagnetic radiation
(b) an electron revolving round nucleus
(c) a negatively charged particle from nucleus
(d) a neutral particle.

29. Radioactivity was first discovered by

(a) Becquerel (b) Madam Curie
(c) J.J. Thomson (d) Rutherford.

30. In nuclear disintegration, the conservation law which is no followed is

(a) Law of conservation of mass
(b) Law of conservation of energy
(c) Law of conservation of momentum
(d) Law of conservation of angular momentum.

31. Activity of a radioactive substance at anytime is 10^3 disint./s. Its half life is 1 s. Its disintegration rate after 3 s is

(a) 250 disint./s (b) 15 disint./s
(c) 500 disint./s (d) 62.5 disint./s.

32. Half life of radioactive radon is 3.8 days. The time after which 1/20 part of radon will be active is

(a) 3.8 days (b) 16.5 days
(c) 33 days (d) 76 days.

33. Percentage of the active part of a radioactive substance left after 5 half lives is

(a) 1 (b) 3
(c) 5 (d) 20.

34. The reciprocal of the time in which 37% the active substance is left is called

(a) disintegration constant (λ)
(b) Curie
(c) half life
(d) mean life.

35. Which of the following is the wrong description of the binding energy of a nucleus?

(a) It is the energy required to break a nucleus into its constituent nucleons

(b) It is the energy made available when free nucleons combine to form a nucleus
(c) It is the sum of rest mass energies of nucleons minus the rest mass energy of nucleus
(d) It is the sum of kinetic energies of all nucleons inside the nucleus.

36. As the mass number (A) of nucleus increases, which of the following quantities related to nucleus do not change?
(a) Mass (b) Volume
(c) Density (d) Binding energy.

37. A free neutron decays to a proton but a free proton does not decay into a neutron. This is because
(a) neutron is a composite particle made of a proton and an electron, whereas proton is a fundamental particle
(b) neutron is an uncharged particle whereas proton is a charged particle
(c) neutron has larger rest mass than the proton
(d) weak force can operate in a neutron but not in a proton.

38. Consider a sample of a pure beta-active material
(a) All the beta-particles emitted have the same energy
(b) The β-particles originally exist inside the nucleus and are ejected at the time of decay.
(c) The anti neutrino emitted in a β-decay has zero mass and hence zero momentum
(d) The active nucleus changes to one of its isobars after the β-decay.

39. In which of the following decays, the element does not change?
(a) α-decay (b) β^+-decay
(c) β^--decay (d) γ-decay.

40. A radioactive nucleus is being produced at constant rate A per second. If its decay constant is λ and N_0 are the number of active nuclei at $t = 0$, then maximum number of nuclei possible are
(a) A/λ (b) $N_0 + A/\lambda$
(c) N_0 (d) $\frac{\lambda}{A} + N_0$.

41. In a sample of a radioactive substance what fraction of the initial number of nuclei will remain undecayed after a time $t = T/2$, where T is half life of radioactive substance?
(a) $\frac{1}{\sqrt{2}}$ (b) $\frac{1}{2\sqrt{2}}$
(c) $\frac{1}{4}$ (d) $\frac{1}{\sqrt{2}-1}$.

42. The activity of a radioactive substance is R_1 at time t_1 and R_2 at time t_2 (> t_1). Its decay constant is λ. Then
(a) $R_1 t_1 = R_2 t_2$ (b) $R_2 = R_1 e^{\lambda(t_1 - t_2)}$
(c) $\frac{R_1 - R_2}{t_2 - t_1} = \text{const}$ (d) $R_2 = R_1 e^{\lambda(t_2 - t_1)}$.

43. In above problem number of atoms decayed during time interval t_1 to t_2 are
(a) $\frac{\log 2}{\lambda}(R_1 - R_2)$ (b) $R_1 e^{-\lambda t_1} - R_2 e^{-\lambda t_2}$
(c) $\lambda(R_1 - R_2)$ (d) $\frac{R_1 - R_2}{\lambda}$.

44. When a radioactive nucleus disintegrates emitting γ-ray photon, as a result of this
(a) the atomic number of element changes
(b) mass number of element changes
(c) both atomic number and mass number change
(d) both atomic number and mass number do no change.

45. Two samples of a radioactive element are found to emit N_1 and N_2 β-particles per second respectively. The ratio of active atoms in the two samples is
(a) $\frac{N_1}{N_2}$ (b) $\frac{N_2}{N_1}$
(c) $\frac{N_2^2}{N_1^2}$ (d) $\frac{N_1^2}{N_2^2}$

46. In a nuclear reaction
${}^{A}_{Z}X \rightarrow {}_{z+1}^{A}Y \rightarrow {}^{A-4}_{z-1}R \rightarrow {}^{A-4}_{z-1}R$ the emitted particles (or radiations) are respectively

(a) α, β, γ (b) β, γ, α
(c) γ, α, β (d) β, α, γ.

47. For a radioactive element, the graph between number of disintegrated atoms (N) and time (t) is

(a) 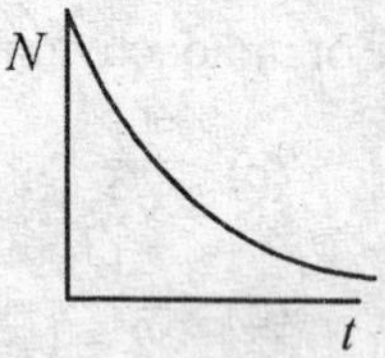 (b) 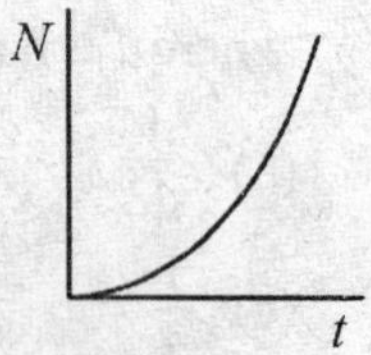

(c) 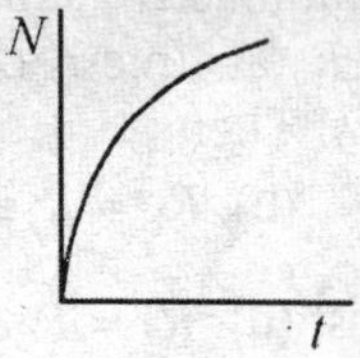 (d) 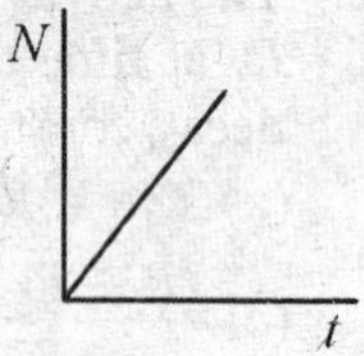

48. When an element emits a positron (β^+), then proton neutron ratio of the daughter element
(a) increases
(b) decreases
(c) remains unchanged
(d) in some elements it increases and in others decreases.

49. Choose the correct statement.
(a) Half life (T) depends on the mass of the active substance
(b) The rate of disintegration of a radioactive element can be increased by heating the sample
(c) After time $2T$ (T is half life), all the nuclei of a radioactive element will disintegrate
(d) The rate of disintegration depends on the number of active atoms.

50. Atomic number of a radioactive element is 42. Its isotopes have mass number 87, 88 and 89. Then find the incorrect statement.
(a) All the isotopes have same chemical properties
(b) All the isotopes have 42 protons
(c) Some of the isotopes may have 46 neutrons
(d) Some of the isotopes may have 42 neutrons.

51. In the following disintegration, X is
$^{239}_{93}\text{Np} \rightarrow {}^{239}_{92}\text{U} + X + \nu$
(a) a proton (b) a positron
(c) an electron (d) a neutron.

52. The binding energies of nuclei X and Y are E_1 and E_2 respectively. The two atoms of X fuse to give one atom of Y and an energy Q is released. Then
(a) $Q = 2E_1 - E_2$ (b) $Q = E_2 - 2E_1$
(c) $Q < 2E_1 - E_2$ (d) $Q > E_2 - 2E_1$.

53. N_1 atoms of a radioactive element emit N_2, β-particles per second. The decay constant of the element is (in s^{-1})
(a) $\dfrac{N_1}{N_2}$ (b) $\dfrac{N_2}{N_1}$
(c) $N_1 \log 2$ (d) $N_2 \log 2$.

54. The ratio of molecular mass of two radioactive substances is 3/2 and the ratio of their decay constant is 4 /3. Then ratio of their initial activity per mole will be
(a) 2 (b) 8/9
(c) 4/3 (d) 9/8.

55. Which of the following processes represents a gamma decay?
(a) $^{A}_{Z}X + \gamma \rightarrow {}^{A}_{z-1}X + a + b$
(b) $^{A}_{Z}X + {}^{1}_{0}\text{n} \rightarrow {}^{A-3}_{z-2}X + c$
(c) $^{A}_{Z}X \rightarrow {}^{A}_{Z}X + f$
(d) $^{A}_{Z}X + e \rightarrow {}^{A}_{z-1}X + g$.

56. The half life of ^{215}At is 100 μs. The time taken for the radioactivity of a sample of ^{215}At to decay to $1/16^{th}$ of its initial value is
(a) 400 μs (b) 6.3 μs
(c) 40 μs (d) 300 μs.

57. A nucleus of mass 220 initially at rest emits an α-particle. If the Q value of the reaction is 5.5 MeV; the energy of the emitted α-particle will be [IIT 2003]

(a) 4.8 MeV (b) 5.4 MeV
(c) 6.0 MeV (d) 6.8 MeV.

58. Half life of a radioactive substance for α-decay is 40 yrs and for β-decay is 20 yrs. What fraction of the sample will decay in a time 80 yrs if the sample is decaying both by α and β-emission simultaneously?

(a) $\frac{63}{64}$ (b) $\frac{31}{32}$
(c) $\frac{15}{16}$ (d) $\frac{7}{8}$.

59. In the nuclear reaction

$${}^{2}_{1}H + {}^{3}_{1}H \rightarrow {}^{4}_{2}He + {}^{1}_{0}n$$

if the binding energies of ${}^{2}_{1}H$, ${}^{3}_{1}H$ and ${}^{4}_{2}He$ are respectively a, b and c (in MeV), then the energy (in MeV) released in this reaction is [CBSE PMT 2005]

(a) $a + b - c$ (b) $c + a - b$
(c) $c - a - b$ (d) $a + b + c$.

60. Nuclei of which one of the following pairs are isotones?

(a) ${}^{74}_{34}Se$, ${}^{71}_{31}Ga$ (b) ${}^{84}_{38}Sr$, ${}^{86}_{38}Sr$
(c) ${}^{92}_{42}Mo$, ${}^{92}_{41}Zr$ (d) ${}^{40}_{20}Ca$, ${}^{32}_{16}Sr$.

61. Half life of a radioactive substance is 1600 yrs. The actual life time of atoms of this element can be from zero to infinity. Mean life then is

(a) infinity (b) 3200 yrs
(c) 2309 yrs (d) 1600 yrs.

62. What percentage of radioactive atoms will disintegrate in time equal to its mean life time?

(a) 100% (b) 63%
(c) 50% (d) 37%.

63. For a radioactive substance half life is 1 month and activity is R. After 3 months its activity will be

(a) R (b) $R/3$
(c) $R/4$ (d) $R/8$.

64. For a radioactive substance graph between $\log R$ (R is its instantaneous activity) and time (t) is

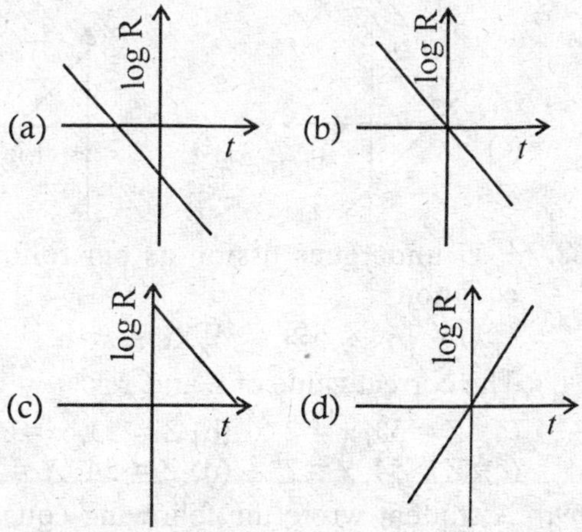

65. ${}^{226}Ra$ has half life 1600 years. The activity of 1 mg of ${}^{226}Ra$ is

(a) micro Curie (b) mili Curie
(c) 0.1 Curie (d) 1 Curie.

66. According to Geiger-Nattal law, the graph between $\log\lambda$ (λ is decay constant) and $\log R$ (R-range of α-particles) is

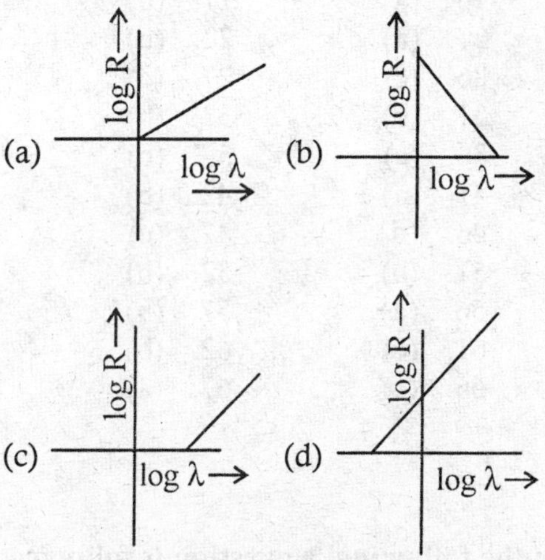

67. In α-decay if energy of α-particles is E and the range is R then correct graph is

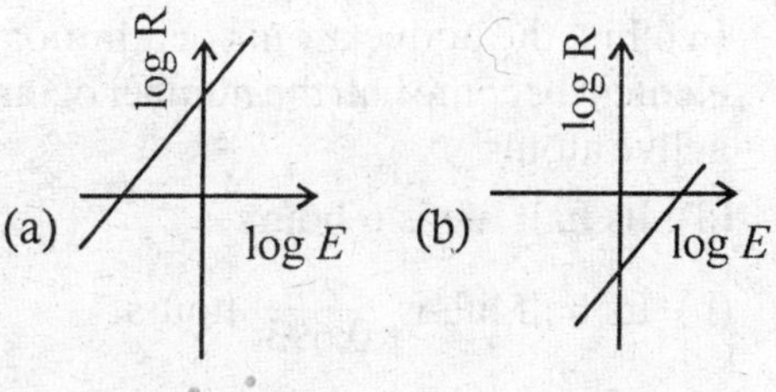

(c) 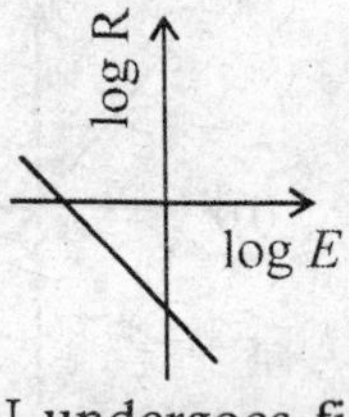

(d) 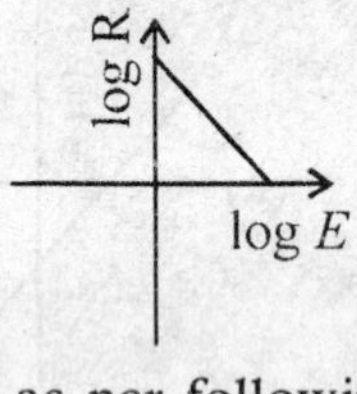

68. $^{235}_{92}U$ undergoes fission as per following equation

$^{235}_{92}U + ^{1}_{0}n \rightarrow ^{91}_{38}Sr + ^{143}_{Z}Xe + x\,^{1}_{0}n$

The correct value of Z and x is

(a) $Z = 53, x = 2$ (b) $Z = 53, x = 3$
(c) $Z = 54, x = 2$ (d) $Z = 54, X = 3$.

69. A student wrote the following equations for proton-proton cycle

(*i*) $^{1}_{1}H + ^{1}_{1}H \rightarrow ^{2}_{1}H + e^{+} + \nu$
(*ii*) $^{1}_{1}H + ^{2}_{1}H \rightarrow ^{3}_{2}He + e^{+} + \nu$
(*iii*) $^{3}_{2}He + ^{3}_{2}He \rightarrow ^{2}_{2}He + ^{1}_{1}H + ^{1}_{1}H$

correct equations out of these are

(a) only (*i*) and (*iii*) (b) only (*ii*)
(c) only (*i*) and (*ii*) (d) only (*iii*).

70. The increasing order of specific charge of a electron (e) proton (ρ) neutron (n) and α-particle (α) respectively is

(a) e, p, n, α (b) n, p, e, α
(c) n, p, α, e (d) n, α, p, e.

ANSWERS

1. (c)	2. (b)	3. (b)	4. (c)	5. (a)
6. (b)	7. (c)	8. (c)	9. (a)	10. (a)
11. (b)	12. (d)	13. (c)	14. (c)	15. (c)
16. (a)	17. (b)	18. (a)	19. (d)	20. (b)
21. (a)	22. (b)	23. (b)	24. (b)	25. (b)
26. (c)	27. (c)	28. (c)	29. (a)	30. (a)
31. (b)	32. (b)	33. (b)	34. (a)	35. (d)
36. (c)	37. (c)	38. (d)	39. (d)	40. (a)
41. (a)	42. (b)	43. (d)	44. (d)	45. (a)
46. (d)	47. (c)	48. (b)	49. (d)	50. (d)
51. (b)	52. (b)	53. (b)	54. (c)	55. (c)
56. (a)	57. (b)	58. (a)	59. (a)	60. (a)
61. (c)	62. (b)	63. (d)	64. (c)	65. (b)
66. (c)	67. (a)	68. (c)	69. (a)	70. (c).

PART II

In the following, a question is followed by four alternatives, of them one or more are correct. Candidate has to select the correct choice(s) from amongst the alternatives.

1. In 6 hrs, the active atoms of a radioactive element become half the number of initially active atoms

(a) Its half life is 6 hours

(b) Its half life is $\frac{6}{0.693}$ hours

(c) Its mean life is 6 hours

(d) Its mean life is $\frac{6}{0.693}$ hours.

2. A radioactive element emits

(a) α-particle

(b) electron

(c) proton
(d) electromagnetic radiation.

3. Particles emitted by a radioactive element are found to be deflected by magnetic field, they can be
(a) electron (b) α-particles
(c) γ-rays (d) protons.

4. Which of the following is the correct description of binding energy of a nucleus?
(a) It is the energy required to break a nucleus into its constituent nucleons
(b) It is the energy made available when free nucleons combine to form a nucleus
(c) It is the sum of the rest mass energies of its nucleons minus the rest mass energy of the nucleus
(d) It is the sum of the kinetic energies of all the nucleons in the nucleus.

5. The heavier nuclei tend to have a larger N/Z ratio because
(a) a neutron is heavier than a proton
(b) a neutron is an unstable particle
(c) a neutron does not exert electric repulsion
(d) Coulomb forces have longer range compared to the nuclear forces.

6. In which of the following decays, the atomic number decreases?
(a) α-decay (b) β^+-decay
(c) β^--decay (d) γ-decay.

7. For nuclei with $A > 100$
(a) the binding energy of the nucleus increases on an average as A increases
(b) the binding energy per nucleon decreases on an average as A increases
(c) if the nucleus breaks into two roughly equal parts energy is released
(d) if two nuclei fuse to from a bigger nucleus, energy is released.

8. Atomic number of a radioactive element is 42. Its isotopes have mass number 87, 88 and 89. Then correct statement is
(a) all the isotopes have same chemical properties
(b) all the isotopes have 42 protons
(c) some of the isotopes may have 46 neutrons
(d) some of the isotopes may have 42 protons.

9. After α-decay the ratio of Z/A for daughter nucleus
(a) decreases if $A > 2Z$
(b) increases if $A > 2Z$
(c) remains unchanged if $A = 2Z$
(d) increases if $A < 2Z$.

10. True statements are
(a) β-particles are fast electrons
(b) γ-rays are same in nature as X-rays
(c) α-particles are helium atoms
(d) positrons are also β-particles.

11. In an atom of atomic no. 11 and mass number 24, there are
(a) 11 electrons (b) 11 neutrons
(c) 13 neutrons (d) 13 protons.

12. The phenomenon of nuclear fission can be carried out both in a controlled and in an uncontrolled way. Out of the following the correct statement *vis-a-vis* this phenomenon are
(a) the fission energy released per reaction is much more than the conventional nuclear reactions and one of the products of reaction is that vary particle which initiates the reaction
(b) it is the surface to volume ratio of the sample of nuclear fuel used which determines whether or not the reaction would sustain itself as a chain reaction
(c) the control rods in a nuclear reactor must be made up of a material that absorbs neutrons effectively
(d) the energy released per fission as well as the energy released per unit mass of the fuel in nuclear fission are both greater than the corresponding quantities for nuclear fusion.

13. Mass number of a nucleus is

(a) always less than its atomic number
(b) always more than its atomic number
(c) sometimes equal to its atomic number
(d) sometimes more than and sometimes equal to its atomic number.

14. $^{235}_{92}U$ is α-active. Then in the bulk of element

(a) the probability of nucleus disintegrating during one second is lower in the first half life and greater in the fifth half life
(b) the probability of a nucleus disintegration during 1 sec remains constant for all times
(c) every nucleus must disintegrate by the lapse of the average life of population
(d) quite an appreciable quantity of U^{235} will remain even after the average life.

15. If has been found that nuclides with 2, 8, 20, 50, 82 and 126 protons or neutrons exceptionally stable. These numbers are referred to as magic numbers and the existence of these has led us to

(a) the idea of periodicity in nuclear properties similar to the periodicity of chemical elements in the periodic table
(b) have a conventional explanation of nuclear fission
(c) have a conventional explanation of nuclear fusion
(d) the so called shell model of the nucleus.

16. In the α-decay process occurring in different types of nuclei at rest

(a) the kinetic energy of the daughter nucleus is always greater than the kinetic energy of α-particles
(b) the kinetic energy of daughter nucleus is always less than the kinetic energy of α-particles
(c) the magnitudes of the linear momenta of the α-particle and the daughter nucleus are always equal
(d) the daughter nucleus is always in a stable state.

17. A nitrogen nucleus $^{14}_{7}N$ absorbs a neutron and can transform into lithium nucleus ($^{7}_{3}Li$) under suitable conditions, after emitting

(a) 4 protons and 4 neutrons
(b) 2 α-particles and 2 gamma photony
(c) 4 protons and 3 neutrons
(d) 1 α-particles, 4 protons and 2 β-particles.

18. Choose the correct statements from the following.

(a) Like other light nuclei, the $^{4}_{2}He$ nuclei also have a low value of the binding energy per nucleon
(b) The binding energy per nucleon decreases for nuclei with small as well as large atomic numbers
(c) The energy required to remove one neutron from $^{7}_{3}Li$ to transform into the isotope $^{6}_{3}Li$ is 5.6 MeV. Which is the same as binding energy per nucleon of $^{6}_{3}Li$
(d) When two deuteron nuclei fuse together, they give rise to tritium nucleus accompanied by a release of energy.

19. From the following equations pick out the possible nuclear fusion reactions.

(a) $^{13}_{5}C + ^{1}_{1}H \rightarrow ^{14}_{6}C + 4.3$ MeV
(b) $^{235}_{92}U + ^{1}_{0}n \rightarrow ^{140}_{54}Xe + ^{94}_{38}Sr + 2\,^{1}_{0}n + \gamma +$ 200 MeV
(c) $^{14}_{7}N + ^{1}_{1}H \rightarrow ^{15}_{8}O + 7.3$ MeV
(d) $^{12}_{6}C + ^{1}_{1}H \rightarrow ^{13}_{7}N + 2$ MeV.

20. Radon decays to polonium according to the decay scheme

$^{222}_{88}Rn \rightarrow ^{218}_{84}Po + ^{4}_{2}He$

Take half life of this decay to be 3.8 days, and state which of the following statements are true?

(a) 60% of the radon will decay approximately in 5 days
(b) Activity of radon after 7.6 days will be $(5.3 \times 10^{-7} N_0)\ s^{-1}$, where N_0 is the original number of atoms of the radon
(c) If original mass of radon was 1 mg the amount of radon disintegrated in 11.4 days will be 0.125 mg.
(d) The decay constant of radon is $2.1 \times 10^{-6}\ s^{-1}$.

ANSWERS

1.	(a, d)	2.	(a, b, d)	3.	(a, b)	4.	(a, b, c)	5.	(c, d)
6.	(a, b)	7.	(b, c)	8.	(a, b, c)	9.	(a, c)	10.	(a, b, c)
11.	(a, c)	12.	(a, b, c)	13.	(c, d)	14.	(b, d)	15.	(a, d)
16.	(b, c)	17.	(a, b, d)	18.	(b, c)	19.	(c, d)	20.	(a, b, d)

29

Semiconductors and Devices

29.1 CLASSIFICATION OF SOLIDS

If an electric field $\boldsymbol{E}$ is established inside a conductor, the free electrons experience force due to the field and acquire a drift speed. This results in an electric current. The conductivity σ is defined in terms of electric field $\boldsymbol{E}$ existing in the conductor and the resulting current density $\boldsymbol{J}$. The relation is

$$\boldsymbol{J} = \sigma \boldsymbol{E} \qquad \ldots(29.1)$$

Larger the conductivity σ, better is the material as a conductor. On the basis of the relative values of electrical conductivity (σ) or resistivity $\left(\rho = \frac{1}{\sigma}\right)$, the solids are classified into three categories.

(1) Conductors: These are the substances whose electric conductivity is very large or resistivity (specific resistance) is very small ($\sim 10^{-8}$ Ω-m). In such substances the free electron concentration is very large ($\approx 10^{29}$ m^{-3}). On the basis of band theory, the valance band (V.B.) and conduction band (C.B.) overlap. Their resistance increases with temperature, i.e., for them the temperature coefficient for resistance is positive. The conductivity of the conductors decreases on adding the impurities. Examples of conductors are, Cu, Ag, Au, Na, Pt, Hg, etc. In general, metals are the good conductors.

(2) Insulators: They have very low conductivity and high resistivity (of the order of 10^{15} Ω-m). In these substances there are no free electrons and this conductivity is practically independent of the temperature. The gap between valance band and conduction band is large (~3 eV to 6 eV). Example of insulators are, wood, plastic mica, glass, etc.

(3) Semiconductors: The conductivity of these substances lies between conductors and insulator. Their resistivity lies between 10^{-1} to 10^{+1} Ω-m. Fourth group elements such as carbon, silicon, germanium are examples of semiconductors. In these substances the valance electrons are neither tightly bound to nucleus nor are loose to be treated as free electrons. *Their resistance decreases with temperature, i.e. the temperature coefficient of the resistance is negative. At zero Kelvin a semiconductor behaves like a perfect insulator.* The energy gap between valance band and conduction band is less than 3 eV. (For germanium $E_g = 0.72$ eV, for Si $E_g = 1.1$, Ga As it is 1.43 eV, etc.)

On the basis of chemical composition, the semiconductors are broadly classified into two classes.

(*a*) **Elemental Semiconductors:** They are pure elements, e.g. silicon, germanium, etc.

(*b*) **Compound Semiconductors:** They are chemical compounds.

Inorganic Compounds: Such as, CdS, GaAs, CdSe, InP, etc.

Organic Compounds: Like anthracine, doped pthalocyanines, etc.

Organic Polymers: Polypyrole, polyaniline, polythiophene, etc. Here we will restrict our study only to the inorganic semiconductors particularly to elemental semiconductors like Ge and Si. The general concepts introduced here for discussing the elemental semiconductors by and large apply to most of the compound semiconductors as well.

Besides the above classification of semiconductors, they are also classified on the basis of the source and nature of charge carriers. This scheme divides the semiconductors into *intrinsic* and *extrinsic* semiconductors.

Intrinsic Semiconductor

These are pure semiconductors with impurity less than 1 part in 10^{10}. They possess mobile charge carriers like electrons and holes. The presence of mobile charge carriers is the intrinsic property of the material. These charge carriers are obtained as a result of thermal excitation and rupturing of the bonds. The electrical conductivity is due to these mobile electrons and holes (we will introduce the concept of hole as a positive charge carrier later in this chapter). In the intrinsic semiconductor, the number density of electrons (n_e) is equal to number density of holes (n_h).

Extrinsic Semiconductor

These are obtained by adding to the pure semiconductor material with small amounts of certain specific impurities with valance other than that of the atoms of the parent material. The process of adding the impurities is called *dopping*. As a result of doping, the concentration (number density) of mobile electrons (or holes) changes drastically. So the electrical conductivity in these materials is essentially due to the foreign atoms or due to extrinsic nature.

29.2 ENERGY BANDS IN SOLIDS: BAND THEORY

In an isolated atom, the energy of an electron is decided by the orbit in which it is moving. It means the electron is in discrete energy states such as 1*s*, 2*s*, 2*p*, ..., etc. They occupy the quantum states available in those energy states. The occupation of these quantum states is governed by the *Pauli's exclusion principle*. In solids, the electron energies will be different because of the presence of many atoms placed close to each other. Because of periodicity of lattice each atom is in electric field of neighbouring atoms. Consequently the energy levels of an individual atom lose their significance. If the crystal contains N-atoms, then due to interaction between the atoms each discrete energy level of an individual atom splits into N-close sub levels. As N is very large, the separation between sub levels is very small ($\sim 10^{-19}$ eV) so that these levels are almost continuous and are said to form an *energy band*. These energy bands are in general separated by regions called *forbidden bands* or *band gaps*.

According to *Pauli's exclusion principle* each energy level can be occupied by two electrons with opposite spins. Hence normally only lower energy bands will be occupied by the electrons. The highest energy band containing electrons is called *valence band*. This band contains electrons which are responsible for forming bonds between the atoms, i.e. the electrons of the outermost orbits of the atoms. These electrons do not take part in the flow of current.

In solids there is a band containing electrons of maximum permissible energies. This band is called *conduction band*. This is also minimum energy vacant band. In case of conductors, this band is partially filled and in semiconductors, at zero Kelvin it is totally vacant but with the increase in temperature, the number of electrons in this band increases resulting increase in the conductivity. The electrons in this band are responsible for the flow of the current in the solids. Conduction band is also called *first allowed band* or *first excited band*.

If there are no electrons in conduction band, no electric conduction can take place and we call them insulators. Semiconductor as mentioned above behave like insulators at zero Kelvin.

In between valance band and conduction band, there is an energy gap called *forbidden energy band or energy gap*. This band is totally empty and there are no electrons in this band.

The minimum energy required for an electron to go from valence band to conduction band is called *band gap* (ΔE_g). As already mentioned, this band gap distinguishes the three types of solids, i.e. conductors, insulators and semiconductors.

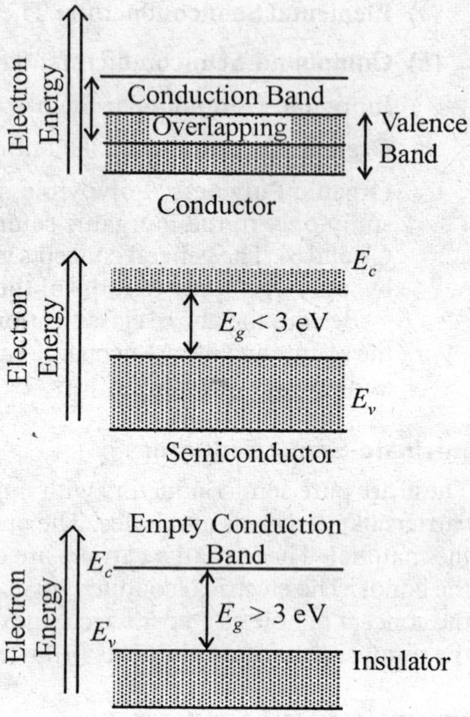

Fig. 29.1

The band contains maximum possible electrons is called *completely filled energy band* and if it contains less number of the electrons than maximum possible number are called *partially filled energy bands*.

The different types of bands with energy gaps are shown in Fig. 29.1.

29.3 INTRINSIC AND EXTRINSIC SEMICONDUCTORS

The semiconductors are classified as (*a*) *intrinsic semiconductors* (*b*) *extrinsic semiconductors*.

Intrinsic Semiconductors

The pure semiconductors without impurities are called intrinsic semiconductors. Their conductivity is due to the presence of intrinsic charge carriers (electrons and holes) and not due to any impurity or foreign atom. The well known intrinsic semiconductors are germanium (Ge) and silicon (Si). Their electronic configuration is

$$\text{Ge (32)} = 1s^2\ 2s^2\ 2p^6\ 3s^2\ 3p^6\ 3d^{10}\ 4s^2\ 4p^2$$

$$\text{Si (14)} = 1s^2\ 2s^2\ 2p^6\ 3s^2\ 3p^6$$

The valance of both Ge and Si is 4. This means that 4 valence electrons of these atoms tightly bound with positive core (nucleus and tightly bound inner electrons) with +4e charge. In Germanium and Silicon crystals, the atoms are arranged in an ordered array such that each atoms lies at the corners of a regular tetrahedron (Fig. 29.2). All the four valance electrons of Ge (and Si) form covalent bonds with neighbouring electrons of other 4 Ge atoms. Thus no free electrons are left for conduction.

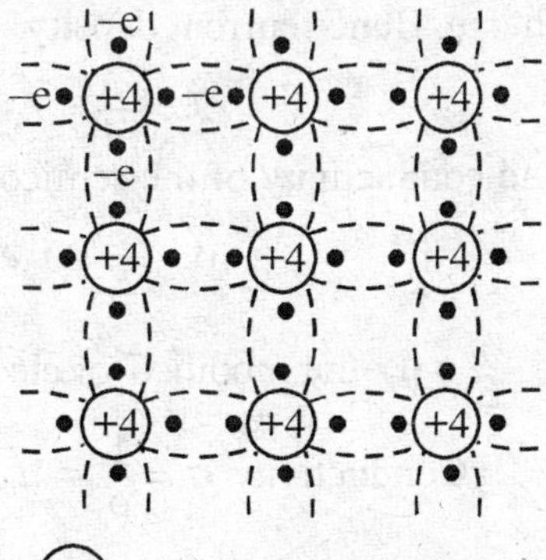

Fig. 29.2

Around 0 K temperature in pure germanium (or silicon) all the valence electrons are rigidly bound to the core of the atom. Thus no free electrons are available for conduction. As such they behave like insulators. But with the rise in temperature, some of the co-valent bonds rupture and some of the valence electrons become free. The bond energy so released and thermal energy are sufficient to carry the electron to conduction band. The vacancy created in the bond (valence band) is called a *hole.* The hole is assumed to possess a positive charge equal in magnitude to that on an electron and is treated as a positive charge carrier in conduction through the semiconductor. Figure 29.3 shows the site of creation of a hole when an electron is released due to rupture of bond.

The process of setting free an electron from a covalent and simultaneously creating a hole requires certain minimum energy, which is equal to the band gap energy (= ΔE_g). The number of electrons (n_e) set free at absolute temperature T is given by

$$n_e = Ce^{-\Delta E_g/2k_BT} \qquad ...(29.2)$$

Where C is a constant and k_B is Boltzmann's constant. As temperature (T) increases n_e increases because more and more bonds rupture with rise in temperature.

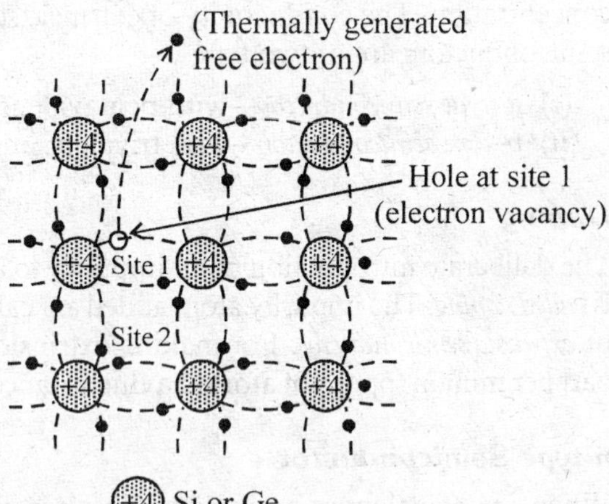

Fig. 29.3

In intrinsic semiconductors the number density of free electrons (n_e) and that of holes (n_h) are equal and each is equal to the intrinsic charge carrier concentration (n_i). Thus

$$n_e = n_h = n_i \qquad ...(29.3)$$

At normal temperatures the charge carrier concentration (n_i) for Ge is 2.4×10^{19} m^{-3} and for Si it is $1.5 \times 10^{+16}$ m^{-3}.

When an electric field is applied across an intrinsic semiconductor, the free electrons move in the direction opposite to the electric field and holes in the direction of the electric field (Fig. 29.4). Thus under the influence of electric field, electrons in conduction band and holes in valence band move in opposite direction creating current in the semiconductor. In terms of drift velocity v_e for electron and v_h for holes, the current in semiconductors is given by

$$i = i_e + i_h = n_e eAv_e + n_h eAv_h$$

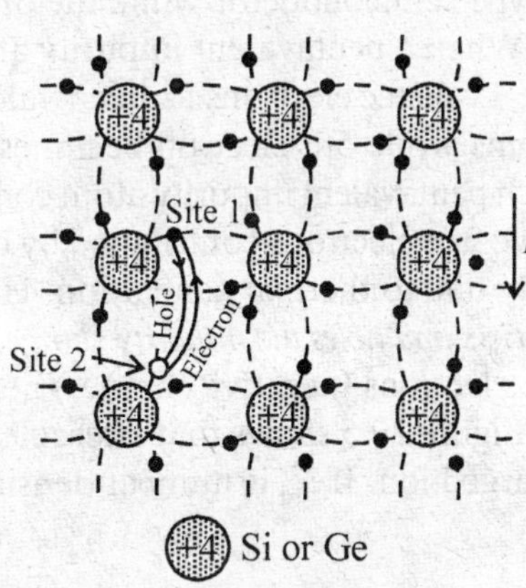

Fig. 29.4

where A is area of cross-section of the semiconductor perpendicular to the applied electric field: and e is electronic

charge. Hence current density

$$J = \frac{i}{A} = n_i e(v_e + v_h) \text{ where } n_i = n_e = n_h$$

and conductivity of the semiconductor using Ohm's law is

$$\sigma = \frac{J}{E} = n_i e\left(\frac{v_e}{E} + \frac{v_h}{E}\right)$$

$\frac{v_e}{E} = \mu_e$ the mobility* of electron and $\frac{v_h}{E} = \mu_h$ mobility of holes.

$$\therefore \quad \text{Conductivity } \sigma = \frac{1}{\rho} = n_i e(\mu_e + \mu_h) \qquad \ldots(29.4)$$

Extrinsic Semiconductor

The conductivity of intrinsic semiconductors is very small. The conductivity of a pure semiconductor can be increased by introducing the impurities. Such an impure semiconductor is called extrinsic semiconductor. The impurities such as pentavalent atoms (As, P, etc.) increases the electron concentration whereas the impurities such as trivalent atoms (Al, B, etc.) increases the holes concentration. The conductivity of extrinsic semiconductor is due to impurities. Thus the extrinsic semiconductors are of two types.

(*i*) *n–type semiconductor*—with pentavalent impurity and
(*ii*) *p–type semiconductor*—with trivalent impurity.

Doping

The deliberate introduction of an impurity to an otherwise pure crystal (like germanium or silicon) is *called doping*. The impurity atom added are called *dopants*. The crystal is called doped semiconductor or *extrinsic semiconductor*. For making extrinsic semiconductor generally impurity of the order of 1 part per million (ppm), of atoms having valance different from 4 is added to the pure semiconductor.

n-type Semiconductor

When a pentavalent impurity like phosphorous, antimony, arsenic is doped in pure germanium or silicon, the conductivity of crystal increases due to surplus electrons and such a crystal is said to be *n*-type semiconductor while the impurity atoms are *called donors*.

When a pentavalent impurity atom replaces the tetravalent atom (germanium or silicon atoms) the 4 valence electrons form co-valent bonds with electrons of neighbouring germanium (or silicon) atoms while 5th electron becomes free. This electron is free to move to cause conduction. Thus each pentavalent impurity atom contributes one extra electron for conduction. These semiconductors have free electrons contributed by donors and generated by the thermal process, while the holes are only due to thermal generation. Hence in these semiconductors, *the electrons are the majority charge carriers and holes the minority charge carriers*. Since majority charge carriers in these semiconductors are electrons (negative charges) hence they are called *n*-type semiconductors.

Figure 29.5 shows that after releasing an electron, the donor atom becomes stationary positively charged ion. If N_d is number density of stationary donor ions, then total number of free electron is

$$n_e = n'_e + n''_e \approx N_d \qquad \ldots(29.5)$$

*Mobility of charge carrier is its drift velocity per unit applied electric field. Its SI units are Amp(m) N^{-1}.

where n'_e is the electron concentration due to donor atoms and n''_e the electron concentration due to temperature; $n''_e << n'_e$.

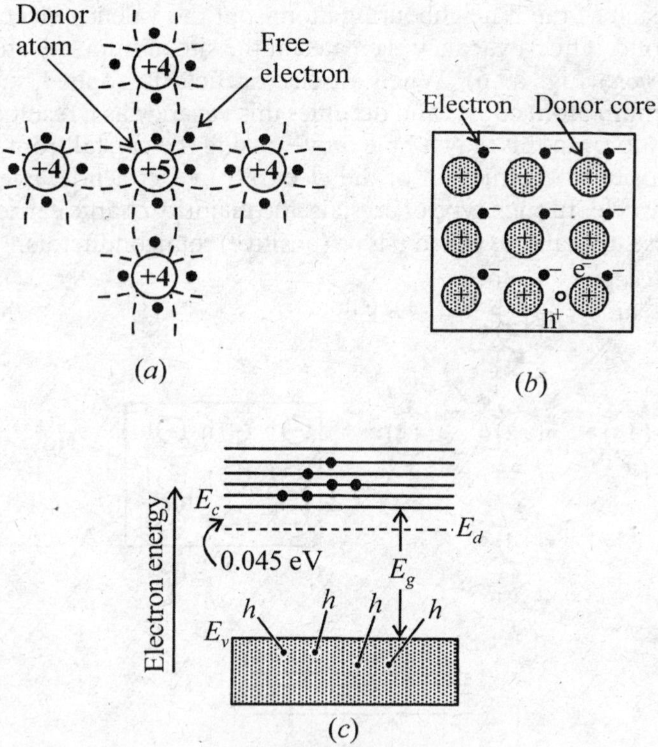

Fig. 29.5

In n-type semiconductor n_e is electron concentration in conduction band and n_h hole concentration in valence band and n_i intrinsic charge carrier concentration, then

$$n_e n_h = n_i^2 \quad \text{...(29.6)}$$

This is called *law of mass action.*

The donor level, i.e., the energy state of donated electrons is near conduction band and away from valence band. As such even at room temperature, the electrons go from donor level to conduction band and the conductivity increases.

In n-type semiconductor, the conductivity is given by

$$\sigma = e(n_e\mu_e + n_h\mu_h) \quad \text{...(29.7)}$$

Since $n_h << n_e$ and $n_e = N_d$ thus

$$\sigma = eN_e\mu_e \quad \text{...(29.8)}$$

here μ is mobility of electrons and μ_h mobility of holes.

Overall the crystal is electrically neutral.

p-type Semiconductors

When a trivalent impurity, like aluminium, indium, boron, etc. is doped in a pure tetravalent germanium (or silicon) the conductivity of crystal increases due to deficiency of electrons, i.e. holes. Such a crystal is said to be p-type semiconductor because of the majority charge carriers being positive charges, i.e. holes. The impurity atoms are called acceptors.

Each trivalent impurity atom (such as, Al, B, etc.) doped replaces one of the germanium (or silicon) atom. The three valence electrons of this impurity atom form co-valent bonds with one valence electron of each of the 3 neighbouring atoms but the valency electron of the 4th atom is unable to form the bond. Thus a vacancy is created at the site of impurity atom in the crystal. This vacancy is called a *hole* (Fig. 29.6). When an electric field is applied, a bound electron from neighbouring germanium atom comes and occupies this vacancy as a result a hole is created at the site of neighbouring atom. In this way a hole moves inside the crystal from one place to the other in the direction opposite to the motion of the electron. In fact a hole is equivalent to a virtual positively charged particle. In such type of crystals, the majority charge carriers are holes which are positive. As such these crystals are called *p*-type (positive) semiconductors.

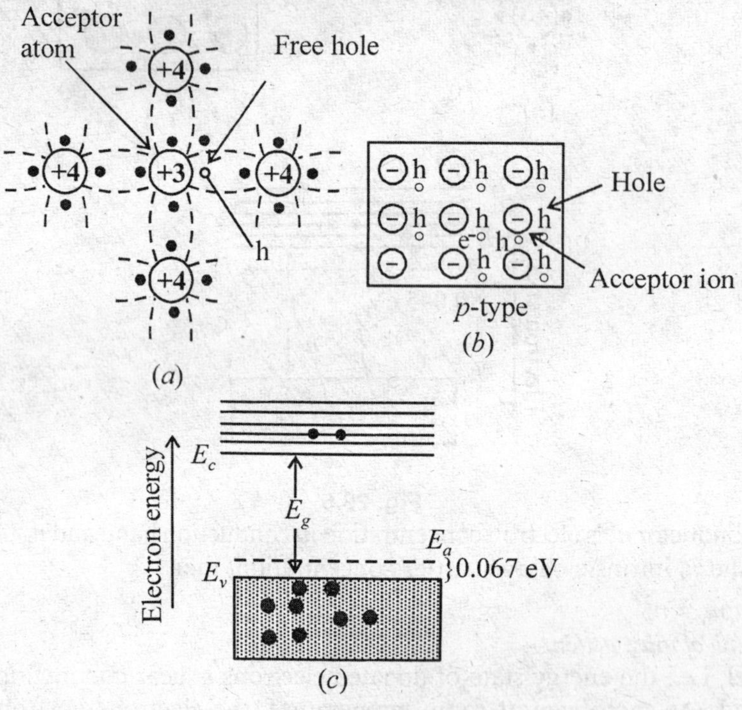

Fig. 29.6

In *p*-type semiconductor majority charge carriers are positive holes and there is equal number of stationary negative acceptor ions. Thus the crystal as a whole is electrically neutral.

If stationary negative acceptor ion concentration is N_a then hole concentration in valence band is

$$n_h = n'_h + n''_h \approx N_a \quad \text{...(29.9)}$$

n''_h is the hole concentration due to thermal agitation and $n'_h = N_a$

The law of mass action, *i.e.*

$$n_e n_h = n_i^2$$

also holds true.

The energy state of holes of acceptor atoms is near valence band and far off the conduction band. Because of this even at room temperature some of the electrons from valence band reach in this acceptor level creating holes in valence band. As a result of this, conductivity increases.

Conductivity in *p*-type semiconductors is given by

$$\sigma = eN_a\mu_h \quad \text{...(29.10)}$$

where μ_h is the mobility of holes. The mobility of holes in *p*-type semiconductors is less than the mobility of electrons in *n*-type semiconductors.

29.4 p–n JUNCTION (JUNCTION DIODE)

A semiconductor having *p*-type impurity at one end and *n*-type impurity at other end is known as *p-n* junction diode. The junction (boundary) when the *p*-type and *n*-type semiconductors combine is called *p-n* junction. To construct a *p-n* junction, one may diffuse a donor impurity to a pure semiconductor so that the entire sample becomes *n*-type. The acceptor impurity may then be diffused in higher concentration from one side to make that side *p*-type.

As the junction is formed, the concentration of majority charge carriers (holes on *p*-side and electrons on *n*-side) is different. As such the diffusion of these majority charge carriers takes place across the junction, *i.e.* the electrons from *n*-side will diffuse towards the *p*-side of junction while holes from *p*-side will go to the *n*-side. While crossing *p-n* boundary, these electrons and holes will recombine with each other because of their opposite charges. These electrons or holes have come from donor or acceptor impurity atom cores, as such donor or acceptor atoms will get depleted of their associated electrons or holes and subsequently will be left with a charged ion core in the layer near the junction boundary as shown in Fig. 29.7 (*a*). Thus a layer called depletion layer is formed at the junction such that on *n*-side near the junction, there is a layer of charged donor atom cores (effective +ve charge) while on the *p*-side there are acceptor charged atom cores with effective negative charge. The charge distribution is shown in Fig. 29.7 (*b*). In the regions far off from the junction, no net charges exists since cores still have their electrons or holes with them.

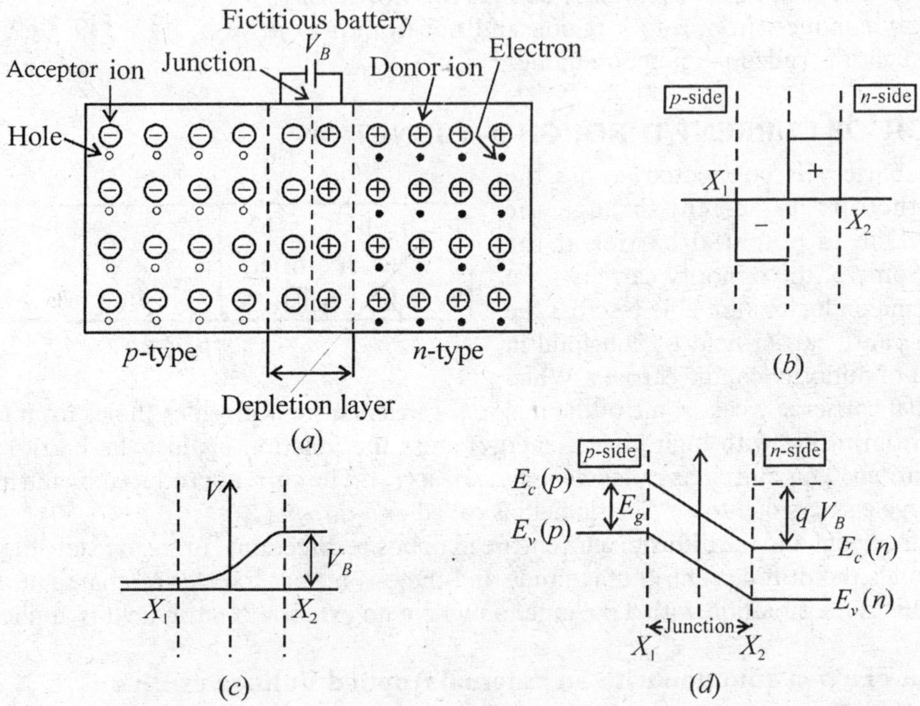

Fig. 29.7

The small region in the vicinity of the junction which is depleted of free charge carriers and has only immobile core ions is called *the depletion layer*. The accumulation of negative charges in the *p*-region and positive charges in the *n*-region sets up a potential difference across the junction, this acts as a barrier and is called potential barriers (V_B) and opposes the further diffusion of electrons and holes across the junction. The junction thus behaves as if a reverse battery with –ve terminal to *p*-region and +ve terminal to *n*-region is connected. The distribution of this barrier voltage as a function of the distance is shown in Fig. 29.7 (*c*) whereas Fig. 29.7 (*b*) shows the distribution of charge near the junction. This barrier voltage (V_B) is also called *Knee voltage*. This voltage barrier sets up a barrier field (*i.e.*, an electric field $\boldsymbol{E_B}$) in the direction from *n*-region to *p*-region.

The barrier voltage (V_B) depends on (*i*) the nature of the semiconductor (*ii*) temperature (*iii*) the amount of doping. The value of barrier potential is 0.7 V for silicon and 0.3 V for germanium. This field is very high of the order of 10^5 to 10^6 V/m because width of the depletion layer is of the order of 1 μm.

To cross this potential barrier in depletion layer an electron of *n*-region must be given an energy eV_B (which is the difference between the energies of electrons on *n*-side and *p*-side). This energy band diagram for a *p-n* junction is shown in Fig. 29.7 (*d*).

If doping concentration is small, the width of depletion layer is large and the barrier field is small because free path for holes and electrons is large. On the other hand if doping concentration are large, the depletion layer width is small and barrier field strength is large because free path for electrons and holes is small.

Symbolically *p-n* junction is represented as shown in Fig. 29.8. The direction of arrow is from *p* to *n* region. The arrow indicates the direction in which the conventional current can flow easily. Since it only conducts from *p* to *n* region and not from *n* to *p* region as such it is called *p-n* junction diode.

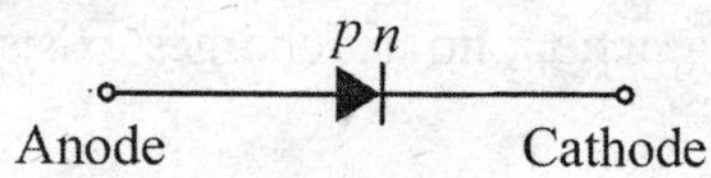

Fig. 29.8

29.5 FLOW OF CURRENT THROUGH *p–n* JUNCTION

When no battery is connected across the junction there is no current through the junction. Due to potential barrier at the junction. Some of the minority carriers from *p* and *n* semiconductors are able to cross the junction because barrier field E_B is helpful in movement of minority charge carriers. When the potential barrier is weak, some of the majority carriers from both sides (holes from *p*-side and electrons from *n*-side with high kinetic energy) cross the junction against the barrier field and produce current. This current is called *diffusion current* (I_{df}). The current produced by the movement of the charge carriers due to an electric field is called *drift current* (I_{dr}).

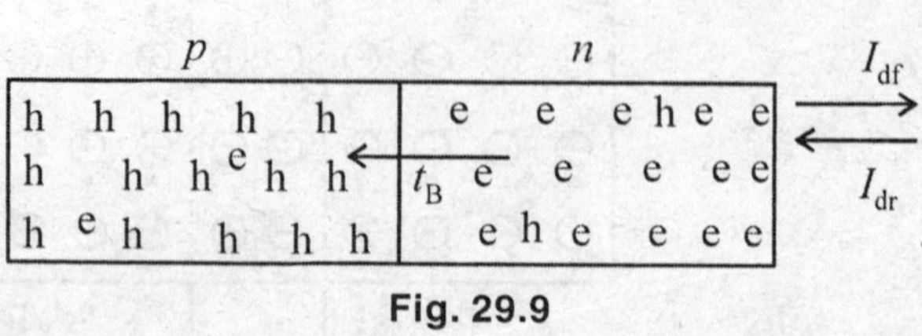

Fig. 29.9

The drift current and the diffusion current are in opposite directions. In steady state the diffusion current equals the drift current in magnitude and there is no net transfer of charge at any cross-section. This is the situation with a *p-n* junction where no external electric field is applied.

Behaviour of a *p–n* Junction with an External Applied Voltage or Bias

A *p-n* junction can be connected in two ways by an external battery making it (*a*) forward bias (*b*) reverse bias.

Forward Bias

If the positive terminal of the battery is connected to the *p*-side and the negative terminal to the *n*-side, the *p-n* junction is said to be forward biased [Fig. 29.10 (*a*)]. In this situation an external electric field E is established across the junction from p to n region. This field E is stronger than the barrier field E_B in the depletion layer. Thus holes (+ve) from p region and electrons n-region start moving towards junction. Holes move in the direction of the field and electrons in the opposite direction of the field. These electrons and holes recombine on reaching near the junction and disappear. Corresponding to each electron-hole combination, a covalent bond ruptures near the positive terminal of battery in *p*-region. This creates an electron and hole pair. The hole moves towards the junction and the electron enters the positive terminal of the battery via connecting wire. At this moment an electron from negative terminal of the battery is freed and enters n region of the junction and replaces the electron disappeared earlier due to combination with hole in the junction. In this way due to motion of majority charge carriers an electric current is established in junction diode. This current is called forward current and is due to the diffusion of the charge carriers (I_{df}).

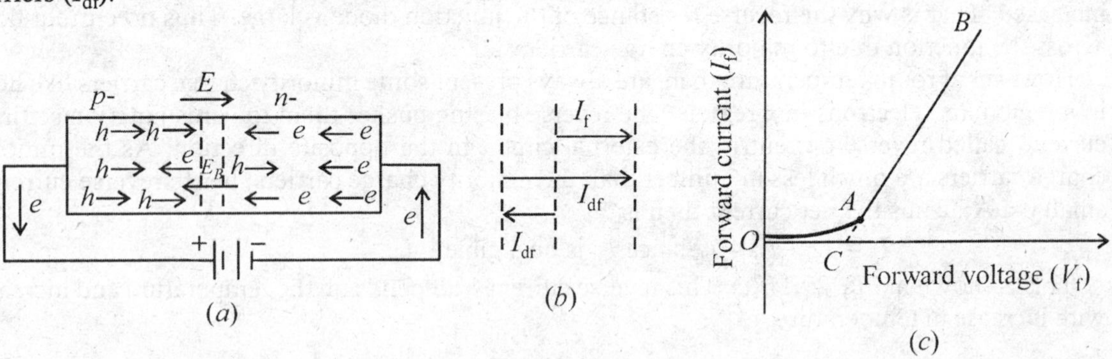

Fig. 29.10

In addition to this current a reverse current due to motion of the minority carriers is also established. This is called drift current (I_{dr}). This current is practically negligible. Thus net current $I_F = I_{df} - I_{dr}$. This is shown in Fig. 29.10 (*b*). Thus the current inside the crystal is due to diffusion of majority charge carriers and outside is due to drift of the electrons only.

In forward biased junction the external applied field E is stronger than the barrier field E_B, as such majority charge carriers (holes in *p*-region and electrons in *n*-region) are attracted towards the junction. This results in decrease in the width of the depletion layer. This is why the forward resistance of junction diode is small. The dynamic resistance of *p-n* junction diode is defined as

$$R = \frac{\Delta V}{\Delta i}$$

where ΔV denotes a small change in the applied potential difference and Δi denotes the corresponding small change in the current. The dynamic resistance is a function of the operating potential difference. It is equal to the reciprocal of the slope of i–V characteristic shown in Fig. 29.10 (*c*).

The graph between the forward voltage applied across the junction and the current is shown in Fig. 29.10 (*c*). Initially due to potential barrier (for Ge it is 0.3 V and for Si it is 0.7 V) the current in the diode is practically zero. As the applied voltage is increased, the current increases slowly but non-linearly till the applied voltage becomes large in comparison to potential barrier. The non-linear part is shown by line OA in the graph. On further increasing the voltage, the current increases rapidly and almost linearly (in fact it increases exponentially). Under this voltage this junction diode acts like a normal conductor. This is shown by straight line curve *AB*. If line *AB* is extended backward, then this line intercepts the voltage axis at a point *C*, which gives the values of potential barrier or this is called *knee voltage.* This is the forward voltage beyond which the forward current increases rapidly. Net current through the junction, under external potential difference *V* is given by

$$I = I_0\left(e^{+qV/k_BT} - 1\right) \quad \text{...(29.11)}$$

where q is the charge on the carrier (i.e., charge on the electron or hole) k_B—Botzmann's constant and T, temperature and I_0 is maximum current in reverse bias.

Reverse Biasing

If the positive terminal of a battery is connected to the *n*-side and negative terminal to the *p*-side, then the *p-n* junction is said to be reversed biased. As shown in Fig. 29.11 (*a*), the applied field E and the barrier field E_B in this case are in the same direction. Now holes in *p*-region and electrons in *n*-region move away from the junction and we can say that the width of the depletion layer has increased. That is why the reverse resistance of the junction diode is large. Thus no current flows across the junction due to majority charge carriers.

However at room temperature there are always present some minority charge carriers like holes in *n*-region and electrons in *p*-region. The reverse biasing pushes them towards junction setting a current, called reverse current in the external circuit in the opposite direction. As the minority charge carriers are much less in number than the majority charge carriers, hence reverse current is small (~μA). Thus the net current then is

$$I_B = I_{dr} - I_{df} = I_{dr} \text{ since } I_{df} \text{ is negligible}$$

This is shown in Fig. 29.11 (*c*). This reverse current I_B depends on the temperature and increases with increase in temperature.

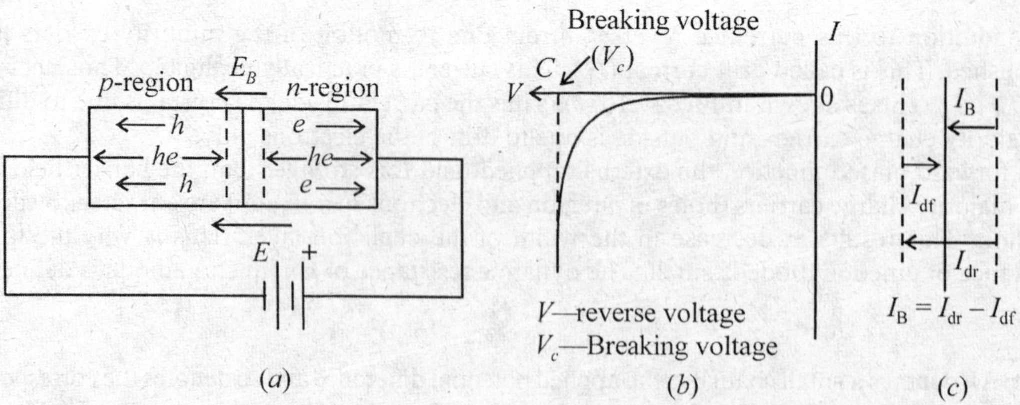

Fig. 29.11

The graph between the applied reverse voltage across the junction and obtained current is shown in Fig. 29.11 (*b*). Since the internal barrier field E_B is also helpful in movement of minority charge carriers as such all of them cross the junction. This is why the reverse current is nearly constant for a vide range of reverse applied voltage. This is shown by part *OC* of the curve in Fig. 29.11 (*b*). If reverse voltage is increased beyond *C*, the current becomes very large. The voltage is called *breakdown voltage* (V_c) and the phenomenon is called *breakdown of the diode.* At this voltage, the rate of creation of hole-electron pairs increases leading to the increased current.

There are two main processes by which breakdown may occur. The holes in the *n*-side and the conduction electrons in the *p*-side are accelerated due to reverse voltage. If these minority carriers acquire sufficient kinetic energy from the electric field and collide with a valance electron, the bond will be broken and valance electron will be taken to the conduction band. Thus a electron hole pair will be created. Breakdown occurring in this manner is called *avalanche breakdown.* Breakdown may also be produced by direct breaking of valance bonds to high electric field. When breakdown occurs in this manner it is called *zener breakdown.*

Zener Diode

A diode meant to operate in the breakdown region is called *an avalanche diode or zener diode* depending upon the mechanism of breakdown. Once the breakdown occurs, the potential difference across the diode does not increase even if potential of the battery applied is increased. Such diodes are used to obtain constant voltage output. A typical circuit which gives constant voltage across load resistance when zener diode is used is given in Fig. 29.12 (*a*) and (*i-V*) characteristic of zener diode including the breakdown voltage is shown in Fig. 29.12 (*b*). Even if there is a small change in the input voltage V_i, the current through R_L remains almost the same. The current through the diode changes but the voltage across it remains essentially the same. Symbol used for zener diode is shown in Fig. 29.12 (*a*).

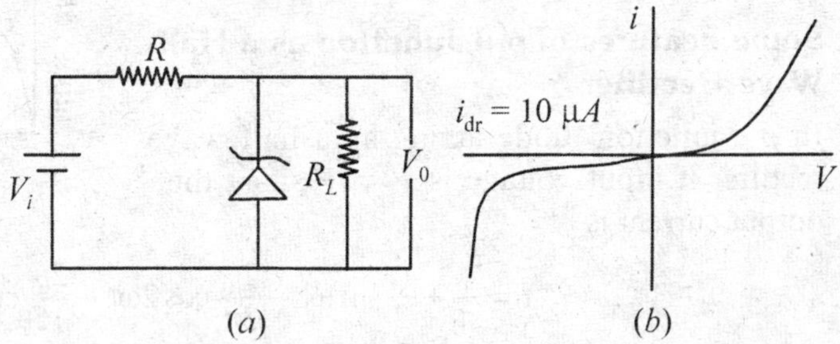

Fig. 29.12

29.6 *p–n* JUNCTION DIODE AS A HALF WAVE RECTIFIER

We have seen that a *p-n* junction diode conducts in forward bias and it does not conduct in reverse bias. In this way, it allows current to flow only in one direction, i.e., from *p*-side to *n*-side. As such if an alternating voltage is applied across the junction diode, the current through the junction flows only for positive half cycle of AC and during negative half cycle the current through junction does not pass. In this way the junction diode acts like a half wave rectifier.

The circuit diagram for *p-n* junction as a half wave rectifier is shown in Fig. 29.13 (*a*) and input-output voltage wave form is shown in Fig. 29.13 (*b*). The input-alternating voltage is connected to the primary (P_1P_2) of a transformer. One end S_1 of the secondary of the transformer is connected to *p*-side of the junction diode. Other end S_2 of the transformer secondary is connected to the *n*-side of the junction through load resistance R_L. DC output voltage is obtained across the load R_L.

Working

For alternating input voltage, when end S_1 of the secondary is positive (and S_2 negative) the junction diode is forward biased. Thus the junction diode conducts for this half cycle of the input voltage and the current through load resistance (R_L) flows as shown by arrow. For next half cycle of the input voltage end S_1 becomes negative (and S_2 positive). Now the junction is reverse biased and does not conduct and the current through load resistance (R_L) becomes zero. This process is repeated after every half cycle (time $T/2$, T is period) of the input voltage. Thus in half wave rectifier, the output current is *Unidirectional but intermittent and pulsating*. Figure 29.13 (*b*) shows the waveform of input and output voltage (or current). Since output current corresponds to only the half part of the input voltage as such it is called half wave rectifier.

The function of the transformer is to provide voltage to the rectifier. If *DC* is to be obtained at high voltage, then step up transformer is used and when DC is required at low voltage step down transformer is used.

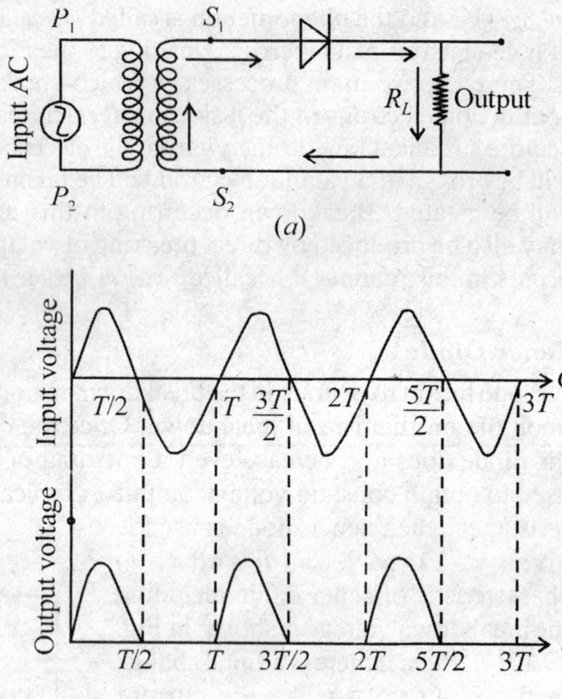

Fig. 29.13

Some Features of *p-n* Junction as a Half Wave Rectifier

In *p-n* junction diode acting as a half wave rectifier if input voltage is $V = V_0 \sin \omega t$ then output current is

$$i = \frac{i_0}{\pi} + \frac{i_0}{2} \sin \omega t - \frac{2i_0}{3\pi} \cos 2\omega t - \frac{2i_0}{15} \cos 4\omega t$$

This result shows that the output is superposition of DC part $\frac{i_0}{\pi}$ and AC part whose fundamental has frequency same as that of the input voltage (= ω). Thus $i_{\text{rms}} = \frac{i_0}{2}$ and $i_{\text{DC}} = \frac{i_0}{\pi}$, where i_0 is the peak value of the current.

$$\text{Ripple factor} \quad \gamma = \sqrt{\left(\frac{I_{\text{rms}}}{i_{\text{DC}}}\right)^2 - 1} = \sqrt{\frac{\pi^2}{4} - 1} = 1.21$$

$$\text{Form factor } F = \frac{i_{\text{rms}}}{i_{\text{DC}}} = \frac{i_0 \pi}{2 i_0} = \frac{\pi}{2}$$

$$\text{Efficiency } \eta = \frac{\text{Power output}}{\text{Power input}} = \frac{i_{\text{DC}}^2 R_L}{i_{\text{rms}}^2 (R_L + R_P)}$$

where R_P is resistance of secondary of transformer and *p-n* junction when it conducts. R_L is load resistance

$$\therefore \quad \eta = \frac{4/\pi^2}{(1 + R_P/R_L)} = \frac{0.405}{1 + R_P/R_L} = \frac{40.5}{1 + R_P/R_L}\%$$

Maximum power output is obtained when $R_P = R_L$ as such the maximum efficiency of half wave rectifier is 20.25%. Peak inverse voltage (PIV) is the potential difference across the junction when it does not conduct. For half wave rectifier PIV is equal to V_0.

29.7 *p-n* JUNCTION DIODE AS A FULL WAVE RECTIFIER

In full wave rectifier output current is obtained during both half cycles (negative as well as positive) of input voltage. In this two diodes are used in such a way that during positive half cycle of input voltage one diode is forward biased and conducts. During the negative half cycle of input voltage second diode becomes forward biased and conducts. During the half cycle one diode conducts the other is reverse biased and does not conduct, i.e., at one time only one diode conducts. Figure 29.14 (*a*) shows the circuit diagram for full wave rectifier and Fig. 29.14 (*b*) gives the input and output wave shape.

The input AC voltage is connected to the primary of P_1, P_2 of the transformer. The two terminals of the secondary S_1, S_2 are connected to *p*-sides of the two diodes (1) and (2). The *n*-sides of both the diodes are connected together. The load is connected to the central tapping *T* of the transformer and *n*-sides of the two diodes.

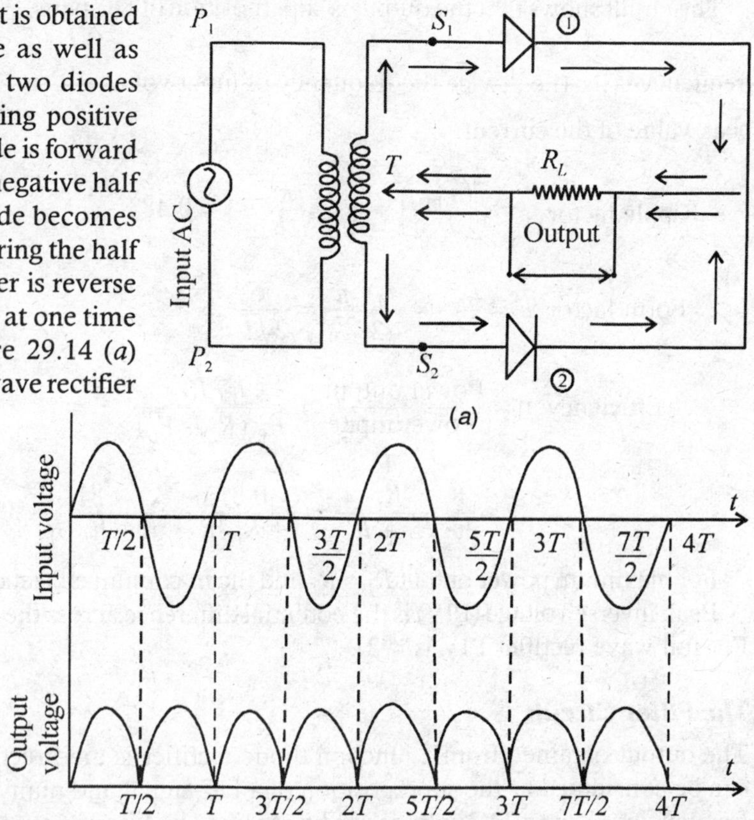

(*b*)

Fig. 29.14

Working

When S_1 terminal of the secondary is positive with respect to central tapping *T*, i.e. S_2 is negative, the diode (1) is forward biased and conducts whereas diode (2) is reverse biased and does not conduct. During this time the current through load and diode (1) flows as shown by arrows. During next half cycle of the input S_2 becomes positive and S_1 negative as such diode (2) becomes forward biased and conducts. The diode (1) becomes reverse biased and does not conduct. The current through load flows as shown in the diagram by arrow. Thus alternately the diodes (1) and (2) conduct and the current through load, during both the cycles of input voltage flows in one and the same direction. Thus in full wave rectifier output current is in the form of *continuous series of unidirectional pulses.* Figure 29.14 (*b*) shows the output current waveform. This shows that corresponding to one cycle of input voltage we have two pulses of output current. This output current is made practically steady DC by using smoothing filters.

Some Features of *p-n* Junction as a Full Wave Rectifier

In *p-n* junction diode acting as a full wave rectifier if input voltage is $V = V_0 \sin\omega t$ then output current is

$$i = \frac{2i_0}{\pi} - \frac{4i_0}{3\pi}\cos 2\omega t - \frac{4i_0}{15\pi}\cos 4\omega t$$

This result shows that the output is superposition of DC part $\frac{2i_0}{\pi}$ and AC part, whose fundamental frequency is 2ω (i.e., twice the frequency of input voltage). $i_{rms} = \frac{i_0}{\sqrt{2}}$ and $i_{DC} = \frac{2i_0}{\pi}$ where i_0 is the peak value of the current.

$$\text{Ripple factor } \gamma = \sqrt{\left(\frac{i_{rms}}{i_{DC}}\right)^2 - 1} = \sqrt{\frac{\pi^2}{8} - 1} = 0.483$$

$$\text{Form factor } F = \frac{i_{rms}}{i_{DC}} = \frac{i_0}{\sqrt{2}} \frac{\pi}{2i_0} = \frac{\pi}{2\sqrt{2}}$$

$$\text{Efficiency } \eta = \frac{\text{Power output}}{\text{Power input}} = \frac{i_{DC}^2 R_L}{i_{rms}^2 (R_L + R_p)}$$

$$= \frac{8}{\pi^2} \frac{R_L}{R_L + R_P} = \frac{0.81}{1 + R_P/R_L} = \frac{81}{1 + R_P/R_L} \%$$

For maximum power output $R_P = R_L$ and the maximum efficiency of the full wave rectifier is 40.5%.

Peak inverse voltage (PIV) is the potential difference across the junction when it does not conduct. For full wave rectifier PIV is = $2V_0$.

The Filter Circuit

The output obtained from a junction diode rectifier is unidirectional but pulsating. Such a signal can be considered as the superposition of a DC signal and many AC signals of different harmonic frequencies. A pure DC voltage can be obtained by filtering out the AC components. Two simplest filter circuits are described here.

(*i*) **Shunt Capacitor Filter:** Figure 29.15 shows a shunt capacitor filter circuit. In this case, a capacitor of appropriate capacity is connected in parallel with load resistance R_L of the rectifier. As already studied a capacitor offers an infinite reactance to DC and a low reactance to high frequency AC ($X_C = 1/\omega t$). Thus the AC component is bipassed through capacitor and a smooth DC appears at the load resistance.

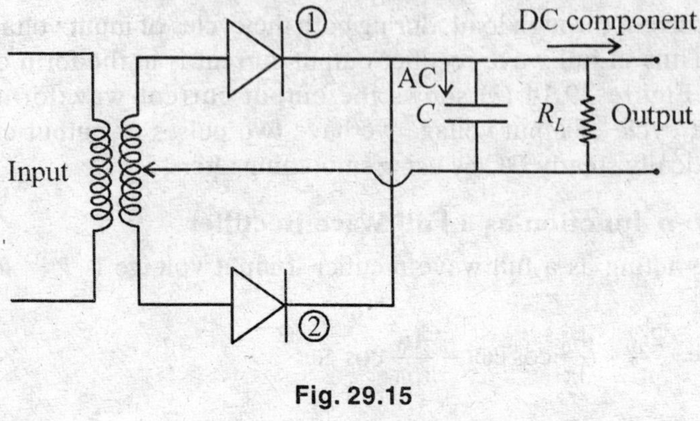

Fig. 29.15

(*ii*) **Series Inductor Filter:** Figure 29.16 shows a circuit of a full wave rectifier with an inductor of inductance L connected in series with its load resistance. An inductor L offers reactance ωL to the flow of current through it. It offers a high reactance to flow of high frequency AC and allows low frequency DC component to pass through it. Thus a smooth DC voltage appears across the load resistance.

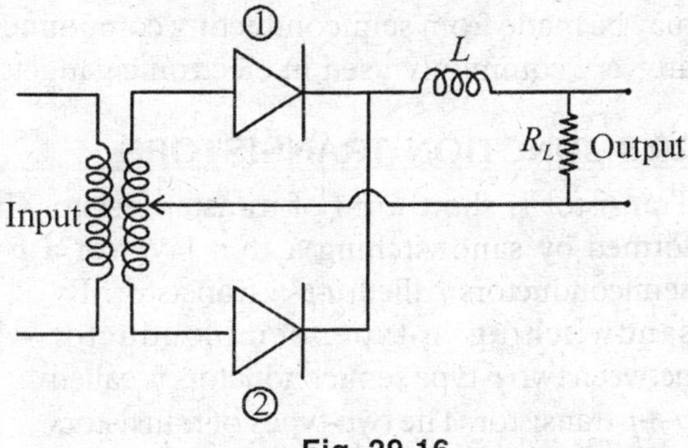

Fig. 29.16

Full Wave Bridge Rectifier

In full wave bridge rectifier, 4, *p-n* junction diodes are connected as shown in Fig. 29.17. The directions of output current through load resistance R_L during positive half cycle and negative half cycle of the input AC voltage is shown respectively in Fig. 29.17 (*a*) and (*b*). This is full wave rectifier and all its parameters are similar to a full wave rectifier.

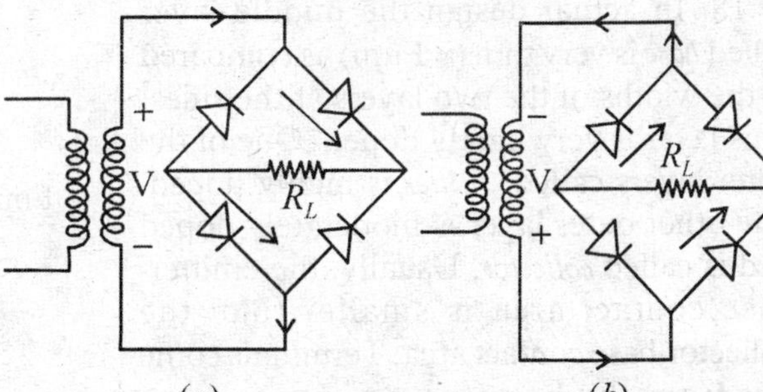

(*a*) (*b*)

Fig. 29.17

Advantages of a Bridge Rectifier

1. It does not require centrally tapped transformer.
2. For a given input voltage, the bridge rectifier has twice power out than the output power of full wave diode rectifier.
3. Peak inverse voltage in this case is half the peak inverse voltage when centrally tapped transformer is used.

Photo Diodes

A photo diode is a *p-n* junction made from a photosensitive semiconducting material in such a way that light can fall on its junction. Its function is controlled by the light allowed to fall on it.

Suppose the wavelength (λ) of light falling on junction is such that the energy of its photon $\left(= \frac{hc}{\lambda}\right)$ is sufficient to break a valance bond. When such light falls on the junction, new electron-hole pair is created. The number of charge carriers increases and hence the conductivity of the junction increases. If the junction is connected in the circuit, the current in the circuit is controlled by the intensity of light.

Light Emitting Diode (LED)

It is a forward biased *p-n* junction which spontaneously converts the biasing electrical energy into light energy in infrared or visible region.

In this a conduction electron makes a transition to valance band to fill up a hole in a *p-n* junction. The excess energy is emitted as a photon. If the wavelength of this photon is in visible region (380 nm–780 nm), one can see the emitted light. Such a *p-n* junction is known as *light emitting diode* (*LED*). For silicon or germanium, this wavelength falls in the infrared region. LED's may be made from semiconducting compounds like gallium, arsenide or indium phosphide. LED's are very commonly used in electronic gadgets as indicator lights.

29.8 JUNCTION TRANSISTORS

Transistor is short form of transfer resistor. It is a three element device. A junction transistor is formed by sandwitching a thin layer of a *p*-type semiconductor between two layers of *n*-type semiconductors called *n-p-n* transistor. By sandwitching *n*-type semiconductor between two *p*-type semiconductors is called *p-n-p* transistor. The two types of transistors with their circuit symbol is shown in Fig. 29.18. In actual design the middle layer called *base* is very thin (≈ 1 μm) as compared to the widths of the two layers at the sides. This layer is very lightly doped. One of the outer layers called *emitter* is highly doped. The other outer layer is moderately doped and is called *collector*. Usually, the emitter-base contact area is smaller than the collector-base contact area. Terminals come out from the base the emitter and the collector for the external connections. Thus a transistor is three element and three terminal device.

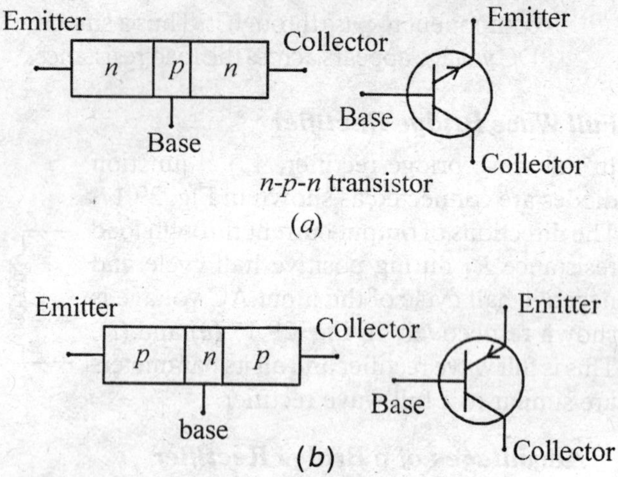

Fig. 29.18

In transistor there are two junctions emitter—base junction and base collector junction. For normal operation of transistor the emitter base junction is always made forward biased and base collector junction is always made reverse biased. The arrow on the emitter line shows the direction of the current through the emitter-base junction.

To understand the action of a transistor let us see the behaviour of depletion layer formed at the emitter-base *p-n* junction when forward biased and base-collector junction is reversed biased. We will see how the charge carriers move across these junctions when proper voltages are applied at its terminals.

Figure 29.19 shows the two depletion formed at the emitter-base junction and the collector-base junction. Since the emitter base junction is forward biased (as well as due to heavy doping of emitter), this depletion layer will be narrow while the collector-base junction being reverse biased will be relatively wider. Forward bias V_{eb} is small (~0.5 to 1 V) whereas the reverse bias voltage V_{cb} is considerably high (~5 to 15 V).

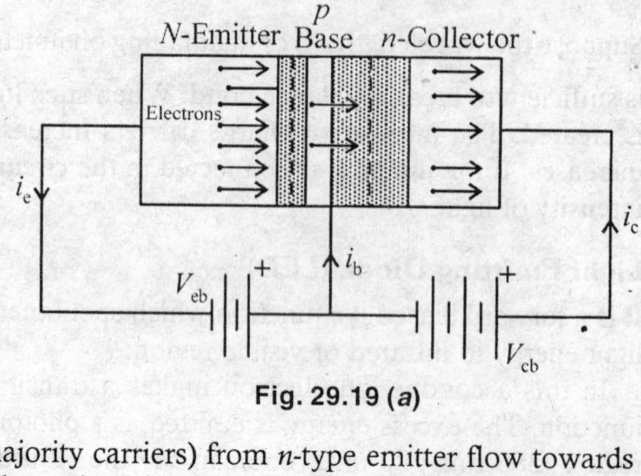

Fig. 29.19 (*a*)

Consider the biased *n-p-n* transistor (Fig. 29.19 (*a*)). Since the emitter base junction is forward biased, a large number of electrons (majority carriers) from *n*-type emitter flow towards the base. These constitute the current through the emitter i_e. This current is from base to emitter.

These electrons have tendency to combine with the holes in p-region of the base. Since base is lightly doped as such a small number of electrons are able to combine with the holes. This gives a small base current (i_b). The base current is very small due to the fact that the base is lightly doped and very thin. Most of the electrons coming from the emitter are able to diffuse through the base region to the collector region. In the collector region, since reverse field is applied hence electrons easily reach the collector terminal and constitute collector current (i_c). It is apparent that

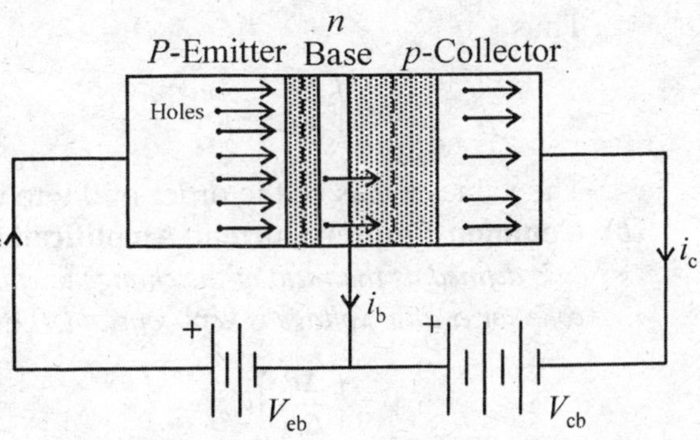

Fig. 29.19 (*b*)

$$i_e = i_c + i_b \text{ with } i_c >> i_b \qquad \ldots(29.12)$$

Similar discussion can be made for a biased p-n-p transistor (Fig. 29.19 (*b*)). Here holes instead of electrons are the majority carriers from p-type emitter region which cross the thin n-base region and are able to reach the collector to give collector current i_c. In this case also

$$i_e = i_c + i_b \text{ with } i_c >> i_b$$

Biasing: Three Configurations of a Transistor

A proper potential difference should be applied across the two junctions to operate the transistor. This is called biasing the transistor. A transistor can be operated in three different modes (configurations). They are

(*i*) Common base (CB) configuration (or ground base mode)
(*ii*) Common emitter (CE) configuration (or grounded emitter mode)
(*iii*) Common collector (CC) configuration (or grounded collector mode)

In Fig. 29.20 the three types of circuit arrangements are shown for an n-p-n transistor. In each case the emitter base junction is forward biased while the collector-base junction is reverse biased.

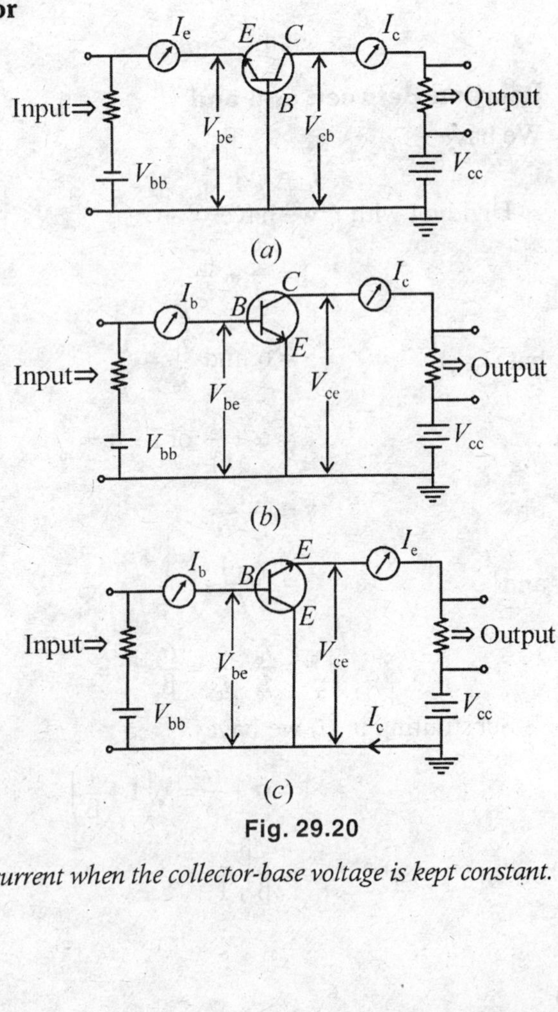

Fig. 29.20

Transistor Constants and their Relation

There are three current gains for a transistor, α, β and γ. They are called the constants of a transistor.

(*a*) **Common Base Current Amplification Factor (α)**

It is defined as the ratio of the change in the collector to the corresponding change in emitter current when the collector-base voltage is kept constant.

Thus

$$\alpha = \left(\frac{\Delta i_c}{\Delta l_e}\right) V_{cb} \qquad \ldots(29.13)$$

The value of α is of the order of 0.9 to 0.99.

(b) Common Emitter Current Amplification Factor (β)

It is defined as the ratio of the change in collector current to the change in the base current when the collector emitter voltage is kept constant. Thus

$$\beta = \left(\frac{\Delta i_c}{\Delta i_b}\right) V_{ec} \qquad \ldots(29.14)$$

The value of β >> 1. This is why the amplification in common emitter amplifier is more than the amplification in common base amplifier.

(c) Common Collector Amplification Factor (γ)

It is defined as the ratio of the change in emitter current to the corresponding change in base current, when emitter-collector voltage is kept constant. Thus

$$\gamma = \left(\frac{\Delta i_c}{\Delta i_b}\right) V_{ec} \qquad \ldots(29.15)$$

$$\gamma \approx \beta >> \alpha$$

Relation Between α, β and γ

We have

$$i_e = i_c + i_b$$

Dividing with i_e we have

$$1 = \frac{i_c}{i_e} + \frac{i_b}{i_c} \qquad \ldots(i)$$

but $$\frac{i_c}{i_e} = \alpha \text{ and } \frac{i_b}{i_e} = \frac{1}{\gamma}$$

∴ $$1 = \alpha + \frac{1}{\gamma} \text{ or } 1 - \alpha = \frac{1}{\gamma}$$

or $$\gamma = \frac{1}{1-\alpha} \qquad \ldots(29.16)$$

and $$\alpha = \frac{\gamma - 1}{\gamma} \qquad \ldots(29.16a)$$

$$\frac{i_b}{i_e} = \frac{i_b}{i_c} \cdot \frac{i_c}{i_e} = \frac{\alpha}{\beta} = \gamma \qquad \ldots(29.17)$$

Substituting in (i) we have

$$1 = \alpha + \frac{\alpha}{\beta} = \alpha\left(1 + \frac{1}{\beta}\right)$$

or $$\alpha = \frac{\beta}{\beta + 1} \qquad \ldots(29.18)$$

and $$\beta = \frac{\alpha}{1-\alpha} \qquad \ldots(29.19)$$

29.9 TRANSISTOR CHARACTERISTICS

Common Emitter Characteristics

Let us consider an *n-p-n* transistor in common emitter configuration. A sample circuit for drawing the input and output characteristics of an *n-p-n* transistor is shown in Fig. 29.21. We can view this circuit as made up of an input section and an output section. The input section contains the base-emitter junction and the voltage source there. Whereas the output section contains the base collector junction and the voltage source there.

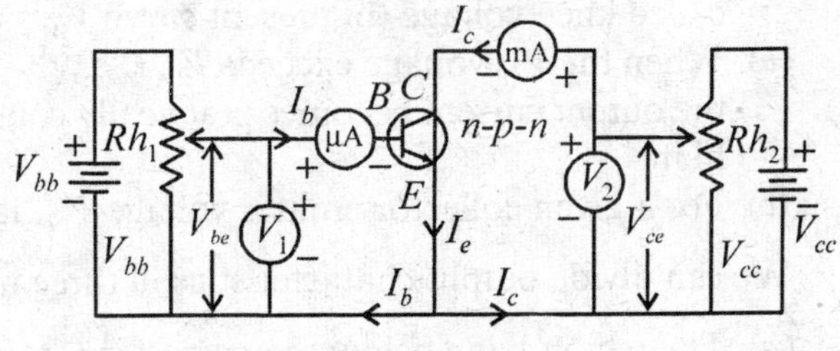

Fig. 29.21

To draw the input characteristics, a curve is plotted between the base current (i_b) and the base emitter voltage (V_{be}). Keeping emitter collector voltage (V_{ce}) constant. The current (i_b) may then be called the input current and the current (i_c) the output current. Figure 29.22 (*a*) shows the input characteristics.

The study of these curves reveals that

(*a*) As long as V_{be} is less than the barrier voltage the base current i_b is small.

(*b*) When the base emitter voltage V_{be} exceeds the barrier voltage, the base current (i_b) increases sharply. These curves resemble the forward bias characteristics of a *p-n* junction diode.

The value of i_b is much smaller than that in a normal diode.

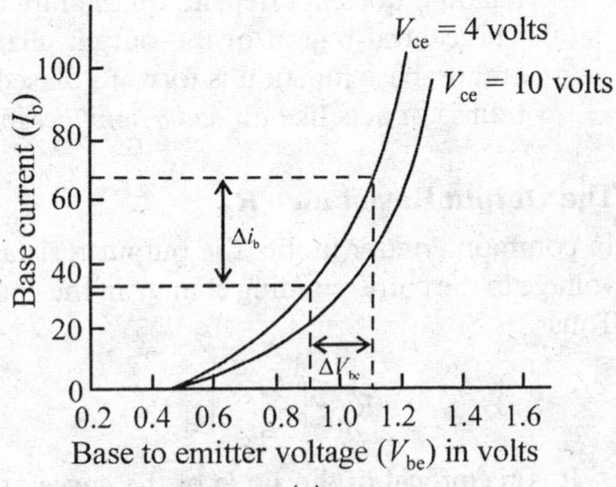

(*a*)

Input Resistance R_i

Input resistance (R_i) of a common emitter mode is defined as the ratio of the change in base emitter voltage to the corresponding change in the base current, when the collector-emitter voltage is kept constant. Thus

$$R_i = \left(\frac{\Delta V_{eb}}{\Delta i_b}\right)_{V_{ce}} \qquad \ldots(29.20)$$

Input characteristics is non-linear as such R_i varies. R_i is at any point of the curve, R_i is reciprocal of the slope of the tangent to the curve.

A graphical relation between the collector current (i_c) and collector-emitter voltage (V_{ce}) at

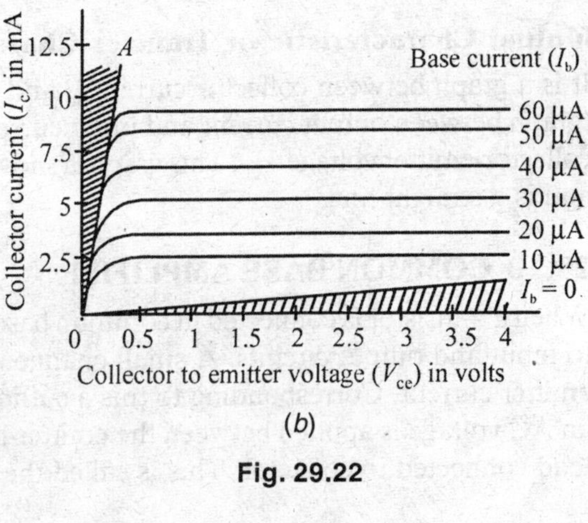

(*b*)

Fig. 29.22

constant base current i_b is called output characteristics. Figure 29.22 (*b*) shows these curves. These curves have the following features.

(*a*) When the collector-emitter voltage (V_{ce}) increases from 0 to V_0 (say), the collector current i_c increases rapidly. The value of collector emitter voltage up to which i_c increases rapidly is called knee voltage (in present curve $V_0 = 0.5$ V).

(*b*) When the V_{ce} voltage exceeds V_{be} (i.e., the collector base junction becomes reverse biased), the output current becomes practically constant. The output resistance in this region is very high.

(*c*) For a given collector emitter voltage V_{ce}, larger the value of i_b larger is the value of i_e.

We can divide output characteristics in three main regions.

(*a*) The region lying below the curve for $i_b = 0$.
This is called *cut-off region*. In this region both the junctions are reverse biased as such $i_c = 0$. In this region (shown shaded) the transistor acts as a *switch. It turns over rapidly from OFF state for which $i_c = 0$ (cut off) to ON-state for which i_c is maximum* (*saturation state*).

(*b*) Region left of line *OA*. This region is called *saturation region* and the line *OA* is called *saturation line.* In this region $V_{ce} < V_{be}$ both the junctions are forward biased. As such in this region i_c does not depend on the input current i_b.

(*c*) The central region of the output characteristics is called *active region.* In this region the emitter-base junction is forward biased and the collector-base junction is reverse biased. A transistor acts like an *audio amplifier* in this region.

The Output Resistance R_0

In common emitter mode, the output resistance R_0 is defined as the change in collector-emitter voltage to the corresponding change in the collector current when the base current is kept constant. Thus

$$R_0 = \left(\frac{\Delta V_{ce}}{\Delta i_c}\right)_{i_b} \quad \text{...(29.21)}$$

It is reciprocal of the slope of the curve and in general is very large.

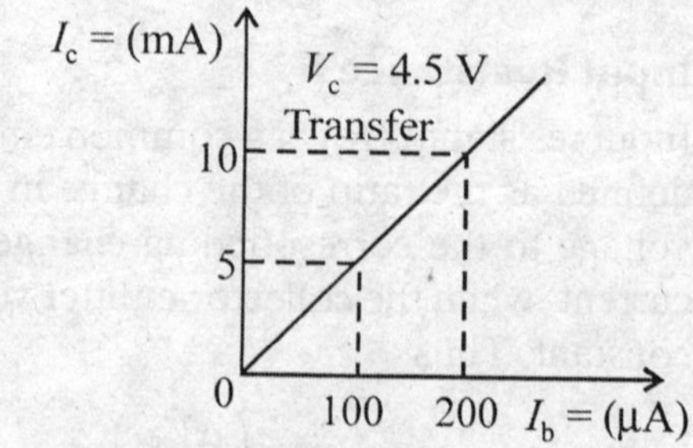

Fig. 29.23

Mutual Characteristic or Transfer Characteristics

It is a graph between collector current i_c and base current i_b (graph between output current and input current) at constant collector-emitter voltage V_{cc}. This graph is shown in Fig. 29.23 and is a straight line.

29.10 COMMON BASE AMPLIFIER

When a transistor is connected in common base configuration, the base of transistor is then common to input and output circuits. A small change in emitter base voltage causes a large change in the emitter current. Corresponding to this a sufficient change in collector current is obtained. Thus if an AC voltage is applied between the emitter-base, then a high alternate voltage is obtained across load connected to collector. This is called the amplifying action of transistor.

A common base amplifier circuit for *n-p-n* transistor is shown in Fig. 29.24. The emitter-base input circuit is made forward biased by using low voltage battery (V_{EB}) as such its resistance becomes small (≈ 100 Ω). The output collector-base circuit is made reverse biased using high voltage battery (V_{cc}). The output circuit has a high resistance (≈ 10000 Ω). A load resistance R_L is connected in base collector circuit across which the output voltage is obtained. An alternating low voltage signal is applied to emitter-base input circuit and an amplified output alternating voltage signal is obtained across the load. A capacitor *C* is connected in the output circuit to obtain pure AC across load.

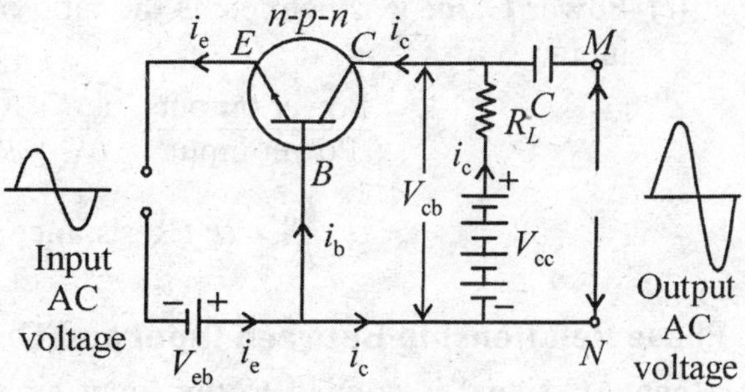

Fig. 29.24

If i_e, i_b and i_c are the emitter current, base current and collector current respectively, then we have

$$i_e = i_b + i_c \quad \ldots(i)$$

From diagram using Kirchhoff's law, we have

$$V_{eb} = V_{cc} - i_c R_L \quad \ldots(ii)$$

where V_{eb} is the collector to base voltage and V_{cc} is the battery voltage in output circuit. This voltage remains constant.

When an *AC* input signal is applied across emitter-base junction, the biasing across emitter-base (V_{EB}) changes. This causes change in emitter current i_e and collector current i_c. Thus base-collector voltage (V_{eb}) changes. This voltage appears as amplified output AC voltage.

Current, Voltage and Power Gains of Common Base Amplifier

(*a*) **Current Gain:** It is the ratio of the change in collector current Δi_c to the corresponding change in emitter current Δi_e when collector base voltage is kept constant.

$$\text{AC current gain } \alpha = \left(\frac{\Delta i_c}{\Delta i_e}\right) V_{cb} \quad \ldots(29.22)$$

This is called AC current gain.

The ratio of the instantaneous collector current i_c and the corresponding instantaneous emitter i_b at constant base-collector voltage is called DC current gain

$$\text{DC current gain } \alpha = \left(\frac{i_c}{i_b}\right) V_{cb} \quad (29.22a)$$

(*b*) **Voltage Gain:** It is the ratio of the change in output voltage to the change in input voltage

$$A_v = \frac{\Delta V_{CB}}{\Delta V_{EB}} \quad \ldots(29.23)$$

$\Delta V_{CB} = R_0 . \Delta i_c$, R_0 is output resistance

and $\Delta V_{EB} = R_i \Delta i_e$, R_i is input resistance

$$A_v = \frac{R_0 \Delta i_c}{R_i \Delta i_e} = \alpha \frac{R_0}{R_i} = \text{Current gain} \times \text{Resistance gain}$$

(c) **Power Gain:** It is defined as the ratio of change in power output to the change in power input.

$$A_P = \frac{\text{Power output}}{\text{Power input}} = \frac{(\Delta i_c)^2 R_0}{(\Delta i_e)^2 R_i}$$

$$= \alpha^2 \frac{R_0}{R_i} = \alpha^2 \text{(Resistance gain)} \quad ...(29.24)$$

Phase Relationship Between Input and Output Signals

When AC signal is applied to the input circuit, the bias between emitter and base changes. Corresponding to the positive half cycle of the signal bias between emitter and base decreases. This decreases the emitter current (i_e) and hence the collector current (i_c). The potential drop across the load resistance ($i_c R_L$) decreases (equation *ii*). This increases the collector-base voltage V_{cb}. As the collector is connected to the positive terminal of the battery V_{ec}, the increase in collector voltage means that the collector becomes more positive. Thus as the input AC signal goes through its positive half cycle, the output signal also goes through the positive half cycle. Similarly corresponding to negative half cycle the output signal also goes through the negative half cycle. Hence in common base amplifier the input and output voltages are in same phase.

Note: A discuss of similar type can be made for *p-n-p* transistor. The results are same.

The common base amplifier is used to amplify high frequency signals and to match low source impedance to high output impedance.

29.11 COMMON-EMITTER AMPLIFIER

The circuit diagram for *n-p-n* transistor as a common emitter amplifier is shown in Fig. 29.25. Emitter is common to both input and output circuits. The emitter base junction is forward biased by the battery V_{be} and the collector is reverse biased by battery V_{cc}. Due to this the input resistance (R_i) become small and output resistance (R_o) becomes large. A load resistance R_L is also connected in emitter-collector circuit as shown in Fig. 29.25. A low AC input signal is applied to the forward biased base-emitter junction.

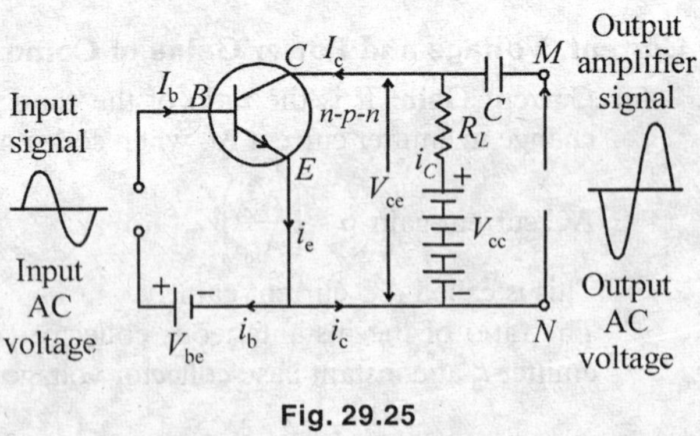

Fig. 29.25

An amplified output alternating signal is obtained between the collector and ground. A capacitor *C* is connected in collector-emitter circuit to obtain a pure AC output signal.

When signal is applied, let i_e, i_c and i_b be the emitter current, collector current and base current respectively then

$$i_e = i_c + i_b \quad ...(i)$$

and the collector to emitter voltage V_{ce} is

$$V_{ce} = V_{cc} - i_c R_L \quad ...(ii)$$

When AC signal is applied between emitter-base junction, the base-emitter voltage varies. Thus changing the collector-emitter voltage as per equation (*ii*). On applying signal in the input circuit, we get amplified variations in the output voltage across collector-emitter.

Current, Voltage and Power Gains of a Common Emitter Amplifier

(*a*) **Current Gain:** AC current gain is defined as the ratio of the change in the collector current (Δi_c) to the change in base current (Δi_b) keeping the collector-emitter voltage (V_{ce}) constant. This is denoted by β.

$$\therefore \qquad \beta = \left(\frac{\Delta i_c}{\Delta i_b}\right)V_{ce} \qquad \text{...(29.25)}$$

DC current gain is defined as the ratio of collector current to the corresponding base current, keeping collector-emitter voltage constant.

$$\beta = \left(\frac{i_c}{i_b}\right)V_{ce} \qquad \text{...(29.25}a\text{)}$$

In linear region of the characteristics the β AC and β DC are usually equal.

(*b*) **Voltage Gain:** Voltage gain is defined as the change in output voltage (ΔV_{ce}) to the change in input voltage (ΔV_{be}). Thus

$$A_v = \frac{\Delta V_{ce}}{\Delta V_{be}} \qquad \text{...(29.26)}$$

But $\qquad \Delta V_{be} = \Delta i_b R_i$ and $\Delta V_{ce} = \Delta i_c R_o$

where R_i and R_o are the input resistance and output resistance respectively.

$$\therefore \text{Voltage gain } A_v = \frac{\Delta i_c R_o}{\Delta i_b R_i} = \beta \cdot \frac{R_o}{R_i}$$

$$\text{or } A_v = \beta \cdot \frac{R_o}{R_i} \text{ current gain} \times \text{resistance gain} \qquad \text{...(29.27)}$$

(*c*) **Power Gain:** It is the ratio of change in output power to the corresponding change in the input power

$$\text{Power gain } A_P = \frac{\text{Change in power output}}{\text{Change in power input}}$$

$$= \frac{R_o(\Delta i_c)^2}{R_i(\Delta i_b)^2} = \beta^2 \cdot \frac{R_o}{R_i} \qquad \text{...(29.28)}$$

$$A_P = \beta^2 \times \text{resistance gain} \qquad \text{...(29.28}a\text{)}$$

Since β >> α, the power gain and voltage gain of a common emitter amplifier is much large than that in common base amplifier.

Transconductance or Transfer Conductance

To have a large amplification, a small change in the collector current i_c is necessary. This property is measured by a quantity called transconductance g_m. Thus

$$g_m = \frac{\Delta i_c}{\Delta V_{be}} \qquad \text{...(29.29)}$$

It is also known as transfer conductance.

Phase Relationship Between Input Signal and Output Signal

When an AC signal is applied to the input circuit, corresponding to positive half cycle of the signal the emitter-base junction bias increases (it becomes more positive). This increases the emitter current. This in turn increases the collector current. This increase in collector current increases the potential difference across load (i.e. increases $i_e R_L$). This makes collector-emitter voltage less positive, i.e. more negative. Thus as the input signal goes through the positive half cycle, the output amplified signal goes through its negative half cycle. Similarly as the input signal goes through its negative half cycle, the amplified output signal goes through its positive half cycle. Hence in common emitter amplifier, the input and output voltages are 180° out of phase or are in opposite phase.

Comparative Study of CB, CE and CC Amplifiers

The Table 29.1 shows the comparative study of the CB, CE and CC amplifiers.

Table 29.1

S.No.	*Parameter*	*CB Amplifier*	*CE Amplifier*	*CC Amplifier*
1.	Input resistance R_i	Minimum $R_i \sim (50\text{–}200)\ \Omega$	R_i greater than CB but less than CC, $R_i \sim (1\text{–}2)\ k\Omega$	Maximum $R_i \sim (150 \sim 800)\ k\Omega$
2.	Output resistance R_o	Ouput resistance is maximum $R_o \sim 450\ k\Omega$	R_o is greater than CC but less than CB, $R_o \sim 50\ k\Omega$	R_o is minimum $R_o \sim 10$ to $100\ \Omega$
3.	Current gain α, β, γ	α is minimum $\alpha < 1$ $\alpha \sim 0.9$ to $.99$	$\beta > \alpha$ $\beta < \gamma$ $\beta \sim 10 - 200$	Gain is maximum $\gamma = \beta + 1 = \frac{1}{1-\alpha}$ $\gamma \sim 10\text{–}200$
4.	Voltage gain A_v	Voltage gain is greater than that in CC but less than CE $A_v \sim 150$	Voltage gain is maximum $A_v \sim 500$	Voltage gain is minimum $A_v < 1$ $A_v \sim <1$
5.	Power gain A_P	Greater than CC but less than CE $A_P \sim 20\text{–}30$	Power gain is maximum $A_P \sim 30\text{–}40$	Power gain is minimum $A_P \sim 10$
6.	Used in which amplifier	Used at high frequency for current amplification	Used at audio frequency as power amplifier	Used as impedance matching device
7.	Phase ϕ	Output and Input are in same phase $\theta = 0$	Input and output are in opposite phase $\theta = \pi$ or $180°$	Zero

29.12 TRANSISTOR AS AN OSCILLATOR

An oscillator is an electronic device which produces electric oscillations of constant frequency and amplitude, without requiring any external input signal. It converts DC energy into an AC energy in some oscillatory circuits.

Figure 29.26 shows a block diagram of an oscillatory circuit. The basic parts of this circuit are (*a*) an amplifier and (*b*) an LC network.

The amplifier section is just a transistor used in common-emitter configuration. The LC network consists of an inductor and a capacitance. This network resonates at a frequency

$$f_0 = \frac{1}{2\pi\sqrt{LC}} \quad \text{...(29.30)}$$

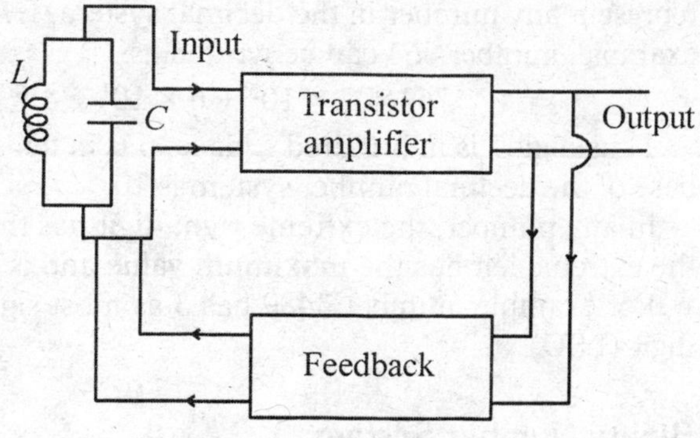

Fig. 29.26

DC batteries are used to bias the transistor and no external input signal is fed to the amplifying section. A part of the output signal is fed back to the input section after going through the LC network. This signal is amplified by the transistor and a part is again fed back to its input section. Thus, it is a self sustained device. The component with the proper frequency f_0 gets reasonably amplified and the output acts as a source of alternating voltage of frequency f_0. The frequency can be varied by varying L or C.

We must note following points for transistor as an oscillator:

(*a*) Oscillator converts DC into AC.
(*b*) Oscillator is a positive feedback amplifier of infinite gain.
(*c*) Barrhansen criterion for oscillation is $\beta A \geq 1$, where β is feedback ratio and A is amplifier gain.
(*d*) An inductance L and condenser C are connected in parallel. This circuit is called a *tank circuit*.
(*e*) Once condenser is charged and allowed to discharge through inductance, the given electrical energy oscillates between capacitance and inductance with frequency

$$f_0 = \frac{1}{2\pi\sqrt{LC}}$$

(*f*) The circuit diagram of *n–p–n* transistor in common emitter configuration is shown in Fig. 29.27.

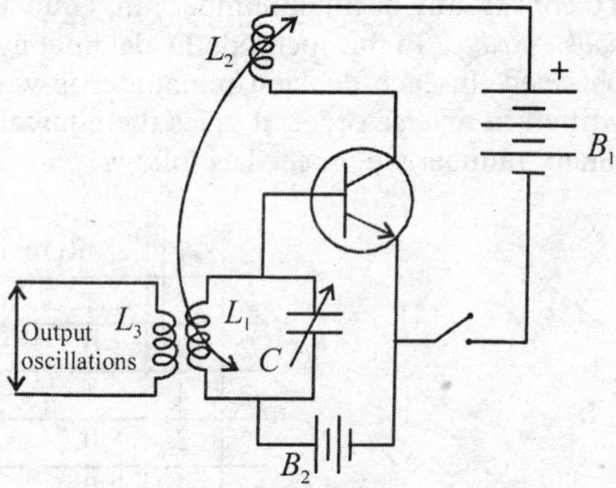

Fig. 29.27

29.13 NUMBER SYSTEMS

A number of number systems are used in mathematics. Some of the main systems are: (*i*) Decimal number system, (*ii*) Binary number system, (*iii*) Hexadecimal number system (*iv*) Octal number system.

Out of all these, here we will discuss only decimal number system and binary number system. It is because in ordinary mathematics, we use decimal number system and the digital electronic devices are based on binary number system.

Decimal Number System

We are familiar with the decimal number system and use this system in day today work. It has ten discrete states corresponding to the ten digits 0, 1, 2, 3, 4, 5, 6, 7, 8, and 9. These are used to represent any number in the decimal system. Every digit in a given number has a place value. For example, number 764 can be written as

$$764 = 7 \times 10^2 + 6 \times 10^1 + 4 \times 10^0 = 700 + 60 + 4$$

Thus digit 7 is in hundred's place, 6 is in ten's place and 4 is in the unit's place. Obviously, the base of the decimal number system is 10.

In any number, the extreme right digit has the least value and is called *least significant digit* and the extreme left has the maximum value and is called *most significant digit*.

For example, number 3489 has 3 as most significant digit (MSD) and 9 is the least significant digit (LSD).

Binary Number System

The binary number system uses only two digits, 0 and 1. Each binary digit is called a *bit*. The base of binary number system is 2. It is a *positional weighted system*. All the decimal numbers in binary number system are the chains of 0 and 1 bits.

In binary number, the place value of each bit corresponds to a definite power of 2. Here also the left end number is most significant bit (MSB) and extreme right number is called least significant bit (LSB).

Conversion of Decimal Number to Binary Number

To convert any decimal number into equivalent binary number, the method used is called *double dabble method*. In this method, the decimal number is repeatedly divided with 2 till zero quotient is obtained. In each division remainder is written in a separate column. When this remainder is written in reverse order, it gives the equivalent binary number. For example, to convert 37 into binary number we proceed as follows:

		Remainder	
2	37		
2	18	1	LSB
2	9	0	
2	4	1	
2	2	0	
2	1	0	
	0	1	MSB

Remainder written in reverse order 100101

The equivalent binary number of decimal number 37 is

100101 *i.e.*, $(37)_{10} = (100101)_2$

In this way 64 in binary number system is 1000000 and 49 is 110001 *i.e.*, $(64)_{10} = (1000000)_2$ and $(49)_{10} = (110001)_2$.

This method of conversion of decimal number into equivalent binary number is applicable to whole numbers only.

Method for Fractions (Numbers Less than 1)

To convert a fraction from decimal number system to equivalent binary number system fraction, the decimal fraction is repeatedly multiplied by 2. Each time whole number (0 or 1) obtained in product is written separately and the fraction part is multiplied by 2. This is done till the fraction part of the product becomes zero. If the product does not become zero, then after few products the fraction is left out.

For example, Decimal fraction 0.125 can be converted into equivalent binary fraction as follows:

Decimal fraction	Binary No.
$0.125 \times 2 = 0.25$	0 MSB
$0.25 \times 2 = 0.5$	0
$0.5 \times 2 = 1.0$	1 LSB

Thus
$(0.125)_{10} = (0.001)_2$

Similarly equivalent binary fraction of decimal fraction.

$$(0.15)_{10} = (0.0010011)_2$$

and $$(0.75)_{10} = (0.110)_2$$

If decimal number contains whole number as well as a fraction (for example 150.75), then binary equivalent numbers of whole number and fraction are separately determined using above method and then equivalent binary number is computed. Here

$$(105)_{10} = (1101001)_2 \text{ and } (0.75)_{10} = (0.110)_2$$

Thus $$(105.75)_{10} = (1101001.110)_2$$

Conversion from Binary Number System to Decimal Number System

(*i*) **Method for Whole Numbers:** As the base in the binary number system is 2, therefore, we multiply least significant bit (LSB) by 2^0, the next bit by 2^1, next by 2^2 and so on till the most significant bit (MSB) is reached. On adding all these products of multiplication gives the equivalent decimal system number

For example to convert 10101, we write as

$$1 \times 2^0 + 0 \times 2^1 + 1 \times 2^2 + 0 \times 2^3 + 1 \times 2^4 = 1 + 0 + 4 + 0 + 16 = 21$$

$$\therefore \quad (10101)_2 = (21)_{10}$$

(ii) **Method to Convert a Binary Fraction into Equivalent Decimal Fraction:** Multiply the successive bits after decimal with 2^{-1}, 2^{-2}, 2^{-3} and so on till we reach to the last bit of the fraction. The addition of all these multiplications will give the fractional decimal number equivalent to decimal binary number.

For example to convert 0.101 into equivalent decimal fraction, we proceed as follows:

$$0.101 = 1 \times 2^{-1} + 0 \times 2^{-2} + 1 \times 2^{-3} = \frac{1}{2} + 0 + \frac{1}{8} = \frac{5}{8} = 0.625$$

Thus $(0.101)_2 = (0.625)_{10}$

If a binary number contains both whole number as well as fraction, then equivalent decimal number and decimal fraction are separately determined as per above method and then final equivalent decimal number is computed.

Note: Positive number in binary system is represented by writing zero to the left of MSB. Thus

$$+43 \rightarrow 0,\ 101011$$

Similarly, a negative binary is represented by writing 1 to the left of MSB. Thus

$$-51 \rightarrow 1,\ 110011$$

29.14 ANALOG AND DIGITAL CIRCUITS

Analog Signal

A continuous time varying current or voltage signal is called analog signal. A sinusoidal varying alternate voltage shown in Fig. 29.28 represents the simplest analog signal. *The electronic circuits which process analog signals are called analog circuits.* The electronic circuits such as amplifier, oscillator, etc. are grouped together as analog circuits.

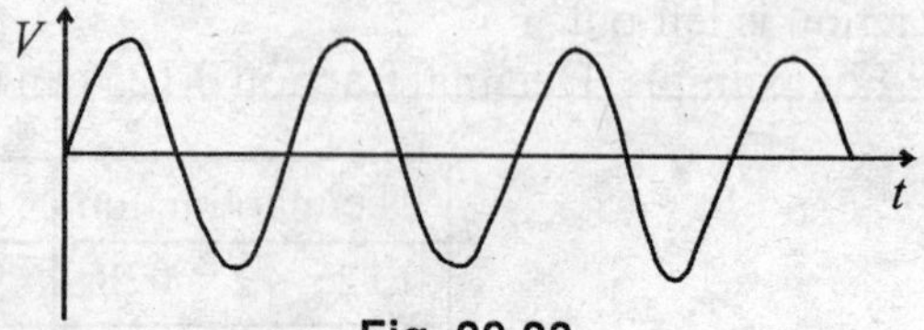

Fig. 29.28

Digital Signal

A pulse-wave form in which only discrete values of voltages are possible is shown in Fig. 29.29. *A signal in which current or voltage can take only two discrete values is called a digital signal.* The high level is termed as '1' while the low level as '0'. Thus digital signals are in the form of pulses of equal level. The electronic circuits which process digital signals are called *digital circuits*. The digital circuits make use of binary number system according to which high level of fixed voltage (say 5 volts) is taken as 1 and low level (zero voltage) is taken as zero. Thus in these signals only two values of input and of output voltages are possible.

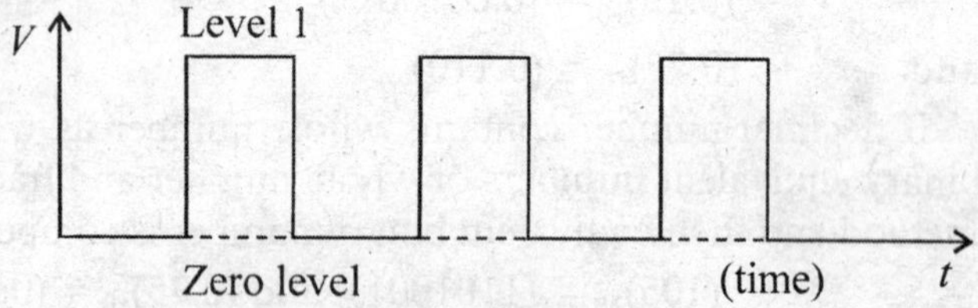

Fig. 29.29

The devices like pocket calculators, electronic watches, modern computers, robots, etc. are all digital circuits. Digital electronics is the study of the circuits which handle the digital data consisting of 0 and 1 level.

29.15 BOOLEAN ALGEBRA

In logic, we have two valued questions *i.e.*, questions whose answer is either yes or no. For example:

1. An electric circuit is open or closed.
2. A statement is true or false.
3. Will India win the match.
4. Action is write or wrong.

It was English mathematician George Boole who thought of relation between mathematics and logic and developed an algebra in 1854 for such two valued questions, which is now known as Boolean algebra. The logical statements are known as Boolean variables. A Boolean variable can have either a true value or a false value. The true value is represented by 1 and false value by 0 (binary numbers).

Just as in ordinary algebra, we have mathematical operators like addition, subtraction and multiplication, similarly in Boolean algebra, we have three basic operators like OR, AND and NOT operators.

The OR Operator

Figure 29.30 shows circuit diagram for operator OR of Boolean algebra. It consists of two parallel switches connected in series with a bulb and a battery. The bulb will not glow when both the switches are off but will glow if either or both the switches are on. This action of switch is called OR operation. Now let S_1 be called input A and S_2 input B and the state of the bulb as output Y. Let off position of switch or bulb be represented by 0 and on position as 1, then result of the OR operator can be written in table form as

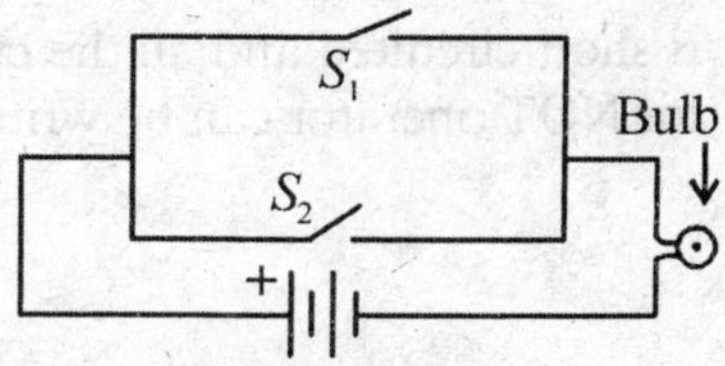

Fig. 29.30

Truth Table

Input A	Input B	Output Y
0	0	0
1	0	1
0	1	1
1	1	1

This table is called *truth table* of the OR operator. In Boolean algebra, it is written as

$$A + B = Y$$

and is read as A or B equals Y.

Obviously output of operation OR is 1 when either or both the two inputs is 1.

The AND Operator

Figure 29.31 shows circuit diagram for AND operator of Boolean algebra. It consists of two switches S_1 and S_2, battery and a bulb all in series. It is obvious that the bulb can glow only when both the switches S_1 and S_2 are closed. This operation can be tabulated in terms of input variables A and B and output Y as follows:

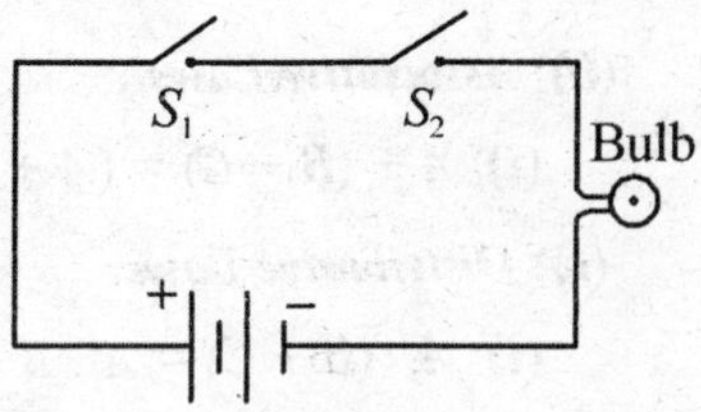

Fig. 29.31

Truth Table

Input A	Input B	Output Y
0	0	0
0	1	0
1	0	0
1	1	1

This table is called truth table of the AND operator. In Boolean algebra it is written as

$$A \cdot B = Y$$

and is read A and B equals Y.

Thus operator AND has output 1 only when both the inputs are 1.

The NOT Operator

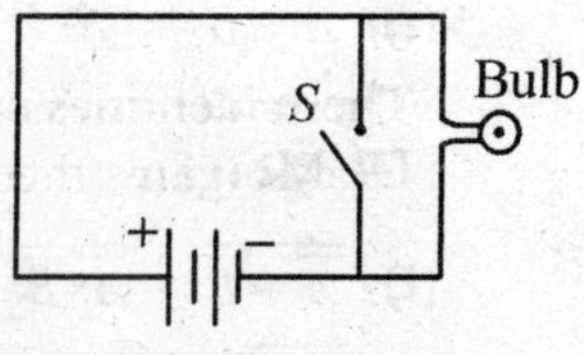

Fig. 29.32

Figure 29.32 shows the circuit diagram of the NOT operator of the Boolean algebra. In this case, bulb is connected to a battery and is short circuited by switch 'S'. When switch is open current flows through the bulb and it glows *i.e.*, output is 1. But when the switch is closed the bulb is short circuited and all the current passes through the switch and output is zero. The truth table for NOT operator can be written as

Truth Table

Input A	Output Y
0	1
1	0

In Boolean algebra NOT operator is written as

$$\overline{A} = Y$$

and is read as A inverse equals Y. In fact NOT operator is an *inverter*. It operates on one Boolean variable and gives its negation or complement as its output. Since it operates on only one input variable as such NOT operator is a unary operator. OR and AND operators are binary operators because they connect two input variables A and B to give a single output.

Some Useful Laws of Boolean Algebra

(a) *Commutative Laws*

(*i*) $A + B = B + A$ (*ii*) $A \cdot B = B \cdot A$.

(b) *Associative Laws*

(*i*) $A + (B + C) = (A + B) + C$ (*ii*) $A \cdot (B \cdot C) = (A \cdot B) \cdot C$.

(c) *Distributive Laws*

(*i*) $A \cdot (B + C) = A \cdot B + A \cdot C$ (*ii*) $(A + B) \cdot (A + C) = A + BC$

(d) *Absorption Laws*

(*i*) $A + A \cdot B = A$ (*ii*) $A \cdot (A + B) = A$ (*iii*) $\overline{A} \cdot (A + B) = \overline{A} \cdot B$

(e) *Double Complement Function*

$$\overline{\overline{A}} = A$$

(f) *Boolean Identities*

(*i*) $A \cdot (\overline{A} + B) = A \cdot B$. (*ii*) $A + \overline{A} \cdot B = A + B$

(*iii*) $A + B \cdot C = (A + B) \cdot (A + C)$

(*iv*) $(\overline{A} + B) \cdot (A + C) = \overline{A} \cdot C + A \cdot B$.

(g) *De-Morgan's Theorem*: It states that the complement of the whole sum is equal to the product of individual complements and *vice versa*, *i.e.*,

(*i*) $\overline{A + B} = \overline{A} \cdot \overline{B}$ (*ii*) $\overline{A \cdot B} = \overline{A} + \overline{B}$

These identities can be proved using the truth table for each operation.

De-Morgan's theorem can also be stated as follows:

(*i*) $\overline{\overline{A} + \overline{B}} = \overline{\overline{A}} \cdot \overline{\overline{B}} = A \cdot B$.

(*ii*) $\overline{\overline{A} \cdot \overline{B}} = \overline{\overline{A}} + \overline{\overline{B}} = A + B$.

Basic OR and AND Relations

OR	AND
(*i*) $A + 0 = A$	(*i*) $A \cdot 0 = 0$
(*ii*) $A + 1 = 1$	(*i*) $A \cdot 1 = 1$
(*iii*) $A + A = A$	(*i*) $A \cdot A = A$
(*iv*) $A + \overline{A} = 1$	(*i*) $A \cdot \overline{A} = 0.$

29.16 LOGIC GATES

A logic gate is an electronic device which may have one or more input but output is always one. It works according to some logical relationship between input and output voltage. Thus it is a digital circuit designed to perform a particular logical operation. In logic gate an output pulse is obtained only for a special combination of input pulses, otherwise there is no output pulse. In logic gate output signal is obtained only for a few input signal combination that means a logic gate may permit a signal to pass through it or may stop it.

There are three basic logic gates: (*i*) OR gate, (*ii*) AND gate, and (*iii*) NOT gate. Besides these, the other gates formed by their combination are (*i*) NAND gate, (*ii*) NOR gate, (*iii*) XOR (exclusive OR) gate, (*iv*) XNOR (exclusive NOR) gate, etc.

NAND gate and NOR gate are called universal gates. Each logic gate is represented by a symbol *called logical symbol* and its function is defined either by truth table or by a Boolean operator (expression).

Truth table of a logic gate is a table which shows the output, corresponding to each possible combination of input.

Negative and Positive Logic

In digital electronics, there are only, two states high and low, OFF and ON, closed and open, etc. The binary numbers 1 and 0 are represented by these two states. In a system in which high state represents 1 and low state zero, is called positive logic. On the other hand, a system in which high state represent 0 and low state 1, is called negative logic.

OR Gate

An OR gate can have two or more than two inputs but output is only one. It gives output (1) if either input A or B are high (1). When both input A and B are zero, the output Y is zero. Figure 29.33 (*a*) shows the circuit diagram for OR gate using two diodes. Figure 29.33 (*b*) shows the logic symbol. This circuit evaluates Y, A OR B, i.e.,

$$Y = A + B$$

read as Y equals A or B. This is Boolean expression for OR gate (same as for OR operator). If $A = B = 0$ both diodes are connected to earth. As such both the diodes do not conduct and output is zero. (*i.e.*, $Y = 0$)

If $A = 1$ and $B = 0$, the diode D_1 is forward biased (since the potential at A is

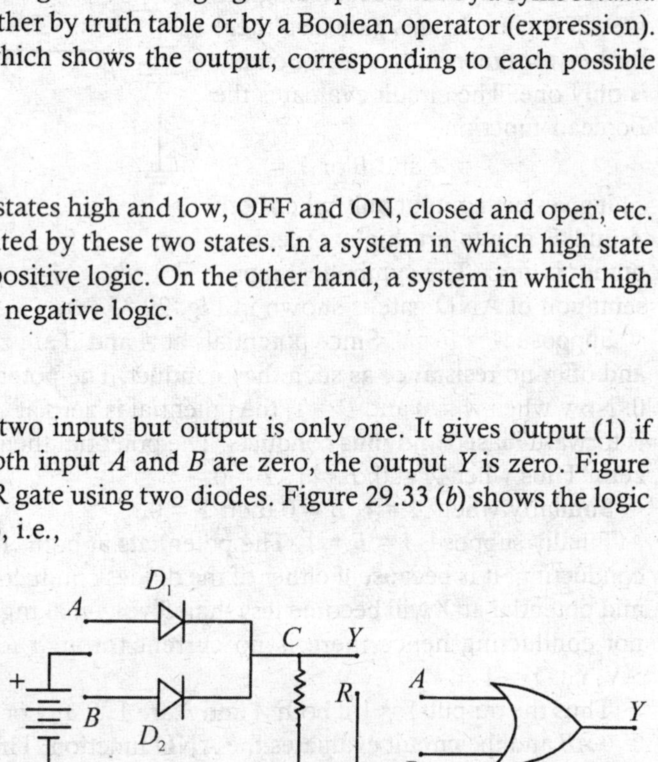

(a) (b)

Fig. 29.33

high say 5 V) and offers no resistance. Thus potential at Y is equal to the potential at A (*i.e.*, 5 V), thus $Y = 1$. Similarly, if $A = 0$ and $B = 1$, $Y = 1$ also. If both A and B are 1, both diodes are forward biased and potential at Y is the same as the common potential at A and B, *i.e.*, 5 V. This also gives $Y = 1$. Hence the circuit evaluates $Y = A$ or $B = A + B$.

The truth table of this gate is same as that of OR operator already discussed.

A typical input and output wave form for an OR gate is shown in Fig. 29.34. The figure shows that for time

$$t < t_1 \; A = 0, B = 0 \text{ and } Y = 0$$
$$t_1 < t < t_2 \; A = 1 \; B = 0 \text{ and } Y = 1$$
$$t_2 < t < t_3 \; A = 1 \; B = 1 \text{ and } Y = 1$$
$$t_3 < t < t_4 \; A = 0 \; B = 1 \text{ and } Y = 1$$
$$t_4 < t < t_5 \; A = 0 \; B = 0 \text{ and } Y = 0$$

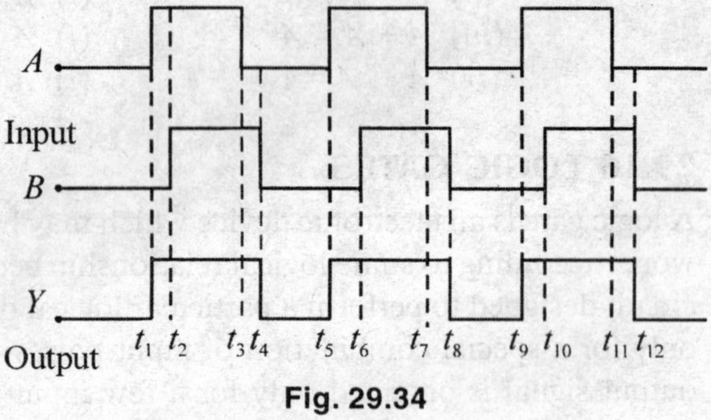

Fig. 29.34

In this way we can get output for entire input wave.

The AND Gate

Figure 29.35 (*a*) shows circuit diagram for AND gate. In this gate also there can be two or more inputs but the output is only one. The circuit evaluates the Boolean function

$$Y = A \text{ and } B \text{ or } Y = AB$$

It gives high output only when both A and B inputs are high, otherwise output is zero. The symbolical representation of AND gate is shown in Fig. 29.35 (*a*).

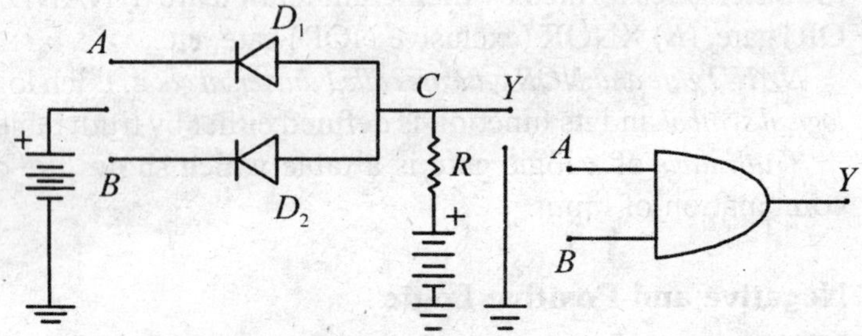

Fig. 29.35

Suppose $A = B = 0$. Since potentials at A and B are zero, so both the diodes are forward biased and offer no resistance as such they conduct. The potential at Y is same as that at A or B and $Y = 0$. Now when $A = 0$ and $B = 1$, the potential is zero at A and that at B is 5 V. So that the diode D_1 is forward biased and thus conducts. The potential then at Y is equal to the potential at A which is zero. Thus when $A = 0$, $B = 1$, $Y = 0$.

Similarly when $A = 1$, $B = 0$ then $Y = 0$.

Finally suppose $A = B = 1$. The potentials at both A and B are 5 V. So neither of the diodes is conducting. It is because if either of the diodes conducts, a current will pass through the resistor R and potential at Y will become less than 5 volts making the diode reverse biased. Since diodes are not conducting hence there is no current through R and the potential at Y will be equal to 5 V, i.e. $Y = 1$.

Thus the output $Y = 1$ if both A and B are 1. If any of the inputs is zero, the output $Y = 0$. Hence $Y = AB$ and the circuit evaluates the AND function. Thus the truth table for an AND gate is same as that of AND operator in Boolean algebra. Figure 29.36 gives a typical input output wave form of an AND gate. The figure show that for time

$$t < t_1 \quad A = 0 \; B = 0 \text{ and } Y = 0$$

$t_1 < t < t_2$ $A = 1$ $B = 0$ and $Y = 0$

$t_2 < t < t_3$ $A = 1$ $B = 1$ and $Y = 1$

$t_3 < t < t_4$ $A = 0$ $B = 1$ and $Y = 0$

$t_4 < t < t_5$ $A = 0$ $B = 0$ and $Y = 0$

$t_5 < t < t_6$ $A = 1$ $B = 0$ and $Y = 0$

$t_6 < t < t_7$ $A = 1$ $B = 1$ and $Y = 1$

In this way a whole wave pattern for output can be obtained corresponding to given two inputs. Here also the truth table is same as for Boolean AND operator.

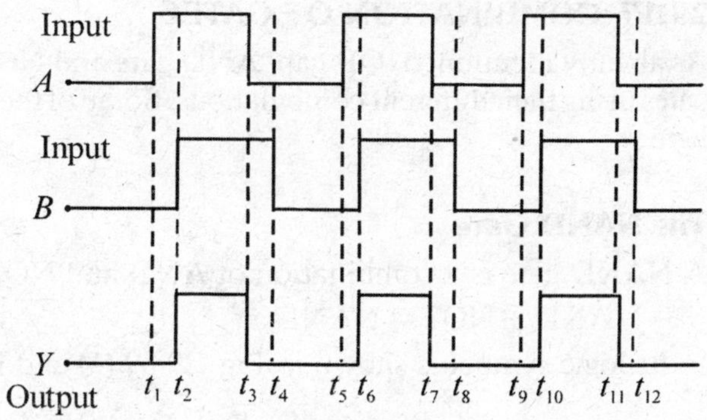

Fig. 29.36

The NOT Gate

Not gate is the simplest gate with one input and one output. Whatever the input may be, it inverts it *i.e.*, it gives a high output $Y (= 1)$ if the input A (= 0) and *vice versa.* A NOT gate cannot be constructed with diodes. Figure 29.37 (*a*) shows a circuit using an *n-p-n* transistor to evaluate the NOT function. Figure 29.37 (*b*) shows the logic symbol of NOT gate.

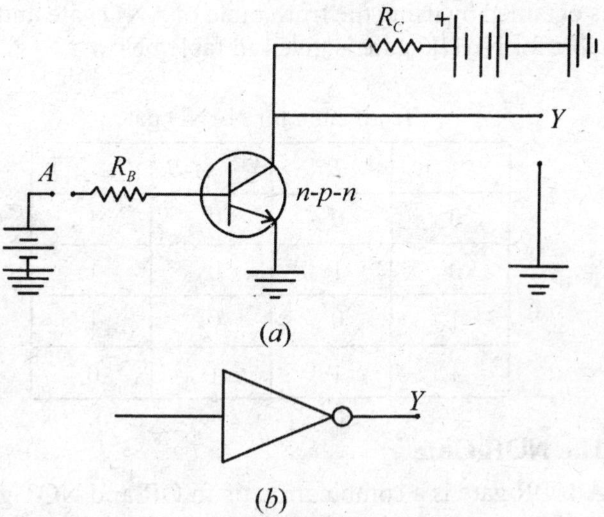

Fig. 29.37

If $A = 0$ the emitter base junction is unbiased and there is no current through it. Correspondingly there is no current through the resistance R_C. The potential at Y is equal to the potential at the positive terminal of the battery, *i.e.*, 5 V. Thus if $A = 0$ $Y = 1$. We observe that the collector base junction is also reverse biased which is consistent with the fact that there is no current in the circuit.

When the potential at A is 5 V the base-emitter junction is forward-biased and there is a large current in the circuit. The direction of the current in the resistance R_C is from right to left. The potential drops across R and its value at Y becomes zero. Thus if $A = 1$ $Y = 0$. The NOT gate is described by the Boolean expression as $\overline{A} = Y$.

An input output wave form for a not gate is shown in Figure 29.38. As we have seen that output of a NOT gate is high when input is low and *vice versa.* Thus for time

$t_1 < t < t_2$ $A = 1$ $Y = 0$

$t_2 < t < t_3$ $A = 0$ $Y = 1$

$t_3 < t < t_4$ $A = 0$ $Y = 0$

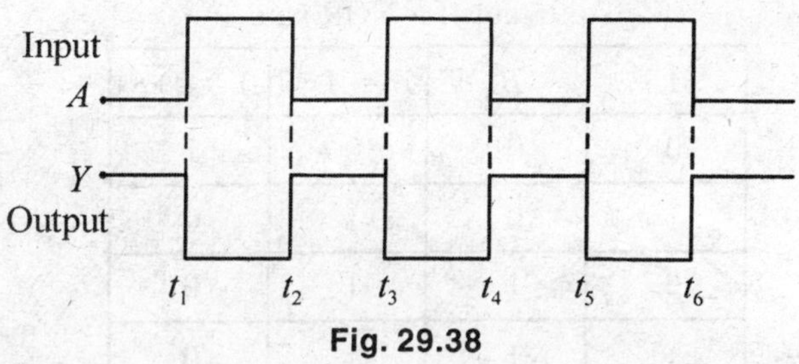

Fig. 29.38

In this way a whole wave pattern for output corresponding to a given input can be obtained. The truth table is same as that for Boolean operator NOT.

29.17 COMBINATION OF GATES

As already mentioned, OR gate AND gate and NOT gate are the basic gates. We can obtain other gates using their different combinations. Some of the gates formed by their combination are discussed here.

The NAND Gate

A NAND gate is a combination of AND and NOT gate (Fig. 29.39 (*a*)) *i.e.*,

AND + NOT = NAND

Its logic symbol is shown in Fig. 29.39 (*b*) and Boolean expression is

$$\overline{A \cdot B} = Y = \overline{A} + \overline{B}$$

It is obtained by connecting the output of an AND gate to the input of a NOT gate. Its truth table is obtained by using the truth table of AND gate and then finding the negation of its output. Truth table for NAND gate is given in table below.

Truth table for NAND gate

A	B	$Y' = A \cdot B$	$Y = \overline{A.B}$
0	0	0	1
0	1	0	1
1	0	0	1
1	1	1	0

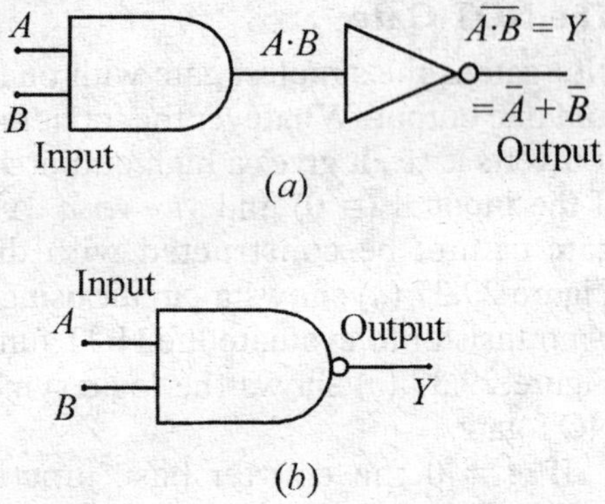

Fig. 29.39

The NOR Gate

A NOR gate is a combination of an OR and NOT gate. It is obtained by connecting the output of an OR gate to the input of a NOT gate as shown in Fig. 29.40 (*a*) *i.e.*,

OR + NOT = NOR

Boolean expression is

$$\overline{A + B} = Y = \overline{A} \cdot \overline{B}$$

Its logical symbol is in shown Fig. 29.40 (*b*). Its truth table is obtained by using the truth table of OR gate and then finding the negation of its output. Truth table for NOR gate is given below.

Truth table for NOR gate

A	B	$Y' = A+B$	$Y = \overline{A+B}$
0	0	0	1
1	0	1	0
0	1	1	0
1	1	1	0

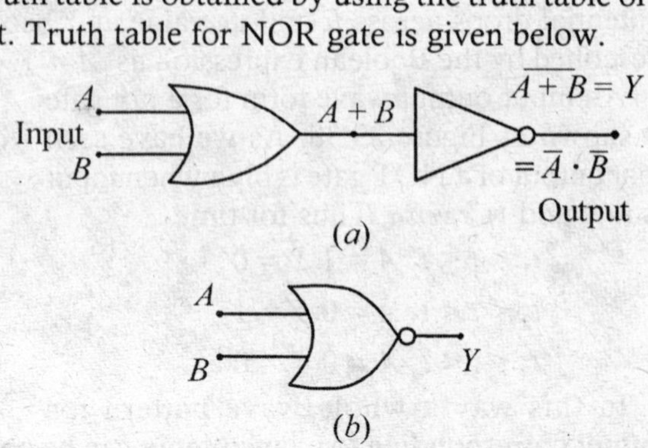

Fig. 29.40

The XOR Gate (Exclusive OR Gate)

XOR gate generally called (exclusive OR gate) evaluates two logical variables A and B to 1 if one of the two variables is zero and the other is 1. If both the variables are 0 or both are 1, the function is zero.

An XOR gate can be constructed with AND, OR and NOT gates as shown in Fig. 29.41 (*a*). Its Boolean symbol is shown in Fig. 29.41 (*b*).

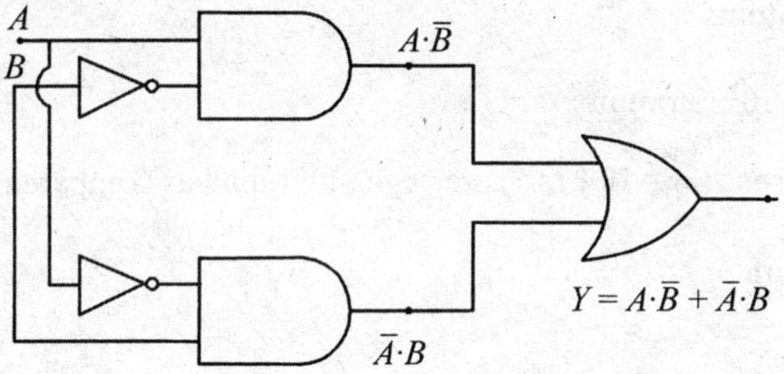

(*a*)

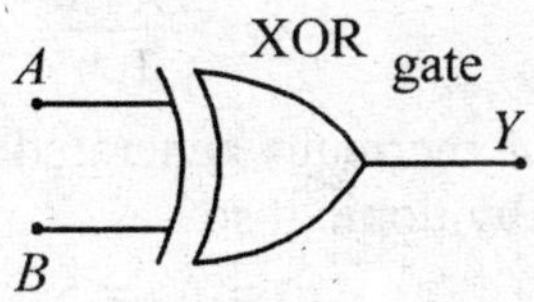

(*b*)

Fig. 29.41

Truth table for this gate is given below.

Truth Table

A	B	OR	XOR
0	0	0	0
1	0	1	1
0	1	1	1
1	1	1	0

The XNOR Gate (Exclusive NOR Gate)

XNOR gate is the combination of XOR and NOT gates, *i.e.*,

XNOR = XOR + NOT

Its logic symbol is shown in Fig. 29.42. Its truth table is shown below.

Truth Table

A	B	XOR	XNOR
0	0	0	1
1	0	1	0
0	1	1	0
1	1	0	1

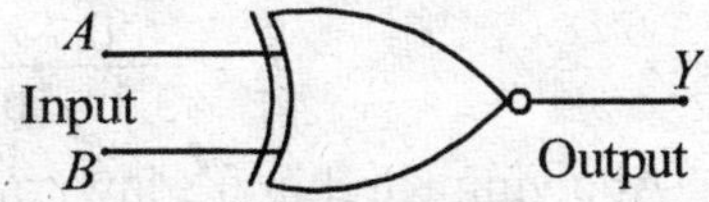

Fig. 29.42

SOLVED EXAMPLES

Example 29.1: A pure Si crystal has 5×10^{28} atoms m^{-3}. It is doped by 1 ppm concentration of pentavalent As. Calculate the number of electrons and holes. (Given that $n_i = 1.5 \times 10^{16}\ m^{-3}$)

Solution: Doping concentration of pentavalent As atom is = 1 ppm

$\therefore$ No density of pentavalent As atoms

$$N_0 = \frac{5 \times 10^{28}}{10^6} = 5 \times 10^{22} \text{ atom m}^{-3}$$

Now the thermally generated electrons ($n_i = 10^{16}\ m^{-3}$) are negligibly small as compared to those produced by doping, so

$$n_e = N_d = 5 \times 10^{22}\ m^{-3}$$

Also $$n_e \cdot n_h = n_i^2$$

$$\therefore \quad n_h = \frac{n_i^2}{n_e} = \frac{(1.5 \times 10^{16})^2}{5 \times 10^{22}} = \frac{2.25}{5} \times 10^{10} = 4.5 \times 10^9\ m^{-3}$$

Example 29.2: In the circuit shown in Fig. 29.43 (*a*). Find the currents i_1, i_2 and i_{D_2}. Given knee voltage for Si diode to be equal to 0.7 V.

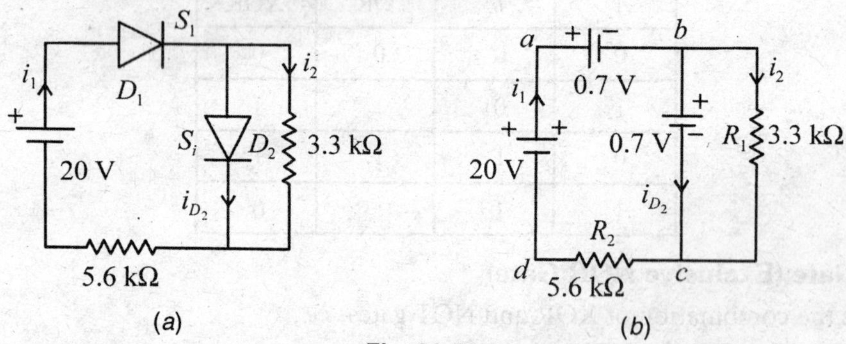

Fig. 29.43

Solution: The direction of current in different branches is shown in Fig. 29.43 (*a*). Equivalent diagram is shown Fig. 29.43 (*b*). Resistance R_1 (= 3.3 kΩ) is in parallel to diode D_2. The potential difference across D_2 is 0.7 V. Hence current i_2 through R_1 is

$$i_2 = \frac{0.7}{3.3 \times 10^3} = 0.212 \times 10^{-3} \text{ A} = 0.212 \text{ mA}$$

Applying Kirchhoff's law to mesh *abcd* we have

$$0.7 + 0.7 + V = 20$$

where V is potential difference across resistance R_2

$$\therefore \quad V = 20 - 1.4 = 18.6 \text{ V}$$

and current i_1 through R_2 is

$$i_1 = \frac{18.6}{5.6 \times 10^3} \text{ A} = 3.32 \text{ mA}$$

At the junction C, we have

$$i_{D_2} + i_2 = i_1$$

$$\therefore \quad i_{D_2} = i_1 - i_2 = 3.32 - 0.212 = 3.108 \text{ mA}$$

Example 29.3: In the circuit shown in Fig. 29.44 determine the output wave form, output DC level and required peak inverse voltage for each diode.

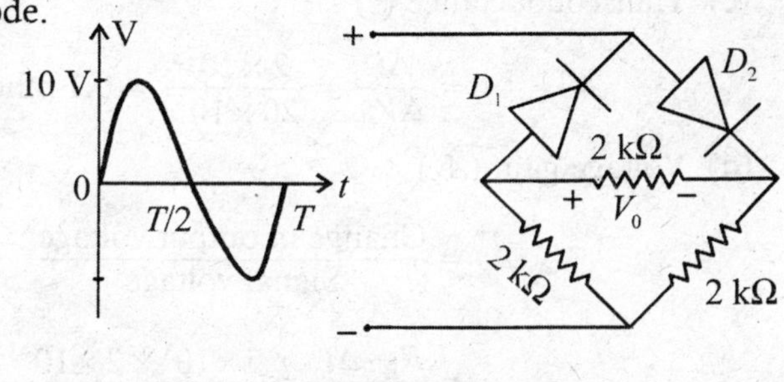

Fig. 29.44

Solution: For positive half cycle of input, diode D_1 is reverse biased and does not conduct whereas D_2 is forward biased and conducts. The circuit is redrawn in Fig. 29.45 (*d*). Figure 29.45 (*b*) shows that $V_0 = 4/2V$. Hence

$$(V_0)_{\max} = \frac{1}{2}(V)_{\max} = 5 \text{ V}$$

Its variation with time is shown in Fig. 29.45 (*c*).

During negative half cycle, *i.e.*, for $t = T/2$ to $t = T$, diode D_1 is forward biased and conducts whereas diode D_2 is reverse biased and does not conduct. The situation again is same. For time $t = T/2$ to $t = T$ the output varies as shown in Fig. 29.45 (*d*). This Figure also shows the full wave output.

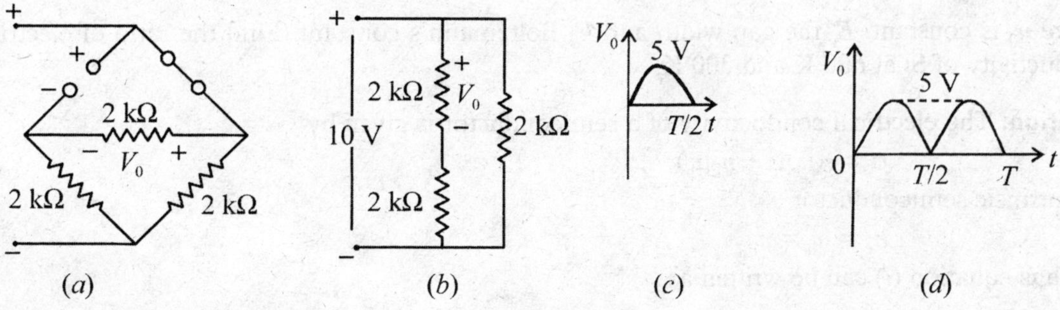

Fig. 29.45

For full wave rectification

$$V_{DC} = \frac{2}{\pi}(V)_{\max} = \frac{2}{\pi} \times 5 = 3.18 \text{ V}$$

For a full wave rectifier $PIV \geq (V)_{\max}$ *i.e.*, peak inverse voltage is the voltage across diode when it is not conducting which in present case is 5 V.

Example 29.4: A transistor is used in common emitter configuration in an amplifier circuit. When a signal of 20 mV is added to the base-emitter voltage, the base current changes by 20 μA and the collector current changes by 2 mA. The load resistance is 5 kΩ. Calculate (*a*) the factor β (*b*) the input resistance (*c*) the transconductance and (*d*) the voltage gain.

Solution:

(*a*) $\beta = \frac{\Delta i_c}{\Delta i_b} = \frac{2 \times 10^{-3}}{20 \times 10^{-6}} = 100$

(*b*) The input resistance (R_i)

$$R_i = \frac{\Delta V_{be}}{\Delta i_b} = \frac{20 \times 10^{-3}}{20 \times 10^{-6}} = 1\ k\Omega$$

(c) Transconductance (g)

$$g = \frac{\Delta i_c}{\Delta V_{eb}} = \frac{2 \times 10^{-3}}{20 \times 10^{-3}} = 0.1 \text{ mho (Siemen)}$$

(d) Voltage gain (A_V)

$$A_V = \frac{\text{Change in output voltage}}{\text{Signal voltage}}$$

$$= \frac{R_L \cdot \Delta i_c}{\Delta V_{be}} = \frac{5 \times 10^3 \times 2 \times 10^{-3}}{20 \times 10^{-3}} = 500$$

Example 29.5: The energy gap of pure Si is 1.1 eV. The mobilities of electrons and holes are respectively 0.135 $m^2V^{-1}s^{-1}$ and 0.048 $m^2V^{-1}s^{-1}$ and can be taken as independent of temperature. The intrinsic carrier concentration is given by

$$n_i = n_0 e^{-E_g/2k_BT}$$

where n_0 is constant, E_g the gap width and k_B Boltzmann's constant. Find the ratio of electrical conductivity of Si at 600 K and 300 K.

Solution: The electrical conductivity of a semiconductor is given by

$$\sigma = e(n_e\mu_e + n_h\mu_h) \qquad \ldots(i)$$

for intrinsic semiconductor

$$n_e = n_h = n_i$$

Thus equation (*i*) can be written as

$$\sigma = e\,n_i(\mu_e + \mu_h) = e(\mu_e + \mu_h)n_0 e^{-E_g/2k_BT}$$

Since the mobilities μ_e and μ_h are independent of temperature as such they can be regarded as constants. Hence the ratio of the conductivities at 600 K and 300 K is

$$\frac{\sigma_{600}}{\sigma_{300}} = \frac{e(\mu_e + \mu_h)\,n_0 e^{-E_g/2k_B \cdot 600}}{e(\mu_e + \mu_h)\,n_0 e^{-E_g/2k_B \cdot 300}} = e^{E_g/k_B \cdot 1200}$$

$$\therefore \quad \frac{\sigma_{600}}{\sigma_{300}} = e^{E_g/120k_B}$$

But $E_g = 1.1 \text{ eV} = 1.1 \times 1.6 \times 10^{-19}$ J and $k_B = 1.38 \times 10^{-23}$ J K^{-1}

$$\therefore \quad \frac{E_g}{1200\,k_B} = \frac{1.1 \times 1.6 \times 10^{-19}}{1200 \times 1.38 \times 10^{-23}} = 10.63$$

$$\therefore \quad \frac{\sigma_{600}}{\sigma_{300}} = e^{10.63} = 4.13 \times 10^4$$

Example 29.6: Find the current produced at room temperature in a pure germanium plate of area 2×10^{-4} m^2 and of thickness 1.2×10^{-3} m. When a potential difference of 5 V is applied across the faces. Concentration of carriers in germanium at room temperature is 1.6×106 m^{-3}. The mobilities of electrons and holes are 0.4 m^{2}V^{-1}s^{-1} and 0.2 m^{2}V^{-1}s^{-1} respectively. How much heat is generated in the plate in 100 seconds?

Solution: The conductivity of germanium is

$$\sigma = e(n_e\mu_c + n_h\mu_h)$$

Since semiconductor is pure hence $n_e = n_h = n_i$.

$$\therefore \quad \sigma = en_i(\mu_e + \mu_h) \qquad \ldots(i)$$

Hence current produced is

$$i = JA = \sigma E \cdot A = \sigma\left(\frac{V}{d}\right)A$$

$$= en_i(\mu_e + \mu_h)\left(\frac{V}{d}\right)A$$

$$= 1.6 \times 10^{-19} \times 1.6 \times 10^6 (0.4 + 0.2)\left(\frac{5}{1.2 \times 10^{-3}}\right) \times 2 \times 10^{-4}$$

$$i = 1.28 \times 10^{-13}\ \text{A}$$

Heat generated is

$$H = Vit = 5 \times 1.28 \times 10^{-13} \times 100 = 6.4 \times 10^{-11}\ \text{J}.$$

Example 29.7: The saturation current density of a p–n junction germanium diode is 250 mA/m^2 at 300 K. Find the voltage that would have to be applied across the junction to cause a forward current density of 10^5 A/m^2.

Solution: Current through a junction under bias V is

$$I = I_s\left(e^{eV/k_BT} - 1\right)$$

where I_s is saturation current in reverse bias.

Dividing with Area of cross-section, we have

$$J = \frac{I}{A} = J_s\left(e^{eV/k_BT} - 1\right) \quad \text{where } J_s = \frac{I_s}{A}$$

where J_s is saturation current density.

Since $e^{eV/kT} >> 1$, hence neglecting 1, we get

$$e^{eV/k_BT} = \frac{J}{J_s} = \frac{10^5}{250 \times 10^{-3}} = 4 \times 10^5$$

$\therefore \qquad \dfrac{eV}{k_B T} = \log(4 \times 10^5) = 12.9$

or $\qquad V = \dfrac{12.9 \times k_B T}{e} = \dfrac{12.9 \times 1.38 \times 10^{-23} \times 300}{1.6 \times 10^{-19}}$

$= 0.33 \text{ V}$

Example 29.8: A *p–n–p* transistor is used in common-emitter mode in an amplifier circuit. A change of 40 μA in the base current brings a change of 2 mA in collector current and 0.04 V in base-emitter voltage. Find the (*a*) input resistance (R_i) and (*b*) the base current amplification factor (β). If a load of 6 kΩ is used, then also find the voltage gain of amplifier.

Solution:

(*a*) Input resistance

$$R_i = \frac{\Delta V_{eb}}{\Delta i_b} = \frac{0.04}{40 \times 10^{-6}} = 10^3\ \Omega = 1\ \text{k}\Omega$$

(*b*) $$\beta = \frac{\Delta i_e}{\Delta i_b} = \frac{2 \times 10^{-3}}{40 \times 10^{-6}} = 50$$

(*c*) Voltage gain $$A_V = \frac{\text{Output voltage}}{\text{Input voltage}} = \frac{i_e R_L}{i_b \cdot R_i}$$

$$= \beta \cdot \frac{R_L}{R_i} = \frac{50 \times 6 \times 10^3}{10^3} = 300$$

Example 29.9: A silicon transistor amplifier circuit is shown in Fig. 29.46. If the current amplification factor β = 100, determine (*i*) base current i_b (*ii*) collector current i_c (*iii*) collector-emitter voltage (*iv*) collector-base voltage. Take potential drop across base-emitter junction = 0.7 V.

Solution:

(*i*) Figure 29.46 represents an *n–p–n* transistor in operation. Applying Kirchhoff's law to base-emitter closed circuit, we have

$$i_b R_1 + V_{eb} = 5$$

or $\qquad i_b R_1 = 5 - V_{eb} = 5 - 0.7 = 4.3 \text{ V}$

$\therefore \qquad i_b = \dfrac{4.3}{R_1} = \dfrac{4.3}{8.6 \times 10^3} = 0.5 \text{ mA}$

(*ii*) $\dfrac{i_c}{i_b} = \beta \therefore i_c = \beta i_b = 100 \times 0.5 \times 10^{-3} = 0.05 \text{ A}$

or $\qquad i_c = 50 \text{ mA}$

(*iii*) Applying Kirchhoff's law to collector-emitter closed circuit, we have

$$R_2 i_c + V_{ce} = 10$$

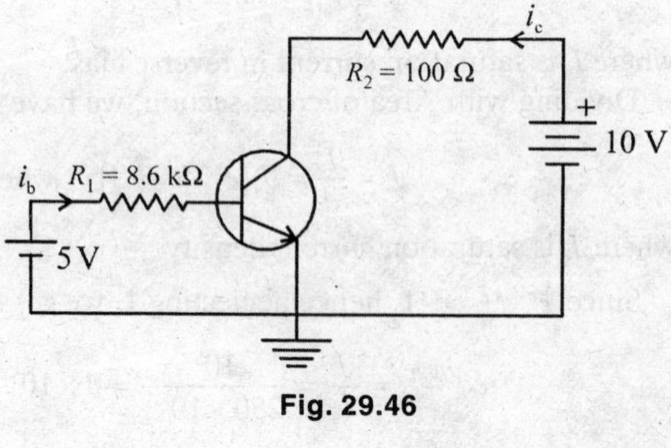

Fig. 29.46

$\therefore \quad V_{ce} = 10 - R_2 i_c = 10 - 100 \times 50 \times 10^{-3}$

$= 10 - 5 = 5 \text{ V}$

(*iv*) From circuit diagram, we have

$V_{ec} = V_{eb} + V_{bc}$

$\therefore \quad V_{bc} = V_{ec} - V_{eb} = 5 - 0.7 = 4.3 \text{ V}$

Example 29.10: A *n–p–n* transistor in a common emitter mode is used as a simple voltage amplifier with a collector current of 4 mA. The terminal of a 8 V battery is connected to the collector through a load resistance R_l and to the base through a resistance R_b. The collector emitter voltage V_{ce} = 4 V, base emitter voltage V_{be}= 0.6 V, and the base current amplification factor β = 100. Calculate the value of R_l and R_b.

Solution: From definition of current amplification in common emitter configuration, we have

$$\beta = \frac{i_c}{i_b} \qquad i_c = 4 \text{ mA and } \beta = 100$$

$$\therefore \quad i_b = \frac{4 \times 10^{-3}}{100} = 0.04 \text{ mA}$$

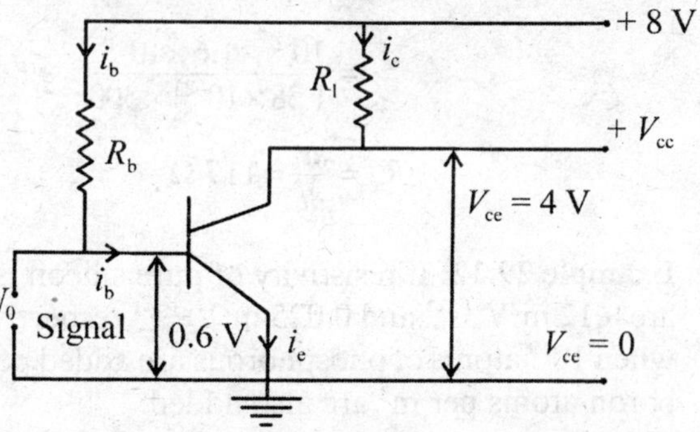

Fig. 29.47

From Fig. 29.47, we have

$$V_{cc} = V_{cb} + V_{eb} = R_b i_b + V_{eb}$$

or $\quad i_b R_b = V_{ec} - V_{eb} = 8 - 0.6 = 7.4$

or $\quad R_b = \frac{7.4}{i_b} = \frac{7.4}{0.04 \times 10^{-3}} = 1.85 \times 10^{+5} \ \Omega = 185 \text{ k}\Omega$

Also from the same figure, we have

$$i_c R_l + V_{ce} = V_{cc}$$

or $\quad i_c R_l = V_{cc} - V_{ce}$

or $\quad R_l = \frac{V_{cc} - V_{ce}}{i_c} = \frac{8-4}{4 \times 10^{-3}} = 1 \text{ k}\Omega$

Example 29.11: Find the static and dynamic resistances of a *p–n* junction germanium diode if the temperature is 27 °C and saturation current I_s = 1 μA for an applied forward bias of 0.2 V.

Solution: Forward current for germanium is

$$i = i_e\left(e^{eV/k_BT} - 1\right)$$

$$= i_s e^{eV/k_BT} \qquad \text{Neglecting 1}$$

$$\frac{eV}{k_BT} = \frac{1.6 \times 10^{-15} \times 0.2}{1.38 \times 10^{-23} \times 300} = 7.7$$

$\therefore$ $$i = 10^{-6} \times e^{7.7} = 2.2 \text{ mA}$$

Static resistance

$$R = \frac{V}{i} = \frac{0.2}{2.2 \times 10^{-3}} = 90 \ \Omega$$

Dynamic resistance is

$$R_d = \frac{dV}{di}$$

Since $$i = i_s(e^{eV/k_BT} - 1)$$

$\therefore$ $$\frac{di}{dV} \equiv \frac{i_s \cdot e}{K_BT} e^{eV/K_BT}$$

$$= \frac{10^{-6} \times 1.6 \times 10^{-19}}{1.38 \times 10^{-23} \times 300} \times e^{7.7}$$

$\therefore$ $$R_d = \frac{dr}{dt} = 11.7 \ \Omega.$$

Example 29.12: If resistivity of pure silicon is 3000 Ω–m and the mobilities of electron and holes are 0.12 $m^2V^{-1}s^{-1}$ and 0.025 $m^2V^{-1}s^{-1}$ respectively. Find (*i*) resistivity of a specimen of the material when 10^{19} atoms of phosphorous are added per m^3, (*ii*) the resistivity of specimen if further 2×10^{19} boron atoms per m^3 are also added.

Solution:

(*i*) The conductivity of pure silicon is

$$\sigma_i = n_i \cdot e(\mu_e + \mu_h)$$

$\therefore$ $$n_i = \frac{\sigma_i}{e(\mu_e + \mu_h)} \qquad \therefore \ \sigma_i = \frac{1}{\rho_i} \qquad (\rho_i \text{ is resistivity})$$

$\therefore$ $$n_i = \frac{1}{3000 \times 1.6 \times 10^{-19}(0.12 + 0.025)}$$

$$= 1.437 \times 10^{16} \text{ charge carriers/m}^3$$

When 10^{19} atoms of phosphorous (donor atoms) are added per m^3. Now $n_e >> n_i$ or $n_e >> n_h$. This means now the conductivity is due to majority carriers, i.e., electrons. Thus,

$$\sigma = \frac{1}{\rho} = n_e e \mu_e$$

$$\rho = \frac{1}{n_e e \mu_e} = \frac{1}{10^{19} \times 1.6 \times 10^{-19} \times 0.12} = 5.2 \ \Omega\text{–m}$$

(*ii*) When 2×10^{19} boron (acceptor) atoms are also added per m^3, we get

$$n_h - n_e = N_a - N_d = 2 \times 10^{19} - 10^{19} = 10^{19}$$

$$n_h \approx 10^{19} \text{ since } n_e << n_h$$

$$\rho = \frac{1}{n_h e \mu_h} = \frac{1}{10^{19} \times 1.6 \times 10^{-19} \times 0.025} = 25 \ \Omega\text{–m}$$

Example 29.13: A silicon diode is connected to a load resistance R_L as shown in Fig. 29.48. If V_{in} = 15 V and R_L = 10 kΩ, (*a*) If barrier voltage V_b = 0.7 V, then calculate (*i*) output voltage across R_L, (*ii*) current in diode (*iii*) forward resistance; (*b*) If diode is assumed ideal, then what will be (*i*) output voltage and (*ii*) output current in diode?

Solution:

(*i*) $V_{in} - V_b = V_{out}$

$$\therefore \quad V_{out} = 15 - 0.7 = 14.3 \text{ V}$$

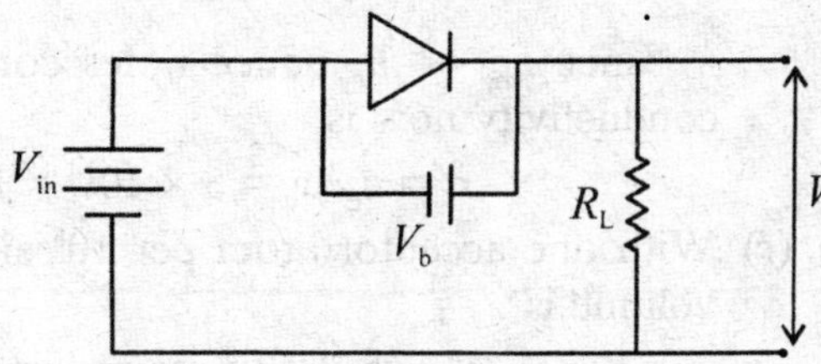

Fig. 29.48

(*ii*) Current

$$i = \frac{V_{out}}{R_L} = \frac{14.3}{10 \times 10^3} = 1.43 \text{ mA}$$

(*iii*) $R_f = \dfrac{V_b}{i}$

$$\text{or} \quad R_f = \frac{0.7}{1.43 \times 10^{-3}} = 490 \ \Omega$$

(*b*) If junction diode is ideal, then

$$R_F \to 0 \ V_b = 0$$

(*i*) V_{out} = 15 V

(*ii*) $i = \dfrac{V_{out}}{R_L} = \dfrac{15}{10 \times 10^3} = 1.5 \text{ mA}$

Example 29.14

(*a*) Find the conductivity of intrinsic silicon at 300 K. It is given that n_i at 300 K in silicon is 1.5 × 10^{10}/cm^3 and the mobilities of electrons and holes in silicon are 1300 cm^2/V-s and 500 cm^2/V-s respectively.

(*b*) If donor type impurity is added to the extent of 1 impurity atom in 10^8 silicon atoms, find its conductivity.

(*c*) If the acceptor impurity is added to the extent of 1 impurity atom in 10^8 silicon atoms, find the conductivity.

Solution:

(*a*) Conductivity in an intrinsic semiconductor is given by

$$\sigma = n_i e\,(\mu_e + \mu_h) = 1.5 \times 10^{10} \times 1.6 \times 10^{-19}\,(1300 + 500)$$
$$= 4.3 \times 10^{-6} \text{ mhos/cm}$$

(*b*) If there is 1 donor atom per 10^8 silicon atoms, then*

$$\text{No. of donor atoms, } N_d = \frac{5 \times 10^{22}}{10^8} = 5 \times 10^{14}/\text{cm}^3$$

Corresponding to each donor atom, we have 1 electron in conduction band.

---

*Number of atoms per unit volume of silicon = $\dfrac{n_0 d}{A} \approx 5 \times 10^{22}/\text{cm}^3$.

Hence electron concentration $n_e \approx N_d = 5 \times 10^{14}/\text{cm}^3$
If n_h is hole concentration, then

$$n_h = \frac{n_i^2}{n_e} = \frac{(1.5\times10^{10})^2}{5\times10^{14}} = 4.5\times10^5/\text{cm}^3$$

Since $n_h << n_e$ hence holes contribution to conductivity can be neglected. Hence conductivity now is

$$\sigma = n_e e\mu_e = 5 \times 10^{14} \times 1.6 \times 10^{-19} \times 1300 = 0.104 \text{ mhos/cm}$$

(*c*) With one acceptor atom per 10^8 silicon atoms, the number of acceptor atoms per unit volume is

$$N_a = 5 \times 10^{14} \text{ atoms/cm}^3$$

The hole concentration is $5 \times 10^{14}/\text{m}^3$. Compared to this the electron concentration is $4.5 \times 10^5/\text{cm}^3$. Hence contribution of electrons to conductivity can be neglected. Thus conductivity now is

$$\sigma = n_h e\mu_h = 5 \times 10^{14} \times 1.6 \times 10^{-19} \times 500$$
$$= 0.04 \text{ mho/cm.}$$

Example 29.15: A potential difference of 0.3 V exists across a *p–n* junction (*a*) If the depletion region is 1 μm wide, what is the intensity of electric field in this region? (*b*) An electron with speed 5×10^5 m/s approaches this *p–n* junction from *n*–side, what will be its speed entering the *p*–side?

Solution:

(*a*) Electric field in the depletion layer is

$$E = \frac{V}{d} = \frac{0.3}{10^{-6}} = 3\times10^5 \text{ V/m}$$

The field is directed from *n* to *p* region.

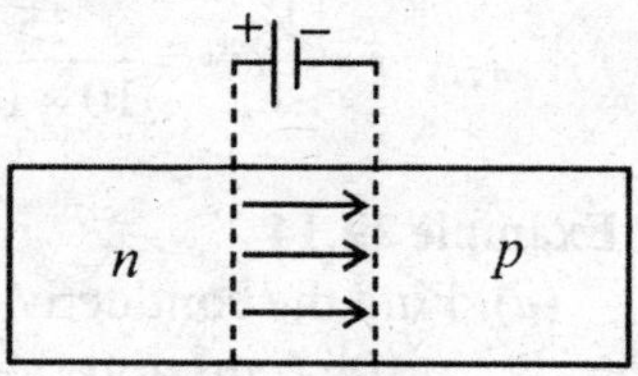

Fig. 29.49

(*b*) Electron in depletion layer is retarded. Retardation is

$$a = \frac{Ee}{m} = \frac{3\times10^5\times1.6\times10^{-19}}{9.1\times10^{-31}} = 5.3\times10^{16} \text{ m/s}^2$$

Thus final speed on reaching *p*–region is (v), then

$$v^2 = v_0^2 - 2as = (5 \times 10^5)^2 - 2 \times 5.3 \times 10^{16} \times 10^{-6}$$
$$= 25 \times 10^{10} - 10.6 \times 10^{10} = 14.4 \times 10^{10}$$

or $\quad v = 3.8 \times 10^5$ m/s.

Example 29.16: An *n*–type silicon sample of width 4×10^{-3} m thickness and length 6×10^{-2} m carries a current of 4.8 mA. When the voltage is applied across the length of the sample. What is the current density? If the free electron density is 10^{22} m^{-3}, then find how much time it takes for the electrons to travel the full length of the sample.

Solution: The current density J is given by

$$J = \frac{I}{A} = \frac{4.8\times10^{-3}}{(4\times10^{-3})\times(6\times10^{-2})} = 20 \text{ A/m}^2$$

$\therefore$ Drift velocity of electron is

$$v_d = \frac{J}{ne} = \frac{20}{10^{22} \times 1.6 \times 10^{-19}} = 1.25 \times 10^{-2} = 0.125 \text{ m/s}$$

Thus time taken to travel length l (= 6×10^{-2} m) is

$$l = vt$$

or
$$t = \frac{l}{v} = \frac{6 \times 10^{-2}}{1.25 \times 10^{-2}} = 4.8 \text{ sec.}$$

Example 29.17: In the circuit shown in Fig. 29.50, the values given are R_1 =15.8 kΩ, R_2 = 2.2 kΩ, R_L = 500 Ω i_b = 55 μA, β = 200, V_{ce} = 9 V. Then find (*i*) V_Q, (*ii*) input resistance, (*iii*) V_{ce} and V_L across R_L, (*iv*) power output P_L.

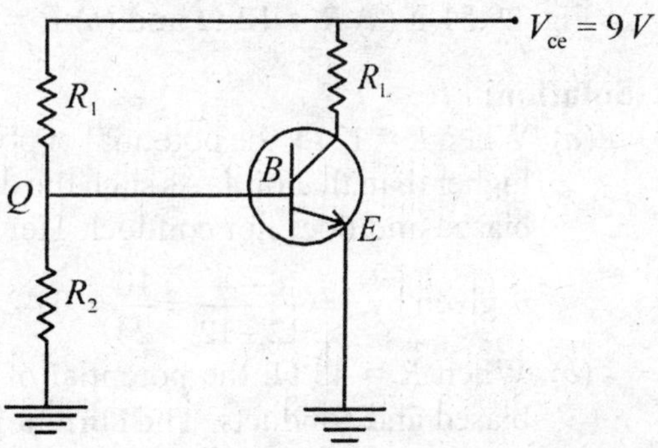

Fig. 29.50

Solution:

(*i*) Potential at Q (= V_Q) is equal to potential at B (= V_B). Hence

$$V_Q = V_B = \frac{V_{ce}}{R_1 + R_2} \cdot R_2 = \frac{2.2 \times 9}{18} = 1.1 \text{ V}$$

(*ii*) Since base current is 55 μA. Hence input resistance at the base is

$$R_i = \frac{V}{i_b} = \frac{1.1}{55 \times 10^{-6}} = 200 \text{ k}\Omega$$

(*iii*) β = 200, then collector current

$$i_c = \beta i_b = 200 \times 55 \times 10^{-6} = 11 \text{ mA}$$

$$V_L = i_c R_L = 11 \times 10^{-3} \times 500 = 5.5 \text{ V}$$

$$V_{ce} = V_{ce} - V_L = 9 - 5.5 = 3.5 \text{ V}$$

(*iv*) $P_L = i_c^2 R_L = (11 \times 10^{-3})^2 \times 500 = 60.5$ mW.

Example 29.18: Let $X = A\overline{BC} + B\overline{CA} + C\overline{BA}$. Evaluate X for (*a*) $A = 1, B = 0, C = 1$ (*b*) $A = B = C = 1$ and (*c*) $A = B = C = 0$.

Solution:

(*a*) $BC = 0$ for $B = 0$ and $C = 1$ since its is B and C

$\therefore$ $\overline{BC} = 1$ and $A\overline{BC} = 1$ since it is A and $\overline{BC}$

$CA = 1$ for $C = 1$ and $A = 1$ $\therefore$ $\overline{CA} = 0$ and $B\overline{CA} = 0$

$BA = 0$ for $B = 0$ and $A = 1$ $\therefore$ $\overline{BA} = 1$ and $C\overline{BA} = 1$

$\therefore$ $X = A\overline{BC} + B\overline{CA} + C\overline{BA} = 1 + 0 + 1 = 1$

(*b*) $A = B = C = 1$

$BC = 1$ $\therefore$ $\overline{BC} = 0$ and $A\overline{BC} = 0$

$CA = 1$ $\therefore$ $\overline{CA} = 0$ and $B\overline{CA} = 0$

$$BA = 1 \therefore \overline{BA} = 0 \text{ and } C\overline{BA} = 0$$

$$\therefore \quad X = \overline{ABC} + \overline{BCA} + \overline{CBA} = 0 + 0 + 0 = 0$$

(*c*) $A = B = C = 0$

$$BC = 0 \therefore \overline{BC} = 1 \text{ and } A\overline{BC} = 0$$

$$CA = 0 \therefore \overline{CA} = 1 \text{ and } B\overline{CA} = 0$$

$$BA = 0 \therefore \overline{BA} = 1 \text{ and } C\overline{BA} = 0$$

$$\therefore \quad X = A\overline{BC} + B\overline{CA} + C\overline{BA} = 0 + 0 + 0 = 0.$$

Example 29.19: Find the current through the resistor R in Fig. 29.51 if (*a*) $R = 12\ \Omega$ and (*b*) $R = 48\ \Omega$.

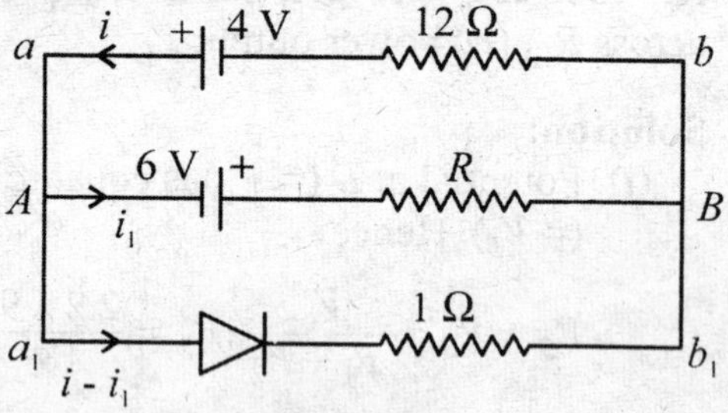

Fig. 29.51

Solution:

(*a*) When $R = 12\ \Omega$, the potential of point B becomes higher than that of A. As such the diode is reverse biased and does not conduct. Hence the current is given by $i = \frac{6+4}{12+12} = \frac{10}{24} = 0.42\text{ A}$

(*b*) When $R = 48\ \Omega$, the potential of A becomes higher than B as such the diode is forward biased and conducts. The current in different branches is shown in Fig. 29.51.

Applying Kirchhoff's law to mesh $aABba$, we have

$$-6 + 48i_1 + 12i = 4$$

or $$48i_1 + 12i = 10 \qquad \ldots(i)$$

Applying Kirchhoff's law to mesh aa_1b_1ba, we get

$$1(i - i_1) + 12i = 4$$

or $$13i - i_1 = 4 \qquad \ldots(ii)$$

or $$13 \times 48i - 48i = 4 \times 48 \qquad \ldots(iii)$$

(*i*) + (*iii*) gives

$$13 \times 48i + 12i = 202$$

or $$53 \times 12i = 202 \text{ or } i = \frac{202}{53 \times 12}$$

and $$i_1 = 13i - 4 = \frac{13 \times 202}{53 \times 12} - 4 = 4.128 - 4$$

$$= 0.128\text{ A}.$$

Example 29.20: Show that $AB + \overline{AB}$ is always 1.

Solution: This can be proved by preparing truth table of the Boolean function. This is given below.

A	B	AB	$\overline{AB}$	$AB+\overline{AB}$
0	0	0	1	1
1	0	0	1	1
0	1	0	1	1
1	1	1	0	1

hence proved.

Example 29.21: Design a logical circuit using AND, OR and NOT gates to evaluate $A\overline{BC} + B\overline{AC}$. Write its truth table.

Solution: Logical circuit is shown in Fig. 29.52. The truth table is given below.

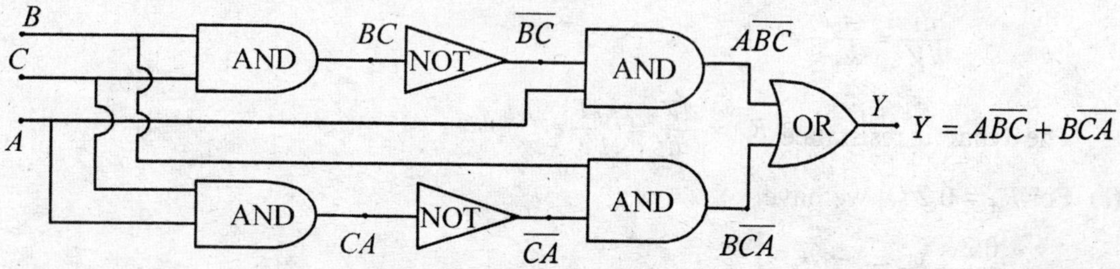

Fig. 29.52

Truth Table

$Y = A\overline{BC} + B\overline{CA}$

A	B	C	BC	$\overline{BC}$	CA	$\overline{CA}$	$A\overline{BC}$	$B\overline{CA}$	Y
0	0	0	0	1	0	1	0	0	0
1	0	0	0	1	0	1	1	0	1
0	1	0	0	1	0	1	0	1	1
0	0	1	0	1	0	1	0	0	0
1	1	0	0	1	0	1	1	1	1
1	0	1	0	1	1	0	1	0	1
0	1	1	1	0	0	1	0	1	1
1	1	1	1	0	1	0	0	0	0

Example 29.22: The current-voltage characteristic of an ideal *p–n* junction diode is given by

$$i = i_0(e^{eV/k_BT} - 1)$$

When the drift current i_0 equals 10 μA. Take the temperature T to be 300 K (*a*) Find the voltage V_0 for which $e^{eV/k_BT} = 100$. One can neglect the term 1 for voltages greater than this value. (*b*) Find an expression for the dynamic resistance of diode as a function of V for $V > V_0$. (*c*) Find the voltage for which the dynamic resistance would be 0.2 Ω.

Solution:

(*a*) For voltage V_0 we have

$$e^{eV_0/k_BT} = 100$$

Taking logs, we get

$$\frac{eV_0}{k_BT} = \log_e 100 = 2.303 \log_{10} 100 = 4.606$$

or $$V_0 = \frac{k_BT}{e} \times 4.604 = \frac{1.38 \times 10^{-23}}{1.6 \times 10^{-19}} \times 300 \times 4.604$$

$$= 0.12 \text{ V}$$

(*b*) $i = i_0(e^{eV/k_BT} - 1)$

$$\therefore \quad \frac{di}{dV} = \frac{i_0 e}{k_BT} e^{eV/k_BT}$$

The dynamic resistance $R_d = \dfrac{dV}{di} = \dfrac{k_BT}{i_0 e} e^{-eV/k_BT}$

(*c*) For $R_d = 0.2\ \Omega$, we have

$$\frac{0.2 \times i_0 e}{k_BT} = e^{-eV/k_BT}$$

or $$e^{+eV/k_BT} = \frac{k_BT}{0.2\, i_0 e} = \frac{1.38 \times 10^{-23} \times 300}{0.2 \times 10 \times 10^{-6} \times 1.6 \times 10^{-19}}$$

$$= 1.3 \times 10^4$$

$$\therefore \quad \frac{eV}{k_BT} = \log_e(1.3 \times 10^4) = 9.5$$

or $$V = \frac{9.5 \times k_BT}{e} = \frac{9.5 \times 1.38 \times 10^{-23} \times 300}{1.6 \times 10^{-19}}$$

$$= 0.25 \text{ V.}$$

Example 29.23: The concentration of hole–electron pairs in pure silicon at $T = 300$ K is 7×10^{15} per cubic metre. Antimony is doped into silicon in a proportion of 1 atom in 10^7 atoms. Assuming that half of the impurity atoms contribute electrons in the conduction band, calculate the factor by which the number of charge carriers increase due to doping. The number of silicon atoms per cubic meter is 5×10^{28}.

Solution: Before doping the electron and hole concentration is same. Hence total number of charge carriers before doping is equal to number of holes plus number of electrons.

$$= 2 \times 7 \times 10^{15} = 1.4 \times 10^{16} \text{ m}^{-3} \quad \ldots(i)$$

Antimony is doped in the ratio 1 in 10^7. Hence number of Sb atoms per cubic metre $= 5 \times 10^{28} \times 10^{-7} = 5 \times 10^{21}\ \text{m}^{-3}$.

Since half of these atoms contribute electrons to the conduction band, the number of extra conduction electrons produced is

$$= 2.5 \times 10^{+21} \text{ m}^{-3}$$

Thus the number of charge carriers per cubic metre after doping is

$$= 2.5 \times 10^{21} + 1.4 \times 10^{16} \approx 2.5 \times 10^{21} \text{ m}^{-3}$$

The factor by which the number of charge carriers is increased is

$$= \frac{2.5 \times 10^{21}}{1.4 \times 10^{16}} = 1.8 \times 10^{5}$$

In fact, as the n–type impurity is doped, the number of holes will decrease. This is because the product of the concentration of holes and conduction electrons remains almost the same. However, this does not affect our result as the number of holes in any way are too small as compared to the number of conduction electrons.

Example 29.24: The number of silicon atoms per m^3 is 5×10^{28}. This is doped simultaneously with 5×10^{22} atoms per m^3 of arsenic and 5×10^{20} m^{-3} atoms of Indium. Calculate the number of electrons and holes. Given that $n_i = 1.5 \times 10^{16}$ m^{-3}. Is the material n–type or p–type.

Solution: If n_e is electron concentration and n_h holes concentration, then

$$n_e n_h = n_i^2 \qquad \ldots(i)$$

If N_d is the number of donor atoms and N_a is the number of acceptor atoms, then charge neutrality requires.

$$N_d - N_a = n_e - n_h \qquad \ldots(ii)$$

Equation (*i*) gives $n_h = \dfrac{n_i^2}{n_e}$

$$\therefore \qquad N_d - N_a = n_e - \frac{n_i^2}{n_e}$$

or

$$n_e^2 - (N_d - N_a)\, n_e - n_i^2 = 0$$

or

$$n_e = \frac{1}{2}\left[(N_d - N_a) + \sqrt{(N_d - N_a)^2 + 4n_i^2}\right]$$

Substituting the values, we get

$$n_e = \frac{1}{2}\left[(5 \times 10^{22} - 5 \times 10^{20}) + \sqrt{(5 \times 10^{22} - 5 \times 10^{20})^2 + 4\,(1.5 \times 10^{16})^2}\right]$$

Since $N_d >> N_a$ and $N_d^2 >> n_i^2$ we can approximately write

$$n_e = \frac{1}{2}[N_d + N_d] = 5 \times 10^{22} \text{ m}^{-3}$$

and

$$n_h = \frac{n_i^2}{n_e} = \frac{(1.5 \times 10^{16})^2}{5 \times 10^{22}} = \frac{2.25}{5} \times 10^{10} = 4.5 \times 10^{9} \text{ m}^{-3}$$

Since majority carriers are electrons, hence the material is n–type.

Example 29.25: The number of electron-hole pairs in an intrinsic semiconductor is 2×10^{19} m^{-3} at 27° C and E_g is 1 eV. Calculate the number of electron-hole pairs at 227° C. Given that Boltzmann's constant is 8.65×10^{-5} eV/Kelvin.

Solution: In an intrinsic semiconductor, the electron-hole pairs at any temperature T is given by

$$n_i = Ce^{-E_g/2k_BT}$$

where C is constant, E_g is band gap.

$$n_1 = Ce^{-E_g/2k_BT_1} \text{ and } n_2 = Ce^{-E_g/2k_BT_2}$$

$$\therefore \quad \frac{n_1}{n_2} = e^{-E_g/2k_B(1/T_1 - 1/T_2)}$$

or $$\log_e \frac{n_1}{n_2} = -\frac{E_g}{2k_B}\left[\frac{1}{T_1} - \frac{1}{T_2}\right]$$ where $T_1 = 300$ K and $T_2 = 500$ K

or $$\log_e \frac{n_1}{n_2} = -\frac{1}{2\times 8.65\times 10^{-5}}\left[\frac{1}{300} - \frac{1}{500}\right]$$

$$= -7.7$$

or $$\log_e \frac{n_2}{n_1} = 7.7$$

and $$\frac{n_2}{n_1} = 2208$$

or $$n_2 = 2208 \times n_1 = 2208 \times 2 \times 10^{19}$$

$$= 4.42 \times 10^{22} \text{ m}^{-3}.$$

Example 29.26: Assume that the silicon diode in the circuit shown in Fig. 29.53 requires a minimum current of 1 mA to be above the knee point (0.7 V) of the *I–V* characteristics. Also assume that the voltage across the diode is independent of current above the knee point.

(*a*) If $V_b = 5$ V what should be the maximum value of R so that the voltage is above the knee point?

(*b*) If $V_b = 5$ V, what must be the value of R to establish the current of 5 mA in the circuit?

(*c*) What is the power dissipated in the resistance R and in the diode when a current of 5 mA flows in the circuit at $V_b = 6$ V?

(*d*) If $R = 1$ kΩ, what is the minimum voltage V_b required to keep the diode above the knee point?

Solution:

(*a*) When a current of 1 mA passes through the circuit, the potential drop of 0.7 V occurs across the diode for a forward bias of 5 V.

Thus the potential drop across $R = 5 - 0.7 = 4.3$ V

Hence maximum value of R is

$$R_{max} = \frac{4.3}{10^{-3}} = 4.3\times 10^3\ \Omega = 4.3\ \text{k}\Omega$$

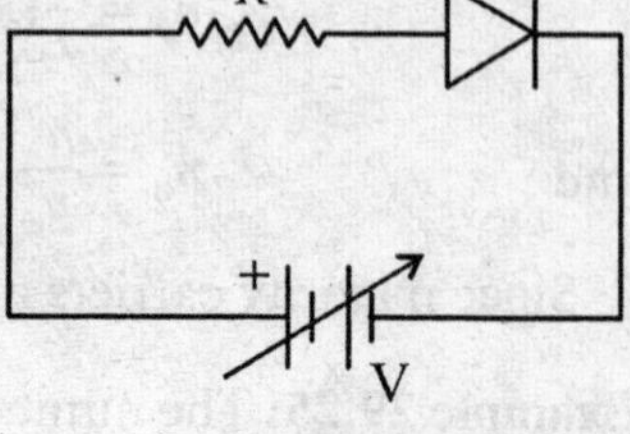

Fig. 29.53

(*b*) To establish a current of 5 mA, the required resistance is

$$R = \frac{4.3}{5\times 10^{-3}} = 860\ \Omega$$

Since knee voltage does not depend on current above knee point.

(c) When $V_b = 6$ volt, the potential difference across resistance R is

$$V_R = 6 - 0.7 = 5.3 \text{ V}$$

and current is 5 mA. Hence power dissipated in resistance R is

$$P_R = V_R \cdot I = 5.3 \times 5 \times 10^{-3} = 26.5 \times 10^{-3} \text{ W} = 26.5 \text{ mW}$$

Power dissipated in diode $P_d = V_d I = 0.7 \times 5 \times 10^{-3} = 3.5 \times 10^{-3}$ W = 3.5 mW

(d) When $R = 1$ kΩ; for minimum current of 1 mA, the potential difference across R is

$$V_R = 1 \times 10^3 \times 10^{-3} = 1 \text{ V}$$

Hence required bias voltage is

$$V_b = V_R + V_d = 1 + 0.7 = 1.7 \text{ V}.$$

Example 29.27: In Fig. 29.54 what is the voltage needed to maintain 15 V across the load resistance $R_L = 2$ kΩ. Assume that the series resistance $R = 200$ Ω and the zener requires a minimum current of 10 mA to work satisfactorily? What is the zener rating required.

Solution: The current distribution in different branches is shown in Fig. 29.54.

Voltage required across R_L (load resistance) $V_L = 15$ volts

Hence current required is

$$i_L = \frac{V_L}{R_L} = \frac{15}{2 \times 10^3} = 7.5 \text{ mA}$$

Zener current is $i_z = 10$ mA

Hence the through resistance R is

$$i_R = i_z = i_L = 10 + 7.5 = 17.5 \text{ mA}$$

Hence voltage across R is

$$V_R = i_R \times R = 17.5 \times 10^{-3} \times 200 = 3.5 \text{ V}$$

Thus input voltage needed is $V_R + V_Z = 18.5$ V

The zener diode should have a current rating of 17.5 mA and break down voltage of 15 V.

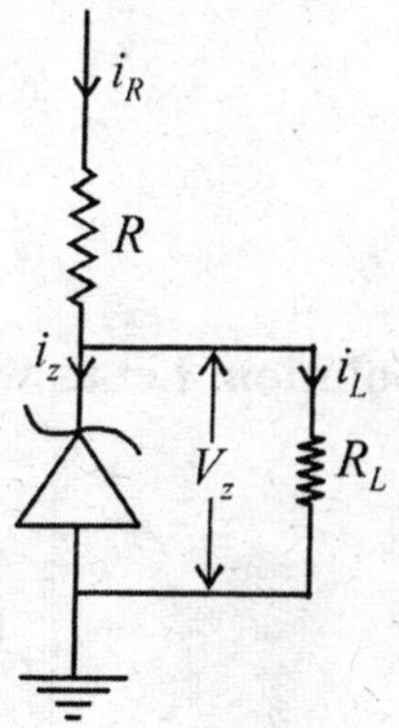

Fig. 29.54

Example 29.28: In the circuit shown in Fig. 29.55, the value of β is 100. Find i_B V_{CE}, V_{BE} and V_{BC} when $i_C = 1.5$ mA. The transistor is in active, cut off or saturation state?

Solution: From the figure, we have

$$R_L i_C + V_{CE} = V_{CC}$$

$$\therefore V_{CE} = V_{CC} - R_L i_E = 24 - 4.7 \times 10^3 \times 1.5 \times 10^{-3}$$

$$= 24 - 7.05 = 16.95 \text{ V}$$

Similarly from base emitter circuit, we have

$$R_B i_b + V_{BE} = V_{CC}$$

$$\therefore \quad V_{BE} = V_{CC} - R_B i_b$$

We know $\beta = \frac{i_c}{i_b}$ or $i_b = \frac{i_c}{\beta}$

$$\therefore \quad V_{BE} = V_{CC} - R_B \frac{i_c}{\beta} = 24 - 220 \times 10^{+3} \frac{1.5 \times 10^{-3}}{100}$$

$$= 24 - 3.3 = 20.7 \text{ V}$$

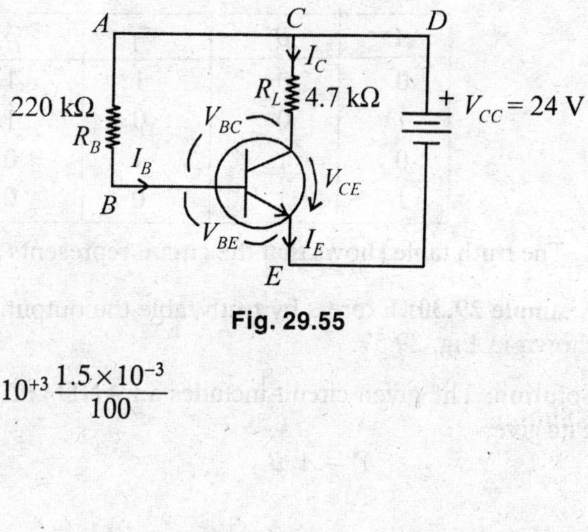

Fig. 29.55

Going along mesh ABCA, we get

$$i_b R_B + V_{BC} - i_c R_C = 0$$

$$\therefore \quad V_{BC} = i_c R_C - i_b R_B = 7.05 - 3.3 = 4.75 \text{ V}.$$

Since $V_{CE} < V_{BE}$ both the junctions are forward biased as such the transistor is in saturation state.

Example 29.29: You are given the two circuits as shown in Fig. 29.56 (*a*) and (*b*). Show that the circuit (*a*) acts as OR gate while the circuit (*b*) acts as AND gate.

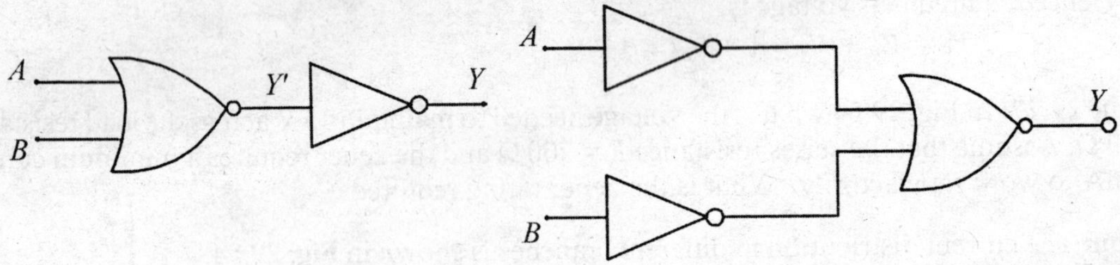

(*a*) (*b*)

Fig. 29.56

Solution: Let us write the truth tables for the two circuits.

(*a*)

A	B	$A+B$	$Y' = \overline{A+B}$	$Y = \overline{Y'}$
0	0	0	1	0
1	0	1	0	1
0	1	1	0	1
1	1	1	0	1

In Fig. 29.56 (*a*), first gate is NOR gate which is combined to NOT gate. This gives $Y' = \overline{A+B}$ and $\overline{Y'} = Y = A + B$. Thus the circuit is representing OR gate. The truth table also shows the OR gate.

(*b*)

A	B	$\overline{A}$	$\overline{B}$	$\overline{A}+\overline{B}$	$Y = \overline{\overline{A}+\overline{B}}$
0	0	1	1	1	0
1	0	0	1	1	0
0	1	1	0	1	0
1	1	0	0	0	1

The truth table shows that the circuit represents AND gate.

Example 29.30: Express by truth table the output Y for all possible inputs A and B in the circuit shown in Fig. 29.57.

Solution: The given circuit includes an AND and an OR gate. Input A and B when fed to AND gate gives

$$Y' = A \cdot B$$

Then A and Y' are the inputs for OR gate. Hence

$$Y = A + Y' = A + A \cdot B$$

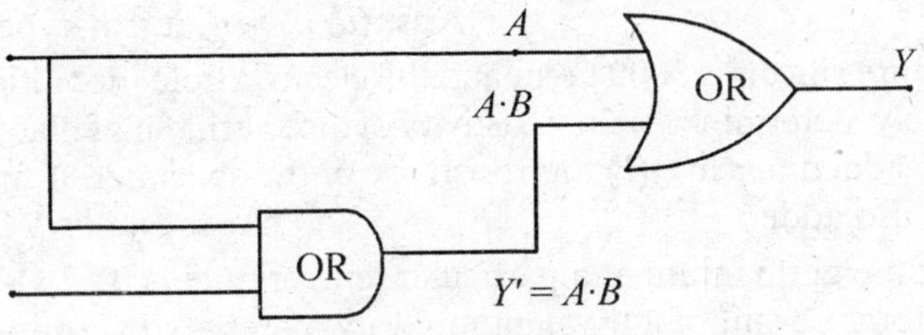

Fig. 29.57

AND Gate			OR Gate		
A	B	$Y' = A + B$	A	Y'	$Y = A + Y'$
0	0	0	0	0	0
0	1	0	0	0	0
1	0	0	1	0	1
1	1	1	1	1	1

This is the truth table of the circuit given in Fig. 29.57.

PROBLEMS

1. The energy gap E_g of pure silicon semiconductor is 1.1 eV. Its hole and electron mobilities respectively are $\mu_h = 0.048\ m^2V^{-1}s^{-1}$ and $\mu_e = 0.135\ m^2V^{-1}s^{-1}$ and are independent to temperature. Find the ratio between its conductivity at 400 K and 300 K. Assume that the temperature dependence of intrinsic carrier concentration given by $n_i = n_0 e^{-E_g/2k_BT}$, where n_0 is a constant, T is absolute temperature and k_B is Boltzmann's constant.

$$\left[\textbf{Ans.}\ \frac{\sigma_{400}}{\sigma_{300}} = 280\right]$$

2. Germanium is doped 1 part in 10^6 with indium at room temperature. Find the conductivity of doped germanium. Given, concentration of germanium atoms is $4.4 \times 10^{28}\ m^{-3}$, intrinsic carrier concentration $n_i = 2.4 \times 10^{19}\ m^{-3}$, $\mu_e = 0.39\ m^2V^{-1}s^{-1}$ and $\mu_h = 0.19\ m^2V^{-1}s^{-1}$.

[**Ans.** $\sigma = 1.34 \times 10^3\ (\Omega\, m)^{-1}$]

3. A block of intrinsic silicon at 300 K has a length of 10 cm and an area of $1.0 \times 10^{-4}\ m^2$. If a battery of emf 2 V is connected across it. Find the electron and hole currents and the total current. The mobilities of electrons and holes respectively are $0.135\ m^2V^{-1}s^{-1}$ and $0.048\ m^2V^{-1}s^{-1}$ and their number densities are equal and each being $1.5 \times 10^{16}\ m^3$.

[**Ans.** $I_e = 6.48 \times 10^{-7}$ A, $I_h = 2.304 \times 10^{-7}$ A and $I = I_c + I_h = 8.787 \times 10^{-7}$ A]

4. A semiconductor is known to have an electron concentration of 8×10^{13} cm^3 and a hole concentration of 5×10^{12} cm^3 (*a*) Is the semiconductor *n*–type or *p*–type? (*b*) What is the resistivity of the sample if the electron mobility is 23000 cm^2/V-s and hole mobility 100 cm^2/V-s.

[**Ans.** (*a*) $n_e > n_h$ it is *n*–type (*b*) $\rho = 3.395$ Ω–m]

5. The resistivity of pure silicon is 3000 Ω–m and the electron hole mobilities are 0.12 and 0.045 m^{2}V^{-1}s^{-1} respectively, determine (*a*) the resistivity of the specimen of material when 10^{19} atoms of phosphorus are added per m^3, (*b*) the resistivity of the specimen if further 2×10^{19} boron atoms per m^3 are also added. [**Ans.** (*a*) $\rho = 5.21$ Ω–m (*b*) 25 Ω–m]

6. The resistivity of pure germanium at a particular temperature is 0.52 Ω–m. If the material is doped with 10^{20} atoms per m^3 of a trivalent impurity material, determine the new resistivity. The electron and hole mobilities are given to be 0.2 and 0.4 m^{2}V^{-1}s^{-1}.

[**Ans.** $\rho = 0.195$ Ω–m]

7. A 10 V zener diode along with a series resistance is connected across a 40 V supply. Calculate the minimum value of the resistance required, if the maximum zener current is 50 mA.

[**Ans.** $R = 600$ Ω]

8. A potential barrier of 0.50 V exists in a *p–n* junction (*i*) If the depletion region is 5.0×10^{-7} m wide, what is the electric field in this region? (*ii*) If an electron approaches the *p–n* junction from the *n*–side with a speed of 5×10^5 m/s with what velocity will it enter the *p*–side?

[**Ans.** (*i*) $E = 10^6$ V/m. (*ii*) $v = 2.7 \times 10^5$ m/s]

9. In the circuit shown in Fig. 29.58, if we assume that when the input voltage of the base resistance is 5 V, V_{BE} is zero and V_{CE} is also zero. What is I_B, I_C and β? When the input is zero, the I_B is zero. What will be the output wave form if the input wave form is as shown in Fig. 29.58?

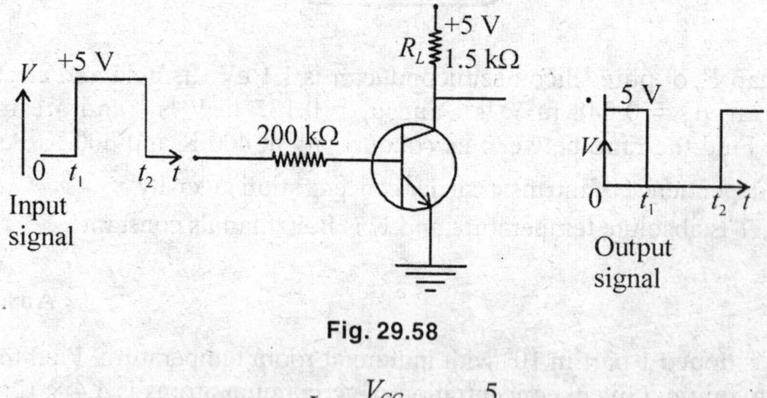

Fig. 29.58

[**Hint:** $V_{CC} = I_C R_L + V_{CE} = I_C R_L$ or $I_C = \dfrac{V_{CC}}{R_L} = \dfrac{5}{1.5 \times 10^3} = 3.3\,\text{mA}$ $V_{CB} = 5$ V, $R_B = 200$ kΩ

$I_B = \dfrac{5}{200 \times 10^3} = 25\,\mu\text{A}$ and $\beta = \dfrac{I_C}{I_B} = 133.3$. The output signal is shown in Fig. 29.58]

10. In the circuit shown in Fig. 29.59, the base current is $I_b = 5.0$ μA, base resistor $R_b = 1.0$ MΩ, collector resistance $R_c = 1.1$ kΩ. The collector current $I_c = 5.0$ mA and the DC voltage in the collector circuit $V_{cc} = 6.0$ V. (*i*) Can this circuit be used as an amplifier? (*ii*) What happens if the resistance R_c is 400 Ω and I_b, I_c and R_b remains same?

[**Hint:** (*i*) $V_{be} = V_{cc} - I_b R_b = 6 - 5 \times 10^{-6} \times 10^6 = 1.0\text{V}$
$V_{ce} = V_{cc} - I_c R_c = 6 - 1.1 \times 10^3 \times 5 \times 10^{-3} = 6 - 5.5 = 0.5$ V. The collector is at +0.5 V with respect to emitter and base is 1 V with respect to emitter. Hence base is 0.5 V with respect to collector. Thus both junctions are forward biased. Thus circuit cannot act like amplifier. (*ii*) When $R_c = 400\ \Omega$ then $V_{ce} = V_{cc} - I_c R_c = 6 - 5 \times 10^{-3} \times 400 = 4$ V.
Now emitter-base junction is forward biased and base collector junction is reverse biased and can work as amplifier]

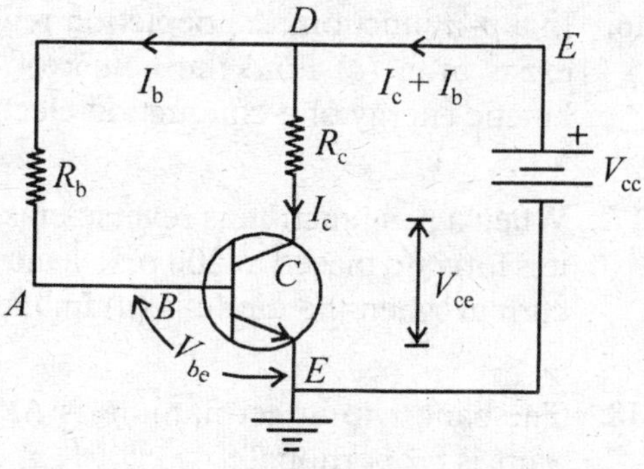

Fig. 29.59

11. A change of 0.2 mA in the base current causes a change of 5 mA in the collector current for a common emitter amplifier. (*i*) Find the AC current gain of the transistor (*ii*) If input resistance is 2 kΩ and its voltage gain is 75, calculate the load resistance used in the circuit. [**Ans.** (*i*) 25 (*ii*) 6 kΩ]

12. Write the truth table for circuit given in Fig. 29.60 consisting of NOR gates only. Identify the logic operations performed by the two circuits.
[**Ans.** (*a*) $Y = \overline{A + B} = \overline{A}$. It is a NOR gate with both input terminals connected acts like a NOT gate. (*b*) $Y = \overline{\overline{A} + \overline{B}} = A \cdot B$ and acts like a AND gate]

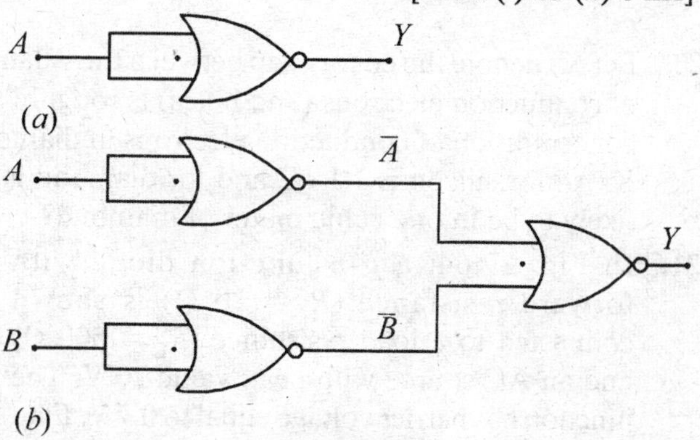

Fig. 29.60

13. Write the truth table for the circuit shown in Fig. 29.61. Show that it represents a XOR gate.

14. Write the Boolean equation and truth table for the circuit given in Fig. 29.62. [**Ans.** $Y = A(B + C)$]

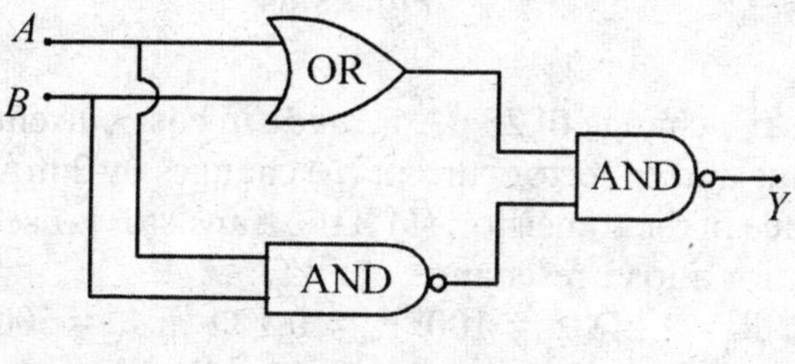

Fig. 29.61

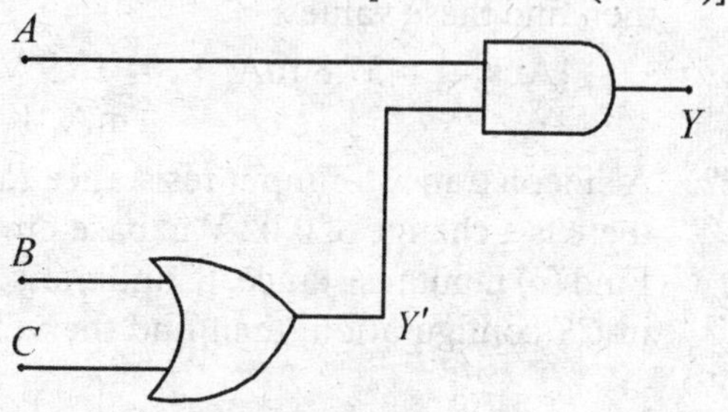

Fig. 29.62

15. Find the output Y of the circuit given in Fig. 29.63 if the inputs are $A = 0, B = 0$; $A = 0, B = 1$; $A = 1, B = 0$ and $A = 1, B = 1$.
[**Ans.** $A = B = 0, Y = 0$; $A = 0, B = 1, Y = 0$, $A = 1, B = 0, Y = 0$, $A = 1, B = 1, Y = 1$. It acts like AND gate]

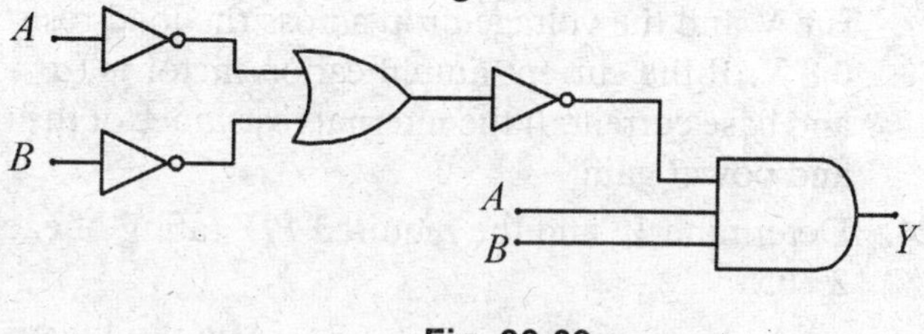

Fig. 29.63

16. In a p–n junction, the depletion region is 400 nm wide and an electric field of 5×10^5 V/m exists in it. (*a*) Find the height of the potential barrier (*b*) What should be the minimum kinetic energy of a conduction electron which can diffuse from the n–side to the p–side?

[**Ans.** (*a*) 0.2 V (*b*) 0.2 eV]

17. When a p–n junction is reverse-biased, the current becomes almost constant at 25 μA. When it is forward biased at 200 mV, a current of 75 μA is obtained. Find the magnitude of diffusion current when the diode is (*a*) unbiased (*b*) reverse biased (*c*) forward biased at 200 mV.

[**Ans.** (*a*) 25 μA (*b*) Zero (*c*) 100 μA]

18. The band gap in germanium is $\Delta E = 0.68$ eV. Assuming that the number of hole-electron pairs is proportional to $e^{-\Delta E/k_BT}$, find the percentage increase in the number of charge carriers in pure germanium as the temperature is increased from 300 K to 320 K. [**Ans.** $f = 127$]

19. Consider a p–n junction diode having the characteristic $i = i_0(e^{eV/k_BT} - 1)$ where $i_0 = 20$ μA. The diode is operated at $T = 300$ K. (*a*) Find the current through the diode when a voltage of 300 mV is applied across it in forward bias. (*b*) At what voltage does the current double?

[**Ans.** (*a*) $i = 2$ A (*b*) 318 V]

20. Let ΔE denote the energy gap between the valance band and conduction band. The population of conduction electrons (and holes) is roughly proportional to $e^{-\Delta E/2k_BT}$. Find the ratio of the concentration of conduction electrons in diamond to that in silicon at room temperature 300 K. ΔE for silicon is 1.1 eV and for diamond is 6.0 eV. How many conduction electrons are likely to be in one cubic meter of diamond? [**Ans.** (*a*) 2.3×10^{-33} (*b*) almost zero]

21. In Fig. 29.64, a p–n junction diode with forward resistance ($R_f = 10\ \Omega$) is shown connected to a load resistance ($R_L = 500\ \Omega$) and an AC source with peak value 20 V. The junction has barrier voltage equal to 0.7 V. Find the peak current through diode and peak voltage across load. If diode is assumed to be ideal, then find these values.

[**Ans.** $i_0 = 37.8$ mA, $V_L = 18.9$ V $i_0 = 4.0$ mA, $V_f = 20$ V]

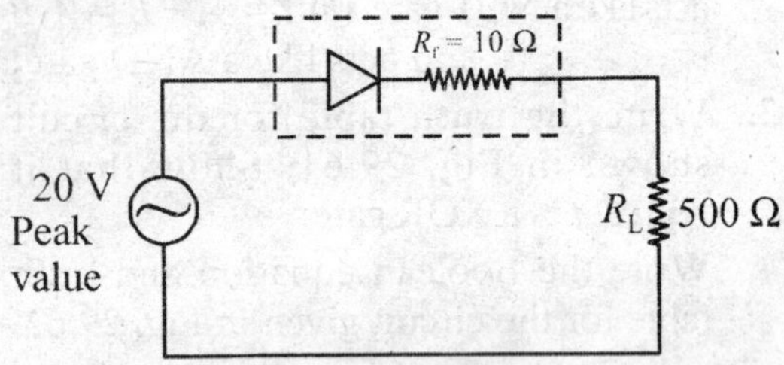

Fig. 29.64

22. A silicon transistor input resistance $R_i = 665\ \Omega$. If a change of 20 μA is made in base current, there is a change of 0.02 V in base-emitter voltage and collector current (i_c) changes by 2 mA. Find (*a*) input resistance, β, and transconductance of the transistor. (*b*) If the transistor is used in *CE* configuration, then find the voltage gain for a load resistance of 5 kΩ.

[**Ans.** $R_i = 1$ kΩ, $\beta = 100$, $g_m = 0.1\ \Omega^{-1}$, $A_V = 500$]

23. An n–p–n transistor is connected in common emitter configuration in which collector supply is 8 V and the voltage drop across the load resistance 800 Ω connected in collector circuit is 0.8 V. If the current amplification factor is ($\alpha = 25/26$) determine collector-emitter voltage and base current. If the internal resistance of the transistor is 200 Ω, calculate the voltage gain and power gain. [**Ans.** $V_{CE} = 7.2$ V, $A_V = 3.846$, $A_P = 3.698$]

24. Determine V_0 and the required *PIV* rating of each diode for the configuration shown in Fig. 29.65.

[**Ans.** Full wave rectified form Peak = 100 V, *PIV* = 100 V]

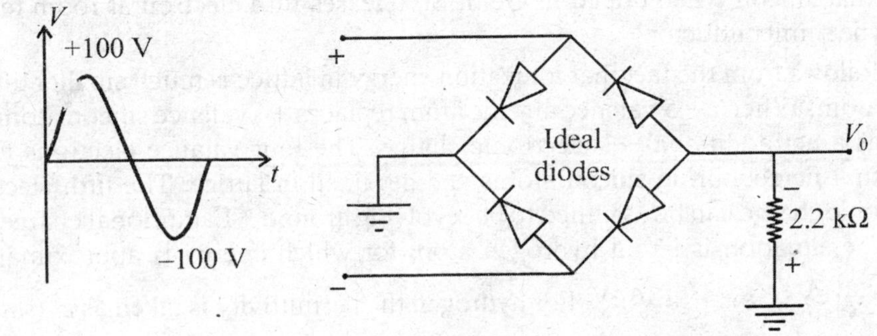

Fig. 29.65

25. (*a*) Given P_{max} = 14 mW for each diode (Fig. 29.66). Determine the maximum current rating of each diode. The knee voltage for diodes is 0.7 V. (*b*) Determine I_{max} for $(V_i)_{max}$ = 160 V. (*c*) Determine the current through each diode at $(V_i)_{max}$ using the results of part (*b*). (*d*) Is the current determined in part (*c*) less than the maximum rating determined in part (*a*)? (*e*) If only one diode was present, determine the diode current and compare it to the maximum rating.

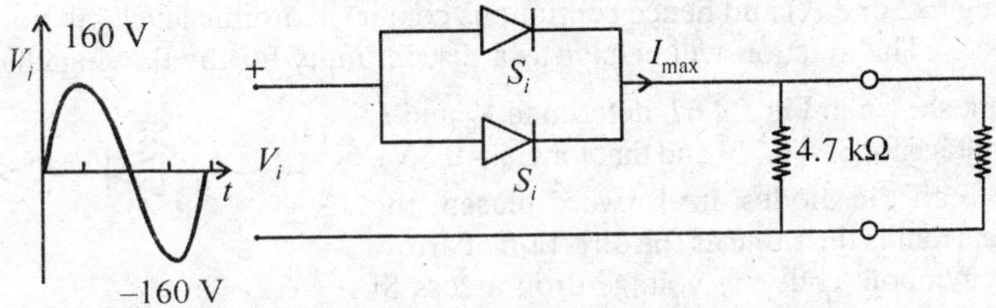

Fig. 29.66

[**Ans.** (*a*) $(I_D)_{max}$ = 20 mA. (*b*) I_{max} = 36.71 mA (*c*) I_D = 18.36 mA (*d*) Yes (*e*) I_D = 36.71 mA > $(I_D)_{max}$ = 20 mA]

SHORT ANSWER TYPE QUESTIONS

1. C, Si and Ge have same lattice structure. Why is C insulator while Si and Ge intrinsic semiconductors?

[**Ans.** The 4 valance electrons (bonding electrons) of C, Si and Ge lie respectively, in the second, third and fourth orbit. Hence energy required to take out an electron from these atoms (*i.e.*, ionization energy) is least for Ge followed by Si and highest for C. Hence number of free electrons for conduction in Ge, and Si are significant but negligibly small for C]

2. Is the ionization energy of an isolated free atom different from the ionization energy E_g for the atoms in a crystalline lattice?

[**Ans.** The ionization energy of an isolated atom is different from its value in crystalline lattice. In crystalline lattice, each bond electron is influenced by large number of atoms in the surrounding crystalline lattice]

3. The ionization energy of an isolated As (a pentavalent atom) is very large. Then how it is possible that silicon when doped in Ge or Si, releases fifth electron at room temperature to form *n*-type semiconductor?

[**Ans.** It follows from the fact that ionization energy in lattice is much smaller than that for an isolated atom. When a +5 valence arsenic atom replaces +4 valence silicon atom. This means that it has a net additional +1 electronic charge. The four valance electrons form covalent bond with 4 neighbouring silicon atoms and get fixed in lattice. The fifth electron with net –1 electronic charge can be assumed to be revolving around +1 additional charge as explained above. The situation is like a hydrogen atom for which energy is approximately given by $E_n = -\frac{13.6}{n^2}$ eV or $E_1 = -13.6$ eV. For hydrogen the permittivity is taken as ε_0 (same as that for vacuum). Whereas if the medium has a permittivity ε_r relative to ε_0, then the above expression can be written as $E_1 = -\frac{13.6}{\varepsilon_r^2}$ eV. For Si $\varepsilon_r = 12$. Hence, E_1 becomes 144 times less ≈ 0.1 eV. Thus ionization is possible]

4. Can we take one slab of *p*–type semiconductor and physically join it to another *n*–type semiconductor to get a *p*–*n* junction?

[**Ans.** No. Any slab, however flat, will have roughness much larger than the inter-atomic crystal spacing (~2 or 3 Å) and hence continuous contact at atomic level will not be possible. The junction will behave as a discontinuity for the flowing charge carriers]

5. In the network shown in Fig. 29.67, determine V_0 and I_d (Take knee voltage for Si = 0.7 V and that for Ge = 0.3 V).

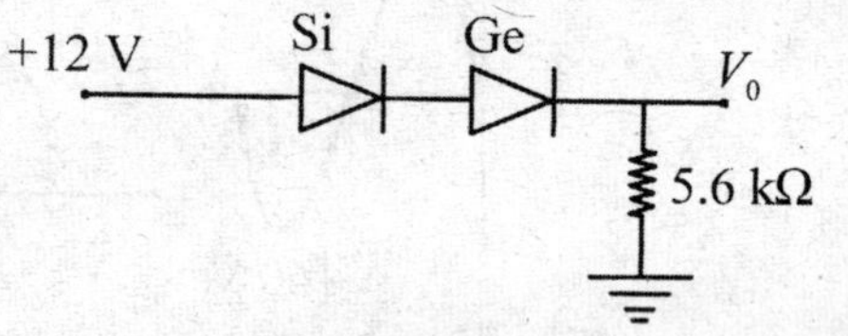

Fig. 29.67

[**Ans.** Since both the diodes are forward biased, the direction of current is the same as the direction of arrow in the diode symbol. Thus the voltage drop across Si diode ($V_1 = 0.7$ V) and that across G diode ($V_2 = 0.3$ V). Here $E = 12$ V > (0.7 + 0.3 V). Hence $V_0 = E - (V_1 + V_2)$

$$= 12 - 1 = 11 \text{ V and current } I_d = \frac{V_0}{R} = \frac{11}{5.6 \times 10^3} = 1.96 \text{ mA}\ ]$$

6. In the circuit shown in Fig. 29.68, determine the voltage V_0 for the circuit net work. Given knee voltage for Si diode as 0.7 V and that for Ge diode as 0.3 V.

[**Ans.** Since the two diodes are in parallel hence the potential difference across the two diodes should be same. When the potential difference becomes 0.3 the Ge diode starts conducting. This will not allow the potential difference to exceed beyond 0.3 V. Hence Si diode will not conduct because the potential difference across Si diode will remain also 0.3 V. Hence $V_0 = 12 - 0.3 = 11.7$ V and the current through resistance is $i = \frac{V_0}{R} = \frac{11.7}{2.2 \times 10^3} = 5.35 \times 10^{-3}$ A

$$= 5.35 \text{ mA}\ ]$$

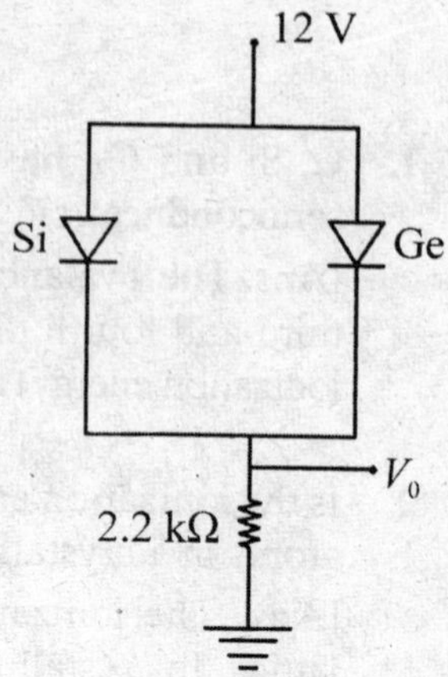

Fig. 29.68

7. The current in the forward bias is known to be more (~mA) than the current in reverse bias (~μA). What is the reason then, to operate the photodiodes in reverse bias?

[**Ans.** Let us consider the case of *p*–type semiconductor. The majority carrier density (of holes *p*) in this is considerably large than the minority electron density (*n*). This means $p >> n$. On irradiating it the number of both types of carriers would increase equally $\Delta n = \Delta p$. Then the fractional change in majority carriers $\left(\frac{\Delta p}{p}\right)$ would be much less than that in the minority carriers $\left(\frac{\Delta n}{n}\right)$. Thus the fractional change due to the photoeffects on the minority carriers dominated reverse bias current is more easily measurable than the fractional change in the forward bias current. Hence photodiodes are preferably used in reverse bias conditions for measuring light intensity]

8. A *p*–*n* junction diode can withstand current up to 10 mA. Under forward bias, the diode has a potential drop of 0.5 V across it which is assumed to be independent of current. What is the maximum voltage of the battery used to forward bias, the diode when a resistance of 200 Ω is connected in series with it?

[**Ans.** As the potential drop across diode is 0.5 V and a current of 10 mA passes through it. Hence its resistance is $R = \frac{0.5}{10 \times 10^{-3}} = 50\,\Omega$. Since a resistance of 200 Ω is connected in series with it, hence total resistance of circuit is 200 + 50 = 250 Ω. Hence maximum voltage of the battery for a current of 10 mA is $V = 250 \times 10 \times 10^{-3} = 2.5$ V]

9. A semiconductor has electron concentration of 0.45×10^{12} m^{-3} and hole concentration of 5.0×10^{20} m^{-3}. Calculate its conductivity. Given electron mobility (μ_e) = 0.135 m^{2}V^{-1}s^{-1}, and hole mobility (μ_h) = 0.048 m^{2}V^{-1}s^{-1}.

[**Ans.** The conductivity of a semiconductor is given by $\sigma = e(\mu_e n_e + \mu_h n_h) = 1.6 \times 10^{-19}\,(0.135 \times 0.45 \times 10^{12} + 0.048 \times 5.0 \times 10^{20})$. Since $n_e << n_h$ hence the contribution of electrons to conductivity can be neglected. Hence $\sigma = 1.6 \times 10^{-19} \times 0.048 \times 5 \times 10^{20} \doteq 3.84$ mho sm^{-1}, mhos stand for Ω^{-1}, this is also called Siemen]

10. The electrical conductivity of a semiconductor increases when electromagnetic radiation of wavelength shorter than 2480 nm is incident on it. Find the band gap of the semiconductor. Given $h = 6.63 \times 10^{-34}$ J–s, $c = 3.0 \times 10^8$ m/s and 1 eV = 1.6×10^{-19} J.

[**Ans.** The band gap is the minimum energy required to push an electron from valance band to the conduction band. Thus $\Delta E = h\nu = \frac{hc}{\lambda} = \frac{6.63 \times 10^{-34} \times 3 \times 10^8}{2480 \times 10^{-9}} = 8.02 \times 10^{-20}$ J

$$\equiv \frac{8.02 \times 10^{-28}}{1.6 \times 10^{-19}} = 0.5 \text{ eV}$$]

11. A transistor has a current amplification factor of 50 in a *CE*–amplifier circuit. The collector resistance is chosen as 5 kΩ and the input resistance is 1 kΩ. Calculate the output voltage if input voltage is 0.01 V.

[**Ans.** In amplifier voltage gain is $A_V = \frac{V_0}{V_i} = \frac{R_c i_c}{R_b \cdot i_b} = \beta \frac{R_c}{R_b} \therefore V_0 = \beta \frac{R_c}{R_b} V_i = \frac{50 \times 5 \times 10^3 \times .01}{10^3}$

$$= 2.5 \text{ V}$$]

12. Construct the truth table for the function *Y* of *A* and *B* represented by Fig. 29.69.

[**Ans.** This is combination of OR and AND gate. Clearly $X = A + B$ and the AND gate receives A and $A + B$ as input. Thus output is $Y = A \cdot (A + B)$.

The truth table is follows

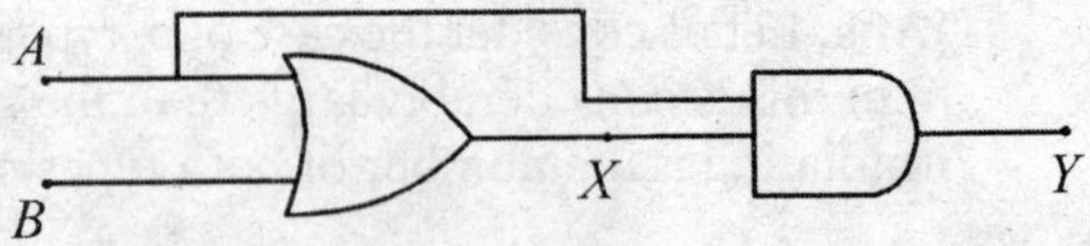

Fig. 29.69

A	B	$X = A + B$	$Y = A(A + B)$
0	0	0	0
0	1	1	0
1	0	1	1
1	1	1	1

A	B	Y
0	0	0
0	1	0
1	0	1
1	1	1

13. Determine the number density of donor atoms which have to be added to an intrinsic germanium semiconductor to produce an n–type semiconductor of conductivity 5 mho/cm. Given that the mobility of electrons in n–type semiconductor is 3850 cm²/Volt-sec.

[**Ans.** Let the concentration of donor atom (N_d) is equal to the concentration of free electrons (n_e) we have conductivity $\sigma = n_e e \mu_e \therefore n_e = \dfrac{\sigma}{e\mu_e}$. Given is σ = 5 mho/cm = 500 mho m^{-1}, μ_e = 3850 cm²/V-sec = 0.385 m^{2}V^{-1}s^{-1} and $e = 1.6 \times 10^{-19}$ $\therefore n_e = N_d = \dfrac{500}{1.6 \times 10^{-19} \times 0.385} = 8.1 \times 10^{21}$ m^{-3}]

14. An intrinsic sample of germanium crystal has a hole density of 10^{13} cm^{-3} at the room temperature. When doped with antimony the whole density is decreased to 10^{11} cm^{-3} at the same temperature. Find the number density of majority charge carriers.

$$\left[\textbf{Ans.}\text{ In semiconductor } n_e n_h = n_i^2. \text{ or } n_e \equiv \frac{n_i^2}{n_h} = \frac{(10^{13})^2}{10^{11}} = 10^{15}\text{ cm}^{-3}\right]$$

15. The circuit shown in Fig. 29.70 contains two diodes each with a forward resistance of 50 Ω and with infinite reverse resistance. If the battery voltage is 6 V, find the current through 100 Ω resistance.

[**Ans.** From the circuit diagram it is apparent that diode to D_2 is reverse biased hence will not conduct. Thus the net resistance of the circuit is R = 150 + 100 = 250 Ω. Hence the current $i = \dfrac{6}{250} = 0.024$ amp. This is the current which passes through the resistance of 100 Ω]

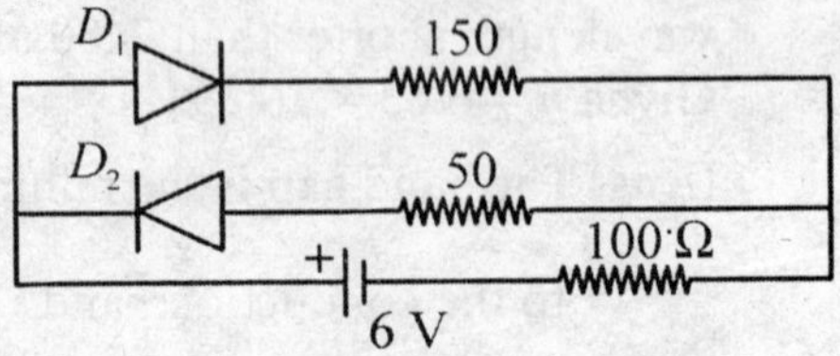

Fig. 29.70

16. In an n–p–n transistor 10^{10} electrons enter the emitter in 10^{-6} s. 2% of the electrons are lost in the base. Calculate the current transfer ratio and current amplification factor.

[**Ans.** Current through emitter $i_e = \dfrac{10^{10} \times 1.6 \times 10^{-19}}{10^{-6}} = 1.6$ mA. The base current is $i_b = 0.02\, i_e$ (2%) = 0.02 × 1.6 = 0.032 mA. In transistor, we have $I_e = i_c + i_b \therefore i_c = i_e - i_b = 1.6 - 0.032$ = 1.568 mA $\therefore$ Current transfer ratio $= \dfrac{i_c}{i_e} = \dfrac{1.568}{1.6} = 0.98$ and current amplification factor

$$= \frac{I_c}{I_b} = \frac{1.568}{0.032} = 49 \Big]$$

17. A transistor is connected in common emitter (CE) configuration. The collector supply is 8 V and the voltage drop across a resistor of 800 Ω in the collector circuit is 0.5 V. If the current gain factor α is 0.96, find the base current.

[**Ans.** $\beta = \frac{\alpha}{1-\alpha} = \frac{0.96}{1-0.96} = 24$, Collector current $i_c = \frac{V_c}{R} = \frac{0.5}{800} = 0.625 \times 10^{-3}$ A $\therefore\ i_b = \frac{i_c}{\beta}$

$= \frac{0.625 \times 10^{-3}}{24} = 26\,\mu\text{A}$]

18. The reverse biased current of a particular *p–n* junction diode increases when it is exposed to light of wavelength less than or equal to 600 nm. Assume that the increase in carrier concentration takes place due to the creation of new hole-electron pairs by the light. Find the band gap.

[**Ans.** The reverse biased current is caused mainly due to the drift current. The drift current in a *p–n* junction is caused by the formation of a new hole-electron pairs and their subsequent motions in the depletion layer. When junction is exposed to light, it may absorb energy from the light photons. If this energy supplied by a photon is greater than (or equal to) the band gap, a hole-electron pair may be formed. Thus the reverse biased current will increase if the light photons have energy greater than (or equal to) the band gap. Hence band gap is equal to the energy of a photon of 600 nm light. This is $E = \frac{hc}{\lambda} = \frac{6.6 \times 10^{-34} \times 3 \times 10^{8}}{600 \times 10^{-9}} = 4.4 \times 10^{-19}$J

$= \frac{3.3 \times 10^{-19}}{1.6 \times 10^{-19}} = 2.06$ eV]

19. A *p–n* photodiode is fabricated from a semiconductor with a band gap of 2.8 eV. Can it detect a wavelength of 600 nm?

[**Ans.** The energy required by electron to go from valence band to conduction band is $E_g = 2.8$ eV. The photon of wavelength λ has energy $E = \frac{hc}{\lambda}$. For photon of λ = 600 nm. $E = \frac{6.6 \times 10^{-34} \times 3 \times 10^{8}}{6 \times 10^{-7}} = 3.3 \times 10^{-19}$ J = 2.07 eV. Since $E < E_g$ the *p–n* junction cannot detect the radiation of wavelength 600 nm]

20. In *p–n–p* transistor circuit, the collector current is 10 mA. If 90% of the holes reach the collector, find the emitter and base current.

[**Ans.** Collector current i_c = 10 mA. Only 90% of the holes reach the collector. This means that emitter current is $i_e = \frac{100}{90} \times 10 = 11$ mA and base current $i_b = i_e - i_c = 11 - 10 = 1$ mA]

21. For a transistor connected in common emitter mode, the voltage drop across the collector is 2 V and β is 50. Find the base current if $R_c = 2$ kΩ.

[**Ans.** The collector current $I_c = \frac{V_{CE}}{R_c}$ $\therefore\ I_c = \frac{2}{2 \times 10^{3}} = 1$ mA. Since β = 50, hence $i_b = \frac{i_c}{\beta}$

$= \frac{1}{50} = 2 \times 10^{-2}$ mA = 20 μA]

22. Why does a semiconductor get damaged when a heavy current flows through it?

[**Ans.** When a heavy current flows, the semiconductor gets heated up. Many covalent bonds break liberating a large number of free electrons. The semiconductor loses its property of controlled conduction. It becomes an ordinary conductor]

23. What is doping in a semiconductor? Why is it done?

[**Ans.** Doping is the process of addition of small amounts of certain specific impurity atoms having valance different from that of host atoms to a pure semiconductor. Doping is done to increase the number of mobile electrons/holes and hence to increase the conductivity of a semiconductor]

24. Why does the width of depletion layer of a *p–n* junction increase in reverse biasing?

[**Ans.** During reverse biasing, the positive terminal of the external battery attracts electrons from *n*–region and its negative terminal attracts holes from the *p*–region *i.e.*, the majority charge carriers move away from the junction. This increases the width of depletion layer.]

25. In the working of a transistor, the emitter-base (EB) junction is forward biased while collector-base (CB) junction is reverse biased? Why?

[**Ans.** Only forward biased emitter-base junction can send the majority charge carriers from emitter to base and only reverse biased collector can collect these majority charge carriers from the base region. If the emitter is reverse biased, no charge carriers will flow towards the collector and hence no current will flow through the transistor]

26. By increasing the load resistance, can we increase the voltage gain of an amplifier indefinitely?

[**Ans.** Initially, the voltage gain increases as voltage gain $A_V = -g_m R_L$. Where g_m is the transconductance $\left(g_m = \frac{\Delta I_c}{\Delta V_B}\right)$. But gain of an amplifier cannot be increased independently by increasing R_L. We know that $V_{CE} = V_{CC} - I_C R_L$. If we increase R_L, V_{CE} decreases and if V_{CE} becomes less than V_{BE} both the junctions become forward biased and saturation starts.]

27. What happens in a transistor when both the emitter and collector are reverse biased? What is this condition known as?

[**Ans.** When both the emitter and collector are reverse biased, no current flows through the transistor as there is no conduction due to majority charge carriers across the emitter-base or collector-base junction. This condition is known as cut-off state.]

28. Why a transistor cannot be used as a rectifier?

[**Ans.** To use a transistor as a rectifier, either its emitter-base portion or collector-base portion has to be used. As base is thin and lightly doped, either of the two portions will not work as a *p–n* junction. As such transistor cannot be used as a rectifier]

29. Is the voltage gain of a CE amplifier constant, irrespective of the magnitude of the input signal? Give reasons.

[**Ans.** No. The voltage is constant for a limited range of input signal V_i. During the positive half cycle, base current I_b increases which increases collector current I_c. If the increase in I_c is large such that V_{CE} becomes less than V_{BE} the transistor goes into saturation state. Thus when $\beta I_b \geq I_c$ the transistor does not work as an amplifier]

30. Which one of the transistors *p–n–p* and *n–p–n* is more useful and why?

[**Ans.** *n–p–n* transistor is more useful than *p–n–p* transistor. In *n–p–n* transistor, electrons are the main charge carriers while in *p–n–p* transistor holes are the main charge carriers. The electrons have higher mobility than holes as such *n–p–n* transistors are more commonly used]

31. Why do semiconductors obey Ohm's law for only low fields?

[**Ans.** The drift velocity of a charge carrier is proportional to electric field E. $v = \frac{eE}{m}\tau$. But v cannot be increased indefinitely by increasing E. At high speeds, due to high collision frequency τ, the relaxation time becomes very small, and the drift-velocity saturates at the thermal velocity (v_{th} = 10 m/s) and becomes independent of electric field at higher E values. At 300 K $v_{th} = \left(\frac{3k_BT}{m}\right)^{1/2} \approx 10^5$ m/*s* and $\tau = 10^{-12}$ sec. An electric field of 10^6 V/m causes saturation of drift velocity. Hence semiconductor obey Ohm's law for low electric fields ($E < 10^6$ V/m) and above this field I becomes independent of V.]

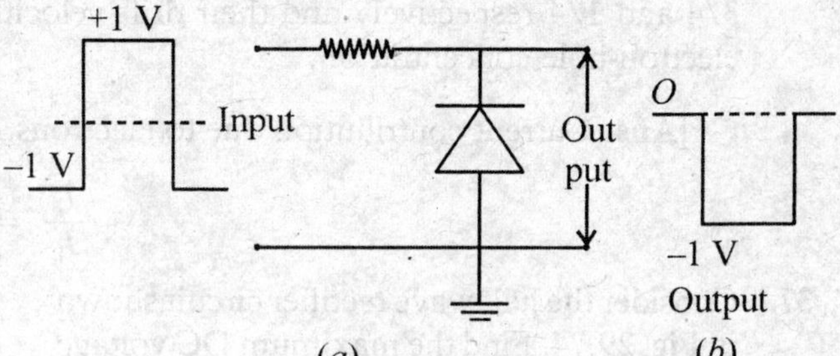

Fig. 29.71

32. A square wave (–1 V to +1 V) is applied to p–n junction diode as shown in Fig. 29.71 (*a*). Draw the output wave forms.

[**Ans.** The p–side of the diode is earthed as such the potential of p–side is zero. So the diode conducts when input is –1 V and does not conduct when input is +1 V. Since diode is ideal, the out-put across it will be either 0 or –1 V]

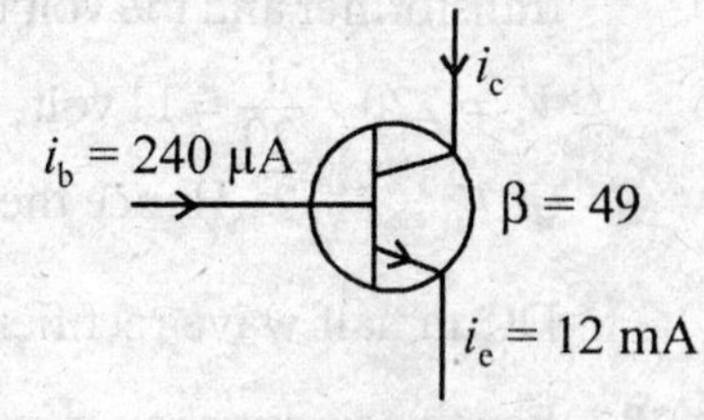

Fig. 29.72

33. In Fig. 29.72 is shown a circuit. Find α and i_c.

[**Ans.** $\alpha = \frac{\beta}{1+\beta} = \frac{49}{50} = 0.98$ and $\alpha = \frac{i_c}{i_e}$ or $i_c = \alpha\, i_e = 0.98 \times 12$ or $i_c = 11.76$ mA]

34. An amplifier is represented by the circuit shown in Fig. 29.73. R_i is the input resistance of the amplifier and voltage V_i is appearing across it. This voltage is amplified by a factor of A_V and appearing across the load as voltage V_0. An external voltage V_s is applied at the input terminals of the amplifier via a series resistance R_s. What will be the apparent gain $A'_V = V_0/V_s$ of the amplifier? Express the gains in terms of A_V, R_s and r_i.

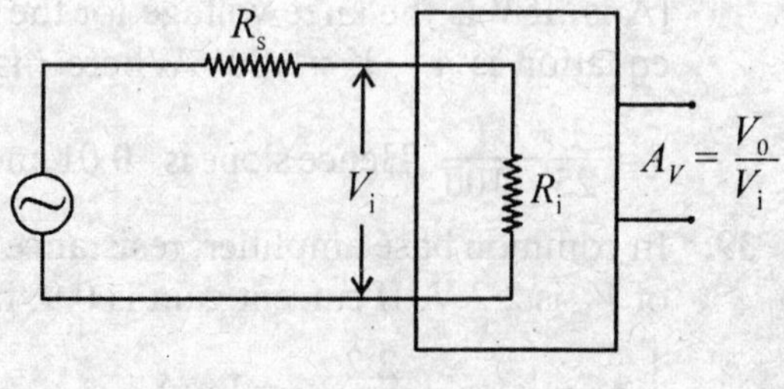

Fig. 29.73

[**Ans.** Voltage gain $A'_V = \frac{V_0}{V_s}$. Since V_s is the source voltage, hence the voltage across the input terminals of the amplifier is $V_i = \frac{V_S R_i}{R_i + R_s}$ and $A_V = \frac{V_0}{V_i} = \frac{V_0(R_i + R_S)}{V_S R_i}$. Hence voltage gain $A'_V = \frac{V_0}{V_S} = \frac{A_V R_i}{R_i + R_S}$]

35. A p-type semiconductor has acceptor levels 57 meV above the valance band. Find the maximum wavelength of light which can create a hole.

[**Ans.** To create a hole in valence band, an electron from the valence band should be given sufficient energy to go into one of the acceptor levels. Since acceptor levels are 57 meV above the valence band, at least 57 meV is needed to create a hole. If λ is wavelength of light, its photon will have an energy hc/λ. Thus to create a hole $\frac{hc}{\lambda} \geq 57$ meV or $\lambda \leq \frac{hc}{57}$

$$= \frac{6.63 \times 10^{-34} \times 3 \times 10^{8}}{57 \times 10^{-3} \times 1.6 \times 10^{-19}} \text{ or } \lambda \leq 2.18 \times 10^{-5} \text{ m}$$]

36. For some semiconductor the fraction of current contributions due to electrons and holes are 3/4 and 1/4 respectively and their drift velocities are in the ratio of 5/2. Find the ratio of electron-hole concentration.

[**Ans.** Current contribution due to electrons $I_e = n_e \cdot \mu_e \cdot Ae$ and that of holes is $I_h = n_h \mu_h Ae$

$$\therefore \frac{I_e}{I_h} = \frac{n_e \cdot \mu_e}{n_h \mu_h} \text{ or } \frac{n_e}{n_h} = \frac{I_e}{I_h} \times \frac{\mu_h}{\mu_e} = \frac{3}{4} \times \frac{4}{1} \times \frac{2}{5} = \frac{6}{5}$$]

37. Consider the half wave rectifier circuit shown in Fig. 29.74. Find the maximum DC voltage available across the load.

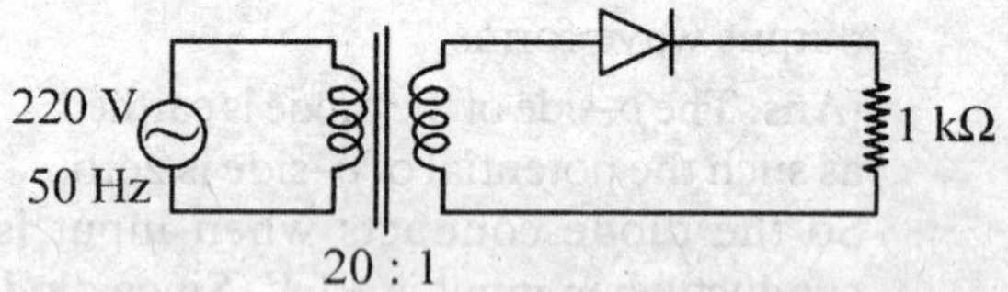

Fig. 29.74

[**Ans.** The transformer is an step down transformer and the voltage across secondary $V_s = 220 \times \frac{1}{20} = 11$ volt. It has a peak value of $V_0 = 11\sqrt{2}$. Hence the maximum value of DC in half wave rectifier is $V_{DC} = \frac{V_0}{\pi} = \frac{11\sqrt{2}}{\pi} = 4.95$ V]

38. For the circuit shown in Fig. 29.75, find the magnitude of the slope of load line on $V \cdot I$ characteristics of the diode.

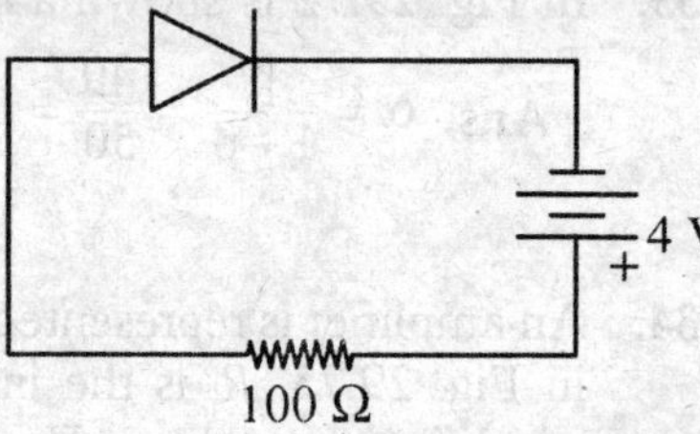

Fig. 29.75

[**Ans.** If V is the knee voltage for the diode then load line equation is $4 - V = 100i$. Where i is the current. Hence $i = \frac{1}{25} - \frac{V}{100}$. Hence slope is –0.01 and magnitude is 0.01]

39. In common base amplifier, resistance is 2.2 kΩ (R_L, which is connected to collector) the value of V_L is 2.2 V. If current gain is 0.9, find value of base current.

[**Ans.** $i_c = \frac{2.2}{2.2 \times 10^3} = 1$ mA. $\alpha = \frac{i_c}{i_e} = 0.9 \therefore i_e = \frac{i_c}{0.9} = \frac{1}{0.9} = 1.1$ mA $\therefore i_b = i_e - i_c = 1.1 - 1$ = 0.1 mA]

40. If the resistivity of copper is 1.7×10^{-6} Ω-cm, then find the mobility of electrons in copper, if each atom of copper contributes one free electron for conduction. Take atomic weight of copper 63.5 and its density 9 g/cc.

[**Ans.** Conductivity in terms of mobility is $\sigma = \frac{1}{\rho} = ne\mu$ where ρ is resistivity $\therefore \mu = \frac{1}{ne\rho}$. If A is atomic weight and n_0 is Avogadro number, the number of free electrons per unit volume is $n = \frac{n_0 d}{A}$. Hence mobility $\mu = \frac{A}{n_0 de\rho} = \frac{63.5}{6.02 \times 10^{23} \times 9 \times 1.6 \times 10^{-19} \times 1.7 \times 10^{-6}} = 43$ cm²/V-s]

41. Find the input resistance of a common emitter transistor or amplifier, if the output resistance is 500 kΩ. The current gain $\alpha = 0.98$ and power gain is 6.0625×10^6.

Ans. Power gain in CE configuration is $A_P = \beta^2\left(\frac{R_0}{R_i}\right)$ and $\beta = \frac{\alpha}{1-\alpha} = \frac{0.98}{1-0.98} = 49$

$$\therefore R_i = \frac{\beta^2 \cdot R_0}{A_P} = \frac{(49)^2 \times 500 \times 10^3}{6.0625 \times 10^6} = 198\,\Omega$$

MULTIPLE CHOICE QUESTIONS

PART I

In the following, a question is followed by four alternatives, of them only one is correct. Therefore, the candidate has to select the correct choice from amongst the alternatives.

1. A semiconducting device is connected in a series circuit with a battery and a resistance. A current is found to pass through the circuit. If the polarity of the battery is reversed, the current drops to almost zero. The device may be

(a) an intrinsic semiconductor
(b) a p–type semiconductor
(c) an n–type semiconductor
(d) a p–n junction.

2. The circuit shown in Fig. 29.76 has two diodes each having forward resistance 50 Ω and reverse resistance ∞. A battery (emf = 6 V) and a resistance 100 Ω is connected. The current through 100 Ω resistor is

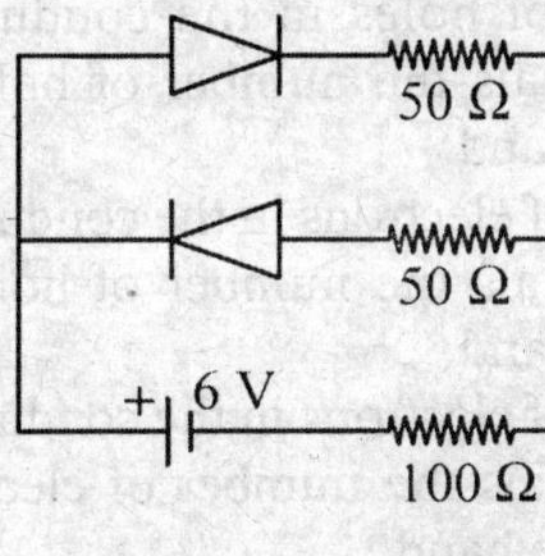

Fig. 29.76

(a) zero (b) 0.02 A
(c) 0.03 A (d) 0.048 A.

3. For given combination of gates (Fig. 29.77). If input at A, B, C is $A = B = C = 0$ and $A = B = 1$ and $C = 0$, the output at 0 respectively is

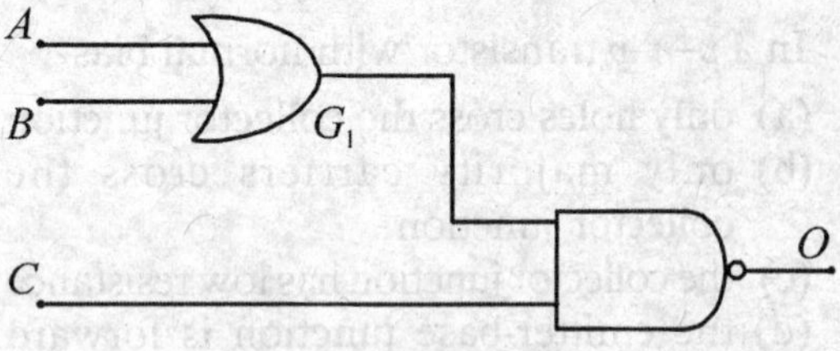

Fig. 29.77

(a) 0, 0 (b) 0, 1
(c) 1, 0 (d) 1, 1.

4. Majority charge carriers are

(a) electrons in p–type semiconductors and holes in n–type semiconductors
(b) electrons in n–type semiconductors and holes in p–type semiconductors
(c) electrons in both p– and n–type semiconductor
(d) holes in both p– and n–type semiconductors.

5. Choose the only incorrect statement from the following.

(a) In conductors, the valence and conduction bands may overlap.

(b) Substances with an energy gap of the order of 10 eV are insulators
(c) The resistivity of a semiconductor increases with increase in temperature
(d) The conductivity of a semiconductor increases with increase in temperature.

6. Two identical capacitors A and B are charged to the same potential V and are connected in two circuits at $t = 0$ as shown in Fig. 29.78. The charges on the capacitors at a time $t = RC$ are respectively

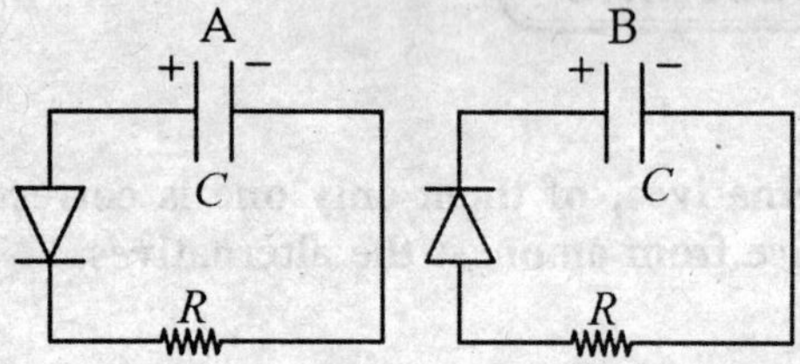

Fig. 29.78

(a) VC, VC (b) $\frac{VC}{e}, VC$
(c) $VC, VC/e$ (d) $\frac{VC}{2}, \frac{VC}{2}$.

7. In a p–n–p transistor with normal bias
(a) only holes cross the collector junction
(b) only majority carriers cross the collector junction
(c) the collector junction has low resistance
(d) the emitter-base junction is forward biased and the collector-base junction is reverse biased.

8. Mass action law in semiconductor is given by
(a) $n = n_i^2$ (b) $\frac{n}{p} = n_i^2$
(c) $np = n_i^2$ (c) $\frac{p}{n} = n_i^2$.

9. Two identical p–n junctions may be connected in series with a battery in three ways (Fig. 29.79). The potential difference across the two p–n junctions are equal in

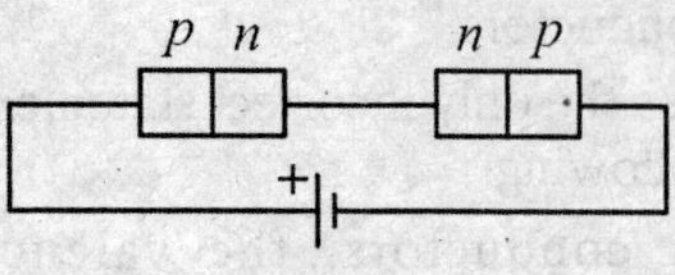

Circuit (i)

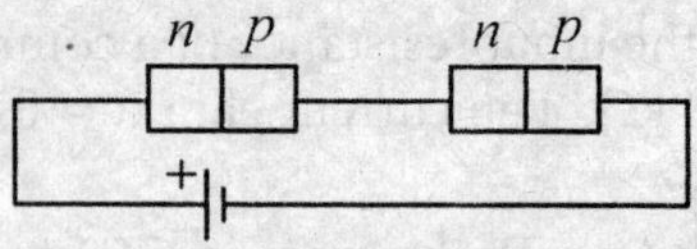

(ii)

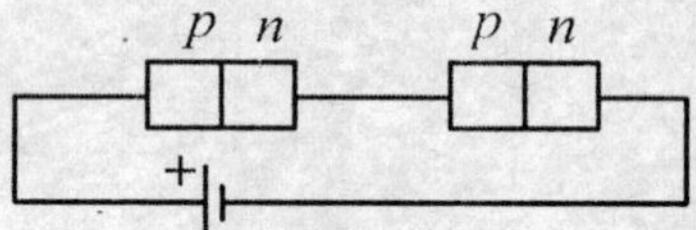

(iii)

Fig. 29.79

(a) circuit (i) and (ii)
(b) circuit (ii) and (iii)
(c) circuit (iii) and (i)
(d) circuit (i) only.

10. The difference between a p–type germanium crystal and ordinary germanium crystal is
(a) the p–type crystal is positively charged due to deficiency of electrons
(b) the p–type crystal is negatively charged due to surplus electrons
(c) the p–type crystal contains impurity atoms of higher valence electrons than germanium
(d) the p–type crystal contains impurity atoms of lower valance electrons than germanium.

11. In an intrinsic semiconductor
(a) number of holes in the conduction band equals the number of electrons in valence band
(b) number of holes in the conduction band equals the number of holes in valence band
(c) number of electrons in the conduction band equals the number of holes in valence band
(d) number of electrons in the conduction band equals the number of electrons in valence band.

12. The drift current in a p–n junction is
(a) from the n–side to the p–side
(b) from the p–side to the n–side

(c) from the n–side to the p–side if the junction is forward biased and in opposite direction if it is reverse biased
(d) from the p–side to the n–side if the junction is forward biased and in opposite direction if it is reverse biased.

13. A hole diffuses from the p–side to the n–side in a p–n junction. This means
(a) a bond is broken on the n–side and the electron freed from the bond jumps to the conduction band
(b) a conduction electron on the p–side jumps to a broken bond to complete it
(c) a bond is broken on the n–side and the electron freed from the bond jumps to a broken bond on the p–side to complete it
(d) a bond is broken on the p–side and the electron freed from the bond jumps to a broken bond on the n–side to complete it.

14. The circuit has two oppositely connected diodes in parallel what is the current flowing in the circuit?

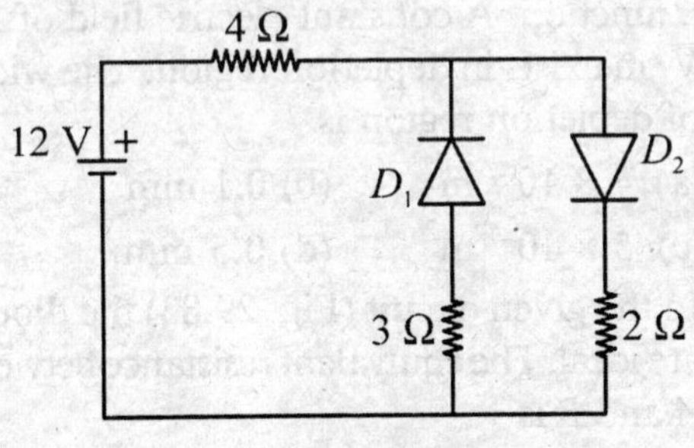

Fig. 29.80

(a) 2.31 A (b) 1.33 A
(c) 1.71 A (d) 2.0 A.

15. Out of the following circuits, which of the diodes is reverse biased?

(a)
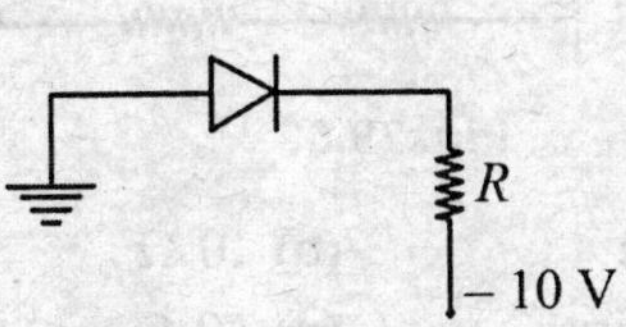

(b)
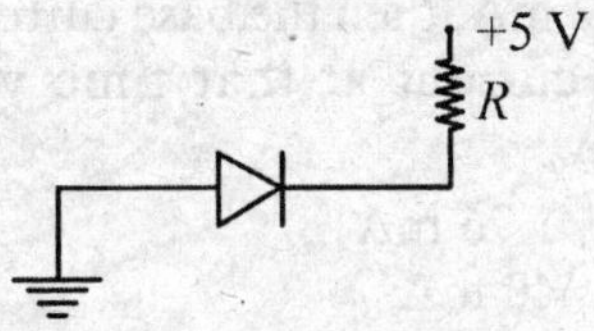

(c)
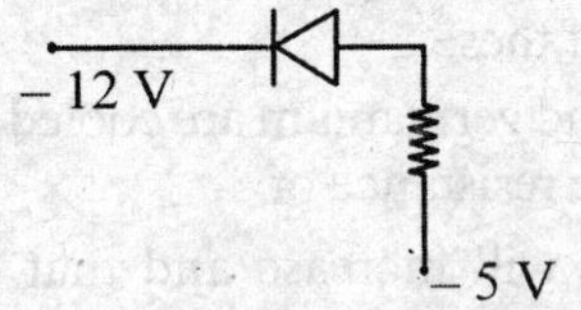

(d)
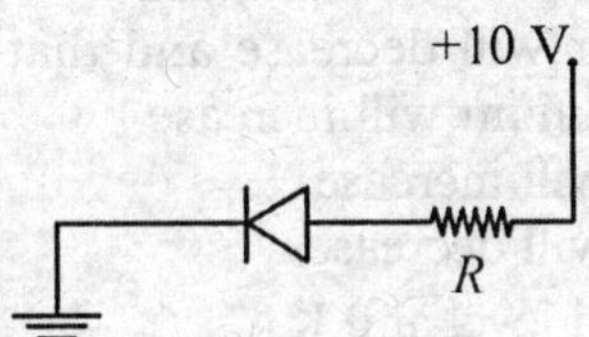

16. In the circuit shown in Fig. 29.81 the current in 5 kΩ resistor is

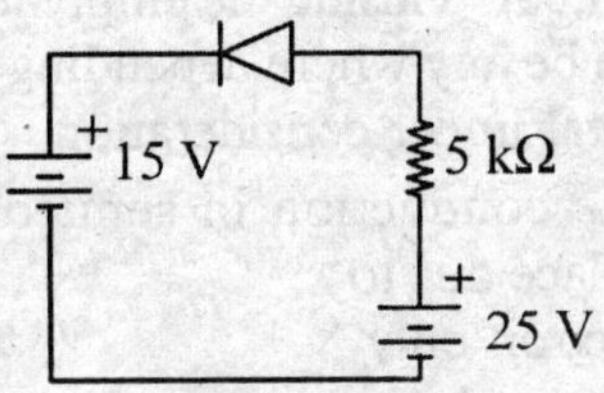

Fig. 29.81

(a) 8 mA (b) 4 mA
(c) 2 mA (d) 0.

17. Assuming the ideal behaviour of the diode the values of V_{out} and V_D for positive half cycle of input will be

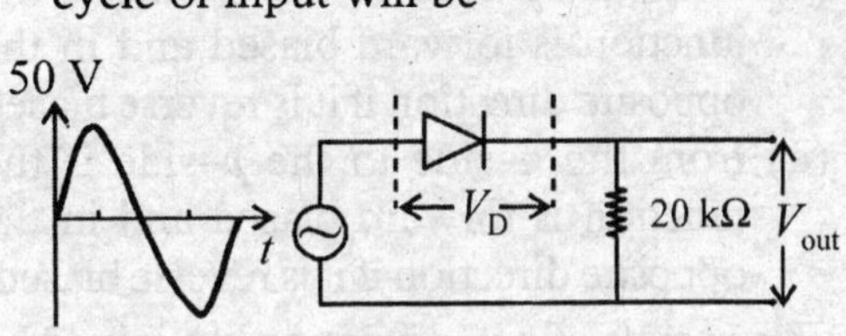

Fig. 29.82

(a) $V_D = 50$ V, $V_{out} = 0$V
(b) $V_D = 0$, $V_{out} = 50$ V
(c) $V_D = 50$ V, $V_{out} = 50$ V
(d) $V_D = 0$, $V_{out} = 0$

18. A transistor has a common base current gain $\alpha = 0.98$. If at some instant the emitter current is 2 mA, then the base current and collector current at that time will be respectively

(a) 40 μA, 1.96 mA
(b) 40 μA, 40 mA
(c) 1.96 μA, 1.96 mA
(d) none of these.

19. If copper and germanium are cooled to 100 K, then the resistance of

(a) copper will increase and that of the germanium will decrease
(b) copper will decrease and that of the germanium will increase
(c) both will increase
(d) both will decrease.

20. The Fermi level at 0 K

(a) separates the empty filled levels
(b) is situated inside the filled level
(c) is situated inside the empty levels
(d) can be any where depending upon the metal under consideration.

21. Electric conduction in semiconductors takes place due to

(a) electron only
(b) holes only
(c) both electrons and holes
(d) neither electrons nor holes.

22. The diffusion current in a *p–n* junction is

(a) from the *n*–side to the *p*–side
(b) from the *p*–side to the *n*–side
(c) from the *p*–side to the *n*–side if the junction is forward biased and in the opposite direction if it is reverse biased
(d) from the *n*–side to the *p*–side if the junction is forward biased and in the opposite direction if it is reverse biased.

23. If two ends of a *p–n* junction are joined by a wire

(a) there will be a steady current from the *n*–side to the *p*–side
(b) there will not be a steady current in the circuit
(c) there will be a steady current from the *p*–side to the *n*–side
(d) there may or may not be a current depending upon the resistance of the connecting wire.

24. Under reverse bias, the depletion layer is crossed

(a) by holes only
(b) by electrons only
(c) by both electrons and holes while moving in same direction
(d) by both electrons and holes when these are moving in opposite directions.

25. A *p*–type silicon semiconductor is made by adding one atom of indium per 5×10^7 atoms of silicon. [atom number density of silicon is $5 \times 10^{28}\ m^{-3}$]. The number of acceptor atoms per cc of silicon is

(a) 2×10^{30} (b) 10×10^{13}
(c) 1.0×10^{15} (d) 2.5×10^{18}.

26. If the intensity of electric field is E, then current density is directly proportional to

(a) E (b) $1/E$
(c) E^2 (d) $1/E^2$.

27. A potential barrier of 0.4 exists across a *p–n* junction. A constant electric field of 10^6 V/m exists in depletion region. The width of depletion region is

(a) 4×10^{-7} m (b) 0.1 mm
(c) 5×10^{-7} m (d) 0.5 mm.

28. In the given circuit (Fig. 29.83) the diodes are ideal. The equivalent resistance between A and B is

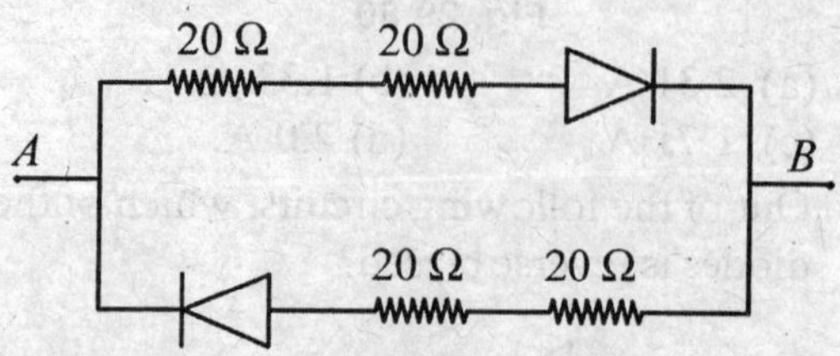

Fig. 29.83

(a) 20 Ω (b) 10 Ω
(c) infinity (d) 40 Ω.

29. A signal of 20 mV is applied to common emitter transistor amplifier circuit. Due to this, the change in base current and the change in collector current are 20 μA and 2 mA. The load resistance is 10 kΩ. The voltage gain is

(a) 20 (b) 10
(c) 50 (d) none of these.

30. In the operation of n–p–n transistor compared to that of a triode the p–base acts as

(a) cathode (b) grid
(c) plate (d) filament.

31. A transistor, when connected in common emitter mode has

(a) high input resistance and low output resistance
(b) medium input resistance and high output resistance
(c) very low input resistance and low output resistance
(d) high input resistance and high output resistance.

32. The depletion layer in a p–n junction diode consists of layer of

(a) positively charged donors on the p–side and negatively charged acceptors on the n–side
(b) negatively charged donors on the p–side and positively charged acceptors on the n–side
(c) positively charged donors on the n–side and negatively charged acceptors on the p–side
(d) negatively charged donors on the n–side and positively charged acceptors on the p–side.

33. The correct symbol for n–p–n transistor is

(a) 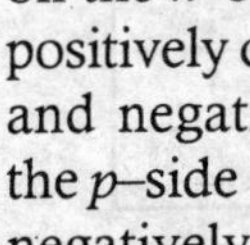
(b) 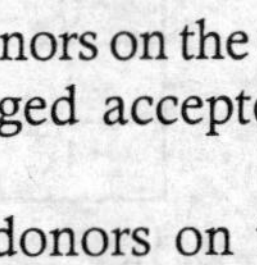
(c) 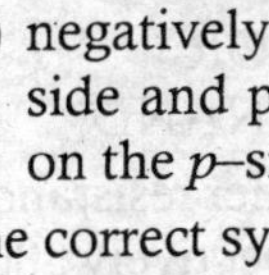
(d) 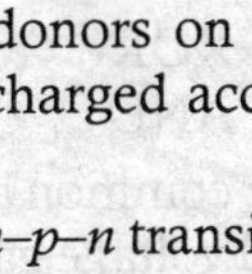

34. For a p–n junction the characteristic V-i graph is

(a) 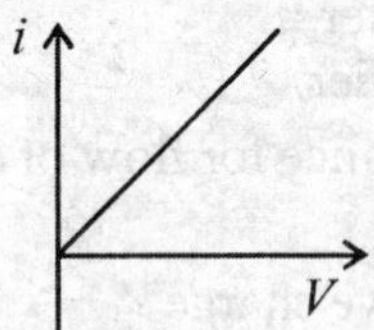

(b) 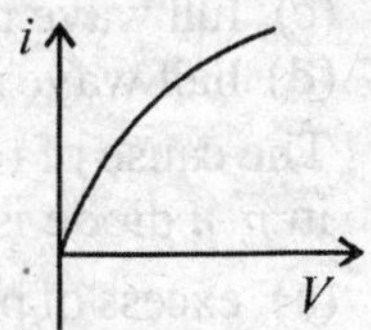

(c) 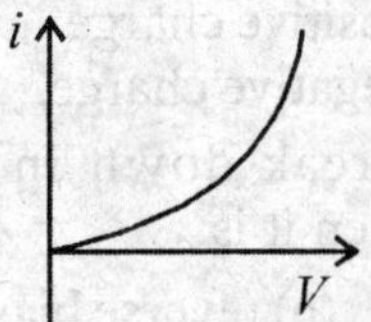

(c) 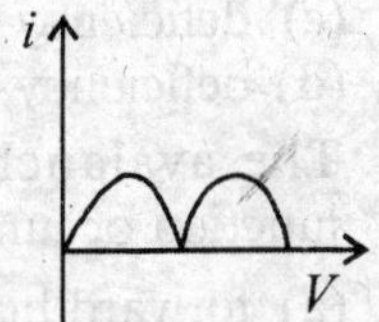

35. Which of the following diagrams correctly represent the direction of motion of current carriers in case of forward biased p–n junction?

(a) 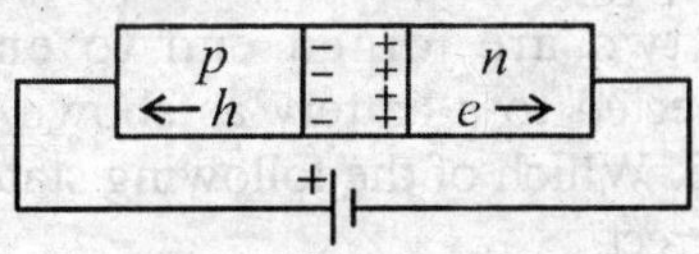

(b) 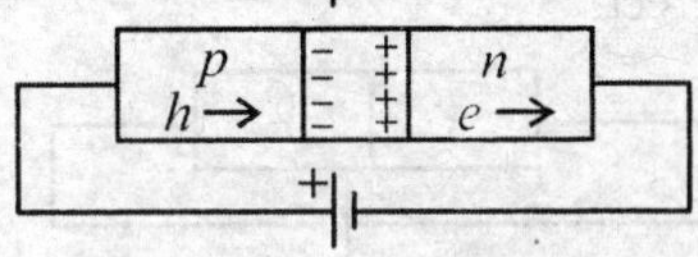

(c) 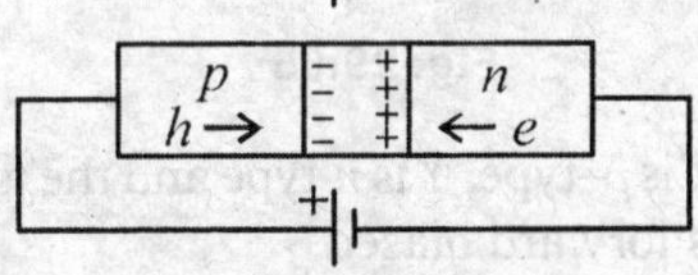

(d) 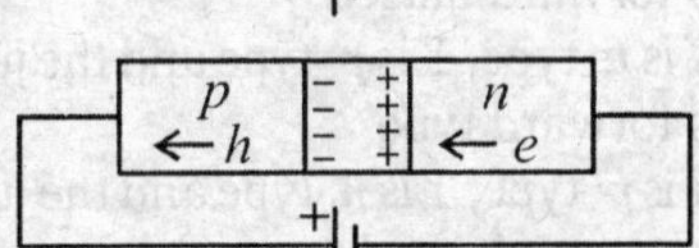

36. In Fig. 29.84 input is applied across AC and output is obtained across B and D. Then the output is

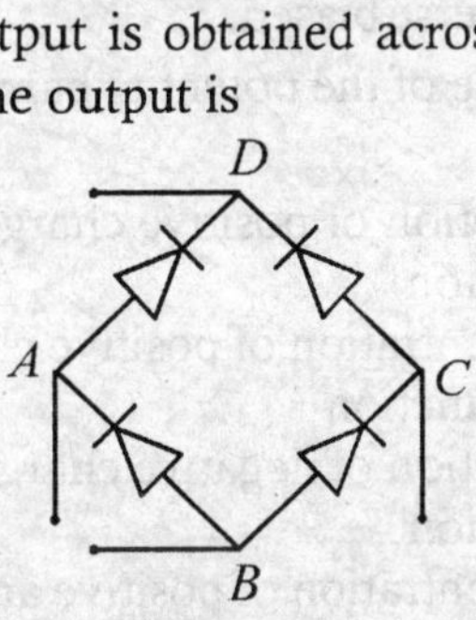

Fig. 29.84

(a) zero
(b) same as input
(c) full wave rectifier
(d) half wave rectifier.

37. The cause of resistance for flow of current in *p–n* diode is
(a) excess of positive charge
(b) excess of positive and negative charge
(c) deficiency of positive charge
(d) deficiency of negative charge.

38. The avalanche break down in a *p–n* junction occurs when it is
(a) forward biased (b) reverse biased
(c) zero biased (d) highly biased.

39. A semiconductor X is made by doping a germanium crystal with arsenic ($Z = 33$). A second semiconductor Y is made by doping germanium with indium ($Z = 49$). The two are joined end to end and connected to a battery as shown in Fig. 29.85. Which of the following statements is correct?

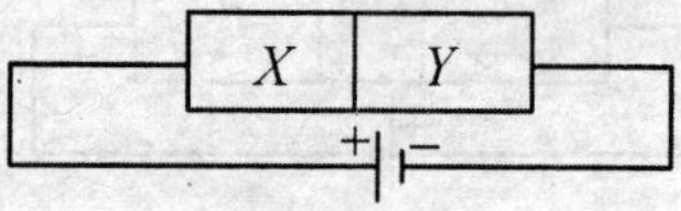

Fig. 29.85

(a) X is *p*–type, Y is *n*-type and the junction is forward biased
(b) X is *n*–type, Y is *p*–type and the junction is forward type
(c) X is *p*–type, Y is *n*–type and the junction is reverse biased
(d) X is *n*–type, Y is *p*–type and the junction is reverse biased.

40. The cause of the potential barrier in a *p–n* diode is
(a) depletion of positive charges near the junction
(b) concentration of positive charges near the junction
(c) depletion of negative charges near the junction
(d) concentration of positive and negative static charges near the junction.

41. A *n–p–n* transistor is biased to work as an amplifier. Which of the following statement is not correct?
(a) The electrons go from base region to collector region
(b) The electrons go from emitter region to base region
(c) The electrons go from base region to emitter region
(d) The electrons go from collector region to base region.

42. Which gate is represented by Fig. 29.86?

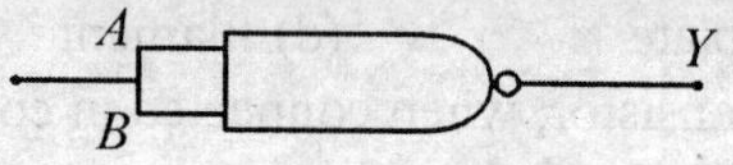

Fig. 29.86

(a) NAND (b) NOR
(c) NOT (d) AND.

43. Which of the following gates will give an output of 1?

(a) 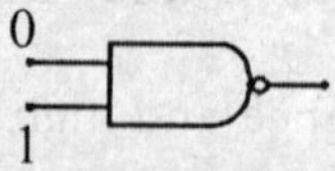

(b) 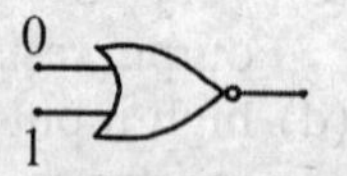

(c) 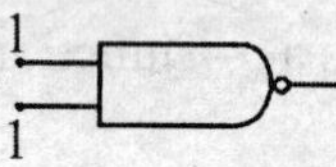

(d) 

44. The truth table shown here is for which of the following gates?

A	B	C
0	0	1
0	1	1
1	0	1
1	1	0

(a) XOR (b) OR
(c) AND (d) NAND.

45. For common base amplifier resistance gain and voltage gain are respectively 3000 and 2500. Current gain is
(a) 0.93 (b) 0.83
(c) 0.73 (d) 0.63.

46. If α of a transistor changes from 0.98 to 0.99 then percentage change in β is
(a) 100% (b) 200%
(c) 300% (d) 400%.

47. When 75 current gain transistor amplifier is given input signal $V_i = 2\sin(50\pi t + \pi/2)$, the output signal obtained is $V_0 = 200\sin(50\pi t + 3\pi/2)$. The transistor is connected
(a) as a simple common collector amplifier
(b) as a simple common base amplifier
(c) as a simple common emitter amplifier
(d) as an oscillator.

48. In Fig. 29.87 an AC source of 50 Hz is connected to a transformer coil by a filter. P and Q ends of the secondary coil are connected to a CRO. Choose the correct statement from the following which describes what we get between terminals of P and Q.

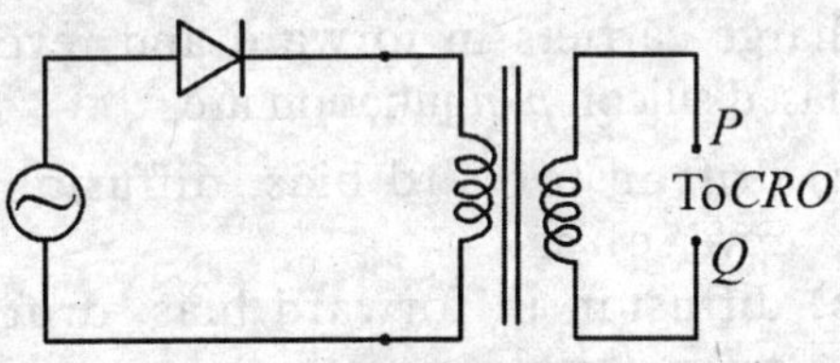

Fig. 29.87

(a) There is no potential difference
(b) There is alternating voltage
(c) There is fluctuating DC between terminals P and Q and minimum value of it is zero
(d) There is a constant DC between P and Q.

49. The region of transistor in which extra impurity is doped to obtain a large number of majority carriers is called as
(a) emitter
(b) base
(c) collector
(d) any one of these depending upon transistor.

50. An oscillator is nothing but an amplifier with
(a) positive feedback
(b) high gain
(c) no feedback
(d) negative feedback.

51. Combination shown in Fig. 29.88 can be shown by the following gate.

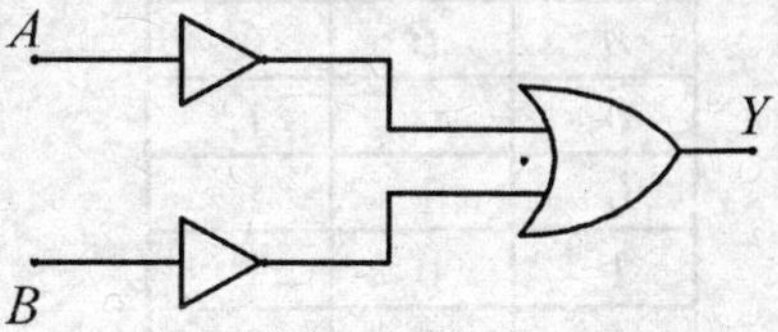

Fig. 29.88

(a) NAND gate (b) AND gate
(c) NOR gate (d) XOR gate.

52. Combination given in Fig. 29.89 can be shown by one of the following gate.

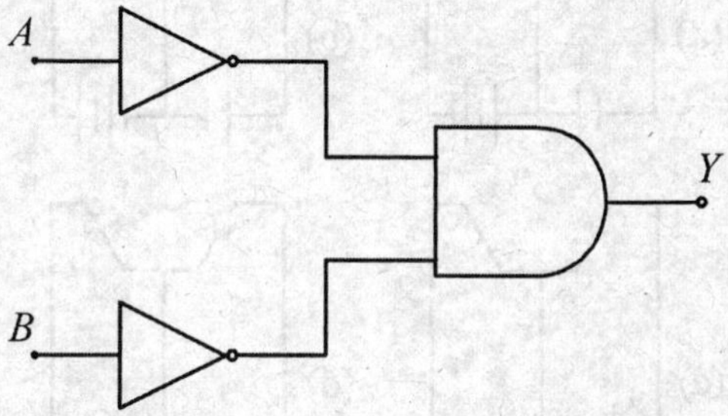

Fig. 29.89

(a) AND gate (b) NOR gate
(c) NAND gate (d) XOR gate.

53. A logic gate has two inputs A and B and one output Y. The Fig. 29.90 shows input and output wave form of the gate. The logic circuit is

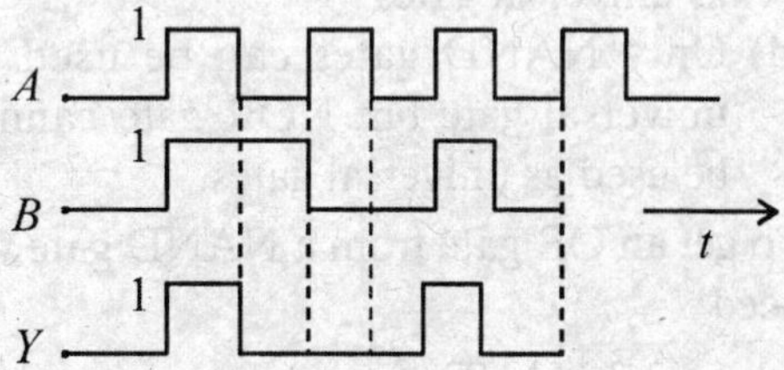

Fig. 29.90

(a) OR gate (b) NAND gate
(c) AND gate (d) NOR gate.

54. Truth table for a logic gate is given below. Where A and B are inputs and Y is output, logic gate is
(a) AND gate (b) OR gate
(c) NOR gate (d) NAND gate.

A	*B*	*C*
0	0	1
0	1	1
1	0	1
1	1	0

55. Correct transistor biasing is

(a) 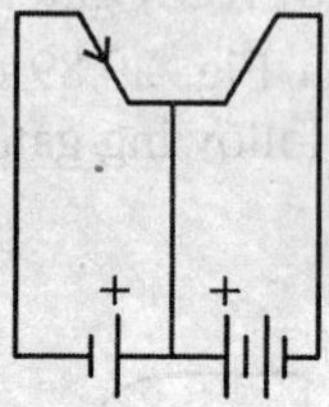

(b) 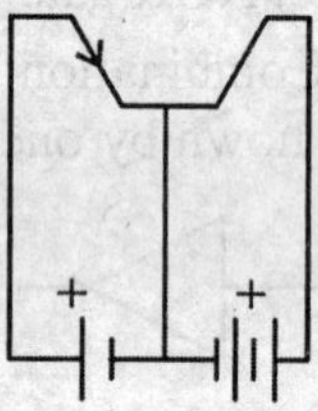

(c) 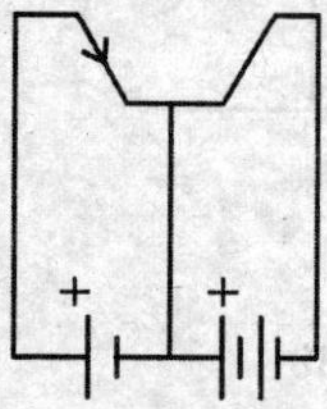

(d) 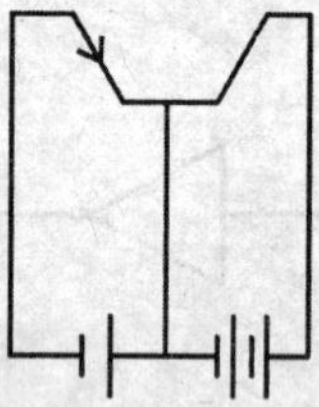

56. Mark the correct statement.

(a) Both XOR and NAND gates can be used as universal gates

(b) Both NAND and NOR gates can be used as universal gates

(c) Both NOR and XOR gates can be used as universal gates

(d) Only NAND gates can be used as universal gate but NOR gate cannot be used as universal gates.

57. To get an OR gate from a NAND gate we need

(a) only 3 NAND gates

(b) two NOT gates obtained from NAND gates and one NAND gates

(c) 4 NAND gates and 2 AND gates obtained from NAND gates

(d) 3 NAND gates and 3 NOT gates obtained from NAND gates.

58. To get AND gate from NAND gate, we need

(a) two AND gates

(b) two NOT gates obtained from NAND gates

(c) one NAND gate and one NOT gate obtained from NAND gate

(d) 3 NAND gates and one NOT gate obtained from NAND gate.

59. The dominant mechanism for motion of charge carriers in forward and reverse biased silicon *p–n* junction are

(a) drift in forward bias, diffusion in reverse bias

(b) diffusion in forward bias, drift in reverse bias

(c) diffusion in both forward bias and reverse bias

(d) drift in both forward and reverse bias.

60. In a common emitter amplifier, the output resistance is 5000 Ω and the input resistance is 2000 Ω. If the peak value of the signal voltage is 10 mV and $\beta = 50$, then the peak value of the output voltage is

(a) 5×10^{-6} (b) 2.5×10^{-4} V

(c) 1.25 V (d) 125 V.

61. A full wave rectifier circuit along with the output is shown in Fig. 29.91. The contribution(s) from the diode (1) is/are

(a) C (b) A, C

(c) B, D (d) A, B, C, D.

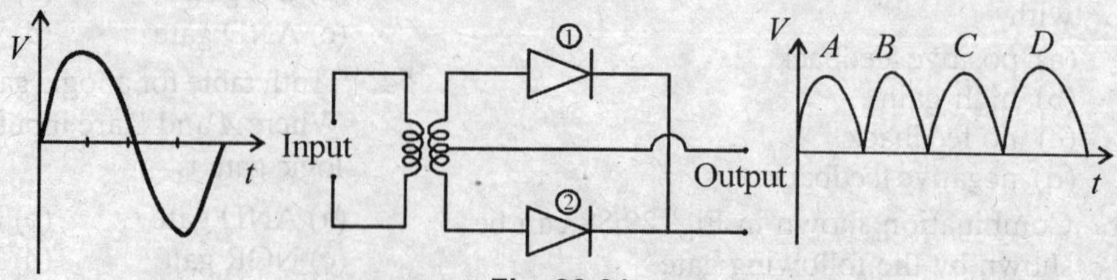

Fig. 29.91

ANSWERS

1. (d)	2. (d)	3. (d)	4. (b)	5. (c)
6. (b)	7. (d)	8. (c)	9. (b)	10. (d)
11. (c)	12. (a)	13. (c)	14. (d)	15. (b)
16. (c)	17. (b)	18. (a)	19. (b)	20. (a)
21. (c)	22. (b)	23. (b)	24. (d)	25. (c)
26. (a)	27. (a)	28. (d)	29. (d)	30. (b)
31. (c)	32. (c)	33. (a)	34. (c)	35. (c)
36. (c)	37. (b)	38. (b)	39. (d)	40. (d)
41. (d)	42. (c)	43. (a)	44. (d)	45. (b)
46. (a)	47. (c)	48. (c)	49. (a)	50. (a)
51. (a)	52. (b)	53. (c)	54. (d)	55. (c)
56. (b)	57. (b)	58. (c)	59. (b)	60. (c)
61. (c).				

MULTIPLE CHOICE QUESTIONS

PART II

In the following, a question is followed by four alternatives, of them one or more are correct. Candidate has to select the correct choice(s) from amongst the alternatives.

1. In a semiconductor diode the field in the depletion layer offers opposition to
(a) majority charge carriers in both regions
(b) minority charge carriers in both regions
(c) free electrons in the n–region
(d) holes in the p–region.

2. Holes are charge carriers in
(a) intrinsic semiconductors
(b) ionic solids
(c) p–type semiconductors
(d) metals.

3. In semiconductors
(a) there are no free electrons at 0 K
(b) there are no free electrons at any temperature
(c) the number of free electrons increase with temperature
(d) the number of free electrons are less than that in a conductor.

4. The impurity atoms with which pure silicon should be doped to make a p–type semiconductor are those of
(a) phosphorous (b) antimony
(c) boron (d) aluminium.

5. Which of the following statements concerning the depletion layer of an unbiased p–n junction are true?
(a) The width of the zone is independent of the densities of the dopants (impurities)
(b) The width of the zone is dependent on the densities of the dopants
(c) The electric field in the zone is provided by the electrons in the conduction band and the holes in the valance band
(d) The electric field in the zone is produced by the ionized dopant atoms.

6. In a p–n junction
(a) new holes and conduction electrons are produced continuously throughout the material

(b) new holes and conduction electrons are produced continuously throughout the material except in the depletion layer
(c) holes and conduction electrons recombine continuously throughout the material
(d) holes and conduction electrons recombine continuously throughout the material except in the depletion layer.

7. An AND gate can be prepared by repetitive use of
(a) NOT gate (b) OR gate
(c) NAND gate (d) NOR gate.

8. A semiconductor is doped with a donor impurity
(a) the hole concentration increases
(b) the hole concentration decreases
(c) the electron concentration increases
(d) the electron concentration decreases.

9. The equivalent resistance of the network shown in Fig. 29.92 between points A and B is (assume diode to be ideal)

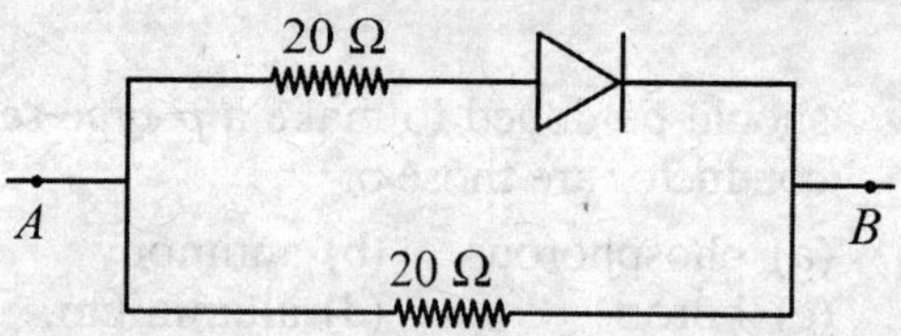

Fig. 29.92

(a) 20 Ω if $V_A > V_B$
(b) 20 Ω if $V_A < V_B$
(c) 10 Ω if $V_A > V_B$
(d) 10 Ω if $V_A < V_B$.

10. In the circuit shown in Fig. 29.93 (assumes diodes to be ideal) current

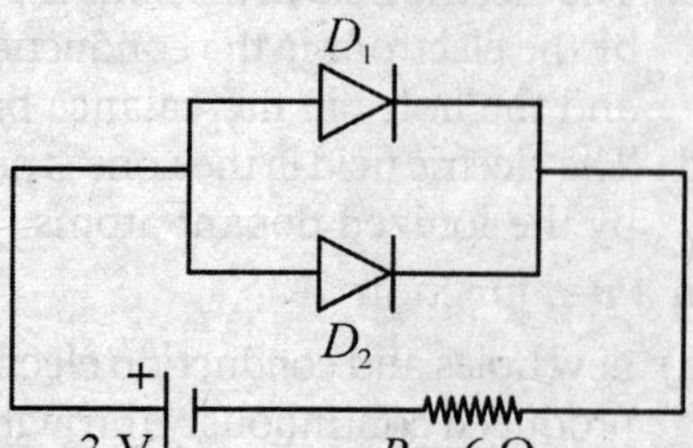

Fig. 29.93

(a) through each diodes is 0.5 A
(b) through resistance is 0.5 A
(c) through each diode is 0.25 A
(d) through battery is 0.5 A.

11. Which of the following statements is/are true?
(a) The resistance of intrinsic semiconductors decreases with increase in temperature
(b) Doping pure silicon with trivalent impurities gives p–type semiconductors
(c) The majority carriers in p–type semiconductors are holes
(d) The majority carriers in n–type semiconductors are holes.

12. The electrical conductivity of pure germanium can be increased by
(a) increasing the temperature
(b) doping acceptor impurities
(c) doping donor impurities
(d) irradiating ultra violet light on it.

13. In a normal operation of a transistor
(a) the base emitter junction is forward-biased
(b) the base collector junction is forward-biased
(c) the base emitter junction is reverse-biased
(d) the base collector junction is reverse-biased.

14. The true statement for semiconductors are
(a) its conductivity is less than conductors and more than that of non conductors
(b) its temperature coefficients of resistance is negative
(c) it works at absolute zero temperature as an ideal non conductor
(d) in pure state, silicon and germanium are called intrinsic semiconductors.

15. In a p–n junction with open ends
(a) there is no systematic motion of charge carriers
(b) holes and conduction electrons systematically go from the p-side to the n–

side and from n–side to the p–side respectively
(c) there is no net charge transfer between the two sides
(d) there is a constant electric field at the junction.

16. A semiconductor is doped with an acceptor impurity
(a) the hole concentration increases
(b) the hole concentration decreases
(c) the electron concentration increases
(d) the electron concentration decreases.

17. Let i_e, i_c and i_b represent the emitter current, the collector current and the base current respectively in a transistor. Then
(a) i_c is slightly smaller than i_e
(b) i_c is slightly greater than i_e
(c) i_b is much smaller than i_e
(d) i_b is much greater than i_e.

18. In the circuit shown in Fig. 29.94

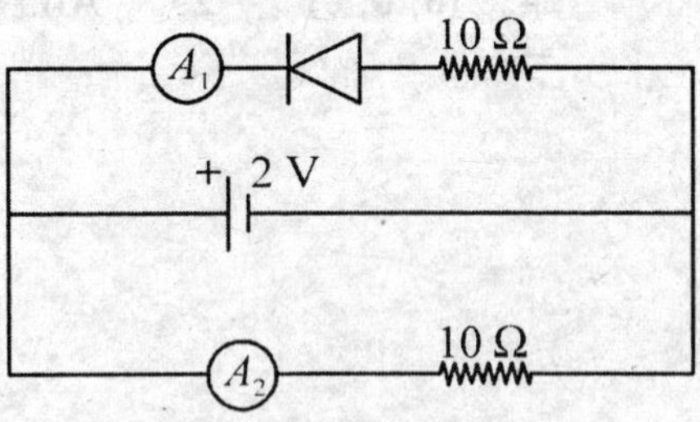

Fig. 29.94

(a) A_1 reads 0.2 A and A_2 also reads 0.2 A
(b) A_1 reads zero and A_2 reads 0.2 A
If polarity of cell is changed
(c) A_1 reads 0.2 A and A_2 also reads 0.2 A
(d) A_1 reads zero and A_2 reads 0.2 A.

19. In the circuit shown in Fig. 29.95 (assume knee voltage for diode = 0.7 V).

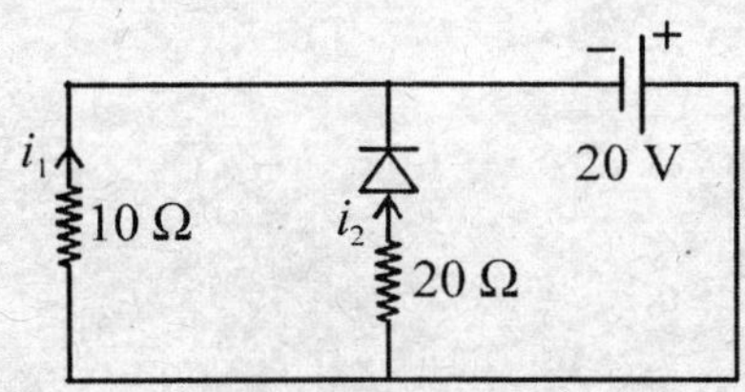

Fig. 29.95

(a) current i_1 = 2.0 A
(b) current i_2 = 1 A
(c) current through battery is 3 A
(d) current i_2 = 0.965 A.

20. In the circuit shown in Fig. 29.96 if knee voltage for diodes be 0.7 V, then

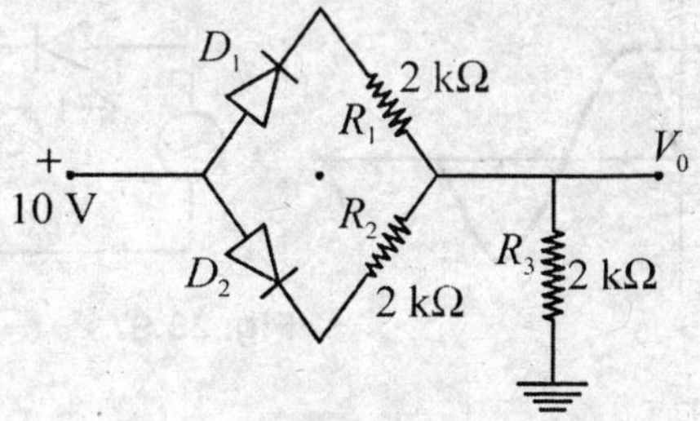

Fig. 29.96

(a) V_0 = 6.2 V
(b) current through each diode is 1.5 mA
(c) current through R_3 = 1.5 mA
(d) current through R_3 is 3 mA.

21. For transistor action
(a) the collector region must be more heavily doped than the base region
(b) the collector base junction must be reverse biased
(c) the base region must be narrow
(d) the base region must be of n–type material.

22. In a p–n junction if I_{dr} is drift current and I_{df} is diffusion current, then
(a) $I_{dr} > I_{df}$ if junction is forward biased
(b) $I_{dr} < I_{df}$ if junction is forward biased
(c) $I_{dr} < I_{df}$ if junction is reverse biased
(d) $I_{dr} > I_{df}$ if junction is reverse biased.

23. For a common emitter amplifier the load resistance is 5 kΩ and input resistance 1 kΩ. If the peak value of signal is 100 mA and current gain β = 50, then
(a) voltage gain is 250
(b) voltage gain is 12500
(c) power gain is 250
(d) power gain is 12500.

24. Assume the behaviour for diode given in Fig. 29.97 to be ideal, then

(a) $V_D = 10$ V, $V_{out} = 0$ for negative half cycle of input
(b) $V_D = 0$ V, $V_{out} = 10$ for positive half cycle of input
(c) DC voltage across 5 kΩ load is 3.2 V
(d) DC voltage across 5 kΩ load is 10 V.

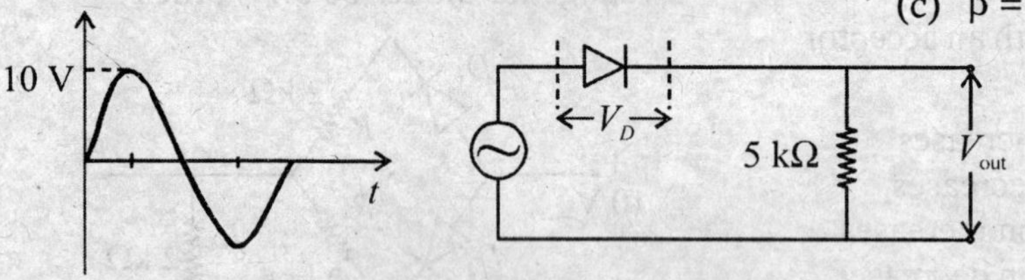

Fig. 29.97

25. If α, β and γ are the current amplification for CB, CE and CC respectively then correct relations are

(a) $\gamma = \beta + 1$ (b) $\alpha = \dfrac{\gamma - 1}{\gamma}$

(c) $\beta = \dfrac{\alpha}{1-\alpha}$ (d) $\beta = \alpha\gamma$

ANSWERS

1. (a, c, d)	2. (a, c)	3. (a, c, d)	4. (c, d)	5. (b, d)
6. (a, d)	7. (c, d)	8. (b, c)	9. (b, c)	10. (b, c d)
11. (a, b, c)	12. (a, c, d)	13. (a, d)	14. All.	15. (b, c, d)
16. (a, b, d)	17. (a, c)	18. (b, c)	19. (a, d)	20. (a, d)
21. (a, b, c)	22. (b, d)	23. (a, d)	24. (a, b, c)	25. All.

30 Principles of Communication

30.1 INTRODUCTION

The exchange of information faithfully between a sender and a receiver is called communication. The set up used for this purpose is called a communication system. A modern communication system requires, sorting, processing and storing information before its transmission. The communication through electrical signals has made communication more wide spread because the electrical signal can be transmitted over much longer distances at extremely high speed (= 3×10^8 m/s).

A communication system is a setup used to transmit information from one point to the other. The communication system has advanced over the years through the use of transistor, integrated circuits, satellites and fibre optics. In principle, a communication system consists of three main parts: (*i*) Transmitter (*ii*) Communication channel (*iii*) Receiver.

(*i*) **Transmitter:** It is a setup that transmits the message to the receiving end through a communication channel. Its main function is to modify the information signal in a form suitable for transmission over the channel and to transmit it.

The transmitter first converts information into electrical oscillations. It then processes and encodes information so as to make it suitable for transmission and subsequently reception. The transmitter also mixes encoded information with a high frequency carrier wave. This process is called modulation. This modulated signal in then amplified and fed to transmitting center. Schematic arrangement for transmitting signal is shown in Fig. 30.1.

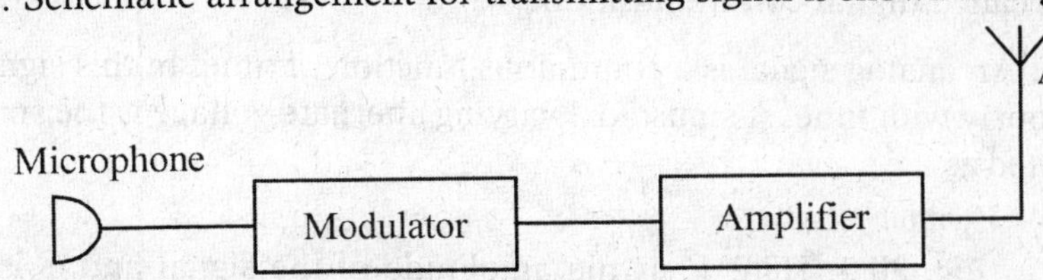

Fig. 30.1

(*ii*) **Communication Channel or Transmission Medium:** The system channel refers to the frequency range allocated to a particular transmission or service. It also includes the medium or the physical path that connects a transmitter to receiver. A modulated wave from transmitter is carried to the receiver through this communication channel. The channel may be a transmission line, an optical fibre or free space.

(*iii*) **Receiver:** This receives the signal sent by transmitter through communication channel and process it and finally converts into original message. The most important function of receiver is demodulation and sometimes decoding as well which is reverse of modulation in transmitter. The output of the receiver may be fed to loud a speaker, video displaying, teleprinter, T.V. picture-tube or computer. The transmitter and receiver must be in synchronization with the modulation and coding methods used. Figure 30.2. shows block diagram of receiver.

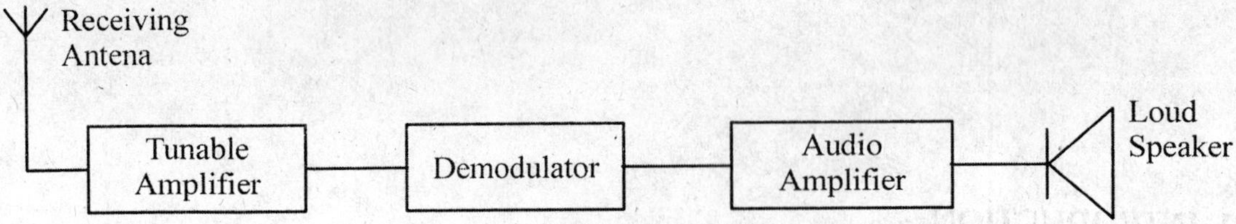

Fig. 30.2

Antenna

In communication system, antenna plays a vital role. It is employed for both transmitting and receiving the radio frequency signal. The transmitting antenna converts electrical energy into electromagnetic waves whereas the receiving (or pick-up) antenna converts electromagnetic waves into electrical energy.

A Hertz antenna is a straight conductor of length equal to half wavelength ($\lambda/2$) of the signal to be transmitted or received. A Maroni antenna is quarter wave ($\lambda/4$) long. It is placed vertically with its one end touching the ground. The ground causes reflection of the voltage and current distributions set up in the antenna.

30.2 ANALOG AND DIGITAL SIGNAL

In communication system, a signal means a time varying voltage or current containing information. The different signals have their own characteristics, such as amplitude, frequency, etc. The signals are of two types: (*i*) Analog Signal (*ii*) Digital Signal.

Thus we have two types of communication systems, depending on the type of signal used.

A communication system using analog signal is called analog communication system and using digital signal is called digital communication.

(*i*) **Analog Signal:** An analog signal is a continuous function of time. In this signal amplitude varies continuously with time. A sinusoidal varying alternate voltage is the simplest analog signal represented as

$$V = V_0 \sin\omega t \qquad \ldots(30.1)$$

and is shown in Fig. 30.3. Here V_0 is the amplitude of the signal and ω is the angular frequency.

The frequency of analog signals associated with speech or music varies over a range between 20 Hz to 20,000 Hz. The range over which the frequencies of a signal vary is called its *bandwidth*. This information carrying signal is also called *baseband signal*. The term base-band stands for the band of frequencies representing the signal supplied by the source of information.

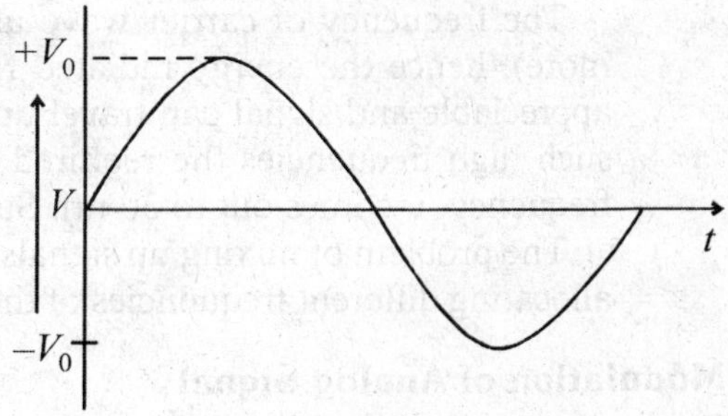

Fig. 30.3

(*ii*) **Digital Signal:** A digital signal is discontinuous function of time. In digital signal, current or voltage have discrete values. It has only two voltage levels, *i.e.*, either high (some constant value of voltage) or low (zero). The two levels in a digital signal are coded usually in binary code. When voltage is low; it is represented by binary digit zero (= 0) and when it is high it is represented by 1. Thus the digital signals are in the form of pulses of equal levels. One such pulsating signal is shown in Fig. 30.4.

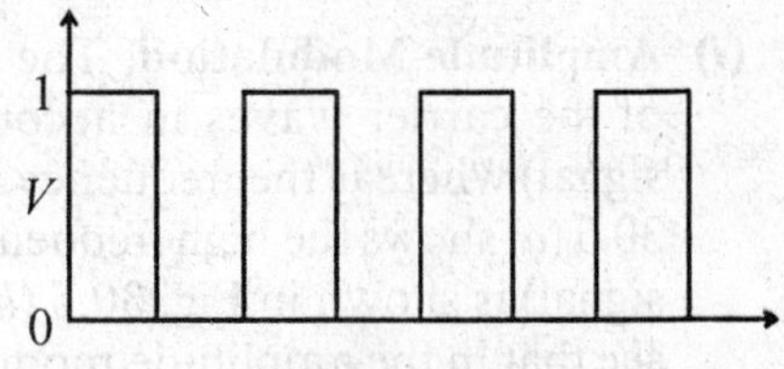

Fig. 30.4

The digital signals can be used as effective commands. Digital signal zero refers to an open circuit whereas 1 refers to closed circuit.

Digits 0 and 1 are called bits. A group of bits is called a *binary word or a byte*. A byte made of 3 bits can give 8 code combination (000, 001, 010, 100, 011, 101, 110 and 111). In general if x is the number of bits in a byte, then the number of code combination will be

$$N = 2^x \qquad \text{...(30.2)}$$

30.3 MODULATION

The audio frequency (frequency range 20 Hz to 20,000 Hz) signals cannot be transmitted directly from the transmitter because of the following reasons:

(*i*) When the frequency of the signal is below 15 kHz; the energy radiated from an antenna is practically zero. This makes the direct transmission of audio signal impracticable.

(*ii*) Since all audio signals from different sources possess frequencies in the range 20 Hz to 20 kHz as such these signals from different transmitting stations will be inseparably mixed up.

(*iii*) For efficient transmission and reception, the transmitting as well as receiving antenna must have a length equal to quarter wavelength ($\lambda/4.$) of the signal. For a frequency of 20 kHz of audio signal the length of antenna required is

$$l = \frac{\lambda}{4} = \frac{c}{4\nu} = \frac{1}{4} \times \frac{3 \times 10^8}{20 \times 10^3} = 3750 \text{ m} = 3.75 \text{ km}$$

and to setup an antenna of this length is practically impossible.

To avoid these difficulties in transmitting audio frequency signals, the signals are *modulated.* For this purpose high frequency carrier wave is used to carry the audio signal to large distances. The modulation involves the super imposition of electrical audio signal on high frequency carrier wave. In the process of modulation some characteristics, usually amplitude, frequency or phase angle of high frequency carrier waves are varied in accordance with the instantaneous value of the audio frequency signals. The resultant waves so obtained are known as *modulated waves.*

The frequency of carrier wave usually chosen is very high (of the order of 1 MHz or more), hence the energy radiated from the antenna at this frequency of 1 MHz is quite appreciable and signal can travel up to a large distance before it dies out. Besides this for such high frequencies the required antenna length is considerably reduced. (For 1 MHz frequency, it comes out to 30 m). Such high antennas can be easily set-up.

The problem of mixing up signals from the different transmitting stations is overcome by allocating different frequencies of the carrier waves to different transmitting stations.

Modulation of Analog Signal

For analog signal there are three types of modulation: Amplitude modulation, frequency modulation and phase modulation. We will discuss here briefly these types of modulation.

(*i*) **Amplitude Modulation:** The amplitude modulation is produced by varying the amplitude of the carrier waves in accordance with the amplitude of the modulating wave (audio signal) whereas the frequency is kept fixed equal to the frequency of the carrier wave. Figure 30.5 (*a*) shows the high frequency carrier wave, the audio frequency signal (called modulating signal) is shown in Fig. 30.5 (*b*) and the modulated signal is shown in Fig. 30.5 (*c*). One can see that in the amplitude modulated wave (Fig. 30.5 (*c*)) the amplitude is not constant when the instantaneous amplitude of modulating signal increases, the amplitude of the carrier wave increases and *vice versa.* Thus the amplitude of the modulated carrier wave is not constant but its envelope has similar sinusoidal variation as that of the low frequency modulating signal. Thus we can say that the carrier modulated wave is loaded with the message contained in the low frequency message signal with amplitude varying between certain maximum value (E_{max}) and minimum value (E_{min}) where

$$E_{max} = E_c + E_m \quad \text{and} \quad E_{min} = E_c - E_m \qquad ...(30.3)$$

Here E_c is amplitude of carrier wave and E_m is amplitude of modulating audio frequency signal.

Here modulation index (or modulation factor) m_a for amplitude modulation is defined as

$$m_a = \frac{E_{max} - E_{min}}{E_{max} + E_{min}}$$

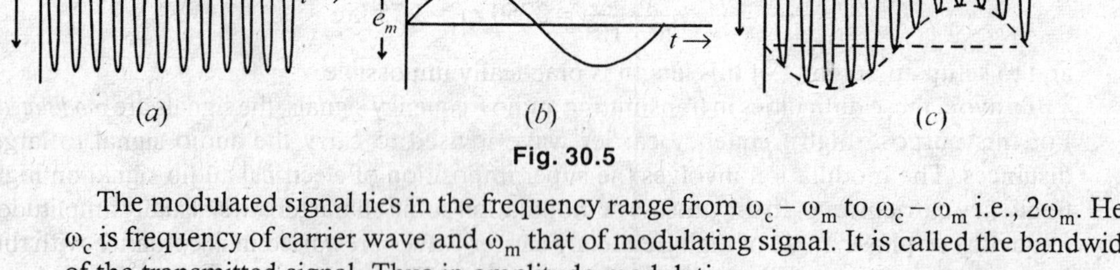

Fig. 30.5

The modulated signal lies in the frequency range from $\omega_c - \omega_m$ to $\omega_c - \omega_m$ i.e., $2\omega_m$. Here ω_c is frequency of carrier wave and ω_m that of modulating signal. It is called the bandwidth of the transmitted signal. Thus in amplitude modulation

$$\text{band with} = 2 \times \text{frequency of audio signal} \qquad ...(30.4)$$

(*ii*) **Frequency Modulation:** The frequency modulation is obtained by varying the frequency of the carrier waves in accordance to the instantaneous amplitude of the modulating audio frequency signal whereas the amplitude is kept constant equal to the amplitude of the carrier wave. Figure 30.6 (*a*) shows the carrier high frequency wave and Fig. 30.6 (*b*) shows the audio frequency modulating wave. The modulated signal is shown in Fig. 30.6 (*c*). The frequency of the carrier wave varies in accordance with the modulating signal. The amplitude of the carrier waves is kept fixed. When the instantaneous amplitude of the modulating signal is large, the instantaneous carrier frequency is higher. It is lower when amplitude of modulating voltage is small.

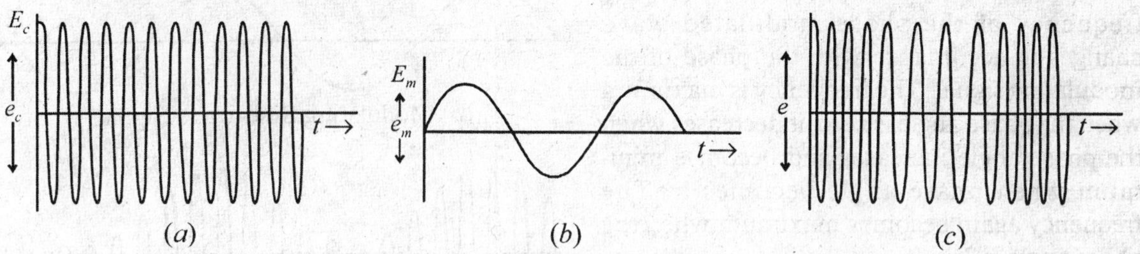

Fig. 30.6

If f is the instantaneous frequency of carrier wave at time t, then deviation in frequency is

$$\delta = f - f_c \qquad \text{...(30.5)}$$

The deviation in frequency is proportional to amplitude of the modulating signal. Thus the maximum deviation is

$$\delta_{max} = \pm k \cdot E_m$$

where k is constant.

The ratio of frequency swing to the frequency of modulating signal (f_m) is called *modulation index* for frequency modulation. It is denoted by m_f

$$\therefore \quad m_f = \frac{\delta_{max}}{f_m} = \frac{f_{max} - f_c}{f_m} = \frac{f_c - f_{min}}{f_m} \qquad \text{...(30.6)}$$

The frequency spectrum of the F.M. wave is much more complex than that of AM wave and has the following features:

(*i*) F.M. modulated signal consists of carrier and an apparently infinite number of side bands, whose frequencies are $(f_c \pm f_m)$, $(f_c \pm 2f_m)$, $(f_c \pm 3f_m)$, ...

(*ii*) The number of side bands with significant amplitudes depends on modulation index m_f.

(*iii*) The number of side bands increases, when the frequency deviation is increased by keeping the frequency of modulating signal fixed. On the other hand, the number of side bands decreases, when the frequency of modulating signal is increased by keeping deviation constant.

(*iv*) The separation of the side bands from the carrier frequency f_c is f_m, $2f_m$, $3f_m$,..., i.e., they have recurrence frequency of f_m.

(*v*) The side bands are disposed symmetrically about the carrier and the side bands at equal distances from the carrier have equal amplitudes. The amplitude of the side bands separated largely from the carrier becomes negligible. Due to this, the number of significant side band pairs is limited.

(*vi*) Bandwidth in frequency modulated signal is

$$\text{bandwidth} = 2n \times \text{frequency of modulating signal} \qquad \ldots(30.7)$$

where n is the number of significant side band pairs.

Phase Modulation

In this, the phase angle ϕ of the wave varies as the instantaneous amplitude of the modulating audio signal. Figure 30.7 gives the phase modulated carrier wave. It is observed that the frequency of the phase modulated wave changes in accordance with the phase of the modulating signal. The frequency is maximum when the phase angle is zero, it decreases when the phase angle increases and becomes minimum when phase angle becomes π. The frequency again becomes maximum when the phase angle is 2π.

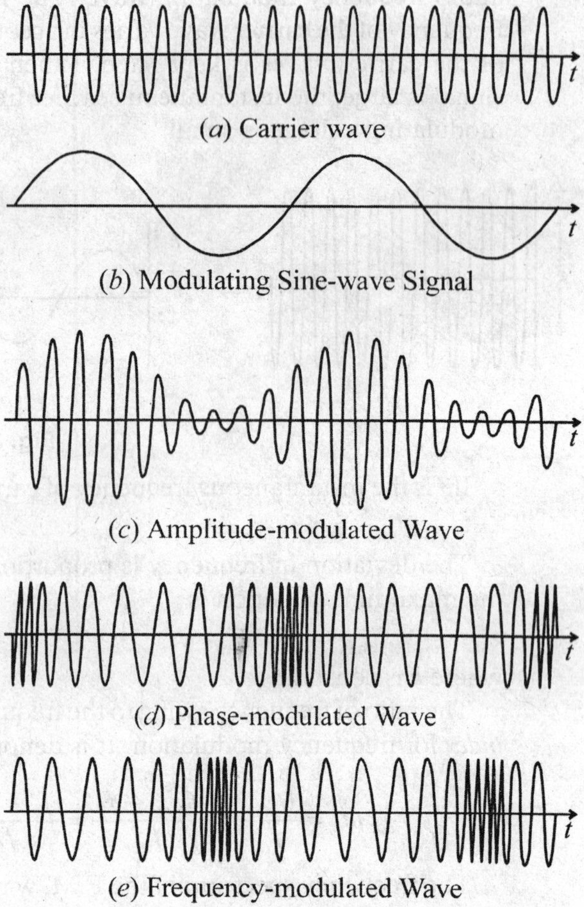

Fig. 30.7

Effective Power Radiated by an Antenna

Power radiated from a linear antenna of length l can be calculated theoretically. It is found that

$$\text{Power radiated} \propto l/\lambda^2$$

This means that for same antenna length, the power radiated by short wavelength or high frequency signals would be very large. Hence the effective power radiated by long wavelength baseband signal would be small. For a good transmission, we require high powers which in turn require high frequency transmission.

Advantages of Frequency Modulation

Some of the advantages of frequency modulation over amplitude modulation are given below:

1. Frequency modulation is highly efficient as compared to A.M. In F.M. all the radiated power is useful whereas in A.M. transmission most of the power is wasted in transmitting the carrier which contains no useful information.
2. F.M. reception is quite immune to noise* as compared to A.M. reception. In F.M. receivers, the noise can be reduced by increasing the frequency deviation or by making use of amplitude limiters.
3. Due to large number of side bands, F.M. transmission can be used for the stereo sound transmission.

*Noise is a form of amplitude variations in the transmitteed signal due to atmosphere, industries, etc.

4. In F.M. standard frequency allocations provide a guard band between commercial F.M. stations. Thus reduces inter channel interference.
5. In the VHF or UHF bands when F.M. operates, there is less noise than in H.F. or M.F. bands where A.M. operates.

Disadvantages

1. In F.M. transmission much wider frequency channel is required. The bandwidth in F.M. transmission is about 10 times as large as that is required in A.M. transmission.
2. F.M. reception is limited to line-of-sight as such the reception area for F.M. is much smaller than that for A.M.
3. F.M. receivers and transmitters are very complex and costly.

The table below gives the range of frequency allotted for F.M. radio and TV broadcasts.

Type of Broadcast	*Frequency Band*
AM Radio	0.5–20 MHz
FM Radio	88–108 MHz
VHF TV	47–230 MH
UHF TV	470–960 MH

30.4 PRODUCTION AND DETECTION OF AMPLITUDE MODULATED WAVE PRODUCTION

The modulation is carried out by superposition of baseband signal and carrier signal. The information signal is fed to the message signal generator. The message signal generator is just a transducer which converts original signal into an electrical signal called baseband signal. A high frequency oscillator generates the carrier wave. The modulator superimposes the baseband signal over the carrier wave and gives out the modulated wave. The block diagram with simple CE amplifier circuit to carry out this modulation is shown in Fig. 30.8.

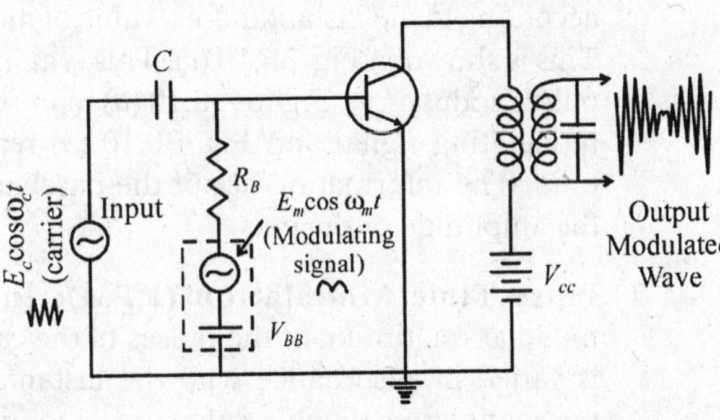

Fig. 30.8

Detection

The process of detection or demodulation is the inverse of modulation. In this process, audio signal is separated from modulated signal. This is done by using a diode and a suitable capacitor filter as shown in Fig. 30.9. The separated audio signal is then fed to the loudspeaker or headphone, which reproduces the original speech or music. The action of a diode as detector is basically the same as that of a half-wave rectifier.

The value of RC' (the time constant of the circuit) is chosen so that the frequency of carrier wave f_c is very large in comparison to reciprocal of time constant, *i.e.*

$$\frac{1}{f_c} << RC' \quad \text{...(30.8)}$$

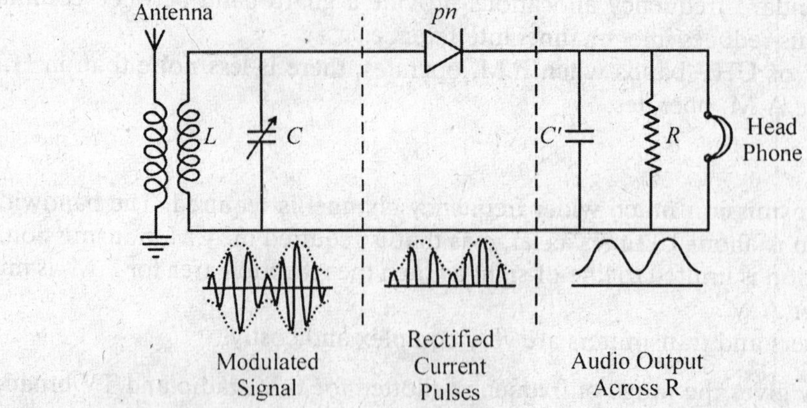

Fig. 30.9

30.5 PULSE MODULATION

In pulse modulation the carriers is no longer continuous sinusoidal wave but is in the form of pulses. In this process, some parameter of a pulse train is varied in accordance with the instantaneous amplitude of the modulating signal. We have two basic types of pulse modulations.

1. **Pulse Amplitude Modulation (PAM):** In the process of modulation, amplitude of the pulse of the carrier is varied according to the instantaneous value of modulating signal. This is shown in Fig. 30.10 (*c*). This is analog to amplitude pulse modulation. Figure 30.10 (*a*) represents a sinusoidal modulating signal and Fig. 30.10 (*b*) represents a pulse train. The information about the baseband signal lies in the amplitude of the pulse.

2. **Pulse Time Modulation (PTM):** In this mode of modulation, timing of the pulses of the carrier pulse train is varied in accordance with the instantaneous value of the modulating signal. Pulse time modulation is further of the two types.

   (*a*) *Pulse Width Modulation* (*PWM*): Here the starting time and amplitude of each pulse are constants but width of duration of each pulse varies proportional to the instantaneous value of the analog signal amplitude. The pulse width is large when the amplitude is large and *vice versa*. Figure 30.10 (*d*) shows the PWM signal. The information about the baseband signal lies in the traiting edge of the pulse.

   (*b*) *Pulse Position Modulation* (*PPM*): In this process of modulation, the position of the pulse relative to its unmodulated time of occurrence is varied according to

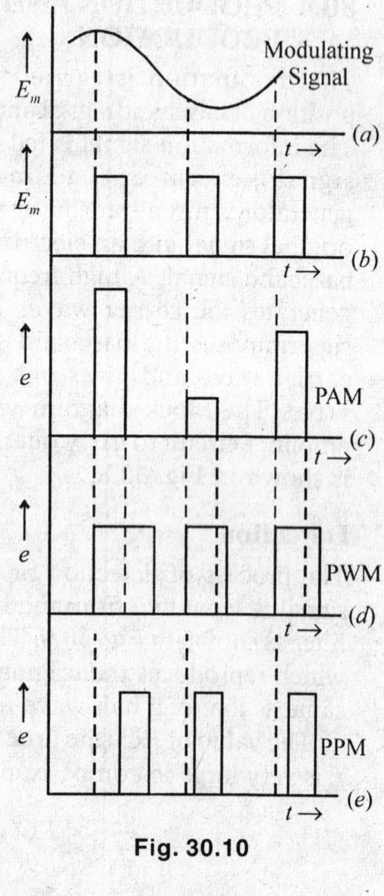

Fig. 30.10

the instantaneous value of the amplitude of modulating signal. Figure 30.10 (*e*) shows PPM wave. Here the shift in position of each pulse is according to the amplitude of the modulating signal at the sampling instant. The information about the baseband signal lies in the leading edge of the pulse.

Pulse Code Modulation

The various pulse modulations discussed above are not completely digital. It is due to the fact that the amplitude, width or position of the pulse transmitted may vary continuously according to the variation in the baseband signal. A *pulse code modulation* (*PCM*) gives a completely digital modulation. It is a process in which the message signal is first *sampled. Sampling is the process of generating pulse of zero duration (width) and of amplitude equal to the instantaneous amplitude of analog signal.* The number of pulses generated per second is called sampling rate. It is then *quantised. Quantisation is the process of dividing the maximum amplitude of the analog voltage signal into a fixed number of levels. These levels are called quantisation levels.* The amplitude of each pulse is then approximated to nearest quantum level. The pulses so obtained are called *quantised pulses*. Each quantised pulse is then coded according to some code. The coding process involves assigning some digits to each of the pulse. For this purpose binary numbers are used. The size of the binary number, i.e., how many number of bits it has which depends on the number of quantisation levels. If p is the number of quantisation levels and n is the number of bits in binary code, then

$$p = 2^n \qquad \text{...(30.9)}$$

An analog signal is described by its frequency range. Instead, a digital signal is described by the *bit rate, i.e.*, number of bits per second (binary digits per second). Then

$$\text{Bit rate} = \text{Sampling rate} \times \text{number of bits per sample} \qquad \text{...(30.10)}$$

30.6 DATA TRANSMISSION AND RETRIEVAL

The pulse code modulated signal is a series of 1 s and 0 s. If they are to be transmitted through cable, they can be directly transmitted as two voltage levels corresponding to 1 V and 0 V. But if they are to be transmitted through space, then same form of modulation has to be employed just as in case analog signal modulation.

The common modulation techniques employed for the purpose of PCM signal are

(*a*) Amplitude Shift Keying (ASK)
(*b*) Frequency Shift Keying (FSK)
(*c*) Phase Shift Keying (PSK)

(*a*) Amplitude Shift Keying (ASK): The information to be transmitted is in the form of a binary code. In ASK the amplitude of the carrier changes according to the information signal. In binary code the amplitude values are 1 and 0. Hence in ASK the carrier amplitude is either zero or has the original value of the carrier. For 1 the original amplitude of constant value, of carrier is present while it is absent as shown in Fig. 30.11 (*b*).

(*b*) Frequency Shift Keying (FSK): In this process the binary code of the PCM signal are represented by two frequencies which are offset from carrier frequency f_c by equal and opposite amounts f and $-f$. The frequency $f_c + f$ represents binary number 1 while $f_c - f$ represents binary value zero. This is less immune to errors than ASK. It is commonly used for HF radio transmission. Fig. 30.11 (*c*) gives FSK modulated signal.

(*c*) **Phase Shift Keying (PSK):** In this process of modulation, binary data are represented by causing shift in the phase of the carrier signal. Here binary value of 0 corresponds to zero phase shift and the binary value 1 corresponds to the phase shift of π. This is shown in Fig. 30.11 (*d*).

Out of all these methods for modulating PCM signal FSK method is preferred as it gives optimum performance in the presence of noise. The ASK method is rarely used as it does not give satisfactory results.

The block diagrams for transmission and reception of PCM communication are shown in Fig. 30.12 (*a*) and (*b*) respectively.

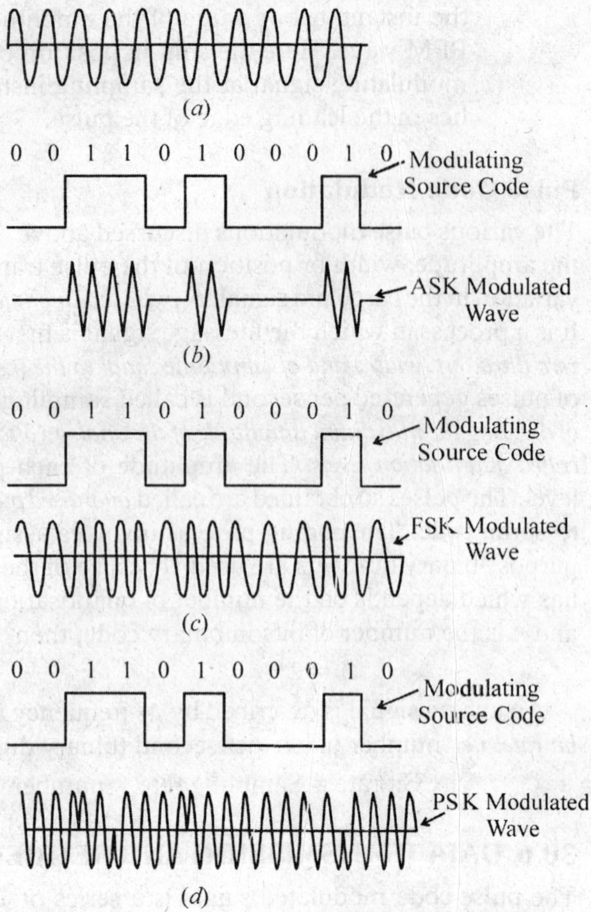

Fig. 30.11

Merits of Digital Communication

The digital communication has the following merits over analog communications.

1. Digital signals are easy to receive because the receiver has only to detect whether the pulse is high or low.
2. AM and FM signals get corrupted over a very short distance as such they cannot be reproduced faithfully. Whereas in case of digital signals as long as the receiver can distinguish between high and low, the original signal can be reproduced accurately.

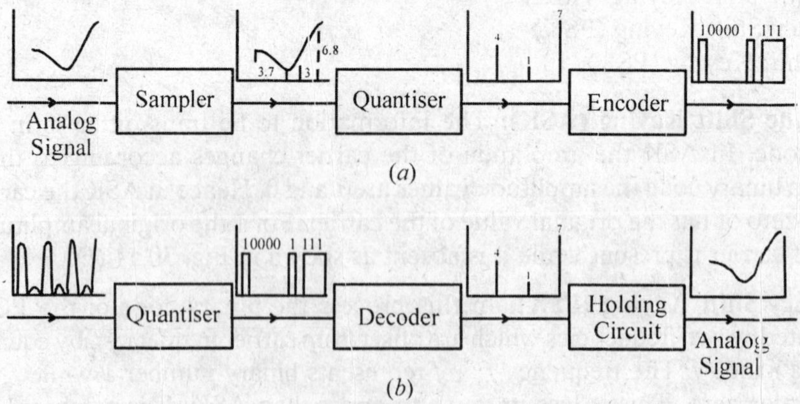

Fig. 30.12

3. As the signal propagates it gets attenuated (*i.e.*, it loses power). Hence when AM and FM signals are amplified, the noise also gets amplified. Thus the quality of the signal is lowered. On the other hand, the digital signal can be cleaned up amplified at different stages by using regenerators to restore quality and power.
4. The noise does not distort the digital signal. The noise at the most can change shape of the pulse but not the pattern of pulse because the information in digital signal is stored in pattern and not in shape.
5. AM and FM signals can be received with suitable receiver whereas the digital signal can be coded so that only the person, they are intended for can receive them.
6. The digital signals can be stored in the absence of the person at the receiving end. Whereas the AM and FM signal are in real time system, *i.e.*, the signal can be received only at the time it reaches the receiving end. It can not be stored.
7. The digital signal can be stored as digital data or used to produce a display in a computer monitor or converted back into analog signal to drive a loudspeaker.

Modem

In modem computer age, the data transmission from one machine to another has become very important. Modems are used to interface two digital sources (or receivers). The name modem is a contraction of the term *modulator* and *demodulator*. As the name suggests, a modem performs functions of both modulator and demodulator. In transmitting mode, modem acts as a modulator and in the receiving mode it acts as a demodulator.

30.7 COMMUNICATION CHANNELS

For establishing a communication between transmitting and receiving station, we need a physical medium through which the signals may propagate. The physical medium between the transmitter and receiver is called the communication channel.

The transmission media are divided broadly into two categories:

(*i*) **Guided Media:** In this communication channel the signals launched are guided along conducting physical path (or line) to the receiver. This mode of communication is commonly referred to as the *line communication*. Important examples of guided media are: (*a*) twisted pair (*b*) coaxial cable (*c*) optical fibre.

(*ii*) **Unguided Media:** Free space is an example of unguided medium. In this, the signal is transmitted freely in space by using an antenna and received at the other end by intercepting the signal with the help of another antenna. This mode of communication is called *space communication*. In space communication the frequency band of the signal produced by the transmitting antenna is more important than the medium in determining the quality of transmission.

Space Communication

If the space around earth is used as communication channel for transmitting the signal, it is called space communication. In this, the information is carried from one place to another by electromagnetic waves of high frequency called carrier waves. The frequencies used are in the range 10^4 Hz to 10^{11} Hz and the corresponding wavelengths range between 1 mm to 30 km. These waves travel through the open space with speed of light. Radio television and satellite communications are all space communications.

There are three modes of propagation used for space communication as mentioned below:

1. **Ground Wave Propagation (also called Ground Waves):** This mode of propagation is possible only when the transmitting and receiving antenne are very close to the surface of earth. A radio wave that can travel directly from transmitting end to the receiving end following the surface of earth is called ground wave. It is guided by the earth and bends round its curvature because there is electrical discontinuity between the earth and the atmosphere.

   As the ground wave propagates over the surface of earth, its energy is absorbed by earth and it is attenuated (*i.e.*, it is weakened). Due to this the ground waves are not suitable for very long communication. These losses are also frequency dependent and are higher for higher frequencies. Hence ground waves propagation is suitable for low frequencies (500 kHz to 1500 kHz), *i.e.*, for broadcasting at long wavelength > 200 m. This region of AM is called medium wave band.

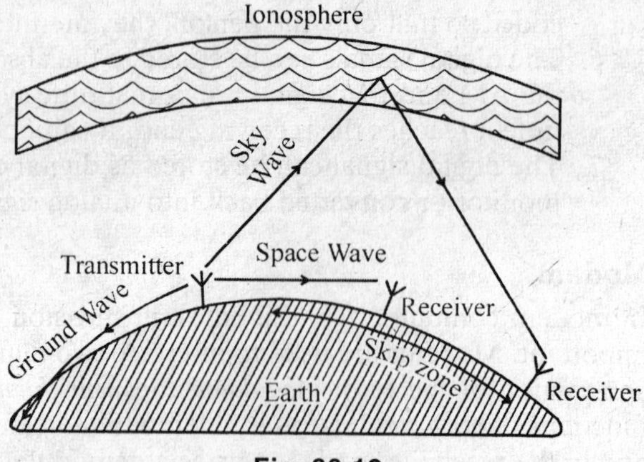

Fig. 30.13

2. **Space Wave Propagation or Tropospheric Wave Propagation:** When the radio waves from transmitting antenna reach the receiving antenna either directly or after reflection from ground or atmosphere, the wave propagation is called space wave propagation.

   If a radio wave transmitted from an antenna, travelling in a straight line directly reaches these receiving antenna, it is called *surface wave or space wave*. It can be seen that due to the finite curvature of the earth, such waves cannot be seen beyond certain points Fig. 30.14. The range of space wave propagation is limited due to finite curvature of earth. The effective reception range of broadcast is essentially the region from P_1 to P_2, *i.e.*, the distance covered by the line of sight. The range $P_1 P_2$ (or $P_1Q = P_2Q = d$ which is half the total range) can be easily calculated by using geometry of the Fig. 30.14. If h is the weight of antenna tower and r the radius of earth, then

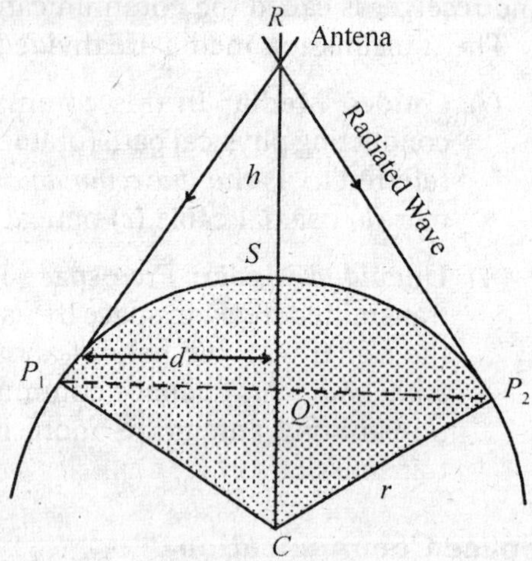

Fig. 30.14

$$(r + h)^2 = r^2 + (RP_1)^2 \quad \ldots(i)$$

Because of large radius of earth its curvature is small. As such point Q is very close to S. Hence from ΔRQP_1, we have

$$(RP_1)^2 = d^2 + h \quad \ldots(ii)$$

Substituting in (*i*), we get

$$(r + h)^2 = r^2 + d^2 + h^2$$

or $\quad 2rh = d^2 \quad$ or $\quad d = \sqrt{2rh}$ ...(30.11)

Thus a TV tower of height 100 m has a range of 36 km.

Besides short-range of this mode, the power radiated by the antenna dissipates as the inverse square of the distance as such the signal becomes weaker with the increasing distance. Hence for long distance transmission through this mode either height of the antenna be made large or repeater transmitting stations should be used.

3. **Sky Wave Propagation or Ionosphere Propagation:** When the radio waves from the transmitting antenna reach the receiving antenna after reflection in the ionosphere, the wave propagation is called sky wave propagation. This mode of propagation is shown in Fig. 30.13. Waves of frequency 2 MHz to 20 MHz can be reflected by the ionosphere. This region of AM band is called short wave band.

The ionosphere extends from 80 to 300 km above the surface of earth. The reflection of electromagnetic waves from ionosphere can be understood on the basis of total internal reflection. The oscillating electric field of electromagnetic waves (frequency ω) changes the velocity of the electrons in the ionosphere. This changes the effective dieletric constant $\varepsilon'$and hence refractive index n'. If n_0 and ε_0 are the value of refractive index and dielectric constant of free space, then

$$n' = n_0\left(1 - \frac{Ne^2}{\varepsilon_0 m\omega^2}\right)^{1/2} \qquad ...(30.12)$$

where N is electron density in the ionosphere.

The equation (30.12) shows that the refractive index of ionosphere is less than free space as such it behaves like a rarer medium. As we go deep inside the ionosphere, N—the electron density goes on increasing and refractive index goes on decreasing. The refraction or bending of the beam will continue till it reaches critical angle after which it will be reflected back. Equation 30.12 shows that different frequencies (ω) will be reflected from different regions of ionosphere having different values of N. Thus different points on earth receive signals reflected from different depths of ionosphere.

If the frequency ω is too high, then after a certain value the density N may not be so high to cause enough bending for reaching critical angle required for total internal reflection. This frequency is called *critical frequency*. If the maximum electron density of the ionosphere is N_{max}/m^3 then the critical frequency is given by.

$$f_c = 9\,(N_{max})^{1/2} \qquad ...(30.13)$$

The value of f_c varies from 5 to 10 MHz. Waves of frequency higher than f_c cross the ionosphere and do not return back to earth.

30.8 SATELLITE COMMUNICATION

The present day requirements of information technology need a very large number of communication channels and hence large bandwidth. This is possible only if the carrier frequency is high. If high frequency is used for carrier waves ($f > 30$ MHz), these waves cannot propagate by conventional methods discussed in last article. As such a new mode of communication, *i.e.*, communication through a satellite is developed which has revolutionised communication technology.

In this mode of communication, microwaves which have a frequency range of 1 GHz to 300 GHz (wavelength of the order of few millimetres to few centimetres) are used as carrier waves and the communicating signal is sent from transmitter to the receiver through a satellite. In short, a satellite communication is just like line of sight microwave communication.

In this mode of communication, a modulated microwave beam (signal) is sent directly towards the satellite from transmitter located on earth. This process is called *uplinking of signal.* The signals received by the satellite are processed, amplified and retransmitted by a device called *transponder* fitted in the satellite. The retransmission of signal by the satellite station to the ground station is called *down-linking signal.* To avoid confusion the up-link and down-link frequencies are kept different. Both these frequencies are in the microwave (UHF) region and can easily cross the ionosphere and reach satellite located well above the atmosphere. The height of geostationary satellite* is ~36000 km.

Thus a communication satellite acts as a microwave link repeater in the sky.

Because of great height of geostationary satellite, it can link communication with a large area of earth. But a single satellite cannot cover whole of the globe because of the curvature of the earth a large part of the earth remains out of site. The communication link over the entire earth can be established by putting three geostationary satellite in synchronous orbits 120° apart from each other (Fig. 30.15).

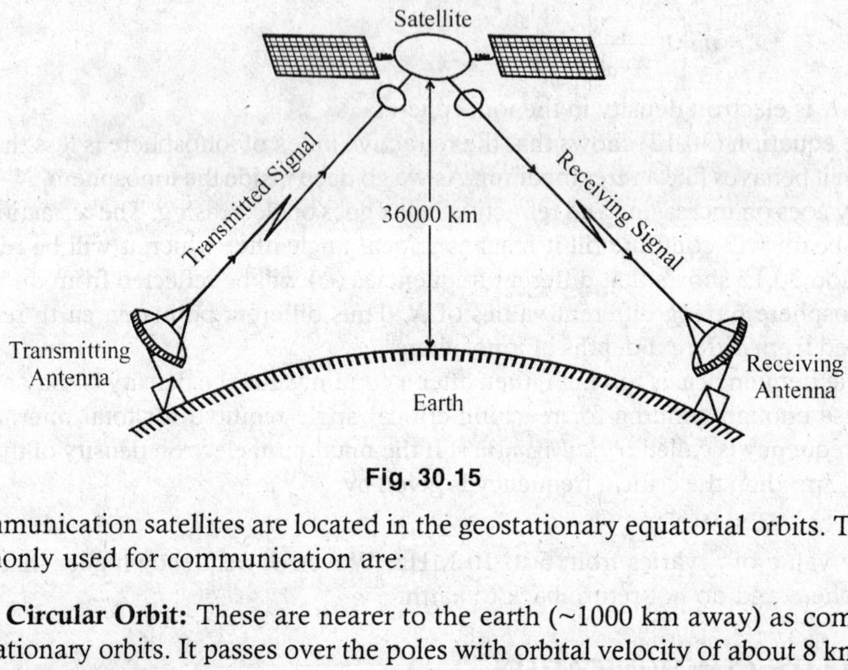

Fig. 30.15

Most communication satellites are located in the geostationary equatorial orbits. The other two orbits commonly used for communication are:

(*a*) **Polar Circular Orbit:** These are nearer to the earth (~1000 km away) as compared to the geostationary orbits. It passes over the poles with orbital velocity of about 8 km/s and time period of about 90 minutes. These satellites are used for *remote sensing,* i.e., obtaining information about an object, area or phenomenon, acquired by sensor that is not indirect contact with the target of investigation. It is basically linked to aerial photography. Now the more advanced technology of *satellite imagery* is used.

---

*A satellite which appears stationary to an observer on the earth is called geostationary satellite. See chapter on gravitation.

(*b*) **Inclined Highly Elliptical Orbits:** These orbits are used for communication in high altitude regions. The preferred inclination in 63° as such they are also called as being in 63° slot.

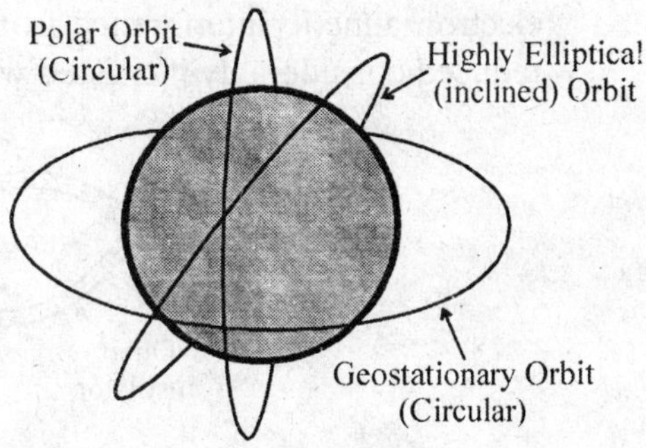

Fig. 30.16

30.9 LINE COMMUNICATION

In line communication, the response of transmission medium towards the signal to be transmitted plays an important role. The transmission characteristics of line communication depends upon the nature of transmission medium and nature of signal.

A communication system involving point-to-point contact between the transmitter and the receiver is called line communication.

The principal types of such communication channels are:

(*i*) Two wire transmission lines
(*ii*) Coaxials cables
(*iii*) Optical fibres.

(*i*) **Two Wire Transmission Lines:** In a parallel wire line, two metallic wires run parallel to each other inside a protective insulating PVC coating. The metallic wires may be hard or flexible depending on the power to be handled. For high power transmission hard wires are used.

The parallel-wire line is employed, when balanced properties of the communication system are the main consideration factors. For this reason such a line is called *balanced line.* When TV is connected to antenna dipole, the impedance of the parallel-wire must match that of the TV. As such a two wire transmission line is most suitable for this purpose. We know that the high frequency signals travel as a wave in the transmission line. Each portion of the transmission line can be considered as a small inductor, resistor and capacitor as shown in Fig. 30.17. Such inductors, resistors and capacitors are distributed throughout the transmission line. As such each length of the transmission line has a characteristic impedance. When the impedance of the detecting device at the receiving end (load) is balanced (*i.e.*, made equal) to the characteristic impedance, then power transferred to the load is maximum.

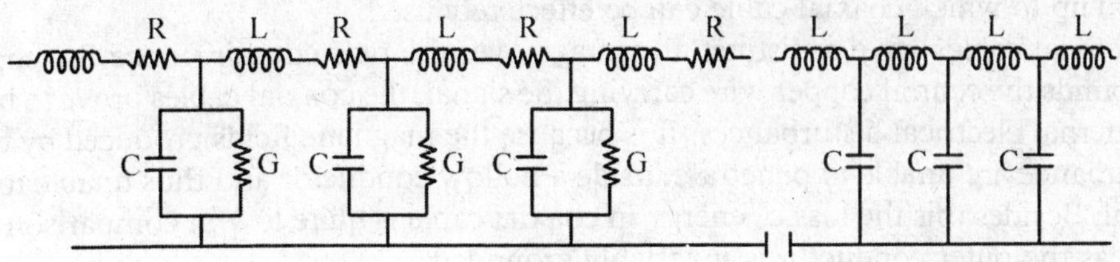

Fig. 30.17

As the frequencies to be transmitted increase towards RF or more, the impedance effects ωL or $1/\omega c$ or even R (R increases with frequency due to skin effect) become significant. As such, the transmission lines become very lossy. Even more serious problem arises because at such high frequencies the transmission line starts radiating energy. Additionally electromagnetic interference from other radiating sources also occurs. The latter can be reduced considerably if twisted wires are used. The geometry of this is shown in Fig. 30.18.

Fig. 30.18

(ii) **Coaxial Cable:** Coaxial cable consists of a central copper wire surrounded by a PVC insulation over which a slieve of copper mesh is placed. The copper mesh is covered with an outer shield of thick PVC material as shown in Fig. 30.19. The signal is transmitted by the inner copper wire which is electrically shielded by the outer copper mesh. The outer copper mesh is always grounded. The use of coaxial cable does not necessitate the matching of the impedance of the different parts of a communication system. For this reason this is also called *unbalanced line.*

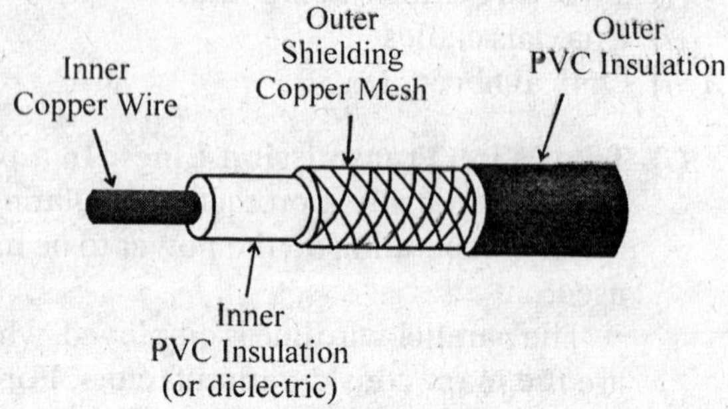

Fig. 30.19

Unlike a two wire-line, in coaxial cables most power is transmitted through electromagnetic waves and only a relatively small amount is transmitted in the form of conduction electrons in the central conductor. As such very high frequencies can be transmitted through coaxial cables before the attenuation becomes too large.

In some coaxial cables the inner PVC insulation is replaced by a dielectric. This dielectric is good insulator for DC or low frequencies and plays an important role in power transmission. As the frequency increases, the dielectric becomes more and more conducting and its resistance begins to contribute towards power loss which is called *dielectric loss.* The dielectric loss becomes quite high after a certain frequency. This puts an upper limit ($\approx$ 20 MHz) up to which coaxial cable can be effectively used.

Coaxial cables have a distinct advantage over the twisted pair. Since the copper mesh surrounds the central copper wire carrying the signal, the coaxial cables prove to be immune to external electrical disturbances. It is because the magnetic fields produced by the electric disturbance are unable to penetrate inside a hollow conductor and thus unable to affect the signal. Besides this the loss of energy in coaxial cable is quite low in comparison to twisted pair, as the outer conductor is invariably grounded.

30.10 OPTICAL COMMUNICATION

The use of light carrier waves (usually called optical wave) for transmission of information from one place to other is called optical communication.

The optical carrier waves have frequencies in the range of 10^{12} to 10^{16} Hz. This is very high as compared to the radio frequencies (10^6 to 10^8 Hz) or microwaves (10^9 to 10^{11} Hz). Due to high frequencies of carrier waves, the transmission bandwidth (which is about 10% of carrier wave frequency) is very large. Hence information carrying capacity (which is proportional to the bandwidth) of an optical communication system is much greater than that of radio or microwave communication.

Recently availability of optical fibres of small diameter ($\sim 10^{-5}$ m) through which light can be transmitted, has made optical communication easier and technology is developing very fast.

The inherent advantages of optical communication over conventional two wire or cable electronic communication system are:

(*i*) Wide channel bandwidth and large channel carrying capacity. It is due to use of higher frequencies $\sim 10^{14}$ Hz as compared to the electronic communication links.

(*ii*) Low transmission losses: In optical fibre power losses per km are less.

(*iii*) Signal security and not accessible to interference. The optical signal is confined to the inside of fibre and cannot be tempered easily. As such secret information is more secure.

Basic Optical Communication Link

The optical transmission requires a point to point link having transmitter at one end and a receiver at the other end. The most important components of an optical communication link are:

(*i*) Optical source or modulator

(*ii*) Communication channel

(*iii*) Optical detector or photo-detector

The components to be used in any optical communication system depend upon the system requirements such as desired transmission distance, channel bandwidth, cost, etc. For example, for a short distance transmission, one can use LED or diode laser of wavelength (λ) of the order of 800 nm. For large distance transmission, we have to use a wavelength for which power loss (attenuation) is small, *i.e.*, of the order of 1300 to 1550 nm. Similarly for obtaining strong optical signals at the receiving end a *p–n* photo-diode may be sufficient, but for detecting weak optical signals one may prefer to use highly sensitive avalanche photo-diode.

Figure 30.20 shows the block diagram for an optical fibre communication system. The information source provides an electrical signal to an electrical transmitter which drives an optical source to give modulated light wave. The optical source which provides the electrical-optical conversion, may be either a semiconductor laser or light emitting diode (LED). The optical fibre cable is used as a transmission medium for optical signal. At the receiving end, the optical signal is detected by optical detector, which converts the optical signal into an electrical signal. The optical detector may be photo-diode or a photo-transistor or a photo-conductor. The electrical signal is further demodulated by electric receiver so as to produce the original information signal, which is then received at the destination.

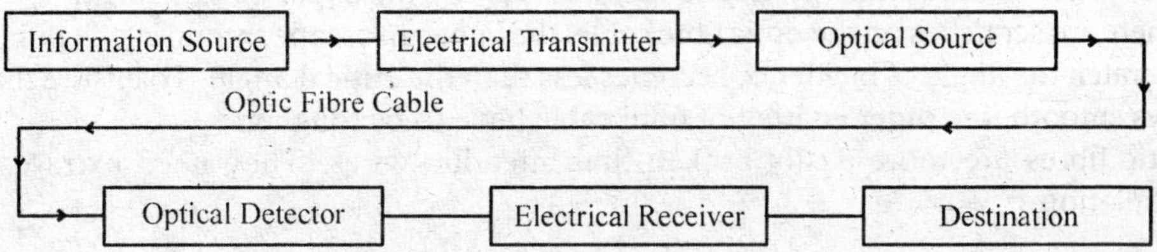

Fig. 30.20

Advantages of Optical Fibre Communication

As compared to the conventional electrical communication (parallel wire and twisted wire), the optical fibre communication is far more advantageous. The main advantages are given below:

1. **Large Transmission Bandwidth:** Due to high frequencies ($\sim 10^{13}$ to 10^{16} Hz) of optical carrier waves, the transmission bandwidth (approximately 10% of the carrier frequency) is large. This gives large information carrying capacity to the system as compared to the coaxial cable.
2. **Immunity to Electrical Disturbance:** Optical fibres serve as dielectric wave guide, so they are free from electromagnetic interference and radio frequency interference. They being insulators thus optical fibres are free from spark and short circuits.
3. **Low Transmission Loss:** Optical fibres have low losses per kilometre as compared to ordinary cables or microwave links. The optic fibres exhibit very low attenuation of the order of 0.2 dB km^{-1}.
4. **Signal Security:** An optical fibre signal is confined to the inside of optical fibres and cannot be tempered easily. So secret information is more secure.
5. **Small Size and Weight:** Smaller size and weight of the system make fibre optics more suitable in space and aeronautical applications.
6. **Ideal for Multiplexing:** Optical fibres can carry a large number of signals at the same time with sufficiently high frequency.
7. **System Reliability:** Due to low intensity loss property of optical fibre cables, the requirement of immediate repeaters or line amplifiers to boost the transmitting signal is much reduced. Thus the reliability of optical signals is enhanced to a greater extent. An optical fibre communication system possesses a predicted life time of the order of 20 to 30 yrs.
8. **High Suitability:** Optical fibres are best suited for digital transmission and switching systems. Laser modulation methods are ideal for such applications.
9. **No Signal Jamming:** Radio signals usually get jammed, when signals are transmitted at the same frequency of radio-waves. Optical communication is free from jamming of signals. They are also noise free.

Disadvantages

1. Manufacturing cost of optical fibres and lasers used is very high.
2. The optical fibre technology can be handled only by the experts in this field.
3. If there are scratches on the optic fibre cable, the light can escape through such points because at scratch the angle of incidence becomes less than the critical angle. To ensure that the core stays smooth, the outer coating of fibre cable have to be tougher.
4. Optic fibres are more easily broken than metallic wires. They need extra care during installation.

SOLVED EXAMPLES

Example 30.1: In a diode amplitude modulated detector, the output circuit consists of $R = 1\ k\Omega$ and $C = 10$ pF. A carrier signal of 100 kHz is to be detected. Is it good? If yes, then explain why? If not what value would you suggest?

Solution: For demodulation, the time constant of the circuit (RC) should be very large in comparison to time period of carrier wave.

i.e. $$\frac{1}{f_c} << RC$$

Here $$\frac{1}{f_c} = \frac{1}{100 \times 10^3} = 10^{-5} \text{ sec}$$

whereas $$RC = 10^3 \times 10 \times 10^{-12} = 10^{-8} \text{ s}$$

In present case, the required condition is not satisfied as $RC << \frac{1}{f_c}$. As such this is not good.

For satisfactory circuit

$$10^{-5} < RC \quad \text{or} \quad C > \frac{10^{-5}}{R} \quad \text{or} \quad C > \frac{10^{-5}}{10^3}$$

or $$C > 10^{-8}$$

for $RC >> \frac{1}{f_c}$ C should be greater than 1 μF.

Example 30.2: A ground receiver station is receiving a signal at (*a*) 5 MHz and (*b*) 100 MHz transmitted from a ground transmitter at a height of 300 m located at a distance of 100 km. Identify whether it is coming via space wave or sky wave propagation or satellite transponder. Radius of earth is 6.4×10^6 m and maximum electron density in ionosphere is $10^{12}\ m^{-3}$.

Solution: Maximum distance covered by space wave communication is

$$d = \sqrt{2rh} = \sqrt{2 \times 6.5 \times 10^6 \times 300} = 62 \text{ km}$$

here h is height of antenna.

Since receiver transmitter distance is 100 km. This is ruled out for both (*a*) and (*b*).

Further f_c for ionosphere propagation

$$f_c = 9\sqrt{N_{max}} \qquad \text{when } N_{max} \text{ is electron concentration}$$

$$= 9\sqrt{10^{32}} = 9 \text{ MHz}$$

So the signal (*a*) of 5 MHz ($<f_c$) comes via ionosphere mode whereas (*b*) signal of 100 MHz comes via satellite mode.

Example 30.3: Consider an optical communication system operating at $\lambda \sim 800$ nm. Suppose only 1% of the optical source frequency is the available channel bandwidth for optical communication. How many channels can be accommodated for transmitting (*a*) audio signals requiring bandwidth of 10 kHz and (*b*) Video TV signals requiring an approximate bandwidth of 5 MHz.

Solution: Optical source frequency $f = c/\lambda$

$$f = \frac{3 \times 10^8}{800 \times 10^{-9}} = 3.75 \times 10^{14} \text{ Hz}$$

Bandwidth is 1% the frequency. Hence bandwidth

$$\Delta f = 3.75 \times 10^{12} \text{ Hz}$$

$$\text{No. of channels} = \frac{\text{Total bandwidth of channel}}{\text{Bandwidth needed per channel}}$$

(*a*) Number of channels for audio signal of bandwidth 8×10^3 Hz is

$$n = \frac{3.75 \times 10^{12}}{8 \times 10^3} = 4.7 \times 10^8$$

(*b*) The number of channels for video TV signal requiring bandwidth 4.5×10^6 Hz

$$n = \frac{3.75 \times 10^{12}}{4.5 \times 10^6} = 8.3 \times 10^5$$

Example 30.4: (*a*) A TV tower has height of 80 m. Find the radius of the circle within which the transmission will be observed, if radius of earth is 6.4×10^6 m. (*b*) How much population is covered by the TV broadcast, if the average population density around tower is 800 km^{-2}?

Solution:

(*a*) If h is the height of the tower and r radius of earth, then radius of the circular area covered by transmission from tower is

$$d = \sqrt{2rh} = \sqrt{2 \times 6.4 \times 10^6 \times 80}$$
$$= 3.2 \times 10^4 \text{ m}$$

(*b*) Area in which transmission can be viewed is

$$A = \pi d^2 = \pi \times (3.2 \times 10^4)^2$$

Since average population density is 800/km^2. Then population covered by transmission is

$$= A \times \text{population density}$$

$$= \frac{800 \times \pi \times (3.2 \times 10^4)^2}{(10^3)^2} = 2.57 \times 10^6$$

Example 30.5: Calculate the time delay between the reception of the ground wave and the sky wave for a receiver at a distance of 200 km from the transmitter given that the reflecting ionosphere layer is effective up to a height of 100 km.

Solution: Let T and R be the transmitter and receiver antennae respectively. Hence $TR = 200$ km (Fig. 30.21). Hence time taken by the ground wave to reach receiver t_1 is

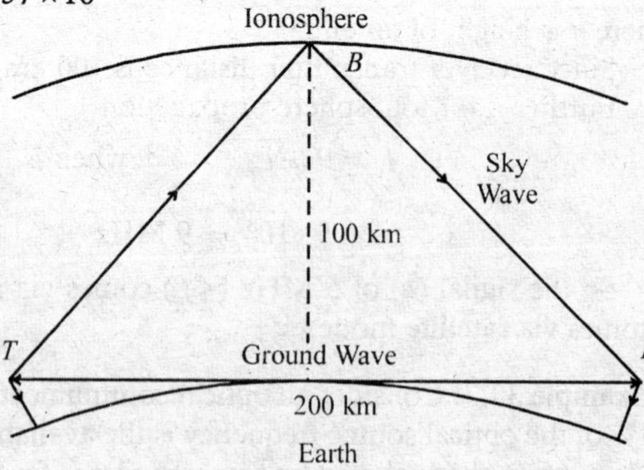

Fig. 30.21

$$t_1 = \frac{200 \times 10^3}{3 \times 10^8} = 6.67 \times 10^{-4} \text{ sec} \qquad \text{(Signal travels with velocity of light)}$$

The sky wave follows the path *TB* and then *TR* to reach the receiver. Hence distance travelled by sky wave is *TB* + *BR*. But

$$TB = TR = \sqrt{(100)^2 + (100)^2} = 100\sqrt{2}$$

$$\therefore \qquad TB + BR = 2 \times 100\sqrt{2} = 283 \text{ km}$$

Hence time taken by sky wave to reach the receiver is

$$t_2 = \frac{383 \times 10^3}{3 \times 10^8} = 9.43 \times 10^{-4} \text{ s}$$

Hence time delay is = $9.4 \times 10^{-4} - 6.67 \times 10^{-4} = 2.76 \times 10^{-4}$ s.

Example 30.6: In an amplitude modulator, the tank circuit consists of a coil of inductance 3.6 mH and a capacitor of capacitance 2.5 pF. If an audio-signal of frequency 15 kHz modulates the carrier generated by the tank circuit, find the frequency of the side bands.

Solution: The frequency of the carrier generated by the tank circuit is

$$f_c = \frac{1}{2\pi\sqrt{LC}} = \frac{1}{2\pi\sqrt{3.6 \times 10^{-3} \times 2.5 \times 10^{-12}}} = 1678 \text{ kHz}$$

The frequency of audio signal

$$f_m = 15 \text{ kHz}$$

Hence frequency *F* of upper side band is

$$f_c + f_m = 1678 + 15 = 1693 \text{ kHz}$$

The frequency of the lower side band is

$$F_c + f_m = 1678 - 15 = 1663 \text{ kHz.}$$

Example 30.7: An amplitude modulated wave is represented as

$$e = 5\,(1 + 0.6\cos 6280t)\,\sin 211 \times 10^4 t \text{ V}$$

(*i*) What are minimum and maximum amplitudes of the AM wave?
(*ii*) What frequency components are contained in the modulated wave.
(*iii*) What are the amplitudes of components.

Solution: Standard amplitude AM wave equation is

$$e = E_c\,(1 + m\cos\omega_m t)\sin\omega_c t$$

when E_c is amplitude of carrier wave and *m* is modulation index

Comparing with given equation

$$e = 5(1 + 0.6\cos 6280t)\,\sin 211 \times 10^4 t \text{ V}$$

we have

$$E_c = 5 \text{ V} \qquad m = 0.6.$$

(*i*) Minimum amplitude of AM wave is

$$e_{min} = 5 - 5 \times 0.6 = 2 \text{ V.}$$

Maximum amplitude of AM wave is

$$e_{max} = 5 + 5 \times .6 = 8 \text{ V}$$

(*ii*) Frequency components of AM modulated wave are

(*a*) $f_c - f_m = \frac{\omega_c}{2\pi} - \frac{\omega_m}{2\pi} = \frac{211 \times 10^4}{2\pi} - \frac{6280}{2\pi}$

$= 336 \times 10^3 - 10^3 = 335$ kH. Lower side band

(*b*) $f_c = 336$ kHz carrier frequency

(*c*) $f_c + f_m = 336 \times 10^3 + 10^3 = 337$ kHz upper side band

(*iii*) Amplitudes of three components are

(*a*) $\frac{mE_c}{2} = \frac{0.6 \times 5}{2} = 1.5$ V

(*b*) $E_c = 5$ V and

(*c*) $\frac{mE_c}{2} = \frac{0.6 \times 5}{2} = 1.5$ V.

Example 30.8: Calculate the length of half wave dipole at (*i*) 30 MHz (*ii*) 300 MHz and (*iii*) 3000 MHz. What information do you draw from the result?

Solution:

(*i*) $f = 30 \times 10^6$ Hz $\therefore$ $\lambda = \frac{c}{f} = \frac{3 \times 10^8}{30 \times 10^6} = 10$ m

Hence length of dipole $\frac{10 \text{ m}}{2} = 5$ m

(*ii*) $f = 300 \times 10^6$ Hz $\lambda = \frac{c}{f} = \frac{3 \times 10^8}{300 \times 10^6} = 1$ m

Hence length of dipole = 1/2 m = 50 cm.

(*iii*) $f = 3000 \times 10^6$ Hz $\lambda = \frac{c}{f} = \frac{3 \times 10^8}{3000 \times 10^6} = 0.1$ m = 10 cm

Hence length of dipole = 1/2 × 10 = 5 cm.

Clearly the length of dipole decreases with the increase in frequency of carrier waves.

Example 30.9: A pulse train is transmitted along a transmission line which is 200 m long. A pulse train consists of pulses with a duration of 30 ns each and separated by 45 ns. How many pulses can be on the line at any given time? Assume the speed of em waves to be the same as in free space.

Solution: Total length of transmission line $d = 200$ m

Pulse time $t_p = 30 \times 10^{-9}$ s

Pulse separation $t_s = 45 \times 10^{-9}$ s

Hence total time $t = (30 + 45) \times 10^{-9} = 75 \times 10^{-9}$ s

Hence space required for one pulse is

$$S = ct = 3 \times 10^8 \times 75 \times 10^{-9} = 22.5 \text{ m}$$

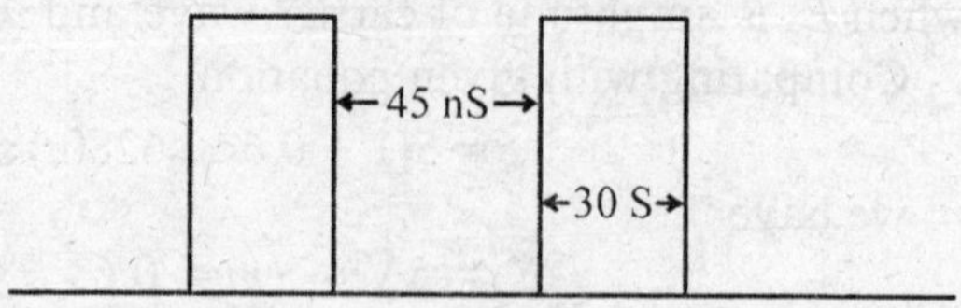

Fig. 30.22

Hence total number of pulses

$$n = \frac{d}{s} = \frac{200}{22.5} = 8.9$$

Hence total 9 pulses can be on the line at any time.

Example 30.10: A particular cable has a capacitance of 40 pF/m and a characteristic impedance of 70 Ω. (*a*) What is the inductance per foot of this cable? (*b*) Determine the impedance of an infinitely long section of such cable?

Solution:

(*a*) We know that characteristic impedance is

$$Z_0 = \sqrt{\frac{L}{C}}$$

$$\therefore \quad 70 = \sqrt{\frac{L}{40 \times 10^{-12}}} \quad \text{or} \quad L = (70)^2 \times 40 \times 10^{-12} = 0.196\ \mu\text{H/m}$$

Inductance per foot = 0.196 × 0.3 = 0.0588 μH/ft

(*b*) The characteristic impedance of transmission line is the impedance that an infinite length of the line would present to power supply at the input end of the line. As such in present case

$$Z_\infty = Z_0 = 70\ \Omega.$$

Example 30.11: Velocity of light in the core of step index fibre is 2×10^8 m/s and critical angle at the core-cladding interface is 80°. Find the numerical aperture and the angle of acceptance for fibre in air. The velocity of light in vacuum is 3×10^8 m/s.

Solution: The refractive index of the core of fibre is

$$\mu_1 = \frac{3 \times 10^8}{2 \times 10^8} = 1.5$$

If μ_2 is the refractive index of cladding, then critical angle at cladding core surface c is given by

$$\sin c = \frac{\mu_2}{\mu_1} \qquad c = 80°.$$

$$\therefore \quad \mu_2 = \mu_1 \sin c = 1.5 \times \sin 80° = 1.48$$

If i is the angle of acceptance of fibre in air, then

$$\sin i = \sqrt{\mu_1^2 - \mu_2^2} = \sqrt{(1.5)^2 - (1.48)^2} = 0.244$$

Hence numerical aperture is 0.244
and angle of acceptance $i = \sin^{-1}(0.244) = 14.12°$.

Example 30.12: Light from a source located in a medium (refractive index = μ_0) enters and optical fibre with core refractive index of μ_1 and clad refractive index of μ_2. Show that the maximum value of angle of acceptance (θ_0) which would undergo total internal reflection in the fibre is

$$\theta_0 = \sin^{-1}\left(\frac{\mu_1^2 - \mu_2^2}{\mu_0^2}\right)^{1/2}$$

Solution: For refraction from medium of refractive index μ_0 to fibre, the angle of refraction should be 90–c for total internal reflection to take place in the fibre at cladding surface. Hence from Fig. 30.23,

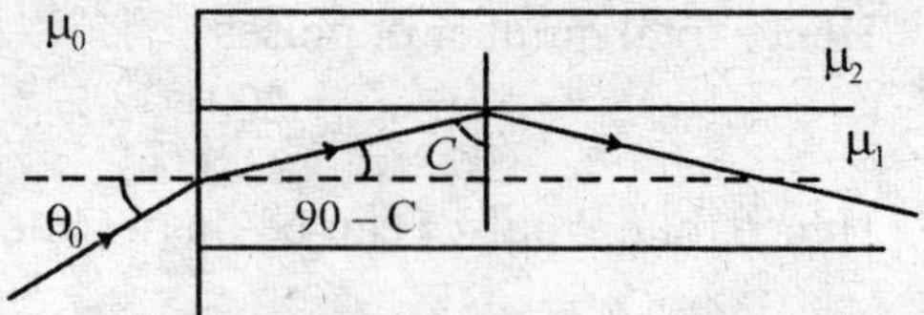

Fig. 30.23

$$\frac{\sin\theta_0}{\sin(90-c)} = \frac{\mu_1}{\mu_0}$$

i.e. $$\sin\theta_0 = \frac{\mu_1}{\mu_2}\cos c \qquad \ldots(i)$$

For refraction at core-cladding surface, we have

$$\sin c = \frac{\mu_2}{\mu_1} \qquad \ldots(ii)$$

$$\therefore \quad \cos c = \sqrt{1-\sin^2 c} = \sqrt{1-\frac{\mu_2^2}{\mu_1^2}} = \frac{1}{\mu_1}\sqrt{\mu_1^2-\mu_2^2}$$

Substituting in (*i*) gives

$$\sin\theta_0 = \frac{\mu_1}{\mu_0}\cos c = \sqrt{\frac{\mu_1^2-\mu_2^2}{\mu_0}}$$

$$\therefore \quad \theta_0 = \sin^{-1}\frac{\sqrt{\mu_1^2-\mu_2^2}}{\mu_0}$$

PROBLEMS

1. An audio signal 15 sin2π (1500 *t*) amplitude modulates a carrier 60 sin2π (100,000 *t*).
   (*a*) Sketch audio signal
   (*b*) Sketch the carrier
   (*c*) Construct the modulated wave
   (*d*) Determine the modulation factor and percent modulation.
   (*e*) What are the frequencies of audio signal and the carrier?
   (*f*) What frequencies would show up in spectrum analyses of the modulated wave?

   [**Ans.** (*d*) $m = \dfrac{\text{Audio amplitude}}{\text{Carrier amplitude}} = 0.25$ and percent modulation is 25% (*e*) $f_a = 1500$ Hz, $f_c = 100000$ Hz (*f*) $f_c + f_a = 101500$ Hz, $f_c - f_a = 98500$ Hz and $f_c = 100000$ Hz]

2. If the maximum and minimum voltage of an AM wave are V_{max} and V_{min} respectively, then show that the modulation factor m is given by

$$m = \frac{V_{max}+V_{min}}{V_{max}-V_{min}}$$

3. A 75 MHz carrier having an amplitude of 50 V is modulated by a 3 kHz audio signal having an amplitude of 20 V.

(*a*) Sketch audio signal
(*b*) Sketch the carrier
(*c*) Construct the modulated wave
(*d*) Determine the modulation factor and percent modulation.
(*e*) What frequencies would show up in spectrum analysis of the modulated wave?
(*f*) Write equations for the carrier and the modulating waves.

[**Ans.** (*d*) m =0.4 (*e*) $f_c + f_a$ = 75003 kHz, f_c = 75 MHz, $f_c - f_a$ = 74997 kHz (*f*) $V_a = 20 \sin 6000\pi t$, $V_c = 50 \sin 150 \times 10^6 \pi t$]

4. A loss-less coaxial cable has a capacitance of 7×10^{-11} F and an inductance of 0.39 μH. Calculate the characteristic impedance of the cable.

$$\left[\textbf{Ans. } Z_0 = \sqrt{\frac{L}{C}} = 75\ \Omega\right]$$

5. A TV tower has a height of 100 m. How much population is covered by the TV broadcast if the average population density around the tower is 1500 $(\text{km})^{-2}$? Radius of earth is 6.4×10^6 m.

[**Ans.** 6×10^6]

6. When modulation percentage is 75, an AM transmitter produces 10 kW. What would be % power saving if the carrier and one of the side bands are suppressed before transmission takes place?

[**Ans.** 89.1%]

[**Hint:** Power dissipated in carrier wave $P_c = \frac{(V_c/\sqrt{2})^2}{R} = \frac{V_c^2}{2R}$. Power dissipated in 2 side bands $P_s = 2\left(\frac{mV_c}{2\sqrt{2}}\right)^2 \frac{1}{R} = \frac{mV_c^2}{4R}$. Total power dissipated $= P_c + P_s = \frac{V_c^2}{2R}\left(1 + \frac{m^2}{2}\right)$. Fraction of the total power carried by side bands $= \frac{P_s}{P_c + P_s} = \frac{m^2}{2 + m^2}$ and fraction of the power carried by one side band $= \frac{m^2}{2(2+m^2)}$. Since m = 0.75, hence power carried by side band $= \frac{m^2}{2(2+m^2)} = 0.1098$. Hence power saved $= 1 - 0.1098 = 0.8902$. Hence % power saved = 89.02%]

7. A ground receiver station is receiving signals at (*i*) 6 MHz and (*ii*) 120 MHz transmitted from a ground transmitter at a height of 300 m, located at a distance of 100 km. Identify whether the signals are coming via space wave or sky wave propagation or through satellite transponder. Radius of earth is 6.4×10^6 m. Maximum electron density in ionosphere is 10^{12} $^{-3}$ m.

[**Ans.** (*i*) Sky wave propagation (*ii*) Satellite communication]

8. A TV transmission tower at a particular station has a height of 160 m. (*a*) What is its coverage range? (*b*) How much population is covered by transmission, if the average population density around the tower is 1200 km^{-2}? (*c*) By how much the height should be increased to double its coverage? Given radius of earth as 6400 km.

[**Ans.** (*a*) d = 45 km (*b*) 76.37 lakhs (*c*) Δh = 480 m]

9. A microwave telephone link operating at the central frequency of 10 GHz has been established. If 2% of this is available for microwave communication channel, then how many telephone channels can be simultaneously granted if each telephone is allotted a bandwidth of 8 kHz. **[Ans.** 2.5×10^4]

10. A step index fibre has a relative refractive index difference of 0.9%. Estimate the critical angle at core cladding interface. When the core index is 1.46. Find the numerical aperture, if source to fibre medium is air:

[**Hint:** $\frac{\mu_1 - \mu_2}{\mu_1} = .009$ or $\mu_1 = 0.991 \times 1.46 = 1.45$. Hence numerical aperture $\sin i = \sqrt{\mu_2^2 - \mu_1^2}$ $= 0.17$ and angle of acceptance $= 9.8°$. If critical angle is c then $\sin c = \frac{\mu_1}{\mu_2} = \frac{1.45}{1.46} = 0.993$ and $c = 83°$]

11. The dielectric constant of the insulation used in a transmission line is 2.6. Calculate the velocity factor of the line. [**Hint:** Velocity factor $= \frac{1}{\sqrt{k}} = 0.62$]

12. Consider an optical communication system operating at $\lambda \sim 800$ nm. Suppose only 1% of the optical source frequency is available as channel band width for optical communical, how many channels can be accommodated for transmitting (*a*) audio signals requiring a bandwidth of 8 kHz and (*b*) Video TV signals requiring an approximate bandwidth of 4.5 MHz.

[**Ans.** (*a*) 4.8×10^8 (b) 8.4×10^5]

13. A radar has a power of 1 kW and is operating at a frequency of 10 GHz. It is located on a steep mountain top of height 600 m. What is the maximum distance upto which it can detect an object located on the surrounding earth's surface? [**Hint:** $d = \sqrt{2rh} = 87.6$ km]

14. Determine the required pulse duration of a pulse so that when pulse travels on a 25 m line, the trailing edge occurs at the generator end of the line just as the leading edge reaches the load. Assume that the speed of the pulse on the line is same as its free space velocity $c = 3 \times 10^8$ m/s.

[**Ans.** $t_p = 0.0833$ μs].

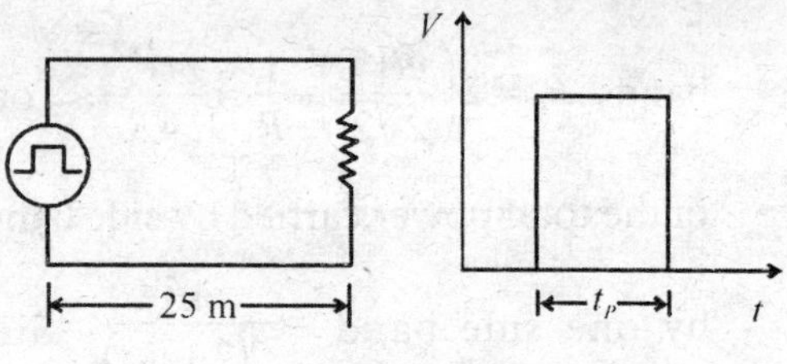

Fig. 30.24

15. How many 600 MHz, waves can be on a 5 mile transmission line simultaneously?

[**Hint:** Wavelength $\lambda = \frac{c}{\nu}$ and $n\lambda = d$ or $n = \frac{d}{\lambda} = \frac{d}{c}\nu$ $\therefore n = \frac{5 \times 600 \times 10^3}{186000}$ $\because c = 186000$ miles/sec $\therefore n = 16.13$]

16. An optical fibre has a numerical aperture of 0.2 and a cladding refractive index of 1.59. Determine (*a*) the acceptance angle for fibre in water which has refractive index of 1.33, (*b*) critical angle at the core-cladding interface.

[**Hint:** Numerical aperture $\sin\theta_0 = \sqrt{\frac{\mu_1^2 - \mu_2^2}{\mu_0^2}}$ or $\sqrt{\mu_1^2 - \mu_2^2} = 1.33 \times .2 \therefore \mu_1{}^2 = 0.071 + (1.59)^2$ or $\mu = 1.612$. Hence angle of acceptance is $i = \sin^{-1}\left(\frac{.2}{1.33}\right) = 8.6°$. Critical angle at interface is $c = \sin^{-1}\left(\frac{\mu_2}{\mu_1}\right) = 80.5°$]

MULTIPLE CHOICE QUESTIONS

In the following, a question is followed by four alternatives, of them one or more are correct. Candidate has to select the correct choice(s) from amongst the alternatives.

1. A 400 Watt carrier is modulated to a depth of 75%. The total power in the modulated wave is

(a) 512.5 Watt (b) 287.5 Watt
(c) 456.25 Watt (d) 443.75 Watt.

2. The characteristic impedance of a loss-less transmission line is

(a) $Z_0 = \sqrt{LC}$ (b) $Z_0 = \sqrt{L/C}$

(c) $Z_0 = \sqrt{\frac{C}{L}}$ (d) $Z_0 = \frac{1}{\sqrt{LC}}$.

3. The power in a two wire transmission line travels

(a) inside the conductors
(b) outside the conductors
(c) both inside and outside the conductor
(d) inside the conductor or outside, depends on the frequency of the carrier waves.

4. Ozone layer above the earth's atmosphere,

(a) prevents infrared radiation from sun reaching earth
(b) prevents infrared rays reflected from earth from escaping earths atmosphere
(c) prevents ultraviolet rays from sun to enter atmosphere
(d) reflects back radiowaves.

5. Digital signals (*i*) do not provide a continuous set of values, (*ii*) represent values as discrete steps (*iii*) can utilize only binary system and (*iv*) can utilize decimal as well as binary system. Which of the following options is true?

(a) Only (*i*) and (*ii*)
(b) Only (*ii*) and (*iii*)
(c) All the above (*i*) to (*iv*)
(d) Only (*i*), (*ii*) and (*iii*) but not (*iv*).

6. A microwave telephone link operating at central frequency of 10 GHz has been established. If 2% of this is available for microwave communication channel, then number of telephone channels with band width 8 kHz can be simultaneously granted is

(a) 2.5×10^7 (b) 2.5×10^6
(c) 2.5×10^5 (d) 2.5×10^4.

7. Modulation is used to

(a) reduce the bandwidth used
(b) separate the transmissions of different users
(c) ensure that intelligence may be transmitted to long distance
(d) allow the use of practical antenna.

8. At which of the following frequencies/ frequency, the communication will not be reliable for a receiver situated beyond horizontal.

(a) 100 kHz (b) 1 MHz
(c) 1 GHz (d) 100 GHz.

9. The audio signals cannot be transmitted directly into space because of

(a) their low frequency range they die out quickly
(b) the height of antenna required is impractically large
(c) signals transmitted from different transmitting stations get inseparably mixed up
(d) their immunity to noise.

10. AM is used for broadcasting because

(a) it is more noise immune than other modulation systems

(b) it requires less power for transmission compared to the other systems
(c) its use avoids receiver complexity
(d) no other system can provide the necessary bandwidth faithful transmission.

11. Frequencies in the UHF range normally propagate by means of

(a) ground waves (b) sky waves
(c) surface waves (d) space waves.

12. Choose the correct statement/statements with regards AM and and FM transmission.

(a) Efficiency of AM transmission is quite high
(b) Noise level in FM transmission is very low
(c) FM transmission gives high fidelity reception due to the presence of a large number of side bands
(d) The adjacent channel interference is high in AM.

13. Choose correct statement with regards to digital communication.

(a) Digital signals can be reproduced more accurately
(b) Digital signals get distorted by noise
(c) Digital signals can be coded
(d) Digital signals cannot be received easily.

14. An antenna is

(a) inductive
(b) capacitive
(c) resistive above its resonance frequency
(d) none of the above.

15. Long distance short-wave radio broadcasting uses

(a) ground wave (b) ionosphere waves
(c) direct wave (d) sky wave.

ANSWERS

1. (a)	2. (b)	3. (b)	4. (a, b, c)	5. (d)
6. (d)	7. (a)	8. (b, c)	9. (a, b, c)	10. (c)
11. (d)	12. (b, c, d)	13. (a, c)	14. (a)	15. (b)

Model Test Paper I

SECTION I

In this section, a question is followed by four alternatives, of them only one is correct.

1. Two metal sphere with radii $r_1 = 1$ cm and $r_2 = 2$ cm are separated by $R = 1$ m and connected to a battery of emf 3000 V. The force between them is

(a) 4.4×10^{-8} N (b) 4×10^{-7} N
(c) 5×10^{-6} N (d) 44×10^{-8} N.

2. Two convex lenses of focal length f_1 and f_2 are separated by a distance d. The combination will work as a glass slab if

(a) $d > f_1 + f_2$ (b) $d < f_1 + f_2$
(c) $d = f_1 + f_2$ (d) never.

3. A voltmeter of resistance 9800 Ω has a least count of 0.2 V and range (0 – 10) Volts. Instead of connecting the voltmeter in parallel with one of the 100 Ω resistor by mistake, a student connects it in series as shown in Fig. MTP-1.1. Which of the following statement is correct?

(a) The voltmeter will get damaged
(b) The voltmeter will read zero
(c) The voltmeter will read 9 Volts
(d) The voltmeter will read 8.8 Volts.

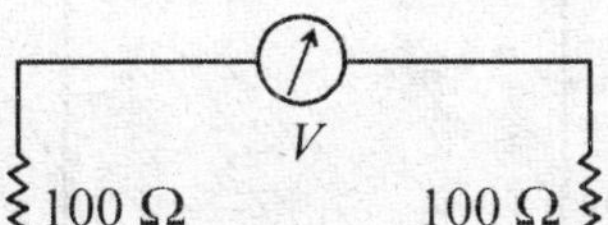

Fig. MTP-1.1

4. The upper half of an inclined plane with inclination ϕ is a perfectly smooth while the lower half is rough. A body starting from rest at the top will again come to rest at the bottom if the co-efficient of friction for the lower half is given by

(a) $\tan\phi$ (b) $2\tan\phi$
(c) $2\cos\phi$ (d) $2\sin\phi$.

5. An annular ring with inner and outer radii R_1 and R_2 is rolling without slipping with a uniform angular speed. The ratio of the forces experienced by the two particles situated on the inner and outer parts of the ring $\frac{F_1}{F_2}$ is

(a) $\frac{R_1}{R_2}$ (b) 1

(c) $\left(\frac{R_1^2}{R_2}\right)$ (d) $\frac{R_2}{R_1}$.

6. The intensity of gamma radiation from a given source is I. On passing through 36 mm of lead it is reduced to $I/8$. The thickness of lead which will reduce the intensity to $I/2$ will be

(a) 12 mm (b) 18 mm
(c) 9 mm (d) 6 mm.

7. The temperature entropy diagram of a reversible engine cycle is given in the Fig. MTP-1.2. Its efficiency is

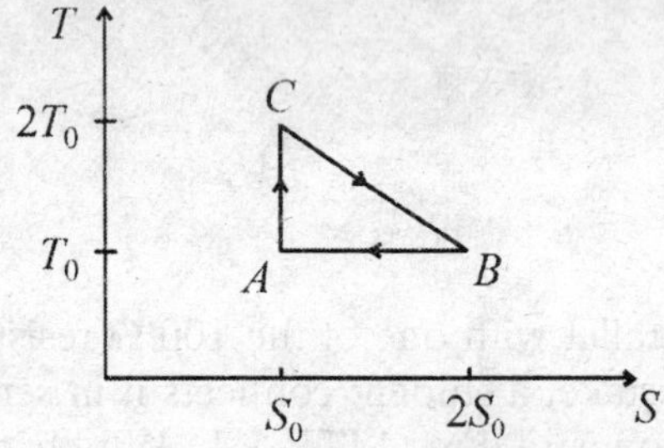

Fig. MTP-1.2

(a) 2/3 (b) 1/3
(c) 1/4 (d) 1/2.

8. A nuclear transformation is denoted by $X(n, \alpha)\,^{7}_{3}\text{Li}$. Which of the following is the nucleus of element X?

(a) $^{11}_{4}$Be (b) $^{9}_{5}$B
(c) $^{10}_{5}$B (d) $^{12}_{6}$C.

9. A uniform non-conducting rod of length l is hinged at an end O, so that it can oscillate freely in vertical plane. A charge q is placed on the other end. If the bar is released from horizontal position, the magnetic induction at hinge when rod becomes vertical is

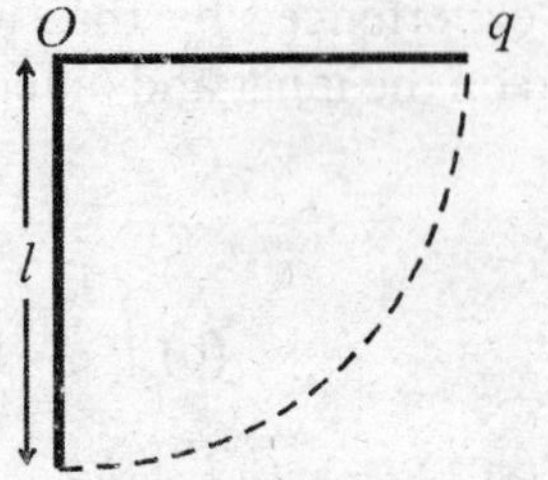

Fig. MTP-1.3

(a) $\frac{\mu_0 q}{4\pi l}\sqrt{\frac{3g}{l}}$ (b) $\frac{\mu_0 q}{2\pi l}\sqrt{\frac{3g}{l}}$

(c) $\frac{\mu_0 q}{4\pi l}\sqrt{3gl}$ (d) $\frac{\mu_0 q}{2\pi}\sqrt{3gl}$.

10. The wavelength of K_α, X-rays of an element having atomic number 31 is λ. The wavelength of K_α X-rays of another element of atomic number Z' is 4λ. Then Z' is

(a) 16 (b) 62
(c) 44 (d) 60.

11. The velocity displacement graph of a particle moving along a straight line is shown in Fig. MTP-1.4. The most suitable acceleration displacement graph is

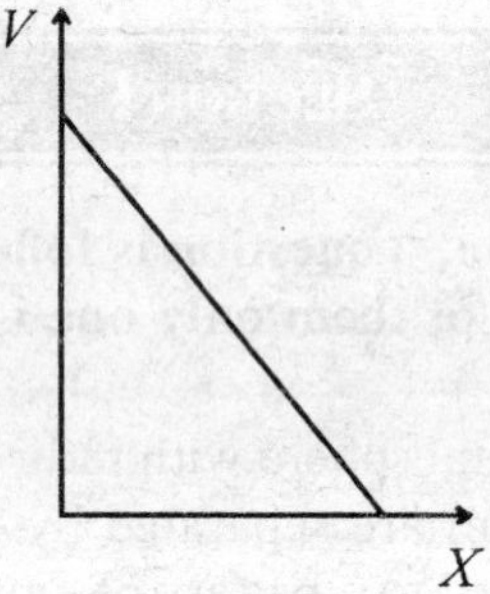

Fig. MTP-1.4

(a) 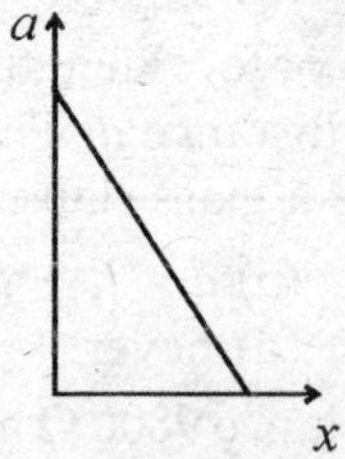 (b) 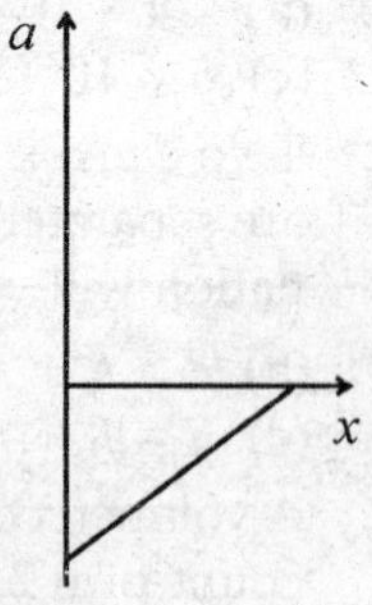

(c) 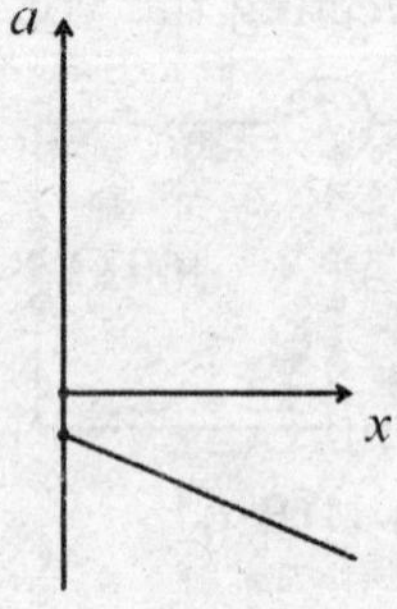 (d) 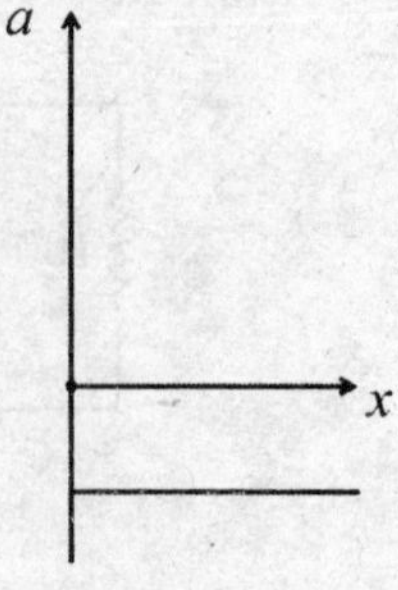

12. The work of 146 J is performed in order to compress one mole of a gas adiabatically and in this process the temperature of the gas increases by 7° C. The gas is ($R = 8.3$ J mole^{-1} K^{-1})
(a) a mixture of monoatomic and diatomic
(b) monoatomic
(c) diatomic
(d) triatomic.

SECTION II

In this section, a question is followed by four alternatives, of them one or more are correct.

13. A particle is moving along an expanding spiral (Fig. MTP-1.5) such that the particle's normal acceleration remains constant. If R is the radius of curvature of the spiral and V its linear velocity and ω its angular velocity, then

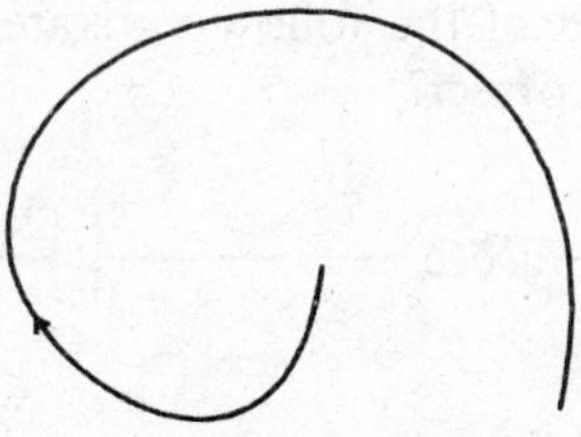

Fig. MTP-1.5

(a) $V \propto \sqrt{R}$ (b) $V \propto R^2$
(c) $\omega \propto 1/\sqrt{R}$ (c) $\omega \propto 1/R^2$

14. Given $\boldsymbol{a} + \boldsymbol{b} + \boldsymbol{c} + \boldsymbol{d} = 0$. Which of the following statements is/(are) true?
(a) $\boldsymbol{a}$, $\boldsymbol{b}$, $\boldsymbol{c}$, and $\boldsymbol{d}$ must each be a null vector
(b) the magnitude of $\boldsymbol{a} + \boldsymbol{c}$ equals the magnitude of $\boldsymbol{b} + \boldsymbol{d}$
(c) The magnitude of $\boldsymbol{a}$ can never be greater than the sum of the magnitudes of $\boldsymbol{b}$, $\boldsymbol{c}$ and $\boldsymbol{d}$.
(d) $\boldsymbol{b} + \boldsymbol{c}$ must be in the plane of $\boldsymbol{a}$ and $\boldsymbol{d}$ if $\boldsymbol{a}$ and $\boldsymbol{d}$ are not collinear and in the line of $\boldsymbol{a}$ and $\boldsymbol{d}$ if they are collinear.

15. A uniform rod of length $6a$ and mass $8m$ lies on a smooth horizontal table. Two particles of mass m and $2m$ moving with velocity $2v$ and v respectively strike the rod normally (Fig. MTP-1.6) and stick to the rod. Denoting angular velocity (about the centre of mass of rod), total energy and translational velocity of centre of mass of rod by ω, E and v_c respectively after the collision

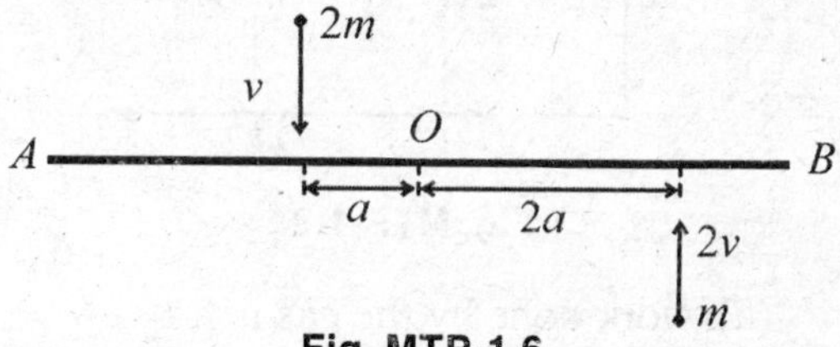

Fig. MTP-1.6

(a) $v_c = 0$ (b) $\omega = \dfrac{3v}{5a}$
(c) $E = \dfrac{3mv^2}{5}$ (d) $\omega = \dfrac{v}{5a}$.

16. A block of mass m attached to a massless spring of force constant K; the other end of which is fixed to the wall of a truck as shown is in Fig. MTP-1.7. The block is placed over a smooth surface and the spring initially is unstretched. Suddenly the truck starts moving towards left with a constant acceleration a_0. As seen from the truck

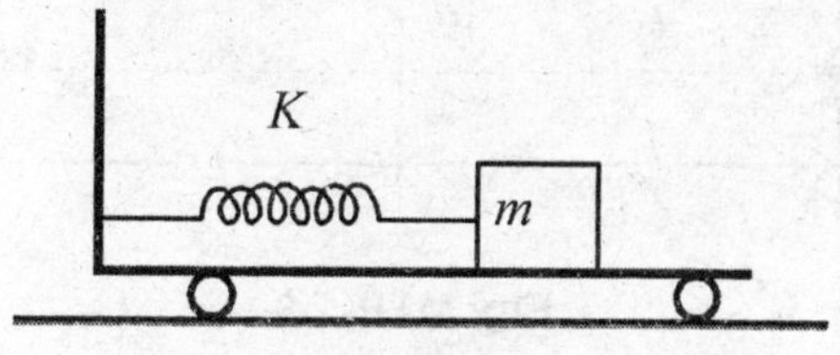

Fig. MTP-1.7

(a) the block will execute S.H.M.
(b) the time period of oscillation will be $2\pi\sqrt{\dfrac{m}{K}}$
(c) the amplitude of oscillation will be $\dfrac{ma_0}{K}$

(d) The energy oscillation will be $\frac{m^2a_0^2}{2K}$.

17. One mole of an ideal monoatomic gas is taken from A to C along the path ABC. The temperature of the gas at A is T_0. For process ABC (Fig. MTP-1.8) (R is universal gas constant)

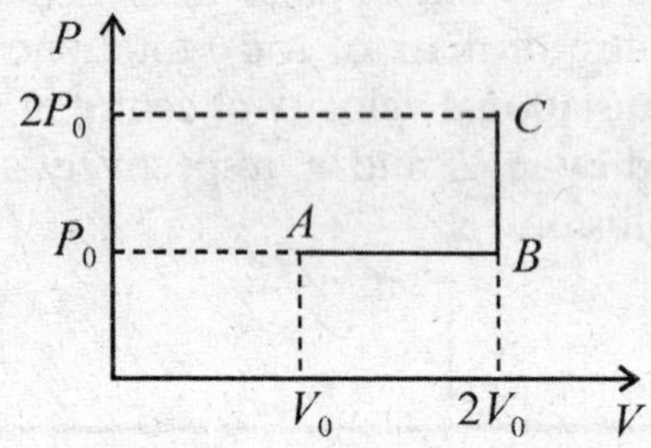

Fig. MTP-1.8

(a) work done by the gas is RT_0

(b) change in internal energy of the gas is $\frac{11}{2}RT_0$

(c) heat absorbed by the gas is $\frac{11}{2}RT_0$

(d) heat absorbed by the gas is $\frac{13}{2}RT_0$.

18. Consider the arrangement shown in Fig. MTP-1.9. The distance D is very large compared to the separation d between the slits

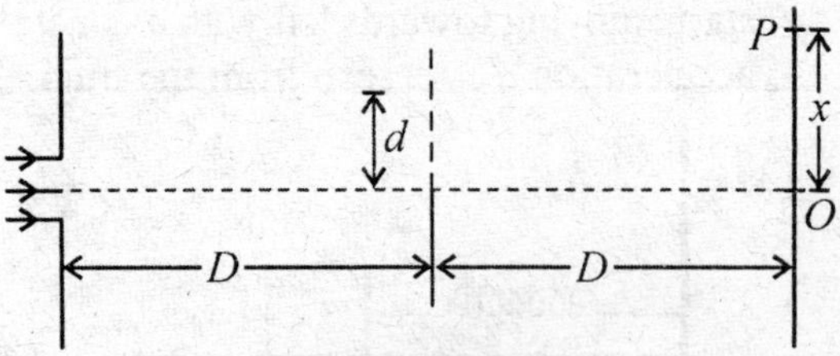

Fig. MTP-1.9

(a) The minimum value of d so that there is dark fringe at O is $\sqrt{\frac{\lambda D}{2}}$

(b) The minimum value of d is $\sqrt{\lambda D}$ so that there is dark fringe at O, suppose d has this small value

(c) The fringe width is $2d$

(d) The next bright fringe is formed at P such that $x = d$.

19. A uniform rope of length l and mass m hangs vertically from rigid support. A wave pulse is generated at the bottom end B (Fig. MTP-1.10) which travels along the rope to A

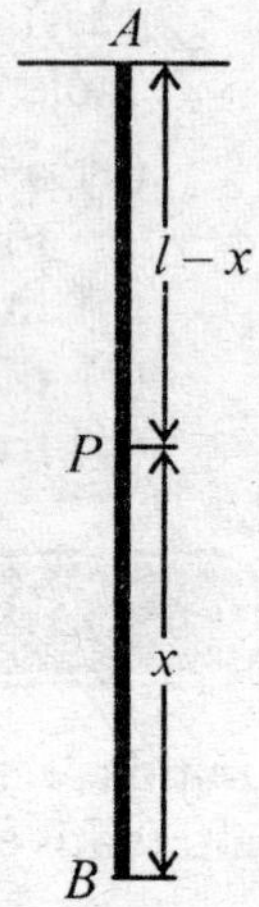

Fig. MTP-1.10

(a) the speed v of the wave pulse when at P, is $v \propto \sqrt{x}$

(b) the speed v of the wave pulse when at P is $v \propto x$

(c) the time taken by the pulse to reach A is $2\sqrt{l/g}$

(d) the time taken by the pulse to reach A is $\sqrt{l/g}$.

20. A capacitor is charged to a potential V_0. It is connected with an inductor through a switch S, the switch is closed at time $t = 0$. Which of the following statement(s) is/(are) correct?

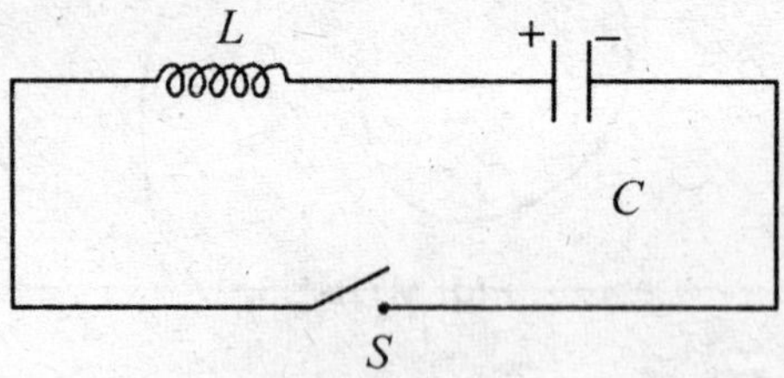

Fig. MTP-1.11

(a) The maximum current in the circuit is $V_0\sqrt{C/L}$

(b) Potential across capacitor becomes zero for the first time at time $t = \pi\sqrt{LC}$

(c) The energy stored in the inductor at time $t = \frac{\pi}{2}\sqrt{LC}$ is $\frac{1}{4}CV_0^2$

(d) The maximum energy stored in the inductor is $\frac{1}{2}CV_0^2$.

SECTION III

In this section, there are four comprehensions, each followed by three multiple choice questions.

Passage 1: A thin rod *AB* of length *a* has variable mass per unit length $\rho_0(a+x)/a$, where x is the distance measured from *A* and ρ_0 is constant.

21. The mass *M* of the rod is

(a) $\frac{3a\rho_0}{2}$ (b) $a\rho_0$

(c) $\frac{5}{2}a\rho_0$ (d) $\frac{3\rho_0}{2a}$.

22. The centre of mass is at the point x where

(a) $x = \frac{4a}{5}$ (b) $x = \frac{5a}{9}$

(c) $x = \frac{a}{3}$ (d) $x = \frac{4a}{9}$.

23. The moment of inertia of rod about an axis through *A* perpendicular to *AB* is

(a) $\frac{2Ma^2}{9}$ (b) $\frac{5}{18}Ma^2$

(c) $\frac{7}{18}Ma^2$ (d) $\frac{7}{9}Ma^2$.

Passage 2: A mica strip and a polysterene strip are fitted on the two slits of double slit experiment. The thickness of the strips is 0.5 mm and separation between the slits is 0.12 cm. The refractive index of mica and polysterene are 1.58 and 1.55 respectively for the light of wavelength 590 nm, which is used in the experiment. The interference is observed on a screen a distance one metre away.

24. The fringe width is

(a) 4.9×10^{-5} m (b) 4.9×10^{-4} m

(c) 1.5×10^{-6} m (d) 1.5×10^{-5} m.

25. The first maximum will be located at distance x from the centre where x is

(a) $x = 0$

(b) $x = 0.021$ cm on one side and 0.028 cm on the other side

(c) $x = 0.024$ cm on one side and $x = 0.025$ cm on the other side

(d) $x = 0.018$ cm on one side and $x = 0.31$ cm on the other side.

26. The first minimum will be located at distance y from the centre, then y is

(a) $y = 0$ (b) $y = 0.0035$ cm

(c) $y = 0.0245$ cm (d) $y = 0.007$ cm.

Passage 3: Two particles *A* and *B* each having a charge *Q*, are held a distance *d* apart. A particle of mass *m* and charge *q* is placed on the perpendicular bisector of *AB*.

27. The distance x from the mid-point of *AB* where this charge q experiences maximum force is

(a) $x = d/2$ (b) $x = \frac{d}{\sqrt{2}}$

(c) $x = \frac{d}{2\sqrt{2}}$ (d) $x = \sqrt{2}d$.

28. The magnitude of this maximum force is

(a) $\frac{4}{3\sqrt{3}\pi}\frac{qQ}{\varepsilon_0 d^2}$ (b) $\frac{2qQ}{3\sqrt{3}\varepsilon_0 d^2}$

(c) $\frac{qQ}{3\sqrt{3}\varepsilon_0 d^2}$ (d) $\frac{qQ}{\pi\varepsilon_0 d^2}$.

29. If this charge q is released from this distance, its subsequent motion will be

(a) an oscillartory motion but not simple harmonic

(b) a simple harmonic motion

(c) a straight line motion away from the mid-point of *AB*

(d) a straight line motion towards the mid-point of *AB*.

Passage 4: Two particles each having a mass *m* are placed at a separation *d* in uniform magnetic field *B* (Fig. MTP-1.12). At time $t = 0$, the particles are projected towards each other, each with speed *v*. Neglect the Coulomb force between the charges.

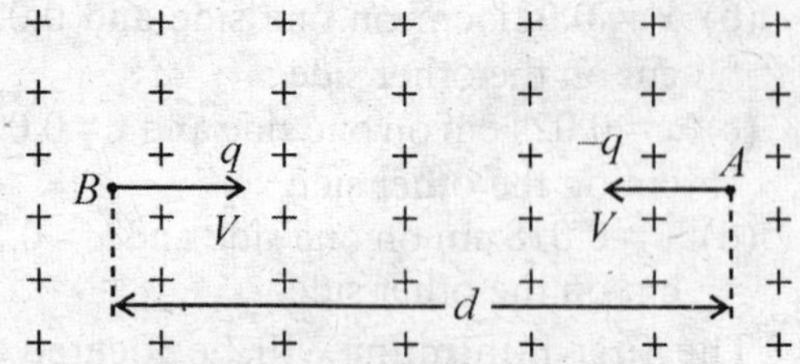

Fig. MTP-1.12

30. The maximum value of the projection speed v_m so that the two particles do not collide is

(a) $\frac{qBd}{m}$ (b) $\frac{2qBd}{m}$

(c) $\frac{qBd}{2m}$ (d) $\frac{qBd}{4m}$.

31. If the projection speed $v = v_m/2$, the maximum and minimum separation between the particles respectively is

(a) $\frac{d}{2}, d$ (b) $\frac{d}{4}, \frac{3d}{2}$

(c) $\frac{d}{2}, \frac{5d}{4}$ (d) $\frac{d}{2}, \frac{3d}{2}$.

32. If $v = 2v_m$ the collision will occur after time

(a) $\frac{\pi}{6}\frac{m}{qB}$ (b) $\frac{\pi}{3}\frac{m}{qB}$

(c) $\frac{\pi}{4}\frac{m}{qB}$ (d) $\frac{\pi}{2}\frac{m}{qB}$.

SECTION IV

In this section, there are four numerical questions. The answers to these questions are to be given in between 0000 to 9999 in the form of nearest integer.

33. A cylinder of mass M and radius R lies over block of mass M (=1 kg) friction between the cylinder and block is sufficient to prevent slipping. While there is no friction between the block and the ground. A light massless string is wrapped around the cylinder and fixed to vertical wall (Fig. MTP-1.13). Now horizontal force F(=11 N) is applied on the block. Find the acceleration of the block and the cylinder.

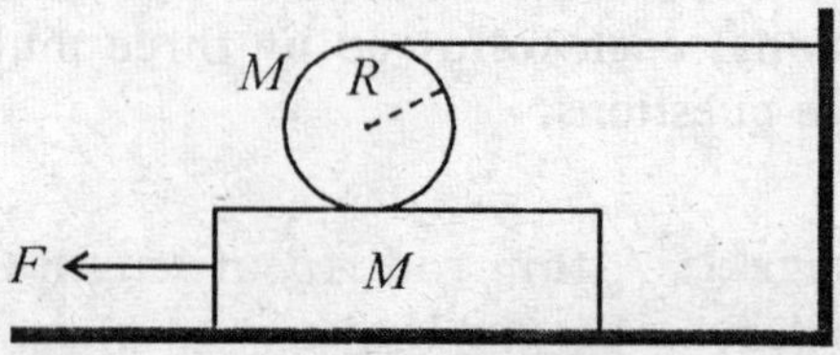

Fig. MTP-1.13

34. One stormy day Ram asked Shyam to go to his friend's house which is towards north from his house. He asked him to go directly towards north. After sometime Ram observed Shyam going with a velocity of 72 km/hr towards 37° west of north as wind is blowing at 36 km/hr towards west. So after 15 minutes Ram blew a whistle so as to give signal to Shyam to come back.

(a) Find the distance S between Ram's house and Shyam when he heard the singnal.

(b) If natural frequency of whistle was 400 Hz, find the frequency as heard by Shyam. Speed of sound in air is 330 m/s.

35. For a X-ray tube the ratio of cut off wave length to characteristic wavelength of k_α line is found to be in the ratio 1 : 2. If the tube is operated at voltage 16.25 kV. Find the atomic number of target metal using Moseley's law.

36. A ray is incident normally on right angled prism whose μ is $\sqrt{3}$ and prism angle α =

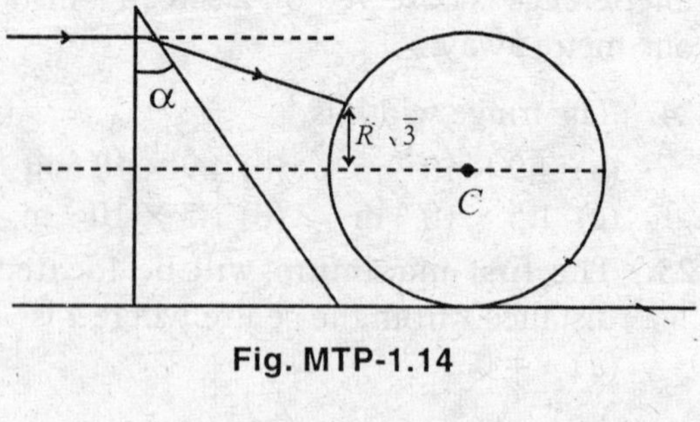

Fig. MTP-1.14

30°. After crossing prism, ray passes through glass sphere. It strikes the glass sphere at $\frac{R}{\sqrt{3}}$ from principal axis (Fig. MTP-1.14). Sphere is half polished. Find the angle of deviation of incident ray.

SECTION V

In this section, there are four questions. Each question has two columns with 4 entries in each column. Entries of Column I are to be matched with Column II. One entry of Column I may have more than one matching in Column II

37.

Column I	Column II
(A) Kepler's laws	(P) Gravitation
(B) Escape velocity	(Q) Communication
(C) Satellites	(R) Areal velocity
(D) Planetary motion	(S) Elliptical orbits

38.

Column I	Column II
(A) Power factor	(P) Alternating currents
(B) Q-factor	(Q) Inductance
(C) rms Value	(R) Capacitance
(D) Resonance	(S) Angular frequency

39. An ideal gas is taken through a cyclic process as shown in Fig. MTP-1.15. If ΔQ is the heat absorbed and ΔU the change in internal energy, then

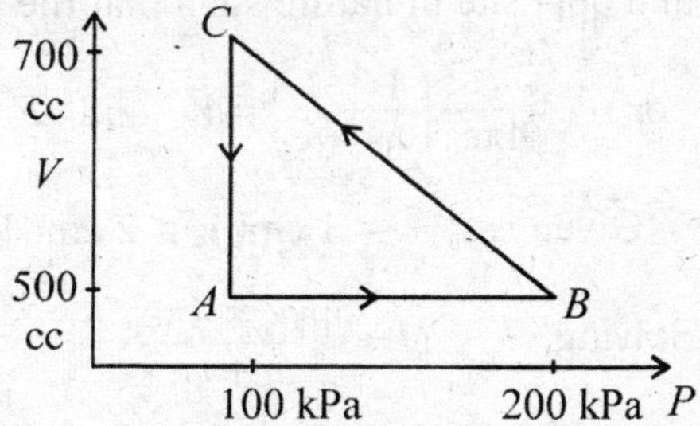

Fig. MTP-1.15

Column I	Column II
(A) AB	(P) $\Delta Q > 0$
(B) BC	(Q) $\Delta Q < 0$
(C) CA	(R) $\Delta U > 0$
(D) $ABCA$	(S) $\Delta U < 0$.

40.

Column I	Column II
(A) Electromagnetic induction	(P) Magnetic field
(B) Gauss theorem	(Q) Electric field
(C) A current carrying straight conductor	(R) Magnetic dipole
(D) A current carrying closed loop	(S) Flux.

ANSWERS

1. (a)	2. (c)	3. (d)	4. (b)	5. (a)
6. (a)	7. (b)	8. (c)	9. (a)	10. (a)
11. (b)	12. (c)	13. (a, c)	14. (b, c)	15. (a, c, d)
16. (a, b, c, d)	17 (a, c)	18 (a, c, d)	19 (a, c)	20 (a, d)
21. (a)	22. (b)	23. (c)	24. (b)	25. (b)
26. (b)	27. (c)	28 (a)	29. (c)	30. (c)
31. (d)	32. (a)			

	A	B	C	D
37.	P-R-S	P	P-Q	P-R-S
38.	P-Q R-S	P-Q-R	P	P-Q R-S
39.	P-R	Q-S	Q-S	P
40.	P-S	P-Q-S	P	P-R-S

SOLUTIONS

1. On connecting the two spheres by a battery, the charge on the spheres will be equal in magnitude and opposite in nature such that the potential difference is 3000 V (i.e. $Q_1 = -Q_2 = Q$)

or $$\frac{Q}{4\pi\varepsilon_0}\left(\frac{1}{r_1}+\frac{1}{r_2}\right) = V \text{ and } F = \frac{1}{4\pi\varepsilon_0}\frac{Q^2}{R^2}$$

Given is $r_1 = 1$ cm, $r_2 = 2$ cm, $V = 3000$ V, and $R = 1$ m

Solving, $$Q = \frac{r_1 r_2 V \times 4\pi\varepsilon_0}{r_1 + r_2}$$

$$\therefore \quad F = \frac{1}{4\pi\varepsilon_0}\frac{(r_1 r_2)^2 V^2 (4\pi\varepsilon_0)^2}{(r_1 + r_2)^2} = 4.4 \times 10^{-8}\text{ N}$$ (a)

2. Equivalent focal length is given by

$$\frac{1}{F} = \frac{1}{f_1} + \frac{1}{f_2} - \frac{d}{f_1 f_2}. \text{ For } F = \infty, \quad d = f_1 + f_2.$$ (b)

3. Total resistance of the circuit is 9800 + 200 = 1000 Ω. Current is

$$i = \frac{9}{10^4} = 9 \times 10^{-4}\text{ amp} \quad \therefore V = 9800 \times 9 \times 10^{-4} = 8.8\text{ V}.$$ (d)

4. Acceleration in upper half is $g\sin\theta$. Hence velocity after distance S is $v^2 = 2g\sin\theta\, S$ and retardation in lower half is $(\mu g\cos\theta - g\sin\theta)$. Hence $v^2 = 2g(\mu\cos\theta - \sin\theta)S$. Equating the two we have $2\tan\theta = \mu$. (b)

5. Centrifugal force on two particles is $F_1 = m\omega^2 R_1$ and $F_2 = m\omega^2 R_2$

or $$\frac{F_1}{F_2} = \frac{R_1}{R_2}$$ (a)

6. Intensity after passing through thickness x of Pb is given by $I = I_0 e^{-\mu x}$ where μ is absorption co-efficient

$$\therefore \quad (2)^{-3} = e^{-\mu x_1} \text{ or } 3\log 2 = \mu x_1 \text{ and } (2)^{-1} = e^{-\mu x_2}$$

$$\therefore \quad \log 2 = \mu x_2 \quad \therefore \quad 3 = \frac{x_1}{x_2} \quad \text{or} \quad x_2 = \frac{x_1}{3} = \frac{36}{3} = 12\text{ mm}$$ (a)

7. Heat received by the system during process CB = area under S-T diagram.

$$\therefore \quad Q_1 = \frac{1}{2}(2T_0 + T_0)S_0 = \frac{3T_0}{2}S_0$$

Heat rejected during process $BA = Q_2 = T_0 S_0$

Efficiency $$\eta = \frac{Q_1 - Q_2}{Q_1} = \frac{1}{3}.$$ (b)

8. The reaction is ${}^A_Z X + {}^1_0\text{n} = {}^7_3\text{Li} + {}^4_2\text{He}$. Conservation of charge gives $Z = 5$ and conservation of nucleons gives $A = 10$. Thus $X = {}^{10}_5\text{B}$. (c)

9. When rod becomes vertical its angular velocity ω is $\frac{1}{2}I\omega^2 = mgl$ or $\frac{1}{2}\cdot\frac{1}{3}ml^2\omega^2 = \frac{1}{2}mgl$ or $\omega = \sqrt{3g/l}$ and linear velocity of charge $v = \omega l = \sqrt{3gl}$. Hence magnetic induction at O is (using Biot Savart's law) $B = \frac{\mu_0 (idl)\sin\theta}{4\pi r^2} = \frac{\mu_0 Vq}{4\pi l^2} = \frac{\mu_0 q}{4\pi l^2}\sqrt{3gl} = \frac{\mu_0 q}{4\pi l}\sqrt{\frac{3g}{l}}$. (a)

10. Moseley's law gives $\sqrt{\nu} \propto (Z-1)$ or $\sqrt{C/\lambda} = K(31-1) = K30$ and $\sqrt{\frac{C}{4\lambda}} = K(Z'-1)$ or $2(Z'-1) = 30$ or $Z' = 16$. (a)

11. Acceleration $a = \frac{dv}{dt} = \frac{dv}{dx} \times \frac{dx}{dt} = -k\frac{dx}{dt}, \frac{dv}{dx}$ is slope of v–x curve or $a = -kv = -k(-kx + c) = k^2x - ck$. A straight line with positive slope with –ve intercepts on acceleration axis and +ve intercepts on distance axis. (b)

12. In adiabatic process, work done $W = \frac{\mu R(T_2 - T_1)}{-\gamma + 1}$ work is done on the system and is –ve. Solving $\gamma = 1.4$. (c)

13. Normal acceleration $a = \frac{v^2}{R} = \omega^2 R$

$\therefore \quad v \propto \sqrt{R}$ and $\omega \propto \frac{1}{\sqrt{R}}$. (b, c)

14. $\mathbf{a} + \mathbf{c} = -(\mathbf{b} + \mathbf{d})$ $\quad \therefore |\mathbf{a} + \mathbf{c}| = |\mathbf{b} + \mathbf{d}|$ (b)

$\mathbf{a} = -(\mathbf{b} + \mathbf{c} + \mathbf{d})$ $\quad \therefore |\mathbf{a}| \le |\mathbf{b}| + |\mathbf{c}| + |\mathbf{d}|$ (c)

15. Conservation of linear momentum gives $v_c = 0$. (a)

Conservation of angular momentum gives $I\omega = 2mva + m2v(2a) = 6mva$

$$I = \frac{1}{12}8m(6a)^2 + 2ma^2 + m(2a)^2 = 30ma^2. \quad \omega = \frac{v}{5a}$$ (d)

$$\text{K.E.} = \frac{1}{2}I\omega^2 = \frac{1}{2}30ma^2\frac{v^2}{25a^2} = \frac{3mv^2}{5}.$$ (c)

16. Free particle diagram is shown in Fig. MTP-1.16. It is equivalent to mass spring system oscillating in vertical plane. It executes S.H.M. (a)

The period is $T = 2\pi\sqrt{\frac{m}{k}}$ (b)

Amplitude a then $ka = ma_0$ or $a = \frac{ma_0}{k}$ (c)

The energy of oscillation $E = \frac{1}{2}ka^2 = \frac{1}{2}\frac{m^2a_0^2}{k}$ (d)

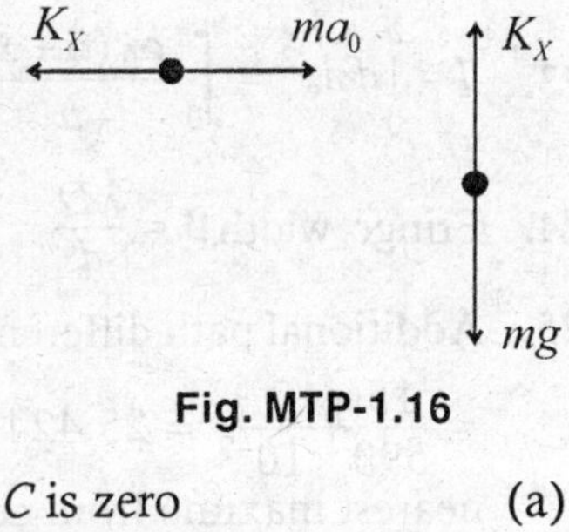

Fig. MTP-1.16

17. Work done by the gas from A to $B = P_0V_0 = RT_0$ and from B to C is zero (a)

Temperature at B is $2T_0$ and at C, $4T_0$. Hence heat absorbed

$$= C_P(2T_0 - T_0) + C_V(4T_0 - 2T_0)$$

$$= \frac{5}{2}RT_0 + \frac{3}{2}R(2T_0) = \frac{11RT_0}{2}$$ (c)

18. Total path difference at $O = 2\left[\left(D^2 + d^2\right)^{1/2} - D\right] = \frac{d^2}{D} = \lambda/2$

$\therefore \quad d = \sqrt{\frac{\lambda D}{2}}$ (a)

Fringe width $= \frac{\lambda D}{d} = \sqrt{2} \,.\, \sqrt{\lambda D} = 2\sqrt{\frac{\lambda D}{2}} = 2d$ (c)

Next bright fringe will appear for path difference $\lambda/2$. At P path difference $\frac{xd}{D} = \lambda/2$

or $\quad x = \frac{\lambda D}{2d} = \frac{d^2}{d} = d$ (d)

19. If λ is mass per unit length then tension at $P = \lambda xg$. Hence pulse velocity $v = \sqrt{\frac{\lambda xg}{\lambda}} = \sqrt{xg}$

or $\quad v \propto \sqrt{x}$ (a)

$\frac{dx}{dt} = \sqrt{xg}$ or $\frac{dx}{\sqrt{xg}} = dt \quad \therefore \int_0^l \frac{dx}{\sqrt{xg}} = t$ or $t = 2\sqrt{\frac{l}{g}}$ (c)

20. The charge oscillates simple harmonically as $q = q_0 \cos \omega t$ where $q_0 = CV_0$ and $\omega = \frac{1}{\sqrt{LC}}$.

Hence current $i = \frac{dq}{dt} = -q_0 \omega \sin \omega t$, $i_{max} = q_0 \omega = V_0 i \frac{1}{\sqrt{LC}} = V_0 \sqrt{\frac{C}{L}}$ (a)

Maximum energy stored in inductor is equal to the maximum enegy stored in capacitor

$= \frac{1}{2} CV_0^2$ (d)

21. Mass of rod $M = \int_0^a \rho_0 \frac{(a+x)}{a} dx = \frac{\rho_0}{a}\left[ax + \frac{x^2}{2}\right]_0^a = \frac{3\rho_0 a}{2}$. (a)

22. $X_{cm} = \frac{\int_0^a \frac{\rho_0 (a+x)}{a} x dx}{M} = \frac{\rho_0}{aM}\left[\frac{ax^2}{2} + \frac{x^3}{3}\right]_0^a = \frac{5a}{9}$ (b)

23. $I = \int_0^a dmx^2 = \int_0^a \frac{\rho_0 (a+x)}{a} x^2 dx = \frac{7}{18} Ma^2$ (c)

24. Fringe width $\beta = \frac{\lambda D}{d} = \frac{590 \times 10^{-9} \times 1}{12 \times 10^{-2}} = 4.9 \times 10^{-4}$ m (b)

25. Additional path difference due to films $\Delta = t(\mu_1 - \mu_2) = 15 \times 10^{-6}$ m. Hence shift in the fringes $= \frac{15 \times 10^{-6}}{590 \times 10^{-9}} = 25.424$. The central fringes is shifted to 25.424 order bright fringe. Hence the nearest maxima from centre is $0.424 \times 4.9 \times 10^{-4}$ m $= 2.1 \times 10^{-4}$ m $= 0.021$ cm towards slit with mica sheet. On the other side of centre there is maxima at 0.028 cm. (b)

26. In between these two maximas there is minima which is at distance $\frac{0.049}{2} - 0.021 = 0.0035$ cm from centre towards slit with Polystrene flim. (b)

27. Force on change q due to two charges Q each at A and B is $F = \frac{1}{4\pi\varepsilon_0}\frac{Qq}{(a^2 + x^2)}$ (Fig. MTP-1.17). Hence net force $f = 2F\cos\theta = \frac{2Qq}{4\pi\varepsilon_0}\frac{x}{(a^2 + x^2)^{3/2}}$. For force to be maximum $\frac{df}{dx} = 0$ which gives $x = \frac{a}{\sqrt{2}} = \frac{d}{2\sqrt{2}}$ (c)

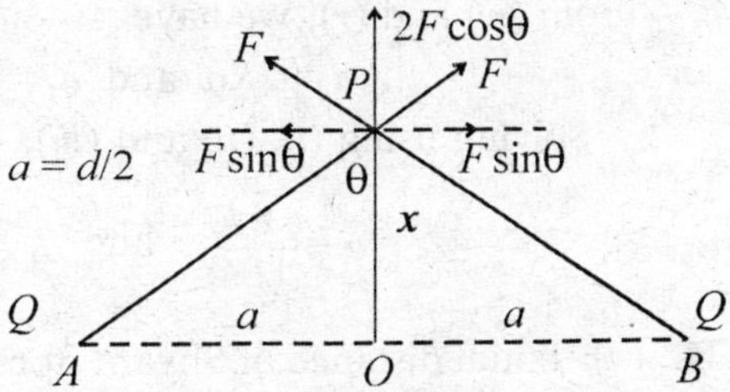

Fig. MTP-1.17

28. Substituting value of x in f, we get $f_{max} = \frac{4Qq}{3\sqrt{3}\pi\varepsilon_0 d^2}$ (a)

29. The force acts on the particle away form O. Hence motion is a straight line away from O the mid point of AB. (c)

30. For no collision, the radius of circular path should be $d/2$. $r = \frac{mv}{qB}$

or $\frac{d}{2} = \frac{mv_{max}}{qB}$ or $v_{max} = \frac{qBd}{2m}$ (c)

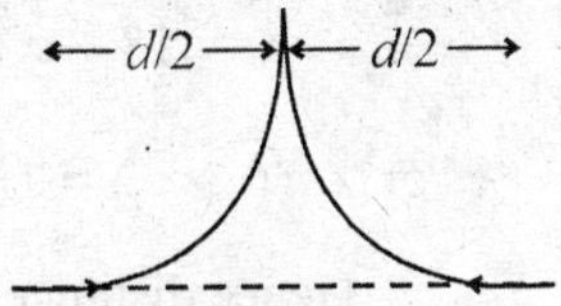

Fig. MTP-1.18

31. When $V = V_m/2$. $r = d/4$. Hence minimum distance is $d/2$ and maximum is $d + d/4 + d/4 = \frac{3d}{2}$ (d)

32. If $V = 2V_m$, $r = \frac{m}{qB} \times \frac{2qBd}{2m} = d$. Hence they will collide after rotating through 30° (Fig. MTP-1.19). Frequency of rotation is $f = \frac{2V_m}{2\pi r} = \frac{qBd}{2m\pi d} = \frac{qB}{2m\pi}$. Time period $= \frac{2m\pi}{qB}$. Hence time for 1/12th of rotation is $t = \frac{2m\pi}{12qB} = \frac{m\pi}{6qB}$. (a)

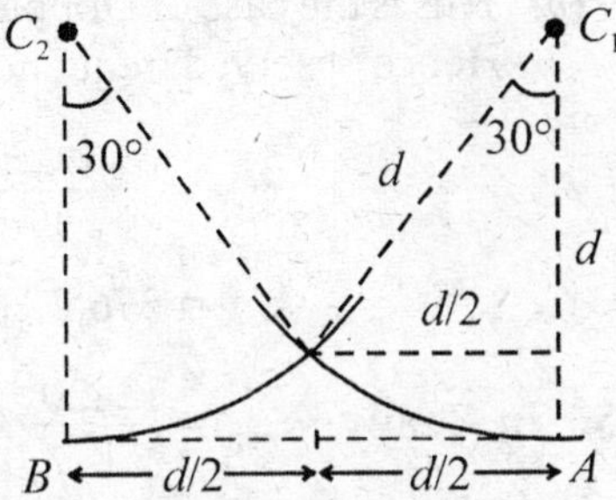

Fig. MTP-1.19

33. Free body diagram for block and cylinder is shown in Fig. MTP-1.20 (*a*) and (*b*) respectively. Let a be the acceleration of block and a_1 be the acceleration of cylinder with respect to block and α is the angular acceleration of cylinder

For block, we have

$$F - f = Ma \qquad ...(i)$$

f is friction force

For cylinder

$$T - f = Ma - Ma_1 \qquad ...(ii)$$

and $$(T + f)R = I\alpha = \frac{MR^2}{2}\alpha \qquad ...(iii)$$

For point P, $R\alpha - a_1 = 0$ ...(*iv*)

and for jpoint Q

$$a_1 - a + R\alpha = 0 \qquad ...(v)$$

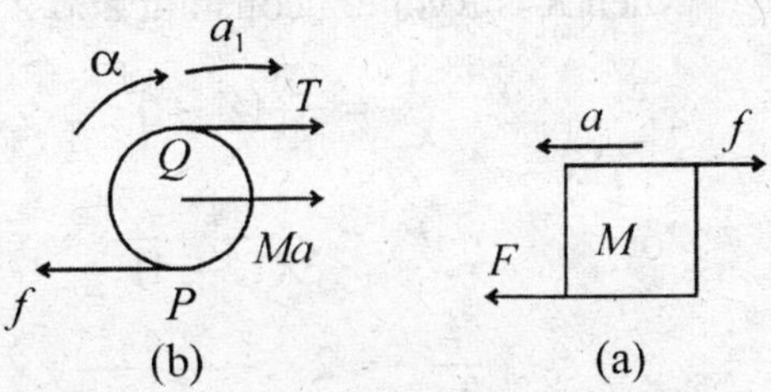

Fig. MTP-1.20

From (*iv*) and (*v*), we have

$$a = 2R\alpha \text{ and } a_1 = R\alpha = a/2$$

Solving using (*i*), (*ii*) and (*iii*), we get

$$a = \frac{8F}{11M} \text{ and } a_1 = \frac{4F}{11M}$$

34. (*a*) Initial distance of Shyam from Ram $= \frac{72}{4} = 18$ km. Sound velocity in still air be v_s then sound velocity in the direction Shyam is

$$v = v_s + v_w \sin 37° = \frac{330 \times 18}{5} + 36 \times 0.6$$

$$= 1188 + 21.6 = 1209.6 \text{ km/hr}$$

If signal reaches Shyam after time t, then

$$vt = 18 + v_1 t \text{ where } v_1 = 72 \text{ km/hr Shyam's velocity}$$

or $\quad t(v - v_1) = 18$ or $t = \frac{18}{1209.6 - 72} = 0.016$ hr

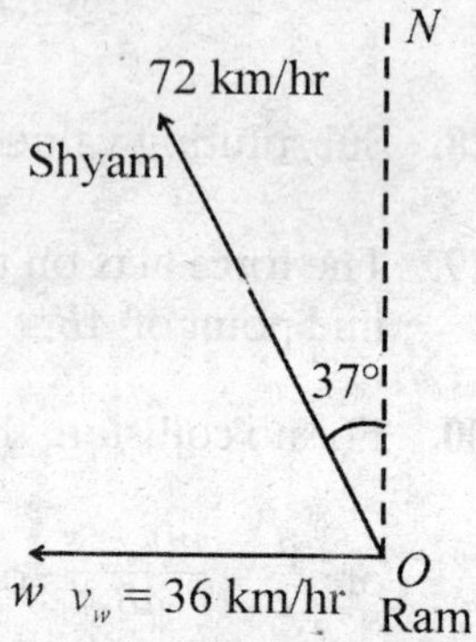

Fig. MTP-1.21

Hence distance of Shyam from Ram's house when he hears the signal

$$S = vt = 1209.6 \times 0.016 = 19.35 \text{ km} = 0019 \text{ km}.$$

(*b*) This is the case of Doppler effect in which source is stationary and observer is moving away. Hence observed frequency is

$$\nu = \frac{\nu_0 (v_s + v_w \sin\theta - v_1)}{v_s + v_w \sin\theta} = \frac{400(1209.6 - 72)}{1209.6}$$

$$= 376.2 \text{ Hz} = 0376 \text{ Hz}.$$

35. In *X*-ray $\quad \lambda_{min} = \frac{12420}{V} \text{ Å} = \frac{12420}{V} \times 10^{-10}$ m

$$\lambda_{min} = 7.6 \times 10^{-11} \text{ m} \quad \ldots(i)$$

where V is the applied accelerating voltage

wavelength K_α lines is given by

$$\frac{1}{\lambda_\alpha} = R\left(\frac{1}{n_1^2} - \frac{1}{n_2^2}\right)(Z-1)^2 \text{ m}^{-1} \quad \ldots(ii)$$

when R is Rydberg constant and Z atomic no. For K_α line $n_1 = 1$ and $n_2 = 2$

$$\therefore \quad \frac{1}{\lambda_\alpha} = \frac{3R}{4}(Z-1)^2$$

or

$$\lambda_\alpha = \frac{4}{3R(Z-1)^2} \text{ m} \quad \ldots(iii)$$

$$\therefore \quad \frac{\lambda_\alpha}{\lambda_{min}} = 2 = \frac{4}{3R(Z-1)^2} \times \frac{1}{7.6 \times 10^{-11}} \qquad \text{But } R = 1.09 \times 10^7 \text{ m}^{-1}$$

$$\therefore \quad Z - 1 = 28 \text{ or } Z = 29.$$

36. Rays through prism and sphere are shown in Fig. MTP-1.22. Angle of refraction δ at the second surface is

$$\sqrt{3}\sin\alpha = \sin\delta \qquad \alpha = 30°$$

$$\therefore \qquad \delta = 60°$$

Hence the refracted ray *BDC* makes an angle of 30° with horizontal diameter *PQ*. Since $DE = \frac{R}{\sqrt{3}}$ then $EC = R$ means ray passes through the centre of sphere, i.e. ray falls normally on the sphere hence it retraces its path and angle of deviation is 180° = 0180.

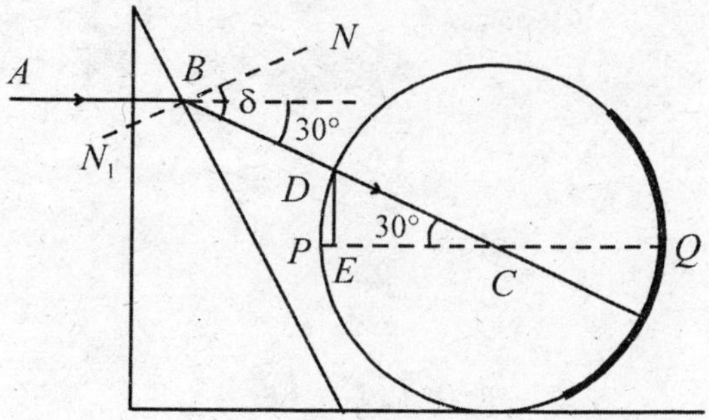

Fig. MTP-1.22

Model Test Paper II

SECTION I

In this section, a question is followed by four alternatives, of them only one is correct.

1. An insect is crawling upon a concave surface of a fixed hemispherical bowl of radius R. If the co-efficient of friction is 1/3 then height upto which the insect crawls is nearly

(a) 5% of R (b) 6% of R
(c) 6.5% of R (d) 7.5% of R.

2. Three concentric charged metallic spherical shells A, B, C have radii a, b, c, charge densities σ, $-\sigma$ and σ and the potentials V_a, V_b, V_c respectively. Then the correct relation is

(a) $V_a = (a + b + c)\sigma/\varepsilon_0$
(b) $(a^2/b - b + c)\sigma/\varepsilon_0$
(c) $(a^2 + b^2 + c^2)\sigma/C\varepsilon_0$
(d) $V_a - V_b = V_c = (a + b + c)\sigma/\varepsilon_0$.

3. In the circuit shown in Fig. MTP-2.1 when R is removed, an additional resistance of 72 Ωm must be introduced in series with the battery in order to keep the current through 30 Ω resistance unaltered. Hence R is

(a) 15 Ω (b) 18 Ω
(c) 20 Ω (d) 21 Ω.

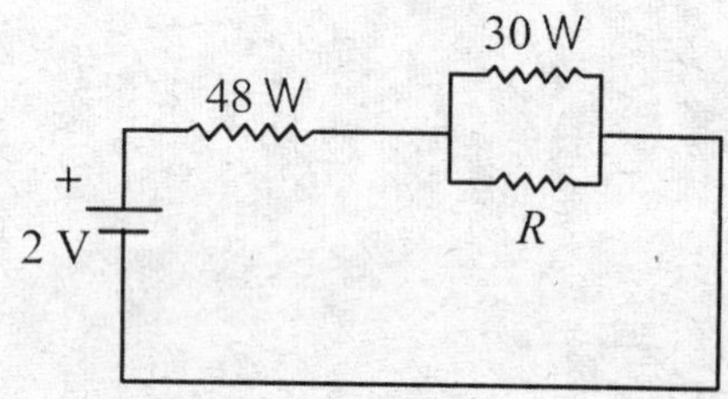

Fig. MTP-2.1

4. Which of the following combinations of three different physical quantities, can never be a meaningful quantity?

(a) $PQ - R$ (b) $\dfrac{PQ}{R}$
(c) $(P - Q)/R$ (d) $(PR - Q^2)/QR$.

5. A point charge q is at a distance 'd' from a large metal sheet. The amount of work required to move this charge to infinity (in units of $q^2/4\pi\varepsilon_0$) is

(a) $\dfrac{1}{d}$ (b) $\dfrac{1}{2d}$
(c) $\dfrac{1}{3d}$ (d) $\dfrac{1}{4d}$.

6. A body of mass m is raised through a height $h = R/5$ from the surface of earth. (R is radius of earth). The increase in potential energy is (g is acceleration due to gravity on surface of earth)

(a) mgh (b) $4mgh/5$
(c) $5mgh/6$ (d) $6mgh/7$.

7. An inductor is connected to a DC source of emf 10 V draws a steady current of 5 A. Now the inductor is connected to a capacitor and an AC source of rms voltage 20 V. What is the rms current in the circuit? It is found that the current and voltage of the source are in same phase.

(a) 10 A
(b) 5 A
(c) 2.4 A
(d) Given data are insufficient and can not be determined.

8. A concave mirror is placed in a medium of refractive index $\mu = 1.2$. An object placed at 30 cm from pole of the mirror and its image coincides. Now a liquid of refractive index μ is placed in the convave mirror in the shaded region and it is found that object needs to be shifted by 7.5 cm below O so the object and image to coincide. The refractive index μ of the liquid is

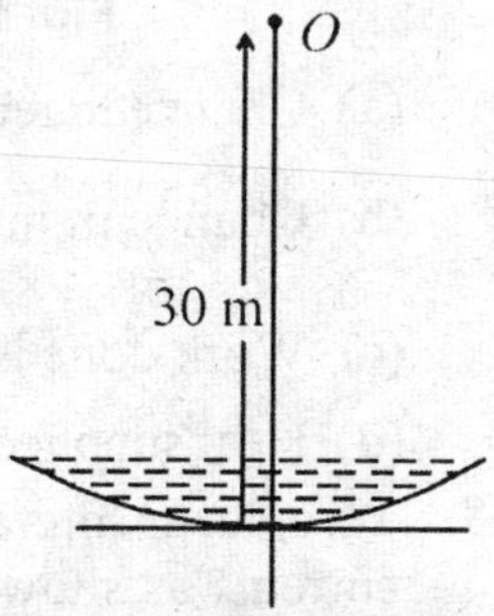

Fig. MTP-2.2

(a) $\frac{4}{3}$ (b) 1.2
(c) 1.6 (d) None of the above.

9. Which of the following is equal to the minimum kinetic energy needed by a hydrogen atom in ground state to produce a photon after collision with another identical hydrogen atom at rest?

(a) 6.8 eV (b) 3.4 eV
(c) 20.4 eV (d) 2.04 eV.

10. In a standard meter bridge experiment to measure the specific resistance of a wire, the following data are found: length $(L) = (1 \pm 0.01)$ m measured by a meter scale. Radius of wire $(r) = (1 \pm 0.01)$ mm measured by screwgauge, resistance of wire $(R) = (5 \pm 0.01)\ \Omega$.

The maximum possible error in the measurement of specific resistance is

(a) 0.1 μΩ-m (b) 0.2 μΩ-m
(c) 0.5024 μΩ-m (d) 0.6024 μΩ-m.

11. When N_0 molecules of a radioactive nuclide are taken activity will be 'A' in time t_0. When $3N_0$ molecules of the same nuclide are taken, activity will become $3Ae$ in time equal to

(a) t_0 (b) $t_0 + 1/\lambda$
(c) $t_0 - 1/\lambda$ (d) $3t_0$.

12. In an adiabatic expansion of air the volume decrease by 5%. Then the percentage change in pressure is

(a) +7% (b) +3%
(c) −5% (d) −7%.

SECTION II

In this section, a question is followed by four alternatives, of them one or more are correct.

13. A long plank P of mass 5 kg is placed on a smooth floor. On P is placed a block Q of mass 2 kg. The co-efficient of friction between P and Q is 0.5. If a horizontal force 15 N is applied to Q as shown in Fig. MTP-2.3 and $g = 10\ m/s^2$.

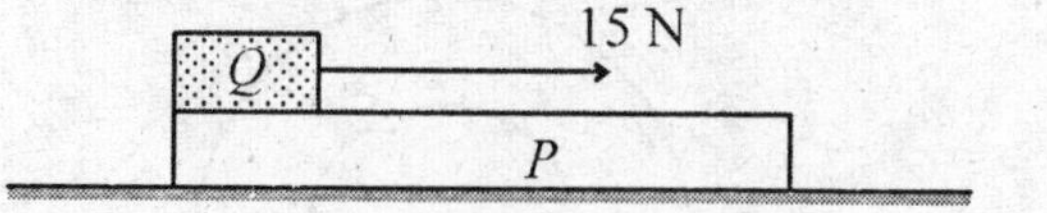

Fig. MTP-2.3

(a) The reaction force on Q due to P is 10 N
(b) The acceleration of Q relative to P is 2.5 m/s^2
(c) The acceleration of P relative to floor is 2.0 m/s^2
(d) The acceleration of centre of mass of P and Q system relative to the floor is 15/7 m/s^2.

14. Which of the following is/are true about vectors?

(a) If $\boldsymbol{A} = \boldsymbol{B} \times \boldsymbol{C}$ and you are given $\boldsymbol{A}$ and $\boldsymbol{C}$, then $\boldsymbol{B}$ is uniquely defined
(b) $\boldsymbol{A} \times (\boldsymbol{B} \times \boldsymbol{C}) = (\boldsymbol{A} \times \boldsymbol{B}) \times \boldsymbol{C}$
(c) If $\boldsymbol{A} \times \boldsymbol{B}$ is zero $\boldsymbol{A}$ is parallel to $\boldsymbol{B}$
(d) If $\boldsymbol{A}, \boldsymbol{B}, \boldsymbol{C}$ are three sides of a parallelopiped, the volume of the parallelopiped is given by $V = \boldsymbol{A} \cdot (\boldsymbol{B} \times \boldsymbol{C})$.

15. Light carries momentum given by $P = E/c$. Where E is energy and c is speed of light. Now which of the following is true?

(a) If an atom emits a photon of energy E the difference between the two energy levels involved in emissioin, E' will be smaller than E
(b) The difference between E and E' is more for γ-ray emission than for emission in visible region
(c) The pressure due to a given beam of light is more on an absorber surface than on a reflector surface
(d) A 10 mW laser beam may cause greater presure on a surface than a 1000 W ordinary lamp placed at the same distance.

16. Two concentric shells have radii R and $2R$, charges q_A and q_B and potentials $2V$ and $(3/2V)$ respectively. Now shell B is earthed and let charges on them become q'_A and q'_B. Then

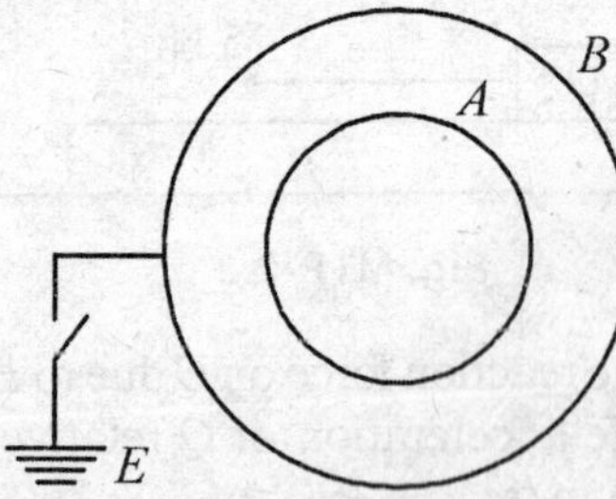

Fig. MTP-2.4

(a) $q_A/q_B = 1/2$
(b) $q'_A/q'_B = 1$
(c) potential of A after earthing becomes $3/2V$
(d) potential difference between A and B after earthing becomes $V/2$.

17. Figure MTP-2.5 gives the P-V indicator diagram for monoatomic gas taken from state A to state B. The units for P and V are standard pressure and molar volume respectively. The temperature at A is T_0 K

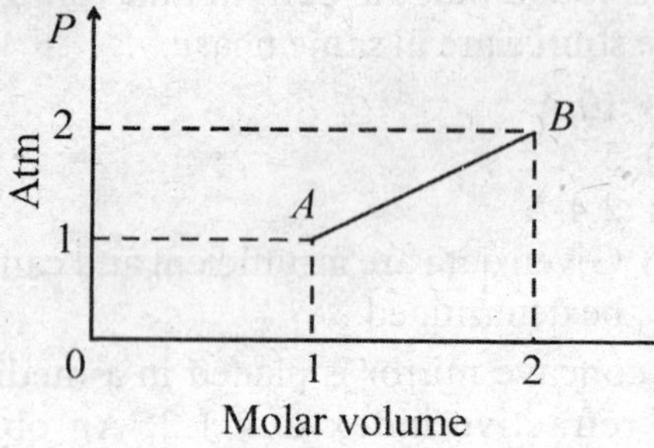

Fig. MTP-2.5

(a) Change in temperature is $4\,T_0$
(b) Change in internal energy is $\frac{9}{2} RT_0$
(c) Work done is $\frac{1}{2} RT_0$
(d) Heat supplied is $6RT_0$.

18. Two large metal sheets carry surface currents as shown in Fig. MTP-2.6. The current through a strip of width dl is Kdl. where K is a constant. Then

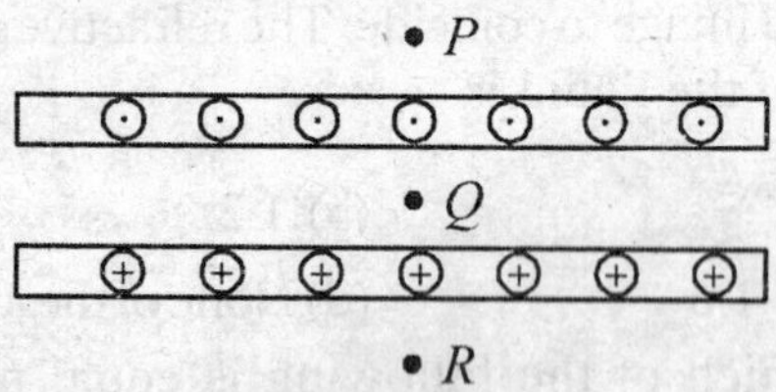

Fig. MTP-2.6

(a) magnetic field at P and R is zero
(b) magnetic field at Q is $\mu_0 K$ towards right
(c) magnetic field at Q is $\mu_0 K$ towards left
(d) magnetic field at Q is zero but at P and R is non-zero.

19. The tension in a stretched string fixed at both ends is changed by 2%. The fundamental frequencey is found to get changed by 15 Hz. Select the correct statements.

(a) Wavelength of the string of fundamental mode does not change
(b) Velocity of propagation of wave changes by 2%
(c) Velocity of propagation changes by 1%
(d) Original frequency is 1500 Hz.

20. The magnetic field perpendicular to the plane of a conducting ring of radius a changes at the rate of α. Then
(a) all the points on the ring are at the same potential
(b) the emf induced in the ring is $\pi a^2\alpha$
(c) electric field intensity E at any point on the ring is zero
(d) $E = \frac{a\alpha}{2}$.

SECTION III

In this section, there are four comprehensions, each followed by three multiple choice questions.

Passage 1: Two conducting plates having area A each are placed parallel at a separation D between them. Initially plates were charged by Q_0 and $2Q_0$. The plates are connected through a cell of emf E (Fig. MTP-2.7).

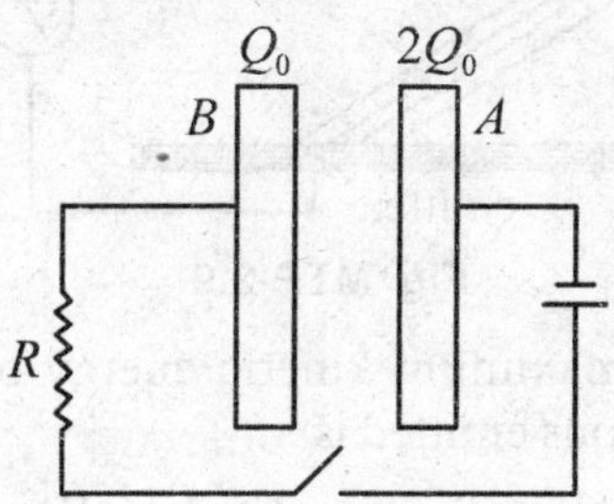

Fig. MTP-2.7

21. The charge distribution on the plates A and B when the key is pressed is
(a) On A: outer surface $\frac{3Q_0}{2}$ and inner surface $\frac{\varepsilon_0 AE}{D}$ and on B outer surface $\frac{Q_0}{2}$ and inner surface, $\frac{\varepsilon_0 AE}{D}$
(b) On A: outer surface $\frac{3Q_0}{2}$ and inner surface $-\frac{\varepsilon_0 AE}{D}$ and on B: outer surface $\frac{3Q_0}{2}$ and inner surface $+\frac{\varepsilon_0 AE}{D}$.
(c) On A: outer surface $-\frac{Q_0}{2}$ and inner surface $+\frac{\varepsilon_0 AE}{D}$ and on B: outer surface $+\frac{Q_0}{2}$ inner surface $-\frac{\varepsilon_0 AE}{D}$
(d) On A: outer surface $-\frac{3Q_0}{2}$ and inner surface $+\frac{\varepsilon_0 AE}{D}$ and on B: outer surface $+\frac{3Q_0}{2}$ and inner surface $-\frac{\varepsilon_0 AE}{D}$.

22. When switch is closed the charge flow through the cell is
(a) $\frac{\varepsilon_0 AE}{D} + \frac{Q_0}{2}$ (b) $\frac{\varepsilon_0 AE}{D} - \frac{Q_0}{2}$
(c) $\frac{\varepsilon_0 AE}{D} - \frac{3Q_0}{2}$ (d) $\frac{\varepsilon_0 AE}{D} + \frac{3Q_0}{2}$.

23. If switch is closed at time $t = 0$, then charge on inner surface of B at any time t is $\left(C = \frac{\varepsilon_0 A}{D} \text{ and } q_0 = CE\right)$
(a) $q = q_0\left(1 - e^{-t/RC}\right) - \frac{Q_0}{2}$
(b) $q = q_0\, e^{-t/RC} + \frac{Q_0}{2}$
(c) $q = q_0\left(1 - e^{-t/RC}\right) + \frac{Q_0}{2}$
(d) none of the above.

Passage 2: A glass slab of thickness 5 cm and refractive index 1.5 has its one face plane and other convex of radius 30 (Fig. MTP-2.8 (*a*)).

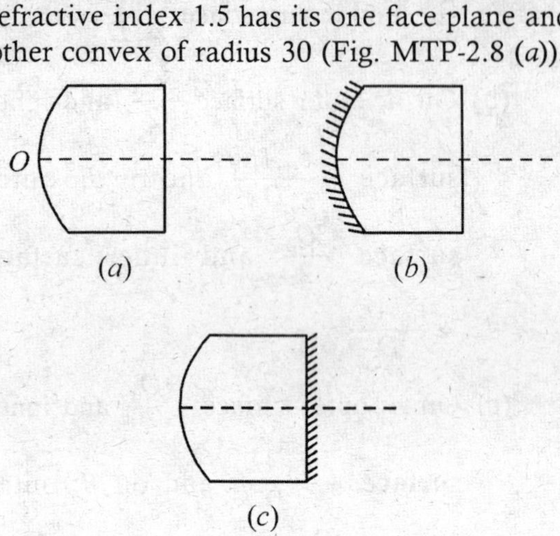

Fig. MTP-2.8

24. A parallel beam of light falls on the plane surface normally, the final image will be formed at
(a) inifinity
(b) 60 cm from O towards plane surface
(c) 60 cm from O on the other side of convex surface
(d) the centre of the slab.

25. If its convex surface is silvered, the position of the object so that its image coincides with the object, then
(a) 16.7 cm
(b) 24.3 cm
(c) 60 cm
(d) there is no such position.

26. If its plane surface is silvered, the position of the object so that its image coincides with the object is
(a) 16.7 cm
(b) 24.3 cm
(c) 60 cm
(d) there is no such position.

Passage 3: Two vessels A and B thermally insulated contain an ideal monoatomic gas. A small tube fitted with a valve connects these vessels. Initially the vessel A has 2 litres of gas at 350 K and 4×10^5 Nm^{-2} pressure. The valve is now opened and the system reaches equilibrium in pressure and temperature.

27. Then the final temperature of the system is
(a) 300 K (b) 323.8 K
(c) 338.76 K (d) 350 K.

28. The final pressure of the system is
(a) 3×10^5 N/m^2 (b) $\frac{10}{3} \times 10^5$ N/m^2
(c) $\frac{6}{5} \times 10^5$ N/m^2 (d) $\frac{8}{3} \times 10^5$ N/m^2.

29. Ratio of the number of molecules in vessel A and B before and after connecting respectively is
(a) $\frac{24}{7}, \frac{1}{2}$ (b) $\frac{7}{24}, 2$
(c) $\frac{7}{24}, \frac{1}{2}$ (d) $\frac{24}{7}, 2$.

Passage 4: In a photoelectric cell collector and emitter plates are separated by a distance 15 cm. The photo cell is connected with ammeter in series (Fig. MTP-2.9). Light of wavelength between 400 and 700 nm fall on the plate whose work function is 2.1 eV.

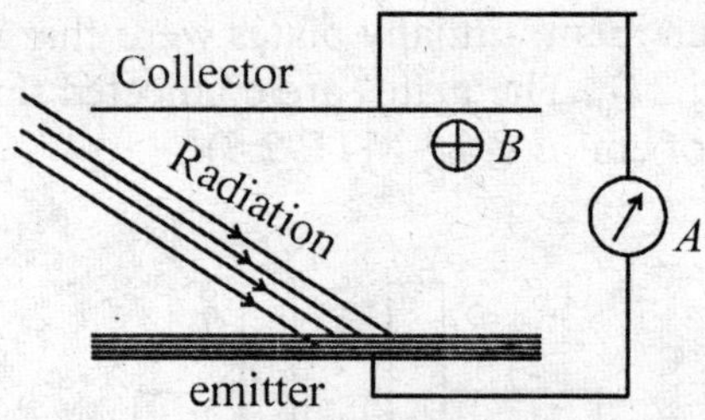

Fig. MTP-2.9

30. The maximum kinetic energy of photo-electrons emitted is
(a) 2 eV (b) 1.5 eV
(c) 1 eV (d) 0.5 eV.

31. The minimum value of magnetic field and its direction present between the plates of the cell so that the ammeter reading falls to zero is
(a) 2.25×10^{-4} T perpendicular to the velocity of electron

(b) 2.25×10^{-5} T parallel to the velocity of electron
(c) 2.25×10^{-5} T perpendicular to the velocity of electron
(d) 2.25×10^{-4} T parallel to the velocity of electron.

32. A monochromatic light of 400 nm and of 1 mW power falls on the emitter and emitter efficiency is 0.2%. The current shown by the ammeter is

(a) 0.65 μA (b) 0.325 μA
(c) 1.3 μA (d) 2.6 μA.

SECTION IV

In this section, there are four numerical questions. The answers to these questions are to be given in between 0000 to 9999 in the form of nearest integer.

33. A thin ring of mass 0.1 kg and diameter 0.5 m carrying charge q is free to rotate about an axis as shown in Fig. MTP-2.10. At $t = 0$, time dependent magnetic field of $8t$ tesla is applied vertically down on the ring. If the ring is at rest before applying magnetic field and induced magnetic effects due to rotation of the ring are neglected, the power developed by the force on acting the ring after 10 sec is found to be 6.4 mW. Find the charge 'q' that the ring is carrying?

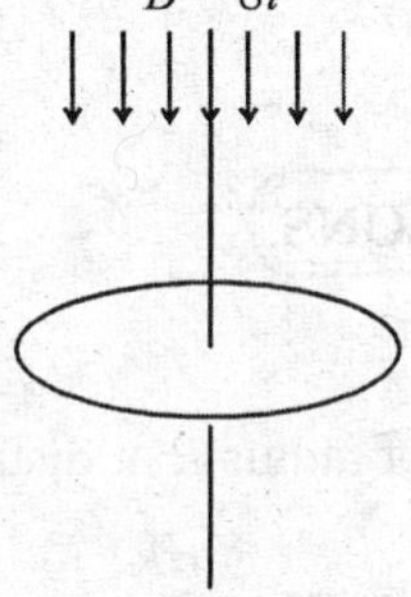

Fig. MTP-2.10

34. A neutron moving with a speed v makes a head-on collision with hydrogen atom in ground state kept at rest. Find the minimum kinetic energy of neutron for which inelastic collision will take place. (assume that mass of proton is nearly equal to the mass of neutron).

35. A fixed pulley carries a weightless thread with mass m_1 and m_2 at its ends. There is friction between the thread and pulley. It is such that thread starts slipping when the ratio $\frac{m_2}{m_1} = \eta_0$. Show that co-efficient of friction is $\mu \frac{1}{\pi} \log(1/\eta_0)$. Find the acceleration of mass if $\frac{m_2}{m_1} = \eta > \eta_0$. Take $\eta_0 =$ 0.5 and $\eta = 4$.

36. Calculate the increase of pressure when a sound wave produce an energy flow of 10^{-7} Wcm^{-2}. Take velocity of sound 340 m/s and density of air 0.001293 g/cc.

SECTION V

In this section, there are four questions. Each question has two columns with 4 entries in each column. Entries of Column I are to be matched with Column II. One entry of Column I may have more than one matching in Column II

37.

Column I	Column II
(A) Sound waves	(P) Dispersion
(B) Chromatic abberation	(Q) Deviation
(C) Direct vision spectroscope	(R) Interference
(D) Electromagnetic waves	(S) Polarisation

38.

Column I	Column II
(A) Characteristic X-rays	(P) Moseley's law
(B) Continous X-rays	(Q) Quantisation of levels
(C) Bohr atom model	(R) Population inversion
(D) LASER	(S) Metastable states.

39.

Column I	Column II
(A) Reactance	(P) Inductance
(B) Impedence	(Q) Frequency
(C) Power factor	(R) Capacitance
(D) Ohm's law	(S) Resistance.

40. An indicator V-T diagram is shown in Fig. MTP-2.11. ΔQ is the change in heat contents of the system, ΔU is the change in internal energy of the system and ΔW is the work done by the system.

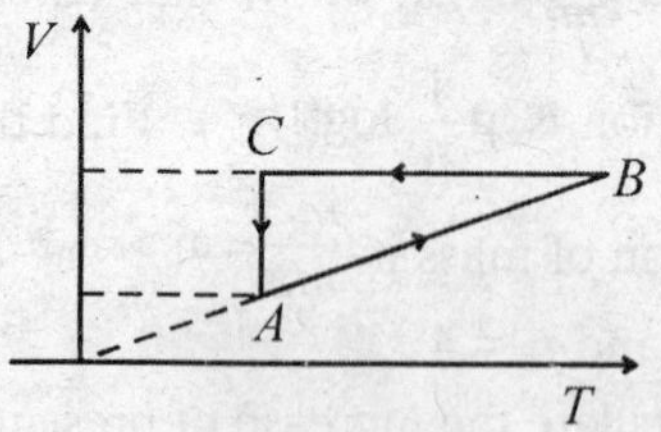

Fig. MTP-2.11

Column I	Column II
(A) AB	(P) $\Delta Q > 0$
(B) BC	(Q) $\Delta U < 0$
(C) CA	(R) $\Delta W > 0$
(D) $ABCA$	(S) $\Delta U = 0$

ANSWERS

1. (a)	2. (b)	3. (c)	4. (c)	5. (b)
6. (c)	7. (a)	8. (a)	9. (c)	10. (c)
11. (c)	12. (a)	13. (c, d)	14. (c, d)	15. (b, d)
16. (a, d)	17 (b, d)	18 (a, b)	19 (a, c, d)	20. (a, b, d)
21. (b)	22. (a)	23. (d)	24. (c)	25. (a)
26. (c)	27. (c)	28 (b)	29. (c)	30. (c)
31. (c)	32. (a)			

SOLUTIONS

1. $\mu = \tan\theta = \frac{1}{3}$. $h = R(1 - \cos\theta) = 0.05R$. (a)

2. Potential outside the charged spherical shell of radius R at distance r from centre is

$$V = \frac{Q}{4\pi\varepsilon_0 r} \text{ and inside every where is } \frac{\sigma R}{\varepsilon_0}.$$

The potential of B is

$$V_b = \frac{1}{4\pi\varepsilon_0}\frac{q}{b} - \frac{\sigma b}{\varepsilon_0} + \frac{\sigma c}{\varepsilon_0} = \frac{\sigma a^2}{\varepsilon_0 b} - \frac{\sigma b}{\varepsilon_0} + \frac{\sigma c}{\varepsilon_0} = \left(a^2/b - b + c\right)\sigma/\varepsilon_0.$$ (b)

3. With 72 Ω in series, the current throughout 30 Ω resistance is

$$i = \frac{2}{48 + 72 + 30} = \frac{2}{150} = \frac{1}{75} \text{ A}$$

Hence potential difference across 30 Ω resistance is $= 30 \times \frac{1}{75}$. P.d. = 0.4 V. Hence potential difference across 48 Ω resistance is = 2 – 0.4 = 1.6 V

$\therefore$ Total current $i = \frac{1.6}{48} = \frac{1}{30}$ A and current through $R = \frac{1}{30} - \frac{1}{75} = \frac{3}{150} = \frac{1}{50}$ A

$\therefore$ P.d. across $R = 0.4 \Big/ \frac{1}{50} = 20\ \Omega$. (c)

4. Physical quantities with same dimensions can only be added or subtracted. (c)

5. According to *uniqueness theorem* and *theorem of electrical images*, the system is equivalent to two charges $+q$ and $-q$ at distance $2d$ (Fig. MTP-2.12).

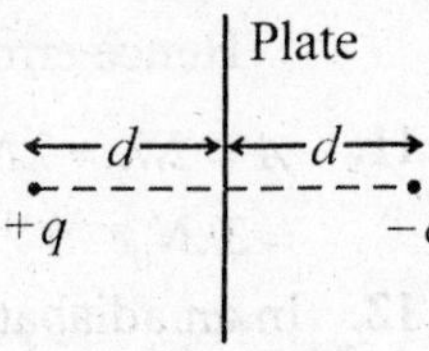

Fig. MTP-2.12

P.E. of system is $-\frac{q^2}{4\pi\varepsilon_0 2d}$. Hence work done in taking charge q to infinity is

$$W = \frac{q^2}{4\pi\varepsilon_0} \times \frac{1}{2d}.$$ (b)

6. P.E. on surface of earth $U_1 = -\frac{GMm}{R}$.

P.E. at height $h = R/5$ is $U_2 = -\frac{GMm}{R + R/5} = -\frac{5GMm}{6R}$

Hence change in P.E. $\Delta U = U_2 - U_1 = -\frac{5GMm}{6R} + \frac{GMm}{R}$

or $$\Delta U = \frac{GMm}{6R} = \frac{GM}{R^2} \cdot \frac{mR}{6} = \frac{mg\,5h}{6} = \frac{5m\,gh}{6}$$ (c)

7. For DC, inductance acts like a short circuit. Hence resistance of inductor is $R = \frac{10}{5} = 2\,\Omega$.

In AC, since current and voltage are in the same phase, i.e. net reactance is zero and impendence $Z = R = 2\ \Omega$ $\quad\therefore\ i = \frac{20}{2} = 10\,\text{A}$. (a)

8. Since medium surrounding is same. Hence surrounding medium ($\mu = 1.2$) does not effect

and $$\mu = \frac{\text{Real radius of curvature}}{\text{Apparent radius of curvature}}$$

or $$\mu = \frac{30}{30 - 7.5} = \frac{4}{3}.$$ (a)

9. If E is the initial K.E. of hydrogen atom, then K.E. after inelastic collisioin is $\frac{E}{2}$, Hence

$E - \frac{E}{2}$ = energy required for 1st excited state = 13.6 – 3.4 eV = 10.2 eV

$\therefore$ $$E = 20.4\ \text{eV}$$ (c)

10. Error in measurement of ρ, the specific resistance is $\left(\rho = \frac{\pi r^2 R}{L}\right)$

$$\frac{d\rho}{\rho}=\frac{2\,dr}{r}+\frac{dR}{R}+\left(\frac{dL}{L}\right)=2\times 0.01+\frac{0.01}{5}+\frac{0.01}{1}$$

$$= 0.02 + 0.002 + 0.01 = 0.032$$

and $$\rho=\frac{\pi\times(10^{-5})\times 5}{1}=15.7\times 10^{-6}\,\Omega-\text{m}.$$

Hence error is $d\rho = 0.032 \times 15.7 \times 10^{-6} = 0.5024\ \mu\Omega$-m. (c)

11. $A = N\lambda = \lambda N_0 e^{-\lambda t_0}$ Activity of $3N_0$ molecules after time t is $= 3N_0\lambda e^{-\lambda t} = 3Ae = 3\lambda N_0 e e^{-\lambda t_0}$ $= 3\lambda N_0 e^{-\lambda t_0+1}$ or $\lambda t = \lambda t_0 - 1$ or $t = t_0 - 1/\lambda$. (c)

12. In an adiabatic expansion PV^γ = const.

or $$\frac{\Delta P}{P}=-\gamma\frac{\Delta V}{V}\quad \frac{\Delta V}{V}=-5\%\ \ \gamma \text{ for air} = 1.4$$

$$\therefore\quad \frac{\Delta P}{P}=(1.4)(5)=7\%.$$ (a)

13. For Q the force of P on Q is 20 N upward (a) wrong

Net force on Q = 15 – 0.5 × 20 = 5 N, acceleration of $Q = \frac{5}{2} = 2.5$ m/s² relative to floor (b) wrong.

The only force on P is 10 N towards right hence its acceleration relative to floor is $\frac{10}{5} = 2$ m/s². (c)

For the system as a whole total force = 15 N total mass 5 + 2 = 7 kg. Hence acceleration of centre of mass $= \frac{15}{7}$ m/s² relative to floor. (d)

14. (a) No. $\mathbf{B} \times \mathbf{C}$ gives area and with given C, B can have many different values

(b) No. If $\mathbf{B}$ is parallel to $\mathbf{C}$, left hand side is zero, right hand side may not be zero

(c) $\mathbf{A} \times \mathbf{B} = AB\sin\theta$ Since $\theta = 0$ $\mathbf{A} \times \mathbf{B} = 0$ (c)

(d) $\mathbf{A} \cdot (\mathbf{B} \times \mathbf{C})$ gives volume of the parallelopiped. (d)

15. (a) No. Since E' will be larger than E and recoil atom carries the rest $E'' = E' - E$

(b) E'' increases as P^2, i.e. as E^2, whereas the fraction $\frac{E''}{E}$ increases as E. Which is far larger in gammas. (b)

(c) No. In absorption momentum transfer is E/C, while in reflection, it is $(1 + r)$ times where r is fraction reflected

(d) Since pressure varies as 1/area, and area for a laser beam may be smaller by a factor much larger than ratio 10^5 given for radial power. (d)

16. (a) $\frac{kq_a}{R}+\frac{kq_b}{2R}=2V$ and $\frac{kq_a}{2R}+\frac{kq_b}{2R}=3/2V$ gives $\frac{q_a}{q_b}=\frac{1}{2}$ (a)

(b) No. After B is earthed potential of B is zero, i.e.

$$q_a' = -q_b' \quad \text{gives} \quad \frac{q_a'}{q_b'} = -1$$

(c) No. After earthing $V_a - V_b = kq_a'\left(\frac{1}{R}-\frac{1}{2R}\right)=\frac{kq_a'}{2R}$ but $q_a' = q_a$. Since charge on inner sphere is same and from (a) $\frac{kq_a}{R} = V$

(d) $V_a - V_b = \frac{kq_a'}{2R} = V/2$. (d)

17. (a) No. Initial temperature T_0, final temperature $4T_0$. $\Delta T = 3T_0$

(b) Change in internal energy $= C_V \cdot \Delta T = \frac{3}{2}R \times 3T_0 = \frac{9RT_0}{2}$ (b)

(c) No. work done is equal to area under indicator diagram

$$W = \frac{1}{2}(P+2P)V = \frac{3}{2}PV = \frac{3}{2}RT_0$$

(d) Heat supplied $= \Delta U + \Delta V = \frac{9RT_0}{2} + \frac{3RT_0}{2} = 6RT_0$. (d)

18. Consider two strips at AB symmetrically situated from P. Magnetic field at P due to these elements is towards left and using Ampere's law net B at P is $\frac{1}{2}\mu_0 K$ towards left

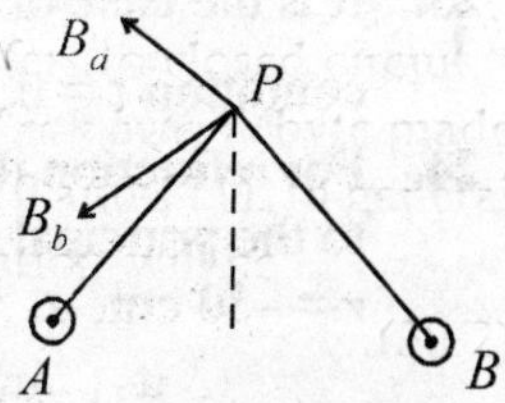

Fig. MTP-2.13

(a) Magnetic field at P and R due to upper and lower plate have same magnitude but opposite direction. Hence it is zero (a)

(b) At Q the field due to upper plate is $\frac{1}{2}\mu_0 K$ towards right and due to lower plate also it is $\frac{1}{2}\mu_0 K$ towards right. Hence net field is $\mu_0 K$ towards right (b)

19. (a) Wavelength depends on length which is fixed. Hence wavelength does not change (a)

(b) No

(c) $v = \sqrt{\frac{T}{m}}$. $v \propto \sqrt{T}$ If T changes by two percent, v then changes by $\frac{1}{2}$ (% of change in T) (c)

(d) Since speed changes by 1%. Hence frequency also changes by 1%. Since frequency changes by 15 Hz. Hence original frequency is 1500 Hz. (d)

20. Flux ϕ linked with the ring is $\phi = \pi a^2 B$ and induced emf

$$e = \frac{d\phi}{dt} = \pi a^2 \frac{dB}{dt} = \pi a^2 \alpha$$ (b)

Let R be the resistance of ring then current in the ring in e/R. Now consider a small element dl of the ring. Its resistance $= \frac{R}{2\pi a}dl$ and P.d.$= \frac{iR}{2\pi a}dl = \frac{edl}{2\pi a}$ and induced emf in this element is $\frac{e}{2\pi a}dl$.

Hence potential difference across dl = emf induced – P.D. due to current = 0. Hence all the points are at the same potential. (a)

Induced electric field is given by $\oint E \cdot dl = \frac{d\phi}{dt}$ or $E\,2\pi a = \pi a^2\alpha$ or $E = \frac{a\alpha}{2}$. (d)

21. The charge distribution on the plates before key is pressed is shown in Fig. MTP-2.14 (*a*) and after key is pressed is shown in Fig. MTP-2.14 (*b*). (b)

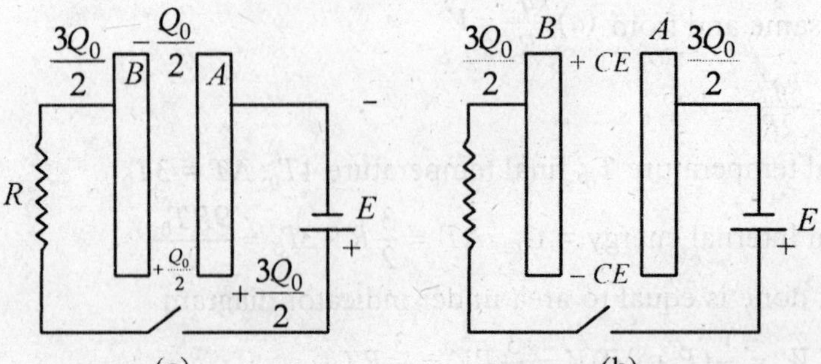

Fig. MTP-2.14

22. When switch is closed the charge flown through the cell is $CE + Q_0/2 = \frac{\varepsilon_0 AE}{D} + Q_0/2$ (d)

23. It is the case of charging of capacitor *R-C* circuit. The equation is $\frac{q}{C} + R\frac{dq}{dt} = E$ with initial condition $t = 0$, $q = -Q_0/2$. Hence $q = q_0 - (q_0 + Q_0/2)\, e^{-t/RC}$ (d)

24. For refraction through lens the ray diagram is shown in Fig. MTP-2.15 (*a*). Rays fall parallel to the principal axis on convex surface from refraction through convex surface $U = \infty$, $V = ?$, $r = -30$ cm, $\mu_1 = 1.5$ and $\mu_2 = 1$ using formula

$$\frac{\mu_2}{V} - \frac{\mu_1}{U} = \frac{\mu_2 - \mu_1}{r}$$

$$\therefore \quad \frac{1}{V} = \frac{1 - 1.5}{-30} = \frac{1}{60} \quad \text{or } V = 60. \qquad \text{(c)}$$

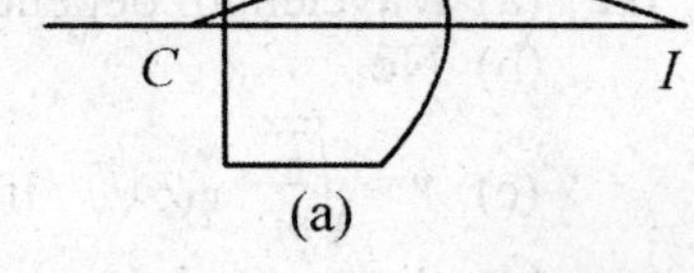

(a)

25. Ray diagram is shown in Fig. MTP-2.15 (*b*). When rays after refraction fall normally on the other surface they are retraced. Hence refraction through plane surface must form image at *C* the centre of curvature *C*. Hence $V = -(30 - 5) = -25$, $U = ?$,

$$\mu_1 = 1,\ \mu_2 = 1.5,\ r = \infty,\ \frac{1.5}{(-25)} - \frac{1}{U} = \frac{1.5 - 1}{-\alpha} \quad \text{or} \quad U = -\frac{25}{1.5}$$

$= -16.7$ cm, i.e. position is 16.7 cm in front of lens. (a)

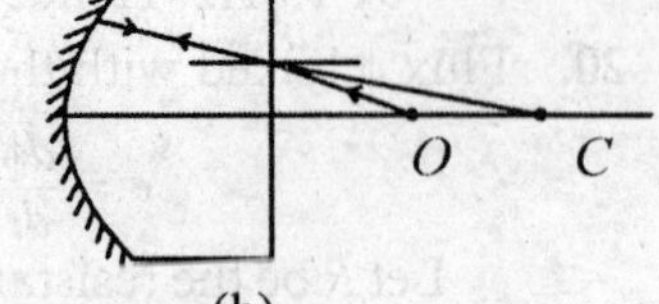

(b)

26. Ray diagram for this system is shown in Fig. MTP-2.15 (*c*). The rays after refraction through convex surface must go parallel to the principal axes. Hence $U = ?$, $V = \infty$, $\mu_1 = 1$, $\mu_2 = 1.5$, $r = +30$ cm.

$$\therefore \quad -\frac{1}{U} = \frac{1.5 - 1}{30} \quad \text{or } U = -60 \text{ cm}$$

Object should be placed at distance 60 cm from lens. (*c*)

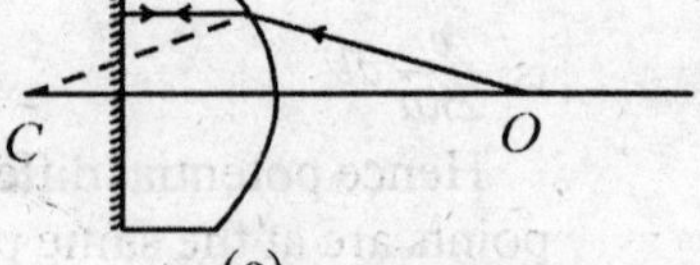

(c)

Fig. MTP-2.15

27. The final temperature is given by

$$T = \frac{n_1 T_1 + n_2 T_2}{n_1 + n_2} = \frac{P_1 V_1 / R + P_2 V_2 / R}{\frac{P_1 V_1}{R T_1} + P_2 V_2 / R T_2} = \frac{1 + 4}{\frac{1}{300} + \frac{4}{350}}$$

$$= \frac{5 \times 300 \times 350}{1550} = 338.76 \text{ K}$$ (c)

28. Total number of molecules will remain conserved. If final pressure is P', then

$$\frac{P_1V_1}{RT_1} + \frac{P_2V_2}{RT_2} = \frac{P'V_1 + P'V_2}{RT}$$

$$\frac{P}{R}\left(\frac{1}{300} + \frac{4}{350}\right) = \frac{P'}{R}\frac{(1+2)}{338.76} \quad \text{where } P = 2 \times 10^5 \text{ Nm}^{-2}.$$

$$\therefore \quad P' = \frac{5 \times 338.76}{3 \times 300 \times 350} = \frac{10}{3} \times 10^5 \text{ Nm}^{-2}.$$ (b)

29. Ratio of n_1 and n_2 before connecting is

$$\frac{n_1}{n_2} = \frac{P_1V_1}{T_1R} \times \frac{T_2R}{P_2V_2} = \frac{350}{300 \times 4} = \frac{7}{24}$$

and after connecting is $\frac{n_1}{n_2} = \frac{V_1}{V_2} = \frac{1}{2}$ (c)

30. Maximum energy of incident photons is of those whose wavelength is 400 nm. Their energy is $E = \frac{hc}{\lambda} = \frac{12.4 \times 10^3}{\lambda(\text{Å})} \text{eV} = 3.1 \text{ eV}$.

Hence maximum K.E. of the photoelectrons emitted = 3.1 – 2.1 = 1 eV. (c)

31. In magnetic field, the electron describes circular motion. If radius of the circular path is equal to the separation between plates the electrons emitted will not strike the collector and ammeter current falls to zero

$$B = \sqrt{\frac{2\,mE}{er^2}} \quad \text{where } E \text{ is electron energy in eV}$$

$$= \sqrt{\frac{2 \times 9.1 \times 10^{-31} \times 1}{1.6 \times 10^{-19} \times 225 \times 10^{-4}}} = 2.25 \times 10^{-5} \text{ T}$$

The field should be perpendicular to velocity. (c)

32. No. of photons falling per second on the emitter $= \frac{P}{h\nu}$. Since efficiency = 0.2%. Hence no. of photoelectrons emitted per second is $N = \frac{P}{h\nu} \times .2 \times 10^{-2} = \frac{P}{h\nu} \times 2 \times 10^{-3}$. Hence current

$$i = Ne = \frac{P}{h\nu} \times 2 \times 10^{-3} \times e$$

$$= \frac{10^{-3}}{3.1} \times 2 \times 10^{-3} = 0.65\ \mu\text{A}.$$ (a)

33. Electric field E induced in ring is

$$E\,(2\pi R) = \pi R^2 \frac{dB}{dt} \quad \text{or} \quad E = \frac{R}{2} B_0, \quad \text{where } B_0 = 8,\ \ B = 8t,\ \ \frac{dB}{dt} = 8.$$

Hence force on ring $F = qE = \frac{qRB_0}{2}$

Torque on ring $\tau = RF = \frac{1}{2}qR^2B_0$

Angular acceleration $\alpha = \frac{\tau}{I} = \frac{qR^2B_0}{2MR^2} = \frac{qB_0}{2M}$

Power generated $P = \frac{d}{dt}(\text{K.E.}) = \frac{d}{dt}\left(\frac{1}{2}I\omega^2\right) = I\omega\frac{d\omega}{dt} = I\omega\alpha$

But $\omega = \alpha t$ $\quad\therefore\quad P = I\alpha^2 t = MR^2.\left(\frac{qB_0}{2M}\right)^2 t$

or $$q = \sqrt{\frac{4PM}{R^2B_0^2t}} = \sqrt{\frac{4\times 6.4\times 10^{-3}\times 10^{-1}}{(0.25)^2\times(8)^2\times 10}} = 8\times 10^{-3}\ \text{Coulomb}$$

34. Let v be speed of neutron before collision, v_1 be the speed of neutron after collision and v_2 be the speed of proton or bydrogen atom after collision and ΔE is the energy of excitation.

Conservation of linear momentum gives $mv = mv_1 + mv_2$ ...(*i*)

Conservation of energy gives $\frac{1}{2}mv^2 = \frac{1}{2}mv_1^2 + \frac{1}{2}mv_2^2 + \Delta E$ ...(*ii*)

From (*i*) we get $v^2 = v_1^2 + v_2^2 + 2v_1v_2$ and comparision with (*ii*) gives

$$2v_1v_2 = \frac{2\Delta E}{m}$$

and $(v_1 - v_2)^2 = (v_1 + v_2)^2 - 4v_1v_2 = v^2 - \frac{4\Delta E}{m}$

v must be real means $v^2 > \frac{4\Delta E}{m}$

or $\frac{1}{2}mv^2 = 2\Delta E$

The minimum energy that can be absorbed by hydrogen atom in ground state to go into excited state is 10.2 eV. Hence minimum kenetic enegy needed is = 2 × 10.2 eV = 20.4 eV. In proper significant digits answer is 0020 eV.

35. Figure MTP-2.16 shows the different forces acting on pully. Components of tensiion parallel to PO are $T\sin d\theta$ and $(T + dT)\sin d\theta$. Since $d\theta$ and dT are small, the net component of tension along PO = $2Td\theta$. Hence normal reaction $dN = 2Td\theta$. If μ is co-efficient of frictioin, then frictional force $f = \mu dN = 2T\mu d\theta = dT$ the difference in tension.

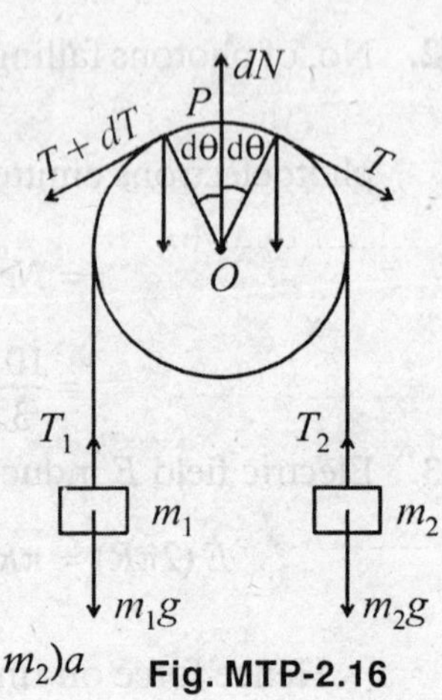

Fig. MTP-2.16

or $2\mu d\theta = \frac{dT}{T}$ or $2\mu\int_0^{\pi/2} d\theta = \int_{T_2}^{T_1}\frac{dT}{T}$ $\quad T_2 = m_2g$ and $T_1 = m_1g$

or $2\mu\frac{\pi}{2} = \log\frac{T_1}{T_2}$ or $\mu = \frac{1}{\pi}\log\frac{m_1}{m_2} = \frac{1}{\pi}\log\frac{1}{\eta_0}$

when $\eta > \eta_0$ then equations of motion for two masses are

$m_2g - T_2 = m_2a$ and $T_1 - m_1g = m_1a$ or $(m_2 - m_1)g - (T_2 - T_1) = (m_1 + m_2)a$

but difference in tension is limiting force of friction or $T_2 - T_1 = g(m_2 - m_1) = m_1 g(m_2/m_1 - 1)$ or $T_2 - T_1 = m_1 g\,(\eta_0 - 1)$

$\therefore \quad (m_2 - m_1)g - m_1 g\,(\eta_0 - 1) = (m_1 + m_2)a$

or $\quad m_1\left(\dfrac{m_2}{m_1} - 1\right)g - m_1 g\,(\eta_0 - 1) = m_1(m_2/m_1 + 1)\,a$ with $\dfrac{m_2}{m_1} = \eta$. Now

$$(\eta - \eta_0)g = (\eta + 1)a \quad \text{or} \quad a = \frac{\eta - \eta_0}{\eta + 1}g$$

for $\quad \eta = 4$ and $\eta_0 = 0.5$, we have

$$a = \frac{4 - 05}{4 + 1} \times 10 = 7 \text{ m/s}^2 \qquad \text{Ans. } 0007 \text{ m/s}^2.$$

36. Intensity I of sound, i.e. energy crossing unit area per second is $I = \frac{1}{2}\omega^2 a^2 \rho v$. Taking wave equation as $y = a\sin(\omega t - kx)$ the pressure changes are given by $\Delta P = E\dfrac{dy}{dx} = Eak\cos(\omega t - kx)$.

Amplitude of pressure changes is $\Delta P_0 = Eak = Ea\dfrac{\omega}{v}$ $\left(\because v = \dfrac{\omega}{k}\right)$ and $v = \sqrt{\dfrac{E}{\rho}}$. And $\Delta P_0 = v^2 \rho a \dfrac{\omega}{v} = a\omega v\rho$ or $a\omega = \dfrac{\Delta P_0}{v\rho}$.

Hence intensity $I = \dfrac{1}{2}\dfrac{(\Delta P_0)^2 \rho v}{v^2 \rho^2}$ and $(\Delta P_0)^2 = 2IvP$.

$$\therefore \quad \Delta P_0 = \sqrt{2Iv\rho} = \sqrt{2 \times 10^{-3} \times 340 \times 1.293} = 0.9376 \text{ N/m}^2 = 0001 \text{ Nm}^{-2}$$

$I = 10^{-7}$ W.cm^{-2} = 10^{-3} Wm^{-2}. $\rho = 0.001293$ g/cc = 1.293 kg/m^3.

37. (A) ⟶ Q – R
(B) ⟶ P – Q
(C) ⟶ P – Q
(D) ⟶ P – Q – R – S

38. (A) ⟶ P – Q
(B) ⟶ None
(C) ⟶ Q – S
(D) ⟶ Q – R – S

39. (A) ⟶ P – Q – R
(B) ⟶ P – Q – R – S
(C) ⟶ P – Q – R – S
(D) ⟶ S

40. (A) ⟶ P – R
(B) ⟶ Q
(C) ⟶ S
(D) ⟶ S

Model Test Paper III

SECTION I

In this section, a question is followed by four alternatives, of them only one is correct.

1. The momentum $\left(P = \frac{mdx}{dt}\right)$ of a point particle executing S.H.M. is plotted against its displacement $x = A\sin\omega t$, the curve is

   (a) straight line (b) sine curve
   (c) parabola (d) ellipse.

2. Ten resistances of equal value R are connected between 5 vertices of a regular pentagon (each vertex to be joined with remaining four) (Fig. MTP-3.1). The equivalent resistance between any two adjacent vertices is

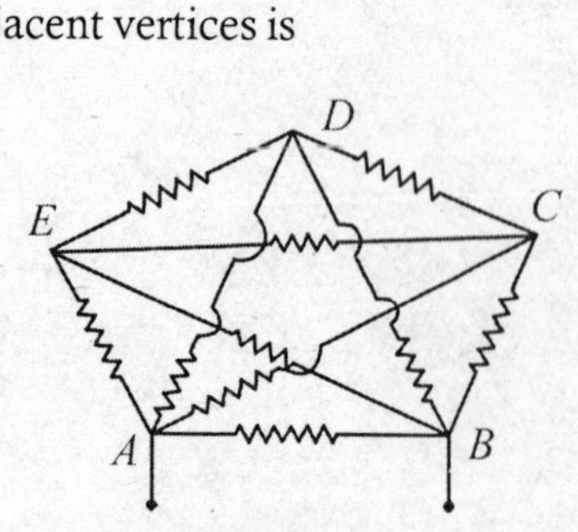

Fig. MTP-3.1

   (a) $R/10$ (b) $2R/10$
   (c) $4R/10$ (d) $5R/10$.

3. Three liquids L_1, L_2, L_3, are at temperatures 30° C, 20° C and 10° C respectively. When equal masses of L_1 and L_2 are mixed, the temperatue of mixture is 26° C and when equal masses of L_1 and L_3 are mixed, the temperature of the mixture is 15° C. If equal masses of L_2 and L_3 are mixed, the temperature of the mixture will be

   (a) 11.8° C (b) 13.5° C
   (c) 16.7° C (d) 18° C.

4. Two coherent point sources S_1 and S_2 are separated by a small distance d (Fig. MTP-3.2). The fringes obtained on the screen will be

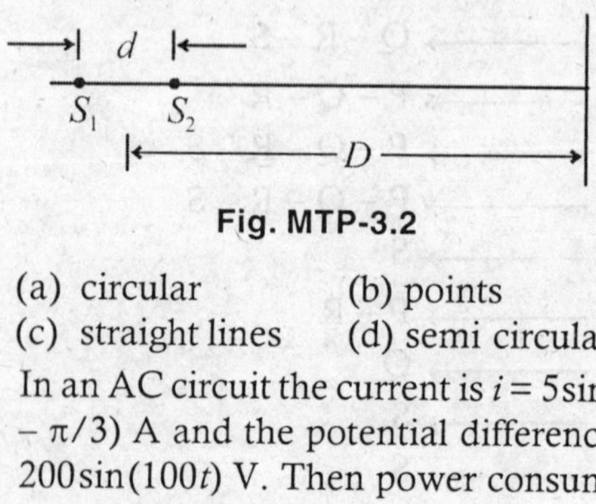

Fig. MTP-3.2

   (a) circular (b) points
   (c) straight lines (d) semi circular.

5. In an AC circuit the current is $i = 5\sin(100t - \pi/3)$ A and the potential difference $V = 200\sin(100t)$ V. Then power consumed is

(a) 100 W (b) 500 W
(c) 250 W (d) 125 W.

6. Three identical resistances each of 10 Ω are connected as shown in Fig. MTP-3.3. The maximum power that can be consumed by individual resistor is 20 Watt. Then the maximum power consumed by the combination is

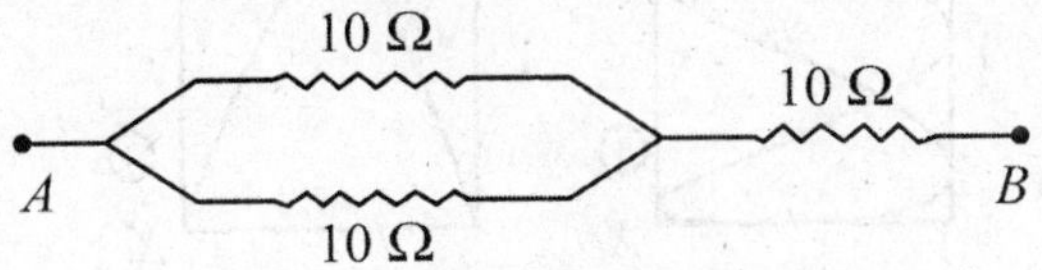

Fig. MTP-3.3

(a) 60 Watt (b) 15 Watt
(c) 30 Watt (d) 40 Watt.

7. A photon having wavelength 6000 Å moves from vacuum to medium of repractive index 1.5. The value of photon energy in vaccum and in the medium are
(a) 2.06 eV and 3.09 eV
(b) 2.06 eV and 1.37 eV
(c) 2.06 eV and 2.06 eV
(d) 1.37 eV and 1.37 eV.

8. The focal length of a concave mirror is f and the distance of the object from the principal focus is P. The magnification of the image is
(a) f/P (b) P/f
(c) fP (d) $\sqrt{P/f}$.

9. An isolated charge q is travelling between plates A and B which are connected through a galvanometer (Fig. MTP-3.4). Choose the correct statement.

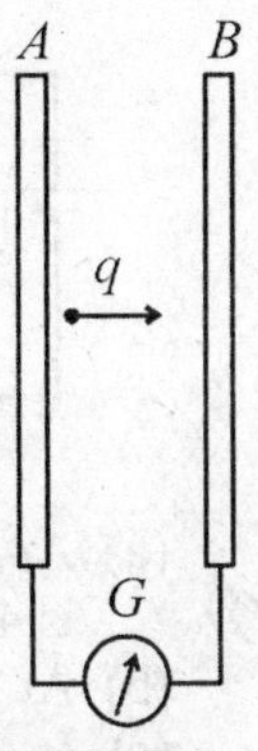

Fig. MTP-3.4

(a) Current flows through the galvanometer from A to B
(b) Current flows through the galvanometer from B to A
(c) No current flows through the galvanometer
(d) When charge is near q a current flows from A to B and when it is near B it flows from B to A.

10. A projectile is fired so as to land at horizontal distance 1.0 km but at the highest point it explodes into two equal parts. One of them falls vertically down with initial speed zero. The distance of the other part from the launching pad where it will land is
(a) 0.75 km (b) 1.0 km
(c) 1.25 km (d) 1.5 km.

11. A point positive charge is moving nonrelativistically along Y-axis. It passes the origin at $t = 0$. At that instant the electric and magnetic fields due to this charge at a point P on the X-axis will point respectively along
(a) $\hat{i}$ and $\hat{j}$ (b) $\hat{j}$ and $\hat{i}$
(c) $\hat{i}$ and $\hat{k}$ (d) $\hat{i}$ and $-\hat{k}$.

12. Figure MTP-3.5 shows the displacement y versus distance x curves of a longitudinal wave at a certain instant. The displacement is taken positive when it is along the positive direction of distance x. The maxima and minima of pressure at this instant are respectively at

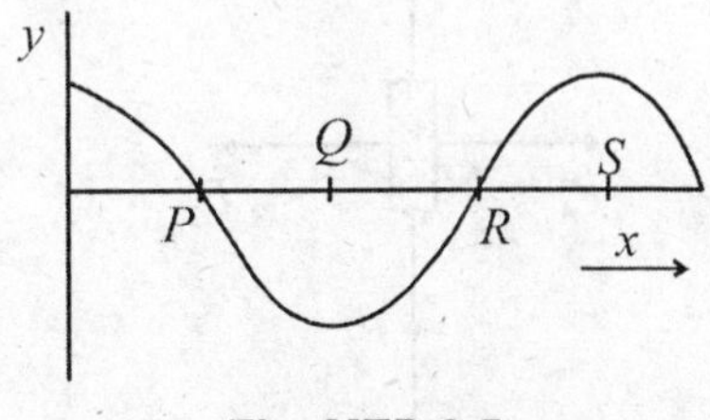

Fig. MTP-3.5

(a) P and R (b) Q and P
(c) Q and S (d) R and P.

SECTION II

In this section, a question is followed by four alternatives, of them one or more are correct.

13. Mark the correct statement

(a) 3.33 cm + 4.71 × 10^2 cm = 4.74 × 10^2 cm

(b) $\frac{11.04 \text{ cm}}{3.0 \text{ sec}} = 3.7$ cm/s

(c) A measured value of 5.1 m can be expresed as 5.1 × 10^3 mm but not 5100 mm

(d) The length of the rod has been measured to 1% accuracy. It is 10 cm.

14. Two closely situated stars are just resolved by the objective of a telescope. The separation between the images formed by the objective would depend upon

(a) aparture of the objective
(b) focal length of the objective
(c) wavelength of light used
(d) brightness of the stars.

15. A bar magnet is pivoted freely in a horizontal plane as shown in the Fig. MTP-3.6. Let B_1, B_2, B_3, B_4 be the net magnetic inductions respectively at points P_1, P_2, P_3, P_4 equidistance from centre O of the magnet. Then

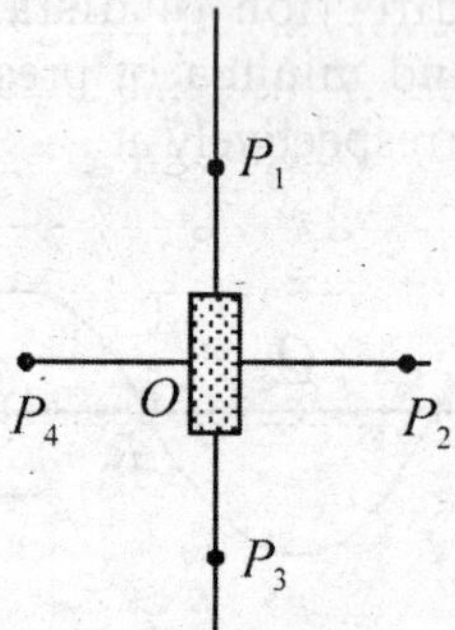

Fig. MTP-3.6

(a) B_1 and B_2 can have the same direction
(d) B_1 and B_3 have the same direction
(c) B_2 and B_4 have the same direction
(d) B_1 and B_2 can have opposite direction.

16. In Fig. MTP-3.7 is shown schematics of some Lissajou's figures formed on cathode ray oscilloscope. The ratio of frequency for vibrations along the vertical to that against the horizontal is

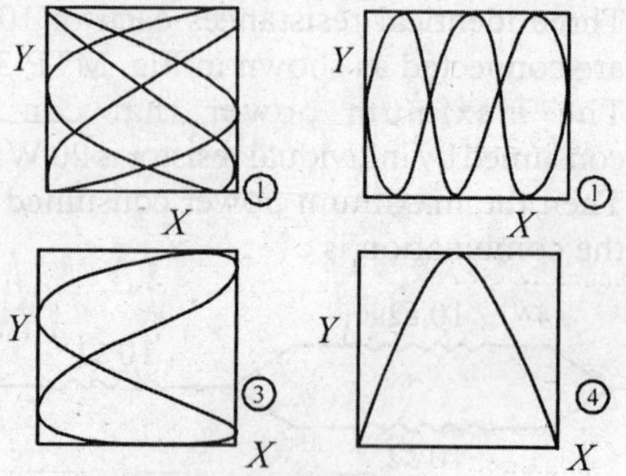

Fig. MTP-3.7

(a) 2/3 in Fig. (1) and 3/1 in Fig. (2)
(b) 2/3 in Fig. (1) and 2/1 in Fig. (4)
(c) 2/1 in Fig. (3) and 1/2 in Fig. (4)
(d) 1/3 in Fig. (2) and 1/2 in Fig (3).

17. Two bodies of masses m_1 and m_2 make a head on collision. During the collisioin, the interaction is all repulsive. At the instant when m_1 and m_2 are closest together

(a) the momenta of m_1 and m_2 will be equal
(b) the velocities of m_1 and m_2 will be equal
(c) m_2 will have gained less kinetic energy than what m_1 will have lost
(d) m_2 will have gained the same momentum as m_1 will have lost.

18. In the circuit shown below, the switch is tunned on at t = 0, then

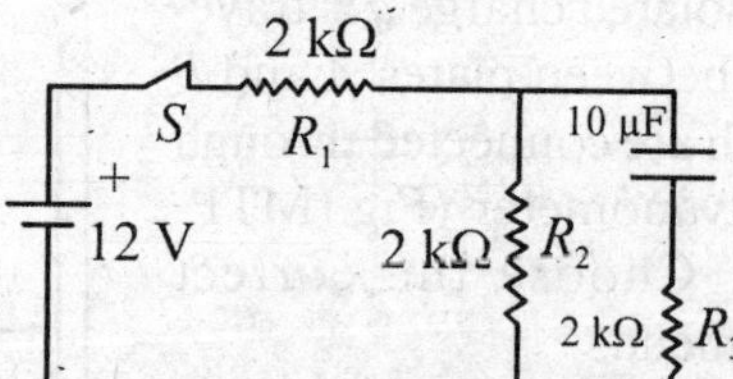

Fig. MTP-3.8

(a) At t = 0, current supplied by the battery is 4 mA
(b) At t = 0 current in R_3 is 2 mA
(c) In steady state current supplied by battery is 3 mA

(d) In steady state current in R_3 is zero.

19. The SI unit of inductance 'henry' is the same as

(a) Weber/ampere (b) Volt-sec/ampere
(c) Joule/(amp)^2 (d) Ohm-second.

20. A charge Q_1 is situated inside an uncharged conducting closed surface. Another charge Q_2 is situated outside the surface. Then

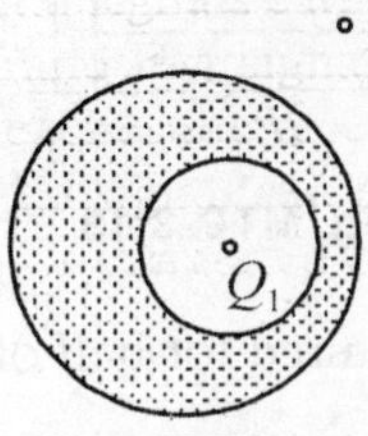

Fig. MTP-3.9

(a) Q_1 and Q_2 experience equal and opposite force
(b) Q_2 experiences a force but Q_1 does not
(c) there is apparent violation of Newton's third law
(d) Force on Q_2 is twice the force on Q_1.

SECTION III

In this section, there are four comprehensions, each followed by three multiple choice questions.

Passage 1: Figure MTP-3.10 shows two long conducting wires placed. They carry currents as indicated. Assume that 1 amp current in a long wire produces magnetic field B at a distance d from the wire

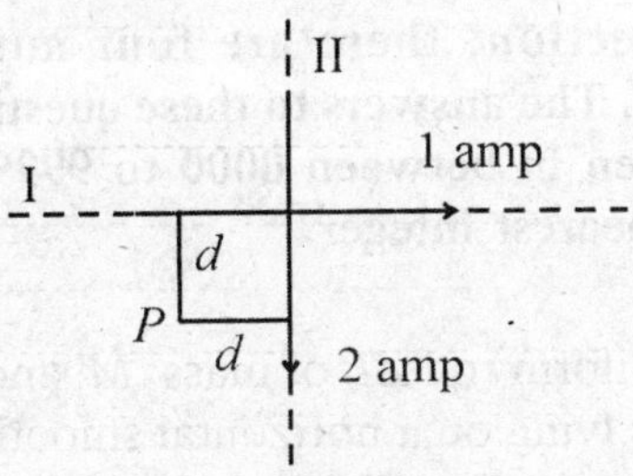

Fig. MTP-3.10

21. The strength and direction of magnetic field at point P in the Fig. MTP-3.10 will be

(a) $3B$ into the plane of paper
(b) B into the plane of paper
(c) B out of the plane of paper
(d) $3B$ out of the plane of paper.

22. Suppose wire II were free to rotate about its point of of intersection with wire I. The wire II will have stable position

(a) as it is
(b) after turning 180° in plane of paper
(c) after turning 90° clockwise in plane of paper
(d) after turning 90° clockwise in plane of paper.

23. If wire II is kept parallel to I and above it at a distance of 1 m, it is found that wire II is in equilibrium. The linear density of wire II is

(a) 2×10^{-8} kg/m (b) 4×10^{-8} kg/m
(c) 2×10^{-7} kg/m (d) 4×10^{-7} kg/m.

Passage 2: In Fig. MTP-3.11 S_1 and S_2 are two small indentical speakers radiating isotropically. They radiate in phase and frequency of sound emitted is 150 Hz. Take speed of sound in air as 300 m/s.

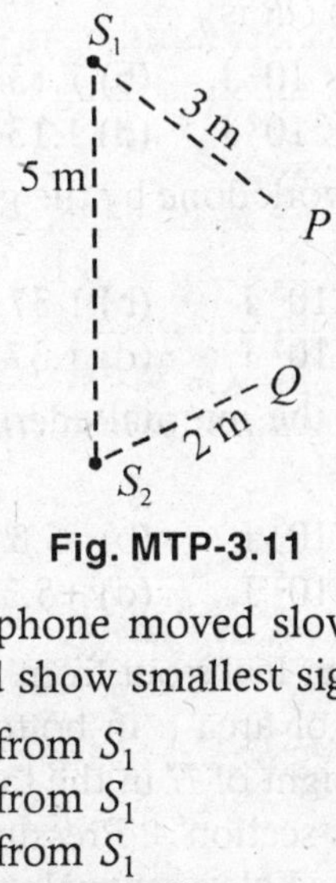

Fig. MTP-3.11

24. A microphone moved slowly from S_1 to S_2 would show smallest signal at

(a) 1 m from S_1
(b) 2 m from S_1
(c) 4 m from S_1
(d) all the above distances.

25. If point P is a nodal point then S_2P could be

(a) 3 m (b) 5 m
(c) 6 m (d) 7 m.

26. If point Q is antinodal point then S_1Q could be

(a) 2 m (b) 3 m
(c) 4 m (d) 5 m.

Passage 3: Three moles of an ideal monoatomic gas perform a cycle shown in Fig. MTP-3.12. The temperature of the gas in different states marked as 1, 2, 3, and 4 shown in the Fig. are 400 K, 700 K, 2500 K and 1100 K respectively

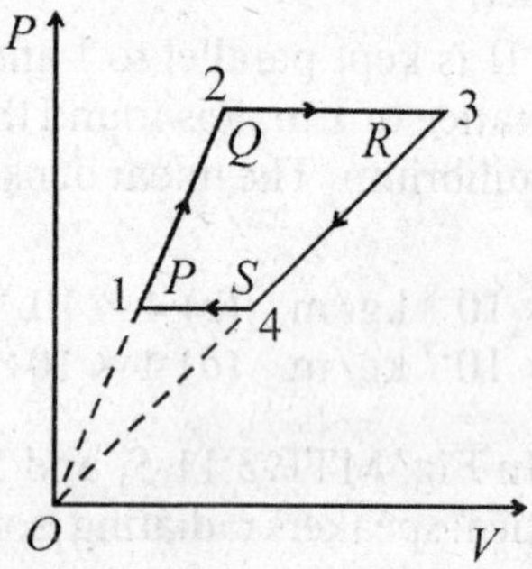

Fig. MTP-3.12

27. Total heat supplied to the system during the process QR is

(a) 1.134×10^2 J (b) 1.134×10^3 J
(c) 1.134×10^4 J (d) 1.134×10^5 J.

28. The total work done by the gas during the process is

(a) 1.37×10^3 J (b) 1.37×10^4 J
(c) 1.37×10^5 J (d) 1.37×10^6 J.

29. Change in the internal energy during the process RS is

(a) -5.3×10^3 J (b) -5.3×10^4 J
(c) -5.3×10^5 J (d) $+5.3 \times 10^4$ J.

Passage 4: Water leaks out from an open tank through a hole of area a in bottom. Water is filled up to a height of H in the tank. The tank has area of cross-section A. Pressure at the open surface and at the hole are equal to atomspheric pressure (= P_0). Neglect the small velocity of the water near the open surface in the tank.

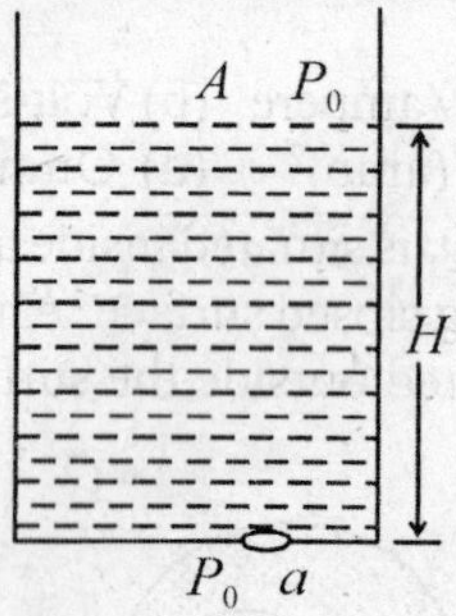

Fig. MTP-3.13

30. The time required for half of the water to leak out is

(a) $\frac{A}{a}\sqrt{\frac{H}{g}}(\sqrt{2}-1)$ (b) $\frac{a}{A}\sqrt{\frac{H}{g}}\left(1-\frac{1}{\sqrt{2}}\right)$

(c) $\frac{A}{a}\sqrt{\frac{H}{g}}\left(1-\frac{1}{\sqrt{2}}\right)$ (d) $\frac{A}{a}\sqrt{\frac{H}{g}}(\sqrt{2}-1)$.

31. The speed of water coming out when half of water has leaked out is

(a) $\sqrt{gH/2}$ (b) $\sqrt{2gH}$
(c) $\sqrt{gH}$ (d) $2\sqrt{gH}$.

32. When water level is h, the decrease in level dh in time interval dt in terms of h is

(a) $\frac{a}{A}\sqrt{2gh}\,dt$ (b) $\frac{A}{a}\sqrt{2gh}\,dt$

(c) $\frac{a}{2A}\sqrt{gh}\,dt$ (d) $\frac{A}{a}\sqrt{gh}\,dt$.

SECTION IV

In this section, there are four numerical questions. The answers to these questions are to be given in between 0000 to 9999 in the form of nearest integer.

33. A uniform rod AB of mass 'M' and length $2L$ is lying on a horizontal smooth plane. A nail at P is fixed to the ground (Fig.

MTP-3.14). A particle of mass $M/2$ moving with velocity u (=16 m/s) collides with rod at a distance $L/2$ from the centre of mass of rod and comes to rest. End B of rod also comes to instantaneous rest after colliding with nail at P. Find the velocity of the centre of mass of rod immediately after end B of rod strikes nail at P. (Initially there is a small gap between nail and rod).

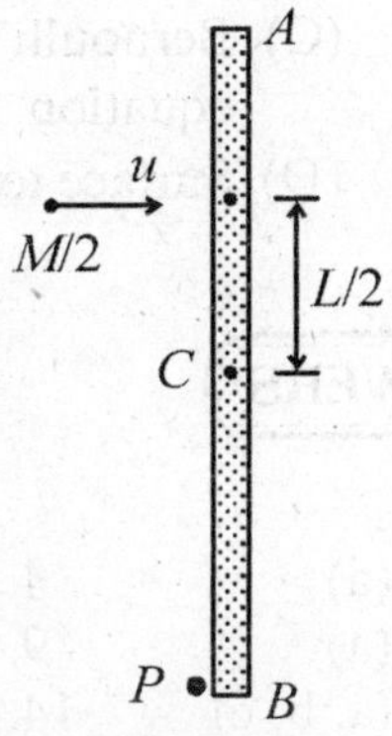

Fig. MTP-3.14

34. A spherical shell placed in an open atomosphere has initial temperature 2.2 T_0 (T_0 is room temperature) and shell is radiating a wavelength $\lambda_1\left(=\frac{hc}{108.8 \text{ eV}} \text{ m}\right)$ corresponding to maximum intensity of radiation. A point source of constant power operating inside the spherical shell is giving it a constant power $P_0 = 0.4\sigma S\varepsilon T_0^4$ (where σ, S and ε are Stefan's constant, surface area and emissivity of material of shell respectively). When the shell attains the steady state, the wavelength of radiation corresponding to maximum intensity, falls over hydrogen like atom, and gets ionised. Assuming the electron is ejected from ground state, find atomic number Z for the atom. (Assume temperature difference to be very small in comparision to T_0).

35. The path of photoelectrons emitted due to electromagnetic radiations incident on a sample of metal A, is found to have maximum bending radius of 0.1 m in a magnetic field of $\frac{\sqrt{2}}{3} \times 10^{-4}$ T. When the radiation is incident normally on double slit having slit saparation of 1 mm. It is observed that there are 10 fringes in a width of 0.31 cm on a screen placed at a distance of 1 m from the double slit. Find the work funtion of the material and the corresponding threshold wavelength. What should be the wavelength of incident light so that the bending radius is one half of what it was before?

36. A current of 1.0 A is established in a tightly wound solenoid of radius 2 cm having 1000 turns/m. Find the magnetic energy stored in each meter of solenoid.

SECTION V

In this section, there are four questions. Each question has two columns with 4 entries in each column. Entries of Column I are to be matched with Column II. One entry of Column I may have more than one matching in Column II

37.

Column I	Column II
(A) Isothermal process	(P) Constant entropy
(B) Adiabatic process	(Q) Constant temperature
(C) Carnot cycle	(R) Reversible process
(D) Latent heat	(S) $\gamma = c_P/c_v$

38.

Column I	Column II
(A) Gauss law	(P) Electric field
(B) Biot Savart's law	(Q) Magnetic induction
(C) A current carrying coil	(R) Magnetic dipole
(D) Ampere's law	(S) Displacement current.

39.

Column I	Column II
(A) Resolving power	(P) Wavelength λ
(B) Critical angle	(Q) Refractive index of medium μ
(C) Interference	(R) Velocity of light in vacuum
(D) Chromatic aberration	(S) Frequency.

40.

Column I	Column II		
(A) Upthrust	(P) Fluids	(C) Bernoulli's equation	(R) Conservation of mechanical energy
(B) Stoke's law	(Q) Molecular forces	(D) Surface tension	(S) Surface energy

ANSWERS

1. (d)	2. (c)	3. (a)	4. (a)	5. (c)
6. (c)	7. (c)	8. (a)	9. (b)	10. (d)
11. (d)	12. (a)	13. (a, b, c)	14. (a, b, c)	15. (a, b, c, d)
16. (a, b)	17. (b, c)	18. (a, b, c, d)	19. (a, b, c, d)	20. (b, c)
21. (a)	22. (d)	23. (b)	24. (d)	25. (c)
26. (c)	27. (d)	28. (b)	29. (b)	30. (a)
31. (c)	32. (a)			

SOLUTIONS

1. $X = A\sin\omega t$, when $v = a\cos\omega t = \omega\sqrt{A^2 - x^2}$ or $\frac{v^2}{\omega^2 A^2} + \frac{x^2}{A^2} = 1$. Hence plot of v vs x is an ellipse. (d)

2. The points C, D, E are equipotential points hence resistances connected in these points do not carry any current and contribute nothing to the effective resistance. Thus the remaining net work is combination of three resistances of each value $2R$ each and one resistance of value R. Hence

$$\frac{1}{R_{eq}} = \frac{1}{2R} + \frac{1}{2R} + \frac{1}{2R} + \frac{1}{R} = \frac{5}{2R} \quad \text{or} \quad R_{eq} = \frac{2R}{5} = \frac{4R}{10}.$$ (c)

3. Law of mixtures Heat lost = Heat gained

when L_1 and L_2 are mixed $\frac{S_1}{S_2} = \frac{3}{2}$.

when L_1 an L_3 are mixed $\frac{S_1}{S_3} = \frac{1}{3}$. This gives $\frac{S_2}{S_3} = \frac{2}{9}$

$$\therefore \quad \frac{20 - Q}{Q - 10} = \frac{9}{2} \quad \text{or} \quad Q = 11.8° \text{ C}.$$ (a)

4. Rays move along lateral surfaces of cones with fixed semi-vertical angles and meeting on the screen form concentric circles. (a)

5. Power consumed in an $AC = I_{rms} \cdot V_{rms} \cdot \cos\theta = \frac{5}{\sqrt{2}} \times \frac{200}{\sqrt{2}} \cos\pi/3$ or $P = 250$ W. (c)

6. Each resistor can carry a maximum current $\sqrt{2}$ A to dissipate maxmum power. Each resistor in parallel combination then must carry $\left(\frac{\sqrt{2}}{2}\right) = \frac{1}{\sqrt{2}}$ A each. Hence total power consummed by the combination is $10\left(\frac{1}{\sqrt{2}}\right)^2 + 10\left(\frac{1}{\sqrt{2}}\right)^2 + 20 = 30$ Watt. (c)

7. In vacuum the photon has energy $= \frac{hc}{\lambda} = 2.06$ eV. On refraction it is wavelength and velocity which change where as frequency remains same, i.e. energy of photon is 2.06 eV. (c)

8. Using Newtons formula we have $x_1 x_2 = f^2$ and $x_1 = P$ and $x_2 = \frac{f^2}{P}$ or $U = f + P$ and $V = \frac{f^2}{P} + f$ and $m = \frac{V}{U} = \frac{f}{P}$. (a)

9. Charge q induces negative charges q_1 and q_2 on the electrodes A and B respectively such that $|q_1| + |q_2| = |q|$. When q is nearer A $|q_1| > |q_2|$ as it moves towards B, q_2 grows at the expense of q_1, i.e. negative charge flows through galvanometer from A to B or the current flows from B to A. (b)

10. After explosion, all momentum is carried by the second half. Hence its horizontal speed is doubled. Time of fall is same, hence travels $2 \times 0.5 = 1$ km from explosion stage. Hence distance from launching pad $= 1 + 0.5 = 1.5$ km. (d)

11. Electric field is always along the line joining the point charge and the point. Hence electric field is along $\hat{i}$.

Magnetic field is given by right hand thumb rule. Which gives direction for B as $-\hat{k}$. (d)

12. At P left molecules displaced towards right and right molecules to the left causing high pressure. At R the situation is reverse causing low pressure. (a)

13. From the definition of significant figures correct options are a, b, c. (a, b, c)

14. Images are just resolved if angular seperatin is $1.22\lambda/D$. Linear separation is obtained by multiplying with focal length. (a, b, c)

15. Magnetic field due to a bar magnet on points on its axis are parallel to the dipole moment and also parallel to horizontal component of earth's magnetic field. Thus B_1 and B_3 are always parallel. (b)

Magnetic field at equitorial points due to magnet are in direction opposite to the magnetic dipole moment and in direction opposite to the horizontal component of earth's magnetic field. As such B_2 and B_4 also have the same direction. (c)

If point is very close to magnet, then magnetic field due to magnet may be greater than the horizontal component of earth's magnetic field. Thus net field at B_1 and B_2 have opposite direction and if the magnetic field due to magnet is less in magnitude than horizontal component of earth's magnetic field, then net field is parallel to horizontal component of earth's magnetic then B_1 and B_2 are in same direction. (a, d)

16. The intercepts on axes are inversely proportional to the frequencies. Then $f_{rv} : f_{nh}$ are Fig. (*i*) 2 : 3, Fig. (*ii*) 3 : 1, Fig. (*iii*) 1 : 2, Fig. (*iv*) 2 : 1. (a, b)

17. Closest approach corresponds to bodies in relative rest. Hence same velocity but not momentum. (b)

During collision part of energy is potential. Hence kinetic energy gained by m_2 is less than that lost by m_1. (c)

18. At $t = 0$ the capacitor acts as a short circuit. Hence equivalent resistance is 3 Ω and current through battery is $\frac{12}{3} = 4$ amp (a)

Then current through R_3 is 2 amp. (b)

In the steady state capacitor acts as open circuit as such current throug R_3 is zero. (d)

In steady state equivalent resistance is 4 Ω and current supplied by battery is $\frac{12}{4} = 3$ amp. (c)

19. Flux ϕ linked with coil when current i passes through it is $\phi = Li$ or $L = \phi/i =$ Weber/ampere (a)

Induced emf $e = \frac{Ldi}{dt}$ or $L = e\frac{dt}{di} =$ Volt - sec/Amp. (b)

When current passes through inductor, energy stored is $\frac{1}{2}LI^2$

or $L = \frac{2E}{I^2} =$ Joule/(Amp)^2 (c)

Inductive reactance $X_L = \omega L$ or $L = \frac{X_L}{\omega}$ Ω-sec. (d)

20. Q_1 does not experience a force as it is inside a conductor and electric field due to Q_2 inside cavity is zero.

Q_2 experience force due to field of Q_1 (b)

Yes there is an apparent violation of Newton's third law. (c)

21. With given direction of current in two wires and the quadrant in which the point P lies, the fields and are into the plane of paper $B + 2B = 3B$. (a)

22. Parallel currents attract. This rule will make wires II turn 90° anticlockwise to reach more stable position. (d)

23. Force between two parallel current carrying coils is $\frac{\mu_0 i_1 i_2}{2\pi d}$ per meter length of wire. Hence

$\frac{\mu_0 i_1 i_2}{2\pi d} = \lambda g$ or $\lambda = \frac{\mu_0 i_1 i_2}{2\pi dg} = 4 \times 10^{-8}$ kg/m. (b)

24. Wavelength $\lambda = \frac{v}{\nu} = \frac{300}{150} = 2$ m. Path distance between waves from S_1 and S_2 at 1 m, 2 m and 4 m from S_1 is respectively 3 m, 1 m and 3 m respectively i.e. odd multiple of $\lambda/2$. Hence these points correspond to destructive interference. (d)

25. If P is nodal point then $S_1P \sim S_1P =$ odd multiple of λ, $6 - 3 = 3 = 1.5\ \lambda/2$. (c)

26. If Q is antinodal point then $S_2Q \sim S_1Q =$ even multiple of λ, $4 - 2 = 2 = 2 \times \lambda/2$. 2 m is not admissible as per diagram. (c)

27. Work done during the process $QR = P(V_3 - V_2) = nR(T_3 - T_2)$

Change in internal energy $= nc_v(T_3 - T_2) = n\ 3/2R(T_3 - T_2)$

Total heat supplied $= 5/2 \times nR(T_3 - T_2) = 5/2 \times 3 \times 8.4(2500 - 700)$

or $\Delta Q = 1.134 \times 10^5$ J. (d)

28. Total work done = area $PQRS$ = area ΔOQR – area ΔOPS

or $W = \frac{1}{2}QR \times P_2 - \frac{1}{2}PS \times P_1 = \frac{1}{2}(V_3 - V_2)P_2 - \frac{1}{2}(V_4 - V_1)P_1$ using $PV = nRT$

$W = \frac{3}{2}R(T_3 - T_2 - T_4 + T_1) = \frac{3R}{2} \times 1100 = 1.37 \times 10^4$ J. (b)

29. Change in internal energy $= nc_v \Delta T$

$$= 3 \times \frac{3}{2} R(1100 - 2500) = -1400 \times \frac{9}{2} \times 8.4 = -5.3 \times 10^4 \text{ J}. \qquad \text{(b)}$$

30. When water level in tank is Y, the velocity of water coming out of hole is obtained using Bernoulli's equation

$$P_0 + g\rho Y = P_0 + \frac{1}{2}\rho v^2 \quad \text{or} \quad v = \sqrt{2gY}$$

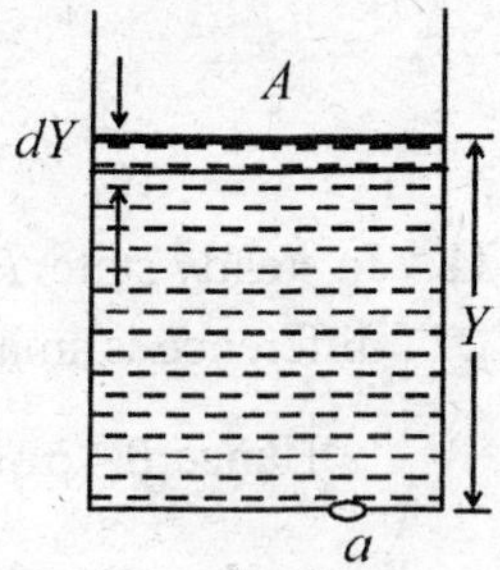

Fig. MTP-3.15

Then volume of the water coming out of hole per second $= av$.

If water level falls at the rate $-\frac{dY}{dt}$ then volume coming out of hole in unit time $= -\frac{A\,dY}{dt}$.

$$\text{or} \qquad -\frac{A\,dY}{dt} = a\sqrt{2gY} \quad \text{or} \quad dt = -\frac{A}{a}\frac{dY}{\sqrt{2gY}} = dt$$

$$\text{integration gives } t = -\frac{A}{a}\int_H^{H/2} \frac{dY}{\sqrt{2gY}} = -\frac{A}{a}\frac{2}{\sqrt{2g}}\sqrt{Y}\Big|_H^{H/2} = \frac{A}{a}\sqrt{\frac{2}{g}}\left[\sqrt{H} - \sqrt{\frac{H}{2}}\right]$$

$$\text{or} \qquad t = \frac{A}{a}\sqrt{\frac{H}{g}}\left(\sqrt{2} - 1\right). \qquad \text{(a)}$$

31. Speed of water coming out of hole when water level is Y is

$$v = \sqrt{2gY}. \text{ Hence for } Y = \frac{H}{2} \quad v = \sqrt{gH}. \qquad \text{(c)}$$

32. When water level is Y, the decrease in level Y in time dt is

$$dY = \frac{a}{A}\sqrt{2gY} \cdot dt \qquad \text{For } Y = h$$

$$dh = \frac{a}{A}\sqrt{2gh} \cdot dt. \qquad \text{(a)}$$

33. After collision of particle with rod let v_{cm} be the velocity of centre of mass of rod and ω its angular velocity about an axis passing throught the centre of mass. Since no external force acts. Conservation of linear momentum gives

$$\frac{M}{2}u = Mv_{\text{cm}} \quad \text{or} \quad v_{\text{cm}} = \frac{u}{2}$$

Conservation of angular momentum gives

$$\frac{M}{2}u\frac{1}{2} = I\omega = \frac{M(2L)^2}{12}\omega \quad \text{or} \quad \omega = \frac{3u}{4L}$$

For second collision at nail, if I_m is the impulse imparted by the nail, then

$$I_m = Mv_0 - Mv_{\text{cm}}$$

where v_0 is the velocity of the centre of mass after second collision and $\quad -I_m L = I\left(\frac{v_0}{L} - \omega\right)$

Substituting values of I_m, ω and v_{cm} and solving for v_0, we get

$v_0 = \frac{9u}{16}$. Since $u = 16$ m/s. Hence

$v_0 = 9$ m/s. **Ans.** 0009 m/s.

34. In steady state $P = \sigma S\varepsilon T^4 - \sigma S\varepsilon T_0^4$. Where T is the temperature of shell. Since temperature difference is small. Let $T = T_0 + \Delta T$

Hence $0.4\sigma S\varepsilon T_0^4 = \sigma S\varepsilon\left[(T_0 + \Delta T)^4 - T_0^4\right]$

or $\quad 0.4T_0^4 = T_0^3 4\Delta T$ or $\Delta T = 0.1$ and $T = 1.1T_0$

Using Wien's displacement law, we have

$\lambda_1 \times 2.2 \times T_0 = \lambda \times 1.1T_0$ or $\lambda = 2\lambda_1$

Hence photon corresponding to wavelength $\lambda = 2\lambda_1$ is 54.4 eV. For isonisation of hydrogen like atom of atomic number Z energy required is $\frac{13.6Z^2}{n^2}$ eV

Equating $\quad \frac{13.6Z^2}{n^2} = 54.4 \quad \frac{Z^2}{n^2} = 4$ or $\frac{Z}{n} = 2$

Hence minimum value of $Z = 2$ **Ans.** 0002.

35. Fringe width is $\beta = \frac{\lambda D}{d}$ or $\lambda = \frac{\beta d}{D}$

$$\beta = \frac{0.31}{10} \text{ cm} = 31 \times 10^{-5} \text{ m}, \; d = 1 \text{ mm} = 10^{-3} \text{ m} \quad \text{and} \quad D = 1 \text{ m}$$

or $$\lambda = \frac{31 \times 10^{-5} \times 10^{-3}}{1} = 31 \times 10^{-8} \text{ m} = 3100 \text{ Å}$$

Hence photon energy $E = \frac{hc}{\lambda} = \frac{124000 \text{ eV}}{3100 \text{ Å}} = 4$ eV

For an electron of energy E the radius of circular path in magnetic field is

$$r = \frac{\sqrt{2mE}}{2Bq} \quad \text{or} \quad E = \frac{r^2B^2q^2}{2m}$$

or $$E = \frac{(0.1)^2\left(\frac{\sqrt{2}}{3}\times 10^{-4}\right)^2 \times (1.6\times 10^{-19})^2}{2\times 9.1\times 10^{-31}} \text{ J}$$

$$= 10^{-2} \times \frac{2}{9} \times \frac{10^{-8} \times 1.6 \times 10^{-19}}{2 \times 9.1 \times 10^{-31}} \text{ eV} = 2 \text{ eV}$$

Hence work function of metal surface = 4 – 2 = 2 eV **Ans.** 0002 eV

Corresponding wavelength $\lambda = \frac{12400}{2} = 6200$ Å **Ans.** 6200 Å.

Radius (r) of the path of electron in magnetic field is proportional to $\sqrt{E}$ (where E is electron energy) or $r^2 \propto E$ and to make radius half, energy of electron should be 1/4. Hence energy of incident photon should be 2 + 0.5 = 2.5 eV. Hence wavelength of incident radiation should be

$$\lambda = \frac{12400}{2.5} = 4960 \text{ Å}$$ **Ans.** 4960 Å.

36. Inductance of a solenoid is givenly $L = \pi \mu_0 n^2 a^2 l$

when n is number of turns per unit length, a, the radius of solenoid and l, its length. For l = 1 m, $L = \pi \mu_0 n^2 a^2$.

Magnetic energy stored in an inductor $E = 1/2\, Li^2 = 1/2\, \pi a^2 \mu_0 n^2 i^2$

or $E = 2\pi^2 \times 10^{-7} \times (2 \times 10^{-2})^2\,(1000)^2\,(1)^2$

$= 7.90 \times 10^{-4}$ J = 790 mJ **Ans.** 0790 mJ.

37. (A) ⟶ Q – R
(B) ⟶ P – R – S
(C) ⟶ P – Q – R – S
(D) ⟶ Q

38. (A) ⟶ P – Q
(B) ⟶ Q
(C) ⟶ Q – R
(D) ⟶ Q – S

39. (A) ⟶ P – Q
(B) ⟶ P – Q – R
(C) ⟶ P – Q – S
(D) ⟶ P – Q

40. (A) ⟶ P
(B) ⟶ P
(C) ⟶ P – R
(D) ⟶ P – Q – S.

Model Test Paper for AIEEE Examination (I)

1. Densities of three liquids are D, $2D$ and $3D$. What will be the density of the resulting mixture if equal volumes of three liquids are mixed?

(a) $6D$ (b) $1.4D$
(c) $2D$ (d) $3D$.

2. Two masses m and M are suspended by means of two strings from two fixed walls as shown in Fig. 1. In equilibrium the value of $\tan\theta$ is

(a) $\left(\frac{2M}{m}-1\right)$

(b) $\left(\frac{2M}{m}+1\right)$

(c) $\left(1-\frac{2M}{m}\right)$

(d) $\left(\frac{2m}{M}-1\right)$.

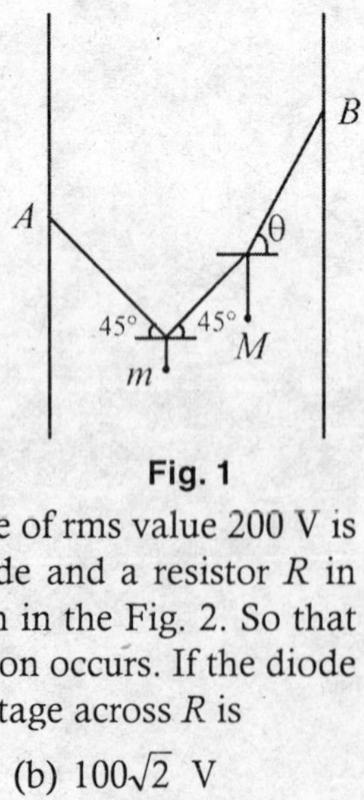

Fig. 1

3. A sinusoidal voltage of rms value 200 V is connected to a diode and a resistor R in the circuit as shown in the Fig. 2. So that half wave rectification occurs. If the diode is ideal; the rms voltage across R is

(a) $200\sqrt{2}$ V (b) $100\sqrt{2}$ V
(c) 200 V (d) 100 V.

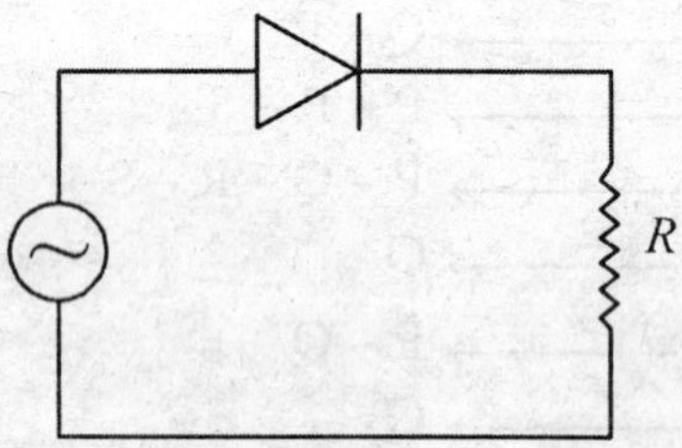

Fig. 2

4. A right prism is to be made by selecting a proper material and the angles as shown in Fig. 3. It is desired that a ray of light incident on AB emerges parallel to the incident direction a after two internal reflections. The minimum refractive index n for this is

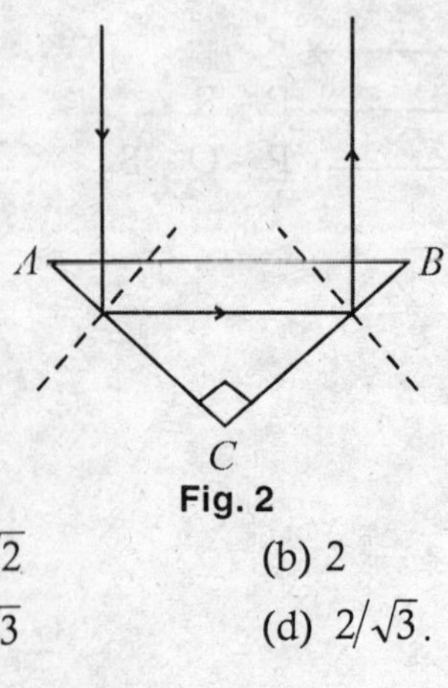

Fig. 2

(a) $\sqrt{2}$ (b) 2
(c) $\sqrt{3}$ (d) $2/\sqrt{3}$.

5. A one litre flask contains some mercury. It is found that at all temperatures the volume of air inside the flask remains the same. If coefficient of linear expansion of glass is $9 \times 10^{-6}/°C$, and the coefficient of volume expansion of mercury is 1.8×10^{-4} /°C, the volume of mercury inside the flask is

(a) 50 cc (b) 150 cc
(c) 450 cc (d) 18.75 cc.

6. A particle of mass m is projected from O with velocity $v_0 \cos\theta \hat{i} + v_0 \sin\theta \hat{j}$. At the same instant another particle of mass $2m$ is projected from point P with $v_0 \hat{k}$. The velocity of 2nd particle with respect to first particle after time $t = \frac{v_0 \sin\theta}{g}$ is

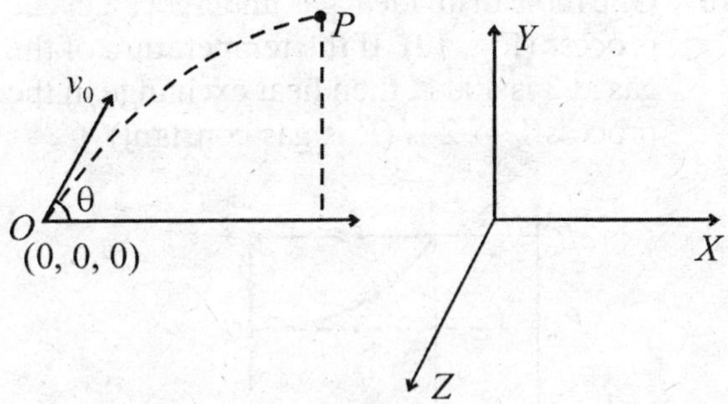

Fig. 4

(a) $v_0\hat{k} + gt\hat{j} + v_0 \cos\theta \hat{i}$
(b) $v_0\hat{k} - gt\hat{j} + v_0 \cos\theta \hat{i}$
(c) $v_0\hat{k} - v_0 \cos\theta \hat{i}$
(d) none of the above.

7. In the position shown in Fig. 5, the spring is at its natural length. The block of mass m is given a velocity v_0 towards the vertical support at $t = 0$. The coefficient of friction between block and the surface is given by $\mu = \alpha x$, where α is a positive constant and x is position of the block from its starting position. The block comes to rest for the first time at $x =$

(a) $v_0\sqrt{\frac{m}{k + \alpha mg}}$ (b) $v_0\sqrt{\frac{m}{k}}$

(c) $v_0\sqrt{\frac{m}{k - \alpha mg}}$ (d) $v_0\sqrt{\frac{m}{\alpha g}}$.

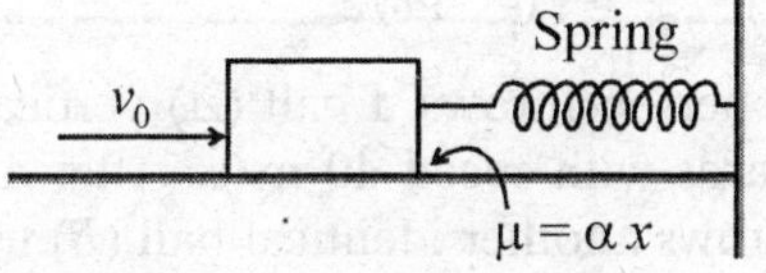

Fig. 5

8. A thin biconvex lens of focal length 30 cm is kept 90 cm away from a plane mirror kept parallel to the plane of the lens. A point object is placed on the optical axis of lens 60 cm away from lens focus of the (Fig. 6). Now if the point object is moved slowly towards the lens with a small speed, the image of the object in mirror

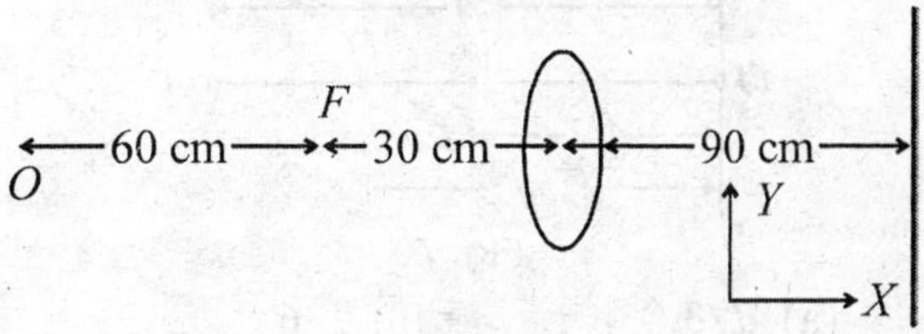

Fig. 6

(a) moves in the +ve X-direction
(b) moves in –ve X-direction
(c) does not move at all
(d) data insufficient.

9. In a region, uniform electric field $E = [10\hat{i} + 10\hat{j}]$ N/C exists. If potential of origin is 0 volts the potential of point (10 m, 10 m, 10 m) is

(a) 20 V (b) –200 V
(c) 10 V (d) +200 V.

10. A spherical drop of radius a and density ρ is falling in air with terminal velocity. The density of air is ρ_0 and its coefficient of viscosity η. The power developed by gravity is

(a) $\frac{\pi a^5 \rho(\rho - \rho_0)}{\eta} g^2$

(b) $\frac{\pi a^5}{2\eta} \rho(\rho - \rho_0) g^2$

(c) $\frac{27}{8\eta}a^5\rho(\rho-\rho_0)g^2$

(d) $\frac{8\pi}{27\eta}a^5\rho(\rho-\rho_0)g^2$.

11. A person throws a ball (A) vertically upwards with speed 40 m/s. After 4 sec he throws another identical ball (B) upwards with same speed. If the collision between balls is head-on perfectly elastic the speed of ball (A) just after collision is

(a) 0 (b) 10 m/s
(c) 20 m/s (d) 40 m/s.

12. A particle is performing S.H.M. with amplitude A and time period T. Time taken by a particle in going from point P to Q is

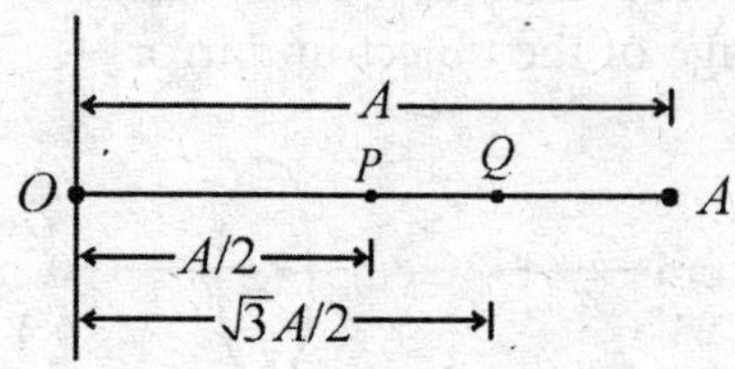

Fig. 7

(a) $T/3$ (b) $T/6$
(c) $T/12$ (d) $T/18$.

13. A plank of length h and width b perpendicular to plane of paper is hinged at one end. The upper end is attached with a spring of spring constant k and water is filled up to height h (Fig. 8). If plank is in equilibrium in vertical position, then potential energy stored in spring is (neglect effect of atmospheric pressure)

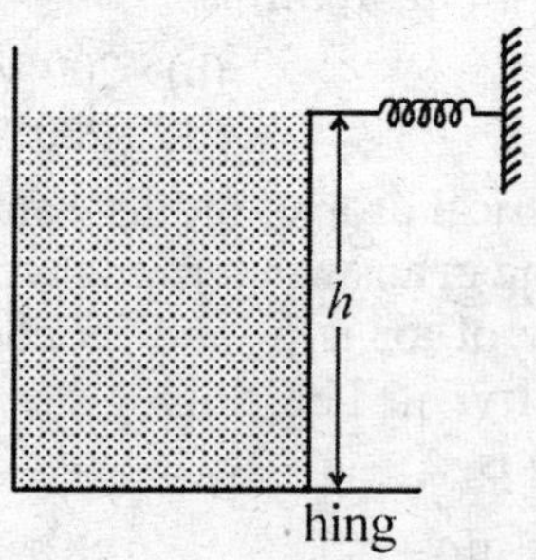

Fig. 8

(a) $\frac{b^2\rho^2g^2h^4}{18k}$ (b) $\frac{b^2\rho^2g^2h^4}{36k}$

(c) $\frac{b^2\rho^2g^2h^4}{72k}$ (d) $\frac{b^2\rho^2g^2h^4}{128k}$.

14. The potential across the resistor as measured by a voltmeter is (60 ± 1.5) V and current through the resistor as measured by an ammeter is (2 ± 0.01) A. The maximum % error in the measurement of R is

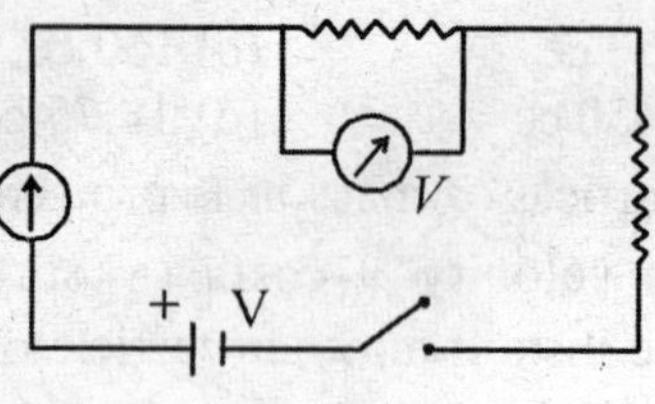

Fig. 9

(a) 3% (b) 1%
(c) 2.5% (d) 0.5%.

15. One mole of an ideal gas undergoes a cyclic process (Fig. 10). If the temperature of the gas at 3 is 400 K then heat exchange in the process 1 → 2 is (R is gas constant)

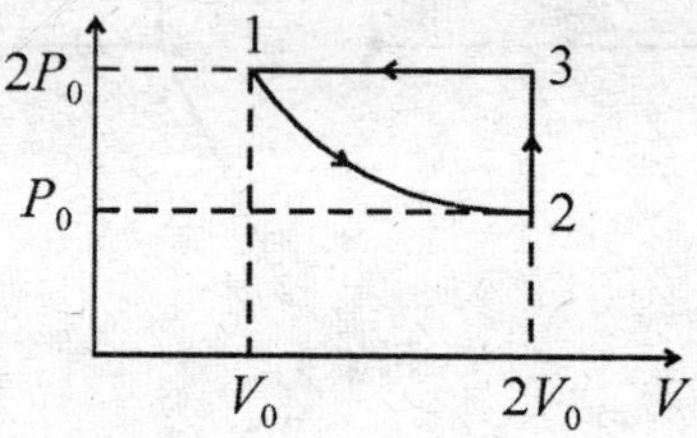

Fig. 10

(a) $200\log 2$ (b) $200R$
(c) $400R\log 2$ (d) $200R\log 2$.

16. 10 g of ice at a temperature of –10° C and 20 g of water at a temperature of 20° C are mixed. Choose the correct alternative regarding the composition and the final temperature of mixture

(a) temperature of mixture is 0° C and quantity of water is less than 20 g
(b) temperature mixture is 0° C and quantity of water is more than 20 g
(c) temperature mixture is slightly less than 0° C and the quantity of water is slightly less than 20 g

(d) temperature of mixture is slightly more than 0° C and the quantity of water is slightly more than 20 g.

17. Figure 11 shows a system of two concentric spheres of radii r_1 and r_2 and kept at temperatures T_1 and T_2 respectively. The radial rate of flow of heat in a substance between the concentric spheres is proportional to

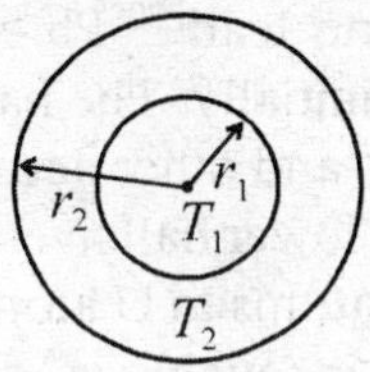

Fig. 11

(a) $(r_2 - r_1)$ (b) $\frac{r_1 r_2}{r_2 - r_1}$

(c) $\log\left(\frac{r_2}{r_1}\right)$ (d) $\frac{r_2 - r_1}{r_1 r_2}$.

18. Two white point dots are 1 mm apart on a black paper. They are viewed by pupil of diameter 3 mm. Approximately what is the maximum distance at which these dots can be resolved by the eye (wavelength of light used is 500 nm)?

(a) 3 m (b) 6 m
(c) 1 m (d) 5 m.

19. Two sources of equal emf are connected to an external resistance R. The internal resistances of the two sources are r_1 and r_2 $(r_2 > r_1)$. If the potential difference across the source having internal resistance r_2 is zero, then

(a) $R = \frac{r_1 r_2}{r_2 - r_1}$

(b) $R = \frac{r_1 r_2}{r_2 + r_1}$

(c) $R = r_2 - r_1$

(d) $R = r_2(r_1 + r_2)/r_2 - r_1$.

20. A magnetic needle is kept in a non-uniform magnetic field. It experiences

(a) a force but not a torque
(b) a force and a torque
(c) neither a force nor a torque
(d) a torque but not a force.

21. The phase difference between the alternating current and emf is $\pi/2$. Which of the following cannot be the constituent of the circuit?

(a) L alone (b) L, C
(c) R, L (d) C alone.

22. The wavelength of K_α-X-ray of an element having atomic number $Z = 11$ is λ. The wavelength of K_α-X-ray of another element of atomic number Z' is 4λ. Then Z' is

(a) 11 (b) 10
(c) 6 (d) 4.

23. A photon of 10.2 eV energy collides with a hydrogen atom in ground state inelastically. Within micro seconds one more photon of energy 15 eV collides with the same hydrogen atom. Particles ejected as a result of collision (they can the detected by suitable detector) are

(a) one photon of energy 10.2 eV and an electron of energy 1.4 eV
(b) 2 photons of energy 10.2 eV
(c) 2 photons of energy 3.4 eV
(d) 1 photon of energy 3.4 eV and one electron of 1.4 eV.

24. A simple pendulum has time period T_1. When the point of suspension moves vertically downwards according to equation $y = kt^2$. Where $k = 1$ m/s^2 and t is time, then time period of the pendulum becomes T_2. Then $(T_1/T_2)^2$ is

(a) $\frac{5}{6}$ (b) $\frac{5}{4}$

(c) $\frac{6}{5}$ (d) $\frac{4}{5}$.

25. A cyclic process $ABCD$ on P-V diagram is shown in the Fig. 12. The same process in a V-T diagram will look like

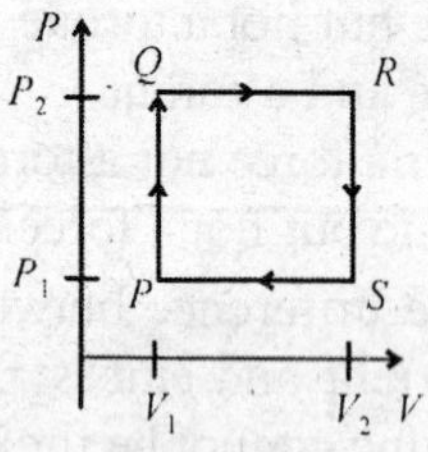

Fig. 12

(a) 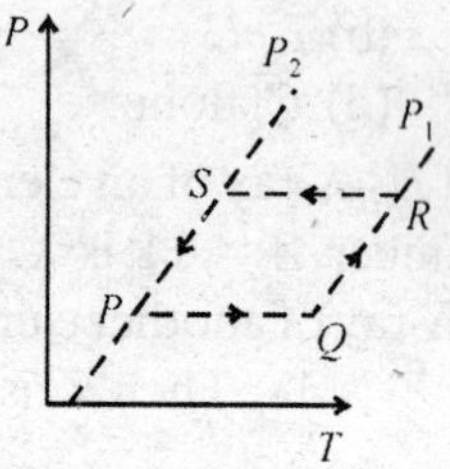 (b) 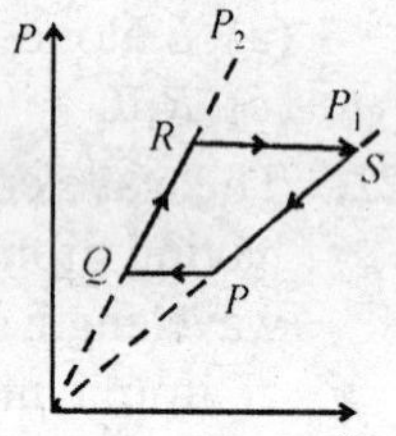

(c) 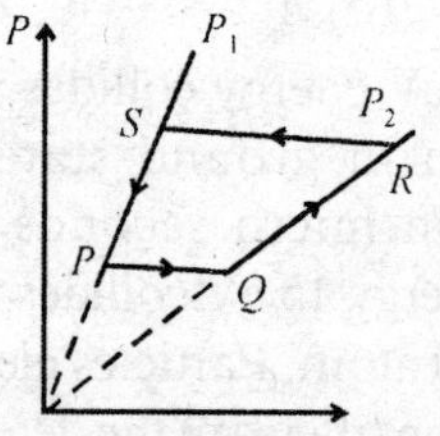 (d) 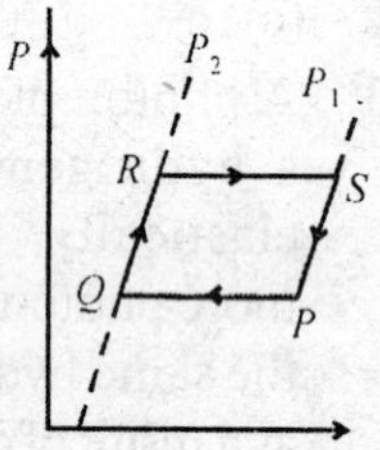

26. A small cube of side L is placed on the axis of spherical mirror. If m is the lateral magnification then the volume of the image will be

(a) mL^3 (b) m^2L^3
(c) m^3L^3 (d) m^4L^3.

27. Light from a point source S on the sphere emerges as a parallel beam (Fig. 13). The region outside the sphere is vacuum. The refractive index of the material of the sphere is

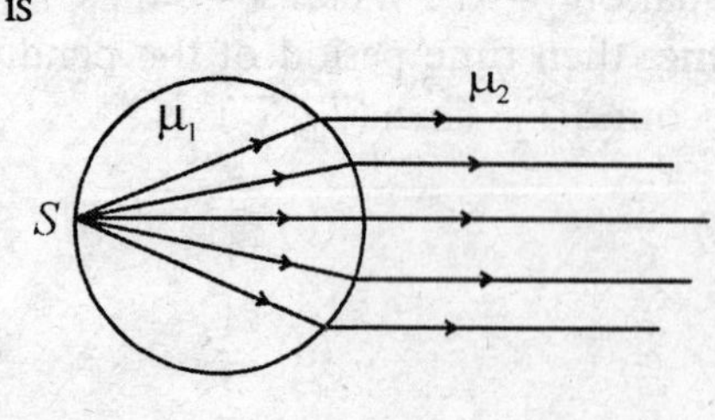

Fig. 13

(a) 3/2 (b) $\sqrt{2}$
(c) $2\sqrt{2}$ (d) 2.

28. The number of images that an observer can see oneself in a rectangular room, the ceiling and two adjacent walls of which are plane mirrors is

(a) 6 (b) 9
(c) 7 (d) 27.

29. A rock solidified some time is part contained no lead ^{208}Pb but did contain some ^{238}U initially. The half life of ^{238}U is 45×10^9 yrs and other decays in the series have much lower half lives. At present the ratio of Pb atoms to U atoms is 7. The age of the rock is roughly is

(a) $7 \times 4.5 \times 10^9$ yrs
(b) $\log(7 \times 4.5 \times 10^9)$ yrs
(c) 9×10^9 yrs
(d) 13.5×10^9 yrs.

30. Two parallel long vertical rails have an inductance connected at their ends as shown in Fig. 14. A copper rod is allowed to slide without friction along the rails. The magnetic induction B is as shown. Then

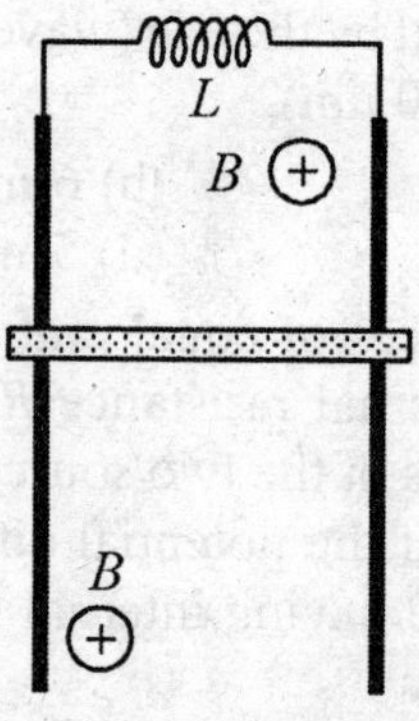

Fig. 14

(a) the motion of the rod is S.H.M.
(b) the rod accelerates and soon attains a terminal velocity
(c) the rod accelerates but acceleration is less than g
(d) the rod accelerates initially then retards and stops.

31. In a series resonance L-C-R circuit with quality factor $Q > 1$

(a) energy is not dissipated
(b) the voltage across the capacitor is more than supply voltage
(c) the voltage across resistor is equal to the voltage across L and C
(d) the voltage across coil is exactly equal to the voltage across capacitor.

32. In arithmetic $20.3 \times 3.9523 = 80.23169$. But as a result of experimental measurement the best way to express the product is

(a) 80.2317 (b) 80.23169
(c) 80.2 (d) 80.232.

33. Figure 15 gives the displacement (S) vs time (t) for two waves A and B in the same medium. Their intensity ratio I_A/I_B is

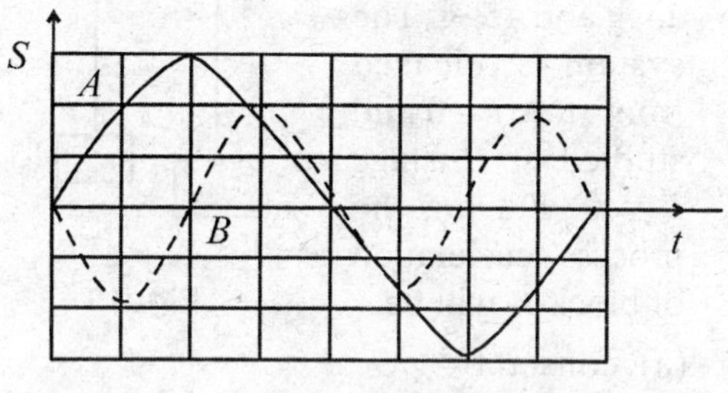

Fig. 15

(a) $\frac{9}{16}$ (b) $\frac{3}{4}$
(c) $\frac{9}{4}$ (d) $\frac{3}{2}$.

34. Figure 16 shows coherent sources S_1 and S_2 emitting at wavelength λ. The separation $S_1S_2 = 1.5\lambda$ and S_1 is ahead of S_2 in phase by $\pi/2$. The of maxima must occur in directions given by $\sin^{-1}$

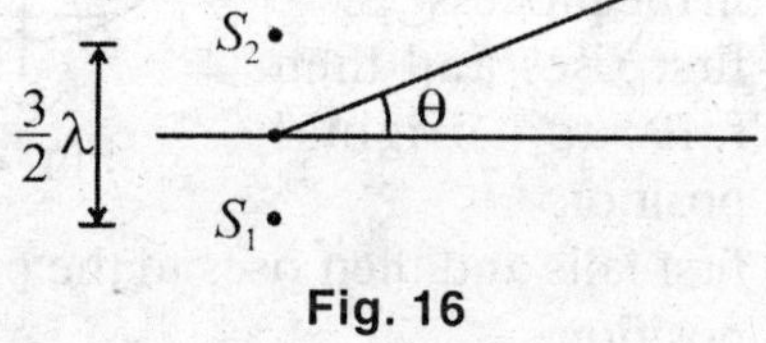

Fig. 16

(a) 0, 1/2 (b) 1, 1/2
(c) 1/6, –1/2 (d) 1/2, 1/6.

35. At identical temperatures the rms speed of hydrogen molecules is 4 times that for oxygen molecules. In a mixture of H_2 and O_2 in mass ratio $H_2 : O_2 = 1 : 8$, the rms speed of all molecules is n fold the rms speed of O_2 molecules where n is

(a) 3 (b) $\frac{4}{3}$
(c) $\sqrt{\frac{8}{3}}$ (d) $\sqrt{11}$.

36. The work required to turn an electric dipole by 60° from equilibrium in a uniform electric field is W. The torque required to hold it in that position will be

(a) $W/2$ (b) $W\sqrt{3}$
(c) $\frac{\sqrt{3}}{2}W$ (d) W.

37. A uniform ring of radius r and mass per unit length ρ is spun about its axis with an angular velocity ω. The increase in tension due to stretching of the ring is

(a) $\rho r^2\omega^2$ (b) $\pi\rho r^2\omega^2$
(c) $\frac{1}{2}\pi\rho r^2\omega^2$ (d) $\frac{1}{2}\rho r^2\omega^2$.

38. Figure 17 shows a potentiometer wire AB with two cells each having zero internal resistance. Instead of a galvanometer a voltmeter V with central zero is connected to read the potential difference. As the jockey point P is slided from end A to end B, the voltmeter reading will vary from

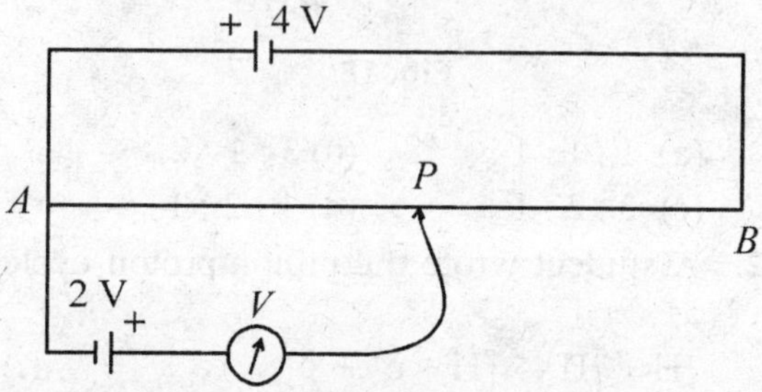

Fig. 17

(a) +2 V to 0 and back to +2 V
(b) +2 V to 6 V

(c) 0 to 4 V
(d) +2 to –2 V.

39. Hydrogen (H), deuterium (D), singly ionised helium (He^+) and doubly ionised lithium (Li^{++}) all have one electron around the nucleus. For a given transition ($n = 2$ to $n = 1$) if wavelengths in four cases be λ_1, λ_2, λ_3, λ_4 respectively, then approximate relation is

(a) $\lambda_1 = \lambda_2 = 2\lambda_3 = 3\lambda_4$
(b) $\lambda_1 = \lambda_2 = 4\lambda_3 = 9\lambda_4$
(c) $\lambda_1 = \frac{1}{2}\lambda_2 = \frac{1}{2}\lambda_3 = \frac{1}{4}\lambda_4$
(d) $\lambda_1 = 2\lambda_2 = 2\sqrt{2}\,\lambda_3 = 3\sqrt{2}\,\lambda_4$.

40. A quantity P is measured using formula $P = Kx^2yz^{-3}$ where K is constant. The quantities x, y and z are measured with percentage accuracy 2, 1, 3 respectively. The percentage accuracy of P is

(a) 14% (b) 8%
(c) 6% (d) 4%.

41. Three resistors, shown in Fig. 18 satisfy the relations $R_2 = 2R_3$ and $R_2^{-1} + R_3^{-1} = R_1^{-1}$. If P_i stands for power dissipated in the ith resistor then $P_1 : P_2 : P_3$ will have the ratio

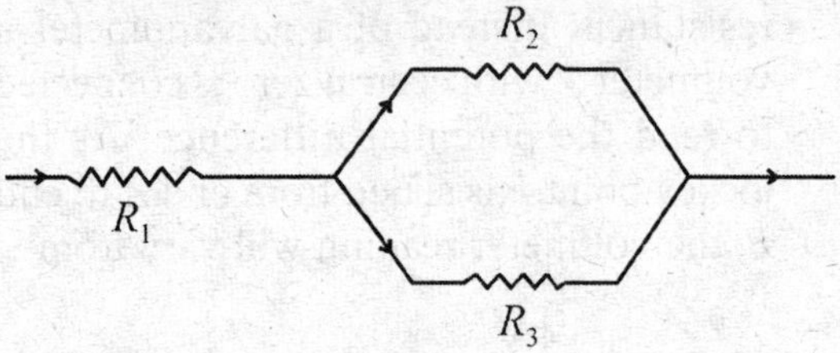

Fig. 18

(a) 1 : 1 : 1 (b) 3 : 1 : 2
(c) 3 : 2 : 1 (d) 1 : 2 : 1.

42. A student wrote the proton-proton cycle as

$^1_1H + ^1_1H \rightarrow ^2_1H + e^+ + \nu$ ...(*i*)
$^1_1H + ^2_1H \rightarrow ^3_2He + e^+ + \nu$ ...(*ii*)
$^3_2He + ^3_2He \rightarrow ^4_2He + ^1_1H + ^1_1H$ ...(*iii*)

out of these equations without error are

(a) (*i*) and (*iii*) only
(b) (*ii*) only
(c) (*i*) and (*ii*) only
(d) (*iii*) only.

43. A radioactive material has decay constant λ, half life T and mean life τ. The correct relation is

(a) $T = \frac{\lambda}{0.693}$ (b) $\tau = 0.693\,T$
(c) $T = \tau \log 2$ (d) $\tau = 2.3\,T$.

44. In the system shown in Fig. 19, the pulley is smooth and massless; and the string has a total mass 5 g. The two suspended blocks have masses 25 g and 15 g. The system is released from state $l = 0$ and studied up to stage $l' = 0$. During the process acceleration of block A will be

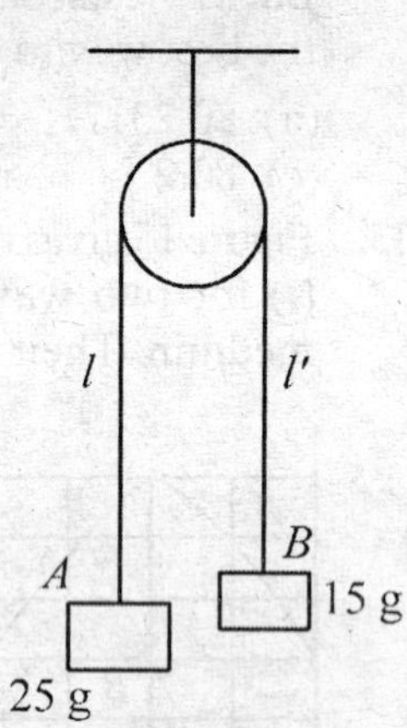

Fig. 19

(a) constant = $g/9$
(b) constant = $g/4$
(c) increases by factor 3
(d) increases by factor 2.

45. A hollow sphere is held suspended (Fig. 20). Sand is poured into it in stages. The centre of gravity of the sphere with the sand

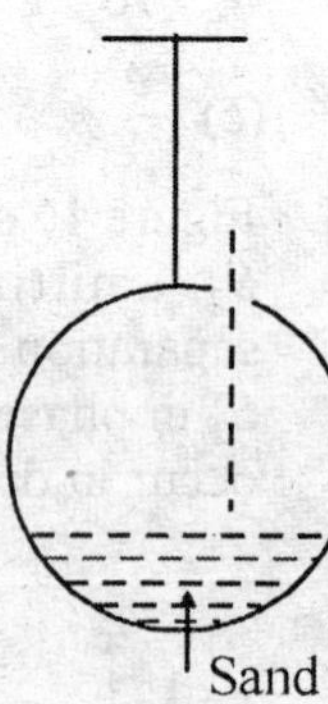

Fig. 20

(a) rises continuously
(b) remains unchanged in the process
(c) first rises and then fall to original position
(d) first falls and then rises to the original position.

46. For a perfect diatomic gas, four indicator diagrams are shown here nearly to scale.

One of them corresponds to isothermal change and another to adiabatic change, these are respectively

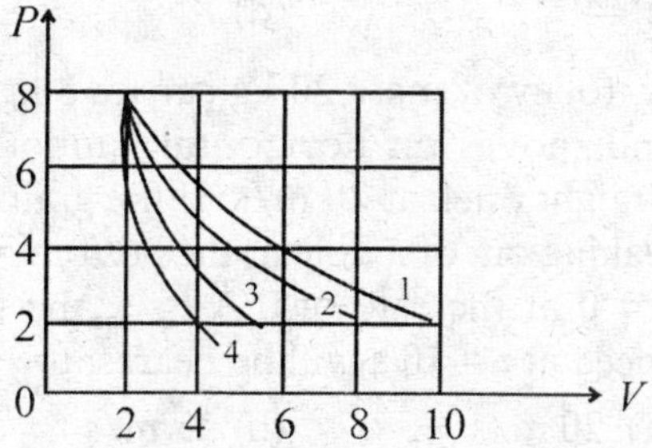

Fig. 21

(a) 1 and 4 (b) 2 and 1
(c) 3 and 2 (d) 2 and 3.

47. Figure 22 shows P and Q as two equally intense coherent sources emitting radiations of wavelength 20 m. The separation PQ is 5.0 m and phase of P is ahead of the phase of Q by 90°. A, B, and C are three distant points of observation, equidistant from the midpoint of PQ. The intensity of radiation at A, B, C will bear the ratio

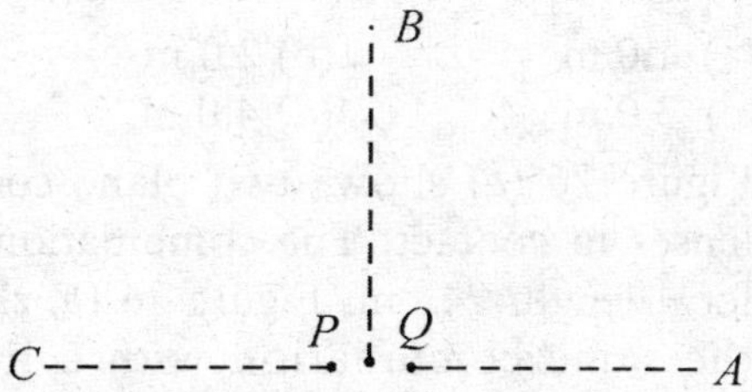

Fig. 22

(a) 0 : 1 : 4 (b) 4 : 1 : 0
(c) 0 : 1 : 2 (d) 2 : 1 : 0.

48. From a steel wire of density ρ is suspended a brass block of density ρ_1. The extension of steel wire comes to e. If the brass block is fully immersed in a liquid of density ρ_0 the extension becomes e'. The ratio e/e' will be

(a) $\frac{\rho_1}{\rho_1 - \rho_0}$ (b) $\frac{\rho_1 - \rho_0}{\rho_1}$

(c) $\frac{\rho_1 - \rho}{\rho_0 - \rho}$ (d) $\frac{\rho_0}{\rho_1 - \rho_0}$.

49. A block of wood is sliding on the floor of a train. The block comes to rest after sliding 5 m when the train is stationary. When the train is in motion, the block comes to rest after sliding 4 m, We can conclude that the train is

(a) moving at a uniform speed in the direction of the sliding block
(b) moving at a uniform speed opposite to the direction of the sliding block
(c) accelerating in the direction of the sliding block
(d) accelerating in the direction opposite to the direction of the sliding block.

50. A 25 W lamp is enclosed in a metal case. When room temperature is 25° C, the metal case is at temperature of 45° C. If the 25 W lamp is replaced with a 50 W lamp, the temperature of metal case is

(a) 45° C (d) 65° C
(c) 70° C (d) 65° C.

51. A ray of light is incident at an angle of 60° on a face of prism of angle 30°. The emergent ray makes an angles of 30° with the incident ray. The emergent ray will

(a) graze the surface from which it emerges
(b) be normal to the surface from which it emerges
(c) make an angle of 60° with the surface from which it emerges
(d) emerges from the same surface.

52. The energy of an electron in excited hydrogen atom is –3.4 eV. Then according to Bohr's theory, the angular momentum of this electron in J-s is

(a) 2.11×10^{-34} (b) 3×10^{-34}
(c) 2×10^{-34} (d) 0.5×10^{-34}.

53. A steel tape is calibrated at 30° C. A reading of 50 m is observed for a certain length when the temperature was 18° C. α for steel is 1.2×10^{-5}/°C. The correct length is

(a) 49.9928 m (b) 50.0118 m
(c) 51.18 m (d) 48.82 m.

54. A 160 W, lamp is connected to 60 V supply. The number of electrons passing through the lamp in one minute is

(a) 10^{21} (b) 10^{19}
(c) 1.6×10^{19} (d) 1.4×10^{20}.

55. Equivalent resistance between *AB* in Fig. 23 is

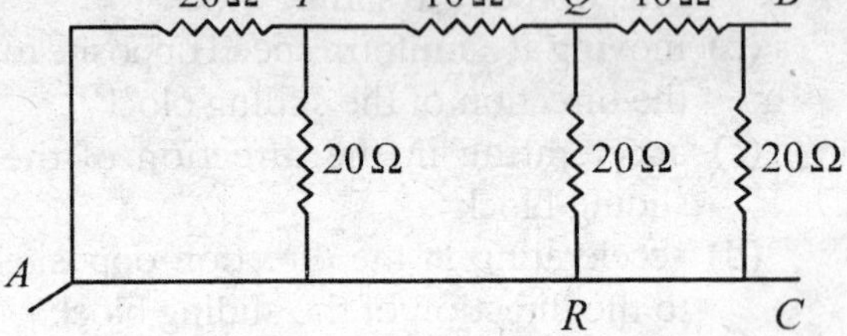

Fig. 23

(a) 10 Ω (b) 20 Ω
(c) 40 Ω (d) 80 Ω.

56. In the Fig. 24, all the capacitors are identical, capable of withstanding 100 V maximum. The maximum voltage which can be applied across terminals *A* and *B* is

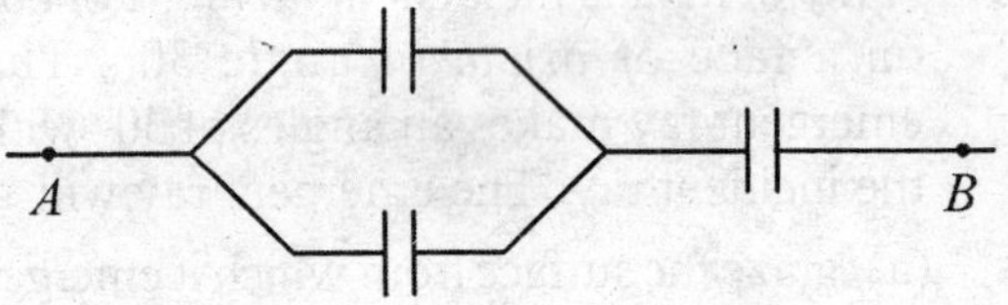

Fig. 24

(a) 100 V (b) 150 V
(c) 200 V (d) 300 V.

57. A collimated beam of light has intensity 5 W/m^2. The wavelength 5×10^{-7} m. The amplitude of electric oscillating field is close to

(a) 0.7 V/m (b) 40 V/m
(c) 60 V/m (d) 10^6 V/m.

58. The pressure across a capillary for a given flow rate of a liquid is proportional to L/r^4, where L is length and r is radius of capillary. If two capillaries in series have $r_1 = 2r_1$ and $L_2 = 4L_1$ and pressure across the capillaries of radii r_1 and r_2 are P_1 and P_2 respectively, then ratio P_1/P_2 will be

(a) 1/4 (b) 1
(c) 4 (d) 8.

59. A trolley of mass 20 kg carries 5 kg grain and moves on horizontal, smooth and straight track at 20 m/s. If the grain starts leaking out of a hole at the bottom at time $t = 0$ at the rate of 0.2 kg/s, the trolley speed at $t = 10$ s will be nearest to

(a) 20 m/s (b) 25 m/s
(c) 15 m/s (d) 13.3 m/s.

60. Figure 25 shows two sources emitting sound at wavelength 0.8 m. An observer walking along a transverse line *AB* at 50 m distance will hear successive maximum sound at spacings

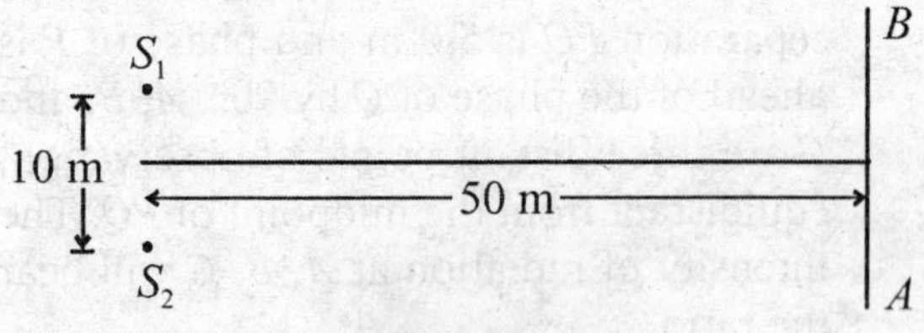

Fig. 25

(a) 1.0 m (b) 2.0 m
(c) 3.0 m (d) 4.0 m.

61. Figure 26 (*a*) shows two plano-convex lenses in contact. The combination has focal length 24 cm. Figure 26 (*b*) shows the same combination with a liquid introduced between them. If refractive index of glass of lenses is 1.5 and that of liquid 1.60. The focal length of system *B* is

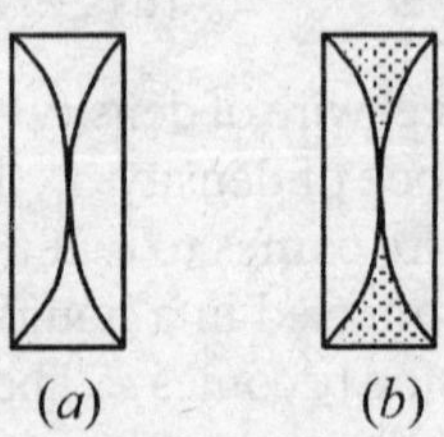

Fig. 26

(a) −120 cm (b) 120 cm
(c) −24 cm (d) 24 cm.

62. In a radioactive decay, if ΔN is the number of nuclei decaying in interval Δt and N is the number of mean active nuclei during that interval, the graph between the quantity $\gamma = \frac{1}{N}\frac{\Delta N}{\Delta t}$ versus time t will have shape

(a) 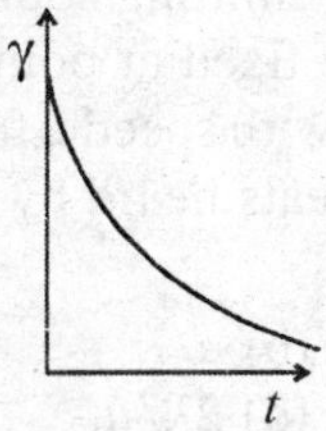

(b) 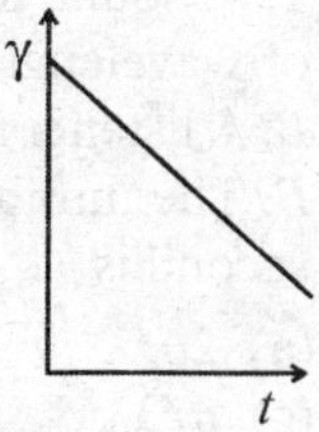

(c) 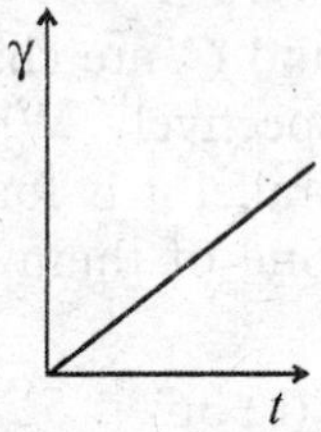

(d) 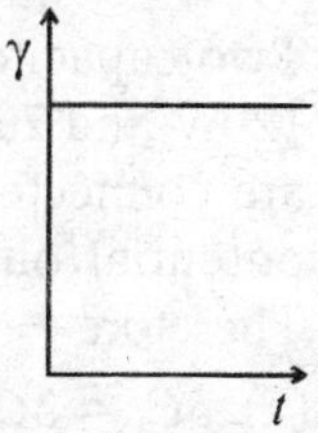

63. A nucleus of mass 24 amu emits a gamma photon of energy 8 MeV. If the emission is assumed to occur when the nucleus is free and at rest in lab frame, the daughter nucleus will have kinetic energy nearest to

(a) 140 keV (b) 14 keV
(c) 1.4 keV (d) 140 eV.

64. Consider the hypothetical situation where negative masses exist in the universe and Newton's laws of gravitation and motion were to hold true. Now consider a positive mass and negative mass of same magnitude separated by a distance

(a) the two masses will attract each other and collide
(b) the two masses will repel and move away from each other
(c) the two masses will initially be attracted and then will slow down and stop at a certain distance between them
(d) two masses will move together in the same direction never colliding.

65. A particle is subjected simultaneously to two perpendicular harmonic motions $X = a \sin \omega t$ and $Y = b \cos \omega t$, $a < b$.

The maximum distance of the particle form the origin will be

(a) $a + b$ (b) a
(c) $\sqrt{a^2 + b^2}$ (d) b.

66. Which of the following temperatures is the highest?

(a) 100 K (b) –13° F
(c) –20° C (d) –23° C.

67. One mole of an ideal gas at temperature T was cooled isochorically till the gas pressure fell from P to P/n. Then, by an isobaric process, the gas was restored to the initial temperature. The net amount of heat absorbed by the gas in the process is

(a) nRT (b) $\frac{RT}{n}$
(c) $RT\left(1 - \frac{1}{n}\right)$ (d) $RT(n - 1)$.

68. A gas has volume 3 litre when the temperature reads 300 K. If the thermometer stands at 400 K the volume of the gas will

(a) 4 litre (b) $\frac{9}{4}$ litre
(c) 1 litre (d) any value.

69. In the circuit of Fig. 27 the steady state voltages V_1 and V_2 respectively are

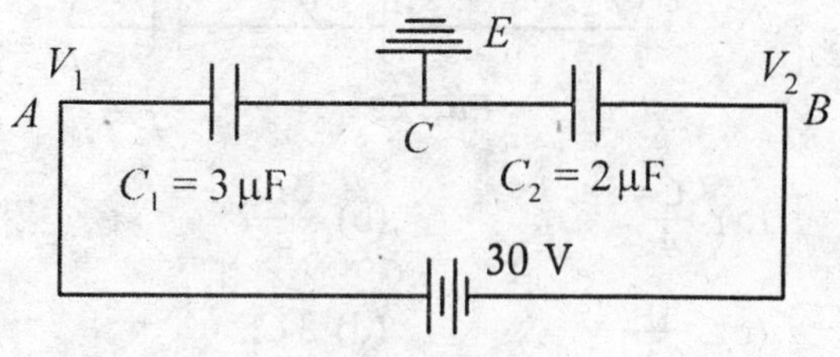

Fig. 27

(a) +12 V, –18 V
(b) –18 V, +12 V
(c) +15 V, –15 V
(d) +30 V, –30 V.

70. Which of the four resistances (Fig. 28) generates the greatest amount of heat when a current flows from A to B?

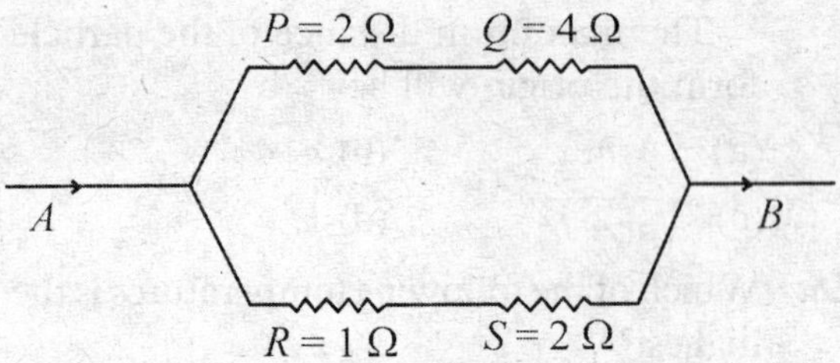

Fig. 28

(a) P (b) Q
(c) R (d) S.

71. A thin rod is bent in the shape of small circle of radius r. If the charge per unit length of the rod is σ and if the circle is rotated about its axis at a rate of 'n' rotations per second, the magnetic induction at a point on the axis at a large distance y from the centre is

(a) $\mu_0 \pi r^3 n\sigma/y^3$ (b) $2\mu_0 \pi r^3 n\sigma/y^3$
(c) $(\mu_0/4\pi) r^3 n\sigma/y^3$ (d) $(\mu_0/2\pi) r^3 n\sigma/y^3$.

72. A network of six identical capacitors each of value C is made as shown in Fig. 29. Equivalent capacitance between points A and B is

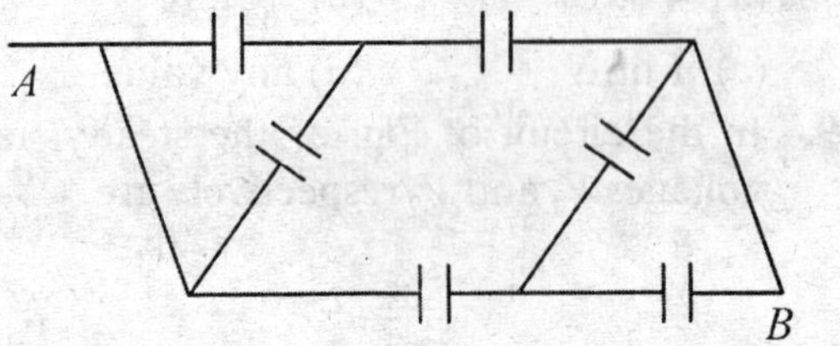

Fig. 29

(a) $\frac{C}{4}$ (b) $\frac{3C}{4}$
(c) $\frac{4C}{3}$ (d) $3\,C$.

73. The radioactive emission which produce an isotope of the original nucleus are

(a) one alpha and four beta
(b) one alpha and two beta
(c) one alpha and one beta
(d) two alpha and one beta.

74. A pipe closed at one end and open at the other resonates with sound of frequency 135 Hz and also 165 Hz but not at any other frequency intermediate between these two. Then the frequency of the fundamental note of the pipe is

(a) 30 Hz (b) 15 Hz
(c) 60 Hz (d) 7.5 Hz.

75. Two sound sources emitting sound, each of wavelength λ are fixed at point A and B. A listener moves with speed u from A to B. The number of beats heard by him per second is

(a) $2u/\lambda$ (b) u/λ
(c) $u/3\lambda$ (d) $2\lambda/u$.

76. Two capacitors C_1 and C_2 are charged to 120 V and 200 V respectively. When they are connected in parallel it is found that potential on each one of them is zero. Therefore

(a) $5C_1 = 3C_2$ (b) $3C_1 = 5C_2$
(c) $3C_1 + 5C_2 = 0$ (d) $9C_1 = 4C_2$.

77. A chain of 5 links each of mass 0.1 kg is lifted vertically with a constant acceleration 1.2 m/s^2. The force of interaction between the top link and the one immediately lower is

(a) 5.5 N (b) 4.4 N
(c) 3.04 N (d) 7.6 N.

78. According to Bohr's theory the energy of an electron in hydrogen atom in states corresponding to $n = 1, 2, 3, 4$ is respectively -13.6 eV, -3.4 eV, -1.51 eV and -0.85 eV. In order to obtain emission of H_β line of Balmer series, the energy that has to be given to a normal hydrogen atom is

(a) 12.75 eV (b) 14.45 eV
(c) 2.55 eV (d) 4.25 eV.

79. A plano-convex lens has curved surface having radius of curvature 60 cm and its material has refractive index 1.5. When the plane surface is silvered, the system will work as a concave mirror of focal length

(a) 20 cm (b) 30 cm
(c) 45 cm (d) 60 cm.

80. Tension in stretched string fixed at both ends is changed by 20%. The fundamental frequency is found to change by 15 Hz. The incorrect statement is
(a) Original frequency is 150 Hz
(b) Velocity of propagation of the transverse wave along the string changes by 50%
(c) Velocity of propagation of transverse wave along the string changes by 10%
(d) Fundamental wavelength of the string does not change.

81. Four nuclei P, Q, R and S participate in process involving α and β decay as shown in Fig. 30. The mass number and atomic number of P are 210 and 83 respectively. Which of the following is correct?

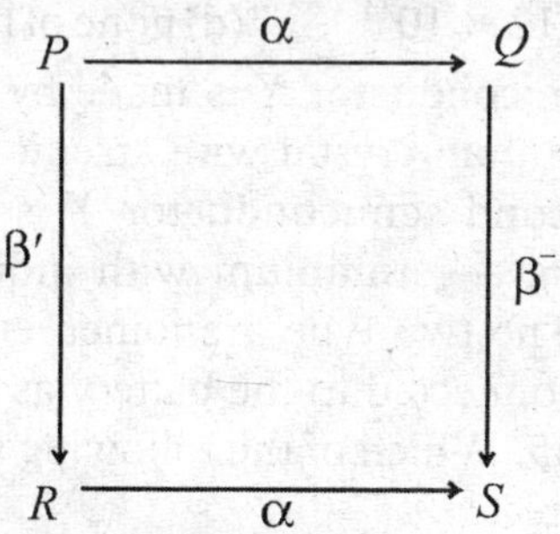

Fig. 30

(a) Mass no. of R is 210
(b) Mass no. of Q is 208
(c) Mass no. of S is 206 and its atomic number is 82
(d) The energy released in resultant process $P \rightarrow S$ via Q and via R is the same.

82. An object of height 0.5 cm is placed laterally in the medium of refractive index $n_1 = 1.5$ at a distance 100 cm away from the pole. The radius of curvature of the surface is 20 cm. The height of the image formed after refraction is

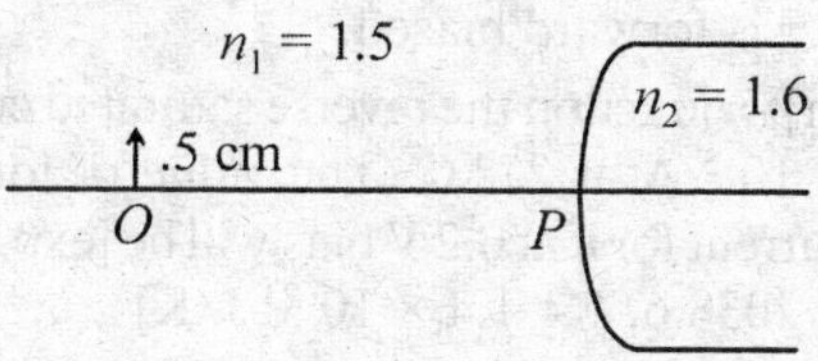

Fig. 31

(a) 0.75 cm (b) 1 cm
(c) 0.25 cm (d) 1.5 cm.

83. A small electric dipole is placed at origin with its dipole moment directed along positive X-axis. The direction of electric field at a point $(2, 2\sqrt{2}, 0)$ is
(a) along Z-axis
(b) along Y-axis
(c) along negative Y-axis
(d) along negative Z-axis.

84. A point charge q is placed inside a conducting spherical shell of inner radius $2R$ and outer radius $3R$ at a distance R from the centre of the shell. The electric potential at the centre of shell will be $\frac{1}{4\pi\varepsilon_0}$ times

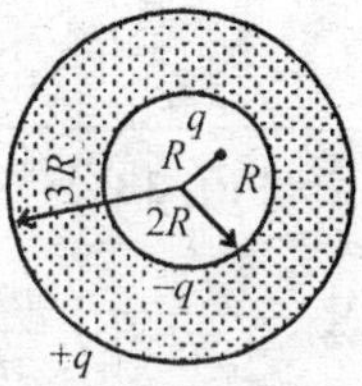

Fig. 32

(a) $\frac{q}{2R}$ (b) $\frac{4q}{3R}$
(c) $\frac{5q}{6R}$ (d) $\frac{2q}{3R}$.

85. A capacitor of capacitance 10 μF is charged to a potential of 50 V with a battery. The battery is now disconnected and an additional charge 200 μC is given to the positive plate of capacitor. Potential difference across the capacitor will be
(a) 50 V (b) 80 V
(c) 100 V (d) 60 V.

86. Two very long straight parallel wires carry steady currents i and $2i$ in opposite direction. The distance between the wires is d. At a certain instant of time a point charge q is at a point equidistant from the two wires in the plane of the wires. Its instantaneous velocity $\boldsymbol{v}$ is perpendicular to this plane. The magnitude of the force due to magnetic field acting on the charge at this instant is

(a) $\frac{\mu_0 iqv}{2\pi d}$ (b) $\frac{\mu_0 iqv}{\pi d}$

(c) $\frac{3\mu_0 iqv}{2\pi d}$ (d) zero.

87. In Fig. 33, a sphere of mass m and radius R is shown with a coil of single turn wound on it. The plane of the coil is parallel to the plane. Current through the coil is i. For the sphere to be in equilibrium, the value of B is

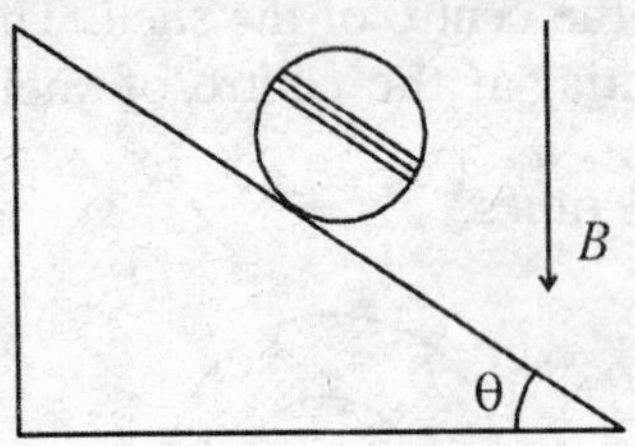

Fig. 33

(a) $\frac{mg \cos\theta}{\pi iR}$ (b) $\frac{mg}{\pi iR}$

(c) $\frac{mg \tan\theta}{\pi iR}$ (d) $\frac{mg \sin\theta}{\pi iR}$.

88. A conducting circular loop of radius a and resistance R is kept on a horizontal plane. A vertical varying magnetic field $B = 2t$ is switched on at time $t = 0$. Then

(a) power generated in the coil at any time t is constant

(b) flow of charge per unit time from any section of the coil is constant

(c) total charge passed through any section between time $t = 0$ to $t = 2$ s is $\frac{4\pi a^2}{R}$

(d) all the above.

89. Radius of a circular conducting loop (Fig. 34) R. Magnetic field is decreasing at a constant rate α. Resistance per unit length of loop is ρ. The current in the wire AB is (AB is diameter of the loop)

(a) $\frac{R\alpha}{2\rho}$ from A to B (b) $\frac{R\alpha}{2\rho}$ from B to A

(c) $\frac{2R\alpha}{\rho}$ from A to B (d) zero.

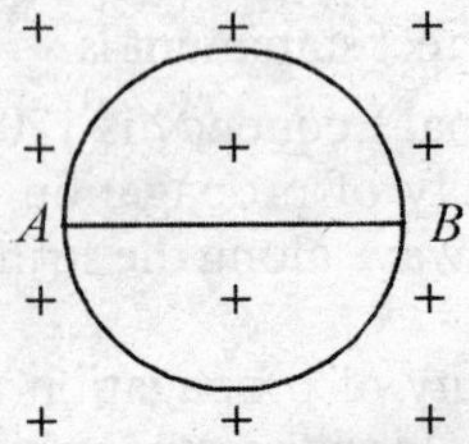

Fig. 34

90. The carrier concentration in pure silicon at room temperature is 1.6×10^{16} m^{-3}. If the electron and hole mobilities are 0.15 m^2 V^{-1}s^{-1} and 0.05 m^{2}V^{-1}s^{-1} respectively, then conductivity of silicon is (in Simen/meter)

(a) 1.28×10^{-4} (b) 2.84×10^{-4}

(c) 5.12×10^{-4} (d) none of the above.

91. A semiconductor X is made by doping a germanium crystal with arsenic ($Z = 33$). A second semiconductor Y is made by doping a germanium with indium ($Z = 49$). The two type are joined end to end and connected to the battery as shown in Fig. 35. Which of the following statement is correct?

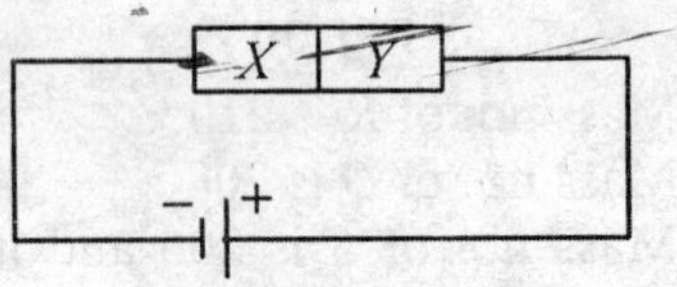

Fig. 35

(a) X is p-type, Y is n-type and the junction is forward biased

(b) X is p-type Y is n-type and the junction is reverse biased

(c) X is n-type, Y is p-type and the junction is reverse biased

(d) X is n-type, Y is p-type and the junction is forward biased.

92. In p-n junction the reverse saturated current is 10^{-5} A at 27° C. The value of forward current for it at 0.2 V bias will be [exp (7.62) = 2038.6, $k = 1.4 \times 10^{-23}$ J/K]

(a) 2037.6×10^{-3} A

(b) 203.76×10^{-3} A
(c) 20.376×10^{-3} A
(d) 2.0376×10^{-3} A.

93. The truth table for function X of A and B represented by Fig. 36

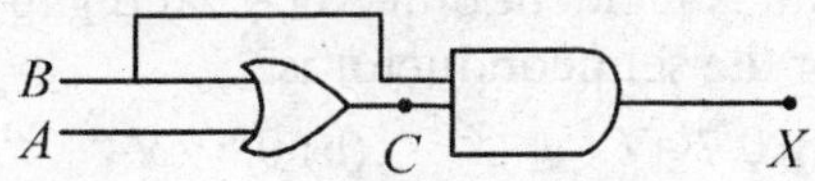

Fig. 36

(a)

A	B	X
0	0	0
0	1	1
1	0	0
1	1	1

(b)

A	B	X
0	0	0
0	1	0
1	0	1
1	1	1

(c)

A	B	X
0	0	0
0	1	0
1	0	0
1	1	1

(d)

A	B	X
0	0	0
0	1	1
1	0	1
1	1	1

94. Let $X = A\overline{BC} + B\overline{CA} + C\overline{AB}$, then value of X for $A = 1$, $B = 0$ and $C = 1$ and $A = B = C = 1$ respectively is

(a) 1, 1 (b) 1, 0
(c) 0, 1 (d) 0, 0.

95. The equivalent resistance of network shown in Fig. 37 between point A and B is

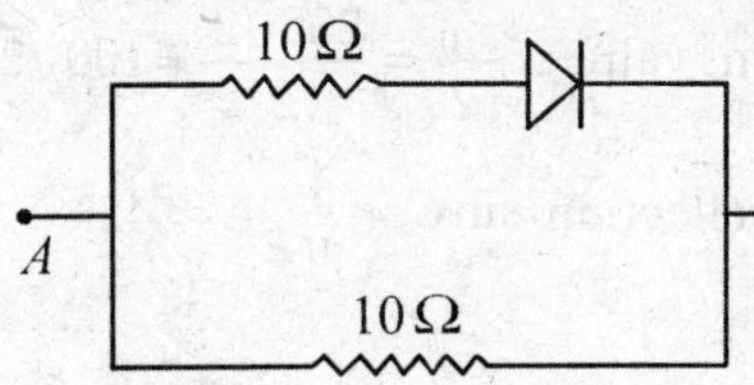

Fig. 37

(a) 10 Ω if $V_A > V_B$
(b) 5 Ω if $V_A > V_B$
(c) zero if $V_A > V_B$
(d) α if $V_A < V_B$.

96. For common emitter transistor amplifier, the audio signal voltage across the collector resistance of 2 kΩ is 2 V. If current amplification factor of transistor is 100, then input signal voltage across base resistance 1 kΩ is

(a) 0.01 V (b) 0.1 V
(c) 1 V (d) none of the above.

97. Which of the following diodes is reverse biased?

(a)

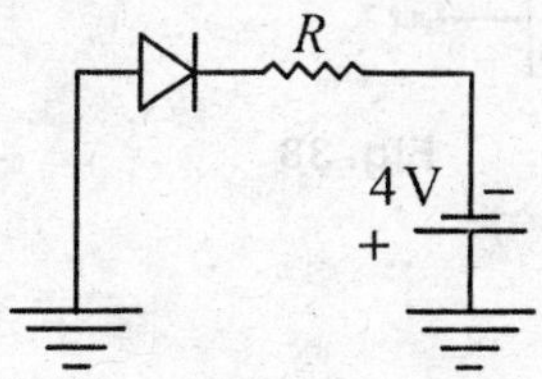

(b)

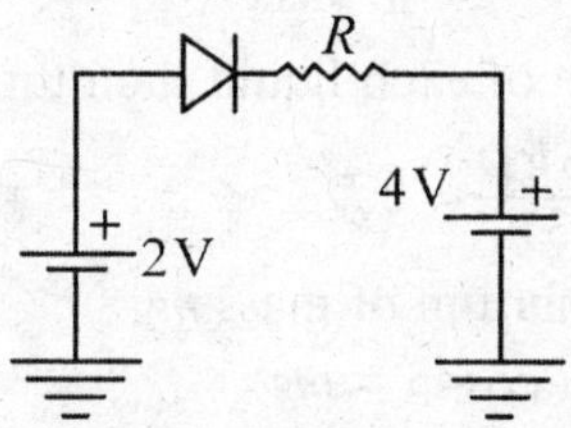

(c)

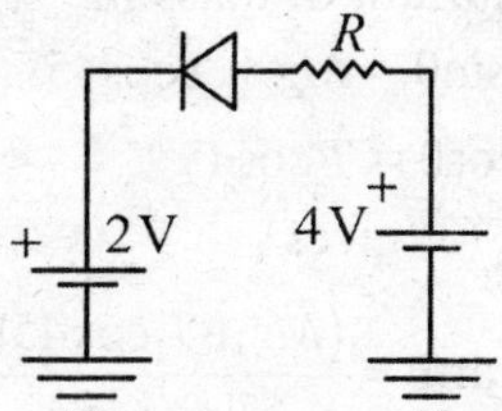

(d)

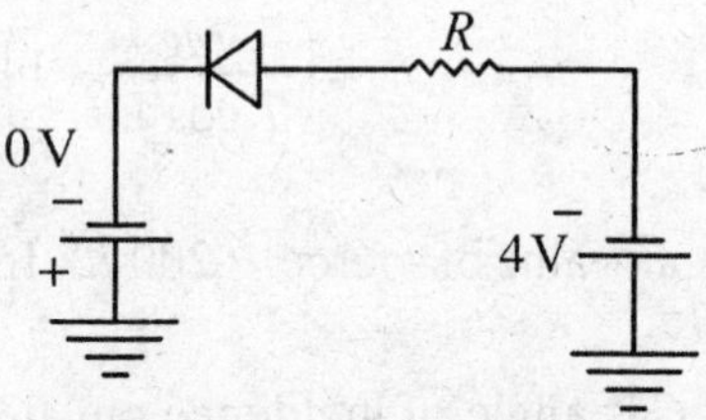

98. For a semiconductor the fraction of current contributions due to electrons and holes are 3/4 and 1/4 respectively. Their drift velocities are in the ratio 5/2, then the ratio of electron hole concentration is

(a) 3 : 1 (b) 5 : 6
(c) 6 : 5 (d) 1 : 3.

99. In Fig. 38 one logic circuit is shown which has two inputs A and B and one output C. Adjacent figure shows voltages A, B and C. Then logic circuit is

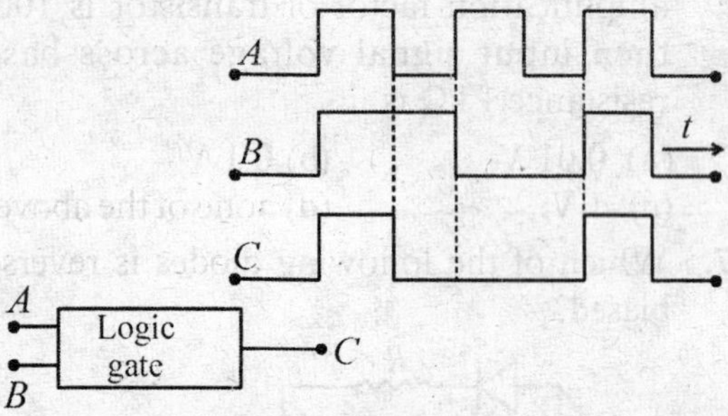

Fig. 38

(a) OR gate (b) NAND gate
(c) AND gate (d) NOR gate.

100. The electrical conductivity of a semiconductor increases when electromagnetic radiation of wavelength shorter than 2480 mm is incident on it. The band gap in eV for the semiconductor is

(a) 0.7 eV (b) 0.5 eV
(c) 2.5 eV (d) 1.1 eV.

SOLUTIONS

1. V volume of each liquid then total mass of liquid $M = VD + 2VD + 3VD$. Hence resulting density $\dfrac{6VD}{3V} = 2D$. (c)

2. For equilibrium of mass m

$$2T\cos 45 = mg \quad \ldots(1)$$

For equilibrium of mass M

$$T_1 \sin\theta = Mg + T\cos 45 \quad \ldots(2)$$

$$T_1 \cos\theta = T\cos 45 \quad \ldots(3)$$

Eqs. (2) and (3) gives

$$\tan\theta = \frac{(Mg + T\cos 45)}{T\cos 45}$$

$$= \left(\frac{Mg}{T\cos 45} + 1\right) = \left(\frac{2M}{m} + 1\right) \quad \text{(b)}$$

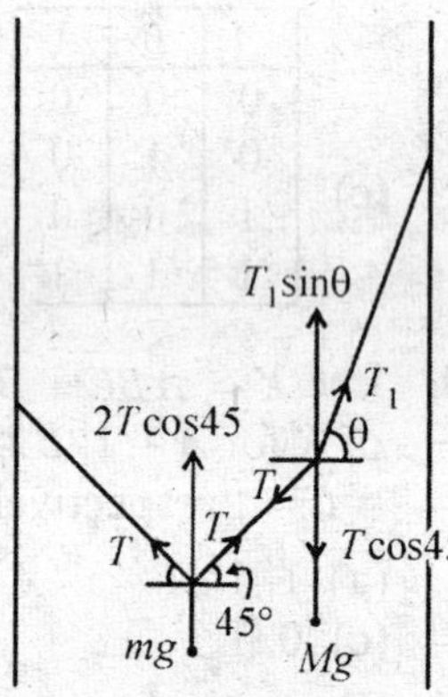

Fig. 39

3. Peak value of source $= 200\sqrt{2}$. In half wave rectifier the rms value is $\dfrac{V_0}{2} = \dfrac{200\sqrt{2}}{2} = 100\sqrt{2}V$.(a)

4. If C is angle of incidence on AC then for total internal reflection $\sin C = \dfrac{1}{n}$, $C = 45°$

$$\therefore \quad n = \frac{1}{\sin C} = \frac{1}{\sin 45} = \sqrt{2} \quad \text{(a)}$$

5. Volume of air = Vol. of glass – Vol. of Hg = Constant at all temperatures

$$[V_0(1+\gamma t)]_{\text{glass}} - [V_0'(1+\gamma t)]_{\text{Hg}} = V_0 - V_0'$$

Given is $\gamma_g = 3 \times 9 \times 10^{-6}/°\text{C}$ and $\gamma_{\text{Hg}} = 1.8 \times 10^{-4}$

Solving $V_0' = \dfrac{V_0\gamma_g}{\gamma_{\text{Hg}}} = \dfrac{1000 \times 3 \times 9 \times 10^{-6}}{1.8 \times 10^{-4}} = 150$ cc. (b)

6. Velocity of projection of particle (1) is $v_0 \sin\theta \hat{j} + v_0 \cos\theta \hat{i}$. After time $t = \dfrac{v_0 \sin\theta}{g}$ its velocity is $\boldsymbol{v}_1 = v_0 \cos\theta \hat{i}$. Velocity of particle (2) after time t is $\boldsymbol{v}_2 = v_0 \hat{k} - gt\hat{j}$. Hence relative velocity is $\boldsymbol{v}_2 - \boldsymbol{v}_1 = v_0 \hat{k} - gt\hat{j} - v_0 \cos\theta i$. (d)

7. When spring is at position x, the net force on it is

$$F = -kx - \mu mg = -kx - \alpha mgx = -(k + \alpha mg)x$$

Hence acceleration

$$\frac{dv}{dt} = \frac{F}{m} = -\frac{(k + \alpha mg)x}{m}$$

or $$vdv = -\frac{k + \alpha mg}{m} x\, dx$$

integration gives $\dfrac{v^2}{2} = -\dfrac{k + \alpha mg}{m}\dfrac{x^2}{2} + C$. C is constant of integration.

When $x = 0$, $v = v_0$ and $C = v_0^2/2$.

Hence equation is $v^2 = v_0^2 - \dfrac{k + \alpha mg}{m} x^2$

and when $v = 0$, $x = v_0\sqrt{\dfrac{m}{k + \alpha mg}}$. (a)

8. Object is at a distance $x > 2f$ (f is focal length of lens). Image is formed at distance between f and $2f$. As object is brought nearer the lens, image moves away from lens, i.e. goes closer to plane mirror. Hence image in plane mirror will come closer to plane mirror, i.e. moves in –ve X-direction. (b)

9. $\dfrac{dV}{dr} = -E$ or $dV = -\mathbf{E} \cdot \mathbf{dr}$

or $$dV = -(10\hat{i} + 10\hat{j}).(dx\hat{i} + dy\hat{j} + dz\hat{k}) = -(10\,dx + 10\,dy)$$

integration given

$$\int_0^V dV = -\int (10dx + 10dy) = -\left[10x\big|_0^{10} + 10y\big|_0^{10}\right] = -200\text{ V}.$$ (b)

10. Terminal velocity $v = \dfrac{2}{9}\dfrac{a^2(\rho - \rho_0)}{\eta} g$. $F = \dfrac{4}{3}\pi a^3 \rho g$

Hence power $P = vF = \dfrac{8\pi}{27\eta} a^5 \rho(\rho - \rho_0) g^2$. (d)

11. Since the collision is elastic and masses same, they will exchange their velocities, their velocities will be same in magnitude. They will collide after time t sec after first particle is projected such that $S = 40t - \dfrac{1}{2}gt^2 = 40(t - 4) - \dfrac{1}{2}g(t - 4)^2$. This gives $t = 6$ sec. Velocity of second particle after 2 sec is $v = 40 - 2g = 20$ m/s upwards. Hence after collision first particle will have speed 20 m/s upwards. (c)

12. Equation for S.H.M is $x = A\sin\omega t$ when $x = A/2$ $\omega t = \pi/6$ or $t = T/12$. When $x = \frac{\sqrt{3}A}{2}$ $\omega t = \pi/3$ or $t = T/6$. Required time $T/6 - T/12 = T/12$. (c)

13. Torque on plank due to water and spring force keep it in equilibrium, i.e.

$$\int_0^h b\rho g y(h-y)\,dy = kxh \quad \text{gives} \quad x = \frac{b\rho g h^2}{6k}$$

$$\text{Energy stored in spring} = \frac{1}{2}kx^2 = \frac{b^2\rho^2 g^2 h^4}{72k}$$ (c)

14. $R = \frac{V}{I}$, $\log R = \log V - \log I$ or $\frac{dR}{R} = \frac{dV}{V} + \left|\frac{dI}{I}\right|$

or $\frac{dR}{R} = \frac{1.5}{60} + \frac{0.01}{2} = \frac{1.8}{60}$ or percentage error = 3%. (a)

15. Process 3 → 1 is isobaric. Hence temperature at 1 is 200 K. Process 2 → 3 is isochoric hence temperature at 2 is also 200 K. For process 1 → 2 PV is constant hence process is isothermal. Hence heat exchanged is equal to work done

$$W = \mu RT \log\frac{V_2}{V_1} = 1 \times R \times 200 \log 2 = 200R\log 2.$$ (d)

16. Heat required to reduce the temperature of 20 g water ($ms\Delta\theta = 20 \times 20 \times 1 = 400$ cal) is more than the heat released when ice temperature increases to 0° C ($10 \times 10 \times 0.5 = 50$ cal) and less than heat released when whole ice melts ($= mL = 10 \times 80 = 800$ cal). Hence final temperature becomes 0° C and some ice melts as such mass of water becomes more than 20 g. (b)

17. Consider a spherical shell of radius r. Let temperature gradient here be $\frac{d\theta}{dr}$. Hence rate of flow of heat through this surface is

$$R = \text{K.A. Temperature Gradient} = -k.4\pi r^2\frac{d\theta}{dr}$$

or

$$R\frac{dr}{r^2} = -4\pi k\,d\theta$$

Integration gives

$$R\int_{r_1}^{r_2}\frac{dr}{r^2} = -4\pi k\int_{T_1}^{T_2} d\theta$$

$$R\left(-\frac{1}{r}\right)_{r_1}^{r_2} = -4\pi k(\theta)_{T_1}^{T_2}$$

$$R\left(\frac{1}{r_1} - \frac{1}{r_2}\right) = 4\pi k(T_1 - T_2) \quad \text{or} \quad R = 4\pi k\frac{r_1 r_2}{r_2 - r_1}(T_1 - T_2).$$ (b)

18. For diffraction through circular aperture, we have $d\sin\theta = 1.22\lambda$. If two objects x apart are placed at distance D then $\tan\theta = \frac{x}{D} \approx \sin\theta$

Then $\frac{3 \times 10^{-3} \times 1 \times 10^{-3}}{D} = 1.22 \times 500 \times 10^{-9}$ $\quad D = 5$ m. (d)

19. Circuit is shown in Fig. 40 current i using Kirchhoff's law is

$$i = \frac{2E}{r_1 + r_2 + R} \text{ and terminal voltage across cell } B \text{ is } V = E - ir_2$$

$$= E - \frac{2Er_2}{r_1 + r_2 + R} = 0 \text{ gives } R = r_2 + r_1$$ (c)

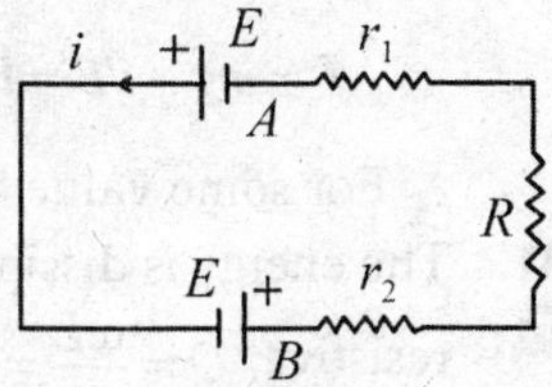

Fig. 40

20. In non uniform magnetic field force on N-pole is different than on S-pole. Hence force always acts.

If needle is placed parallel to the magnetic field no torque acts $\tau = \boldsymbol{M} \times \boldsymbol{B} = 0$. But if needle is not parallel to magnetic field a torque also acts. (b)

21. If phase difference is θ, then $\tan\theta = \frac{X}{R}$ for $\theta = \frac{\pi}{2}$ $R = 0$. (c)

22. Using Moseley's law, we have $\sqrt{\nu} \propto (Z-1)$

or $$\sqrt{\frac{\nu}{\nu'}} = \frac{(Z-1)}{Z'-1} = \sqrt{\frac{\lambda'}{\lambda}} = 2, \; Z = 11 \text{ and } Z' = 6.$$ (c)

23. Excited hydrogen atom on de-excitation gives photon of energy 10.2 V. Now the electron in ground state on absorbing energy of 15 eV comes out with K.E. = 15 – 13.6 eV = 1.4 eV. (a)

24. Acceleration of frame of reference is $a = 2K = 2\,\text{m/s}^2$

$$T_1 = 2\pi\sqrt{\frac{L}{g}} = 2\pi\sqrt{\frac{L}{10}} \quad \text{and} \quad T_2 = 2\pi\sqrt{\frac{L}{g-a}} = 2\pi\sqrt{\frac{L}{8}}$$

or $$\left(\frac{T_1}{T_2}\right)^2 = \frac{4}{5}.$$ (d)

25. The process contains two isochoric processes PQ and RS and two isobaric processes QR and SP.

On V-T graph isobars are straight lines passing through origin with smaller slope for higher pressure and isochores are lines parallel to T-axis. (c)

26. Lateral (Transverse) magnification = $V/U = m$, longitudinal magnification $= \frac{dV}{dU} = \frac{V^2}{U^2} = m^2$.
Hence volume of the image of a cube of side $L = (mL)(mL)(m^2L) = m^4L^3$. (d)

27. For refraction through spherical surfaces $\frac{\mu_2}{V} - \frac{\mu_1}{U} = \frac{\mu_2 - 1}{r}$, $V = \alpha$, $U = -2r$, $r = -r$ and $\mu_2 = 1$ gives $\mu_1 = 2$.

28. Walls are mirrors perpendicular to each other will give three images in a plane parallel to roof. Roof mirror will give 3 image of these images and one directly of the object. Total images 7. (c)

29. U^{238} decays to Pb^{208} with half life of 4.5×10^9 yrs, Pb : U = 7 : 1 means 7/8 of U has disintegrated and 1/8 remains means 3 half lives have elapsed = $3 \times 4.5 \times 10^9$. (d)

30. Copper rod moving in magnetic field, emf induced $e = -\frac{d\phi}{dt} = -BlV$, $-BlV = -L\frac{di}{dt}$, $Ldi = Bldx$

or $Li = Blx$, $i = \frac{Blx}{L}$. Force due to induced current $F = ilB$. Hence net force on rod is

$$f = mg - ilB = mg - \left(\frac{i^2B^2x}{L}\right)$$

For some value of x, $f = 0$, the body attains terminal velocity. (b)

31. The energy is dissipated in R and no losses in L and C. At resonance the impedance is purely resistive, $Q = \frac{\omega L}{R} = \frac{1}{\omega CR} > 1 \quad \frac{1}{\omega C} < R$ or $V_C > V$. C is also wrong and $V_C = -V_L$. (b)

32. Result can be correct to 3 significant places (due to 20.3). (c)

33. Intensity of sound wave $I \propto f^2A^2$, where A is amplitude and f is frequency. Hence

$$\frac{I_a}{I_b} = \frac{(1\times 3)^2}{(2\times 2)^2} = \frac{9}{16}.$$ (a)

34. Path difference = $1.5\lambda\sin\theta$ in addition to this there is initial phase difference of $\pi/2$ or path difference $\lambda/4$. Hence net path difference is

$$\Delta = 1.5\lambda\sin\theta - \lambda/4$$

For maximum intensity we have $\Delta = n\lambda$ or $1.5\,\lambda\sin\theta - \lambda/4 = n\lambda$.
For $n = 0$, $\sin\theta = 1/6$, For $n = -1$, $\sin\theta = -1/2$. (c)

35. Molecule number ratio in mixture is $H_2 : O_2 = 2 : 1$

This gives $C_{rms}^2 = \frac{1}{3}\left(2V_H^2 + 1V_O^2\right) = \frac{V_O^2}{3}\left\{2\left(\frac{V_H}{V_O}\right)^2 + 1\right\} = V_O^2 \times 11$ and $C_{rms} = \sqrt{11}V_O$. (d)

36. Work done = $EP(1 - \cos\theta)$. Thus $W = \frac{EP}{2}$. Torque required to hold it in this position is $C = EP\sin\theta = 2\omega\sin 60 = \sqrt{3}\omega$. (b)

37. Net force on small element AB of ring towards the centre is $F = 2T\sin\theta \approx 2T\theta$, since θ is very small. This provides the necessary centripetal force to element AB. Mass of this element $m = 2\theta \cdot r \cdot \rho$. Hence $m\omega^2 r = 2T\theta$

$$2\theta \cdot r\rho\omega^2 r = 2T\theta \quad \text{or} \quad T = \rho r^2\omega^2.$$ (a)

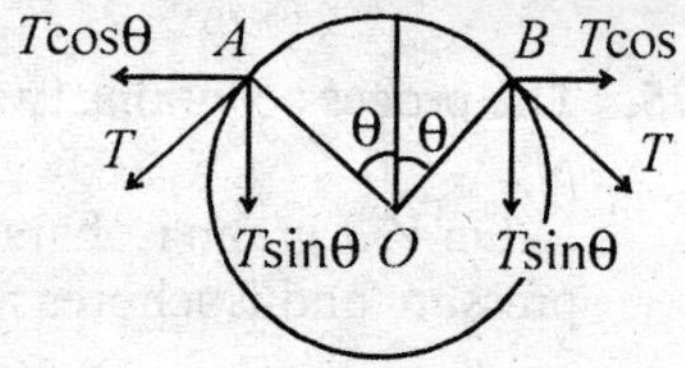

Fig. 41

38. When P is at A, voltmeter reads 2 V and when at B its one terminal is connected to –ve of 4 volt battery and +ve to 2 V cell reading is 6 V. (b)

39. As an approximation $\lambda \propto Z^{-2}$. We expect $\lambda_2 = \lambda_1$, $\lambda_3 = \frac{\lambda_1}{4}$, $\lambda_4 = \frac{\lambda_1}{9}$.

Hence ratio is 1 : 1 : 4 : 9. (b)

40. $P = kx^2yz^{-3}$ taking logs we have $\log P = \log k + 2\log x + \log y - 3\log z$. Hence error is

$$\frac{dP}{P} = 2\frac{dx}{x} + \frac{dy}{y} + 3\frac{dz}{z} = 2\times 2 + 1 + 3\times 3 = 14\%.$$ (a)

41. Power for given voltage is proportional to $1/R$. Hence $P_2 : P_3 = 1 : 2$, $P_1 = P_2 + P_3$ gives

$$P_1 : P_2 : P_3 = 3 : 1 : 2.$$ (b)

42. Mass balance right in all three, but charge balance wrong is 2. (a)

43. $T = \frac{0.693}{\lambda} = \frac{\log 2}{\lambda}$ and $\tau = \frac{1}{\lambda}$, Thus $T = \tau\log 2$. (a)

44. Acceleration $a = \frac{m_1 - m_2}{m_1 + m_2} g$. when $l = 0$, $m_1 = 25$, $m_2 = 20$ and $a = g/9$ when $l' = 0$ $m_1 = 30$, $m_2 = 15$ and $a' = g/3$ and $a' = 3a$. (c)

45. Initially c.g. is at centre when mass is added it lowers the centre of gravity. Then it goes on rising and finally comes to the centre when filled. (d)

46. *PV* constant decides curve 2 for isothermal. Adiabatic has larger slope. Hence 2 and 3 option correct. (d)

47. *P* is ahead of *Q* in phase by 90°, i.e. path difference $\lambda/4$. In direction *A* it is compensated by spacing $PQ(= \lambda/4)$. Hence path difference at *A* is zero and intensity is maximum $(= 4I)$. Along *B* phase difference is 90°. Hence intensity is $2I$. Along *C* net path difference is $\lambda/4 + \lambda/4 = \lambda/2$, i.e. we get destructive interference and zero intensity. (d)

48. Extension $\propto$ weight, i.e. $e \propto W$ and weight in air is proportional to density of block, i.e. $e \propto \rho_1$ and weight in liquid is proportional to $(\rho_1 - \rho_0)$

$\therefore \quad e' \propto (\rho_1 - \rho_0) \qquad \therefore \quad e/e' = \frac{\rho_1}{\rho_1 - \rho_0}$. (a)

49. In addition to the frictional force an inertial force $-$***ma*** would also act on the block. (c)

50. In steady state, the power input is equal to the radiation loss, which is proportional to difference of temperature $(\Delta\theta)$ over surroundings. Doubling the power doubles the temperature difference. Thus temperature becomes 25 + 40 = 65° C. (b)

51. $A + \delta = i + e$ where *e* is angle of emergence. 30 + 30 = 60 + *e* gives $e = 0$, i.e. ray emerges normally from the surface. (b)

52. $E_n = -3.4 \text{ eV} = -\frac{13.6}{n^2}$ gives $n = 2$. Angular momentum of electron in 2nd orbit $= 2h/2\pi = h/\pi = 2.11 \times 10^{-34}$ Js. (a)

53. $\Delta L = l_0 \propto \Delta\theta = 50 \times 1.2 \times 10^{-5} \times 12 = 7.2 \times 10^{-3}$ m.

The tape contracts, correct length = 50 – 0.0072 = 49.9928 m. (a)

54. $i = \frac{W}{V} = \frac{160}{60} = \frac{8}{3}$ amp. If *n* is no. of electrons crossing in 1 sec, then $i = ne$ or $n = i/e$

$= \frac{8}{3 \times 1.6 \times 10^{-19}} = \frac{1}{6} \times 10^{20}$. Hence electrons crossing per minute $= \frac{1}{6} \times 10^{20} \times 60 = 10^{21}$. (a)

55. Point A can be shown on right as *C* (Fig. 23)

From left, 20 Ω and 20 Ω in parallel and thus = 10 Ω.

This 10 Ω is in series with 10 Ω *PQ* resistance = 20 Ω.

This 20 Ω and 20 Ω at *R* are in parallel = 10 Ω.

This 10 in series with 10 Ω (*QB*) = 20 Ω.

This 20 Ω and 20 Ω at S in parallel = 10 Ω. (a)

56. Maximum charge on capacitor can be $CV (V = 100 \text{ V})$. It divides equally on two capacitors in parallel. Hence potential difference across these two is $V/2$. Hence total potential which can be applied across *AB* is $V + V/2 = 3/2 \text{ V} = 150$ V. (b)

57. Energy density of electromagnetic wave $U = \frac{1}{2}\varepsilon_0 E^2$ and intensity $I = UC = \frac{1}{2}\varepsilon_0 C E_0^2$ or

$$E_0 = \sqrt{\frac{25}{\varepsilon_0 C}} = 61.4 \qquad \text{(c)}$$

58. $Q = \frac{\pi P_1 r_1^4}{8\eta L_1} = \frac{\pi P_2 r_2^4}{8\eta L_2}$ or $\frac{P_1}{P_2} = \frac{L_1}{L_2}\left(\frac{r_2}{r_1}\right)^4 = \frac{1}{4}(2)^4 = 4$. (c)

59. At $t = 0$ mass m and velocity v momentum mv. Velocity of released grain relative to trolly is zero, i.e. no force acts and momentum remains conserved. After time Δt mass released be Δm and let trolley has speed v' then

$$mv = (m - \Delta m)v' + \Delta mv \quad \text{gives } v' = v = 20 \text{ m/s}.$$ (a)

60. Fringe width $\beta = \frac{\lambda D}{d} = \frac{0.8 \times 50}{10} = 4$ m. (d)

61. Focal length of combination in Fig. *a* is 24 cm, hence each lens has focal length 48 cm. For glass lens $\frac{1}{f} = (\mu - 1)\left(\frac{1}{r_1} - \frac{1}{r_2}\right)$ or $\frac{1}{48} = (1.5 - 1)\left(\frac{1}{r} - \frac{1}{\infty}\right)$ or $r = 24$ cm. Hence focal length of liquid lens is $\frac{1}{f_1} = (1.6 - 1)\left(-\frac{1}{24} - \frac{1}{24}\right)$ or $f_1 = -20$ cm. Focal length of combination is $\frac{1}{f} = \frac{1}{24} - \frac{1}{20} = -\frac{1}{120}$ or $f = -120$ cm. (a)

62. For radioactive decay $N = N_0 e^{-\lambda t}$ or $\frac{\Delta N}{\Delta t} = -N_0 \lambda e^{-\lambda t}$ and $\frac{1}{N}\frac{\Delta N}{\Delta t} = \gamma = -\lambda$. Thus γ is independent of time. (d)

63. Momentum of recoil nucleus is equal to that of photon, i.e. $P = \frac{E_p}{c}$ and energy of recoil nucleus $E = \frac{P^2}{2m}$ or $E = \frac{E_p^2}{2mc^2} = \frac{64 \times 10^6}{2 \times 24 \times 931} = 1.4 \times 10^3$ eV. (c)

64. Gravitational force on each mass will be repulsive. The acceleration of positive mass $= \frac{F}{m}$ towards left. Acceleration of negative mass $= \frac{F}{m}$ also towards left. (d)

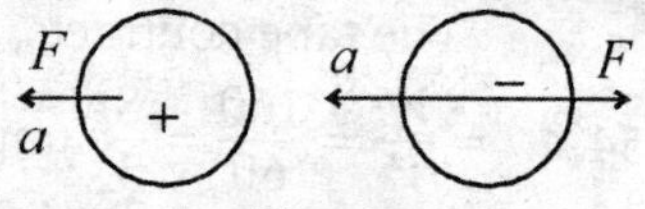

Fig. 42

65. Resulting motion will be an ellipse $\frac{x^2}{a^2} + \frac{y^2}{b^2} = 1$ with semi major axis equal to b. Hence maximum distance is b. (d)

66. 100 K = −173° F, $-13°\text{C} = -\frac{5}{9}(32 + 13) = -\frac{5}{9} \times 45 = -25°\text{C}$. (c)

67. At the end of whole process temperature is same, i.e. $\Delta U = 0$. In process I (isochoric) $\Delta W = 0$. In second process (isobasic) pressure constant P/n volume rises from V to nV. Hence work done $= p/n\,(nV - V) = PV(1 - 1/n)$. (c)

68. Nothing is stated about pressure. So volume can have any value. (d)

69. Two capacitors in series. Potential difference across them inversely proportional to capacity. Hence potential difference across $C_1 = 12$ V and that across $C_2 = 18$ V. Potential of point $C = 0$. Hence $V_1 = +12$ and $V_2 = -18$. (a)

70. If current in P and $Q = i$, then current in R and $S = 2i$. Heat generated $= i^2R$ is greatest in S. (d)

71. Current $i = 2\pi r\sigma n$. $B = \dfrac{\mu_0 nir^2}{2(r^2 + y^2)^{3/2}} \approx \dfrac{\mu_0 nir^2}{2y^3} = \dfrac{\mu_0 \pi r^3 \sigma n}{y^3}$. (a)

72. The net work is equivalent to that shown in Fig. 43. Therefore equivalent capacitance is $2C$ and C series and two such circuits in parallel

$$= 2\frac{2C \times C}{2C + C} = \frac{4}{3}C.$$ (c)

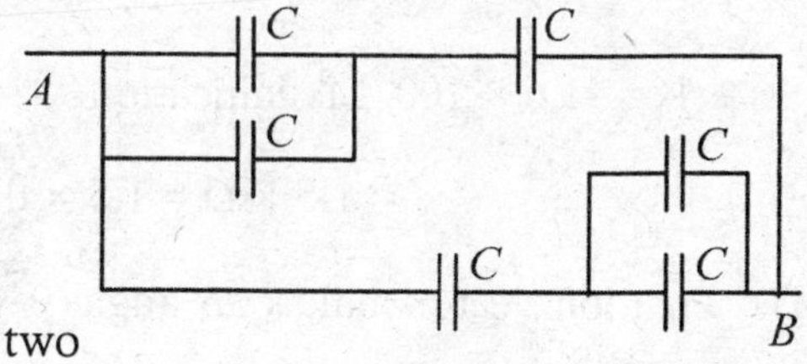

Fig. 43

73. Emission of one alpha reduces atomic number by two whereas that of one beta increases it by one. (b)

74. Only odd harmonics are present. Hence $(2m - 1)n = 135$ and $(2m + 1)n = 165$ gives $m = 5$ and $n = 15$. (b)

75. From one source observer is moving away and frequency observed $f_1 = f_0\dfrac{(c - U)}{c}$ and towards other he is moving, hence observed frequency $f_2 = f_0\dfrac{(c - U)}{c}$. No. of beats observed

$$= f_2 - f_1 = \frac{24f_0}{c} = \frac{2U}{\lambda}.$$ (a)

76. The charges on the capacitors are $120\,C_1$ and $200\,C_2$ respectively. Since the potential becomes zero, the net charge on the combination must be zero. This requires that the magnitude of the charges on each capacitor must be equal, i.e. $120\,C_1 = 200\,C_2$ or $3\,C_1 = 5\,C_2$. (b)

77. If F is the upward force and a is net acceleration upward then $F - 5mg = 5ma$ or $F = 5m(g + a)$. When m is mass of each link.

If T is tension between top link and the one in immediately below, then

$F - mg - T = ma$

or $T = F - m(g + a) = 4m(g + a) = 4 \times 0.1 \times 11 = 4.4$ N. (b)

78. Hydrogen atom in ground state ($E_1 = -13.6$ eV) must be excited to energy level with $n = 4$ ($E_4 = -0.85$ eV) so as to have H_β line. (a)

79. A ray of light passes twice through the lens and gets reflected once from the plan surface. If f_1, f_2 and f represent focal lengths of lens, mirror and combination respectively, then $\dfrac{1}{f} = \dfrac{2}{f_1} + \dfrac{1}{f_2}$,

$f_2 = \infty$ and $\dfrac{1}{f_1} = \dfrac{(\mu - 1)}{r} = \dfrac{1}{120}$. This gives $f = 60$ cm. (d)

80. $n \propto \sqrt{T}$ and $n' \propto \sqrt{T'}$ or $\dfrac{n'}{n} = \sqrt{\dfrac{T'}{T}} = \sqrt{1.2} = 1.1$

or $\dfrac{n' - n}{n} = 0.1$ or $n = \dfrac{15}{0.1} = 150$ Hz (Correct).

$v = n\lambda$, Fundamental wavelength does not change. Since frequency changes by 10% hence velocity changes by 10%. Correct option. (b)

81. With β-emission atomic number increases by 1 and mass no. remains same (a) is correct.

With α-emission mass no. changes by 4 (b) is wrong.

Thus mass number of S is 206 and atomic number 83 (c) is wrong.

The energy released in two processes is different (d) also wrong.

82. After refraction image is formed at V such that

$$\frac{n_2}{V}-\frac{n_1}{U}=\frac{n_2-n_1}{r},\ U=-100\text{ cm. } \frac{1.6}{V}+\frac{1.5}{100}=\frac{1.6-1.5}{20}\text{ gives}$$

$V = -1.6 \times 100$. Magnification $m=\frac{I}{O}=\frac{\mu_1 V}{\mu_2 U}=\frac{1.5\times1.6\times100}{1.6\times100}=1.5$

$I = 1.5 \times O = 1.5 \times 0.5 = 0.75.$ (a)

83. Position vector makes an angle θ with dipole moment such that $\tan\theta=\frac{2\sqrt{2}}{2}=\sqrt{2}$. The electric field at this point makes an angle β with vector $\boldsymbol{r}$ such that $\tan\beta=\frac{1}{2}\tan\theta=\frac{1}{\sqrt{2}}$, and the angle that electric field makes with dipole moment is $(\theta+\beta)$.

$$\tan(\theta+\beta)=\frac{\tan\theta+\tan\beta}{1-\tan\theta\tan\beta}=\infty \quad \text{or} \quad \theta+\beta=90.$$ (b)

84. Induced charges are $-q$ and $+q$ on the inner and outer surfaces of shell. Hence potential at the centre is

$$V=\frac{1}{4\pi\varepsilon_0}\left[\frac{q}{R}-\frac{q}{2R}+\frac{q}{3R}\right]=\frac{q}{4\pi\varepsilon_0}\frac{5}{6R}.$$ (c)

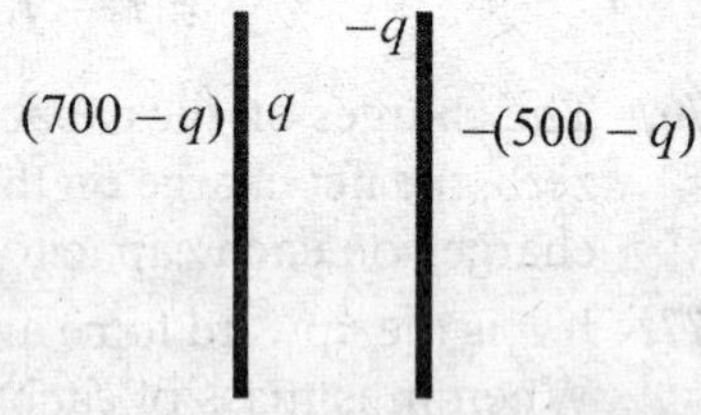

Fig. 44

85. Charge acquired by the plates of capacitor $q_0 = CV = 500\ \mu C$. Let the new distribution be as shown in Fig. 44. The charge on outer surfaces of both plates is same. Hence

$700 - q = -500 + q$ or $2q = 1200$ and $q = 600\ \mu C$

This gives $V = 60$ V. (d)

86. Magnetic field at P (Fig. 45) is perpendicular to the paper inwards due to both wires. Charge particle is projected parallel to the magnetic field. Hence force

$\boldsymbol{F} = q\boldsymbol{V}\times\boldsymbol{B} = 0.$ (d)

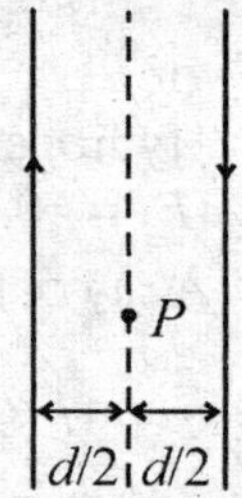

Fig. 45

87. For rotational equilibrium $(mg\sin\theta)R = M(B\sin\theta)$. When $M = i\pi R^2$ is magnetic moment of current carrying coil and is perpendicular to the plane of coil

$$B=\frac{mg}{\pi iR}.$$ (b)

88. Induced emf $e=(\pi a^2)\frac{dB}{dt}=\pi a^2\times2=2\pi a^2$.

Hence current $i=\frac{e}{R}=\frac{2\pi a^2}{R}$ = constant (b) is correct

Charge passed in time $t = 0$ to $t = 2$, $i=\frac{dq}{dt}$ or $q=it=\frac{4\pi a^2}{R}$ (c) is correct

Power $P=i^2R=\left(\frac{2\pi a^2}{R}\right)^2R=\frac{4\pi^2a^4}{R}$ = constant (a) is correct.

89. According to Lenz's law emfs of same magnitude in the clockwise directions are induced in two loops. So the current is induced in the clockwise direction in the outer boundary but no current in AB. (d)

90. $\sigma = e(\mu_n + \mu_h) n_i = 1.6 \times 10^{-19} (0.15 + 0.05) \times 1.6 \times 10^{16} = 5.12 \times 10^{-4}$. (c)

91. Arsenic is 5th group element and has 5 electrons in valance orbit hence X is n-type. Indium is 3rd group element with 3 electrons in valance orbit as such Y is p-type and the circuit is reverse biased. (c)

92. $i = i_s\ e \times p\left(\frac{+eV}{kT}\right) = 10^{-5}\ \exp\left(\frac{1.6 \times 10^{-19} \times 0.2}{1.4 \times 10^{-23} \times 300}\right) = 10^{-5}\ \exp(7.62)$, i.e. $i = 2037.6 \times 10^{-5} = 20.376 \times 10^{-3}$. (c)

93.

Truth table

A	B	$C = A + B$	$X = A(A + B)$
0	0	0	0
0	1	1	0
1	0	1	1
1	1	1	1

(b)

94. For $A = 1$, $B = 0$ and $C = 1$ $\overline{BC} = 1$ and $A\overline{BC} = 1$. Whereas $\overline{CA} = 0$ and $B\overline{CA} = 0$. Similarly $C\overline{AB} = 0$. Hence $X = A\overline{BC} + B\overline{CA} + C\overline{AB} = 1$. For $A = B = C = 1$ $\overline{BC} = 0$ and $A\overline{BC} = 0$. Similarly $B\overline{CA} = C\overline{AB} = 0$

Hence $X = A\overline{BC} + B\overline{CA} + C\overline{AB} = 0 + 0 + 0 = 0.$ (b)

95. If $V_a > V_b$ diode is forward biased. Two resistors each of 10 Ω are in parallel. Equivalent resistance = 5 Ω. (b)

96. $i_c = \frac{V_c}{R_0} = \frac{2}{2 \times 10^3} = 10^{-3}$ amp $\beta = \frac{i_c}{i_b}$ $i_b = \frac{i_i}{\beta} = \frac{10^{-3}}{100} = 10^{-5}$ A

$$V_i = i_b R_i = 10^{-5} \times 10^3 = 10^{-2} \text{ V}$$ (a)

97. For diode to be reverse biased n-side potential should be more than that on p-side. (b)

98. $I_e = en_e AV_e$ and $i_h = en_h AV_h$ or $\frac{I_e}{i_h} = \frac{n_e V_e}{n_h V_h}$ or $\frac{h_e}{n_h} = \frac{i_e V_h}{i_n V_e}$

$$\frac{n_e}{n_h} = \frac{3}{4} \times 4 \times \frac{2}{5} = \frac{6}{5}.$$ (c)

99. Truth table of the wave form is

A	B	X
1	1	1
0	1	0
1	0	0
0	0	0

AND gate (c)

100. Band gap $= \frac{hc}{\lambda} = \frac{12400}{\lambda(\text{Å})}$ eV $= \frac{12400}{24800} = 0.5$ eV. (b)

Model Test Paper for AIEEE Examination (II)

1. A cubical clock of side a is moving on a smooth plane with velocity v. During its motion it collides with a linear obstacle O (Fig. 1). After collision with O the angular speed of block will be

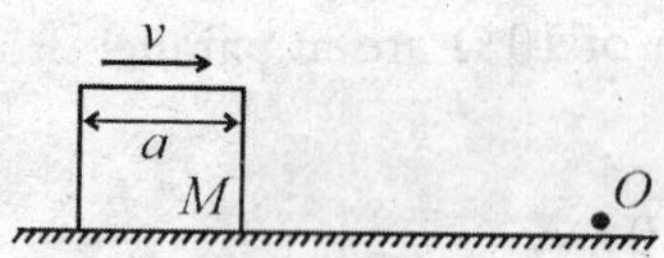

Fig. 1

(a) $\frac{3v}{4a}$ (b) $\frac{3v}{2a}$

(c) $\sqrt{\frac{3}{2}}\frac{v}{a}$ (d) zero.

2. Ratio of speed of sound in nitrogen and helium at 300 K is

(a) $\sqrt{\frac{2}{7}}$ (b) $\sqrt{\frac{1}{7}}$

(c) $\frac{\sqrt{3}}{5}$ (d) $\frac{\sqrt{6}}{5}$.

3. A disc of mass M and radius R rolls on a smooth horizontal surface (Fig. 2). Angular momentum of disc about origin if its angular speed is ω, is

(a) $\frac{1}{2}MR^2\omega$ (b) $MR^2\omega$

(c) $\frac{3}{2}MR^2\omega$ (d) $2MR^2\omega$.

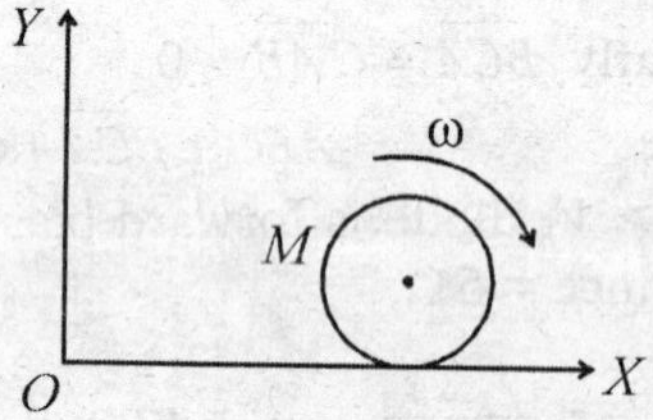

Fig. 2

4. A spring of constant K is divided in two parts such that one is of twice the length than other one. The spring constant of longer one is

(a) $\frac{2}{3}K$ (b) $\frac{3}{2}K$

(c) $3K$ (d) $6K$.

5. Two identical metallic plates are given charge Q_1 and $Q_2 (< Q_1)$. If these plates are brought near to each other to form a capacitor of capacity C, then potential difference between the plates will be

(a) $\frac{Q_1 + Q_2}{2C}$ (b) $\frac{Q_1 + Q_2}{C}$

(c) $\frac{Q_1 - Q_2}{C}$ (d) $\frac{Q_1 - Q_2}{2C}$.

6. For circuit shown in Fig. 3, the correct statement is

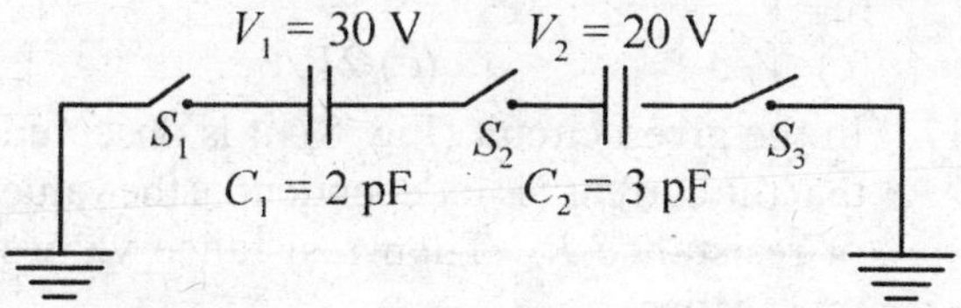

Fig. 3

(a) On closing S_1, $V_1 = 15$ V, $V_2 = 20$ V
(b) On closing S_2, $V_1 = V_2 = 25$ V
(c) On closing S_1 and S_2, $V_1 = V_2 = 0$
(d) On closing both S_1 and S_3, $V_1 = 30$ V, $V_2 = 20$ V.

7. On absorbing energy ^{22}Ne nucleus emits two α-particles and disintegrates into an unknown nucleus. This unknown nucleus is

(a) nitrogen (b) carbon
(c) boron (d) oxygen.

8. A quantity X is given by $\varepsilon_0 L \frac{\Delta V}{\Delta t}$ where ε_0 is the permittivity of free space, L is length, ΔV is a potential difference and Δt is a time interval. The dimensional formula for X is the same as that of

(a) resistance (b) charge
(c) voltage (d) current.

9. Two particles of masses m_1 and m_2 in projectile motion have velocities v_1 and v_2 respectively at $t = 0$. They collide at time t_0. Their velocities become v_1' and v_2' at time $2t_0$ while still in air. The value of $\left|(m_1 v_1' + m_2 v_2') - (m_1 v_1 + m_2 v_2)\right|$ is

(a) zero (b) $(m_1 + m_2)gt_0$
(c) $2(m_1 + m_2)gt_0$ (d) $\frac{1}{2}(m_1 + m_2)gt_0$.

10. One quarter sector is cut from a uniform circular disc of radius r (Fig. 4). This sector has mass m. It is made to rotate about a line perpendicular to its plane and passing through the centre of the original disc. Its moment of inertia about the axis of rotation is

(a) $\frac{1}{2}mr^2$ (b) $\frac{1}{4}mr^2$
(c) $\frac{1}{8}mr^2$ (d) $\sqrt{2}mr^2$.

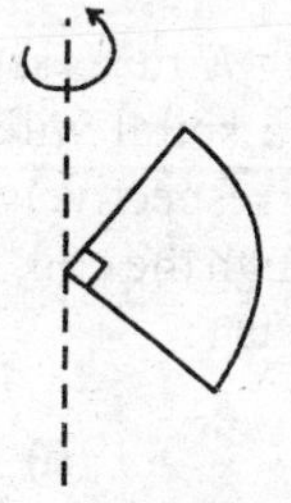

Fig. 4

11. A hemispherical portion of radius r is removed from the bottom of a cylinder of radius r. The volume of the remaining cylinder is V and its mass is m. It is suspended by a string in a liquid of density ρ where it stays vertical. The upper surface of the cylinder is at depth h below the liquid surface. The force on the bottom of the cylinder by the liquid is

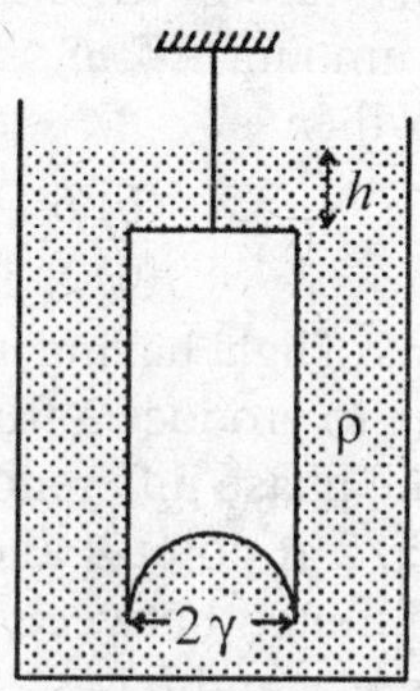

Fig. 5

(a) mg (b) $mg - V\rho g$
(c) $mg + \pi r^2 h g \rho$ (d) $\rho g(V + \pi r^2 h)$.

12. P-V plots for two gases during adiabatic processes are shown in Fig. 6. Plots (1) and (2) should correspond respectively to

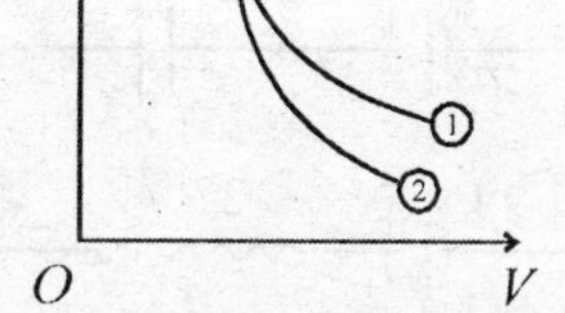

Fig. 6

(a) He and O_2 (b) O_2 and He
(c) He and Ar (d) O_2 and N_2.

13. A coil having N turns is wound tightly in the form of a spiral with inner and outer radii a and b respectively. When a current I passes through the coil, the magnetic field at the centre is

(a) $\frac{\mu_0 NI}{b}$ (b) $\frac{2\mu_0 NI}{a}$

(c) $\frac{\mu_0 NI}{2(b-a)}\log\frac{b}{a}$ (d) $\frac{\mu_0 I^N}{2(b-a)}\log\frac{b}{a}$.

14. A wire of length L and three identical cells of negligible internal resistance are connected in series. Due to the current, the temperature of the wire is raised by ΔT in time t. A number N of similar cells is now connected in series with a wire of the same material and cross-section but of length $2L$. The temperature of the wire is raised by the same amount ΔT in same time. The value of N then is

(a) 4 (b) 6
(c) 8 (d) 9.

15. Two beams of light having intensities I and $4I$ interfere to produce a fringe pattern on screen. The phase difference between the beams is $\pi/2$ at point A and π at point B. Then difference between the resultant intensities at A and B is

(a) $2I$ (d) $4I$
(c) $5I$ (d) $7I$.

16. In the given circuit (Fig. 7), with steady state, the potential drop across the capacitor must be

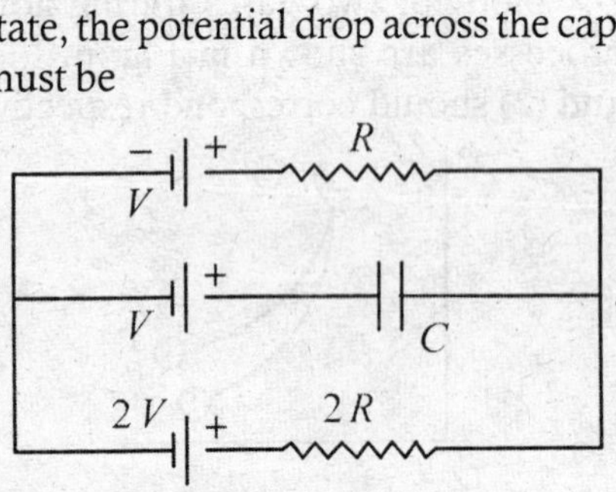

Fig. 7

(a) V (b) $V/2$
(c) $V/3$ (d) $2V/3$.

17. In the given circuit (Fig. 8), it is observed that the current I is independent of the value of resistance R_6. Then resistance values must satisfy

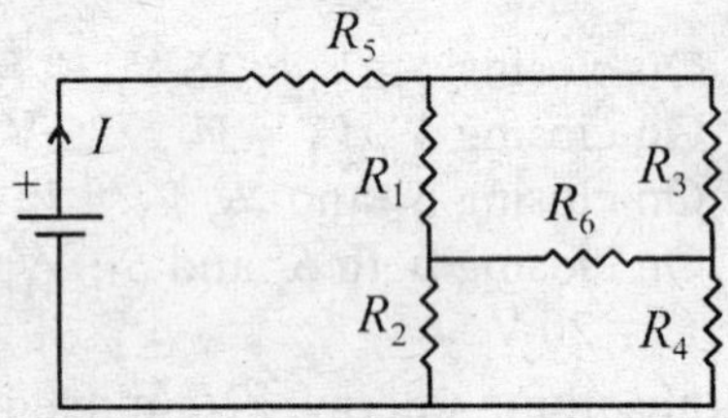

Fig. 8

(a) $R_1R_2R_5 = R_3R_4R_6$

(b) $\frac{1}{R_5}+\frac{1}{R_6}=\frac{1}{R_1+R_2}+\frac{1}{R_3+R_4}$

(c) $R_1R_4 = R_2R_3$

(d) $R_1R_3 = R_2R_4 = R_4R_6$.

18. A non-planer loop of conducting wire carrying a current I is placed as shown in Fig. 9. Each of the straight sections of the loop is of length $2a$. The magnetic field due to this loop at point $P(a, 0, a)$ point in the direction

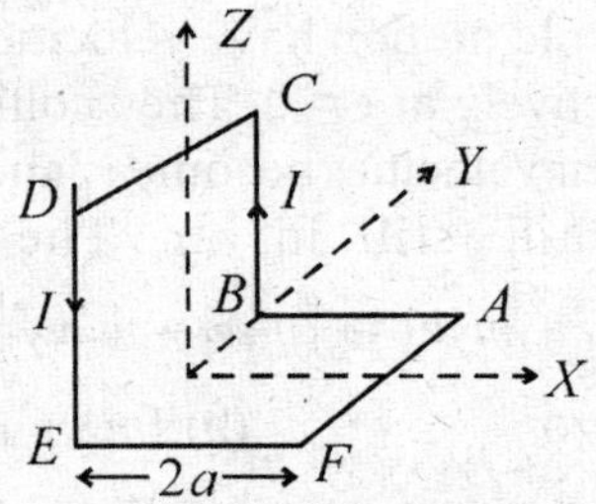

Fig. 9

(a) $\frac{1}{\sqrt{2}}(-\hat{j}+\hat{k})$ (b) $\frac{1}{\sqrt{3}}(-\hat{j}+\hat{k}+\hat{i})$

(c) $\frac{1}{\sqrt{3}}(\hat{i}+\hat{j}+\hat{k})$ (d) $\frac{1}{\sqrt{2}}(\hat{i}+\hat{k})$.

19. Two blocks of mass 10 kg and 4 kg are connected by a massless spring. The system is placed on a frictionless surface. An impulse of 14 m/s is given to heavier mass in the direction towards the lighter mass. The velocity of the centre of mass then is

(a) 30 m/s (b) 20 m/s
(c) 10 m/s (d) 5 m/s.

20. Two equal point changes are placed on X-axis at $x = -a$ and $x = +a$. Another charge Q is placed on the origin. If charge Q is slightly displaced through x along x-axis, its electrical potential energy changes which is approximately proportional to ($x << a$)

(a) x (b) x^2
(c) x^3 (d) $1/x$.

21. Two capacitors each of capacity C are charged to potentials V_1 and V_2 respectively. The negative plates of two capacitors are connected together when the positive plates of the two are connected together, then loss in electrostatic potential energy of the system is

(a) $\frac{1}{4}C(V_1^2 - V_2^2)$ (b) $\frac{1}{4}C(V_1^2 + V_2^2)$
(c) $\frac{1}{4}C(V_1 - V_2)^2$ (d) $\frac{1}{4}C(V_1 + V_2)^2$.

22. An ideal gas is taken round a cyclic process $A \to B \to C \to A$ (Fig. 10). If total heat given to the system is 5 J, then work done during the process CA is

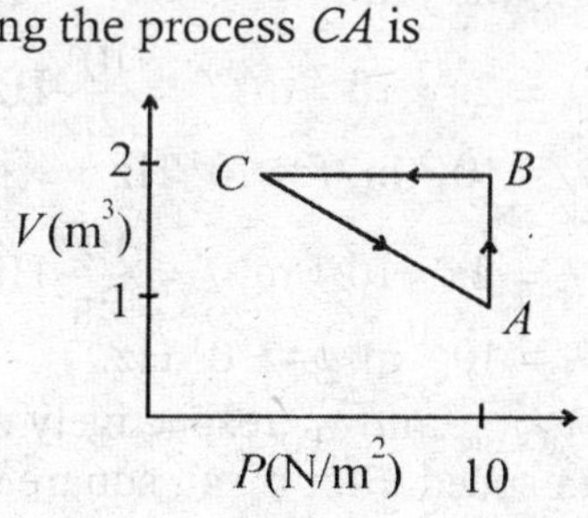

Fig. 10

(a) –5 J (b) –10 J
(c) –15 J (d) –20 J.

23. A block of mass 0.1 kg is held against a wall by applying a horizontal force of 5 N on the block. If the co-efficient of friction between the block and the wall is 0.5, the magnitude of the frictional force acting on the block is

(a) 2.5 N (b) 0.98 N
(c) 4.9 N (d) 0.49 N.

24. If $|\boldsymbol{A} \times \boldsymbol{B}| = \sqrt{3}\, \boldsymbol{A} \cdot \boldsymbol{B}$, then the value of $|\boldsymbol{A} + \boldsymbol{B}|$ is

(a) $(A^2 + B^2 + AB)^{1/2}$
(b) $\left(A^2 + B^2 + \frac{AB}{\sqrt{3}}\right)^{1/2}$
(c) $A + B$
(d) $\left(A^2 + B^2 + \sqrt{3}AB\right)^{1/2}$.

25. A long horizontal rod has a bead which can slide along its length, and initially placed at a distance L from one end A of rod. The rod is set in angular motion about A with constant angular acceleration α. If the coefficient of friction between the rod and the bead is μ and gravity is neglected, then time after which the bead starts slipping is

(a) $\sqrt{\frac{\mu}{\alpha}}$
(b) $\frac{\mu}{\sqrt{\alpha}}$
(c) $\frac{1}{\sqrt{\mu\alpha}}$
(d) infinitesimal small.

26. An infinite long conductor PQR is bent to form a right angle (Fig. 11). A current I flows through PQR. The magnetic field due to this current at the point M is B_1. Now another infinitely long straight conductor QS is connected at Q so that the current is $I/2$ in QR as well as in QS. The current in PQ remains unchanged. The magnetic field at M is now B_2. The ratio B_1/B_2 is

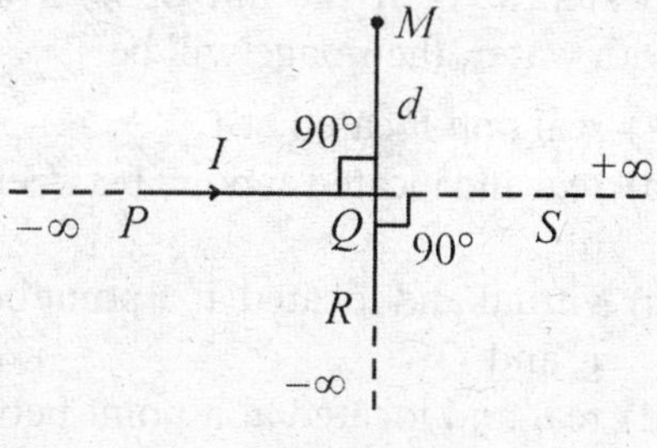

Fig. 11

(a) 1/2 (b) 1
(c) 2/3 (d) 2.

27. Two radioactive materials X_1 and X_2 have decay constants 10λ and λ respectively. If initially they have the same number of nuclei then the ratio of number of nuclei of X_1 to that of X_2 will be $1/e$ after time

(a) $1/10\lambda$ (b) $1/(11\lambda)$
(c) $11/(10\lambda)$ (d) $1/(9\lambda)$.

28. The electron in a hydrogen atom makes a transition $n_1 \rightarrow n_2$ where n_1 and n_2 are principal quantum numbers of the two states. Assume the Bohr model to be valid, the time period of the electron in the initial state is eight times that of the final state. The possible values of n_1 and n_2 are

(a) $n_1 = 4,\ n_2 = 2$ (b) $n_1 = 4,\ n_2 = 1$
(c) $n_1 = 8,\ n_2 = 6$ (d) $n_1 = 6,\ n_2 = 4$.

29. Two identical containers A and B with frictionless piston contain the same ideal gas at the same temperature and the same volume V. The mass of the gas in A is m_A and that in B is m_B. The gas in each cylinder is now allowed to expand isothermally to the same volume $2V$. The changes in the pressures in A and B are found to be ΔP and $1.5\Delta P$ respectively. Then

(a) $4m_A = 9m_B$ (b) $2m_A = 3m_B$
(c) $3m_A = 2m_B$ (d) $9m_A = 4m_B$.

30. A concave mirror is placed on a horizontal table, with its axis directed vertically upwards. Let O be the pole of the mirror and C its centre of curvature. A point object is placed at C. It has the real image also located at C. If the mirror is now filled with water, the image will be

(a) real and located at C
(b) real and located at point between C and α
(c) virtual and located at a point between C and O
(d) real and located at a point between C and D.

31. In a *p-n* junction diode not connected to any circuit

(a) the potential is the same every where
(b) the *p*-type side is at a higher potential then the *n*-type side
(c) there is an electric field at the junction directed from the *n*-type side to the *p*-type side
(d) there is an electric field at the junction directed from the *p*-type side to the *n*-type side.

32. A black body is at temperature 2880 K. The energy of radiation emitted by this object with wavelength between 499 nm and 500 nm is U_1, between 999 nm and 1000 nm is U_2 and between 1499 nm and 1500 nm is U_3. The Wien constant $b = 2.88 \times 10^6$ nm-K. Then

(a) $U_1 = 0$ (b) $U_3 = 0$
(c) $U_1 > U_2$ (d) $U_2 > U_1$.

33. A transverse sinusoidal wave of amplitude a, wavelength λ and frequency f is travelling on a stretched string. The maximum speed of any point on the string is $v/10$, where v is speed of propagation of the wave. If $a = 10^{-3}$ m and $v = 10^{-1}$ m/s, then λ and f are

(a) $\lambda = 2\pi \times 10^{-2}$ m, $f = \dfrac{10^3}{2\pi}$ Hz
(b) $\lambda = 10^{-3}$ m, $f = 10^4$ Hz
(c) $\lambda = 2\pi \times 10^{-1}$ m, $f = \dfrac{10^2}{2\pi}$ Hz
(d) $\lambda = 10^{-2}$ m, $f = 10^3$ Hz.

34. Let $\bar{v}$, v_{rms} and v_P respectively denote the mean speed, root mean square speed and most probable speed of the molecules in an ideal monoatomic gas at absolute temperature T. The mass of molecule is m. Then

(a) no molecule can have speed greater than $\sqrt{2}\, v_{rms}$
(b) no molecule can have speed less then $\dfrac{1}{\sqrt{2}} v_P$

(c) $\bar{v} < v_P < v_{rms}$
(d) The average kinetic energy of molecule is $\frac{3}{4} m v_P^2$.

35. Two cylinders A and B fitted with piston contain equal amounts of an ideal diatomic gas at 300 K. The piston of A is free to move while that of B is held fixed. Same amount of heat is given to the gas in each cylinder. If the rise in temperature of the gas in A is 30 K then rise in temperature of the gas in B is

(a) 30 K (b) 18 K
(c) 50 K (d) 42 K.

36. A string of length 0.4 m and mass 10^{-2} kg is tightly clamped at its ends. The tension is the string is 1.6 N. Identical wave pulses are produced at one end at equal intervals of time Δt. The minimum value of Δt which allows constructive interference between successive pulses is

(a) 0.05 s (b) 0.10 s
(c) 0.20 s (d) 0.40 s.

37. A spherical surface of radius of curvature R, separates air (refractive index 1.0) from glass (refractive index 1.5). The centre of curvature is in the glass. A point object P, placed in air is found to have a real image Q in glass. The line PQ cuts the surface at a point O and $PO = OQ$. The distance PO is

(a) $5R$ (b) $3R$
(c) $2R$ (d) $1.5R$.

38. A parallel monochromatic beam of light is incident normally on a narrow slit. A diffraction pattern is formed on a screen placed perpendicular to the direction of incident beam. At the first minimum of the diffraction pattern, the phase difference between the rays coming from the two edges of the slit is

(a) 0 (b) $\pi/2$
(c) π (d) 2π.

39. The maximum value of force F such that the block in Fig. 12 does not move is

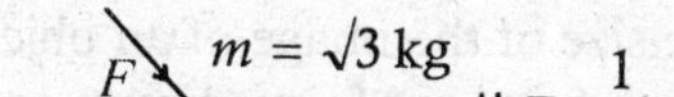

Fig. 12

(a) 20 N (b) 10 N
(c) 12 N (d) 15 N.

40. In Fig. 13 CP represents wavefront and AO and BP the corresponding two rays. For constructive interference at P between the ray BP and reflected OP, the correct condition is

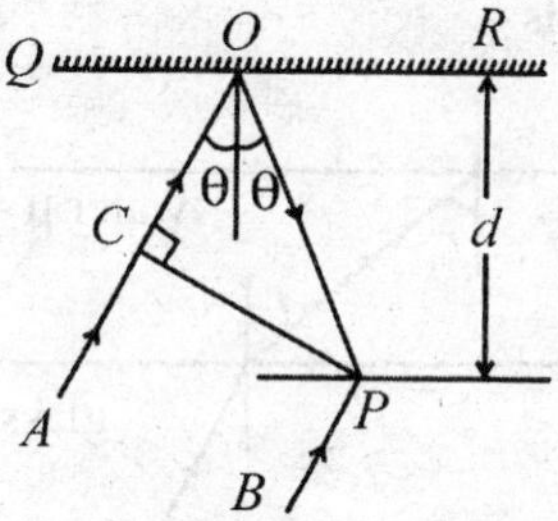

Fig. 13

(a) $\cos\theta = \frac{3\lambda}{2d}$

(b) $\cos\theta = \frac{\lambda}{4d}$

(c) $\sec\theta - \cos\theta = \frac{\lambda}{d}$

(d) $\sec\theta - \cos\theta = \frac{4\lambda}{d}$.

41. Two rods, one of aluminium and the other made of steel, having initial length l_1 and l_2 are connected together to form a single rod of length $l_1 + l_2$. The coefficients of linear expansion for aluminium and steel are α_a and α_s respectively. If the length of each rod increases by the same amount when their temperatures are raised by t° C, then ratio $l_1/(l_1 + l_2)$ is

(a) $\frac{\alpha_s}{\alpha_a}$ (b) $\frac{\alpha_a}{\alpha_s}$

(c) $\frac{\alpha_a}{\alpha_a + \alpha_s}$ (d) $\frac{\alpha_s}{\alpha_a + \alpha_s}$.

42. The size of the image of an object, which is at infinity, as formed by a convex lens of focal length 30 cm is 2 cm. If a concave lens of focal length 20 cm is placed between the convex lens and the image at a distance of 26 cm from the convex lens, then the size of final image is

(a) 1.25 cm (b) 2.5 cm
(c) 1.05 cm (d) 2 cm.

43. A ray of light is incident at the glass-water interface at an angle i, it emerges finally parallel to the surface of water, then value of μ_g is

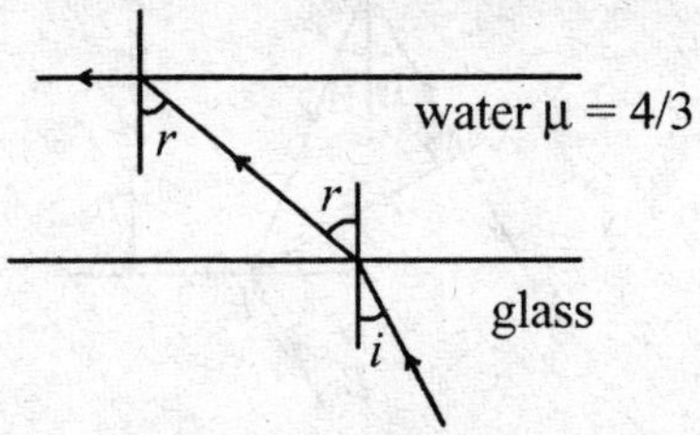

Fig. 14

(a) $\frac{4}{3}\sin i$ (b) $\frac{1}{\sin i}$
(c) $\frac{4}{3}$ (d) 1.

44. Electric potential energy between proton an electron is given by $U = U_0 \log \frac{r}{r_0}$, when r_0 is a constant. Assuming Bohr's model to be applicable, the variation of r_n with n is (n being the principal quantum number)

(a) $r_n \propto n$ (b) $r_n \propto 1/n$
(c) $r_n \propto n^2$ (d) $r_n \propto \frac{1}{n^2}$.

45. A nucleus with mass number 220 initially at rest emits an α-particle. If Q-value of the reaction is 5.5 MeV, the kinetic energy of the α particle is

(a) 4.4 MeV (b) 5.4 MeV
(c) 5.6 MeV (d) 6.5 MeV.

46. Rate of cooling of a body at temperature of 28° C in the surrounding at 20° C is R. To make its cooling rate $4R$ its temperature should be raised to

(a) 112° C (b) 80° C
(c) 60° C (d) 52° C.

47. Half life of a radioactive substance is 5 minutes. The mass disintegrated in 20 minutes is

(a) 93.7% (b) 75%
(c) 25% (d) 6.25%.

48. Two waves of intensity I_1 and I_2 interfere. The maximum intensity produced to the minimum intensity produced are in the ratio α. Then ratio I_1/I_2 is

(a) $\frac{\alpha - 1}{\alpha + 1}$ if $I_1 > I_2$

(b) $\frac{1+\alpha}{\alpha - 1}$ if $I_1 > I_2$

(c) $\left(\frac{1+\sqrt{\alpha}}{\sqrt{\alpha}-1}\right)^2$ if $I_1 > I_2$

(d) $\left(\frac{\sqrt{\alpha}-1}{1+\sqrt{\alpha}}\right)^2$ if $I_1 > I_2$.

49. An inductor connected to a DC source of emf 10 V draws a steady current of 5 A. Now the inductor is connected to a capacitor and an AC source of rms voltage 20 V. It is found that current and voltage of the source are in phase. The rms current in the circuit is

(a) 10 A (b) 5 A
(c) 2.4 A (d) data insufficient.

50. Monochromatic light of wavelength 4800 Å is used in a Young's double slit experiment. When one slit is covered by a transparent sheet of thickness 20 μm and refractive index 1.48, the fringe pattern shifts by n number of fringes. Here n is

(a) 2 (b) zero
(c) 20 (d) 48.

51. A satellite of mass m is orbiting round earth (Mass—M, radius—R) at a distance $2R$ from the centre of earth. The amount of energy that must be given to the satellite in order to make its radius $6R$ is

(a) less then $\frac{GMm}{3R}$

(b) equal to $\frac{GMm}{6R}$

(c) greater than $\frac{GMm}{4R}$

(d) less than $\frac{GMm}{4R}$.

52. In the given arrangement (Fig. 15) two capillaries of glass having radii r and $2r$ are dipped in pure water of density ρ, the surface tension of water is

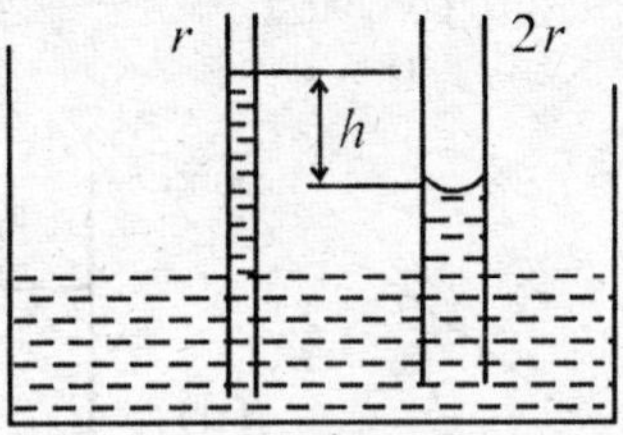

Fig. 15

(a) $2\rho hgr$ (b) $\frac{\rho hgr}{2}$
(c) ρhgr (d) none of these.

53. An object is placed at 30 cm from a convex lens of focal length 15 cm. On the other side of the lens a convex mirror of focal length 12 cm is placed so that the principal axis of both coincide. It is observed that the object and image coincide. The separation between lens and mirror is

(a) 6 cm (b) 30 cm
(c) 15 cm (d) 12 cm.

54. A mass m is attached to one end of a light rod. This assembly is placed on the origin (Fig. 16). If all surfaces are smooth, then the position of lower end when the rod becomes horizontal after its upper end has been disturbed gently towards right is

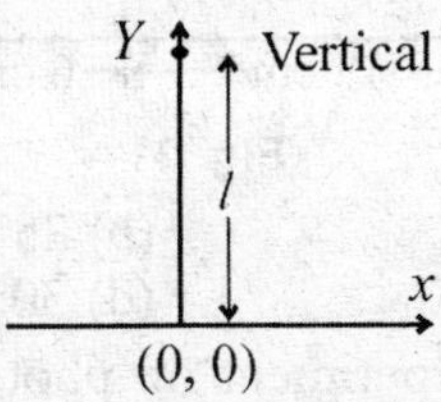

Fig. 16

(a) $(l, 0)$ (b) $(l/2, 0)$
(c) $(-l, 0)$ (d) $(-l/2, 0)$.

55. A particle is projected horizontally from a tower with velocity v_0. Wind is blowing in opposite direction and is providing a constant horizontal acceleration a_0. If particle strikes the ground vertically, then height of the building is

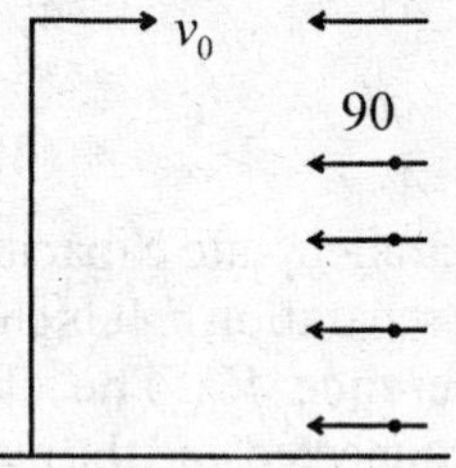

Fig. 17

(a) $\frac{v_0^2 g}{a_0^2}$ (b) $\frac{v_0^2 g}{2a_0^2}$

(c) $\frac{2v_0^2 g}{a_0^2}$ (d) $\frac{v_0^2}{g}$.

56. A uniform disc of mass m is resting on a rough horizontal surface. A force F is applied at the center of disc at $t = 0$. If disc performs pure rolling then power developed by F at any time t is

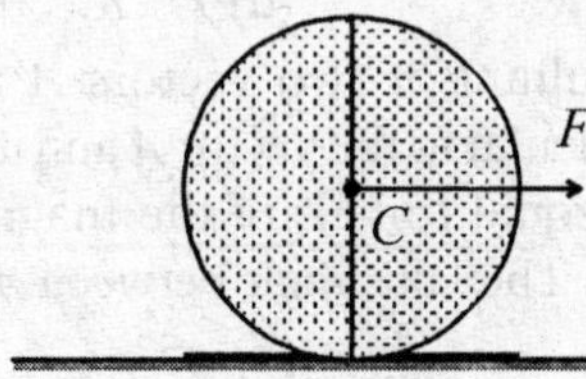

Fig. 18

(a) $\frac{3F^2t}{2m}$ (b) $\frac{4F^2t}{3m}$

(c) $\frac{3F^2t}{4m}$ (d) $\frac{2F^2t}{3m}$.

57. Three similar plates are arranged 'd' meters apart (Fig. 19). If area of each plate is A and a charge q is given to middle plate then potential of middle plate is

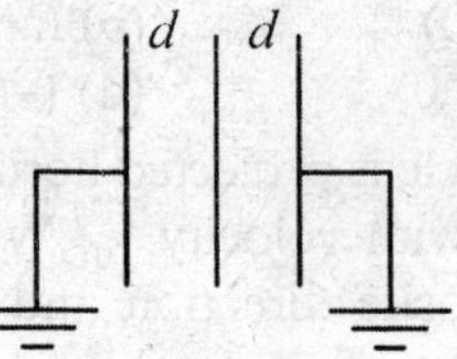

Fig. 19

(a) $\frac{2qd}{A\varepsilon_0}$ (b) $\frac{qd}{4A\varepsilon_0}$

(c) $\frac{qd}{A\varepsilon_0}$ (d) $\frac{qd}{2A\varepsilon_0}$.

58. A parallel plate capacitor has plate area A and separation d. It is charged to a potential difference V_0. The charging battery is disconnected and the plates are pulled apart to three times the initial separation. The work required to separate the plates is

(a) $\frac{3\varepsilon_0 A V_0^2}{d}$ (b) $\frac{\varepsilon_0 A V_0^2}{2d}$

(c) $\frac{\varepsilon_0 A V_0^2}{3d}$ (d) $\frac{\varepsilon_0 A V_0^2}{d}$.

59. Resistances P, Q, S and R are arranged in cyclic order to form a balanced Wheatstone's bridge network. The ratio of power consumed in branches $P + Q$ and $R + S$ is

(a) 1 : 1 (b) $R : P$

(c) $P^2 : R^2$ (d) $P : R$.

60. The resultant of two vectors $\boldsymbol{A}$ and $\boldsymbol{B}$ is perpendicular to the vector $\boldsymbol{A}$ and its magnitude is equal to half of the magnitude of vector $\boldsymbol{B}$. Then the angle between $\boldsymbol{A}$ and $\boldsymbol{B}$ is

(a) 30° (b) 45°

(c) 150° (d) 120°.

61. Two parallel light rays are incident at one surface of a prism (Fig. 20). The prism is made of glass of refractive index 1.5. The angle between the rays as they emerge is nearly

(a) 19° (b) 37°

(c) 45° (d) 49°.

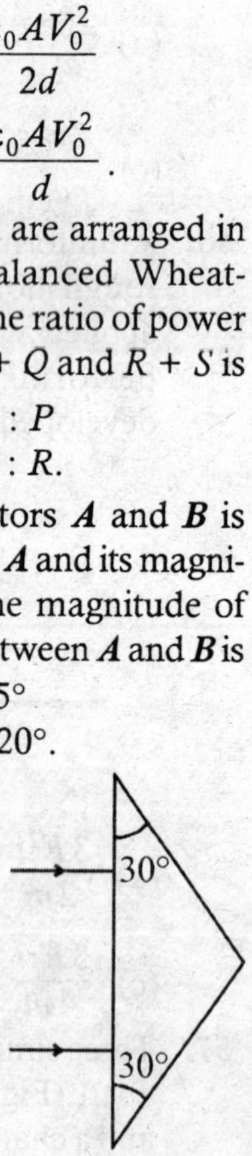

Fig. 20

62. A block is placed on a rough plane whose inclination to the horizontal θ can be varied. The angle of repose is 30°. The graph that correctly indicates the variation of the frictional force (f) between the block and the plane with θ is

(a) 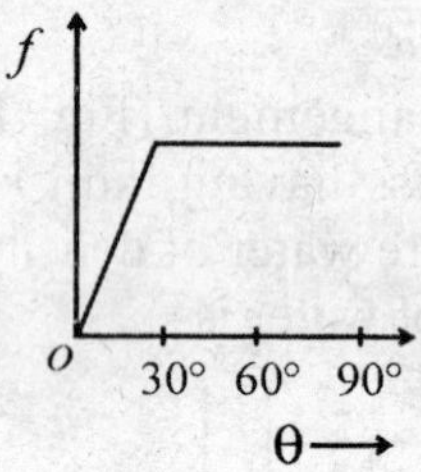

(b) 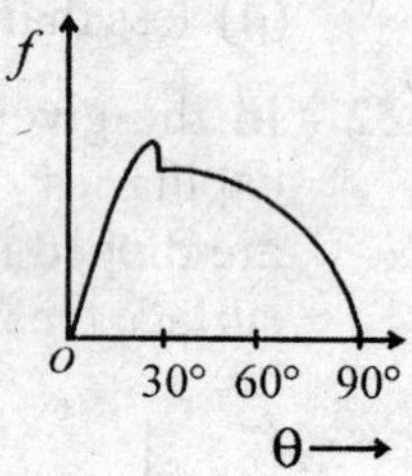

(c) 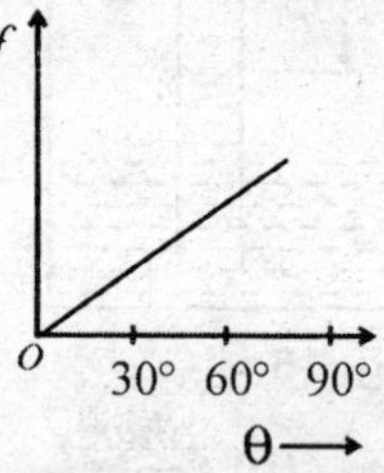

(d) 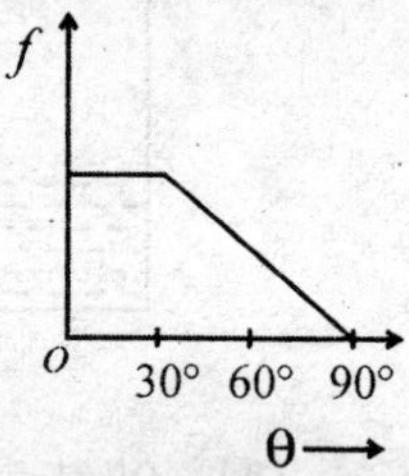

63. A particle of mass m is located in one dimensional potential field where potential energy is given by $U(x) = A(1 - \cos Px)$, where A and P are constants. The period of small oscillations of the particle is

(a) $2\pi\sqrt{\frac{m}{AP^2}}$ (b) $2\pi\sqrt{\frac{m}{AP}}$

(c) $2\pi\sqrt{\frac{m}{A}}$ (d) $2\pi\sqrt{\frac{AP}{m}}$.

64. The coefficient of thermal conductivity of copper may be assumed to be 9 times that of steel. In the composite cylindrical bar shown in Fig. 21, the temperature at the junction of copper and steel is

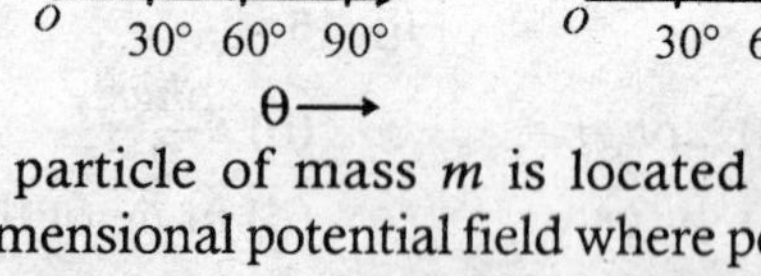

Fig. 21

(a) 75° (b) 25° C

(c) 67° C (d) 30° C.

65. In an experiment on photoelectric effect stopping potential is 1.0 V, when light of wavelength 6520 Å is incident on the emitting surface. The stopping potential is 2.9 V for light of wavelength 3260 Å. The work function of the metal is

(a) 0.9 eV
(b) 1.9 eV
(c) 5.8 eV
(d) cannot be determined from given data.

66. Pick up the correct statement.

(a) As the voltage applied to the X-ray tube increases, the intensity of X-rays increases
(b) All the β-active nuclei emit electrons
(c) Infrared radiation is used in the study of molecular structure
(d) Moderator is used in a reactor to reduce the rate of reaction.

67. A vernier callipers has 20 divisions on the vernier scale which coincide with 19 on the main scale. The least count of the instrument is 0.1 mm. The main scale division are of

(a) 0.5 mm (b) 1 mm
(c) 2 mm (d) 0.25 mm.

68. A particle moves in a circle of radius 4 cm clockwise at constant speed 2 cm/s. If $\hat{i}$ and $\hat{j}$ are unit vectors along X and Y respectively, the acceleration of the particle at the instant half way between P and Q is (in cm/s^2)

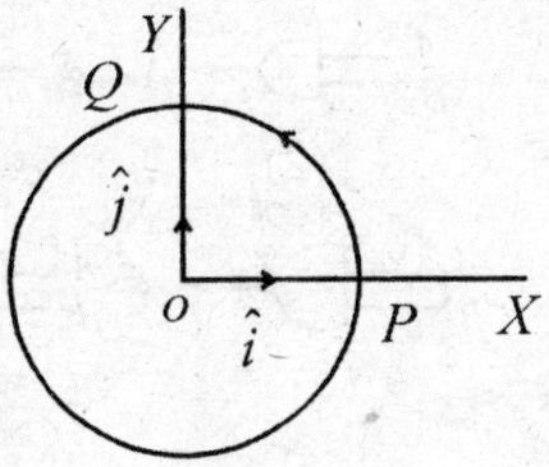

Fig. 22

(a) $-4(\hat{i}+\hat{j})$ (b) $4(\hat{i}+\hat{j})$
(c) $-\frac{1}{\sqrt{2}}(\hat{i}+\hat{j})$ (d) $\frac{1}{4}(\hat{i}-\hat{j})$.

69. Which of the following statements about the velocity of sound is not true?

(a) For same pressure, v in different gases is proportional to (density)$^{-1/2}$ only
(b) v in any gas is smaller than v in any solid or liquid
(c) v in gas is independent of pressure
(d) v is proportional to $T^{1/2}$ for gases.

70. Consider a parallelogram $ABCD$ with angle at $B = 120°$. A charge $+Q$ is placed at corner A produces field E and Potential V at corner D. If we now add charges $-2Q$ and $+Q$ at corners B and C respectively, the magnitude of field and potential at D will become respectively

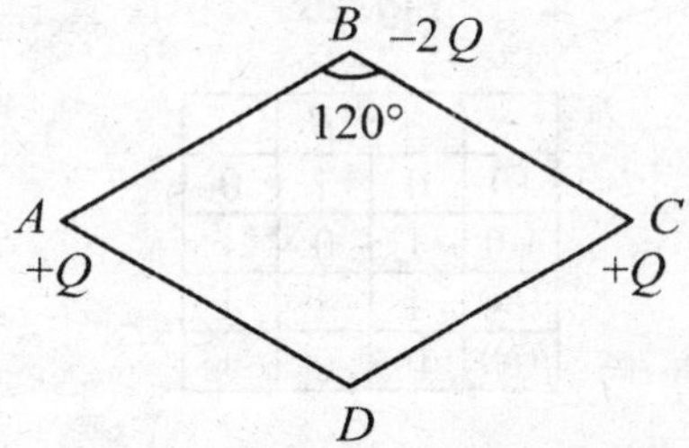

Fig. 23

(a) E and 0
(b) 0 and V
(c) $E\sqrt{2}$ and $V/\sqrt{2}$
(d) $E/\sqrt{2}$ and $V/\sqrt{2}$.

71. In the circuit (Fig. 24) capacitors A and B have identical geometry but a material of dielectric constant 3 is present between plates of B. The potential difference across A and B are respectively

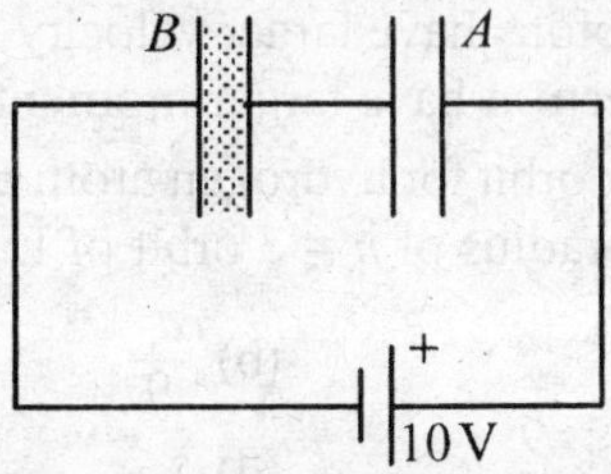

Fig. 24

(a) 2.5 V, 7.5 V (b) 2 V, 8 V
(c) 7.5 V, 2.5 V (d) 8 V, 2 V.

72. A pair of coil of turns n_1 and n_2 are kept close together. If a current passing through the first is reduced at rate r, an emf of 2 mV is developed across the other coil. If the second coil carries current which is then reduced at the rate $3r$, the emf produced across the first coil will be $(n_1/n_2 = 1.5)$

(a) 9 mV (b) 4 mV
(c) 6 mV (d) 1.5 mV.

73. Consider a two-input AND gate (Fig. 25). Out of the four entries in the truth table given below, the correct entries are

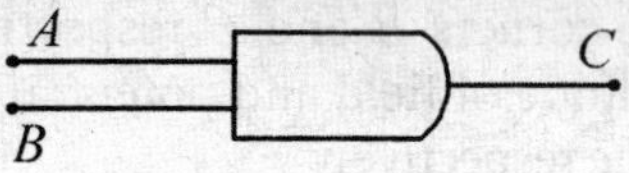

Fig. 25

	A	B	C
(*i*)	0	1	0
(*ii*)	1	0	1
(*iii*)	1	1	1
(*iv*)	0	0	1

(a) All
(b) (*i*) and (*ii*) only
(c) (*i*) and (*iii*) only
(d) (*i*), (*ii*) and (*iii*) only.

74. Compare electrons accelerated through certain potential difference and protons accelerated through the same potential difference. If initial velocities are negligible, the emergent.

(a) protons have larger momentum
(b) electrons have small kinetic energy
(c) protons have larger velocity
(d) electron have larger momentum.

75. If Bohr orbit for hydrogen atom has radius r_0, the radius of $n = 3$ orbit of Li^{++} ion is

(a) a_0 (b) $\frac{a_0}{9}$
(c) $a_0 \times \frac{9}{4}$ (d) $3a_0$.

76. Three capillaries have lengths 10, 5 and 2 cm, radii 1.0, 0.5 and 0.2 mm respectively. For streamline flow, their assembly in series is equivalent to one capillary of radius 0.5 mm length L. The value of L (in cms) is nearest to

(a) 17 (b) 38
(c) 54 (d) 84.

77. Figure 26 shows a battery of emf 120 V connected through a 4.0 M Ω resistor to a 0.50 μF capacitor. A neon bulb N which strikes at 100 V is connected as shown. It shows negative glow. The period of this glow is nearest to

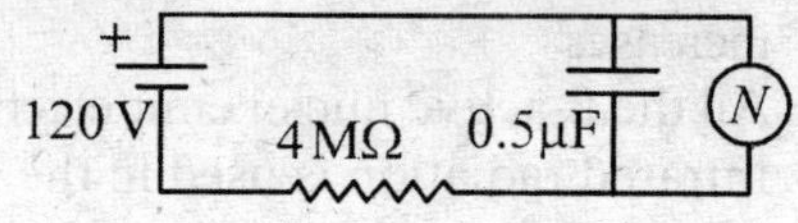

Fig. 26

(a) $\frac{1}{2}\log 6/5$ sec (b) 2log6 sec
(c) $\frac{1}{2}\log 5$ sec (d) 2log6/5 sec.

78. The mean temperature of a planet may be estimated from the balance between radiant energy received from the sun and that emitted by the surface of the planet. If d is distance of the planet from sun, planet temperature is proportional to

(a) d^1 (b) $d^{-1/2}$
(c) d^2 (d) d^{-4}.

79. Given in Fig. 27 are logic symbols for same gates which of the following identification is correct?

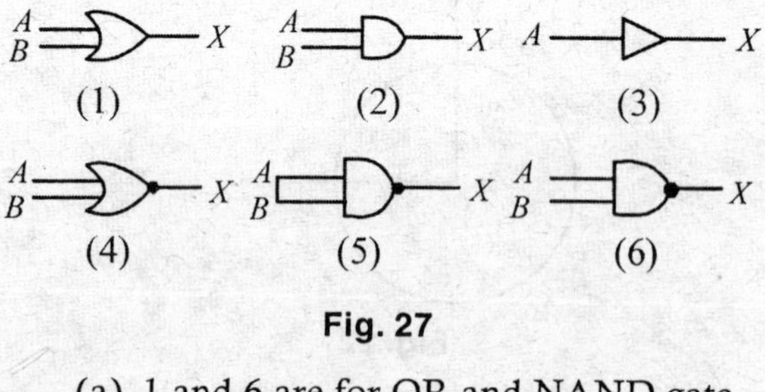

Fig. 27

(a) 1 and 6 are for OR and NAND gate
(b) 2 and 3 are for AND and NOR gate
(c) 4 and 5 are for NAND and NOR gate
(d) 2 and 5 are for NOR and NAND gate.

80. In resonance column, the first three resonances are observed at lengths L_1, L_2 and L_3. These are related through

(a) $L_3 = 2L_2 = 4L_1$

(b) $L_3 = \frac{5}{3}L_2 = 5L_1$

(c) $L_3 - L_2 = L_2 - L_1$

(d) $L_3 - L_2 = 2(L_2 - L_3)$.

81. In the circuit shown in Fig. 28, the steady state voltage across capacitor C is a fraction f of the battery emf. The fraction is decided by

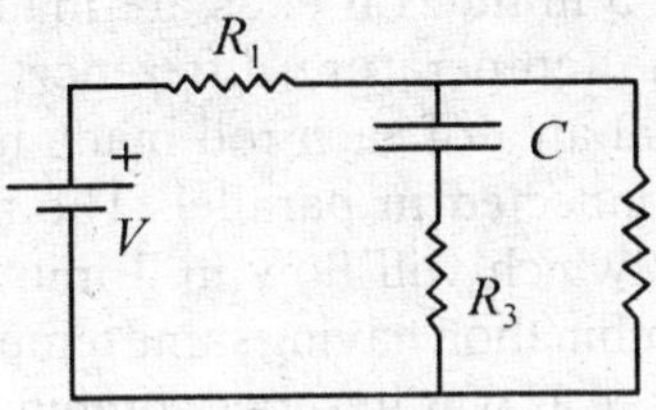

Fig. 28

(a) R_1 only (b) R_1 and R_3 only

(c) R_1 and R_2 only (d) R_1, R_2 and R_3.

82. In Fig. 29, A and B are two horizontal metal plates separated by insulating plugs. The region between A and B is evacuated. An electron of mass m remains at rest inside the chamber. If a proton of mass M replaces electron under similar conditions, then the proton

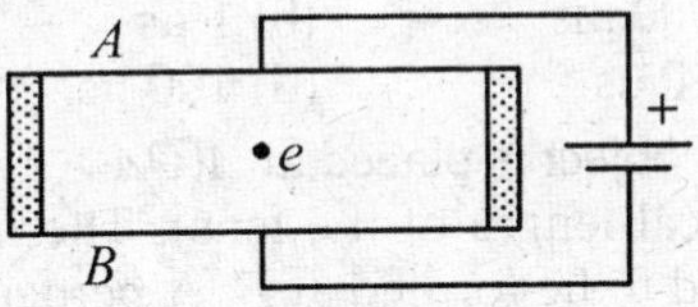

Fig. 29

(a) Continues to remain at rest

(b) acquires an acceleration of $\frac{eE}{M}$ in the upward direction

(c) acquires an acceleration $(Mg - eE)/M$ in the downward direction

(d) acquires an acceleration $\frac{(m+M)}{M}g$ in the downward direction.

83. An uncharged conducting sphere of radius R is at a distance r from point charge Q. The potential at the point P inside the sphere at distance r_1 from the centre of sphere is

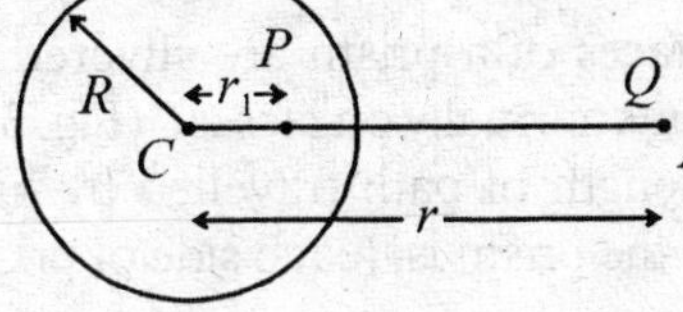

Fig. 30

(a) zero (b) $\frac{Q}{4\pi\varepsilon_0 R}$

(c) $\frac{Q}{4\pi\varepsilon_0 r}$ (d) $-\frac{Q}{4\pi\varepsilon_0 R}$.

84. Power factor for the circuit shown in Fig. 31 is

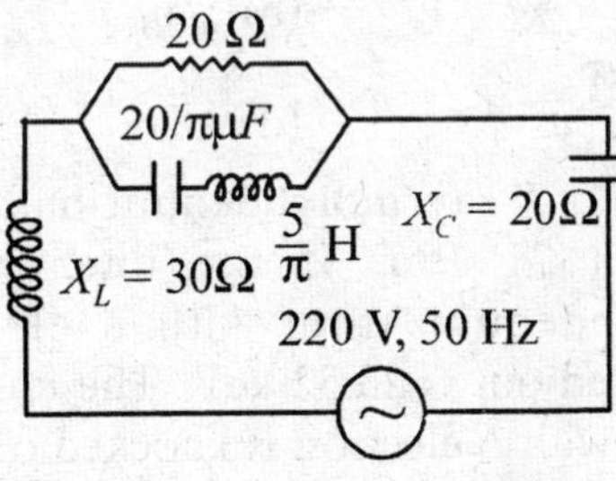

Fig. 31

(a) zero (b) $\frac{1}{\sqrt{5}}$

(c) $\frac{2}{\sqrt{5}}$ (d) $\frac{1}{2}$.

85. Two blocks having mass 6 kg and 5 kg are placed on a rough surface. Both blocks are connected by a light string (Fig. 32). String (L = 50 cm) vibrates in fundamental mode of natural frequency 100 Hz. Coefficient of friction between both block and surface is $\mu = 0.5$, $F = 30$ N and mass per unit length of string is 1 g/m. If frictional force on 5 kg and 6 kg block is f_1 and f_2 respectively, then

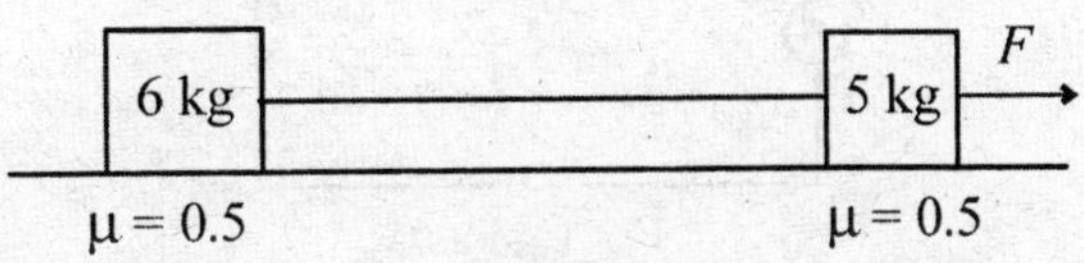

Fig. 32

(a) $f_1 = 25$ N, $f_2 = 30$ N
(b) $f_1 = 10$ N, $f = 20$ N
(c) $f_1 = 20$ N, $f_2 = 10$ N
(d) $f_1 = 30$ N, $f_2 = 25$ N.

86. Two faces of a prism are silvered. A ray incidents normally on face AC (Fig. 33). The total length of path travelled by light ray inside the prism is, (each side of prism is a)

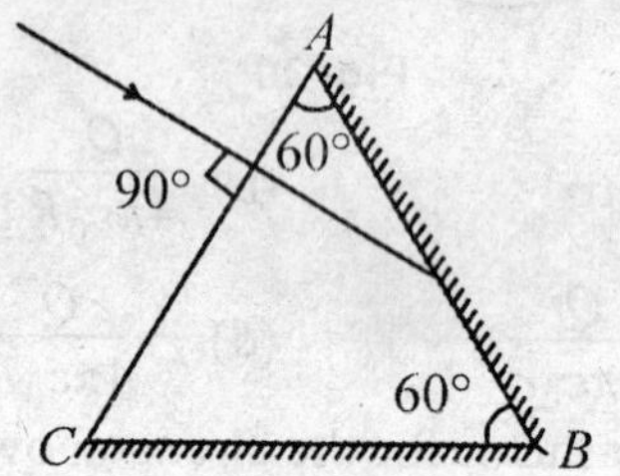

Fig. 33

(a) a (b) $\sqrt{3}a$
(c) $\frac{\sqrt{3}}{2}a$ (d) $\sqrt{2}a$.

87. The K_α X-ray molybdenum has wavelength 71 Pm. If the energy of a molybdenum atom with K-electrons knocked out is 23.32 keV. The energy of atom with L electron is knocked out is

(a) 17.5 keV (b) 12.42 keV
(c) 5.82 keV (d) 3.4 keV.

88. An AC source $V = V_0 \sin\omega t$ is connected to a circuit (Fig. 34). The reading of the ammeter when $\omega = \frac{1}{\sqrt{LC}}$ is

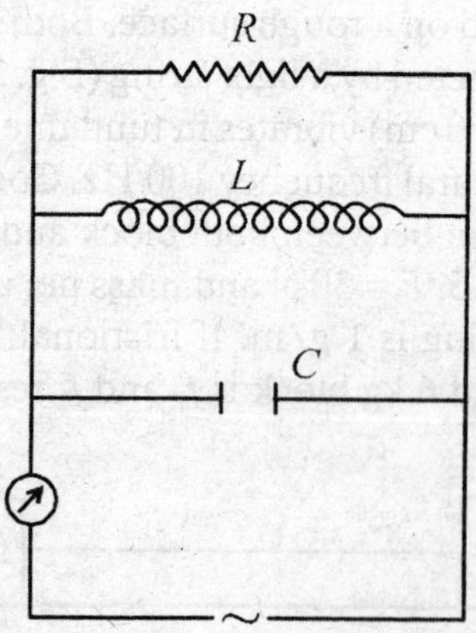

Fig. 34

(a) $\frac{V_0}{R}$

(b) $\frac{V_0}{\sqrt{\omega^2L^2 - 1/\omega^2C^2}}$

(c) $\frac{V_0}{\sqrt{2}R}$

(d) $\frac{V_0}{\sqrt{2(\omega^2L^2 - 1/\omega^2C^2)}}$.

89. Two identical rods are connected in series. The combination allows to flow 30 J of heat in 3 minutes if ends are maintained at some temperature difference. Cross-sectional area of each rod made half and now connected in parallel. The amount of heat which will flow in 3 minutes for this combination having same temperature difference as was in series combination is

(a) 15 J (b) 30 J
(c) 60 J (d) 120 J.

90. A radioactive source in the form of a metal sphere of radius 8×10^{-3} m emits β-particles at constant rate of 5.6×10^{10} particles per second. If the source is electrically insulated, how long will it take for its potential to increase by 0.8 V. Assuming that 80% of the emitted β-particles escape from the surface.

(a) 100 ms (b) 1 μs
(c) 10 μs (d) 100 μs.

91. If an object is placed at $A(OA > f$, where f is focal length of the lens). The image is found to be formed at B. A perpendicular is erected at O and C is chosen on it such that the $\angle ACB$ is right angle. Then the value of f will be

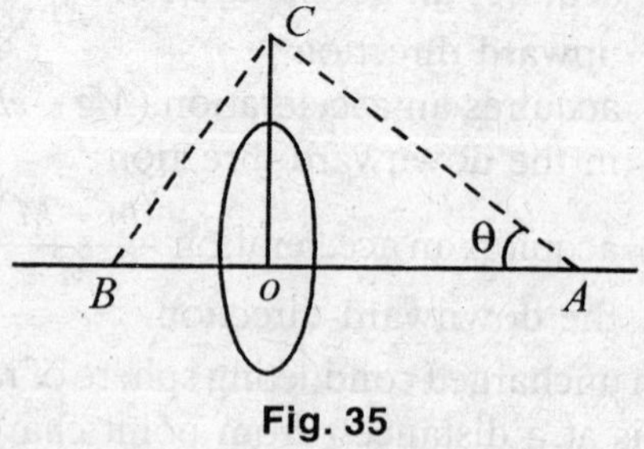

Fig. 35

(a) $(AB)/(OC)^2$
(b) $(AC)(BC)/(OC)$
(c) $(OC)^2/AB$
(d) $(OC)(AB)/(AC + BC)$.

92. A particle executing SHM crosses points P and Q with same velocity. Having taken 2 s in passing from P to Q, it returns to Q after another 2 s. The periodic time is
(a) 2 s (b) 4 s
(c) 6 s (d) 8 s.

93. An electric field is applied to a semi-conductor. Let the number of charge carriers be n and the average drift speed be v. If the temperature is increased
(a) both n and v will increase
(b) n will increase and v will decrease
(c) v will increase but n will decrease
(d) both n and v will decrease.

94. An AND gate can be obtained by repetitive use of
(a) NOT gate (b) OR gate
(c) NOR gate (d) none of the above.

95. In a p-n junction the depletion layer is 400 nm wide and an electric field of 5×10^5 V/m exist in it. The minimum kinetic energy of a electron which can diffuse from n-side to p-side is
(a) 0.2 eV (b) 0.125 eV
(c) 0.8 eV (d) 0.02 eV

96. Suppose energy liberated in the recombination of a hole-electron pair is converted into electromagnetic radiation. If the maximum wavelength emitted is 820 nm, then the band gap is
(a) 0.5 eV (b) 1 eV
(c) 1.5 eV (d) 2 eV.

97. n-p-n transistor circuit is shown in Fig. 36. This is
(a) a common base amplifier circuit
(b) a common emitter amplifier circuit
(c) a common collector amplifier circuit
(d) it is not an amplifier circuit.

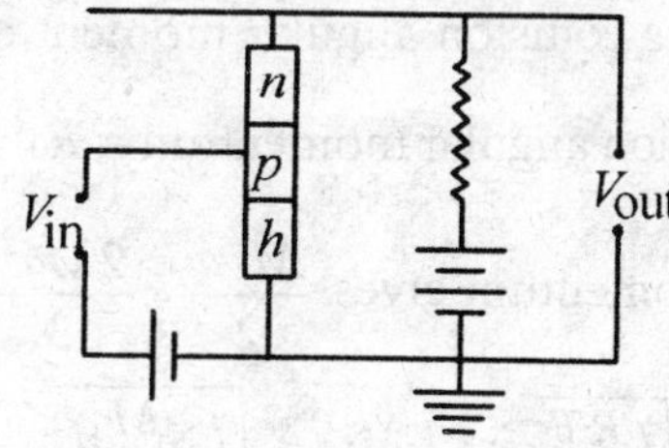

Fig. 36

98. The diode used in the circuit shown in Fig. 37 has a constant voltage drop of 0.5 V at all currents and a maximum power rating of 100 milli-Watts. What should be the value of the resistor R connected in series with diode for obtaining maximum current?

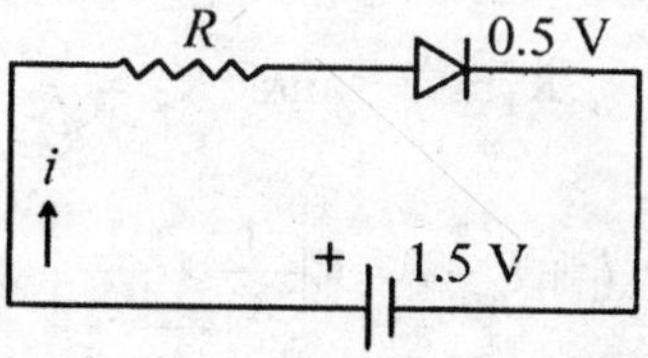

Fig. 37

(a) 5 Ω (b) 5.6 Ω
(c) 6.76 Ω (d) 20 Ω.

99. In a sample of a radioactive substance, the fraction of initial number of nuclei will remain undecayed after a time $t = T/2$, where T is half life of a radioactive substance is
(a) $\frac{1}{\sqrt{2}}$ (b) $\frac{1}{2\sqrt{2}}$
(c) $\frac{1}{4}$ (d) $\frac{1}{\sqrt{2}-1}$.

100. The de Broglie wavelength of electron in ground state of hydrogen atom is
(a) 1.06 Å (b) 1.52 Å
(c) 0.53 Å (d) 3.33 Å.

SOLUTIONS

1. Before collision angular moment of block about horizontal axis through $O = Mv\,a/2$. After collision angular momentum $= I\omega$. I about its one edge $= \frac{Ma^2}{6} + \frac{Ma^2}{2} = \frac{2Ma^2}{3}$. Conservation of momentum gives $\frac{Mv\,a}{2} = \frac{2Ma^2}{3}\omega$ or $\omega = \frac{3}{4}\frac{v}{a}$. (a)

2. $v = \sqrt{\frac{\gamma RT}{M}}$ and $\frac{v_N}{v_H} = \sqrt{\frac{\gamma_N M_H}{\gamma_H M_N}}$ $\gamma_N = \frac{7}{5}$, $\gamma_H = \frac{5}{3}$, $M_H = 4$, $M_N = 28$.

or $\frac{v_N}{v_H} = \sqrt{\frac{7 \times 4 \times 3}{5 \times 5 \times 28}} = \frac{\sqrt{3}}{5}$. (c)

3. Disc has angular speed ω and its centre of mass is moving with speed $v = \omega R$. Hence angular momentum $L = I\omega + MvR$

$$L = \frac{1}{2}MR^2\omega + M\omega R^2 = \frac{3}{2}MR^2\omega.$$ (c)

4. $K = \frac{C}{l}$, $K_1 = \frac{C}{l_1}$ and $K_2 = \frac{C}{l_2}$ or $\frac{K_1}{K_2} = \frac{l_2}{l_1} = 2$. $K_1 = 2K_2$

$$l_1 + l_2 + l = C\left[\frac{1}{K_1} + \frac{1}{K_2}\right] = \frac{C \times 3}{2K_2} = \frac{C}{K} \quad \text{or} \quad K_2 = \frac{3}{2}K$$ (b)

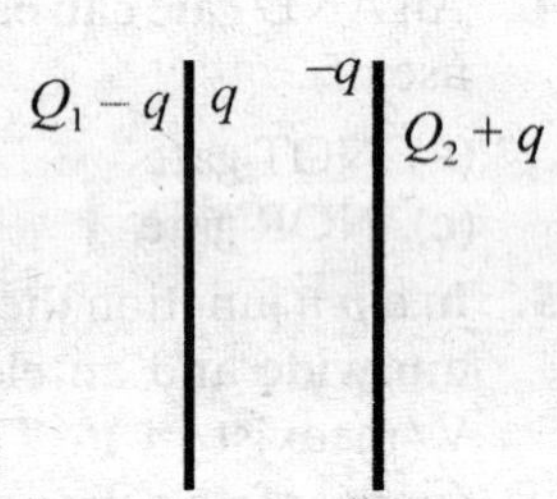

Fig. 38

5. When the plates are brought closer, the distribution of charge is as shown in Fig. 38. Such that $Q_1 - q = Q_2 + q$

or $q = \frac{Q_1 - Q_2}{2}$ and $V = \frac{q}{C} = \frac{Q_1 - Q_2}{2C}$. (d)

6. When S_1 is closed $V_1 = 30$ V and $V_2 = 20$ V. When S_3 is closed the two capacitors are in series and each have charge 60 pC as such $V_1 = 30$ and $V_2 = 20$ V. When S_1 and S_2 are closed the P.d. across C_1 and C_2 remains same. But 1st plate of C_1 comes to 0 potential and second plate has potential –30 V. The second plate of capacitor C_2 is at potential –50 V. When S_1 and S_3 are closed $V_1 = 30$ V and $V_2 = 20$ V with first plate of C_1 at 0 potential and second plate of C_2 at zero potential. (d)

7. Disintegration equation is ${}^{22}_{10}\text{N} \rightarrow {}^{14}_{6}\text{C} + {}^{4}_{2}\text{He} + {}^{4}_{2}\text{He} - Q$. (b)

8. Dimensions of $\varepsilon_0 = \text{M}^{-1}\text{L}^{-3}\text{T}^4\text{I}^{-2}$ and $V = \frac{\text{Joule}}{\text{Coul}} = \text{M}^1\text{L}^2\text{T}^{-3}\text{I}^{-1}$. Hence dimensions of

$$X = \frac{\text{M}^{-1}\text{L}^{-3}\text{T}^4\text{I}^2\text{L}^1\text{M}^1\text{L}^2\text{T}^{-3}\text{I}^{-1}}{\text{T}} = \text{I}.$$ (d)

9. At the time of collision the velocites of colliding particles are $(\boldsymbol{v}_1 - gt_0\hat{j})$ and $(\boldsymbol{v}_2 - gt_0\hat{j})$. Let their velocities after collision be $\boldsymbol{v}$ and $\boldsymbol{v}'$. Then conservation of momentum gives

$$m_1(\boldsymbol{v}_1 - gt_0\hat{j}) + m_2(\boldsymbol{v}_2 - gt_0\hat{j}) = m_1\boldsymbol{v} + \ldots\ m_2\boldsymbol{v}'$$ (i)

After time t_0 after collision their velocities are v_1' and v_2'. Hence $v_1' = v - gt_0\hat{j}$ and $v_2' = v' - gt_0\hat{j}$. Substituting in (*i*), we get

$$m_1\left(v_1 - gt_0\hat{j}\right) + m_2\left(v_2 - gt_0\hat{j}\right) = m_1\left(v_1' + gt_0\hat{j}\right) + m_2\left(v_2' + gt_0\hat{j}\right)$$

or $$\left(m_1 v_1 + m_2 v_2\right) - \left(m_1 v_1' + m_2 v_2'\right) = 2(m_1 + m_2)gt_0\hat{j}$$

and $$\left|\left(m_1 v_1 + m_2 v_2\right) - \left(m_1 v_1' + m_2 v_2'\right) = 2(m_1 + m_2)gt_0\right|$$ (c)

10. Since m is mass of quarter sector of disc hence total mass of disc is $4m$ and moment of inertia of disc $= \frac{1}{2}Mr^2 = \frac{1}{2}(4m)r^2$. Thus quarter sector has moment of inertia $\frac{1}{4} \times \frac{1}{2}(4mr^2) = \frac{1}{2}mr^2$. (a)

11. Force at the bottom of cylinder is (*i*) due to upthrust = $V\rho g$ and (*ii*) due to pressure of liquid above the cylinder $hg\rho \cdot \pi r^2$. Thus net force is $F = V\rho g + hg\rho \cdot \pi r^2 = \rho g(V + \pi r^2 h)$. (d)

12. For adiabatic process PV^γ = constant and slope of the curve $\frac{dP}{dV} = -\gamma\frac{P}{V}$ greater γ, greater slope. Hence curve (1) is for diatomic and (2) for monoatomic. (b)

13. Consider a ring of spiral with inner radius r and $r + dr$. No. of turns in this ring are $\frac{N}{b-a}dr$.

Magnetic field at the centre due to current I through this ring is $dB = \frac{\mu_0 iNdr}{2(b-a)r}$. Integration from $r = a$ to $r = b$ gives

$$B = \frac{\mu_0 NI}{2(b-a)}\log b/a\,.$$ (c)

14. Initially total emf = $3e$. Hence heat generated $= \frac{(3e)^2}{R}t$ and $\frac{(3e)^2}{R}t = mS\Delta t$ with N cells heat generated $= \frac{(Ne)^2}{2R}t$ and $\frac{(Ne)^2}{2R}t = 2mS\Delta t$. Solving, we get $N = 6$. (b)

15. At a point where the phase difference between two waves is θ intensity is given by

$$I = I_1 + I_2 + 2\sqrt{I_1 I_2}\cos\theta\,.$$

For point A, $\theta = \pi/2$ hence $I_a = I + 4I = 5I$. For point B, $\theta = \pi$ hence $I_b = I + 4I - 4I = I$. Thus $I_a - I_b = 4I$. (b)

16. In steady state, using Kirchhoff's law current in circuit is $i = \frac{V}{3R}$. Hence potential difference across ab is $\frac{4V}{3}$. Hence potential difference across capacitor $V_C = \frac{4}{3}V - V = \frac{V}{3}$. (c)

17. Resistances R_1, R_2, R_3 and R_4 form a balanced Wheatstone's bridge. Hence $\frac{R_1}{R_2} = \frac{R_3}{R_4}$ or $R_1R_4 = R_2R_3$. (c)

18. Contribution to magnetic field at $P(a, 0, a)$ due to ABC and DEF sections are equal in magnitude and opposite in direction. Hence net contribution is zero.

Contribution from FA is proportional to $(Idl\hat{j})\times(2a\hat{k}) = 2aIdl\hat{i}$ and contribution from CD is proportional to $(-Idl\hat{j})\times(2a\hat{i}) = 2aIdl\hat{k}$. Hence unit vector in the direction of magnetic field is $\frac{1}{\sqrt{2}}(\hat{i}+\hat{k})$. (d)

19. $c_{cm} = \frac{m_1v_1 + m_2v_2}{m_1 + m_2} = \frac{10\times 14 + 4\times 0}{10+4} = 10$ m/s . (c)

20. Initial $P.E. = \frac{2kQq}{a}$. Final $P.E. = \frac{kQq}{a+x} + \frac{kQq}{a-x} = \frac{2kQqa}{a^2-x^2}$ and change in

$$P.E. = 2kQq\left(\frac{a}{a^2-x^2} - \frac{1}{a}\right) = \frac{2kQqx^2}{(a^2-x^2)a} \approx \frac{2kQqx^2}{a^3}.$$ (b)

21. Loss in electric potential energy is $\frac{1}{2}\frac{C_1C_2}{C_1+C_2}(V_1-V_2)^2$, $C_1 = C_2$. (c)

22. Process AB is isobaric, hence work done $= P\Delta V = 10 \times 1 = 10$ J. BC is isochoric and hence work done is zero. If W is work done in process CA then total work done $= W + 10$. Change in internal energy is zero. Hence $W + 10 = 5$ or $W = -5$ J. (a)

23. Limiting force of friction is $= 5 \times 0.5 = 2.5$ N.
But force on block $= mg = 0.1 \times 9.8 = 0.98$ N.
Hence frictional force on the block is 0.98 N. (b)

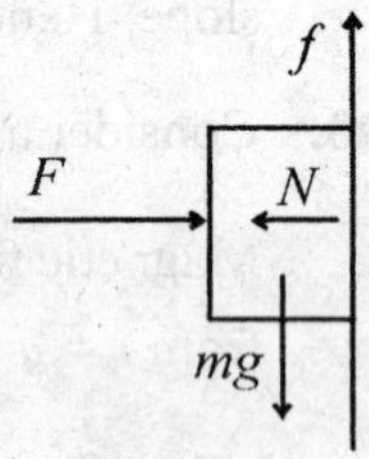

Fig. 39

24. $|\boldsymbol{A}+\boldsymbol{B}| = AB\sin\theta = \sqrt{3}AB\cos\theta$ given $\theta = 60°$.
$|\boldsymbol{A}+\boldsymbol{B}| = [A^2 + B^2 + 2AB\cos 60°]^{1/2}$ $\cos 60 = 1/2$. (a)

25. The bead has tangential acceleration $= \alpha L$. Hence tangential force $= m\alpha L$ the rod gives reaction $= m\alpha L$ and frictional force along rod $= \mu m\alpha L$. When this becomes equal to centrepetal force $= m\omega^2 L$, the head starts moving. Thus

$$m\omega^2 L = \mu m\alpha L \text{ or } \omega = \sqrt{\mu\alpha} = \alpha t \text{ or } t = \sqrt{\frac{\mu}{\alpha}}.$$ (a)

26. Magnetic field at M due to portion $PQ = \frac{\mu_0 I}{4\pi d}$ and due to $QR = 0$ and $B_1 = \frac{\mu_0 I}{4\pi d}$. Magnetic field due to $QS = \frac{\mu_0 I}{8\pi d}$ and $B_2 = \frac{3\mu_0 I}{8\pi d}$, $\frac{B_1}{B_2} = \frac{2}{3}$. (c)

27. $N_1 = N_0e^{-10\lambda t}$, $N_2 = N_0e^{-\lambda t}$ or $\frac{N_1}{N_2} = e^{-1} = e^{-9\lambda t}$ or $t = 1/9\lambda$. (d)

28. Time period T of electron is an orbit of quantum number n is $T \propto n^3$. This gives $\frac{n_1}{n_2} = 2$. (a)

29. For gas (A) $P_aV = \frac{m_aRT}{M}$ and $(P_a + \Delta P)2V = \frac{m_aRT}{M}$ or $\Delta P = \frac{m_aRT}{2MV}$. Similarly for gas B $1.5\Delta P = \frac{m_bRT}{2MV}$. This gives $\frac{m_b}{1.5} = m_a$ or $3m_a = 2m_b$. (c)

30. The refraction on water surface will form virtual image beyond C which will act like object for mirror and the final image is formed by reflection between focus and centre of curvature, i.e. between C and O and is real image. (d)

31. In *p-n* junction diode there is depletion layer at the junction. In this layer there is electric field which stops diffusion of holes from p to n and electrons from n to p. Its direction is from n to p. (c)

32. According to Prevost theory everybody irrespective of its temperature emits radiation means $U_1 \neq 0$ and $U_3 \neq 0$. Wien's law gives the wavelength corresponding to maximum energy emitted is $\lambda = \frac{b}{T}$, $\lambda = 10^3$ nm gives $U_2 > U_1$. (c)

33. Maximum particle velocity $= a\omega = \frac{v}{10}$ or $\omega = \frac{v}{10a} = \frac{10}{10 \times 10^{-3}} = 10^3$ and $f = \frac{10^3}{2\pi}$ Hz, $\lambda f = v$,

$\lambda = \frac{10}{10^3/2\pi} = 2\pi \times 10^{-2}$ m. (a)

34. Postulates of kinetic theory give molecules all possible velocities from 0 to ∞. This gives (*a*) and (*b*) are wrong.

$$\bar{v} = \sqrt{\frac{8RT}{\pi M}},\ v_{rms} = \sqrt{\frac{3RT}{M}} \text{ and } v_P = \sqrt{\frac{2RT}{M}} \text{ gives } v_P < \bar{v} < v_{rms}$$

Average kinetic energy of a molecule is $\frac{1}{2}mv_{rms}^2 = \frac{3}{2}m\frac{RT}{M} = \frac{3m}{4}v_P^2$. (d)

35. Gas A is heated at constant pressure. Hence $\Delta Q = \mu C_P \Delta T$ or $\Delta Q = \mu C_P \times 30$. Whereas gas B is heated at constant volume and $\Delta Q = \mu C_V \Delta T$ or $\Delta T = 30 \times \frac{C_P}{C_V} = 30 \cdot \gamma \cdot \gamma = 7/5$ for diatomic gas. $\Delta T = 30 \times \frac{7}{5} = 42$ K. (d)

36. For constructive interference path difference is λ or time lag equal to time period. In stetched string $\lambda = 2l = 0.8$ m. $v = \sqrt{\frac{T}{m}} = 8$ m/s. Hence frequency $f = \frac{v}{\lambda} = 10$ Hz and time period $= \Delta t \frac{1}{10} = 0.10$ s. (b)

37. For refraction through curved surfaces $\frac{\mu_2}{V} - \frac{\mu_1}{U} = \frac{\mu_2 - \mu_1}{R}$. Given is $\mu_2 = 1.5$, $\mu_1 = 1$, $V = +X$, $U = -X$ and R is +ve. This $\frac{1.5}{X} + \frac{1}{X} = \frac{1.5-1}{R}$ gives $X = 5R$. (a)

38. In single slit Fraunhoffer diffraction, the path difference between rays from two edges of slit with slit width 'a' is $a\sin\theta$, where θ is angle of diffraction. In the direction in which $a\sin\theta = \lambda$. We get first minima. Hence phase difference is 2π. (d)

39. For no motion $F\cos 60° = \mu(mg + F\sin 60°)$. Solving gives $F = 20$ N. (a)

40. For constructine interference at P, the resultant path difference $\Delta = n\lambda$. Δ at P is $\Delta = (CO + OP) + \lambda/2$ (due to reflection at O). From geometry $CO = OP\cos 2\theta$ and $OP = d/\cos\theta$. Thus $CO + OP = OP(1 + \cos 2\theta) = 2\cos^2\theta \frac{d}{\cos\theta} = 2d\cos\theta$. Thus $2d\cos + \lambda/2 = n\lambda$. For $n = 1$ $2d\cos\theta = \lambda/2$ or $\cos\theta = \frac{\lambda}{4d}$. (b)

41. $\alpha_a l_1 t = \alpha_s l_2 t$ or $\dfrac{l_2}{l_1} = \dfrac{\alpha_a}{\alpha_s}$ or $\dfrac{l_2 + l_1}{l_1} = \dfrac{\alpha_a + \alpha_S}{\alpha_S}$. (c)

42. Convex lens forms image at distance 30 cm and is of length 2 cm. This serves as an object for concave lens. Then for concave lens $U = +4$ cm, $f = -20$ cm, $\dfrac{1}{V} - \dfrac{1}{U} = \dfrac{1}{f}$ or $\dfrac{1}{V} - \dfrac{1}{4} = -\dfrac{1}{20}$ gives $V = 5$ cm hence size of image $= \dfrac{5}{4} \times 2 = 2.5$ cm. (b)

43. $\dfrac{\mu_w}{\mu_g} = \dfrac{\sin i}{\sin r}$ and $\mu_w = \dfrac{1}{\sin r}$ gives $\mu_g = 1/\sin i$. (b)

44. $F = \left|-\dfrac{dU}{dr}\right| = \dfrac{U_0}{r} = \dfrac{mv^2}{r}$ or $v = \sqrt{\dfrac{U_0}{m}} =$ constant.

Quantum condition is $mv_n r_n = \dfrac{nh}{2\pi}$ or $r_n \propto n$. (*a*)

45. Momentum P of α particle and daugher nucleus will be same. Hence $\dfrac{P^2}{2}\left[\dfrac{1}{m} + \dfrac{1}{M}\right] = Q$. This gives $\dfrac{P^2}{2m} = \dfrac{QM}{m + M} = \dfrac{5.5 \times 216}{220} = 5.4$ MeV. (b)

46. Rate of cooling is proportional to excess temperature over surroundings. Excess temperature should be four times, i.e. temperature 20 + 32 = 52° C. (d)

47. 20 minutes means 4 half lives. Hence remaining active substance is 1/16 or decay fraction is $\dfrac{15}{16} = \dfrac{15}{16} \times 100 = 93.75\%$. (a)

48. $\alpha = \left(\dfrac{\sqrt{I_1} + \sqrt{I_2}}{\sqrt{I_1} - \sqrt{I_2}}\right)^2 = \left(\dfrac{x+1}{x-1}\right)^2$. Where $x = \sqrt{\dfrac{I_1}{I_2}}$ if $I_1 > I_2$, $x > 1$ then $\sqrt{\alpha} = \dfrac{x+1}{x-1}$ gives $\sqrt{\alpha}x - \sqrt{\alpha} = x + 1$ or $x = \dfrac{1 + \sqrt{\alpha}}{\sqrt{\alpha} - 1}$ gives $\dfrac{I_1}{I_2} = x^2 = \left[\dfrac{1 + \sqrt{\alpha}}{\sqrt{\alpha} - 1}\right]^2$. (c)

49. Resistance of inductor $R = \dfrac{10}{5} = 2\,\Omega$. When connected to AC along with capacitor, the voltage and current are in same phase means circuit is resonance circuit. All voltage comes across-resistance. Hence $i = \dfrac{20}{2} = 10$ A. (a)

50. Path difference introduced is $= t(\mu - 1) = 20 \times 10^{-6} \times 0.48$. A path difference of wavelength (λ) shifts fringe pattern by 1 fringe. The number of fringes shifted is

$$n = \frac{20 \times 10^{-6} \times 0.48}{4800 \times 10^{-10}} = 20. \quad \text{(c)}$$

51. Total energy of satellite in orbit of radius $2R$ is $E_i = -\dfrac{GMm}{4R}$. Total energy in orbit of radius

$6R$ is $E_f = -\frac{GMm}{12R}$. Hence required energy $E = E_f - E_i$,

$$E = -\frac{GMm}{12R} + \frac{GMm}{4R} = \frac{GMm}{4R}\left(-\frac{1}{3}+1\right) = \frac{GMm}{6R}. \quad \text{(b)}$$

52. If h_1 is height of liquid level in bigger tube, then $h_1 = \frac{2T}{g\rho(2r)} = \frac{T}{g\rho r}$ and for second capillary $h + h_1 = \frac{2T}{\rho gr}$. This gives $h = \frac{T}{g\rho r}$ or $T = hg\rho r$.

53. Object is placed at $2f$ distance from lens. Hence image is formed at $2f$ distance (= 30 cm) from lens. This serves as object for mirror. In order to form final image at the object the rays must fall normally on mirror. That is the distance between mirror and image formed by lens must be twice the focal length of mirror = 24. This gives the distance between lens and mirror as 6 cm. (a)

54. The only force on the particle is gravitational force mg acting vertically downwards. Hence when rod is horizontal mass reaches origin (0, 0) and lower end has co-ordinates $(-l, 0)$. (c)

55. The particle reaches the ground in time $t = \sqrt{\frac{2h}{g}}$. In this time if the horizontal velocity becomes zero, the particle will strike the ground vertically. This gives $v_0 = a_0\sqrt{\frac{2h}{g}}$ or $h = \frac{v_0^2 g}{2a_0^2}$. (b)

56. Pure rolling is equivalent to rotation about O. Then torque $\tau = Fr = I\alpha$ or $\alpha = \frac{Fr}{I}$ and angular velocity after time t is $\omega = \alpha t = \frac{Fr.t}{3/2\, mr^2}$ and $v = \omega r = \frac{2}{3}\frac{Ft}{m}$. Power $Fv = \frac{2}{3}\frac{Ft^2}{m}$. (d)

57. Figure 19 shows two capacitors each of capacity $C = \frac{\varepsilon_0 A}{d}$ connected in parallel and charge q is given. Potential $V = \frac{q}{2C} = \frac{qd}{2\varepsilon_0 A}$. (d)

58. Initially charge is q and energy stored is $\frac{q^2}{2C}$. When separation is made 3 times capacity becomes $C/3$ and charge is conserved. Then energy stored $= \frac{3q^2}{2C}$. Hence work done $= \frac{3q^2}{2C} - \frac{q^2}{2C} = \frac{q^2}{C} = V^2C$. $C = \frac{\varepsilon_0 A}{d}$ and $q = CV$ gives $W = \frac{V^2\varepsilon_0 A}{d}$. (d)

59. For balanced bridge $P/Q = R/S$ or $\frac{P+Q}{R+S} = \frac{Q}{S} = \frac{P}{R}$. Power dissipated $\frac{V^2}{R}$. i.e. $\frac{P_1}{P_2} = \frac{R+S}{P+Q} = \frac{R}{P}$. (b)

60. In Fig. 40 given is $B\sin\theta = B/2$ or $\theta = 30°$. This gives the angle between two vectors = 180 – 30 = 150°. (c)

61. Angle of incidence on second surface is 30° and angle of refraction r is $\sin r = \mu \sin i = 0.75$, $r \approx 48°$. From Fig. 41. $\phi = 90 - r + 30° = 72°$. Hence $\theta = 180 - 2(72) = 36°$. (b)

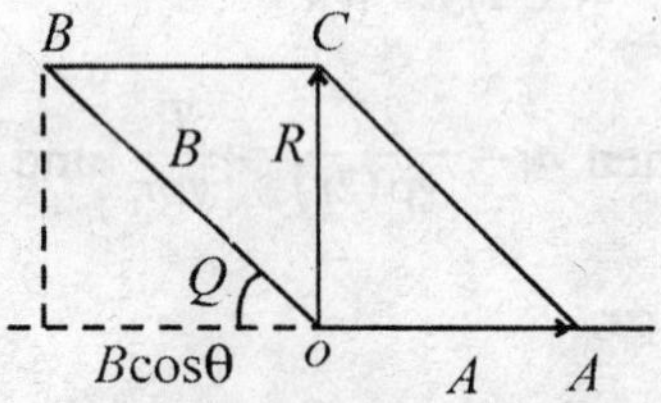

Fig. 40

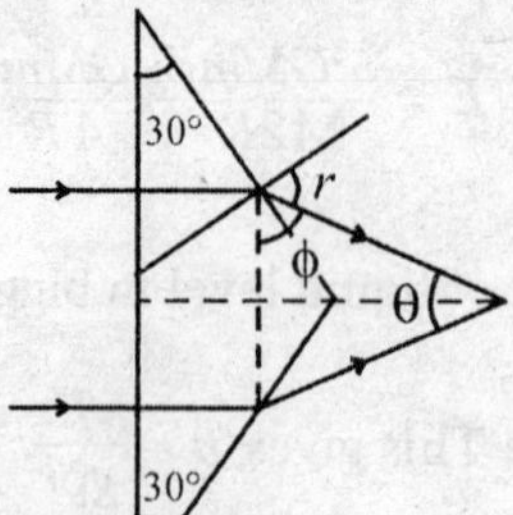

Fig. 41

62. Upto the angle of repose, the static frictional force is $(mg\sin\theta)$ and for angle greater than angle of repose, the force of sliding friction is given by $\mu mg\cos\theta$. (b)

63. Force $F = -\frac{du}{dx} = -AP\sin Px$. For small angles $\sin Px \approx Px$ force $F = -AP^2x$ and acceleration $= \frac{F}{m} = -\frac{AP^2}{m}x$. The particle executes S.H.M. with time period $T = 2\pi\sqrt{\frac{m}{AP^2}}$. (a)

64. $\frac{9KA(100-\theta)}{18} = \frac{KA(\theta - 0)}{6}$ same heat flow through each, $\theta = 75°$ C. (a)

65. $h\nu_1 - \phi = 1$ and $h\nu_2 - \phi = 2.9$, , $\nu_2 = 2\nu_1$, gives $2h\nu_1 - \phi = 2.9$, $\phi + 2.9 = 2\phi + 2$ or $\phi = 0.9$ eV. (a)

66. Wavelength and not the intensity depends on voltage. β-decay means β^+ (positron) emission and *K*-capture. Moderator does not slow down the rate of chain reaction. (c)

67. Difference of divisions 1 in 20 least count 0.1 mm gives main scale div = 0.1 × 20 = 2 mm. (c)

68. Centreptal acceleration at P is $a = \frac{v^2}{r} = 1$ along OP gives $\mathbf{a} = -\frac{1}{\sqrt{2}}(\hat{i} + \hat{j})$. (c)

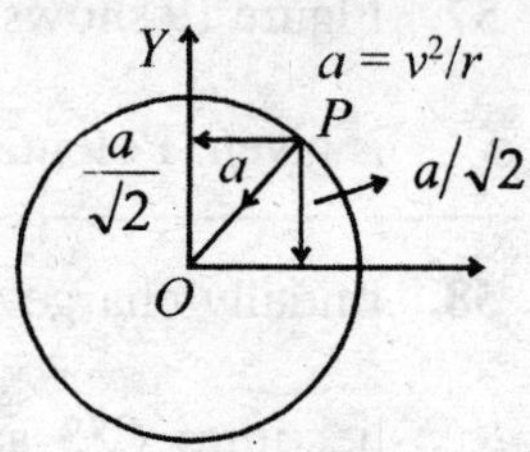

Fig. 42

69. None of the four appears wrong. But different γ identify (*a*) as untrue. (a)

70. Figure shows $BD = AD$. So V becomes zero. This gives (*a*) we find electric field is E towards B. (a)

71. Two capacities are in series with capacity ratio 3 : 1. Same charge hence potential ratio is 1 : 3. This gives $V_a = 7.5$ V and $V_b = 2.5$ V. (c)

72. Mutual inductance is same. Hence emf is proportional to $\left|\frac{di}{dt}\right|$ only. Hence $e = 6$ mV. (c)

73. AND gate means C is 1 only when $A = B = 1$ and zero otherwise. Correct entry (*i*) and (*iii*) only. (c)

74. Energy same $P = \sqrt{2mE}$ means greater mass greater momentum. (a)

75. $a_n \propto Z^{-1}n^2$, $a_3 \propto \frac{(3)^2}{3}$ or $a_3 = 3a_0$. (d)

76. If P is pressure difference, then rate of flow Q through the tube is given by $Q = \frac{\pi P r^4}{8\eta l}$ or $P = \frac{Q 8\eta}{\pi} \frac{l}{r^4}$. For three tubes in series rate of flow is same. Thus $\frac{L}{R^4} = \frac{l_1}{r_1^4} + \frac{l_2}{r_2^4} + \frac{l_3}{r_3^4}$ which gives $16L = 10 + 80 + 1250$ or $L = \frac{1340}{16} = 83.75$. (d)

77. Time constant of circuit $RC = 2$ sec. Voltage $V = V_0(1 - e^{-t/RC})$, $100 = 120(1 - e^{-t/2})$ or $e^{-t/2} = \frac{1}{6}$
Thus $t = 2\log 6$. The period is close to this because discharge takes negligible time. (b)

78. Radiation received $\propto d^{-2}$, emitted $\propto T^4$ so $T \propto d^{-1/2}$. (b)

79. Gates 1 to 6 are OR, AND, NOT, NOR, NOT and NAND gates. (a)

80. If e is end correction then $L_3 + e = \frac{5\lambda}{4}$ and $L_2 + e = \frac{3\lambda}{4}$, $L_1 + e = \frac{\lambda}{4}$ which gives $L_3 - L_2 = L_2 - L_1$. (c)

81. R_1 and R_2 divide the potential acting across C, R_3 does not influence the P.d. though it affects rate of charging. (c)

82. Electron at rest $Ee = mg$. Downward force on proton

$$F = Ee + Mg = mg + Mg. \quad a = \frac{(m+M)}{M} g.$$ (c)

83. Potential at all points in the sphere is same. The potential at the centre of sphere is equal to the potential due to point charge, plus the potential by charges induced on the surface of sphere. The total charge induced on the sphere is zero. Hence potential at centre $= \frac{Q}{4\pi\varepsilon_0 r}$ = Potential at P. (c)

84. At this frequency $C = \frac{20}{\pi}\,\mu F$ and $L = \frac{5}{\pi} H$ provide zero impedence. Hence resistance 20 Ω is short circuited. Thus the total resistance is zero and power factor $= R/Z = 0$. (a)

85. $\frac{1}{2L}\sqrt{\frac{T}{m}} = 100$ gives $T = 10$ N.
Friction of 6 kg $= T = f_2 = 10$ N and $f_1 + T = 30$ or $f_1 = 20$ N. (c)

86. Path of the ray through prism is shown in Fig. 43. Total path length in prism is $2(RP + PQ) = 2(AP\cos 30 + PB\cos 30)$ or

$$\text{Path length} = 2\cos 30(AP + PB) = 2a\frac{\sqrt{3}}{2} = a\sqrt{3}.$$ (b)

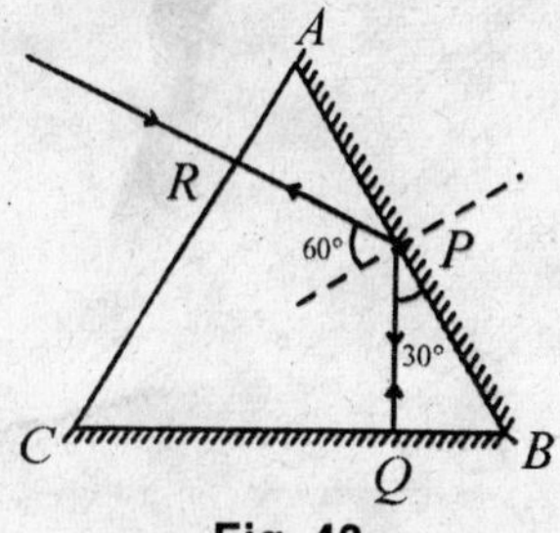

Fig. 43

87. Energy of K_α X-ray photon is $\frac{1242}{71 \times 10^{-3}} = 17.5$ keV,
$E_k - E_L = 17.5$ keV or $E_L = 23.32 - 17.5 = 5.82$ keV. (c)

88. L-C circuit is a rejector circuit and no current passes through this branch. Hence ammeter shows $I_{rms} = \frac{V_0}{\sqrt{2}R}$. (c)

89. In series combination rate of heat flow $Q = \frac{KA(\theta_1 - \theta_2)}{2l}$ rods are identical. In parallel combination $Q' = 2\frac{KA(\theta_1 - \theta_2)}{2 \times l} = 2Q = 60$ J. (c)

90. If charge on sphere is q, potential $V = \frac{kq}{r}$ and q = (net × 0.8) or $V = \frac{k.net \times 0.8}{r}$

or $t = \frac{Vr}{kne \times 0.8} = \frac{0.8 \times 8 \times 10^{-3}}{9 \times 10^9 \times 5.6 \times 10^{10} \times 1.6 \times 10^{-19} \times 0.8}$ or $t = 10^{-4}$ sec. (d)

91. $\frac{1}{OA} + \frac{1}{OB} = \frac{1}{f}$ and $f = \frac{OA \cdot OB}{OA + OB} = \frac{OC^2}{AB}$ since $\tan\theta = \frac{OC}{OA} = \frac{OB}{OC}$. (c)

92. As shown in Fig. 44 time period is 8 s. (d)

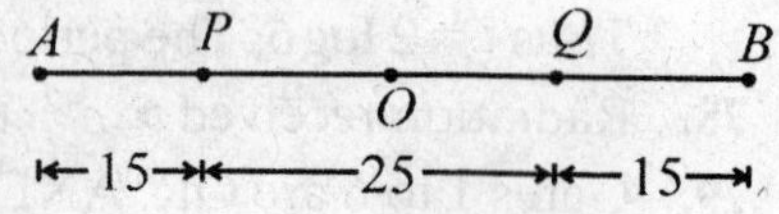

Fig. 44

93. With rise in temperature number of charge carriers increase. The collision frequecy also increases. This decreases average drift speed. (b)

94. AND gate is obtained by repetitive use of NOR gate. (c)

95. Potential barrier across depletion layer is = 5 × 10^5 × 400 × 10^{-9} = 0.2 V. Hence required K.E. is 0.2 eV. (a)

96. $\Delta E = h\nu = \frac{hc}{\lambda} = \frac{12.4 \times 1000}{\lambda(\text{Å})} = \frac{12.4 \times 10^3}{8200} = 1.5$ eV. (c)

97. It is common emitter amplifier. (b)

98. $i = \frac{P}{V} = \frac{100 \times 10^{-3}}{0.5} = 0.2$ A. P.d. across $R = 1.5 - 0.5 = 1$ V gives $R = \frac{V}{C} = \frac{1}{0.2} = 5\ \Omega$. (a)

99. $N = N_0 e^{-\lambda t} = N_0 (2)^{-t/T}$. Hence $t = T/2$, $\frac{N}{N_0} = (2)^{-1/2} = \frac{1}{\sqrt{2}}$. (a)

100. Radius of first orbit is 0.53 Å. Hence de Broglie wavelength should be equal to the length of orbit, i.e. $\lambda = 2\pi r = 2\pi \times 0.53 = 3.33$ Å. (d)